PERIODIC TABLE OF THE ELEMENTS

Group	1 IA	2 IIA	3 IIIB	4 IVB	5 VB	6 VIB	7 VIIB	8	9 VIIIB	10	11 IB	12 IIB	13 III IIIA	14 IV IVA	15 V VA	16 VI VIA	17 VII VIIA	18 VIII VIIA
Period 1	1 H hydrogen 1.0080 1s¹																	2 He helium 4.00 1s²
2	3 Li lithium 6.94 2s¹	4 Be beryllium 9.01 2s²											5 B boron 10.81 2s²2p¹	6 C carbon 12.01 2s²2p²	7 N nitrogen 14.01 2s²2p³	8 O oxygen 16.00 2s²2p⁴	9 F fluorine 19.00 2s²2p⁵	10 Ne neon 20.18 2s²2p⁶
3	11 Na sodium 22.99 3s¹	12 Mg magnesium 24.31 3s²											13 Al aluminium 26.98 3s²3p¹	14 Si silicon 28.09 3s²3p²	15 P phosphorus 30.97 3s²3p³	16 S sulfur 32.06 3s²3p⁴	17 Cl chlorine 35.45 3s²3p⁵	18 Ar argon 39.95 3s²3p⁶
4	19 K potassium 39.10 4s¹	20 Ca calcium 40.08 4s²	21 Sc scandium 44.96 3d¹4s²	22 Ti titanium 47.87 3d²4s²	23 V vanadium 50.94 3d³4s²	24 Cr chromium 52.00 3d⁵4s¹	25 Mn manganese 54.95 3d⁵4s²	26 Fe iron 55.84 3d⁶4s²	27 Co cobalt 58.93 3d⁷4s²	28 Ni nickel 58.69 3d⁸4s²	29 Cu copper 63.55 3d¹⁰4s¹	30 Zn zinc 65.38 3d¹⁰4s²	31 Ga gallium 69.72 4s²4p¹	32 Ge germanium 72.63 4s²4p²	33 As arsenic 74.92 4s²4p³	34 Se selenium 78.97 4s²4p⁴	35 Br bromine 79.90 4s²4p⁵	36 Kr krypton 83.80 4s²4p⁶
5	37 Rb rubidium 85.47 5s¹	38 Sr strontium 87.62 5s²	39 Y yttrium 88.91 4d¹5s²	40 Zr zirconium 91.22 4d²5s²	41 Nb niobium 92.91 4d⁴5s¹	42 Mo molybdenum 95.95 4d⁵5s¹	43 Tc technetium (97) 4d⁵5s²	44 Ru ruthenium 101.07 4d⁷5s¹	45 Rh rhodium 102.91 4d⁸5s¹	46 Pd palladium 106.42 4d¹⁰	47 Ag silver 107.87 4d¹⁰5s¹	48 Cd cadmium 112.41 4d¹⁰5s²	49 In indium 114.82 5s²5p¹	50 Sn tin 118.71 5s²5p²	51 Sb antimony 121.76 5s²5p³	52 Te tellurium 127.60 5s²5p⁴	53 I iodine 126.90 5s²5p⁵	54 Xe xenon 131.29 5s²5p⁶
6	55 Cs caesium 132.91 6s¹	56 Ba barium 137.33 6s²	57 La lanthanum 138.91 5d¹6s²	72 Hf hafnium 178.49 5d²6s²	73 Ta tantalum 180.95 5d³6s²	74 W tungsten 183.84 5d⁴6s²	75 Re rhenium 186.21 5d⁵6s²	76 Os osmium 190.23 5d⁶6s²	77 Ir iridium 192.22 5d⁷6s²	78 Pt platinum 195.08 5d⁹6s¹	79 Au gold 196.97 5d¹⁰6s¹	80 Hg mercury 200.59 5d¹⁰6s²	81 Tl thallium 204.38 6s²6p¹	82 Pb lead 207.2 6s²6p²	83 Bi bismuth 208.98 6s²6p³	84 Po polonium (209) 6s²6p⁴	85 At astatine (210) 6s²6p⁵	86 Rn radon (222) 6s²6p⁶
7	87 Fr francium (223) 7s¹	88 Ra radium (226) 7s²	89 Ac actinium (227) 6d¹7s²	104 Rf rutherfordium (267) 6d²7s²	105 Db dubnium (268) 6d³7s²	106 Sg seaborgium (269) 6d⁴7s²	107 Bh bohrium (270) 6d⁵7s²	108 Hs hassium (269) 6d⁶7s²	109 Mt meitnerium (277) 6d⁷7s²	110 Ds darmstadtium (281) 6d⁸7s²	111 Rg roentgenium (282) 6d¹⁰7s¹	112 Cn copernicium (285) 6d¹⁰7s²	113 Nh nihonium (286) 7s²7p¹	114 Fl flerovium (290) 7s²7p²	115 Mc moscovium (290) 7s²7p³	116 Lv livermorium (293) 7s²7p⁴	117 Ts tennessine (294) 7s²7p⁵	118 Og oganesson (294) 7s²7p⁶

Lanthanoids (lanthanides)

58 Ce cerium 140.12 4f¹5d¹6s²	59 Pr praseodymium 140.91 4f³6s²	60 Nd neodymium 144.24 4f⁴6s²	61 Pm promethium (145) 4f⁵6s²	62 Sm samarium 150.36 4f⁶6s²	63 Eu europium 151.96 4f⁷6s²	64 Gd gadolinium 157.25 4f⁷5d¹6s²	65 Tb terbium 158.93 4f⁹6s²	66 Dy dysprosium 162.50 4f¹⁰6s²	67 Ho holmium 164.93 4f¹¹6s²	68 Er erbium 167.26 4f¹²6s²	69 Tm thulium 168.93 4f¹³6s²	70 Yb ytterbium 173.05 4f¹⁴6s²	71 Lu lutetium 174.97 5d¹6s²

Actinoids (actinides)

90 Th thorium 232.04 6d²7s²	91 Pa protactinium 231.04 5f²6d¹7s²	92 U uranium 238.03 5f³6d¹7s²	93 Np neptunium (237) 5f⁴6d¹7s²	94 Pu plutonium (244) 5f⁶7s²	95 Am americium (243) 5f⁷7s²	96 Cm curium (247) 5f⁷6d¹7s²	97 Bk berkelium (247) 5f⁹7s²	98 Cf californium (251) 5f¹⁰7s²	99 Es einsteinium (252) 5f¹¹7s²	100 Fm fermium (257) 5f¹²7s²	101 Md mendelevium (258) 5f¹³7s²	102 No nobelium (259) 5f¹⁴7s²	103 Lr lawrencium (262) 6d¹7s²

Molar masses (atomic weights) quoted to the number of significant figures given here can be regarded as typical of most naturally occurring samples.

The elements

Name	Symbol	Atomic number	Molar mass (g mol^{-1})
Actinium	Ac	89	227
Aluminium (aluminum)	Al	13	26.982
Americium	Am	95	243
Antimony	Sb	51	121.76
Argon	Ar	18	39.95
Arsenic	As	33	74.922
Astatine	At	85	210
Barium	Ba	56	137.33
Berkelium	Bk	97	247
Beryllium	Be	4	9.0122
Bismuth	Bi	83	208.98
Bohrium	Bh	107	270
Boron	B	5	10.81
Bromine	Br	35	79.904
Cadmium	Cd	48	112.41
Caesium (cesium)	Cs	55	132.91
Calcium	Ca	20	40.078
Californium	Cf	98	251
Carbon	C	6	12.011
Cerium	Ce	58	140.12
Chlorine	Cl	17	35.45
Chromium	Cr	24	51.996
Cobalt	Co	27	58.933
Copernicium	Cn	112	285
Copper	Cu	29	63.546
Curium	Cm	96	247
Darmstadtium	Ds	110	281
Dubnium	Db	105	268
Dysprosium	Dy	66	162.50
Einsteinium	Es	99	252
Erbium	Er	68	167.26
Europium	Eu	63	151.96
Fermium	Fm	100	257
Flerovium	Fl	114	290
Fluorine	F	9	18.998
Francium	Fr	87	223
Gadolinium	Gd	64	157.25
Gallium	Ga	31	69.723
Germanium	Ge	32	72.630
Gold	Au	79	196.97
Hafnium	Hf	72	178.49
Hassium	Hs	108	269
Helium	He	2	4.0026
Holmium	Ho	67	164.93
Hydrogen	H	1	1.0080
Indium	In	49	114.82
Iodine	I	53	126.90
Iridium	Ir	77	192.22
Iron	Fe	26	55.845
Krypton	Kr	36	83.798
Lanthanum	La	57	138.91
Lawrencium	Lr	103	262
Lead	Pb	82	207.2
Lithium	Li	3	6.94
Livermorium	Lv	116	293
Lutetium	Lu	71	174.97
Magnesium	Mg	12	24.305
Manganese	Mn	25	54.938
Meitnerium	Mt	109	277
Mendelevium	Md	101	258
Mercury	Hg	80	200.59
Molybdenum	Mo	42	95.95
Moscovium	Mc	115	290
Neodymium	Nd	60	144.24
Neon	Ne	10	20.180
Neptunium	Np	93	237
Nickel	Ni	28	58.693
Nihonium	Nh	113	286
Niobium	Nb	41	92.906
Nitrogen	N	7	14.007
Nobelium	No	102	259
Oganesson	Og	118	294
Osmium	Os	76	190.23
Oxygen	O	8	15.999
Palladium	Pd	46	106.42
Phosphorus	P	15	30.974
Platinum	Pt	78	195.08
Plutonium	Pu	94	244
Polonium	Po	84	209
Potassium	K	19	39.098
Praseodymium	Pr	59	140.91
Promethium	Pm	61	145
Protactinium	Pa	91	231.04
Radium	Ra	88	226
Radon	Rn	86	222
Rhenium	Re	75	186.21
Rhodium	Rh	45	102.91
Roentgenium	Rg	111	282
Rubidium	Rb	37	85.468
Ruthenium	Ru	44	101.07
Rutherfordium	Rf	104	267
Samarium	Sm	62	150.36
Scandium	Sc	21	44.956
Seaborgium	Sg	106	269
Selenium	Se	34	78.971
Silicon	Si	14	28.085
Silver	Ag	47	107.87
Sodium	Na	11	22.990
Strontium	Sr	38	87.62
Sulfur	S	16	32.06
Tantalum	Ta	73	180.95
Technetium	Tc	43	97
Tellurium	Te	52	127.60
Tennessine	Ts	117	294
Terbium	Tb	65	158.93
Thallium	Tl	81	204.38
Thorium	Th	90	232.04
Thulium	Tm	69	168.93
Tin	Sn	50	118.71
Titanium	Ti	22	47.867
Tungsten	W	74	183.84
Uranium	U	92	238.03
Vanadium	V	23	50.942
Xenon	Xe	54	131.29
Ytterbium	Yb	70	173.05
Yttrium	Y	39	88.906
Zinc	Zn	30	65.38
Zirconium	Zr	40	91.224

INORGANIC CHEMISTRY

8th edition

MARK WELLER
Cardiff University

JONATHAN ROURKE
Cardiff University

FRASER ARMSTRONG
University of Oxford

SIMON LANCASTER
University of East Anglia

TINA OVERTON

OXFORD
UNIVERSITY PRESS

Great Clarendon Street, Oxford, OX2 6DP,
United Kingdom

Oxford University Press is a department of the University of Oxford.
It furthers the University's objective of excellence in research, scholarship,
and education by publishing worldwide. Oxford is a registered trade mark of
Oxford University Press in the UK and in certain other countries

© Oxford University Press 2025

The moral rights of the authors have been asserted

Seventh Edition 2018
Sixth Edition 2014
Fifth Edition 2010

All rights reserved. No part of this publication may be reproduced, stored in a retrieval system, transmitted, used for text and data mining, or used for training artificial intelligence, in any form or by any means, without the prior permission in writing of Oxford University Press, or as expressly permitted by law, by licence or under terms agreed with the appropriate reprographics rights organization. Enquiries concerning reproduction outside the scope of the above should be sent to the Rights Department, Oxford University Press, at the address above.

You must not circulate this work in any other form
and you must impose this same condition on any acquirer

Published in the United States of America by Oxford University Press
198 Madison Avenue, New York, NY 10016, United States of America

British Library Cataloguing in Publication Data

Data available

Library of Congress Control Number: 2024939982

ISBN 978-0-19-886691-6

Printed in the UK by
Bell & Bain Ltd., Glasgow

Links to third party websites are provided by Oxford in good faith and
for information only. Oxford disclaims any responsibility for the materials
contained in any third party website referenced in this work.

The manufacturer's authorised representative in the EU for product safety is Oxford University Press España S.A. of El Parque Empresarial San Fernando de Henares, Avenida de Castilla, 2 – 28830 Madrid (www.oup.es/en or product.safety@oup.com). OUP España S.A. also acts as importer into Spain of products made by the manufacturer.

Preface

Introducing *Inorganic Chemistry*
Our aim in the eighth edition of *Inorganic Chemistry* is to provide a comprehensive, fully updated, and contemporary introduction to the diverse and fascinating discipline of inorganic chemistry. Inorganic chemistry deals with the properties of all of the elements in the periodic table. Those classified as metallic range from the highly reactive sodium and barium to the noble metals, such as gold and platinum. The nonmetals include solids, liquids, and gases, and their properties encompass those of the aggressive, highly oxidizing fluorine and the unreactive gases such as helium. Although this variety and diversity are features of any study of inorganic chemistry, there are underlying patterns and trends which enrich and enhance our understanding of the subject. These trends in reactivity, structure, and properties of the elements and their compounds provide an insight into the landscape of the periodic table and provide the foundation on which to build a deeper understanding of the chemistry of the elements and their compounds.

Inorganic compounds vary from ionic solids to covalent compounds and metals. The text builds on simple bonding models, such as the classical electrostatic model of ionic solids, which should already be familiar from introductory chemistry courses. We proceed to rationalize and interpret the properties of many elements and compounds by using qualitative models that are based on quantum mechanics, including the interaction of atomic orbitals to form molecular orbitals and the band structures of solids.

Although qualitative models of bonding and reactivity clarify and systematize the subject, inorganic chemistry is essentially an experimental subject, and lies at the heart of many of the most important recent advances in chemistry. New, often unusual, inorganic compounds and materials are constantly being synthesized and identified. Modern inorganic syntheses continue to enrich the field with compounds that give us fresh perspectives on structure, bonding, and reactivity.

Inorganic chemistry has considerable impact on our everyday lives and on other scientific disciplines. The chemical industry is highly dependent on inorganic chemistry: it is essential for forming and improving the modern materials and compounds used as catalysts, energy storage materials, semiconductors, optoelectronics, superconductors, and advanced ceramics. The environmental, biological, and medical impacts of inorganic chemistry on our lives are enormous. Current topics in industrial, materials, biological, and environmental chemistry are highlighted throughout the early sections of the book to illustrate their importance and encourage the reader to explore further. These aspects of inorganic chemistry are then developed more thoroughly later in the text, particularly in Part 4, which explores the current applications of inorganic chemistry in more detail.

Part 4 also describes the horizons of inorganic chemistry focusing on exciting developments across catalysis, energy materials, and medicine with the aim of enthusing the reader about the vibrant and contemporary nature of the subject.

Throughout the book there is extensive cross-referencing so that the fundamental and descriptive chemistry of Parts 1–3 is directly linked to applications and research in Part 4. One aim of this is to demonstrate to the reader how their understanding of the core subject underlies the most contemporary aspects of inorganic chemistry.

What is new to this edition?
In this new edition we have continued to refine the presentation, organization, and visual representation. The book has been extensively revised, much has been rewritten, and there is some completely new material.

- A new four-part structure features a new Part 3, which focuses on complex solids and biological systems, and a refocused Part 4, which showcases a range of the current applications of inorganic chemistry, including sustainable chemistry, nanochemistry, catalysis, and inorganic chemistry in medicine.
- A new Chapter 10, 'Themes across the periodic table', includes a section on relativistic effects and the oxo wall concept.
- The relativistic concept is now introduced in Chapter 1.
- Coverage of the VSEPR model in Chapter 2, 'Molecular structure, bonding, and energetics', has been rewritten to provide greater clarity and more examples.
- The introduction to catalysis in Chapter 2 has been extended to include the Sabatier principle.
- The different mechanisms of acid–base catalysis—general, specific, and Lewis—are now defined in Chapter 5, 'Acids and bases'.
- A new section on electrocatalysis has been introduced in Chapter 6, 'Oxidation and reduction'.
- Coordination compounds are now introduced earlier in the book, as the subject of Chapter 7, 'An introduction to coordination compounds'.
- The discussion of amphoterism now appears in Chapter 9, 'Periodic trends'.

- The chemistry of the 'superheavy' 6d elements is now introduced in Chapter 20, 'The d-block elements'.
- The chemistry of the actinoids in Chapter 21, 'The f-block elements', has been expanded, including a new section on the heavy actinoids.
- Chapter 26, 'Biological inorganic chemistry', features new material on non-heme oxygenases, cobalt enzymes, and Fe(IV) chemistry.
- A dedicated Chapter, 27, on the increasingly important area of green chemistry framed in terms of the 12 principles.
- Chapters 28 and 29 focus in turn on the advances and derived applications of nanostructured inorganic materials and porous solids, including metal–organic frameworks.
- A new Chapter 30, 'Energy materials', that draws together research and developments on functional inorganic materials involved in energy generation and transmission, including photovoltaics, photocatalysts, superconductors, rechargeable battery components, and hydrogen storage materials.
- A revamped Chapter 31, 'Catalysis', brings all aspects of this important topic together in one place, with new and updated examples from across inorganic chemistry.
- An extensively reviewed and revised Chapter 32, 'Inorganic chemistry in medicine', includes new material on metal complexes as antiviral agents, and the mechanisms by which drugs become active once administered.

In addition, many of the approximately 2000 illustrations and the marginal structures have been redrawn in line with current accessibility standards, especially to support readers with colour-vision deficiency.

How is the content organized?

The topics in Part 1, 'Foundations', have been revised to make them more accessible to the reader, with additional qualitative explanation accompanying the more mathematical treatments. The material has been reorganized to allow a more coherent progression through the topics of symmetry and bonding and to introduce the important topic of catalysis early in the text.

Part 2, 'The elements and their compounds', has been thoroughly updated, building on the improvements made in earlier editions. Chapter 9 draws together periodic trends, while Chapter 10 considers cross-cutting themes that explain the richness of the patterns in the periodic table. Chapters 9 and 10 provide extensive cross-references heralding detailed treatment in the subsequent descriptive chapters. These chapters start with hydrogen and proceed across the periodic table, taking in the s-block metals and the diverse elements of the p block, before ending with extensive coverage of the d- and f-block elements.

Each of these chapters is organized in two sections: 'Essentials' describes the fundamental chemistry of the elements, and 'Detail' provides a more extensive account. The chemical properties of each group of elements and their compounds are further enriched with descriptions of current applications and recent advances made in inorganic chemistry. The patterns and trends that emerge are rationalized by drawing on the principles introduced in Part 1 and Chapters 9 and 10.

Part 3, 'Inorganic chemistry of complex solids and biological systems', extends the fundamental concepts and theories introduced in Part 1, with a focus on a deeper exploration of complex solid materials and biological inorganic chemistry. The concepts and ideas introduced here are central to many important advances and applications of inorganic chemistry that we discuss in Part 4.

Part 4, 'Expanding our horizons: advances and applications', takes the reader to the edge of knowledge in several areas of current research. These chapters explore specialized vibrant topics that are of importance to industry and biology, and include green chemistry, catalysis, materials chemistry including nanomaterials, energy materials, and an introduction to the different ways in which inorganic compounds are used in medicine.

Extensive cross-referencing is a feature of the book. This enables the reader to understand the importance of the subject fundamentals, covered in Parts 1 and 2, in the advanced and applied inorganic chemistry, covered in Parts 3 and 4.

About the authors

Mark Weller is an honorary Professor of Chemistry at Cardiff University. His research interests cover a wide range of synthetic and structural inorganic chemistry, including photovoltaic compounds, zeolites, battery materials, and specialist pigments; he is the author of over 350 primary literature publications in these fields. Mark has taught both inorganic chemistry and physical chemistry methods at undergraduate and postgraduate levels for more than 40 years. He is a co-author of OUP's *Characterisation Methods in Inorganic Chemistry* and an OUP Primer (23) on *Inorganic Materials Chemistry*.

Jonathan Rourke is Reader in Chemistry at Cardiff University and Honorary Professor of Chemistry at the University of Warwick. He received his PhD at the University of Sheffield on organometallic polymers and liquid crystals, followed by postdoctoral work in Canada and back in Britain. His initial independent research career began at Bristol University and then at Warwick, before he moved to Cardiff. Over the years, Dr Rourke has taught most aspects of inorganic chemistry, all the way from basic bonding, through symmetry analysis, to advanced transition-metal chemistry.

Fraser Armstrong is an Emeritus Research Fellow of St John's College, University of Oxford, where he was tutor in inorganic chemistry for 28 years. His interests span the fields of renewable energy, hydrogen, enzymology, biological inorganic chemistry, and electrocatalysis by enzymes. He was an Associate Professor at the University of California, Irvine, before joining the Department of Chemistry at Oxford in 1993. In 2008 he was elected as a Fellow of the Royal Society of London. His awards include the 2012 Davy Medal of the Royal Society.

Simon Lancaster is a Professor of Chemistry Education at the University of East Anglia. He won a National Teaching Fellowship and an RSC Higher Education Teaching Award in 2013. He was selected as one of RSC's 175 faces of diversity in 2016 and President of the Education Division from 2017 to 2020. As a passionate advocate for evidence-based teaching innovation, he has led the adoption of tailored active blended learning approaches and extended reality technologies at UEA.

Tina Overton was Director of the Leeds Institute for Teaching Excellence at the University of Leeds, with previous appointments at Monash University and the University of Hull. Tina has co-authored several textbooks on inorganic chemistry and skills development. She has been awarded the Royal Society of Chemistry's HE Teaching Award, Tertiary Education Award, and Nyholm Prize.

Acknowledgements

We have taken care to ensure that the text is free of errors. This is difficult in a rapidly changing field, where today's knowledge is soon replaced by tomorrow's. We thank all those colleagues who so willingly gave their time and expertise to a careful reading of a variety of draft chapters, and who shaped the content of this edition in other ways.

MTW would like to thank Hollie for her support, inspiration, and considerable forbearance during the writing of this text.

Chris Amodio, *University of Surrey*
Paul Anastas, *Yale University*
Frances H. Arnold, *California Institute of Technology*
Rolf Berger, *University of Uppsala, Sweden*
Harry Bitter, *University of Utrecht, The Netherlands*
Richard Blair, *University of Central Florida*
Andrew Bond, *University of Southern Denmark, Denmark*
Josef T. Boronsky, *Department of Chemistry and St John's College, Oxford*
Darren Bradshaw, *University of Liverpool*
Paul Brandt, *North Central College*
Karen Brewer, *Hamilton College*
Andrew Ballantyne, *University of Northampton*
Katherine Blundell, *University of Oxford*
George Britovsek, *Imperial College, London*
Scott Bunge, *Kent State University*
Jenny Burnham, *University of Sheffield*
David Cardin, *University of Reading*
Claire Carmalt, *University College London*
Carl Carrano, *San Diego State University*
Brian Chalmers, *University of St Andrews*
Aldrik Velders, *Wageningen University*
Neil Champness, *University of Nottingham*
Ferman Chavez, *Oakland University*
Ann Chippindale, *University of Reading*
Karl Coleman, *University of Durham*
Simon Collison, *University of Nottingham*
Bill Connick, *University of Cincinnati*
Stephen Daff, *University of Edinburgh*
Sandra Dann, *University of Loughborough*
Nancy Dervisi, *University of Cardiff*
Richard Douthwaite, *University of York*
Simon Duckett, *University of York*
Peter P. Edwards, *University of Oxford*
A.W. Ehlers, *Free University of Amsterdam, The Netherlands*
Oliver Einsle, *Albert-Ludwigs University of Freiburg*
Anders Eriksson, *University of Uppsala, Sweden*
Andrew Fogg, *University of Liverpool*
Dr Anna-Marie Fuller, *University of East Anglia*
Kyle Galloway, *University of Nottingham*
Margaret Geselbracht, *Reed College*
Martin Gill, *Swansea University*
Mario Henrique Gonzalez, *São Paulo State University*

Gregory Grant, *University of Tennessee*
Harry B. Gray, *California Institute of Technology*
Yurii Gun'ko, *Trinity College Dublin*
Simon Hall, *University of Bristol*
Justin Hargreaves, *University of Glasgow*
Bill Harrison, *University of Aberdeen*
Richard Henderson, *University of Newcastle*
Eva Hervia, *University of Strathclyde*
Michael S. Hill, *University of Bath*
Jan Philipp Hofmann, *Eindhoven University*
Brendan Howlin, *University of Surrey*
Carol Hua, *University of Melbourne*
Songping Huang, *Kent State University*
Carl Hultman, *Gannon University*
Stephanie Hurst, *Northern Arizona University*
Jon Iggo, *University of Liverpool*
S. Jackson, *University of Glasgow*
Guy Jameson, *University of Melbourne*
Michael Jensen, *Ohio University*
Pavel Karen, *University of Oslo, Norway*
Terry Kee, *University of Leeds*
Paul King, *Birbeck, University of London*
Rachael Kipp, *Suffolk University*
Caroline Kirk, *University of Loughborough*
Lars Kloo, *KTH Royal Institute of Technology, Sweden*
Randolph Kohn, *University of Bath*
Hanno Kossen, *Newcastle University*
Paul Lickiss, *Imperial College, London*
Sven Lindin, *University of Stockholm, Sweden*
Paul Loeffler, *Sam Houston State University*
Paul Low, *University of Durham*
Astrid Lund Ramstad, *The Norwegian University of Science and Technology, Norway*
Jason Lynam, *University of York*
Joel Mague, *Tulane University*
Mary F. Mahon, *University of Bath*
Francis Mair, *University of Manchester*
Mikhail Maliarik, *University of Uppsala, Sweden*
David E. Marx, *University of Scranton*
John McGrady, *University of Oxford*
Sankar Meenakshisundaram, *Cardiff University*
Roland Meier, *Friedrich-Alexander-Universität*
Katrina Miranda, *University of Arizona*
Grace Morgan, *University College Dublin*
Ebbe Nordlander, *University of Lund, Sweden*

Nick Norman, *University of Bristol*
Lars Öhrström, *Chalmers (Goteborg), Sweden*
Ivan Parkin, *University College London*
Dan Price, *University of Glasgow*
Paul Pringle, *University of Bristol*
T. B. Rauchfuss, *University of Illinois*
Jennifer Readman, *University of Central Lancashire*
Jan Reedijk, *University of Leiden, The Netherlands*
David Richens, *St Andrews University*
Denise Rooney, *National University of Ireland, Maynooth*
Julia Sarju, *University of York*
Graham Saunders, *Queens University Belfast*
Ian Shannon, *University of Birmingham*
P. Shiv Halasyamani, *University of Houston*
Stephen Skinner, *Imperial College, London*
Bob Slade, *University of Surrey*
Peter Slater, *University of Surrey*
LeGrande Slaughter, *Oklahoma State University*
Martin B. Smith, *University of Loughborough*
Sheila Smith, *University of Michigan*
Jake Soper, *Georgia Institute of Technology*
David M. Stanbury, *Auburn University*
Jonathan Steed, *University of Durham*
Hongzhe Sun, *Hong Kong University*
Gunnar Svensson, *University of Stockholm, Sweden*
Hernando A. Trujillo, *Wilkes University*
Andrei Verdernikov, *University of Maryland*
Ramon Vilar, *Imperial College, London*
Adelina Voutchkova, *George Washington University*
Keith Walters, *Northern Kentucky University*
Robert Wang, *Salem State College*
David Weatherburn, *University of Victoria, Wellington*
Adam Weller, *Laureate Academy, Hemel Hempstead*
Michael K. Whittlesey, *University of Bath*
Paul Wilson, *University of Bath*
Joseph A. Wright, *University of East Anglia*
Vivian Yam, *Hong Kong University*
Nigel A. Young, *University of Hull*
Jingdong Zhang, *Denmark Technical University*

Many of the figures in Chapter 26 and 32 were produced using PyMOL software; for more information, see W. L. DeLano, *The PyMOL Molecular Graphics System* (2002), DeLano Scientific, San Carlos, CA, USA.

We would also like to acknowledge the inspirational role and major contributions of Peter Atkins, whose early editions of *Inorganic Chemistry* formed the foundations of this text.

Learning with this book

Inorganic Chemistry provides numerous learning features to help you master this wide-ranging subject. In addition, it has been designed so that you can either work through the chapters chronologically, or dip in at an appropriate point in your studies.

Organizing the information

Key points
The key points outline the main take-home message(s) of the section that follows. These will help you to focus on the principal ideas being introduced in the text.

> **KEY POINT** Regions where wavefunctions pass through zero are called nodes. Inorganic chemists generally find it adequate to use visual representations of atomic orbitals rather than mathematical expressions. However, we need to be aware of the mathematical expressions that underlie these representations.
>
> Because the potential energy of an electron in the field of a

Context boxes
Context boxes demonstrate the diversity of inorganic chemistry and its wide-ranging applications to, for example, advanced materials, industrial processes, environmental chemistry, and everyday life.

> **BOX 1.3** Technetium—what is a synthetic element?
>
> A synthetic element is one that does not occur naturally on Earth but that can be artificially generated by nuclear reactions. The first synthetic element was technetium (Tc, $Z = 43$), named from the Greek word for 'artificial'. Its discovery—or more

Notes on good practice
In some areas of inorganic chemistry, the nomenclature commonly in use can be confusing or archaic. To address this, we have included brief 'Notes on good practice' to help you avoid making common mistakes.

> **A NOTE ON GOOD PRACTICE**
>
> An electronvolt is the amount of kinetic energy gained by an electron as it accelerates through a potential of one volt. It is a useful, but non-SI, unit. In chemistry, kinetic energy gained by a mole of electrons passing through a potential of one volt is 96.485 kJ mol^{-1}. The approximation 1 eV ≈ 100 kJ mol^{-1} is worth

Further reading
Each chapter lists sources where further information can be found. We have tried to ensure that these sources are easily available and have indicated the type of information each one provides.

> **FURTHER READING**
>
> E. Scerri, *The Periodic Table: Its Story and its Significance*. 2nd ed. Oxford University Press (2019). A lively account of the development of the periodic system from early discoveries of elements to the impact of quantum theory.

Resource section
You will find a comprehensive collection of resources, including an extensive data section and information relating to group theory and spectroscopy, at the end of the book.

> **Resource section 1**
> **Selected ionic radii**
>
> Ionic radii are given (in picometres, pm) for the most common oxidation states and coordination geometries. The coordination number is given in parentheses: (4) refers to tetrahedral and (4SP) refers to square planar. All d-block species are low-spin unless labelled with †, in which case values for high-spin are quoted. Most data are taken from R. D. Shannon, *Acta Crystallogr.*, 1976, **A32**, 751, where values for other coordination geometries can be found. Where Shannon values are not available, Pauling ionic radii are quoted and are indicated by*.

Visualizing inorganic chemistry

Interactive 3D figures
Interactive, three-dimensional versions of many of the figures and structures that appear throughout this book have been developed by Professor Nick Greeves and can be accessed by visiting www.chemtube3d.com.

You can use the Search function on ChemTube3D.com (using a search based on the figure caption) or the ChemTube3D menus to locate the required pages.

Problem solving

Brief illustrations
'A brief illustration' shows you how to use equations or concepts that have just been introduced in the main text and will help you to understand how to manipulate data correctly.

> **A BRIEF ILLUSTRATION**
>
> Difluorine, F_2, has the configuration $1\sigma_g^2 \sigma_u^2 2\sigma_g^2 1\pi_u^4 1\pi_g^4$ and, because $1\sigma_g$, $1\pi_u$, and $2\sigma_g$ orbitals are bonding but $1\sigma_u$ and $1\pi_g$ are antibonding, $b = \frac{1}{2}(2+2+4-2-4) = 1$. The bond order of F_2 is 1, which is consistent with the structure F–F and the conventional description of the molecule as having a single

Worked examples and self-tests
Numerous worked 'Examples' provide a more detailed illustration of the application of the material being discussed. Each one demonstrates an important aspect of the topic under discussion or provides practice with calculations and problems. Each 'Example' is followed by a 'Self-test' designed to help you monitor your progress.

> **EXAMPLE 2.5 Accounting for the structure of a heteronuclear diatomic molecule**
>
> The halogens form compounds among themselves. One of these 'interhalogen' compounds is iodine monochloride, ICl, in which the order of orbitals is 1σ, 2σ, 1π, 3σ, 2π, 4σ (from calculation). What is the ground-state electron configuration of ICl?

Exercises
There are many brief 'Exercises' at the end of each chapter, which can be used to check your understanding and gain experience and practice in tasks such as balancing equations, predicting and drawing structures, and manipulating data.

> **EXERCISES**
>
> 2.1 Draw feasible Lewis structures for (a) NO^+, (b) ClO^-, (c) H_2O_2, (d) CCl_4, (e) HSO_3^-.
>
> 2.2 Draw the resonance structures for CO_3^{2-}.
>
> 2.3 What shapes would you expect for the species (a) H_2Se, (b) BF_4^-, (c) NH_4^+?

Tutorial problems
The 'Tutorial problems' are more demanding in content and style than the 'Exercises' and are often based on a research paper or other additional source of information. 'Tutorial problems' generally require a discursive response, and there may not be only a single correct answer. They may be used as essay-type questions or for classroom discussion.

> **TUTORIAL PROBLEMS**
>
> 2.1 In valence bond theory, hypervalence is usually explained in terms of d-orbital participation in bonding. In the paper 'On the role of orbital hybridisation' (*J. Chem. Educ.*, 2007, **84**, 783) the author argues that this is not the case. Give a concise summary of

Teaching materials for registered adopters of the text

Lecturer resources are available only to registered adopters of the textbook. To register, simply visit **www.oup.com/he/weller8e/** and follow the appropriate links.

Figures and tables from the book

The artwork and tables from the book are available in ready-to-download format. These can be used for educational purposes without charge (but not for commercial purposes without specific permission).

Solutions to self-tests and end-of-chapter exercises

While the final answers to all self-tests and end-of-chapter exercises are available to students, the full solutions, prepared by Dr Alen Hadzovic, University of Toronto, are available to lecturers only.

Test bank

Prepared by Dr Alen Hadzovic, University of Toronto, the test bank contains multiple-choice questions for each chapter, with formative feedback provided for each question.

Summary of contents

PART 1 Foundations — 1

1. Atomic structure — 3
2. Molecular structure, bonding, and energetics — 35
3. Molecular symmetry — 72
4. The structures and energetics of simple solids — 100
5. Acids and bases — 145
6. Oxidation and reduction — 185
7. An introduction to coordination compounds — 217
8. Physical techniques in inorganic chemistry — 262

PART 2 The elements and their compounds — 307

9. Periodic trends — 309
10. Themes across the periodic table — 325
11. Hydrogen — 338
12. The Group 1 elements — 362
13. The Group 2 elements — 385
14. The Group 13 elements — 409
15. The Group 14 elements — 441
16. The Group 15 elements — 471
17. The Group 16 elements — 501
18. The Group 17 elements — 525
19. The Group 18 elements — 550
20. The d-block elements — 562
21. The f-block elements — 595
22. d- and f-Metal complexes: electronic and magnetic properties — 621
23. Coordination chemistry: reactions of complexes — 646
24. d-Metal organometallic chemistry — 675

PART 3 Inorganic chemistry of complex solids and biological systems — 719

25. Advanced solid-state chemistry: structures, compounds, and electronic properties — 721
26. Biological inorganic chemistry — 776

PART 4 Expanding our horizons: advances and applications — 845

27. Green chemistry — 847
28. Nanoparticle inorganic chemistry — 859
29. Framework and porous materials — 882
30. Energy materials — 899
31. Catalysis — 927
32. Inorganic chemistry in medicine — 955

Resource section 1 Selected ionic radii — 975
Resource section 2 Electronic properties of the elements — 978

Resource section 3	Standard potentials	980
Resource section 4	Character tables	993
Resource section 5	Symmetry-adapted orbitals	998
Resource section 6	Tanabe–Sugano diagrams	1003
Resource section 7	Useful data: conversion factors, fundamental constants, and other information	1006
Index		*1009*

Detailed contents

Glossary of chemical abbreviations	xxvi

PART 1 Foundations 1

1 Atomic structure 3
The structures of hydrogenic atoms 7
- 1.1 Spectroscopic information 7
- 1.2 Some principles of quantum mechanics 8
- 1.3 Atomic orbitals 9

Many-electron atoms 15
- 1.4 Penetration and shielding 15
- 1.5 The building-up principle 18
- 1.6 The classification of the elements 20
- 1.7 Atomic properties 23

FURTHER READING 33
EXERCISES 33
TUTORIAL PROBLEMS 34

2 Molecular structure, bonding, and energetics 35
Lewis structures 35
- 2.1 The octet rule 35
- 2.2 Resonance 37
- 2.3 The VSEPR model 37

Valence bond theory 42
- 2.4 The hydrogen molecule 42
- 2.5 Homonuclear diatomic molecules 43
- 2.6 Polyatomic molecules 44

Molecular orbital theory 46
- 2.7 An introduction to the theory 46
- 2.8 Homonuclear diatomic molecules 49
- 2.9 Heteronuclear diatomic molecules 53
- 2.10 Bond properties 56
- 2.11 Bands formed from extended overlapping atomic orbitals 58

Bond properties, reaction enthalpies, and kinetics 62
- 2.12 Bond length 62
- 2.13 Bond strength and reaction enthalpies 63
- 2.14 Electronegativity and bond enthalpy 64
- 2.15 An introduction to catalysis 66

FURTHER READING 69
EXERCISES 70
TUTORIAL PROBLEMS 71

3 Molecular symmetry 72
An introduction to symmetry analysis 72
- 3.1 Symmetry operations, elements, and point groups 72
- 3.2 Character tables 79

Applications of symmetry 81
- 3.3 Polar molecules 81
- 3.4 Chiral molecules 81
- 3.5 Molecular vibrations 82

The symmetries of molecular orbitals 86
- 3.6 Symmetry-adapted linear combinations 86
- 3.7 The construction of molecular orbitals 88
- 3.8 The vibrational analogy 90

Representations 91
- 3.9 The reduction of a representation 91
- 3.10 Projection operators 92
- 3.11 Polyatomic molecules 93

FURTHER READING 98
EXERCISES 98
TUTORIAL PROBLEMS 99

4 The structures and energetics of simple solids 100
The description of the structures of solids 101
- 4.1 Unit cells and the description of crystal structures 101
- 4.2 The close packing of spheres 104
- 4.3 Holes in close-packed structures 108

The structures of metals and simple alloys 111
- 4.4 Polytypism 112
- 4.5 Nonclose-packed structures 112
- 4.6 Polymorphism of metals 113
- 4.7 Metallic radii of metals 114
- 4.8 The structures of simple alloys and interstitials 115

Ionic solids 117
- 4.9 Characteristic structures of ionic solids 117
- 4.10 The rationalization of structures 127

The energetics of ionic bonding 130
- 4.11 Lattice enthalpy and the Born–Haber cycle 131
- 4.12 The calculation of lattice enthalpies 132
- 4.13 Comparison of experimental and theoretical values 134
- 4.14 The Kapustinskii equation 136
- 4.15 Consequences of lattice enthalpies 137

Further information: the Born–Mayer equation — 141
FURTHER READING — 141
EXERCISES — 142
TUTORIAL PROBLEMS — 144

5 Acids and bases — 145

Brønsted acidity — 146
- 5.1 Proton transfer equilibria in water — 147

Characteristics of oxygen-containing Brønsted acids — 153
- 5.2 Periodic trends in aqua acid strength — 154
- 5.3 Simple oxoacids — 155
- 5.4 Anhydrous oxides — 157
- 5.5 Polyoxo compound formation — 158

Lewis acidity — 159
- 5.6 Identifying Lewis acids and bases — 160
- 5.7 The fundamental types of reaction — 161
- 5.8 Properties of main group Lewis acids — 162
- 5.9 Hydrogen bonding — 166
- 5.10 Factors governing interactions between Lewis acids and bases — 169
- 5.11 Thermodynamic Lewis acidity parameters — 171

Nonaqueous solvents — 173
- 5.12 Solvent levelling — 173
- 5.13 The Hammett acidity function and its application to strong, concentrated acids — 174
- 5.14 The solvent system definition of acids and bases — 175
- 5.15 Solvents as acids and bases — 175

Applications of acid–base chemistry — 178
- 5.16 Acid and base catalysis — 178
- 5.17 Superacids and superbases — 179
- 5.18 Heterogeneous acid–base reactions — 180

FURTHER READING — 181
EXERCISES — 182
TUTORIAL PROBLEMS — 184

6 Oxidation and reduction — 185

Reduction potentials — 186
- 6.1 Redox half-reactions — 186
- 6.2 Standard potentials and spontaneity — 187
- 6.3 Trends in standard potentials — 191
- 6.4 The electrochemical series — 191
- 6.5 The Nernst equation — 192

Redox stability — 193
- 6.6 The influence of pH — 194
- 6.7 Reactions with water — 194
- 6.8 Oxidation by atmospheric oxygen — 196
- 6.9 Disproportionation and comproportionation — 196
- 6.10 The influence of complexation — 197
- 6.11 The relation between solubility and standard potentials — 198

Useful presentation of potential data — 199
- 6.12 Latimer diagrams — 199
- 6.13 Frost diagrams — 200
- 6.14 Proton-coupled electron transfer: Pourbaix diagrams — 204

Applications — 205
- 6.15 Environmental chemistry: natural waters — 205
- 6.16 Electrocatalysis — 206
- 6.17 Extraction of the elements by chemical reduction — 207
- 6.18 Extraction of the elements by chemical oxidation — 211
- 6.19 Electrochemical extraction — 212

FURTHER READING — 213
EXERCISES — 213
TUTORIAL PROBLEMS — 216

7 An introduction to coordination compounds — 217

The language of coordination chemistry — 218
- 7.1 Representative ligands — 219
- 7.2 Nomenclature — 222

Constitution and geometry — 223
- 7.3 Low coordination numbers — 223
- 7.4 Intermediate coordination numbers — 224
- 7.5 Higher coordination numbers — 227
- 7.6 Polymetallic complexes — 228

Isomerism and chirality — 229
- 7.7 Square-planar complexes — 230
- 7.8 Tetrahedral complexes — 232
- 7.9 Trigonal-bipyramidal and square-pyramidal complexes — 232
- 7.10 Octahedral complexes — 233
- 7.11 Ligand chirality — 237

The thermodynamics of complex formation — 239
- 7.12 Formation constants — 239
- 7.13 Trends in successive formation constants — 240
- 7.14 The chelate and macrocyclic effects — 241
- 7.15 Steric effects and electron delocalization — 242

Electronic structure — 244
- 7.16 Crystal-field theory — 244
- 7.17 Ligand-field theory — 254

FURTHER READING — 258
EXERCISES — 258
TUTORIAL PROBLEMS — 260

8 Physical techniques in inorganic chemistry — 262

Diffraction methods — 263
- 8.1 X-ray diffraction — 263
- 8.2 Neutron diffraction — 267

Absorption and emission spectroscopies — 269
- 8.3 Ultraviolet–visible spectroscopy — 270
- 8.4 Fluorescence or emission spectroscopy — 274
- 8.5 Infrared and Raman spectroscopy — 276

Resonance techniques — 279
- 8.6 Nuclear magnetic resonance — 279
- 8.7 Electron paramagnetic resonance — 285
- 8.8 Mössbauer spectroscopy — 287

Ionization-based techniques — 289
- 8.9 Photoelectron spectroscopy — 289
- 8.10 X-ray absorption spectroscopy — 291
- 8.11 Mass spectrometry — 292

Chemical analysis — 294
- 8.12 Atomic absorption spectroscopy — 294
- 8.13 CHN analysis — 295
- 8.14 X-ray fluorescence elemental analysis — 296
- 8.15 Thermal analysis — 296

Magnetometry and magnetic susceptibility — 298
Electrochemical techniques — 299
Microscopy — 302
- 8.16 Scanning probe microscopy — 302
- 8.17 Electron microscopy — 302

FURTHER READING — 303
EXERCISES — 304
TUTORIAL PROBLEMS — 306

PART 2 The elements and their compounds — 307

9 Periodic trends — 309

Trends in distribution — 309
- 9.1 Occurrence of the elements — 309

Trends in properties — 310
- 9.2 Atomic parameters — 310
- 9.3 Metallic character — 314
- 9.4 Oxidation states — 315

Trends in compounds — 318
- 9.5 Amphoterism — 318
- 9.6 Coordination numbers — 319
- 9.7 Bond enthalpy trends — 319

Where trends collide — 320
- 9.8 Wider aspects of periodicity — 320

FURTHER READING — 323
EXERCISES — 323
TUTORIAL PROBLEMS — 324

10 Themes across the periodic table — 325

Causes — 325
- 10.1 Presence of unpaired electrons — 325
- 10.2 Anomalous nature of the first member of each group — 326
- 10.3 Diagonal relationships — 328
- 10.4 The alternation effect — 329
- 10.5 The lanthanoid contraction — 330
- 10.6 Relativistic effects — 331

Consequences — 332
- 10.7 The inert pair effect — 332
- 10.8 The oxo wall — 334
- 10.9 Group n versus Group $n + 10$ — 336

FURTHER READING — 337
EXERCISES — 337
TUTORIAL PROBLEM — 337

11 Hydrogen — 338

Part A: The essentials — 338
- 11.1 The element — 340
- 11.2 Simple compounds — 342

Part B: The detail — 344
- 11.3 Nuclear properties — 344
- 11.4 Production of dihydrogen — 345
- 11.5 Reactions of dihydrogen — 348
- 11.6 Compounds of hydrogen — 351
- 11.7 General methods for synthesis of binary hydrogen compounds — 358

FURTHER READING — 359
EXERCISES — 360
TUTORIAL PROBLEMS — 361

12 The Group 1 elements — 362

Part A: The essentials — 362
- 12.1 The elements — 363
- 12.2 Simple compounds — 365
- 12.3 The atypical properties of lithium — 366

Part B: The detail — 367
- 12.4 Occurrence and extraction — 367
- 12.5 Uses of the elements and their compounds — 369
- 12.6 Hydrides — 372
- 12.7 Halides — 372
- 12.8 Oxides and related compounds — 374
- 12.9 Sulfides, selenides, and tellurides — 375

12.10	Hydroxides	375
12.11	Compounds of oxoacids	376
12.12	Nitrides and carbides	377
12.13	Solubility and hydration	378
12.14	Solutions in liquid ammonia	379
12.15	Zintl phases containing alkali metals	379
12.16	Coordination compounds	380
12.17	Organometallic compounds	381

FURTHER READING 383
EXERCISES 383
TUTORIAL PROBLEMS 383

13 The Group 2 elements — 385

Part A: The essentials — 386

13.1	The elements	386
13.2	Simple compounds	387
13.3	The anomalous properties of beryllium	389

Part B: The detail — 389

13.4	Occurrence and extraction	389
13.5	Uses of the elements and their compounds	390
13.6	Hydrides	392
13.7	Halides	393
13.8	Oxides, sulfides, and hydroxides	395
13.9	Nitrides and carbides	398
13.10	Salts of oxoacids	398
13.11	Solubility, hydration, and beryllates	402
13.12	Coordination compounds	403
13.13	Organometallic compounds	404
13.14	Lower oxidation state Group 2 compounds	406

FURTHER READING 406
EXERCISES 407
TUTORIAL PROBLEMS 407

14 The Group 13 elements — 409

Part A: The essentials — 410

14.1	The elements	410
14.2	Compounds	411
14.3	Boron clusters and borides	414

Part B: The detail — 415

14.4	Occurrence and recovery	415
14.5	Uses of the elements and their compounds	416
14.6	Simple hydrides of boron	416
14.7	Boron trihalides	420
14.8	Boron–oxygen compounds	422
14.9	Compounds of boron with nitrogen	423
14.10	Metal borides	425
14.11	Higher boranes and borohydrides	426
14.12	Metalaboranes and carboranes	431
14.13	The hydrides of aluminium, gallium, indium, and thallium	433
14.14	Trihalides of aluminium, gallium, indium, and thallium	434
14.15	Low oxidation state halides of aluminium, gallium, indium, and thallium	434
14.16	Oxo compounds of aluminium, gallium, indium, and thallium	435
14.17	Sulfides of gallium, indium, and thallium	436
14.18	Compounds with Group 15 elements	436
14.19	Zintl phases	437
14.20	Organometallic compounds	437

FURTHER READING 439
EXERCISES 439
TUTORIAL PROBLEMS 440

15 The Group 14 elements — 441

Part A: The essentials — 441

15.1	The elements	441
15.2	Simple compounds	444
15.3	Extended silicon–oxygen compounds and their derivatives	446

Part B: The detail — 448

15.4	Occurrence and recovery	448
15.5	Diamond and graphite	448
15.6	Other forms of carbon	449
15.7	Hydrides	451
15.8	Compounds with halogens	453
15.9	Compounds of carbon with oxygen and sulfur	455
15.10	Compounds of silicon with oxygen	458
15.11	Oxides of germanium, tin, and lead	459
15.12	Compounds with nitrogen	459
15.13	Carbides	460
15.14	Silicides	463
15.15	Extended silicon–oxygen materials	463
15.16	Compounds of heavier Group 14 elements with organic substituents	465

FURTHER READING 468
EXERCISES 468
TUTORIAL PROBLEMS 469

16 The Group 15 elements — 471

Part A: The essentials — 472

16.1	The elements	472
16.2	Simple compounds	473
16.3	Oxides and oxanions of nitrogen	476

Part B: The detail	**476**
16.4 Occurrence and recovery	476
16.5 Uses	477
16.6 Nitrogen activation	479
16.7 Nitrides and azides	480
16.8 Phosphides	481
16.9 Arsenides, antimonides, and bismuthides	482
16.10 Hydrides	482
16.11 Halides	485
16.12 Oxohalides	486
16.13 Oxides and oxoanions of nitrogen	487
16.14 Oxides of phosphorus, arsenic, antimony, and bismuth	492
16.15 Oxoanions of phosphorus, arsenic, antimony, and bismuth	492
16.16 Condensed phosphates	493
16.17 Phosphazenes	494
16.18 Organometallic compounds of arsenic, antimony, and bismuth	495
FURTHER READING	498
EXERCISES	498
TUTORIAL PROBLEMS	500

17 The Group 16 elements — 501

Part A: The essentials	**502**
17.1 The elements	502
17.2 Simple compounds	503
Part B: The detail	**505**
17.3 Oxygen	505
17.4 Sulfur	508
17.5 Selenium, tellurium, and polonium	509
17.6 Compounds formed with hydrogen	509
17.7 Halides	513
17.8 Metal oxides	513
17.9 Metal sulfides, selenides, tellurides, and polonides	514
17.10 Oxides of sulfur, selenium, and tellurium	515
17.11 Oxoacids of sulfur	517
17.12 Polyanions of sulfur, selenium, and tellurium	520
17.13 Polycations of sulfur, selenium, and tellurium	521
17.14 Sulfur–nitrogen compounds	521
FURTHER READING	523
EXERCISES	523
TUTORIAL PROBLEMS	524

18 The Group 17 elements — 525

Part A: The essentials	**526**
18.1 The elements	526
18.2 Simple compounds	527
18.3 The interhalogens	528
Part B: The detail	**530**
18.4 Occurrence, recovery, and uses	530
18.5 Molecular structure and properties	532
18.6 Reactivity trends	534
18.7 Pseudohalogens	535
18.8 Special properties of fluorine compounds	536
18.9 Structural features	536
18.10 The interhalogens	537
18.11 Halogen oxides	541
18.12 Oxoacids and oxoanions	542
18.13 Thermodynamic aspects of oxoanion redox reactions	542
18.14 Trends in rates of oxoanion redox reactions	544
18.15 Redox properties of individual oxidation states	545
18.16 Fluorocarbons	546
FURTHER READING	547
EXERCISES	547
TUTORIAL PROBLEMS	548

19 The Group 18 elements — 550

Part A: The essentials	**551**
19.1 The elements	551
19.2 Simple compounds	551
Part B: The detail	**552**
19.3 Occurrence and recovery	552
19.4 Uses	553
19.5 Synthesis and structure of xenon fluorides	554
19.6 Reactions of xenon fluorides	555
19.7 Xenon-oxygen compounds	556
19.8 Insertion compounds	557
19.9 Organoxenon compounds	558
19.10 Coordination compounds	558
19.11 Other compounds of noble gases	559
FURTHER READING	559
EXERCISES	560
TUTORIAL PROBLEMS	560

20 The d-block elements — 562

Part A: The essentials	**563**
20.1 Occurrence and recovery	563
20.2 Chemical and physical properties	563
Part B: The detail	**566**
20.3 Group 3: scandium, yttrium, and lanthanum	566
20.4 Group 4: titanium, zirconium, and hafnium	568
20.5 Group 5: vanadium, niobium, and tantalum	570

20.6	Group 6: chromium, molybdenum, and tungsten	573
20.7	Group 7: manganese, technetium, and rhenium	579
20.8	Group 8: iron, ruthenium, and osmium	581
20.9	Group 9: cobalt, rhodium, and iridium	583
20.10	Group 10: nickel, palladium, and platinum	584
20.11	Group 11: copper, silver, and gold	586
20.12	Group 12: zinc, cadmium, and mercury	588
20.13	The 6d elements	592

FURTHER READING 593
EXERCISES 593
TUTORIAL PROBLEMS 593

21 The f-block elements — 595

The elements 596
- 21.1 The valence orbitals — 596
- 21.2 Occurrence and recovery — 597
- 21.3 Physical properties and applications — 598

Lanthanoid chemistry 599
- 21.4 General trends — 599
- 21.5 Binary ionic compounds — 603
- 21.6 Ternary and complex oxides — 605
- 21.7 Coordination compounds — 606
- 21.8 Organometallic compounds — 608

Actinoid chemistry 611
- 21.9 General trends — 612
- 21.10 Thorium and uranium — 615
- 21.11 Neptunium, plutonium, and americium — 617
- 21.12 The heavy actinoids — 618

FURTHER READING 619
EXERCISES 619
TUTORIAL PROBLEMS 620

22 d- and f-Metal complexes: electronic and magnetic properties — 621

Electronic spectra 621
- 22.1 Electronic spectra of atoms — 622
- 22.2 Electronic spectra of d-metal complexes — 626
- 22.3 Charge-transfer bands — 631
- 22.4 Selection rules and intensities — 633
- 22.5 Luminescence — 635
- 22.6 Electronic spectra of the lanthanoids — 636
- 22.7 Electronic spectra of the actinoids — 639

Magnetism 640
- 22.8 Cooperative magnetism — 640
- 22.9 Spin-crossover complexes — 642
- 22.10 Magnetic properties of the lanthanoids — 643

FURTHER READING 643
EXERCISES 644
TUTORIAL PROBLEMS 645

23 Coordination chemistry: reactions of complexes — 646

Ligand substitution reactions 647
- 23.1 Rates of ligand substitution — 647
- 23.2 The classification of mechanisms — 648

Ligand substitution in square-planar complexes 652
- 23.3 The nucleophilicity of the entering group — 652
- 23.4 The shape of the transition state — 653

Ligand substitution in octahedral complexes 656
- 23.5 Rate laws and their interpretation — 656
- 23.6 The activation of octahedral complexes — 657
- 23.7 Base hydrolysis — 661
- 23.8 Stereochemistry — 661
- 23.9 Isomerization reactions — 662

Redox reactions 663
- 23.10 The classification of redox reactions — 663
- 23.11 The inner-sphere mechanism — 664
- 23.12 The outer-sphere mechanism — 666

Photochemical reactions 670
- 23.13 Prompt and delayed reactions — 670
- 23.14 d–d and charge-transfer reactions — 670
- 23.15 Transitions in metal–metal bonded systems — 672

FURTHER READING 672
EXERCISES 673
TUTORIAL PROBLEMS 674

24 d-Metal organometallic chemistry — 675

Bonding 677
- 24.1 Stable electron configurations — 677
- 24.2 Electron-count preference — 678
- 24.3 Electron counting and oxidation states — 679
- 24.4 Nomenclature — 681

Ligands 682
- 24.5 Carbon monoxide — 682
- 24.6 Phosphines — 684
- 24.7 Hydrides and dihydrogen complexes — 685
- 24.8 η^1-Alkyl, -alkenyl, -alkynyl, and -aryl ligands — 686
- 24.9 η^2-Alkene and -alkyne ligands — 687
- 24.10 Nonconjugated diene and polyene ligands — 688
- 24.11 Butadiene, cyclobutadiene, and cyclooctatetraene — 688
- 24.12 Benzene and other arenes — 690

24.13	The allyl ligand	691
24.14	Cyclopentadiene and cycloheptatriene	692
24.15	Carbenes	694
24.16	Alkanes, agostic hydrogens, and noble gases	695
24.17	Dinitrogen and nitrogen monoxide	696

Compounds 696

24.18	d-Block carbonyls	696
24.19	Metallocenes	702
24.20	Metal–metal bonding and metal clusters	706

Reactions 710

24.21	Ligand substitution	710
24.22	Oxidative addition and reductive elimination	712
24.23	σ-Bond metathesis	713
24.24	1,1-Migratory insertion reactions	714
24.25	1,2-Insertions and β-hydride elimination	715
24.26	α-, γ-, and δ-Hydride eliminations and cyclometallations	715

FURTHER READING 716
EXERCISES 716
TUTORIAL PROBLEMS 717

PART 3 Inorganic chemistry of complex solids and biological systems 719

25 Advanced solid-state chemistry: structures, compounds, and electronic properties 721

Defects and nonstoichiometry 722

25.1	The origins and types of defects	722
25.2	Nonstoichiometric compounds and solid solutions	726
25.3	Complex alloys and interstitials	728
25.4	Extended defects	731

The electronic structures of solids 733

25.5	The conductivities of inorganic solids	733
25.6	Semiconductors: Si analogues and extrinsic semiconductors	734

Ion transport in solids 737

25.7	Atom and ion diffusion and conduction	737
25.8	Solid electrolytes and electrodes	738

Synthesis of materials 743

25.9	The formation of bulk materials	743

Metal oxides, nitrides, and fluorides 747

25.10	Monoxides of the 3d metals	747
25.11	Higher oxides and complex oxides	749
25.12	Oxide glasses	755
25.13	Nitrides, fluorides, and mixed-anion phases	757

Sulfides, intercalation compounds, and metal-rich phases 759

25.14	Metal sulfides, selenides, and tellurides	759
25.15	Metal hydrides and solids for hydrogen storage	761

Framework structures 763

25.16	Structures based on tetrahedral oxoanions	764
25.17	Structures based on linked polyhedral metal centres	765

Molecular inorganic materials and fullerides 769

25.18	Fullerides	769
25.19	Molecular inorganic materials chemistry	770

FURTHER READING 773
EXERCISES 774
TUTORIAL PROBLEMS 775

26 Biological inorganic chemistry 776

The organization of cells 777

26.1	The physical structure of cells	777
26.2	The inorganic composition of living organisms	777
26.3	Biological metal-coordination sites	780

Metal ions in transport and communication 785

26.4	Sodium and potassium transport	785
26.5	Calcium signalling proteins	787
26.6	Selective transport and storage of iron	789
26.7	Oxygen transport and storage	791
26.8	Electron transfer	795

Catalytic processes 801

26.9	Acid–base catalysis	801
26.10	Enzymes dealing with H_2O_2 and O_2	808
26.11	Enzymes dealing with radical rearrangements and alkyl group transfer	821
26.12	Oxygen atom transfer by molybdenum and tungsten enzymes	825
26.13	Hydrogenases, enzymes that activate H_2	827
26.14	The nitrogen cycle	828

Metals in gene regulation 832

26.15	Transcription factors and the role of Zn	832
26.16	Iron proteins as sensors	833
26.17	Proteins that sense Cu and Zn levels	836
26.18	Biomineralization	837

Perspectives 838

26.19	The contributions of individual elements	838
26.20	Future directions	840

FURTHER READING 843
EXERCISES 844
TUTORIAL PROBLEMS 844

PART 4 Expanding our horizons: advances and applications — 845

27 Green chemistry — 847
Twelve principles — 848
- 27.1 Prevention — 848
- 27.2 Atom economy — 849
- 27.3 Less hazardous chemical species — 850
- 27.4 Designing safer chemicals — 851
- 27.5 Safer solvents and auxiliaries — 851
- 27.6 Design for energy efficiency — 853
- 27.7 Use of renewable feedstocks — 854
- 27.8 Minimize derivatives — 854
- 27.9 Catalysis — 855
- 27.10 Design for degradation — 855
- 27.11 Real-time analysis for pollution prevention — 855
- 27.12 Inherently safer chemistry for accident prevention — 856

FURTHER READING — 857
EXERCISES — 857
TUTORIAL PROBLEMS — 857

28 Nanoparticle inorganic chemistry — 859
Introduction to nanomaterials and their synthesis — 860
- 28.1 Nanomaterial terminology and history — 860
- 28.2 Solution-based synthesis of nanoparticles — 861
- 28.3 Vapour-phase synthesis of nanoparticles via solutions or solids — 863
- 28.4 Templated synthesis of nanomaterials using frameworks, supports, and substrates — 864
- 28.5 Characterization and formation of nanomaterials using microscopy — 866

Nanostructures and mesoporous materials: properties and applications — 866
- 28.6 Zero-dimensional (0D) control, nanoparticles of carbon, metals, metal oxides, and metal chalcogenides — 866
- 28.7 Special optical properties of 0D nanoparticles and quantum dots — 868
- 28.8 One-dimensional control: carbon nanotubes and inorganic nanowires — 870
- 28.9 Two-dimensional control: graphene, quantum wells, and solid-state superlattices — 873
- 28.10 Three-dimensional control of nanostructural units: mesoporous materials and composites — 876

FURTHER READING — 880
EXERCISES — 881
TUTORIAL PROBLEMS — 881

29 Framework and porous materials — 882
Zeolites — 882
- 29.1 Advances in zeolite design and synthesis — 882
- 29.2 Applications of zeolites — 885

Zeotypes and metal–organic frameworks — 889
- 29.3 Inorganic structures based on linked polyhedral metal centres — 889
- 29.4 Metal–organic frameworks — 891

FURTHER READING — 897
EXERCISES — 898
TUTORIAL PROBLEMS — 898

30 Energy materials — 899
Energy harvesting, conversion, optical and magnetic materials — 899
- 30.1 Pigments, coatings, and phosphors — 899
- 30.2 Photovoltaics and solar cell materials — 905
- 30.3 Photocatalysts and water-splitting — 908
- 30.4 Thermal solar farms; piezoelectric and thermoelectric energy materials — 910
- 30.5 Fuel cell materials — 912
- 30.6 Strong permanent magnets and magnetoresistive compounds — 913

Energy transmission and storage — 915
- 30.7 Superconductors — 916
- 30.8 Hydrogen storage — 919
- 30.9 Rechargeable battery materials — 921

FURTHER READING — 925
EXERCISES — 925
TUTORIAL PROBLEMS — 926

31 Catalysis — 927
Homogeneous catalysis — 928
- 31.1 Alkene metathesis — 929
- 31.2 Hydrogenation of alkenes — 930
- 31.3 Hydroformylation — 931
- 31.4 Wacker oxidation of alkenes — 933
- 31.5 Asymmetric oxidations — 934
- 31.6 Palladium-catalysed C–C bond-forming reactions — 934
- 31.7 Methanol carbonylation: ethanoic acid synthesis — 936

Heterogeneous and nanoparticle catalysis — 936
- 31.8 The nature of heterogeneous catalysts — 937
- 31.9 Ammonia synthesis — 940
- 31.10 Sulfur dioxide oxidation — 941
- 31.11 Fischer–Tropsch synthesis — 942

31.12	Reactions involving heterogeneous nanoparticle catalysts	942
31.13	Zeolites and microporous structures in heterogeneous catalysis	943

Heterogenized homogeneous and hybrid catalysis 946

31.14	Alkene oligomerization and polymerization	946
31.15	Tethered catalysts	950
31.16	Biphasic systems	951

FURTHER READING 952
EXERCISES 952
TUTORIAL PROBLEMS 953

32 Inorganic chemistry in medicine 955

The chemistry of elements in medicine 955

32.1	Inorganic compounds in cancer treatment	958
32.2	Anti-arthritis drugs	963
32.3	Bismuth pharmaceuticals	964
32.4	Lithium in the treatment of bipolar disorders	966
32.5	Involvement of metal complexes in the cause and treatment of malaria	966
32.6	Metal complexes as antiviral agents	966
32.7	Metal drugs that slowly release CO: an agent against post-operative stress	968
32.8	Chelation therapy	969
32.9	Imaging agents	970
32.10	Nanoparticles in directed drug delivery	972
32.11	Outlook	972

FURTHER READING 973
EXERCISES 974
TUTORIAL PROBLEMS 974

Resource section 1	Selected ionic radii	975
Resource section 2	Electronic properties of the elements	978
Resource section 3	Standard potentials	980
Resource section 4	Character tables	993
Resource section 5	Symmetry-adapted orbitals	998
Resource section 6	Tanabe–Sugano diagrams	1003
Resource section 7	Useful data: conversion factors, fundamental constants, and other information	1006

Index 1009

Glossary of chemical abbreviations

Ac	acetyl, CH_3CO
acac	acetylacetonato
aq	aqueous solution species
bpy	2,2′-bipyridine
cod	1,5-cyclooctadiene
cot	cyclooctatetraene
Cy	cyclohexyl
Cp	cyclopentadienyl
Cp*	pentamethylcyclopentadienyl
cyclam	tetraazacyclotetradecane
dien	diethylenetriamine
DMSO	dimethylsulfoxide
DMF	dimethylformamide
h	hapticity
edta	ethylenediaminetetraacetato
en	ethylenediamine (1,2-diaminoethane)
Et	ethyl
gly	glycinato
Hal	halide
^{i}Pr	isopropyl
KCP	$K_2Pt(CN)_4Br_{0.3} \cdot 3H_2O$
L	a ligand
μ	signifies a bridging ligand
M	a metal
Me	methyl
mes	mesityl, 2,4,6-trimethylphenyl
Ox	an oxidized species
ox	oxalato
Ph	phenyl
phen	phenanthroline
py	pyridine
Sol	solvent, or a solvent molecule
soln	nonaqueous solution species
^{t}Bu	tertiary butyl
THF	tetrahydrofuran
TMEDA	N,N,N',N'-tetramethylethylenediamine
trien	2,2′,2″-triaminotriethylene
X	generally halogen, also a leaving group or an anion
Y	an entering group

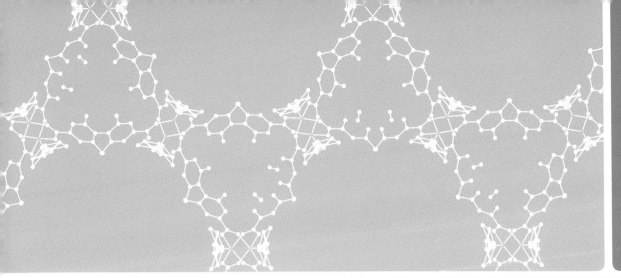

PART 1

Foundations

The eight chapters in this part of the book lay the foundations of inorganic chemistry. The first four chapters develop an understanding of the structures of atoms, molecules, and solids. Chapter 1 introduces the structure of atoms in terms of quantum theory and describes and explains important fundamental properties. Chapter 2 develops molecular structure in terms of increasingly sophisticated models of covalent bonding and introduces band theory of solids. Chapter 3 shows how a systematic consideration of the symmetry of molecules can be used to enhance our understanding of the bonding and structure of molecules and help interpret data from some of the techniques described in Chapter 8. Chapter 4 describes ionic bonding, the structures and properties of a range of typical solids. Chapter 5 explains how acid–base properties are defined, measured, and applied across a wide area of chemistry. Chapter 6 describes oxidation and reduction and demonstrates how electrochemical data can be used to predict and explain the outcomes of reactions in which electrons are transferred between molecules. Chapter 7 describes the coordination compounds of the elements. We discuss geometry, bonding, electronic structure, and reactions of complexes, and see how symmetry considerations can provide useful insight into this important class of compounds. Chapter 8 provides a toolbox for inorganic chemistry: it describes a wide range of the instrumental techniques that are used to identify and determine the structures and compositions of inorganic compounds.

Atomic structure

The structures of hydrogenic atoms
1.1 **Spectroscopic information**
1.2 **Some principles of quantum mechanics**
1.3 **Atomic orbitals**

Many-electron atoms
1.4 **Penetration and shielding**
1.5 **The building-up principle**
1.6 **The classification of the elements**
1.7 **Atomic properties**

Further reading

Exercises

Tutorial problems

This chapter lays the foundations for the explanation of the trends in the physical and chemical properties of all elements and inorganic compounds. To understand the behaviour of molecules and solids we need to understand atoms: our study of inorganic chemistry must therefore begin with a review of their structures and properties. We start with a discussion of the origin of matter in the solar system and then consider the development of our understanding of atomic structure and the behaviour of electrons in atoms. We introduce quantum theory qualitatively and use the results to rationalize properties such as atomic radii, ionization energy, electron affinity, and electronegativity. A knowledge of these properties allows us to begin to understand the diverse chemical properties of nearly 120 elements known today.

The observation that the universe is expanding has led to the current view that about 14 billion years ago the currently visible universe was concentrated into a point-like region that exploded in an event called the **Big Bang**. With initial temperatures immediately after the Big Bang of about 10^9 K, the fundamental particles produced in the explosion had too much kinetic energy to bind together in the forms we know today. However, the universe cooled as it expanded, the particles moved more slowly, and they soon began to adhere together under the influence of a variety of forces. In particular, the **strong force**, a short-range but powerful attractive force between nucleons (protons and neutrons), bound these particles together into nuclei. As the temperature fell still further, the **electromagnetic force**, a relatively weak but long-range force between electric charges, bound electrons to nuclei to form atoms, and the universe acquired the potential for complex chemistry and the existence of life (Box 1.1).

About two hours after the start of the universe, the temperature had fallen so much that most of the matter was in the form of H atoms (89%) and He atoms (11%). In one sense, not much has happened since then for, as Fig. 1.1 shows, hydrogen and helium remain overwhelmingly the most abundant elements in the universe. However, nuclear reactions have formed dozens of other elements and have immeasurably enriched the variety of matter in the universe, and thus given rise to the whole area of chemistry (Boxes 1.2 and 1.3).

Table 1.1 summarizes the properties of the subatomic particles that we need to consider in chemistry. All the known elements—by 2016 all up to 118 had been confirmed—that are formed from these subatomic particles are distinguished by their **atomic number**, Z, the number of protons in the nucleus of an atom of the element. Many elements have a number of **isotopes**, which are atoms with the same atomic number but different atomic masses. These isotopes are distinguished by the **mass number**, A, which is the total number of protons and neutrons in the nucleus. The mass number is also sometimes termed the *nucleon number*. Hydrogen, for instance, has three isotopes. In each case $Z = 1$, indicating that the nucleus contains one proton. The most abundant isotope has $A = 1$, denoted ^{1}H, its nucleus consisting of a

BOX 1.1 How are elements created?

The earliest stars resulted from the gravitational condensation of clouds of H and He atoms. This gave rise to high temperatures and densities within the clouds, and fusion reactions began as nuclei merged together.

Energy is released when light nuclei fuse together to give elements of higher atomic number. Nuclear reactions are very much more energetic than normal chemical reactions because the **strong force** which binds protons and neutrons together is much stronger than the electromagnetic force that binds electrons to nuclei. Whereas a typical chemical reaction might release about 10^3 kJ mol^{-1}, a nuclear reaction typically releases a million times more energy, about 10^9 kJ mol^{-1}.

Elements up to $Z = 26$ (iron) are formed inside stars. These elements are the products of the nuclear fusion reactions referred to as 'nuclear burning'. The burning reactions, which should not be confused with chemical combustion, involve H and He nuclei and a complicated fusion cycle catalysed by C nuclei. The stars that formed in the earliest stages of the evolution of the cosmos lacked C nuclei and used non-catalysed H-burning. Nucleosynthesis reactions are rapid at temperatures of $5–10 \times 10^6$ K. Here we have another contrast between chemical and nuclear reactions, because chemical reactions take place at temperatures a hundred thousand times lower. Moderately energetic collisions between atoms or molecules can result in chemical change, but only highly vigorous collisions can provide the energy required to bring about most nuclear transformations.

The elements beyond iron ($Z > 26$) are produced in significant quantities when hydrogen burning is complete and the collapse of the star's core raises its density to 10^8 kg m^{-3} (about 10^5 times the density of water) and the temperature to 10^8 K. Under these extreme conditions, a star will become a red giant and helium burning can occur.

The high abundance of iron and nickel in the universe is consistent with these elements having the most stable of all nuclei. This stability is expressed in terms of the **binding energy**, which is the difference in energy between the nucleus itself and the same numbers of individual protons and neutrons. This binding energy is often presented in terms of the difference in mass between the nucleus and its individual protons and neutrons because, according to Einstein's theory of relativity, mass and energy are related by $E = mc^2$, where c is the speed of light. Therefore, if the mass of a nucleus differs from the total mass of its components by $\Delta m = m_{nucleons} - m_{nucleus}$, then its binding energy is $E_{bind} = (\Delta m)c^2$. The binding energy of ^{56}Fe, for example, is the difference in energy between the ^{56}Fe nucleus and 26 protons and 30 neutrons. A positive binding energy corresponds to a nucleus that has a lower, more favourable, energy (and lower mass) than its constituent nucleons.

Figure B1.1 shows the binding energy per nucleon, E_{bind}/A (obtained by dividing the total binding energy by the number of nucleons), for the elements up to $Z = 100$. Iron and nickel occur at the maximum of the curve, showing that their nucleons are bound together more strongly than in any other nuclide. Harder to see from the graph is an alternation of binding energies as the atomic number varies from even to odd, with even-Z nuclides slightly more stable than their odd-Z neighbours. There is a corresponding alternation in cosmic abundances, with nuclides of even atomic number being marginally more abundant than those of odd atomic number. This stability of even-Z nuclides is attributed to the lowering of energy by pairing nucleons in the nucleus.

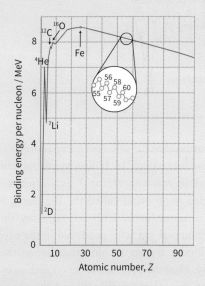

FIGURE B1.1 Nuclear binding energies. The greater the binding energy, the more stable the nucleus. Note the alternation in stability shown in the inset.

single proton. Far less abundant (only 1 atom in 6000) is deuterium, with $A = 2$. This mass number indicates that, in addition to a proton, the nucleus contains one neutron. The formal designation of deuterium is ^{2}H, but it is commonly denoted D. The third, short-lived, radioactive isotope of hydrogen is tritium, ^{3}H or T. Its nucleus consists of one proton and two neutrons. In certain cases it is helpful to display the atomic number of the element as a left suffix; so the three isotopes of hydrogen would then be denoted $^1_1$H, $^2_1$H, and $^3_1$H. Hydrogen is the only element for which there are such significant chemical distinctions between the isotopes that the isotopes warrant individual names.

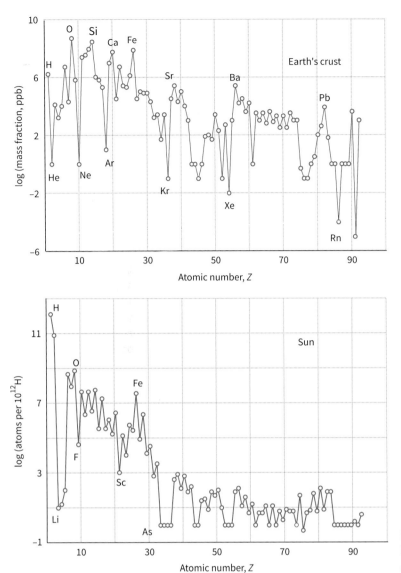

FIGURE 1.1 The abundances of the elements in the Earth's crust and the Sun. Elements with odd Z are less stable than their neighbours with even Z.

BOX 1.2 What are nuclear fusion and nuclear fission?

If two nuclei with mass numbers lower than 56 merge to produce a new nucleus with a larger nuclear binding energy, then excess energy is released. This process is called **fusion**. For example, two neon-20 nuclei may fuse to give a calcium-40 nucleus:

$$2\,^{20}_{10}\text{Ne} \rightarrow {}^{40}_{20}\text{Ca}$$

The value of the binding energy per nucleon, E_{bind}/A, for ^{20}Ne is approximately 8.0 MeV. Therefore, the total binding energy of the species on the left-hand side of the equation is $2 \times 20 \times 8.0$ MeV = 320 MeV. The value of E_{bind}/A for ^{40}Ca is close to 8.6 MeV, and so the total energy of the species on the right-hand side is 40×8.6 MeV = 344 MeV. The difference in the binding energies of the products and reactants is therefore 24 MeV.

For nuclei with $A > 56$, binding energy can be released when they split into lighter products with higher values of E_{bind}/A. This process is called **fission**. For example, uranium-236 can undergo fission into (among many other modes) xenon-140 and strontium-93 nuclei:

$$^{236}_{92}\text{U} \rightarrow {}^{140}_{54}\text{Xe} + {}^{93}_{38}\text{Sr} + 3\,^{1}_{0}\text{n}$$

The values of E_{bind}/A for ^{236}U, ^{140}Xe, and ^{93}Sr nuclei are 7.6, 8.4, and 8.7 MeV, respectively. Therefore, the energy released in this reaction is $(140 \times 8.4) + (93 \times 8.7) - (236 \times 7.6)$ MeV = 191.5 MeV for the fission of each ^{236}U nucleus.

Fission can also be induced by bombarding heavy elements with neutrons:

$$^{235}_{92}\text{U} + {}^{1}_{0}\text{n} \rightarrow \text{fission products} + \text{neutrons}$$

The kinetic energy of fission products from ^{235}U is about 165 MeV, that of the neutrons is about 5 MeV, and the γ-rays produced have an energy of about 7 MeV. The fission products are themselves radioactive and decay by β-, γ-, and X-radiation, releasing about 23 MeV. In a nuclear fission reactor the neutrons that are not consumed by fission are captured with the release of about 10 MeV, which escapes from the reactor as radiation, and about 1 MeV which remains as undecayed fission products in the spent fuel. Therefore, the total energy produced for one fission event is about 200 MeV, or 32 pJ. It follows that about 1 W of reactor heat (where $1\,W = 1\,J\,s^{-1}$) corresponds to about 3.1×10^{10} fission events per second. A nuclear reactor producing 3 GW has an electrical output of approximately 1 GW and corresponds to the fission of 3 kg of ^{235}U per day.

The use of nuclear power is controversial in large part on account of the risks associated with the highly radioactive, long-lived spent fuel. The declining stocks of fossil fuels, however, make nuclear power attractive as it is estimated that stocks of uranium could last for hundreds of years. The cost of uranium ores is currently low, and 100 g of uranium oxide generates as much energy as sixty barrels of oil or 20 tonnes of coal. The increased use of nuclear power would also drastically reduce the rate of emission of greenhouse gases into the atmosphere. The environmental drawback with nuclear power is the storage and disposal of radioactive waste, and the public are nervous about possible nuclear accidents, such as that in Chernobyl in 1986 and Fukushima in 2011, and the misuse of nuclear capabilities in pursuit of political ambitions.

BOX 1.3 Technetium—what is a synthetic element?

A synthetic element is one that does not occur naturally on Earth but that can be artificially generated by nuclear reactions. The first synthetic element was technetium (Tc, $Z = 43$), named from the Greek word for 'artificial'. Its discovery—or more precisely, its preparation—filled a gap in the periodic table and its properties matched those predicted by Mendeleev and others. The longest-lived isotope of technetium (^{98}Tc) has a half-life of 4.2 million years, so any produced when the Earth was formed has long since decayed. Technetium is produced in red-giant stars.

The most widely used isotope of technetium is ^{99m}Tc, where the 'm' indicates a metastable isotope. Technetium-99m emits high-energy γ-rays but has a relatively short half-life of 6.01 hours. These properties make the isotope particularly attractive for use *in vivo* as the γ-ray energy is sufficient for it to be detected outside the body and its half-life means that most of it will have decayed within 24 hours. Consequently, ^{99m}Tc is widely used in nuclear medicine—for example, in radiopharmaceuticals for imaging and in functional studies of the brain, bones, blood, lungs, liver, heart, thyroid gland, and kidneys (Section 32.9). Technetium-99m is generated through nuclear fission in nuclear power plants, but a more useful laboratory source of the isotope is a technetium generator, which uses the decay of ^{99}Mo to ^{99m}Tc.

The half-life of ^{99}Mo is 66 hours, which makes it more convenient for transport and storage than ^{99m}Tc itself. Most commercial generators are based on ^{99}Mo (made from stable ^{98}Mo by neutron capture) in the form of the molybdate ion, MoO_4^{2-}, adsorbed on Al_2O_3. The $^{99}MoO_4^{2-}$ ion decays by beta emission to the pertechnetate ion, $^{99m}TcO_4^{2-}$, which is less tightly bound to the alumina (Section 32.9).

$$^{99}Mo \rightarrow \,^{99m}Tc + \,^{0}_{-1}\beta$$

Sterile saline solution is washed through a column of the immobilized ^{99}Mo, and the ^{99m}Tc solution is collected.

TABLE 1.1 Subatomic particles of relevance to chemistry

Particle	Symbol	Mass/m_u*	Mass number	Charge/e†	Spin
Electron	e^-	5.486×10^{-4}	0	−1	½
Proton	p	1.0073	1	+1	½
Neutron	n	1.0087	1	0	½
Photon	γ	0	0	0	1
Neutrino	ν	c. 0	0	0	½
Positron	e^+	5.486×10^{-4}	0	+1	½
α particle	α	[$^4_2He^{2+}$ nucleus]	4	+2	0
β particle	β	[e^- ejected from nucleus]	0	−1	½
γ photon	γ	[electromagnetic radiation from nucleus]	0	0	1

* Masses are expressed relative to the atomic mass constant, $m_u = 1.6605 \times 10^{-27}$ kg.
† The elementary charge is $e = 1.602 \times 10^{-19}$ C.

The structures of hydrogenic atoms

So far we have discussed the nuclear properties of the elements. As chemists we are much more interested in the electronic structure of atoms and the organization of the periodic table which is a direct consequence of periodic variations in the electronic structure of atoms. Initially, we consider hydrogen-like or **hydrogenic atoms**, which have only one electron and so are free of the complicating effects of electron–electron repulsions. Hydrogenic atoms include ions such as He^+ and C^{5+} (found in the interiors of stars) as well as the hydrogen atom itself. Then we use the concepts that these atoms introduce to build up an approximate description of the structures of **many-electron atoms** (or *polyelectron atoms*).

1.1 Spectroscopic information

KEY POINTS Spectroscopic observations on gaseous hydrogen atoms suggest that an electron can occupy only certain energy levels and that the emission of discrete frequencies of electromagnetic radiation occurs when an electron makes a transition between these levels.

When an electric discharge is applied to hydrogen gas, excited hydrogen atoms are generated and electromagnetic radiation is emitted. When passed through a prism or diffraction grating, this radiation is found to consist of a series of components: one in the ultraviolet region, one in the visible region, and several in the infrared region of the electromagnetic spectrum (Fig. 1.2; Box 1.4). The nineteenth-century spectroscopist Johann Rydberg found that all the wavelengths (λ, lambda) can be described by the expression

$$\frac{1}{\lambda} = R\left(\frac{1}{n_1^2} - \frac{1}{n_2^2}\right) \tag{1.1}$$

where R is the **Rydberg constant**, an empirical constant with the value 1.097×10^7 m^{-1}. The n are integers, with $n_1 = 1, 2, \ldots$ and $n_2 = n_1 + 1, n_1 + 2, \ldots$. The series with $n_1 = 1$ is called the *Lyman series* and lies in the ultraviolet region. The series with $n_1 = 2$ lies in the visible region and is called the *Balmer series*. The infrared series include the *Paschen series* ($n_1 = 3$) and the *Brackett series* ($n_1 = 4$).

The energy of a photon is given by the equation $E = h\nu$, where h is Planck's constant, 6.626×10^{-34} J s, and ν is frequency, the number of times per second that a wave travels through a complete cycle, expressed in units of hertz, where 1 Hz = 1 s^{-1}. We can use this expression and the equation $\nu = c/\lambda$, where c is the speed of light (2.998×10^8 m s^{-1}) and λ is wavelength in metres, to derive the expression $E = hc/\lambda$. The quantity $1/\lambda$ is referred to as the wavenumber $\tilde{\nu}$, and gives the number of wavelengths in a given distance: it is directly proportional to the energy of the photon.

The structure of the spectrum is explained if it is supposed that the emission of radiation takes place when an electron in an atom makes a transition from a state of energy $-hcR/n_2^2$ to a state of energy $-hcR/n_1^2$ and that the energy difference, which is equal to $hcR(1/n_1^2 - 1/n_2^2)$, is carried away as a photon of energy, $E = hc/\lambda$. By equating $E = hcR(1/n_1^2 - 1/n_2^2)$ and $E = hc/\lambda$, and cancelling hc, we obtain eqn. 1.1.

> **A NOTE ON GOOD PRACTICE**
>
> Although wavelength is usually expressed in nano- or picometres, wavenumbers are usually expressed in cm^{-1}, or reciprocal centimetres. A wavenumber of 1 cm^{-1} denotes one complete wavelength in a distance of 1 cm. 1 cm^{-1} is equivalent to 11.96 J mol^{-1}.

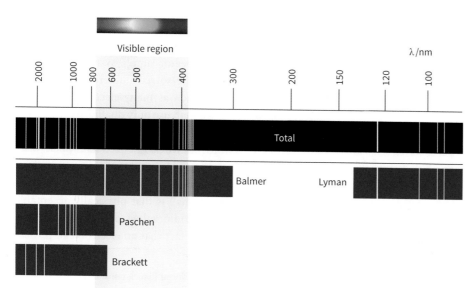

FIGURE 1.2 The spectrum of atomic hydrogen and its analysis into series.

BOX 1.4 How do sodium atoms light our streets?

The emission of light when atoms are excited is put to good use in lighting streets in many parts of the world. The widely used yellow street lamps are based on the emission of light from excited sodium atoms.

Low pressure sodium (LPS) lamps consist of a glass tube coated with indium tin oxide (ITO). The indium tin oxide reflects infrared light and transmits visible light. Two inner glass tubes hold solid sodium and a small amount of neon and argon—the same mixture as found in neon lights. When the lamp is turned on, the neon and argon emit a red glow which heats the sodium metal. Within a few minutes, the sodium starts to vaporize, the electrical discharge excites electrons in the atoms to a high energy level, and they re-emit the energy as yellow light.

One advantage of these lamps over other types of street lighting is that they do not lose light output as they age. They do, however, use more energy towards the end of their life, which may make them less attractive from environmental and economic perspectives. Increasingly, sodium street lights are being replaced by energy efficient LED lighting (Section 30.1).

The question these observations raise is why the energy of the electron in the atom is limited to the values $-hcR/n^2$ and why R has the value observed. An initial attempt to explain these features was made by Niels Bohr in 1913 using an early form of quantum theory in which he supposed that the electron could exist in only certain circular orbits about the atom. Although he obtained the correct value of R, his model was later shown to be untenable as it conflicted with the version of quantum theory developed by Erwin Schrödinger and Werner Heisenberg in 1926.

EXAMPLE 1.1 Predicting the wavelength of lines in the atomic spectrum of hydrogen

Predict the wavelengths of the first three lines in the Balmer series.

Answer For the Balmer series, $n_1 = 2$ and $n_2 = 3, 4, 5, 6$. So if we substitute into eqn. 1.1 we obtain $\frac{1}{\lambda} = R\left(\frac{1}{2^2} - \frac{1}{3^2}\right)$ for the first line which gives $1\,513\,888\,\text{m}^{-1}$ or 661 nm. Using values of $n_2 = 4$ and 5 for the next two lines give values of λ of 486 and 434 nm, respectively, see Fig. 1.2.

Self-test 1.1 (a) Predict the wavenumber and wavelength of the second line in the Paschen series. (b) Calculate the values of n_1 and n_2 for the line in the Lyman series with a wavelength of 103 nm.

1.2 Some principles of quantum mechanics

KEY POINTS Electrons can behave as particles or as waves; solution of the Schrödinger equation gives wavefunctions, which describe the location and properties of electrons in atoms. The probability of finding an electron at a given location is proportional to the square of the wavefunction. Wavefunctions generally have regions of positive and negative amplitude and may undergo constructive or destructive interference with one another.

In 1924, Louis de Broglie suggested that because electromagnetic radiation could be considered to consist of particles called photons yet at the same time exhibit wave-like properties, such as interference and diffraction, then the same might be true of electrons. This dual nature is called **wave–particle duality**. An immediate consequence of duality is that it is impossible to know the linear momentum (the product of mass and velocity) and the location of an electron (and any particle) simultaneously. This restriction is called the **Heisenberg uncertainty principle** which states that the product of the uncertainty in momentum and the uncertainty in position cannot be less than a quantity of the order of Planck's constant (specifically, ½ℏ, where $\hbar = h/2\pi$).[1]

Schrödinger formulated an equation that took account of wave–particle duality and accounted for the motion of electrons in atoms. To do so, he introduced the **wavefunction**, ψ (psi), a mathematical function of the position coordinates x, y, and z, which describes the behaviour of an electron. The **Schrödinger equation** for an electron free to move in one dimension (the x axis) is

$$\underbrace{-\frac{\hbar^2}{2m_e}\frac{d^2\psi}{dx^2}}_{\text{Kinetic energy contribution}} + \underbrace{V(x)\psi(x)}_{\text{Potential energy contribution}} = \underbrace{E\psi(x)}_{\text{Total energy}} \quad (1.2)$$

where m_e is the mass of an electron, V is the potential energy of the electron, and E is its total energy. The Schrödinger equation is a second-order differential equation that can be solved exactly for a number of simple systems (such as a hydrogen atom) and can be solved numerically for many more complex systems (such as many-electron atoms and molecules). However, we shall use only qualitative aspects of its solutions. The generalization of eqn. 1.2 to three dimensions is straightforward, but we do not need its explicit form.

One crucial feature of eqn. 1.2 and its analogues in three dimensions and the imposition of certain requirements (called 'boundary conditions') is that physically acceptable solutions exist only for certain values of E. Therefore, the **quantization** of energy, the fact that an electron can possess

---

[1] $\hbar$ (pronounced h-bar) is the reduced Planck constant. It is used when angular frequency in radians per second is more appropriate than cycles per second.

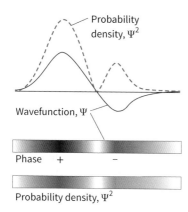

FIGURE 1.3 The Born interpretation of the wavefunction is that its square is a probability density. There is zero probability density at a node.

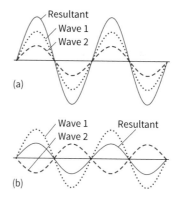

FIGURE 1.4 Wavefunctions interfere where they spread into the same region of space. (a) If they have the same sign in a region, they interfere constructively and the total wavefunction has an enhanced amplitude in the region. (b) If the wavefunctions have opposite signs, then they interfere destructively, and the resulting superposition has a reduced amplitude.

only certain discrete energies in an atom, follows naturally from the Schrödinger equation.

A wavefunction contains all the dynamical information possible about the electron, including where it is and how fast it is travelling. As Heisenberg's uncertainty principle means it is impossible to know all this information simultaneously, this leads naturally to the concept of the probability of finding an electron at a given location. Specifically, the probability of finding an electron at a given location is proportional to the square of the wavefunction at that point, ψ^2. According to this interpretation, there is a high probability of finding the electron where ψ^2 is large, and the electron will not be found where ψ^2 is zero (Fig. 1.3). The quantity ψ^2 is called the **probability density** of the electron. It is a 'density' in the sense that the product of ψ^2 and the infinitesimal volume element $d\tau = dxdydz$ (where τ is tau) is proportional to the probability of finding the electron in that volume. The probability is *equal* to $\psi^2 d\tau$ if the wavefunction is 'normalized'. A normalized wavefunction is one that is scaled so that the total probability of finding the electron somewhere is 1. The wavefunction of an electron in an atom is called an **atomic orbital**.

Like other waves, wavefunctions in general have regions of positive and negative amplitude, or sign. To help keep track of the relative signs of different regions of a wavefunction, or atomic orbital, in illustrations we label regions of opposite sign with dark and light shading corresponding to + and − signs, respectively. The sign of the wavefunction is of crucial importance when two wavefunctions spread into the same region of space and interact. Then a positive region of one wavefunction may add to a positive region of the other wavefunction to give a region of enhanced amplitude. This enhancement is called **constructive interference** (Fig. 1.4(a)). It means that, where the two wavefunctions spread into the same region of space, such as occurs when two atoms are close together, there may be a significantly enhanced probability of finding the electrons in that region. Conversely, a positive region of one wavefunction may be cancelled by a negative region of the second wavefunction (Fig. 1.4(b)). This **destructive interference** between wavefunctions reduces the probability that an electron will be found in that region. As we shall see, the interference of wavefunctions is of great importance in the explanation of chemical bonding.

1.3 Atomic orbitals

Chemists use hydrogenic atomic orbitals to develop models that are central to the interpretation of inorganic chemistry, and we shall spend some time describing their shapes and significance.

(a) Hydrogenic energy levels

KEY POINTS The energy of the bound electron is determined by n, the principal quantum number; in addition, l specifies the magnitude of the orbital angular momentum and m_l specifies the orientation of that angular momentum.

Each of the wavefunctions obtained by solving the Schrödinger equation for a hydrogenic atom is uniquely labelled by a set of three integers called **quantum numbers**. These quantum numbers are designated n, l, and m_l: n is called the **principal quantum number**, l is the **orbital angular momentum quantum number** (formerly the 'azimuthal quantum number'), and m_l is called the **magnetic quantum number**. Each quantum number specifies a physical property of the electron: n specifies the energy, l labels the magnitude of the orbital angular momentum, and m_l labels the orientation of that angular momentum. The value of n also indicates the size of the orbital, with larger n, high-energy orbitals, more diffuse than low n, compact, tightly bound, low-energy orbitals. The value of l also indicates the angular shape of the orbital, with the number of lobes increasing as l increases. The value of m_l indicates the orientation of these lobes.

The allowed energies are specified by the principal quantum number, n. For a hydrogenic atom of atomic number Z, they are given by

$$E_n = -\frac{hcRZ^2}{n^2} \tag{1.3}$$

with $n = 1, 2, 3, \ldots$ and

$$R = \frac{m_e e^4}{8h^3 c \varepsilon_0^2} \tag{1.4}$$

(The fundamental constants in this expression are given inside the back cover.) The calculated numerical value of R is 1.097×10^7 m^{-1}, in excellent agreement with the empirical value determined spectroscopically by Rydberg. For future reference, the value of hcR corresponds to 13.6 eV or 1312 kJ mol^{-1}.

A NOTE ON GOOD PRACTICE

An electronvolt is the amount of kinetic energy gained by an electron as it accelerates through a potential of one volt. It is a useful, but non-SI, unit. In chemistry, kinetic energy gained by a mole of electrons passing through a potential of one volt is 96.485 kJ mol^{-1}. The approximation 1 eV ≈ 100 kJ mol^{-1} is worth remembering. The Faraday constant, F, the electric charge of a mole of electrons, is 96 485 C mol^{-1}.

The energies given by eqn. 1.3 are all negative, signifying that the energy of the electron in a bound state is lower than a widely separated stationary electron and nucleus. The zero of energy (at $n = \infty$) corresponds to the electron and nucleus being widely separated and stationary. Positive values of the energy correspond to unbound states of the electron in which it may travel with any velocity and hence possess any energy. Finally, because the energy is proportional to $1/n^2$, the energy levels in the bound state converge as the energy increases (becomes less negative, Fig. 1.5).

The value of l specifies the magnitude of the orbital angular momentum through $\{l(l+1)\}^{1/2}\hbar$, with $l = 0, 1, 2, \ldots$. We can think of l as indicating the momentum with which the electron circulates around the nucleus via the lobes of the orbital. As we shall see shortly, the third quantum number m_l specifies the orientation of this momentum—for instance, whether the circulation is clockwise or anticlockwise.

(b) Shells, subshells, and orbitals

KEY POINTS All orbitals with a given value of n belong to the same shell, all orbitals of a given shell with the same value of l belong to the same subshell, and individual orbitals are distinguished by the value of m_l.

In a hydrogenic atom, all orbitals with the same value of n have the same energy and are said to be **degenerate**. The principal quantum number therefore defines a series of **shells** of the atom or sets of orbitals with the same value of n and hence

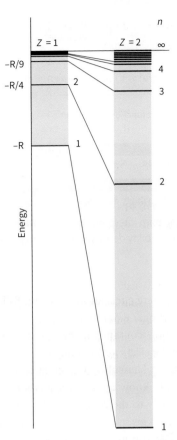

FIGURE 1.5 The quantized energy levels of an H atom ($Z = 1$) and an He$^+$ ion ($Z = 2$). The energy levels of a hydrogenic atom are proportional to Z^2.

with the same energy and approximately the same radial extent. Shells with $n = 1, 2, 3, \ldots$ are sometimes referred to as K, L, M, ... shells—for example, when electronic transitions between these shells are referred to in X-ray spectroscopy.

The orbitals belonging to each shell are classified into **subshells** distinguished by a quantum number l. For a given value of n, the quantum number l can have the values $l = 0, 1, \ldots, n - 1$, giving n different values in all. For example, the shell with $n = 1$ consists of just one subshell with $l = 0$, the shell with $n = 2$ consists of two subshells, one with $l = 0$ and the other with $l = 1$, the shell with $n = 3$ consists of three subshells, with values of l of 0, 1, and 2. It is common practice to refer to each subshell by a letter:

Value of l	0	1	2	3	4	...
Subshell designation	s	p	d	f	g	...

For most purposes in chemistry we need consider only s, p, d, and f subshells.[2]

A subshell with quantum number l consists of $2l + 1$ individual orbitals. These orbitals are distinguished by the

---

[2] The orbital labels s, p, d, and f come from terms used to describe groups of lines in the atomic spectra. They stand for sharp, principal, diffuse, and fundamental, respectively.

magnetic quantum number, m_l, which can have the $2l + 1$ integer values from $+l$ down to $-l$. This quantum number specifies the component of orbital angular momentum around an arbitrary axis (commonly designated z) passing through the nucleus. So, for example, a d subshell of an atom ($l = 2$) consists of five individual atomic orbitals that are distinguished by the values $m_l = +2, +1, 0, -1, -2$. An f subshell ($l = 3$) consists of seven individual atomic orbitals with the values $m_l = +3, +2, +1, 0, -1, -2, -3$.

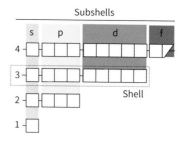

FIGURE 1.6 The classification of orbitals into subshells (same value of l) and shells (same value of n).

A NOTE ON GOOD PRACTICE

Write the sign of m_l even when it is positive. Thus, we write $m_l = +2$, not $m_l = 2$.

The practical conclusion for chemistry from these rules is that there is only one orbital in an s subshell ($l = 0$), the one with $m_l = 0$: this orbital is called an **s orbital**. There are three orbitals in a p subshell ($l = 1$), with quantum numbers $m_l = +1, 0, -1$; they are called **p orbitals**. The five orbitals of a d subshell ($l = 2$) are called **d orbitals**, with quantum numbers $m_l = +2, +1, 0, -1, -2$, and so on (Fig. 1.6).

EXAMPLE 1.2 **Identifying orbitals from quantum numbers**

Which set of orbitals is defined by $n = 4$ and $l = 1$? How many orbitals are there in this set?

Answer We need to remember that the principal quantum number n identifies the shell and that the orbital quantum number l identifies the subshell. The subshell with $l = 1$ consists of p orbitals. The allowed values of $m_l = +l, \ldots, -l$ give the number of orbitals of that type. In this case, $m_l = +1, 0$, and -1. There are therefore three 4p orbitals.

Self-test 1.2 (a) Which set of orbitals is defined by the quantum numbers $n = 3$ and $l = 2$? How many orbitals are there in this set? (b) What are the quantum numbers n and l that define a 5f orbital? How many orbitals are there in this set?

(c) Electron spin

KEY POINTS The intrinsic spin angular momentum of an electron is defined by the two quantum numbers s and m_s. Four quantum numbers are needed to define the state of an electron in a hydrogenic atom.

In addition to the three quantum numbers required to specify the spatial distribution of an electron in a hydrogenic atom, two more quantum numbers are needed to define the state of an electron. These additional quantum numbers relate to the intrinsic angular momentum of an electron, its **spin**. This evocative name suggests that an electron can be regarded as having an angular momentum arising from a spinning motion, rather like the daily rotation of the Earth as it travels in its annual orbit around the Sun. However, spin is a quantum-mechanical property, and this analogy must be viewed with great caution.

Spin is described by two quantum numbers, s and m_s. The former is the analogue of l for orbital motion but, for an electron it is restricted to the single, unchangeable value $s = \frac{1}{2}$. The magnitude of the spin angular momentum is given by the expression $\{s(s + 1)\}^{1/2}\hbar$, so when we substitute for $s = \frac{1}{2}$ we find that this magnitude is fixed at $\frac{1}{2}\sqrt{3}\hbar$ for any electron. The second quantum number, the **spin magnetic quantum number**, m_s, may take only two values, $+\frac{1}{2}$ (anticlockwise spin, imagined from above) and $-\frac{1}{2}$ (clockwise spin). The two states are often represented by the two arrows ↑ ('spin-up', $m_s = +\frac{1}{2}$) and ↓ ('spin-down', $m_s = -\frac{1}{2}$) or by the Greek letters α and β, respectively.

Because the spin state of an electron must be specified if the state of the atom is to be specified fully, it is common to say that the state of an electron in a hydrogenic atom is characterized by four quantum numbers, n, l, m_l, and m_s.

(d) Nodes

KEY POINT Regions where wavefunctions pass through zero are called nodes. Inorganic chemists generally find it adequate to use visual representations of atomic orbitals rather than mathematical expressions. However, we need to be aware of the mathematical expressions that underlie these representations.

Because the potential energy of an electron in the field of a nucleus is spherically symmetric (it is proportional to Z/r and independent of orientation relative to the nucleus), the orbitals are best expressed in terms of the spherical polar coordinates defined in Fig. 1.7, rather than the Cartesian coordinates, x, y, and z. In these coordinates, the orbitals all have the form

$$\psi_{nlm_l} = \underbrace{R_{nl}(r)}_{\text{Variation with radius}} \times \underbrace{Y_{lm_l}(\theta, \phi)}_{\text{Variation with angle}} \quad (1.5)$$

This expression reflects the simple idea that a hydrogenic orbital can be written as the product of a function $R(r)$ of the radius (the distance the electron is from the nucleus) and a function $Y(\theta, \phi)$ of the angular coordinates. The positions where either component of the wavefunction passes

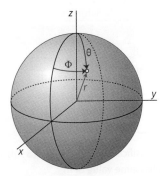

FIGURE 1.7 Spherical polar coordinates: r is the radius, θ (theta) the colatitude, and ϕ (phi) the azimuth.

through zero are called **nodes**. Consequently, there are two types of node. **Radial nodes** occur where the radial component of the wavefunction passes through zero, and **angular nodes** occur where the angular component of the wavefunction passes through zero. The numbers of both types of node increase with increasing energy and are related to the quantum numbers n and l. The total number of radial and angular nodes for any orbital is equal to $n - 1$.

(e) The radial variation of atomic orbitals

KEY POINT An s orbital has nonzero amplitude at the nucleus; all other orbitals (those with $l > 0$) vanish at the nucleus.

Figs. 1.8 and 1.9 show the radial variations of some atomic orbitals. A 1s orbital, the wavefunction with $n = 1$, $l = 0$, and $m_l = 0$, decays exponentially with distance from the nucleus and never passes through zero (it has no nodes). All orbitals decay exponentially at sufficiently great distances from the nucleus and this distance increases as n increases.

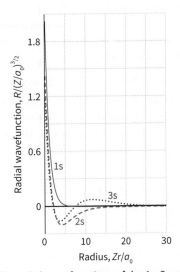

FIGURE 1.8 The radial wavefunctions of the 1s, 2s, and 3s hydrogenic orbitals. Note that the number of radial nodes is 0, 1, and 2, respectively. Each orbital has a nonzero amplitude at the nucleus (at $r = 0$).

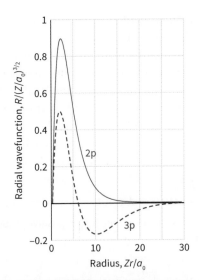

FIGURE 1.9 The radial wavefunctions of the 2p and 3p hydrogenic orbitals. Note that the number of radial nodes is 0 and 1, respectively. Each orbital has zero amplitude at the nucleus (at $r = 0$).

Some orbitals oscillate through zero close to the nucleus and thus have one or more radial nodes before beginning their final exponential decay. As the principal quantum number of an electron increases, it is likely to be found further away from the nucleus and its energy increases.

An orbital with quantum numbers n and l has $n - l - 1$ radial nodes. This oscillation is evident in the 2s orbital, the orbital with $n = 2$, $l = 0$, and $m_l = 0$, which passes through zero once and hence has one radial node. A 3s orbital passes through zero twice and so has two radial nodes (Figs. 1.8 and 1.9). A 2p orbital (one of the three orbitals with $n = 2$ and $l = 1$) has no radial node because its radial wavefunction does not pass through zero anywhere. For any series of the same type of orbital, the first occurrence has no radial node, the second has one radial node, the third has two, and so on (Fig. 1.10).

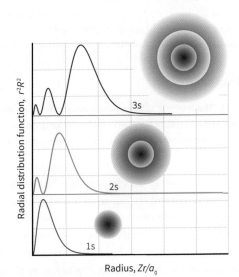

FIGURE 1.10 The 1s, 2s, and 3s orbitals showing the radial nodes.

Although an electron in an s orbital may be found at the nucleus, an electron in any other type of orbital will not be found there. We shall soon see that this apparently minor detail, which is a consequence of the absence of orbital angular momentum when $l = 0$, is one of the key concepts for understanding the layout of the periodic table and the chemistry of the elements.

EXAMPLE 1.3 Predicting numbers of radial nodes

How many radial nodes do 3p, 3d, and 4f orbitals each have?

Answer We need to make use of the fact that the number of radial nodes is given by the expression $n - l - 1$ and use it to find the number of radial nodes using values of n and l. The 3p orbitals have $n = 3$ and $l = 1$, and so the number of radial nodes is $n - l - 1 = 1$. The 3d orbitals have $n = 3$ and $l = 2$. Therefore, the number of radial nodes is $n - l - 1 = 0$. The 4f orbitals have $n = 4$ and $l = 3$ and the number of radial nodes is $n - l - 1 = 0$. The 3d and 4f orbitals are the first occurrence of the d and f orbitals, so this also indicates that they will have no radial node. The 3p orbitals are the second occurrence of the p orbitals and so we would expect them to have one radial node.

Self-test 1.3 (a) How many radial nodes does a 5s orbital have? (b) Which p orbital has two radial nodes?

(f) The radial distribution function

KEY POINT A radial distribution function gives the probability that an electron will be found at a given distance from the nucleus, regardless of the direction.

The Coulombic (electrostatic) force that binds the electron is centred on the nucleus, so it is often of interest to know the probability of finding an electron at a given distance from the nucleus, regardless of its direction. This information enables us to judge how tightly the electron is bound. The total probability of finding the electron in a spherical shell of radius r and thickness dr is the integral of $\psi^2 d\tau$ over all angles. This result is written $P(r)dr$, where $P(r)$ is called the **radial distribution function**. In general,

$$P(r) = r^2 \psi(r)^2 \tag{1.6}$$

(For s orbitals, this expression is the same as $P = 4\pi r^2 \psi^2$.) If we know the value of P at some radius r, then we can state the probability of finding the electron somewhere in a shell of thickness dr at that radius simply by multiplying P by dr.

Because the wavefunction of a 1s orbital decreases exponentially with distance from the nucleus and the factor r^2 in eqn. 1.6 increases, the radial distribution function of a 1s orbital goes through a maximum (Fig. 1.11). Therefore, there is a distance at which the electron is most likely to

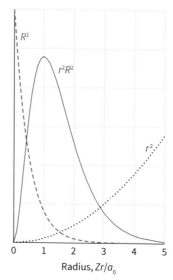

FIGURE 1.11 The radial distribution function, r^2R^2, of a hydrogenic 1s orbital. r^2R^2 is the product of r^2 (which increases as r increases) and the square of the radial component of the wavefunction Ψ (labelled R^2 in the figure and which decreases exponentially). This distance increases as nuclear charge increases and passes through a maximum at $r = a_0/Z$.

be found. In general, this most probable distance decreases as the nuclear charge increases (because the electron is attracted more strongly to the nucleus), and specifically

$$r_{max} = \frac{a_0}{Z} \tag{1.7}$$

where a_0 is the **Bohr radius**

$$a_0 = \frac{4\pi\varepsilon_0 \hbar^2}{m_e e^2} \tag{1.8}$$

a quantity that appeared in Bohr's formulation of his model of the hydrogen atom; its numerical value is 52.9 pm.

EXAMPLE 1.4 Interpreting radial distribution functions

Figure 1.12 shows the radial distribution functions for 2s and 2p hydrogenic orbitals. Which orbital gives the electron a greater probability of close approach to the nucleus?

Answer By examining Fig. 1.12 we can see that the radial distribution function of a 2p orbital approaches zero near the nucleus faster than a 2s electron does. This difference is a consequence of the fact that a 2p orbital has zero amplitude at the nucleus on account of its orbital angular momentum. The 2s electron has a greater probability of close approach to the nucleus indicated by the inner maximum. Note that the 2s orbital extends further into space.

Self-test 1.4 Which orbital, 3p or 3d, gives an electron a greater probability of being found close to the nucleus?

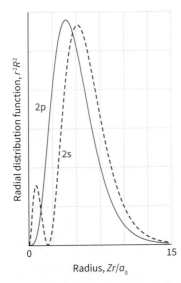

FIGURE 1.12 The radial distribution functions of hydrogenic orbitals. Although the 2p orbital is *on average* closer to the nucleus (note where its maximum lies), the 2s orbital has a high probability of being close to the nucleus on account of the inner maximum.

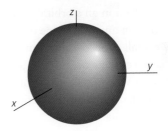

FIGURE 1.13 The spherical boundary surface of an s orbital.

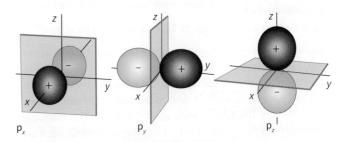

FIGURE 1.14 The representation of the boundary surfaces of the p orbitals. Each orbital has one nodal plane running through the nucleus. For example, the nodal plane of the p_z orbital is the *xy* plane. The darkly shaded lobe (labelled '+') has a positive amplitude, the more lightly shaded one (labelled '−') is negative.

(g) The angular variation of atomic orbitals

KEY POINTS The boundary surface of an orbital indicates the region of space within which the electron is most likely to be found; orbitals with the quantum number l have l nodal planes.

The angular wavefunction expresses the variation of angle around the nucleus, and this describes the orbital's angular shape. An s orbital has the same amplitude at a given distance from the nucleus whatever the angular coordinates of the point of interest: that is, an s orbital is spherically symmetrical. The orbital is normally represented by a spherical surface with the nucleus at its centre. The surface is called the **boundary surface** of the orbital and defines the region of space within which there is a high (typically 90%) probability of finding the electron. This boundary surface is what chemists draw to represent the shape of an orbital. The planes on which the angular wavefunction passes through zero are called **angular nodes** or **nodal planes**. An electron will not be found anywhere on a nodal plane. A nodal plane cuts through the nucleus and separates the regions of positive and negative sign of the wavefunction.

In general, an orbital with the quantum number l has l nodal planes. An s orbital, with $l = 0$, has no nodal plane and the boundary surface of the orbital is spherical (Fig. 1.13).

All orbitals with $l > 0$ have amplitudes that vary with angle and, for p orbitals, m_l values of +1, 0, and −1. In the most common graphical representation, the boundary surfaces of the three p orbitals of a given shell are identical apart from the fact that their axes lie parallel to each of the three different Cartesian axes centred on the nucleus, and each one possesses a nodal plane passing through the nucleus (Fig. 1.14). In the diagrammatic representation of the orbitals, the two lobes are shaded differently (dark and light respectively) or labelled '+' and '−' to indicate that one has a positive and one has a negative amplitude. This representation is the origin of the labels p_x, p_y, and p_z. Each p orbital, with $l = 1$, has a single nodal plane.

The boundary surfaces and labels we use for the d and f orbitals are shown in Figs. 1.15 and 1.16, respectively. The d_{z^2} orbital looks different from the remaining d orbitals. There are in fact six possible combinations of double

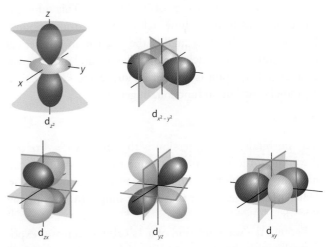

FIGURE 1.15 One representation of the boundary surfaces of the d orbitals. Four of the orbitals have two perpendicular nodal planes that intersect in a line passing through the nucleus. In the d_{z^2} orbital, the nodal surface forms two cones that meet at the nucleus.

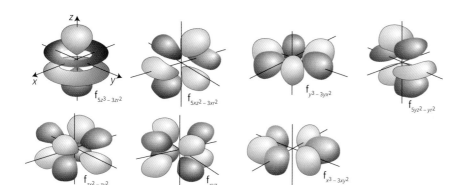

FIGURE 1.16 One representation of the boundary surfaces of the f orbitals. Other representations (with different shapes) are sometimes encountered.

dumb-bell-shaped orbitals around three axes: three with lobes between the axes, as in d_{xy}, d_{yz}, and d_{zx}, and three with lobes along the axis. However, only five d orbitals are allowed. One of these orbitals is assigned $d_{x^2-y^2}$ and lies along the x and y axes. The remaining orbital is the $d_{2z^2-x^2-y^2}$ from the algebra which is simplified to d_{z^2} and can be thought of as the superposition of the remaining two combinations, the $d_{z^2-y^2}$ and the $d_{z^2-x^2}$. Note that a d orbital (with $l = 2$) has two nodal planes that intersect at the nucleus; a typical f orbital ($l = 3$) has three nodal planes.

Many-electron atoms

As we have remarked, a 'many-electron atom' is an atom with more than one electron, so even He, with two electrons, is technically a many-electron atom. The exact solution of the Schrödinger equation for an atom with N electrons would be a function of the $3N$ coordinates of all the electrons. It would be extremely challenging to find exact formulas for such complicated functions; however, it is straightforward to perform numerical computations by using widely available software to obtain precise energies and probability densities, see Further Reading. This software can also generate graphical representations of the resulting orbitals that can assist in the interpretation of the properties of the atom. For most of inorganic chemistry we rely on the **orbital approximation**, in which each electron occupies an atomic orbital that resembles those found in hydrogenic atoms. When we say that an electron 'occupies' an atomic orbital, we mean that it is described by the corresponding wavefunction and set of quantum numbers.

1.4 Penetration and shielding

KEY POINTS The ground-state electron configuration is a specification of the orbital occupation of an atom in its lowest energy state. The exclusion principle forbids more than two electrons to occupy a single orbital. The nuclear charge experienced by an electron is reduced by shielding by other electrons, including those in the same shell. Trends in effective nuclear charge can be used to rationalize the trends in many properties. As a result of the combined effects of penetration and shielding, the order of energy levels in a shell of a many-electron atom is s < p < d < f.

It is quite easy to account for the electronic structure of the helium atom in its **ground state**, its state of lowest energy. According to the orbital approximation, we suppose that both electrons occupy an atomic orbital that has the same spherical shape as a hydrogenic 1s orbital. However, the orbital will be more compact because, as the nuclear charge of helium is greater than that of hydrogen, the electrons are drawn in towards the nucleus more closely than is the one electron of an H atom. The ground-state **configuration** of an atom is a statement of the orbitals its electrons occupy in the ground state. For helium, with two electrons in the 1s orbital, the ground-state configuration is denoted $1s^2$ (read as 'one s two').

As soon as we come to the next atom in the periodic table, lithium ($Z = 3$), we encounter several major new features. The configuration $1s^3$ is forbidden by a fundamental feature of nature known as the **Pauli exclusion principle**:

No more than two electrons may occupy a single orbital and, if two do occupy a single orbital, then their spins must be paired.

By 'paired' we mean that one electron spin must be ↑ ($m_s = +½$) and the other ↓ ($m_s = -½$); the pair is denoted ↑↓. Another way of expressing the principle is to note that, because an electron in an atom is described by four variable quantum numbers, n, l, m_l, and m_s, no two electrons can have the same four quantum numbers. The Pauli principle was introduced originally to account for the absence of certain transitions in the spectrum of atomic helium.

Because the configuration 1s³ is forbidden by the Pauli exclusion principle, the third electron must occupy an orbital of the next higher shell, the shell with $n = 2$. The question that now arises is whether the third electron occupies a 2s orbital or one of the three 2p orbitals. To answer this question, we need to examine the energies of the two subshells and the effect of the other electrons in the atom. Although 2s and 2p orbitals have the same energy in a hydrogenic atom, spectroscopic data and calculations show that this is not the case in a many-electron atom.

In the orbital approximation, we treat the repulsion between electrons in an approximate manner by supposing that the electronic charge is distributed spherically around the nucleus. Then each electron moves in the attractive field of the nucleus and also experiences an average repulsive charge from the other electrons. According to classical electrostatics, the field that arises from a spherical distribution of charge is equivalent to the field generated by a single point charge at the centre of the distribution (Fig. 1.17). This negative charge reduces the actual charge of the nucleus, Z, to Z_{eff}, where Z_{eff} is called the **effective nuclear charge**.

This effective nuclear charge depends on the values of n and l of the electron of interest because electrons in different shells and subshells approach the nucleus to different extents. The reduction of the true nuclear charge to the effective nuclear charge by the other electrons is called **shielding**. The effective nuclear charge is sometimes expressed in terms of the true nuclear charge and an empirical **shielding constant**, σ, by writing $Z_{eff} = Z - \sigma$. The shielding constant can be determined by fitting hydrogenic orbitals to those computed numerically. It can also be approximated by using an empirical set of rules, the most widely used are **Slater's rules**.

Slater's rules attribute a numerical contribution to electrons in an atom in the following way:

Write out the electron configuration of the atom and group orbitals together in the form

(1s)(2s2p)(3s3p)(3d)(4s4p)(4d)(4f)(5s5p) etc.

If the outermost electron is in an s or p orbital:

Each of the other electrons in the (ns np) grouping contributes 0.35 to σ.

Each electron in the $n - 1$ shell contributes 0.85 to σ.

Each electron in lower shells contributes 1.0 to σ.

If the outermost electron is in a d or f orbital:

Each of the other electrons in the (nd) or (nf) grouping contributes 0.35 to σ.

Each electron in lower shells or earlier groupings contributes 1.0 to σ.

For example, to calculate the shielding constant for the outermost electron, and hence the effective nuclear charge of fluorine, F, we first write down the electron configuration with appropriate groupings:

$(1s^2)(2s^2 2p^5)$

Then $\sigma = (6 \times 0.35) + (2 \times 0.85) = 3.80$ and, therefore, $Z_{eff} = Z - \sigma = 9 - 3.80 = 5.20$. The values of Z_{eff} calculated this way are not the same as those given in Table 1.2 although they do follow the same pattern. The Slater model is an approximation and does not take into account the difference between s and p orbitals or the effects of the various interactions between electrons in the same orbital, such as spin correlation (which is discussed in Section 1.5(a)).

EXAMPLE 1.5 Calculating shielding constants

Calculate the shielding constants for the outermost electron in Mg.

Answer We need to write down the electron configuration of the atom and group the orbitals as described above: Mg $(1s^2)(2s^2 2p^6)(3s^2)$. We can now calculate the shielding constant by assigning values to each electron other than the outermost one. So for Mg we have $\sigma = (1 \times 0.35) + (8 \times 0.85) + (2 \times 1.0) = 9.15$.

Self-test 1.5 (a) Calculate the shielding constant for the outermost electron in Si. (b) Calculate the effective nuclear charge on the outermost electron in Cl.

The closer to the nucleus that an electron can approach, the closer the value of Z_{eff} is to Z itself because the electron is repelled less by the other electrons present in the atom. With this point in mind, consider a 2s electron in the Li atom. There is a nonzero probability that the 2s electron can be found inside the 1s shell and experience the full nuclear charge (Fig. 1.18). The potential for the presence of an electron inside shells of other electrons is called **penetration**. A 2p electron does not penetrate so effectively through the **core**, the filled inner shells of electrons, because

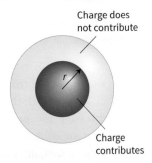

FIGURE 1.17 The electron at the r radius experiences a repulsion from the total charge within the sphere of radius r; charge outside that radius has no net effect.

TABLE 1.2 Computed effective nuclear charge, Z_{eff}

	H							He
Z	1							2
1s	1.00							1.69
	Li	Be	B	C	N	O	F	Ne
Z	3	4	5	6	7	8	9	10
1s	2.69	3.68	4.68	5.67	6.66	7.66	8.65	9.64
2s	1.28	1.91	2.58	3.22	3.85	4.49	5.13	5.76
2p			2.42	3.14	3.83	4.45	5.10	5.76
	Na	Mg	Al	Si	P	S	Cl	Ar
Z	11	12	13	14	15	16	17	18
1s	10.63	11.61	12.59	13.57	14.56	15.54	16.52	17.51
2s	6.57	7.39	8.21	9.02	9.82	10.63	11.43	12.23
2p	6.80	7.83	8.96	9.94	10.96	11.98	12.99	14.01
3s	2.51	3.31	4.12	4.90	5.64	6.37	7.07	7.76
3p			4.07	4.29	4.89	5.48	6.12	6.76

feature of many-electron atoms. This pattern can be seen from Table 1.2, which gives the calculated values of Z_{eff} for all atomic orbitals in the ground-state electron configuration of atoms. The typical trend in effective nuclear charge is an increase across a period, for in most cases the increase in the positive nuclear charge in successive elements is not fully cancelled by the additional electron. The values in the table also confirm that an s electron in the outermost shell of the atom is generally less shielded than a p electron of that shell. So, for example, $Z_{eff} = 5.13$ for a 2s electron in an F atom, whereas for a 2p electron $Z_{eff} = 5.10$, a lower value. Similarly, the effective nuclear charge is larger for an electron in an np orbital than for one in an nd orbital.

As a result of penetration and shielding, the order of energies in many-electron atoms is typically ns, np, nd, nf, because, in a given shell, s orbitals are the most penetrating and f orbitals are the least penetrating. The overall effect of penetration and shielding is depicted in the energy-level diagram for a neutral atom shown in Fig. 1.19.

Figure 1.20 summarizes the energies of the orbitals through the periodic table. The effects are quite subtle, and the order of the orbitals depends strongly on the numbers of electrons present in the atom and may change on ionization. For example, the effects of penetration are very pronounced for 4s electrons in K and Ca, and in these atoms the 4s orbitals lie lower in energy than the 3d orbitals. However, from Sc through Zn, the 3d orbitals in the neutral atoms lie close to but lower than the 4s orbitals. In atoms from Ga ($Z \geq 31$) onwards, the 3d orbitals lie well below the 4s orbital in energy, and the outermost electrons are unambiguously those of the 4s and 4p subshells.

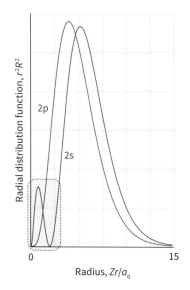

FIGURE 1.18 The penetration of a 2s electron through the inner core is greater than that of a 2p electron because the latter vanishes at the nucleus. Therefore, the 2s electrons are less shielded than the 2p electrons.

its wavefunction goes to zero at the nucleus. As a consequence, it is more fully shielded from the nucleus by the core electrons. We can conclude that in a many-electron atom a 2s electron has a lower energy (is bound more tightly) than a 2p electron, and therefore that the 2s orbital will be occupied before the 2p orbitals, giving a ground-state electron configuration for Li of $1s^22s^1$. This configuration is commonly denoted [He]$2s^1$, where [He] denotes the atom's helium-like $1s^2$ core.

The pattern of orbital energies in lithium, with 2s lower than 2p, and normally ns lower than np, is a general

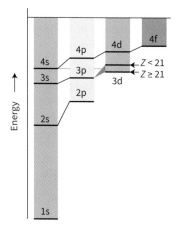

FIGURE 1.19 A schematic diagram of the energy levels of a many-electron atom with $Z < 21$ (as far as calcium). There is a change in order for $Z \geq 21$ (from scandium onwards). This is the diagram that justifies the building-up principle, with up to two electrons being allowed to occupy each orbital.

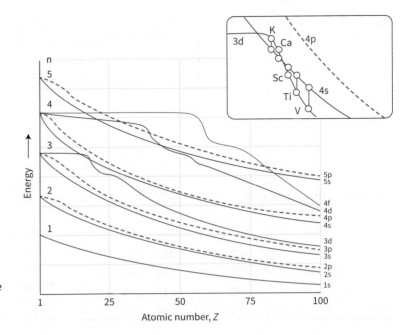

FIGURE 1.20 A more detailed portrayal of the energy levels of many-electron atoms in the periodic table. The inset shows a magnified view of the order near $Z = 20$, where the 3d series of elements begins.

EXAMPLE 1.6 Accounting for trends in effective nuclear charge

From Table 1.2, the increase in Z_{eff} (2p) between C and N is 0.69, whereas the increase between N and O (2p) is only 0.62. Suggest a reason why the increase in Z_{eff} for a 2p electron is smaller between N and O than between C and N given the configurations of the atoms listed above.

Answer We need to identify the general trend and then think about an additional effect that might modify it. In this case, we expect to see an increase in effective nuclear charge across a period, and indeed we do. However, on going from C to N, the additional electron occupies an empty 2p orbital, whereas on going from N to O, the additional electron must occupy a 2p orbital that is already occupied by one electron. It therefore experiences stronger electron–electron repulsion. Electron–electron repulsion contributes to the overall shielding effect and so the increase in Z_{eff} is not as great.

Self-test 1.6 (a) Account for the larger increase in effective nuclear charge for a 2p electron on going from B to C compared with a 2s electron on going from Li to Be. (b) Account for the increase in effective nuclear charge on going from B to Al.

1.5 The building-up principle

The ground-state electron configurations of many-electron atoms are determined experimentally by spectroscopy and are summarized in Resource section 2. To account for them, we need to consider both the effects of penetration and shielding on the energies of the orbitals and the role of the Pauli exclusion principle. The **building-up principle** (which is also known as the *Aufbau principle* and is described below) is a procedure that leads to plausible ground-state configurations. It is not infallible, but it is an excellent starting point for the discussion. Moreover, as we shall see, it provides a theoretical framework for understanding the structure and implications of the periodic table.

(a) Ground-state electron configurations

KEY POINTS The order of occupation of atomic orbitals follows the order 1s, 2s, 2p, 3s, 3p, 4s, 3d, 4p, ... Degenerate orbitals are occupied singly before being doubly occupied; certain modifications of the order of occupation occur for d and f orbitals.

According to the building-up principle, orbitals of neutral atoms are treated as being occupied in the order determined in part by the principal quantum number and in part by penetration and shielding:

Order of occupation: 1s 2s 2p 3s 3p 4s 3d 4p…

Each orbital can accommodate up to two electrons. Thus, the three orbitals in a p subshell can accommodate a total of six electrons and the five orbitals in a d subshell can accommodate up to ten electrons. The ground-state configurations of the first five elements are therefore expected to be

H	He	Li	Be	B
$1s^1$	$1s^2$	$1s^2 2s^1$	$1s^2 2s^2$	$1s^2 2s^2 2p^1$

This order agrees with experiment. When more than one orbital of the same energy is available for occupation, such as when the 2p orbitals begin to be filled in boron and carbon, we adopt **Hund's rule**:

When more than one orbital has the same energy, electrons occupy separate orbitals and do so with parallel spins (↑↑).

The occupation of separate orbitals of the same value of l (such as a p_x orbital and a p_y orbital) can be understood in terms of the weaker repulsive interactions that exist between electrons occupying different regions of space (electrons in different orbitals) than between those occupying the same region of space (electrons in the same orbital). The requirement of parallel spins for electrons that do occupy different degenerate orbitals is a consequence of a quantum-mechanical effect called **spin correlation**, the tendency for two electrons with parallel spins to stay apart from one another. The motions of two such electrons are strongly correlated to avoid simultaneous occupancy of the same space, which would result in a force aptly known as **Pauli repulsion**.

An additional, purely quantum mechanical, factor that stabilizes arrangements of electrons with parallel spins is the **exchange energy**. The exchange energy is the extra stability that a parallel spin configuration (↑↑) gains because the electrons are indistinguishable and interchangeable. When two electrons have parallel spins in degenerate orbitals they very effectively avoid each other and can swap positions leading to a reduction in the energy of the system relative to configurations with antiparallel spins. If one of the electrons of a pair with parallel spins is removed then this exchange energy is lost. Consequently, arrangements of electrons in degenerate orbitals with large numbers of parallel spins are stabilized relative to those without.

Obvious consequences of exchange energy arise with electron configurations having a maximum number of parallel spins. A conspicuous example is that **half-filled shells**, such as p^3, d^5, and f^7, have enhanced stability, as removing an electron from these configurations is accompanied by (local) maximum loss of exchange energy. Removing one electron from the d^5 configuration (↑↑↑↑↑) to give (↑↑↑↑) reduces the number of pairs of electrons with parallel spins, and hence the possible number of electron exchanges, from 10 to 6 if each possible pair of electrons is considered. One result of this preference for arrangements with half-filled shells is that the ground state of the chromium atom is $4s^1 3d^5$ rather than $4s^2 3d^4$ as the former maximizes the exchange energy.

It is arbitrary which of the p orbitals of a subshell is occupied first because they are degenerate, but it is common to adopt the alphabetical order p_x, p_y, p_z. It then follows from the building-up principle that the ground-state configuration of C is $1s^2 2s^2 2p_x^1 2p_y^1$ or, more simply, $1s^2 2s^2 2p^2$. If we recognize the helium-like core ($1s^2$), an even briefer notation is [He]$2s^2 2p^2$, and we can think of the electronic valence structure of the atom as consisting of two paired 2s electrons and two parallel 2p electrons surrounding a closed helium-like core. The electron configurations of the remaining elements in the period are similarly

C	N	O	F	Ne
[He]$2s^2 2p^2$	[He]$2s^2 2p^3$	[He]$2s^2 2p^4$	[He]$2s^2 2p^5$	[He]$2s^2 2p^6$

The $2s^2 2p^6$ configuration of neon is another example of a **closed shell**, a shell with its full complement of electrons. The configuration $1s^2 2s^2 2p^6$ is denoted [Ne] when it occurs as a core.

The ground-state configuration of Na is obtained by adding one more electron to a neon-like core, and is [Ne]$3s^1$, showing that it consists of a single electron outside a completely filled $1s^2 2s^2 2p^6$ core. Now a similar sequence of filling subshells begins again, with the 3s and 3p orbitals complete at argon, with configuration [Ne]$3s^2 3p^6$, which can be denoted [Ar]. Because the 3d orbitals are so much higher in energy, this configuration is effectively closed. Moreover, the 4s orbital is next in line for occupation, so the configuration of K is analogous to that of Na, with a single electron outside a noble-gas core: specifically, it is [Ar]$4s^1$. The next electron, for Ca, also enters the 4s orbital, giving [Ar]$4s^2$, which is the analogue of Mg. However, in the next element, Sc, the added electron occupies a 3d orbital, and filling of the d orbitals begins.

(b) Exceptions

KEY POINTS Inter-electronic repulsion can result in higher-energy orbitals being filled before lower-energy orbitals; the additional energy associated with a half-filled or fully filled orbital can result in unexpected orbital occupancies.

The energy levels in Figs. 1.19 and 1.20 are for individual atomic orbitals and do not fully take into account repulsion between electrons. For elements with an incompletely filled d subshell, the determination of actual ground states by spectroscopy and calculation shows that it is advantageous to occupy orbitals predicted to be *higher* in energy (the 4s orbitals). The explanation for this order is that the occupation of the 4s orbitals can result in a reduction in the repulsions between electrons that would occur if the 3d orbitals were occupied. It is essential when assessing the total energy of the electrons to consider all contributions to the energy of a configuration and particularly electron repulsions, not merely the one-electron orbital energies. Spectroscopic data show that the ground-state configurations of these atoms, the first row transition metals, are mostly of the form $3d^n 4s^2$, with the 4s orbitals fully occupied despite individual 3d orbitals being lower in energy.

An additional feature, another consequence of spin correlation and exchange energies, is that in some cases a lower total energy may be obtained by forming a half-filled or filled d subshell, even though that may mean moving an s electron into the d subshell. Therefore, as a half-filled d shell is approached, the ground-state configuration is likely to be d^5s^1 and not d^4s^2 (as for Cr). As a full d subshell is approached, the configuration is likely to be $d^{10}s^1$ rather than d^9s^2 (as for Cu) or $d^{10}s^0$ rather than d^8s^2 (as for Pd). A similar effect occurs where f orbitals are being occupied, and a d electron may move into the f subshell so as to achieve an f^7 or an f^{14} configuration, with a net lowering of energy.

For cations and complexes of the d-block elements the removal of electrons reduces the effect of electron–electron repulsions and the 3d orbital energies fall well below that of the 4s orbitals. Consequently, all d-block cations and complexes have d^n configurations and no electron in the outermost s orbital. For example, the configuration of an Fe atom is $[Ar]3d^64s^2$ whereas in $[Fe(CO)_5]$ the configuration is $[Ar]3d^8$ and Fe^{2+} has the configuration $[Ar]3d^6$. For the purposes of chemistry, the electron configurations of the d-block ions are more important than those of the neutral atoms. In later chapters (starting in Chapter 20) we shall see the great significance of the configurations of the d-metal ions, for the subtle modulations of their energies provide the basis for the explanations of important properties of their compounds.

EXAMPLE 1.7 Deriving an electron configuration

Predict the ground-state electron configurations of (a) P, (b) Ti, and (c) Ti^{3+}.

Answer We need to use the building-up principle and Hund's rule to populate atomic orbitals with electrons. (a) For the P atom, for which $Z = 15$, we must add 15 electrons in the order specified above, with no more than two electrons in any one orbital. This procedure results in the configuration $[Ne]3s^23p^3$ with the three 3p electrons each in a different p orbital with parallel spins. (b) For the neutral Ti atom, for which $Z = 22$, we must add 22 electrons in the order specified above, with no more than two electrons in any one orbital. This results in the configuration $[Ar]4s^23d^2$, with the two 3d electrons in different orbitals with parallel spins. However, because the 3d orbitals lie below the 4s orbitals for elements beyond Ca, it is usual to reverse the order in which they are written. The configuration is therefore reported as $[Ar]3d^24s^2$. (c) The Ti^{3+} cation has 19 electrons. We should fill the orbitals in the order specified above remembering, however, that the cation will have a d^n configuration and no electrons in the s orbital. The configuration of Ti^{3+} is therefore $[Ar]3d^1$.

Self-test 1.7 (a) Predict the ground-state electron configurations of Ni and Ni^{2+}. (b) Predict the ground-state electron configurations of Cu, Cu^+, and Cu^{2+}.

1.6 The classification of the elements

KEY POINTS The elements are broadly divided into metals, nonmetals, and metalloids according to their physical and chemical properties; the organization of elements into the form resembling the modern periodic table is accredited to Mendeleev.

A useful broad division of elements is into **metals** and **nonmetals**. Metallic elements (such as iron and copper) are typically lustrous, malleable, ductile, electrically conducting solids at about room temperature. Nonmetals are often gases (oxygen), liquids (bromine), or solids that do not conduct electricity appreciably (sulfur). The chemical implications of this classification should already be clear from introductory chemistry:

1. Metallic elements combine with nonmetallic elements to give compounds that are typically hard, nonvolatile solids (for example sodium chloride).

2. When combined with each other, the nonmetals often form volatile molecular compounds (for example, carbon dioxide).

3. When metals combine (or simply mix together) they produce alloys that have most of the physical characteristics of metals (for example, brass from copper and zinc).

Some elements have properties that make it difficult to classify them as metals or nonmetals. These elements are called **metalloids**. Examples of metalloids are silicon, germanium, arsenic, and tellurium.

A NOTE ON GOOD PRACTICE

You will sometimes see metalloids referred to as 'semimetals'. This name is best avoided because a semimetal has a well-defined and quite distinct meaning in solid-state physics (see Section 2.11).

(a) The periodic table

A more detailed classification of the elements is the one devised by Dmitri Mendeleev in 1869; this scheme is familiar to every chemist, and many non-chemists, as the **periodic table**. Mendeleev arranged the known elements in order of increasing atomic weight (molar mass). This arrangement resulted in families of elements with similar chemical properties, which he arranged into the groups of the periodic table. For example, the fact that C, Si, Ge, and Sn all form hydrides of the general formula EH_4 suggests that they belong to the same group. That N, P, As, and Sb all form hydrides with the general formula EH_3 suggests that they belong to a different group. Other compounds of these elements show family similarities, as in the formulas CF_4 and SiF_4 in the first group, and NF_3 and PF_3 in the second.

Mendeleev concentrated on the chemical properties of the elements. At about the same time, Lothar Meyer in Germany was investigating their physical properties, and found that similar values repeated periodically with increasing molar mass. Figure 1.21 shows a classic example, where the molar volume of the element (its volume per mole of atoms) at 1 bar and 298 K is plotted against atomic number.

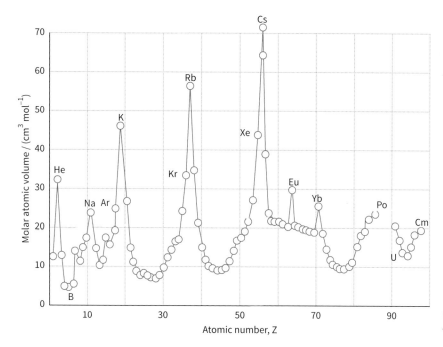

FIGURE 1.21 The periodic variation of molar volume with atomic number.

Mendeleev provided a spectacular demonstration of the usefulness of the periodic table by predicting the general chemical properties, such as the numbers of bonds they form, of unknown elements such as gallium, germanium, and scandium corresponding to gaps in his original periodic table. (He also predicted elements that we now know cannot exist, and denied the presence of elements that we now know do exist, but that is overshadowed by his positive achievement.) The same process of inference from periodic trends is still used by inorganic chemists to rationalize trends in the physical and chemical properties of compounds and to suggest the synthesis of previously unknown compounds. For instance, by recognizing that carbon and silicon are in the same family, the existence of alkenes $R_2C=CR_2$ suggests that $R_2Si=SiR_2$ ought to exist too. Compounds with silicon–silicon double bonds (di-silaethenes) do indeed exist, but it was not until 1981 that chemists succeeded in isolating one. The periodic trends in the properties of the elements are explored further in Chapter 9. The periodic table is an iconic and evolving representation of the known elements and is still being added to today (Box 1.5).

> **BOX 1.5** How far can the periodic table expand?
>
> In January 2016 the International Union of Pure and Applied Chemistry (IUPAC) confirmed the discovery of four new elements, so completing the seventh row of the periodic table. These elements had atomic numbers 113, 115, 117, and 118 and were initially given the temporary, and rather uninspiring, names ununtrium, Uut, ununpentium, UUp, ununseptimum, Uus, and ununoctium, Uuo, though they have since been named as nihonium, Nh, muscovium, Mc, tennessine, Ts, and organesson, Og. The procedures for naming new elements are clearly laid out by IUPAC: the claim for the discovery of a new element is verified by IUPAC and the International Union of Pure and Applied Physics (IUPAP) and then formally assigned to a group of scientists or a laboratory. The IUPAC Inorganic Chemistry Committee then invites the discoverers to propose a name and a symbol, which must fall within strict guidelines. The name can be based on a mythological concept, a mineral, a person, a place, or a property, and should have an ending that fits in with other elements in the appropriate part of the periodic table. For example, elements in groups 1 to 16 must end in 'ium', those in group 17 would end in 'ine', and those in group 18 must end in 'on'.
>
> Element 113 was synthesized in Japan, and its discoverers suggested the name nihonium to reflect that. Nihon is one of two ways to say 'Japan' and means 'land of the rising Sun'. Nihonium was the first element to be discovered in Asia. Similarly, elements 114 and 117 were named in honour of the geographical regions in which their discoverers carried out their science, those being Moscow and Tennessee, USA, respectively. The discoverers of element 118 were based in Moscow and the USA and, in proposing organesson, chose to honour scientist Yuri Organessian for his contribution to the chemistry and physics of the superheavy elements.
>
> The synthesis of new elements is not really within the remit of typical chemists, being more the domain of physicists. The

interest in new elements does, however, excite both groups of scientists who are extremely interested is knowing how far the periodic table can extend, and whether there is limit to elemental size. Based on nuclear stability, some calculations predict that a nucleus cannot exist beyond Z = 126. Another consideration is electron velocity: the average velocity of an electron is proportion to Z (eqn. 1.9). It is predicted that at Z = 137 electron velocity will reach that of the speed of light, providing a different predicted limit for atomic number.

Up to atomic number 100 (fermium), unstable elements are normally synthesized via neutron capture by one element, which is followed by β decay, resulting in an increase in atomic number of one. For the very heavy nuclei, it is necessary to fuse two lighter nuclei together. The four new elements reported in 2016 were produced by accelerating ^{48}Ca nuclei and impacting them on californium or berkelium targets. This method is incredibly inefficient and the total number of atoms produced of each of the new elements is in single figures. Even these few atoms of the new superheavy element were very unstable and decayed almost immediately, producing new isotopes of known lighter elements.

Although the very short half-lives of the heavy elements precludes many synthetic procedures, innovative techniques have been applied, and complexes of einsteinium (Section 21.11), dubnium, and seaborgium (Section 20.13) have all been characterized. More compounds will surely follow.

(b) The format of the periodic table

KEY POINTS The blocks of the periodic table reflect the identity of the orbitals that are occupied last in the building-up process. The period number is the principal quantum number of the valence shell. The group number is related to the number of valence electrons.

The layout of the periodic table reflects the electronic structure of the atoms of the elements (Fig. 1.22). We can now see, for instance, that a **block** of the table indicates the type of subshell currently being occupied according to the building-up principle. Each **period**, or row, of the table corresponds to the completion of the s, p, d, and f subshells of a given shell. The period number is the value of the principal quantum number n of the shell which according to the building-up principle is currently being occupied in the main groups of the table. For example, Period 2 corresponds to the $n = 2$ shell and the filling of the 2s and 2p subshells.

The group numbers, G, are closely related to the number of electrons in the **valence shell**, the outermost shell of the atom. In the '1–18' numbering system recommended by IUPAC:

Block:	s	p	d
Number of electrons in valence shell:	G	$G - 10$	G

For the purpose of this expression, the 'valence shell' of a d-block element consists of the ns and $(n - 1)$d orbitals, so a Sc atom has three valence electrons (two 4s and one 3d electron). The number of valence electrons for the p-block element Se (Group 16) is $16 - 10 = 6$, which corresponds to the configuration s^2p^4.

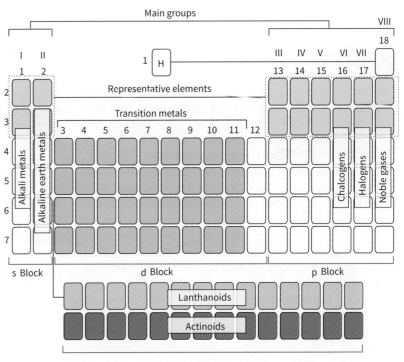

FIGURE 1.22 The general structure of the periodic table. Compare this template with the complete table at the front of the book for the identities of the elements that belong to each block.

EXAMPLE 1.8 Placing elements within the periodic table

Without consulting a periodic table, state to which period, group, and block of the periodic table the element with the electron configuration $1s^2 2s^2 2p^6 3s^2 3p^4$ belongs. Identify the element.

Answer We need to remember that the period number is given by the principal quantum number, n, that the group number can be found from the number of valence electrons, and that the identity of the block is given by the type of orbital last occupied according to the building-up principle. The valence electrons have $n = 3$, therefore the element is in Period 3 of the periodic table. The six valence electrons identify the element as a member of Group 16. The electron added last is a p electron, so the element is in the p block. The element is sulfur.

Self-test 1.8 (a) To which period, group, and block of the periodic table will the element with the electron configuration $1s^2 2s^2 2p^6 3s^2 3p^6 4s^2$ belong? Identify the element. (b) To which period, group, and block of the periodic table will the element with the electron configuration $1s^2 2s^2 2p^6 3s^2 3p^6 4s^2 3d^{10} 4p^6 5s^1 4d^5$ belong? Identify the element.

1.7 Atomic properties

Certain characteristic properties of atoms, particularly their radii and the energies associated with the removal and addition of electrons, show regular periodic variations with atomic number. These atomic properties are of considerable importance for understanding the chemical properties of the elements and are discussed further in Chapters 9 and 10. A knowledge of these trends enables chemists to rationalize observations and predict likely chemical and structural behaviour without having to refer to tabulated data for each element.

(a) Atomic and ionic radii

KEY POINTS Atomic radii increase down a group and, within the s and p blocks, decrease from left to right across a period. The lanthanoid contraction results in a decrease in atomic radius for elements following the f block. All monatomic anions are larger than their parent atoms and all monatomic cations are smaller.

One of the most useful atomic characteristics of an element is the size of its atoms and ions. As we shall see in later chapters, geometrical considerations are central to explaining the structures of many solids and individual molecules. In addition, the average distance of an electron from the nucleus of an atom correlates with the energy needed to remove it in the process of forming a cation.

An atom does not have a precise radius because far from the nucleus the electron density falls off exponentially (but sharply). However, we can expect atoms with numerous electrons to be larger, in some sense, than atoms that have only a few electrons. Such considerations have led chemists to propose a variety of definitions of atomic radius on the basis of empirical considerations.

The **metallic radius** of a metallic element is defined as half the experimentally determined distance between the centres of nearest-neighbour atoms in the solid (Fig. 1.23(a), but see Section 4.7 for a refinement of this definition). The **covalent radius** of a nonmetallic element is similarly defined as half the internuclear distance between neighbouring atoms of the same element in a molecule (Fig. 1.23(b)).

Metallic and covalent radii are frequently both referred to as **atomic radii** but, because of their different origins, they are often different. In this chapter we shall restrict our discussion to covalent radii, as these values permit comparisons across the whole of the periodic table; in later chapters we will make reference to metallic radii, as appropriate. The periodic trends in covalent radii are illustrated diagrammatically in Figs. 1.24 and 1.25.

Covalent radii are very useful in predicting the separation of differing atoms in covalently bonded molecules, i.e. the bond length, but there are also many compounds that contain ions, where an atom has gained or lost electrons. Thus, a further useful concept in inorganic chemistry is a measurement of the size an ion, its **ionic radius**.

As with atomic radii there is no abrupt extent of an electron cloud in an ion so the ionic radius (Fig. 1.23(c)) of an element can be related to the distance between the centres of neighbouring cations and anions in an ionic compound. An arbitrary decision has to be taken on how to apportion the cation–anion distance between the two ions. There have been many suggestions: in one common scheme, the radius of the O^{2-} ion is taken to be 140 pm (Table 1.3; see Section 4.10 for a refinement of this definition). For example,

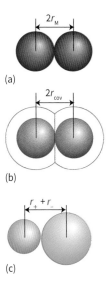

FIGURE 1.23 A representation of (a) metallic radius, (b) covalent radius, and (c) ionic radius.

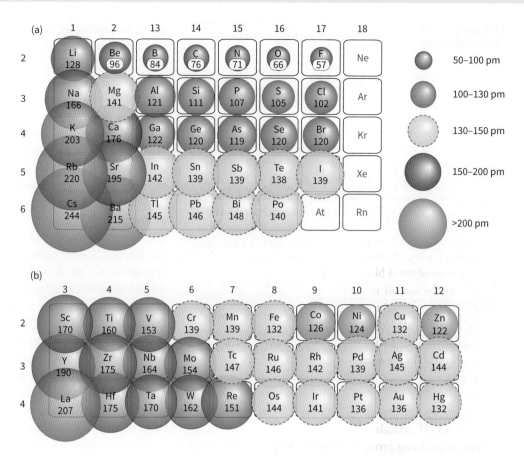

FIGURE 1.24 The variation of covalent radii (in picometres) within the periodic table: (a) main group; (b) d block.

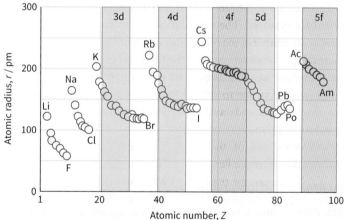

FIGURE 1.25 The variation of atomic radii through the periodic table. Note the contraction of radii following the lanthanoids in Period 6.

TABLE 1.3 Ionic radii, r/pm* (For more values, see Resource section 1)

Li⁺	Be²⁺	B³⁺			N³⁻	O²⁻	F⁻
59(4)	27(4)	11(4)			146(4)	140(6)	133(6)
Na⁺	Mg²⁺	Al³⁺			P³⁻	S²⁻	Cl⁻
102(6)	72(6)	53(6)			192(4)	184(6)	181(6)
K⁺	Ca²⁺	Ga³⁺			As³⁻	Se²⁻	Br⁻
138(6)	100(6)	62(6)			222(4)	198(6)	196(6)
Rb⁺	Sr²⁺	In³⁺	Sn²⁺	Sn⁴⁺		Te²⁻	I⁻
152(6)	118(6)	80(6)	102(6)	69(6)		221(6)	220(6)
Cs⁺	Ba²⁺	Tl³⁺	Pb²⁺	Pb⁴⁺			
167(6)	135(6)	89(6)	119(6)	78(6)			

* Numbers in parentheses are the coordination number of the ion.

the ionic radius of Mg^{2+} is obtained by subtracting 140 pm from the internuclear distance between adjacent Mg^{2+} and O^{2-} ions in solid MgO of 212 pm; this gives the ionic radius of Mg^{2+} as 72 pm. See Resource section 1 for other values of ionic radii.

The data in Figs. 1.24 and 1.25 show that *atomic radii increase down a group*, and that they *decrease from left to right across a period*. These trends are readily interpreted in terms of the electronic structure of the atoms. On descending a group, the valence electrons are found in orbitals of successively higher principal quantum number. The atoms within the group have a greater number of completed shells of electrons in successive periods and hence their radii normally increase down the group. Across a period, the valence electrons enter orbitals of the same shell; however, the increase in effective nuclear charge across the period draws in the electrons and results in progressively more compact atoms. The general increase in radius down a group and decrease across a period should be remembered as they correlate well with trends in many chemical properties.

The increase in atomic radii on descending a group is relatively small for the elements that follow the d block due to the presence of the poorly shielding d electrons. For example, although there is a substantial increase in atomic radius between C and Si (76 and 111 pm, respectively), the atomic radius of Ge (120 pm) is only slightly greater than that of Si (111 pm). An effect with a similar origin is seen with period 6 (Cs, Ba, ...). We see from Fig. 1.24 that the atomic radii in the third row of the d block (Hf, Ta, W, ...) are very similar to those in the second row (Zr, Nb, Mo, ...), and not significantly larger as might be expected given their considerably greater numbers of electrons. The reduction of radius below that expected on the basis of a simple extrapolation down the group is a result of the so-called **lanthanoid contraction**.

The name points to the origin of the effect. The elements in the third row of the d block (Period 6) are preceded by the elements of the first row of the f block, the lanthanoids, in which the 4f orbitals are being occupied. These orbitals have poor shielding properties and so the valence electrons experience more attraction from the nuclear charge than might be expected. The repulsions between electrons being added on crossing the f block fail to compensate for the increasing nuclear charge, so Z_{eff} increases between La and Lu. The dominating effect of the latter is to draw in all the electrons and hence to result in a more compact atom for the later lanthanoids and the third row d-block elements that follow them.

The properties of the heavier elements, notably those within and following Period 6, are increasingly influenced by relativistic effects, the reasons for which are easily explained, albeit only at a qualitative level. Electrons in s and p orbitals approach closely to the nucleus and may experience strong accelerations. As the average velocity of an electron, v, is directly proportional to Z, for large values of Z, v approaches the speed of light (predicted to occur at Z = 137).

$$v = \frac{Ze^2}{2\varepsilon_0 nh} \quad (1.9)$$

According to the theory of relativity, the mass of an electron increases above its rest mass m_e as its velocity approaches the speed of light.

$$m = \frac{m_e}{\sqrt{1-(v/c)^2}} \quad (1.10)$$

The increase in mass causes a contraction in the radii of s orbitals (eqn. 1.8) and to a lesser extent p orbitals, resulting in their stabilization. The less penetrating d and f orbitals, now more effectively screened by the contracted s and p shells, expand, becoming more diffuse, and their energies increase. For light elements, relativistic effects can be neglected, but for the heavier elements with high atomic numbers they become significant; for instance, with Hg (Z = 80) the 1s orbital is contracted by 23% compared to that expected before allowing for relativity. Relativistic effects (Section 10.6) underpin many familiar observations—the colour of gold, the liquid state of mercury, the inert pair effect (Section 10.7) being a few examples—and they are essential for understanding the properties of actinoids.

Ionic radii show similar trends to atomic radii, and it is immediately apparent from Table 1.3 that all monatomic anions are larger than their parent atoms, and all monatomic cations are smaller than their parent atoms (in some cases markedly so). The increase in radius of an atom on anion formation is a result of the greater electron–electron repulsions that occur when an additional electron is added to form an anion. There is also an associated decrease in the value of Z_{eff}. The smaller radius of a cation compared with its parent atom is a consequence not only of the reduction in electron–electron repulsions that follow electron loss, but also of the fact that cation formation typically results in the loss of the valence electrons and an increase in Z_{eff}. That loss often leaves behind only the much more compact closed shells of electrons. Once these gross differences are taken into account, the variation in ionic radii through the periodic table mirrors that of the atoms.

Although small variations in atomic radii may seem of little importance, in fact atomic radius plays a central role in the chemical properties of the elements. Small changes can have profound consequences, as we shall see in Chapters 9 and 10.

(b) Enthalpies of atomization

KEY POINT Across each row, enthalpies of atomization increase as bonding orbitals are filled, then decrease as antibonding orbitals become occupied.

The enthalpy of atomization of an element, $\Delta_a H^\ominus$, is a measure of the energy required to form gaseous atoms. For solids, the enthalpy of atomization is the enthalpy change associated with the atomization of the solid; for molecular species, it is the enthalpy of dissociation of the molecules. As can be seen in Table 1.4, enthalpies of atomization first increase and then decrease across Periods 2 and 3, reaching a maximum at C in Period 2 and Si in Period 3. The values decrease between C and N, and between Si and P: even though N and P each have five valence electrons, two of these electrons form a lone pair and only three are involved in bonding. A similar effect is seen between N and O, where O has six valence electrons, of which four form lone pairs and only two are involved in bonding. These trends are shown in Fig. 1.26.

TABLE 1.4 Enthalpies of atomization, $\Delta_a H^\ominus$ (kJ mol^{-1})

Li	Be											B	C	N	O	F
159	324											563	717	473	249	79
Na	Mg											Al	Si	P	S	Cl
107	146											326	456	315	279	122
K	Ca	Sc	Ti	V	Cr	Mn	Fe	Co	Ni	Cu	Zn	Ga	Ge	As	Se	Br
89	178	378	471	515	397	281	415	426	431	338	131	277	377	302	227	112
Rb	Sr	Y	Zr	Nb	Mo	Tc	Ru	Rh	Pd	Ag	Cd	In	Sn	Sb	Te	I
81	164	425	605	733	659	661	652	556	377	285	112	243	302	262	197	107
Cs	Ba	La	Hf	Ta	W	Re	Os	Ir	Pt	Au	Hg	Tl	Pb	Bi	Po	
76	182	431	621	782	860	776	789	671	565	368	64	182	195	207	142	

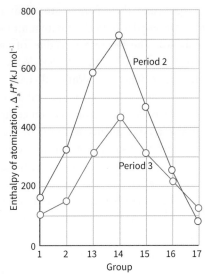

FIGURE 1.26 The variation of the enthalpy of atomization in the s- and p-block elements.

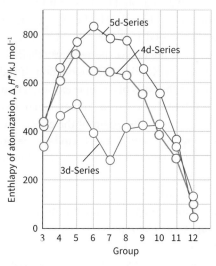

FIGURE 1.27 The variation of the enthalpy of atomization in the d-block elements.

The enthalpies of atomization of the d-block elements are higher than those of the s- and p-block elements, in line with their greater number of valence electrons and consequently stronger bonding. The values reach a maximum at Groups 5 and 6 (Fig. 1.27), where there is a maximum number of unpaired electrons available to form bonds. The middle of each row shows an irregularity due to the exchange energy optimization (Section 1.5(a)), which favours a half-filled d shell for the free atom. This effect is more pronounced for compact orbitals and is particularly evident for the 3d series, in which Cr (3d^{5}4s^1) and Mn (3d^{5}4s^2) have significantly lower atomization energies than expected from a simple consideration of their number of valence electrons.

The enthalpy of atomization decreases down a group in the s and p blocks but increases down a group in the d block. Thus, s and p orbitals become less effective at forming bonds as the period number increases, whereas d orbitals become more effective. These trends are attributed to the expansion of p orbitals on descending a group from optimal for overlap to too diffuse for extensive overlap and, in contrast, d orbitals expanding in size from too contracted to optimal for overlap.

The same trends can be seen in the melting points (Table 1.5) of the elements, where a greater number of valence electrons leads to greater binding energy and a higher melting temperature. The melting points of the Group 15–17 elements are influenced more by intermolecular interactions than valence electrons.

(c) Ionization energy

KEY POINTS First ionization energies are lowest at the lower left of the periodic table (near caesium) and greatest near the upper right (near helium). Successive ionizations of a species require higher energies.

The ease with which an electron can be removed from an atom is measured by its **ionization energy**, I, the minimum energy needed to remove an electron from a gas-phase atom:

$$A(g) \rightarrow A^+(g) + e^-(g) \quad I = E(A^+, g) - E(A, g) \quad (1.11)$$

The **first ionization energy**, I_1, is the energy required to remove the least tightly bound electron from the neutral atom, the **second ionization energy**, I_2, is the energy required to remove the least tightly bound electron from the resulting

TABLE 1.5 Normal melting points of the elements, θ_{mp}/°C

Li	Be										B	C	N	O	F
180	1280										2300	3730	−210	−218	−220
Na	Mg										Al	Si	P	S	Cl
97.8	650										660	1410	44*	113	−101
K	Ca	Sc	Ti	V	Cr	Mn	Fe	Co	Ni	Cu	Zn	Ga	Ge	As†	Se	Br
63.7	850	1540	1675	1900	1890	1240	1535	1492	1453	1083	420	29.8	937	817	217	−7.2
Rb	Sr	Y	Zr	Nb	Mo	Tc	Ru	Rh	Pd	Ag	Cd	In	Sn	Sb	Te	I
38.9	768	1500	1850	2470	2610	2200	2500	1970	1550	961	321	157	232	630	450	114
Cs	Ba	La	Hf	Ta	W	Re	Os	Ir	Pt	Au	Hg	Tl	Pb	Bi	Po	
28.7	714	920	2220	3000	3410	3180	3000	2440	1769	1063	13.6	304	327	271	254	

* White allotrope.
† Grey allotrope under 28 atm

cation, and so on. Ionization energies are often expressed in electronvolts (eV) but are easily converted into kilojoules per mole by using $1\,\text{eV} = 96.485\,\text{kJ mol}^{-1}$. The ionization energy of the H atom is 13.6 eV, so to remove an electron from an H atom is equivalent to dragging the electron through a potential difference of 13.6 V. This equates to $1312.196\,\text{kJ mol}^{-1}$.

> **A NOTE ON GOOD PRACTICE**
>
> In thermodynamic calculations it is often more appropriate to use the **ionization enthalpy**, the standard enthalpy of the process in eqn. 1.11, typically at 298 K. The molar ionization enthalpy is larger by $\tfrac{5}{2}RT$ than the ionization energy. This difference stems from the change from $T=0$ (assumed implicitly for I) to the temperature T (typically 298 K) to which the enthalpy value refers, and the replacement of 1 mol of gas particles by 2 mol of gaseous ions plus electrons. However, because RT is only $2.5\,\text{kJ mol}^{-1}$ (corresponding to 0.026 eV) at room temperature and ionization energies are of the order of 10^2–$10^3\,\text{kJ mol}^{-1}$ (1–10 eV), the difference between ionization energy and enthalpy can often be ignored.

To a large extent, the first ionization energy of an element is determined by the energy of the highest occupied orbital of its ground-state atom. First ionization energies vary systematically through the periodic table (Table 1.6, Fig. 1.28), being smallest at the lower left (near Cs) and greatest near the upper right (near He). The variation follows the pattern of effective nuclear charge including some subtle modulations arising from the effect of electron–electron repulsions within the same subshell. A useful approximation is that for an electron from a shell with principal quantum number n:

$$I \propto \frac{Z_{\text{eff}}^2}{n^2}$$

Ionization energies also correlate strongly with atomic radii, and elements that have small atomic radii generally have high ionization energies. The explanation of the correlation is that in a small atom an electron is close to the nucleus and experiences a strong Coulombic attraction, making it difficult to remove. Therefore, as the atomic radius increases down a group, the ionization energy decreases and

TABLE 1.6 First, second, and third (and some fourth) ionization energies of the elements, $I/(\text{kJ mol}^{-1})$

H							He
1312							2373
							5259
Li	Be	B	C	N	O	F	Ne
513	899	801	1086	1402	1314	1681	2080
7297	1757	2426	2352	2855	3386	3375	3952
11 809	14 844	3660	4619	4577	5300	6050	6122
		25 018					
Na	Mg	Al	Si	P	S	Cl	Ar
495	737	577	786	1011	1000	1251	1520
4562	1476	1816	1577	1903	2251	2296	2665
6911	7732	2744	3231	2911	3361	3826	3928
		11 574					
K	Ca	Ga	Ge	As	Se	Br	Kr
419	589	579	762	947	941	1139	1351
3051	1145	1979	1537	1798	2044	2103	3314
4410	4910	2963	3302	2734	2974	3500	3565
Rb	Sr	In	Sn	Sb	Te	I	Xe
403	549	558	708	834	869	1008	1170
2632	1064	1821	1412	1794	1795	1846	2045
3900	4210	2704	2943	2443	2698	3197	3097
Cs	Ba	Tl	Pb	Bi	Po	At	Rn
375	502	590	716	704	812	926	1036
2420	965	1971	1450	1610	1800	1600	
3400	3619	2878	3080	2466	2700	2900	

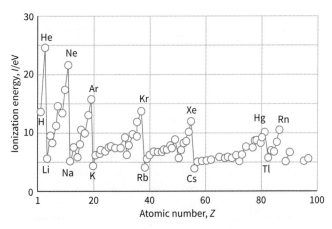

FIGURE 1.28 The periodic variation of first ionization energies.

the decrease in radius across a period is accompanied by a gradual increase in ionization energy.

There are some deviations from this general trend in ionization energy and these can be explained quite readily. One example is the observation that the first ionization energy of boron (801 kJ mol⁻¹) is smaller than that of beryllium (899 kJ mol⁻¹), despite the former's higher nuclear charge. This anomaly is readily explained by noting that, for boron, the outermost electron occupies a 2p orbital and hence is less strongly bound than if it had occupied a 2s orbital. As a result, the value of I_1 decreases from Be to B. A similar effect can be seen with magnesium and aluminium.

Another anomaly is the decrease between nitrogen (1402 kJ mol⁻¹) and oxygen (1314 kJ mol⁻¹), and this has a slightly different explanation. The configurations of the two atoms are

N [He]$2s^2 2p_x^1 2p_y^1 2p_z^1$ O [He]$2s^2 2p_x^2 2p_y^1 2p_z^1$

We see that, in an O atom, two electrons are present in a single 2p orbital and they repel each other strongly. This strong repulsion offsets the greater nuclear charge, and the ionization energy of this electron is reduced. The ionization of an electron from the $2s^2 2p^3$ configuration of nitrogen involves a significant loss of exchange energy from breaking into the half-filled shell, resulting in a higher ionization energy, whereas removing an electron from an O atom to produce an O⁺ ion does not involve any reduction in the exchange energy, as the ionized electron is the only one with the ↓ spin orientation. These two factors result in the ionization energy for oxygen being somewhat smaller than for nitrogen. A similar effect is seen lower down the two groups; for example, with phosphorus and sulfur (Example 1.9).

When considering F and Ne on the right of Period 2, the highest-energy electrons are in orbitals that are already half full and continue the trend from O towards higher ionization energy. The higher values of the ionization energies of these two elements reflect the high value of Z_{eff} and the increase in exchange energy. The value of I_1 falls back sharply from Ne to Na as the outermost electron occupies the next shell with an increased principal quantum number and is therefore further from the nucleus.

EXAMPLE 1.9 Accounting for a variation in ionization energy

Account for the decrease in first ionization energy between phosphorus and sulfur.

Answer We approach this question by considering the groundstate configurations of the two atoms:

P [Ne]$3s^2 3p_x^1 3p_y^1 3p_z^1$ S [Ne]$3s^2 3p_x^2 3p_y^1 3p_z^1$

As in the analogous case of N and O, in the ground state of S, two electrons are present in a single 3p orbital. They are so close together that they repel each other strongly, and this increased repulsion offsets the effect of the greater nuclear charge of S compared with P.

Self-test 1.9 (a) Account for the decrease in first ionization energy between fluorine and chlorine. (b) Explain the decrease in ionization energy between magnesium and aluminium.

Another deviation from the simple trend of decrease down the group is seen in Group 13 where the ionization energy of Ga is greater than that of Al. This is due to the intervention of the 3d subshell earlier in Period 4 leading to an increase in Z_{eff} and a smaller atomic radius for Ga than would be expected in comparison with the trend between Al and In. These deviations are manifestations of the **alternation effect** which is explored further in Section 10.4. The lanthanoid contraction affects the ionization energies of the 5d-series elements, making them higher than expected on the basis of a straightforward extrapolation from the 3d- and 4d-series elements. Some of the metals—specifically Au, Pt, Ir, and Os—have such high ionization energies that they are unreactive except when subjected to extreme conditions.

Another important pattern is that successive ionizations of an element require increasingly higher energies (Table 1.6). Thus, the second ionization energy of an element E (the energy needed to remove an electron from the cation E⁺) is higher than its first ionization energy, and its third ionization energy (the energy needed to remove an electron from E²⁺) is higher still. The explanation is that the higher the positive charge of a species, the greater the electrostatic attraction experienced by the electron being removed, that is, there is a higher proton:electron ratio. When an electron is removed, Z_{eff} increases and the orbitals contract, and it is then even more difficult to remove an electron from this smaller, more compact, cation. The difference in ionization energy is greatly magnified when the electron is removed from a closed shell of the atom (as is the case for the second ionization

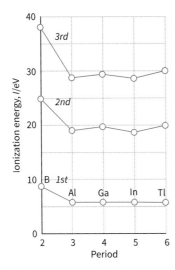

FIGURE 1.29 The first, second, and third ionization energies of the elements of Group 13. Successive ionization energies increase, but there is no clear pattern of ionization energies down the group.

energy of Li and any of its congeners) because the electron must then be extracted from a compact orbital in which it interacts strongly with the nucleus. The first ionization energy of Li, for instance, is 513 kJ mol⁻¹, but its second ionization energy is 7297 kJ mol⁻¹, more than ten times greater.

The pattern of successive ionization energies down a group is far from simple. Figure 1.29 shows the first, second, and third ionization energies of the members of Group 13. Although they lie in the expected order $I_1 < I_2 < I_3$, there is no simple trend. The lesson to be drawn is that whenever an argument hangs on trends in small differences in ionization energies, it is always best to refer to actual numerical values rather than to guess a likely outcome.

EXAMPLE 1.10 **Accounting for values of successive energies of ionization**

Rationalize the following values for successive ionization energies of boron, where $\Delta_{ion}H(N)$ is the Nth enthalpy of ionization:

N	1	2	3	4	5
$\Delta_{ion}H(N)/$(kJ mol⁻¹)	801	2426	3660	25 018	32 834

Answer When considering trends in ionization energy, a sensible starting point is the electron configurations of the atoms. The electron configuration of B is 1s²2s²2p¹. The first ionization energy corresponds to removal of the electron in the 2p orbital. This electron is shielded from nuclear charge by the core and the full 2s orbital. The second value corresponds to removal of a 2s electron from the B⁺ cation. This electron is more difficult to remove on account of the increased effective nuclear charge. The effective nuclear charge increases further on removal of this electron, resulting in an increase between $\Delta_{ion}H(2)$ and $\Delta_{ion}H(3)$. There is a large increase between $\Delta_{ion}H(3)$ and $\Delta_{ion}H(4)$ because the 1s shell lies at very low energy as it experiences almost the full nuclear charge and also has $n=1$. The final electron to be removed experiences no shielding of nuclear charge so $\Delta_{ion}H(5)$

is very high, and is given by $hcRZ^2$ with $Z = 5$, corresponding to (13.6 eV) × 25 = 340 eV (32.8 MJ mol⁻¹).

Self-test 1.10 Study the values listed below of the first five ionization energies of an element and deduce to which group of the periodic table the element belongs. Give your reasoning.

N	1	2	3	4	5
$\Delta_{ion}H(N)/$(kJ mol⁻¹)	1086	2352	4619	6229	37 838

(d) Electron affinity

KEY POINT Electron affinities are highest for elements near fluorine in the periodic table.

The **electron-gain enthalpy**, $\Delta_{eg}H^\ominus$, is the change in standard molar enthalpy when a gaseous atom gains an electron:

$$A(g) + e^-(g) \rightarrow A^-(g)$$

Electron gain may be either exothermic or endothermic. Although the electron-gain enthalpy is the thermodynamically appropriate term, much of inorganic chemistry is discussed in terms of a closely related property, the **electron affinity**, E_a, of an element (Table 1.7), which is the difference in energy between the gaseous atoms and the gaseous ions at $T = 0$.

$$E_a = E(A, g) - E(A^-, g) \quad (1.12)$$

The two terms are often, and wrongly, used interchangeably. Although the precise relation is $\Delta_{eg}H^\ominus = -E_a - \tfrac{5}{2}RT$, the contribution $\tfrac{5}{2}RT$ is commonly ignored. A positive electron affinity indicates that the ion A⁻ has a lower energy than the neutral atom, A, whereas a negative one indicates the A⁻ anion has a higher energy than the neutral atom. The second electron affinity for the attachment of a second electron is invariably negative because it represents adding a second electron to an already negatively charged anion and this repulsion outweighs any nuclear attraction.

TABLE 1.7 Electron affinities of the main-group elements, $E_a/$(kJ mol⁻¹)*

H							He
72							−48
Li	Be	B	C	N	O	F	Ne
60	≤0	27	122	−8	141	328	−116
					−780		
Na	Mg	Al	Si	P	S	Cl	Ar
53	≤0	43	134	72	200	349	−96
					−492		
K	Ca	Ga	Ge	As	Se	Br	Kr
48	2	29	116	78	195	325	−96
Rb	Sr	In	Sn	Sb	Te	I	Xe
47	5	29	116	103	190	295	−77

*The first values refer to the formation of the ion X⁻ from the neutral atom; the second value to the formation of X²⁻ from X⁻.

The electron affinity of an element is largely determined by the energy of the *lowest unfilled* (or half-filled) orbital of the ground-state atom. This orbital is one of the two **frontier orbitals** of an atom, the other one being the *highest filled* atomic orbital. The frontier orbitals are the sites of many of the changes in electron distributions when bonds form, and we shall see more of their importance as the text progresses.

An element has a high electron affinity if the additional electron can enter a shell where it experiences a strong effective nuclear charge. This is the case for elements towards the top right of the periodic table, as we have already explained. Therefore, elements close to fluorine (specifically O and Cl, but not the noble gases) can be expected to have the highest electron affinities as their Z_{eff} is large and it is possible to add electrons to the valence shell. Nitrogen has a very low electron affinity for the same reason it has a high ionization energy: the neutral atom has a half-filled p shell which has significant exchange energy. Thus there is a high electron-electron repulsion when the incoming electron enters an orbital that is already half full and there is no gain in exchange energy as the additional electron has a spin anti-parallel to those of the other 2p electrons.

A NOTE ON GOOD PRACTICE

Be alert to the fact that some people use the terms 'electron affinity' and 'electron-gain enthalpy' interchangeably. In such cases, a positive 'electron affinity' could incorrectly indicate that energy is given out on forming A⁻ from A.

EXAMPLE 1.11 Accounting for the variation in electron affinity

Account for the large decrease in electron affinity between Li and Be despite the increase in nuclear charge.

Answer When considering trends in electron affinities, as in the case of ionization energies, a sensible starting point is the electron configurations of the atoms. The electron configurations of Li and Be are [He]2s^1 and [He]2s^2, respectively. The additional electron enters the 2s orbital of Li but enters the 2p orbital of Be, and hence is much less tightly bound. In fact, the nuclear charge is so well shielded in Be that electron gain is endothermic.

Self-test 1.11 (a) Account for the decrease in electron affinity between C and N. (b) Explain why the values of electron affinity for the Group 18 elements are all negative.

(e) Electronegativity

KEY POINTS The electronegativity of an element is the power of an atom of the element to attract electrons when it is part of a compound; there is a general increase in electronegativity across a period and a general decrease down a group.

The **electronegativity**, χ (chi), of an element was originally defined by Linus Pauling as the power of an atom of the element in a molecule to attract electrons to itself. If an atom has a strong tendency to acquire electrons, it is said to be highly electronegative (like the elements close to fluorine). Electronegativity is a very useful concept in chemistry and has numerous applications, which include a rationalization of bond energies and the types of reaction that substances undergo and the prediction of the polarities of bonds and molecules (Chapter 2).

Quantitative measures of electronegativity have been defined in many different ways. Linus Pauling's original formulation (which results in the values denoted χ_P in Table 1.8) draws on concepts relating to the energetics of bond formation, which will be dealt with in Chapter 2. A later definition more in the spirit of this chapter, in the sense that it is based on the properties of individual atoms, was proposed by Robert Mulliken. He observed that if an atom has a high ionization energy, I, and a high electron affinity, E_a, then it will be likely to acquire rather than lose electrons when it is part of a compound, and hence be classified as highly electronegative. Conversely, if its ionization energy and electron affinity are both low, then the atom will in its compounds still donate electrons rather than gain them, and hence be classified as electropositive. These observations provide the basis of the definition of the **Mulliken electronegativity**, χ_M, as the average value of the ionization energy and the electron affinity of the element (both expressed in electronvolts):

$$\chi_M = \frac{1}{2}(I + E_a) \qquad (1.13)$$

The hidden complication in the apparently simple definition of the Mulliken electronegativity is that the ionization energy and electron affinity in the definition relate to the **valence state**, the electron configuration the atom is supposed to have when it is part of a molecule. Hence, some calculation is required because the ionization energy and electron affinity to be used in calculating χ_M are mixtures of values for various actual spectroscopically observable states of the atom. We need not go into the calculation, but the resulting values given in Table 1.8 may be compared with the Pauling values (Fig. 1.30). The two scales give similar values and show the same trends. One reasonably reliable conversion between the two is

$$\chi_P = 1.35\chi_M^{1/2} - 1.37 \qquad (1.14)$$

Because the elements near F (other than the noble gases) have high ionization energies and appreciable electron affinities, these elements have the highest Mulliken electronegativities. Because χ_M depends on atomic energy levels—and in particular on the location of the highest filled and lowest empty orbitals—the electronegativity of an element is high if the two frontier orbitals of its atoms are low in energy.

TABLE 1.8 Pauling, χ_P, Mulliken, χ_M, and Allred–Rochow, χ_{AR}, electronegativities

	H							He
Pauling, χ_P	2.20							
Mulliken, χ_M	3.06							3.5
Allred–Rochow, χ_{AR}	2.20							5.5
	Li	Be	B	C	N	O	F	Ne
Pauling, χ_P	0.98	1.57	2.04	2.55	3.04	3.44	3.98	
Mulliken, χ_M	1.28	1.99	1.83	2.67	3.08	3.22	4.43	4.60
Allred–Rochow, χ_{AR}	0.97	1.47	2.01	2.50	3.07	3.50	4.10	5.10
	Na	Mg	Al	Si	P	S	Cl	Ar
Pauling, χ_P	0.93	1.31	1.61	1.90	2.19	2.58	3.16	
Mulliken, χ_M	1.21	1.63	1.37	2.03	2.39	2.65	3.54	3.36
Allred–Rochow, χ_{AR}	1.01	1.23	1.47	1.74	2.06	2.44	2.83	3.30
	K	Ca	Ga	Ge	As	Se	Br	Kr
Pauling, χ_P	0.82	1.00	1.81	2.01	2.18	2.55	2.96	3.0
Mulliken, χ_M	1.03	1.30	1.34	1.95	2.26	2.51	3.24	2.98
Allred–Rochow, χ_{AR}	0.91	1.04	1.82	2.02	2.20	2.48	2.74	3.10
	Rb	Sr	In	Sn	Sb	Te	I	Xe
Pauling, χ_P	0.82	0.95	1.78	1.96	2.05	2.10	2.66	2.6
Mulliken, χ_M	0.99	1.21	1.30	1.83	2.06	2.34	2.88	2.59
Allred–Rochow, χ_{AR}	0.89	0.99	1.49	1.72	1.82	2.01	2.21	2.40
	Cs	Ba	Tl	Pb	Bi			
Pauling, χ_P	0.79	0.89	2.04	2.33	2.02			
Mulliken, χ_M	0.70	0.90	1.80	1.90	1.90			
Allred–Rochow, χ_{AR}	0.86	0.97	1.44	1.55	1.67			

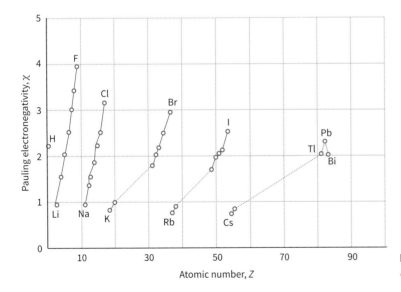

FIGURE 1.30 The periodic variation of Pauling electronegativities.

Various alternative 'atomic' definitions of electronegativity have been proposed. A widely used scale, suggested by A. L. Allred and E. Rochow, is based on the view that electronegativity is determined by the electric field at the surface of an atom. As we have seen, an electron in an atom experiences an effective nuclear charge Z_{eff}. The Coulombic potential at the surface of such an atom is proportional to Z_{eff}/r, and the electric field there is proportional to Z_{eff}/r^2. In the **Allred–Rochow definition** of electronegativity, χ_{AR} is assumed to be proportional to this field, with r taken to be the covalent radius of the atom:

$$\chi_{AR} = 0.744 + \frac{35.90 Z_{eff}}{(r/\text{pm})^2} \qquad (1.15)$$

The numerical constants have been chosen to give values comparable to Pauling electronegativities. According to the Allred–Rochow definition, elements with high electronegativity are those with high effective nuclear charge and a small covalent radius; such elements lie close to F. The Allred–Rochow values parallel closely those of the Pauling electronegativities and are useful for discussing the electron distributions in compounds.

Note that the two definitions of electronegativity that are based on atoms—the Mulliken and the Allred–Rochow—allow values of electronegativity to be calculated for the noble gases that do not readily form stable compounds, i.e. He, Ne, and Ar.

Periodic trends in electronegativity can be predicted by trends in the size of the atoms and electron configuration, even though electronegativity refers to atoms within compounds. If an atom is small and has an almost closed shell of electrons, then it is more likely to have a high electronegativity. Consequently, the electronegativities of the elements typically increase left to right across a period and decrease down a group.

There are some exceptions to this general trend, as can be seen from the following electronegativities:

Al	Si
1.61	1.90
Ga	Ge
1.81	2.01
In	Sn
1.78	1.96
Tl	Pb
2.04	2.33

This departure from a smooth decrease down the group between Al and Ga and between Si and Ge is due to the intervention of the 3d subshell. There is also an increase in electronegativity values for Tl and Pb, and this is due to the presence of the 4f subshell. The **alternation effect** (Section 10.4) is partly a manifestation of these exceptions.

> **A NOTE ON GOOD PRACTICE**
>
> Care should be taken to avoid incorrect use of the electronegativity concept when referring to stable discrete anions formed from that element. Whereas a fluorine atom is very electronegative the fluoride ion, F⁻, is not.

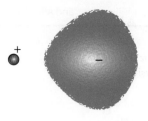

FIGURE 1.31 A schematic representation of the polarization of an electron cloud on an anion by an adjacent cation.

(f) Polarizability

KEY POINTS A polarizable atom or ion is one with frontier orbitals that lie close in energy; large, heavy atoms and ions tend to be highly polarizable.

The **polarizability**, α, of an atom is its ability to be distorted by an electric field (such as that of a neighbouring ion). An atom or ion is highly **polarizable** if its electron distribution can be distorted readily. This is most likely to occur for large anions with a low charge density and low effective nuclear charge. Species that effectively distort the electron distribution of a neighbouring atom or anion are described as having **polarizing ability**. These are typically small, highly charged cations with a high charge density (Fig. 1.31).

We shall see the consequences of polarizability when considering the nature of bonding in Section 2.2, but it is appropriate to anticipate here that extensive polarization leads to covalency. **Fajan's rules** summarize the factors that affect polarization:

- Small, highly charged cations have high polarizing ability.
- Large, highly charged anions are easily polarized.
- Cations that do not have a noble-gas electron configuration are easily polarized.

The last rule is particularly important for the d-block elements.

EXAMPLE 1.12 Identifying polarizable species

Which would be the more polarizable, an F⁻ ion or an I⁻ ion?

Answer We can make use of the fact that polarizable anions are typically large and highly charged. An F⁻ ion is small and singly charged. An I⁻ ion has the same charge but is large. Therefore, an I⁻ ion is likely to be the more polarizable.

Self-test 1.12 (a) Which would be more polarizing, Na⁺ or Cs⁺? (b) Which compound would exhibit most covalent character, NaF or NaI?

FURTHER READING

E. Scerri, *The Periodic Table: Its Story and its Significance*. 2nd ed. Oxford University Press (2019). A lively account of the development of the periodic system from early discoveries of elements to the impact of quantum theory.

H. Aldersley-Williams, *Periodic Tales: The Curious Lives of the Elements*. Viking (2011). Not an academic book, but provides social and cultural background to the use or discovery of many elements.

M. Laing, The different periodic tables of Dmitri Mendeleev. *J. Chem. Educ.*, 2008, 85, 63.

P. Ball, Whose periodic table is it anyway? *Chemistry World*, 2019.

M. W. Cronyn, The proper place for hydrogen in the periodic table. *J. Chem. Educ.*, 2003, 80, 947.

P. A. Cox, *Introduction to Quantum Theory and Atomic Structure*. Oxford University Press (1996). An introduction to the subject.

P. Atkins, J. de Paula, and J. Keeler, *Atkins' Physical Chemistry*. Oxford University Press (2022). Chapters 7 and 8 give an account of quantum theory and atomic structure.

J. Emsley, *Nature's Building Blocks: An A–Z Guide to the Elements*. Oxford University Press (2011). An interesting guide to the elements.

D. M. P. Mingos, *Essential trends in inorganic chemistry*. Oxford University Press (1998). Includes a detailed discussion of the important horizontal, vertical, and diagonal trends in the properties of the atoms.

P. A. Cox, *The Elements: Their Origin, Abundance, and Distribution*. Oxford University Press (1989). Examines the origin of the elements, the factors controlling their widely differing abundances, and their distributions in the Earth, the solar system, and the universe.

The Orbitron, http://winter.group.shef.ac.uk/orbitron/ This resource is an illustrated gallery of atomic and molecular orbitals.

P. Pyykkö, Relativistic effects in chemistry: More common than you thought. *Ann. Rev. Phys. Chem.*, 2012, 63, 45.

EXERCISES

1.1 What is the ratio of the energy of an electronic ground-state He^+ ion to that of a Be^{3+} ion?

1.2 According to the Born interpretation, the probability of finding an electron in a volume element $d\tau$ is proportional to $\psi^2 d\tau$. (a) What is the most probable location of an electron in an H atom in its ground state? (b) What is its most probable distance from the nucleus, and why is this different? (c) What is the most probable distance of a 2s electron from the nucleus?

1.3 The ionization energy of H is 13.6 eV. What is the difference in energy between the $n = 1$ and $n = 6$ levels?

1.4 The ionization energies of rubidium and silver are 4.18 and 7.57 eV, respectively. Calculate the ionization energies of an H atom with its electron in the same outermost orbitals as in these two atoms, and account for the differences in values in these different elements.

1.5 When 58.4 nm radiation from a helium discharge lamp is directed on a sample of krypton, electrons are ejected with a velocity of 1.59×10^6 m s^{-1}. The same radiation ejects electrons from Rb atoms with a velocity of 2.45×10^6 m s^{-1}. What are the ionization energies (in electronvolts, eV) of the two elements?

1.6 Calculate the wavelength of the line in the atomic spectrum of hydrogen in which $n_1 = 1$ and $n_2 = 3$. What is the energy change for this transition?

1.7 Calculate the wavenumber ($\tilde{\nu} = 1/\lambda$) and wavelength of the first transition in the visible region of the atomic spectrum of hydrogen.

1.8 Show that the following four lines in the Lyman series can be predicted from eqn. 1.1: 91.127, 97.202, 102.52, and 121.57 nm.

1.9 What is the relation of the possible angular momentum quantum numbers to the principal quantum number?

1.10 How many orbitals are there in a shell of principal quantum number n? (Hint: begin with $n = 1, 2$, and 3, and see if you can recognize the pattern.)

1.11 Complete the following table:

n	l	m_l	Orbital designation	Number of orbitals
2			2p	
3	2			
			4s	
4		+3, +2, …, −3		

1.12 What are the values of the n, l, and m_l quantum numbers that describe the 5f orbitals?

1.13 Use sketches of 2s and 2p orbitals to distinguish between (a) the radial wavefunction and (b) the radial distribution function.

1.14 Sketch the radial distribution functions for the 2p, 3p, and 3d orbitals and, with reference to your diagrams, explain why a 3p orbital is lower in energy than a 3d orbital.

1.15 Predict how many nodes and how many nodal planes a 4p orbital will have.

1.16 Draw pictures of the two d orbitals in the xy-plane as flat projections in the plane of the paper. Label each drawing with the appropriate mathematical function, and include a labelled pair of Cartesian coordinate axes. Label the orbital lobes correctly with + and − signs.

1.17 Consider the process of shielding in atoms, using Be as an example. What is being shielded? What is it shielded from? What is doing the shielding?

1.18 Calculate the screening constants for the outermost electron in the elements Li to F. Comment on the values you obtain.

1.19 In general, ionization energies increase across a period from left to right. Explain why the second ionization energy of Cr is higher, not lower, than that of Mn.

1.20 Compare the first ionization energy of Ca with that of Zn. Explain the difference in terms of the balance between shielding with increasing numbers of d electrons and the effect of increasing nuclear charge.

1.21 Compare the first ionization energies of Sr, Ba, and Ra. Relate the irregularity to the lanthanoid contraction.

1.22 The second ionization energies of some Period 4 elements are

Ca	Sc	Ti	V	Cr	Mn
1145	1235	1310	1365	1592	1509 kJ mol^{-1}

Identify the orbital from which ionization occurs, and account for the trend in values.

1.23 Give the ground-state electron configurations of (a) C, (b) F, (c) Ca, (d) Ga^{3+}, (e) Bi, (f) Pb^{2+}.

1.24 Give the ground-state electron configurations of (a) Sc, (b) V^{3+}, (c) Mn^{2+}, (d) Cr^{2+}, (e) Co^{3+}, (f) Cr^{6+}, (g) Cu, (h) Gd^{3+}.

1.25 Give the ground-state electron configurations of (a) W, (b) Rh^{3+}, (c) Eu^{3+}, (d) Eu^{2+}, (e) V^{5+}, (f) Mo^{4+}.

1.26 Identify the elements that have the ground-state electron configurations: (a) [Ne]3s^{2}3p^4, (b) [Kr]5s^2, (c) [Ar]4s^{2}3d^3, (d) [Kr]5s^{2}4d^5, (e) [Kr]5s^{2}4d^{10}5p^1, (f) [Xe]6s^{2}4f^6.

1.27 Without consulting reference material, draw the form of the periodic table with the numbers of the groups and the periods, and identify the s, p, and d blocks. Identify as many elements as you can. (As you progress through your study of inorganic chemistry, you should learn the positions of all the s-, p-, and d-block elements and associate their positions in the periodic table with their chemical properties.)

1.28 Account for the trends across Period 3 in (a) ionization energy, (b) electron affinity, (c) electronegativity.

1.29 Account for the fact that the two Group 4 elements zirconium (Period 5) and hafnium (Period 6) have the same covalent radius.

1.30 Identify the frontier orbitals of a Be atom in its ground state.

1.31 Use the data in Tables 1.6 and 1.7 to test Mulliken's proposition that electronegativity values are proportional to $I + E_a$.

TUTORIAL PROBLEMS

1.1 In the paper 'What can the Bohr–Sommerfeld model show students of chemistry in the 21st century?' (M. Niaz and L. Cardellini, *J. Chem. Educ.*, 2011, 88, 240) the authors use the development of models of atomic structure to deliberate on the nature of science. What were the shortcomings of the Bohr model of the atom? How did Sommerfeld refine Bohr's model? How did Pauli resolve some of the new model's shortcomings? Discuss what these developments teach us about the nature of science.

1.2 Survey the early and modern proposals for the construction of the periodic table. You should consider attempts to arrange the elements on helices and cones as well as the more practical two-dimensional surfaces. What, in your judgement, are the advantages and disadvantages of the various arrangements?

1.3 In their 2009 paper 'Icon of chemistry: The periodic system of chemical elements in the new century' (*Angew. Chem. Int. Ed.*, 2009, 48, 3404), S. Wang and W. Schwarz claim that the periodic system of elements has suffered from the 'invention of scientific facts'. Summarize their key arguments in relation to the electron configurations of the transition elements.

1.4 The decision about which elements should be identified as belonging to the f block has been a matter of some controversy. A view has been expressed by W. B. Jensen (*J. Chem. Educ.*, 1982, 59, 635). Summarize the controversy and Jensen's arguments. An alternative view has been expressed by L. Lavalle (*J. Chem. Educ.*, 2008, 85, 1482). Summarize the controversy and the arguments.

1.5 During 1999 several papers appeared in the scientific literature claiming that d orbitals of Cu$_2$O had been observed experimentally. In his paper 'Have orbitals really been observed?' (*J. Chem. Educ.*, 2000, 77, 1494), Eric Scerri reviews these claims and discusses whether orbitals can be observed physically. Summarize his arguments briefly.

1.6 At various times the following two sequences have been proposed for the elements to be included in Group 3: (a) Sc, Y, La, Ac, (b) Sc, Y, Lu, Lr. Because ionic radii strongly influence the chemical properties of the metallic elements, it might be thought that ionic radii could be used as one criterion for the periodic arrangement of the elements. Use this criterion to describe which of these sequences is preferred.

1.7 In the paper 'Ionization energies of atoms and atomic ions' (P. F. Lang and B. C. Smith, *J. Chem. Educ.*, 2003, 80, 938) the authors discuss the apparent irregularities in the first and second ionization energies of d- and f-block elements. Describe how these inconsistencies are rationalized.

1.8 The electron configuration of the transition metals is described by W. H. E. Schwarz in his paper 'The full story of the electron configurations of the transition elements' (*J. Chem. Educ.*, 2010, 87, 444). Schwarz discusses five features that must be considered to fully understand the electron configurations of these elements. Discuss each of these five features and summarize the impact of each on our understanding of the electronic configurations.

Molecular structure, bonding, and energetics

2

Lewis structures
2.1 The octet rule
2.2 Resonance
2.3 The VSEPR model

Valence bond theory
2.4 The hydrogen molecule
2.5 Homonuclear diatomic molecules
2.6 Polyatomic molecules

Molecular orbital theory
2.7 An introduction to the theory
2.8 Homonuclear diatomic molecules
2.9 Heteronuclear diatomic molecules
2.10 Bond properties
2.11 Bands formed from extended overlapping atomic orbitals

Bond properties, reaction enthalpies, and kinetics
2.12 Bond length
2.13 Bond strength and reaction enthalpies
2.14 Electronegativity and bond enthalpy
2.15 An introduction to catalysis

Further reading

Exercises

Tutorial problems

The interpretation of structures and reactions in inorganic chemistry is often based on semiquantitative models. In this chapter we examine the development of models of molecular structure in terms of the concepts of valence bond and molecular orbital theory. In addition, we review methods for predicting the shapes of molecules. This chapter introduces concepts that will be used throughout the text to explain the structures and reactions of a wide variety of species. The chapter also illustrates the importance of the interplay between qualitative models, experiment, and calculation.

Lewis structures

In 1916 G. N. Lewis proposed that a **covalent bond** is formed when two neighbouring atoms share an electron pair. A **single bond**, a shared electron pair (A:B), is denoted A–B; likewise, a **double bond**, two shared electron pairs (A::B), is denoted A=B, and a **triple bond**, three shared pairs of electrons (A:::B), is denoted A≡B. An unshared pair of valence electrons on an atom (A:) is called a **lone pair** or **nonbonding pair**. In this model, lone pairs do not contribute directly to the bonding but they can influence the shape of the molecule and play an important role in its properties.

2.1 The octet rule

KEY POINT The 'octet rule' is widely used to explain the formation of many compounds of main-group elements: it states that atoms share electron pairs until they have acquired eight valence electrons.

Lewis found that he could account for the existence and stability of a wide range of molecules by proposing the **octet rule**:

> *Each atom shares electrons with neighbouring atoms to achieve a total of eight valence electrons (an 'octet').*

As we saw in Section 1.5, a closed-shell, noble-gas configuration is achieved when eight electrons occupy the s and p subshells of the valence shell. One exception is the hydrogen atom, which fills its valence shell, the 1s orbital, with two electrons (a 'duplet').

The octet rule provides a simple way of constructing a **Lewis structure**, a diagram that shows the pattern of bonds and lone pairs in a molecule. In most cases we can construct a Lewis structure in three steps.

1. Decide on the number of electrons that are to be included in the structure by adding together the numbers of all the valence electrons provided by the atoms.

Each atom provides all its valence electrons (thus, H provides one electron, and O, with the configuration $[He]2s^22p^4$, provides six). Each negative charge on an ion corresponds to an additional electron; each positive charge corresponds to one less electron.

2. Write the chemical symbols of the atoms in the arrangement that shows which atoms are bonded together.

In most cases we know the arrangement or can make an informed guess. The less electronegative element is usually the central atom of a molecule, as in CO_2 and SO_4^{2-}, but there are many well-known exceptions (H_2O and NH_3 among them).

3. Distribute the electrons in pairs so that there is one pair of electrons forming a single bond between each pair of atoms bonded together, and then supply electron pairs (to form lone pairs or multiple bonds) until each atom has an octet.

Each bonding pair (:) is then represented by a single line (−). The net charge of a polyatomic ion is supposed to be possessed by the ion as a whole, not by a particular individual atom.

EXAMPLE 2.1 Drawing a Lewis structure

Draw a Lewis structure for the BF_4^- ion.

Answer We need to consider the total number of electrons supplied and how they are shared to complete an octet around each atom. The atoms supply $3 + (4 \times 7) = 31$ valence electrons; the single negative charge of the ion signifies the presence of an additional electron. We must therefore accommodate 32 electrons in 16 pairs around the five atoms. One solution, with the less electronegative boron at the centre, is (**1**). The negative charge is ascribed to the ion as a whole, not to a particular individual atom.

1 BF_4^-

Self-test 2.1 Draw a Lewis structure for (a) the PCl_3 molecule and (b) the BF_3 molecule.

Table 2.1 gives examples of Lewis structures of some common molecules and ions. Except in simple cases, a Lewis structure does not portray the shape of the species, but only the pattern of bonds and lone pairs: it shows the number of the links, not the geometry of the molecule. For example, the BF_4^- ion is actually tetrahedral (**2**), not planar, and PF_3 is trigonal-bipyramidal (**3**).

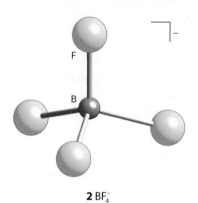

2 BF_4^-

TABLE 2.1 Lewis structures of some simple molecules*

H—H	:N≡N:	:C≡O:
:Ö=C=Ö:	:N≡N—Ö:	[:N≡C—Ö:]⁻
:Ö=Ö—Ö:	[:Ö—N=Ö:]⁻	:Ö=S—Ö:
H—N(H)—H (with lone pair)	:F—P(:F:)—F:	[:Ö=S(:Ö:)=Ö:]²⁻
[:Ö—P(=Ö)(—Ö:)—Ö:]³⁻	[:Ö—S(=Ö)(=Ö)—Ö:]²⁻	[:Ö=Cl(=Ö)—Ö:]⁻

*Only representative resonance structures are given. Shapes are indicated for triatomic molecules.

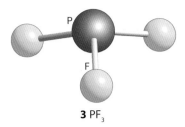

3 PF$_3$

2.2 Resonance

KEY POINTS Resonance between Lewis structures lowers the energy of the molecule and distributes the bonding character of electrons over the molecule; Lewis structures with similar energies provide the greatest resonance stabilization.

A single Lewis structure is often an inadequate description of the molecule: as an example, we consider ozone, O$_3$ (**4**), the shape of which is explained using the model described in Section 2.3. The Lewis structure suggests incorrectly that one O–O bond is different from the other, whereas in fact they have identical lengths (128 pm) intermediate between those of typical single O–O and double O–O bonds (148 pm and 121 pm, respectively). This deficiency of the Lewis description is overcome by introducing the concept of **resonance**, in which the actual structure of the molecule is taken to be a superposition, or average, of all the feasible Lewis structures corresponding to a given atomic arrangement.

4 O$_3$

Resonance is indicated by a double-headed arrow, as in

Resonance should be pictured as a *blending* of structures, not a flickering alternation between them. In quantum-mechanical terms, the electron distribution of each structure is represented by a wavefunction, and the actual wavefunction, ψ, of the molecule is the superposition of the individual wavefunctions for each contributing structure:[1]

$$\psi = \psi(O-O=O) + \psi(O=O-O)$$

The overall wavefunction is written as a superposition with equal contributions from both structures because the two structures have identical energies. The *blended* structure of two or more Lewis structures is called a **resonance hybrid**. Note that resonance occurs between structures that differ only in the allocation of electrons; resonance does not occur between structures in which the atoms themselves lie in different positions. For instance, there is no resonance between the structures SOO and OSO.

Resonance has two main effects:

- Resonance averages the bond characteristics over the molecule.
- The energy of a resonance hybrid structure is lower than that of any single contributing structure.

The energy of the O$_3$ resonance hybrid, for instance, is lower than that of either individual structure alone. Resonance is most important when there are several structures of identical energy that can be written to describe the molecule, as for O$_3$. In such cases, all the structures of the same energy contribute equally to the overall structure.

Structures with different energies may also contribute to an overall resonance hybrid but, in general, the greater the energy difference between two Lewis structures, the smaller the contribution of the higher-energy structure. The BF$_3$ molecule, for instance, could be regarded as a resonance hybrid of the structures shown in (**5**), but the first structure dominates even though the octet is incomplete. Consequently, BF$_3$ is regarded *primarily* as having that structure with a small admixture of double-bond character. In contrast, for the NO$_3^-$ ion (**6**), the last three structures dominate, and we treat the ion as having considerable double-bond character.

5 BF$_3$ resonance hybrids

6 NO$_3^-$ resonance hybrids

2.3 The VSEPR model

There is no simple method for predicting the numerical value of bond angles even in simple molecules, except where the shape is governed by symmetry. However, the **valence shell electron pair repulsion (VSEPR)** model of molecular shape, sometimes known as the Gillespie–Nyholm rules, is extremely useful in predicting the shapes of molecules, particularly those formed from elements belonging to the main groups.

---

[1] This wavefunction is not normalized (Section 1.2). We shall often omit normalization constants from linear combinations in order to clarify their structure. The wavefunctions themselves are formulated in the valence bond theory, which is described later.

(a) The basic shapes

KEY POINTS In the VSEPR model, valence electrons are arranged into domains on the surface of a sphere representing the central atom. These domains, comprised of bonding or nonbonding electrons, take up positions as far apart as possible. The shape of the molecule is identified by referring to the locations of the peripheral atoms, associated with the bonding electron pairs, in the resulting structure.

The primary assumption of the VSEPR model is that regions of enhanced electron density, by which we mean bonding pairs, lone (nonbonding) pairs, or the concentrations of electrons associated with multiple bonds, take up positions as far apart as possible so that the repulsions between them are minimized. For instance, four such regions of electron density will lie at the corners of a regular tetrahedron, five will lie at the corners of a trigonal bipyramid, and so on (Table 2.2).

Each domain may correspond to a single bond or lone pair (two electrons), a double bond (two pairs, four electrons), or a triple bond (three pairs, six electrons). The size of each domain depends on the number of electrons contained within it and the degree to which bond formation allows it to move away from the central atom (and contract in size) or (lacking attraction to a second nucleus) spread out onto its surface, as occurs with a lone pair. The steric influence of domains is thus expected to decrease in the following order: lone pair > triple bond > double bond > single bond. As explained in Box 2.1, the repulsion between electron pairs is mainly quantum-mechanical in origin.

It is convenient to represent the framework by the generic notation AX_pE_q where A is the central atom, X_p represents the number (p) of peripheral atoms and E represents the number (q) of lone pairs. Although the arrangement of domains in the framework governs the molecular shape, the *name* of the shape is determined by the resulting arrangement of the atoms alone (Table 2.3). For instance, the NH_3 molecule has a AX_3E framework: it has four domains that are arranged tetrahedrally, but as one of them is a lone pair the molecule has a trigonal-pyramidal shape with the apex occupied by the lone pair. The greater steric influence of the lone pair domain results in a distortion of the regular tetrahedron, decreasing the H–N–H bond angle from 109.5° to 107.3°. Similarly, H_2O (AX_2E_2) has four domains in a tetrahedral

TABLE 2.2 The basic arrangement of electron domains according to the VSEPR model

Number of domains	Framework
2	Linear
3	Trigonal-planar
4	Tetrahedral
5	Trigonal-bipyramidal
6	Octahedral

BOX 2.1 The basis of valence shell electron pair repulsion

A consequence of the Pauli exclusion principle introduced in Sections 1.4 and 1.5 is that two or more electrons having the same spin (all ↑ or all ↓) have zero probability of occupying the same space: the outcome is a strong force that acts to keep these electrons apart. **Pauli repulsion**, which is entirely quantum-mechanical in origin and distinct from the Coulombic repulsion that acts between all electrons regardless of spin, forms the basis of why molecules adopt particular shapes. The principle can be explained by placing the central atom of a molecule at the centre of a cube and introducing a valence shell octet in two stages, according to electron spin (Figure B2.1).

For the first four electrons (having spin ↑), Pauli repulsion is minimized if their positions of maximum probability lie at alternate corners of the cube (a): the remaining electrons (spin ↓) might then be expected (through electrostatic repulsion) to favour the remaining four corners (b), but their location is determined instead by their attraction to the nuclei of the peripheral atoms to form bonds, in this case resulting in four **domains** (c) situated as far apart as possible and thus arranged tetrahedrally on the surface of the central atom (d). Analogous arrangements—the **framework**—are

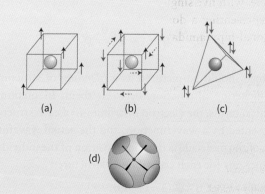

FIGURE B2.1 The basis for understanding the VSEPR model follows from the Pauli exclusion principle, which keeps electrons with the same spin as far apart as possible.

generated for two, three, five, and six electron pairs. Residual electron pairs that are not used to form a bond will remain closer to the central nucleus and at a greater distance from each other (hence the reason that a lone pair takes up more space on the surface of the central atom).

arrangement, but as two correspond to lone pairs the molecule is classified as angular (or 'bent'): the resulting H–O–H bond angle is contracted even further, to 104.5°. A deficiency of the VSEPR model, however, is that it cannot be used to predict the actual bond angles adopted by a molecule.

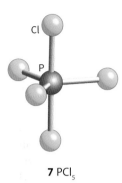

7 PCl$_5$

To apply the VSEPR model systematically, we first draw the Lewis structure for the molecule or ion and identify the central atom A. Next, we count the number of peripheral atoms X and lone pairs E carried by that atom. It is important to note that each peripheral atom (regardless of whether it is singly or multiply bonded to the central atom) and each lone pair count as a single domain. To achieve the lowest energy, the domains take up positions as far apart as possible, so we can identify the framework they adopt by referring to Table 2.2. Finally, we take account of the composition of each domain and identify the shape of the molecule from Table 2.3. Thus, a PCl$_5$ molecule, with five single bonds around the central atom, each representing a domain, is predicted (and found) to be trigonal-bipyramidal (**7**).

TABLE 2.3 The description of molecular shapes

Shape	Framework	Examples
Linear	AX$_2$, AX$_2$E$_3$	CO$_2$, HCN, XeF$_2$
Angular (bent)	AX$_2$E, AX$_2$E$_2$	H$_2$O, O$_3$, NO$_2^-$
Trigonal-planar	AX$_3$	BF$_3$, SO$_3$, NO$_3^-$, CO$_3^{2-}$
Trigonal-pyramidal	AX$_3$E	NH$_3$, SO$_3^{2-}$, SOCl$_2$
T-shape	AX$_3$E$_2$	ClF$_3$
Tetrahedral	AX$_4$	CH$_4$, SO$_4^{2-}$
Disphenoidal ('see-saw')	AX$_4$E	SF$_4$
Square-planar	AX$_4$E$_2$	XeF$_4$
Trigonal-bipyramidal	AX$_5$	PCl$_5$(g), SF$_4$O
Square-pyramidal	AX$_5$E	IF$_5$
Octahedral	AX$_6$	SF$_6$, PCl$_6^-$, IO(OH)$_5^+$
Pentagonal	AX$_5$E$_2$	XeF$_5^-$
Pentagonal-bipyramidal	AX$_7$	IF$_7$

EXAMPLE 2.2 Using the VSEPR model to predict shapes

Predict the shape of (a) a BF$_3$ molecule, (b) an SO$_3^{2-}$ ion, (c) a PCl$_4^+$ ion.

Answer We begin by drawing the Lewis structure of each species and then consider the number of bonding and lone pairs of electrons and how they are arranged around the central atom. (a) The Lewis structure of BF$_3$ is shown in (**5**). The central boron atom forms bonds to three F atoms and there is no lone pair. There are therefore three equivalent domains giving an AX$_3$ molecule having a trigonal-planar shape (**8**). (b) The SO$_3^{2-}$ ion has three important resonance forms (**9**) each having three bonds to oxygen (with 67% single and 33% double bond character) and one lone pair. The four domains form a AX$_3$E tetrahedron, and the resulting molecular shape is trigonal-pyramidal with bond angles contracted to 107° (**10**). (c) Phosphorus has five valence electrons, four of which are used to form bonds to the four Cl atoms. In PCl$_4^+$, the fifth electron is removed to give the +1 charge on the ion, so all the electrons supplied by the P atom are used in bonding and there is no lone pair, The ion is of the AX$_4$ category with four equivalent bonding domains, so it is tetrahedral (**11**).

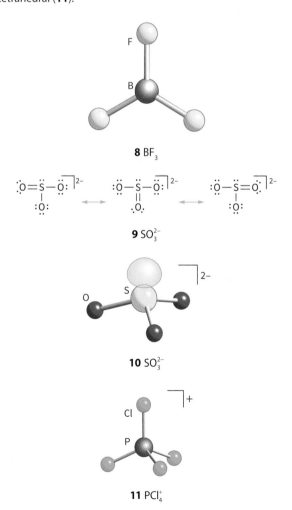

8 BF$_3$

9 SO$_3^{2-}$

10 SO$_3^{2-}$

11 PCl$_4^+$

Self-test 2.2 Predict the shape of (a) an H$_2$S molecule, (b) a CO$_2$ molecule, (c) an XeO$_4$ molecule.

(b) Modifications of the basic shapes

KEY POINT Lone pairs repel other pairs more strongly than bonding pairs do.

Once the basic shape of a AX_pE_q molecule has been identified, adjustments are made by taking into account how the size of each domain determines the degree to which it repels neighbouring domains. For domains comprised of a single electron pair, the ordering is lone-pair/lone-pair > lone-pair/bonding pair > bonding pair/bonding pair. Further adjustments are needed when the peripheral atoms differ in identity (X, Y etc) and bond type. Molecules of the type AX_3E, AX_2E_2 have tetrahedral domain frameworks and trigonal-pyramidal or angular geometries, respectively. For many other molecules, such as those having five domains (AX_4E or AX_3E_2) or six domains (AX_4E_2), assignment of the molecular shape requires further consideration.

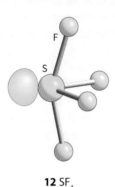

12 SF_4

The first question is where to place each lone pair, which require more space than any bonding domain. In a five-domain trigonal-bipyramidal framework for which the choice is between an axial and an equatorial site, a lone pair occupies an equatorial site. This preference occurs because in an equatorial site the lone pair is repelled by only two bonding domains at 90° (Fig. 2.1), whereas in an axial position the lone pair is repelled by three bonding domains at 90°. An example is SF_4, which has a 'see-saw' shape (**12**).

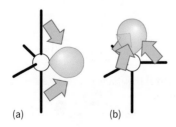

FIGURE 2.1 In the VSEPR model a lone pair in the equatorial position (a) of a trigonal-bipyramidal arrangement interacts strongly with two bonding pairs, but in an axial position (b) it interacts strongly with three bonding pairs. The former arrangement is generally lower in energy.

In an octahedral framework, a single lone pair can occupy any position as they are all equivalent, but (to avoid being next to each other) *two* lone-pairs will adopt a *trans* arrangement, which results in a square-planar structure. An example is XeF_4 (**13**). In this section, lone pairs are shown in flattened form to emphasize their greater spatial influence.

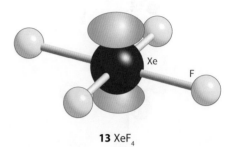

13 XeF_4

The second question is how to predict the positions of double and single bonds in molecules of the type AX_2Y_2E or AX_3YE, where each X is now attached to A via a double bond (two electron pairs), and Y is now introduced to represent the peripheral atom(s) attached by a single bond. Since a double bond domain has a greater steric influence than a single bond domain, we expect that X should (like a lone pair) occupy equatorial sites that minimize the number of neighbours at 90°. Examples are XeO_3F_2 (**14**), SOF_4 (**15**), and the see-saw shaped molecule XeO_2F_2 (**16**). The trigonal-bipyramidal frameworks in the latter two are distorted in such a way that the bond angles O-S-F and O-Xe-F are opened up slightly, reflecting the steric dominance of a single double bond over two single bonds and two double bonds over a single lone pair.

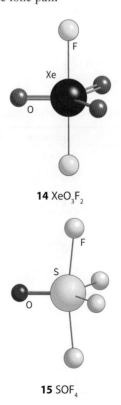

14 XeO_3F_2

15 SOF_4

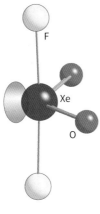

16 XeO$_2$F$_2$

A third question is how to assign the positions of different peripheral atoms where choices exist. As mentioned previously, the size of a domain depends on the degree to which electrons are drawn away from the central atom, so we expect that the more electronegative a peripheral atom (relative to A), the smaller will be its domain size. Trigonal-bipyramidal molecules of the type AX$_4$Y, where X is more electronegative than Y, provide good examples. The compounds PF$_4$H (**17**) and PF$_4$(CH$_3$) (**18**) each adopt structures in which the less electronegative Y atom occupies an equatorial site. Although it could be argued that a methyl group prefers an equatorial site because of its greater steric demand, this explanation cannot apply for hydrogen: indeed, in the related compound PF$_3$H$_2$ (**19**) the second hydrogen atom also occupies an equatorial position.

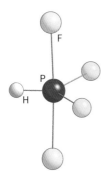

17 PF$_4$H

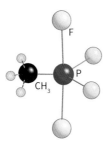

18 PF$_4$(CH$_3$)

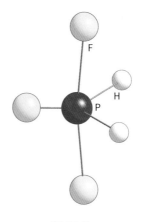

19 PF$_3$H$_2$

EXAMPLE 2.3 Accounting for the effect of lone pairs on molecular shape

Predict the shape of a ClF$_3$ molecule.

Answer We begin by drawing the Lewis structure of the molecule and identify the number of bonding and lone pairs of electrons; then we identify the shape of the molecule and finally consider any modifications due to the presence of lone pairs. The Lewis structure of ClF$_3$ is shown in (**20**). The central Cl atom has three F atoms attached to it and two lone pairs. The framework adopted by these five regions is a trigonal bipyramid.

$$:\ddot{\underset{..}{F}}-\ddot{\underset{|}{Cl}}-\ddot{\underset{..}{F}}:$$
$$:\ddot{\underset{..}{F}}:$$

20 Lewis structure of ClF$_3$

To predict the molecular shape we need to minimize the number of unfavourable ($\leq 90°$) interactions experienced by the lone pairs, as they occupy more space. Consequently, the potential energy is minimized if the lone pairs occupy equatorial sites (where they experience only two perpendicular interactions compared to the three they would experience in axial positions): this results in a distorted T-shaped molecule (**21**) with the two axial F atoms bent away slightly from the two lone pairs.

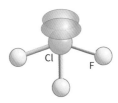

21 ClF$_3$

Self-test 2.3 Predict the shape of (a) an XeF$_2$ molecule and (b) an ICl$_2^+$ molecular ion.

The VSEPR model is highly successful, but sometimes runs into difficulty when there is more than one basic shape of similar energy. For example, with five regions of electron density around the central atom, a square-pyramidal arrangement is only slightly higher in energy than a trigonal-bipyramidal arrangement, and there are several examples of both structures being found, one being BrF$_5$ (22, 23).

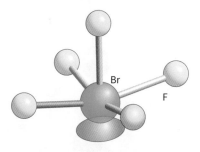

22 BrF$_5$ (square-pyramidal structure)

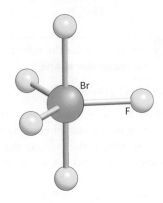

23 BrF$_5$ (trigonal-bipyramidal structure where lone pair is not influential)

Similarly, the basic shapes for seven regions of electron density are not easily predicted, partly because so many different conformations have similar energies. However, in the p block, seven-coordination is dominated by pentagonal-bipyramidal structures. For example, IF$_7$ (**24**) is pentagonal-bipyramidal, and XeF$_5^-$ (**25**) with five bonds and two lone pairs is based on a pentagonal-bipyramidal framework, but with two lone pairs the molecule has pentagonal-planar geometry.

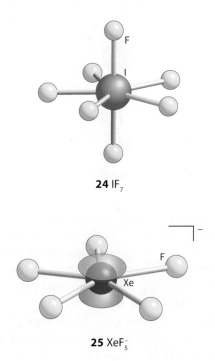

24 IF$_7$

25 XeF$_5^-$

Lone pairs are stereochemically less influential when they belong to heavy p-block elements. The SeF$_6^{2-}$ and TeCl$_6^{2-}$ ions, for instance, with six bonding pairs of electrons and one lone pair on the Se and Te atoms, are octahedral despite the presence of the lone pair. Lone pairs that do not influence the molecular geometry are said to be **stereochemically inert** and are usually present in the non-directional s orbitals. The VSEPR model is generally less reliable for d-block molecules where structures are determined by other factors.

Valence bond theory

Valence bond (VB) theory was the first quantum-mechanical theory of bonding to be developed. Valence bond theory considers the interaction of atomic orbitals on separate atoms as they are brought together to form a molecule. Although the computational techniques involved have been largely superseded by molecular orbital theory, much of the language and some of the concepts of VB theory still remain and are used throughout chemistry.

2.4 The hydrogen molecule

KEY POINTS In valence bond theory, the wavefunction of an electron pair is formed by superimposing the wavefunctions for the separated fragments of the molecule; a molecular potential energy curve shows the variation of the molecular energy with internuclear separation.

As we saw in Chapter 1, quantum mechanics and the concept of a wavefunction arises as a consequence of particle–wave duality and the Heisenberg uncertainty principle. The two-electron wavefunction for two widely separated H atoms, H$_A$ and H$_B$, is $\psi = \chi_A(1)\chi_B(2)$, where χ_A and χ_B are H1s orbitals on atoms A and B, electron 1 is on atom H$_A$ and electron 2 is on H$_B$.[2] When the atoms are close, it is not possible to know whether it is electron 1 or electron 2 that is on H$_A$. An equally valid description is therefore $\psi = \chi_A(2)\chi_B(1)$, in which electron 2 is on H$_A$ and electron 1 is on H$_B$. When two outcomes are equally probable,

---

[2] Although χ, chi, is also used for electronegativity, the context makes it unlikely that the two usages will be confused: χ is commonly used to denote an atomic orbital in computational chemistry.

quantum mechanics instructs us to describe the true state of the system as a superposition of the wavefunctions for each possibility, so a better description of the molecule than either wavefunction alone is the linear combination of the two possibilities:

$$\psi = \chi_A(1)\chi_B(2) + \chi_A(2)\chi_B(1) \tag{2.1}$$

This function is the (unnormalized) VB wavefunction for an H–H bond. The formation of the bond can be pictured as being due to the high probability that the two electrons will be found between the two nuclei and hence will bind them together (Fig. 2.2(a)). More formally, the wave pattern represented by the term $\chi_A(1)\chi_B(2)$ interferes constructively with the wave pattern represented by the contribution $\chi_A(2)\chi_B(1)$ and there is an enhancement in the amplitude of the wavefunction in the internuclear region. For technical reasons stemming from the Pauli exclusion principle, only electrons with paired spins can be described by a wavefunction of the type written in eqn. 2.1, so only paired electrons can contribute to a bond in VB theory. We say, therefore, that a VB wavefunction is formed by **spin pairing** of the electrons in the two contributing atomic orbitals. The electron distribution described by the wavefunction in eqn. 2.1 is called a σ bond. As shown in Fig. 2.2, a σ bond has cylindrical symmetry around the internuclear axis, and the electrons in it have zero orbital angular momentum about that axis.

The **molecular potential energy curve** for H_2, a graph showing the variation of the energy of the molecule with internuclear separation, is calculated by changing the internuclear separation R and evaluating the energy at each selected separation (Fig. 2.3). The energy is found to fall below that of two separated H atoms as the two atoms are brought within bonding distance of each other and each electron becomes free to migrate to the other atom. However, the resulting lowering of energy is counteracted by an increase in energy from the Coulombic (electrostatic) repulsion between the two positively charged nuclei. This positive contribution to the energy becomes large as the atoms approach each other and R becomes small. Consequently, the total potential

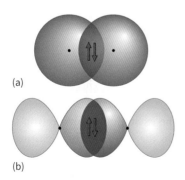

FIGURE 2.2 The formation of a σ bond from (a) s orbital overlap, (b) p orbital overlap. A σ bond has cylindrical symmetry around the internuclear axis.

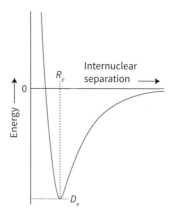

FIGURE 2.3 A molecular potential energy curve showing how the total energy of a molecule varies as the internuclear separation is changed.

energy curve passes through a minimum and then climbs to a strongly positive value at small internuclear separations. The depth of the minimum of the curve, at the internuclear separation R_e, is denoted D_e. The deeper the minimum, the more strongly the atoms are bonded together. The steepness of the well shows how rapidly the energy of the molecule rises as the bond is stretched or compressed. The steepness of the curve, an indication of the *stiffness* of the bond, therefore governs the vibrational frequency of the molecule (Section 8.5).

2.5 Homonuclear diatomic molecules

KEY POINT Electrons in atomic orbitals of the same symmetry but on neighbouring atoms are paired to form σ and π bonds.

A similar description can be applied to more complex molecules, and we begin by considering **homonuclear diatomic molecules**—diatomic molecules in which both atoms belong to the same element (dinitrogen, N_2, is an example). To construct the VB description of N_2, we consider the valence electron configuration of each atom, which from Section 1.5 we know to be $2s^2 2p_z^1 2p_y^1 2p_x^1$. It is conventional to take the z-axis to be the internuclear axis, so we can imagine each atom as having a $2p_z$ orbital pointing towards a $2p_z$ orbital on the other atom, with the $2p_x$ and $2p_y$ orbitals perpendicular to this axis. A σ bond is then formed by spin pairing between the two electrons in the opposing $2p_z$ orbitals (Fig. 2.2(b)). Its spatial wavefunction is still given by eqn. 2.1, but now χ_A and χ_B stand for the two $2p_z$ orbitals. A simple way of identifying a σ bond is to envisage rotation of the bond around the internuclear axis: if the wavefunction remains unchanged, the bond is classified as σ.

The remaining 2p orbitals cannot merge to give σ bonds as they do not have cylindrical symmetry around the internuclear axis. Instead, the orbitals merge to form two π bonds. A π bond arises from the spin pairing of electrons in two p orbitals that approach side by side (Fig. 2.4).

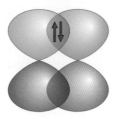

FIGURE 2.4 The formation of a π bond.

The bond is so called because, viewed along the internuclear axis, it resembles a pair of electrons in a p orbital. More precisely, an electron in a π bond has one unit of orbital angular momentum about the internuclear axis. A simple way of identifying a π bond is to envisage rotation of the bond through 180° around the internuclear axis. If the signs (as indicated by the shading) of the lobes of the orbital are interchanged, then the bond is classified as π.

There are two π bonds in N_2; one is formed by spin pairing in two neighbouring $2p_x$ orbitals, and the other by spin pairing in two neighbouring $2p_y$ orbitals. The overall bonding pattern in N_2 is therefore a σ bond plus two π bonds (Fig. 2.5), which is consistent with the structure N≡N. Analysis of the total electron density in a triple bond shows that it has cylindrical symmetry around the internuclear axis, with the four electrons in the two π bonds forming a ring of electron density around the central σ bond.

2.6 Polyatomic molecules

KEY POINTS Each σ bond in a polyatomic molecule is formed by the spin pairing of electrons in any neighbouring atomic orbitals with cylindrical symmetry about the relevant internuclear axis; π bonds are formed by pairing electrons that occupy neighbouring atomic orbitals of the appropriate symmetry.

To introduce polyatomic molecules we consider the VB description of H_2O. The valence electron configuration of a hydrogen atom is $1s^1$ and that of an O atom is $2s^2 2p_z^2 2p_y^1 2p_x^1$.

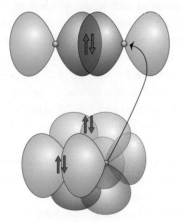

FIGURE 2.5 The VB description of N_2. Two electrons form a σ bond and another two pairs form two π bonds. In linear molecules, where the x- and y-axes are not specified, the electron density of π bonds is cylindrically symmetrical around the internuclear axis.

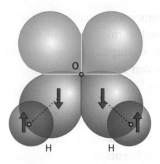

FIGURE 2.6 The VB description of H_2O. There are two σ bonds formed by pairing electrons in O2p and H1s orbitals. This model predicts a bond angle of 90°.

The two unpaired electrons in the O2p orbitals can each pair with an electron in an H1s orbital, and each combination results in the formation of a σ bond (each bond has cylindrical symmetry about the respective O–H internuclear axis). Because the $2p_y$ and $2p_x$ orbitals lie at 90° to each other, the two σ bonds also lie at 90° to each other (Fig. 2.6). We can predict, therefore, that H_2O should be an angular molecule, which it is. However, the theory predicts a bond angle of 90° whereas the actual bond angle is 104.5°. Similarly, to predict the structure of an ammonia molecule, NH_3, we start by noting that the valence electron configuration of an N atom given previously suggests that three H atoms can form bonds by spin pairing with the electrons in the three half-filled 2p orbitals. The latter are perpendicular to each other, so we predict a trigonal-pyramidal molecule with a bond angle of 90°. An NH_3 molecule is indeed trigonal-pyramidal, but the experimental bond angle is 107°.

Another deficiency of the VB theory presented so far is its inability to account for the tetravalence of carbon; that is, its familiar ability to form four bonds as exemplified in methane, CH_4, which is tetrahedral, like PCl_4^+ (**11**). The ground-state configuration of C is $2s^2 2p_z^1 2p_y^1$, which suggests that a C atom should be capable of forming only two bonds, not four. Clearly, something is missing from the VB approach.

These two deficiencies—the failure to account for bond angles, and the valence of carbon—are overcome by introducing two new features: *promotion* and *hybridization*.

(a) Promotion

KEY POINT Promotion of electrons may occur if the outcome is to achieve more or stronger bonds and a lower overall energy.

Promotion is the excitation of an electron to an orbital of higher energy in the course of bond formation. Although electron promotion requires an investment of energy, that investment is worthwhile if the energy can be more than recovered from the greater strength or number of bonds that it allows to be formed. Promotion is not a 'real' process in which an atom somehow becomes excited and then forms bonds: it is a contribution to the overall energy change that occurs when bonds form.

In carbon, for example, the promotion of a 2s electron to a 2p orbital can be thought of as leading to the configuration $2s^1 2p_z^1 2p_y^1 2p_x^1$, with four unpaired electrons in separate orbitals. These electrons may pair with four electrons in orbitals provided by four other atoms, such as four H1s orbitals if the molecule is CH_4, and hence form four σ bonds. Although energy was required to promote the electron, it is more than recovered by the atom's ability to form four bonds in place of the two bonds of the unpromoted atom. Promotion, and the formation of four bonds, is a characteristic feature of carbon and of its congeners in Group 14 (Chapter 15) because the promotion energy is quite small: the promoted electron leaves a doubly occupied ns orbital and enters a vacant np orbital, hence significantly relieving the electron–electron repulsion it experiences in the ground state. As the group is descended, bond energies increasingly fail to compensate for the promotion energy: divalent molecular compounds are common for tin and lead (Section 15.8).

(b) Hypervalence

KEY POINT Hypervalence and octet expansion occur for elements following Period 2.

The elements of Period 2, Li through Ne, obey the octet rule quite well, but elements of later periods show deviations from it. For example, the bonding in PCl_5 (**7**) requires the P atom to have 10 electrons in its valence shell, one pair for each P–Cl bond. Similarly, in SF_6 the S atom must have 12 electrons if each F atom is to be bound to the central S atom by an electron pair (**26**). Species of this kind, which in terms of Lewis structures demand the presence of more than an octet of electrons around at least one atom, are called **hypervalent**.

26 Lewis structure of SF_6

One explanation of hypervalence invokes the availability of low-lying unfilled d orbitals, which can accommodate the additional electrons. According to this explanation, a P atom can accommodate more than eight electrons if it uses its vacant 3d orbitals. In PCl_5, with its five pairs of bonding electrons, at least one 3d orbital must be used in addition to the four 3s and 3p orbitals of the valence shell. The rarity of hypervalence in Period 2 is then ascribed to the absence of 2d orbitals. However, a more obvious reason for the rarity of hypervalence in Period 2 may be the geometrical difficulty of packing more than four atoms around a small central atom and may in fact have little to do with the availability of d orbitals. The molecular orbital theory of bonding, which is described in Section 2.7, describes the bonding in hypervalent compounds without invoking participation of d orbitals.

(c) Hybridization

KEY POINTS Hybrid orbitals are formed when atomic orbitals on the same atom interfere; specific hybridization schemes correspond to each local molecular geometry.

The description of the bonding in AB_4 molecules of Group 14 is still incomplete because it appears to imply the presence of three σ bonds of one type (formed from χ_B and χ_{A2p} orbitals) and a fourth σ bond of a distinctly different character (formed from χ_B and χ_{A2s}), whereas all the experimental evidence (bond lengths and strengths) points to the equivalence of all four A–B bonds, as in CH_4, for example.

This problem is overcome by realizing that the electron density distribution in the promoted atom is equivalent to the electron density in which each electron occupies a **hybrid orbital** formed by interference, or 'mixing', between the A2s and the A2p orbitals. The origin of the hybridization can be appreciated by thinking of the four atomic orbitals, which are waves centred on a nucleus, as being like ripples spreading from a single point on the surface of a lake: the waves interfere destructively and constructively in different regions, and give rise to four new shapes.

The specific linear combinations that give rise to four equivalent hybrid orbitals are

$$h_1 = s + p_x + p_y + p_z \quad h_2 = s - p_x - p_y + p_z$$
$$h_3 = s - p_x + p_y - p_z \quad h_4 = s + p_x - p_y - p_z \quad (2.2)$$

As a result of the interference between the component orbitals, each hybrid orbital consists of a large lobe pointing in the direction of one corner of a regular tetrahedron and a smaller lobe pointing in the opposite direction (Fig. 2.7). The angle between the axes of the hybrid orbitals is the tetrahedral angle, 109.47°. Because each hybrid

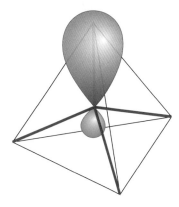

FIGURE 2.7 One of the four equivalent sp³ hybrid orbitals. Each one points towards a different vertex of a regular tetrahedron.

is built from one s orbital and three p orbitals, it is called an **sp³ hybrid orbital**.

It is now easy to see how the VB description of a CH_4 molecule is consistent with a tetrahedral shape with four equivalent C–H bonds. Each hybrid orbital of the promoted carbon atom contains a single unpaired electron; an electron in χ_{H1s} can pair with each one, giving rise to a σ bond pointing in a tetrahedral direction. Because each sp³ hybrid orbital has the same composition, all four σ bonds are identical apart from their orientation in space.

A further feature of hybridization is that a hybrid orbital has pronounced directional character, in the sense that it has enhanced amplitude in the internuclear region. This directional character arises from the constructive interference between the s orbital and the positive lobes of the p orbitals. As a result of the enhanced amplitude in the internuclear region, the bond strength is greater than for an s or p orbital alone. This increased bond strength is another factor that helps to repay the promotion energy.

Hybrid orbitals of different compositions are used to match different molecular geometries and to provide a basis for their VB description. For example, sp² hybridization, arising from one s and two p orbitals, is used to reproduce the electron distribution needed for trigonal-planar species, such as on B in BF_3 and N in NO_3^-, whereas sp hybridization,

TABLE 2.4 Some hybridization schemes

Coordination number	Arrangement	Composition
2	Linear	sp, pd, sd
	Angular	sd
3	Trigonal-planar	sp², p²d
	Unsymmetrical planar	spd
	Trigonal-pyramidal	pd²
4	Tetrahedral	sp³, sd³
	Irregular tetrahedral	spd², p³d, pd³
	Square-planar	p²d², sp²d
5	Trigonal-bipyramidal	sp³d, spd³
	Square-pyramidal	sp²d², sd⁴, pd⁴, p³d²
	Pentagonal-planar	p²d³
6	Octahedral	sp³d²
	Trigonal-prismatic	spd⁴, pd⁵
	Trigonal-antiprismatic	p³d³

arising from one s and one p orbital, reproduces a linear distribution. Table 2.4 gives the hybrids needed to match the geometries of a variety of electron distributions and includes hybridization schemes that include d orbitals, thus accounting for hypervalence as discussed in Section 3.11.

Molecular orbital theory

We have seen that VB theory provides a reasonable description of bonding in simple molecules. However, it does not handle polyatomic molecules very elegantly. **Molecular orbital (MO) theory** is a more sophisticated model of bonding that can be applied equally successfully to simple and complex molecules. In MO theory, we extend the *atomic* orbital description of atoms in a very natural way to describe **molecular orbitals** of molecules in which electrons spread over *all* the atoms in a molecule and bind them all together. In the spirit of this chapter, we continue to treat the concepts qualitatively and to provide a sense of how inorganic chemists discuss the electronic structures of molecules by using MO theory. Almost all qualitative discussions and calculations on inorganic molecules and ions are now carried out within the framework of MO theory.

2.7 An introduction to the theory

We begin by considering homonuclear diatomic molecules and diatomic ions formed by two atoms of the same element. The concepts these species introduce are readily extended to heteronuclear diatomic molecules formed between two atoms or ions of different elements. They are also easily extended to polyatomic molecules and solids composed of huge numbers of atoms and ions. However, for polyatomic molecules it is important to consider the symmetry of the orbitals involved. Therefore, we will discuss the molecular orbital treatment of polyatomic molecules in Chapter 3 after we have discussed molecular symmetry.

(a) The approximations of the theory

KEY POINTS Molecular orbitals are constructed as linear combinations of atomic orbitals; there is a high probability of finding electrons in atomic orbitals that have large coefficients in the linear combination; each molecular orbital can be occupied by up to two electrons.

As in the description of the electronic structures of atoms, we set out by making the **orbital approximation**, in which we assume that the wavefunction, ψ, of the N_e electrons in the molecule can be written as a product of one-electron wavefunctions: $\psi = \psi(1)\,\psi(2)\ldots\psi(N_e)$. The interpretation of this expression is that electron 1 is described by the wavefunction ψ(1), electron 2 by the wavefunction ψ(2), and so on. These one-electron wavefunctions are the **molecular orbitals** of the theory. As for atoms, the square of a one-electron wavefunction gives the probability distribution for that electron in the molecule: an electron in a molecular orbital is likely to be found where the orbital has a large amplitude, and will not be found at all at any of its nodes.

The next approximation is motivated by noting that, when an electron is close to the nucleus of one atom, its wavefunction closely resembles an atomic orbital of that atom. For instance, when an electron is close to the nucleus of an H atom in a molecule, its wavefunction is like a 1s orbital of that atom. Therefore, we may suspect that we can construct a reasonable first approximation to the molecular orbital by superimposing atomic orbitals contributed by each atom. This modelling of a molecular orbital in terms of contributing atomic orbitals is called the **linear combination of atomic orbitals** (LCAO) approximation. A 'linear combination' is a sum with various weighting coefficients. In simple terms, we combine the atomic orbitals of contributing atoms to produce molecular orbitals that extend over the entire molecule.

In the most elementary form of MO theory, only the valence shell atomic orbitals are used to form molecular orbitals. Thus, the molecular orbitals of H_2 are approximated by using two hydrogen 1s orbitals, one from each atom:

$$\psi = c_A \chi_A + c_B \chi_B \tag{2.3}$$

In this case the **basis set**, the atomic orbitals χ from which the molecular orbital is built, consists of two H1s orbitals, one on atom A and the other on atom B. The principle is exactly the same for more complex molecules. For example, the basis set for the methane molecule consists of the 2s and 2p orbitals on carbon and four 1s orbitals on the hydrogen atoms. The coefficients c in the linear combination show the extent to which each atomic orbital contributes to the molecular orbital: the greater the value of c, the greater the contribution of that atomic orbital to the molecular orbital. To interpret the coefficients in eqn. 2.3 we note that c_A^2 is the probability that the electron will be found in the orbital χ_A and c_B^2 is the probability that the electron will be found in the orbital χ_B. The fact that both atomic orbitals contribute to the molecular orbital implies that there is interference between them where their amplitudes are nonzero, with the probability distribution being given by

$$\psi^2 = c_A^2 \chi_A^2 + 2c_A c_B \chi_A \chi_B + c_B^2 \chi_B^2 \tag{2.4}$$

The term $2c_A c_B \chi_A \chi_B$ represents the contribution to the probability density arising from this interference.

Because H_2 is a homonuclear diatomic molecule, its electrons are equally likely to be found near each nucleus, so the linear combination that gives the lowest energy will have equal contributions from each 1s orbital ($c_A^2 = c_B^2$), leaving open the possibility that $c_A = +c_B$ or $c_A = -c_B$. Thus, ignoring normalization, the two molecular orbitals are

$$\psi_\pm = \chi_A \pm \chi_B \tag{2.5}$$

The relative signs of coefficients in LCAOs play a very important role in determining the energies of the orbitals. As we shall see, they determine whether atomic orbitals interfere constructively or destructively where they spread into the same region and hence lead to an accumulation or a reduction of electron density in those regions.

Two more preliminary points should be noted. We see from this discussion that *two* molecular orbitals may be constructed from *two* atomic orbitals. In due course, we shall see the importance of the general point that N molecular orbitals can be constructed from a basis set of N atomic orbitals. For example, if we use all four valence orbitals on each O atom in O_2, then from the total of eight atomic orbitals we can construct eight molecular orbitals. In addition, as in atoms, the Pauli exclusion principle implies that each molecular orbital may be occupied by up to two electrons; if two electrons are present in that orbital, then their spins must be paired (antiparallel). Thus, in a diatomic molecule constructed from two Period 2 atoms and in which there are eight molecular orbitals available for occupation, up to 16 electrons may be accommodated before all the molecular orbitals are full. The same rules that are used for filling atomic orbitals with electrons (the building-up principle and Hund's rule, Section 1.5) apply to filling molecular orbitals with electrons.

The general pattern of the energies of N molecular orbitals formed from N atomic orbitals is that one molecular orbital lies below that of the parent atomic energy levels, one lies higher in energy than they do, and the remainder are distributed between these two extremes.

(b) Bonding and antibonding orbitals

KEY POINTS The wave-like interference between orbitals of neighbouring atoms determines the spaces that electrons can occupy in a molecule. A bonding orbital arises from the constructive interference of neighbouring atomic orbitals, leading to enhanced probability of electrons being found in the internuclear region where they hold the positively charged nuclei together by electrostatic interactions. Conversely, an antibonding orbital arises from their destructive interference, as indicated by a node between the atoms, and realized by the lowered probability of finding an electron in the internuclear region. Enhancement of the probability of electrons being found in the internuclear region is taken to be the origin of the strength of bonds.

The orbital ψ_+ is an example of a **bonding orbital**. It is so called because the energy of the molecule is lowered relative to that of the separated atoms if this orbital is occupied by electrons. To understand the bonding character of ψ_+ we note that it arises from constructive interference between the two atomic orbitals, which produces an enhanced amplitude in the internuclear space (Fig. 2.8). As the probability of an electron occupying a particular location is proportional to the square of the wavefunction at that point, one that occupies ψ_+ has an enhanced probability of being

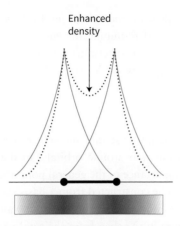

FIGURE 2.8 The enhancement of electron density in the internuclear region arising from the constructive interference between the atomic orbitals on neighbouring atoms.

found in the internuclear region, and can therefore interact strongly with both nuclei instead of just one. Hence orbital overlap, the spreading of one orbital into the region occupied by another, leading to enhanced probability of electrons being found in the internuclear region, is taken to be the origin of the strength of bonds.

The orbital ψ_- is an example of an **antibonding orbital**. It is so called because, if it is occupied, the energy of the molecule is higher than for the two separated atoms. The greater energy of an electron in this orbital arises from the destructive interference between the two atomic orbitals, which cancels their amplitudes and gives rise to a nodal plane between the two nuclei (Fig. 2.9). As a result, electrons that occupy ψ_- are largely excluded from the internuclear region. An antibonding orbital is slightly more antibonding than its partner bonding orbital is bonding: the asymmetry can be interpreted in terms of the nucleus-nucleus repulsion, which raises the energies of both bonding and antibonding orbitals relative to the isolated atoms.[3] The asymmetry is greatest for molecular orbitals involving small atoms and where overlap is strong. The energetic cost of placing excess electrons in antibonding orbitals (which for Lewis structures amounts to increasing the number of nonbonding electron repulsions) accounts partly for the weak bonds formed between electron-rich 2p elements.

The energies of the two molecular orbitals in H_2 are depicted in Fig. 2.10, which is an example of a **molecular orbital energy-level diagram** depicting the relative energies of molecular orbitals. The two electrons occupy the lower-energy molecular orbital. An indication of the size of the energy gap between the two molecular orbitals is the observation of a spectroscopic absorption in H_2 at 11.4 eV (in the ultraviolet at 109 nm), which can be ascribed to the transition of an electron from the bonding orbital to the antibonding orbital. The dissociation energy of H_2 is 4.5 eV (436 kJ mol^{-1}), which gives an indication of the location of the bonding orbital relative to the separated atoms.

The Pauli exclusion principle limits to two the number of electrons that can occupy any molecular orbital and requires that those two electrons be paired ($\uparrow\downarrow$). The exclusion principle is the origin of the importance of the pairing of the electrons in bond formation in MO theory just as it is in VB theory: in the context of MO theory, two is the maximum number of electrons that can occupy an orbital that contributes to the stability of the molecule. The H_2 molecule, for example, has a lower energy than that of the separated atoms because two electrons can occupy the orbital ψ_+, and both can contribute to the lowering of its energy (as shown in Fig. 2.10). A weaker bond can be expected if only one electron is present in a bonding orbital, but, nevertheless, H_2^+ is

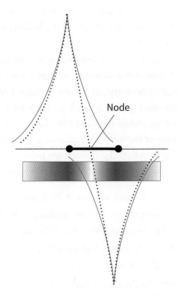

FIGURE 2.9 The destructive interference that arises if the overlapping orbitals have opposite signs. This interference leads to a nodal surface in an antibonding molecular orbital.

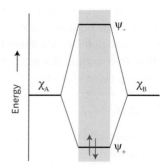

FIGURE 2.10 The molecular orbital energy-level diagram for H_2 and analogous molecules.

[3] The asymmetric energy distribution of bonding and antibonding orbitals is explained further in P. W. Atkins and R. Friedman, *Molecular Quantum Mechanics*. Oxford University Press (2010).

known as a transient gas-phase ion: its dissociation energy is 2.6 eV (250.8 kJ mol^{-1}). Three electrons (as in H$_2^-$) are less effective than two electrons because the third electron must occupy the antibonding orbital ψ_- and hence destabilize the molecule. With four electrons, the antibonding effect of two electrons in ψ_- easily overcomes the bonding effect of two electrons in ψ_+. There is then no net bonding. It follows that a four-electron molecule with only 1s orbitals available for bond formation, such as He$_2$, is not expected to be stable relative to dissociation into its atoms.

So far, we have discussed interactions of atomic orbitals that give rise to molecular orbitals that are lower in energy (bonding) and higher in energy (antibonding) than the separated atoms. In addition, it is possible to generate a molecular orbital that has the same energy as the initial atomic orbitals. In this case, occupation of this orbital neither stabilizes nor destabilizes the molecule and so is described as a **nonbonding orbital**. Typically, a nonbonding orbital is a molecular orbital that consists of a single orbital on one atom, perhaps because there is no atomic orbital of the correct symmetry on a neighbouring atom with which to form a bonding orbital.

(c) The influence of Pauli repulsion

The Pauli exclusion principle, which states that no two electrons can have the same set of quantum numbers, does not apply when those electrons are on two different atoms separated at sufficient distance, but results in a strong resistance when the atoms approach each other to form a bond and their wavefunctions overlap. Pauli repulsion contributes to all bonds except that formed between two hydrogen atoms, and is particularly important for compounds formed between 2p elements for which atoms must approach each other closely.

2.8 Homonuclear diatomic molecules

Although the structures of diatomic molecules can be calculated easily by using commercial software packages, the validity of any such calculations must, at some point, be confirmed by experimental data. Moreover, a wide variety of experimental information can be combined (Chapter 8). One of the most direct portrayals of electronic structure is obtained from ultraviolet photoelectron spectroscopy (UPS, Section 8.9), in which electrons are ejected from the orbitals they occupy in molecules and their energies determined. Because the peaks in a photoelectron spectrum correspond to the various kinetic energies of photoelectrons ejected from different orbitals of the molecule, the spectrum gives a vivid portrayal of the molecular orbital energy levels of a molecule (Fig. 2.11).

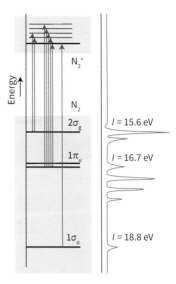

FIGURE 2.11 The UV photoelectron spectrum of N$_2$. The fine structure in the spectrum arises from excitation of vibrations in the cation formed by photoejection of an electron.

(a) The orbitals

KEY POINTS Molecular orbitals are classified as σ, π, or δ according to their rotational symmetry about the internuclear axis, and (in centrosymmetric species) as g or u according to their symmetry with respect to inversion.

Our task is to see how MO theory can account for the features revealed by photoelectron spectroscopy and the other techniques, principally absorption spectroscopy, that are used to study molecules. We are concerned predominantly with outer-shell valence orbitals, rather than core orbitals. As with H$_2$, the starting point in the theoretical discussion is the **minimal basis set**, the smallest set of atomic orbitals from which useful molecular orbitals can be built. In Period 2 diatomic molecules, the minimal basis set consists of the one valence s orbital and three valence p orbitals on each atom, giving eight atomic orbitals in all. We shall now see how the minimal basis set of eight valence shell atomic orbitals (four from each atom, one s and three p) is used to construct eight molecular orbitals. Then we shall use the Pauli principle to predict the ground-state electron configurations of the molecules.

The energies of the atomic orbitals (the valence orbital energies) that form the basis set are indicated on either side of the molecular orbital diagram in Fig. 2.12, which is appropriate for O$_2$ and F$_2$. We form **σ orbitals** by allowing overlap between atomic orbitals that have cylindrical symmetry around the internuclear axis, which (as remarked earlier) is conventionally labelled z. Symmetry and its many applications to MO theory will be discussed in more detail in Chapter 3, but for diatomic molecules the concept is straightforward.

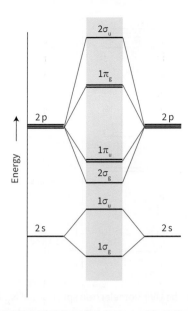

FIGURE 2.12 The molecular orbital energy-level diagram for the later Period 2 homonuclear diatomic molecules. This diagram should be used for O_2 and F_2.

The notation σ signifies that the orbital has cylindrical symmetry; atomic orbitals that can form σ orbitals include the 2s and $2p_z$ orbitals on the two atoms (Fig. 2.13). From these four orbitals (the 2s and the $2p_z$ orbitals on atom A and the corresponding orbitals on atom B) with cylindrical symmetry we can construct four σ molecular orbitals, two of which arise predominantly from interaction of the 2s orbitals and two from interaction of the $2p_z$ orbitals. These molecular orbitals are labelled $1\sigma_g$, $1\sigma_u$, $2\sigma_g$, and $2\sigma_u$, respectively.

The remaining two 2p orbitals on each atom, which have a nodal plane containing the z-axis, overlap to give π **orbitals** (Fig. 2.14). Bonding and antibonding π orbitals can be formed from the mutual overlap of the two $2p_x$ orbitals, and also from the mutual overlap of the two $2p_y$ orbitals. This pattern of overlap gives rise to the two pairs of doubly degenerate energy levels (two energy levels of the same energy) shown in Fig. 2.12 and labelled $1\pi_u$ and $1\pi_g$.

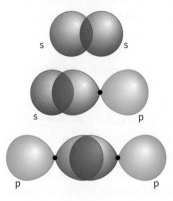

FIGURE 2.13 A σ orbital can be formed in several ways, including s,s overlap, s,p overlap, and p,p overlap, with the p orbitals directed along the internuclear axis.

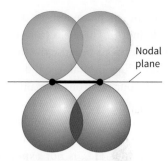

FIGURE 2.14 Two p orbitals can overlap to form a π orbital. The orbital has a nodal plane passing through the internuclear axis, shown here from the side.

For homonuclear diatomics, it is sometimes convenient (particularly for spectroscopic discussions) to signify the symmetry of the molecular orbitals with respect to their behaviour under inversion through the centre of the molecule. The operation of **inversion** consists of starting at an arbitrary point in the molecule, travelling in a straight line to the centre of the molecule, and then continuing an equal distance out on the other side of the centre. This procedure is indicated by the arrows in Figs. 2.15 and 2.16. The orbital is designated g (for *gerade*, even) if it is identical under inversion, and u (for *ungerade*, odd) if it changes sign. Thus, a bonding σ orbital is g and an antibonding σ orbital is u (Fig. 2.15). On the other hand, a bonding π orbital is u and an antibonding π orbital is g (Fig. 2.16). Note that the σ orbitals are numbered separately from the π orbitals.

The procedure can be summarized as follows:

1. From a basis set of four atomic orbitals on each atom, eight molecular orbitals are constructed.
2. Four of these eight molecular orbitals are σ orbitals and four are π orbitals.
3. The four σ orbitals span a range of energies, one being strongly bonding and another strongly antibonding; the remaining two lie between these extremes.
4. The four π orbitals form one doubly degenerate pair of bonding orbitals and one doubly degenerate pair of antibonding orbitals.

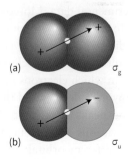

FIGURE 2.15 (a) Bonding and (b) antibonding σ interactions, with the arrow indicating the inversions.

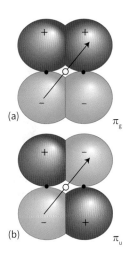

FIGURE 2.16 (a) Bonding and (b) antibonding π interactions, with the arrow indicating the inversions.

The most bonding orbital is usually a σ orbital, as electrons are concentrated between the two atomic nuclei. To establish the actual location of the energy levels, it is necessary to use electronic absorption spectroscopy, photoelectron spectroscopy, or detailed computation.

Photoelectron spectroscopy and detailed computation (the numerical solution of the Schrödinger equation for the molecules) enable us to build the orbital energy schemes shown in Fig. 2.17. As we see there, from Li_2 to N_2 the arrangement of orbitals is that shown in Fig. 2.18, whereas for O_2 and F_2 the order of the $2\sigma_g$ and $1\pi_u$ orbitals is reversed and the array is that shown in Fig. 2.12. The reversal of order can be traced to the increasing energy separation of the 2s and 2p orbitals that occurs on going to the right across Period 2. A general principle of quantum mechanics is that the mixing of wavefunctions is strongest if their energies are similar; mixing is not important if their energies differ by more than about 10 eV. When the s-p energy separation is small, each σ molecular orbital is a mixture of s and p character on each atom. For Li_2 to N_2 the energy separation of the 2s and 2p orbitals is small so we see mixing, and the resulting molecular orbitals have both s and p character. As the s and p energy separation increases as we move across the row to O and F, there is less s–p mixing and the molecular orbitals become more purely s-like or p-like. The decrease in s–p mixing also affects the energies of the molecular orbitals, as can be seen in Fig. 2.17. Additionally, a feature of quantum mechanics, often known as the 'noncrossing rule' and mentioned later in Section 22.2, tells us that molecular orbitals that have the same symmetry, for instance $1\sigma_g$ and $2\sigma_g$, and are not too widely separated in energy, will interact in such a way as to increase that energy separation; thus a $1\sigma_g$ orbital is somewhat stabilized while the $2\sigma_g$ orbital is destabilized.

When considering species containing two neighbouring d-block atoms, as in Hg_2^{2+} and $[Cl_4ReReCl_4]^{2-}$, we should also allow for the possibility of forming bonds from d orbitals. A d_{z^2} orbital has cylindrical symmetry with respect to

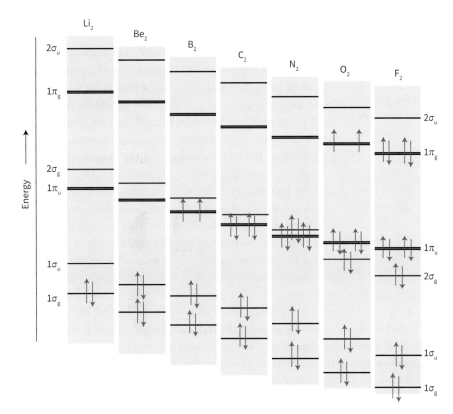

FIGURE 2.17 The variation of orbital energies for Period 2 homonuclear diatomic molecules from Li_2 to F_2.

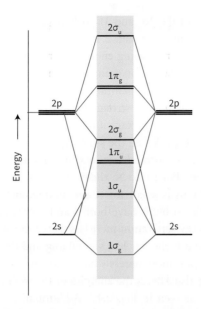

FIGURE 2.18 The molecular orbital energy-level diagram for Period 2 homonuclear diatomic molecules from Li_2 to N_2.

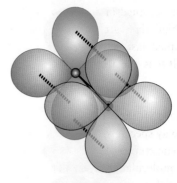

FIGURE 2.19 The formation of δ orbitals by d-orbital overlap. The orbital has two mutually perpendicular nodal planes that intersect along the internuclear axis.

the internuclear (z) axis, and hence can contribute to the σ orbitals that are formed from s and p_z orbitals. The d_{yz} and d_{zx} orbitals both look like p orbitals when viewed along the internuclear axis, and hence can contribute to the π orbitals formed from p_x and p_y. The new feature is the role of $d_{x^2-y^2}$ and d_{xy}, which have no counterpart in the orbitals discussed up to now. These two orbitals can overlap with matching orbitals on the other atom to give rise to doubly degenerate pairs of bonding and antibonding δ orbitals (Fig. 2.19). As we shall see in Chapter 20, δ orbitals are important for the discussion of bonds between d-metal atoms, in d-metal complexes, and in many organometallic compounds.

(b) The building-up principle for molecules

KEY POINTS The building-up principle is used to predict the ground-state electron configurations by accommodating electrons in the array of molecular orbitals summarized in Fig. 2.12 or Fig. 2.18 and recognizing the constraints of the Pauli principle.

We use the building-up principle in conjunction with the molecular orbital energy-level diagram in the same way as for atoms. The order of occupation of the orbitals is the order of increasing energy, as depicted in Fig. 2.12 or Fig. 2.18. Each orbital can accommodate up to two electrons. If more than one orbital is available for occupation (because they happen to have identical energies, as in the case of pairs of π orbitals), then the orbitals are occupied separately. In that case, the electrons in the half-filled orbitals adopt parallel spins (↑↑) with one electron in each orbital, just as is required by Hund's rule for atoms (Section 1.5(a)). With very few exceptions, these rules lead to the actual ground-state configuration of the Period 2 diatomic molecules. Molecular orbital configurations are written like those for atoms: the orbitals are listed in order of increasing energy, and the number of electrons in each one is indicated by a superscript. Note that $π^4$ is shorthand for the complete occupation of two different, energetically equivalent π orbitals.

Fig. 2.17 shows the molecular orbital energy levels filled with electrons for the diatomic molecules Li_2 to F_2. The Li_2 molecule is predicted to have a single Li–Li bond, and indeed such molecules are observed in the gas phase; the electron configuration of Li_2 is $1\sigma_g^2$. Be_2 is predicted to be an unstable species as there are equal numbers of bonding and antibonding electrons. The electron configuration of Be_2 is $1\sigma_g^2 1\sigma_u^2$. B_2 molecules are observed in the gas phase, and the molecular orbital description supports this observation as there are more electrons in bonding than in antibonding orbitals, with an electron configuration of $1\sigma_g^2 \sigma_u^2 1\pi_u^2$. The molecular orbital description of C_2, with the electron configuration $1\sigma_g^2 1\sigma_u^2 1\pi_u^4$, predicts the existence of the C=C species with two π bonds. C_2 is in fact very uncommon as tetravalent carbon is much more favoured. The electron configuration of N_2 is $1\sigma_g^2 1\sigma_u^2 1\pi_u^4 2\sigma_g^2$ and predicts an N≡N triple bond comprised of one σ and two π bonds, in agreement with the stability of the N_2 species. When we reach O_2 we see the effect of the increased energy gap between the s and p orbitals on the relative energies of the molecular orbitals. The electron configuration of O_2 is thus $1\sigma_g^2 1\sigma_u^2 2\sigma_g^2 1\pi_u^4 1\pi_g^2$, and the fact that there are more bonding than antibonding electrons predicts that the molecule is stable, as we know it is. The molecular orbital treatment also predicts that it has two unpaired electrons in the $1\pi_u$ orbitals. Thus, O_2 is predicted to be paramagnetic. This explanation of the observed paramagnetism of oxygen is one of the great successes of molecular orbital theory over the valence bond model. The electron configuration of F_2 is $1\sigma_g^2 1\sigma_u^2 2\sigma_g^2 1\pi_u^4 1\pi_g^4$ and indicates that it should be stable as there are more bonding than antibonding electrons. Adding two more electrons to fill the strongly antibonding $2\sigma_u$ orbital would result in strong repulsion between the two atoms; thus, as expected, the neon molecule Ne_2 does not exist.

The **highest occupied molecular orbital** (HOMO) is the molecular orbital that, according to the building-up principle, is occupied last. The **lowest unoccupied molecular orbital** (LUMO) is the next higher molecular orbital. In Fig. 2.17, the HOMO of F_2 is $1\pi_g$ and its LUMO is $2\sigma_u$; for N_2 the HOMO is $2\sigma_g$ and the LUMO is $1\pi_g$. We shall increasingly see that these **frontier orbitals**, the LUMO and the HOMO, play special roles in the interpretation of structural and kinetic studies. The term SOMO, denoting a **singly occupied molecular orbital**, is sometimes encountered and is of crucial importance for the properties of radical species. Radicals may dimerize by forming a bond between the atoms that contribute most to the SOMO. The frontier orbitals form the basis of the **isolobal analogy** mentioned later in the book and described in more detail in Section 24.20. In short, the isolobal analogy is a concept applied to make intuitive predictions about the bond-making abilities of atoms and molecules (molecular 'fragments') that might have little obvious chemical similarity but share a common identity in terms of the symmetries and electron occupancy of their frontier orbitals.

2.9 Heteronuclear diatomic molecules

The molecular orbitals of heteronuclear diatomic molecules differ from those of homonuclear diatomic molecules in having unequal contributions from each atomic orbital. Each molecular orbital has the form

$$\psi = c_A \chi_A + c_B \chi_B + \cdots \tag{2.6}$$

The unwritten orbitals include all the other orbitals of the correct symmetry for forming σ or π bonds but which typically make a smaller contribution than the two valence shell orbitals we are considering. In contrast to orbitals for homonuclear species, the coefficients c_A and c_B are not necessarily equal in magnitude. If $c_A^2 > c_B^2$ the orbital is composed principally of χ_A and an electron that occupies the molecular orbital is more likely to be found near atom A than atom B. The opposite is true for a molecular orbital in which $c_A^2 < c_B^2$.

(a) Heteronuclear molecular orbitals

KEY POINTS Heteronuclear diatomic molecules are polar; bonding electrons tend to be found closer to the more electronegative atom and antibonding electrons closer to the less electronegative atom.

Since the energy of an atomic orbital decreases (is stabilized) as Z_{eff} increases, the relative energies of the two sets of atomic orbitals reflect the respective electronegativities (or Z_{eff}) of each atom. Consequently, the greater contribution to a bonding molecular orbital normally comes from the more electronegative atom: the bonding electrons are then likely to be found close to that atom and hence be in an energetically favourable location. The less electronegative atom normally contributes more to an antibonding orbital (Fig. 2.20); that is, antibonding electrons are more likely to be found in an energetically unfavourable location, closer to the less electronegative atom. A heteronuclear diatomic molecule is thus inherently polar, the covalent bonding

---

EXAMPLE 2.4 **Predicting the electron configurations of diatomic molecules**

Predict the ground-state electron configurations of the oxygen molecule, O_2, the superoxide ion, O_2^-, and the peroxide ion, O_2^{2-}.

Answer We need to determine the number of valence electrons and then populate the molecular orbitals with them in accord with the building-up principle. An O_2 molecule has 12 valence electrons. The first 10 electrons recreate the N_2 configuration except for the reversal of the order of the $1\pi_u$ and $2\sigma_g$ orbitals (see Fig. 2.17). Next in line for occupation are the doubly degenerate $1\pi_g$ orbitals. The last two electrons enter these orbitals separately and have parallel spins. The configuration is therefore

$O_2: 1\sigma_g^2 1\sigma_u^2 2\sigma_g^2 1\pi_u^4 1\pi_g^2$

The O_2 molecule is interesting because the lowest energy configuration has two unpaired electrons in different π orbitals. Hence, O_2 is paramagnetic (tends to be attracted into a magnetic field). The next two electrons can be accommodated in the $1\pi_g$ orbitals, giving

$O_2^-: 1\sigma_g^2 1\sigma_u^2 2\sigma_g^2 1\pi_u^4 1\pi_g^3$
$O_2^{2-}: 1\sigma_g^2 1\sigma_u^2 2\sigma_g^2 1\pi_u^4 1\pi_g^4$

We are assuming that the orbital order remains that shown in Fig. 2.17, although this need not be the case.

Self-test 2.4 (a) Determine the number of unpaired electrons on O_2, O_2^-, and O_2^{2-}. (b) Write the valence electron configuration you would expect for S_2^{2-} and S_2^{2+}.

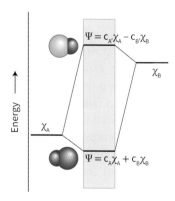

FIGURE 2.20 The molecular orbital energy-level diagram arising from the interaction of two atomic orbitals with different energies. The lower molecular orbital is primarily composed of the lower-energy atomic orbital, and vice versa. The shift in energies of the two levels is less than if the atomic orbitals had the same energy.

involving electrons that are unequally shared by the two atoms. Both bonding electrons may be provided by one of the atoms: this is referred to as a **dative** (or **coordinate**) bond which is often written as A←:B to emphasize that an electron pair is 'donated' from B (the donor) to A (the acceptor). Such bonding is important in describing Lewis acids and bases (Chapter 5) and coordination compounds (Chapter 7). In the extreme case, corresponding to a large difference in electronegativity, one atom gains complete control over all the valence electrons, resulting in an ionic bond.

We have already remarked that two wavefunctions interact less strongly as their energies diverge. This dependence on energy separation implies that the stabilization of a bonding molecular orbital (lowering of its energy relative to that of the atomic orbitals) should be lower in a heteronuclear molecule than for a homonuclear molecule in which the atomic orbitals have the same energies. However, we should not conclude that A–B bonds are weaker than A–A bonds because other factors (including orbital size and closeness of approach) are also important. In Section 2.14 we will mention the Pauling electronegativity scale which assumes that A–B bonds benefit from an ionic contribution that does not depend on orbital overlap. Notably, the heteronuclear CO molecule has an even higher bond enthalpy (1070 kJ mol^{-1}) than its isoelectronic counterpart N$_2$ (946 kJ mol^{-1}).

(b) Hydrogen fluoride

KEY POINT In hydrogen fluoride the bonding orbital is more concentrated on the F atom and the antibonding orbital is more concentrated on the H atom.

As an illustration of these general points, consider a simple heteronuclear diatomic molecule, HF. The five valence orbitals available for molecular orbital formation are the 1s orbital of H and the 2s and 2p orbitals of F; there are 1 + 7 = 8 valence electrons to accommodate in the five molecular orbitals that can be constructed from the five basis orbitals.

The σ orbitals of HF can be constructed by allowing an H1s orbital to overlap with the F2s and F2p$_z$ orbitals (z being the internuclear axis). These three atomic orbitals combine to give three σ molecular orbitals of the form $\psi = c_1 \chi_{H1s} + c_2 \chi_{F2s} + c_3 \chi_{F2p}$. This procedure leaves the F2p$_x$ and F2p$_y$ orbitals unaffected as they have π symmetry and there is no valence H orbital of that symmetry. These π orbitals are therefore examples of the nonbonding orbitals mentioned previously, and are molecular orbitals confined to a single atom. Note that because there is no centre of inversion in a heteronuclear diatomic molecule, we do not use the g, u classification for its molecular orbitals.

Fig. 2.21 shows the resulting energy-level diagram. The 1σ bonding orbital is predominantly F2s in character as the energy difference between it and the H1s orbital is large.

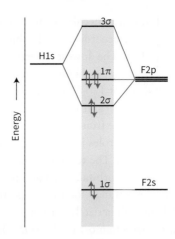

FIGURE 2.21 The molecular orbital energy-level diagram for HF. The relative positions of the atomic orbitals reflect the ionization energies of the atoms.

It is, therefore, confined mainly to the F atom and essentially nonbonding. The 2σ orbital is much more bonding than the 1σ orbital and has both H1s and F2p character. The 3σ orbital is antibonding, and principally H1s in character: the H1s orbital has a relatively high energy (compared with the fluorine orbitals) and hence contributes predominantly to the high-energy antibonding molecular orbital.

Of the eight valence electrons in HF, two fill the nonbonding 1σ orbital localized on F, two fill the strongly bonding 2σ orbital, and the remaining four fill the two 1π orbitals that are confined to the F atom and nonbonding because the net overlap with H 1s is zero (Fig. 2.22). The resulting arrangement is consistent with the conventional model of three lone pairs on the fluorine atom. All the electrons are now accommodated, so the configuration of the molecule is 1σ^{2}2σ^{2}1π^4. One important feature to note is that all the electrons occupy orbitals that are predominantly on the F atom. It follows that we can expect the HF molecule to be polar, with a partial negative charge on the F atom, which is found experimentally.

(c) Carbon monoxide

KEY POINTS The HOMO of a carbon monoxide molecule is an almost nonbonding σ orbital largely localized on C; the LUMO is an antibonding π orbital.

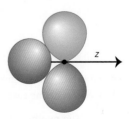

FIGURE 2.22 Interaction along a z-axis between an s orbital on one atom and p$_x$ or p$_y$ orbitals on the other atom results in no net overlap and the molecular orbital is nonbonding.

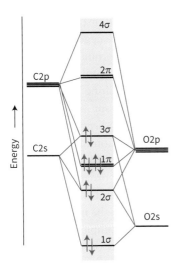

FIGURE 2.23 The molecular orbital energy-level diagram for CO.

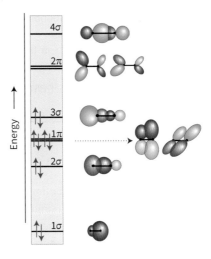

FIGURE 2.24 A schematic illustration of the molecular orbitals of CO, with the size of the atomic orbital indicating the magnitude of its contribution to the molecular orbital.

The molecular orbital energy-level diagram for carbon monoxide is a somewhat more complicated example than HF because both atoms have 2s and 2p orbitals that can participate in the formation of σ and π orbitals. The energy-level diagram is shown in Fig. 2.23. The ground-state configuration is

$$\text{CO}: 1\sigma^2 2\sigma^2 1\pi^4 3\sigma^2$$

As the ladder is ascended, orbitals increasingly take on the character of the more electropositive atom. Thus, for the eight MOs of CO, the lower four tend to have more O character, and the upper four have more C character. The 1σ orbital is localized mostly on the O atom and therefore essentially nonbonding or weakly bonding. The 2σ orbital is bonding. The 1π orbitals constitute the doubly degenerate pair of bonding π orbitals, with mainly C2p orbital character.

The HOMO in CO is 3σ, which is predominantly C2p_z in character. The LUMO is the doubly degenerate pair of antibonding π orbitals, with mainly O2p orbital character (Fig. 2.24). This combination of frontier orbitals—a full σ orbital largely localized on C and a pair of empty π orbitals—is one reason why so many compounds are known in which CO is bonded to a d metal atom. In the so-called d-metal carbonyls, the HOMO lone pair orbital of CO participates in the formation of a σ bond to the metal atom and the LUMO antibonding π orbital participates in the formation of π bonds with the metal atom (Section 24.5).

Although the difference in electronegativity between C and O is large, the experimental value of the electric dipole moment of the CO molecule (0.1 D, where D is a unit of dipole moment, the debye) is small. Moreover, the negative end of the dipole is on the C atom despite that being the less electronegative atom. This odd situation stems from the fact that the lone pairs and bonding pairs have a complex distribution. It is wrong to conclude that because the bonding electrons are mainly on the more electronegative O atom,

O is the negative end of the dipole, as this ignores the balancing effect of the lone pair on the C atom. The inference of polarity from electronegativity is particularly unreliable when antibonding orbitals are occupied as these orbitals are more likely to have the character of the more electropositive atom.

EXAMPLE 2.5 Accounting for the structure of a heteronuclear diatomic molecule

The halogens form compounds among themselves. One of these 'interhalogen' compounds is iodine monochloride, ICl, in which the order of orbitals is 1σ, 2σ, 1π, 3σ, 2π, 4σ (from calculation). What is the ground-state electron configuration of ICl?

Answer First, we identify the atomic orbitals that are to be used to construct molecular orbitals: these are the 3s and 3p valence shell orbitals of Cl and the 5s and 5p valence shell orbitals of I. As for Period 2 elements, an array of σ and π orbitals can be constructed, and is shown in Fig. 2.25. The bonding orbitals are predominantly Cl in character (because that is the more

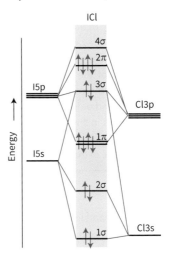

FIGURE 2.25 A schematic illustration of the energies of the molecular orbitals of ICl.

electronegative element) and the antibonding orbitals are predominantly I in character. There are 7+7=14 valence electrons to accommodate, which results in the ground-state electron configuration $1\sigma^2 2\sigma^2 1\pi^4 3\sigma^2 2\pi^4$.

Self-test 2.5 Predict the ground-state electron configuration of the hypochlorite ion, ClO^-.

2.10 Bond properties

We have already seen the origin of the importance of the electron pair: two electrons is the maximum number that can occupy a bonding orbital and hence contribute to a chemical bond. We now extend this idea by introducing the concept of 'bond order'.

(a) Bond order

KEY POINTS The bond order assesses the net number of bonds between two atoms in the molecular orbital formalism; the greater the bond order between a given pair of atoms, the greater the bond strength.

The **bond order**, b, identifies a shared electron pair as counting as a 'bond' and an electron pair in an antibonding orbital as an 'antibond' between two atoms. More precisely, the bond order is defined as

$$b = \frac{1}{2}(n - n^*) \quad (2.7)$$

where n is the number of electrons in bonding orbitals and n^* is the number in antibonding orbitals. Nonbonding electrons are ignored when calculating bond order.

A BRIEF ILLUSTRATION

Difluorine, F_2, has the configuration $1\sigma_g^2 \sigma_u^2 2\sigma_g^2 1\pi_u^4 1\pi_g^4$ and, because $1\sigma_g$, $1\pi_u$, and $2\sigma_g$ orbitals are bonding but $1\sigma_u$ and $1\pi_g$ are antibonding, $b = \frac{1}{2}(2+2+4-2-4) = 1$. The bond order of F_2 is 1, which is consistent with the structure F–F and the conventional description of the molecule as having a single bond. Dinitrogen, N_2, has the configuration $1\sigma_g^2 1\sigma_u^2 1\pi_u^4 2\sigma_g^2$ and $b = \frac{1}{2}(2+4+2-2) = 3$. A bond order of 3 corresponds to a triply bonded molecule, which is in line with the structure N≡N. The high bond order is reflected in the high bond enthalpy of the molecule (946 kJ mol^{-1}), one of the highest for any molecule.

Isoelectronic molecules and ions have the same bond order, so F_2 and O_2^{2-} both have bond order 1. The bond order of the CO molecule, like that of the isoelectronic molecule N_2, is 3, in accord with the analogous structure C≡O. However, this method of assessing bonding is primitive, especially for heteronuclear species. For instance, inspection of the computed molecular orbitals suggests that 1σ and 3σ are best regarded as nonbonding orbitals largely localized on O and C respectively, and hence should really be disregarded in the calculation of b. The resulting bond order is unchanged by this modification. The lesson is that the definition of bond order provides a useful indication of the multiplicity of the bond, but any interpretation of contributions to b needs to be made in the light of guidance from the composition of computed orbitals. Bond order is a semiquantitative guide to bond strength: it is useful when comparing the strengths of different bonds that can be formed between atoms of two specified elements; for example, C–O vs C=O.

The definition of bond order allows for the possibility that an orbital is only singly occupied. Electron loss from N_2 leads to the formation of the transient species N_2^+ in which the bond order is reduced from 3 to 2.5. This reduction in bond order is accompanied by a corresponding decrease in bond strength (from 946 to 855 kJ mol^{-1}), and an increase in the bond length from 109 pm for N_2 to 112 pm for N_2^+.

EXAMPLE 2.6 Determining bond order

Determine the bond order of the oxygen molecule, O_2, the superoxide ion, O_2^-, and the peroxide ion, O_2^{2-}.

Answer We must determine the number of valence electrons, use them to populate the molecular orbitals, and then use eqn. 2.7 to calculate b. The species O_2, O_2^-, and O_2^{2-} have 12, 13, and 14 valence electrons, respectively. Recall, from Example 2.4, that their configurations are

$O_2: 1\sigma_g^2 1\sigma_u^2 2\sigma_g^2 1\pi_u^4 1\pi_g^2$

$O_2^-: 1\sigma_g^2 1\sigma_u^2 2\sigma_g^2 1\pi_u^4 1\pi_g^3$

$O_2^{2-}: 1\sigma_g^2 1\sigma_u^2 2\sigma_g^2 1\pi_u^4 1\pi_g^4$

The $1\sigma_g$, $1\pi_u$, and $2\sigma_g$ orbitals are bonding and the $1\sigma_u$ and $1\pi_g$ orbitals are antibonding. Therefore, the bond orders are

$O_2: b = \frac{1}{2}(2-2+4+2-2) = 2$

$O_2^-: b = \frac{1}{2}(2-2+4+2-3) = 1.5$

$O_2^{2-}: b = \frac{1}{2}(2-2+4+2-4) = 1$

Self-test 2.6 Predict the bond order of (a) the dicarbide anion, C_2^{2-}, and (b) Ne_2.

(b) Bond correlations

KEY POINT For a given pair of elements, bond strength increases and bond length decreases as bond order increases.

The strengths and lengths of bonds correlate quite well with each other and with the bond order. For a given pair of atoms,

Bond enthalpy increases as bond order increases.
Bond length decreases as bond order increases.

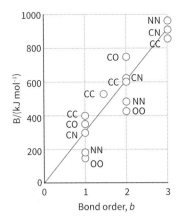

FIGURE 2.26 The correlation between bond enthalpy (B) and bond order.

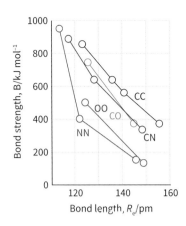

FIGURE 2.28 The correlation between bond enthalpy (B) and bond length.

These trends are illustrated in Figs. 2.26 and 2.27. The strength of the dependence varies with the elements. In Period 2 the correlation is not linear for carbon–carbon bonds, with the result that a C=C double bond is less than twice as strong as a C–C single bond (586 kJ mol^{-1} in ethene vs 331 kJ mol^{-1} in ethane). This difference has profound consequences in organic chemistry, particularly for the reactions of unsaturated compounds. It implies, for example, that it is energetically favourable (but slow in the absence of a catalyst) for ethene and ethyne to polymerize: in this process, C–C single bonds form at the expense of the appropriate numbers of multiple bonds.

Familiarity with carbon's properties, however, must not be extrapolated without caution to the bonds between other elements, particularly as non-bonded electron repulsion increases across Period 2. An N=N double bond (409 kJ mol^{-1}) is more than twice as strong as an N–N single bond (163 kJ mol^{-1}), and an N≡N triple bond (946 kJ mol^{-1}) is more than five times as strong. It is on account of this trend that NN multiply bonded compounds are stable relative to polymers or three-dimensional compounds having only single bonds. The same is not true of phosphorus,

where the P–P, P=P, and P≡P bond enthalpies are 201, 310, and 490 kJ mol^{-1}, respectively. For phosphorus, single bonds are stable relative to the matching number of multiple bonds. Thus, phosphorus exists in a variety of solid forms in which P–P single bonds are present, including the tetrahedral P$_4$ molecules of white phosphorus (Section 16.1). Diphosphorus molecules, P$_2$, are species generated at high temperatures and low pressures.

The two correlations with bond order taken together imply that, for a given pair of elements,

Bond enthalpy increases as bond length decreases.

This correlation is illustrated in Fig. 2.28: it is a useful feature to bear in mind when considering the stabilities of molecules because bond lengths may be readily available from independent sources.

EXAMPLE 2.7 Predicting correlations between bond order, bond length, and bond strength

Use the bond orders of the oxygen molecule, O$_2$, the superoxide ion, O$_2^-$, and the peroxide ion, O$_2^{2-}$, calculated in Example 2.6 to predict the relative bond lengths and strengths of the species.

Answer We need to remember that bond enthalpy increases as bond order increases. The bond orders of O$_2$, O$_2^-$, and O$_2^{2-}$ are 2, 1.5, and 1, respectively. Therefore, we expect the bond enthalpies to increase in the order O$_2^{2-}$ < O$_2^-$ < O$_2$. Indeed, the gas-phase bond enthalpy determined for the O–O bond in H$_2$O$_2$ is just 146 kJ mol^{-1}, much lower than that of O$_2$ (497 kJ mol^{-1}). Bond length decreases as the bond enthalpy increases, so bond lengths should follow the opposite trend: i.e. O$_2^{2-}$ > O$_2^-$ > O$_2$. These predictions are supported by bond lengths of 149, 128, and 121 pm, for O$_2^{2-}$ (in BaO$_2$), O$_2^-$ (in KO$_2$), and O$_2$, respectively.

Self-test 2.7 Predict the order of bond enthalpies and bond lengths for C–N, C=N, and C≡N bonds.

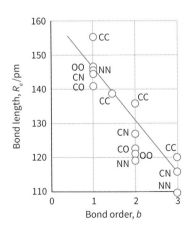

FIGURE 2.27 The correlation between bond length and bond order.

2.11 Bands formed from extended overlapping atomic orbitals

An important extension of MO theory is used to explain the electronic properties of solid materials. Here, valence electrons supplied by the atoms may spread through the entire structure, which is now treated like an indefinitely large molecule. We begin by showing how metals are described in terms of molecular orbitals. Then we go on to show that the same principles can be applied, but with a different outcome, to ionic and molecular solids.

(a) Band formation by extended orbital overlap

KEY POINT The overlap of atomic orbitals in solids gives rise to bands of energy levels separated by energy gaps.

The extension of MO theory to solid materials can be illustrated by considering a line of atoms, and supposing that each atom has an s orbital that overlaps the s orbitals on its immediate neighbours (Fig. 2.29).

When the line consists of only two atoms, there is a bonding and an antibonding molecular orbital. When a third atom joins them, there are three molecular orbitals. The central orbital of the set is nonbonding and the outer two are at low energy and high energy, respectively. As more atoms are added, each one contributes an atomic orbital, and hence one more molecular orbital is formed. When there are N atoms in the line, there are N molecular orbitals. The orbital of lowest energy has no nodes between neighbouring atoms, and the orbital of highest energy has a node between every pair of neighbours. The remaining orbitals have successively $1, 2, \ldots$, internuclear nodes and a corresponding range of energies between the two extremes (Fig. 2.30).

The overlap of a large number of atomic orbitals in a solid can be viewed as leading to a very large number of molecular orbitals that are extremely closely spaced in energy and so form an almost continuous **band** of energy levels (Fig. 2.31).

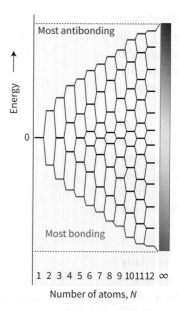

FIGURE 2.30 The energies of the orbitals that are formed when N atoms are brought up to form a one-dimensional array.

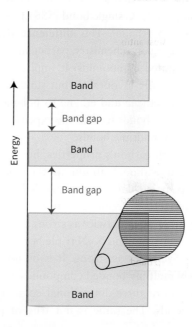

FIGURE 2.31 The electronic structure of a solid is characterized by a series of bands of orbitals separated by gaps at energies where orbitals do not occur.

The total width of the band (its energy spread) which remains finite even as N approaches infinity (as shown in Fig. 2.30) depends on the strength of the interaction between neighbouring atoms. Just as for homonuclear diatomic molecules, where strong orbital overlap leads to a greater separation of the bonding and antibonding energy levels, a higher degree of orbital overlap between neighbours in the solid produces a larger energy separation of the nonnode orbital and the all-node orbital. However, whatever the number of atomic orbitals used to form the molecular orbitals, there is only a finite spread of orbital energies (both

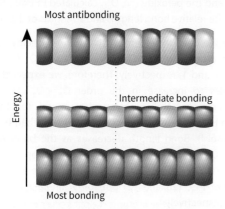

FIGURE 2.29 A band can be thought of as formed by bringing up atoms successively to form a line of atoms. N atomic orbitals give rise to N molecular orbitals.

the lowest and highest energies converge on limiting values, as depicted in Fig. 2.30). It follows that the separation in energy between neighbouring orbitals must approach zero as N approaches infinity, otherwise the range of orbital energies could not be finite. That is, a band consists of a near-continuum of energy levels.

The band just described is built from s orbitals and is called an **s band**. If there are p orbitals available, a **p band** can be constructed from their overlap, as shown in Fig. 2.32. Because p orbitals lie higher in energy than s orbitals of the same valence shell, there is often an energy gap between the s band and the p band (Fig. 2.33). However, if the bands span a wide range of energy and the atomic s and p energies are similar (as is often the case), then the two bands overlap. The **d band** is similarly constructed from the overlap of d orbitals. The formation of bands is not restricted to one type of atomic orbital and bands may be formed in compounds by combinations of different orbital types; for example, the d orbitals of a metal atom may overlap the p orbitals of neighbouring O atoms.

In general, a band structure diagram can be produced for any solid and is constructed using the frontier orbitals on all the atoms present. The energies of these bands, and whether they overlap, will depend on the energies of the contributing atomic orbitals, and bands may be empty, full, or partially filled depending on the total number of electrons in the system. As with heteronuclear diatomic molecules, the more electronegative element in an ionic solid contributes mostly to the lower energy bands (analogous to the bonding orbital) while the more electropositive element contributes mostly to the higher energy bands (analogous to the antibonding orbital).

(b) The Fermi level and electronic conductivity

KEY POINTS The Fermi level is the highest occupied energy level in a solid at $T = 0$. When the Fermi level lies within a band, electrons are highly mobile and the material conducts electricity

At $T = 0$, electrons occupy the individual molecular orbitals forming the bands in accordance with the building-up principle. If each atom, in an array of N atoms forming a band from s orbitals alone, supplies one s electron, then at $T = 0$ the lowest N orbitals are occupied; as each contributing s orbital can accommodate two electrons the band can accommodate $2N$ electrons and is 'half-filled'. The highest occupied orbital at $T = 0$ is called the **Fermi level**, and in this case (a band formed from s orbitals and each atom having one s electron) it lies at the centre of the band (Fig. 2.34). When the band is not completely full the electrons close to the Fermi level can easily be promoted to nearby empty levels, the electron occupancy across the range of these levels being known as the **Fermi distribution**. As a result, they are mobile and can

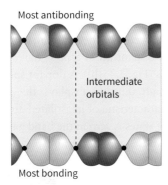

FIGURE 2.32 An example of a p band in a one-dimensional solid.

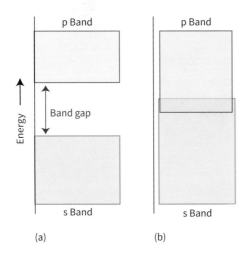

FIGURE 2.33 (a) The s and p bands of a solid and the gap between them. Whether or not there is in fact a gap depends on the separation of the s and p orbitals of the atoms and the strength of the interaction between them in the solid. (b) If the interaction is strong, the bands are wide and may overlap.

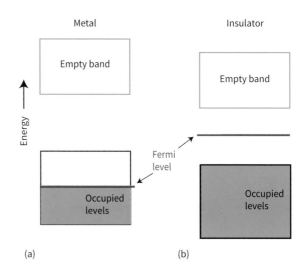

FIGURE 2.34 (a) Typical band structure for a metal showing the Fermi level; if each of the N atoms supplies one s electron, then at $T = 0$ the lower ½N orbitals are occupied and the Fermi level lies near the centre of the band. (b) Typical band structure for an insulator with the Fermi level midway in the band gap.

move relatively freely through the solid, and the substance can conduct electricity. The solid is in fact a *metallic* conductor. An example of a material with a half-filled band formed from s-orbitals is sodium which exhibits a conductivity at room temperature of 2.1×10^7 Sm^{-1}. Electronic conductivity is described in detail in Chapter 25.

(c) Densities of states and width of bands

KEY POINT The density of states is not uniform across a band: in most cases, the states are densest close to the centre of the band.

The number of energy levels in an energy range divided by the width of the range is called the **density of states**, ρ (Fig. 2.35(a)). The density of states is not uniform across a band because the energy levels are packed together more closely at some energies than at others. In three dimensions, the variation of density of states is like that shown in Fig. 2.36, with the greatest density of states near the centre of the band and the lowest density at the edges. The reason for this behaviour can be traced to the number of ways of producing a particular linear combination of atomic orbitals. There is only one way of forming a fully bonding molecular orbital (the lower edge of the band) and only one way of forming a fully antibonding orbital (the upper edge). However, there are many ways (in a three-dimensional array of atoms) of forming a molecular orbital with an energy corresponding to the interior of a band.

The number of orbitals contributing to a band determines the total number of states within it; that is, the area enclosed by the density of states curve. Large numbers of contributing atomic orbitals which have strong overlap produce broad (in energy terms) bands with a high density of states. If only a relatively few atoms contribute to a band and these are well separated in the solid, as is the case for a species (a dopant) introduced into the material, then the band associated with this dopant atom type is narrow and contains only a few states (Fig. 2.35(b)). A dopant species introduced at extremely high dilution displays the discrete atomic energy levels characteristic of its isolated atomic state.

The density of states is zero in the band gap itself—there is no energy level in the gap. In certain special cases, however, a full band and an empty band might coincide in energy but with a zero density of states at their conjunction (Fig. 2.37). Solids with this band structure are called **semimetals**. One important example is graphite (Section 15.5), which is a semimetal in directions parallel to the sheets of carbon atoms.

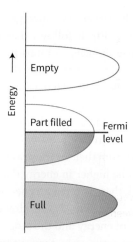

FIGURE 2.36 Typical density of states diagram for a three-dimensional metal. The density of states is represented by the horizontal direction.

> **A NOTE ON GOOD PRACTICE**
> This use of the term 'semimetal' should be distinguished from its other use as a synonym for metalloid. In this text we avoid the latter usage.

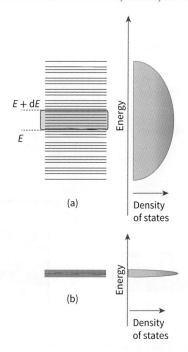

FIGURE 2.35 (a) The density of states in a metal is the number of energy levels in an infinitesimal range of energies between E and $E + dE$. (b) The density of states associated with a low concentration of dopant.

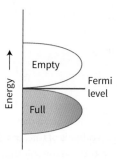

FIGURE 2.37 The density of states in a semimetal.

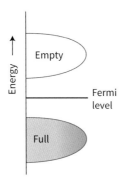

FIGURE 2.38 The structure of a typical insulator: there is a significant gap between the filled and empty bands.

(d) Insulators

KEY POINT An insulator is a solid with a large band gap.

A solid is an insulator if enough electrons are present to fill a band completely and there is a considerable energy gap before an empty orbital and its associated band become available (Fig. 2.38). In a sodium chloride crystal, for instance, the N Cl$^-$ ions are nearly in contact and their 3s and three 3p valence orbitals overlap to form a narrow band consisting of $4N$ levels. The Na$^+$ ions are also nearly in contact and also form a band. The electronegativity of chlorine is so much greater than that of sodium that the chlorine band lies well below the sodium band, and the band gap (E_g) is about 7 eV. A total of $8N$ electrons are to be accommodated (seven from each Cl atom, and one from each Na atom). These $8N$ electrons enter the lower chlorine band, fill it, and leave the sodium band empty. Because the energy of thermal motion available at room temperature is $kT \approx 0.03$ eV (k is Boltzmann's constant), very few electrons have enough energy to occupy the orbitals of the sodium band.

In an insulator the band of highest energy that contains electrons (at $T = 0$) is normally termed the **valence band**. The next-higher band (which is empty at $T = 0$) is called the **conduction band**. In NaCl the band derived from the Cl orbitals is the valence band, and the band derived from the Na orbitals is the conduction band.

We normally think of an ionic or molecular solid as consisting of discrete ions or molecules. According to the picture just described, however, they can be regarded as having a band structure. The two pictures can be reconciled because it is possible to show that a full band is equivalent to a sum of localized electron densities. In sodium chloride, for example, a full band built from Cl orbitals is equivalent to a collection of discrete Cl$^-$ ions, and an empty band built from Na orbitals is equivalent to the array of Na$^+$ ions.

(e) Intrinsic semiconductors

KEY POINT The band gap in a semiconductor controls the temperature dependence of the conductivity through an Arrhenius-like expression.

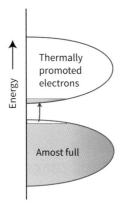

FIGURE 2.39 In an intrinsic semiconductor, the band gap is so small that the Fermi distribution results in the population of some orbitals in the upper band.

The characteristic physical property of a semiconductor is that its electrical conductivity increases with increasing temperature.

In an **intrinsic semiconductor**, the band gap is relatively small, 0.5–2 eV, so that the thermal energy results in some electrons from the valence band populating the empty upper band (Fig. 2.39). This occupation of the conduction band introduces **positive holes**, equivalent to an absence of electrons, into the lower band, and as a result the solid is conducting because both the holes and the promoted electrons can move. At room temperature, the conductivities of semiconductors are typically intermediate between those of metals and insulators because only very few electrons and holes can act as charge carriers. The strong, increasing with temperature, dependence of the conductivity follows from the exponential Boltzmann-like temperature dependence of the electron population in the upper band. The dividing line between insulators and semiconductors is a matter of the size of the band gap (Table 2.5); the conductivity itself is an unreliable criterion because, as the temperature is increased, a given substance may have in succession a low, intermediate, and high conductivity. The values of the band gap and conductivity that are taken as indicating semiconduction rather than insulation depend on the application being considered.

TABLE 2.5 Some typical band gaps (E_g) at 298 K

Material	E_g/eV
Carbon (diamond)	5.47
Silicon carbide	3.00
Silicon	1.11
Germanium	0.66
Gallium arsenide	1.35
Indium arsenide	0.36

It follows from the exponential form of the population of the conduction band that the conductivity of a semiconductor should show an Arrhenius-like temperature dependence of the form

$$\sigma = \sigma_0 e^{-E_g/2kT} \tag{2.8}$$

The conductivity of a semiconductor is therefore expected to be Arrhenius-like with an activation energy equal to half the band gap, $E_a \approx \frac{1}{2} E_g$. This is found to be the case in practice.

Bond properties, reaction enthalpies, and kinetics

Certain properties of bonds are approximately the same in different compounds of the elements. Thus, if we know the strength of an O–H bond in H_2O, then with some confidence we can use the same value for the O–H bond in CH_3OH. At this stage we confine our attention to two of the most important characteristics of a bond: its length and its strength. We also extend our understanding of bonds to predict the shapes of simple inorganic molecules.

2.12 Bond length

KEY POINTS The equilibrium bond length in a molecule is the separation of the centres of the two bonded atoms; covalent radii vary through the periodic table in much the same way as metallic and ionic radii.

The **equilibrium bond length** in a molecule is the distance between the centres of the two bonded atoms. A wealth of useful and accurate information about bond lengths is available in the literature, most of it obtained by X-ray diffraction on solids (Section 8.1). Equilibrium bond lengths of molecules in the gas phase are usually determined by infrared or microwave spectroscopy, or more directly by electron diffraction. Some typical values are given in Table 2.6.

To a reasonable first approximation, equilibrium bond lengths can be partitioned into contributions from each atom of the bonded pair. The contribution of an atom to a covalent bond is called the **covalent radius** of the element (**27**). We can use the covalent radii in Table 2.7 to predict, for example, that the length of a P–N bond is 107 pm + 71 pm = 178 pm; experimentally, this bond length is close to 180 pm in a number of compounds. Experimental bond lengths should be used whenever possible, but covalent radii are useful for making cautious estimates when experimental data are not available.

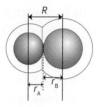

27 Covalent radius

Covalent radii vary through the periodic table in much the same way as metallic and ionic radii (Section 1.7(a)), for the same reasons, and are smallest close to F. Covalent radii are approximately equal to the separation of nuclei when the cores of the two atoms are in contact: the valence electrons draw

TABLE 2.6 Equilibrium bond lengths, R_e/pm

H_2^+	106
H_2	74
HF	92
HCl	127
HBr	141
HI	160
N_2	109
O_2	121
F_2	144
Cl_2	199
I_2	267

TABLE 2.7 Covalent radii, r/pm*

H			
37			
C	N	O	F
76 (1)	71 (1)	66 (1)	57
67 (2)	65 (2)	57 (2)	
60 (3)	54 (3)		
70 (a)			
Si	P	S	Cl
111	107	105 (1)	102
		95 (2)	
Ge	As	Se	Br
120	119	120	120
	Sb	Te	I
	139	138	139

* Values are for single bonds except where otherwise stated (in parentheses); (a) denotes aromatic.

the two atoms together until the repulsion between the cores starts to dominate. A covalent radius expresses the closeness of approach of *bonded* atoms; the closeness of approach of *nonbonded* atoms in neighbouring molecules that are in contact is expressed in terms of the **van der Waals radius** of the element, which is the internuclear separation when the *valence* shells of the two atoms are in nonbonding contact (28). van der Waals radii are of paramount importance for understanding the packing of molecular compounds in crystals, the conformations adopted by small but flexible molecules, and the shapes of biological macromolecules (Chapter 26).

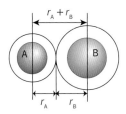

28 van der Waals radius

2.13 Bond strength and reaction enthalpies

KEY POINTS The strength of a bond is measured by its dissociation enthalpy; mean bond enthalpies are used to make estimates of reaction enthalpies.

A convenient thermodynamic measure of the strength of an AB bond is the **bond dissociation enthalpy**, $\Delta H^\ominus$ (A–B), the standard reaction enthalpy for the process

$$AB(g) \rightarrow A(g) + B(g)$$

Bond dissociation enthalpy is always positive as energy is required to break bonds. The **mean bond enthalpy**, B, is the average bond dissociation enthalpy taken over a series of A–B bonds in different molecules (Table 2.8).

Mean bond enthalpies can be used to estimate reaction enthalpies. However, thermodynamic data on actual species should be used whenever possible in preference to mean values, because the latter can be misleading. For instance, the Si–Si bond enthalpy ranges from $226\,kJ\,mol^{-1}$ in Si_2H_6 to $322\,kJ\,mol^{-1}$ in $Si_2(CH_3)_6$. Homonuclear bond energies are sometimes determined from enthalpies of atomization of the element, taking into consideration the overall number of bonds broken in the process; for instance, $456/2 = 226$ in the case of silicon. The values in Table 2.8 are best considered as data of last resort: they may be used to make rough estimates of reaction enthalpies when enthalpies of formation or actual bond enthalpies are unavailable.

Bond enthalpy data can be used to understand why some reactions occur. In general, strong bonds in the product favour its formation. In any reaction we need to consider both enthalpy and entropy. A detailed discussion of

TABLE 2.8 Mean bond enthalpies, $B/(kJ\,mol^{-1})$*

	H	C	N	O	F	Cl	Br	I	S	P	Si
H	436										
C	412	348 (1)									
		612 (2)									
		837 (3)									
		518 (a)									
N	388	305 (1)	163 (1)								
		613 (2)	409 (2)								
		890 (3)	946 (3)								
O	463	360 (1)	157	146 (1)							
		743 (2)		497 (2)							
				1070 (3)							
F	565	484	270	185	159						
Cl	431	338	200	203	254	242					
Br	366	276				219	193				
I	299	238				210	178	151			
S	338	259	464	523	343	250	212		264		
P	322 (1)			407						201	
				560 (2)						490 (3)	
Si	318			466							226

* Values are for single bonds except where otherwise stated (in parentheses); (a) denotes aromatic.

thermodynamics is the realm of physical chemistry and will not be attempted here, but we can use bond enthalpy arguments to rationalize the outcomes of some simple reactions. For example, when HCl is formed from hydrogen and chlorine the entropy change will be small, as there are the same number of gaseous molecules on each side of the equation:

$$H_2(g) + Cl_2(g) \rightarrow 2\ HCl(g)$$

As the entropy change is small, we would expect the outcome of the reaction to be indicated by the change in enthalpy. From the data in Table 2.8 we can see that the H–Cl bond enthalpy (431 kJ mol^{-1}) is similar to the H–H bond (436 kJ mol^{-1}) but greater than that of the Cl–Cl bond (242 kJ mol^{-1}). The reaction enthalpy can be calculated from the difference between the sum of the bond enthalpies of bonds broken and the sum of the bond enthalpies for the bonds that are formed. Therefore, the enthalpy for this reaction will be $(436 + 242) - (2 \times 431) = -184$ kJ mol^{-1}, and the reaction is exothermic. If we consider the following reactions, we would expect to see a reduction in entropy as we produce two moles of gaseous product from four moles of gaseous reactants:

$$N_2(g) + 3\ F_2(g) \rightarrow 2\ NF_3(g)$$
$$N_2(g) + 3\ Cl_2(g) \rightarrow 2\ NCl_3(g)$$

The enthalpy for the formation of NF$_3$ is $(946 + 3 \times 159) - (6 \times 270) = -197$ kJ mol^{-1}, and the molecule is stable and not very reactive. In contrast, the formation of NCl$_3$ is endothermic with an enthalpy of formation of $(946 + 3 \times 242) - (6 \times 200) = 472$ kJ mol^{-1}, and the molecule is unstable and explosive. Bond enthalpies are once again important in understanding the differences in these superficially similar reactions. If we examine the bond enthalpy data in Table 2.8 we rationalize that it is the lower value for the F–F bond compared to the Cl–Cl bond coupled with the stronger N–F bond (270 kJ mol^{-1}) compared to the N–Cl bond (200 kJ mol^{-1}) that accounts for the greater stability of NF$_3$. Weak X–X bonding is an important contributor to the reactivity of common compounds such as F$_2$ or H$_2$O$_2$. With such small electron-rich atoms, weak bonds arise due to Pauli repulsion, which restricts effective σ overlap, and the Coulombic repulsion between lone-pairs (or occupancy of strongly antibonding π* orbitals).

A related reaction is the formation of NH$_3$:

$$N_2(g) + 3\ H_2(g) \rightarrow 2\ NH_3(g)$$

Once again we might expect entropy to decrease on the formation of two moles of product from four moles of reactants. Both experimental measurements and bond enthalpy data show that the reaction is exothermic (−74 kJ mol^{-1} from bond enthalpy data in Table 2.8).

However, the reaction proceeds very slowly and requires high temperature, high pressure, and a catalyst to proceed at a measurable rate. This is an example of the influence of kinetics on the outcome of a reaction, and demonstrates that thermodynamic considerations have to be used with caution (Section 2.15).

EXAMPLE 2.8 **Making estimates using mean bond enthalpies**

Estimate the reaction enthalpy for the production of SF$_6$(g) from SF$_4$(g) given that the mean bond enthalpies of F$_2$, SF$_4$, and SF$_6$ are 159, 343, and 327 kJ mol^{-1}, respectively, at 25°C.

Answer We make use of the fact that the enthalpy of a reaction is equal to the difference between the sum of the bond enthalpies for broken bonds and the sum of the enthalpies of the bonds that are formed. The reaction is

$$SF_4(g) + F_2(g) \rightarrow SF_6(g)$$

In this reaction, 1 mol F–F bonds and 4 mol S–F bonds (in SF$_4$) must be broken, corresponding to an enthalpy change of 159 kJ + (4 × 343 kJ) = +1531 kJ. This enthalpy change is positive because energy will be used in breaking bonds. Then 6 mol S–F bonds (in SF$_6$) must be formed, corresponding to an enthalpy change of 6 × (−327 kJ) = −1962 kJ. This enthalpy change is negative because energy is released when the bonds are formed. The net enthalpy change is therefore

$$\Delta H^\ominus = +1531\ kJ - 1962\ kJ = -431\ kJ$$

Hence, the reaction is strongly exothermic. The experimental value for the reaction is −434 kJ, which is in excellent agreement with the estimated value. The weakness of the F–F bond is seen to be a major factor favouring this reaction.

Self-test 2.8 Estimate the enthalpy of formation of (a) HF from H$_2$ and F$_2$ and (b) H$_2$S from S$_8$ (a cyclic molecule) and H$_2$.

2.14 Electronegativity and bond enthalpy

The concept of electronegativity was introduced in Section 1.7(e), where it was defined as the power of an atom of the element to attract electrons to itself when it is part of a compound. The greater the difference in electronegativity between two elements A and B, the greater the ionic character of the A–B bond.

Linus Pauling's original formulation of electronegativity drew on concepts relating to the energetics of bond formation. For example, in the formation of AB from the diatomic A$_2$ and B$_2$ molecules,

$$A_2(g) + B_2(g) \rightarrow 2\ AB(g)$$

he argued that the excess energy, ΔE, of the A–B bond over the average energy of A–A and B–B bonds can be attributed to the presence of ionic contributions to the bonding. He defined the difference in electronegativity as

$$|\chi_p(A) - \chi_p(B)| = 0.102(\Delta E/\text{kJ mol}^{-1})^{1/2} \quad (2.9a)$$

where

$$\Delta E = B(A-B) - \tfrac{1}{2}[B(A-A) + B(B-B)] \quad (2.9b)$$

with $B(A-B)$ the mean A–B bond enthalpy. Thus, if the A–B bond enthalpy is significantly greater than the average of the nonpolar A–A and B–B bonds, then it is presumed that there is a substantial ionic contribution to the bonding and hence a large difference in electronegativity between the two atoms. Pauling electronegativities increase with increasing oxidation number of the element, and the values given in Table 1.8 are for the most common oxidation state.

A NOTE ON GOOD PRACTICE

The oxidation number, N_{ox},[4] is a parameter obtained by exaggerating the *ionic* character of a bond. It can be regarded as the charge that an atom would have if the more electronegative atom in a bond acquired the two electrons of the bond completely. The oxidation state is the physical state of the element corresponding to its oxidation number. Thus, an atom may *be assigned* an oxidation number and be *in* the corresponding oxidation state.[5]

A BRIEF ILLUSTRATION

Oxidation numbers are assigned by applying a set of simple rules (Table 2.9). These rules reflect the consequences of electronegativity for the 'exaggerated ionic' structures of compounds and match the increase in the degree of oxidation that we would expect as the number of oxygen atoms in a compound increases (as in going from NO to NO_3^-). This aspect of oxidation number is taken further in Chapter 6. Many elements, for example nitrogen, the halogens, and the d-block elements, can exist in a variety of oxidation states (Table 2.9). To determine an oxidation number, work through the rules in the order given. Stop as soon as the oxidation number has been assigned. For example, to determine the oxidation number of Cr in CrO_4^{2-} we have a total charge of -2, and -2 for each of four oxygens. So $-2 = Cr_{ox.no.} + 4(-2)$ and so the oxidation number of Cr must be $+6$. These rules are not exhaustive, but they are applicable to a wide range of common compounds.

---

[4] There is no formally agreed symbol for oxidation number.
[5] In practice, inorganic chemists use the terms 'oxidation number' and 'oxidation state' interchangeably, but in this text we shall preserve the distinction.

TABLE 2.9 The determination of oxidation number

	Oxidation number
1. The sum of the oxidation numbers of all the atoms in the species is equal to its total charge	
2. For atoms in their elemental form	0
3. For atoms of Group 1	+1
For atoms of Group 2	+2
For atoms of Group 13 (except B)	+3(EX_3), +1(EX)
For atoms of Group 14 (except C, Si)	+4(EX_4), +2(EX_2)
4. For hydrogen	+1 in combination with nonmetals
	−1 in combination with metals
5. For fluorine	−1 in all its compounds
6. For oxygen	−2 unless combined with F
	−1 in peroxides (O_2^{2-})
	$-\tfrac{1}{2}$ in superoxides (O_2^-)
	$-\tfrac{1}{3}$ in ozonides (O_3^-)
7. For halogens	−1 in most compounds, unless the other elements include oxygen or more electronegative halogens

Electronegativity values are useful for estimating the enthalpies of bonds between elements of different electronegativity and for making qualitative assessments of the polarities of bonds. Binary compounds in which the difference in electronegativity between the two elements is greater than about 1.7 can generally be regarded as being predominantly ionic. However, this crude distinction was refined by Anton van Arkel and Jan Ketelaar in the 1940s, when they proposed a triangle with vertices representing ionic, covalent, and metallic bonding. The **Ketelaar triangle** (more appropriately, the *van Arkel–Ketelaar triangle*) has been elaborated by Gordon Sproul, who constructed a triangle based on the difference in electronegativities ($\Delta\chi$) of the elements in a binary compound and their average electronegativity (χ_{mean}) (Fig. 2.40). The Ketelaar triangle can be used to classify a wide range of compounds of different kinds.

Ionic bonding is characterized by a large difference in electronegativity. Because a large difference indicates that the electronegativity of one element is high and that of the other is low, the average electronegativity must be intermediate in value. The compound CsF, for instance, with $\Delta\chi_p = 3.19$ and $\chi_{mean} = 2.38$, lies at the 'ionic' apex of the triangle. Highly ionic solids have large band gaps and are insulators.

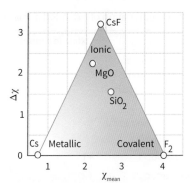

FIGURE 2.40 A Ketelaar triangle, showing how a plot of average electronegativity against electronegativity difference can be used to classify the bond type for binary compounds.

Covalent bonding is characterized by a small difference in electronegativities. Such compounds lie at the base of the triangle. Binary compounds that are predominantly covalently bonded are typically formed between nonmetals, which commonly have high electronegativities: they tend to be semiconductors. It follows that the covalent region of the triangle is the lower, right-hand corner. This corner of the triangle is occupied by F_2, which has $\Delta\chi = 0$ and $\chi_{mean} = 3.98$ (the maximum value of any Pauling electronegativity).

Metallic bonding is also characterized by a small electronegativity difference, and also lies towards the base of the triangle. In metallic bonding, however, electronegativities are low, the average values are therefore also low, and consequently metallic bonding occupies the lower, left-hand corner of the triangle. The outermost position is occupied by Cs, which has $\Delta\chi = 0$ and $\chi_{mean} = 0.79$ (the lowest value of Pauling electronegativity).

Aside from its conceptual significance, the advantage of using a Ketelaar triangle over simple electronegativity difference is that it allows us to distinguish between covalent and metallic bonding, which are both indicated by a small electronegativity difference.

EXAMPLE 2.9 Using the Ketelaar triangle to classify binary compounds

Use Fig. 2.40 and the data in Table 1.8 to predict which type of bonding will dominate in (a) MgO and (b) SiO_2.

Answer (a) The Pauling electronegativity values for Mg and O are 1.31 and 3.44, respectively. Therefore, $\Delta\chi_p = 3.44 - 1.31 = 2.13$ and $\chi_{mean} = 2.38$. These values place MgO in the ionic region of the triangle. (b) The Pauling electronegativity values for Si and O are 1.90 and 3.44, respectively, and therefore $\Delta\chi_p = 3.44 - 1.90 = 1.54$ and $\chi_{mean} = 2.67$. These values place SiO_2 lower on the triangle compared to MgO and in the covalent bonding region.

Self-test 2.9 Use Fig. 2.40 and the data in Table 1.8 to predict the type of bonding that will dominate in (a) BeF_2 and (b) NO.

2.15 An introduction to catalysis

We have seen in Section 2.13 that thermodynamic considerations are crucial in predicting the stabilities of molecules and the outcomes of reactions. However, thermodynamically feasible reactions may take place too slowly to be useful. Thus we have to consider the kinetics of a reaction as well as the thermodynamics. For example, graphite is the thermodynamically stable form of carbon, but the conversion of diamond into graphite, although thermodynamically favoured, occurs at too slow a rate to be important. Kinetic considerations play an essential role in determining the useful outcomes of a simple reaction. This kinetic barrier can be overcome by increasing the energy in the system to match that of bonds that must be broken, or by using a **catalyst**. In general, a catalyst is a substance that increases the rate of a reaction but is not itself consumed. In this section we discuss the basic principles of catalysis. Specific examples and applications of catalysis are discussed in Chapters 5, 6, 24, 25, 26, and 31.

(a) Energetics

KEY POINTS A catalyst increases the rates of processes by introducing new pathways with lower Gibbs energies of activation; the reaction profile contains no high peaks and no deep troughs.

A catalyst increases the rates of processes by introducing new pathways with lower Gibbs energies of activation, $\Delta^{\ddagger}G$. We need to focus on the Gibbs energy profile of a catalytic reaction, not just the enthalpy or energy profile, because the new elementary steps that occur in the catalysed process are likely to have quite different entropies of activation. A catalyst does not affect the Gibbs energy of the overall reaction, $\Delta_r G^{\ominus}$, because G is a state function.[6] The difference is illustrated in the **reaction coordinates** shown in Fig. 2.41, where the overall reaction Gibbs energy is the same in all cases. Reactions that are thermodynamically unfavourable cannot be made favourable by a catalyst.

Figure 2.41 also shows that the Gibbs energy profile of a catalysed reaction contains no high peaks and no deep troughs. The new pathway introduced by the catalyst changes the mechanism of the reaction to one with a very different shape and with lower maxima. However, an equally important point is that stable or nonlabile catalytic intermediates do not occur in the cycle. Similarly, the product must be released in a thermodynamically favourable step.

---

[6] That is, G depends only on the current state of the system and not on the path that led to the state.

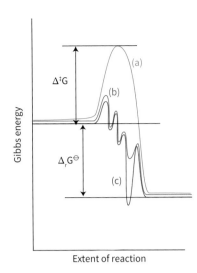

FIGURE 2.41 A representation (known as a reaction coordinate) of the energetics of a catalytic cycle. The uncatalysed reaction (a) has a higher $\Delta^{\ddagger}G$ than a step in the catalysed reaction (b). The Gibbs energy of the overall reaction, $\Delta_r G^{\ominus}$, is the same for routes (a) and (b). The curve (c) shows the profile for a reaction mechanism with an intermediate that is more stable than the product.

If, as shown by line (c) in Fig. 2.41, a stable complex were formed with the catalyst, it would turn out to be the product of the reaction and the cycle would terminate. Similarly, impurities may suppress catalysis, by coordinating strongly to catalytically active sites, and act as catalyst poisons.

(b) Catalytic cycles

KEY POINT A catalytic cycle is a sequence of reactions that consumes the reactants and forms products, with the catalytic species being regenerated after the cycle.

The essence of catalysis is a cycle of reactions in which the reactants are consumed, the products are formed, and the catalytic species is regenerated. A simple example of a catalytic cycle involving a homogeneous catalyst is the isomerization of prop-2-en-1-ol (allyl alcohol, $CH_2=CHCH_2OH$) to prop-1-en-1-ol ($CH_3CH=CHOH$) catalysed by the compound $[Co(CO)_4H]$ (Fig. 2.42). The active form of the catalyst is $[Co(CO)_3H]$ which is formed upon reversible dissociation of a CO ligand, which generates a vacant coordination site. The dashed line black square and the solid line blue square, respectively, denote the unreactive precursor (often referred to as the 'resting' state) and initial active form of the catalyst. The first step is the coordination of the reactant to the catalyst at the vacant site; this is followed by reorganization of the bonding within the coordination sphere of the catalyst and, finally, release of the product and regeneration of the active catalyst. Once released, the prop-1-en-1-ol

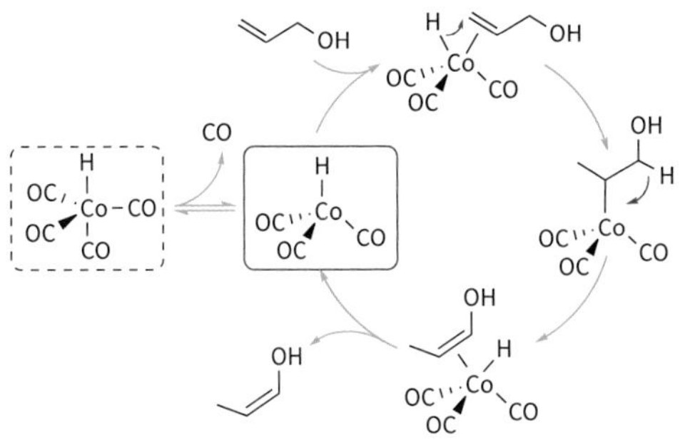

FIGURE 2.42 The catalytic cycle for the isomerization of prop-2-en-1-ol to prop-1-en-1-ol. The initial active state of the catalyst is shown in the blue square; it is formed from the precursor (dashed line black square) by dissociation of a CO ligand to generate a vacant site for binding the reactant.

tautomerizes to propanal (CH_3CH_2CHO). As with all mechanisms, this cycle has been proposed on the basis of a range of information. The elucidation of catalytic mechanisms is complicated by the occurrence of several delicately balanced reactions, which often cannot be studied in isolation.

(c) Catalytic efficiency and lifetime

KEY POINTS A highly active catalyst—one that results in a fast reaction even in low concentrations—has a large turnover frequency. A catalyst must be able to survive a large number of catalytic cycles if it is to be useful in a synthesis process.

The **turnover frequency**, **TOF**, is the number of catalytic cycles completed per unit time and is often used to express the efficiency of a catalyst. A highly active catalyst—one that results in a fast reaction even in low concentrations—has a high turnover frequency. The slowest step in a catalytic cycle is called the **rate-determining step, RDS**.

The **turnover number, TON**, is the number of cycles for which a catalyst survives. If it is to be economically viable, a catalyst must have a large turnover number. However, it may be destroyed by side reactions to the main catalytic cycle or by the presence of small amounts of impurities in the starting materials (the *feedstock*). For example, many alkene polymerization catalysts are destroyed by O_2, so in the synthesis of polyethene (polyethylene) and polypropene (polypropylene) the concentration of O_2 in the ethene or propene feedstock should be no more than a few parts per billion.

Some catalysts can be regenerated quite readily. For example, the supported metal catalysts used in the reforming reactions that convert hydrocarbons to high-octane gasoline become covered with carbon because the catalytic reaction is accompanied by a small amount of dehydrogenation. These supported metal particles can be cleaned by interrupting the catalytic process periodically and burning off the accumulated carbon.

(d) Selectivity

KEY POINT A selective catalyst yields a high proportion of the desired product with minimum amounts of side products.

In industry, there is considerable economic incentive to develop **selective catalysts**, which yield a high proportion of the desired product with minimum amounts of side products. For instance, the simplest epoxide, oxirane (**29**), a major industrial chemical known commonly as ethene oxide, is produced by reacting ethene with O_2 using metallic Ag particles supported on Al_2O_3 as the catalyst.

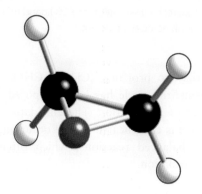

29 Oxirane (ethene oxide)

The economic viability of the process depends on minimizing two competing reactions, direct combustion of ethene or that of the oxirane product to form CO_2 and H_2O, each of which are far more favoured thermodynamically ($\Delta_r H^\ominus$ values indicated in the scheme shown). Much research is therefore being carried out to establish the mechanism, which involves the adsorption and dissociation of O_2 to form Ag–O species on particular crystal faces of Ag.

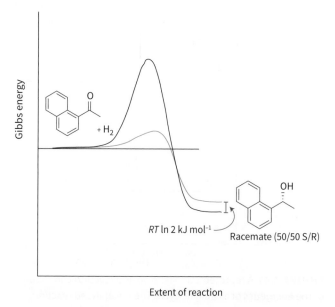

FIGURE 2.43 The energetics of a catalytic reaction in which the product is a pure enantiomer and therefore less favoured thermodynamically.

Selectivity can be ignored in only a very few simple inorganic reactions, where there is essentially only one thermodynamically favourable product, as in the formation of NH_3 from H_2 and N_2.

One area where selectivity is of considerable and growing importance is asymmetric synthesis, where only one enantiomer of a particular compound is required and catalysts are designed to produce one chiral form in preference to any others. As with oxirane production, enantioselective catalysts must favour the kinetic rather than the thermodynamic product: in this case a pure enantiomer is only slightly more stable than the corresponding racemic mixture, by the entropy factor $-RT \ln 2$ (Fig. 2.43). Examples of such catalysts include the chiral organometallic complexes pioneered by Ryoji Noyori, a recipient of the Nobel Prize for Chemistry in 2001.

(e) Homogeneous and heterogeneous catalysts

KEY POINTS Homogeneous catalysts are present in the same phase as the reagents, and are often well defined; heterogeneous catalysts are present in a different phase from the reagents.

As described in detail in Chapter 31, catalysts are classified as **homogeneous** if they are present in the same phase as the reagents; this normally means that they are present as solutes in liquid reaction mixtures. Catalysts are **heterogeneous** if they are present in a different phase from that of the reactants; this normally means that they are present as solids with the reactants present either as gases or in solution. From a practical standpoint, homogeneous catalysis is attractive because it is often highly selective towards the formation of a desired product. In large-scale industrial processes, homogeneous catalysts are preferred for exothermic reactions because it is easier to dissipate heat from a solution than from the solid bed of a heterogeneous catalyst. In principle, every homogeneous catalyst molecule in solution is accessible to reagents, potentially leading to very high activities. It should also be borne in mind that the mechanism of homogeneous catalysis is more accessible to

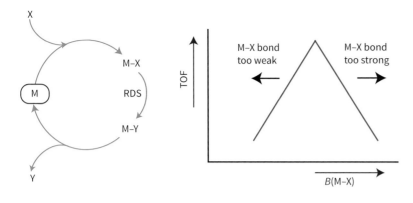

FIGURE 2.44 Illustration of the Sabatier principle which links the turnover frequency of a catalyst to the strength of the bond it forms with a reactant. The optimal bond strength B(M–X) is one that promotes formation of intermediate M–X but does not render it so stable that the subsequent conversion to M–Y (leading to release of product Y) is too slow.

detailed investigation than that of heterogeneous catalysis, as species in solution are often easier to characterize than those on a surface and because the interpretation of rate data is frequently easier. The major disadvantage of homogeneous catalysts is that a separation step is required.

Heterogeneous catalysts are used very extensively in industry and have a much greater economic impact than homogeneous catalysts. One attractive feature is that many of these solid catalysts are robust at high temperatures and therefore tolerate a wide range of operating conditions. Reactions are faster at high temperatures, so at high temperatures solid catalysts generally produce higher outputs for a given amount of catalyst and reaction time than homogeneous catalysts operating at lower temperatures in solutions. Another reason for their widespread use is that extra steps are not needed to separate the product from the catalyst, resulting in efficient and more environmentally friendly processes (Chapter 27). Typically, gaseous or liquid reactants enter a tubular reactor at one end and pass over a bed of the catalyst, and products are collected at the other end. This same simplicity of design applies to the catalytic converter used to oxidize CO and hydrocarbons and reduce nitrogen oxides in automobile exhausts.

An important guideline for rationalizing the efficacy of a catalyst is the Sabatier principle, named after French chemist Paul Sabatier (Nobel laureate, 1912), which states that the bond formed between a catalyst and a reactant should be neither too weak or too strong. Consider a reaction, illustrated in Fig. 2.44 (*left*) in which X is converted to Y on a catalyst M (typically but not always a metal): a weak interaction between M and X raises the energy of the intermediate M–X whereas a strong bond stabilizes it against the subsequent reaction that is the rate-determining step. A volcano diagram, also called a plot (*right*) is often used to show how the turnover frequency depends on the strength of interaction between catalyst and reactant.

FURTHER READING

M. J. Winter, *Chemical Bonding*. Oxford University Press (2016). This short text introduces the concepts of chemical bonding in a descriptive and non-mathematical way.

R. J. Gillespie and I. Hargittai, *The VSEPR Model of Molecular Geometry*. Dover Publications (2013). An excellent introduction to attitudes to VSEPR theory.

R. J. Gillespie and P. L. A. Popelier, *Chemical Bonding and Molecular Geometry: from Lewis to Electron Densities*. Oxford University Press (2001). A comprehensive survey of modern theories of chemical bonding and geometry.

T. Albright, *Orbital Interactions in Chemistry*. Wiley-Blackwell (2013). This text covers the application of molecular orbital theory to organic, organometallic, inorganic, and solid-state chemistry.

D. M. P. Mingos, *Essential Trends in Inorganic Chemistry*. Oxford University Press (2004). An overview of inorganic chemistry from the perspective of structure and bonding.

I. D. Brown, *The Chemical Bond in Inorganic Chemistry*. Oxford University Press (2006).

J. Barratt, *Structure and Bonding*. RSC Publishing (2001).

D. O. Hayward, *Quantum Mechanics*. RSC Publishing (2002).

D. Rodríguez-Padrón, A. R. Puente-Santiago, A. M. Balu, M. J. Muñoz-Batista, and R. Luque, Environmental catalysis: Present and future. *ChemCatChem.*, 2018, **11**, 18. A comprehensive review of directions being taken to advance the impact of catalysis in sustainability.

A. B. Laursen, A. F. Varela, F. Dionigi, H. Fanchiu, C. Miller, O. L. Trinhammer, J. Rossmeisl, and S. Dahl, Electrochemical hydrogen evolution: Sabatier's principle and the volcano plot. *J. Chem. Educ.*, 2012, **89**, 1595. This article describes the role of the Sabatier principle in understanding a reaction of huge societal importance.

M. D. Wodrich, B. Sawatlon, M. Busch, and C. Corminboeuf, The genesis of molecular volcano plots, *Accounts of Chemical Research*, 2021, **54**, 1107. This account explains how the Sabatier principle is applied in homogeneous catalysis.

M. J. Sahre, G. F. von Rudorff, and O. A. von Lilienfeld, Quantum alchemy based bonding trends and their link to Hammett's equation and Pauling's electronegativity model. *J. Amer. Chem. Soc.*, 2023, **145**, 5899. This article compares methods for estimating bond energies and introduces a new approach.

EXERCISES

2.1 Draw feasible Lewis structures for (a) NO$^+$, (b) ClO$^-$, (c) H$_2$O$_2$, (d) CCl$_4$, (e) HSO$_3^-$.

2.2 Draw the resonance structures for CO$_3^{2-}$.

2.3 What shapes would you expect for the species (a) H$_2$Se, (b) BF$_4^-$, (c) NH$_4^+$?

2.4 What shapes would you expect for the species (a) SO$_3$, (b) SO$_3^{2-}$, (c) IF$_5$?

2.5 What shapes would you expect for the species (a) IF$_6^+$, (b) IF$_3$, (c) XeOF$_4$?

2.6 What shapes would you expect for the species (a) ClF$_3$, (b) ICl$_4^-$, (c) I$_3^-$?

2.7 In which of the species ICl$_6^-$ and SF$_4$ is the bond angle closest to that predicted by the VSEPR model?

2.8 Solid phosphorus pentachloride is an ionic solid composed of PCl$_4^+$ cations and PCl$_6^-$ anions, but the vapour is molecular. What are the shapes of the ions in the solid?

2.9 Use the covalent radii in Table 2.7 to calculate the bond lengths in (a) CCl$_4$ (177 pm), (b) SiCl$_4$ (201 pm), (c) GeCl$_4$ (210 pm). (The values in parentheses are experimental bond lengths and are included for comparison.)

2.10 Use the concepts from Chapter 1, particularly the effects of penetration and shielding on the radial wavefunction, to account for the variation of single-bond covalent radii with position in the periodic table.

2.11 Given that B(Si–O) = 640 kJ mol^{-1}, show that bond enthalpy considerations predict that silicon–oxygen compounds are likely to contain networks of tetrahedra with Si–O single bonds and not discrete molecules with Si–O double bonds.

2.12 The common forms of nitrogen and phosphorus are N$_2$(g) and P$_4$(s), respectively. Account for the difference in terms of the single and multiple bond enthalpies.

2.13 Use the data in Table 2.8 to calculate the standard enthalpy of the reaction 2 H$_2$(g) + O$_2$(g) → 2 H$_2$O(g). The experimental value is –484 kJ. Account for the difference between the estimated and experimental values.

2.14 Rationalize the bond dissociation energy (D) and bond length data of the gaseous diatomic species given in the following table and highlight the atoms that obey the octet rule.

	D/(kJ mol^{-1})	Bond length/pm
C$_2$	607	124.3
BN	389	128.1
O$_2$	498	120.7
NF	343	131.7
BeO	435	133.1

2.15 Predict the standard enthalpies of the reactions

$$S_2^{2-}(g) + \tfrac{1}{4}S_8(g) \rightarrow S_4^{2-}(g)$$
$$O_2^{2-}(g) + O_2(g) \rightarrow O_4^{2-}(g)$$

by using mean bond enthalpy data. Assume that the unknown species O$_4^{2-}$ is a singly bonded chain analogue of S$_4^{2-}$.

2.16 Determine the oxidation states of the element emboldened in each of the following species: (a) **S**O$_3^{2-}$, (b) **N**O$^+$, (c) **Cr**$_2$O$_7^{2-}$, (d) **V**$_2$O$_5$, (e) **P**Cl$_5$.

2.17 Four elements arbitrarily labelled A, B, C, and D have electronegativities of 3.8, 3.3, 2.8, and 1.3, respectively. Place the compounds AB, AD, BD, and AC in order of increasing covalent character.

2.18 Use the Ketelaar triangle in Fig. 2.40 and the electronegativity values in Table 1.8 to predict what type of bonding is likely to dominate in (a) BCl$_3$, (b) KCl, (c) BeO.

2.19 Predict the hybridization of orbitals required in (a) BCl$_3$, (b) NH$_4^+$, (c) SF$_4$, (d) XeF$_4$.

2.20 Use molecular orbital diagrams to determine the number of unpaired electrons in (a) O$_2^-$, (b) O$_2^+$, (c) BN, (d) NO$_2$.

2.21 Use Fig. 2.17 to write the electron configurations of (a) Be$_2^+$, (b) B$_2^-$, (c) C$_2^-$, (d) F$_2^+$ and sketch the form of the HOMO in each case.

2.22 At –152°C, NO condenses to a liquid which contains (NO)$_2$ dimers. By considering the frontier orbitals of NO, predict the structure of the dominant dimeric form. (The ordering of molecular orbitals in NO is the same as for O$_2$.)

2.23 When acetylene (ethyne) is passed through a solution of copper(I) chloride a red precipitate of copper acetylide, CuC$_2$, is formed. This is a common test for the presence of acetylene. Describe the bonding in the C$_2^{2-}$ ion in terms of molecular orbital theory and compare the bond order to that of C$_2$.

2.24 Draw and label a molecular orbital energy-level diagram for the gaseous homonuclear diatomic molecule dicarbon, C$_2$. Annotate the diagram with pictorial representations of the molecular orbitals involved. What is the bond order of C$_2$?

2.25 Draw a molecular orbital energy-level diagram for the gaseous heteronuclear diatomic molecule boron nitride, BN. How does it differ from that for C$_2$?

2.26 Using the following values for the valence orbital energies of boron and oxygen, construct a molecular orbital diagram for BO and explain why the anion BO$^-$ forms a complex with metal ions by bonding through the B atom.

B(2s), – 14.0 eV; B(2p), – 8.3 eV; O(2s), – 32.2 eV; O(2p), – 15.8 eV

2.27 Assume that the MO diagram of IBr is analogous to that of ICl (Fig. 2.25). (a) What basis set of atomic orbitals would be used to generate the IBr molecular orbitals? (b) Calculate the bond order of IBr.

2.28 Explain how a Ketelaar triangle can be used to categorize binary solids AB as insulators, semiconductors, and conductors.

2.29 Account for the bond dissociation energies (kJ mol^{-1}) determined for the following diatomic molecules. C$_2$, 612; N$_2$, 946; O$_2$, 497; BN, 389; NF, 321; CO, 1070; F$_2$, 159; Cl$_2$, 242.

2.30 Determine the bond orders of (a) S$_2$, (b) Cl$_2$, and (c) NO$^+$ from their molecular orbital configurations and compare the values with the bond orders determined from Lewis structures.

2.31 What are the expected changes in bond order and bond distance that accompany the following ionization processes?

(a) $O_2 \rightarrow O_2^+ + e^-$; (b) $N_2 + e^- \rightarrow N_2^-$; (c) $NO \rightarrow NO^+ + e^-$

2.32 From a consideration of their respective molecular orbital diagrams, sketch the forms of the frontier orbitals of C_2^{2-}, CO, NO, and O_2 and predict any likely consequences for chemical properties.

2.33 Assign the lines in the UV photoelectron spectrum of CO shown in Fig. 2.45 and predict the appearance of the UV photoelectron spectrum of the SO molecule (see Section 8.9).

2.34 When an He atom absorbs a photon to form the excited configuration $1s^1 2s^1$ (here called He*) a weak bond forms with another He atom to give the diatomic molecule HeHe*. Construct a molecular orbital description of the bonding in this species.

2.35 Construct and label molecular orbital diagrams for N_2, NO, and O_2 showing the principal linear combinations of atomic orbitals being used. Comment on the following bond lengths: N_2 110 pm, NO 115 pm, O_2 121 pm.

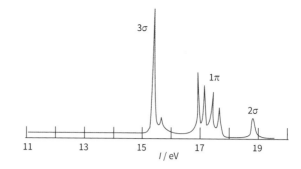

FIGURE 2.45 The ultraviolet photoelectron spectrum of CO obtained using 21 eV radiation.

2.36 Do the hypothetical species (a) square H_4^{2+}, (b) angular O_3^{2-} have a duplet or octet of electrons? Explain your answer and decide whether either of them is likely to exist.

2.37 Describe the difference between a semiconductor and a semimetal.

TUTORIAL PROBLEMS

2.1 In valence bond theory, hypervalence is usually explained in terms of d-orbital participation in bonding. In the paper 'On the role of orbital hybridisation' (*J. Chem. Educ.*, 2007, **84**, 783) the author argues that this is not the case. Give a concise summary of the method used and the author's reasoning.

2.2 Develop an argument based on bond enthalpies for the importance of Si–O bonds, in preference to Si–Si or Si–H bonds, in substances common in the Earth's crust. How and why does the behaviour of silicon differ from that of carbon?

2.3 The van Arkel–Ketelaar triangle has been in use since the 1940s. A quantitative treatment of the triangle was carried out by Gordon Sproul in 1994 (*J. Phys. Chem.*, 1994, **98**, 6699). How many scales of electronegativity and how many compounds did Sproul investigate? What criteria were used to select compounds for the study? Which two electronegativity scales were found to give the best separation between areas of the triangle? What were the theoretical bases of these two scales?

2.4 In their short article 'In defense of the hybrid atomic orbitals' (P. C. Hiberty, F. Volatron, and S. Shaik, *J. Chem. Educ.*, 2012, **89**, 575), the authors defend the continuing use of the concept of the hybrid atomic orbital. Summarize the criticisms that they are addressing and present an outline of their arguments in favour of hybrid orbitals.

2.5 In their article 'Some observations on molecular orbital theory' (J. F. Harrison and D. Lawson, *J. Chem. Educ.*, 2005, **82**, 1205) the authors discuss several limitations of the theory. What are these limitations? Sketch the MO diagram for Li_2 given in the paper. Why do you think this version does not appear in textbooks? Use the data given in the paper to construct MO diagrams for B_2 and C_2. Do these versions differ from those in Fig. 2.18 in this textbook? Discuss any variations.

2.6 (a) Use a molecular orbital program or input and output from software supplied by your instructor to construct a molecular orbital energy-level diagram to correlate the MO (from the output) and AO (from the input) energies and indicate the occupancy of the MOs (in the manner of Fig. 2.21) for one of the following molecules: HF (bond length 92 pm), HCl (127 pm), or CS (153 pm). (b) Use the output to sketch the form of the occupied orbitals, showing the signs of the AO lobes by shading and their amplitudes by means of size of the orbital.

2.7 In the article 'Interatomic Repulsion and the Pauli Principle' (*J. Chem. Educ.*, 2021, **98**, 2912–2918) David McMillin considers the diatomic molecule HHe, which has a bond order of ½. Explain why this compound appears not to exist.

2.8 The effects of the nonbonding lone pair in the tin and lead anions in compounds such as $Sr(MX_3)_2 \cdot 5H_2O$ (M = Sn or Pb, X = Cl or Br) have been studied by crystallography and by electronic structure calculations (I. Abrahams et al., *Polyhedron*, 2006, **25**, 996). Briefly outline the synthetic method used to prepare the compounds and indicate which compound could not be prepared. Explain how this compound was handled in the electronic structure calculations as no experimental structural data were available. State the shape of the $[MX_3]^-$ anions and describe how the effect of the nonbonding lone pair varies between Sn and Pb in the gas phase and the solid phase.

2.9 In their article 'Overview of selective oxidation of ethylene to ethylene oxide by Ag catalysts', *ACS Catalysis*, 2019, **9**, 10727, Pu and co-workers give an overview of efforts being made to understand the detailed mechanism of an extremely important industrial process. Summarize the principles that are discussed and the mechanistic options that are open.

3 Molecular symmetry

An introduction to symmetry analysis
3.1 **Symmetry operations, elements, and point groups**
3.2 **Character tables**

Applications of symmetry
3.3 **Polar molecules**
3.4 **Chiral molecules**
3.5 **Molecular vibrations**

The symmetries of molecular orbitals
3.6 **Symmetry-adapted linear combinations**
3.7 **The construction of molecular orbitals**
3.8 **The vibrational analogy**

Representations
3.9 **The reduction of a representation**
3.10 **Projection operators**
3.11 **Polyatomic molecules**

Further reading

Exercises

Tutorial problems

Symmetry and bonding of molecules are intimately linked. In this chapter we explore some of the consequences of molecular symmetry and introduce the systematic arguments of group theory. We shall see that symmetry considerations are essential for constructing molecular orbitals and analysing molecular vibrations, particularly where these are not immediately obvious. They also enable us to extract information about molecular and electronic structure from spectroscopic data.

The systematic treatment of symmetry makes use of a branch of mathematics called **group theory**. Group theory is a rich and powerful subject, but we shall confine our use of it at this stage to the classification of molecules in terms of their symmetry properties, the construction of molecular orbitals, and the analysis of molecular vibrations and the selection rules that govern their excitation. We shall also see that it is possible to draw some general conclusions about the properties of molecules, such as polarity and chirality, without doing any calculations at all.

An introduction to symmetry analysis

That some molecules are 'more symmetrical' than others is intuitively obvious. Our aim, though, is to define the symmetries of individual molecules precisely, not just intuitively, and to provide a scheme for specifying and reporting these symmetries. It will become clear in later chapters that symmetry analysis is one of the most pervasive techniques in inorganic chemistry.

3.1 Symmetry operations, elements, and point groups

KEY POINTS Symmetry operations are actions that leave the molecule apparently unchanged; each symmetry operation is associated with a symmetry element. The point group of a molecule is identified by noting its symmetry elements and comparing these elements with the elements that define each group.

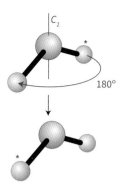

FIGURE 3.1 An H$_2$O molecule may be rotated through any angle about the bisector of the HOH bond angle, but only a rotation of 180° (the C_2 operation) leaves it apparently unchanged.

A fundamental concept of the chemical application of group theory is the **symmetry operation**, an action, such as rotation through a certain angle, that leaves the molecule apparently unchanged. An example is the rotation of an H$_2$O molecule by 180° around the bisector of the HOH angle (Fig. 3.1). Associated with each symmetry operation there is a **symmetry element**—a point, line, or plane with respect to which the symmetry operation is performed. Table 3.1 lists the most important symmetry operations and their corresponding elements. All these operations leave at least one point unchanged, and hence they are referred to as the operations of **point-group symmetry**.

The **identity operation**, E, consists of doing nothing to the molecule. Every molecule has at least this operation and some have only this operation, so we need it if we are to classify all molecules according to their symmetry.

The rotation of an H$_2$O molecule by 180° around a line bisecting the HOH angle (as in Fig. 3.1) is a symmetry operation, denoted C_2. In general, an **n-fold rotation** is a symmetry operation if the molecule appears unchanged after rotation by 360°/n. The corresponding symmetry element is a line, an **n-fold rotation axis**, C_n, about which

TABLE 3.1 Symmetry operations and symmetry elements

Symmetry operation	Symmetry element	Symbol
Identity	'whole of space'	E
Rotation by 360°/n	n-fold symmetry axis	C_n
Reflection	mirror plane	σ
Inversion	centre of inversion	i
Rotation by 360°/n followed by reflection in a plane perpendicular to the rotation axis	n-fold axis of improper rotation*	S_n

*Note the equivalences $S_1 = \sigma$ and $S_2 = i$.

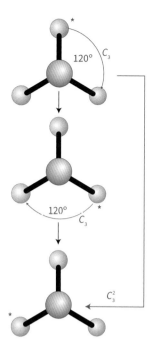

FIGURE 3.2 A three-fold rotation and the corresponding C_3 axis in NH$_3$. There are two rotations associated with this axis: one through 120° (C_3) and one through 240° (C_3^2).

the rotation is performed. So for the H$_2$O molecule a two-fold rotation leaves the molecule unchanged after rotation by 360°/2 or 180°. There is only one rotation operation associated with a C_2 axis (as in H$_2$O) because clockwise and anticlockwise rotations by 180° are identical. The trigonal-pyramidal NH$_3$ molecule has a three-fold rotation axis, denoted C_3, on rotation of the molecule through 360°/3 or 120°. There are now two operations associated with this axis: a clockwise rotation by 120° and an anticlockwise rotation by 120° (Fig. 3.2). The two operations are denoted C_3 and C_3^2 (because two successive clockwise rotations by 120° are equivalent to an anticlockwise rotation by 120°), respectively.

The square-planar molecule XeF$_4$ has a four-fold axis, C_4, but in addition it also has two pairs of two-fold rotation axes that are perpendicular to the C_4 axis: one pair (C_2') passes through each *trans*-FXeF unit and the other pair (C_2'') passes through the bisectors of the FXeF angles (Fig. 3.3). By convention, the highest-order rotational axis, which is called the **principal axis**, defines the z-axis (and is typically drawn vertically). For XeF$_4$ the C_4 axis is the principal axis. The C_4^2 operation is equivalent to a C_2 rotation, and this is normally listed separately from the C_4 operation as '$C_2(=C_4^2)$'.

The reflection of an H$_2$O molecule in either of the two planes shown in Fig. 3.4 is a symmetry operation; the corresponding symmetry element is a **mirror plane**, σ. The H$_2$O molecule has two mirror planes that intersect at the bisector of the HOH angle. Because the planes are 'vertical', in

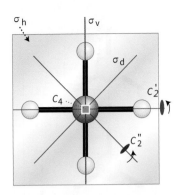

FIGURE 3.3 Some of the symmetry elements of a square-planar molecule such as XeF$_4$.

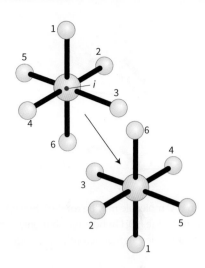

FIGURE 3.5 The inversion operation and the centre of inversion i in SF$_6$.

the sense of containing the rotational (z) axis of the molecule, they are labelled with a subscript v, as in σ_v and σ'_v. The XeF$_4$ molecule in Fig. 3.3 has a mirror plane σ_h in the plane of the molecule. The subscript h signifies that the plane is 'horizontal' in the sense that the vertical principal rotational axis of the molecule is perpendicular to it. This molecule also has two more sets of two mirror planes that intersect the four-fold axis. The symmetry elements (and the associated operations) are denoted σ_v for the planes that pass through the F atoms and σ_d for the planes that bisect the angle between the F atoms. The v denotes that the plane is 'vertical', and the d denotes 'dihedral' and signifies that the plane bisects the angle between two C'_2 axes (the FXeF axes).

To understand the **inversion operation**, i, we need to imagine that each atom is projected in a straight line through a single point located at the centre of the molecule and then out to an equal distance on the other side (Fig. 3.5). In an octahedral molecule such as SF$_6$, with the point at the centre of the molecule, diametrically opposite pairs of atoms at the corners of the octahedron are interchanged. In general, under inversion, an atom with coordinates (x, y, z) moves to ($-x$, $-y$, $-z$). The symmetry element, the point through which the projections are made, is called the **centre of inversion**, i.

For SF$_6$, the centre of inversion lies at the nucleus of the S atom. Likewise, the molecule CO$_2$ has an inversion centre at the C nucleus. However, there need not be an atom at the centre of inversion: an N$_2$ molecule has a centre of inversion midway between the two nitrogen nuclei, and the S$_4^{2+}$ ion (**1**) has a centre of inversion in the middle of the square ion. An H$_2$O molecule does not possess a centre of inversion, and no tetrahedral molecule can have a centre of inversion.

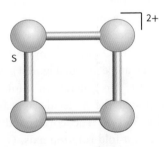

1 The S$_4^{2+}$ cation

Although an inversion and a two-fold rotation may sometimes achieve the same effect, that is not the case in general and the two operations must be distinguished (Fig. 3.6).

An **improper rotation** consists of a rotation of the molecule through a certain angle around an axis followed by a reflection in the plane perpendicular to that axis (Fig. 3.7). The illustration shows a four-fold improper rotation of a CH$_4$ molecule. In this case, the operation consists of a 90° (i.e. 360°/4) rotation about an axis bisecting two HCH bond angles, followed by a reflection through a plane perpendicular to the rotation axis. Neither the 90° (C_4) operation nor the reflection alone is a symmetry operation for CH$_4$, but

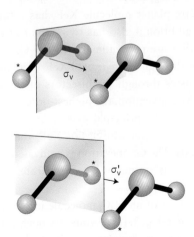

FIGURE 3.4 The two vertical mirror planes σ_v and σ'_v in H$_2$O and the corresponding operations. Both planes cut through the C_2 axis.

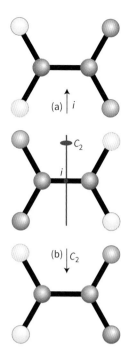

FIGURE 3.6 Care must be taken not to confuse (a) an inversion operation with (b) a two-fold rotation. Although the two operations may sometimes appear to have the same effect, that is not the case in general, as can be seen when the four terminal atoms of the same element are coloured differently.

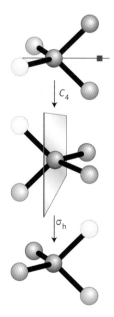

FIGURE 3.7 A four-fold axis of improper rotation S_4 in the CH_4 molecule. The four terminal atoms of the same element are coloured differently to help track their movement.

their overall effect is a symmetry operation. A four-fold improper rotation is denoted S_4. The symmetry element, the **improper-rotation axis**, S_n (S_4 in the example), is the corresponding combination of an n-fold rotational axis and a perpendicular mirror plane.

An S_1 axis, a rotation through 360° followed by a reflection in the perpendicular plane, is equivalent to a reflection

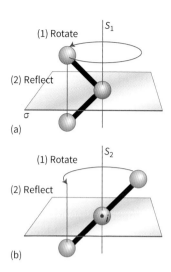

FIGURE 3.8 (a) An S_1 axis is equivalent to a mirror plane and (b) an S_2 axis is equivalent to a centre of inversion.

alone, so S_1 and σ_h are the same; the symbol σ_h is used rather than S_1. Similarly, an S_2 axis, a rotation through 180° followed by a reflection in the perpendicular plane, is equivalent to an inversion, i (Fig. 3.8); the symbol i is employed rather than S_2.

By identifying the symmetry elements of the molecule, and referring to Table 3.2, we can assign a molecule to its **point group**. In practice, the shapes in the table give a very good clue to the identity of the group to which the molecule belongs, at least in simple cases. The decision tree in Fig. 3.9 can also be used to assign most common point groups systematically by answering the questions at each decision point. The name of the point group is normally its **Schoenflies symbol**, such as C_{3v} for an ammonia molecule.

It is very useful to be able to recognize immediately the point groups of some common molecules. Linear molecules with a centre of symmetry, such as H_2, CO_2 (**2**), and HC≡CH belong to $D_{\infty h}$. A molecule that is linear but has no centre of symmetry, such as HCl or OCS (**3**), belongs to $C_{\infty v}$.

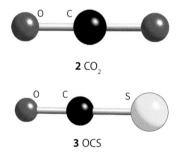

2 CO_2

3 OCS

Tetrahedral (T_d) and octahedral (O_h) molecules have more than one principal axis of symmetry (Fig. 3.10): a tetrahedral CH_4 molecule, for instance, has four C_3 axes, one along each CH bond. The O_h and T_d point groups are known as **cubic groups** because they are closely related to the symmetry of a cube. A closely related group, the icosahedral group, I_h, characteristic of the icosahedron, has 12 five-fold axes (Fig. 3.11). The icosahedral group contains the regular dodecahedron and is important for boron compounds (Section 14.11) and the C_{60} fullerene molecule (Section 15.6).

The distribution of molecules among the various point groups is very uneven. Some of the most common groups for molecules are the low-symmetry groups C_1 and C_s. There are many examples of molecules in groups C_{2v} (such as SO_2) and C_{3v} (such as NH_3). There are many linear molecules, which belong to the groups $C_{\infty v}$ (HCl, OCS) and $D_{\infty h}$ (Cl_2 and CO_2), and a number of planar-trigonal molecules (such as BF_3, **4**), which are D_{3h}; trigonal-bipyramidal molecules (such as PCl_5, **5**), which are also D_{3h}; and square-planar molecules, which are D_{4h} (**6**).

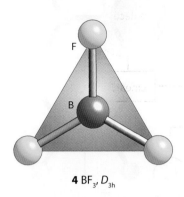

4 BF_3, D_{3h}

TABLE 3.2 The composition of some common groups

Point group	Symmetry elements	Shape	Examples
C_1	E		SiHClBrF
C_2	$E\ C_2$		H_2O_2
C_s	$E\ \sigma$		NHF_2
C_{2v}	$E\ C_2\ \sigma_v\ \sigma_v'$		SO_2Cl_2, H_2O
C_{3v}	$E\ 2C_3\ 3\sigma_v$		NH_3, PCl_3, $POCl_3$
$C_{\infty v}$	$E\ 2C_\infty\ \infty\sigma_v$		OCS, CO, HCl
D_{2h}	$E\ 3C_2\ i\ 3\sigma$		N_2O_4, C_2H_4
D_{3h}	$E\ 2C_3\ 3C_2\ \sigma_h\ 2S_3\ 3\sigma_v$		BF_3, PCl_5
D_{4h}	$E\ 2C_4\ C_2\ C_2'\ 2C_2''\ i\ 2S_4\ \sigma_h\ 2\sigma_v\ 2\sigma_d$		XeF_4, trans-[MA_4B_2]
$D_{\infty h}$	$E\ \infty C_2'\ 2C_\infty\ i\ \infty\sigma_v\ 2S_\infty$		CO_2, H_2, C_2H_2
T_d	$E\ 8C_3\ 3C_2\ 6S_4\ 6\sigma_d$		CH_4, $SiCl_4$
O_h	$E\ 8C_3\ 6C_2\ 6C_4\ 3C_2\ i\ 6S_4\ 8S_6\ 3\sigma_h\ \sigma_d$		SF_6

An introduction to symmetry analysis 77

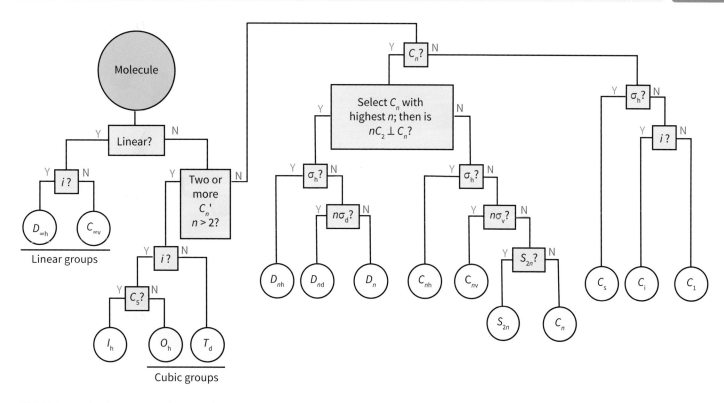

FIGURE 3.9 The decision tree for identifying a molecular point group. The symbols of each point refer to the symmetry elements.

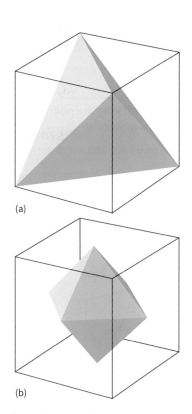

(a)

(b)

FIGURE 3.10 Shapes having cubic symmetry: (a) the tetrahedron, point group T_d; (b) the octahedron, point group O_h.

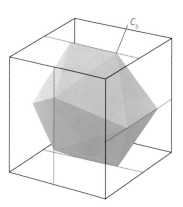

FIGURE 3.11 The regular icosahedron, point group I_h, and its relation to a cube.

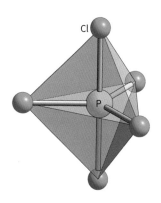

5 PCl_5, D_{3h}

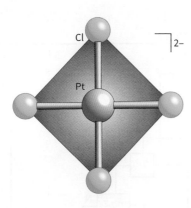

6 [PtCl$_4$]$^{2-}$, D_{4h}

An 'octahedral molecule' belongs to the octahedral point group O_h only if all six groups and the lengths of their bonds to the central atom are identical and all angles are 90°. For instance, 'octahedral' molecules with two identical substituents opposite each other, as in (7), are actually D_{4h}. The last example shows that the point-group classification of a molecule is more precise than the casual use of the terms 'octahedral' or 'tetrahedral' that indicate molecular geometry but say little about symmetry.

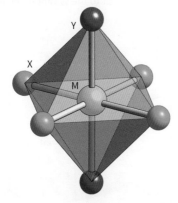

7 trans-[MX$_4$Y$_2$], D_{4h}

EXAMPLE 3.1 Identifying symmetry elements

Identify the symmetry elements in the eclipsed conformation of an ethane molecule.

Answer We need to identify the rotations, reflections, and inversions that leave the molecule apparently unchanged. Do not forget that the identity is a symmetry operation. By inspection of the molecular models, we see that the eclipsed conformation of a CH$_3$CH$_3$ molecule (**8**) has the elements E (do nothing), C_3 (a three-fold rotation axis), $3C_2$ (three two-fold rotation axes running through the C–C bond), σ_h (a horizontal mirror plane bisecting the C–C bond), $3\sigma_v$ (three separate vertical mirror planes running along each C–H bond), and S_3 (an improper rotation on rotating around the three-fold axis of symmetry followed by reflection in the plane perpendicular to it). We can see that the staggered conformation (**9**) additionally has the elements i (inversion) and S_6 (an improper rotation around the six-fold axis of symmetry arising from the six staggered H atoms).

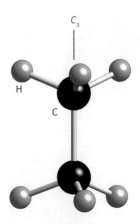

8 A C_3 axis

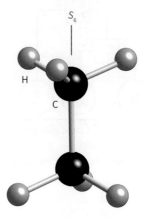

9 An S_6 axis

Self-test 3.1 Sketch the S_4 axis of an NH$_4^+$ ion. How many of these axes does the ion possess?

EXAMPLE 3.2 Identifying the point group of a molecule

To what point groups do H$_2$O and XeF$_4$ belong?

Answer We need to either work through Table 3.2 or use Fig. 3.9. (a) The symmetry elements of H$_2$O are shown in Fig. 3.12. H$_2$O possesses the identity (E), a two-fold rotation axis (C_2), and two vertical mirror planes (σ_v and σ_v'). The set of elements

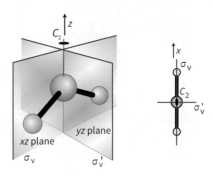

FIGURE 3.12 The symmetry elements of H$_2$O. The diagram on the right is the view from above and summarizes the diagram on the left.

(E, C_2, σ_v, σ_v') corresponds to those of the group C_{2v} listed in Table 3.2. Alternatively we can work through Fig. 3.9: the molecule is not linear; does not possess two or more C_n with $n > 2$; does possess a C_n (a C_2 axis); does not have 2$C_2 \perp$ to the C_2; does not have σ_h; does not have 2σ_v; it is therefore C_{2v}. (b) The symmetry elements of XeF$_4$ are shown in Fig. 3.3. XeF$_4$ possesses the identity (E), a four-fold axis (C_4), two pairs of two-fold rotation axes that are perpendicular to the principal C_4 axis, a horizontal reflection plane σ_h in the plane of the paper, and two sets of two vertical reflection planes, σ_v and σ_d. Using Table 3.2, we can see that this set of elements identifies the point group as D_{4h}. Alternatively we can work through Fig. 3.9: the molecule is not linear; does not possess two or more C_n with $n > 2$; does possess a C_n (a C_4 axis); does have 4$C_2 \perp$ to the C_4; and does have σ_h; it is therefore D_{4h}.

Self-test 3.2 Identify the point groups of (a) BF$_3$, a trigonal-planar molecule, and (b) the tetrahedral SO$_4^{2-}$ ion.

3.2 Character tables

KEY POINT The systematic analysis of the symmetry properties of molecules is carried out using character tables.

We have seen how the symmetry properties of a molecule define its point group, and how that point group is labelled by its Schoenflies symbol. Associated with each point group is a **character table**. A character table displays all the symmetry elements of the point group together with a description of how various objects or mathematical functions transform under the corresponding symmetry operations. In simple terms, it summarizes how each of the symmetry elements transforms the molecule. A character table is complete: every possible object or mathematical function relating to the molecule belonging to a particular point group must transform like one of the rows in the character table of that point group.

The structure of a typical character table is shown in Table 3.3. The entries in the main part of the table are called **characters**, χ (chi). Each character shows how an object or mathematical function, such as an atomic orbital, is affected by the corresponding symmetry operation of the group. Thus:

Character	Significance
1	The orbital is unchanged
−1	The orbital changes sign
0	The orbital undergoes a more complicated change, or is the sum of changes of degenerate orbitals

For instance, the rotation of a p$_z$ orbital about the z axis leaves it apparently unchanged (hence its character is 1); a reflection of a p$_z$ orbital in the xy-plane changes its sign (character −1). In some character tables, numbers such as 2 and 3 appear as characters: this feature is explained later.

The **class** of an operation is a specific grouping of symmetry operations of the same geometrical type: the two (clockwise and anticlockwise) three-fold rotations about an axis form one class, reflections in a mirror plane form another, and so on. The number of members of each class is shown in the heading of each column of the table, as in 2C_3, denoting that there are two members of the class of three-fold rotations. All operations of the same class have the same character. The **order**, h, of the group is the total number of symmetry operations that can be carried out.

Each row of characters corresponds to a particular **irreducible representation** of the group. An irreducible representation has a technical meaning in group theory but, broadly speaking, it is a fundamental type of symmetry in the group. The label in the first column is the **symmetry species** of that irreducible representation. The two columns on the right contain examples of functions that exhibit the characteristics of each symmetry species. One column contains functions defined by a single axis, such as translations (x,y,z), p orbitals (p$_x$,p$_y$,p$_z$), or rotations around an axis (R_x,R_y,R_z), and the other column contains quadratic functions such as those that represent d orbitals (xy, etc.). The letter A used to label a symmetry species in the group C_{2v} means that the function to which it refers is symmetric with respect to rotation about the two-fold axis (i.e. its character is 1). The label B indicates that the function changes sign under that rotation (the character is −1). The subscript 1 on A$_1$ means that the function to which it refers is also symmetric with respect to reflection in the principal vertical plane (for H$_2$O this is the plane that contains all three atoms). A subscript 2 is used to denote that the function changes sign under this reflection.

Character tables for a selection of common point groups are given in Resource section 4.

Now consider the slightly more complex example of NH$_3$, which belongs to the point group C_{3v} (Table 3.4). An NH$_3$ molecule is described as having higher symmetry than H$_2$O. This higher symmetry is apparent by noting the **order**, h, of the group, the total number of symmetry operations that can be carried out. For H$_2$O, $h = 4$, and for NH$_3$, $h = 6$.

TABLE 3.3 The components of a character table

Name of point group*	Symmetry operations R arranged by class (E, C_n, etc.)	Functions	Further functions	Order of group, h
Symmetry species (Γ)	Characters (χ)	Translations and components of dipole moments (x, y, z) of relevance to IR activity; rotations	Quadratic functions such as z^2, xy, etc., of relevance to Raman activity	

* Schoenflies symbol.

3 Molecular symmetry

TABLE 3.4 The C_{3v} character table

C_{3v}	E	$2C_3$	$3\sigma_v$	$h=6$	
A_1	1	1	1	z	x^2+y^2, z^2
A_2	1	1	−1	R_z	
E	2	−1	0	$(R_x, R_y)(x,y)$	$(zx, yz)(x^2-y^2, xy)$

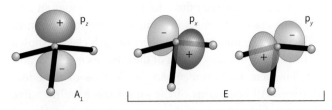

FIGURE 3.13 The nitrogen $2p_z$ orbital in ammonia is symmetric under all operations of the C_{3v} point group and therefore has A_1 symmetry. The $2p_x$ and $2p_y$ orbitals behave identically under all operations (they cannot be distinguished) and are given the symmetry label E.

For highly symmetrical molecules, h is large; for example, $h = 48$ for the point group O_h.

Inspection of the NH$_3$ molecule (Fig. 3.13) shows that the N$2p_z$ orbital remains unchanged under the E, $2C_3$, and $3\sigma_v$ operations, giving the characters 1, 1, 1 and therefore A_1 symmetry. In contrast the N$2p_x$ and N$2p_y$ do change under the $2C_3$ and $3\sigma_v$ operations; since both orbitals have the same symmetry characteristics they are degenerate and must be treated together. The N$2p_x$ and N$2p_y$ orbitals have symmetry representation E, and the degeneracy is indicated by the appearance of 2 in the column under E.

In general, the characters in the column headed by the identity operation E give the degeneracy of the orbitals:

Symmetry label	Degeneracy
A, B	1
E	2
T	3

So, for NH$_3$ there is one orbital with A_1 symmetry and two orbitals with E symmetry. Be careful to distinguish between the italic E for the operation and the roman E for the symmetry label: all operations are italic and all labels are roman.

Degenerate irreducible representations also contain zero values for some operations because the character is the sum of the characters for the two or more orbitals of the set, and if one orbital changes sign but the other does not, then the total character is 0. For example, the reflection through the vertical mirror plane containing the y-axis in NH$_3$ results in no change of the p_y orbital (1), but an inversion of the p_x orbital (−1).

EXAMPLE 3.3 Identifying the symmetry species of orbitals

Identify the symmetry species of each of the oxygen valence-shell atomic orbitals in an H$_2$O molecule, which has C_{2v} symmetry.

Answer The symmetry elements of the H$_2$O molecule are shown in Fig. 3.12, and the character table for C_{2v} is given in Table 3.5. We need to see how the orbitals behave under these symmetry operations. An s orbital on the O atom is unchanged by all four operations, so its characters are (1,1,1,1) and thus it has symmetry species A_1. Likewise, the $2p_z$ orbital on the O atom is unchanged by all operations of the point group and is thus totally symmetric under C_{2v}: it therefore has symmetry species A_1. The character of the O$2p_x$ orbital under C_2 is −1, which means simply that it changes sign under a two-fold rotation. A p_x orbital also changes sign (and therefore has character −1) when reflected in the yz-plane (σ'_v), but is unchanged (character 1) when reflected in the xz-plane (σ_v). It follows that the characters of an O$2p_x$ orbital are (1,−1,1,−1) and therefore that its symmetry species is B_1. The

TABLE 3.5 The C_{2v} character table.

C_{2v}	E	C_2	$\sigma_v(xz)$	$\sigma'_v(yz)$	$h=4$	
A_1	1	1	1	1	z	x^2, y^2, z^2
A_2	1	1	−1	−1	R_z	xy
B_1	1	−1	1	−1	x, R_y	zx
B_2	1	−1	−1	1	y, R_x	yz

character of the O$2p_y$ orbital under C_2 is −1, as it is when reflected in the xz-plane (σ_v). The O$2p_y$ is unchanged (character 1) when reflected in the yz-plane (σ'_v). It follows that the characters of an O$2p_y$ orbital are (1,−1,−1,1) and therefore that its symmetry species is B_2.

Self-test 3.3 Identify the symmetry species of all five d orbitals of the central S atom in H$_2$S.

EXAMPLE 3.4 Determining degeneracy

Determine whether there are triply degenerate orbitals in BF$_3$.

Answer To decide if there can be triply degenerate orbitals in BF$_3$ we note that the point group of the molecule is D_{3h}. Reference to the character table for this group (Resource section 4) shows that, because no character exceeds 2 in the column headed E, the maximum degeneracy is 2. Therefore, none of its orbitals can be triply degenerate. This is confirmed by the appearance of only A and E symmetry labels in the character table.

Self-test 3.4 The SF$_6$ molecule is octahedral. What is the maximum possible degree of degeneracy of its orbitals?

Applications of symmetry

Important applications of symmetry in inorganic chemistry include the construction and labelling of molecular orbitals and the interpretation of spectroscopic data to determine structure. However, there are several simpler applications as some molecular properties, such as polarity and chirality, can be deduced with only the knowledge of the point group to which a molecule belongs. Other properties, such as the classification of molecular vibrations and the identification of their IR and Raman activity, require us to know the detailed structure of the character table. We illustrate both applications in this section.

3.3 Polar molecules

KEY POINT A molecule cannot be polar if it belongs to any group that includes a centre of inversion, any of the groups D and their derivatives, the cubic groups (T, O), the icosahedral group (I), or their modifications.

A **polar molecule** is a molecule that has a permanent electric dipole moment. A molecule cannot be polar if it has a centre of inversion. Inversion implies that a molecule has matching charge distributions at all diametrically opposite points about a centre, which rules out a dipole moment. For similar reasons, a dipole moment cannot lie perpendicular to any mirror plane or axis of rotation that the molecule may possess. For example, a mirror plane demands identical atoms on either side of the plane, so there can be no dipole moment across the plane. Similarly, a symmetry axis implies the presence of identical atoms at points related by the corresponding rotation, which rules out a dipole moment perpendicular to the axis.

In summary:

- A molecule cannot be polar if it has a centre of inversion.
- A molecule cannot have an electric dipole moment perpendicular to any mirror plane.
- A molecule cannot have an electric dipole moment perpendicular to any axis of rotation.

Some molecules have a symmetry axis that rules out a dipole moment in one plane and another symmetry axis or mirror plane that rules it out in another direction. The two or more symmetry elements jointly forbid the presence of a dipole moment in any direction. Thus, any molecule that has a C_n axis and a C_2 axis perpendicular to that C_n axis (as do all molecules belonging to a D point group) cannot have a dipole moment in any direction. For example, the BF$_3$ molecule (D_{3h}) is nonpolar. Likewise, molecules belonging to the tetrahedral, octahedral, and icosahedral groups have several perpendicular rotation axes that rule out dipoles in all three directions, so such molecules must be nonpolar; hence SF$_6$ (O_h) and CCl$_4$ (T_d) are nonpolar.

EXAMPLE 3.5 **Judging whether or not a molecule can be polar**

The ruthenocene molecule (**10**) is a pentagonal prism with the Ru atom sandwiched between two C$_5$H$_5$ rings. Predict whether it is polar.

Answer We should decide whether the point group is D or cubic because in neither case can it have a permanent electric dipole. Reference to Fig. 3.9 shows that a pentagonal prism belongs to the point group D_{5h}. Therefore, the molecule must be nonpolar.

Self-test 3.5 A conformation of the ruthenocene molecule that lies above the lowest energy conformation is a pentagonal antiprism (**11**). Determine the point group and predict whether the molecule is polar.

10 11

3.4 Chiral molecules

KEY POINT A molecule cannot be chiral if it possesses an improper rotation axis (S_n).

A **chiral molecule** (from the Greek word for 'hand') is a molecule that cannot be superimposed on its own mirror image. An actual hand is chiral in the sense that the mirror image of a left hand is a right hand, and the two hands cannot be superimposed. A chiral molecule and its mirror image partner are called **enantiomers** (from the Greek word for 'both parts'). Chiral molecules that do not interconvert rapidly between enantiomeric forms are **optically active** in the sense that they can rotate the plane of polarized light. Enantiomeric pairs of molecules rotate the plane of polarization of light by equal amounts in opposite directions.

A molecule with a mirror plane is obviously not chiral. However, a small number of molecules without mirror planes are not chiral either. In fact, the crucial condition is that a molecule with an improper rotation axis, S_n, cannot be chiral. A mirror plane is an S_1 axis of improper rotation and a centre of inversion is equivalent to an S_2 axis; therefore, molecules with either a mirror plane or a centre of inversion have axes of improper rotation and cannot be chiral. Groups in which S_n is present include D_{nh}, D_{nd}, and some of the cubic groups (specifically, T_d and O_h). Therefore, molecules such as CH$_4$ and [Ni(CO)$_4$] that belong to the group T_d are not chiral. That a 'tetrahedral' carbon atom leads to optical activity (as in CHClFBr) should serve as another reminder that group theory is stricter in its terminology than casual conversation. Thus CHClFBr (**12**) belongs to the group C_1, not to the group T_d; it has tetrahedral geometry but not tetrahedral symmetry.

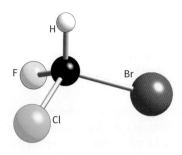

12 CHClFBr, C_1

When judging chirality, it is important to be alert for axes of improper rotation that might not be immediately apparent. Molecules with neither a centre of inversion nor a mirror plane (and hence with no S_1 or S_2 axes) are usually chiral, but it is important to verify that a higher-order improper-rotation axis is not also present. For instance, the quaternary ammonium ion (**13**) has neither a mirror plane (S_1) nor an inversion centre (S_2), but it does have an S_4 axis and so it is not chiral.

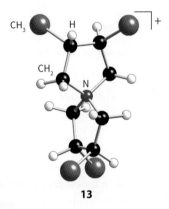

13

EXAMPLE 3.6 **Judging whether or not a molecule is chiral**

The complex [Mn(acac)$_3$], where acac denotes the acetylacetonato ligand (CH$_3$COCHCOCH$_3^-$), has the structure shown as (**14**). Predict whether it is chiral.

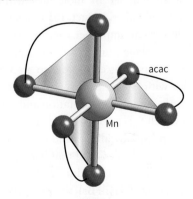

14 [Mn(acac)$_3$]

Answer We begin by identifying the point group in order to judge whether it contains an improper-rotation axis either explicitly or in a disguised form. The chart in Fig. 3.9 shows that the complex belongs to the point group D_3, which consists of the elements (E, C_3, $3C_2$) and hence does not contain an S_n axis either explicitly or in a disguised form. The complex is chiral and hence, because it is long-lived, optically active.

Self-test 3.6 Is the conformation of H_2O_2 shown in (**15**) chiral? The molecule can usually rotate freely about the O–O bond: comment on the possibility of observing optically active H_2O_2.

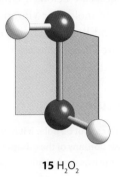

15 H_2O_2

3.5 Molecular vibrations

KEY POINTS If a molecule has a centre of inversion, none of its vibrations can be both IR and Raman active; a vibrational mode is IR active if it has the same symmetry as a component of the electric dipole vector; a vibrational mode is Raman active if it has the same symmetry as a component of the molecular polarizability.

A knowledge of the symmetry of a molecule can assist and greatly simplify the analysis of infrared (IR) and Raman spectra (Section 8.5). It is convenient to consider two aspects of symmetry. One is the information that can be obtained directly by knowing to which point group a molecule as a whole belongs. The other is the additional information that comes from knowing the symmetry species of each vibrational mode. All we need to know at this stage is that the absorption of infrared radiation can occur when a vibration results in a change in the electric dipole moment of a molecule; a Raman transition can occur when the polarizability of a molecule changes during a vibration.

For a molecule of N atoms there are $3N$ displacements to consider as the atoms move in the three orthogonal directions, x, y, and z. For a nonlinear molecule, three of these displacements correspond to translational motion of the molecule as a whole (in each of the x, y, and z directions), and three correspond to an overall rotation of the molecule (about each of the x, y, and z axes). Thus the remaining $3N - 6$ atomic displacements must correspond to molecular deformations or vibrations. There is no rotation around the molecular axis, z, if the molecule is linear, only around the x and y axes. So linear molecules have only two rotational degrees of freedom instead of three, leaving $3N - 5$ vibrational displacements.

(a) The exclusion rule

The three-atom nonlinear molecule H_2O has $(3 \times 3) - 6 = 3$ vibrational modes (Fig. 3.14). All three vibrational displacements lead to a change in the dipole moment (Fig. 3.15), and this can be confirmed by group theory. It follows that all three modes of this C_{2v} molecule are IR active. It is

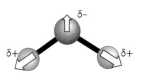

Symmetric stretch ν_1

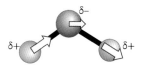

Antisymmetric stretch ν_3

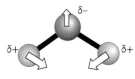

Bend ν_2

FIGURE 3.15 The vibrations of an H_2O molecule all change the dipole moment.

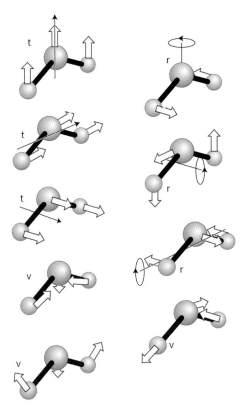

FIGURE 3.14 An illustration of the counting procedure for displacements of the atoms in a nonlinear molecule.

difficult to judge intuitively whether or not a vibrational mode is Raman active because it is hard to know whether a particular distortion of a molecule results in a change of polarizability (although modes that result in a change in volume, and thus the electron density of the molecule, such as the symmetric stretch (A_{1g}) of SF_6 (O_h), are good prospects). This difficulty is partly overcome by the **exclusion rule**, which is sometimes helpful:

If a molecule has a centre of inversion, none of its modes can be both IR and Raman active. A mode may be inactive in both.

EXAMPLE 3.7 Using the exclusion rule

There are four vibration modes of the linear triatomic CO_2 molecule (Fig. 3.16). Which of these are IR or Raman active?

Answer To establish whether or not a stretch is IR active, we need to consider its effect on the dipole moment of the molecule. If we consider the symmetric stretch, ν_1, we can see it leaves the electric dipole moment unchanged at zero and so it is IR inactive: it may therefore be Raman active (and is). In contrast, for the antisymmetric stretch, ν_3, the C atom moves in the opposite direction relative to that of the two O atoms: as a result, the electric dipole moment changes from zero in the course of the vibration and the mode is IR active. Because the CO_2 molecule has a centre of inversion, it follows from the exclusion rule that this mode cannot be Raman active. Both bending modes cause a departure of the dipole moment from zero and are therefore IR active. It follows from the exclusion rule that the two bending modes (they are degenerate) are Raman inactive.

Self-test 3.7 The bending mode of linear N_2O is active in the IR. Predict whether it is also Raman active.

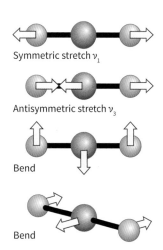

FIGURE 3.16 The stretches and bends of a CO_2 molecule.

(b) Information from the symmetries of normal modes

It is often intuitively obvious whether a vibrational mode gives rise to a changing electric dipole and is therefore IR active. When intuition is unreliable, perhaps because the molecule is complex or the mode of vibration is difficult to visualize, a symmetry analysis can be used instead. We shall illustrate the procedure by considering the two square-planar palladium species, (**16**) and (**17**). The Pt analogues of these species and the distinction between them are of considerable social and practical significance because the *cis* isomer is used as a chemotherapeutic agent against certain cancers, whereas the *trans* isomer is therapeutically inactive (Section 32.1).

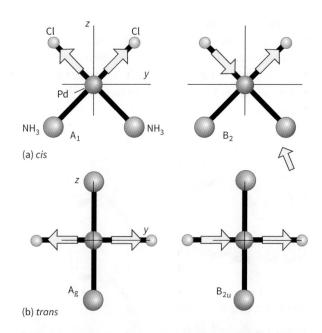

FIGURE 3.17 The Pd–Cl stretching modes of *cis* (a) and *trans* (b) forms of [PdCl$_2$(NH$_3$)$_2$]. The motion of the Pd atom (which preserves the centre of mass of the molecule) is not shown.

16 *cis*-[PdCl$_2$(NH$_3$)$_2$]

17 *trans*-[PdCl$_2$(NH$_3$)$_2$]

First, we note that the *cis* isomer (**16**) has C_{2v} symmetry, whereas the *trans* isomer (**17**) is D_{2h}. Both species have IR absorption bands in the Pd–Cl stretching region between 200 and 400 cm^{-1}, and these are the only bands we are going to consider. If we think of the PdCl$_2$ fragment in isolation, and compare the *trans* form with CO$_2$ (Fig. 3.16), we can see that there are two stretching modes; similarly, the *cis* form also has one symmetric and one asymmetric stretch. We know immediately from the exclusion rule that the two modes of the *trans* isomer (which has a centre of symmetry) cannot be active in both IR and Raman. However, to decide which modes are IR active and which are Raman active we consider the characters of the modes themselves. It follows from the symmetry properties of dipole moments and polarizabilities (which we do not verify here) that:

> The symmetry species of the vibration must be the same as that of *x*, *y*, or *z* in the character table for the vibration to be IR active, and the same as that of a quadratic function, such as *xy* or *x*^{2}, for it to be Raman active.

Our first task, therefore, is to classify the normal modes according to their symmetry species, and then to identify which of these modes have the same symmetry species as *x*, etc., and *xy*, etc., by referring to the final columns of the character table of the molecular point group.

Fig. 3.17 shows the symmetric (left) and antisymmetric (right) stretches of the Pd–Cl bonds for each isomer, where the NH$_3$ group is treated as a single mass point. The arrows in the diagram show the vibration or, more formally, they show the displacement vectors representing the vibration. To classify them according to their symmetry species in their respective point groups we use an approach similar to the symmetry analysis of molecular orbitals that we will use for determining SALCs (Section 3.6).

Consider the *cis* isomer and its point group C_{2v} (Table 3.5), and note that we represent the vibration as an arrow. For the symmetric stretch, we see that the pair of displacement vectors representing the vibration is apparently unchanged by each operation of the group. For example, the two-fold rotation simply interchanges two equivalent displacement vectors. It follows that the character of each operation is 1:

E	C_2	σ_v	σ_v'
1	1	1	1

The symmetry of this vibration is therefore A_1. For the antisymmetric stretch, the identity E leaves the displacement vectors unchanged and the same is true of σ_v' which lies in the plane containing the two Cl atoms. However, both C_2 and σ_v interchange the two oppositely directed displacement vectors, and so convert the overall displacement into -1 times itself. The characters are therefore

E	C_2	σ_v	σ_v'
1	−1	−1	1

The C_{2v} character table identifies the symmetry species of this mode as B_2. So for the *cis* isomer we have the A_1 and B_2 symmetry species.

A similar analysis of the *trans* isomer, but using the D_{2h} group (Resource section 4), results in the A_g and B_{2u} symmetry species for the symmetric and antisymmetric Pd–Cl stretches, respectively, as demonstrated in the following example.

EXAMPLE 3.8 Identifying the symmetry species of vibrational displacements

The *trans* isomer in Fig. 3.17 has D_{2h} symmetry. Verify that the symmetry species of the antisymmetric Pd–Cl stretches is B_{2u}.

Answer We need to start by considering the effect of the various elements of the group on the displacement vectors of the Cl⁻ ligands, noting that the molecule is in the *yz*-plane. The elements of D_{2h} are E, $C_2(x)$, $C_2(y)$, $C_2(z)$, i, $\sigma(xy)$, $\sigma(yz)$, and $\sigma(zx)$. Of these, E, $C_2(y)$, $\sigma(xy)$, and $\sigma(yz)$ leave the displacement vectors unchanged and so have characters 1. The remaining operations reverse the directions of the vectors, so giving characters of −1:

E	$C_2(x)$	$C_2(y)$	$C_2(z)$	i	$\sigma(xy)$	$\sigma(yz)$	$\sigma(zx)$
1	−1	1	−1	−1	1	1	−1

We now compare this set of characters with the D_{2h} character table and establish that the symmetry species is B_{2u}.

Self-test 3.8 Confirm that the symmetry species of the symmetric mode of the Pd–Cl stretches in the *trans* isomer is A_g.

As we have remarked, a vibrational mode is IR active if it has the same symmetry species as the displacements *x*, *y*, or *z*. To identify whether either of the two vibrational modes of the *cis* isomer are IR active we inspect the C_{2v} character table (Table 3.5). The last two columns show that *z* is A_1 and *y* is B_2. Therefore, both A_1 and B_2 vibrations of the *cis* isomer are IR active. For the *trans* isomer we inspect the D_{2h} character table. The last two columns show that the A_g symmetry species has no *x*, *y*, or *z* and that B_{2u} is *y* and therefore only the antisymmetric Pd–Cl stretch of the *trans* isomer with symmetry B_{2u} is IR active. The symmetric A_g mode of the *trans* isomer is not IR active.

To determine the Raman activity, we note that in the C_{2v} character table the quadratic forms *xy*, etc. transform as A_1, A_2, B_1, and B_2 and therefore in the *cis* isomer the modes of symmetry A_1 and B_2 are Raman active. In the D_{2h} character table the quadratic forms transform to A_g, B_{1g}, B_{2g}, and B_{3g}. Therefore, in the *trans* isomer the A_g symmetry is Raman active.

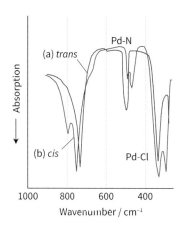

FIGURE 3.18 The IR spectra of *trans* (a) and *cis* (b) forms of [PdCl$_2$(NH$_3$)$_2$]. (R. Layton, D. W. Sink, and J. R. Durig, *J. Inorg. Nucl. Chem.*, 1966, **28**, 1965.)

The experimental distinction between the *cis* and *trans* isomers now emerges. In the Pd–Cl stretching region, the *cis* (C_{2v}) isomer has two bands in both the Raman and IR spectra. By contrast, the *trans* (D_{2h}) isomer has one band at a different frequency in each of the IR and Raman spectra. The IR spectra of the two isomers are shown in Fig. 3.18.

(c) The assignment of molecular symmetry from vibrational spectra

An important application of vibrational spectra is the identification of molecular symmetry and hence shape and structure. An especially important example arises in metal carbonyls, in which CO molecules are bound to a metal atom. Vibrational spectra are especially useful because the CO stretch is responsible for very strong characteristic absorptions between 1850 and 2200 cm⁻¹ (Section 24.7).

When we consider a set of vibrations, the characters obtained by considering the symmetries of the displacements of atoms are often found not to correspond to any one particular row in the character table. However, because the character table is a complete summary of the symmetry properties of an object, the characters that have been determined must correspond to a sum of two or more of the rows in the table. In such cases we say that the displacements span a **reducible representation**. Our task is to find the **irreducible representations** that they span. To do so, we identify the rows in the character table that must be added together to reproduce the set of characters that we have obtained. This process is called **reducing a representation**. In some cases the reduction is obvious; in others it may be carried out systematically by using a procedure explained in Section 3.9 and Example 3.9.

EXAMPLE 3.9 Reducing a representation

One of the first metal carbonyls to be characterized was the tetrahedral (T_d) molecule [Ni(CO)$_4$]. The vibrational modes of the molecule that arise from stretching motions of the CO groups are four combinations of the four CO displacement vectors. Which modes are IR or Raman active? The CO displacements of [Ni(CO)$_4$] are shown in Fig. 3.19.

Answer We need to consider the motion of the four CO displacement vectors, consider how many of them remain unchanged, and then consult the character table for T_d (Table 3.6).

Under the E operation all four vectors remain unchanged, under a C_3 operation only one remains the same, under both C_2 and S_4 none of the vectors remains unchanged, and under σ_d two remain the same. The characters are therefore:

E	$8C_3$	$3C_2$	$6S_4$	$6\sigma_d$
4	1	0	0	2

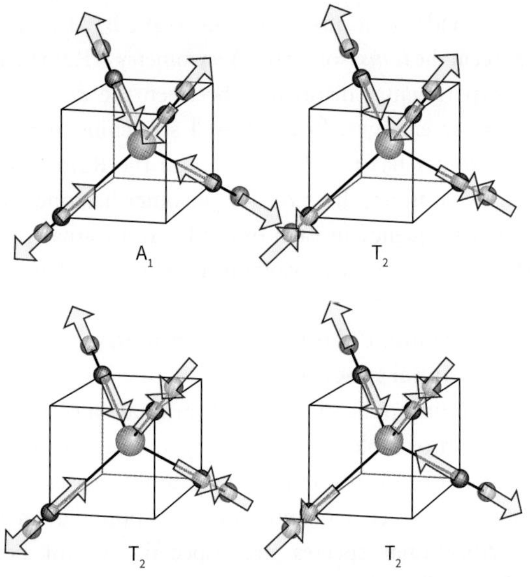

TABLE 3.6 The T_d character table

T_d	E	$8C_3$	$3C_2$	$6S_4$	$6\sigma_d$	$h = 24$	
A_1	1	1	1	1	1		$x^2 + y^2 + z^2$
A_2	1	1	1	−1	−1		
E	2	−1	2	0	0		$(2z^2 − x^2 − y^2, x^2 − y^2)$
T_1	3	0	−1	1	−1	(R_x, R_y, R_z)	
T_2	3	0	−1	−1	1	(x, y, z)	(xy, yz, zx)

This set of characters does not correspond to any one symmetry species. However, it does correspond to the sum of the characters of symmetry species A_1 and T_2:

	E	$8C_3$	$3C_2$	$6S_4$	$6\sigma_d$
A_1	1	1	1	1	1
T_2	3	0	−1	−1	1
$A_1 + T_2$	4	1	0	0	2

It follows that the CO displacement vectors transform as $A_1 + T_2$. By consulting the character table for T_d, we see that the combination labelled A_1 transforms like $x^2 + y^2 + z^2$, indicating that it is Raman active but not IR active. By contrast, x, y, and z and the products xy, yz, and zx transform as T_2, so the T_2 modes are both Raman and IR active. Consequently, a tetrahedral carbonyl molecule is recognized by one IR band and two Raman bands in the CO stretching region.

Self-test 3.9 Show that the four CO stretches in the square-planar (D_{4h}) [Pt(CO)$_4$]$^{2+}$ cation transform as $A_{1g} + B_{1g} + E_u$. How many bands would you expect in the IR and Raman spectra for the [Pt(CO)$_4$]$^{2+}$ cation?

FIGURE 3.19 The modes of [Ni(CO)$_4$] that correspond to the stretching of CO bonds.

The symmetries of molecular orbitals

We shall now see in more detail the significance of the labels used for molecular orbitals introduced in Sections 2.7–2.9 and gain more insight into their construction. At this stage the discussion will continue to be informal and pictorial, our aim being to provide an introduction to group theory but not the details of the calculations involved. The specific objective here is to show how to identify the symmetry label of a molecular orbital from a drawing like those in Resource section 5 and, conversely, to appreciate the significance of a symmetry label. The arguments later in the book are all based on simply 'reading' molecular orbital diagrams qualitatively.

3.6 Symmetry-adapted linear combinations

KEY POINT Symmetry-adapted linear combinations of orbitals are combinations of atomic orbitals that conform to the symmetry of a molecule and are used to construct molecular orbitals of a given symmetry species.

A fundamental principle of the MO theory of diatomic molecules (Section 2.7) is that molecular orbitals are constructed from atomic orbitals of the same symmetry. Thus, in a diatomic molecule, an s orbital may have a nonzero overlap integral with another s orbital or with a p_z orbital on the second atom (where z is the internuclear

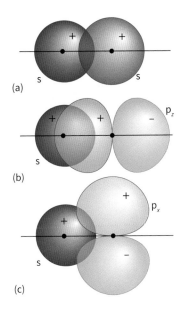

FIGURE 3.20 An s orbital can overlap (a) an s or (b) a p_z orbital on a second atom with constructive interference. (c) An s orbital has zero net overlap with a p_x or p_y orbital because the constructive interference between the parts of the atomic orbitals with the same sign exactly matches the destructive interference between the parts with opposite signs.

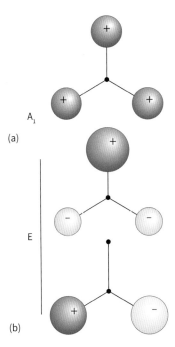

FIGURE 3.21 The (a) A_1 and (b) E symmetry-adapted linear combinations of H1s orbitals in NH_3.

direction; Fig. 3.20), but not with a p_x or p_y orbital. Formally, whereas the p_z orbital of the second atom has the same rotational symmetry as the s orbital of the first atom and the same symmetry with respect to reflection in a mirror plane containing the internuclear axis, the p_x and p_y orbitals do not. The restriction that σ, π, or δ bonds can be formed from atomic orbitals of the same symmetry species stems from the requirement that all components of the molecular orbital must behave identically under any transformation (e.g. reflection, rotation) if they are to have nonzero overlap.

Exactly the same principle applies in polyatomic molecules, where the symmetry considerations may be more complex and require us to use the systematic procedures provided by group theory. The general procedure is to group atomic orbitals, such as the three H1s orbitals of NH_3, together to form combinations of a particular symmetry and then to build molecular orbitals by allowing combinations of the same symmetry on different atoms to overlap, such as a N2s orbital and the appropriate combination of the three H1s orbitals. Specific combinations of atomic orbitals that are used to build molecular orbitals of a given symmetry are called **symmetry-adapted linear combinations** (SALCs). A collection of commonly encountered SALCs of orbitals is shown in Resource section 5; it is usually simple to identify the symmetry of a combination of orbitals by comparing it with the diagrams provided there.

The generation of SALCs of a given symmetry is a task for group theory, as we explain in Section 3.10. However, they often have an intuitively obvious form. For instance, the fully symmetric A_1 SALC of the H1s orbitals of NH_3 (Fig. 3.21) is

$$\phi_1 = \psi_{A1s} + \psi_{B1s} + \psi_{C1s}$$

where A, B, and C are the three H atoms.

To verify that this SALC is indeed of symmetry A_1 we note that it remains unchanged under the identity E, each C_3 rotation, and any of the three vertical reflections, so its characters are (1,1,1) and hence it spans the fully symmetric irreducible representation of C_{3v}. The E SALCs are less obvious, but, as we shall see, are

$$\phi_2 = 2\psi_{A1s} - \psi_{B1s} - \psi_{C1s}$$
$$\phi_3 = \psi_{B1s} - \psi_{C1s}$$

EXAMPLE 3.10 Identifying the symmetry species of a SALC

Identify the symmetry species of the SALCs that may be constructed from the H1s orbitals of NH_3.

Answer We start by establishing how the set of H1s orbitals transforms under the operations of the appropriate symmetry group of the molecule. An NH_3 molecule has symmetry C_{3v} and the three H1s orbitals all remain unchanged under the identity operation E. None of the H1s orbitals remains unchanged under a C_3 rotation, and only one remains unchanged under a vertical reflection σ_v. As a set they therefore span a representation with the characters

E	$2C_3$	$3\sigma_v$
3	0	1

We now need to reduce this set of characters, and by inspection of the character table in Table 3.4 we can see that they correspond to $A_1 + E$ (1,1,1 and 2,−1,0). It follows that the three H1s orbitals contribute two SALCs, one with A_1 symmetry and the other with E symmetry. The C_{3v} character table has both x and y components in the fourth column for the E symmetry. Therefore, the SALC with E symmetry has two members of the same energy (Fig. 3.21). In more complicated examples the reduction might not be obvious, and we use the systematic procedure discussed in Section 3.10.

Self-test 3.10 What is the symmetry label of the SALC $\phi = \psi_{A1s} + \psi_{B1s} + \psi_{C1s} + \psi_{D1s}$ in CH_4, where ψ_{J1s} is an H1s orbital on atom J?

3.7 The construction of molecular orbitals

KEY POINT Molecular orbitals are constructed from SALCs and atomic orbitals of the same symmetry species.

We have seen in Example 3.10 that the SALC ϕ_1 of H1s orbitals in NH_3 has A_1 symmetry. The N2s and N2p_z orbitals also have A_1 symmetry in this molecule, so all three can contribute to the same molecular orbitals. The symmetry species of these molecular orbitals will be A_1, like their components, and they are called a_1 **orbitals**. Note that the labels for molecular orbitals are lower-case versions of the symmetry species of the orbital. Three such molecular orbitals are possible, each of the form

$$\psi = c_1\psi_{N2s} + c_2\psi_{N2p_z} + c_3\phi_1$$

with c_i coefficients that are found by computational methods and can be positive or negative in sign. They are labelled $1a_1$, $2a_1$, and $3a_1$ in order of increasing energy (the order of increasing number of internuclear nodes), and correspond to bonding, nonbonding, and antibonding combinations (Fig. 3.22).

We have also seen (and can confirm by referring to Resource section 5) that in a C_{3v} molecule the SALCs ϕ_2 and ϕ_3 of the H1s orbitals have E symmetry. The C_{3v} character table shows that the same is true of the N2p_x and N2p_y orbitals (Fig. 3.23). It follows that ϕ_2 and ϕ_3 can combine with these two N2p orbitals to give doubly degenerate bonding and antibonding orbitals of the form

$$\psi = c_4\psi_{N2p_x} + c_5\phi_2 \text{ and } c_6\psi_{N2p_y} + c_7\phi_3$$

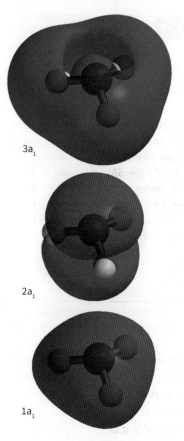

FIGURE 3.22 The three a_1 molecular orbitals of NH_3 as computed by molecular modelling software.

These molecular orbitals have E symmetry and are therefore called **e orbitals**. The pair of lower energy, denoted 1e, are bonding (the coefficients have the same sign), and the upper pair, 2e, are antibonding (the coefficients have the opposite sign).

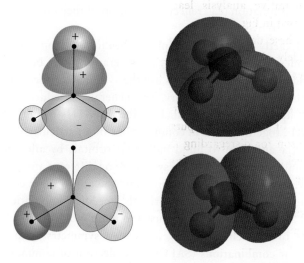

FIGURE 3.23 The two bonding e orbitals of NH_3 as schematic diagrams and as computed by molecular modelling software.

> **EXAMPLE 3.11** **Identifying the symmetry species of SALCs**
>
> Identify the symmetry species of the SALC $\phi = \psi_O' - \psi_O''$ in the C_{2v} molecule NO_2, where ψ_O' is a $2p_x$ orbital on one O atom and ψ_O'' is a $2p_x$ orbital on the other O atom.
>
> **Answer** To establish the symmetry species of a SALC we need to see how it transforms under the symmetry operations of the group. A picture of the SALC is shown in Fig. 3.24, and we can see that under C_2, the SALC, ϕ, changes into itself, implying a character of 1. Under σ_v, both atomic orbitals change sign, so ϕ is transformed into $-\phi$, implying a character of -1. The SALC also changes sign under σ_v' so the character for this operation is also -1. The characters are therefore
>
> | E | C_2 | σ_v | σ_v' |
> |---|---|---|---|
> | 1 | 1 | -1 | -1 |
>
>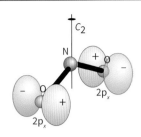
>
> **FIGURE 3.24** The combination of O $2p_x$ orbitals referred to in Example 3.11.
>
> Inspection of the character table for C_{2v} shows that these characters correspond to symmetry species A_2.
>
> **Self-test 3.11** Identify the symmetry species of the combination $\phi = \psi_{A1s} - \psi_{B1s} + \psi_{C1s} - \psi_{D1s}$ for a square-planar (D_{4h}) array of H atoms A, B, C, D.

A symmetry analysis has nothing to say about the energies of orbitals other than to identify degeneracies. To calculate the energies, and even to arrange the orbitals in order, it is necessary to use quantum mechanics; to assess them experimentally it is necessary to use techniques such as photoelectron spectroscopy. In simple cases, however, we can use the general rules set out in Section 2.8 to judge the relative energies of the orbitals. For example, in NH_3, the $1a_1$ orbital, containing the low-lying N2s orbital, can be expected to lie lowest in energy, and its antibonding partner, $3a_1$, will probably lie highest, with the nonbonding $2a_1$ approximately halfway between. The 1e bonding orbital is next-higher in energy after $1a_1$, and the 2e correspondingly lower in energy than the $3a_1$ orbital. This qualitative analysis leads to the energy-level scheme shown in Fig. 3.25.

These days, there is no difficulty in using one of the widely available software packages to calculate the energies of the orbitals directly by either an *ab initio* or a semi-empirical procedure; the relative energies shown in Fig. 3.25 have in fact been calculated in this way. Nevertheless, the ease of achieving computed values should not be seen as a reason for disregarding the understanding of the energy-level order that comes from investigating the structures of the orbitals.

The general procedure for constructing a molecular orbital scheme for a reasonably simple molecule can now be summarized as follows:

1. Assign a point group to the molecule.
2. Look up the shapes of the SALCs in Resource section 5.

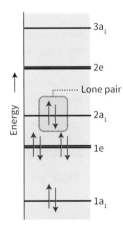

FIGURE 3.25 A schematic molecular orbital energy-level diagram for NH_3 and an indication of its ground-state electron configuration.

3. Arrange the SALCs of each molecular fragment in increasing order of energy, first noting whether they stem from s, p, or d orbitals (and putting them in the order s < p < d), and then their number of internuclear nodes.
4. Combine SALCs of the same symmetry type from the two fragments, and from N SALCs form N molecular orbitals.
5. Estimate the relative energies of the molecular orbitals from considerations of overlap and relative energies of the parent orbitals, and draw the levels on a molecular orbital energy-level diagram (showing the origin of the orbitals).
6. Confirm, correct, and revise this qualitative order by carrying out a molecular orbital calculation by using appropriate software.

EXAMPLE 3.12 Constructing molecular orbitals from SALCs

The two SALCs of H1s orbitals in the C_{2v} molecule H$_2$O are $\phi_1 = \psi_{A1s} + \psi_{B1s}$ (18) and $\phi_2 = \psi_{A1s} - \psi_{B1s}$ (19). Which oxygen orbitals can be used to form molecular orbitals with them?

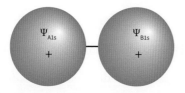

18 $\phi_1 = \psi_{A1s} + \psi_{B1s}$

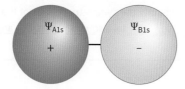

19 $\phi_2 = \psi_{A1s} - \psi_{B1s}$

Answer We start by establishing how the SALCs transform under the symmetry operations of the group (C_{2v}). Under E neither SALC changes sign, so their characters are 1. Under C_2, ψ_1 does not change sign but ψ_2 does; their characters are therefore 1 and −1, respectively. Under σ_v the combination ψ_1 does not change sign but ψ_2 does change sign, so their characters are again +1 and −1, respectively. Under the reflection σ_v' neither SALC changes sign, so their characters are 1. The characters are therefore

	E	C_2	σ_v	σ_v'
ψ_1	1	1	1	1
ψ_2	1	−1	−1	1

We now consult the character table and identify their symmetry labels as A$_1$ and B$_2$, respectively. The same conclusion could have been obtained more directly by referring to Resource section 5. According to the entries on the right of the character table, the O2s and O2p$_z$ orbitals also have A$_1$ symmetry; O2p$_y$ has B$_2$ symmetry. The linear combinations that can be formed are therefore

a$_1$ $\quad \psi = c_1\psi_{O2s} + c_2\psi_{O2p_z} + c_3\phi_1$

b$_2$ $\quad \psi = c_4\psi_{O2p_y} + c_5\phi_2$

The three a$_1$ orbitals are bonding, intermediate, and antibonding in character according to the relative signs of the coefficients c_1, c_2, and c_3, respectively. Similarly, depending on the relative signs of the coefficients c_4 and c_5, one of the two b$_2$ orbitals is bonding and the other is antibonding.

Self-test 3.12 The four SALCs built from Cl3s orbitals in the square-planar (D_{4h}) [PtCl$_4$]$^{2-}$ anion have symmetry species A$_{1g}$, B$_{1g}$, and E$_u$. Which Pt atomic orbitals can combine with which of these SALCs?

3.8 The vibrational analogy

KEY POINT The shapes of SALCs are analogous to stretching displacements.

One of the great strengths of group theory is that it enables disparate phenomena to be treated analogously. We have already seen how symmetry arguments can be applied to molecular vibrations, so it should come as no surprise that SALCs have analogies in the normal modes of molecules. In fact, the illustrations of SALCs in Resource section 5 can be interpreted as contributions to the normal vibrational modes of molecules. The following example illustrates how this is done.

EXAMPLE 3.13 Predicting the IR and Raman bands of an octahedral molecule

Consider an AB$_6$ molecule, such as SF$_6$, that belongs to the O_h point group. Sketch the normal modes of A–B stretches and comment on their activities in IR or Raman spectroscopy.

Answer We argue by analogy with the shapes of SALCs and identify the SALCs that can be constructed from s orbitals in an octahedral arrangement (Resource section 5). These orbitals are the analogues of the stretching displacements of the A–B bonds, and the signs represent their relative phases. The SALCs that can be constructed from s orbitals have the symmetry species A$_{1g}$, E$_g$, and T$_{1u}$. The resulting linear combinations of stretches are illustrated in Fig. 3.26.

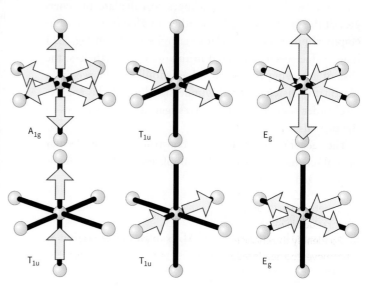

FIGURE 3.26 The A$_{1g}$, E$_g$, and T$_{1u}$ M–L stretching modes of an octahedral ML$_6$ complex. The motion of the central metal atom M, which preserves the centre of mass of the molecule, is not shown (it is stationary in both the A$_{1g}$ and E$_g$ modes).

By inspecting the final column of the character table (Resource section 4) we can see that the A_{1g} (totally symmetric) and E_g modes are Raman active and the T_{1u} mode is IR active. Note that an O_h molecule has a centre of inversion, and we would not expect modes to be active in both the IR and the Raman.

Self-test 3.13 Predict how the IR and Raman spectra of D_{4h} trans-SF_4Cl_2 differ from those of SF_6, considering only bands due to S–F stretching vibrations.

Representations

We now move on to a more quantitative treatment and introduce two topics that are important for applying symmetry arguments in the treatment of molecular orbitals and spectroscopy systematically.

3.9 The reduction of a representation

KEY POINT A reducible representation can be resolved into its constituent irreducible representations by using the reduction formula.

We have seen that the three H1s orbitals of NH_3 give rise to—the technical term is 'span'—two irreducible representations in C_{3v}, one of symmetry species A_1 and the other of symmetry species E. Here we present a systematic way of arriving at the identification of the symmetry species spanned by a set of orbitals or atom displacements.

The fact the three H1s orbitals of NH_3 span two particular irreducible representations is expressed formally by writing $\Gamma = A_1 + E$, where Γ (upper-case gamma) denotes the symmetry species of the reducible representation. In general, we write

$$\Gamma = c_1\Gamma_1 + c_2\Gamma_2 + \cdots \tag{3.1}$$

where the Γ_i denote the various symmetry species of the group and the c_i tell us how many times each symmetry species appears in the reduction. More advanced group theory (see Further Reading) provides an explicit formula for calculating the coefficients c_i in terms of the characters χ_i of the irreducible representation Γ_i and the corresponding characters χ of the original reducible representation Γ:

$$c_i = \frac{1}{h}\sum_C g(C)\chi_i(R)\chi(R) \tag{3.2}$$

Here h is the order of the point group (the number of symmetry operations; it is given in the top row of the character table) and the sum is over each class C of the group, with $g(C)$ the number of elements in that class. How this expression is used is illustrated by the following example.

Many of the modes of cis-$[PdCl_2(NH_3)_2]$ found in Example 3.14 are complex motions that are not easy to visualize: they include Pd–N stretches and various buckling motions of the plane. However, even without being able to visualize them easily, we can infer at once that the A_1, B_1, and B_2 modes are IR active (because the functions x, y, and z, and hence the components of the electric dipole, span these symmetry species) and all the modes are Raman active (because the quadratic forms span all four species), as can be seen from inspection of the right-hand columns of the character tables.

EXAMPLE 3.14 Using the reduction formula

Consider the molecule cis-$[PdCl_2(NH_3)_2]$, which, if we ignore the hydrogen atoms, belongs to the point group C_{2v}. What are the symmetry species spanned by the displacements of the atoms?

Answer To analyse this problem we consider the 15 displacements of the five nonhydrogen atoms, that is five atoms displaced along each of three axes, x, y, and z (Fig. 3.27), and obtain the characters of what will turn out to be a reducible representation Γ by examining what happens when we apply the symmetry operations of the group. Then we use eqn. 3.2 to identify the symmetry species of the irreducible representations into which that reducible representation can be reduced.

To identify the characters of Γ we note that each displacement that moves to a new location under a particular symmetry operation contributes 0 to the character of that operation; those that remain

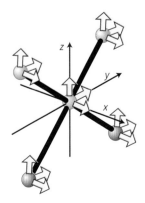

FIGURE 3.27 The atomic displacements in cis-$[PdCl_2(NH_3)_3]$ with the H atoms ignored.

the same contribute 1; those that are reversed contribute −1. The analysis is simplified by considering only the displacements of atoms that have not changed their equilibrium position under the symmetry operation being considered. Thus, the first stage is to determine the number of unchanged atoms for each class of operation, and the second stage is to multiply that number by the characteristic contribution for that operation. Thus, because all five atoms and each of three displacements of those atoms remain unmoved under the identity, $\chi(E) = 5 \times 3 = 15$.

A C_2 rotation leaves only one atom (the Pd) unchanged; it leaves the z displacement of that atom unchanged (contributing 1) and reverses the x and y displacements on that atom (contributing −2), so $\chi(C_2) = 1 \times (1 - 2) = -1$. Under the reflection σ_v again only the Pd is unchanged, with the z and x displacements on the Pd unchanged (contributing 2) and the y displacement being reversed (contributing −1), so $\chi(\sigma_v) = 1 \times (2 - 1) = 1$. Finally, for any reflection in a vertical plane passing through the plane of the atoms, all five atoms remain unmoved; on each of these five atoms, the z displacements remain the same (contributing 1) and so do the y displacements (another 1), but the x displacements are reversed (contributing −1); therefore $\chi(\sigma'_v) = 5 \times (1+1-1) = 5$. The characters of Γ are therefore

E	C_2	σ_v	σ'_v
15	−1	1	5

Now we use eqn. 3.2, noting that $h = 4$ for this group, and noting that $g(C) = 1$ for all C. To find how many times the symmetry species A_1 (1,1,1,1) appears in the reducible representation, we write

$$c_1 = \tfrac{1}{4}[(1 \times 15) + (1 \times (-1)) + (1 \times 1) + (1 \times 5)] = 5$$

So A_1 appears five times, $5A_1$. If we repeat the procedure for A_2 (1,1,−1,−1) we obtain

$$c_2 = \tfrac{1}{4}[(1 \times 15) + (1 \times (-1)) + (-1 \times 1) + (-1 \times 5)] = 2$$

and A_2 appears twice in the reducible representation, $2A_2$. By repeating this procedure for the other species, we find

$$\Gamma = 5A_1 + 2A_2 + 3B_1 + 5B_2$$

For C_{2v} the translations of the entire molecule span $A_1 + B_1 + B_2$ (as given by the functions x, y, and z in the final column in the character table) and the rotations span $A_2 + B_1 + B_2$ (as given by the functions R_x, R_y, and R_z in the final column in the character table). By subtracting these symmetry species from the ones we have just found ($5A_1 + 2A_2 + 3B_1 + 5B_2 − A_1 − B_1 − B_2 − A_2 − B_1 − B_2$), we can conclude that the vibrations of the molecule span $4A_1 + A_2 + B_1 + 3B_2$.

Self-test 3.14 Determine the symmetries of all the vibration modes of $[PdCl_4]^{2-}$, a D_{4h} molecule.

3.10 Projection operators

KEY POINT A projection operator is used to generate SALCs from a basis of orbitals.

To generate an unnormalized SALC of a particular symmetry species from an arbitrary set of basis atomic orbitals, we select any one of the set and form the following sum:

$$\varphi = \sum_R \chi_i(R) R \psi \qquad (3.3)$$

where $\chi_i(R)$ is the character of the operation R for the symmetry species of the SALC we want to generate for the atomic orbitals ψ. Once again, the best way to illustrate the use of this expression is with an example.

EXAMPLE 3.15 Generating a SALC

Generate the SALC of Clσ orbitals for $[PtCl_4]^{2-}$. The basis orbitals are denoted ψ_1, ψ_2, ψ_3, and ψ_4 and are shown in Fig. 3.28(a).

Answer To implement eqn. 3.3 we start with one of the basis orbitals and subject it to all the symmetry operations of the D_{4h} point group, writing down the basis function $R\psi$ into which it is transformed. For example, the operation C_4 moves ψ_1 into the position occupied by ψ_2, C_2 moves it to ψ_3, and C_4^3 moves it to ψ_4. Continuing for all operations we obtain

Operation R:	E	C_4	C_4^3	C_2	C'_2	C'_2	C''_2	C''_2	i	S_4	S_4^3	σ_h	σ_v	σ_v	σ_d	σ_d
$R\psi_1$	ψ_1	ψ_2	ψ_4	ψ_3	ψ_1	ψ_3	ψ_2	ψ_4	ψ_3	ψ_2	ψ_4	ψ_1	ψ_1	ψ_3	ψ_2	ψ_4

We now add together all the new basis functions and find $4\psi_1 + 4\psi_2 + 4\psi_3 + 4\psi_4$, and for each class of operation we multiply by the character $\chi_i(R)$ for the irreducible representation in which we are interested. Thus, for A_{1g} (as all characters are 1) we obtain $4\psi_1 + 4\psi_2 + 4\psi_3 + 4\psi_4$. The (unnormalized) SALC is therefore

$$\phi(A_{1g}) = 4(\psi_1 + \psi_2 + \psi_3 + \psi_4)$$

with the normalized version being

$$\phi(A_{1g}) = \tfrac{1}{2}(\psi_1 + \psi_2 + \psi_3 + \psi_4)$$

As we continue down the character table using the various symmetry species, the SALCs emerge as follows:

$$\phi(B_{1g}) = \tfrac{1}{2}(\psi_1 - \psi_2 + \psi_3 - \psi_4)$$
$$\phi(E_u) = \tfrac{1}{\sqrt{2}}(\psi_1 - \psi_3)$$

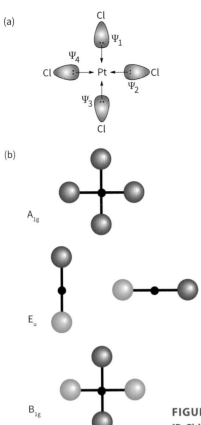

Under all other irreducible representations the projection operators vanish (thus no SALCs exist of those symmetries). We then continue by using ψ_2 as our basis function, whereupon we obtain the same SALCs except for

$$\phi(B_{1g}) = \tfrac{1}{2}(\psi_2 - \psi_1 + \psi_4 - \psi_3)$$

$$\phi(E_u) = \tfrac{1}{\sqrt{2}}(\psi_2 - \psi_4)$$

Completing the process with ψ_3 and ψ_4 gives similar SALCs (only the signs of some of the component orbitals change). The forms of the SALCs are therefore $A_{1g} + B_{1g} + E_u$ (Fig. 3.28(b)).

Self-test 3.15 Use projection operators in SF_6 to determine the SALCs for σ bonding in an octahedral complex.

FIGURE 3.28 (a) The Cl orbital basis used to construct SALCs in $[PtCl_4]^{2-}$, and (b) the SALCs constructed for $[PtCl_4]^{2-}$.

Using an analysis like the one above, it is possible to construct SALCs for any molecule we wish to consider. Resource section 5 contains diagrammatic representations of SALCs for the most useful point groups, including those necessary for both σ and π bonding interactions.

3.11 Polyatomic molecules

In Chapter 2 we discussed the application of molecular orbital theory to simple homonuclear and heteronuclear diatomic molecules. The symmetry of the contributing atomic orbitals is more important for polyatomic molecules, so it is appropriate for us to consider polyatomics here. We saw in Section 2.8 that the general structure of molecular orbital energy-level diagrams can be derived by grouping the orbitals into different sets, the σ and π orbitals, according to their shapes. The same procedure is used in the discussion of the molecular orbitals of polyatomic molecules. However, because their shapes are more complex than diatomic molecules, we need a more powerful approach.

Molecular orbital theory can be used to discuss in a uniform manner the electronic structures of triatomic molecules, finite groups of atoms, and the almost infinite arrays of atoms in solids. In each case the molecular orbitals resemble those of diatomic molecules, the only important difference being that the orbitals are built from a more extensive basis set of atomic orbitals. A key point to bear in mind is that from N atomic orbitals it is possible to construct N molecular orbitals.

The photoelectron spectrum of NH_3 (Fig. 3.29) indicates some of the features that a theory of the structure of polyatomic molecules must elucidate. The spectrum shows two bands. The one with the lower ionization

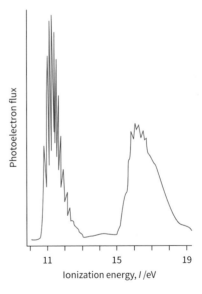

FIGURE 3.29 The UV photoelectron spectrum of NH_3, obtained using He 21 eV radiation.

energy (in the region of 11 eV) has considerable vibrational structure, an indication (see later) that the orbital from which the electron is ejected plays a considerable role in determining the molecule's shape. The broad band in the region of 16 eV arises from electrons that are bound more tightly.

(a) Polyatomic molecular orbitals

KEY POINTS Molecular orbitals are formed from linear combinations of atomic orbitals of the same symmetry; their energies can be determined experimentally from gas-phase photoelectron spectra and interpreted in terms of the pattern of orbital overlap.

The features that were discussed in Chapter 2 in connection with diatomic molecules are present in all polyatomic molecules. In each case, we write the molecular orbital of a given symmetry (such as the σ orbitals of a linear molecule) as a sum of *all* the atomic orbitals that can overlap to form orbitals of that symmetry:

$$\psi = \sum_i c_i \chi_i \tag{3.4}$$

In this linear combination, the χ_i are atomic orbitals (usually the valence orbitals of each atom in the molecule) and the index i runs over all the atomic orbitals that have the appropriate symmetry. From N atomic orbitals we can construct N molecular orbitals. Then,

- The greater the number of nodes in a molecular orbital, the greater the antibonding character and the higher the orbital energy.
- Orbitals constructed from lower-energy atomic orbitals lie lower in energy (so atomic s orbitals typically produce lower-energy molecular orbitals than atomic p orbitals of the same shell).
- Interactions between nonnearest-neighbour atoms are weakly bonding (lower the energy slightly) if the orbital lobes on these atoms have the same sign (and interfere constructively). They are weakly antibonding if the signs are opposite (and interfere destructively).

A BRIEF ILLUSTRATION

To account for the features in the photoelectron spectrum of NH_3, we need to build molecular orbitals that will accommodate the eight valence electrons in the molecule. Each molecular orbital is a combination of seven atomic orbitals: the three H1s orbitals, the N2s orbital, and the three N2p orbitals. It is possible to construct seven molecular orbitals from these seven atomic orbitals (Fig. 3.30).

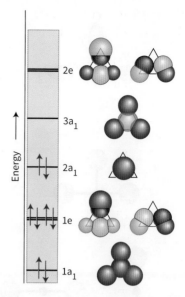

FIGURE 3.30 A schematic illustration of the molecular orbitals of NH_3 with the size of the atomic orbital indicating the magnitude of its contribution to the molecular orbital. The view is along the z-axis.

As was seen in Section 3.7, the molecular orbitals of NH_3 are constructed from SALCs and atomic orbitals of the same symmetry. The molecular orbitals constructed in this way are labelled according to their symmetry in the following way:

- a, b orbitals have A or B symmetry and are nondegenerate
- e orbitals have E symmetry and are doubly degenerate (two orbitals of the same energy)
- t orbitals have T symmetry and are triply degenerate (three orbitals of the same energy).

As we have already seen, subscripts and superscripts are sometimes added to these letters, as in a_1, b_2, e_g, and t_2 because it is sometimes necessary to distinguish different a, b, e, and t orbitals according to a more detailed analysis of their symmetries.

The formal rules for the construction of the orbitals were described in Section 3.7, but it is possible to obtain a sense of their origin by imagining viewing the NH_3 molecule along its three-fold axis (designated z). The N2p_z and N2s orbitals both have cylindrical symmetry about that axis. If the three H1s orbitals are superimposed with the same sign relative to each other (i.e. so that all have the same size and tint in the diagram, Fig. 3.30), then they match this cylindrical symmetry. It follows that we can form molecular orbitals of the form

$$\psi = c_1 \chi_{N2s} + c_2 \chi_{N2p_z} + c_3 [\chi_{H1sA} + \chi_{H1sB} + \chi_{H1sC}] \tag{3.5}$$

From these *three* basis orbitals (the specific combination of H1s orbitals counts as a single 'symmetry adapted' basis orbital), it is possible to construct three molecular orbitals (with different values of the coefficients c). The orbital with no nodes between the N and H atoms is the lowest in energy, that with a node between all the NH neighbours is the highest in energy, and the third orbital lies between the two. The three orbitals are nondegenerate and are labelled $1a_1$, $2a_1$, and $3a_1$ in order of increasing energy.

The $N2p_x$ and $N2p_y$ orbitals have π symmetry with respect to the z-axis, and can be used to form orbitals with combinations of the H1s orbitals that have a matching symmetry. For example, one such superposition will have the form

$$\psi = c_1\chi_{N2p_x} + c_2[\chi_{H1sA} + \chi_{H1sB}] \quad (3.6)$$

As can be seen from Fig. 3.30, the signs of this H1s orbital combination match those of the $N2p_x$ orbital. The N2s orbital cannot contribute to this superposition, so only *two* combinations can be formed—one without a node between the N and H orbitals, and the other with a node. The two orbitals differ in energy, the former being lower. A similar combination of orbitals can be formed with the $N2p_y$ orbital, and it turns out by symmetry arguments that the two orbitals are degenerate with the two we have just described. The combinations are examples of e orbitals (because they form doubly degenerate pairs), and are labelled 1e and 2e in order of increasing energy.

The general form of the molecular orbital energy-level diagram is shown in Fig. 3.31. The actual location of the orbitals (particularly the relative positions of the a and the e sets) can be found only by detailed computation or by identifying the orbitals responsible for the photoelectron spectrum. We have indicated the probable assignment of the 11 eV and 16 eV peaks, which fixes the locations of two

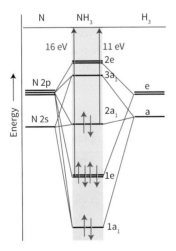

FIGURE 3.31 The molecular orbital energy-level diagram for NH_3 when the molecule has the observed bond angle (107°) and bond length; excitation energies are also shown.

of the occupied orbitals. The third occupied orbital is out of range of the 21 eV radiation used to obtain the spectrum.

The photoelectron spectrum is consistent with the need to accommodate eight electrons in the orbitals. The electrons enter the molecular orbitals in increasing order of energy, starting with the orbital of lowest energy, and taking note of the requirement of the exclusion principle that no more than two electrons can occupy any one orbital. The first two electrons enter $1a_1$ and fill it. The next four enter the doubly degenerate 1e orbitals and fill them. The last two enter the $2a_1$ orbital, which calculations show is almost nonbonding and localized on the N atom. The resulting overall ground-state electron configuration is therefore $1a_1^2 1e^4 2a_1^2$. No antibonding orbitals are occupied, so the molecule has a lower energy than the separated atoms.

The conventional description of NH_3 as a molecule with a lone pair is also mirrored in the configuration: the HOMO is $2a_1$, which is largely confined to the N atom and makes only a small contribution to the bonding. We saw in Section 2.3 that lone-pair electrons play a considerable role in determining the shapes of molecules. The extensive vibrational structure in the 11 eV band of the photoelectron spectrum is consistent with this observation, as photoejection of a $2a_1$ electron removes the spatial influence of the lone pair and the shape of the ionized molecule is considerably different from that of NH_3 itself. Photoionization therefore results in extensive vibrational structure in the spectrum.

(b) Hypervalence in the context of molecular orbitals

KEY POINT The delocalization of molecular orbitals means that an electron pair can contribute to the bonding of more than two atoms.

In Section 2.6 we used valence bond theory to explain hypervalence by using d orbitals to allow the valence shell of an atom to accommodate more than eight electrons. Molecular orbital theory explains it rather more elegantly.

We consider SF_6, which has six S-F bonds and hence 12 electrons involved in forming bonds and is therefore hypervalent. The simple basis set of atomic orbitals that are used to construct the molecular orbitals consists of the valence shell s and p orbitals of the sulfur atom and one p orbital of each of the six F atoms and pointing towards the sulfur atom. We use the F2p orbitals rather than the F2s orbitals because they match the sulfur orbitals more closely in energy. From these 10 atomic orbitals it is possible to construct 10 molecular orbitals. Calculations indicate that four of the orbitals are bonding and four are antibonding; the two remaining orbitals are nonbonding (Fig. 3.32).

There are 12 electrons to accommodate. The first two enter $1a_1$ and the next six enter $1t_1$. The remaining four fill

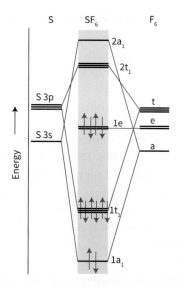

FIGURE 3.32 A schematic molecular orbital energy-level diagram for SF$_6$.

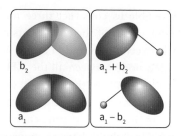

FIGURE 3.33 The two occupied 1a$_1$ and 1b$_2$ orbitals of the H$_2$O molecule and their sum 1a$_1$+1b$_2$ and difference 1a$_1$−1b$_2$. In each case we form an almost fully localized orbital between a pair of atoms.

the nonbonding pair of orbitals, resulting in the configuration $1a_1^2 1t_1^6 1e^4$. As we see, none of the antibonding orbitals (2a$_1$ and 2t$_1$) is occupied. Molecular orbital theory, therefore, accounts for the formation of SF$_6$, with four bonding orbitals and two nonbonding orbitals occupied, and does not need to invoke S3d orbitals and octet expansion required for the valence bond approach. This does not mean that d orbitals cannot participate in the bonding, but it does show that they are not *necessary* for bonding six F atoms to the central sulfur atom. The limitation of valence bond theory is the assumption that each atomic orbital on the central atom can participate in the formation of only one bond. Molecular orbital theory takes hypervalence in its stride by having available plenty of orbitals, not all of which are antibonding. Therefore, the question of when hypervalence can occur appears to depend on factors other than d-orbital availability, such as the ability of small atoms to pack around a large atom.

(c) Localization

KEY POINT Localized and delocalized descriptions of bonds are mathematically equivalent, but one description may be more suitable for a particular property, as summarized in Table 3.7.

A striking feature of the VB approach to chemical bonding is its accord with chemical instinct, as it identifies something that can be called 'an A-B bond'. Both OH bonds in H$_2$O, for instance, are treated as localized, equivalent structures because each one consists of an electron pair shared between O and H. This feature appears to be absent from MO theory because molecular orbitals are delocalized and the electrons that occupy them bind all the atoms together, not just a specific pair of neighbouring atoms. The concept of an A-B bond as existing independently of other bonds in the molecule, and of being transferable from one molecule to another, seems to have been lost. However, we shall now show that the molecular orbital description is mathematically almost equivalent to the overall electron distribution as described by individual bonds. The demonstration hinges on the fact that linear combinations of molecular orbitals can be formed that result in the same overall electron distribution, but the individual orbitals are distinctly different.

Consider the H$_2$O molecule. The two occupied bonding orbitals of the delocalized description, 1a$_1$ and 1b$_2$, are shown in Fig. 3.33. If we form the sum 1a$_1$ + 1b$_2$, the negative half of 1b$_2$ cancels half the 1a$_1$ orbital almost completely, leaving a localized orbital between O and the other H. Likewise, when we form the difference 1a$_1$ − 1b$_2$, the other half of the 1a$_1$ orbital is cancelled almost completely, so leaving a localized orbital between the other pair of atoms. Therefore, by taking sums and differences of delocalized orbitals, localized orbitals are created (and vice versa). Because these are two equivalent ways of describing the same overall electron population, one description cannot be said to be better than the other.

Table 3.7 suggests when it is appropriate to select a delocalized description or a localized description. In general,

TABLE 3.7 A general indication of the properties for which localized and delocalized descriptions are appropriate

Localized appropriate	Delocalized appropriate
Bond strengths	Electronic spectra
Force constants	Photoionization
Bond lengths	Electron attachment
Brønsted acidity*	Magnetism
VSEPR description	Standard potentials†

* Chapter 5.
† Chapter 6.

a delocalized description is needed for dealing with global properties of the entire molecule. Such properties include electronic spectra (UV and visible transitions, Section 8.3), photoionization spectra, ionization and electron gain enthalpies (Section 1.7), and reduction potentials (Section 6.2). In contrast, a localized description is most appropriate for dealing with properties of a fragment of a total molecule. Such properties include bond strength, bond length, bond force constant, and some aspects of reactions (such as acid–base character): in these aspects the localized description is more appropriate because it focuses attention on the distribution of electrons in and around a particular bond.

(d) Localized bonds and hybridization

KEY POINT Hybrid atomic orbitals are sometimes used in the discussion of localized molecular orbitals.

The localized molecular orbital description of bonding can be taken a stage further by invoking the concept of hybridization. Strictly speaking, hybridization belongs to VB theory, but it is commonly invoked in simple qualitative descriptions of molecular orbitals.

We have seen that in general a molecular orbital is constructed from all atomic orbitals of the appropriate symmetry. However, it is sometimes convenient to form a mixture of orbitals on one atom (the O atom in H_2O, for instance), and then to use these hybrid orbitals to construct localized molecular orbitals. In H_2O, for instance, each OH bond can be regarded as formed by the overlap of an H1s orbital and a hybrid orbital composed of O2s and O2p orbitals (Fig. 3.34).

We have already seen that the mixing of s and p orbitals on a given atom results in hybrid orbitals that have a definite direction in space, as in the formation of tetrahedral hybrids. Once the hybrid orbitals have been selected, a localized molecular orbital description can be constructed. For example, four bonds in CF_4 can be formed by building bonding and antibonding localized orbitals by overlap of each hybrid and one F2p orbital directed towards it. Similarly, to describe the electron distribution of BF_3, we could consider each localized BF σ orbital as formed by the overlap of an sp^2 hybrid with an F2p orbital. A localized orbital description of a PCl_5 molecule would be in terms of five P–Cl σ bonds formed by overlap of each of the five trigonal-bipyramidal sp^3d hybrid orbitals with a 2p orbital of a Cl atom. Similarly, where we wanted to form six localized orbitals in a regular octahedral arrangement (e.g. in SF_6), we would need two d-orbitals: the resulting six sp^3d^2 hybrids point in the required directions.

(e) Electron deficiency

KEY POINT The existence of electron-deficient species is explained by the delocalization of the bonding influence of electrons over several atoms.

The VB model of bonding fails to account for the existence of **electron-deficient compounds**, which are compounds for which, according to Lewis's approach, there are not enough electrons to form the required number of bonds. This point can be illustrated most easily with diborane, B_2H_6 (20). There are only 12 valence electrons but, according to Lewis's approach, at least eight electron pairs (16 electrons) are needed to bind eight atoms together.

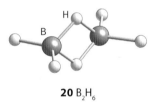

20 B_2H_6

The formation of molecular orbitals by combining several atomic orbitals accounts effortlessly for the existence of these compounds. The eight atoms of this molecule contribute a total of 14 valence orbitals (three p and one s orbital from each B atom, making eight, and one s orbital each from the six H atoms). These 14 atomic orbitals can be used to construct 14 molecular orbitals. About seven of these molecular orbitals will be bonding or nonbonding, which is more than enough to accommodate the 12 valence electrons provided by the atoms.

The bonding can be best understood if we consider that the MOs produced are associated with either the terminal BH fragments or with the bridging BHB fragments. The localized MOs associated with the terminal BH bonds are constructed simply from atomic orbitals on two atoms (the H1s and a $B2s2p^n$ hybrid). The molecular orbitals associated with the two BHB fragments are linear

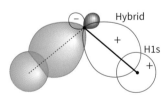

FIGURE 3.34 The formation of localized O–H orbitals in H_2O by the overlap of hybrid orbitals on the O atom and H1s orbitals. The hybrid orbitals are a close approximation to the sp^3 hybrids shown in Fig. 2.7.

combinations of the B2s2p^n hybrids on each of the two B atoms and an H1s orbital of the H atom lying between them (Fig. 3.35). Three molecular orbitals are formed from these three atomic orbitals: one is bonding, one is nonbonding, and the third is antibonding. The bonding orbital can accommodate two electrons and hold the BHB fragment together. The same remark applies to the second BHB fragment, and the two occupied 'bridging' bonding molecular orbitals hold the molecule together. Thus, overall, 12 electrons account for the stability of the molecule because their influence is spread over more than six pairs of atoms.

Electron deficiency is well developed not only in boron (where it was first clearly recognized) but also in carbocations and a variety of other classes of compounds that we encounter later in the text.

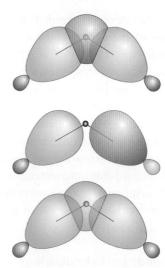

FIGURE 3.35 The molecular orbital formed between two B atoms and one H atom lying between them, as in B_2H_6. Two electrons occupy the bonding combination and hold all three atoms together.

FURTHER READING

P. Atkins, J. de Paula, and J. Keeler, *Atkins' Physical Chemistry*. Oxford University Press (2022). An account of the generation and use of character tables without too much mathematical background.

For more rigorous introductions, see:

J. S. Ogden, *Introduction to Molecular Symmetry*. Oxford University Press (2001).

P. Atkins and R. Friedman, *Molecular Quantum Mechanics*. Oxford University Press (2010).

M. T. Weller and N. A. Young, *Characterisation Methods in Inorganic Chemistry*. Oxford University Press (2017).

EXERCISES

3.1 Use the VSEPR model to predict the structures of the following molecules or ions:

BrF_4^- $TeCl_4$ I_2Cl_6 (contains no I–I bonds) ClO_3^-

In each case show the principal rotation axis and indicate any horizontal or vertical mirror planes. Which of the molecules have an inversion centre?

3.2 Draw sketches to identify the following symmetry elements: (a) a C_3 axis and a σ_v plane in the NH_3 molecule, (b) a C_4 axis and a σ_h plane in the square-planar $[PtCl_4]^{2-}$ ion.

3.3 Which of the following molecules and ions has (a) a centre of inversion, (b) an S_4 axis: (i) CO_2, (ii) C_2H_2, (iii) BF_3, (iv) SO_4^{2-}?

3.4 Determine the symmetry elements and assign the point group of (a) NH_2Cl, (b) CO_3^{2-}, (c) SiF_4, (d) HCN, (e) SiFClBrI, (f) BF_4^-.

3.5 How many planes of symmetry does a benzene molecule possess? Which chloro-substituted benzene of formula $C_6H_nCl_{6-n}$ has exactly four planes of symmetry?

3.6 Determine the symmetry elements of objects with the same shape as the boundary surface of (a) a p orbital, (b) a d_{xy} orbital, (c) a d_{z^2} orbital.

3.7 (a) Determine the point group of an SO_3^{2-} ion. (b) What is the maximum degeneracy of a molecular orbital in this ion? (c) If the sulfur orbitals are 3s and 3p, which of them can contribute to molecular orbitals of this maximum degeneracy?

3.8 (a) Determine the point group of the PF_5 molecule. (Use VSEPR, if necessary, to assign geometry.) (b) What is the maximum degeneracy of its molecular orbitals? (c) Which P3p orbitals contribute to a molecular orbital of this degeneracy?

3.9 Consider the displacements of the atoms in the trigonal-bipyramidal molecule PF_5. Calculate the number and symmetry of its vibrational modes.

3.10 How many vibrational modes does an SO_3 molecule have (a) in the plane of the nuclei, (b) perpendicular to the molecular plane?

3.11 What are the symmetry species of the vibrations of (a) SF_6, (b) BF_3 that are both IR and Raman active?

3.12 Consider CH_4. Use the projection operator method to construct the SALCs of $A_1 + T_2$ symmetry that derive from the four H1s orbitals. With which atomic orbitals on C would it be possible to form MOs with the H1s SALCs?

3.13 Use the projection operator method to determine the SALCs required for formation of σ bonds in (a) BF_3, (b) PF_5.

3.14 (a) Based on the MO discussion of NH_3 in the text, find the average NH bond order in NH_3 by calculating the net number of bonds and dividing by the number of NH groups.

3.15 From the relative atomic orbital and molecular orbital energies depicted in Fig. 3.32, describe the character as mainly F or mainly S for the frontier orbitals e (the HOMO) and 2t (the LUMO) in SF_6. Explain your reasoning.

3.16 Draw the linear combinations of H1s orbitals for linear and square forms of the hypothetical H_4 molecule and construct the respective MO diagrams to demonstrate that one of these forms should be paramagnetic (contain unpaired electrons).

3.17 (a) Construct the form of each molecular orbital in linear $[HHeH]^{2+}$ using 1s basis atomic orbitals on each atom and considering successive nodal surfaces. (b) Arrange the MOs in increasing energy. (c) Indicate the electron population of the MOs. (d) Should $[HHeH]^{2+}$ be stable in isolation or in solution? Explain your reasoning.

TUTORIAL PROBLEMS

3.1 Consider a molecule IF_3O_2 (with I as the central atom). How many isomers are possible? Assign point group designations to each isomer.

3.2 How many isomers are there for 'octahedral' molecules with the formula MA_3B_3, where A and B are monoatomic ligands? What is the point group of each isomer? Are any of the isomers chiral? Repeat this exercise for molecules with the formula $MA_2B_2C_2$.

3.3 Group theory is often used by chemists as an aid in the interpretation of IR spectra. For example, there are four NH bonds in NH_4^+ and four stretching modes are possible. There is the possibility that several vibrational modes occur at the same frequency, and hence are degenerate. A quick glance at the character table will tell if degeneracy is possible. (a) In the case of the tetrahedral NH_4^+ ion, is it necessary to consider the possibility of degeneracies? (b) Are degeneracies possible in any of the vibrational modes of $NH_2D_2^+$?

3.4 Determine whether the number of IR and Raman active stretching modes could be used to determine uniquely whether a sample of gas is BF_3, NF_3, or ClF_3.

3.5 Reaction of $AsCl_3$ with Cl_2 at low temperature yields a product, believed to be $AsCl_5$, which shows Raman bands at 437, 369, 295, 220, 213, and 83 cm^{-1}. Detailed analysis of the 369 and 295 cm^{-1} bands shows them to arise from totally symmetric modes. Show that the Raman spectrum is consistent with a trigonal-bipyramidal geometry.

3.6 Show how you would use Raman and infrared spectroscopy to distinguish between regular octahedral and regular trigonal prismatic geometries of a six-coordinate species, ML_6. Discuss the possible distortions that could occur in either case (you do not need to determine the symmetries of vibrations for the distorted states).

3.7 Consider the p orbitals on the four Cl atoms of tetrahedral $[CoCl_4]^{2-}$, with one p orbital on each Cl pointing directly at the central metal atom. (a) Confirm that the four p orbitals which point at the metal transform in an identical manner to the four s orbitals on the Cl atoms. How might these p orbitals contribute to the bonding of the complex? (b) Take the remaining eight p orbitals and determine how they transform. Reduce the representation you derive to determine the symmetry of the SALCs these orbitals contribute to. Which metal orbitals can these SALCs bond with? (c) Generate the SALCs referred to in (b).

3.8 Consider all 12 of the p orbitals on the four Cl atoms of a square-planar complex like $[PtCl_4]^{2-}$. (a) Determine how these p orbitals transform under D_{4h} and reduce the representation. (b) Which metal orbitals can these SALCs bond with? (c) Which SALCs and which metal orbitals contribute to σ bonds? (d) Which SALCs and which metal orbitals contribute to the in-plane π bonds? (e) Which SALCs and which metal orbitals contribute to the out-of-plane π bonds?

3.9 Consider an octahedral metal complex and construct all the σ- and π-bonding SALCs.

3.10 Construct an approximate molecular orbital energy diagram for a hypothetical planar form of NH_3. You may refer to Resource sections 4 and 5 to determine the form of the appropriate orbitals on the central N atom and on the triangle of H_3 atoms. From a consideration of the atomic energy levels, place the N and H_3 orbitals on either side of a molecular orbital energy-level diagram. Then use your judgement about the effect of bonding and antibonding interactions and energies of the parent orbitals to construct the molecular orbital energy levels in the centre of your diagram, and draw lines indicating the contributions of the atomic orbitals to each molecular orbital. Ionization energies are $I(H1s) = 13.6$ eV, $I(N2s) = 26.0$ eV, and $I(N2p) = 13.4$ eV.

4 The structures and energetics of simple solids

The description of the structures of solids

4.1 Unit cells and the description of crystal structures
4.2 The close packing of spheres
4.3 Holes in close-packed structures

The structures of metals and simple alloys

4.4 Polytypism
4.5 Nonclose-packed structures
4.6 Polymorphism of metals
4.7 Metallic radii of metals
4.8 The structures of simple alloys and interstitials

Ionic solids

4.9 Characteristic structures of ionic solids
4.10 The rationalization of structures

The energetics of ionic bonding

4.11 Lattice enthalpy and the Born–Haber cycle
4.12 The calculation of lattice enthalpies
4.13 Comparison of experimental and theoretical values
4.14 The Kapustinskii equation
4.15 Consequences of lattice enthalpies

Further information: the Born–Mayer equation

Further reading

Exercises

Tutorial problems

An understanding of the chemistry of compounds in the solid state is central to the study of many important inorganic materials, including alloys, metal salts, graphene, rechargeable battery components, photovoltaic materials, nanomaterials, zeolites, and high-temperature superconductors. This chapter surveys the structures adopted by atoms and ions in simple solids and explores why one arrangement may be preferred to another. We begin with the simplest model, in which atoms are represented by hard spheres and the structure of the solid is the outcome of stacking these spheres tightly together. This 'close-packed' arrangement provides a good description of many metals and alloys and is a useful starting point for the discussion of numerous ionic solids. These simple solid structures can then be considered as the building blocks for the construction of more complex inorganic materials, including their nanoparticulate forms. Introduction of partial covalent character into the bonding influences the choice of structure and thus trends in the adopted structural type correlate with the electronegativities of the constituent atoms. The chapter also describes some of the energy considerations that can be used to rationalize the trends in structure and reactivity. These arguments also systematize the discussion of the thermal stabilities and solubilities of ionic solids formed by the elements of Groups 1 and 2.

The majority of elements and inorganic compounds exist as solids and comprise ordered arrays of atoms, ions, or molecules. The structures of most metals can be described in terms of a regular, space-filling arrangement of the metal atoms. These metal centres interact through **metallic bonding**, where the electrons are delocalized throughout the solid; that is, the electrons are not associated with a particular atom or bond. One model that can be used to describe a metal is an enormous molecule with a multitude of atomic orbitals which overlap to produce molecular orbitals that extend throughout the sample (Section 2.11).

Metallic bonding is characteristic of elements with low ionization energies, such as those on the left of the periodic table, through the d block and into part of the p block close to the d block. Most of the elements are metals, but metallic bonding also occurs in many other solids; for example, the oxides and sulfides of some of the d metals. The lustrous-red rhenium oxide, ReO_3, and shiny 'fool's gold' (iron pyrites, FeS_2), illustrate the occurrence of metallic bonding, and delocalized electrons, in compounds. The familiar properties of an elemental metal stem from the characteristics of its bonding and in particular the delocalization of electrons throughout the solid. Thus, metals are malleable (easily deformed by the application of pressure) and ductile (able to be drawn into a wire) because the electrons can adjust rapidly to any relocation of the metal atom nuclei caused by an external force and there is no directionality in the bonding. They are lustrous because the electrons can respond almost freely to an incident wave of electromagnetic radiation and reflect it.

In **ionic bonding**, ions of different elements are held together in rigid, symmetrical arrays as a result of the attraction between their opposite charges. As ionic bonding depends upon electrostatic interactions between oppositely charged ions it is commonly found in compounds between cationic metal ions and the anions of electronegative elements. There are many other compound types that exhibit ionic bonding; for example, some compounds of nonmetals (such as ammonium nitrate) contain features of ionic bonding, between the molecular ions, $[NO_3]^-$ and $[NH_4]^+$, as well as covalent interactions, within the molecular ions (the N–O and N–H bonds). There are also materials that exhibit features of both ionic and metallic bonding.

Both ionic and metallic bonding are nondirectional, so structures where these types of bonding occur are most easily understood in terms of space-filling models that maximize, for example, the number and strength of the electrostatic interactions between the ions. The regular arrays of atoms, ions, or molecules in solids that produce these structures are best represented using a repeating unit that is produced as a result of the efficient methods of filling space, known as the unit cell. In bulk inorganic solids these units will repeat thousands of times, typically forming crystallites or crystals of a micron (10^{-3} cm) or more in size; in nanoparticles (Chapter 28), with dimensions of only 10–100 nm, these units may repeat only a few times.

The description of the structures of solids

The arrangement of atoms or ions in simple solid structures can often be represented by different arrangements of hard spheres. The spheres used to describe metallic solids represent neutral atoms because each cation can still be considered as surrounded by its full complement of electrons. The spheres used to describe ionic solids represent the cations and anions.

4.1 Unit cells and the description of crystal structures

A crystal of an element or compound can be regarded as constructed from regularly repeating structural elements, which may be atoms, molecules, or ions. The 'crystal lattice' is the geometric pattern formed by the points that represent the positions of these repeating structural elements.

(a) Lattices and unit cells

KEY POINTS The lattice defines a network of identical points that has the translational symmetry of a structure. A unit cell is a subdivision of a crystal that when stacked together following translations reproduces the crystal.

A **lattice** is a three-dimensional, infinite array of points, the **lattice points**, each of which is surrounded in an identical way by neighbouring points. The lattice defines the repeating nature of the crystal. The **crystal structure** itself is obtained by associating one or more identical structural units, such as atoms, ions, or molecules, with each lattice point. In many cases the structural unit may be centred on the lattice point, but that is not necessary.

A **unit cell** of a three-dimensional crystal is an imaginary parallel-sided region (a 'parallelepiped') from which the entire crystal can be built up by purely translational displacements;[1] unit cells so generated fit perfectly together with no space excluded. Unit cells may be chosen in a variety of ways but it is generally preferable to choose the smallest cell that exhibits the greatest symmetry. Thus, in the **two-dimensional pattern** in Fig. 4.1, a variety of unit cells (a parallelogram in two dimensions) may be chosen, each of which repeats the contents of the box under translational displacements. Two possible choices of repeating unit are shown, but (b) would be preferred to (a) because it is smaller.

The relationship between the lattice parameters in three dimensions as a result of the symmetry of the structure gives rise to the seven **crystal systems** (Table 4.1 and Fig. 4.2). All ordered structures adopted by compounds belong to one of these crystal systems; most of those described in this chapter, which deals with simple compositions and stoichiometries,

---

[1] A translation exists where it is possible to move an original figure or motif in a defined direction by a certain distance to produce an exact image. In this case a unit cell reproduces itself exactly by translation parallel to a unit cell edge by a distance equal to the unit cell parameter.

4 The structures and energetics of simple solids

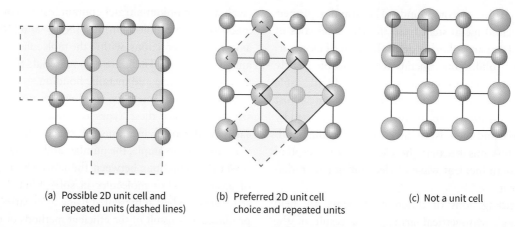

(a) Possible 2D unit cell and repeated units (dashed lines)

(b) Preferred 2D unit cell choice and repeated units

(c) Not a unit cell

FIGURE 4.1 A two-dimensional solid and two choices of a unit cell. The entire 2D crystal is produced by translational displacements of either unit cell, but (b) is generally preferred to (a) because it is smaller. Note that this is a two-dimensional unit cell, and expanding this description and choice of unit cell to three-dimensions would require all layers to sit directly above each other.

TABLE 4.1 The seven crystal systems

System	Relationships between lattice parameters	Unit cell defined by	Essential symmetries
Triclinic	$a \neq b \neq c, \alpha \neq \beta \neq \gamma \neq 90°$	$a\,b\,c\,\alpha\,\beta\,\gamma$	None
Monoclinic	$a \neq b \neq c, \alpha = \gamma = 90°, \beta \neq 90°$	$a\,b\,c\,\beta$	One two-fold rotation axis and/or a mirror plane
Orthorhombic	$a \neq b \neq c, \alpha = \beta = \gamma = 90°$	$a\,b\,c$	Three perpendicular two-fold axes and/or mirror planes
Rhombohedral	$a = b = c, \alpha = \beta = \gamma \neq 90°$	$a\,\alpha$	One three-fold rotation axis
Tetragonal	$a = b \neq c, \alpha = \beta = \gamma = 90°$	$a\,c$	One four-fold rotation axis
Hexagonal	$a = b \neq c, \alpha = \beta = 90°, \gamma = 120°$	$a\,c$	One six-fold rotation axis
Cubic	$a = b = c, \alpha = \beta = \gamma = 90°$	a	Four three-fold rotation axes tetrahedrally arranged

belong to the higher symmetry cubic and hexagonal systems. The angles (α, β, γ) and lengths (a, b, c) used to define the size and shape of a unit cell, relative to an origin, are the **unit cell parameters** (the 'lattice parameters'); the angle between a and b is denoted γ, that between b and c is α, and that between a and c is β; a general triclinic unit cell is illustrated in Fig. 4.2.

A **primitive** unit cell (denoted by the symbol P) has just one lattice point in the unit cell (Fig. 4.3), and the translational symmetry present is just that on the repeating unit cell.

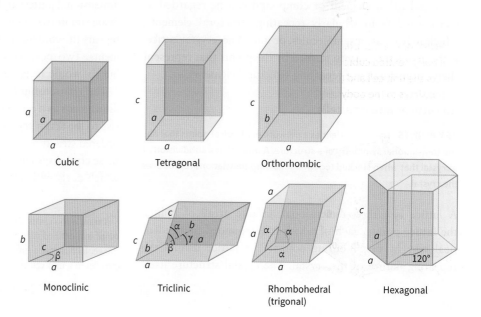

FIGURE 4.2 The seven crystal systems.

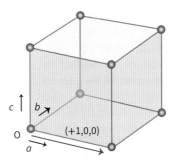

FIGURE 4.3 Lattice points describing the translational symmetry of a primitive cubic unit cell. The translational symmetry is just that of the unit cell; for example, the α lattice point at the origin, O, translates by (+1,0,0) to another corner of the unit cell.

More complex lattice types are **body-centred** (I, from the German word *innenzentriert*, referring to the lattice point at the unit cell centre) and **face-centred** (F) with two and four lattice points in each unit cell, respectively, and additional translational symmetry beyond that of the unit cell (Figs. 4.4 and 4.5(a)). For orthorhombic unit cells it is possible to have just have face-centring of one of the cell faces with the additional translational symmetry $(+\frac{1}{2},+\frac{1}{2},0)$, Fig. 4.5(b), which is normally denoted **C-centred**.

The additional translational symmetry in the **body-centred cubic** (bcc) lattice, equivalent to the displacement $(+\frac{1}{2},+\frac{1}{2},+\frac{1}{2})$ from the unit cell origin at (0,0,0), produces a lattice point at the unit cell centre; note that the surroundings of each lattice point are identical, consisting of eight other lattice points at the corners of a cube. Centred lattices are usually preferred to primitive (although it is always possible to use a primitive lattice for any structure), for with them the full structural symmetry of the cell is more apparent.

We use the following rules to work out the number of lattice points in a three-dimensional unit cell. The same process can be used to count the number of atoms, ions, or molecules that the unit cell contains (Section 4.9).

- A lattice point within the body of a cell belongs entirely to that cell and counts as 1.
- A lattice point on a face is shared by two cells and contributes ½ to the cell.
- A lattice point on an edge is shared by four cells and hence contributes ¼.
- A lattice point at a corner is shared by eight cells that share the corner, and so contributes ⅛.

Thus, for the face-centred cubic lattice depicted in Fig. 4.5 the total number of lattice points in the unit cell is $(8 \times \frac{1}{8}) + (6 \times \frac{1}{2}) = 4$. For the body-centred cubic lattice depicted in Fig. 4.4, the number of lattice points is $(1 \times 1) + (8 \times \frac{1}{8}) = 2$.

(b) Fractional atomic coordinates and projections

KEY POINT Structures may be drawn in projection, with atom positions denoted by fractional coordinates.

The position of an atom in a unit cell is normally described in terms of **fractional coordinates**—coordinates expressed as a fraction of the length of a side of the unit cell. Thus, the position of an atom, relative to an origin (0,0,0), located at xa parallel to a, yb parallel to b, and zc parallel to c is denoted (x,y,z), with $0 \leq x, y, z \leq 1$. Three-dimensional representations of complex structures are often difficult to draw and to interpret in two dimensions. A clearer method of representing three-dimensional structures on a two-dimensional surface is to draw the structure in projection by viewing the unit cell down one direction, typically one of the axes of the

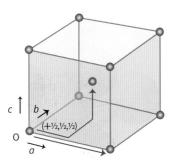

FIGURE 4.4 Lattice points describing the translational symmetry of a body-centred cubic unit cell. The translational symmetry is that of the unit cell and (+½,+½,+½), so a lattice point at the origin, O, translates to the body centre of the unit cell.

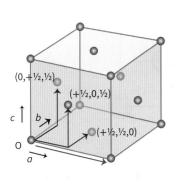

(a)

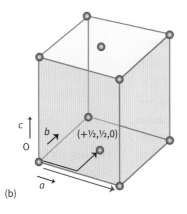

(b)

FIGURE 4.5 (a) Lattice points describing the translational symmetry of a face-centred cubic unit cell (F-centred). The translational symmetry is that of the unit cell and (+½,+½,0), (+½,0,+½), and (0,+½,+½) so a lattice point at the origin, O, translates to points in the centres of each of the faces. (b) Lattice points describing the translational symmetry of a C-centred orthorhombic unit cell; the translational symmetry is that of the unit cell and (+½,+½,0).

EXAMPLE 4.1   Identifying lattice types

Determine the translational symmetry present in the structure of cubic ZnS (Fig. 4.6) and identify the lattice type to which this structure belongs.

Answer We need to identify the displacements that, when applied to the entire cell, result in every atom arriving at an equivalent location (same atom type with the same coordination environment). In this case, the displacements $(0,+½,+½)$, $(+½,+½,0)$, and $(+½,0,+½)$, where $+½$ in the x, y, or z coordinate represents a translation along the appropriate cell direction by a distance of $a/2$, $b/2$, or $c/2$ respectively, have this effect. For example, starting at the labelled Zn^{2+} ion towards the near bottom left-hand corner of the unit cell (the origin), which is surrounded by four S^{2-} ions at the corners of a tetrahedron, and applying the translation $(+½,0,+½)$, we arrive at the Zn^{2+} ion towards the top front right-hand corner of the unit cell, which has the same tetrahedral coordination to sulfur. Identical translational symmetry exists for all the ions in the structure. These translations correspond to those of the face-centred lattice, so the lattice type is F.

Self-test 4.1 Determine the lattice type of CsCl (Fig. 4.7).

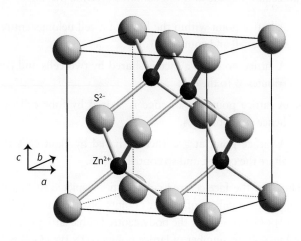

FIGURE 4.6 The cubic ZnS structure.

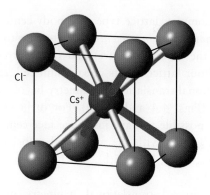

FIGURE 4.7 The cubic CsCl structure.

unit cell. The positions of the atoms relative to the projection plane are denoted by the fractional coordinate above the base plane and written next to the symbol defining the atom in the projection. If two atoms lie above each other, then both fractional coordinates are noted in parentheses. For example, the structure of body-centred tungsten, shown in three dimensions in Fig. 4.8(a), is represented in projection in Fig. 4.8(b).

4.2 The close packing of spheres

KEY POINT The close packing of identical spheres can result in a variety of polytypes, of which hexagonal and cubic close-packed structures are the most common.

Many metallic and ionic solids can be regarded as constructed from atoms and ions represented as hard spheres. If there is no directional covalent bonding, these spheres are free to pack together as closely as geometry allows and hence adopt a **close-packed structure**, a structure in which there is least unfilled space.

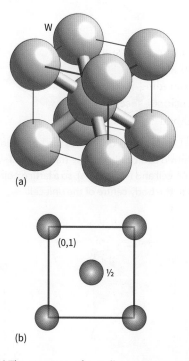

FIGURE 4.8 (a) The structure of metallic tungsten and (b) its projection representation.

EXAMPLE 4.2 Drawing a three-dimensional representation in projection

Convert the face-centred cubic lattice shown in Fig. 4.5(a) into a projection diagram.

Answer We need to identify the locations of the lattice points by viewing the cell from a position perpendicular to one of its faces. The faces of the cubic unit cell are square, so the projection diagram viewed from directly above the unit cell is a square. There is a lattice point at each corner of the unit cell, so the points at the corners of the square projection are labelled (0,1). There is a lattice point on each vertical face, which projects to points at fractional coordinate ½ on each edge of the projection square. There are lattice points on the lower and on the upper horizontal face of the unit cell, which project to two points at the centre of the square at 0 and 1, respectively, so we place a final point in the centre of a square and label it (0,1). The resulting projection is shown in Fig. 4.9.

Self-test 4.2 Convert the projection diagram of the unit cell of the SiS$_2$ structure shown in Fig. 4.10 into a three-dimensional representation.

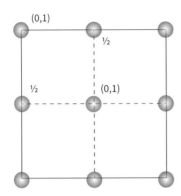

FIGURE 4.9 The projection representation of a face-centred cubic unit cell.

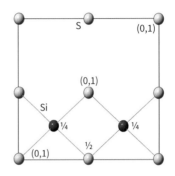

FIGURE 4.10 The structure of silicon sulfide (SiS$_2$).

To determine the arrangement which fits identical spheres as closely together as possible in three dimensions we start by considering two dimensions and packing spheres in sheets so that there is the least space between them. The greatest number of immediate neighbours that can be packed around one sphere is six in a regular hexagon, Figs. 4.11(a) and 4.11(b).[2] This pattern, where each sphere has six surrounding it, can be continued infinitely so that a two-dimensional sheet is formed with every sphere having six neighbouring spheres, Fig. 4.11(c). There is only one way of constructing this close-packed layer. Note that the environment of each sphere in the layer is identical, with six others placed around it in a hexagonal pattern.

The single layer of identical spheres (Fig. 4.11(c)) now needs to be extended into three dimensions. A second close-packed layer of spheres is formed by placing spheres in the dips between the spheres of the first layer so that each sphere in this second layer touches three spheres in the layer below (Figs. 4.12(a), 4.12(b), and 4.12(c)). (Note that only half the dips in the original layer are occupied, as there is insufficient space to place spheres into all the dips, Fig. 4.12(c); spheres will not fit in the dips marked X around the original sphere but only at the places marked O.) Figures 4.12(d) and 4.12(e) show how the spheres in this second layer are added sequentially, slowly building up the complete close-packed layer. The arrangement of spheres in this second layer is identical to that in the first, each with six nearest neighbours; the pattern is just slightly displaced horizontally.

The third close-packed layer, while identical in its arrangement of spheres within the layer, can be positioned (relative to the first two layers) in two different ways. Remember only half the dips from one close-packed layer can be occupied by spheres in the second close-packed layer. These two differing arrangements give rise to one of two **polytypes**, or structures, that are the same in two dimensions (in this case, in the layers of close-packed spheres) but different in the relative position of the third. Later we shall see that many different polytypes can be formed, but those described here are two very important special cases. We use the nomenclature A, B, C to denote the relative positions of the close-packed layers relative to one another; the first layer position is given the symbol A

---

[2] A good way of showing this yourself is to get a number of identical coins and push them together on a flat surface; the most efficient arrangement for covering the area is with six coins around each coin. This simple modelling approach can be extended to three dimensions by using any collection of identical spherical objects such as balls, oranges, Maltesers® (which can be stuck together by partially melting the chocolate coating), or marbles.

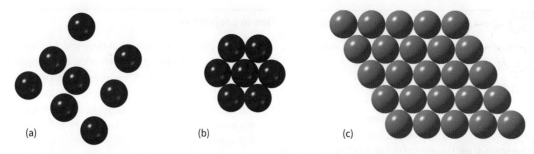

FIGURE 4.11 Close-packing of hard spheres to form a close-packed layer (a) nonclose packed spheres with large voids between them, (b) six spheres close packing around a central sphere in a hexagonal arrangement minimizing the empty space, (c) a close-packed layer where each sphere has six neighbours.

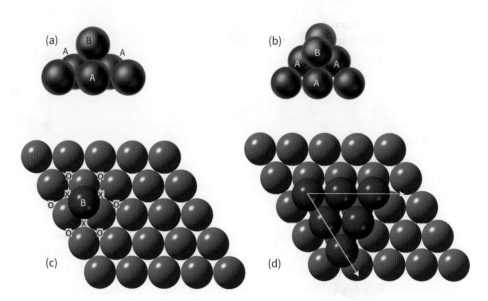

FIGURE 4.12 Constructing and positioning a second close-packed layer. (a) and (b). A sphere in a second layer (A) nestles in the dip created between three spheres (B) in the first layer. (c) A second sphere in the second layer can only be placed in the dips marked O ; there is insufficient room to place the second sphere in the dips marked X. (d) Placing more spheres in the second layer builds up a second close-packed layer slightly-shifted positionally from the first close-packed layer (blue spheres).

and any layer positionally directly above that is also given the symbol A. The other positions for the close-packed layers, corresponding to the dip positions O and X in Fig. 4.12(c), are denoted B and C.

In one polytype, the spheres of the third layer lie directly above the spheres of the first and each sphere in the second layer gains three more neighbours in this layer above it. This ABAB . . . pattern of layers, where A denotes layers that have spheres directly above each other and likewise for B, gives a structure with a hexagonal unit cell and hence is said to be **hexagonally close-packed** (often written in the shortened form hcp) (Figs. 4.13(c) and 4.14).

In the second polytype, the spheres of the third layer are placed above the dips that were not occupied in the first layer. The second layer fits into half the dips in the first layer and the third layer lies above the remaining dips. This arrangement results in an ABCABC . . . pattern, where C denotes a layer that has spheres that are not directly above spheres of the A or the B layer positions (but will be directly above another C-type layer). This pattern corresponds to a structure with a cubic unit cell and hence is termed **cubic close-packed** (shortened to ccp) (Figs. 4.13(d) and 4.15).

Because each ccp unit cell has a sphere at one corner and one at the centre of each face, a ccp unit cell is sometimes referred to as **face-centred cubic** (fcc). The **coordination number** (CN) of a sphere in a close-packed arrangement (the 'number of nearest neighbours') is 12, formed from six touching spheres in the original close-packed layer and three from each layer above and below it. This is the greatest number that geometry in three dimensions allows.[3] When directional bonding is important, as with structures described later in this chapter, the resulting structures are no longer close-packed and the coordination number will be less than the maximum of 12.

[3] That this arrangement, where each sphere has 12 nearest-neighbours, is the highest possible density of packing spheres was conjectured by Johannes Kepler in 1611; the proof was found only in 1998.

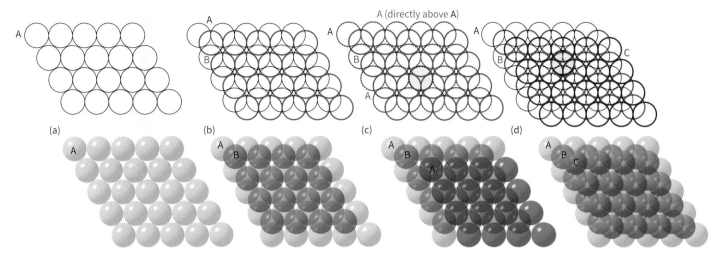

FIGURE 4.13 The formation of two close-packed polytypes. (a) A single close-packed layer, A. (b) The second close-packed layer, B, lies in dips above A. (c) The third layer reproduces the first to give an ABA arrangement structure (hcp). (d) The third layer lies above the gaps in the first layer, giving an ABC arrangement (ccp). The different colours (labelled A, B, and C) identify the different layers of identical spheres.

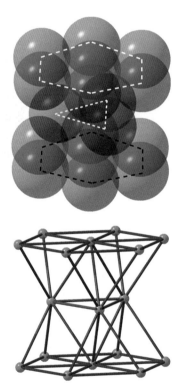

FIGURE 4.14 The hexagonal close-packed (hcp) unit cell of the ABAB... polytype. The colours of the spheres correspond to the layers in Fig. 4.13(c).

A NOTE ON GOOD PRACTICE

The descriptions ccp and fcc are often used interchangeably, although strictly ccp refers only to the close-packed arrangement, whereas fcc refers to the face-centred cubic lattice type which is a common representation of ccp, Fig. 4.15. Throughout this text the term ccp will be used to describe the cubic close-packing arrangement. It will be drawn using the cubic unit cell, with the fcc lattice type, as this representation is easiest to visualize.

The occupied space in a close-packed structure amounts to 74% of the total volume (see Example 4.3). However, the remaining unoccupied space, 26%, is not empty in a real solid because electron density of an atom does not end as abruptly as the hard-sphere model suggests. The type and distribution of these spaces between the close-packed spheres are known as 'holes'. They are important because many structures, including those of some alloys and many ionic compounds, can be regarded as formed from an expanded close-packed arrangement in which additional atoms or ions occupy all or some of the holes.

The ccp and hcp arrangements are the most efficient simple ways of filling space with identical spheres. They differ only in the stacking sequence of the close-packed layers, and other, more complex, close-packed layer sequences may be

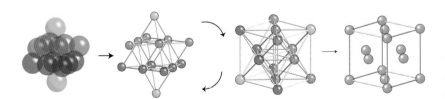

FIGURE 4.15 The cubic close-packed (ccp) unit cell of the ABC... polytype. The colours of the spheres correspond to the layers in Fig. 4.13(d).

EXAMPLE 4.3 Calculating the occupied space in a close-packed array

Calculate the percentage of unoccupied space in a close-packed arrangement of identical spheres.

Answer Because the space occupied by hard spheres is the same in the ccp and hcp arrays, we can choose the geometrically simpler structure, ccp, for the calculation. Consider Fig. 4.16. The spheres of radius r are in contact across the face of the cube and so the length of this diagonal is $r + 2r + r = 4r$. The side of such a cell is $\sqrt{8}r$ from Pythagoras' theorem (the square of the length of the diagonal, $(4r)^2$, equals the sum of the squares of the two sides of length a, so $2 \times a^2 = (4r)^2$, giving $a = \sqrt{8}r$), so the cell volume is $(\sqrt{8}r)^3 = 8^{3/2}r^3$. The unit cell contains $\tfrac{1}{8}$ of a sphere at each corner (for $8 \times \tfrac{1}{8} = 1$ in all) and half a sphere on each face (for $8 \times \tfrac{1}{2} = 3$ in all), for a total of 4. Because the volume of each sphere is $\tfrac{4}{3}\pi r^3$, the total volume occupied by the spheres themselves is $4 \times \tfrac{4}{3}\pi r^3 = \tfrac{16}{3}\pi r^3$. The occupied fraction is therefore $(\tfrac{16}{3}\pi r^3)/(8^{3/2}r^3) = \tfrac{16}{3}\pi/8^{3/2}$, which evaluates to 0.740.

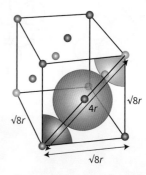

FIGURE 4.16 The dimensions involved in the calculation of the packing fraction in a close-packed arrangement of identical spheres of radius r.

Self-test 4.3 Calculate the fraction of space occupied by identical spheres in (a) a primitive cubic cell and (b) a body-centred cubic unit cell. Comment on your answers in comparison with the value obtained for close-packed structures.

---

formed by locating successive planes in different positions (A, B, and C) relative to their neighbours (Section 4.4). Any collection of identical atoms, such as those in the simple picture of an elemental metal, or of approximately spherical molecules, is likely to adopt one of these close-packed structures unless there are additional energetic reasons, such as covalent bonding interactions, for adopting an alternative arrangement. Indeed, many metals adopt such close-packed structures (Section 4.4), as do the solid forms of the noble gases (which are ccp). Almost-spherical molecules, such as fullerene, C_{60}, in the solid state, also adopt the ccp arrangement (Fig. 4.17), and so do many small molecules that rotate around their centres in the solid state and thus appear spherical, such as H_2, F_2, and one form of solid oxygen, O_2.

4.3 Holes in close-packed structures

KEY POINTS The structures of many solids can be discussed in terms of close-packed arrangements of one atom type in which the tetrahedral or octahedral holes are occupied by other atoms or ions. The ratio of spheres to octahedral holes to tetrahedral holes in a close-packed structure is 1:1:2.

The feature of a close-packed structure that enables us to extend the concept to describe structures more complicated than elemental metals is the existence of two types of **hole**, or unoccupied space between the close-packed spheres. An **octahedral hole** lies between two triangles of spheres on adjoining layers (Fig. 4.18(a)); note that two triangles rotated by 60° relative to each other define two opposite faces of an octahedron, so a point at the centre of this hole has six nearest-neighbour spheres at the corners of an octahedron. Octahedral holes lie halfway between two parallel close-packed layers at all the points where the two, staggered triangles are directly adjacent in neighbouring layers.

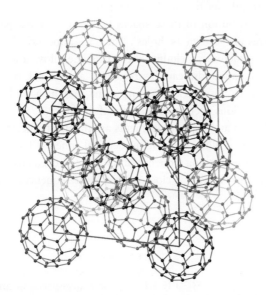

FIGURE 4.17 The structure of solid C_{60} showing the packing of C_{60} polyhedra in an fcc unit cell.

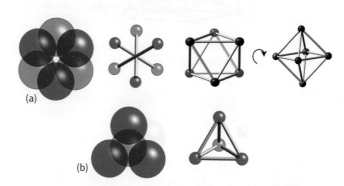

FIGURE 4.18 (a) An octahedral hole and (b) a tetrahedral hole formed in an arrangement of close-packed spheres.

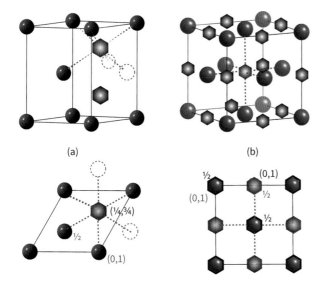

The size of the largest sphere that can fit on to the octahedral site and be in contact with the six nearest-neighbour spheres can be calculated, see Example 4.4. If each hard close-packed sphere has radius r, and if those close-packed spheres are to remain in contact, then each octahedral hole can accommodate a hard sphere representing another type of atom with a radius no larger than $0.414r$.

A **tetrahedral hole** (Figs. 4.18(b) and 4.12(b)) is formed by a planar triangle of touching spheres in a close-packed layer capped by a single sphere of an adjacent close-packed layer which lies in the dip between them. This capping atom may be in the close-packed layer above or below the three atoms in the original close-packed plane. As a result, the tetrahedral holes in any close-packed solid can be divided into two sets: in one the apex of the tetrahedron is directed up (T) and in the other the apex points down (T'). In an arrangement of N close-packed spheres there are N tetrahedral holes of each set and $2N$ tetrahedral holes in all. The position of the tetrahedral holes between the close-packed layers lies closer to the layer from which three atoms are chosen to define the limits of the tetrahedral hole. The arrangement of close-packed layers and holes can be summarized as the repeating sequence PTOT'PTOT'P . . . , where P is a close-packed layer and O and T are the octahedral and tetrahedral holes.

In a close-packed structure of spheres of radius r, a tetrahedral hole can accommodate another hard sphere of radius no greater than $0.225r$ (see Self-test 4.4). The location of tetrahedral holes, and the four nearest-neighbour spheres for one hole, is shown in Fig. 4.20(a) for an hcp arrangement and in Fig. 4.20(b) for a ccp arrangement.

FIGURE 4.19 (a) The location (represented by a hexagon) of the two octahedral holes in the hcp unit cell and (b) the locations (represented by hexagons) of the octahedral holes in the ccp unit cell. Positions of close-packed spheres in neighbouring unit cells are shown as dotted circles in the hcp case to illustrate the octahedral coordination; dotted lines show the coordination geometry for one octahedral hole in each structure type.

For a crystal consisting of N spheres in a close-packed structure, there are N octahedral holes. The positions and distribution of these holes in an hcp unit cell are shown in Fig. 4.19(a) and those in a ccp unit cell in Fig. 4.19(b). These illustrations also show that every hole of this type has local octahedral symmetry in the sense that it is surrounded by six nearest-neighbour spheres with their centres at the corners of an octahedron.

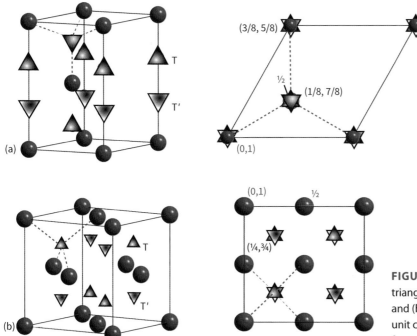

FIGURE 4.20 (a) The locations (represented by triangles) of the tetrahedral holes in the hcp unit cell and (b) the locations of the tetrahedral holes in the ccp unit cell. Dotted lines show the coordination geometry for one tetrahedral hole in each structure type.

Individual tetrahedral holes in ccp and hcp structures are identical. In the ccp arrangement of close-packed spheres, the T and T' sites are well separated and both can be occupied by a sphere no bigger than $0.225r$. However, in the hcp arrangement the neighbouring T and T' holes are just above and below a close-packed layer; they share a common tetrahedral face and are so close together that they are never occupied simultaneously.

These concepts based around the close-packing of spheres and the existence and occupation of the holes between these close-packed spheres are useful in describing many simple structures. As we have seen, close-packed spheres describe the solid-state structures of many metals and spherical molecules, and the structures of many alloys and interstitial compounds, such as steels (Box 4.2), can be represented by adding atoms to the hole sites. Where two types of sphere of different radius pack together (for instance, when cations and anions form a structure together), the larger spheres (normally the anions) can be considered to form a close-packed array, and the smaller spheres (usually the cations) can occupy the octahedral or tetrahedral holes. This representation allows many simple ionic structures to be described in terms of the occupation of holes in close-packed arrays (Section 4.9).

EXAMPLE 4.4 Calculating the size of an octahedral hole

Calculate the maximum radius of a sphere that may be accommodated in an octahedral hole in a close-packed solid composed of spheres of radius r.

Answer The structure of a hole is shown in Fig. 4.21(a). If the radius of a sphere is r and that of the hole is r_h, it follows from Pythagoras's theorem that $(r + r_h)^2 + (r + r_h)^2 = (2r)^2$ and therefore that $(r + r_h)^2 = 2r^2$, which implies that $r + r_h = \sqrt{2}r$. That is, $r_h = (\sqrt{2} - 1)r$, which evaluates to $0.414r$. Note that this is the permitted maximum size subject to keeping the close-packed spheres in contact; if the spheres are allowed to separate slightly while maintaining their relative positions, then the hole can accommodate a larger sphere.

Self-test 4.4 Show that the maximum radius of a sphere that can fit into a tetrahedral hole is $r_h = 0.225r$: base your calculation on Fig. 4.21(b). Note that a tetrahedron may be inscribed inside a cube using four non-adjacent vertices and the centre of the hole has the same central point as that of the tetrahedron and the encompassing cube.

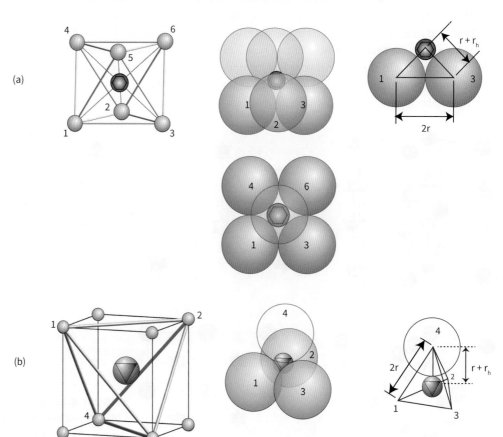

FIGURE 4.21 The distances used to calculate the size of (a) an octahedral hole and (b) a tetrahedral hole.

EXAMPLE 4.5 Demonstrating that the ratio of close-packed spheres to octahedral holes in ccp is 1:1

Determine the number of close-packed spheres and octahedral holes in the ccp arrangement and hence show the ratio is 1 sphere:1 hole.

Answer The ccp unit cell, with the positions of the octahedral holes marked, is shown in Fig. 4.19(b). For the close-packed spheres the calculation of the number in the unit cell follows that given in Section 4.1 for lattice points in the F-centred lattice, as there is one sphere associated with each lattice point. The number of close-packed spheres in the unit cell is therefore $(8 \times \frac{1}{8}) + (6 \times \frac{1}{2}) = 4$. The octahedral holes are sited along each edge of the cube (12 edges in total), shared between four unit cells, with a further hole at the centre of the cube which is not shared between unit cells. So the total of octahedral holes in the unit cell is $(6 + \frac{1}{2}) + 1 = 4$. The ratio of close-packed spheres to holes in the unit cell is 4:4, equivalent to 1:1. As the unit cell is the repeating unit for the whole structure, this result applies to the complete close-packed array, and it is often quoted that 'for N close-packed spheres there are N octahedral holes'.

Self-test 4.5 Show that the ratio of close-packed spheres to tetrahedral holes in ccp is 1:2.

The structures of metals and simple alloys

X-ray diffraction studies (Section 8.1) reveal that many metallic elements have close-packed structures, indicating that the bonds between the atoms have little directional covalent character (Table 4.2, Fig. 4.22). One consequence of this close packing is that metals often have high densities because the most mass is packed into the smallest volume. Indeed, the elements deep in the d block, near iridium and osmium, include the densest solids known under normal

TABLE 4.2 The crystal structures adopted by metals under normal conditions

Crystal structure	Element
Hexagonal close-packed (hcp)	Be, Co, Mg, Ti, Zn
Cubic close-packed (ccp)	Ag, Al, Au, Ca, Cd, Cu, Ni, Pb, Pt
Body-centred cubic (bcc)	Ba, Cr, Fe, W, alkali metals
Primitive cubic (cubic-P)	Po

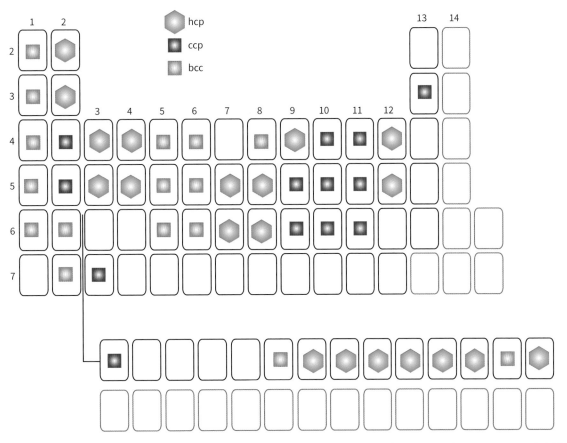

FIGURE 4.22 The structures of the metallic elements at room temperature. Elements with more complex structures are left blank.

conditions of temperature and pressure. Osmium has the highest density of all the elements, measured by X-ray crystallography at 22.59 g cm⁻³, and the density of tungsten, 19.25 g cm⁻³, which is almost twice that of lead (11.3 g cm⁻³), results in its use as weighting material in fishing equipment and as ballast in high-performance cars.

> **EXAMPLE 4.6** Calculating the density of a substance from a structure
>
> Calculate the density of gold, with a cubic close-packed array of atoms of molar mass $M = 196.97$ g mol⁻¹ and a cubic lattice parameter $a = 409$ pm.
>
> **Answer** Density is an intensive property (a property of matter that does not depend on the amount); therefore, the density of the unit cell is the same as the density of any macroscopic sample. We represent the ccp arrangement as a face-centred lattice with a sphere at each lattice point; there are four spheres associated with the unit cell. The mass of each atom is M/N_A, where N_A is Avogadro's constant, and the total mass of the unit cell containing four gold atoms is $4M/N_A$. The volume of the cubic unit cell is a^3. The mass density of the cell is $\rho = 4M/N_A a^3$. At this point we insert the data:
>
> $$\rho = \frac{4 \times (196.97 \times 10^{-3}\,\text{kg mol}^{-1})}{(6.022 \times 10^{23}\,\text{mol}^{-1}) \times (409 \times 10^{-12}\,\text{m})^3} = 1.91 \times 10^4\,\text{kg m}^{-3}$$
>
> That is, the density of the unit cell, and therefore of the bulk metal, is 19.1 g cm⁻³. The experimental value is 19.2 g cm⁻³, in good agreement with this calculated value.
>
> **A NOTE ON GOOD PRACTICE**
>
> It is always best to proceed symbolically with a calculation for as long as possible: this reduces the risk of numerical error and gives an expression that can be used in other circumstances.
>
> **Self-test 4.6** Calculate the lattice parameter of silver assuming that it has the same structure as elemental gold but a density of 10.5 g cm⁻³.

4.4 Polytypism

KEY POINT Polytypes involving complex stacking arrangements of close-packed layers occur for some metals.

Which of the common close-packed polytypes, hcp or ccp, a metal adopts depends on the details of the electronic structure of its atoms, the extent of interaction between second-nearest-neighbours, and the potential for some directional character in the bonding. It has been observed that softer, more malleable metals, such as copper and gold, adopt the ccp arrangement while metals with hcp, such as cobalt and magnesium, are harder and more brittle. This behaviour is related to how easily planes of atoms can slip past each other. In hcp only the neighbouring close-packed planes, A and B, can move easily relative to each other, but consideration of the ccp structure, when drawn as in Fig. 4.14, shows that the planes ABC can be chosen in different orthogonal directions, allowing the closed-packed layers of atoms to move easily in multiple directions.

A close-packed structure need not be either of the common ABAB . . . or ABCABC . . . polytypes. An infinite range of close-packed polytypes can in fact occur, as the layers may stack in a more complex repetition of A, B, and C layers or even in some permissible random sequence. The stacking cannot be a completely random choice of A, B, and C sequences, however, because adjacent layers cannot have exactly the same sphere positions; for instance, AA, BB, and CC cannot occur because spheres in one layer must occupy dips in the adjacent layer. Cobalt is an example of a metal that displays this more complex polytypism. Above 500°C, cobalt is ccp but it undergoes a transition when cooled. The structure that results is a nearly randomly stacked set (for instance, ABACBABABC . . .) of close-packed layers of Co atoms. In some samples of cobalt the polytypism is not random, as the sequence of planes of atoms repeats after several hundred layers. The long-range repeat may be a consequence of a spiral growth of the crystal that requires several hundred turns before a stacking pattern is repeated.

4.5 Nonclose-packed structures

KEY POINTS A common nonclose-packed metal structure is body-centred cubic; a primitive cubic structure is occasionally encountered. Metals that have structures more complex than those described so far can sometimes be regarded as slightly distorted versions of simple structures.

Not all elemental metals have structure based on close packing and some other packing patterns use space nearly as efficiently. One commonly adopted arrangement has the translational symmetry of the body-centred cubic lattice and is known as the **body-centred cubic** structure (cubic-I or bcc), in which a sphere is at the centre of a cube with spheres at each corner (Fig. 4.23). Metals with this structure have a coordination number of 8 because the central atom is in contact with all the atoms at the corners of the unit cell. Although a bcc structure is less closely packed than the ccp and hcp structures (for which the coordination number is 12), the difference is not very great because the central atom has six second-nearest-neighbours, at the centres of the adjacent unit cells, only 15% further away. This arrangement leaves 32% of the space unfilled compared with 26% in the close-packed structures (see Example 4.3). A bcc structure

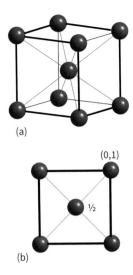

FIGURE 4.23 (a) A bcc structure unit cell and (b) its projection representation.

is adopted by 15 of the elements under standard conditions, including all the alkali metals and the metals in Groups 5 and 6. Accordingly, this simple arrangement of atoms is sometimes referred to as the 'tungsten type'.

The least common metallic structure is the **primitive cubic** (cubic-P) structure (Fig. 4.24), in which spheres are located at the lattice points of a primitive cubic lattice, taken as the corners of the cube. The coordination number of a cubic-P structure is 6.

One form of polonium (α-Po) is the only example of this structure among the elements under normal conditions, though bismuth also adopts this structure under pressure. Solid mercury (α-Hg), however, has a closely related structure: it is obtained from the cubic-P arrangement by stretching the cube along one of its body diagonals (Fig. 4.25(a)); a second form of solid mercury (β-Hg) has a structure based

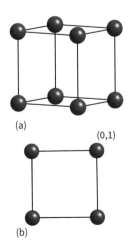

FIGURE 4.24 (a) A primitive cubic unit cell and (b) its projection representation.

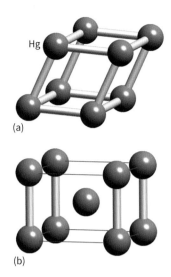

FIGURE 4.25 The structures of (a) α-mercury and (b) β-mercury that are closely related to the unit cells with primitive cubic and body-centred cubic lattices, respectively.

on the bcc arrangement but compressed along one cell direction (Fig. 4.25(b)).

Metals that have structures more complex than those described so far can sometimes be regarded, like solid mercury, as having slightly distorted versions of simple structures. Zinc and cadmium, for instance, have structures that are almost hcp, but the planes of close-packed atoms are separated by a slightly greater distance than in perfect hcp.

4.6 Polymorphism of metals

KEY POINTS Polymorphism is a common consequence of the low directionality of metallic bonding. At high temperatures a bcc structure is common for metals that are close-packed at low temperatures on account of the increased amplitude of atomic vibrations.

The lack of directionality in the interactions between metal atoms accounts for the wide occurrence of **polymorphism**, the ability to adopt different crystal forms under different conditions of pressure and temperature. It is often, but not universally, found that the most closely packed phases are thermodynamically favoured at low temperatures and the less closely packed structures are favoured at high temperatures. Similarly, the application of high pressure leads to structures with higher packing densities, such as ccp and hcp.

The polymorphs of metals are generally labelled α, β, γ, … with increasing temperature. Some metals revert to a low-temperature form at higher temperatures. Iron, for example, shows several solid-solid phase transitions; α-Fe, which is bcc, occurs up to 906°C; γ-Fe, which is ccp, occurs up to 1401°C; and then α-Fe occurs again up to the melting point at 1530°C. The hcp polymorph, β-Fe, is formed at

BOX 4.1 What happens to metals under pressure?

The Earth has an innermost core about 1200 km in diameter that consists of solid iron and is responsible for generating the planet's powerful magnetic field. The pressure at the centre of the Earth has been calculated to be around 370 GPa (about 4.7 million atm) at a temperature between 5000°C and 6500°C. The polymorph of iron that exists under these conditions has been much debated, with information from theoretical calculations and measurements using seismology. The current thinking is that the iron core consists of the body-centred cubic polymorph. It has been proposed that this exists either as a giant crystal or a large number of oriented crystals such that the long diagonal of the bcc unit cell aligns along the Earth's axis of rotation (Fig. B4.1).

The study of the structures and polymorphism of elements and compounds under high-pressure conditions goes beyond the study of the Earth's core. Hydrogen, when subjected to pressures similar to those at the Earth's core, is predicted to become a metallic solid, similar to the alkali metals, and the cores of planets such as Jupiter have been hypothesized to contain

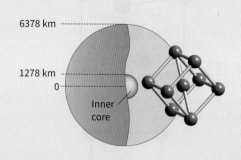

FIGURE B4.1

hydrogen in this form. Recent X-ray diffraction measurements indicate that hydrogen molecules start to dissociate into an hcp arrangement of hydrogen atoms under pressures around 250 GPa, and that at pressures above 425 GPa this material then shows the metallic reflectivity of a metal. When pressures of over 55 GPa are applied to iodine, the I_2 molecules dissociate and adopt the simple face-centred cubic structure; the element becomes metallic and is a superconductor below 1.2 K.

---

high pressures and was believed to be the form that exists at the Earth's core, but recent studies indicate that a bcc polymorph is more likely (Box 4.1).

The bcc structure is common at high temperatures for metals that are close-packed at low temperatures because the increased amplitude of atomic vibrations in the hotter solid results in a less close-packed structure. For many metals (among them Ca, Ti, and Mn) the transition temperature is above room temperature; for others (among them Li and Na) the transition temperature is below room temperature (lithium transforms from a bcc structure to hcp on cooling below 70 K). It is also found empirically that a bcc structure is favoured by metals with a small number of valence electrons per orbital, so the alkali metals (ns^1 electronic structure) and many early-mid transition metals (with nd^m and $m \leq 5$ electronic structures) adopt bcc structures.

4.7 Metallic radii of metals

KEY POINT The Goldschmidt correction converts metallic radii of metals to the value they would have in a close-packed structure with 12-fold coordination.

A definition of the metallic radius of a metal was given in Section 1.7 as half the distance between the centres of adjacent atoms in the solid. Note that values given for the atomic or covalent radius of a metal can be different from metallic radii as they represent the size of metal atoms when covalently bonded in a compound. Values of atomic or metallic radii are given in Table 9.2.

Many metals exhibit polymorphism, adopting structures with different coordination numbers, and metallic radii are found to increase with the coordination number of the lattice. The same atom in structures with different coordination numbers may therefore appear to have different radii, and an atom of a metal with coordination number 12 appears bigger than one with coordination number 8. In an extensive study of internuclear separations in a wide variety of polymorphic elements and alloys, V. Goldschmidt found that the average relative radii are related, as shown in Table 4.3.

TABLE 4.3 The variation of metallic radius with coordination number

Coordination number	Relative radius
12	1
8	0.97
6	0.96
4	0.88

A BRIEF ILLUSTRATION

The empirical metallic radius of Na is 185 pm, but that is for the bcc structure in which the coordination number is 8. To adjust to 12-coordination we multiply this radius by 1/0.97 = 1.03 and obtain 191 pm as the radius that a Na atom would have if it were in a close-packed structure.

The essential features of atomic radii in Section 1.7 apply here to metallic radii and Goldschmidt-corrected (for CN 12) metallic radii generally increase down a group and decrease from left to right across a period.

EXAMPLE 4.7 Calculating a metallic radius

The cubic unit cell parameter, a, of primitive cubic polonium (α-Po) is 335 pm. Use the Goldschmidt correction to calculate a metallic radius for this element.

Answer We need to infer the radius of the atoms from the dimensions of the unit cell and the coordination number, and then apply a correction to coordination number 12. Because the Po atoms of radius r are in contact along the unit cell edges, the length of the primitive cubic unit cell is $2r$. Thus, the metallic radius of 6-coordination Po is $a/2$ with $a = 335$ pm. The conversion factor from 6-fold to 12-fold coordination from Table 4.3 (1/0.960) gives the metallic radius of Po as $\frac{1}{2} \times 335\,\text{pm} \times 1/0.960 = 174\,\text{pm}$.

Self-test 4.7 Predict the lattice parameter for Po when it adopts a bcc structure.

4.8 The structures of simple alloys and interstitials

An **alloy** is a blend of metallic elements prepared by mixing the molten components and then cooling the mixture to produce a metallic solid. Alloys may be homogeneous solid solutions in which the atoms of one metal are distributed randomly among the atoms of the other, or they may be compounds with a definite composition and internal structure. Alloys typically form from two electropositive metals, so they are likely to be located towards the bottom left-hand corner of a Ketelaar triangle (Fig. 4.26).

The majority of simple alloys can be classified as either 'substitutional' or 'interstitial'. A **substitutional alloy or solid solution** is a solution in which atoms of the solute metal replace some of the parent pure metal atoms, Fig. 4.27.

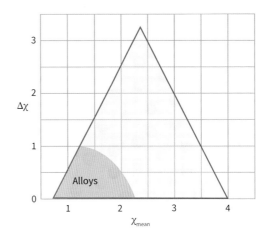

FIGURE 4.26 The approximate locations of alloys in a Ketelaar triangle.

Some of the classic examples of alloys are brass (up to 38 atom per cent Zn in Cu), bronze (a metal other than Zn or Ni in Cu; casting bronze, for instance, is 10 atom per cent Sn and 5 atom per cent Pb), and stainless steel (over 12 atom per cent Cr in Fe). The structures of these substitutional metal alloys can usually be discussed in terms of close-packing of spheres (or near close-packing bcc) as one metal atom is randomly replaced in the structure by one of a similar metallic radius. Thus the structure of a brass with the composition 3 parts Cu and 1 part Zn ($Cu_{0.75}Zn_{0.25}$) has the ccp structure adopted by pure copper with zinc replacing copper randomly on a quarter of the sites, Fig. 4.28.

Interstitial solid solutions are often formed between metals and small atoms (such as boron, carbon, and nitrogen) that can occupy interstices, such as octahedral and tetrahedral holes, at low levels in a parent metal while maintaining its crystal structure. Examples include the carbon steels, such as **austenite**, where low levels of carbon (typically ≤2%) occupy some of the octahedral holes in a ccp arrangement of iron metal atoms, Fig. 4.29, Box 4.2. Substitutional and interstitial alloys are described more fully in Section 25.3.

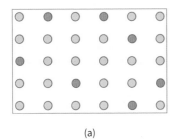
(a)

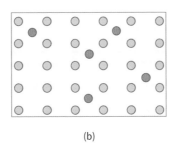
(b)

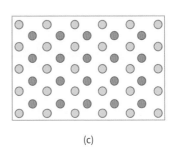
(c)

FIGURE 4.27 (a) Substitutional and (b) interstitial alloys. (c) A regular arrangement of interstitial atoms can lead to a new structure.

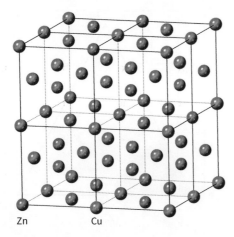

FIGURE 4.28 The structure of brass of the composition $Cu_{0.75}Zn_{0.25}$ with zinc atoms randomly replacing copper atoms in a ccp arrangement.

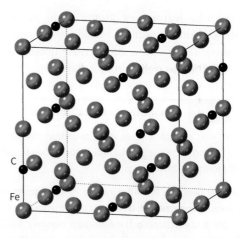

FIGURE 4.29 The structure of the carbon steel austenite with a ccp arrangement of iron atoms with carbon atoms in a small proportion of the octahedral holes.

BOX 4.2 How do alloying and interstitials turn iron into steels?

Steels are alloys of iron, carbon, and other elements. They are classified as mild, medium-carbon, or high-carbon steels according to the percentage of carbon they contain. Mild steels contain up to 0.25 atom per cent C, medium-carbon steels contain 0.25–0.45 atom per cent, and high-carbon steels contain 0.45–1.50 atom per cent. Increasing the carbon content of steels up to 0.8% improves their hardness and strength as the presence of the interstitial carbon atom restricts the ability of iron layers to move relative to each other on the application of an external force. At carbon levels above 0.8%, which includes cast iron with approximately 3% carbon content, the carbon atoms in the steel tend to congregate at the surfaces of the individual micron-sized particles forming the steel; the strength of the steels decrease and they become more brittle. The brittleness of cast iron makes it unsuitable for large structural purposes; for example, bridges and early uses in railway bridges resulted in structural failures and collapse after less than a year's use.

The addition of other metals to these carbon steels can have a major effect on the structure, properties, and therefore applications of the steel. Examples of metals added to carbon steels, so forming 'stainless steels', are listed in Table B4.2. Stainless steels are also classified by their crystalline structures, which are controlled by factors such as the rate of cooling following their formation in a furnace and the type of added metal. Thus, pure iron adopts different polymorphs (Section 4.6) depending on temperature, and some of these high-temperature structures can be stabilized at room temperature in steels by quenching (cooling very rapidly).

Stainless steels with the **austenite** structure comprise over 70% of total stainless steel production. Austenite is a solid solution of carbon and iron that exists in steel above 723°C and is ccp iron with about 2% of the octahedral holes filled with carbon. As it cools, it breaks down into other materials, including **ferrite** and martensite, as the solubility of carbon in the iron drops to below 1 atom per cent. The rate of cooling determines the relative proportions of these materials and therefore the mechanical properties (e.g. hardness and tensile strength) of the steel. The addition of certain other metals, such as Mn, Ni, and Cr, can allow the austenitic structure to survive cooling to room temperature. These steels contain a maximum of 0.15 atom per cent C and typically 10–20 atom per cent Cr plus Ni or Mn as a substitutional solid solution; they can retain an austenitic structure at all temperatures from the cryogenic region to the melting point of the alloy. A typical composition of 18 atom per cent Cr and 8 atom per cent Ni in addition to iron is known as *18/8 stainless*.

Ferrite is α-Fe with only a very small level of carbon, less than 0.1 atom per cent, with a bcc iron crystal structure. Ferritic stainless steels are highly corrosion-resistant, but far less durable than austenitic grades. They contain between 10.5 and 27 atom per cent Cr and some Mo, Al, or W. Martensitic stainless steels are not as corrosion-resistant as the other two classes, but are strong and tough as well as highly machineable, and can be hardened by heat treatment. They contain 11.5–18 atom per cent Cr and 1–2 atom per cent C which is trapped in the iron structure as a result of quenching compositions with the austenite structure type. The martensitic crystal structure is closely related to that of ferrite, but the unit cell is tetragonal rather than cubic.

TABLE B4.2 Metal composition of some common stainless steels

Metal	Atom percentage added	Effect on properties
Copper	0.2–1.5	Improves atmospheric corrosion resistance
Nickel	0.1–1	Benefits surface quality
Niobium	0.02	Increases tensile strength and yield point
Nitrogen	0.003–0.012	Improves strength
Manganese	0.2–1.6	Improves strength
Vanadium	Up to 0.12	Increases strength

Ionic solids

KEY POINTS The ionic model treats a solid as an assembly of oppositely charged spheres that interact by nondirectional electrostatic forces; if the thermodynamic properties of the solid calculated on this model agree with experiment, then the compound is normally considered to be ionic.

Ionic solids, such as NaCl, SrO, and KNO_3, are often recognized by their brittleness because the electrons made available by cation formation are localized on a neighbouring anion; striking the solid can cause ions of the same charge to shift to adjacent positions and the resultant repulsion causes it to fracture. Ionic solids also commonly have high melting points, because of the strong Coulombic forces between oppositely charged ions that have to be overcome in producing the molten state, and many are soluble in polar solvents, particularly water, where the ions become strongly solvated. However, there are exceptions especially for higher-charged ions where the solubility can be low: CaF_2, for example, is a high-melting ionic solid but it is insoluble in water. Ammonium nitrate, NH_4NO_3, is ionic in terms of the interactions between the ammonium and nitrate ions, but melts at 170°C. Binary ionic materials are typical of elements with large electronegativity differences, typically $\Delta\chi > 3$, and such compounds are, therefore, likely to be found at the top corner of a Ketelaar triangle (Fig. 4.30).

The classification of a solid as ionic is based on comparison of its properties with those of the **ionic model**, which treats the solid as an assembly of oppositely charged, hard spheres that interact primarily by nondirectional electrostatic forces (Coulombic forces). If the thermodynamic properties of the solid calculated using this model agree with experiment, then the solid may be ionic. However, it should be noted that many examples of coincidental agreement with the ionic model are known, so numerical agreement alone does not imply ionic bonding. The nondirectional nature of electrostatic interactions between ions in an ionic solid contrasts with those present in a covalent solid, where the symmetries of the atomic orbitals play a strong role in determining the geometry of the structure. However, the assumption that ions can be treated as perfectly hard spheres (of fixed radius for a particular ion type) that have no directionality in their bonding is far from true for real ions. For example, with halide anions some directionality might be expected in their bonding as a result of the orientations of their p orbitals, and large ions such as Cs^+ and I^- are easily polarizable so do not behave as hard spheres. Even so, the ionic model is a useful starting point for describing many simple structures.

We start by describing some common ionic structures in terms of the packing of hard spheres of different sizes and opposite charges. After that, we see how to rationalize the structures in terms of the energetics of crystal formation. The structures described have been obtained by using X-ray diffraction (Section 8.1), and were among the first substances to be examined in this way.

4.9 Characteristic structures of ionic solids

The ionic structures described in this section are prototypes of a wide range of solids. For instance, although the rock-salt structure takes its name from a mineral form of NaCl, it is characteristic of numerous other solids (Table 4.4).

Many, but not all, of the structures can be regarded as derived from arrays in which the larger of the ions, usually the anions, stack together in ccp or hcp patterns and the smaller counter-ions (usually the cations) occupy the octahedral or tetrahedral holes in the lattice (Table 4.5).

Throughout the following discussion it will be helpful to refer back to Figs. 4.19 and 4.21 to see how the structure being described is related to the hole patterns shown there. Throughout this discussion it should be remembered that the ratio of close-packed spheres to octahedral holes to tetrahedral holes is 1:1:2 for both cubic and hexagonal close packing. Filling all the octahedral holes by an atom or ion of type A in a cubic close-packed array of spheres of type X will give a ratio of X to A of 1:1, that is A_1X_1; therefore, this arrangement gives rise to a compound of the stoichiometry AX.

The holes in a close-packed array are much smaller than the spheres (of radius r) that form the close-packed structure; a maximum of $0.414r$ for the octahedral holes and $0.225r$ for the tetrahedral holes, see Example 4.4. Therefore, the close-packed layers usually need to expand to accommodate the counter-ions, but this expansion is often a minor perturbation of the anion arrangement, which will still be referred to as

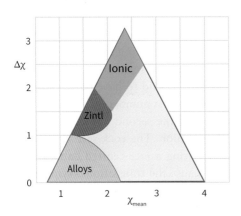

FIGURE 4.30 The approximate locations of ionic and Zintl phases (see Chapter 25) in a Ketelaar triangle.

TABLE 4.4 The crystal structures of compounds, at standard conditions unless otherwise stated

Crystal structure	Examples*
Antifluorite	K_2O, K_2S, Li_2O, Na_2O, Na_2Se, Na_2S
Caesium chloride	**CsCl**, TlI (low T), CsAu, CsCN, CuZn, NbO
Fluorite	**CaF$_2$**, UO_2, HgF_2, LaH_2, PbO_2 (high pressure, >6 GPa)
Nickel arsenide	**NiAs**, NiS, FeS, PtSn, CoS
Perovskite	**CaTiO$_3$** (distorted), $SrTiO_3$, $PbZrO_3$, $LaFeO_3$, $LiSrH_3$, $KMnF_3$
Rock-salt	**NaCl**, KBr, RbI, AgCl, AgBr, MgO, SrO, TiO, FeO, NiO, SnAs, UC, ScN
Rutile	**TiO$_2$** (one polymorph), MnO_2, SnO_2, WO_2, MgF_2, NiF_2
Sphalerite (zinc blende, cubic)	**ZnS** (one polymorph), CuCl, CdS (Hawleyite polymorph), HgS, GaP, AgI (at high pressure, >6 GPa, transforms to rock-salt structure), InA, ZnO (high pressure, >6 GPa)
Spinel	**MgAl$_2$O$_4$**, $ZnFe_2O_4$, $ZnCr_2S_4$
Wurtzite (hexagonal)	**ZnS** (one polymorph), ZnO, BeO, AgI (one polymorph, iodargyrite), AlN, SiC, NH_4F, CdS (Greenockite polymorph)

* A substance in bold type is the one that gives its name to the structure.

TABLE 4.5 The relation of structure to the filling of holes

Close-packing type	Hole-filling	Structure type (exemplar)
Cubic (ccp)	All octahedral	Rock salt (NaCl)
	All tetrahedral	Fluorite (CaF$_2$)
	Half octahedral	CdCl$_2$
	Half tetrahedral	Sphalerite (ZnS)
Hexagonal (hcp)	All octahedral	Nickel arsenide (NiAs); with some distortion from perfect hcp (CdI$_2$)
	Half octahedral	Rutile (TiO$_2$); with distortion from perfect hcp
	All tetrahedral	No structure exists: tetrahedral holes share faces
	Half tetrahedral	Wurtzite (ZnS)

ccp and hcp. This expansion avoids some of the strong repulsion between the identically charged ions (usually the anions) and also allows larger species to be inserted into the holes between the close-packed ions. Overall, examining the opportunities for hole-filling in a close-packed array of the larger ion type provides an excellent starting point for the descriptions of many simple ionic structures.

(a) Binary phases, AX$_n$

KEY POINTS Important structures that can be expressed in terms of the occupation of holes include the rock-salt, caesium-chloride, sphalerite, fluorite, wurtzite, nickel-arsenide, and rutile structures.

The simplest ionic compounds contain just one type of cation (A) and one type of anion (X), present in various ratios covering compositions, such as AX and AX$_2$. Several different structures may exist for each of these compositions, depending on the relative sizes of the cations and anions, and which hole types are filled and to what degree in the close-packed array (Table 4.5). We start by considering compositions AX with equal numbers of cations and anions and then consider AX$_2$, the other commonly found stoichiometry.

The **rock-salt structure** is based on a ccp array of the larger anions with cations in all the octahedral holes (Fig. 4.31). Alternatively, it can be viewed as a structure in which the anions occupy all the octahedral holes in a ccp array of cations; the two descriptions are equivalent because the arrangement of octahedral holes is the same as a close-packed sphere arrangement. As the number of octahedral holes in a close-packed array is equal to the number of ions forming the array (the X ions), then filling them all with A ions yields the stoichiometry AX. Because each ion is surrounded by an octahedron of six counter-ions, the coordination number of each type of ion is 6 and the structure is said to have **6:6 coordination**. In this notation **N:N**, the first number is the coordination number of the cation and the second number is the coordination number of the anion. The rock-salt structure can still be described as having a face-centred cubic lattice after this hole-filling because the translational symmetry demanded by this lattice type is preserved when all the octahedral sites are occupied.

To visualize the local environment of an ion in the rock-salt structure, we should note that the six nearest neighbours

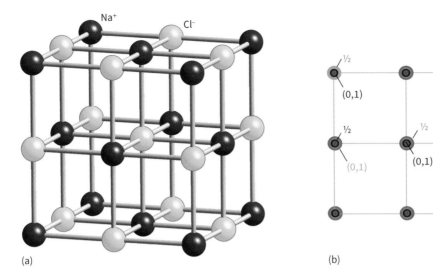

FIGURE 4.31 (a) The rock-salt structure and (b) its projection representation. Note the relation of this structure to the fcc structure in Fig. 4.19 with an atom in each octahedral hole.

of the central ion of the cell shown in Fig. 4.31 lie at the centres of the faces of the cell and form an octahedron around the central ion. All six neighbours have a charge opposite to that of the central ion. The 12 second-nearest neighbours of the central ion are at the centres of the edges of the cell, and all have the same charge as the central ion. The eight third-nearest neighbours are at the corners of the unit cell, and have a charge opposite to that of the central ion. We can use the rules described in Section 4.1 to determine the composition of the unit cell and the number of atoms or ions of each type present.

A BRIEF ILLUSTRATION

In the unit cell shown in Fig. 4.31, there are the equivalent of $(8 \times \tfrac{1}{8}) + (6 \times \tfrac{1}{2}) = 4\,\text{Na}^+$ ions and $(12 \times \tfrac{1}{4}) + 1 = 4\,\text{Cl}^-$ ions. Hence, each unit cell contains four NaCl formula units. The number of formula units present in the unit cell is commonly denoted Z, so in this case $Z = 4$.

The rock-salt structure is found widely for ionic compounds of the stoichiometry AX, and the several examples given in Table 4.4 include many Group 1 halides and Group 2 oxides. The rock-salt arrangement is not just formed for simple monatomic species such as M^+ and X^-, but also for many 1:1 ionic compounds, AX, in which the ions A^{n+} and X^{n-} are complex units, such as $[\text{Co(NH}_3)_6][\text{TlCl}_6]$. The structure of this compound can be considered as an array of close-packed octahedral $[\text{TlCl}_6]^{3-}$ anions with $[\text{Co(NH}_3)_6]^{3+}$ cations in all the octahedral holes. Similarly, compounds such as CaC_2, CsO_2, KCN, and FeS_2 all adopt structures closely related to the rock-salt structure, with alternating cations and complex anions (C_2^{2-}, O_2^-, CN^-, and S_2^{2-}, respectively) in three orthogonal directions (Fig. 4.32(a) and (b)). However, the orientation of these (nonspherical) linear diatomic species can lead to elongation of the unit cell and elimination of the cubic symmetry in CaC_2 (Fig. 4.32(a)).

Further compositional flexibility, but retaining a rock-salt type of structure, can come from having more than one

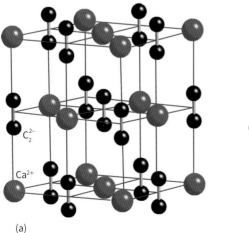

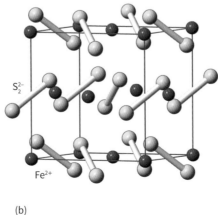

FIGURE 4.32 (a) The structure of CaC_2 is based on the rock-salt structure but is elongated in the direction parallel to the axes of the C_2^{2-} ions giving a tetragonal unit cell. (b) The structure of FeS_2 has S_2^{2-} anions orientated in different directions producing a cubic unit cell based on the rock-salt structure type.

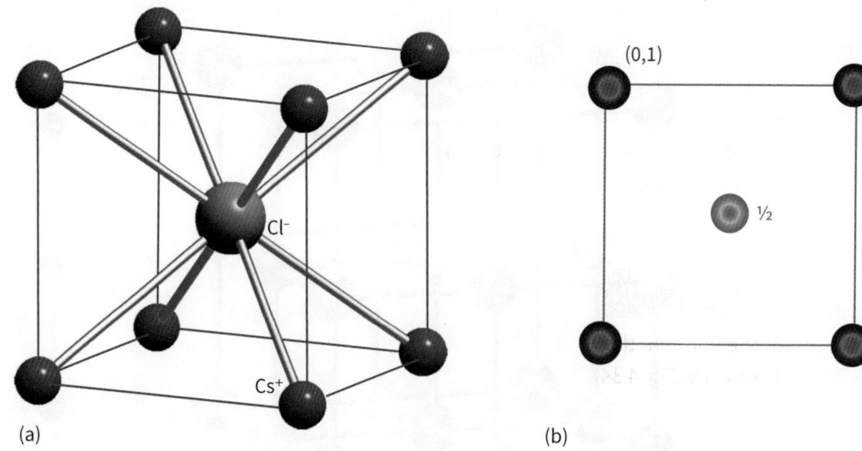

FIGURE 4.33 (a) The caesium-chloride structure. The corner lattice points, which are shared by eight neighbouring cells, are surrounded by eight nearest-neighbour lattice points. The anion occupies a cubic hole in a primitive cubic lattice; (b) its projection.

cation or anion type while maintaining the overall 1:1 ratio between ions of opposite charge. Thus, filling half of the A sites in the rock-salt structure type as Li^+ and half as Ni^{3+} gives rise to the formula $(Li_{1/2}Ni_{1/2})O$, which is normally written as $LiNiO_2$, and the known compound of this stoichiometry adopts this structure type.

Much less common than the rock-salt structure for compounds of stoichiometry AX is the **caesium-chloride structure** (Fig. 4.33), which is possessed by CsCl, CsBr, and CsI, as well as some other compounds formed of ions of similar radii to these, including TlI (see Table 4.4). The caesium-chloride structure is not based on close-packing but has a primitive cubic unit cell, with each corner occupied by an anion, and a cation occupying the 'cubic hole' at the cell centre (or vice versa); as a result, $Z = 1$. An alternative view of this structure is as two interpenetrating primitive cubic cells, one of Cs^+ and the other of Cl^-. The coordination number of both types of ion is 8, so the structure is described as having 8:8 coordination.

Note that NH_4Cl also forms this structure despite the relatively small size of the NH_4^+ ion because the cation can form hydrogen bonds with four of the Cl^- ions at the corners of the cube (Fig. 4.34). Many simple 1:1 alloys, such as AlFe and CuZn, have a caesium-chloride arrangement of the two metal atom types.

The **sphalerite structure** (Fig. 4.35), which is also known as the **zinc-blende structure**, takes its name from one of the mineral forms of ZnS. Like the rock-salt structure, it is based on an expanded ccp anion arrangement, but now the cations occupy one type of tetrahedral hole, one-half the tetrahedral holes present in a close-packed structure. Each ion is surrounded by four neighbours and so the structure has 4:4 coordination and $Z = 4$. This structure is adopted by a number of the Group 12 chalcogenides (e.g. one polymorph of ZnS, CdSe, HgS), Group 11 halides (CuCl, AgI

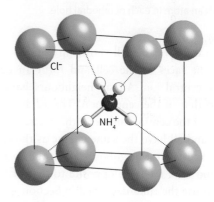

FIGURE 4.34 The structure of ammonium chloride, NH_4Cl, reflects the ability of the tetrahedral NH_4^+ ion to form hydrogen bonds to the tetrahedral array of Cl^- ions around it.

above 147°C), and Group 13 pnictides (GaP, BN). Different cations may fill the tetrahedral holes as in $LiNbAs_2$, with Li^+ and Nb^{5+} randomly placed across the half-filled tetrahedral

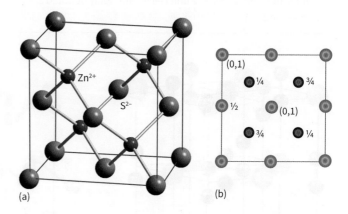

FIGURE 4.35 (a) The sphalerite (zinc blende) structure and (b) its projection representation. Note its relation to the ccp lattice in Fig. 4.20(a), with half the tetrahedral holes occupied by Zn^{2+} ions.

holes, and CuZn$_2$InSe$_4$, that has Cu$^+$, Zn^{2+} and In^{3+} distributed on the tetrahedral site in a close-packed array of Se^{2-} anions. Note that the tetrahedral coordination geometry present for both cations and anions in this structure type reflects a greater directionality in the bonding and its formation with more polarizable species.

A BRIEF ILLUSTRATION

To count the ions in the unit cell shown in the sphalerite structure shown in Fig. 4.34, we draw up the following table:

Location (share)	Number of cations	Number of anions	Contribution
Centre (1)	4 × 1	0	4
Face (½)	0	6 × ½	3
Edge (¼)	0	0	0
Vertex (⅛)	0	8 × ⅛	1
Total	4	4	8

There are four cations and four anions in the unit cell. This ratio is consistent with the chemical formula ZnS, with $Z = 4$.

The **wurtzite structure** (Fig. 4.36) takes its name from another polymorph of zinc sulfide that occurs naturally as a mineral. It differs from the sphalerite structure in being derived from an expanded hcp anion array rather than a ccp array, but as in sphalerite the cations occupy half the tetrahedral holes; that is just one of the two types (either T or T' as discussed in Section 4.3). This structure, which has 4:4 coordination, is adopted by ZnO, one form of AgI, and one polymorph of SiC, as well as several other compounds (Table 4.4). The local symmetries of the cations and anions are identical with respect to their nearest neighbours in wurtzite and sphalerite, but differ at the second-nearest neighbours. Many compounds show polymorphism exhibiting both sphalerite and wurtzite structure types; the choice depends on the conditions under which they crystallize or the temperature and pressure to which they are subjected.

The **nickel-arsenide structure** (NiAs, Fig. 4.37) is also based on an expanded, distorted hcp anion array, but the Ni atoms now occupy the octahedral holes, and each As atom lies at the centre of a trigonal prism of Ni atoms. This structure is adopted by NiS, FeS, and a number of other chalcogenides including FeSe, CrSe, VS, and PdTe.

The nickel-arsenide structure is typical of MX compounds that contain polarizable ions and are formed from elements with smaller electronegativity differences than elements that, as ions, adopt the rock-salt structure. Compounds that form this structure type lie in the 'polarized ionic salt area' of a Ketelaar triangle (Fig. 4.38). There is also potential for some degree of metal–metal bonding between metal atoms in adjacent layers (see Fig. 4.37(c)) and this structure type (or distorted forms of it) is also common for a large number of alloys based on d- and p-block elements including PtSn, BiRh, and IrPb.

A common AX$_2$ structural type is the **fluorite structure**, which takes its name from its exemplar, the naturally occurring mineral fluorite, CaF$_2$. In fluorite, the Ca^{2+} ions lie in an expanded ccp array and the F$^-$ ions occupy all the tetrahedral holes (Fig. 4.39). In this description it is the cations that are close-packed because the F$^-$ anions are small. The lattice has 8:4 coordination, which is consistent with there being twice as many anions as cations. The anions in their tetrahedral holes have four nearest neighbours and the cation site is surrounded by a cubic array of eight anions. The fluorite structure is adopted by compounds that have a relatively large cation and small anion (for example, O^{2-}, F$^-$ and H$^-$) including SrF$_2$, ThO$_2$, TiH$_2$, and PbF$_2$. Mixtures of both cations and anions are possible provided the overall

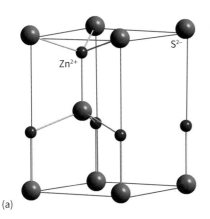
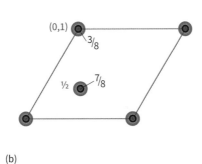

(a) (b)

FIGURE 4.36 (a) The wurtzite structure and (b) its projection representation.

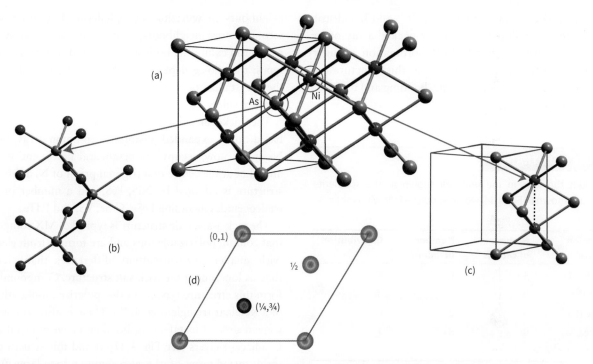

FIGURE 4.37 (a) The nickel-arsenide structure; (b) and (c) show the six-fold coordination geometries of As (trigonal-prismatic) and Ni (octahedral), respectively, and (d) is the projection representation of the unit cell. The short M–M interaction is shown as a dotted line in (c).

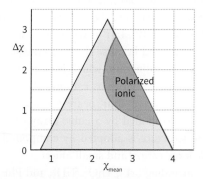

FIGURE 4.38 The location of polarized Ionic salts in a Ketelaar triangle.

stoichiometry remains AX_2, and more complex compound stoichiometries such as CfOF and $BaSnF_4$ adopt the fluorite structure type.

The **antifluorite** structure is the inverse of the fluorite structure in the sense that the locations of cations and anions are reversed; this reflects the fact that the structure is adopted in compounds with the smallest cations such as Li^+ (r = 59 pm in four-fold coordination). The structure is shown by some alkali metal oxides, including Li_2O. In it the cations (which are twice as numerous as the anions) occupy all the tetrahedral holes of a ccp array of anions. The coordination is 4:8 rather than the 8:4 of fluorite itself. Again,

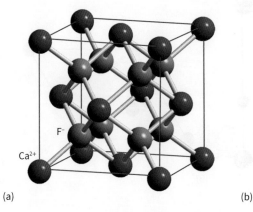 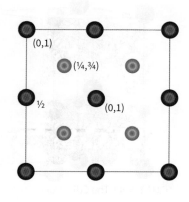

FIGURE 4.39 (a) The fluorite structure and (b) its projection representation. This structure has a ccp array of cations and all the tetrahedral holes are occupied by anions.

a single A or X species may be replaced by a mixture of cations or anions respectively and the compound Li_7NbP_4 adopts the antifluorite structure with P^{3-} anions forming a cubic close-packed array with $\frac{7}{8}$ of the tetrahedral holes filled by Li^+ and $\frac{1}{8}$ by Nb^{5+}.

The **rutile structure** (Fig. 4.40) takes its name from rutile, a mineral form of titanium(IV) oxide, TiO_2. The structure can also be considered an example of hole-filling in an hcp anion arrangement, but now the cations occupy only half the octahedral holes and there is considerable buckling of the close-packed anion layers. Each Ti^{4+} atom is surrounded by six O atoms, though the Ti-O distances are not identical and fall into two sets, so its coordination is more accurately described as (4+2). Each O atom is surrounded by three Ti^{4+} ions and hence the rutile structure has 6:3 coordination. The principal ore of tin, cassiterite SnO_2, has the rutile structure, and many transition metal difluorides adopt this structure type (Table 4.4).

In the **cadmium-iodide structure** (as in CdI_2, Fig. 4.41), the octahedral holes between every other pair of hcp layers of I^- ions (that is, half of the total number of octahedral holes) are filled by Cd^{2+} ions. The CdI_2 structure is often referred to as a 'layer structure', as the repeating layers of atoms perpendicular to the close-packed layers form the sequence I–Cd–I···I–Cd–I···I–Cd–I with weak van der Waals interactions between the iodine atoms in adjacent layers. The structure has (6,3) coordination, being octahedral for the cation and trigonal-pyramidal for the anion. The structure type is found commonly for many d-metal halides and chalcogenides (e.g. $FeBr_2$, MnI_2, ZrS_2, and $NiTe_2$).

The **cadmium-chloride structure** (as in $CdCl_2$, Fig. 4.42) is analogous to the CdI_2 structure but with a ccp arrangement of anions; half the octahedral sites, those between alternate anion layers, are occupied. This layer structure has identical coordination numbers (6,3) and geometries for the ions to those found for the CdI_2 structure type, although it is preferred for a number of d-metal dichlorides, such as $MnCl_2$ and $NiCl_2$.

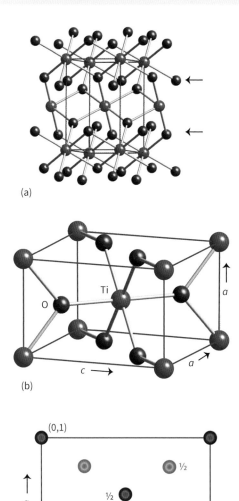

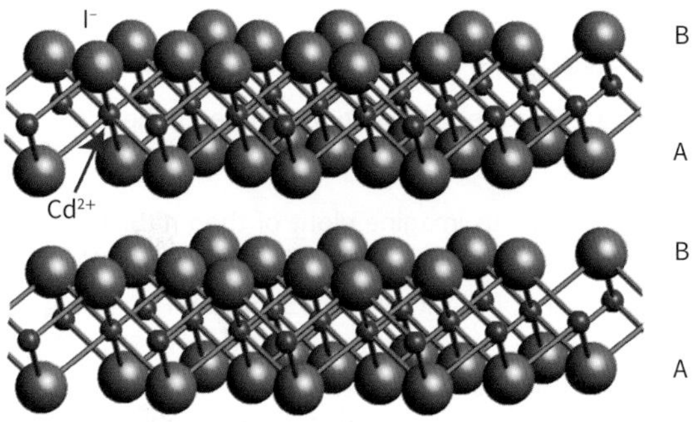

FIGURE 4.40 The rutile structure adopted by one polymorph of TiO_2: (a) the buckled close-packed layers of oxide ions, arrowed, with titanium cations in half the octahedral holes (the unit cell is outlined); (b) the unit cell, showing the titanium coordination to oxide ions; and (c) its projection representation.

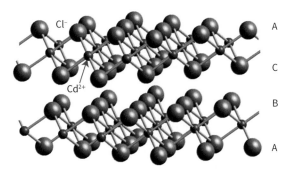

FIGURE 4.41 The CdI_2 structure; the iodide layers have hcp stacking.

FIGURE 4.42 The $CdCl_2$ structure; the chloride layers have ccp stacking.

EXAMPLE 4.8 Determining the stoichiometry of a hole-filled structure

Identify the stoichiometries of the following structures based on hole-filling using a cation, A, in close-packed arrays of anions, X: (a) an hcp array in which one-third of the octahedral sites are filled; (b) a ccp array in which all the tetrahedral and all the octahedral sites are filled.

Answer We need to be aware that in an array of N close-packed spheres there are $2N$ tetrahedral holes and N octahedral holes (Section 4.3). Therefore, filling all the octahedral holes in a closed-packed array of anions X with cations A would produce a structure in which cations and anions were in the ratio 1:1, corresponding to the stoichiometry AX. (a) As only one-third of the holes are occupied, the A:X ratio is $\frac{1}{3}:1$, corresponding to the stoichiometry AX_3. An example of this type of structure is BiI_3. (b) The total number of A species is $2N + N$ with N X species. The A:X ratio is therefore 3:1, corresponding to the stoichiometry A_3X. An example of this type of structure is Li_3Bi.

Self-test 4.8 Determine the stoichiometry of an hcp array with two-thirds of the octahedral sites occupied.

EXAMPLE 4.9 Predicting a possible structure-type for simple and complex ion compositions

Predict possible structure types for the following compounds: (a) PuO_2, (b) EuO, (c) $CsPF_6$, (d) Li_2TiO_3.

Answer We can use Table 4.4 and the known structures of compounds with similar compositions and ion sizes (Resource section 1) to propose structures for the simple binary compositions; for the compounds that contain complex ions we need also to consider the ratio of anions to cations. Possible structures for (a) are fluorite and rutile and Pu^{4+} is a similar size to U^{4+}, so from Table 4.4 we would predict the fluorite structure. (b) Eu^{2+} is a very similar size to Sr^{2+} and SrO has a rock-salt structure type. (c) We can write this compound as $Cs^+[PF_6]^-$, so likely structure types for this AX composition are rock salt and CsCl; in fact, this material adopts the rock-salt structure type with alternating Cs^+ and $[PF_6]^-$ ions. (d) This composition can be rewritten as $[AX]_3$ with the A sites filled by two thirds Li and one third Ti. This analysis is similar to the case of $LiNiO_2$, but now the A cation sites in the rock-salt structure type are filled randomly with 2/3 Li:1/3 Ti.

Self-test 4.9 Describe possible structures for (a) PrO_2, (b) CrO_2, (c) $CrTaO_4$, (d) AcOF, (e) Li_2TiF_6.

(b) Ternary phases, $A_aB_bX_n$

KEY POINT The perovskite and spinel structures are adopted by many compounds with the stoichiometries ABO_3 and AB_2O_4, respectively.

Structural possibilities increase very rapidly once the compositional complexity is increased to three ionic species. Unlike binary compounds, it is difficult to predict the most likely structure type based on the three different ion sizes and preferred coordination numbers. This section describes two important structures formed by ternary oxides and some ternary halides; the O^{2-} ion is the most common anion, so oxide chemistry is central to a significant part of solid-state chemistry.

The mineral perovskite, $CaTiO_3$, is the structural prototype of many ABX_3 solids (Table 4.4), particularly oxides. In its ideal form, the **perovskite structure** is cubic with each A cation surrounded by 12 X anions and each B cation surrounded by 6 X anions (Fig. 4.43). In fact, the perovskite structure may also be described as a close-packed array of A cations and O^{2-} anions (arranged such that each A cation is surrounded by 12 O^{2-} anions from the original close-packed layers; Fig. 4.43(d)), with B cations in all the octahedral holes that are formed from six of the O^{2-} ions, giving $B_{n/4}[AO_3]_{n/4}$, which is equivalent to ABO_3.

In oxides, X = O and the sum of the charges on the A and B ions must be +6. That sum can be achieved in several ways ($A^{2+}B^{4+}$ and $A^{3+}B^{3+}$ among them), including the possibility of mixed oxides of formula $A(B_{0.5}B'_{0.5})O_3$, as in $La(Ni_{0.5}Ir_{0.5})O_3$. The A-type cation in perovskites is therefore usually a large ion (of radius greater than 110 pm) of lower charge, such as Ba^{2+} or La^{3+}, and the B cation is a small ion (of radius less than 100 pm, typically 60–70 pm) of higher charge, such as Ti^{4+}, Nb^{5+}, or Fe^{3+}. In halide perovskites, X = F, Cl, Br, and I and the sum of the charges on the A and B cations must be +3. Example phases include $RbMgCl_3$, $KMnF_3$, and $CsPbI_3$. With X = I and B = Pb the A cation site is very large and the perovskite structure only forms with the largest inorganic cation, Cs^+ with an ionic radius of >180 pm. Alternatively, large organic cations, such as methylammonium, can adopt the A-site position and the material $[(CH_3)NH_3]PbI_3$ has recently been studied as a photovoltaic (Chapter 15, Box 15.4).

Materials adopting the perovskite structure often show interesting and useful electrical properties, such as piezoelectricity, ferroelectricity, and high-temperature superconductivity (Section 25.11(d) and Chapter 30).

EXAMPLE 4.10 Determining coordination numbers

Demonstrate that the coordination number of the Ti^{4+} ion in the perovskite $SrTiO_3$ is 6.

Answer We need to imagine eight of the unit cells shown in Fig. 4.43 stacked together with a Ti atom shared by them all. A local fragment of the structure is shown in Fig. 4.44; it shows that there are six O^{2-} ions around the central Ti^{4+} ion, so the coordination number of Ti in perovskite is 6. An alternative way of viewing the perovskite structure is as BO_6 octahedra sharing all vertices in three orthogonal directions with the A cations at the centres of the cubes so formed (Fig. 4.43(c)).

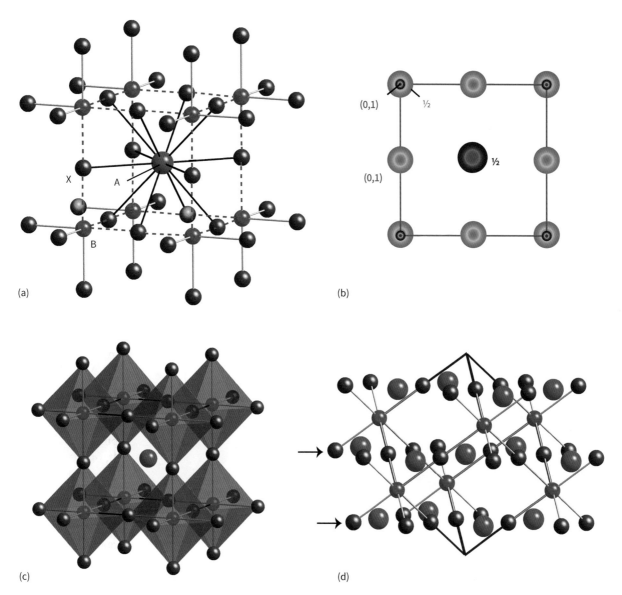

FIGURE 4.43 The perovskite structure, ABX_3: (a) the cubic unit cell outlined in dashed blue lines, emphasizing the coordination geometry of the A (12-fold) and B (6-fold octahedral) cations to X; (b) the projection representation of the unit cell; (c) the same structure, emphasizing the octahedral coordination of the B sites and description of the structure as linked BX_6 octahedra; (d) the relationship of the perovskite structure to a close-packed arrangement of A and X (arrowed) with B in octahedral holes; the unit cell outlined is the same as that in (a).

Self-test 4.10 What is the coordination environment of the O^{2-} site in $SrTiO_3$?

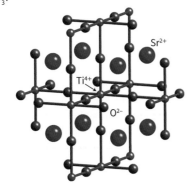

FIGURE 4.44 The local coordination environment of a Ti atom in perovskite.

Spinel itself is $MgAl_2O_4$, and oxide spinels, in general, have the formula AB_2O_4. The **spinel structure** consists of a ccp array of O^{2-} ions in which the A cations occupy one-eighth of the tetrahedral holes and the B cations occupy half the octahedral holes (Fig. 4.45). Spinel formulas are sometimes written $A[B_2]O_4$, the square brackets denoting the cation type (normally the smaller, higher-charged ion of A and B) that occupies the octahedral holes. So, for example, $ZnAl_2O_4$ can be written $Zn[Al_2]O_4$ to show that all the Al^{3+} cations occupy octahedral sites. Examples of compounds that have spinel structures include many ternary oxides with the stoichiometry AB_2O_4 that contain a 3d-series metal, such as $NiCr_2O_4$ and $ZnFe_2O_4$, and some simple binary d-block oxides, such as Fe_3O_4, Co_3O_4, and Mn_3O_4; note that in these structures A and B are the same element but in different

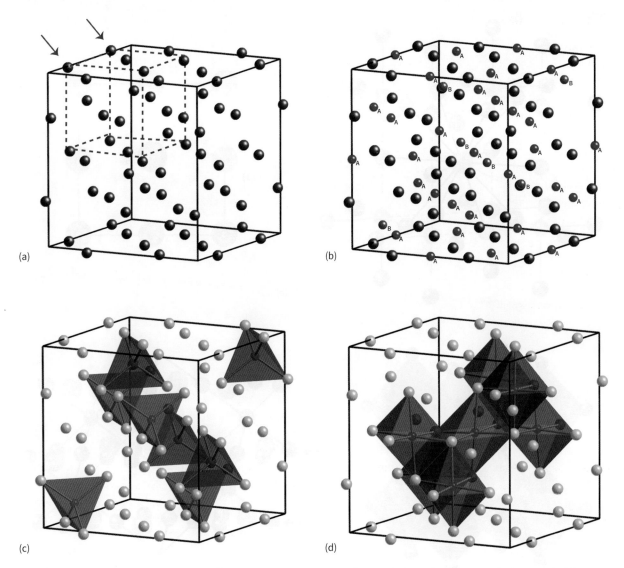

FIGURE 4.45 The spinel structure AB_2O_4: (a) the close-packed arrangement of anions (O^{2-}) in the full unit cell (close-packed layers are arrowed); the smaller simple ccp unit cell is outlined by dashed lines; (b) the arrangement of cations and anions within the full unit cell with the A cations shown in magenta and B cations in red; (c) and (d) the polyhedral coordination of A (tetrahedral) and B (octahedral) by oxide for cation sites fully within the unit cell.

oxidation states, as in $Mn^{2+}[Mn^{3+}]_2O_4$. There are also a number of compositions termed **inverse spinels**, in which the cation distribution is B[AB]O_4 and in which the more abundant cation is distributed over both tetrahedral and octahedral sites. Spinels and inverse spinels are discussed again in Sections 7.16(h), 22.1, and 25.11.

EXAMPLE 4.11 Predicting possible ternary phases

What ternary oxides with the perovskite or spinel structure might it be possible to synthesize that contain the cations Ti^{4+}, Zn^{2+}, In^{3+}, and Pb^{2+}? Use the ionic radii given in Resource section 1.

Answer For each of the possible cation pairings we need to consider whether the sizes of the ions permit the occurrence of the two structures. Starting with Zn^{2+} and Ti^{4+} we can predict that $ZnTiO_3$ does not exist as a perovskite as the Zn^{2+} ion is too small for the A-type site with B as Ti^{4+}; likewise, $PbIn_2O_4$ does not adopt the spinel structure as the Pb^{2+} cation is too large for the tetrahedral sites. We conclude that the permitted structures are $PbTiO_3$ (perovskite), $TiZn_2O_4$ (spinel), and $ZnIn_2O_4$ (spinel).

Self-test 4.11 What additional oxide perovskite composition(s) might be obtained if La^{3+} is added to this list of cations?

4.10 The rationalization of structures

The thermodynamic stabilities and structures of ionic solids can be treated very simply using the **ionic model** where ions are treated purely as hard, charged spheres. However, a model of a solid in terms of charged spheres interacting only electrostatically is crude and we should expect significant departures from its predictions because many solids involve some covalent bonding. Even conventional 'good' ionic solids, such as the alkali metal halides, have some covalent character. Nevertheless, the ionic model provides an attractively simple and effective scheme for correlating and understanding many of the properties exhibited by ionic solids.

(a) Ionic radii

KEY POINT The sizes of ions—ionic radii—generally increase down a group, decrease across a period, increase with coordination number, and decrease with increasing charge number.

A difficulty that confronts us at the outset is the meaning of the term 'ionic radius'. As remarked in Section 1.7, it is necessary to apportion the single internuclear separation of nearest-neighbour ions between the two different species (e.g. an Na^+ ion and a Cl^- ion in contact). The most direct way to solve the problem is to make an assumption about the radius of one ion, and then to use that value to compile a set of self-consistent values for all other ions. The O^{2-} ion has the advantage of being found in combination with a wide range of elements. It is also reasonably unpolarizable, so its size does not vary much as the identity and electropositivity of an accompanying cation is changed. In many compilations the values are based on $r(O^{2-}) = 140$ pm. However, this value is by no means sacrosanct: a set of values compiled by Goldschmidt was based on $r(O^{2-}) = 132$ pm, and other value sets use the F^- ion as the basis.

For certain purposes (such as predicting the sizes of unit cells) ionic radii can be helpful, but they are reliable only if they are all based on the same fundamental choice (such as the value 140 pm for O^{2-}). If values of ionic radii are used from different sources, it is essential to verify that they are based on the same convention. An additional complication, first noted by Goldschmidt, is that, as we have already seen for metals, apparent ionic radii increase with coordination number (Fig. 4.46, see also Resource section 1). This is because, when fitting larger numbers of oppositely charged ions around a central ion, the repulsions between the former will push them further away from the central ion, increasing its apparent radius. Hence, when comparing ionic radii, we should compare like with like, and use values for a single coordination number (typically 6).

The problems of the early workers have been resolved only partly by developments in X-ray diffraction (Section 8.1).

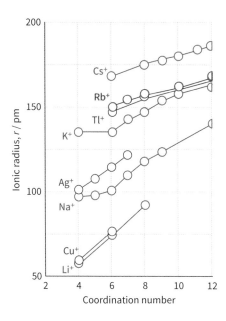

FIGURE 4.46 The variation of ionic radius with coordination number.

It is now possible to measure the electron density between two neighbouring ions and identify the minimum as the boundary between them. However, as can be seen from Fig. 4.47, the electron density passes through a very broad minimum, and its exact location may be very sensitive to experimental uncertainties and to the identities of the two neighbours. By analysing X-ray diffraction data on thousands of compounds, particularly oxides and fluorides, very extensive lists of self-consistent values have been compiled, and some are given in Table 1.4 and Resource section 1. These values for ionic radii are used throughout this chapter and the rest of the book.

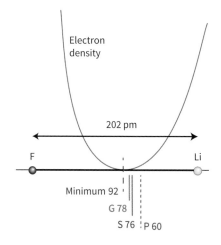

FIGURE 4.47 The variation in electron density along the Li–F axis in LiF. The point P indicates the Pauling radii of the two ions, G the original (1927) Goldschmidt radii, and S the Shannon radii (Resource section 1).

The general trends observed for ionic radii are the same as for atomic radii. Thus:

- Ionic radii increase down a group. (The lanthanoid contraction, discussed in Section 1.7(a), restricts the increase between the 4d- and 5d-series metal ions.)
- The radii of ions of the same charge decrease across a period.
- If an element can form cations with different charge numbers, then for a given coordination number its ionic radius decreases with increasing charge number.
- Because a positive charge indicates a reduced number of electrons, and hence a more dominant nuclear attraction, cations are smaller than anions for elements with similar atomic numbers.
- When an ion can occur in environments with different coordination numbers, the observed radius, as measured by considering the average distances to the nearest neighbours, increases as the coordination number increases. This increase reflects the fact that the repulsions between the surrounding ions are reduced if they move apart, so leaving more room for the central ion.

(b) The radius ratio

KEY POINT The radius ratio indicates the likely coordination numbers of the ions in a binary compound.

A parameter that figures widely in the literature of inorganic chemistry, particularly in introductory texts, is the **radius ratio**, γ (gamma), of the ions. This is the ratio of the radius of the smaller ion (r_{small}) to that of the larger (r_{large}):

$$\gamma = \frac{r_{small}}{r_{large}} \tag{4.1}$$

In most cases, r_{small} is the cation radius and r_{large} is the anion radius, though there are exceptions to this for the smallest anions (e.g. O^{2-} and F^-) with the larger cations (e.g. Cs^+, Ba^{2+}). The minimum radius ratio that can support a given coordination number is then calculated by considering the geometrical problem of packing together spheres of different sizes (Table 4.6).

It is argued that, if the radius ratio falls below the minimum given, then ions of opposite charge will not be in contact and ions of like charge will touch. According to a simple electrostatic argument, a lower coordination number, in which the contact of oppositely charged ions is restored, then becomes favourable. Another way of looking at this argument is that, as the radius of the M^+ ion increases, more anions can pack around it, so giving a larger number of favourable Coulombic interactions. Note that these simple electrostatic arguments only consider nearest-neighbour interactions, and a better model for predicting the packing ions requires more detailed calculations taking into account the whole ion array; these are undertaken in Section 4.12.

We can use our previous calculations of hole size (Example 4.4) to put these ideas on a firmer footing. A cation of radius $0.225r$ or below will fit into (and one below $0.225r$ will rattle around in) a tetrahedral hole. Note that $0.225r$ represents the size of the largest cation that will fit in a tetrahedral hole, and cations of radius between $0.225r$ and $0.414r$ will push the anions slightly apart. Therefore, a cation of radius between $0.225r$ and $0.414r$ will only occupy a tetrahedral hole in a slightly expanded close-packed array of anions of radius r, but this slight expansion of the anions remains the most energetically favourable arrangement. However, once the radius of a cation reaches $0.414r$, the anions are forced so far apart that octahedral coordination becomes possible and more favourable on electrostatic grounds with the greater number of cation to anion interactions. This will continue to be the most favourable arrangement until it becomes possible to fit eight anions around the cation when its radius reaches $0.732r$.

In summary, the coordination number will not increase to 6, with good contacts between cation and anions, until the radius goes above $0.414r$, and 6-coordination will be the preferred arrangement for $0.414 < \gamma < 0.732$. Similar arguments apply for the tetrahedral holes that can be filled by smaller ions with sizes between $0.225r$ and $0.414r$.

These concepts of ion-packing based on radius ratios can often be used to predict which structure is most likely for any particular choice of cation and anion (Table 4.6). In practice, the radius ratio is most reliable when the cation coordination number is 8, and less reliable with 6- and

TABLE 4.6 The correlation of structural type with radius ratio

Radius ratio (γ)	CNs for 1:1 and 1:2 stoichiometries*	Binary AB structure type	Binary AB_2 structure type
1	12	None known	None known
0.732–1	8:8 and 8:4	CsCl	CaF_2
0.414–0.732	6:6 and 6:3	NaCl (ccp), NiAs (hcp)	TiO_2
0.225–0.414	4:4	ZnS (ccp and hcp)	

* CN denotes coordination number.

4-coordination cations because directional covalent bonding becomes more important for these lower coordination numbers. Polarization effects are also important for larger ions. These factors, which depend upon the electronegativity and polarizability of ions, are considered in more detail in Section 4.10(c) as refinements to the simple ionic model.

A BRIEF ILLUSTRATION

To predict the crystal structure of TlCl we note that the ionic radii are $r(Tl^+)$ = 159 pm and $r(Cl^-)$ = 181 pm, giving γ = 0.88. As 0.732 < γ < 1 we can predict 8:8 coordination (that is the Tl^+ ionic radius is large enough that 8 chloride ions can fit around it) and that TlCl is likely to adopt a caesium-chloride structure. That is the structure found in practice.

EXAMPLE 4.12 Predicting structures

Predict structures for the ionic compounds RbI, BeO, and PbF_2 using radius-ratio rules and the ionic radii for six-fold coordination from Resource section 1.

Answer For each compound we need to calculate the radius ratio, γ, and then use Table 4.6 to select the most likely structure type. For RbI the ionic radii are Rb^+ = 148 pm and I^- = 220 pm, so γ = 0.672. This value falls in the range 0.414–0.732, so we would predict CNs 6:6 and the rock-salt structure (NiAs is a possibility on packing considerations but is normally only found where there is a degree of covalency in the bonding). Similar calculations for BeO and PbF_2 give γ = 0.321 and 0.894, respectively. BeO would thus be predicted to have one of the structures with coordination numbers 4:4 (in practice it adopts the zinc sulfide (wurtzite) structure with 4:4 coordination). PbF_2, as an AB_2 compound, would be predicted to have the fluorite structure (8:8 CNs), and again experimentally one form of this compound is found to adopt this structure type.

Self-test 4.12 Predict structures for CaO and BkO_2 (Bk = berkelium, an actinoid) using radius-ratio rules and the ionic radii for six-fold coordination from Resource section 1.

The ionic radii used in these calculations are those obtained by consideration of structures under normal ambient conditions. At high pressures, different structures may be preferred, especially those with higher coordination numbers and greater density. Thus many simple compounds transform between the common 4:4-, 6:6-, and 8:8-coordination structures under pressure. Examples of this behaviour include most of the lighter alkali metal halides, which change from a 6:6-coordination rock-salt structure to an 8:8-coordination caesium-chloride structure at 5 kbar (the rubidium halides) or 10–20 kbar (the sodium and potassium halides). The ability to predict the structures of compounds under pressure is important for understanding the behaviour of ionic compounds under such conditions; for example, in geochemistry. Calcium oxide, for instance, is predicted to transform from the rock-salt to the caesium-chloride structure at around 600 kbar, the pressure in the Earth's lower mantle.

Similar arguments involving the relative ionic radii of cations and anions and their preferred coordination numbers (that is, preferences for octahedral, tetrahedral, or cubic geometries) can be applied throughout structural solid-state chemistry and aid the prediction of which ions might be incorporated into a particular structure type. For more complex stoichiometries, such as the ternary compounds with perovskite and spinel structure types, the ability to predict which combinations of cations and anions will yield a specific structure type has proved very useful. One example is that, for the high-temperature superconducting cuprates (Section 30.7), the design of a particular structure feature, such as Cu^{2+} in octahedral coordination to oxygen, can be achieved using ionic-radii considerations. Similarly, for the lead iodide perovskites $APbI_3$ (Section 30.2) that have been investigated as photovoltaic materials, the size of the cation A that can adopt 12 co-ordination to I^- can be estimated; only Cs^+ of the simple metal cations is large enough.

(c) Structure maps

KEY POINT A structure map is a representation of the variation in crystal structure with the character of the bonding.

The use of radius-ratio rules is not totally reliable (it only predicts the experimental structure for about 50% of compounds). This fallibility is often a result of additional, non-ionic interactions between the ions including some degree of directional covalent bonding and polarization effects. However, it is possible to rationalize the choice of structure of a compound further by collecting enough information empirically and looking for patterns. This approach has motivated the compilation of **structure maps**. An example of a structure map is an empirically compiled map that depicts the dependence of crystal structure on the electronegativity difference between the elements present and the average principal quantum number of the valence shells of the two atoms. As such, a structure map can be regarded as an extension of the ideas introduced in Chapter 2 in relation to the Ketelaar triangle. As we have seen, binary ionic salts are formed for large differences in electronegativity, $\Delta\chi$, but as this difference is reduced, polarized ionic salts and more covalently bonded networks become preferred. Now we can focus on this region of the triangle and explore how small changes in electronegativity and polarizability affect the choice of ion arrangement in addition to ionic-radii considerations.

The ionic character of a bond increases with $\Delta\chi$, so moving from left to right along the horizontal axis of a structure

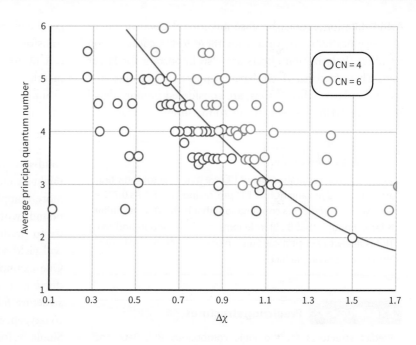

FIGURE 4.48 A structure map for compounds of formula MX. A point is defined by the electronegativity difference ($\Delta\chi$) between M and X and their average principal quantum number n. The location on the map indicates the coordination number expected for that pair of properties. (Based on E. Mooser and W. B. Pearson, *Acta Crystallogr.*, 1959, **12**, 1015. Reproduced with permission of the International Union of Crystallography.)

map (Fig. 4.48) correlates with an increase in ionic character in the bonding. The principal quantum number is an indication of the radius of an ion, so moving up the vertical axis corresponds to an increase in the average radius of the ions. Because atomic energy levels become closer as the atom expands, the polarizability of the atom also increases (Section 1.7(f)). Consequently, the vertical axis of a structure map corresponds to increasing size and polarizability of the bonded atoms.

Figure 4.48 shows an example of a structure map for MX compounds. We see that the structures we have been discussing for MX compounds fall into distinct regions of the map. Elements with large $\Delta\chi$ have 6:6 coordination, such as is found in the rock-salt structure; elements with small $\Delta\chi$ (and hence where there is the expectation of covalence) have lower coordination numbers. In terms of a structure-map representation, GaN is in a more covalent region of Fig. 4.48 than ZnO because $\Delta\chi$ is appreciably smaller.

A BRIEF ILLUSTRATION

To predict the type of crystal structure that should be expected for magnesium sulfide, MgS, we note that the electronegativities of magnesium and sulfur are 1.3 and 2.6, respectively, so $\Delta\chi = 1.3$. The average principal quantum number is 3 (both elements are in Period 3). The point $\Delta\chi = 1.3$, $n = 3$ lies just in the six-fold coordination region of the structure map in Fig. 4.48. This location is consistent with the observed rock-salt structure of MgS.

The energetics of ionic bonding

A compound tends to adopt the crystal structure that corresponds to the lowest Gibbs energy. Therefore, if for the process

$$M^+(g) + X^-(g) \to MX(s)$$

the change in standard reaction Gibbs energy, $\Delta_r G^\ominus$, is more negative for the formation of a structure A rather than B, then the transition from B to A is spontaneous under the prevailing conditions, and we can expect the solid to be found with structure A.

The process of solid formation from the gas of ions is so exothermic that at and near room temperature the contribution of the entropy to the change in Gibbs energy (as in $\Delta_r G^\ominus = \Delta H^\ominus - T\Delta S^\ominus$) may be neglected; this neglect is rigorously true at $T = 0$. Hence, discussions of the thermodynamic properties of solids normally focus, initially at least, on changes in enthalpy. That being so, we look for the structure that is formed most exothermically and identify it as the thermodynamically most stable form. Some typical values of lattice enthalpies are given in Table 4.7 for a number of simple ionic compounds.

TABLE 4.7 Lattice enthalpies of some simple inorganic solids

Compound	Structure type	$\Delta_L H^{exp}/(kJ\ mol^{-1})$
LiF	Rock salt	1030
LiI	Rock salt	757
NaF	Rock salt	923
NaCl	Rock salt	786
NaBr	Rock salt	747
NaI	Rock salt	704
KCl	Rock salt	719
KI	Rock salt	659
CsF	Rock salt	744
CsCl	Caesium chloride	657
CsBr	Caesium chloride	632
CsI	Caesium chloride	600
MgF_2	Rutile	2922
CaF_2	Fluorite	2597
$SrCl_2$	Fluorite	2125
LiH	Rock salt	858
NaH	Rock salt	782
KH	Rock salt	699
RbH	Rock salt	674
CsH	Rock salt	648
BeO	Wurtzite	4293
MgO	Rock salt	3795
CaO	Rock salt	3414
SrO	Rock salt	3217
BaO	Rock salt	3029
Li_2O	Antifluorite	2799
TiO_2	Rutile	12150
CeO_2	Fluorite	9627

4.11 Lattice enthalpy and the Born–Haber cycle

KEY POINTS Lattice enthalpies are determined from enthalpy data by using a Born–Haber cycle; the most stable crystal structure of the compound is commonly the structure with the greatest lattice enthalpy under the prevailing conditions.

The **lattice enthalpy**, $\Delta_L H^\ominus$, is the standard molar enthalpy change accompanying the formation of a gas of ions from the solid:

$$MX(s) \rightarrow M^+(g) + X^-(g) \quad \Delta_L H^\ominus$$

A NOTE ON GOOD PRACTICE

The definition of lattice enthalpy as an endothermic (positive) term corresponding to the break-up of the lattice is correct but contrary to many school and college texts, where it is defined with respect to lattice formation (and listed as a negative quantity).

A NOTE ON GOOD PRACTICE

The terms 'lattice enthalpy' and 'lattice energy' are often used interchangeably, though because of variations in thermodynamic functions that define these quantities under standard conditions (such as the work, $P\Delta V$, involved in forming the gaseous ions) they differ by a few kJ mol⁻¹. This difference is, however, negligible compared with errors in experimental and theoretical determined values, hence the accepted use of either term.

Because lattice disruption is always endothermic, lattice enthalpies are always positive and their positive signs are normally omitted from their numerical values. As remarked above, if entropy considerations are neglected, then the most stable crystal structure of the compound is the structure with the greatest lattice enthalpy under the prevailing conditions.

Lattice enthalpies are determined from empirical enthalpy data by using a **Born–Haber cycle**, a closed path of steps that includes lattice formation as one stage, such as that shown in Fig. 4.49. The standard enthalpy of decomposition of a compound into its elements in their reference states (their most stable states under the prevailing conditions) is the negative of its standard enthalpy of formation, $\Delta_f H^\ominus$:

$$M(s) + X(s,l,g) \rightarrow MX(s) \quad \Delta_f H^\ominus$$

Likewise, the standard enthalpy of lattice formation from the gaseous ions is the negative of the lattice enthalpy as specified above. For a solid element, the standard enthalpy

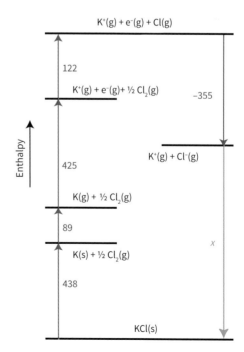

FIGURE 4.49 The Born–Haber cycle for KCl. The lattice enthalpy is equal to $-x$. All numerical values are in kJ mol⁻¹.

of atomization, $\Delta_{atom}H^\ominus$, is the standard enthalpy of sublimation, as in the process

$$M(s) \rightarrow M(g) \quad \Delta_{atom}H^\ominus$$

For a gaseous element, the standard enthalpy of atomization is the standard enthalpy of dissociation, $\Delta_{dis}H^\ominus$, as in

$$X_2(g) \rightarrow 2X(g) \quad \Delta_{dis}H^\ominus$$

The standard enthalpy of formation of ions from their neutral atoms is the enthalpy of ionization (for the formation of cations, $\Delta_{ion}H^\ominus$) and the electron-gain enthalpy (for anions, $\Delta_{eg}H^\ominus$):

$$M(g) \rightarrow M^+(g) + e^-(g) \quad \Delta_{ion}H^\ominus$$
$$X(g) + e^-(g) \rightarrow X^-(g) \quad \Delta_{eg}H^\ominus$$

The value of the lattice enthalpy—the only unknown in a well-chosen cycle—is found from the requirement that the sum of the enthalpy changes round a complete cycle is zero (because enthalpy is a state property).[4] The value of the lattice enthalpy obtained from a Born–Haber cycle depends on the accuracy of all the measurements being combined, and as a result there can be significant variations, typically ±10 kJ mol⁻¹, in tabulated values.

EXAMPLE 4.13 Using a Born–Haber cycle to determine a lattice enthalpy

Calculate the lattice enthalpy of KCl(s) using a Born–Haber cycle and the information shown in the following table:

	$\Delta H^\ominus$/(kJ mol⁻¹)
Sublimation of K(s)	+89
Ionization of K(g)	+425
Dissociation of Cl$_2$(g)	+244
Electron gain by Cl(g)	−355
Formation of KCl(s)	−438

Answer The required cycle is shown in Fig. 4.49. The sum of the enthalpy changes around the cycle is zero, so

$$\Delta_L H^\ominus = 438 + 425 + 89 + (244/2) - 355 = 719 \text{ kJ mol}^{-1}$$

Note that the calculation becomes more obvious if we draw an energy-level diagram showing the signs of the various steps of the cycle; all lattice enthalpies are positive. Also, as only one Cl atom from Cl$_2$(g) is required to produce KCl, half the dissociation energy of Cl$_2$, $\frac{1}{2} \times 244$ kJ mol⁻¹, is used in the calculation.

---

[4] Note that when the lattice enthalpy is known from calculation (Section 4.12), a Born–Haber cycle may be used to determine the value of another elusive quantity, the electron-gain enthalpy (and hence the electron affinity).

Self-test 4.13 Calculate the lattice enthalpy of magnesium bromide from the data shown in the following table:

	$\Delta H^\ominus$/(kJ mol⁻¹)
Sublimation of Mg(s)	+148
Ionization of Mg(g)	+2187 to Mg^{2+}(g)
Vaporization of Br$_2$(l)	+31
Dissociation of Br$_2$(g)	+193
Electron gain by Br(g)	−331
Formation of MgBr$_2$(s)	−524

4.12 The calculation of lattice enthalpies

Once the lattice enthalpy is known, it can be used to judge the character of the bonding in the solid. If the value calculated on the assumption that the lattice consists of ions interacting electrostatically is in good agreement with the measured value, then it may be appropriate to adopt a largely ionic model of the compound. A discrepancy indicates that a purely ionic model of the interactions is inadequate and a degree of covalence exists in the bonding. As mentioned earlier, it is important to remember that numerical coincidences can be misleading in this assessment.

(a) The Born–Mayer equation

KEY POINTS The Born–Mayer equation is used to estimate the lattice enthalpy for an ionic lattice. The Madelung constant reflects the effect of the geometry of the lattice on the strength of the net Coulombic interaction.

To calculate the lattice enthalpy of a supposedly ionic solid we need to take into account several contributions, including the Coulombic attractions and repulsions between the ions and the repulsive interactions that occur when the electron densities of the ions overlap. This calculation yields the **Born–Mayer equation** for the lattice enthalpy at $T = 0$:

$$\Delta_L H^\ominus = \frac{N_A |z_A z_B| e^2}{4\pi\varepsilon_0 d}\left(1 - \frac{d^*}{d}\right) A \quad (4.2)$$

where $d = r_1 + r_2$ is the distance between centres of neighbouring cations and anions, and hence a measure of the 'scale' of the unit cell (for the derivation, see Further Information: the Born–Mayer Equation). In this expression N_A is Avogadro's constant, z_A and z_B the charge numbers of the cation and anion, e the fundamental charge, ε_0 the vacuum permittivity, and d^* a constant (typically 34.5 pm) used to represent the repulsion between ions at short range. The quantity A is called the **Madelung constant**, and depends on the structure (specifically, on the relative distribution of ions, Table 4.8);

TABLE 4.8 Madelung constants

Structural type	A
Caesium chloride	1.763
Fluorite	2.519
Rock salt	1.748
Rutile	2.408
Sphalerite	1.638
Wurtzite	1.641

see Box 4.3. The Born–Mayer equation in fact gives the lattice energy as distinct from the lattice enthalpy, but the two are identical at $T = 0$ and the difference may be disregarded in practice at normal temperatures.

A BRIEF ILLUSTRATION

To estimate the lattice enthalpy of sodium chloride, we use $z(\text{Na}^+) = +1$, $z(\text{Cl}^-) = -1$, $A = 1.748$ from Table 4.8, and $d = r_{\text{Na}^+} + r_{\text{Cl}^-} = 283$ from Table 1.4; hence (using fundamental constants) and ensuring that the units of d are appropriate to each part of the equation

$$\Delta_L H^\ominus = \frac{(6.022 \times 10^{23}\,\text{mol}^{-1}) \times (+1) \times (-1) \times (1.602 \times 10^{-19}\,\text{C})^2}{4\pi \times (8.854 \times 10^{-12}\,\text{J}^{-1}\text{C}^2\text{m}^{-1}) \times (2.83 \times 10^{-10}\,\text{m})}$$

$$\times \left(1 - \frac{34.5\,\text{pm}}{283\,\text{pm}}\right) \times 1.748$$

$$= 7.56 \times 10^5\,\text{J mol}^{-1}$$

or 756 kJ mol^{-1}. This value compares reasonably well with the experimental value from the Born–Haber cycle, 788 kJ mol^{-1}.

BOX 4.3 How the Madelung constant is calculated

The calculation of the total Coulombic energy of a crystal involves summing all the individual Coulomb potential terms of the form

$$V_{AB} = \frac{(z_A e) \times (z_B e)}{4\pi\varepsilon_0 r_{AB}}$$

for ions of charge number z_A and z_B (ensuring that the charges on cations and anions are given the correct signs) separated by a distance r_{AB}. This summation can be carried out for any arrangement of ions or structure type, but in practice it converges very slowly. This is because while r_{AB} increases (with a decreasing contribution to V_{AB}) the number of pairs of ions at that distance in a crystal also increases; note also that every other 'shell' of ions around a central point is normally of opposite charge and so they alternate between positive and negative contributions to V_{AB}.

The calculation of the Madelung constant can be illustrated by consideration of a uniformly spaced one-dimensional line of alternating cations and anions (Fig. B4.2). The shortest two equivalent interactions contribute a Coulombic potential energy proportional to $\frac{-2z^2}{d}$, the second pair $\frac{+2z^2}{2d}$, and so on, giving

$$\frac{4\pi\varepsilon_0 V}{e^2} = -\frac{2z^2}{d} + \frac{2z^2}{2d} - \frac{2z^2}{3d} + \frac{2z^2}{4d} - \frac{2z^2}{5d}$$

$$= \frac{-2z^2}{d}\left(1 - \frac{1}{2} + \frac{1}{3} - \frac{1}{4} + \frac{1}{5}\cdots\right)$$

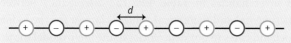

FIGURE B4.2

The slowly converging series may be shown to equal ln 2, so that for the general case of a row of ions of alternating charge z, separated by a distance d,

$$V = \frac{e^2}{4\pi\varepsilon_0} \times \frac{z^2}{d} \times 2\ln 2$$

2 ln 2 = 1.386, which is the Madelung constant for this array of ions. Similar computations can be carried out for all structure types to yield the values given in Table 4.8. The basis of the calculation for the rock-salt structure type can be seen from Fig. B4.3, where the series has the terms

$$\frac{4\pi\varepsilon_0 V}{e^2} = -\frac{6z^2}{d} + \frac{12z^2}{\sqrt{2}d} - \frac{8z^2}{\sqrt{3}d} + \frac{6z^2}{2d}\cdots$$

for 6 anions around a central cation at a distance d (the sum of the two ionic radii), 12 cations at a distance $\sqrt{2}d$, 8 anions at a distance $\sqrt{3}d$, and so on. In this case, the series sums to give the Madelung constant for the rock-salt structure of 1.748.

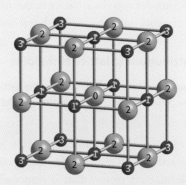

FIGURE B4.3 Basis for the calculation of the Madelung constant for the rock-salt structure type, showing the shells of ions at different distances from the central ion, labelled 0.

The form of the Born–Mayer equation for lattice enthalpies allows us to account for their dependence on the charges and radii of the ions in the solid. Thus, the heart of the equation is

$$\Delta_L H^\ominus \propto \frac{|z_A z_B|}{d}$$

Therefore, a large value of d (corresponding to large cations and anions) results in a low lattice enthalpy, whereas high ionic charges result in a high lattice enthalpy. This dependence is seen in some of the values given in Table 4.7. For the alkali metal halides, the lattice enthalpies decrease from LiF to LiI and from LiF to CsF as the halide and alkali metal ion radii increase, respectively. We also note that the lattice enthalpy of MgO ($|z_A z_B| = 4$) is just over four times that of NaCl ($|z_A z_B| = 1$), because of the increased charges on the ions for a similar value of d, noting that the Madelung constant is the same.

The Madelung constant typically increases slightly with coordination number. For instance, $A = 1.748$ for the 6:6-coordination rock-salt structure, but 1.763 for the 8:8-coordination caesium-chloride structure and 1.638 for the 4:4-coordination sphalerite structure. This dependence reflects the fact that a large contribution comes from nearest neighbours, and such neighbours are more numerous when the coordination number is large. However, a high coordination number does not necessarily mean that the interactions are stronger in the caesium-chloride structure because the potential energy also depends on the scale of the lattice. Thus, d may be so large in lattices with ions big enough to adopt eight-fold coordination that the separation of the ions reverses the effect of the small increase in the Madelung constant and results in a smaller lattice enthalpy. One manifestation of the larger Madelung constant for higher coordination numbers is that such a high coordination number structure is often adopted when simple radius-ratio rules (Section 4.10(b)) would predict the lower coordination number structure. Thus LiI, which has a radius ratio $\gamma = 0.34$, predicting 4:4 coordination, is found from experiment to adopt the rock-salt structure with 6:6 coordination.

(b) Other contributions to lattice enthalpies

KEY POINT Nonelectrostatic contributions to the lattice enthalpy include van der Waals interactions, particularly the dispersion interaction.

Another contribution to the lattice enthalpy is the **van der Waals interaction** between the ions and molecules, the weak intermolecular interaction that is responsible for the formation of condensed phases of electrically neutral species. An important and sometimes dominant contribution of this kind is the **dispersion interaction** (sometimes also termed the 'London interaction'). The dispersion interaction arises from the transient fluctuations in electron density (and, consequently, instantaneous electric dipole moment) on one molecule driving a fluctuation in electron density (and dipole moment) on a neighbouring molecule, and the attractive interaction between these two instantaneous electric dipoles. The molar potential energy of this interaction, V, is expected to vary as

$$V = -\frac{N_A C}{d^6} \quad (4.3)$$

The constant C depends on the substance. For ions of low polarizability, this contribution is only about one per cent of the electrostatic contribution and is ignored in elementary lattice enthalpy calculations of ionic solids. However, for highly polarizable ions such as Tl^+ and I^-, such terms can make significant contributions of several per cent. Thus, the dispersion interaction for compounds such as LiF and CsBr is estimated to contribute 16 kJ mol^{-1} and 50 kJ mol^{-1}, respectively.

4.13 Comparison of experimental and theoretical values

KEY POINTS For compounds formed from elements with $\Delta\chi > 2$, the ionic model is generally valid and lattice enthalpy values derived using the Born–Mayer equation and Born–Haber cycles are similar. For structures formed with small electronegativity differences and polarizable ions there may be additional, nonionic contributions to the bonding. Lattice enthalpy calculations may also be used to predict the stability or otherwise of unknown compounds.

The agreement between the experimental lattice enthalpy and the value calculated using the ionic model of the solid (in practice, from the Born–Mayer equation) provides a measure of the extent to which the solid is ionic. Table 4.9 lists some calculated and measured lattice enthalpies together with electronegativity differences. The ionic model is reasonably valid if $\Delta\chi > 2$, but the bonding becomes increasingly covalent if $\Delta\chi < 2$. However, it should be remembered that the electronegativity criterion ignores the role of polarizability of the ions. Thus, experimental lattice enthalpies for the alkali metal halides all show very good or fairly good agreement with the ionic model: the best with the least polarizable halide ions (F^-) formed from the highly electronegative F atom, and the worst with the highly polarizable halide ions (I^-) formed from the less electronegative I atom.

This trend is also seen in the lattice enthalpy data for the silver halides in Table 4.9. The discrepancy between experimental and theoretical values is largest for the iodide, which indicates major deficiencies in the ionic model for this

The energetics of ionic bonding

TABLE 4.9 Comparison of experimental and theoretical lattice enthalpies for rock-salt structures

	$\Delta_L H^{calc}$/(kJ mol^{-1})	$\Delta_L H^{exp}$/(kJ mol^{-1})	$(\Delta_L H^{exp} - \Delta_L H^{calc})$/(kJ mol^{-1}), percentage difference	$\Delta\chi$
LiF	1029	1030	1, 0.1%	3.00
LiCl	834	853	19, 2.2%	2.18
LiBr	788	807	19, 2.4%	1.98
LiI	730	757	27, 3.6%	1.68
AgF	920	953	33, 3.5%	2.05
AgCl	832	903	71, 7.9%	1.23
AgBr	815	895	80, 8.9%	1.03
AgI	777	882	105, 11.9%	0.73

compound. Overall the agreement is much poorer with Ag than with Li, as the electronegativity of silver ($\chi = 1.93$) is much higher than that of lithium ($\chi = 0.98$) and significant covalent character in the bonding would be expected.

It is not always clear whether it is the electronegativity of the atoms or the polarizability of the resultant ions that should be used as a criterion. The worst agreement with the ionic model is for polarizable-cation/polarizable-anion combinations that are substantially covalent. Here again, though, the difference between the electronegativities of the parent elements is small and it is not clear whether electronegativity or polarizability provides the better criterion.

EXAMPLE 4.14 Using the Born–Mayer equation to decide the theoretical stability of unknown compounds

Decide whether solid ArCl is likely to exist.

Answer The answer hinges on whether the enthalpy of formation of ArCl is significantly positive or negative: if it is significantly positive (endothermic), the compound is unlikely to be stable (there are, of course, exceptions). Consideration of a Born–Haber cycle for the synthesis of ArCl would show two unknowns: the enthalpy of formation of ArCl and its lattice enthalpy. We can estimate the lattice enthalpy of a purely ionic ArCl by using the Born–Mayer cycle and associated equation, assuming the radius of Ar$^+$ to be midway between that of Na$^+$ and K$^+$. That is, the lattice enthalpy is somewhere between the values for NaCl and KCl at about 745 kJ mol^{-1}. A cycle similar to that used for the formation of KCl(s), Fig. 4.49, can be used for ArCl, Fig. 4.50. We need to produce 1 mol of Cl atoms so, taking half the enthalpy of dissociation of Cl$_2$, 122 kJ mol^{-1}, the ionization enthalpy of Ar as 1524 kJ mol^{-1}, and the electron affinity of Cl as 355 kJ mol^{-1} gives $\Delta_f H(\text{ArCl, s}) = 1524 - 745 - 355 + 122$ kJ mol^{-1} = +546 kJ mol^{-1}. That is, the compound is predicted to be very unstable with respect to its elements, mainly because the large ionization enthalpy of Ar is not compensated by the lattice enthalpy.

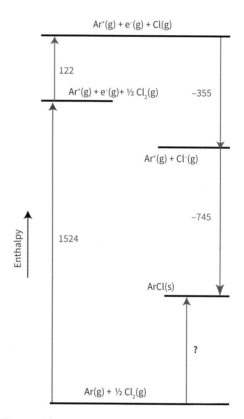

FIGURE 4.50 The Born–Haber cycle for the hypothetical ArCl; note that the lattice enthalpy is a negative quantity for the process shown. All numerical values are in kilojoules per mole.

Self-test 4.14 Predict whether CsCl$_2$ with the fluorite structure is likely to exist.

Calculations such as that in Example 4.14 were used to predict the stability of the first noble gas compounds. The ionic compound O$_2^+$PtF$_6^-$ had been obtained from the reaction of oxygen with PtF$_6$. Consideration of the ionization energies of O$_2$ (1176 kJ mol^{-1}) and Xe (1169 kJ mol^{-1}) showed them to be almost identical and the sizes of Xe$^+$ and

O_2^+ would be expected to be similar, implying similar lattice enthalpies for their compounds. Hence, once O_2 had been found to react with platinum hexafluoride, it could be predicted that Xe should too, as indeed it does, to give an ionic compound which is believed to contain XeF^+ and PtF_6^- ions.

Similar calculations may be used to predict the stability or otherwise of a wide variety of compounds; for example, alkaline earth monohalides such as MgCl. Calculations based on Born–Mayer lattice enthalpies and Born–Haber cycles, Fig. 4.51, show that such a compound would be expected to disproportionate into Mg and $MgCl_2$. By considering the reaction

$$2\ MgCl(s) \rightarrow MgCl_2(s) + Mg(s)$$

and estimating the lattice enthalpy of MgCl to be similar to that of NaCl at +786 kJmol^{-1}, the enthalpy change for the reaction can be estimated as

$$\begin{aligned}\Delta_{disprop}H &= +(2 \times 786)(\Delta_L H \text{ of MgCl}) - 373(1^{st} \text{ IE of Mg}) \\ &\quad +1451(2^{nd} \text{ IE of Mg}) - 148(\Delta_{sub}H \text{ of Mg}) \\ &\quad -2526(\Delta_L H \text{ of MgCl}_2) \\ &= -388\ \text{kJ mol}^{-1}\end{aligned}$$

so the reaction would be expected to proceed to the right-hand side. Note that calculations of this type provide only an estimate of the enthalpies of formation of ionic compounds, and hence some idea of the thermodynamic stability of a compound. It may still be possible to isolate a thermodynamically unstable compound if its decomposition is very slow. Indeed, a compound containing Mg(I) was reported in 2007 (Section 13.14), though the bonding in this compound is mainly covalent.

4.14 The Kapustinskii equation

KEY POINT The Kapustinskii equation is used to estimate lattice enthalpies of ionic compounds and to give a measure of the thermochemical radii of the constituent ions.

A. F. Kapustinskii observed that, if the Madelung constants for a number of structures are divided by the number of ions per formula unit, N_{ion}, then approximately the same value is obtained for them all; N_{ion} is 2 for AX structures, 3 for AX_2 etc. He also noted that the value so obtained increases with the coordination number. Therefore, because ionic radius also increases with coordination number, the variation in $A/(N_{ion}d)$ from one structure to another can be expected to be fairly small. This observation led Kapustinskii to propose that there exists a hypothetical rock-salt structure that is energetically equivalent to the true structure of any ionic solid. Consequently, the lattice enthalpy can be calculated by using the rock-salt Madelung constant, N_{ion}, and the appropriate ionic radii for 6:6 coordination. The resulting expression, derived from eqn. 4.2, is called the **Kapustinskii equation**:

$$\Delta_L H^\ominus = -\frac{N_{ion}|z_A z_B|}{d}\left(1 - \frac{d^*}{d}\right)\kappa \quad (4.4)$$

In this equation, $\kappa = 1.21 \times 10^5$ kJ pm mol^{-1}, d (given in pm) is the sum of the ionic radii ($r_+ + r_-$), and d^* is taken, as before, as 34.5 pm. If the repulsive energy term, d^*/d, has a value appropriate to the NaCl structure, at around 10% of the Coulombic attractive energy—that is, 0.1—then the Kapustinskii equation can be simplified further to

$$\Delta_L H^\ominus = \frac{N_{ion}|z_A z_B|}{r_+ + r_-}\kappa' \quad (4.5)$$

where $\kappa' = 1.08 \times 10^5$ kJ pm mol^{-1}. Eqn. 4.5 can be used to estimate the lattice enthalpy for any crystalline solid

A BRIEF ILLUSTRATION

To estimate the lattice enthalpy of radium fluoride, RaF_2, we need the number of ions per formula unit ($N_{ion} = 3$), their charge numbers ($z(Ra^{2+}) = +2$, $z(F^-) = -1$), and the sum of their ionic radii (148 pm + 133 pm = 281 pm). Then, with $d^* = 34.5$ pm, using eqn. 4.4,

$$\Delta_L H^\ominus = \frac{3|(+2)(-1)|}{281\,\text{pm}} \times \left(1 - \frac{34.5\,\text{pm}}{281\,\text{pm}}\right) \times (1.21 \times 10^5)\,\text{kJ pm mol}^{-1}$$
$$= 2266\,\text{kJ mol}^{-1}$$

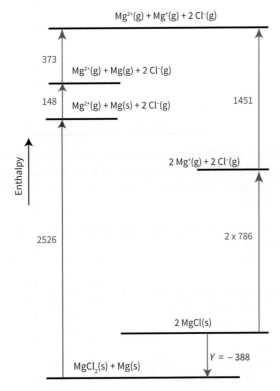

FIGURE 4.51 The Born–Haber cycle for the disproportionation of MgCl(s); all numerical values are in kilojoules per mole.

TABLE 4.10 The thermochemical radii of ions, r/pm

Main-group elements

BeF_4^{2-}	BF_4^-	CO_3^{2-}	NO_3^-	OH^-	
(245)	232	178	179	133	
		CN^-	NO_3^-	O_2^{2-}	
		191	(189)	173	
			PO_4^{3-}	SO_4^{2-}	ClO_4^-
			(238)	258	240
			AsO_4^{3-}	SeO_4^{2-}	BrO_3^-
			(248)	249	154
			SbO_4^{3-}	TeO_4^{2-}	IO_3^-
			(260)	(254)	182

Complex ions				d-Metal oxoanions		
$[TiF_6]^{2-}$	$[PtCl_6]^{2-}$	$[SiF_6]^{2-}$	$[SnCl_6]^{2-}$	$[CrO_4]^{2-}$		$[MnO_4]^-$
289	313	269	326	(256)		(240)
$[TiCl_6]^{2-}$	$[PtBr_6]^{2-}$	$[GeF_6]^{2-}$	$[SnBr_6]^{2-}$	$[MoO_4]^{2-}$		
331	342	265	363	(254)		
$[ZrCl_6]^{2-}$			$[PbCl_6]^{2-}$			
358			348			

Sources: H. D. B. Jenkins and K. P. Thakur, *J. Chem. Educ.*, 1979, **56**, 576; A. F. Kapustinskii, *Q. Rev. Chem. Soc.*, 1956, **10**, 283 (values in parentheses).

containing one cation type and one anion type whose ionic radii are known.

The Kapustinskii equation can also be used to ascribe numerical values to the 'radii' of nonspherical molecular ions, as their values can be adjusted until the calculated value of the lattice enthalpy matches that obtained experimentally from the Born–Haber cycle. The self-consistent parameters obtained in this way are called **thermochemical radii** (Table 4.10). Once tabulated, these values can be used to estimate lattice enthalpies, and hence enthalpies of formation, of a wide range of compounds without needing to know the structure, assuming that the bonding is essentially ionic.

A BRIEF ILLUSTRATION

To estimate the lattice enthalpy of potassium nitrate, KNO_3, we need the number of ions per formula unit ($N_{ion} = 2$), their charge numbers ($z(K^+) = +1, z(NO_3^-) = -1$), and the sum of their thermochemical radii (138 pm + 189 pm = 327 pm). Then, with $d^* = 34.5$ pm, using eqn. 4.4,

$$\Delta_L H^\ominus = \frac{2|(+1)(-1)|}{327\,\text{pm}} \times \left(1 - \frac{34.5\,\text{pm}}{327\,\text{pm}}\right) \times (1.21 \times 10^5)\,\text{kJ pm mol}^{-1}$$
$$= 622\,\text{kJ mol}^{-1}$$

This calculated value compares reasonably well with the experimental value of 685 kJ mol^{-1}.

EXAMPLE 4.15 Using the Kapustinskii equation

Calculate a value for the lattice enthalpy of barium pernitride, BaN_2, assuming that the thermochemical radius of the pernitride anion, $[N_2]^{2-}$, is similar to that of peroxide in Table 4.10. The ionic radius of Ba^{2+} is 135 pm. Use the simplified Kapustinskii equation (eqn. 4.5).

Answer We substitute the correct values into the Kapustinskii equation, noting that the anion in this compound is the diatomic pernitride ion, $[N_2]^{2-}$, so N_{ion} is 2 and $z_A = z_B = 2$.

$$\Delta_L H^\ominus = \frac{2|(+2)(-2)|}{(135+173)\,\text{pm}} \times (1.08 \times 10^5)\,\text{kJ pm mol}^{-1}$$
$$= 2805\,\text{kJ mol}^{-1}$$

Self-test 4.15 Calculate a value for the lattice enthalpy of lithium pernitride, Li_2N_2.

4.15 Consequences of lattice enthalpies

The Born–Mayer equation shows that, for a given lattice type (a given value of A), the lattice enthalpy increases with increasing ion charge numbers (as $|z_A z_B|$). The lattice enthalpy also increases as the ions come closer together and the scale of the lattice decreases. Energies that vary as the **electrostatic parameter**, ζ (xi),

$$\zeta = \left|\frac{z_A z_B}{d}\right| \tag{4.6}$$

(which is often written more succinctly as $\zeta = z^2/d$) are widely adopted in inorganic chemistry as indicative that an ionic model is appropriate. In this section we consider three consequences of lattice enthalpy and its relation to the electrostatic parameter.

(a) Thermal stabilities of ionic solids

KEY POINT Lattice enthalpies may be used to explain the chemical properties of many ionic solids, including their thermal decomposition.

The particular aspect we consider here is the temperature needed to bring about thermal decomposition of Group 2 carbonates (although the arguments can easily be extended to many inorganic solids):

$$MCO_3(s) \rightarrow MO(s) + CO_2(g)$$

Magnesium carbonate, for instance, decomposes in air when heated to about 300°C, whereas calcium carbonate decomposes only if the temperature is raised to over 800°C. The decomposition temperatures of thermally unstable compounds (such as carbonates) increase with cation radius (Table 4.11). In general, large cations stabilize large anions to decomposition (and vice versa). The lower temperature for the decomposition of $MgCO_3$ versus $CaCO_3$ is an important consideration in assessing and mitigating the environmental impact of cement manufacture; see Section 13.5. If magnesium-based cements can be developed then the amount of fossil fuel required to heat MCO_3 to produce MO would be much reduced (compared to current calcium-based materials).

The stabilizing influence of a large cation on an unstable anion, that can break up and form a smaller anion, can be explained in terms of trends in lattice enthalpies. First, we note that the decomposition temperatures of solid inorganic compounds can be discussed in terms of their Gibbs energies of decomposition into specified products. The standard Gibbs energy for the decomposition of a solid, $\Delta G^\ominus = \Delta H^\ominus - T\Delta S^\ominus$, becomes negative when the second term on the right exceeds the first, which is when the temperature exceeds

$$T = \frac{\Delta H^\ominus}{\Delta S^\ominus} \tag{4.7}$$

TABLE 4.11 Decomposition data for carbonates

	$MgCO_3$	$CaCO_3$	$SrCO_3$	$BaCO_3$
$\Delta G^\ominus$ / (kJ mol^{-1})*	+48.3	+130.4	+183.8	+218.1
$\Delta H^\ominus$ / (kJ mol^{-1})	+100.6	+178.3	+234.6	+269.3
$\Delta S^\ominus$ / (J K^{-1} mol^{-1})	+175.0	+160.6	+171.0	+172.1
θ_{decomp} /°C	300	840	1100	1300

* Data are for the reaction $MCO_3(s) \rightarrow MO(s) + CO_2(g)$ at 298 K. θ is the temperature required to reach $p(CO_2) = 1$ bar, and has been estimated from the thermodynamic data at 298 K.

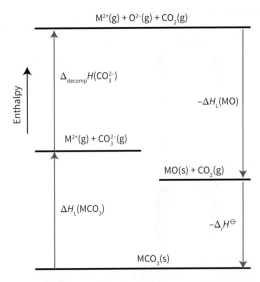

FIGURE 4.52 A thermodynamic cycle showing the enthalpy changes involved in the decomposition of a solid carbonate MCO_3.

In many cases it is sufficient to consider only trends in the reaction enthalpy, as the reaction entropy is essentially independent of M because it is dominated by the formation of gaseous product, in the case of carbonates CO_2. The standard enthalpy of decomposition of the solid is then given by

$$\Delta H^\ominus = \Delta_{decomp} H^\ominus + \Delta_L H^\ominus(MCO_3, s) - \Delta_L H^\ominus(MO, s)$$

where $\Delta_{decomp} H^\ominus$ is the standard enthalpy of decomposition of CO_3^{2-} in the gas phase (Fig. 4.52):

$$CO_3^{2-}(g) \rightarrow O^{2-}(g) + CO_2(g)$$

Because $\Delta_{decomp} H^\ominus$ is large and positive, the overall reaction enthalpy is positive (decomposition is endothermic), but it is less strongly positive if the lattice enthalpy of the oxide is markedly greater than that of the carbonate because then $\Delta_L H^\ominus(MCO_3, s) - \Delta_L H^\ominus(MO, s)$ is negative. It follows that the decomposition temperature will be low for oxides that have relatively high lattice enthalpies compared with their parent carbonates. The compounds for which this is true are composed of small, highly charged cations, such as Mg^{2+}; remembering that $\Delta_L H^\ominus \propto |z_A z_B|/d$ also supports this statement as d will be smaller for Mg^{2+}.

Fig. 4.53 illustrates why a small cation has a more significant influence on the change in the lattice enthalpy as the anion size is varied. The change in separation, d, is relatively small when the parent compound has a large cation initially as the cation contribution to d (= $r_+ + r_-$) is more significant. As the illustration shows in an exaggerated way, when the cation is very big, the change in size of the anion barely affects the scale of the lattice. Therefore, with a given unstable polyatomic anion, the lattice enthalpy difference is more significant and favourable to decomposition when the cation is small than when it is large.

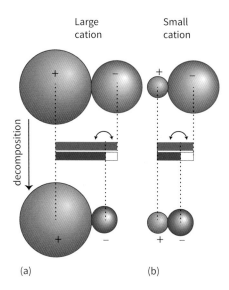

FIGURE 4.53 A greatly exaggerated representation of the relative change in lattice parameter d for cations of different radii. (a) When the anion changes size (as when CO_3^{2-} decomposes into O^{2-} and CO_2, for instance) and the cation is large, the lattice parameter changes by a comparatively small amount. (b) If the cation is small, however, the relative change in lattice parameter is large, resulting in a larger increase in lattice enthalpy, and decomposition is thermodynamically more favourable.

The difference in lattice enthalpy between MO_n and $M(CO_3)_n$ is also magnified by a larger charge on the cation, as $\Delta_L H^\ominus \propto |z_A z_B|/d$, so the thermal decomposition of a carbonate will occur at lower temperatures if it contains a higher-charged cation. This dependence on cation charge is observed in the fact that alkaline earth carbonates (M^{2+}) tend to decompose at lower temperatures than the corresponding alkali metal carbonates (M^+).

EXAMPLE 4.16 Assessing the dependence of compound stability on ionic radius

Present an argument to account for the fact that when they burn in oxygen, lithium forms the oxide Li_2O but sodium forms the peroxide Na_2O_2.

Answer We need to consider the role of the relative sizes of cations and anions in determining the stability of a compound. Because the small Li^+ ion results in Li_2O having a more favourable lattice enthalpy (in comparison with M_2O_2) than Na_2O, the decomposition reaction $M_2O_2(s) \rightarrow M_2O(s) + O_2(g)$ is thermodynamically more favourable for Li_2O_2 than for Na_2O_2.

Self-test 4.16 Predict the order of decomposition temperatures of alkaline earth metal sulfates in the reaction $MSO_4(s) \rightarrow MO(s) + SO_3(g)$.

The use of a large cation to stabilize a large anion that is otherwise susceptible to decomposition, forming a smaller anionic species, is widely used by inorganic chemists to prepare compounds that are otherwise thermodynamically unstable towards decomposition. For example, the interhalogen anions, such as ICl_4^-, are obtained by the oxidation of I^- ions by Cl_2 but are susceptible to decomposition to iodine monochloride and Cl^-:

$$MI(s) + 2Cl_2(g) \rightarrow MICl_4(s) \rightarrow MCl(s) + ICl(g) + Cl_2(g)$$

To disfavour the decomposition, a large cation is used to reduce the lattice enthalpy difference between $MICl_4$ and MCl/MI. The larger alkali metal cations such as K^+, Rb^+, and Cs^+ can be used in some cases, but it is even better to use a really bulky alkylammonium ion, such as $N^tBu_4^+$.

(b) The stabilities of oxidation states

KEY POINT The relative stabilities of different oxidation states in solids can often be predicted from consideration of lattice enthalpies.

A similar argument can be used to account for the general observation that high metal oxidation states are stabilized by small anions. In particular, F has a greater ability than the other halogens to stabilize the high oxidation states of metals. Thus, the only known binary halides of Ag(II), Co(III), and Mn(IV) are the fluorides. Another sign of the decrease in stability of the heavier halides of metals in high oxidation states is that the iodides of Cu(II) and Fe(III) decompose on standing at room temperature (to CuI and FeI_2). Oxygen is also a very effective species for stabilizing the highest oxidation states of elements because of the high charge and small size of the O^{2-} ion.

To explain these observations, consider the reaction

$$MX(s) + \tfrac{1}{2}X_2(g) \rightarrow MX_2(s)$$

where X is a halogen. The aim is to show why this reaction is most strongly spontaneous for X = F. If we ignore entropy contributions we must show that the reaction is most exothermic for fluorine. One contribution to the reaction enthalpy is the conversion of $\tfrac{1}{2}X_2$ to X^-. Despite F having a lower electron affinity than Cl, this step is more exothermic for X = F than for X = Cl because the bond enthalpy of F_2 is lower than that of Cl_2. The lattice enthalpies, however, play the major role. In the conversion of MX to MX_2, the charge number of the cation increases from +1 to +2, so the lattice enthalpy increases. As the radius of the anion increases, however, this difference in the two lattice enthalpies diminishes, and the exothermic contribution to the overall reaction decreases too. Hence, both the lattice enthalpy and the X^- formation enthalpy lead to a less exothermic reaction as the halogen changes from F to I.

Provided entropy factors are similar, which is plausible, we expect an increase in thermodynamic stability of MX relative to MX_2 on going from X = F to X = I down Group 17. Thus many iodides do not exist for metals in their higher oxidation states, and compounds such as $Cu^{2+}(I^-)_2$, $Tl^{3+}(I^-)_3$, and VI_5 are unknown, whereas the corresponding fluorides CuF_2, TlF_3, and VF_5 are easily obtained. In effect, even if these iodides could be formed the thermodynamics shows that the high-oxidation-state metals would oxidize I^- ions to I_2, leading to formation of a lower metal oxidation state, such as Cu(I), Tl(I), and V(III), in the iodides of these metals.

(c) Solubility

KEY POINT The solubilities of salts in water can be rationalized by considering lattice and hydration enthalpies.

Lattice enthalpies play a role in solubilities, as the dissolution involves breaking up the lattice, but the trend is much more difficult to analyse than for decomposition reactions. One rule that is reasonably well obeyed is that *compounds that contain ions with widely different radii are soluble in water*. Conversely, the least water-soluble salts are those of ions with similar radii. That is, in general, *difference in size favours solubility in water*. It is found empirically that an ionic compound MX tends to be very soluble when the radius of M^+ is smaller than that of X^- by about 80 pm.

Two familiar series of compounds illustrate these trends. In gravimetric analysis, Ba^{2+} is used to precipitate SO_4^{2-}, and the solubilities of the Group 2 sulfates decrease from $MgSO_4$ to $BaSO_4$. In contrast, the solubility of the Group 2 hydroxides increases down the group: $Mg(OH)_2$ is the sparingly soluble 'milk of magnesia', but $Ba(OH)_2$ can be used as a soluble hydroxide for preparation of solutions of OH^-. The first case shows that a large anion requires a large cation for precipitation. The second case shows that a small anion requires a small cation for precipitation.

Before attempting to rationalize the observations, we should note that the solubility of an ionic compound depends on the standard reaction Gibbs energy for

$$MX(s) \rightarrow M^+(aq) + X^-(aq)$$

In this process, the interactions responsible for the lattice enthalpy of MX are replaced by hydration (and by solvation in general) of the ions. However, the exact balance of enthalpy and entropy effects is delicate and difficult to assess, particularly because the entropy change also depends on the degree of order of the solvent molecules that is brought about by the presence of the dissolved solute. The data in Fig. 4.54 suggest that enthalpy considerations are important in some cases at least, as the graph shows that there is a correlation between the enthalpy of solution of a salt and the difference in hydration enthalpies of the two ions. If the cation has a larger hydration enthalpy than its

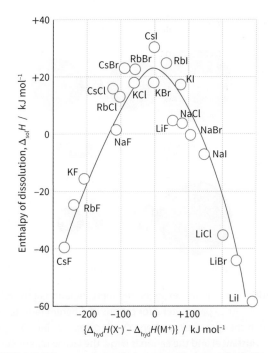

FIGURE 4.54 The correlation between enthalpies of solution of halides in H_2O and the differences between the hydration enthalpies of the ions. Dissolution is most exothermic when the difference is large.

anion partner (reflecting the difference in their sizes) or vice versa, then the dissolution of the salt is exothermic (reflecting the favourable solubility equilibrium).

The variation in enthalpy can be explained by using the ionic model. The lattice enthalpy is inversely proportional to the distance between the centres of the ions:

$$\Delta_L H^\ominus \propto \frac{1}{r_+ + r_-}$$

However, the hydration enthalpy, with each ion being hydrated individually, is the sum of individual ion contributions:

$$\Delta_{hyd} H \propto \frac{1}{r_+} + \frac{1}{r_-}$$

If the radius of one ion is small, the term in the hydration enthalpy for that ion will be large. However, in the expression for the lattice enthalpy, one small ion cannot make the denominator of the expression small by itself. Thus, one small ion can result in a large hydration enthalpy but not necessarily lead to a high lattice enthalpy, so ion size asymmetry can result in exothermic dissolution. If both ions are small, then both the lattice enthalpy and the hydration enthalpy may be large, and dissolution might not be very exothermic.

EXAMPLE 4.17 Accounting for trends in the solubility of s-block compounds

What is the trend in the solubilities of the Group 2 metal carbonates (Mg to Ra)?

Answer We need to consider the role of the relative sizes of cations and anions. The CO_3^{2-} anion has a large radius (Table 4.10) and has the same magnitude of charge as the cations M^{2+} of the Group 2 elements. The least soluble carbonate of the group is predicted to be that of the largest cation, Ra^{2+}. The most soluble is expected to be the carbonate of the smallest cation, Mg^{2+}. Although magnesium carbonate is more soluble than radium carbonate, it is still only sparingly soluble: its solubility constant (its solubility product, K_{sp}) is only 3×10^{-8}.

Self-test 4.17 Which can be expected to be more soluble in water, $NaClO_4$ or $KClO_4$?

Further information: the Born–Mayer equation

As we have seen in Box 4.3, the Coulomb potential energy of a single cation in a one-dimensional line of alternating cations A and anions B of charges $+e$ and $-e$ separated by a distance d is given by

$$V = \frac{2e^2 \ln 2}{4\pi\varepsilon_0 d}$$

The total molar contribution of all the ions is this potential energy multiplied by Avogadro's constant, N_A (to convert to a molar value), and divided by 2 (to avoid counting each interaction twice).

The total molar potential energy also needs to include the repulsive interaction between the ions. We can model that by a short-range exponential function of the form Be^{-d/d^*}, with d^* a constant that defines the range of the repulsive interaction, and B a constant that defines its magnitude. The total molar potential energy of interaction is therefore

$$V = -\frac{N_A e^2}{4\pi\varepsilon_0 d} A + Be^{-d/d^*}$$

This potential energy passes through a minimum when $dV/dd = 0$, which occurs at

$$\frac{dV}{dd} = \frac{N_A e^2}{4\pi\varepsilon} A - \frac{B}{d^*} e^{-d/d^*} = 0$$

It follows that, at the minimum,

$$Be^{-d^*/d} = \frac{N_A e^2 d^*}{4\pi\varepsilon_0 d^2} A$$

This relation can be substituted into the expression for V to give

$$V = -\frac{N_A e^2}{4\pi\varepsilon_0 d}\left(1 - \frac{d^*}{d}\right) A$$

On identifying $-V$ with the lattice enthalpy (more precisely, with the lattice energy at $T = 0$), we obtain the Born–Mayer equation (eqn. 4.2) for the special case of singly charged ions. The generalization to other charge types is straightforward.

If a different expression for the repulsive interaction between the ions is used, then this expression will be modified. One alternative is to use an expression such as $1/r^n$ with a large n, typically $6 \le n \le 12$, which then gives rise to a slightly different expression for V known as the **Born–Landé equation**:

$$V = -\frac{N_A e^2}{4\pi\varepsilon_0 d}\left(1 - \frac{1}{n}\right) A$$

The semiempirical Born–Mayer expression, with $d^* = 34.5$ pm determined from the best agreement with experimental data, is generally preferred to the Born–Landé equation.

FURTHER READING

R. D. Shannon, in *Encyclopaedia of Inorganic Chemistry*, ed. R. B. King. John Wiley & Sons (2005). A survey of ionic radii and their determination.

A. F. Wells, *Structural Inorganic Chemistry*. Oxford University Press (1985). The standard reference book, which surveys the structures of a huge number of inorganic solids.

J. K. Burdett, *Chemical Bonding in Solids*. Oxford University Press (1995). Further details of the electronic structures of solids.

Some introductory texts on solid-state inorganic chemistry are:

U. Müller, *Inorganic Structural Chemistry*. John Wiley & Sons (1993).

A. R. West, *Solid State Chemistry and its Applications, Student Edition*, 2nd ed. John Wiley & Sons (2014).

S. E. Dann, *Reactions and Characterization of Solids*. Royal Society of Chemistry (2000).

L. E. Smart and E. A. Moore, *Solid State Chemistry: An Introduction*, 5th ed. CRC Press (2020).

P. A. Cox, *The Electronic Structure and Chemistry of Solids*. Oxford University Press (1987).

D. K. Chakrabarty, *Solid State Chemistry*, 2nd ed. New Age Science Ltd (2010).

P. M. Woodward, P. Karen, J. S. O. Evans, and Thomas Vogt, *Solid State Materials Chemistry*. Cambridge University Press (2021).

Two very useful texts on the application of thermodynamic arguments to inorganic chemistry are:

W. E. Dasent, *Inorganic Energetics*. Cambridge University Press (1982).

D. A. Johnson, *Some Thermodynamic Aspects of Inorganic Chemistry*. Cambridge University Press (1982).

EXERCISES

4.1 What are the relationships between the unit cell parameters in the monoclinic crystal system? Draw a projection of a monoclinic cell viewed down the *b* axis and hence show that the packing of monoclinic unit cells will completely fill three-dimensional space.

4.2 Show that regular pentagons cannot completely fill a two-dimensional space.

4.3 Draw a tetragonal unit cell and mark on it a set of points that would define (a) a face-centred lattice and (b) a body-centred lattice. Demonstrate, by considering two adjacent unit cells, that a tetragonal face-centred lattice of dimensions *a* and *c* can always be redrawn as a body-centred tetragonal lattice with dimensions $a/\sqrt{2}$ and *c*.

4.4 (a) Draw a cubic unit cell with lattice points at (½,0,0), (0,½,0), (0,0,½), and (½,½,½). What is the lattice type of this unit cell and how many lattice points are contained within the unit cell? (b) Draw an orthorhombic unit cell with lattice points at (0,0,0) and (½,½,0). What is the lattice type and how many lattice points are there in the unit cell?

4.5 Which of the following schemes for the repeating pattern of close-packed planes are not ways of generating close-packed lattices? (a) CBACBA . . ., (b) ABA . . ., (c) ABCBA . . ., (d) ABCBC . . ., (e) ABABC . . ., (f) ABCBCA . . . .

4.6 Determine the formulas of the compounds produced by (a) filling a quarter of the tetrahedral holes with cations M in a hexagonal close-packed array of anions X; (b) filling half the octahedral holes with cations M in a cubic close-packed array of anions X; (c) filling a sixth of the tetrahedral holes in a hcp array of M with X?

4.7 Metallic copper adopts an fcc structure with density 8960 kg m^{-3}. Draw the unit cell of copper and mark the shortest copper-atom to copper-atom distance. How many copper atoms are there in the unit cell? Use the density and mass of the unit cell (determined from the atomic mass of a copper atom and the unit cell content) to calculate the lattice parameter for copper metal, and hence calculate a value for the metallic radius of copper.

4.8 (a) Potassium reacts with C_{60} (Fig. 4.17) to produce a compound in which all the octahedral and tetrahedral holes are filled by potassium ions. Derive a stoichiometry for this compound. (b) The reaction of C_{60} with excess potassium yields a material that has a body-centred arrangement of C_{60} molecules with two potassium ions along each unit cell edge; calculate the stoichiometry of this compound.

4.9 In the structure of MoS_2, the S atoms are arranged in close-packed layers that repeat themselves in the sequence AAA . . . . The Mo atoms occupy holes with coordination number 6. Show that each Mo atom is surrounded by a trigonal prism of S atoms.

4.10 Metallic lithium adopts a bcc structure with density of 535 kg m^{-3}. What is the length of the edge of the unit cell?

4.11 An alloy of copper and gold has the structure shown in Fig. 4.55. Calculate the composition of this unit cell. What is the lattice type of this structure? Given that 24-carat gold is pure gold, what carat gold does this alloy represent?

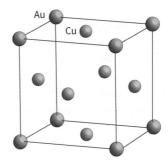

FIGURE 4.55 The structure of Cu_3Au.

4.12 Bronze is a compound of tin and copper in the ratio 85% Cu and 15% Sn. Electronegativities of Cu and Sn are $\chi(Cu) = 1.95$ and $\chi(Sn) = 1.96$. Would you classify bronze as an alloy?

4.13 Alloy bicycle frames use aluminium in combination with other light metals. What other low-density metals are likely to form an alloy with aluminium?

4.14 Depending on temperature, RbCl can exist in either the rock-salt or caesium-chloride structure. (a) What is the coordination number of the cation and anion in each of these structures? (b) In which of these structures will Rb have the larger apparent ionic radius?

4.15 Consider the structure of caesium chloride. How many Cs$^+$ ions occupy second-nearest-neighbour locations of a Cs$^+$ ion? How many Cl$^-$ ions occupy the third-nearest-neighbour sites?

4.16 The ReO_3 structure is cubic with a rhenium atom at each corner of the unit cell and one oxygen atom on each unit cell edge midway between the rhenium atoms. Sketch this unit cell and determine (a) the coordination numbers of the ions and (b) the identity of the structure type that would be generated if a cation were inserted in the centre of each ReO$_3$ unit cell.

4.17 Describe the coordination around the oxide ions in the perovskite structure, ABO$_3$, in terms of coordination to the A- and B-type cations.

4.18 If the A-type cation in a perovskite of composition ABO$_3$ is replaced by an oxide anion to produce the stoichiometry BO$_4$, the arrangement of oxide ions is the same as that in cubic close-packing. What type of hole is filled by the B-type cation, and what proportion of this hole-type is filled?

4.19 Obtain formulae (MX$_n$ or M$_n$X) for structures derived from hole-filling in close-packed arrays with (a) half the octahedral holes filled, (b) one-quarter of the tetrahedral holes filled, (c) two-thirds of the octahedral holes filled. What are the average coordination numbers of M and X in (a) and (b)?

4.20 Use radius-ratio rules and the ionic radii given in Resource section 1 to predict structures of (a) UO$_2$, (b) FrI, (c) BeS, (d) InP.

4.21 Given the following data for the length of a side of the unit cell for compounds that crystallize in the rock-salt structure, determine the cation radii: MgSe (545 pm), CaSe (591 pm), SrSe (623 pm), BaSe (662 pm). (Hint: To determine the radius of Se^{2-}, assume that the Se^{2-} ions are in contact in MgSe.)

4.22 Use the structure map in Fig. 4.48 to predict the coordination numbers of the cations and anions in (a) LiF, (b) RbBr, (c) SrS, (d) BeO. The observed coordination numbers are (6,6) for LiF, RbBr, and SrS, and (4,4) for BeO. Propose a possible reason for any discrepancies.

4.23 Describe how the structures of the following compounds could be described in terms of the simple structure types of Table 4.4 but with complex ions replacing simple monoatomic cations and anions. (a) K$_2$PtCl$_6$, (b) [Ni(H$_2$O)$_6$][SiF$_6$], (c) CsCN, (d) CsPF$_6$.

4.24 The structure of calcite (CaCO$_3$) is shown in Fig. 4.56. Describe how this structure is related to that of NaCl.

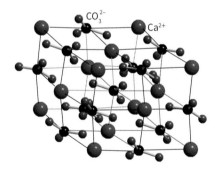

FIGURE 4.56 The structure of CaCO$_3$.

4.25 The projection representation of a unit cell of an ionic crystal of the binary compound A$_p$B$_q$ is shown in Fig. 4.57. Determine the stoichiometry and the number of formula units in the unit cell. State the coordination number and coordination geometry of ions A and B. Give an example of a compound that adopts this structure, and comment on how a derivative of this structure could exist with the formula A$_p$B$_{2q}$.

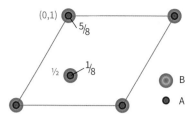

FIGURE 4.57 Compound A$_p$B$_q$.

4.26 What are the most significant terms in the Born–Haber cycle for the formation of Ca$_3$N$_2$?

4.27 Explain, by considering the magnitude of the key terms in the Born–Haber cycle that involve the halogen or halide ion, why AgF$_2$ exists but AgCl$_2$ does not.

4.28 On the basis of the factors that contribute to lattice enthalpies, place LiF, SrO, RbCl, AlP, NiO, and CsI, all of which adopt the rock-salt structure, in order of increasing lattice enthalpy.

4.29 By considering the parameters that change in the Born–Mayer expression, estimate lattice enthalpies for MgO and AlN, given that MgO and AlN adopt the rock-salt structure with very similar lattice parameters to NaCl and $\Delta_L H^\ominus$(NaCl) = 786 kJ mol^{-1}.

4.30 (a) Calculate the enthalpy of formation of the hypothetical compound KF$_2$ assuming a CaF$_2$ structure. Use the Born–Mayer equation to obtain the lattice enthalpy and estimate the radius of K^{2+} by extrapolation of trends in Table 1.4 and Resource section 1. Ionization enthalpies and electron gain enthalpies are given in Tables 1.6 and 1.7. (b) What factor prevents the formation of this compound despite the favourable lattice enthalpy?

4.31 The common oxidation number for an alkaline earth metal is +2. Using the Born–Mayer equation and a Born–Haber cycle, show that CaCl is an exothermic compound. Use a suitable analogy to estimate an ionic radius for Ca$^+$. The sublimation enthalpy of Ca(s) is 176 kJ mol^{-1}. Show that an explanation for the non-existence of CaCl can be found in the enthalpy change for the reaction 2 CaCl(s) → Ca(s) + CaCl$_2$(s).

4.32 Determine the first four terms in the expression for the Madelung constant calculation for the CsCl structure.

4.33 (a) Explain why lattice energy calculations based on the Born–Mayer equation reproduce the experimentally determined values to within 1% for LiCl but only 10% for AgCl, given that both compounds have the rock-salt structure. (b) Identify a pair of compounds containing M^{2+} ions that might be expected to show similar behaviour.

4.34 Use the Kapustinskii equation and the ionic and thermochemical radii given in Resource section 1 and Table 4.10, and r(Bk^{4+}) = 96 pm to calculate lattice enthalpies of (a) BkO$_2$, (b) K$_2$SiF$_6$, and (c) LiClO$_4$.

4.35 Which member of each pair is likely to be more soluble in water: (a) BaSeO$_4$ or CaSeO$_4$, (b) NaF or NaBF$_4$?

4.36 Infrared spectrometers can have optical windows made of NaCl, KBr, or CsI. What is the disadvantage of using CsI optics?

4.37 Recommend a specific cation for the quantitative precipitation of selenate ion, SeO$_4^{2-}$, from water. Suggest two different cations, one that would have a water-soluble phosphate (PO_4^{3-}) and the other a highly insoluble phosphate.

4.38 Which of the following pairs of isostructural compounds are likely to undergo thermal decomposition at lower temperature? Give your reasoning. (a) $MgCO_3$ and $CaCO_3$ (decomposition products $MO + CO_2$). (b) CsI_3 and $N(CH_3)_4I_3$ (both compounds contain the $[I_3]^-$ anion; decomposition products $MI + I_2$).

4.39 The anion $[ICl_4]^-$ may be synthesized by bubbling $Cl_2(g)$ through a solution of iodide ions. This complex anionic species can undergo decomposition, losing Cl_2, to form initially $[ICl_2]^-$ and eventually regenerating I^-. Suggest a counter-ion that could be used to crystallize $[ICl_4]^-$ and prevent its decomposition.

TUTORIAL PROBLEMS

4.1 Both ccp and hcp have identical close-packed volumes, but the calculated Gibbs free energy of the two arrangements shows that ccp is more stable by 0.002–0.012 kJ mol^{-1} at room temperature (L.V. Woodcock, *Nature*, 1997, **385**, 141; P. G. Bolhuis, D. Frenkel, S.-C. Mau and D. A. Huse, *Nature*, 1997, **388**, 235). Determine independently the number of neighbours in the first five coordination shells for these two close-packed structures and, thereby, describe the origin of this difference in free energy.

4.2 Calculate the number of atoms present in a cubic gold nanoparticle of dimensions 2 × 2 × 2 nm^3; gold adopts an fcc unit cell with lattice parameter of 400 pm. Discuss how the size of silver and gold nanoparticles affects their colour and how this has led to their historical application in stained glass.

4.3 The relative positons of tetrahedral and octahedral holes between close-packed layers may be described using the **PTOT** system, where **P** represents a close-packed layer, and **T** and **O** the tetrahedral and octahedral holes, respectively. Use this description and the formation of 'stuffed' **PTOT** arrangements (B. E. Douglas, *J. Chem. Educ.*, 2007, **84**, 1846) to describe the structures of (a) Li_3Bi, (b) $CdCl_2$, (c) MoS_2, and (d) $BeAl_2O_4$.

4.4 By considering the rock-salt structure and the distances and charges around one central ion, show that the first six terms of the Na$^+$ Madelung series are

$$\frac{6}{\sqrt{1}} - \frac{12}{\sqrt{2}} + \frac{8}{\sqrt{3}} - \frac{6}{\sqrt{4}} + \frac{24}{\sqrt{5}} - \frac{24}{\sqrt{6}}$$

Discuss methods for showing that this series converges to 1.748 by reference to R. P. Grosso, J. T. Fermann, and W. J. Vining, *J. Chem. Educ.*, 2001, **78**, 1198.

4.5 Nanocrystals with dimensions of 1–10 nm are of increasing importance in technological applications (see Chapter 28). The calculation of the Madelung constant for a particular structure type requires the summation of an infinite series of terms, and this method is not applicable to a crystal of nanometre dimensions. Discuss how the concept of a Madelung factor, A^*, may be used to describe the total ionic interactions in nanocrystals. How does A^* vary as a function of nanocrystal size for (a) the NaCl structure type and (b) the CsCl structure type? How might the properties of a nanocrystal differ from the bulk solid as a result? See M. D. Baker and A. D. Baker, *J. Chem. Educ.*, 2010, **87**, 280.

4.6 Discuss the thermodynamic factors behind the observation that CuO is a well-known, stable oxide of Cu(II) but AgO is a mixed-valence oxide, Ag(I)Ag(III)O$_2$. See D. Tudela, *J. Chem. Educ.*, 2008, **85**, 863. AgF_2 is a stable compound of Ag(II); discuss what factors might contribute to the stability of Ag(II) in combination with fluoride.

4.7 The Kapustinskii equation shows that lattice enthalpies are inversely proportional to the interionic distances. Later work has shown that further simplification of the Kapustinskii equation allows lattice enthalpies to be estimated from the molecular (formula) unit volume (the unit cell volume divided by the number of formula units, Z, that it contains) or the mass density (see, for example, H. D. B. Jenkins and D. Tudela, *J. Chem. Educ.*, 2003, **80**, 1482). How would you expect the lattice enthalpy to vary as a function of (a) the molecular unit volume and (b) the mass density? Given the following unit cell volumes (all in cubic angstroms, Å^3; 1 Å = 10^{-10} m) for the alkaline earth carbonates MCO_3 and oxides, predict the observed decomposition behaviour of the carbonates.

MgCO$_3$	CaCO$_3$	SrCO$_3$	BaCO$_3$
47	61	64	76
MgO	CaO	SrO	BaO
19	28	34	42

4.8 The following linear correlation, sometimes referred to as Bartlett's relationship, has been found between lattice enthalpy (kJ mol^{-1}) and the inverse cube root of the formula unit volume V (in nm^3) for 1:1 MX salts.

$$\Delta H_L = \frac{232.8}{\sqrt[3]{V}} + 110$$

Show that this expression is related to the Kapustinskii equation. Formula unit cell volumes are readily obtainable from X-ray diffraction studies of crystalline, ionic MX structures, and therefore Bartlett's relationship may be used where individual thermochemical radii are not available. Discuss how application of this relationship has been used to account for the inability to synthesize $[I_2]^+[AlCl_4]^-$ and the instability of $[I_3]^+[AsF_6]^-$ (H. D. B. Jenkins, H. K. Roobottom, J. Passmore, and L. Glasser, *Inorg. Chem.*, 1999, **38**, 3609).

4.9 The analysis undertaken in Problem 4.8 may be extended to give the 'isomegethic rule', which states that 'The formula unit volumes, V_m, of isomeric ionic salts are approximately the same'; see H. D. B. Jenkins et al., *Inorg. Chem.*, 2004, **43**, 6238; L. Glasser, *J. Chem. Educ.*, 2011, **88**, 581. Discuss the basis for this 'rule' and its applications in solid-state chemistry.

Acids and bases

Brønsted acidity

5.1 **Proton transfer equilibria in water**

Characteristics of oxygen-containing Brønsted acids

5.2 **Periodic trends in aqua acid strength**
5.3 **Simple oxoacids**
5.4 **Anhydrous oxides**
5.5 **Polyoxo compound formation**

Lewis acidity

5.6 **Identifying Lewis acids and bases**
5.7 **The fundamental types of reaction**
5.8 **Properties of main group Lewis acids**
5.9 **Hydrogen bonding**
5.10 **Factors governing interactions between Lewis acids and bases**
5.11 **Thermodynamic Lewis acidity parameters**

Nonaqueous solvents

5.12 **Solvent levelling**
5.13 **The Hammett acidity function and its application to strong, concentrated acids**
5.14 **The solvent system definition of acids and bases**
5.15 **Solvents as acids and bases**

Applications of acid–base chemistry

5.16 **Acid and base catalysis**
5.17 **Superacids and superbases**
5.18 **Heterogeneous acid–base reactions**

Further reading

Exercises

Tutorial problems

This chapter focuses on the wide variety of species that are classified as acids and bases and processes known as acid–base reactions. The first part describes the Brønsted definition, in which an acid is a proton donor and a base is a proton acceptor. Proton-transfer equilibria can be discussed quantitatively in terms of acidity constants, which are a measure of the tendency for species to donate protons. In the second part, we introduce the Lewis definition of acids and bases which deals with reactions involving electron-pair sharing between a donor (the base) and an acceptor (the acid). This broadening enables us to extend our discussion of acids and bases to include species that do not contain protons and to reactions in nonprotic media. We also introduce a molecular orbital description of hydrogen bonding, which has a very special place in chemistry and biology. Because of the greater diversity of Lewis species, a single scale of strength is not appropriate, and two approaches are taken: in one, acids and bases are classified as 'hard' or 'soft'; in the other, thermochemical data are used to obtain a set of parameters characteristic of each species. The final sections introduce nonaqueous solvents and describe some of the most important applications of acid–base chemistry.

The original distinction between acids and bases was based, hazardously, on criteria of taste and feel: acids were regarded as sour, and bases felt soapy. A deeper chemical understanding of their properties emerged from Arrhenius's (1884) conception of an acid as a compound that produces (positively charged) hydrogen ions in water. The modern definitions that we consider in this chapter are based on a broader range of chemical reactions. The definition due to Brønsted and Lowry focuses on proton transfer, and that due to Lewis is based on dative bond formation—the interaction of electron pair acceptor and electron pair donor molecules and ions. The latter was introduced in Section 2.9.

Acid–base reactions are very common, although we do not always immediately recognize them as such, especially if they involve more subtle definitions of what it is to be an acid or base. For instance, production of harmful acid rain begins with a very simple reaction between the industrial pollutant sulfur dioxide (the acid) and water (the base):

$$2SO_2(g) + H_2O(l) \rightarrow HOSO_2^-(aq) + H^+(aq)$$

Likewise, increasing acidification of the oceans due to the reaction of atmospheric CO_2 with water (Sections 5.3 and 15.9) threatens the existence of coral reefs and other sensitive ecosystems. On the other hand, saponification, the process used in making soap, is an industrial-scale acid–base reaction of obvious importance for human wellbeing:

$$NaOH(aq) + RCOOR'(aq) \rightarrow RCO_2Na(aq) + R'OH(aq)$$

Many simple inorganic compounds which are discrete molecules in the gas phase, such as $BeCl_2$, condense as polymers because they contain both Lewis acid (Be) and Lewis base (Cl) centres that can form intermolecular dative bonds.

$$nBeCl_2(g) \rightarrow (BeCl_2)_n(s)$$

There are many such reactions, and in due course we shall see why they are regarded as reactions between acids and bases.

Brønsted acidity

KEY POINTS A Brønsted acid is a proton donor, and a Brønsted base is a proton acceptor. A proton has no separate existence in chemistry, and it is always associated with other species. A simple representation of a hydrogen ion in water is as the hydronium ion, H_3O^+. The acidity of an aqueous solution is measured by its pH value. The strength of a Brønsted acid in water (and by convention, the strength of its conjugate Brønsted base) is given by the acidity constant pK_a.

Johannes Brønsted in Denmark and Thomas Lowry in England proposed (in 1923) that the essential feature of an acid–base reaction is the transfer of a hydrogen ion, H^+, from one species to another. In the context of this definition, a hydrogen ion is often referred to as a proton. They suggested that any substance that acts as a proton donor should be classified as an acid, and any substance that acts as a proton acceptor should be classified as a base. Substances that act in this way are now called 'Brønsted acids' and 'Brønsted bases', respectively:

A **Brønsted acid** is a proton donor.

A **Brønsted base** is a proton acceptor.

The definitions make no reference to the environment in which proton transfer occurs, so they apply to proton-transfer behaviour in any solvent and even in no solvent at all.

An example of a Brønsted acid is hydrogen fluoride, HF, which can donate a proton to another molecule, such as H_2O, when it dissolves in water:

$$HF(g) + H_2O(l) \rightarrow H_3O^+(aq) + F^-(aq)$$

An example of a Brønsted base is ammonia, NH_3, which can accept a proton from a proton donor:

$$H_2O(l) + NH_3(aq) \rightarrow NH_4^+(aq) + OH^-(aq)$$

As these two examples show, water is an example of an **amphiprotic** substance, a substance that can act as both a Brønsted acid and a Brønsted base.

When an acid donates a proton to a water molecule, the latter is converted into a **hydronium ion**, H_3O^+ (1; the dimensions are taken from the crystal structure of $H_3O^+ClO_4^-$). However, the entity H_3O^+ is almost certainly an oversimplified description of the proton in water, for it participates in extensive **hydrogen bonding** (Section 5.9), and a better representation is $H_9O_4^+$, known as an Eigen complex (2). Gas-phase studies of water clusters using mass spectrometry suggest further that a cage of H_2O molecules can condense around one H_3O^+ ion in a regular pentagonal dodecahedral arrangement, resulting in the formation of the species $H^+(H_2O)_{21}$. As these structures indicate, the most appropriate description of a proton in water varies according to the environment and the experiment under consideration; for simplicity, we shall use the representation H_3O^+ throughout.

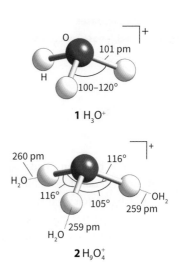

1 H_3O^+

2 $H_9O_4^+$

5.1 Proton transfer equilibria in water

Proton transfer reactions are commonly represented as equilibria

$$HF(aq) + H_2O(l) \rightleftharpoons H_3O^+(aq) + F^-(aq)$$

$$H_2O(l) + NH_3(aq) \rightleftharpoons NH_4^+(aq) + OH^-(aq)$$

As discussed further in Section 5.9, protons are highly mobile in water; accordingly, the central feature of Brønsted acid–base chemistry in aqueous solution is that equilibria are attained rapidly, and we concentrate on this aspect.

(a) Conjugate acids and bases

KEY POINTS When a species donates a proton, it becomes the conjugate base; when a species gains a proton, it becomes the conjugate acid. Conjugate acids and bases are in equilibrium in solution.

The form of the two forward and reverse reactions given above, both of which depend on the transfer of a proton from an acid to a base, is expressed by writing the general Brønsted equilibrium as:

$$Acid_1 + Base_2 \rightleftharpoons Acid_2 + Base_1$$

The species $Base_1$ is called the **conjugate base** of $Acid_1$, and $Acid_2$ is the **conjugate acid** of $Base_2$. The conjugate base of an acid is the species remaining after a proton is lost. The conjugate acid of a base is the species formed when a proton is gained. Thus, F^- is the conjugate base of HF, and H_3O^+ is the conjugate acid of H_2O. There is no *fundamental* distinction between an acid and a conjugate acid or a base and a conjugate base: a conjugate acid is just another acid and a conjugate base is just another base.

EXAMPLE 5.1 Identifying acids and bases

Identify the Brønsted acid and its conjugate base in the forward reaction of the following equilibria:

(a) $HSO_4^-(aq) + OH^-(aq) \rightleftharpoons H_2O(l) + SO_4^{2-}(aq)$

(b) $PO_4^{3-}(aq) + H_2O(l) \rightleftharpoons HPO_4^{2-}(aq) + OH^-$

Answer We need to identify the species that loses a proton and its conjugate partner. (a) The hydrogensulfate ion, HSO_4^-, transfers a proton to hydroxide; it is therefore the acid and the SO_4^{2-} ion produced is its conjugate base. (b) The H_2O molecule transfers a proton to the phosphate ion acting as a base; thus H_2O is the acid and the OH^- ion is its conjugate base.

Self-test 5.1 Identify the acid, base, conjugate acid, and conjugate base in the following reactions:

(a) $HNO_3(aq) + H_2O(l) \rightarrow H_3O^+(aq) + NO_3^-(aq)$

(b) $CO_3^{2-}(aq) + H_2O(l) \rightarrow HCO_3^-(aq) + OH^-(aq)$

(c) $NH_3(aq) + H_2S(aq) \rightarrow NH_4^+(aq) + HS^-(aq)$.

(b) The strengths of Brønsted acids

KEY POINTS The Brønsted acidity of a solution is described by its pH value. The strength of a Brønsted acid is measured by its acidity constant, and the strength of a Brønsted base is measured by its basicity constant; the stronger the base, the weaker is its conjugate acid.

Throughout this discussion, we shall apply the concept of pH, which we assume to be familiar from introductory chemistry. The pH value of a solution is defined by the Sørensen equation (5.1(a)):

$$pH = -\log a(H^+) \qquad (5.1(a))$$

in which the *activity* of the proton $a(H^+)$ is its effective thermodynamic concentration. Activity is a product of the real concentration and the activity constant f, which is unity for an ideal solution (one in which solute ions or molecules do not interact with each other, a condition that is approached at sufficiently high dilution). Assuming $f = 1$ we can consider pH to represent proton concentration through eqn. 5.1(b):

$$pH = -\log[H_3O^+] \qquad (5.1(b))$$

The strength of a Brønsted acid, such as HF, in aqueous solution is expressed by its **acidity constant** (or 'acid ionization constant'), K_a:

$$HF(aq) + H_2O(l) \rightleftharpoons H_3O^+(aq) + F^-(aq)$$

$$K_a = \frac{[H_3O^+][F^-]}{[HF]} \qquad (5.2(a))$$

More generally, adopting, for Brønsted acids, the letter X to denote the conjugate base:

$$HX(aq) + H_2O(l) \rightleftharpoons H_3O^+(aq) + X^-(aq)$$

$$K_a = \frac{[H_3O^+][X^-]}{[HX]} \qquad (5.2(b))$$

In this definition, $[X^-]$ denotes the numerical value of the molar concentration of the species X^- (so, if the molar concentration of HF molecules is $0.001\,mol\,dm^{-3}$, then $[HF] = 0.001$). A value $K_a \ll 1$ implies that $[HX]$ is large with respect to $[X^-]$, and so proton retention by the acid is favoured. The experimental value of K_a for hydrogen fluoride in water is 3.5×10^{-4}, indicating that under normal conditions, only a very small fraction of HF molecules are deprotonated. The actual fraction deprotonated can be calculated as a function of acid concentration from the numerical value of K_a.

> **EXAMPLE 5.2 Calculating acidity constants**
>
> The pH of 0.145 M $CH_3COOH(aq)$ is 2.80. Calculate K_a for ethanoic acid.
>
> **Answer** To calculate K_a we need to calculate the concentrations of H_3O^+, $CH_3CO_2^-$, and CH_3COOH in the solution. The concentration of H_3O^+ is obtained from the pH by writing $[H_3O^+] = 10^{-pH}$, so in a solution of pH = 2.80, the molar concentration of H_3O^+ is $1.6 \times 10^{-3}\,mol\,dm^{-3}$. Each deprotonation event produces one H_3O^+ ion and one $CH_3CO_2^-$ ion, so the concentration of $CH_3CO_2^-$ is the same as that of the H_3O^+ ions (provided the autoprotolysis of water can be neglected). The molar concentration of the remaining acid is $0.145 - 0.0016\,mol\,dm^{-3} = 0.143\,mol\,dm^{-3}$. Therefore
>
> $$K_a = \frac{(1.6 \times 10^{-3})^2}{0.143} = 1.7 \times 10^{-5}$$
>
> This value corresponds to $pK_a = 4.77$.
>
> **Self-test 5.2** For hydrofluoric acid $K_a = 3.5 \times 10^{-4}$. Calculate the pH of 0.10 M HF(aq).

Likewise, the proton-transfer equilibrium characteristic of a base, such as NH_3, in water can be expressed in terms of a **basicity constant**, K_b:

$$NH_3(aq) + H_2O(l) \rightleftharpoons NH_4^+(aq) + OH^-(aq) \quad K_b = \frac{[NH_4^+][OH^-]}{[NH_3]}$$

More generally:

$$X(aq) + H_2O(l) \rightleftharpoons HX^+(aq) + OH^-(aq) \quad K_b = \frac{[HX^+][OH^-]}{[X]}$$

(5.3)

If $K_b \ll 1$, then $[HX^+] \ll [X]$ at typical concentrations of X and only a small fraction of X molecules are protonated. Therefore, the base is a weak proton acceptor and its conjugate acid is present in low concentration. The experimental value of K_b for ammonia in water is 1.8×10^{-5}, indicating that under normal conditions only a very small fraction of NH_3 molecules is protonated. As for the acid calculation, the actual fraction of base protonated can be determined from the value of K_b.

Because water is amphiprotic, a proton-transfer equilibrium exists even in the absence of added acids or bases. The proton transfer from one water molecule to another is called **autoprotolysis** (or 'autoionization'). Proton transfer in water is very fast because it involves the interchange of weak hydrogen bonds between neighbouring molecules (Section 5.9). The extent of autoprotolysis and the composition of the solution at equilibrium is described by the **autoprotolysis constant** (or 'autoionization constant') of water:

$$2H_2O(l) \rightleftharpoons H_3O^+(aq) + OH^-(aq) \quad K_w = [H_3O^+][OH^-]$$

The experimental value of K_w is 1.00×10^{-14} at 25°C, indicating that only a very tiny fraction of water molecules are present as ions in pure water. Indeed, we know that because the pH of pure water is 7.00, and $[H_3O^+] = [OH^-]$, then $[H_3O^+] = 1.0 \times 10^{-7}\,mol\,dm^{-3}$. Tap water and bottled water have a pH slightly below 7 due to dissolved CO_2 (Sections 5.8(c) and 15.9).

An important role for the autoprotolysis constant of a solvent is that it enables us to relate the strength of a base to the strength of its conjugate acid, thereby allowing use of a single constant to express both acidity and basicity strength. Thus, the value of K_b for the ammonia equilibrium in which NH_3 acts as a base is related to the value of K_a for the equilibrium:

$$NH_4^+(aq) + H_2O(l) \rightleftharpoons H_3O^+(aq) + NH_3(aq)$$

in which its conjugate acid acts as an acid by

$$K_a K_b = K_w \tag{5.4}$$

The implication of eqn. 5.4 is that the larger the value of K_b, the smaller the value of K_a. That is, the stronger the base, the weaker its conjugate acid. It is conventional to report the strength of a base in terms of the acidity constant K_a of its conjugate acid rather than use its K_b value.

> **A BRIEF ILLUSTRATION**
>
> The K_b of ammonia in water is 1.8×10^{-5}. It follows that K_a of the conjugate acid NH_4^+ is
>
> $$K_a = \frac{K_w}{K_b} = \frac{1 \times 10^{-14}}{1.8 \times 10^{-5}} = 5.6 \times 10^{-10}$$

Because, like molar concentrations, acidity constants span many orders of magnitude, it is convenient to report them as their common logarithms (logarithms to the base 10) like pH by using

$$pK = -\log K \tag{5.5}$$

where K may be any of the constants we have introduced. At 25°C, for instance, $pK_w = 14.00$. It follows from this definition and the relation in eqn. 5.4 that

$$pK_a + pK_b = pK_w \tag{5.6}$$

A similar expression relates the strengths of conjugate acids and bases in any solvent, with pK_w replaced by the appropriate autoprotolysis constant of the solvent, pK_{sol}.

(c) Strong and weak acids and bases

KEY POINTS An acid or base is classified as either weak or strong depending on the size of its acidity constant.

Table 5.1 lists the acidity constants of some common acids and conjugate acids of some bases in water. A substance is classified as a **strong acid** if the proton-transfer equilibrium lies strongly in favour of donation of a proton to the solvent. Thus, a substance with $pK_a < 0$ (corresponding to $K_a > 1$ and usually to $K_a \gg 1$) is a strong acid. Such acids are commonly regarded as being fully deprotonated in solution (but it should never be forgotten that this situation is only an approximation). For example, hydrochloric acid is regarded as a solution of H_3O^+ and Cl^- ions, and a negligible concentration of HCl molecules. A substance with $pK_a > 0$ (corresponding to $K_a < 1$) is classified as a **weak acid**; for such species, the proton-transfer equilibrium lies in favour of nonionized acid. Hydrogen fluoride is a weak acid in water, and hydrofluoric acid consists of hydronium ions, fluoride ions, and a high proportion of HF molecules. The hydrate of CO_2, commonly referred to as carbonic acid (H_2CO_3), is another weak acid.

A **strong base** is a species that is almost fully protonated in water. An example is the oxide ion, O^{2-}, which is immediately converted into OH^- ions in water. A **weak base** is only partially protonated in water, an example being NH_3. The conjugate base of any strong acid is a weak base, because it is thermodynamically unfavourable for such a base to accept a proton.

TABLE 5.1 Acidity constants for species in aqueous solution at 25°C

Acid	HX (familiar formula)	X⁻	K_a	pK_a
Hydriodic	HI	I⁻	10^{11}	−11
Perchloric	$HClO_4$	ClO_4^-	10^{10}	−10
Hydrobromic	HBr	Br⁻	10^9	−9
Hydrochloric	HCl	Cl⁻	10^7	−7
Sulfuric	H_2SO_4	HSO_4^-	10^2	−2
Nitric	HNO_3	NO_3^-	10^2	−2
Hydronium ion	H_3O^+	H_2O	1	0.0
Chloric	$HClO_3$	ClO_3^-	10^{-1}	1
Sulfurous	H_2SO_3	HSO_3^-	1.5×10^{-2}	1.81
Hydrogensulfate ion	HSO_4^-	SO_4^{2-}	1.2×10^{-2}	1.92
Phosphoric	H_3PO_4	$H_2PO_4^-$	7.5×10^{-3}	2.12
Hydrofluoric	HF	F⁻	3.5×10^{-4}	3.45
Formic	HCOOH	HCO_2^-	1.8×10^{-4}	3.75
Ethanoic	CH_3COOH	$CH_3CO_2^-$	1.74×10^{-5}	4.76
Pyridinium ion	$HC_5H_5N^+$	C_5H_5N	5.6×10^{-6}	5.25
Carbonic	H_2CO_3	HCO_3^-	4.3×10^{-7}	6.37
Hydrogen sulfide	H_2S	HS⁻	9.1×10^{-8}	7.04
Dihydrogenphosphate ion	$H_2PO_4^-$	HPO_4^{2-}	6.2×10^{-8}	7.21
Boric acid*	$B(OH)_3$	$B(OH)_4^-$	7.2×10^{-10}	9.14
Ammonium ion	NH_4^+	NH_3	5.6×10^{-10}	9.25
Hydrocyanic	HCN	CN⁻	4.9×10^{-10}	9.31
Hydrogencarbonate ion	HCO_3^-	CO_3^{2-}	4.8×10^{-11}	10.32
Hydrogenarsenate ion	$HAsO_4^{2-}$	AsO_4^{3-}	3.0×10^{-12}	11.53
Hydrogenphosphate ion	HPO_4^{2-}	PO_4^{3-}	2.2×10^{-13}	12.67
Hydrogensulfide ion	HS⁻	S^{2-}	1.1×10^{-19}	19

* The proton-transfer equilibrium is $B(OH)_3(aq) + 2H_2O(l) \rightleftharpoons H_3O^+(aq) + B(OH)_4^-(aq)$.

The pH of a dilute solution containing a simple mixture of acid HX and its conjugate base X^- is given by

$$pH = pK_a + \log \frac{[X^-]}{[HX]} \quad (5.7)$$

(d) Polyprotic acids

KEY POINTS A polyprotic acid releases protons in succession, and successive deprotonations are progressively less favourable; a distribution diagram summarizes how the fraction of each species present depends on the pH of the solution.

A **polyprotic acid** is a substance that can donate more than one proton. An example is hydrogen sulfide, H_2S, a diprotic acid. For a diprotic acid, there are two successive proton donations and two acidity constants:

$$H_2S(aq) + H_2O(l) \rightleftharpoons HS^-(aq) + H_3O^+(aq)$$

$$HS^-(aq) + H_2O(l) \rightleftharpoons S^{2-}(aq) + H_3O^+(aq)$$

$$K_{a1} = \frac{[H_3O^+][HS^-]}{[H_2S]}$$

$$K_{a2} = \frac{[H_3O^+][S^{2-}]}{[HS^-]}$$

From Table 5.1, $K_{a1} = 9.1 \times 10^{-8}$ ($pK_{a1} = 7.04$) and $K_{a2} \approx 1.1 \times 10^{-19}$ ($pK_{a2} = 19$). The second acidity constant, K_{a2}, is always smaller than K_{a1} (and hence pK_{a2} is larger than pK_{a1}). The decrease in K_a is consistent with an electrostatic model of the acid in which, in the second deprotonation, a proton must separate from a centre with one more negative charge than in the first deprotonation. Because additional electrostatic work must be done to remove the positively charged proton, the deprotonation is less favourable.

EXAMPLE 5.3 Calculating concentrations of ions in polyprotic acids

Calculate the concentration of carbonate ions in a 0.10 M aqueous solution of CO_2. The hydrated form is represented as H_2CO_3; K_{a1} is given in Table 5.1; $K_{a2} = 4.6 \times 10^{-11}$.

Answer We need to consider the equilibria for the successive deprotonation steps with their acidity constants:

$$H_2CO_3(aq) + H_2O(l) \rightleftharpoons HCO_3^-(aq) + H_3O^+(aq)$$

$$HCO_3^-(aq) + H_2O(l) \rightleftharpoons CO_3^{2-}(aq) + H_3O^+(aq)$$

$$K_{a1} = \frac{[H_3O^+][HCO_3^-]}{[H_2CO_3]}$$

$$K_{a2} = \frac{[H_3O^+][CO_3^{2-}]}{[HCO_3^-]}$$

We assume that the second deprotonation is so slight that it has no effect on the value of $[H_3O^+]$ arising from the first deprotonation, in which case we can write $[H_3O^+] = [HCO_3^-]$. These two terms therefore cancel in the expression for K_{a2}, which results in

$$K_{a2} = [CO_3^{2-}]$$

that is, K_{a2} is independent of the initial concentration of the acid. It follows that the concentration of carbonate ions in the solution is 4.6×10^{-11} mol dm^{-3}.

Self-test 5.3 Calculate the pH of 0.20 M $H_2C_4H_4O_6$(aq) (tartaric acid), given $K_{a1} = 1.0 \times 10^{-3}$ and $K_{a2} = 4.6 \times 10^{-5}$.

The clearest representation of the concentrations of the species that are formed in the successive proton-transfer equilibria of polyprotic acids is a **distribution diagram**, showing the fraction of solute present as a specified species X, $f(X)$, plotted against the pH. Consider, for instance, the triprotic acid H_3PO_4, which releases three protons in succession to give $H_2PO_4^-$, HPO_4^{2-}, and PO_4^{3-}. The fraction of solute present as intact H_3PO_4 molecules is

$$f(H_3PO_4) = \frac{[H_3PO_4]}{[H_3PO_4] + [H_2PO_4^-] + [HPO_4^{2-}] + [PO_4^{3-}]} \quad (5.8)$$

The concentration of each solute at a given pH can be calculated from the pK_a values.[1] Fig. 5.1 shows the fraction of all four solute species as a function of pH and hence summarizes the relative importance of each acid and its conjugate base at each pH. Conversely, the diagram indicates the pH of the solution that contains a particular fraction

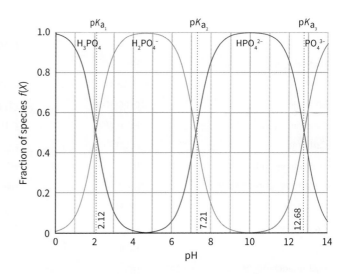

FIGURE 5.1 Distribution diagram for the various forms of the triprotic acid phosphoric acid in water, as a function of pH.

---

[1] For the calculations involved, see P. W. Atkins, L. Jones, L. Laverman, *Chemical Principles: the Quest for Insight*. MacMillan (2016).

of the species. We see, for instance, that if pH < pK_{a1}, corresponding to high hydronium ion concentrations, then the dominant species is the fully protonated H_3PO_4 molecule. However, if pH > pK_{a3}, corresponding to low hydronium ion concentrations, then the dominant species is the fully deprotonated PO_4^{3-} ion. The intermediate species are dominant when pH values lie between the relevant pK_a values.

(e) Factors governing the strengths of Brønsted acids and bases

KEY POINTS Proton affinity is the negative of the gas-phase proton-gain enthalpy. The proton affinities of p-block conjugate bases decrease to the right along a period and down a group. Proton affinities (and accordingly, the strengths of bases) are influenced by solvation, which stabilizes species carrying a charge.

A quantitative understanding of the relative acidities of X–H protons can be obtained by considering the enthalpy changes accompanying proton transfer. We shall consider gas-phase proton transfer reactions first, and then consider the effects of the solvent.

The simplest reaction of a proton is its attachment to a base, X^- (which, although denoted here as a negatively charged species, could be a neutral molecule, such as NH_3), in the gas phase:

$$X^-(g) + H^+(g) \rightarrow HX(g)$$

The standard enthalpy of this reaction is the **proton-gain enthalpy**, $\Delta_{pg}H^\ominus$. The negative of this quantity is often reported as the **proton affinity**, A_p (Table 5.2). When $\Delta_{pg}H^\ominus$ is large and negative, corresponding to an exothermic proton attachment, the proton affinity is high, indicating strongly basic character in the gas phase. If the proton-gain enthalpy is only slightly negative, then the proton affinity is low, indicating a weaker basic (or more acidic) character.

TABLE 5.2 Gas-phase and solution proton affinities*

Conjugate acid	Base (X)	A_p/kJ mol^{-1}	A'_p/kJ mol^{-1}
HF	F$^-$	1554	1150
HCl	Cl$^-$	1395	1090
HBr	Br$^-$	1354	1079
HI	I$^-$	1315	1068
H$_2$O	OH$^-$	1645	1188
HCN	CN$^-$	1476	1183
H$_3$O$^+$	H$_2$O	723	1130
NH$_4^+$	NH$_3$	865	1182

* A_p is the gas-phase proton affinity, A'_p is the effective proton affinity for the base in water.

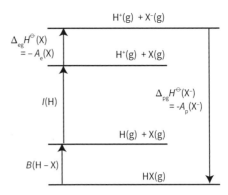

FIGURE 5.2 Thermodynamic cycle for a proton gain reaction.

Thus, HF is a stronger acid than H_2O, and HI is the strongest acid of the hydrogen halides. In other words, the order of proton affinities of their conjugate bases is I$^-$ < OH$^-$ < F$^-$. These trends can be explained by using a thermodynamic cycle such as that shown in Fig. 5.2 in which proton gain can be considered as the outcome of three steps:

Electron loss from X^-: $X^-(g) \rightarrow X(g) + e^-(g)$
$$-\Delta_{eg}H^\ominus(X) = A_e(X)$$

(the reverse of electron gain by X)

Electron gain by H^+: $H^+(g) + e^-(g) \rightarrow H(g)$
$$-\Delta_i H^\ominus(H) = -I(H)$$

(the reverse of the ionization of H)

Combination of H and X: $H(g) + X(g) \rightarrow HX(g)$
$$-B(H-X)$$

(the reverse of homolytic H–X bond dissociation)

The proton-gain enthalpy of the conjugate base X^- is the sum of these enthalpy changes:

Overall: $H^+(g) + X^-(g) \rightarrow HX(g)$
$$\Delta_{pg}H^\ominus(X^-) = A_e(X) - I(H) - B(H-X)$$

Therefore, the proton affinity of base X^- is

$$A_p(X^-) = B(H-X) + I(H) - A_e(X) \quad (5.9)$$

The resulting trends for the simplest stable hydrogen-containing compounds of p-block elements are represented in Fig. 5.3. The proton affinities of the conjugate bases of p-block binary acids H_nX decrease to the right along a period and down a group, thereby indicating an increase in gas-phase acidity. The dominant factor in the variation in proton affinity across a period is the trend in electron affinity of X, which increases from left to right and hence lowers

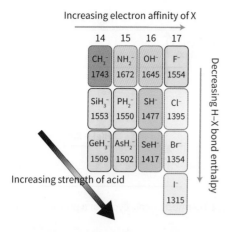

FIGURE 5.3 Gas-phase proton affinities (A_p/kJ mol^{-1}) of anions derived from stable binary hydrogen compounds of p-block elements H_nX.

the proton affinity of X^-. Thus, because the proton affinity of X^- decreases, the gas-phase acidity of HX *increases* across a period as the electron affinity of X increases. Because increasing electron affinity correlates with increasing electronegativity (Section 1.7), the gas-phase acidity of HX also increases as the electronegativity of A increases. The dominant factor when descending a group is the decrease in the H–X bond dissociation enthalpy, which lowers the proton affinity of X^- and therefore increases the gas-phase acid strength of H_nX.

The overall result of these effects is a decrease in gas-phase proton affinity of X^-, and therefore an increase in the gas-phase acidity of HX, from the top left to bottom right of the p-block. On this basis we see that HI is a much stronger acid than CH_4.

The correlation we have described is modified when a solvent (typically water) is present. The gas-phase process $X^-(g) + H^+(g) \rightarrow HX(g)$ becomes

$$X^-(aq) + H^+(aq) \rightarrow HX(aq)$$

and the negative of the accompanying proton-gain enthalpy is called the **effective proton affinity**, A'_p, of $X^-(aq)$.

If the species X^- denotes H_2O itself, the effective proton affinity of H_2O is the enthalpy change accompanying the process

$$H_2O(l) + H^+(aq) \rightarrow H_3O^+(aq)$$

The energy released as water molecules are attached to a proton in the gas phase, as in the process

$$nH_2O(g) + H^+(g) \rightarrow H^+(H_2O)_n(g)$$

can be measured by mass spectrometry and used to assess the energy change for the hydration process in solution. It is found that the energy released passes through a maximum value of 1130 kJ mol^{-1} as n increases, and this value is taken to be the effective proton affinity of H_2O in bulk water.

The effective proton affinity of the ion OH^- in water is simply the negative of the enthalpy of the reaction

$$OH^-(g) + H^+(aq) \rightarrow H_2O(l)$$

which can be measured by conventional means (such as the temperature dependence of its equilibrium constant, K_w). The value found is 1188 kJ mol^{-1}.

The reaction

$$HX(aq) + H_2O(l) \rightarrow H_3O^+(aq) + X^-(aq)$$

is exothermic if the effective proton affinity of $X^-(aq)$ is lower than that of $H_2O(l)$ (less than 1130 kJ mol^{-1}) and (provided entropy changes are negligible and so enthalpy changes are a guide to spontaneity) HX will give up protons to the water and be strongly acidic. Likewise, the reaction

$$X^-(aq) + H_2O(l) \rightarrow HX(aq) + OH^-(aq)$$

is exothermic if the effective proton affinity of $X^-(aq)$ is higher than that of $OH^-(aq)$ (1188 kJ mol^{-1}). Provided enthalpy changes are a guide to spontaneity, $X^-(aq)$ will accept protons and will act as a strong base.

A BRIEF ILLUSTRATION

The effective proton affinity of I^- in water is 1068 kJ mol^{-1} compared to 1315 kJ mol^{-1} in the gas phase, showing that the I^- ion is stabilized by hydration. The effective proton affinity is also smaller than the effective proton affinity of water (1130 kJ mol^{-1}), which is consistent with the fact that HI is a strong acid in water. All the halide ions except F^- have effective proton affinities smaller than that of water, which is consistent with all the hydrogen halides except HF being strong acids in water.

The effects of solvation can be rationalized in terms of an electrostatic model in which the solvent is treated as a continuous dielectric medium. The solvation of a gas-phase ion is always strongly exothermic. The magnitude of the enthalpy of solvation $\Delta_{solv}H^\ominus$ (the enthalpy of hydration in water, $\Delta_{hyd}H^\ominus$) depends on the radius of the ions, the relative permittivity of the solvent, and the possibility of specific bonding (especially hydrogen bonding) between the ions and the solvent.

When considering the gas phase we assume that entropy contributions for the proton transfer process are small and so $\Delta G^\ominus \approx \Delta H^\ominus$. In solution, entropy effects cannot be ignored and we must use $\Delta G^\ominus$. The Gibbs energy of solvation of an ion can be identified as the energy involved in transferring the anion from a vacuum into a solvent of relative permittivity ε_r. The **Born equation** can be derived using this model:[2]

$$\Delta_{solv}G^\ominus = -\frac{N_A z^2 e^2}{8\pi\varepsilon_0 r}\left(1 - \frac{1}{\varepsilon_r}\right) \qquad (5.10)$$

where z is the charge number of the ion, r its effective radius which includes part of the radii of solvent molecules, N_A is Avogadro's constant, ε_0 is the vacuum permittivity, and ε_r is the relative permittivity (the dielectric constant).

The Gibbs energy of solvation is proportional to z^2/r (known also as the electrostatic parameter, ξ), and so small, highly charged ions are stabilized in polar solvents (Fig. 5.4). The Born equation also shows that the larger the relative permittivity the more negative the value of $\Delta_{solv}G^\ominus$. This stabilization is particularly important for water, for which $\varepsilon_r = 80$ (and the term in parentheses is close to 1), compared with nonpolar solvents for which ε_r may be as low as 2 (and the term in parentheses is close to 0.5).

Because $\Delta_{solv}G^\ominus$ is the change in molar Gibbs energy when an ion is transferred from the gas phase into aqueous solution, a large, negative value of $\Delta_{solv}G^\ominus$ favours the formation of ions in solution compared with the gas phase (Fig. 5.4). The interaction of the charged ion with the polar solvent molecules stabilizes the conjugate base X^- relative to the parent acid HX, and as a result the acidity of HX is enhanced by the polar solvent. On the other hand, the effective proton affinity of a neutral base X is higher than in the gas phase because the conjugate acid HX^+ is stabilized by solvation. Because a cationic acid, such as NH_4^+, is stabilized by solvation, the effective proton affinity of its conjugate base (NH_3) is higher than in the gas phase. The acidity of cationic acids is therefore lowered by a polar solvent.

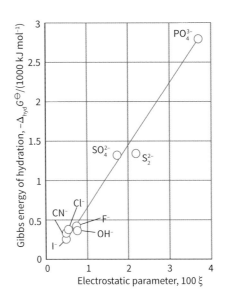

FIGURE 5.4 Correlation between $\Delta_{solv}G^\ominus$ and the dimensionless electrostatic parameter ξ ($= z^2/(r/pm)$) for selected anions.

The Born equation ascribes stabilization by solvation to Coulombic interactions. However, hydrogen bonding (Section 5.9) is an important factor in protic solvents such as water and leads to the formation of hydrogen-bonded clusters around some solutes. As a result, water has a greater stabilizing effect on small, highly charged ions than the Born equation predicts. This stabilizing effect is particularly great for F^-, OH^-, and Cl^-, with their high charge densities and for which water acts as a hydrogen-bond donor. Because a water molecule has two additional pairs of electrons on O, it can also be a hydrogen-bond acceptor. Acidic ions such as NH_4^+ are stabilized by hydrogen bonding and consequently have a lower acidity than predicted by the Born equation.

Characteristics of oxygen-containing Brønsted acids

KEY POINT Aqua acids, hydroxoacids, and oxoacids are typical of specific regions of the periodic table.

We have focused the discussion so far on acids of the type HX. However, the largest class of acids in water consists of species that donate protons from an –OH group attached to a central atom. A donatable proton of this kind is called an **acidic proton** to distinguish it from other protons that may be present in the molecule, such as the nonacidic methyl protons in CH_3COOH.

There are three classes of acids to consider:

1. An **aqua acid**, in which the acidic proton is on a water molecule coordinated to a central metal ion.

    $E(OH_2)(aq) + H_2O(l) \rightleftharpoons E(OH)^-(aq) + H_3O^+(aq)$

An example is

$[Fe(OH_2)_6]^{3+}(aq) + H_2O(l) \rightleftharpoons [Fe(OH_2)_5OH]^{2+}(aq) + H_3O^+(aq)$

The aqua acid, the hexaaquairon(III) ion, is shown as (3).

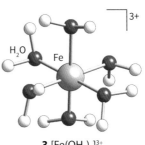

3 $[Fe(OH_2)_6]^{3+}$

---

[2] For the derivation of the Born equation, see P. Atkins, J. de Paula, and J. Keeler, *Physical Chemistry*, Oxford University Press and W.H. Freeman & Co. (2022).

2. A **hydroxoacid**, in which the acidic proton is on a hydroxyl group without a neighbouring oxo group (=O). An example is Te(OH)$_6$ (4).

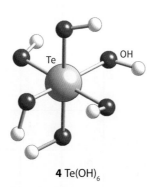

4 Te(OH)$_6$

3. An **oxoacid**, in which the acidic proton is on a hydroxyl group with an oxo group attached to the same atom. Sulfuric acid, H$_2$SO$_4$ (SO$_2$(OH)$_2$; 5), is an example of an oxoacid.

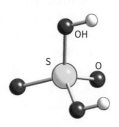

5 SO$_2$(OH)$_2$, H$_2$SO$_4$

The three classes of acid can be regarded as successive stages in the deprotonation of an aqua acid:

$$\text{aqua acid} \xrightarrow{-H^+} \text{hydroxoacid} \xrightarrow{-H^+} \text{oxoacid}$$

An example of these successive stages is provided by a complex of a d-block metal in an intermediate oxidation state, such as Ru(IV):

Aqua acids are characteristic of central atoms in low oxidation states, of s- and d-block metals, and of metals on the left of the p block. Oxoacids are commonly found where the central element is in a high oxidation state. An element from the right of the p block may also produce an oxoacid in one of its intermediate oxidation states (HClO$_2$ is an example).

5.2 Periodic trends in aqua acid strength

KEY POINTS The strengths of aqua acids typically increase with increasing positive charge of the central metal ion and with decreasing ionic radius; exceptions are indicative of greater covalence in the metal-OH$_2$ bond.

The strengths of aqua acids typically increase with increasing positive charge of the central metal ion and with decreasing ionic radius. This variation can be rationalized to some extent in terms of the ionic model, in which the metal cation is represented by a sphere of radius r_+ carrying z positive charges. Because protons are more easily removed from the vicinity of cations of high charge and small radius, the model predicts that the acidity should increase with increasing z and with decreasing r_+.

The validity of the ionic model of acid strengths can be judged from Fig. 5.5. Aqua ions of elements that form ionic solids (principally those from the s block) have pK_a values that are quite well described by the ionic model. Several d-block ions (such as Fe^{2+} and Cr^{3+}) lie reasonably near the same straight line, but many ions (particularly those with low pK_a, corresponding to high acid strength) deviate markedly from it. This deviation indicates that the metal ions repel the departing proton more strongly than is predicted by the ionic model. This enhanced repulsion can be rationalized by supposing that the positive charge of the cation is not confined to the central ion but is delocalized over the ligands and hence is closer to the departing proton. The delocalization is equivalent to attributing covalence to the element–oxygen bond. Indeed, the correlation is worst

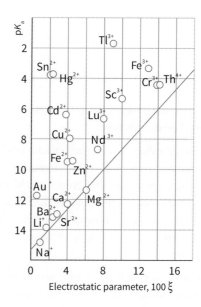

FIGURE 5.5 The correlation between acidity constant pK_a and the dimensionless electrostatic parameter ξ (= $z^2/(r/\text{pm})$) of aqua ions. The line indicates the trend expected on the basis of the ionic model.

for ions that are disposed to form more covalent M–OH$_2$ bonds.

For the later d- and the p-block metal ions (such as Cu^{2+} and Sn^{2+}, respectively), the strengths of the aqua acids are much greater than the ionic model predicts. For these species, covalent bonding is more important than ionic bonding and the ionic model is unrealistic. The overlap between metal d orbitals and the orbitals of an oxygen ligand increases down a group, so aqua ions of 2nd and 3rd-row d-block metals tend to be stronger acids than aqua ions of the 1st row.

EXAMPLE 5.4 Accounting for trends in aqua acid strength

Account for the trend in acidity [Fe(OH$_2$)$_6$]$^{2+}$ < [Fe(OH$_2$)$_6$]$^{3+}$ < [Al(OH$_2$)$_6$]$^{3+}$ ≈ [Hg(OH$_2$)]$^{2+}$.

Answer We need to consider the charge density on the metal centre and its effect on the ease with which the H$_2$O ligands can be deprotonated. The weakest acid is the Fe^{2+} complex, on account of its relatively large ionic radius and low charge. The increase of charge to +3 increases the acid strength. The greater acidity of Al^{3+} can be explained by its smaller radius. The anomalous ion in the series is the Hg^{2+} complex. This complex reflects the failure of the ionic model, because in the complex there is a substantial transfer of positive charge to oxygen as a result of covalent bonding.

Self-test 5.4 Arrange [K(OH$_2$)$_6$]$^+$, [Sc(OH$_2$)$_6$]$^{3+}$, [Ca(OH$_2$)$_6$]$^{2+}$, and [Ni(OH$_2$)$_6$]$^{2+}$ in order of increasing acidity.

5.3 Simple oxoacids

The simplest oxoacids are the **mononuclear acids**, which contain one atom of the parent element. They include H$_2$CO$_3$, HNO$_3$, H$_3$PO$_4$, and H$_2$SO$_4$.[3] These oxoacids are formed by the electronegative elements at the upper right of the periodic table and by other elements in high oxidation states (Table 5.3). One interesting feature in the table is the occurrence of planar H$_2$CO$_3$ and HNO$_3$ molecules but not their analogues in later periods. As we saw in Chapter 2, pπ–pπ bonding is more important among the Period 2 elements, so their atoms are more likely to be constrained to lie in a plane.

(a) Substituted oxoacids

KEY POINTS Substituted oxoacids have strengths that may be rationalized in terms of the electron-withdrawing power of the substituent; in a few cases, a nonacidic H atom is attached directly to the central atom of an oxoacid.

One or more –OH groups of an oxoacid may be replaced by other groups to give a series of substituted oxoacids, which include fluorosulfuric acid, SO$_2$F(OH), and aminosulfuric

[3] These acids are more helpfully written as CO(HO)$_2$, NO$_2$(OH), PO(OH)$_3$, and SO$_2$(HO)$_2$, and boric acid written as B(OH)$_3$ rather than H$_3$BO$_3$. In this text we use both forms of notation, the *familiar formula* or the *structural formula*, depending upon the properties being explained—the familiar formulas being adopted wherever structural details are not of concern.

TABLE 5.3 The structure and pK_a values of oxoacids*

p = 0	p = 1		p = 2	p = 3
HO—Cl	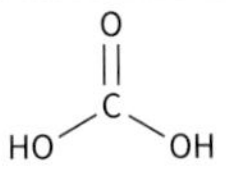 HO–C(=O)–OH		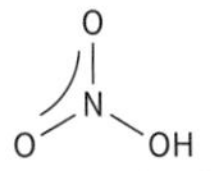 O=N(–O)–OH	
7.2	3.6		−1.4	
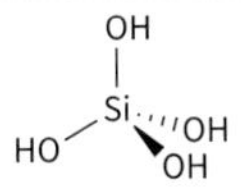 HO–Si(OH)(OH)–OH	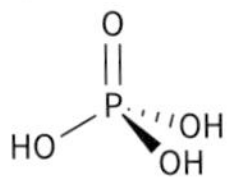 HO–P(=O)(OH)–OH	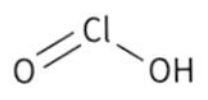 O=Cl–OH	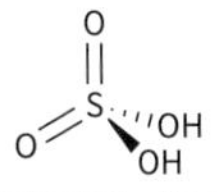 O=S(=O)(OH)–OH	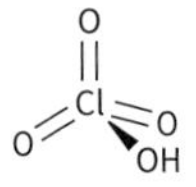 O=Cl(=O)(=O)–OH
10	2.1, 7.4, 12.7	2.0	−1.9	−10
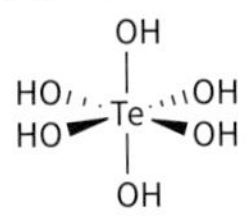 HO–Te(OH)(OH)(OH)–OH	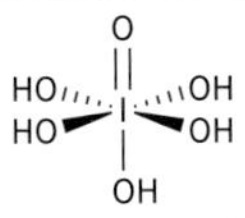 HO–I(=O)(OH)(OH)–OH	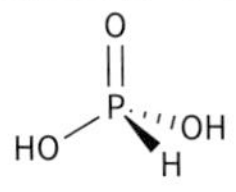 HO–P(=O)(H)–OH	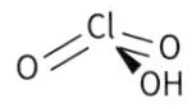 O=Cl(=O)–OH	
7.8, 11.2	1.6, 7.0	1.8, 6.6	−1.0	
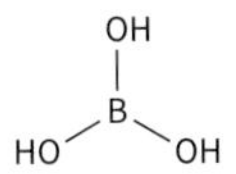 HO–B(OH)–OH	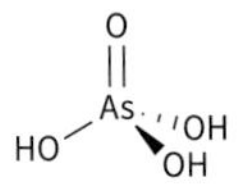 HO–As(=O)(OH)–OH	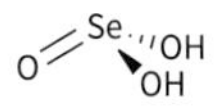 O=Se(OH)–OH		
9.1†	2.3, 6.9, 11.5	2.6, 8.0		

* p is the number of nonprotonated O atoms.
† Boric acid is a special case; see Section 14.8.

acid, $SO_2(NH_2)(OH)$ (**6**). Because fluorine is highly electronegative, it withdraws electrons from the central S atom and confers on S a higher effective positive charge. As a result, the substituted acid is stronger than $SO_2(OH)_2$. Another electron acceptor substituent is $-CF_3$, as in the strong acid trifluoromethylsulfonic acid, CF_3SO_3H (i.e. $SO_2(CF_3)(OH)$). By contrast, the $-NH_2$ group, which has a lone pair, can donate electron density to S by π bonding. This transfer of charge reduces the positive charge of the central atom and weakens the acid.

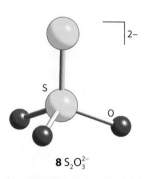

8 $S_2O_3^{2-}$

A NOTE ON GOOD PRACTICE

The structures of oxoacids are drawn with double bonds to oxo groups, =O. This representation indicates the connectivity of the O atom to the central atom, but in reality resonance lowers the calculated energy of the molecule and distributes the bonding character of electrons over the molecule.

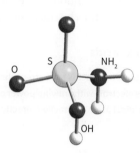

6 $SO_2(NH_2)(OH)$

A trap for the unwary is that not all oxoacids follow the familiar structural pattern of a central atom surrounded by OH and O groups. Occasionally an H atom is attached directly to the central atom, as in phosphonic (phosphorous) acid, H_3PO_3. Phosphonic acid is in fact only a *di*protic acid, as the substitution of two OH groups leaves an H–P bond (**7**) and consequently a nonacidic proton. This structure is consistent with ^{1}H- and ^{31}P-NMR (Section 8.6) and vibrational spectra (Section 8.5), and the structural formula, $HPO(OH)_2$, emphasizes the presence of the H–P bond. The nonacidity of the H–P bond mainly reflects the much lower electron-withdrawing ability of the central P atom compared to O.

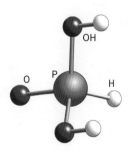

7 $HPO(OH)_2$, H_3PO_3

Substitution for an oxo group (as distinct from a hydroxyl group) is another example of a structural change that can occur. An important example is the thiosulfate ion, $S_2O_3^{2-}$ (**8**), in which an S atom replaces an O atom of a sulfate ion.

(b) Pauling's rules

KEY POINT The strengths of a series of oxoacids containing a specific central atom with a variable number of oxo and hydroxyl groups are summarized by Pauling's rules.

For a series of mononuclear oxoacids of an element E, the strength of the acids increases with increasing number of O atoms. This trend can be explained qualitatively by considering the electron-withdrawing properties of oxygen. The electronegative O atoms withdraw electrons, so making each O–H bond weaker. Consequently, protons are more readily released. In general, for any series of oxoacids, the one with the most O atoms is the strongest. For example, the acid strengths of the oxoacids of chlorine decrease in the order $HClO_4 > HClO_3 > HClO_2 > HClO$. Similarly, HNO_3 is a stronger acid than HNO_2.

Another important factor is the degree to which differing numbers of terminal oxo groups stabilize the deprotonated (conjugate) base by resonance. For example, the conjugate base of H_2SO_4, the HSO_4^- anion, can be described as a resonance hybrid of three contributions (**9**), whereas the conjugate base of H_2SO_3, the HSO_3^- anion, has only two resonance contributions (**10**). Consequently, H_2SO_4 is a stronger acid than H_2SO_3. An interesting corollary of this comparison, relating to the nature of H_2SO_3, is discussed later.

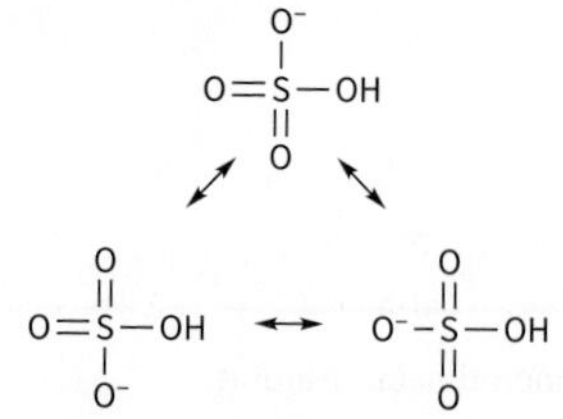

9

$$O=S\begin{smallmatrix}OH\\\\O^-\end{smallmatrix} \longleftrightarrow O^--S\begin{smallmatrix}OH\\\\O\end{smallmatrix}$$

10

The trends can be systematized semiquantitatively by using two empirical rules devised by Linus Pauling, where p is the number of oxo groups and q is the number of hydroxyl groups:

Rule 1. For the oxoacid $EO_p(OH)_q$, $pK_a \approx 8 - 5p$.

Rule 2. The successive pK_a values of polyprotic acids (those with $q > 1$), increase by 5 units for each successive proton transfer.

Rule 1 predicts that neutral hydroxoacids with $p = 0$ have $pK_a \approx 8$, acids with one oxo group have $pK_a \approx 3$, and acids with two oxo groups have $pK_a \approx -2$. For example, sulfuric acid, $SO_2(OH)_2$, has $p = 2$ and $q = 2$, and $pK_{a1} \approx -2$ (signifying a strong acid). Similarly, pK_{a2} is predicted to be +3, although comparison with the experimental value of 1.9 reminds us that these rules are only approximations.

The success of Pauling's rules may be gauged by inspection of Table 5.3, in which acids are grouped according to p. The variation in strengths down a group is not large, and the complicated and perhaps cancelling effects of changing structures allow the rules to work moderately well. The more important variation across the periodic table from left to right and the effect of change of oxidation number are taken into account by the number of oxo groups. In Group 15 the oxidation number +5 requires one oxo group (as in $PO(OH)_3$), whereas in Group 16 the oxidation number +6 requires two (as in $SO_2(OH)_2$).

(c) Structural anomalies

KEY POINT In certain cases, a simple molecular formula misrepresents the composition of aqueous solutions of nonmetal oxides.

An interesting use of Pauling's rules is to detect structural anomalies. For example, hydrated CO_2, which is referred to as carbonic acid, $CO(OH)_2$, is commonly reported as having $pK_{a1} = 6.4$, but the rules predict $pK_{a1} = 3$. The anomalously low acidity indicated by the experimental value is the result of treating the concentration of dissolved CO_2 as if it were all $CO(OH)_2$. However, in the equilibrium

$$CO_2(aq) + H_2O(l) \rightleftharpoons CO(OH)_2(aq)$$

only about 1% of the dissolved CO_2 is present as $CO(OH)_2$, so the actual concentration of acid is much less than the concentration of dissolved CO_2. When this difference is taken into account, the true pK_{a1} of carbonic acid is about 3.6, as Pauling's rules predict. The equilibrium between CO_2 and $CO(OH)_2$ explains why CO_2 acidifies water, and is mentioned later when we discuss Lewis acidity.

The experimental value $pK_{a1} = 1.8$ reported for sulfurous acid, H_2SO_3, suggests another anomaly, in this case acting in the opposite way. In fact, spectroscopic studies have failed to detect the molecule $SO(OH)_2$ in solution, and the equilibrium constant for

$$SO_2(aq) + H_2O(l) \rightleftharpoons SO(OH)_2(aq)$$

is less than 10^{-9}. The equilibria of dissolved SO_2 are complex, and a simple analysis is inappropriate. The ions that have been detected include $SO_2(OH)^-$ and $S_2O_5^{2-}$, and there is evidence for an SH bond in the solid salts of the hydrogensulfite ion.

This discussion of the composition of aqueous solutions of CO_2 and SO_2 calls attention to the important point that not all nonmetal oxides react fully with water to form acids. Carbon monoxide is another example: although it is formally the anhydride of methanoic acid, HCOOH, carbon monoxide does not in fact react with water at room temperature to produce the acid. The same is true of some metal oxides: OsO_4, for example, can exist as dissolved neutral molecules.

EXAMPLE 5.5 **Using Pauling's rules**

Identify the structural formulas that are consistent with the following pK_a values: H_3PO_4, 2.12; H_3PO_3, 1.80; H_3PO_2, 2.0.

Answer We can apply Pauling's rules to use the pK_a values to predict the number of oxo groups. All three values are in the range that Pauling's first rule associates with one oxo group. This observation suggests the formulas $PO(OH)_3$, $HPO(OH)_2$, and $H_2PO(OH)$. The second and the third formulas are derived from the first by replacement of –OH by H bound to P (as in structure **7**).

Self-test 5.5 Predict the pK_a values of (a) H_3PO_4, (b) $H_2PO_4^-$, (c) HPO_4^{2-}.

5.4 Anhydrous oxides

We have treated oxoacids as derived by deprotonation of their parent aqua acids. It is also useful to take the opposite viewpoint and to consider aqua acids and oxoacids as derived by hydration of the oxides of the central atom. This approach emphasizes the acid and base properties of oxides and their correlation with the location of the element in the periodic table.

(a) Acidic, basic, and amphoteric oxides

KEY POINTS Acidic oxides (those that react with bases) are typically formed by nonmetallic elements, while basic oxides (those that react with acids) are typically formed by metallic elements. The term 'amphoteric' is used to describe oxides that react with both acids and bases.

An **acidic oxide** is an oxide that, on dissolution in water, binds an H$_2$O molecule and releases a proton to the surrounding solvent:

$$CO_2(g) + H_2O(l) \rightleftharpoons CO(OH)_2(aq)$$
$$CO(OH)_2(aq) + H_2O(l) \rightleftharpoons H_3O^+(aq) + CO_2(OH)^-(aq)$$

An equivalent interpretation is that an acidic oxide is an oxide that reacts with an aqueous base (as occurs with OH$^-$ in an alkaline solution):

$$CO_2(g) + OH^-(aq) \rightarrow CO_2(OH)^-(aq)$$

A **basic oxide** is an oxide to which a proton is transferred when it dissolves in water:

$$BaO(s) + H_2O(l) \rightarrow Ba^{2+}(aq) + 2OH^-(aq)$$

The equivalent interpretation in this case is that a basic oxide is an oxide that reacts with an acid:

$$BaO(s) + 2H_3O^+(aq) \rightarrow Ba^{2+}(aq) + 3H_2O(l)$$

Because acidic and basic oxide character often correlates with other chemical properties, a wide range of properties can be predicted from a knowledge of the character of oxides. In a number of cases the correlations follow from the basic oxides being largely ionic and of acidic oxides being largely covalent. For instance, an element that forms an acidic oxide is likely to form volatile, covalent halides. By contrast, an element that forms a basic oxide is likely to form solid, ionic halides. In short, the acidic or basic character of an oxide is an indication of whether an element should be regarded as a metal or a nonmetal. Generally, nonmetals form acidic oxides and metals form basic oxides—exceptions being when the metal is in a high oxidation state.

An **amphoteric oxide** is an oxide that reacts with both acids and bases.[4] Thus, aluminium oxide reacts with acids and alkalis:

$$Al_2O_3(s) + 6H_3O^+(aq) + 3H_2O(l) \rightarrow 2[Al(OH_2)_6]^{3+}(aq)$$
$$Al_2O_3(s) + 2OH^-(aq) + 3H_2O(l) \rightarrow 2[Al(OH)_4]^-(aq)$$

The occurrence of amphoterism among the elements of the periodic table is discussed in Section 9.5.

5.5 Polyoxo compound formation

KEY POINTS Metal aqua ions tend to form polymers as the pH is raised, whereas oxoanions tend to form polymers as the pH is lowered. Polyoxoanions account for most of the mass of oxygen in the Earth's crust.

As the pH of a solution is increased, the aqua ions of metals that have basic or amphoteric oxides generally undergo polymerization and precipitation. Because the precipitation occurs quantitatively at a pH characteristic of each metal, one application of this behaviour is the separation of metal ions.

With the exception of Be^{2+} (which is amphoteric), the elements of Groups 1 and 2 have no important solution species beyond the aqua ions M$^+$(aq) and M^{2+}(aq). By contrast, the solution chemistry of the elements becomes very rich as the amphoteric region of the periodic table is approached (Section 9.5). The two most common examples are polymers formed by Fe(III) and Al(III), both of which are abundant in the Earth's crust. In acidic solutions, both form octahedral hexaaqua ions, [Al(OH$_2$)$_6$]$^{3+}$ and [Fe(OH$_2$)$_6$]$^{3+}$. In solutions of pH > 4, both precipitate as gelatinous hydrated hydroxides:

$$[Fe(OH_2)_6]^{3+}(aq) + (3+n) H_2O(l) \rightarrow Fe(OH)_3 \cdot nH_2O(s)$$
$$+ 3H_3O^+(aq)$$
$$[Al(OH_2)_6]^{3+}(aq) + (3+n) H_2O(l) \rightarrow Al(OH)_3 \cdot nH_2O(s)$$
$$+ 3H_3O^+(aq)$$

The precipitated polymers, which are often of colloidal dimensions (between 1 nm and 1 μm), slowly crystallize to stable mineral forms. The extensive network structure of aluminium polymers, which are neatly packed in three dimensions, contrasts with the linear polymers of their iron analogues.

Polyoxoanion formation from oxoanions (condensation) occurs by protonation of an O atom (which is formally in the 2- oxidation state) and its departure as H$_2$O. An example of the simplest condensation reaction, starting with the orthophosphate ion, PO$_4^{3-}$, is formation of the diphosphate ion, P$_2$O$_7^{4-}$.

$$2\,PO_4^{3-} + 2\,H_3O^+ \longrightarrow \text{[structure]} + 3\,H_2O$$

The elimination of water consumes protons and decreases the average charge number of each P atom to −2. If each phosphate group is represented as a tetrahedron with the O atoms located at the corners, P$_2$O$_7^{4-}$ can be drawn as connected polyhedra (**11**).

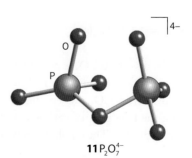

11 P$_2$O$_7^{4-}$

---

[4] The word 'amphoteric' is derived from the Greek word for 'both'.

Phosphoric acid can be prepared by hydrolysis of the solid phosphorus(V) oxide, P_4O_{10}. An initial step using a limited amount of water produces a metaphosphate ion with the formula $P_4O_{12}^{4-}$ (12).

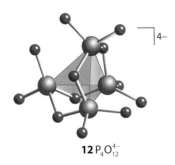

12 $P_4O_{12}^{4-}$

This reaction is only the simplest among many, and the separation of products from the hydrolysis of phosphorus(V) oxide by chromatography reveals the presence of chain species containing one to nine P atoms. Chain polymers of formula P_n with $n = 10$–50 can be isolated as mixed amorphous glasses analogous to those formed by silicates (Section 16.15).

The polyphosphates are biologically important. At physiological pH (close to 7.4), the P–O–P entity is unstable with respect to hydrolysis. Consequently, its hydrolysis can serve as a mechanism for providing the energy to drive a reaction (the Gibbs energy). Similarly, the formation of the P–O–P bond is a means of storing Gibbs energy. The key to energy exchange in metabolism is the hydrolysis of adenosine triphosphate, ATP (13a), to adenosine diphosphate, ADP (13b) as shown here, or to adenosine monophosphate. In living cells, ATP and ADP are present as their Mg^{2+} complexes (Section 26.3).

$[MgATP]^{2-} + H_2O \rightarrow [MgADP]^- + H_2PO_4^-$
$\Delta_r G^\ominus = -31\,kJ\,mol^{-1}$

13b ADP^{3-}

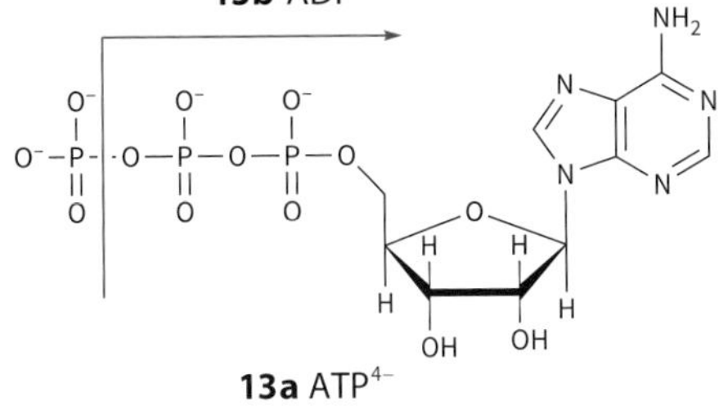

13a ATP^{4-}

Energy flow in metabolism depends on the subtle construction of pathways to condense ATP from ADP and AMP (and phosphate units) and hydrolyse it back when required. A thermodynamically unfavourable reaction can be selectively driven by tightly coupling it to ATP hydrolysis, the two processes occurring simultaneously on a specific enzyme catalyst.

Reactant ⇌ Product

$[MgATP]^{2-}$ + Reactant ⇌(enzyme) $[MgADP]^-$ + Product

Polyoxoanions account for most of the mass of oxygen in the Earth's crust, as they include almost all silicate minerals. Higher nuclearity species formed by condensation of nonmetal oxoanions are typically found as rings and chains. The silicates are very important examples of polymeric oxoanions, and we discuss them in detail in Sections 15.10, 25.16, and 29.1. One example of a polysilicate mineral is $MgSiO_3$, which contains a virtually infinite chain of SiO_3^{2-} units.

The formation of polyoxoanions is important for early d-block ions in their highest oxidation states, particularly V(V), Mo(VI), W(VI), and (to a lesser extent) Nb(V), Ta(V), and Cr(VI), for which the term **polyoxometallates** (POMs) is used. Polyoxometallates form a variety of three-dimensional framework structures that are stable to oxidation and are finding increasing application in catalysis and analytical chemistry. Different metals and nonmetals such as P can be incorporated to produce compounds known as heteropolyoxometallates. We describe polyoxometallates in more detail in Chapter 20 and Box 20.1.

Lewis acidity

KEY POINTS A Lewis acid is an electron pair acceptor; a Lewis base is an electron pair donor.

The Brønsted–Lowry theory of acids and bases focuses on the transfer of a proton between species. A more general theory of acid–base reactions was introduced by G. N. Lewis in the same year as Brønsted and Lowry introduced theirs (1923). However, Lewis's approach became influential only in the 1930s.

A **Lewis acid** is a substance that acts as an electron pair acceptor. A **Lewis base** is a substance that acts as an electron pair donor. In this section we denote a Lewis acid by A and a Lewis base by :B, usually omitting any other lone electron pairs ('lone' in this case meaning not directly involved in bonding) that may be present. The fundamental reaction of Lewis acids and bases is the formation of a **complex** (or adduct), A–B, in which A and :B form a bond by sharing the electron pair supplied by the base. Such a bond, often called a **dative** or **coordinate** bond, was introduced briefly in Section 2.9. To encourage familiarity we will use all of these terms interchangeably throughout the following text. Lewis acid–base interactions are responsible for the structures adopted by many compounds and for specific reactivities that are important in catalysis.

> **A NOTE ON GOOD PRACTICE**
>
> The terms Lewis *acid* and *base* are used in discussions of the equilibrium (thermodynamic) properties of reactions. In the context of reaction rates (kinetics) an electron-pair donor is called a *nucleophile* and an electron-pair acceptor is called an *electrophile*.

The bonding between a Lewis acid and Lewis base can be viewed from a molecular orbital perspective, as illustrated in Fig. 5.6. The Lewis acid provides an empty orbital that is usually the LUMO of that component, and the Lewis base provides a full orbital that is usually the respective HOMO. The newly formed bonding orbital is populated by the two electrons supplied by the base, whereas the newly formed antibonding orbital is left unoccupied. As a result, there is a net lowering of energy when the bond forms.

5.6 Identifying Lewis acids and bases

KEY POINTS Brønsted acids and bases *exhibit* Lewis acidity and basicity respectively; the Lewis definition can be applied to aprotic systems.

A proton is a Lewis acid because it can attach to an electron pair, as in the formation of NH_4^+ from NH_3. It follows that any Brønsted acid, as it provides protons, serves as an agent for Lewis acidity. Note that the Brønsted acid HX that we considered previously in this chapter is actually the complex formed by the Lewis acid H^+ (A) with the Lewis base B^-. We say that a Brønsted acid *exhibits* Lewis acidity rather than that a Brønsted acid *is* a Lewis acid.

All Brønsted bases are Lewis bases, because a proton acceptor is also an electron pair donor: an NH_3 molecule, for instance, is a Lewis base as well as a Brønsted base. Therefore, the whole of the material presented in the preceding sections of this chapter can be regarded as a special case of Lewis's approach.

We will focus mainly on Lewis acids. Because the proton is not essential to the definition of a Lewis acid or base, a much wider range of substances can be classified as acids in the Lewis scheme than can be classified in the Brønsted scheme. One obvious factor is that the stability of a complex may be strongly influenced by adverse steric interactions between the acid and the base. A further factor is that the practical application of Lewis acids depends very much on their stability, ease of handling, and toxicity.

We meet many examples of Lewis acids later, but we should be alert to the following possibilities:

1. A molecule with an incomplete octet of valence electrons can complete its octet by accepting an electron pair donated by the base. Good examples are compounds of Group 13, such as BF_3, which can accept the lone pair of NH_3 and other donors.

2. A molecule or ion with a complete octet may be able to rearrange its valence electrons and accept an additional electron pair. For example, CO_2 acts as a Lewis acid when it forms HCO_3^- (more accurately $CO_2(OH)^-$) by accepting an electron pair from an O atom in an OH^- ion.

3. A molecule or ion may be able to expand its valence shell to accept another electron pair, which may be used either in σ or π bonding. An example is the formation of the complex $[SiF_6]^{2-}$ when two F^- ions (the Lewis bases) bond to SiF_4 (the acid). This type of Lewis acidity is common for compounds of the heavier p-block elements, such as SiX_4, AsX_3, and PX_5 (with X being a halogen).

4. A metal cation can accept an electron pair supplied by the base in a coordination compound. This aspect of Lewis acids and bases is treated at length in Chapters 7 and 23. Hydration is a particularly familiar reaction; for instance, when $CoCl_2$ is dissolved in water, lone pairs of H_2O molecules (acting as a Lewis base) donate to the central cation to produce $[Co(OH_2)_6]^{2+}$. The anhydrous Co^{2+} cation (which would exist in the gas phase) is therefore the Lewis acid.

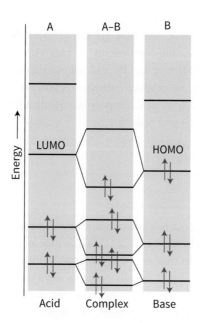

FIGURE 5.6 Molecular orbital representation of the orbital interactions responsible for formation of a complex between a Lewis acid A and a Lewis base :B.

EXAMPLE 5.6 Identifying Lewis acids and bases

Identify the Lewis acids and bases in the reactions (a) $BrF_3 + F^- \rightarrow BrF_4^-$, (b) $KH + H_2O \rightarrow KOH + H_2$ (occurring in the gas phase).

Answer We need to identify the electron-pair acceptor (the acid) and the electron-pair donor (the base). (a) The acceptor BrF_3 accepts a pair of electrons from the donor F^-. Therefore, BrF_3 is a Lewis acid and F^- is a Lewis base. (b) The salt KH provides H^- (the hydride anion) which displaces H^+ from water to give H_2 and OH^-. The net reaction is

$$H^- + H_2O \rightarrow H_2 + OH^-$$

If we think of this reaction as

$$H{:}^- + H^+ {-} {:}OH^- \rightarrow H{-}H + {:}OH^-$$

we see that H^- provides an electron pair and is therefore a Lewis base. It reacts with H_2O to drive out OH^-, another Lewis base.

Self-test 5.6 Identify the acids and bases in the reactions (a) $FeCl_3 + Cl^- \rightarrow FeCl_4^-$, (b) $I^- + I_2 \rightarrow I_3^-$.

5.7 The fundamental types of reaction

Lewis acids and bases undergo a variety of characteristic reactions.

(a) Complex formation

KEY POINT In complex formation, a free Lewis acid and free Lewis base become linked by a coordinate (dative) bond.

The simplest Lewis acid–base reaction in the gas phase (or a noncoordinating solvent) is **complex formation**:

$$A + {:}B \rightarrow A{-}B$$

Two examples are

Both reactions involve Lewis acids and bases that are independently stable in the gas phase or in solvents that do not form complexes with them. Consequently, the individual species (as well as the complexes) may be studied experimentally. Complex formation is an exothermic process: as we saw from Fig. 5.6, the electrons occupying the HOMO of the resulting complex are at a lower energy than in the HOMO of the Lewis base from which the complex was formed.

(b) Displacement reactions

KEY POINT In a displacement reaction, an acid or base drives out another acid or base from a Lewis complex. A double displacement reaction is known as metathesis.

A **displacement** of one Lewis partner by another is a reaction of the form

$$A{-}B + {:}B' \rightarrow B{:} + A{-}B' \quad \text{or} \quad A{-}B + A' \rightarrow A + A'{-}B$$

An example is

All Brønsted proton-transfer reactions in water are displacement reactions, as in

$$HS^-(aq) + H_2O(l) \rightarrow S^{2-}(aq) + H_3O^+(aq)$$

Here, the Lewis base H_2O displaces the Lewis base S^{2-} from its complex with the Lewis acid H^+.

Reactions that alter the nuclearity of oxoacids, mentioned in Section 5.5, are displacement reactions. For instance, the condensation of two PO_4^{3-} ions to form $P_2O_7^{4-}$ involves the replacement (at a single PO_4^{3-} ion) of one Lewis base (O^{2-}) by another (written for clarity as $P(O)O_3^{3-}$).

In the context of complexes of d- and f-block metal ions, a displacement reaction in which one ligand is driven out of the complex and is replaced by another is called a **substitution reaction** (Section 23.1). For instance, the complex ion $[Co(NH_3)_5Cl]^{2+}$ undergoes hydrolysis in aqueous solution.

$$[Co(NH_3)_5Cl]^{2+} + H_2O \rightarrow [Co(NH_3)_5(H_2O)]^{3+} + Cl^-$$

The toxicity of carbon monoxide to animals is an example of a Lewis acid–base reaction that depends on strong π-bonding. Normally, the oxygen molecule forms a bond to the Fe(II) atom of haemoglobin and does so reversibly (Section 26.7), however, carbon monoxide displaces O_2 in an irreversible process that has potentially lethal consequences.

$$Hb - Fe^{II}O_2 + CO \rightarrow Hb - Fe^{II}CO + O_2$$

A double displacement (an interchange of both acid and base partners) is commonly known as **metathesis**:[5]

$$A{-}B + A'{-}B' \rightarrow A{-}B' + A'{-}B$$

---

[5] The name metathesis comes from the Greek word for exchange.

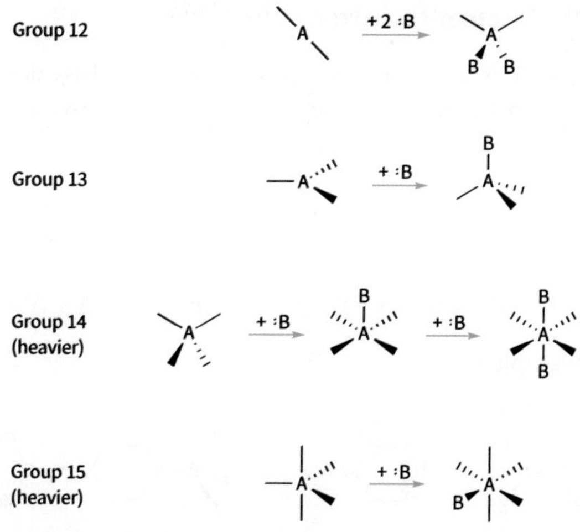

FIGURE 5.7 Reactions of Lewis acid centres (A) with Lewis bases (:B) showing typical stereochemical outcomes. Substituents on A are not shown.

The displacement of the base :B by a different base :B' is assisted by the extraction of :B by the acid A'. An example is the reaction

Here the base Br⁻ displaces I⁻, and the extraction is assisted by the formation of the less soluble AgI.

5.8 Properties of main group Lewis acids

An understanding of the trends in Lewis acidity and basicity enables us to predict the outcome of many reactions of the s- and p-block elements. The typical reactions of neutral (uncharged) Lewis acids of Groups 12–16 involve the stereochemical changes shown in Fig. 5.7. The shapes of the acids and their resulting complexes are predicted by the VSEPR model (Section 2.3).

(a) Lewis acids and bases of the s-block elements

KEY POINTS The most common example of the Lewis acidity of s-block elements is hydration of their cations, which is largely electrostatic in nature. The one-dimensional polymeric structure of beryllium chloride in the solid state arises from interactions between monomeric $BeCl_2$ units and demonstrates the importance of Lewis acidity and basicity in influencing structure.

The observation that metal ions in water are present as aqua complexes shows that they behave as Lewis acids, with H_2O molecules serving as Lewis bases. With the exception of the lightest s-block elements, individual donor-acceptor interactions are weak. The Be atom in beryllium dihalides acts as a Lewis acid by forming a polymeric chain structure in the solid state (**14**). In this structure, new σ bonds are formed when each of the chlorine atoms on Cl-Be-Cl acts as a Lewis base and donates an electron pair to an adjacent Be atom (the resultant Be orbitals becoming sp^3 hybrids). The Lewis acidity of beryllium chloride is also demonstrated by the ready formation of tetrahedral complexes such as $BeCl_4^{2-}$.

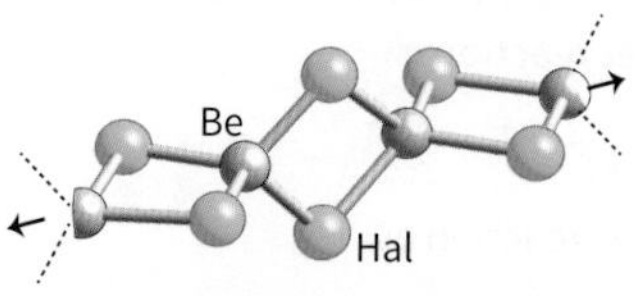

14 Be(Hal)$_2$

(b) Group 13 Lewis acids

KEY POINTS Boron and aluminium form numerous compounds that are strong Lewis acids: many are useful in synthesis and for the theoretical insight they provide.

The basis for the Lewis acidity of boron and aluminium compounds is that planar molecules represented as BX_3 and AlX_3 have incomplete octets, and the vacant p orbital perpendicular to the plane can accept an electron pair from a Lewis base, the molecule becoming pyramidal in the process (Fig. 5.7).

Compounds of boron containing large electron-withdrawing groups such as $B(OTeF_5)_3$ (**15**) and $B(OTf)_3$ (**16**) (OTf (triflate) = trifluoromethane sulfonate, OSO_2CF_3) are among the strongest neutral Lewis acids known.

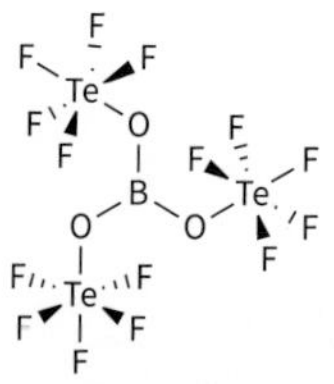

15 B(OTeF$_5$)$_3$

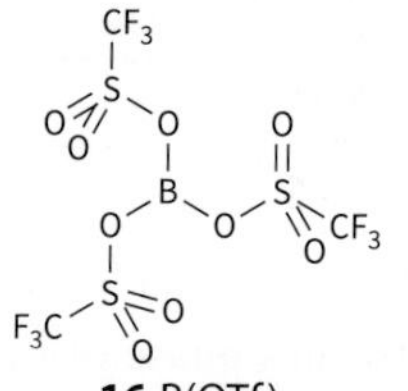

16 B(OTf)$_3$

Despite being a much weaker Lewis acid, BF_3 is widely used as an industrial catalyst. It is a stable gas at room temperature and pressure, and dissolves in diethyl ether to give a solution that is convenient to use, the commercial material being the 1:1 adduct (**17**). This dissolution is also an aspect of Lewis acid character because, as BF_3 dissolves, it forms a complex with the O atom of a solvent molecule.

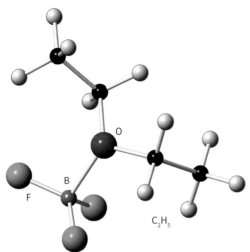

17 Adduct of BF_3 with diethyl ether

The strength of $B(OTeF_5)_3$ and $B(OTf)_3$ as Lewis acids is ascribed to the highly electron-withdrawing properties of the bulky X substituents, which will confer greater electron deficiency on the boron atom. However, when X is a simple halogen, the order of thermodynamic stability of complexes formed with trimethylamine, $:N(CH_3)_3$, is $BF_3 < BCl_3 < BBr_3$, opposite to that expected on the basis of the relative electronegativities of the halogens.

The simplest and most widely accepted explanation hinges on the fact that the planarity of the BX_3 molecule is lost when a complex is formed and the B–X bonds are forced to bend away from their new neighbour. The planar structure is expected on the basis of the VSEPR model (Section 2.3) but is supported further by π-bonding between filled p_z orbitals on X and the empty $B2p_z$ orbital that is to serve as acceptor orbital (**18**).

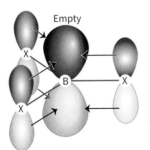

18 p–p π bonding in BX_3

The small F atom forms the strongest π bonds with the $B2p$ orbital: recall that p–p π bonding is strongest for Period 2 elements, largely on account of the small atomic radii of these elements and the significant overlap of their compact $2p$ orbitals (Section 2.5). As a result, the BF_3 molecule has the strongest π bond to be broken when the amine forms an N–B bond, and distortion of the planar structure becomes the dominant factor.

As with $BeCl_2$, Lewis acidity is important in determining structure. Aluminium halides even form dimers in the gas phase: aluminium chloride, which contains sheets of 6-coordinate Al ions in the solid state exists as Al_2Cl_6 molecules in the vapour (**19**). Each Al atom acts as an acid towards a Cl atom initially belonging to the other Al atom. Like BF_3, aluminium halides are important Lewis acid catalysts in organic reactions (Section 5.16).

19 Al_2Cl_6

(c) Group 14 Lewis acids

KEY POINTS Carbon dioxide is among the most familiar of Lewis acids: its reactions involve major rearrangement of its orbitals. Group 14 elements other than carbon exhibit hypervalence and act as Lewis acids by becoming five or six-coordinate; compounds of tin(II) that possess a lone pair on Sn may also act as Lewis bases.

The interconversion between CO_2 and carbonate ions occurring in solids or solutions is one of the most important reactions on Earth: it is essential in biology and a central topic in combating climate change. In the forward reaction, the Lewis base, an oxide (or hydroxide) ion, attacks the carbon atom on which the empty 2π antibonding orbitals are largely localized.

The reverse reaction occurs in the production of cement (Box 5.1) and is responsible for the release of large amounts of CO_2 into the atmosphere.

Unlike carbon, the larger Group 14 atoms can expand their valence shell to become hypervalent. Stable structures with five-coordinate trigonal-bipyramidal geometry can be isolated, and a six-coordinate adduct is formed when the Lewis acid SiF_4 reacts with two F^- ions:

BOX 5.1 Acid–base chemistry in the industrial release and capture of CO_2

Cement is made by grinding together limestone ($CaCO_3$) and a source of aluminosilicates (Section 15.3), such as clay, shale, or sand, which are then heated to 1500°C in a rotary cement kiln. The limestone decomposes with release of CO_2 to form lime (CaO) which reacts with the aluminosilicates to form molten calcium silicates and aluminates with various empirical formulae Ca_2SiO_4, Ca_3SiO_5, and $Ca_3Al_2O_6$. The formulas are usually written to indicate the persistence of the reactive CaO species in the product.

$$n\,CaCO_3 + SiO_2 \rightarrow (CaO)_n SiO_2 + n\,CO_2$$

The compounds cool to form a material known as *clinker*, which is ground to a fine powder and stored under anhydrous conditions. The adhesive properties of cement stem from complex acid–base reactions that occur when clinker is mixed with water, giving rise to anionic chains of SiO_4 or AlO_4 tetrahedra linked by Ca^{2+} ions. The decomposition of carbonate to yield CO_2 and oxide exemplifies the role of CO_2 as a Lewis acid, O^{2-} as a Lewis base, and carbonate as adduct. The formation of silicate or aluminate chains exemplifies the importance of Lewis acid–base chemistry in cohesion.

Cement production is responsible for a significant proportion of the CO_2 released annually (up to 8% in 2018), mainly by the direct release of CO_2 but also through use of fossil fuels to generate the heat required. Much of the CO_2 that is produced by centralized industrial installations (such as is formed during combustion of fossil fuels) can be selectively removed from flue gas using liquid amine scrubbers.

$$2\,RNH_2(aq) + CO_2(g) + H_2O(l) \rightarrow (RNH_3)_2CO_3(s)$$

The solid product is easily decomposed to release the CO_2 which is then pressurized and incarcerated in subterranean wells—an obvious and useful site being the vast cavities produced in spent oil wells which are otherwise filled with water. Such a process is known generally as **Carbon Capture** (see Box 15.6) and is one of the key tools under development in the battle to decrease greenhouse gas emissions.

Because the Lewis base F^-, aided by a proton, can displace O^{2-} from silicates, hydrofluoric acid is corrosive towards glass (which is derived from SiO_2). With most bases, the trend in acidity for tetrahalides SiX_4 follows the order $SiF_4 > SiCl_4 > SiBr_4 > SiI_4$, which correlates with the decrease in the electron-withdrawing power of the halogen from F to I. Germanium and tin fluorides can react similarly. In place of halogens, other electron-withdrawing groups can be placed on the Si atom to produce even stronger Lewis acids, such as bis(perchlorocatacholate silane, $Si(O_2C_6Cl_4)_2$, which can bind one or two moles of Lewis base.

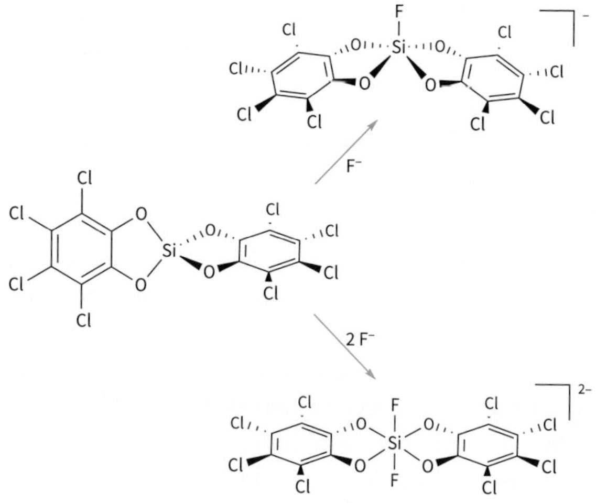

Lower oxidation states become more common as Group 14 is descended. The tin(II) salt $SnCl_2$ is also a Lewis acid as it combines with Cl^- ions to form the $SnCl_3^-$ anion (20) which (through its lone pair (not shown)) acts as a base to form complexes with transition metals, such as $(CO)_5Mn-SnCl_3$ (21).

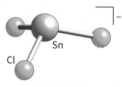

20 $SnCl_3^-$

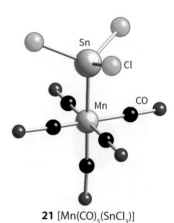

21 $[Mn(CO)_5(SnCl_3)]$

EXAMPLE 5.7 Predicting the relative Lewis basicity of compounds

Rationalize the following relative Lewis basicities: (a) $(H_3Si)_2O < (H_3C)_2O$; (b) $(H_3Si)_3N < (H_3C)_3N$.

Answer Nonmetallic elements in Period 3 and below can expand their valence shells by delocalizing the additional electrons that can be donated by O or N lone pairs, thereby creating multiple bonds (O and N are thus acting as π electron donors). The silyl ether and silyl amine are able to form additional internal dative bonds in this way, and are the weaker Lewis bases in each pair.

Self-test 5.7 Given that π bonding between Si and the lone pair of N is important, what difference in structure between $(H_3Si)_3N$ and $(H_3C)_3N$ do you expect?

(d) Group 15 Lewis acids

KEY POINTS Compounds of the heavier Group 15 elements act as Lewis acids; those of antimony (V) have important applications in catalysis.

The heavier elements of Group 15 form very important Lewis acids. Phosphorus pentafluoride is a strong Lewis acid and forms complexes with ethers and amines. Antimony pentafluoride, SbF_5, is one of the most widely studied compounds, despite being highly corrosive and toxic, and is capable of generating carbocations.

$$(CH_3)_3C-F + SbF_5 \rightarrow (CH_3)_3C^+ + SbF_6^-$$

As with $BeCl_2$ and $AlCl_3$, SbF_5 undergoes internal acid–base reactions, and exists as various aggregates and as a tetramer Sb_4F_{20} in the solid state. The $(Sb_2F_5)_n$ aggregates are even stronger Lewis acids than the monomer. Structurally characterized adducts include the acetonitrile complex (**22**).

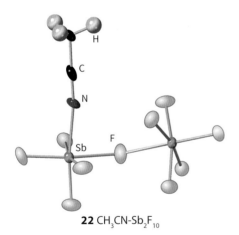

22 $CH_3CN \cdot Sb_2F_{10}$

The reaction of SbF_5 with HF produces a Brønsted-Lewis **superacid** by generating H_2F^+ (Section 5.17).

(e) Group 16 Lewis acids

KEY POINTS Sulfur dioxide can act as a Lewis acid by accepting an electron pair at the S atom; to act as a Lewis base, the SO_2 molecule can donate either its S or its O lone pair to a Lewis acid.

Sulfur dioxide is both a Lewis acid and a Lewis base. Its Lewis acidity is illustrated by the formation of a complex with a trialkylamine acting as a Lewis base:

To act as a Lewis base, the SO_2 molecule can donate either its S or its O lone pair to a Lewis acid. When SbF_5 is the acid, the O atom of SO_2 acts as the electron-pair donor, but when Ru(II) is the acid, the S atom acts as the donor (**23**).

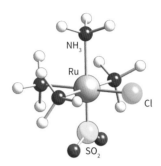

23 $[RuCl(NH_3)_4(SO_2)]^+$

Sulfur trioxide is a strong Lewis acid and a very weak (O donor) Lewis base. Its acidity is illustrated by the reaction

A classic aspect of the acidity of SO_3 is its highly exothermic reaction with water in the formation of sulfuric acid. The resulting problem of having to remove large quantities of heat from the reactor used for the commercial production of sulfuric acid is alleviated by exploiting the Lewis acidity of sulfur trioxide further to carry out the hydration by a two-stage process (Section 17.11). Before dilution, sulfur trioxide is dissolved in sulfuric acid to form the mixture known as *oleum*. This reaction is an example of Lewis acid–base complex formation:

The resulting $H_2S_2O_7$ can then be hydrolysed in a less exothermic reaction:

$$H_2S_2O_7 + H_2O \rightarrow 2H_2SO_4$$

(f) Group 17 Lewis acids

KEY POINTS Bromine and iodine molecules act as mild Lewis acids. A weak interaction known as halogen bonding is a subtle yet significant influence on structure and reactivity.

Lewis acidity is expressed in an interesting and subtle way by Br_2 and I_2, which are both strongly coloured. The strong visible absorption spectra of Br_2 and I_2 arise from transitions to

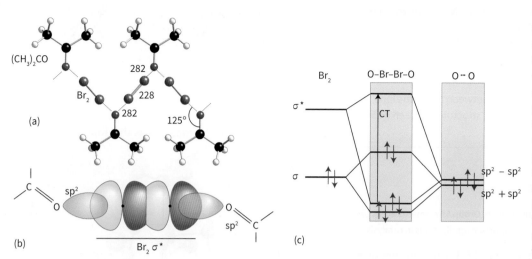

FIGURE 5.8 Interaction of Br₂ with the carbonyl group of propanone. (a) Structure of $(CH_3)_2CO \rightarrow Br_2$ determined by X-ray diffraction. Bond lengths given in pm. (b) The orbital overlap responsible for complex formation. (c) A partial molecular orbital energy level diagram for the σ and σ* orbitals of Br₂ with appropriate combinations of the sp² orbitals on the two O atoms. The charge-transfer transition is labelled CT.

low-lying unfilled antibonding orbitals. The colours of the species therefore suggest that the empty orbitals may be low enough in energy to serve as acceptor orbitals in Lewis acid–base complex formation.[6] Iodine is violet in the solid and gas phases and in non-donor solvents such as trichloromethane. However, when dissolved in water, propanone (acetone), or ethanol, all of which are Lewis bases, iodine is brown. The colour changes because a solvent–solute complex is formed using a lone pair on the donor molecule O atoms and a low-lying σ* orbital of the dihalogen.

The weak interaction between halogen atoms and Lewis bases is known as **halogen bonding**. Use of the empty σ* orbital on the halogen in compound AX (A is typically alkyl or halogen) leads to a tendency for linear arrangement along an axis A–X–B. Like hydrogen bonding that is discussed next, intermolecular halogen bonding is responsible for the arrangement of molecules as they appear in crystals, such as the infinite chains formed by pyrazine and iodine (**24**). Halogen bonding may also be important in the microbiological degradation of environmentally harmful halocarbons by Co-containing enzymes (Section 26.11).

24 Linear chains formed by I₂ and pyrazine

The interaction of Br₂ with the carbonyl group of propanone is shown in Fig. 5.8. The illustration also shows the electronic transition responsible for the new absorption band observed when a complex is formed. The orbital from which the electron originates in the transition is predominantly a lone pair on the carbonyl-O atom. The orbital to which the transition occurs is predominantly the LUMO of the acid (the dihalogen). Thus, to a first approximation, the transition transfers an electron from the base to the acid and is therefore called a **charge-transfer transition**.

The triiodide ion, I_3^-, is another example of a complex between a halogen acid (I₂) and a halide base (I⁻). One of the applications of its formation is to render molecular iodine soluble in water so that it can be used as a titration reagent:

$$I_2(s) + I^-(aq) \rightarrow I_3^-(aq) \qquad K = 725$$

The triiodide ion is one example of a large class of polyhalide ions (Section 18.10).

5.9 Hydrogen bonding

Hydrogen bonding has a special place in chemistry. A hydrogen bond is a Lewis acid–base interaction between a hydrogen atom that is covalently attached to an electronegative atom (H–A) and a second electronegative atom (B), H–A and B acting, respectively, as acid and base. Each bond, represented as the dashed section in A–H·····B, is the result of a weak interaction between a hydrogen atom that is covalently bound to an electronegative atom and a lone pair on an electronegative atom that is positioned in close proximity.

(a) Hydrogen bonding in structure stabilization

KEY POINTS Hydrogen bonds are much weaker than normal covalent bonds; but acting collectively, they are responsible for stabilizing intermolecular structures of many liquids and solids, and for determining the internal structures of large molecules like proteins and nucleic acids.

The significance of hydrogen bonding lies in the way that a very large number of weak, specific bonds can act **collectively**, hence water exists as a liquid at ambient temperature and pressure, and protein molecules interact internally and

---
[6] The terms *donor–acceptor complex* and *charge-transfer complex* were at one time used to denote these complexes. However, the distinction between these complexes and the more familiar Lewis acid–base complexes is arbitrary, and in the current literature the terms are used more or less interchangeably.

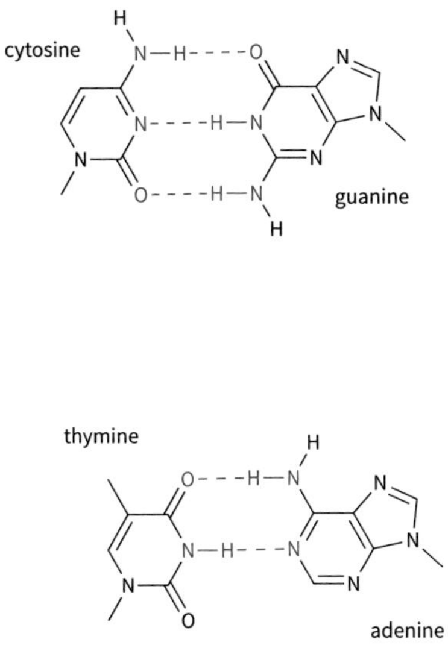

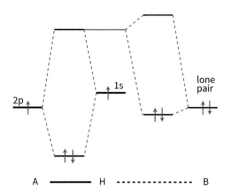

FIGURE 5.9 The specificity of collective hydrogen bonding forms the basis for genetic information storage by DNA.

with other macromolecules in highly specific ways. Among numerous examples, hydrogen bonding is responsible for the structure and information-storing properties of DNA (Fig. 5.9) and the strength of special materials like Kevlar® (**25**)—the fabric incorporated in bullet-proof vests. The strong interstrand hydrogen bonding in these structures resembles the way a zipper binds materials together while retaining flexibility. In proteins, hydrogen bonds between peptide chains are responsible for the so-called secondary structure that gives rise to characteristic folds and shapes (Section 26.3).

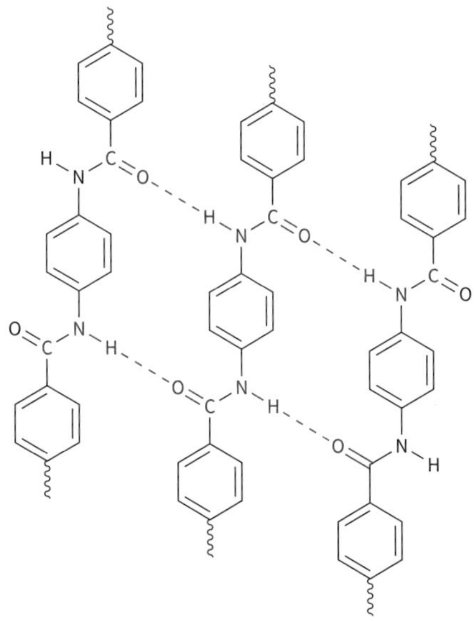

25 Kevlar®

Hydrogen bonds are much weaker than normal covalent bonds, with typical dissociation energies averaging only about 20 kJ mol^{-1} but spanning a wide range (10–60 kJ mol^{-1}). The bond length is usually taken as the distance between A and B, as it is usually difficult to define the exact position of the H atom. Long, weak hydrogen bonds tend to be electrostatic in nature and thus nondirectional, whereas stronger, shorter hydrogen bonds are more covalent and can be described by a simple MO model. An MO diagram for a fairly strong H-bond is shown (Fig. 5.10). Note the directionality that is implicit when covalence is dominant, because specific orbitals on the donor and acceptor are used.

The fact that solid HCN consists of linear chains (Fig. 5.11) may be traced to the collinear arrangement of the 3σ orbital on the carbon atom that is the H-bond donor and the empty 4σ orbital on the N atom that acts as acceptor.

The structures of hydrogen-bonded complexes determined in the gas phase by microwave spectroscopy (Fig. 5.12) show further the importance of lone-pair orientation as implied by VSEPR theory (Section 2.3).

It is important to note that hydrogen bonding is usually unsymmetrical in that the H atom is not midway between the two nuclei, even when the heavier linked atoms are identical. For example, the [ClHCl]$^-$ ion is linear but the H atom is not midway between the Cl atoms (Fig. 5.13). By contrast, the bifluoride ion, [FHF]$^-$, displays symmetric H-bonding: the H atom lies midway between the F atoms, and the F–F separation (226 pm) is significantly less than twice the van der Waals radius of the F atom (2 × 135 pm).

FIGURE 5.10 A molecular orbital representation of hydrogen bonding.

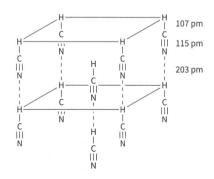

FIGURE 5.11 The structure of solid HCN, showing the linear chains formed by hydrogen bonding.

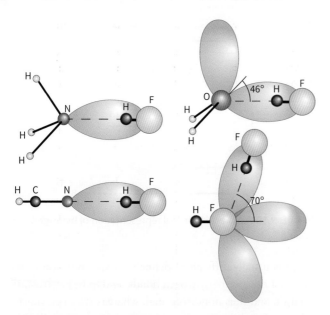

FIGURE 5.12 The H-bonding interactions between gaseous HF and other small molecules, showing how their lone-pairs dictate the shapes of the adducts formed. The HF molecule (shown as enlarged spheres) is oriented along the three-fold axis of NH_3, collinear with HCN, out of the HOH plane in its complex with H_2O, and off the HF axis in the (HF)$_2$ dimer.

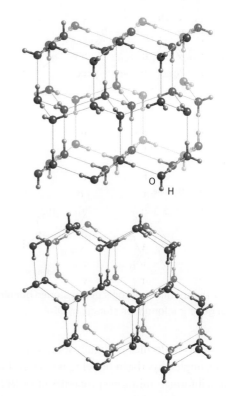

FIGURE 5.14 The structure of the hexagonal form of ice (I_h) shown in two orientations.

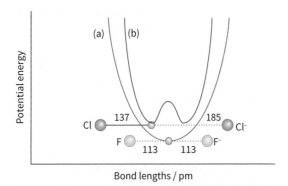

FIGURE 5.13 The variation of the potential energy with the position of the proton between two atoms in a hydrogen bond (lengths given in pm). (a) The double minimum potential characteristic of a weak hydrogen bond. (b) The single minimum potential characteristic of a strong hydrogen bond.

One of the most interesting manifestations of hydrogen bonding is the structure of ice. There are at least ten different phases of ice, but only one is stable under ordinary conditions. The familiar low-pressure phase of ice, ice-I_h, crystallizes in a hexagonal unit cell with each O atom surrounded tetrahedrally by four others (as shown in Fig. 5.14). These O atoms are held together by hydrogen bonds with O–H⋯O and O⋯H–O bonds largely randomly distributed through the solid. The resulting structure is quite open, which accounts for the density of ice being lower than that of water. When ice melts, the network of hydrogen bonds partially collapses.

Water molecules held together by hydrogen bonds may form molecular cages known as **clathrates**, which can encapsulate other molecules that are able to fit inside. Clathrate compounds are **host-guest complexes**, those with water as host being called **clathrate hydrates**. A structurally characterized example is the clathrate hydrate of composition $Xe_2CCl_4\cdot(H_2O)_{17}$ (Fig. 5.15), in which the cages consist of 14-faced and 12-faced polyhedra in the ratio 3:2. The guest molecules occupy the interiors of the polyhedra. Aside from their interesting structures, which illustrate the organization that can be enforced by hydrogen bonding, clathrate hydrates are often used as models for the way in which water

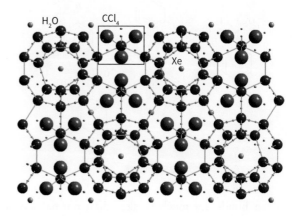

FIGURE 5.15 The cages of water molecules in clathrate hydrates that host foreign species; in this case, Xe atoms in the compound formulated as $Xe_2(CCl_4)_4(H_2O)_{17}$.

appears to become organized around nonpolar groups, such as those in proteins. Methane clathrate hydrates occur in the Earth at high pressures, and it is estimated that huge quantities of CH_4 are trapped in these formations (Box 15.3).

(b) Proton transfer in water

KEY POINT The high mobility of protons in water is explained by combining Brønsted and Lewis concepts.

Water acts as a fast 'proton conductor' both in bulk liquid and in the narrow hydrated pores that occur in biological macromolecules. The extremely high mobility of protons in water and other hydrogen-bonded liquids is accounted for by a hopping mechanism, first proposed by Theodor Grotthuss in 1806. Although details are still debated by theoreticians, the reaction step shown in Fig. 5.16 is widely accepted as a consensus for understanding how H^+ transfer occurs between molecules in liquid water. The mechanism involves two types of hydrogen-bonded cationic water cluster, two Eigen cations $H_9O_4^+$ sharing a H_2O (shown here as $(H_2O)_3–H^+–H_2O$ (left) and $H_2O–H^+–(H_2O)_3$ (right)), and the slightly less stable Zundel cation $H_5O_2^+$ ($H_2O–H^+–H_2O$) (shown in oval) which is considered to be the transition state.

5.10 Factors governing interactions between Lewis acids and bases

The proton (H^+) was the key electron pair acceptor in the discussion of Brønsted acid and base strengths. When considering Lewis acids and bases we must allow for a greater variety of acceptors and hence more factors that influence the interactions between electron pair donors and acceptors in general.

(a) The 'hard'/'soft' classification of acids and bases

KEY POINTS Hard and soft acids and bases are identified empirically by the trends in stabilities of the complexes that they form: hard acids tend to bind to hard bases and soft acids tend to bind to soft bases.

It proves helpful when considering the interactions of Lewis acids and bases containing elements drawn from throughout the periodic table to consider at least two main classes of substance. The classification of substances as 'hard' and 'soft' acids and bases was introduced by R. G. Pearson in the 1960s; it is a generalization—and a more evocative renaming—of the distinction between two types of behaviour that were originally named simply 'class a' and 'class b', respectively, by S. Ahrland, J. Chatt, and N. R. Davies.

The two classes are identified empirically by the opposite order of strengths (as measured by the equilibrium constant, K_f, for the formation of the complex) with which they form complexes with halide ion bases:

- Hard acids form complexes with stabilities in the order: $I^- < Br^- < Cl^- < F^-$
- Soft acids form complexes with stabilities in the order: $F^- < Cl^- < Br^- < I^-$

Fig. 5.17 shows the trends in K_f for complex formation with a variety of halide ion bases. The equilibrium constants increase steeply from F^- to I^- when the acid is Hg^{2+}, indicating that Hg^{2+} is a soft acid. The trend is less steep but in the same direction for Pb^{2+}, which indicates that this ion is a borderline soft acid. The trend is in the opposite direction

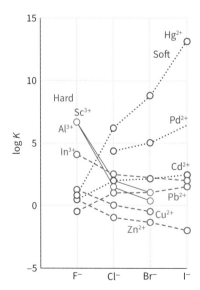

FIGURE 5.17 Trends in stability constants for complex formation by hard and soft metal ions with a variety of halide ion bases. Hard ions are indicated by the solid blue lines, soft ions by the dotted red lines. Borderline hard or borderline soft ions are indicated by dashed green lines.

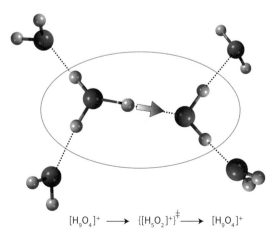

$[H_9O_4]^+ \longrightarrow \{[H_5O_2]^+\}^‡ \longrightarrow [H_9O_4]^+$

FIGURE 5.16 A mechanism for proton transfer in water involving two Eigen cations $H_9O_4^+$ as initial and final states, and a Zundel cation $H_5O_2^+$ (represented by the oval) as transition state.

for Zn^{2+}, so this ion is a borderline hard acid. The steep downward slope for Al^{3+} indicates that it is a hard acid. A useful rule of thumb is that small cations, which are not easily polarized, are hard and form complexes with small anions. Large cations are more polarizable and are soft.

For Al^{3+}, the binding strength increases as the electrostatic parameter (z^2/r) of the anion increases, which is consistent with an ionic model of the bonding. For Hg^{2+}, the binding strength increases with increasing polarizability of the anion. These two correlations suggest that hard acid cations form complexes in which simple Coulombic, or ionic, interactions are dominant, and that soft acid cations form more complexes in which covalent bonding is important.

A similar classification can be applied to neutral molecular acids and bases. For example, the Lewis acid phenol forms a more stable complex by hydrogen bonding to $(C_2H_5)_2O$: than to $(C_2H_5)_2S$:. This behaviour is analogous to the preference of Al^{3+} for F^- over Cl^-. By contrast, the Lewis acid I_2 forms a more stable complex with $(C_2H_5)_2S$:. We can conclude that phenol is hard whereas I_2 is soft.

In general, acids are identified as hard or soft by the thermodynamic stability of the complexes they form, as set out for the halide ions above and for other species as follows:

- Hard acids form complexes with stabilities in the order: $R_3P \ll R_3N, R_2S \ll R_2O$
- Soft acids form complexes with stabilities in the order: $R_2O \ll R_2S, R_2N \ll R_3P$

Bases can also be defined as soft or hard. Bases such as halides and oxoanions are classified as hard because ionic contributions to the bonding will be predominant in most of the complexes they form. Many soft bases bond through a carbon atom, such as CO and CN^-. In addition to donating electron density to the metal through a σ interaction, these small multiply bonded ligands are able to accept electron density through the low-lying empty π orbitals (the LUMO) present on the base (see Chapter 2). The bonding is, consequently, predominantly covalent in character. As these soft bases are able to accept electron density into π orbitals they are known as π **acids**. The nature of this bonding is described using ligand-field theory, introduced in Section 7.17.

It follows from the definition of hardness that

- Hard acids tend to bind to hard bases.
- Soft acids tend to bind to soft bases.

When species are analysed with these rules in mind, it is possible to identify the classification summarized in Table 5.4.

(b) Interpretation of hardness

KEY POINTS Hard acid–base interactions are predominantly electrostatic; soft acid–base interactions are predominantly covalent.

TABLE 5.4 The classification of Lewis acids and bases

Hard	Borderline	Soft
Acids		
H^+, Li^+, Na^+, K^+	$Fe^{2+}, Co^{2+}, Ni^{2+}$	$Cu^+, Au^+, Ag^+, Tl^+, Hg_2^{2+}$
$Be^{2+}, Mg^{2+}, Ca^{2+}$	$Cu^{2+}, Zn^{2+}, Pb^{2+}$	$Pd^{2+}, Cd^{2+}, Pt^{2+}, Hg^{2+}$
$Cr^{2+}, Cr^{3+}, Al^{3+}$	SO_2, BBr_3	BH_3
SO_3, BF_3		
Bases		
F^-, OH^-, H_2O, NH_3	NO_2^-, SO_3^{2-}, Br^-	H^-, R^-, CN^-, CO, I^-
$CO_3^{2-}, NO_3^-, O^{2-}$,	N_3^-, N_2	$\underline{S}CN^-\ast, R_3P, C_6H_5^-$
$SO_4^{2-}, PO_4^{3-}, ClO_4^-$	$C_6H_5N, S\underline{C}N^-\ast$	R_2S

* The underlined element is the site of attachment to which the classification refers.

The bonding between hard acids and bases can be described approximately in terms of ionic or dipole–dipole interactions. Soft acids and bases are more polarizable than hard acids and bases, so the acid–base interaction has a more pronounced covalent character.

It is important to note that although we associate soft acid/soft base interactions with covalent bonding, the bond itself may be surprisingly weak. This point is illustrated by reactions involving Hg^{2+}, a representative soft acid. The metathesis reaction

$$BeI_2 + HgF_2 \rightarrow BeF_2 + HgI_2$$

is exothermic, as predicted by the hard–soft rule. The bond dissociation energies (in kJ mol^{-1}) measured for these molecules in the gas phase are.

| Be–F | 632 | Hg–F | 268 |
| Be–I | 289 | Hg–I | 145 |

Therefore, it is not a large Hg–I bond energy that ensures that the reaction is exothermic but the especially strong bond between Be and F, which is an example of a hard–hard interaction. In fact, an Hg atom forms only weak bonds to any other atom. In aqueous solution, the reason why Hg^{2+} forms a much more stable complex with iodide ions compared to chloride ions is the much more favourable hydration energy of Cl^- compared to I^-.

(c) Chemical consequences of hardness

KEY POINTS Hard–hard and soft–soft interactions help to systematize complex formation but must be considered in the light of other possible influences on bonding.

The concepts of hardness and softness help to rationalize a great deal of inorganic chemistry. For instance, they are useful for choosing preparative conditions and predicting the directions of reactions, and they help to rationalize the outcome of metathesis reactions. However, the concepts must always be used with due regard for other factors that may affect the outcome of reactions. This deeper understanding of chemical reactions will grow during the course of the rest of this book. For the time being we shall limit the discussion to a few straightforward examples.

The classification of molecules and ions as hard or soft acids and bases helps to clarify the terrestrial distribution of the elements described in Chapter 1. Hard cations such as Li^+, Mg^{2+}, Ti^{4+}, and Cr^{3+} are found in association with the hard base O^{2-}. The soft cations Cd^{2+}, Pb^{2+}, and Sn^{2+} are found in association with soft anions, particularly S^{2-}, Se^{2-}, and Te^{2-}. The consequences of this correlation are discussed in more detail in Section 9.1.

Polyatomic anions may contain two or more donor atoms differing in their hard–soft character. For example, the SCN^- ion is a base that comprises both the harder N atom and the softer S atom. The ion tends to bind hard acids through N. However, with a soft acid, such as a metal ion in a low oxidation state, the ion bonds through the S atom. Platinum(II), for example, forms four Pt–SCN bonds in the complex $[Pt(SCN)_4]^{2-}$.

(d) Other contributions to complex formation

Although the type of bond formation is a major reason for the distinction between the two classes, there are other contributions to the Gibbs energy of complex formation and hence to the equilibrium constant. Important factors are:

1. Competition with the solvent when reactions occur in solution. The solvent may be a Lewis acid, a Lewis base, or both.
2. Rearrangement of the substituents of the acid and base that may be necessary to permit formation of the complex; for instance, a major change in structure occurs when CO_2 reacts with OH^- to form $HOCO_2^-$.
3. Steric repulsion between substituents on the acid and the base, which can also give rise to interactions that depend on chirality (Section 7.11). The compound 'oxazaborolidine' (**26**) is an important catalyst for selective enantiomeric reduction of ketones.

26 Oxazaborolidine

Bifunctional compounds such as that shown in Fig. 5.18 exemplify the concept of the Frustrated Lewis Pair (FLP), which is described in more detail in Box 14.4. The phosphorus atom possessing a lone pair is a strong Lewis base, while the boron atom with its vacant orbital is a strong Lewis acid. Interactions

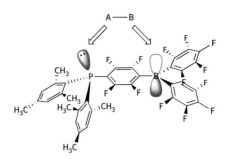

FIGURE 5.18 An example of a Frustrated Lewis Pair (FLP), $(C_6H_2Me_3)_2P(C_6F_4)B(C_6F_5)_2$. The spatially-separated phosphorus (Lewis base) and boron (Lewis acid) sites on the molecule cleave the bond of the incoming reactant A–B.

between the acid and base centres in these phosphinoborane compounds are weak due to steric restraints, but they react with molecular H_2 by cleaving it heterolytically to form a boron-hydride adduct that can react further with aldehydes.

As pointed out by Fontaine and Stephan (see Further Reading) the 'dual-attack' mode forms the basis for catalysis by a host of enzymes, and the FLP concept is, in fact, a very ancient trick.

5.11 Thermodynamic Lewis acidity parameters

KEY POINTS The standard enthalpies of complex formation are reproduced by the E and C parameters of the Drago–Wayland equation that reflect, in part, the ionic and covalent contributions to the bond in the complex.

A more quantitative alternative to the hard–soft classification of acids and bases makes use of an approach in which electronic, structural rearrangement, and steric effects are incorporated into a small set of parameters. The standard reaction enthalpies of complex formation in the gas phase

$$A(g) + B(g) \rightarrow A\text{–}B(g) \quad \Delta_r H^\ominus(A\text{–}B)$$

can be predicted empirically by the Drago–Wayland equation:

$$-\Delta_r H^\ominus(A\text{–}B)/\text{kJ mol}^{-1} = E_A E_B + C_A C_B \quad (5.11)$$

TABLE 5.5 Drago–Wayland parameters for some acids and bases*

	E	C
Acids		
Antimony pentachloride	15.1	10.5
Boron trifluoride	20.2	3.31
Iodine	2.05	2.05
Iodine monochloride	10.4	1.70
Phenol	8.86	0.90
Sulfur dioxide	1.88	1.65
Trichloromethane	6.18	0.32
Trimethylboron	12.6	3.48
Bases		
Acetone	2.02	4.67
Ammonia	2.78	7.08
Benzene	0.57	1.21
Dimethyl sulfide	0.70	15.26
Dimethyl sulfoxide	2.76	5.83
Methylamine	2.66	12.00
p-Dioxane	2.23	4.87
Pyridine	2.39	13.10
Trimethylphosphine	17.2	13.40

* E and C parameters are often reported to give ΔH in kcal mol^{-1}; we have multiplied both by $\sqrt{(4.184)}$ to obtain ΔH in kJ mol^{-1}.

The parameters E and C were introduced with the idea that they represent 'electrostatic' and 'covalent' factors, respectively, but in fact they must accommodate all factors except solvation. The compounds for which the parameters are listed in Table 5.5 satisfy the equation with an error of less than ±3 kJ mol^{-1}, as do a much larger number of examples in the original papers.

> **A BRIEF ILLUSTRATION**
>
> From Table 5.5 we find $E = 21.2$ and $C = 3.31$ for BF$_3$, and $E = 2.78$ and $C = 7.98$ for NH$_3$. Applying the Drago–Wayland equation to the formation of the adduct H$_3$N:BF$_3$ gives $\Delta_r H^\ominus = -[(21.1 \times 2.79) + (3.31 \times 7.98)] = -85.28$ kJ mol^{-1}, indicating an exothermic reaction.

The Drago–Wayland equation (eqn. 5.11) is semi-empirical but very successful and useful across a wide range of reactions. In addition to providing estimates of the enthalpies of complex formation for over 1500 complexes, these enthalpies can be combined to calculate the enthalpies of displacement and metathesis reactions, revealing that many hard–soft interactions may also be strong. Moreover, the equation is useful for reactions of acids and bases in non-polar, noncoordinating solvents as well as for reactions in the gas phase. The major limitation is that the equation is restricted to substances that can conveniently be studied in the gas phase or in noncoordinating solvents; hence, in the main it is limited to neutral molecules. It also fails to take into account steric restrictions.

More quantitative measurements of Lewis acidity are obtained in various ways, but the complicating distinction between hard and soft interactions must be taken into consideration when interpreting data. The simplest and most widely adopted methods use NMR spectroscopy (Section 8.6) to measure the strength of the acid–base interaction through the change in electron density (observed as 'chemical shift') experienced at a specific atom on a standard base. Standard bases are Et$_3$PO, for which ^{31}P NMR is used, or *trans*-crotonaldehyde, CH$_3$C$\underline{\text{H}}$=CH–CHO, where ^{1}H NMR is used and the chemical shift of H3 (underlined) is measured. A method providing direct thermodynamic insight into the reactivity towards small Lewis bases involves measuring or calculating the enthalpy change when a Lewis acid reacts with fluoride ion in the gas phase.

$$A(g) + F^-(g) \xrightarrow{\Delta_r H^\ominus} AF^-(g)$$

The value obtained is known as the Fluoride Ion Affinity (FIA), and some examples are given in Fig. 5.19. Although FIA is widely used, it is rather restricted to hard acids; thus an alternative scale, Hydride Ion Affinity (HIA) is often quoted when soft acids are compared.

As expected, cationic Lewis acids are far more acidic than their closest neutral counterparts. The trifluoroselenium cation SiF$_3^+$ has an FIA of 1294 kJ mol^{-1}, whereas isoelectronic BF$_3$ has a value of just 343 kJ mol^{-1}.

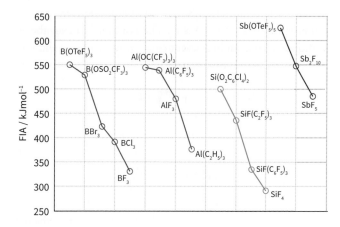

FIGURE 5.19 Strengths of some Lewis acids based on Fluoride Ion Affinity (FIA).

Nonaqueous solvents

In this section we explore how the properties of acids and bases are altered by using nonaqueous solvents.

5.12 Solvent levelling

KEY POINT A solvent with a large autoprotolysis constant can be used to discriminate between a wide range of acid and base strengths.

A Brønsted acid that is weak in water may appear strong in a solvent that is a more effective proton acceptor, and vice versa. Indeed, in sufficiently basic solvents (such as liquid ammonia), it may not be possible to discriminate between their strengths, because all of them will be fully deprotonated. Similarly, Brønsted bases that are weak in water may appear strong in a more strongly proton-donating solvent (such as anhydrous acetic acid). It may not be possible to arrange a series of bases according to strength, for all of them will be effectively fully protonated in acidic solvents. We shall now see that the autoprotolysis constant of a solvent plays a crucial role in determining the range of Brønsted acid or base strengths that can be distinguished for species dissolved in it.

To understand how an acid or base will behave in a particular solvent we need to return to the concept of proton affinity introduced in Section 5.1(e). Any Brønsted acid stronger than H_3O^+ in water donates a proton to H_2O and forms H_3O^+. Consequently, no acid significantly stronger than H_3O^+ (i.e. the A'_p value of its conjugate base is lower than that of H_2O) can remain protonated in water. No experiment conducted in water can tell us which of HBr and HI is the stronger acid because both transfer their protons essentially completely to give H_3O^+. In effect, solutions of the strong acids HX and HY behave as though they are solutions of H_3O^+ ions regardless of whether HX is intrinsically stronger than HY. Water is therefore said to have a **levelling effect** that brings all stronger acids down to the acidity of H_3O^+. The strengths of such acids can be distinguished by using a less basic solvent. For instance, although HBr and HI have indistinguishable acid strengths in water, in acetic acid HBr and HI behave as weak acids and their strengths can be distinguished: in this way it is found that HI is a stronger proton donor than HBr.

The levelling effect can be expressed in terms of the pK_a of the acid. An acid such as HCN dissolved in a protic solvent, HSol, is classified as strong if $pK_a < 0$, where K_a is the acidity constant of the acid in the solvent Sol:

$$HCN(sol) + HSol(l) \rightleftharpoons H_2Sol^+(sol) + CN^-(sol)$$

$$K_a = \frac{[H_2Sol^+][CN^-]}{[HCN]}$$

That is, all acids with $pK_a < 0$ (corresponding to $K_a > 1$) display the acidity of H_2Sol^+ when they are dissolved in the solvent HSol.

An analogous effect can be found for bases introduced to water. Any base that is strong enough to undergo complete protonation by water (i.e. its A'_p value is higher than that of OH^-) produces an OH^- ion for each molecule of base added. The solution behaves as though it contains OH^- ions. Therefore, we cannot distinguish the proton-accepting power of such bases, and we say that they are levelled to a common strength. Indeed, the OH^- ion is the strongest base that can exist in water because any species that is a stronger proton acceptor immediately forms OH^- ions by proton transfer from water. For this reason, we cannot study NH_2^- or CH_3^- in water by dissolving alkali metal amides or methides because both anions generate OH^- ions and are fully protonated to NH_3 and CH_4:

$$KNH_2(s) + H_2O(l) \rightarrow K^+(aq) + OH^-(aq) + NH_3(aq)$$

$$Li_4(CH_3)_4(s) + 4H_2O(l) \rightarrow 4Li^+(aq) + 4OH^-(aq) + 4CH_4(g)$$

The base levelling effect can also be expressed in terms of the pK_b of the base. A base dissolved in HSol is classified as strong if $pK_b < 0$, where K_b is the basicity constant of the base in HSol:

$$NH_3(sol) + HSol(l) \rightleftharpoons NH_4^+(sol) + Sol^-(sol)$$

$$K_a = \frac{[NH_4^+][Sol^-]}{[NH_3]}$$

That is, all bases with $pK_b < 0$ (corresponding to $K_b > 1$) display the basicity of Sol^- in the solvent HSol. Because $pK_{sol} = pK_a + pK_b$, this criterion for levelling may be expressed as follows: all bases with $pK_a > pK_{sol}$ give a negative value for pK_b and behave like Sol^- in the solvent HSol.

It follows from this discussion of acids and bases in a common solvent HSol that because any acid is levelled if $pK_a < 0$ in HSol and any base is levelled if $pK_a > pK_{sol}$ in the same solvent, the window of strengths that are not levelled in the solvent ranges from $pK_a = 0$ to pK_{sol}. For water, $pK_w = 14$. For liquid ammonia, the autoprotolysis equilibrium is

$$2NH_3(l) \rightleftharpoons NH_4^+(sol) + NH_2^-(sol) \qquad pK_{NH_3} = 33$$

Consequently, acids and bases are discriminated much less in water than they are in ammonia. The discrimination windows of a number of solvents are shown in Fig. 5.20. The window for dimethyl sulfoxide (DMSO, $(CH_3)_2SO$) is much wider because $pK_{DMSO} = 37$. Consequently, DMSO can be used to study a larger range of acids (from H_2SO_4 to PH_3). Water has a narrow window compared to some of the other solvents shown in the illustration. One fundamental reason

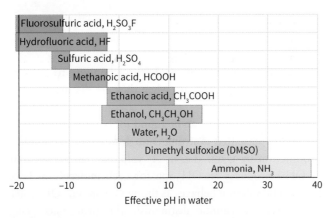

FIGURE 5.20 Acid–base discrimination window for a variety of solvents. The width of each window is proportional to the autoprotolysis constant of the solvent.

is the high relative permittivity of water (ε = 80), which favours the formation (and separation) of H_3O^+ and OH^- ions (eqn. 5.10). Permittivity is a measure of the ability of a material to resist the formation of an electric field within it.

EXAMPLE 5.8 Differentiating acidities in different solvents

Which of the solvents shown in Fig. 5.20 could be used to differentiate the acidities of HCl ($pK_a \approx -6$) and HBr ($pK_a \approx -9$)?

Answer We need to look for a solvent with an acid–base discrimination window between −6 and −9. The only solvents in the table for which the window covers the range −6 to −9 are methanoic (formic) acid, HCOOH, and hydrofluoric acid, HF.

Self-test 5.8 Which of the solvents given in Fig. 5.20 could be used to discriminate the acidities of PH_3 ($pK_a \approx 27$) and GeH_4 ($pK_a \approx 25$)?

Nonaqueous solvents can be selected for reactions of molecules that are readily hydrolysed, to avoid levelling by water, or to enhance the solubility of a solute. Nonaqueous solvents are often selected on the basis of their liquid range and relative permittivity. The physical properties of some common nonaqueous solvents are shown in Table 5.6.

5.13 The Hammett acidity function and its application to strong, concentrated acids

KEY POINT The Hammett acidity function is used to compare the protonating power of strong acids as concentrated aqueous solutions, in nonaqueous solvents, or as neat substances.

Concentrated solutions of strong acids have important applications, but the pH scale (which is based on an ideal thermodynamic quantity, proton *activity*) is no longer applicable for comparing their acidities, which are scaled instead using the **Hammett acidity function** H_0 defined as

$$H_0 = pK + \log \frac{[X]}{[HX]} \quad (5.12)$$

The Hammett acidity function is applicable for nonaqueous solvents as water can be avoided altogether, and is particularly useful for comparing the protonation power of pure, undiluted ('neat') acids. Table 5.7 lists Hammett functions for some common pure acids and mixtures. The larger the value of $-H_0$, the greater the protonating ability of the acid.

Eqn. 5.12 resembles eqn. 5.7 in that $pK = -\log K$, where K is the acidity constant for HX, except that HX is a very strong acid in high concentration and activities are no longer meaningful. Thus, although H_0 replaces pH, it cannot be regarded in an analogous way: most obviously, Hammett constants may well exceed the concentration of H^+ that is physically possible in a liquid.

TABLE 5.6 Physical properties of some nonaqueous solvents

Solvent	Melting point/°C	Boiling point/°C	Relative permittivity*
Liquid ammonia	−77.7	−33.3	23.9 (at −33.3°C)
Glacial acetic acid	16.7	117.9	6.15
Sulfuric acid	10.4	290 (decomposes)	100
Hydrogen fluoride	−83.4	19.5	84
Ethanol	−114.5	78.3	24.55
Dinitrogen tetroxide	−11.2	21.1	2.42
Bromine trifluoride	8.8	125.8	107
Dimethyl sulfoxide (DMSO)	18.5	189	46.45

*These values are temperature dependent.

TABLE 5.7 Hammett acidity functions H_0 for some strong acids, either as pure substances or as mixtures

Acid	H_0
$HSbF_6$ (50% HF, 50% SbF_5)	−31.3
'Magic acid' (25% SbF_5, 75% HSO_3F)	−21.5
HSO_3F	−15.1
CF_3SO_3H	−14.1
$H_2S_2O_7$	−14.1
$HClO_4$	−13.0
H_2SO_4	−12.0
HF	−11

5.14 The solvent system definition of acids and bases

KEY POINT The solvent system definition of acids and bases extends the Brønsted–Lowry definition to include species that do not participate in proton transfer.

The Brønsted–Lowry definition of acids and bases describes acids and bases in terms of the proton. This system can be extended to species that cannot participate in proton transfer by recognizing an analogy with the autoprotolysis reaction of water:

$$2H_2O(l) \rightleftharpoons H_3O^+(aq) + OH^-(aq)$$

An acid increases the concentration of H_3O^+ ions, and a base increases the concentration of OH^- ions. We can recognize a similar structure in the autoionization reaction of some aprotic solvents, such as bromine trifluoride, BrF_3:

$$2BrF_3(l) \rightleftharpoons BrF_2^+(sol) + BrF_4^-(sol)$$

where sol denotes solution (i.e. that species is dissolved) in the nonionized species (BrF_3 in this case). In the **solvent-system definition**, any solute that increases the concentration of the cation generated by autoionization of the solvent is defined as an acid, and any that increases the concentration of the corresponding anion is defined as a base. The solvent system definition can be applied to any solvent that autoionizes and applies to both protic and aprotic nonaqueous solvents.

EXAMPLE 5.9 Identifying acids and bases using the solvent system method

The salt BrF_2AsF_6 is soluble in BrF_3. Is it an acid or base in this solvent?

Answer We need to identify the autoionization products of the solvent and then decide whether the solute increases the concentration of the cation (an acid) or the anion (a base). The autoionization products of BrF_3 are BrF_2^+ and BrF_4^-. The solute produces BrF_2^+ and AsF_6^- ions when it dissolves. As the salt increases the concentration of the cations it is defined as an acid in the solvent system.

Self-test 5.9 Is $KBrF_4$ an acid or a base in BrF_3?

5.15 Solvents as acids and bases

The solvent system definition of acids and bases allows solutes to be defined as acids and bases by considering the autoionization products of the solvent. Most solvents are also either electron pair acceptors or donors and hence are either Lewis acids or bases. The chemical consequences of solvent acidity and basicity are considerable, as they help to account for the differences between reactions in aqueous and nonaqueous media. It follows that a displacement reaction often occurs when a solute dissolves in a solvent, and that the subsequent reactions of the solution are also usually either displacements or metatheses. For example, when antimony pentafluoride dissolves in bromine trifluoride, the following displacement reaction occurs:

$$SbF_5(s) + BrF_3(l) \rightarrow BrF_2^+(sol) + SbF_6^-(sol)$$

In the reaction, the strong Lewis acid SbF_5 abstracts F^- from BrF_3. A more familiar example of the solvent as a participant in a reaction is in Brønsted theory, where the Lewis acid (H^+) is always regarded as complexed with the solvent (as in H_3O^+ if the solvent is water), and reactions are treated as the transfer of the acid, the proton, from a solvent molecule to another base. Only the saturated hydrocarbons among common solvents lack significant Lewis acid or base character.

Solvents with Lewis base character are common. Most of the well-known polar solvents, including water, alcohols, ethers, amines, dimethyl sulfoxide (DMSO, $(CH_3)_2SO$), dimethylformamide (DMF, $(CH_3)_2NCHO$), and acetonitrile (CH_3CN) are hard Lewis bases. Dimethyl sulfoxide is an interesting example of a solvent that is hard on account of its O donor atom, and soft on account of its S donor atom. Reactions of acids and bases in these solvents are generally displacements:

Among Lewis acids, liquid sulfur dioxide is a good soft solvent for dissolving the soft base benzene. Unsaturated hydrocarbons may act as acids or bases by using their π or π^* orbitals as frontier orbitals. Alkanes with electronegative substituents, such as haloalkanes (e.g. $CHCl_3$), are significantly acidic at the hydrogen atom, although saturated fluorocarbon solvents lack Lewis acid and base properties.

(a) Liquid ammonia

KEY POINTS Liquid ammonia is a useful nonaqueous solvent. Many reactions in liquid ammonia are analogous to those in water.

Liquid ammonia is widely used as a nonaqueous solvent. It boils at $-33.3°C$ at 1 atm, and despite a somewhat lower relative permittivity ($\varepsilon_r = 24$) than that of water, it is a good solvent for inorganic compounds such as ammonium salts, nitrates, cyanides, and thiocyanides, and organic compounds such as amines, alcohols, and esters. It closely resembles the aqueous system as can be seen from the autoionization

$$2NH_3(l) \rightleftharpoons NH_4^+(sol) + NH_2^-(sol)$$

Solutes that increase the concentration of NH_4^+, the solvated proton, are acids. Solutes that decrease the concentration of

NH_4^+ or increase the concentration of NH_2^- are defined as bases. Thus, ammonium salts are acids in liquid ammonia and amines are bases.

Liquid ammonia is a more basic solvent than water and enhances the acidity of many compounds that are weak acids in water. For example, acetic acid is almost completely ionized in liquid ammonia:

$$CH_3COOH(sol) + NH_3(l) \rightarrow NH_4^+(sol) + CH_3COO^-(sol)$$

Many reactions in liquid ammonia are analogous to those in water. The following acid–base neutralization can be carried out:

$$NH_4Cl(sol) + NaNH_2(sol) \rightarrow NaCl(sol) + 2NH_3(l)$$

Liquid ammonia is a very good solvent for alkali and alkali earth metals, with the exception of beryllium. The alkali metals are particularly soluble, and as much as 336 g of caesium can be dissolved in 100 g of liquid ammonia at −50°C. The metals can be recovered by evaporating the ammonia. These solutions are very conducting, and are blue when dilute and bronze when concentrated. Electron paramagnetic resonance spectra (Section 8.7) show that the solutions contain unpaired electrons. The blue colour typical of the solutions is the outcome of a very broad optical absorption band in the near IR with a maximum near 1500 nm. The metal is ionized in ammonia solution to produce 'solvated electrons':

$$Na(s) + NH_3(l) \rightarrow Na^+(sol) + e^-(sol)$$

The blue solutions survive for long times at low temperature but decompose slowly to produce hydrogen and sodium amide, $NaNH_2$. The exploitation of the blue solutions to produce compounds called 'electrides' is discussed in Section 12.17.

(b) Hydrogen fluoride

KEY POINT Hydrogen fluoride is a reactive toxic solvent that is highly acidic.

Liquid hydrogen fluoride (b.p. 19.5°C) is an acidic solvent with a relative permittivity (ε_r = 84 at 0°C) comparable to that of water (ε_r = 78 at 25°C). It is a good solvent for ionic substances. However, as it is both highly reactive and toxic, it presents handling problems, including its ability to etch glass. In practice, liquid hydrogen fluoride is usually contained in polytetrafluoroethylene and polychlorotrifluoroethylene vessels. Hydrogen fluoride is particularly hazardous because it penetrates tissue rapidly and interferes with nerve function. Consequently, burns may go undetected and treatment may be delayed. It can also etch bone and reacts with calcium in the blood.

Liquid hydrogen fluoride is a highly acidic solvent as it has a high autoprotolysis constant and produces solvated protons very readily:

$$3HF(l) \rightarrow H_2F^+(sol) + HF_2^-(sol)$$

Although the conjugate base of HF is formally F^-, the ability of HF to form a strong hydrogen bond to F^- means that the conjugate base is better regarded as the bifluoride ion, HF_2^-. Only very strong acids are able to donate protons and function as acids in HF; for example, fluorosulfonic acid:

$$HSO_3F(sol) + HF(l) \rightleftharpoons H_2F^+(sol) + SO_3F^-(sol)$$

Organic compounds such as acids, alcohols, ethers, and ketones can accept a proton and act as bases in HF(l). Other bases increase the concentration of H_2F^- to produce basic solutions:

$$CH_3COOH(l) + 2HF(l) \rightleftharpoons CH_3COOH_2^+(sol) + H_2F^-(sol)$$

In this reaction acetic acid, an acid in water, is acting as a base.

Many fluorides are soluble in liquid HF as a result of the formation of the HF_2^- ion; for example:

$$LiF(s) + HF(l) \rightarrow Li^+(sol) + HF_2^-(sol)$$

(c) Anhydrous sulfuric acid

KEY POINT The autoionization of anhydrous sulfuric acid is complex, with several competing side reactions.

Anhydrous sulfuric acid is an acidic solvent. It has a high relative permittivity and is viscous because of extensive hydrogen bonding. Despite this association the solvent is appreciably autoionized at room temperature. The major autoionization is

$$2H_2SO_4(l) \rightleftharpoons H_3SO_4^+(sol) + HSO_4^-(sol)$$

However, there are secondary autoionizations and other equilibria, such as

$$H_2SO_4(l) \rightleftharpoons H_2O(sol) + SO_3(sol)$$
$$H_2O(sol) + H_2SO_4(l) \rightleftharpoons H_3O^+(sol) + HSO_4^-(sol)$$
$$SO_3(sol) + H_2SO_4(l) \rightleftharpoons H_2S_2O_7(sol)$$
$$H_2S_2O_7(sol) + H_2SO_4(l) \rightleftharpoons H_3SO_4^+(sol) + HS_2O_7^-(sol)$$

The high viscosity and high level of association through hydrogen bonding would usually lead to low ion mobilities. However, the mobilities of $H_3SO_4^+$ and HSO_4^- are comparable to those of H_3O^+ and OH^- in water, indicating that similar proton transfer mechanisms are taking place. The main species taking part are $H_3SO_4^+$ and HSO_4^-:

Most strong oxo acids accept a proton in anhydrous sulfuric acid and are thus bases:

$$H_3PO_4(sol) + H_2SO_4(l) \rightleftharpoons H_4PO_4^+(sol) + HSO_4^-(sol)$$

An important reaction is that of nitric acid with sulfuric acid to generate the nitronium ion, NO_2^+, which is the active species in aromatic nitration reactions:

$$HNO_3(sol) + 2H_2SO_4(l) \rightleftharpoons NO_2^+(sol) + H_3O^+(sol) + 2HSO_4^-(sol)$$

Some acids that are very strong in water act as weak acids in anhydrous sulfuric acids; for example, perchloric acid, $HClO_4$, and fluorosulfuric acid, $HFSO_3$.

(d) Dinitrogen tetroxide

KEY POINTS Dinitrogen tetroxide autoionizes by two reactions. The preferred route can be enhanced by addition of electron-pair donors or acceptors.

Dinitrogen tetroxide, N_2O_4, has a narrow liquid range with a freezing point at −11.2°C and boiling point of 21.2°C. Two autoionization reactions occur:

$$N_2O_4(l) \rightleftharpoons NO^+(sol) + NO_3^-(sol)$$
$$N_2O_4(l) \rightleftharpoons NO_2^+(sol) + NO_2^-(sol)$$

The first autoionization is enhanced by addition of Lewis bases, such as diethyl ether:

$$N_2O_4(l) + :X \rightleftharpoons XNO^+(sol) + NO_3^-(sol)$$

Lewis acids such as BF_3 enhance the second autoionization reaction:

$$N_2O_4(l) + BF_3(sol) \rightleftharpoons NO_2^+(sol) + F_3BNO_2^-(sol)$$

Dinitrogen tetroxide has a low relative permittivity and is not a very useful solvent for inorganic compounds. It is however a good solvent for many esters, carboxylic acids, halides, and organic nitro compounds.

(e) Ionic liquids

KEY POINT Ionic liquids are polar, nonvolatile solvents able to provide very high concentrations of Lewis acids or bases as catalysts for many reactions.

Ionic liquids are salts with low melting points, usually below 100°C, which typically comprise asymmetric quaternary (alkyl) ammonium cations and complex anions such as $AlCl_4^-$ and carboxylates of various chain lengths. Ionic liquids are characterized by low volatility, high thermal stability, inherent inertness over a wide range of electrode potentials, and high conductivity, making possible numerous applications such as providing alternative solvents for organic syntheses and electrolytes for batteries, although care is often needed to avoid moisture. The ability of these ionic compounds to exist as liquids under ambient conditions is attributed to the large size and conformational flexibility of one or both of the ions—properties that give rise to small lattice energies and large entropy increases accompanying melting.

Reactions of dialkyl imidazolium chlorides with $AlCl_3$ produce the corresponding dialkyl imidazolium heptachlorodialuminate salts—among a growing class of promising electrolytes for high-energy-density batteries (Section 30.9).

The cation and/or anion may be selected to impart chirality and specific acid–base properties to the solvent. The anion of ionic liquids is often a Lewis base, such as the dicyanamide ion $((NC)_2N^-)$ which is a catalyst for acetylation reactions. In some cases the cation may possess a basic group, an example being 1-alkyl-1,4-diazabicyclo[2.2.2]octane, known as $[C_n\text{dabco}]^+$, which contains a tertiary-N atom capable of forming hydrogen bonds and conferring water solubility. Salts of $[C_n\text{dabco}]^+$ with the bis(trifluoromethane)sulfonimide anion, TFSI⁻ (**27**), are miscible with water for $n = 2$ but immiscible with water for $n = 8$, while the melting point also drops with increasing n. Salts such as $Cu(NO_3)_2$ are usually insoluble in ionic liquids but dissolve in $[C_n\text{dabco}]^+$ TFSI⁻ because Cu^{2+} is complexed by the tertiary-N donors.

27 $[C_n\text{dabco}]^+$ TFSI⁻

(f) Supercritical fluids

KEY POINT Supercritical fluids have special properties as solvents and are finding increasing use in environmentally benign industrial processes.

A supercritical (sc) fluid is a state of matter where the liquid and vapour phases are indistinguishable: it has low viscosity combined with high dissolving capability for many solutes, and many gases are completely miscible. Supercritical fluids are produced by applying a combination of temperature and pressure that exceed the critical point (Fig. 5.21).

The most important example is supercritical carbon dioxide ($scCO_2$) which has a critical point at $P_c = 72.8$ atm and $T_c = 30.95$°C. The CO_2 molecule can act as a Lewis base

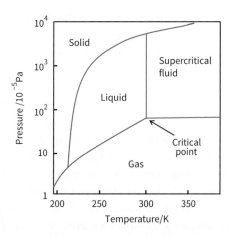

FIGURE 5.21 Pressure–temperature phase diagram for carbon dioxide showing the conditions under which it behaves as a supercritical fluid (1 atm = 1.01 × 10⁵ Pa).

(**28**) or a Lewis acid; indeed, both types of interaction can occur at a single CO_2 molecule (**29**). As a solvent, $scCO_2$ has some important applications, such as decaffeination of coffee and an increasing number of green industrial processes where $scCO_2$ is replacing organic solvents that cause so much environmental concern. Unlike organic solvents, $scCO_2$ can be removed at the end of a process by depressurization and then recycled (Section 27.5).

28 $CO_2{:}AlCl_3$

29 $CO_2{:}OC(R)CH_3$

Compared to normal water, supercritical water (scH_2O) is an excellent solvent for organic compounds and a poor solvent for ions. Close to the critical conditions (P_c = 218 atm, T_c = 374°C) water undergoes a remarkable change in properties, from having a very high extent of autoprotolysis as it approaches the critical point (pK_w is approximately 11, compared to 14), to having a much lower value above the critical point (pK_w is approximately 20 at 600°C and 250 atm). Pressure and temperature can therefore be used to optimize the solvent for specific chemical reactions.

A particularly important application of scH_2O is the destruction of organic waste materials—a process that exploits the complete miscibility of both organic compounds and O_2 in this solvent. The process, known as supercritical water oxidation (SCWO), results in very rapid oxidation of the material to form CO_2 and H_2O (Section 27.5). Much of the heat generated in the highly exothermic reaction is recovered to help sustain the high temperature required in the reactor.

$$\text{organic waste} \xrightarrow[>374°C/>218\text{atm}]{scH_2O} CO_2 + H_2O$$

EXAMPLE 5.10 **Accounting for properties in terms of the Lewis basicity of solvents**

Silver perchlorate, $AgClO_4$, is significantly more soluble in benzene than in alkane solvents. Account for this observation in terms of Lewis acid–base properties.

Answer We need to consider how the solvent interacts with the solute. The π electrons of benzene, a soft base, are available for complex formation with the empty orbitals of the cation Ag^+, a soft acid. The Ag^+ ion is thus solvated favourably by benzene. The species $[Ag—C_6H_6]^+$ is the complex of the acid Ag^+ with π electrons of the weak base benzene.

Self-test 5.10 Predict the structures of the complex cation formed in each case when anhydrous NaF is dissolved in (a) CH_3CN, (b) HF.

Applications of acid–base chemistry

The Brønsted and Lewis definitions of acids and bases do not have to be considered separately from each other. In fact, many applications of acid–base chemistry utilize both Lewis and Brønsted acids or bases simultaneously.

5.16 Acid and base catalysis

We introduced catalysis in Section 2.15. Acids and bases play an important role in accelerating the rates of reactions without necessarily being included in the products. Catalysis by Brønsted acids and bases is usually discussed in terms of two mechanisms.

General acid catalysis occurs when any species capable of donating a proton accelerates the reaction rate. Likewise, **general base catalysis** occurs when any base accelerates the rate. The direction of the reaction depends, naturally, on thermodynamic factors, but in either case transfer of the proton occurs in the transition state. Such catalysis is common in enzyme reactions, where amino acid side chains function as acids or bases. The example shown in Fig. 5.22

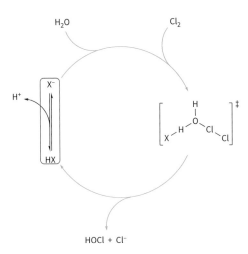

FIGURE 5.22 An example of general acid–base catalysis. The hydrolysis of Cl_2 to form hypochlorite and chloride, catalysed by a general base X^- that is required in the transition state. The reverse reaction would involve the general acid XH.

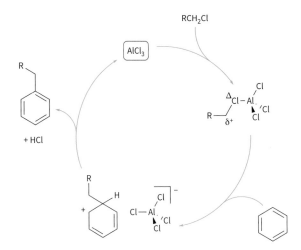

FIGURE 5.24 An example of Lewis acid catalysis. A Friedel–Crafts alkylation reaction catalysed by $AlCl_3$.

is the base-catalysed reaction of Cl_2 with water, X^- typically being the conjugate base of a weak acid such as ethanoate.

In **specific acid catalysis**, protonated solvent, H^+_{solv} is the catalyst: the Brønsted acid must be sufficiently strong to produce H^+_{solv} at a high concentration, so it is typically one having a very negative pK or H_0 value. The reactant is protonated to form a reactive intermediate before the transition state is reached. An example, shown in Fig. 5.23, is the decarboxylation of pyrrole-2-carboxylic acid catalysed by a strong acid, in which H^+ must first be transferred to the pyrrole ring.

Lewis acids serve as catalysts for many organic reactions. The classic example of **Lewis acid catalysis** is Friedel–Crafts alkylation (the substitution of an H atom in an aromatic ring by an alkyl group) and acylation (the substitution of

H by RCO) using $AlCl_3$ as catalyst. Alkylation of benzene is believed to proceed as shown in Fig. 5.24. The first step is the reaction between the Lewis acid and the alkyl halide, which generates a polarized intermediate bearing a positive charge on the alkyl group: such a cationic species is electrophilic and attacks the benzene ring to generate a transient aryl cation radical that is deprotonated by $AlCl_4^-$ to yield HCl and $AlCl_3$.

5.17 Superacids and superbases

KEY POINTS Superacids are defined as being stronger proton donors than anhydrous sulfuric acid. Superbases are stronger proton acceptors than the hydroxide ion.

A **superacid** is a substance that is a stronger proton donor than pure (anhydrous) H_2SO_4. They are typically viscous, corrosive liquids and can be up to 10^{18} times more acidic than H_2SO_4 itself, their strengths being expressed by their large (negative) Hammett acidity functions (Table 5.7). Most superacids are formed *in situ* by dissolving a very strong Lewis acid in a strong anhydrous Brønsted acid, common examples being when SbF_5 is dissolved in fluorosulfonic acid, HSO_3F, or anhydrous HF. A mixture of SbF_5 and HSO_3F is known as 'magic acid', so named because of its ability to dissolve candle wax. The enhanced acidity is due to the displacement, by SbF_5, of the proton on HSO_3F, which is thus taken up by a second HSO_3F.

$$SbF_5(l) + 2HSO_3F(l) \rightarrow H_2SO_3F^+(sol) + SbF_5SO_3F^-(sol)$$

An even stronger superacid is formed when SbF_5 is added to anhydrous HF:

$$SbF_5(l) + 2HF(l) \rightarrow H_2F^+(sol) + SbF_6^-(sol)$$

Other pentafluorides also form superacids in HSO_3F and HF, and the acidity of these compounds decreases in the order $SbF_5 > AsF_5 > TaF_5 > NbF_5 > PF_5$.

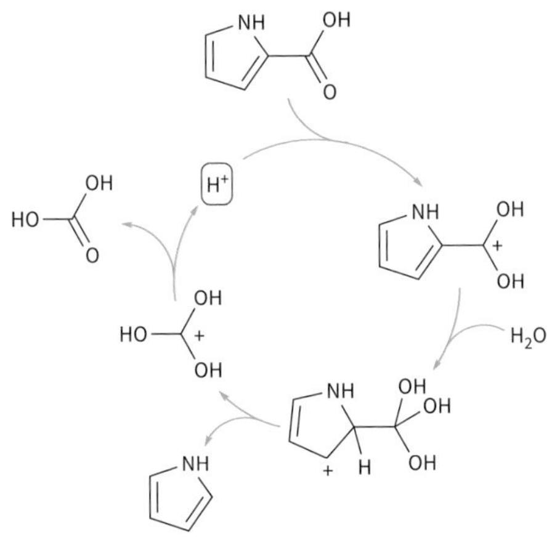

FIGURE 5.23 An example of specific acid catalysis: the catalyst is the solvated proton. Hydrolytic decarboxylation of the carboxylic acid is initiated by a step that requires a very strong proton donor. This kind of catalysis is common for strong acids in polar solvents, including water, and is important in catalysis by superacids (Section 5.17).

Superacids are known that can protonate almost any organic compound. In the 1960s, George Olah and his colleagues found that carbonium ions were stabilized when hydrocarbons were dissolved in superacids.[7] In inorganic chemistry, superacids have been used to observe a wide variety of reactive cations such as S_8^{2+}, $H_3O_2^+$, Xe_2^+, and HCO^+, some of which have been isolated for structural characterization. Even carbon nanotubes (see Section 15.6) can be protonated, rendering them soluble in polar solvents.

Superacids are known that do not involve a Lewis acid; moreover, some can be synthesized and stored as crystalline materials that may be handled under anhydrous conditions. The fluorinated carborane cluster compound H-[CHB$_{11}$F$_{11}$] is recognized as the strongest Brønsted acid known. The structure of the anion is shown (30).

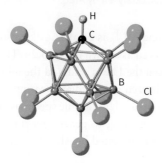

30 [CHB$_{11}$F$_{11}$]$^-$

Superbases are compounds which have extremely high basicity. Superbases are usually compounds of Group 1 or Group 2 elements, and they react with water to produce OH$^-$ (the solvent levelling effect) so are used in nonpolar or highly basic solvents. For example, lithium nitride, Li$_3$N, formed by the direct combination of Li metal and N$_2$, reacts violently with water to produce ammonia:

$$Li_3N(s) + 3H_2O(l) \rightarrow 3Li(OH)(aq) + NH_3(g)$$

The nitride anion is a stronger base than the hydride ion and deprotonates hydrogen in a reaction that is reversible at 270°C.

$$Li_3N(s) + 2H_2(g) \rightleftharpoons LiNH_2(s) + 2LiH(s)$$

Sodium hydride is a superbase that is used in organic chemistry to deprotonate carboxylic acids, alcohols, phenols, and thiols. Calcium hydride reacts with water to liberate hydrogen:

$$CaH_2(s) + 2H_2O(l) \rightarrow Ca(OH)_2(s) + 2H_2(g)$$

---

[7] Carbocations could not be studied before Olah's experiments, and he won the 1994 Nobel Prize for Chemistry for this work.

Calcium hydride is used as a desiccant, to inflate weather balloons, and as a laboratory source of pure hydrogen.

The superbase anion may be organic, as represented by the widely used compound lithium diisopropylamide (known as LDA) (31). The sterically hindered anion is an extremely strong base for the selective deprotonation of alkyl groups in reactions where it is necessary to achieve kinetic control by favouring formation of a single transient intermediate.

31 Lithium diisopropylamide

5.18 Heterogeneous acid–base reactions

KEY POINT The surfaces of many catalytic materials and minerals have Brønsted and Lewis acid sites.

Some of the most important reactions involving the Lewis and Brønsted acidity of inorganic compounds occur at solid surfaces. For example, **surface acids**, which are solids with a high surface area and Lewis acid sites, are used as catalysts in the petrochemical industry for the interconversion of hydrocarbons. The surfaces of many materials that are important in the chemistry of soil and natural waters also have Brønsted and Lewis acid sites.

Silica surfaces do not readily produce Lewis acid sites because –OH groups remain tenaciously attached at the surface of SiO$_2$ derivatives; as a result, Brønsted acidity is dominant. The Brønsted acidity of silica surfaces themselves is only moderate (and comparable to that of acetic acid). However, aluminosilicates, in which some tetrahedral Si sites (SiO$_4$ tetrahedra) are replaced by Coulombically equivalent Al sites (AlO$_3$OH/AlO$_4^-$), display strong Brønsted acidity. When surface OH groups are removed by heat treatment, the aluminosilicate surface possesses strong Lewis acid sites.

The best-known class of aluminosilicates is the **zeolites** (Sections 15.15, 25.16, 29.1, 29.2) which are widely used as environmentally benign heterogeneous catalysts (Section 31.13). The catalytic activity of zeolites arises from their acidic nature and they are known as **solid acids**. Selectivity is conferred by the sizes of the cavities. Other solid acids include supported heteropoly acids and acidic clays. Some reactions occurring at these catalysts are very sensitive to the presence of Brønsted or Lewis acid sites. For example, toluene can be subjected to Friedel–Crafts alkylation at various positions over a bentonite clay catalyst:

When the reagent is benzyl chloride, Lewis acid sites are involved in the reaction and when the reagent is benzyl alcohol, Brønsted sites are involved.

Surface reactions carried out using the Brønsted acid sites of silica gels are used to prepare thin coatings of a wide variety of organic groups using surface modification reactions such as

$$\equiv\!Si\!-\!OH + HOSiR_3 \longrightarrow \equiv\!Si\!-\!OSiR_3 + H_2O$$

$$\equiv\!Si\!-\!OH + ClSiR_3 \longrightarrow \equiv\!Si\!-\!OSiR_3 + HCl$$

Thus, silica gel surfaces can be modified to have affinities for specific classes of molecules. This procedure greatly expands the range of stationary phases that can be used for chromatography. The surface –OH groups on glass can be modified similarly, and glassware treated in this manner is sometimes used in the laboratory when proton-sensitive compounds are being studied.

Solid acids are finding new applications in green chemistry (Chapter 27). Traditional industrial processes generate large volumes of hazardous waste during the final stages of the process when the product is separated from the reagents and by-products. Solid catalysts are easily separated from liquid products, and reactions can often operate under milder conditions and provide greater selectivity.

FURTHER READING

D. Himmel, V. Radtke, B. Butschke, and I. Krossing, Basic Remarks on Acidity. *Angew. Chem. Int. Ed.*, 2018, **57**, 4386–4411.

W. Stumm and J. J. Morgan, *Aquatic Chemistry: Chemical Equilibria and Rates in Natural Waters*. Wiley, New York (1995). The classic text on the chemistry of natural waters.

J. Burgess, *Ions in Solution: Basic Principles of Chemical Interactions*. Ellis Horwood, Chichester (1999).

T. Akiyama, Stronger Brønsted acids. *Chem. Rev.*, 2007, **107**, 5744.

G. Cavallo, P. Metrangolo, R. Milani, T. Pilati, A. Priimagi, G. Resnati, and G. Terraneo, The halogen bond. *Chem. Rev.*, 2016, **116**, 2478.

A. Staubitz, A. P. M. Robertson, M. E. Sloan, and I. Manners, Amine- and phosphine borane adducts: New interest in old molecules. *Chem. Rev.*, 2010, **110**, 4023.

P. Erdmann and L. Greb, What Distinguishes the Strength and the Effect of a Lewis Acid: Analysis of the Gutmann-Beckett Method. *Angew. Chem. Int. Ed.*, 2022, **61**, e202114550.

T. P. Silverstein, The real reason why oil and water don't mix. *Journal of Chemical Education*, 1988, **75**, 116–118.

D. W. Stephan and G. Erker, Frustrated Lewis Pairs: Metal-free hydrogen activation and more. *Angew. Chem. Int. Ed.*, 2010, **49**, 46.

F.-G. Fontaine and D. W. Stephan, On the concept of Frustrated Lewis Pairs. *Philosophical Transactions of the Royal Society A*, 2017, **375**, 20170004. http://dx.doi.org/10.1098/rsta.2017.0004

A. R. Jupp and D. W. Stephan, New directions for Frustrated Lewis Pair chemistry. *Trends in Chemistry*, 2019, **1**, 35.

S. J. Grabowski, What is the covalency of hydrogen bonding? *Chem. Rev.*, 2011, **111**, 2597.

E. J. Corey, Enantioselective catalysis based on cationic oxazaborolidines. *Angew. Chem. Int. Ed.*, 2009, **48**, 2100.

L. Greb, Lewis superacids: Classifications, candidates, and applications. *Chem. Eur. J.*, 2018, **24**, 17881.

P. Raveendran, Y. Ikushima, and S. L. Wallen, Polar attributes of supercritical carbon dioxide. *Acc. Chem. Res.*, 2005, **38**, 478.

F. Jutz, J.-M. Andanson, and A. Baiker, Ionic liquids and dense carbon dioxide: A beneficial biphasic system for catalysis. *Chem. Rev.*, 2011, **111**, 322.

D. R. MacFarlane, J. M. Pringle, K. M. Johansson, S. A. Forsyth, and M. Forsyth, Lewis base ionic liquids. *Chem. Commun.*, 2006, 1905.

T. Welton, Ionic liquids: A brief history. *Biophysical Reviews*, 2019, **19**, 691.

J. M. Gomes, S. S. Silva, and R. L. Reis, Biocompatible ionic liquids: Fundamental behaviours and applications. *Chem. Soc. Rev.*, 2019, **48**, 4317.

I. Krossing, J. M. Slattery, C. Daguenet, P. J. Dyson, A. Oleinikova, and H. Weingärtner, Why are ionic liquids liquid? A simple explanation based on lattice and solvation energies. *J. Am. Chem. Soc.*, 2006, **128**, 13427.

N. Corcoran, *Chemistry in Non-Aqueous Solvents*. Kluwer Academic Publishers (2003). A comprehensive account.

G. A. Olah, G. K. Prakash, and J. Sommer, *Superacids*. Wiley, New York (1985).

S. Ramesh, L. M. Ericson, V. A. Davis, R. K. Saini, C. Kittrell, M. Pasquali, W. E. Billups, W. W. Adams, R. H. Hauge, and R. E. Smalley, Dissolution of pristine single walled carbon nanotubes in superacids by direct protonation. *J. Phys. Chem. B.*, 2004, **108**, 8794.

H. F. T. Klare and Martin Oestreich, The power of the proton: From superacidic media to superelectrophile catalysis. *J. Am. Chem. Soc.*, 2021, **143**, 15490.

EXERCISES

5.1 Sketch an outline of the s and p blocks of the periodic table and indicate on it the elements that form (a) strongly acidic oxides and (b) strongly basic oxides, and (c) show the regions for which amphoterism is common.

5.2 Identify the conjugate bases corresponding to the following acids: $[Co(NH_3)_5(OH_2)]^{3+}$, HSO_4^-, CH_3OH, $H_2PO_4^-$, $Si(OH)_4$, HS^-.

5.3 Identify the conjugate acids of the bases C_5H_5N (pyridine), HPO_4^{2-}, O^{2-}, CH_3COOH, $[Co(CO)_4]^-$, CN^-.

5.4 Calculate the equilibrium concentration of H_3O^+ in a 0.10 M solution of butanoic acid ($K_a = 1.86 \times 10^{-5}$). What is the pH of this solution?

5.5 The K_a of ethanoic acid, CH_3COOH, in water is 1.8×10^{-5}. Calculate K_b of the conjugate base, $CH_3CO_2^-$.

5.6 Write balanced equations for the main reaction occurring when (a) H_3PO_4 and Na_2HPO_4 and (b) CO_2 and $CaCO_3$ are mixed in aqueous media.

5.7 Explain why hydrogen selenide is a stronger acid than hydrogen sulfide.

5.8 The effective proton affinity A'_p of F^- in water is $1150\,kJ\,mol^{-1}$. Predict whether it will behave as an acid or a base in water.

5.9 Although NH_3 is a stronger Brønsted base than $(CH_3)_3N$ in water, it is a weaker Brønsted base in the gas phase. Account for this observation.

5.10 The gas-phase proton affinities of PH_2^- and F^- are very similar (1550 and $1554\,kJ\,mol^{-1}$ respectively), but PH_2^- is unable to exist in aqueous solution (the pK_a of $PH_3 = 27$) whereas HF behaves as a weak acid (pK_a for HF = 3.2). Account for this apparent discrepancy.

5.11 Draw the structures of chloric acid and chlorous acid and predict their pK_a values using Pauling's rules.

5.12 Aided by Fig. 5.5 (taking solvent levelling into account), identify which bases from the following lists are (a) too strong to be studied experimentally, (b) too weak to be studied experimentally, or (c) of directly measurable base strength. (i) CO_3^{2-}, O^{2-}, ClO_4^-, and NO_3^- in water; (ii) HSO_4^-, NO_3^-, ClO_4^- in H_2SO_4.

5.13 The aqueous solution pK_a values for HOCN, H_2NCN, and CH_3CN are approximately 4, 10.5, and 20 (estimated), respectively. Explain the trend in these cyano derivatives of binary acids and compare them with H_2O, NH_3, and CH_4. Is the CN group electron donating or withdrawing?

5.14 Use Pauling's rules to place the following acids in order of increasing acid strength: HNO_2, H_2SO_4, $HBrO_3$, and $HClO_4$ in a nonlevelling solvent.

5.15 The acids H_3PO_4, H_3PO_3, and H_3PO_2 all have a pK_a value of 2, but the pK_a values of HOCl, $HClO_2$, and $HClO_3$ are 7.5, 2.0, and −3.0, respectively. Explain this observation.

5.16 Explain why metal ions in aqueous solution can show acidic properties.

5.17 Arrange the following ions in order of increasing acidity in aqueous solution:

Fe^{3+}, Na^+, Mn^{2+}, Ca^{2+}, Al^{3+}, Sr^{2+}.

5.18 Which member of the following pairs is the stronger acid? Give reasons for your choice. (a) $[Fe(OH_2)_6]^{3+}$ or $[Fe(OH_2)_6]^{2+}$, (b) $[Al(OH_2)_6]^{3+}$ or $[Ga(OH_2)_6]^{3+}$, (c) $Si(OH)_4$ or $Ge(OH)_4$, (d) $HClO_3$ or $HClO_4$, (e) H_2CrO_4 or $HMnO_4$, (f) H_3PO_4 or H_2SO_4.

5.19 The ions Na^+ and Ag^+ have similar radii. Which aqua ion is the stronger acid? Why?

5.20 Arrange the oxides Al_2O_3, B_2O_3, BaO, CO_2, Cl_2O_7, SO_3 in order from the most acidic through amphoteric to the most basic.

5.21 Arrange the acids HSO_4^-, H_3O^+, H_4SiO_4, CH_3GeH_3, NH_3, HSO_3F in order of increasing acid strength.

5.22 When a pair of aqua cations forms an M–O–M bridge with the elimination of water, what is the general rule for the change in charge per M atom on the ion?

5.23 List the considerations needed to predict how global warming will influence the pH of the oceans given that the solubility of CO_2 in seawater decreases as the temperature rises and hydration of CO_2 is an exothermic reaction.

5.24 For each of the following processes, identify the acids and bases involved and characterize the process as complex formation or acid–base displacement. Identify the species that exhibit Brønsted acidity as well as Lewis acidity.

(a) $SO_3 + H_2O \rightarrow HSO_4^- + H^+$

(b) $CH_3[B_{12}] + Hg^{2+} \rightarrow [B_{12}]^+ + CH_3Hg^+$; $[B_{12}]$ designates the Co-porphyrin, vitamin B_{12} (Section 26.11).

(c) $KCl + SnCl_2 \rightarrow K^+ + [SnCl_3]^-$

(d) $AsF_3(g) + SbF_5(l) \rightarrow [AsF_2]^+[SbF_6]^-(s)$

(e) Ethanol dissolves in pyridine to produce a nonconducting solution.

5.25 Aluminium sulfide, Al_2S_3, gives off a foul odour characteristic of hydrogen sulfide when it becomes damp. Write a balanced chemical equation for the reaction and discuss it in terms of acid–base concepts.

5.26 Explain why the following metathesis reaction is favourable:

$HgF_2(aq) + 2\,I^-(aq) \rightarrow HgI_2(aq) + 2F^-(aq)$

5.27 Using hard–soft concepts, which of the following reactions are predicted to have an equilibrium constant greater than 1? Unless otherwise stated, assume gas-phase or hydrocarbon solution and 25°C.

(a) $R_3PBBr_3 + R_3NBF_3 \rightleftharpoons R_3PRF_3 + R_3NBBr_3$

(b) $SO_2 + (C_6H_5)_3PHOC(CH_3)_3 \rightleftharpoons (C_6H_5)_3PSO_2 + HOC(CH_3)_3$

(c) $CH_3HgI + HCl \rightleftharpoons CH_3HgCl + HI$

(d) $[AgCl_2]^-(aq) + 2CN^-(aq) \rightleftharpoons [Ag(CN)_2]^-(aq) + 2Cl^-(aq)$

5.28 Identify the products from the reaction between the following pairs of reagents. In each case identify the species which are acting as a Lewis acid or a Lewis base in the reactions.

(a) $CsF + BrF_3$

(b) $ClF_3 + SbF_5$

(c) $B(OH)_3 + H_2O$

(d) $B_2H_6 + PMe_3$

5.29 The enthalpies of reaction of trimethylboron with NH_3, CH_3NH_2, $(CH_3)_2NH$, and $(CH_3)_3N$ are −58, −74, −81, and −75 kJ mol^{-1}, respectively. Why is trimethylamine out of line?

5.30 Explain why $N(SiH_3)_3$ and NF_3 are much weaker Lewis bases than NMe_3. (Unlike carbon, silicon can expand its octet.)

5.31 Explain the following enthalpy changes (kJ mol^{-1}) for adduct formation:

(a) $BEt_3 + NMe_3$,	−72.3
(b) $B(OMe)_3 + NMe_3$,	−31.5
(c) $BMe_3 + NMe_3$,	−75.2
(d) $BMe_3 + N(SiH_3)_3$,	+4

5.32 In the gas phase, the base strength of amines increases regularly along the series $NH_3 < CH_3NH_2 < (CH_3)_2NH < (CH_3)_3N$. Consider the role (if any) of steric effects and the electron-donating ability of CH_3 in determining this order for Brønsted and Lewis basicity. In aqueous solution, the order is reversed. What solvation effect is likely to be responsible?

5.33 An electrically conducting solution is produced when $AlCl_3$ is dissolved in the basic polar solvent CH_3CN. Give formulas for the most probable conducting species and describe their formation using Lewis acid–base concepts.

5.34 Predict whether the equilibrium constants for the following reactions should be greater than 1 or less than 1:

(a) $CdI_2(s) + CaF_2(s) \rightleftharpoons CdF_2(s) + CaI_2(s)$
(b) $[CuI_4]^{2-}(aq) + [CuCl_4]^{3-}(aq) \rightleftharpoons [CuCl_4]^{2-}(aq) + [CuI_4]^{3-}(aq)$
(c) $NH_2^-(aq) + H_2O(l) \rightleftharpoons NH_3(aq) + OH^-(aq)$

5.35 Pyridine forms a stronger Lewis acid–base complex with SO_3 than with SO_2. However, pyridine forms a weaker complex with SF_6 than with SF_4. Explain the difference.

5.36 Although AlF_3 is insoluble in liquid HF it dissolves if NaF is present. When BF_3 is added to the resulting solution AlF_3 precipitates. Explain these observations.

5.37 Describe the solvent properties that would (a) favour displacement of Cl$^-$ by I$^-$ from an acid centre, (b) favour basicity of R_3As over R_3N, (c) favour acidity of Ag$^+$ over Al^{3+}, (d) promote the reaction $2\,FeCl_3 + ZnCl_2 \rightarrow Zn^{2+} + 2\,[FeCl_4]^-$. In each case, suggest a specific solvent that might be suitable.

5.38 Use acid–base concepts to comment on the fact that the only important ore of mercury is cinnabar, HgS, whereas zinc occurs in nature as sulfides, silicates, carbonates, and oxides.

5.39 With the aid of the table of *E* and *C* values (Table 5.5), discuss the relative basicities of (a) acetone and dimethyl sulfoxide, (b) dimethyl sulfide and dimethyl sulfoxide. Comment on a possible ambiguity for dimethyl sulfoxide.

5.40 Use the data in Table 5.5 to calculate the enthalpy change for the reaction of iodine with phenol.

5.41 Explain why it is not possible to determine by experiment whether HBr or HI is a stronger acid in H_2O, and thus suggest a solvent in which their acidity can be differentiated.

5.42 Write balanced Brønsted acid–base equations for the dissolution of the following compounds in liquid hydrogen fluoride: (a) CH_3CH_2OH, (b) NH_3, (c) C_6H_5COOH.

5.43 Give the equation for the dissolution of SiO_2 glass by HF and interpret the reaction in terms of Lewis and Brønsted acid–base concepts.

5.44 The f-block elements are found as M^{3+} cations in silicate minerals. What does this indicate about their hardness?

5.45 The hydroxoacid $Si(OH)_4$ is weaker than H_2CO_3. Write balanced equations to show how dissolving a solid M_2SiO_4 can lead to a reduction in the pressure of CO_2 over an aqueous solution. Explain why silicates in ocean sediments might limit the increase of CO_2 in the atmosphere.

5.46 The precipitation of $Fe(OH)_3$ discussed in the chapter is used to clarify waste waters, because the gelatinous hydrous oxide is very efficient at the co-precipitation of some contaminants and the entrapment of others. The solubility constant of $Fe(OH)_3$ is $K_s = [Fe^{3+}][OH^-]^3 \approx 1.0 \times 10^{-38}$. As the autoprotolysis constant of water links $[H_3O^+]$ to $[OH^-]$ by $K_w = [H_3O^+][OH^-] = 1.0 \times 10^{-14}$, we can rewrite the solubility constant by substitution as $[Fe^{3+}]/[H^+]^3 = 1.0 \times 10^4$. (a) Balance the chemical equation for the precipitation of $Fe(OH)_3$ when iron(III) nitrate is added to water. (b) If 6.6 kg of $Fe(NO_3)_3 \cdot 9H_2O$ is added to 100 dm^3 of water, what is the final pH of the solution and the molar concentration of Fe^{3+}, neglecting other forms of dissolved Fe(III)? Provide formulas for two Fe(III) species that have been neglected in this calculation.

5.47 Explain the following observations: (a) aluminium kitchen utensils are discoloured by cooking mildly alkaline foods such as potatoes but are cleaned by cooking acidic foods such as citrus fruit; (b) silver cutlery becomes tarnished in atmospheres containing traces of hydrogen sulfide.

5.48 The compounds SO_2 and $SOCl_2$ can undergo an exchange of radioactively labelled sulfur. The exchange is catalysed by Cl$^-$ and $SbCl_5$. Suggest mechanisms for these two exchange reactions with the first step being the formation of an appropriate complex.

5.49 For parts (a), (b), and (c), state which of the two solutions has the lower pH:

(a) 0.1 M $Fe(ClO_4)_2$(aq) or 0.1 M $Fe(ClO_4)_3$(aq)
(b) 0.1 M $Ca(NO_3)_2$(aq) or 0.1 M $Mg(NO_3)_2$(aq)
(c) 0.1 M $Hg(NO_3)_2$(aq) or 0.1 M $Zn(NO_3)_2$(aq)

5.50 Why are strongly acidic solvents (e.g. SbF_5/HSO_3F) used in the preparation of cations such as I_2^+ and Se_8^{2+}, whereas strongly basic solvents are needed to stabilize anionic species such as S_4^{2-} and Pb_9^{4-}?

5.51 A standard procedure for improving the detection of the stoichiometric point in titrations of weak bases with strong acids is to use acetic acid as a solvent. Explain the basis of this approach.

5.52 Write equations to account for the acidity of CO_2 and boric acid ($B(OH)_3$), each dissolved in water.

5.53 Catalysis of the acylation of aryl compounds by the Lewis acid $AlCl_3$ was described in Section 5.16. Propose a mechanism for a similar reaction catalysed by an alumina surface.

TUTORIAL PROBLEMS

5.1 In their paper 'The strengths of the hydrohalic acids' (*J. Chem. Educ.*, 2001, **78**, 116), R. Schmid and A. Miah discuss the validity of literature values of the pK_a values for HF, HCl, HBr, and HI. (a) On what basis have the literature values been estimated? (b) To what is the low acid strength of HF relative to HCl usually attributed? (c) What reason do the authors suggest for the high acid strength of HCl?

5.2 Despite a century passing since Brønsted and Lowry introduced their proton-based theories of acidity, scientists are still divided on how to understand and apply many of the fundamental principles that apply to all kinds of compounds and the different environments/solvents in which they are studied. In 'Basic remarks on acidity' (*Angew. Chem. Int. Ed.* 2018, **57**, 4386), Ingo Krossing and co-workers attempt to unify the field. Summarize the arguments that they use to present reference states for solution acidity (pH) and acid strength (pK), and explain how they extend the principles to strong acids and a variety of different solvents.

5.3 In a review article (*Angew. Chem. Int. Ed.*, 2009, **48**, 2100), E. J. Corey describes asymmetric catalysis by chiral boranes. Show how these chiral boranes, which are Lewis acids, are also able to direct Brønsted acidity.

5.4 The long-taught explanation that the ordering of Lewis acidities of BX_3 ($BCl_3 > BBr_3 > BF_3$) arises because stronger X-B π-backbonding resists adoption of a pyramidal geometry, is discussed critically by Popelier and co-workers in their article 'Electrostatics explains the reverse Lewis acidity of BH_3' (*J. Phys. Chem. A.*, 2021, **125**, 8615). Summarize the various explanations listed in their introduction, and draw diagrams to represent their alternative model.

5.5 Stephan and and co-workers have advocated an alternative approach for scaling Lewis acidity called the Global Electrophilicity Index. In their article 'The global electrophilicity index as a metric for Lewis acidity', by A. R. Jupp, T. C. Johnstone, and D. W. Stephan (*Dalton Trans.*, 2018, **47**, 7029), they explain the GIA and compare it with FIA and NMR-based methods for Lewis scaling. What are the advantages and disadvantages of the GIA approach?

5.6 'Super' Lewis acids are defined and discussed in a review by Lutz Greb, 'Lewis superacids: classifications, candidates and applications' (*Chem. Eur. Journal*, 2018, **24**, 17881). How does he define 'super Lewis acids'? How are they scaled? What are the criteria required for making these compounds useful in the laboratory or in industry?

5.7 An article by Poliakoff and co-workers describes how a green industrial chemical process was initiated, in which conventional solvents were replaced by $scCO_2$ (*Green Chem.*, 2003, **5**, 99). Explain the advantages and challenges of introducing such changes to traditional technology.

5.8 The reversible reaction of CO_2 gas with aqueous emulsions of long-chain alkyl amidine compounds has important practical applications. Describe the chemistry that is involved in this demonstration of 'switchable surfactants' (*Science*, 2006, **313**, 958).

5.9 An article by Krossing and co-workers (*J. Am. Chem. Soc.*, 2006, **128**, 13427) explains the behaviour of ionic liquids in terms of a thermodynamic cycle approach. Describe the principles that are applied, and summarize the predictions that are made.

5.10 In 'The power of the proton: From superacidic media to superelectrophile catalysis' (*J. Am. Chem. Soc.* 2021, **143**, 15490), Klare and Oestreich discuss some of the legacy of G. Olah's Nobel Prize-winning research on superacids. Explain the ways in which these compounds have been 'tamed' to facilitate chemical reactions that would otherwise be very difficult to achieve.

6 Oxidation and reduction

Reduction potentials

6.1 **Redox half-reactions**
6.2 **Standard potentials and spontaneity**
6.3 **Trends in standard potentials**
6.4 **The electrochemical series**
6.5 **The Nernst equation**

Redox stability

6.6 **The influence of pH**
6.7 **Reactions with water**
6.8 **Oxidation by atmospheric oxygen**
6.9 **Disproportionation and comproportionation**
6.10 **The influence of complexation**
6.11 **The relation between solubility and standard potentials**

Useful presentation of potential data

6.12 **Latimer diagrams**
6.13 **Frost diagrams**
6.14 **Proton-coupled electron transfer: Pourbaix diagrams**

Applications

6.15 **Environmental chemistry: natural waters**
6.16 **Electrocatalysis**
6.17 **Extraction of the elements by chemical reduction**
6.18 **Extraction of the elements by chemical oxidation**
6.19 **Electrochemical extraction**

Further reading

Exercises

Tutorial problems

Oxidation is the removal of electrons from a species; reduction is the addition of electrons. Almost all elements and their compounds can undergo oxidation and reduction reactions, and the element is said to exhibit one or more different oxidation states. In this chapter we present examples of this 'redox' chemistry and develop concepts for understanding why oxidation and reduction reactions occur, considering mainly their thermodynamic aspects. We discuss the procedures for analysing redox reactions in solution and see that the electrode potentials of electrochemically active species provide data that are useful for determining and understanding the stability of species and solubility of salts. We describe procedures for displaying trends in the stabilities of various oxidation states, including the influence of pH. The chapter concludes with discussions of applications, first in environmental chemistry and electrocatalysis, then with examples of some major industrial processes, including the extraction of metals from their ores.

A large class of reactions of inorganic compounds can be regarded as occurring by the transfer of electrons from one species to another. Electron gain is called **reduction** and electron loss is called **oxidation**; the joint process is called a **redox reaction**. The species that supplies electrons is the **reducing agent** (or 'reductant') and the species that removes electrons is the **oxidizing agent** (or 'oxidant'). Many redox reactions release a great deal of energy and they are exploited in combustion or battery technologies as well as biology.

Many redox reactions occur between reactants in the same physical state. Some examples are:

In gases:

$$2\,NO(g) + O_2(g) \rightarrow 2\,NO_2(g)$$
$$2\,C_4H_{10}(g) + 13\,O_2(g) \rightarrow 8\,CO_2(g) + 10\,H_2O(g)$$

In solution:

$$Fe^{3+}(aq) + Cr^{2+}(aq) \rightarrow Fe^{2+}(aq) + Cr^{3+}(aq)$$
$$3\,CH_3CH_2OH(aq) + 2\,CrO_4^{2-}(aq) + 10\,H^+(aq) \rightarrow$$
$$3\,CH_3CHO(aq) + 2\,Cr^{3+}(aq) + 8\,H_2O(l)$$

In biological systems:

$$\text{'Mn}_4\text{'(V,IV,IV,IV)} + 2\,H_2O(l) \rightarrow \text{'Mn}_4\text{'(IV,III,III,III)} +$$
$$4\,H^+(aq) + O_2(g)$$

In solids:

$$LiCoO_2(s) + 6\,C(s) \rightarrow LiC_6(s) + CoO_2(s)$$
$$CeO_2(s) \xrightarrow{heat} CeO_{2-\delta}(s) + \delta/2\,O_2(g)$$

followed by

$$CeO_{2-\delta}(s) + \delta H_2O(g) \rightarrow CeO_2(s) + \delta H_2(g)$$

The biological example refers to the production of O_2 from water by a Mn_4CaO_5 cofactor contained in one of the photosynthetic complexes of plants (Section 26.10). In the first solid-state example Li_6C represents a compound where Li^+ ion has penetrated between the sheets of carbon atoms in graphite to form an **intercalation compound** (Chapter 25) The reaction takes place in a lithium-ion battery during charging, and its reverse takes place during discharge (Section 30.9). The second solid-state example is a thermal cycle that converts water into H_2 and O_2 using heat that can be supplied by concentrating solar radiation (Box 11.4). Redox reactions can also occur at interfaces (phase boundaries), such as a gas/solid or a solid/liquid interface: examples include the dissolution of a metal and reactions occurring at an electrode.

Because of the diversity of redox reactions, it is often convenient to analyse them by applying a set of formal rules expressed in terms of oxidation numbers (Section 2.14) and not to think in terms of actual electron transfers. Oxidation then corresponds to an increase in the oxidation number of an element, and reduction corresponds to a decrease in its oxidation number. If no element in a reaction undergoes a change in oxidation number, then the reaction is not redox. We shall adopt this approach when we judge it appropriate.

A BRIEF ILLUSTRATION

The simplest redox reactions involve the formation of cations and anions from the elements. Examples include the oxidation of lithium to Li^+ ions when it burns in air to form Li_2O, and the reduction of chlorine to Cl^- when it reacts with calcium to form $CaCl_2$. For the Group 1 and 2 elements the only oxidation numbers commonly encountered are those of the element (0) and of the ions, +1 and +2, respectively. However, many of the other elements form compounds in more than one oxidation state. Thus lead is commonly found in its compounds as Pb(II), as in PbO, and as Pb(IV), as in PbO_2.

The ability to exhibit multiple oxidation numbers is seen at its fullest in Groups 5, 6, 7, 8, and 9 for the d block and in Groups 15, 16, and 17 for the p block. Osmium, for instance, forms compounds that span oxidation numbers between −2, as in $[Os(CO)_4]^{2-}$, and +8, as in OsO_4. Likewise, oxidation numbers for sulfur span between −2 in H_2S, and +6 in H_2SO_4. Compounds of 3d-elements are most likely to exhibit consecutive oxidation numbers—that is, +2, +3, +4—and thus be suited to undergo one-electron transfers. Because the oxidation state of an element is often reflected in the properties of its compounds, the ability to express the tendency of an element to form a compound in a particular oxidation state quantitatively is very useful in inorganic chemistry.

Reduction potentials

Because electrons are transferred between species in redox reactions, electrochemical methods (using electrodes to measure electron transfer reactions under controlled thermodynamic conditions) are of major importance and lead to a scale of 'standard potentials' for inorganic species. The tendency of an electron to migrate from one species to another is expressed in terms of the differences between their standard potentials. Techniques such as cyclic voltammetry (Chapter 8) are used to measure reduction potentials under different conditions, and extract information about the reversibility and kinetics of electron-transfer reactions and influences of pH and complexation.

6.1 Redox half-reactions

KEY POINT A redox reaction can be expressed as the difference of two reduction half-reactions.

It is convenient to think of a redox reaction as the combination of two conceptual **half-reactions** in which the electron loss (oxidation) and gain (reduction) are displayed explicitly. In a reduction half-reaction, a substance gains electrons, as in

$$2H^+(aq) + 2e^- \rightarrow H_2(g)$$

In an oxidation half-reaction, a substance loses electrons, as in

$$Zn(s) \rightarrow Zn^{2+}(aq) + 2e^-$$

Electrons are not ascribed a state in the equation of a half-reaction: they are 'in transit'. The oxidized and reduced species in a half-reaction constitute a **redox couple**. A couple is written with the oxidized species before the reduced, as in H^+/H_2 and Zn^{2+}/Zn, and typically the phases are not shown.

For reasons that will soon become clear, it is useful to represent oxidation half-reactions by the corresponding reduction half-reaction. To do so, we simply reverse the equation for the oxidation half-reaction. Thus, the reduction half-reaction associated with the oxidation of zinc is written

$$Zn^{2+}(aq) + 2e^- \rightarrow Zn(s)$$

A redox reaction in which zinc is oxidized by hydrogen ions,

$$Zn(s) + 2H^+(aq) \rightarrow Zn^{2+}(aq) + H_2(g)$$

is then written as the *difference* of the two reduction half-reactions. In some cases it may be necessary to multiply each half-reaction by a factor to ensure that the numbers of electrons released and used match.

6.2 Standard potentials and spontaneity

KEY POINT A reaction is thermodynamically favourable (spontaneous) in the sense $K > 1$, if $E^\ominus > 0$, where $E^\ominus$ is the difference of the standard potentials corresponding to the half-reactions into which the overall reaction may be divided.

Thermodynamic arguments can be used to identify which reactions are spontaneous (i.e. have a natural tendency to occur). The thermodynamic criterion of spontaneity is that, at constant temperature and pressure, the reaction Gibbs energy change, $\Delta_r G$, is negative. It is usually sufficient to consider the standard reaction Gibbs energy, $\Delta_r G^\ominus$, which is related to the equilibrium constant through

$$\Delta_r G^\ominus = -RT \ln K \quad (6.1)$$

A negative value of $\Delta_r G^\ominus$ corresponds to $K > 1$ and therefore to a 'favourable' reaction in the sense that the products dominate the reactants at equilibrium. It is important to realize, however, that $\Delta_r G$ depends on the composition and that all reactions ultimately become spontaneous (i.e. have $\Delta_r G < 0$) under appropriate conditions. This is another way of saying that no equilibrium constant has an *infinite* value.

EXAMPLE 6.1 Combining half-reactions

Write a balanced equation for the oxidation of Fe^{2+} by permanganate ions (MnO_4^-) in acid solution.

Answer Balancing redox reactions often requires additional attention to detail because species other than products and reactants, such as electrons and hydrogen ions, often need to be considered. A systematic approach is as follows:

(1) Write the unbalanced half-reactions for the two species as reductions.
(2) Balance the elements other than hydrogen.
(3) Balance O atoms by adding H_2O to the other side of the arrow.
(4) If the solution is acidic, balance the H atoms by adding H^+; if the solution is basic, balance the H atoms by adding OH^- to one side and H_2O to the other.
(5) Balance the charge by adding e^-.
(6) Multiply each half-reaction by a factor to ensure that the numbers of e^- match.
(7) Combine the two half-reactions by subtracting the one containing the most reducing species from that containing the most oxidizing species. Cancel redundant terms.

The half-reaction for the reduction of Fe^{3+} is straightforward as it involves only the balance of charge:

$$Fe^{3+}(aq) + e^- \rightarrow Fe^{2+}(aq)$$

The unbalanced half-reaction for the reduction of MnO_4^- is

$$MnO_4^-(aq) \rightarrow Mn^{2+}(aq)$$

Balance the O with H_2O:

$$MnO_4^-(aq) \rightarrow Mn^{2+}(aq) + 4H_2O(l)$$

Balance the H with $H^+(aq)$:

$$MnO_4^-(aq) + 8H^+(aq) \rightarrow Mn^{2+}(aq) + 4H_2O(l)$$

Balance the charge with e^-:

$$MnO_4^-(aq) + 8H^+(aq) + 5e^- \rightarrow Mn^{2+}(aq) + 4H_2O(l)$$

To balance the number of electrons in the two half-reactions, the first is multiplied by 5. Then subtracting the iron half-reaction from the permanganate half-reaction and rearranging so that all stoichiometric coefficients are positive gives

$$MnO_4^-(aq) + 8H^+(aq) + 5Fe^{2+}(aq) \rightarrow Mn^{2+}(aq) + 5Fe^{3+}(aq) + 4H_2O(l)$$

Self-test 6.1 Use reduction half-reactions to write a balanced equation for the oxidation of zinc metal by permanganate ions in acid solution.

A BRIEF COMMENT

Standard conditions are all substances at 100 kPa pressure (1 bar) and unit activity. For reactions involving H^+ ions, standard conditions correspond to pH = 0, which is approximately 1 mol dm^{-3} [H^+]. Pure solids and liquids have unit activity. Although we use ν (nu) for the stoichiometric coefficient of the electron, electrochemical equations in inorganic chemistry are also commonly written with n in its place; we use ν to emphasize that it is a dimensionless number, not an amount in moles.

Because the overall chemical equation is the difference of two reduction half-reactions, the standard Gibbs energy of the overall reaction is the difference of the standard Gibbs energies of the two half-reactions. However, because reduction half-reactions always occur in pairs in any actual chemical reaction, only the difference in their standard Gibbs energies has any significance. Therefore, we can choose one half-reaction to have $\Delta_r G^\ominus = 0$, and report all other values relative to it. By convention, the specially chosen half-reaction is the reduction of hydrogen ions at pH = 0, 1 bar H_2:

$$H^+(aq) + e^- \rightarrow \tfrac{1}{2} H_2(g) \quad \Delta_r G^\ominus = 0$$

at all temperatures.

A BRIEF ILLUSTRATION

The standard Gibbs energy for the reduction of Zn^{2+} ions is found by determining experimentally that

$$Zn^{2+}(aq) + H_2(g) \rightarrow Zn(s) + 2H^+(aq) \quad \Delta_r G^\ominus = +147 \text{ kJ mol}^{-1}$$

Then, because the H^+ reduction half-reaction makes zero contribution to the reaction Gibbs energy (according to our convention), it follows that

$$Zn^{2+}(aq) + 2e^- \rightarrow Zn(s) \quad \Delta_r G^\ominus = +147 \text{ kJ mol}^{-1}$$

Standard reaction Gibbs energies may be measured by setting up a **galvanic cell**—an electrochemical cell in which a chemical reaction is used to generate an electric current, in which the reaction driving the electric current through the external circuit is the reaction of interest (Fig. 6.1). The potential difference between the two electrodes is then measured. The **cathode** is the electrode at which reduction occurs and the **anode** is the site of oxidation. In practice, we must ensure that the cell is acting reversibly in a thermodynamic sense, which means that the potential difference must be measured with no current flowing. If desired, the measured potential difference can be converted to a reaction Gibbs energy by using $\Delta_r G = -\nu F E$, where ν is the stoichiometric coefficient of the electrons transferred when the half-reactions are combined and F is Faraday's constant ($F = 96{,}484$ C mol^{-1}). Tabulated values, normally for standard conditions, are

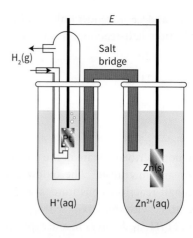

FIGURE 6.1 A schematic diagram of a galvanic cell. The standard potential, $E_{cell}^\ominus$, is the potential difference when the cell is not generating current and all the substances are in their standard states.

usually kept in the units in which they were measured; namely, volts (V).

The potential that corresponds to the $\Delta_r G^\ominus$ of a half-reaction is written $E^\ominus$, with

$$\Delta_r G^\ominus = -\nu F E^\ominus \tag{6.2}$$

in which $\Delta_r G^\ominus$ is expressed in J mol^{-1} for $F = 96{,}485$ C mol^{-1}. The potential $E^\ominus$ is called the **standard potential** (or 'standard reduction potential', to emphasize that by convention, the half-reaction is a reduction and is written with the oxidized species and electrons on the left). Because $\Delta_r G^\ominus$ for the reduction of H^+ is arbitrarily set at zero, the standard potential of the H^+/H_2 couple is also zero at all temperatures:

$$H^+(aq) + e^- \rightarrow \tfrac{1}{2} H_2(g) \quad E^\ominus(H^+, H_2) = 0$$

The standard potential of the H^+/H_2 couple, adopted by convention as the reference potential for other redox couples, is referred to as the **standard hydrogen electrode** (SHE). Although this standard reference is often assumed when displaying data, in practice other reference electrodes are used for experiments (a real H^+/H_2 electrode is rarely practical) and potential data presented in publications are defined further by stipulating 'V vs SHE' (or 'V vs X') X being a more convenient reference, to provide clarity.

A BRIEF ILLUSTRATION

For the Zn^{2+}/Zn couple, for which $\nu = 2$, it follows from the measured value of $\Delta_r G^\ominus$ that at 25°C:

$$Zn^{2+}(aq) + 2e^- \rightarrow Zn(s)$$
$$E^\ominus(Zn^{2+}, Zn) = -[147/(2 \times 96.484)] \text{ V} = -0.76 \text{ V}$$

Because the standard reaction Gibbs energy is the difference of the $\Delta_r G^\ominus$ values for the two contributing half-reactions, $E_{cell}^\ominus$ for an overall reaction is also the difference of the two standard potentials of the reduction half-reactions into which the overall reaction can be divided. Thus, from the half-reactions given above it follows that the difference is

$$2H^+(aq) + Zn(s) \rightarrow Zn^{2+}(aq) + H_2(g) \quad E^\ominus_{cell} = +0.76\,V$$

Note that the $E^\ominus$ values for couples (and their half-reactions) are called standard potentials and that their difference is denoted $E^\ominus_{cell}$ and called the **standard cell potential**. The consequence of the negative sign in eqn. 6.2 is that a reaction is favourable (in the sense $K > 1$) if the corresponding standard cell potential is positive. Because $E^\ominus > 0$ for the reaction in the illustration ($E^\ominus = +0.76\,V$), we know that zinc has a thermodynamic tendency to reduce H$^+$ ions under standard conditions (aqueous, pH = 0, and Zn^{2+} at unit activity); that is, zinc metal dissolves in acids. The same is true for any metal that has a couple with a negative standard potential, although the reactivity depends also on kinetic factors.

A NOTE ON GOOD PRACTICE

The cell potential used to be called (and in practice is still widely called) the electromotive force (emf). However, a potential is not a force, and IUPAC favours the name 'cell potential'.

EXAMPLE 6.2 Calculating a standard cell potential

Use the following standard potentials to calculate the standard potential of a copper–zinc cell.

$$Cu^{2+}(aq) + 2e^- \rightarrow Cu(s) \quad E^\ominus(Cu^{2+},Cu) = +0.34\,V$$

$$Zn^{2+}(aq) + 2e^- \rightarrow Zn(s) \quad E^\ominus(Zn^{2+},Zn) = -0.76\,V$$

Answer For this calculation we note from the standard potentials that Cu^{2+} is the more oxidizing species (the couple with the higher potential), and will be reduced by the species with the lower potential (Zn in this case). The spontaneous reaction is therefore Cu^{2+}(aq) + Zn(s) → Zn^{2+}(aq) + Cu(s), and the cell potential is the difference of the two half-reactions (copper − zinc),

$$E^\ominus_{cell} = E^\ominus(Cu^{2+},Cu) - E^\ominus(Zn^{2+},Zn)$$
$$= +0.34\,V - (-0.76\,V) = +1.10\,V$$

The cell will produce a potential difference of 1.1 V (under standard conditions).

Self-test 6.2 Is copper metal expected to react with dilute hydrochloric acid?

A battery based on the Cu-Zn cell is limited to single use because the products are not easily restored to their original state. **Rechargeable batteries** are devices in which after discharge, the two half-reactions can be driven in the opposite directions to restore the original energized state, using mains electricity or solar energy. A familiar example is the nickel cadmium (NiCad) battery which harnesses the energy released when Cd metal (acting as one electrode) is oxidized by Ni(O)OH (layered on Ni metal) according to the two half-reactions

$$Cd(s) + 2\,OH^-(aq) \rightarrow Cd(OH)_2(s) + 2e^-$$

$$Ni(O)OH(s) + 2\,H_2O(l) + 2e^- \rightarrow Ni(OH)_2(s) + 2\,OH^-(aq)$$

A NiCad battery produces a voltage of approximately 1.5 V. The recharging reactions are simply the reverse of those shown and driven by connecting a low-voltage DC source for a sufficient period of time.

A BRIEF COMMENT

Although *anode* and *cathode* are used to indicate electrodes carrying out oxidation and reduction, an ambiguity arises for rechargeable batteries since the roles are reversed for discharging and recharging cycles. To avoid confusion, *anode* and *cathode* refer to the roles of each electrode during normal use (discharge): thus in a NiCad battery, the Cd electrode is regarded as the anode while the Ni-based electrode is the cathode.

Combustion is a familiar type of redox reaction, and the energy that is released can be exploited in heat engines. A **fuel cell** converts a chemical fuel directly into electrical power (Box 6.1).

BOX 6.1 What are fuel cells?

A fuel cell converts a chemical fuel, such as hydrogen (used for larger power requirements) or methanol (a convenient fuel for small applications), directly into electrical power, using O_2 or air as the oxidant. As power sources, fuel cells offer several advantages over rechargeable batteries or combustion engines, and their use is steadily increasing. The materials used in fuel cells are described in Section 30.5. Compared to batteries, which have to be replaced or recharged over a significant period of time, a fuel cell operates as long as fuel is supplied. Furthermore, a fuel cell does not contain large amounts of environmental contaminants such as Ni and Cd, although relatively small amounts of Pt and other metals are required as electrocatalysts. The operation of a fuel cell is more efficient than combustion devices, with near-quantitative conversion of fuel to H_2O and (for methanol) CO_2. Fuel cells are also much less polluting, because nitrogen oxides are not produced at the relatively low temperatures that are used. Because an individual cell potential is less than about 1 V, fuel cells are connected in series known as 'stacks' in order to produce a useful voltage.

For H_2, important classes of fuel cell are the **proton-exchange membrane fuel cell** (PEMFC), the **alkaline fuel cell** (AFC), and the **solid oxide fuel cell** (SOFC) (Section 30.5), which differ in their mode of electrode reactions, chemical charge transfer, and operational temperature. These fuel cells have direct counterparts in electrolyser cells that perform the opposite reaction and are described in more detail in Section 11.4. For methanol, the standard example is the **direct methanol fuel cell** (DMFC). Details are provided in the table.

Fuel cell	Reaction at anode	Electrolyte	Transfer ion	Reaction at cathode	Temp. range/°C	Pressure/ atm	Efficiency/ per cent
PEMFC	$H_2 \rightarrow 2H^+ + 2e^-$	H^+-conducting polymer (PEM)	H^+	$2H^+ + \tfrac{1}{2}O_2 + 2e^- \rightarrow H_2O$	80–100	1–8	35–40
AFC	$H_2 \rightarrow 2H^+ + 2e^-$	Aqueous alkali	OH^-	$H_2O + \tfrac{1}{2}O_2 + 2e^- \rightarrow 2OH^-$	80–250	1–10	50–60
SOFC	$H_2 + O^{2-} \rightarrow H_2O + 2e^-$	Solid oxide	O^{2-}	$\tfrac{1}{2}O_2 + 2e^- \rightarrow O^{2-}$	600–1000	1	50–60
DMFC	$CH_3OH + H_2O \rightarrow CO_2 + 6H^+ + 6e^-$	H^+-conducting polymer	H^+	$2H^+ + \tfrac{1}{2}O_2 + 2e^- \rightarrow H_2O$	0–40	1	20–40

The basic principles of fuel cells are illustrated by a PEMFC (Fig. B6.1), which operates at modest temperatures (80–100°C) and is suited as an on-board power supply for road vehicles. At the anode, a continuous supply of H_2 is oxidized and the resulting H^+ ions, the chemical charge carriers, pass through a membrane to the cathode at which O_2 is reduced to H_2O; this process produces a flow of electrons from anode to cathode (the current) which is directed through the load (which is typically an electric motor). The anode (the site of H_2 oxidation) and the cathode (the site of O_2 reduction) are both loaded with a Pt catalyst to obtain efficient electrochemical conversions of fuel and oxidant.

The major factor limiting the efficiency of PEM and other fuel cells is the sluggish reduction of O_2 at the cathode, which involves expenditure of a significant *overpotential* (see Section 6.16) to drive this reaction at a practical rate. The operating voltage is usually about 0.7 V. The membrane is composed of a H^+-conducting polymer, sodium perfluorosulfonate (invented by Du Pont and known commercially as Nafion®).

An AFC is more efficient than a PEMFC because the reduction of O_2 at the Pt cathode is much easier under alkaline conditions. Hence the operating voltage is typically greater than about 0.8 V. The membrane of the PEMFC is now replaced by a pumped flow of hot aqueous alkali between the two electrodes. A long-established technology, AFCs were used to provide power for the pioneering Apollo spacecraft Moon missions. An SOFC operates at much higher temperatures (600–1000°C) and is used to provide electricity and heating in buildings (in the arrangement called 'combined heat and power', CHP).

Methanol is used as a fuel in either of two ways. One exploits methanol as an 'H_2 carrier', because the reforming reaction (see Section 11.4) is used to generate H_2 which is then supplied *in situ* to a normal hydrogen fuel cell as mentioned above. This indirect method avoids the need to store H_2 under pressure. Alternatively, methanol is oxidized directly in a DMFC which incorporates anode and cathode, each loaded with Pt or a Pt alloy, and a PEM. The methanol is supplied to the anode as an aqueous solution (at 1 mol dm^{-3}).

The DMFC is particularly suited for small low-power devices such as mobile phones and portable electronic processors, and it provides a promising alternative to the Li-ion battery. The principal disadvantage of the DMFC is its relatively low efficiency. This inefficiency arises from two factors that lower the operating voltage: the sluggish kinetics at the anode (oxidation of CH_3OH to CO_2 and H_2O) in addition to the poor cathode kinetics already mentioned, and transfer of methanol across the membrane to the cathode ('crossover') which occurs because methanol permeates the hydrophilic PEM easily. A 50/50 Pt/Ru mixture supported on carbon is used as the anode catalyst to improve the rate of methanol oxidation.

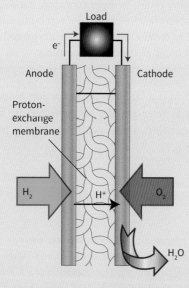

FIGURE B6.1 Schematic diagram of a proton-exchange membrane (PEM) fuel cell. The anode and cathode are loaded with a catalyst (Pt) to convert fuel (H_2) and oxidant (O_2) into H^+ and H_2O, respectively. The membrane (usually a material called Nafion®) allows the H^+ ions produced at the anode to be transferred to the cathode.

6.3 Trends in standard potentials

KEY POINT The atomization and ionization of a metal and the hydration enthalpy of its ions all contribute to the value of the standard potential.

The factors that contribute to the standard potential of the couple M^+/M can be identified by consideration of a thermodynamic cycle and the corresponding changes in Gibbs energy that contribute to the overall reaction

$$M^+(aq) + \tfrac{1}{2}H_2(g) \rightarrow H^+(aq) + M(s)$$

The thermodynamic cycle shown in Fig. 6.2 has been simplified by ignoring the reaction entropy, which is largely independent of the identity of M. The entropy contribution $T\Delta S^\ominus$ lies in the region of -20 to $-40\,\text{kJ}\,\text{mol}^{-1}$, which is small in comparison with the reaction enthalpy, the difference between the standard enthalpies of formation of $H^+(aq)$ and $M^+(aq)$. In this analysis we use the absolute values of the enthalpies of formation of M^+ and H^+, not the values based on the convention $\Delta_f H^\ominus(H^+,aq) = 0$. Thus, we use $\Delta_f H^\ominus(H^+,aq) = +445\,\text{kJ}\,\text{mol}^{-1}$, which is obtained by considering the formation of an H atom from $\tfrac{1}{2}H_2(g)$ ($+218\,\text{kJ}\,\text{mol}^{-1}$), ionization to $H^+(g)$ ($+1312\,\text{kJ}\,\text{mol}^{-1}$), and hydration of $H^+(g)$ (approximately $-1085\,\text{kJ}\,\text{mol}^{-1}$).

The analysis of the cell potential into its thermodynamic contributions allows us to account for trends in the standard potentials. For instance, the variation of standard potential down Group 1 seems contrary to expectation based on electronegativities insofar as Cs^+/Cs ($\chi = 0.79$, $E^\ominus = -2.94\,\text{V}$) has a less negative standard potential than Li^+/Li ($\chi = 0.98$, $E^\ominus = -3.04\,\text{V}$), despite Li having a higher electronegativity than Cs. Lithium has a higher enthalpy of sublimation and ionization energy than Cs, and in isolation this difference would imply a less negative standard potential as formation of the ion is less favourable. However, Li^+ has a large negative

TABLE 6.1 Thermodynamic contributions to $E^\ominus$ for a selection of metals at 298 K

	Li	Na	Cs	Ag
$\Delta_{sub}H^\ominus/(\text{kJ}\,\text{mol}^{-1})$	+161	+109	+79	+284
$I/(\text{kJ}\,\text{mol}^{-1})$	526	502	382	735
$\Delta_{hyd}H^\ominus/(\text{kJ}\,\text{mol}^{-1})$	−520	−406	−264	−468
$\Delta_f H^\ominus(M^+,aq)/(\text{kJ}\,\text{mol}^{-1})$	+167	+206	+197	+551
$\Delta_r H^\ominus/(\text{kJ}\,\text{mol}^{-1})$	+278	+240	+248	−106
$T\Delta_r S^\ominus/(\text{kJ}\,\text{mol}^{-1})$	−16	−22	−34	−29
$\Delta_r G^\ominus/(\text{kJ}\,\text{mol}^{-1})$	+294	+262	+282	−77
$E^\ominus/\text{V}$	−3.04	−2.71	−2.92	+0.80

$\Delta_f H^\ominus(H^+,aq) = +445\,\text{kJ}\,\text{mol}^{-1}$

enthalpy of hydration, which results from its small size (its ionic radius is 90 pm) compared with Cs^+ (181 pm) and its consequent strong electrostatic interaction with water molecules. Overall, the favourable enthalpy of hydration of Li^+ outweighs terms relating to the formation of $Li^+(g)$ and gives rise to a more negative standard potential. The less negative standard potential for Na^+/Na ($-2.71\,\text{V}$) in comparison with the rest of Group 1 (close to $-2.9\,\text{V}$) can be explained in terms of a combination of a fairly high sublimation enthalpy and moderate hydration enthalpy (Table 6.1).

The value of $E^\ominus(Na^+, Na) = -2.71\,\text{V}$ may also be compared with that for $E^\ominus(Ag^+, Ag) = +0.80\,\text{V}$. The (6-coordinate) ionic radii of these ions ($r_{Na^+} = 116\,\text{pm}$ and $r_{Ag^+} = 129\,\text{pm}$) are similar, and consequently their ionic hydration enthalpies are similar too. However, the much higher enthalpy of sublimation of silver and its higher ionization energy result in a positive standard potential. This difference is reflected in the very different behaviour of the metals when treated with a dilute acid: sodium reacts and dissolves vigorously producing H_2, whereas silver is unreactive. Similar arguments can be used to explain many of the trends observed in the standard potentials given in Table 6.2. For instance, the positive potentials characteristic of the noble metals result in large part from their very high sublimation enthalpies.

A NOTE ON GOOD PRACTICE

Always include the sign of a reduction potential, even when it is positive.

6.4 The electrochemical series

KEY POINTS The oxidized member of a couple is a strong oxidizing agent if $E^\ominus$ is positive and large; the reduced member is a strong reducing agent if $E^\ominus$ is negative and large.

A negative standard potential ($E^\ominus < 0$) signifies a couple in which the reduced species (the Zn in Zn^{2+}/Zn) is a reducing

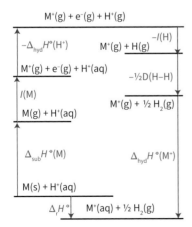

FIGURE 6.2 A thermodynamic cycle showing the properties that contribute to the standard potential of a metal couple. Endothermic processes are drawn with upward-pointing arrows, and exothermic contributions with downward pointing arrows.

TABLE 6.2 Selected standard potentials at 298 K; further values are included in Resource section 3

Couple	$E^\ominus$/V
$F_2(g) + 2\,e^- \rightarrow 2\,F^-(aq)$	+2.87
$Ce^{4+}(aq) + e^- \rightarrow Ce^{3+}(aq)$	+1.76
$MnO_4^-(aq) + 8\,H^+(aq) + 5\,e^- \rightarrow Mn^{2+}(aq) + 4\,H_2O(l)$	+1.51
$Cl_2(g) + 2\,e^- \rightarrow 2\,Cl^-(aq)$	+1.36
$O_2(g) + 4\,H^+(aq) + 4\,e^- \rightarrow 2\,H_2O(l)$	+1.23
$[IrCl_6]^{2-}(aq) + e^- \rightarrow [IrCl_6]^{3-}(aq)$	+0.87
$Fe^{3+}(aq) + e^- \rightarrow Fe^{2+}(aq)$	+0.77
$[PtCl_4]^{2-}(aq) + 2\,e^- \rightarrow Pt(s) + 4\,Cl^-(aq)$	+0.60
$I_3^-(aq) + 2\,e^- \rightarrow 3\,I^-(aq)$	+0.54
$[Fe(CN)_6]^{3-}(aq) + e^- \rightarrow [Fe(CN)_6]^{4-}(aq)$	+0.36
$AgCl(s) + e^- \rightarrow Ag(s) + Cl^-(aq)$	+0.22
$2\,H^+(aq) + 2\,e^- \rightarrow H_2(g)$	0
$AgI(s) + e^- \rightarrow Ag(s) + I^-(aq)$	−0.15
$Zn^{2+}(aq) + 2\,e^- \rightarrow Zn(s)$	−0.76
$Al^{3+}(aq) + 3\,e^- \rightarrow Al(s)$	−1.68
$Ca^{2+}(aq) + 2\,e^- \rightarrow Ca(s)$	−2.84
$Li^+(aq) + e^- \rightarrow Li(s)$	−3.04

agent for H^+ ions under standard conditions in aqueous solution. That is, if $E^\ominus$ (Ox, Red) < 0, then the substance 'Red' is a strong enough reducing agent to reduce H^+ ions (in the sense that $K > 1$ for the reaction).

Ox / Red couple with strongly positive $E^\ominus$

[Ox is strongly oxidizing]

⋮

Ox / Red couple with strongly negative $E^\ominus$

[Red is strongly reducing]

The list of $E^\ominus$ values given in Table 6.2 is arranged in the order of the **electrochemical series**. An important feature of the electrochemical series is that the reduced member of a couple has a thermodynamic tendency to reduce the oxidized member of any couple that lies above it in the series. Note that the classification refers only to the thermodynamic aspect of the reaction—its spontaneity under standard conditions and the value of K, not its rate. Thus even reactions that are found to be thermodynamically favourable from the electrochemical series may not progress, or progress only extremely slowly, if the kinetics of the process are unfavourable.

EXAMPLE 6.3 Using the electrochemical series

Among the couples in Table 6.2 is the permanganate ion, MnO_4^-, the common analytical reagent used in redox titrations of iron. Which of the ions Fe^{2+}, Cl^-, and Ce^{3+} can permanganate oxidize in acidic solution?

Answer We need to note that a reagent that is prone to oxidation by MnO_4^- ions must be the reduced form of a redox couple having a more negative standard potential than the couple MnO_4^-/Mn^{2+}. The standard potential of the couple MnO_4^-/Mn^{2+} in acidic solution is +1.51 V. The standard potentials of Fe^{3+}/Fe^{2+}, Cl_2/Cl^-, and Ce^{4+}/Ce^{3+} are +0.77, +1.36, and +1.76 V, respectively. It follows that MnO_4^- ions are sufficiently strong oxidizing agents in acidic solution (pH = 0) to oxidize Fe^{2+} and Cl^-, which have less positive standard potentials. Permanganate ions cannot oxidize Ce^{3+}, which has a more positive standard potential. It should be noted that the presence of other species in the solution can modify the potentials and the conclusions: this variation with conditions is particularly important in the case of H^+ ions, and the influence of pH is discussed in Section 6.6. The ability of MnO_4^- ions to oxidize Cl^- means that HCl cannot be used to acidify redox reactions involving permanganate, but instead H_2SO_4 is used.

Self-test 6.3 Another common analytical oxidizing agent is an acidic solution of dichromate ions, $Cr_2O_7^{2-}$, for which $E^\ominus(Cr_2O_7^{2-}, Cr^{3+}) = +1.38$ V. Is the solution useful for a redox titration of Fe^{2+} to Fe^{3+}? Could there be a side reaction when Cl^- is present?

6.5 The Nernst equation

KEY POINT The cell potential at an arbitrary composition of the reaction mixture is given by the Nernst equation.

To judge the tendency of a reaction to run in a particular direction at an arbitrary composition, we need to know the sign and value of $\Delta_r G$ at that composition. For this information we use the thermodynamic result that

$$\Delta_r G = \Delta_r G^\ominus + RT \ln Q \qquad (6.3a)$$

where Q is the reaction quotient:[1]

$$a\mathrm{Ox}_A + b\mathrm{Red}_B \rightarrow a'\mathrm{Red}_A + b'\mathrm{Ox}_B \quad Q = \frac{[\mathrm{Red}_A]^{a'}[\mathrm{Ox}_B]^{b'}}{[\mathrm{Ox}_A]^{a}[\mathrm{Red}_B]^{b}}$$

(6.3b)

The reaction quotient has the same form as the equilibrium constant K, but the concentrations refer to an arbitrary stage of the reaction; at equilibrium, $Q = K$. When evaluating Q and K, the quantities in square brackets are to be interpreted as the numerical values of the molar concentrations. Both Q and K are therefore dimensionless quantities. The reaction is spontaneous at an arbitrary stage if $\Delta_r G < 0$. This criterion can be expressed in terms of the potential of the corresponding cell by substituting $E_{cell} = -\Delta_r G/\nu F$ and $E_{cell}^\ominus = -\Delta_r G^\ominus/\nu F$ into eqn. 6.3a, which gives the **Nernst equation**:

$$E_{cell} = E_{cell}^\ominus - \frac{RT}{\nu F} \ln Q \qquad (6.4)$$

---

[1] For reactions involving gas-phase species, the molar concentrations of the latter are replaced by partial pressures relative to $p^\ominus = 1$ bar.

TABLE 6.3 The relation between K and $E^\ominus$

$E^\ominus/V$	K
+2	10^{34}
+1	10^{17}
0	1
−1	10^{-17}
−2	10^{-34}

A reaction is spontaneous if, under the prevailing conditions, $E_{cell} > 0$, for then $\Delta_r G < 0$. At equilibrium $E_{cell} = 0$ and $Q = K$, so eqn. 6.4 implies the following very important relation between the standard potential of a cell and the equilibrium constant of the cell reaction at a temperature T:

$$\ln K = \frac{\nu F E^\ominus_{cell}}{RT} \tag{6.5}$$

Table 6.3 lists the values of K that correspond to cell potentials in the range −2 to +2 V, with $\nu = 1$ and at 25°C. The table shows that, although electrochemical data are often compressed into the range −2 to +2 V, that narrow range corresponds to 68 orders of magnitude in the value of the equilibrium constant for $\nu = 1$.

If we regard the cell potential E_{cell} as the difference of two reduction potentials, just as $E^\ominus_{cell}$ is the difference of two *standard* reduction potentials, then the potential of each couple, E, that contributes to the cell reaction can be written like eqn. 6.4,

$$E = E^\ominus - \frac{RT}{\nu F} \ln Q \tag{6.6a}$$

but with

$$a\text{Ox} + \nu e^- \rightarrow a'\text{Red} \quad Q = \frac{[\text{Red}]^{a'}}{[\text{Ox}]^a} \tag{6.6b}$$

By convention, the electrons do not appear in the expression for Q.

The temperature dependence of a standard cell potential (eqns. 6.6a and 6.6b) provides a straightforward way to determine the standard entropy of many redox reactions. From eqn. 6.2, we can write

$$-\nu F E^\ominus_{cell} = \Delta_r G^\ominus = \Delta_r H^\ominus - T\Delta_r S^\ominus \tag{6.7a}$$

Then, if we suppose that $\Delta_r H^\ominus$ and $\Delta_r S^\ominus$ are independent of temperature over the small range usually of interest, it follows that

$$-\nu F E^\ominus_{cell}(T_2) - \{-\nu F E^\ominus_{cell}(T_1)\} = -(T_2 - T_1)\Delta_r S^\ominus$$

and therefore that

$$\Delta_r S^\ominus = \frac{-\nu F\{E^\ominus_{cell}(T_2) - E^\ominus_{cell}(T_1)\}}{T_2 - T_1} \tag{6.7b}$$

In other words, $\Delta_r S^\ominus$ is proportional to the slope of a graph of a plot of the standard cell potential against temperature.

The standard reaction entropy change $\Delta_r S^\ominus$ often reflects the change in solvation accompanying a redox reaction: for each half-reaction a positive entropy contribution is expected when the corresponding reduction results in a decrease in electric charge (solvent molecules are less tightly bound and more disordered). Conversely, a negative contribution is expected when there is an increase in charge. As mentioned in Section 6.3, entropy contributions to standard potentials are usually very similar when comparing redox couples involving the same change in charge.

EXAMPLE 6.4 **The potential generated by a fuel cell**

Calculate the cell potential (measured using an electrical load of such high resistance that negligible current flows) produced by a fuel cell in which the reaction is $2\,H_2(g) + O_2(g) \rightarrow 2\,H_2O(l)$ with H_2 and O_2 each at 25°C and a pressure of 100 kPa. (Note that in a working proton exchange membrane (PEM) fuel cell the temperature is usually 80–100°C to improve performance.)

Answer We note that under zero-current conditions, the cell potential is given by the difference of standard potentials of the two redox couples. For the reaction as stated, we write

Right: $O_2(g) + 4H^+(aq) + 4e^- \rightarrow 2H_2O(l)$ $\quad E^\ominus = +1.23\,V$
Left: $2H^+(aq) + 2e^- \rightarrow H_2(g)$ $\quad E^\ominus = 0$
Overall (right − left): $2H_2(g) + O_2(g) \rightarrow 2H_2O(l)$

The standard potential of the cell is therefore

$E^\ominus_{cell} = (+1.23\,V) - 0 = +1.23\,V$

The reaction is spontaneous as written, and the right-hand electrode is the cathode (the site of reduction).

Self-test 6.4 What potential difference would be produced in a fuel cell operating with oxygen and hydrogen with both gases at 5.0 bar?

Redox stability

When assessing the thermodynamic stability of a species in solution, we must bear in mind all possible reactants: the solvent, other solutes, the species itself, and dissolved oxygen. In the following discussion we focus on the types of reaction that result from the thermodynamic instability of a solute. We also comment briefly on kinetic factors, but the trends they show are generally less systematic than those shown by stabilities.

6.6 The influence of pH

KEY POINT Many redox reactions in aqueous solution involve transfer of H^+ as well as electrons, and the electrode potential therefore depends on the pH.

For many reactions in aqueous solution the electrode potential varies with pH because reduced species of a redox couple are usually much stronger Brønsted bases than the oxidized species. For a redox couple in which there is transfer of v_e electrons and v_H protons, it follows from eqn. 6.6b that

$$Ox + v_e e^- + v_H H^+ \rightleftharpoons RedH_{v_H} \quad Q = \frac{[RedH_{v_H}]}{[Ox][H^+]^{v_H}}$$

and

$$E = E^\ominus - \frac{RT}{v_e F} \ln \frac{[RedH_{v_H}]}{[Ox][H^+]^{v_H}}$$

$$= E^\ominus - \frac{RT}{v_e F} \ln \frac{[RedH_{v_H}]}{[Ox]} + \frac{v_H RT}{v_e F} \ln[H^+]$$

If the concentrations of Red and Ox are combined with $E^\ominus$ we define E' as

$$E' = E^\ominus - \frac{RT}{v_e F} \ln \frac{[RedH_{v_H}]}{[Ox]}$$

And by using $\ln[H^+] = \ln 10 \times \log[H^+]$ with $pH = -\log[H^+]$, the potential of the electrode can be written

$$E = E' - \frac{v_H RT \ln 10}{v_e F} pH \tag{6.8a}$$

At 25°C,

$$E = E' - \frac{(0.059 \text{ V})v_H}{v_e} pH \tag{6.8b}$$

That is, the potential decreases (becoming more negative) as the pH increases and the solution becomes more basic.

A BRIEF ILLUSTRATION

The half-reaction for the perchlorate/chlorate (ClO_4^-/ClO_3^-) couple is

$$ClO_4^-(aq) + 2H^+(aq) + 2e^- \rightarrow ClO_3^-(aq) + H_2O(l)$$

Therefore, whereas at pH = 0, $E^\ominus$ = +1.201 V, at pH = 7 the reduction potential for the ClO_4^-/ClO_3^- couple is 1.201 − (2/2)(7 × 0.059) V = +0.788 V. The perchlorate anion is a stronger oxidant under acid conditions.

The reduction potential for a half-reaction under a specified, non-standard set of conditions (typically pH ≠ 0) is known as the *formal* reduction potential (or simply formal potential), denoted $E^{\ominus\prime}$. Formal potentials are particularly useful in biochemical discussions because cell fluids are buffered near pH = 7. The condition pH = 7 (with unit activity for the other electroactive species present) corresponds to the so-called **biological standard state**; in biochemical contexts the formal potential is sometimes denoted E_{m7}, the 'm7' denoting the 'midpoint' potential at pH = 7.

A BRIEF ILLUSTRATION

To determine the formal potential of the H^+/H_2 couple at pH = 7.0, the other species being present in their standard states, we note that $E' = E^\ominus (H^+, H_2) = 0$. The reduction half-reaction is $2H^+(aq) + 2e^- \rightarrow H_2(g)$, so $v_e = 2$ and $v_H = 2$. The formal potential for the H^+/H_2 couple at pH = 7.0 is therefore

$$E' = 0 - (2/2)(0.059 \text{ V}) \times 7.0 = -0.41 \text{ V}$$

6.7 Reactions with water

Water may act as an oxidizing agent, when it is reduced to H_2:

$$H_2O(l) + e^- \rightarrow \tfrac{1}{2} H_2(g) + OH^-(aq)$$

For the equivalent reduction of hydronium ions in water at any pH (and partial pressure of H_2 of 1 bar) we have seen that the Nernst equation gives

$$H^+(aq) + e^- \rightarrow \tfrac{1}{2} H_2(g) \quad E = -0.059 \text{ V} \times pH \tag{6.9}$$

This is the reaction that chemists typically have in mind when they refer to 'the reduction of water'. Water may also act as a reducing agent, when it is oxidized to O_2:

$$2H_2O(l) \rightarrow O_2(g) + 4H^+(aq) + 4e^-$$

When the partial pressure of O_2 is 1 bar, the Nernst equation for the $O_2, 4H^+/2H_2O$ half-reaction becomes

$$E = 1.23 \text{ V} - (0.059 \text{ V} \times pH) \tag{6.10}$$

because $v_H/v_e = 4/4 = 1$. Both H^+ and O_2 therefore have the same pH dependence for their reduction half-reactions. The variation of these two potentials with pH is shown in Fig. 6.3.

(a) Oxidation by water

KEY POINT For metals with large, negative standard potentials, reaction with aqueous acids leads to the production of H_2 unless a passivating oxide layer is formed.

The reaction of a metal with water or aqueous acid is in fact the oxidation of the metal by water or hydrogen ions (see Section 6.4), because the overall reaction is one of the following processes (and their analogues for more highly charged metal ions):

$$M(s) + H_2O(l) \rightarrow M^+(aq) + \tfrac{1}{2} H_2(g) + OH^-(aq)$$

$$M(s) + H^+(aq) \rightarrow M^+(aq) + \tfrac{1}{2} H_2(g)$$

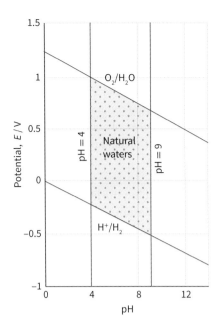

FIGURE 6.3 The variation of the reduction potentials of water with pH. The sloping lines defining the upper and lower limits of thermodynamic water stability are the potentials for the O_2/H_2O and H^+/H_2 couples, respectively. The central zone represents the stability range of natural waters.

These reactions are thermodynamically favourable when M is an s-block metal, a 3d-series metal from Group 3 to Group 10 (Sc, Ti, V, Cr, Mn, Fe, Co, Ni), or a lanthanoid. An example from Group 3 is

$$2Sc(s) + 6H^+(aq) \rightarrow 2Sc^{3+}(aq) + 3H_2(g)$$

When the potential for the reduction of a metal ion to the metal is negative, the metal should undergo oxidation in 1 M acid with the evolution of hydrogen.

Although the reactions of magnesium and aluminium with moist air are spontaneous, both metals can be used for years in the presence of water and oxygen. They survive because they are **passivated**, or protected against reaction, by an impervious film of oxide. Magnesium oxide and aluminium oxide both form a protective skin on the parent metal. A similar passivation occurs with iron, copper, and zinc. The process of 'anodizing' a metal, in which the metal is made an anode in an electrolytic cell, is one in which partial oxidation produces a smooth, hard passivating film on its surface. Anodization, using chromic or sulfuric acid as the electrolyte, is especially effective for the protection of aluminium or titanium by the formation of an inert and cohesive oxide layer.

Production of H_2 by electrolysis or photolysis of water is widely viewed as one of the renewable energy solutions for the future, and is discussed in more detail in Section 11.4.

(b) Reduction by water

KEY POINT Water can act as a reducing agent; that is, be oxidized by other species.

The strongly positive potential of the O_2, 4 H^+/2 H_2O couple (eqn. 6.10) shows that acidified water is a poor reducing agent except towards strong oxidizing agents. An example of the latter is $Co^{3+}(aq)$, for which $E^\ominus$ (Co^{3+}, Co^{2+}) = +1.92 V. It is reduced by water with the evolution of O_2, and Co^{3+} does not survive in aqueous solution:

$$4Co^{3+}(aq) + 2H_2O(l) \rightarrow 4Co^{2+}(aq) + O_2(g) + 4H^+(aq)$$
$$E^\ominus_{cell} = +0.69\,V$$

Because H^+ ions are produced in the reaction, lower acidity (higher pH) favours the evolution of O_2.

Only a few oxidizing agents (such as $Co^{3+}(aq)$) can oxidize water rapidly enough to give appreciable rates of O_2 evolution. Standard potentials higher than +1.23 V occur for several redox couples that are regularly used in aqueous solution, including Ce^{4+}/Ce^{3+} ($E^\ominus$ = +1.76 V), the acidified dichromate ion couple $Cr_2O_7^{2-}/Cr^{3+}$ ($E^\ominus$ = +1.38 V), and the acidified permanganate couple MnO_4^-/Mn^{2+}($E^\ominus$ = +1.51 V). The origin of the barrier to reaction is a kinetic one, stemming from the need to transfer four electrons and to form an O–O bond (the single bond that is initially formed being weak).

Given that the rates of redox reactions are often controlled by the slow rate at which an O–O bond can be formed, it remains a challenge for inorganic chemists to develop good catalysts for O_2 evolution in electrolytic or photoelectrolytic devices (Section 30.3). The importance of this process is not due to any large-scale commercial demand for O_2, but because of the desire to generate H_2 (a 'green' fuel) from water by electrolysis or photolysis. Sustaining life in a spacecraft or any other extraterrestrial environment also depends upon the efficient recycling of H_2O, H_2, and O_2 using electricity supplied by solar panels or a nuclear reactor. Existing catalysts include the relatively poorly understood coatings that are used in the anodes of cells for the commercial electrolysis of water. They also include the enzyme system found in the O_2 evolution apparatus of the plant photosynthetic centre. This system is based on a special cofactor containing four Mn atoms and one Ca atom (Section 26.10). Although Nature is elegant and efficient, it is also complex, and the photosynthetic process is only slowly being elucidated by biochemists and bioinorganic chemists. Significant progress in mimicking Nature's efficiency is being made using Ru, Ir, and Co complexes (Box 17.2).

(c) The stability field of water

KEY POINT The stability field of water shows the region of pH and reduction potential where couples are neither oxidized by nor reduce hydrogen ions.

A reducing agent that can reduce water to H_2 rapidly, or an oxidizing agent that can oxidize water to O_2 rapidly, cannot survive in aqueous solution. The **stability field** of water, which is shown in Fig. 6.3, is the range of values of potential and pH for which water is thermodynamically stable towards both oxidation and reduction.

The upper and lower boundaries of the stability field are identified by finding the dependence of E on pH for the relevant half-reactions. As we have seen above, both oxidation (to O_2) and reduction of water have the same pH dependence (a slope of -0.059 V when E is plotted against pH at 25°C), and the stability field is confined within the boundaries of a pair of parallel lines of that slope. Any species with a potential more negative than that given in eqn. 6.9 can reduce water (specifically, can reduce H^+) with the production of H_2; hence the lower line defines the low-potential boundary of the stability field. Similarly, any species with a potential more positive than that given in eqn. 6.10 can liberate O_2 from water, and the upper line gives the high-potential boundary. Couples that are thermodynamically unstable in water lie outside (above or below) the limits defined by the sloping lines in Fig. 6.3: species that are oxidized by water have potentials lying below the H_2 production line and species that are reduced by water have potentials lying above the O_2 production line.

The stability field in 'natural' water is represented by the addition of two vertical lines at pH = 4 and pH = 9, which mark the limits on pH that are commonly found in lakes and streams. A diagram like that shown in the illustration is known as a **Pourbaix diagram** and is widely used in environmental chemistry, as we shall see in Section 6.14.

6.8 Oxidation by atmospheric oxygen

KEY POINT The O_2 present in air and dissolved in water can oxidize metals and metal ions in solution.

The possibility of reaction between the solutes and dissolved O_2 must be considered when a solution is contained in an open beaker or is otherwise exposed to air. As an example, consider an aqueous solution containing Fe^{2+} in contact with an inert atmosphere such as N_2. Because $E^\ominus$ (Fe^{3+}, Fe^{2+}) = +0.77 V, which lies within the stability field of water, we expect Fe^{2+} to survive in water. Moreover, we can also infer that the oxidation of metallic iron by $H^+(aq)$ should not proceed beyond Fe(II), because further oxidation to Fe(III) is unfavourable (by 0.77 V) under standard conditions. However, the picture changes considerably in the presence of O_2. Many elements occur naturally as oxidized species, either as soluble oxoanions such as SO_4^{2-}, NO_3^-, and MoO_4^{2-} or as ores such as Fe_2O_3. In fact, Fe(III) is the most common form of iron in the Earth's crust, and most iron in sediments that have been deposited from aqueous environments is present as Fe(III). The reaction

$$4Fe^{2+}(aq) + O_2(g) + 4H^+(aq) \rightarrow 4Fe^{3+}(aq) + 2H_2O(l)$$

is the difference of the following two half-reactions:

$O_2(g) + 4H^+(aq) + 4e^- \rightarrow 2H_2O \quad E^\ominus = +1.23\,V$

$Fe^{3+}(aq) + e^- \rightarrow Fe^{2+}(aq) \quad E^\ominus = +0.77\,V$

which implies that $E^\ominus_{cell}$ = +0.46 V at pH = 0. The oxidation of Fe^{2+}(aq) by O_2 is therefore spontaneous (in the sense $K > 1$) at pH = 0 and also at higher pH, although Fe(III) aqua species are hydrolysed and are precipitated as 'rust' (Section 6.14).

EXAMPLE 6.5 Judging the importance of atmospheric oxidation

The oxidation of copper roofs to a green substance known as verdigris (typically 'basic copper carbonate') is an example of atmospheric oxidation in a damp environment. Estimate the potential for oxidation of copper metal by atmospheric O_2 in acid-to-neutral aqueous solution. Cu^{2+}(aq) is not deprotonated between pH = 0 and 7, so we may assume no H^+ ions are involved in the half-reaction.

Answer We need to consider the reaction between Cu metal and atmospheric O_2 in terms of the two relevant reduction half-reactions:

$O_2(g) + 4H^+(aq) + 4e^- \rightarrow 2H_2O \quad E = +1.23\,V - (0.059\,V) \times pH$

$Cu^{2+}(aq) + 2e^- \rightarrow Cu(s) \quad E^\ominus = +0.34\,V$

The difference is

$E_{cell} = 0.89\,V - (0.059\,V) \times pH$

Therefore, E_{cell} = +0.89 V at pH = 0 and +0.48 V at pH = 7, so atmospheric oxidation by the reaction

$2Cu(s) + O_2(g) + 4H^+(aq) \rightarrow 2Cu^{2+}(aq) + 2H_2O(l)$

has $K > 1$ in both neutral and acid environments. The familiar green surface is a passive layer of an almost impenetrable hydrated copper(II) carbonate, sulfate or, near the sea, chloride. These compounds are formed from oxidation in the presence of atmospheric CO_2, SO_2, or salt water, and the anion is also involved in the redox chemistry.

Self-test 6.5 The standard potential for the conversion of sulfate ions, SO_4^{2-}, to SO_2(aq) by the reaction $SO_4^{2-}(aq) + 4H^+(aq) + 2e^- \rightarrow SO_2(aq) + 2H_2O(l)$ is +0.16 V. What is the thermodynamically expected fate of SO_2 emitted into fog or clouds?

6.9 Disproportionation and comproportionation

KEY POINT Standard potentials can be used to define the inherent stability and instability of different oxidation states in terms of disproportionation and comproportionation.

Because $E^\ominus$ (Cu^+,Cu) = +0.52 V and $E^\ominus$ (Cu^{2+},Cu^+) = +0.16 V, and both potentials lie within the stability field of water, Cu^+ ions neither oxidize nor reduce water. Nevertheless, Cu(I) is not stable in aqueous solution because it can undergo **disproportionation**—a redox reaction in which the oxidation number of an element is simultaneously raised and lowered. In other words, the element undergoing disproportionation serves as its own oxidizing and reducing agent:

$$2Cu^+(aq) \rightarrow Cu^{2+}(aq) + Cu(s)$$

This reaction is the difference of the following two half-reactions:

$$Cu^+(aq) + e^- \rightarrow Cu(s) \quad E^\ominus = +0.52\,V$$
$$Cu^{2+}(aq) + e^- \rightarrow Cu^+(aq) \quad E^\ominus = +0.16\,V$$

Because $E^\ominus_{cell} = 0.52\,V - 0.16\,V = +0.36\,V$ for the disproportionation reaction, $K = 1.3 \times 10^6$ at 298 K, so the reaction is highly favourable; i.e. the Cu^+ aqua ion is *inherently* unstable. Hypochlorous acid also undergoes disproportionation:

$$5HClO(aq) \rightarrow 2Cl_2(g) + ClO_3^-(aq) + 2H_2O(l) + H^+(aq)$$

This redox reaction is the difference of the following two half-reactions:

$$4HClO(aq) + 4H^+(aq) + 4e^- \rightarrow 2Cl_2(g) + 4H_2O(l)$$
$$E^\ominus = +1.63\,V$$
$$ClO_3^-(aq) + 5H^+(aq) + 4e^- \rightarrow HClO(aq) + 2H_2O(l)$$
$$E^\ominus = +1.43\,V$$

So overall $E^\ominus_{cell} = 1.63\,V - 1.43\,V = +0.20\,V$, and $K = 3 \times 10^{13}$ at 298 K.

EXAMPLE 6.6 Assessing the likelihood of disproportionation

Show that Mn(VI) is unstable with respect to disproportionation into Mn(VII) and Mn(II) in acidic aqueous solution.

Answer To answer this question we need to consider the two half-reactions, one an oxidation, the other a reduction, that involve the species Mn(VI). The overall reaction (noting, from Pauling's rules, Section 5.3, that the Mn(VI) oxoanion MnO_4^{2-} should be protonated at pH = 0)

$$5HMnO_4^-(aq) + 3H^+(aq) \rightarrow 4MnO_4^-(aq) + Mn^{2+}(aq) + 4H_2O(l)$$

is the difference of the following two half-reactions

$$HMnO_4^-(aq) + 7H^+(aq) + 4e^- \rightarrow Mn^{2+}(aq) + 4H_2O(l)$$
$$E^\ominus = +1.63\,V$$
$$4MnO_4^-(aq) + 4H^+(aq) + 4e^- \rightarrow 4HMnO_4^-(aq)$$
$$E^\ominus = +0.90\,V$$

The difference of the standard potentials is +0.73 V, so the disproportionation is essentially complete ($K = 10^{50}$ at 298 K). A practical consequence of the disproportionation is that high concentrations of $HMnO_4^-$ ions cannot be obtained in acidic solution, although MnO_4^{2-} is stabilized in basic solution.

Self-test 6.6 The standard potentials for the couples Fe^{2+}/Fe and Fe^{3+}/Fe^{2+} are −0.41 V and +0.77 V, respectively. Should we expect Fe^{2+} to disproportionate in aqueous solution?

In **comproportionation**, the reverse of disproportionation, two species with the same element in different oxidation states form a product in which the element is in an intermediate oxidation state. An example is

$$Ag^{2+}(aq) + Ag(s) \rightarrow 2Ag^+(aq) \quad E^\ominus_{cell} = +1.18\,V$$

The large positive potential indicates that Ag(II) and Ag(0) are completely converted to Ag(I) in aqueous solution ($K = 1 \times 10^{20}$ at 298 K).

6.10 The influence of complexation

KEY POINTS The formation of a more thermodynamically stable complex when the metal is in the higher oxidation state of a couple favours its oxidation and makes the standard potential more negative; the formation of a more stable complex when the metal is in the lower oxidation state of the couple favours its reduction and the standard potential becomes more positive.

The formation of metal complexes (see Chapter 7) affects standard potentials because the ability of a complex (ML) formed by coordination of a ligand (L) to accept or release an electron differs from that of the corresponding aqua ion (M).

$$M^{v+}(aq) + e^- \rightarrow M^{(v-1)+}(aq) \quad E^\ominus(M)$$
$$ML^{v+}(aq) + e^- \rightarrow ML^{(v-1)+}(aq) \quad E^\ominus(ML)$$

The change in standard potential for the ML redox couple relative to that of M reflects the degree to which the ligand L coordinates more strongly to the oxidized or reduced form of M. In certain cases the standard potential associated with particular oxidation states may be varied over more than two volts depending on the choice of ligand. For instance, the standard potential for one-electron reduction of Fe(III) complexes ranges between $E > +1\,V$ for L = bpy (**1**) to $E < -1\,V$ when L is the naturally occurring ligand known as enterobactin (Section 26.6). Complexes of Ru containing bpy-like ligands are used in dye-sensitized photovoltaic cells, and their reduction potentials can be tuned by placing different substituents on the organic rings.

1 2,2′-bipyridine (bpy)

A change in standard potential due to complexation is analysed by considering a generic thermodynamic cycle such as that shown in Fig. 6.4. Because the sum of reaction Gibbs energies round the cycle is zero, we can write

$$-FE^\ominus(M) - RT\ln K^{red} + FE^\ominus(ML) + RT\ln K^{ox} = 0 \quad (6.11)$$

where K^{ox} and K^{red} are equilibrium constants for L binding to M^{v+} and $M^{(v-1)+}$, respectively (of the form $K = [ML]/[M][L]$),

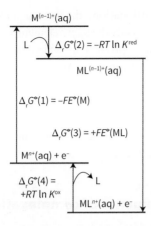

FIGURE 6.4 Thermodynamic cycle showing how the standard potential of the couple $M^{v+}/M^{(v-1)+}$ is altered by the presence of a ligand L.

and we have used $\Delta_r G^\circ = -RT \ln K$ in each case. This expression rearranges to

$$E^\ominus(M) - E^\ominus(ML) = \frac{RT}{F} \ln \frac{K^{ox}}{K^{red}} \quad (6.12a)$$

At 25°C, and with $\ln x = \ln 10 \log x$,

$$E^\ominus(M) - E^\ominus(ML) = (0.059\,V) \log \frac{K^{ox}}{K^{red}} \quad (6.12b)$$

Thus, every ten-fold increase in the equilibrium constant for ligand binding to M^{v+} compared to $M^{(v-1)+}$ decreases the reduction potential by 0.059 V.

A BRIEF ILLUSTRATION

The standard potential for the half-reaction $[Fe(CN)_6]^{3-}(aq) + e^- \rightarrow [Fe(CN)_6]^{4-}(aq)$ is 0.36 V; that is, 0.41 V more negative than that of the aqua redox couple $[Fe(OH_2)_6]^{3+}(aq) + e^- \rightarrow [Fe(OH_2)_6]^{2+}(aq)$. This equates to CN⁻ having a 10^7-fold greater affinity (in the sense $K^{ox} \approx 10^7 K^{red}$) for Fe(III) compared to Fe(II).

EXAMPLE 6.7 Interpreting potential data to identify bonding trends in complexes

Ruthenium is located immediately below iron in the periodic table. The following reduction potentials have been measured for species of Ru in aqueous solution. What do these values suggest when compared to their Fe counterparts?

$[Ru(OH_2)_6]^{3+} + e^- \rightarrow [Ru(OH_2)_6]^{2+}$ $E^\ominus = +0.25\,V$

$[Ru(CN)_6]^{3-} + e^- \rightarrow [Ru(CN)_6]^{4-}$ $E^\ominus = +0.80\,V$

Answer We can answer this question by noting that if complexation by a certain ligand causes the reduction potential of a metal ion to shift in a *positive* direction, then the new ligand must be stabilizing the reduced metal ion. In this case we see that CN⁻ stabilizes Ru(II) with respect to Ru(III). This behaviour is in stark contrast to the behaviour of Fe (see the preceding Brief

Illustration) where we noted that CN⁻ (an excellent σ-donor ligand) stabilizes Fe(III). As we will see later, the greater radial extension of 4d orbitals compared to 3d orbitals means that the ability of CN⁻ to use its empty 2π orbital and act as a π-acceptor ligand becomes more important when it is bound to Ru, in this case favouring the more electron-rich Ru(II) state.

Self-test 6.7 The ligand bpy (**1**) forms complexes with Ru(III) and Ru(II). The standard potential of the $[Ru(bpy)_3]^{3+}/[Ru(bpy)_3]^{2+}$ couple is +1.26 V. Does bpy bind preferentially to Ru(III) or Ru(II)? By how many orders of magnitude is the binding of three bpy to Ru(III) enhanced or decreased relative to the binding of six water molecules to Ru(III)?

6.11 The relation between solubility and standard potentials

KEY POINTS The standard cell potential can be used to determine the solubility product.

The solubility of sparingly soluble compounds is expressed by an equilibrium constant known as the **solubility product**, K_{sp}. The approach is analogous to that introduced above for relating complexation equilibria to standard potentials. For a compound MX_v that dissolves in water to give metal ions $M^{v+}(aq)$ and anions $X^-(aq)$, we write

$$M^{v+}(aq) + vX^-(aq) \rightleftharpoons MX_v(s) \quad K_{sp} = [M^{v+}][X^-]^v \quad (6.13)$$

To generate the overall (non-redox) solubility reaction we use the difference of the two reduction half-reactions

$M^{v+}(aq) + ve^- \rightarrow M(s)$ $E^\ominus(M^{v+}/M)$

$MX_v(s) + ve^- \rightarrow M(s) + v\,X^-(aq)$ $E^\ominus(MX_v/M, X^-)$

From which it follows that

$$\ln K_{sp} = \frac{vF\{E^\ominus(MX_v/M, X^-) - E^\ominus(M^{v+}/M)\}}{RT} \quad (6.14)$$

EXAMPLE 6.8 Determining a solubility product from standard potentials

The possibility of plutonium waste leaking from nuclear facilities is a serious environmental problem. Calculate the solubility product of $Pu(OH)_4$, based upon the following potentials measured in acid or basic solution. Hence, comment on the consequences of Pu(IV) waste leaking into environments of low pH as compared to high pH.

$Pu^{4+}(aq) + 4e^- \rightarrow Pu(s)$ $E^\ominus = -1.28\,V$

$Pu(OH)_4(s) + 4e^- \rightarrow Pu(s) + 4OH^-(aq)$ $E^\ominus = -2.06\,V$ at pH = 14

Answer We need to consider a thermodynamic cycle that combines the changes in Gibbs energy for the electrode reactions at pH = 0 and 14 using the potentials given, and the standard Gibbs energy for the reaction $Pu^{4+}(aq)$ with $OH^-(aq)$. The solubility product for $Pu(OH)_4$ is $K_{sp} = [Pu^{4+}][OH^-]^4$, so the corresponding Gibbs energy term is $-RT \ln K_{sp}$. For the thermodynamic cycle $\Delta G = 0$, so we obtain

$$-RT \ln K_{sp} = 4FE^{\ominus}(Pu^{4+}/Pu) - 4FE^{\ominus}(Pu(OH)_4/Pu)$$

and therefore

$$\ln K_{sp} = \frac{4F\{(-2.06\,V) - (-1.28\,V)\}}{RT}$$

It follows that $K_{sp} = 1.7 \times 10^{-53}$: $Pu(OH)_4$ is so insoluble that under basic conditions very little Pu^{4+} will escape into the environment.

Self-test 6.8 Given that the standard potential for the Ag^+/Ag couple is $+0.80\,V$, calculate the potential of the $AgCl/Ag,Cl^-$ couple under conditions of $[Cl^-] = 1.0\,mol\,dm^{-3}$, given that $K_{sp} = 1.77 \times 10^{-10}$.

Useful presentation of potential data

There are several useful diagrammatic summaries of the relative stabilities of different oxidation states in aqueous solution. 'Latimer diagrams' are useful for summarizing quantitative data for individual elements. 'Frost diagrams' are useful for the qualitative portrayal of the relative and inherent stabilities of oxidation states of a range of elements. Pourbaix (E–pH) diagrams display how reduction potentials depend on pH and are useful for predicting the predominant species existing under a particular set of conditions.

6.12 Latimer diagrams

In a **Latimer diagram** (also known as a *reduction potential diagram*) for an element, the numerical value of the standard potential (in volts) is written over a horizontal line (or arrow) connecting species with the element in different oxidation states. The most highly oxidized form of the element is on the left, and in species to the right the element is in successively lower oxidation states. Latimer diagrams summarize a great deal of information in a compact form and (as we explain) show the relationships between the various species in a particularly clear manner. They are used to provide standard potential data in Resource section 3.

(a) Construction

KEY POINTS In a Latimer diagram, oxidation numbers decrease from left to right and the numerical values of $E^{\ominus}$ in volts are written above the line joining the species involved in the couple.

The Latimer diagram for chlorine in acidic solution, for instance, is

$$ClO_4^- \xrightarrow{+1.20} ClO_3^- \xrightarrow{+1.18} HClO_2 \xrightarrow{+1.67} $$
$$+7 \qquad\qquad +5 \qquad\qquad +3$$
$$HClO \xrightarrow{+1.63} Cl_2 \xrightarrow{+1.36} Cl^-$$
$$+1 \qquad\qquad 0 \qquad\qquad -1$$

As in this example, oxidation numbers are sometimes written under (or over) the species. Conversion of a Latimer diagram to a half-reaction equation requires careful consideration of all species involved in the reaction, some of which are not included in the Latimer diagram (H^+ and H_2O). The procedure for balancing redox equations was shown in Section 6.1. The standard state for this couple includes the condition that pH = 0. For instance, the notation

$$HClO \xrightarrow{+1.63} Cl_2$$

denotes

$$2HClO(aq) + 2H^+(aq) + 2e^- \rightarrow Cl_2(g) + 2H_2O(l)$$
$$E^{\ominus} = +1.63\,V$$

Similarly,

$$ClO_4^- \xrightarrow{+1.20} ClO_3^-$$

denotes

$$ClO_4^-(aq) + 2H^+(aq) + 2e^- \rightarrow ClO_3^-(aq) + H_2O(l)$$
$$E^{\ominus} = +1.20\,V$$

Note that both of these half-reactions involve H^+ ions, and therefore the potentials depend on pH.

In basic aqueous solution (corresponding to pOH = 0 and therefore pH = 14), the Latimer diagram for chlorine is

$$ClO_4^- \xrightarrow{+0.37} ClO_3^- \xrightarrow{+0.30} ClO_2^- \xrightarrow{+0.68}$$
$$+7 \qquad\qquad +5 \qquad\qquad +3$$
$$ClO^- \xrightarrow{+0.42} Cl_2 \xrightarrow{+1.36} Cl^-$$
$$+1 \qquad\qquad 0 \qquad\qquad -1$$

The value for the Cl_2/Cl^- couple is the same as in acidic solution because its half-reaction does not involve the transfer of protons.

(b) Nonadjacent species

KEY POINT The standard potential of a couple that is the combination of two other couples is obtained by combining the standard Gibbs energies, not the standard potentials, of the half-reactions.

To derive the standard potential of a nonadjacent couple that involves two or more electrons, we cannot in general just add their standard potentials but must make use of eqn. 6.2 ($\Delta_r G^{\ominus} = -vFE^{\ominus}$) and the fact that the overall $\Delta_r G^{\ominus}$ for two successive steps a and b is the sum of the individual values:

$$\Delta_r G^{\ominus}(a+b) = \Delta_r G^{\ominus}(a) + \Delta_r G^{\ominus}(b)$$

To find the standard potential of the composite process, we convert the individual $E^{\ominus}$ values to $\Delta_r G^{\ominus}$ through multiplication by the relevant factor $-vF$, add them together, and then convert the sum back to $E^{\ominus}$ for the nonadjacent couple by division by $-vF$ for the overall electron transfer:

$$-vFE^{\ominus}(a+b) = -v(a)FE^{\ominus}(a) - v(b)FE^{\ominus}(b)$$

Because the factors $-F$ cancel and $v = v(a) + v(b)$, the net result is

$$E^{\ominus}(a+b) = \frac{v(a)E^{\ominus}(a) + v(b)E^{\ominus}(b)}{v(a) + v(b)} \quad (6.15)$$

A BRIEF ILLUSTRATION

To use the Latimer diagram to calculate the value of $E^{\ominus}$ for the ClO_2^-/Cl_2 couple in basic aqueous solution we note the following two standard potentials:

$$ClO_2^-(aq) + H_2O(l) + 2e^- \rightarrow ClO^-(aq) + 2OH^-(aq) \quad E^{\ominus}(a) = +0.68\,V$$
$$ClO^-(aq) + H_2O(l) + e^- \rightarrow \tfrac{1}{2}Cl_2(aq) + 2OH^-(aq) \quad E^{\ominus}(b) = +0.42\,V$$

Their sum,

$$ClO_2^-(aq) + 2H_2O(l) + 3e^- \rightarrow \tfrac{1}{2}Cl_2(g) + 4OH^-(aq)$$

is the half-reaction for the couple we require. We see that $v(a) = 2$ and $v(b) = 1$. It follows from eqn. 6.15 that the standard potential of the ClO_2^-/Cl^- couple is

$$E^{\ominus} = \frac{(2)(0.68\,V) + (1)(0.42\,V)}{3} = +0.59\,V$$

EXAMPLE 6.9 Identifying a tendency to disproportionate

A version of the Latimer diagram for oxygen at pH 0 is

$$O_2 \xrightarrow{+0.70} H_2O_2 \xrightarrow{+1.76} H_2O$$

Does hydrogen peroxide have a tendency to disproportionate in acid solution?

Answer We can approach this question by reasoning that if H_2O_2 is a stronger oxidant than O_2, then it should react with itself to produce O_2 by oxidation and 2 H_2O by reduction. The potential to the right of H_2O_2 is higher than that to its left, so we anticipate that H_2O_2 should disproportionate into its two neighbours under acid conditions. From the two half-reactions

$$2H^+(aq) + 2e^- + H_2O_2(aq) \rightarrow 2H_2O(l) \quad E^{\ominus} = +1.76\,V$$
$$O_2(g) + 2H^+(aq) + 2e^- \rightarrow H_2O_2(aq) \quad E^{\ominus} = +0.70\,V$$

we conclude that for the overall reaction

$$2H_2O_2(aq) \rightarrow 2H_2O(l) + O_2(g) \quad E^{\ominus} = +1.06\,V$$

and is spontaneous (in the sense $K > 1$).

Self-test 6.9 Use the following Latimer diagram (acid solution) to discuss whether (a) Pu(IV) disproportionates to Pu(III) and Pu(V) in aqueous solution; (b) Pu(V) disproportionates into Pu(VI) and Pu(IV).

$$\underset{+6}{PuO_2^{2+}} \xrightarrow{+1.02} \underset{+5}{PuO_2^+} \xrightarrow{+1.04} \underset{+4}{Pu^{4+}} \xrightarrow{+1.01} \underset{+3}{Pu^{3+}}$$

(c) Disproportionation

KEY POINT A species has a tendency to disproportionate into its two neighbours if the potential on the right of the species in a Latimer diagram is more positive than that on the left.

Consider the disproportionation

$$2M^+(aq) \rightarrow M(s) + M^{2+}(aq)$$

This reaction has $K > 1$ if $E^{\ominus} > 0$. To analyse this criterion in terms of a Latimer diagram, we express the overall reaction as the difference of two half-reactions:

$$M^+(aq) + e^- \rightarrow M(s) \quad E^{\ominus}(R)$$
$$M^{2+}(aq) + e^- \rightarrow M^+(aq) \quad E^{\ominus}(L)$$

The designations L and R refer to the relative positions, left and right respectively, of the couples in a Latimer diagram (recall that the more highly oxidized species lies to the left). The standard potential for the overall reaction is $E^{\ominus} = E^{\ominus}(R) - E^{\ominus}(L)$, which is positive if $E^{\ominus}(R) > E^{\ominus}(L)$. We can conclude that a species is inherently unstable (that is, it has a tendency to disproportionate into its two neighbours) if the potential on the right of the species is more positive than the potential on the left.

6.13 Frost diagrams

A **Frost diagram** (also known as an *oxidation state diagram*) of an element X is a plot of vE for the couple X(N)/X(0) against the oxidation number, N, of the element (v is the net number of electrons that is transferred to form each oxidation state, starting from $N = 0$). The general form of a Frost diagram is given in Fig. 6.5. Frost diagrams depict whether a particular species X(N) is a good oxidizing agent or reducing agent. They also provide an important guide for identifying oxidation states of an element that are inherently stable or unstable.

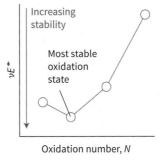

FIGURE 6.5 Oxidation state stability as viewed in a Frost diagram.

(a) Gibbs energies of formation for different oxidation states

KEY POINTS A Frost diagram shows how the Gibbs energies of formation of different oxidation states of an element vary with oxidation number. The most stable oxidation state of an element corresponds to the species that lies lowest in its Frost diagram. Frost diagrams are conveniently constructed by using electrode potential data.

For a half-reaction in which a species X with oxidation number N is converted to its elemental form, the reduction half-reaction is written

$$X(N) + v e^- \rightarrow X(0)$$

Because $vE^\ominus$ is proportional to the standard reaction Gibbs energy for the conversion of the species $X(N)$ to the element (explicitly, $vE^\ominus = -\Delta_r G^\ominus/F$, where $\Delta_r G^\ominus$ is the standard reaction Gibbs energy for the half-reaction given above), a Frost diagram can also be regarded as a plot of standard reaction Gibbs energy (divided by F) against oxidation number. Consequently, the most stable states of an element in aqueous solution correspond to species that lie lowest in its Frost diagram. Any condition can be represented by using the appropriate formal potentials, $E^{\ominus'}$.[2]

The example given in Fig. 6.6 shows data for nitrogen species formed in aqueous solution at pH = 0 and pH = 14. Only $NH_4^+(aq)$ is exergonic ($\Delta_f G^\ominus < 0$); all other species are endergonic ($\Delta_f G^\ominus > 0$). The diagram shows that the higher oxides and oxoacids are highly endergonic in acid solution but relatively stabilized in basic solution. The opposite is generally true for species with $N < 0$ except that hydroxylamine is particularly unstable regardless of pH.

EXAMPLE 6.10 Constructing a Frost diagram

Construct a Frost diagram for oxygen from the Latimer diagram in Example 6.9.

Answer We begin by placing the element in its zero oxidation state (O_2) at the origin for the vE and N axes. For the reduction of O_2 to H_2O_2 (for which $N = -1$), $E^\ominus = +0.70$ V, so $vE^\ominus = -0.70$ V. Because the oxidation number of O in H_2O is -2 and $E^\ominus$ for the O_2/H_2O couple is $+1.23$ V, $vE^\ominus$ at $N = -2 = -2.46$ V. These results are plotted in Fig. 6.7, along with data (see Resource section 3) processed in the same way for pH 14.

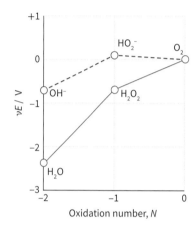

FIGURE 6.7 The Frost diagram for oxygen in acidic solution (solid red line, pH = 0) and alkaline solution (dashed blue line, pH = 14).

Self-test 6.10 Construct a Frost diagram from the Latimer diagram for Tl:

$$Tl^{3+} \xrightarrow{+1.25} Tl^+ \xrightarrow{-0.34} Tl$$

(b) Interpretation

KEY POINTS Frost diagrams may be used to gauge the inherent stabilities of different oxidation states of an element and to decide whether particular species are good oxidizing or reducing agents. The slope of a line connecting two species having different oxidation numbers is the reduction potential for that redox couple.

To interpret the qualitative information contained in a Frost diagram it is important to note (Fig. 6.8) that the slope of the line connecting two species having oxidation numbers N'' and N' is $vE^\ominus/(N' - N'') = E^\ominus$ (since $v = N' - N''$).

This simple rule leads to the following features, which should be considered alongside Fig. 6.9.

1. The steeper the line joining two points (left to right) in a Frost diagram, the more positive the standard potential of the corresponding couple (Fig. 6.9(a)).

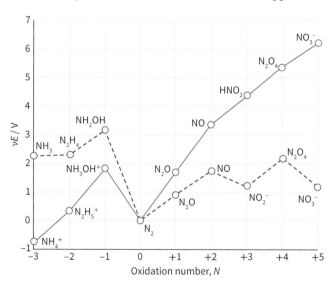

FIGURE 6.6 The Frost diagram for nitrogen: the steeper the slope of a line, the higher the standard potential for the redox couple. The solid red line refers to standard (acid) conditions (pH = 0), the dashed blue line refers to pH = 14. Note that because HNO_3 is a strong acid, it is present as its conjugate base NO_3^- even at pH = 0. The N(IV) species is shown as its dimer N_2O_4.

---

[2] The superscripted $\ominus$ is often omitted from the y-axis label when data for different conditions are displayed together for comparison.

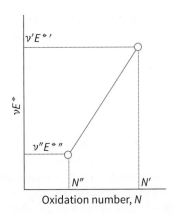

FIGURE 6.8 The general structure of a region of a Frost diagram used to establish the relationship between the slope of a line connecting species having different oxidation numbers and the standard potential of the corresponding redox couple.

2. The oxidizing agent in the couple with the more positive slope (the more positive $E^{\ominus}$) is liable to undergo reduction (Fig. 6.9(b)).

3. The reducing agent of the couple with the less positive slope (the most negative $E^{\ominus}$) is liable to undergo oxidation (Fig. 6.9(b)).

For instance, the steep slope connecting NO_3^- to lower oxidation numbers in Fig. 6.6 shows that nitrate is a good oxidizing agent under standard conditions.

We saw in the discussion of Latimer diagrams that a species is liable to undergo disproportionation if the potential for its reduction from $X(N)$ to $X(N-1)$ is greater than its potential for oxidation from $X(N)$ to $X(N+1)$. The same criterion can be expressed in terms of a Frost diagram (Fig. 6.9(c)).

4. A species in a Frost diagram is unstable with respect to disproportionation (it is *inherently* unstable) if its point lies above the line connecting the two adjacent species.

When this criterion is satisfied, the standard potential for the couple to the left of the species is greater than that for the couple on the right. A specific example is NH_2OH; as can be seen in Fig. 6.6, this compound is unstable with respect to disproportionation into NH_3 and N_2. The origin of this rule is illustrated in Fig. 6.9(d), where we show geometrically that the reaction Gibbs energy of a species with intermediate oxidation number lies above the average value for the two species on either side. As a result, there is a tendency for the intermediate species to disproportionate into the two other species.

The criterion for comproportionation to be spontaneous can be stated analogously (Fig. 6.9(e)):

5. Two species will tend to comproportionate into an intermediate species that lies below the straight line joining the terminal species.

A substance that lies below the line connecting its neighbours in a Frost diagram is inherently more stable than they are because their average molar Gibbs energy is higher (Fig. 6.9(f)) and hence comproportionation is thermodynamically

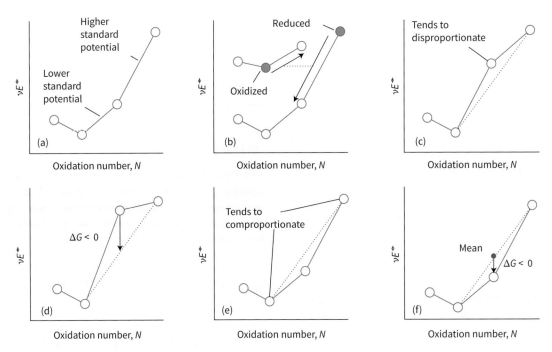

FIGURE 6.9 The interpretation of a Frost diagram to gauge (a) reduction potential, (b) tendency towards oxidation and reduction, (c, d) disproportionation, and (e, f) comproportionation.

favourable. The nitrogen in NH_4NO_3, for instance, has two ions with oxidation numbers $-3(NH_4^+)$ and $+5 (NO_3^-)$. Because N_2O lies below a line directly joining NH_4^+ to NO_3^-, their comproportionation is spontaneous:

$$NH_4^+(aq) + NO_3^-(aq) \rightarrow N_2O(g) + 2H_2O(l)$$

However, although the reaction is expected to be spontaneous on thermodynamic grounds under standard conditions, the reaction is kinetically inhibited in solution and does not ordinarily occur. The corresponding reaction

$$NH_4NO_3(s) \rightarrow N_2O(g) + 2H_2O(g)$$

in the solid state is both thermodynamically spontaneous ($\Delta_r G^\ominus = -168\,\text{kJ}\,\text{mol}^{-1}$) and, once initiated by a detonation, explosively fast. Indeed, ammonium nitrate is often used in place of dynamite for blasting rocks.

A BRIEF ILLUSTRATION

Refer to the oxygen diagram in Fig. 6.7. At the point corresponding to $N = -1$ (for H_2O_2), $(-1) \times E^\ominus = -0.70\,\text{V}$, and at $N = -2$ (for H_2O), $(-2) \times E^\ominus = -2.46\,\text{V}$. The difference of the two values is $-1.76\,\text{V}$. The change in oxidation number of oxygen on going from H_2O_2 to H_2O is -1. Therefore, the slope of the line is $(-1.76\,\text{V})/(-1) = +1.76\,\text{V}$, in accord with the value for the H_2O_2/H_2O couple in the Latimer diagram.

We saw that Frost diagrams can equally well be constructed for other conditions. The formal potentials at pH = 14 are often denoted $E_B^\ominus$, and the dotted blue line in Fig. 6.6 is a '*basic* Frost diagram' for nitrogen. The important difference from the behaviour in acidic solution is the stabilization of NO_2^- against disproportionation: its point in the basic Frost diagram no longer lies above the line connecting its neighbours. The practical outcome is that metal nitrites are stable in neutral and basic solutions and can be isolated, whereas HNO_2 cannot (although solutions of HNO_2 have some short-term stability as their decomposition is kinetically slow). In some cases, there are marked differences between strongly acidic and basic solutions, as for the phosphorus oxoanions. This example illustrates an important general point about oxoanions: when their reduction requires removal of oxygen, the reaction consumes H^+ ions, and all oxoanions are stronger oxidizing agents in acidic than in basic solution.

EXAMPLE 6.11 Using a Frost diagram to judge the thermodynamic stability of ions in solution

Fig. 6.10 shows the Frost diagram for manganese. Comment on the stability of Mn^{3+} in acidic aqueous solution.

Answer We approach this question by inspecting how the $\nu E^\ominus$ value for Mn^{3+} ($N = +3$) compares with the values for species on either side ($N < +3$, $N > +3$). Because Mn^{3+} lies *above* the line joining Mn^{2+} to MnO_2, it should disproportionate into these two species. The chemical reaction is

$$2Mn^{3+}(aq) + 2H_2O(l) \rightarrow Mn^{2+}(aq) + MnO_2(s) + 4H^+(aq)$$

Self-test 6.11 What is the oxidation number of Mn in the product when MnO_4^- is used as an oxidizing agent in aqueous acid?

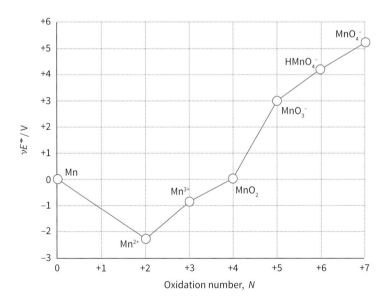

FIGURE 6.10 The Frost diagram for manganese in acidic solution (pH = 0). Note that $HMnO_4$, H_2MnO_4, and $HMnO_3$ are strong acids and exist as their conjugate bases at pH = 0.

> **EXAMPLE 6.12** Application of Frost diagrams at different pH
>
> Potassium nitrite (containing N(III)) is stable in basic solution, but when the solution is acidified a gas is evolved that turns brown on exposure to air. What is the reaction?
>
> **Answer** To answer this we use the Frost diagram (Fig. 6.6) to compare the inherent stabilities of N(III) in acid and basic solutions. The point representing NO_2^- ion in basic solution lies below the line joining NO to NO_3^-; the ion therefore is not liable to disproportionation. On acidification, the HNO_2 point rises and the straightness of the line through NO, HNO_2, and N_2O_4 (dimeric NO_2) implies that all three species are present at equilibrium. The brown gas is NO_2 formed from the reaction of NO evolved from the solution with air. In solution, the species of oxidation number +2 (NO) tends to disproportionate. However, the escape of NO from the solution prevents its disproportionation to N_2O and HNO_2.
>
> **Self-test 6.12** By reference to Fig. 6.6, compare the strength of NO_3^- as an oxidizing agent in acidic and basic solution.

6.14 Proton-coupled electron transfer: Pourbaix diagrams

KEY POINTS A Pourbaix diagram is a map of the conditions of potential and pH under which species are stable in water. A horizontal line separates species related by electron transfer only, a vertical line separates species related by proton transfer only, and sloped lines separate species related by both electron and proton transfer.

A **Pourbaix diagram** (also known as an **E–pH diagram**) indicates the conditions of pH and potential under which a species is thermodynamically stable. Pourbaix diagrams are used to represent and analyse proton-coupled electron-transfer reactions. The diagrams were introduced by Marcel Pourbaix in 1938 as a convenient way of discussing the chemical properties of species in natural waters and they are applied in environmental and corrosion science, as well as in the interpretation of electrochemical kinetics and electrocatalysis.

Iron is essential for almost all life forms and the problem of its uptake from the environment is discussed further in Section 26.6. Fig. 6.11 shows a simplified Pourbaix diagram for iron, omitting such low concentration species as oxygen-bridged Fe(III) dimers and minerals such as magnetite, Fe_3O_4, containing both Fe^{3+} and Fe^{2+}. This diagram is useful for the discussion of iron species in natural waters (see Section 6.15) because the total iron concentration is low; at high concentrations, complex multinuclear iron species can form. We can see how the diagram has been constructed by considering some of the reactions involved.

The reduction half-reaction

$$Fe^{3+}(aq) + e^- \rightarrow Fe^{2+}(aq) \quad E^\ominus = +0.77\,V$$

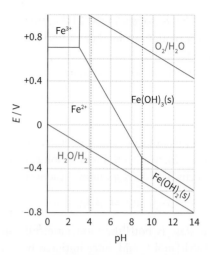

FIGURE 6.11 A simplified Pourbaix diagram for some important naturally occurring aqua-species of iron.

does not involve H^+ ions, so its potential is independent of pH and hence corresponds to a *horizontal* line on the diagram. If the environment contains a couple with a potential above this line (a more positive, oxidizing couple), then the oxidized species, Fe^{3+}, will be the major species. Hence, the horizontal line towards the top left of the diagram is a boundary that separates the regions where Fe^{3+} and Fe^{2+} dominate.

Another reaction to consider is

$$Fe^{3+}(aq) + 3H_2O(l) \rightarrow Fe(OH)_3(s) + 3H^+(aq)$$

This reaction is not a redox reaction (there is no change in oxidation number of any element), so it is insensitive to the electrochemical potential in its environment and therefore is represented by a *vertical* line on the diagram. However, this boundary does depend on pH, with $Fe^{3+}(aq)$ favoured by low pH and $Fe(OH)_3(s)$ favoured by high pH. We adopt the convention that Fe^{3+} is the dominant species in the solution if its concentration exceeds 10 µmol dm^{-3} (a typical freshwater value). The equilibrium concentration of Fe^{3+} varies with pH, and the vertical boundary at pH = 3 represents the pH at which Fe^{3+} becomes dominant according to this definition. In general, a vertical line in a Pourbaix diagram does not involve a redox reaction but signifies a pH-dependent change of state of either the oxidized or reduced form.

As the pH is increased, the Pourbaix diagram includes reactions such as

$$Fe(OH)_3(s) + 3H^+(aq) + e^- \rightarrow Fe^{2+}(aq) + 3H_2O(l)$$

(for which the slope of potential against pH, according to eqn. 6.8b, is $v_H/v_e = -3(0.059\,V)$) and eventually $Fe^{2+}(aq)$ is also precipitated as $Fe(OH)_2$. Inclusion of the metal dissolution couple ($Fe^{2+}/Fe(s)$) would complete construction of the Pourbaix diagram for well-known aqua species of iron.

The lines representing the O_2/H_2O and H_2O/H_2 redox couples are parallel since v_H/v_e is 1 in both cases, deviation occurring only as the pK_a of water is approached. In applications that rely on comparing data for H_2 redox chemistry with that of O_2 or other oxidants (such as fuel cells and electrolysers) over a range of pH, an additional reference scale, the **reversible hydrogen electrode (RHE)**, is often used. For any given pH, the potential of the oxidant couple is measured against the corresponding potential for H_2O/H_2 at 1 bar H_2. Thus E_{RHE} for the four-electron reduction of O_2 to H_2O is +1.23 V throughout the entire pH range of Fig. 6.11.

Applications

In the following sections we discuss some applications of the redox chemistry described above. Regardless of whether we are dealing with the environment, sustainability, or industry, it is difficult to avoid reference to oxidation and reduction reactions and the processes that depend on them. Redox reactions form the basis of a major part of the chemical industry.

6.15 Environmental chemistry: natural waters

KEY POINTS The quality of a natural water system, freshwater or marine, is typically gauged by its oxygen content and pH, which in turn determine the availability of dissolved substances, both nutrients and pollutants. Pourbaix diagrams are useful tools, for instance, in predicting the availability of dissolved metal ions such as Fe^{2+} in different environments.

The chemistry of natural waters can be rationalized by using Pourbaix diagrams of the kind we have just constructed. Thus, where fresh water is in contact with the atmosphere, it is saturated with O_2, and many species may be oxidized by this strong oxidizing agent. More fully reduced forms are found in the absence of oxygen, especially where there is organic matter to act as a reducing agent. The major acid system that controls the pH of the medium is $CO_2/H_2CO_3/HCO_3^-/CO_3^{2-}$, where atmospheric CO_2 provides the acid and dissolved carbonate minerals provide the base. Biological activity is also important, because respiration releases CO_2. This acidic oxide lowers the pH and hence makes the potential more positive. The reverse process, photosynthesis, consumes CO_2, thus raising the pH and making the potential more negative. The condition of typical natural waters—their pH and the potentials of the redox couples they contain—is summarized in Fig. 6.12.

From Fig. 6.11 we see that the simple cation $Fe^{3+}(aq)$ can exist in water only if the environment is oxidizing; that is, where O_2 is plentiful and the pH is low (below 3). Because few natural waters are so acidic, $Fe^{3+}(aq)$ is not found in the environment. The iron in insoluble Fe_2O_3 or insoluble hydrated forms such as FeO(OH) can enter solution as Fe^{2+} if it is reduced, which occurs when the condition of the water lies below the sloping boundary in the diagram. We should observe that, as the pH rises, Fe^{2+} can form only if there are strong reducing couples present, and its formation is very unlikely in oxygen-rich water. Comparison with Fig. 6.12 shows that iron will be reduced and dissolved in the form of Fe^{2+} in both bog waters and organic-rich waterlogged soils (at pH near 4.5 in both cases and with corresponding E values near +0.03 V and −0.1 V, respectively).

It is instructive to analyse a Pourbaix diagram in conjunction with an understanding of the physical processes that occur in water. As an example, consider a lake where the temperature gradient, cool at the bottom and warmer above, tends to prevent vertical mixing. At the surface, the water is fully oxygenated and the iron must be present in particles of the insoluble FeO(OH); these particles tend to settle. At greater depth, the O_2 content is low. If the organic content or other sources of reducing agents are sufficient, the oxide will be reduced and iron will dissolve as Fe^{2+}. The Fe(II) ions will then diffuse towards the surface where they encounter O_2 and be oxidized to insoluble FeO(OH) again.

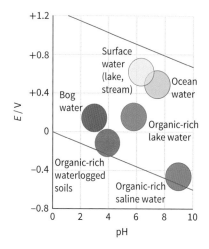

FIGURE 6.12 The stability field of water showing regions typical of various natural waters.

EXAMPLE 6.13 Using a Pourbaix diagram

Fig. 6.13 shows part of a Pourbaix diagram for manganese. Identify the environment in which the solid MnO_2 or its corresponding hydrous oxides are important. Is Mn(III) formed under any conditions?

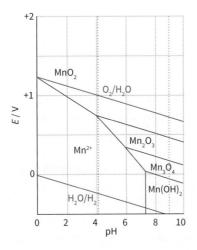

FIGURE 6.13 A section of the Pourbaix diagram for manganese. The broken black vertical lines represent the normal pH range (pH 4–9) in natural waters.

Answer We approach this problem by locating the zone of stability for MnO_2 on the Pourbaix diagram and inspecting its position relative to the boundary between O_2 and H_2O. Manganese dioxide is the thermodynamically favoured state in well-oxygenated water under all pH conditions, with the exception of strong acid (pH < 1). Under mildly reducing conditions, in waters having neutral-to-acidic pH, the stable species is Mn^{2+}(aq). Manganese(III) species are stabilized only in oxygenated waters at higher pH.

Self-test 6.13 Use Figs. 6.11 and 6.12 to evaluate the possibility of finding $Fe(OH)_3$(s) in a waterlogged soil.

6.16 Electrocatalysis

KEY POINT An electrocatalyst increases the rate of a reaction occurring at an electrode.

A reaction occurring at an electrode is often too slow to be useful, whether for investigating the reaction itself or exploiting it for synthesis or energy conversion. One such example is the four-electron interconversion between O_2 and H_2O that is relevant in fuel cells (Box 6.1) and electrolysers (Section 11.4). The reaction rates in both directions, forward (the oxygen reduction reaction, ORR) and reverse (the oxygen evolution reaction, OER) are negligible unless a substantially more negative (or positive) electrode potential is applied.

$$O_2(aq) + 4H^+(aq) + 4e^- \rightleftharpoons 2H_2O(l)$$

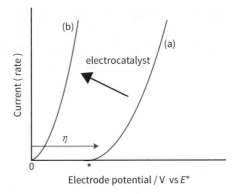

FIGURE 6.14 The effect of an electrocatalyst in facilitating an electrode reaction. The uncatalysed reaction (a) does not occur until a sizeable overpotential has been applied to the electrode (onset potential is marked by asterisk). An electrocatalyst increases the rate and lowers the overpotential requirement (b) so that the reaction commences at a potential close to $E^\ominus$.

The dependence of the rate of the electron-transfer reaction on electrode potential appears in current–voltage plots known as voltammograms. Voltammetry, as the general method is known, is mentioned in more detail in Chapter 8. The two voltammograms shown in Fig. 6.14 represent two scenarios that apply to H_2O oxidation to O_2. Trace (a) shows how the rate depends on increasing electrode potential E when the reaction is extremely sluggish: no current is apparent until a far more positive oxidizing potential, indicated by *, is applied. The difference between the onset potential * and $E^\ominus$ is the **overpotential** required to drive the reaction at a measurable rate. In contrast, trace (b) shows the potential dependence that should be obtained when the overpotential requirement is decreased so much so that the oxidation reaction proceeds at $E^\ominus$ and accelerates rapidly, often with an exponential dependence on the difference $E - E^\ominus$.

Such an enhancement can be achieved by lowering the activation energy through an **electrocatalyst**—a species that is either inherently present on the electrode surface or is coated on to it. An electrocatalytic cycle (Fig. 6.15)

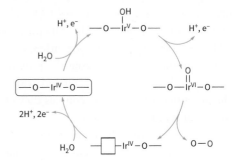

FIGURE 6.15 Electrocatalytic cycle showing a mechanism proposed for the O_2 evolution reaction (OER) catalysed at an IrO_2 electrode. The cycle involves successive proton-coupled electron transfers that lower the activation energy.

is constructed in the same way as other catalytic cycles (Sections 2.15 and 5.16) except that the individual steps now include electron transfers. In this proposal for the four-electron oxidation of water catalysed at an IrO_2 surface, the reaction is facilitated by step-wise, one-electron increases in Ir oxidation state that are coupled to successive deprotonations of H_2O, and direct involvement of occupied (-O-) and vacant (-☐-)-oxygen atom sites.

6.17 Extraction of the elements by chemical reduction

Only a few metals, such as gold, occur in nature as their elements. Most metals are found as their oxides, such as Fe_2O_3, or as ternary compounds, such as $FeTiO_3$. Sulfides are also common, particularly in mineral veins where deposition occurred under water-free and oxygen-poor conditions. Prehistoric humans slowly learned how to transform ores to produce metals for making tools and weapons. Copper could be extracted from its ores by aerial oxidation at temperatures attainable in the primitive hearths that became available about 6000 years ago:

$$2Cu_2S(s) + 3O_2(g) \rightarrow 2Cu_2O(s) + 2SO_2(g)$$
$$2Cu_2O(s) + Cu_2S(s) \rightarrow 6Cu(s) + SO_2(g)$$

This was probably the first example of a **pyrometallurgical** process.

It was not until nearly 3000 years ago that sufficiently high temperatures could be reached to extract metals from more stable ores, thereby leading to the Iron Age. Iron and other elements were produced by heating the ore to its molten state with a reducing agent such as carbon, in a process known as **smelting**. Carbon remained the dominant reducing agent until the end of the nineteenth century, and metals that needed higher temperatures for their production remained unavailable even though their ores were reasonably abundant.

The availability of electric power expanded the scope of carbon reduction, because electric furnaces can reach much higher temperatures than carbon-combustion furnaces, such as the blast furnace. Thus, magnesium was a metal of the twentieth century because one of its modes of recovery, the **Pidgeon process**, involves the very high temperature, electrothermal reduction of the oxide by carbon:

$$MgO(s) + C(s) \xrightarrow{\Delta} Mg(l) + CO(g)$$

Note that the carbon is oxidized only to carbon monoxide, the product favoured thermodynamically at the very high reaction temperatures used.

The technological breakthrough in the nineteenth century that resulted in the conversion of aluminium from a rarity into a major construction metal was the introduction of electrolysis, the driving of a nonspontaneous reaction (including the reduction of ores) by electrical energy involving the passage of an electric current.

(a) Thermodynamic aspects

KEY POINTS An Ellingham diagram summarizes the temperature dependence of the standard Gibbs energies of formation of metal oxides and is used to identify the temperature at which reduction by carbon or carbon monoxide becomes spontaneous.

As we have seen, the standard reaction Gibbs energy, $\Delta_r G^\ominus$, is related to the equilibrium constant, K, through $\Delta_r G^\ominus = -RT \ln K$, and a negative value of $\Delta_r G^\ominus$ corresponds to $K > 1$. It should be noted that equilibrium is rarely attained in commercial processes as many such systems involve dynamic stages where, for instance, reactants and products are in contact only for short times. Furthermore, even a process at equilibrium for which $K < 1$ can be viable if the product (particularly a gas) is swept out of the reaction chamber and the reaction continues to chase the ever-vanishing equilibrium composition. In principle, we also need to consider rates when judging whether a reaction is feasible in practice, but reactions are often fast at high temperature and thermodynamically favourable reactions are likely to occur. A fluid phase (typically a gas or solvent) is usually required to facilitate what would otherwise be a sluggish reaction between coarse particles.

To achieve a negative $\Delta_r G^\ominus$ for the reduction of a metal oxide with carbon or carbon monoxide, one of the following reactions

(a) $C(s) + \frac{1}{2}O_2(g) \rightarrow CO(g)$ $\Delta_r G^\ominus(C, CO)$
(b) $\frac{1}{2}C(s) + \frac{1}{2}O_2(g) \rightarrow \frac{1}{2}CO_2(g)$ $\Delta_r G^\ominus(C, CO_2)$
(c) $CO(g) + \frac{1}{2}O_2(g) \rightarrow CO_2(g)$ $\Delta_r G^\ominus(CO, CO_2)$

must have a more negative $\Delta_r G^\ominus$ than a reaction of the form

(d) $x\, M(s\text{ or } l) + \frac{1}{2}O_2(g) \rightarrow M_xO(s)$ $\Delta_r G^\ominus(M, M_xO)$

under the same reaction conditions. If that is so, then one of the reactions

(a – d) $M_xO(s) + C(s) \rightarrow x\, M(s\text{ or } l) + CO(g)$
$\Delta_r G^\ominus(C, CO) - \Delta_r G^\ominus(M, M_xO)$

(b – d) $M_xO(s) + \frac{1}{2}C(s) \rightarrow x\, M(s\text{ or } l) + \frac{1}{2}CO_2(g)$
$\Delta_r G^\ominus(C, CO_2) - \Delta_r G^\ominus(M, M_xO)$

(c – d) $M_xO(s) + CO(g) \rightarrow x\, M(s\text{ or } l) + CO_2(g)$
$\Delta_r G^\ominus(CO, CO_2) - \Delta_r G^\ominus(M, M_xO)$

will have a negative standard reaction Gibbs energy, and therefore have $K > 1$. The procedure followed here is similar

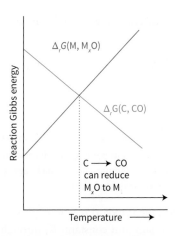

FIGURE 6.16 The variation of the standard reaction Gibbs energies for the formation of a metal oxide and carbon monoxide with temperature. The formation of carbon monoxide from carbon can reduce the metal oxide to the metal at temperatures higher than the point of intersection of the two lines. More specifically, at the intersection the equilibrium constant changes from $K < 1$ to $K > 1$. This type of display is an example of an Ellingham diagram.

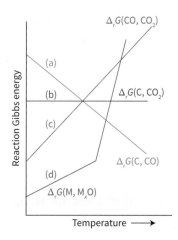

FIGURE 6.17 Part of an Ellingham diagram showing the standard Gibbs energy for the formation of a metal oxide and the three carbon oxidation Gibbs energies. The slopes of the lines are determined largely by whether or not there is net gas formation or consumption in the reaction. A phase change generally results in a kink in the graph (because the entropy of the substance changes).

to that adopted with half-reactions in aqueous solution (Section 6.1), but now all the reactions are written as oxidations with ½ O_2 in place of 2 e⁻, and the overall reaction is the difference of reactions with matching numbers of oxygen atoms. The relevant information is commonly summarized in an **Ellingham diagram** (Fig. 6.16), which is a graph of $\Delta_r G^\ominus$ against temperature.

We can understand the appearance of an Ellingham diagram by noting that $\Delta_r G^\ominus = \Delta_r H^\ominus - T\Delta_r S^\ominus$ and using the fact that the enthalpy and entropy of reaction are, to a reasonable approximation, independent of temperature. That being so, the slope of a line in an Ellingham diagram should therefore be equal to $-\Delta_r S^\ominus$ for the relevant reaction.

Because the standard molar entropies of gases are much larger than those of solids, the reaction entropy of (a), in which there is a net formation of gas (because 1 mol CO replaces ½ mol O_2), is positive, and its line therefore has a negative slope. The standard reaction entropy of (b) is close to zero as there is no net change in the amount of gas, so its line is horizontal. Reaction (c) has a negative reaction entropy because $\frac{3}{2}$ mol of gas molecules is replaced by 1 mol CO_2; hence the line in the diagram has a positive slope. The standard reaction entropy of (d), in which there is a net consumption of gas, is negative, and hence the plot has a positive slope (Fig. 6.17). The kinks in the lines, where the slope of the metal oxidation line changes, are where the metal undergoes a phase change, particularly melting, and the reaction entropy changes accordingly. At temperatures for which the C/CO line (a) lies above the metal oxide line (d), $\Delta_r G^\ominus(M,M_xO)$ is more negative than $\Delta_r G^\ominus(C,CO)$. At these temperatures, $\Delta_r G^\ominus(C,CO) - \Delta_r G^\ominus(M,M_xO)$ is positive, so the reaction (a – d) has $K < 1$. However, for temperatures for which the C/CO line lies below the metal oxide line, the reduction of the metal oxide by carbon has $K > 1$. Similar remarks apply to the temperatures at which the other two carbon oxidation lines (b) and (c) lie above or below the metal oxide line. In summary:

- For temperatures at which the C/CO line lies below the metal oxide line, carbon can be used to reduce the metal oxide and itself is oxidized to carbon monoxide.

- For temperatures at which the C/CO_2 line lies below the metal oxide line, carbon can be used to achieve the reduction, but is oxidized to carbon dioxide.

- For temperatures at which the CO/CO_2 line lies below the metal oxide line, carbon monoxide can reduce the metal oxide to the metal and is oxidized to carbon dioxide.

Fig. 6.18 shows an approximate Ellingham diagram for a selection of common metal oxides. In principle, production of all the metals from their oxides, even calcium and aluminium, could be accomplished by heating with carbon at a sufficiently high temperature. However, there are severe practical limitations. Efforts to produce aluminium in this way (most notably in Japan, where electricity is expensive) were frustrated by the volatility of Al_2O_3 at the very high temperatures required (around 3000°C). A difficulty of a different kind is encountered in the pyrometallurgical extraction of titanium, where titanium carbide, TiC, is formed instead of the metal. In practice, pyrometallurgical extraction of metals is confined principally to magnesium, iron, cobalt, nickel, zinc, and a variety of ferroalloys (alloys with iron).

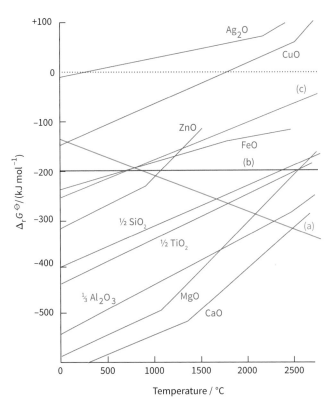

FIGURE 6.18 An approximate Ellingham diagram for the reduction of metal oxides.

EXAMPLE 6.14 Using an Ellingham diagram

What is the lowest temperature at which ZnO can be reduced to zinc metal by carbon? What is the overall reaction at this temperature?

Answer To answer this question we examine the Ellingham diagram in Fig. 6.18 and estimate the temperature at which the ZnO line crosses the C,CO line. The C,CO line (a) lies below the ZnO line at approximately 1200°C; above this temperature, reduction of the metal oxide is spontaneous. The contributing reactions are reaction (a) and the reverse of

$$Zn(g) + \tfrac{1}{2}O_2(g) \rightarrow ZnO(s)$$

so the overall reaction is the difference, or

$$C(s) + ZnO(s) \rightarrow CO(g) + Zn(g)$$

The physical state of zinc is given as a gas because the element boils at 907°C, corresponding approximately to the change in the gradient of the ZnO line.

Self-test 6.14 What is the minimum temperature for reduction of MgO by carbon?

Similar principles apply to reductions using other reducing agents. For instance, an Ellingham diagram can be used to explore whether a metal M′ can be used to reduce the oxide of another metal M. In this case, we note from the diagram whether at a temperature of interest the M′, M′O line lies below the M,MO line, as M′ is now taking the place of C. When

$$\Delta_r G^\ominus = \Delta_r G^\ominus(M', M'O) - \Delta_r G^\ominus(M, MO)$$

is negative, where the Gibbs energies refer to the reactions

(a) $M'(s \text{ or } l) + \tfrac{1}{2}O_2(g) \rightarrow M'O(s)$ $\Delta_r G^\ominus(M', M'O)$

(b) $M(s \text{ or } l) + \tfrac{1}{2}O_2(g) \rightarrow MO(s)$ $\Delta_r G^\ominus(M, MO)$

the reaction

(a − b) $MO(s) + M'(s \text{ or } l) \rightarrow M(s \text{ or } l) + M'O(s)$

(and its analogues for MO_2, and so on) is feasible (in the sense $K > 1$). For instance, because in Fig. 6.18 the line for MgO remains below the line for SiO_2 at temperatures below 2500°C, magnesium may be used to reduce SiO_2 below that temperature. This reaction has in fact been used to produce low-grade silicon, as discussed in the following section.

(b) Survey of processes

KEY POINTS A blast furnace produces the conditions required to reduce iron oxides with carbon; electrolysis may be used to bring about an otherwise nonspontaneous reduction as required for the extraction of aluminium from its oxide. Metal oxides are being used as intermediates (carriers) in two-stage processes known as chemical looping that facilitate clean combustion and extraction of O_2 from air.

Industrial processes for achieving the reductive extraction of metals show a greater variety than the thermodynamic analysis might suggest. An important factor is that the ore and carbon are both solids, and a reaction between two solids is rarely fast. Most processes exploit gas/solid or liquid/solid heterogeneous reactions. Current industrial processes are varied in the strategies they adopt to ensure economical rates, exploit materials, and avoid environmental problems. We can explore these strategies by considering three important examples that reflect low, moderate, and extreme difficulty of reduction.

The least difficult reductions include those of copper ores. Roasting and smelting are still widely used in the pyrometallurgical extraction of copper. However, some recent techniques seek to avoid the major environmental problems caused by the production of the large quantity of SO_2, released to the atmosphere, that accompanies roasting. One promising development is the **hydrometallurgical extraction** of copper, the extraction of the metal by reduction of aqueous solutions of its ions, using H_2 or scrap iron as the reducing agent. In this process, Cu^{2+} ions, leached from low-grade ores by acid or bacterial action, are reduced by hydrogen in the reaction (or by a similar reduction using iron):

$$Cu^{2+}(aq) + H_2(g) \rightarrow Cu(s) + 2H^+(aq)$$

This process is less harmful to the environment provided the acid by-product is reused or neutralized locally rather than contributing to acidic atmospheric pollutants. It also allows economic exploitation of lower grade ores.

That extraction of iron is of intermediate difficulty is shown by the fact that the Iron Age followed the Bronze Age. In economic terms, iron ore reduction is the most important application of carbon pyrometallurgy. In a blast furnace (Fig. 6.19), which is still the major source of the element, the mixture of iron ores (Fe_2O_3, Fe_3O_4), coke (C), and limestone ($CaCO_3$) is heated with a blast of hot air. Combustion of coke in this blast raises the temperature to 2000°C, and the carbon burns to carbon monoxide in the lower part of the furnace. The supply of Fe_2O_3 from the top of the furnace meets the hot CO rising from below. The iron(III) oxide is reduced, first to Fe_3O_4 and then to FeO at 500–700°C, and the CO is oxidized to CO_2. The final reduction to iron, from FeO, by carbon monoxide occurs between 1000 and 1200°C in the central region of the furnace. Thus overall

$$Fe_2O_3(s) + 3CO(g) \rightarrow 2Fe(l) + 3CO_2(g)$$

The function of the lime, CaO, formed by the thermal decomposition of calcium carbonate, is to combine with the silicates present in the ore to form a molten layer of calcium silicates (slag) in the hottest (lowest) part of the furnace. Slag is less dense than molten iron, so it separates and can

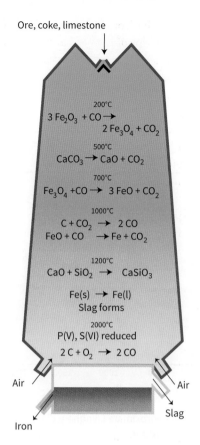

FIGURE 6.19 A schematic diagram of a blast furnace showing the typical composition and temperature profile.

BOX 6.2 Hydrogen in steel production

The current method of producing iron using carbon monoxide derived from coal as the reducing agent is not sustainable. As well as their role in the direct chemical reduction of iron ore, fossil fuels are used to generate the high temperatures, 2000°C, that are needed during the extraction and refinement processes. In the early 2020s, the CO_2 released during the manufacture of iron and steel accounted for 5–7% of anthropological emissions. Renewable energy is continually becoming cheaper and more abundant, directing attention to the use of H_2 to replace CO in steel making. Iron ore (mainly Fe_2O_3) is pelleted and fed into a fluidized bed reactor where it meets a flow of H_2 at 800 °C. The following reactions occur as the ore, intermediates, and final product pass down the reactor, indicated in Fig. B6.2 by their respective colour coding: orange, brown, green, and grey.

$$3Fe_2O_3(s) + H_2(g) \rightarrow 2Fe_3O_4(s) + H_2O(g) \quad \Delta_rH^\ominus = -6.02\,kJ\,mol^{-1}$$

$$Fe_3O_4(s) + H_2(g) \rightarrow 3FeO(s) + H_2O(g) \quad \Delta_rH^\ominus = +46.6\,kJ\,mol^{-1}$$

$$FeO(s) + H_2(g) \rightarrow Fe(s) + H_2O(g) \quad \Delta_rH^\ominus = +16.4\,kJ\,mol^{-1}$$

Apart from the reduction of Fe_2O_3 to Fe_3O_4, the reactions are endothermic, so the process is favoured by high temperatures and by the removal of H_2O, which is the only gaseous product.

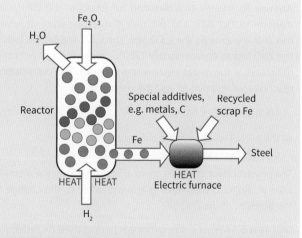

FIGURE B6.2 Diagram showing steel production from iron ore using renewable H_2 as the chemical reducing agent. Electricity (ideally from a renewable source) is used for heating.

be drained away. The iron formed melts at about 400°C below the melting point of the pure metal on account of the dissolved carbon it contains. The impure iron, the densest phase, settles to the bottom and is drawn off to solidify into 'pig iron', in which the carbon content is high (about 4% by mass). The manufacture of steel is then a series of reactions in which the carbon content is reduced and other metals are used to form alloys with the iron. Hydrogen offers a more sustainable route for steel production (Box 6.2).

More difficult than the extraction of either copper or iron is the extraction of silicon from its oxide: indeed, silicon is very much an element of the twentieth century. Silicon of 96–99% purity is prepared by reduction of quartzite or sand (SiO_2) with high purity coke. The Ellingham diagram (Fig. 6.18) shows that the reduction is feasible only at temperatures in excess of about 1700°C. This high temperature is achieved in an electric arc furnace in the presence of excess silica (to prevent the accumulation of SiC):

$$SiO_2(l) + 2C(s) \xrightarrow{1500°C} Si(l) + 2CO(g)$$
$$2SiC(s) + SiO_2(l) \rightarrow 3Si(l) + 2CO(g)$$

Very pure silicon (for semiconductors) is made by converting crude silicon to volatile compounds, such as silanes or $SiCl_4$ (Section 15.4). These compounds are purified by exhaustive fractional distillation and then reduced to silicon with pure hydrogen. The resulting semiconductor-grade silicon is melted, and large single crystals are pulled slowly from the cooled surface of a melt: this procedure is called the **Czochralski process**.

Most titanium is produced by the **Kroll process**, in which a titanium ore such as $FeTiO_3$ is first reacted with Cl_2 and carbon to produce $TiCl_4$. The purified $TiCl_4$ is then reduced with magnesium metal.

$$2FeTiO_3 + 7Cl_2 + 6C \xrightarrow{900°C} 2TiCl_4 + 2FeCl_3 + 6CO$$
$$TiCl_4 + 2Mg \xrightarrow{1100°C} Ti + 2MgCl_2$$

As we have remarked, the Ellingham diagram shows that the direct reduction of Al_2O_3 with carbon becomes feasible only above 2300°C, which makes it uneconomically expensive and wasteful in terms of any fossil fuels used to heat the system. However, the reduction can be brought about **electrochemically** (Section 6.19).

The information represented in Ellingham diagrams may be applied in technologies known as chemical looping, in which a reactive metal oxide (MO_x) is used as a carrier to separate two chemical reactions. Examples include a two-stage combustion system that facilitates the capture of CO_2 by releasing it in a concentrated form (Fig. 6.20) and a process for extracting pure O_2 from air (Fig. 6.21).

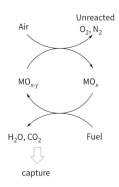

FIGURE 6.20 A chemical loop process for two-stage combustion, in which use of a metal oxide as intermediate leads to a concentrated CO_2 product, hence facilitating its capture.

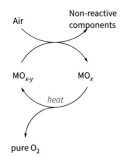

FIGURE 6.21 A chemical loop process for extracting pure O_2 from air.

6.18 Extraction of the elements by chemical oxidation

KEY POINT Elements obtained by chemical oxidation include the heavier halogens, sulfur, and (in the course of their purification) certain noble metals.

As elemental oxygen is available from fractional distillation of air, chemical methods for its production are not necessary. Elemental sulfur is either mined or produced by oxidation of the H_2S that is removed from 'sour' natural gas and crude oil. The oxidation is accomplished by the **Claus process**, which consists of two stages. In the first stage, some hydrogen sulfide is oxidized to sulfur dioxide at high temperature:

$$2H_2S + 3O_2 \rightarrow 2SO_2 + 2H_2O$$

In the second stage, this sulfur dioxide is allowed to react in the presence of a catalyst with more hydrogen sulfide:

$$2H_2S + SO_2 \xrightarrow{\text{Oxide catalyst, 300°C}} 3S + 2H_2O$$

The catalyst is typically Fe_2O_3 or Al_2O_3. The Claus process is environmentally benign, otherwise it would be necessary to burn the toxic hydrogen sulfide to polluting sulfur dioxide.

The only important metals extracted in a process using an oxidation stage are those that occur in native form (i.e. as the element). Gold is an example, because it is difficult to separate the granules of metal in low-grade ores by simple 'panning'. The dissolution of gold depends on oxidation, which is favoured by complexation with CN^- ions:

$$Au(s) + 2CN^-(aq) \rightarrow [Au(CN)_2]^-(aq) + e^-$$

This complex is then reduced to the metal by reaction with another reactive metal, such as zinc:

$$2[Au(CN)_2]^-(aq) + Zn(s) \rightarrow 2Au(s) + [Zn(CN)_4]^{2-}(aq)$$

However, because of the toxicity of cyanide, alternative methods of extracting gold have been used. One of these involves the use of sulfur-cycle bacteria (Box 17.4) that can release gold from sulfidic ores.

The lighter, strongly oxidizing halogens are extracted electrochemically, as described in Section 6.19. The more readily oxidizable halogens, Br_2 and I_2, are obtained by chemical oxidation of the aqueous halides with chlorine, for example,

$$2NaBr(aq) + Cl_2(g) \rightarrow 2NaCl(aq) + Br_2(l)$$

6.19 Electrochemical extraction

KEY POINTS Elements obtained by electrochemical reduction include aluminium; those obtained by electrochemical oxidation include chlorine.

The extraction of metals from ores by electrochemical means is confined mainly to the more electropositive elements, as discussed in the case of aluminium later in this section. For other metals produced in bulk quantities, such as iron and copper, the more energy-efficient and cleaner routes used by industry (using chemical reduction methods) were described in Section 6.17(b). In some specialist cases, electrochemical reduction is used to isolate small quantities of platinum group metals. So, for instance, treatment of spent catalytic converters with acids under oxidizing conditions produces a solution containing complexes of Pt(II) and other platinum group metals, which can then be reduced electrochemically. The metals are deposited at the cathode with an overall 80% efficient extraction from the ceramic catalytic converter.

As we saw in Section 6.17, an Ellingham diagram shows that the reduction of Al_2O_3 with carbon becomes feasible only above 2400°C, which is uneconomically expensive. However, the reduction can be brought about **electrolytically**, and all modern production uses the electrochemical **Hall–Héroult process**, which was invented in 1886 independently by Charles Hall and Paul Héroult.

The process requires pure aluminium hydroxide that is extracted from aluminium ores by using the **Bayer process**. In this process the bauxite ore used as a source of aluminium is a mixture of the acidic oxide SiO_2 and amphoteric oxides and hydroxides, such as Al_2O_3, AlOOH, and Fe_2O_3. The Al_2O_3 is dissolved in hot aqueous sodium hydroxide, which separates the aluminium from much of the less soluble Fe_2O_3, although silicates are also rendered soluble in these strongly basic conditions. Cooling the sodium aluminate solution results in the precipitation of $Al(OH)_3$, leaving the silicates in solution.

For the final stage, in the Hall–Héroult process, the aluminium hydroxide is dissolved in molten cryolite (Na_3AlF_6) and the melt is reduced electrolytically at a steel cathode with graphite anodes. The latter participate in the electrochemical reaction by reacting with the evolved oxygen atoms,

$$Al_2O_3 + 3C \rightarrow 2Al + 3CO$$

although the oxidation reaction at the graphite anode results mainly in the production of CO_2 and the overall process is best written as

$$2Al_2O_3 + 3C \rightarrow 4Al + 3CO_2$$

As the power consumption of a typical plant is huge, aluminium is usually produced where electricity is cheap (e.g. from hydroelectric sources in Canada) and not where bauxite is mined (e.g. in Jamaica).

The lighter halogens are the most important elements extracted by electrochemical oxidation. The standard reaction Gibbs energy for the oxidation of Cl^- ions in neutral water

$$2Cl^-(aq) + 2H_2O(l) \rightarrow 2OH^-(aq) + H_2(g) + Cl_2(g)$$
$$\Delta_r G^\ominus = +422 \, kJ \, mol^{-1}$$

is strongly positive, which suggests that electrolysis is required. The minimum potential difference that can achieve the oxidation of Cl^- is about 2.2 V (from $\Delta_r G^\ominus = -\nu FE^\ominus$ and $\nu = 2$).

It may appear that there is a problem with the competing OER reaction (Section 6.16)

$$2H_2O(l) \rightarrow 2H_2(g) + O_2(g) \quad \Delta_r G^\ominus = +414 \, kJ \, mol^{-1}$$

which can be driven forward by a potential difference of only 1.2 V (in this reaction, $\nu = 4$). However, the OER is sluggish and a substantial overpotential is needed to drive it. Consequently, the electrolysis of brine produces Cl_2, H_2, and aqueous NaOH, but not much O_2.

Oxygen, not fluorine, is produced if aqueous solutions of fluorides are electrolysed. Therefore, F_2 is prepared by the electrolysis of an anhydrous mixture of potassium fluoride and hydrogen fluoride—an ionic conductor that is molten above 72°C.

FURTHER READING

A. J. Bard, M. Stratmann, F. Scholtz, and C. J. Pickett, *Encyclopedia of Electrochemistry: Inorganic Chemistry*, vol. 7b. John Wiley & Sons (2006).

J.-M. Savéant, *Elements of Molecular and Biomolecular Electrochemistry: An Electrochemical Approach to Electron-Transfer Chemistry*. John Wiley & Sons (2006).

R. M. Dell and D. A. J. Rand, *Understanding Batteries*. Royal Society of Chemistry (2001).

A. J. Bard, R. Parsons, and R. Jordan, *Standard Potentials in Aqueous Solution*. M. Dekker (1985). A collection of cell potential data with discussion.

A. Wieckowski and J. Nørskov (eds.), *Fuel Cell Science: Theory, Fundamentals, and Biocatalysis*. John Wiley & Sons (2010).

R. O'Hayre, S-K. Cha, W. Colella, and F. B. Prinz, *Fuel Cell Fundamentals*, 3rd ed. John Wiley & Sons (2016).

I. Barin, *Thermochemical Data of Pure Substances*, vols. 1 and 2. VCH Weinheim (1989). A comprehensive source of thermodynamic data for inorganic substances.

J. Emsley, *The Elements*. Oxford University Press (1998). An excellent source of data on the elements including standard potentials.

A. G. Howard, *Aquatic Environmental Chemistry*. Oxford University Press (1998). Discussion of the compositions of freshwater and marine systems explaining the effects of oxidation and reduction processes.

M. Pourbaix, *Atlas of Electrochemical Equilibria in Aqueous Solution*. Pergamon Press (1966). The original and still good source of Pourbaix diagrams.

A. Minguzzi, F.-R. F. Fan, A. Vertova, S. Rondinini, and A. J. Bard, Dynamic potential–pH diagrams application to electrocatalysts for water oxidation. *Chem. Sci.*, 2012, **3**, 217. A discussion of the application of Pourbaix diagrams in electrocatalysis relevant to renewable energy.

W. Stumm and J. J. Morgan, *Aquatic Chemistry*. John Wiley & Sons (1996). A standard reference on natural water chemistry.

P. Zanello and F. Fabrizi de Biani, *Inorganic Electrochemistry: Theory, Practice and Applications*, 2nd ed. Royal Society of Chemistry (2011). An introduction to electrochemical investigations.

P. G. Tratnyek, T. J. Grundl, and S. B. Haderlien (eds.), *Aquatic Redox Chemistry*. American Chemical Society Symposium Series, vol. 1071 (2011). A collection of articles describing recent developments in the field.

K. Shah, B. Moghtaderi, and T. Wall, Selection of suitable oxygen carriers for chemical looping air separation: A thermodynamic approach. *Energy Fuels*, 2012, **26**, 2038.

W. Liu, H. Zuo, J. Wang, Q. Xue, B. Ren, and F. Yang, The production and application of hydrogen in steel industry. *Int. J. Hydrogen Energy*, 2021, **46**, 10548.

EXERCISES

6.1 Assign oxidation numbers for each of the elements participating in the following reactions.

$2NO(g) + O_2(g) \rightarrow 2NO_2(g)$

$2Mn^{3+}(aq) + 2H_2O(l) \rightarrow MnO_2(s) + Mn^{2+}(aq) + 4H^+(aq)$

$LiCoO_2(s) + C(s) \rightarrow LiC_6(s) + CoO_2(s)$

$Ca(s) + H_2(g) \rightarrow CaH_2(s)$

6.2 Use data from Resource section 3 to suggest chemical reagents that would be suitable for carrying out the following transformations, and write balanced equations for the reactions: (a) oxidation of HCl to chlorine gas, (b) reduction of Cr(III)(aq) to Cr(II)(aq), (c) reduction of Ag$^+$ to Ag(s), (d) reduction of I$_2$ to I$^-$.

6.3 Use standard potential data from Resource section 3 as a guide to write balanced equations for the reactions that each of the following species might undergo in aerated aqueous acid. If the species is stable, write 'no reaction'. (a) Cr^{2+}, (b) Fe^{2+}, (c) Cl$^-$, (d) HClO, (e) Zn(s).

6.4 Use the information in Resource section 3 to write balanced equations for the reactions, including disproportionations, that can be expected for each of the following species in aerated acidic aqueous solution: (a) Fe^{2+}, (b) Ru^{2+}, (c) HClO$_2$, (d) Br$_2$.

6.5 Explain why the standard potentials for the half-reactions

$[Ru(NH_3)_6]^{3+}(aq) + e^- \rightarrow [Ru(NH_3)_6]^{2+}(aq)$

$[Fe(CN)_6]^{3-}(aq) + e^- \rightarrow [Fe(CN)_6]^{4-}(aq)$

vary with temperature in opposite directions.

6.6 Balance the following redox reaction in acid solution: $MnO_4^- + H_2SO_3 \rightarrow Mn^{2+} + HSO_4^-$. Predict the qualitative pH dependence on the net potential for this reaction (i.e. increases, decreases, remains the same).

6.7 Using the data shown in the table, and given that the standard reduction potential for the Fe^{3+}/Fe^{2+} couple in aqueous solution is +0.77 V vs SHE, calculate the standard reduction potentials for the remaining redox couples. Compare your answers with values given in Resource section 3, and comment on your results.

	Ti	V	Cr	Mn	Fe	Co
$\Delta_{hyd}H^\ominus(3+)/(kJ\,mol^{-1})$	4154	4375	4560	4544	4430	4651
$\Delta_{hyd}H^\ominus(2+)/(kJ\,mol^{-1})$	1882	1918	1904	1841	1946	1996
I_3 (kJ mol^{-1})	2652	2828	2987	3247	2957	3232

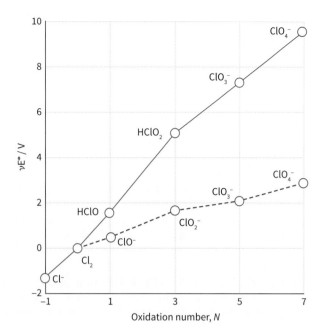

FIGURE 6.22 A Frost diagram for chlorine. The solid red line refers to acid conditions (pH = 0) and the dashed blue line to pH = 14.

6.8 Write the Nernst equation for

(a) the reduction of $O_2(g)$: $O_2(g) + 4\,H^+(aq) + 4\,e^- \rightarrow 2\,H_2O(l)$

(b) the reduction of $Fe_2O_3(s)$:

$$Fe_2O_3(s) + 6\,H^+(aq) + 6\,e^- \rightarrow 2\,Fe(s) + 3\,H_2O(l)$$

In each case express the formula in terms of pH. What is the potential for the reduction of O_2 at pH = 7 and $p(O_2) = 0.20$ bar (the partial pressure of oxygen in air)?

6.9 Answer the following questions using the Frost diagram in Fig. 6.22. (a) What are the consequences of dissolving Cl_2 in aqueous basic solution? (b) What are the consequences of dissolving Cl_2 in aqueous acid? (c) Is the failure of $HClO_2$ to disproportionate in aqueous solution a thermodynamic or a kinetic phenomenon?

6.10 Use standard potentials as a guide to write equations for the main net reaction that you would predict in the following experiments: (a) N_2O is bubbled into aqueous NaOH solution, (b) zinc metal is added to aqueous sodium triiodide, (c) I_2 is added to excess aqueous $HClO_3$.

6.11 Adding NaOH to an aqueous solution containing Ni^{2+} results in precipitation of $Ni(OH)_2$. The standard potential for the Ni^{2+}/Ni couple is -0.25 V and the solubility product $K_{sp} = [Ni^{2+}][OH^-]^2 = 1.5 \times 10^{-16}$. Calculate the electrode potential at pH = 14.

6.12 Characterize the condition of acidity or basicity that would most favour the following transformations in aqueous solution:

(a) $Mn^{2+} \rightarrow MnO_4^-$, (b) $ClO_4^- \rightarrow ClO_3^-$, (c) $H_2O_2 \rightarrow O_2$, (d) $I_2 \rightarrow 2I^-$.

6.13 From the following Latimer diagram (which does not correspond to standard conditions), calculate the value of $E^\ominus$ for the reaction $2\,HO_2(aq) \rightarrow O_2(g) + H_2O_2(aq)$.

$$O_2 \xrightarrow{-0.05} HO_2 \xrightarrow{+1.44} H_2O_2$$

Comment on the thermodynamic tendency of HO_2 to undergo disproportionation.

6.14 Use the Latimer diagram for chlorine to determine the potential for reduction of ClO_4^- to Cl_2. Write a balanced equation for this half-reaction.

6.15 Using the following Latimer diagram, which shows the standard potentials for sulfur species in acid solution (pH = 0), construct a Frost diagram and calculate the standard potential for the $HSO_4^-/S_8(s)$ couple.

$$S_2O_8^{2-} \xrightarrow{+1.96} HSO_4^- \xrightarrow{+0.16} H_2SO_3 \xrightarrow{+0.40} S_2O_3^{2-}$$
$$\xrightarrow{+0.60} S \xrightarrow{+0.14} H_2S$$

6.16 Calculate the reduction potential at 25°C for the conversion of MnO_4^- to $MnO_2(s)$ in aqueous solution at pH = 9.00 and 1 M $MnO_4^-(aq)$ given that $E^\ominus(MnO_4^-/MnO_2) = +1.69$ V.

6.17 Using the following aqueous acid solution reduction potentials $E^\ominus(Pd^{2+},Pd) = +0.915$ V and $E^\ominus([PdCl_4]^{2-},Pd) = +0.60$ V, calculate the equilibrium constant for the reaction $Pd^{2+}(aq) + 4\,Cl^-(aq) \rightleftharpoons [PdCl_4]^{2-}(aq)$ in 1 M HCl(aq).

6.18 Calculate the equilibrium constant of the reaction $Au^+(aq) + 2\,CN^-(aq) \rightleftharpoons [Au(CN)_2]^-(aq)$ from the standard potentials

$$Au^+(aq) + e^- \rightarrow Au(s) \qquad E^\ominus = +1.68\,V$$
$$[Au(CN)_2]^-(aq) + e^- \rightarrow Au(s) + 2\,CN^-(aq) \qquad E^\ominus = -0.6\,V$$

6.19 The ligand edta forms stable complexes with hard acid centres. How will complexation with edta affect the reduction of M^{2+} to the metal in the 3d series?

6.20 Draw a Frost diagram for mercury in acid solution, given the following Latimer diagram:

$$Hg^{2+} \xrightarrow{0.911} Hg_2^{2+} \xrightarrow{0.796} Hg$$

Comment on the tendency of any of the species to act as an oxidizing agent, a reducing agent, or to undergo disproportionation.

6.21 Use Fig. 6.12 to find the approximate potential of an aerated lake at pH = 6. With this information and Latimer diagrams from Resource section 3, predict the species at equilibrium for the elements (a) iron, (b) manganese, (c) sulfur.

6.22 Explain why water with high concentrations of dissolved carbon dioxide and open to atmospheric oxygen is very corrosive towards iron.

6.23 The species Fe^{2+} and H_2S are important at the bottom of a lake where O_2 is scarce. If the pH = 6, what is the maximum value of E characterizing the environment?

6.24 In Fig. 6.11, which of the boundaries depend on the choice of Fe^{2+} concentration as 10^{-5} mol dm^{-3}?

6.25 Consult the Ellingham diagram in Fig. 6.18 and determine if there are any conditions under which aluminium might be expected to reduce MgO. Comment on these conditions.

6.26 The standard potentials for phosphorus species in aqueous solution at pH = 0 and pH = 14 are provided in Resource section 3.

(a) Account for the difference in reduction potentials between pH 0 and pH 14.

(b) Construct a single Frost diagram showing both sets of data.

(c) Phosphine (PH_3) can be prepared by heating phosphorus with aqueous alkali. Discuss the reactions that are feasible and estimate their equilibrium constants.

6.27 Given the following standard potentials in basic solution

$CrO_4^{2-}(aq) + 4H_2O(l) + 3e^- \rightarrow Cr(OH)_3(s) + 5OH^-(aq)$ $E^\ominus = -0.11$ V
$[Cu(NH_3)_2]^+(aq) + e^- \rightarrow Cu(s) + 2NH_3(aq)$ $E^\ominus = -0.10$ V

and assuming that a reversible reaction can be established on a suitable catalyst, calculate $E^\ominus$, $\Delta_r G^\ominus$, and K for the reductions of (a) CrO_4^{2-} and (b) $[Cu(NH_3)_2]^+$ in basic solution. Comment on why $\Delta_r G^\ominus$ and K are so different between the two cases despite the values of $E^\ominus$ being so similar.

6.28 Many of the tabulated data for standard potentials have been determined from thermochemical data rather than direct electrochemical measurements of cell potentials. Carry out a calculation to illustrate this approach for the half-reaction $Sc_2O_3(s) + 3H_2O(l) + 6e^- \rightarrow 2Sc(s) + 6OH^-(aq)$.

	Sc^{3+}(aq)	OH^-(aq)	H_2O(l)	Sc_2O_3(s)	Sc(s)
$\Delta_f H^\ominus$/(kJ mol^{-1})	−614.2	−230.0	−286.8	−1908.7	0
$S_m^\ominus$/(JK^{-1} mol^{-1})	−255.2	−10.75	69.91	77.0	34.76

6.29 Standard potentials at 25°C for indium and thallium in aqueous solution (pH = 0) are given below.

$In^{3+}(aq) + 3e^- \rightarrow In(s)$ $E^\ominus = -0.338$ V
$In^+(aq) + e^- \rightarrow In(s)$ $E^\ominus = -0.126$ V
$Tl^{3+}(aq) + 3e^- \rightarrow Tl(s)$ $E^\ominus = +0.721$ V
$Tl^+(aq) + e^- \rightarrow Tl(s)$ $E^\ominus = -0.336$ V

Use the data to construct a Frost diagram for the two elements and discuss the relative stabilities of the species.

6.30 Data shown below refer to redox couples for the Group 8 elements Fe and Ru.

$Fe^{2+}(aq) + 2e^- \rightarrow Fe(s)$ $E^\ominus = -0.44$ V
$Fe^{3+}(aq) + e^- \rightarrow Fe^{2+}(aq)$ $E^\ominus = +0.77$ V
$Ru^{2+}(aq) + 2e^- \rightarrow Ru(s)$ $E^\ominus = +0.80$ V
$Ru^{3+}(aq) + 2e^- \rightarrow Ru^{2+}(aq)$ $E^\ominus = +0.25$ V

(a) Comment on the relative stability of Fe^{2+} and Ru^{2+} in acidic aqueous solution.

(b) Give a balanced equation for the reaction that would occur if iron filings were to be added to an acidified aqueous solution of an Fe^{3+} salt.

(c) Calculate the equilibrium constant for the reaction between Fe^{2+}(aq) and an acidified solution of potassium permanganate under standard conditions (use Resource section 3).

6.31 The Latimer diagram for vanadium species in acidic (pH = 0) solution is:

$V(OH)_4^+ \xrightarrow{+1.00} VO_2^{2+} \xrightarrow{+0.34} V^{3+} \xrightarrow{-0.26} V^{2+} \xrightarrow{-1.13} V$

Using these data:

(a) Calculate the potential for the reduction of VO^{2+}(aq) to V(s) and write a balanced chemical equation for this process.

(b) Construct a Frost diagram for vanadium at pH = 0.

(c) Calculate the equilibrium constant for the disproportionation of V^{3+}(aq) to V^{2+}(aq) and VO^{2+}(aq) at pH = 0. Hence comment on the stability of V^{3+}(aq) in acidic (pH = 0) solution.

(d) How might the potential for the reduction of VO^{2+}(aq) to V^{3+}(aq) be affected by increasing the pH of the solution?

6.32 The following Latimer diagrams show the standard reduction potentials $E^\ominus$/V for some oxidation states of iron in acid and alkaline solution:

pH = 0 $Fe^{3+} \xrightarrow{+0.77} Fe^{2+} \xrightarrow{-0.44} Fe$
pH = 14 $FeO_4^{2-} \xrightarrow{+0.55} FeO_2^- \xrightarrow{-0.69} Fe(O)OH^- \xrightarrow{-0.80} Fe$

(a) Plot a Frost diagram showing the states of Fe under acid and alkaline conditions.

(b) Calculate the standard potential for the Fe^{3+}/Fe couple in acid solution.

(c) Explain why the potentials for the Fe(III)/Fe(II) reduction are very different in acid and alkaline conditions, and comment on the fact that FeO_4^{2-} is unknown in acid solution.

(d) FeO_4^{2-} can be prepared by oxidation of Fe(III) by ClO^- in alkaline solution. Making use of the standard reduction potentials given below, write an equation for the reaction occurring in this preparation.

Standard potentials $E^\ominus$/V at pH = 14:

ClO^-/Cl_2 +0.42 V Cl_2/Cl^- +1.36 V.

6.33 Reduction potentials for half-reactions of simple carbon species measured in aqueous solution at pH 7.0, 25°C, are as follows.

$CO_2(g) + 4H^+ + 4e^- \rightarrow HCHO(aq) + H_2O$ $E^\ominus = -0.51$ V
$CO_2(g) + e^- \rightarrow CO_2^-(aq)$ $E^\ominus = -1.90$ V
$CO_2(g) + 2H^+ + 2e^- \rightarrow CO(g) + H_2O$ $E^\ominus = -0.52$ V
$CO_2(g) + H^+ + 2e^- \rightarrow HCO_2^-(aq)$ $E^\ominus = -0.43$ V
$CO_2(g) + 6H^+ + 6e^- \rightarrow CH_3OH(aq) + H_2O$ $E^\ominus = -0.38$ V
$CO_2(g) + 8H^+ + 8e^- \rightarrow CH_4(g) + 2H_2O$ $E^\ominus = -0.24$ V

(a) Assign carbon oxidation numbers to each of the species involved.

(b) Construct a Frost diagram, making HCHO the reference point at the origin.

(c) Identify the redox couple having the most positive reduction potential.

(d) Is hydration of CO to give formate (HCO_2^-) favourable at pH 7.0?

(e) At pH 7.0, 25°C, the reduction potential for the hydrogen half-reaction is given by:

$2H^+(aq) + 2e^- \rightarrow H_2(g)$ $E^\ominus = -0.41$ V

Calculate the equilibrium constant for the water-gas shift reaction

$CO_2(g) + H_2(g) \rightleftharpoons CO(g) + H_2O(l)$

in aqueous solution, under these conditions.

TUTORIAL PROBLEMS

6.1 Use standard potential data to suggest why permanganate (MnO_4^-) is not a suitable oxidizing agent for the quantitative estimation of Fe^{2+} in the presence of HCl, but becomes so if sufficient Mn^{2+} and phosphate ion are added to the solution. (Hint: Phosphate forms complexes with Fe^{3+}, thereby stabilizing it.)

6.2 Explain the significance of reduction potentials in inorganic chemistry, highlighting their applications in investigations of stability, solubility, and reactivity in water.

6.3 Using Resource sections 1–3, and data for atomization of the elements $\Delta_a H^\ominus = 397\,kJ\,mol^{-1}$ (Cr) and $664\,kJ\,mol^{-1}$ (Mo), construct thermodynamic cycles for the reactions of Cr or Mo with dilute acids and thus consider the importance of metallic bonding in determining the standard reduction potentials for formation of cations from metals.

6.4 Enterobactin (Ent) is a special ligand secreted by some bacteria to sequester Fe from the environment (Fe is an essential nutrient for almost all living species, see Section 26.6). The equilibrium constant for formation of [Fe(III)(Ent)] ($10^{52}\,mol^{-1}\,dm^3$) is at least forty orders of magnitude higher than the equilibrium constant for the corresponding Fe(II) complex. Determine the feasibility of releasing Fe from [Fe(III)(Ent)] by its reduction to Fe(II) under conditions of neutral pH, noting that the strongest reducing agent normally available to bacteria is H_2.

6.5 The reduction potential of an ion such as OH^- can be strongly influenced by the solvent. (a) From the review article by D. T. Sawyer and J. L. Roberts (*Acc. Chem. Res.*, 1988, **21**, 469), describe the magnitude of the change in the potential of the OH/OH^- couple on changing the solvent from water to acetonitrile, CH_3CN. (b) Suggest a qualitative interpretation of the difference in solvation of the OH^- ion in these two solvents.

6.6 Construct an Ellingham diagram for the thermal splitting of water ($H_2O(g) \rightarrow H_2(g) + \tfrac{1}{2}O_2(g)$) by using $\Delta_r H^\ominus = +260\,kJ\,mol^{-1}$; $\Delta_r S^\ominus = +60\,J\,K^{-1}\,mol^{-1}$ (you may assume that $\Delta_r H^\ominus$ is independent of temperature). Hence, calculate the temperature at which H_2 may be obtained by spontaneous decomposition of water (*Chem. Rev.*, 2007, **107**, 4048). Comment on the feasibility of producing H_2 by this means.

6.7 By reference to the paper by K. Shah, B. Moghtaderi, and T. Wall (*Energy Fuels*, 2012, **26**, 2038), discuss how the information represented in Ellingham diagrams can be used to identify metal oxides suitable for the production of pure O_2 from air.

6.8 Discuss how the equilibrium $Cu^+(aq) + Cu(s) \rightleftharpoons 2Cu^+(aq)$ can be shifted by complexation with chloride ions (see J. Malyyszko and M. Kaczor, *J. Chem. Educ.*, 2003, **80**, 1048).

6.9 In their article 'Variability of the cell potential of a given chemical reaction' (*J. Chem. Educ.*, 2004, **81**, 84), L. H. Berka and I. Fishtik conclude that $E^\ominus$ for a chemical reaction is not a state function because the half-reactions are arbitrarily chosen and may contain different numbers of transferred electrons. Discuss this objection.

6.10 In their article 'A thermochemical study of ceria: Exploiting an old material for new modes of energy conversion and CO_2 mitigation' (*Philos. Trans. R. Soc. London*, 2010, **368**, 3269), Chueh and Haile describe a way to convert water into H_2 and O_2 using CeO_2. Explain the chemical and thermodynamic principles of this innovation.

6.11 In their article 'Enzymes and bio-inspired electrocatalysts in solar fuel devices' (*Energy Environ. Sci.*, 2012, **5**, 7470), Woolerton et al. use a generic Frost diagram (Fig. 6.23) to illustrate the concept of chemical energy storage and release. So-called 'energy-rich substances' are compounds (fuels and oxidants) in which chemical energy is stored: fuels (such as hydrocarbons, boranes, metal hydrides) lie towards the upper left, whereas strong oxidants such as O_2 lie towards the upper right. 'Energy-poor substances' are stable, reduced compounds (e.g. H_2O) lying towards the lower left and oxidized compounds (e.g. CO_2, components of ash) lying towards the lower right. Energy release (whether by combustion or in a fuel cell or battery) occurs when a species from the upper left reacts with a species from the upper right to give products positioned diagonally down their respective arrows. During energy storage (such as photosynthesis or charging a battery), compounds from the lower left and lower right are transformed using sunlight or electricity into products that are positioned diagonally upwards along their respective arrows. Using data from Resource section 3, assess the usefulness of this concept for a methanol/oxygen system and a lead–acid battery.

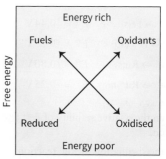

FIGURE 6.23 A generic Frost diagram depicting the relationships between fuels, oxidants, and energy-poor substances such as water, carbon dioxide, and ash.

An introduction to coordination compounds

The language of coordination chemistry

7.1 **Representative ligands**

7.2 **Nomenclature**

Constitution and geometry

7.3 **Low coordination numbers**

7.4 **Intermediate coordination numbers**

7.5 **Higher coordination numbers**

7.6 **Polymetallic complexes**

Isomerism and chirality

7.7 **Square-planar complexes**

7.8 **Tetrahedral complexes**

7.9 **Trigonal-bipyramidal and square-pyramidal complexes**

7.10 **Octahedral complexes**

7.11 **Ligand chirality**

The thermodynamics of complex formation

7.12 **Formation constants**

7.13 **Trends in successive formation constants**

7.14 **The chelate and macrocyclic effects**

7.15 **Steric effects and electron delocalization**

Electronic structure

7.16 **Crystal-field theory**

7.17 **Ligand-field theory**

Further reading

Exercises

Tutorial problems

Metal complexes, in which a metal atom or ion is surrounded by several ligands, play an important role in inorganic chemistry, especially for elements of the d block. In this chapter we introduce the common structural arrangements for ligands around a single metal atom and the isomeric forms that are possible. We then introduce a discussion on the nature of ligand–metal bonding in transition metal complexes in terms of two theoretical models. We start with the simple but useful crystal-field theory, which is based on an electrostatic model of the bonding, and then progress to the more sophisticated ligand-field theory. Both theories invoke a parameter, the ligand-field splitting parameter, to correlate spectroscopic and magnetic properties.

In the context of metal coordination chemistry, the term **complex** means a central metal atom or ion surrounded by a set of ligands. A **ligand** is an ion or molecule that can have an independent existence. Two examples of complexes are $[Co(NH_3)_6]^{3+}$, in which the Co^{3+} ion is surrounded by six NH_3 ligands, and $[Na(OH_2)_6]^+$, in which the Na^+ ion is surrounded by six H_2O ligands. We shall use the term **coordination compound** to mean a neutral complex or an ionic compound in which at least one of the ions is a complex. Thus, $[Ni(CO)_4]$ (**1**) and $[Co(NH_3)_6]Cl_3$ (**2**) are both coordination compounds.

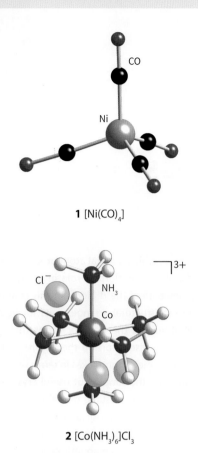

1 [Ni(CO)$_4$]

2 [Co(NH$_3$)$_6$]Cl$_3$

A complex is a combination of a Lewis acid (the central metal atom) with a number of Lewis bases (the ligands). The atom in the Lewis base ligand that forms the bond to the central atom is called the **donor atom** because it donates the electrons used in bond formation. Thus, N is the donor atom when NH$_3$ acts as a ligand, and O is the donor atom when H$_2$O acts as a ligand. The metal atom or ion, the Lewis acid in the complex, is the **acceptor** atom. All metals, from all blocks of the periodic table, form complexes.

The principal features of the geometrical structures of metal complexes were identified in the late nineteenth and early twentieth centuries by Alfred Werner, whose training was in organic stereochemistry. Werner combined the interpretation of optical and geometrical isomerism, patterns of reactions, and conductance data in work that remains a model of how to use physical and chemical evidence effectively and imaginatively. The striking colours of many d- and f-metal coordination compounds, which reflect their electronic structures, were a mystery to Werner. This characteristic was clarified only when the description of electronic structure in terms of orbitals was applied to the problem in the period from 1930 to 1960. We look at the electronic structure of d- and f-metal complexes in Chapter 22.

The geometrical structures of metal complexes can now be determined in many more ways than Werner had at his disposal. When single crystals of a compound can be grown, X-ray diffraction (Section 8.1) gives precise shapes, bond distances, and angles. Nuclear magnetic resonance (Section 8.6) can be used to study complexes with lifetimes longer than microseconds. Very short-lived complexes—those with lifetimes comparable to diffusional encounters in solution (a few nanoseconds)—can be studied by vibrational and electronic spectroscopy. It is possible to infer the geometries of complexes with long lifetimes in solution (such as the classic complexes of Co(III), Cr(III), and Pt(II) and many organometallic compounds) by analysing patterns of reactions and isomerism. This method was originally exploited by Werner, and it still teaches us much about the synthetic chemistry of the compounds as well as helping to establish their structures.

The language of coordination chemistry

KEY POINTS In an inner-sphere complex, the ligands are attached directly to a central metal ion; outer-sphere complexes occur where cation and anion associate in solution or an ionic solid.

In what we normally understand as a complex, more precisely an **inner-sphere complex**, the ligands are attached directly to the central metal atom or ion. These ligands form the **primary coordination sphere** of the complex and their number is called the **coordination number** of the central metal atom. As in solids, a wide range of coordination numbers can occur, and the origin of the structural richness and chemical diversity of complexes is the ability of the coordination number to range up to 12.

Although we shall concentrate on inner-sphere complexes throughout this chapter, we should keep in mind that complex cations can associate electrostatically with anionic ligands (and, by other weak interactions, with solvent molecules) without displacement of the ligands already present. The product of this association is called an **outer-sphere complex**. In an aqueous solution of [Mn(OH$_2$)$_6$]$^{2+}$ and SO$_4^{2-}$ ions, for instance, the equilibrium concentration of the outer-sphere complex {[Mn(OH$_2$)$_6$]$^{2+}$SO$_4^{2-}$} (**3**) can, depending on the concentration, exceed that of the inner-sphere complex [Mn(OH$_2$)$_5$SO$_4$] in which the ligand SO$_4^{2-}$ is directly attached to the metal ion.

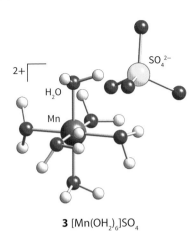

3 [Mn(OH$_2$)$_6$]SO$_4$

It is worth remembering that most methods of measuring complex formation equilibria do not distinguish outer-sphere from inner-sphere complex formation but simply detect the sum of all bound ligands. Within crystalline solids, it is possible to have both inner-sphere and outer-sphere coordination of oppositely charged anions and of molecules of neutral solvent or ligand. Water molecules that are not directly coordinated to the complex cation, equivalent to outer-sphere ligand molecules, are termed 'water of crystallization'. In solid manganese(II) sulfate pentahydrate, {[Mn(OH$_2$)$_4$]SO$_4$}·H$_2$O, each manganese ion is coordinated to two sulfate anions and four water molecules; the remaining noncoordinated water molecule is a water of crystallization. In crystalline iron(II) sulfate heptahydrate, {[Fe(OH$_2$)$_6$]$^{2+}$}·SO$_4^{2-}$·H$_2$O, the iron cation is only coordinated to water molecules; the sulfate anion is outer-sphere and there is one water of crystallization.

A large number of molecules and ions can behave as ligands, and a large number of metal ions form complexes. We now introduce some representative ligands and consider the basics of naming complexes.

7.1 Representative ligands

KEY POINTS Polydentate ligands can form chelates; a bidentate ligand with a small bite angle can result in distortions from standard structures.

Table 7.1 gives the names and formulas of a number of common simple ligands, and Table 7.2 gives the common prefixes used. Some of these ligands have only a single donor pair of electrons and will have only one point of attachment to the metal: such ligands are classified as **monodentate** (from the Latin meaning 'one-toothed'). Ligands that have more than one point of attachment are classified as **polydentate**. Ligands that specifically have two points of attachment are known as **bidentate**, those with three **tridentate**, and so on.

Ambidentate ligands have more than one different potential donor atom. An example is the thiocyanate ion (NCS$^-$), which can attach to a metal atom either by the N atom, to give thiocyanato-κN complexes, or by the S atom, to give thiocyanato-κS complexes. Another example of an ambidentate ligand is NO$_2^-$: as M–NO$_2^-$ (**4**) the ligand is nitrito-κN, and as M–ONO (**5**) it is nitrito-κO.

TABLE 7.1 Typical ligands and their names

Name	Formula	Abbreviation	Donor atoms	Number of donors
Acetylacetonato	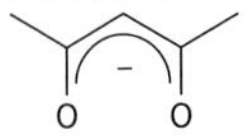 (structure)	acac$^-$	O	2
Ammine	NH$_3$		N	1
Aqua	H$_2$O		O	1
2,2-Bipyridine	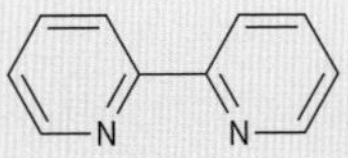 (structure)	bpy	N	2
Bromido	Br$^-$		Br	1
Carbonato	CO$_3^{2-}$		O	1 or 2
Carbonyl	CO		C	1
Chlorido	Cl$^-$		Cl	1
1,4,7,10,13-Hexaoxa-cyclooctadecane	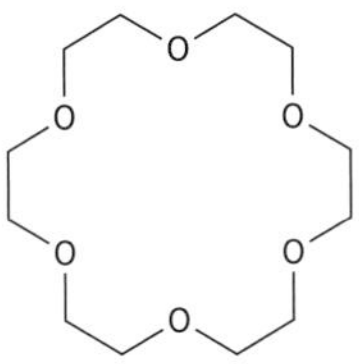 (structure)	18-crown-6	O	6

(Continued)

TABLE 7.1 (Continued)

Name	Formula	Abbreviation	Donor atoms	Number of donors
4,7,13,16,21-Pentaoxa-1,10-diaza-bicyclo[8.8.5]tricosane	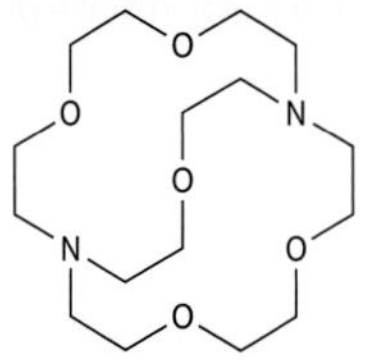 (cryptand structure)	2.2.1 crypt	N, O	2N, 5O
Cyanido	CN⁻		C	1
Diethylenetriamine	NH(CH$_2$CH$_2$NH$_2$)$_2$	dien	N	3
Bis(diphenylphosphino)ethane	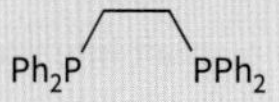 Ph$_2$P–CH$_2$CH$_2$–PPh$_2$	dppe	P	2
Bis(diphenylphosphino)methane	 Ph$_2$P–CH$_2$–PPh$_2$	dppm	P	2
Cyclopentadienyl	C$_5$H$_5^-$	Cp⁻	C	5
1,2-Diaminoethane	NH$_2$CH$_2$CH$_2$NH$_2$	en	N	2
Ethylenediaminetetraacetato	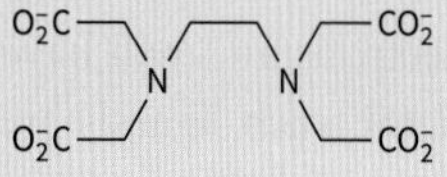 (O$_2^-$C-CH$_2$)$_2$N-CH$_2$CH$_2$-N(CH$_2$-CO$_2^-$)$_2$	edta⁴⁻	N, O	2N, 4O
Fluorido	F⁻		F	1
Glycinato	NH$_2$CH$_2$CO$_2^-$	gly⁻	N, O	1N, 1O
Hydrido	H⁻		H	1
Hydroxido	OH⁻		O	1
Iodido	I⁻		I	1
Nitrato	NO$_3^-$		O	1 or 2
Nitrito-κN	NO$_2^-$		N	1
Nitrito-κO	ONO⁻		O	1
Oxido	O^{2-}		O	1
Oxalato	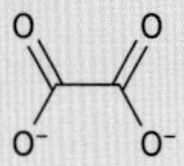 (oxalate structure)	ox²⁻	O	2
Pyridine	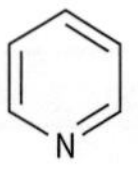 (pyridine structure)	py	N	1
Sulfido	S^{2-}		S	1
Tetraazacyclotetradecane	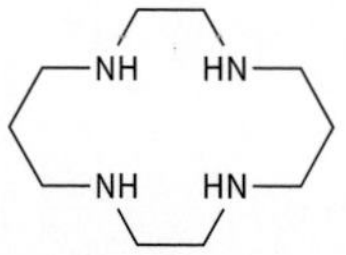 (cyclam structure)	cyclam	N	4
Thiocyanato-κN	NCS⁻		N	1
Thiocyanato-κS	SCN⁻		S	1
Thiolato	RS⁻		S	1
Triaminotriethylamine	N(CH$_2$CH$_2$NH$_2$)$_3$	tren	N	4
Tricyclohexylphosphine*	P(C$_6$H$_{11}$)$_3$	PCy$_3$	P	1
Trimethylphosphine*	P(CH$_3$)$_3$	PMe$_3$	P	1
Triphenylphosphine*	P(C$_6$H$_5$)$_3$	PPh$_3$	P	1

*The PR$_3$ ligand is formally a substituted phosphane, but the older nomenclature based on the name phosphine is widely used.

TABLE 7.2 Prefixes used for naming complexes

Prefix	Meaning
mono-	1
di-, bis-	2
tri-, tris-	3
tetra-, tetrakis-	4
penta-	5
hexa-	6
hepta-	7
octa-	8
nona-	9
deca-	10
undeca-	11
dodeca-	12

forms a five-membered ring when both N atoms attach to the same metal atom (**6**).

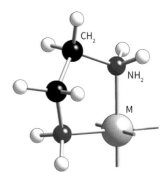

6 1,2-Diaminoethane (en) ligand attached to M

It is important to note that normal chelating ligands will attach to the metal only at two adjacent coordination sites, in a *cis* fashion. The hexadentate ligand ethylenediaminetetraacetic acid, as its anion (edta^{4-}), can attach at six points (at two N atoms and four O atoms) and can form an elaborate complex with five five-membered rings (**7**). This ligand is used to trap metal ions, such as Ca^{2+} ions, in 'hard' water. Complexes of chelating ligands often have additional stability over those of nonchelating ligands; the origin of this so-called **chelate effect** is discussed later in this chapter (Section 7.14). Table 7.1 includes some of the most common chelating ligands.

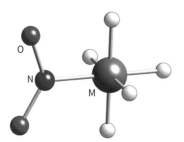

4 Nitrito-κ*N* ligand

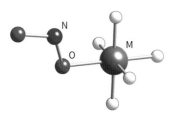

5 Nitrito-κ*O* ligand

A NOTE ON GOOD PRACTICE

Although the 'κ terminology' in which the letter κ (kappa) is used to indicate the atom of ligation has been around for some time, the old names isothiocyanato, indicating attachment of the SCN$^-$ ligand by the N atom, and thiocyanato, indicating attachment by the S atom, are still widely encountered. Similarly, the old names nitro, indicating attachment of the NO$_2^-$ ligand by the N atom, and nitrito, indicating attachment by the O atom, are also still widely encountered.

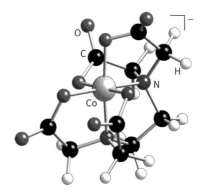

7 [Co(edta)]$^-$

In a chelate formed from a saturated organic ligand, such as 1,2-diaminoethane, a five-membered ring can fold into a conformation that preserves the tetrahedral angles within the ligand and yet still achieve an L–M–L angle of 90°, the angle typical of octahedral complexes. Six-membered rings may be favoured sterically or by electron delocalization through their π orbitals. The bidentate β-diketones, for example, coordinate as the anions of their enols in six-membered ring structures (**8**).

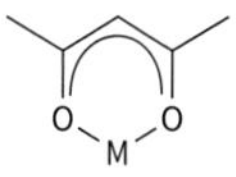

Polydentate ligands can produce a **chelate** (from the Greek for 'claw'), a complex in which a ligand forms a ring that includes the metal atom. An example is the bidentate ligand 1,2-diaminoethane (en, NH$_2$CH$_2$CH$_2$NH$_2$), which

8

An important example is the acetylacetonato anion (acac⁻, **9**).

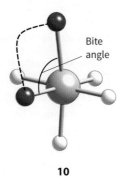

9

Because biochemically important amino acids can form five- or six-membered rings, they also chelate readily. The degree of strain in a chelating ligand is often expressed in terms of the **bite angle**, the L–M–L angle in the chelate ring (**10**).

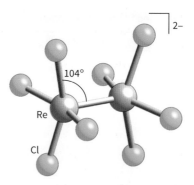

11 $[Re_2Cl_8]^{2-}$

10

7.2 Nomenclature

KEY POINTS *The cation and anion of a complex are named according to a set of rules; cations are named first, and ligands are named in alphabetical order.*

Detailed guidance on nomenclature is beyond the scope of this book, and only a general introduction is presented here. In fact, the names of complexes often become so cumbersome that inorganic chemists often prefer to give the formula rather than spell out the entire name.

For compounds that consist of one or more ions, the cation is named first, followed by the anion (as for simple ionic compounds), regardless of which ion is complex. Complex ions are named with their ligands in alphabetical order (ignoring any numerical prefixes). The ligand names are followed by the name of the metal with either its oxidation number in parentheses, as in hexaamminecobalt(III) for $[Co(NH_3)_6]^{3+}$, or with the overall charge on the complex specified in parentheses, as in hexaamminecobalt(3+). The suffix -ate is added to the name of the metal if the complex is an anion, as in tetrachloridoplatinate(II) for $[PtCl_4]^{2-}$. Some metals, such as iron, copper, silver, gold, tin, and lead, have anion names that derive from the Latin form of the element name (i.e. ferrate, cuprate, argentate, aurate, stannate, and plumbate, respectively).

The number of a particular type of ligand in a complex is indicated by the prefixes mono-, di-, tri-, and tetra-. The same prefixes are used to state the number of metal atoms if more than one is present in a complex, as in octachloridodirhenate(III), $[Re_2Cl_8]^{2-}$ (**11**).

Where confusion with the names of ligands is likely—perhaps because the name already includes a prefix, as with 1,2-diaminoethane—the alternative prefixes bis-, tris-, and tetrakis- are used, with the ligand name in parentheses. For example, dichlorido- is unambiguous, but tris(1,2-diaminoethane) shows more clearly that there are three 1,2-diaminoethane ligands, as in tris(1,2-diaminoethane)cobalt(II), $[Co(en)_3]^{2+}$. Ligands that bridge two metal centres are denoted by a prefix μ (mu) added to the name of the relevant ligand, as in μ-oxido-bis(pentamminecobalt(III)) (**12**). If the number of centres bridged is greater than two, a subscript is used to indicate the number; for instance, a hydride ligand bridging three metal atoms is denoted μ_3-H.

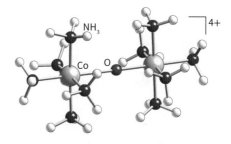

12 $[(H_3N)_5CrOCr(NH_3)_5]^{4+}$

> **A NOTE ON GOOD PRACTICE**
>
> The letter κ can be superscripted with a number that indicates the number of points of attachment: thus a bidentate 1,2-diaminoethane ligand bound through both N atoms is indicated as $\kappa^2 N$. The letter η (eta) is used with superscripts to indicate bonding modes of certain ligands in organometallic chemistry (Section 24.4).

Square brackets are used to indicate which groups are bound to a metal atom, and should be used whether the complex is charged or not. The metal symbol is given first, then the ligands in alphabetical order (the earlier rule that anionic ligands precede neutral ligands has been superseded), as in $[CoCl_2(NH_3)_4]^+$ for tetraamminedichloridocobalt(III). This order is sometimes varied to clarify which ligand is involved in a reaction. Polyatomic ligand formulas are sometimes written in an unfamiliar sequence (as for OH$_2$ in [Fe(OH$_2$)$_6$]$^{2+}$ for hexaaquairon(II)) to place the donor atom adjacent to the metal atom and so help to make the structure of the complex clear. The donor atom of an ambidentate ligand is sometimes indicated by underlining it; for, example [Fe(OH$_2$)$_5$(NCS)]$^{2+}$. Note that, somewhat confusingly, the ligands in the formula are in alphabetical order of binding element, and thus, as in the cobalt example above, the formula and name of the complex may differ in the order in which the ligands appear.

EXAMPLE 7.1 Naming complexes

Name the complexes (a) [PtCl$_2$(NH$_3$)$_4$]$^{2+}$; (b) [Ni(CO)$_3$(py)]; (c) [Cr(edta)]$^-$; (d) [CoCl$_2$(en)$_2$]$^+$; (e) [Rh(CO)$_2$I$_2$]$^-$.

Answer To name a complex, we start by working out the oxidation number of the central metal atom and then add the names of the ligands in alphabetical order. (a) The complex has two anionic ligands (Cl$^-$), four neutral ligands (NH$_3$), and an overall charge of +2; hence the oxidation number of platinum must be +4. According to the alphabetical order rules, the name of the complex is tetraamminedichloridoplatinum(IV). (b) The ligands CO and py (pyridine) are neutral, so the oxidation number of nickel must be 0. It follows that the name of the complex is tricarbonylpyridinenickel(0). (c) This complex contains the hexadentate edta^{4-} ion as the sole ligand. The four negative charges of the ligand result in a complex with a single negative charge if the central metal ion is Cr^{3+}. The complex is therefore ethylenediaminetetraacetatochromate(III). (d) This complex contains two anionic chloride ligands and two neutral en ligands. The overall charge of +1 must be the result of the cobalt having oxidation number +3. The complex is therefore dichloridobis(1,2-diaminoethane)cobalt(III). (e) This complex contains two anionic I$^-$ (iodido) ligands and two neutral CO ligands. The overall charge of −1 must be the result of the rhodium having oxidation number +1. The complex is therefore dicarbonyldiiodidorhodate(I).

Self-test 7.1 Write the formulas of the following complexes: (a) diaquadichloridoplatinum(II); (b) diamminetetra(thiocyanato-κN)chromate(III); (c) tris(1,2-diaminoethane)rhodium(III); (d) bromidopentacarbonylmanganese(I); (e) chloridotris(triphenylphosphine)rhodium(I).

Constitution and geometry

KEY POINTS The number of ligands in a complex depends on the size of the metal atom, the identity of the ligands, and the electronic interactions.

The coordination number of a metal atom or ion is not always evident from the composition of the solid, as solvent molecules and species that are potentially ligands may simply fill spaces within the structure and not have any direct bonds to the metal ion. For example, X-ray diffraction shows that CoCl$_2$·6H$_2$O contains the neutral complex [CoCl$_2$(OH$_2$)$_4$] and two uncoordinated (outer-sphere) H$_2$O molecules occupying well-defined positions in the crystal. Such additional solvent molecules are called **solvent of crystallization**.

Three factors govern the coordination number of a complex:

- the size of the central atom or ion
- the steric interactions between the ligands
- electronic interactions between the central atom or ion and the ligands.

In general, the large radii of atoms and ions in the later periods favour higher coordination numbers. For similar steric reasons, bulky ligands often result in low coordination numbers, especially if the ligands are also charged (when unfavourable electrostatic interactions also come into play). High coordination numbers are also most common on the left of a period, where the ions have larger radii. They are especially common when the metal ion has only a few electrons, because a small number of valence electrons means that the metal ion can accept more electrons from Lewis bases; one example is [Mo(CN)$_8$]$^{4-}$.

Lower coordination numbers are found on the right of the d block where atomic radii are smallest, particularly if the ions are rich in electrons, as they are less able to accept additional electrons; an example is [PtCl$_4$]$^{2-}$. Low coordination numbers occur if the ligands can form multiple bonds with the central metal, as in [MnO$_4$]$^-$ and [CrO$_4$]$^{2-}$, as now the electrons provided by each ligand tend to exclude the attachment of more ligands. We consider these coordination number preferences in more detail in Section 7.16.

7.3 Low coordination numbers

KEY POINTS Two-coordinate complexes are found for Cu$^+$ and Ag$^+$; these complexes often accommodate more ligands if they are available. Complexes may have coordination numbers higher than their empirical formulas suggest.

The best-known complexes of metals with coordination number 2 that are formed in solution under ordinary laboratory conditions are linear species of the Group 11 and

12 ions. Linear two-coordinate complexes with two identical, symmetrical ligands have $D_{\infty h}$ symmetry. The complex [AgCl$_2$]$^-$, which is responsible for the dissolution of solid silver chloride in aqueous solutions containing excess Cl$^-$ ions, is one example; dimethyl mercury, Me–Hg–Me, is another. A series of linear Au(I) complexes of formula LAuX, where X is a halogen and L is a neutral Lewis base such as a substituted phosphine, R$_3$P, or thioether, R$_2$S, are also known. Two-coordinate complexes sometimes gain additional ligands, if extra ligands are available, to form three- or four-coordinate complexes.

A formula that suggests a certain coordination number in a solid compound might conceal a polymeric chain with a higher coordination number. For example, CuCN appears to have coordination number 1, but it in fact exists as linear –Cu–CN–Cu–CN– chains in which the coordination number of copper is 2.

Three-coordination is relatively rare among metal complexes, but is found with bulky ligands such as tricyclohexylphosphine, as in [Pt(PCy$_3$)$_3$] (**13**) (Cy = cyclohexyl, –C$_6$H$_{11}$) with its trigonal arrangement of the ligands. MX$_3$ compounds, where X is a halogen, are usually chains or networks with a higher coordination number and shared ligands. Three-coordinate complexes with three identical, symmetrical ligands normally have D_{3h} symmetry.

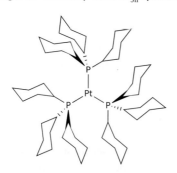

13 [Pt(PCy$_3$)], Cy=*cyclo*-C$_6$H$_{11}$

7.4 Intermediate coordination numbers

Complexes of metal ions with the intermediate coordination numbers 4, 5, and 6 are the most important class of complex. They include the vast majority of complexes that exist in solution and almost all the biologically important complexes.

(a) Four-coordination

KEY POINTS Tetrahedral complexes are favoured over higher-coordinate complexes if the central atom is small or the ligands large; square-planar complexes are typically observed for metals with d^8 configurations.

Four-coordination is found in an enormous number of compounds. Tetrahedral complexes of approximately T_d symmetry (**14**) are favoured over higher coordination numbers when the central atom is small and the ligands are large (such as Cl$^-$, Br$^-$, and I$^-$), because then ligand–ligand repulsions override the energy advantage of forming more metal–ligand bonds. Four-coordinate s- and p-block complexes with no lone pair on the central atom, such as [BeCl$_4$]$^{2-}$, [AlBr$_4$]$^-$, and [AsCl$_4$]$^+$, are almost always tetrahedral, and tetrahedral complexes are common for oxoanions of metal atoms on the left of the d block in high oxidation states, such as [MoO$_4$]$^{2-}$. Examples of tetrahedral complexes from Groups 5–11 are [VO$_4$]$^{3-}$, [CrO$_4$]$^{2-}$, [MnO$_4$]$^-$, [FeCl$_4$]$^{2-}$, [CoCl$_4$]$^{2-}$, [NiBr$_4$]$^{2-}$, and [CuBr$_4$]$^{2-}$.

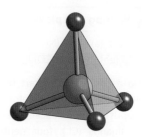

14 Tetrahedral complex, T_d

Another type of four-coordinate complex is also found: those where the four ligands surround the central metal in a square-planar arrangement (**15**). Complexes of this type were originally identified because they can lead to different isomers when the complex has the formula MX$_2$L$_2$. We discuss this isomerism in Section 7.7. Square-planar complexes with four identical, symmetrical ligands have D_{4h} symmetry.

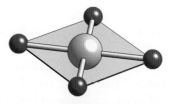

15 Square-planar complex, D_{4h}

Square-planar complexes are rarely found for s- and p-block complexes but are abundant for d^8 complexes of the elements belonging to the 4d- and 5d-series metals such as Rh$^+$, Ir$^+$, Pd^{2+} Pt^{2+}, and Au^{3+}, which are almost invariably square-planar. For 3d metals with d^8 configurations (for example, Ni^{2+}), square-planar geometry is favoured by ligands that can form π bonds by accepting electrons from the metal atom, as in [Ni(CN)$_4$]$^{2-}$. Examples of square-planar complexes from Groups 9, 10, and 11 are [RhCl(PPh$_3$)$_3$], *trans*-[Ir(CO)Cl(PMe$_3$)$_2$], [Ni(CN)$_4$]$^{2-}$, [PdCl$_4$]$^{2-}$, [Pt(NH$_3$)$_4$]$^{2+}$, and [AuCl$_4$]$^-$. Square-planar geometry can also be forced on a central atom by complexation with a ligand that contains a rigid ring of four donor atoms, such as in the formation of a porphyrin complex (**16**).

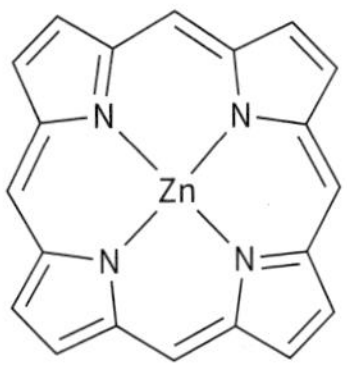

16 Zinc porphyrin complex

Section 7.16 gives a more detailed explanation of the factors that help to stabilize square-planar complexes.

(b) Five-coordination

KEY POINTS *In the absence of polydentate ligands that enforce the geometry, the energies of the various geometries of five-coordinate complexes differ little from one another, and such complexes are often fluxional.*

Five-coordinate complexes, which are less common than four- or six-coordinate complexes, are normally either square-pyramidal or trigonal-pyramidal. A square-pyramidal complex would have C_{4v} symmetry if all the ligands were identical and the trigonal-bipyramidal complex (**17**) would have D_{3h} symmetry with identical ligands.

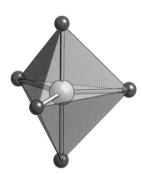

17 Trigonal-bipyramidal complex, D_{3h}

Distortions from these ideal geometries are common, and only small changes in bond angles are often necessary to change a structure from one ideal to the other or to an intermediate configuration. A trigonal-bipyramidal shape minimizes ligand–ligand repulsions, but steric constraints on ligands that can bond through more than one site to a metal atom can favour a square-pyramidal structure. For instance, square-pyramidal five-coordination is found among the biologically important porphyrins, where the ligand ring enforces a square-planar structure and a fifth ligand attaches above the plane. Structure (**18**) shows part of the active centre of myoglobin, the oxygen transport protein; the oxygen molecule binds in the vacant coordination site below the Fe (Section 26.7).

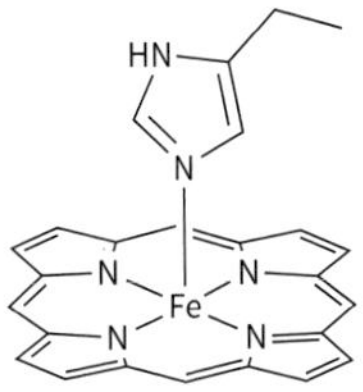

18

In some cases, five-coordination is induced by a polydentate ligand containing a donor atom that can bind to an axial location of a trigonal bipyramid, with its remaining donor atoms reaching down to the three equatorial positions (**19**). Ligands that force a trigonal-bipyramidal structure in this fashion are called **tripodal**.

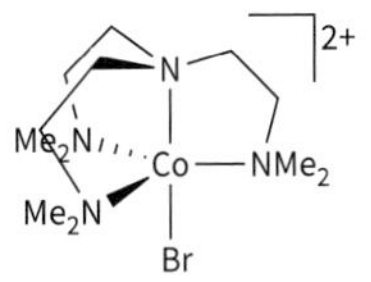

19 [CoBrN(CH$_2$CH$_2$NMe$_2$)$_3$]$^{2+}$

(c) Six-coordination

KEY POINT *The overwhelming majority of six-coordinate complexes are octahedral or have shapes that are small distortions of octahedral.*

Six-coordination is the most common arrangement for metal complexes and is found in s-, p-, d-, and (more rarely) f-metal coordination compounds. Almost all six-coordinate complexes are octahedral (**20**), at least if we consider the ligands as represented by structureless points.

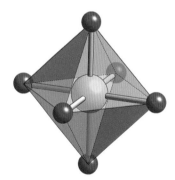

20 Octahedral complex, O_h

A *regular* octahedral (O_h) arrangement of ligands is highly symmetrical (Fig. 7.1). It is especially important, not only because it is found for many complexes of formula ML$_6$ but also because it is the starting point for discussions of complexes of lower symmetry, such as those shown in Fig. 7.2.

The simplest deviation from O_h symmetry is tetragonal (D_{4h}), and occurs when two ligands along one axis differ from the other four; these two ligands, which are *trans* to each other,

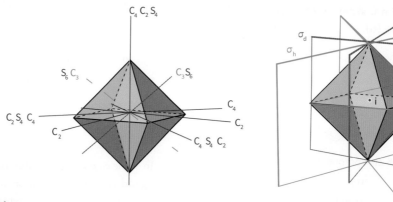

FIGURE 7.1 The highly symmetric octahedral arrangement of six ligands around a central metal and the corresponding symmetry elements of an octahedron. Note that not all the symmetry elements are shown.

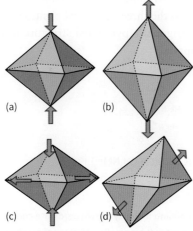

FIGURE 7.2 Distortions of a regular octahedron: (a) and (b) tetragonal distortions, (c) rhombic distortion, (d) trigonal distortion.

might be closer in than the other four or, more commonly, further away. For the d^9 configuration (particularly for Cu^{2+} complexes), a tetragonal distortion may occur even when all ligands are identical because of an inherent effect known as the Jahn–Teller distortion (Section 7.16). Rhombic (D_{2h}) distortions, in which a *trans* pair of ligands are close in and another *trans* pair are further out, can occur. Trigonal (D_{3d}) distortions occur when two opposite faces of the octahedron move away and give rise to a large family of structures that are intermediate between regular octahedral and trigonal-prismatic (**21**); such structures are sometimes referred to as rhombohedral.

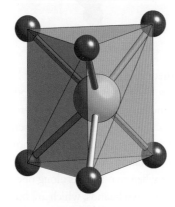

21 Trigonal prism, D_{3h}

Trigonal-prismatic (D_{3h}) complexes are rare, but have been found in solid MoS_2 and WS_2; the trigonal prism is also the shape of several complexes of formula $[M(S_2C_2R_2)_3]$ (**22**).

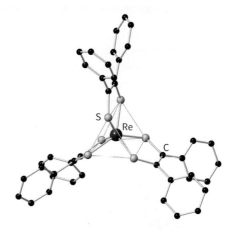

22 $[Re(SCPh=CPhS)_3]$

Trigonal-prismatic d^0 complexes such as $[Zr(CH_3)_6]^{2-}$ have also been isolated. Such structures require either very small **σ-donor ligands**, ligands that bind by forming a σ bond to the central atom, or favourable ligand–ligand interactions that can constrain the complex into a trigonal prism; such ligand–ligand interactions are often provided by ligands that contain sulfur atoms, which can form long, weak covalent bonds to each other. A chelating ligand that permits only a small bite angle can cause distortion from octahedral towards trigonal-prismatic geometry in six-coordinate complexes (Fig. 7.3).

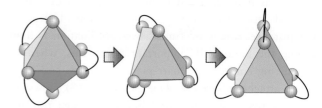

FIGURE 7.3 A chelating ligand that permits only a small bite angle can distort an octahedral complex into trigonal-prismatic geometry.

7.5 Higher coordination numbers

KEY POINTS Larger atoms and ions tend to form complexes with high coordination numbers; nine-coordination is particularly important in the f block.

Seven-coordination is encountered for the larger Group 2 metals, a few 3d complexes, and many more 4d and 5d complexes, where the larger central atom can accommodate more than six ligands. Seven-coordination resembles five-coordination in the similarity in energy of its various geometries. These limiting 'ideal' geometries include the pentagonal bipyramid (**23**), a capped octahedron (**24**), and a capped trigonal prism (**25**); in each of the latter two, the seventh capping ligand occupies one face.

and $[ReCl_6O]^{2-}$ from the d block, and $[UO_2(OH_2)_5]^{2+}$ from the f block.

A method to force seven- rather than six-coordination on the lighter elements is to synthesize a ring of five donor atoms (**26**) that then occupy the equatorial positions, leaving the axial positions free to accommodate two more ligands.

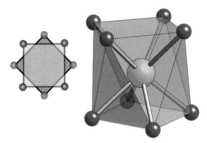

26

Stereochemical nonrigidity is also shown in eight-coordination; such complexes may be square-antiprismatic (**27**) in one crystal but dodecahedral (**28**) in another.

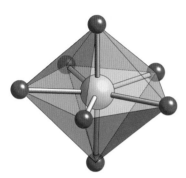

23 Pentagonal bipyramid, D_{5h}

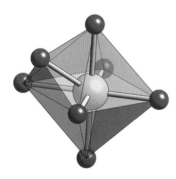

24 Capped octahedron

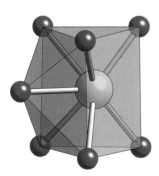

25 Capped trigonal prism

There are a number of intermediate structures, and interconversion between them is often rapid at room temperature. Examples include $[Mo(CNR)_7]^{2+}$, $[ZrF_7]^{3-}$, $[TaCl_4(PR_3)_3]$,

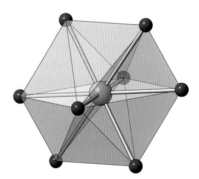

27 Square antiprism, D_4

28 Dodecahedron D_{2d}

Two examples of complexes with these geometries are shown as (**29**) and (**30**), respectively.

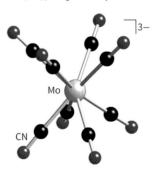

29 $[Mo(CN)_8]^{3-}$

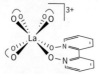

30 [Zr(ox)$_4$]$^{4-}$

Cubic geometry (**31**) is rare, but an example comes from the complex of lanthanum with four bipyridinedioxide ligands (**32**).

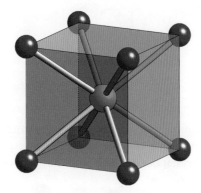

31 Cube

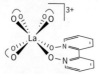

32 [La(bpyO$_2$)$_4$]$^{3+}$

Nine-coordination is important in the structures of f-block elements; their relatively large ions can act as host to a large number of ligands. A simple example of a nine-coordinate lanthanoid complex is [Nd(OH$_2$)$_9$]$^{3+}$, which has a tri-capped trigonal-prismatic structure (**33**).

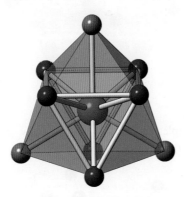

33 Tricapped trigonal prism, D_{3h}

More complex examples arise with the MCl$_3$ solids, with M ranging from La to Gd, where a coordination number of 9 is achieved through metal–halide–metal bridges (Section 21.5). An example of nine-coordination in the d block is [ReH$_9$]$^{2-}$ (**34**), which has small enough ligands for this coordination number to be feasible; the geometry can be thought of as a capped square antiprism.

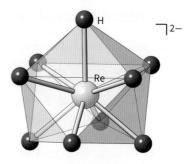

34 [ReH$_9$]$^{2-}$

Ten-, eleven- and twelve-coordination are encountered in complexes of the f-block M^{3+} ions. An example of a 10-coordinate complex is [Th(ox)$_4$(OH$_2$)$_2$]$^{4-}$, in which each oxalate ion ligand (ox^{2-}, C$_2$O$_4^{2-}$) provides two O donor atoms; an example of an 11-coordinate complex is thorium nitrate [Th(NO$_3$)$_4$(OH$_2$)$_3$], where each NO$_3^-$ binds by two oxygen atoms. Finally, an example of a 12-coordinate complex is [Ce(NO$_3$)$_6$]$^{2-}$ (**35**), which is formed in the reaction of Ce(IV) salts with nitric acid. Each NO$_3^-$ ligand is bonded to the metal atom by two O atoms. These high coordination numbers are rare with s-, p-, and d-block ions.

35 [Ce(NO$_3$)$_6$]$^{2-}$

7.6 Polymetallic complexes

KEY POINT Polymetallic complexes are classified as metal clusters if they contain M—M bonds or as cage complexes if they contain ligand-bridged metal atoms.

Polymetallic complexes are complexes that contain more than one metal atom. In some cases, the metal atoms are held together by bridging ligands; in others there are direct metal–metal bonds; in yet others there are both types of link. The term **metal cluster** is usually reserved for polymetallic complexes in which there are direct metal–metal bonds that form triangular or larger closed structures. This rigorous definition, however, would exclude linear M—M compounds, and is normally relaxed. We shall consider any M—M bonded system to be a metal cluster.

Polymetallic complexes may be formed with a wide variety of anionic ligands. For example, two Cu^{2+} ions can be held together with acetate-ion bridges (**36**). Structure (**37**) is an example of a cubic structure formed from four Fe atoms bridged by RS$^-$ ligands. This type of structure is of great biological importance as it is involved in a number of biochemical redox reactions (Section 26.8).

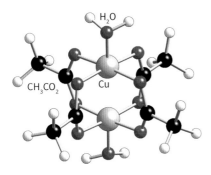

36 [(H$_2$O)Cu(μ-CH$_3$CO$_2$)$_4$Cu(OH$_2$)]

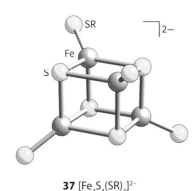

37 [Fe$_4$S$_4$(SR)$_4$]$^{2-}$

With the advent of modern structural techniques, such as automated X-ray diffractometers and multinuclear NMR,

many polymetallic clusters containing metal–metals bonds have been discovered and have given rise to an active area of research. A simple example is the mercury(I) cation Hg$_2^{2+}$ and complexes derived from it, such as [Hg$_2$(Cl)$_2$] (**38**), which is commonly written simply Hg$_2$Cl$_2$. A metal cluster containing ten CO ligands and two Mn atoms is shown as (**39**).

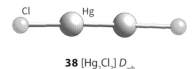

38 [Hg$_2$Cl$_2$] $D_{\infty h}$

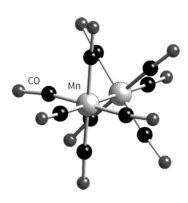

39 [(OC)$_5$Mn–Mn(CO)$_5$]

Isomerism and chirality

KEY POINT A molecular formula may not be sufficient to identify a coordination compound: linkage, ionization, hydrate, and coordination isomerism are all possible for coordination compounds.

A molecular formula often does not provide enough information to identify a compound unambiguously. We have already noted that the existence of ambidentate ligands gives rise to the possibility of **linkage isomerism**, in which the same ligand may link through different atoms. This type of isomerism accounts for the red and yellow isomers of the formula [Co(NH$_3$)$_5$(NO$_2$)]$^{2+}$. The red compound has a nitrito-κO O–Co link (**5**); the yellow isomer, which forms from the unstable red form over time, has a nitrito-κN N–Co link (**4**). Linkage isomerism is a type of **constitutional** or **structural isomerism**, and we will consider three further types of constitutional isomerism briefly before looking at geometric and optical isomerism in more depth.

Ionization isomerism occurs when ligand and counterion in one compound exchange places. An example is [PtCl$_2$(NH$_3$)$_4$]Br$_2$ and [PtBr$_2$(NH$_3$)$_4$]Cl$_2$. If the compounds are soluble, the two isomers exist as different ionic species in solution (in this example, with free Br$^-$ and Cl$^-$ ions, respectively).

Very similar to ionization isomerism is **hydrate isomerism**, which arises when one of the ligands is water; for example, there are three differently coloured hydration isomers of a compound with molecular formula CrCl$_3$·6H$_2$O: the violet [Cr(OH$_2$)$_6$]Cl$_3$, the pale green [CrCl(OH$_2$)$_5$]Cl$_2$·H$_2$O, and the dark green [CrCl$_2$(OH$_2$)$_4$]Cl·2H$_2$O.

Coordination isomerism arises when there are different complex ions that can form from the same molecular formula, as in [Co(NH$_3$)$_6$][Cr(CN)$_6$] and [Cr(NH$_3$)$_6$][Co(CN)$_6$].

Once we have established which ligands bind to which metals, and through which donor atoms, we can consider how to arrange these ligands in space. The three-dimensional character of metal complexes can result in a multitude of possible arrangements of the ligands. We now explore these varieties of isomerism by considering the permutations of ligand arrangement for each of the common complex geometries: this type of isomerism is known as **stereo isomerism**.

Complexes with coordination numbers greater than six have the potential for a great number of stereoisomers, both **geometrical** and **optical**. As these complexes are often stereochemically nonrigid, the isomers are usually not separable and we do not consider them further.

EXAMPLE 7.2 Isomerism in metal complexes

What types of isomerism are possible for complexes with the following molecular formulas: (a) [Pt(PEt$_3$)$_3$SCN]$^+$, (b) CoBr(NH$_3$)$_5$SO$_4$, (c) FeCl$_2$(H$_2$O)$_6$?

Answer (a) The complex contains the ambidentate thiocyanate ligand, SCN$^-$, which can bind through either the S or the N atom to give rise to two linkage isomers: [Pt(<u>S</u>CN)(PEt$_3$)$_3$]$^+$ and [Pt(<u>N</u>CS)(PEt$_3$)$_3$]$^+$. (b) With an octahedral geometry and five coordinated ammonia ligands, it is possible to have two ionization isomers: [Co(NH$_3$)$_5$SO$_4$]Br and [CoBr(NH$_3$)$_5$]SO$_4$. (c) Hydrate isomerism occurs as complexes of formula [Fe(OH$_2$)$_6$]Cl$_2$, [FeCl(OH$_2$)$_5$]Cl·H$_2$O, and [FeCl$_2$(OH$_2$)$_4$]·2H$_2$O are possible.

Self-test 7.2 Two types of isomerism are possible for the six-coordinate complex with the molecular formula Cr(NO$_2$)$_2$(H$_2$O)$_6$. Identify all isomers.

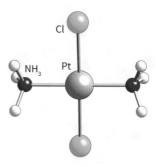

41 *trans*-[PtCl$_2$(NH$_3$)$_2$]

In the simple case of two sets of two different monodentate ligands, as in [MA$_2$B$_2$], there is only the case of *cis/trans* isomerism to consider, (**42**) and (**43**).

7.7 Square-planar complexes

KEY POINT *The only simple isomers of square-planar complexes are cis–trans isomers.*

Werner studied a series of four-coordinate Pt(II) complexes formed by the reactions of PtCl$_2$ with NH$_3$ and HCl. For a complex of formula MX$_2$L$_2$, only one isomer is expected if the species is tetrahedral, but two isomers are expected if the species is square-planar (**40**, **41**). Because Werner was able to isolate two complexes of formula [PtCl$_2$(NH$_3$)$_2$], he concluded that they could not be tetrahedral and were, in fact, square-planar. The complex with like ligands on adjacent corners of the square is called a *cis* isomer (**40**, point group C_{2v}) and the complex with like ligands opposite is the *trans* isomer (**41**, D_{2h}). Geometric isomerism is far from being of only academic interest: platinum complexes are used in cancer chemotherapy, and it is found that only *cis*-Pt(II) complexes can bind to the bases of DNA for long enough to be effective (Section 32.1).

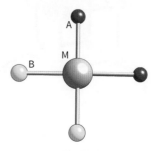

42 *cis*-[MA$_2$B$_2$]

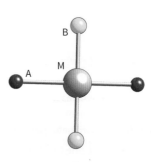

43 *trans*-[MA$_2$B$_2$]

With three different ligands, as in [MA$_2$BC], the locations of the two A ligands also allow us to distinguish the geometric isomers as *cis* and *trans*, (**44**) and (**45**).

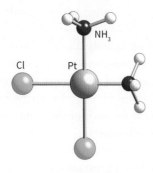

40 *cis*-[PtCl$_2$(NH$_3$)$_2$]

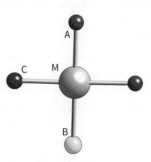

44 *cis*-[MA$_2$BC]

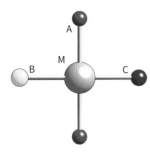

45 *trans*-[MA₂BC]

When there are four different ligands, as in [MABCD], there are three different isomers and we have to specify the geometry more explicitly, as in (46), (47), and (48).

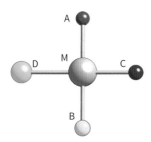

46 [MABCD], A *trans* to B

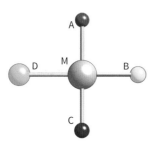

47 [MABCD], A *trans* to C

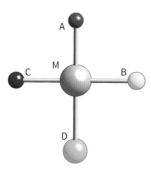

48 [MABCD], A *trans* to D

Bidentate ligands with different endgroups, as in [M(AB)₂], can also give rise to geometrical isomers that can be classified as *cis* (49) and *trans* (50).

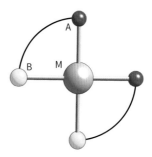

49 *cis*-[M(AB)₂]

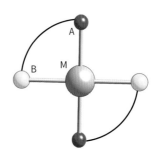

50 *trans*-[M(AB)₂]

EXAMPLE 7.3 Identifying isomers from chemical evidence

How would the differing reactivity of the two isomers of diamminedichloridoplatinum(II) with 1,2-diaminoethane allow them to be distinguished?

Answer The *cis* diamminedichlorido isomer reacts with one equivalent of 1,2-diaminoethane (en, **6**), replacing two NH₃ ligands with one bidentate en at adjacent positions. The *trans* isomer cannot displace the two NH₃ ligands with only one en ligand, Fig. 7.4. A reasonable explanation is that the en ligand cannot reach across the square plane to bond to the two *trans* positions. This conclusion is supported by X-ray crystallography; the driving force for the reaction of the *cis* isomer is the favourable entropic change associated with the chelate effect (Section 7.14).

FIGURE 7.4 The differing reactivity of *cis*- and *trans*-diamminedichloroplatinum(II) provides a chemical method for distinguishing the isomers.

Self-test 7.3 The two square-planar isomers of [PtBrCl(PR$_3$)$_2$] (where PR$_3$ is a trialkylphosphine) have different ^{31}P-NMR spectra (Fig. 7.5). For the sake of this exercise we ignore coupling to ^{195}Pt ($I = \frac{1}{2}$ at 33% abundance), Section 8.6. One isomer (A) shows a single ^{31}P resonance; the other (B) shows two ^{31}P resonances, each of which is split into a doublet by the second ^{31}P nucleus. Which isomer is *cis* and which is *trans*?

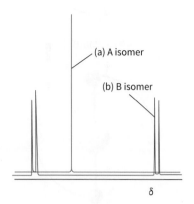

FIGURE 7.5 Idealized ^{31}P NMR spectra of two isomers of [PtBrCl(PR$_3$)$_2$]. The fine structure due to Pt is not shown.

7.8 Tetrahedral complexes

KEY POINT The only simple isomers of tetrahedral complexes are optical isomers.

The only isomers of tetrahedral complexes normally encountered are those where either all four ligands are different or where there are two unsymmetrical bidentate chelating ligands. In both cases, (51) and (52), the molecules are **chiral**, not superimposable on their mirror image (Section 3.4).

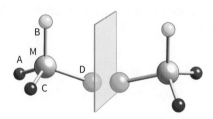

51 [MABCD] enantiomers

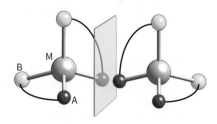

52 [M(AB)$_2$] enantiomers

Two mirror-image isomers jointly make up an **enantiomeric pair**. The existence of a pair of chiral complexes that are each other's mirror image (like a right hand and a left hand), and that have lifetimes that are long enough for them to be separable, is called **optical isomerism**. Optical isomers are so called because they are **optically active**, in the sense that one enantiomer rotates the plane of polarized light in one direction and the other rotates it through an equal angle in the opposite direction.

7.9 Trigonal-bipyramidal and square-pyramidal complexes

KEY POINTS Five-coordinate complexes are not stereochemically rigid; two chemically distinct coordination sites exist within both trigonal-bipyramidal and square-pyramidal complexes.

The energies of the various geometries of five-coordinate complexes often differ little from one another. The delicacy of this balance is underlined by the fact that [Ni(CN)$_5$]$^{3-}$ can exist as both square-pyramidal (53) and trigonal-bipyramidal (54) conformations in the same crystal.

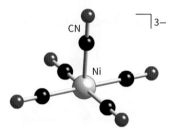

53 [Ni(CN)$_5$]$^{3-}$, square-pyramidal

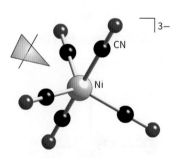

54 [Ni(CN)$_5$]$^{3-}$, trigonal-bipyramidal

In solution, trigonal-bipyramidal complexes with monodentate ligands are often highly fluxional (that is, able to twist into different shapes), so a ligand that is axial at one moment becomes equatorial at the next moment: the conversion from one stereochemistry to another may occur by a **Berry pseudorotation** (Fig. 7.6). Thus, although isomers of five-coordinate complexes do exist, they are commonly not separable.

It is important to be aware that both trigonal-bipyramidal and square-pyramidal complexes have two chemically distinct

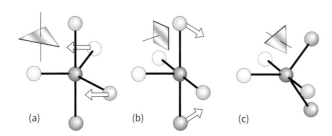

FIGURE 7.6 A Berry pseudorotation in which (a) a trigonal-bipyramidal Fe(CO)$_5$ complex distorts into (b) a square-pyramidal isomer and then (c) becomes trigonal-bipyramidal again, but with the two initially axial ligands now equatorial.

sites: axial (a) and equatorial (e) for the trigonal bipyramid (**55**) and axial (a) and basal (b) for the square pyramid (**56**). Certain ligands have preferences for the different sites because of their steric and electronic requirements.

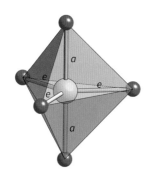

55 [ML$_5$], trigonal bipyramid

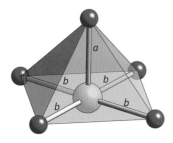

56 [ML$_5$], square pyramid

7.10 Octahedral complexes

There are huge numbers of complexes with nominally octahedral geometry, where in this context the nominal structure '[ML$_6$]' is taken to mean a central metal atom surrounded by six ligands, not all of which are necessarily the same.

(a) Geometrical isomerism

KEY POINTS *Cis* and *trans* isomers exist for octahedral complexes of formula [MA$_4$B$_2$], and *mer* and *fac* isomers are possible for complexes of formula [MA$_3$B$_3$]. More complicated ligand sets lead to further isomers.

Whereas there is only one way of arranging the ligands in octahedral complexes of general formula [MA$_6$] or [MA$_5$B], the two B ligands of an [MA$_4$B$_2$] complex may be placed on adjacent octahedral positions to give a *cis* isomer (**57**) or on diametrically opposite positions to give a *trans* isomer (**58**). Provided we treat the ligands as structureless points, the *trans* isomer has D_{4h} symmetry and the *cis* isomer has C_{2v} symmetry.

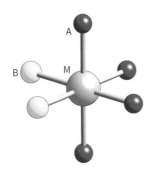

57 *cis*-[MA$_4$B$_2$]

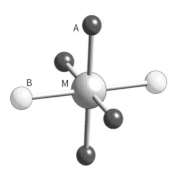

58 *trans*-[MA$_4$B$_2$]

There are two ways of arranging the ligands in [MA$_3$B$_3$] complexes. In one isomer, three A ligands lie in one plane and three B ligands lie in a perpendicular plane (**59**). This complex is designated the *mer* isomer (for meridional) because each set of ligands can be regarded as lying on a meridian of a sphere.

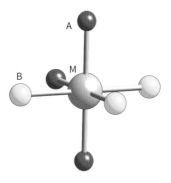

59 *mer*-[MA$_3$B$_3$]

In the second isomer, all three A (and B) ligands are adjacent and occupy the corners of one triangular face of the octahedron (**60**); this complex is designated the *fac* isomer (for facial) because the ligands sit on the corners of one face of an octahedron. Provided we treat the ligands as structureless points, the *mer* isomer has C_{2v} symmetry and the *fac* isomer has C_{3v} symmetry.

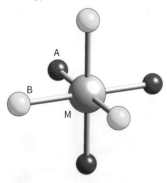

60 *fac*-[MA$_3$B$_3$]

For a complex of composition [MA$_2$B$_2$C$_2$], there are five different geometrical isomers: an all-*trans* isomer (**61**); three different isomers where one pair of ligands is *trans* while the other two are *cis*, as in (**62**), (**63**), and (**64**); and an enantiomeric pair of all-*cis* isomers (**65**).

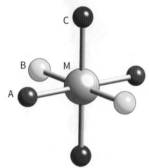

61 [MA$_2$B$_2$C$_2$]

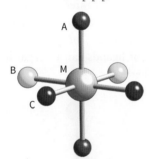

62 [MA$_2$B$_2$C$_2$]

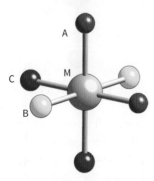

63 [MA$_2$B$_2$C$_2$]

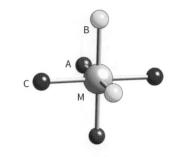

64 [MA$_2$B$_2$C$_2$]

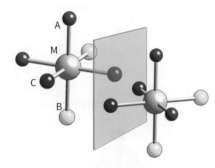

65 [MA$_2$B$_2$C$_2$] enantiomers

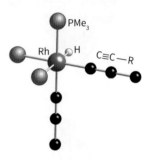

66 *fac*-[RhH(C≡CR)$_2$(PMe$_3$)$_3$]

More complicated compositions, such as [MA$_2$B$_2$CD] or [MA$_3$B$_2$C], result in more extensive geometrical isomerism. For instance, the rhodium compound [RhH(C≡CR)$_2$(PMe$_3$)$_3$] exists as three different isomers: *fac* (**66**), *mer-trans* (**67**), and *mer-cis* (**68**). Although octahedral complexes are normally stereochemically rigid, isomerization reactions do sometimes occur (Section 23.9).

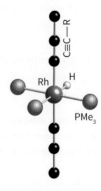

67 *mer-trans*-[RhH(C≡CR)$_2$(PMe$_3$)$_3$]

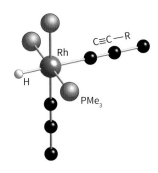

68 *mer-cis*-[RhH(C≡CR)$_2$(PMe$_3$)$_3$]

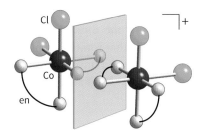

70 *cis*-[CoCl$_2$(en)$_2$]$^+$ enantiomers

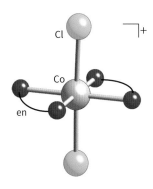

71 *trans*-[CoCl$_2$(en)$_2$]$^+$

(b) Chirality and optical isomerism

KEY POINTS A number of ligand arrangements at an octahedral centre give rise to chiral compounds; isomers are designated Δ or Λ depending on their configuration.

In addition to the many examples of geometrical isomerism shown by octahedral compounds, some are also chiral. A very simple example is [Mn(acac)$_3$] (**69**), where three bidentate acetylacetonato (acac) ligands result in the existence of enantiomers. One way of looking at the optical isomers that arise in complexes of this nature is to imagine looking down one of the three-fold axes and seeing the ligand arrangement as a propeller or screw thread.

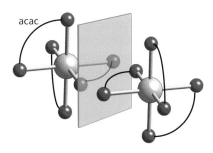

69 [Mn(acac)$_3$] enantiomers

Chirality can also exist for complexes of formula [MA$_2$B$_2$C$_2$] when the ligands of each pair are *cis* to each other (**65**). In fact, many examples of optical isomerism are known for octahedral complexes with both monodentate and polydentate ligands, and we must always be alert to the possibility of optical isomerism.

As a further example of optical isomerism, consider the products of the reaction of cobalt(III) chloride and 1,2-diaminoethane in a 1:2 mole ratio. The product includes a pair of dichlorido complexes, one of which is violet (**70**) and the other green (**71**); they are, respectively, the *cis* and *trans* isomers of dichloridobis(1,2-diaminoethane)cobalt(III), [CoCl$_2$(en)$_2$]$^+$. As can be seen from their structures, the *cis* isomer cannot be superimposed on its mirror image. It is therefore chiral and hence (because the complexes are long-lived) optically active. The *trans* isomer has a mirror plane and can be superimposed on its mirror image; it is achiral and optically inactive.

The absolute configuration of a chiral octahedral complex is described by imagining a view along a three-fold rotation axis of the regular octahedron and noting the handedness of the helix formed by the ligands (Fig. 7.7). Clockwise rotation of the helix is then designated Δ (delta) whereas the anticlockwise rotation is designated Λ (lambda). The designation of the configuration must be distinguished from the experimentally determined direction in which an isomer rotates polarized light: some Λ

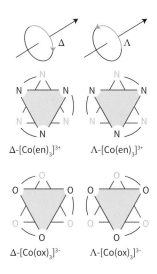

FIGURE 7.7 Absolute configurations of M(L–L)$_3$ complexes. Δ is used to indicate clockwise rotation of the helix and Λ to indicate anticlockwise rotation.

compounds rotate in one direction, others rotate in the opposite direction, and the direction may change with wavelength. The isomer that rotates the plane of polarization clockwise (when viewed into the oncoming beam) at a specified wavelength is designated the *d*-isomer, or the (+)-**isomer**; the one rotating the plane anticlockwise is designated the *l*-isomer, or the (−)-**isomer**.

Box 7.1 describes how the specific isomers of a complex might be synthesized and how enantiomers of metal complexes may be separated.

BOX 7.1 How can specific isomers be synthesised and separated?

The synthesis of specific isomers often requires subtle changes in synthetic conditions. For example, the most stable Co(II) complex in ammoniacal solutions of Co(II) salts, $[Co(NH_3)_6]^{2+}$, is only slowly oxidized. As a result, a variety of Co(III) complexes containing other ligands as well as NH_3 can be prepared by bubbling air through a solution containing ammonia and a Co(II) salt. Starting with ammonium carbonate yields $[Co(CO_3)(NH_3)_4]^+$, in which CO_3^{2-} is a bidentate ligand that occupies two adjacent coordination positions. The complex *cis*-$[CoL_2(NH_3)_4]^+$ can be prepared by displacement of the CO_3^{2-} ligand in acidic solution. When concentrated hydrochloric acid is used, the violet *cis*-$[CoCl_2(NH_3)_4]Cl$ compound (**B1**) can be isolated:

$$[Co(CO_3)(NH_3)_4]^+(aq) + 2H^+(aq) + 3Cl^-(aq) \rightarrow$$
$$cis\text{-}[CoCl_2(NH_3)_4]Cl(s) + H_2CO_3(aq)$$

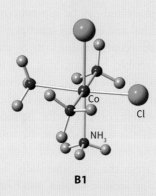

B1

By contrast, reaction of $[Co(NH_3)_6]^{2+}$ directly with a mixture of HCl and H_2SO_4 in air gives the bright green *trans*-$[CoCl_2(NH_3)_4]$Cl isomer (**B2**).

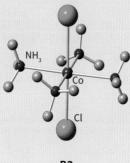

B2

Optical activity is the only physical manifestation of chirality for a compound with a single chiral centre. However, as soon as more than one chiral centre is present, other physical properties, such as solubility and melting point, are affected because they depend on the strengths of intermolecular forces, which are different between different isomers (just as there are different forces between a given nut and bolts with left- and right-handed threads). One method of separating a pair of enantiomers into the individual isomers is therefore to prepare **diastereomers**. As far as we need be concerned, diastereomers are isomeric compounds that contain two chiral centres, one being of the same absolute configuration in both components and the other being enantiomeric between the two components. An example of diastereomers is provided by the two salts of an enantiomeric pair of cations, A, with an optically pure anion, B, and hence of composition [Δ-A][Δ-B] and [Λ-A][Δ-B]. Because diastereomers differ in physical properties (such as solubility), they are separable by conventional techniques.

A classical chiral resolution procedure begins with the isolation of a naturally optically active species from a biochemical source (many naturally occurring compounds are chiral). A convenient compound is *d*-tartaric acid (**B3**), a carboxylic acid obtained from grapes. This molecule is a chelating ligand for complexation of antimony, so a convenient resolving agent is the potassium salt of the singly charged antimony *d*-tartrate anion. This anion is used for the resolution of $[Co(en)_2(NO_2)_2]^+$ as follows:

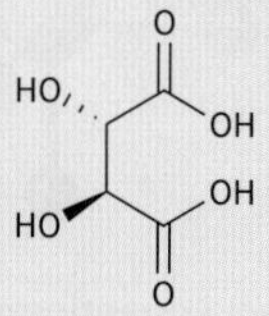

B3

The enantionmeric mixture of the cobalt(III) complex is dissolved in warm water, and a solution of potassium antimony *d*-tartrate is added. The mixture is cooled immediately to induce crystallization. The less soluble diastereomer {*l*-[Co(en)$_2$(NO$_2$)$_2$]}{*d*-[SbOC$_4$H$_4$O$_6$]} separates as fine yellow crystals. The filtrate is reserved for isolation of the *d*-enantiomer. The solid diastereomer is ground with water and sodium iodide. The sparingly soluble compound *l*-[Co(en)$_2$(NO$_2$)$_2$]I separates, leaving sodium antimony tartrate in the solution. The *d*-isomer is obtained from the filtrate by precipitation of the bromide salt.

Further reading

A. von Zelewsky, *Stereochemistry of Coordination Compounds*. John Wiley & Sons (1996).

W. L. Jolly, *The Synthesis and Characterization of Inorganic Compounds*. Waveland Press (1991).

EXAMPLE 7.4 Identifying types of isomerism

When the four-coordinate square-planar complex [IrCl(PMe$_3$)$_3$] (where PMe$_3$ is trimethylphosphine) reacts with Cl$_2$, two six-coordinate products of formula [IrCl$_3$(PMe$_3$)$_3$] are formed. ^{31}P-NMR spectra indicate one P environment in one of these isomers, and two in the other. What isomers are possible?

Answer Because the complexes have the formula [MA$_3$B$_3$], we expect meridional and facial isomers. Structures (**72**) and (**73**) show the arrangement of the three Cl$^-$ ions in the *fac* and *mer* isomers, respectively. All P atoms are equivalent in the *fac* isomer and two environments exist in the *mer* isomer.

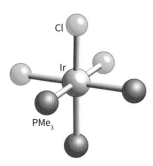

72 *fac*-[IrCl$_3$(PMe$_3$)$_3$]

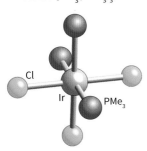

73 *mer*-[IrCl$_3$(PMe$_3$)$_3$]

Self-test 7.4 When the glycinate anion, H$_2$NCH$_2$CO$_2^-$ (gly$^-$), reacts with cobalt(III) oxide, both the N and an O atom of gly$^-$ coordinate and two Co(III) nonelectrolyte *mer* and *fac* isomers of [Co(gly)$_3$] are formed. Sketch the two isomers. Sketch the mirror images of the two isomers: are they superimposable?

7.11 Ligand chirality

KEY POINT Coordination to a metal can stop a ligand inverting and hence lock it into a chiral configuration.

In certain cases, achiral ligands can become chiral on coordination to a metal, leading to a complex that is chiral. Usually, the achiral ligand contains a donor that rapidly inverts as a free ligand, but becomes locked in one configuration on coordination. An example is MeNHCH$_2$CH$_2$NHMe, where the two N atoms become chiral centres on coordination to a metal atom. For a square-planar complex, this imposed chirality results in four isomers: one pair of chiral enantiomers (**74**) and two complexes that are not chiral (**75**) and (**76**).

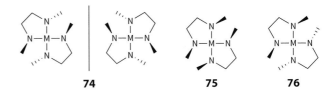

74 **75** **76**

EXAMPLE 7.5 Recognizing chirality

Which of the complexes (a) [Cr(edta)]$^-$, (b) [Ru(en)$_3$]$^{2+}$, (c) [Pt(dien)Cl]$^+$ are chiral?

Answer If a complex has either a mirror plane or centre of inversion, it cannot be chiral. If we look at the schematic complexes drawn in (**77**), (**78**), and (**79**), we can see that neither (**77**) nor (**78**) has a mirror plane or a centre of inversion; so both are chiral (they also have no higher S_n axis). Conversely, (**79**) has a plane of symmetry and hence is achiral. (Although the CH$_2$ groups in a dien ligand are not in the mirror plane, they oscillate rapidly above and below it.)

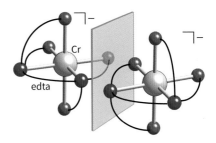

77 [Cr(edta)]$^-$ enantiomers

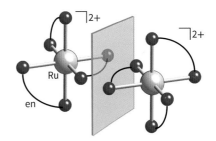

78 [Ru(en)$_3$]$^{2+}$ enantiomers

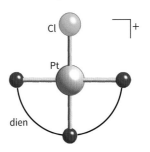

79 [PtCl(dien)]$^+$

Self-test 7.5 Which of the complexes (a) *cis*-[CrCl$_2$(ox)$_2$]$^{3-}$, (b) *trans*-[CrCl$_2$(ox)$_2$]$^{3-}$, (c) *cis*-[RhH(CO)(PR$_3$)$_2$] are chiral?

Metal complexes of all shapes and sizes have roles in biology and medicine (Box 7.2).

BOX 7.2 Where are metal complexes found in biology and medicine?

Coordination complexes have a role in many of the most important biological processes known. Familiar examples include magnesium at the centre of plant photosynthesis in chlorophyll (**B4**), and iron at the heart of oxygen transport in haemoglobin (**B5**).

B4

B5

Recent estimates suggest that around 20% of all enzymes contain a coordinated metal at the active site. Many enzymes contain more than one active centre, and these may contain different metals, such as the copper and iron centres in the synthetic model of cytochrome *c* oxidase (**B6**). Other multi-metallic enzymes include hydrogenases such as (**B7**), which contains six iron centres, together with a variety of ligand types.

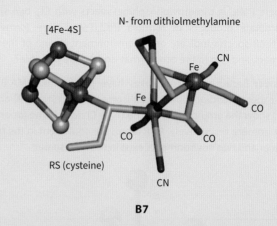

B7

Metals complexes also have significant uses in medicine. The use of *cis* platin (**B8**) as a treatment for some types of cancer is well known, but other metals are also widely used. Thus gallium complexes (**B9**) are under investigation as anti-cancer drugs, gold complexes (**B10**) are effective against arthritis, and gadolinium (**B11**) and technetium complexes (**B12**) are used to assist imaging.

B8

B9

B10

B11

B6

B12

Chapters 26 and 32 discuss these complexes, alongside many others, in more detail.

The thermodynamics of complex formation

When assessing chemical reactions we need to consider both thermodynamic and kinetic aspects because, although a reaction may be thermodynamically feasible, there might be kinetic constraints.

7.12 Formation constants

KEY POINTS A formation constant expresses the interaction strength of a ligand relative to the interaction strength of the solvent molecules (usually H_2O) as a ligand; a stepwise formation constant is the formation constant for each individual solvent replacement in the synthesis of the complex; an overall formation constant is the product of the stepwise formation constants.

Consider the reaction of Fe(III) with SCN^- to give $[Fe(SCN)(OH_2)_5]^{2+}$, a red complex used to detect either iron(III) or the thiocyanate ion:

$$[Fe(OH_2)_6]^{3+}(aq) + SCN^-(aq) \rightleftharpoons$$
$$[Fe(SCN)(OH_2)_5]^{2+}(aq) + H_2O(l)$$

$$K_f = \frac{[Fe(SCN)(OH_2)_5^{2+}]}{[Fe(OH_2)_6^{3+}][SCN^-]}$$

TABLE 7.3 Formation constants for the reaction
$[M(OH_2)_n]^{m+} + L \rightarrow [M(L)(OH_2)_{n-1}]^{m+} + H_2O$

Ion	Ligand	K_f	log K_f
Mg^{2+}	NH_3	1.7	0.23
Ca^{2+}	NH_3	0.64	−0.2
Ni^{2+}	NH_3	525	2.72
Cu^+	NH_3	8.50×10^5	5.93
Cu^{2+}	NH_3	2.0×10^4	4.31
Hg^{2+}	NH_3	6.3×10^8	8.8
Rb^+	Cl^-	0.17	−0.77
Mg^{2+}	Cl^-	4.17	0.62
Cr^{3+}	Cl^-	7.24	0.86
Co^{2+}	Cl^-	4.90	0.69
Pd^{2+}	Cl^-	1.25×10^5	6.1
Na^+	SCN^-	1.2×10^4	4.08
Cr^{3+}	SCN^-	1.2×10^3	3.08
Fe^{3+}	SCN^-	234	2.37
Co^{2+}	SCN^-	11.5	1.06
Fe^{2+}	pyridine	5.13	0.71
Zn^{2+}	pyridine	8.91	0.95
Cu^{2+}	pyridine	331	2.52
Ag^+	pyridine	93	1.97

The equilibrium constant, K_f, of this reaction is called the **formation constant** of the complex. The concentration of solvent (normally H_2O) does not appear in the expression because it is taken to be constant in dilute solution and ascribed unit activity. The value of K_f indicates the strength of binding of the ligand relative to H_2O: if K_f is large, the incoming ligand binds more strongly than the solvent, H_2O; if K_f is small, the incoming ligand binds more weakly than H_2O. Because the values of K_f can vary over a huge range (Table 7.3) they are often expressed as their logarithms, log K_f.

> **A NOTE ON GOOD PRACTICE**
>
> In expressions for equilibrium constants and rate equations, we omit the brackets that are part of the chemical formula of the complex; the surviving square brackets denote molar concentration of a species (with the units mol dm^{-3} removed).

The discussion of stabilities is more involved when more than one ligand may be replaced. For instance, in the reaction of $[Ni(OH_2)_6]^{2+}$ to give $[Ni(NH_3)_6]^{2+}$,

$$[Ni(OH_2)_6]^{2+}(aq) + 6NH_3(aq) \rightleftharpoons$$
$$[Ni(NH_3)_6]^{2+}(aq) + 6H_2O(l)$$

there are at least six steps, even if *cis–trans* isomerization is ignored. For the general case of the complex ML_n, for which the overall reaction is $M + nL \rightarrow ML_n$, the **stepwise formation constants** are

$$M + L \rightleftharpoons ML \quad K_{f1} = \frac{[ML]}{[M][L]}$$

$$ML + L \rightleftharpoons ML_2 \quad K_{f2} = \frac{[ML_2]}{[ML][L]}$$

and so on, and in general

$$ML_{n-1} + L \rightleftharpoons ML_n \quad K_{fn} = \frac{[ML_n]}{[ML_{n-1}][L]}$$

These stepwise constants are the ones to consider when seeking to understand the relationships between structure and reactivity.

When we want to calculate the concentration of the final product (the complex ML_n) we use the **overall formation constant**, β_n:

$$M + nL \rightleftharpoons ML_n \quad \beta_n = \frac{[ML_n]}{[M][L]^n}$$

As may be verified by multiplying together the individual stepwise constants, the overall formation constant is the product of the stepwise constants

$$\beta_n = K_{f1} K_{f2} \ldots K_{fn}$$

The inverse of each K_f, the **dissociation constant**, K_d, is also sometimes useful, and is often preferred when we are interested in the concentration of ligand that is required to give a certain concentration of complex:

$$ML \rightleftharpoons M + L \quad K_{d1} = \frac{[M][L]}{[ML]} = \frac{1}{K_{f1}}$$

For a 1:1 reaction, like the one above, when half the metal ions are complexed and half are not, so that [M] = [ML], then $K_{d1} = [L]$. In practice, if initially [L] ≫ [M], so that there is an insignificant change in the concentration of L when M is added and undergoes complexation, K_d is the ligand concentration required to obtain 50% complexation.

Because K_d has the same form as K_a for acids, with L taking the place of H^+, its use facilitates comparisons between metal complexes and Brønsted acids. The values of K_d and K_a can be tabulated together if the proton is considered to be simply another cation. For instance, HF can be considered as the complex formed from the Lewis acid H^+, with the Lewis base F^- playing the role of a ligand.

7.13 Trends in successive formation constants

KEY POINTS Stepwise formation constants typically lie in the order $K_{fn} > K_{fn+1}$, as expected statistically; deviations from this order indicate a major change in structure.

The magnitude of the formation constant is a direct reflection of the sign and magnitude of the standard Gibbs energy of formation (because $\Delta_r G^\ominus = -RT \ln K_f$). It is commonly observed that stepwise formation constants lie in the order $K_{f1} > K_{f2} > \ldots > K_{fn}$. This general trend can be explained quite simply by considering the decrease in the number of the ligand H_2O molecules available for replacement in the formation step, as in

$$[M(OH_2)_5L](aq) + L(aq) \rightleftharpoons [M(OH_2)_4L_2](aq) + H_2O(l)$$

compared with

$$[M(OH_2)_4L_2](aq) + L(aq) \rightleftharpoons [M(OH_2)_3L_3](aq) + H_2O(l)$$

The decrease in the stepwise formation constants reflects the diminishing statistical factor as successive ligands are replaced, coupled with the fact that an increase in the number of bound ligands increases the likelihood of the reverse reaction. That such a simple explanation is more or less correct is illustrated by data for the successive complexes in

TABLE 7.4 Formation constants of Ni(II) ammines, $[Ni(NH_3)_n(OH_2)_{6-n}]^{2+}$

n	K_f	log K_f	K_n/K_{n-1} (experimental)	K_n/K_{n-1} (statistical)*
1	525	2.72		
2	148	2.17	0.28	0.42
3	45.7	1.66	0.31	0.53
4	13.2	1.12	0.29	0.56
5	4.7	0.63	0.35	0.53
6	1.1	0.03	0.23	0.42

* Based on ratios of numbers of ligands available for replacement, with the reaction enthalpy assumed constant.

the series from $[Ni(OH_2)_6]^{2+}$ to $[Ni(NH_3)_6]^{2+}$ (Table 7.4). The reaction enthalpies for the six successive steps are known to vary by less than $2\,kJ\,mol^{-1}$.

A reversal of the relation $K_{fn} > K_{fn+1}$ is usually an indication of a major change in the electronic structure of the complex as more ligands are added. An example is the observation that the tris(bipyridine) complex of Fe(II), $[Fe(bpy)_3]^{2+}$, is strikingly stable compared with the bis complex, $[Fe(bpy)_2(OH_2)_2]^{2+}$. This observation can be correlated with the change in electronic configuration from a high-spin (weak-field) $t_{2g}^4 e_g^2$ configuration in the bis complex (note the presence of weak-field H_2O ligands) to a low-spin (strong-field) t_{2g}^6 configuration in the tris complex, where there is a considerable increase in the ligand-field stabilization energy (LFSE) (see Sections 7.16 and 7.17).

$$[Fe(OH_2)_6]^{2+}(aq) + bpy(aq) \rightleftharpoons$$
$$[Fe(bpy)(OH_2)_4]^{2+}(aq) + 2H_2O(l) \quad \log K_{f1} = 4.2$$

$$[Fe(bpy)(OH_2)_4]^{2+}(aq) + bpy(aq) \rightleftharpoons$$
$$[Fe(bpy)_2(OH_2)_2]^{2+}(aq) + 2H_2O(l) \quad \log K_{f2} = 3.7$$

$$[Fe(bpy)_2(OH_2)_2]^{2+}(aq) + bpy(aq) \rightleftharpoons$$
$$[Fe(bpy)_3]^{2+}(aq) + 2H_2O(l) \quad \log K_{f3} = 9.3$$

A contrasting example is the halide complexes of Hg(II), where K_{f3} is anomalously low compared with K_{f2}:

$$[Hg(OH_2)_6]^{2+}(aq) + Cl^-(aq) \rightleftharpoons$$
$$[HgCl(OH_2)_5]^+(aq) + H_2O(l) \quad \log K_{f1} = 6.74$$

$$[HgCl(OH_2)_5]^+(aq) + Cl^-(aq) \rightleftharpoons$$
$$[HgCl_2(OH_2)_4](aq) + H_2O(l) \quad \log K_{f2} = 6.48$$

$$[HgCl_2(OH_2)_4](aq) + Cl^-(aq) \rightleftharpoons$$
$$[HgCl_3(OH_2)]^-(aq) + 3H_2O(l) \quad \log K_{f3} = 0.95$$

The decrease between the second and third values is too large to be explained statistically and suggests a major change in the nature of the complex, such as the onset of four-coordination:

$$[Hg(OH_2)_3Cl_2] + Cl^- \longrightarrow [HgCl_3(OH_2)]^- + 3H_2O$$

EXAMPLE 7.6 Interpreting irregular successive formation constants

The successive formation constants for complexes of cadmium with Br^- are $K_{f1} = 36.3$, $K_{f2} = 3.47$, $K_{f3} = 1.15$, $K_{f4} = 2.34$. Suggest an explanation of why $K_{f4} > K_{f3}$.

Answer The anomaly suggests a structural change, so we need to consider what it might be. Aqua complexes are usually six-coordinate whereas halogeno complexes of M^{2+} ions are commonly tetrahedral. The reaction of the complex with three Br^- groups to add the fourth is

$$[CdBr_3(OH_2)_3]^-(aq) + Br^-(aq) \rightleftharpoons [CdBr_4]^{2-}(aq) + 3H_2O(l)$$

This step is favoured by the release of three H_2O molecules from the relatively restricted coordination sphere environment. The result is an increase in K_f.

Self-test 7.6 Assuming the displacement of a water by a ligand were so favoured that the back reaction could be ignored, calculate all the stepwise formation constants you would expect in the formation of $[ML_6]^{2+}$ from $[M(OH_2)_6]^{2+}$, and the overall formation constant, given that $K_{f1} = 1 \times 10^5$.

7.14 The chelate and macrocyclic effects

KEY POINTS The chelate and macrocyclic effects are the greater stability of complexes containing co-ordinated polydentate ligands compared with a complex containing the equivalent number of analogous monodentate ligands; the chelate effect is largely an entropic effect; the macrocyclic effect has an additional enthalpic contribution.

When K_{f1} for the formation of a complex with a bidentate chelate ligand, such as 1,2-diaminoethane (en), is compared with the value of β_2 for the corresponding bis(ammine) complex, it is found that the former is generally larger:

$$[Cd(OH_2)_6]^{2+}(aq) + en(aq) \rightleftharpoons$$
$$[Cd(en)(OH_2)_4]^{2+}(aq) + 2H_2O(l)$$

$\log K_{f1} = 5.84 \quad \Delta_r H^\ominus = -29.4 \text{ kJ mol}^{-1}$
$\Delta_r S^\ominus = +13.0 \text{ J K}^{-1} \text{ mol}^{-1}$

$$[Cd(OH_2)_6]^{2+}(aq) + 2NH_3(aq) \rightleftharpoons$$
$$[Cd(NH_3)_2(OH_2)_4]^{2+}(aq) + 2H_2O(l)$$

$\log \beta_2 = 4.95 \quad \Delta_r H^\ominus = -29.8 \text{ kJ mol}^{-1}$
$\Delta_r S^\ominus = -5.2 \text{ J K}^{-1} \text{ mol}^{-1}$

Two similar Cd–N bonds are formed in each case, yet the formation of the chelate-containing complex is distinctly more favourable. This greater stability of chelated complexes compared with their nonchelated analogues is called the **chelate effect**.

The chelate effect can be traced primarily to differences in reaction entropy between chelated and nonchelated complexes in dilute solutions. The chelation reaction results in an increase in the number of independent molecules in solution, and hence disorder. By contrast, the nonchelating reaction produces no net change (compare the two chemical equations above). The former therefore has the more positive reaction entropy and hence is the more favourable process. The reaction entropies measured in dilute solution support this interpretation.

The entropy advantage of chelation extends beyond bidentate ligands, and applies, in principle, to any polydentate ligand. In fact, the greater the number of donor sites the multidentate ligand has, the greater is the entropic advantage of displacing monodentate ligands. Macrocyclic ligands, where multiple donor atoms are held in a cyclic array, such as crown ethers or phthalocyanin (80), give complexes of even greater stability than might otherwise be expected. This so-called **macrocyclic effect** is thought to be a combination of the entropic effect seen in the chelate effect, together with an additional energetic contribution that comes from the pre-organized nature of the ligating groups (that is, no additional strains are introduced to the ligand on coordination).

80

The chelate and macrocyclic effects are of great practical importance. The majority of reagents used in complexometric titrations in analytical chemistry are polydentate chelates like $edta^{4-}$, and most biochemical metal binding sites are chelating or macrocyclic ligands. A formation constant as high as 10^{12} to 10^{25} is generally a sign that the chelate or macrocyclic effect is in operation.

In addition to the thermodynamic rationalization for the chelate effect we have described, there is an additional role in the chelate effect for kinetics. Once one ligating group of a polydentate ligand has bound to a metal ion, it becomes more likely that its other ligating groups will bind, as they are now constrained to be in close proximity to the metal ion; thus chelate complexes are favoured kinetically too.

7.15 Steric effects and electron delocalization

KEY POINT The stability of chelate complexes of d metals involving diimine ligands is a result of the chelate effect in conjunction with the ability of the ligands to act as π acceptors as well as σ donors.

Steric effects have an important influence on formation constants. They are particularly important in chelate formation because ring completion may be difficult geometrically. Chelate rings with five members are generally very stable because their bond angles are near ideal in the sense of there being no ring strain. Six-membered rings are reasonably stable and may be favoured if their formation results in electron delocalization. Three-, four-, and seven-membered (and larger) chelate rings are found only rarely because they normally result in distortions of bond angles and unfavourable steric interactions.

Complexes containing chelating ligands with delocalized electronic structures may be stabilized by electronic effects in addition to the entropy advantages of chelation. For example, diimine ligands (**81**), such as bipyridine (**82**) and phenanthroline (**83**), are constrained to form five-membered rings with the metal atom.

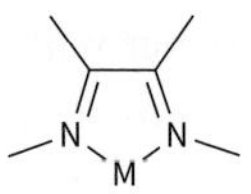

81 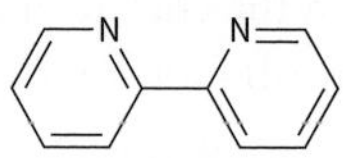 **82** bpy 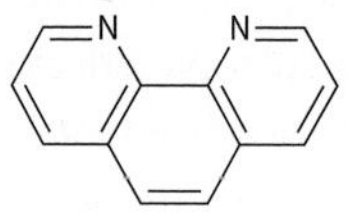 **83** phen

The great stability of their complexes with d metals is probably a result of their ability to act as π acceptors as well as σ donors and to form π bonds by overlap of the full metal d orbitals and the empty ring π* orbitals (Section 7.17). This bond formation is favoured by electron population in the metal t_{2g} orbitals, which allows the metal atom to act as a π donor and transfer electron density to the ligand rings. An example is the complex $[Ru(bpy)_3]^{2+}$ (**84**). In some cases the chelate ring that forms can have appreciable aromatic character, which stabilizes the chelate ring even more.

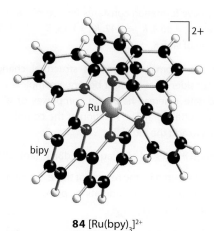

84 $[Ru(bpy)_3]^{2+}$

Box 7.3 describes how complicated chelating and macrocyclic ligands might be synthesized.

BOX 7.3 How can we make rings and knots?

A metal ion such as Ni(II) can be used to assemble a group of ligands that then undergo a reaction among themselves to form a **macrocyclic ligand**, a cyclic molecule with several donor atoms. A simple example is

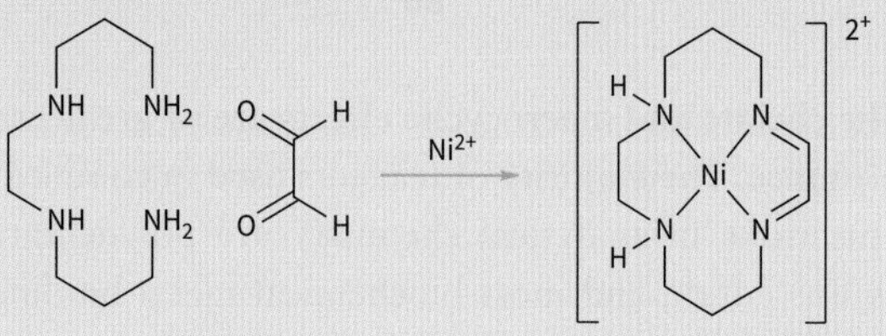

This phenomenon, which is called the **template effect**, can be applied to produce a surprising variety of macrocyclic ligands. The reaction shown above is an example of a **condensation reaction**, in which a bond is formed between two molecules, and a small molecule (in this case H_2O) is eliminated. If the metal ion had not been present, the condensation reaction of the component ligands would have been an ill-defined polymeric mixture, not a macrocycle. Once the macrocycle has been formed, it is normally stable on its own, and the metal ion may be removed to leave a multidentate ligand that can be used to complex other metal ions.

A wide variety of macrocyclic ligands can be synthesized by the template approach. Two more complicated ligand syntheses are shown.

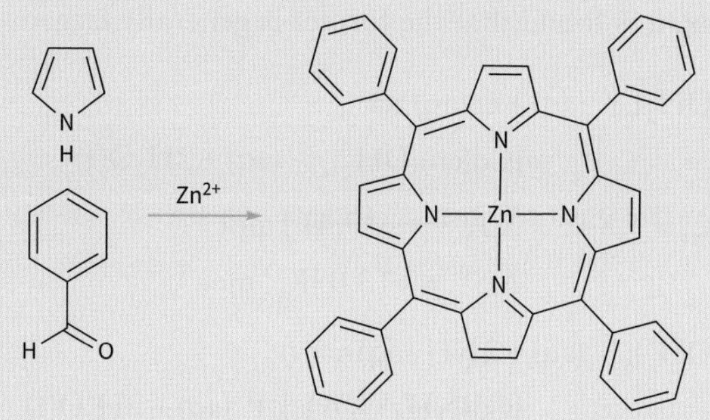

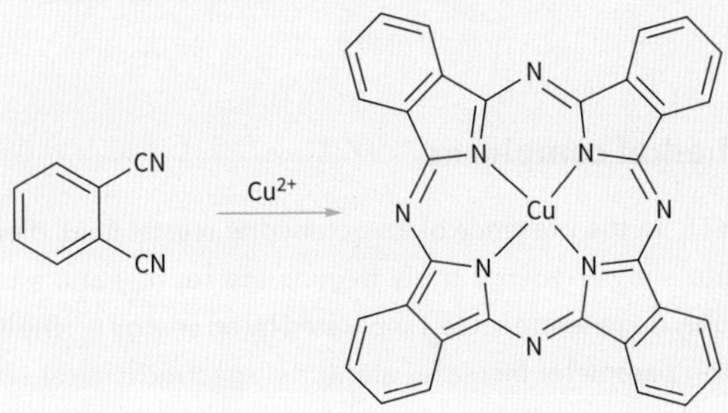

The origin of the template effect may be either kinetic or thermodynamic. For example, the condensation may stem either from the increase in the rate of the reaction between coordinated ligands (on account of their proximity or electronic effects) or from the added stability of the chelated ring product.

More complicated template syntheses can be used to construct topologically complex molecules, such as the chain-like *catenanes*, molecules that consist of interlinked rings. An example of the synthesis of a catenane containing two rings is shown below.

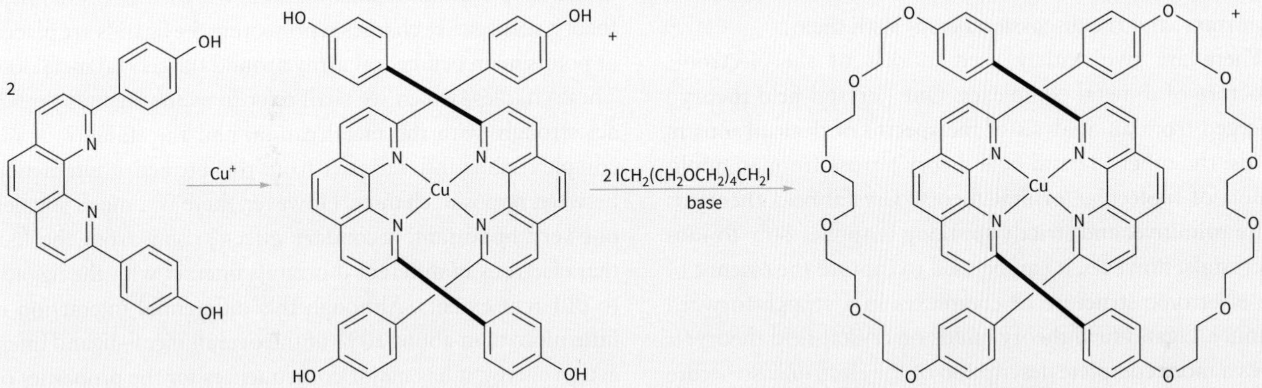

Here, two bipyridine-based ligands are coordinated to a copper ion, and then the ends of each ligand are joined by a flexible linkage. The metal ion can then be removed to give a **catenand** (catenane ligand), which can be used to complex other metal ions.

Even more complicated systems, equivalent to knots and links,[1] can be constructed with multiple metals. For instance, the following synthesis gives rise to a single molecular strand tied in a trefoil knot:

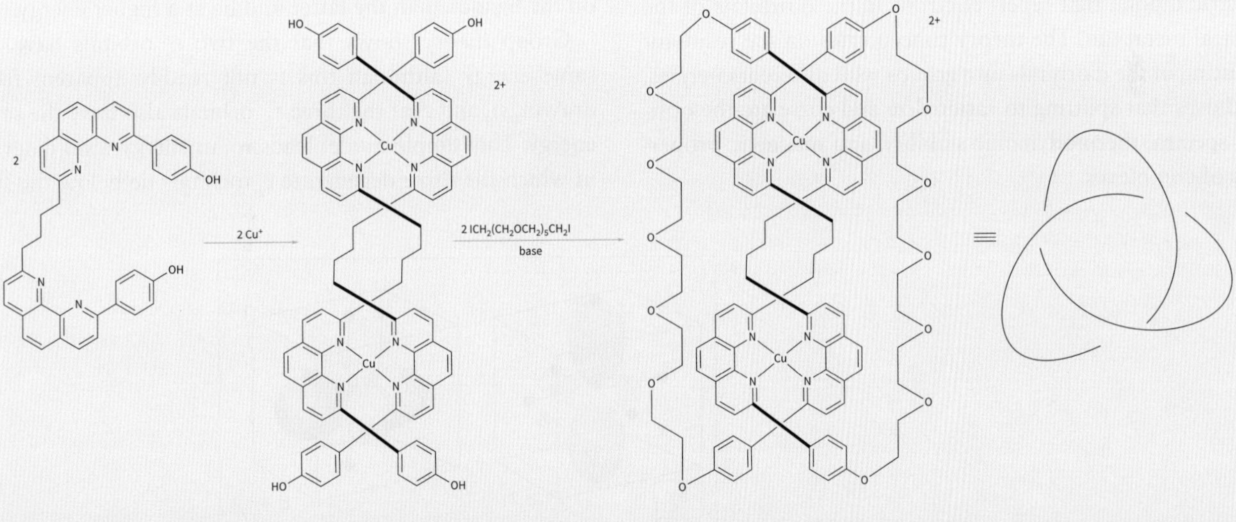

Work on these systems was rewarded in 2016 with the award of the Nobel Prize for Chemistry to J.-P. Sauvage, J. F. Stoddart and B. Feringa 'for the design and synthesis of molecular machines'.

---

[1] Knotted and linked systems are far from being purely of academic interest, and many proteins exist in these forms: see P. Dabrowski-Tumanski and J. I. Sulkowska, Topological knots and links in proteins, *Proceedings of the National Academy of Sciences* 2017, **114**, 3415.

Electronic structure

We now examine in detail the bonding, electronic structure, electronic spectra, and magnetic properties of the d-metal complexes. The striking colours of many d-metal complexes were a mystery to Werner when he elucidated their structures, and the origin of the colours was clarified only when the description of electronic structure in terms of orbitals was applied to the problem in the period from 1930 to 1960. Tetrahedral and octahedral complexes are the most important, and the discussion begins with them.

There are two widely used models of the electronic structure of d-metal complexes. One ('crystal-field theory') emerged from an analysis of the spectra of d-metal ions in solids; the other ('ligand-field theory') arose from an application of molecular orbital theory. Crystal-field theory is more primitive, and strictly speaking it applies only to ions in crystals; however, it can be used to capture the essence of the electronic structure of complexes in a straightforward manner. Ligand-field theory builds on crystal-field theory: it gives a more complete description of the electronic structure of complexes and accounts for a wider range of properties.

7.16 Crystal-field theory

In **crystal-field theory**, a ligand lone pair is modelled as a point negative charge (or as the partial negative charge of an electric dipole) that repels electrons in the d orbitals of the central metal ion. The theory concentrates on the resultant splitting of the d orbitals into groups with different energies, and uses that splitting to rationalize and correlate the optical spectra, thermodynamic stability, and magnetic properties of complexes.

(a) Octahedral complexes

KEY POINTS In the presence of an octahedral crystal field, d orbitals are split into a lower-energy triply degenerate set (t_{2g}) and a higher-energy doubly degenerate set (e_g) separated by an energy Δ_O; the ligand-field splitting parameter increases along the spectrochemical series of ligands and also varies with the identity and charge of the metal ion.

In the model of an octahedral complex used in crystal-field theory, six negative charges representing the ligands are placed at points in an octahedral array around the central metal ion. These charges (which we shall refer to as the 'ligands') interact strongly with the metal cation, and the stability of the complex stems in large part from this attractive interaction between opposite charges. However, there is a much smaller, but very important, secondary effect arising from the fact that electrons in different d orbitals interact with the ligands to different extents. Although this differential interaction is little more than about 10% of the overall metal–ligand interaction energy, it has major consequences for the properties of the complex and is the principal focus of this section.

Electrons in d_{z^2} and $d_{x^2-y^2}$ orbitals (which are of symmetry type e_g in O_h; Section 3.11) are concentrated close to the ligands, along the axes, whereas electrons in d_{xy}, d_{yz}, and d_{zx} orbitals (which are of symmetry type t_{2g}) are concentrated in regions that lie between the ligands (Fig. 7.8). As a result, the former are repelled more strongly by the negative charge on the ligands than the latter and lie at a higher energy.

Group theory shows that the two e_g orbitals have the same energy (although this is not readily apparent from drawings), and that the three t_{2g} orbitals also have the same energy. This simple model leads to an energy-level diagram in which the three degenerate t_{2g} orbitals lie below the two

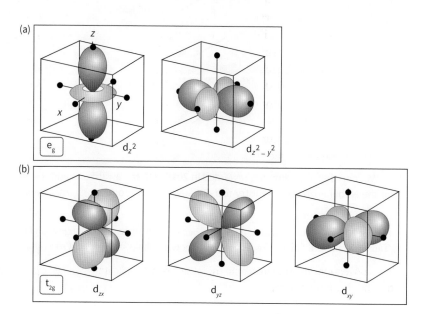

FIGURE 7.8 The orientation of the five d orbitals with respect to the ligands of an octahedral complex: the degenerate (a) e_g and (b) t_{2g} orbitals.

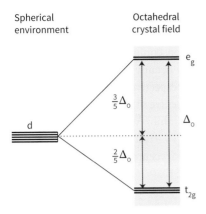

FIGURE 7.9 The energies of the d orbitals in an octahedral crystal field. Note that the mean energy remains unchanged relative to the energy of the d orbitals in a spherically symmetric environment (such as in a free atom).

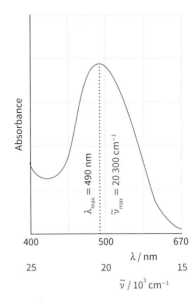

FIGURE 7.10 The optical absorption spectrum of $[Ti(OH_2)_6]^{3+}$.

degenerate e_g orbitals (Fig. 7.9). The energy separation of the two sets of orbitals is called the **ligand-field splitting parameter**, Δ_O (where the subscript 'o' signifies an octahedral crystal field).

> **A NOTE ON GOOD PRACTICE**
>
> In the context of crystal-field theory, the ligand-field splitting parameter should be called the *crystal-field splitting parameter*, but we use ligand-field splitting parameter to avoid a proliferation of names.

The energy level that corresponds to the hypothetical spherically symmetrical environment (in which the negative charge due to the ligands is evenly distributed over a sphere instead of being localized at six points) defines the **barycentre** of the array of levels, with the two e_g orbitals lying at $\tfrac{3}{5}\Delta_O$ above the barycentre and the three t_{2g} orbitals lying at $\tfrac{2}{5}\Delta_O$ below it. As in the representation of the configurations of atoms, a superscript is used to indicate the number of electrons in each set, for example t_{2g}^2.

The simplest property that can be interpreted by crystal-field theory is the absorption spectrum of a one-electron complex. Fig. 7.10 shows the optical absorption spectrum of the d^1 hexaaquatitanium(III) ion, $[Ti(OH_2)_6]^{3+}$. Crystal-field theory assigns the first absorption maximum at 493 nm (20 300 cm^{-1}) to the transition $e_g \leftarrow t_{2g}$ and identifies 20 300 cm^{-1} with Δ_O for the complex.

> **A NOTE ON GOOD PRACTICE**
>
> The convention in spectroscopic notation is to indicate transitions as [upper state] ← [lower state].

It is not so straightforward to obtain values of Δ_O for complexes with more than one d electron because the energy of a transition then depends not only on orbital energies but also on the electron–electron repulsion energies. This aspect is treated more fully in Section 22.1, and the results from the analyses described there have been used to obtain the values of Δ_O in Table 7.5.

The ligand-field splitting parameter, Δ_O, varies systematically with the identity of the ligand. For instance, in the series of complexes $[CoX(NH_3)_5]^{n+}$ with X = I$^-$, Br$^-$, Cl$^-$, H$_2$O, and NH$_3$, the colours range from purple (for X = I$^-$) through pink (for Cl$^-$) to yellow (with NH$_3$). This sequence indicates that the energy of the lowest-energy electronic transition (and therefore Δ_O) increases as the ligands are varied along the series. The same order is followed regardless of the identity of the metal ion. Thus ligands can be arranged in a **spectrochemical series**, in which the members

TABLE 7.5 Ligand-field splitting parameters Δ_O of ML$_6$ complexes*

	Ions	Ligands				
		Cl$^-$	H$_2$O	NH$_3$	en	CN$^-$
d^3	Cr^{3+}	13 700	17 400	21 500	21 900	26 600
d^5	Mn^{2+}	7500	8500		10 100	30 000
	Fe^{3+}	11 000	14 300			(35 000)
d^6	Fe^{2+}		10 400			(32 800)
	Co^{3+}		(20 700)	(22 900)	(23 200)	(34 800)
	Rh^{3+}	(20 400)	(27 000)	(34 000)	(34 600)	(45 500)
d^8	Ni^{2+}	7500	8500	10 800	11 500	

* Values are in cm^{-1}; entries in parentheses are for low-spin complexes.

are arranged in order of increasing energy of transitions that occur when they are present in a complex:

$I^- < Br^- < S^{2-} < \underline{S}CN^- < Cl^- < N\underline{O}_2^- < N^{3-} < F^- < \underline{O}H^-$
$< C_2O_4^{2-} < O^{2-} < H_2O < \underline{N}CS^- < CH_3C \equiv N < py < NH_3$
$< en < bpy < phen < \underline{N}O_2^- < PPh_3 < \underline{C}N^- < CO$

(The donor atom in an ambidentate ligand is underlined.) Thus, the series indicates that, for the same metal, the optical absorption of the cyano complex will occur at higher energy than that of the corresponding chlorido complex. A ligand that gives rise to a high-energy transition (such as CO) is referred to as a **strong-field ligand**, whereas one that gives rise to a low-energy transition (such as Br⁻) is referred to as a **weak-field ligand**. Crystal-field theory alone cannot explain these strengths, but ligand-field theory can, as we shall see in Section 7.17.

The ligand-field strength also depends on the identity of the central metal ion, the order being approximately

$Mn^{2+} < Ni^{2+} < Co^{2+} < Fe^{2+} < V^{2+} < Fe^{3+} < Co^{3+}$
$< Mo^{3+} < Rh^{3+} < Ru^{3+} < Pd^{4+} < Ir^{3+} < Pt^{4+}$

The value of Δ_O increases with increasing oxidation state of the central metal ion (compare the two entries for Fe and Co) and also increases down a group (compare, for instance, the locations of Co, Rh, and Ir). The variation with oxidation state reflects the smaller size of more highly charged ions and the consequently shorter metal–ligand distances and stronger interaction energies. The increase down a group reflects the larger size of the 4d and 5d orbitals compared with the compact 3d orbitals and the consequent stronger interactions with the ligands.

(b) Ligand-field stabilization energies

KEY POINTS The ground-state configuration of a complex reflects the relative values of the ligand-field splitting parameter and the pairing energy. For octahedral 3d^n species with $n = 4 - 7$, high-spin and low-spin complexes occur in the weak-field and strong-field cases, respectively. Octahedral complexes of 4d- and 5d-series metals are typically low-spin.

Because the d orbitals in an octahedral complex do not all have the same energy, the ground-state electron configuration of a complex is not immediately obvious. To predict it, we use the d-orbital energy-level diagram shown in Fig. 7.9 as a basis for applying the building-up principle. That is, we identify the lowest-energy configuration subject to the Pauli exclusion principle (a maximum of two electrons in an orbital) and (if more than one degenerate orbital is available) to the requirement that electrons first occupy separate orbitals and do so with parallel spins.

First, we consider complexes formed by the 3d-series elements. In an octahedral complex, the first three d electrons of a 3d^n complex occupy separate t_{2g} nonbonding orbitals, and do so with parallel spins. For example, the ions Ti^{2+} and V^{2+} have electron configurations 3d^2 and 3d^3, respectively. The d electrons occupy the lower t_{2g} orbitals, as shown in (85) and (86), respectively. The energy of a t_{2g} orbital relative to the barycentre of an octahedral ion is $-0.4\Delta_O$, and the complexes are stabilized by $2 \times (-0.4\Delta_O) = -0.8\Delta_O$ (for Ti^{2+}) and $3 \times (-0.4\Delta_O) = -1.2\Delta_O$ (for V^{2+}). This additional stability, relative to the barycentre is called the **ligand-field stabilization energy** (LFSE).

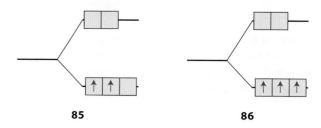

85   86

> **A NOTE ON GOOD PRACTICE**
>
> The term *crystal-field stabilization energy* (CFSE) is widely used in place of LFSE, but strictly speaking the term is appropriate only for ions in crystals.

The next electron needed for the 3d^4 ion Cr^{2+} may enter one of the t_{2g} orbitals and pair with the electron already there (87). However, if it does so, it experiences a strong Coulombic repulsion, which is called the **pairing energy**, P. Alternatively, the electron may occupy one of the e_g orbitals (88).

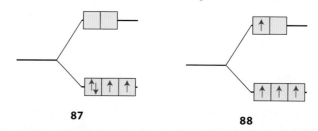

87   88

Although the pairing penalty is now avoided, the orbital energy is higher by Δ_O. In the first case (t_{2g}^4), there is a stabilization of $-1.6\Delta_O$, countered by the pairing energy of P, giving a net LFSE of $-1.6\Delta_O + P$. In the second case ($t_{2g}^3 e_g^1$), the LFSE is $3 \times (-0.4\Delta_O) + 0.6\Delta_O = -0.6\Delta_O$, as there is no pairing energy to consider. Which configuration is adopted depends on which of $(-1.60\Delta_O + P)$ and $(-0.60\Delta_O)$ is the larger in magnitude.

If $\Delta_O < P$, which is called the **weak-field case**, a lower energy is achieved when the upper orbital is occupied to give the configuration $t_{2g}^3 e_g^1$. If $\Delta_O > P$, which is called the **strong-field case**, a lower energy is achieved by occupying only the lower orbitals despite the cost of the pairing energy. The resulting configuration is now t_{2g}^4. For example, [Cr(OH$_2$)$_6$]$^{2+}$ has the ground-state configuration $t_{2g}^3 e_g^1$, whereas [Cr(CN)$_6$]$^{4-}$, with relatively strong-field ligands (as indicated by the spectrochemical series), has the configuration t_{2g}^4. In the weak-field case all the electrons occupy different orbitals and have parallel spins. The resulting spin-correlation effect (the tendency of electrons of the same spin to avoid each other) helps to offset the cost of occupying orbitals of higher energy.

The ground-state electron configurations of $3d^1$, $3d^2$, and $3d^3$ octahedral complexes are unambiguous because there is no competition between the additional stabilization achieved by occupying the t_{2g} orbitals and the pairing energy: the configurations are t_{2g}^1, t_{2g}^2, and t_{2g}^3, respectively, with each electron in a separate orbital. As remarked above, there are two possible configurations for $3d^4$ complexes; the same is true of $3d^n$ complexes in which $n = 5$, 6, or 7. In the strong-field case the lower orbitals are occupied preferentially, and in the weak-field case electrons avoid the pairing energy by occupying the upper orbitals.

When alternative configurations are possible, the species with the smaller number of parallel electron spins is called a **low-spin complex**, and the species with the greater number of parallel electron spins is called a **high-spin complex**. As we have noted, an octahedral $3d^4$ complex is likely to be low-spin if the crystal field is strong, but high-spin if the field is weak (Fig. 7.11); the same applies to $3d^5$, $3d^6$, and $3d^7$ complexes:

	Weak-field ligands		Strong-field ligands	
	Configuration	Unpaired electrons	Configuration	Unpaired electrons
$3d^4$	$t_{2g}^3 e_g^1$	4	t_{2g}^4	2
$3d^5$	$t_{2g}^3 e_g^2$	5	t_{2g}^5	1
$3d^6$	$t_{2g}^4 e_g^2$	4	t_{2g}^6	0
$3d^7$	$t_{2g}^5 e_g^2$	3	$t_{2g}^6 e_g^1$	1

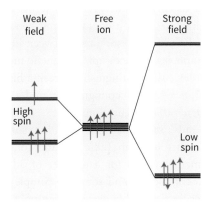

FIGURE 7.11 The effect of weak and strong ligand fields on the occupation of orbitals for a d^4 complex. The former results in a high-spin configuration and the latter in a low-spin configuration.

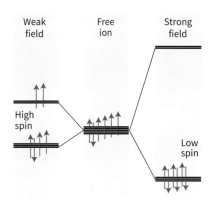

FIGURE 7.12 The effect of weak and strong ligand fields on the occupation of orbitals for a d^6 complex. The former results in a high-spin configuration and the latter in a low-spin configuration.

The ground-state electron configurations of $3d^8$, $3d^9$, and $3d^{10}$ complexes are unambiguous and the configurations are, respectively, $t_{2g}^6 e_g^2$, $t_{2g}^6 e_g^3$, and $t_{2g}^6 e_g^4$.

In general, the net energy of a $t_{2g}^x e_g^y$ configuration relative to the barycentre, without taking the pairing energy into account, is $(-0.4x + 0.6y)\Delta_O$. Pairing energies need to be taken into account only for pairing that is additional to the pairing that occurs in a spherical field. Fig. 7.12 shows the case of a d^6 ion. In both the free ion and the high-spin complex two electrons are paired, whereas in the low-spin case all six electrons occur as three pairs. Thus, we do not need to consider the pairing energy in the high-spin case, as there is no additional pairing. There are two additional pairings in the low-spin case, so two pairing energy contributions must be taken into account.

High-spin complexes always have the same number of unpaired electrons as in a spherical field (free ion), and we therefore do not need to consider pairing energies for high-spin complexes. Table 7.6 lists the values for the LFSE of the

TABLE 7.6 Ligand-field stabilization energies for octahedral complexes*

d^n	Example	N (high spin)	LFSE/Δ_O	N (low spin)	LFSE
d^0		0	0		
d^1	Ti^{3+}	1	-0.4		
d^2	V^{3+}	2	-0.8		
d^3	Cr^{3+}, V^{2+}	3	-1.2		
d^4	Cr^{2+}, Mn^{3+}	4	-0.6	2	$-1.6\Delta_O + P$
d^5	Mn^{2+}, Fe^{3+}	5	0	1	$-2.0\Delta_O + 2P$
d^6	Fe^{2+}, Co^{3+}	4	-0.4	0	$-2.4\Delta_O + 2P$
d^7	Co^{2+}	3	-0.8	1	$-1.8\Delta_O + P$
d^8	Ni^{2+}	2	-1.2		
d^9	Cu^{2+}	1	-0.6		
d^{10}	Cu^+, Zn^{2+}	0	0		

* N is the number of unpaired electrons.

various configurations of octahedral ions, with the appropriate pairing energies taken into account for the low-spin complexes. Remember that the LFSE is generally only a small fraction of the overall interaction between the metal atom and the ligands.

The strength of the crystal field (as measured by the value of Δ_O) and the spin-pairing energy (as measured by P) depend on the identity of both the metal and the ligand, so it is not possible to specify a universal point in the spectrochemical series at which a complex changes from high spin to low spin. For 3d-metal ions, low-spin complexes commonly occur for ligands that are high in the spectrochemical series (such as CN^-) and high-spin complexes are common for ligands that are low in the series (such as F^-). For octahedral d^n complexes with $n = 1 - 3$ and $8 - 10$ there is no ambiguity about the configuration (see Table 7.6), and the designations 'high-spin' and 'low-spin' are not used.

As we have seen, the values of Δ_O for complexes of 4d- and 5d-series metals are typically higher than for the 3d-series metals. Pairing energies for the 4d- and 5d-series metals tend to be lower than for the 3d-series metals because the orbitals are less compact and electron–electron repulsions correspondingly weaker. Consequently, complexes of these metals generally have electron configurations that are characteristic of strong crystal fields and typically have low spin. An example is the $4d^4$ complex $[RuCl_6]^{2-}$, which has a t_{2g}^4 configuration, which is typical of a strong crystal field despite Cl^- being low in the spectrochemical series. Likewise, $[Ru(ox)_3]^{3-}$ has the low-spin configuration t_{2g}^5 whereas $[Fe(ox)_3]^{3-}$ has the high-spin configuration $t_{2g}^3 e_g^2$.

EXAMPLE 7.7 Calculating the LFSE

Determine the LFSE for the following octahedral ions from first principles and confirm the value matches those in Table 7.6: (a) d^3, (b) high-spin d^5, (c) high-spin d^6, (d) low-spin d^6, (e) d^9.

Answer We need to consider the total orbital energy in each case and, when appropriate, the pairing energy. (a) A d^3 ion has configuration t_{2g}^3 (no pairing of electrons) and therefore LFSE = 3 × (−0.4Δ_O) = −1.2Δ_O. (b) A high spin d^5 ion has configuration $t_{2g}^3 e_g^2$ (no pairing of electrons), therefore LFSE = (3 × −0.4 + 2 × 0.6)Δ_O = 0. (c) A high-spin d^6 ion has configuration $t_{2g}^4 e_g^2$ with the pairing of two electrons. However, since those two electrons would be paired in a spherical field there is no additional pairing energy to be concerned with. Therefore LFSE = (4 × −0.4 + 2 × 0.6)Δ_O = −0.4Δ_O. (d) A low-spin d^6 ion has configuration t_{2g}^6 with the pairing of three pairs of electrons. Since one pair of electrons would be paired in a spherical field, the additional pairing energy is $2P$. Therefore LFSE = 6 × (−0.4Δ_O) + 2P = −2.4Δ_O + 2P. (e) A d^9 ion has configuration $t_{2g}^6 e_g^3$ with the pairing of four pairs of electrons. Since all four pairs of electrons would be paired in a spherical field there is no additional pairing energy. Therefore LFSE = (6 × −0.4 + 3 × 0.6)Δ_O = −0.6Δ_O.

Self-test 7.7 What is the LFSE for both high- and low-spin d^7 configurations?

(c) Magnetic measurements

KEY POINTS Magnetic measurements are used to determine the number of unpaired spins in a complex and hence to identify its ground-state configuration. A spin-only calculation may fail for low-spin d^5 and for high-spin $3d^6$ and $3d^7$ complexes.

The experimental distinction between high-spin and low-spin octahedral complexes is based on the determination of their magnetic properties. Compounds are classified as **diamagnetic** if they are repelled by a magnetic field and **paramagnetic** if they are attracted by a magnetic field. The two classes are distinguished experimentally by magnetometry (Chapter 8). The magnitude of the paramagnetism of a complex is commonly reported in terms of the magnetic dipole moment it possesses: the higher the magnetic dipole moment of the complex, the greater the paramagnetism of the sample.

In a free atom or ion, both the orbital and the spin angular momenta give rise to a magnetic moment and contribute to the paramagnetism. When the atom or ion is part of a complex, any orbital angular momentum is normally **quenched**, or suppressed, as a result of the interactions of the electrons with their nonspherical environment. However, if any electrons are unpaired the net electron spin angular momentum survives and gives rise to **spin-only paramagnetism**, which is characteristic of many d-metal complexes. The spin-only magnetic moment, μ, of a complex with total spin quantum number S is

$$\mu = 2\sqrt{S(S+1)}\mu_B \quad (7.1)$$

where μ_B is the **Bohr magneton**, $\mu_B = e\hbar/2m_e$, the magnetic moment of a single electron, with the value 9.274×10^{-24} J T^{-1}. Because $S = \frac{1}{2} N$, where N is the number of unpaired electrons, each with spin $s = \frac{1}{2}$,

$$\mu = \sqrt{N(N+2)}\mu_B \quad (7.2)$$

A measurement of the magnetic moment of a d-block complex can usually be interpreted in terms of the number of unpaired electrons it contains, and hence the measurement can be used to distinguish between high-spin and low-spin complexes. For example, magnetic measurements on a d^6 complex easily distinguish between a high-spin $t_{2g}^4 e_g^2$ ($N = 4$, $S = 2$, $\mu = 4.90\ \mu_B$) configuration and a low-spin t_{2g}^6 ($N = 0$, $S = 0$, $\mu = 0$) configuration.

The spin-only magnetic moments for some electron configurations are listed in Table 7.7 and compared there with experimental values for a number of 3d complexes. For most 3d complexes (and some 4d complexes), experimental values lie reasonably close to spin-only predictions, so it becomes possible to identify correctly the number of unpaired electrons and assign the ground-state configuration. For instance, $[Fe(OH_2)_6]^{3+}$ is paramagnetic with a

TABLE 7.7 Calculated spin-only magnetic moments

Ion	Electron configuration	N	S	μ/μ_B Calculated	Experimental
Ti^{3+}	t_{2g}^1	1	$\frac{1}{2}$	1.73	1.7–1.8
V^{3+}	t_{2g}^2	2	1	2.83	2.7–2.9
Cr^{3+}	t_{2g}^3	3	$\frac{3}{2}$	3.87	3.8
Mn^{3+}	$t_{2g}^3 e_g^1$	4	2	4.90	4.8–4.9
Fe^{3+}	$t_{2g}^3 e_g^2$	5	$\frac{5}{2}$	5.92	5.9

magnetic moment of 5.9 μ_B. As shown in Table 7.7, this value is consistent with there being five unpaired electrons (N = 5 and S = 5/2), which implies a high-spin $t_{2g}^3 e_g^2$ configuration.

The interpretation of magnetic measurements is sometimes less straightforward than this example might suggest. For example, the potassium salt of $[Fe(CN)_6]^{3-}$ has $\mu = 2.3$ μ_B, which is between the spin-only values for one and two unpaired electrons (1.7 μ_B and 2.8 μ_B, respectively). In this case, the spin-only assumption has failed because the orbital contribution to the magnetic moment is substantial; however, it is still possible to use this value to distinguish between the two possibilities for the d^5 Fe^{3+} ion: a low-spin complex would have a single unpaired electron (1.7 μ_B) whereas a high-spin complex would have five unpaired electrons (5.9 μ_B).

For orbital angular momentum to contribute, and hence for the paramagnetism to differ significantly from the spin-only value, there must be one or more unfilled or half-filled orbitals similar in energy to the orbitals occupied by the unpaired spins and of the appropriate symmetry (one that is related to the occupied orbital by rotation round the direction of the applied field). If that is so, the applied magnetic field can force the electrons to circulate around the metal ion by using the low-lying orbitals and hence it generates orbital angular momentum and a corresponding orbital contribution to the total magnetic moment (Fig. 7.13).

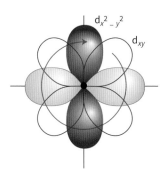

FIGURE 7.13 If there is a low-lying orbital of the correct symmetry, the applied field may induce the circulation of the electrons in a complex and hence generate orbital angular momentum. This diagram shows the way in which circulation may arise when the field is applied perpendicular to the xy-plane (perpendicular to the page or screen you are viewing it on).

Departure from spin-only values is generally large for low-spin d^5 and for high-spin $3d^6$ and $3d^7$ complexes. It is also possible for the electronic state of the metal ion to change (for example, with temperature), leading to a change from high-spin to low-spin and a change in the magnetic moment. Such complexes are referred to as **spin-crossover** complexes and are discussed in more detail, together with the effects of cooperative magnetism, in Sections 22.8 and 30.6.

EXAMPLE 7.8 Inferring an electron configuration from a magnetic moment

The magnetic moment of a certain octahedral Co(II) complex is 4.0 μ_B. What is its d-electron configuration?

Answer We need to match the possible electron configurations of the complex with the observed magnetic moment. A Co(II) complex is d^7. The two possible configurations are $t_{2g}^5 e_g^2$ (high spin, N = 3, S = $\frac{3}{2}$) with three unpaired electrons or $t_{2g}^6 e_g^1$ (low spin, N = 1, S = $\frac{1}{2}$) with one unpaired electron. The spin-only magnetic moments are 3.87 μ_B and 1.73 μ_B, respectively (see Table 7.7). Therefore, the only consistent assignment is the high-spin configuration $t_{2g}^5 e_g^2$.

Self-test 7.8 The magnetic moment of the complex $[Mn(NCS)_6]^{4-}$ is 6.06 μ_B. What is its electron configuration?

(d) Thermochemical correlations

KEY POINT The experimental variation in hydration enthalpies reflects a combination of the variation in radii of the ions (the linear trend) and the variation in LFSE (the saw-tooth variation).

The concept of ligand-field stabilization energy helps to explain the double-humped variation in the hydration enthalpies of the high-spin octahedral 3d-metal M^{2+} ions (Fig. 7.14).

The hydration enthalpy for the 2+ ions is the enthalpy change for the reaction

$$M^{2+}(g) + 6H_2O \rightarrow [M(H_2O)_6]^{2+}$$

We would expect a nearly linear increase in the magnitude of the hydration enthalpy across the period as the ionic radii of the central metal ion decreases from left to right across the period, resulting in an increase in the strength of the bonding to the H_2O ligands; this is shown as the filled circles in Fig. 7.14. The deviation of hydration enthalpies from this straight line arises from the additional ligand-field stabilization energies in the octahedral complexes formed from the free ion. As Table 7.6 shows, the LFSE increases from d^1 to d^3, decreases again to d^5, then rises to d^8, before dropping back to zero for d^{10}.

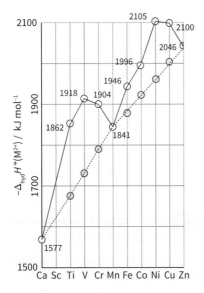

FIGURE 7.14 The hydration enthalpy of M^{2+} ions of the first row of the d block. The straight dotted lines show the expected trend without considering ligand-field stabilization energy. Note the general trend to greater hydration enthalpy (more exothermic hydration) on crossing the period from left to right.

EXAMPLE 7.9 **Using the LFSE to account for thermochemical properties**

The following oxides of formula MO have the lattice enthalpies given, when in the rock-salt structure which gives octahedral coordination of the metal ion:

MnO	FeO	CoO	NiO	ZnO
3810	3921	3988	4071	4032 kJ mol^{-1}

Account for the trends in terms of the LFSE.

Answer We need to consider the simple trend that would be expected on the basis of trends in ionic radii, and then deviations that can be traced to the LFSE. The general trend across the d block is the increase in lattice enthalpy from MnO (d^5) to ZnO (d^{10}) as the ionic radii of the metals decrease (recall that lattice enthalpy is proportional to $1/(r_1 + r_2)$, Section 4.12). The Mn^{2+} ion, being high spin (O^{2-} is a weak-field ligand), has an LFSE of zero, as does the d^{10} Zn^{2+} ion. For a linear increase in lattice enthalpy from manganese to zinc oxides we would expect the lattice enthalpies to increase by (4032 − 3810)/5 kJ mol^{-1} from Mn^{2+} to Fe^{2+} to Co^{2+} to Ni^{2+} to Cu^{2+} to Zn^{2+}. We would therefore expect FeO, CoO, and NiO to have lattice enthalpies of 3854, 3899, and 3943 kJ mol^{-1}, respectively. In fact, FeO (d^6) has a lattice enthalpy of 3921 kJ mol^{-1} and we can ascribe this difference of 67 kJ mol^{-1} to an LFSE of −0.4Δ_O. Likewise the actual lattice enthalpy of 3988 kJ mol^{-1} for CoO (d^7) is 89 kJ mol^{-1} greater than predicted, with this difference arising from an LFSE of −0.8Δ_O and for NiO (d^8) 4071 kJ mol^{-1} is 128 kJ mol^{-1} greater than predicted, with this difference arising from an LFSE of −1.2Δ_O.

Self-test 7.9 Account for the variation in lattice enthalpy of the solid fluorides in which each metal ion is surrounded by an octahedral array of F$^-$ ions: MnF$_2$ (2780 kJ mol^{-1}), FeF$_2$ (2926 kJ mol^{-1}), CoF$_2$ (2976 kJ mol^{-1}), NiF$_2$ (3060 kJ mol^{-1}), and ZnF$_2$ (2985 kJ mol^{-1}).

(e) Tetrahedral complexes

KEY POINTS In a tetrahedral complex, the e orbitals lie below the t$_2$ orbitals; only the high-spin case need be considered.

Four-coordinate tetrahedral complexes are second only in abundance to octahedral complexes for the 3d metals. The same kind of arguments based on crystal-field theory can be applied to these species as we used for octahedral complexes.

A tetrahedral crystal field splits d orbitals into two sets but with the two e orbitals (the d$_{x^2-y^2}$ and the d$_{z^2}$) lower in energy than the three t$_2$ orbitals (the d$_{xy}$, the d$_{yz}$, and the d$_{zx}$) (Fig. 7.15).[2] The fact that the e orbitals lie below the t$_2$ orbitals can be understood from a consideration of the spatial arrangement of the orbitals: the e orbitals point between the positions of the ligands and their partial negative charges, whereas the t$_2$ orbitals point more directly towards the ligands (Fig. 7.16). A second difference is that the ligand-field splitting parameter in a tetrahedral complex, Δ_T, is less than Δ_O, as should be expected for complexes with fewer ligands, none of which is oriented directly at the d orbitals (in fact, $\Delta_T \approx \frac{4}{9}\Delta_O$). The pairing energy is invariably more unfavourable than Δ_T, and normally only high-spin tetrahedral complexes are encountered.

Ligand-field stabilization energies can be calculated in exactly the same way as for octahedral complexes. Since tetrahedral complexes are always high-spin, there is never any need to consider the pairing energy in the LFSE, and the only differences compared with octahedral complexes are the order of occupation (e before t$_2$) and the contribution of

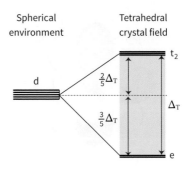

FIGURE 7.15 The orbital energy level diagram used in the application of the building-up principle in a crystal-field analysis of a tetrahedral complex.

---

[2] Because there is no centre of inversion in a tetrahedral complex, the orbital designation does not include the parity label g or u.

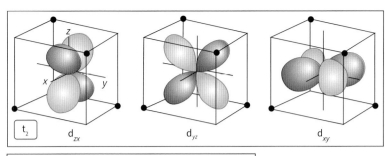
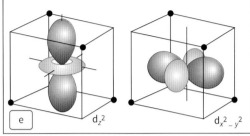

FIGURE 7.16 The effect of a tetrahedral crystal field on a set of d orbitals is to split them into two sets; the e pair (which point less directly at the ligands) lie lower in energy than the t_2 triplet.

TABLE 7.8 Ligand-field stabilization energies for tetrahedral complexes*

d^n	Config.	N	LFSE/Δ_T
d^0		0	0
d^1	e^1	1	−0.6
d^2	e^2	2	−1.2
d^3	$e^2 t_2^1$	3	−0.8
d^4	$e^2 t_2^2$	4	−0.4
d^5	$e^2 t_2^3$	5	0
d^6	$e^3 t_2^3$	4	−0.6
d^7	$e^4 t_2^3$	3	−1.2
d^8	$e^4 t_2^4$	2	−0.8
d^9	$e^4 t_2^5$	1	−0.4
d^{10}	$e^4 t_2^6$	0	0

* N is the number of unpaired electrons.

each orbital to the total energy ($-\tfrac{3}{5}\Delta_T$ for an e orbital and $+\tfrac{2}{5}\Delta_T$ for a t_2 orbital). Table 7.8 lists the configurations of tetrahedral d^n complexes together with the calculated values of the LFSE, and Table 7.9 lists some experimental values of Δ_T for a number of complexes.

TABLE 7.9 Values of Δ_T for representative tetrahedral complexes

Complex	Δ_T/cm^{-1}
[VCl$_4$]	9010
[CoCl$_4$]$^{2-}$	3300
[CoBr$_4$]$^{2-}$	2900
[CoI$_4$]$^{2-}$	2700
[Co(NCS)$_4$]$^{2-}$	4700

(f) Square-planar complexes

KEY POINTS A d^8 configuration, coupled with a strong crystal field, favours the formation of square-planar complexes. This tendency is enhanced with the 4d and 5d metals because of their larger size and the greater ease of electron pairing.

Although a tetrahedral arrangement of four ligands is the least sterically demanding arrangement, some complexes exist with four ligands in an apparently higher-energy square-planar arrangement. If only electrostatic interactions are considered, a square-planar arrangement of ligands gives the d-orbital splitting shown in Fig. 7.17, with $d_{x^2-y^2}$ raised above all the others. This arrangement may become energetically favourable when there are eight d electrons and the crystal field is strong enough to favour the low-spin $d_{yz}^2 d_{zx}^2 d_{z^2}^2 d_{xy}^2$ configuration. In this configuration the electronic stabilization energy can more than compensate for any unfavourable steric interactions. Thus, many square-planar complexes are found for complexes of the large 4d^8

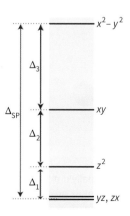

FIGURE 7.17 The orbital splitting parameters for a square-planar complex.

and 5d⁸ Rh(I), Ir(I), Pd(II), Pt(II), and Au(III) ions, in which unfavourable steric constraints have less effect and there is a large ligand-field splitting associated with the 4d- and 5d-series metals.

By contrast, small 3d-series metal complexes such as $[NiX_4]^{2-}$, with X a halide, are generally tetrahedral because the ligand-field splitting parameter is generally quite small and will not compensate sufficiently for the unfavourable steric interactions. Only when the ligand is high in the spectrochemical series is the LFSE large enough to result in the formation of a square-planar complex, as, for example, with $[Ni(CN)_4]^{2-}$. We have already noted that pairing energies for the 4d- and 5d-series metals tend to be lower than for the 3d-series metals, and this difference provides a further factor that favours the formation of low-spin square-planar complexes with these metals.

(g) Tetragonally distorted complexes: the Jahn–Teller effect

KEY POINTS A tetragonal distortion can be expected when the ground electronic configuration of a complex is orbitally degenerate; the complex will distort so as to remove the degeneracy and achieve a lower energy.

Six-coordinate d⁹ complexes of copper(II) usually depart considerably from octahedral geometry and show pronounced tetragonal distortions (Fig. 7.18). High-spin d⁴ (for instance, Cr^{2+} and Mn^{3+}) and low-spin d⁷ six-coordinate complexes (for instance, Ni^{3+}) may show a similar distortion, but complexes of these ions are less common, and the distortions are less pronounced than those in copper(II). These distortions are manifestations of the **Jahn–Teller effect**: if the ground electronic configuration of a nonlinear complex is orbitally degenerate, and asymmetrically filled, then the complex distorts so as to remove the degeneracy and achieve a lower energy.

The physical origin of the effect is quite easy to identify. Thus, a tetragonal distortion of a regular octahedron,

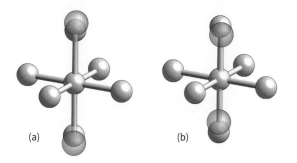

FIGURE 7.18 (a) A tetragonally distorted complex where two of the ligands have moved further away from the central ion; (b) a tetragonally distorted complex where two of the ligands have moved closer towards the central ion.

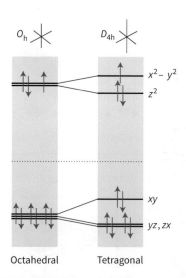

FIGURE 7.19 The effect of tetragonal distortions (compression along x and y and extension along z) on the energies of d orbitals. The orbital occupation is for a d⁹ complex.

corresponding to extension along the z-axis and compression on the x- and y-axes, lowers the energy of the $e_g(d_{z^2})$ orbital and increases the energy of the $e_g(d_{x^2-y^2})$ orbital (Fig. 7.19). Therefore, if one or three electrons occupy the e_g orbitals (as in high-spin d⁴, low-spin d⁷, and d⁹ complexes) a tetragonal distortion may be energetically advantageous. For example, in a d⁹ complex (with configuration that would be $t_{2g}^6 e_g^3$ in O_h), such a distortion leaves two electrons in the d_{z^2} orbital with a lower energy and one in the $d_{x^2-y^2}$ orbital with a higher energy.

The Jahn–Teller effect identifies an unstable geometry (a nonlinear complex with an orbitally degenerate ground state); it does not predict the preferred distortion. For instance, with an octahedral complex, instead of axial elongation and equatorial compression the degeneracy can also be removed by axial compression and equatorial elongation. Which distortion occurs in practice is a matter of energetics, not symmetry. However, because axial elongation weakens two bonds but equatorial elongation weakens four, axial elongation is more common than axial compression.

A Jahn–Teller effect is possible for other electron configurations of octahedral complexes (the d¹, d², low-spin d⁴ and d⁵, high-spin d⁶, and d⁷ configurations) and for tetrahedral complexes (the d¹, d³, d⁴, d⁶, d⁸, and d⁹ configurations). However, as neither the t_{2g} orbitals in an octahedral complex nor any of the d orbitals in a tetrahedral complex point directly at the ligands, the effect is normally much smaller, as is any measurable distortion. Tetrahedral Cu^{2+} compounds often show a slightly 'flattened' tetrahedral geometry as seen for the $[CuCl_4]^{2-}$ anion (89) in Cs_2CuCl_4.

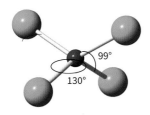

89 The flattened $[CuCl_4]^{2-}$ anion

A Jahn–Teller distortion can hop from one orientation to another and give rise to the **dynamic Jahn–Teller effect**. For example, below 20 K the EPR spectrum of $[Cu(OH_2)_6]^{2+}$ shows a static distortion (more precisely, one that is effectively stationary on the timescale of the resonance experiment). However, above 20 K the distortion disappears because it hops more rapidly than the timescale of the EPR observation.

(h) Octahedral versus tetrahedral coordination

KEY POINTS Consideration of the LFSE predicts that d^3 and d^8 ions strongly prefer an octahedral geometry over a tetrahedral one; for other configurations the preference is less pronounced; LFSE has no bearing on the geometry of d^0, high-spin d^5, and d^{10} ions.

An octahedral complex has six M–L bonding interactions and, in the absence of significant steric and electronic effects, this arrangement will have a lower energy than a tetrahedral complex with just four M–L bonding interactions. We have already discussed the effects of steric bulk on a complex (Section 7.3), and have just seen the electronic reasons that favour a square-planar complex. We can now complete the discussion by considering the electronic effects that favour an octahedral complex over a tetrahedral one.

Fig. 7.20 illustrates the variation of the LFSE for tetrahedral and high-spin octahedral complexes for all electronic configurations. It is apparent that, in terms of LFSE, octahedral geometries are strongly preferred over tetrahedral for d^3 and d^8 complexes: chromium(III) (d^3) and nickel(II) (d^8) do indeed show an exceptional preference for octahedral geometries. Similarly, d^4 and d^9 configurations show a preference for octahedral complexes (for example Mn(III) and Cu(II); note that the Jahn–Teller effect enhances this preference), whereas tetrahedral complexes of d^1, d^2, d^6, and d^7 ions will not be too disfavoured; thus V(II) (d^2) and Co(II) (d^7) form tetrahedral complexes ($[MX_4]^{2-}$) with chloride, bromide, and iodide ligands. The geometry of complexes of ions with d^0, d^5, and d^{10} configurations will not be affected by the number of d electrons, as there is no LFSE for these species.

Because the size of the d-orbital splitting, and hence the LFSE, depends on the ligand, it follows that a preference for octahedral coordination will be least pronounced for weak-field ligands. With strong-field ligands, low-spin complexes might be preferred and, although the situation is complicated by the pairing energy, the LFSE of a low-spin octahedral complex will be greater than that of a high-spin complex. There will thus be a correspondingly greater preference for octahedral over tetrahedral coordination when the octahedral complex is low-spin.

This preference for octahedral over tetrahedral coordination plays an important role in the solid state by influencing the structures that are adopted by d-metal compounds. This influence is demonstrated by the ways in which the different metal ions A and B in spinels (of formula AB_2O_4, Sections 4.9(b) and 25.11) occupy the octahedral or tetrahedral sites. Thus, Co_3O_4 is a normal spinel because the low-spin d^6 Co(III) ion strongly favours octahedral coordination, resulting in $(Co^{2+})_T(2Co^{3+})_OO_4$, whereas Fe_3O_4 (magnetite) is an inverse spinel because Fe(II), but not Fe(III), can acquire greater LFSE by occupying an octahedral site. Thus, magnetite is formulated as $(Fe^{3+})_T(Fe^{2+}Fe^{3+})_OO_4$.

(i) The Irving–Williams series

KEY POINT The Irving–Williams series summarizes the relative stabilities of complexes formed by M^{2+} ions, and reflects a combination of electrostatic effects and LFSE.

Fig. 7.21 shows log K_f values (Section 7.12) for complexes of the octahedral M^{2+} ions of the group 2 metals and 3d series. The variation in formation constants shown there is summarized by the **Irving–Williams series**:

$$Ba^{2+} < Sr^{2+} < Ca^{2+} < Mg^{2+} < Mn^{2+} < Fe^{2+} < Co^{2+} < Ni^{2+} < Cu^{2+} > Zn^{2+}$$

The order is relatively insensitive to the choice of ligands, but, since formation constants refer to forming complexes from aqua ions, the ligand concerned has to be able to displace water at a metal centre.

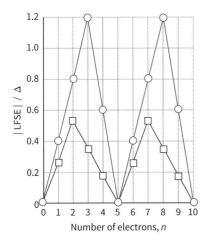

FIGURE 7.20 The ligand-field stabilization energy for d^n complexes in octahedral (high-spin, circles) and tetrahedral (squares) complexes. The LFSE is shown in terms of Δ_O, by applying the relation $D_T = \tfrac{4}{9}D_O$.

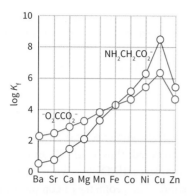

FIGURE 7.21 The variation of formation constants for the M^{2+} ions of the Irving–Williams series.

In general, the increase in stability correlates inversely with ionic radius, which suggests that the Irving–Williams series reflects electrostatic effects. However, beyond Mn^{2+} there is a sharp increase in the value of K_f for d^6 Fe(II), d^7 Co(II), d^8 Ni(II), and d^9 Cu(II), with strong-field ligands. These ions experience an additional stabilization proportional to the extra ligand-field stabilization energies in the complexes (compared to that produced by the displaced water molecules, Table 7.6). There is one important exception: the stability of Cu(II) complexes is greater than that of Ni(II) even though Cu(II) has an additional e_g electron. This anomaly is a consequence of the stabilizing influence of the Jahn–Teller effect, which results in strong binding of four of the ligands in the plane of the tetragonally distorted Cu(II) complex, and that stabilization enhances the value of K_f. The formation constants of Zn(II) complexes, which are not enhanced by either Jahn–Teller distortions or LFSE considerations, are normally greater than those of Mn(II) and Fe(II), but less than those of Co(II), Ni(II), and Cu(II).

7.17 Ligand-field theory

Crystal-field theory provides a simple conceptual model that can be used to interpret magnetic, spectroscopic, and thermochemical data by using empirical values of Δ_O. However, the theory is defective because it treats ligands as point charges or dipoles and does not take into account the overlap of ligand and metal-atom orbitals. One consequence of this oversimplification is that crystal-field theory cannot account for the ligand spectrochemical series. **Ligand-field theory**, which is an application of molecular orbital theory that concentrates on the d orbitals of the central metal atom, provides a more substantial framework for understanding the origins of Δ_O.

The strategy for describing the molecular orbitals of a d-metal complex follows procedures similar to those described in Chapter 2 for bonding in polyatomic molecules: the valence orbitals on the metal and ligand are used to form symmetry-adapted linear combinations (SALCs; Section 3.6), and then the relative energies of the molecular orbitals are estimated by using empirical energy and overlap considerations. These relative energies can be verified and positioned more precisely by comparison with experimental data (particularly UV–visible absorption and photoelectron spectroscopy).

We shall first consider octahedral complexes, initially taking into account only the metal–ligand σ bonding. We then consider the effect of π bonding, and see that it is essential for understanding Δ_O (which is one reason why crystal-field theory cannot explain the spectrochemical series). Finally, we consider complexes with different symmetries, and see that similar arguments apply to them. Later, in Chapter 22, we shall see how information from optical spectroscopy is used to refine the discussion and provide quantitative data on the ligand-field splitting parameter and electron–electron repulsion energies.

(a) σ Bonding

KEY POINT In ligand-field theory, the building-up principle is used in conjunction with a molecular orbital energy-level diagram constructed from metal-atom orbitals and symmetry-adapted linear combinations of ligand orbitals.

We begin by considering an octahedral complex in which each ligand (L) has a single valence orbital directed towards the central metal atom (M); each of these orbitals has local σ symmetry with respect to the M–L axis. Examples of such ligands include the NH_3 molecule and the F^- ion.

In an octahedral (O_h) environment, the orbitals of the central metal atom divide by symmetry into four sets (Fig. 7.22 and Resource section 4):

Metal orbital	Symmetry label	Degeneracy
s	a_{1g}	1
p_x, p_y, p_z	t_{1u}	3
$d_{x^2-y^2}, d_{z^2}$	e_g	2
d_{xy}, d_{yz}, d_{zx}	t_{2g}	3

Six symmetry-adapted linear combinations of the six ligand σ orbitals can also be formed as explained in Section 3.8. These combinations can be taken from Resource section 5 and are also shown in Fig. 7.22.

One (unnormalized) SALC has symmetry a_{1g}:

$$a_{1g}: \sigma_1 + \sigma_2 + \sigma_3 + \sigma_4 + \sigma_5 + \sigma_6$$

where σ_i denotes a σ orbital on ligand i. There are three SALCs of symmetry t_{1u}:

$$t_{1u}: \sigma_1 - \sigma_3, \sigma_2 - \sigma_4, \sigma_5 - \sigma_6$$

and two SALCs of symmetry e_g:

$$e_g: \sigma_1 - \sigma_2 + \sigma_3 - \sigma_4, 2\sigma_6 + 2\sigma_5 - \sigma_1 - \sigma_2 - \sigma_3 - \sigma_4$$

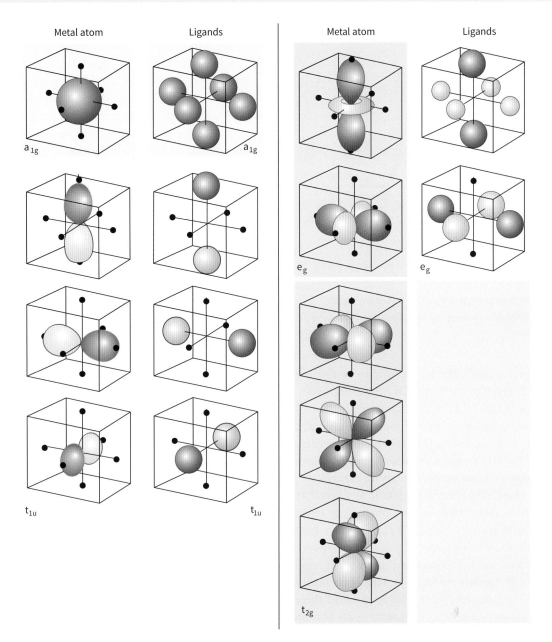

FIGURE 7.22 Symmetry-adapted combinations of ligand σ orbitals (represented here by spheres) in an octahedral complex. For symmetry-adapted orbitals in other point groups, see Resource section 5.

These six SALCs account for all the ligand orbitals of σ symmetry: there is no combination of ligand σ orbitals that has the symmetry of the metal t_{2g} orbitals, so the latter do not participate in σ bonding.[3]

Molecular orbitals are formed by combining SALCs and metal-atom orbitals of the same symmetry. For example, the (unnormalized) form of an a_{1g} molecular orbital is $c_M\psi_{Ms} + c_L\psi_{La_{1g}}$, where ψ_{Ms} is the s orbital on the metal atom M and $\psi_{La_{1g}}$ is the ligand SALC of symmetry a_{1g}. The metal s orbital and ligand a_{1g} SALC overlap to give two molecular orbitals, one bonding and one antibonding.

Similarly, the doubly degenerate metal e_g orbitals and the ligand e_g SALCs overlap to give four molecular orbitals (two degenerate bonding, two degenerate antibonding), and the triply degenerate metal t_{1u} orbitals and the three t_{1u} SALCs overlap to give six molecular orbitals (three degenerate bonding, three degenerate antibonding). There are therefore six bonding combinations in all, and six antibonding combinations. The three triply degenerate metal t_{2g} orbitals remain nonbonding and fully localized on the metal atom. Calculations of the resulting energies (adjusted to agree with

---

[3] The normalization constants (with overlap neglected) are $N(a_{1g}) = \left(\frac{1}{6}\right)^{1/2}$, $N(t_{1u}) = \left(\frac{1}{2}\right)^{1/2}$ for all three orbitals, and $N(e_g) = \frac{1}{2}$ and $\left(\frac{1}{12}\right)^{1/2}$, respectively.

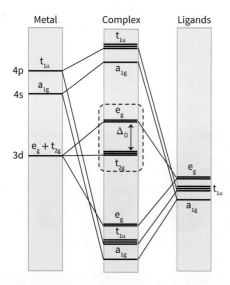

FIGURE 7.23 Molecular orbital energy levels of a typical octahedral complex. The frontier orbitals are inside the box with the dotted outline.

a variety of spectroscopic data of the kind to be discussed in Section 22.2) result in the molecular orbital energy-level diagram shown in Fig. 7.23.

The greatest contribution to the molecular orbital of lowest energy is from atomic orbitals of lowest energy (Section 2.9). For NH_3, F^-, and most other ligands, the ligand σ orbitals are derived from atomic orbitals with energies that lie well below those of the metal d orbitals. As a result, the six bonding molecular orbitals of the complex are mainly ligand-orbital in character (that is, $c_L^2 > c_M^2$). These six bonding orbitals can accommodate the 12 electrons provided by the six ligand lone pairs. The electrons that we can regard as provided by the ligands are therefore largely confined to the ligands in the complex, just as the crystal-field theory presumes. However, because the coefficients c_M are nonzero, the bonding molecular orbitals do have some d-orbital character and the 'ligand electrons' are partly delocalized on to the central metal atom.

The total number of electrons to accommodate, in addition to those supplied by the ligands, now depends on the number of d electrons, n, supplied by the metal atom. These additional electrons enter the orbitals next in line for occupation, which are the nonbonding d orbitals (the t_{2g} orbitals) and the antibonding combination (the upper e_g orbitals) of the d orbitals and ligand orbitals. The t_{2g} orbitals are wholly confined (in the present approximation) to the metal atom, and the antibonding e_g orbitals are largely metal-atom in character too, so the n electrons supplied by the central atom remain largely on that atom. The frontier orbitals of the complex are, therefore, the nonbonding entirely metal t_{2g} orbitals and the antibonding, mainly metal e_g orbitals. Thus, we have arrived at an arrangement that is qualitatively the same as in crystal-field theory. In the ligand-field

approach the octahedral ligand-field splitting parameter, Δ_O, is the separation between the molecular orbitals largely, but not completely, confined to the metal atom (Fig. 7.23).

With the molecular orbital energy-level diagram established, we use the building-up principle to construct the ground-state electron configuration of the complex. For a six-coordinate d^n complex, there are $12 + n$ electrons to accommodate. The six bonding molecular orbitals accommodate the 12 electrons supplied by the ligands. The remaining n electrons are accommodated in the nonbonding t_{2g} orbitals and the antibonding e_g orbitals. Now the picture is essentially the same as for crystal-field theory, the types of complexes that are obtained (high-spin or low-spin, for instance) depending on the relative values of Δ_O and the pairing energy P. The principal difference from the crystal-field discussion is that ligand-field theory gives deeper insight into the origin of the ligand-field splitting, and we can begin to understand why some ligands are strong and others are weak. For instance, a good σ-donor ligand should result in strong metal–ligand overlap, hence a more strongly antibonding e_g set and consequently a larger value of Δ_O. However, before drawing further conclusions we must go on to consider what crystal-field theory ignores completely: the role of π bonding.

(b) π Bonding

KEY POINTS π-Donor ligands decrease Δ_O whereas π-acceptor ligands increase Δ_O; the spectrochemical series is largely a consequence of the effects of π bonding when such bonding is feasible.

If the ligands in a complex have orbitals with local π symmetry with respect to the M–L axis (as two of the p orbitals of a halide ligand have), they may form bonding and antibonding π orbitals with the metal orbitals (Fig. 7.24). For an octahedral complex, the combinations that can be formed from the ligand π orbitals include SALCs of t_{2g} symmetry. These ligand combinations have net overlap with the metal t_{2g} orbitals, which are therefore no longer purely nonbonding on the metal atom. Depending on the relative energies of the ligand and metal orbitals, the energies of the now molecular t_{2g} orbitals lie above or below the energies they

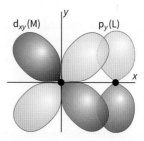

FIGURE 7.24 The p overlap that may occur between a ligand p orbital perpendicular to the M–L axis and a metal d_{xy} orbital.

had as nonbonding atomic orbitals, so Δ_O is decreased or increased, respectively.

To explore the role of π bonding in more detail, we need two of the general principles described in Chapter 2. First, we shall make use of the idea that when atomic orbitals overlap effectively they mix strongly: the resulting bonding molecular orbitals are significantly lower in energy, and the antibonding molecular orbitals are significantly higher in energy than the atomic orbitals. Second, we note that atomic orbitals with similar energies interact strongly, whereas those of very different energies mix only slightly, even if their overlap is large.

A **π-donor ligand** is a ligand that, before any bonding is considered, has filled orbitals of π symmetry around the M–L axis. Such ligands include Cl^-, Br^-, OH^-, O^{2-}, and even H_2O. In Lewis acid–base terminology (Section 5.6), a π-donor ligand is a π **base**. The energies of the full π orbitals on the ligands will not normally be higher than their σ-donor orbitals (HOMO) and must therefore also be lower in energy than the metal d orbitals. Because the full π orbitals of π-donor ligands lie lower in energy than the partially filled d orbitals of the metal, when they form molecular orbitals with the metal t_{2g} orbitals, the bonding combination lies lower than the ligand orbitals and the antibonding combination lies above the energy of the d orbitals of the free metal atom (Fig. 7.25). The electrons supplied by the ligand π orbitals occupy and fill the bonding combinations, leaving the electrons originally in the d orbitals of the central metal atom to occupy the antibonding t_{2g} orbitals. The net effect is that the previously nonbonding metal t_{2g} orbitals become antibonding and hence are raised closer in energy to the antibonding e_g orbitals. It follows that π-donor ligands *decrease* Δ_O.

A **π-acceptor ligand** is a ligand that has empty π orbitals that are available for occupation. In Lewis acid–base

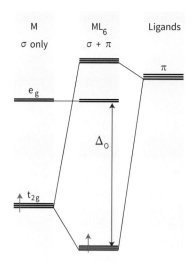

FIGURE 7.26 Ligands that act as π acceptors increase Δ_O. Only the π orbitals of the ligands are shown.

terminology, a π-acceptor ligand is a π **acid**. Typically, the π-acceptor orbitals are vacant antibonding orbitals on the ligand (usually the LUMO), as in CO and N_2, which are higher in energy than the metal d orbitals. The two π* orbitals of CO, for instance, have their largest amplitude on the C atom and have the correct symmetry for overlap with the metal t_{2g} orbitals, so CO can act as a π-acceptor ligand (Section 24.5). Phosphines (PR_3) are also able to accept π-electron density and also act as π acceptors (Section 24.6).

Because the π-acceptor orbitals on most ligands are higher in energy than the metal d orbitals, they form molecular orbitals in which the bonding t_{2g} combinations are largely of metal d-orbital character (Fig. 7.26). These bonding combinations lie lower in energy than the d orbitals themselves. The net result is that π-acceptor ligands *increase* Δ_O.

We can now put the role of π bonding in perspective. The order of ligands in the spectrochemical series is partly that of the strengths with which they can participate in M–L σ bonding. For example, both CH_3^- and H^- are high in the spectrochemical series (similar to NCS^-) because they are very strong σ donors. However, when π bonding is significant, it has a strong influence on Δ_O: π-donor ligands decrease Δ_O and π-acceptor ligands increase Δ_O. This effect is responsible for CO (a strong π acceptor) being high on the spectrochemical series and for OH^- (a strong π donor) being low in the series. The overall order of the spectrochemical series may be interpreted in broad terms as dominated by π effects (with a few important exceptions), and in general the series can be interpreted as follows:

— increasing Δ_O →

π-donor,   weak π-donor,   no π effects,   π-acceptor

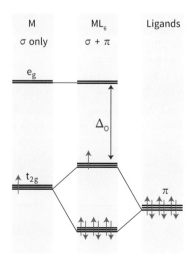

FIGURE 7.25 The effect of p bonding on the ligand-field splitting parameter. Ligands that act as π donors decrease Δ_O. Only the π orbitals of the ligands are shown.

Representative ligands that match these classes are

π donor	weak π donor	no π effects	π acceptor
I$^-$, Br$^-$, Cl$^-$, F$^-$	H$_2$O	NH$_3$	PR$_3$, CO

Notable examples of where the effect of σ bonding dominates include amines (NR$_3$), CH$_3^-$, and H$^-$, none of which has orbitals of π symmetry of an appropriate energy and thus are neither π-donor nor π-acceptor ligands. It is important to note that the classification of a ligand as strong-field or weak-field does not provide any guide as to the strength of the M–L bond.

The effect of π bonding on geometries other than octahedral is qualitatively similar, though we should note that in the tetrahedral geometry it is the e orbitals that form the π interactions. In the case of square-planar complexes the order of some of the notionally metal d orbitals changes compared with the pure crystal-field picture. Fig. 7.27 shows the arrangement of metal d orbitals with π interactions considered and is worth comparing with Fig. 7.17, where only electrostatic interactions with a crystal field were considered. We see that, even though most of the d orbitals have different energies in the

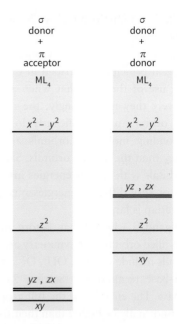

FIGURE 7.27 The orbital splitting pattern for a square-planar complex with π interactions taken into account.

different cases, in all cases it is the $d_{x^2-y^2}$ orbital that is highest in energy, and therefore unoccupied, in a d^8-metal complex ion.

FURTHER READING

G. B. Kauffman, *Inorganic Coordination Compounds*. John Wiley & Sons (1981). A fascinating account of the history of structural coordination chemistry.

G. B. Kauffman, *Classics in Coordination Chemistry: I. Selected Papers of Alfred Werner*. Dover (1968). Provides translations of Werner's key papers.

G. J. Leigh and N. Winterbottom (eds.), *Modern Coordination Chemistry: The Legacy of Joseph Chatt*. Royal Society of Chemistry (2002). A readable historical discussion of this area.

A. von Zelewsky, *Stereochemistry of Coordination Compounds*. John Wiley & Sons (1996). A readable book that covers chirality in detail.

E. C. Constable, G. Parkin, and L. Que (eds.), *Comprehensive Coordination Chemistry III*. Elsevier (2019).

N. G. Connelly, T. Damhus, R. M. Hartshorn, and A. T. Hutton, *Nomenclature of Inorganic Chemistry: IUPAC Recommendations 2005*. Royal Society of Chemistry (2005).

Also known as 'The IUPAC red book', the definitive guide to naming inorganic compounds.

R. A. Marusak, K. Doan, and S. D. Cummings, *Integrated Approach to Coordination Chemistry: An Inorganic Laboratory Guide*. John Wiley & Sons (2007). This unusual textbook describes the concepts of coordination chemistry and illustrates these concepts through well-explained experimental projects.

J.-M. Lehn (ed.), *Transition Metals in Supramolecular Chemistry, Vol. 5 of Perspectives in Supramolecular Chemistry*. John Wiley & Sons (2007). Inspiring accounts of developments and applications in coordination chemistry.

J. W. Steed and J. L. Atwood (eds.), *Supramolecular Chemistry*, 3rd ed. Wiley (2022).

M. T. Weller and N. A. Young, *Characterisation Methods in Inorganic Chemistry*. Oxford University Press (2017). Contains a more detailed analysis of electronic spectroscopy of transition metals.

EXERCISES

7.1 Name and draw structures of the following complexes: (a) [Ni(CN)$_4$]$^{2-}$, (b) [CoCl$_4$]$^{2-}$, (c) [Mn(NH$_3$)$_6$]$^{2+}$.

7.2 Give formulas for (a) chloridopentaamminecobalt(III) chloride, (b) hexaaquairon(3+) nitrate, (c) cis-dichloridobis (1,2-diaminoethane)ruthenium(II), (d) μ-hydroxidobis(pentaamminechromium(III)) chloride.

7.3 Name the octahedral complex ions (a) cis-[CrCl$_2$(NH$_3$)$_4$]$^+$, (b) trans-[Cr(NH$_3$)$_2$(κN-NCS)$_4$]$^-$, (c) [Co(C$_2$O$_4$)(en)$_2$]$^+$.

7.4 (a) Sketch the two structures that describe most four-coordinate complexes. (b) In which structure are isomers possible for complexes of formula MA_2B_2?

7.5 Sketch the two structures that describe most five-coordinate complexes. Label the two different sites in each structure.

7.6 (a) Sketch the two structures that describe most six-coordinate complexes. (b) Which one of these is rare?

7.7 Explain the meaning of the terms *monodentate*, *bidentate*, and *tetradentate*.

7.8 What type of isomerism can arise with ambidentate ligands? Give two examples.

7.9 What is the denticity of the following molecules? Which could act as bridging ligands? Which could act as chelating ligands?

7.10 Draw the structures of representative complexes that contain the ligands (a) en, (b) ox^{2-}, (c) phen, (d) 12-crown-4, (e) tren, (f) terpy, (g) $edta^{4-}$.

7.11 What types of isomer are the two compounds $[RuBr(NH_3)_5]Cl$ and $[RuCl(NH_3)_5]Br$?

7.12 For which of the following tetrahedral complexes are isomers possible? Draw all the isomers. $[CoBr_2Cl_2]^-$, $[CoBrCl_2(OH_2)]$, $[CoBrClI(OH_2)]$.

7.13 For which of the following square-planar complexes are isomers possible? Draw all the isomers. $[Pt(NH_3)_2(ox)]$, $[PdBrCl(PEt_3)_2]$, $[IrH(CO)(PR_3)_2]$, $[Pd(gly)_2]$.

7.14 For which of the following octahedral complexes are isomers possible? Draw all the isomers. $[FeCl(OH_2)_5]^{2+}$, $[IrCl_3(PEt_3)_3]$, $[Ru(bpy)_3]^{2+}$, $[CoCl_2(en)(NH_3)_2]^+$, $[W(CO)_4(py)_2]$.

7.15 How many isomers are possible for an octahedral complex of general formula $[MA_2BCDE]$? Draw all that are possible.

7.16 Which of the following complexes are chiral? (a) $[Cr(ox)_3]^{3-}$, (b) *cis*-$[PtCl_2(en)]$, (c) *cis*-$[RhCl_2(NH_3)_4]^+$, (d) $[Ru(bpy)_3]^{2+}$, (e) *fac*-$[Co(NO_2)_3(dien)]$, (f) *mer*-$[Co(NO_2)_3(dien)]$. Draw the enantiomers of the complexes identified as chiral, and identify the plane of symmetry in the structures of the achiral complexes.

7.17 Which isomer is the following tris(acac) complex?

7.18 Draw and label both Λ and Δ isomers of the $[Ru(en)_3]^{2+}$ cation.

7.19 The stepwise formation constants for complexes of NH_3 with $[Cu(OH_2)_6]^{2+}$(aq) are log K_{f1} = 4.15, log K_{f2} = 3.50, log K_{f3} = 2.89, log K_{f4} = 2.13, and log K_{f5} = −0.52. Suggest a reason why K_{f5} is so different.

7.20 The stepwise formation constants for complexes of $NH_2CH_2CH_2NH_2$ (en) with $[Cu(OH_2)_6]^{2+}$(aq) are log K_{f1} = 10.72 and log K_{f2} = 9.31. Compare these values with those of ammonia given in Exercise 7.19, and suggest why they are different.

7.21 Determine the configuration (in the form $t_{2g}^x e_g^y$ or $e^x t_2^y$, as appropriate), the number of unpaired electrons, and the ligand-field stabilization energy in terms of Δ_O or Δ_T and P for each of the following complexes. Use the spectrochemical series to decide, where relevant, which are likely to be high-spin and which low-spin. (a) $[Co(NH_3)_6]^{3+}$, (b) $[Fe(OH_2)_6]^{2+}$, (c) $[Fe(CN)_6]^{3-}$, (d) $[Cr(NH_3)_6]^{3+}$, (e) $[W(CO)_6]$, (f) tetrahedral $[FeCl_4]^{2-}$, (g) tetrahedral $[NiCl_4]^{2-}$.

7.22 Both H^- and $P(C_6H_5)_3$ are ligands of similar field strength, high in the spectrochemical series. Recalling that phosphines act as π acceptors, is π-acceptor character required for strong-field behaviour? What orbital factors account for the field strength of each ligand?

7.23 Estimate the spin-only contribution to the magnetic moment for each complex in Exercise 7.21.

7.24 Match each of the magnetic moments, 3.8, 0, 1.8, 5.9 (in μ_B), with one of the species: $[Fe(CN)_6]^{3-}$, $[Fe(H_2O)_6]^{3+}$, $[CrO_4]^{2-}$, $[Cr(H_2O)_6]^{3+}$.

7.25 Solutions of the complexes $[Co(NH_3)_6]^{2+}$, $[Co(OH_2)_6]^{2+}$ (both O_h), and $[CoCl_4]^{2-}$ are coloured. One is pink (absorbs blue light), another is yellow (absorbs violet light), and the third is blue (absorbs red light). Considering the spectrochemical series and the relative magnitudes of Δ_T and Δ_O, assign each colour to one of the complexes.

7.26 For each of the following pairs of complexes, identify the one that has the larger LFSE:

(a) $[Cr(OH_2)_6]^{2+}$ or $[Mn(OH_2)_6]^{2+}$
(b) $[Fe(OH_2)_6]^{2+}$ or $[Fe(OH_2)_6]^{3+}$
(c) $[Fe(OH_2)_6]^{3+}$ or $[Fe(CN)_6]^{3-}$
(d) $[Fe(CN)_6]^{3-}$ or $[Ru(CN)_6]^{3-}$
(e) tetrahedral $[FeCl_4]^{2-}$ or tetrahedral $[CoCl_4]^{2-}$

7.27 Interpret the variation, including the overall trend across the 3d series, of the following values of oxide lattice enthalpies (in kJ mol^{-1}). All the compounds have the rock-salt structure: CaO (3460), TiO (3878), VO (3913), MnO (3810), FeO (3921), CoO (3988), NiO (4071).

7.28 Enthalpies of hydration and values of the ligand-field splitting parameter, Δ_O, are given for some octahedrally coordinated ions. (a) Plot the enthalpies of hydration against the number of d electrons. (b) Calculate LFSE in terms of Δ_O for high-spin configuration. Use the given values of Δ_O to find LFSE in kJ mol^{-1} for each ion. (c) Apply this energy as a correction term to the enthalpies of hydration and plot the estimated ΔH in the absence of ligand-field effects. Comment on this plot. (1 kJ mol^{-1} = 83.7 cm^{-1}.)

Ion	$\Delta_{hyd}H$/kJ mol^{-1}	Δ_O/cm^{-1}
Ca^{2+}	1 577	0
V^{2+}	1 918	12 600
Cr^{2+}	1 904	13 900
Mn^{2+}	1 841	7 800
Fe^{2+}	1 946	10 400
Co^{2+}	1 996	9 300
Ni^{2+}	2 105	8 300
Cu^{2+}	2 100	12 600
Zn^{2+}	2 046	0

7.29 A neutral macrocyclic ligand with four donor atoms produces a red diamagnetic low-spin d^8 complex of Ni(II) if the anion is the weakly coordinating perchlorate ion. When perchlorate is replaced by two thiocyanate ions, SCN^-, the complex turns violet and is high-spin with two unpaired electrons. Interpret the change in terms of structure.

7.30 Bearing in mind the Jahn–Teller effect, predict the structure of $[Cr(OH_2)_6]^{2+}$.

7.31 The spectrum of d^1 Ti^{3+}(aq) is attributed to a single electronic transition $e_g \leftarrow t_{2g}$. The band shown in Fig. 7.10 is not symmetric and suggests that more than one state is involved. Suggest how to explain this observation using the Jahn–Teller effect.

TUTORIAL PROBLEMS

7.1 The compound Na_2IrCl_6 reacts with triphenylphosphine in diethyleneglycol in an atmosphere of CO to give trans-$[IrCl(CO)(PPh_3)_2]$, known as 'Vaska's compound'. Excess CO gives a five-coordinate species, and treatment with $NaBH_4$ in ethanol gives $[IrH(CO)_2(PPh_3)_2]$. Derive a formal name for Vaska's compound. Draw and name all isomers of the two five-coordinate complexes.

7.2 A pink solid has the formula $CoCl_3 \cdot 5NH_3 \cdot H_2O$. A solution of this salt is also pink and rapidly gives 3 mol AgCl on titration with silver nitrate solution. When the pink solid is heated, it loses 1 mol H_2O to produce a purple solid with the same ratio of NH_3:Cl:Co. The purple solid, on dissolution and titration with $AgNO_3$, releases two of its chlorides rapidly. Deduce the structures of the two octahedral complexes, and draw and name them.

7.3 The hydrated chromium chloride that is available commercially has the overall composition $CrCl_3 \cdot 6H_2O$. On boiling a solution, it becomes violet and has a molar electrical conductivity similar to that of $[Co(NH_3)_6]Cl_3$. By contrast, another form of $CrCl_3 \cdot 5H_2O$ is green and has a lower molar conductivity in solution. If a dilute acidified solution of the green complex is allowed to stand for several hours, it turns violet. Interpret these observations with structural diagrams.

7.4 The complex first denoted β-$[PtCl_2(NH_3)_2]$ was identified as the trans isomer. (The cis isomer was denoted α.) It reacts slowly with solid Ag_2O to produce $[Pt(NH_3)_2(OH_2)_2]^{2+}$. This complex does not react with 1,2-diaminoethane to produce a chelated complex. Name and draw the structure of the diaqua complex. A third isomer of composition $PtCl_2 \cdot 2NH_3$ is an insoluble solid that, when ground with $AgNO_3$, produces a mixture containing $[Pt(NH_3)_4](NO_3)_2$ and a new solid phase of composition $Ag_2[PtCl_4]$. Give the structures and names of each of the three Pt(II) compounds.

7.5 Air oxidation of Co(II) carbonate and aqueous ammonium chloride gives a pink chloride salt with a ratio of $4NH_3$:Co. On addition of HCl to a solution of this salt, a gas is rapidly evolved and the solution slowly turns violet on heating. Complete evaporation of the violet solution yields $CoCl_3 \cdot 4NH_3$. When this is heated in concentrated HCl, a green salt can be isolated with composition $CoCl_3 \cdot 4NH_3 \cdot HCl$. Write balanced equations for all the transformations occurring after the air oxidation. Give as much information as possible concerning the isomerism occurring and the basis of your reasoning. Is it helpful to know that the form of $[Co(Cl)_2(en)_2]^+$ that is resolvable into enantiomers is violet?

7.6 When cobalt(II) salts are oxidized by air in a solution containing ammonia and sodium nitrite, a yellow solid, $[Co(NO_2)_3(NH_3)_3]$, can be isolated. In solution it is nonconducting; treatment with HCl produces a complex that, after a series of further reactions, can be identified as trans-$[Co(Cl)_2(NH_3)_3(OH_2)]^+$. It requires an entirely different route to prepare cis-$[Co(Cl)_2(NH_3)_3(OH_2)]^+$. Is the yellow substance fac or mer? What assumption must you make to arrive at a conclusion?

7.7 The reaction of $[ZrCl_4(dppe)]$ with $Mg(CH_3)_2$ gives $[Zr(CH_3)_4(dppe)]$. NMR spectra indicate that all methyl groups are equivalent. Draw octahedral and trigonal prism structures for the complex, and show how the conclusion from NMR supports the trigonal prism assignment (P. M. Morse and G. S. Girolami, *J. Am. Chem. Soc.*, 1989, **111**, 4114).

7.8 The resolving agent d-cis$[Co(NO_2)_2(en)_2]$Br can be converted to the soluble nitrate by grinding in water with $AgNO_3$. Outline the use of this species for resolving a racemic mixture of the d and l enantiomers of $K[Co(edta)]$. (The l-$[Co(edta)]^-$ enantiomer forms the less soluble diastereomer (F. P. Dwyer and F. L. Garvan, *Inorg. Synth.*, 1965, **6**, 192).)

7.9 Show how the coordination of two $MeHNCH_2CH_2NH_2$ ligands to a metal atom in a square-planar complex results in not only cis and trans but also optical isomers. Identify the mirror planes in the isomers that are not chiral.

7.10 Use a group-theory analysis to assign point groups to the all cis and all trans isomers of $[MA_2B_2C_2]$; use the character tables associated with each to determine whether they are chiral or not.

7.11 BINAP is a chelating diphosphine ligand shown below. Discuss the reasons for the observed chirality of BINAP, and of its complexes.

7.12 The equilibrium constants for the successive reactions of 1,2-diaminoethane with Co^{2+}, Ni^{2+}, and Cu^{2+} are as follows:

$$[M(OH_2)_6]^{2+} + en \rightleftharpoons [M(en)(OH_2)_4]^{2+} + 2H_2O \quad K_1$$

$$[M(en)(OH_2)_4]^{2+} + en \rightleftharpoons [M(en)_2(OH_2)_2]^{2+} + 2H_2O \quad K_2$$

$$[M(en)_2(OH_2)_2]^{2+} + en \rightleftharpoons [M(en)_3]^{2+} + 2H_2O \quad K_3$$

Ion	log K_1	log K_2	log K_3
Co^{2+}	5.89	4.83	3.10
Ni^{2+}	7.52	6.28	4.26
Cu^{2+}	10.72	9.31	−1.0

Discuss whether these data support the generalizations in the text about successive formation constants. Discuss the effect of the metal on the formation constant. How might the Irving–Williams series provide insight into these formation constants? In particular, how can you account for the very low value of K_3 for Cu^{2+}?

7.13 What are *rotaxanes*, and how do they differ from *pseudorotaxanes*? Discuss how coordination chemistry might be used to synthesize such molecules.

7.14 In a fused magma liquid from which silicate minerals crystallize, the metal ions can be four-coordinate. In olivine crystals, the M(II) co-ordination sites are octahedral. Partition coefficients, which are defined as $K_p = [M(II)]_{olivine}/[M(II)]_{melt}$, follow the order Ni(II) > Co(II) > Fe(II) > Mn(II). Account for this in terms of ligand-field theory. (See I. M. Dale and P. Henderson, *24th Int. Geol. Congress, Sect.*, 1972, **10**, 105.)

7.15 By considering the splitting of the octahedral orbitals as the symmetry is lowered, draw the symmetry-adapted linear combinations and the molecular orbital energy-level diagram for σ bonding in a *trans*-[ML$_4$X$_2$] complex. Assume that the ligand X is lower in the spectrochemical series than L.

7.16 Referring to Resource section 5, draw the appropriate symmetry-adapted linear combinations and the molecular orbital diagram for σ bonding in a square-planar complex. The point group is D_{4h}. Take note of the small overlap of the ligand with the d_{z^2} orbital. What is the effect of π bonding?

7.17 Starting with Fig. 7.19, show how, using a crystal-field approach, an extreme tetragonal distortion leads to the orbital energy-level diagram in Fig. 7.17. Perform a similar analysis for two-coordinate linear complexes.

7.18 Consider a trigonal-prismatic six-coordinate ML$_6$ complex with D_{3h} symmetry. Use the D_{3h} character table (Resource section 4) to divide the d orbitals of the metal into sets of defined symmetry type. Assume that the ligands are at the same angle relative to the *xy*-plane as in a tetrahedral complex.

7.19 In a trigonal-bipyramidal complex the axial and equatorial sites have different steric and electronic interactions with the central metal ion. Consider a range of some common ligands and decide which coordination site in a trigonal-bipyramidal complex they would favour. (See A. R. Rossi and R. Hoffmann, *Inorg. Chem.*, 1975, **14**, 365.)

8 Physical techniques in inorganic chemistry

Diffraction methods
- 8.1 **X-ray diffraction**
- 8.2 **Neutron diffraction**

Absorption and emission spectroscopies
- 8.3 **Ultraviolet–visible spectroscopy**
- 8.4 **Fluorescence or emission spectroscopy**
- 8.5 **Infrared and Raman spectroscopy**

Resonance techniques
- 8.6 **Nuclear magnetic resonance**
- 8.7 **Electron paramagnetic resonance**
- 8.8 **Mössbauer spectroscopy**

Ionization-based techniques
- 8.9 **Photoelectron spectroscopy**
- 8.10 **X-ray absorption spectroscopy**
- 8.11 **Mass spectrometry**

Chemical analysis
- 8.12 **Atomic absorption spectroscopy**
- 8.13 **CHN analysis**
- 8.14 **X-ray fluorescence elemental analysis**
- 8.15 **Thermal analysis**

Magnetometry and magnetic susceptibility

Electrochemical techniques

Microscopy
- 8.16 **Scanning probe microscopy**
- 8.17 **Electron microscopy**

Further reading

Exercises

Tutorial problems

All the structures of the molecules and materials to be covered in this book have been determined experimentally by applying one or more physical techniques. The characterization techniques and instruments available vary greatly in complexity, and all the methods produce data that help to determine the structure, composition, or properties of a compound. Many of the physical techniques used in contemporary inorganic research rely on the interaction of electromagnetic radiation with matter. The full range of energies present in the electromagnetic spectrum, from gamma rays to long-wavelength radio waves, is used to investigate the nature of inorganic compounds. In this chapter we introduce the most important physical techniques that are used to investigate the atomic and electronic structures of inorganic compounds. As well as studying structures and properties, physical techniques can also be used to investigate the reactions and real-time processes involving inorganic compounds and solids. These *in situ* or *in operando* techniques, common to many of the experimental methods described in this chapter, are employed to study, for example, inorganic reaction mechanisms, catalytic processes, and the phase changes occurring in rechargeable batteries.

Diffraction methods

Diffraction techniques, particularly those using X-rays, are the most important methods available to the inorganic chemist for the determination of structures. X-ray diffraction has been used to establish the structures of at least a million different substances, including hundreds of thousands of purely inorganic compounds and many organometallic compounds. The method is used to locate the positions of the atoms and ions that make up a solid compound, and hence provides a description of structures in terms of features such as bond lengths, bond angles, and the relative positions of ions and molecules in a unit cell. This structural information has been interpreted in terms of atomic and ionic radii, which then allow chemists to predict structure and explain trends in many properties. Diffraction methods are normally nondestructive in the sense that the sample remains unchanged and may be analysed further by using a different technique.

8.1 X-ray diffraction

KEY POINTS The scattering of radiation with wavelengths of about 100 pm from crystals gives rise to diffraction; the interpretation of the diffraction patterns provides quantitative structural information and in many cases the complete molecular or ionic structure.

Diffraction is the interference between waves that occurs as a result of an object in their path. X-rays are scattered elastically (with no change in energy) by the electrons in atoms, and diffraction can occur for a periodic array of scattering centres separated by distances similar to the wavelength of the radiation (in the region 50–250 pm), such as exist in a crystal (typical bond lengths are 100–300 pm). If we think of scattering as equivalent to reflection from two adjacent parallel planes of atoms separated by a distance d (Fig. 8.1),

then the angle at which constructive interference occurs (to produce a diffraction intensity maximum) between waves of wavelength λ is given by **Bragg's equation**:

$$2d\sin\theta = n\lambda \qquad (8.1)$$

where n is an integer, normally set to unity. Thus an X-ray beam impinging on a crystalline compound with an ordered array of atoms will produce a set of diffraction maxima, termed a **diffraction pattern**, with each maximum, or **reflection**, occurring at an angle θ corresponding to a different separation of planes of atoms, d, in the crystal.

> **A NOTE ON GOOD (AND CONVENTIONAL) PRACTICE**
> Crystallographers still generally use the angstrom (1 Å = 10^{-10} m = 10^{-8} cm = 10^{-2} pm) as a unit of measurement. This unit is convenient because bond lengths typically lie between 1 and 3 Å and the X-ray wavelengths used to derive them usually lie between 0.5 and 2.5 Å.

An atom or ion scatters X-rays in proportion to the number of electrons it possesses, and the intensities of the measured diffraction maxima are proportional to the square of that number. Thus the diffraction pattern produced is characteristic of the positions and types (in terms of their number of electrons) of atom present in the crystalline compound, and the measurement of X-ray diffraction angles and intensities provides structural information. Because of its dependence on the number of electrons, X-ray diffraction is particularly sensitive to any electron-rich atoms in a compound. Thus, X-ray diffraction by $NaNO_3$ probes and derives information on the three nearly isoelectronic ion and atoms to a similar degree, but for $Pb(OH)_2$ the scattering and structural information is dominated by the lead atom.

There are two principal X-ray techniques: the **powder method**, in which the materials being studied are in polycrystalline form with crystallite sizes around a few micrometres or less, and **single-crystal diffraction**, in which the sample is a single crystal of dimensions of several tens or hundreds of micrometres.

(a) Powder X-ray diffraction

KEY POINT Powder X-ray diffraction is used mainly for phase identification and the determination of lattice parameters and lattice type.

A powdered (polycrystalline) sample contains an enormous number of very small crystallites, typically 0.1–10 μm in dimension and orientated at random. An X-ray beam striking a polycrystalline sample is scattered in all directions; at

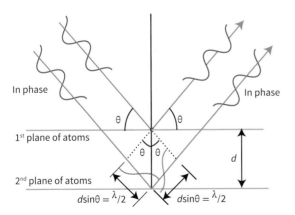

FIGURE 8.1 Bragg's equation is derived by treating layers of atoms as reflecting planes. X-rays interfere constructively when the additional path length, $2d\sin\theta$, is equal to an integer multiple of the wavelength λ.

some angles, those given by Bragg's equation, constructive interference occurs. As a result, each set of planes of atoms separated by the lattice spacing d gives rise to a cone of diffraction intensity: as there are many different choices for the planes of atoms (and associated d) a large number of diffraction cones is observed. Each cone consists of a set of closely-spaced diffracted rays or diffraction spots, each one of which represents diffraction from a single crystallite within the powder sample (Fig. 8.2). With a very large number of crystallites these diffracted spots merge together to form the observed diffraction cone. A **powder diffractometer** (Fig. 8.3(a)) uses an electronic detector to measure the angles of the diffracted beams. Scanning the detector around the sample along the circumference of a circle cuts through the diffraction cones at the various diffraction maxima, and the intensity of the diffracted X-rays is recorded as a function of the detector angle (Fig. 8.3(b)). Note that the angle of the detector relative to the incident beam is measured at 2θ, while the Bragg angle is θ.

The number and positions of the reflections depend on the cell parameters, crystal system, lattice type, and wavelength used to collect the data; the peak intensities depend on the types of atoms present and their positions. Nearly all crystalline solids produce a unique set of powder X-ray diffraction maxima, normally described as a powder X-ray diffraction 'pattern', in terms of the various angles of the reflections and their intensities. In mixtures of compounds, each crystalline phase present contributes to the powder diffraction pattern

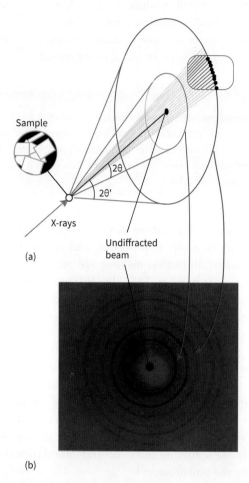

FIGURE 8.2 (a) The cones of diffraction that result from X-ray scattering by a powdered sample. The cone consists of thousands of individual diffraction spots, from individual crystallites, that merge together. (b) A photographic image of the diffraction pattern from a powdered sample. The undiffracted straight-through beam is at the centre of the image and the various diffraction cones, corresponding to different d-spacings, are observed as concentric circles. Using the Bragg equation, eqn. 8.1, it is possible, from the diffraction angles, to determine the d-spacing associated with each diffraction maximum. Computational analysis of all these experimentally determined d-spacings allows identification of which planes of atoms in the crystal gave rise to the diffraction maxima. These planes are identified using three Miller Indices, denoted h, k, and l, related to the orientation of the plane of atoms with respect to the unit cell, Section 4.1.

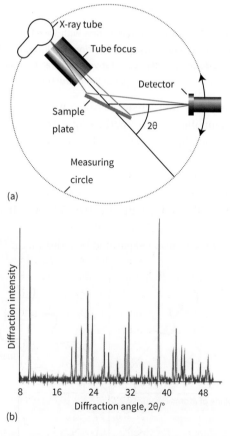

FIGURE 8.3 (a) Schematic diagram of a powder diffractometer operating in reflection mode, in which the X-ray scattering occurs from a sample mounted as a flat plate. For weakly absorbing compounds the samples may be mounted in a capillary and the diffraction data collected in transmission mode. (b) The form of a typical powder diffraction pattern, showing a series of reflections as a function of measured diffraction angle, 2θ.

TABLE 8.1 Applications of powder X-ray diffraction

Application	Typical use and information extracted
Identification of unknown materials	Rapid (<30 minutes) identification of most crystalline phases
Determination of sample purity	Qualitative analysis, quality control in manufacturing. Monitoring the progress of a chemical reaction occurring in the solid state
Determination and refinement of lattice parameters	Phase identification and monitoring structure as a function of composition
Investigation of phase diagrams/new materials	Mapping out phase behaviour and structure as a function of composition
Determination of crystallite size/stress	Particle size measurement, particularly of nanoparticles, and uses in metallurgy
Structure refinement	Extraction of detailed crystallographic data, such as atomic positions, for a known structure type
Ab initio structure determination	Structure determination is possible in some cases without initial knowledge of the crystal structure
Phase changes/expansion coefficients	Studies as a function of temperature (cooling or heating typically in the range 100–1200 K). Expansion coefficients. Observation of structural transitions
In situ and *in operando* studies	Investigating a real-time chemical process, e.g. discharge of a lithium battery, formation of an excited state

its own unique set of reflection angles and intensities. Typically, the method is sensitive enough to detect a small level (5–10% by mass) of a particular crystalline component in a mixture.

The effectiveness of powder X-ray diffraction has led to its becoming the major technique for the characterization of polycrystalline inorganic materials (Table 8.1). Many of the powder diffraction data sets collected from inorganic, organometallic, and organic compounds have been compiled into a database by the International Centre for Diffraction Data (ICDD). This database, which contains over a million unique powder X-ray diffraction patterns, can be used like a fingerprint library to identify an unknown material from its powder pattern alone. Powder X-ray diffraction is used routinely in the investigation of phase formation and changes in structures of solids. The synthesis of a metal oxide can be verified by collecting a powder diffraction pattern and demonstrating that the data are consistent with a single pure phase of that material. Indeed, the progress of a chemical reaction is often monitored by observing the formation of the product phase at the expense of the reactants.

Basic crystallographic information, such as lattice parameters, can often be extracted from powder X-ray diffraction (PXRD) data, usually with very high precision. The determination of the lattice parameters and unit cell symmetry (Section 4.1) from PXRD data is easily accomplished for small high symmetry unit cells, such as cubic or orthorhombic, but requires very-high-quality data and computational methods for triclinic systems. The presence or absence of certain reflections in the diffraction pattern permits the determination of the lattice type. In recent years the technique of fitting the intensities of the peaks in the diffraction pattern has become a popular method of extracting structural information, including atomic positions. The analysis, which is known as the **Rietveld method**, involves fitting a calculated diffraction pattern to the experimental trace. The technique is not as powerful as the single-crystal method, for it gives less accurate atomic positions and can normally only be used for relatively simple structures where a starting structural model is known, but has the advantage of not requiring the growth of a single crystal. Increasing computing power coupled with very high quality PXRD data is allowing the *ab initio* (without a trial starting model) structure determination of some simple inorganic compound structures.

Powder diffraction methods, including both X-ray and neutron (Section 8.2), maybe be used to study chemical processes happening in real time or *in operando*. Examples include synthesis reactions involving crystalline solids and the chemical processes occurring during the charging and discharging of a lithium-ion battery. In the latter example, PXRD patterns are collected from a lithium-ion battery cell as a current flows through it over a number of hours. The various diffraction patterns collected show the formation or destruction of different crystalline materials during these charging and discharging processes.

EXAMPLE 8.1 Using powder X-ray diffraction

Titanium dioxide exists as several polymorphs, the most common of which are anatase, rutile, and brookite. The experimental diffraction angles 2θ for the six strongest reflections collected from each of these different polymorphs are summarized in the following Table. The powder X-ray diffraction pattern collected using 154 pm X-rays from a sample of white paint, known to contain TiO_2 in one or more of these polymorphic forms, shows the diffraction pattern in Fig. 8.4. Identify the TiO_2 polymorph(s) present and, approximately, their relative amounts.

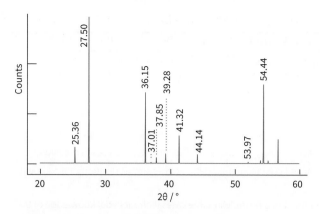

FIGURE 8.4 A powder diffraction pattern obtained from a paint containing a mixture of TiO_2 polymorphs.

Rutile	Anatase	Brookite
27.50	25.36	19.34
36.15	37.01	25.36
39.28	37.85	25.71
41.32	38.64	30.83
44.14	48.15	32.85
54.44	53.97	34.90

Answer We need to identify the polymorph that has a diffraction pattern that matches the one observed. The lines closely match those of rutile (the strongest reflections) and anatase (a few weak reflections), so the paint contains these phases with rutile as the major TiO_2 phase.

Self-test 8.1 Chromium(IV) oxide also adopts the rutile structure. By consideration of Bragg's equation and the ionic radii of Ti^{4+} and Cr^{4+} (Resource section 1), predict in outline the main features of the CrO_2 powder X-ray diffraction pattern.

(b) Single-crystal X-ray diffraction

KEY POINT Analysis of the X-ray diffraction patterns obtained from single crystals allows the structure of molecules and extended lattices to be determined, although hydrogen atoms are often not directly located for inorganic compounds.

Analysis of the diffraction data obtained from single crystals is the most important method of obtaining the structures of inorganic solids. Provided a compound can be grown as a crystal of sufficient size and quality, the data provide definitive information about molecular and extended lattice structures.

The collection of diffraction data from a single crystal can be carried out using a single crystal diffractometer (Fig. 8.5) where the crystal is rotated around three orthogonal directions, denoted ω, ϕ, and χ, in an X-ray beam A **four-circle diffractometer** uses a scintillation detector to measure the diffracted X-ray beam intensity as a function of the diffraction angle, 2θ.

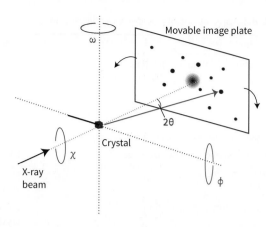

FIGURE 8.5 The layout of an image plate single crystal X-ray diffractometer. A computer controls the location of the detector as the four angles are changed systematically.

Most contemporary diffractometers use an **area-detector** or **image plate** that is sensitive to X-rays: these detectors, which may also be rotated around the crystal, measure a large number of diffraction maxima simultaneously so that full data sets can typically be collected in just a few hours (Fig. 8.6).

Analysis of the diffraction data from single crystals is a complex process, extracting structural information from the locations and intensities of many thousands of reflections. But with the recent increases in computational power a skilled crystallographer can complete the structure determination of even complex inorganic molecules in under an hour. Single-crystal X-ray diffraction can be used to determine the structures of the vast majority of inorganic compounds when they can be obtained as crystals with dimensions of about $20 \times 20 \times 20\,\mu m^3$ or larger. Positions for most atoms, including C, N, O, and metals, in most inorganic compounds can be determined with sufficient accuracy that bond lengths

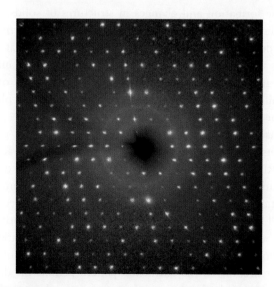

FIGURE 8.6 Part of a single-crystal X-ray diffraction pattern. Individual spots arise by diffraction of X-rays scattered from different planes of atoms within the crystal.

can be defined to within a fraction of a picometre. As an example, the S–S bond length in monoclinic sulfur has been reported, with the estimated standard deviation (esd) given in brackets, as 204.7(3) pm or 2.047(3)Å.

> **A NOTE ON GOOD PRACTICE**
>
> The S–S bond length is reported with an estimated standard deviation of 0.3 pm, which is calculated on the basis of the sensitivity of the exact experimental method used, the quality of the data obtained, and the sensitivity of the data to the extracted quantity. With high-quality single-crystal diffraction data and structures of moderate complexity, as defined by the number of different atom sites the structure contains, bond length esds are typically reported with values of 0.1–0.5 pm. When comparing structural information, the size of the reported esd should be borne in mind so as to gauge the significance of any trends or differences.

The positions of hydrogen atoms can be determined for inorganic compounds that contain only light atoms (Z less than about 18, argon), but their locations in many inorganic compounds that also contain heavy atoms, such as the 4d and 5d-series elements, can be difficult or impossible. The problem lies with the small number of electrons on a hydrogen atom (just one); this electron density is often reduced even further when hydrogen forms bonds to other atoms. Furthermore, as the one electron is normally part of a bond, the X-ray-determined position of this electron density is often displaced along the bond direction; this produces bond lengths shortened compared to the true internuclear distance. Other techniques, such as neutron diffraction (Section 8.2), can often be applied to determine the positions of hydrogen atoms in inorganic compounds.

Molecular structures obtained by the analysis of single-crystal X-ray diffraction data are often represented in ORTEP diagrams (Fig. 8.7; the acronym stands for Oak Ridge Thermal Ellipsoid Plot). In an ORTEP diagram an ellipsoid is used to represent the volume within which the scattering electron density most probably lies, more correctly called the displacement ellipsoid, taking into account its thermal motion. The size of the ellipsoid increases with temperature and, as a result, so does the imprecision of the bond lengths extracted from the data and hence the reported error.

(c) X-ray diffraction at synchrotron sources and using free electron lasers

KEY POINT High-intensity X-ray beams generated by synchrotron sources allow the structures of very complex molecules to be determined.

Much more intense X-ray beams, than are available from laboratory sources, can be obtained by using **synchrotron radiation**. Synchrotron radiation is produced by electrons circulating close to the speed of light in a storage ring and is typically several orders of magnitude more intense than laboratory sources. Because of their size, synchrotron X-ray sources are normally national or international facilities. Diffraction equipment located at such an X-ray source permits the study of much smaller samples, and crystals as small as $1 \times 1 \times 1\,\mu m^3$ have been used. Furthermore, on larger crystals the collection of high-quality data can be undertaken much more rapidly, and more complex structures, such as those of enzymes, can be determined more easily. It is also possible to study the structures of the excited states of some molecules by irradiating a single crystal in the synchrotron X-ray beam (with a high-power laser) to generate the excited state of some of the molecules *in situ* while the diffraction data are collected.

X-ray free electron lasers (XFEL) are powerful pulsed X-ray sources that can be used for studying proteins and metalloenzymes. Continuous high-power X-ray beams from synchrotrons often cause damage to a protein structure—a particular problem with many metalloenzymes where ionized electrons can cause changes to the oxidation state and coordination of the important metal centre (Chapter 26). An XFEL applies a femtosecond intense pulse of X-rays which is diffracted by a nanocrystal before the crystal is destroyed. By repeatedly introducing a new nanocrystal to the pulsed beam, a full diffraction pattern can be obtained from the complex molecules that can only be crystallized as extremely small crystals. An XFEL can also be used for time-resolved studies of crystals providing dynamic information about unstable intermediates.

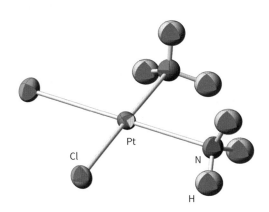

FIGURE 8.7 An ORTEP diagram of cisplatin, *cis*-[Pt(NH$_3$)$_2$Cl$_2$] (Section 32.1). The ellipsoids correspond to a 90% probability of locating the electron density associated with the atoms.

8.2 Neutron diffraction

KEY POINT The scattering of neutrons by crystals yields diffraction data that provide additional information on structure, particularly the positions of light atoms.

Diffraction occurs from crystals for any particle with a velocity such that its associated wavelength (through the de Broglie relation, $\lambda = h/mv$) is comparable to the separations of the atoms or ions in the crystal. Neutrons and electrons (used in electron microscopes, Section 8.17) travelling at suitable velocities have wavelengths of the order of 100–200 pm and thus undergo diffraction by crystalline inorganic compounds.

Neutron beams of the appropriate wavelength are generated by 'moderating' (slowing down) neutrons produced in nuclear reactors or through a process known as **spallation**, in which neutrons are chipped off the nuclei of heavy elements by accelerated beams of protons. The instrumentation used for collecting data and analysing single-crystal or powder neutron diffraction patterns is often similar to that used for X-ray diffraction. The scale is much larger, however, because neutron-beam fluxes are much lower than laboratory X-ray sources. Furthermore, whereas many chemistry laboratories have X-ray diffraction equipment for structure characterization, neutron diffraction can be undertaken only at a few specialist sources worldwide. The investigation of an inorganic compound with this technique is, therefore, much less routine, and its application is essentially limited to systems where X-ray diffraction fails to determine an important part of the structure containing hydrogen or other light atoms, such as Li. With the growing importance of hydrogen-containing compounds in relation to energy storage and generation (Chapters 11 and 30), locating hydrogen atoms in inorganic compounds is becoming increasingly vital. Similarly, many batteries for mobile applications use lithium-ion transport in transition metal oxides, and the determination of the Li^+ positions in these complex inorganic materials has become highly important for the development of a new generation of battery materials.

The advantages of neutron diffraction stem from the fact that neutrons are scattered by nuclei rather than by the surrounding electrons. As a result, neutrons are sensitive to structural parameters that often complement those for X-rays. In particular, the neutron scattering is not dominated by heavy elements, which can be a problem with the X-ray diffraction technique for many inorganic compounds. For example, locating the position of a light element, such as H or Li, in a material that also contains Pb can be impossible with X-ray diffraction, as almost all the electron density is associated with the Pb atoms. With neutrons, in contrast, the scattering from light atoms is often similar to that of heavy elements, so the light atoms contribute significantly to the intensities in the diffraction pattern. Thus neutron diffraction is frequently used in conjunction with X-ray diffraction techniques to define an inorganic structure more accurately in terms of atoms such as H, Li, and O when they are in the presence of heavier, electron-rich metal atoms.

Typical applications include studies of complex metal oxides, such as the high-temperature superconductors (where accurate oxide ion positions are required in the presence of metals such as Ba and Tl), systems where H atom positions are of interest such as metal hydrides studied as potential hydrogen storage materials, hybrid organic-inorganic lead perovskite photovoltaics, and Li-ion battery materials.

Another use for neutron diffraction is to distinguish nearly isoelectronic species. In X-ray scattering, pairs of neighbouring elements in a period of the periodic table, such as O and N or Cl and S, are nearly isoelectronic and scatter X-rays to about the same extent, so they are hard to tell apart in a crystal structure that contains both of them. However, the atoms of these pairs do scatter neutrons to very different extents, N 50% more strongly than O, and Cl about four times better than S, so the clear designation between these atom pairs is much easier than with X-ray diffraction data.

One area where distinguishing neighbouring elements has become important recently is in the study of new photovoltaic materials based on metals of Groups 11, 12, 13, and 14 in combination with chalcogenides (S, Se, Te) and halides; neutron diffraction allows the distribution of near isoelectronic species, to be investigated. One example is the study of Cu_2ZnSnS_4 (CZTS), where the locations of copper and zinc atoms can be distinguished.

Because of the penetrating nature of neutrons (a result of having zero charge and not interacting with the majority of electrons in solid matter) it is possible to build specialist apparatus and environments around the sample and for the neutron beam not to be absorbed or scattered. This allows the study of samples under very high pressures, including in diamond pressure cells, and in intense magnetic fields. Materials can also be studied *in operando*, and the study of lithium compounds inside a battery during charge and discharge cycles (to obtain structural information about the location of the light lithium ion) has been an important success of the technique.

A further property of neutrons, which have $I = \frac{1}{2}$, is that they are also scattered by unpaired electrons, each having $S = \frac{1}{2}$, as well as by the nucleus. In inorganic compounds that have ordered arrangements of unpaired electrons, including ferromagnetic and antiferromagnetic materials (Section 22.10), this scattering will produce additional diffraction intensity. This effect is called magnetic scattering to distinguish it from the nuclear scattering (Fig. 8.8). Analysis of this magnetic scattering provides detailed information on the way in which the electron magnetic moments are ordered, the so-called **magnetic structure**.

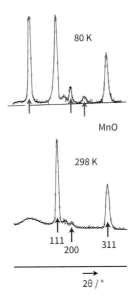

FIGURE 8.8 Powder neutron diffraction patterns collected from MnO showing the additional magnetic reflections that derive from the ordered antiferromagnetic arrangement of electron spins that occurs at low temperature, 80 K. MnO is paramagnetic at room temperature, with no long-range ordering of magnetic moments, so no magnetic reflection is seen at 298 K.

EXAMPLE 8.2 Choosing a suitable diffraction technique for a structural problem in inorganic chemistry

Which experimental diffraction technique would you use to obtain the following information?

(i) Accurate Se–Se bond lengths, with errors less than 0.3 pm, in K_2Se_5.
(ii) Accurate hydrogen positions in $\{(CpY)_4(\mu\text{-}H)_7)(\mu\text{-}H)_4WCp^*(PMe_3)\}$.
(iii) The distribution of copper and zinc in a γ-brass, Cu_5Zn_8.
(iv) The structure of BO(OH).

Answer We need to relate the information needed to the sensitivities of different diffraction techniques.

For (i) both potassium and selenium are 'heavy atoms' and will scatter X-rays strongly so that, provided suitable crystals can be obtained, single-crystal X-ray diffraction will provide very accurate bond lengths. The structure of this compound has been reported with Se–Se bond lengths in the range 2.335–2.366 ± 0.002 Å using such data.

For (ii) the presence of heavy atoms such as Y and W will mean that they will dominate the scattering in an X-ray diffraction experiment. The structure of this compound, with accurate hydrogen positions, was reported in 2011 following a single-crystal neutron diffraction experiment on a crystal of volume 9 mm³.

(iii) Because copper ($Z = 29$) and zinc ($Z = 30$) are near neighbours in the periodic table and near isoelectronic, they will 'look' identical to X-rays. They can, however, be distinguished using neutrons as their relative scattering powers, Cu = 7.72 and Zn = 5.68, are very different. Therefore either powder neutron diffraction or, if large single crystals are available, single crystal neutron diffraction could be employed.

(iv) While BO(OH) contains hydrogen, all the other atoms are 'light' atoms (B $Z = 5$ and O $Z = 8$), so the hydrogen position could still be determined to a reasonable accuracy level using X-rays. Provided crystals of sufficient size can be grown, then single crystal X-ray diffraction would be the structure determination method of choice.

Self-test 8.2 (a) If K_2Se_5 could only be obtained as crystals with dimensions 5 × 10 × 20 μm³ how would this modify the experimental technique employed? (b) Why might neutron diffraction give a slightly different position for the hydrogen atom in BO(OH) in comparison with that obtained from X-ray diffraction methods?

Absorption and emission spectroscopies

The majority of physical techniques used to investigate inorganic compounds involve the absorption and sometimes the re-emission of electromagnetic radiation. The frequency of the radiation absorbed provides useful information on the energy levels of an inorganic compound, and the intensity of the absorption can often be used to provide quantitative analytical information. Absorption spectroscopy techniques are normally nondestructive as after the measurement the sample can be recovered for further analysis.

The spectrum of electromagnetic radiation used in chemistry ranges from the short wavelengths associated with γ-rays and X-rays (about 1 nm), to radio waves with wavelengths of several metres (Fig. 8.9). This spectrum covers the full range of atomic and molecular energies associated with characteristic phenomena such as ionization, vibration, rotation, and nuclear reorientation. Thus, X- and ultraviolet (UV) radiation can be used to determine the electronic structures of atoms and molecules, and infrared (IR) radiation can be used to examine their vibrational behaviour. Radiofrequency (RF) radiation, as used in nuclear magnetic resonance (NMR), explores the energies associated with reorientations of the nucleus in a magnetic field, and those energies are sensitive to the

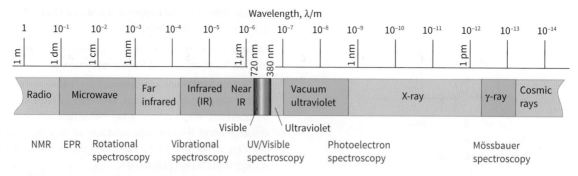

FIGURE 8.9 The electromagnetic spectrum, with wavelengths and techniques that make use of the different regions.

chemical environment of the nucleus. In general, absorption spectroscopic methods make use of the absorption of electromagnetic radiation by a molecule or material at a characteristic frequency corresponding to the energy of a transition between the relevant energy levels. The intensity is related to the probability of the transition, which in turn is determined in part by symmetry rules, such as those described in Chapter 3 for vibrational spectroscopy.

The various spectroscopic techniques involving electromagnetic radiation have different associated timescales, and this variation can influence the structural information that is extracted. When a photon interacts with an atom or molecule we need to consider factors such as the lifetime of any excited state and how a molecule may change during that interval. Table 8.2 summarizes the timescales associated with various spectroscopic techniques discussed in this section. We see that IR spectroscopy takes a much faster snapshot of the molecular structure than NMR; thus if a molecule can reorientate or change its shape within a nanosecond the different states are distinguished by IR but not by NMR, which instead sees a time-averaged structure of the molecule. Such a species is said to be 'fluxional' on the NMR timescale. The temperature at which data are collected should also be taken into account as molecular reorientation rates increase with increasing temperature.

TABLE 8.2 Typical timescales of some common characterization methods

X-ray diffraction	10^{-17} s
Mössbauer	10^{-18} s
Electronic spectroscopy UV–visible	10^{-15} s
Vibrational spectroscopy IR/Raman	10^{-10}–10^{-12} s
EPR	10^{-6} s
NMR	$c.10^{-3}$–10^{-6} s

8.3 Ultraviolet–visible spectroscopy

KEY POINTS The energies and intensities of electronic transitions provide information on electronic structure and chemical environment; changes in spectral properties are used to monitor the progress of reactions.

Ultraviolet–visible spectroscopy (UV–visible spectroscopy) is the observation of the absorption of electromagnetic radiation in the UV and visible regions of the spectrum. It is sometimes known as **electronic spectroscopy** because the energy is used to excite electrons to higher energy levels. UV–visible spectroscopy is among the most widely used techniques for studying inorganic compounds, including light-absorbing compounds (Box 8.1) and their reactions, and most teaching laboratories possess a UV–visible spectrophotometer (Fig. 8.10).

Sometimes electronic transitions occur between energy levels that are close in energy, particularly those involving d and f electrons, and these can occur beyond the visible portion of the electromagnetic spectrum and in the near-infrared region (NIR), λ = 800–2000 nm. This section describes only basic principles; these are elaborated in later chapters, particularly Chapter 22.

(a) Measuring a spectrum

The sample for a UV–visible spectrum determination is usually a solution but may also be a gas or a solid. A gas or liquid is contained in a cell (a 'cuvette') constructed of an optically transparent material such as glass or, for UV spectra at wavelengths below 320 nm, pure silica. Usually,

A BRIEF ILLUSTRATION

Iron pentacarbonyl, [Fe(CO)$_5$], illustrates why consideration of such timescales is important when analysing the spectra to obtain structural information. Infrared spectroscopy suggests that [Fe(CO)$_5$] has D_{3h} symmetry, with distinct axial and equatorial carbonyl groups; whereas an NMR spectrum obtained at room temperature suggests that all the carbonyl groups are equivalent because the axial and equatorial carbonyl groups exchange positions faster than 10^6 times per second, within the NMR timescale.

BOX 8.1 How can solar cell materials be characterized?

The Sun is an abundant source of clean energy, and the development of high-efficiency solar cells, which convert sunlight to electrical energy, is a major current research topic. The majority of commercial photovoltaic cells use the semiconductor silicon to absorb sunlight, but many alternative materials are being investigated with the aim of developing cheaper solar cells that could, for example, be printed as thin films directly onto any surface. Examples of alternative semiconductors being researched include CdTe, CIGS (copper indium gallium selenide, $Cu(In,Ga)Se_2$), CZTS (copper zinc tin sulfide, Cu_2ZnSnS_4), and lead halide perovskites, such as methylammonium lead iodide (MAPI), $(CH_3NH_3)PbI_3$. A key property of these photovoltaic semiconducting materials is that their band gaps are around 1.5–1.8 eV, and they therefore absorb (sun)light efficiently across the whole of the ultraviolet and visible regions (up to wavelengths of 780 nm) and ideally also into the near-infrared (up to 1200 nm). Potential new photovoltaic materials are therefore characterized using UV–visible spectroscopy, as shown in Figure B8.1.

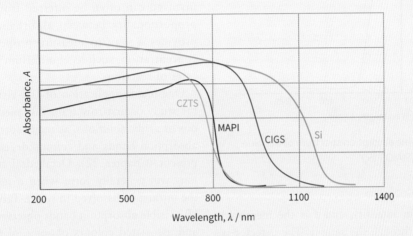

FIGURE B8.1 UV–visible absorption spectra of materials (Si, CIGS, CZTS, and MAPI) used in photovoltaic cells.

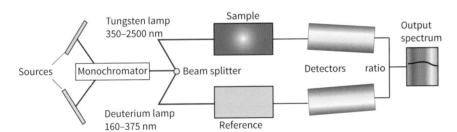

FIGURE 8.10 The layout of a typical UV–visible absorption spectrometer.

the beam of incident radiation is split into two, one passing through the sample and the other passing through a cell that is identical except for the absence of the sample. The emerging beams are compared at the detector (a photodiode), and the absorption is obtained as a function of wavelength.

Conventional spectrometers sweep the wavelength of the incident beam by changing the angle of a diffraction grating, but it is now more common for the entire spectrum to be recorded at once using a diode array detector. For solid samples, the intensity of UV–visible radiation *reflected* from the sample is more easily measured than the small amount that can be transmitted through a solid. The absorption spectrum can be obtained by subtraction of the reflected intensity from the intensity of the incident radiation and subsequent application of various wavelength-dependent corrections (Fig. 8.11).

The intensity of absorption is measured as the **absorbance**, A, defined as

$$A = \log_{10}\left(\frac{I_0}{I}\right) \tag{8.2}$$

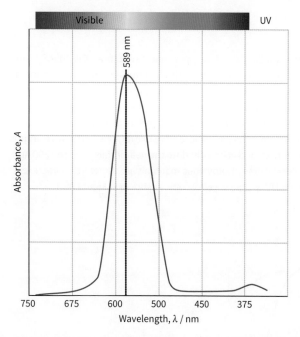

FIGURE 8.11 The UV–visible absorption spectrum of the blue pigment ultramarine, $Na_7[SiAlO_4]_6(S_3)$.

where I_0 is the incident intensity, and I is the measured intensity after passing through the sample. The detector is the limiting factor for strongly absorbing species because the measurement of low photon flux is unreliable; in such cases the sample is often diluted.

A BRIEF ILLUSTRATION

A sample that attenuates the light intensity by 10% (so $I_0/I = 100/90$) has an absorbance of 0.05, one that attenuates it by 90% (so $I_0/I = 100/10$) has an absorbance of 1.0, one that attenuates it by 99% (so $I_0/I = 100/1$) has an absorbance of 2.0, and so on.

The empirical **Beer–Lambert law** is used to relate the absorbance to the molar concentration, $[J]$, of the absorbing species J and the optical path length, L:

$$A = \varepsilon[J]L \tag{8.3}$$

where ε (epsilon) is the **molar absorption coefficient** (still commonly referred to as the 'extinction coefficient' and sometimes the 'molar absorptivity'). Values of ε range from above $10^5 \text{ dm}^3\text{ mol}^{-1}\text{ cm}^{-1}$ for fully allowed transitions; for example, an electron transferring from the 3d to 4p energy levels in an atom ($\Delta l = 1$), to less than 1 $\text{dm}^3\text{ mol}^{-1}\text{ cm}^{-1}$ for 'forbidden' atomic transitions, such as those with $\Delta l = 0$. In molecules these selection rules extend to transitions between states based on molecular orbitals, although they may be relaxed, particularly by vibrations that alter the symmetries of the orbitals (Chapter 22). For small molar absorption coefficients, the absorbing species may be difficult to observe unless the concentration or path length is increased accordingly.

Fig. 8.12 shows a typical solution UV–visible spectrum obtained from a d-metal compound, in this case Ti(III), which has a d^1 configuration. From the wavelength of the radiation absorbed, the energy levels of the compound, including the effect of the ligand environment on a d-metal atom, can be derived. The type of transition involved can often be inferred from the value of ε. The proportionality between absorbance and concentration provides a way to measure properties that depend on concentration, such as equilibrium compositions and the rates of reaction.

One major application of UV–visible spectroscopy is in the study of first-row transition metal complexes in aqueous or acidic solutions, as many species have characteristic absorption bands and colours depending on the complex present (Chapter 22). Octahedral hexaaqua $[M(H_2O)_6]^{n+}$ ions will generally form when a metal salt, for example the nitrate, is dissolved in water. Table 8.3 summarizes the UV–visible absorption bands observed in the visible region, and associated colours of some of the common first-row transition metal hexaaqua ions, $[M(H_2O)_6]^{n+}$. The colours of these species are usually the result of d-d electronic transitions, Chapter 22.

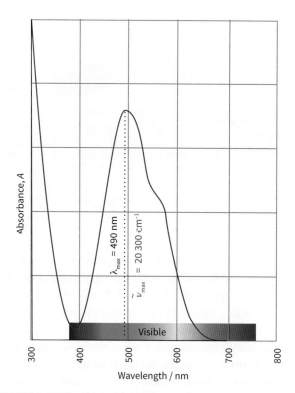

FIGURE 8.12 The UV–visible spectrum of $[Ti(OH_2)_6]^{3+}$ (aq).

TABLE 8.3 Summary of absorption bands and colours associated with different $[M(H_2O)_6]^{n+}$ ions

Ion	Typical octahedral example	Main absorption band(s)/nm	Colour
Ti^{3+}	$[Ti(H_2O)_6]^{3+}$	450–650 (Fig. 8.12)	purple, blue
V^{4+}	$[VO(H_2O)_n]^{2+}$	600–800	blue
V^{3+}	$[V(H_2O)_6]^{3+}$	350–450, 550–650	green
V^{2+}	$[V(H_2O)_6]^{2+}$	250–400, 550–600	purple/lilac
Cr^{3+}	$[Cr(H_2O)_6]^{3+}$	350–450, 550–650	blue-purple
Mn^{2+}	$[Mn(H_2O)_6]^{2+}$	350, 400, 425, 530 (all weak)	very pale pink
Fe^{2+}	$[Fe(H_2O)_6]^{2+}$	650–800 (weak)	pale green–blue
Co^{2+}	$[Co(H_2O)_6]^{2+}$	450–560	pink–red
Ni^{2+}	$[Ni(H_2O)_6]^{2+}$	380–440, 600–800	green
Cu^{2+}	$[Cu(H_2O)_6]^{2+}$	620–800	Blue

EXAMPLE 8.3 Relating UV–visible spectra and colour

Fig. 8.13 shows the UV–visible absorption spectrum of $PbCrO_4$. What colour would you expect $PbCrO_4$ to be?

Answer We need to be aware that the removal of light of a particular wavelength from incident white light results in the remaining light being perceived as having its complementary

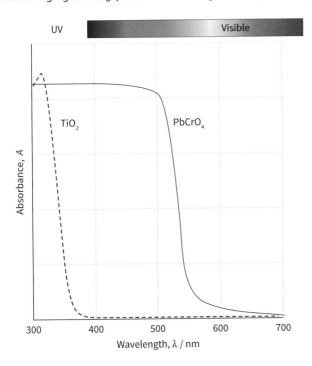

FIGURE 8.13 The UV–visible spectra of $PbCrO_4$ (solid red) and TiO_2 (dashed blue). Absorbance is given as a function of wavelength.

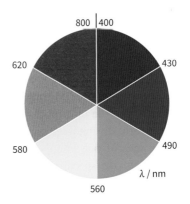

FIGURE 8.14 An artist's colour wheel: complementary colours lie opposite each other across a diameter.

colour. Complementary colours are diametrically opposite each other on an artist's colour wheel (Fig. 8.14). The only absorption from $PbCrO_4$ in the visible region is in the blue region of the spectrum. The rest of the light is scattered into the eye, which, according to Fig. 8.14, perceives the complementary colour yellow. Because the electronic transition involves an allowed transition, from a p orbital on oxygen to a d orbital on chromium, the absorption coefficient is very large and $PbCrO_4$ is an intense yellow colour and has been used as a pigment.

Self-test 8.3 Explain why TiO_2, for which the UV–visible absorption spectrum is also given in Fig. 8.13, is widely used in sunscreens to protect against harmful UVA radiation (UV radiation with wavelengths in the range 320–360 nm). What colour is TiO_2?

(b) Spectroscopic monitoring of titrations and kinetics

When the emphasis is on measurement of the intensities rather than the energies of transitions, the spectroscopic investigation is usually called **spectrophotometry**. Provided at least one of the species involved has a suitable absorption band, it is usually straightforward to carry out a 'spectrophotometric titration' in which the extent of reaction is monitored by measuring the concentrations of the components present in the mixture. The measurement of UV–visible absorption spectra of species in solution also provides a method for monitoring the progress of reactions and determining rate constants.

Techniques that use UV–visible spectral monitoring range from those measuring reactions with half-lives of picoseconds (photochemically initiated by an ultrafast laser pulse) to the monitoring of slow reactions with half-lives of hours or even days. The **stopped-flow technique** (Fig. 8.15) is commonly used to study reactions with half-lives of between 5 ms and 10 s and which can be initiated by mixing. Two solutions, each containing one of the reactants, are mixed rapidly by a pneumatic impulse, then the flowing, reacting solution is brought to an abrupt stop by filling a 'stop-syringe' chamber, which triggers the monitoring of absorbance. The reaction can be monitored at a single wavelength, or successive spectra can be measured very rapidly by using a diode array detector.

The spectral changes that occur during a titration or the course of a reaction provide information about the number of species that form during its progress. An important case is the appearance of one or more **isosbestic points**, wavelengths at which two species have equal values for their molar absorption coefficients (Fig. 8.16; the name comes from the Greek for 'equal extinguishing'). The detection of isosbestic points in a titration or during the course of a reaction is evidence for there being only two dominant species (reactant and product) in the solution. It is extremely unlikely that three coloured species (hence an intermediate as well as reactant and product) could have the same molar absorption coefficient at one or more wavelengths.

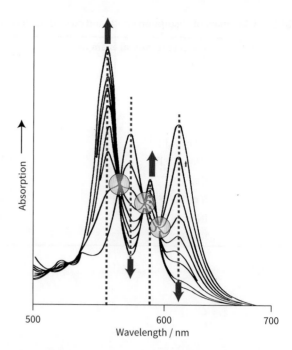

FIGURE 8.16 Isosbestic points (circled) observed in the absorption spectral changes during the reaction of HgTPP (TPP = tetraphenylporphyrin) with Zn^{2+} in which Zn replaces Hg in the macrocycle. The initial and final spectra are those of the reactant and product, which indicates that the free TPP ligand does not reach a detectable concentration during the reaction. Adapted from 'Kinetics of electrophilic substitution reactions involving metal ions in metalloporphyrins', C. Grant and P. Hambright. *J. Am. Chem. Soc.* **91**, 4195. Copyright 1969, American Chemical Society.

8.4 Fluorescence or emission spectroscopy

Fluorescence or **emission spectroscopy**, also sometimes referred to as fluorimetry or spectrofluorimetry, concerns the emitted electromagnetic radiation, usually in the visible or near-infrared regions of the spectrum, from a compound that has been electronically excited, normally with UV radiation. The emitted photon is normally at a lower energy than the exciting radiation because of losses to nonradiative processes, such as vibrational modes, in the compound under investigation. The origins and relationships between absorption and emission spectra of a compound are summarized in Fig. 8.17.

The emission or fluorescence spectrum depends on the energy of the exciting radiation, and experimentally it may be collected for a variety of incident wavelengths. Electronic absorption and emission spectra are often collected and analysed together. The luminescence spectrometer (Fig. 8.18) typically uses a xenon lamp as a source of excitation, and a monochromator selects an excitation wavelength from the light produced in the range 200–800 nm. This monochromatic beam is directed onto the compound under investigation and its emission spectrum analysed, using a second monochromator, for photon energies between 200 and 900 nm.

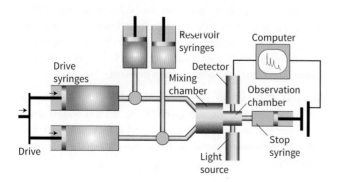

FIGURE 8.15 The structure of a stopped-flow instrument for studying fast reactions in solution.

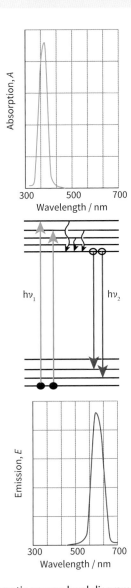

FIGURE 8.17 Schematic energy-level diagram showing the origins of the UV–visible and emission spectra of a typical inorganic compound.

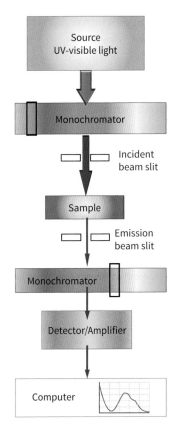

FIGURE 8.18 The layout of a typical fluorescence spectrometer.

blue-emitting LED is used in combination with a Ce^{3+} doped oxide that has a fluorescent emission of yellow light: the combination of residual blue light and fluorescent yellow light is seen as white light by the human eye.

Fig. 8.19 shows the emission spectrum collected from nanoparticles of CdSe/ZnS, known as quantum dots

In inorganic chemistry, the experimental collection of emission spectra is of particular interest for materials that are used as phosphors, in low-energy lighting and in displays, where UV radiation, produced by a light-emitting diode (LED) or by a discharge in a mercury vapour, is converted into visible light. Alternatively, high-energy visible light, for example 460 nm blue light, can induce fluorescence emission at lower-energy visible or infrared wavelengths.

As with UV–visible spectroscopy, transitions between the electronic energy levels of d-block and f-block elements, which contain unpaired electrons, frequently produce emission in the visible and near-infrared regions of the electromagnetic spectrum. Therefore, compounds containing low levels of lanthanoid ions, such as Eu^{3+} and Tm^{3+} (Sections 22.6 and 30.1), and some transition metal-doped sulfides, for example, Mn^{2+}-doped ZnS, are normally studied for use in these applications. In some types of low-energy lighting, a

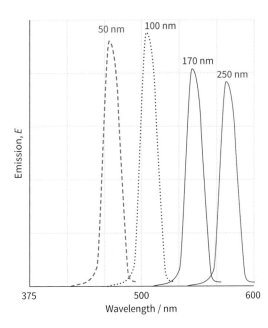

FIGURE 8.19 The emission spectrum of CdSe/ZnS nanoparticles as a function of size, using an excitation wavelength of 320 nm.

(Box 28.1), as a function of particle size. The smaller the particles become, the shorter the wavelength emitted as the energy separation of the excited and ground states increases.

The analysis of the absorption and emission spectra of gas phase atomic species, atomic absorption, and flame emission spectroscopies, is employed for the determination of the chemical compositions of inorganic compounds. These techniques are described more fully in Section 8.12.

8.5 Infrared and Raman spectroscopy

KEY POINTS Infrared and Raman spectroscopy are often complementary in that a particular type of vibration may be observed in one method but not in the other; the information is used in many ways, ranging from the identification of functional groups to complete molecular geometry determination.

Vibrational spectroscopy is used to characterize compounds in terms of the strength, stiffness, and number of bonds that are present. It is also used to detect the presence of known compounds (fingerprinting), to monitor changes in the concentration of a species during a reaction, to determine the components of an unknown compound (such as the presence of CO ligands), to determine a likely structure for a compound, and to measure properties of bonds (their force constants). There are two main experimental techniques used to obtain vibrational spectra: **infrared spectroscopy** and **Raman spectroscopy**.

(a) The energies of molecular vibrations

A bond in a molecule behaves like a spring: stretching it through a distance x produces a restoring force F. For small displacements, the restoring force is proportional to the displacement and $F = -kx$, where k is the **force constant** of the bond: the stiffer the bond, the larger the force constant. Such a system is known as a harmonic oscillator, and solution of the Schrödinger equation gives the energies

$$E_v = \left(v + \tfrac{1}{2}\right)\hbar\omega \qquad (8.4a)$$

where $\omega = (k/\mu)^{1/2}$, $v = 0, 1, 2, \ldots$, and μ is the **effective mass** of the oscillator. For a diatomic molecule composed of atoms of masses m_A and m_B,

$$\mu = \frac{m_A m_B}{m_A + m_B} \qquad (8.4b)$$

This effective mass is different for isotopomers (molecules composed of different isotopes of an element), which in turn leads to changes in E_v. If $m_A \gg m_B$, then $\mu \approx m_B$ and only atom B moves appreciably during the vibration: in this case the vibrational energy levels are determined largely by m_B, the mass of the lighter atom. The frequency v is therefore high when the force constant is large (a stiff bond) and the effective mass of the oscillator is low (only light atoms are moved during the vibration). Vibrational energies are usually expressed in terms of the wavenumber $\tilde{v} = \omega/2\pi c$; typical values of $\tilde{v}$ lie in the range 300–3800 cm^{-1} (Table 8.4).

TABLE 8.4 Characteristic fundamental stretching wavenumbers of some common molecular species as free molecules or ions, or coordinated to a metal centre

Species	Range/cm^{-1}
OH	3400–3600
NH	3200–3400
CH	2900–3200
BH	2600–2800
CN$^-$	2000–2200
CO (terminal)	1900–2100
CO (bridging)	1800–1900
R$_2$C=O	1600–1760
NO	1675–1870
O$_2^-$	920–1120
O$_2^{2-}$	800–900
Si–O	900–1100
Metal–Cl	250–500
Metal–metal bonds	120–400

> **A NOTE ON GOOD PRACTICE**
>
> μ is commonly referred to as the 'reduced mass', for the same term appears in the separation of internal motion from translational motion. However, in polyatomic molecules each vibrational mode corresponds to the motion of different quantities of mass, which depends on the individual masses in a much more complicated way, and the 'effective mass' is the more general term for use when discussing vibrational modes.

A molecule consisting of N atoms can vibrate in $3N - 6$ different, independent ways if it is nonlinear, and $3N - 5$ different ways if it is linear. These different, independent vibrations are called **normal modes**. For instance, a CO$_2$ molecule has four normal modes of vibration (as was shown in Fig. 3.16)—two corresponding to stretching the bonds and two corresponding to bending the molecule in two perpendicular planes.

Bending modes typically occur at lower frequencies than stretching modes, and their effective masses, and therefore their frequencies, depend on the masses of the atoms in a complicated way that reflects the extent to which the various atoms move in each mode. The modes are labelled v_1, v_2, etc., and are sometimes given descriptive names such as 'symmetric stretch', 'antisymmetric stretch', and in-plane/out-of-plane bends. Only normal modes that correspond to a changing electric dipole moment can absorb infrared radiation, so only these modes are **IR active** and contribute to an IR spectrum. A normal mode is **Raman active** if it corresponds to a change in polarizability of the molecule. As we saw in Chapter 3, group theory is a powerful tool for predicting the IR and Raman activities of molecular vibrations.

The lowest level ($v = 0$) of any normal mode corresponds to $E_0 = \tfrac{1}{2}\hbar\omega$, the so-called **zero-point energy**, which is the lowest vibrational energy that a bond can possess. In addition to the fundamental transitions with $\Delta v = +1$, vibrational spectra may also show bands arising from double quanta ($\Delta v = +2$) at $2\tilde{v}$, known as **overtones**, and **combinations** of two different vibrational modes (e.g. $v_1 + v_2$). These special transitions may be helpful, as they often arise even when the fundamental transition is not allowed by the selection rules.

(b) The techniques

The IR vibrational spectrum of a compound is obtained by exposing the sample to infrared radiation and recording the variation of the absorbance with frequency, wavenumber, or wavelength.

> **A NOTE ON GOOD (AND CONVENTIONAL) PRACTICE**
>
> An absorption is often stated as occurring at, say, 'a frequency of 1000 wavenumbers'. This widespread practice should be noted and used with caution; 'wavenumber' is the name of a physical observable related to frequency v by $\tilde{v} = v/c$, and wavenumber is not a unit. The dimensions of wavenumber are 1/length, and it is commonly reported in inverse centimetres (cm^{-1}).

In early IR spectrometers, the transmission was measured as the frequency of the radiation was swept between two limits. Nowadays the spectrum is extracted from an interferogram by **Fourier transformation**, which converts information in the time domain (based on the interference of waves travelling along paths of different lengths) to the frequency domain. The sample must be contained in a material that does not absorb IR radiation, which means that standard glass cannot be used and aqueous solutions are unsuitable unless the spectral bands of interest occur at frequencies not absorbed by water. Optical windows are typically constructed from CsI or CaF_2. Traditional procedures of sample preparation include KBr pellets (where the sample is diluted with dried KBr and then pressed into a translucent disc) and paraffin mulls (where the sample is produced as a suspension that is then placed as a droplet between the optical windows). These methods are still widely used, but now becoming standard in laboratories are **attenuated total reflectance** (ATR) devices in which the sample is simply placed on top of a diamond window. An IR beam is passed through the diamond, and a small part of the wave escapes from the surface as an evanescent wave which can be absorbed by the sample.

A typical range of an IR spectrum is between 4000 and 250 cm^{-1}, which corresponds to a range of wavelengths between 2.5 and 40 μm; this range covers many important vibrational modes of inorganic bonds. Fig. 8.20 shows a typical IR spectrum.

In Raman spectroscopy the sample is exposed to intense laser radiation in the visible region of the spectrum. Most of the photons are scattered elastically (with no change of frequency) but some are scattered inelastically, having given up some of their energy to excite vibrations; it is also possible for photons to gain energy from vibrations. The latter photons have frequencies that differ from that of the incident radiation (v_0) by amounts equivalent to vibrational frequencies (v_i) of the molecule. Those that lose energy and excite vibrations give rise to **Stokes lines** at $v_0 - v_i$, and those that gain energy produce **anti-Stokes lines** at $v_0 + v_i$. A disadvantage of Raman compared to IR is that linewidths are usually much greater. Conventional Raman spectroscopy involves the photon causing a transition to a 'virtual' excited state which then collapses back to a real lower state,

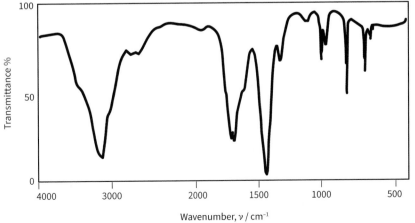

FIGURE 8.20 The IR spectrum of ammonium carbonate, NH_4CO_3, showing characteristic absorptions due to the carbonate ion (C-O stretches near 1450 and 1080 cm^{-1} and a bending mode at 712 cm^{-1}) and the ammonium ion (N-H symmetric stretch near 3200 cm^{-1} and NH_4^+ deformation near 1400 cm^{-1}).

emitting the detected photon in the process. The technique is not very sensitive, though with modern laser technology Raman spectra are relatively easy to obtain. In Raman microspectroscopy the laser can be focused onto an area less than 1 μm in diameter and scanned across a sample to obtain the Raman spectrum as a function of position.

Great signal enhancement is achieved if the species under investigation is coloured and the excitation laser is tuned to a real electronic transition. The latter technique is known as **resonance Raman spectroscopy** and is particularly valuable for studying the environment of d-metal atoms in enzymes (Chapter 26) because only vibrations close to the electronic chromophore (the group primarily responsible for the electronic excitation) are excited and the many thousands of bonds in the rest of the molecule are silent. In **surface enhanced Raman spectroscopy** (**SERS**) the Raman signal is greatly enhanced (up to a factor of 10^{11}) when a molecule is absorbed onto a metal surface, such as a gold nanoparticle.

Raman spectra may be obtained over a similar range to IR spectra (200–4000 cm^{-1}), and Fig. 8.21 shows a typical spectrum. Raman spectroscopy is often complementary to IR spectroscopy, as the two techniques probe vibrational modes with different activities: one mode might correspond to a change in dipole moment and thus be detectable in the IR spectrum; another mode might correspond to a change in polarizability (a change in the electron distribution in a molecule caused by an applied electric field) and be seen in the Raman spectrum. We saw in Section 3.5 that for group theoretical reasons, no mode can be both IR and Raman active in a molecule that has a centre of inversion (the exclusion rule).

(c) Applications of IR and Raman spectroscopy

One important application of vibrational spectroscopy is in the determination of the shape of an inorganic molecule. A five-coordinate structure, AX_5 (where X is a single element) for instance, may adopt a square-pyramidal or trigonal-bipyramidal geometry having C_{4v} or D_{3h} symmetry, respectively. A normal-mode analysis of these geometries of the kind explained in Section 3.5 reveals that a trigonal-bipyramidal AX_5 molecule has five stretching modes (of symmetry species $2A_1' + A_2'' + E'$, E' represents a pair of doubly degenerate vibrations) of which three are IR active ($A_2'' + E'$, corresponding to two absorption bands allowing for the degeneracy of the E' modes), and four are Raman active ($2A_1' + E'$, corresponding to three bands). A similar analysis of square-pyramidal geometry shows it has four IR active stretching modes ($2A_1 + E$, three observed bands) and five Raman active stretching modes ($2A_1 + B_1 + E$, four observed bands). The vibrational spectra of BrF_5 reveal three IR and four Raman Br-F stretching bands, showing that this molecule is square-pyramidal, just as expected from VSEPR theory (Section 2.3).

EXAMPLE 8.4 Identifying molecular geometry

Fig. 8.22 shows the IR and Raman spectra obtained from XeF_4 over the same range of wavenumbers. Does XeF_4 adopt a square-planar or tetrahedral geometry?

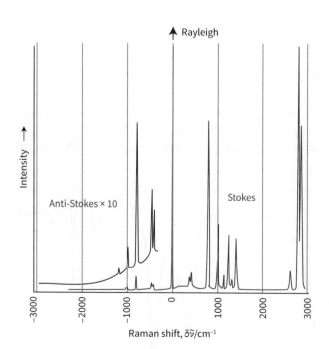

FIGURE 8.21 A typical Raman spectrum showing Rayleigh scattering (scattering of the laser light with no change in wavelength that occurs at the origin) and Stokes and anti-Stokes lines.

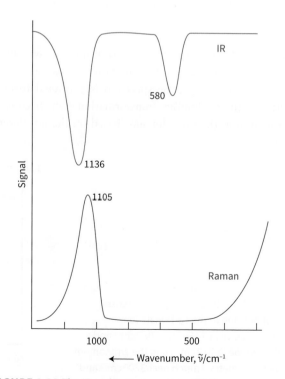

FIGURE 8.22 The IR and Raman spectra of XeF_4.

Answer An AB_4 molecule may be tetrahedral (T_d), square-planar (D_{4h}), the base of a square pyramid (C_{4v}), or see-saw (C_{2v}). The spectra have no absorption energy in common and, therefore, it is likely that the molecule has a centre of symmetry. Of the symmetry groups available, only square-planar geometry, D_{4h}, has a centre of symmetry.

Self-test 8.4 Use VSEPR theory to predict a molecular shape for XeF_2 and hence determine the total number of vibrational modes expected to be observed in its IR and Raman spectra. Would any of these absorptions occur at the same frequency in both the Raman and IR spectra?

A major use of IR and Raman spectroscopy is the characterization of the many compounds of the d block that contain one or more carbonyl ligands. Carbonyl groups give rise to intense vibrational absorption bands in a region where few other molecules produce absorptions. Free CO absorbs at 2143 cm^{-1}, but when coordinated to a metal atom in a compound the stretching frequency (and correspondingly the wavenumber) is lowered by an amount that depends on the extent to which electron density is transferred into the two π antibonding orbitals (the LUMOs) by back donation from the metal atom (Section 24.5). The CO stretching absorption also allows terminal and bridging ligands to be distinguished, with bridging ligands occurring at lower frequencies. Isotopically labelled compounds show a shift in the absorption bands (by about 40 cm^{-1} to lower wavenumbers when ^{13}C replaces ^{12}C in a CO group) through a change in effective mass (eqn. 8.4b), and this effect can be used to assign spectra and probe reaction mechanisms involving this ligand.

A BRIEF ILLUSTRATION

Deuteration, replacing H by D in a compound, has a large effect on its vibrational spectrum due to the large mass ratio of these isotopes. In water the O–H stretch occurs at 3550 cm^{-1} while the O–D stretch in D_2O occurs at 2440 cm^{-1}; this is consistent with the change in effective mass of the system, which means that vibrations involving deuterium are shifted to lower wavenumbers by a factor of approximately $1/\sqrt{(m_D/m_H)} = 1/\sqrt{2}$ compared with H.

The speed of data acquisition possible with Fourier-transform IR (FTIR) spectroscopy has meant that it can be incorporated into rapid kinetic techniques, including ultra-fast laser photolysis and stopped-flow methods.

Raman and IR spectroscopy are excellent methods for studying molecules that are formed and trapped in inert matrices, in the technique known as **matrix isolation** (Box 19.3). The principle of matrix isolation is that highly unstable species that would not normally exist under ambient conditions can be generated in an inert matrix at low temperatures, such as solid xenon. The unstable molecules SiH_3, FeF_4, and AuF_2 have all been matrix isolated and studied using vibrational spectroscopy.

Resonance techniques

Several techniques of structural investigation depend on bringing energy-level separations into resonance with electromagnetic radiation, with the separations in some cases controlled by the application of a magnetic field. Two of these techniques involve **magnetic resonance**: in one, the energy levels are those of magnetic nuclei (nuclei with nonzero spin, $I > 0$); in the other, they are the energy levels of unpaired electrons.

8.6 Nuclear magnetic resonance

KEY POINTS Nuclear magnetic resonance is suitable for studying compounds containing elements with magnetic nuclei, especially hydrogen. The technique gives information on molecular structure, including chemical environment, connectivity, and internuclear separations. It also probes molecular dynamics and is an important tool for investigating rearrangement reactions occurring on a millisecond timescale.

Nuclear magnetic resonance (NMR) is the most powerful and widely used spectroscopic method for the determination of molecular structures in solution and pure liquids. In many cases it provides information about shape and symmetry with greater certainty than is possible with other spectroscopic techniques, such as IR and Raman spectroscopy. It also provides information about the rate and nature of the interchange of ligands in fluxional molecules and can be used to monitor the course of reactions, in many cases providing great mechanistic detail. The technique has been used to obtain the structures of protein molecules with molar masses of up to 30 kg mol^{-1} (corresponding to a molecular mass of 30 kDa) and complements the more static descriptions obtained with X-ray single-crystal diffraction. However, unlike X-ray diffraction, NMR studies of molecules in solution generally cannot provide very precise bond distances and angles, although they can provide some information on internuclear separations. It is usually a nondestructive technique because the sample can be recovered from solution after the resonance spectrum has been collected.

The sensitivity of NMR depends on several factors, including the abundance of the isotope and the size of its nuclear magnetic moment. For example, ^1H, with 99.98%

TABLE 8.5 Nuclear spin characteristics of common nuclei used in NMR

Nucleus	Natural abundance/per cent	Sensitivity*	Spin	NMR frequency/MHz[†]
^{1}H	99.98	5680	$\frac{1}{2}$	100.000
^{2}H	0.015	0.00821	1	15.351
^{7}Li	92.58	1540	$\frac{3}{2}$	38.863
^{11}B	80.42	754	$\frac{3}{2}$	32.072
^{13}C	1.11	1.00	$\frac{1}{2}$	25.145
^{15}N	0.37	0.0219	$\frac{1}{2}$	10.137
^{17}O	0.037	0.0611	$\frac{5}{2}$	13.556
^{19}F	100	4730	$\frac{1}{2}$	94.094
^{23}Na	100	525	$\frac{3}{2}$	26.452
^{29}Si	4.7	2.09	$\frac{1}{2}$	19.867
^{31}P	100	377	$\frac{1}{2}$	40.481
^{89}Y	100	0.668	$\frac{1}{2}$	4.900
^{103}Rh	100	0.177	$\frac{1}{2}$	3.185
^{109}Ag	48.18	0.276	$\frac{1}{2}$	4.654
^{119}Sn	8.58	28.7	$\frac{1}{2}$	37.272
^{183}W	14.4	0.0589	$\frac{1}{2}$	4.166
^{195}Pt	33.8	19.1	$\frac{1}{2}$	21.462
^{199}Hg	16.84	5.42	$\frac{1}{2}$	17.911

* Sensitivity is relative to ^{13}C = 1 and is the product of the relative sensitivity of the isotope and the natural abundance.
[†] At 2.349 T (a '100 MHz spectrometer' for ^{1}H). Most modern spectrometers operate with larger magnetic fields and hence higher frequencies, typically 200–600 MHz. Values for NMR frequencies for these spectrometers may be easily calculated from the 100 MHz value by multiplying the given frequency by the ratio (spectrometer frequency)/(100 MHz).

natural abundance and a large magnetic moment, is easier to observe than ^{13}C, which has a smaller magnetic moment and only 1.1% natural abundance. With modern multinuclear NMR techniques it is particularly easy to observe spectra for ^{1}H, ^{19}F, and ^{31}P, and useful spectra can also be obtained for many other isotopes. Table 8.5 lists a selection of important NMR nuclei and their sensitivities. A common limitation for exotic nuclei is the presence of a nuclear quadrupole moment, a nonuniform distribution of electric charge (which is present for all nuclei with nuclear spin quantum number $I > \frac{1}{2}$), which broadens signals and degrades spectral information. Most nuclei with even atomic numbers and even mass numbers (such as ^{12}C and ^{16}O) have zero spin and are invisible in NMR.

(a) Observation of the spectrum

A nucleus of spin I can take up $2I + 1$ orientations relative to the direction of an applied magnetic field. Each orientation has a different energy (Fig. 8.23), with the lowest level the most highly populated. The energy separation of the two states $m_I = +\frac{1}{2}$ and $m_I = -\frac{1}{2}$ of a nucleus with spin $\frac{1}{2}$ (such as ^{1}H or ^{13}C) is

$$\Delta E = \hbar\gamma B_0 \tag{8.5}$$

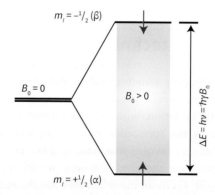

FIGURE 8.23 When a nucleus with spin $I > 0$ is in a magnetic field, its $2I + 1$ orientations (designated m_I) have different energies. This diagram shows the two energy levels of a nucleus with $I = \frac{1}{2}$ (as for ^{1}H, ^{13}C, ^{31}P).

where B_0 is the magnitude of the applied magnetic field (more precisely, the magnetic induction in tesla, 1 T = 1 kg s^{-2} A^{-1}) and γ is the gyromagnetic ratio of the nucleus; that is, the ratio of its magnetic moment to its spin angular momentum. With modern superconducting magnets producing fields of 5–23 T, resonance is achieved with electromagnetic radiation in the range 200–1200 MHz. Because the difference in energy of the $m_I = +\frac{1}{2}$ and $m_I = -\frac{1}{2}$ states in the

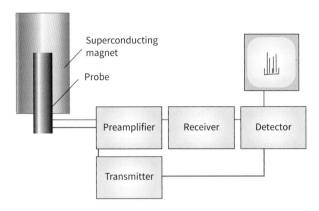

FIGURE 8.24 The layout of a typical NMR spectrometer. The link between transmitter and detector is arranged so that only low-frequency signals are processed.

applied magnetic field is small, the population of the lower level is only marginally greater than that of the higher level. Consequently, the sensitivity of NMR is low but can be increased by using a stronger magnetic field, which increases the energy difference and therefore the population difference and the signal intensity.

Spectra were originally obtained in a continuous wave (CW) mode in which resonances are encountered by subjecting the sample to a constant radiofrequency while increasing the magnetic field, or holding the field constant while sweeping the radiofrequency. In contemporary spectrometers, the energy separations are identified by exciting nuclei in the sample with a sequence of radiofrequency pulses and then observing the return of the nuclear magnetization to equilibrium. Fourier transformation then converts the time-domain data to the frequency domain with peaks appearing at frequencies corresponding to transitions between the different nuclear energy levels. Fig. 8.24 shows the experimental arrangement of an NMR spectrometer.

(b) Chemical shifts

The frequency of an NMR transition depends on the local magnetic field experienced by the nucleus and is expressed in terms of the **chemical shift**, δ, the difference between the resonance frequency of nuclei (v) in the sample and that of a reference compound ($v°$):

$$\delta = \frac{v - v°}{v°} \times 10^6 \qquad (8.6)$$

A NOTE ON GOOD PRACTICE

The chemical shift δ is dimensionless. However, common practice is to report it as 'parts per million' (ppm) in acknowledgement of the factor of 10^6 in the definition. This practice is unnecessary.

A common standard for ^{1}H, ^{13}C, or ^{29}Si spectra is tetramethylsilane (TMS), Si(CH$_3$)$_4$. When $\delta < 0$ the nucleus is said to be **shielded** (with a resonance that is referred to as occurring at 'low frequency') relative to the standard; $\delta > 0$ corresponds to a nucleus that is **deshielded** (with a resonance that is said to occur at 'high frequency') with respect to the reference. An H atom bound to a low-oxidation-state, d-block element from Groups 6 to 10 (such as [HCo(CO)$_4$]) is generally found to be highly shielded whereas in an oxoacid (such as H$_2$SO$_4$) it is deshielded. From these examples it might be supposed that the higher the electron density around a nucleus, the greater its shielding. However, as several factors contribute to the shielding, a simple physical interpretation of chemical shifts in terms of electron density is generally not possible (Section 11.3).

The chemical shifts of ^{1}H and other nuclei in various chemical environments are tabulated, so empirical correlations can often be used to identify compounds or the element to which the resonant nucleus is bound. For example, the H chemical shift in CH$_4$ is only 0.1 because the H nuclei are in an environment similar to that in the standard, tetramethylsilane, but the H chemical shift in GeH$_4$ is $\delta = 3.1$ (Fig. 8.25). The number of resonances with different chemical shifts can be used to determine the number of different (inequivalent) nuclei that are present in a molecule. For instance, in ClF$_3$ the chemical shift of the unique ^{19}F nucleus is separated by $\Delta\delta = 120$ from that of the other two F nuclei (Fig. 8.26).

Changes in the chemical shift are caused by the introduction of a paramagnetic species into the solution due to the

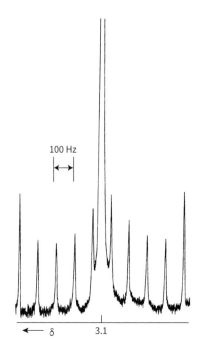

FIGURE 8.25 The ^{1}H-NMR spectrum of GeH$_4$. The main resonance is at $\delta = 3.1$. Satellites produced through spin–spin coupling (Section 8.6(c)) are given by $J(^1\text{H}-^{73}\text{Ge})$.

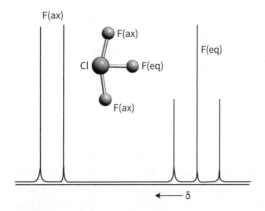

FIGURE 8.26 The ^{19}F-NMR spectrum of ClF_3.

A BRIEF ILLUSTRATION

The ^{19}F NMR spectrum of ClF_3 is shown in Fig. 8.26. The signal ascribed to the two axial F nuclei (each with $I = \frac{1}{2}$) is split into a doublet by the single equatorial ^{19}F nucleus, and the signal from the latter is split into a triplet by the two axial ^{19}F nuclei (^{19}F is in 100% abundance). Thus, the pattern of ^{19}F resonances readily distinguishes this unsymmetrical structure from trigonal-planar and trigonal-pyramidal structures, both of which would have equivalent F nuclei and hence a single ^{19}F resonance.

---

local magnetic fields they produce. Measurement of this difference in chemical shift may be used to determine the number of unpaired electrons in the paramagnetic species (see the later section, Magnetometry and Magnetic Susceptibility).

(c) Spin–spin coupling

Structural assignment is often helped by the observation of the **spin–spin coupling**, which gives rise to multiplets in the spectrum due to interactions between nuclear spins. Spin–spin coupling arises when the orientation of the spin of a nearby nucleus affects the energy levels of the nucleus being probed, and thereby causes small changes in the latter's resonance frequency. The strength of spin–spin coupling, which is reported as the spin–spin coupling constant, J (in Hertz, Hz), and is independent of the applied magnetic field, decreases rapidly with distance through chemical bonds, and in many cases is greatest when the two atoms are directly bonded to each other. In **first-order spectra**, which are being considered here, the coupling constant is equal to the separation of adjacent lines in a multiplet. As can be seen in Fig. 8.25, $J(^1H{-}^{73}Ge) \approx 100$ Hz. The resonances of chemically equivalent nuclei do not display the effects of their mutual spin–spin coupling. Thus, a single ^{1}H signal is observed for the CH_3I molecule even though there is coupling between the H nuclei.

A multiplet of $2I + 1$ lines is obtained when a nucleus of spin $\frac{1}{2}$ (or a set of symmetry-related $I = \frac{1}{2}$ nuclei) is coupled to a nucleus of spin I. In the spectrum of GeH_4 shown in Fig. 8.25, the single central line arises from the four equivalent H nuclei in GeH_4 molecules that contain Ge isotopes with $I = 0$. This central line is flanked by ten evenly spaced but less intense lines that arise from a small fraction of GeH_4 that contains ^{73}Ge, for which $I = \frac{9}{2}$; the four ^{1}H nuclei are coupled to the ^{73}Ge nucleus to yield a 10-line multiplet $(2 \times \frac{9}{2} + 1 = 10)$.

The coupling of the nuclear spins of different isotopes is called **heteronuclear coupling**; the Ge–H coupling just discussed is an example. **Homonuclear coupling** between nuclei of the same isotope is detectable when the nuclei are in chemically inequivalent locations.

The sizes of ^{1}H–^{1}H homonuclear coupling constants in organic molecules are typically 18 Hz or less. By contrast, ^{1}H–X heteronuclear coupling constants can be several hundred hertz. Homonuclear and heteronuclear coupling between nuclei other than ^{1}H can lead to coupling constants of many kilohertz. The sizes of coupling constants can often be related to the geometry of a molecule by noting empirical trends. In square-planar Pt(II) complexes, J(Pt–P) is sensitive to the group *trans* to a phosphine ligand, and the value of J(Pt–P) increases in the following order of *trans* ligands:

$$R^-, H^-, PR_3, NH_3, Br^-, Cl^-$$

For example, *cis*-$[PtCl_2(PEt_3)_2]$, where Cl^- is *trans* to P, has J(Pt–P) = 3.5 kHz, whereas *trans*-$[PtCl_2(PEt_3)_2]$, with P *trans* to P, has J(Pt–P) = 2.4 kHz. These systematic variations allow *cis* and *trans* isomers to be distinguished quite readily. The rationalization for the variation in the sizes of the coupling constants above stems from the fact that a ligand that exerts a large *trans* influence (Section 23.4) substantially weakens the bond *trans* to itself, causing a reduction in the NMR coupling between the nuclei.

In complex molecules, coupling between nuclei can give rise to very large numbers of overlapping resonances in the NMR spectrum, making their assignment to a specific nucleus in the molecule difficult. In such cases the NMR spectrum can sometimes be simplified by applying a technique known as **broad band decoupling**. In this technique, irradiation of the sample with a wide range of frequencies corresponding to the resonance frequency range of a coupled nucleus (denoted X) causes it to reorient rapidly; this averages the possible interactions with the NMR nucleus under study (denoted A) and removes the spin–spin coupling between the **A** and **X** nuclei, producing a much simpler A-nucleus NMR spectrum. NMR spectra of nucleus **A** collected in this manner are denoted using the terminology A{X}. Thus, a proton-decoupled phosphorus NMR spectrum would be described as the ^{31}P{^{1}H} NMR spectrum. The ^{31}P-NMR spectrum of phosphine, PH_3, is a quartet due to $J(^1H{-}^{31}P)$ spin–spin coupling, but the ^{31}P{^{1}H} NMR spectrum shows just a single resonance.

(d) Intensities

The integrated intensity of a signal arising from a group of chemically equivalent nuclei is proportional to the number of nuclei in the group. Provided sufficient time is allowed during spectrum acquisition for the full relaxation of the observed nucleus, integrated intensities can be used to aid spectral assignment for most nuclei. (However, for nuclei with low sensitivities (such as ^{13}C), allowing sufficient time for full relaxation may not be realistic, so quantitative information from signal intensities is difficult to obtain.) For instance, in the spectrum of ClF$_3$ the integrated ^{19}F intensities are in the ratio 2:1 (for the doublet and triplet, respectively). The pattern of ^{19}F resonances indicates the presence of two equivalent F nuclei and one inequivalent F nucleus and distinguishes this less symmetrical structure from a trigonal-planar structure, D_{3h}, which would have a single resonance with all F environments equivalent.

The relative intensities of the $2N + 1$ lines in a multiplet that arises from coupling to N equivalent $I = \frac{1}{2}$ nuclei are given by Pascal's triangle (**1**); thus three equivalent protons give a 1:3:3:1 quartet. Groups of nuclei with higher spin quantum numbers give different patterns. The ^{1}H-NMR spectrum of HD, for instance, consists of three lines of equal intensity as a result of coupling to the ^{2}H nucleus ($I = 1$, with $2I + 1 = 3$ orientations).

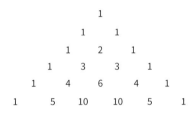

1 Pascal's triangle

EXAMPLE 8.5 Interpreting an NMR spectrum

Explain why (i) the ^{19}F-NMR spectrum of SF$_4$ consists of two 1:2:1 triplets of equal intensity and (ii) the ^{77}Se-NMR spectrum of SeF$_4$ consists of a triplet of triplets (^{77}Se, $I = \frac{1}{2}$).

Answer (i) We need to recall that an SF$_4$ molecule (**2**) has a 'see-saw' structure based on a trigonal-bipyramidal arrangement of electron pairs with a lone pair occupying one of the equatorial sites. The two axial F nuclei are chemically different from the two equatorial F nuclei and give two signals of equal intensity. The signals are in fact 1:2:1 triplets, as each ^{19}F nucleus couples to the two chemically distinct ^{19}F nuclei. (ii) Noting that SeF$_4$ has the same molecular geometry as SF$_4$ the selenium nucleus will couple to two environmentally different fluorine nuclei, equatorial and axial; coupling to one type with J(Se–F$_{eq}$) will produce a triplet and each of these resonances will be split into a triplet with J(Se–F$_{ax}$), producing a triplet of triplets.

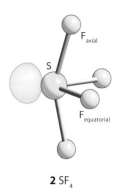

2 SF$_4$

Self-test 8.5 (a) Describe the expected form of the ^{19}F NMR spectrum of BrF$_5$. (b) Use the isotope information in Table 8.5 to explain why the ^{1}H resonance of the hydrido (H$^-$) ligand in cis-[Rh(CO)H(PMe$_3$)$_2$] consists of eight lines of equal intensity.

(e) Fluxionality

The timescale of NMR is slow in the sense that structures can be resolved provided their lifetime is not less than a few milliseconds. For example, [Fe(CO)$_5$] shows just one ^{13}C resonance, indicating that, on the NMR timescale, all five CO groups are equivalent. However, the IR spectrum (of timescale about 1 ps) shows distinct axial and equatorial CO groups, and by implication a trigonal-bipyramidal structure. The observed ^{13}C-NMR spectrum of [Fe(CO)$_5$] is the weighted average of these separate resonances.

Because the temperature at which NMR spectra are recorded can easily be changed, samples can often be cooled down to a temperature at which the rate of interconversion becomes slow enough for separate resonances to be observed. Fig. 8.27, for instance, shows the idealized ^{31}P-NMR spectra of [RhMe(PMe$_3$)$_4$] (**3**) at room temperature and at −80°C. At low temperatures the spectrum consists of a doublet of doublets of relative intensity 3 near $\delta = -24$, which arises from the equatorial P atoms (coupled to ^{103}Rh and the single axial ^{31}P), and a quartet of doublets of intensity 1, derived from the axial P atom (coupled to ^{103}Rh and the three equatorial ^{31}P atoms). At room temperature the scrambling of the PMe$_3$ groups makes them all equivalent and a doublet is observed (from coupling to ^{103}Rh).

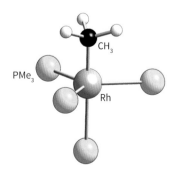

3 [RhMe(PMe$_3$)$_4$]

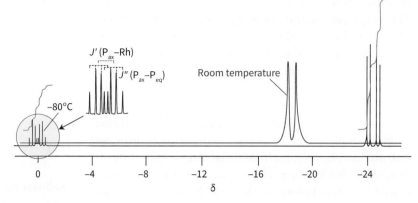

FIGURE 8.27 The ^{31}P-NMR spectra of [RhMe(PMe$_3$)$_4$] (**3**) at room temperature and at −80°C.

Collection of spectra as a function of temperature allows the determination of the point at which the spectrum changes from the high-temperature form to the low-temperature form ('the coalescence temperature'). Further detailed analysis of these variable-temperature NMR data allows extraction of a value for the energy barrier to interconversion.

(f) Solid-state NMR

The NMR spectra of solids rarely show the high resolution that can be obtained from solution NMR. This difference is mainly due to anisotropic interactions, such as dipolar magnetic couplings between nuclei where the individual magnetic moments on neighbouring nuclei (having different fixed positions with respect to an external magnetic field) modify the local magnetic field experienced by the nuclei being probed. Note that these anisotropic interactions are averaged in solution due to rapid molecular tumbling on the NMR timescale. These effects mean that in the solid state chemically equivalent nuclei usually experience a variety of magnetic environments and so have different resonance frequencies. A typical result of this range of additional couplings is to produce an NMR spectrum that has very broad resonances, often more than 10 kHz wide.

To average out anisotropic interactions, samples are spun at very high speeds (typically 10–25 kHz) at the 'magic angle' (54.7°) with respect to the field axis; at this angle the anisotropic interactions, which vary as $(1 - 3\cos^2\theta)$, are averaged to zero. This so-called **magic-angle spinning** (MAS) reduces the effect of anisotropy substantially and removes dipolar (spin–spin) couplings. Resonance signals are still significantly broadened compared to those obtained in solution NMR spectra. The broadening of signals can sometimes be so great that signal widths are comparable to the chemical shift range for some nuclei. This broadening is a particular problem for ^{1}H, which has a small chemical shift range of $\Delta\delta = 10$. Broad signals are less of a problem for nuclei such as ^{195}Pt, for which the range in chemical shifts is $\Delta\delta = 16\,000$, although this large range can be reflected in large anisotropic linewidths. Quadrupolar nuclei (those with $I > \tfrac{1}{2}$) present additional problems as the peak position becomes field-dependent, and unless the nucleus is in a high-symmetry environment, for example, tetrahedral or octahedral, the peak can no longer be identified with the chemical shift.

Despite these difficulties, developments in the technique have made possible the observation of high-resolution NMR spectra for solids and are of far-reaching importance in many areas of chemistry. An example is the use of ^{29}Si-MAS-NMR to determine the environments of Si atoms in natural and synthetic silicates and aluminosilicates, such as zeolites (Fig. 8.28).

The techniques of homonuclear and heteronuclear 'decoupling' enhance the resolution of spectra and the use of multiple pulse sequences has allowed the observation of spectra with some difficult samples. The high-resolution technique CPMAS-NMR, a combination of MAS with **cross-polarization**

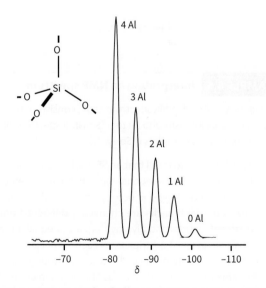

FIGURE 8.28 An example of a ^{29}Si-MAS-NMR spectrum obtained from the aluminosilicate zeolite analcine; each resonance represents a different silicon environment in the zeolite structure Si(OAl)$_{4-n}$(OSi)$_n$ as n varies from 0 to 4.

(CP), usually with heteronuclear decoupling, has been used to study many compounds containing ^{13}C, ^{31}P, and ^{29}Si. In cross-polarization an abundant, high-sensitivity NMR nucleus is excited by an initial radiofrequency pulse and some of the magnetization energy is transferred to a neighbouring low sensitivity nucleus. As a result the observation of the NMR relaxation of this low-abundance nucleus becomes more facile. The technique is often used to study molecular compounds in the solid state. For example, the ^{13}C-CPMAS spectrum of [Fe$_2$(C$_8$H$_8$)(CO)$_5$] at −160°C indicates that all C atoms in the C$_8$ ring are equivalent on the timescale of the experiment. The interpretation of this experimental observation is that the molecule is fluxional even in the solid state.

EXAMPLE 8.6 Interpreting an MAS-NMR spectrum

The ^{29}Si-MAS-NMR spectrum of the mineral bentonite, Ba^{2+}Ti^{4+}[Si$_3$O$_9$]$^{6-}$, shows a single resonance while that of (Mg^{2+})$_4$[Si$_3$O$_{10}$]$^{8-}$ consists of two resonances with intensity ratio 2:1. Propose structures for the [Si$_3$O$_9$]$^{6-}$ and [Si$_3$O$_{10}$]$^{8-}$ anions consistent with these observations. (^{29}Si, $I = \tfrac{1}{2}$, 5% abundant.)

Answer MAS-NMR does not observe dipolar couplings so the number of resonances in the spectrum will tell us how many different environments are present for the nucleus under investigation. For the [Si$_3$O$_9$]$^{6-}$ anion we need to propose a structure that has all of the silicon nuclei in the anion in identical environments; this can only be achieved in the cyclic species (**4**). For the [Si$_3$O$_{10}$]$^{8-}$ anion with the three silicon nuclei, the resonance data tell us that two of the nuclei are equivalent and one is in a unique environment; this can be achieved for the linear species (**5**).

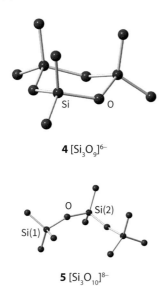

4 [Si$_3$O$_9$]$^{6-}$

5 [Si$_3$O$_{10}$]$^{8-}$

Self-test 8.6 Predict the ^{29}Si-MAS-NMR spectrum of (a) Tm$_4$(SiO$_4$)(Si$_3$O$_{10}$), where the trisilicate anion is linear, and (b) K$_2$Ta$_2$Si$_4$O$_{14}$, which contains the cyclic [Si$_4$O$_{12}$]$^{8-}$ anion.

8.7 Electron paramagnetic resonance

KEY POINTS Electron paramagnetic resonance spectroscopy is used to study compounds possessing unpaired electrons, particularly those containing a d-block element; it is often the technique of choice for identifying and studying metals such as Fe and Cu at the active sites of metalloenzymes.

Electron paramagnetic resonance (EPR; or electron spin resonance, ESR) spectroscopy, the observation of resonant absorption by unpaired electrons in a magnetic field, is a technique for studying paramagnetic species such as organic and main-group radicals. Its principal importance in inorganic chemistry is for characterizing compounds containing the d- and f-block elements.

The simplest case is for a species having one unpaired electron ($S = \tfrac{1}{2}$): by analogy with NMR, the application of an external magnetic field B_0 produces a difference in energy between the $m_s = +\tfrac{1}{2}$ and $m_s = -\tfrac{1}{2}$ states of the electron, with

$$\Delta E = g\mu_B B_0 \qquad (8.7)$$

where μ_B is the Bohr magneton and g is a numerical factor known simply as the **g value** (Fig. 8.29).

The conventional method of recording EPR spectra is to use a continuous-wave (CW) spectrometer (Fig. 8.30), in which the sample is irradiated with a constant microwave frequency and the applied magnetic field is varied. The resonance frequency for most spectrometers is approximately 9 GHz, and the instrument is then known as an 'X-band spectrometer'. The magnetic field in an X-band spectrometer is about 0.3 T. Laboratories specializing in EPR spectroscopy often have a range of instruments, each operating at different fields. Thus an S-band (resonant frequency 3 GHz) spectrometer and particularly those operating at high fields, Q-band (35 GHz) and W-band (95 GHz) spectrometers, are used to complement the information gained with an X-band spectrometer.

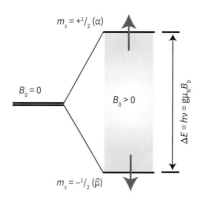

FIGURE 8.29 When an unpaired electron is in a magnetic field, its two orientations (α, $m_s = +\tfrac{1}{2}$, and β, $m_s = -\tfrac{1}{2}$) have different energies. Resonance is achieved when the energy separation matches the energy of the incident microwave photons.

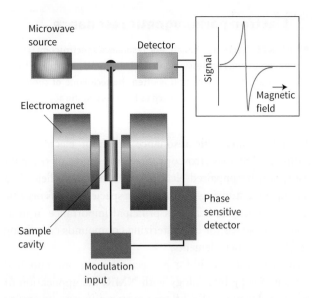

FIGURE 8.30 The layout of a typical continuous wave EPR spectrometer.

Pulsed EPR spectrometers are increasingly in use and offer new opportunities analogous to the way that pulsed Fourier-transform techniques have revolutionized NMR. Pulsed-EPR techniques provide time resolution, making it possible to measure the dynamic properties of paramagnetic systems.

(a) The g value

For a free electron $g = 2.0023$ but in compounds this value is altered by spin–orbit coupling which changes the local magnetic field experienced by the electron. For many species, particularly d-metal complexes, g values may be highly anisotropic so that the resonance condition depends on the angle that the paramagnetic species makes with the applied field (Fig. 8.31). The illustration shows the EPR spectra of frozen solutions or 'glasses', expected for isotropic (all three g values the same along perpendicular axes), axial (two the same, $g_\perp \neq g_\parallel$), and rhombic (all three different, $g_{zz} \neq g_{yy} \neq g_{xx}$) spin systems.

The sample (which is usually contained in a quartz tube) comprises the paramagnetic species in dilute form, either in the solid state (doped crystals or powders) or in solution. Relaxation is so efficient for d-metal ions that the spectra are often too broad to detect; consequently, liquid nitrogen and sometimes liquid helium are used to cool the sample. Frozen solutions behave as amorphous powders, so resonances are observed at all the g values of the compound, somewhat analogous to powder X-ray diffraction (Section 8.1). More detailed studies can be made with oriented single crystals. Provided relaxation is slow, EPR spectra can also be observed at room temperature in liquids.

Spectra can also be obtained for systems having more than one unpaired electron, such as triplet states, but the theoretical background is much more complicated. Whereas species

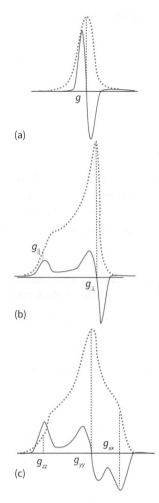

FIGURE 8.31 The forms of EPR powder (frozen solution) spectra expected for different types of g-value anisotropy (a), (b), and (c). The dotted green line is the absorption and the solid red line the first derivative of the absorption (its slope). For technical reasons related to the CW instrument detection technique, the first derivative is normally observed in EPR spectrometers.

having an odd number of electrons are usually detectable, it may be very difficult to observe spectra for systems having an even number of electrons. Table 8.6 shows the suitability of common paramagnetic species for EPR detection.

(b) Hyperfine coupling

The **hyperfine structure** of an EPR spectrum, the multiplet structure of the resonance lines, is due to the coupling of the electron spin to any magnetic nuclei present. A nucleus with spin I splits an EPR line into $2I + 1$ lines of the same intensity (Fig. 8.32).

A distinction is sometimes made between the hyperfine structure due to coupling to the nucleus of the atom on which the unpaired electron is primarily located (Fig. 8.33) and the 'superhyperfine coupling', the coupling to ligand nuclei. Superhyperfine coupling to ligand nuclei is used to measure the extent of electron delocalization and covalence in metal complexes.

TABLE 8.6 EPR detectability of common d-metal ions

Usually easy to study		Usually difficult to study or diamagnetic	
Species	S	Species	S
Ti(III)	$\tfrac{1}{2}$	Ti(II)	1
Cr(III)	$\tfrac{3}{2}$	Ti(IV)	0
V(IV)	$\tfrac{1}{2}$	Cr(II)	2
Fe(III)	$\tfrac{1}{2}, \tfrac{5}{2}$	V(III)	1
Co(II)	$\tfrac{3}{2}, \tfrac{1}{2}$	V(V)	0
Ni(III)	$\tfrac{3}{2}, \tfrac{1}{2}$	Fe(II)	2, 1, 0
Ni(I)	$\tfrac{1}{2}$	Co(III)	0
Cu(II)	$\tfrac{1}{2}$	Co(I)	0
Mo(V)	$\tfrac{1}{2}$	Ni(II)	0, 1
W(V)	$\tfrac{1}{2}$	Cu(I)	0
		Mo(VI)	0
		Mo(IV)	1, 0
		W(VI)	0

EXAMPLE 8.7 Interpreting superhyperfine coupling

Use the data provided in Table 8.5 to suggest how the EPR spectrum of a Co(II) complex might change when a single OH^- ligand is replaced by an F^- ligand.

Answer We need to note that ^{19}F (100% abundance) has nuclear spin $I = \tfrac{1}{2}$ whereas ^{16}O (close to 100%) has $I = 0$. Consequently, any part of the EPR spectrum might be split into two lines.

Self-test 8.7 Predict how you might show that an EPR signal that is characteristic of a new material arises from tungsten sites.

By using spectrometers operating at different fields (X, Q, W, etc.) it is possible to distinguish between spectral features that are due to g-value anisotropy (which becomes more spread out at higher field) and features that are due to hyperfine coupling (which becomes relatively less significant at higher field).

8.8 Mössbauer spectroscopy

KEY POINT Mössbauer spectroscopy is based on the resonant absorption of γ-radiation by nuclei and exploits the fact that nuclear energies are sensitive to the electronic and magnetic environment.

The **Mössbauer effect** makes use of recoil-less absorption and emission of γ-radiation by a nucleus. To understand what is involved, consider a radioactive ^{57}Co nucleus that decays by electron capture to produce an excited state of ^{57}Fe, denoted $^{57}Fe^{**}$ (Fig. 8.34). This nuclide decays to another

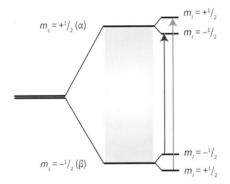

FIGURE 8.32 When a magnetic nucleus is present, its $2I + 1$ orientations give rise to a local magnetic field that splits each spin state of an electron into $2I + 1$ levels. The allowed transitions ($\Delta m_s = +1$, $\Delta m_I = 0$) give rise to the hyperfine structure of an EPR spectrum.

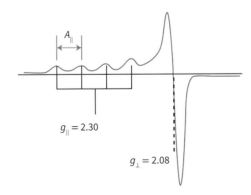

FIGURE 8.33 The EPR spectrum of Cu^{2+} (d^9, one unpaired electron) in frozen aqueous solution. The tetragonally distorted Cu^{2+} ion shows an axially symmetrical spectrum in which hyperfine coupling to Cu ($I = \tfrac{3}{2}$, four hyperfine lines) is clearly evident for the $g_\parallel$ component.

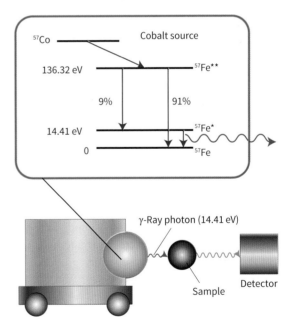

FIGURE 8.34 The layout of a Mössbauer spectrometer. The speed of the carriage is adjusted until the Doppler-shifted frequency of the emitted γ-ray matches the corresponding nuclear transition in the sample. The inset shows the nuclear transitions responsible for the emission of the γ-ray.

excited state, denoted ^{57}Fe*, that lies 14.41 eV above the ground state and which emits a γ-ray of energy 14.41 eV as it decays. If the emitting nucleus (the source) is pinned down in a rigid lattice, the nucleus does not recoil and the radiation produced is highly monochromatic. Mössbauer spectroscopy is increasingly being carried out using synchrotron radiation in place of the Co source.

If a sample containing ^{57}Fe (which occurs with 2% natural abundance) is placed close to the source, the monochromatic γ-ray emitted by ^{57}Fe* can be expected to be absorbed resonantly. However, because changes in the exact electronic and magnetic environment affect the nuclear energy levels to a small degree, resonant absorption occurs only if the environment of the sample ^{57}Fe is chemically identical to that of the emitting ^{57}Fe* nucleus. It might seem that the energy of the γ-ray cannot easily be varied, but the Doppler effect can be exploited for moving the transmitting nucleus at a speed v relative to the sample and a frequency shift is induced of magnitude $\Delta\nu = (v/c)\nu_\gamma$. This shift is sufficient even for velocities of a few millimetres per second, to match the absorption frequency to the transmitted frequency. A **Mössbauer spectrum** is a portrayal of the resonant absorption peaks (vertical axis, intensity of absorption) that occur as the velocity of the source is changed (horizontal axis in mm s^{-1}).

The Mössbauer spectrum of a sample containing iron in a single chemical environment can be expected to consist of a single line due to absorption of radiation at the energy required (ΔE) to excite the nucleus from its ground state to the excited state. The difference between ΔE of the sample and that of metallic ^{57}Fe is called the **isomer shift** and is expressed in terms of the velocity that is required to achieve resonance by the Doppler shift. The value of ΔE depends on the magnitude of the electron density at the nucleus, and although this effect is primarily due to s electrons (because their wavefunctions are nonzero at the nucleus), shielding effects cause ΔE to be sensitive to the number of p and d electrons too. As a result, different oxidation states, such as Fe(II), Fe(III), and Fe(IV), can be distinguished as well as ionic and covalent bonding.

The element most suited for study by Mössbauer spectroscopy is iron, as a result of a number of factors which make the experiment relatively easy to undertake and the data obtained useful. These factors include a half-life of ^{57}Co of 272 days, so the γ-ray intensity produced by the source is relatively strong for over a year, and the absorption peaks are reasonably sharp and well resolved. Iron has great importance in chemistry and is common in minerals, oxides, alloys, and biological samples, for which many other characterization techniques—for example, NMR—are less effective. Mössbauer spectroscopy is also used to study some other nuclei, including ^{119}Sn, ^{129}I, and ^{197}Au, which have suitable nuclear energy levels and γ-emission half-lives.

A ^{57}Fe* nucleus has $I = \frac{3}{2}$ and so possesses an electric quadrupole moment (a nonspherical distribution of electric charge) that interacts with electric field gradients produced by the charge distribution around the nucleus. As a result, and providing the electronic environment of the nucleus is not isotropic, the Mössbauer spectrum splits into two lines of separation ΔE_Q (Fig. 8.35). The splitting is a good guide as to the state of Fe in proteins and minerals as it depends on the oxidation state and the distribution of d-electron density. Magnetic fields, produced by a large magnet or internally in some ferromagnetic materials, also cause changes in the energies of the various spin orientation states of an ^{57}Fe* nucleus. As a result, the Mössbauer spectrum splits into six lines, the allowed transitions between the levels derived from an $I = \frac{3}{2}$ state in a magnetic field ($m_I = +\frac{3}{2}, +\frac{1}{2}, -\frac{1}{2}, -\frac{3}{2}$) and an $I = \frac{1}{2}$ state ($m_I = +\frac{1}{2}, -\frac{1}{2}$) with $\Delta I = 1$, Fig. 8.35(c).

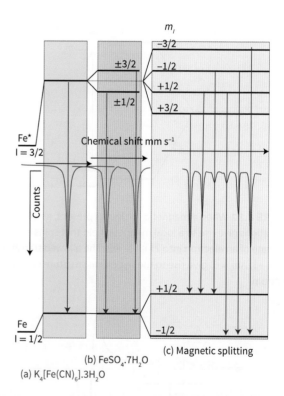

(a) K$_4$[Fe(CN)$_6$]·3H$_2$O (b) FeSO$_4$·7H$_2$O (c) Magnetic splitting

FIGURE 8.35 The effects of an electric field gradient and a magnetic field on the energy levels involved in the Mössbauer technique for Fe samples. The derived spectra show, from left to right, the origins of the isomer shift, quadrupole coupling, and magnetic hyperfine coupling (but no quadrupole splitting). The spectrum of (a) K$_4$Fe(CN)$_6$·3H$_2$O, octahedral low spin d^6, is representative of a highly symmetrical environment in [Fe(CN)$_6$]$^{4-}$ and shows a single peak with an isomer shift. The spectrum of (b) FeSO$_4$·7H$_2$O, d^6 in a nonsymmetrical environment, shows quadrupolar splitting. (c) The form of the spectrum of FeSO$_4$·7H$_2$O, produced in a magnetic field.

EXAMPLE 8.8 Interpreting a Mössbauer spectrum

The isomer shifts of Fe(II) compounds relative to metallic iron, Fe(0), are generally in the range +1 to +1.5 mm s^{-1}, whereas isomer shifts for Fe(III) compounds lie in the range +0.2 to +0.5 mm s^{-1}. Explain these values in terms of the electronic configurations of Fe(0), Fe(II), and Fe(III).

Answer The outermost electron configurations of Fe(0), Fe(II), and Fe(III) are (formally) $3d^64s^2$, $3d^6$, and $3d^5$, respectively. The s-electron density at the nucleus is decreased in Fe(II) compared with Fe(0), producing a large positive isomer shift. When a 3d electron is removed to produce Fe(III) from Fe(II), there is a small increase in s-electron density at the nucleus, as any 3d electron present partly screens the nucleus from the remaining inner s electrons (1s, 2s, and 3s; see Chapter 1). The associated reduction in s-electron density at the nucleus means that the isomer shift becomes slightly less positive for Fe(III) compared to Fe(II).

Self-test 8.8 Predict a likely isomer shift for iron in Sr_2FeO_4.

Ionization-based techniques

Ionization-based techniques measure the energies of products, electrons, or molecular fragments generated when a sample is ionized by bombardment with high-energy radiation or particles.

8.9 Photoelectron spectroscopy

KEY POINT Photoelectron spectroscopy is used to determine the energies and energy ordering of orbitals in molecules and solids by analysing the kinetic energies of photoejected electrons.

The basis of **photoelectron spectroscopy** (PES) is the measurement of the kinetic energies of electrons (photoelectrons) emitted by ionization of a sample that is irradiated with monochromatic, high-energy radiation (Fig. 8.36). It follows from the law of conservation of energy that the kinetic energy of the ejected photoelectrons, E_k, is related to the ionization energies, E_i, from their orbitals by the relation

$$E_k = h\nu - E_i \qquad (8.8)$$

where ν is the frequency of the incident radiation. **Koopmans' theorem** states that the ionization energy is equal to the negative of the orbital energy, so the kinetic energies of photoelectrons can be used to determine orbital energies. The theorem assumes that the energy involved in electron reorganization after ionization is offset by the increase in electron–electron repulsion energy as the orbital contracts. This approximation is usually taken as reasonably valid. Corrections are also often required to allow for the additional energy, the work function φ, required for an electron to escape from the surface of the material under investigation.

There are two major types of photoionization technique: **X-ray photoelectron spectroscopy** (XPS) and **ultraviolet photoelectron spectroscopy** (UPS). Although much more intense sources can be obtained using synchrotron beam lines, the standard laboratory source for XPS is usually a magnesium or aluminium anode that is bombarded by a high-energy electron beam. This bombardment results in radiation at 1.254 and 1.486 keV, respectively, due to the transition of a 2p electron into a vacancy in the 1s orbital caused by ejection of an electron. These energetic photons cause ionizations from core orbitals in other elements that are present in the sample; the ionization energies are characteristic of the element and its oxidation state. Because the linewidth is high (usually 1–2 eV), XPS is not suitable for probing fine details of valence orbitals but can be used to study the band structures of solids. The mean free path of electrons in a solid is only about 1 nm, so XPS is suitable for surface elemental analysis, and in this application it is commonly known as **electron spectroscopy for chemical analysis** (ESCA).

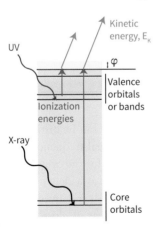

FIGURE 8.36 In photoelectron spectroscopy, high-energy electromagnetic radiation (UV for the ejection of valence and valence band electrons, X-ray for core electrons) expels an electron from its orbital, and the kinetic energy of the photoelectron is equal to the difference between the photon energy and the ionization energy of the electron allowing for corrections due to the work function, φ.

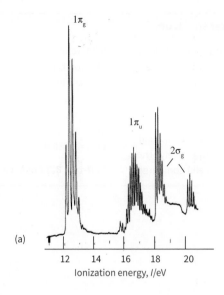

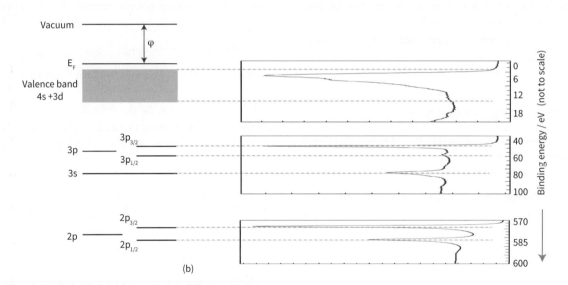

FIGURE 8.37 (a) The UV photoelectron spectrum of O_2. Loss of an electron from $2\sigma_g$ (see the MO energy-level diagram, Fig. 2.12) gives rise to two bands because the unpaired electron that remains can be parallel or antiparallel to the two unpaired electrons in the $1\pi_g$ orbitals. (b) A schematic combined UV and X-ray photoelectron spectrum from Cr metal and its band structure showing the origin of the observed PES bands.

The source for UPS is typically a helium discharge lamp that emits He(I) radiation (21.22 eV) or He(II) radiation (40.8 eV). Linewidths are much smaller than in XPS, so the resolution is far greater. The technique is used to study valence-shell energy levels, and the vibrational fine structure often provides important information on the bonding or antibonding character of the orbitals from which electrons are ejected (Fig. 8.37(a)). When the electron is removed from a nonbonding orbital the product is formed in its vibrational ground state, and a narrow line is observed. However, when the electron is removed from a bonding or antibonding orbital the resulting ion is formed in several different vibrational states and extensive fine structure is observed. Bonding and antibonding orbitals can be distinguished by determining whether the vibrational frequencies in the resulting ion are higher or lower than for the original molecule.

The PES technique is also widely used for the study of the electronic structure of solids and investigation of their valence bands. Valence bands lie at or close to the Fermi level, E_F, in metals and many semiconductors, and excitations from these bands are observed as broad features in the PES spectrum at low binding energies, Fig. 8.37(b). The surfaces of solids, which often differ from the bulk material forming—for example, oxides—are also readily studied by the PES method and angle-resolved PES techniques allow depth profiling of the electronic levels within the surface region.

Another useful aid is the comparison of photoelectron intensities for a sample irradiated with He(I) and He(II). The higher-energy source preferentially ejects electrons from d or f orbitals, allowing these contributions to be distinguished from s and p orbitals, for which He(I) causes higher intensities. The origin of this effect lies in differences in absorption cross-sections (see Further Reading).

8.10 X-ray absorption spectroscopy

KEY POINT X-ray absorption spectra can be used to determine the oxidation state of an element in a compound and to investigate its local environment.

As mentioned in Section 8.9, the intense X-ray radiation from synchrotron sources may be used to eject electrons from the cores of elements present in a compound. **X-ray absorption spectra** (XAS) are obtained by varying the photon energy across a range of energies at which electrons in the various atoms present in a compound can be excited and ionized (typically between 0.1 and 100 keV). The characteristic absorption energies correspond to the binding energies of different inner-shell electrons of the various elements present. Thus the frequency of an X-ray beam may be swept across an absorption edge of a selected element, and information on the oxidation state and neighbourhood of this chosen chemical element obtained.

Fig. 8.38 shows a typical X-ray absorption spectrum. Each region of the spectrum can provide different useful information on the chemical environment of the element under investigation:

1. Just prior to the absorption edge is the 'pre-edge', where core electrons are excited to higher empty orbitals but not ejected. This 'pre-edge structure' can provide information on the energies of excited electronic states and also on the local symmetry of the atom. It is more intense for low-symmetry environments such as those that do not have a centre of symmetry.

2. In the edge region, where the photon energy, E, is between E_i and $E_i + 10$ eV, where E_i is the ionization energy, the 'X-ray absorption near-edge structure' (XANES) is observed. Information that can be extracted from the XANES region includes oxidation state and the coordination environment, including any subtle geometrical distortions. The near-edge structure can also be used as a 'fingerprint', as it is characteristic of a specific environment and valence state.

3. The 'near-edge X-ray absorption fine structure' (NEXAFS) region lies between $E_i + 10$ eV and $E_i + 50$ eV. It has particular application to chemisorbed molecules on surfaces because it is possible to infer information about the orientation of the adsorbed species.

4. The 'extended X-ray absorption fine structure' (EXAFS) region lies at energies greater than $E_i + 50$ eV. The photoelectrons ejected from a particular atom by absorption of X-ray photons with energies in this region may be backscattered by any adjacent atoms. This effect can result in an interference pattern that is detected as periodic variations in intensity at energies just above the absorption edge. In EXAFS these variations are analysed to reveal the nature (in terms of their electron density) and number of nearby atoms and the distance between the absorbing atom and the scattering atom. An advantage of this method is that it can provide bond lengths in amorphous samples and for species in solutions—systems that are not amenable to study using diffraction techniques.

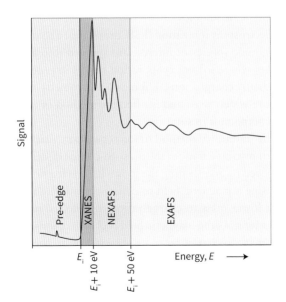

FIGURE 8.38 A typical X-ray absorption-edge spectrum defining the various regions discussed in the text.

EXAMPLE 8.9 Information from pre-edge and edge features in XAS spectra

Interpret the following data which give the energy (eV) of the main pre-edge feature in Mn K-edge XAS (the K-edge represents ionization of the 1s electron):

Mn(II)	Mn(III)	Mn(IV)	Mn(V)	Mn(VI)	Mn(VII)
6540.6	6541.0	6541.5	6542.1	6542.5	6543.8

An oxide containing manganese showed a pre-edge feature in its XAS spectrum consisting of peaks at 6540.6 eV (intensity 1) and 6540.9 eV (intensity 2). Explain the observed variation in the energy of the pre-edge feature and propose a formula for the manganese oxide.

Answer We need to understand how the energies of the orbitals involved in producing the pre-edge feature change with increasing manganese oxidation state. The pre-edge feature at the K-edge results from excitation of the 1s electron to a higher-energy orbital such as an empty 3d orbital (there may be several overlapping features corresponding to different excited states). The higher the oxidation state of manganese, the greater the effective nuclear charge experienced by the 1s electron and the greater the energy needed to excite or ionize it. This relationship is reflected in the experimentally observed monotonic increase in pre-edge feature energy from Mn(II) to Mn(VII). The manganese oxide shows pre-edge features at energies corresponding to those of Mn(II) and Mn(III) with the intensity ratio 1:2, so the oxide is likely to be $MnO:Mn_2O_3 = Mn_3O_4$.

Self-test 8.9 Describe the expected trend in the energy of the XAS K-edge for sulfur in compounds with oxidation states from S^{2-} (sulfide) to SO_4^{2-} (sulfate, S(VI)).

8.11 Mass spectrometry

KEY POINT Mass spectrometry is a technique for determining the mass of a molecule and of its fragments.

Mass spectrometry measures the mass-to-charge ratio of gaseous ions. The ions can be either positively or negatively charged, and it is normally trivial to infer the actual charge on an ion and hence the mass of a species. It is a destructive analytical technique because the sample cannot be recovered for further analysis.

The precision of measurement of the mass of the ions varies according to the use being made of the spectrometer (Fig. 8.39). If all that is required is a crude measure of the mass, for instance to within $\pm 1 m_u$ (where m_u is the atomic mass constant, 1.66054×10^{-27} kg), then the resolution of the mass spectrometer need be of the order of only 1 part in 10^4. In contrast, to determine the mass of individual atoms so that the **mass defect** (the difference between the atomic mass and the mass of the constituent protons and neutrons due to the nuclear binding energy) can be determined, the precision must approach 1 part in 10^{10}. With a mass spectrometer of this precision, molecules of nominally the same mass such as $^{12}C^{16}O$ (of mass $27.9949 m_u$) can be distinguished from $^{14}N_2$ (of mass $28.0061 m_u$), and the elemental and isotopic composition of ions of nominal mass less than $1000 m_u$ may be determined unambiguously.

(a) Ionization and detection methods

The major practical challenge with mass spectrometry is the conversion of a sample into gaseous ions. Typically, less than a milligram of compound is used. Many different experimental arrangements have been devised to produce gas-phase ions but all suffer from a tendency to fragment the compound of interest.

Electron impact ionization (EI) relies on bombarding a sample with high-energy electrons to cause both vaporization and ionization. A disadvantage is that EI tends to induce considerable decomposition in larger molecules.

Fast atom bombardment (FAB) is similar to EI, except that bombardment of the sample with fast neutral atoms is used to vaporize and ionize the sample; it induces less fragmentation than EI.

Matrix-assisted laser desorption/ionization (MALDI) is similar to EI, except that a short laser pulse is used to the same effect; this technique is particularly effective with solid and polymeric samples.

In **electrospray ionization** (ESI), droplets of a solution containing ionic species of interest are sprayed into a vacuum chamber where solvent evaporation results in generation of individually charged ions; ESI mass spectrometry is becoming more widely used and is often the method of choice for ionic compounds in solution.

In the traditional method of ion separation, rarely used in modern instrumentation, the mass-to-charge discrimination is achieved by accelerating the ions with an electric field and then using a magnetic field to deflect the moving ions: ions with a lower mass-to-charge ratio are deflected more than heavier ions. As the magnetic field is changed, ions with different mass-to-charge ratio are directed on to the detector (Fig. 8.39(a)). A **quadrupole mass filter** employs an oscillating electric field between four parallel rods to separate the ions; ions can only pass through the field and onto the detector at a specific applied field which can be swept rapidly to record all the ion m/z values present (Fig. 8.39(b)).

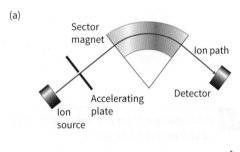

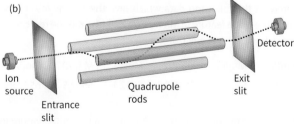

FIGURE 8.39 (a) A magnetic sector mass spectrometer. The molecular fragments are deflected according to their mass-to-charge ratio, allowing for separation at the detector. (b) A quadrupole mass spectrometer where ions are selected to reach the detector by adjusting the electric field produced by four charged rods.

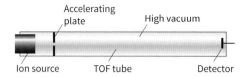

FIGURE 8.40 A time-of-flight (TOF) mass spectrometer. The molecular fragments are accelerated to different speeds by the potential difference and arrive at different times at the detector.

In a **time-of-flight** (TOF) mass spectrometer, the ions from a sample are accelerated by an electric field for a fixed time and then allowed to fly freely (Fig. 8.40). Because the force on all the ions of the same charge is the same, the lighter ions are accelerated to higher speeds than the heavier ions and strike a detector sooner.

In an **ion cyclotron resonance** (ICR) mass spectrometer (often denoted FTICR, for Fourier transform-ICR), ions are collected in a small cyclotron cell inside a strong magnetic field. The ions circle round in the magnetic field, effectively behaving as an electric current. Because an accelerated current generates electromagnetic radiation, the signal generated by the ions can be detected and used to establish their mass-to-charge ratio.

Mass spectrometry is most widely used in organic chemistry but is also very useful for the analysis of many classes of inorganic compounds such as organometallics and metal cofactors in enzymes. However, inorganic compounds with ionic structures or covalently bonded networks (for example, SiO_2), are not volatile and do not fragment into molecular ion units (even with the MALDI technique), so cannot generally be analysed by this method. Conversely, the weaker bonding in some inorganic coordination compounds means that they fragment very easily in the mass spectrometer which often makes determination of the parent molecular ion species problematic.

(b) Interpretation

Fig. 8.41 shows a typical mass spectrum. To interpret a spectrum, it is helpful to detect a peak corresponding to the singly charged, intact molecular ion (parent ion). Sometimes a peak occurs at half the molecular mass and is then ascribed to a doubly charged ion. Peaks from multiply charged ions are usually easy to identify because the separation between the peaks from the different isotopomers is no longer m_u but fractions of that mass. For instance, in a doubly charged ion, isotopic peaks are $\frac{1}{2}m_u$ apart, in a triply charged ion they are $\frac{1}{3}m_u$ apart, and so on.

In addition to indicating the mass of the molecule or ion that is being studied (and hence its molar mass), a mass spectrum also provides information about fragmentation pathways of molecules. This information can be used to confirm structural assignments. For example, complex ions

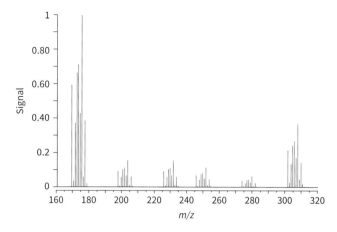

FIGURE 8.41 The mass spectrum of $[Mo(\eta^6\text{-}C_6H_6)(CO)_2PMe_3]$. See Example 8.10 for the detailed interpretation of the peaks in this spectrum.

often lose ligands and peaks are observed that correspond to the complete ion less one or more ligands.

Multiple peaks are observed when an element is present as a number of isotopes (for instance, chlorine is 75.5% ^{35}Cl and 24.5% ^{37}Cl). Thus, for a molecule containing chlorine, the mass spectrum will show two peaks $2m_u$ apart in an intensity ratio of about 3:1. Different patterns of peaks are obtained for elements with a more complex isotopic composition and can be used to identify the presence of an element in compounds of unknown composition. An Hg atom, for instance, has six isotopes in significant abundance (Fig. 8.42). The actual proportion of isotopes of an element varies according to its geographic source, and this subtle aspect is easily identified with high-resolution mass spectrometers. Thus, the precise determination of the proportions of isotopes can be used to determine the source of a sample.

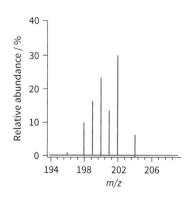

FIGURE 8.42 The mass spectrum of a sample containing mercury, showing the isotopic composition of the atoms.

EXAMPLE 8.10 Interpreting a mass spectrum

Fig. 8.41 shows part of the mass spectrum of $[Mo(\eta^6\text{-}C_6H_6)(CO)_2PMe_3]$. Assign the main peaks.

Answer The complex has an average molecular mass of $306m_u$, but a simple molecular ion is not seen because Mo has a large number of isotopes. Ten peaks centred on $306m_u$ are detected. The most abundant isotope of Mo is ^{98}Mo (24%) and the ion that contains this isotope has the highest intensity of the peaks of the molecular ion. In addition to the peaks representing the molecular ion, peaks at $M^+ - 28$, $M^+ - 56$, $M^+ - 76$, $M^+ - 104$, and $M^+ - 132$ are seen. These peaks represent the loss of one CO, two CO, PMe_3, PMe_3 + CO, and PMe_3 + 2CO ligands, respectively, from the parent compound.

Self-test 8.10 Explain why the mass spectrum of $ClBr_3$ (Fig. 8.43) consists of five peaks separated by $2m_u$.

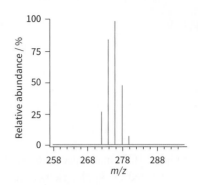

FIGURE 8.43 The mass spectrum of $ClBr_3$.

Chemical analysis

A classic application of physical techniques is the determination of the elemental composition of compounds. The techniques now available are highly sophisticated and in many cases can be automated to achieve rapid, reliable results. In this section we include thermal techniques that can be used to follow the phase changes of substances without change of composition, as well as processes that result in changes of composition. In each of these techniques the compound is usually destroyed during the analysis.

8.12 Atomic absorption spectroscopy

KEY POINT Almost every metallic element can be determined quantitatively by using the spectral absorption characteristics of atoms.

The principles of atomic absorption spectroscopy (AAS) are similar to those of UV–visible spectroscopy except that the absorbing species are free atoms, either neutral or charged. Unlike molecules, atoms do not have rotational or vibrational energy levels, and the only transitions that occur are between electronic energy levels. Consequently, atomic absorption spectra consist of sharply defined lines rather than the broad bands typical of molecular spectroscopy.

Fig. 8.44 shows the basic components of an atomic absorption spectrophotometer. The gaseous sample is exposed to radiation of a specific wavelength from a 'hollow cathode' lamp, which consists of a cathode constructed of a particular element and a tungsten anode in a sealed tube filled with neon. If a particular element is present in the sample, the radiation emitted by the lamp characteristic for that element is reduced in intensity at the detector because it stimulates an electronic transition in the sample and is partly absorbed. By determining the level of absorption

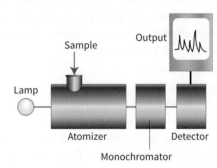

FIGURE 8.44 The layout of a typical atomic absorption spectrophotometer.

relative to standard materials, a quantitative measurement can be made of the amount of the element present. A different lamp is required for each element that is to be analysed. Because the ionization process may produce spectral lines from other components of the analyte (the substance being analysed), a monochromator is placed after the atomizer to isolate the desired wavelength for passage to the detector.

The major differences in instrumentation arise from the different methods used to convert the analyte to free, unbound atoms or ions. In **flame atomization** the analyte solution is mixed with the fuel in a 'nebulizer', which creates an aerosol. The aerosol enters the burner where it passes into a fuel–oxidant flame. Typical fuel–oxidant mixtures are acetylene–air, which produces flame temperatures of up to 2500 K, and acetylene–nitrous oxide, which generates temperatures of up to 3000 K. A common type of 'electrothermal atomizer' is the graphite furnace. The temperatures reached in the furnace are comparable to those attained in a flame atomizer, but detection limits can be 1000 times better. The increased sensitivity is due to the ability to generate atoms quickly and keep them in the optical path for longer. Another advantage of the graphite furnace is that solid samples may be used.

Almost every metallic element can be analysed using AAS, although not all with high sensitivity nor a usefully low detection limit. For example, the detection limit for Cd in a flame ionizer is 1 part per billion (1 ppb = 1 in 10^9), whereas that for Hg is only 500 ppb. Limits of detection using a graphite furnace can be as low as 1 part in 10^{15} for some elements. Direct determination is possible for any element for which hollow cathode lamp sources are available. Other species can be determined by indirect procedures. For example, PO_4^{3-} reacts with MoO_4^{2-} in acid conditions to form $H_3PMo_{12}O_{40}$, which can be extracted into an organic solvent and analysed for molybdenum. To analyse for a particular element, a set of calibration standards is prepared in a similar matrix to the sample, and the standards and the sample are analysed under the same conditions.

In **flame emission spectroscopy** (sometimes called atomic emission spectroscopy, AES) and **flame photometry** the intensity of light given out by ions excited in a high-temperature flame is analysed. In essence, these techniques are quantified flame tests and are often applied to the analysis of the Group 1 and Group 2 metal ions which impart characteristic colours to flames.

8.13 CHN analysis

KEY POINT The carbon, hydrogen, nitrogen, oxygen, and sulfur content of a sample can be determined by high-temperature decomposition.

Instruments are available that allow automated analysis of C, H, N, O, and S. Fig. 8.45 shows the arrangement for an instrument that analyses for C, H, and N, sometimes referred to as **CHN analysis**.

During this analysis, the sample is heated to 900°C in oxygen, and a mixture of carbon dioxide, carbon monoxide, water, nitrogen, and nitrogen oxides is produced. A stream of helium sweeps the products into a tube furnace at 750°C, where copper reduces nitrogen oxides to nitrogen and removes excess oxygen. Copper oxide converts carbon monoxide to carbon dioxide. The resulting mixture of H_2O, CO_2, and N_2 is analysed by passing it through a series of three thermal conductivity detector pairs. The first cell of the first detector pair measures the total conductivity of the gas mixture, water is then removed in a trap, and the thermal conductivity is measured again. The difference in the two conductivity values corresponds to the amount of water in the gas and thus the hydrogen content of the sample. A second detector pair separated by a carbon dioxide trap yields the carbon content, and the remaining nitrogen is measured at the third detector. The data obtained from this technique are reported as mass percentages of C, H, and N.

Oxygen may be analysed if the reaction tube is replaced with a quartz tube filled with carbon that has been coated with catalytic platinum. When the gaseous products are swept through this tube, the oxygen is converted to carbon monoxide, which is then converted to carbon dioxide by passage over hot copper oxide. The rest of the procedure is the same as described above. Sulfur can be measured if the sample is oxidized in a tube filled with copper oxide. Water is removed by trapping in a cool tube, and the sulfur dioxide is determined at what is normally the hydrogen detector.

EXAMPLE 8.11 Interpreting CHN analytical data

A CHN analysis of a compound of iron gave the following mass percentages for the elements present: C 64.54, N 0, and H 5.42, with the residual mass being iron. Determine the empirical formula of the compound.

Answer The molar masses of C, H, and Fe are 12.01, 1.008, and 55.85 g mol^{-1}, respectively. The mass of each element in exactly 100 g of the sample is 64.54 g C, 5.42 g H, and (the difference from 100 g) 30.04 g Fe. The amounts present are therefore

$$n(C) = \frac{64.54\,g}{12.01\,g\,mol^{-1}} = 5.37\,mol$$

$$n(H) = \frac{5.42\,g}{1.008\,g\,mol^{-1}} = 5.38\,mol$$

$$n(Fe) = \frac{30.04\,g}{55.85\,g\,mol^{-1}} = 0.538\,mol$$

These amounts are in the ratio 5.37:5.38:0.538 ≈ 10:10:1. The empirical formula of the compound is therefore $C_{10}H_{10}Fe$, corresponding to the molecular formula [Fe(C_5H_5)$_2$].

Self-test 8.11 Assuming that samples are weighed to the same accuracy, why are the percentages of hydrogen in 5d-series compounds as determined by CHN methods less accurate than those determined for the analogous 3d-series compounds?

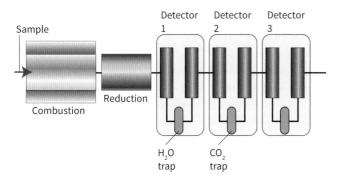

FIGURE 8.45 The layout of the apparatus used for CHN analysis.

8.14 X-ray fluorescence elemental analysis

KEY POINT Qualitative and quantitative information on the elements present in a compound may be obtained by exciting and analysing X-ray emission spectra.

As discussed in Section 8.9, ionization of core electrons can occur when a material is exposed to short-wavelength X-rays. When an electron is ejected in this way, an electron from a higher-energy orbital can take its place and the difference in energy is released in the form of a photon, which is also typically in the X-ray region, with an energy characteristic of the atoms present. This fluorescent radiation can be analysed by either energy-dispersive analysis or wavelength-dispersive analysis. By matching the peaks in the spectrum with the characteristic values of the elements it is possible to identify the presence of a particular element. This is the basis of the X-ray fluorescence (XRF) technique. The intensity of the characteristic radiation is also directly related to the amount of each element in the material. Once the instrument is calibrated with appropriate standards it can be used to determine quantitatively most elements with $Z > 8$ (oxygen). Fig. 8.46 shows a typical XRF energy-dispersive spectrum.

A technique similar to XRF is used in electron microscopes (Section 8.17), where the method is known as energy-dispersive analysis of X-rays (EDAX) or energy-dispersive spectroscopy (EDS). Here the X-rays generated by bombardment of a sample by energetic electrons result in the ejection of core electrons, and X-ray emission occurs as the outer electrons fall into the vacancies in the core levels. These X-rays are characteristic of the elements present, and their intensities are representative of the amounts present. The spectrum can be analysed to determine qualitatively and quantitatively (using appropriate standards) the presence and amount of most elements (generally those with $Z > 8$) in the material. Quantitative information is not highly accurate; typically errors on determined percentages are at least a few per cent even with careful use of standards, and the accuracy is lowered further if the X-ray spectra of the elements being investigated overlap.

EXAMPLE 8.12 Interpreting EDAX analytical data

The EDAX spectrum of a sample investigated using a scanning electron microscope showed the following atomic percentages: Ca 29.5%, Ti 35.2%, O 35.3%. Determine the composition of the sample.

Answer The molar masses of Ca, Ti, and O are 40.08, 47.87, and 16.00 g mol^{-1}, respectively. The amounts present are therefore

$$n(\text{Ca}) = \frac{29.5\,\text{g}}{40.08\,\text{g mol}^{-1}} = 0.736\,\text{mol}$$

$$n(\text{Ti}) = \frac{35.2\,\text{g}}{47.87\,\text{g mol}^{-1}} = 0.735\,\text{mol}$$

$$n(\text{O}) = \frac{35.3\,\text{g}}{16.00\,\text{g mol}^{-1}} = 2.206\,\text{mol}$$

These amounts are in the ratio 0.736:0.735:2.206 ≈ 1:1:3. The empirical formula of the compound is therefore CaTiO$_3$ (a complex oxide with the perovskite structure; see Chapter 4).

Self-test 8.12 Explain why EDAX analysis of a magnesium aluminium silicate would give poor quantitative information.

8.15 Thermal analysis

KEY POINT Thermal methods include thermogravimetric analysis, differential thermal analysis, and differential scanning calorimetry.

Thermal analysis is the analysis of a change in a property of a sample induced by heating. The sample is usually a solid and the changes that occur include melting, phase transition, sublimation, and decomposition.

The analysis of the change in the mass of a sample on heating is known as **thermogravimetric analysis** (TGA). The measurements are carried out using a *thermobalance*, which consists of an electronic microbalance, a temperature-programmable furnace, and a controller, which enables the sample to be simultaneously heated and weighed (Fig. 8.47).

The sample is weighed into a sample holder and then suspended from the balance within the furnace. The temperature of the furnace is usually increased linearly, but more complex heating schemes, isothermal heating (heating that maintains constant temperature at a phase transition), and cooling protocols can also be used. The balance and furnace are situated within an enclosed system so that the atmosphere can be controlled. That atmosphere may be

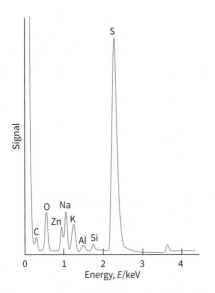

FIGURE 8.46 An XRF spectrum obtained from a metal silicate sample, showing the presence of various elements by their characteristic X-ray emission lines.

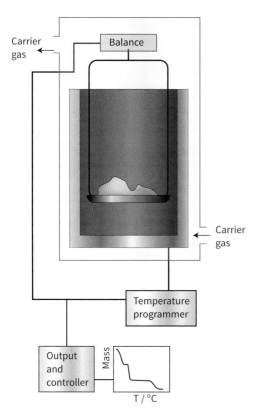

FIGURE 8.47 A thermogravimetric analyser: the mass of the sample is monitored as the temperature is raised.

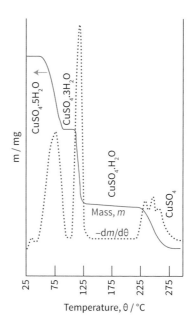

FIGURE 8.48 The thermogravimetric curve obtained for $CuSO_4 \cdot 5H_2O$ as the temperature is raised from 20°C to 500°C. The solid red line is the mass of the sample, and the dotted blue line is its first derivative (the slope of the solid red line).

inert or reactive, depending on the nature of the investigation, and can be static or flowing. A flowing atmosphere has the advantage of carrying away any volatile or corrosive species and prevents the condensation of reaction products. In addition, any species produced can be fed into a mass spectrometer for identification.

TGA is most useful for desorption, decomposition, dehydration, and oxidation processes. For example, the thermogravimetric curve for $CuSO_4 \cdot 5H_2O$ from room temperature to 300°C shows three stepwise mass losses (Fig. 8.48), corresponding to the three stages in the dehydration to form first $CuSO_4 \cdot 3H_2O$, then $CuSO_4 \cdot H_2O$, and finally $CuSO_4$.

Another widely used thermal method of analysis is **differential thermal analysis** (DTA). In this technique the temperature of the sample is compared to that of a reference material while they are both subjected to the same heating procedure. In a DTA instrument, the sample and reference are placed in low thermal conductivity sample holders that are then located within cavities in a block in the furnace. Common reference samples for the analysis of inorganic compounds are alumina (Al_2O_3) and carborundum (SiC). The temperature of the furnace is increased linearly, and the difference in temperature between the sample and the reference is plotted against the furnace temperature. If an endothermic event takes place within the sample, the temperature of the sample lags behind that of the reference and a minimum is observed on the DTA curve. If an exothermic event takes place, the temperature of the sample rises above that of the reference and a maximum is observed on the curve. The area under the endotherm or exotherm (the resulting curve in each case) is related to the enthalpy change accompanying the thermal event.

DTA is most useful for the investigation of processes such as phase changes, where one form of a solid changes to another and there is no weight change observed in the TGA experiment. Examples include the crystallization of amorphous glasses and the transition from one structure type to another (e.g. the transition of TlI from a rock-salt structure type to a CsCl structure type that occurs on heating to 175°C).

A technique closely related to DTA is **differential scanning calorimetry** (DSC). In DSC, the sample and the reference are maintained at the same temperature throughout the heating procedure by using separate power supplies to the sample and reference holders. Any difference between the power supplied to the sample and reference is recorded against the furnace temperature. Thermal events appear as deviations from the DSC baseline as either endotherms or exotherms, depending on whether more or less power has to be supplied to the sample relative to the reference. In DSC, endothermic reactions are usually represented as positive deviations from the baseline, corresponding to increased power supplied to the sample. Exothermic events are represented as negative deviations from the baseline.

The information obtained from DTA and DSC is very similar. The former can be used up to higher temperatures, although the quantitative data, such as the enthalpy of a phase change, obtained from DSC are more reliable. Both DTA and DSC are used to obtain a 'fingerprint' comparison of the results obtained from a sample relative to a reference material.

EXAMPLE 8.13 Interpreting thermal analysis data

When a sample of bismuth nitrate hydrate, $Bi(NO_3)_3 \cdot nH_2O$, of mass 100 mg was heated to 500°C and dryness, the loss in mass observed was 18.56 mg. Determine n.

Answer We need to make a stoichiometric analysis of the decomposition, $Bi(NO_3)_3 \cdot nH_2O \rightarrow Bi(NO_3)_3 + nH_2O$, to determine the value of n. The molar mass of $Bi(NO_3)_3 \cdot nH_2O$ is $(395.01 + 18.02n)$ g mol^{-1}, so the initial amount of $Bi(NO_3)_3 \cdot nH_2O$ present is (100 mg)/$(395.01 + 18.02n$ g mol^{-1}). As each formula unit of $Bi(NO_3)_3 \cdot nH_2O$ contains n mol of H_2O, the amount of H_2O present in the solid is n times this amount, or $n(100$ mg$)/(395.01 + 18.02n$ g mol$^{-1}) = 100n/(395.01 + 18.02n)$ mmol. The mass loss is 18.56 mg. This loss is entirely due to the loss of water, so the amount of H_2O lost is (18.56 mg)/(18.02 g mol^{-1}) = 1.030 mmol. We equate this amount to the amount of H_2O in the solid initially:

$$\frac{100n}{395.01+18.02n} = 1.030$$

(The units mmol have cancelled.) It follows that $n = 5$, and that the solid is $Bi(NO_3)_3 \cdot 5H_2O$.

Self-test 8.13 Reduction of a 10.000 mg sample of an oxide of tin in hydrogen at 600°C resulted in the formation of 7.673 mg of tin metal. Determine the stoichiometry of the tin oxide.

Magnetometry and magnetic susceptibility

KEY POINTS Magnetometry is used to determine the characteristic response of a sample to an applied magnetic field. The magnetic susceptibility provides information on the number of unpaired electrons in a metal complex.

The classic way of monitoring the magnetic properties of a sample is to measure the attraction into or repulsion out of an inhomogeneous magnetic field by monitoring the change in the apparent weight of a sample when the magnetic field is applied by using equipment known as a **Gouy balance** (Fig. 8.49(a)). The sample is hung on one side of a balance by a fine thread so that one end of it lies in the field of a powerful electromagnet and the other end of the sample is in just the Earth's magnetic field. The sample is weighed with the electromagnet field applied and then with it turned off. From the change in apparent weight, the force acting on the sample as a result of the application of the field can be determined and, from this—with a knowledge of various instrumental constants, sample volume, and the molar mass—the molar susceptibility. The effective magnetic moment of a d-metal or f-block ion present in a material may be deduced from the magnetic susceptibility and used to infer the number of unpaired electrons and the spin state (Sections 7.16 and 22.8).

In a **Faraday balance** a magnetic field gradient is generated between two curved magnets; this technique yields precise susceptibility measurements and also allows collection of magnetization data as a function of the magnitude and direction of the applied field. In the related **Evans balance** method the sample under investigation is placed between a pair of suspended electromagnets. The change in current supplied to the electromagnets required to maintain their position relative to the sample versus an empty sample tube is measured.

The more modern **vibrating sample magnetometer** (VSM, Fig. 8.49(b)) measures the magnetic properties of a material

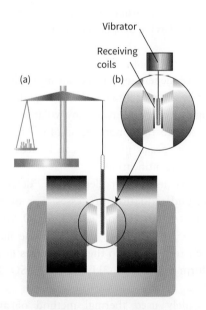

FIGURE 8.49 A schematic diagram of (a) a Gouy balance and (b, insert) a schematic diagram of a vibrating sample magnetometer modification.

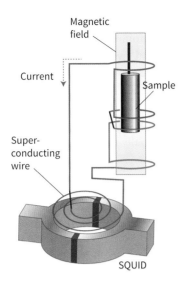

FIGURE 8.50 The magnetic susceptibility of a sample is measured by using a SQUID: the sample, which is exposed to a magnetic field, is moved through the loops in small increments and the potential difference across the SQUID is monitored.

Measurements of magnetic properties are now made more routinely using a **superconducting quantum interference device** (SQUID, Fig. 8.50). A SQUID makes use of the quantization of magnetic flux and the property of current loops in superconductors that are part of the circuit. The current that flows in the loop in a magnetic field is determined by the value of the magnetic flux and hence the magnetic susceptibility of the sample.

The **Evans NMR method** can be used to determine the magnetic moment of the paramagnetic species *in solution* and thus the number of unpaired electrons. The technique uses the change in the NMR chemical shift of a dissolved species (Section 8.6) that results from adding the paramagnetic species to the solution. This observed shift in resonance frequency, $\Delta \nu$ (in Hz), is related to the specific susceptibility χ_g of the solute (cm³ g⁻¹) through the expression

$$\chi_g = \frac{-3\Delta\nu}{4\pi\nu m} + \chi_0 + \frac{\chi_0(d_0 - d_s)}{m}$$

where ν is the spectrometer frequency (Hz), χ_0 is the mass susceptibility of the solvent (cm³ g⁻¹), m is the mass of substance dissolved per cm³ of solution, and d_0 and d_s are the densities of, respectively, the solvent and the solution (g cm⁻³). Values derived for the susceptibility of a transition-metal species in solution may differ from those obtained for the solid phase using a magnetic balance because the coordination geometry, and therefore the number of unpaired electrons, may change on dissolution.

by using a modified version of a Gouy balance. The sample is placed in a uniform magnetic field, which induces a net magnetization. As the sample is vibrated, an electrical signal is induced in suitably placed pick-up coils. The signal has the same frequency of vibration and its amplitude is proportional to the induced magnetization. The vibrating sample may be cooled or heated, allowing a study of magnetic properties as a function of temperature.

Electrochemical techniques

KEY POINT Cyclic voltammetry is used to measure reduction potentials and investigate coupled chemical reactions of redox-active species.

In **cyclic voltammetry** the current flowing due to interfacial electron transfer between an electrode and a species in solution is measured while the potential difference applied to the electrode is varied linearly with time, back and forth between two potential limits. Cyclic voltammetry provides direct information on reduction potentials and reactions coupled to redox chemistry, such as catalysis, and stabilities of products of oxidation or reduction. The technique provides rapid qualitative insight into the redox properties of an electroactive compound and reliable quantitative information on its thermodynamic and kinetic properties. The solution containing the species of interest, usually a transition metal complex, is contained in an electrochemical cell into which three electrodes are placed (Fig. 8.51).

The 'working electrode' at which the electrochemical reaction of interest occurs is usually constructed from platinum, silver, gold, or graphite. The reference electrode is normally a silver/silver chloride electrode, and the counter electrode is normally a piece of platinum. The role of the counter electrode is to complete the electrical circuit, ensuring that electrons do not need to pass across the reference–solution interface. The concentration of electroactive species is usually quite low (less than 0.001 mol dm⁻³) and the solution contains a relatively high concentration of inert 'supporting' electrolyte (at concentrations greater than about 0.1 mol dm⁻³) to provide conductivity. The potential difference is applied between the working electrode and the reference electrode and is scanned back and forth between two limits, tracing out a triangular waveform. The potential sweep rate can be varied between 1 mV s⁻¹ and 100 V s⁻¹, depending on the instrument and size of electrode.

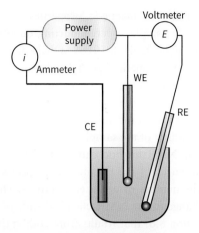

FIGURE 8.51 An electrochemical cell into which three electrodes are placed. The half-cell reaction of interest occurs at the working electrode (WE), the potential of which is controlled with respect to the reference electrode (RE). No current flows between the WE and the RE: current flows instead between the WE and the counter electrode (CE), at which reactions such as solvent or electrolyte breakdown are driven in order to balance the charge passed across the WE.

To understand what is involved, consider the simple redox couple $[Fe(CN)_6]^{3-}/[Fe(CN)_6]^{4-}$, where, initially, only the reduced form (the Fe(II) complex) is present (Fig. 8.52). No current flows while the electrode potential is sufficiently negative relative to the reduction potential. As the potential is increased and approaches the reduction potential of the Fe(III)/Fe(II) couple, the Fe(II) must be oxidized to preserve Nernstian equilibrium close to the working electrode, and a current starts to flow. This current rises to a peak, then decreases steadily because Fe(II) becomes depleted close to the electrode (the solution is unstirred) and must be supplemented by species diffusing from increasingly distant regions of the solution. Once the upper potential limit is reached, the potential sweep is reversed. Initially, Fe(II) diffusing to the electrode continues to be oxidized, but eventually the potential difference becomes sufficiently negative to reduce the Fe(III) that has been formed and accumulated close to the electrode surface; the current reaches a peak, then decreases gradually to zero as the lower potential limit is reached.

The average of the two peak potentials is a good approximation to the reduction potential E under the prevailing conditions (E is not in general the standard potential because nonstandard conditions are usually adopted). In the ideal case, the oxidation and reduction peaks are similar in magnitude and separated by a small potential increment, which is usually $(59\,\text{mV})/v_e$ at 25°C, where v_e is the number of electrons transferred during the reaction at the electrode (see also Section 6.5). This case is an example of a reversible redox reaction, with electron transfer at the electrode sufficiently fast that equilibrium (the concentrations of species close to the electrode surface are as expected from the Nernst equation) is always maintained throughout the potential sweep. In such a case the current is usually limited by diffusion of electroactive species to the electrode.

Slow kinetic processes at the electrode result in a large separation of reduction and oxidation peaks that increases with increasing scan rate. This separation arises because an overpotential (effectively a driving force) is required to overcome barriers to electron transfer in each direction. Moreover, the peak due to a reduction or an oxidation in the initial part of the cycle process is often not matched by a corresponding peak in the reverse direction. This absence occurs because the species that is initially generated undergoes a further chemical reaction during the cycle and produces either a species with a different reduction potential or one that is not electroactive within the range of potential scanned. Inorganic chemists often refer to this behaviour as 'irreversible'.

An electrochemical reaction followed by a chemical reaction is known as an **EC process**. By analogy, a **CE process** is a reaction in which the species able to undergo the electrochemical (E) reaction must first be generated by a chemical (C) reaction. Thus, for a molecule that is suspected of decomposing upon oxidation, it may be possible to observe the initial unstable species formed by the E process provided the scan rate is sufficiently fast to re-reduce it before it undergoes further reaction. Consequently, by varying the scan rate, the kinetics of the chemical reaction can be determined.

The chemical step 'C' is often a proton transfer reaction that is fast compared to electron transfer so that its

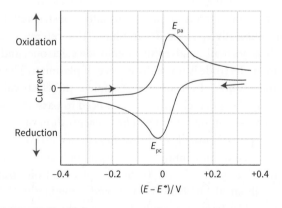

FIGURE 8.52 The cyclic voltammogram for an electroactive species present in solution as the oxidized form, and displaying a reversible one-electron reaction at an electrode. The peak potentials E_{pa} and E_{pc} for oxidation and reduction, respectively, are separated by 0.06 V. The reduction potential is the mean of E_{pa} and E_{pc}.

kinetics cannot be determined. Even so, cyclic voltammetry provides a great deal of information on the thermodynamics of proton-coupled electron transfer that were described in Sections 6.6 and 6.14. Fig. 8.53(a) shows the cyclic voltammetry of an Os(II) complex containing a single H_2O ligand; two one-electron processes are revealed, generating first Os(III), then Os(IV). The separation between the potentials at which each process occurs, and the fact that the return sweep regenerates first Os(III) and then Os(II), shows that all species are stable. Carrying out the experiment over a wide pH range (0–10) allows a Pourbaix diagram (Section 6.14) to be constructed, as shown in Fig. 8.53(b). The interpretation is that over the pH range 2–9, oxidation of Os(II) to Os(III) results in loss of one proton (the slope is −59 mV/pH unit), and subsequent oxidation to Os(IV) results in loss of both protons and formation of an oxoligand. Bends in the data lines for Os(III)/Os(II) at pH 1.9 and pH 9.2 are ascribed to the pK values for the first deprotonations of Os(III) and Os(II), respectively, but no such deviations are observed in the data line for Os(IV)/Os(III), showing that the pK values for the second deprotonations of Os(III) or protonation of Os(IV) are out of range.

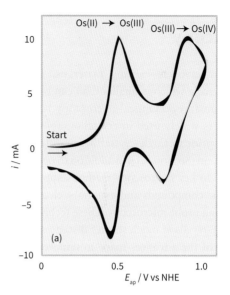

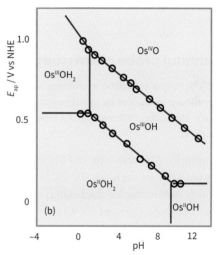

EXAMPLE 8.14 Kinetic resolution of proton and electron transfer reactions using cyclic voltammetry

Figure 8.53c shows the possible elementary reactions that the Os(II) complex [Os^{II}(bpy)$_2$py(OH$_2$)]$^{2+}$ can undergo during its oxidation at an electrode. With a solution at pH 3.1, when the sweep rate is increased to a very large value, the voltammetry (shown in Fig. 8.53(a)) changes and eventually consists of a pair of widely spaced reduction and oxidation peaks centred at the potential $E_{III/II(2H)}$. What does this tell us about the rates of proton transfer? Hint: The rate of an E process increases with increasing potential difference, whereas the rate of a C process should not.

Answer To answer this question we note that horizontal arrows in Fig. 8.53(c) denote E processes, whereas vertical arrows denote C processes. Increasing the potential sweep rate will disfavour the formation of species for which a C step must occur first. The ultimate observation of a single couple at $E_{III/II(2H)}$ means that deprotonation of the coordinated water ligand is slow compared to interfacial electron transfer.

Self-test 8.14 If interfacial electron-transfer rates involving the Os series are always faster than proton transfer, predict the appearance of the voltammograms observed at very high sweep rate when starting with solutions (pH 3.1) of the Os(III) or Os(IV) complexes.

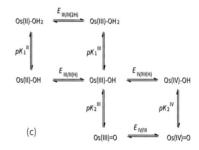

FIGURE 8.53 (a) Cyclic voltammetry of an Os(II) complex [Os^{II}(bpy)$_2$py(OH$_2$)]$^{2+}$ (abbreviated as Os-OH$_2$) at pH 3.1, showing two one-electron processes generating first Os(III) and then Os(IV) species. Sweep rate is 0.2 V/s. By carrying out the experiment over a wide pH range (0–10) the Pourbaix diagram (b) can be constructed; (c) shows the possible elementary reactions that the Os(II) complex [Os^{II}(bpy)$_2$py(OH$_2$)]$^{2+}$ can undergo during its oxidation at an electrode. Adapted from an original kindly supplied by Prof. J. M. Saveant. Article published in *Proc. Natl. Acad. Sci. USA*. **106**, 11,829-11,836 (2009).

Microscopy

Microscopy involves viewing materials, usually in solid form, on a scale which is smaller than can be resolved by the unaided human eye. Optical microscopy using visible light and lenses is widely used in crystallography to assess crystal quality prior to single-crystal X-ray diffraction structure determination (Section 8.1). In addition to optical microscopy there are two key microscopy methods applied to the study of inorganic compounds and materials: **electron microscopy** and **scanning probe microscopy**. These techniques are able to visualize materials at much shorter distances, down to 1 nm, and many of the advances made in nanoscience and nanotechnology (Chapter 28) would not have occurred without the ability to characterize the nanoscale structural, chemical, and physical properties of materials.

8.16 Scanning probe microscopy

KEY POINTS Scanning tunnelling microscopy uses the tunnelling current from a sharp conductive tip to image and characterize a conductive surface; atomic force microscopy uses intermolecular forces to image the surface.

Scanning tunnelling microscopy (STM) and **atomic force microscopy** (AFM) were the first to be discovered and are now the most widely used examples of **scanning probe microscopies** (SPM). This family of techniques allows the three-dimensional imaging of the surface of materials by using a sharply pointed probe brought into close proximity (or in contact) with the specimen. As the probe is moved across the surface an image is constructed by monitoring the spatial variation in the value of a physical parameter such as potential difference, electric current, magnetic field, or mechanical force.

In STM, an atomically sharp conductive tip is scanned at about 0.3–10 nm above the surface of the sample (Fig. 8.54(a)). Electrons in the conductive tip, which is held either at a constant potential or at a constant height above the sample, can tunnel through the gap with a probability that is related exponentially to the distance from the surface. As a result, the electron tunnelling current reflects the distance between the sample and the tip. If the tip is scanned over the surface at a constant height, then the current represents the change in the topography of the surface. The movement of the tip at a constant height with great precision is accomplished by using piezoelectric ceramics for the displacement of the tip. Fig. 8.54(b) shows the results of an STM study of a graphite substrate. In AFM, the atoms at the tip of the probe interact with the surface atoms of the sample through intermolecular forces (such as van der Waals interactions). The cantilever holding the probe bends up and down in response to the forces, and the extent of deflection is monitored with a reflected laser beam.

Variations of AFM include the following analytical methods:

- **Frictional force microscopy** measures variations in the lateral forces on the tip that are based on chemical variations on the surface.
- **Magnetic force microscopy** uses a magnetic tip to image the surface magnetic structure of a material.
- **Electrostatic force microscopy** senses variations in electric fields.
- **Molecular recognition AFM** can be undertaken where the tip is functionalized with specific ligands and the interaction between the decorated tip and surface is measured. Such microscopes can provide resolution of the chemical properties on the surface.

8.17 Electron microscopy

KEY POINT Transmission and scanning electron microscopes use electrons to image the sample in a similar fashion to optical microscopes but at much higher resolution.

An imaging electron microscope operates like a conventional optical microscope, but instead of imaging photons,

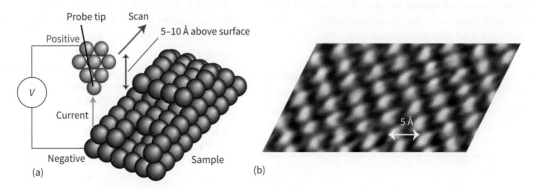

FIGURE 8.54 (a) Schematic of the operation of a scanning tunnelling microscope (STM); (b) STM image of a graphite surface.

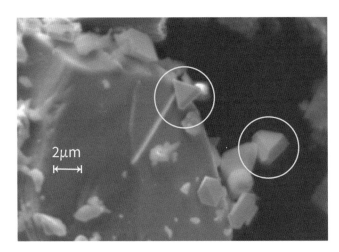

FIGURE 8.55 SEM image of gold crystals (circled) with dimensions of ~2 μm, on the surface of a larger crystal.

as in the visible microscope, electron microscopes use electrons. In these instruments, electron beams are accelerated through 1–200 kV and electric and magnetic fields are used to focus the electrons. In **transmission electron microscopy** (TEM) the electron beam passes through the thin sample being examined and is imaged on a phosphorescent screen. In **scanning electron microscopy** (SEM) the beam is scanned over the object and the reflected (scattered) beam is then imaged by the detector (Fig. 8.55).

The ultimate resolution of SEM depends on how sharply the incident beam is focused on the sample, how it is moved over the sample, and how much the beam spreads out into the sample before it is reflected, but features with dimensions of 1 μm or less are typically resolved. In both microscopes, the electron probes cause the production of X-rays with energies characteristic of the elemental composition of the material. As such, **energy-dispersive spectroscopy** of these characteristic X-rays is used to quantify the chemical make-up of materials by using electron microscopes (Section 8.14).

The primary advantage of SEM over TEM is that it can form images of electron-opaque samples without the need for difficult specimen preparations: SEM is therefore the electron microscopy method of choice for the straightforward characterization of materials. However, SEM samples need to be conductive; otherwise electrons collect on the sample and interact with the electron beam itself, resulting in blurring. Nonconductive samples must therefore be coated with a thin layer of a conducting material, usually gold or graphitic carbon.

FURTHER READING

Although this chapter has introduced many of the methods used by chemists to characterize inorganic compounds, it is not exhaustive. Other techniques used for investigating the structures and properties of solids and solutions include nuclear quadrupole resonance and inelastic neutron scattering, to name just two. The following references are a source of information on these techniques and a greater depth of coverage on the major techniques introduced here.

M. T. Weller and N. A. Young, *Characterisation Methods in Inorganic Chemistry*. Oxford University Press (2017).

A. K. Brisdon, *Inorganic Spectroscopic Methods*. Oxford Science Publications (1998).

R. P. Wayne, *Chemical Instrumentation*. Oxford Science Publications (1994).

D. A. Skoog, F. J. Holler, and T. A. Nieman, *Principles of Instrumental Analysis*. Brooks Cole (1997).

R. S. Drago, *Physical Methods for Chemists*. Saunders (1992).

F. Rouessac and A. Rouessac, *Chemical Analysis: Modern Instrumentation and Techniques*, 2nd ed. Wiley-Blackwell (2007).

S. K. Chatterjee, *X-Ray Diffraction: Its Theory and Applications*. Prentice Hall of India (2004).

B. D. Cullity and S. R. Stock, *Elements of X-Ray Diffraction*. Prentice Hall (2003).

B. Henderson and G. F. Imbusch, *Optical Spectroscopy of Inorganic Solids. Monographs on the Physics & Chemistry of Materials*. Oxford University Press (2006).

E. I. Solomon and A. B. P. Lever, *Inorganic Electronic Structure and Spectroscopy. Vol. 1. Methodology*. John Wiley & Sons (2006).

E. I. Solomon and A. B. P. Lever, *Inorganic Electronic Structure and Spectroscopy. Vol. 2. Applications and Case Studies*. John Wiley & Sons (2006).

J. S. Ogden, *Introduction to Molecular Symmetry*. Oxford University Press (2001).

F. Siebert and P. Hildebrandt, *Vibrational Spectroscopy in Life Science*. Wiley-VCH (2007).

J. R. Ferraro and K. Nakamoto, *Introductory Raman Spectroscopy*. Academic Press (1994).

K. Nakamoto, *Infrared and Raman Spectra of Inorganic and Coordination Compounds*. Wiley-Interscience (1997).

J. K. M. Saunders and B. K. Hunter, *Modern NMR Spectroscopy: A Guide for Chemists*. Oxford University Press (1993).

J. A. Iggo, *NMR Spectroscopy in Inorganic Chemistry*. Oxford University Press (1999).

J. W. Akitt and B. E. Mann, *NMR and Chemistry*. Stanley Thornes (2000).

K. J. D. MacKenzie and M. E. Smith, *Multinuclear Solid-State Nuclear Magnetic Resonance of Inorganic Materials*. Pergamon Press (2004).

D. P. E. Dickson and F. J. Berry, *Mössbauer Spectroscopy*. Cambridge University Press (2005).

M. E. Brown, *Introduction to Thermal Analysis*. Kluwer Academic Press (2001).

P. J. Haines, *Principles of Thermal Analysis and Calorimetry*. Royal Society of Chemistry (2002).

A. J. Bard and L. R. Faulkner, *Electrochemical Methods: Fundamentals and Applications*, 2nd ed. John Wiley & Sons (2001).

O. Kahn, *Molecular Magnetism*. VCH (1993).

R. G. Compton and C. E. Banks. *Understanding Voltammetry*. World Scientific Publishing (2007).

EXERCISES

8.1 The synthesis of the spinel Mg_2TiO_4 can be achieved by heating TiO_2 with MgO at 1400°C over a period of several days. How would powder X-ray diffraction be able to demonstrate that the reaction had gone to completion?

8.2 A student forgot to label their samples of MgO and CaO; both are white, insoluble powders. How could they identify which sample is which without destroying them?

8.3 The minimum size of crystal that can typically be studied using a laboratory single-crystal diffractometer is $50 \times 50 \times 50$ μm³. The X-ray flux from a synchrotron source is expected to be 10^6 times the intensity of a laboratory source. Calculate the minimum size of a cubic crystal that could be studied on a diffractometer by using this source. A neutron flux is 10^3 times weaker. Calculate the minimum size of crystal that could be studied by single-crystal neutron diffraction.

8.4 Why are the reported errors on the N–H bond lengths determined in $(NH_4)_2SeO_4$ by single-crystal X-ray diffraction much larger than those on the Se–O bond lengths?

8.5 Strong hydrogen bonds can either be symmetric, as in the [F–H–F]⁻ anion (with both F–H distances equal at 1.14 Å) or asymmetric as in KH_2PO_4 (where the O–H bond distance is 1.05 Å and H–O hydrogen bonding separation is 1.51 Å). How could you determine whether the hydrogen bond in the following compounds was symmetric or asymmetric: (a) BO(OH) (b) CrO(OH)?

8.6 (a) Calculate the wavelength associated with a neutron moving at 2.60 km s⁻¹. Is this wavelength suitable for diffraction studies ($m_n = 1.675 \times 10^{-27}$ kg)? (b) For magnetic neutron diffraction studies, long neutron wavelengths are often needed; calculate the velocity of a neutron that has an associated wavelength of 5 Å.

8.7 Discuss why the length of an O–H bond obtained from X-ray diffraction experiments averages 85 pm, whereas that obtained in neutron diffraction experiments averages 96 pm. Would you expect to see similar effects with C–H bond lengths measured by these techniques?

8.8 Explain why a Raman band assigned to the symmetric N–C stretching mode in $N(CH_3)_3$ shows a shift to lower frequency when ^{14}N is substituted by ^{15}N but no such shift is observed for the N–Si symmetric stretch in $N(SiH_3)_3$.

8.9 The ambidentate thiocyanate ligand (SCN⁻) can coordinate to a metal centre either through sulfur (SCN) or nitrogen (NCS), Chapter 7. Explain the following observed vibrational frequencies for the C–S bond ($\nu(CS)$): $K_2[Co(SCN)_4]\cdot 4H_2O$, 850 cm⁻¹; KSCN, 746 cm⁻¹; $K_2[Pd(SCN)_4]$, 730 cm⁻¹. $RbBi(SCN)_4$ shows an absorption in its IR spectrum at 722 cm⁻¹; what is the likely mode of bonding of the thiocyanate group to bismuth in this compound?

8.10 Calculate the expected wavenumber for an N–D stretch given that N–H stretches are normally observed at 3400 cm⁻¹.

8.11 Suggest reasons for the order of stretching frequencies observed with isoelectronic diatomic species: CN⁻ > CO > NO⁺. Predict a position for the carbide anion $[C_2]^{2-}$, which is found in calcium carbide CaC_2, in this series.

8.12 Use the data in Table 8.4 to estimate the O–O stretching wavenumber expected for a compound believed to contain the oxygenyl species O_2^+. Would you expect to observe this stretching vibration in (a) the IR spectrum or (b) the Raman spectrum?

8.13 Predict the form of the ¹⁹F-NMR and the ⁷⁷Se-NMR spectra of ⁷⁷SeF₄. For ⁷⁷Se, $I = \frac{1}{2}$.

8.14 The ³¹P-NMR spectrum of PF_3Cl_2 consists of a doublet of triplets. Explain this observation.

8.15 Explain the observation that the ¹⁹F-NMR spectrum of the $[XeF_5]^-$ anion consists of a central peak symmetrically flanked by two peaks, each of which is roughly one-sixth of the intensity of the central peak. Predict the form of the ¹²⁹Xe-NMR spectrum for this ion. ¹²⁹Xe, $I = \frac{1}{2}$, 26.5% abundant.

8.16 The ¹²⁹Xe- and ¹⁹F-NMR spectra of $[XeF_3]^+$ are shown in Fig. 8.56. Interpret these spectra in light of the shape of $[XeF_3]^+$ predicted from VSEPR theory.

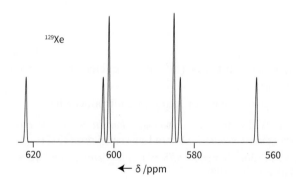

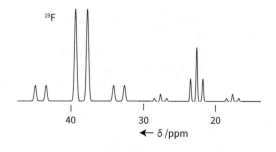

FIGURE 8.56

8.17 Explain why the ^{13}C-NMR spectrum of [Fe$_3$(CO)$_{12}$] shows only a single peak at 212.5 ppm at room temperature, while the IR spectrum shows the presence of both terminal and bridging carbonyl groups. How might the ^{13}C-NMR spectrum change on cooling?

8.18 The ^{31}P-MAS-NMR spectrum of solid PCl$_5$ shows two resonances, one of which has a chemical shift similar to that found for ^{31}P in the salt CsPCl$_6$. Explain.

8.19 The solution ^{31}P-NMR spectrum of P$_4$S$_3$ consists of a doublet and quartet with intensities in the ratio 3:1. Suggest a structure consistent with this pattern.

8.20 Sketch the forms of the following solution phase NMR spectra (abundances shown as per cent):

(a) The ^{1}H-NMR spectrum of KBH$_4$

(b) The ^{1}H- and ^{195}Pt-NMR spectra of *cis*-[Pt(CO)$_2$(H)Cl]

^{1}H, $I = \frac{1}{2}$, 100%; ^{195}Pt, $I = \frac{1}{2}$, 34%;

^{10}B, $I = 3$; 20%; ^{11}B, $I = \frac{3}{2}$; 80%; no other nuclei contribute to these NMR spectra.

8.21 Determine the g values of the EPR spectrum shown in Fig. 8.57, measured for a frozen sample using a microwave frequency of 9.43 GHz.

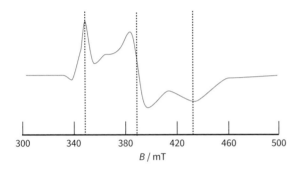

FIGURE 8.57

8.22 Which technique is sensitive to the fastest processes, NMR or EPR?

8.23 The EPR spectrum of bis[tris(2-pyridyl)methane]copper(II) dinitrate is isotropic at 293 K, but is axial with $g_\parallel > g_\perp$ at 150 K. Explain these observations by references to the EPR timescale and dynamic Jahn–Teller distortions that can occur in copper complexes.

8.24 For a paramagnetic compound of a d-metal having one unpaired electron, outline the main difference you would expect between an EPR spectrum measured in aqueous solution at room temperature and one recorded for a frozen solution.

8.25 Predict a value for the isomer shift for iron in the Mössbauer spectrum of Na$_3$Fe(V)O$_4$.

8.26 Some formally Fe(IV) compounds, such as CaFeO$_3$, are believed to contain the iron as a mixture of Fe(III) and Fe(V). Predict the key features of the ^{57}Fe Mössbauer spectra of this compound when described as (a) CaFe(IV)O$_3$ and (b) CaFe(III)$_{0.5}$Fe(V)$_{0.5}$O$_3$, assuming iron is in a regular octahedral coordination to oxygen in both descriptions. Could Mössbauer spectroscopy be used to distinguish these two descriptions? CaFeO$_3$ orders antiferromagnetically at low temperatures, and the Mössbauer spectrum at 4 K consists of two overlying, magnetic hyperfine six-line patterns. Which is the best description of the iron oxidation states in CaFeO$_3$ at this temperature?

8.27 How could the analysis of the XANES spectrum at the Fe K-edge help identify the iron oxidation state(s) present in CaFeO$_3$ as discussed in Exercise 8.26?

8.28 Addition of concentrated HCl to an aqueous solution of CoSO$_4$.7H$_2$O results in major changes to the Co K-edge EXAFS spectrum. Explain this observation.

8.29 Explain why, even though the average atomic mass of silver is 107.9m_u, no peak at 108m_u is observed in the mass spectrum of pure silver. Describe the expected form of the molecular ion peak in the mass spectra of (a) methyl silver, CH$_3$Ag, and (b) methylsilver tetramer, (CH$_3$Ag)$_4$.

8.30 What peaks should you expect to observe in the mass spectrum of [Mo(C$_6$H$_6$)(CO)$_3$]?

8.31 Thermogravimetric analysis of a zeolite of composition CaAl$_2$Si$_6$O$_{16}$.nH$_2$O shows a mass loss of 25% on heating to dryness. Determine n.

8.32 LiBH$_4$ has been proposed as a possible hydrogen storage material. Heating LiBH$_4$ on a thermobalance to 1000°C results in a weight loss of 18.5%. Is all the hydrogen released during this decomposition?

8.33 Interpret the cyclic voltammogram shown in Fig. 8.58, which has been recorded for an Fe(III) complex in aqueous solution.

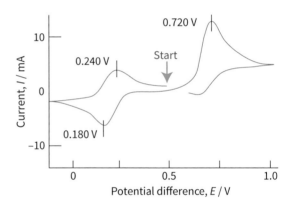

FIGURE 8.58

8.34 Sodium carbonate, boron oxide, and silicon dioxide when heated together and quickly cooled produce a borosilicate glass. Explain why the powder diffraction pattern of this product shows no diffraction maxima. Heating the borosilicate glass in a DTA instrument shows an exothermic event at 500°C, and the powder X-ray diffraction pattern observed from the product shows diffraction maxima. Explain these observations.

8.35 A cobalt(II) salt was dissolved in water and reacted with excess acetylacetone (2,4-pentanedione, CH$_3$COCH$_2$COCH$_3$) and hydrogen peroxide. A green solid was formed that gave the following results for elemental analysis: C, 50.4%; H, 6.2%; Co, 16.5% (all by mass). Determine the ratio of cobalt to acetylacetonate ion in the product.

8.36 How would the cyclic voltammetry shown in Fig. 8.53(a) differ if (a) the Os(IV) complex decomposed rapidly, (b) Os(III) is oxidized in a single, rapid two-electron step to Os(V)?

8.37 Particle sizes in the range 50–1000 nm can be determined using powder X-ray diffraction as the particle size in this range is inversely correlated with diffraction peak width. How else could a particle size in this range be measured? What technique could be used to measure the size of particles in the range 1–10 nm?

8.38 How would you determine the composition of a crystal obtained from the reaction of CuS, SnS, and ZnS in a sealed tube, without destroying it? The analysis shows the following atomic weight percentages, Cu 26.74%, Zn 15.83%, Sn 10.90%, and S 46.54%. Determine the composition of the compound. The crystal was left in air for a week and then analysed again; how might the determined composition change?

TUTORIAL PROBLEMS

8.1 Discuss the importance of X-ray crystallography in inorganic chemistry. See, for example, W. P. Jensen, G. J. Palenik, and I.-H. Suh, The history of molecular structure determination viewed through the Nobel Prizes, *J. Chem. Educ.*, 2003, **80**, 753.

8.2 How is single-crystal neutron diffraction of use in inorganic chemistry?

8.3 How may metal hydride materials, of interest for hydrogen storage applications, be characterized using the various analytical methods discussed in this chapter?

8.4 Discuss the information available from the following techniques in the analysis of inorganic pigments used in antique oil paintings: (a) powder X-ray diffraction, (b) infrared and Raman spectroscopies, (c) UV–visible spectroscopy, (d) X-ray fluorescence.

8.5 Discuss how hydrogen bonding in inorganic compounds can be characterized using infrared spectroscopy. How does the strength of a hydrogen bond affect (a) OH stretching modes, (b) OH bending modes?

8.6 How may a consideration of ^{31}P chemical shifts and ^{1}H coupling constants be used to distinguish the isomers of the octahedral complex $[Rh(CCR)_2H(PMe_3)_3]$? (See J. P. Rourke, G. Stringer, D. S. Yufit, J. A. K. Howard, and T. B. Marder, *Organometallics*, 2002, **21**, 429.)

8.7 Discuss the importance of NMR spectroscopy in the development of Group 18 chemistry. (See M. Gerken and G. J. Schrobilgen, *Coord. Chem. Rev.*, 2000, **197**, 335.)

8.8 Ionization techniques in mass spectrometry frequently induce fragmentation and other undesirable reactions. However, under some circumstances these reactions can be helpful. Using fullerenes as an example, discuss this phenomenon. (See M. M. Boorum, Y. V. Vasil'ev, T. Drewello, and L. T. Scott, *Science*, 2001, **294**, 828.)

8.9 Discuss how EXAFS can be used to follow the isomerization of the spiked triangular cluster $[Ru_3Pt(\mu\text{-}H)(\mu_4\text{-}\eta^2\text{-}CCBu^t)(CO)_9(dppe)]$ (dppe = 1,2-bis(diphenylphosphino)ethane) to the butterfly cluster $[Ru_3Pt(\mu_4\text{-}\eta^2\text{-}C{=}CHBu^t)(CO)_9(dppe)]$. (See A. J. Dent, L. J. Farrugia, A. G. Orpen, and S. E. Stratford, *J. Chem. Soc., Chem. Commun.*, 1992, 1456.)

8.10 Discuss how you would carry out the following analyses and determinations: (a) calcium levels in breakfast cereal, (b) mercury in shellfish, (c) the geometry of BrF_5, (d) the number and type of organic ligands in a d-metal complex, (e) the number of coordinated ammonia molecules in $Mg(NH_3)_nCl_2$.

8.11 The iron content of a dietary supplement tablet was determined using atomic absorption spectrometry. A tablet (0.4878 g) was ground to a fine powder, and 0.1123 g was dissolved in dilute sulfuric acid and transferred to a 50 cm^3 volumetric flask. A 10 cm^3 sample of this solution was taken and made up to 100 cm^3 in another volumetric flask. A series of standards was prepared that contained 1.00, 3.00, 5.00, 7.00, and 10.0 ppm iron. The absorptions of the standards and the sample solution were measured at the iron absorption wavelength.

Calculate the mass of iron in the tablet.

Concentration/ppm	Absorbance
1.00	0.095
3.00	0.265
5.00	0.450
7.00	0.632
10.0	0.910
Sample	0.545

8.12 A sample of water from a reservoir was analysed for copper content. The sample was filtered and diluted 10-fold with deionized water. A set of standards was prepared with copper concentrations between 100 and 500 ppm. The standards and the sample were aspirated into an atomic absorption spectrometer and the absorbance was measured at the copper absorption wavelength. The results obtained are given below. Calculate the concentration of copper in the reservoir.

Concentration/ppm	Absorbance
100	0.152
200	0.388
300	0.590
400	0.718
500	0.865
Sample	0.751

8.13 A sample from an effluent stream was analysed for phosphate levels. Dilute hydrochloric acid and excess sodium molybdate were added to a 50 cm^3 sample of the effluent. The molybdophosphoric acid, $H_3PMo_{12}O_{40}$, that was formed was extracted into two 10 cm^3 portions of an organic solvent. A molybdenum standard with a concentration of 10 ppm was prepared in the same solvent. The combined extract and the standard were aspirated into an atomic absorption spectrometer set up to measure molybdenum. The extracts gave an absorbance of 0.573, and the standard gave an absorbance of 0.222. Calculate the concentration of phosphate in the effluent.

8.14 The particle sizes determined using powder X-ray diffraction data peak widths (using the Scherrer formula which is widely described in the literature) and those obtained from imaging using electron microscopy often differ, with electron microscopy normally yielding a larger value. Determine why this is the case.

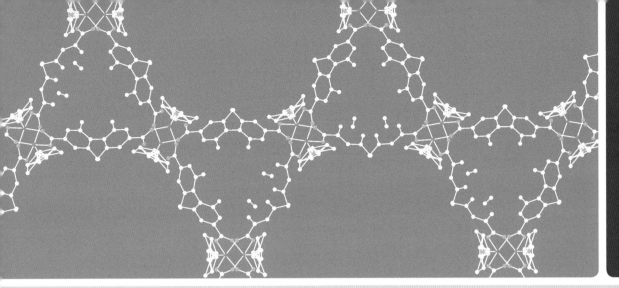

PART 2

The elements and their compounds

This part of the text describes the physical and chemical properties of the elements as they are set out in the periodic table. This 'descriptive chemistry' of the elements reveals a rich tapestry of patterns and trends, many of which can be rationalized and explained by application of the concepts developed in Part 1.

The first chapter of this part of the text, Chapter 9, summarizes the trends and patterns in the properties of the elements and their compounds in the context of the periodic table and the principles described in Part 1. Chapter 10 describes the pivotal themes, such as the lanthanoid contraction, that are much discussed in reference to the periodic table. Chapter 11 deals with the distinctive chemistry of hydrogen. The following eight chapters (Chapters 12–19) proceed systematically across the main groups of the periodic table. The elements of these groups demonstrate the variety, richness, and fascinating nature of inorganic chemistry.

The chemical properties of the d- and f-block elements are so diverse and extensive that the next five chapters of this part are devoted to them. Chapter 20 reviews the descriptive chemistry of the three series of d-block elements. Chapter 21 provides an account of the important and unusual properties of the f-block elements. Chapter 22 describes how electronic structure affects the chemical and physical properties of the d- and f-metal complexes, and Chapter 23 describes their reactions in solution. Chapter 24 deals with the industrially vital d-metal organometallic compounds.

9 Periodic trends

Trends in distribution
  9.1 **Occurrence of the elements**

Trends in properties
  9.2 **Atomic parameters**
  9.3 **Metallic character**
  9.4 **Oxidation states**

Trends in compounds
  9.5 **Amphoterism**
  9.6 **Coordination numbers**
  9.7 **Bond enthalpy trends**

Where trends collide
  9.8 **Wider aspects of periodicity**

Further reading

Exercises

Tutorial problems

The periodic table provides an organizing principle that coordinates and rationalizes the varied physical and chemical properties of the elements. Periodicity is the regular manner in which the physical and chemical properties of the elements vary with atomic number. An understanding of periodicity helps to understand the chemistry of the elements and their compounds.

Although the chemical properties of the elements can seem diverse, the periodic table helps to show that they vary reasonably systematically with atomic number. Once these trends and patterns are recognized and understood, many of the detailed properties of the elements are revealed as consistent and predictable. In this chapter we summarize some of the trends in the physical and chemical properties of the elements and interpret them in terms of the underlying principles presented in Chapter 1.

Trends in distribution

The general structure of the modern periodic table was discussed in Section 1.6. Almost all trends in the physical and chemical properties of the elements can be traced to the electron configuration of the atoms and atomic radii, and their variation with atomic number.

9.1 Occurrence of the elements

KEY POINT Hard–hard and soft–soft interactions help to systematize the terrestrial distribution of the elements.

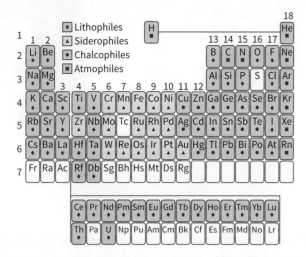

FIGURE 9.1 The Goldschmidt classification of the elements.

Although some elements do occur in their elemental state in nature—for example, the gases nitrogen and oxygen, the non-metal sulfur, and the metals silver and gold—most elements occur naturally only as compounds with other elements.

The concept of hardness and softness (Section 5.10) helps to rationalize a great deal of inorganic chemistry, including the type of compound that the element forms in Nature. Thus, soft acids tend to bond to soft bases and hard acids tend to bond to hard bases. These tendencies explain certain aspects of the **Goldschmidt classification** of the elements into four types (Fig. 9.1), a scheme widely used in geochemistry:

Lithophiles are found primarily in the Earth's crust (the lithosphere) in silicate minerals, and include Li, Mg, Ti, Al, and Cr (as their cations). These cations are hard, and are found in association with the hard base O^{2-}.

Chalcophiles are often found in combination with sulfide (and selenide and telluride) minerals, and include Cd, Pb, Sb, and Bi. These elements (as their cations) are soft, and preferentially form bonds with the soft base S^{2-} (or Se^{2-} and Te^{2-}). Zinc cations are borderline hard, but softer than Al^{3+} and Cr^{3+}, and Zn is also often found as its sulfide.

Siderophiles are intermediate in terms of hardness and softness and show an affinity for both oxygen and sulfur. They occur mainly in their elemental state and include Pt, Pd, Ru, Rh, and Os.

Atmophiles are elements that occur as gases, such as H, N, and Group 18 elements (the noble gases).

EXAMPLE 9.1 Explaining the Goldschmidt classification

The common ores of Ni and Cu are sulfides. By contrast, Al is obtained from a mixture of the oxide and its hydrate, and Ca from the carbonate. Can these observations be explained in terms of hardness?

Answer We need to assess whether the hard–hard and soft–soft rule applies. From Table 5.4 we know that OH^-, O^{2-}, and CO_3^{2-} are hard bases; S^{2-} is a soft base. The table also shows that the cations Ni^{2+} and Cu^{2+} are considerably softer acids than Al^{3+} or Ca^{2+}. Hence the hard–hard and soft–soft rule accounts for the differentiation observed.

Self-test 9.1 (a) Of the metals Cd, Rb, Cr, Pb, Sr, and Pd, which might be expected to be found in aluminosilicate minerals, coordinated by SiO_4^{4-} and AlO_4^{5-}, and which in sulfides? (b) Predict whether thallium occurs naturally as a sulfide or silicate ore.

Trends in properties

9.2 Atomic parameters

Although this part of the text deals with the *chemical* properties of the elements and their compounds, we need to keep in mind that these chemical properties spring from the *physical* characteristics of atoms. As we saw in Chapter 1, these physical characteristics—the radii of atoms and ions, and the energy changes associated with the formation of ions—vary periodically. Here we review these variations.

(a) Atomic radii

KEY POINTS Atomic radii increase down a group and, within the s and p blocks, decrease from left to right across a period. In the d block the atomic radii of the 5d series are similar to those of the 4d series of elements.

As we saw in Section 1.7, atomic radii increase down a group and decrease from left to right across a period. Across a period, as a result of the increasing nuclear charge there is an increase in effective nuclear charge, despite the mitigating effects of penetration and shielding. This increase draws in the electrons and results in a smaller atom. On descending a group, electrons occupy successive shells outside a completed core and the radii increase (Fig. 9.2).

The increase in radii is relatively small for the elements that follow the d block due to the presence of the poorly shielding d electrons. For example, although there is a substantial increase in atomic radius between C and Si (76 and 111 pm, respectively), the atomic radius of Ge (120 pm) is only slightly greater than that of Si (Fig. 9.2(a)). The radii of isolated d-metal atoms and ions generally decrease on moving to the right, reflecting the poor shielding of d electrons and the increasing effective nuclear charge (Section 1.7). The radius of the metal atoms in the solid element is determined by the strength of the metallic bonding and the number of

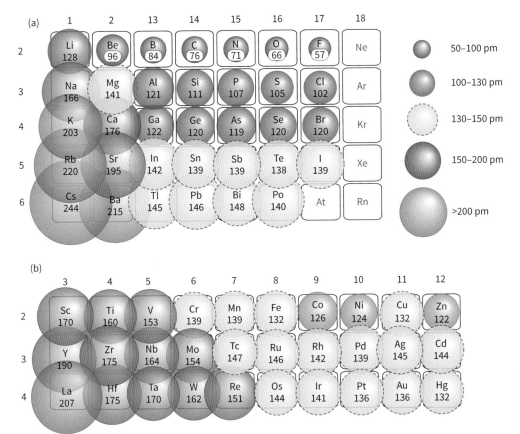

FIGURE 9.2 The variation of atomic radii (in picometres) within the periodic table. (a) Main group (b) d block.

electrons contributing to that. Thus, the separations of the centres of the atoms in the solid generally follow a similar pattern to the melting points: they decrease to the middle of the d block, with the smallest separations occurring where the maximum number of electrons can contribute to metallic bonding. Manganese is exceptional, as discussed later.

An important observation is that despite the increase in principal quantum number, the atomic radii of the elements in the 5d series (Hf, Ta, W, ...) are approximately the same as those of their 4d-series congeners (Zr, Nb, Mo, ...). This interruption can mainly be traced to the **lanthanoid contraction** introduced in Section 1.7(a) and discussed further in Section 10.5.

(b) Ionization energies and electron affinities

KEY POINTS Ionization energy increases across a period and decreases down a group. Electron affinities are highest for elements near fluorine, particularly the halogens.

We need to be aware of the energies needed to form cations and anions of the elements. Ionization energies are relevant to the formation of cations; electron affinities are relevant to the formation of anions.

The ionization energy of an element is the energy required to remove an electron from a gas-phase atom (Section 1.7).

Ionization energies correlate strongly with atomic radii, and elements that have small atomic radii generally have high ionization energies. Therefore, as the atomic radius increases down a group, the ionization energy decreases (Fig. 9.3). An exception is seen in Group 13 where the ionization energy of Ga is greater than that of Al. The increase is due to the intervention of the 3d subshell just before Ga which leads to a step increase in Z_{eff} and a smaller atomic radius for Ga. The **alternation effect** as it is known also manifests itself for the other elements of Period 4, and is discussed further in Section 10.4.

Across each period, higher ionization energies occur when electrons are removed from half-filled or full shells or subshells—a consequence of exchange energy introduced in Section 1.5. Thus, the first ionization energy of nitrogen ($[He]2s^22p^3$) is 1402 kJ mol^{-1}, which is higher than the value for oxygen ($[He]2s^22p^4$, 1314 kJ mol^{-1}). Similarly, the ionization energy of phosphorus (1011 kJ mol^{-1}) is higher than that of sulfur (1000 kJ mol^{-1}). The ionization energies of the 5d-series elements are also higher than expected on the basis of a straightforward extrapolation, some (specifically Au, Pt, Ir, and Os) having such high ionization energies that they are unreactive under normal conditions. The increase is a result of the lanthanoid contraction and the **relativistic influence** that will be described further in Section 10.6.

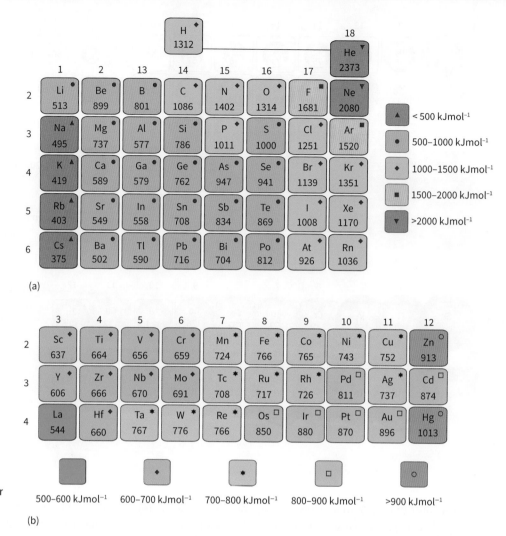

FIGURE 9.3 The variation of first ionization energy (in kilojoules per mole) within the periodic table. (a) Main group, (b) d-block.

The energies of electrons in hydrogenic atoms are proportional to Z^2/n^2. A first approximation to the energies of electrons in many-electron atoms is that they are proportional to Z_{eff}^2/n^2, where Z_{eff} is the effective nuclear charge (Section 1.4), although this proportionality should not be taken definitively. Plots of first ionization energies against Z_{eff}^2/n^2 for the outermost electrons for the elements Li to Ne ($n = 2$) and Hf to Hg ($n = 6$) are shown in Figs. 9.4 and 9.5, respectively. The graphs confirm that the proportionality is broadly followed, especially at high values of n when the outermost electron experiences an interaction with an almost point-like core.

The electron affinity plays a role in assessing the energy required to form an anion. As we saw in Section 1.7(d), an element has a high electron affinity if the additional electron can enter a shell where it experiences a strong effective nuclear charge. Therefore, across a period, elements to the right (other than the noble gases) have the highest electron affinities as Z_{eff} is then large. The addition of an electron to a singly charged anion (as in the formation of O^{2-} from O^-) is invariably unfavourable (that is, the process is endothermic) because it takes energy to push an electron onto a negatively charged species. However, that does not mean it cannot happen, for it is important to assess the overall consequences of ion formation, and it is often the

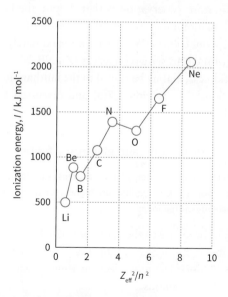

FIGURE 9.4 Plot of first ionization energies against Z_{eff}^2/n^2 for the outermost electrons for the elements lithium to neon ($n = 2$).

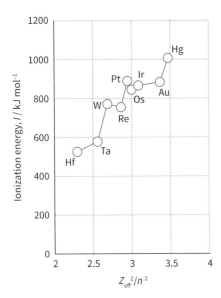

FIGURE 9.5 Plot of first ionization energies against Z_{eff}^2/n^2 for the outermost electrons for the elements hafnium to mercury ($n = 6$).

case that the interaction between highly charged ions in the solid overcomes the additional energy needed to form them. When assessing the energetics of compound formation, it is essential to think globally and not to rule out an overall process simply because an individual step is endothermic. The unfavourable thermodynamics of the transformation of interest are often compensated by the favourable thermodynamics of by-product formation; for example, precipitation of a solid with a high lattice energy.

(c) Electronegativity

KEY POINTS Electronegativity increases across a period and decreases down a group.

We saw in Section 1.7 that the electronegativity, χ, is the power of an atom of the element to attract electrons to itself when it is part of a compound. In general, the electronegativity of main-group elements increases diagonally from the lower left (Cs) to the upper right (F). The trend correlation is most readily understood in terms of the Mulliken definition of electronegativity as the mean of the ionization energy and electron affinity of an element. If an atom has a high ionization energy (so is unlikely to give up electrons) and a high electron affinity (so there are energetic advantages in its gaining electrons), then it is more likely to attract an electron to itself. Although the Mulliken (I and E_a) and

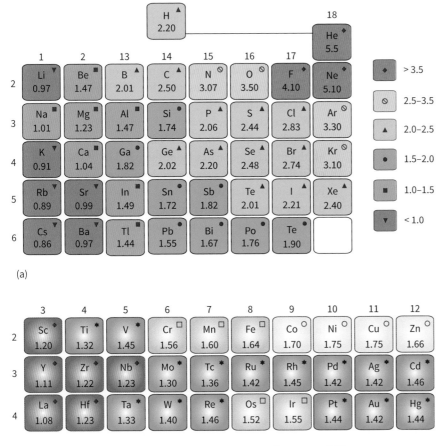

FIGURE 9.6 The variation of Allred–Rochow electronegativity within the periodic table. (a) Main group, (b) d block.

Allred–Rochow (Z_{eff} and covalent radius) (Fig. 9.6) definitions are most useful for understanding the basis for the trends in electronegativity values, the Pauling values, derived empirically from experimental bond energy data, are more commonly used.

There are some exceptions to this general trend. The departure from a smooth decrease down the group between Al and Ga and between Si and Ge is a manifestation of the alternation effect (Section 10.4) due to the intervention of the 3d subshell. We also see an increase in electronegativity values for Tl and Pb, which is attributed to the relativistic stabilization of the valence orbitals and referred to as the **inert pair effect** (Section 10.7).

9.3 Metallic character

KEY POINT The metallic character of the elements decreases across a period and increases down a group.

The chemical properties of the metallic elements can be considered as arising from the ability of the elements to donate their valence electrons into a band and produce metallic bonding (Section 2.11). Consequently, elements with low ionization energies are likely to be metals, and those with high ionization energies are likely to be nonmetals. Thus, as ionization energies decrease down a group the elements become more metallic, and as the ionization energies increase across a row the elements become less metallic (Fig. 9.7).

These trends can also be directly related to the trends in atomic radii, as large atoms typically have low ionization energies and are more metallic in character. This trend is most noticeable within Groups 13–16, where the elements at the head of the group are nonmetals and those at the foot of the group are metals (Sections 14.1, 15.1, 16.1, and 17.1). Within this general trend there are allotropic variations in

TABLE 9.1 Some allotropes of the p-block elements

C		O	
Diamond, graphite, amorphous, fullerenes		Dioxygen, ozone	
	P	S	
	White, red, black	Many catenated rings, chains, amorphous	
	As	Se	
	Yellow, grey (metallic), black	Red (α, β, γ), grey, black	
Sn	Sb		
Grey, white (metallic)	Blue (metallic), yellow, black		
	Bi		
	Amorphous, crystalline (metallic)		

the sense that some elements exist as both metals and nonmetals. An example is Group 15: N and P are nonmetals, As exists as nonmetal, metalloid, and metallic allotropes, and Sb and Bi are metals (Section 16.1). Elements in the p block typically form several allotropes (Table 9.1). All the d-block elements are metallic. Their properties vary from the immensely strong and light titanium, the highly electrically conducting copper, the malleable gold and platinum, and the very dense osmium and iridium. To a large extent these properties derive from the nature of the metallic bonding that binds the atoms together and how that bonding varies between metals.

We introduced the concept of band structure in Chapter 2. Generally speaking, the same band structure is present for all the metals and arises from the overlap of the ns and np orbitals in the main group metals to give an s band and a p band, and overlap of the ns and the $(n-1)d$ orbitals in the d block to give an s band and a d band. The principal difference between the metals is the number of electrons available to occupy these bands: K ($4s^1$) has one bonding electron, Ti ($4s^2 3d^2$) has four bonding electrons, V ($4s^2 3d^3$) five, Cr ($4s^1 3d^5$) six, and so on. The lower, net bonding region of the valence band is, therefore, progressively filled with electrons on going to the right across the block, which results in stronger bonding, until around Group 7 (at Mn, Tc, Re) when the electrons begin to populate the upper, net antibonding part of the band.

This trend in bonding strength is reflected in the increase in melting point from the low-melting alkali metals (effectively only one bonding electron for each atom, resulting in melting points typically less than 100 °C) up to Cr, and its decline thereafter to the low-melting Group 12 metals (Table 9.2). The strength of metallic bonding in tungsten is such that its melting point (3410 °C) is exceeded by that of

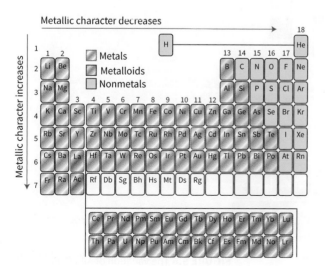

FIGURE 9.7 The variation of metallic character through the periodic table.

TABLE 9.2 Normal melting points of the elements, $\theta_{mp}/°C$

Li	Be										B	C	N	O	F
180	1280										2300	3730	−210	−218	−220
Na	Mg										Al	Si	P	S	Cl
97.8	650										660	1410	44*	113	−110
K	Ca	Sc	Ti	V	Cr	Mn	Fe	Co	Ni	Cu	Zn	Ga	Ge	As†	Se	Br
63.7	850	1540	1675	1900	1890	1240	1535	1492	1453	1083	420	29.8	937	817	217	−7.2
Rb	Sr	Y	Zr	Nb	Mo	Tc	Ru	Rh	Pd	Ag	Cd	In	Sn	Sb	Te	I
38.9	768	1500	1850	2470	2610	2200	2500	1970	1550	961	321	2000	232	630	450	114
Cs	Ba	La	Hf	Ta	W	Re	Os	Ir	Pt	Au	Hg	Tl	Pb	Bi	Po	
28.7	714	920	2220	3000	3410	3180	3000	2440	1769	1063	13.6	304	327	271	254	

* White allotrope.
† Grey allotrope under 28 atm.

only one other element, carbon. The unusually low melting point for mercury is a consequence of the relativistic effect (Section 10.6).

9.4 Oxidation states

The trends in stable oxidation states within the periodic table can be understood to some extent by considering electron configurations.

(a) Main-group elements

KEY POINTS The group oxidation number can be predicted from the electron configuration of an element in the s and p blocks. For the heavier elements there is increasing stability of an oxidation state that is 2 less than the number of valence electrons.

The group oxidation number is the highest possible positive oxidation number based on the group number, and hence number of valence electrons. It has values of N where N is the group number for s- and d-block elements and $N - 10$ for the p-block elements, where it is simpler to consider the number of valence electrons. A noble-gas configuration is achieved in the s and p blocks when eight electrons occupy the s and p subshells of the valence shell. In Groups 1, 2, and 13 the loss of electrons to leave the inner complete shell can be achieved with a relatively small input of energy. Thus, for these elements the oxidation numbers typical of the groups are +1, +2, and +3, respectively.

For the elements at the head of Group 14 to Group 17 it becomes increasingly energetically favourable—provided we consider the overall contributions to the energy, such as the interaction between oppositely charged ions—for the atoms to accept electrons in order to complete the valence shell. Consequently, the observed oxidation numbers are −4, −3, −2, −1 in combination with less electronegative elements. Group 18 elements already have a complete octet of electrons and are neither readily oxidized nor reduced.

The heavier elements of the p block also form compounds where the element has an oxidation number 2 less than the number of valence electrons. The relative stability of an oxidation state in which the oxidation number is 2 less than the number of valence electrons is an example of the **inert pair effect** discussed further in Section 10.7.

(b) The d- and f-block elements

KEY POINTS The group oxidation state can be achieved by elements that lie towards the left of the d block but not by elements on the right. Oxygen is usually more effective than fluorine at bringing out the highest oxidation states because less crowding is involved. Oxidation state M(III) is common to the left of the 3d series and M(II) is common for metals from the middle to the right of the block. The highest oxidation state of an element becomes more stable on descending a group.

The group oxidation number is achieved in the d block only for Groups 3–8, and even then the highly electronegative halides and O are required (Section 20.2). For Group 7 and 8, only O can produce anions or neutral oxides with an element having oxidation numbers +7 and +8, as in the permanganate anion, MnO_4^-, and osmium tetroxide, OsO_4. Oxygen affords the group oxidation state for many elements more readily than does fluorine because fewer O atoms than F atoms are needed to achieve the same oxidation number, thus decreasing steric crowding. The range of observed oxidation states is shown in Table 9.3. As can be seen, up to Mn, all the 3d and 4s electrons can be removed or used in bonding and the maximum oxidation state corresponds to the group number.

The trend in thermodynamic stability of the group oxidation states of the 3d-series elements is illustrated in Fig. 9.8, which shows the Frost diagram (Section 6.13) for species in aqueous acidic solution. We see that the group oxidation states of Sc(III), Ti(V), and V(V) fall in the lower part of the diagram. This location indicates that the element and any species in intermediate oxidation states are readily oxidized to the group oxidation state. By contrast, species in

TABLE 9.3 The range of observed positive oxidation numbers for 3d-series elements

Sc	Ti	V	Cr	Mn	Fe	Co	Ni	Cu	Zn
		+1					+1	+1	
	+2	+2	+2	+2	+2	+2	+2	+2	+2
+3	+3	+3	+3	+3	+3	+3	+3	+3	
	+4	+4	+4	+4	+4	+4	+4	+4	
		+5	+5	+5					
			+6	+6	+6				
				+7					

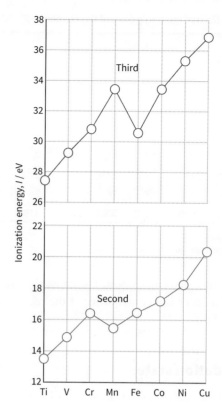

FIGURE 9.9 The second and third ionization energies of the 3d metals.

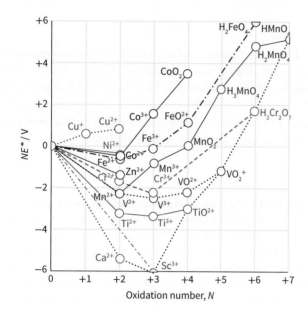

FIGURE 9.8 A Frost diagram for the first series of the d-block elements in acidic solution (pH = 0). The dashed line (----) connects species in their group oxidation states.

the group oxidation state for Cr(VI) and Mn(VII) lie in the upper part of the diagram. This location indicates that they are very susceptible to reduction. The Frost diagram shows that the group oxidation state is not achieved in Groups 8–12 of the 3d series (Fe, Co, Ni, Cu, and Zn), and also shows the oxidation states that are most stable under acidic solution conditions, namely Ti(III), V(III), Cr(III), Mn(II), Fe(II), Co(II), and Ni(II).

Fig. 9.9 shows the second and third ionization energies of the 3d metals, and we can see the expected increase across the period, in line with increasing nuclear charge. The anomalous values for manganese and iron are a result of the very stable d^5 configurations of the Mn^{2+} and Fe^{3+} ions.

As might be expected from the values of the third ionization energy, the M(III) oxidation state is common at the left of the period, and is the only oxidation state normally encountered for scandium. Titanium, vanadium, and chromium all form a wide range of compounds in oxidation state M(III), and under normal conditions the M(III) oxidation state is more stable than the M(II) state. Manganese(II) is especially stable to oxidation due to its half-filled d shell, and relatively few Mn(III) compounds are known. Beyond manganese, many Fe(III) complexes are known, but are often oxidizing. In acid solution, $Co^{3+}(aq)$ is powerfully oxidizing and O_2 is evolved:

$$4Co^{3+}(aq) + 2H_2O(l) \rightarrow 4Co^{2+}(aq) + 4H^+(aq) + O_2(g)$$
$$E^\circ = +0.69V$$

Aqua ions of Ni^{3+} and Cu^{3+} have not been prepared.

In contrast, M(II) becomes increasingly common from left to right across the series. For example, among the early members of the 3d series, $Sc^{2+}(aq)$ is unknown and $Ti^{2+}(aq)$ is formed only upon bombarding solutions of Ti^{3+} with electrons, in the technique known as **pulse radiolysis**. For Groups 5 and 6, $V^{2+}(aq)$ and $Cr^{2+}(aq)$ are also thermodynamically unstable with respect to oxidation by H^+ ions:

$$2V^{2+}(aq) + 2H^+(aq) \rightarrow 2V^{3+}(aq) + H_2(g) \quad E^\circ_{cell} = +0.26V$$

Beyond Cr (for Mn^{2+}, Fe^{2+}, Co^{2+}, Ni^{2+}, and Cu^{2+}), M(II) is stable with respect to reaction with water, and only Fe^{2+} is oxidized by air.

EXAMPLE 9.2 Judging trends in redox stability in the d block

On the basis of trends in the properties of the 3d-series elements, suggest possible M^{2+} aqua ions for use as reducing agents and write a balanced chemical equation for the reaction of one of these ions with O_2 in acidic solution.

Answer We need to identify an element that has an accessible, but oxidizable, M(II) oxidation state. The M(II) state is most stable for the late 3d-series elements, with ions of the metals on the left of the series, such as $V^{2+}(aq)$ and $Cr^{2+}(aq)$, being strong reductants and widely used in mechanistic studies (Chapter 23). As we move to the right, the $Fe^{2+}(aq)$ ion is only weakly reducing. While the $Co^{2+}(aq)$, $Ni^{2+}(aq)$, and $Cu^{2+}(aq)$ ions are not reducing in water. The Latimer diagram (Section 6.12) for iron indicates that Fe^{3+} is the only possible higher oxidation state of iron in acidic solution:

$$Fe^{3+} \xrightarrow{+0.77} Fe^{2+} \xrightarrow{-0.44} Fe$$

The chemical equation for the oxidation is then

$$4Fe^{2+}(aq) + O_2(g) + 4H^+(aq) \rightarrow 4Fe^{3+}(aq) + 2H_2O(l)$$

Self-test 9.2 (a) Refer to the appropriate Latimer diagram in Resource section 3 and identify the oxidation state and formula of the species that is thermodynamically favoured when an acidic aqueous solution of V^{2+} is exposed to oxygen. (b) By examination of the Latimer diagram for Mn in Resource section 3 predict which is the most stable oxidation state in acidic solution.

The stability of high oxidation states increases down a group in Groups 4–12 as the radii increase and higher coordination numbers are more likely. In addition, the high oxidation state compounds are generally halides or oxides in which donation of electrons from the ligands to the metal stabilizes the high oxidation state. In Group 12 the M(II) oxidation state dominates.

The relative stabilities of oxidation states of the 4d and 5d series members of each group are similar, as their atomic radii are very similar (due to the lanthanoid contraction). As already remarked, half-filled shells of electrons with parallel spins are particularly stable due to the exchange energy contribution (Section 1.5). This additional stability has important consequences for the chemistry of the d-block elements with exactly half-filled and completely filled shells.

The importance of spin correlation diminishes as the orbitals become larger because for the more diffuse 4d and 5d orbitals, electron repulsion, which promotes high-spin configurations, is less important. For example, Tc and Re, which are the 4d and 5d counterparts of Mn, do not form M(II) compounds readily. d-Block elements also form stable compounds with the metal in its zero oxidation state. These complexes are typically stabilized by a ligand that acts as a π-acid such as CO. π-Acids, or π-acceptors, are ligands that can accept electron density through the formation of a π bond to the metal (Section 24.5).

TABLE 9.4 Highest oxidation states of the d-block binary halides*

Group							
4	5	6	7	8	9	10	11
TiI_4	VF_5	CrF_5†	MnF_4	$FeBr_3$	CoF_4	NiF_4	$CuBr_2$
ZrI_4	NbI_5	$MoCl_6$	$TcCl_6$	RuF_6	RhF_6	PdF_4	AgF_3
HfI_4	TaI_5	WBr_6	ReF_7	OsF_6	IrF_6	PtF_6	AuF_5

*The formulas show the least electronegative halide that brings out the highest stable oxidation state of the d metal.
† CrF_6 exists for several days at room temperature in an inert vessel.

The increasing stability of high oxidation states for the heavier d metals can be seen in the formulas of their halides (Table 9.4), and the formulas MnF_4, TcF_6, and ReF_7 show a greater ease of oxidizing the 4d- and 5d-series metals than the 3d-series. The hexafluorides of the heavier d metals (as in PtF_6) have been prepared from Group 6 through to Group 10 except for Pd. In keeping with the stability of high oxidation states for the heavier metals, WF_6 is not a significant oxidizing agent. However, the oxidizing character of the hexafluorides increases to the right, and PtF_6 is so potent that it can oxidize O_2 to O_2^+ giving $(O_2)^+[PtF_6]^-$:

$$O_2(g) + PtF_6(s) \rightarrow (O_2)PtF_6(s)$$

Even Xe can be oxidized by PtF_6 (Section 19.5).

Compounds of d metals in low oxidation states often exist as ionic solids, whereas compounds of d metals in high oxidation states tend to take on covalent character: compare OsO_2, which is an ionic solid with the rutile structure, and OsO_4, which is a low-melting-point covalent solid comprised of tetrahedral OsO_4 molecules (Section 20.8). We discussed the effect in Section 1.7(f). Low and zero oxidation state organometallic compounds exist for the d-block elements, and these are discussed in Chapter 24.

The only commonly observed oxidation state for lanthanoid ions in coordination compounds or in solution is the stable Ln(III). Oxidation states other than M(III) occur for the relatively stable empty (f^0), half-filled (f^7), or filled (f^{14}) subshell. Thus, Ce^{3+} is readily oxidized to Ce^{4+} (f^0), and Eu^{3+} can be reduced to Eu^{2+} (f^7). This uniformity in stable oxidation state is reflected in the reduction potentials of the lanthanoids, with values only ranging from −1.99V for $E^{\ominus}(Eu^{3+}/Eu)$ to −2.38V for $E^{\ominus}(La^{3+}/La)$ (an 'honorary' member of the f block, Chapter 21).

The elements from thorium (Th, Z = 90) to lawrencium (Lr, Z = 103) have ground-state electron configurations that involve the filling of the 5f subshell, and in this sense are analogues of the lanthanoids. However, the early members of the actinoid series do not exhibit the chemical uniformity of the lanthanoids and form compounds in a variety of oxidation states up to +6. The +3 oxidation state becomes predominant in Am and beyond. The 5f orbitals of the actinoids lie higher in energy than the 4f and for the early actinoids remain at a similar energy to the 6d and 7s orbitals. As a result, they participate in bonding for elements up to berkelium (Chapter 21).

Trends in compounds

9.5 Amphoterism

KEY POINTS The boundary between metals and nonmetals in the periodic table is characterized by the formation of amphoteric oxides; amphoterism also varies with the oxidation state of the element.

The term 'amphoteric' was introduced in Section 5.4 and is used to describe oxides that react with both acids and bases. Amphoterism is observed for the lighter elements of Groups 2 and 13, as in BeO, Al_2O_3, and Ga_2O_3. It is also observed for some of the d-block elements in high oxidation states, in which the central atom is very electron withdrawing, and some of the heavier elements of Groups 14 and 15, such as SnO_2 and Sb_2O_3.

Figure 9.10 shows the location of elements that in their characteristic main group oxidation states have amphoteric oxides. They lie in the region between acidic and basic oxides, and hence serve as an important guide to the metallic or nonmetallic character of an element. The onset of amphoterism correlates with a significant degree of covalent character in the bonds formed by the elements, either because the metal ion has a high charge density (Section 10.3) and is strongly polarizing (as for Be) or because the metal ion is polarized by the O atom attached to it (as for Sb).

An important issue in the d block is the oxidation number necessary for amphoterism. Figure 9.11 shows the oxidation number for which an element in the first row of the block has a stable amphoteric oxide. We see that on the left of the block, from titanium to manganese, oxidation state +4 is amphoteric (with higher values on the border of acidic and lower values on the border of basic). On the right of the block, amphoterism occurs at lower oxidation numbers: the oxidation states M(III) for cobalt and nickel and M(II) for copper and zinc are fully amphoteric. There is no simple way of predicting the onset of amphoterism. The degree of covalence typically increases with the oxidation number of the metal as the increasingly positively charged cation becomes more strongly polarizing (Section 1.7).

EXAMPLE 9.3 Using oxide acidity in qualitative analysis

In the traditional scheme of qualitative analysis, a solution of metal ions is oxidized and then aqueous ammonia is added to raise the pH. The ions Fe^{3+}, Ce^{3+}, Al^{3+}, Cr^{3+}, and V^{3+} precipitate as hydrated hydroxides. The addition of H_2O_2 and NaOH redissolves the aluminium, chromium, and vanadium oxides. Discuss these steps in terms of the acidities of oxides.

Answer When the oxidation number of the metal is +3, all the metal oxides are sufficiently basic to be insoluble in a solution with pH ≈ 10. Aluminium(III) oxide is amphoteric and redissolves in alkaline solution to give aluminate ions, $[Al(OH)_4]^-$. Vanadium(III) and chromium(III) oxides are oxidized by H_2O_2 to give vanadate ions, $[VO_4]^{3-}$, and chromate ions, $[CrO_4]^{2-}$, which are the anions derived from the acidic oxides V_2O_5 and CrO_3, respectively.

Self-test 9.3 If Ti(IV) ions were present in the sample, how would they behave in their reaction with water?

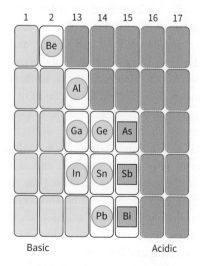

FIGURE 9.10 The location of elements having amphoteric oxides. The circled elements have amphoteric oxides in all oxidation states. The elements in boxes have acidic oxides in the highest oxidation state and amphoteric oxides in lower oxidation states.

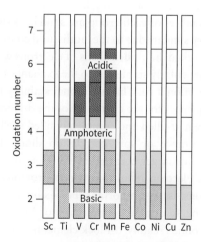

FIGURE 9.11 The oxidation numbers for which elements in the first row of the d-block have amphoteric oxides. Predominantly acidic oxides are shown by the hatched pink-shaded blocks; predominantly basic oxides are shown by the dotted blue-shaded blocks.

9.6 Coordination numbers

KEY POINTS Low coordination numbers generally dominate for small atoms; high coordination numbers are possible as a group is descended. The 4d- and 5d-series elements often exhibit higher coordination numbers than their 3d-series congeners; compounds of d metals in high oxidation states tend to have covalent structures.

The coordination number of an atom in a compound depends very much on the relative sizes of the central atom and the surrounding atoms. In the p block, low coordination numbers are most common for the compounds of Period 2 elements, but higher coordination numbers are observed as each group is descended and the radius of the central atom increases. For example, in Group 15, N forms three-coordinate molecules such as NCl_3 and four-coordinate ions such as NH_4^+, whereas its congener P forms molecules with coordination numbers of 3 and 5, as in PCl_3 and PCl_5, and six-coordinate ionic species such as PCl_6^- (Section 16.11). This higher coordination number in Period 3 elements is an example of hypervalence, and as remarked on in Section 2.6(b), it is likely to be due to the possibility of arranging a greater number of atoms or molecules around the larger central atom and the filling of low-energy bonding molecular orbitals.

In the d block the 4d and 5d series of elements tend to exhibit higher coordination numbers than the 3d series elements because of their larger radii (Section 20.2). With the small F^- ligand the 3d series metals tend to form six-coordinate complexes, but the larger 4d and 5d series metals in the same oxidation state tend to form seven-, eight-, and nine-coordinate complexes. The octacyanidomolybdate complex, $[Mo(CN)_8]^{3-}$, illustrates the tendency towards high coordination numbers with less sterically demanding ligands.

The radii of the lanthanoid ions, Ln^{3+}, contract steadily across the lanthanoid series. This decrease is attributed in part to the increase in Z_{eff} as electrons are added to the 4f subshell, but relativistic effects also start to contribute (Sections 1.7(a) and 10.6). Very high coordination numbers have been observed for the large atoms and ions of the f block. For example, Nd forms the 9-coordinate $[Nd(OH_2)_9]^{3+}$ and Th forms the 10-coordinate $[Th(C_2O_4)_4(OH_2)_2]^{4-}$ ion. Examples of 11- and 12-coordinate species are $[Th(NO_3)_4(OH_2)_3]$ (**19b**, Section 21.10) and $[Ce(NO_3)_6]^{2-}$, in which the NO_3^- is bound to the metal through two O atoms (Section 7.5). Thorium also forms the 15-coordinate $[Th(H_3BNMe_2BH_3)_4]$ and 16-coordinate $[Li(Et_2O)_2][Th(BH_4)_5]$ complexes in which the Th is bonded to 15 and 16 H atoms respectively.

9.7 Bond enthalpy trends

KEY POINTS As a general rule, bond enthalpies decrease down each group for p-block elements and increase down each group for d-block elements: but there are notable exceptions. The overall trend is for bond enthalpies to decrease from left to right across the d block.

Mean E–X bond enthalpies in compounds where E has a lone pair typically decrease down a group in the p block. However, the bond enthalpy for an element at the head of a group in Period 2 is anomalous and is smaller than that of the corresponding element in Period 3:

B/(kJ mol^{-1})		B/(kJ mol^{-1})	
N–N	163	N–Cl	200
P–P	201	P–Cl	319
As–As	180	As–Cl	317

The relative weakness of single bonds between atoms of the later Period 2 elements (nitrogen, oxygen, fluorine) is often ascribed to the electron–electron repulsion that is much more marked for the smaller atoms (Section 10.2). For a p-block element E that has no lone pairs, the E–X bond enthalpy generally decreases down a group, but now the electron–electron repulsion is diminished and the homonuclear bond strength at the head of the group is not weaker than those below it:

B/(kJ mol^{-1})		B/(kJ mol^{-1})	
C–C	348	C–Cl	338
Si–Si	226	Si–Cl	381
Ge–Ge	188	Ge–Cl	342

In the absence of the lone pair/lone pair repulsion the smaller atoms form stronger bonds as the shared electrons are closer to each of the atomic nuclei and orbital overlap is more efficient. The strength of the Si–Cl bond can be attributed to the fact that the atomic orbitals of the two elements have similar energies and efficient overlap. High bond enthalpy values like this are also sometimes attributed to a contribution from π-bonding involving d orbitals but are more appropriately ascribed to an ionic contribution to the bonding (Section 2.14). Thus, if we consider the M–X bonds (X=halogen):

B/(kJ mol^{-1})		B/(kJ mol^{-1})	
C–F	485	Si–F	565
C–Cl	327	Si–Cl	381
C–Br	285	Si–Br	310
C–I	213	Si–I	234

we see that bond strengths decrease as the electronegativity difference decreases and the ionic contribution decreases.

One application of bond enthalpy arguments concerns the existence or otherwise of subvalent compounds—compounds in which fewer bonds are formed than valence rules suggest, such as PH_2. Although this compound is thermodynamically stable with respect to dissociation into the constituent atoms, it is unstable with respect to disproportionation:

$$3PH_2(g) \rightarrow 2PH_3(g) + \tfrac{1}{4}P_4(s)$$

The origin of the spontaneity of this reaction is the strength of the P–P bonds in molecular phosphorus, P_4. There are the same number (six) of P–H bonds in the reactants as there are in the products, but the reactants have no P–P bonds.

In the d block, bond enthalpies generally increase down a group—a trend that is the opposite of the general trend for the p block. For example, consider the following M–H, M–C, and M–O bond strengths:

B/(kJ mol^{-1})		B/(kJ mol^{-1})	
Cr–H	258	Fe–C	390
Mo–H	282	Ru–C	528
W–H	339	Os–C	598

B/(kJ mol^{-1})		B/(kJ mol^{-1})	
Ti–O	667	Zr–O	766
V–O	637	Nb–O	727
Cr–O	461	Mo–O	525
Mn–O	362	Tc–O	548
Fe–O	407	Ru–O	528
Co–O	397	Rh–O	405

As we saw in Section 9.2, d orbitals appear to become more effective at forming bonds down a group as they expand in size, from being too contracted for effective overlap with the 1s, 2s, and 2p orbitals of hydrogen, carbon, and oxygen, to more optimal for overlap.

The periodic trend is for bond enthalpies to decrease from left to right across the d block. This can be attributed to the consequences of increasing effective nuclear charge, both through d orbital contraction, reducing orbital overlap, and the decreasing ionic contribution, as the electronegativity of the metal increases.

Where trends collide

9.8 Wider aspects of periodicity

The differences in chemical properties of elements and compounds result from a complex interplay of periodic trends. In this section we illustrate how these trends compensate, conflict with, and enhance one another.

(a) Ionic chlorides

KEY POINT For ionic compounds, trends in lattice enthalpies, ionization energies, and enthalpies of atomization have significant effects on the enthalpy of formation of ionic halides.

As can be seen from Table 9.5, the values of $\Delta_f H^\ominus$ for the Group 1 halides are reasonably constant on descending a group. The ionization energy and enthalpy of atomization both become less positive down the group as the atomic radii increase but these trends are largely offset by changes in the lattice enthalpy, which becomes less favourable as the cations increase in size (Section 4.11). The values of $\Delta_f H^\ominus$ for Group 2 halides are up to twice the values for the Group 1 halides. The decrease in ionization enthalpy down Group 2 is not offset by the decreasing lattice energy to as great an extent as in Group 1, leading to more favourable enthalpies of formation, especially for the heavier congeners.

The ionization energy and enthalpy of atomization both become more positive across a row in the periodic table. However, the more significant factor is the very large increase in lattice enthalpy as the radii decrease and the charges on the ions increase. The influence of these combined factors can be seen in the values of $\Delta_f H^\ominus$ for KCl, $CaCl_2$, and $ScCl_3$, which are -436, -795, and -925 kJ mol^{-1}, respectively (Sections 12.7 and 13.7).

(b) Covalent halides

KEY POINT Bond enthalpy and entropy effects are the most important factors in determining whether or not Group 16 halides exist.

Compounds formed between sulfur and the halogens can provide some insight into the factors that influence the values of $\Delta_f H^\ominus$ for covalent halides (Section 17.7). Sulfur forms several different compounds with F, most of which are gases. Sulfur hexafluoride, SF_6, sulfur difluoride, SF_2, and sulfur dichloride, SCl_2, exist, whereas

TABLE 9.5 Standard enthalpies of formation of Group 1 and 2 chlorides, $\Delta_f H^\ominus$/kJ mol^{-1}

LiCl	−409	$BeCl_2$	−512
NaCl	−411	$MgCl_2$	−642
KCl	−436	$CaCl_2$	−795
RbCl	−431	$SrCl_2$	−828
CsCl	−433	$BaCl_2$	−860

SCl_6 is not known. The values of $\Delta_f H^\ominus$ calculated from bond enthalpy data are:

	SF_2	SF_6	SCl_2	SCl_6
$\Delta_f H^\ominus/(\text{kJ mol}^{-1})$	−298	−1220	−49	−74

Thus, although the formation of SCl_6 is more exothermic than SCl_2, other factors including steric crowding ensure that SCl_6 cannot be prepared under standard conditions. The explanation can be found by considering the bond enthalpies of sulfur–halogen bonds, along with respective F–F and Cl–Cl bonds:

	F–SF	F–SF$_5$	Cl–SCl	F–F	Cl–Cl
$B/(\text{kJ mol}^{-1})$	367	329	271	155	242

There is a decrease in bond enthalpy between SF_2 and SF_6, possibly due to steric crowding around the S atom and repulsion between crowded F atoms. A similar decrease would be expected between SCl_2 and SCl_6. This weak bonding is one factor in the nonexistence of SCl_6, the other being that a Cl–Cl bond is much stronger than a F–F bond.

This explanation emphasizes the importance of considering *all* the species involved in a possible reaction; thus, when we compare the thermodynamics of decomposition of SCl_6 (which releases Cl_2) with the analogous decomposition of SF_6, we find that the latter is stable also because the F–F bond is nearly 90 kJ mol^{-1} weaker than the Cl–Cl bond. In contrast, compounds containing the PCl_6^- ion are known. Bonding between P and Cl should be stronger than that between S and Cl because P is less electronegative than S. Compounds containing the PCl_6^- ion will also be stabilized by the lattice energy of the ionic solid.

EXAMPLE 9.4 Assessing the factors affecting the formation of a compound

Estimate $\Delta_f H^\ominus$ (SH_6, g) by assuming that B(H–S) is the same for H–SH$_5$ as for H–SH (375 kJ mol^{-1}). The value of B(H–H) is 436 kJ mol^{-1} and $\Delta_{atm} H^\ominus$ (S,s) 218 kJ mol^{-1}. Suggest a way of reconciling your answer with observation.

Answer We estimate $\Delta_f H^\ominus$ (SH_6, g) from the difference between the enthalpies of bonds broken and bonds formed in the reaction

$$\tfrac{1}{8} S_8(s) + 3H_2(g) \rightarrow SH_6(g)$$

The enthalpy change accompanying bond breaking is $\tfrac{1}{8} \times 279$ kJ mol^{-1} + 3 × (436 kJ mol^{-1}) = 1587 kJ mol^{-1}. The enthalpy change accompanying bond making is −6 × (375 kJ mol^{-1}) = −2250 kJ mol^{-1}. Therefore,

$$\Delta_f H^\ominus (SH_6, g) = 1587 \text{ kJ mol}^{-1} - 2250 \text{ kJ mol}^{-1} = -663 \text{ kJ mol}^{-1}$$

indicating that the compound is exothermic and, on the basis of this calculation, can be expected to exist. However, SH_6 does not exist. The reason must lie in the S–H bond being much weaker than that used in the calculation. A further contribution to the Gibbs energy of formation is the unfavourable entropy change due to forming the molecule from three H_2 molecules instead of only one when SH_2 is formed.

Self-test 9.4 (a) Comment on the following $\Delta_f H^\ominus$ values (kJ mol^{-1}):

S(g)	Se(g)	Te(g)	SF$_4$	SeF$_4$	TeF$_4$	SF$_6$	SeF$_6$	TeF$_6$
+279	+227	+197	−762	−850	−1036	−1220	−1030	−1319

(b) Use the data above to calculate a value of B(Te–F) in kJ mol^{-1}.

(c) Ionic oxides

KEY POINT The ionic model is more appropriate for the 3d-element oxides than for the 4d oxides.

The contrasting enthalpies of formation of various metal oxides of formula MO (Table 9.6) provide some interesting insights into different aspects of periodicity (Section 13.8). The high exothermicity of the Group 2 oxide is a result of the relatively low ionization energy and low enthalpy of atomization of the s-block metals. The experimental lattice enthalpies are very close to those calculated from the Kapustinskii equation, indicating that the compound conforms well to the ionic model. Values of $\Delta_f H^\ominus$ (MO) for the 3d-series elements become less negative across the period. There are opposing trends across this series because, although the ionization energy increases, the atomization enthalpy decreases. The experimental lattice enthalpies of the 4d oxides deviate from those calculated from the Kapustinskii equation, indicating that the ionic model is no longer adequate.

TABLE 9.6 Some thermodynamic data (in kJ mol^{-1}) for metal oxides, MO

	$\Delta_{ion(1+2)} H^\ominus$	$\Delta_a H^\ominus$		$\Delta_f H^\ominus$	$\Delta_L H^\ominus$(calc*)	$\Delta_L H^\ominus$(exp†)
Ca	1735	177	CaO	−636	3464	3390
V	2064	514	VO	−431	3728	4037
Ni	2490	430	NiO	−240	4037	4436
Pd	2674	380	PdO	−378	4257	2172

* As calculated from the Kapustinskii equation.
† As calculated from a Born–Haber cycle.

EXAMPLE 9.5 Predicting the thermal stabilities of d-block oxides

Compare the stabilities of V_2O_5 and Nb_2O_5 towards thermal decomposition in the reaction

$$M_2O_5(s) \rightarrow 2MO(s) + \tfrac{3}{2} O_2(g)$$

Use the data in Table 9.6 and the enthalpies of formation of NbO, Nb_2O_5, and V_2O_5 of −406, −1901, and −1552 kJ mol⁻¹, respectively.

Answer We need to consider the reaction enthalpy for each oxide. The enthalpies of reaction can be calculated from the difference between the enthalpies of formation of products and those of the reactants:

For Nb_2O_5:

$$\Delta_r H^\ominus = 2(-406 \text{ kJ mol}^{-1}) - (-1901 \text{ kJ mol}^{-1}) = +1089 \text{ kJ mol}^{-1}$$

For V_2O_5:

$$\Delta_r H^\ominus = 2(-431 \text{ kJ mol}^{-1}) - (-1552 \text{ kJ mol}^{-1}) = +690 \text{ kJ mol}^{-1}$$

The enthalpy of the reaction for V_2O_5 is less endothermic than that of Nb_2O_5. Therefore, V_2O_5 is thermally less stable.

Self-test 9.5 Given that $\Delta_f H^\ominus$ (P_4O_{10}, s) = −3012 kJ mol⁻¹, what further data would be useful when drawing comparisons with the value for V_2O_5?

(d) Noble character

KEY POINT Metals on the right of the d block tend to exist in low oxidation states and form compounds with soft ligands.

With the exception of Group 12, the metals at the lower right of the d block are resistant to oxidation. This resistance is largely due to strong intermetallic bonding and high ionization energies. It is most evident for Ag, Au, and the 4d- and 5d-series metals in Groups 8–10 (Fig. 9.12). The latter are referred to as the **platinum metals** because they occur together in platinum-bearing ores. In recognition of their traditional use, Cu, Ag, and Au are referred to as the **coinage metals**. Gold occurs as the metal; silver, gold, and the platinum metals are also recovered in the electrolytic refining of copper.

Copper, silver, and gold are not susceptible to oxidation by H⁺ under standard conditions, and this noble character accounts for their use, together with platinum, in jewellery and ornaments. Aqua regia, a 3:1 mixture of concentrated hydrochloric and nitric acids, is an old but effective reagent for the oxidation of gold and platinum. Its function is twofold: the NO_3^- ions provide the oxidizing power, and the Cl⁻ ions act as complexing agents. The overall reaction is

$$Au(s) + 4H^+(aq) + NO_3^-(aq) + 4Cl^-(aq) \rightarrow [AuCl_4]^-(aq) + NO(g) + 2H_2O(l)$$

The active species in solution are thought to be Cl_2 and NOCl, which are generated in the reaction

$$3HCl(aq) + HNO_3(aq) \rightarrow Cl_2(aq) + NOCl(aq) + 2H_2O(l)$$

Oxidation number preferences are erratic in Group 11. For Cu, the +1 and +2 states are most common, but for Ag +1 is typical, and for Au +1 and +3 are common. The simple aqua ions Cu⁺(aq) and Au⁺(aq) undergo disproportionation in aqueous solution:

$$2Cu^+(aq) \rightarrow Cu(s) + Cu^{2+}(aq)$$
$$3Au^+(aq) \rightarrow 2Au(s) + Au^{3+}(aq)$$

Complexes of Cu(I), Ag(I), and Au(I) are often linear. For example, $[H_3NAgNH_3]^+$ forms in aqueous solution and linear $[XAgX]^-$ complexes have been identified by X-ray crystallography. The currently preferred explanation for the tendency towards linear coordination is the similarity in energy of the outer nd and $(n + 1)s$ and $(n + 1)p$ orbitals, which permits the formation of collinear spd hybrids (Fig. 9.13).

The soft Lewis acid character of Cu⁺, Ag⁺, and Au⁺ is illustrated by their affinity order, which is I⁻ > Br⁻ > Cl⁻. Complex formation, as in the formation of $[Cu(NH_3)_2]^+$ and $[AuI_2]^-$, provides a means of stabilizing the +1 oxidation state of these metals in aqueous solution. Many tetrahedral complexes are also known for Cu(I), Ag(I), and Au(I) (Section 7.8).

Square-planar complexes are common for the platinum metals and gold in oxidation states that yield the d^8 electron configuration, which include Rh(I), Ir(I), Pd(II), Pt(II), and Au(III) (for example, Section 20.10(d)). An example is $[Pt(NH_3)_4]^{2+}$. Characteristic reactions for these complexes are ligand substitution (Section 23.1) and, except for Au(III) complexes, oxidative addition (Section 24.22).

The noble character that these elements developed across the d block is suddenly lost at Group 12 (Zn, Cd, Hg), where the metals once again become susceptible to atmospheric oxidation. The greater ease of oxidation of the Group 12 metals is due to a decrease in the extent of intermetallic bonding and an abrupt stabilization of the d^{10} subshell.

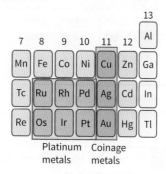

FIGURE 9.12 The location of the platinum and coinage metals in the periodic table.

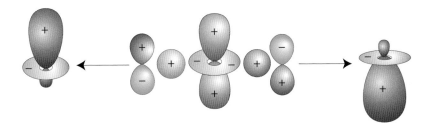

FIGURE 9.13 The hybridization of s, p_z, and d_z^2 with the choice of phases shown here produces a pair of collinear orbitals that can be used to form strong σ bonds.

FURTHER READING

P. Enghag, *Encyclopaedia of the Elements*. John Wiley & Sons (2004).

D. M. P. Mingos, *Essential Trends in Inorganic Chemistry*. Oxford University Press (2004). Includes a detailed discussion of the important horizontal, vertical, and diagonal trends in the properties of the atoms.

N. C. Norman, *Periodicity and the s- and p-Block Elements*. Oxford University Press (1997). Includes coverage of essential trends and features of s-block chemistry.

E. Scerri, *The Periodic Table: Its Story and its Significance*, 2nd ed. Oxford University Press (2019). A lively account of the development of the periodic system from early discoveries of elements to the impact of quantum theory.

C. Benson, *The Periodic Table of the Elements and Their Chemical Properties*. Kindle edition. MindMelder.com (2009).

EXERCISES

9.1 Give the expected maximum stable oxidation state for (a) Ba, (b) As, (c) P, (d) Cl, (e) Ti, (f) Cr.

9.2 The elements vary from metals through metalloids to nonmetals. They form fluorides with oxidation numbers +5 and +3 and the hydrides are all toxic gases. Identify this group of elements.

9.3 The elements are all metals. The most stable oxidation number at the top of the group is +3, the most stable at the bottom is +6. Identify the group of elements.

9.4 Draw a Born–Haber cycle for the formation of the hypothetical compound $NaCl_2$. State which thermochemical step is responsible for the fact that $NaCl_2$ does not exist.

9.5 Summarize the relationship between ionic radii, ionization energy, and metallic character.

9.6 Which one of each of the following pairs has the higher first ionization energy: (a) Be or B, (b) C or Si, (c) Cr or Mn?

9.7 Which one of each of the following pairs is the more electronegative: (a) Na or Cs, (b) Si or O?

9.8 Classify each of the following oxides as acidic, basic, or amphoteric: (a) Na_2O, (b) P_2O_5, (c) ZnO, (d) SiO_2, (e) Al_2O_3, (f) MnO.

9.9 In the traditional scheme for the separation of metal ions from solution that is the basis of qualitative analysis, ions of Au, As, Sb, and Sn precipitate as sulfides but redissolve on addition of excess ammonium polysulfide. By contrast, ions of Cu, Pb, Hg, Bi, and Cd precipitate as sulfides but do not redissolve. In the language of Section 9.5, the first group is amphoteric for reactions involving SH^- in place of OH^-. The second group is less acidic. Locate the amphoteric boundary in the periodic table for sulfides implied by this information. Compare this boundary with the amphoteric boundary for oxides shown in Fig. 9.10. Does this analysis agree with describing S^{2-} as a softer base than O^{2-}?

9.10 Place the following halides in order of increasing covalent character: CrF_2, CrF_3, CrF_6.

9.11 Use the following data to calculate average values of $B(Se–F)$ in SeF_4 and SeF_6. Comment on your answers in view of the corresponding values for $B(S–F)$ in SF_4 (+340 kJ mol^{-1}) and SF_6 (+329 kJ mol^{-1}): $\Delta_a H^\ominus$ (Se) = +227 kJ mol^{-1}, $\Delta_a H^\ominus$ (F) = +159 kJ mol^{-1}, $\Delta_f H^\ominus$ (SeF_6, g) = −1030 kJ mol^{-1}, $\Delta_f H^\ominus$ (SeF_4, g) = −850 kJ mol^{-1}.

9.12 Use the bond enthalpy data in Table 2.7 to calculate the enthalpy of formation for NF_3 and NCl_3. Explain why NF_3 is thermodynamically stable whereas NCl_3 is unstable and reactive.

9.13 Explain why the oxides of N(III), P(III), and As(III) are acidic, whereas Sb(III) oxide is amphoteric and Bi(III) oxide is basic.

TUTORIAL PROBLEMS

9.1 In the paper 'Diffusion cartograms for the display of periodic table data' (*J. Chem. Educ.*, 2011, 88, 1507), M. J. Winter describes a technique commonly used in geography to represent periodic trends. Write a critique of this method by comparing two of the cartograms with two of the figures used in this chapter.

9.2 In the paper 'What and how physics contributes to understanding the periodic law' (*Found. Chem.*, 2001, 3, 145), V. Ostrovsky describes the philosophical and methodological approaches taken by physicists to explain periodicity. Compare and contrast the approaches of physics and chemistry to explaining chemical periodicity.

9.3 Many models of the periodic table have been proposed since the version devised by Mendeleev. Review the more recent versions and discuss the theoretical basis for each one.

9.4 How many electronegativity scales have been proposed? Discuss the relative merits and most appropriate application of each scale.

Themes across the periodic table

10

Causes

10.1 Presence of unpaired electrons

10.2 Anomalous nature of the first member of each group

10.3 Diagonal relationships

10.4 The alternation effect

10.5 The lanthanoid contraction

10.6 Relativistic effects

Consequences

10.7 The inert pair effect

10.8 The oxo wall

10.9 Group n versus Group *n* + 10

Further reading

Exercises

Tutorial problem

Chapters 1 and 9 have described the periodic table and its ability to systematically account for the properties and chemistry of the elements. However, there is significant diversity in the properties of the individual elements that can only be rationalized and understood in terms of their specific electronic structures and valence orbitals. This chapter explains in detail many of the more distinctive (and distinctively named) characteristic behaviours of sets, rows, or regions of elements in the periodic table that are commonly referred to when discussing periodicity.

Causes

10.1 Presence of unpaired electrons

KEY POINTS Compounds with unpaired electrons are most commonly formed with elements in which the 2p, 3d, and 4f orbitals are being filled. These orbitals lack a radial node, thus disfavouring occupancy by two electrons. Radicals are less common for heavier elements in each block.

Radicals, molecules, and ions in which there are unpaired electrons, appear among the various blocks of the periodic table with increasing abundance in the order s < p < d < f (Fig. 10.1). Within each block, the occurrence of radicals also depends on principal quantum number (n), with the 2p, 3d, and 4f rows hosting by far the greater number.

To understand these observations, we need to consider several factors. Firstly, the s-block elements are highly electropositive and, with few exceptions, form ionic compounds in which they attain a noble gas (closed shell) electron configuration. Similar considerations apply to the highly electropositive elements of Group 3, Zr and Hf in Group 4, and more marginally to Nb and Ta in Group 5. Secondly, for the p, d, and f blocks, radicals are most abundant in the first rows: 2p, 3d, and 4f orbitals, being the first occurrence of each orbital type, lack a radial node, leading to compact orbitals and large electron–electron repulsion effects (Sections 1.3 and 1.5). Radial nodes 'push out' the outer maxima of radial distribution functions, and electron–electron repulsions are lowered as the orbital becomes more diffuse.

The strong influence of electron–electron repulsion in orbitals lacking a radial node accounts partly for the characteristic properties of the lighter transition elements and the lanthanoids: the other factor is exchange energy—a quantum-mechanical effect (Section 1.5) which affords additional stability to systems in which pairs of electrons with parallel spins in degenerate orbitals can exchange their position. One consequence of strong electron–electron repulsions is that each successive ionization stabilizes the other electrons and successive ionization energies increase

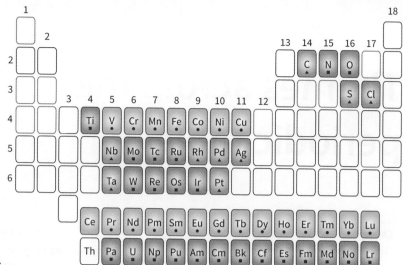

FIGURE 10.1 Occurrence of elements forming compounds having one or more unpaired electrons. Green shading + (●) represents elements normally forming compounds with unpaired electrons, blue shading + (■) represents elements for which stable radicals are common, and red shading + (▲) represents elements for which stable radicals are rare.

steeply, making it more favourable to retain partially filled orbitals than reach the group oxidation state.

In some larger atoms, potential radicals are removed by covalent bond formation; thus, whereas N radicals are common (NO, NO$_2$), compounds of heavier p-block elements, E, in oxidation states that should require them to be radicals are found instead to have E–E bonds. Examples of the latter include Ga–Ga in [Ga$_2$Cl$_6$]$^{2-}$ (Ga(II)) and P–P in P$_2$O$_6^{2-}$ (P(IV)). Sulfur is an interesting borderline case: radicals such as S$_2$O$_4^{2-}$ (the highly reactive dithionite ion) and S$_3^-$ (stabilized in the mineral *lapis lazuli*, and the source of its colour) are common, while the strong disulfide bond RS–SR (S$^-$) determines the structures of many proteins.

10.2 Anomalous nature of the first member of each group

KEY POINTS *The first member of each group within the p block shows differences from the rest of the group that are attributed to smaller size, a larger degree of electron–electron repulsion, and a much greater ability to form multiple bonds. Within the d block, the 3d metals form compounds that tend to be more ionic and in a lower oxidation state than the corresponding 4d and 5d elements. Within the f block, compounds of the 4f elements are almost exclusively ionic compounds containing the Ln^{3+} cation, whereas the 5f elements exhibit much more covalence and variation in oxidation state.*

The first periods of each block are atypical. Most obviously, the chemistry of hydrogen is very different from that of the other s-block elements, as we will see in Chapter 11. Whereas most group 1 compounds are ionic in nature, lithium salts have some covalent character to their bonding. The chemistry of beryllium differs significantly from the other Group 2 elements. The oxide of Be is amphoteric, and reacts with an excess of OH$^-$ ions, to form [Be(OH)$_4$]$^{2-}$; no equivalent chemistry is observed for the basic Mg oxide. Be$_2$C contains the C^{4-} ion and hydrolysis produces methane; the other Group 2 carbides contain the C$_2^{2-}$ ion and yield ethyne on hydrolysis. Be forms covalent hydrides and halides; the analogous compounds of the other Group 2 elements are predominantly ionic.

In the p block, the 2p elements display a number of important differences compared to their congeners. These anomalies are attributable to the small atomic radius and the correlated properties: high ionization energies; high electronegativities; lower coordination numbers and (for the elements to the right of the period) particularly high electron-electron repulsion. For example, in Group 14 carbon forms countless catenated hydrocarbons with strong C–C bonds (Section 15.2). Carbon also forms strong multiple bonds in the alkenes and alkynes. This tendency to catenation is much decreased for its congeners as the E–E bond enthalpy decreases, and the longest silane formed contains just seven Si atoms (Section 15.2). The greater atomic size of silicon means it can exhibit a coordination number of 6, while carbon is normally limited to 4. Nitrogen shows distinct differences from phosphorus and the rest of Group 15 (Section 16.1). Thus, nitrogen commonly exhibits a coordination number of 3, as in NF$_3$, and a coordination number of 4 in species such as NH$_4^+$ and NF$_4^+$, whereas the larger phosphorus can form three- and five-coordinate compounds, such as PF$_3$ and PF$_5$, and six-coordinate species such as PF$_6^-$.

A major difference between 2p and 3p elements is their differing propensity for π bonding. The extent to which p orbitals can interact with each other to form a π bond depends greatly on the internuclear separation (Fig. 10.2); thus, whereas C, N, O, and F atoms are able to form strong p-p π interactions (yielding strongly bonding and

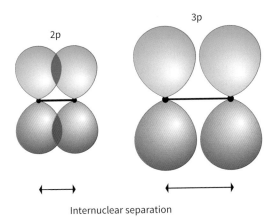

FIGURE 10.2 Comparison between p overlap for elements in Periods 2 and 3.

antibonding orbitals), the same is not true for Si, P, S, and Cl. One or two π bonds may be formed, noting that the overall bond order is governed by the relative occupancies of bonding and antibonding orbitals; indeed, the full occupancy of strongly antibonding orbitals is one reason why the bonding in F_2 is weak compared to that of Cl_2.

The strength of π bonding accounts for: (a) why CO_2 is a gas whereas SiO_2 is an extended solid (see Example 10.1); (b) why N_2 is such a stable molecule; (c) why elemental oxygen exists as O_2 while sulfur occurs as S_8 rings. In each case thermodynamic arguments and cycles can be used to demonstrate that the strong p-p π interactions in the 2p row elements stabilize multiple bonds while weaker p-p π interactions for 3p row elements mean that molecular structures with several single bonds are more stable.

EXAMPLE 10.1 Explaining trends in structures and stabilities

Using the following data, explain why CO_2 occurs as discrete molecules whereas SiO_2 exists as a solid that consists of SiO_4 tetrahedra interconnected by Si–O–Si bridges (Section 15.3). Mean bond enthalpies $B(X-Y)$ in kJ mol^{-1}: C–O, 360; C=O, 743; Si–O, 466; Si=O, 642.

Answer We can use the bond enthalpy data for respective E–O and E=O bonds to calculate the total bond energy for each unit. The tetrahedral orthosilicate (SiO_4) unit present in quartz has four Si–O single bonds, giving a total bond energy (4×466) of 1864 kJ mol^{-1}. If SiO_2 existed as discrete molecules, there would be two Si=O double bonds giving a total bond energy (2×642) of 1284 kJ mol^{-1}. It thus follows that SiO_2 exists as an extended network solid consisting of interconnected orthosilicate units (Section 15.3).

A discrete CO_2 molecule has two C=O double bonds, giving a total bond energy (2×743) = 1486 kJ mol^{-1}: were CO_2 to have the quartz (SiO_2) structure, the analogous 'orthocarbonate' unit would comprise four C–O single bonds, giving a total bond energy (4×360) of 1440 kJ mol^{-1}. A discrete molecular structure is thus favoured for CO_2.

Self-test 10.1 Use the following data to explain why elemental nitrogen exists as a diatomic molecule, whereas the common form of elemental phosphorus is the tetrahedral molecule P_4. Mean bond enthalpies $B(X-Y)$ in kJ mol^{-1}: N–N, 163; N≡N, 946; P–P, 201; P≡P, 480.

The extent of hydrogen bonding (see Section 5.9) is much greater in the compounds of the first member of each group, where the greater electronegativity results in a more polarized E–H bond. For example, the boiling point of ammonia is −33°C, which is higher than that of the other Group 15 hydrides (Section 16.10). Likewise, water and hydrogen fluoride are liquids at room temperature whereas H_2S and HCl are gases (Sections 17.6(d) and 18.2).

The striking differences between the 2p elements and their congeners is also reflected, but less markedly, in the d block. Thus, the properties of the 3d series metals differ from those of the 4d and 5d series. For binary compounds, lower oxidation states tend to be more stable in the 3d series, with the stability of higher oxidation states increasing down each group (Sections 9.4(b) and 20.2). For example, the most stable oxidation state of chromium is Cr(III) whereas it is M(VI) for Mo and W. The degree of covalency and the coordination numbers increase between the compounds of 3d and the 4d and 5d elements. Thus, whereas 3d elements form ionic solids, as in CrF_2 the 4d and 5d-series elements form higher halides such as MoF_6 and WF_6, which are liquids at room temperature. These differences are attributable to the smaller ionic radii of the 3d-series elements, the fact that the radii of the 4d and 5d-series elements are quite similar (due to the lanthanoid contraction), and the ability of halides to stabilize high oxidation state metals by donation of electrons through a π bond that is more readily formed with larger d orbitals. The general trends in bond enthalpies are discussed in Section 9.7.

The properties of the elements in the first row of the f block, the lanthanoids, are significantly different from those of the second row, the actinoids (Chapter 21). The elements Ce through to Lu, generically referred to as Ln, are all highly electropositive, with the $E^{\ominus}(Ln^{3+}/Ln)$ standard potential lying between those of $E^{\ominus}(Li^+/Li)$ and $E^{\ominus}(Mg^{2+}/Mg)$. The elements favour the oxidation state Ln(III) with a uniformity that is unprecedented in the periodic table, and this is ascribed to the 4f orbitals being 'buried' in the core of the atom. The actinoid elements show less uniformity in their properties than the lanthanoid elements because the larger 5f orbitals can participate in bonding (Section 21.9).

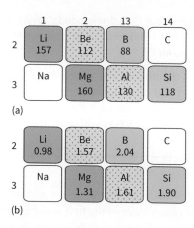

FIGURE 10.3 The diagonal relationship between (a) atomic radii (in picometres) and (b) Pauling electronegativity in Periods 2 and 3.

10.3 Diagonal relationships

KEY POINTS The chemical properties of some Period 2 elements are similar to those of the element positioned diagonally to the lower right in Period 3.

The so-called *diagonal relationships* refer to the observation that pairs of p-block elements located in diagonally neighbouring positions (to the right and down) display similar chemical behaviour. The similarity is largely attributed to compensations in atomic properties: thus, whereas a shift to the next period (2p to 3p) is expected to result in an increase in atomic (Fig. 10.3(a)) and ionic radii, a shift to the next group (1 to 2, 2 to 3, etc.) results in a decrease in atomic and ionic radii of the higher-charged ions: these two effects tend to cancel each other out. For example, the ionic radii of Na^+ and Ca^{2+} are the same to two significant figures at 1.0 Å (Table 1.3). These similarities are reflected in charge density values for ions (ion charge/ionic volume) and values for the 2p, 3p, and 4p elements of Groups 1, 2, 13, and 14, which are summarized in Table 10.1. Note the similar values of Li^+ and Mg^{2+}, Be^{2+} and Al^{3+}, and B^{3+} and Si^{4+}. Also noteworthy is the closeness of the values for Mg^{2+} and Zn^{2+} with that of Sc^{3+}.

The two diagonally related elements also have Pauling electronegativities that are closer in value than with their immediate vertical neighbours (Fig. 10.3(b)): their compounds are thus expected to occupy similar positions on the Ketelaar triangle (Section 2.14). Where the ionic model is appropriate, the two elements are expected to resemble each other in terms of properties that depend on lattice and solvation energies and the degree of departure from the model due to polarization.

Here we describe the resulting similarities from the most significant diagonal relationships in the periodic table.

(a) Li/Mg

A comparison of lithium (Chapter 12) and magnesium (Chapter 13) reveals common physical properties and chemical behaviours:

- Elemental Li and Mg are hard metals and relatively unreactive, compared to the lower congeners in their groups.
- Both Li and Mg metal react with O_2 to form the monoxides.
- They are the only two elements that react directly with N_2, forming the nitrides Li_3N and Mg_3N_2, which release NH_3 upon hydrolysis.
- Their relatively high charge densities mean both lithium and magnesium salts have some covalent character to their bonding.
- Their fluorides and hydroxides are only sparingly soluble in water (Section 4.15(c)).
- Their salts with larger anions, such as sulfate, perchlorate, are very soluble in water.
- The carbonates of lithium and magnesium decompose easily upon heating, giving the oxide (Section 4.15(a)).

(b) Be/Al

The chemistry of Be is described in Chapter 13 and that of Al in Chapter 14, and there are many similarities:

- Be and Al dissolve in both acid and alkali.
- The high charge densities of Be^{2+} and Al^{3+} ions mean they are strongly polarizing and the aqua ions are subject to extensive hydrolysis even under acidic conditions.
- Polarization also means the oxides of Be and Al are amphoteric (Section 9.5).
- In the presence of excess OH^- ions, Be and Al form $[Be(OH)_4]^{2-}$ and $[Al(OH)_4]^-$, respectively.
- Both elements form structures based on linked tetrahedra: Be forms structures built from $[BeO_4]^{n-}$ and $[BeX_4]^{n-}$ tetrahedra (X=halide) and Al forms numerous aluminates and aluminosilicates containing the $[AlO_4]^{n-}$ unit.
- Their carbonates are unstable and decompose, producing the oxide, as a result of the small highly charged cations.
- Both elements form carbides that contain the C^{4-} ion and produce methane on reaction with water.

TABLE 10.1 Charge densities, X, (C mm^{-3}) for selected cations calculated according to the formula $X = ne/(4/3 \pi r^3)$. Ionic radii are used for octahedral coordination except where noted (T) for 4 coordinate tetrahedral.

Li^+	Be^{2+}			B^{3+}	C^{4+}	
98(T)	1108(T)			1663(T)	6265(T)	
Na^+	Mg^{2+}			Al^{3+}	Si^{4+}	P^{5+}
24	120			770(T)	970	1358
K^+	Ca^{2+}			Ga^{3+}	Ge^{4+}	As^{5+}
11	52			261	508	884

- Both Be and Al form covalent hydrides.
- Their polarizing nature means the halides of Be and Al are molecular, and have M–X–M bridges.
- The alkyl compounds of Be and Al are electron-deficient compounds that contain M–C–M bridges.
- Both Be and Al form metal–metal bonded species in their lower oxidation states Be(I) and Al(II).

There are also analogies between the chemical properties of Be and Zn. As well as exhibiting the typical properties of electropositive metals in forming salts, Zn is also amphoteric and dissolves in strong bases to produce zincates. Structures containing linked $[ZnO_4]^{n-}$ and $[BeO_4]^{n-}$ tetrahedra are both known.

(c) B/Si

Although boron has anomalous behaviour with respect to the lower elements in Group 13 (Chapter 14), it has a pronounced diagonal relationship with silicon (Chapter 15):

- Elemental B and Si are both semiconductors (respective bandgaps approximately 1.5 and 1.1 eV) (Section 2.11).
- Boron and silicon form acidic oxides with amorphous polymeric structures that are combined in borosilicate glasses (Sections 14.8 and 15.10).
- Both B and Si form reactive and highly flammable gaseous hydrides.
- The chlorides BCl_3 and $SiCl_4$ are Lewis acids with polar E–Cl bonds that react with water.

(d) C/P

Carbon (Chapter 15) and phosphorus (Chapter 16) share some characteristics:

- Both C and P form extensive series of allotropes (Table 9.2). Despite the presence of a lone pair not present on carbon, the larger size of phosphorus versus nitrogen favours catenation and extended structures, while being smaller than silicon favours multiple bonds.
- The two elements have almost identical Pauling electronegativities.

(e) N/S

By nitrogen (Chapter 16) and sulfur (Chapter 17) the diagonal relationship is becoming less clear-cut, and the general similarities in chemistry are shared by many of the nonmetal elements.

10.4 The alternation effect

KEY POINTS The chemical properties of the 4p elements Ga–Br deviate from the trends expected upon descending each group. This interruption is known as the alternation effect and can be attributed to an increase in Z_{eff} resulting from completion of the first occurrence of a poorly screening d subshell.

The alternation effect is the term used to describe the interruption to the expected trend, that elements become more electropositive (resulting in increased metallic behaviour) as a p block group is descended, between periods 3 and 4, due to the completion of the first set of d orbitals.

The $3d^{10}$ subshell increases the effective nuclear charge Z_{eff} because of the poor shielding ability acting on an electron in the 4p orbitals. Consequently, although atomic radii continue to increase down each group, electronegativity values, according to the Allred–Rochow definition (eqn. 1.15), demonstrate a step increase for the elements Ga to Br (Fig. 10.4).

The alternation effect explains the unusual properties of elemental gallium. Instead of having a typical metallic structure like aluminium, gallium contains Ga_2 dimers that resemble I_2 and even Hg_2. The 'molecule-like' structure accounts for the unusually low melting point (30°C).

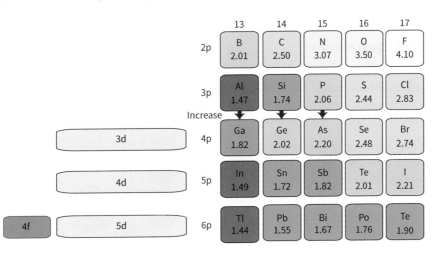

FIGURE 10.4 Filling of the first set of d orbitals results in a step increase in the electronegativities of the 4p elements. Allred–Rochow electronegativities are represented by shade, with light yellow shades representing high electronegativity and darker blue-green shades representing lower (below 2) values.

The step increase in electronegativity has important consequences for compounds, one of which is that bonds formed to hydrogen are more covalent than a simple trend would predict. Thus, whereas aluminium hydride is a non-volatile polymeric solid, gallium resembles boron in forming the discrete molecular hydrides diborane (Section 14.6) and digallane (Section 14.13) respectively.

The higher oxides of the 4p elements are much more reactive than expected. Compounds of arsenic (V), selenium (VI), and bromine (VII) are powerfully oxidizing compared to their corresponding 3p analogues phosphorus (V), sulfur (VI), and chlorine (VII), and even with respect to their 5p counterparts. The interruption in stability is illustrated by the following standard electrode potentials for perchlorate, perbromate, and periodate.

$$ClO_4^- + 2e^- + 2H^+ \rightarrow ClO_3^- + H_2O \quad E^\ominus = 1.20\,V$$
$$BrO_4^- + 2e^- + 2H^+ \rightarrow BrO_3^- + H_2O \quad E^\ominus = 1.85\,V$$
$$H_5IO_6 + 2e^- + H^+ \rightarrow IO_3^- + 3H_2O \quad E^\ominus = 1.60\,V$$

The pentahalides of the Group 15 elements provide another useful demonstration. Shaded cells denote unknown or extremely unstable compounds.

NF_5	NCl_5	NBr_5
PF_5	PCl_5	PBr_5
AsF_5	$AsCl_5$	$AsBr_5$
SbF_5	$SbCl_5$	$SbBr_5$
BiF_5	$BiCl_5$	$BiBr_5$

The absence of higher halides for N is expected because of steric crowding and very unfavourable electron repulsions. The limitation on pentahalides of Bi is a manifestation of the inert pair effect (Section 10.7). The interruption clearly occurs at As ($AsCl_5$ is unstable above −50°C).

10.5 The lanthanoid contraction

KEY POINTS Shielding by f electrons is less effective than that by d electrons, resulting in increasing effective nuclear charge. The higher ionization energies but smaller ions result in little change in the stability of the M(III) oxidation state.

The 4f orbitals start filling after lanthanum, giving the lanthanoid elements (Chapter 21). This means the successive addition of 14 electrons, but electrons in f orbitals provide poor screening of the nuclear charge. Outer electrons feel a greater effective nuclear charge and are drawn closer to the nucleus, resulting in a steady contraction in atomic radius. The contraction in metallic (atomic) radii is generally uniform, apart from in the middle and at the end, due to the half-shell and full-shell effects (Section 21.4).

The increasing effective nuclear charge from left to right across the f block yields the expected increase in ionization energies. The radii of the 3+ ions show a smooth decrease, with little influence from electron configuration, including ligand-field effects. The greater sum of the first three ionization energies is compensated for by an increase in hydration energy for the smaller ions, so little change in three-electron $E^\ominus(Ln^{3+}/Ln)$ standard potentials is observed across the row.

The lanthanoid contraction has an important impact on the properties of the 5d row of elements that follow completion of the 4f subshell, as the metallic (atomic) and 3+ ionic radius of the 14th Ln element Lu are each slightly smaller than for Y, the 'equivalent' element immediately preceding the 4d row. The result is that elements of the 5d row have metallic (atomic) radii very similar to their 4d counterparts, instead of being larger as would be expected from the increase in principal quantum number. Thus, a Hf atom is approximately the same size as Zr rather than being larger, and a Ta atom is the same size as an Nb atom (Fig. 10.5).

The main difference between the 4d and 5d metals is their densities. For example, Hf is twice as dense as Zr, and Ta is twice as dense as Nb (Fig. 10.6).

More significant for the chemistry of the 5d elements is the similarity in ionic radii between 4d and 5d ions, as shown by comparing the radii for the first three 4d/5d counterparts

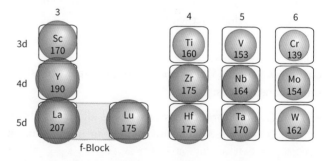

FIGURE 10.5 The impact of the lanthanoid contraction on the metallic (atomic) radii (in picometres) of 5d elements.

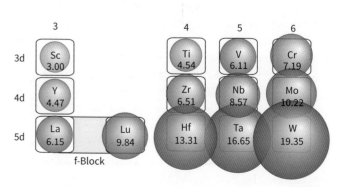

FIGURE 10.6 The impact of the lanthanoid contraction on the densities (in g cm^{-3}) of 5d elements.

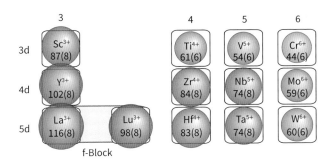

FIGURE 10.7 A comparison of the ionic radii (in picometres) in the group oxidation states for the first three 3d, 4d, and 5d ions. Coordination numbers are given in parentheses.

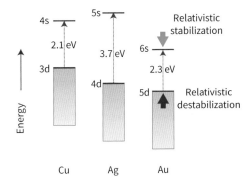

FIGURE 10.8 The frontier orbitals of metallic Cu, Ag, and Au showing how the relativistic stabilization of the 6s orbital and destabilization of the 5d orbitals account for why Au is yellow in colour, resembling Cu more than Ag.

in their group oxidation states, and for the same coordination number (shown in parentheses); Fig. 10.7.

The consequence of there being little change in ionic radii between Hf and Zr and Ta and Nb is that these pairs of elements are very difficult to separate. Thus, Hf has very similar chemistry to Zr, likewise Ta and Nb. As the 5d row is crossed, greater differences with 4d counterparts start to occur, partly because the 5d elements become increasingly influenced by relativistic effects (Section 10.6); Au, for example, is quite easily distinguished from Ag, likewise Pt and Pd.

10.6 Relativistic effects

KEY POINTS Relativistic effects, which act directly to stabilize s (and p) orbitals and indirectly to destabilize d (and f) orbitals, are important in determining the properties of the heavier elements and their compounds. The consequences include the colour of gold, the unique properties of Hg, the 'inert pair' stabilization of Tl(I) and Pb(II), and the rich chemistry of the early actinoids.

We saw in Chapter 1 how relativistic effects influence the energies of atomic orbitals, the primary factor being the marked increase in the mass of an electron as its velocity approaches the speed of light, as given by eqn. 1.10. This mass increase acts directly to contract and stabilize s orbitals, and to some extent p orbitals. The ensuing increased nuclear screening results in the destabilization of d and f orbitals. Relativistic effects govern spin–orbit coupling, which increases dramatically for the heavier elements, enriching (and complicating) physical properties such as magnetism and electronic spectra.

The manifestation of relativistic effects begins in Period 6, towards the end of the third row of the d-block (Hf to Hg), the 5d orbitals of which are otherwise stabilized through the poor screening by the underlying 4f shell (the lanthanoid contraction). Relativistic effects play a major role in determining the chemical properties of the heaviest elements, including some very familiar examples such as the elements Au ([Xe](4f^{14})(5d^{10})(6s^1)) and Hg ([Xe](4f^{14})((5d^{10})(6s^2)), each of which display clearly visible unusual properties. The colour of gold arises because a photon colliding with surface atoms excites an electron from the filled 5d orbitals (which form a band) to the half-filled 6s orbital, the energy of which corresponds to the Fermi level (Section 2.11): this transition occurs in the visible region of the spectrum. The simplified diagram shown in Fig. 10.8 compares the energy of this transition with that of the other Group 11 metals Cu and Ag. The small gap for Cu gives the metal its reddish colour, and the larger gap for Ag lies in the UV region, accounting for its silver lustre, that is typical of most metals. Continuing the trend, we might expect that the gap should be larger for Au; however, the 6s orbital is stabilized and the 5d band is destabilized, narrowing the gap so that the resulting excitation involves the absorption of blue light. Stabilization of the 6s orbital leads to a high electron affinity, and gold is the most electronegative of the metals. Gold forms compounds containing the auride ion Au$^-$ and diatomic species Au$_2$, analogous to the halogens and with a bond energy greater than I$_2$.

The small s–d gap also favours the hybridization that accounts for the preponderance of linear coordination in compounds of Au(I). In the case of mercury, the filled 6s^2 orbital has just as profound an influence on properties. Orbital overlap between the contracted 6s orbitals of Hg atoms is poor, giving rise only to weak metallic bonding; this accounts for its liquid state and poor electronic conductivity relative to other metals. Stable compounds include those of the mercurous ion Hg$_2^{2+}$, isoelectronic with Au$_2$.

The stability of Pb(II), which exemplifies the inert pair effect, is another consequence of the relativistic stabilization of s orbitals. Calculations show that without relativistic stabilization of Pb^{2+} ([Xe](5d^{10})(6s^2)), the standard potential for the $E^{\ominus}$(PbO$_2$/Pb^{2+}) couple would be much less positive, and that for the $E^{\ominus}$(Pb^{2+}/Pb) couple much less negative: the lead–acid battery (Section 15.11) would therefore not be possible. The filled orbital (6s^2) is the inert pair.

Relativistic effects become progressively dominant as the atomic number increases further, the velocity of s electrons varying with Z. In Groups 1 and 2, the first (and second) ionization energies of Fr and Ra are actually higher than for Cs and Ba, abruptly opposing the trends that are otherwise observed as the groups are descended. The actinoids behave very differently to the lanthanoids because the expected expansion of the 5f orbitals compared to the 4f orbitals is further enhanced by their relativistic destabilization. Uranium, plutonium, and other early actinoids thus display a wealth of coordination chemistry and several oxidation states, to an extent that easily rivals the rich chemistry of most d-block elements.

Consequences

10.7 The inert pair effect

KEY POINTS The inert pair effect leads to an increasing stability of an oxidation state that is 2 less than the group oxidation number for the heavier elements.

(a) General description

The heavier elements of the p block also form compounds where the element has an oxidation number 2 less than the group oxidation number. This relative stability of the lower oxidation state is an example of the **inert pair effect** and is a recurring theme within the p block (Sections 14.15, 15.11, and 16.13).

For example, in Group 13 the group oxidation number is +3, whereas the +1 oxidation number increases in stability down the group. In fact, the most common oxidation state of thallium is Tl(I). There is no simple explanation for this effect: it is often ascribed to the large energy that is needed to remove the ns^2 electrons after the np^1 electron has been removed. However, the sum of the first three ionization energies of Tl ($5438\,kJ\,mol^{-1}$) is not higher than the value for Ga ($5521\,kJ\,mol^{-1}$) and is only slightly higher than the value for In ($5083\,kJ\,mol^{-1}$). Given the prevailing trends, we would expect the value to be lower than that of In or Ga. Its relatively high value is related to relativistic stabilization of the 6s orbital (Sections 1.7(a) and 10.6).

Below we explore other factors that contribute to the inert pair effect, including the low M–X bond enthalpies for the heavier p-block elements, and the decreasing lattice energy as the ionic radii increase down a group.

(b) Thermodynamic rationalization of the inert pair effect in ionic compounds

To rationalize the observation of the inert pair effect for Group 13–16 elements in ionic compounds, we can consider the relative thermodynamic stability of $MX_{m-2}(s)$ and $MX_m(s)$, X=halogen, where m is the maximum valence exhibited in the group (equal to the group number minus 10). We illustrate this analysis using Group 13 elements (M=Al, Ga, In, and Tl) and the ionic solids MX(s) and $MX_3(s)$; these arguments can readily be extended to the later groups.

We can consider whether $MX_3(s)$ is stable towards decomposition to MX(s) by determining the enthalpy for the reaction.

$$MX(s) + X_2(g,l,s) \rightarrow MX_3(s) \quad (10.1)$$

The Born–Haber cycle for the formation of MX(s) from the elements has been discussed in Section 4.11 and is included as Fig. 10.9(a). We can now consider the thermodynamic cycle for reaction (eqn.) 10.1, which involves the key stages of breaking up the solid structure of MX(s) ($+\Delta_L H(MX)$) followed by ionization to form M^{3+} and formation of the solid $MX_3(s)$, Fig. 10.9(b).

The important variable terms for the different group elements are highlighted in Fig. 10.9; they are the second and third ionization energy and the lattice enthalpies of MX(s) and $MX_3(s)$. Notice that the lattice enthalpy of MX is much lower (by about an order of magnitude) than that of MX_3 due to the lower number of ions, larger cation, and lower cation charge, so we can ignore this term.

Table 10.2 summarizes the sum of the second and third ionization energy for the Group 13 elements, and Table 10.3 the lattice enthalpies of $MX_3(s)$, $\Delta_L H^{\ominus}$, calculated using the Kapustinskii equation (eqn. 4.4).

As expected, the lattice enthalpy of $MX_3(s)$ decreases with increasing Group 13 cation size and also larger halide ion size. Set against this is the trend in 2nd IE + 3rd IE, which instead of decreasing with increasing atomic mass, as might be expected for increasing principal quantum number, levels off and increases for thallium: note the higher value for gallium as a result of the alternation effect (Section 10.4). The high value of 2nd IE + 3rd IE for thallium is attributable to relativistic effects (Section 10.6).

The thermodynamic stability of $MX_3(s)$ phases versus decomposition to MX(s) and halogen can now be explained as a balance between the endothermic 2nd IE + 3rd IE and the exothermic $-\Delta_L H^{\ominus}$. For the majority of Group 13 elements combined with small halides, this balance means that

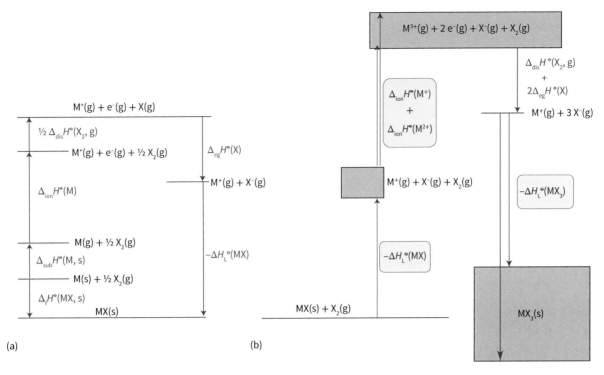

FIGURE 10.9 Thermodynamic cycles for (a) the formation of MX(s) from the elements and (b) reaction (eqn.) 10.1.

TABLE 10.2 The sum of the second and third ionization energy for the Group 13 elements.

Element	B	Al	Ga	In	Tl
Additional ionization enthalpy (kJ mol^{-1}) 2nd IE + 3rd IE	6086	4560	4941	4524	4848

TABLE 10.3 Calculated lattice enthalpies (kJ mol^{-1}) for MX$_3$(s) using the Kapustinskii equation with ionic radii in picometres (Resource section 1) given in parentheses.

	Al^{3+} (53)	Ga^{3+} (62)	In^{3+} (80)	Tl^{3+} (89)
F$^-$ (133)	6968	6646	6085	5838
Cl$^-$ (181)	5538	5333	4966	4800
Br$^-$ (196)	5205	5023	4696	4547
I$^-$ (220)	4747	4596	4320	4194

formation of MX$_3$(s) is thermodynamically favourable with negative enthalpies for reaction (eqn.) 10.1. However, for thallium the increased ionization energy caused by relativistic effects and low lattice enthalpies for large halides result in the enthalpy change for reaction (eqn.) 10.1 being positive. For example, Tl(III)I$_3$(s) is thermodynamically unstable with respect to decomposition to TlI(s) and I$_2$(s).

While these data and analysis are specific to Group 13 halides, similar arguments apply in general to ionic Group 13 compounds and the oxides and halides of the heavier elements in Groups 14, 15, and 16. As such, they provide a thermodynamic rationalization for the inability to synthesize compounds such as PbBr$_4$ and BiI$_5$.

(c) Thermodynamic rationalization of the inert pair effect in covalent compounds

For covalent compounds a thermodynamic rationalization of the inert pair effect can also be made, though the arguments are less rigorous, as specific thermodynamic terms are less well determined.

Let us consider the thermodynamic cycle for the formation of the higher oxidation state halide of a Group 13 element, MX$_3$, from MX. Fig. 10.10 shows that the main varying factor is the average bond enthalpy, M–X, in MX and MX$_3$ and the factors involved in forming three covalent bonds.

For the formation of three covalent M–X bonds in MX$_3$, the ground-state electron configuration ns^2np^1 requires the promotion of one ns electron to the np level; with associated hybridization to form, for example, the sp^2 bonding set. A consideration of the relative energies of the ns and np electrons (Table 10.4) for the Group 13 elements shows that the promotion of an ns electron to an np level will require more energy for the heavy elements, particularly Tl; this is due to relativistic effects having a greater influence in stabilizing the 6s orbital energy of Tl (Section 10.6). This additional electron promotion energy needs to be compensated for by the

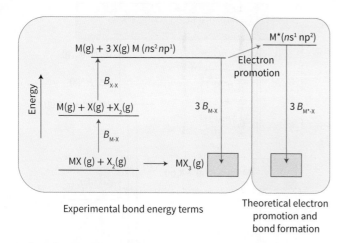

FIGURE 10.10 Thermodynamic cycle for the formation of MX_3 from MX showing the bond enthalpies involved. The formation of MX_3 from M(g) involves the theoretical promotion of an ns electron to an np orbital.

additional bond enthalpies involved in forming three M–X bonds in MX_3 rather than one in MX, to make MX_3 thermodynamically stable versus decomposition to MX and X_2.

Table 10.4 also summarizes the average experimental M–X bond dissociation enthalpies for M–F and M–I bonds; these values decrease down the group as the bond length increases and orbital overlap decreases. Values are much lower for M–I compared to M–F, again due to weaker orbital overlap with 5s and 5p orbitals on iodine.

The origins of the inert pair effect and thermodynamic instability of some covalent compounds displaying the maximum group oxidation state can now be rationalized. The increased promotion energy for 6s to 6p electrons in Tl, Pb, and Bi due to relativistic effects (compared to 3s and 3p) may not be compensated for by the bond enthalpies of two additional M–X bonds, particularly for large, weakly bonding elements such as iodine. Thus $Tl(III)I_3$ has not been synthesized, neither has PbI_4, and $PbCl_4$ is only stable below 0°C. While the inert pair effect is most apparent for row 6 of the periodic table due to relativistic effects, the decreasing bond enthalpies for larger atoms in rows 4 and 5 are also manifested in increasing stability of the lower oxidation state. Numerous further examples are provided in Chapters 14, 15, and 16.

10.8 The oxo wall

KEY POINT The d block is divided into two regions by an oxo wall, which separates metals able to form a multiple M≡O or M=O bond (left) and those that do not (right). Elements lying close to the wall are used in catalysts for oxygen-atom transfer reactions.

The oxo wall refers to the scarcity or nonexistence of d-block complexes containing the entities M≡O or M=O (in which an oxido ligand O^{2-} forms a multiple bond to M) beyond Group 8.[1] Thus whereas complexes of V(IV) contain the vanadyl entity, as in $[VO(H_2O)_4]^{2+}$, no such species are found for elements on the right side of the d block (Fig. 10.11).

We saw in Section 5.2 that the Brønsted acidity of a coordinated water molecule, leading to formation of OH^- and ultimately O^{2-}, depends on the degree to which the M–O bond is polarized by the positive charge on the metal ion. Deprotonation is favoured for high oxidation states. Unlike vanadium, nickel does not form oxido complexes such as $[NiO(H_2O)_4]^{2+}$, and a first impression might be that Ni(IV) is in any case too powerfully oxidizing to exist in aqueous solution. However, the prime consideration is that O^{2-} is a very strong π-donor ligand, able to use all three p orbitals to form double (M=O) and triple (M≡O) bonds (**1**). The overall bonding picture depends on how the metal-based π-symmetry orbitals, which are antibonding, are occupied.

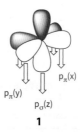

1

We must therefore use ligand-field theory to see how these orbitals are arranged in energy and occupied by electrons.

[1] The oxo wall concept has been developed and popularized by Professor Harry Gray and close colleagues.

TABLE 10.4 The average experimental M–X bond dissociation enthalpies for M–F and M–I bonds and the relative energies of the ns and np electrons for the Group 13 elements.

Element, n	M–F bond dissociation enthalpy/kJ mol^{-1}	M–I bond dissociation enthalpy/kJ mol^{-1}	Valence atomic orbital energies/eV		
			s	p	Δ
B, 2	766	384	−14.05	−8.30	5.75
Al, 3	664	368	−11.32	−5.98	5.34
Ga, 4	577	339	−12.61	−5.93	6.68
In, 5	506	~300	−11.89	−5.60	6.29
Tl, 6	445	272	−13.14	−5.47	7.67

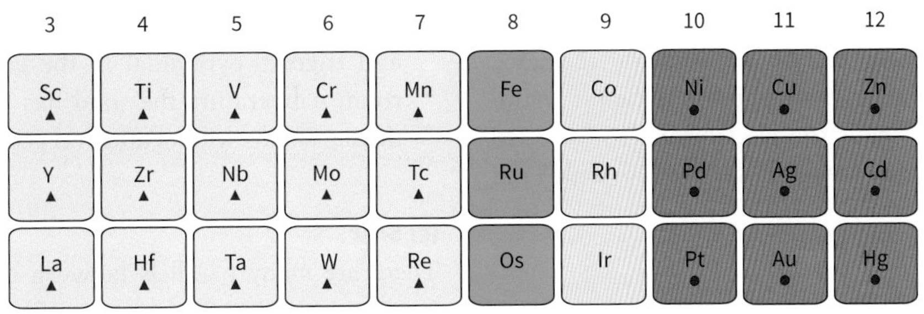

FIGURE 10.11 The oxo wall. Elements on the left side of the d block (pale green, ▲) form many stable oxido complexes, whereas those on the right (red, ●) do not form oxido complexes. Group 8 represents the classical oxo wall, although 'violations' explainable in terms of variations in ligand field and symmetry occur in Group 9. Oxido complexes of metals at the oxo wall are often highly active catalysts for oxygen-atom transfer reactions.

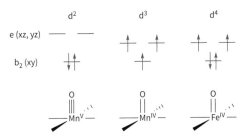

FIGURE 10.12 Ligand-field splittings and d-electron occupancies for tetragonal (C_{4v}) complexes of Mn and Fe. The more strongly antibonding a_1 (z^2) and b_1 (x^2-y^2) orbitals are not shown.

The explanation is found most easily by considering a tetragonal complex MOL_4, having C_{4v} symmetry, noting that the bonding orbitals are occupied by electrons donated by the O^{2-} ligand. The ordering of the nonbonding and antibonding metal d orbitals is derived from the familiar O_h case that we described in Section 7.17, the t_{2g} orbitals becoming b_2 (xy) and e (xz, yz). The b_2 orbital is nonbonding, but the e orbitals, which have π symmetry, are antibonding with respect to the MO bond; these are the orbitals of principal interest. Each electron that occupies one of these e orbitals lowers the π-bond order by ½ from a maximum of 2. The x^2-y^2 (b_1) and z^2 (a_1) orbitals derived from the e_g set and having σ-symmetry lie still higher in energy and need not concern us at this point.

The simplified ligand fields for tetragonal complexes of Mn and Fe, with d^2, d^3, and d^4 electron configurations are shown in Fig. 10.12. The e–b_2 energy gap for d^2 is sufficiently large that both electrons occupy the b_2 orbital, leaving the two e orbitals empty. This configuration is advantageous, giving the maximum possible π-bond order of 2, and (with the strongly antibonding a_1 (z^2) orbital also being unoccupied) the overall total bond order is three: one sigma and two pi. For Mn(IV) and Fe(IV), each e orbital is singly occupied, the π-bond order is reduced to 1, and a double bond results. As the metal–oxygen bond distance increases, the e–b_2 energy gap decreases.

Two consequences of Fig. 10.12 are relevant for certain highly specialized roles that Mn and Fe have in catalysis. The O^{2-} bound to Mn(V) is an extremely weak base ($[M\equiv O]^{3+}$ is the conjugate base of a super acid) but a very strong electrophile: the $[M\equiv O]^{3+}$ entity is thus perfectly suited for attacking a second O^{2-} positioned very close by to form an O–O bond, as it does during the O_2-evolving reaction of photosynthesis (Section 26.10(d)). In contrast, the O^{2-} bound to Fe(IV) is a sufficiently strong base to 'capture a proton', as occurs during the electron-transfer coupled hydrogen atom abstraction reaction responsible for C–H bond activation by oxygenases (Section 26.10(c)).

As well as Mn(V), the powerfully oxidizing d^0 Cr(VI) oxoanion (CrO_4^{2-}) and the d^0 Mo(VI) intermediate in enzyme catalysts are further examples of how high reactivity is found *close to* the oxo wall. In contrast, even in their highest oxidation states, oxido-complexes of the early or 'light' d-block metals further to the left are relatively unreactive.

Beyond Group 8, the requirement that the metal be in a very high oxidation state to leave a space in the antibonding e orbitals that would allow a nonzero π-bond-order becomes very demanding. Examples apparently violating the oxo wall (complexes of Group 9 and beyond) are often reported, but are either disproved or readily explained in terms of how their symmetries and ligand-field splitting 'allow the rules to be relaxed'. The oxo wall can shift to the right if a different structure is adopted. For example, upon changing from tetragonal to trigonal geometry (ideally C_{3v}) the x^2-y^2 orbital becomes accessible to accommodate an additional pair of electrons without their having to be placed in the antibonding π-symmetry orbitals. For instance, in Group 9, an Ir(V) oxido species (Ir=O, d^4) formed on the surface of a IrO_2 electrode is an intermediate in the electrocatalytic O_2-evolving reaction of water electrolysis, mentioned in Section 6.16. It is also necessary to be aware that powerfully oxidizing 'oxido' complexes formed by metals beyond the wall may instead contain a single bond to an oxyl radical (O·).

The nitride (N^{3-}) ion is another very strong π-donor ligand, and a 'nitrido wall' can be described on the same basis as above. Interestingly, there appears to be no analogous 'fluoro wall': fluorine is too electronegative for F^- to be a strong π donor, so the M–F bond is always single (and much more ionic).

10.9 Group n versus Group $n + 10$

KEY POINT For the group oxidation state, there are chemical similarities between p-block elements in Group $n + 10$ (with atomic number Z) and the d-block elements in Group n (with atomic numbers $Z + 8$). The relationship can be explained in terms of electron configurations and atomic radii.

There are similarities between p-block elements in Group $n + 10$ and the d-block element in Group n. This reflects the equal number of valence electrons. More specifically, where elements from Group $n + 10$ (with atomic numbers Z) are compared with the d-block element in Group n (with atomic numbers $Z + 8$), the atomic radii will also be similar. The chemical similarities are observed when the p-block elements are in their group oxidation states and the d-block element has a d^0 configuration (such configurations are often referred to as 'closed' shell).

When dealing with the 5p elements (In–Xe) the relationship is between Z and $Z + 22$, as there are a further 14 lanthanoids in between. For example, Al ($Z = 13$, Group 13) shows similarities to Sc ($Z = 21$, Group 3): both have three valence electrons. Their atomic radii are fairly similar: Al is 143 pm and Sc is 160 pm and the standard potential $E^\ominus(Al^{3+}/Al)$ (−1.66 V) is close to $E^\ominus(Sc^{3+}/Sc)$ (−1.88 V). Similarities also exist between the following pairs.

(a) Si/Ti

A comparison of silicon (group 14, $Z = 14$, Chapter 15) and titanium (Group 4, $Z = 22$, Chapter 20) reveals common physical properties and chemical behaviours:

- $SiCl_4$ and $TiCl_4$ are both volatile covalent liquids with tetrahedral geometries.
- Si and Ti alkoxides hydrolyse and condense in sol-gel reactions to produce silica and titania materials. Extending the n versus $n + 10$ relationship, tetragonal GeO_2 and TiO_2 adopt the rutile structure (Section 4.9).

(b) P/V

The oxygen and fluorine compounds of phosphorus (group 15, $Z = 15$, Chapter 16) and vanadium (group 5, $Z = 23$, Chapter 20) have a number of structural similarities:

- PO_4^{3-} and VO_4^{3-} are respectively weakly and mildly oxidizing tetrahedral anions that undergo condensation upon acidification.

- PF_5 is a trigonal-pyramidal molecule, VF_5 is monomeric and trigonal-pyramidal in the gas phase, albeit below room temperature the solid has bridging fluorides producing chains with octahedral geometry.

(c) S/Cr

There are many parallels between the chemistry of sulfur (group 16, $Z = 16$, Chapter 17) and both chromium (group 6, $Z = 24$, Chapter 20) and molybdenum (group 6, Chapter 20):

- SO_3 is a trigonal-planar molecule and a strong oxidant. It polymerizes to form a trimeric ring or chains of vertex-sharing tetrahedral. CrO_3 forms chains of vertex-sharing tetrahedral.
- SO_4^{2-} is a mildly oxidizing tetrahedral anion. CrO_4^{2-} and MoO_4^{2-} are respectively strongly and very weakly oxidizing tetrahedral anions.
- SF_6 is a very stable octahedral molecule. Although CrF_6 does not exist, MoF_6 is a molecular species.

(d) Cl/Mn

Comparing the oxides of chlorine (group 17, $Z = 17$, Chapter 18) and manganese (group 7, $Z = 25$, Chapter 20) shows common physical and chemical properties:

- Cl_2O_7 and Mn_2O_7 are both liquids, and though chorine(VII) oxide is only weakly oxidizing, the Mn(VII) compound is strongly oxidizing.
- Similarly, ClO_4^- is a mildly oxidizing tetrahedral anion, while the structurally similar MnO_4^- anion is strongly oxidizing.

EXAMPLE 10.2 Predicting the chemical properties of a $Z + 8$ element

The perchlorate ion, ClO_4^- is unstable, and its compounds can detonate on contact or with heat. Predict whether the analogous compound of the $Z + 8$ element would be a suitable replacement for a perchlorate compound in a reaction.

Answer We need to identify the $Z + 8$ element. Because the atomic number of Cl is 17, the $Z + 8$ element is Mn ($Z = 25$). The compound of Mn analogous to ClO_4^- is the permanganate ion, $[MnO_4]^-$, which is in fact an oxidizing agent, but less likely to detonate. Permanganate is likely to be a suitable replacement for perchlorate.

Self-test 10.2 (a) Xenon is very unreactive but forms a few compounds with oxygen and fluorine such as XeO_4. Predict the shape of XeO_4 and identify the $Z + 22$ compound with the same structure. (b) Silicon forms the dioxide SiO_2. Identify the $Z + 8$ element and predict the formula of its oxide.

FURTHER READING

P. Enghag, *Encyclopaedia of the Elements*. John Wiley & Sons (2004).

D. M. P. Mingos, *Essential Trends in Inorganic Chemistry*. Oxford University Press (2004). Includes a detailed discussion of the important horizontal, vertical, and diagonal trends in the properties of the atoms.

N. C. Norman, *Periodicity and the s- and p-Block Elements*. Oxford University Press (1997). Includes coverage of essential trends and features of s-block chemistry.

E. Scerri, *The Periodic Table: Its Story and Its Significance*, 2nd ed. Oxford University Press (2019). A lively account of the development of the periodic system from early discoveries of elements to the impact of quantum theory.

C. Benson, *The Periodic Table of the Elements and Their Chemical Properties*. Kindle edition. MindMelder.com (2009).

J. T. Boronski, Inducing nucleophilic reactivity at beryllium with an aluminyl ligand. *J. Am. Chem. Soc.*, 2023, **145**, 4408. A fascinating account of a compound containing an Al(II)—Be(I) bond, illustrating their diagonal relationship.

L. J. Norrby, Why is mercury a liquid? *J. Chem. Educ.*, 1991, **68**, 110. A very readable article by a critic of the neglect of relativistic effects in understanding the properties of heavy elements.

P. Pyykkö, Relativistic effects in structural chemistry. *Chem. Rev.*, 1998, **88**, 563. A comprehensive account of how relativistic effects influence the structure and properties of heavier elements and their compounds.

P. Pyykkö, Relativistic effects in chemistry: More common than you thought. *Ann. Rev. Phys. Chem.*, 2012, **63**, 45. A short and engaging account that gives some excellent examples.

N. C. Pyper, Relativity and the periodic table. *Phil. Trans. Royal Society A.*, 2020, **378**, 20190305. A very readable account of how relativistic effects influence chemical properties across the periodic table.

H. B. Gray, Elements of life at the oxo wall. *Chem. Int.*, 2019, **41**, 16. A short essay outlining the oxo wall concept and its importance in the catalysis of oxygen-atom transfer reactions.

EXERCISES

10.1 Predict how the properties of PO_2, an unstable species formed by thermolysis of phosphate salts, should resemble and differ from those of NO_2.

10.2 Use the following bond enthalpy data to explain why elemental sulfur forms rings or chains with S–S single bonds, whereas oxygen exists as diatomic molecules. $B(X–Y)$ in kJ mol^{-1}: O–O, 142; O=O, 498; S–S, 263; S=S, 431.

10.3 From the data provided for Exercise 10.2, account for why sulfur forms catenated polysulfides of formula $[S–S–S]^{2-}$ and $[S–S–S–S]^{2-}$, whereas polyoxygen anions beyond O_3^- are unknown.

10.4 Compare and contrast the consequences of the alternation effect with regard to 4p elements and the lanthanoid contraction with regard to 5d elements.

10.5 (a) Identify the $Z + 8$ element for P. Briefly summarize any similarities between the elements. (b) Identify the $Z + 22$ element for tin, and summarize any similarities between them.

10.6 Use the following data to calculate average values of $B(Se–F)$ in SeF_4 and SeF_6. Comment on your answers in view of the corresponding values for $B(S–F)$ in SF_4 (+340 kJ mol^{-1}) and SF_6 (+329 kJ mol^{-1}): $\Delta_aH^\ominus$ (Se) = +227 kJ mol^{-1}, $\Delta_aH^\ominus$ (F) = +159 kJ mol^{-1}, $\Delta_fH^\ominus$ (SeF$_6$, g) = −1030 kJ mol^{-1}, $\Delta_fH^\ominus$ (SeF$_4$, g) = −850 kJ mol^{-1}.

10.7 Account for the following observations made for pentachlorides of Group 15 elements:

NCl_5 is unknown, PCl_5 is thermally stable at 200°C, $AsCl_5$ decomposes above −50°C

$SbCl_5$ is stable up to ~140°C, $BiCl_5$ is unknown.

10.8 Element 112, copernicium (Cn), which lies below mercury in Group 12, is expected to behave as an inert gas. Offer an explanation for this interesting prediction.

10.9 A newly synthesized compound of Cu(III) is reported to contain the *cupryl* entity ($[CuO]^+$). Comment on this discovery.

TUTORIAL PROBLEM

10.1 In January 2016 the International Union of Pure and Applied Chemistry (IUPAC) announced the discovery of four new elements: nihonium (element 113), moscovium (element 115), tennessine (element 117), and oganesson (element 118). These elements are very unstable and form isotopes of more stable elements by radioactive decay. Assuming that samples of these elements could be made that were stable enough to be investigated, use your knowledge of periodic trends to predict some of the expected physical and chemical properties of these elements.

11 Hydrogen

Part A: The essentials
11.1 The element
11.2 Simple compounds

Part B: The detail
11.3 Nuclear properties
11.4 Production of dihydrogen
11.5 Reactions of dihydrogen
11.6 Compounds of hydrogen
11.7 General methods for synthesis of binary hydrogen compounds

Further reading

Exercises

Tutorial problems

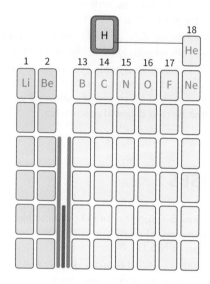

Hydrogen is an element which has a very rich chemistry despite its simple atomic structure. The environmental and societal importance of hydrogen is wide-reaching and obvious in the media where frequent reference is made to the *hydrogen economy*. In this chapter we discuss reactions of hydrogen species that are particularly interesting in terms of fundamental chemistry as well as important applications, including carbon-neutral energy technologies. We describe how H_2 is produced in the laboratory, on an industrial scale from fossil fuels and, looking to the future, by making increasing use of renewable resources. We summarize the synthesis and properties of binary compounds that range from volatile, molecular compounds to salt-like and metallic solids. We shall see that many of the properties of these compounds are understandable in terms of their ability to supply H^- or H^+ ions, and that it is possible to anticipate which one is likely to dominate. We consider how the H_2 molecule is activated by binding to catalysts, the processes that can produce H_2 from water using solar energy, and the efforts being made to make it a more convenient fuel for vehicles.

PART A: The essentials

Hydrogen is the most abundant element in the universe and the tenth most abundant by mass on Earth, where it is found in the oceans, minerals, and in all forms of life. The stable form of elemental hydrogen under normal conditions is *dihydrogen*, H_2, which occurs at trace levels in the Earth's lower atmosphere (0.5 ppm) and is essentially the only component of the extremely thin outer atmosphere. The partial depletion of elemental hydrogen from Earth reflects its volatility during formation of the planet. Dihydrogen is produced naturally by microbial organisms, as a product of some fermentation processes and as a by-product of ammonia biosynthesis (Box 11.1). Some H_2 is also produced in reactions occurring deep underground where reduced Fe minerals come into contact with water. The industrial production and major uses of H_2 are shown in Fig. 11.1.

Although most H_2 is still produced using fossil fuels, the ultimate goal is to obtain it using renewable energy to split water into its elements. Dihydrogen is an essential raw material for the industrial production of ammonia and many chemicals, and is being increasingly used in steel production. As an *energy vector*, H_2 offers a way to store electricity derived from solar energy, complementing the use of batteries. In fuel cells, the clean reaction of H_2 with O_2 in a large domestic device produces electricity and heat, and even potable water, a valuable resource in many parts of the world. However, the low volumetric energy density of H_2 poses challenges for its use as a fuel for transport. By developing a greater understanding of H_2 and its associated reactions, chemists are playing a leading role in bringing forth a more sustainable world.

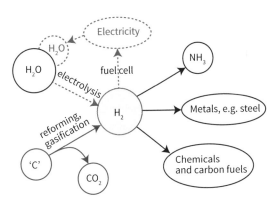

FIGURE 11.1 The production and major uses of H_2. Renewable pathways are shown in dashed green: they include the generation of potable water by domestic fuel cells. Production from fossil fuels 'C' will eventually be phased out.

BOX 11.1 What is the biological hydrogen cycle?

Dihydrogen is cycled by microbial organisms using metalloenzymes (Section 26.13). Although H_2 is present only to the extent of about 0.5 ppm at the Earth's surface, its levels are hundreds of times higher in anaerobic environments, such as wetland soils and sediments at the bottom of deep lakes and hot springs. The H_2 is produced in these O_2-free zones as a waste product by strict anaerobes (fermentative bacteria) which break down organic material (biomass) in processes that effectively use H^+ as an oxidant (it acts as a terminal electron acceptor in primitive forms of respiration). Hydrogen is also generated by thermophilic organisms that derive their carbon and energy entirely from CO, and by nitrogen-fixing bacteria, which yield H_2 as a by-product of ammonia formation. Other microorganisms, some of them aerobic, use H_2 as a 'fuel' like sugars, and they are responsible for the formation of the familiar gases CH_4 (methanogens) and H_2S (*Desulfovibrio*) as well as nitrite and other products. Fig. B11.1 summarizes some of the overall processes occurring in a freshwater environment.

In animals, including humans, the anaerobic environment of the large intestine is host to bacteria that form H_2 by the breakdown of carbohydrates. The mucus layer of the mouse intestine has been found to contain H_2 at levels above 0.04 mmol dm^{-3}, equivalent to a partial atmosphere of 5% H_2. In turn, this H_2 is utilized by methanogens, such as those found in ruminating mammals, to produce CH_4 and by other bacteria, including dangerous pathogens such as those of the *Salmonella* genus and *Helicobacter pylori*, which is responsible for gastric ulcers. High levels of H_2 in the breath have been used to diagnose conditions related to carbohydrate intolerance, and

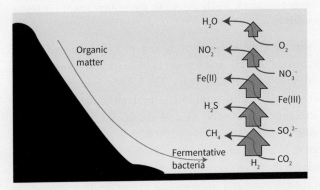

FIGURE B11.1 Some of the processes that contribute to the biological hydrogen cycle in a freshwater environment.

these levels may reach >70 ppm following lactose ingestion by lactose-intolerant patients.

Production of H_2 by microorganisms (*biohydrogen*) has a role to play in renewable 'green' energy alongside other industrial processes mentioned in Section 11.4. There are two different approaches, each of which ultimately uses sunlight as the primary energy input. The first way uses anaerobic organisms to ferment organic matter from sources ranging from cultivated biomass (including seaweed) to domestic waste. The second approach involves the direct manipulation of photosynthetic organisms such as green algae and cyanobacteria to produce H_2 as well as biomass. In either case H_2 can be extracted continuously by gas filters without the interruptions that harvesting requires.

11 Hydrogen

11.1 The element

The hydrogen atom, with ground-state configuration 1s¹, has only one electron so it might be thought that the element's chemical properties will be limited, but this is far from the case. Hydrogen has richly varied chemical properties and forms compounds with nearly every other element. It ranges in character from being a strong Lewis base (the hydride ion, H⁻) to being a strong Lewis acid (as the hydrogen cation, H⁺, the proton; Section 5.1).

Under certain circumstances H atoms can form bonds to more than one other atom simultaneously. The 'hydrogen bond' formed when an H atom bridges two electronegative atoms is fundamental to life on Earth, both in terms of biological chemistry and the surrounding environment: it was discussed in detail in Section 5.9. It is because of intermolecular hydrogen bonding that compounds formed with the light 2p elements N, O, and F have relatively high boiling points—the prevalence of water as a liquid across much of Earth's surface serving as a constant reminder of this important yet weak chemical bond.

(a) The atom and its ions

KEY POINTS The proton, H⁺, is always found in combination with a Lewis base and is highly polarizing; the hydride ion, H⁻, is highly polarizable.

There are three isotopes of hydrogen: hydrogen itself (^{1}H), deuterium (D, ^{2}H), and tritium (T, ^{3}H); tritium is radioactive with a half-life of 12.3 years. The lightest isotope, ^{1}H (very occasionally called protium), is by far the most abundant. Deuterium has variable natural abundance with an average value of about 16 atoms in 100 000. Tritium occurs to the extent of only 1 atom in 10^{21}. The different names and symbols for the three isotopes reflect the large relative differences in their masses and the chemical properties that depend on mass, such as the rates of diffusion and bond-cleavage reactions. The nuclear spin of ^{1}H ($I = ½$) is exploited in NMR spectroscopy (Section 8.6) for identifying hydrogen-containing molecules and determining their structures.

The free hydrogen cation (H⁺, the proton) has a huge charge/radius ratio, and it is not surprising to find that it is an extremely strong, hard Lewis acid. In the gas phase it readily attaches to other molecules and atoms; it even attaches to He to form HeH⁺. In the condensed phase, H⁺ is always found in combination with a Lewis base, and its ability to transfer between Lewis bases gives it the special role in chemistry that was explored in Chapter 5. The molecular cations H_2^+ and H_3^+ (the triangular hydrogenonium ion) are unknown in solution, although H_3^+ is the most abundant molecular cation in interstellar space. In contrast to H⁺, which is highly polarizing, the hydride ion, H⁻, is highly polarizable because two electrons are bound by just one proton: H⁻ is a strong, soft base. The radius of H⁻ varies considerably depending on the atom to which it is attached. The lack of core electrons to scatter X-rays means that bond distances and angles involving an H atom in a compound are difficult to measure by X-ray diffraction: for this reason, neutron diffraction is used when it is crucial to determine the precise positions of H atoms (Section 8.2).

(b) Properties and reactions

KEY POINTS Hydrogen has distinctive atomic properties that place it in a special position in the periodic table. Dihydrogen is quite an inert molecule and its reactions require a catalyst or initiation by radicals.

Hydrogen's special properties distinguish it from all other elements in the periodic table. It is usually placed at the head of Group 1 because like the alkali metals it has only one electron in its valence shell. That position, however, does not truly reflect the chemical or physical properties of the element. Its ionization energy is far higher than the other Group 1 elements, so hydrogen is not a metal under ambient conditions encountered on Earth. However, when subjected to extreme pressures that exist naturally in certain extraterrestrial environments such as the core of Jupiter, liquid H_2 becomes metallic with a conductivity comparable to heavier Group 1 elements. At the other extreme, solitary hydrogen atoms dispersed in outer space (just one or two atoms per m³ between galaxies) emit characteristic radio-frequency radiation, relatively free of interference from cosmic dust, that enables us to explore the universe. The source of the emission is the relaxation of an excited state (F1) in which the proton and electron have the same spin orientation (p↑e↑), to the ground state (F0) in which the spins are opposed (p↑e↓). The F1 state has an extremely long lifetime, so the emission line at 1.4204 GHz (wavelength 21.1 cm) is inherently narrow, like those due to electronic transitions occurring with visible light in the Balmer series (Section 1.1). Such spectral lines allow measurement of the relative velocities of different structures in outer space (for example, rotating galaxies and binary stars) through the Doppler effect, which alters the frequency or width of the signal. Hydrogen atoms thus equip astrophysicists with a sensitive intergalactic speed camera.

In some versions of the periodic table, hydrogen is placed at the head of Group 17 because, like the halogens, it requires only one electron to complete its valence shell. But the electron affinity of hydrogen is far lower than any of the elements of Group 17, and the discrete hydride ion, H⁻, is encountered only in certain compounds. To reflect its unique characteristics, we place H in a special position at the head of the entire table.

Because H_2 has so few electrons, the intermolecular forces between H_2 molecules are weak, and at 1 atm the gas condenses to a liquid only when cooled to 20 K. If an electric discharge is passed through H_2 gas at low pressure, the molecules dissociate, ionize, and recombine, forming a plasma

BOX 11.2 Dihydrogen as a fuel for transport

The use of H_2 as a fuel (an energy carrier or *vector*) has been investigated seriously since the 1970s when oil prices first rose dramatically; interest has increased greatly in more recent times owing to environmental pressures on further use of fossil fuels. As a fuel, H_2 is clean burning and nontoxic, and its production from fully renewable resources is slowly but inexorably replacing its production from fossil carbon feedstocks. Table B11.1 compares the performance data for H_2 and other energy carriers including hydrocarbon fuels and a lithium ion battery. The high specific enthalpy of H_2 suits it for aerospace applications (rockets and aircraft) where lightness is important. However, H_2 has a low volumetric energy density (its standard enthalpy of combustion divided by its volume), and it falls far below hydrocarbon fuels in this respect.

Along with lithium ion batteries, H_2 offers a viable way to power carbon-neutral road transport. Although it can be used in internal combustion engines (with modifications in design and specifications), the most important way of utilizing H_2 in a small vehicle is to react it in a fuel cell (Box 6.1 and Section 30.5) to produce electricity for an electric motor. The main advantage of H_2 over batteries lies with range: whereas it may take just minutes to refuel a vehicle with H_2, recharging a battery imposes a much longer interruption in a long journey.

Challenges facing the ongoing commercial adoption of H_2 include ensuring its safe handling by consumers and allaying some common misconceptions. The safety issue is historic: as a large component of coal gas, which also included toxic CO, H_2 was routinely piped to homes in industrialized countries up to the mid–late twentieth century. Frequent failures of the supply infrastructure may be traced at least in part to its reaction with metals that causes them to become brittle. But there are positive comparisons. Being much lighter than hydrocarbon fuels, H_2 is very quickly dispersed from the source of a leak, thus reducing the possibility of ignition and lowering the intensity of a fire. The radiant heat produced by an H_2 flame is lower than that from fossil fuels, thereby further minimizing the chance of a fire spreading. The risk of an H_2 explosion is also reduced by its high O_2 requirement (20–60%) compared to carbon-based fuels (>3%).

The challenge for using H_2 in road transport, and particularly for small vehicles, lies in overcoming the disadvantage of its low volumetric energy density. Although the situation is greatly improved by compressing H_2 (typically, to a practical standard of 70 MPa), there is considerable interest in developing the advanced H_2 storage materials outlined in Box 11.5 and described in more detail in Section 30.8.

TABLE B11.1 Specific enthalpies and volumetric energy densities of common energy carriers
(1 MJ = 0.278 kWh, 1 MPa = 10 bar)

Fuel	Specific enthalpy/ MJ kg^{-1}	Volumetric energy density/ MJ dm^{-3}
Liquid H_2*	142	8.5
H_2 at 70 MPa	142	5.3
Liquid natural gas	50	20.2
Natural gas at 20 MPa	50	8.3
Petrol (gasoline)	46	34.2
Diesel*	45	38.2
Coal	30	27.4
Ethanol*	27	22.0
Methanol*	20	15.8
Wood*	15	14.4
Lithium battery* (see Box 12.3)	2.0	6.1

(* denotes an energy carrier that can be derived or recharged from renewable resources, including biowaste)

---

containing, in addition to H_2, spectroscopically observable amounts of H, H^+, H_2^+, and H_3^+.

The H_2 molecule has a high bond enthalpy (436 kJ mol^{-1}) and a short bond length (74 pm). The high bond strength results in H_2 being quite an inert molecule, and reactions of H_2 do not occur readily unless a special activation pathway is provided. In the gas phase it is much more difficult to dissociate H_2 heterolytically than homolytically,[1] because the former incurs a large additional energy cost to separate the opposite charges. Heterolytic cleavage is therefore assisted by reagents that form strong bonds to H^+ and H^-:

$$H_2(g) \rightarrow H(g) + H(g) \quad \Delta_r H^\ominus = +436 \text{ kJ mol}^{-1}$$
$$H_2(g) \rightarrow H^+(g) + H^-(g) \quad \Delta_r H^\ominus = +1675 \text{ kJ mol}^{-1}$$

Both homolytic and heterolytic dissociation are catalysed by molecules or active surfaces. In the gas phase, the explosive reaction of H_2 with O_2,

$$2H_2(g) + O_2(g) \rightarrow 2H_2O(g) \quad \Delta_r H^\ominus = -242 \text{ kJ mol}^{-1}$$

proceeds by a complex radical chain mechanism (Section 11.5).

In addition to reactions that result in dissociation of the H–H bond, H_2 can also react reversibly without cleavage to form dihydrogen d-metal complexes (Sections 11.6(d) and 24.7).

Hydrogen is an excellent fuel for large rockets on account of its high specific enthalpy (the standard enthalpy of combustion divided by the mass), which is approximately three times that of a typical hydrocarbon (Box 11.2).

---

[1] In homolytic dissociation of H_2 the H–H bond breaks symmetrically to give two hydrogen atoms, whereas heterolytic dissociation of H_2 produces H^+ and H^-.

11.2 Simple compounds

The nature of the bonding in binary compounds of hydrogen—that is, with other elements E (EH_n)—is largely rationalized by noting that an H atom has a high ionization energy (1312 kJ mol^{-1}) and a low positive electron affinity (77 kJ mol^{-1}). Although binary hydrogen compounds are often known as 'hydrides' (we will use this term throughout this chapter), very few actually contain a discrete H$^-$ anion. The (Pauling) electronegativity of 2.2 (Section 1.7(e)) is intermediate in value, so hydrogen is normally assigned the oxidation number −1 when in combination with metals (as in NaH and AlH$_3$), and +1 when in combination with non-metals (as in H$_2$O and HCl).

(a) Classification of binary compounds

KEY POINTS Compounds formed between hydrogen and other elements vary in their nature and stability. In combination with p-block elements, hydrogen forms compounds having covalent E–H bonds, some of which show Brønsted acidity. Compounds with the most electropositive s-block metals are usually regarded as salts of the hydride ion and are strongly basic. In combination with d- and f-block metals, hydrogen atoms often occupy interstitial sites.

The binary compounds of hydrogen fall into three classes, although there is a range of structural types, and some elements form compounds with hydrogen that do not fall strictly into any one category:

1. **Molecular hydrides** exist as individual, discrete molecules; they are usually formed with p-block elements of similar or higher electronegativity than H. Their E–H bonds are best regarded as covalent.

Familiar examples of molecular hydrides include methane, CH$_4$ (1), ammonia, NH$_3$ (2), and water, H$_2$O (3).

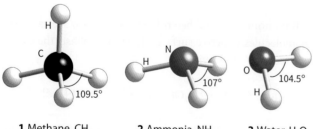

1 Methane, CH$_4$ **2** Ammonia, NH$_3$ **3** Water, H$_2$O

As shown in Section 5.1, molecular hydrides formed with elements located on the far right of the p-block (HCl, HBr, HI) are strong Brønsted acids.

2. **Saline hydrides**, also known as *ionic hydrides*, are formed with the most electropositive elements.

Saline hydrides, such as LiH and CaH$_2$, are nonvolatile crystalline solids that are electrically insulating, although only those in Group 1 and the heavier elements of Group 2

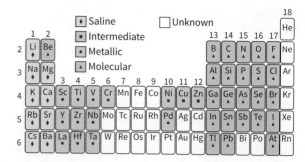

FIGURE 11.2 Classification of the binary hydrogen compounds of the s-, p-, and d-block elements. Although some d-block elements such as iron and ruthenium do not form binary hydrides, they do form metal complexes containing the hydride ligand.

should be regarded as hydride 'salts' containing discrete H$^-$ ions.

3. **Metallic hydrides** are nonstoichiometric, electrically conducting solids with a metallic lustre.

Metallic hydrides are formed with many d- and f-block elements. The H atoms are often regarded as occupying interstitial sites within the metal structure, although this occupation rarely occurs without expansion or phase change and frequently leads to loss of ductility and a tendency to fracture—a process known as *embrittlement*.

Fig. 11.2 summarizes the classification and the distribution of the different classes of binary hydrogen compound throughout the periodic table. It also identifies 'intermediate' hydrides that do not fall strictly into any of these categories, and elements for which binary hydrides have not been characterized.

In addition to binary compounds, hydrogen is found in complex anions of some p-block elements, examples being the BH_4^- ion (tetrahydridoborate, also known as boranate and more traditionally as 'borohydride') in NaBH$_4$ or the AlH_4^- ion (tetrahydridoaluminate, also known as alanate and more traditionally as 'aluminium hydride') in LiAlH$_4$. Both H and H$_2$ are encountered as ligands in certain d-block metal complexes.

(b) Thermodynamic considerations

KEY POINTS In the s and p blocks, strengths of E–H bonds *decrease* down each group. In the d block, strengths of E–H bonds *increase* down each group.

The standard Gibbs energies of formation of the hydrogen compounds of s- and p-block elements reveal a regular variation in stability (Table 11.1). With the possible exception of BeH$_2$ (for which good data are not available) all the s-block hydrides are exergonic ($\Delta_f G^\ominus < 0$) and therefore thermodynamically stable with respect to their elements at room temperature. Although the trend is erratic in Group 13, for all the other groups of the p block, the simple hydrogen

TABLE 11.1 Standard Gibbs energy of formation, $\Delta_f G^\ominus/(\text{kJ mol}^{-1})$ of binary s- and p-block hydrogen compounds at 25°C*

Period	Group 1	Group 2	Group 3	Group 14	Group 15	Group 16	Group 17
2	LiH(s)	BeH$_2$(s)	B$_2$H$_6$(g)	CH$_4$(g)	NH$_3$(g)	H$_2$O(l)	HF(g)
	−68.4	(+20)	+37.2	−50.7	−16.5	−237.1	−273.2
3	NaH(s)	MgH$_2$(s)	AlH$_3$(s)	SiH$_4$	PH$_3$(g)	H$_2$S(g)	HCl(g)
	−33.5	−35.9	+46.4	+56.9	+13.4	−33.6	−95.3
4	KH(s)	CaH$_2$(s)	Ga$_2$H$_6$(s)	GeH$_4$(g)	AsH$_3$(g)	H$_2$Se(g)	HBr(g)
	(−36)	−147.2	>0	+113.4	+68.9	+15.9	−53.5
5	RbH(s)	SrH$_2$(s)		SnH$_4$(g)	SbH$_3$(g)	H$_2$Te(g)	HI(g)
	(−30)	(−141)		+188.3	+147.8	>0	+1.7
6	CsH(s)	BaH$_2$(s)					
	(−32)	(−140)					

*Values in parentheses are estimates.

compounds of the first members of the groups (CH$_4$, NH$_3$, H$_2$O, and HF) are exergonic but the analogous compounds of their congeners become progressively less stable down the groups. The heavier hydrides become more stable on going from Group 14 across to the halogens. For example, SnH$_4$ is strongly endergonic ($\Delta_f G^\ominus \gg 0$) whereas HI is barely so.

The trends can be traced to the variation in E–H bond energies (Fig. 11.3). The H–H bond is the strongest single homonuclear bond known (apart from a ^{2}H–^{2}H or ^{3}H–^{3}H bond; see Section 11.3), and in order for a compound to be exergonic and stable with respect to its elements, it needs to have E–H bonds that are sufficiently strong that the cost of breaking each H–H bond is largely offset. For molecular hydrides of the p-block elements, bonding is strongest with the Period 2 elements and becomes progressively weaker down each group. The weak bonds formed by the heavier p-block elements are due to the poor overlap between the relatively compact H1s orbital and the more diffuse s and p orbitals of their atoms. Although d-block elements do not form binary molecular compounds, many complexes contain one or more hydride ligands. Metal–hydrogen bond strengths in the d block increase down a group because the 3d orbitals are too contracted to overlap well with the H1s orbital and better overlap is afforded by 4d and 5d orbitals.

(c) Reactions of binary compounds

KEY POINT The reactions of binary compounds of hydrogen fall into three classes depending on the polarity of the E–H bond.

In compounds where E and H have similar electronegativities, cleavage of the E–H bond tends to be homolytic, producing, initially, an H atom and an E atom or radical, each of which can go on to combine with other available atoms or radicals.

For $\chi(E) \approx \chi(H)$: $\quad$ E–H $\rightarrow$ E· + H·

Common examples of homonuclear cleavage include the thermolysis and combustion of hydrocarbons.

In compounds where E is more electronegative than H, heterolytic cleavage can occur, releasing a proton.

For $\chi(E) > \chi(H)$: $\quad$ E–H $\rightarrow$ E$^-$ + H$^+$

The compound behaves as a Brønsted acid and is able to transfer H$^+$ to a base. In such compounds the H atom is termed **protonic** (noting from Chapter 5 that the term **protic** is a description of the compound, not the H atom).

Heterolytic bond cleavage also occurs in compounds where E is less electronegative than H, including saline hydrides.

For $\chi(E) < \chi(H)$: $\quad$ E–H $\rightarrow$ E$^+$ + H$^-$

In this case the H atom is **hydridic** and an H$^-$ ion is transferred to a Lewis acid, such as a boron-containing reagent

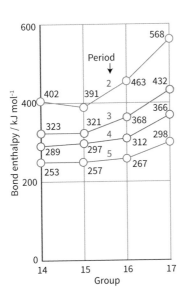

FIGURE 11.3 Average bond energies (kJ mol^{-1}) for binary molecular hydrides of p-block elements.

(Section 5.8). The reducing agents $NaBH_4$ and $LiAlH_4$ used in organic synthesis are examples of hydride-transfer reagents. By analogy with Brønsted acidity, which measures the ability of a species to donate a proton, a **hydridicity** scale (often referred to as a **hydricity** scale) can be compiled that compares the abilities of species to donate a hydride (Section 11.6). The scale may be based on calculations performed for species in the gas phase or experimental data for hydride transfer equilibria in a suitable solvent. Through its ability to exist in both protonic (H^+) and hydridic (H^-) states, a bound hydrogen atom can act as a two-electron redox agent.

EXAMPLE 11.1 **Frontier molecular orbitals of acidic and hydridic compounds**

Using a simple molecular orbital model, compare the properties (protonic or hydridic) expected of molecular diatomic compounds HX, where (a) X is an electronegative element, and (b) X is an electropositive element.

Answer We need only consider the orbitals required to make σ bonds, and we draw the atomic orbitals at appropriate energy levels depending on the electronegativity of X. In case (a) the atomic orbital on X lies at lower energy than that of H. Consequently, the filled bonding MO has mostly X character, whereas the empty antibonding orbital has mostly H character: the H atom is thus protonic. In case (b) the atomic orbital on X lies at higher energy than that of H, and it is the filled bonding orbital that has most H-character, making the H atom hydridic.

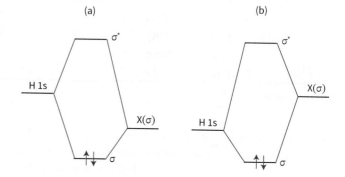

Self-test 11.1 Use the data in Table 1.8 to predict which of the hydrides XH_3 formed by Group 15 elements (N, P, As, Sb, Bi) should contain the most hydridic H atoms.

PART B: The detail

In this part of the chapter we present a more detailed discussion of the chemical properties of hydrogen, identifying and interpreting trends. We explain how dihydrogen is prepared on a small scale in the laboratory and how it is produced industrially from fossil fuels; then we outline methods for its production from water using renewable energy. We describe the reactions that dihydrogen undergoes with other elements and classify the different types of compounds formed. Finally, we present the strategies for synthesizing various hydrogen-containing compounds.

11.3 Nuclear properties

KEY POINT The three hydrogen isotopes H, D, and T have large differences in their atomic masses and different nuclear spins, which give rise to easily observed changes in IR, Raman, and NMR spectra of molecules containing these isotopes.

Neither ^1H nor ^2H (deuterium, D) is radioactive, but ^3H (tritium, T) decays by the loss of a β particle to yield a rare but stable isotope of helium:

$$^3_1H \rightarrow {}^3_2He + \beta^-$$

The half-life for this decay is 12.4 years. Tritium's abundance of 1 in 10^{21} hydrogen atoms in surface water reflects a steady state between its production by bombardment of cosmic rays on the upper atmosphere and its loss by radioactive decay. Tritium can be synthesized by neutron bombardment of ^6Li (exothermic) or ^7Li (endothermic):

$$^1_0n + {}^6_3Li \rightarrow {}^3_1H + {}^4_2He \quad \text{(releases 4.78 MeV energy)}$$
$$^1_0n + {}^7_3Li \rightarrow {}^3_1H + {}^4_2He + {}^1_0n \quad \text{(requires 2.87 MeV energy)}$$

Continuous production of tritium from lithium is a key step in the projected future generation of energy from nuclear fusion rather than nuclear fission. In typical prototype fusion reactors, tritium and deuterium are heated to over 100 MK to give a plasma in which the nuclei react to produce ^4He and a neutron.

$$^2_1H + {}^3_1H \rightarrow {}^4_2He + {}^1_0n \quad \text{(releases 17.6 MeV;}$$
$$\Delta H = -1698 \text{ MJ mol}^{-1})$$

The neutron is used to bombard a lithium blanket that has been enriched in ^6Li to generate further tritium. This process

TABLE 11.2 The effect of deuteration on physical properties

	H_2	D_2	H_2O	D_2O
Normal boiling point/°C	−252.8	−249.7	100.0	101.4
Mean bond enthalpy/(kJ mol^{-1})	436.0	443.3	463.5	470.9

carries far fewer environmental risks than fission of ^{235}U and is essentially renewable: of the two primary fuels required, deuterium is readily available from water and lithium is also widely distributed (Chapter 12).

The physical and chemical properties of **isotopologues**, isotopically substituted molecules, are usually very similar, but not when D is substituted for H, as the mass of the substituted atom is doubled. Table 11.2 shows that the differences in boiling points and bond enthalpies are easily measurable for H_2 and D_2. The difference in boiling point between D_2O and H_2O reflects the greater strength of the O···D–O hydrogen bond compared with that of the O···H–O hydrogen bond, because the zero-point energy (Section 8.5) of the former is lower. The compound D_2O is known as 'heavy water' and is used as a moderator in the nuclear power industry; it slows down emitted neutrons and increases the rate of induced fission.

> **A NOTE ON GOOD PRACTICE**
>
> An *isotopologue* is a molecular entity that differs only in isotopic composition. An *isotopomer* is an isomer having the same number of each isotopic atom but differing in their positions in the molecule.

Reaction rates are often different for processes in which E–H and E–D bonds, where E is another element, are broken, made, or rearranged. The detection and measurement of this **kinetic isotope effect** can often help to support a proposed reaction mechanism. Kinetic isotope effects are frequently observed when an H atom is transferred from one atom to another in an activated complex. For example, the electrochemical reduction of H$^+$(aq) to H_2(g) occurs with a substantial isotope effect, with H_2 being liberated much more rapidly. A practical consequence of the difference in rates of formation of H_2 and D_2 is that D_2O may be concentrated electrolytically, thus facilitating separation of the two isotopes.

In general, reactions involving D_2O occur more slowly than those involving H_2O and, not surprisingly, D_2O and D-substituted foods ingested in large quantities are poisonous for higher organisms. In pharmacology, tetrabenazine, used to treat involuntary movement disorder (associated with Huntington's disease), is deuterated (the resulting product is known as 'deutetrabenazine', (**4**)) to improve the drug's residence time in a patient. As with many 'foreign' substances, its degradation in the body may involve enzymes called oxygenases that insert an O-atom into CH bonds (Section 26.10).

4 'Deutetrabenazine'

Because the frequencies of molecular vibrations depend on the masses of atoms, they are strongly influenced by substitution of D for H. The heavier isotope results in the lower frequency (Section 8.5). The isotope effect can be exploited by observing the IR spectra of isotopologues to determine whether a particular infrared absorption involves significant motion of a hydrogen atom in the molecule.

The distinct properties of the isotopes make them useful as **tracers**. The involvement of H and D through a series of reactions can be followed by infrared (IR, Section 8.5) and mass spectrometry (Section 8.11), as well as by NMR spectroscopy (Section 8.6). Tritium can be detected by its radioactivity, which can be a more sensitive probe than spectroscopy.

Another important property of the hydrogen nucleus is its spin. The nucleus of hydrogen, a proton, has $I = ½$; the nuclear spins of D and T are 1 and ½, respectively. As explained in Section 8.6, proton NMR detects the presence of H nuclei in a compound and is a powerful method for determining structures of molecules—even proteins with molecular masses in excess of 20 kDa. Fig. 11.4 shows the ranges of typical ^{1}H-NMR chemical shifts for some compounds of p- and d-block elements. Although hydrogen atoms bonded to electronegative elements ('protonic' H atoms) tend to display more positive values of chemical shift than hydrogen atoms coordinated to metal ions with incomplete d-subshells (d^n, where $n > 0$), other factors contribute, such as the mass of the atom to which the hydrogen is attached, and the solvent in which the compound is dissolved.

11.4 Production of dihydrogen

Hydrogen is important both as a raw material for the chemical industry and, increasingly, as a fuel. Although H_2 is not present in significant quantities at the Earth's surface, there is a high biological turnover due to the various microorganisms that use H$^+$ as an oxidant or H_2 as a fuel (Box 11.1). Industrially, H_2 has long been produced from fossil fuels, and nowadays most is produced from methane

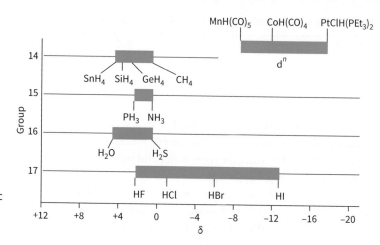

FIGURE 11.4 Typical ^{1}H-NMR chemical shifts for diamagnetic compounds relative to TMS (tetramethylsilane). The tinted boxes show families of elements grouped together.

(natural gas). In 2021, world production of H_2 by all means exceeded 75 Mt. Most H_2 is used close to its site of production, for the synthesis of ammonia by the Haber–Bosch process (Section 16.6), hydrogenation of unsaturated fats, hydrocracking of crude oil, and large-scale manufacture of organic chemicals. In efforts to reduce global CO_2 emissions, H_2 will increasingly be produced from renewable sources, the aim being to use solar energy or electricity derived from it to 'split' water. Reactions of such 'green' H_2 with CO_2 or CO to produce liquid hydrocarbon fuels would amount to a carbon-neutral technology. To assist industries and policy-makers, a terminology is being adopted to classify the different large-scale sources of H_2. Known as the 'colours of hydrogen', the categories are explained in Box 11.3.

(a) Small-scale preparation

KEY POINTS In the laboratory, H_2 is easily generated by the reactions of electropositive elements with aqueous acid or alkali, or by hydrolysis of saline hydrides. It is also produced by electrolysis.

BOX 11.3 The 'colours' of H_2

Sources of commercial H_2 are classified by assigning a colour to represent the degree to which production has avoided the emission of CO_2. The 'spectrum' is shown in Fig. B11.2.

Brown H_2 is produced by gasification of coal: its production releases more CO_2 than any other method.

Grey H_2 is produced by steam reforming and currently accounts for most of the global demand.

Blue H_2 is produced by steam reforming linked to CO_2 capture, although 10–20% is still released.

Turquoise H_2 is produced by hydrocarbon pyrolysis: it is carbon neutral, but economic viability depends on the value of the carbon product. High-demand nanomaterials are promising options.

Pink H_2 is produced by electrolysis using nuclear energy. Nuclear fission is not universally regarded as a completely renewable resource.

Green H_2 is produced from solar energy, by electrolysis using fully renewable electricity, by photolysis, and by biological methods such as fermentation of biomass.

White H_2 is not shown, but it refers to the rare fraction that originates from deep underground deposits. It is formed continuously through the reduction of water by Fe(II)-containing minerals. A variation is **orange H_2**, also not shown,

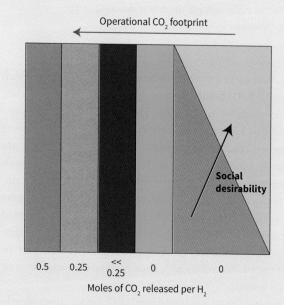

FIGURE B11.2 The 'renewability spectrum' of H_2 sources: renewability and lowering in carbon footprint of the products (as CO_2) increase from left to right.

in which the underground H_2 production is stimulated; for example, by injection of water that may also be enriched with carbon dioxide.

There are many straightforward procedures for preparing small quantities of pure H_2. In the laboratory, H_2 is produced by reaction of Al or Si with hot alkali solution.

$$2\,Al(s) + 2\,OH^-(aq) + 6\,H_2O(l) \rightarrow 2\,Al(OH)_4^-(aq) + 3\,H_2(g)$$
$$Si(s) + 2\,OH^-(aq) + H_2O(l) \rightarrow SiO_3^{2-}(aq) + 2\,H_2(g)$$

or, at room temperature, by reaction of Zn with mineral acids:

$$Zn(s) + 2\,H_3O^+(aq) \rightarrow Zn^{2+}(aq) + H_2(g) + 2\,H_2O(l)$$

The reaction of metal hydrides with water provides a convenient way to obtain small amounts of H_2 outside a laboratory. Calcium dihydride is particularly suited for remote, on-site H_2 production, as it is commercially available and inexpensive and reacts with H_2O at room temperature:

$$CaH_2(s) + 2\,H_2O(l) \rightarrow Ca(OH)_2(s) + 2\,H_2(g)$$

Pure H_2 is also produced in small amounts, using a simple electrolysis cell; electrolysis of heavy water is a convenient way to prepare D_2.

(b) Production of H_2 using carbon sources

KEY POINT Most H_2 for industry is produced by the high-temperature reaction of H_2O with CH_4 or a similar reaction with coke.

Hydrogen is needed in huge quantities to satisfy the needs of industry; in fact, production is often integrated directly (without transport) into chemical processes that require H_2 as a feedstock, such as oil refining and ammonia production. As of 2021, most H_2 is still produced by reacting fossil fuels, i.e. natural gas (methane) or coal, with H_2O (as steam). The catalysed reaction of hydrocarbons with H_2O is known as **hydrocarbon (steam) reforming**. High temperatures are required, and other hydrocarbons apart from methane can be used.

$$CH_4(g) + H_2O(g) \rightarrow CO(g) + 3\,H_2(g) \quad \Delta_r H^\ominus = +206.2\,\text{kJ mol}^{-1}$$

Hydrogen is also produced by the reaction of steam with coal, in a process known as **coal gasification**, which occurs at 1000°C:

$$C(s) + H_2O(g) \rightleftharpoons CO(g) + H_2(g) \quad \Delta_r H^\ominus = +131.4\,\text{kJ mol}^{-1}$$

The mixture of CO and H_2 produced by steam reforming or gasification is known as **synthesis gas** ('syngas'), and further reaction with water (the **water gas shift reaction**) produces more H_2:

$$CO(g) + H_2O(g) \rightleftharpoons CO_2(g) + H_2(g) \quad \Delta_r H^\ominus = -41.2\,\text{kJ mol}^{-1}$$

Overall

$$CH_4(g) + 2\,H_2O(g) \rightleftharpoons CO_2(g) + 4\,H_2(g) \quad \Delta_r H^\ominus = +165.0\,\text{kJ mol}^{-1}$$

or

$$C(s) + 2\,H_2O(g) \rightleftharpoons CO_2(g) + 2\,H_2(g) \quad \Delta_r H^\ominus = +90.2\,\text{kJ mol}^{-1}$$

In a related technology suited for vehicles, H_2 can be produced from methanol using an on-board steam reformer that mixes methanol vapour with H_2O (steam reforming) and with O_2 (air-reforming) in two parallel catalytic processes.

$$CH_3OH(g) + H_2O(g) \xrightleftharpoons{Cu/ZnO} CO_2(g) + 3\,H_2(g)$$
$$\Delta_r H^\ominus = +49.4\,\text{kJ mol}^{-1}$$

$$CH_3OH(g) + \tfrac{1}{2}O_2(g) \xrightleftharpoons{Pd} CO_2(g) + 2\,H_2(g)$$
$$\Delta_r H^\ominus = -192.3\,\text{kJ mol}^{-1}$$

The two reactions, which occur over the temperature range 200–350°C, are controlled to ensure that the heat produced by the exothermic oxidation reaction just offsets that required for (a) the reaction with steam and (b) the vaporization of all components. The CO_2 and H_2 products are separated with a membrane. Despite CO_2 being emitted, the technology is carbon neutral provided the methanol (which can thus be regarded as an H_2 carrier, see Box 11.2) has been produced from a renewable resource.

Hydrogen can also be obtained by hydrocarbon **pyrolysis**—a process that yields elemental carbon instead of CO_2. A longstanding disadvantage has been that the elemental carbon that is normally formed is an amorphous material, known commonly as carbon black or soot, which is described in Section 15.6.

$$CH_4(g) \rightarrow C(s) + 2\,H_2(g) \quad \Delta_r H^\ominus = +74.6\,\text{kJ mol}^{-1}$$

However, the reaction may also offer a route to producing bulk carbon nanomaterials, prompting research to identify catalysts directing the formation of desirable high-value products, the H_2 itself becoming a useful by-product.

(c) Production of H_2 from water alone

KEY POINTS The costs of producing H_2 from water in processes known as 'water splitting' are high compared to the costs of reforming fossil fuels. Environmental pressures are driving technologies to produce H_2 more efficiently by using surplus or renewable electricity, particularly that derived from solar energy.

In the well-established **chloralkali process** (Box 12.5), H_2 is produced as a by-product of NaOH manufacture by electrolysis of brine, the other gaseous product being Cl_2 (as it is more easily evolved at the anode than O_2). The amount of H_2 produced in this way is significant but is determined by the economic demand for NaOH and Cl_2, each of which are caustic substances. Electrolysis of water is increasingly being used to produce H_2 as the intended product, on a much larger scale with benign O_2 as the only by-product. Compared to production from fossil sources, H_2 produced by electrolysis is free from contaminants such as CO, and is therefore better suited for operation in fuel cells where poisoning of electrode surfaces must be avoided. Solar energy, whether from direct radiation or wind, is intermittent, and demand rarely matches supply and vice versa (the same is true of nuclear power, which is best provided at a constant level): thus, it becomes attractive to use surplus electricity to make H_2 which can then be used directly or stored then recycled to produce electricity when it is needed. It is as such a facilitator that the role of H_2 as an **energy vector** becomes very clear.

The water-splitting reaction for liquid H_2O is written:

$$H_2O(l) \rightarrow H_2(g) + \tfrac{1}{2}O_2(g) \quad \Delta_r H^\ominus = +285.9 \text{ kJ mol}^{-1}$$

This reaction can be driven electrochemically under ambient conditions (Section 6.16) by supplying a potential difference of at least 1.49 V between the anode (where the O_2 evolution reaction (OER) occurs) and the cathode (where the H_2 evolution reaction (HER) occurs). The cell is known as an electrolyser, and three types are currently in use and under continuous development. Design principles for electrolysers are very similar to those used for H_2 fuel cells (Box 6.1) where the reaction in the opposite direction is harnessed to produce electrical energy. The **alkaline electrolyser** is the simplest and longest established: it operates at 30–80°C with Ni electrodes immersed in aqueous NaOH or KOH as electrolyte, separated by a diaphragm that allows transport of OH^- ions produced at the cathode. The **solid oxide electrolyser** operates at high temperatures (500–800°C), and uses a Ni-based cathode and a lanthanoid-based ceramic anode. Steam is applied to the cathode, producing H_2 and O^{2-} ions which are transported to the anode, through a solid ZrO_2-based electrolyte, to evolve O_2. The **PEM electrolyser** (Polymer Exchange Membrane) offers high rates of electrolysis at low voltages and temperatures (20–80°C). Water is introduced to the anode, and the protons produced by the OER are transported to the cathode through the PEM, which is usually a commercially available material such as Nafion®. The anode and cathode are coated with very thin layers of electrocatalyst, typically IrO_2 or RuO_2 for the OER, and Pt or Pd for the HER. Platinum is chosen because the HER and HOR are reversible and only a small overpotential (Section 6.16) is needed to drive the reaction at a high rate, the principle being analogous to the PEM fuel cell (Box 6.1). A typical **stack**, the core of a commercial electrolyser, comprises hundreds of individual cells connected in series (Fig. 11.5).

Due in part to the expensive materials required, electrolysis of water is economical and environmentally benign only if the electrical power stems from cheap, renewable resources or if it is surplus to demand. These conditions are already met in countries that have plenty of hydroelectric or nuclear energy, and strides are underway to produce H_2 using electricity produced by solar and wind farms when its

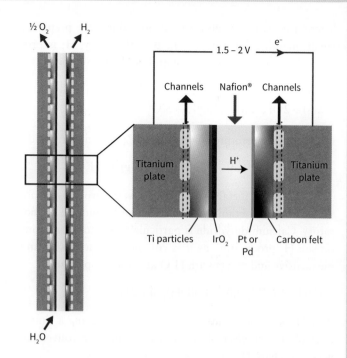

FIGURE 11.5 Diagram showing a single unit of a PEM H_2 electrolyser stack. An enlarged cross-section of the cell is shown at the right.

supply exceeds immediate demand. With suitable adaptation, even seawater can be electrolysed.

As of 2022, less than 4% of the global H_2 demand is produced by electrolysis, including that produced in the chloralkali process, and there is still an overriding reliance on fossil resources. Looking into the future, the production of H_2 from water is increasingly seen as a way of storing the energy derived from sunlight (photovoltaic, solar-thermal, wind) and thus levelling its intermittency, but new technology is required to lower costs and improve rates. Some physical methods for 'solar water-splitting' are outlined in Box 11.4 and Section 30.3.

Hydrogen can also be produced by fermentation, using anaerobic bacteria that use cultivated biomass or biological waste as their energy source (Box 11.1). Biological production could take place in 'hydrogen farms' by nurturing photosynthetic microorganisms that have been modified to produce H_2 as well as organic molecules.

11.5 Reactions of dihydrogen

KEY POINTS Molecular hydrogen is activated by homolytic or heterolytic dissociation on a metal or metal oxide surface or by coordination to a d-block metal. Direct reactions of H_2 with O_2 and halogens involve a radical chain mechanism.

Continuing from the introduction in Section 11.1, we now discuss some specific examples of the types of reaction that H_2 undergoes.

BOX 11.4 How can H_2 be produced directly using solar energy?

The Earth receives about 100 000 TW from the Sun, which is approximately 7000 times greater than the present global rate of energy consumption (15 TW in the early 2020s). As well as photosynthesis which produces all kinds of biomass, solar energy is harnessed to produce electricity, using photovoltaic cells and wind turbines. However, direct sunlight and wind are intermittent energy sources and are often incompatible with society's need for reliable, uninterrupted power. As an energy vector, H_2 offers a solution to rectify this conflict of supply and demand. By using solar energy to generate H_2 from water (water splitting), humans now have the greatest opportunity to end the world's dependence on fossil fuels and help curb global climate change. It is argued that *direct* conversion of solar energy to H_2 in small installations—i.e. without first converting it to electricity—could offer higher efficiency. Two such direct technologies being considered are photocatalytic H_2 production (photolysis) and high-temperature solar H_2 production (thermolysis).

Solar photocatalytic H_2 production ('artificial photosynthesis') combines and adopts principles used in photovoltaic cells and by plants for natural photosynthesis. The basis of a water-splitting system based on a single semiconductor particle is shown in Fig. B11.3, and may be understood from the introduction to band theory given in Section 2.11, and the principles of electrochemistry (Chapter 6). A more detailed description of photocatalytic water-splitting is presented in Section 30.3.

The essentials are as follows.

1. A photon excites an electron in the semiconductor, from the valence band to the conduction band, leaving a hole (h^+) in the valence band.
2. The conduction band potential E_{CB} must be more negative than 0 V vs RHE in order to reduce protons to H_2 (the HER).
3. The valence band potential E_{VB} must be more positive than +1.23 V vs RHE in order to oxidize water to O_2 (the OER).
4. The band gap E_g—the difference in energy between valence band and conduction band—must therefore exceed the minimal electrochemical potential required to split water (>1.23V). Light having a wavelength below 1000 nm is sufficiently energetic.
5. The HER and OER (Section 11.4(c)) are facilitated by respective catalysts (often referred to as co-catalysts) which are typically based on d-block metals, such as Rh, Cr, and Co.
6. To avoid the risk of explosion, the H_2 and O_2 that are formed must be separated, either *in situ* (by designing units that traverse a membrane; see broken line) or subsequently.

The so called 'sunbelt' regions, which include Australia, southern Europe, the Sahara desert, and southwestern states of the US, receive about 1 kW m^{-2} of solar power, representing a substantial quantity of useful thermal energy (Section 30.4). These regions are suitable for high-temperature solar H_2 production using solar-concentrating systems that reflect and focus radiation onto a receiver furnace, producing temperatures in excess of 1500°C. The intense heat can be used indirectly to split water into H_2 and O_2.

Direct, single-step thermolysis of water requires temperatures in excess of 4000°C—well above the threshold readily attainable in a solar concentrator, or compatible with containment materials and engineering. By using a multistep process it is possible to produce H_2 at much lower temperatures. Several systems are under investigation and development, the simplest of which are two-stage processes involving metal oxides, such as the sequence:

$$Fe_3O_4(s) \xrightarrow{\Delta} 3FeO(s) + \tfrac{1}{2}O_2(g) \quad \Delta_r H^\ominus = +319.5 \text{ kJ mol}^{-1}$$

$$H_2O(l) + 3FeO(s) \xrightarrow{cool} Fe_3O_4(s) + H_2(g) \quad \Delta_r H^\ominus = -33.6 \text{ kJ mol}^{-1}$$

although H_2 production by this route still requires temperatures above 2200°C.

A cerium oxide system performs the thermolysis cycle at temperatures below 2000°C.

$$CeO_2(s) \xrightarrow{\Delta} CeO_{2-\delta}(s) + \delta/2\, O_2(g)$$

$$CeO_{2-\delta}(s) + \delta H_2O(g) \xrightarrow{cool} CeO_2(s) + \delta H_2(g)$$

Water splitting at lower temperatures has been achieved with hybrid processes combining thermochemical and electrochemical reactions, such as

$2Cu(s) + 2HCl(g) \rightarrow H_2(g) + 2CuCl(l)$	(at 425°C)
$4CuCl(solv) \rightarrow 2Cu(s) + 2CuCl_2(solv)$	(electrochemical in HCl as solvent-electrolyte)
$2CuCl_2(s) + H_2O(g) \rightarrow Cu_2OCl_2(s) + 2HCl(g)$	(at 325°C)
$Cu_2OCl_2(s) \rightarrow 2CuCl(l) + \tfrac{1}{2}O_2(g)$	(at 550°C)

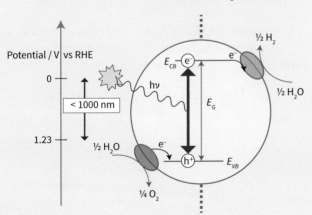

FIGURE B11.3 The principle of a semiconductor-based water-splitting device for solar H_2 generation.

(a) Homolytic dissociation

High temperatures are required to dissociate H_2 directly into isolated atoms, but a catalyst provides an alternative pathway with a substantially lower activation barrier through the simultaneous formation of H-catalyst bonds. Homolytic dissociation on the surface of a metal is represented:

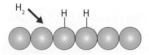

An important example is the reaction of H_2 at finely divided Pd or Ni metal which is used in the catalytic hydrogenation of alkenes, the catalytic reduction of aldehydes to alcohols, and in electrocatalysis (Section 6.16). Platinum is used as the electrocatalyst for H_2 oxidation in PEM electrolysers (Fig. 11.5) and fuel cells that are suitable for transport purposes (Box 6.1). Homolytic dissociation is a key step in one of two mechanisms proposed for electrocatalytic H_2 activation, Fig. 11.6 (left). Following the Sabatier principle (Section 2.15), the Pt–H bond is neither too weak to form nor too strong to be broken, accounting for why Pt is so dominant in H_2 activation. However, Pt is an expensive element and there is much interest in finding alternatives; we mention this aspect again in Section 11.6.

(b) Heterolytic dissociation

Heterolytic dissociation of H_2 depends upon a metal ion (for hydride coordination) and a Brønsted base being in close proximity. Such a condition is provided by the surface of a metal oxide, on which heterolytic dissociation is represented:

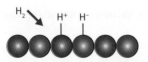

The Brønsted base can also be a solvent water molecule. In an alternative mechanism for electrocatalytic H_2 activation at Pt (Fig. 11.6 right), one H-atom is present as a hydridic Pt–H surface species, while the other is a solvated H+.

Reaction of H_2 with a ZnO surface appears to produce a Zn(II)-bound hydride and an O-bound proton. This reaction is involved in the production of methanol by catalytic hydrogenation of carbon monoxide over $Cu/ZnO/Al_2O_3$:

$$CO(g) + 2H_2(g) \rightarrow CH_3OH(g)$$

Dihydrogen is also dissociated very rapidly into a metal-bound hydride and a base-bound proton during its oxidation at the active site of metalloenzymes known as hydrogenases (Section 26.13).

Beyond these long-established processes, the frustrated Lewis pair (FLP) concept introduced in Section 5.10(d) offers further new possibilities for intricate, atomic-level design of catalysts for heterolytic H_2 dissociation.

(c) Radical chain reactions

Radical chain mechanisms account for the thermally or photochemically initiated reactions between H_2 and the halogens in which atoms are generated that act as radical chain carriers in the propagation reaction. Chain termination occurs when the radicals recombine:

$$H_2 \xrightarrow{X} XH\bullet + H\bullet \xrightarrow{O_2} HOO\bullet \xrightarrow{XH\bullet} 2\,OH\bullet \xrightarrow{H_2} H_2O + H\bullet\ldots\text{etc}$$

Once initiated, the activation energy for radical attack is low because a new bond is formed at the same time as one bond is lost.

Initiation, by heat or light: $\quad Br_2 \rightarrow Br\bullet + Br\bullet$

Propagation: $\quad Br\bullet + H_2 \rightarrow HBr + H\bullet$
$\quad H\bullet + Br_2 \rightarrow HBr + Br\bullet$

Termination: $\quad H\bullet + H\bullet \rightarrow H_2$
$\quad Br\bullet + Br\bullet \rightarrow Br_2$

Combustion of H_2, its highly exothermic reaction with O_2, also occurs by a radical chain mechanism. Certain mixtures explode violently when detonated:

$$2H_2(g) + O_2(g) \rightarrow 2H_2O(g) \quad \Delta_r H^\ominus = -242\,\text{kJ mol}^{-1}$$

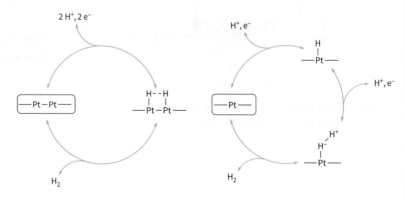

FIGURE 11.6 Homolytic and heterolytic mechanisms for the activation of H_2 at a Pt surface, which occurs during HER and HOR electrocatalysis in PEM electrolysers and fuel cells. The Volmer–Tafel mechanism (left), involves *homolytic* dissociation/recombination of H_2 on adjacent Pt atoms, whereas the Volmer–Heyrovský mechanism (right) involves a *heterolytic* dissociation/recombination step using solvent H_2O as a base.

11.6 Compounds of hydrogen

Hydrogen forms compounds with most of the elements. These compounds are classified into molecular hydrides, saline hydrides (salts of the hydride anion), metallic hydrides (interstitial compounds of d-block elements), and discrete complexes of d-block elements in which hydride or dihydrogen are ligands.

(a) Molecular hydrides

Molecular hydrides are formed with p-block elements and with Be, the most electronegative of the s-block elements. The bonding is covalent but variations in bond polarity (depending on the electronegativity of the atoms to which hydrogen is attached) result in a range of reaction types in which hydrogen is formally transferred as H^+, H^-, or H-atom.

(i) Nomenclature and classification

KEY POINTS Molecular compounds of hydrogen are classified as electron-rich, electron-precise, or electron-deficient. Electron-deficient hydrides provide some of the most intriguing examples of molecular structure and bonding, as their simplest units tend to associate via bridging hydrogen atoms to form dimers and higher polymers.

The systematic names of the molecular hydrogen compounds are formed from the name of the element and the suffix -ane, as in phosphane for PH_3. The more traditional names, however, such as phosphine and hydrogen sulfide (H_2S, sulfane) are still widely used (Table 11.3). The common names ammonia and water are universally used rather than their systematic names azane and oxidane.

Molecular compounds of hydrogen are divided further into three sub-categories:

Electron-precise, in which all valence electrons of the central atom are engaged in bonds.

Electron-rich, in which there are more electron pairs on the central atom than are needed for bond formation (that is, there are lone pairs on the central atom).

Electron-deficient, in which there are too few electrons available to fill the bonding and nonbonding orbitals.

Electron-precise molecular hydrogen compounds include hydrocarbons such as methane and ethane, and their heavier analogues silane, SiH_4, and germane, GeH_4 (Section 15.7). All these molecules are characterized by the presence of two-centre, two-electron bonds (2c,2e bonds) and the absence of lone pairs on the central atom. Electron-rich compounds are formed by the elements in Groups 15–17. Important examples include ammonia, water, and the hydrogen halides. Electron-deficient hydrogen compounds are common for boron and aluminium. The analogous simple hydride of boron, BH_3, is not found as a monomer; instead, it occurs as a dimer, B_2H_6 (diborane, 5), in which the two B atoms are bridged by a pair of H atoms in two three-centre, two-electron bonds (3c,2e bonds).

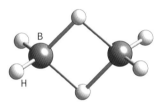

5 Diborane, B_2H_6

The shapes of the molecules of the electron-precise and electron-rich compounds can all be predicted by the VSEPR rules (Section 2.3), which explain why CH_4 is tetrahedral (1), NH_3 is trigonal-pyramidal (2), and H_2O is angular (3).

Electron-deficient compounds provide some of the most interesting and unusual examples of structure and bonding. A Lewis structure for diborane, B_2H_6, would require at least 14 valence electrons to bind the eight atoms together, but the molecule has only 12 valence electrons. The simple explanation of its structure is the presence of BHB three-centre, two-electron bonds (3c,2e; Section 3.11) acting as bridges between the two B atoms, so two electrons contribute to binding three atoms. These bridging B–H bonds are longer and weaker than the terminal B–H bonds. An alternative way of viewing the molecule is that each BH_3 moiety is a strong Lewis acid and gains the share of an electron pair from a B–H bond in the other BH_3 moiety. Being so small, the H atoms pose little or no steric hindrance to dimer

TABLE 11.3 Some molecular hydrogen compounds

Group	Formula	Traditional name	IUPAC name
13	B_2H_6	diborane	diborane[6]
	AlH_3	alane	alane
	Ga_2H_6	digallane	digallane
14	CH_4	methane	methane
	SiH_4	silane	silane
	GeH_4	germane	germane
	SnH_4	stannane	stannane
15	NH_3	ammonia	azane
	PH_3	phosphine	phosphane
	AsH_3	arsine	arsane
	SbH_3	stibine	stibane
16	H_2O	water	oxidane
	H_2S	hydrogen sulfide	sulfane
	H_2Se	hydrogen selenide	selane
	H_2Te	hydrogen telluride	tellane
17	HF	hydrogen fluoride	hydrogen fluoride
	HCl	hydrogen chloride	hydrogen chloride
	HBr	hydrogen bromide	hydrogen bromide
	HI	hydrogen iodide	hydrogen iodide

formation. The structures of boron hydrides are described more fully in Sections 14.2 and 14.6.

As expected, aluminium shows related behaviour that is modified by the larger atomic radius of this Period 3 element. The compound AlH_3 does not exist as a monomer but forms a polymer in which each relatively large Al atom is surrounded octahedrally by six H atoms. Beryllium exhibits a *diagonal relationship* (Section 10.3) with Al, and also forms a polymeric covalent hydride BeH_2. Although BH_3 and AlH_3 do not exist as monomers they do form important complex anions in combination with the hydride anion. The common reagents sodium tetrahydridoborate ($NaBH_4$) and lithium tetrahydridoaluminate ($LiAlH_4$) are examples of adduct formation between BH_3 or AlH_3, each a Lewis acid, and the Lewis base :H⁻.

Interaction between a metal atom and a C–H bond is frequently encountered in organometallic compounds of d-metals. A 3c,2e M–H–C bond known as an **agostic bond** is formed when a vacant orbital of the metal atom accepts electrons from a C–H bond that is present in a coordinated ligand. The normally inert C–H bond is thus weakened, providing a route for the activation of hydrocarbons by transition metal catalysts, as discussed further in Section 24.16.

(ii) Reactions of molecular hydrides

KEY POINTS Homolytic dissociation of an E–H bond to produce a radical (E·) and hydrogen atom H occurs most readily for the hydrides of the heavy p-block elements. Hydrogen attached to an electronegative element has protonic character and the compound is typically a Brønsted acid. Hydrogen attached to a more electropositive element can be transferred to an acceptor as a hydride ion.

As was summarized in Section 11.2, the reactions of binary molecular hydrides are discussed in terms of their ability to undergo homolytic dissociation and, when the dissociation is heterolytic, in terms of their protonic or hydridic character.

Homolytic dissociation occurs readily for the hydrogen compounds of some p-block elements, especially the heavier elements. For example, the use of a radical initiator greatly facilitates the reaction of trialkylstannanes, R_3SnH, with haloalkanes, RX, as a result of the formation of $R_3Sn·$ radicals:

$$R_3SnH + R'X \rightarrow R'H + R_3SnX$$

Thermal decomposition reactions of molecular hydrides yielding H_2 and the element occur by homolytic dissociation. Decomposition temperatures usually correlate with E–H bond energies and inversely with enthalpies of formation. For example, AsH_3 (As–H bond enthalpy 297 kJ mol⁻¹), which is an *endothermic* hydride—that is, its formation from the elements is endothermic—decomposes quantitatively at 250–300°C:

$$AsH_3(g) \rightarrow As(s) + \tfrac{3}{2}H_2(g) \quad \Delta_rH^\ominus = -66.4 \text{ kJ mol}^{-1}$$

In contrast, water (O–H bond enthalpy 464 kJ mol⁻¹), which is a highly *exothermic hydride*—that is, its formation from the elements is exothermic—is only 4% decomposed into H_2 and O_2 at 2200°C:

$$H_2O(g) \rightarrow \tfrac{1}{2}O_2(g) + H_2(g) \quad \Delta_rH^\ominus = +242 \text{ kJ mol}^{-1}$$

Direct thermolysis of water is therefore not a practical solution for H_2 production although step-wise processes are feasible (see Box 11.4).

Compounds reacting by proton donation are said to show protic behaviour: in other words, they are Brønsted acids. We saw in Section 5.1 that Brønsted acid strength increases from left to right across a period in the p block (in the order of increasing electron affinity) and down a group (in the order of decreasing bond energy). One striking example of this trend is the increase in acidity from CH_4 to HF and then from HF to HI. Binary hydrogen compounds of elements on the right of the periodic table typically undergo these reactions.

Molecules in which hydrogen is bound to a more electropositive element can act as hydride ion donors. Important examples are the complex hydrido anions such as BH_4^- and AlH_4^-, which are used to hydrogenate compounds containing a multiple bond. Other examples include numerous compounds of d-block elements, including many catalysts.

Hydride affinities, analogous to proton affinities, can be calculated. For instance, the hydride affinity of a boron compound BX_3 ($-\Delta_HH^\ominus$) is the enthalpy for the reaction:

$$BX_3(g) + H^-(g) \rightarrow HBX_3^-(g)$$

A strong hydride *donor* is associated with a low value of $-\Delta_HH^\ominus$.

A practical **hydridicity** scale may also be determined experimentally by comparing the hydride ion donor abilities of different species HY in a particular aprotic solvent, such as acetonitrile. The scale is derived by considering the following equilibria, which include the heterolytic cleavage of H_2:

$$HA(\text{solv}) + HY(\text{solv}) \rightleftharpoons A^-(\text{solv}) + Y^+(\text{solv}) + H_2(g) \quad (1)$$

$$HA(\text{solv}) \rightleftharpoons H^+(\text{solv}) + A^-(\text{solv}) \quad (2)$$

$$H_2(g) \rightleftharpoons H^+(\text{solv}) + H^-(\text{solv}) \quad (3)$$

$$HY(\text{solv}) \rightleftharpoons H^-(\text{solv}) + Y^+(\text{solv}) \quad (4)$$

Values of equilibrium constants (hence $\Delta G^\ominus$) for reactions (1) and (2) are experimentally measurable, and $\Delta G^\ominus$ for reaction (3) in acetonitrile is set at 317 kJ mol⁻¹ at 298 K. The hydride donor ability of HY ($\Delta_HG^\ominus$) for reaction (4) is thus determined from:

$$\Delta_HG^\ominus = \Delta G^\ominus_{(1)} - 2.3RTpK_{HA} + 317$$

TABLE 11.4 Comparison of covalent E–H bond enthalpies in molecular hydrides with the corresponding intermolecular hydrogen bond enthalpies

E–H	Bond enthalpy/ kJ mol^{-1}	E–H⋯E	Hydrogen bond enthalpy/kJ mol^{-1}
S–H	363	HS–H⋯SH$_2$	7
N–H	386	H$_2$N–H⋯NH$_3$	17
O–H	464	HO–H⋯OH$_2$	22
F–H	565	F–H⋯FH	29
Cl–H	428	HO–H⋯Cl$^-$	55
F–H	565	F⋯H⋯F	465

A strong hydride donor is thus associated with a low value of $\Delta_H G^\ominus$. Hydricity is an important concept for understanding some catalytic reactions of d-metal complexes and is discussed further in Section 11.6(d).

(iii) The influence of hydrogen bonding

KEY POINT Hydrogen bonding influences the physical properties of molecular compounds and is an important determinant of polymolecular structure.

We saw in Section 5.9 that a hydrogen bond is an interaction between a hydrogen atom covalently bonded to an electronegative atom and the lone pair of a neighbouring electronegative atom. Although individual hydrogen bonds are usually much weaker than conventional X–H bonds (Table 11.4), their collective influence can be striking, as is easily demonstrated by the fact that H$_2$O under normal conditions is a liquid and not a gas.

Normal boiling points (Fig. 11.7) are unusually high for strongly hydrogen-bonded molecules: ammonia (containing N–H⋯N bonds), hydrogen fluoride (containing F–H⋯F bonds, leading to zigzag polymeric chains (6)), and, of course, water, for which there are two O–H⋯O bonds per molecule leading to greater three-dimensional cohesion. Compared to the hydrides of the first row, the boiling points

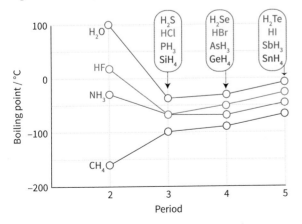

FIGURE 11.7 Normal boiling points of p-block binary hydrogen compounds.

of corresponding second-row compounds PH$_3$, H$_2$S and HCl are much lower, consistent with hydrogen bonds being much weaker. As observed for all hydrides of Group 14, boiling points tend otherwise to increase slightly upon descending each group, due to the increase in van der Waals forces.

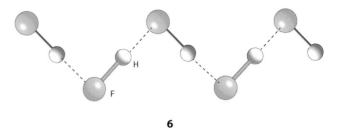

6

(b) Saline hydrides

KEY POINTS Hydrogen compounds of the most electropositive metals may be regarded as ionic hydrides; they liberate H$_2$ upon contact with Brønsted acids and transfer H$^-$ to electrophiles. As direct hydride donors they react with halide compounds to form anionic hydride complexes.

The saline hydrides are ionic solids containing discrete H$^-$ ions and are analogous to corresponding halide salts. The ionic radius of H$^-$ varies from 126 pm in LiH to 154 pm in CsH. This wide variability reflects the poor control that the single charge of the proton has on its two surrounding electrons and the resulting high compressibility and polarizability of H$^-$. Hydrides of Group 1 and 2 elements, with the exception of Be, are ionic compounds. All Group 1 hydrides adopt the rock-salt structure. With the exception of MgH$_2$, which has the rutile structure, the Group 2 (saline) hydrides adopt the fluorite structure at high temperature and the related PbCl$_2$ structure at low temperature (Table 11.5).

The saline hydrides are insoluble in common nonaqueous solvents, but they do dissolve in molten alkali halides and hydroxides, such as NaOH (m.p. 318°C). Electrolysis of these stable molten-salt solutions produces hydrogen gas at the anode (the site of oxidation):

$$2H^-(melt) \rightarrow H_2(g) + 2e^-$$

This reaction provides chemical evidence for the existence of discrete H$^-$ ions. Saline hydrides react, often violently, with water to produce H$_2$:

$$NaH(s) + H_2O(l) \rightarrow NaOH(aq) + H_2(g)$$

TABLE 11.5 Structures of s-block hydrides

Compound	Crystal structure
LiH, NaH, RbH, CsH	Rock salt
MgH$_2$	Rutile
CaH$_2$, SrH$_2$, BaH$_2$	Distorted PbCl$_2$

Alkali metal hydrides are convenient reagents for making other hydride compounds because they are direct providers of H⁻ ions for the following synthetically useful reactions:

1. Metathesis with a halide, such as the reaction of finely divided lithium hydride with silicon tetrachloride dissolved in dry diethyl ether (et):

$$4LiH(s) + SiCl_4(et) \rightarrow 4LiCl(s) + SiH_4(g)$$

2. Addition to a Lewis acid: For example, reaction with a trialkylboron compound yields a hydride complex that is a useful reducing agent and source of hydride ions in organic solvents:

$$NaH(s) + B(C_2H_5)_3(et) \rightarrow Na[HB(C_2H_5)_3](et)$$

3. Reaction with a proton source, to produce H_2:

$$NaH(s) + CH_3OH(et) \rightarrow NaOCH_3(s) + H_2(g)$$

The absence of convenient solvents limits the use of saline hydrides as reagents, but this problem is partially overcome by the availability of commercial dispersions of finely divided NaH in oil. Even more finely divided and reactive alkali metal hydrides can be prepared from the metal alkyl and hydrogen.

Saline hydrides are pyrophoric; indeed, finely divided sodium hydride can ignite simply if it is left exposed to humid air. Such fires are difficult to extinguish because even carbon dioxide is reduced when it comes into contact with hot metal hydrides (water, of course, forms yet more flammable H_2 product), and blanketing with an inert solid such as sand is normally employed.

In addition to the use of calcium dihydride in portable H_2 generators, magnesium dihydride, MgH_2, is one of many compounds under investigation as a hydrogen storage medium for transport purposes (Box 11.5 and Section 30.8). The amount of H atoms in a given volume of MgH_2 is about 50% higher than in the same volume of liquid H_2.

(c) Metallic hydrides

KEY POINTS Most metallic hydrides are nonstoichiometric, electrically conducting solids in which the H-atoms are mobile. No stable binary metal hydrides are known for the metals in Groups 7–9.

Many of the d- and f-block elements react with H_2 to produce metallic hydrides, which are discussed in more detail in Chapters 25 and 30. Most of these compounds (and the hydrides of alloys) have a metallic lustre and are electrically conducting (hence their name). They are less dense than the parent metal and tend to be brittle materials. Long established and familiar to metallurgists, *embrittlement* is a form of corrosion caused when metals are directly exposed to H_2,

---

BOX 11.5 Designing H_2 carriers to meet the challenge of fuelling H_2-powered vehicles

The need to develop practical systems for on-board H_2 storage is considered as a major obstacle to its future use as a fuel for road vehicles. A practical solution is achieved by compression, typically to 70 MPa at which the volumetric energy density is 5.3 MJ dm⁻³. At fuelling stations, the highly compressed H_2 is transferred to a vehicle at −40°C using a sealed valve connector: refuelling a car takes just minutes (a much shorter time than needed to charge a battery). However, the volumetric energy density at 70 MPa is still low compared to petrol (Table B11.1), and larger fuel tanks (the walls of which are constructed from tough, lightweight carbon materials, Section 15.6, that are not degraded by embrittlement) are needed to ensure a useful range of 600 km. The challenge, therefore, is to identify and develop materials, known as hydrogen carriers, which can store a greater density of H_2 without such compression and in a fully reversible manner, undergoing hydrogenation and desorption at high rates under reasonable temperatures and pressure conditions. Both inorganic and organic carriers are under consideration. The structures and properties of hydrogen carriers based on inorganic materials are described in detail in Section 30.8.

An H_2 carrier suitable for transportation must fulfil the following requirements:

- Have a high H_2 storage capacity (energy density, adjusted for packing volume).
- Be based on light elements.
- Have a low desorption temperature.
- Be able to undergo long life cycles of H_2 uptake and H_2 release without decomposition.
- Be compatible with fuel infrastructure.
- Have low production costs.
- Be safe to handle.

Examples of some potential H_2 carriers are shown in Fig. B11.4.

Materials under investigation include hydrides, borohydrides, and amides of the lightest metals, along with corresponding composites (Section 30.8). Liquid ammonia and liquid organic hydrogen carriers (LOHCs) offer the easiest handling for the customer, as hydrogenation (charging) is carried out within the supply infrastructure. At a filling station, liquid ammonia fuel would be transferred directly into the vehicle using a sealed connection, whereas with LOHCs the discharged (dehydrogenated) product would simply be exchanged for hydrogenated fuel.

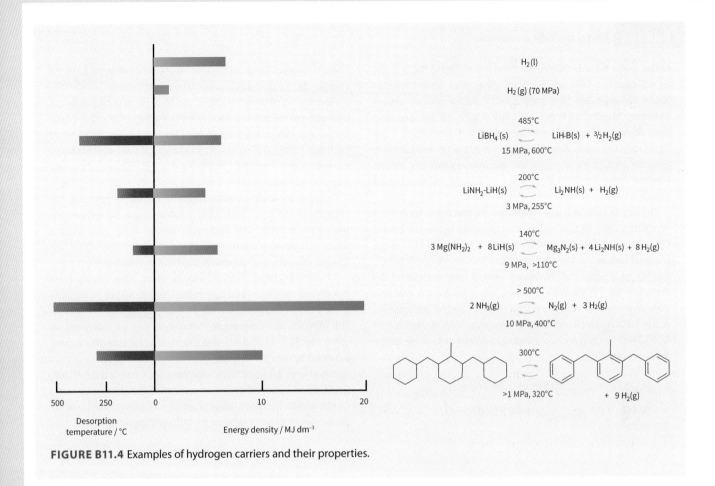

FIGURE B11.4 Examples of hydrogen carriers and their properties.

or to water that may be reduced to H_2 *in situ*: it results in cracks and fractures and poses challenges for the construction of pipes and other infrastructure needed to support a hydrogen economy.

Most metallic hydrides have variable composition (they are nonstoichiometric). For example, at 550°C zirconium hydride exists over a composition range from $ZrH_{1.30}$ to $ZrH_{1.75}$; it has the fluorite structure (Fig. 4.39) with a variable number of anion sites unoccupied. The variable stoichiometry and metallic conductivity of these hydrides can be understood in terms of a model based on bands (Section 2.11) in which the delocalized orbitals responsible for the conductivity accommodate the electrons supplied by additional H atoms.

Metallic hydrides are formed by all the d-block metals of Groups 3, 4, and 5 and by almost all f-block elements (Fig. 11.8). However, the only hydride in Group 6 is CrH, and no hydrides are known for the unalloyed metals of Groups 7, 8, and 9. The region of the periodic table covered by Groups 7 through to 9 is sometimes referred to as the **hydride gap** because few, if any, stable binary metal–hydrogen compounds are formed by these elements. However, these metals are important as hydrogenation catalysts because they can *activate* hydrogen.

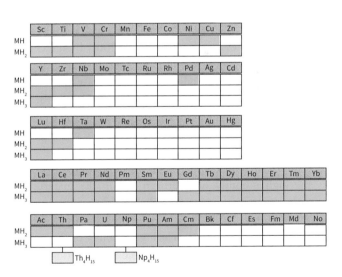

FIGURE 11.8 Hydrides formed by d- and f-block elements. The formulas are limiting stoichiometries based on the structure type.

The Group 10 metals, especially Ni and Pt, are often used as hydrogenation catalysts in which surface hydride formation is thought to be involved. However, somewhat surprisingly, at moderate pressures only Pd forms a stable bulk phase; its composition is PdH_x, with $x < 1$. Nickel forms hydride phases at very high pressures, but Pt does not form

BOX 11.6 Metal hydride batteries

Nickel metal-hydride (NiMH) batteries are related to the popular nickel–cadmium (NiCad) batteries. The main advantages of NiMH batteries are that they can be recharged an almost indefinite number of times in a typical 3–5-year lifespan, and they do not contain the very toxic element Cd, enabling them to be recycled easily. Being less expensive than Li-ion batteries (Section 30.9), they are attractive alternatives for applications ranging from portable electronics to all-electric and hybrid vehicles.

The key feature of a NiMH battery is an electrode comprised of a metal alloy M that is able to intercalate hydrogen atoms. One such alloy is $LaNi_6$, established as a reversible H_2-storage material except it is far too heavy for road vehicle applications. During charging, the hydrogen atoms are generated by reduction of the electrolyte, an aqueous solution containing 30% (by mass) KOH. Common to both NiMH and NiCad batteries is the $Ni(OH)_2$ electrode: during charging, Ni(II) is oxidized to Ni(III). The charging and discharging reactions at each electrode are summarized as

$$M + H_2O + e^- \underset{discharge}{\overset{charge}{\rightleftharpoons}} M-H + OH^-$$

$$Ni(OH)_2 + OH^- \underset{discharge}{\overset{charge}{\rightleftharpoons}} Ni(O)OH + H_2O + e^-$$

There is no net change in the electrolyte concentration over the charge–discharge cycle.

Fine tuning of the strength of the M–H bond is crucial to the operation of the battery, and represents another example of the Sabatier principle introduced in Section 2.15. The ideal bond enthalpy falls in the range 25–50 kJ mol^{-1}. If the bond enthalpy is too low, M–H bonds are not formed and H_2 is evolved instead. If the bond enthalpy is too high, the reaction is not reversible.

Other factors influence the choice of metal. For example, the alloy must not react with KOH solution, must be resistant to oxidation and corrosion, and must tolerate overcharge (during which O_2 is generated at the Ni(O)OH electrode) and overdischarge (during which H_2 is generated at the $Ni(OH)_2$ electrode).

To satisfy these diverse requirements, the alloys have disordered structures and use metals that would not be suitable if used alone, including Li, Mg, Al, Ca, V, Cr, Mn, Fe, Cu, and Zr. The number of H atoms per metal atom can be increased by using Mg, Ti, V, Zr, and Nb, and the M–H bond enthalpy can be adjusted by using V, Mn, and Zr.

The charge and discharge reactions are catalysed by Al, Mn, Co, Fe, and Ni and the corrosion resistance is improved by using Cr, Mo, and W. This wide range of properties allows NiMH battery performance to be optimized for different applications.

---

any at all. Apparently, the Pt–H bond enthalpy is sufficiently high to disrupt the H–H bond, but is not high enough to offset the loss of Pt–Pt metallic bonding, which would occur upon formation of a bulk platinum hydride. In agreement with this interpretation, the enthalpies of sublimation, which reflect the strength of metallic bonding, increase in the order Pd (378 kJ mol^{-1}) < Ni (430 kJ mol^{-1}) < Pt (565 kJ mol^{-1}). The M–H bond enthalpy is a crucial factor in the design of metal hydride batteries, the principles of which are explained in Box 11.6.

Another striking property of many metallic hydrides is the high mobility of hydrogen within the material at slightly elevated temperatures. This mobility is used in the ultrapurification of H_2 by diffusion through a palladium/silver alloy tube (Fig. 11.9). The high mobility of the hydrogen they contain and their variable composition make the metallic hydrides potential hydrogen storage media (Section 30.8), albeit not for transport purposes as they are too heavy. Upon cooling from red heat, palladium absorbs up to 900 times its own volume of H_2, which is given off again upon heating. As a result, palladium is sometimes referred to as a 'hydrogen sponge'. The intermetallic compound $LaNi_5$ forms a hydride phase with a limiting composition $LaNi_5H_6$, and at this composition it contains a greater volumetric density of hydrogen than liquid H_2. The rapid and reversible redox reaction of $LaNi_6$ with water

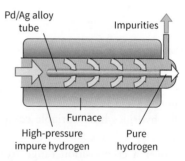

FIGURE 11.9 Schematic diagram of a hydrogen purifier. Because of a pressure differential and the mobility of H atoms in palladium, hydrogen diffuses through the palladium–silver alloy as H atoms but impurities do not.

to produce interstitial H makes it an important battery material (Box 11.6). A less expensive system with the composition $FeTiH_x$ ($x < 1.95$) is commercially available for low-pressure hydrogen storage.

EXAMPLE 11.2 Correlating the classification and properties of hydrogen compounds

Classify the compounds PH_3, CsH, and B_2H_6 and discuss their probable physical properties. For the molecular compounds specify their sub-classification (electron-deficient, electron-precise, or electron-rich).

Answer We need to consider the group to which the element E belongs. The compound CsH is a compound of a Group 1 element, and so it is expected to be a saline hydride, typical of the s-block metals. It has the rock-salt structure. The large difference in electronegativity between Cs and H produces a large band gap, and CsH is an electrical insulator. As with the hydrogen compounds of other p-block elements, the hydrides PH_3 and B_2H_6 are molecular with low molar masses and high volatilities. They are in fact gases under normal conditions. The Lewis structure indicates that PH_3 has a lone pair on the phosphorus atom and that it is therefore an electron-rich molecular compound. On the other hand, diborane, B_2H_6, is an electron-deficient compound.

Self-test 11.2 Give balanced equations (or NR, for no reaction) for (a) $Ca + H_2$, (b) $NH_3 + BF_3$, (c) $LiOH + H_2$.

(d) Hydrido and dihydrogen complexes of d-metals

KEY POINTS A large number of d-block metal complexes are known in which the dihydrogen molecule or hydride anion are ligands. These complexes play important roles in catalysis and hydrogen activation.

The H atom and the H_2 molecule play important roles in organometallic chemistry, particularly homogeneous catalysis involving hydrogenation of alkenes and carbonyl groups (Sections 24.7, 31.2, and 31.3). An individual H atom bound to a metal atom is usually regarded as a H^- (hydrido) ligand—the highly polarizable H^- entity behaving as a soft, two-electron σ-donor. There are a very large number of complexes of d- and f-block elements containing one or more hydride ligands; these complexes include those of elements in the 'hydride gap' that do not form binary metallic hydrides. Hydrido complexes can be synthesized by many routes, such as the reaction of a metal ion or complex with a suitable hydride source:

$$[FeI_2(CO)_4] + 2NaBH_4 \xrightarrow{THF} [Fe(H)_2(CO)_4] + B_2H_6 + 2NaI$$

As with binary compounds of the main group, the coordinated H ligand can be protonic or hydridic depending on the electron-withdrawing or electron-donating character of the metal atom, which in turn depends upon the nature of the other ligands. Electron-withdrawing CO ligands in the coordination shell render the H atom protonic because they can stabilize the conjugate base that results from proton release. As explained in more detail in Section 24.7, compounds such as $[CoH(CO)_4]$ are Brønsted acids that can be ranked in terms of their pK_a values, that for $[CoH(CO)_4]$ being 8.3 when measured in acetonitrile. By contrast, electron-donating ligands tend to confer greater hydricity. Bis(diphosphine) d-metal complexes of Rh are comparable to boron trihalides as hydride-transfer agents:

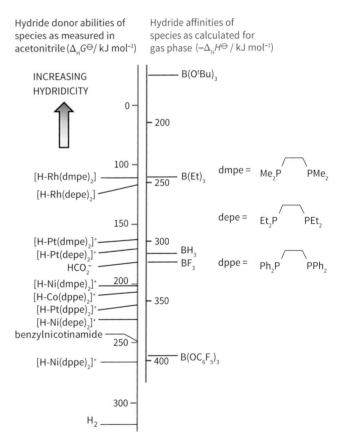

Some trends in hydricities of d-metal hydrido complexes are illustrated in Fig. 11.10. Values of hydride affinities ($-\Delta_H H^\ominus$) calculated for some boron compounds using density functional theory are plotted alongside hydride donor abilities ($\Delta_H G^\ominus$) for bis(diphosphine) d-metal complexes and other species, calculated from experimental data obtained in solution. Hydride donor abilities vary with ligand donor strength (methyl > ethyl > phenyl groups as substituents), group (Group 7 > isoelectronic Group 8), and row (3d < 4d > 5d). Included in Fig. 11.10 are values for formate and benzylnicotinamide, the latter being an analogue

FIGURE 11.10 Scaling hydricity for d-block hydrido complexes and main-group hydrides. The right-hand scale shows values of hydride affinities ($-\Delta_H H^\ominus$) calculated for some boron compounds using density functional theory. The left-hand scale shows hydride donor abilities ($\Delta_H G^\ominus$) for bis(diphosphine) d-metal complexes and other species, calculated from experimental data obtained in solution. The two scales are aligned using the observed equilibrium constant of approximately 1 between $[HRh(dmpe)_2]$ and BEt_3. Adapted with permission from *J. Am. Chem. Soc.* 2009, **131**, 40, 14454–14465. American Chemical Society.

of NADH, the highly specific hydride transfer agent used by biology in enzyme-catalysed reactions.

> **EXAMPLE 11.3** Determining hydridicities of d-metal complexes from hydride transfer equilibria in aprotic solvents
>
> The complex $[Pt(PNP)_2]^{2+}$ ('PNP' = $Et_2PCH_2N(Me)CH_2PEt_2$) equilibrates with H_2 and NEt_3 in acetonitrile yielding a mixture containing $HNEt_3$ and $[PtH(PNP)_2]^{2+}$. From the concentrations of $[Pt(PNP)_2]^{2+}$ and $[PtH(PNP)_2]^+$ present at equilibrium under 1 atm H_2 (determined using ^{31}P NMR), the equilibrium constant for the reaction:
>
> $$[Pt(PNP)_2]^{2+} + H_2 + NEt_3 \rightleftharpoons [HPt(PNP)_2]^+ + HNEt_3^+$$
>
> is 790 atm^{-1}. Given that the pK_a for $HNEt_3^+$ in acetonitrile is 18.8, calculate the hydride donor ability of $[PtH(PNP)_2]^{2+}$ ($\Delta_H G^\ominus$).
>
> **Answer** We use the thermodynamic arguments described in Section 11.6(a)(ii), using the value of 317 kJ mol^{-1} for the heterolysis of H_2 in acetonitrile. We obtain:
>
> $$\Delta_H G^\ominus = 2.3RT \log(790) - 2.3RT(18.8) + 317$$
> $$= 232 \text{ kJ mol}^{-1}$$
>
> **Self-test 11.3** The analogous $[Pd(PNP)_2]^{2+}$ complex does not react with H_2 in the presence of NEt_3, but the equilibrium constant for the hydride exchange reaction:
>
> $$[PdH(PNP)_2]^+ + [Pt(PNP)_2]^{2+} \rightleftharpoons [Pd(PNP)_2]^{2+} + [PtH(PNP)_2]^+$$
>
> is 450 at 298 K. Calculate the hydride donor ability of $[PdH(PNP)_2]^+$.

As with some main-group hydrides, an H atom can also occupy a bridging position (in a 3c,2e bond) between two metal atoms, usually in conjunction with a metal–metal bond. The complex $[(\mu\text{-}H)W_2(CO)_{10}]^-$ (7) provides a rare example of an H atom bridging two metal atoms that are not otherwise bonded.

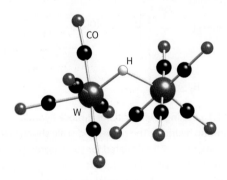

7 $[(\mu\text{-}H)W_2(CO)_{10}]^-$

A **homoleptic complex** is a complex that contains only one type of ligand. Examples of homoleptic hydrido metal complexes are provided by Fe, Rh, and Tc. The dark green compound Mg_2FeH_6, which contains the octahedral $[FeH_6]^{4-}$ complex anion, is obtained by reacting the elements together under pressure. The complex anion $[ReH_9]^{2-}$ (8) is formed by the reaction of perrhenate, $[ReO_4]^-$, with elemental K or Na in ethanol. In the solid state the H atoms form a tricapped trigonal prism around the Re, which is formally in the oxidation state +7. The $[TcH_9]^{2-}$ complex has the same structure.

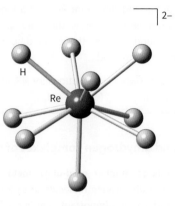

8 $[ReH_9]^{2-}$

The H_2 molecule can also coordinate as an intact molecule, using the $1\sigma_g$ orbital to donate an electron pair and the $1\sigma_u$ orbital to accept an electron pair back from the metal, in what is known as π-back-donation or **synergistic bonding** (Section 24.5). Many d-block complexes have been isolated that contain a relatively stable, intact H_2 ligand. The first such compound to be identified was $[W(CO)_3(H_2)(P^iPr_3)_2]$ (9), where ^iPr denotes isopropyl, $CH(CH_3)_2$.

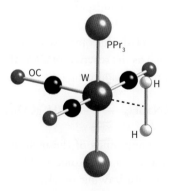

9 $[W(CO)_3(H_2)(P^iPr_3)_2]$

11.7 General methods for synthesis of binary hydrogen compounds

KEY POINTS The general routes to binary hydrogen compounds are direct reaction of H_2 and the element, protonation of nonmetal anions, and metathesis between a hydride source and a halide or pseudohalide.

A negative Gibbs energy of formation is an indicator that the direct combination of hydrogen and an element may be the preferred synthetic route for a hydrogen compound. When a compound is thermodynamically unstable with respect to its elements, an indirect synthetic route from other compounds can often be found, but each step in the indirect route must be thermodynamically favourable.

There are three common methods for synthesizing binary hydrogen compounds:

1. Direct combination of the elements (hydrogenolysis):

   $2E + H_2(g) \rightarrow 2EH$

2. Protonation of a Brønsted basic anion:

   $E^- + H_2O(l) \rightarrow EH + OH^-(aq)$

3. Reaction of an ionic hydride or hydride donor (MH) with a halide (metathesis):

   $MH + EX \rightarrow EH + MX$

In such general equations, the symbol E can also denote an element with higher valence, with corresponding changes of detail in the formulas and stoichiometric coefficients.

Direct combination is used commercially for the synthesis of compounds that have negative Gibbs energies of formation, including NH_3 and the hydrides of lithium, sodium, and calcium. However, in some cases high pressure, high temperature, and a catalyst are necessary to overcome unfavourable kinetic barriers (a notable example is the industrial production of ammonia, Section 31.9). The high temperature used for the lithium reaction is required to melt the metal and break up the surface layer of hydride that would otherwise passivate it. This inconvenience is avoided in many laboratory preparations by adopting one of the alternative synthesis routes, which may also be used to prepare compounds with positive Gibbs energies of formation.

An example of protonation of a Brønsted base, such as a nitride ion, is

$Li_3N(s) + 3H_2O(l) \rightarrow 3LiOH(aq) + NH_3(g)$

Lithium nitride is too expensive for the stoichiometric reaction to be suitable for the commercial production of ammonia, but it is very useful in the laboratory for the preparation of ND_3 (by using D_2O in place of H_2O). Water is a sufficiently strong acid to protonate the very strong base N^{3-}, but a stronger acid, such as H_2SO_4, is required to protonate the weak base Cl^- to form HCl.

$NaCl(s) + H_2SO_4(l) \rightarrow NaHSO_4(s) + HCl(g)$

An example of synthesis by metathesis is the preparation of silane:

$LiAlH_4(s) + SiCl_4(l) \rightarrow LiAlCl_4(s) + SiH_4(g)$

Hydrides of the more electropositive elements (LiH, NaH, and the AlH_4^- anion) are the most active H^- sources. Salts such as $LiAlH_4$ and $NaBH_4$ are soluble in ether solvents that solvate the alkali metal ion. Of the two anionic complexes, AlH_4^- has much the higher hydride donor ability.

EXAMPLE 11.4 Using hydrogen compounds in synthesis

Suggest a procedure for synthesizing lithium tetraethoxyaluminate, $Li[Al(OEt)_4]$, from $LiAlH_4$ and reagents and solvents of your choice.

Answer We need to note that AlH_4^- is a strong H^- donor. Because H^- is an even stronger Brønsted base than ethoxide ($CH_3CH_2O^- = EtO^-$) it should react with ethanol to produce H_2 and yield EtO^- which will thus replace H^-. The reaction of the slightly acidic compound ethanol with the strongly hydridic AlH_4^- should yield the desired alkoxide and hydrogen. The reaction might be carried out by dissolving $LiAlH_4$ in tetrahydrofuran and dropping ethanol into this solution slowly:

$LiAlH_4(thf) + 4C_2H_5OH(l) \rightarrow Li[Al(OEt)_4](thf) + 4H_2(g)$

This type of reaction should be carried out slowly under a stream of inert gas (N_2 or Ar) to dilute the H_2, which is explosively flammable.

Self-test 11.4 Suggest a way of making triethylmethylstannane, $MeEt_3Sn$, from triethylstannane, Et_3SnH, and a reagent of your choice.

FURTHER READING

F. Hensel and P. P. Edwards, Hydrogen, the first alkali metal. *Chem. Eur. J.*, 1996, **2**, 1201.

F. Hensel and P. P. Edwards, Will solid hydrogen ever be a metal? *Nature*, 1997, **388**, 621.

T. I. Sigfusson, Pathways to hydrogen as an energy carrier. *Philos. Trans. R. Soc., A*, 2007, **365**, 1025.

B. Sørensen, *Hydrogen and Fuel Cells*. Elsevier Academic Press (2005).

P. Ball, *H₂O: A Biography of Water*. Phoenix (2004). An entertaining look at the chemistry and physics of water.

M. J. Schultz, T. H. Vu, B. Meyer, and P. Bisson, Water: A responsive small molecule. *Acc. Chem. Res.*, 2012, **45**, 15.

N. P. Brandon and Z. Kurban, Clean energy and the hydrogen economy. *Phil. Trans. Royal Society A*, 2017, **375**: 20160400.

W. Lubitz and W. Tumas (eds.), Hydrogen. *Chem. Rev.* (100th thematic issue), 2007, **107**. There are many interesting articles in this special edition.

P. M. Modisha, C. N. M. Ouma, R. Garidzirai, P. Wasserscheid and D. Bessarabov, The prospect of hydrogen storage using liquid organic hydrogen carriers. *Energy and Fuels*, 2019, **33**, 2778.

M. Flekiewicz and G. Kubica, Hydrogen onboard storage technologies for vehicles. In *Diesel Engines: Current Challenges and Future Perspectives* (H. Koten, ed.) IntechOpen (2023), DOI:10.5772/intechopen.1003040

T. Takata and K. Domen, Particulate photocatalysts for water splitting: Recent advances and future prospects. *ACS Energy Lett.*, 2019, **4**, 542.

T. Kodama and N. Gokon, Thermochemical cycles for high-temperature solar hydrogen production. *Chem. Rev.*, 2007, **107**, 4048.

N. S. Lewis and D. G. Nocera, Powering the planet: Chemical challenges in solar energy utilization. *Proc. Natl. Acad. Sci. USA*, 2006, **103**, 157.

S. Y. Reece, J. A. Hamel, K. Sung, T. D. Jarvi, A. J. Esswein, J. J. H. Pijpers, and D. G. Nocera, Wireless solar water splitting using silicon-based semiconductors and earth-abundant catalysts. *Science*, 2011, **334**, 645.

A. Bocarsly and D. M. P. Mingos (eds.), Fuel cells and hydrogen storage. *Struct. Bonding*, 2011, **141**.

J. R. Norton and J. Sowa (eds.), Metal hydrides. *Chem. Rev.*, 2016, **116**, Issue 15. A series of articles describing cutting-edge aspects of metal hydride chemistry.

M. Pasquali and C. Mesters, We can use carbon to decarbonize—and get hydrogen for free. *Proc. Natl. Acad. Sci. USA*, 2021, **118**, e2112089118; https://doi.org/10.1073/pnas.2112089118.

A. M Oliveira, R. R Beswick, and Y. Yan, A green hydrogen economy for a renewable energy society. *Current Opin. Chem. Eng.*, 2021, **33**: 100701.

F. Osselin, C.Soulaine, C. Fauguerolles, E. C. Gaucher, B. Scaillet, and M. Pichavant, Orange hydrogen is the new green. *Nature Geoscience*, 2022, **15**, 765. Contains an account of the various 'colours' of hydrogen.

EXERCISES

11.1 It has been suggested that hydrogen could be placed in Group 1, Group 14, or Group 17 of the periodic table. Give arguments for and against each of these positions.

11.2 Write balanced chemical equations for three major industrial preparations of hydrogen gas. Propose two different reactions that would be convenient for the preparation of hydrogen in the laboratory.

11.3 Preferably without consulting reference material, construct the periodic table, identify the elements, and (a) indicate positions of salt-like, metallic, and molecular hydrides, (b) add arrows to indicate trends in $\Delta_f G^\ominus$ for the hydrogen compounds of the p-block elements, (c) identify the areas where the molecular hydrides are electron-deficient, electron-precise, and electron-rich.

11.4 Predict the expected physical properties of water in the absence of hydrogen bonding.

11.5 Name and classify the following hydrogen compounds: (a) BaH_2, (b) SiH_4, (c) NH_3, (d) AsH_3, (e) $PdH_{0.9}$, (f) HI.

11.6 Identify the compounds from Exercise 11.5 that provide the most pronounced example of the following chemical characteristics, and produce a balanced equation that illustrates each of the characteristics: (a) hydridic character, (b) Brønsted acidity, (c) variable composition, (d) Lewis basicity.

11.7 Divide the compounds in Exercise 11.5 into those that are solids, liquids, or gases at room temperature and pressure. Are any of the solids likely to be good electrical conductors?

11.8 Comment on the following radii of the H^- ion calculated from structures of ionic compounds.

	LiH	NaH	KH	CsH	MgH_2	CaH_2	BaH_2
Radius/pm	114	129	134	139	109	106	111

11.9 Identify the reaction that is most likely to produce the highest proportion of HD, and give your reasoning: (a) $H_2 + D_2$ equilibrated over a platinum surface, (b) D_2O + NaH, (c) electrolysis of HDO.

11.10 Identify the compound in the following list that is most likely to undergo radical reactions with alkyl halides, and explain the reason for your choice: H_2O, NH_3, $(CH_3)_3SiH$, $(CH_3)_3SnH$.

11.11 What is the trend in hydridic character of BH_4^-, AlH_4^-, and GaH_4^-? Which is the strongest reducing agent? Produce the equations for the reaction of GaH_4^- with excess 1 M HCl(aq).

11.12 Describe the important physical differences and a chemical difference between each of the hydrogen compounds of the p-block elements in Period 2 with their counterparts in Period 3.

11.13 Stibane, SbH_3 ($\Delta_f H^\ominus = +145.1$ kJ mol^{-1}), decomposes above −45°C. Assess the difficulty in preparing a sample of BiH_3 ($\Delta_f H^\ominus = +277.8$ kJ mol^{-1}), and suggest a method for its preparation.

11.14 Dihydrogen is a familiar reducing agent, but it is also an oxidizing agent. Explain this statement, giving examples.

11.15 Comment on the observation that addition of hydrogen gas to the complex *trans*-$[W(CO)_3(PCy_3)_2]$ (Cy = cyclohexyl) results in formation of two complexes that are in equilibrium with one another. One complex has the tungsten centre in a formal W(0) oxidation state, while the other is in a formal W(II) oxidation state. Removal of the H_2 atmosphere regenerates the starting material.

11.16 Correct the faulty statements in the following description of hydrogen compounds. 'Hydrogen, the lightest element, forms thermodynamically stable compounds with all of the nonmetals and most metals. The isotopes of hydrogen have mass numbers of 1, 2, and 3, and the isotope of mass number 2 is radioactive. The structures of the hydrides of the Group 1 and 2 elements are typical of ionic compounds because the H^- ion is compact and has a well-defined radius. The structures of the hydrogen compounds of the nonmetals are adequately described by VSEPR theory. The compound $NaBH_4$ is a versatile reagent because it has greater hydridic character than the simple Group 1 hydrides such as NaH. Heavy element hydrides such as the tin hydrides frequently undergo radical reactions, in part because of the low E–H bond energy. The boron hydrides are called electron-deficient compounds because they are easily reduced by hydrogen.'

11.17 Estimate the expected infrared stretching wavenumber of gaseous $^3H^{35}Cl$, given that the corresponding value for $^1H^{35}Cl$ is 2991 cm^{-1}?

11.18 Consult Chapter 8 and then sketch the qualitative splitting pattern and relative intensities within each set for the ^1H- and ^{31}P-NMR spectra of PH_3.

TUTORIAL PROBLEMS

11.1 In his novel *The Mysterious Island*, published in 1874, the French science-fiction author Jules Verne wrote: 'I believe that water will one day be used as a fuel, that hydrogen and oxygen, which constitute it, used alone or simultaneously, will provide an inexhaustible source of heat and light, of an intensity of which coal is not capable.' How many parts of this statement have proved to be true?

11.2 In their article 'Hydrogen storage in metal–organic frameworks' (*Chem. Soc. Rev.*, 2009, **38**, 1294), Jeffrey Long and co-workers discuss some of the design principles for hydrogen storage materials. List the advantages and disadvantages of storing H_2 by using weak rather than strong binding interactions.

11.3 In his paper 'The proper place for hydrogen' (*J. Chem. Educ.*, 2003, **80**, 947), M. W. Cronyn argues that hydrogen should be placed at the head of Group 14 immediately above carbon. Summarize his reasoning.

11.4 In their article 'Reversible, metal-free hydrogen activation' (*Science*, 2006, **314**, 1124), Douglas Stephan and co-workers describe how H_2 is heterolytically cleaved at a molecule containing only main-group elements. By referring also to Section 5.10, explain the principles of this reaction, the procedures used to investigate the mechanism, and the possible implications for H_2 storage.

11.5 In their article 'Ru-based catalysts for ammonia decomposition: A mini-review', L. Jiang and co-workers (*Energy Fuels*, 2021, **35**, 11693) outline the prospects and challenges for using NH_3 as a H_2 carrier. State the advantages that such technology might bring, and list the principles and priorities in the search for catalysts able to facilitate NH_3 decomposition at low temperature.

12 The Group 1 elements

Part A: The essentials
12.1 The elements
12.2 Simple compounds
12.3 The atypical properties of lithium

Part B: The detail
12.4 Occurrence and extraction
12.5 Uses of the elements and their compounds
12.6 Hydrides
12.7 Halides
12.8 Oxides and related compounds
12.9 Sulfides, selenides, and tellurides
12.10 Hydroxides
12.11 Compounds of oxoacids
12.12 Nitrides and carbides
12.13 Solubility and hydration
12.14 Solutions in liquid ammonia
12.15 Zintl phases containing alkali metals
12.16 Coordination compounds
12.17 Organometallic compounds

Further reading

Exercises

Tutorial problems

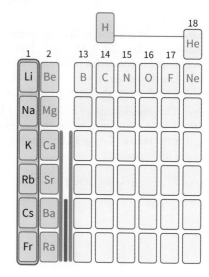

All the Group 1 elements are metallic, but unlike most metals they have low densities and are very reactive. In this chapter we provide a survey of the chemistry of these elements, highlighting the similarities and trends in properties and commenting on the anomalous behaviour of lithium. We then provide a detailed review of the chemistry of the alkali metals, discussing the occurrence of the elements in the natural environment and how they are extracted and used. The chapter also interprets trends in the properties of the simple binary compounds in terms of the ionic model and the nature of complexes and organometallic compounds of the elements.

PART A: The essentials

The Group 1 elements, the **alkali metals**, are lithium, sodium, potassium, rubidium, caesium (cesium), and francium. We shall not discuss francium, which exists naturally only in minute quantities and is highly radioactive. All the elements are metals and form simple ionic compounds, most of which are soluble in water. The elements form a limited number

TABLE 12.1 Selected properties of the Group 1 elements

	Li	Na	K	Rb	Cs
Metallic radius/pm	152	186	231	244	262
Covalent radius/pm	128	166	203	220	244
Ionic radius/pm (coordination number)	59(4) 76(6)	102(6)	138(6) 151(8)	148(6) 160(8)	174(8)
Ionization energy/kJ mol^{-1}	519	494	418	402	376
Standard potential/ V	−3.04	−2.71	−2.94	−2.92	−3.03
Density/g cm^{-3}	0.53	0.97	0.86	1.53	1.90
Melting point/ °C	180	98	64	39	29
$\Delta_{hyd}H^{\ominus}(M^+)$/(kJ mol^{-1})	−519	−406	−322	−301	−276
$\Delta_{sub}H^{\ominus}$/(kJ mol^{-1})	159	107	89	81	76

of complexes and organometallic compounds. In this first section of the chapter we summarize the key features of the chemistry of the Group 1 elements.

12.1 The elements

KEY POINT The trends in the properties of the Group 1 metals and their compounds can be explained in terms of variations in their atomic radii and ionization energies.

Sodium and potassium have high natural abundances, occurring widely as salts such as the chlorides. Lithium is relatively rare, occurring mainly in the mineral spodumene, $LiAlSi_2O_6$. Rubidium and caesium are rarer still, but occur in reasonable concentrations in some minerals, including the zeolite *pollucite*, $Cs_2Al_2Si_4O_{12} \cdot nH_2O$. Sodium and lithium metals are extracted by the electrolysis of molten metal chloride. Potassium is obtained by reacting KCl with sodium metal, and rubidium and caesium by reaction of the metal chloride with calcium or barium.

All the Group 1 elements are metals with the valence electron configuration ns^1. They conduct electricity and heat, are soft, and have low melting points that decrease down the group. Their softness and low melting points stem from the fact that their metallic bonding is weak because each atom contributes only one electron to the valence band (Section 2.11). This softness is particularly evident for Cs, which melts at only 29°C. Liquid sodium and sodium/potassium mixtures, which have melting points as low as −12.6°C, have been used as the coolant in some nuclear power plants and supercomputers because of their excellent thermal conductivities. All the elements adopt a body-centred cubic structure (Section 4.5) and, because that structure-type is not close-packed and their metallic radii are large, they have low densities compared to most metals. The alkali metals readily form alloys among themselves—for example NaK—and with many other metals, such as sodium/mercury amalgam. The simplest 'reaction' ever undertaken is that between one atom of sodium and one atom of caesium; ultracold lasers were used to manipulate individual atoms of Na and Cs to produce one molecule of NaCs. Table 12.1 summarizes some important properties of the alkali metals.

Flame tests are commonly used for the identification of the presence of the alkali metals and their compounds. Electronic transitions occur within the metal atoms and ions formed in the flames with energies that fall in the visible part of the spectrum, which give a characteristic colour to the flame:

Li	Na	K	Rb	Cs
Crimson	Yellow	Purple-pink	Violet	Blue

The intensity of the emission spectrum obtained from an alkali metal salt solution can be measured with a flame photometer to provide a quantitative measurement of the element's concentration in the solution, Section 8.12.

The chemical properties of the Group 1 elements correlate with the trend in their covalent and metallic radii (Fig. 12.1).

The increase in atomic radius from Li to Cs leads to a decrease in first ionization energy down the group because the valence shell is increasingly distant from the nucleus (Fig. 12.2, Section 1.7).

Because their first ionization energies are all low, the metals are reactive and form M$^+$ ions increasingly readily down the group. Their reaction with water,

$$2M(s) + 2H_2O(l) \rightarrow 2MOH(aq) + H_2(g)$$

illustrates this trend:

Li	Na	K	Rb	Cs
gently	vigorously	vigorously with ignition	explosively	explosively

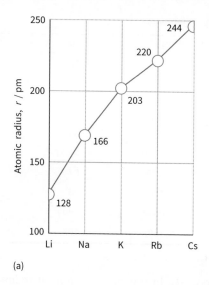
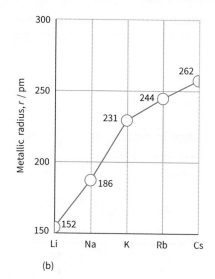

FIGURE 12.1 The variation in (a) the atomic radius and (b) the metallic radius of the elements of Group 1.

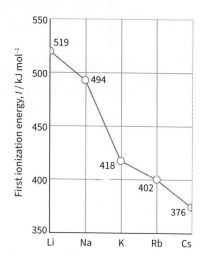

FIGURE 12.2 The variation in first ionization energy of the elements of Group 1.

A part of the reason for the explosive character of the reaction of Rb and Cs with water is that these metals are denser than water, sink below the surface, and the ignition of the hydrogen scatters the water violently. High-speed camera imaging of the reaction shows that electrons move rapidly from the alkali metal surface into the water (forming a blue solution due to the solvated electrons), and the remaining alkali metal cations repel each other on the metal surface, leading to the formation of metal spikes. The propagation of these very high surface area spikes leads to further extremely rapid reactions and the explosion.

The thermodynamic tendency to form M^+ in aqueous conditions is confirmed by the standard potentials of the couples M^+/M, which are all large and negative (Table 12.1), indicating that the metals are readily oxidized. The surprising uniformity of the standard potentials of the alkali metals can be explained by studying the thermodynamic cycle for the reduction half-reaction (Fig. 12.3). The enthalpies of

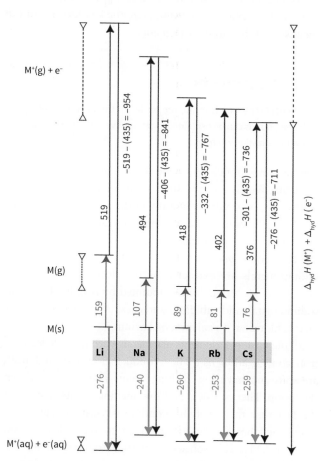

FIGURE 12.3 Thermochemical cycle (standard enthalpy changes in kJ mol^{-1}) for the oxidation half-reaction $M(s) \rightarrow M^+(aq) + e^-(aq)$. A theoretical value of 435 kJ mol^{-1} is included for the enthalpy of hydration of an electron, corresponding to the process $\frac{1}{2}H_2(g) + H_2O \rightarrow H_3O^+(aq) + e^-(aq)$ to allow the derived values for this half-reaction to be compared to the standard electrode potentials in Table 12.1.

sublimation and ionization both decrease down the group (making the formation of the $M^+(g)$ ion easier and the oxidation more favourable); however, this trend is counteracted by a smaller enthalpy of hydration as the radii of the ions increase (making oxidation in aqueous conditions less favourable).

All the elements must be stored under a hydrocarbon oil to prevent their reaction with atmospheric oxygen and moisture, although Li, Na, and K can be handled in air for short periods when only their surfaces tarnish. Rb and Cs must be handled under an inert atmosphere at all times. Lithium is usually handled in an argon-filled glove box as it reacts with N_2.

12.2 Simple compounds

KEY POINT The binary compounds of the alkali metals contain the cations of the elements and exhibit predominantly ionic bonding.

Group 1 elements form ionic (saline) hydrides with the rock-salt structure; the anion present is the hydride ion, H^-. These hydrides were discussed in some detail in Section 11.6(b). All the Group 1 elements form halides, MX. They can be obtained by direct combination of the elements or more normally from solutions; for example, reaction of the metal hydroxide or carbonate with hydrohalic acid (HX, X=F, Cl, Br, I). The halides occur widely, and a litre of seawater contains about 35 g of NaCl. Most of the halides have the 6:6-coordinate rock-salt structure (Fig. 12.4), but CsCl, CsBr, and CsI have the 8:8-coordinate CsCl structure (Fig. 12.5) as the larger caesium ion is able to fit a larger number of halide anions around it (Section 4.9). Most of the Group 1 metals halides and hydroxides crystallize from aqueous solution as the anhydrous salts. There are a few exceptions for the smaller Li and Na ions; for example, $LiX \cdot 3H_2O$ for X=Cl, Br, I, and $LiOH \cdot 8H_2O$. Lithium iodide is deliquescent, absorbing water rapidly from the air to form $LiI \cdot 3H_2O$ initially and then a solution.

> **EXAMPLE 12.1** Predicting structures for the francium halides
>
> Predict a structure for francium bromide, FrBr, using the ionic radii Fr^+ 196 pm and Br^- 196 pm.
>
> **Answer** We need to calculate the radius ratio ($\gamma = r_+/r_-$) and use the predictive rules in Section 4.9 to suggest a structure type. $\gamma = r_+/r_- = 196/196 = 1.00$. Therefore 8:8 coordination would be expected from Table 4.6 and the caesium chloride structure type predicted.
>
> **Self-test 12.1** Predict a structure type for FrF.

Under very high pressures, above 113 GPa, sodium has been found to form a stable compound with helium, Na_2He. Na_2He is an electride, where the formula can be written as $[Na^+]_2He[e^-]_2$. In the solid structure the electrons form the

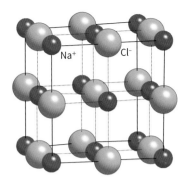

FIGURE 12.4 The rock-salt structure adopted by the majority of Group 1 metal halides.

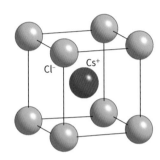

FIGURE 12.5 The CsCl structure adopted by CsCl, CsBr, and CsI under normal conditions.

anionic part of an antifluorite-type crystal structure and are separated from Na^+ cations by helium atoms, Fig. 12.6.

The Group 1 elements react vigorously with oxygen. Only Li reacts directly with excess oxygen to form predominantly the simple oxide containing O^{2-}, Li_2O. Sodium reacts with oxygen to produce the peroxide, Na_2O_2, which contains the peroxide ion, O_2^{2-}, and the other Group 1 elements form the superoxides, which contain the paramagnetic superoxide ion, O_2^-. All the hydroxides are white, translucent solids absorbing water from the atmosphere in an exothermic reaction. Lithium hydroxide, LiOH, forms

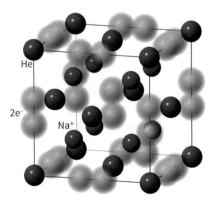

FIGURE 12.6 The structure of the high pressure Na_2He, $[Na^+]_2He[e^-]_2$; sites occupied by two electrons are shown schematically.

the stable hydrate LiOH·8H$_2$O, while the other hydroxides are deliquescent, yielding solutions. The solubility of the hydroxides makes them a ready source of OH$^-$ ions in the laboratory and in industry. The metals react with sulfur to form compounds with the formula M$_2$S$_x$, where x lies in the range 1 to 6. The simple sulfides Na$_2$S and K$_2$S have the antifluorite structure, whereas the polysulfides, with $n \geq 2$, contain S$_n^{2-}$ chains. Lithium readily forms a nitride, Li$_3$N, when it is heated in nitrogen (or more slowly at room temperature), but the other alkali metals do not react with nitrogen gas at ambient pressures. Sodium and potassium nitrides form by reacting the liquid or gaseous metals with nitrogen atoms, N, formed by passing an electric discharge through N$_2$(g).

Only Li reacts directly with graphitic carbon to form a carbide of the stoichiometry Li$_2$C$_2$ which contains the dicarbide (acetylide) anion, C$_2^{2-}$. Similar carbides are formed by the other alkali metals by heating them in ethyne (acetylene, C$_2$H$_2$). Potassium, Rb, and Cs react with graphite to form intercalation compounds, for example C$_8$K (Section 15.5) which has potassium ions between the graphite sheets. In combination with the p-block metals (from Groups 13 to 15) the alkali metals are strongly reducing and often form Zintl phases; the latter contain the alkali metal cation together with a reduced, complex anionic species, such as Ge$_4^{4-}$ in K$_4$Ge$_4$.

Sodium dissolves in liquid ammonia without evolving hydrogen, producing, at low concentrations, deep blue solutions that contain solvated electrons. These solutions survive for long periods at and below the normal boiling point of ammonia (−33°C) and in the absence of air. Concentrated metal–ammonia solutions have a metallic bronze colour and have electrical conductance close to that of a solid metal, about 10^7 S m^{-1}. It is possible to isolate the alkalide anions, M$^-$, from solutions of the metals in amines, which are formed by disproportionation of the element into M$^+$ and M$^-$.

The Group 1 element ions are hard Lewis acids (Section 5.6), and complexes are formed mainly with small, hard donors such as O or N atoms. Their hardness decreases down the group with increasing ionic radius. Interactions with monodentate ligands are weak, and the hydrated species, M(OH$_2$)$_n^+$, readily exchange H$_2$O ligands with the surrounding solvent. Chelating ligands, such as the hexadentate ethylenediaminetetraacetate ion, edta [(O$_2$CCH$_2$)$_2$NCH$_2$CH$_2$N(CH$_2$CO$_2$)$_2$]$^{4-}$, have much higher formation constants. Macrocycles and crown ethers can form strong complexes with the Group 1 elements provided their ions have the correct radius to fit into the ligand coordination environment (Section 7.14).

The lighter Group 1 elements form organometallic compounds which are highly reactive, hydrolysed by water, liberating hydrogen, and pyrophoric (spontaneously igniting) in air. Protic (proton-donating) organic compounds are reduced by the elements to form ionic organometallic compounds; for example, cyclopentadiene reacts with Na metal in the solvent THF to form Na$^+$[C$_5$H$_5$]$^-$. Lithium alkyls and aryls are by far the most important Group 1 organometallic compounds. They are thermally stable, soluble in organic and nonpolar solvents such as THF, and widely used as a source of nucleophilic alkyl or aryl groups in synthesis.

12.3 The atypical properties of lithium

KEY POINT The chemical properties of Li are anomalous because of its small ionic radius and tendency to exhibit covalent bonding.

As we saw in Chapters 9 and 10, most of the trends in the chemical properties of the elements in the periodic table are best discussed in terms of vertical trends within groups or horizontal trends across periods. However, the lightest member of a group, in this case Li, often displays properties that are markedly different from that of its congeners, Section 10.2. This difference can also be described as a diagonal relationship with the element to its lower right in the periodic table (Section 10.3). In the case of Group 1 and Li the following differences can be noted:

- Lithium can exhibit a high degree of covalent character in its bonding. This covalent character is due to the high polarizing power of the Li$^+$ ion associated with high charge density (Section 1.7).
- Lithium forms the normal oxide, Li$_2$O, containing the O^{2-} anion, when burnt in oxygen, whereas other Group 1 elements form peroxides or superoxides.
- Lithium is the only alkali metal to form a nitride, Li$_3$N, directly when heated in nitrogen under atmospheric pressures, and a carbide, Li$_2$C$_2$, when heated with graphite.
- Some lithium salts such as the carbonate, phosphate, and fluoride have very low solubilities in water. Other lithium salts crystallize as hydrates or are hygroscopic.
- Lithium forms many stable organometallic compounds.
- Lithium nitrate decomposes directly to the oxide, whereas the other alkali metals initially form nitrites, MNO$_2$.
- Lithium hydride is stable to heating to 900°C, whereas the other hydrides decompose on heating above 400°C.

The very low molar mass of lithium, which also makes it the least dense metal (0.53 g cm^{-3}), leads to applications where low weight is important. Examples include rechargeable batteries (LiCoO$_2$, LiFePO$_4$, LiC$_6$) and systems for hydrogen storage such as lithium metal hydrides, lithium borohydrides, and lithium amides and imides (Box 11.5).

PART B: The detail

In this section we present a more detailed discussion of the chemistry of the elements of Group 1, interpreting some of the observed properties in thermodynamic terms. As the bonding in the compounds formed by these elements is usually ionic, we apply the concepts of the ionic model.

12.4 Occurrence and extraction

KEY POINT The Group 1 elements can be extracted by electrolysis.

The name 'lithium' comes from the Greek *lithos* for stone. The natural abundance of lithium is low, the most abundant minerals being spodumene, $LiAlSi_2O_6$, from which lithium used to be extracted, and lepidolite, which has the approximate formula $K_2Li_3Al_4Si_7O_{21}(F,OH)_3$. Lithium is now most commonly obtained from brines as lithium carbonate, though because of the increasing demand for lithium in lithium-ion batteries there is an increasing need to discover other exploitable natural resources (Box 12.1) and ensure that lithium can be recycled effectively (Box 12.2). Lithium metal is usually isolated by the electrolysis of a molten LiCl/KCl mixture.

Sodium occurs as the mineral rock salt (NaCl) and in salt lakes, seawater, and the residue of ancient dried-up saline lakes that are often buried underground. Sodium chloride makes up 2.6% by mass of the biosphere, with the oceans containing 4×10^{19} kg of the salt. Subsurface deposits can be mined in the conventional way or water may be pumped

---

BOX 12.1 Where is lithium found, and how is it extracted?

The growing technological importance of lithium—particularly for primary and rechargeable battery applications (see Box 12.3) and specifically for electric vehicles—has focused attention on worldwide reserves and extraction of the element. Lithium use in batteries has increased to over 50% of worldwide demand (overtaking more traditional applications in glasses and ceramics), with over 30 million electric vehicles on the road globally in 2023. Worldwide production of lithium is near 8 100 000 tonnes per annum, and supply security has become a top priority for technology and automotive companies. Lithium is the twenty-fifth most abundant element (at 20 ppm) in the Earth's crust, but it is widely distributed; seawater has a concentration of around 0.20 ppm, equivalent to 230 billion tonnes. Lithium forms a minor part of igneous rocks such as spodumene and petalite, and lepidolite and hectorite clays derived from weathering of igneous rocks; these are all commercially viable sources.

The vast majority of lithium is now extracted from brines, aqueous solutions of alkali metal salts (halides, nitrates, and sulfates), or **caliche**, which is a hardened deposit of these sedimentary salts. These caliches are normally mainly $CaCO_3$, but in some parts of the world they consist of alkali metal salts, and in northern Chile, Argentina, and Bolivia these minerals are relatively rich in lithium and potassium. The caliche rock is dissolved to produce brine, and this is concentrated by solar evaporation. Because of the high solubility of lithium halide salts (Section 12.7) the solutions become enriched in lithium compared with other Group 1 elements. Lithium carbonate is precipitated by the addition of a solution of sodium carbonate to the hot, lithium-rich brine. Lithium can also be concentrated in brines by reverse osmosis, where pressure is applied to the dilute solution, causing water to migrate through a semipermeable membrane, thereby increasing the salt content of the solution. Industries involved in the recycling of lithium from batteries have also sprung up over the past few years, Box 12.2.

Lithium extraction from spodumene, $LiAlSi_2O_6$, involves heating the ore to 1100°C to convert the structure from the α-form to the β-form, which is much more reactive. β-Spodumene is then treated with concentrated sulfuric acid at 225°C to produce Li_2SO_4 and this is then dissolved and converted to Li_2CO_3 by addition of sodium carbonate. This process is very energy-intensive and generates eight times as much solid waste than lithium extraction from brines, so research effort is being directed at more environmentally friendly methods of producing lithium from ores. Effective methods of lithium extraction from low-concentration ores and even sea-water are also being studied.

Identified worldwide, extractable resources of lithium are estimated at 35 million tonnes, which, if used exclusively for car batteries using current technologies, is sufficient for approximately 3 billion cars with 24 kWh batteries—approximately twice the present number of all vehicles. However, the demand for lithium in other industries also needs to be met, including nonautomotive battery applications. The majority of the rest of the extracted lithium (29%) is currently used in the ceramics and glass industry to produce low-melting-point materials, lithium hydroxide-based lubricating greases (12%), and psychopharmalogical agents for the treatment of bipolar disorder, Section 32.4.

A further area of research involves finding alternatives to lithium-based systems for energy storage. For applications where weight is not of high importance—that is, nonmobile applications, such as the storage of intermittently generated wind power—rechargeable batteries based on the more widely available sodium can be used, Box 12.4.

BOX 12.2 How to recycle and reuse lithium

With the recent exponential growth in electric vehicle numbers (EVs) and the limited lifetime of lithium-ion batteries (around 1500 cycles or 8–10 years) there is a growing demand for methods of recycling and reusing the lithium, and other valuable components, from these batteries.

A lithium-ion battery capacity falls to around 70–80% of its initial value after around a decade of typical use, and while this might then become unacceptable for EV use, the battery can still be used for energy storage where specific capacity (energy stored per kg) is of less importance. Historically, lead–acid batteries (Box 15.7) have been used for back-up energy storage—for example, in telecommunication networks and signal towers—but second-life lithium ion batteries are now widely employed in these systems, particularly in China.

Eventually even the reuse of lithium batteries becomes uneconomic as their specific capacity drops still further and they are broken up and disposed of or recycled. Different types of Li-ion battery (Section 30.9) contain a variety of metal components, including cobalt, nickel, copper, and manganese, in addition to lithium, and it is the recovery of these more valuable transition metals that is currently driving Li-ion battery recycling.

The transition metals are extracted by smelting old batteries, giving rise to alloys containing the most valuable Ni, Co, Ni, and Cu, and a slag containing Al and Li (as Li_2O and Li_2CO_3). Some research is focused on separating the various battery components and layers (anode, cathode, and electrolyte) to enable the extraction of the different transition metals more efficiently.

As the demand for, and price of, lithium both rise, extracting it from the waste battery materials is becoming more economically viable. Li may be extracted from the smelted slag by dissolving Li_2CO_3 and recrystallization, though the salt's low solubility (Section 12.11(a)) results in high costs and further waste. Lithium may also be leached from battery waste solids using strong acids and lithium salts subsequently precipitated by addition of NaOH or Na_2CO_3.

Electrochemical extraction routes are also being researched. Waste battery cathode materials will dissolve Li^+ ions into solution versus a lithium metal anode, and these lithium ions are transported electrochemically through a solid electrolyte (Box 12.3, Section 30.9) where they can be deposited as high-purity Li_2CO_3.

---

underground to dissolve the rock salt, which is then pumped out as saturated brine solution. The metal is extracted by Down's process—the electrolysis of molten sodium chloride:

$$2\,NaCl(l) \rightarrow 2\,Na(l) + Cl_2(g)$$

The sodium chloride is kept molten at 600°C—a temperature considerably below its melting point of 808°C—by the addition of calcium chloride. A high potential difference, typically between 4 and 8 V, is applied between a carbon anode and an iron cathode immersed in the molten salt. The electrolysis liberates liquid sodium metal at the cathode, which rises to the surface of the cell where it is collected under an inert atmosphere. This process is also used for the industrial production of chlorine, which is generated at the anode.

Potassium occurs naturally as potash (K_2CO_3) and carnallite ($KCl \cdot MgCl_2 \cdot 6H_2O$). Natural potassium contains 0.012% of the radioactive isotope ^{40}K, which undergoes β-decay, with a half-life of 1.25×10^9 years, to ^{40}Ca and electron capture to ^{40}Ar. The ratio of ^{40}K and ^{40}Ar can be used for dating rocks, specifically the time when the rock solidified, at which point it traps any ^{40}Ar formed. In principle, potassium could be extracted electrolytically, but the high reactivity of the element makes this far too hazardous. Instead, molten sodium and molten potassium chloride are heated together and potassium and sodium chloride are formed:

$$Na(l) + KCl(l) \rightleftharpoons NaCl(l) + K(l)$$

At the temperature of operation, potassium is a vapour, and removing it from the system drives the equilibrium to the right.

Rubidium (from the Latin *rubidus* for deep red) and caesium (from *caesius* for sky blue) were discovered by Robert Bunsen in 1861 and named from the colour their salts impart to a flame. Both elements occur as minor constituents of the mineral lepidolite, which has the composition $(K,Rb,Cs)Li_2Al(Al,Si)_3O_{10}(F,OH)_2$, from which they are obtained as by-products of the extraction of lithium. Prolonged treatment of lepidolite with sulfuric acid forms the alums of the alkali metals, $M_2SO_4 \cdot Al_2(SO_4)_3 \cdot nH_2O$, M=K, Rb, Cs. The alums are separated by multiple fractional crystallizations, then converted to the hydroxide by reaction with $Ba(OH)_2$ and finally to the chloride by ion exchange. The metals are obtained from the molten chloride by reduction with calcium or barium:

$$2\,RbCl(l) + Ca(s) \rightarrow CaCl_2(s) + 2\,Rb(s)$$

Caesium also occurs as the mineral pollucite, $Cs_4Al_4Si_9O_{26} \cdot H_2O$. The element is extracted from the mineral by leaching with sulfuric acid to form the alum $Cs_2SO_4 \cdot Al_2(SO_4)_3 \cdot 24H_2O$, which is then converted to the sulfate by roasting with carbon. The chloride is formed by ion exchange and is then reduced with calcium or barium as described above. Caesium metal can also be obtained by the electrolysis of molten CsCN.

12.5 Uses of the elements and their compounds

KEY POINTS Common uses of lithium are related to its low density; the most widely used compounds of Group 1 are sodium chloride and sodium hydroxide.

The applications of lithium metal are in large part due to its low atomic mass and consequently low density. It is used in alloys where weight is of a premium concern, such as aircraft parts; Al containing around 2% Li has a mass density 6% lower than that of pure Al, and is used, for example, in parts of aircraft wings to reduce the overall weight and thereby improve fuel efficiency. Similar lithium-containing alloys have been used in aerospace applications including as the oxygen and rocket-propellent tanks of the Falcon 9 launch system used by SpaceX.

The low molar mass of lithium (6.94 g mol^{-1}), only 3.3% that of lead, coupled with the strongly negative standard potential of the Li$^+$/Li couple (Table 12.1), make lithium batteries an attractive alternative to lead–acid batteries (Box 12.3). Lithium carbonate is widely used to treat bipolar conditions (manic depression, Section 32.4), where

---

BOX 12.3 What are lithium batteries?

The very negative standard potential and low molar mass of lithium make it an ideal anode material for batteries. These batteries have relatively high specific energy (energy production divided by the mass of the battery) because lithium metal and compounds containing lithium are light in comparison with some other materials used in batteries, such as lead and zinc. Lithium batteries are now extremely common, but there are many types based on different lithium compounds and reactions. Batteries that contain lithium which are used once and then disposed of are termed primary lithium batteries, while rechargeable systems are described as secondary or Li-ion batteries.

Primary (non-rechargeable) lithium batteries

The reaction between lithium and MnO$_2$ is used in the majority of these cells, producing a voltage of 3 V, twice that of a zinc–carbon battery or an alkaline battery (which uses the reaction between the less electropositive zinc and MnO$_2$). The reactions are as follows:

Anode: $Li \rightarrow Li^+ + e^-$
Cathode: $Mn(IV)O_2 + Li^+ + e^- \rightarrow LiMn(III)O_2$

While such batteries are widely used in Japan, accounting for 30% of the primary battery market, they only account for a small but growing proportion in the UK and EU. Their main applications are those that require a long battery life, particularly at low operating temperatures, and high voltages; for example, home security equipment, smoke detectors, tracking devices, and the 'Internet of things'. Lithium–iron sulfide (FeS$_2$) batteries are also produced commercially and have twice the capacity of alkaline batteries, producing a voltage of ~1.5 V. They have low self-discharge rates, leading to a long shelf-life.

Another popular lithium battery uses thionyl chloride, SOCl$_2$. This system produces a light, high-voltage cell with a stable energy output. The overall reaction in the battery is

$4Li(s) + 2SOCl_2(l) \rightarrow 4LiCl(s) + S(s) + SO_2(l)$

The battery requires no additional solvent as both SOCl$_2$ and SO$_2$ are liquids at the internal battery pressure, but this battery is not rechargeable as both sulfur and LiCl are precipitated. It is used in military applications and in spacecraft. Another battery system is based on the reduction of SO$_2$:

$2Li(s) + 2SO_2(l) \rightarrow Li_2S_2O_2(s)$

The system is also not rechargeable as solid Li$_2$S$_2$O$_4$ deposits on the cathode. This battery uses acetonitrile (CH$_3$CN) as a co-solvent, and the handling of this compound and the SO$_2$ present safety hazards. The batteries are hermetically sealed and not available to the general public. They are used in military communications and automated external defibrillators that are used to restore normal heart rhythm.

Rechargeable lithium batteries

Lithium rechargeable batteries, used widely in portable computers, EVs, and mobile phones, mainly use Li$_{1-x}$CoO$_2$ ($x \leq 1$) or a lithium manganese nickel oxide as the cathode, coupled with a lithium/graphite anode, LiC$_6$. Lithium ions are produced at the anode during the battery discharge. To maintain charge balance, Co(IV) is reduced to Co(III) in the form of LiCoO$_2$ at the cathode. The reactions occurring during battery discharge are

Cathode: $Li_{1-x}CoO_2(s) + x\ Li^+(sol) + x\ e^- \rightarrow LiCoO_2(s)$
Anode: $C_6Li \rightarrow 6\ C(graphite) + Li^+(sol) + e^-$

The battery is rechargeable because both the cathode and the anode can act as host for the Li$^+$ ions, which can move back and forth between them when charging and discharging. There are many other lithium batteries using different electrode materials, mainly d-metal compounds that take part in the redox reaction in a similar way to cobalt, Section 30.9. EVs use lithium battery technology rather than lead–acid cells (Box 15.7) and employ a variety of materials as the cathode, depending on the battery costs and desired vehicle range. Very large factories—'Giga factories' producing up to 50 GWh of Li-ion battery capacity annually—have been built to produce the enormous volume of lithium-ion batteries needed for the growing number of EVs.

Lithium–air batteries are also being researched as they could produce very high energy densities of 12 kWh kg^{-1}, about five times that of the Li$_{1-x}$CoO$_2$ system described previously. These rechargeable batteries use atmospheric oxygen as the cathode, where it is reduced to lithium oxide during discharge, and lithium metal as the anode. However, these battery types need considerable further research as many side-reactions can take place, degrading battery performance; for example, the formation of lithium peroxide, carbonate, and hydroxide at the cathode in air and reactions involving the electrolyte separating the anode and cathode. Recent work using a high surface area, composite electrolyte of $Li_{10}GeP_2S_{12}$ nanoparticles embedded in a polyethylene oxide polymer controlled the battery discharge product as almost single-phase Li_2O. This improvement yielded practical energy densities of around $6\,kWh\,kg^{-1}$. If concerns in scaling-up the requirement of these cells for very high surface area materials and interfaces can be overcome, then large-scale lithium–air batteries approaching the theoretical limit of $12\,kWh\,kg^{-1}$ and rechargeable over thousands of cycles may become viable. Rechargeable batteries are discussed further in Chapters 25 and 30.

therapeutic effects arise by it normalizing levels of sodium and calcium, interacting with neurotransmitters and altering cell signalling. Lithium stearate is used as a lubricant in the automotive industry. The high polarizing power of Li^+ means that some complex oxides, such as $LiMO_3$, M=Nb, or Ta, show important nonlinear optical and acousto-optical effects, and are widely used in mobile communication devices.

Sodium and potassium are essential for physiological function (Section 26.4), and a major use of NaCl is in flavouring food. Sodium is used in the extraction of rarer metals, including Ti via titanium(IV) chloride. Other major uses of NaCl are road deicing and production of NaOH in the chloralkali industry (Box 12.5). However, because of concerns over the effects on the environment of distributing large amounts of rock salt as a deicing agent, alternative materials using less NaCl are being sought. For example, a mixture of sodium chloride and molasses has been employed. Sodium hydroxide is one of the 10 most important industrial chemicals in terms of annual tonnage produced.

Other historical applications of sodium and its compounds include the use of the metal in some kinds of street lamp, which produce a distinctive yellow glow when an electrical discharge is passed through sodium vapour (Box 1.4), table salt, baking soda, and caustic soda (NaOH).

Sodium compounds are currently being investigated for use in sodium-ion batteries which may have applications in grid-scale energy storage (Box 12.4). Sodium salts and compounds with ion-exchangable Na^+ are also widely used in water-softening equipment (Box 12.6).

Potassium hydroxide is used in soap manufacture to make 'soft' liquid soaps. Potassium chloride and sulfate are used as fertilizers; the nitrate and chlorate are used in fireworks and explosives, such as gunpowder. Potassium bromide has been used as an antaphrodisiac (a compound that reduces libido). Potassium cyanide is used in the metal extraction and plating industries to obtain or aid in the deposition of copper, silver, and gold.

Rubidium and caesium are often used in the same applications, and one element may often be substituted for the other. The market for these elements is small and highly specialized. Applications include glass for fibre optics in the telecommunications industry, night-vision equipment, and photoelectric cells. The 'caesium clock' (atomic clock) is used for the international standard measure of time and for the definition of the second and the metre. Caesium salts, mainly caesium formate $Cs(HCO_2)$, are also used as high-density drilling fluids: the high density of the solutions arises from the high atomic mass of Cs.

The large size of the caesium cation is frequently used in the inorganic chemistry laboratory to stabilize species in

---

BOX 12.4 Are sodium ion batteries the future for energy storage?

Much of renewable energy, such as that generated by solar and wind farms, is intermittent, and will therefore require a storage system if these green energy sources are to become the principal ones in the future. Large arrays of rechargeable batteries are one possible energy storage solution. For such large energy storage requirements, also known as 'grid-scale' energy storage, the low weight requirements demanded by electronic vehicles and portable electronic goods are no longer relevant, and the expensive lithium in Li-ion batteries (Box 12.3) can be replaced by cheap, ubiquitous sodium. Considerable current research activity is focused on the development of rechargeable sodium-ion batteries. These batteries are similar in terms of their components and chemistry to lithium-ion batteries but with Na^+ replacing Li^+ in each of the anode, electrolyte, and cathode phases. Possible cathode materials include transition-metal oxides, such as $NaCoO_2$, phosphates, and silicates. Sodium metal, sodium metal alloys, such as NaSn, graphitic carbon, and sodium titanium oxides, are being investigated as the anode and salts, such as $NaPF_6$ and $NaClO_4$, as the electrolyte. The first prototype rechargeable Na-ion batteries of this type were launched in 2015. Sodium sulfur batteries which can also be used in grid-scale energy storage are discussed further in Box 12.7.

BOX 12.5 What does the chloralkali industry produce?

The chloralkali industry has its roots in the industrial revolution, when large quantities of alkali were first required for the manufacture of soap, paper, and textiles. Today, sodium hydroxide is one of the top 10 most important inorganic chemicals in terms of quantity produced, and continues to be important in the manufacture of other inorganic chemicals and in the pulp and paper industries. Chlorine and dihydrogen are the gaseous products. Chlorine is very important industrially and is used in the manufacture of PVC, in the extraction of titanium, and in the pulp and paper industries.

The industrial process is based on the electrolysis of aqueous sodium chloride. Water is reduced to hydrogen gas and hydroxide ions at the cathode, and chloride ions are oxidized to chlorine gas at the anode:

$$2H_2O(l) + 2e^- \rightarrow H_2(g) + 2OH^-(aq)$$
$$2Cl^-(aq) \rightarrow Cl_2(g) + 2e^-$$

There are three different types of cells that are used for the electrolysis. In a diaphragm cell there is a diaphragm which prevents the OH^- ions produced at the cathode from coming into contact with the Cl_2 gas produced at the anode. This diaphragm was at one time made of asbestos, but is now made of a polytetrafluoroethylene mesh. During the electrolysis, the solution at the cathode is removed continuously and evaporated in order to crystallize the sodium chloride impurities. The final sodium hydroxide typically contains approximately 1% by mass NaCl.

A *membrane cell* functions like the diaphragm cell except that the anode and cathode solutions are separated by a microporous polymer membrane that is permeable only to Na^+ ions. The sodium hydroxide solution produced using this cell typically contains approximately 50 ppm Cl^-. The disadvantages of this method are that the membrane is very expensive, and can become clogged by trace impurities.

The *mercury cell* uses liquid mercury as the cathode. Chlorine gas is produced at the anode, but sodium metal is produced at the cathode:

$$Na^+(aq) + e^- \rightarrow Na(Hg)$$

The sodium–mercury amalgam is reacted with water on a graphite surface:

$$2Na(Hg) + 2H_2O(l) \rightarrow 2NaOH(aq) + H_2(g)$$

The sodium hydroxide solution produced by this route is very pure and the mercury cell is the preferred source of high-quality solid sodium hydroxide. Unfortunately, the process is accompanied by discharge of mercury into the environment by a number of routes. Consequently, the chloralkali industry is under pressure to move away from the use of mercury electrodes.

BOX 12.6 What are sodium ion-exchange materials used for?

Hard water contains high levels of Ca^{2+} and Mg^{2+} ions, which precipitate from the water on heating (as limescale, which is largely $CaCO_3$) and inhibit the formation of a lather with soap or detergent, so reducing their effectiveness. Domestic water softeners contain zeolites, or ion-exchange resins, which contain Na^+ ions that exchange with the Ca^{2+} and Mg^{2+} ions. Zeolites are often also a component of washing detergents, where they perform the same role.

Zeolites are microporous aluminosilicates that contain weakly held cations and water molecules within their cavities (Sections 15.15 and 29.1). They are often available naturally or are synthesized as their sodium-containing form and denoted 'Na-zeolite'. The ion-exchange reaction that occurs when hard water is exposed to Na-zeolite is

$$Na_2\text{-zeolite}(s) + Ca^{2+}(aq) \rightarrow Ca\text{-zeolite}(s) + 2Na^+(aq)$$

with the result that Ca^{2+} ions are removed from the solution to the solid phase. The softened water contains mainly Na^+ ions as the dissolved cation species. As sodium carbonate and the sodium salts of soap and detergent molecules are very soluble, the softened water is more effective for washing and avoids problems such as the deposition of limescale on the heating elements of kettles, dishwashers, and washing machines.

The reverse reaction, which regenerates the ion-exchanger Na-zeolite, can be performed in a water softener by exposing the exhausted zeolite to high concentrations of Na^+ ions (for example, sodium chloride solution):

$$Ca\text{-zeolite}(s) + 2Na^+(aq, conc) \rightarrow Na_2\text{-zeolite}(s) + Ca^{2+}(aq)$$

In the case of zeolites added to detergents, the Ca-zeolite that is produced exists as a finely divided solid that is flushed away. This procedure is environmentally benign because the effluent contains calcium, silicon, aluminium, oxygen, and water—the same as many natural minerals.

Instead of zeolites, some water softeners contain resins, which are porous polymeric organic compounds formed from cross-linked polystyrene decorated with functional groups such as carboxylates and sulfonates. The charge on these anionic groups is balanced by Na^+ ions on the surface of the resin, which are readily exchanged with Ca^{2+} and Mg^{2+}.

> **BOX 12.7** The sodium–sulfur battery: an alternative to the sodium-ion oxide batteries?
>
> The sodium–sulfur battery is an alternative to metal oxide based Na-ion batteries, described in Box 12.4, and uses the power generated by the reaction of sodium with sulfur. The battery has a high energy density, a good efficiency of charge and discharge (90%), a long cycle life, and is fabricated using inexpensive materials. Molten sodium metal forms the anode and is separated from the cathode (steel in contact with sulfur absorbed into a porous carbon) by a β-alumina solid electrolyte (Section 25.6). Sodium β-alumina is an ionic conductor but a poor electrical conductor, so avoiding self-discharge of the battery. When the battery is discharging Na gives up an electron to the external circuit and the resulting Na$^+$ ions migrate through the sodium β-alumina to the sulfur container. At the cathode, electrons from the external circuit react with sulfur to form sodium polysulfides, S_n^{2-}. The overall battery discharge process is
>
> $$2\ Na(l) + 4\ S(l) \rightarrow Na_2S_4(l) \quad E_{cell} \approx 2.1\ V$$
>
> During charging, the reverse process takes place and small heat losses in the system keep it at the operating temperature of 300–350°C. On account of the high temperature of their operation and the highly corrosive nature of the battery components, such cells are primarily suitable for large-scale static applications rather than transport. Sodium–sulfur batteries therefore offer an energy-storage system that could be used in association with renewable energy plants that only operate at certain periods. At a wind farm the battery would store energy during times of high wind but low power demand, and the stored energy would then be discharged from the batteries during peak load periods. Similarly, solar farm energy generated and stored during daylight hours could be used at night.

ionic compounds that are susceptible to thermal decomposition (Section 4.15(a)). This stabilizing effect coupled with high positivity of caesium can be seen in the reaction of platinum with caesium metal and caesium hydride to produce the unique Pt^{2-} containing compound, $4Cs_2Pt \cdot CsH$. As the largest single-atom cation, caesium is also used in solid-state chemistry to facilitate the synthesis of structures that would not be stable with smaller cations; the existence of $CsPbI_3$, as the only inorganic lead iodide perovskite, demonstrates this.

12.6 Hydrides

KEY POINT The hydrides of the Group 1 elements are ionic and contain the H$^-$ ion.

The Group 1 elements react with hydrogen to form ionic (saline) hydrides with the rock-salt structure; the anion present is the hydride ion, H$^-$. These hydrides were discussed in detail in Section 11.6(b), and LiH, containing 12.6 wt% hydrogen, has been evaluated as a single-use hydrogen storage material; see also Section 30.8.

The hydrides react violently with water:

$$NaH(s) + H_2O(l) \rightarrow NaOH(aq) + H_2(g)$$

Finely divided sodium hydride can even ignite if it is left exposed to humid air. Such fires are difficult to extinguish, because even carbon dioxide is reduced when it comes into contact with hot metal hydrides. Hydrides are useful as non-nucleophilic bases and reductants:

$$NaH(s) + NH_3(l) \rightarrow NaNH_2(am) + H_2(g)$$

where 'am' denotes a solution in ammonia.

EXAMPLE 12.2 Predicting the structures of the alkali metal hydrides

Use the ionic radius values of Li$^+$ and Na$^+$ (from Table 12.1), that of the hydride ion, 140 pm, and radius ratio rules to predict structures for the salts MH.

Answer The values of the radius ratio, γ, can be calculated for both MH compounds, and the values used to predict a structure based on Table 4.6. The values of γ using data in Table 12.1 are LiH, 0.42, and NaH, 0.73. These values lie (just!) within the range that predicts 6:6 coordination and, therefore, for the AX stoichiometry the rocksalt structure type is adopted (as is found experimentally).

Self-test 12.2 The values of γ for the hydrides KH, RbH, and CsH all lie in the range that predicts 8:8 coordination and the CsCl structure type. By considering the polarizability of the hydride ion, propose a reason why these saline hydrides adopt the rocksalt structure type.

12.7 Halides

KEY POINT On descending the group, the enthalpy of formation becomes less negative for the fluorides but more negative for the chlorides, bromides, and iodides.

All the Group 1 elements form halides, MX, by direct combination of the elements. Most of the halides have the 6:6-coordinate rock-salt structure, but CsCl, CsBr, and CsI have the 8:8-coordinate CsCl structure (Section 4.9). The simple radius-ratio arguments presented in Section 4.10 may be used to help rationalize this choice of structure. Table 12.2 summarizes the radius ratio (γ) for the various alkali metal halides. As we saw in Section 4.10, a rock-salt

TABLE 12.2 Radius ratio, γ, for the alkali metal halides.*

	F	Cl	Br	I
Li	0.57	0.42	0.39	0.35
Na	0.77	0.56	0.52	0.46
K	0.96	0.76	0.70	0.63
Rb	0.90	0.82	0.76	0.67
Cs	0.80	0.92	0.85	0.76

* Based on ionic radii for six-coordination. Values in italic type are compounds that adopt the rock-salt structure type.

structure with 6:6 coordination is expected for radius-ratio values between 0.414 and 0.732, and the CsCl structure is expected for larger values; a 4:4-coordinate zinc-sulfide structure is expected for values below 0.414. However, the lattice energies of the CsCl and rock-salt arrangements differ by only a small percentage, and factors such as polarization (Section 4.12) help to stabilize the rock-salt structure over the CsCl structure for most of the alkali metal halides. At 445°C CsCl changes from the CsCl structure-type to the rock-salt structure-type, and on cooling below room temperature RbCl converts to the CsCl structure-type.

EXAMPLE 12.3 Investigating a pressure-induced phase transition using powder X-ray diffraction

The powder X-ray diffraction data collected from RbI at standard conditions show that the lattice type is face-centred cubic and the lattice parameter is 734 pm. Application of 4×10^8 Pa of pressure causes the powder X-ray diffraction pattern to change, showing that the lattice type becomes primitive with a lattice parameter of 446 pm. Interpret these data, given that the ionic radii for Rb^+ and I^- in six-fold (eight-fold in parentheses) coordination are 148 (160) pm and 220 (232) pm, respectively.

Answer On the basis of radius-ratio rules and $\gamma = 0.67$, the predicted structure for rubidium iodide is rock-salt with 6:6 coordination (Section 4.10(b)). The lattice type of this structure is face-centred cubic, in agreement with the diffraction data.

It is also possible to predict the lattice parameter for this structure by using the ionic radii. In the rock-salt structure the lattice parameter is equivalent to the overall Rb–I–Rb distance (Fig. 12.4), which is twice the sum of the anion and cation ionic radii. Thus, the predicted lattice parameter is 736 pm, in good agreement with the experimental value.

Structures under pressure have a thermodynamic tendency to rearrange to denser arrangements. The radius ratio for rubidium iodide lies close to 0.732, above which the CsCl structure is expected. Therefore, under pressure, a phase transformation occurs in RbI. The lattice type of the CsCl structure type is primitive (Fig. 12.5), in agreement with the X-ray diffraction data, and a lattice parameter for this structure type can be calculated by using the ionic radii of Rb^+ and I^- in eight-fold coordination (160 and 232 pm, respectively) as 453 pm (from $2(r_+ + r_-)/3^{1/2}$ as the body diagonal of the unit cell of length a is $\sqrt{3}a$ and corresponds to Rb–I–Rb). This result closely matches the experimental data: the diffraction data are collected at 4 kbar, so the lattice parameter is slightly smaller than the calculated value, which is based on ionic radii evaluated at 1 bar.

Self-test 12.3 X-ray diffraction data show that FrI adopts a structure exhibiting a primitive cubic unit cell with a lattice parameter of 490 pm. Are these experimental data consistent with the predicted structure using the ionic radii in Resource section 1?

The enthalpies of formation of all the halides are large and negative, becoming less negative from fluoride to iodide for each element. The enthalpies of formation become less negative for the fluorides on descending the group, but become more negative for the chlorides, bromides, and iodides (Fig. 12.7).

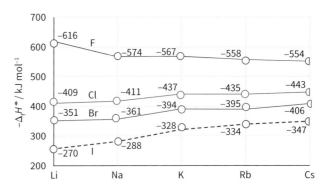

FIGURE 12.7 The standard enthalpies of formation of the halides of Group 1 elements at 298 K.

These trends can be rationalized by considering a Born–Haber cycle for formation of the halides from the elements (Fig. 12.8). Note that these calculations treat the bonding

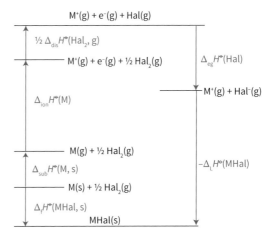

FIGURE 12.8 The Born–Haber cycle for the formation of the Group 1 halides. The sum of the enthalpy changes round the cycle is zero.

in alkali metal compounds as purely ionic; however, for the heavier-metal ions there is increasing contribution from covalence as the ions become larger, more polarizable, and less hard.

Consideration of the Born–Haber cycle (Fig. 12.8) shows that the trend in the enthalpy of formation of the alkali metal halides depends on the relative values of $(\Delta_{sub}H^\ominus + \Delta_{ion}H^\ominus) - \Delta_L H^\ominus$. For the chlorides, bromides, and iodides, the variation in values of the sum $(\Delta_{sub}H^\ominus + \Delta_{ion}H^\ominus)$ is greater than the variation in $\Delta_L H^\ominus$, and the enthalpies of formation become more negative down the group. However, for the fluorides, the small ionic radius of fluorine ensures that the differences in $\Delta_L H^\ominus$ are greater than those in $(\Delta_{sub}H^\ominus + \Delta_{ion}H^\ominus)$, and the enthalpies of formation become less negative down the group.

12.8 Oxides and related compounds

KEY POINTS Only Li forms a normal oxide, containing O^{2-} anions, on direct reaction with excess oxygen; Na forms the peroxide, and the heavier elements form mainly the superoxides.

As mentioned previously, all the Group 1 elements react vigorously with oxygen. Only Li reacts directly with excess oxygen to produce just the oxide, Li_2O, which adopts the antifluorite structure (Section 4.9):

$$4Li(s) + O_2(g) \rightarrow 2Li_2O(s)$$

Sodium reacts with oxygen to give mainly the peroxide, Na_2O_2, which contains the peroxide ion, O_2^{2-}:

$$2Na(s) + O_2(g) \rightarrow Na_2O_2(s)$$

The other Group 1 elements form mixtures of peroxides and, mainly, the superoxides:

$$K(s) + O_2(g) \rightarrow KO_2(s)$$

These compounds contain the paramagnetic superoxide ion, O_2^-, and adopt a CaC_2 type structure that is based on the rock-salt structure (Fig. 4.32; see also Fig. 12.9 for an alternative view).

All the varieties of oxides are basic and react with water to produce the OH^- ion by extraction of H^+ from H_2O in a Brønsted acid–base reaction:

$$Li_2O(s) + H_2O(l) \rightarrow 2Li^+(aq) + 2OH^-(aq)$$

$$Na_2O_2(s) + 2H_2O(l) \rightarrow 2Na^+(aq) + 2OH^-(aq) + H_2O_2(aq)$$

$$2KO_2(s) + 2H_2O(l) \rightarrow 2K^+(aq) + 2OH^-(aq) + H_2O_2(aq) + O_2(g)$$

The oxide and the peroxide react by proton transfer from H_2O. The initial formation of 'hydrogen superoxide', HO_2, by proton transfer to the superoxide ion is followed immediately by the disproportionation of HO_2 into O_2 and H_2O_2.

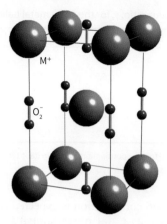

FIGURE 12.9 The structure of the superoxides, MO_2, of the Group 1 elements.

The normal oxides of Na, K, Rb, and Cs can be prepared by heating the metal with a limited amount of oxygen or by thermal decomposition of the peroxide or superoxide:

$$Na_2O_2(s) \rightarrow Na_2O(s) + \tfrac{1}{2}O_2(g)$$

The oxides Na_2O, K_2O, and Rb_2O adopt the antifluorite structure. The stability of the peroxides and superoxides to this decomposition increases down the group, Li_2O_2 being the least stable and Cs_2O_2 the most. The tendency of the peroxide or superoxide to decompose to the oxide can be explained by examining the lattice enthalpies of the compounds as discussed in Section 4.15 and Example 4.16; see also Example 12.4. Sodium peroxide is widely used as an oxidizing agent, as it provides a ready source of oxygen on warming.

EXAMPLE 12.4 Predicting the stabilities of peroxides using thermochemical data

The ionic radii of the O^{2-} and O_2^{2-} ions are 126 and 180 pm, respectively. Use this information to confirm that there is a decreased tendency for the peroxide to decompose on descending the group.

Answer To assess stabilities we need to compare lattice enthalpies. To do so, we use the data in Table 12.1 and the Kapustinskii equation (eqns. 4.4 and 4.5) to calculate the difference in lattice enthalpy between Na_2O and Na_2O_2 and then for Rb_2O and Rb_2O_2. We need to remember that the peroxide formula is $(M^+)_2(O_2^{2-})$, so in the Kapustinskii equation the number of ions is 3 and their charge numbers are +1 and −2. Insertion of the data gives the following values:

Na_2O	Na_2O_2	Rb_2O	Rb_2O_2
2702	2260	2316	1980 kJ mol^{-1}

The difference between the values for Na_2O and Na_2O_2 is 442 kJ mol^{-1}, and the difference for Rb_2O and Rb_2O_2 is 336 kJ mol^{-1}.

These results show that the difference between the values decreases down the group, suggesting that there is a lower thermodynamic tendency for the peroxide to decompose to form the oxide (providing entropy considerations are similar).

Self-test 12.4 All the Group 1 ozonides are unstable and decompose unless kept at low temperature. Predict how the decomposition temperature would be expected to vary down the group.

Potassium superoxide, KO_2, absorbs carbon dioxide, liberating oxygen:

$$4\, KO_2(s) + 2\, CO_2(g) \rightarrow 2\, K_2CO_3(s) + 3\, O_2(g)$$

This reaction is exploited to regenerate air in applications that include breathing apparatus for firefighters and submarine air purification systems. For applications in aerospace—for example, spacesuit life-support units—lithium peroxide is often used instead of KO_2 to reduce weight.

Ozonides, compounds that contain the ozonide ion, O_3^-, exist for all the Group 1 elements. The ozonides of K, Rb, and Cs are obtained by heating the hydroxide, peroxide, or superoxide with ozone. Sodium and lithium ozonides may be prepared by ion exchange of CsO_3 in liquid ammonia. All these ozonide compounds are very unstable and explode violently:

$$2KO_3(s) \rightarrow 2KO_2(s) + O_2(g)$$

Partial oxidation of Rb and Cs yields suboxides of various compositions. Special conditions are needed to form these compounds, in which the elements occur with average oxidation numbers lower than +1; that is, a mixture of +1 and zero oxidation states. These compounds are formed only when air, water, and other oxidizing agents are rigorously excluded. A series of metal-rich oxides are formed by the reaction of Rb or Cs with a limited supply of oxygen. These compounds are dark, highly-reactive, metallic conductors with formulas such as Rb_6O, Rb_9O_2, Cs_4O, and Cs_7O. A clue to the nature of these compounds is that Rb_9O_2 consists of O atoms surrounded by octahedra of six Rb atoms, with two neighbouring octahedra sharing faces (Fig. 12.10). These compounds were some of the earliest metal cluster compounds, containing metal-to-metal bonds, to be synthesized and characterized, although many other systems—for example, the Zintl phases (Section 14.19)—have now been found. The metallic conduction of the compounds suggests that the valence electrons are delocalized beyond the individual Rb_9O_2 clusters.

12.9 Sulfides, selenides, and tellurides

KEY POINT The Group 1 elements form simple sulfides, M_2S, and polysufides in combination with sulfur.

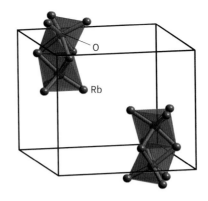

FIGURE 12.10 The structure of Rb_9O_2, showing two units, in which each O atom is surrounded by an octahedron of Rb atoms, and neighbouring octahedra share triangular faces. The unit cell is outlined.

All the alkali metals form a simple sulfide of stoichiometry M_2S; those of the smaller ions, Li^+ to K^+, adopt the antifluorite structure with simple S^{2-} ions. The polysulfides, M_2S_n, with n ranging from 2 to 6, are also known for the heavier alkali metals where the softer acids and larger cations, M^+, stabilize the soft bases S_n^{2-}. For $n \geq 3$, the structures contain polysulfide anions as zigzag chains separated by the alkali metal cations (Fig. 12.11). The sodium/sulfur battery is being studied as a possible stationary energy-storage system for use in combination with wind farms and solar energy plants (Box 12.4). Selenium and tellurium react with the alkali metals to form selenides such as K_2Se, and tellurides, respectively; polyselenides, K_2Se_5, and polytellurides, Cs_2Te_5, are also known.

12.10 Hydroxides

KEY POINT All Group 1 hydroxides are soluble in water and absorb water and carbon dioxide from the atmosphere.

All the hydroxides of Group 1 elements are white, translucent, hygroscopic solids. They absorb water and carbon

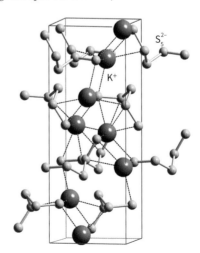

FIGURE 12.11 The structure of K_2S_5.

dioxide from the atmosphere in an exothermic reaction. LiOH forms the stable hydrate LiOH·8H$_2$O, while the heavier alkali metal hydroxides are deliquescent. The solubility of the hydroxides makes them a ready source of OH$^-$ ions in the laboratory and in industry. Potassium hydroxide, KOH, is soluble in ethanol, and this 'ethanolic KOH' is a useful reagent in organic synthesis. Alkali metal hydroxide solutions rapidly absorb carbon dioxide from the air,

$$2\,MOH(aq) + CO_2(g) \rightarrow M_2CO_3(aq) + H_2O(l)$$

and solutions left open to the air rapidly become contaminated with carbonate. For this reason the concentration of an MOH solution to be used in qualitative volumetric analysis should be checked prior to its use. Concentrated MOH solutions also react slowly at room temperature (more rapidly when heated) with silicate glass to produce alkali metal silicates, so reactions involving their use at high temperatures should be undertaken in inert plastic laboratory-ware.

Sodium hydroxide is produced by the chloralkali industry (Box 12.5), and is used as a reagent in the organic chemical industry and in the preparation of other inorganic chemicals. It is also used in the paper-making industry and by the food industry to break down proteins. For example, olives are soaked in sodium hydroxide solution to make the skins soft enough to be edible, and the Norwegian delicacy *lutefisk* has a jelly-like consistency produced by the dissolution of the proteins from dried cod. Domestic applications are based on the action of NaOH on grease, and it is used extensively in oven and drain cleaners. In some 'foaming' drain cleaners it is mixed with aluminium powder which reacts with the aqueous hydroxide ions to liberate hydrogen gas.

12.11 Compounds of oxoacids

The Group 1 elements form salts with most oxoacids. The most industrially important Group 1 salts of oxoacids are sodium carbonate, commonly called soda ash, and sodium hydrogen carbonate, commonly known as sodium bicarbonate.

(a) Carbonates

KEY POINT The Group 1 carbonates are soluble and decompose to the oxide when heated strongly.

The Group 1 elements form the only soluble carbonates (with the exception of the NH$_4^+$ ion), although lithium carbonate is only sparingly soluble.

Sodium carbonate has been produced by the Solvay process for many years. The overall reaction, which uses the commonly available feedstocks of NaCl and CaCO$_3$, can be represented by the equilibrium

$$2\,NaCl(aq) + CaCO_3(s) \rightleftharpoons Na_2CO_3(s) + CaCl_2(aq)$$

However, the equilibrium lies to the left on account of the high lattice energy of CaCO$_3$, and the actual process uses a complex stepwise route involving ammonia. Calcium oxide, produced by the thermal decomposition of calcium carbonate, is reacted with ammonium chloride to generate ammonia

$$2\,NH_4Cl(s) + CaO(s) \rightarrow 2\,NH_3(g) + CaCl_2(s) + H_2O(l)$$

Ammonia and carbon dioxide (from the thermal decomposition of CaCO$_3$ and NaHCO$_3$) are passed into a saturated sodium chloride solution to form a solution of NH$_4^+$, Na$^+$, Cl$^-$, and HCO$_3^-$ ions:

$$NaCl(aq) + CO_2(g) + NH_3(g) + H_2O(l) \rightleftharpoons Na^+(aq)$$
$$+ HCO_3^-(aq) + NH_4^+(aq) + Cl^-(aq)$$

When cooled to below 15°C, NaHCO$_3$ precipitates from the solution, is filtered off, and is then heated to produce the desired Na$_2$CO$_3$, with evolution of CO$_2$ and H$_2$O. The residual NH$_4$Cl is isolated and reused in the initial reaction stage with CaO. The process is energy-intensive and produces large amounts of CaCl$_2$ as a by-product. These problems mean that Na$_2$CO$_3$ is mined wherever sources of the mineral trona, sodium sesquicarbonate, Na$_3$(CO$_3$)(HCO$_3$)·2H$_2$O, exist.

The main uses of sodium carbonate are in glass manufacture, where it is heated with silica to form sodium silicate, Na$_2$O·xSiO$_2$, as a cleaning product 'washing soda', and as a water softener, where it removes Ca^{2+} ions as calcium carbonate, the 'scale' formed in kettles in hard-water areas. Potassium carbonate is produced by treating KOH with carbon dioxide and is used in glass and ceramics manufacture.

Lithium carbonate decomposes when heated above 650°C:

$$Li_2CO_3(s) \xrightarrow{\Delta} Li_2O(s) + CO_2(g)$$

The carbonates of the heavier elements only decompose significantly when heated above 800°C. This stabilizing influence of a large cation on a large anion can be explained in terms of trends in lattice energies, and was discussed in Section 4.15. The differences between the lattice enthalpies of the carbonate and the oxide decrease down the group, suggesting that there is a lower thermodynamic tendency for the carbonate to form the oxide on descending the group (entropy effects being supposed similar). The decomposition temperature also increases: sodium carbonate begins to decompose above 800°C whereas Rb$_2$CO$_3$ requires heating to near 1000°C.

(b) Hydrogencarbonates

KEY POINTS Sodium hydrogencarbonate is less soluble than sodium carbonate and liberates CO$_2$ when heated.

Lithium hydrogencarbonate is only known in solution and readily decomposes to lithium carbonate. Sodium hydrogencarbonate (sodium bicarbonate) is less soluble than sodium carbonate in water and can be prepared by bubbling carbon dioxide through a saturated solution of the carbonate:

$$Na_2CO_3(aq) + CO_2(g) + H_2O(l) \rightarrow 2\,NaHCO_3(s)$$

The reverse of this reaction occurs when the hydrogencarbonate is heated:

$$2\,NaHCO_3(s) \rightarrow Na_2CO_3(s) + CO_2(g) + H_2O(l)$$

This reaction provides the basis for the use of sodium hydrogencarbonate as a fire extinguisher. The powdered salt smothers the flames and decomposes in the heat to liberate carbon dioxide and water, which themselves act as extinguishers. This reaction is also the basis for the use of sodium hydrogencarbonate in baking, when the carbon dioxide and water vapour released during the baking process cause the dough to rise. A more effective raising agent is baking powder, in which sodium hydrogencarbonate is mixed with calcium dihydrogenphosphate:

$$2\,NaHCO_3(s) + Ca(H_2PO_4)_2(s) \rightarrow Na_2HPO_4(s)$$
$$+ CaHPO_4(s) + 2\,CO_2(g) + 2\,H_2O(l)$$

Potassium hydrogencarbonate is used as a buffer in wine production and in water treatment. It is also used as a buffer in low-pH liquid detergents, as an additive in soft drinks, and as an antacid to combat indigestion.

(c) Other oxosalts

KEY POINT The nitrates of Group 1 elements are used as fertilizers and in explosives.

Sodium sulfate, Na_2SO_4, is very soluble and readily forms hydrates. The major commercial source of sodium sulfate is as a by-product of the production of hydrochloric acid from sodium chloride:

$$NaCl(aq) + H_2SO_4(aq) \rightarrow Na_2SO_4(aq) + 2HCl(aq)$$

It is also obtained as a by-product of several other industrial processes, including flue-gas desulfurization and the manufacture of rayon. The principal use of sodium sulfate is in processing wood pulp for making the tough brown paper used in packaging and cardboard. During the process, the sodium sulfate is reduced to sodium sulfite, which dissolves the lignin in the wood. (The lignin is recovered from the pulp and used as an adhesive and binder.) It is also used in glass manufacture, in detergents, and as a mild laxative.

Sodium nitrate, $NaNO_3$, is deliquescent and is used in making other nitrates, fertilizers, and explosives. Potassium nitrate, KNO_3, occurs naturally as the mineral saltpetre. It is slightly soluble in cold water and very soluble in hot water. It has been used extensively in the manufacture of gunpowder since about the twelfth century and is used in explosives, fireworks, matches, and fertilizers.

EXAMPLE 12.5 Applying thermogravimetric analysis to study the decomposition of alkali metal nitrates

When heated above 900°C a sample of lithium nitrate, $LiNO_3$, of mass 100.0 mg loses 71.76% of its mass in a single stage, whereas potassium nitrate heated to the same temperature loses mass in two stages, with respective total mass losses of 15.82% (at 350°C) and 53.42% (above 950°C) of the original sample. Determine the compositions of the various products formed in the decomposition reactions of potassium and lithium nitrates. Use thermodynamic arguments based on changes in lattice enthalpy to interpret your answer.

Answer We need to consider the changes in molar mass that the data represent, and then identify the corresponding empirical formulas (Section 8.15). The molar mass of $LiNO_3$ is 68.95 g mol^{-1}, so 100.0 mg corresponds to (100.0 mg)/(68.95 g mol^{-1}) = 1.450 mmol $LiNO_3$. Because 1 mol $LiNO_3$ produces 1 mol of solid lithium-containing decomposition product **X**, 1.450 mmol $LiNO_3$ produces 1.450 mmol **X**. However, we know that the mass of **X** produced is 28.24 mg. Therefore, its molar mass is (28.24 mg)/(1.450 mmol) = 19.48 g mol^{-1}. This molar mass corresponds to the empirical formula $LiO_{0.5}$ (or Li_2O) and results from the loss of $NO_2(g)$ and $O_2(g)$ with the overall equation for the decomposition of lithium nitrate as

$$LiNO_3(s) \rightarrow \tfrac{1}{2}Li_2O(s) + NO_2(g) + \tfrac{1}{4}O_2(g)$$

Similar calculations for KNO_3 show the initial mass loss corresponds to the formation of KNO_2 (potassium nitrite) at 350°C and K_2O at 450°C with the sequential reactions

$$KNO_3(s) \rightarrow KNO_2(s) + \tfrac{1}{2}O_2(g)$$
$$2KNO_2(s) \rightarrow K_2O(s) + 2NO(g) + \tfrac{1}{2}O_2(g)$$

The decompositions of lithium and potassium nitrates to their oxides proceed by different routes, which is another example of the atypical behaviour of lithium for this group. The larger alkali metal cations stabilize the NO_2^- ion against its immediate decomposition to oxide. A similar difference in decomposition route and temperature occurs for lithium carbonate, which is the only alkali metal carbonate to decompose readily on heating.

Self-test 12.5 Use similar arguments to rationalize the different decomposition temperatures of the two alkali metal nitrates to the final product.

12.12 Nitrides and carbides

KEY POINT Only Li forms a nitride and a carbide by direct reaction with nitrogen and carbon, respectively.

Although Li is the least reactive of the Group 1 metals, it is the only one that forms a nitride (which is normally red) by direct reaction with nitrogen:

$$6\,Li(s) + N_2(g) \rightarrow 2\,Li_3N(s)$$

FIGURE 12.12 The structure of Li_3N.

The structure of lithium nitride (Fig. 12.12) consists of sheets of composition $[Li_2N]^-$, containing six-coordinate N^{3-} ions, that are separated by further Li^+ ions. The Li^+ ions in solid lithium nitride are highly mobile, as there are vacant sites in the structures on to which lithium ions can hop, and it is, therefore, classified as a 'fast ion conductor' (Sections 25.8, 30.5, and 30.9). It is being studied as a solid electrolyte and as a possible anode material for use in rechargeable batteries.

Lithium nitride also shows potential as a hydrogen storage material (Box 11.5, Section 30.8). It stores up to 11.5% by mass of hydrogen when exposed to hydrogen gas at elevated temperatures and pressures. The Li_3N reacts with hydrogen to form $LiNH_2$ and LiH in a reversible reaction:

$$Li_3N(s) + 2H_2(g) \leftrightarrow LiNH_2(s) + 2LiH(s)$$

When heated to 170°C the $LiNH_2$ and LiH react together to form Li_3N and liberate hydrogen.

Sodium nitride has recently been synthesized by the deposition of Na and N atoms on a cooled sapphire surface at liquid nitrogen temperatures or by reaction of liquid Na with a nitrogen atom plasma. Its structure is analogous to the ReO_3 structure type (Section 25.4), with N^{3-} replacing Re(VI) and Na^+ replacing O^{2-}; the material is metallic as the sodium 3s electrons are not fully transferred to nitrogen and conduction pathways exist through voids in the solid. K_3N has been synthesized in a similar manner and adopts a hexagonal unit cell. The other Group 1 elements do not form nitrides, although the azides, which contain the N_3^- ion, can be obtained by the reaction

$$2MNH_2(s) + N_2O(g) \rightarrow MN_3(s) + MOH(s) + NH_3(g)$$

Sodium azide, NaN_3, is used in vehicle safety air bags where its rapid decomposition, following the application of an electrical discharge, produces large volumes of nitrogen gas in under 30 ms. Azides are poisonous in the same way as cyanides as they inhibit the enzyme cytochrome oxidase and the production of cell energy from ATP.

Lithium reacts directly with graphitic carbon at high temperatures to form a carbide of the stoichiometry Li_2C_2, which contains the dicarbide (acetylide) anion, C_2^{2-}. The other alkali metals do not form carbides by direct reaction of the elements, although ionic compounds of the stoichiometry M_2C_2 are obtained by heating the metal in ethyne. Potassium, Rb, and Cs react with graphite at low temperatures to form intercalation compounds such as C_8K (Section 15.5). Lithium may be inserted into graphite electrochemically to produce LiC_6, which has an important role as the anode in rechargeable lithium battery systems (Box 12.3). The alkali metals Na to Cs also react with fullerene, C_{60}, to form fullerides such as Na_2C_{60}, Cs_3C_{60}, and K_6C_{60}, which contain the alkali metal cation and a fulleride anion, C_{60}^{n-}. The structure of K_3C_{60} is described in Section 15.6 and contains K^+ ions in all the octahedral and tetrahedral holes of a cubic close-packed array of C_{60}^{3-} anions; this material becomes superconducting below 30 K.

EXAMPLE 12.6 Applying NMR to study Group 1 compounds

All the Group 1 elements have quadrupolar nuclei, for instance $I(^{23}Na) = \frac{3}{2}$ and $I(^{133}Cs) = \frac{7}{2}$. However, NMR spectra, including solid-state MASNMR spectra (Section 8.6), can be obtained for such nuclei, particularly if they are in high-symmetry environments. The ^{23}Na-NMR spectrum of the fulleride Na_3C_{60} (which is obtained by reacting sodium metal with the fullerene C_{60} forming $[C_{60}]^{3-}$, fulleride, anions) shows two resonances. Interpret this information and describe how the structure of Na_3C_{60} is related to that of solid C_{60} (Section 4.9).

Answer The two resonances in the low-temperature spectrum indicate that the compound contains two different environments for Na. We know that C_{60} adopts a structure formed by cubic close-packing of the C_{60} molecules (Section 4.9). In the reaction with sodium metal, the C_{60} molecules are reduced to anions; the small Na^+ cations can occupy all the available tetrahedral and octahedral holes of a slightly expanded, but still close-packed, array of C_{60}^{3-} anions. Each type of hole corresponds to one of the environments detected by NMR.

Self-test 12.6 Predict the ^{7}Li-NMR spectrum of Li_3N (Fig. 12.12) at high and low temperatures, assuming that a high-resolution spectrum can be obtained for this nucleus.

12.13 Solubility and hydration

KEY POINTS There is wide variation in the solubility of the common salts; only Li and Na form hydrated salts.

All the common salts of the Group 1 elements are soluble in water. The solubilities cover a wide range of values, some of the most soluble being those for which there is the greatest difference between the radii of the cation and anion. Thus, the solubilities of the Li halides increase from the fluoride to the bromide, whereas for Cs the trend is reversed. The explanation for these trends was discussed in Section 4.15.

Not all alkali metal salts occur as their hydrates. The lattice enthalpies of hydrated salts are lower than for the anhydrous salt because the radius of the cation is effectively increased by the hydration sphere and is further from its surrounding anions. The hydrated salt will be favoured if this decrease in lattice enthalpy is offset by the hydration enthalpy. The hydration enthalpy depends on the ion-dipole interaction between the cation and the polar water molecule. This interaction is greatest when the cation has high charge density. The Group 1 metal cations have low charge density on account of their large radii and low charge, and, consequently, most of their salts are anhydrous. There are a few exceptions for the smaller Li^+ and Na^+ ions; for example, $LiOH \cdot 8H_2O$ and $Na_2SO_4 \cdot 10H_2O$ (historically known as Glauber's salt).

12.14 Solutions in liquid ammonia

KEY POINTS Sodium dissolves in liquid ammonia to produce a solution that is blue when dilute and bronze when concentrated.

Sodium dissolves in pure anhydrous liquid ammonia (without hydrogen evolution) to produce solutions that are deep blue when dilute. The colour of these **metal–ammonia solutions** originates from the tail of a strong absorption band that peaks in the near infrared.[1] The dissolution of sodium in liquid ammonia to produce a very dilute solution is represented by the equation

$$Na(s) \rightarrow Na^+(am) + e^-(am)$$

These solutions survive for long periods at the temperature of boiling ammonia (−33°C) and in the absence of air. However, they are only metastable, and their decomposition is catalysed by some d-block compounds:

$$Na^+(am) + e^-(am) + NH_3(l) \rightarrow NaNH_2(am) + \tfrac{1}{2}H_2(g)$$

Concentrated metal–ammonia solutions have a metallic bronze colour and an electrical conductance close to that of a metal. These solutions have been described as 'expanded metals' in which $e^-(am)$ associates with the ammoniated cation. This description is supported by the fact that, in saturated solutions, the ammonia-to-metal ratio is between 5 and 10, which corresponds to a reasonable coordination number for the metal.

The blue metal–ammonia solutions are excellent reducing agents. For example, the Ni(I) complex $[Ni_2(CN)_6]^{4-}$, in which nickel is in an unusually low oxidation state, Ni(I), may be prepared by the reduction of Ni(II) with potassium in liquid ammonia:

$$2K_2[Ni(CN)_4] + 2K^+(am) + 2e^-(am) \rightarrow$$
$$K_4[Ni_2(CN)_6](am) + 2KCN$$

---

[1] Other electropositive metals with low enthalpies of sublimation, including Ca and Eu, dissolve in liquid ammonia to produce solutions with a blue colour that is independent of the metal.

The reaction is performed in the absence of air in a vessel cooled to the boiling point of ammonia. Other reactions of M(am) as a strong reducing reagent include the formation of graphitic intercalates (Section 15.5), fullerides (Section 15.6), and Zintl phases (Section 14.19) by, for example, the following reactions:

$$8C(graphite) + K^+(am) + e^-(am) \rightarrow [K(am)]^+[C_8]^-(s)$$
$$C_{60}(s) + 3Rb^+(am) + 3e^-(am) \rightarrow [Rb(am)]_3C_{60} \xrightarrow{\Delta} Rb_3C_{60}(s)$$

The alkali metals also dissolve in ethers and alkylamines to give solutions with absorption spectra that depend on the identity of the metal. The dependence on the metal suggests that the spectrum is associated with charge transfer from an *alkalide ion*, M^- (such as a sodide ion, Na^-), to the solvent. When ethylenediamine (1,2-diaminoethane, en) is used as a solvent the dissolution equation is written as

$$2Na(s) \rightarrow Na^+(en) + Na^-(en)$$

Further evidence for the occurrence of alkalide ions is the diamagnetism associated with the species assigned as M^-, which would have the spin-paired ns^2 valence-electron configuration. Another observation in agreement with this interpretation is that, when sodium/potassium alloy is dissolved,

$$NaK(l) \rightarrow K^+(en) + Na^-(en)$$

the metal-dependent absorption band is the same as for solutions of Na itself.

12.15 Zintl phases containing alkali metals

KEY POINT The alkali metals reduce the Groups 13–16 metals to produce Zintl phases containing polymeric anions.

Zintl phases are formed when a Group 1 element is combined with a p-block metal from Groups 13 to 16. Alkali metal solutions in liquid ammonia are strong reducing agents and react with the metal to form such phases. Alternatively, Zintl phases can be obtained by direct reaction of the Group 1 element and the p-block element at high temperatures. Group 1 Zintl phases are ionic compounds in which electrons are transferred from the alkali metal atom to a cluster of the p-block atoms to form a polyanion; these compounds are normally diamagnetic, semiconducting or poor conductors, and brittle.

With the Group 14 elements (E), compounds of stoichiometry M_4E_4, which contain a tetrahedral E_4^{4-} anion (Fig. 12.13), and M_4E_9, for example Cs_4Ge_9 containing the monocapped square antiprismatic Ge_9^{4-} anion, can be obtained. For Group 13, compounds such as Rb_2In_3 (containing $[In_6]^{4-}$ octahedral anions) and KGa (with $[Ga_8]^-$ polyhedral anions) are known. The compound Cs_5Bi_4 contains tetrameric chains of stoichiometry Bi_4^{5-}. Even more exotic Zintl phases obtained with

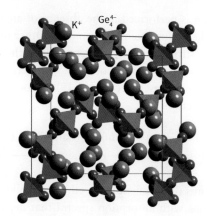

FIGURE 12.13 The structure of K_4Ge_4.

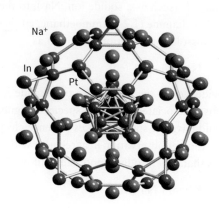

FIGURE 12.14 Part of the structure of $Na_{172}In_{192}Pt_2$ showing the complex fulleride-like networks formed by the In atoms around Na ions.

Group 1 elements include the fullerene-type structures of $Na_{96}In_{91}M_2$ and $Na_{172}In_{192}M_2$, M=Ni, Pd, Pt (Fig. 12.14).

12.16 Coordination compounds

KEY POINT The Group 1 elements form stable complexes with polydentate ligands.

The Group 1 ions, particularly Li⁺ to K⁺, are hard Lewis acids (Section 5.6). Therefore, most of the complexes they form arise from Coulombic interactions with small, hard donors, such as those possessing O or N atoms. Monodentate ligands are only weakly bound on account of the weak Coulombic interactions and lack of significant covalent bonding by these ions. However, a number of factors (such as the formation of peroxides and ozonides, rather than oxides, by the heavier alkali metals and the insolubility of their perchlorates) indicate that the metals become less hard as the group is descended.

The H_2O ligands in $M(OH_2)_n^+$ species readily exchange with the surrounding H_2O molecules of the solvent, although this is slowest for the very hard Li⁺ ion and faster for the increasingly less hard Rb⁺ and Cs⁺ ions. Chelating ligands

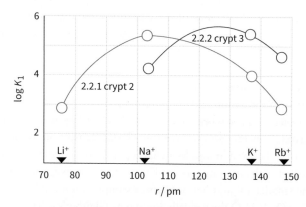

FIGURE 12.15 The formation constants of complexes of Group 1 metals with cryptand ligands plotted against cation radius (6 coordination). Note that the smaller 2.2.1 crypt favours complex formation with Na⁺ and the larger 2.2.2 crypt favours K⁺.

such as the ethylenediaminetetraacetate ion, $[(O_2CCH_2)_2NCH_2CH_2N(CH_2CO_2)_2]^{4-}$, have much higher formation constants, particularly with the larger alkali metal cations. Macrocycles and related ligands form the most stable complexes. Crown ethers such as 18-crown-6 (**1**) form complexes with alkali metal ions that are reasonably stable in nonaqueous solution. Bicyclic cryptand ligands, such as 2.2.1-crypt (**2**) and 2.2.2-crypt (**3**), form complexes with alkali metals that are even more stable, and they can survive even in aqueous solution (**4**). These ligands are selective for a particular metal ion, the dominant factor being the fit between the cation and the cavity in the ligand that accommodates it (Fig. 12.15).

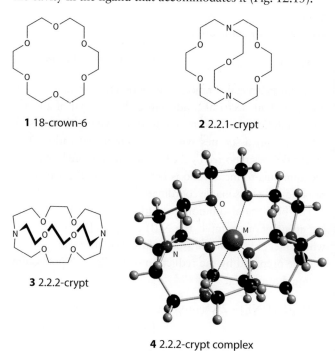

Another example of this fit between cation and ligand cavity is found in the potassium channel, responsible for the selective transport of Na⁺ and K⁺ ions across cell membranes

(Section 26.3). The ions cross the hydrophobic cell membrane by means of embedded protein molecules that contain cavities lined with donor atoms. The donor atoms are arranged to form a cavity, the size of which determines whether Na⁺ or K⁺ is bound. Such *ion channels* modulate the Na⁺/K⁺ concentration differential across the cell membrane that is essential for particular functions of the cell. The naturally occurring molecule valinomycin (5) is an antibiotic that selectively coordinates K⁺: the resulting hydrophobic 1:1 complex transports K⁺ through a bacterial cell membrane, depolarizing the ion differential and resulting in cell death.

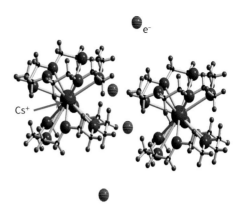

FIGURE 12.16 The crystal structure of [Cs(18-crown-6)₂]⁺e⁻. The striped blue spheres mark the sites of highest electron density and so indicate the locations of the 'anion' e⁻.
Source: M. Faber, K. Moeggenborg, et al. Structure of K+(cryptand[2.2.2]) electride and evidence for trapped electron pairs. *Nature* **331**, 599–601 (1988).

5 Valinomycin

The complexation of sodium with a cryptand can be used to prepare solid sodides, such as [Na(2.2.2)]⁺Na⁻, where (2.2.2) denotes the cryptand ligand. X-ray structure determination reveals the presence of [Na(2.2.2)]⁺ and Na⁻ ions, with the latter located in a cavity of the crystal with an apparent radius larger than that of I⁻. The precise nature of the products of this reaction varies with the ratio of sodium to cryptand. It is also possible to crystallize solids containing solvated electrons, the so-called **electrides**, and to obtain their X-ray crystal structures. Fig. 12.16, for example, shows the inferred position of the maxima of the electron density in such a solid. The preparation of sodides and other alkalides demonstrates the powerful influence of solvents and complexing agents on the chemical properties of metals.

A further example of these influences is the ability of crown ethers to produce reactive Cl⁻ ions in organic solvents. When an aqueous solution of NaCl is shaken in a separating funnel with a solution of 18-crown-6 in organic solvent, the coordinated Na⁺ ions cross into the organic phase, drawing Cl⁻ ions with them. The poorly solvated Cl⁻ ions are highly reactive.

12.17 Organometallic compounds

KEY POINT The organometallic compounds of the Group 1 elements react rapidly with water and are pyrophoric.

Group 1 elements form a number of organometallic compounds that are unstable in the presence of water and are pyrophoric in air. They are prepared in organic solvents such as tetrahydrofuran (THF). Protic (proton-donating) organic compounds form ionic organometallic compounds with the Group 1 metals. For example, cyclopentadiene reacts with sodium metal in THF:

$$Na(s) + C_5H_6(l) \rightarrow Na^+[C_5H_5]^-(sol) + \tfrac{1}{2}H_2(g)$$

The resulting cyclopentadienide anion is an important intermediate in the synthesis of d-block organometallic compounds (Chapter 24).

Lithium, sodium, and potassium form intensely coloured compounds with aromatic species. The oxidation of the metal results in transfer of an electron to the aromatic system to produce a **radical anion**—an anion that possesses an unpaired electron:

Na + (naphthalene) ⟶ Na⁺[C₁₀H₈]⁻

Sodium and potassium alkyls are colourless solids that are insoluble in organic solvents and, when stable, have fairly high melting temperatures. They are produced by a **transmetallation reaction**, which involves breaking a metal–carbon bond and forming a metal–carbon bond to a different metal. Alkylmercury compounds are often the starting materials in these reactions. For example, methylsodium is produced in the reaction between sodium metal and dimethylmercury in a hydrocarbon solvent:

$$Hg(CH_3)_2 + 2\,Na \rightarrow 2\,NaCH_3 + Hg$$

Organolithiums are by far the most important Group 1 organometallic compounds. They are liquids or low-melting solids, are the most thermally stable of the entire group, and are soluble in organic and nonpolar solvents such as THF. They can be synthesized from an alkyl halide and lithium metal, or by reacting the organic species with butyllithium, $Li(C_4H_9)$, commonly abbreviated to BuLi:

$$BuCl(sol) + 2\,Li(s) \rightarrow BuLi(sol) + LiCl(s)$$
$$BuLi(sol) + C_6H_6(l) \rightarrow Li(C_6H_5)(sol) + C_4H_{10}(g)$$

A feature of many main-group organometallic compounds is the presence of bridging alkyl groups. When ethers are the solvent, methyl lithium exists as $Li_4(CH_3)_4$, with a tetrahedron of Li atoms and bridging CH_3 groups (6). In hydrocarbon solvents, $Li_6(CH_3)_6$ (7) is formed; its structure is based on an octahedral arrangement of Li atoms. Other alkyllithiums adopt similar structures except when the alkyl groups become very bulky, as in the case of t-butyl, $-C(CH_3)_3$, when tetramers are the largest species formed. Many of these alkyllithiums are electron-deficient compounds and contain the $3c,2e$ and $4c,2e$ bonds characteristic of such compounds (Section 3.11).

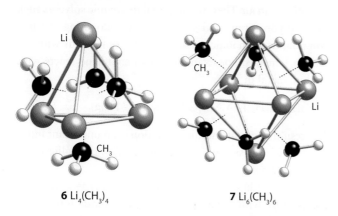

6 $Li_4(CH_3)_4$ **7** $Li_6(CH_3)_6$

Clusters of lithium atoms in organolithium compounds may also encapsulate anions; for example, the oxide anion in the octahedral unit $[Li_6O]^{4+}$ cation which exists in $\{(c\text{-}C_4H_9)N(H)\}_{12}Li_{14}O$, and the cubic ($\mu_8$) hydride centred moiety $[Li_8H]^{7+}$ found in the $[(hpp)_6HLi_8]^+$ cation, where hppH is guanidine 1,3,4,6,7,8-hexahydro-2H-pyrimido[1,2-a]pyrimidine.

Organolithium compounds are very important in organic synthesis, the most important reactions being those in which they act as nucleophiles and attack, for example, a carbonyl group:

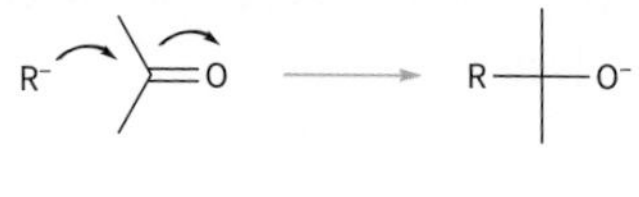

Organolithium compounds are also used to convert p-block halides to organoelement compounds, as we see in later chapters. For example, boron trichloride reacts with butyllithium in THF to produce an organoboron compound:

$$BCl_3(sol) + 3\,BuLi(sol) \rightarrow Bu_3B(sol) + 3\,LiCl(s)$$

The driving force for this and many other reactions of s- and p-block organometallic compounds is the formation of the insoluble halide compound of the more electropositive metal.

Alkyllithiums are important industrially in the stereospecific polymerization of alkenes to form synthetic rubber. Butyllithium is used as an initiator in solution polymerization to produce a wide range of elastomers and polymers. Organolithium compounds are also used in the synthesis of a range of pharmaceuticals, including vitamins A and D, analgesics, antihistamines, antidepressants, and anticoagulants. Alkyllithiums can be used in the synthesis of other organometallic compounds. For example, they can be used to introduce alkyl groups into d-metal organometallic compounds (Section 23.8):

$$(C_5H_5)_2MoCl_2 + 2\,CH_3Li \rightarrow (C_5H_5)_2Mo(CH_3)_2 + 2\,LiCl$$

The reactivity and solubility of alkyllithiums are enhanced by adding a chelating ligand such as tetramethylethylenediamine, TMEDA (8), which breaks up any tetramers to produce complexes such as $[BuLi(TMEDA)]_2$.

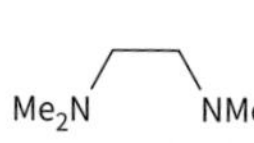

8 TMEDA

Reaction of the alkali metals with polyarenes, such as corannulene, $C_{20}H_{10}$, leads to the formation of highly reduced corannulene anions; for example, $[C_{20}H_{20}]^{4-}$. These species can coordinate and form sandwich-like compounds with multiple alkali metal cations that are situated between the polyaromatic rings. Example species include $[Li_5(C_{20}H_{10}^{4-})_2]^{3-}$ (9) and $[Li_3K_3(C_{20}H_{10}^{4-})_2]^{2-}$.

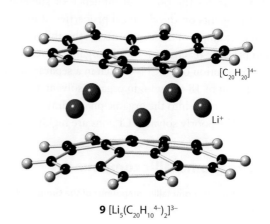

9 $[Li_5(C_{20}H_{10}^{4-})_2]^{3-}$

FURTHER READING

R. B. King, *Inorganic Chemistry of the Main Group Elements*. John Wiley & Sons (2005).

P. Enghag, *Encyclopedia of the Elements*. John Wiley & Sons (2004).

D. M. P. Mingos, *Essential Trends in Inorganic Chemistry*. Oxford University Press (2004). A survey of inorganic chemistry from the perspective of structure and bonding.

V. K. Grigorovich, *The Metallic Bond and the Structure of Metals*. Nova Science Publishers (1989).

N. C. Norman, *Periodicity and the s- and p-Block Elements*. Oxford University Press (1997). Includes coverage of essential trends and features of s-block chemistry.

A. Sapse and P. V. Schleyer (eds.), *Lithium Chemistry: A Theoretical and Experimental Overview*. John Wiley & Sons (1995).

W.-P. Leung and Y.-C. Chan, *Alkali Metal Inorganic Chemistry*, in *Encyclopedia of Inorganic and Bioinorganic Chemistry*. John Wiley and Sons (1999–2014).

H. Bae and Y. Kim, Technologies of lithium recycling from waste lithium ion batteries: A review. *Materials Advances*, 2021, **2**, 3234.

EXERCISES

12.1 Why are Group 1 elements (a) strong reducing agents, (b) poor complexing agents?

12.2 Describe the processes involved in extracting caesium metal from natural minerals.

12.3 The auride anion, Au^-, has a similar ionic radius to Br^- at 196 pm. Predict structures for the ionic compounds CsAu and RbAu.

12.4 Use the data given in Tables 12.1 and 12.2 to calculate the enthalpies of formation for the Group 1 fluorides and chlorides. Plot the data, and comment on the trends observed.

12.5 Describe the origin of the diagonal relationship between Li and Mg.

12.6 Explain why the standard electrode potentials of Li and Cs are almost identical despite the much lower first ionization energy of caesium. How might the standard electrode potential $Li(s)/Li^+$ change in (a) dimethylsulfoxide, a strongly coordinating solvent, (b) a weakly coordinating ionic liquid?

12.7 Which of the following pairs is most likely to form the desired compound? Describe the periodic trend and the physical basis for your answer in each case. (a) Ethanoate ion or edta ion to react with Cs^+. (b) Li^+ or K^+, to form a complex with 2.2.2-crypt.

12.8 Identify the metal-containing compounds **A**, **B**, **C**, and **D** in the following array of reactions when (a) M is Li, (b) M is Rb.

$$A \xleftarrow{H_2O} M \xrightarrow{O_2} B \xrightarrow{\Delta} C$$
$$\downarrow NH_3(l)$$
$$D$$

12.9 Describe how you would prepare a pure sample of caesium ozonide starting from caesium metal, including its recrystallization from a suitable solvent.

12.10 Use the Kapustinskii equation (eqns. 4.4 and 4.5) to calculate lattice enthalpies for LiF and CsF. Account for the fact that LiF and CsI have low solubility in water, whereas LiI and CsF are very soluble.

12.11 Which halide salts of francium would be the least soluble? How could francium be precipitated and isolated from solution that also contains sodium ions?

12.12 Explain why LiH has greater thermal stability than the other Group 1 hydrides, whereas solid $LiHCO_3$ cannot be isolated, unlike the other Group 1 hydrogencarbonates.

12.13 Which of lithium sulfite (Li_2SO_3, lithium sulfate(IV)) or potassium sulfite would you expect to decompose at the higher temperature?

12.14 Draw the structures of NaCl and CsCl, and give the coordination number of the metal in each case. Explain why the compounds adopt different structures.

12.15 What would you expect to happen to the structure of RbBr under the application of very high pressures?

12.16 Predict the products of the following reactions:

(a) $CH_3Br + Li \rightarrow$

(b) $MgCl_2 + LiC_2H_5 \rightarrow$

(c) $C_2H_5Li + C_6H_6 \rightarrow$

12.17 Explain how the nature of the alkyl group affects the structure of lithium alkyls.

TUTORIAL PROBLEMS

12.1 It has been argued that a better diagonal relationship exists between Li and Ca than between Li and Mg (T. P. Hanusa, *J. Chem. Ed.*, 1967, **64**, 686–7). Summarize the arguments for and against this viewpoint.

12.2 Francium has been described as 'the last discovered natural element' (J.-P. Adloff and G. B. Kauffman, *The Chemical Educator*, 2005, **10**, 387–94). Discuss the justification for this statement. Summarize the evidence and claims for the first discovery of francium.

12.3 Under ambient conditions lithium and sodium adopt simple bcc structures. Under high pressures these alkali metals undergo a series of complex phase transitions to fcc and then lower-symmetry structures (M. I. McMahon et al., *Proc. Natl. Acad. Sci. U.S.A.*, 2007, **104**(44) 17297; B. Rousseau et al., *Eur. Phys. J. B.*, 2011, **81**, 1–14). Discuss these phase transitions and the changes in electronic properties that accompany them.

12.4 Discuss the industrial uses of lithium and likely future demand for compounds of the metal. How are these demands likely to be met? A useful resource is the United States Geological Survey at https://minerals.usgs.gov/minerals/pubs/commodity/lithium/. Accessed May 2016.

12.5 By 2030 it has been projected that European gigafactory capacity will be around 1350 GWh (Faraday Institution Report "*UK electric vehicle and battery production potential to 2040*" (September 2024)). Discuss the materials and processes that have led to the reduction of Li-ion battery costs over the last decade.

12.6 Summarize the chemistry of sodium that is being researched for the development of rechargeable sodium-ion batteries (M. D. Slater, D. Kim, E. Lee, and C. S. Johnson, *Adv. Func. Mater.*, 2013, **23**, 947–58).

12.7 Identify the incorrect phrase in each of the following statements; explain your answer in each case. (a) Sodium dissolves in ammonia and amines to produce the sodium cation and solvated electrons or the sodide ion. (b) Sodium dissolved in liquid ammonia will not react with NH_4^+ because of strong hydrogen bonding with the solvent.

12.8 Z. Jedlinski and M. Sokol describe the solubility of alkali metals in nonaqueous supramolecular systems (*Pure Appl. Chem.*, 1995, **67**, 587). They dissolved the metals in THF containing crown ethers or cryptands. Sketch the structure of the 18-crown-6 ligand. Give the equations proposed for the dissolution process. Outline the two methods used to prepare the alkali metal solutions. What factors affect the stability of the solutions?

12.9 Alkali metal halides can be extracted from aqueous solution by solid-phase ditopic salt receptors (J. M. Mahoney, A. M. Beatty, and B. D. Smith, *Inorg. Chem.*, 2004, **43**, 7617). (a) What is a ditopic receptor? (b) What is the order of selectivity to extraction of the alkali metal ions in aqueous solution? (c) What is the order of selectivity to extraction from the solid phase? (d) Explain the observed order of selectivity.

12.10 The molecular geometries of crown ether derivatives play an important role in capturing and transporting alkali metal ions. K. Okano and co-workers (K. Okano, H. Tsukube, and K. Hori, *Tetrahedron*, 2004, **60**, 10877) studied stable conformations of 12-crown-O_3N and its Li^+ complex in aqueous and acetonitrile solutions. (a) Which three programs did the authors use in their study, and what did each program calculate? (b) Which Li^+ complex was found to be most stable in (i) aqueous and (ii) acetonitrile solutions?

13 The Group 2 elements

Part A: The essentials
13.1 The elements
13.2 Simple compounds
13.3 The anomalous properties of beryllium

Part B: The detail
13.4 Occurrence and extraction
13.5 Uses of the elements and their compounds
13.6 Hydrides
13.7 Halides
13.8 Oxides, sulfides, and hydroxides
13.9 Nitrides and carbides
13.10 Salts of oxoacids
13.11 Solubility, hydration, and beryllates
13.12 Coordination compounds
13.13 Organometallic compounds
13.14 Lower oxidation state Group 2 compounds

Further reading

Exercises

Tutorial problems

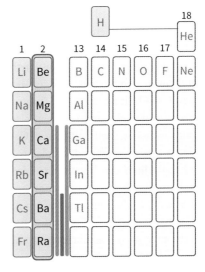

In this chapter we describe the occurrence and isolation of the Group 2 elements and study the chemical properties of their simple compounds, complexes, and organometallic compounds. Throughout the chapter comparisons are drawn with the elements of Group 1, and we show how the chemical properties of beryllium differ markedly from those of the other Group 2 elements. We discuss how the insolubility of some of the calcium compounds leads to the existence of many inorganic minerals that provide the raw materials for the infrastructure of our built environment and the basis of many rigid biological structures.

The elements calcium, strontium, barium, and radium are known as the **alkaline earth metals**, but the term is often applied to the whole of Group 2. All the elements are silvery white metals, and the bonding in their compounds is normally described in terms of the ionic model (Section 4.9). Some aspects of the chemical properties of beryllium are more like those of a metalloid with a degree of covalence in its bonding. The elements are denser, harder, and less reactive than the elements of Group 1 but are still more reactive than many typical metals. The lighter elements beryllium and magnesium form a number of complexes and organometallic compounds.

PART A: The essentials

In this first section of the chapter we summarize the key features of the chemistry of the Group 2 elements.

13.1 The elements

KEY POINT The most important factors influencing the chemical properties of the Group 2 elements are their ionization energies and ionic radii.

Beryllium occurs naturally as the semiprecious mineral beryl, $Be_3Al_2(SiO_3)_6$. Magnesium is the eighth most abundant element in the Earth's crust and the third most abundant element dissolved in seawater; it is commercially extracted from seawater and from the mineral dolomite, $CaCO_3 \cdot MgCO_3$. Calcium is the fifth most abundant element in the Earth's crust, but only the seventh most common in seawater because of the low solubility of $CaCO_3$; it occurs widely in its carbonate as limestone, marble, and chalk, and is a major component of biominerals, such as bones, shells, and coral. Calcium, strontium, and barium are all extracted by electrolysis of their molten chlorides. Radium can be extracted from uranium-bearing minerals, although all its isotopes are radioactive.

The greater mechanical hardness and higher melting points of the Group 2 compared with the Group 1 elements indicates an increase in the strength of metallic bonding on going from Group 1 to Group 2, which can be attributed to the increased number of electrons available (Section 2.11). The metallic and covalent radii of the Group 2 elements are smaller than those of Group 1 and demonstrate the trend that these radii increase down the Group (Table 13.1, Fig. 13.1).

The reduction in atom size between the neighbouring groups is responsible for their higher densities and ionization energies. The ionization energies of the elements decrease down the group as the radius increases (Fig. 13.2), and the elements become more reactive and more electropositive as it becomes easier to form the +2 ions. This decrease in ionization energy is reflected in the trend in standard potentials for the M^{2+}/M couples, which become more negative down the group.

Calcium, strontium, barium, and radium react readily with cold water and with increasing vigour; magnesium reacts only with hot water (it forms an insoluble passivating layer of the hydroxide with cold water):

$$M(s) + 2H_2O(l) \rightarrow M(OH)_2(aq) + H_2(g)$$

Beryllium does not react with water or steam beyond the formation of a passivating layer of oxide or hydroxide.

Beryllium and magnesium occur as hexagonal closed-packed structures. Calcium and strontium adopt cubic close-packed structures at room temperature, which undergo a phase transition (710 K, Ca, and 820 K, Sr) to the more open body-centred cubic structure before melting. Barium and radium also adopt the 8-coordinate body-centred cubic structure. The density decreases from Be to Mg to Ca (in contrast to the lighter Group 1 elements) as a result of strong metallic bonding in the Group 2 elements; this leads to very short metal–metal distances in the lightest elements (only 225 pm in beryllium metal, for instance, Table 13.1) and, as a result, small unit cells with higher densities despite the low atomic masses. The change in structure from 12-coordinate hcp (in Be and Mg) to 8-coordinate bcc (in Ca, Sr, Ba, and Ra at just below their melting points) probably accounts for the somewhat anomalous trend in melting points down the group; the increase between Mg and Ca is a result of the relatively stronger metallic bonding in the 8-coordinate bcc structure.

TABLE 13.1 Selected properties of the Group 2 elements and their divalent cations

	Be	Mg	Ca	Sr	Ba	Ra
Metallic radius/pm	114	160	197	214	218	223
Covalent radius / pm	96	141	176	195	215	221
Ionic radius, $r(M^{2+})$/pm (coordination number)	27(4)	72(6)	100(6)	126(8)	142(8)	170(12)
First ionization energy, I/kJ mol^{-1}	900	736	590	548	502	510
Second ionization energy, I/kJ mol^{-1}	1757	1451	1145	1064	965	979
Density, ρ/g cm^{-3}	1.85	1.74	1.54	2.62	3.51	5.00
Melting point/°C	1280	650	850	768	714	700
$E^{\ominus}(M^{2+},M)$/V	−1.85	−2.38	−2.87	−2.89	−2.90	−2.92
$\Delta_{hyd}H^{\ominus}(M^{2+})$/kJ mol^{-1}	−2500	−1920	−1650	−1480	−1360	–
$\Delta_{atm}H^{\ominus}$/kJ mol^{-1}	324	148	178	164	180	~130

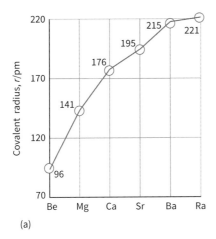
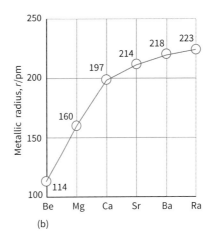

FIGURE 13.1 Trends in (a) covalent and (b) metallic radii for Group 2.

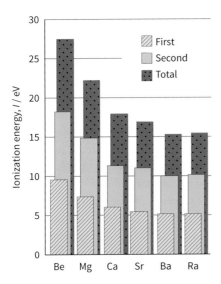

FIGURE 13.2 The variation of first, second, and total (first plus second) ionization energies in Group 2.

Beryllium is inert in air as its surface is passivated by the formation of a thin layer of BeO. Magnesium and calcium metals tarnish in air with the formation of an oxide layer, but will burn completely to their oxides and nitrides when heated. Strontium and barium, especially in powdered forms, ignite in air and are stored under hydrocarbon oils.

EXAMPLE 13.1 Determining how the enthalpy of a reaction informs about magnesium chemistry

Determine the enthalpy of reaction of magnesium metal with carbon dioxide, and comment on whether a CO_2 fire extinguisher could be used to extinguish a fire from magnesium burning in oxygen. The standard enthalpies of formation of MgO(s) and CO_2(g) are −602 and −394 kJ mol^{-1} respectively.

Answer For the reaction

$$2\,Mg(s) + CO_2(g) \rightarrow 2\,MgO(s) + C(s)$$

we can calculate $\Delta_{react}H = 2 \times (-602) - (-394) = -810$ kJ mol^{-1}, as magnesium and carbon are in their elemental standard states. As this reaction is highly exothermic, magnesium metal would continue to burn in a CO_2 atmosphere.

Self-test 13.1 Could water be used to extinguish a magnesium fire? Suggest which gases could be used to extinguish a magnesium fire.

As with the Group 1 elements (Section 12.1), flame tests are commonly used for the identification of the presence of the heavier Group 2 elements and their compounds:

Ca	Sr	Ba	Ra
Orange-red	Crimson	Yellowish-green	Deep red

Compounds of the Group 2 elements, particularly those of strontium and barium, are used to colour fireworks.

13.2 Simple compounds

KEY POINT The binary compounds of the Group 2 metals contain the cations of the elements and exhibit predominantly ionic bonding.

All the elements occur as M(II) in their simple compounds, which is consistent with their ns^2 valence-electron configuration. Apart from Be, their compounds are predominantly ionic. With the exception of beryllium, the Group 2 elements form ionic (saline) **hydrides** containing the hydride anion, H$^-$. In contrast, beryllium hydride adopts a three-dimensional network of covalently linked BeH$_4$ tetrahedra. Magnesium hydride, MgH$_2$, loses hydrogen when heated above 250°C and is being studied as a hydrogen storage material (Box 13.4 and Section 30.8). The hydrides react with water to produce hydrogen gas.

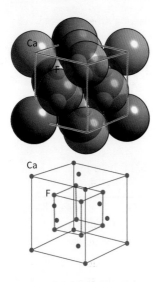

FIGURE 13.3 The fluorite structure adopted by CaF_2, SrF_2, BaF_2, and $SrCl_2$.

All the elements form **halides**, MX_2, by direct combination of the elements. The halides of the elements other than Be, however, are normally formed from solution, such as by reaction of the metal hydroxide or carbonate with HX(aq), X=Cl, Br, I, followed by dehydration of the resulting hydrous salt. The fluorides of the larger cations (from Ca to Ba) adopt the 8:4-coordinate fluorite structure (Fig. 13.3), while MgF_2 crystallizes with a rutile structure with 6:3 coordination for the smaller Mg^{2+} ion. The beryllium halides form covalently bonded networks of edge- or corner-linked tetrahedra.

EXAMPLE 13.2 Predicting Lewis acid strengths of Group 2 halides

Which of the Group 2 halides would you expect to be the strongest Lewis acid?

Answer Lewis acids are generally molecules that accept a pair of electrons from a Lewis base (Section 5.6). The bonding in beryllium halides is predominantly covalent in nature (compared with Mg–Ba where it is predominantly ionic), and the electron density on beryllium can be decreased (making it a stronger Lewis acid) through bonding with the most electronegative element, fluorine. Therefore, BeF_2 would be expected to be the strongest Lewis acid, and this is exemplified by its chemistry in reactions where it forms adducts with amines and ethers (Lewis bases).

Self-test 13.2 Predict the reaction product when $BeCl_2$ is dissolved in diethyl ether.

Beryllium oxide, BeO, is a white, insoluble solid adopting the wurtzite structure with 4:4 coordination, as expected for the small Be^{2+} ion; the oxides of the other Group 2 elements all adopt the rock-salt structure with 6:6 coordination. The strongly insulating properties of BeO and its high thermal conductivity resulting from the interlinked, short Be-O bonds leads to its use in electronics as a heat-dissipating spacer in integrated circuits. Magnesium oxide is insoluble but reacts slowly with water to form $Mg(OH)_2$; likewise CaO reacts with water to form the partially soluble $Ca(OH)_2$. The **oxides** of Sr and Ba, SrO and BaO, dissolve in water to form strongly basic hydroxide solutions:

$$BaO(s) + H_2O(l) \rightarrow Ba^{2+}(aq) + 2OH^-(aq)$$

Magnesium hydroxide, $Mg(OH)_2$, is basic but only very sparingly soluble; beryllium hydroxide, $Be(OH)_2$, is amphoteric and in strongly basic solutions it forms the tetrahydroxyberyllate ion, $[Be(OH)_4]^{2-}$:

$$Be(OH)_2(s) + 2OH^-(aq) \rightarrow [Be(OH)_4]^{2-}(aq)$$

which has been isolated in the salt $SrBe(OH)_4$.

The sulfides can be prepared by direct reaction of the elements, and adopt the rock-salt structure for all the elements except Be, which has a sphalerite structure (Fig. 4.35). Beryllium carbide, Be_2C, has an antifluorite structure formally containing Be^{2+} and C^{4-} (methide) ions. The carbides of the other members of the group have the formula MC_2 and contain the dicarbide (acetylide) anion, C_2^{2-}; they react with water to generate ethyne (acetylene), C_2H_2. The elements from Mg to Ra react directly with nitrogen when heated to produce the nitrides, M_3N_2, which react with water to produce ammonia.

With the exception of the fluorides, the salts of singly charged anions are usually soluble in water, although beryllium salts—on account of the highly polarizing nature of the Be^{2+} ion—often hydrolyse in aqueous solutions with the formation of $[Be(OH_2)_3(OH)]^+$ and H_3O^+. The radium halides are the least soluble of the group halides, and this property is used to extract radium by using fractional crystallization. In general, the salts of the Group 2 elements are much less soluble in water than those of Group 1 on account of the higher lattice enthalpies of solid structures containing doubly charged cations. This is especially true when they are combined with more highly charged divalent or trivalent anions, such as carbonate, sulfate, or phosphate. The Group 2 salts of these anions are insoluble or only sparingly soluble.

The carbonates and sulfates of the Group 2 elements have important roles in natural water systems and rock formation (for example, limestone and chalk), and as materials for forming hard structures. The carbonates and sulfates are insoluble, as a result of the high lattice energy of structure formed from 2+ and 2– ions. The solubility of calcium carbonate increases if CO_2 is dissolved in the water, as in rainwater, owing to the formation of HCO_3^- with its lower charge. 'Temporary hardness' of water is caused by the presence of magnesium and calcium hydrogencarbonates; the cations are precipitated as carbonates on boiling solutions containing the hydrogencarbonates.

Calcium carbonate is widely used by living organisms in the construction of hard structural biomaterials such as shells, bones, and teeth (Section 26.18). Limestone, which forms from the skeletal remains of such organisms, is a rock extensively used in construction, and famous buildings that contain limestone include the Great Pyramid, the Tower of London, and the Empire State Building.

When heated, an alkaline earth carbonate decomposes to the oxide, although for Sr and Ba this decomposition process requires temperatures above 800°C. Calcium sulfate is used extensively in the construction industry (plaster) and occurs naturally as gypsum, which is the dihydrate, $CaSO_4 \cdot 2H_2O$.

The Group 2 cations form complexes with charged polydentate ligands, such as the analytically important ethylenediaminetetraacetate ion (edta^{4-}; see Table 7.1) and crown and crypt ligands. The most important macrocyclic complexes are the chlorophylls, which are porphyrin complexes of Mg and are involved in photosynthesis; Sections 26.3 and 26.10(d).

Beryllium forms an extensive series of organometallic compounds. Alkyl- and arylmagnesium halides are very well known as Grignard reagents and are widely used in synthetic organic chemistry, where they behave as a source of alkyl and aryl anions. In organometallic compounds, beryllium has been obtained in the zero oxidation state, and magnesium as Mg(I).

13.3 The anomalous properties of beryllium

KEY POINTS The compounds of beryllium show a high level of covalency, and beryllium shows a strong diagonal relationship with aluminium.

Section 10.2 described the anomalous chemistry frequently observed for the lightest member of a group in the periodic table. The small size of Be^{2+} (ionic radius 27 pm) and its consequent high charge density and polarizing power results in the compounds of Be being largely covalent; the ion is a strong Lewis acid. The coordination number most commonly observed for this small atom is 4 and the local geometry tetrahedral. Group 2 elements heavier than beryllium typically have coordination numbers of 6 or more. Some consequences of these properties of the lightest Group 2 element are:

- A significant covalent contribution to the bonding in compounds such as the beryllium halides $BeCl_2$, $BeBr_2$, and BeI_2 and the hydride, BeH_2. BeF_2 and $BeCl_2$ are strong Lewis acids.
- A greater tendency to form complexes, with the formation of molecular compounds including $Be_4O(O_2CCH_3)_6$.
- Hydrolysis (deprotonation) of beryllium salts in aqueous solution, forming species such as $[Be(OH_2)_3OH]^+$ and acidic solutions. Hydrated beryllium salts tend to decompose by hydrolysis reactions, in which beryllium oxo- or hydroxo-salts are formed, rather than by the simple loss of water.
- The oxide and other chalcogenides of Be adopt structures with the more directional 4:4-coordination structures.
- Beryllium forms many stable organometallic compounds, including methylberyllium ($Be(CH_3)_2$), ethylberyllium, t-butylberyllium, and beryllocene (($C_5H_5)_2Be$).

Another important general feature of Be is its strong diagonal relationship with Al, as described in Section 10.3. These commonalities in chemistry include the amphoteric behaviours of BeO and Al_2O_3 (Section 9.5) and the formation of covalent hydrides and halides by both elements. There are also analogies between the chemical properties of Be and Zn (Section 10.3); for example, both metals form tetrahedral $[MO_4]^{2-}$ anions on dissolution in strongly basic solutions.

PART B: The detail

In this section we present a more detailed discussion of the chemistry of the elements of Group 2 and their compounds. Because the bonding in the compounds formed by these elements is usually ionic (as always, bearing in mind the individuality of Be) we can usually interpret many of their properties in terms of the ionic model.

13.4 Occurrence and extraction

KEY POINTS Magnesium is the only Group 2 element extracted on an industrial scale; magnesium, calcium, strontium, and barium can be obtained by electrolysis of their molten chlorides.

Beryllium occurs naturally as the semiprecious mineral beryl, $Be_3Al_2(SiO_3)_6$, from which its name is taken. Beryl is the basis of several gemstones, including emerald, in which a small fraction of Al^{3+} is replaced by Cr^{3+}, yielding the green colour, aquamarine (pale blue green), heliodor (yellow), and morganite (pale pink-brown). The other main ore of beryllium is betrandite, $Be_4Si_2O_7(OH)_2$. Beryllium is extracted by heating beryl with sodium hexafluorosilicate, Na_2SiF_6, to produce BeF_2, which is then reduced to the element by magnesium.

Magnesium is the eighth most abundant element in the Earth's crust. It occurs naturally in a number of minerals such as dolomite, $CaCO_3 \cdot MgCO_3$, and magnesite, $MgCO_3$, and is the third most abundant element dissolved in seawater (after Na and Cl), from which it is commercially extracted. A litre of seawater contains more than 1 g of magnesium ions. The extraction from seawater relies on the fact that magnesium hydroxide is less soluble than calcium hydroxide because the solubility of the salts of mononegative anions increases down the group (Section 13.11). Either CaO (quicklime) or $Ca(OH)_2$ (slaked lime) is added to seawater and $Mg(OH)_2$ precipitates. The hydroxide is converted to the chloride by treatment with hydrochloric acid:

$$CaO(s) + H_2O(l) \rightarrow Ca^{2+}(aq) + 2OH^-(aq)$$
$$Mg^{2+}(aq) + 2OH^-(aq) \rightarrow Mg(OH)_2(s)$$
$$Mg(OH)_2(s) + 2HCl(aq) \rightarrow MgCl_2(aq) + 2H_2O(l)$$

The magnesium is then extracted by electrolysis of molten magnesium chloride.

Cathode: $Mg^{2+}(l) + 2e^- \rightarrow Mg(s)$
Anode: $2Cl^-(l) \rightarrow Cl_2(g) + 2e^-$

Magnesium is also extracted from dolomite, a mixed metal carbonate, $MgCO_3 \cdot CaCO_3$, which is initially heated in air to produce magnesium and calcium oxides. This mixture is heated with ferrosilicon (FeSi), which forms calcium silicate (Ca_2SiO_4), iron, and magnesium. Magnesium is a liquid at the high operating temperatures used in the process, and can be removed by distillation.

A major problem in the production of magnesium is its high reactivity towards water, oxygen, and moist air. Nitrogen, which is commonly used to provide an inert atmosphere for the production of many other reactive metals, cannot be used for magnesium because it reacts to form the nitride, Mg_3N_2. Sulfur hexafluoride or sulfur dioxide added to dry air are used as alternatives to nitrogen, as these gases inhibit the formation of MgO. Although hot and liquid magnesium are very reactive towards oxygen and water, the metal can be handled safely under ambient conditions on account of the presence of an inert passivating oxide film on its surface.

Calcium is the fifth most abundant element in the Earth's crust and occurs widely as limestone, $CaCO_3$. The name 'calcium' comes from the Latin *calx*, which means 'lime'. Calcium concentrations in seawater are lower than those of magnesium because of the lower solubility of $CaCO_3$ compared with $MgCO_3$ and the greater use of calcium by marine organisms. The element is a major component of biominerals such as bone, shells, and teeth (Section 26.18), and is central to cell signalling processes including hormonal functions and electrical activation of enzymes in higher organisms (Section 26.5). The average adult human contains approximately 1 kg of calcium. Calcium binds strongly to oxalate ions to form insoluble $Ca(C_2O_4)$; kidney stones are formed when this reaction occurs in the kidneys.

Calcium is extracted by electrolysis of the molten chloride, which is itself obtained as a by-product of the Solvay process for the production of sodium carbonate (Section 12.11). Calcium tarnishes in air and ignites on heating with the formation of calcium oxide and nitride.

Strontium is named after the Scottish village of Strontian, where the strontium-containing ores celestine, $SrSO_4$, and strontianite, $SrCO_3$, were first found. It is extracted by electrolysis of molten $SrCl_2$ or by reduction of SrO with Al:

$$6SrO(s) + 2Al(s) \rightarrow 3Sr(s) + Sr_3Al_2O_6(s)$$

The metal reacts vigorously with water, and as a finely divided powder it ignites in air; initially the product is SrO but, once burning, the nitride, Sr_3N_2, is also formed. Barium is extracted by electrolysis of the molten chloride or by reduction of BaO with Al. It reacts very vigorously with water and ignites readily in air.

All the isotopes of radium are radioactive. They undergo α, β, and γ decay with half-lives that vary from 42 minutes to 1599 years. Radium was discovered by Marie and Pierre Curie in 1898 after painstaking extraction from the uranium-bearing mineral pitchblende. Pitchblende is a complex mineral containing many elements: it contains approximately 1 g of Ra in 10 t of ore, and it took the Curies three years to isolate 0.1 g of $RaCl_2$.

13.5 Uses of the elements and their compounds

KEY POINTS Magnesium and its compounds have major applications in pyrotechnics, alloys, and common medicines; calcium compounds are widely used in the construction industry, and magnesium and calcium are very important for biological function.

Beryllium is unreactive in air on account of a passivating layer of an inert oxide film on its surface, which makes it very resistant to corrosion. This inertness, combined with the fact that it is one of the lightest metals, results in its use in alloys to make precision instruments, aircraft, and missiles. The James Webb Space Telescope has eighteen lightweight beryllium mirrors. Beryllium is highly transparent to X-rays because of its low atomic number (and thus electron count), and is used for X-ray tube windows. An alloy of beryllium with copper and aluminium has excellent fatigue or failure resistance to spring and stressed functions; applications include automobile suspensions, electromechanical devices, and the springs in computer keyboards and printers. Beryllium is also used as a moderator for nuclear reactions (where it slows down fast-moving neutrons through inelastic collisions) because the beryllium nucleus is a very weak absorber of neutrons and the metal has a high melting point.

Most of the applications of elemental magnesium are based on the formation of light alloys, especially with aluminium,

> **BOX 13.1** How are magnesium alloys used in applications that require lightweight materials?
>
> Magnesium alloys are the lightest alloys used for structural purposes with densities of around $1.8\,\text{g cm}^{-3}$ compared with $2.7\,\text{g cm}^{-3}$ for aluminium, $4.5\,\text{g cm}^{-3}$ for titanium, and $7.7\,\text{g cm}^{-3}$ for steels. Mg-based alloys demonstrate a high strength-to-weight ratio leading to their use in aerospace, portable electronic devices, automotive parts (particularly in electric vehicles, EVs) and other applications where a strong, lightweight material is required. Pure magnesium metal is reactive and brittle and is hard to form into structural objects, hence the use of Mg alloys which overcome these deficiencies. Mg-based alloys can be cast, where the alloy is poured into a mould, or wrought, where the alloy is subjected to mechanical working to produce wires, tubes, or sheets.
>
> Magnesium–aluminium alloys are the most widely used alloys of this type, as the addition of Al results in a high strength and hard material suitable for casting into structural objects.
>
> Addition of manganese to Mg improves its resistance to salt water corrosion, and Mg–Be alloys are less prone to surface oxidation during casting and welding processes. Recently, the addition of just 0.1% of Ca to magnesium has been found to markedly reduce corrosion by saltwater.
>
> Specific applications of magnesium alloys include the cases of mobile phones and laptops, lightweight luggage, and high-speed fabrication machinery used, for example, in the textile industry. Many aerospace and rocket parts are now made from Mg alloys; for example, aeroplane thrust reversers and transmission cases. With the growth in numbers of EVs there is a need to keep the weight of the vehicle as low as possible to improve vehicle range. Mg alloys have been used in steering column brackets, supports for brakes and clutches, instrument panels, and wheels.

that are widely used in construction where weight is an issue; for example, in the aerospace industry (Box 13.1).

As beryllium oxide is extremely toxic and carcinogenic by inhalation and soluble beryllium salts are mildly poisonous, the industrial applications of beryllium compounds are limited. BeO is used in high-power electronic devices ranging in size from the most powerful nanoscale, electronic chips to large-scale power distribution, and communication structures. The thermal conductivity of BeO of $300\,\text{W m}^{-1}\,\text{K}^{-1}$ (similar to aluminium metal) is only second to that of diamond—a property that derives from the short Be-O bond length, 165 pm, and the highly interconnected tetrahedral coordination of beryllium in its wurtzite structure (Section 4.9). As the size of silicon semiconductor nodes has fallen towards 10 nm in the 2020s, the use of BeO as the insulator and heat dissipater between these nodes (and with other metal components in electronics) has increased. The low dielectric constant of BeO, 6.7, permits its use in high-frequency circuitry. Grid-scale power distribution—for example, transformers and telecommunication radiofrequency transmitters—also require an insulating material with a high thermal conductivity and low dielectric constant, and BeO is again the material of choice for these applications.

Various medical applications of magnesium compounds include 'Milk of Magnesia', $Mg(OH)_2$, which is a common remedy for indigestion, and 'Epsom Salts', $MgSO_4 \cdot 7H_2O$, which is used for a variety of health treatments, including as a treatment for constipation, a purgative, and a soak for sprains and bruises. Magnesium oxide, MgO, is used as a refractory lining for furnaces. Sorel cement is a mixture of $MgCl_2 \cdot nH_2O$ and MgO that has been used in hard artificial stones, and originally as an ivory substitute. Organomagnesium compounds are widely used in organic synthesis, and are known as Grignard reagents (Section 13.13).

The compounds of calcium are much more useful than the element itself. Calcium oxide (as lime or quicklime) is a major component of mortar and cement (Box 13.2). It is also employed in steel manufacture and paper-making. Calcium sulfate dihydrate, $CaSO_4 \cdot 2H_2O$, is widely used in building materials, including plasterboard, and anhydrous $CaSO_4$ is a common drying agent. Calcium carbonate is used in the Solvay process (Section 12.11) for the production of sodium carbonate (except in the United States, where sodium carbonate occurs naturally and is mined as trona) and as the raw material for production of CaO. Calcium fluoride is insoluble and transparent over a wide range of wavelengths; it is used to make cells and windows for infrared and ultraviolet spectrometers.

Strontium is used in pyrotechnics (Box 13.3) and in phosphors; for example, $SrAl_2O_4$ doped with Eu^{2+} or Dy^{3+}, Section 30.1. Insoluble barium compounds, taking advantage of the large number of electrons of each Ba^{2+} ion, are very effective at absorbing X-rays: they have been used as 'barium meals' and 'barium enemas' to investigate the intestinal tract. Barium is highly toxic, so the totally insoluble sulfate is used in this application. Barium carbonate is used in glass-making and as a flux to aid the flow of glazes and enamels. It is also used as rat poison. The sulfide has been used as a depilatory to remove unwanted body hair. Barium sulfate is colourless, with no absorption in the visible region of the electromagnetic spectrum, and it is used as a reference standard in UV–visible spectroscopy (Section 8.3).

Soon after its discovery, radium was used to treat malignant tumours; its compounds are still used as precursors for radon used in similar applications. Luminous radium paint was once widely used on clock and watch faces but has been replaced by less hazardous phosphorescent compounds.

BOX 13.2 Which calcium compounds are formed in cement and concrete and how can concrete repair itself?

Cement is made by grinding together limestone and a source of aluminosilicates, such as clay, shale, or sand, and then heating the mixture to 1500°C in a rotary cement kiln. The first important reaction to occur in the lower-temperature portion of the kiln (900°C) is the calcining of limestone (the process of heating to a high temperature to oxidize or decompose a substance and convert it to a powder), when calcium carbonate (limestone) decomposes to calcium oxide (lime) and carbon dioxide is driven off. At higher temperatures the calcium oxide reacts with the aluminosilicates and silicates to form molten Ca_2SiO_4, Ca_3SiO_5, and $Ca_3Al_2O_6$. The relative proportions of these compounds determine the properties of the final cement. As the compounds cool, they solidify into a form called clinker. The clinker is ground to a fine powder, and a small amount of calcium sulfate (gypsum) is added to form Portland cement.

Concrete is produced by mixing cement with sand, gravel, or crushed stone and water. Often small amounts of additives are added to achieve particular properties. For example, flow and dispersion are improved by adding polymeric materials such as phenolic resins, and resistance to frost damage is improved by adding surfactants. When the water is added to the cement, complex hydration reactions occur, producing the hydrated phases $Ca_3Si_2O_7 \cdot H_2O$, $Ca_3Si_2O_7 \cdot 3H_2O$, and $Ca(OH)_2$:

$$2Ca_2SiO_4(s) + 2H_2O(l) \rightarrow Ca_3Si_2O_7 \cdot H_2O(s) + Ca(OH)_2(aq)$$

$$2Ca_2SiO_4(s) + 4H_2O(l) \rightarrow Ca_3Si_2O_7 \cdot 3H_2O(s) + Ca(OH)_2(aq)$$

These hydrates form a gel or slurry that coats the surfaces of the sand or aggregates and fills the voids to form the solid concrete. The properties of concrete are determined by the relative proportions of calcium silicates and calcium aluminosilicates in the cement used, the additives, and the amount of water, which determines the degree of hydration.

The raw materials for cement manufacture often contain traces of sodium and potassium sulfates, and sodium and potassium hydroxides are formed during the hydration process. These hydroxides are responsible for the cracking, swelling, and distortion of many ageing concrete structures. The hydroxides take part in a complex series of reactions with the aggregate material to form an alkali silicate gel. This gel is hygroscopic and expands as it absorbs water, producing stress in the concrete, which leads to cracking and deformation. The susceptibility of concrete to this 'alkali silicate reaction' is now monitored by calculating the total alkali levels in the concrete produced, and strategies are in place to minimize its effects. For example, adding 'fly ash' (a waste product from coal-fired power stations, containing SiO_2, Al_2O_3, and CaO) to the mix can reduce the problem.

Concrete has a low tensile strength in comparison with other building materials, and it can break if it is stressed through stretching or pulling. This results in the formation of cracks in concrete, especially over time in high-stress applications; for example, vehicle bridges and skyscrapers. Once cracks form in concrete it is susceptible to further degradation as water and air borne pollutants enter the structure; any reinforcement inside the concrete in the form of steel bars may also rust.

Many concretes are self-healing as a result of their innate chemical composition. Levels of clinker remain in the concrete after the hydration reactions, and water that enters through cracks will react with this, forming new calcium silicate hydrates that will seal a small crack. In moist air the presence of any lime, CaO, remaining from the clinker formation will react to form $Ca(OH)_2$ and then $CaCO_3$, which also helps fill and seal cracks formed in concrete. These self-healing processes can be enhanced by the addition of encapsulated compounds to the concrete; both microscopic (capsules under 1mm) and macroscopic encapsulation methods have been used. These polymer capsules can contain a variety of reagents ranging from organic glues and bacterial spores to nanoparticulate silica and sodium silicate. As a crack develops, these capsules come in contact with air and water and release their contents which directly fill the pore, or undergo a chemical reaction, forming a material to fill the pore. Calcium carbonate can also be deposited by bacteria (see Section 26.18), and incorporation of carbonatogenesic microbes in the concrete leads to self-healing concretes that also remove CO_2 from the air.

Magnesium and calcium are of great biological importance. Magnesium is a component of chlorophyll, but is also coordinated by many other biologically important ligands, including ATP (adenosine triphosphate, Section 26.3). It is essential for human health, being responsible for the activity of many enzymes. The recommended adult human dose is approximately 0.3 g per day, and the average adult contains about 25 g of magnesium. The bioinorganic chemistry of calcium is discussed in detail in Section 26.5.

13.6 Hydrides

KEY POINT All the Group 2 elements form ionic hydrides with the exception of beryllium, which forms a polymeric covalent compound.

Like the Group 1 elements, the Group 2 elements, with the exception of Be, form ionic, saline hydrides that contain the H^- ion. They can be prepared by direct reaction between the metal and hydrogen. Beryllium hydride is covalent and must be prepared from alkylberyllium (Section 13.13).

BOX 13.3 What produces the colour in fireworks and flares?

Fireworks use exothermic reactions to produce heat, light, and sound. Common oxidants are nitrates and perchlorates, which decompose when heated to liberate oxygen. Common fuels are carbon, sulfur, powdered aluminium or magnesium, and organic materials, such as poly(vinyl chloride) (PVC), starch, and gums (which also bind the mixture together). The most common constituent of fireworks is gunpowder or black powder, a mixture of potassium nitrate, sulfur, and charcoal combining an oxidant and fuels. Special effects, such as colours, flashes, smoke, and noises are provided by additives to the firework mixture. The Group 2 elements are used in fireworks to provide colour.

Barium compounds are added to fireworks to produce green flames. The species responsible for the colour is $BaCl^+$, which is produced when Ba^{2+} ions combine with Cl^- ions. The Cl^- ions are produced during decomposition of the perchlorate oxidant or during combustion of the PVC fuel:

$KClO_4(s) \rightarrow KCl(s) + 2O_2(g)$
$KCl(s) \rightarrow K^+(g) + Cl^-(g)$
$Ba^{2+}(g) + Cl^-(g) \rightarrow BaCl^+(g)$

Barium chlorate, $Ba(ClO_3)_2$, has been used instead of $KClO_4$ and a separate barium compound, but is very unstable to shock and friction. Similarly, strontium nitrate and carbonate are used to produce a red colour on formation of $SrCl^+$. Strontium chlorate and perchlorate are effective at producing the red colour, but for routine use they are too unstable to shock and friction.

Distress flares also use strontium compounds. Strontium nitrate is mixed with sawdust, waxes, sulfur, and $KClO_4$ and packed into a waterproof tube. When ignited, the flares burn with an intense red flame for up to 30 minutes.

As well as being used as a fuel, powdered magnesium is added to fireworks and flares to maximize light output. Burning the magnesium produces an intense white light, and the illumination is increased by the incandescence from high-temperature MgO particles that are produced in the oxidation reaction.

It has a three-dimensional network structure with bridging H atoms (Fig. 13.4); the long-held view that it is a linear chain is incorrect.

The ionic hydrides of the heavier elements react violently with water to produce hydrogen:

$MgH_2(s) + 2H_2O(l) \rightarrow Mg(OH)_2(s) + 2H_2(g)$

This reaction is not as violent as that for the Group 1 elements, and can be used as a source of hydrogen in fuel cells. For hydrogen storage, a reversible reaction involving uptake of hydrogen near room temperature is needed. Magnesium hydride loses hydrogen on heating above 250°C so the process

$Mg(s) + H_2(g) \rightarrow MgH_2(s)$

is reversible, and the low molar mass of Mg (24.3 g mol^{-1}) makes MgH_2 potentially an excellent hydrogen storage material. Attempts to reduce the decomposition temperature to closer to room temperature involve making complex magnesium hydrides doped with other metals, fabricating nanoparticulate forms (Chapter 28), and forming molecular complexes with magnesium hydride cores (Box 13.4).

CaH_2, SrH_2, and BaH_2 adopt the $PbCl_2$ structure-type with nine-fold coordination of the Group 2 cation to hydride anions. Calcium hydride is used as a desiccant for amine and halocarbon solvents, removing water through the formation of $Ca(OH)_2$ and hydrogen gas. This reaction with water is also a convenient source of hydrogen gas used for inflating weather balloons and life rafts.

13.7 Halides

KEY POINTS The halides of beryllium are covalent; all the fluorides, except BeF_2, are insoluble in water; all other halides are soluble.

All the beryllium halides are covalent. Beryllium fluoride is prepared from the thermal decomposition of $(NH_4)_2BeF_4$ and is a glassy solid that exists in several temperature-dependent phases similar to those of SiO_2 (Section 14.10). It is soluble in water, forming the hydrated ion $[Be(OH_2)_4]^{2+}$ under acidic conditions and hydrolysed species, e.g. $[BeOH(OH_2)_3]^+$, at higher pH. Beryllium chloride, $BeCl_2$, can be made from the oxide:

$BeO(s) + C(s) + Cl_2(g) \rightarrow BeCl_2(s) + CO(g)$

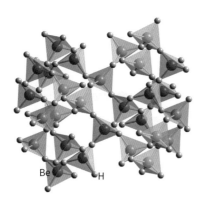

FIGURE 13.4 The structure of BeH_2.

BOX 13.4 How and why is MgH$_2$ used as a hydrogen storage material?

The development of materials that can reversibly store hydrogen is of considerable importance for practical transport applications (Box 11.5). MgH$_2$ contains 7.7 weight percent hydrogen and displays reversible hydrogen release and uptake above 250°C, but with slow kinetics:

$$MgH_2(s) \rightleftharpoons Mg(s) + H_2(g)$$

The kinetics can be greatly improved by doping MgH$_2$ with transition metals, particularly Ti as TiH$_2$, by ball milling the solid to decrease particle size, or through treatment with fluoride-containing solutions. Consideration of the thermodynamics of the MgH$_2$ decomposition reaction show a positive enthalpy for this reaction of 74.4 kJ mol^{-1}, which derives from the very high lattice enthalpy of MgH$_2$(s) ($\Delta_L H = 2718$ kJ mol^{-1}) relative to atomization of Mg(s) ($\Delta_{atm} H = 147$ kJ mol^{-1}). While the entropy change for hydrogen release is strongly favourable ($\Delta S = 135$ J mol^{-1} K^{-1}), these values mean that bulk MgH$_2$ will only decompose at significantly above room temperature. Consequently, bulk MgH$_2$ cannot be used for the reversible storage of hydrogen under, or near, ambient conditions.

Theoretical calculations demonstrate that for subnanometer-sized solid MgH$_2$ or (MgH$_2$)$_n$ clusters ($n < 20$), the enthalpy of decomposition decreases sharply because of the reduced lattice enthalpy of a high-surface-area particle; surface ions have low coordination numbers and thus a reduced contribution to the lattice enthalpy in the Born–Mayer equation (Section 4.12). A decomposition temperature of about 200°C has been estimated for very small clusters of (MgH$_2$)$_n$. It has been found experimentally that nanoparticles of MgH$_2$ with sizes near 1–10 nm demonstrate a small reduction in H$_2$ evolution temperatures compared with the bulk material, where crystallite sizes are of the order of 1 μm.

An alternative approach to the production of 'magnesium hydride' particles in the subnanometre range involves a 'bottom-up' approach in which a molecule with a central unit consisting of magnesium and hydride ions is synthesized, with the aim that hydrogen will be reversibly released from this subnanometre-sized unit at, or near to, room temperature. Complexes with a variety of magnesium hydride cores of the type [Mg$_n$H$_{2n-x}$]$^{x+}$, including [Mg$_4$H$_4$]$^{4+}$, [Mg$_8$H$_{10}$]$^{6+}$, and [Mg$_{13}$H$_{18}$]$^{8+}$ (**B1**), show decomposition temperatures near and below 200°C—considerably lower temperatures than solid, crystalline MgH$_2$. (D. Mukherjee and J. Okuda, *Molecular magnesium hydrides*, Angew. Chem. Int. Ed., 2018, **57**, 1458.)

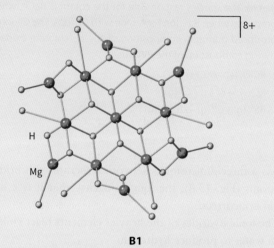

B1

The [Mg$_{13}$H$_{18}$]$^{8+}$ core of exakis(μ3-Hydrido)-hexakis(1,4,7-trimethyl-1,4,7,10-tetra-azacyclododecane)-dodecakis(μ^2-hydrido)-trideca-magnesium bis(tetrakis(3,5-bis(trifluoromethyl)phenyl)borate) n-pentane tetrahydrofuran solvate.

The chloride, as well as BeBr$_2$ and BeI$_2$, can also be prepared from the direct reaction of the elements at elevated temperatures.

The structure of solid BeCl$_2$ is a polymeric chain (**1**).

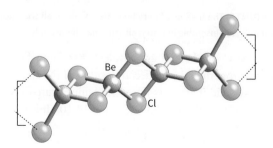

1 Solid phase (BeCl$_2$)$_n$

The local structure is an almost regular tetrahedron around the Be atom, and the bonding can be considered to be based on sp^3 hybridization. In BeCl$_2$ the chloride ion has sufficient valence electrons for 2c,2e covalent bonding to take place. Beryllium fluoride and chloride are Lewis acids, readily forming adducts with electron-pair donors such as diethyl ether (**2**).

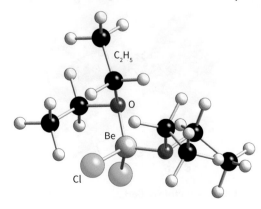

2 BeCl$_2$(O(C$_2$H$_5$)$_2$)$_2$

In the vapour phase the compound tends to form a dimer based on sp² hydridization (3), and when the temperature is above 900°C linear monomers are formed, consistent with sp hybridization (4).

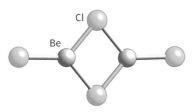

3 Vapour phase (BeCl$_2$)$_2$, T < ~900°C

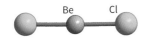

4 Vapour phase BeCl$_2$, T > ~900°C

The anhydrous halides of magnesium can be prepared by the direct combination of the elements or, for strontium and barium, dehydration of the hydrated halide salt; preparation from aqueous solutions yields the hydrated salts, which, for the lighter elements, are partially hydrolysed on heating. All the fluorides except BeF$_2$ are sparingly soluble, although the solubility increases slightly down the group. As the radius of the cation increases from Be to Ba, the coordination number in the fluorides increases from 4 to 8; BeF$_2$ forms structures analogous to those of SiO$_2$ (such as quartz, 4:2 coordination), MgF$_2$ adopts the rutile structure (6:3), and CaF$_2$, SrF$_2$, and BaF$_2$ adopt the fluorite structure (8:4) (Section 4.9). The other halides of Group 2 form layer structures, reflecting the increasing polarizability of the halide ions. Magnesium chloride adopts the CdCl$_2$ layered structure, in which the layers are arranged so that the Cl⁻ ions are cubic close-packed (Fig. 13.5). Both MgI$_2$ and CaI$_2$ adopt the closely related cadmium-iodide structure in which the layers of I⁻ ions are hexagonal close-packed (Section 4.9, Figure 4.41).

The most important fluoride of the group is CaF$_2$. Its mineral form, fluorite or fluorspar, is the only large-scale source of fluorine. Anhydrous hydrogen fluoride is prepared by the action of concentrated sulfuric acid on fluorspar:

$$CaF_2(s) + H_2SO_4(l) \rightarrow CaSO_4(s) + 2HF(l)$$

All the chlorides are deliquescent and form hydrates; they have lower melting points than the fluorides. Magnesium chloride is the most important chloride for industry and applications. It is extracted from seawater and then used in the production of magnesium metal. Calcium chloride is produced on a massive scale industrially. Its hygroscopic character leads to its widespread use as a laboratory drying agent. Both MgCl$_2$ and CaCl$_2$ are used to deice roads, where they are more effective than NaCl for two reasons. First, the dissolution is very exothermic:

$$CaCl_2(s) \rightarrow Ca^{2+}(aq) + 2Cl^-(aq) \quad \Delta_{sol}H^\ominus = -82\,\mathrm{kJ\,mol^{-1}}$$

The heat generated helps to melt the ice. Second, the minimum freezing mixture of CaCl$_2$ in water has a freezing point of −55°C, compared to −18°C for that of NaCl in water. Calcium and magnesium chlorides are also less toxic than NaCl to plant life surrounding roads, and are less corrosive to iron and steel. The exothermic dissolution of the chlorides also leads to applications in instant-heating packs and self-heating drink containers.

Radium has the least soluble halides as a result of the low hydration enthalpy of the large Ra^{2+} ion; the solubilities of RaCl$_2$ and BaCl$_2$ at 20°C are ~200 g dm^{-3} and ~350 g dm^{-3}, respectively. This property is used to separate Ra^{2+} from Ba^{2+} by fractional crystallization of the chloride or bromide.

EXAMPLE 13.3 Predicting the nature of the halides

Use the data in Table 1.8 and Fig. 2.40 to predict whether CaF$_2$ is predominantly ionic or covalent.

Answer One approach is to identify the electronegativities of the two elements in the compound and then to refer to the Ketelaar triangle (Section 2.14) to judge the type of bonding present. The Pauling electronegativity values of Ca and F are 1.00 and 3.98, respectively. The average electronegativity is therefore 2.49 and the difference is 2.98. These values on the Ketelaar triangle in Fig. 2.40 indicate that CaF$_2$ should be ionic.

Self-test 13.3 Predict whether (a) BeCl$_2$ and (b) BaF$_2$ are predominantly ionic or covalent. Discuss the structures adopted by these two compounds in reference to the prediction made.

13.8 Oxides, sulfides, and hydroxides

The Group 2 elements react with O$_2$ to form the oxides, [M^{2+}][O^{2-}]. All the elements except Be also form unstable peroxides. The oxides of Mg to Ra react with water to form the basic hydroxides; BeO and Be(OH)$_2$ are amphoteric.

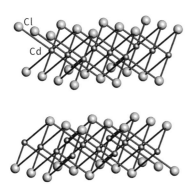

FIGURE 13.5 The CdCl$_2$ structure adopted by MgCl$_2$.

(a) Oxides, peroxides, and complex oxides

KEY POINTS All the Group 2 elements form normal oxides by reaction with oxygen, except Ba, which forms the peroxide; all the peroxides decompose to the oxides, their stabilities increasing down the group.

Beryllium oxide is obtained by ignition of the metal in oxygen. It is a white, insoluble solid with the wurtzite structure (Section 4.9). Its high melting point (2570°C), low reactivity, and excellent thermal conductivity, the highest of any oxide, lead to its use as a refractory material. It is highly toxic on inhalation, leading to chronic beryllosis (a disease of the lungs) and cancer. This problem is exacerbated by its relatively low density (3.0 g cm^{-3}), as dust particles remain airborne for long periods, but BeO is safe for many applications when used as a sintered monolith or in electronic components. In combination with electropositive metals, beryllium forms complex beryllates, such as K_2BeO_2 and $La_2Be_2O_5$, which contain BeO_4 tetrahedra and are structurally analogous to silicates.

The oxides of the other Group 2 elements can be obtained by direct combination of the elements (except Ba, which forms the peroxide) but they are more commonly obtained by decomposition of the carbonates:

$$MCO_3(s) \xrightarrow{\Delta} MO(s) + CO_2(g)$$

The oxides of the elements from Mg to Ba all adopt the rock-salt structure (Section 4.9). Their melting points decrease down the group as the lattice enthalpies decrease with increasing cation radius. Magnesium oxide only melts at 2852°C, and is used as a refractory lining in industrial furnaces. Like BeO, MgO has a high thermal conductivity coupled with a low electrical conductivity. This combination of properties leads to its use as an electrically insulating material and in electronics, Section 13.5.

Calcium oxide (as lime or quicklime) is used in large quantities in the steel industry to remove P, Si, and S. When heated, CaO is thermoluminescent and emits a bright white light (hence 'limelight'). Calcium oxide is also used as a water softener to remove hardness by reacting with soluble carbonates and hydrogencarbonates to form the insoluble $CaCO_3$. It reacts with water to form $Ca(OH)_2$, which is sometimes known as slaked lime and is used to neutralize acidic soils.

The peroxides SrO_2 and BaO_2 can be made by direct reaction of the elements, while the insoluble peroxides of Mg and Ca are obtained by addition of sodium peroxide, NaO_2, to aqueous solutions of these metals. All the peroxides are strong oxidizing agents and decompose to the oxide:

$$MO_2(s) \rightarrow MO(s) + \tfrac{1}{2}O_2(g)$$

The thermal stability of the peroxides increases down the group as the radius of the cation increases. This trend is explained by considering the lattice enthalpies of the peroxide and the oxide, and their dependence on the relative radii of the cations and anions. As O^{2-} is smaller than O_2^{2-}, the lattice enthalpy of the oxide is greater than that of the corresponding peroxide. The difference between the two lattice enthalpies decreases down the group as both values become smaller with increasing cation radius, so the tendency to decompose decreases. Magnesium peroxide, MgO_2, is consequently the least stable peroxide and is used as an *in situ* source of oxygen in a range of applications, including bioremediation to clean up polluted waterways. CaO_2 has similar applications in disinfecting water and as a flour-bleaching agent.

EXAMPLE 13.4 Explaining the thermal stabilities of peroxides

Estimate the difference between the lattice enthalpies of the peroxide and oxide of Mg and Ba, and comment on the values obtained.

Answer The Kapustinskii equation (eqns. 4.4 and 4.5) can be used to estimate lattice enthalpies, using the ionic radii in Table 13.1 and the radii of the oxide and peroxide ions (136 and 180 pm, respectively). Remembering that peroxide is a single anion, O_2^{2-}, substitution of the values gives

	MgO	MgO$_2$	BaO	BaO$_2$
$\Delta_L H$/kJ mol^{-1}	4037	3315	3147	2684
Difference/kJ mol^{-1}	722		463	

This calculation confirms that the difference between the lattice enthalpies of the oxide and peroxide decreases down the group. As a result, the heavier group members are found to form the most thermally stable peroxides.

Self-test 13.4 Estimate the lattice enthalpies for CaO and CaO$_2$ and check that the above trend is confirmed.

The heavier Group 2 elements also form a number of complex oxides such as perovskite, $SrTiO_3$, and spinel, $MgAl_2O_4$ (Section 4.9). The range of ionic radii available, from Mg^{2+} (72 pm for coordination number 6) to Ba^{2+} (142 pm and larger for coordination number 8 and higher) means that complex oxides can be synthesized that contain these cations in a large variety of different structure types. Important examples of such complex oxides include the ferroelectric perovskite $BaTiO_3$, the phosphor $SrAl_2O_4$:Eu^{2+}, and many of the high-temperature superconductors; for example, $YBa_2Cu_3O_7$ and $Bi_2Sr_2CaCu_2O_8$ (Section 30.7). The coordination number preferences of the cations are important in solid-state chemistry, where they can be used to control the structures of many complex oxides. If a doubly charged ion is required to occupy the A site of a perovskite structure (Section 4.9) with a coordination number of 12, then Sr^{2+} or Ba^{2+} is normally selected (as in $SrTiO_3$). On the other hand, for the spinel structure (general formula AB_2O_4, with six-coordinate B-type sites) Mg^{2+} is a good choice (as in $GeMg_2O_4$).

A BRIEF ILLUSTRATION

The control of the local coordination number by the Group 2 ions is illustrated by considering the structure of the superconducting phase $Tl_2Ba_2Ca_2Cu_3O_{10}$; the larger Ba^{2+} cations demand a higher coordination number to O than does Ca^{2+}, so the former are found exclusively on the nine-coordinate sites and the latter on eight-coordinate sites. It would be impossible to synthesize $Tl_2Ca_2Ca_2Cu_3O_{10}$ with Ca^{2+} replacing Ba^{2+} on the higher coordinate site, as that location would be thermodynamically unfavourable.

(b) Sulfides

KEY POINT The sulfides mostly adopt the rock-salt structure and have applications as phosphors.

Beryllium sulfide adopts a zinc-blende structure, whereas the sulfides of the heavier elements all crystallize with the rock-salt structure. Rechargeable magnesium sulfur batteries are being developed for EV applications where the cell reaction

$$8Mg + S_8(s) \rightleftharpoons 8MgS(s)$$

occurs between a magnesium foil anode and a sulfur cathode separated by a magnesium salt electrolyte, such as magnesium tetrakis(hexafluoroisopropyloxy) borate $Mg[B(hfip)_4]_2$ (hfip = $OC(H)(CF_3)_2$). This battery develops a voltage in the range 1.65–1.0 V, lower than a Li-ion battery, but the theoretical volumetric capacity of 3850 mAh cm^{-3} is almost twice that of a Li-ion cell.

Barium sulfide, produced by reducing the naturally occurring barytes, $BaSO_4$, with coke,

$$BaSO_4(s) + 2C(s) \rightarrow BaS(s) + 2CO_2(g)$$

displays strong phosphoresence and was the first synthetic phosphor produced around AD 1600. A mixed calcium/strontium sulfide doped with bismuth is a long-lifetime phosphor and has been used in glow-in-the-dark pigments.

EXAMPLE 13.5 Predicting and identifying the structure type of a Group 2 chalcogenide

Analysis of the powder X-ray diffraction pattern obtained from CaSe showed that the lattice type was face-centred with lattice parameter 592 pm. Predict a structure type for this compound by using the ionic radii for Ca^{2+} and Se^{2-} as 100 and 184 pm, respectively. Does your prediction agree with the experimental observation?

Answer First, we need to be aware that a binary compound such as CaSe is likely to adopt one of the simple AX structure types described in Section 4.9; then we can use the radius-ratio rule to guide us to the likely structural type. Two of the AX structure types have face-centred lattice types; namely, rock-salt and zinc-blende. The radius ratio is (100 pm)/(194 pm) = 0.52, which according to Table 4.6 suggests that the rock-salt structure is preferred. The lattice parameter for the rock-salt structure is calculated by considering the unit cell in Fig. 4.32, as it is equal to the length of the side of the unit cell. As can be seen from the figure, this length is twice the sum $r(Ca^{2+}) + r(Se^{2-})$. Thus, from the ionic radii the lattice parameter is predicted to be 2 × (100 + 194) pm = 588 pm, in good agreement with the X-ray diffraction value, which therefore demonstrates that CaSe adopts the rock-salt structure.

Self-test 13.5 Use ionic radii to predict the structure type of BeSe.

(c) Hydroxides

KEY POINT The solubility of the hydroxides increases down the group.

Beryllium hydroxide, $Be(OH)_2$, is amphoteric and is precipitated by the addition of sodium hydroxide to an aqueous solution of Be^{2+}. Its structure consists of an infinite network of $Be(OH)_4$ tetrahedra linked through a shared hydroxide ion. The remainder of the Group 2 hydroxides are formed by reaction of the oxides with water. The hydroxides become apparently more basic down the group because their solubility increases from $Mg(OH)_2$ to $Ba(OH)_2$. Magnesium hydroxide, $Mg(OH)_2$, is sparingly soluble and forms a mildly basic solution because a saturated solution contains a low concentration of the OH$^-$ ions. Calcium hydroxide, $Ca(OH)_2$, is more soluble than $Mg(OH)_2$, so a saturated solution has a higher concentration of OH$^-$ ions and the solution is described as moderately basic. A saturated solution of $Ca(OH)_2$ is called limewater and is used to test for the presence of CO_2. If CO_2 is bubbled through the limewater, a white precipitate of $CaCO_3$ is formed, which then disappears on further reaction with CO_2 to form the hydrogencarbonate ion:

$$Ca(OH)_2(aq) + CO_2(g) \rightarrow CaCO_3(s) + H_2O(l)$$
$$CaCO_3(s) + CO_2(g) + H_2O(l) \rightarrow Ca^{2+}(aq) + 2HCO_3^-(aq)$$

Barium hydroxide, $Ba(OH)_2$, is soluble, and aqueous solutions are strongly basic.

A BRIEF ILLUSTRATION

The molar solubility of $Mg(OH)_2$ is 1.54×10^{-4} mol dm^{-3}. Therefore, the concentration of OH$^-$ ions in the saturated solution is 3.08×10^{-4} mol dm^{-3}. We know from Section 5.1(b) that $K_w = 1.0 \times 10^{-14}$, so (ignoring deviations from ideality) $[H_3O^+] = K_w/[OH^-] = 3.25 \times 10^{-11}$ mol dm^{-3}. Provided we can identify activities with molar concentrations, this concentration corresponds to pH = 10.5.

EXAMPLE 13.6 Amphoteric behaviour of beryllium hydroxide

Write balanced equations for the dissolution of $Be(OH)_2$ in (i) dilute sulfuric acid and (ii) strontium hydroxide solution. For (ii) comment on the acid–base characteristics shown by the two Group 2 elements.

Answer

(i) $Be(OH)_2(s) + H_2SO_4(aq) + 2H_2O \rightarrow Be^{2+}(aq) + [SO_4]^{2-}(aq) + 4H_2O(l)$

Here beryllium hydroxide acts as a base and the sulfate salt is formed. The product collected after crystallization is beryllium sulfate tetrahydrate, $BeSO_4 \cdot 4H_2O$, with four water molecules solvating the Be^{2+} ion in the solid structure (consisting of alternating layers of $[Be(H_2O)_4]^{2+}$ and $[SO_4]^{2-}$ ions).

(ii) $Be(OH)_2(s) + Sr(OH)_2(aq) \rightarrow SrBe(OH)_4(s)$

In this reaction, strontium hydroxide acts as a strong base, reacting to form the salt Sr^{2+} (conjugate acid of $Sr(OH)_2$) and the tetrahydroxyberyllate anion $[Be(OH)_4]^{2-}$. This reaction shows the increasing basic behaviour of the Group 2 metal oxides and hydroxides as the group is descended.

Self-test 13.6 What would be formed in the case of reaction (i) and reaction (ii) by heating the products to 500°C with the loss of water?

13.9 Nitrides and carbides

KEY POINT The Group 2 nitrides and carbides react with water to produce ammonia and either methane or ethyne, respectively.

All the elements form nitrides of composition M_3N_2 when heated in nitrogen. These react with water to form ammonia and the metal hydroxide:

$M_3N_2(s) + 6H_2O(l) \rightarrow 3M(OH)_2(s, aq) + 2NH_3(g)$

Magnesium burns in nitrogen to form greenish-yellow Mg_3N_2, which has been used as a catalyst for preparing cubic BN (Section 14.9). Calcium nitride, Ca_3N_2, reacts with hydrogen gas at 400°C to produce CaNH and CaH_2. Beryllium nitride only melts above 2200°C and is therefore used as a refractory material.

All the elements also form carbides. Beryllium carbide, Be_2C, formally contains the methide ion, C^{4-}, although some covalency would be expected in the bonding in this compound; it is a crystalline solid with the antifluorite structure (Fig. 13.6). The carbides of Mg, Ca, Sr, and Ba have the formula MC_2 and contain the dicarbide (acetylide) anion, C_2^{2-}. The carbides of Ca, Sr, and Ba are prepared by heating the oxide or carbonate with carbon in a furnace at 2000°C:

$MO(s) + 3C(s) \rightarrow MC_2(s) + CO(g)$
$MCO_3(s) + 4C(s) \rightarrow MC_2(s) + 3CO(g)$

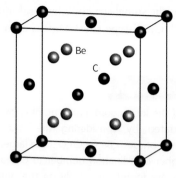

FIGURE 13.6 The antifluorite structure adopted by Be_2C.

All the carbides react with water to produce the hydrocarbon corresponding to the carbon ion present: beryllium carbide produces methane, whereas the other elements produce ethyne (acetylene):

$Be_2C(s) + 4H_2O(l) \rightarrow 2Be(OH)_2(s) + CH_4(g)$
$CaC_2(s) + 2H_2O(l) \rightarrow Ca(OH)_2(s) + C_2H_2(g)$

Methane and ethyne are flammable, and the latter burns with a bright light from the incandescent carbon particles that form in the flame. When this reaction was discovered in the late nineteenth century, calcium carbide found widespread use in vehicle lights, enabling safe night-time driving for the first time, and in miners' and potholers' lamps.

13.10 Salts of oxoacids

The most important oxo compounds of the Group 2 elements are the carbonates, hydrogencarbonates, and sulfates.

(a) Carbonates and hydrogencarbonates

KEY POINTS All the carbonates are sparingly soluble in water, with the exception of $BeCO_3$; the carbonates decompose to the oxide on heating, most readily for the lighter elements in the group. The hydrogencarbonates are more soluble than the carbonates.

Beryllium carbonate readily decomposes upon contact with water, forming CO_2 and $[Be(OH_2)_4]^{2+}$ which is immediately hydrolyzed due to the high charge density on the Be^{2+} ion and its polarization of the O–H bond of a hydrating H_2O molecule:

$[Be(OH_2)_4]^{2+}(aq) + H_2O(l) \rightarrow [Be(OH_2)_3(OH)]^+(aq) + H_3O^+(aq)$

The carbonates of the other elements are all sparingly soluble and are decomposed to the oxide on heating:

$MCO_3(s) \xrightarrow{\Delta} MO(s) + CO_2(g)$

The temperature at which this decomposition occurs increases from 350°C for Mg to 1360°C for Ba (Fig. 13.7).

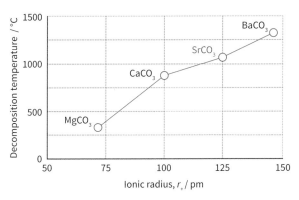

FIGURE 13.7 The variation of the decomposition temperature of the Group 2 carbonates with cationic radius.

The Group 2 carbonates demonstrate a similar thermal stability pattern to that shown by the Group 1 carbonates, with stability increasing down the group. As discussed in Section 4.15, these trends can be explained in terms of trends in lattice enthalpies, and hence, more fundamentally, in terms of trends in ionic radii. **Calcium-looping**, Box 13.5, is being investigated as a part of a method of carbon capture and storage.

Calcium carbonate is the most important oxo compound of the elements. It occurs widely in nature as limestone, chalk, marble, and dolomite (with magnesium), and as coral, pearl, and seashell. Calcium carbonate crystallizes in a variety of polymorphs. The most common forms are calcite, aragonite, and vaterite (Fig. 13.8; Box 13.6). Calcium carbonate is an important biomineral and a major constituent of bones and shells (Box 13.7). It also has widespread use in construction and road-building, and is used as an antacid, an abrasive in toothpaste, in chewing gum, and as a health supplement where it is taken to maintain bone density. Powdered limestone is known as agricultural lime and is used to neutralize acidic soil:

$$CaCO_3(s) + 2H^+(aq) \rightarrow Ca^{2+}(aq) + CO_2(g) + H_2O(l)$$

BOX 13.5 How can CaO be used to capture CO_2 and help reduce global warming?

Anthropogenic contributions to global warming through CO_2 emissions from the burning of fossil fuels are a major environmental concern. One way to tackle these CO_2 emissions is through Carbon Capture and Storage (CCS), with the aim of capturing CO_2 where it would otherwise be emitted from large polluting sources, including power stations and cement factories. CO_2 captured in this way could then potentially be stored in, for example, old expended oil wells and gas reservoirs. One alternative, but related, technology is to chemically capture CO_2 already present in the atmosphere, which could remove (in the same land area) far more CO_2 than can trees and plants. However, this is not a practical solution.

One of the best CO_2 capture technologies is 'calcium looping', which uses CaO to remove CO_2 and which can then be heated, using renewable energy, to release it in concentrated form for easier storage. CaO will also absorb CO_2 from low-concentration environments, such as air, so could be used to stabilize and reduce current global atmospheric CO_2 levels, so-called air remediation. Industrial processes that produce high concentrations of CO_2 are the burning of fossil fuel, especially coal, and the cement industry that produces CaO though the thermal decomposition of $CaCO_3$.

Calcium looping usually employs a dual fluidized bed system where very small solid particles of CaO act like a fluid due to the flue gas of CO_2 passing through them. This ensures rapid uptake of CO_2 into the solid. The uptake of CO_2 by CaO is exothermic:

$$CaO(s) + CO_2 \rightarrow CaCO_3(s) \quad \Delta H = -178 \text{ kJ mol}^{-1}$$

and some of this thermal energy can be recovered to drive the reverse process. Solar thermal energy could also be used to drive the decomposition of $CaCO_3$ above 800°C.

Calcium-looping is a cheap, low-toxicity technology compared with amine scrubbing, which typically uses liquid 2-aminoethanol to absorb CO_2 reversibly.

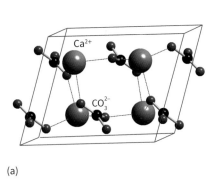

(a)

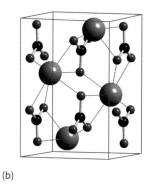

(b)

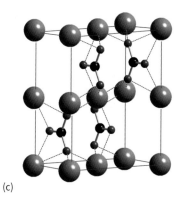

(c)

FIGURE 13.8 The structures of the (a) calcite, (b) aragonite, and (c) vaterite polymorphs of $CaCO_3$.

Calcium carbonate is sparingly soluble in water, but its solubility is increased if CO_2 is dissolved in the water, as in rainwater. Thus, caverns have been eroded in limestone rock by the reaction to form the more soluble hydrogencarbonate:

$$CaCO_3(s) + H_2O(l) + CO_2(g) \rightarrow Ca^{2+}(aq) + 2HCO_3^-(aq)$$

This reaction is reversible, and over time calcium carbonate stalactites and stalagmites are formed following very slow evaporation of water and precipitation of $CaCO_3$.

Because HCO_3^- has a lower charge than CO_3^{2-}, Group 2 hydrogencarbonates do not precipitate from solutions containing these ions. Temporary hardness of water (hardness

BOX 13.6 Where is calcium carbonate in rocks and minerals, and what uses does it have?

Calcium carbonate occurs as vast rock deposits. The most common and stable form is the hexagonal calcite which comprises about 4% by mass of the Earth's crust. Calcite is formed in many different geological environments and constitutes a significant part of all three major rock classification types: igneous, sedimentary, and metamorphic. Calcite is a major component in the igneous rock called carbonatite, and forms a portion of many hydrothermal veins. Limestone is the sedimentary form of calcite and is normally formed from the fossilized remains of marine creatures. Limestone metamorphoses to marble under heat and pressure, which increase the density of the rock and destroy any texture. Pure white marble is the result of metamorphism of very pure limestone, while the characteristic swirls and veins of many coloured marble varieties are usually due to the presence of mineral impurities, including clay, silt, sand, and iron oxides. Iceland spar is a form of transparent, colourless calcite originally found in Iceland. This crystalline mineral form exhibits birefringence, also known as double refraction, which is the division of a ray of light into two separate rays, depending on the polarization of the light. The phenomenon is the result of the different polarizations of light having different refractive indices in the crystalline material. As the two beams exit a crystal they are bent into two different angles of refraction.

Aragonite is a less abundant polymorph of calcite. It is orthorhombic, with three crystal forms. It is less stable than calcite and converts into it at 400°C, and given enough time, aragonite in the environment will convert to calcite. Most bivalve animals, such as oysters, clams, mussels, and corals, secrete aragonite, and their shells and pearls are composed mostly of aragonite. The pearlization and iridescent colours in seashells, such as abalone, are due to multiple layers of aragonite, of less than a micron thick, producing interference effects with visible light. Other natural sources of aragonite include hot springs and cavities in volcanic rocks.

Synthetic aragonite can be formed, and is used as a filler in the paper industry, where its fine texture, whiteness, and absorbent properties add quality to the product. Powdered limestone is heated in a kiln to form CaO (lime), which is then slurried with water to form milk of lime. Carbon dioxide is then bubbled through the slurry until aragonite is formed:

$$CaCO_3(s) \xrightarrow{\Delta} CaO(s) + CO_2(g)$$
$$CaO(s) + H_2O(l) \rightarrow Ca(OH)_2(s)$$
$$Ca(OH)_2(s) + CO_2(g) \rightarrow CaCO_3(s) + H_2O(l)$$

The conditions, such as temperature and flow rate of carbon dioxide, determine the final particle size distribution and crystal type.

Vaterite is an even rarer polymorph of calcium carbonate, also adopting a hexagonal unit cell. It is more soluble than calcite or aragonite and converts slowly to these forms on contact with water. It is deposited from some mineral springs in cold climates; gallstones are often composed of this form of $CaCO_3$. Travertine is a white, naturally occurring, very hard form of calcium carbonate. It is deposited from the water of hot mineral springs or streams containing calcium ions.

The two main polymorphs of $CaCO_3$ may be distinguished using infrared spectroscopy (Section 8.5). In calcite and aragonite the CO_3^{2-} anions are surrounded by Ca^{2+} ions to produce distinct local environments. As a result, the vibrational modes of CO_3^{2-} observed in their infrared spectra occur at slightly different frequencies. Moreover, due to differences in the local symmetry, certain vibrational modes are detected for each polymorph, as summarized in the following table, and infrared or Raman spectra may be used to rapidly distinguish the two forms. For example, the infrared spectrum obtained from a piece of snail shell is shown in Fig. B13.1. It shows a strong feature at around 1078 cm^{-1} that is characteristic of the aragonite form of calcium carbonate.

Wavenumber/cm^{-1}	
Calcite	Aragonite
714	698
876	857
	1080
1420 broad	1480 broad
1800	1785

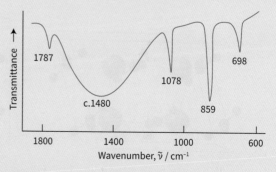

FIGURE B13.1 The IR spectrum of a piece of snail shell.

> **BOX 13.7** How do organisms use calcium carbonate as a biomineral?
>
> Biomineralization involves the production by organisms of inorganic solids, and over 50 biominerals have been identified, of which calcium carbonate is one of the most common. Nature is adept at controlling mineralization processes to produce single crystals and polycrystalline and amorphous structures with remarkable morphologies and mechanical properties (Section 26.18). The biomineral often fulfils a structural role in the organism; for example, as teeth, bones, or shells. Many shells are composed mostly of calcium carbonate with only up to about 2% protein.
>
> Most of the calcium carbonate mineral deposits found today were formed by sea creatures that manufactured shells and skeletons of calcium carbonate. When these animals died, their shells settled on the sea floor and were compressed to form limestone and chalk. The White Cliffs of Dover in England are formed from chalk made from the shells of microscopic sea creatures called *Foraminifera* that lived about 136 million years ago.
>
> In pearls and mother-of-pearl, very small $CaCO_3$ crystals are deposited to form smooth, hard, lustrous alternating layers of calcite and aragonite. The different structural forms of the calcium carbonate layers make the shell very strong, as they have different preferred directions of fracture and the interleaved layers resist breaking under pressure. The thickness of the layers is similar to the wavelength of light, which produces the interference effects and the observed pearlescence.
>
> Many attempts are being made to replicate these complicated crystal growth patterns of $CaCO_3$ in the laboratory. Crystals can sometimes be grown around a template, which is then removed to produce a porous structure; or the addition of certain chemicals can cause different shapes of crystals to grow from a particular solution. Similarly, the growth of bone, which is mainly calcium hydroxyapatite, $Ca_5(PO_4)_3(OH)$, is being investigated, including the development of synthetic bone materials. Biominerals are discussed further in Section 26.18.

which is removed by boiling the water) is caused by the presence of magnesium and calcium hydrogencarbonates in solution. These ions are precipitated as carbonate on boiling, when the equilibrium in the following reaction is moved to the right:

$$Ca(HCO_3)_2(aq) \rightleftharpoons CaCO_3(s) + CO_2(g) + H_2O(l)$$

Temporary hardness may also be removed by adding $Ca(OH)_2$, which also precipitates the carbonate:

$$Ca(HCO_3)_2(aq) + Ca(OH)_2(aq) \rightarrow 2CaCO_3(s) + 2H_2O(l)$$

If temporary hard water is not treated to remove the Ca^{2+} and Mg^{2+} ions, these ions can react with soap (sodium stearate, $NaC_{17}H_{35}CO_2$) or detergent molecules to form an insoluble precipitate, scum, which reduces the effectiveness of the detergent:

$$2NaC_{17}H_{35}CO_2(aq) + Ca^{2+}(aq) \rightarrow Ca(C_{17}H_{35}CO_2)_2(s) + 2Na^+(aq)$$

(b) Sulfates and nitrates

KEY POINTS Calcium sulfate, which occurs naturally as gypsum and alabaster, is used in building materials as plaster.

'Permanent hardness' (so called because the hardness is not removed by boiling) is caused by magnesium and calcium sulfates. In this case the water is softened by passing it through an ion-exchange resin, which replaces the Mg^{2+} and Ca^{2+} ions with Na^+ ions.

Calcium sulfate is the most important of the Group 2 sulfates. It occurs naturally as gypsum, which is the dihydrate, $CaSO_4 \cdot 2H_2O$ (Fig. 13.9). Alabaster is a dense, fine-grained

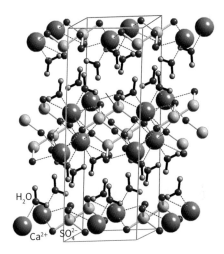

FIGURE 13.9 The structure of $CaSO_4 \cdot 2H_2O$, highlighting the water molecules that are mainly lost on heating above 150°C.

form of the same composition; it resembles marble and can be carved into sculptures. When the dihydrate is heated above 150°C it loses water to form the hemihydrate, $CaSO_4 \cdot \frac{1}{2}H_2O$, which is also known as building plaster or plaster of Paris as it was first mined in the Montmartre district of the city. When mixed with water, plaster expands as it forms the dihydrate, so providing a strong structure that is used to make casts for broken limbs and as a wall covering. Gypsum is mined and used directly as a building material. One application is in fireproof wallboard; in the event of fire, the dihydrate will dehydrate to form the hemihydrate and release water vapour:

$$2CaSO_4 \cdot 2H_2O(s) \rightarrow 2CaSO_4 \cdot \tfrac{1}{2}H_2O(s) + \tfrac{3}{2}H_2O(g)$$

The reaction is endothermic ($\Delta_r H^\ominus = +117\,\text{kJ mol}^{-1}$), so it absorbs heat from the fire. In addition, the water produced absorbs heat and evaporates, and then the gaseous water provides an inert barrier, reducing the supply of oxygen to the fire.

The insolubility of $BaSO_4$ coupled to the strong X-ray absorbing properties of barium, due to its high atomic number (56), lead to its use in X-ray imaging of the digestive tract. The white pigment lithopone is a mixture of $BaSO_4$ and ZnS, and has excellent chemical stability, being inert to attack by sulfides, unlike lead white, $PbCO_3$. Barium sulfate, with its high density of 4.5 g cm^{-3}, is a component of many drilling muds, which clean and cool the drill bit and then carry the drilled rock away.

Hydrated nitrates, for example $Ca(NO_3)_2 \cdot 4H_2O$, can be obtained by treating the oxides, hydroxides, and carbonates with nitric acid and crystallizing the salt from the resulting aqueous solution. For Mg to Ba the anhydrous salts are easily obtained by thermal dehydration. Heating hydrated beryllium nitrate, $Be(NO_3)_2 \cdot 4H_2O$, results in its decomposition and the evolution of NO_2. Anhydrous beryllium nitrate, $Be(NO_3)_2$, can be obtained by dissolving $BeCl_2$ in N_2O_4 and heating the resultant solvate $Be(NO_3)_2 \cdot 2N_2O_4$ gently to drive off NO_2. Heating $Be(NO_3)_2$ further produces the basic nitrate $Be_4O(NO_3)_6$, which contains a central Be_4O tetrahedral unit with edge-bridging nitrate groups with a structure similar to that of the basic acetate (ethanoate) (5).

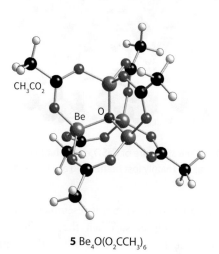

5 $Be_4O(O_2CCH_3)_6$

13.11 Solubility, hydration, and beryllates

KEY POINTS The large negative hydration enthalpies of the salts of mono-negative ions ensure that they are soluble. For salts of dinegative ions, the lattice enthalpies are more influential and the salts are insoluble.

Compounds of the Group 2 elements are generally much less soluble in water than those of Group 1 elements, even though the hydration enthalpies of the divalent cations are more negative:

	Na$^+$	K$^+$	Mg^{2+}	Ca^{2+}
$\Delta_{hyd}H^\ominus$/kJ mol^{-1}	−406	−322	−1920	−1650

With the exception of the fluorides, the salts of singly charged anions are usually soluble in water, and those of doubly charged anions (such as oxides, sulfates, and oxalates) are usually only sparingly soluble. For the latter, including the carbonates and the sulfates, the high lattice enthalpy arising from the high charge of the anion is the deciding factor in making dissolution unfavourable (remembering the lattice enthalpy is proportional to the product of the ion charges, Section 4.12), outweighing the influence of the enthalpy of hydration. This insolubility is responsible for the enormous natural deposits of magnesium- and calcium-containing minerals that include limestone, gypsum, and dolomite widely exploited in the construction industry. With the exception of BeF_2, the fluorides are all insoluble in water because the small size of F$^-$ leads to a high lattice enthalpy. Beryllium fluoride has a very high hydration enthalpy due to the high charge density on the small cation: in this case, hydration enthalpy rather than lattice enthalpy is the dominant factor:

$$BeF_2(s) + 4H_2O(l) \rightarrow [Be(OH_2)_4]^{2+}(aq) + 2F^-(aq)$$
$$\Delta_r H^\ominus = -250\,\text{kJ mol}^{-1}$$

The H$_2$O molecules directly coordinated to Be^{2+} are very strongly held in aqueous solution, exchanging only very slowly with free water. The hydrated Be^{2+} cation, $[Be(OH_2)_4]^{2+}$, acts as an acid in water:

$$[Be(OH_2)_4]^{2+}(aq) + H_2O(l) \rightarrow [Be(OH_2)_3(OH)]^+(aq) + H_3O^+(aq)$$

This reaction can be traced to the high polarizing power of the small, doubly charged cation. Solutions of the hydrated salts of the heavier elements are neutral.

Trends in Group 2 salt solubility and hydrolysis can also be explained in hard/soft acid–base terms. As the group is descended the ions become less hard; the fluorides and hydroxides (small, hard anions) are insoluble for the harder Be^{2+} and Mg^{2+}, but more soluble for Ba^{2+}. Be^{2+} is most strongly hydrated by the hard O atom in H$_2$O, and the relatively soft Ba^{2+} much more weakly.

The amphoteric nature of Be, leading to the formation of $[Be(OH)_4]^{2-}$ under strongly basic conditions, means that the element forms an extensive series of beryllates that are built from BeO$_4$ tetrahedra. The tetrahedral anion, $[Be(OH)_4]^{2-}$, has been isolated in the salt $SrBe(OH)_4$. The beryl family of minerals, $Be_3Al_2(SiO_3)_6$, which includes emerald, aquamarine, and morganite, contains this unit as do a number of other complex beryllates such as phenakite, Be_2SiO_4, and the zeolite nabesite, $Na_2BeSi_4O_{10}\cdot 4H_2O$. The compound $BeAl_2O_4$ occurs naturally as the mineral chrysoberyl, which, in one form, alexandrite, is doped with chromium and changes in colour from green in daylight to purple under incandescent lighting.

EXAMPLE 13.7 Assessing the factors affecting solubility

Estimate the lattice enthalpies of $MgCl_2$ and $MgCO_3$. Comment on the likely implications for their solubilities.

Answer Once again we can use the data in Table 13.1 and the Kapustinskii equation (eqns. 4.4 and 4.5); we also need to know that the ionic and thermochemical ionic radii of Cl^- and CO_3^{2-} are 167 and 178 pm, respectively (Table 4.10). Substitution of the data gives the lattice enthalpies of $MgCO_3$ and $MgCl_2$ as 3260 and 2478 kJ mol^{-1}, respectively. As the value for $MgCO_3$ is larger, it is more likely to offset the hydration enthalpy and the solubility will be lower than that of $MgCl_2$.

Self-test 13.7 Estimate the lattice enthalpies of MgF_2, $MgBr_2$, and MgI_2, and comment on how solubilities of Group 2 halides vary with increasing halide ion size.

13.12 Coordination compounds

KEY POINTS Only beryllium forms coordination compounds with simple ligands including halides; the most stable complexes are formed with polydentate chelating ligands such as the edta anion.

Compounds of Be show properties consistent with a greater covalent character than those of its congeners, and some of its complexes with ordinary ligands are stable. The complexes are usually tetrahedral, although the coordination number of Be can fall to 3 or 2 if the ligands are bulky. The least labile complexes are formed with halide or chelating O-donor ligands, such as oxalate, alkoxides, and diketonates. For example, basic beryllium acetate (beryllium oxoethanoate, $Be_4O(O_2CCH_3)_6$) consists of a central O atom surrounded by a tetrahedron of four Be atoms, which in turn are bridged by ethanoate ions (5). It can be prepared by the reaction of ethanoic (acetic) acid with beryllium carbonate:

$$4BeCO_3(s) + 6CH_3COOH(l) \rightarrow 4CO_2(g) + 3H_2O(l) + Be_4O(O_2CCH_3)_6(s)$$

Basic beryllium acetate is a colourless, sublimable, molecular compound; it is soluble in chloroform, from which it can be recrystallized.

Dihydroxymagnesium carboxylates were identified for the first time through the study of nonterrestrial material (in meteorites). Soluble organic matter from meteorites was analysed by mass spectrometry and found to contain species of the type $[(OH)_2Mg(O_2CR)]^-$, where R is a long chain alkyl group C_nH_{2n+1} with $n = 11$–18. These compounds exhibit high thermal stability and may have existed during the early phases of planetary formation.

Group 2 cations form complexes with crown and cryptand ligands. The least labile of these complexes are formed with the larger Sr^{2+} and Ba^{2+} cations. All the complexes are more stable than those of the smaller, lower-charge, Group 1 cations. The most stable complexes are formed with charged polydentate ligands, such as the analytically important ethylenediaminetetraacetate ion (edta^{4-}). The formation constants of edta^{4-} complexes lie in the order Ca^{2+} > Mg^{2+} > Sr^{2+} > Ba^{2+}. In the solid state, the structure of the Mg^{2+} edta complex is seven-coordinate (6), with H_2O at one coordination site.

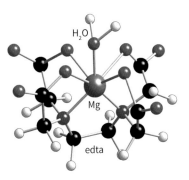

6 $[Mg(edta)(OH_2)]^{2-}$

In crystal structures of $[Ca^{2+}(edta)^{4-}]^{2-}$ complexes, calcium is eight-coordinate with two additional H_2O molecules serving as ligands (7). Spectroscopic studies also support the existence of similar eight-coordinate Ca^{2+} ions in aqueous edta^{4-} solutions.

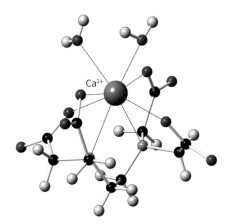

7 $[Ca^{2+}(edta)^{4-}(OH_2)_2]^{2-}$

Many complexes of Ca^{2+} and Mg^{2+} occur naturally in biological systems. The most important macrocyclic complexes are the chlorophylls (8), which are porphyrin complexes of Mg and are central to photosynthesis (Section 26.10(d)).

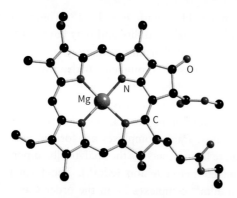

8 Chlorophyll fragment (Mg–C–N–O skeleton)

Magnesium is involved in biological catalysis, including phosphate transfer and carbohydrate metabolism. Calcium is a component of biominerals and is also coordinated by proteins, notably those involved in cell signalling and muscle action (Section 26.5). The development of Group 2 metal hydride complexes such as [(DIPP-nacnac)CaH$_2$(THF)]$_2$ (where DIPP-nacnac = CH{(CMe)(2,6-iPr$_2$C$_6$H$_3$N)}$_2$), containing two calcium atoms separated by two bridging hydride ions, is of importance with respect to the development of hydrogen storage materials (Box 11.5 and Chapter 30).

13.13 Organometallic compounds

KEY POINTS Alkylberyllium compounds polymerize in the solid phase; Grignard reagents are some of the most important main-group organometallic compounds.

Organometallic compounds of Be are pyrophoric in air and unstable in water. Methylberyllium can be prepared by transmetallation from methylmercury in a hydrocarbon solvent:

$$Hg(CH_3)_2(sol) + Be(s) \rightarrow Be(CH_3)_2(sol) + Hg(l)$$

Another synthetic route is by halogen exchange or metathesis reactions in which a beryllium halide reacts with an alkyllithium compound. The products are the lithium halide and an alkylberyllium compound. In this way, the halogen and organic groups are transferred between the two metal atoms. The driving force for this and similar reactions is the formation of the insoluble halide of the more electropositive metal:

$$2n\text{-BuLi}(sol) + BeCl_2(sol) \rightarrow (n\text{-Bu})_2Be(sol) + 2LiCl(s)$$

Grignard reagents in ether can also be used in the synthesis of organoberyllium compounds:

$$2RMgCl(sol) + BeCl_2(sol) \rightarrow R_2Be(sol) + 2MgCl_2(s)$$

Methylberyllium, Be(CH$_3$)$_2$, is predominantly a monomer in the vapour phase and in hydrocarbon solvents, where it adopts a linear structure, as expected from the VSEPR model. In the solid it forms polymeric chains in which the bridging CH$_3$ groups form 3c,2e bridging bonds (Section 3.11(e); **9**).

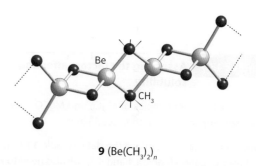

9 (Be(CH$_3$)$_2$)$_n$

Bulkier alkyl groups lead to a lower degree of polymerization; ethylberyllium (**10**) is a dimer and *t*-butylberyllium (**11**) is a monomer.

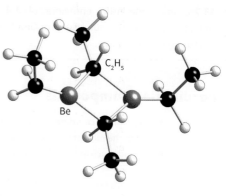

10 (BeEt$_2$)$_2$

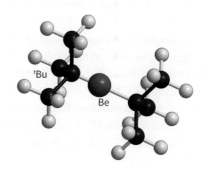

11 Be^{t}Bu$_2$ (^{t}Bu = (CH$_3$)$_3$C)

An interesting organoberyllium compound is 'beryllocene', (C$_5$H$_5$)$_2$Be, which, although the formula suggests an analogy with ferrocene (Section 24.19), in fact has a different structure in the crystalline state, with the Be atom positioned directly above the centre of one cyclopentadienyl ring and below a single C atom on the other ring (**12**). However, the low-temperature (−135°C) NMR spectrum of this compound in solution suggests that the two rings are equivalent, indicating that even at this low temperature the Be atom and the C_5H_5 rings are rearranging rapidly in the molecular form.

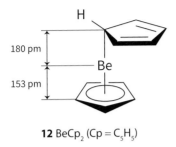

12 $BeCp_2$ (Cp = C_5H_5)

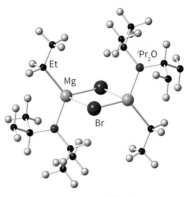

14 $[EtMgBr(O^iPr_2)]_2$

Alkyl- and arylmagnesium halides are very well known as **Grignard reagents** and are widely used in synthetic organic chemistry, where they behave as a source of R⁻. They are prepared from magnesium metal and an organohalide. As the surface of magnesium is covered by a passivating oxide film, it has to be activated before the reaction can proceed. A trace of iodine is usually added to the reactants, forming magnesium iodide; this compound is soluble in the solvent used, and dissolves to expose an activated magnesium surface. Alternatively, a highly active, finely powdered form of magnesium can be generated by reducing $MgCl_2$ with potassium in THF. The reaction to produce the Grignard reagents is carried out in diethyl ether or tetrahydrofuran:

$Mg(s) + RBr(sol) \rightarrow RMgBr(sol)$

The structural chemistry of Grignard reagents is far from simple, and is not generally represented by the simple empirical formula 'RMgX'. This is due to the formation of dimers and coordination by Lewis basic solvent molecules under their preparative conditions. When EtMgBr is formed in, and crystallized from, diethylether, the monomeric complex (**13**), with tetracoordinated magnesium, is formed; if the more bulky diisopropylether is used as a solvent, then dimeric $[EtMgBr(O^iPr_2)]_2$ (**14**) is produced, containing three coordinate pyramidal Mg.

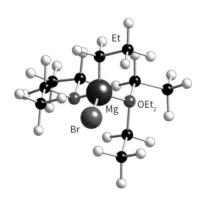

13 $MgEtBr(OEt_2)_2$

In addition, complex equilibria in solution, known as **Schlenk equilibria**, lead to the presence of several species, the exact nature of which depend on temperature, concentration, and solvent. For example, R_2Mg, RMgX, and MgX_2 have all been detected:

$2RMgX(sol) \rightarrow R_2Mg(sol) + MgX_2(sol)$

Grignard reagents are widely used in the synthesis of organometallic compounds of other metals, as in the formation of alkylberyllium compounds mentioned above. They are also widely used in organic synthesis. One reaction is **organomagnesiation**, which involves addition of the Grignard reagent to an unsaturated bond:

$R^1MgX(sol) + R^2R^3C=CR^4R^5(sol) \rightarrow$
$\qquad R^1R^2R^3CCR^4R^5MgX(sol)$

Grignard reagents undergo side reactions such as **Wurtz coupling** to form a carbon–carbon bond:

$R^1MgX(sol) + R^2X(sol) \rightarrow R^1R^2(sol) + MgX_2(sol)$

The compound $^iPrMgCl\cdot LiCl$ is known as a **turbo Grignard reagent** in which the postulated formation of a four-membered ring $^iPr-Mg(Cl)(Cl)Li$ or $[^iPrMgCl_2]^- Li^+$, accelerates reaction rates and increases selectivity, probably due to the generation of an increased negative charge on magnesium complex.

The organometallic compounds of Ca, Sr, and Ba are generally ionic and very unstable. They all form analogues of Grignard reagents by direct interaction of the finely divided metal with organohalide.

Complexes of the heavier alkaline earth metals (Mg, Ca, Sr, Ba) elements are becoming widely used in alkene and alkyne hydroamination and hydrophosphination catalysis allowing the synthesis of a wide variety of acyclic and heterocyclic small molecules. These Group 2 catalysts have significant advantages over the more commonly used transition metal organometallic complexes due to much lower costs and inherent lower toxicity. The heavier alkaline earth hydride complexes are used as pre-catalysts and intermediates in multiple bond hydrogenation, hydrosilylation, and hydroboration reactions and in a variety of dehydrocoupling processes that allow facile catalytic construction of Si–C, Si–N and B–N bonds.

13.14 Lower oxidation state Group 2 compounds

KEY POINT The lighter elements of Group 2 have been stabilized in covalent organometallic compounds with the oxidation states Mg(I) and Be(0).

Although the members of Group 2 occur almost exclusively in oxidation state +2 in their compounds, the reduction of Mg(II) to Mg(I) by potassium has been achieved in the synthesis of the compounds LMg–MgL, where L = [Ar(NC)(N^{i}Pr$_2$)N(Ar)]$^-$ and Ar = 2,6-diisopropylphenyl. The structures contain a central Mg–Mg bond of length 285 pm (**15**), which is shorter than the Mg–Mg distance in magnesium metal (320 pm).

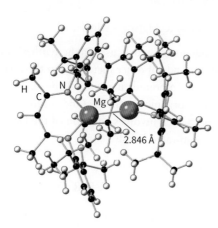

15 [Mg(DIPP-nacnac)]$_2$, [DIPP-nacnac] = (ArNCMe)$_2$CH, Ar = 2,6-^{i}Pr$_2$C$_6$H$_3$]

The chemistry of Mg(I) is being increasingly developed with significant advances in C–H bond activation, including those in arenes. For example, photoactivation of dimagnesium(I) compounds with blue or UV light allows reduction of benzene to cyclohexadienediyl-bridged dimagnesium(II) compounds. The reduction of beryllocene (**12**) using these Mg(I) compounds results in the formation of diberyllocene (**16**), CpBeBeCp, which contains Be(I) and a Be–Be bond of length 205 pm. Diberyllocene acts as a reductant that can be used to synthesize compounds with Be–Al and Be–Zn bonds.

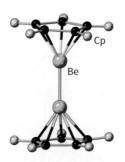

16 Cp-Be-Be-Cp

Zero valent beryllium compounds have also been synthesized using cyclic carbene ligands to stabilize this oxidation state. Brightly coloured compounds, such as the purple bis(1-(2,6-diisopropylphenyl)-3,3,5,5-tetramethylpyrrolidin-2-ylidene)-beryllium (**17**), contain Be(0) which is linearly coordinated to the two carbene ligands via two very short Be–C bonds of 1.664 Å, indicating possible double bond character; that is, a three centre–two electron π bond in addition to the Be–C sigma bond. Such low and zerovalent compounds are electron rich and may have important application in catalysis activating C–H and H–H bonds.

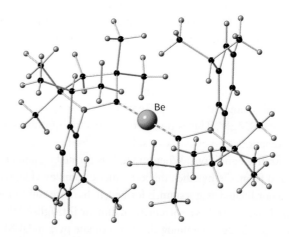

17 Bis(1-(2,6-diisopropylphenyl)-3,3,5,5-tetramethylpyrrolidin-2-ylidene)-beryllium(0)

FURTHER READING

R. B. King, *Inorganic Chemistry of the Main Group Elements*. John Wiley & Sons (1994).

P. Enghag, *Encyclopedia of the Elements*. John Wiley & Sons (2004).

D. M. P. Mingos, *Essential Trends in Inorganic Chemistry*. Oxford University Press (2004). A survey of inorganic chemistry from the perspective of structure and bonding.

N. C. Norman, *Periodicity and the s- and p-Block Elements*. Oxford University Press (1997). Includes coverage of essential trends and features of s-block chemistry.

J. A. H. Oates, *Lime and Limestone: Chemistry and Technology, Production and Uses*. John Wiley & Sons (1998).

M. S. Hill, D. J. Liptrot, and C. Weetman, Alkaline earths as main group reagents and molecular catalysts. *Chem. Soc Rev.* 2016, **45**, 972–88.

EXERCISES

13.1 Explain why compounds of beryllium are mainly covalent whereas those of the other Group 2 elements are predominantly ionic.

13.2 Why are the properties of beryllium more similar to aluminium and zinc than to magnesium?

13.3 The melting points (°C) of the Group 2 chlorides are as follows: $BeCl_2$, 405; $MgCl_2$, 714; $CaCl_2$, 772; $SrCl_2$, 874; and $BaCl_2$, 962. Account for the significantly lower temperature for $BeCl_2$.

13.4 Using electronegativity values of 1.57 for Be and 0.79 for Cs and a Ketelaar triangle (Fig. 2.40), predict what type of compound might form between these elements.

13.5 Identify the metal-containing compounds **A**, **B**, **C**, and **D** in this scheme:

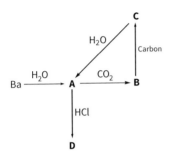

13.6 Why does beryllium fluoride form a glass when cooled from a melt?

13.7 Predict the product of the reaction

$$BeF_2(s) + nNH_3(l) \rightarrow$$

13.8 Dissolution of $MgCl_2 \cdot 6H_2O$ in water gives a weakly acidic solution while dissolution of $BaCl_2 \cdot 2H_2O$ at the same concentration gives a neutral pH solution. Explain these observations.

13.9 Calculate the weight percent of hydrogen in each of the Group 2 hydrides. Why is MgH_2 being investigated as a hydrogen storage material while BeH_2 is not?

13.10 Explain why Group 1 hydroxides are much more corrosive to metals than those of most Group 2 hydroxides.

13.11 What are the advantages of using $Ba(OH)_2$ over NaOH as a base in analytical acid–base titrations, especially when the solutions have to be handled in air?

13.12 Which of the salts, $MgSeO_4$ or $BaSeO_4$, would be expected to be more soluble in water?

13.13 2.4% of all global CO_2 emissions (36 gigatonnes in 2021) derive from cement production. Assuming all this CO_2 is derived from the decomposition of $CaCO_3$ calculate the annual level of CaO production for the cement industry

13.14 How can Ra be separated from solutions containing the other Group 2 metal cations?

13.15 Which Group 2 salts are used as drying agents and why?

13.16 Predict structures for BeTe and BaTe using an ionic radius for Te^{2-} of 207 pm.

13.17 Use the data in Table 1.7 and the Ketelaar triangle in Fig. 2.40 to predict the nature of the bonding in $BeBr_2$, $MgBr_2$, and $BaBr_2$.

13.18 The two Grignard compounds C_2H_5MgBr and $2,4,6\text{-}(CH_3)_3C_6H_2MgBr$ both dissolve in THF. What differences would be expected in the structures of the species formed in these solutions?

13.19 Predict the products of the following reactions:

(a) $MgCl_2 + LiC_2H_5 \rightarrow$

(b) $Mg + (C_2H_5)_2Hg \rightarrow$

(c) $Mg + C_2H_5HgCl \rightarrow$

TUTORIAL PROBLEMS

13.1 Marble and limestone buildings are eroded by contact with acid rain. Define the term 'acid rain' and discuss the origins of the acidity. Describe the processes by which the marble and limestone are attacked.

13.2 Discuss the advantages and disadvantages of calcium looping for CO_2 capture in comparison with amine scrubbing.

13.3 In their paper 'Noncovalent interaction of chemical bonding between alkaline earth cations and benzene?' (*Chem. Phys. Lett.*, 2001, **349**, 113), X. J. Tan and co-workers carried out theoretical calculations of complexes formed between beryllium, magnesium, and calcium ions and benzene. To which orbital interactions were the binding of the alkali metal to benzene attributed? How was the C–C bond length in benzene affected by this interaction? Place the M–C bonds in order of increasing bond enthalpy. How did the strength of the bonds compare to those formed between Group 1 elements and benzene? Sketch the geometry of the metal–benzene complexes.

13.4 Discuss beryllium fluoride glasses, noting how the chemistry of BeF_2 is analogous to that of SiO_2.

13.5 P. C. Junk and J. W. Steed (*J. Chem. Soc., Dalton. Trans.*, 1999, 407) prepared crown ether complexes from the nitrates of Mg, Ca, Sr, and Ba. Outline the general procedure that was used for the syntheses. Sketch the structures of the two crown ethers used. Comment on the structures of the complexes and how they change with different cations.

13.6 Use your knowledge of chemical trends in Group 2 chemistry and data from Table 13.1 to predict the chemistry of radium. Compare your predictions with experimental observations. See, for example, H. W. Kirby and M. L. Salutsky, *The radiochemistry of radium*. *Nuclear Science Series, National Academy of Sciences, National Research Council. National Bureau of Standards, US Department of Commerce* (1964). http://library.lanl.gov/cgi-bin/getfile?rc000041.pdf

13.7 The aluminosilicates form a large group of minerals based on linked AlO_4 and SiO_4 tetrahedra which includes many clay and zeolite minerals whose compositions may be written $M^{x+}[Al_xSi_{1-x}]O_4]^{x-}$. Discuss the occurrence of the BeO_4 unit in natural minerals. To what extent has it proved possible to produce beryllophosphates based on linked BeO_4 and PO_4 tetrahedral units that are structural and compositional analogues of $M^{x+}[Al_xSi_{1-x}]O_4]^{x-}$ and $M^{x+}[Be_xP_{1-x}]O_4]^{x-}$?

13.8 An experiment was carried out to determine the hardness of domestic water. A few drops of pH = 10 buffer were added to a 100 cm³ sample of the water. This sample was titrated against 0.01 M edta(aq) using Eriochrome Black T indicator and gave a titre of 33.8 cm³. Under these circumstances both Mg^{2+} and Ca^{2+} ions react with edta. A second 100 cm³ sample was titrated against the edta solution after 5.0 cm³ of 0.1 M NaOH(aq) and a few drops of murexide indicator had been added. Under these conditions only the Ca^{2+} ions react with the edta and a titre of 27.5 cm³ was obtained. Determine the hardness of the water sample in terms of the concentration of the Mg^{2+} and of the Ca^{2+} ions.

13.9 The synthesis of a Mg(I) compound has been reported (*Science*, 2007, **318**, 1754). Describe how this synthesis was carried out and how this unusual Group 2 oxidation state was stabilized. Discuss how Mg(I) compounds are being developed for the activation of C–H bonds in inert arenes (*Angew. Chem. Int. Ed.*, 2021, **60**, 7087–92).

13.10 Discuss the bonding in the C-Be-C unit of zero valent beryllium compounds (*Nature Chemistry*, 2016, **8**, 890–4).

13.11 'Nature is adept at controlling mineralisation processes to produce single crystals and polycrystalline and amorphous structures with remarkable morphologies and mechanical properties.' (F. Meldrum Int. *Mater. Rev.*, 2003, **48**, 187; F. Meldrum and H Cölfen *Chem. Rev.*, 2008, **108**, 4332–432). Calcium carbonate is one of the most common biominerals. Summarize the chemistry involved in the biomimetic engineering of $CaCO_3$.

The Group 13 elements

Part A: The essentials
14.1 **The elements**
14.2 **Compounds**
14.3 **Boron clusters and borides**

Part B: The detail
14.4 **Occurrence and recovery**
14.5 **Uses of the elements and their compounds**
14.6 **Simple hydrides of boron**
14.7 **Boron trihalides**
14.8 **Boron–oxygen compounds**
14.9 **Compounds of boron with nitrogen**
14.10 **Metal borides**
14.11 **Higher boranes and borohydrides**
14.12 **Metallaboranes and carboranes**
14.13 **The hydrides of aluminium, gallium, indium, and thallium**
14.14 **Trihalides of aluminium, gallium, indium, and thallium**
14.15 **Low oxidation state halides of aluminium, gallium, indium, and thallium**
14.16 **Oxo compounds of aluminium, gallium, indium, and thallium**
14.17 **Sulfides of gallium, indium, and thallium**
14.18 **Compounds with Group 15 elements**
14.19 **Zintl phases**
14.20 **Organometallic compounds**

Further reading

Exercises

Tutorial problems

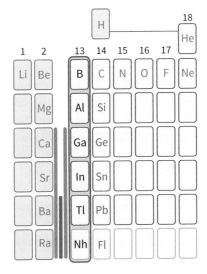

There are clear trends in the chemical properties of the Group 13 elements, such as oxidation number and amphoteric character, that are repeated in the other groups of the p block. In this chapter we look at the occurrence and isolation of each element in Group 13 and consider the chemical properties of the elements and their simple compounds, the role of the inert pair effect, coordination compounds, and organometallic compounds.

The stable elements of Group 13—boron, aluminium, gallium, indium, and thallium—have diverse physical and chemical properties. The first member of the group, boron, is essentially non-metallic, whereas the heavier members of the group are distinctly metallic. Aluminium is the most important element commercially and is produced on a massive scale for a wide range of applications requiring a strong, light metal. Boron forms a large number of cluster compounds involving hydrogen, metals, and carbon. Gallium and indium in alloys and compounds have important electronic and optical properties. Element 113, with the name nihonium was discovered in 2004; its most stable isotope has a half-life of only 20 s and the chemistry of this element is not known.

PART A: The essentials

In this section we discuss the essential features of the chemistry of the Group 13 elements.

14.1 The elements

KEY POINTS Boron is the only nonmetal in the group. Aluminium is the most abundant Group 13 element.

The elements of Group 13 show a wide variation in abundance in crustal rocks, the oceans, and the atmosphere. Aluminium is abundant, but the low cosmic and terrestrial abundance of boron, like that of lithium and beryllium, reflects how some of the lightest elements are sidestepped in nucleosynthesis (Box 1.1). The low abundance of heavier members of the group is in keeping with the progressive decrease in nuclear stability of the elements that follow iron. Boron occurs naturally as borax, $Na_2B_4O_5(OH)_4 \cdot 8H_2O$, and kernite, $Na_2B_4O_5(OH)_4 \cdot 2H_2O$, from which the impure element is obtained. Aluminium occurs in numerous clays and aluminosilicate minerals but the commercially most important mineral is bauxite, a complex mixture of hydrated aluminium hydroxide and aluminium oxide, from which it is extracted on an immense scale. Gallium oxide occurs as an impurity in bauxite and is normally recovered as a by-product of the manufacture of aluminium. Indium and thallium occur in trace amounts in many metal sulfide minerals, including iron pyrites FeS_2.

Whereas the elements of the s and d blocks are all metallic, the elements of the p block range from nonmetals, often through metalloids, to metals. This variety results in a diversity of chemical properties and some distinctive trends (Chapters 9 and 10). There is an increase in metallic character from B to Tl: B is a nonmetal; Al is essentially metallic, although it can be classed as a metalloid on account of its amphoteric character; Ga, In, and Tl are metals, with Ga and In showing some amphoteric chemistry forming, for example, gallates and thioindates, $[InS_4]^{5-}$, in strongly basic conditions. Associated with this trend is a variation from predominantly covalent to ionic bonding in the compounds of the elements that can be rationalized in terms of the increase in atomic radius and related decrease in ionization energy down the group (Table 14.1). Because the ionization energies of the heavier elements are low, the metals form cations increasingly readily down the group. In contrast to the expected trend in electronegativity (Section 1.7(e)), Ga demonstrates the alternation effect (Section 10.4) and is more electronegative than Al.

As discussed in Section 10.2, the first member of each group differs from its congeners on account of its small atomic radius. This difference is particularly evident in Group 13, where the chemical properties of B are very distinct from those of the rest of the group. However, B does have a pronounced diagonal relationship with Si in Group 14, Section 10.3.

The valence electron configuration of the Group 13 elements is ns^2np^1 and, as this configuration suggests, all the elements adopt the +3 oxidation state in their compounds. However, the heavier elements of the group demonstrate very clearly the 'inert pair effect', Section 10.7, and also form compounds with the metal in the +1 oxidation state that have increasing stability down the Group. In fact, the most common oxidation state exhibited by Tl is Tl(I). This trend is particularly evident with the halides, with the relative stability of MX increasing relative to MX_3. Thallium(I) is intensely poisonous because its ionic radius is very similar to that of potassium ions: it enters cells and disrupts the mechanisms of potassium and sodium transport (Section 26.4, Example 26.3).

TABLE 14.1 Selected properties of the elements

	B	Al	Ga	In	Tl
Covalent radius/pm	84	121	122	142	145
Metallic radius/pm	–	143	153	167	171
Ionic radius, $r(M^{3+})$/pm*	27	53	62	80	89
Melting point/°C	2300	660	30	157	304
Boiling point/°C	3930	2470	2403	2072	1473
First ionization energy, I_1/kJ mol^{-1}	799	577	577	556	590
Second ionization energy, I_2/kJ mol^{-1}	2427	1817	1979	1821	1971
Third ionization energy, I_3/kJ mol^{-1}	3660	2745	2963	2704	2878
Electron affinity, E_a/kJ mol^{-1}	26.7	42.5	28.9	28.9	19.2
Pauling electronegativity	2.04	1.61	1.81	1.78	2.04
$E^{\ominus}$ (M^{3+}, M)/V	−0.89	−1.68	−0.53	−0.34	+0.72

* For coordination number 6.

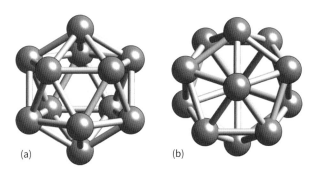

FIGURE 14.1 A view of the B_{12} icosahedron in rhombohedral boron (a) along and (b) perpendicular to the three-fold axis of the crystal. The individual icosahedra are linked by 3c,2e bonds.

Boron exists in several allotropes. Amorphous boron is a brown powder but the hard and refractory crystalline B forms shiny black crystals. The three solid phases for which crystal structures are available contain the icosahedral (20-faced) B_{12} unit as a building block (Fig. 14.1). This icosahedral unit is a recurring motif in boron chemistry and is found in the structures of metal borides and boron hydrides. The icosahedral unit is also found in some intermetallic compounds of other Group 13 elements, such as Al_5CuLi_3, $RbGa_7$, and K_3Ga_{13}. Boron is inert, and under normal conditions finely divided B is attacked only by F_2 and HNO_3.

Even though Al is an electropositive metal it is very inert on account of the presence of a passivating surface oxide film. If this film is removed, then Al is rapidly oxidized by air. Aluminium has a high reflectance, which is maintained in the powdered form, making it a useful component of silver-coloured paints. It is a good thermal and electrical conductor. Gallium is brittle at low temperatures but melts at 30°C. Its low melting point is attributed to its crystal structure, in which each Ga atom has only one nearest neighbour and six next-nearest neighbours: thus, solid Ga requires little thermal energy (ΔH_{fus} = 5.59 kJ mol^{-1}) to dissociate and form the Ga–Ga pairs present in the liquid state. Gallium has a wide liquid range (30–2403°C) and wets glass and skin, making it difficult to handle. Gallium readily forms alloys with other metals and diffuses into their lattices, making them brittle. Indium forms a distorted ccp lattice and Tl is hexagonal close-packed.

14.2 Compounds

KEY POINTS All of the elements form hydrides, oxides, and halides in the +3 oxidation state. The +1 oxidation state becomes more stable down the group and is the most stable oxidation state for compounds of thallium.

The Group 13 elements have an ns^2np^1 valence electron configuration, which yields a maximum of six electrons in the valence shell when three covalent bonds are formed by electron sharing. As a result, many of their compounds have an incomplete octet and act as Lewis acids, being able to complete their octet by accepting a pair of electrons from a donor. Moreover, as is typical of an element at the head of its group, the chemical properties of B and its compounds are strikingly different from those of its congeners.

> **A NOTE ON GOOD PRACTICE**
>
> Be careful to distinguish electron deficiency from the possession of an incomplete octet. The former refers to the lack of sufficient electrons to account for the connections between atoms as normal covalent bonds; the latter is the possession of fewer than eight electrons in a valence shell.

The binary hydrogen compounds of boron are called boranes. The simplest member of the series, diborane, B_2H_6 (**1**), is electron-deficient and its structure is commonly described in terms of 2c,2e and 3c,2e bonds (Section 3.11(e)): bridging 3c,2e bonds are a recurring theme in borane chemistry.

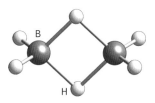

1 Diborane, B_2H_6

All the boron hydrides burn with a characteristic green flame, and several of them ignite explosively on contact with air. Alkali metal tetrahydridoborates, $NaBH_4$ and $LiBH_4$, are very useful in the laboratory as general reducing agents and as precursors for most boron–hydrogen compounds. Alkali and alkaline earth metal tetrahydridoborates and ammonia borane, NH_3BH_3, have been investigated as hydrogen storage materials (Box 14.1).

Boron trihalides consist of trigonal-planar BX_3 molecules. Unlike the halides of the other elements in the group, they are monomeric in the gas, liquid, and solid states. Boron trifluoride and boron trichloride are gases, the tribromide is a volatile liquid, and the triiodide is a solid (Table 14.2).

This trend in volatility is consistent with the increase in strength of dispersion forces with the number of electrons in the molecules. Boron trihalides have an incomplete octet and are Lewis acids. The order of Lewis acidity is $BF_3 < BCl_3 \leq BBr_3$ and is contrary to the order that might be expected from consideration of the electronegativity of the attached halogens (Section 5.8(b)). The electron deficiency is partially removed by X–B π bonding between the halogen atoms and the B atom, giving rise to the occupation of the vacant p orbital on the B atom by electrons donated by the halogen atoms (Fig. 14.2). The trend in Lewis acidity stems

BOX 14.1 How are the compounds of the Group 13 elements used for hydrogen storage?

Hydrogen fuel cells are seen as an alternative to carbon-based fuels and are starting to find applications in mobile technologies and motor vehicles. Efficient fuel cells demand an effective source of hydrogen, and many methodologies for storing hydrogen have been investigated. These include using high pressure and porous materials, but others focus on chemical compounds that generate H_2 on heating or on reaction with water. The boron and aluminium hydrides come into this last category. Compounds of interest for commercial applications have a high mass percentage hydrogen content. The values for $LiBH_4$, $NaBH_4$, $LiAlH_4$, and AlH_3 are approximately 18, 11, 11, and 10 mass percent, respectively.

Sodium tetrahydridoborate, $NaBH_4$, reacts with water to generate hydrogen gas in an exothermic reaction.

$$NaBH_4(aq) + 4H_2O(l) \rightarrow 4H_2(g) + NaB(OH)_4(aq)$$

$$\Delta_r H^\ominus = -2300 \text{ kJ mol}^{-1}$$

The reaction requires a nickel or platinum catalyst and rapidly produces moist hydrogen for the engine or fuel cell.

The $NaBH_4$ is used as a 30 mass percent solution in water, and the fuel is thus a nonvolatile, nonflammable liquid at atmospheric pressure. There are no side reactions or volatile by-products, and the borate product can be recycled, though not directly.

Ammonia borane, BH_3NH_3, with a hydrogen content of 21 mass percent, has also been investigated for hydrogen generation. It was also investigated as a rocket fuel in the 1950s, but the studies were abandoned. Ammonia borane decomposes to liberate hydrogen when heated to 500°C. The residue is boron nitride, which cannot be easily recycled. Recent studies have investigated the hydrogen storage potential of the ammonia complex of magnesium borohydride, $Mg(BH_4)_2 \cdot 2NH_3$ (Section 30.8). The complex contains 16 mass percent hydrogen, which is released when a solution of the complex flows over a ruthenium catalyst. The complex begins to decompose at 150°C, with a maximum hydrogen release rate at 205°C, making it competitive with ammonia borane, BH_3NH_3, as a hydrogen storage material, Section 30.8.

TABLE 14.2 Properties of the boron trihalides

	BF_3	BCl_3	BBr_3	BI_3
Melting point/°C	−127	−107	−46	50
Boiling point/°C	−100	13	91	210
Bond length/pm	130	175	187	210
$\Delta_f G^\ominus$/kJ mol^{-1}	−1112	−339	−232	+21

from more efficient X–B π bonding for the lighter, smaller halogens, the F–B bond being one of the strongest formally single bonds known.

The most important oxide of B, B_2O_3, is prepared by dehydration of boric acid.

$$4B(OH)_3(s) \xrightarrow{\Delta} 2B_2O_3(s) + 6H_2O(l)$$

The vitreous form of the oxide consists of a network of partially ordered trigonal BO_3 units. Crystalline B_2O_3 consists of an ordered network of BO_3 units joined through O atoms. Metal oxides dissolve in molten B_2O_3 to produce coloured glasses. Boron oxide and silica are the main constituents of borosilicate glass, which, because of the low thermal expansivity of the glass due to the strong B–O bonds, is used to make heat-resistant laboratory glassware.

There are many molecular compounds that contain BN bonds, and many of them are analogous to carbon compounds. The similarities between compounds containing BN and CC units can be explained by the fact that these units are isoelectronic. The simplest compound of B and N, boron nitride, BN, is easily synthesized by heating boron oxide with a nitrogen compound (Box 14.2):

$$B_2O_3(l) + 2NH_3(g) \xrightarrow{1200°C} 2BN(s) + 3H_2O(g)$$

The structure of one form of boron nitride consists of planar sheets of atoms like those in graphite (Section 15.5), and some of the physical properties of BN are similar to those of graphite. For example, both graphite and BN have a slippery feel and are used as lubricants. However, BN is a white, nonconducting solid, not a black, metallic conductor. Apart from layered boron nitride, the best-known unsaturated compound of B and N is borazine, $B_3N_3H_6$ (2), which is isoelectronic and isostructural with benzene and, like benzene, is a colourless liquid (b.p. 55°C).

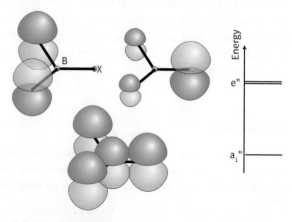

FIGURE 14.2 The bonding π orbitals of boron trihalides are largely localized on the electronegative halogen atoms, but overlap with the p_z orbital of boron is significant in the a_1'' orbital.

| BOX 14.2 | **What properties of boron nitride lead to applications?** |

Hexagonal boron nitride was first developed to meet the needs of the aerospace industry. It is stable in oxygen and is not attacked by steam below 900°C. It is a good thermal insulator, has low thermal expansion, and is resistant to thermal shock. These applications have led to its use in industry to make high-temperature crucibles. The powder is used as a mould-release and thermal insulator. Boron nitride nanotubes have been formed by depositing boron and nitrogen on a tungsten surface under high vacuum. These nanotubes could be suitable for high-temperature conditions under which carbon nanotubes would burn. BN nanotubes also offer the possibility of ambient-temperature hydrogen storage as they have been found to take up 2.6 weight percent of H_2, at elevated pressure.

The softness and sheen of powdered boron nitride has led to its widest application, in the cosmetics and personal-care industries. It is nontoxic and presents no known hazard, and is added to many products up to around 10%. It adds a pearlescent sheen to products such as nail polishes and lipsticks, and is added to foundations to hide wrinkles. Its light-reflective properties scatter the light, making wrinkles less noticeable.

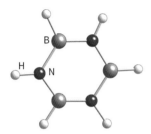

2 Borazine, $B_3N_3H_6$

The elements Al, Ga, In, and Tl are metals with many similarities in their chemical properties. Like B, they form electron-deficient compounds that act as Lewis acids. Aluminium forms alloys with many other metals and produces light, corrosion-resistant materials. When Al is alloyed with Ga, the Ga prevents the formation of the tightly held passivating oxide film on the Al. When the alloy is dropped in water the Al reacts with the water, forming aluminium hydroxide and liberating hydrogen.

Aluminium hydride, AlH_3, is a polymeric solid formed from AlH_6 octahedra linked through hydrogen bridges. Unlike CaH_2 and NaH, which are more readily available commercially, AlH_3 has few applications in the laboratory, but $NaAlH_4$ is a widely used reducing agent. The alkyl-aluminium hydrides, such as $Al_2(C_2H_5)_4H_2$, are well-known molecular compounds and contain Al–H–Al $3c,2e$ bonds (Section 3.11(e)). Gallium hydride can be synthesized as the unstable gas phase species digallane, Ga_2H_6, which when trapped at low temperature exhibits an analogous structure to diborane. InH_3 has only been isolated in an inert matrix at low temperature; TlH_3 has not been isolated.

All the elements form trihalides with the metal in its +3 oxidation state. However, as expected from the inert-pair effect (Section 10.7), the +1 oxidation state becomes more common on descending the group, and Tl forms stable monohalides. Gallium, In, and Tl also form mixed I/III oxidation state halides. Because the F^- ion is so small, the trifluorides are mechanically hard ionic solids that have much higher melting points and sublimation enthalpies than the other halides. Their high lattice enthalpies also result in their having very limited solubility in most solvents, and they do not act as Lewis acids to simple donor molecules. The heavier trihalides of Al, Ga, and In are soluble in a wide variety of polar solvents and are excellent Lewis acids. The trigonal-planar MX_3 monomer occurs only at elevated temperatures in the gas phase. Otherwise, the trihalides exist as M_2X_6 dimers in the vapour phase and in solution. The solid forms of the halides MX_3 mainly contain dimeric units and are volatile; an exception is $AlCl_3$, which has a six-coordinate layer structure in the solid phase that converts to four-coordinate molecular dimers (3) at its low melting point of 192°C. TlF and TlI have distorted rocksalt-type structures, and TlCl and TlBr adopt the CsCl structure type.

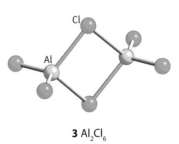

3 Al_2Cl_6

The most stable form of Al_2O_3, α-alumina, is a very hard, refractory, and amphoteric material. Dehydration of aluminium hydroxide at temperatures below 900°C leads to the formation of γ-alumina, which is a metastable polycrystalline form with a defect spinel structure (Section 4.9(b)) and a very high surface area. The α and γ forms of Ga_2O_3 have the same structures as their Al analogues. Indium and thallium form In_2O_3 and Tl_2O_3. Thallium also forms the Tl(I) oxide and peroxide, Tl_2O and Tl_2O_2, respectively.

14 The Group 13 elements

EXAMPLE 14.1   Understanding the stabilities of oxides

Explain why In_2O_3 is stable to over 2000°C while Tl_2O_3 readily decomposes to Tl_2O and O_2 on heating to 600°C in nitrogen.

Answer This is a manifestation of the inert pair effect (Section 10.7). Additional energy is required to form the Group 13 element in its higher oxidation state through ionization from M^+ to M^{3+}. As the group is descended, the lattice enthalpies of the oxide in the higher oxidation state will be decreased due to the larger ionic radius as $\Delta_L H \propto 1/(r_+ + r_-)$. For the thallium oxides there is a fine balance between this decreased lattice enthalpy and increased ionization enthalpy, such that thallium(III) oxide decomposes to Tl_2O on heating if the oxygen is removed. For indium, the thermodynamic balance lies strongly in favour of In_2O_3. Note that the heats of formation of In_2O_3 and Tl_2O_3 are -923 kJ mol^{-1} and -94 kJ mol^{-1}, respectively, demonstrating the greater stability of In_2O_3.

Self-test 14.1 Which oxide would you expect to be formed by heating nihonium in air?

The most important oxo salts of Group 13 are the alums, $MAl(SO_4)_2 \cdot 12H_2O$, where M is a univalent cation such as Na^+, K^+, Rb^+, Cs^+, Tl^+, or NH_4^+. Gallium and In can also form analogous series of salts of this type, but B and Tl do not. The alums can be thought of as double salts containing the hydrated trivalent cation $[Al(OH_2)_6]^{3+}$. The remaining water molecules form hydrogen bonds between the cations and sulfate ions. The mineral alum, $KAl(SO_4)_2 \cdot 12H_2O$, from which aluminium takes its name, is the only common, water-soluble, aluminium-bearing mineral. It has been used since ancient times as a mordant to fix dyes to textiles. The mordant forms a coordination complex with the dye, which then attaches to the fabric, preventing it from being washed out. The term 'alum' is used widely to describe other compounds with the general formula $M(I)M'(III)(SO_4)_2 \cdot 12H_2O$, where M' is often a d-block metal, such as Fe in 'ferric alum', $KFe(SO_4)_2 \cdot 12H_2O$.

14.3 Boron clusters and borides

KEY POINT Boron forms an extensive range of polymeric, cage-like compounds which include the borohydrides, metallaboranes, and carboranes.

In addition to the simple hydrides, B forms several series of neutral and anionic polymeric cage-like boron–hydrogen compounds. Borohydrides are formed with up to 12 B atoms and fall into three classes called *closo*, *nido*, and *arachno*.

The borohydrides with the formula $[B_nH_n]^{2-}$ have a *closo* structure—a name derived from the Greek for 'cage'. This series of anions is known for $n = 5$ to 12, and examples include the trigonal-bipyramidal $[B_5H_5]^{2-}$ ion (4), the octahedral $[B_6H_6]^{2-}$ ion (5), and the icosahedral $[B_{12}H_{12}]^{2-}$ ion (6).

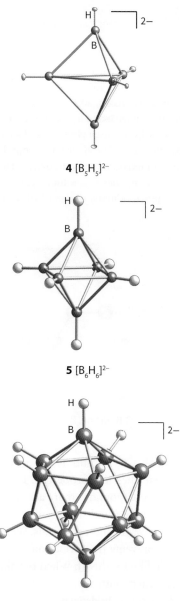

4 $[B_5H_5]^{2-}$

5 $[B_6H_6]^{2-}$

6 $[B_{12}H_{12}]^{2-}$

When boron clusters have the formula B_nH_{n+4} they adopt the *nido* structure—a name derived from the Latin for 'nest'. An example is B_5H_9 (7). Clusters of formula $[B_nH_{n+6}]$ have an *arachno* structure, from the Greek for 'spider' (as they resemble untidy spiders' webs). One example is pentaborane(11) (B_5H_{11}, 8).

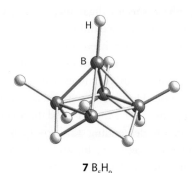

7 B_5H_9

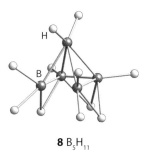

8 B_5H_{11}

Boron forms many metal-containing clusters called **metallaboranes**. In some cases the metal is attached to a borohydride ion through hydrogen bridges. A more common and generally more robust group of metallaboranes have direct M–B bonds. Closely related to the polyhedral boranes and borohydrides are the **carboranes** (more formally, the carbaboranes)—a large family of clusters that contain both B and C atoms. An analogue of $[B_6H_6]^{2-}$ (**5**) is the neutral carborane $B_4C_2H_6$ (**9**). Other heteroatoms such as N, P, and As can also be introduced into boranes.

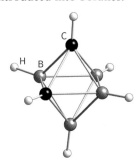

9 *closo*-1,2-$B_4C_2H_6$

In combination with less electronegative elements, such as metals, boron forms borides. These compounds contain a wide variety of boron-containing structural features ranging from single atoms, through B_2 pairs, chains, polyhedra, such as icosahedra, to three-dimensional networks. Many of these compounds are hard, refractory materials having melting points in excess of 2000°C.

> **EXAMPLE 14.2** Representing the structures of metal borides
>
> Draw the structure of calcium boride CaB_6, which is based on the CsCl structure with calcium replacing Cs and octahedral $[B_6]^{2-}$ ions replacing Cl. What is the coordination number of calcium to boron in this structure?
>
> **Answer** The CsCl structure type is simple cubic with Cs on the unit cell corners and Cl at the unit cell centre. Replacing the atoms as indicated produces the structure shown in Fig. 14.3. The face of the $[B_6]$ octahedron at the centre of the cube in Fig. 14.3 is directed at the calcium ion, and each calcium ion is surrounded by eight $[B_6]$ octahedra, so the total coordination number of calcium to boron is 3 × 8 = 24.
>
>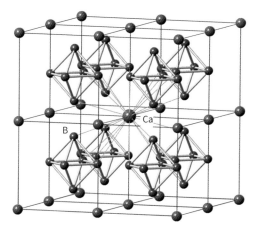
>
> **FIGURE 14.3** The structure of CaB_6 highlighting the $[B_6]^{2-}$ octahedron and coordination of the central Ca^{2+} cation.
>
> **Self-test 14.2** Draw a schematic representation of the structure of YB_{12} which adopts a rock-salt arrangement of Y^{3+} cations and cuboctahedral $[B_{12}]^{3-}$ anions.

PART B: The detail

In this section a more detailed discussion of the chemistry of the elements of Group 13 is presented. This discussion interprets the trend from nonmetallic to metallic character that occurs down the group and the impact of the incomplete octet and associated Lewis acidity on the properties of the Group 13 compounds. The contrasting properties of boron are dealt with in several sections.

14.4 Occurrence and recovery

KEY POINTS Aluminium is highly abundant; gallium, indium, and thallium are rare, and are obtained as by-products in the extraction of more common metals.

Boron exists in several hard and refractory allotropes. The three phases for which crystal structures, under ambient conditions, have been determined contain the icosahedral (20-faced) B_{12} unit as a building block (Fig. 14.1). This icosahedral unit is a recurring motif in boron chemistry, and we shall meet it again in the structures of metal borides and boron hydrides. A high-pressure phase of boron contains a mixture of B_{12} icosahedra and B_2 pairs.

Boron occurs naturally as borax, $Na_2B_4O_5(OH)_4 \cdot 8H_2O$, and kernite, $Na_2B_4O_5(OH)_4 \cdot 2H_2O$, from which the impure element is obtained. The borax is converted to boric acid, $B(OH)_3$, and then to boron oxide, B_2O_3. The oxide is reduced with magnesium and washed with alkali and then hydrofluoric acid. Pure boron is produced by reduction of BBr$_3$ vapour with H$_2$:

$$2BBr_3(g) + 3H_2(g) \rightarrow 2B(s) + 6HBr(g)$$

Aluminium is the most abundant metallic element in the Earth's crust and makes up approximately 8% by mass of crustal rocks. It occurs in numerous clays and aluminosilicate minerals, but the commercially most important mineral is bauxite, a complex mixture of hydrated aluminium hydroxide and aluminium oxide, from which it is extracted by the Hall–Héroult process on an immense scale (Section 6.19). In this process the Al$_2$O$_3$ is dissolved in molten cryolite, Na$_3$AlF$_6$, the mixture is electrolysed, and aluminium is deposited at the cathode. The process is very energy-intensive, but the expense is offset by the scale of production, the low cost of the raw material, and frequently the use of hydroelectric power. The oxide, alumina, occurs naturally as ruby, sapphire, corundum, and as a constituent of emery. Aluminium metal adopts a face-centred cubic unit cell of dimension 4.041 Å.

Gallium oxide occurs as an impurity in bauxite and is normally recovered as a by-product of the manufacture of aluminium. The process results in the concentration of gallium in the residues, from which it is extracted by electrolysis.

Indium is produced as a by-product of the extraction of lead and zinc, and is isolated by electrolysis. Thallium compounds are found in flue dust, the fine particles emitted with the gases produced during smelting. The dust is dissolved in dilute sulfuric acid; hydrochloric acid is then added to precipitate thallium(I) chloride, and the metal is extracted by electrolysis.

14.5 Uses of the elements and their compounds

KEY POINTS The most useful compound of boron is borax; the most commercially important element is aluminium.

The main use of boron is in borosilicate glasses. Borax, Na$_2$B$_4$O$_5$(OH)$_4$·8H$_2$O, used to have many domestic applications, for example, as a water softener, cleaner, and mild pesticide; however, the UK and EU have banned these uses due to concerns about reproductive toxicity. Amorphous brown boron is used in pyrotechnics to impart a bright green colour. Boron is an essential micronutrient in plants. Lightweight, strong boron filaments are used in composite materials for the aerospace industry and in sports equipment.

Many compounds of boron, the borides, are superhard materials, having hardness approaching that of diamond. Cubic boron nitride is synthesized at high pressures, which makes it expensive. Rhenium diboride does not require high pressures and so production is relatively cheap, but Re is an expensive metal. The material known as 'heterodiamond', sometimes labelled BCN, is formed from diamond and boron nitride by explosive shock synthesis. These compounds are used as substitutes for diamond in cutting tools and blades.

Sodium perborate, NaBO$_3$·H$_2$O, which exists as the dimer Na$_2$B$_2$O$_4$(OH)$_4$, is used as a chlorine-free bleach in laundry products; its use as a tooth whitener has now been banned in Europe due to safety concerns. It is less aggressive to textiles than chlorine bleaches and active at low temperatures (40°C) when mixed with an activator such as tetraacetylethylenediamine (TAED). Sodium tetrahydridoborate, NaBH$_4$, is used on a massive scale for bleaching wood pulp. Boranes used to be popular as rocket fuels but were found to be too pyrophoric to be handled safely. Boranes are being investigated as possible hydrogen storage materials, with the hydrogen stored as the ammonia–borane complex H$_3$N:BH$_3$ (Box 11.5 and Section 30.8).

Aluminium is the most widely used nonferrous metal. The technological uses of aluminium metal exploit its lightness, resistance to corrosion, and the fact that it is easily recycled. It is used in cans, foils, utensils, construction, and aircraft alloys (Box 14.3). Many Al compounds are used as mordants, in water and sewage treatment, in paper production, as food additives, and for waterproofing textiles. Aluminium chloride and chlorohydride are used in antiperspirants, and the hydroxide is used as an antacid. Sodium tetrahydridoaluminate, NaAlH$_4$, doped with TiF$_3$ is used as a hydrogen storage material.

Because the melting point of Ga is just above room temperature (30°C) it is used in high-temperature thermometers. Gallium and indium form a low-melting-point alloy that is used as the safety device in sprinkler systems. Both elements are deposited on glass surfaces to form corrosion-resistant mirrors, and In$_2$O$_3$ doped with Sn is used as a transparent, conducting coating for electronic displays and as a heat-reflective coating for light bulbs. Gallium nitride is used in blue LEDs and laser diodes and is the basis of Blu-ray® technology. It is insensitive to ionizing radiation, and is used in solar cells in satellites. Gallium arsenide is a semiconductor and is used in integrated circuits, light-emitting diodes, and solar cells. Thallium compounds were once used to treat ringworm and as a rat and ant poison. However, this application has been banned because of their very high toxicity, which arises from the transport of Tl$^+$ ions across cell membranes together with K$^+$ ions (Section 26.4, Example 26.3). Thallium is absorbed more preferentially by tumour cells, and has been used in nuclear medicine as an imaging agent.

14.6 Simple hydrides of boron

The simplest hydride of boron is gaseous diborane, B$_2$H$_6$. Higher boranes exist and can be liquids such as B$_5$H$_9$ and solids such as B$_{10}$H$_{14}$. The boranes are cleaved by Lewis bases.

BOX 14.3 How and why is aluminium recycled?

Aluminium production is energy-intensive and expensive, and for these reasons aluminium recycling is very attractive. Aluminium has been recycled for many decades, but the popularity of aluminium beverage cans and the focus on domestic recycling has increased the activity enormously. The aluminium economy is a good example of a circular economy, where a waste product becomes the feedstock of another process, minimizing the impact on the environment and natural resources. The aluminium is not consumed during a product's lifetime, but has the potential to be recycled and reused many times, as the process has no detrimental effects on the metal's physical or chemical properties.

The cost of recycling aluminium is 5% of the cost of extracting aluminium from bauxite, even when the costs of collection and separation are taken into account. There are added benefits associated with the reduced use of landfill and the cost of shipping new aluminium or bauxite. The process is fairly straightforward. Cans are separated from other waste and shredded into small pieces which are cleaned and formed into blocks. The formation of blocks minimizes oxidation. The blocks are then heated in a furnace at 750°C to produce molten aluminium. Solid waste is removed, and any dissolved hydrogen is degassed by adding ammonium perchlorate, which decomposes to liberate chlorine that reacts with the hydrogen, nitrogen, and oxygen. Additives can be added to alter the properties of the final alloy. The aluminium is then cast into ingots. Recycling the aluminium from motor vehicles is also well established and follows the same basic process but with more complex separation stages in order to separate out other metals, polymers, textiles, etc.

The global recycling rate for aluminium beverage cans is over 75%, with Brazil leading the way with 98% of all cans being recycled. The global recycling rate for aluminium from building and transport uses is approaching 90%, and recycled aluminium plays an increasing role in meeting society's demands for the metal.

(a) Boranes

KEY POINTS Diborane can be synthesized by metathesis between a boron halide and a hydride source; many of the higher boranes can be prepared by the partial pyrolysis of diborane; all the boron hydrides are flammable, sometimes explosively, and many of them are susceptible to hydrolysis.

Diborane, B_2H_6, can be prepared in the laboratory by metathesis of a boron halide with either $LiAlH_4$ or $LiBH_4$ in ether:

$$3\,LiBH_4(et) + 4\,BF_3(et) \rightarrow 3\,LiBF_4(et) + 2\,B_2H_6(g)$$

That this reaction is a metathesis (an exchange of partners) can be seen by writing it in the simplified form

$$\tfrac{3}{4}BH_4^-(et) + BF_3(et) + \tfrac{3}{4}BF_4^-(et) + BH_3(g)$$

Both $LiBH_4$ and $LiAlH_4$, like LiH, are good reagents for the transfer of H^-, but they are generally preferred over LiH and NaH because they are soluble in ethers. The synthesis is carried out with the strict exclusion of air (typically under an inert atmosphere) because diborane ignites on contact with air. Diborane decomposes very slowly at room temperature, forming higher boron hydrides and a nonvolatile and insoluble yellow solid that consists of $B_{10}H_{14}$ and the polymeric species BH_n.

The compounds fall into two classes. One class has the formula B_nH_{n+4}, and the other, which is richer in hydrogen and less stable, has the formula B_nH_{n+6}. Examples include pentaborane(11), B_5H_{11} (8), tetraborane(10), B_4H_{10} (10), and pentaborane(9), B_5H_9 (7). The IUPAC nomenclature, in which the number of boron atoms is specified by a prefix and the number of hydrogen atoms is given in parentheses, should be noted. Thus, the systematic name for diborane is diborane(6); however, as there is no diborane(8), the simpler term 'diborane' is almost always used.

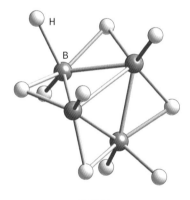

10 B_4H_{10}

All the boranes are colourless and diamagnetic. They range from gases (B_2H_6 and B_4H_8), through volatile liquids (B_5H_9 and B_6H_{10} hydrides), to the sublimable solid $B_{10}H_{14}$. All the boron hydrides are flammable, and several of the lighter ones, including diborane, react spontaneously with air, often with explosive violence and a green flash (an emission from an excited state of the reaction intermediate BO). The final product of the reaction is boric acid:

$$B_2H_6(g) + 3\,O_2(g) \rightarrow 2\,B(OH)_3(s)$$

Boranes are readily hydrolysed by water to produce boric acid and hydrogen:

$$B_2H_6(g) + 6\,H_2O(l) \rightarrow 2\,B(OH)_3(aq) + 6\,H_2(g)$$

As described below, B_2H_6 is a Lewis acid, and the mechanism of this hydrolysis reaction involves coordination of H_2O acting as a Lewis base. Molecular hydrogen then forms as a result of the combination of the partially positively charged H atom on O with the partially negatively charged H atom on B.

(b) Lewis acidity of diborane and alkylboranes

KEY POINTS Soft and bulky Lewis bases cleave diborane symmetrically; more compact and hard Lewis bases cleave the hydrogen bridge unsymmetrically; although it reacts with many hard Lewis bases, diborane is best regarded as a soft Lewis acid.

As implied by the mechanism of hydrolysis, diborane and many other light boron hydrides act as Lewis acids and are cleaved by reaction with Lewis bases. Two different cleavage patterns have been observed; namely, symmetrical cleavage and unsymmetrical cleavage. In **symmetrical cleavage**, B_2H_6 is broken symmetrically into two BH_3 fragments, each of which forms a complex with a Lewis base:

Many complexes of this kind exist, and are isoelectronic with hydrocarbons. For instance, the product of this reaction is isoelectronic with 2,2-dimethylpropane, $C(CH_3)_4$. Stability trends indicate that BH_3 is a soft Lewis acid, as illustrated by the reaction

$$H_3B-N(CH_3)_3 + F_3B-S(CH_3)_2 \rightarrow H_3B-S(CH_3)_2 + F_3B-N(CH_3)_3$$

in which BH_3 transfers to the soft S donor atom, and the harder Lewis acid, BF_3, combines with the hard N donor atom.

The direct reaction of diborane and ammonia results in **unsymmetrical cleavage**, which is cleavage leading to an ionic product:

Unsymmetrical cleavage of this kind is generally observed when diborane and a few other boron hydrides react with strong, sterically uncrowded bases at low temperatures. Two ligands can attack one B atom in the course of the reaction only if they are small and steric repulsion is avoided.

EXAMPLE 14.3 Using NMR to identify reaction products

Explain how ^{11}B-NMR could be used to determine whether cleavage of diborane with an NMR inactive Lewis base is symmetrical or unsymmetrical (Section 8.6).

Answer We need to identify the possible products of the two reactions and then decide how the features of their NMR spectra will differ. Symmetrical cleavage of B_2H_6 with L yields $BH_3L + BH_3L$ and unsymmetrical cleavage yields $BH_2L_2^+$ and BH_4^-. In the former, ^{11}B is coupled to three equivalent ^{1}H nuclei, and we would therefore observe a quartet in the NMR spectrum. In unsymmetrical cleavage the first product has ^{11}B coupled to two equivalent ^{1}H nuclei, which would produce a triplet. The second product has ^{11}B coupled to four equivalent nuclei, which would produce a quintuplet.

Self-test 14.3 ^{11}B nuclei have $I = \frac{3}{2}$. Predict the number of lines and their relative intensities in the ^{1}H-NMR spectrum of BH_4^-.

Substitution of hydrogen on diborane by alkyl groups (R) leads to compounds such as, in the case of the methyl group, 1-methyldiborane, 1,1-dimethyldiborane ... 1,1,2,2-tetramethyldiborane, where any alkyl groups adopt terminal rather than bridging positions. With very bulky R-groups the ability of alkylborane to act as a Lewis base is impeded and leads to the recently important area of 'frustrated Lewis pair' chemistry and catalysis (Box 14.4).

(c) Hydroboration

KEY POINT Hydroboration, the reaction of diborane with alkenes in ether solvent, produces organoboranes that are useful intermediates in synthetic organic chemistry.

An important component of a synthetic chemist's repertoire of reactions is hydroboration, the addition of HB across a multiple bond:

$$H_3B-OR_2 + H_2C=CH_2 \xrightarrow{\Delta, ether} CH_3CH_2BH_2 + R_2O$$

From the viewpoint of an organic chemist, the C–B bond in the primary product of hydroboration is an intermediate stage in the stereospecific formation of C–H or C–OH bonds, to which it can be converted. From the viewpoint of the inorganic chemist, the reaction is a convenient method for the preparation of a wide variety of organoboranes. The hydroboration reaction is one of a class of reactions in which EH adds across the multiple bond; hydrosilylation (Section 15.7(b)) is another important example. Hydroboration of alkene reactions follows an 'anti-Markovnikov' rule for regioselectivity in that the hydrogen becomes bonded to the more substituted carbon atom of the original alkene.

BOX 14.4 What is a 'frustrated Lewis pair'?

As we have seen, many simple boron hydrides—for example, diborane—act as Lewis acids and form adducts with electron pair Lewis acids such as phosphines. The formation of the adduct requires a close approach of the Lewis acid and base, and the B–N distance in ammonia borane adduct, H_3BNH_3, is 1.58(2) Å. Obstruction of the boron atom—for instance, by bulky substituents—inhibits the formation of a Lewis adduct. Thus, solutions of $(t\text{-}Bu)_3P$ and $B(C_6F_5)_3$ in toluene show no evidence for the formation of an adduct, even on cooling to $-50°C$. Systems of this type are known as a **frustrated Lewis pair** or **FLP** (see also Section 5.10(d)).

The main interest in FLP chemistry derives from their ability to cleave bonds in small molecules and in particular that of dihydrogen, H_2 (which normally requires a metal catalyst). Bubbling hydrogen gas through a solution of $(t\text{-}Bu)_3P$ and $B(C_6F_5)_3$ in toluene affords the white salt $[(t\text{-}Bu)_3PH]^+[HB(C_6F_5)_3]^-$ where the hydrogen has been heterolytically cleaved. In some systems, such as those with the phosphine 1,8-bis(diphenylphosphino) naphthalene (**B1**) in combination with $B(C_6F_5)_3$, the uptake of hydrogen at room temperature and 1.5 bar is reversible on heating the salt formed to 60°C.

B1

Some FLPs consist of groups within a single molecule that are prevented from approaching each other by the constraints imposed by the bonding in the molecule, and such systems can be particularly active:

$Mes_2P\text{- - - -}B(C_6H_5)_2 \xrightarrow{H_2} Mes_2P^+\text{-}...\text{-}B(C_6H_5)_2$ (with H's)

One of the main areas of interest for the application of FLP chemistry is in hydrogen catalysis. Addition of a polar reducible substrate, such as an alkene, alkyne, imine, enamine, polyaromatic, or ketone, to an FLP in the presence of hydrogen, leads to facile hydrogen addition across the unsaturated bond. FLPs of chiral alkenylboranes can be used for enantioselective imine hydrogenation and asymmetric hydrosilylations. FLP compounds can also capture and activate other small molecules, including CO_2, SO_2, and N_2O; the ability to capture CO_2 and release it (on heating) has been proposed as a carbon capture technology.

(d) The tetrahydridoborate ion

KEY POINT The tetrahydridoborate ion is a useful intermediate for the preparation of metal hydride complexes and borane adducts.

Diborane reacts with alkali metal hydrides to produce salts containing the tetrahydridoborate ion, BH_4^-. Because of the sensitivity of diborane and LiH to water and oxygen, the synthesis must be carried out in the absence of air and in a nonaqueous solvent such as the short-chain polyether $CH_3OCH_2CH_2OCH_3$ (denoted here 'polyet'):

$$B_2H_6(\text{polyet}) + 2\,LiH(\text{polyet}) \rightarrow 2\,LiBH_4(\text{polyet})$$

We can view this reaction as another example of the Lewis acidity of BH_3 towards the strong Lewis basicity of H^-. The BH_4^- ion is isoelectronic with CH_4 and NH_4^+, and the three species show the following variation in chemical properties as the electronegativity of the central atom increases:

	BH_4^-	CH_4	NH_4^+
Character:	hydridic	–	protic

where 'protic' denotes Brønsted acid (proton-donating) character; CH_4 is neither acidic nor basic under the conditions prevailing in aqueous solution.

Alkali metal tetrahydridoborates are very useful laboratory and commercial reagents. They are often used as a mild source of H^- ions, as general reducing agents, and as precursors for most boron–hydrogen compounds, and they are utilized as hydrogen storage materials (Box 11.5 and Section 30.8). Most of these reactions are carried out in polar nonaqueous solvents. The preparation of diborane, mentioned previously,

$$3\,NaBH_4(s) + 4\,BF_3(g) \rightarrow 2\,B_2H_6(g) + 3\,NaBF_4(s)$$

is one example, and $NaBH_4$ in tetrahydrofuran (THF) is used to reduce aldehydes and ketones to alcohols. Although BH_4^- is thermodynamically unstable with respect to hydrolysis, the reaction is very slow at high pH, and some synthetic applications have been devised in water. For example, germane (GeH_4) can be prepared by dissolving GeO_2 and KBH_4 in aqueous potassium hydroxide and then acidifying the solution:

$$HGeO_3^-(aq) + BH_4^-(aq) + 2\,H^+(aq) \rightarrow GeH_4(g) + B(OH)_3(aq)$$

Aqueous BH_4^- can also serve as a simple reducing agent, as in the reduction of aqua ions such as $Ni^{2+}(aq)$ or $Cu^{2+}(aq)$ to the metal or metal boride. With halogen complexes of 4d and 5d elements that also have stabilizing ligands, such as phosphines, tetrahydridoborate ions can be used to introduce a hydride ligand by a metathesis reaction in a non-aqueous solvent:

$$[RuCl_2(PPh_3)_3] + NaBH_4 + PPh_3 \xrightarrow{\Delta, \text{benzene/alcohol}} RuH(PPh_3)_4 + \text{other products}$$

It is probable that many of these metathesis reactions proceed through a transient BH_4^- complex. Indeed, tetrahydridoborate complexes are known, especially with highly electropositive metals: they include $[Al(BH_4)_3]$ (**11**), which contains a diborane-like double hydride bridge, and $[Zr(BH_4)_4]$ (**12**), in which triple hydride bridges are present. The bonding in these compounds can be described in terms of $3c,2e$ bonds (Section 3.11(e)).

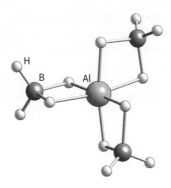

11 $Al(BH_4)_3$

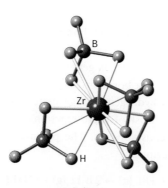

12 $Zr(BH_4)_4$

EXAMPLE 14.4 **Predicting the reactions of boron–hydrogen compounds**

By means of a chemical equation, indicate the products resulting from the interaction of equal amounts of $[HN(CH_3)_3]Cl$ with $LiBH_4$ in tetrahydrofuran (THF).

Answer We should expect LiCl, with its high lattice enthalpy, to be a likely product. If this is the case we shall be left with BH_4^- and $[HN(CH_3)_3]^+$. The interaction of the hydridic BH_4^- ion with the protic $[HN(CH_3)_3]^+$ ion will evolve hydrogen to produce trimethylamine and BH_3. In the absence of other Lewis bases, the BH_3 molecule would coordinate to THF; however, the stronger Lewis base trimethylamine is produced in the initial reactions, so the overall reaction will be

$$[HN(CH_3)_3]Cl + LiBH_4 \rightarrow H_2 + H_3BN(CH_3)_3 + LiCl$$

Self-test 14.4 Write an equation for the reaction of B_2H_6 with propene in ether solvent and a 1:2 stoichiometry, and another equation for its reaction with ammonium chloride in THF with the same stoichiometry.

14.7 Boron trihalides

KEY POINTS Boron trihalides are useful Lewis acids, with BCl_3 stronger than BF_3, and important electrophiles for the formation of boron–element bonds; subhalides with B–B bonds, such as B_2Cl_4, are also known.

All the boron trihalides except BI_3 may be prepared by direct reaction between the elements. However, the preferred method for BF_3 is the reaction of B_2O_3 with CaF_2 in H_2SO_4. This reaction is driven in part by production of HF from the reaction of the H_2SO_4 with CaF_2 and the stability of the $CaSO_4$:

$$B_2O_3(s) + 3CaF_2(s) + 6H_2SO_4(l) \rightarrow$$
$$2BF_3(g) + 3[H_3O][HSO_4](\text{soln}) + 3CaSO_4(s)$$

All the boron trihalides form simple Lewis complexes with suitable bases, as in the reaction

$$BF_3(g) + :NH_3(g) \rightarrow F_3B-NH_3(s)$$

However, boron chlorides, bromides, and iodides are susceptible to protolysis by mild proton sources such as water, alcohols, and even amines. As shown in Fig. 14.4, this reaction, together with metathesis reactions, is very useful in preparative chemistry. An example is the rapid hydrolysis of BCl_3 to produce boric acid, $B(OH)_3$:

$$BCl_3(g) + 3H_2O(l) \rightarrow B(OH)_3(aq) + 3HCl(aq)$$

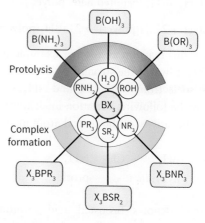

FIGURE 14.4 The reactions of boron–halogen compounds (X = halogen).

> **EXAMPLE 14.5** Predicting the products of reactions of the boron trihalides
>
> Predict the likely products of the following reactions and write the balanced chemical equations: (a) BF_3 and excess NaF in acidic aqueous solution, (b) BCl_3 and excess NaCl in acidic aqueous solution, (c) BBr_3 and excess $NH(CH_3)_2$ in a hydrocarbon solvent.
>
> **Answer** We need to consider whether the B–X bond is susceptible to hydrolysis. (a) The F^- ion is a chemically hard and fairly strong base; BF_3 is a hard and strong Lewis acid with a high affinity for the F^- ion. Hence, the reaction should result in a complex:
>
> $$BF_3(g) + F^-(aq) \rightarrow BF_4^-(aq)$$
>
> Excess F^- and acid prevent the formation of hydrolysis products such as BF_3OH^-, which are formed at high pH. (b) Unlike B–F bonds, which are very strong and only mildly susceptible to hydrolysis, the other boron–halogen bonds are hydrolysed vigorously by water. We can anticipate that BCl_3 will undergo hydrolysis rather than coordinate to aqueous Cl^-:
>
> $$BCl_3(g) + 3H_2O(l) \rightarrow B(OH)_3(aq) + 3HCl(aq)$$
>
> (c) Boron tribromide will undergo protolysis with formation of a B–N bond:
>
> $$BBr_3(g) + 6NH(CH_3)_2(solv) \rightarrow B[N(CH_3)_2]_3(solv) + 3[NH_2(CH_3)_2]Br(solv)$$
>
> In this reaction the HBr produced by the protolysis protonates excess dimethylamine.
>
> **Self-test 14.5** Write and justify balanced equations for plausible reactions between (a) BCl_3 and ethanol, (b) BCl_3 and pyridine in hydrocarbon solution, (c) BBr_3 and $F_3BN(CH_3)_3$.

It is probable that a first step in this reaction is the formation of the complex Cl_3B-OH_2, which then eliminates HCl and reacts further with water.

The tetrafluoridoborate anion, BF_4^-, mentioned in Example 14.5, is used in preparative chemistry when a relatively large weakly coordinating anion is needed. The tetrahalidoborate anions BCl_4^- and BBr_4^- can be prepared in nonaqueous solvents. However, because of the ease with which B–Cl and B–Br bonds undergo solvolysis, they are stable in neither water nor alcohols.

Boron halides are the starting point for the synthesis of many boron–carbon and boron–pseudohalogen compounds (Section 18.7).[1] Examples include the formation of alkylboron and arylboron compounds, such as trimethylboron, by the reaction of boron trifluoride with a methyl Grignard reagent in ether solution:

$$BF_3 + 3CH_3MgI \rightarrow B(CH_3)_3 + \text{magnesium halides}$$

---

[1] Pseudohalogens are species that resemble the halogens in their chemical properties. Cyanogen, $(CN)_2$, is a pseudohalogen, and the cyanide ion, CN^-, is a pseudohalide.

When an excess of the Grignard (or organolithium) reagent is present, tetraalkyl- or tetraarylborates are formed:

$$BF_3 + Li_4(CH_3)_4 \rightarrow Li[B(CH_3)_4] + 3LiF$$

Boron halides containing B–B bonds have been prepared. The best-known of these compounds have the formula B_2X_4, with X = F, Cl, or Br, and the tetrahedral cluster compound B_4Cl_4. The B_2Cl_4 molecules are planar (**13**) in the solid state, with efficient packing of this geometry, but staggered (**14**) in the gas phase. This conformational difference suggests that rotation about the B–B bond is quite easy, as is expected for a single bond.

13 B_2Cl_4, D_{2h}

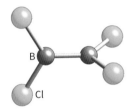

14 B_2Cl_4, D_{2d}

One route to B_2Cl_4 is to pass an electric discharge through BCl_3 gas in the presence of a Cl atom scavenger, such as mercury vapour. Metathesis reactions can be used to make B_2X_4 derivatives from B_2Cl_4. The thermal stability of these derivatives increases with increasing tendency of the X group to form a π bond with boron:

$$B_2Cl_4 < B_2F_4 < B_2(OR)_4 \ll B_2(NR_2)_4$$

It was thought for a long time that X groups with lone pairs were essential for the existence of B_2X_4 compounds, but diboron compounds with alkyl or aryl groups have been prepared. Compounds that survive at room temperature can be obtained when the groups are bulky, as in $B_2({}^tBu)_4$.

A secondary product in the synthesis of B_2Cl_4 is B_4Cl_4, a pale yellow solid composed of molecules with the four boron atoms forming a tetrahedron (**15**). Like B_2Cl_4, B_4Cl_4 does not have a formula analogous to those of the boranes (such as B_2H_6) discussed above. This difference may lie in the tendency of halogens to form π bonds with boron by the donation of lone electron pairs on the halide into the otherwise vacant p orbital on B, as in Fig. 14.2 (Section 5.8(b)).

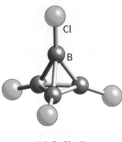

15 B_4Cl_4, T_d

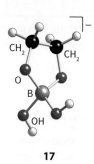

17

14.8 Boron–oxygen compounds

KEY POINT Boron forms boric acid, B_2O_3, polyborates, and borosilicate glasses.

Boric acid, $B(OH)_3$, is a very weak Brønsted acid in aqueous solution. However, the equilibria are more complicated than the simple Brønsted proton transfer reactions characteristic of the later p-block oxoacids. Boric acid is in fact primarily a weak Lewis acid, and the complex it forms with H_2O, $H_2OB(OH)_3$, is the actual source of protons:

$$B(OH)_3(aq) + 2\,H_2O(l) \rightleftharpoons H_3O^+(aq) + [B(OH)_4]^-(aq)$$
$$pK_a = 9.2$$

As is typical of many of the lighter elements of the p block, there is a tendency for the anion to polymerize by condensation, with the loss of H_2O. Thus, in concentrated neutral or basic solution, equilibria such as

$$3\,B(OH)_3(aq) \rightleftharpoons [B_3O_3(OH)_4]^-(aq) + H^+(aq) + 2\,H_2O(l)$$
$$pK_a = 0.85$$

occur to yield polynuclear anions (**16**).

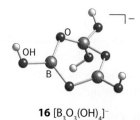

16 $[B_3O_3(OH)_4]^-$

The reaction of boric acid with an alcohol in the presence of sulfuric acid leads to the formation of simple borate esters, which are compounds of the form $B(OR)_3$:

$$B(OH)_3 + 3\,CH_3OH \xrightarrow{H_2SO_4} B(OCH_3)_3 + 3\,H_2O$$

Borate esters are much weaker Lewis acids than the boron trihalides, presumably because the oxygen atom acts as an intramolecular π donor, like the F atom in BF_3 (Section 5.8(b)), and donates electron density to the p orbital of the B atom. Hence, judging from Lewis acidity, an O atom is more effective than an F atom as a π donor towards B. 1,2-Diols have a particularly strong tendency to form cyclic borate esters (**17**), including with sugars, on account of the chelate effect (Section 7.14).

As with silicates and aluminates, there are many polynuclear borates, and both cyclic and chain species are known. An example is the cyclic polyborate anion, $[B_3O_6]^{3-}$ (**18**). A notable feature of borate formation is the possibility of both three-coordinate B atoms, as in (**18**), and four-coordinate B atoms, as in $[B(OH)_4]^-$. The mineral borax contains the $[B_4O_5(OH)_4]^{2-}$ anion (**19**), which has both three- and four-coordinate B atoms. Polyborates form by sharing one O atom with a neighbouring B atom, as in (**18**); structures in which two adjacent B atoms share two or three O atoms are unknown.

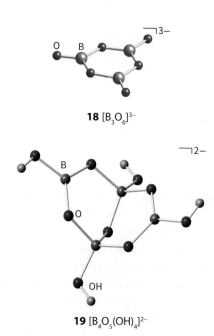

18 $[B_3O_6]^{3-}$

19 $[B_4O_5(OH)_4]^{2-}$

Boron oxide, B_2O_3, is acidic and is prepared by dehydration of boric acid:

$$2\,B(OH)_3(s) \xrightarrow{\Delta} B_2O_3(s) + 3\,H_2O(g)$$

The rapid cooling of molten B_2O_3 or metal borates often leads to the formation of borate glasses. Although these glasses themselves have little technological significance, the fusion of sodium borate with silica leads to the formation of borosilicate glasses (such as Pyrex®). Borosilicate glasses are resistant to thermal shock and can be heated over a flame or other source of direct heat. Boron also forms the suboxide B_6O, by reaction of boron with ZnO at 1400°C.

Sodium perborate is used as a bleach in laundry powders, automatic dishwasher powders, and, until it was banned in the EU, whitening toothpastes. Although the formula is often given as $NaBO_3 \cdot H_2O$ or $NaBO_3 \cdot 4H_2O$, the compound contains the peroxide anion, O_2^{2-} and is more accurately described as $Na_2[B_2(O_2)_2(OH)_4] \cdot 6H_2O$.

14.9 Compounds of boron with nitrogen

KEY POINTS Compounds containing BN, which is isoelectronic with CC, include the ethane analogue ammonia borane, H_3NBH_3, the benzene analogue, $N_3B_3H_6$, and BN analogues of graphite and diamond.

The thermodynamically stable phase of boron nitride, BN, consists of planar sheets of atoms like those in graphite (Section 15.5). The planar sheets of alternating B and N atoms consist of edge-shared hexagons and, as in graphite, the B–N distance within the sheet (145 pm) is much shorter than the distance between the sheets (333 pm; Fig. 14.5). The difference between the structures of graphite and boron nitride, however, lies in the register of the atoms of neighbouring sheets: in BN, the hexagonal rings are stacked directly over each other, with B and N atoms alternating in successive layers; in graphite, the hexagons are staggered. Molecular orbital calculations suggest that the stacking in BN stems from a partial positive charge on B and a partial negative charge on N. This charge distribution is consistent with the electronegativity difference of the two elements ($\chi^P(B) = 2.04$, $\chi^P(N) = 3.04$).

As with impure graphite, layered boron nitride is a slippery material that is used as a lubricant. Unlike graphite, however, it is a colourless electrical insulator, as there is a large energy gap between the filled and vacant π bands. The size of the band gap is consistent with its high electrical resistivity and lack of absorption in the visible spectrum. In keeping with this large band gap, BN forms a much smaller number of intercalation compounds than graphite (Section 15.5). In contrast to graphite, layered boron nitride is stable in air up to 1000°C, making it a useful refractory material.

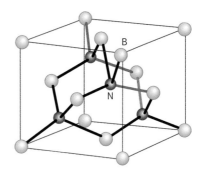

FIGURE 14.6 The sphalerite structure of cubic boron nitride.

Layered boron nitride changes into a denser cubic phase at high pressures and temperatures (60 kbar and 2000°C; Fig. 14.6). This phase is a hard crystalline analogue of diamond but, as it has a lower lattice enthalpy, it has a slightly lower mechanical hardness (Fig. 14.7). Cubic boron nitride is manufactured and used as an abrasive for certain high-temperature applications in which diamond cannot be used because it forms carbides with the material being ground.

The fact that BN and CC are isoelectronic suggests that there might be analogies between these compounds and hydrocarbons. Many **amine–boranes**, the boron–nitrogen analogues of saturated hydrocarbons, can be synthesized by reaction between a nitrogen Lewis base and a boron Lewis acid:

$$\tfrac{1}{2}B_2H_6 + N(CH_3)_3 \rightarrow H_3BN(CH_3)_3$$

However, although amine–boranes are isoelectronic with hydrocarbons, their properties are significantly different, in large part due to the difference in electronegativities of B and N. For example, whereas ammonia–borane, H_3NBH_3, (20) is a solid at room temperature with a vapour pressure of a few pascals, its analogue ethane, H_3CCH_3, is a gas that condenses at −89°C. This difference can be traced to the difference in polarity of the two molecules: ethane is nonpolar, whereas ammonia–borane has a large dipole moment of 5.2 D (20).

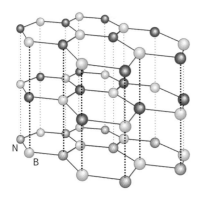

FIGURE 14.5 The structure of layered hexagonal boron nitride. Note that the rings are in register between layers.

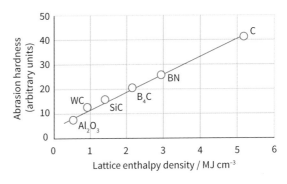

FIGURE 14.7 The correlation of hardness with lattice enthalpy density (the lattice enthalpy divided by the molar volume of the substance). The point for carbon represents diamond; that for boron nitride represents the diamond-like sphalerite structure.

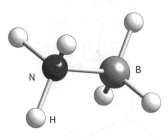

20 NH_3BH_3

Several BN analogues of the amino acids have been prepared, including ammonia–carboxyborane, H_3NBH_2COOH, the analogue of glycine, NH_2CH_2COOH. These compounds display significant physiological activity, including tumour inhibition and reduction of serum cholesterol.

The simplest unsaturated boron–nitrogen compound is aminoborane, H_2NBH_2, which is isoelectronic with ethene. It has only a transient existence in the gas phase because it readily forms cyclic ring compounds such as a cyclohexane analogue (**21**).

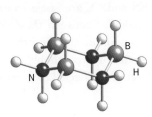

21 $N_3B_3H_{12}$

However, the aminoboranes do survive as monomers when the double bond is shielded from reaction by bulky alkyl groups on the N atom and by Cl atoms on the B atom (**22**).

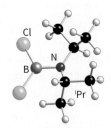

22 $Cl_2B-N(^iPr)_2$, $^iPr = (CH_3)_2CH$

For instance, monomeric aminoboranes can be synthesized readily by the reaction of a dialkylamine and a boron halide:

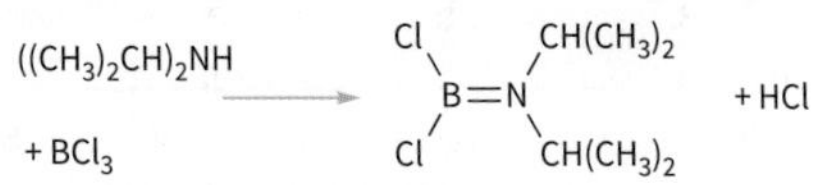

$$((CH_3)_2CH)_2NH + BCl_3 \longrightarrow \underset{Cl}{\overset{Cl}{B}}=N\underset{CH(CH_3)_2}{\overset{CH(CH_3)_2}{}} + HCl$$

The reaction also occurs with xylyl (2,4,6-trimethylphenyl) groups in place of isopropyl groups.

Apart from layered boron nitride, the best-known unsaturated compound of boron and nitrogen is borazine, $B_3N_3H_6$ (**2**), which is isoelectronic and isostructural with benzene. Borazine was first prepared by Alfred Stock in 1926 by the reaction between diborane and ammonia. Since then, many symmetrically trisubstituted derivatives have been made by procedures that depend on the protolysis of BCl bonds of BCl_3 by an ammonium salt (**23**):

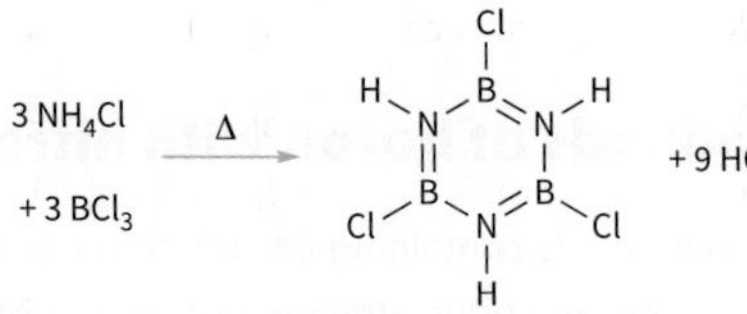

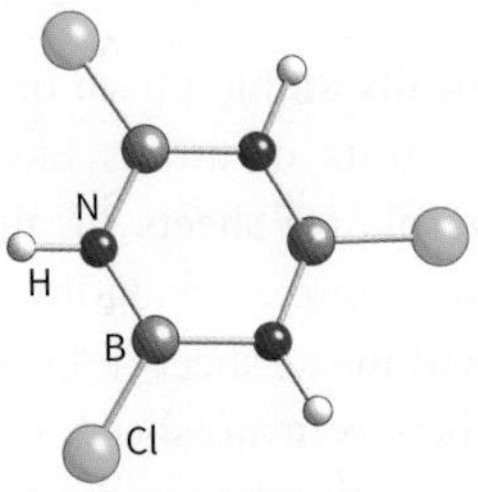

$$3\,NH_4Cl + 3\,BCl_3 \xrightarrow{\Delta} \text{(B}_3\text{N}_3\text{H}_3\text{Cl}_3\text{)} + 9\,HCl$$

23 $B_3N_3H_3Cl_3$

The use of an alkylammonium chloride yields *N*-alkyl substituted *B,B′,B″*-trichloroborazines.

Despite their structural resemblance, there is little chemical resemblance between borazine and benzene. Once again, the difference in the electronegativities of boron and nitrogen is influential, and B–Cl bonds in trichloroborazine are much more labile than the C–Cl bonds in trichlorobenzene. In the borazine compound, the π electrons are concentrated on the N atoms, and there is a partial positive charge on the B atoms that leaves them open to nucleophilic attack. A sign of the difference is that the reaction of a chloroborazine with a Grignard reagent or hydride source results in the substitution of Cl by alkyl, aryl, or hydride groups. Another example of the difference is the ready addition of HCl to borazine to produce a trichlorocyclohexane analogue (**24**):

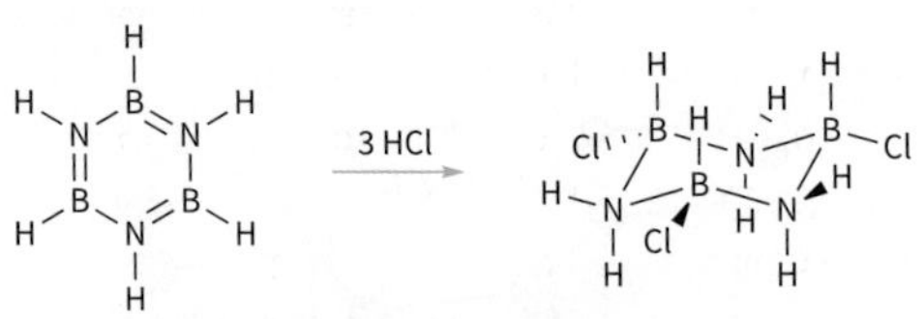

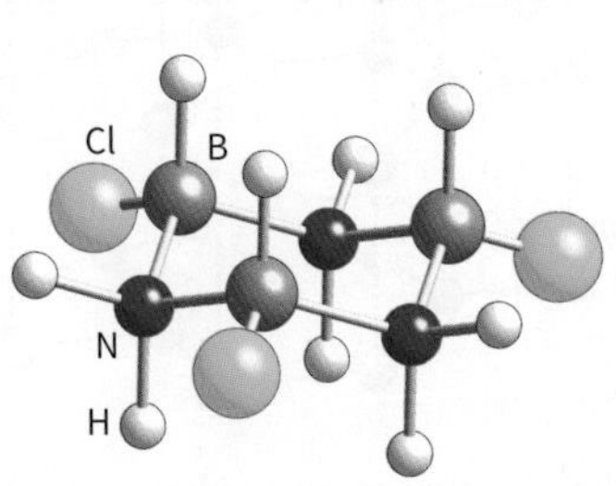

$$B_3N_3H_6 + 3\,HCl \longrightarrow B_3N_3H_9Cl_3$$

24 $B_3N_3H_9Cl_3$

The electrophile, H⁺, in this reaction attaches to the partially negative N atom, and the nucleophile, Cl⁻, attaches to the partially positive boron atom.

EXAMPLE 14.6 Preparing borazine derivatives

Give balanced chemical equations for the synthesis of borazine starting with NH_4Cl, BCl_3, and other reagents of your choice.

Answer The first step will be the protolysis of the B–Cl bond in BCl_3 by the ammonium ion. Therefore, reaction of NH_4Cl with BCl_3 will yield

$$3NH_4Cl + 3BCl_3 \rightarrow H_3N_3B_3Cl_3 + 9HCl$$

The Cl atoms in B,B',B''-trichloroborazine can then be displaced by hydride ions from reagents such as $LiBH_4$, to yield borazine:

$$3LiBH_4 + H_3N_3B_3Cl_3 \xrightarrow{THF} H_3B_3N_3H_3 + 3LiCl + 3THF \cdot BH_3$$

Self-test 14.6 Suggest a reaction or series of reactions for the preparation of N,N',N''-trimethyl-B,B',B''-trimethylborazine starting with methylamine and boron trichloride.

14.10 Metal borides

KEY POINT Metal borides include boron anions as isolated B atoms, linked *closo*-boron polyhedra, and hexagonal boron networks.

The direct reaction of elemental boron and a metal at high temperatures provides a useful route to many metal borides. An example is the reaction of Ca and some other highly electropositive metals with B to produce a phase of composition MB_6:

$$Ca(l) + 6B(s) \rightarrow CaB_6(s)$$

Metal borides are found with a wide range of compositions, as B can occur in numerous types of structural unit, including isolated boron atoms, chains, planar and puckered nets, and clusters. The simplest metal borides are metal-rich compounds that contain isolated B^{3-} ions. The most common examples of these compounds have the formula M_2B, where M may be one of the middle to late 3d metals (Mn to Ni) in low oxidation states. Another important class of metal borides contains planar or puckered hexagonal nets that have the composition MB_2 (Fig. 14.8). These compounds are formed primarily by electropositive metals, including Mg (Box 14.5), Al, the early d metals (from Sc to Mn, for instance, in Period 4), and U.

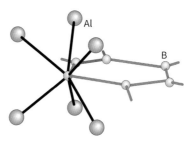

FIGURE 14.8 The AlB_2 structure. To give a clear picture of the hexagonal layer, B atoms outside the unit cell are displayed.

The boron-rich borides, typically MB_6 and MB_{12}, where M is an electropositive metal, are of even greater structural interest. In them, the B atoms link to form an intricate network of interconnecting cages. In MB_6 compounds (which are formed by the electropositive s-block metals, such as Na, K, Ca, Sr, and Ba, and f-block metals), the B_6 octahedra are linked by their vertices to form a cubic framework (Fig. 14.9). The linked B_6 clusters bear a charge of –1, –2, or –3, depending on the cation with which they are associated. In the MB_{12} compounds, such as YB_{12}, the B-atom networks are based on linked cuboctahedra (25) rather than the more familiar icosahedron. This type of compound is formed by some of the heavier electropositive metals, particularly those of the f block.

BOX 14.5 What makes magnesium diboride a 38 K superconductor?

Magnesium diboride, MgB_2, is an inexpensive compound that has been known in the laboratory for more than fifty years. In 2001 this simple compound was found to have superconducting properties (Sections 25.5 and 30.7). Jun Akimitsu and his co-workers discovered by chance that MgB_2 loses its electrical resistance when cooled. At the time, they were characterizing materials used to enhance the performance of known high-temperature superconductors. The discovery led to a major flurry of research around the world on this new superconductor.

The transition temperature of MgB_2 is 38 K and is exceeded only by a few compounds such as the much more complicated perovskite cuprate structures (Section 30.7). Many of the first measurements were made using MgB_2 powder straight from the commercial bottle. High-quality MgB_2 can be synthesized by heating finely divided boron and magnesium powders together at around 950°C under pressure. Thin films, wires, and tapes have since been formed that have potential for applications in superconducting magnets, microwave communications, and power applications.

Magnesium diboride has a simple structure in which the B atoms are arranged in graphite-like planes with alternating layers of Mg ions. The Mg atoms donate their two valence electrons to the network of B atoms producing $\{[B_2]^{2-}\}_n$ sheets. Varying the number of electrons donated to the boron conduction bands can dramatically affect the transition temperature. The transition temperature of the compound falls if some of the Mg atoms are replaced by Al and increases when doped with Cu.

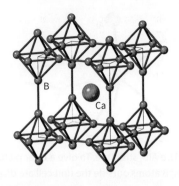

FIGURE 14.9 The CaB$_6$ structure. Note that the B$_6$ octahedra are connected by a bond between vertices of adjacent B$_6$ octahedra. The crystal is a simple cube analogue of CsCl. Thus, eight connected B$_6$ octahedra surround a central Ca ion.

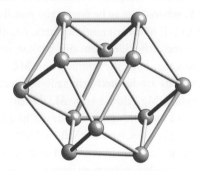

25 B$_{12}$ cuboctahedron

14.11 Higher boranes and borohydrides

KEY POINT The bonding in boron hydrides and polyhedral borohydride ions can be approximated by conventional 2c,2e bonds together with 3c,2e bonds.

In this section we describe the structures and properties of the cage-like boranes and borohydrides, which include Stock's series B$_n$H$_{n+4}$ and B$_n$H$_{n+6}$ as well as the more recently discovered [B$_n$H$_n$]$^{2-}$ closed polyhedra. The borohydrides have been studied for many years as an interesting class of compounds and have more recently found applications (Box 14.6).

Boron-cluster compounds are best considered from the standpoint of fully delocalized molecular orbitals containing electrons that contribute to the stability of the entire molecule. However, it is sometimes fruitful to identify groups of three atoms and to regard them as bonded together by versions of the 3c,2e bonds of the kind that occur in diborane itself (**1**). In the more complex boranes, the three centres of the 3c,2e bonds may be BHB bridge bonds, but they may also be bonds in which three B atoms lie at the corners of an equilateral triangle with their sp^3 hybrid orbitals overlapping at its centre (**26**). To reduce the complexity of the structural diagrams, the illustrations that follow will not in general indicate the 3c,2e bonds in the structures.

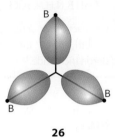

26

(a) Wade's rules

KEY POINTS Wade's rules can be used to predict the structures of polyhedral borohydrides; boron hydride structures include simple polyhedral *closo* compounds and the progressively more open *nido* and *arachno* structures.

A correlation between the number of electrons (counted in a specific way), the formula, and the shape of the molecule was established by Kenneth Wade in the 1970s. These so-called **Wade's rules** apply to a class of polyhedra called **deltahedra** (because they are made up of triangular faces resembling Greek deltas, Δ) and can be used in two ways. For molecular

---

BOX 14.6 Which physical and chemical properties of boron lead to its use in cancer treatments?

A promising new form of radiotherapy for brain, head, and neck tumours involves the irradiation of boron compounds with low-energy neutrons. Boron neutron-capture therapy (BNCT) involves injecting the patient with a ^{10}B-enriched (^{10}B has a 20% natural abundance) boron compound which can bind to tumour cells. When irradiated with neutrons, the ^{10}B nucleus (which strongly absorbs neutrons unlike carbon, hydrogen, and oxygen) undergoes fission and produces a helium nucleus (an alpha particle) and a ^{7}Li nucleus, and liberates approximately 2.3 MeV of energy:

$$^{10}_{5}\text{B} + ^{1}_{0}\text{n} \rightarrow ^{4}_{2}\text{He} + ^{7}_{3}\text{Li}$$

The highly energetic alpha particle and ^{7}Li$^+$ nucleus cause high levels of local (5–10 μm) ionization and tumour cell death.

The most promising boron-containing compounds for this application have been polyhedral borohydrides, and Na$_2$B$_{12}$H$_{11}$SH (sodium borocaptate) has also been used clinically. The factor limiting progress is the amount of boron that can be introduced into a tumour cell without causing toxicity to normal cells. A breakthrough may come with the recent development of boron carbide nanoparticles. The nanoparticles are introduced into a sample of the patient's own T-cells, which are then injected back into the patient, where they travel to the tumour and deliver the nanoparticles. The nanoparticles have also been coated with a peptide that improves cell uptake, and labelled with a fluorescent dye that enables the nanoparticles to be tracked within the body.

and anionic boranes, they enable us to predict the general shape of the molecule or anion from its formula. However, because the rules are also expressed in terms of the number of electrons, we can extend them to analogous species in which there are atoms other than boron, including carboranes and other p-block clusters. Here we concentrate on the boron clusters, where knowing the formula is sufficient for predicting the shape. However, so that we can cope with other clusters we shall show how to count the framework electrons too.

The building block from which the deltahedron is constructed is assumed to be one BH group (27) that contributes two electrons. The electrons in the B–H bond are ignored in the counting procedure, but all others are included whether or not it is obvious that they help to hold the skeleton together. The 'skeleton' refers to the framework of the cluster with each BH group counted as a unit. If a B atom happens to carry two H atoms, only one of the B–H bonds is treated as a unit. The second B–H bond lies within the same spherical surface as the B atoms and is included in the skeleton electron count. For instance, in B_5H_{11}, one of the B atoms has two 'terminal' H atoms, but only one BH entity is treated as a unit, the other pair of electrons being treated as part of the skeleton and hence referred to as 'skeletal electrons'. A BH group makes two electrons available to the skeleton (the B atom provides three electrons and the H atom provides one, but of these four, two are used for the B–H bond). Writing the formula as $(BH)_5H_6$ yields five BH electron pairs and six additional electrons, eight pairs. This corresponds to an $n+3$ *arachno* cluster.

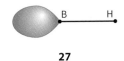

27

A BRIEF ILLUSTRATION

To count the number of skeletal electrons in B_4H_{10} (10) we consider the number of BH units and the number of H atoms. There are four BH units, which contribute $4 \times 2 = 8$ electrons, and the six additional H atoms, which contribute a further six electrons, giving 14 in all. The resulting seven pairs are distributed as shown in (28): two are used for the additional terminal B–H bonds, four are used for the four BHB bridges, and one is used for the central B–B bond.

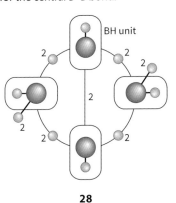

28

TABLE 14.3 Classification of boron hydrides

Type	Formula*	Skeletal electrons	Examples
Closo	$[B_nH_n]^{2-}$	$n+1$	$[B_5H_5]^{2-}$ to $[B_{12}H_{12}]^{2-}$
Nido	B_nH_{n+4}	$n+2$	B_2H_6, B_5H_9, B_6H_{10}
Arachno	B_nH_{n+6}	$n+3$	B_4H_{10}, B_5H_{11}
Hypho†	B_nH_{n+8}	$n+4$	None‡

* In some cases, protons can be removed; thus $[B_5H_8]^-$ arises from the deprotonation of B_5H_9.
† The name comes from the Greek for 'net'.
‡ Some derivatives are known.

According to Wade's rules (Table 14.3), species of formula $[B_nH_n]^{2-}$ and $(n + 1)$ pairs of skeletal electrons have a *closo* structure, with a B atom at each corner of a closed deltahedron having n vertices and no B–H–B bonds. This series of anions would have n pairs of skeletal electrons from the n BH groups, plus two electrons from the 2− charge. The series is known for $n = 5$ to 12, and examples include the trigonal-bipyramidal $[B_5H_5]^{2-}$ ion, the octahedral $[B_6H_6]^{2-}$ ion, and the icosahedral $[B_{12}H_{12}]^{2-}$ ion. The *closo*-borohydrides and their carborane analogues (Section 14.12) are typically thermally stable and moderately unreactive.

Boron clusters of formula B_nH_{n+4} have the *nido* structure. They can be regarded as derived from a *closo*-borane that has lost one vertex but have B–H–B bonds as well as B–B bonds. An example is B_5H_9, in which there are $(5 \times 2) + 4 = 14$ skeletal electrons, or 7 pairs. The $(n + 1)$ rule (Table 14.3) states that the structure will be based on an n-vertex deltahedron. In this case $n = 6$, so, as there are only five B atoms, the cluster is based on an octahedron with one vertex removed (7). In general, the thermal stability of the *nido*-boranes is intermediate between that of *closo*- and *arachno*-boranes.

A NOTE ON GOOD PRACTICE

Recognize that the variable n is used in two different contexts here. The general formulas of borohydrides use n, as in, for example, B_nH_{n+4}. However, when we calculate the number of cluster electron pairs we designate them as n as well.

Clusters of formula B_nH_{n+6} have an *arachno* structure. They can be regarded as *closo*-borane polyhedra less two vertices (and must have B–H–B bonds). One example of an *arachno*-borane is pentaborane(11), B_5H_{11}, which has $(5 \times 2) + 6 = 16$ skeletal electrons or 8 skeletal electron pairs. According to the $(n + 1)$ rule, $n = 7$ and the structure will be based on a seven-vertex deltahedron with two vertices removed (8). As with most *arachno*-boranes, pentaborane(11) is thermally unstable at room temperature and is highly reactive.

> **EXAMPLE 14.7** Using Wade's rules to predict a borohydride structure
>
> Infer the structure of $[B_6H_6]^{2-}$ from its formula and from its electron count.
>
> **Answer** We should note that the formula $[B_6H_6]^{2-}$ belongs to a class of borohydrides having the formula $[B_nH_n]^{2-}$, which is characteristic of a *closo* species. Alternatively, we can count the number of skeletal electron pairs and from that deduce the structural type. Assuming one B–H bond per B atom, there are six BH units to take into account and therefore 12 skeletal electrons plus two from the overall charge of -2: $(6 \times 2) + 2 = 14$, or seven electron pairs, which is $(n + 1)$ with $n = 6$. Therefore, the cluster is based on an octahedron with no missing vertices and is a *closo* cluster (**5**).
>
> **Self-test 14.7** (a) How many framework electron pairs are present in B_4H_{10}, and to what structural category does it belong? Sketch its structure. (b) Predict the structure of $[B_5H_8]^-$.

(b) The origin of Wade's rules

KEY POINT *The molecular orbitals in a closo-borane can be constructed from BH units, each of which contributes one radial atomic orbital pointing towards the centre of the cluster and two perpendicular p orbitals that are tangential to the polyhedron.*

Wade's rules have been justified by molecular orbital calculations. We shall indicate the kind of reasoning involved by considering the first of them, the $(n + 1)$ rule. In particular, we shall show that $[B_6H_6]^{2-}$ has a low energy if it has an octahedral *closo* structure, as predicted by the rules.

A B–H bond uses one electron and one orbital of the B atom, leaving three orbitals and two electrons for the skeletal bonding. One of these orbitals, called a **radial orbital**, can be considered to be a boron sp hybrid pointing towards the interior of the fragment (as in **26**). The remaining two boron p orbitals, the **tangential orbitals**, are perpendicular to the radial orbital (**29**).

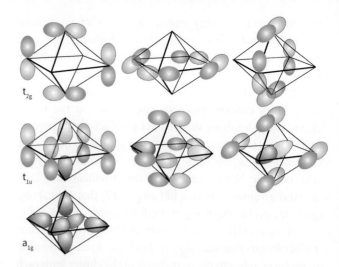

FIGURE 14.10 Radial and tangential bonding molecular orbitals for $[B_6H_6]^{2-}$. The relative energies are $a_{1g} < t_{1u} < t_{2g}$.

The lowest-energy molecular orbital is totally symmetric (a_{1g}) and arises from in-phase contributions from all the radial orbitals. Calculations show that the next-higher orbitals are the t_{1u} orbitals, each of which is a combination of four tangential and two radial orbitals. Above these three degenerate orbitals lie another three t_{2g} orbitals, which are tangential in character, giving seven bonding orbitals in all. Hence, there are seven orbitals with net bonding character delocalized over the skeleton, and they are separated by a considerable gap from the remaining 11 largely antibonding orbitals (Fig. 14.11).

There are seven electron pairs to accommodate, one pair from each of the six B atoms and one pair from the overall charge (-2). These seven pairs can all enter and fill the seven bonding skeleton orbitals, and hence give rise to a stable structure, in accord with the $(n + 1)$ rule. Note that the unknown neutral octahedral B_6H_6 molecule would have too few electrons to fill the t_{2g} bonding orbitals. Similar arguments can be used for all *closo* structures.

29

The shapes of the 18 symmetry-adapted linear combinations of these 18 orbitals in an octahedral B_6H_6 cluster can be inferred from the drawings in Resource section 4, and we show the ones with net bonding character in Fig. 14.10.

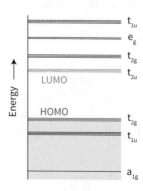

FIGURE 14.11 Schematic molecular orbital energy levels of the B-atom skeleton of $[B_6H_6]^{2-}$. The form of the bonding orbitals is shown in Fig. 14.10.

(c) Structural correlations

KEY POINT Conceptually, the *closo*, *nido*, and *arachno* structures are related by the successive removal of a BH fragment and addition of H or electrons.

A very useful structural correlation between *closo*, *nido*, and *arachno* species is based on the observation that clusters with the same numbers of skeletal electrons are related by removal of successive BH groups and the addition of the appropriate numbers of electrons and H atoms. This conceptual process provides a good way to think about the structures of the various boron clusters but does not represent how they are interconverted chemically.

The idea is amplified in Fig. 14.12, where the removal of a BH unit and two electrons and the addition of four H atoms converts the octahedral *closo*-$[B_6H_6]^{2-}$ anion to the square-pyramidal *nido*-B_5H_9 borane. A similar process (removal of a BH unit and addition of two H atoms) converts *nido*-B_5H_9 into a butterfly-shaped *arachno*-B_4H_{10} borane. Each of these three boranes has 14 skeletal electrons, but as the number of skeletal electrons per B atom increases, the structure becomes more open.

A more systematic correlation of this type is indicated for many different boranes in Fig. 14.13.

(d) Synthesis of higher boranes and borohydrides

KEY POINT Pyrolysis followed by rapid quenching provides one method of converting small boranes to larger boranes.

As discovered by Stock and perfected by many subsequent workers, the controlled pyrolysis of B_2H_6 in the gas phase provides a route to most of the higher boranes and borohydrides, including B_4H_{10}, B_5H_9, and $B_{10}H_{14}$. A key first step in the proposed mechanism is the dissociation of B_2H_6 and the condensation of the resulting BH_3 fragment with borane fragments. For example, the mechanism of the formation of tetraborane(10) by the pyrolysis of diborane appears to be

$$B_2H_6 \rightarrow BH_3 + BH_3$$
$$B_2H_6 + BH_3 \rightarrow B_3H_7 + H_2$$
$$BH_3 + B_3H_7 \rightarrow B_4H_{10}$$

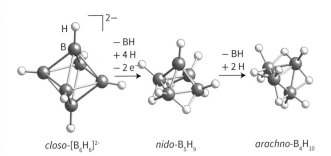

FIGURE 14.12 Structural correlations between a B_6 *closo* octahedral structure, a B_5 *nido* square pyramid, and a B_4 *arachno* butterfly.

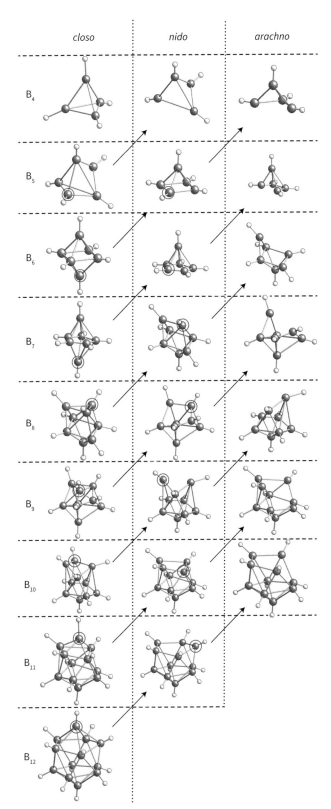

FIGURE 14.13 Structural relations between *closo*, *nido*, and *arachno* boranes and heteroatomic boranes. Diagonal lines connect species that have the same number of skeletal electrons. Hydrogen atoms beyond those in the B–H framework and charges have been omitted. The encircled atom shows which one is removed to produce the structure to its upper right. (Based on *Acc. Chem. Res.*, 1976, **9**, 12, 446–52, with permission, American Chemical Society.)

The synthesis of tetraborane(10), B_4H_{10}, is particularly difficult because it is highly unstable, in keeping with the instability of the B_nH_{n+6} (*arachno*) series. To improve the yield, the product that emerges from the hot reactor is immediately quenched on a cold surface. Pyrolytic syntheses to form species belonging to the more stable B_nH_{n+4} (*nido*) series proceed in higher yield, without the need for a rapid quench. Thus, B_5H_9 and $B_{10}H_{14}$ are readily prepared by the pyrolysis reaction. More recently, these brute-force methods of pyrolysis have given way to more specific methods that are described below.

(e) Characteristic reactions of boranes and borohydrides

KEY POINTS Characteristic reactions of boranes are cleavage of a BH_2 group from diborane and tetraborane by NH_3, deprotonation of large boron hydrides by bases, reaction of a boron hydride with a borohydride ion to produce a larger borohydride anion, and Friedel–Crafts type substitution of an alkyl group for hydrogen in pentaborane and some larger boron hydrides.

The characteristic reactions of boron clusters with a Lewis base range from cleavage of BH_n from the cluster to deprotonation of the cluster, cluster enlargement, and abstraction of one or more protons. All the boranes are reactive, sensitive to air and moisture, and susceptible to hydrolysis. These reactions with water produce boric acid and hydrogen, and the outcome of this reaction can be used to establish the stoichiometry of the borane:

$$B_nH_m + 3nH_2O \rightarrow nB(OH)_3 + \frac{3n+m}{2}H_2$$

Lewis base cleavage reactions have already been introduced in Section 14.6(b) in connection with diborane. With the robust higher borane B_4H_{10}, cleavage may break some B–H–B bonds, leading to partial fragmentation of the cluster:

EXAMPLE 14.8 Determining the stoichiometry of a borane or borohydride from its hydrolysis reaction

Hydrolysis of a mole of a borohydride yields 11 mol of H_2 and 4 mol of $B(OH)_3$. Determine its stoichiometry.

Answer The hydrolysis reaction is $B_nH_m + 3nH_2O \rightarrow n\,B(OH)_3 + \frac{3n+m}{2}H_2$ and so $n = 4$ and $\frac{3n+m}{2} = 11$, which gives $m = 10$. The compound is B_4H_{10}.

Self-test 14.8 Hydrolysis of a mole of a borohydride yields 12 mol of H_2 and 5 mol of $B(OH)_3$. Identify the compound and suggest a structure.

Deprotonation, rather than cleavage, occurs readily with the large borane $B_{10}H_{14}$:

$$B_{10}H_{14} + N(CH_3)_3 \rightarrow [HN(CH_3)_3]^+[B_{10}H_{13}]^-$$

The structure of the product anion indicates that deprotonation occurs from a 3c,2e BHB bridge, leaving the electron count on the boron cluster unchanged. This deprotonation of a BHB 3c,2e bond to yield a 2c,2e bond occurs without major disruption of the bonding:

The Brønsted acidity of boron hydrides increases approximately with size: $B_4H_{10} < B_5H_9 < B_{10}H_{14}$. This trend correlates with the greater delocalization of charge in the larger clusters, in much the same way that delocalization accounts for the greater acidity of phenol than methanol. The variation in acidity is illustrated by the observation that, as shown above, the weak base trimethylamine deprotonates decaborane(14), but the much stronger base methyllithium is required to deprotonate B_5H_9:

Hydridic character is most characteristic of small anionic borohydrides. As an illustration, whereas BH_4^- readily surrenders an H^- ion in the reaction

$$BH_4^- + H^+ \rightarrow \tfrac{1}{2}B_2H_6 + H_2$$

the $[B_{10}H_{10}]^{2-}$ ion survives even in strongly acidic solution. Indeed, the hydronium salt $[(H_3O)^+]_2\,[B_{10}H_{10}]^{2-}$ can even be crystallized.

The cluster-building reaction between a borane and a borohydride provides a convenient route to higher borohydride ions:

$$5\,K[B_9H_{14}] + 2\,B_5H_9 \xrightarrow{\text{dimethoxyethane, 85°C}} 5\,K[B_{11}H_{14}] + 9\,H_2$$

Similar reactions are used to prepare other borohydrides, such as $[B_{10}H_{10}]^{2-}$. This type of reaction has been used to synthesize a wide range of polynuclear borohydrides. ^{11}B-NMR spectroscopy reveals that the boron skeleton in $[B_{11}H_{14}]^-$ consists of an icosahedron with a missing vertex (Fig. 14.14).

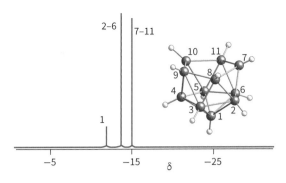

FIGURE 14.14 The proton-decoupled ^{11}B-NMR spectrum of $[B_{11}H_{14}]^-$. The *nido* structure (a truncated icosahedron) is indicated by the 1:5:5 intensity pattern.

The electrophilic displacement of H$^+$ provides a route to alkylated and halogenated species. As with Friedel–Crafts reactions, the electrophilic displacement of H is catalysed by a Lewis acid, such as aluminium chloride, and the substitution generally occurs on the closed portion of the boron clusters:

14.12 Metallaboranes and carboranes

KEY POINTS Main-group and d-block metals may be incorporated into boron hydrides through B–H–M bridges or more robust B–M bonds. When CH is introduced in place of BH in a polyhedral boron hydride, the charge of the resulting carboranes is one unit more positive; carborane anions are useful precursors of boron-containing organometallic compounds.

Metallaboranes are metal-containing boron clusters. In some cases the metal is appended to a borohydride ion through hydrogen bridges. A more common and generally more robust group of metallaboranes have direct metal–boron bonds. An example of a main-group metallaborane with an icosahedral framework is *closo*-$[B_{11}H_{11}AlCH_3]^{2-}$ (**30**). It is prepared by interaction of the acidic hydrogens in Na$_2$[B$_{11}$H$_{13}$] with trimethylaluminium:

$$2[B_{11}H_{13}]^{2-} + Al_2(CH_3)_6 \xrightarrow{\Delta} 2[B_{11}H_{11}AlCH_3]^{2-} + 4CH_4$$

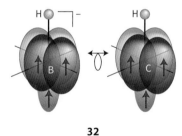

30 *closo*-$[B_{11}H_{11}AlCH_3]^{2-}$

When B$_5$H$_9$ is heated with Fe(CO)$_5$, a metallated analogue of pentaborane is formed (**31**).

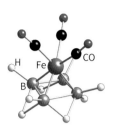

31 [Fe(CO)$_3$B$_4$H$_8$]

Generally, boranes are quite reactive to metal reagents and attack can occur at several points on the polyhedral cage. Therefore, reactions produce complex mixtures of metallaboranes from which individual species can be isolated.

Closely related to the polyhedral boranes and borohydrides are the **carboranes**—a large family of clusters that contain both B and C atoms. Now we begin to see the full generality of Wade's electron-counting rules, as BH$^-$ is isoelectronic and isolobal with CH (**32**), and we can expect the polyhedral borohydrides and carboranes to be related. For example, C$_2$B$_3$H$_5$ has (5 × 2) electrons from each of the B–H or C–H bonds and an additional electron from each C, giving a total of 12 cluster electrons, or six pairs.

32

The ($n + 1$) rule predicts that the molecule will be based on a five-vertex polyhedron, or trigonal bipyramid (**33**). Wade's rules do not enable us to predict the positions of the carbon atoms. Spectroscopic techniques have to be used to further elucidate the structure.

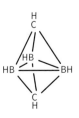

33

EXAMPLE 14.9 Using Wade's rules to predict the structure of a carborane

Predict the structure of $C_2B_5H_7$.

Answer The number of skeletal electrons is $(7 \times 2) + 2 = 16$, or 8 skeletal electron pairs. The $(n + 1)$ rule predicts that the shape is based on a seven-vertex polyhedron, a pentagonal bipyramid. As there are seven vertex atoms, this is a *closo* structure (**34**).

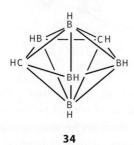

34

Self-test 14.9 Predict the structure of $C_2B_4H_6$.

The carboranes are often prepared by reacting boranes with ethyne:

$$B_5H_9 + C_2H_2 \xrightarrow{C_2H_2, 500–600°C} 1,5\text{-}C_2B_3H_5 + 1,6\text{-}C_2B_4H_6 + 2,4\text{-}C_2B_5H_7$$

An interesting reaction is the conversion of decaborane(14) to *closo*-1,2-$B_{10}C_2H_{12}$ (**35**).

35 *closo*-1,2-$B_{10}C_2H_{12}$

The first reaction in this preparation is the displacement of an H_2 molecule from decaborane by a thioether:

$$B_{10}H_{14} + SEt_2 \rightarrow B_{10}H_{12}(SEt_2) + H_2$$

The loss of two H atoms in this reaction is compensated by the donation of electron pairs by the added thioether, so the electron count is unchanged. The product of the reaction is then converted to the carborane by the addition of an alkyne:

$$B_{10}H_{12}(SEt_2) + C_2H_2 \rightarrow B_{10}C_2H_{12} + SEt_2 + H_2$$

The four π electrons of ethyne displace one thioether molecule (two-electron donor) and an H_2 molecule (which leaves two additional electrons). The net loss of two electrons correlates with the change in structure from a *nido* starting material to the *closo* product. The C atoms are in adjacent (1,2) positions, reflecting their origin from ethyne. This *closo*-carborane survives in air and can be heated without decomposition. At 500°C in an inert atmosphere it undergoes isomerization into 1,7-$B_{10}C_2H_{12}$ (**36**), which in turn isomerizes at 700°C to the 1,12-isomer (**37**).

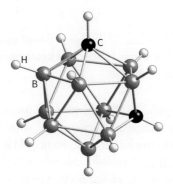

36 *closo*-1,7-$B_{10}C_2H_{12}$

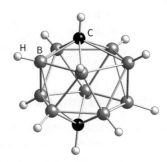

37 *closo*-1,12-$B_{10}C_2H_{12}$

The H atoms attached to carbon in *closo*-$B_{10}C_2H_{12}$ are very mildly acidic, so it is possible to lithiate these compounds with butyllithium:

$$B_{10}C_2H_{12} + 2\,LiC_4H_9 \rightarrow B_{10}C_2H_{10}Li_2 + 2\,C_4H_{10}$$

These dilithiocarboranes are good nucleophiles and undergo many of the reactions characteristic of organolithium reagents (Section 12.17). Thus a wide range of carborane derivatives can be synthesized. For example, reaction with CO_2 gives a carborane dicarboxylic acid:

$$B_{10}C_2H_{10}Li_2 \xrightarrow{(1)\,2\,CO_2;\,(2)\,2\,H_2O} B_{10}C_2H_{10}(COOH)_2$$

Similarly, I_2 leads to the diiodocarborane and NOCl yields $B_{10}C_2H_{10}(NO)_2$.

Although 1,2-$B_{10}C_2H_{12}$ is very stable, the cluster can be partially fragmented in strong base, and then deprotonated with NaH to yield *nido*-$[B_9C_2H_{11}]^{2-}$:

$$B_{10}C_2H_{12} + EtO^- + 2\,EtOH \rightarrow [B_9C_2H_{12}]^- + B(OEt)_3 + H_2$$

$$Na[B_9C_2H_{12}] + NaH \rightarrow Na_2[B_9C_2H_{11}] + H_2$$

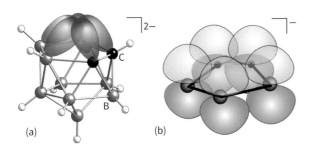

FIGURE 14.15 The isolobal relation between (a) $[B_9C_2H_{11}]^{2-}$ and (b) $[C_5H_5]^-$. The H atoms have been omitted for clarity.

The importance of these reactions is that $nido$-$[B_9C_2H_{11}]^{2-}$ (Fig. 14.15(a)) is an excellent ligand. In this role it mimics the cyclopentadienyl ligand ($[C_5H_5]^-$; Fig. 14.15(b)), which is widely used in organometallic chemistry (Section 24.14):

$$2\,Na_2[B_9C_2H_{11}] + FeCl_2 \xrightarrow{THF} 2\,NaCl + Na_2[Fe(B_9C_2H_{11})_2]$$

$$2\,Na[C_5H_5] + FeCl_2 \xrightarrow{THF} 2\,NaCl + [Fe(C_5H_5)_2]$$

Although we shall not enter into the details of their synthesis, a wide range of metal-coordinated carboranes can be synthesized. A notable feature is the ease of formation of multi-decker sandwich compounds containing carborane ligands (**38** and **39**). The highly negative $[B_3C_2H_5]^{4-}$ and $[B_2C_3H_5]^{3-}$ ligands have a much greater tendency to form stacked sandwich compounds than the less negative and therefore poorer donor $[C_5H_5]^-$.

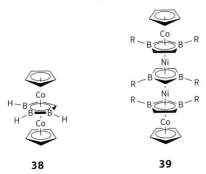

38   **39**

14.13 The hydrides of aluminium, gallium, indium, and thallium

KEY POINTS $LiAlH_4$ and $LiGaH_4$ are useful precursors of MH_3L_2 complexes; $LiAlH_4$ is also used as a source of H^- ions in the preparation of metalloid hydrides, such as SiH_4. The alkylaluminium hydrides are used to couple alkenes.

Aluminium hydride, AlH_3, is a polymeric solid formed from AlH_6 octahedra linked through bent Al–H–Al bridges which melts at 150°C. The alkylaluminium hydrides, including $Al_2(C_2H_5)_4H_2$, are well-known molecular compounds and contain Al–H–Al $3c,2e$ bonds (Section 3.11(e)). Hydrides of this kind are used to couple alkenes, the initial step being the addition of the AlH entity across the C=C double bond, as in hydroboration (Section 14.6(c)). Pure gaseous digallane, Ga_2H_6, was prepared for the first time in 1989; it liquefies below –50°C, and decomposes above –30°C. Matrix isolation studies of digallane molecules show a structure analogous to diborane with terminal Ga–H bonds of 152 pm and bridging Ga–H at 171 pm. The solid adopts a polymeric structure linked through Ga–H–Ga bridges. The hydrides of indium and thallium are very unstable, although InH_3 may be isolated at low temperature. Adducts, where InH_3 acts as a Lewis acid, such as in $[InH_3\{P(C_6H_{11})_3\}]$, are stable to above 50°C.

The metathesis of the halides with LiH leads to lithium tetrahydridoaluminate, $LiAlH_4$, or the analogous $LiGaH_4$:

$$4\,LiH + ECl_3 \xrightarrow{\Delta,\,ether} LiEH_4 + 3\,LiCl \quad (E = Al, Ga)$$

The direct reaction of Li, Al, and H_2 leads to the formation of either $LiAlH_4$ or Li_3AlH_6, depending on the conditions of the reaction. Their formal analogy with halide complexes such as $AlCl_4^-$ and AlF_6^{3-} should be noted.

The AlH_4^- and GaH_4^- ions are tetrahedral, and are much more hydridic than BH_4^- (Section 11.6(d)). Their hydridic character is consistent with the higher electronegativity of B compared with Al and Ga, and the fact that BH_4^- is more covalent than AlH_4^- and GaH_4^-. For example, $NaAlH_4$ reacts violently with water but, as we saw earlier, basic aqueous solutions of $NaBH_4$ are useful in synthetic chemistry. They are also much stronger reducing agents; $LiAlH_4$ is commercially available and is widely used as a strong hydride source and as a reducing agent.

With the halides of many nonmetallic elements AlH_4^- serves as a hydride source in metathesis reactions; for example, in the reaction of lithium tetrahydridoaluminate with silicon tetrachloride in tetrahydrofuran solution to produce silane:

$$LiAlH_4 + SiCl_4 \xrightarrow{THF} LiAlCl_4 + SiH_4$$

The general rule in this important type of reaction is that H^- migrates from the element of lower electronegativity (Al in the example) to the element of greater electronegativity (Si).

Under conditions of controlled protolysis, both AlH_4^- and GaH_4^- lead to complexes of aluminium or gallium hydride:

$$LiEH_4 + [(CH_3)_3NH]Cl \rightarrow (CH_3)_3N-EH_3 + LiCl + H_2$$
$$(E = Al, Ga)$$

In striking contrast to BH_3 complexes, these complexes will add a second molecule of base to form five-coordinate complexes of aluminium or gallium hydride:

$$(CH_3)_3N-EH_3 + N(CH_3)_3 \rightarrow ((CH_3)N)_2EH_3 \quad (E = Al, Ga)$$

This behaviour is consistent with the trend for Period 3 and heavier p-block elements to form five- and six-coordinate hypervalent compounds (Section 3.11(b)).

14.14 Trihalides of aluminium, gallium, indium, and thallium

KEY POINTS Aluminium, gallium, and indium all favour the +3 oxidation state, and their trihalides are Lewis acids. Thallium trihalides are less stable than those of their congeners.

Although direct reaction of Al, Ga, or In with a halogen yields a halide, these electropositive metals also react with HCl or HBr gas, and the latter is usually a more convenient route:

$$2\,Al(s) + 6\,HCl(g) \xrightarrow{100°C} 2\,AlCl_3(s) + 3\,H_2(g)$$

AlF_3 and GaF_3 form salts of the type Na_3AlF_6 (cryolite) and Na_3GaF_6, which contain octahedral $[MF_6]^{3-}$ complex ions. Cryolite occurred naturally at Ivigtût in Greenland, but this one commercially viable deposit was mined to exhaustion in the 1980s. Molten synthetic cryolite, manufactured by reaction of sodium aluminate, $NaAlO_2$, with HF, is now used as a solvent for Al_2O_3 in the industrial extraction of aluminium.

The Lewis acidities of the trihalides reflect the relative chemical hardness of the Group 13 elements. Thus, towards a hard Lewis base (such as ethyl acetate, which is hard because of its O donor atoms), the Lewis acidities of the halides weaken as the softness of the acceptor element increases, so the Lewis acidities fall in the order $BCl_3 > AlCl_3 > GaCl_3$. By contrast, towards a soft Lewis base (such as dimethylsulfane, Me_2S, which is soft because of its S atom), the Lewis acidities strengthen as the softness of the acceptor element increases: $GaX_3 > AlX_3 > BX_3$ (X = Cl or Br).

Aluminium trichloride is a useful starting material for the synthesis of Al organometallic compounds:

$$AlCl_3(sol) + 3\,LiR(sol) \rightarrow AlR_3(sol) + 3\,LiCl(s)$$

This reaction is an example of **transmetallation**, where the halide formed is that of the more electronegative element and the high lattice enthalpy of that compound can be regarded as the 'driving force' of the reaction. The main industrial application of $AlCl_3$ is as a Friedel–Crafts catalyst in organic synthesis.

The thallium trihalides are much less stable than those of their lighter congeners. A trap for the unwary is that thallium triiodide is a compound of Tl(I) rather than Tl(III), as it contains the I_3^- ion, not I^-. This is confirmed by considering the standard potentials, which indicate that Tl(III) is rapidly reduced to Tl(I) by iodide:

$$Tl^{3+}(aq) + 2e^- \rightarrow Tl^+(aq) \quad E^\ominus = +1.25\,V$$
$$I_3^-(aq) + 2e^- \rightarrow 3I^-(aq) \quad E^\ominus = +0.55\,V$$

However, in excess iodide Tl(III) is stabilized by the formation of a complex:

$$TlI_3(s) + I^-(aq) \rightarrow [TlI_4]^-(aq)$$

In keeping with the general tendency towards higher coordination numbers for the larger atoms of heavier p-block elements, halides of Al and its heavier congeners may form coordinate bonds with more than one Lewis base:

$$AlCl_3 + N(CH_3)_3 \rightarrow Cl_3AlN(CH_3)_3$$
$$Cl_3AlN(CH_3)_3 + N(CH_3)_3 \rightarrow Cl_3Al(N(CH_3)_3)_2$$

14.15 Low oxidation state halides of aluminium, gallium, indium, and thallium

KEY POINT The +1 oxidation state becomes progressively more stable from aluminium to thallium.

The increasing stability of the M(I) oxidation state as Group 13 is descended can be understood in terms of the inert pair effect, Section 10.7. All the AlX compounds and both GaF and InF are unstable, gaseous species that disproportionate in the solid phase:

$$2\,AlX(s) \rightarrow 2\,Al(s) + AlX_3(s)$$

The other monohalides of Ga, In, and Tl are more stable. Gallium monohalides are formed by reacting GaX_3 with the metal in a 1:2 ratio:

$$GaX_3(s) + 2\,Ga(s) \rightarrow 3\,GaX(s) \quad (X = Cl, Br, or\ I)$$

The stability increases from the chloride to the iodide. The stability of the +1 oxidation state is increased by the formation of compounds such as $Ga[AlX_4]$. The apparently divalent GaX_2 can be prepared by heating GaX_3 with gallium metal in a 2:1 ratio:

$$2\,GaX_3(s) + Ga(s) \xrightarrow{\Delta} 3\,GaX_2(s) \quad (X = Cl, Br, or\ I)$$

The formula GaX_2 is deceiving, as this solid and most other apparently divalent salts do not contain Ga(II); instead, they are mixed-oxidation-state compounds containing equimolar mixtures of Ga(I) and Ga(III), as in $Ga(I)[Ga(III)Cl_4]$. Mixed-oxidation-state halogen compounds are also known for the heavier metals, such as $InCl_2$ and $TlBr_2$. The presence of M^{3+} ions is indicated by the existence of MX_4^- complexes in these salts with short M–X distances, and the presence of M^+ ions is indicated by the longer and less regular separation from the halide ions representative of an ionic interaction. There is in fact only a fine line between the formation of a mixed-oxidation-state ionic compound and the formation of a Ga(II) compound that contains M–M bonds (Section 10.1). For example, mixing $GaCl_2$ with a solution of [N(CH$_3$)$_4$]Cl in a nonaqueous solvent yields the compound [N(CH$_3$)$_4$]$_2$[Cl$_3$Ga–GaCl$_3$], in which the anion has an ethane-like structure with a Ga–Ga bond.

Indium monohalides are prepared by direct interaction of the elements or by heating the metal with HgX$_2$. The stability increases from the chloride to the iodide and is enhanced by the formation of compounds such as In[AlX$_4$]. Gallium(I) and In(I) halides both disproportionate when dissolved in water:

$$3\,MX(s) \rightarrow 2\,M(s) + M^{3+}(aq) + 3\,X^-(aq)$$
$$(M = Ga, In; X = Cl, Br, I)$$

Thallium(I) is stable with respect to disproportionation in water and thallium(I) halides can be prepared by the action of HX on an acidified solution of a soluble Tl(I) salt. Thallium(I) fluoride has a distorted rock-salt structure whereas TlCl and TlBr have the caesium-chloride structure (Section 4.9). Yellow TlI has an orthorhombic layer structure in which the inert pair manifests itself structurally as a stereoactive lone pair, but when pressure is applied it is converted to red TlI with a caesium-chloride structure. Thallium(I) iodide is used in photomultiplier tubes to detect ionizing radiation. Other low-oxidation-state halides of indium and thallium are known: TlX$_2$ is actually Tl(I)[Tl(III)X$_4$], Tl$_2$X$_3$ is Tl(I)$_3$[Tl(III)X$_6$], and In$_4$Br$_6$ is In(I)$_3$[In(III)Br$_6$].

EXAMPLE 14.10 Proposing reactions of Group 13 halides

Propose chemical equations (or indicate no reaction) for reactions between (a) AlCl$_3$ and (C$_2$H$_5$)$_3$NGaCl$_3$ in toluene, (b) (C$_2$H$_5$)$_3$NGaCl$_3$ and GaF$_3$ in toluene, (c) TlCl and NaI in water.

Answer (a) We need to note that the trichlorides are excellent Lewis acids and that Al(III) is a stronger and harder Lewis acid than Ga(III). Therefore, the following reaction can be expected:

$$AlCl_3 + (C_2H_5)_3NGaCl_3 \rightarrow (C_2H_5)_3NAlCl_3 + GaCl_3$$

(b) In this case we need to note that the fluorides are ionic, so GaF$_3$ has a very high lattice enthalpy and is not a good Lewis acid. There is no reaction. (c) Now we note that Tl(I) is a chemically borderline soft Lewis acid, so it combines with the softer I$^-$ ion rather than Cl$^-$:

$$TlCl(s) + NaI(aq) \rightarrow TlI(s) + NaCl(aq)$$

Like silver halides, Tl(I) halides have low solubility in water, so the reaction will probably proceed very slowly.

Self-test 14.10 Propose, with reasons, the chemical equation (or indicate no reaction) for reactions between (a) (CH$_3$)$_2$SAlCl$_3$ and GaBr$_3$, (b) TlCl$_3$ and formaldehyde (HCHO) in acidic aqueous solution. (Hint: Formaldehyde is easily oxidized to CO$_2$ and H$^+$.)

14.16 Oxo compounds of aluminium, gallium, indium, and thallium

KEY POINTS Aluminium and gallium form α and β forms of the oxide in which the elements are in their +3 oxidation state; thallium forms oxides exhibiting the +1 and +3 oxidation states and a peroxide.

The most stable form of Al$_2$O$_3$, α-alumina, is a very hard and refractory material. In its mineral form it is known as corundum, and as a gemstone it is sapphire or ruby, depending on the metal-ion impurities. The blue of sapphire arises from a charge-transfer transition from Fe^{2+} to Ti^{4+} ion impurities (Section 25.1(b), Box 25.1). Ruby is α-alumina in which a small fraction of the Al^{3+} ions is replaced by Cr^{3+}. The structure of α-alumina and gallia, Ga$_2$O$_3$, consists of an hcp array of O^{2-} ions with the metal ions occupying two thirds of the octahedral holes in an ordered array.

Dehydration of aluminium hydroxide at temperatures below 900°C leads to the formation of γ-alumina, which is a metastable polycrystalline form with a defect spinel structure (Section 4.9(b)) and a very high surface area. Partly because of its surface acid and base sites, this material is used as a solid phase in chromatography and as a heterogeneous catalyst and catalyst support (Section 31.8).

The α and γ forms of Ga$_2$O$_3$ have the same structures as their Al analogues. The metastable form is β-Ga$_2$O$_3$, which has a ccp structure with Ga(III) in distorted octahedral and tetrahedral sites. Half the Ga(III) ions are therefore four-coordinate despite its large radius (compared to Al(III)). This coordination may be due to the effect of the filled 3d^{10} shell of electrons, as remarked previously. Indium and Tl form In$_2$O$_3$ and Tl$_2$O$_3$, respectively. Thallium also forms the Tl(I) oxide, Tl$_2$O (adopting the cadmium iodide structure type with the cations and anions reversed, the anti-cadmium iodide structure), and a peroxide, Tl$_2$O$_2$, showing the similarity of the chemistry of Tl(I) to that of the alkali metals. Indium tin oxide (ITO) is In$_2$O$_3$ doped with 10% by mass SnO$_2$ to form an n-type semiconductor, which has applications as a transparent conducting oxide in touch-screens (Box 14.7) and in heated aircraft windows. ITO is also used in infrared-reflecting coatings and mirrors and as an antireflective coating on binoculars, telescopes, and spectacles. The melting point of ITO is 1900°C, which makes ITO thin film strain gauges very useful in harsh environments; for example, jet engines and gas turbines.

Thallium is an important component of many high-temperature superconducting complex oxides, including those that demonstrate some of the higher critical temperatures (the temperature below which they become superconducting, T_c). Thallium barium calcium copper

BOX 14.7 How does my touchscreen work?

Indium oxide is a white solid adopting the bixbyite structure, in which indium adopts a distorted octahedral coordination. Replacing around 10% of the indium sites with tin produces a pale yellow-grey solid, of the approximate composition $In_{1.8}Sn_{0.2}O_3$, known as indium tin oxide or ITO. ITO is close to transparent in the visible region and transmits visible light well, particularly when it is deposited as a thin film (around 100 nm thick) onto glass. It is also a reasonable electrical conductor with a 100-nm thick film having a surface resistivity of around only 15 Ω, Section 30.1(a).

ITO can be deposited in thin films on surfaces by a variety of methods such as physical vapour deposition and ion beam sputtering, Section 28.3. The main use of these films is as transparent conducting coatings for liquid crystal and plasma displays, touch panels, solar cells, and organic light-emitting diodes (OLEDs). Light from the device will pass through the ITO coating placed on the back of the glass screen.

In a capacitive touchscreen, used on most modern devices, a small voltage is applied to the ITO film etched into a grid-like pattern. When any conducting object, including the human finger, is placed on the insulating front of the glass screen a change in the capacitance occurs at that point. The position of this capacitance change can be determined electronically (using the grid pattern), and this can also be achieved for several points (or finger touches) on the screen simultaneously.

---

oxides include the series of phases of the general formula $Tl_mBa_2Ca_{n-1}Cu_nO_{2n+m+2}$ with $m = 1, 2$ and $n = 1–5$. Of these, when $m = 2$ and $n = 3$ the composition $Tl_2Ba_2Ca_2Cu_3O_{10}$ is obtained which shows a T_C of 125 K. These phases can be synthesized by the reaction of the component oxides, including Tl_2O_3, at around 900°C.

14.17 Sulfides of gallium, indium, and thallium

KEY POINT Gallium, indium, and thallium form many sulfides with a wide range of structures.

The only sulfide of Al is Al_2S_3, which is prepared by direct reaction of the elements at elevated temperatures:

$$2\,Al(s) + 3\,S(s) \xrightarrow{\Delta} Al_2S_3(s)$$

It is rapidly hydrolysed in aqueous solution:

$$Al_2S_3(s) + 6\,H_2O(l) \rightarrow 2\,Al(OH)_3(s) + 3\,H_2S(g)$$

Aluminium sulfide exists in α, β, and γ forms. The structures of the α and β forms are based on the wurtzite structure (Section 4.9): in α-Al_2S_3 the S^{2-} ions are hcp and the Al^{3+} ions occupy two thirds of the tetrahedral sites in an ordered fashion; in β-Al_2S_3 the Al^{3+} ions occupy two thirds of the tetrahedral sites randomly. The γ form adopts the same structure as γ-Al_2O_3.

The sulfides of Ga, In, and Tl are more numerous and varied than those of Al and adopt many different structural types. Some examples are given in Table 14.4. Many of the sulfides are semiconductors, photoconductors, or light emitters, and are used in electronic devices.

TABLE 14.4 Selected sulfides of gallium, indium, and thallium

Sulfide	Structure
GaS	Layer structure with Ga–Ga bonds
α-Ga_2S_3	Defect wurtzite structure (hexagonal)
γ-Ga_2S_3	Defect sphalerite structure (cubic)
InS	Layer structure with In–In bonds
β-In_2S_3	Defect spinel (as γ-Al_2O_3)
TlS	Structure containing Tl(I)S_6 and Tl(III)S_6 octahedra
Tl_4S_3	Chains of [Tl(III)S_4] and [Tl(I)S_6] and [Tl(I)S_5] polyhedra

14.18 Compounds with Group 15 elements

KEY POINT Aluminium, gallium, and indium react with phosphorus, arsenic, and antimony to form materials that act as semiconductors.

The compounds formed between Group 13 and Group 15 elements (the pnictogens) are important commercially and technologically, as they are isoelectronic with Si and Ge and act as semiconductors (Sections 2.11, 25.6(a), and 30.2(a)). The nitrides adopt the wurtzite structure, and the phosphides, arsenides, and stibnides all adopt the zinc-blende (sphalerite) structure (Section 4.9). All the Group 13/15 (still commonly 'Group III/V') binary compounds can be prepared by direct reaction of the elements at high temperature and pressure:

$$Ga(s) + As(s) \rightarrow GaAs(s)$$

The most widely used Group 13/15 semiconductor is gallium arsenide, GaAs, which is used to make devices such as integrated circuits, light-emitting diodes, and laser diodes.

TABLE 14.5 Band gap at 298 K

	E_g/eV
GaN	3.40
GaAs	1.35
GaSb	0.67
InAs	0.36
InSb	0.16
Si	1.11

Its band gap is similar to that of Si and larger than those of other Group 13/15 compounds (Table 14.5) apart from GaN. Gallium arsenide is superior to Si for such applications because it has higher electron mobility, allowing it to function at frequencies in excess of 250 GHz. Gallium arsenide devices also generate less electronic noise than silicon devices.

One disadvantage of Group 13/15 semiconductors is that the compounds decompose in moist air and must be kept under an inert atmosphere, usually nitrogen, or be completely encapsulated. Gallium nitride's band gap of 3.40 eV corresponds to violet-blue light, and since the 1990s this compound has been widely used to produce blue light in light-emitting diodes and in the lasers of Blu-ray® DVD players, Box 25.4.

14.19 Zintl phases

KEY POINT Group 13 elements form Zintl phases with Group 1 and Group 2 elements which are poor conductors and diamagnetic.

The Group 13 elements form Zintl phases (Sections 4.8 and 25.3(c)) with Group 1 or Group 2 metals. Zintl phases are compounds of two metals which are brittle, diamagnetic, and poor conductors. They are therefore quite different from alloys. Zintl phases are formed between a very electropositive Group 1 or 2 element and a moderately electronegative p-block metal or metalloid. They are ionic, with electrons being transferred from the Group 1 or 2 metal to the more electronegative element. The anion, referred to as a 'Zintl ion', has a complete octet of valence electrons and is polymeric; the cations are located within the anionic lattice. The structure of NaTl consists of a polymeric anion in a covalent diamond structure with Na$^+$ ions fitted into the anionic lattice. In Na$_2$Tl the polymeric anion is tetrahedral Tl$_4^{8-}$. The Zintl anions can be isolated by reacting with salts containing the tetraalkylammonium ion, which substitutes for the Group 1 or 2 metal ion, or by encapsulation within a cryptand. Some compounds appear to be Zintl phases but are conducting and paramagnetic. For example, K$_8$In$_{11}$ contains the In$_{11}^{8-}$ anion, which has one delocalized electron per formula unit.

14.20 Organometallic compounds

The most important organometallic compounds of the Group 13 elements are those of B and Al. Organoboron compounds are commonly treated as organometallic compounds, even though B is not a metal.

(a) Organoboron compounds

KEY POINTS Organoboron compounds are electron-deficient and act as Lewis acids; tetraphenylborate is an important anion.

Organoboranes of the type BR$_3$ can be prepared by hydroboration of an alkene with diborane.

$$B_2H_6 + 6\,CH_2{=}CH_2 \rightarrow 2\,B(CH_2CH_3)_3$$

Alternatively, they can be produced from a Grignard reagent (Section 13.13):

$$(C_2H_5)_2O{:}BF_3 + 3\,RMgX \rightarrow BR_3 + 3\,MgXF + (C_2H_5)_2O$$

Alkylboranes are not hydrolysed but are pyrophoric. The aryl species are more stable. They are all monomeric and planar. Like other B compounds, the organoboron species are electron-deficient and consequently act as Lewis acids and form adducts easily.

An important anion is the tetraphenylborate ion, [B(C$_6$H$_5$)$_4$]$^-$, more commonly written BPh$_4^-$, analogous to the tetrahydridoborate ion, BH$_4^-$ (Section 14.6). The sodium salt can be obtained by a simple addition reaction:

$$BPh_3 + NaPh \rightarrow Na^+[BPh_4]^-$$

The Na salt is soluble in water, but the salts of most large, monopositive ions are insoluble. Consequently, the anion is useful as a precipitating agent and can be used in gravimetric analysis.

(b) Organoaluminium compounds

KEY POINTS Methyl- and ethyl-aluminium are dimers; bulky alkyl groups result in monomeric species.

Alkylaluminium compounds can be prepared on a laboratory scale by transmetallation of a mercury compound:

$$2\,Al + 3\,Hg(CH_3)_2 \rightarrow Al_2(CH_3)_6 + 3\,Hg$$

Trimethylaluminium is prepared commercially by the reaction of Al metal with chloromethane to give Al$_2$Cl$_2$(CH$_3$)$_4$. This intermediate is then reduced with Na, and the Al$_2$(CH$_3$)$_6$ (**40**) is removed by fractional distillation.

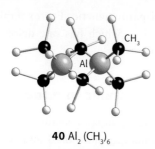

40 Al$_2$(CH$_3$)$_6$

Alkylaluminium dimers are similar in structure to the analogous dimeric halides, but the bonding is different. In the halides, the bridging Al–Cl–Al bonds are $2c,2e$ bonds; that is, each Al–Cl bond involves an electron pair. In the alkylaluminium dimers the Al–C–Al bonds are longer than the terminal Al–C bonds, which suggests that they are $3c,2e$ bonds, with one bonding pair shared across the Al–C–Al unit, somewhat analogous to the bonding in diborane, B$_2$H$_6$ (Section 14.6).

Triethylaluminium and higher alkyl compounds are prepared from the metal, an appropriate alkene, and hydrogen gas at elevated temperatures and pressures:

$$2\,Al + 3\,H_2 + 6\,CH_2=CH_2 \xrightarrow{60-110°C,\ 10-20\ MPa} Al_2(CH_2CH_3)_6$$

This route is relatively cost-effective and, as a result, alkylaluminium compounds have found many commercial applications. Triethylaluminium, often written as the monomer Al(C$_2$H$_5$)$_3$, is an organometallic complex of Al of major industrial importance. It is used in the Ziegler–Natta polymerization catalyst (Section 31.14).

Steric factors have a powerful effect on the structures of alkylaluminiums. Where dimers are formed, the long weak bridging bonds are easily broken. This tendency increases with the bulkiness of the ligand. So, for example, triphenylaluminium is a dimer, but trimesitylaluminium, with the very bulky trimesityl 2,4,6-(CH$_3$)$_3$C$_6$H$_2$– group, is a monomer.

(c) Organometallic compounds of Ga, In, and Tl

KEY POINT The only +1 oxidation state organocompounds are formed with the cyclopentadienyl ligand.

The trigonal-planar Ga(III), In(III), and Tl(III) organocompounds, R$_3$Tl, where R = Me, Et, Ph, are reactive, air-sensitive compounds that are soluble in organic solvents such as THF and ether. Ga(III) and In(III) organocompounds can be prepared by direct interaction between the metal and R$_2$Hg. This synthesis requires no solvent so R$_3$Ga or R$_3$In are easily isolated. The Tl compounds can be prepared from the monohalides, R$_2$TlX, which are stable to air and water and are insoluble in organic solvents:

$$R_2TlX + R'Li \rightarrow R_2TlR' + LiX$$

The R$_3$Tl compounds are useful in carbon–carbon bond formation:

$$R_3Tl + R'COCl \rightarrow R_2TlCl + R'COR \quad R = Me, Et, Ph$$
$$R' = \text{alkyl or aryl group}$$

Me$_3$Ga and Me$_3$In are monomeric trigonal-planar molecules in the vapour phase but exist as tetramers in the solid phase. Ph$_3$Ga and Ph$_3$In consist of stacked trigonal-planar units with the Ga or In lying between phenyl rings of the units above and below. The monohalide compounds, R$_2$GaX and R$_2$InX, form stacked arrangements in the solid phase (**41**). The fluorides Me$_2$GaF and Et$_2$GaF adopt a six-membered ring structure (**42**).

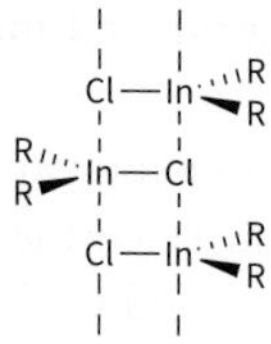

41 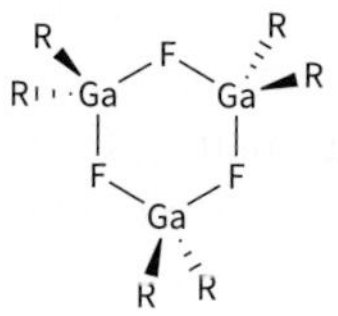   **42**

The only stable Ga(I), In(I), and Tl(I) organocompounds are those with the cyclopentadienyl ligand, C$_5$H$_5^-$ (**43**). Compounds of this type are useful sources of the cyclopentadienyl ligand in the synthesis of other organometallic compounds:

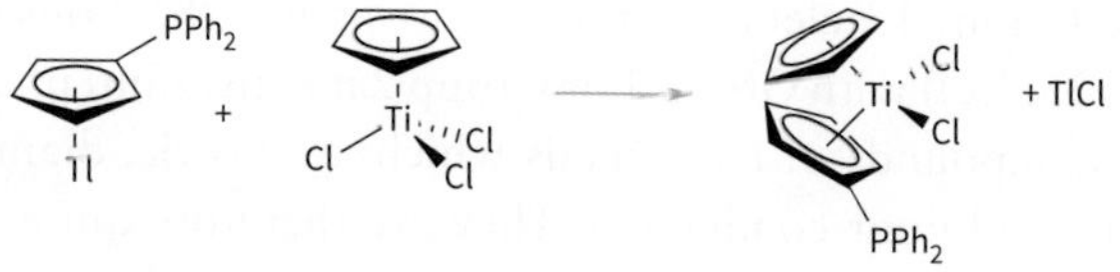

43

C$_5$H$_5$In is made via an In(III) intermediate:

$$InCl_3 + 3\,NaC_5H_5 \rightarrow (C_5H_5)_3In + 3\,NaCl$$
$$(C_5H_5)_3In \rightarrow (C_5H_5)In + (C_5H_5)_2$$

It is the only stable and soluble In(I) organometallic compound and produces C$_5$H$_6$ and InX with acids, HX.

FURTHER READING

D. M. P. Mingos, *Essential Trends in Inorganic Chemistry*. Oxford University Press (2004). A survey of inorganic chemistry from the perspective of structure and bonding.

N. C. Norman, *Periodicity and the s- and p-Block Elements*. Oxford University Press (1997). Includes coverage of essential trends and features of s-block chemistry.

R. B. King (ed.), *Encyclopedia of Inorganic Chemistry*. John Wiley & Sons (2005).

C. E. Housecroft, *Boranes and Metalloboranes*. Ellis Horwood (2005). An introduction to borane chemistry.

C. Benson, *The Periodic Table of the Elements and Their Chemical Properties*. Kindle ed. MindMelder.com (2009).

EXERCISES

14.1 Give a balanced chemical equation and conditions for the recovery of boron.

14.2 Explain the important role of cryolite, Na_3AlF_6, in the extraction of aluminium metal.

14.3 Why does a droplet of mercury cause aluminium to react rapidly with air?

14.4 Describe the bonding in (a) BF_3, (b) $AlCl_3$, (c) B_2H_6, (d) TlF.

14.5 Arrange the following in order of increasing Lewis acidity: BF_3, BCl_3, $AlCl_3$. In the light of this order, write balanced chemical reactions (or no reaction) for:

(a) $BF_3N(CH_3)_3 + BCl_3 \rightarrow$

(b) $BH_3CO + BBr_3 \rightarrow$

14.6 Thallium tribromide (1.11 g) reacts quantitatively with 0.257 g of NaBr to form a product A. Deduce the formula of A. Identify the cation and anion.

14.7 Identify the B-containing compounds **A**, **B**, and **C**.

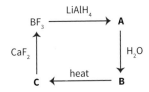

14.8 Does B_2H_6 survive in air? If not, write the equation for the reaction.

14.9 Predict how many different boron environments would be apparent in the proton-decoupled ^{11}B-NMR of (a) B_5H_{11}, (b) B_4H_{10}.

14.10 Predict the products from the hydroboration of (a) $(CH_3)_2C=CH_2$, (b) CH≡CH.

14.11 Diborane has been used as a rocket propellant. Calculate the energy released from 1.00 kg of diborane given the following values of $\Delta_f H^\ominus$/kJ mol^{-1}: B_2H_6 = 31, H_2O = −242, B_2O_3 = −1264. The combustion reaction is $B_2H_6(g) + 3 O_2(g) \rightarrow 3 H_2O(g) + B_2O_3(s)$. What would be the problem with diborane as a fuel?

14.12 How would you expect the Group 13 element-to-hydrogen bond enthalpy to change as the group is descended? Hence explain the nonexistence of TlH_3.

14.13 Using BCl_3 as a starting material and other reagents of your choice, devise a synthesis for the Lewis acid chelating agent, $F_2B-C_2H_4-BF_2$.

14.14 Determine the point group of a B_2Cl_4 molecule in (a) the gas phase (staggered) and (b) the solid (planar) structures.

14.15 Give the IUPAC names of (a) $B_{10}H_{14}$, (b) $[B_{12}H_{12}]^{2-}$, and (c) $[B_6H_7]^-$.

14.16 Give balanced chemical equations for the synthesis of 1,2-$B_{10}C_2H_{10}(Si(CH_3)_3)_2$, starting with decaborane(14) and other reagents of your choice.

14.17 Given $NaBH_4$, a hydrocarbon of your choice, and appropriate ancillary reagents and solvents, give formulas and conditions for the synthesis of (a) $B(C_2H_5)_3$, (b) Et_3NBH_3.

14.18 Draw the B_{12} unit that is a common motif of boron structures; take a viewpoint along a C_2 axis.

14.19 Which boron hydride would you expect to be more thermally stable, B_6H_{10} or B_6H_{12}? Give a generalization by which the thermal stability of a borane can be judged.

14.20 How many skeletal electrons are present in B_5H_9?

14.21 (a) From its formula, classify $B_{10}H_{14}$ as *closo*, *nido*, or *arachno*. (b) Use Wade's rules to determine the number of framework electron pairs for decaborane(14). (c) Verify by detailed accounting of valence electrons that the number of cluster valence electrons of $B_{10}H_{14}$ is the same as that determined in (b).

14.22 Use Wade's rules to predict a structure for (a) B_5H_{11}, (b) B_5H_9, (c) $B_4H_7^-$, (d) $[B_6H_6]^{2-}$.

14.23 Hydrolysis of 1 mol of a borohydride yields 15 mol of H_2 and 6 mol of $B(OH)_3$. Identify the compound and suggest a structure.

14.24 Starting with $B_{10}H_{14}$ and other reagents of your choice, give the equations for the synthesis of $[Fe(nido-B_9C_2H_{11})_2]^{2-}$, and sketch the structure of this species.

14.25 ^{11}B-NMR is an excellent spectroscopic tool for inferring the structures of boron compounds. With ^{11}B–^{11}B coupling ignored, it is possible to determine the number of attached H atoms by the multiplicity of a resonance: BH gives a doublet, BH_2 a triplet, and BH_3 a quartet. B atoms on the closed side of *nido* and *arachno* clusters are generally more shielded than those on the open face. Assuming no B–B or B–H–B coupling, predict the general pattern of the ^{11}B-NMR spectra of (a) BH_3CO and (b) $[B_{12}H_{12}]^{2-}$.

14.26 The structure of B_2O_3 is formed from vertex-linked BO_3 planar-triangular units, while in Al_2O_3 aluminium ions adopt solely octahedral coordination to oxygen. Explain these observations on the preferred Group 13 element coordination geometries.

14.27 Predict the product of the reactions of BCl_3 with (a) water and (b) dimethyl sulfide $(CH_3)_2S$.

14.28 BF_3 reacts with CsF to form $CsBF_4$ while there is no reaction between BBr_3 and CsBr. Explain these observations.

14.29 (a) What are the similarities and differences in the structures of layered BN and graphite (Section 14.5)? (b) Contrast their reactivity with Na, air, and Br_2. (c) Suggest a rationalization for the differences in structure and reactivity.

14.30 Under what conditions could you convert hexagonal BN to cubic BN?

14.31 Devise a synthesis for the borazines (a) $Ph_3N_3B_3Cl_3$ and (b) $Me_3N_3B_3H_3$, starting with BCl_3 and other reagents of your choice. Draw the structures of the products.

14.32 Identify the incorrect statements in the following description of Group 13 chemistry, and provide corrections along with explanations of the principle or chemical generalization that applies:

(a) All the elements in Group 13 are nonmetals.

(b) The increase in chemical hardness on going down the group is illustrated by greater oxophilicity and fluorophilicity for the heavier elements.

(c) The Lewis acidity increases for BX_3 from X = F to Br, and this may be explained by stronger Br–B π bonding.

(d) *Arachno*-boron hydrides have a $2(n + 3)$ skeletal electron count, and they are more stable than *nido*-boron hydrides.

(e) In a series of *nido*-boron hydrides acidity increases with increasing size.

(f) Layered boron nitride is similar in structure to graphite and, because it has a small separation between HOMO and LUMO, it is a good electrical conductor.

14.33 Indium forms a chloride of the stoichiometry In_5Cl_9. Rewrite this formula in terms of the oxidation states commonly adopted by indium. Describe the bonding in this compound given that it has been stated that it adopts the same structure type as $Cs_3Tl_2Cl_9$.

14.34 Indium tin oxide has important applications as a transparent conducting oxide. Which oxide(s) could be doped with Al_2O_3 to produce a material with similar properties?

14.35 Use the data in Table 14.5 to estimate a band gap for the semiconductor GaP. Use this to calculate the wavelength of light that is emitted when an electron drops from the conduction band to the valence band. Is this consistent with the observed green light of gallium phosphide LEDs?

14.36 Determine the position of the compound Rb_2In_3 on a Ketelaar triangle, and hence predict the type of bonding in this phase. By comparison with known structural motifs from boride cluster chemistry suggest a possible indium-containing unit if the formula is rewritten as Rb_4In_6.

TUTORIAL PROBLEMS

14.1 Discuss the similarities in the chemistry of Tl(I) to those of the alkali metals and particularly that of potassium.

14.2 An article in *Chemistry World*, 'The fifth element' (*Chemistry World*, 8 April 2019) opens with the sentence: 'Boron's chemistry is as much defined by what it isn't—carbon, or a metal—as by what it is.' Discuss this statement with reference to recent discoveries in boron chemistry.

14.3 In his paper 'Covalent and ionic molecules: Why are BeF_2 and AlF_3 high melting point solids whereas BF_3 and SiF_4 are gases?' (*J. Chem. Educ.*, 1998, **75**, 923), R. J. Gillespie makes a case for the classification of the bonding in BF_3 and SiF_4 as predominantly ionic. Summarize his arguments and describe how these differ from the conventional view of bonding in gaseous molecules.

14.4 Nanotubes of BN have been synthesized by C. Colliex et al. (*Science*, 1997, **278**, 653). (a) What are the advantageous properties of these nanotubes over carbon analogues? (b) Outline the method used for the preparation of these compounds. (c) What was the main structural feature of the nanotubes, and how could this be exploited in applications?

14.5 Discuss the unusual electronic properties of $Ir_6In_{32}S_{21}$ made in a liquid indium solvent (*Chem. Sci.*, 2020, **11**, 870–8).

14.6 M. Montiverde discusses 'Pressure dependence of the superconducting temperature of MgB_2' (*Science*, 2001, **292**, 75).

(a) Describe the bases of the two theories that have been postulated to explain the superconductivity of MgB_2. (b) How does the T_c of MgB_2 vary with pressure? What insight does this provide into the superconductivity?

14.7 Use the references in Z. W. Pan, Z. R. Dai, and Z. L. Wang (*Science*, 2001, **291**, 1947) as a starting point to write a review of wire-like nanomaterials of Group 13 elements. Indicate how In_2O_3 nanobelts were prepared, and give the dimensions of a typical nanobelt.

14.8 Shuji Nakamura was awarded the 2014 Nobel Prize for Physics, in part for his invention of blue LEDs based on gallium nitride. Discuss the chemistry of gallium nitride, including its synthesis, structures adopted, and semiconducting chemistry (including the doped forms).

14.9 In their paper 'New structural motifs in metallaborane chemistry: Synthesis, characterization, and solid-state structures of $[(Cp^*W)_3(\mu-H)B_8H_8]$, $[(Cp^*W)_2B_7H_9]$, and $[(Cp^*Re)_2B_7H_7]$ $(Cp^* = \eta^5-C_5Me_5)$' (*Organometallics*, 1999, **18**, 853), A. S. Weller, M. Shang, and T. P. Fehlner discuss the synthesis and characterization of some novel boron-rich metallaboranes. Sketch and explain the ^{11}B- and ^{1}H-NMR spectra of $(Cp^*W)_2B_7H_9$.

14.10 Discuss how some boranates can act as both a nucleophile and an electrophile. C. F. Lee et al., *Nat. Chem.*, 2018, DOI: 10.1038/s41557-018-0097-5.

15 The Group 14 elements

Part A: The essentials
15.1 The elements
15.2 Simple compounds
15.3 Extended silicon–oxygen compounds and their derivatives

Part B: The detail
15.4 Occurrence and recovery
15.5 Diamond and graphite
15.6 Other forms of carbon
15.7 Hydrides
15.8 Compounds with halogens
15.9 Compounds of carbon with oxygen and sulfur
15.10 Compounds of silicon with oxygen
15.11 Oxides of germanium, tin, and lead
15.12 Compounds with nitrogen
15.13 Carbides
15.14 Silicides
15.15 Extended silicon–oxygen materials
15.16 Compounds of heavier Group 14 elements with organic substituents

Further reading

Exercises

Tutorial problems

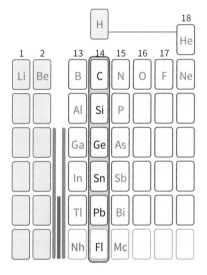

The elements of Group 14—carbon, silicon, germanium, tin, and lead—show great diversity in their chemical and physical properties. Carbon, the building block of complex organic chemistry, has taken on immense geopolitical significance as nations debate how to curb global increases in atmospheric levels of carbon dioxide and methane, each of which are potent greenhouse gases. Silicon is widely distributed across the Earth's surface, silicate-based minerals being a major component of crustal rocks, while germanium, tin, and lead are mined and find important applications in technology.

PART A: The essentials

The elements of Group 14 (the carbon group) are of fundamental importance in industry and nature. We discuss carbon in many contexts throughout this text, including organometallic compounds in Chapter 24, materials in Chapters 25, 28, and 29, and catalysis in Chapter 31. This section focuses on the general properties of Group 14 elements, and how and why these vary as we descend the group from carbon to lead.

15.1 The elements

KEY POINTS The lightest elements of the group are nonmetals; tin and lead are metals. All the elements except lead exist as several allotropes.

The lightest members of the group, carbon and silicon, are nonmetals, germanium is a metalloid, and tin and lead are metals. This increase in metallic properties on descending

15 The Group 14 elements

TABLE 15.1 Selected properties of the Group 14 elements

	C	Si	Ge	Sn	Pb
Melting point/°C	3730 (graphite sublimes)	1410	937	232	327
Atomic radius/pm	76	111	120	139	146
Ionic radius, $r(M^{n+})$/pm			73 (+2)	93 (+2)	119 (+2)
			53 (+4)	69 (+4)	78 (+4)
First ionization energy, I/kJ mol^{-1}	1090	786	762	707	716
Pauling electronegativity	2.5	1.9	2.0	1.9	2.3
Electron affinity, E_a/kJ mol^{-1}	122	134	116	116	35
$E^\ominus(M^{4+},M^{2+})$/V				+0.15	+1.46
$E^\ominus(M^{2+},M)$/V				−0.14	−0.13

a group is a striking feature of the p block, and can be understood in terms of the increasing atomic radius and associated decrease in ionization energy down the group (Table 15.1). The elements form discrete cations increasingly readily down the group. The artificial member flerovium, element 114, was first synthesized in 1998, and the most stable isotope ^{289}Fl has a half-life of less than three seconds so is not discussed further in this text.

As the valence configuration ns^2np^2 suggests, the +4 oxidation state is dominant in the compounds of the elements. The major exception is lead, for which the most common oxidation state is +2. The relative stability of the low oxidation state is an example of the inert pair effect (Section 10.7), a prominent feature of the heaviest p-block elements that arises mainly from the relativistic stabilization of the 6s orbital. Carbon is particularly noted for catenation, the formation of extensive chains and networks. Carbon and silicon are strong **oxophiles** and **fluorophiles**, in the sense that they have high affinities for the hard anions O^{2-} and F^-, respectively (Section 5.10). Their oxophilic character is evident in the existence of an extensive series of oxoanions, the carbonates and silicates. In contrast, Pb^{2+} forms more stable compounds with soft anions, such as I^- and S^{2-}, than with hard anions, and is therefore classified as chemically soft.

Two almost pure crystalline forms of carbon, diamond and graphite, are mined and can also be synthesized. There are other pure forms of carbon as well as many less pure forms, such as coke, which is made by the pyrolysis of coal, and lampblack, the sooty product of incomplete combustion of hydrocarbons. Silicon is widely distributed in the natural environment and constitutes 26% by mass of the Earth's crust. It occurs as sand, quartz, amethyst, agate, and opal, and is also found in asbestos, feldspar, clays, and micas. Germanium is low in abundance and occurs naturally in the ore germanite, $Cu_{13}Fe_2Ge_2S_{16}$, as an impurity in zinc ores, particularly sphalerite, ZnS, and in coal. Tin occurs as the mineral cassiterite, SnO_2, and lead occurs as galena, PbS.

Diamond and graphite differ remarkably: diamond is an electrical insulator, whereas graphite is a good conductor; diamond is the hardest known natural substance, whereas graphite is soft; diamond is transparent, and graphite is black. The origin of these widely different physical properties can be traced to the structures and bonding in the two allotropes.

In diamond, each C atom forms single bonds of length 154 pm with four adjacent C atoms at the corners of a regular tetrahedron (Fig. 15.1); the result is a rigid, covalent, three-dimensional framework. The cubic diamond structure is also adopted by silicon, germanium, and tin.

Graphite consists of stacks of planar layers within which each C atom has three nearest-neighbours at 142 pm (Fig. 15.2) producing a hexagonal network. The σ bonds between neighbours within the sheets are formed from the overlap of sp^2 hybrid orbitals, and the remaining perpendicular p orbitals overlap to form π bonds that are delocalized over the plane. The ready cleavage of graphite parallel to the planes of atoms (which is enhanced due to the presence of impurities) accounts for its slipperiness and its use as a lubricant. In contrast, as is well known, diamond is one of the hardest materials to cut, its brittleness causing it to fracture easily.

Other allotropes of carbon have come to the fore since the 1980s. The first molecular species to be identified, known as fullerenes (and informally as 'buckyballs', (**1**)),

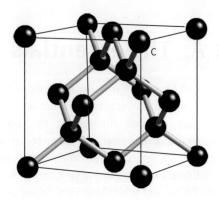

FIGURE 15.1 The cubic diamond structure.

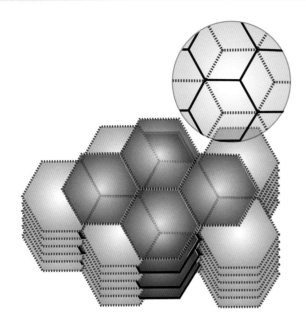

FIGURE 15.2 The structure of graphite. The rings are in register in alternate planes, not adjacent planes.

have been joined by several other technologically important forms of carbon, contributing immensely to the field known as nanotechnology.

1

Single two-dimensional atomic layers of graphite can be isolated—the discovery of graphene (Box 15.1) by Andre Geim and Kostya Novoselov earning them the 2010 Nobel Prize in Physics. Carbon nanotubes, discovered in the early 1990s, are composed of graphene-like tubes with hemispherical buckyball-like caps.

All the elements of the group except lead have at least one solid phase with a diamond structure (Fig. 15.1). At room temperature the stable phase of tin is white tin (tetragonal β-Sn), in which the Sn atoms have six nearest-neighbours in a highly distorted octahedral array. When white tin is cooled to below 13.2°C, it converts to grey tin (α-Sn), which has a cubic, diamond-type structure. The effects of this transformation, known as tin pest or disease, were first recognized on organ pipes in medieval European cathedrals. Legend has it that Napoleon's armies were defeated in Russia because, as the temperature fell, the white tin buttons on the soldiers' uniforms were converted to grey tin, the increase in atomic volume causing the material to break up and crumble.

The gap between the valence and conduction bands (Section 2.11) decreases steadily from diamond (which is classed as a wide-band-gap semiconductor but is commonly regarded as an insulator) to tin, which behaves like a metal above its transition temperature.

Elemental carbon in the form of coal or coke is used as a fuel and reducing agent in the recovery of metals from their ores. Graphite is used as a lubricant and in pencils, and diamond is used in industrial cutting tools. The band gap and consequent semiconductivity of silicon leads to its many applications in integrated circuits, computer chips, solar cells, and other electronic solid-state devices. Silica (SiO_2) is

---

BOX 15.1 What is graphene?

Graphene (discussed further in Section 28.9) is monolayer graphite: a single layer of hexagonally arranged carbon atoms completely separated from other such layers (Fig. B15.1). Graphene is the strongest known material, with a breaking strength of approximately $40\,N\,m^{-1}$, which is 200 times greater than structural steel. It has the highest recorded thermal conductivity and shows more elasticity than any other crystal, stretching by up to 20%. Graphene also exhibits other intriguing properties. For example, it shrinks with increasing temperature, and exhibits simultaneously high pliability with brittleness, so it can be folded but will shatter like glass under high strain. It is also impermeable to gases. But it is graphene's high electronic conductivity that excites most interest: it has no band gap, and so is permanently conducting and cannot be switched off, but its electronic properties are modified in subtle ways by 'twisting' it.

Advances in graphene technology are hampered by the lack of large-scale production of pure graphene sheets. The original

FIGURE B15.1 The structure of graphene.

method used to produce graphene is exfoliation—a 'top-down' approach by which the surface is mechanically prised away from a graphite crystal. Often referred to as the 'Scotch® tape method', exfoliation is achieved quite simply by using sticky tape, but separating the useful thin flakes from the graphite debris is time-consuming and inefficient. A 'bottom-up' approach—chemical vapour deposition using a hydrocarbon precursor (normally methane)—yields good-quality graphene, but, like the Scotch® tape method, it only produces single layers at a time. Other methods are being explored, including epitaxial growth on single-crystal silicon carbide.

the major raw material used to make glass. Germanium was the first widely used material for the construction of transistors because it was easier to purify than silicon and, having a smaller band gap than silicon (0.72 eV for Ge, 1.11 eV for Si), is a better intrinsic semiconductor.

Tin has widespread applications, generally in combination with other elements. Resistant to corrosion and having low toxicity, it is used to plate steel for use in tin cans. Bronze is an alloy of tin and copper that typically contains less than 12% by mass of tin; bronze with higher tin content is used to make bells. Solder, an alloy of tin and lead, has been in use since Roman times. The glass used for making windows contains a thin coating of tin that gives it a very flat surface. Ternary oxides of tin such as indium tin oxide combine electronic conductivity with optical transparency and are widely used in optoelectronics. Trialkyl- and triaryltin compounds are used as fungicides and biocides.

The softness and malleability of lead has resulted in its use in plumbing (from the Latin name *plumbum* for lead) and its high density (11.34 g cm^{-3}) leads to its use in ammunition. Many of these uses are now being phased out because of concerns over lead poisoning. Lead is used to make shields to protect against ionizing radiation. Lead oxide was added to glass to raise its refractive index and form 'lead' or 'crystal' glass. The rechargeable lead–acid battery (see Box 15.7) invented in 1859 and improved continually to improve performance and minimize environmental issues, remains very important despite huge recent growth in alternative battery technologies.

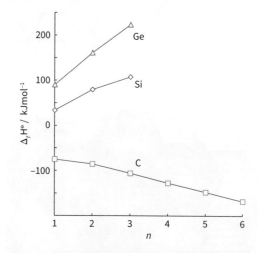

FIGURE 15.3 Variation in the enthalpies of formation of linear alkanes, silanes, and germanes (E_nH_{2n+2}) with increasing chain length.

15.2 Simple compounds

KEY POINTS As Group 14 is descended, the dominant oxidation number changes from 4 to 2 and the stabilities of simple binary compounds with hydrogen and the halogens decrease. Oxides of silicon and carbon vary greatly in their structures and properties and include some of the most abundant compounds on Earth.

All the Group 14 elements form tetravalent hydrides, EH_4, while carbon, silicon, and germanium also form catenated molecular hydrides, general formula E_nH_{2n+2}, containing chains of E–E bonds. All the hydrides react with O_2, the combustion of alkanes being particularly familiar. Their formation is exothermic for carbon but endothermic for the heavier members (Fig. 15.3); thus, whereas alkanes are stable fuels, silanes are used as precursors for the thermal deposition of pure silicon. With increasing chain length, $\Delta_fH^\ominus$ values become more negative for alkanes and more positive for silanes and germanes, reflecting the tendency for catenation to decrease down the group. Thus, while alkane chains may be extremely long, as in polyethylene, the longest silane is heptasilane, Si_7H_{16}. Plumbane (PbH_4) is extremely unstable.

Silanes, with their greater number of electrons and stronger intermolecular forces, are less volatile than their hydrocarbon analogues. Thus, whereas propane, C_3H_8, is a gas under normal conditions, its silicon analogue, trisilane, Si_3H_8, is a liquid that boils at 53°C. Carbon also forms strong multiple bonds in the unsaturated alkenes and alkynes (Table 15.2). The strength of the C–C bond and the ability to form multiple bonds are largely responsible for the diversity and stability of carbon compounds that form the basis of organic chemistry. Multiple bonding is much

TABLE 15.2 Selected mean bond enthalpies, $B(X–Y)/\text{kJ mol}^{-1}$

C–H	412	Si–H	318	Ge–H	288	Sn–H	250	Pb–H	<157
C–O	360	Si–O	466	Ge–O	350				
C=O	743	Si=O	642						
C–C	348	Si–Si	226	Ge–Ge	186	Sn–Sn	150	Pb–Pb	87
C=C	612	Si=Si	270						
C≡C	837								
C–F	486	Si–F	584	Ge–F	466				
C–Cl	322	Si–Cl	390	Ge–Cl	344	Sn–Cl	320	Pb–Cl	301

TABLE 15.3 Properties of CO and CO_2

Oxide	m.p./°C	b.p./°C	Vibrational frequencies/cm^{-1}		Bond length (CO)/pm
CO	−199	−192	2145 (s)		113
CO_2	sublimes	−78	2349* $\overrightarrow{O=C}\overrightarrow{=O}$	1333 $\overleftarrow{O=C}\overrightarrow{=O}$	116
			671* (bend) $O=C=O$ ↑↓ ↑↓		

* Vibrations responsible for the absorption of infrared radiation.

less common for the heavier members, and accounts for why elemental silicon has no graphite analogue (Section 10.2).

The tetrahalomethanes, the simplest halocarbons, vary from the highly stable and volatile CF_4 to the thermally unstable solid CI_4. The full range of tetrahalides is known for silicon and germanium: all of them are volatile molecular compounds. Tetrahalides become increasingly unstable as the group is descended and +2 oxidation state becomes increasingly dominant, as seen, for example, in the stable, well-known compounds $SnCl_2$ and PbF_2.

The two familiar oxides of carbon are CO and CO_2, data for which are summarized in Table 15.3. The bond in CO is short and very strong, in accordance with the Lewis structure :C≡O: (Section 2.9). Carbon dioxide, CO_2, is a linear molecule with longer bonds compared to CO, which is consistent with the bonds being double rather than triple. Absorption of infrared radiation by CO_2 is an important factor contributing to the greenhouse effect: CO_2 has two IR-active modes (Fig. 3.16). Other carbon oxides contain carbon–carbon bonds, the simplest being carbon suboxide, O=C=C=C=O, formed upon dehydration of malonic acid ($HOOC$-CH_2-$COOH$). Graphite oxide, formed when graphite is oxidized with concentrated sulfuric acid and potassium chlorate, contains epoxide and hydroxyl groups located not only on the surface but between the layers. The graphite layers cleave easily giving two-dimensional sheets of graphene oxide that can be reduced to produce further graphene-like materials.

Silicon monoxide, SiO, can be synthesized as a surface-oxidizable polymeric solid, but it is not well characterized. The vapour phase contains diatomic Si≡O molecules, which have been detected in interstellar space—as expected, given estimated cosmic abundance rankings for silicon and oxygen (O, third; Si, eighth). On Earth, silicon is found as silica, SiO_2, which occurs in many forms and derivatives: all of these are based upon the orthosilicate unit $[SiO_4]^{4-}$ (**2**), which is often represented as a simple tetrahedron within which the Si atom is assumed to lie at the centre and O atoms at the vertices.

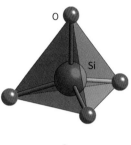

2

The most familiar form of silica is quartz: its high-temperature polymorph, β-cristabolite, has a cubic structure analogous to diamond except that each carbon atom is replaced by a SiO_4 tetrahedron (Fig. 15.4). Silica and many silicates crystallize slowly. Amorphous solids known as **glasses** can be obtained instead of crystals by cooling the melt at an appropriate rate. In some respects these glasses resemble liquids, as their structures are ordered over distances of only a few interatomic spacings (such as within a single SiO_4 tetrahedron). Unlike liquids, however, their viscosities are very high, and for most practical purposes they behave like solids.

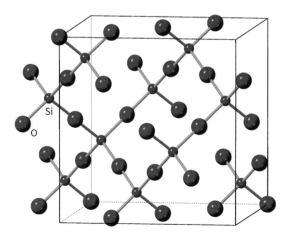

FIGURE 15.4 The structure of β-cristabolite, a high-temperature form of quartz. The silicon atoms occupy the same positions as carbon atoms in diamond.

The Si–O bond is among the strongest 'single bonds' known, leading to a diverse group of compounds known as silicates. Their diverse structures are often easier to comprehend if we consider that each terminal O atom of the tetrahedral $[SiO_4]^{4-}$ unit contributes -1 to the charge of the unit, but each shared O atom contributes 0. Thus, two orthosilicate units joined at a vertex give disilicate $[O_3SiOSiO_3]^{6-}$ (**3**).

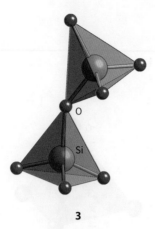

3

Germanium(IV) oxide, GeO_2, is stable and resembles silica, while germanium(II) oxide, GeO, is very unstable and disproportionates readily to Ge and GeO_2. Tin forms two stable oxides: tin(IV) oxide, SnO_2, has the rutile structure, and tin(II) oxide, SnO, exists as blue-black and red polymorphs, both of which are readily oxidized to SnO_2 when heated in air. Lead(IV) oxide, PbO_2, is a brown, oxidizing compound which decomposes upon heating to produce a mixed valence oxide, Pb_3O_4, containing both Pb(IV) and Pb(II) (this is known as 'red lead'), and ultimately lead(II) oxide. The increasing stability of the $n-2$ oxidation state exemplifies the inert pair effect (Sections 10.6 and 10.7) and the relativistic stabilization of the $(6s^2)$ sub-shell in Pb(II).

Carbon forms hydrogen cyanide, HCN, ionic cyanides containing the CN^- ion, and the gas cyanogen $(CN)_2$. These substances are all extremely toxic. The direct reaction of silicon and nitrogen gas at high temperatures produces silicon nitride, Si_3N_4, a very hard and inert material that is used in high-temperature ceramics.

Carbon forms numerous binary carbides with metals and metalloids. Group 1 and 2 metals form ionic saline carbides, d-block metals form metallic carbides, and boron and silicon form covalent solids. Silicon carbide, SiC, is an important abrasive (carborundum).

EXAMPLE 15.1 **Explaining why alkanes form long chains whereas silanes do not**

Use thermodynamic arguments and data provided (e.g. in Tables 2.8 and 15.2) to account for why alkanes can form infinite chains of $-CH_2-$ units, whereas the chain length is very limited for silanes.

Answer We write the general reaction as

$$nX(s) + (n+1)H_2(g) \rightarrow X_nH_{2n+2}(g)$$

Each time the chain is extended by one XH_2 unit, we require the atomization of one mole of X from the solid and dissociation of one H_2 molecule into its atoms. In turn, the product contains one additional X–X bond and two additional X–H bonds.

For alkanes, the overall cost in terms of reactant bonds broken is $717 + 436 = 1153\,kJ/mol$, whereas the gain in terms of new bonds made in the product is $348 + (2 \times 412) = 1172\,kJ/mol$. The incremental enthalpy change is therefore $-19\,kJ$ per $-CH_2-$ unit. The result explains the negative slope for the alkane data shown in Fig. 15.3. Long chain alkanes are stable.

For silanes, the overall cost in terms of reactant bonds broken is $456 + 436 = 892\,kJ/mol$, whereas the gain in terms of new bonds made in the product is $226 + (2 \times 318) = 862\,kJ/mol$. The incremental enthalpy change is therefore $+30\,kJ$ per $-SiH_2-$ unit. The result is consistent with the positive slope for the silane data shown in Fig. 15.3. Silane chain length can thus be sustained only up to a limited degree. The origin of the thermodynamic stability of long-chain alkanes over silanes is a combination of the weakness of the Si–Si bond relative to C–C and the weakness of the two Si–H bonds relative to two C–H bonds, that accompany each unit.

Self-test 15.1 Use data tables to explain why the reaction

$$2X_2H_6(g) \rightarrow X_4H_{10}(g) + H_2(g)$$

is endothermic for X = C but exothermic for X = Si

15.3 Extended silicon–oxygen compounds and their derivatives

KEY POINTS The orthosilicate unit is a building block for a vast range of extended network solids, including ones in which silicon atoms are replaced by aluminium, resulting in spaces that contain mobile cations. These materials have important applications in industry.

The strong single Si–O bond strength ($466\,kJ\,mol^{-1}$) accounts for the existence of a vast number of silicate-based minerals and synthetic silicon–oxygen compounds. The long-range structures of these materials are comprised of limited 3D sheets or infinite 3D networks (Fig. 15.5) of orthosilicate tetrahedra linked via their vertices (**3**) or edges (**4**).

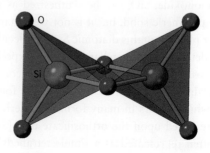

4

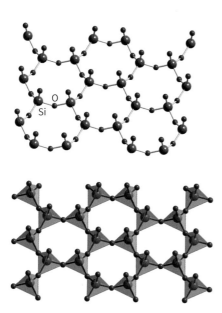

FIGURE 15.5 Part of a network of orthosilicate units. Shown below is the same network displayed as vertex-linked tetrahedra.

In the mineral talc, $Mg_3(OH)_2Si_4O_{10}$, Mg^{2+} and OH^- ions are sandwiched between layers of $[Si_4O_{10}]^{4-}$ anions. The arrangement is electrically neutral, and as a result, talc readily cleaves between the layers and accounts for its familiar slippery feel.

Aluminosilicates are formed when Al atoms replace some of the Si atoms in a silicate, and occur naturally as clays, minerals, and rocks. Zeolite aluminosilicates are widely used as molecular sieves, microporous catalysts, and catalyst support materials (Sections 25.16(a), 29.1, and 29.2). Because Al occurs as Al(III), its presence in place of Si(IV) in an aluminosilicate renders the overall charge more negative by one unit. An additional cation, such as H^+, Na^+, or $\frac{1}{2}Ca^{2+}$ is therefore required for each Al atom that replaces a Si atom (5). These additional cations have a profound effect on the properties of the materials.

5

Numerous minerals feature varieties of layered aluminosilicates that also contain metals such as lithium, magnesium, and iron: they include clays and micas. An example of a simple layered aluminosilicate is kaolinite, $Al_2(OH)_4Si_2O_5$, which is used commercially as china clay and in medical applications. Muscovite mica, $KAl_2(OH)_2Si_3AlO_{10}$ (Fig. 15.6(a)), has charged layers because one Al(III) atom substitutes for one Si(IV) atom, and the resulting negative charge is compensated by a K^+ ion that lies between the repeating layers. Because of this electrostatic cohesion, muscovite is not soft like talc but it is readily cleaved into sheets. Many minerals are based on a three-dimensional aluminosilicate framework. The feldspars, typically $XAl_{1+x}Si_{3-x}O_8$, X = Na, K, Ca, are the most important class of rock-forming minerals.

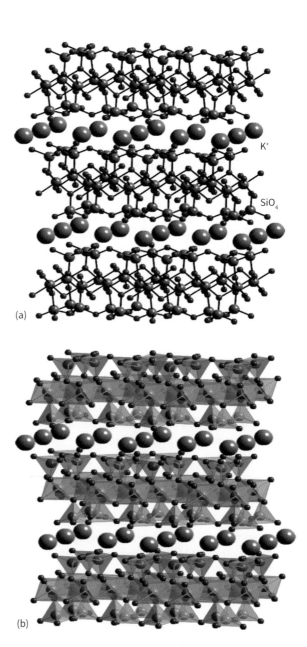

FIGURE 15.6 (a) The structure of 2:1 clay minerals such as muscovite mica $KAl_2(OH)_2Si_3AlO_{10}$, in which K^+ ions reside between the charged layers (exchangeable cation sites), Si^{4+} resides in sites of coordination number 4, and Al^{3+} in sites of coordination number 6. (b) The polyhedral representation. In talc, Mg^{2+} ions occupy the octahedral sites and O atoms on the top and bottom are replaced by OH groups and the K^+ sites are vacant.

PART B: The detail

In this section we discuss the detailed chemistry of the Group 14 elements, exploring the decreasing stability of higher oxidation states and increasing metallic character down the group.

15.4 Occurrence and recovery

KEY POINTS Elemental carbon is mined as graphite and diamond; elemental silicon is recovered from SiO_2 by carbon-arc reduction; tin and lead are obtained from their respective oxide and sulfide ores; germanium is found in zinc ores.

Elemental carbon occurs as diamond and graphite, and in several forms of low crystallinity. As compounds, it occurs as carbon dioxide in the atmosphere and dissolved in natural waters, as the insoluble carbonates of calcium and magnesium, and in forms derived from extended biological action, such as coal, oil, and natural gas.

Elemental silicon is produced from silica, SiO_2, by high-temperature reduction with carbon in an electric arc furnace:

$$SiO_2(s) + 2C(s) \rightarrow Si(s) + 2CO(g)$$

Further refinement of silicon, required for electronics and solar panels, is carried out by the thermal decomposition of silanes or $SiCl_4$. Germanium is much lower in abundance and generally not concentrated in nature. It is obtained by the reduction of GeO_2 with carbon monoxide or hydrogen (Section 6.17). Tin is produced by reducing the mineral cassiterite, SnO_2, with coke in an electric furnace. Lead is obtained from its sulfide ores, which are converted to oxide and reduced by carbon in a blast furnace.

15.5 Diamond and graphite

KEY POINTS Diamond has a cubic structure. Graphite consists of stacked two-dimensional carbon sheets; oxidizing agents or reducing agents may be intercalated between these sheets with concomitant electron transfer.

Diamond has the highest known thermal conductivity of any 3D solid material because its structure (shown in Fig. 15.1) distributes thermal motion in three dimensions very efficiently. The measurement of thermal conductivity is used to identify fake diamonds. Because of its durability, clarity, and high refractive index, diamond is one of the most highly prized gemstones.

The ready cleavage of graphite parallel to the planes of atoms (as evident from Fig. 15.2) is largely due to the presence of impurities and accounts for its slipperiness. These graphene planes are widely separated from each other (at 335 pm), indicating that there are weaker forces between them. These forces are sometimes, but not very appropriately, called 'van der Waals forces'. Unlike diamond, graphite is soft and black with a slightly metallic lustre; it is neither durable nor particularly attractive.

The conversion of diamond to graphite at room temperature and pressure is spontaneous ($\Delta_{trs}G^{\ominus} = -2.90 \text{ kJ mol}^{-1}$) but does not occur at an observable rate under ordinary conditions: diamonds older than the solar system have been isolated from meteorites. Diamond is the denser phase (3.51 g cm^{-3} instead of 2.26 g cm^{-3}), so it is favoured by high pressures, and large quantities of diamond abrasive are manufactured commercially by a d-metal-catalysed high-temperature, high-pressure process (Box 15.2). Synthetic multi-carat (>200 mg) diamonds produced in this way are replacing natural diamonds in jewellery. Thin films of boron-doped diamond are piezoresistive (their electrical resistance changes when pressure is applied) and are deposited on silica surfaces for use as high-temperature pressure sensors.

The electrical conductivity and many of the chemical properties of graphite are closely related to the delocalized π bonding. The electrical conductivity perpendicular to the planes is low (5 S cm^{-1} at 25°C) and increases with temperature, signifying that graphite is a semiconductor in that direction (Section 2.11). The electrical conductivity is much higher parallel to the planes (30 kS cm^{-1} at 25°C) but decreases as the temperature is raised, indicating that graphite behaves as a metal in that direction; in fact it is classed as a semimetal (Figure 2.37). This effect is most striking in pyrolytic graphite, manufactured by decomposing a hydrocarbon gas at high temperature in a vacuum furnace. The resulting graphite is of very high purity with desirable mechanical, thermal, and electrical properties. Pyrolytic graphite is used in ion beam grids, thermal insulators, rocket nozzles, heater elements, and as an electrode material.

Graphite can act as either an electron donor or an electron acceptor towards atoms and ions that penetrate between its sheets and give rise to an **intercalation compound**. Thus, K atoms reduce graphite by donating their valence electron to the empty conduction band formed from the overlapping π* orbitals and the resulting K⁺ ions penetrate between the layers (Fig. 15.7). The electrons introduced to the conduction band are mobile, and therefore alkali metal–graphite intercalates have high electrical conductivity. The stoichiometry of the compound depends on the quantity of alkali metal and the reaction conditions. The different stoichiometries are associated with an interesting series of structures, where the alkali metal ion may insert between neighbouring layers of C atoms, every other layer, and so on in a process known as 'staging'.

Graphite reacts with fluorine to produce 'graphite fluoride', a nonstoichiometric species with formula CF_x ($0 < x < 1.25$).

BOX 15.2 How are diamonds synthesized?

Diamonds were first synthesized in 1955 after many failed attempts, using graphite and a d-metal heated to 1500–2000 K and subjected to 7 GPa pressure. The graphite and metal must both be molten for diamond to be produced. The d metal (typically nickel) dissolves the graphite and the less soluble diamond phase crystallizes out. Low-temperature synthesis produces dark, impure crystals. High-temperature synthesis produces paler, purer crystals. Common impurities are species that can be accommodated into the lattice with minimum distortion. The diamonds are often contaminated with graphite or the metal catalyst. The lattice dimensions of nickel are similar to those of diamond, and crystallites of nickel may be included in the diamond lattice.

Large, high-quality diamonds are formed by seeding with diamond crystals positioned in a cooler part of the apparatus, inducing crystallization in a slow, controlled way. Diamonds can be synthesized directly from graphite without a metal catalyst if the temperature and pressure are high enough. The shock synthesis method (the Du Pont method) exposes graphite to the intense pressure generated by a high-explosive charge. The graphite reaches a temperature of 1000 K and a pressure of 30 GPa for a few milliseconds and some of it is converted to diamond. In the static pressure method, graphite is heated electrically at high pressure. Polycrystalline lumps of diamond are formed at 3300–4500 K and 13 GPa. Aliphatic hydrocarbons may also be used as the carbon source in this method, but aromatic compounds such as naphthalene and anthracene produce graphite.

Microscopic diamond crystals mixed with graphite can be formed by depositing C atoms on a hot surface in the absence of air. The C atoms are produced by the pyrolysis of methane, and the atomic hydrogen also produced in the pyrolysis plays an important role in favouring diamond over graphite. Atomic hydrogen reacts more rapidly with the graphite than with diamond to produce volatile hydrocarbons, so the unwanted graphite is removed. Applications of synthetic diamond films range from the hardening of surfaces subjected to wear, such as cutting tools and drills, to the construction of electronic devices. Boron-doped diamond films are electronically conducting and are used as electrodes in electrochemistry.

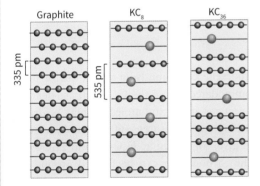
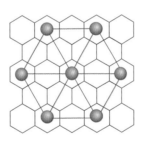

FIGURE 15.7 Potassium graphite compounds showing two types of alternation of intercalated atoms. The top-down view of KC_8 (right) shows the regular array of K⁺ ions.

This compound is black when x is low in its range, and colourless when $x > 1$. It is used as a lubricant in high-vacuum applications and as the cathode in lithium batteries. The other halogens show an alternation effect (Section 10.4) in their tendency to form intercalation compounds with graphite. Chlorine reacts slowly to form C_8Cl, and iodine does not react at all. By contrast, bromine intercalates readily to give C_8Br, $C_{16}Br$, and $C_{20}Br$ in another example of staging. Like KC_8 the halogen intercalation compounds are electronically conducting, but the reason now is that electrons are removed from the valence (π) band.

15.6 Other forms of carbon

Carbon also exists in several less crystalline forms, as well as the fullerenes and related compounds.

(a) Carbon clusters

KEY POINT Fullerenes are formed when an electric arc is discharged between carbon electrodes in an inert atmosphere.

Metal and nonmetal cluster compounds have been known for decades, but the discovery of the soccer-ball-shaped C_{60} cluster in the 1980s by Richard Smalley, Robert Curl, and Harold Kroto (which led to their award of the 1996 Nobel Prize in Chemistry) created great excitement because carbon is a common element and there had seemed little likelihood that new molecular carbon structures would be found.

When an electric arc is struck between carbon electrodes in an inert atmosphere, a large quantity of soot is formed together with significant quantities of C_{60} (**1**) and much smaller quantities of related **fullerenes**, such as C_{70}, C_{76}, and C_{84}, Section 25.18. The fullerenes can be dissolved in a hydrocarbon or halogenated hydrocarbon and separated by chromatography on an alumina column. The structure of C_{60} has been determined by X-ray crystallography on the solid at low temperature and electron diffraction in the gas phase. The molecule consists of five- and six-membered carbon rings, and the overall symmetry is icosahedral in the gas phase.

Fullerenes can be reduced to form [60]fulleride salts, C_{60}^{n-} ($n = 1-12$) which are discussed in more detail in Section 25.18. Fullerides of alkali metals are solids having compositions such as K_3C_{60} and are electronically conductive. The structure of K_3C_{60} consists of a face-centred cubic array of C_{60} ions in which K^+ ions occupy the one octahedral and two tetrahedral sites available to each C ion (Fig. 25.63). At temperatures below 40K, K_3C_{60} becomes superconducting.

(b) Fullerene–metal complexes

KEY POINT The polyhedral fullerenes undergo reversible multi-electron reduction and form complexes with d-metal organometallic compounds.

Methods of synthesizing different fullerenes have been developed and their redox and coordination chemistry have been extensively investigated. The observation that C_{60} undergoes electrochemically reversible electron-transfer steps in nonaqueous solvents (Fig. 15.8) suggests that the fullerenes ought to serve as either electrophiles or nucleophiles. One illustration of this ability is the attack of electron-rich Pt(0) phosphine complexes on C_{60}, yielding **exohedral** fullerene complexes such as (6), in which the Pt atom spans a pair of C atoms in the fullerene molecule. This reaction is analogous to the coordination of alkenes described in Section 24.9.

6

Although analogy with η^6-benzenechromium complexes (Section 24.19) suggests that a metal atom might coordinate to a six-fold face of C_{60}, that such hexahapto complexes do not in fact form is attributed to the radial arrangement of the C2pπ orbitals (7), which results in them having a poor overlap with d orbitals of a metal atom centred above a six-fold face of the molecule.

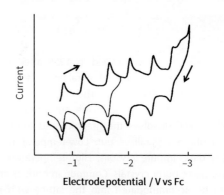

FIGURE 15.8 Cyclic voltammograms of C_{60} in DMF/toluene, recorded at low temperature, emphasizing the reversibility and stability of species reduced by 4–6 electrons. The reference electrode is ferrocene (Fc, Fe^{3+}/Fe^{2+}).

7

Endohedral fullerene compounds, in which one or more atoms are accommodated *inside* the C_{60} shell, are denoted M@C_{60}. Small inert gas atoms and molecules may be driven inside the cage at high temperatures (>600°C) and pressures (>2000 atm) to give, for example, H_2@C_{60} and H_2O@C_{60}. Alternatively, the carbon cage can be formed around the endohedral atom by using a metal-doped carbon rod in an electric arc. Larger shells, such as La@C_{82} and La_3@C_{106}, are often formed.

(c) Carbon nanotubes

Carbon nanotubes, closely related to both fullerenes and graphene, consist of one or more concentric cylindrical tubes conceptually formed by rolling graphene sheets. The ends of the nanotubes are often capped by hemispherical fullerene-like caps containing six five-membered rings of atoms (Fig. 15.9). Nanotubes formed from a single graphene sheet are known as single-walled nanotubes (SWNT). Their properties are determined by the diameter (in the nm range) and length of the tube. Multiwalled nanotubes (MWNT) consist of concentric tubes of graphene. The MWNT can be made up of concentric cylinders of graphene sheets, in the so-called 'Russian-doll' model, or from a single graphene sheet rolled around itself (the 'parchment' model). **Nanobuds** combine both carbon nanotubes and fullerenes: they have a

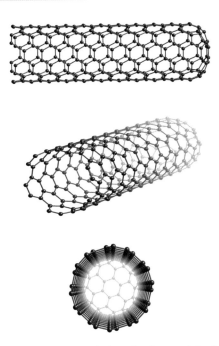

FIGURE 15.9 The structure of a length of capped single-walled carbon nanotube.

fullerene covalently bonded to the outer wall of a nanotube which can act as an anchor, reducing the amount by which the nanotubes slide over each other. Carbon nanomaterials are covered more fully in Chapters 25 and 28.

(d) Partially crystalline carbon

KEY POINTS Amorphous and partially crystalline carbons in the form of small particles are used on a large scale as adsorbents and as strengthening agents for rubber. Carbon fibres impart strength to polymeric materials.

There are many forms of carbon that have a low degree of crystallinity. These partially crystalline materials have considerable commercial importance; they include carbon black, activated carbon, and carbon fibres. Because single crystals suitable for complete X-ray analyses of these materials are not available, their structures are uncertain. However, what information there is suggests that their structures are similar to that of graphite and multiple layer graphenes, but the extent of the crystallinity and shapes of the particles differ.

Carbon black, containing a mixture of nanoparticulate forms, is prepared (on a scale that exceeds 8 Mt annually) by the combustion of hydrocarbons under oxygen-deficient conditions. It is used on a huge scale as a pigment, in printer's ink (as on this page), and as a filler for rubber goods, including tyres, where it greatly improves the strength and wear resistance of the rubber and helps to protect it from degradation by sunlight.

Activated carbon is prepared by controlled pyrolysis of organic material, including coconut shells. It has a high surface area (in some cases exceeding $1000\,\text{m}^2\,\text{g}^{-1}$) that arises from the small particle size. It is therefore a very efficient adsorbent for molecules, including organic pollutants from drinking water, noxious gases from the air, and impurities from reaction mixtures. Parts of the surface defined by the edges of the hexagonal sheets are covered with oxidation products, including carboxyl and hydroxyl groups (8).

8

Carbon fibres are made by the controlled pyrolysis of asphalt fibres or synthetic fibres and are incorporated into a variety of high-strength plastic products, such as tennis rackets and aircraft components. Their structure bears a resemblance to that of graphite, but in place of the extended sheets the layers consist of ribbons parallel to the axis of the fibre. The strong in-plane bonds (which resemble those in graphite) give the fibre its very high tensile strength.

EXAMPLE 15.2 **Comparing the bonding in diamond and boron**

Each B atom in elemental boron is bonded to five other B atoms but each C atom in diamond is bonded to four nearest-neighbours. Suggest an explanation for this difference.

Answer We need to consider the valence electrons on each atom and the orbitals available for bond formation. The B and C atoms both have four orbitals available for bonding (one s and three p). However, a C atom has four valence electrons, one for each orbital, and it can therefore use all its electrons and orbitals to form 2c, 2e bonds with the four neighbouring C atoms. By contrast, B has only three electrons, so to use all four orbitals it forms 3c, 2e bonds. The formation of these three-centre bonds brings another B atom into bonding distance.

Self-test 15.2 Describe how the electronic structure of graphite is altered when it reacts with (a) potassium, (b) bromine.

15.7 Hydrides

The Group 14 elements form tetravalent hydrides, EH_4, with hydrogen, and carbon and silicon form catenated molecular hydrides.

(a) Hydrocarbons

KEY POINT Saturated hydrocarbons ranging from methane to long chain alkanes are stable and rather unreactive. Methane, formed naturally by biological action, is a greenhouse gas.

As seen from Fig. 15.3, the alkanes are exothermic compounds relative to carbon and hydrogen. Those up to butane, C_4H_{10} (b.p. $-1°C$) are gases; those containing from 5 to 17 carbon atoms are liquids, and the heavier hydrocarbons are solids. All are flammable.

Methane, CH_4, is the simplest hydrocarbon. It is formed naturally through the action of anaerobic bacteria known as methanogens, and found in large natural underground deposits from which it is extracted as natural gas and used as domestic and industrial fuel:

$$CH_4(g) + 2O_2(g) \rightarrow CO_2(g) + 2H_2O(g)$$
$$\Delta_{comb}H^\ominus = -882\,kJ\,mol^{-1}$$

Apart from this combustion reaction, methane is not very reactive. It is not hydrolysed by water (Box 15.3) and reacts with halogens only when exposed to ultraviolet radiation:

$$CH_4(g) + Cl_2(g) \xrightarrow{h\nu} CH_3Cl(g) + HCl(g)$$

Methane is a greenhouse gas, as some of its IR-active t_2 vibrations are strongly absorbing. Fortunately, its low molecular mass ultimately allows it to escape Earth's atmosphere, but it still accounts for >10% of greenhouse gas emissions. Its leakage into the atmosphere is being closely monitored, whether from natural sources such as methane clathrates (Box 15.3) and the digestive action of livestock, or through leaking infrastructure of oil and gas industries.

(b) Silanes

KEY POINTS Silanes are much more reactive than alkanes—a property that is largely related to the increased size and lower electronegativity of silicon compared to carbon.

Silane, SiH_4, is prepared commercially by the reduction of SiO_2 with Al under a high pressure of H_2 in a molten salt mixture of NaCl and $AlCl_3$. An idealized equation for this reaction is

$$6H_2(g) + 3SiO_2(g) + 4Al(s) \rightarrow 3SiH_4(g) + 2Al_2O_3(s)$$

Linear chain silanes exist for the range Si_nH_{2n+2}, $n = 2-7$; branched as well as cyclic silanes, such as the liquid cyclohexasilane, Si_6H_{12}, are stable compounds. Silane itself, SiH_4, is a reducing agent: it is spontaneously flammable in air and reacts violently with halogens. Silanes are much more reactive than the corresponding alkanes and their inherent stability decreases with increasing chain length (Fig. 15.3). Their instability is exploited in the thermal deposition of silicon as an atomic-level layer for electronic devices and solar panels

---

BOX 15.3 Where can methane be found on the ocean floor?

Methane clathrates (methane hydrates) are crystalline solids formed at low temperatures when ice crystallizes around CH_4 molecules (Fig. B15.2). Often containing other small, gaseous molecules such as ethane and propane, their formation can cause major problems by clogging gas pipelines in cold climates. Several different clathrate structures are known. The unit cell of the most common one, known as Structure I, contains 46 H_2O molecules and up to eight CH_4 molecules. Since 1 m^3 of clathrate liberates up to 164 m^3 of methane gas, they are good sources of energy but a contributing factor to greenhouse gas emissions.

Methane clathrates are thought to form by migration of 'fossil' methane from beneath the ocean floor along geological faults, followed by crystallization on contact with cold seawater. The methane is also generated continually by bacterial degradation of organic matter (catalysed by Ni-containing enzymes) in low-oxygen environments at the ocean floor (Section 26.11). Below the zone of solid clathrates, large volumes of methane may occur as bubbles of free gas in the sediment. Methane hydrates are stable at low temperatures and high pressures. Because of these conditions and the need for relatively large amounts of organic matter for bacterial methanogenesis, clathrates are mainly restricted to high latitudes and along continental margins in the oceans. On the continental margins the supply of organic material is high enough to generate enough methane and water temperatures are close to freezing. In polar regions, the gas hydrates are commonly linked to the occurrence of permafrost. The permafrost reservoir of methane has been estimated at about 400 Gt of carbon in the Arctic, and perhaps higher still in Antarctic reservoirs. The oceanic reservoir has been estimated to be about 5 Tt of carbon.

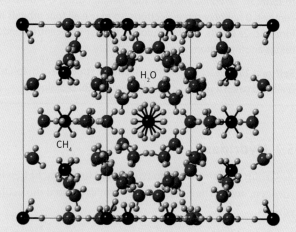

FIGURE B15.2 The structure of methane hydrate, $(CH_4)_4(H_2O)_{23}$; note that both the water and methane molecules are depicted showing their orientational disorder.

(Section 30.2). This instability compared to the alkanes is often attributed to the lower bond enthalpies of Si–Si and Si–H compared to C–C and C–H (Table 15.2), but the higher reactivity is more likely to be kinetic in origin, stemming from the larger size and lower electronegativity of silicon atoms. The larger atomic radius of Si and the greater polarity of the Si–H bond leaves it open to attack by nucleophiles: these, as well as the possible availability of low-lying d orbitals, may facilitate the formation of intermediate adducts.

Bonds between silicon and hydrogen are not readily hydrolysed in neutral water, but the reaction is rapid in strong acid or in the presence of traces of base. Similarly, alcoholysis is accelerated by catalytic amounts of alkoxide:

$$SiH_4 + 4ROH \xrightarrow{\Delta, OR^-} Si(OR)_4 + 4H_2$$

Kinetic studies indicate that the reaction proceeds through a structure in which OR^- attacks the Si atom while H_2 is being formed via an H···H hydrogen bond between hydridic and protic H atoms.

The silicon analogue of hydroboration (Section 14.6(c)) is **hydrosilylation**, the addition of SiH across the multiple bonds of alkenes and alkynes. This reaction, which is used in both industrial and laboratory syntheses, can be carried out under conditions (300°C or ultraviolet irradiation) that produce a radical intermediate. In practice, it is usually performed under far milder conditions by using a platinum complex as catalyst:

$$CH_2=CH_2 + SiH_4 \xrightarrow{\Delta, H_2PtCl_6, \text{isopropanol}} CH_3CH_2SiH_3$$

The current view is that this reaction proceeds through an intermediate in which both the alkene and silane are attached to the Pt atom.

While double bonds between carbon atoms are a common and important feature of the organic chemistry of carbon, the increasing weakness of p–p overlap and of π bonds on descending a group (Section 10.2) means that E=E double bonds become increasingly rarer for the heavier elements. The first compound containing an Si=Si double bond was not isolated until 1978, when tetramesityldisilene (mesityl = 1,3,5-trimethylbenzene) was synthesized. Stabilized with bulky organic substituents, a number of compounds of the heavier elements can be synthesized that are analogues of alkenes and alkynes. Their structures and stability raise interesting questions about bonding, and they are discussed in Section 15.16.

(c) Germane, stannane, and plumbane

KEY POINT Thermal stability decreases rapidly from germane to plumbane.

Germane (GeH_4) and stannane (SnH_4) can be synthesized by the reaction of the appropriate tetrachloride with $LiAlH_4$ in tetrahydrofuran solution. Plumbane (PbH_4) has been synthesized in trace amounts by the protolysis of a magnesium/lead alloy, but it is extremely unstable. The presence of alkyl or aryl groups stabilizes the hydrides of all three elements. For example, trimethylplumbane, $(CH_3)_3PbH$, begins to decompose at 230°C, but it can survive for several hours at room temperature.

15.8 Compounds with halogens

Silicon, germanium, and tin react with all the halogens to form tetrahalides. Carbon reacts only with fluorine, and lead forms stable dihalides.

(a) Halides of carbon

KEY POINTS Nucleophiles displace halogens in carbon–halogen bonds; organometallic nucleophiles produce new M–C bonds; mixtures of polyhalocarbons and alkali metals are explosion hazards.

Carbon tetrafluoride is a colourless gas, CCl_4 is a dense liquid, CBr_4 is a pale yellow solid, and CI_4 is a red solid. Carbon tetrachloride was once used widely as a laboratory solvent and as a dry-cleaning fluid, a refrigerant, and in fire extinguishers, but its use has declined steeply owing to its toxicity and the availability of more benign alternatives. The stabilities of the tetrahalomethanes decrease from CF_4 to CI_4 (Table 15.4).

Carbon tetrafluoride is produced by burning any carbon-containing compound, including elemental carbon, in fluorine. The other tetrahalomethanes are prepared from methane and the halogen:

$$CH_4(g) + 4Cl_2(g) \rightarrow CCl_4(l) + 4HCl(g)$$

Tetrahalomethanes are thermodynamically unstable with respect to hydrolysis:

$$CX_4(l \text{ or } g) + 2H_2O(l) \rightarrow CO_2(g) + 4HX(aq)$$

However, the reaction for C–F bonds is very slow, and fluorocarbon polymers such as poly(tetrafluoroethene) are highly resistant to attack by water.

Partially halogenated alkanes are commonly prepared from alcohols:

$$CH_3OH(l) + HI(solv) \rightarrow CH_3I(l) + H_2O(l)$$

TABLE 15.4 Properties of tetrahalomethanes

	CF_4	CCl_4	CBr_4	CI_4
Melting point/°C	−187	−23	90	171 *dec**
Boiling point/°C	−128	77	190	*sub**
$\Delta_f G^\ominus$/kJ mol^{-1}	−879	−65	148	>0

* *dec*, decomposes; *sub*, sublimes.

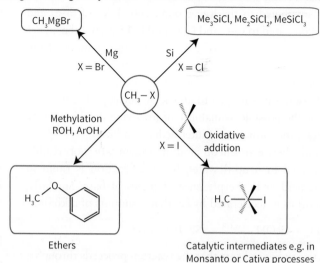

FIGURE 15.10 Some important reactions of carbon–halogen bonds (X = halogen).

Halogenated alkanes provide a route to a wide variety of derivatives, mainly by nucleophilic displacement of one or more halogen atoms. Some useful reactions are outlined in Fig. 15.10. The rates increase greatly from fluorine to iodine, and lie in the order F << Cl < Br < I.

The **carbonyl halides** (Table 15.5) are planar molecules and useful chemical intermediates. The simplest of these compounds, $COCl_2$, phosgene (9), is an extremely toxic gas. It is prepared on a large scale by the reaction of chlorine with carbon monoxide:

$$CO(g) + Cl_2(g) \xrightarrow{200°C, charcoal} COCl_2(g)$$

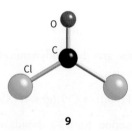

9

The main utility of phosgene lies in the ease of nucleophilic displacement of Cl to produce carbonyl compounds, isocyanates, and polycarbonates, which are important thermoplastic polymers containing carbonate ester linkages.

(b) Compounds of silicon and germanium with halogens

KEY POINT Substitution reactions of silicon halides occur more readily than those of carbon halides because silicon can form intermediates having a higher coordination number than four, whereas carbon cannot.

Among the silicon tetrahalides, the most important is the tetrachloride, which is prepared by direct reaction of the elements or by chlorination of silica in the presence of carbon:

$$Si(s) + 2Cl_2(g) \rightarrow SiCl_4(l)$$
$$SiO_2(s) + 2Cl_2(g) + 2C \xrightarrow{\Delta} SiCl_4(l) + 2CO(g)$$

Silicon and germanium halides are mild Lewis acids and add one or two ligands to yield five- or six-coordinate complexes:

$$SiF_4(g) + 2F^-(aq) \rightarrow [SiF_6]^{2-}(aq)$$
$$GeCl_4(l) + N\equiv CCH_3(l) \rightarrow Cl_4GeN\equiv CCH_3(s)$$

Hydrolysis of the Si and Ge tetrahalides is fast, and can be represented schematically as

$$EX_4 + 2H_2O \rightarrow EX_4(OH_2)_2 \rightarrow EO_2 + 4HX$$
(E = Si or Ge, X = halogen)

Hydrolysis of $SiCl_4$ and alkylchlorosilanes are important processes for making silicas and 'silicone' polymers (Section 15.16). The corresponding carbon tetrahalides and chloroalkanes are kinetically more resistant to hydrolysis because of the lack of access to the sterically shrouded C atom to form an intermediate aqua complex. Fully halogenated silanes and cyclosilanes, such as dodecachlorocyclohexasilane, Si_6Cl_{12} (**10**), are known.

TABLE 15.5 Properties of carbonyl halides

	OCF_2	$OCCl_2$	$OCBr_2$
Melting point/°C	−114	−128	−12
Boiling point/°C	−83	8	65
$\Delta_f G^\ominus$/kJ mol^{-1}	−619	−205	−111

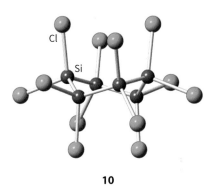

10

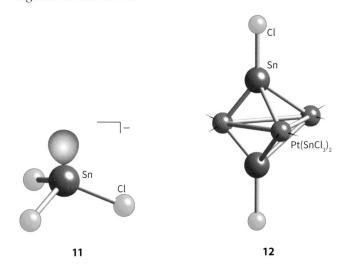

11   **12**

$Pt_3Sn_8Cl_{20}$, which is trigonal-bipyramidal (**12**). The $SnCl_3^-$ ligands are not drawn.

The substitution reactions are more facile than for their carbon analogues because a Si atom can expand its coordination sphere to accommodate the incoming nucleophile. The stereochemistry suggests that a five-coordinate intermediate is formed, with the most electronegative substituents adopting the axial position. Moreover, substituents leave from the axial position. The H^- ion is a poor leaving group, and alkyl groups are even poorer:

Complex halides containing a Group IV metal and one or more electropositive metal cation—for instance, an alkali metal or alkaline earth metal—are a large family of compounds. Examples include those adopting the perovskite structure type, such as $CsSnCl_3$ and $CsPbI_3$ (at high temperature), and K_2SnBr_6 which contains $[SnBr_6]^{2-}$ octahedra. Hybrid lead iodide perovskites (phases that contain a mixture of inorganic and organic species) are of great interest as solar cell absorber materials (Box 15.4 and Section 30.2).

(c) Tin and lead halides

KEY POINTS Tin forms stable dihalides and tetrahalides, whereas for lead, only the dihalides are stable.

Aqueous and nonaqueous solutions of Sn(II) salts are useful mild reducing agents (see Resource section 3), but they must be stored under an inert atmosphere because air oxidation is spontaneous and rapid:

$$Sn^{2+}(aq) + H_2O + \tfrac{1}{2}O_2(g) \rightarrow SnO_2(s) + 2H^+(aq)$$

Tin dihalides and tetrahalides are both well known. The tetrachloride, tetrabromide, and tetraiodide are molecular compounds, but the tetrafluoride has a solid-state structure formed from SnF_6 octahedra sharing four vertices to form infinite sheets. Mixed valence halides, such as Sn_2F_6 (Sn(II) Sn(IV)F$_6$), are also known. Lead tetrafluoride can be considered as an ionic solid, but $PbCl_4$ is an unstable, covalent, yellow oil that decomposes into $PbCl_2$ and Cl_2 at room temperature. Lead tetrabromide and tetraiodide are unknown.

Both Sn(IV) and Sn(II) form a variety of complexes. Thus, $SnCl_4$ forms complex ions such as $[SnCl_5]^-$ and $[SnCl_6]^{2-}$ in acidic solution. In nonaqueous solution, a variety of donors interact with the moderately strong Lewis acid $SnCl_4$ to form complexes like cis-$[SnCl_4(OPMe_3)_2]$. In aqueous and nonaqueous solutions Sn(II) forms trihalo complexes, such as $[SnCl_3]^-$ where the pyramidal structure indicates the presence of a stereochemically active lone pair (**11**). The $[SnCl_3]^-$ ion can act as a soft donor to d-metal ions. One unusual example of this ability is the red cluster compound

15.9 Compounds of carbon with oxygen and sulfur

KEY POINTS Carbon monoxide is a common ligand in d-metal chemistry; carbon dioxide, available in the atmosphere at low levels, is the ultimate source of carbon for almost all organic molecules; the sulfur compounds CS and CS_2 have similar structures to their oxygen analogues.

Carbon forms two major molecular oxides: CO and CO_2 (Table 15.3). The uses of CO include the reduction of metal oxides in a blast furnace (Section 6.17), and the water gas shift reaction (Section 11.4) for the production of H_2:

$$CO(g) + H_2O(g) \rightleftharpoons CO_2(g) + H_2(g)$$

Catalytic reactions which convert carbon monoxide to acetic acid and aldehydes are discussed in Chapter 31. The CO molecule has very low Brønsted basicity and negligible Lewis acidity towards neutral electron-pair donors. Despite its weak Lewis acidity, however, CO is attacked by strong Lewis bases at high pressure and somewhat elevated temperatures. Thus, the reaction with OH^- ions yields the formate ion, HCO_2^-:

$$CO(g) + OH^-(g) \rightarrow HCO_2^-(s)$$

Similarly, the reaction with methoxide ions (CH_3O^-) yields the acetate ion, $CH_3CO_2^-$.

> **BOX 15.4** Can lead iodide perovskites replace silicon in solar cells?
>
> The majority of commercial solar cells (photovoltaic cells) use semiconducting silicon as the absorber layer for sunlight (Section 30.2). However, these cells remain relatively expensive to manufacture due to the costs of producing high-purity crystalline silicon—a process that requires very high temperatures and clean environments. Alternative semiconducting materials, particularly those that could be made into inks and printed onto any surface, are the goal of much current research on photovoltaics. An exciting prototype of these materials is methylammonium lead iodide, $[CH_3NH_3]^+[PbI_3]^-$ (MAPI). Solar cells based on MAPI show efficiencies of over 20%, rivalling those of silicon-based devices.
>
> MAPI has the perovskite structure (Section 4.9) in which Pb^{2+} occupies the B-type cation site, X is I^-, and the A-type cation is the large methylammonium ion (Fig. B15.3). Because of the large sizes of Pb^{2+} and I^-, the cation occupying the A-site has to be very large, and while Cs^+ ($r_+ = 174$ pm) can fulfil this role in the high temperature form of $CsPbI_3$, no simple inorganic cation is large enough to do so at room temperature. The thermochemical radius (Section 4.14) of the methylammonium cation has been estimated as 217 pm.
>
> MAPI can be precipitated from an aqueous solution of HI containing $(CH_3NH_3)I$ and PbI_2, thus making it suitable for room-temperature solar cells printed as thin films. It is a shiny, grey
>
>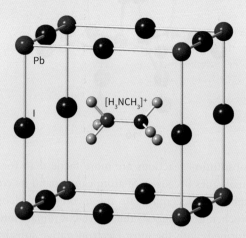
>
> **FIGURE B15.3** The perovskite structure of $[CH_3NH_3]^+[PbI_3]^-$.
>
> semiconducting solid with a band gap of 1.6 eV. However, MAPI is susceptible to decomposition in moist atmospheres, and devices have yet to show the long-term stability required of a commercial photovoltaic device. The design and synthesis of new, stable perovskite-type semiconducting materials, inspired by MAPI, is one of great current activity (Section 30.2).

Carbon monoxide is an excellent ligand towards d-metal atoms in low oxidation states (Section 24.5). Its well-known toxicity is an example of this behaviour: it binds very tightly to the Fe atom in haemoglobin (Section 26.7) thereby disabling the uptake of O_2.

Carbon dioxide is a weak Lewis acid, the reaction with water (which has great environmental importance as it affects the pH of oceans) normally being written as:

$$CO_2(g) + H_2O(l) \rightleftharpoons HOCO_2^-(aq) + H^+(aq)$$

The uncatalysed reaction is very slow, but rapid equilibrium is essential for life: a Zn-containing enzyme, carbon dioxide hydratase (carbonic anhydrase, Section 26.9), accelerates the reaction by a factor of about 10^9.

Carbon dioxide is one of several small molecules that are implicated in the **greenhouse effect** through molecular vibrations that are activated by infrared radiation (Sections 3.5 and 8.5). The effect is to block the release of heat into space, like a 'blanket'. There has been a significant increase in atmospheric CO_2 since the industrialization of society (from 280 ppm to over 420 ppm currently). In the past, nature has managed to stabilize the concentration of atmospheric CO_2, in part by precipitation of calcium carbonate in the deep oceans, but the process is too slow to compensate for the increased influx of CO_2 into the atmosphere (Box 15.5).

One strategy to slow the rate of increase in atmospheric CO_2 is **carbon capture and storage**, CCS (Box 15.6). In CCS, CO_2 is captured from industrial flues by reaction with amines, then after release and liquefaction by compression, it is pumped underground ('buried'), often back into gas or oil wells in order to drive out further oil or gas. Alternatively, and of increasing importance, the CO_2 is used to synthesize useful products, a process known as **carbon capture and re-utilization (CCU)**.

Reactions of the CO_2 molecule generally proceed via nucleophilic attack on the carbon atom, often accompanied by electrophilic attack on one or both oxygen atoms. From an economic perspective, an important reaction is with ammonia to yield ammonium carbonate, $(NH_4)_2CO_3$, which upon heating is converted directly to urea, $CO(NH_2)_2$, a fertilizer and chemical intermediate. In the soft-drinks industry, CO_2 is dissolved under pressure to produce carbonic acid, H_2CO_3 (which has a pleasant acidic taste), and forms bubbles when the pressure is released. In organic chemistry, a common synthetic reaction is that between CO_2 and carbanion reagents to produce carboxylic acids. With suitable catalysts, CO_2 is now replacing toxic phosgene in the industrial synthesis of polycarbonates such as polypropenecarbonate, PPC.

BOX 15.5 The fast carbon cycle

The dangers of climate change have raised scientists' attention on the fast carbon cycle. There is also a slow carbon cycle which involves the formation and destruction of carbonate-containing rocks; while carbon moves much more slowly in this cycle it accounts for by far the greatest amount of carbon on Earth.

Oxygen was not present when the Earth first cooled and liquid water first became available: CO_2 was the principal atmospheric gas. The first photosynthetic organisms used very simple, non-O_2-evolving forms of photosynthesis to assimilate CO_2 into organic molecules. Some of these non-oxygenic processes persist today in anaerobic microorganisms that use molecules such as sulfate and organic acids instead of H_2O as electron acceptors. Compared to water, these compounds are in limited supply, and non-O_2-evolving photosynthesis is capable of reducing only a small amount of CO_2.

The emergence of oxygenic photosynthesis approximately 2.4 billion years ago changed the Earth in what has often been termed 'the second big bang'. Once oxygenic photosynthesis evolved, planetary biomass could be produced and sustained at levels two to three orders of magnitude larger than previously, and CO_2 levels decreased. In the crucial biological process known as the Calvin–Benson–Bassham (CBB) cycle (commonly abbreviated to Calvin cycle) CO_2 is 'fixed' (to the extent of 100 Gt per year) into organic molecules.

The simplified biological cycle shown in green in Fig. B15.4 includes processes that depend on metal catalysts in the form of enzymes, which are described in Chapter 26. Biomass is taken to mean the total product of photosynthetic CO_2 reduction: it includes also the uptake of atmospheric N_2 through biological nitrogen fixation by microorganisms possessing an Fe- and Mo-containing enzyme known as nitrogenase. With each cycle, some of the biomass becomes unavailable for rapid biological oxidation (respiration, using Fe and Cu enzymes) because organisms die, and much of the organic material may become buried in sediments. Over geological time scales this buried matter accumulates and is converted into the coal, shale, oil, and natural gas that constitute our fossil fuel reserves.

The same process that created our fossil fuel reserves also formed the O_2 of the atmosphere (using a Mn-containing enzyme) and eventually decreased the high level of CO_2 that had previously prevailed The rise of O_2 was initially slow on account of the vast reservoir of iron(II) present in the oceans, as well as sulfide minerals. Only once the iron(II) and reduced sulfur were consumed did O_2 begin to accumulate in the atmosphere. The insoluble iron(III) products precipitated, giving banded iron(III) formations that are well known to geologists.

Currently, we are extracting and burning fossil fuels on a geologically very short time scale, building up CO_2 (the red pathway) much faster than the biological cycle (the green pathway) can cope with.

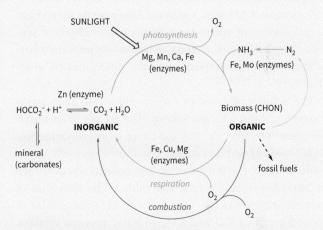

FIGURE B15.4 The fast carbon cycle, showing the principal elements involved.

BOX 15.6 Carbon capture

Finding ways to minimize increases in the levels of atmospheric CO_2 is one of the biggest challenges of the twenty-first century. Our reliance on carbon-based fossil fuels could be lowered by using energy more efficiently and reducing our consumption. Alternatively, our use of low-carbon fuels such as nuclear and renewable energies could be increased.

A feasible way of managing atmospheric CO_2 levels is by capturing it where it is produced at high levels, then either storing it or using it—strategies known as carbon capture and storage, **CCS**, or carbon capture and re-utilization, **CCU**. A typical new 1 GW coal-fired power station produces around 6 Mt of CO_2 annually. Adding CO_2 capture scrubbers to remove CO_2 from flue gas can significantly reduce these emissions. One process uses aqueous solutions of amines, which react with the CO_2 to form solid ammonium carbamate, NH_2COONH_4. A problem with this process is that the aqueous phase evaporates in the gas stream: a new approach uses ionic liquids having an attached amine group. These low-temperature molten salts react reversibly with CO_2, do not need water to function, and can be recycled. An alternative strategy is based on the use of CaO (Box 13.5) and metal-organic frameworks (Section 29.4) are also being considered.

Carbon dioxide sequestration lowers the power output of power plants by 25–40%, raising the cost of energy production. It is thus important to find ways of using CO_2 as an industrial resource rather than 'burying' it. Ultimately, CCU may bridge the gap between environmental and economic demands.

Metal complexes of CO_2 are known (**13**), but they are rare and far less important than the metal carbonyls. The neutral CO_2 molecule acts as a Lewis acid, and the bonding is dominated by electron donation from a low-oxidation state metal atom into a carbon-centred antibonding π orbital of CO_2. The side-on bonding to the metal atom resembles that between an alkene and an electron-rich metal centre (Section 24.9).

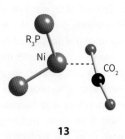

13

An important use of supercritical fluid CO_2 (i.e. highly compressed carbon dioxide but above its critical temperature, Section 5.15) is as a solvent. Applications range from decaffeination of coffee beans to its use in chemical synthesis in place of conventional solvents—an important part of the strategy for implementing the procedures of 'green chemistry'.

The sulfur analogues of carbon monoxide and carbon dioxide, CS and CS_2, are known. The former is an unstable transient molecule and the latter is very endothermic ($\Delta_f H^\ominus = +89\,\mathrm{kJ\,mol^{-1}}$). Carbon disulfide (b.p. 46°C) is a useful solvent for nonpolar or low polarity species, such as phosphorus, sulfur, fats, and rubber. The structures of metal complexes of CS (**14**) and CS_2 (**15**) are similar to those formed by CO and CO_2. In basic aqueous solution, CS_2 undergoes hydrolysis and yields a mixture of CO_3^{2-} and trithiocarbonate (CS_3^{2-}) ions,

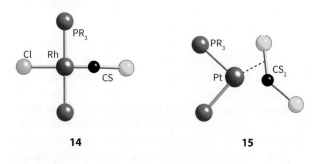

14   **15**

EXAMPLE 15.3 Proposing a synthesis that uses the reactions of carbon monoxide

Propose a synthesis of $CH_3^{13}CO_2^-$ that uses ^{13}CO, a primary starting material for many carbon-13 labelled compounds.

Answer We should bear in mind that CO_2 is readily attacked by strong nucleophiles such as $LiCH_3$ to produce acetate ions. Therefore, an appropriate procedure would be to oxidize ^{13}CO to $^{13}CO_2$ and then to react the latter with $LiCH_3$. A strong oxidizing agent, such as solid MnO_2, can be used in the first step to avoid the problem of excess O_2 in the direct oxidation.

$$^{13}CO(g) + 2MnO_2(s) \xrightarrow{\Delta} {}^{13}CO_2(g) + Mn_2O_3(s)$$
$$4\,{}^{13}CO_2(g) + Li_4(CH_3)_4(et) \rightarrow 4Li[CH_3{}^{13}CO_2](et)$$

where 'et' denotes solution in ether. (Another method involves the reaction of $[Rh(I)_2(CO)_2]^-$ with ^{13}CO. The basis of this reaction is discussed in Sections 24.24 and 31.7.)

Self-test 15.3 Propose a synthesis of $D^{13}CO_2^-$ starting from ^{13}CO.

15.10 Compounds of silicon with oxygen

KEY POINT The Si–O–Si link is present in silica, a wide range of metal silicate minerals, and silicone polymers.

Following from Section 15.3, it should be clear that an endless single-stranded chain or a ring of SiO_4 units, which has two shared O atoms for each Si atom, will have the formula and charge $[(SiO_3)^{2-}]_n$. A compound containing such a cyclic metasilicate ion is the mineral beryl, $Be_3Al_2Si_6O_{18}$, which contains the $[Si_6O_{18}]^{12-}$ ion (**16**). A chain metasilicate (**17**) is present in the mineral jadeite, $NaAl(SiO_3)_2$, one of two minerals sold as jade, the green colour arising from trace iron impurities. In addition to other configurations for the single chain, there are double-chain silicates, which include the family of minerals known commercially as asbestos.

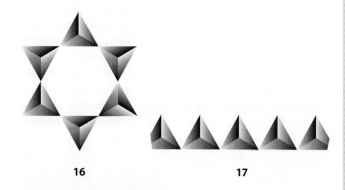

16   **17**

A BRIEF ILLUSTRATION

The cyclic silicate anion $[Si_3O_9]^{n-}$ is a six-membered ring with alternating Si and O atoms and six terminal O atoms, two on each Si atom. Because each terminal O atom contributes −1 to the charge, the overall charge is −6. From another perspective, the oxidation numbers of silicon and oxygen, +4 and −2, respectively, also indicate a charge of −6 for the anion.

White asbestos, or chrysotile, $Mg_3Si_2O_5(OH)_4$, has a sheet silicate structure (Fig. 15.11). When viewed under a microscope the long flexible fibres appear as a bunched mass. Brown asbestos, amosite, and blue asbestos, crocidolite, have a double-chain silicate structure and appear as

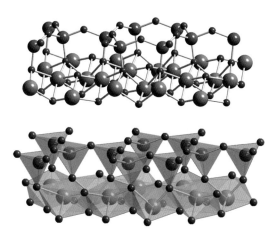

FIGURE 15.11 A double layer of the sheet silicate structure of chrysotile (white asbestos).

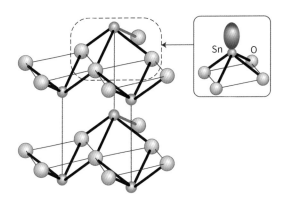

FIGURE 15.12 The structure of blue-black SnO, showing parallel layers of square-based pyramidal SnO_4 units, and the stereochemically active lone pair.

individual needles. All these forms have important properties that include thermal stability, heat resistance, nonbiodegradability, resistance to a wide range of chemicals, and low electrical conductivity. However, prolonged inhalation of asbestos particles, particularly the long, thin needles of blue and brown forms, presents a severe health hazard. Only white asbestos continues to be used, as it presents a lower health risk.

In silicate glasses, $[SiO_4]$ tetrahedra link together without any long-range order. The composition of silicate glasses has a strong influence on their physical properties. For example, fused quartz (amorphous SiO_2) softens at about 1300°C, borosilicate glass (which contains a proportion of boron oxide; Section 14.8) softens at about 800°C, and soda-lime glass softens at even lower temperatures. The variation in softening point can be understood by appreciating that the Si–O–Si links in silicate glasses form the framework that imparts rigidity. When basic oxides such as Na_2O and CaO are incorporated (as in soda-lime glass), they react with the SiO_2 melt and convert Si–O–Si links into terminal SiO$^-$ groups, and hence, by lowering the connectivity, lower its softening temperature. Very different properties are found for the –Si–O–Si– backbone of silicone polymers, which are described later in this chapter.

15.11 Oxides of germanium, tin, and lead

KEY POINT The 2+ oxidation state becomes increasingly important as Group 14 is descended, and is the central feature of the chemistry of lead.

Germanium(II) oxide, GeO, is a reducing agent and disproportionates to Ge and GeO_2. Germanium dioxide, GeO_2, exists as two crystalline polymorphs. One form has the same structure type as quartz, SiO_2, based on tetrahedral four-coordinate GeO_4 units, and the other has a rutile-like structure with six-coordinate germanium. A vitreous form of GeO_2 that resembles fused silica can also be prepared by rapidly cooling molten GeO_2 (m.p. 1115°C). Germanium analogues of silicates and aluminosilicates (Section 15.15) are also known.

In SnO the four O^{2-} ions around the Sn(II) lie in a square to one side, producing a flattened pyramidal coordination (Fig. 15.12). This geometry is rationalized by the presence of a stereochemically active lone pair on the Sn atom, and SnO can be regarded as having the fluorite structure (Section 4.9) with alternate layers of anions missing.

When heated in the absence of air, SnO disproportionates into Sn and SnO_2. The latter occurs naturally as the mineral cassiterite and has a rutile structure (Section 4.9). It has low solubility in glasses and glazes, and is used in large quantities as an opacifier and pigment carrier in ceramic glazes to make them less transparent.

The oxides of lead are very interesting structurally. The red form of PbO has the same structure as blue-black SnO, with a stereochemically active lone pair (Fig. 15.12). Lead also forms mixed oxidation state oxides, such as 'red lead', Pb_3O_4, which contains Pb(IV) in an octahedral environment and Pb(II) in an irregular six-coordinate environment. The assignment of different oxidation numbers to the lead in these two sites is based on the shorter PbO distances for the atom identified as Pb(IV). The maroon form of lead(IV) oxide, PbO_2, crystallizes in the rutile structure and is unlike all the earlier members of Group 14; it is strongly oxidizing. The marked stabilization of the 2+ state, now fully in evidence, is a manifestation of the inert pair effect that arises from the relativistic stabilization of the $6s^2$ sub-shell (Sections 10.6 and 10.7). This oxide is a component of the cathode of a lead–acid battery (Box 15.7).

15.12 Compounds with nitrogen

KEY POINTS The cyanide ion, CN$^-$, forms complexes with many d-metal ions; its coordination to the active sites of enzymes such as cytochrome c oxidase accounts for its high toxicity.

BOX 15.7 How does a lead–acid battery work?

The chemistry of the lead–acid battery is noteworthy: as well as being the most successful rechargeable battery, it illustrates the role of both kinetics and thermodynamics in the operation of cells.

In its fully charged state, the active material on the cathode is PbO_2, and at the anode it is lead metal; the electrolyte is dilute sulfuric acid. One feature of this arrangement is that the lead-containing reactants and products at both electrodes are insoluble. When the cell is producing current (during discharge), the reaction at the cathode is the reduction of PbO_2 to Pb(II), which in the presence of sulfuric acid is deposited as insoluble $PbSO_4$:

$$PbO_2(s) + HSO_4^-(aq) + 3H^+(aq) + 2e^- \rightarrow PbSO_4(s) + 2H_2O(l)$$
$$E^\ominus = +1.70 \text{ V}$$

At the anode, lead is oxidized to Pb(II), which is also deposited as the sulfate. The half-cell reaction is:

$$PbSO_4(s) + 2e^- \rightarrow Pb(s) + SO_4^{2-}(aq) \quad E^\ominus = -0.36 \text{ V}$$

The overall cycle, i.e. the processes of discharging and charging, is thus written

$$PbO_2(s) + Pb(s) + H_2SO_4 \xrightleftharpoons[charge]{discharge} 2PbSO_4(s) + 2H_2O(l)$$

The lead–acid battery delivers a cell potential of about 2.1 V, exceeding by far the potential for the oxidation of water to O_2, which is 1.23 V. The success of the battery thus hinges on the high overpotentials (and hence low rates) of oxidation of H_2O on PbO_2 and of reduction of H_2O on lead. The battery owes its existence to the large relativistic stabilization of Pb(II) relative to Pb(IV); a similar battery based on tin, based on SnO_2, SnO and Sn, would deliver a negligible voltage.

Hydrogen cyanide, HCN, is produced in large amounts by the high-temperature catalytic partial oxidation of methane and ammonia, and is used as an intermediate in the synthesis of many common polymers, such as poly(methylmethacrylate) and poly(acrylonitrile). It is highly volatile (b.p. 26°C) and highly poisonous.

Unlike the neutral ligand CO, the CN^- ion is a reasonably strong Brønsted base ($pK_a = 9.4$) and a much poorer Lewis acid π acceptor. Consequently, the coordination chemistry of CN^- is more often associated with metal ions in higher oxidation states, an example being that whereas CO stabilizes Ni(0), as with $[Ni(CO)_4]$, CN^- forms a stable Ni(II) complex $[Ni(CN)_4]^{2-}$ and even stabilizes Fe(III) relative to Fe(II) (the electrode potential for the $[Fe(CN)_6]^{3-}/[Fe(CN)_6]^{4-}$ couple is more negative than for the $Fe^{3+}(aq)/Fe^{2+}(aq)$ couple). The extreme toxicity of cyanide is due to its binding to an Fe(III) intermediate of cytochrome c oxidase, the enzyme responsible for reducing O_2 and generating energy (Section 26.10).

The toxic, flammable gas cyanogen, $(CN)_2$ (18), dissociates to give CN radicals and forms interpseudohalogen compounds, such as FCN and ClCN. Cyanogen is known as a **pseudohalogen** because of its similarity to a halogen (Section 18.7).

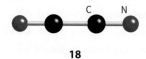

18

The direct reaction of Si and N_2 at high temperatures produces silicon nitride, Si_3N_4, a very hard and inert compound used in high-temperature ceramic materials. Suitable organosilicon–nitrogen compounds are being sought, that undergo pyrolysis to yield silicon nitride fibres and other shapes. Trisilylamine, $(H_3Si)_3N$, the silicon analogue of trimethylamine, has very low Lewis basicity. It has a planar structure, or is fluxional with a very low barrier to inversion. The low basicity and planar structure have traditionally been attributed to d-orbital participation in bonding, allowing sp^2 hybridization around the N atom and delocalization of the lone pair through π bonding. An alternative rationale is that the lower electronegativity of Si compared to C renders the Si–N bond more polar than the C–N bond. This difference leads to long-range electrostatic repulsion between the three $H_3Si^{\delta+}$- groups in trisilylamine and hence a planar structure.

15.13 Carbides

The numerous binary compounds of carbon with metals and metalloids, the *carbides*, are classified as follows:

- **Saline carbides**, which are largely ionic solids; they are formed by the elements of Groups 1 and 2 and by aluminium.

- **Metallic carbides**, which have a metallic conductivity and lustre; they are formed by the p- and d-block elements.

- **Metalloid carbides**, which are hard covalent solids formed by boron and silicon.

Fig. 15.13 summarizes the distribution of the different carbide types in the periodic table; it also includes binary molecular compounds of carbon with electronegative elements, which are not normally regarded as carbides. This classification is useful for correlating chemical and physical properties, but (as so often in inorganic chemistry) the borderlines are imprecise.

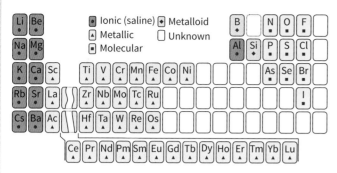

FIGURE 15.13 The distribution of carbides in the periodic table. Molecular compounds of carbon are included for completeness, but are not carbides.

(a) Saline carbides

KEY POINTS Metal–carbon compounds of highly electropositive metals are saline; nonmetal carbides are mechanically hard and are semiconductors.

Saline carbides of the Group 1 and 2 metals may be divided into three subcategories: **graphite intercalation compounds**, such as KC_8, **dicarbides** (or 'acetylides'), which contain the C_2^{2-} anion, and **methides**, which formally contain the C^{4-} anion.

Graphite intercalation compounds were mentioned earlier in Section 15.5. A series of alkali metal–graphite intercalation compounds can be prepared with different metal:carbon ratios, including KC_8 and KC_{16} (Fig. 15.7).

The dicarbides are formed by a broad range of electropositive metals, including those from Groups 1 and 2 (Sections 12.12 and 13.9) and the lanthanoids. The C_2^{2-} ion has a very short CC distance in some dicarbides (e.g. 119 pm in CaC_2), which is consistent with it having a triply bonded $[C\equiv C]^{2-}$ ion isoelectronic with $[C\equiv N]^-$ and $N\equiv N$. Some dicarbides have a structure related to rock salt, but replacement of the spherical Cl^- ion by the elongated $[C\equiv C]^{2-}$ ion leads to an elongation of the crystal along one axis and tetragonal symmetry (Fig. 15.14). The CC bond

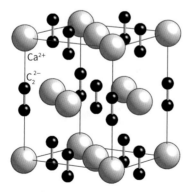

FIGURE 15.14 The calcium-carbide structure. Note that this structure bears a similarity to the rock-salt structure. Because C_2^{2-} is not spherical, the cell is elongated along one axis. This crystal is therefore tetragonal rather than cubic.

is significantly longer in the lanthanoid dicarbides, suggesting that the simple triple-bonded structure is not a good approximation for those compounds.

EXAMPLE 15.4 Predicting the bond order of the dicarbide anion

Use a molecular orbital approach to predict the bond order of the C_2^{2-} anion.

Answer We use the molecular orbital energy-level diagram in Fig. 2.18 and populate it with 10 electrons. This gives the $1\sigma_g^2 1\sigma_u^2 1\pi_u^4 2\sigma_g^2$ configuration. The bond order, b, is given by $\frac{1}{2}(n-n^*) = \frac{1}{2}(8-2) = 3$.

Self-test 15.4 Predict what will happen to the bond length and bond strength of C_2^{2-} if it is oxidized to C_2^-.

Methides, also known as methanides, such as Be_2C and Al_4C_3, are borderline between saline and metalloid, and the isolated C ion is only formally C^{4-}. As expected, methide compounds tend to occur with smaller metal cations. The existence of directional bonding to the C atom in methides (as distinct from the nondirectional character expected of purely ionic bonding) is indicated by the crystal structures of methides, which are not those expected for the simple packing of spherical ions. The methide ion is a central component of the active site of nitrogenase, the enzyme responsible for converting N_2 into useful NH_3. The structure of the [Mo7Fe-9S,C] cofactor is discussed in Section 26.14.

The principal synthetic routes to the saline dicarbides of Groups 1 and 2 are very straightforward:

Direct reaction of the elements at high temperatures:

$$Ca(g) + 2C(s) \xrightarrow{>2000°C} CaC_2(s)$$

The formation of graphite intercalation compounds is another example of a direct reaction, but is carried out at much lower temperatures. The intercalation reaction is more facile because no CC covalent bonds are broken when an ion slips between the graphite layers.

Reaction of a metal oxide and carbon at a high temperature:

$$CaO(s) + 3C(s) \xrightarrow{2000°C} CaC_2(s) + CO(g)$$

Crude calcium dicarbide is prepared in electric arc furnaces by this method. The carbon serves both as a reducing agent and as a source of carbon to form the dicarbide.

Reaction of ethyne (acetylene) with a metal–ammonia solution:

$$2Na(am) + C_2H_2(g) \rightarrow Na_2C_2(s) + H_2(g)$$

This reaction occurs under mild conditions and leaves the carbon–carbon bonds of the starting material intact.

The saline dicarbides and methides are readily oxidized and protonated. For example, calcium dicarbide reacts with water to produce ethyne:

$$CaC_2(s) + 2H_2O(l) \rightarrow Ca(OH)_2(s) + HC\equiv CH(g)$$

This reaction is used to generate ethyne in facilities where calcium dicarbide is available, as it is cheaper and easier than producing it from oil. Similarly, the hydrolysis of aluminium methide liberates methane:

$$Al_4C_3(s) + 12H_2O(l) \rightarrow 4Al(OH)_3(s) + 3CH_4(g)$$

The controlled hydrolysis or oxidation of the graphite intercalation compound KC_8 restores the graphite and produces a hydroxide or oxide of the metal:

$$2KC_8(s) + 2H_2O(g) \rightarrow graphite + 2KOH(s) + H_2(g)$$

(b) Metallic carbides

KEY POINT d-Metal carbides are often hard materials with the carbon atom octahedrally surrounded by metal atoms.

The d metals provide the largest class of carbides. Examples are Co_6Mo_6C and Fe_3Mo_3C. They are sometimes referred to as **interstitial carbides** because it was long thought that the structures were the same as those of the metals and that they were formed by the insertion of C atoms in octahedral holes. In fact, the structure of the metal and the metal carbide often differ. For example, tungsten metal has a body-centred structure whereas tungsten carbide (WC) is hexagonal close-packed. The name 'interstitial carbide' gives the erroneous impression that the metallic carbides are not legitimate compounds. In fact, the hardness and other properties of metallic carbides demonstrate that strong metal–carbon bonding is present. Carbides are economically and technologically useful materials. Tungsten carbide (WC), for example, is used for cutting tools and high-pressure apparatus such as that used to produce diamond. Cementite, Fe_3C, is a major constituent of steel and cast iron.

Metallic carbides of composition MC have an fcc or hcp arrangement of metal atoms with the C atoms in the octahedral holes. The C atoms in carbides of composition M_2C occupy only half the octahedral holes between the close-packed metal atoms. A C atom in an octahedral hole is formally **hypercoordinate** (i.e. has an untypically high coordination number) because it is surrounded by six metal atoms. However, the bonding can be expressed in terms of delocalized molecular orbitals formed from the C2s and C2p orbitals and the d orbitals (and perhaps other valence orbitals) of the surrounding metal atoms.

It has been found empirically that the formation of simple compounds in which the C atom resides in an octahedral hole of a close-packed structure occurs when $r_C/r_M < 0.59$, where r_C is the covalent radius of C and r_M is the metallic radius of M. This relationship also applies to metal compounds containing nitrogen or oxygen.

(c) Metalloid carbides

KEY POINT Boron and silicon form the extremely hard materials B_4C and SiC

Silicon and boron form metalloid carbides. Boron carbide is an extremely hard ceramic material that is used in tank armour, bulletproof vests, and many industrial applications such as cutting tools and wear-resistant coatings. It is also used as a neutron absorber in nuclear reactors. Boron carbide is prepared by reducing B_2O_3 with carbon in an electric arc furnace:

$$2B_2O_3(s) + 7C(s) \rightarrow B_4C(s) + 6CO(g)$$

Its formula is usually written as B_4C, but a better representation is $B_{12}C_3$. The complex structure contains icosohedral B_{12} units linked directly by B–B bonds and by linear C–C–C chains (Fig. 15.15).

When SiO_2 is heated with carbon, CO is evolved and silicon carbide, SiC, forms. This very hard material is widely used as the abrasive carborundum. Grains of SiC can be bonded together by high-temperature treatment to form hard ceramics that are used in car brakes, car clutches, and in bulletproof vests. Electronic applications of silicon carbide include light-emitting diodes and high-temperature and high-voltage semiconductors. Silicon carbide exists in over 200 different crystalline forms. The most common polymorph is α-SiC, which has a hexagonal wurtzite structure (Fig. 4.36) and is formed at temperatures above 1700°C. β-SiC is formed below 1700°C and has the cubic zinc-blende structure (Fig. 4.35).

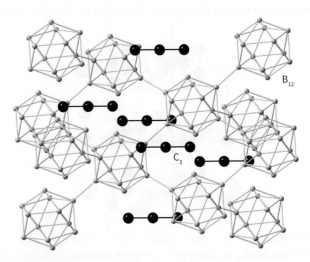

FIGURE 15.15 The structure of boron carbide, showing the icosahedral B_{12} units and the C–C–C chains that link them.

15.14 Silicides

KEY POINTS Silicon–metal compounds (silicides) contain isolated Si, tetrahedral Si_4 units, or hexagonal nets of Si atoms.

Silicon, like its neighbours boron and carbon, forms a wide variety of binary compounds with metals. Some of these **silicides** contain isolated Si atoms. The structure of ferrosilicon, Fe_3Si, for instance, which plays an important role in steel manufacture, can be viewed as an fcc array of Fe atoms with some atoms replaced by Si. Compounds such as K_4Si_4 contain isolated tetrahedral cluster anions $[Si_4]^{4-}$ that are isoelectronic with P_4. Many of the f-block elements form compounds with the formula MSi_2 that adopt the AlB_2 structure shown in Fig. 14.8.

15.15 Extended silicon–oxygen materials

As well as forming simple binary compounds with oxygen, silicon forms a range of extended network solids that have numerous applications in industry. Aluminosilicates occur naturally as clays and rocks. Zeolite aluminosilicates are widely used as molecular sieves, catalysts, and catalyst support materials. These compounds are discussed in more detail in Sections 25.16, 29.1, and 29.2.

(a) Aluminosilicates

KEY POINTS Aluminium may replace silicon in a silicate framework to form an aluminosilicate. Layered aluminosilicates are the primary constituents of clay and some common minerals.

Aluminosilicates are largely responsible for the rich variety of the mineral world. We have already seen that in γ-alumina, Al^{3+} ions are present in both octahedral and tetrahedral holes (Section 14.16). This versatility carries over into the aluminosilicates, where Al may substitute for Si in tetrahedral sites (5); it may also enter an octahedral environment external to the silicate framework, or, more rarely, occur with other coordination numbers. As we mentioned earlier, substitution of Al for Si renders the overall charge of that site negative by one unit, and an additional cation, such as H^+, Na^+, or half as many Ca^{2+}, is therefore required for each Al atom that replaces a Si atom. As we shall see, these additional cations have a profound effect on the properties of the materials.

(b) Microporous solids

KEY POINT Zeolite aluminosilicates have large open cavities or channels, giving rise to useful properties such as ion exchange and molecular absorption.

The molecular sieves are crystalline aluminosilicates having open structures with apertures of molecular dimensions.

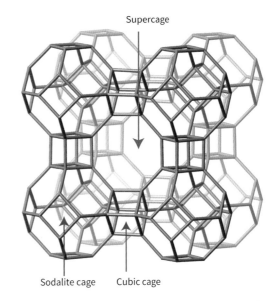

FIGURE 15.16 Framework representation of a type-A zeolite. Note the sodalite cages (truncated octahedral), the small cubic cages, and the central supercage.

The name 'molecular sieve' is prompted by the observation that these materials absorb only molecules that are smaller than the aperture dimensions and so can be used to separate molecules of different sizes. These 'microporous' substances represent a major triumph of solid-state chemistry, for their synthesis and our understanding of their properties combine challenging determinations of structures, imaginative synthetic chemistry, and important practical applications. The cages are defined by the crystal structure, so they are highly regular and of precise size. Consequently, molecular sieves capture molecules with greater selectivity than high-surface-area solids such as silica gel or activated carbon, where molecules may be caught in irregular voids between the small particles.

The zeolites[1] have an aluminosilicate framework with cations (typically from Groups 1 or 2) trapped inside tunnels or cages (Fig. 15.16). In addition to their function as molecular sieves, zeolites are used as ion-exchange resins as they can exchange their ions for those in a surrounding solution.

Zeolites are used for shape-selective heterogeneous catalysis (Section 31.13). For example, the molecular sieve ZSM-5 is used to synthesize 1,4-dimethylbenzene (o-xylene) for use as an octane booster in petrol. The other xylenes are not produced because the catalytic process is controlled by the size and shape of the zeolite cages and tunnels. This and other applications are summarized in Table 15.6 and discussed in Sections 25.17 and 31.13.

---

[1] The name 'zeolite' is derived from the Greek for 'boiling stone'. Geologists found that certain rocks gave off steam and seemed to boil when subjected to the flame of a blowpipe.

TABLE 15.6 Some uses of zeolites

Function	Application
Ion exchange	Water softeners in detergents
Absorption of molecules	Selective gas separation
	Gas chromatography
Solid acid	Cracking high molar mass hydrocarbons for fuel and petrochemical intermediates
	Shape-selective alkylation and isomerization of aromatics for petroleum and polymer intermediates

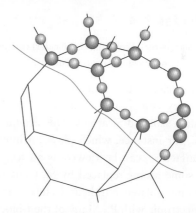

FIGURE 15.17 Framework representation of a truncated octahedron (truncation perpendicular to the four-fold axes of the octahedron) and the relationship of Si and O atoms to the framework. Note that a Si atom is at each vertex of the truncated octahedron and an O atom is approximately halfway along each edge.

Synthetic procedures have added to the many naturally occurring zeolite varieties and have produced zeolites that have specific cage sizes and specific chemical properties within the cages. These synthetic zeolites are usually produced in a high-pressure autoclave. Their open structures seem to form around hydrated cations or other large cations such as NR_4^+ ions introduced into the reaction mixture. For example, a synthesis may be performed by heating colloidal silica to 100–200°C in an autoclave with an aqueous solution of tetrapropylammonium hydroxide. The microcrystalline product, which has the typical composition $[N(C_3H_7)_4]OH(SiO_2)_{48}$, is converted into the zeolite by burning away the C, H, and N of the quaternary ammonium cation at 500°C in air. Aluminosilicate zeolites are made by including high-surface-area alumina in the starting materials.

A wide range of zeolites has been prepared with varying cage and bottleneck sizes (Table 15.7). Their structures are based on approximately tetrahedral MO_4 units, which in the great majority of cases are SiO_4 and AlO_4. Because the structures involve many such tetrahedral units, it is common practice to abandon the polyhedral representation in favour of one that emphasizes the position of the Si and Al atoms. In this scheme, the Si or Al atom lies at the intersection of four line segments and the O atom bridge lies on the line segment (Fig. 15.17).

This **framework representation** has the advantage of giving a clear impression of the shapes of the cages and channels in the zeolite. Some examples are illustrated in Fig. 15.18.

The important zeolites have structures that are based on the 'sodalite cage' (Fig. 15.16), which can also be viewed as a truncated octahedron formed by slicing off each vertex of an octahedron (**19**). The truncation leaves a square face in the place of each vertex and the triangular faces of the octahedron are transformed into regular hexagons.

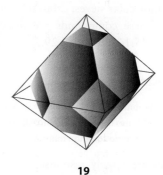

19

The substance known as 'zeolite type A' is based on sodalite cages that are joined by O bridges between the square faces. Eight such sodalite cages are linked in a cubic pattern with a large central cavity called an α-cage. The α-cages share octagonal faces, with an open diameter of 420 pm. Thus H_2O or other small molecules can fill them and diffuse through octagonal faces. However, these faces are too small to permit the entry of molecules with van der Waals diameters larger than 420 pm.

TABLE 15.7 Composition and properties of some molecular sieves

Molecular sieve	Composition	Bottleneck diameter/pm	Chemical properties
A	$Na_{12}[(AlO_2)_{12}(SiO_2)_{12}]\cdot xH_2O$	400	Absorbs small molecules; ion exchanger, hydrophilic
X	$Na_{86}[(AlO_2)_{86}(SiO_2)_{106}]\cdot xH_2O$	800	Absorbs medium-sized molecules; ion exchanger, hydrophilic
Chabazite	$Ca_2[(AlO_2)_4(SiO_2)_8]\cdot xH_2O$	400–500	Absorbs small molecules; ion exchanger, hydrophilic; acid catalyst
ZSM-5	$Na_3[(AlO_2)_3(SiO_2)_{93}]\cdot xH_2O$	550	Moderately hydrophilic
ALPO-5	$AlPO_4\cdot xH_2O$	800	Moderately hydrophobic
Silicalite	SiO_2	600	Hydrophobic

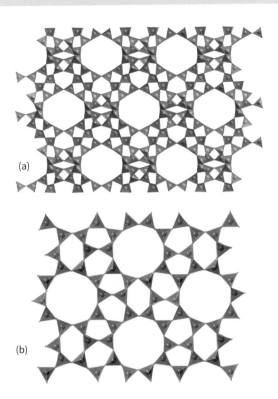

FIGURE 15.18 Two zeolite framework structures: (a) Zeolite-X, and (b) ZSM-5. In each case only the SiO_4 tetrahedra that form the framework are shown; nonframework atoms such as charge-balancing cations and water molecules are omitted.

A BRIEF ILLUSTRATION

To identify the four-fold and six-fold axes in the truncated octahedral polyhedron used to describe the sodalite cage (**19**), we note that there is one four-fold axis running through each pair of opposite square faces, for a total of three four-fold axes. Similarly, a set of four six-fold axes runs through opposite six-fold faces.

The charge on the aluminosilicate zeolite framework is neutralized by cations lying within the cages. In the type-A zeolite, Na^+ ions are present and the formula is $Na_{12}(AlO_2)_{12}(SiO_2)_{12} \cdot xH_2O$. Numerous other ions, including d-block cations and NH_4^+, can be introduced by ion exchange with aqueous solutions. Zeolites are therefore used for water softening and as a component of laundry detergent to remove the di- and tripositive ions that decrease the effectiveness of the surfactant.

In addition to the control of properties by selecting a zeolite with the appropriate cage and bottleneck size, a zeolite can be chosen for its affinity for molecules according to their polarity (Table 15.7). The aluminosilicate zeolites, which always contain charge-compensating ions, have high affinities for polar molecules such as H_2O and NH_3. By contrast, the nearly pure silica molecular sieves bear no net charge and are nonpolar to the point of being hydrophobic. Another group of hydrophobic zeolites is based on the aluminium phosphate frameworks; $AlPO_4$ is isoelectronic with Si_2O_4 and the framework is similarly uncharged.

Another interesting aspect of zeolite chemistry is that large molecules can be synthesized from smaller molecules inside the zeolite cage. Like a 'ship in a bottle', once assembled, the molecule is too large to escape. For example, Na^+ ions in a type-Y zeolite may be replaced by Fe^{2+} ions (by ion exchange). The resulting Fe^{2+}–Y zeolite is heated with phthalonitrile, which diffuses into the zeolite and condenses around the Fe^{2+} ion to form iron phthalocyanine, which remains imprisoned in the cage (Fig. 15.19). The entrapped Fe complex is a catalyst for oxidation of hydrocarbons, resembling the action of some enzymes (Section 26.10).

15.16 Compounds of heavier Group 14 elements with organic substituents

KEY POINTS In contrast to silanes and the heavier Group 14 hydrides, silicon, germanium, tin, and even lead form a wide range of stable compounds in which hydrogen is substituted by bulkier alkyl and aryl groups. Reactivity is conferred by substituting one or more alkyl groups by a halogen. Methylchlorosilanes are important starting materials for the manufacture of silicone polymers.

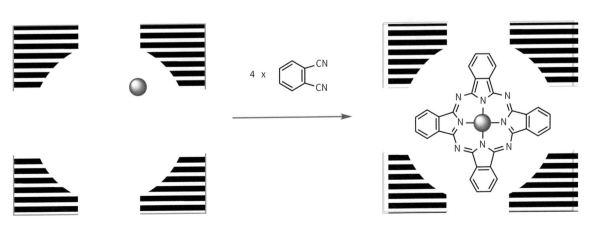

FIGURE 15.19 Synthesis of an Fe phthalocyanine complex inside a zeolite cage.

Tetraalkyl and tetraaryl analogues of silanes, germanes, and stannanes are prepared by reacting the tetrahalides with appropriate organolithium or Grignard (Mg) reagents:

$SiCl_4 + 4RLi \rightarrow SiR_4 + 4LiCl$

$SnCl_4 + 4RMgI \rightarrow SnR_4 + 4MgICl$

The less stable lead alkyls are made starting from $PbCl_2$:

$2PbCl_2 + 4RLi \rightarrow R_4Pb + 4LiCl + Pb$

$2PbCl_2 + 4RMgBr \rightarrow R_4Pb + Pb + 4MgBrCl$

In general, these compounds have low polarity bonds and their stability decreases down the group. Tetraethyl lead used to be added to petrol (gasoline) to prevent premature combustion and improve fuel economy, but this use has largely been discontinued, as lead poses a dangerous environmental hazard. Tetramethyl- and tetraethylgermanium are used in the microelectronics industry as precursors for GeO_2 chemical deposition.

Substitution of one or more alkyl groups by chlorine atoms affords compounds that are precursors for a very wide range of important products. One such compound is chlorotrimethyl tin, $(CH_3)_3SnCl$, which is used to produce stabilizers for PVC plastics: it is produced by the Kocheshkov redistribution reaction:

$SnCl_4 + 3Sn(CH_3)_4 \rightarrow 4(CH_3)_3SnCl$

Alkyl tin compounds are also widely used as fungicides, wood preservatives, and antifouling coatings on ships, although the latter use has become restricted because of environmental concerns for marine life. The trialkyl halides R_3SnX can polymerize via Sn–X–Sn bridges and form chain structures having five-coordinate Sn. The presence of bulky R groups may affect the shape. For example, in $(SnFMe_3)_n$ (20), the Sn–F–Sn backbone is in a zigzag arrangement: by way of contrast, in Ph_3SnF the chain has straightened, and $(Me_3SiC)Ph_2SnF$ is a monomer.

a Cu_2O catalyst at 300°C. The products are separated by distillation, the major product being dimethyldichlorosilane:

$nMeCl + Si \rightarrow Me_nSiCl_{4-n}$

Methylchlorosilanes react with water to produce oligomers containing alkylated silicon atoms and oxygen atoms connected by Si–O–Si bridges. Thus Me_3SiCl yields a simple oligomer:

$Me_3SiCl + H_2O \rightarrow Me_3SiOH + HCl$

$2Me_3SiOH \rightarrow Me_3SiOSiMe_3 + H_2O$

More importantly though, hydrolysis of Me_2SiCl_2 produces chains or rings, and the hydrolysis of $MeSiCl_3$ produces a cross-linked polymer (Fig. 15.20). It is interesting to note that most silicone polymers are based on a Si–O–Si backbone, whereas carbon polymers are generally based on a C–C backbone, reflecting the strengths of the Si–O and C–C bonds (Table 15.2).

Silicone polymers have a range of structures and uses. Their properties depend on the degree of polymerization and cross-linking, which are influenced by the choice and mix of reactants, and the use of dehydrating agents such as sulfuric acid and elevated temperatures. The liquid silicones are thermally more stable than hydrocarbon oils. Moreover, unlike hydrocarbons, their viscosity changes only slightly with temperature. Consequently, silicones are used as lubricants and wherever inert fluids are needed, for instance, in hydraulic braking systems.

Silicones are very hydrophobic and are used in water-repellent sprays for shoes and other items. The lower molar mass silicones are essential in personal-care products such as shampoos, conditioners, shaving foams, hair gels, and toothpastes, and impart to them a 'silky' feel. Decamethylcyclopentasiloxane (21) is used as a hair conditioner and is becoming increasingly popular as an

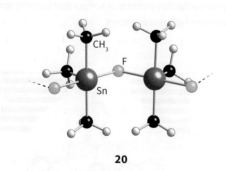

20

Chlorine-substituted methylsilanes, represented as Me_nSiCl_{4-n}, are important starting materials for the manufacture of silicon polymers known as silicones. They are prepared by reacting elemental silicon with chloromethane over

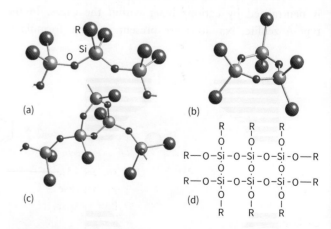

FIGURE 15.20 The structure of (a) a chain, (b) a ring, and (c) cross-linked silicone; (d) the chemical formula of a cross-linked fragment.

environmentally friendly dry-cleaning fluid. It is odourless, nontoxic, and decomposes to SiO_2 with traces of H_2O and CO_2. 'Bouncy putty' or 'silly putty' is a silicone polymer made from a borate cross-linked dimethyl silicone oil, $\{(CH_3)_2SiO\}_n$. At the other end of the spectrum of practical usefulness, silicone greases, oils, and resins are used as sealants, lubricants, varnishes, waterproofing, synthetic rubbers, prosthetic implants, and hydraulic fluids.

21

Analogues of carbenes (R_2C) are known for Si, Ge, Sn, and Pb. Whereas the carbene unit is a triplet (**22**), the heavier analogues (collectively known as tetralenes) are singlets (**23**).

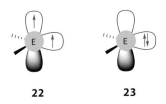

22   **23**

The germylenes, R_2Ge, and other analogues are stabilized by very bulky R groups; thus, dimethylgermylene, $(CH_3)_2Ge$, is very unstable whereas bis(2,4,6-tritertiarybutylphenyl) germylene (**24**) is stable. They are synthesized by reduction of the corresponding tetravalent dichloro compounds:

$$nR_2GeCl_2 + 2nLi \rightarrow (R_2Ge)_n + 2nLiCl$$

24

Many of the reactions of the heavier tetralenes are analogous to those of the carbenes (Section 24.15), and they are useful in organometallic chemistry because they will insert into bonds such as carbon–halogen and metal–carbon:

$$RX + R_2Ge \rightarrow R_3GeX$$

$$R_2Ge \xrightarrow{R'Li} R_2R'GeLi \xrightarrow{R'Li} R_2R'_2Ge$$

FIGURE 15.21 Bonding models for ditetralenes: (a) valence bond model with resonating nonbonding electron pairs; (b) molecular orbital model with Jahn-Teller interaction.

The ER_2 compounds dimerize to produce analogues of alkenes, R_2EER_2, collectively known as ditetralenes. From E = Si downwards, the R groups do not lie in the E–E plane, as for alkenes, but are displaced to give trans-oriented bent structures, while the E–E distances become increasingly longer than expected for a double bond. Noting that two singlet R_2E: fragments can be combined using internal Lewis acid–base interactions as explained in Chapter 5, the structures of ditetralenes illustrate some of the principles described in Chapter 2 (Fig. 15.21).

The VB/VSEPR approach explains the bending in terms of repulsions between resonance-stabilized nonbonding electron pairs on each E atom. The MO model considers the stabilization arising from the mixing of two orbitals of the same symmetry that are generated upon bending—an effect known as quasi or 2nd-order Jahn-Teller mixing. Trans bending lowers the D_{2h} symmetry to C_{2h}, causing the original π orbital to adopt the same symmetry as the σ* orbital (b_u), the result being to raise the energy of the σ* orbital and stabilize the π-derived nonbonding orbital. This mixing increases down Group 14 as orbitals become increasingly close in energy, thus the bending is greatest for Pb and least for Si.

Ditetralynes are analogues of ethynes which are dimers of the carbyne group RC. Many compounds of the type REER are known for E = Si–Pb, with R being a bulky alkyl or aryl group, and apart from Pb, E–E distances are shorter than expected for a single bond. All have bent structures.

FURTHER READING

R. A. Layfield, Highlights in low-coordinate Group 14 organometallic chemistry. *Organomet. Chem.*, 2011, **37**, 133. This review summarizes the key advances that have been made in the organoelement chemistry of silicon, germanium, tin, and lead.

M. C. McCarthy, C. A. Gottlieb, and P. Thaddeus, Silicon molecules in space and in the laboratory. *Molecular Physics*, 2009, **101**, 697. A short account describing the discovery of simple Si compounds in cosmic dust.

N. Notman, The diamond synthesisers. *Chemistry World*, 2023, **20**, 29. A very helpful summary of the applications of synthetic diamonds.

M. A. Pitt and D. W. Johnson, Main group supramolecular chemistry. *Chem. Soc. Rev.*, 2007, **36**, 1441.

A. Schnepf, Metalloid Group 14 cluster compounds: An introduction and perspectives on this novel group of cluster compounds. *Chem. Soc. Rev.*, 2007, **36**, 745.

H. Berke, The invention of blue and purple pigments in ancient times. *Chem. Soc. Rev.*, 2007, **36**, 15. An interesting account of the uses of silicate pigments.

D. M. P. Mingos, *Essential Trends in Inorganic Chemistry*. Oxford University Press (2004). A survey of inorganic chemistry from the perspective of structure and bonding.

N. C. Norman, *Periodicity and the s- and p-Block Elements*. Oxford University Press (1997). Includes coverage of essential trends and features of p-block chemistry.

J. Baggot, *Perfect Symmetry: The Accidental Discovery of Buckminsterfullerene*. Oxford University Press (1994). A general account of the story of the discovery of the fullerenes.

P. J. F. Harris, *Carbon Nanotube Science: Synthesis, Properties and Applications*. Cambridge University Press (2011).

P. W. Fowler and D. W. Manolopoulos, *An Atlas of Fullerenes*. Dover Publications (2007).

R. Ahuja et al., Relativity and the lead–acid battery. *Physical Review Letters*, 2011, **106**, 018301.

P. P. Power, Bonding and reactivity of heavier group 14 element alkyne analogues. *Organometallics*, 2007, **26**, 4362.

C. Hepburn et al., The technological and economic prospects for CO_2 utilization and removal. *Nature*, 2019, **575**, 87. This review article assesses practical pathways for the utilization of captured carbon dioxide (part of the CCU strategy).

EXERCISES

15.1 Correct any inaccuracies in the following descriptions of Group 14 chemistry:

(a) At very high pressures, diamond is a thermodynamically stable phase of carbon.

(b) Both CO_2 and CS_2 are weak Lewis acids and the hardness increases from CO_2 to CS_2.

(c) Zeolites are layered materials exclusively composed of aluminosilicates.

(d) The reaction of calcium carbide with water yields ethyne, and this product reflects the presence of a highly basic C_2^{2-} ion in calcium carbide.

(e) SnO_2 is a stronger oxidizing agent than PbO_2 because it has a different structure.

15.2 The lightest p-block elements often display different physical and chemical properties from the heavier members. Discuss the similarities and differences by comparison of:

(a) The structures and electrical properties of carbon and silicon.

(b) The physical properties and structures of the oxides of carbon and silicon.

(c) The Lewis acid–base properties of the tetrahalides of carbon and silicon.

15.3 Silicon forms the chlorofluorides $SiCl_3F$, $SiCl_2F_2$, and $SiClF_3$. Sketch the structures of these molecules. What are their point groups?

15.4 Explain why CH_4 burns in air whereas CF_4 does not. The enthalpy of combustion of CH_4 is $-888 \, kJ \, mol^{-1}$ and the C–H and C–F bond enthalpies are 412 and 486 $kJ \, mol^{-1}$, respectively.

15.5 The data shown in Fig. 15.3 indicate that the enthalpies of formation for alkanes and silanes, X_nH_{2n+2}, vary smoothly with n from n = 2, but values for n = 1 are more negative. Suggest an explanation for this observation.

15.6 Describe the types of bonding in the structures of SiF_4 and PbF_2. Explain why these are so different.

15.7 SiF_4 reacts with $(CH_3)_4NF$ to form $[(CH_3)_4N][SiF_5]$. (a) Use the VSEPR rules to determine the shape of the cation and anion in the product. (b) Account for the fact that the ^{19}F-NMR spectrum shows two fluorine environments.

15.8 Predict the appearance of (a) the ^{119}Sn-NMR spectrum and (b) the ^{31}P-NMR spectrum of $(CH_3)_3SnPH_2$, (trimethylstannyl)phosphine.

15.9 Use the data in Table 15.2 and the additional bond enthalpy data given here to calculate the enthalpy of hydrolysis of CCl_4 and CBr_4. Bond enthalpies/kJ mol^{-1}: O–H = 463, H–Cl = 431, H–Br = 366.

15.10 Identify the compounds **A** to **F**.

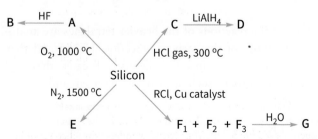

A is a material having crystalline and amorphous forms.

B is a colourless gas, molecular mass 104.

C is a colourless liquid, b.p. 57.6 °C.

D is a flammable gas.

E is a hard, inert ceramic material.

F_1, F_2, F_3 are products containing 1, 2, and three R groups per Si.

G is a polymer: chain length and branching are determined by the F precursor inserted.

15.11 (a) Summarize the trends in relative stabilities of the oxidation states of the elements of Group 14, and indicate the elements that display the inert pair effect. (b) With this information in mind, write balanced chemical reactions or NR (for no reaction) for the following combinations, and explain how the answer fits the trends:

(i) $Sn^{2+}(aq) + PbO_2(s)(excess) \xrightarrow{\text{air excluded}}$

(ii) $Sn^{2+}(aq) + O_2(air) \rightarrow$

15.12 Explain why PbI_4 is unknown while reaction of Pb with fluorine yields PbF_4.

15.13 Why does silicon occur naturally in silicates while the major ore of lead is the sulfide galena, PbS?

15.14 Give balanced chemical equations and conditions for the recovery of silicon and germanium from their ores.

15.15 (a) Describe the trend in band gap energy, E_g, for the elements carbon (diamond) to tin (grey). (b) Explain whether the electrical conductivity of silicon increases or decreases when its temperature is changed from 20°C to 40°C.

15.16 Describe the preparation, structure, and classification of (a) KC_8, (b) CaC_2, (c) K_3C_{60}.

15.17 Explain why silicon can exhibit six-fold coordination in $[SiF_6]^{2-}$ but the $[SiCl_6]^{2-}$ anion is unknown.

15.18 Describe in general terms the nature of the $[SiO_3]_n^{2n-}$ ion present in jadeite.

15.19 Three SiO_4 units can link together to form a short chain of composition $[Si_3O_{10}]^{8-}$ or a cyclic unit of stoichiometry $[Si_3O_9]^{6-}$. How could ^{29}Si NMR be used to distinguish these two polymeric anions?

15.20 Describe the physical properties of pyrophyllite and muscovite mica and explain how these properties arise from the composition and structures of these closely related aluminosilicates.

15.21 There are major commercial applications for semicrystalline and amorphous solids, many of which are formed by Group 14 elements or their compounds. List four different examples of amorphous or partially crystalline solids described in this chapter and briefly state their useful properties.

15.22 The layered silicate compound $CaAl_2(Al_2Si_2)O_{10}(OH)_2$ contains a double aluminosilicate layer, with both Si and Al in four-coordinate sites. Sketch an edge-on view of a reasonable structure for the double layer, involving only vertex sharing between the SiO_4 and AlO_4 units. Discuss the likely sites occupied by Ca^{2+} in relation to the silica–alumina double layer.

15.23 Describe the key compositional and structural features of 'silly putty'.

15.24 Discuss the data shown below for the following compounds of Group 14 elements E:

Compound	Bond dissociation energy (E–E)/kJmol^{-1}	Bond length (E–E)/pm
$H_3C–CH_3$	368	154
$H_2C=CH_2$	710	134
$H_3Si–SiH_3$	310	233
$H_2Si=SiH_2$	270	214
$H_3Ge–GeH_3$	276	244
$H_2Ge=GeH_2$	155	229

TUTORIAL PROBLEMS

15.1 The chemical properties of flerovium are predicted to be significantly different from those of the other members of Group 14 (P. Schwerdtfeger and M. Seth, *J. Nucl. Radiochem. Sci.*, 2002, **3**, 133). Specifically, the only oxidation state exhibited by Fl will be +2 and the element will be chemically highly inert. Do these predictions follow the observed trends in Group 14? Your discussion should include the role of relativistic effects on the energies of the 7s and 7p orbitals in the superheavy elements.

15.2 Discuss the solid-state chemistry of silicon in silicates with reference to how the various structures can be built up from SiO_4 tetrahedra linked into polymeric anions, chains, rings, sheets, and three-dimensional networks.

15.3 One of your friends is studying English and is taking a course in science fiction. A common theme in the sources is silicon-based life forms. Your friend wonders why silicon should be chosen and why all life is carbon-based. Prepare a short article that presents arguments for and against silicon-based life.

15.4 Karl Marx remarked in *Das Kapital* that 'If we could succeed, at a small expenditure of labour, in converting carbon into diamonds, their value might fall below that of bricks.' Review current methods of producing synethetic diamonds and comment on the degree to which they have displaced natural diamonds.

15.5 In their paper 'Mesoporous silica nanoparticles in biomedical applications', Li Zongxi et al. (*Chem. Soc. Rev.*, 2012, **41**, 2590) discuss the utility of mesoporous silica nanoparticles (MSNPs) as supports for the delivery of therapeutic drugs in the body. What are the properties of MSNPs that make them particularly suitable for these applications? Describe how the rate of drug release is controlled. Outline the synthetic methods used to produce the MSNPs and identify the chemists who first synthesized MSNPs.

15.6 In the paper 'Developing drug molecules for therapy with carbon monoxide' (*Chem. Soc. Rev.*, 2012, **41**, 3571), the authors discuss the use of carbon monoxide as a therapeutic agent for treating diseased tissue. Outline the problems associated with using CO as a therapeutic agent and how the authors overcame some of these.

15.7 One way of addressing the effects of CO_2 emissions on the atmosphere is to use it as a chemical feedstock in the production of useful chemicals, fuels, and polymers (M. Aresta and A. Dibenedetto, *Dalton Trans.*, 2007, 2975; C.-H. Huang and C.-S. Tan, *Aerosol Air Qual. Res.*, 2014, **14**, 480). Summarize the various approaches to using CO_2 as a feedstock and the chemistry involved in the production of organic compounds.

15.8 The paper 'Metallacarboranes and their interactions: Theoretical insights and their applicability' (P. Farràs et al., *Chem. Soc. Rev.*, 2012, **41**, 3445) discusses the properties of metallacarboranes and how they can be analysed using computational methods. Outline the properties of metallacarboranes that were investigated and describe the principles of the computational methods used to model them.

15.9 The combination of mesoporosity with semiconductivity would produce materials with interesting properties. A synthesis of such a material was discussed in 'Hexagonal mesoporous germanium' by S. Gerasimo et al. (*Science*, 2006, **313**, 5788). Summarize the expected advantages of such a material and describe how the mesoporous germanium was synthesized.

15.10 Many potentially useful semiconducting materials for photovoltaic applications contain Group 14 elements in sulfides. Examples include Cu_2ZnSnS_4 (CZTS) and SnS, which have been studied because they contain nontoxic elements that are relatively abundant on Earth. Describe the structures and properties of these sulfide materials and include a discussion of the sustainability of these Group 14 element-based materials in comparison with CdTe and CIGS (copper indium gallium sulfide).

15.11 Hybrid organic–inorganic lead and tin iodide perovskites are potential low-cost, solution-synthesized semiconductors for solar cell applications (T. M. Brenner, D. A. Egger, L. Kronik, G. Hodes, and D. Cahen, *Nat. Rev. Mater.*, 2016, **1**, Article 15007). Summarize the recent advances made with these materials focusing on their synthesis, structures, and chemical properties.

15.12 In 'An atomic seesaw switch formed by tilted asymmetric Sn–Ge dimers on a Ge (001) surface' (*Science*, 2007, **315**, 1696), K. Tomatsu et al. describe the synthesis and operation of a molecular switch. Describe how such molecular switches operate, and summarize their current and potential applications.

15.13 In the paper 'High-temperature carbon dioxide capture in a porous material with terminal zinc-hydride sites' (*Science*, 2024, **386**, 814) R. C. Rohde et al. describe the reversible capture of CO_2 from dilute sources at high temperature. Explain the basis of this carbon capture process and why temperature is an important consideration.

The Group 15 elements

Part A: The essentials
16.1 The elements
16.2 Simple compounds
16.3 Oxides and oxanions of nitrogen

Part B: The detail
16.4 Occurrence and recovery
16.5 Uses
16.6 Nitrogen activation
16.7 Nitrides and azides
16.8 Phosphides
16.9 Arsenides, antimonides, and bismuthides
16.10 Hydrides
16.11 Halides
16.12 Oxohalides
16.13 Oxides and oxoanions of nitrogen
16.14 Oxides of phosphorus, arsenic, antimony, and bismuth
16.15 Oxoanions of phosphorus, arsenic, antimony, and bismuth
16.16 Condensed phosphates
16.17 Phosphazenes
16.18 Organometallic compounds of arsenic, antimony, and bismuth

Further reading

Exercises

Tutorial problems

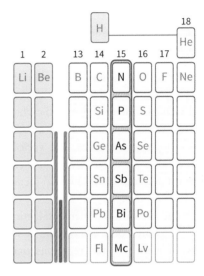

The chemical properties of the Group 15 elements are very diverse. Although the simple trends that we observed for Groups 13 and 14 are still apparent, they are complicated by the fact that the Group 15 elements exhibit a wide range of oxidation states and form many complex compounds with oxygen. Nitrogen makes up a large proportion of the atmosphere and is widely distributed in the biosphere. Phosphorus is essential for both plant and animal life. In stark contrast, arsenic is a well-known poison.

The Group 15 elements—nitrogen, phosphorus, arsenic, antimony, bismuth, and moscovium—are some of the most important elements for life, geology, and industry. They range from gaseous nitrogen to metallic bismuth. All isotopes of moscovium have very short half-lives. The members of this group, the 'nitrogen group', are sometimes referred to collectively as the **pnictogens** (from the Greek for 'to stifle', a property of nitrogen). This name is neither widely used nor officially sanctioned. As in the rest of the p block, the element at the head of Group 15, nitrogen, differs significantly from its congeners (Section 10.2). Its coordination number is generally lower and it is the only member of the group to exist as a gaseous, diatomic molecule under normal conditions.

PART A: The essentials

The properties of the Group 15 elements are diverse and more difficult to rationalize in terms of atomic radii and electron configuration than the p-block elements encountered so far. The usual trends of increasing metallic character down a group and stability of low oxidation states at the foot of the group are still evident, but they are complicated by the wide range of oxidation states available.

16.1 The elements

KEY POINTS Nitrogen is a gas; the heavier elements are all solids that exist in several allotropes.

All the members of the group other than N are solids under normal conditions and exist in several allotropes. The expected trend to increasing metallic character down the group is not clear-cut (Table 16.1) as semiconductivity depends so much on structure. The various allotropes existing for each element differ widely in their band gaps, and even bismuth does not have a normal metallic structure under ambient conditions.

Under ambient conditions, elemental nitrogen exists solely as N_2: the triply-bonded molecule is so stable that N_2 is used as an inert gas, and reactions that produce it often do so with explosive force. Inorganic chemists have been seeking to produce other allotropes of nitrogen that could serve to store the massive amount of energy that is released upon decomposition to N_2, but the research is technically challenging as well as dangerous. A solid allotrope, N_6, related to black phosphorus mentioned below, is produced under extremely high pressure (140 GPa) and high temperature, but while this and other 'polynitrogen' compounds (Section 16.7) have potential applications as high energy density materials, the overriding obstacle is how to produce a practical technology.

In contrast, the heavier elements of Group 15 exist as several allotropes under ambient conditions. Like the gaseous N_2 molecule, (the high temperature allotrope) P_2 has a formal triple bond and a short bond length (189 pm). The strength of π bonds formed by Period 3 elements is weak relative to those of Period 2, so the allotrope P_2 is much less favoured than N_2.

White phosphorus is a waxy solid consisting of tetrahedral P_4 molecules (**1**). Despite the small P–P–P angle (60°), the molecules persist in the vapour up to about 800°C, but above that temperature the equilibrium concentration of P_2 becomes appreciable. White phosphorus is very reactive and bursts into flame in air to yield P_4O_{10}.

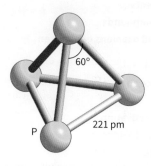

1 P_4, T_d

Red phosphorus can be obtained by heating white phosphorus at 300°C in an inert atmosphere for several days. It is normally obtained as an amorphous solid, but crystalline materials can be prepared that have very complex three-dimensional network structures. By trapping white phosphorus molecules in carbon nanotubes, the transition to red phosphorus can be restricted to the P_4 molecules linking into chains—a material which has been dubbed 'pink phosphorus'. Unlike white phosphorus, red phosphorus does not ignite readily in air and is relatively inert. Red phosphorus can be converted to the more useful polyphosphide salts. When potassium ethoxide in an organic solvent is passed through red phosphorus with mild heating the polyphosphides KP_5, K_2P_{16}, and K_3P_{21} are obtained.

TABLE 16.1 Selected properties of the Group 15 elements

	N (N_2)	P	As (grey)	Sb (metallic)	Bi (metallic)
Melting point/°C	−210	44 (white) 590 (red)	817	630	271
Atomic radius/pm	71	107	119	139	148
First ionization energy/kJ mol^{-1}	1402	1011	947	834	704
Electrical conductivity/10^6 S m^{-1}		10 (α-black)	3.33	2.50	0.77
Pauling electronegativity	3.0	2.2	2.2	2.05	2.0
Electron affinity/kJ mol^{-1}	−8	72	78	103	105
B(E–H)/kJ mol^{-1}	390	322	297	254	

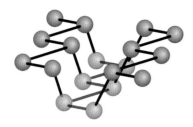

FIGURE 16.1 A section of one of the puckered layers of black phosphorus. Note the trigonal-pyramidal coordination of the atoms.

When phosphorus is heated under high pressure, a series of phases of black phosphorus are formed, the thermodynamically most stable form below 550°C. One of the phases consists of puckered layers composed of pyramidal three-coordinate P atoms (Fig. 16.1). Two-dimensional sheets of black phosphorus can be synthesized which are termed phosphorene, by analogy with graphene. Phosphorene has high tensile strength, and the band gap can be modulated by strain, giving it useful applications in electronics and energy storage. In contrast to the usual practice of choosing the most stable ambient phase of an element as the reference phase for thermodynamic calculations, white phosphorus is adopted because it is more accessible, more uniform, and therefore better characterized than the other forms.

Arsenic exists in two solid forms: yellow arsenic and grey or metallic arsenic. Yellow arsenic and gaseous arsenic both consist of tetrahedral As_4 molecules. Yellow arsenic is transformed into the more stable metallic arsenic by exposure to light. The structure at room temperature of metallic As, and of Sb and Bi, is built from puckered hexagonal layers in which each atom has three nearest neighbours. The layers stack in a way that gives three more distant neighbours in the adjacent net (Fig. 16.2). That even bismuth[1] favours an extended molecular structure, accounts for why it is among the most brittle of metals.

Nitrogen is readily available as dinitrogen, N_2, since it makes up 78% by mass of the atmosphere. The principal raw material for the production of elemental phosphorus is phosphate rock, the insoluble, crushed, and compacted remains of ancient organisms, which consists primarily of the minerals fluorapatite, $Ca_5(PO_4)_3F$, and hydroxyapatite, $Ca_5(PO_4)_3OH$. The chemically softer elements As, Sb, and Bi are often found in sulfide ores. Arsenic is found in the ores realgar, As_4S_4, orpiment, As_2S_3, arsenolite, As_2O_3, and arsenopyrite, FeAsS. Antimony occurs naturally as the minerals stibnite, Sb_2S_3, and ullmanite, NiSbS.

---

[1] The only naturally occurring isotope of bismuth (209) was long believed to be stable towards radioactive decay. However, it actually undergoes α emission with a half-life of 1.9×10^{19} years, which is about a billion times longer than the current age of the universe.

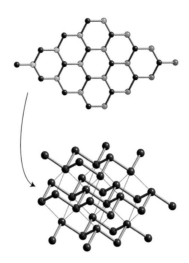

FIGURE 16.2 The puckered network structure of bismuth. Each Bi atom has three nearest neighbours; the dark shaded atoms indicate next-nearest-neighbours from an adjacent puckered sheet. Lightest atoms are furthest from the viewer.

EXAMPLE 16.1 Using bond enthalpies to predict stabilities

The enthalpy of the reaction $P_4(g) \rightarrow 2\,P_2(g)$ is $+217$ kJ mol^{-1}. If the P–P bond enthalpy is 201 kJ mol^{-1} calculate the energy of the P≡P bond.

Answer The enthalpy of any reaction is equal to the difference between the sum of the bond enthalpies for broken bonds and the sum of the enthalpies of the bonds that are formed. In this reaction, 6 mol P–P bonds in P_4 (1) must be broken, corresponding to an enthalpy change of 6×201 kJ $= +1206$ kJ mol^{-1}. Then 2 mol P≡P bonds in P_2 must be formed. Therefore, $217 = 1206 - 2 \times B(P≡P)$ and thus $B(P≡P) = 494$ kJ mol^{-1} which is in good agreement with the value given in Table 2.8 (490 kJ mol^{-1}).

Self-test 16.1 Use the bond energy data in Table 2.8 to calculate the combustion enthalpy for white phosphorus to form P_4O_{10}.

16.2 Simple compounds

KEY POINTS The Group 15 elements form binary compounds on direct interaction with many elements. Nitrogen achieves oxidation number +5 only with oxygen and fluorine. Oxidation state E(V) is common for phosphorus, arsenic, and antimony but rare for bismuth, for which the E(III) state is the more stable.

The wide variety of possible oxidation states of the Group 15 elements can be understood to a large extent by considering the valence-electron configuration of the elements, which is ns^2np^3. This configuration suggests that the highest oxidation state should be E(V), as is indeed the case (Section 9.4(a)). Consistent with the inert pair effect (Section 10.7), the E(III) oxidation state is more stable for Bi.

Nitrogen has a very high electronegativity (significantly exceeded by only O and F) and in many compounds—for example, the nitrides, which contain the N^{3-} ion, and ammonia, NH_3—nitrogen is in a negative oxidation state. Nitrogen achieves positive oxidation states only in compounds with the more electronegative elements O and F. Nitrogen does achieve the group oxidation state (N(V)), but only under much stronger oxidizing conditions than are necessary to achieve this state for the other elements of the group. Nitrogen plays a crucial role in biology as it is a constituent of amino acids, nucleic acids, and proteins, and the nitrogen cycle is one of the most important processes in the ecosystem (Box 16.1 and Section 26.14).

The distinctive nature of nitrogen is due in large part to its high electronegativity, its small atomic radius, and the absence of accessible d orbitals. Thus, N seldom has coordination numbers greater than 4 in simple molecular compounds, but the heavier elements frequently reach coordination numbers of 5 and 6, as in PCl_5 and AsF_6^-.

Nitrogen forms binary compounds, the nitrides, with almost all the elements. Depending on the second element, the nitrides are saline, covalent, or interstitial. Nitrogen also forms the azides, which contain the N_3^- ion. Like N, P forms compounds with almost every element in the periodic table. There are many varieties of phosphides, with formulas ranging from M_4P to MP_{15}. The P atoms may be arranged in rings, chains, or cages,

BOX 16.1 How is nitrogen recycled?

Most of the molecules used by biological systems—including proteins, nucleic acids, chlorophyll, various enzymes and vitamins, and many other cellular constituents—contain nitrogen. In all these compounds, nitrogen is in its reduced form with an oxidation number of −3. Although N_2 is the most abundant constituent of the Earth's atmosphere, its bioavailability is limited by its unreactivity, and the requirements of the biosphere come from the process of nitrogen fixation. Therefore, a major challenge to biology (and technology) involves the reduction of N_2 for incorporation into essential nitrogen compounds.

The nitrogen cycle shown in Fig. B16.1 integrates the biological and industrial reactions that link the reactions of simple nitrogen species. The inner cycle (blue) represents the biological cycle, in which ammonia plays a central role in the incorporation (assimilation) of nitrogen atoms into organic molecules (the release of simple nitrogen species from organic molecules is known as dissimilation). Some of the enzymes of the nitrogen cycle are discussed in Section 26.14. When organisms die and biomass decays, organonitrogen compounds decompose and release nitrogen, mainly as ammonia, back to the local environment. The natural primary source of ammonia is atmospheric N_2, which is 'fixed' by microbes present in the soil or living 'symbiotically' in specialized root nodules present in certain plants. The enzyme catalysing ammonia biosynthesis from N_2 and H_2O is known as nitrogenase and has long been a source of inspiration as the reaction occurs at ambient temperature and pressure.

Agriculture, needing to feed an expanding human population and do so economically, makes massive use of synthetic fertilizers to supply plants with ammonia and other sources of fixed nitrogen (nitrate, urea) along with essential trace elements. Industrial ammonia synthesis is carried out by the catalytic Haber–Bosch process (Sections 16.6 and 31.9), although this currently entails considerable cost in terms of carbon footprint as, unlike biological nitrogen fixation, high pressures and temperatures are required. Industrial ammonia is also converted to nitrate salts by the Ostwald process (Section 16.13), and

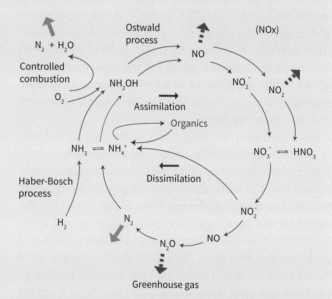

FIGURE B16.1 The nitrogen cycle. The inner cycle (blue) shows the pathways taken by simple nitrogen species occurring within a living cell, the production of ammonia from N_2 being limited to microbes. The outer cycle (orange) shows in a parallel fashion how ammonia is produced and used to make nitrates by industrial routes, resulting in fertilizers to augment the biological cycle. The ammonia may also be used as a fuel and hydrogen carrier. Block arrows show escape of harmful (patterned red) or benign (solid green) gases.

both ammonia and nitrates enter the biological nitrogen cycle, usually raising growth rates and crop yields far beyond those achievable without the industrial intervention. A downside of synthetic fertilizers is that natural reservoirs are inadequate sinks for excess input: ammonia is toxic, and nitrate or nitrite may accumulate as an undesirable component of groundwater or produce eutrophication in lakes, wetlands, river deltas, and coastal areas.

for example P_7^{3-} (**2**), $(P_8^{2-})_n$ (**3**), and P_{11}^{3-} (**4**). The arsenides and antimonides of the Group 13 elements In and Ga are semiconductors (Section 25.6).

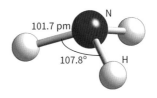

5 Ammonia, NH_3, C_{3v}

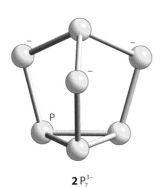

2 P_7^{3-}

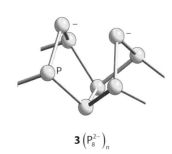

3 $(P_8^{2-})_n$

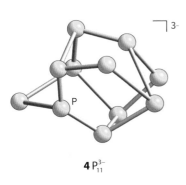

4 P_{11}^{3-}

All the elements form simple hydrides (Section 11.6). Ammonia NH_3 (**5**) is a pungent gas that is toxic at high levels of exposure. Ammonia is an excellent solvent for the Group 1 metals; for instance, it is possible to dissolve 330 g of Cs in 100 g of liquid ammonia at −50°C. These highly coloured, electrically conducting solutions contain solvated electrons (Section 12.14). The chemical properties of ammonium salts are similar to those of the Group 1 ions, especially K^+ and Rb^+. Ammonium salts decompose on heating and ammonium nitrate is a component of some explosives: it is also widely used as a fertilizer. Nitrogen also forms the colourless liquid hydrazine, N_2H_4. The other hydrides of Group 15 are phosphine (formally phosphane, PH_3), arsine (arsane, AsH_3), and stibine (stibane, SbH_3), which are all poisonous gases. Bismuthane is unstable, decomposing below 0°C to the constituent elements.

A NOTE ON GOOD PRACTICE

Although phosphane is the correct formal name of phosphine, the latter name is widely used and we adopt it here. However, we shall use the formal names arsane and stibane for these less common compounds and their derivatives.

Halogen compounds of P, As, and Sb are numerous and important in synthetic chemistry. Trihalides are known for all the Group 15 elements. However, whereas pentafluorides are known for all members of the group from P to Bi, pentachlorides are known only for P, As, and Sb, and the pentabromide is known only for P. Nitrogen does not reach its group oxidation number (N(V)) in neutral binary halogen compounds, but it does achieve it in NF_4^+. Presumably, an N atom is too small for NF_5 to be sterically feasible. Bismuth (Z = 83) is a relativistic element, and the difficulty of oxidizing Bi(III) to Bi(V) is an example of the inert pair effect (Sections 10.6 and 10.7). Bismuth pentafluoride, BiF_5, exists, but $BiCl_5$ and $BiBr_5$ do not.

Nitrogen forms many oxides and oxoanions, which are treated separately in Section 16.3. Phosphorus, As, Sb, and Bi form oxides and oxoanions in a range of oxidation numbers from +5 to +1. The most common oxidation state is E(V), but the E(III) state becomes increasingly more important by bismuth.

The complete combustion of phosphorus yields phosphorus(V) oxide, P_4O_{10}. Each P_4O_{10} molecule has a cage structure in which a tetrahedron of P atoms is held together by bridging O atoms, and each P atom has a terminal O atom (**6**). Combustion in a limited supply of oxygen results in the formation of phosphorus(III) oxide, P_4O_6; this molecule has the same O-bridged framework as P_4O_{10}, but lacks the terminal O atoms (**7**). Arsenic, Sb, and Bi form As_2O_3, Sb_2O_3, and Bi_2O_3 (Section 16.14).

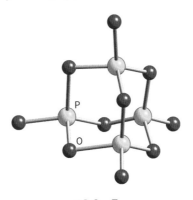

6 P_4O_{10}, T_d

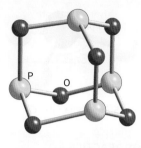

7 P_4O_6, T_d

16.3 Oxides and oxanions of nitrogen

KEY POINTS The nitrate ion is a strong but sluggish oxidizing agent. The intermediate oxidation states of nitrogen are often susceptible to disproportionation. Dinitrogen oxide is unreactive.

Nitrogen forms oxo compounds and oxoanions in all oxidation states from N(V) to N(I). Nitrogen is in the oxidation state N(V) in nitric acid, HNO_3, which is a major industrial chemical used in the production of fertilizers, explosives, and a wide variety of nitrogen-containing chemicals. The nitrate ion, NO_3^-, is a strong but sluggish oxidizing agent. When concentrated nitric acid is mixed with concentrated hydrochloric acid, the orange, fuming aqua regia is formed, which is one of the few reagents that are able to dissolve platinum and gold. The anhydride of nitric acid is N_2O_5. It is a crystalline solid of composition $[NO_2^+][NO_3^-]$.

Nitrogen(IV) oxide, which is commonly called nitrogen dioxide, exists as an equilibrium mixture of the brown NO_2 radical and its colourless dimer, N_2O_4 (dinitrogen tetroxide). The equilibrium constant for the dimerization

$$2\,NO_2(g) \rightleftharpoons N_2O_4(g)$$

is $K = 8.7$ at 25°C

In nitrous acid, HNO_2, nitrogen is present as N(III). Nitrous acid is a strong oxidizing agent. Dinitrogen trioxide, N_2O_3, the anhydride of nitrous acid, is a blue solid that melts above −100°C to give a blue liquid that dissociates into NO and NO_2.

Nitrogen(II) oxide, more commonly known as nitric oxide, NO, is another example of a radical (Section 10.1). However, unlike NO_2, it does not form a stable dimer in the gas phase because the odd electron is distributed almost equally over both atoms and not, as in NO_2, largely confined to the N atom. Until the late 1980s, no beneficial biological roles were known for NO. However, since then it has been found that NO is generated *in vivo* (Section 26.14), and that it performs functions such as the reduction of blood pressure, vasodilation, neurotransmission, and the destruction of microbes.

Dinitrogen oxide, N_2O (specifically NNO), is a colourless, unreactive gas that is released into the atmosphere by both natural and anthropogenic activities. One application of its inertness is that N_2O has been used as the propellant gas for instant whipping cream. Similarly, N_2O was used for many years as a mild anaesthetic; however, this practice has been discontinued because of undesirable physiological side effects, particularly mild hysteria, indicated by its common name of 'laughing gas'. It is still used in a 50:50 mixture with oxygen as an analgesic in childbirth and in clinical procedures, such as wound suturing. Significantly, N_2O is not only a potent greenhouse gas, but is now acknowledged to be an ozone-depleting agent equivalent to chlorofluorocarbons.

EXAMPLE 16.2 Predicting the shapes of nitrogen oxides

N_2O_5 dissociates into NO_2^+ and NO_3^- ions. Use the VSEPR model to predict the shape of these ions.

Answer The Lewis structure for NO_2^+ is shown in (**8**). The ion has two bonding regions and no lone pairs around the central N atom and is therefore linear. One possible Lewis structure for NO_3^- is shown in (**9**) and shows that there are three bonding regions around the central N atom. The ion is therefore trigonal-planar.

$$[\mathrm{\ddot{O}=N=\ddot{O}}]^+$$

8 NO_2^+

9 NO_3^-

Self-test 16.2 Predict the shapes of (a) N_2O and (b) NO_2^-.

PART B: The detail

In this section we review the detailed chemistry of the Group 15 elements. We shall see the wide variety of oxidation states achieved by the elements, particularly nitrogen and phosphorus.

16.4 Occurrence and recovery

KEY POINTS Nitrogen is recovered by distillation from liquid air; it is used as an inert gas and in the production of ammonia. Elemental phosphorus is recovered from the minerals fluorapatite and hydroxyapatite by carbon arc reduction; the resulting white phosphorus is a molecular solid, P_4. Treatment of apatite with sulfuric acid yields phosphoric acid, which is converted to fertilizers and other chemicals.

Nitrogen is obtained on a massive scale by the distillation of liquid air. Liquid nitrogen is a very convenient cryogen as well as a way of storing and handling N_2 in the laboratory. Membrane materials that are more permeable to O_2 than to N_2 are used in laboratory-scale separations from air at room temperature (Fig. 16.3).

Phosphorus was first isolated by Hennig Brandt in 1669. Brandt, misinterpreting their colour, was trying to extract gold from urine and sand, and instead isolated a white solid, which glowed in the dark. This element was called phosphorus after the Greek for 'light bearer'. Today, phosphorus is produced by the action of concentrated sulfuric acid on the mineral fluorapatite to generate phosphoric acid, from which elemental phosphorous is subsequently obtained:

$$Ca_5(PO_4)_3F(s) + 5H_2SO_4(l) \rightarrow 3H_3PO_4(l) + 5CaSO_4(s) + HF(g)$$

The potential pollutant HF is scavenged by reaction with silicates to yield the less reactive SiF_6^{2-} complex ion.

The crude product of the treatment of phosphate-containing minerals with acid contains d-metal contaminants that are difficult to remove completely, so its use is largely confined to fertilizers and metal treatment. Most pure phosphoric acid and phosphorus compounds are still produced from the element because it can be purified by sublimation. The production of elemental phosphorus starts with crude calcium phosphate, which is reduced with carbon in an electric arc furnace. Silica is added (as sand) to produce a slag of calcium silicate:

$$2Ca_3(PO_4)_2(s) + 6SiO_2(s) + 10C(s) \xrightarrow{1500°C} 6CaSiO_3(l) + 10CO(g) + P_4(g)$$

The slag is molten at these high temperatures and so can easily be removed from the furnace. The phosphorus vaporizes and is condensed to the solid, which is stored under

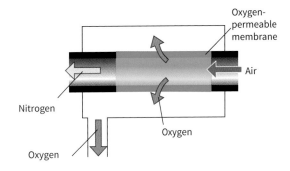

FIGURE 16.3 Schematic diagram of a membrane separator for nitrogen and oxygen.

water to protect it from reaction with air. Most phosphorus produced in this way is burned to form P_4O_{10}, which is then hydrated to yield pure phosphoric acid.

Arsenic is usually extracted from the flue dust of copper and lead smelters (Box 16.2). However, it is also obtained by heating the ores in the absence of oxygen:

$$FeAsS \xrightarrow{700°C} FeS(s) + As(g)$$

Antimony is extracted by heating the ore stibnite with iron, which produces the metal and iron sulfide:

$$Sb_2S_3(s) + 3Fe(s) \rightarrow 2Sb(s) + 3FeS(s)$$

Bismuth occurs as bismite, Bi_2O_3, and bismuthinite, Bi_2S_3. Its ores commonly occur with those of copper, tin, lead, and zinc and it is extracted by reduction as a by-product of the processing of those elements.

16.5 Uses

KEY POINTS Nitrogen is essential for the industrial production of ammonia and nitric acid; the major use of phosphorus is in the manufacture of fertilizers.

Nitrogen gas is used extensively as an inert atmosphere in metal processing, petroleum refining, and food processing:

BOX 16.2 Is environmental arsenic toxic?

The environmental toxicity of arsenic is a problem of groundwater contamination. The worst occurrence of arsenic pollution is in Bangladesh and the neighbouring Indian province of West Bengal, where hundreds of thousands of people have been diagnosed with arsenicosis. Three major rivers drain into this region, bringing iron-laden sediments from the mountains. The fertile delta is heavily farmed, and organic matter leaches into the shallow aquifer, creating reducing conditions. Arsenic levels are correlated with iron levels in groundwater, and the arsenic is thought to be released on the dissolution of iron oxides and hydroxides from the ores.

Arsenicosis develops over a period of up to twenty years. The first symptoms are keratoses of the skin, which develop into cancers; the liver and kidneys also deteriorate. The early stage is reversible if arsenic ingestion is discontinued, but once cancers develop, effective treatment becomes more difficult. The biochemistry of these effects is uncertain. Arsenate is reduced to As(III) complexes in the body, which probably act by binding sulfhydryl (thiol) groups. A plausible link to cancer is suggested by the laboratory finding that low levels of arsenic inhibit hormone receptors that turn on cancer suppressor genes.

it is also used to provide an inert atmosphere in the laboratory, and liquid nitrogen (b.p. −196°C, 77 K) is an important refrigerant. The major industrial use of nitrogen is in the production of ammonia by the Haber–Bosch process (Section 16.6) and conversion to nitric acid by the Ostwald process (Section 16.13).

Ammonia provides a route to a wide range of nitrogen compounds, which include fertilizers, plastics, and explosives (Fig. 16.4). The Haber–Bosch process is costly in terms of energy, and global production is currently responsible for nearly 2% of CO_2 emissions. As with hydrogen, the green credentials of ammonia are represented by 'colours' (Box 11.3): brown NH_3 is produced using H_2 generated by steam reforming of fossil fuels, blue NH_3 is produced using H_2 generated by reforming but implementing carbon capture, and green NH_3 uses H_2 generated from electrolysis of water using renewable energy. Ammonia is a hydrogen carrier (Box 11.5) and can be used directly as a 'green' fuel (energy density > 5 MWh/tonne) in diesel and jet engines that have been modified to curb NOx emissions.

Phosphorus is used in pyrotechnics, smoke bombs, steel-making, and alloys. Red phosphorus mixed with silica is used as the striking strip on matchboxes. The friction involved in striking a match produces enough heat to convert some of the red phosphorus to white phosphorus, which then ignites. Sodium phosphate is used as a cleaning agent, a water softener, and to prevent scaling in boilers and pipes. Condensed phosphates are added to detergents as builders that enhance the detergency by forming complexes with metal ions, softening the water. In the natural environment, phosphorus is usually present as phosphate ions.

Phosphorus (together with N and K) is an essential plant nutrient. However, as a result of the low solubility of many metal phosphates it is often depleted in soil, and hence hydrogenphosphates are important components of balanced fertilizers. Approximately 85% of the phosphoric acid produced goes into fertilizer manufacture.

Phosphorus is also an important constituent of bones and teeth (which are predominantly calcium phosphate), cell membranes (phosphate esters of fatty acids), and nucleic acids, including DNA, RNA, and adenosine triphosphate (ATP), the energy-transfer unit of living organisms (Section 26.2). The phosphines, PX_3, are widely used ligands (Table 7.1, Section 24.6).

Arsenic is used as a dopant in solid-state devices such as integrated circuits and lasers. GaAs is a 13/15 (III/V) semiconductor (Section 25.6) which has better electron mobility and heat stability than silicon. It is used in mobile-phone and satellite applications, solar cells, and optical windows. Arsenic compounds are highly poisonous (Box 16.2). In forensic science, advent of the Marsh test enabled arsenic to be detected for the first time. Gaseous arsane is produced when the oxide is reacted with nitric acid and zinc:

$$As_2O_3 + 6Zn + 6H_2SO_4 \rightarrow 2AsH_3 + 6ZnSO_4 + 3H_2O$$

Ignition of AsH_3 produces arsenic, which can be observed as a black powder.

As with many elements known to be highly poisonous, the toxicity of arsenic is exploited for human benefit. Used under highly controlled conditions, As_2O_3 is an effective drug for treating acute leukaemia (Section 32.1). Arsenilic acid, $C_6H_8AsNO_3$, and sodium arsenilate, $NaAsC_6H_8$, are used as antimicrobial agents in animal and poultry feed to prevent the growth of moulds. Arsenicals are also used as insecticides and herbicides. Monosodium methylarsenate (MSMA) is used to control weeds in cotton and turf crops and in domestic lawn treatments. The first arsenic-containing insecticide was Paris green, $Cu(CH_3CO_2)_2 \cdot 3Cu(AsO_2)_2$, which was manufactured in 1865 for the treatment of the Colorado potato beetle. Sodium arsenite, $NaAsO_2$, is used in poison baits to control grasshoppers and as a dip to prevent parasites in livestock.

Antimony is used in semiconductor technologies to produce infrared detectors and light-emitting diodes. It is used in alloys, where it leads to stronger and harder products. Antimony oxide is used to increase the activity of chlorinated hydrocarbon flame retardants, where it enhances the release of halogenated radicals. Sodium stibogluconate, a compound of Sb(V), is used to treat Leishmaniasis, a tropical skin disease.

Bismuth is much less poisonous than As or Sb and has long been used in some orally administered medicines (Section 32.3). Bismuth subsalicylate, $HOC_6H_4CO_2BiO$, is used in conjunction with antibiotics and as a treatment for peptic ulcers. Bismuth is finding uses in solar energy conversion and thermoelectrics (Chapter 30), as it offers an alternative to lead iodide perovskites (Section 30.2).

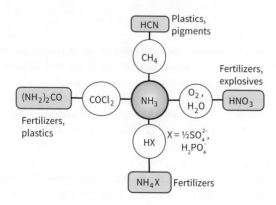

FIGURE 16.4 The industrial uses of ammonia.

EXAMPLE 16.3 Examining the electronic structure and chemistry of P₄

Draw the Lewis structure of P₄ and discuss its possible role as a ligand.

Answer We use the rules described in Section 2.1 to develop the Lewis structure. There is a total of $4 \times 5 = 20$ valence electrons. If each P atom forms a bond to each of the other three P atoms, then 12 electrons will be accounted for, leaving eight electrons, or one lone pair on each P atom (**10**).

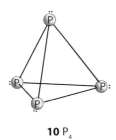

10 P₄

This structure, together with the fact that the electronegativity of P is moderate ($\chi_P = 2.06$), suggests that P₄ might be a moderately good donor ligand. Indeed, P₄ complexes with d-block metals are known.

Protonation of white phosphorus might therefore be expected to yield a compound in which the hydrogen is bonded to an apex. In practice, on treatment with super Brønsted acids, reaction with H⁺ sees one of the P–P edge bonds being broken and the hydrogen bridging between two phosphorus atoms (**11**).

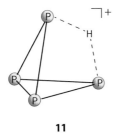

11

Self-test 16.3 (a) Consider the Lewis structure of a segment of the structure of bismuth shown in Fig. 16.2. Is this puckered structure consistent with the VSEPR model? (b) Use the Lewis model to predict the nature of the bonding in N₂ and use this to explain the properties of nitrogen.

16.6 Nitrogen activation

KEY POINT Nitrogen is unreactive. The commercial Haber–Bosch process requires high temperatures and pressures to yield ammonia, which is a major ingredient in fertilizers and an important chemical intermediate. Considerable effort is being invested in developing approaches to activating nitrogen.

Nitrogen occurs in many compounds, but N₂ itself, with a triple bond between the two atoms, is strikingly unreactive. A few strong reducing agents can transfer electrons to the N₂ molecule at room temperature, leading to scission of the N≡N bond, but usually the reaction needs a very strong reducing agent and extreme conditions. The reduction approach is exemplified by the slow reaction with lithium metal at room temperature, which yields Li₃N. Similarly, when Mg (the diagonal neighbour of Li, Section 10.3) burns in air it forms the nitride as well as the oxide.

The slowness of the reactions of N₂ is the result of at least three factors. The first is the strength of the N≡N triple bond and hence the high activation energy required for breaking it. (The strength of this bond also accounts for the lack of nitrogen allotropes stable under ambient conditions.) The second is the relatively large size of the HOMO–LUMO gap in N₂ (Section 2.8(b)), which makes the molecule resistant to simple electron-transfer redox processes. The third factor is the low polarizability of N₂, which does not encourage the formation of the highly polar transition states that are often involved in electrophilic and nucleophilic displacement reactions.

Cheap methods of nitrogen activation—its conversion into useful compounds—are highly desirable because they would have a profound effect on the economy. In the Haber–Bosch process for the production of ammonia, H₂ and N₂ are combined at high temperatures and pressures over an Fe catalyst, as we discuss in detail in Section 16.10. Much of the recent research aimed at achieving more economical ways of activating N₂ has been inspired by the way in which bacteria carry out the transformation at room temperature. Catalytic conversion of nitrogen to NH₄⁺ involves the metalloenzyme nitrogenase, which occurs in nitrogen-fixing bacteria such as those found in the root nodules of legumes. The mechanism by which nitrogenase carries out this reaction, at an active site containing Fe, Mo, and S, is the topic of considerable research. Dinitrogen complexes of metals were discovered in 1965, at about the same time that it was realized that nitrogenase contains Mo (Section 26.14). These developments led to optimism that efficient homogeneous catalysts might be developed in which metal ions would coordinate to N₂ and promote its reduction. Many N₂ complexes have in fact been prepared, and in some cases the preparation is as simple as bubbling N₂ through an aqueous solution of a complex:

$$[Ru(NH_3)_5(OH_2)]^{2+}(aq) + N_2(g) \rightarrow$$
$$[Ru(NH_3)_5(N_2)]^{2+}(aq) + H_2O(l)$$

As with the isoelectronic CO molecule, end-on bonding is typical of N₂ when it acts as a ligand (**12**; Section 24.17). The N–N bond length in the Ru(II) complex is only slightly altered from that in the free molecule. However, when N₂ is coordinated to a more strongly reducing metal centre, this bond is considerably lengthened by back-donation of electron density into the π* orbitals of N₂.

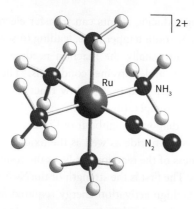

12 $[Ru(NH_3)_5(N_2)]^{2+}$

The early work demonstrated that N_2 can be activated through its coordination in a metal complex, indicating a way to generate ammonia from N_2 and a proton source under ambient conditions. Much research has since been carried out, aimed particularly at electrocatalysis by complexes of abundant metals such as Fe, at which N_2 is activated in nitrogenase. The trigonal Fe complex **13** undergoes several catalytic ($N_2 \rightarrow NH_3$) turnovers when operated as an electrocatalyst.

13

16.7 Nitrides and azides

Nitrogen forms simple binary compounds with other elements; they are classified as nitrides or azides.

(a) Nitrides

KEY POINT Nitrides are classified as saline, covalent, or interstitial.

The nitrides of metals can be prepared by direct interaction of the element with nitrogen or ammonia or by thermal decomposition of an amide:

$6\,Li(s) + N_2(g) \rightarrow 2\,Li_3N(s)$

$3\,Ca(s) + 2\,NH_3(l) \rightarrow Ca_3N_2(s) + 3\,H_2(g)$

$3\,Zn(NH_2)_2(s) \rightarrow Zn_3N_2(s) + 4\,NH_3(g)$

The compounds of N with H, O, and the halogens are treated separately.

The **saline nitrides** can be regarded as containing the nitride ion, N^{3-}. However, the high negative charge of this ion means that it is highly polarizable (Section 1.7(f)) and even saline nitrides are likely to have considerable covalent character. Saline nitrides occur for lithium, Li_3N, and the Group 2 elements, M_3N_2.

The **covalent nitrides**, in which the E–N bond is covalent, possess a wide range of properties depending on the element to which N is bonded. Some examples of covalent nitrides are boron nitride, BN, cyanogen, $(CN)_2$, tetrasulfur tetranitride, S_4N_4, and disulfur dinitride, S_2N_2. These compounds are discussed in the context of the other elements. Phosphorus nitride, P_3N_5, exists in two polymorphs; α-P_3N_5 exists at pressures up to 6 GPa above which it converts to γ-P_3N_5. PN has been observed in the interstellar medium and in the atmospheres of Jupiter and Saturn.

The largest category of nitrides consists of the **interstitial nitrides** of the d-block elements with formulas MN, M_2N, or M_4N. The N atom occupies some or all of the octahedral sites within the cubic or hexagonal close-packed lattice of metal atoms. The compounds are hard and inert, with a metallic lustre and conductivity. They are widely used as refractory materials and find applications as crucibles, high-temperature reaction vessels, and thermocouple sheaths.

The nitride ion, N^{3-}, is often found as a ligand in d-metal complexes. Its high negative charge, small size, and ability to serve as a good π-donor as well as a σ-donor means that it can stabilize metals in high oxidation states. The short coordinate bond between the ion and the metal atom is often represented as M=N. An example is the complex $[Os(N)(NH_3)_5]^{2+}$ (**14**).

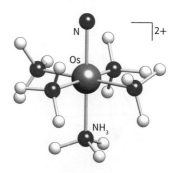

14 $[Os(NH_3)_5N]^{2+}$

(b) Azides and polynitrogen compounds

KEY POINTS Azide salts are used as detonators in explosives. Polynitrogen compounds are very unstable.

Azides, in which nitrogen is present as N_3^-, may be synthesized by the oxidation of sodium amide with either NO_3^- ions or N_2O at elevated temperatures:

$3\,NH_2^- + NO_3^- \xrightarrow{175°C} N_3^- + 3\,OH^- + NH_3$

$2\,NH_2^- + N_2O \xrightarrow{190°C} N_3^- + OH^- + NH_3$

The ion is isoelectronic with both dinitrogen oxide, N_2O, and CO_2, and thus, like these two molecules, is linear. It is a reasonably strong Brønsted base, the pK_a of its conjugate acid, hydrazoic acid, HN_3, being 4.75. It is also a good ligand towards d-block ions. However, heavy-metal complexes or salts, such as $Pb(N_3)_2$ and $Hg(N_3)_2$, are shock-sensitive detonators and decompose to produce the metal and nitrogen:

$$Pb(N_3)_2(s) \rightarrow Pb(s) + 3N_2(g)$$

Ionic azides such as NaN_3 are thermodynamically unstable but kinetically inert; they can be handled at room temperature. Sodium azide is toxic and is used as a chemical preservative and in pest control. When alkali metal azides are heated or detonated by impact they explode, liberating N_2; this reaction is used in the inflation of air bags in cars, in which the heating of the azide is electrical.

A BRIEF ILLUSTRATION

A typical airbag contains approximately 50 g of NaN_3. To estimate the volume of nitrogen produced when the azide is detonated at room temperature and pressure (20°C and 1 atm), we need to consider the amount (in moles) of N_2 molecules produced in the decomposition reaction $2\,NaN_3(s) \rightarrow 2\,Na(s) + 3\,N_2(g)$. Because 50 g of NaN_3 contains 0.77 mol NaN_3, this reaction alone liberates 1.2 mol N_2 (an amount occupying 26 dm^3 at 20°C and 1 atm). The Na produced reacts immediately with KNO_3 to produce even more N_2. As the airbag is restricted in volume, the pressure of nitrogen in the airbag will be high, so providing protection to the driver.

Polynitrogen compounds featuring more than three N atoms, in which the N–N bonding is very weak, were introduced in Section 16.1. Pentazole, HN_5, is very unstable but the cyclic anion has been stabilized. A section of the crystal structure of the salt, formulated as $(H_3O)_3(NH_4)_4(N_5)_6Cl$, is shown (**15**).

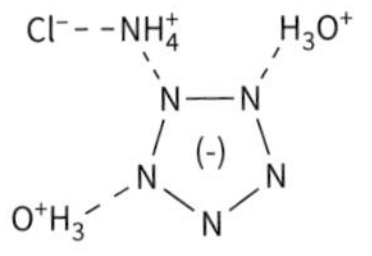

15

Compounds containing the polynitrogen cation, N_5^+ (**16**) have been synthesized from species containing N_3^- and N_2F^+ ions. For example, N_5AsF_6 is prepared from N_2FAsF_6 and HN_3 in anhydrous HF solvent:

$$[N_2F][AsF_6](sol) + HN_3(sol) \rightarrow [N_5][AsF_6](sol) + HF(l)$$

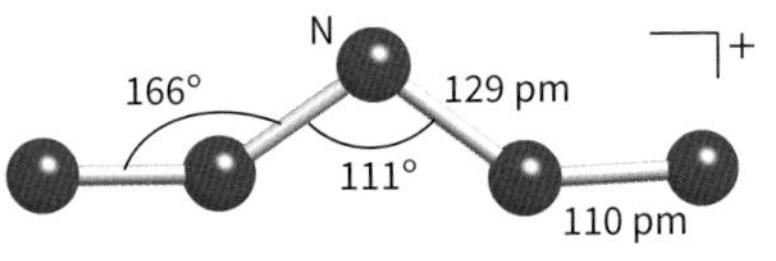

16 N_5^+

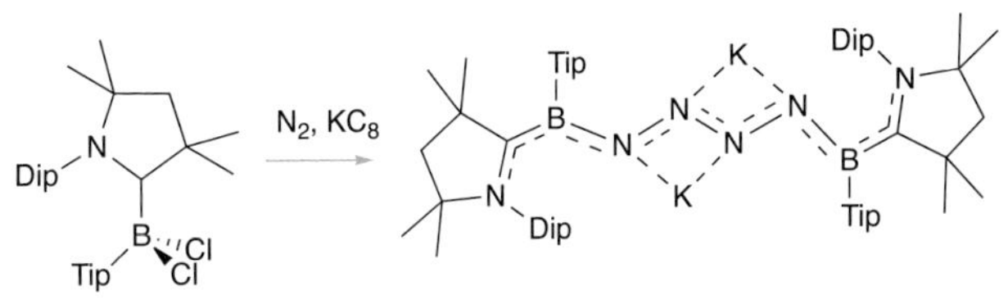

FIGURE 16.5 The borylene mediated coupling of dinitrogen to make a four-nitrogen chain.

The compound is a white solid that decomposes explosively above 60°C. It is a powerful oxidizing agent and ignites organic material even at low temperatures. Salts of dinegative anions can be prepared by metathesis with the salts of mononegative anions in anhydrous HF:

$$2[N_5]SbF_6(sol) + Cs_2SnF_6(sol) \rightarrow [N_5]_2SnF_6(sol) + 2\,CsSbF_6(s)$$

The product is a white solid that is friction-sensitive and decomposes to N_5SnF_6 above 25°C. This product, N_5SnF_6, is stable up to 50°C.

The formation of a four-nitrogen chain directly from dinitrogen has been achieved through reduction of the dihaloborane precursor [(CAAC)BCl$_2$Tip] (Tip, 2,4,6-triisopropylphenyl) with potassium graphite in the presence of nitrogen (Fig. 16.5). Instead of forming end-on bonded N_2 complexes like **12**, the product is a bis-borylene complex of a [N_4K_2] moiety, {[(CAAC)TipB]$_2$(μ_2-N_4K_2)}, which arises from the linking of two N_2 molecules mediated by a transient borylene. Carbene complexes are discussed in Section 24.15.

16.8 Phosphides

KEY POINT Phosphides may be metal-rich or phosphorus-rich.

The compounds of phosphorus with hydrogen, oxygen, and the halogens are discussed separately. The phosphides of other elements can be prepared by heating the appropriate element with red phosphorus in an inert atmosphere:

$$n\,M + m\,P \rightarrow M_nP_m$$

There are many varieties of phosphides, with formulas ranging from M_4P to MP_{15}. They include metal-rich phosphides, in which M:P > 1, monophosphides, in which M:P = 1, and phosphorus-rich phosphides, in which M:P < 1. Metal-rich phosphides are usually very inert, hard, brittle, refractory materials and resemble the parent metal in having high electrical and thermal conductivities. The structures have a trigonal-prismatic arrangement of six, seven, eight, or nine metal ions around a P atom (**17**).

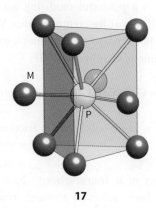

FIGURE 16.6 The preparation and synthetic application of the bis(trichlorosilyl)phosphide anion.

17

Monophosphides adopt a variety of structures depending on the relative size of the other atom. For example, AlP adopts the zinc-blende structure, SnP adopts the rock-salt structure, and VP adopts the nickel-arsenide structure (Section 4.9). Phosphorus-rich phosphides have lower melting points and are less stable than metal-rich phosphides and monophosphides. They are semiconductors rather than conductors.

The synthetic chemistry of phosphorus compounds normally begins with white phosphorus. However, white phosphorus is toxic, pyrophoric, and in relatively short supply. Bis(trichlorosilyl)phosphide can be prepared from the trimetaphosphate salt, itself the product of phosphate condensation (Section 16.16), by reaction with trichlorosilane (Fig. 16.6). The salt can be hydrolysed to form phosphine. It also acts as a nucleophile towards organic electrophiles, allowing the preparation of organophosphorus compounds, without the preparation of white phosphorous or its oxidation with chlorine.

16.9 Arsenides, antimonides, and bismuthides

KEY POINT Indium and gallium arsenides and antimonides are semiconductors.

The compounds formed between metals and arsenic, antimony, and bismuth can be prepared by direct reaction of the elements:

$$Ni(s) + As(s) \rightarrow NiAs(s)$$

The arsenides and antimonides of the Group 13 elements In and Ga are semiconductors. Gallium arsenide (GaAs) is the more important and is used to make devices such as integrated circuits, light-emitting diodes, and laser diodes. Its band gap is similar to that of silicon and larger than those of other Group 13/15 semiconductors (Table 14.5; Section 25.6). Gallium arsenide is superior to Si for such applications because it has higher electron mobility, and the devices produce less electronic noise. Silicon still has major advantages over GaAs in the sense that silicon is cheap, and the wafers are stronger than those of GaAs, so processing is easier. Silicon is also less of an environmental problem than GaAs. Gallium arsenide integrated circuits are commonly used in mobile phones, satellite communications, and some radar systems.

16.10 Hydrides

All the Group 15 elements form binary compounds with hydrogen. All the EH_3 hydrides are toxic. Nitrogen also forms a catenated hydride, hydrazine, N_2H_4.

(a) Ammonia

KEY POINTS Ammonia is produced by the Haber–Bosch process; it is used to manufacture fertilizers and many other useful nitrogen-containing chemicals.

Ammonia is produced in huge quantities worldwide for use as a fertilizer and as a primary source of nitrogen in the production of many chemicals. As already mentioned, the Haber–Bosch process is used for the entire global production. In this process, N_2 and H_2 combine directly at high temperature (450°C) and pressure (100 atm) over a promoted Fe catalyst:

$$N_2(g) + 3H_2(g) \rightleftharpoons 2NH_3$$

The promoters (compounds that enhance the catalyst's activity) include SiO_2, MgO, and other oxides (Section 31.9). The high temperature and catalyst are required to overcome the kinetic inertness of N_2, and the high pressure is needed to overcome the thermodynamic effect of an unfavourable equilibrium constant at the operating temperature.

So novel and great were the chemical and engineering problems arising from the then (early twentieth century) uncharted area of large-scale, high-pressure technology, that two Nobel Prizes were awarded in connection with the process. One went to Fritz Haber (in 1918), who developed the chemical process. The other went to Carl Bosch (in 1931), the chemical engineer who designed the first plants to realize Haber's process. The technology is also referred to as the Haber–Bosch process in recognition of Bosch's contribution. It has had a major impact on civilization because ammonia is the primary source of most nitrogen-containing compounds, including fertilizers and most commercially important compounds of nitrogen. In the early twentieth century there were predictions of widespread starvation across Europe—predictions that were never realized because of the widespread introduction of nitrogen-based fertilizers. Before the development of the Haber–Bosch process the main sources of nitrogen for fertilizers were guano (bird droppings) and saltpetre, which had to be mined and transported from South America.

The boiling point of ammonia is −33°C, which is higher than that of the hydrides of the other elements in the group and indicates the influence of extensive hydrogen bonding. Liquid ammonia is a useful nonaqueous solvent for solutes such as alcohols, amines, ammonium salts, amides, and cyanides. Reactions in liquid ammonia closely resemble those in aqueous solution, as indicated by the following autoprotolysis equilibria:

$2H_2O(l) \rightleftharpoons H_3O^+(aq) + OH^-(aq) \quad pK_w = 14.00$ at $25°C$
$2NH_3(l) \rightleftharpoons NH_4^+(am) + NH_2^-(am) \quad pK_{am} = 34.00$ at $-33°C$

Many of the reactions are analogous to those carried out in water. For example, simple acid–base neutralization reactions can be carried out:

$NH_4Cl(am) + NaNH_2(am) \rightarrow NaCl(am) + 2NH_3(l)$

Ammonia is a water-soluble weak base. The chemical properties of the conjugate acid, ammonium salts are very similar to those of Group 1 salts, especially of K^+ and Rb^+. They are soluble in water and solutions such as NH_4Cl, are acidic because of the equilibrium

$NH_4^+(aq) + H_2O(l) \rightleftharpoons H_3O^+(aq) + NH_3(aq) \quad pK_a = 9.25$

Ammonium salts decompose readily on heating and, for many salts, such as the halides, carbonate, and sulfate, ammonia is evolved:

$NH_4Cl(s) \rightarrow NH_3(g) + HCl(g)$
$(NH_4)_2SO_4(s) \rightarrow 2NH_3(g) + H_2SO_4(l)$

When the anion is oxidizing, as in the case of NO_3^-, ClO_4^-, and $Cr_2O_7^{2-}$, the NH_4^+ is oxidized to N_2 or N_2O:

$NH_4NO_3(s) \rightarrow N_2O(g) + 2H_2O(g)$

When ammonium nitrate is heated strongly or detonated, the decomposition of 2 mol $NH_4NO_3(s)$ in the reaction

$2NH_4NO_3(s) \rightarrow 2N_2(g) + O_2(g) + 4H_2O(g)$

produces 7 mol of gaseous molecules, corresponding to an increase in volume from about 200 cm³ to about 140 dm³, a factor of 700. This feature leads to the use of ammonium nitrate as an explosive, and nitrate fertilizers are often mixed with materials such as calcium carbonate or ammonium sulfate to make them more stable. Ammonium sulfate and the ammonium hydrogenphosphates, $NH_4H_2PO_4$ and $(NH_4)_2HPO_4$, are also used as fertilizers because phosphate is a plant nutrient. Ammonium perchlorate is used as the oxidizing agent in solid-fuel rocket propellants.

EXAMPLE 16.4 Explaining bond angles of the hydrides

The bond angles of the hydrides of the Group 15 elements are: NH_3, 107.8°; PH_3, 93.6°; AsH_3, 91.8°; SbH_3, 91.3°. Use valence bond considerations to explain this trend.

Answer Valence bond theory (Section 2.6) explains the bonding in NH_3 in terms of the formation of a set of four sp³ hybrid orbitals, giving rise to the tetrahedral molecule and the bond angle close to that of the perfect tetrahedron of 109°. The bond angle is slightly smaller than 109° because of the presence of the lone pair on the N atom. Hybridization is reliant on efficient mixing of s and p orbitals on the central atom. For N the s and p orbitals are close in energy. However, as we descend the group, the energy gap between s and p orbitals increases and the mixing is less efficient and bond angles deviate from the ideal tetrahedral angle.

Self-test 16.4 (a) The basicity of the Group 15 hydrides decreases down the group; $NH_3 > PH_3 > AsH_3 > SbH_3 > BH_3$. Explain this trend. (b) Explain why NF_3 is stable whereas NCl_3 decomposes explosively.

(b) Hydrazine and hydroxylamine

KEY POINT Hydrazine is a weaker base than ammonia and forms two series of salts.

Hydrazine, N_2H_4, is a fuming, colourless liquid with an odour like that of ammonia. It has a liquid range similar to that of water (2–114°C), and like water there are two hydrogen bond acceptors and two donors per molecule. In the liquid phase, hydrazine adopts a *gauche* conformation around the N–N bond (**18**).

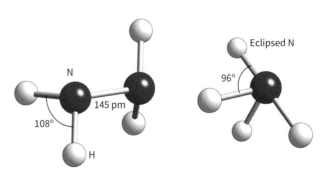

18 Hydrazine, N_2H_4

Hydrazine is manufactured by the Raschig process, in which ammonia and sodium hypochlorite react in dilute aqueous solution. The reaction proceeds through several steps, which can be simplified to

$$NH_3(aq) + NaOCl(aq) \rightarrow NH_2Cl(aq) + NaOH(aq)$$
$$2\,NH_3(aq) + NH_2Cl(aq) \rightarrow N_2H_4(aq) + NH_4Cl(aq)$$

There is a competing side-reaction that is catalysed by d-metal ions:

$$N_2H_4(aq) + 2\,NH_2Cl(aq) \rightarrow N_2(g) + 2\,NH_4Cl(aq)$$

Gelatine is therefore added to the reaction mixture to form a complex with the d-metal ions and remove them. The resulting dilute aqueous solution of hydrazine is converted to a concentrated solution of hydrazine hydrate, $N_2H_4 \cdot H_2O$, by distillation. This product is often preferred commercially as it is cheaper than hydrazine and has a wider liquid range. Hydrazine is produced by distillation of the hydrate in the presence of a drying agent such as solid NaOH or KOH.

Hydrazine is a much weaker base than ammonia:

$$N_2H_4(aq) + H_2O(l) \rightleftharpoons N_2H_5^+(aq) + OH^-(aq) \quad pk_{a1} = 6.07$$
$$N_2H_5^+(aq) + H_2O(l) \rightleftharpoons N_2H_6^{2+}(aq) + OH^-(aq) \quad pk_{a2} = -1.05$$

It reacts with acids, HX, to form two series of salts, N_2H_5X and $N_2H_6X_2$.

The major use of hydrazine and its methyl derivatives, CH_3NHNH_2 and $(CH_3)_2NNH_2$, is as a rocket fuel. Hydrazine is also used as a foam-blowing agent and as a treatment in boiler water to scavenge dissolved oxygen and prevent oxidation of pipes. Both N_2H_4 and $N_2H_5^+$ are reducing agents and are used in the recovery of precious metals.

EXAMPLE 16.5 Evaluating rocket fuels

Hydrazine, N_2H_4, and dimethylhydrazine, $N_2H_2(CH_3)_2$, are used as rocket fuels. Given the following data, suggest which would be the more efficient fuel in terms of its thermodynamic properties:

	$\Delta_f H^\ominus$ /kJ mol^{-1}
$N_2H_4(l)$	+50.6
$N_2H_2(CH_3)_2(l)$	+42.0
$CO_2(g)$	−394
$H_2O(g)$	−242

Answer We must evaluate which combustion reactions release the most heat by calculating their (standard) enthalpies of combustion. The combustion reactions are

$$N_2H_4(l) + O_2(g) \rightarrow N_2(g) + 2H_2O(g)$$
$$N_2H_2(CH_3)_2(l) + O_2(g) \rightarrow N_2(g) + 4H_2O(g) + 2CO_2(g)$$

The enthalpy of reaction (in this case, combustion) is calculated from

$$\Delta_c H^\ominus = \sum_{products} \Delta_f H^\ominus - \sum_{reactants} \Delta_f H^\ominus$$

We find −535 kJ mol^{-1} for N_2H_4 and −1798 kJ mol^{-1} for $N_2H_2(CH_3)_2$. An important factor for selecting rocket fuels is the specific enthalpy (the enthalpy of combustion divided by the mass of fuel), which for these fuels has values −16.7 and −29.9 kJ g^{-1}, respectively, indicating that $N_2H_2(CH_3)_2$ is the better fuel even when mass is significant.

Self-test 16.5 Refined hydrocarbons and liquid hydrogen are also used as rocket fuels. What are the advantages of dimethylhydrazine over these fuels?

Hydroxylamine, NH_2OH (**19**), is a colourless, hygroscopic solid with a low melting point (32°C). It is usually available as one of its salts or in aqueous solution. It is a weaker base than either ammonia or hydrazine:

$$NH_2OH(aq) + H_2O(l) \rightleftharpoons NH_3OH^+(aq) + OH^-(aq)$$
$$pK_a = 5.97$$

19 Hydroxylamine, NH_2OH

Anhydrous hydroxylamine can be prepared by adding sodium butoxide, NaC_4H_9O (NaOBu), to a solution of hydroxylamine hydrochloride in 1-butanol. The NaCl produced is filtered off and the hydroxylamine precipitated by the addition of ether:

$$[NH_3OH]Cl(sol) + NaOBu \rightarrow NH_2OH(sol) + NaCl(s)$$
$$+ BuOH(l)$$

The major commercial use of hydroxylamine is in the synthesis of caprolactam, which is an intermediate in the manufacture of nylon.

(c) Phosphine, arsane, and stibane

KEY POINTS Unlike liquid ammonia, liquid phosphine, arsane, and stibane do not associate through hydrogen bonding; their much more stable alkyl and aryl analogues are useful soft ligands.

In contrast to the commanding role that ammonia plays in nitrogen chemistry, the highly poisonous hydrides of the heavier nonmetallic elements of Group 15 (particularly phosphine, PH_3, and arsane, AsH_3) are of minor importance in the chemistry of their respective elements. Both phosphine and arsane are used in the semiconductor industry to dope Si or to prepare other semiconductor compounds, such as GaAs, by chemical vapour deposition (Section 28.4). These thermal decomposition reactions reflect the positive Gibbs energy of formation of these hydrides.

The commercial synthesis of PH$_3$ uses the disproportionation of white phosphorus in basic solution:

$$P_4(s) + 3OH^-(aq) + 3H_2O(l) \rightarrow PH_3(g) + 3H_2PO_2^-(aq)$$

Arsane and stibane may be prepared by the protolysis of compounds that contain an electropositive metal in combination with arsenic or antimony:

$$Zn_3E_2(s) + 6H_3O^+(l) \rightarrow 2EH_3(g) + 3Zn^{2+}(aq) + 6H_2O(l)$$
$$(E = As, Sb)$$

Phosphine and arsane are poisonous gases that readily ignite in air, but the much more stable organic derivatives PR$_3$ and AsR$_3$ (R = alkyl or aryl groups) are widely used as ligands in metal coordination chemistry. In contrast with the hard-donor properties of ammonia and alkylamine ligands, the organophosphines and organoarsanes, such as P(C$_2$H$_5$)$_3$ and As(C$_6$H$_5$)$_3$, are soft ligands and are therefore often incorporated into metal complexes having central metal atoms in low oxidation states. The stability of these complexes correlates with the soft-acceptor nature of the metals in low oxidation states, and the stability of soft-donor–soft-acceptor combinations (Section 5.10).

All the Group 15 hydrides are pyramidal, but the bond angle decreases down the group:

NH$_3$ 107.8°   PH$_3$ 93.6°   AsH$_3$ 91.8°   SbH$_3$ 91.3°

The large change in bond angle was attributed in Example 16.4 to a decrease in the extent of sp^3 hybridization from NH$_3$ to SbH$_3$, but steric effects could also play a part. The E–H bonding pairs of electrons will repel each other. This repulsion is greatest when the central element, E, is small, as in NH$_3$, and the H atoms will be as far away from each other as possible in a near-tetrahedral arrangement. As the size of the central atom is increased down the series, the repulsion between bonding pairs decreases and the bond angle is close to 90°.

It is evident from the boiling points plotted in Fig. 11.7 that PH$_3$, AsH$_3$, and SbH$_3$ are subject to little, if any, hydrogen bonding with themselves; however, PH$_3$ and AsH$_3$ can be protonated by strong acids, such as HI, to form phosphonium and arsonium ions, PH$_4^+$ and AsH$_4^+$, respectively.

16.11 Halides

All the elements form a trihalide with at least one halogen. Phosphorus, arsenic, and antimony form stable pentahalides.

(a) Nitrogen halides

KEY POINT Except for NF$_3$ which is stable, nitrogen trihalides have limited stability and are explosive.

Nitrogen trifluoride, NF$_3$, is the only exergonic binary halogen compound of nitrogen. This pyramidal molecule is not very reactive. Thus, unlike NH$_3$, it is only a very weak Lewis base because the strongly electronegative F atoms render the lone pair of electrons much less available: whereas the polarity of the N–H bond in NH$_3$ is $^{\delta-}$N–H$^{\delta+}$, that of the N–F bond in NF$_3$ is $^{\delta+}$N–F$^{\delta-}$. Nitrogen trifluoride can be converted into the N(V) species NF$_4^+$ by the reaction

$$NF_3(l) + 2F_2(g) + SbF_3(l) \rightarrow [NF_4^+][SbF_6^-](sol)$$

Nitrogen trichloride, NCl$_3$, is a highly endergonic, explosive, yellow oil. It is prepared commercially by the electrolysis of an aqueous solution of ammonium chloride and was once used as an oxidizing bleach for flour. The electronegativities of nitrogen and chlorine are similar, and the N–Cl bond is not very polar. Nitrogen tribromide, NBr$_3$, is an explosive, deep-red oil. Nitrogen triiodide, NI$_3$, is an explosive solid. Nitrogen is more electronegative than both bromine and iodine, and so the N–X bonds are polar in the sense $^{\delta-}$N–X$^{\delta+}$ and, formally, the oxidation numbers are −3 for the N and +1 for each halogen.

(b) Halides of the heavy elements

KEY POINTS Whereas the halides of nitrogen have limited stability, their heavier congeners form an extensive series of compounds; the trihalides and pentahalides are useful starting materials for the synthesis of derivatives by metathetical replacement of the halide.

The trihalides and pentahalides of Group 15 elements other than nitrogen are used extensively in synthetic chemistry, and their simple empirical formulas conceal an interesting and varied structural chemistry.

The trihalides range from gases and volatile liquids, such as PF$_3$ (b.p. −102°C) and AsF$_3$ (b.p. 63°C), to solids, such as BiF$_3$ (m.p. 649°C). A common method of preparation is direct reaction of the element and halogen. For phosphorus, the trifluoride is prepared by metathesis of the trichloride and a metal fluoride:

$$2PCl_3(l) + 3ZnF_2(s) \rightarrow 2PF_3(g) + 3ZnCl_2(s)$$

The trichlorides PCl$_3$, AsCl$_3$, and SbCl$_3$ are useful starting materials for the preparation of a variety of alkyl, aryl, alkoxy, and amino derivatives because they are susceptible to protolysis and metathesis:

$$ECl_3(sol) + 3EtOH(l) \rightarrow$$
$$E(OEt)_3(sol) + 3HCl(sol)(E = P, As, Sb)$$
$$ECl_3(sol) + 6Me_2NH(sol) \rightarrow$$
$$E(NMe_2)_3(sol) + 3[Me_2NH_2]Cl(sol)(E = P, As, Sb)$$

Phosphorus trifluoride, PF$_3$, is an interesting ligand because in some respects it resembles CO. Like CO, it is a weak σ donor but a strong π acceptor, and complexes of PF$_3$, such as [Ni(PF$_3$)$_4$], are analogous to [Ni(CO)$_4$] (Section 24.6). The π acceptor character is attributed to a P–F antibonding

LUMO, which has mainly P p-orbital character. The trihalides also act as mild Lewis acids towards Lewis bases such as trialkylamines and halides. Many halide complexes have been isolated, such as the simple mononuclear species $AsCl_4^-$ (20) and SbF_5^{2-} (21). More complex dinuclear and polynuclear anions linked by halide bridges, such as the polymeric chain $([BiBr_3]^{2-})_n$ in which Bi(I) is surrounded by a distorted octahedron of Br atoms, are also known.

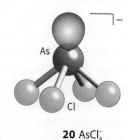

20 $AsCl_4^-$

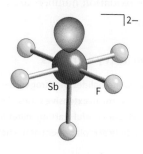

21 SbF_5^{2-}

The pentahalides vary from gases, such as PF_5 (b.p. −85°C) and AsF_5 (b.p. −53°C), to solids, such as PCl_5 (sublimes at 162°C) and BiF_5 (m.p. 154°C). The five-coordinate gas-phase molecules are trigonal-bipyramidal. In contrast to PF_5 and AsF_5, SbF_5 is a highly viscous liquid in which the molecules are associated through F-atom bridges. In solid SbF_5 these bridges result in a cyclic tetramer (22), which reflects the tendency of Sb(V) to achieve a coordination number of 6.

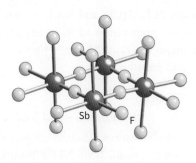

22 $(SbF_5)_4$

A related phenomenon occurs with PCl_5, which in the solid state exists as $[PCl_4^+][PCl_6^-]$. In this case, the ionic contribution to the lattice enthalpy provides the driving force for the transfer of a Cl^- ion from one PCl_5 molecule to

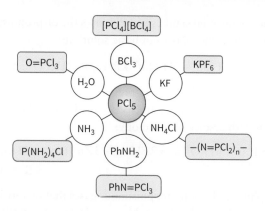

FIGURE 16.7 The uses of phosphorus pentachloride.

another. The pentafluorides of P, As, Sb, and Bi are strong Lewis acids (Section 5.6). SbF_5 is a very strong Lewis acid; it is much stronger, for example, than the aluminium halides. When SbF_5 or AsF_5 is added to anhydrous HF, a superacid is formed (see Section 5.17):

$$SbF_5(l) + 2HF(l) \rightarrow H_2F^+(sol) + SbF_6^-(sol)$$

Of the pentachlorides, PCl_5 and $SbCl_5$ are stable, whereas $AsCl_5$ is very unstable. This difference is a manifestation of the alternation effect (Section 10.4). The pentahalides of P and Sb are very useful in syntheses. Phosphorus pentachloride, PCl_5, is widely used in the laboratory and in industry as a starting material, and some of its characteristic reactions are shown in Fig. 16.7. Note, for example, that reaction of PCl_5 with Lewis acids yields PCl_4^+ salts, and simple Lewis bases like F^- produce six-coordinate complexes such as PF_6^-.[2] Compounds containing the NH_2 group lead to the formation of PN bonds, and the interaction of PCl_5 with either H_2O or P_4O_{10} yields $O=PCl_3$.

16.12 Oxohalides

KEY POINTS Nitrosyl and nitryl halides are useful halogenating agents; phosphoryl halides are important industrially in the synthesis of organophosphorus derivatives.

Nitrogen forms all the nitrosyl halides, NOX, and the nitryl halides, NO_2X. The nitrosyl halides and NO_2F are prepared by direct interaction of the halogen with NO or NO_2, respectively:

$$2NO(g) + Cl_2(g) \rightarrow 2NOCl(g)$$
$$2NO_2(g) + F_2(g) \rightarrow 2NO_2F(g)$$

They are all reactive gases and the oxofluorides and oxochlorides are useful fluorinating and chlorinating agents. NOF_3 is

---

[2] Not only is a Lewis adduct formed, but all the chlorides are exchanged for fluorides.

also known. It has a short N=O bond (116 pm) and long N–F bond (143 pm) suggesting the [O=NF$_2$]$^+$F$^-$ structure.

Phosphorus readily forms the phosphoryl halides POCl$_3$ and POBr$_3$ by the reaction of the trihalides PX$_3$ with O$_2$ at room temperature. The fluorine and iodine analogues are prepared by the reaction of POCl$_3$ with a metal fluoride or iodide:

$$POCl_3(l) + 3\,NaF \rightarrow POF_3(g) + 3\,NaCl(s)$$

All the molecules are tetrahedral and contain a P=O bond (the symmetry being C$_{3v}$): POF$_3$ is gaseous, POCl$_3$ is a colourless liquid, POBr$_3$ is a brown solid, and POI$_3$ is a violet solid. They are all readily hydrolysed, fume in air, and form adducts with Lewis acids. They provide a route to the synthesis of organophosphorus compounds, which are manufactured on a large scale for use as plasticizers, oil additives, pesticides, and surfactants. For example, reaction with alcohols and phenols gives (RO)$_3$PO and Grignard reagents (Section 13.13) yield R$_n$POCl$_{3-n}$:

$$3\,ROH(l) + POCl_3(l) \rightarrow (RO)_3PO(sol) + 3\,HCl(sol)$$
$$n\,RMgBr(sol) + POCl_3(sol) \rightarrow R_nPOCl_{3-n}(sol) + n\,MgBrCl(s)$$

16.13 Oxides and oxoanions of nitrogen

KEY POINT Reactions of nitrogen–oxygen compounds that liberate or consume N$_2$ are generally very slow at normal temperatures and pH = 7.

We can infer the redox properties of the compounds of the elements in Group 15 in acidic aqueous solution from the Frost diagram in Fig. 16.8(a). The steepness of the slopes of the lines on the far right of the diagram show the thermodynamic tendency for reduction of the E(V) oxidation states of the elements. They show, for instance, that Bi$_2$O$_5$ is potentially a very strong oxidizing agent, which is consistent with the inert pair effect and the instability of Bi(V) compared to Bi(III) (Section 10.7). The next-strongest oxidizing agent is NO$_3^-$. Both As(V) and Sb(V) are milder oxidizing agents, and P(V), in the form of phosphoric acid, is a very weak oxidant. The higher oxidation states are stabilized under basic conditions (Fig. 16.8(b)).

The redox properties of nitrogen are important because of its widespread occurrence in the atmosphere, the biosphere, industry, and the laboratory. Nitrogen chemistry is quite complex, partly because of the large number of accessible oxidation states but also because reactions that are thermodynamically favourable are often slow or have rates that depend crucially on the identity of the reactants. As the N$_2$ molecule is kinetically inert, redox reactions that consume N$_2$ are slow. Moreover, the formation of N$_2$ is often slow and may be sidestepped in aqueous solution (Fig. 16.9). As with several other p-block elements, the activation barriers to reaction of high-oxidation-state oxoanions, such as NO$_3^-$, are greater than for low-oxidation-state oxoanions, such as NO$_2^-$.

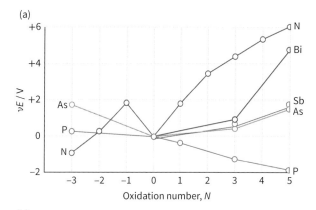

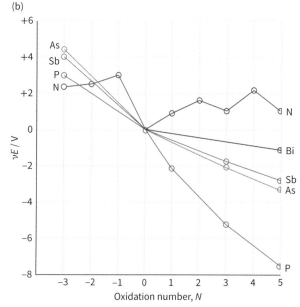

FIGURE 16.8 Frost diagram for the elements of the nitrogen group in (a) acidic solution and (b) basic solution.

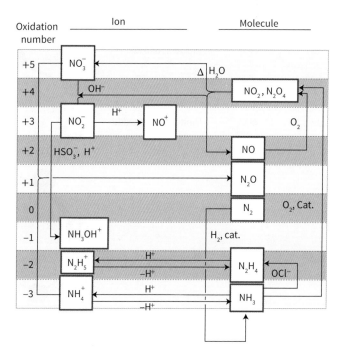

FIGURE 16.9 The interconversion of important nitrogen species.

In addition to destabilizing high oxidations states, low pH also often accelerates their oxidizing reactions by protonation, and this step may facilitate subsequent NO bond breaking.

Table 16.2 summarizes some of the properties of the nitrogen oxides, and Table 16.3 does the same for the nitrogen oxoanions. Both tables will help us to navigate through the details of their properties.

(a) Nitrogen(V) oxides and oxoanions

KEY POINTS The nitrate ion is a strong but slow oxidizing agent at room temperature; strong acid and heating are required to accelerate the reaction.

The most common source of N(V) is nitric acid, HNO_3, which is a major industrial chemical used in the production of fertilizers, explosives, and a wide variety of nitrogen-containing chemicals. It is produced by modern versions of the Ostwald process, which make use of an indirect route from N_2 to the highly oxidized compound HNO_3 via the fully reduced compound NH_3. Thus, after nitrogen has been reduced to the −3 state as NH_3 by the Haber–Bosch process, it is oxidized to the +4 state:

$$4\,NH_3(g) + 7\,O_2(g) \rightarrow 6\,H_2O(g) + 4\,NO_2(g)$$
$$\Delta_r G^\ominus = -308.0\,kJ(mol\,NO_2)^{-1}$$

The NO_2 then undergoes disproportionation into N(II) and N(V) in water at elevated temperatures:

$$3\,NO_2(aq) + H_2O(l) \rightarrow 2\,HNO_3(aq) + NO(g)$$
$$\Delta_r G^\ominus = -5.0\,kJ(mol\,HNO_3)^{-1}$$

TABLE 16.2 Oxides of nitrogen

Oxidation number	Formula	Name	Structure (gas phase)	Comments
+1	N_2O	Nitrous oxide (dinitrogen oxide)	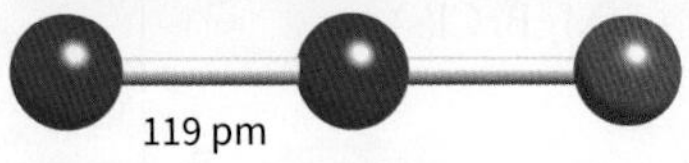 119 pm	Colourless gas, not very reactive
+2	NO	Nitric oxide (nitrogen monoxide)	 115 pm	Colourless, reactive, paramagnetic gas
+3	N_2O_3	Dinitrogen trioxide	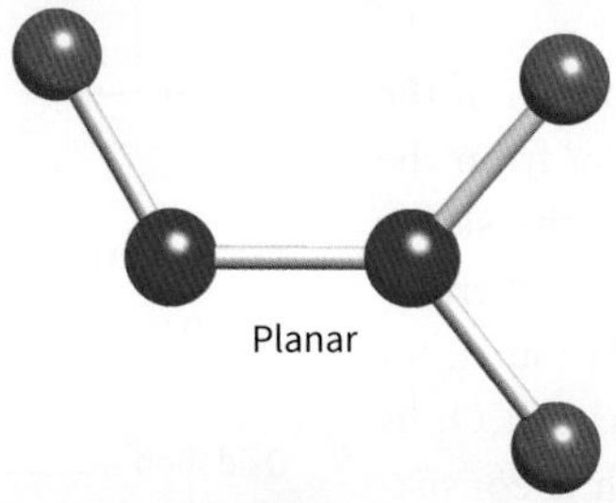 Planar	Blue liquid (m.p. 101°C); dissociates into NO and NO_2 in the gas phase
+4	NO_2	Nitrogen dioxide	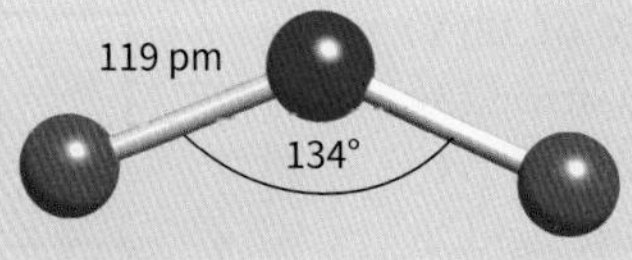 119 pm, 134°	Brown, reactive, paramagnetic gas
+4	N_2O_4	Dinitrogen tetroxide	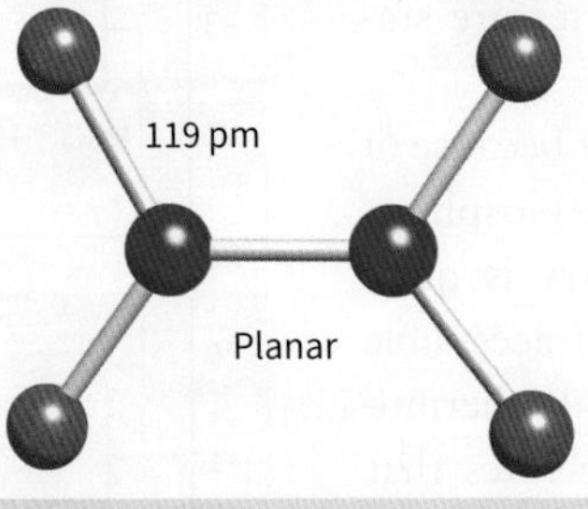 119 pm, Planar	Colourless liquid (m.p. −11°C); in equilibrium with NO_2 in the gas phase
+5	N_2O_5	Dinitrogen pentoxide	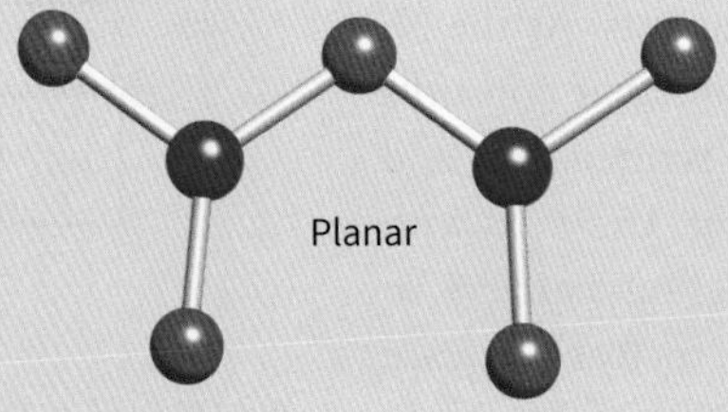 Planar	Colourless, unstable, crystallizes as ionic solid $[NO_2][NO_3]$

TABLE 16.3 Nitrogen–oxygen ions

Oxidation number	Formula	Common name	Structure	Comments
+1	$N_2O_2^{2-}$	Hyponitrite	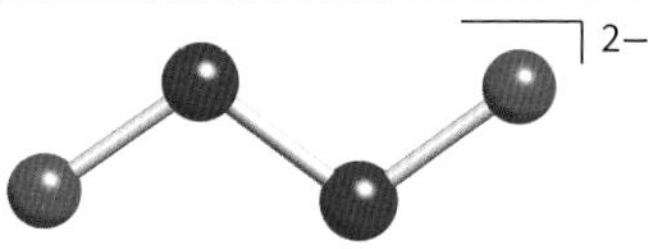	Usually acts as a reducing agent
+3	NO_2^-	Nitrite	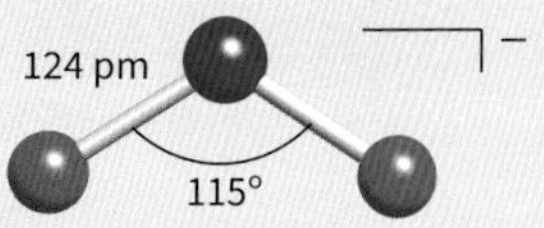 124 pm, 115°	Weak base; acts as an oxidizing agent and a reducing agent
+3	NO^+	Nitrosonium (nitrosyl cation)	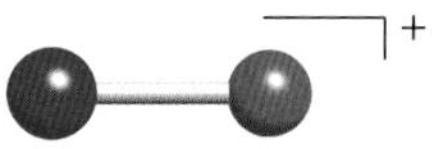	Oxidizing agent and Lewis acid; π-acceptor ligand
+5	NO_3^-	Nitrate	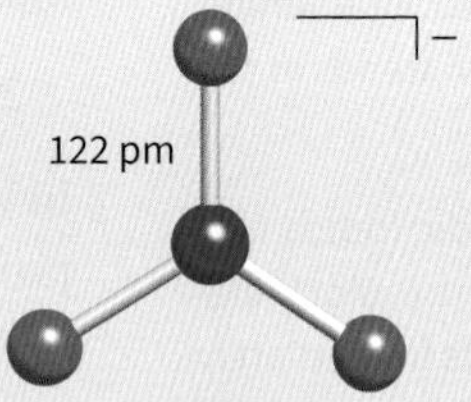 122 pm	Very weak base; an oxidizing agent
+5	NO_2^+	Nitronium (nitryl cation)	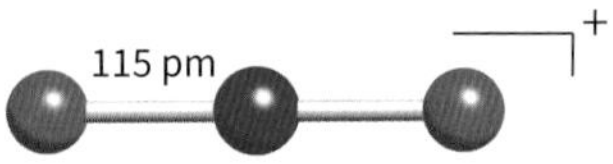 115 pm	Oxidizing agent, nitrating agent, Lewis acid

All the steps are thermodynamically favourable. The by-product NO is oxidized with O_2 to NO_2 and recirculated. Such an indirect route is used because the direct oxidation of N_2 to NO_2 is thermodynamically unfavourable, with $\Delta_r G^\ominus$ (NO_2, g) = +51 kJ mol^{-1}. This endergonic character is largely due to the great strength of the N≡N bond (950 kJ mol^{-1}).

Standard potential data imply that the NO_3^- ion is a moderately strong oxidizing agent. However, its reactions are generally slow in dilute acid solution. Because protonation of an O atom promotes NO bond breaking, concentrated HNO_3 (in which NO_3^- is protonated) undergoes more rapid reactions than the dilute acid (in which HNO_3 is fully deprotonated). It is also a thermodynamically more potent oxidizing agent at low pH. A sign of this oxidizing character is the yellow colour of the concentrated acid, which indicates its instability with respect to decomposition into NO_2:

$$4\,HNO_3(aq) \rightarrow 4\,NO_2(aq) + O_2(g) + 2\,H_2O(l)$$

This decomposition is accelerated by light and heat.

The reduction of NO_3^- ions rarely yields a single product, as so many lower oxidation states of nitrogen are available. For example, a strong reducing agent such as zinc can reduce a substantial proportion of dilute HNO_3 as far as the 3− oxidation state:

$$HNO_3(aq) + 4\,Zn(s) + 9\,H^+(aq) \rightarrow$$
$$NH_4^+(aq) + 3\,H_2O(l) + 4\,Zn^{2+}(aq)$$

A weaker reducing agent, such as copper, proceeds only as far as the 4+ oxidation state in the concentrated acid:

$$2\,HNO_3(aq) + Cu(s) + 2\,H^+(aq) \rightarrow$$
$$2\,NO_2(g) + Cu^{2+}(aq) + 2\,H_2O(l)$$

With the dilute acid, the N(II) oxidation state is favoured, and NO is formed:

$$2\,NO_3^-(aq) + 3\,Cu(s) + 8\,H^+(aq) \rightarrow$$
$$2\,NO(g) + 3\,Cu^{2+}(aq) + 4\,H_2O(l)$$

Aqua regia is a mixture of concentrated nitric acid and concentrated hydrochloric acid: it is yellow because of the presence of the decomposition products NOCl and Cl_2, and loses its potency as these volatile products are formed:

$$HNO_3(aq) + 3\,HCl(aq) \rightarrow NOCl(g) + 2\,H_2O(l)$$

'*Aqua regia*' is Latin for 'royal water' and was so called by alchemists because of its ability to dissolve the noble metals gold and platinum. Gold will dissolve to a very small extent in concentrated nitric acid. In *aqua regia* the Cl$^-$ ions present react immediately with the Au^{3+} ions formed to produce $[AuCl_4]^-$ and move the equilibrium to the right:

$$Au(s) + NO_3^- + 4\,Cl^-(aq) + 4\,H^+(aq) \rightarrow$$
$$[AuCl_4]^-(aq) + NO(g) + 2\,H_2O(l)$$

The anhydride of nitric acid is N_2O_5. It is an ionic crystalline solid more accurately written with the formula

[NO$_2^+$][NO$_3^-$] and can be prepared by dehydration of nitric acid with P$_4$O$_{10}$:

$$4\,HNO_3(l) + P_4O_{10}(s) \rightarrow N_2O_5(s) + 4\,HPO_3(l)^3$$

The solid sublimes at 320°C and the gaseous molecules dissociate to produce NO$_2$ and O$_2$. The compound is a strong oxidizing agent and can be used for the synthesis of anhydrous nitrates:

$$N_2O_5(s) + Na(s) \rightarrow NaNO_3(s) + NO_2(g)$$

EXAMPLE 16.6 **Explaining the relative stabilities of E(V)**

Compounds of N(V), As(V), and Bi(V) are stronger oxidizing agents than the E(V) oxidation states of the two intervening elements, P and Sb. Correlate this observation with trends in the periodic table.

Answer We need to consider some of the periodic trends and themes discussed in Chapters 9 and 10. The light p-block elements are more electronegative than the elements immediately below them in the periodic table; accordingly, these light elements are generally less easily oxidized themselves and are thus good oxidizing agents. Nitrogen is generally a good oxidizing agent in its positive oxidation states. As(V) compounds are much less stable than those of P and Sb on account of the alternation effect (Section 10.4), which arises from the increased Z_{eff} due to poor shielding of the 3d electrons. Bismuth is much less electronegative but favours the M(III) oxidation state in preference to the +5 state—a reflection of the inert pair effect (Section 10.7).

Self-test 16.6 From trends in the periodic table, decide whether phosphorus (V) or sulfur (VI) is likely to be the stronger oxidizing agent.

(b) Nitrogen(IV) and nitrogen(III) oxides and oxoanions

KEY POINT The intermediate oxidation states of nitrogen are often susceptible to disproportionation.

Nitrogen(IV) oxide exists as an equilibrium mixture of the brown NO$_2$ radical and its colourless dimer, N$_2$O$_4$ (dinitrogen tetroxide):

$$N_2O_4(g) \rightleftharpoons 2\,NO_2(g) \quad K = 0.115 \text{ at } 25°C$$

This readiness to dissociate is consistent with the N–N bond in N$_2$O$_4$ (23) being long and weak, and arises because the molecular orbital occupied by the unpaired electron is spread almost equally over all three atoms in NO$_2$ rather than being concentrated on the N atom. This structure is in contrast to the isoelectronic oxalate ion, C$_2$O$_4^{2-}$, where the C–C bond is stronger because in -CO$_2^-$ the electron is more concentrated on the C atom (reflecting the larger gap in electronegativity between carbon and oxygen).

³ HPO$_3$ is a metaphosphoric acid ((27) and (28)).

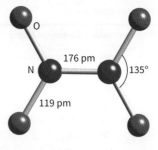

23 N$_2$O$_4$, D_{2h}

Nitrogen(IV) oxide is a poisonous oxidizing agent that is present in low concentrations in the atmosphere, especially in photochemical smog. In basic aqueous solution it disproportionates into N(III) and N(V), forming NO$_2^-$ and NO$_3^-$ ions (Fig. 16.8):

$$2\,NO_2(aq) + 2\,OH^-(aq) \rightarrow NO_2^-(aq) + NO_3^-(aq) + H_2O(l)$$

In acidic solution (as in the Ostwald process) the reaction product is N(II) in place of N(III) because nitrous acid readily disproportionates:

$$3\,HNO_2(aq) \rightleftharpoons NO_3^-(aq) + 2\,NO(g) + H_3O^+(aq)$$
$$E^\ominus = +0.05\,V, K = 50$$

Nitrous acid, HNO$_2$, is a strong oxidizing agent:

$$HNO_2(aq) + H^+(aq) + e^- \rightarrow NO(g) + H_2O(l) \quad E^\ominus = +1.00\,V$$

and its reactions as an oxidizing agent are often more rapid than its disproportionation (Box 16.3).

The reaction kinetics are greatly enhanced under acid conditions as a result of its conversion to the nitrosonium ion, NO$^+$:

$$HNO_2(aq) + H^+(aq) \rightarrow H_2NO_2^+(aq) \rightarrow NO^+(aq) + H_2O(l)$$

The nitrosonium ion is a strong Lewis acid and forms complexes rapidly with anions and other Lewis bases. The resulting species may not themselves be susceptible to oxidation (as in the case of SO$_4^{2-}$ and F$^-$ ions, which form [O$_3$SONO]$^-$ (24) and ONF (25), respectively).

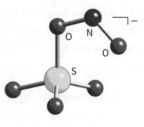

24 O$_3$SONO$^-$

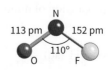

25 ONF

BOX 16.3 What is the role of nitrite in curing meat?

For centuries, meat has been preserved by the use of common salt, which dehydrates the meat, removing moisture essential for the growth of bacteria. One by-product of this process was that some meats took on a red appearance and distinctive taste. This was found to be due to traces of sodium nitrate in the salt, which is reduced to nitrite by bacterial action during the processing. Today, sodium nitrite is used in curing of meats such as bacon, hams, and sausage.

The nitrite delays the onset of botulism, retards the development of rancidity, and preserves the flavours of spices. The nitrite is converted to nitric oxide which binds to myoglobin, the O_2-binding protein responsible for the natural red colour of uncured meat (Section 26.7). The myoglobin–nitric oxide complex is deep red, which results in the bright pink hue typical of cured meats. A reaction of nitrite with myoglobin is also responsible for the green tinge which is sometimes seen in bacon. This is known as 'nitrite burn' and occurs when the haem group in myoglobin is nitrated by the nitrite.

There is good experimental evidence that the reaction of HNO_2 with I^- ions leads to the rapid formation of INO:

$$I^-(aq) + NO^+(aq) \rightarrow INO(aq)$$

followed by the rate-determining second-order reaction between two INO molecules:

$$2\,INO(aq) \rightarrow I_2(aq) + 2\,NO(g)$$

Nitrosonium salts containing poorly coordinating anions, such as $[NO][BF_4]$, are useful reagents in the laboratory as facile oxidizing agents and as a source of NO^+.

Dinitrogen trioxide, N_2O_3, the anhydride of nitrous acid (Table 16.2), is a blue solid that melts above −100°C to produce a blue liquid that dissociates to give NO and NO_2:

$$N_2O_3(l) \rightarrow NO(g) + NO_2(g)$$

The yellow-brown colour of NO_2 means that the liquid becomes progressively more green as the dissociation proceeds.

(c) Nitrogen(II) oxide

KEY POINTS Nitric oxide is a strong π-acceptor ligand and a troublesome pollutant in urban atmospheres; the molecule acts as a neurotransmitter.

Nitric oxide reacts with O_2 to generate NO_2, but in the gas phase the rate law is second-order in NO because a transient dimer, $(NO)_2$, is produced that subsequently reacts with an O_2 molecule. Because the reaction is second-order, atmospheric NO (which is produced in low concentrations by coal-fired power plants and by internal combustion engines) is slow to convert to NO_2.

Since NO is endergonic, it should be possible to find a catalyst to convert the pollutant NO to the natural atmospheric gases N_2 and O_2 before emission. Three classes of catalyst have been developed: (i) metal oxides (including lanthanide oxides), (ii) supported metal oxides (including cobalt oxide), and (iii) copper zeolites (Section 31.13).

(d) Low-oxidation-state nitrogen–oxygen compounds

KEY POINT Dinitrogen oxide is unreactive, for kinetic reasons.

Dinitrogen oxide, N_2O (commonly referred to as 'nitrous oxide'), is a colourless, unreactive gas, and is produced by the comproportionation of molten ammonium nitrate. Care must be taken to avoid an explosion in this reaction, in which the cation is oxidized by the anion:

$$NH_4NO_3(l) \xrightarrow{250°C} N_2O(g) + 2\,H_2O(g)$$

Standard potential data suggest that N_2O should be a strong oxidizing agent in acidic and basic solutions:

$$N_2O(g) + 2\,H^+(aq) + 2e^- \rightarrow N_2(g) + H_2O(l)$$
$$E^\ominus = +1.77\,\text{V at pH} = 0$$

$$N_2O(g) + H_2O(l) + 2e^- \rightarrow N_2(g) + 2\,OH^-(aq)$$
$$E^\ominus = +0.94\,\text{V at pH} = 14$$

However, kinetic considerations are paramount, and the gas is unreactive towards many reagents at room temperature. As noted earlier, it is a potent greenhouse gas.

EXAMPLE 16.7 Comparing the redox properties of nitrogen oxoanions and oxo compounds

Compare (a) NO_3^- and NO_2^- as oxidizing agents, and (b) N_2H_4 and H_2NOH as reducing agents.

Answer We need to refer to the Frost diagram for nitrogen, which is included in Fig. 16.8, and use the interpretation described in Section 6.13. (a) Both NO_3^- and NO_2^- ions are strong oxidizing agents. The reactions of the former are often sluggish but are generally faster in acidic solution. The reactions of NO_2^- ions are generally faster and become even faster in acidic solution, where the NO^+ is a common identifiable intermediate. (b) Hydrazine and hydroxylamine are both good reducing agents. In basic solution, hydrazine becomes a stronger reducing agent.

Self-test 16.7 (a) Compare NO_2, NO, and N_2O with respect to their ease of oxidation in air. (b) Summarize the reactions that are used for the synthesis of hydrazine and hydroxylamine. Are these reactions best described as electron-transfer processes or nucleophilic displacements?

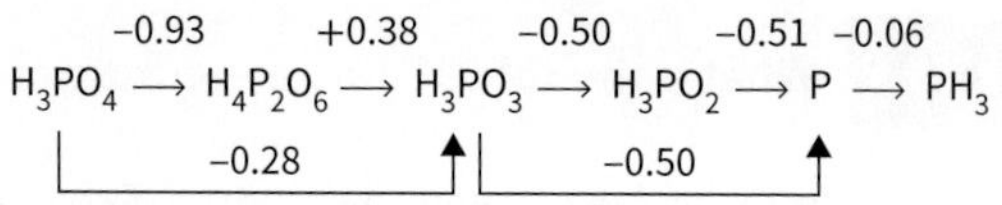

16.14 Oxides of phosphorus, arsenic, antimony, and bismuth

KEY POINTS The oxides of phosphorus include P_4O_6 and P_4O_{10}, both of which are cage compounds with T_d symmetry; on progressing from arsenic to bismuth, the E(V) oxidation state is more readily reduced to E(III).

Phosphorus forms phosphorus(V) oxide, P_4O_{10} (6), and phosphorus(III) oxide, P_4O_6 (7). It is also possible to isolate the intermediate compositions having one, two, or three O atoms terminally attached to the P atoms. Both principal oxides can be hydrated to yield the corresponding acids, the P(V) oxide giving phosphoric acid, H_3PO_4, and the P(III) oxide giving phosphonic acid, H_3PO_3. As remarked in Section 5.3(c), phosphonic acid has one H atom attached directly to the P atom; it is therefore a diprotic acid and better represented as $HPO(OH)_2$.

In contrast to the high stability of phosphorus(V) oxide, arsenic, antimony, and bismuth more readily form oxides with oxidation number +3, specifically As_2O_3, Sb_2O_3, and Bi_2O_3. In the gas phase, the arsenic(III) and antimony(III) oxides have the molecular formula E_4O_6, with the same tetrahedral structure as P_4O_6. Arsenic, Sb, and Bi do form oxides with oxidation number +5, but Bi(V) oxide is unstable and has not been structurally characterized. This is another example of the consequences of the inert pair effect (Section 10.7).

16.15 Oxoanions of phosphorus, arsenic, antimony, and bismuth

KEY POINTS Important oxoanions are the P(V) species phosphate, PO_4^{3-}, the P(III) species phosphite, HPO_3^{2-} and the P(I) species hypophosphite, $H_2PO_2^-$. The existence of P–H bonds and the highly reducing character of the two lower oxidation states is notable. Phosphorus(V) also forms an extensive series of O-bridged polyphosphates. In contrast to N(V), P(V) species are not strongly oxidizing. As(V) is a stronger oxidant than P(V).

$$H_3PO_4 \xrightarrow{-0.93} H_4P_2O_6 \xrightarrow{+0.38} H_3PO_3 \xrightarrow{-0.50} H_3PO_2 \xrightarrow{-0.51} P \xrightarrow{-0.06} PH_3$$
$$\underset{-0.28}{\longleftarrow} \quad \underset{-0.50}{\longleftarrow}$$

FIGURE 16.10 Latimer diagram for phosphorus under acidic conditions.

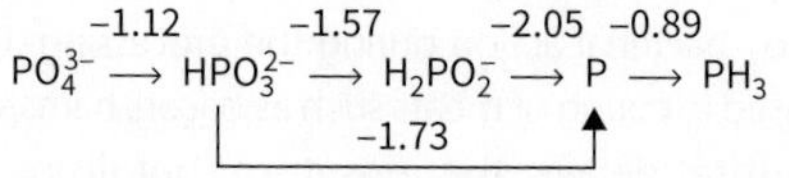

$$PO_4^{3-} \xrightarrow{-1.12} HPO_3^{2-} \xrightarrow{-1.57} H_2PO_2^- \xrightarrow{-2.05} P \xrightarrow{-0.89} PH_3$$
$$\underset{-1.73}{\longleftarrow}$$

FIGURE 16.11 Latimer diagram for phosphorus under basic conditions.

It can be seen from the Latimer diagrams in Figs. 16.10 and 16.11 that elemental P and most of its compounds other than P(V) are strong reducing agents. White phosphorus disproportionates into phosphine, PH_3 (oxidation number –3), and hypophosphite ions (oxidation number +1) in basic solution (Fig. 16.6):

$$P_4(s) + 3\,OH^-(aq) + 3\,H_2O(l) \rightarrow PH_3(g) + 3\,H_2PO_2^-(aq)$$

Table 16.4 lists some common P oxoanions (Box 16.4). The approximately tetrahedral environment of the P atom in their structures should be noted, as should the existence of P–H bonds in the hypophosphite and phosphite anions.

The synthesis of various P(III) oxoacids and oxoanions, including HPO_3^{2-} and alkoxophosphanes, is conveniently performed by reaction with phosphorus(III) chloride under mild conditions, such as in cold tetrachloromethane solution:

$$PCl_3(l) + 3\,H_2O(l) \rightarrow H_3PO_3(sol) + 3\,HCl(sol)$$
$$PCl_3(l) + 3\,ROH(sol) + 3\,N(CH_3)_3(sol) \rightarrow$$
$$P(OR)_3(sol) + 3\,[HN(CH_3)_3]Cl(sol)$$

Reductions with $H_2PO_2^-$ and HPO_3^{2-} are usually fast. One of the commercial applications of this lability is the use of $H_2PO_2^-$ to reduce $Ni^{2+}(aq)$ ions, coating surfaces with metallic Ni in the process called 'electrodeless plating':

$$Ni^{2+}(aq) + 2\,H_2PO_2^-(aq) + 2\,H_2O(l) \rightarrow Ni(s) + 2\,H_2PO_3^-(aq)$$
$$+ H_2(g) + 2\,H^+(aq)$$

TABLE 16.4 Some phosphorus oxoanions

Oxidation number	Formula	Name	Comments
+1	$H_2PO_2^-$	Hypophosphite (dihydrodioxophosphate)	Facile reducing agent
+3	HPO_3^{2-}	Phosphite	Facile reducing agent
+4	$P_2O_6^{4-}$	Hypophosphate	Basic
+5	PO_4^{3-}	Phosphate	Strongly basic
+5	$P_2O_7^{4-}$	Diphosphate	Basic; longer chain

BOX 16.4 How are phosphates useful in the food industry?

Phosphorus in the form of phosphates is essential to life, and phosphate fertilizers in forms such as bone, fish, and guano have been used since ancient times. The phosphate industry started in the mid-nineteenth century when sulfuric acid was used to decompose bones and phosphate minerals to make the phosphate more readily available. The development of more economical routes led to the diversification of the industrial applications of phosphoric acid and phosphate salts.

More than 90% of world production of phosphoric acid is used to make fertilizers but there are several other applications. One of the most important is in the food industry. A dilute solution of phosphoric acid is nontoxic and has an acidic taste and it is used extensively in beverages to give a tart taste, as a buffering agent in jams and jellies, and as a purifying agent in sugar refining. Sodium dihydrogenphosphate, NaH_2PO_4, is added to animal feeds as a dietary supplement. The disodium salt, Na_2HPO_4, is used as an emulsifier for processing cheese as it interacts with the protein casein and prevents separation of the fat and water.

The potassium salts are more soluble and more expensive than the sodium salts. The dipotassium salt, K_2HPO_4, is used as an anticoagulant in coffee creamer. It interacts with the protein and prevents coagulation by the coffee acids. Calcium dihydrogenphosphate monohydrate, $Ca(H_2PO_4)_2 \cdot H_2O$, is used as a raising agent in bread, cake mixes, and self-raising flour. Together with $NaHCO_3$ it produces CO_2 during the baking process but it also reacts with the protein in the flour to control the elasticity and viscosity of the dough or mixture.

The largest use of calcium monohydrogenphosphate, $CaHPO_4 \cdot 2H_2O$, is as a dental polish in nonfluoride toothpaste. Calcium diphosphate, $Ca_2P_2O_7$, is used in fluoride toothpaste. Calcium phosphate, Ca_3PO_4, is added to sugar and salt to improve their flow.

The Frost diagram for the elements, shown in Fig. 16.8, reveals similar trends in aqueous solution, with oxidizing character following the order PO_4^{3-} < AsO_4^{3-} < $Sb(OH)_6^-$ < Bi(V). The thermodynamic tendency and kinetic ease of reducing AsO_4^{3-} is thought to be key to its toxicity towards animals. Thus, As(V) as AsO_4^{3-} readily mimics PO_4^{3-}, and so may be incorporated into cells. There, unlike P, it is reduced to an As(III) species, which is thought to be the actual toxic agent. This toxicity may stem from the affinity of As(III) for sulfur-containing amino acids. The enzyme arsenite oxidase, which contains a Mo cofactor (Section 26.12), is produced by certain bacteria and is used to lower the toxicity of As(III) by converting it to As(V).

16.16 Condensed phosphates

KEY POINT Dehydration of phosphoric acid leads to the formation of chain or ring structures that may be based on many PO_4 units.

When phosphoric acid, H_3PO_4, is heated above 200°C, condensation occurs, resulting in the formation of P–O–P bridges between two neighbouring PO_4^{3-} units (Section 5.5). The extent of this condensation depends upon the temperature and duration of heating.

$$2\,H_3PO_4(l) \rightarrow H_4P_2O_7(l) + H_2O(g)$$
$$H_3PO_4(l) + H_4P_2O_7(l) \rightarrow H_5P_3O_{10}(l) + H_2O(g)$$

The simplest condensed phosphate is thus $H_4P_2O_7$. The most commercially important condensed phosphate is the sodium salt of the triacid, $Na_5P_3O_{10}$ (**26**) which is widely used in detergents for laundry and dishwashers, in other cleaning products, and in water treatment (Box 16.5). Polyphosphates are also used in various ceramics and as food additives. Triphosphates such as adenosine triphosphate (ATP) are of vital importance in living organisms (Section 26.2) where generation of phosphate esters (phosphorylation) plays a central role.

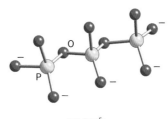

26 $P_3O_{10}^{5-}$

A range of condensed phosphates occurs with chain lengths ranging from those based on two PO_4 units to polyphosphates having chain lengths of several thousand units. Di-, tri-, tetra-, and pentapolyphosphates have been isolated but higher members of the series always contain mixtures. However, the average chain length can be determined by the usual methods used in polymer analysis or by titration. Just as the three successive acidity constants for phosphoric acid differ, so do the acidity constants for the two types of OH group of the polyphosphoric acids. The terminal OH groups, of which there are two per molecule, are weakly acidic. The remaining OH groups, of which there is one per P atom, are strongly acidic because they are situated adjacent to strongly electron-withdrawing =O groups. The ratio of weakly to strongly acidic protons gives an indication of average chain length. The long-chain polyphosphates are viscous liquids or glasses.

If NaH_2PO_4 is heated and the water vapour is allowed to escape, then the tricyclo anion $P_3O_9^{3-}$ (**27**) is formed. If this reaction is carried out in a closed system, the product is Maddrell's salt, a crystalline material that contains long chains of PO_4 units. The tetracyclo anion (**28**) is formed when P_4O_{10} is treated with cold aqueous solutions of NaOH or $NaHCO_3$.

BOX 16.5 Industrial applications of polyphosphates

The most widely used polyphosphate is sodium tripolyphosphate, $Na_5P_3O_{10}$. Its major use is as a 'builder' for synthetic detergents used in domestic laundry products, car shampoos, and industrial cleaners. Its role in these applications is to form stable complexes with the calcium and magnesium ions in hard water, effectively making them unavailable for precipitation. It also acts as a buffer and prevents the flocculation of dirt and the redeposition of soil particles. Potassium tripolyphosphate is more soluble and more expensive than the sodium analogue and is used in liquid detergents. For some applications an effective trade-off between solubility and cost can be achieved by using sodium potassium tripolyphosphate, $Na_3K_2P_3O_{10}$. The use of polyphosphates in detergents has been implicated in excessive algae growth and eutrophication of some natural waters. This has led to restrictions in their use in many countries and the reduction in their use in home laundry products. However, phosphate is still widely used as an essential fertilizer and more of it enters rivers and lakes by run-off from farmland than from detergents.

Food-grade sodium tripolyphosphate is used in the curing of hams and bacon. It interacts with the proteins and leads to good moisture retention during curing. It is also used to improve the quality of processed chicken and seafood products. The technical-grade product is used as a water softener, by sequestration as above, and in the paper pulping and textile industries, where it is used to help break down cellulose.

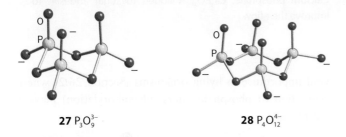

27 $P_3O_9^{3-}$ **28** $P_4O_{12}^{4-}$

29 $P_4(NR)_6$

EXAMPLE 16.8 Determining the chain length of a polyphosphoric acid by titration

A sample of a polyphosphoric acid was dissolved in water and titrated with dilute NaOH(aq). Two stoichiometric points were observed at 16.8 and 28.0 cm³. Determine the chain length of the polyphosphate.

Answer We need to determine the ratio of the two different types of OH group. The strongly acidic OH groups are titrated by the first 16.8 cm³. The two terminal OH groups are titrated by the remaining 28.0 − 16.8 cm³ = 11.2 cm³. Because the concentrations of analyte and titrant are such that each OH group requires 5.6 cm³ of the titrant (because 11.2 cm³ is used to titrate two such groups), we conclude that there are (16.8 cm³)/(5.6 cm³) = 3 strongly acidic OH groups per molecule. A molecule with two terminal OH groups and three further OH groups is a tripolyphosphate.

Self-test 16.8 When titrated against base, a sample of polyphosphate gave end points at 30.4 and 45.6 cm³. What is the chain length?

16.17 Phosphazenes

KEY POINTS The range of PN compounds is extensive, and includes cyclic and polymeric phosphazenes, $(PX_2N)_n$; phosphazenes form highly flexible elastomers.

Many analogues of phosphorus–oxygen compounds exist in which the O atom is replaced by the isolobal (Sections 2.8 and 24.20(c)) NR or NH group, such as $P_4(NR)_6$ (**29**), the analogue of P_4O_6.

Other compounds exist in which OH or OR groups are replaced by the isolobal NH_2 or NR_2 groups. An example is $P(NMe_2)_3$, the analogue of $P(OMe)_3$. Another indication of the scope of PN chemistry, and a useful point to consider, is that PN is structurally equivalent to SiO. For example, various phosphazenes, which are chains and rings containing R_2PN units (**30**), are analogous to the siloxanes (Section 15.16) and their R_2SiO units (**31**).

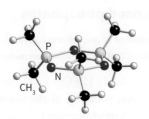

30 $((CH_3)_2PN)_3$

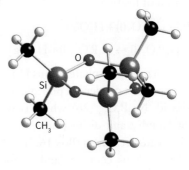

31 $((CH_3)_2SiO)_3$

BOX 16.6 What are the biomedical applications of polyphosphazenes?

Biodegradable polymers are attractive biomedical materials (see also Chapter 32) because they survive for only a limited time *in vivo*. Polyphosphazenes are proving to be very useful in this respect as they degrade to harmless by-products and their physical properties can be tuned by altering the substituents on the P atoms. They are used as bio-inert housing materials for implantation of devices, as structural materials for construction of heart valves and blood vessels, and as biodegradable supports for *in vivo* bone regeneration. The best polyphosphazenes for this last application form fibres in which the P–N backbone consists of alkyoxy groups that form bonds to Ca^{2+} ions. The polymer fibre becomes populated with the patient's osteoblasts (bone-making cells). The polymer degrades as the osteoblasts multiply and fill the space between the fibres. Polyphosphazenes have been designed that hydrolyse at a specific rate and maintain their strength as the erosion process proceeds.

Polyphosphazenes are also used as drug-delivery systems (Chapter 32). The bioactive molecule is trapped within the structure of the polymer or incorporated into the P–N backbone, and the drug is released when the polymer degrades. The rate of degradation can be controlled by altering the structure of the polymer backbone, thus providing control over the rate of drug delivery. Drugs that can be delivered in this way include cisplatin, dopamine, and steroids:

$$(Cl_2PN)_n + 2n\, CF_3CF_2O^- \rightarrow [(CF_3CF_2O)_2PN]_n + 2n\, Cl^-$$

Like silicone rubber, the polyphosphazenes remain rubbery at low temperatures because, like the isoelectronic SiOSi group (Section 15.15), the molecules are helical and the PNP groups are highly flexible.

The cyclic phosphazene dichlorides are good starting materials for the preparation of the more elaborate phosphazenes. They are easily synthesized:

$$nPCl_3 + nNH_4Cl \rightarrow (Cl_2PN)_n + 4nHCl \qquad n = 3 \text{ or } 4$$

Use of a chlorocarbon solvent and temperatures near 130°C results in the cyclic trimer (**32**) and tetramer (**33**), and when the trimer is heated to about 290°C it changes to polyphosphazene (Box 16.6). The Cl atoms in the trimer, tetramer, and polymer are readily displaced by other Lewis bases.

Then, because Cl atoms are readily replaced by strong Lewis bases, such as alkoxides, the chlorophosphazene is used as follows:

$$(Cl_2PN)_4 + 8NaOCH_3 \rightarrow [(CH_3O)_2PN]_4 + 8NaCl$$

16.18 Organometallic compounds of arsenic, antimony, and bismuth

Oxidation numbers +3 and +5 are encountered in many of the organometallic compounds of arsenic, antimony, and bismuth. An example of a compound with an element in the E(III) oxidation state is $As(CH_3)_3$ (**34**), and an example of the E(V) state is $As(C_6H_5)_5$ (**35**). Organoarsenic compounds were once widely used to treat bacterial infections and as herbicides and fungicides. However, because of their high toxicity they no longer have major commercial applications.

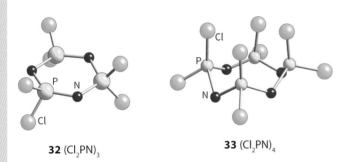

32 $(Cl_2PN)_3$ **33** $(Cl_2PN)_4$

The large bis(triphenylphosphine)iminium cation, $[Ph_3P=N=PPh_3]^+$, which is commonly abbreviated as PPN^+, is very useful for forming crystalline salts of large anions. The salts of this cation are usually soluble in polar aprotic solvents such as HMPA, dimethylformamide, and even dichloromethane.

A BRIEF ILLUSTRATION

To prepare $[NP(OCH_3)_2]_4$ from PCl_5, NH_4Cl, and $NaOCH_3$ the cyclic chlorophosphazene is synthesized first:

$$4PCl_5 + 4NH_4Cl \xrightarrow{130°C} (Cl_2PN)_4 + 16HCl$$

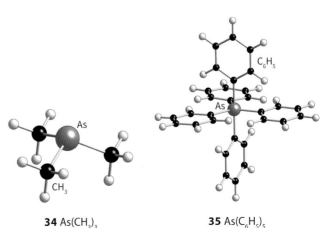

34 $As(CH_3)_3$ **35** $As(C_6H_5)_5$

(a) Oxidation state E(III)

KEY POINTS The stability of the organometallic compounds decreases in the order As > Sb > Bi; the aryl compounds are more stable than the alkyl compounds.

Organometallic compounds of arsenic(III), antimony(III), and bismuth(III) can be prepared in an ether solvent by metathesis using a Grignard reagent, an organolithium compound, or direct synthesis with an organohalide:

$$AsCl_3(et) + 3RMgCl(et) \rightarrow AsR_3(et) + 3MgCl_2(et)$$

$$2As(et) + 3RBr(et) \xrightarrow{Cu/\Delta} AsRBr_2(et) + AsR_2Br(et)$$

$$AsR_2Br(et) + R'Li(et) \rightarrow AsR_2R'(et) + LiBr(et)$$

The compounds are all readily oxidized but are stable to water. The M–C bond strength decreases for a given R group in the order As > Sb > Bi (Section 9.7). Consequently, the stability of the compounds decreases in the same order. In addition, the aryl compounds, such as $(C_6H_5)_3As$, are generally more stable than the alkyl compounds. The halogen-substituted compounds R_nMX_{3-n} are well known.

All the compounds act as Lewis bases and form complexes with d metals. The basicity decreases in the order As > Sb > Bi. Many complexes of alkyl- and arylarsanes have been prepared, but fewer stibane complexes are known. A useful ligand, for example, is the bidentate compound known as diars (**36**).

Because of their soft-donor character, many aryl- and alkylarsane complexes of the soft species Rh(I), Ir(I), Pd(II), and Pt(II) have been prepared. However, hardness criteria are only approximate, so we should not be surprised to see phosphine and arsane complexes of some metals in higher oxidation states. For example, the unusual Pd(IV) oxidation state of palladium is stabilized by the diars ligand (**37**).

The synthesis of diars provides a good illustration of some common reactions in the synthesis of organoarsenic compounds (Fig. 16.12). The starting material is $(CH_3)_2AsI$. This compound is not conveniently prepared by metathesis reaction between AsI_3 and a Grignard or similar carbanion reagent, because that reaction is not selective to partial substitution on the As atom when the organic group is compact. Instead, the compound can be prepared by the direct action of a haloalkane, CH_3I, on elemental arsenic.

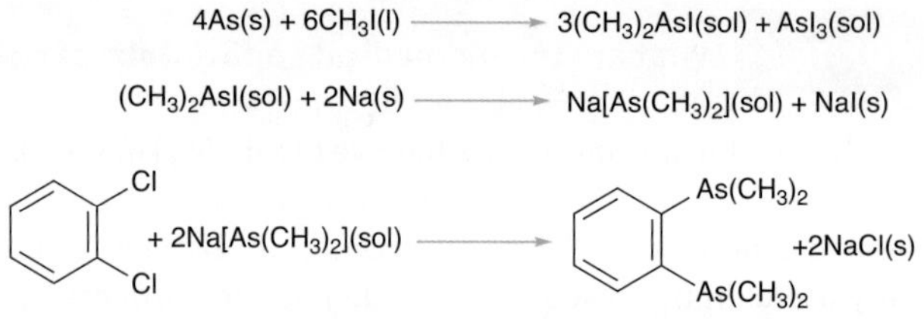

FIGURE 16.12 The synthesis of diars.

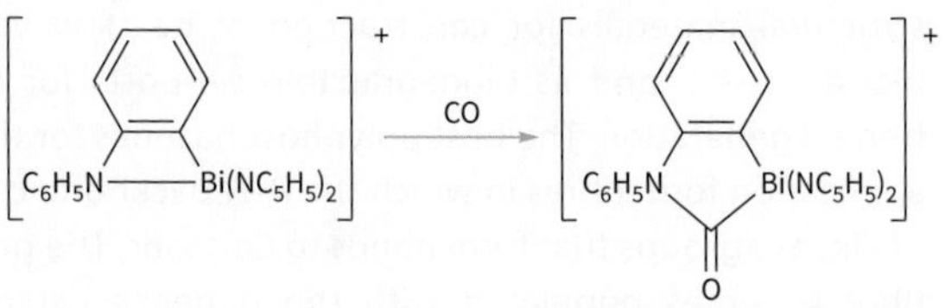

FIGURE 16.13 Insertion of CO into a Bi–N bond.

In the next step, the action of sodium on $(CH_3)_2AsI$ is used to produce $[(CH_3)_2As]^-$. The resulting powerful nucleophile $[(CH_3)_2As]^-$ is then used to displace chlorine from 1,2-dichlorobenzene.

Polyarsane compounds, $(RAs)_n$, can be prepared in ether, by reduction of a pentavalent organometallic compound, R_5As, or by treating an organohaloarsenic compound with Li:

$$nRAsX_2(et) + 2nLi(et) \rightarrow (RAs)_n(et) + 2nLiX(et)$$

The compound R_2AsAsR_2 is very reactive because the As–As bond is readily cleaved. It reacts with oxygen, sulfur, and species containing C=C bonds, and forms complexes with d-metal species in which the As–As bond may be cleaved (for example **38**) or left intact.

Polyarsanes of up to six units have been characterized. Polymethylarsane exists as a yellow, puckered cyclic pentamer (**39**) and as a purple-black ladder-like structure (**40**). The strength of the M–M bond decreases in the order As > Sb > Bi, so whereas arsenic forms several catenated organometallic compounds, only R_2Bi–BiR_2 has been isolated.

In Box 14.4, small molecule activation by main group frustrated Lewis pairs was described. Transition metal-like reactivity is also observed with the cationic bismuth amide shown in Fig. 16.13. The strained Bi–N bond inserts CO yielding a cationic bismuth carbamoyl species.

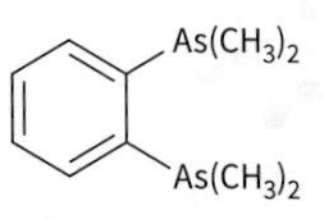

36 $C_6H_4(As(CH_3)_2)_2$, diars

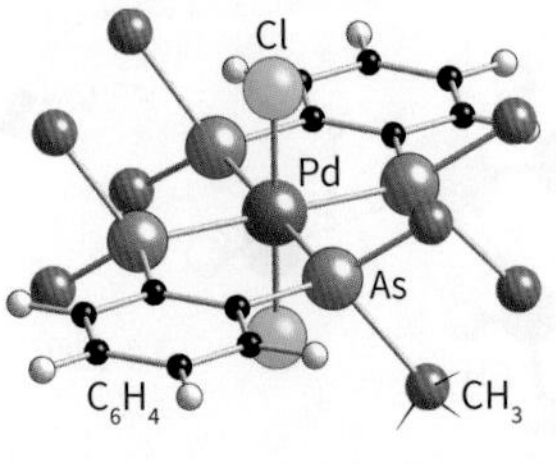

37 $[PdCl_2(diars)_2]^{2+}$

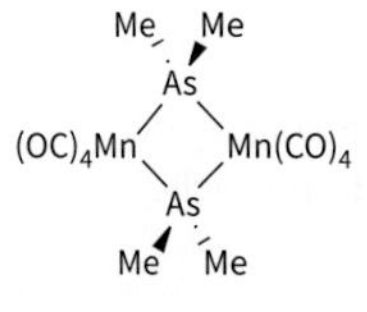

38

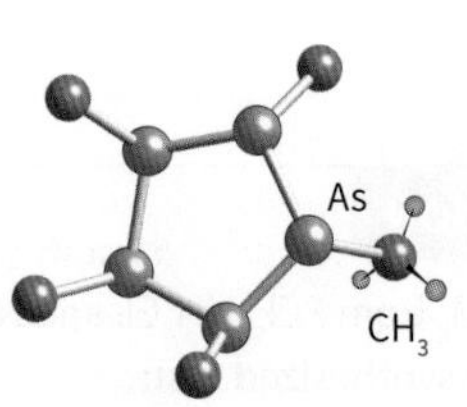

39 $As_5(CH_3)_5$

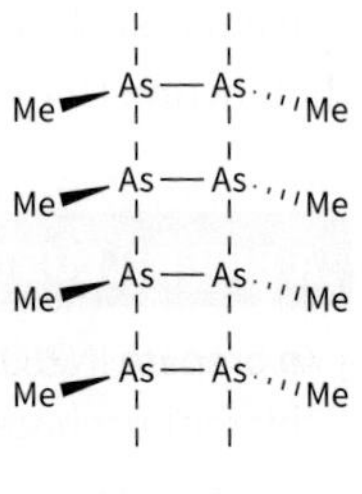

40 $(AsMe)_n$

FIGURE 16.14 The synthesis of a phosphorus dication.

As well as forming single M–C bonds, As, Sb, and Bi also form M=C bonds. A well-studied group of compounds is the arylometals, in which a metal atom forms part of a heterocyclic six-membered benzene-like ring (**41**).

41 C_5H_5M

Analogues of pyridine, C_5H_5N, are known. Arsabenzene, C_5H_5As, is stable up to 200°C, stibabenzene, C_5H_5Sb, can be isolated but readily polymerizes, and bismabenzene, C_5H_5Bi, is very unstable. Introduction of triisopropylsilyl substituents on the adjacent carbon atoms produces a stable derivative (**42**). These compounds exhibit typical aromatic character although arsabenzene is 1000 times more reactive than benzene. A related group of compounds is arsole, stibole, and bismuthole, C_4H_4MH, in which the metal atom forms part of a five-membered ring (**43**).

42   **43** C_4H_5M

While compounds containing phosphorus–carbon bonds are not normally regarded as organometallic, cationic cyclopentadienyl complexes of phosphorus adopt structures reminiscent of metal complexes. Single fluoride abstraction from Cp*PF$_2$ (Cp* = pentamethylcyclopentadienyl) initially gives a η^2-bonded monocation, [η^2-Cp*PF]$^+$. Removal of the second fluoride produces the extremely Lewis acidic dication [η^5-Cp*P]$^{2+}$ (Fig. 16.14). See Section 24.14 for a discussion of transition-metal complexes of the cyclopentadienyl ligand.

(b) Oxidation state E(V)

KEY POINT The tetraphenylarsonium ion is a starting material for the preparation of other As(V) organometallic compounds.

The trialkylarsanes act as nucleophiles towards haloalkanes to produce tetraalkylarsonium salts, which contain As(V):

$$As(CH_3)_3(sol) + CH_3Br(sol) \rightarrow [As(CH_3)_4]Br(sol)$$

This type of reaction cannot be used for the preparation of the tetraphenylarsonium ion, [AsPh$_4$]$^+$, because triphenylarsane is a much weaker nucleophile than trimethylarsane. Instead, a suitable synthetic reaction is:

$$Ph_3AsO(sol) + PhMgBr(sol) \rightarrow [Ph_4As]Br(sol) + MgO(s)$$

Fundamentally, this reaction is simply a metathesis in which the Ph$^-$ anion replaces the formal O^{2-} ion attached to the As atom, resulting in a compound in which the arsenic retains its E(V) oxidation state. The formation of the highly exergonic compound MgO also contributes to the Gibbs free energy of this reaction, making it more thermodynamically favourable.

The tetraphenylarsonium, tetraalkylammonium, and tetraphenylphosphonium cations are used in synthetic inorganic chemistry as bulky cations to stabilize bulky anions. The tetraphenylarsonium ion is also a starting material for the preparation of other As(V) organometallic compounds. For instance, the action of phenyllithium on a tetraphenylarsonium salt produces pentaphenylarsenic (**35**), a compound of As(V):

$$[AsPh_4]Br(sol) + LiPh(sol) \rightarrow AsPh_5(sol) + LiBr(s)$$

Pentaphenylarsenic, AsPh$_5$, is trigonal-bipyramidal, as expected from VSEPR considerations. We have seen (Section 2.3) that a square-pyramidal structure is often close in energy to the trigonal-bipyramidal structure, and the antimony analogue, SbPh$_5$, is in fact square-pyramidal (**44**).

44 Sb(C$_6$H$_5$)$_5$

FURTHER READING

P. Enghag, *Encyclopaedia of the Elements*. John Wiley & Sons (2004).

D. M. P. Mingos, *Essential Trends in Inorganic Chemistry*. Oxford University Press (2004). An overview of inorganic chemistry from the perspective of structure and bonding.

R. B. King (ed.), *Encyclopedia of Inorganic Chemistry*. John Wiley & Sons (2005).

H. R. Allcock, *Chemistry and Applications of Polyphosphazenes*. John Wiley & Sons (2002).

J. Emsley, *The Shocking History of Phosphorus: A Biography of the Devil's Element*. Pan (2001).

W. T. Frankenberger, *The Environmental Chemistry of Arsenic*. Marcel Dekker (2001).

G. J. Leigh, *The World's Greatest Fix: A History of Nitrogen and Agriculture*. Oxford University Press (2004).

C. Benson, *The Periodic Table of the Elements and Their Chemical Properties*. Kindle edition. MindMelder.com (2009).

J. C. Peters, Advancing electrocatalytic nitrogen fixation: Insights from molecular systems. *Faraday Discuss.*, 2023, **243**, 450.

Royal Society, Ammonia: zero-carbon fertiliser, fuel and energy store. 2020. ISBN: 978-1-78252-448-9.

D. Laniel et al., Aromatic hexazine $[N_6]^{4-}$ anion featured in the complex structure of the high-pressure potassium nitrogen compound K_9N_{56}. *Nat. Chem.*, 2023, **15**, 641. One of several accounts of the challenging synthesis of polynitrogen compounds.

EXERCISES

16.1 List the elements in Group 15 and indicate the ones that are (a) diatomic gases, (b) nonmetals, (c) metalloids, (d) true metals. Indicate those elements that display the inert pair effect.

16.2 Predict the chemical properties of element 115, muscovium.

16.3 Write Lewis structures for the following: N_2O, NO_2, Me_3NO.

16.4 How can the following oxides of nitrogen be prepared in the laboratory?

N_2O, NO, NO_2, N_2O_4

16.5 What is the average oxidation number of nitrogen in dinitrogen oxide, N_2O? Draw a Lewis structure and compare this result with the oxidation numbers of the individual nitrogen atoms in the molecule.

16.6 Explain why phosphorus exhibits much more extensive allotropy under ambient conditions than nitrogen.

16.7 Account for the variation in the following bandgaps (E_g/eV) for compounds of Group 15 elements.

BN (cubic) 7.5   GaN 3.37   GaP 2.26   GaAs 1.42   GaSb 0.72

16.8 (a) Give complete and balanced chemical equations for each step in the synthesis of H_3PO_4 from hydroxyapatite to yield (a) high-purity phosphoric acid, and (b) fertilizer-grade phosphoric acid. (c) Account for the large difference in cost between these two methods.

16.9 Ammonia can be prepared by (a) the hydrolysis of Li_3N or (b) the high-temperature, high-pressure reduction of N_2 by H_2. Give balanced chemical equations for each method, starting with N_2, Li, and H_2, as appropriate. (c) Account for the lower cost of the second method.

16.10 Show with an equation why aqueous solutions of NH_4NO_3 are acidic.

16.11 Carbon monoxide is a good ligand and is toxic. Why is the isoelectronic N_2 molecule not toxic?

16.12 Explain why NF_3 has weak electron donor properties whereas PF_3 is a good ligand in metal complexes. How can the existence of NOF_3 be explained?

16.13 Compare and contrast the formulas and stabilities of the oxidation states of the common nitrogen chlorides with the phosphorus chlorides.

16.14 Account for the mean bond enthalpies (in kJ mol^{-1}) for the following Group 15 hydrides and chlorides:

	N	P	As	Sb
EH_3	391	321	297	254
ECl_3	193	322	309	314

16.15 Use the VSEPR model to predict the probable shapes of (a) PCl_4^+, (b) PCl_4^-, (c) $AsCl_5$, (d) SbF_5^{2-}, (e) SbF_6^-.

16.16 Explain why compounds of Bi(III) typically have the metal atom in a distorted geometry.

16.17 Give balanced chemical equations for each of the following reactions: (a) oxidation of P_4 with excess oxygen, (b) reaction of the product from part (a) with excess water, (c) reaction of the product from part (b) with a solution of $CaCl_2$, and name the product.

16.18 Starting with NH_3(g) and other reagents of your choice, give the chemical equations and conditions for the synthesis of (a) HNO_3, (b) NO_2^-, (c) NH_2OH, (d) N_3^-.

16.19 Write the balanced chemical equation corresponding to the standard enthalpy of formation of P_4O_{10}(s). Specify the structure, physical state (s, l, or g), and allotrope of the reactants. Do either of the reactants differ from the usual practice of taking as reference state the most stable form of an element?

16.20 Without reference to the text, sketch the general form of the Frost diagrams for phosphorus (oxidation numbers 0 to +5) and bismuth (0 to +5) in acidic solution and discuss the relative stabilities of the E(III) and E(V) oxidation states of both elements.

16.21 Account for the variation in stabilities of the following compounds, given as their empirical formulae:

P_2O_5: stable, m.p. = 422°C at 5×10^5 Pa

As_2O_5: decomposes at 300°C

Sb_2O_5: decomposes at 380°C

Bi_2O_5: decomposes at 20°C

16.22 Explain why all forms of P_2O_5 contain tetrahedral P, whereas all forms of Sb_2O_5 contain octahedral Sb.

16.23 Are reactions of NO_2^- as an oxidizing agent generally faster or slower when pH is lowered? Give a mechanistic explanation for the pH dependence of NO_2^- oxidations.

16.24 When equal volumes of nitric oxide (NO) and air are mixed at atmospheric pressure a rapid reaction occurs, to form NO_2 and N_2O_4. However, nitric oxide from an automobile exhaust, which is present in the parts per million concentration range, reacts slowly with air. Produce an explanation for this observation in terms of the rate law and the probable mechanism.

16.25 On account of their slow reactions at electrodes, the potentials of most redox reactions of nitrogen compounds cannot be measured in an electrochemical cell. Instead, the values must be determined from other thermodynamic data. Illustrate such a calculation by using $\Delta_fG^\ominus(NH_3,aq) = -26.5$ kJ mol^{-1} to calculate the standard potential of the N_2/NH_3 couple in basic aqueous solution.

16.26 Comment on the factors influencing the following standard potentials (pH = 0):

H_3PO_4/H_3PO_3 −0.28 V
H_3AsO_4/H_3AsO_3 +0.56 V
$Bi_2O_5/BiO^+(aq)$ +2.00 V

16.27 Give balanced chemical equations for the reactions of the following reagents with PCl_5 and indicate the structures of the products: (a) water (1:1), (b) water in excess, (c) $AlCl_3$, (d) NH_4Cl.

16.28 Explain how you could use ^{31}P-NMR to distinguish between PF_3 and POF_3.

16.29 Use data in Resource section 3 to calculate the standard potential of the reaction of H_3PO_2 with Cu^{2+}. Are HPO_2^{2-} and $H_2PO_2^-$ useful as oxidizing or as reducing agents?

16.30 The tetrahedral P_4 molecule may be described in terms of localized 2c,2e bonds. Determine the number of skeletal valence electrons and from this decide whether P_4 is *closo*, *nido*, or *arachno* (these terms are specified in Section 14.11). If it is not *closo*, determine the parent *closo* polyhedron from which the structure of P_4 could be formally derived by the removal of one or more vertices.

16.31 Sketch the two possible geometric isomers of the octahedral $[AsF_4Cl_2]^-$ and explain how they could be distinguished by ^{19}F-NMR.

16.32 Use the Latimer diagrams in Resource section 3 to determine which species of N and P disproportionate in acid conditions.

16.33 Identify the arsenic compounds **A**, **B**, **C**, and **D**.

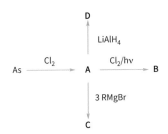

16.34 Identify the nitrogen compounds **A**, **B**, **C**, **D**, and **E**.

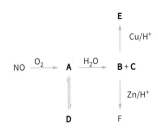

16.35 In the reaction scheme shown below, identify the species **A**–**G** and comment on their properties. Account for the difference between **B** and **F**.

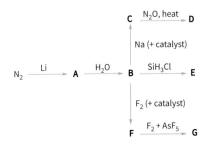

A is a solid having a layer structure.
B is a colourless basic gas, dipole moment = 1.5 Debye.
C is an ionic solid that is a strong base.
D is the Na salt of a linear anion.
E is a liquid and weak Lewis base.
F is a colourless gas, dipole moment = 0.2 Debye.
G is an ionic solid.

16.36 Identify the compounds formed in the following reaction scheme.

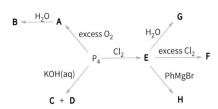

A is a white solid.
B is a tribasic acid.
C is a flammable gas.
D is a monobasic acid.
E is a colourless liquid, b.p. 76°C.
F is a crystalline solid that sublimes at 160°C.
G is a dibasic acid.
H is a white solid that is commonly used as a ligand in organometallic chemistry.

For the sequence leading from P_4 to **E** and **F**, explain why replacing Cl_2 by F_2 would give a different outcome.

16.37 Identify the compounds **A**–**G** formed from arsenic in the following reaction scheme and comment on their properties and the reactions leading to them. Could **B** be obtained directly from arsenic?

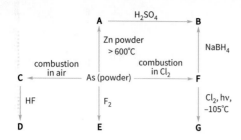

A is a semiconductor with a bandgap of 1.0 V.

B is a colourless gas (b.p. of liquid = −62°C).

C is a white solid that dissolves in water to give an acidic solution with behaviour typical of a dibasic acid.

D is a liquid, b.p. = 57°C.

E is a gas (b.p. of liquid = −53°C).

F is a colourless liquid, b.p. = 130°C.

G is an unstable solid below its decomposition temperature of −50°C: the corresponding compounds of P and Sb are stable.

TUTORIAL PROBLEMS

16.1 Write a short review of the uses of bismuth compounds in the treatment of gastric ulcers.

16.2 A paper published by J. Lee and co-workers (*Scientific Reports*, 2015, **5**, 11512, DOI: 10.1038/srep11512) proposed the existence of two-dimensional honeycomb sheet and ribbon forms of the Group 15 elements. This prediction has now been confirmed experimentally by the groups of D. Laniel and of C. Ji and their respective co-workers (*Phys. Rev. Lett.*, 2020, **124**, 216001 and *Sci. Adv.*, 2020, **6**, aba9206). (a) What methods did the authors use to explore these structures? (b) Which Group 15 element requires the most extreme conditions to form the honeycomb structures? (c) What properties and uses are these structures predicted to have? (d) What is it about these structures of Group 15 elements that makes them potentially more useful than graphene?

16.3 The structure of a butterfly-shaped P_4 unit has been reported by J. Bresien and co-workers (*Dalton Trans.*, 2016, **45**, 1998). (a) Sketch the P_4 unit. (b) Give one example of a bucyclic tetraphosphene synthesized from P_4 with a Lewis acid and one synthesized via self-assembly of P_1 units. (c) Draw the two isomers of Mes*P_4Mes*. (d) Which isomer would give two triplets in the ^{31}P NMR spectrum?

16.4 Nitric oxide has been discovered to play a crucial role in biological systems. A paper published by A. W. Carpenter and M. H. Schoenfisch (*Chem. Soc. Rev.*, 2012, **41**, 3742) discusses therapeutic applications of NO. Outline the main applications and drawbacks of gaseous NO as a therapeutic agent. Discuss how NO-donor molecules have been developed to enable NO therapies for a wide range of medical conditions.

16.4 Describe sewage treatment methods that result in a decrease in phosphate levels in wastewater. Outline a laboratory method that could be used to monitor phosphate levels in water.

16.5 A compound containing five-coordinate nitrogen has been characterized (A. Frohmann, J. Riede, and H. Schmidbaur, *Nature*, 1990, **345**, 140). Describe (a) the synthesis, (b) the structure of the compound, (c) the bonding.

16.6 Two articles (A. Lykknes and L. Kvittingen, Arsenic: not so evil after all?, *J. Chem. Educ.*, 2003, **80**, 497; J. Wang and C. M. Chien, Arsenic in drinking water: a global environmental problem, *J. Chem. Educ.*, 2004, **81**, 207) present opposing perspectives on the toxic nature of arsenic. Use these references to produce a critical assessment of the beneficial and detrimental effects of arsenic.

16.7 A paper published by N. Tokitoh et al. (*Science*, 1997, **277**, 78) gives an account of the synthesis and characterization of a stable bismuthene, containing Bi=Bi double bonds. Give the equations for the synthesis of the compound. Name and sketch the structure of the steric protecting group that was used. Why was the isolation of the product simple? What methods were used to determine the structure of the compound?

16.8 A paper published by Y. Zhang et al. (*Inorg. Chem.*, 2006, **45**, 10446) describes the synthesis of phosphazene cations as precursors for polyphosphazenes. Polyphosphazenes are prepared by ring-opening polymerization of the cyclic $(NPCl_2)_3$, and the reaction is initiated by phosphazene cations. Discuss which Lewis acids were used to produce the cations, and give the reaction scheme for the ring-opening polymerization of $(NPCl_2)_3$.

16.9 In their paper 'Catalytic reduction of dinitrogen to ammonia at single molybdenum center' (*Science*, 2003, **301**, 5629), D. Yandulov and R. Schrock describe the catalytic conversion of nitrogen to ammonia at room temperature and atmospheric pressure. Discuss why this development could be important commercially. Review nonbiological methods of nitrogen activation.

16.10 Nitric oxide is a potential atmospheric pollutant. In their review, 'A review on the catalytic decomposition of NO to N_2 and O_2: catalysts and processes' (*Catal. Sci. Technol.*, 2018, **8**, 4563–75), Wang and co-workers describe a number of catalytic approaches to its decomposition. What are the relative merits of the three general approaches?

The Group 16 elements

17

Part A: The essentials
- 17.1 The elements
- 17.2 Simple compounds

Part B: The detail
- 17.3 Oxygen
- 17.4 Sulfur
- 17.5 Selenium, tellurium, and polonium
- 17.6 Compounds formed with hydrogen
- 17.7 Halides
- 17.8 Metal oxides
- 17.9 Metal sulfides, selenides, tellurides, and polonides
- 17.10 Oxides of sulfur, selenium, and tellurium
- 17.11 Oxoacids of sulfur
- 17.12 Polyanions of sulfur, selenium, and tellurium
- 17.13 Polycations of sulfur, selenium, and tellurium
- 17.14 Sulfur–nitrogen compounds

Further reading

Exercises

Tutorial problems

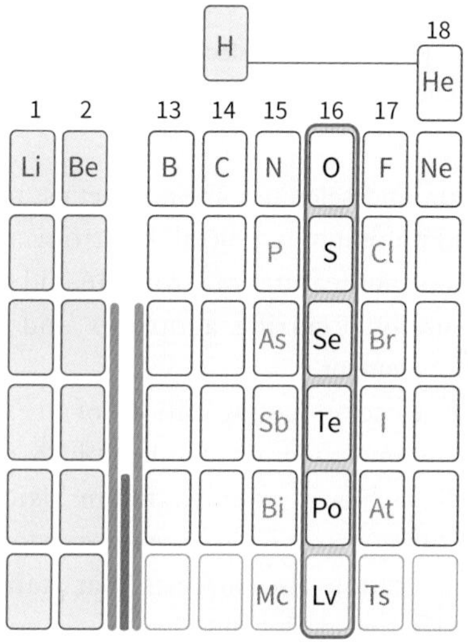

Group 16 elements are all nonmetals, apart from polonium, the heaviest member. The Earth abounds with oxygen, whether it is present in minerals, water, or as atmospheric O_2 that is produced by photosynthesis and required by all higher life forms. Metal oxides are key materials in many electronic and optical applications. Sulfur is essential to all life forms, and selenium is required in trace amounts. Sulfur and selenium exhibit a tendency to catenation and form rings and chains. Single bonds between O-atoms are weak, and thus peroxides are highly reactive. In contrast single bonds between S-atoms are strong, providing the stability required to crosslink regions of protein molecules.

The Group 16 elements oxygen, sulfur, selenium, tellurium, and polonium are often called the **chalcogens**. The name derives from the Greek word for 'bronze' and refers to the association of sulfur and its congeners with copper in this metal's ores. As in the rest of the p block, the element at the head of the group, oxygen, differs significantly from the other members of the group (Section 10.2). The coordination numbers of its compounds are generally lower, with the frequent formation of double bonds, and oxygen is the only member of the group to exist as diatomic molecules under normal conditions.

PART A: The essentials

In this section we discuss the essential features of the chemistry of the Group 16 elements. All the naturally occurring elements other than oxygen exist as solids under normal conditions. The artificial element livermorium has limited stability (the half-life of ^{293}Lv is < 0.1 s) and will not be discussed further. Metallic character generally increases down the group.

17.1 The elements

KEY POINTS Oxygen is the most electronegative element in Group 16 and is the only one to exist normally as a gas; all the elements occur in allotropic forms.

Oxygen, sulfur, and selenium are nonmetals, tellurium is a metalloid, and polonium is a metal. Allotropy and polymorphism are important features of Group 16 and sulfur occurs in more naturally occurring allotropes and polymorphs than any other element.

The group electron configuration of ns^2np^4 suggests a group maximum oxidation number of +6 (Table 17.1). Oxygen never achieves this maximum oxidation state, although the other elements do in many circumstances. The electron configuration also suggests that stability may be achieved with an oxidation number of −2, which is overwhelmingly common for O. The most remarkable feature of S, Se, and Te is that they all form stable compounds with oxidation numbers between −2 and +6.

The very high electronegativity of oxygen has an enormous influence on its chemical properties, as does its small atomic radius and the absence of accessible d orbitals. Thus, O seldom has a coordination number greater than 3 in simple molecular compounds, but its heavier congeners frequently reach coordination numbers of 4 and 6, as in SF_6.

Dioxygen (O_2) reacts with many organic and inorganic compounds under suitable conditions. Only the noble gases He, Ne, and Ar do not form oxides directly. Even though the O=O bond energy of +497 kJ mol^{-1} is high, many exothermic combustion reactions occur because the resulting E–O covalent bond enthalpies or MO_n lattice enthalpies are also high. One of the most important reactions of O_2 is the formation of a labile Fe–O_2 complex in haemoglobin (Section 26.7(b)), which greatly increases its effective solubility in water (blood) and allows its efficient transport from air to tissue.

Oxygen is the most abundant element in the Earth's crust at 46% by mass, and is present in all silicate minerals. It comprises 86% by mass of the oceans and 89% of water. The average human contains two-thirds oxygen by mass. Dioxygen, which is essentially derived completely from the water-splitting action of photosynthetic organisms, makes up 21% by mass of the atmosphere (Box 17.1). Oxygen is also the third most abundant element in the Sun, and the most abundant element on the surface of the Moon (46% by mass). Oxygen also occurs as ozone, O_3, a highly reactive, pungent gas that is crucial to the protection of life on Earth, for it shields the surface from solar ultraviolet radiation.

Sulfur occurs as deposits of the native element, in meteorites, volcanoes, and hot springs, and in minerals ranging from galena, PbS, to barite, $BaSO_4$. It also occurs as H_2S in natural gas and as organosulfur compounds in crude oil. That sulfur can exist in a large number of allotropic forms can be explained by the ability of S atoms to catenate, the high S–S bond energy of 264 kJ mol^{-1} being exceeded only by C–C (348 kJ mol^{-1}) and H–H (436 kJ mol^{-1}). All the crystalline forms of sulfur that can be isolated at room temperature consist of puckered S_n rings, S_8 (**1**) being the most common.

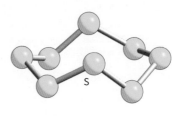

1 S_8

The striking difference between O–O and S–S single bond energies has important consequences. The O–O bond enthalpy is 146 kJ mol^{-1} and peroxides are powerful oxidizing agents; in contrast, the strong S–S bond is exploited in biology to stabilize protein structure by forming permanent linkages (RS–SR) between cysteine residues on different protein strands or different regions of one strand. This difference in bond enthalpies is rationalized most easily in terms of the greater electron–electron repulsion in an O–O bond.

TABLE 17.1 Selected properties of the elements

	O	S	Se	Te	Po
Covalent radius/pm	66	105	120	138	140
Ionic radius/pm	140	184	198	221	
First ionization energy/kJ mol^{-1}	1314	1000	941	869	812
Melting point/°C	−218	113 (α)	217	450	254
Boiling point/°C	−183	445	685	990	960
Pauling electronegativity	3.4	2.6	2.6	2.1	2.0
Electron affinity*/kJ mol^{-1}	141	200	195	190	183
	−844	−532			

* The first value is for $X(g) + e^-(g) \rightarrow X^-(g)$; the second value is for $X^-(g) + e^-(g) \rightarrow X^{2-}(g)$.

BOX 17.1 The Great Oxygenation Event: the impact of biogenic O_2

In the course of the evolution of the Earth's atmosphere, the proliferation of O_2-evolving photosynthesis eventually resulted in its presence in the atmosphere at 21% by volume. The appearance of O_2 some 2.3 billion years ago in what is known as the **Great Oxygenation Event** (GOE) led to the extinction of many species—primitive microorganisms lacking the ability to deal with such a toxic molecule. Some species retreated to habitats deeper in the soil or waters, where anaerobic conditions persisted and where their descendants remain today. Other organisms adapted differently and evolved to exploit the increasingly abundant and powerful oxidant. These organisms were the early *aerobes*, among which some would evolve to become our ancestors. Metaphorically speaking, the advantage that O_2 offered to life would amount to replacing a 0.1 V battery by a 1 V battery—a massive leap forward.

But the atmosphere that we have today appeared only after the first O_2 had decimated much of the huge, mineral-based reducing pool contained in or near the Earth's surface. Sulfur, which was present largely in the form of sulfide in anaerobic waters, was oxidized to sulfate. Metal ion concentrations also changed dramatically. Three of the metals most profoundly affected were molybdenum, copper, and iron.

In Earth's modern oceans, molybdenum is the most abundant d metal (at 0.01 ppm). However, before O_2 became established, molybdenum was locked away as insoluble solids, mainly MoO_2 and MoS_2. Oxidation of these solids produced the soluble molybdate ion:

$$2MoS_2(s) + 7O_2(g) + 2H_2O(l) \rightarrow 2[MoO_4]^{2-}(aq) + 4SO_2(g) + 4H^+(aq)$$

The water-soluble molybdate ion now became readily available to aquatic organisms. Copper, previously trapped as insoluble Cu_2S, was likewise mobilized.

In contrast, iron suffered the opposite fate. In the ancient oceans that were devoid of O_2, the element was present as Fe(II). Iron(II) hydroxide and sulfide are sufficiently soluble that iron would have been readily available to the earliest aquatic organisms. Well before O_2 levels rose in the atmosphere, the oxidation of Fe^{2+} to Fe^{3+} led to the precipitation of iron(III) hydroxides and oxides in what are known as **banded iron formations** (BIFs). Consisting of thin layers of magnetite (Fe_3O_4) or hematite (Fe_2O_3) separated by silica, BIFs are typically several hundred metres deep and cover a wide area: as well as being important sources of iron ore, they are testimony to the precipitation of iron from the oceans that occurred as the GOE got underway. Needing to adapt even further, the early organisms evolved to produce special ligands known as siderophores, capable of sequestering sufficient Fe from the environment to enable their survival (Section 26.6).

The chemically soft elements Se and Te occur in metal sulfide ores, and their principal source is the electrolytic refining of copper. Both elements are important components of thermoelectric materials which can interconvert heat and electrical energy (Section 30.4). Like sulfur, elemental selenium exists in several allotropic and polymorphic forms that include red Se_8 (puckered ring, analogous to yellow S_8), a stable grey hexagonal form based on chains, and a glassy amorphous black form. Tellurium exists as infinite chains arranged in a hexagonal lattice. Polonium, a shiny metal, occurs in 33 known isotopes, all of which are radioactive and extremely toxic. In the gas phase, all the heavier chalcogens exist, like O_2, as diatomic molecules.

As we will see later, the propensity for S, Se, and Te to undergo catenation extends to both anions and cations of these elements.

17.2 Simple compounds

KEY POINT The elements of Group 16 form simple binary compounds with hydrogen, halogens, oxygen, and metals.

Water, H_2O, so vital to life, is the only Group 16 hydride that is not a poisonous, malodorous gas. Its melting and boiling points (0°C and 100°C, respectively) are both very high compared to compounds of similar molecular mass and the analogous molecules in Group 16 (Table 17.2). Oxygen is by far the most electronegative element in Group 16: the high boiling point of H_2O, and of hydrogen peroxide, H_2O_2 (a liquid from −0.4°C to 150°C), is due to strong and extensive hydrogen bonding (O–H···O), as explained in Section 5.9.

Oxygen forms oxides with most metals, and peroxides and superoxides with Group 1 and 2 metals. When the oxidation number of the metal is lower than +4, the oxide is commonly ionic, whereas if the oxidation number of the

TABLE 17.2 Selected properties of the Group 16 hydrides

	H_2O	H_2S	H_2Se	H_2Te	H_2Po
Melting point/°C	0.0	−85.6	−65.7	−51	−36
Boiling point/°C	100.0	−60.3	−41.3	−4	37
$\Delta_f H^\ominus$/kJ mol^{-1}	−285.6 (l)	−20.1	+73.0	+99.6	
Bond length/pm	96	134	146	169	
Bond angle/°	104.5	92.1	91	90	
Acidity constants					
pK_{a1}	14.00	6.89	3.89	2.64	
pK_{a2}		14.15	11	10.80	

metal is greater than +4 the oxide is molecular. Sulfur forms sulfides, S^{2-}, disulfides, S_2^{2-}, and polysulfides, S_n^{2-}, with metals. Such catenated anions persist, albeit to a decreasing extent, for selenium and tellurium.

Sulfur, Se, Te, and Po have a rich chemistry with the halogens, and some of the most common halides are summarized in Table 17.3. Those in which the chalcogen is in an odd oxidation state (+1, +5) contain E–E bonds, reflecting the general instability of radicals below the first row of the p block (Section 10.1). Only F brings out the maximum group oxidation state, and the low-oxidation-state fluorides of Se, Te, and Po are unstable with respect to disproportionation into the element and a higher-oxidation-state fluoride.

The molecules of the two common oxides of S, SO_2 (b.p. −10°C), and SO_3 (b.p. 44.8°C), are angular (bent) (**2**) and trigonal-planar (**3**), respectively, in the gas phase. In the solid, sulfur trioxide exists as cyclic trimers (**4**).

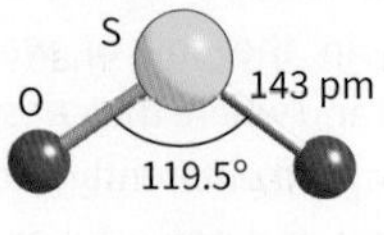

2 SO_2, C_{2v}

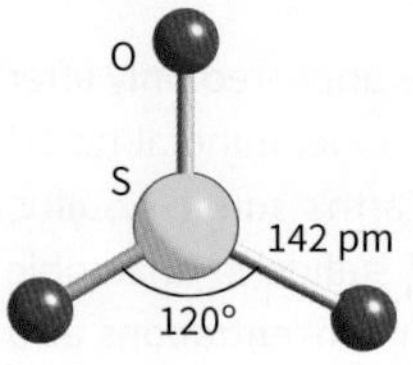

3 SO_3, D_{3h}

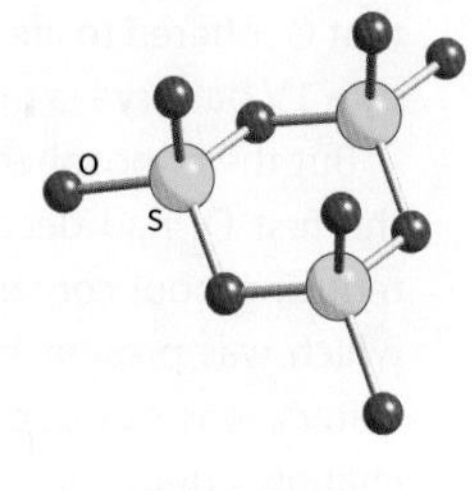

4 $(SO_3)_3$, C_{3v}

Sulfur dioxide is a poisonous gas with a sharp, choking odour. The major use of SO_2 is in the manufacture of sulfuric acid via the contact process, where it is first oxidized to SO_3. It is also used as a bleach, disinfectant, and food preservative. Sulfur trioxide, SO_3, is made on a huge scale by catalytic oxidation of SO_2. It is seldom isolated but immediately converted to sulfuric acid, H_2SO_4. Being extremely corrosive, anhydrous SO_3 is rarely handled in the laboratory, but is available as oleum (also known as fuming sulfuric acid), $H_2S_2O_7$, which is a solution of 25–65% SO_3 by mass in concentrated sulfuric acid. The reaction with metal oxides to produce sulfates is used to scrub undesirable SO_3 from effluent gases from industrial processes. Selenium, Te, and Po all form a dioxide and trioxide.

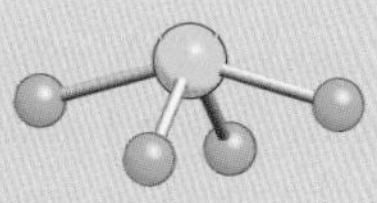

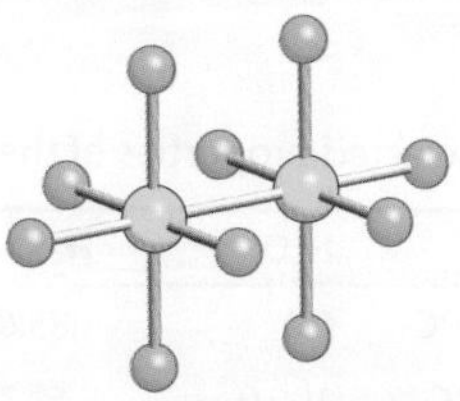

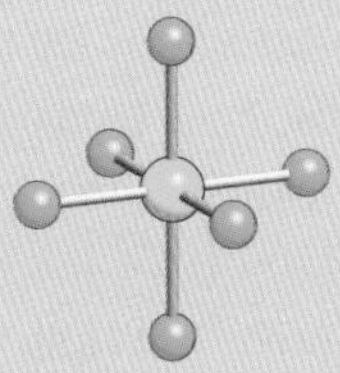

Sulfuric acid, H_2SO_4, is a dense, viscous, and highly corrosive liquid that even extracts water from organic matter to leave a charred, carbonaceous residue. It is a strong acid (for the first deprotonation step), a useful nonaqueous solvent, and exhibits extensive autoprotolysis (Section 5.1). Two series of salts are formed; the sulfates, SO_4^{2-}, and the hydrogensulfates, HSO_4^-. Sulfurous acid, H_2SO_3, has never been isolated. Aqueous solutions of SO_2, which are referred to as 'sulfurous acid', are better considered as hydrates, $SO_2 \cdot nH_2O$. Two series of salts, the sulfites, SO_3^{2-}, and the hydrogensulfites, HSO_3^-, are known and are moderately strong reducing agents, becoming oxidized to sulfates, SO_4^{2-}, or dithionates, $S_2O_6^{2-}$.

TABLE 17.3 Some halides of sulfur, selenium, and tellurium

Oxidation number (average, where relevant)	Formula	Structure	Remarks
$+\frac{1}{2}$	Te_2X (X = Br, I)	Halide bridges	Silver-grey
+1	S_2F_2	Two isomers: 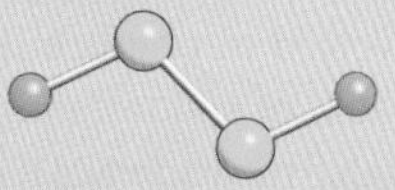 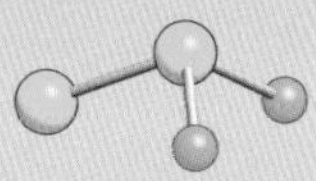	
	S_2Cl_2	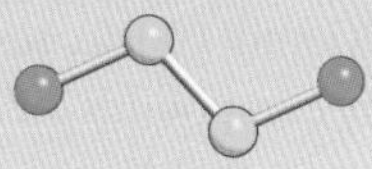	Reactive
+2	SCl_2	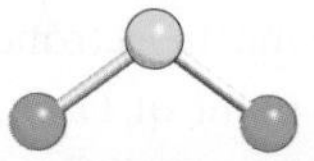	Reactive
+4	SF_4		
	SeX_4 (X = F, Cl, Br)		Gas
	TeX_4 (X = F, Cl, Br, I)		SeF_4 liquid TeF_4 solid
+5	S_2F_{10}		Reactive
	Se_2F_{10}		
+6	SF_6, SeF_6		Colourless gases
	TeF_6		Liquid (b.p. 368°C)

EXAMPLE 17.1 Using bond energies to explain elemental forms

Use the following bond enthalpy data to explain why elemental oxygen occurs as diatomic O_2 molecules whereas elemental sulfur occurs as S_8 rings.

	O–O	O=O	S–S	S=S
B/kJ mol^{-1}	146	497	264	421

Answer We can consider the formation of four moles of S_2 diatomic molecules from one mole of S_8. The enthalpy of this reaction is given by the difference between the bond enthalpies of the bonds that are broken and the bond enthalpies for bonds formed. In this case then $\Delta H^\ominus = 8(265) - 4(421) = +428$ kJ mol^{-1}. This is an endothermic reaction so S_2 will only be formed from S_8 if energy is supplied. The analogous reaction for oxygen gives $\Delta H^\ominus = 8(146) - 4(497) = -820$ kJ mol^{-1}, which is clearly exothermic. So the larger difference between the enthalpies of the single and double bonds in oxygen ensures that O_2 is formed.

Self-test 17.1 Use a molecular orbital diagram to determine the bond order of the S_2^{2-} ion.

PART B: The detail

In this section we discuss the detailed chemistry of the Group 16 elements and observe the rich variation in the structures of the compounds they form.

17.3 Oxygen

KEY POINTS Oxygen has two allotropes, dioxygen and ozone. Dioxygen has a triplet ground state and oxidizes hydrocarbons by a radical chain mechanism. Reaction with an excited state molecule can produce a fairly long-lived singlet state that can react as an electrophile. Ozone is an unstable and highly aggressive oxidizing agent.

Dioxygen is a biogenic gas (i.e. one that has been produced by the action of organisms): almost all of it is the result of photosynthesis, although traces are produced in the upper atmosphere by the action of ultraviolet radiation on water vapour. Dioxygen gas is colourless, odourless, and soluble in water to the extent of 3.08 cm^3 per 100 cm^3 water at 25°C and atmospheric pressure. This solubility falls to below 2.0 cm^3 in sea water but is still sufficient to support aerobic marine life. The solubility of O_2 in organic solvents is approximately 10 times greater than in water. The high solubility of O_2 makes it necessary to purge all solvents used in the synthesis of oxygen-sensitive compounds.

(a) Reactions of O_2

Oxygen is readily available as O_2 from the atmosphere and is obtained on a massive scale by the liquefaction and distillation of liquid air. The main commercial motivation is to recover O_2 for steelmaking, in which it reacts exothermically with coke (carbon) to produce carbon monoxide. About 1 tonne (1 t = 10^3 kg) of O_2 is needed to make 1 tonne of steel. Oxygen is also required to produce the white pigment TiO_2 by the chloride process:

$$TiCl_4(l) + O_2(g) \rightarrow TiO_2(s) + 2Cl_2(g)$$

Small-scale oxygen production, such as in the home for asthma sufferers, is achieved using pressure-swing adsorption, in which air is passed through a zeolite that preferentially adsorbs nitrogen. Oxygen is used in many oxidation processes; for example, the production of oxirane (ethylene oxide) from ethene. Oxygen is also supplied on a large scale for sewage treatment, renewal of polluted waterways, paper-pulp bleaching, and as an artificial atmosphere in medical and submarine applications. Although electrochemical or solar water-splitting is normally intended for the production of 'green' H_2 (Section 11.4), it also represents a crucial means of producing O_2 in an extraterrestrial environment.

Liquid oxygen is very pale blue, and boils at −183°C. Its colour arises from electronic transitions involving pairs of neighbouring molecules: one photon from the red–yellow–green region of the visible spectrum can raise two O_2 molecules to an excited state to form a molecular pair. Under high pressure, the colour of solid oxygen changes from light blue to orange and then to red at approximately 10 GPa.

The molecular orbital description of O_2 implies the existence of a double bond; however, as we saw in Section 2.8, the ground-state electron configuration is $1\sigma_g^2 1\sigma_u^2 2\sigma_g^2 1\pi_u^4 1\pi_g^2$ and the outermost two electrons occupy separate antibonding π orbitals with parallel spins, giving a triplet state and accounting for why the molecule is paramagnetic (Fig. 2.12). Oxygen is a strong oxidant, yet most of its reactions are sluggish (a point first made in connection with overpotentials in Section 6.16). For example, a solution of Fe^{2+} is only slowly oxidized by air, even though the reaction is thermodynamically favourable. More obviously (and fortunately for us), the combustion of organic matter in air—a much more favourable reaction than oxidation of Fe(II)—does not occur unless initiated by a source of intense heat.

Several factors contribute to the appreciable activation energy of many reactions of O_2. The first factor is the high bond energy (497 kJ mol^{-1}), which results in a high activation energy for reactions that depend on homolytic dissociation. This energy is available at the elevated temperatures sustained during highly exothermic combustion reactions that proceed by radical chain mechanisms. In contrast, reactions under milder conditions depend upon the ability of O_2 to form bonds with reaction partners. The triplet ground state of O_2, with both π^* orbitals singly occupied, is neither an effective Lewis acid nor an effective Lewis base, and therefore has little tendency to react with p-block Lewis bases or acids in steps that could initiate thermodynamically favourable two- or four-electron transfer reactions. The reactions of O_2 are thus often referred to as being 'spin-restricted'.

Reactions with d-block ions having one or more unpaired electrons are not restricted in this way, although the simplest such process—a one-electron transfer to O_2 that results in superoxide—is unfavourable thermodynamically and requires a reasonably strong reducing agent to achieve a significant rate:

$$O_2(g) + H^+(aq) + e^- \rightarrow HO_2(aq) \quad E^\ominus = -0.13\,V \text{ at pH}=0$$
$$O_2(g) + e^- \rightarrow O_2^-(aq) \quad E^\ominus = -0.33\,V \text{ at pH}=14$$

Reductions of O_2 by d-block metals lie behind important catalytic reactions, such as the Wacker process for oxidation of ethene (Section 31.4) and its conversion to water at the cathode of fuel cells (Box 6.1). In metalloenzymes (Section 26.10), O_2 is activated by coordination to metals such as Fe and Cu, which leads to very rapid four-electron reduction of O_2 to water or incorporation of one or both O atoms into organic molecules.

Oxygen is the 'unavoidable' by-product of electrochemical H_2 generation from water and there is much interest in facilitating its formation, the sluggish kinetics of which represent the chief obstacle to further progress in renewable H_2 generation as well as overpotential losses in fuel cells. Many recent advances in developing O_2-producing catalysts are based on mimicking the Mn oxide cluster that evolves O_2 in photosynthesis (Box 17.2).

(b) Singlet oxygen

The term symbol for the triplet ground state of O_2 is $^3\Sigma_g^-$.[1] The singlet state, $^1\Sigma_g^+$, with antiparallel electrons in the same two π^* orbitals, is higher in energy by 1.63 eV (158 kJ mol^{-1}), and another singlet state $^1\Delta_g$ ('singlet delta'), with both electrons paired in one π^* orbital, lies between these two terms at 0.98 eV (94 kJ mol^{-1}) above the ground state (Fig. 17.1). Of the two singlet states, the latter has much the longer excited-state lifetime, and $O_2(^1\Delta_g)$ survives long enough to participate in chemical reactions. When it is needed for reactions, $O_2(^1\Delta_g)$ can be generated in solution by energy transfer from a photoexcited molecule. Thus [Ru(bpy)$_3$]$^{2+}$ is excited

---

[1] The symbols Σ, Π, and Δ are used for linear molecules such as dioxygen in place of the symbols S, P, and D used for atoms. The Greek letters represent the magnitude of the total orbital angular momentum around the internuclear axis.

BOX 17.2 Finding efficient electrocatalysts for water oxidation

Oxygen is not only essential as a reactant in combustion or fuel cells: it is also a by-product of 'green' hydrogen generation by electrochemical water-splitting. The oxygen evolution reaction (OER) occurring during water electrolysis is of great industrial and economic importance because it limits the rate that H_2 can be produced.

We have already mentioned the importance of developing efficient catalysts for producing H_2 from water (Section 11.4). The direct oxidation of water to O_2 ($E^\ominus = 1.23\,V$) that must occur simultaneously at the anode is far more challenging because it involves several unstable intermediates; hence the OER is sluggish and a much larger and more wasteful overpotential is needed. To produce 'green' H_2 economically it is not only important that OER catalysts perform with only a small overpotential but also that they are robust and composed of cheap, abundant elements. In terms of performance the best catalyst for the OER is IrO$_2$ (Fig. 6.15), but iridium is among the rarest elements known.

In biology, photosynthetic O_2 evolution occurs at a Mn–O cluster that can produce more than 100 molecules of O_2 per second (see Section 26.10(d)). Some promising nonbiological electrocatalysts for water oxidation are based on Co–O clusters. Electrochemical oxidation of aqueous Co^{2+} ions with boronate or phosphate anions results in the electrodeposition of a cobalt oxide layer that catalyses O_2 evolution at a low overpotential. A plausible mechanism (Fig. B17.1) may be understood by considering how the Brønsted acidity of a coordinated water molecule increases as the oxidation number of the metal ion increases (Section 5.2). The Co ions at the surface undergo oxidation by successive coupled proton-electron transfers resulting in an adjacent pair of Co(III)-coordinated oxyl radicals: O–O bond formation occurs and an O_2 molecule is released. Note that as explained in Section 10.8, a tetragonal Co(IV)-O species should not have a Co=O bond (i.e. a bond order of 2), thus it is unstable with respect to the species Co(III)-O$^-$, which has an oxyl radical.

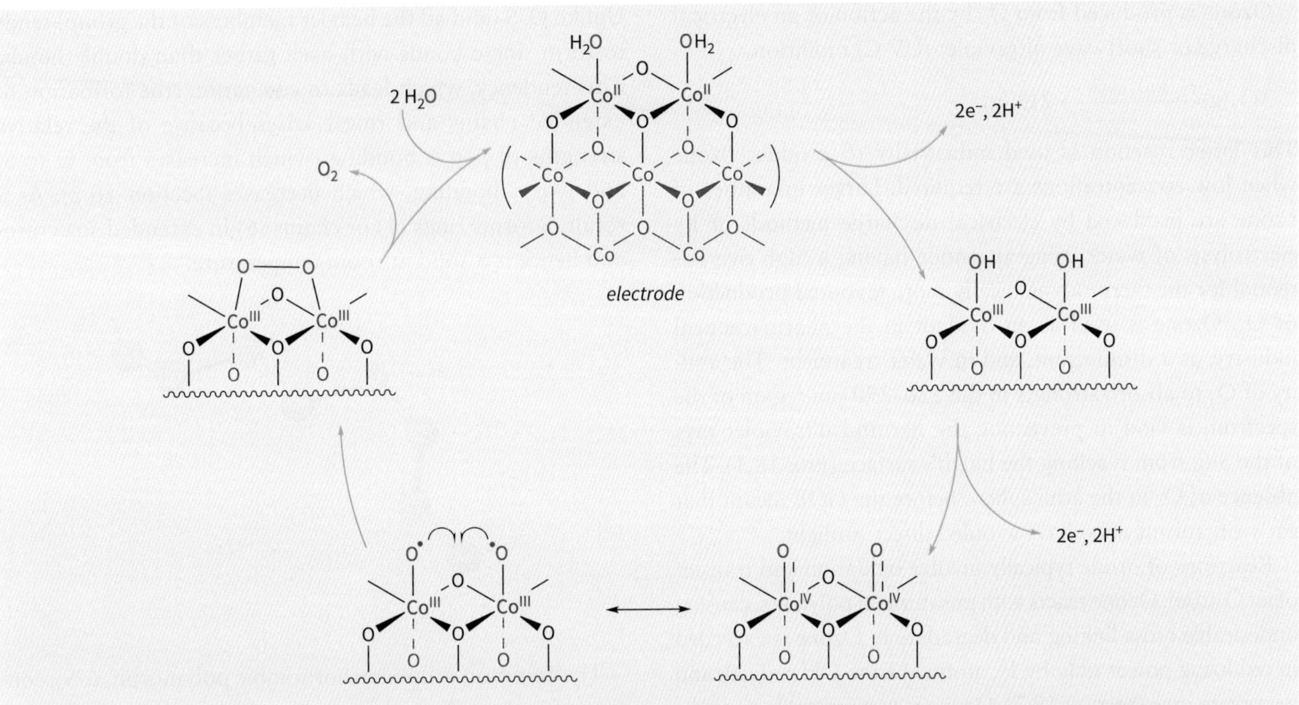

FIGURE B17.1 Possible mechanism for electrocatalytic production of O_2 by a cobalt oxide layer formed on an electrode.

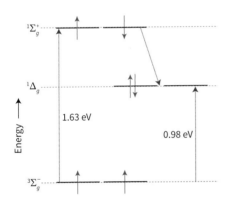

FIGURE 17.1 Occupancies of the antibonding $1\pi_g$ orbitals in triplet (ground state) and singlet excited states of O_2.

by absorption of blue light (452 nm) to give an electronically excited state, denoted $*[Ru(bpy)_3]^{2+}$ (Section 22.5), and this state transfers energy to $O_2(^3\Sigma_g^-)$:

$$*[Ru(bpy)_3]^{2+} + O_2(^3\Sigma_g^-) \rightarrow [Ru(bpy)_3]^{2+} + O_2(^1\Delta_g)$$

Another way to generate $O_2(^1\Delta_g)$ is through the thermal decomposition of an ozonide:

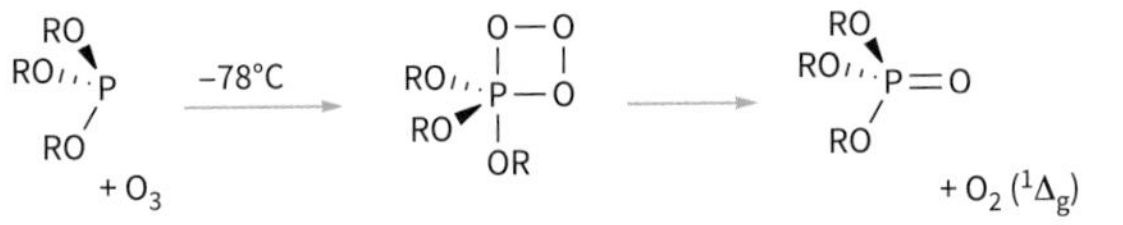

In contrast to the radical character of many $O_2(^3\Sigma_g^-)$ reactions, $O_2(^1\Delta_g)$ reacts as an electrophile. This mode of reaction is feasible because $O_2(^1\Delta_g)$ has an empty π^* orbital, rather than two that are each occupied by a single electron. For example, $O_2(^1\Delta_g)$ adds across a diene, thus mimicking the Diels–Alder reaction of butadiene with an electrophilic alkene:

Singlet oxygen is implicated as one of the biologically hazardous products of photochemical smog. Advantageously, it is one of the destructive agents of programmed cell death (apoptosis) and photodynamic therapy for cancer (Section 32.1).

(c) Ozone

The other allotrope of oxygen, *ozone*, O_3, boils at −112°C and is an explosive and highly reactive endoergic blue gas ($\Delta_f G^\ominus = +163 \text{ kJ mol}^{-1}$). It decomposes into dioxygen:

$$2 O_3(g) \rightarrow 3 O_2(g)$$

but this reaction is slow in the absence of a catalyst or ultraviolet radiation.

Ozone has a pungent odour that resembles that of chlorine; this property is reflected in its name, which is derived from the Greek *ozein*, 'to smell'. The O_3 molecule is angular, in accord with the VSEPR model (5), and has a bond angle of 117°; it is diamagnetic. Gaseous ozone is blue, liquid ozone is blue-black, and solid ozone is violet-black.

5 O_3, C_{2v}

Ozone is produced from O_2 by the action of an electrical discharge or shortwave ultraviolet (UV-C) radiation.

$$3O_2(g) \xrightarrow{h\nu(<240nm)} 2O_3(g)$$

This latter reaction is used industrially to produce ozone when low concentrations are required. Larger quantities of ozone are produced by electrical discharge methods or by electrolysis of water using an anode having a high overpotential for the thermodynamically more favoured production of O_2. Ozone is used as an oxidant in the pharmaceutical industry, as a disinfectant, and in water treatment. The ability of O_3 to absorb strongly in the 220–290 nm region of the spectrum is vital in preventing the harmful ultraviolet rays of the Sun from reaching the Earth's surface (Box 18.1). The absence of O_3 in the atmosphere before the GOE meant that early organisms must have avoided direct sunlight.

Reactions of ozone typically involve oxidation and transfer of an O atom. Ozone reacts with unsaturated polymers, causing undesirable cross-linking and degradation. Ozone is exceeded in oxidizing power only by F_2, atomic O, the OH radical, and perxenate ions (Section 19.7). Ozone is very unstable in acidic solution but is much more stable under basic conditions:

$$O_3(g) + 2H^+(aq) + 2e^- \rightarrow O_2(g) + H_2O(l) \quad E^\ominus = +2.08\,V$$
$$O_3(g) + 2H_2O(l) + 2e^- \rightarrow O_2(g) + 2OH^-(aq) \quad E^\ominus = +1.25\,V$$

Ozone forms ozonides with Group 1 and 2 elements. They are prepared by passing gaseous ozone over the powdered hydroxide, MOH or $M(OH)_2$, at temperatures below −10°C. The ozonides are red-brown solids that decompose on warming:

$$MO_3(s) \rightarrow MO_2(s) + \tfrac{1}{2}O_2(g)$$

The ozonide ion, O_3^-, is angular, like O_3, but with the slightly larger bond angle of 119.5°.

17.4 Sulfur

KEY POINTS Sulfur is extracted as the element from underground deposits. It has many allotropic and polymorphic forms, including a metastable polymer, but its most stable form is the cyclic S_8 molecule.

Sulfur is essential to all life as it occurs in some amino acids and other biologically active molecules. Up to the end of the twentieth century it was obtained by the Frasch process, in which underground deposits are forced to the surface using superheated water and compressed air. This energy-intensive process has largely been overtaken by the Claus process in which H_2S from natural gas and crude oil is first oxidized in air at 1000–1400°C (Section 6.18). This step produces some SO_2 that then reacts with the remaining H_2S at 200–350°C over a catalyst:

$$2H_2S(g) + SO_2(g) \rightarrow 3S(l) + 2H_2O(l)$$

Unlike O, S (and all the heavier members of the group) tends to form single bonds with itself rather than double bonds. This tendency, which leads to catenation (the formation of extended chains and rings) arises because of the relative strengths of p–p σ bonding (which increases from O to S) and p–p π bonding, which decreases (Section 10.2). As a result, S forms rings (**1**) or chains (**6**) in extended structures and hence is a solid at room temperature.

6 S_n

The common yellow orthorhombic polymorph, α-S_8, consists of crown-like eight-membered rings (**1**), and all other forms of S eventually revert to this form. Orthorhombic α-sulfur is an electrical and thermal insulator. When it is heated to 93°C, the packing of the S_8 rings is modified and monoclinic β-S_8 forms. Monoclinic γ-sulfur is formed when molten sulfur that has been heated above 150°C is cooled slowly. This polymorph consists of S_8 rings like the α and β forms, but with more efficient packing, resulting in a higher density.

> **A NOTE ON GOOD PRACTICE**
>
> The various molecular entities that S forms are called allotropes of the element. The various crystalline forms that these entities exist as are termed polymorphs.

It is possible to synthesize and crystallize sulfur rings containing 6–20 S atoms (Table 17.4) which may also exist in several polymorphs, as well as infinite chains. Orthorhombic sulfur melts at 113°C; the yellow liquid darkens above 160°C and becomes more viscous as the sulfur rings break open and polymerize. The resulting helical S_n polymers (**6**) can be drawn from the melt and quenched to form metastable rubber-like materials that slowly revert to α-S_8 at room temperature. The unstable sulfur analogues of dioxygen and ozone, S_2 and S_3, occur naturally in the sulfur vapors released from hot springs and volcanoes. Both are coloured gases: S_2 being violet while S_3 is cherry red. Like O_2, the S_2 molecule has both σ and π bonding to give a triplet ground state and a bond dissociation energy of 421 kJ mol^{-1}.

Sulfur reacts directly with many elements. It ignites in F_2 to form SF_6, reacts rapidly with Cl_2 to form S_2Cl_2, and dissolves in Br_2 to give S_2Br_2, which readily dissociates. It does

TABLE 17.4 Properties of selected sulfur allotropes and polymorphs

Allotrope/polymorph	Melting point/°C	Appearance
S_2	Gas (unstable)	Violet
S_3	Gas (unstable)	Cherry red
S_6	50*d	Orange-red
S_7	39*d	Yellow
α-S_8	113	Yellow
β-S_8	119	Yellow
γ-S_8	107	Pale yellow
S_{12}	148	Pale yellow
S_{20}	124	Pale yellow
S_∞	104	Yellow

d, decomposes.

not react with liquid I_2, which is therefore used as a low-temperature solvent for sulfur.

Elemental sulfur is used in the vulcanization of natural rubber, as a component of gunpowder, and as a fungicide. Most of the sulfur produced industrially is used to manufacture sulfuric acid, H_2SO_4.

17.5 Selenium, tellurium, and polonium

KEY POINTS Selenium and tellurium crystallize in helical chains; polonium crystallizes in a primitive cubic form.

Selenium can be extracted from the waste sludge produced by sulfuric acid plants. Both selenium and tellurium can be extracted from copper sulfide ores, where they occur as the copper selenide or telluride. The extraction method depends on the other compounds or elements present. The first step usually involves oxidation in the presence of sodium carbonate, as in:

$$Cu_2Se(s) + Na_2CO_3(aq) + 2\,O_2(g) \rightarrow$$
$$2\,CuO(s) + Na_2SeO_3(aq) + CO_2(g)$$

The solution containing Na_2SeO_3 and Na_2TeO_3 is acidified with sulfuric acid. The Te precipitates out as the dioxide, leaving selenous acid, H_2SeO_3, in solution. Selenium is recovered by treatment with SO_2:

$$H_2SeO_3(aq) + 2\,SO_2(g) + H_2O(l) \rightarrow Se(s) + 2\,H_2SO_4(aq)$$

Pure tellurium is obtained by dissolving the TeO_2 in aqueous sodium hydroxide, followed by electrolytic reduction:

$$TeO_2(s) + 2\,NaOH(aq) \rightarrow Na_2TeO_3(aq) + H_2O(l) \rightarrow$$
$$Te(s) + 2\,NaOH(aq) + O_2(g)$$

As with S, three polymorphs of Se exist as Se_8 rings (1) and differ only in the packing of the rings to give α, β, and γ forms of red selenium. The most stable form at room temperature is metallic grey selenium, a crystalline material composed of helical chains. The common commercial form of the element is amorphous black selenium; it has a very complex structure comprising rings containing up to 1000 Se atoms. Another amorphous form of Se, obtained by deposition of the vapour, is used as the photoreceptor in the xerographic photocopying process. Selenium is an essential element for humans, but, as with many essential elements, there is only a narrow margin between the minimum daily requirement and toxicity.

Selenium exhibits both photovoltaic character, where light is converted directly into electricity, and photoconductive character. It is a p-type semiconductor (Section 25.6) and is used in electronic and solid-state applications. The photoconductivity of grey selenium arises from the ability of incident light to excite electrons across its small band gap (2.6 eV in the crystalline material, 1.8 eV in the amorphous material). These properties make Se useful in the production of photocells and exposure meters for photography, as well as solar cells. It is also used in photocopier drums and in the glass industry to make red glasses and enamels.

Tellurium crystallizes in a chain structure like that of grey selenium. Polonium crystallizes in a primitive cubic structure, and a closely related higher-temperature form above 36°C. We remarked in Section 4.5 that the primitive cubic structure represents inefficient packing of atoms: Po is the only element that adopts this structure under normal conditions. Tellurium and Po are both highly toxic. The extreme toxicity of Po is enhanced by its intense radioactivity (all 33 isotopes of polonium are radioactive). Having little natural occurrence, Po can be produced in small amounts (gram quantities) through irradiation of ^{209}Bi (atomic number 83) with neutrons, which gives ^{210}Po (atomic number 84) having a half-life of 138 d:

$$^{209}_{83}\text{Bi} + ^{1}_{0}\text{n} \rightarrow ^{210}_{84}\text{Po} + e^-$$

Selenium, Te, and Po combine directly with most elements, although less readily than O or S. The occurrence of multiple bonds is lower than with O or S, as is the tendency towards catenation (compared with S) and the number of allotropes. The unexpected difficulty of oxidizing Se to Se(VI) (Fig. 17.2) is an example of the alternation effect (Section 10.4) that characterizes much of the chemistry of the 4p elements.

17.6 Compounds formed with hydrogen

The impact of hydrogen bonding is seen clearly in the compounds formed between the Group 16 elements and hydrogen. Oxygen forms water and hydrogen peroxide, which are both liquids, while the compounds formed with the heavier elements are all toxic, foul-smelling gases.

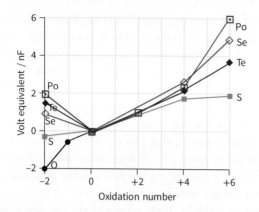

FIGURE 17.2 Frost diagram for the elements of Group 16 in acidic solution. The species with oxidation number −2 are H₂E. For oxidation number −1 the compound is H₂O₂. The positive oxidation numbers refer to the oxoacids or oxoanions.

(a) Water

KEY POINT Hydrogen bonding in water results in a high-boiling point liquid and a highly structured arrangement in the solid state.

At least nine distinct forms of ice have been identified. At 0°C and atmospheric pressure, hexagonal ice, I_h, forms (Fig. 5.14), but between −120°C and −140°C the cubic form, I_c, is produced. At very high pressures, several higher-density polymorphs are formed.

Water is formed by the direct interaction of the elements:

$$H_2(g) + \tfrac{1}{2}O_2(g) \rightarrow H_2O(l) \quad \Delta_f H^\ominus(H_2O,l) = -286\,\text{kJ mol}^{-1}$$

This reaction is very exothermic and provides the basis for the development of the hydrogen economy and hydrogen fuel cells (Fig. B6.1, Box 11.2, and Section 30.3).

Water is the most widely used solvent, not only because it is so widely available but also because of its high relative permittivity (dielectric constant), wide liquid range, and—through a combination of its polar character and ability to form hydrogen bonds—solvating ability. Many anhydrous and hydrated compounds dissolve in water to give hydrated cations and anions. Some predominantly covalent compounds, such as ethanol and ethanoic (acetic) acid, are soluble in water or are miscible with it because of hydrogen-bonded interactions with the solvent. Many other covalent compounds react with water in hydrolysis reactions and examples are discussed in the appropriate chapters.

In addition to simple dissolution and hydrolysis reactions, the importance of aqueous solution chemistry can be seen in redox reactions (Chapter 6) and acid–base reactions (Chapter 5). Water also acts as a Lewis base ligand in metal complexes (Chapter 7). The deprotonated forms, OH^- and particularly the oxide ion O^{2-}, are important ligands for stabilizing higher oxidation states, examples of which are found in the simple oxocations of early d-block elements, such as the vanadyl ion, VO^{2+}. The O^{2-} ion is a very strong π donor and is thus able to form double and triple dative bonds to a metal ion. The so-called 'oxo wall' divides the d block between groups 8 and 9: the former (Fe, Ru, Os) able to form tetragonal complexes in which O^{2-} can form a double bond, the latter (Co, Rh, Ir) unable (Section 10.8).

(b) Hydrogen peroxide and peroxides

KEY POINT Hydrogen peroxide is susceptible to decomposition by disproportionation at elevated temperatures or in the presence of catalysts.

Hydrogen peroxide is a very pale blue, viscous liquid. It has a higher boiling point than water (150°C) and a greater density (1.445 g cm⁻³ at 25°C). It is miscible in water and is usually handled in aqueous solution. The Frost diagram for oxygen (Fig. 17.2) shows that H_2O_2 is unstable with respect to disproportionation:

$$H_2O_2(l) \rightarrow H_2O(l) + \tfrac{1}{2}O_2(g) \quad \Delta_f G^\ominus = -119\,\text{kJ mol}^{-1}$$

This reaction is slow but is explosive when catalysed by a metal surface or alkali dissolved from glass. For this reason, hydrogen peroxide and its solutions are stored in plastic bottles and a stabilizer is added. This reaction can be considered in terms of the reduction half-reactions:

$$\tfrac{1}{2}H_2O_2(aq) + H^+(aq) + e^- \rightarrow H_2O(l) \quad E^\ominus = +1.68\,\text{V}$$
$$H^+(aq) + \tfrac{1}{2}O_2(g) + e^- \rightarrow \tfrac{1}{2}H_2O_2(aq) \quad E^\ominus = +0.70\,\text{V}$$

Consequently, any substance with a standard potential for one-electron oxidation or reduction in the range 0.70–1.68 V that has suitable binding sites can catalyse this reaction.

In acid solution, hydrogen peroxide is a very powerful oxidizing agent:

$$2\,Ce^{3+}(aq) + H_2O_2(aq) + 2\,H^+(aq) \rightarrow 2\,Ce^{4+}(aq) + 2\,H_2O(l)$$

The underlying reason for the oxidizing nature of hydrogen peroxide and its instability with respect to H_2O and O_2 lies in the weakness of the O–O single bond (146 kJ mol⁻¹).

In basic solution, hydrogen peroxide can act as a reducing agent:

$$O_2(g) + H_2O(l) + 2\,e^- \rightarrow HO_2^-(aq) + OH^-(aq) \quad E^\ominus = +0.09\,\text{V}$$

and thus:

$$2\,Ce^{4+}(aq) + H_2O_2(aq) + 2\,OH^-(aq) \rightarrow$$
$$2\,Ce^{3+}(aq) + 2\,H_2O(l) + O_2(g)$$

Hydrogen peroxide reacts with d-metal ions such as Fe^{2+} to form the hydroxyl radical in the Fenton reaction:

$$Fe^{2+}(aq) + H_2O_2(aq) \rightarrow Fe^{3+}(aq) + OH^-(aq) + OH^\bullet(aq) \quad (12)$$

The Fe^{3+} product can react with a second H_2O_2 to regenerate Fe^{2+}, so that the production of hydroxyl radical is catalytic. The hydroxyl radical is one of the strongest oxidizing agents known ($E = +2.85\,\text{V}$), and the reaction is used to oxidize organic matter. In living cells, its reaction with DNA has potentially lethal consequences.

BOX 17.3 How is hydrogen peroxide used as an environmentally friendly bleach?

Hydrogen peroxide is rapidly replacing chlorine and hypochlorite bleaches in industrial applications as it is environmentally benign, generating only water and oxygen as by-products.

The major consumers of hydrogen peroxide bleach are the paper, textile, and wood-pulp industries. A growing market is in de-inking of recycled paper and in the manufacture of kraft paper (strong brown paper). Approximately 85% of all cotton and wool is bleached with hydrogen peroxide. One of its advantages over chlorine-based bleaches is that it does not affect many modern dyes. It is also used to decolour oils and waxes.

Hydrogen peroxide is used to treat domestic and industrial effluent and sewage. Other industrial uses of hydrogen peroxide are the epoxidation of soybean and linseed oils to produce plasticizers and stabilizers for the plastics industry and as a propellant for torpedoes and missiles. Hydrogen peroxide is increasingly being used as a green oxidant. It can be used in aqueous solution if generated *in situ*, and the only by-product is water.

Hydrogen peroxide is a slightly stronger acid than water:

$$H_2O_2(aq) + H_2O(l) \rightleftharpoons H_3O^+(aq) + HO_2^-(aq) \quad pK_a = 11.65$$

Deprotonation occurs in other basic solvents such as liquid ammonia, and NH_4OOH has been isolated and found to consist of NH_4^+ and HO_2^- ions. When solid NH_4OOH melts (at 25°C), the melt contains hydrogen-bonded NH_3 and H_2O_2 molecules.

The oxidizing ability of hydrogen peroxide and the harmless nature of its by-products lead to its many applications. It is used in water treatment to oxidize pollutants, as a mild antiseptic, and as a bleach in the textile, paper, and hair-care industries (Box 17.3).

(c) Hydrogen superoxide and superoxide

One-electron reduction of O_2 produces the superoxide anion O_2^-. In aqueous solution, O_2^- is a weak reductant but quite a potent oxidant, but it rapidly undergoes disproportionation to give O_2 and H_2O_2. Superoxide is easily produced in biological environments, and along with H_2O_2 it is referred to in biochemical and medical circles as a 'reactive oxygen species' (ROS). These agents are held responsible for a host of disorders, ranging from mutagenesis to aging.

EXAMPLE 17.2 The properties of O_2 and its reduction products in water

The Pourbaix diagram shown in Fig. 17.3 describes the redox chemistry of different oxygen species in water. Write the half-cell reactions in each case, taking into consideration the changes indicated by the vertical dashed lines, and identify those labelled A and B.

We note first of all that the line corresponding to the four-electron reduction of O_2 is linear between pH 0 and pH 14, with a slope, −0.06 V/pH unit, as expected for the reaction:

$$O_2(g) + 4H^+(aq) + 4e^- \rightarrow 2H_2O(l)$$

The line corresponding to the two-electron reduction of O_2 occurring at a less positive potential has a slope of −0.06 V/pH unit at pH < 11.7 (the pK_a for H_2O_2 and −0.03 V/pH unit at pH > 11.8). The half-cell reactions for these two regions are:

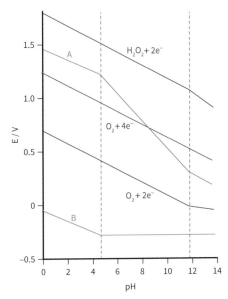

FIGURE 17.3 Combined Pourbaix diagrams for the half-cell reactions of O_2 and its reduction products.

$$O_2(g) + 2H^+(aq) + 2e^- \rightarrow 2H_2O_2(aq) \quad \text{at pH} < 11.7$$
$$O_2(g) + H^+(aq) + 2e^- \rightarrow 2HO_2^-(aq) \quad \text{at pH} > 11.7$$

Likewise, the line corresponding to the two-electron reduction of peroxide which occurs at a more positive potential corresponds to the half-cell reactions:

$$H_2O_2(aq) + 2H^+(aq) + 2e^- \rightarrow 2H_2O(l) \quad \text{at pH} < 11.7$$
$$HO_2^-(aq) + 3H^+(aq) + 2e^- \rightarrow 2H_2O(l) \quad \text{at pH} > 11.7$$

Lines A and B correspond to the one-electron reactions associated with superoxide, assuming a value of 4.8 for the pK of HO_2. Thus for A:

$$HO_2(aq) + H^+(aq) + e^- \rightarrow H_2O_2(aq) \quad \text{at pH} < 4.8$$
$$O_2^-(aq) + 2H^+(aq) + e^- \rightarrow H_2O_2(aq) \quad \text{for pH} > 4.8, < 11.7$$
$$O_2^-(aq) + H^+(aq) + e^- \rightarrow HO_2^-(aq) \quad \text{at pH} > 11.7$$

and for B:

$$O_2(g) + H^+(aq) + e^- \rightarrow HO_2(aq) \quad \text{at pH} < 4.8$$
$$O_2(g) + e^- \rightarrow O_2^-(aq) \quad \text{at pH} > 4.8$$

Self-test 17.2 Comment on the role of pH in determining the inherent stabilities of superoxide and peroxide species in water.

(d) Compounds of sulfur, selenium, and tellurium with hydrogen

KEY POINTS Unlike water, compounds of the type H_2X are gases under normal conditions, a consequence of much weaker hydrogen bonding.

Hydrogen sulfide is produced by volcanoes and by some microorganisms (Box 17.4). It is an impurity in natural gas and must be removed before the gas is used.

Pure H_2S can be prepared by direct combination of the elements above 600°C:

$$H_2(g) + S(g) \rightarrow H_2S(g)$$

Hydrogen sulfide is easily generated in the laboratory by trickling dilute hydrochloric or phosphoric acid on to FeS:

$$FeS(s) + 2HCl(aq) \rightarrow H_2S(g) + FeCl_2(aq)$$

It can also be prepared by hydrolysis of aluminium sulfide, which is easily generated by ignition of a mixture of the elements:

$$2Al(s) + 3S(s) \rightarrow Al_2S_3(s)$$
$$Al_2S_3(s) + 3H_2O(l) \rightarrow Al_2O_3(s) + 3H_2S(g)$$

It is readily soluble in water and is a weak acid:

$$H_2S(g) + H_2O(l) \rightleftharpoons H_3O^+(aq) + HS^-(aq) \quad pK_{a1} = 6.89$$
$$HS^-(aq) + H_2O(l) \rightleftharpoons H_3O^+(aq) + S^{2-}(aq) \quad pK_{a2} = 19.00$$

BOX 17.4 How is sulfur recycled by biology?

Sulfur is essential to all life forms through its presence in the amino acids cysteine and methionine, and in many active sites of enzymes. Moreover, many organisms obtain energy by the oxidation or reduction of inorganic sulfur compounds. The resultant transformations, catalysed by numerous enzymes, constitute the biological sulfur cycle. The simplified version shown in Fig. B17.2 distinguishes the roles and impacts of each half cycle.

Sulfur-reducing bacteria (SRB) such as strains of *Desulfovibrio* generate sulfide (usually releasing H_2S) under anaerobic conditions using sulfate as their oxidant. The role of this half cycle is to generate energy and produce reduced S species for biosynthesis. These bacteria are found in anaerobic environments where both SO_4^{2-} and reduced organic matter are found, for example, in anoxic marine sediments and in the rumen of sheep and cattle. The first stage is the reduction of sulfate to sulfite: this is usually followed by reduction to sulfide which in many microorganisms occurs in a complex eight-electron process catalysed by a single metalloenzyme known as sulfite reductase. The impacts of SRBs are found in sulfide ore formation, H_2S release into the atmosphere, biocorrosion, the souring of petroleum under anaerobic conditions, the Cu–Mo antagonism in ruminants, and many other physiological, ecological, and biogeochemical contexts.

The oxidative half cycle is the province of bacteria such as *Thiobacillus* and *Rhodobacter*. *Thiobacilli* live solely on inorganic materials (they are known as **chemoautotrophs**): they use energy obtained by oxidizing reduced S-species to drive all their cellular reactions, including the fixation of carbon from CO_2. *Rhodobacter* are purple-coloured photosynthetic bacteria that can use H_2S instead of H_2O as their electron source. The oxidation of sulfide to sulfate produces an acidic environment in which some bacteria thrive. Mine drainage water can have a microbially produced pH as low as 1.5, and *Thiobacilli* are used commercially to mobilize metals from sulfide ores. For example, *Thiobacillus ferrooxidans* not only oxidizes the sulfur in iron sulfide deposits, but also oxidizes the Fe(II) to Fe(III) which is soluble at low pH:

$$4FeS_2 + 15O_2 + 2H_2O \rightarrow 4Fe^{3+} + 8SO_4^{2-} + 4H^+$$

The final step, oxidation of sulfite to sulfate, has direct importance for humans, as many foods contain sulfite, either as a preservative or naturally. The reaction is catalysed by sulfite oxidase, a Mo-containing enzyme produced by higher organisms, that is mentioned in Section 26.12.

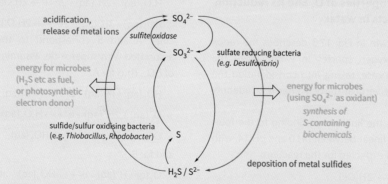

FIGURE B17.2 The biological sulfur cycle, indicating the common bacteria that are involved, the roles of the metabolic processes, and the impacts made on the environment.

Acidic solutions of H_2S are mild reducing agents and deposit elemental S on standing.

In a similar way, H_2Se can be made by direct combination of the elements, by the reaction of FeSe with hydrochloric acid, or by hydrolysis of Al_2Se_3:

$$H_2(g) + Se(s) \rightarrow H_2Se(g)$$
$$FeSe(s) + 2\,HCl(aq) \rightarrow H_2Se(g) + FeCl_2(aq)$$
$$Al_2Se_3(s) + 3\,H_2O(l) \rightarrow Al_2O_3(s) + 3\,H_2Se(g)$$

H_2Te is made by hydrolysis of Al_2Te_3 or by the action of hydrochloric acid on Mg, Zn, or Al tellurides:

$$Al_2Te_3(s) + 6\,H_2O(l) \rightarrow 3\,H_2Te(g) + 2\,Al(OH)_3(aq)$$
$$MgTe(s) + 2\,HCl(aq) \rightarrow H_2Te(g) + MgCl_2(aq)$$

All three gases—H_2S, H_2Se, and H_2Te—dissolve in water to produce acidic solutions. The acidity constants increase from H_2S to H_2Te (Table 17.2). Like their sulfur analogue, aqueous solutions of H_2Se and H_2Te are readily oxidized and deposit elemental selenium and tellurium on standing.

17.7 Halides

KEY POINTS The halides of oxygen have limited stability but its heavier congeners form an extensive series of halogen compounds; typical formulas are EX_2, EX_4, and EX_6.

The oxidation number of O is −2 in all its compounds with the halogens other than F. Oxygen difluoride, OF_2, is the highest fluoride of oxygen and hence contains O in its highest oxidation state (+2).

The structures of the sulfur halides S_2F_2, SF_4, SF_6, and S_2F_{10} (Table 17.3) are all in line with the VSEPR model. Thus, SF_4 has 10 valence electrons around the S atom, two of which form a lone pair in an equatorial position of a trigonal bipyramid. We have already mentioned the theoretical evidence that the molecular orbitals bonding the F atoms to the central atom in SF_6 primarily use the sulfur 4s and 4p orbitals, with the 3d orbitals playing a relatively unimportant role (Section 2.6). The same seems to be true of SF_4 and S_2F_{10}.

Sulfur hexafluoride is a gas at room temperature. It is very unreactive and its inertness stems from strong S–F bonds and the suppression, presumably by steric protection of the central S atom, of thermodynamically favourable reactions, such as hydrolysis

$$SF_6(g) + 4\,H_2O(l) \xrightarrow{\;\;\;\;} 6\,HF(aq) + H_2SO_4(aq)$$

The SeF_6 molecule is easily hydrolysed and is generally more reactive than SF_6. Similarly, the sterically less hindered molecule SF_4 is reactive and undergoes rapid partial hydrolysis:

$$SF_4(g) + H_2O(l) \rightarrow OSF_2(aq) + 2\,HF(aq)$$

Both SF_4 and SeF_4 are selective fluorinating agents for the conversion of –COOH into –CF_3 and C=O and P=O groups into –CF_2 and –PF_2 groups:

$$2\,R_2CO(l) + SF_4(g) \rightarrow 2\,R_2CF_2(sol) + SO_2(g)$$

Sulfur chlorides are commercially important. The reaction of molten S with Cl_2 yields the foul-smelling and toxic substance disulfur dichloride, S_2Cl_2, which is a yellow liquid at room temperature (b.p. 138°C). Disulfur dichloride and its further chlorination product sulfur dichloride, SCl_2, an unstable red liquid, are produced on a large scale for use in the vulcanization of rubber. In this process, S-atom bridges are introduced between polymer chains so the rubber object can retain its shape.

Sulfur iodides are unstable, but the iodides of Te and Po are more robust. A series of catenated subhalides exist for the heavy members. For example, Te_2I and Te_2Br consist of ribbons of edge-shared Te hexagons with halogen bridges (7).

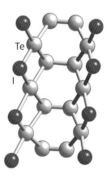

7 Te_2I

17.8 Metal oxides

KEY POINTS The oxides formed by metals include the basic oxides with high oxygen coordination number that are formed with most M^+ and M^{2+} ions. Oxides of metals in intermediate oxidation states often have more complex structures and are amphoteric. Metal peroxides and superoxides are formed between O_2 and alkali metals and alkaline earth metals. Terminal E=O linkages and E–O–E bridges are common with nonmetals and with metals in high oxidation states. Many different oxides of the d-block elements exist, with a wide variety of structures, varying from ionic lattices to covalent molecules.

The O_2 molecule readily removes electrons from metals to form a variety of metal oxides containing the anions O^{2-} (oxide), O_2^- (superoxide), and O_2^{2-} (peroxide). Even though the existence of O^{2-} can be rationalized in terms of a closed-shell noble gas electron configuration, the formation of $O^{2-}(g)$ from $O_2(g)$ is highly endothermic, and the ion is stabilized in the solid state by the very favourable lattice energies that result from its large charge-to-radius ratio (Section 4.12).

Alkali metals and alkaline earth metals often form peroxides or superoxides (Sections 12.8 and 13.8), but peroxides and superoxides of other metals are rare. Among the metals, only some of the noble metals do not form thermodynamically stable oxides. However, even where no bulk oxide phase is formed, an atomically clean metal surface (which can be prepared only in an ultrahigh vacuum) is quickly covered with a surface layer of oxide when it is exposed to traces of oxygen.

Many different oxides are known for the d-block elements, with a number of different structures. Oxygen has the ability to bring out the highest oxidation state for some elements, but oxides also exist for some elements in very low oxidation states: in Cu_2O, copper is present as Cu(I). Monoxides are known for all of the 3d-series metals and have the rock-salt structure characteristic of ionic solids, but their properties, which are discussed in more detail in Chapter 25, indicate significant deviations from the simple ionic $M^{2+}O^{2-}$ model. For example, TiO has metallic conductivity and FeO is always deficient in iron. The early d-block monoxides are strong reducing agents. Thus, TiO is easily oxidized by water or oxygen, and MnO is a convenient oxygen scavenger that is used in the laboratory to remove oxygen impurity in inert gases down to the parts-per-billion range.

Structural trends in the metal oxides are not readily summarized, but for oxides in which the metal has oxidation number +1, +2, or +3, the O^{2-} ion is generally in a site of high coordination number:

M(I): M_2O oxides often have a rutile or antifluorite structure (6:3 and 8:4 coordination, respectively).

M(II): MO oxides usually have the rock-salt structure (6:6 coordination).

M(III): M_2O_3 oxides often have 6:4 coordination.

At the other extreme, higher oxides such as M_2O_7 or MO_4 compounds are molecular: examples being manganese heptoxide, Mn_2O_7, which exists as a volatile, corrosive liquid, and osmium tetroxide, OsO_4. The structures of the oxides with metals in high oxidation states and the oxides of non-metallic elements frequently have multiple-bond character in which the O^{2-} donates an electron pair into a σ bond and uses one or two electron pairs for π bonding. The O^{2-} entity may be terminal, often represented as E=O, or bridging, as in E–O–E.

17.9 Metal sulfides, selenides, tellurides, and polonides

KEY POINTS Monatomic and polyatomic sulfide ions occur as discrete anions and as ligands. Monosulfides with the nickel-arsenide structure are formed by most 3d metals. The 4d- and 5d-series metals often form disulfides with alternating layers of metal ions and sulfide ions; binary disulfides of the early d metals often have a layered structure, whereas Fe^{2+} and many of the later d-metal disulfides contain discrete S_2^{2-} ions.

Many metals occur naturally as their sulfide ores. The ores are roasted in air to form the oxide or the water-soluble sulfate, from which the metals are extracted. The sulfides can be prepared in the laboratory or industrially by several routes; direct combination of the elements, reduction of a sulfate, or precipitation of an insoluble sulfide from solution by addition of H_2S:

$$Fe(s) + S(s) \rightarrow FeS(s)$$
$$MgSO_4(s) + 4C(s) \rightarrow MgS(s) + 4CO(g)$$
$$M^{2+}(aq) + H_2S(g) \rightarrow MS(s) + 2H^+(aq)$$

The solubilities of the metal sulfides vary enormously. The Group 1 and 2 sulfides are soluble, whereas the sulfides of the heavy elements of Group 11 and 12 are among the least soluble compounds known. The wide variation enables selective separation of metals to take place on the basis of the solubilities of the sulfides.

The Group 1 sulfides, M_2S, adopt the antifluorite structure (Section 4.9). The Group 2 elements and some f-block elements form monosulfides, MS, with a rock-salt structure. Monosulfides are common in the 3d series (Table 17.5) and most have the nickel-arsenide structure (Fig. 4.37).

The disulfides of the d metals fall into two classes (Table 17.6). One class consists of layered compounds with either the CdI_2 or the MoS_2 structure; the other consists of compounds containing discrete S_2^{2-} groups, the pyrites and marcasite structures.

The layered disulfides are built from a sulfide layer, a metal layer, and then another sulfide layer (e.g. Fig. 17.4). These sandwiches stack together in the crystal with sulfide layers in one slab adjacent to a sulfide layer in the next. Clearly, this structure is not consistent with a simple ionic model and its formation indicates a large degree of covalence. The metal ion in these layered structures is surrounded by six S atoms. Its coordination environment is octahedral in some cases (such as PtS_2, which adopts the CdI_2 structure) and trigonal-prismatic in others (MoS_2). The layered MoS_2 structure (Fig. 17.4) is favoured by S–S bonding as indicated by short S–S distances within each of the MoS_2 slabs. Some of the layered metal sulfides undergo intercalation reactions in

TABLE 17.5 Structures of d-block MS compounds

	Group					
	4	5	6*	7	8	9	10
Nickel-arsenide structure (shaded)	Ti	V		Mn†	Fe	Co	Ni
Rock-salt structure (unshaded)	Zr	Nb					

* Metal monosulfides of Group 6 are not shown; some of the heavier metals have more complex structures.
† MnS has two polymorphs; one has a rock-salt structure, the other has a wurtzite structure.

TABLE 17.6 Structures of d-block MS$_2$ compounds

	Group*							
	4	5	6	7	8	9	10	11
Layered (shaded)	Ti			Mn	Fe	Co	Ni	Cu
Pyrite or marcasite	Zr	Nb	Mo		Ru	Rh		
(unshaded)	Hf	Ta	W	Re	Os	Ir	Pt	

* Metals not shown do not form disulfides or have disulfides with complex structures.
Adapted from A. F. Wells, *Structural Inorganic Chemistry*. Oxford University Press (1984).

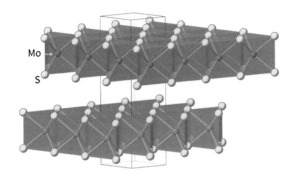

FIGURE 17.4 The structure of MoS$_2$.

which ions or molecules penetrate between adjacent sulfide layers (Section 25.14).

Compounds containing discrete S_2^{2-} ions adopt the pyrite (Fig. 17.5) or marcasite structure. The stability of the formal S_2^{2-} ion in metal sulfides is much greater than that of the O_2^{2-} ion in peroxides.

Simple thiometallate complexes such as $[MoS_4]^{2-}$ can be synthesized easily by passing H$_2$S gas through a strongly basic aqueous solution of molybdate or tungstate ions:

$$[MoO_4]^{2-}(aq) + 4H_2S(g) \rightarrow [MoS_4]^{2-}(aq) + 4H_2O(l)$$

These tetrathiometallate anions are building blocks for the synthesis of complexes containing more metal atoms.

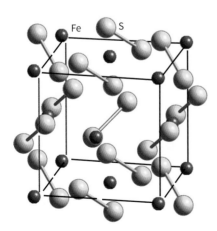

FIGURE 17.5 The structure of pyrite, FeS$_2$.

For example, they will coordinate to many dipositive metal ions, such as Co^{2+} and Zn^{2+}:

$$Co^{2+}(aq) + 2\,[MoS_4]^{2-}(aq) \longrightarrow \left[\begin{array}{c}S\diagdown_{Mo^{VI}}\diagup^S\diagdown_{Co^{II}}\diagup^S\diagdown_{Mo^{VI}}\diagup^S \\ S\diagup\quad\diagdown S\quad S\diagup\quad\diagdown S\end{array}\right]^{2-}(aq)$$

The selenides and tellurides are the most common naturally occurring sources of the elements. Group 1 and 2 selenides, tellurides, and polonides are prepared by direct interaction of the elements in liquid ammonia. They are water-soluble solids that are rapidly oxidized in air to produce the elements, with the exception of the polonides (for which they are among the most stable compounds of the element). The selenides and tellurides of Li, Na, and K adopt the antifluorite structure whereas those of the heavier elements of Group 1 adopt the rock-salt structure. Selenides, tellurides, and polonides of the d metals are also prepared by direct interaction of the elements and are nonstoichiometric. Two examples are compounds of approximate stoichiometry Ti$_2$Se and Ti$_3$Se.

Sulfides, selenides, and tellurides of Group 12 elements form the industrially important Group 12/16 semiconductors, previously known as II/VI semiconductors (Section 25.6). These compounds contain a Group 12 cation and a Group 16 anion and are more ionic than the Group 13/15 semiconductors. Examples are CdS, CdSe, CdTe, and ZnSe, and they are used in optoelectronic applications, such as solar cells and light-emitting diodes.

17.10 Oxides of sulfur, selenium, and tellurium

In this section we concentrate on compounds formed between oxygen and its congeners in Group 16. Oxides of other elements are described in the relevant group chapters.

(a) Sulfur oxides and oxohalides

KEY POINTS Sulfur dioxide is a mild Lewis acid towards p-block bases.

Sulfur dioxide (**2**) and sulfur trioxide (**3**) are both Lewis acids, with the S atom being the acceptor site, but SO$_3$ is much the stronger and harder acid. The high Lewis acidity of SO$_3$ accounts for its occurrence as a cyclic trimeric O-bridged solid at room temperature and pressure (**4**).

Sulfur dioxide is manufactured on a large scale by combustion of sulfur or H$_2$S or by roasting sulfide ores in air:

$$4\,FeS(s) + 7\,O_2(g) \rightarrow 4\,SO_2(g) + 2\,Fe_2O_3(s)$$

The atmospheric release of SO$_2$ has been implicated in the production of acid rain (Box 17.5). Although SO$_2$ dissolves in water to produce a solution commonly referred to as

> **BOX 17.5 What is acid rain?**
>
> The main components of acid rain are nitric and sulfuric acids produced by interaction of the oxides with hydroxyl radicals. The hydroxyl radicals are formed when water reacts with oxygen atoms which are produced by the photodecomposition of ozone:
>
> $HO\cdot + NO_2 \rightleftharpoons NHO_3$
>
> $HO\cdot + SO_2 \rightleftharpoons HSO_3\cdot$
>
> $HSO_3\cdot + O_2 + H_2O \rightleftharpoons H_2SO_4 + HO_2\cdot$
>
> The resulting hydroperoxyl radicals produce additional hydroxyl radicals:
>
> $HO_2\cdot + X \rightleftharpoons XO + HO\cdot \quad X = NO \text{ or } SO_2$
>
> Molecules of sulfuric and nitric acid form strong hydrogen bonds with one another, with metal oxides and gases in the atmosphere, and with water, to form particles. These small particles are a major health threat in polluted air, and increased concentrations of particulate matter in the size range of 2.5 μm or less are linked to increased mortality from respiratory and heart diseases. These particles are small enough to lodge deep in the lungs, and they can carry noxious chemicals on their surface.
>
> These particles affect the wider ecosystem because their acidity is carried into natural rainfall, increasingly leaching alkali metal (Na^+, K^+) and alkaline earth (Ca^{2+}, Mg^{2+}) ions from the soil (where they are normally trapped in ion-exchange sites in clay and humus or in limestone). Depletion of these nutrients limits plant growth. The same chemistry also erodes marble statues and buildings. Granite-lined lakes (which have low buffer capacity) can be acidified, leading to the disappearance of fish and other aquatic life. Because emissions from combustion sources can travel long distances, acid rain is a regional problem, with large areas at risk, particularly those downwind of coal-fired power plants, which have tall smokestacks to disperse the NO_x and SO_2 exhaust gases. These health and environmental effects combine to make NO_x and SO_2 a main focus of air pollution regulatory activity.

sulfurous acid, H_2SO_3, this is in fact a complex mixture of numerous species. Sulfur dioxide forms weak complexes with simple p-block Lewis bases. For example, although it does not form a stable complex with H_2O, it does form stable complexes with stronger Lewis bases, such as trimethylamine and F^- ions. Sulfur dioxide is a useful solvent for acidic substances.

(b) Oxides of selenium and tellurium

KEY POINTS Selenium and tellurium dioxides are polymorphic. Illustrating the alternation effect, both SeO_2 and SeO_3 are thermodynamically less stable than their sulfur and tellurium counterparts.

The dioxides of Se, Te, and Po can be prepared by direct reaction of the elements. Selenium dioxide is a white solid having a polymeric structure, that sublimes at 315°C (**8**). It is used as an oxidizing agent in organic chemistry. It is thermodynamically less stable than SO_2 or TeO_2 and is reduced to selenium on reaction with NH_3, N_2H_4, or aqueous SO_2:

$3 SeO_2(s) + 4 NH_3(l) \rightarrow 3 Se(s) + 2 N_2(g) + 6 H_2O(l)$

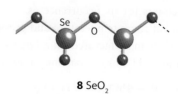

8 SeO_2

Tellurium dioxide occurs naturally as the mineral tellurite, β-TeO_2, which has a layer structure in which TeO_4 units form dimers (**9**). Synthetic α-TeO_2 consists of similar TeO_4 units that share all vertices to form a three-dimensional rutile-like structure (**10**). Illustrating the trend for increasing metal-like behaviour down the group, polonium dioxide exists as the yellow form with the fluorite structure and as the red tetragonal form.

9 A $(TeO_4)_2$ fragment of β-TeO_2 **10** A TeO_4 fragment of α-TeO_2

The fact that selenium trioxide, unlike SO_3 or TeO_3, is thermodynamically less stable than the dioxide (Table 17.7), is a well-known example of the alternation effect (Section 10.4). Selenium trioxide is a white hygroscopic solid which sublimes at 100°C and decomposes at 165°C. In the solid state the structure is based on Se_4O_{12} tetramers (**11**) but it is monomeric in the vapour phase.

11 A $(SeO_3)_4$ (C_{4v}) unit of SeO_3

TABLE 17.7 Standard enthalpies of formation, $\Delta_f H^\ominus/\text{kJ mol}^{-1}$, of sulfur, selenium, and tellurium oxides

SO_2	−297	SO_3	−432
SeO_2	−230	SeO_3	−184
TeO_2	−325	TeO_3	−348

Tellurium trioxide exists as the yellow α-TeO_3, which is prepared by dehydration of $Te(OH)_6$, and the more stable β-TeO_3, which is made by heating α-TeO_3 or $Te(OH)_6$ in oxygen.

(c) Chalcogen oxohalides

KEY POINTS The most important oxohalides are those of sulfur; selenium and tellurium oxofluorides are known, and the 'teflate' ion is a useful ligand.

Many chalcogen oxohalides are known. The most important are the thionyl dihalides, OSX_2, and the sulfuryl dihalides, O_2SX_2. One laboratory application of thionyl dichloride is the dehydration of metal chlorides:

$$MgCl_2 \cdot 6H_2O(s) + 6OSCl_2(l) \rightarrow MgCl_2(s) + 6SO_2(g) + 12HCl(g)$$

The compound $F_5TeOTeF_5$ and its selenium analogue are known. The $OTeF_5^-$ ion, known informally as 'teflate', is a bulky anion in which the oxygen atom can act as an electron-pair donor in a coordinate bond. Teflate is a well-established ligand for high-oxidation-state d-metal and main-group element complexes such as $[Ti(OTeF_5)_6]^{2-}$ (**12**), $[Xe(OTeF_5)_6]$, and $[M(C_5H_5)_2(OTeF_5)_2]$, where M=Ti, Zr, Hf, W, and Mo.

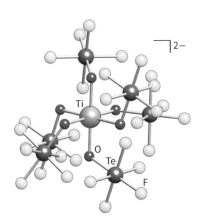

12 $[Ti(OTeF_5)_6]^{2-}$

17.11 Oxoacids of sulfur

Sulfur (like N and P) forms many oxoacids that are important in the laboratory and in industry. These exist in aqueous solution or as the solid salts of the oxoanions (Table 17.8).

(a) Redox properties of the oxoanions

KEY POINTS The oxoanions of sulfur include the sulfite ion, SO_3^{2-}, which is a good reducing agent, the rather unreactive sulfate ion, SO_4^{2-}, and the strongly oxidizing peroxodisulfate ion, $O_3SOOSO_3^{2-}$. As with sulfur, the redox reactions of selenium and tellurium oxoanions are often slow.

Sulfur's common oxidation numbers are −2, 0, +2, +4, and +6, but there are also many S–S bonded species that are assigned odd and fractional average oxidation numbers. A simple example is the thiosulfate ion, $S_2O_3^{2-}$, in which the average oxidation number of S is +2, but in which the environments of the two S atoms are quite different. The thermodynamic relations between the oxidation states are summarized by the Frost diagram (Fig. 17.2).

We saw in Section 6.6 that the pH of a solution has a marked effect on the redox properties of oxoanions. This strong dependence is true for SO_2 and SO_3^{2-} because the former is easily reduced in acidic solution and is therefore an oxidizing agent, whereas the latter in basic solution is primarily a reducing agent:

$$SO_2(aq) + 4H^+(aq) + 4e^- \rightarrow S(s) + 2H_2O(l) \quad E^\ominus = +0.50\,\text{V}$$
$$SO_4^{2-}(aq) + H_2O(l) + 2e^- \rightarrow SO_3^{2-}(aq) + 2OH^-(aq) \quad E^\ominus = -0.94\,\text{V}$$

The principal species present in acidic solution is SO_2, not H_2SO_3. The oxidizing character of SO_2 is exploited for use as a mild disinfectant and as a preservative for dried fruit and wine.

The peroxodisulfate ion, $OSOOSO_3^{2-}$, is a powerful and useful oxidizing agent, although this reactivity reflects the properties of O rather than S because it arises from the weakness of the O–O bond, as we have discussed above for hydrogen peroxide:

$$E^\ominus = +1.96\,\text{V}$$

The oxoanions of selenium and tellurium are a much less diverse and extensive group. Selenic acid is thermodynamically a strong oxidizing acid:

$$SeO_4^{2-}(aq) + 4H^+(aq) + 2e^- \rightarrow H_2SeO_3(aq) + H_2O(l)$$
$$E^\ominus = +1.15\,\text{V}$$

However, like SO_4^{2-}, and in common with the behaviour of oxoanions of other elements in high oxidation states, the reduction of SeO_4^{2-} is generally slow. Telluric acid exists as $Te(OH)_6$ and also as $(HO)_2TeO_2$ in solution. Again, its reduction is thermodynamically favourable but kinetically sluggish.

TABLE 17.8 Some sulfur oxoanions

Oxidation number	Formula	Name	Structure	Remarks
One S atom				
+4	SO_3^{2-}	Sulfite	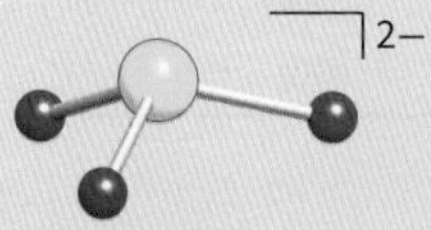	Basic, reducing agent
+6	SO_4^{2-}	Sulfate	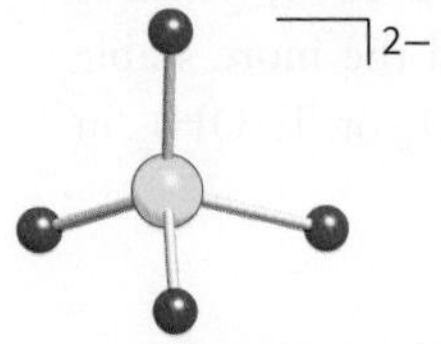	Weakly basic
Two S atoms				
+2 (average)	$S_2O_3^{2-}$	Thiosulfate	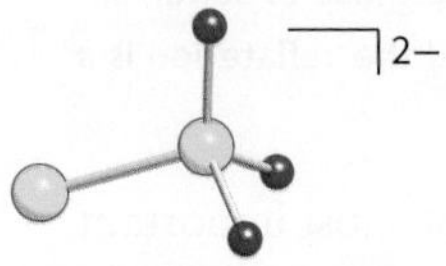	Moderately strong reducing agent
+3	$S_2O_4^{2-}$	Dithionite	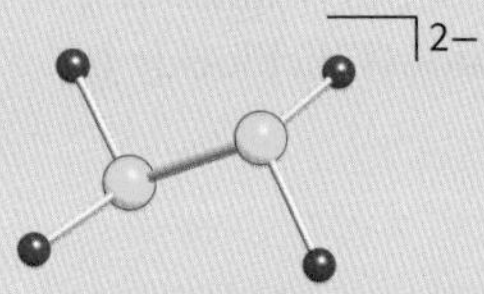	Strong reducing agent
+5	$S_2O_6^{2-}$	Dithionate	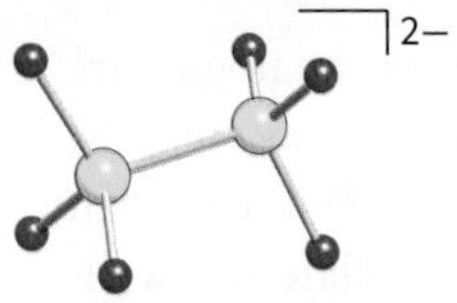	Resists oxidation and reduction
Polysulfur oxoanions				
Variable	$S_nO_{2n+2}^{2-}$ $3 \leq n \leq 20$	$n = 4$, tetrathionate	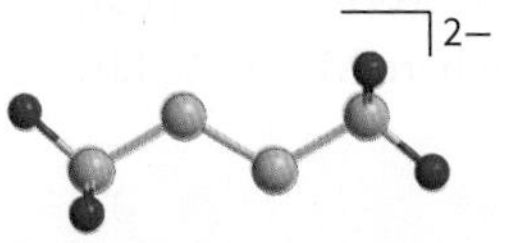	

(b) Sulfuric acid

KEY POINTS Sulfuric acid is a strong acid; it is a useful nonaqueous solvent because of its extensive autoprotolysis.

Sulfuric acid is a dense viscous liquid. It dissolves in water in a highly exothermic reaction:

$$H_2SO_4(l) \rightarrow H_2SO_4(aq) \quad \Delta_r H^\ominus = -880 \text{ kJ mol}^{-1}$$

It is a strong Brønsted acid in water ($pK_{a1} = -2$) but not for its second deprotonation ($pK_{a2} = 1.92$). Anhydrous H_2SO_4 has a very high relative permittivity and high electrical conductivity, consistent with extensive autoprotolysis:

$$2H_2SO_4(l) \rightleftharpoons H_3SO_4^-(sol) + HSO_4^+(sol) \quad K = 2.7 \times 10^{-4}$$

The equilibrium constant for this autoprotolysis (Section 5.1) is greater than that of water by a factor of more than 10^{10}. This property leads to the use of sulfuric acid as a nonaqueous, protic solvent.

As well as undergoing autoprotolysis, H_2SO_4 dissociates into H_2O and SO_3, which react further with H_2SO_4 to give a number of products:

$$SO_3 + H_2SO_4 \rightleftharpoons H_2S_2O_7$$

$$H_2S_2O_7 + H_2SO_4 \rightleftharpoons H_3SO_4^+ + HS_2O_7^-$$

Consequently, rather than being a single substance, anhydrous sulfuric acid is made up of a complex mixture of at least seven characterized species.

Sulfuric acid is one of the most important chemicals produced on an industrial scale. Over 80% of it is used to manufacture phosphate fertilizers from phosphate rock:

$$Ca_5F(PO_4)_3(s) + 5H_2SO_4(aq) + 10H_2O(l) \rightarrow$$
$$5CaSO_4 \cdot 2H_2O(aq) + HF(aq) + 3H_3PO_4(aq)$$

It is also used to remove impurities from petroleum, to 'pickle' (clean) iron and steel before electroplating, as the electrolyte in lead–acid batteries (Box 15.7), and in the manufacture of many other bulk chemicals, such as hydrochloric and nitric acids.

Concentrated sulfuric acid is manufactured by the Contact process (Section 31.10). The first stage is the exothermic oxidation of a sulfur compound to SO_2. Most plants use elemental sulfur, but metal sulfides and H_2S are also used:

$$S(s) + O_2(g) \rightarrow SO_2(g)$$

$$4FeS(s) + 7O_2(g) \rightarrow 2Fe_2O_3(s) + 4SO_2(g)$$

$$2H_2S(g) + 3O_2(g) \rightarrow 2SO_2(g) + 2H_2O(g)$$

The second stage is the oxidation of SO_2 to SO_3. This reaction is carried out at high temperatures and pressures over a V_2O_5 catalyst supported on silica beads:

$$2SO_2(g) + O_2(g) \rightarrow 2SO_3(g)$$

The SO_3 is then passed into the bottom of a packed column and washed with oleum, $H_2S_2O_7$, before being treated with 98% by mass H_2SO_4. The SO_3 reacts with the 2% of water to produce sulfuric acid, H_2SO_4:

$$SO_3(g) + H_2O(sol) \rightarrow H_2SO_4(l)$$

(c) Sulfurous acid and disulfurous acid

KEY POINTS Sulfurous and disulfurous acids have never been isolated. However, salts of both acids exist; sulfites are moderately strong reducing agents and are used as bleaches; disulfites rapidly decompose in acidic conditions.

Although an aqueous solution of SO_2 is referred to as 'sulfurous acid', H_2SO_3 has never been isolated, and the predominant species present are the hydrates, $SO_2 \cdot nH_2O$. The first and second proton donations are therefore best represented as follows:

$$SO_2 \cdot nH_2O(aq) + 2H_2O(l) \rightleftharpoons H_3O^+(aq) + HSO_3^-(aq) + nH_2O(l)$$
$$pK_a = 1.81$$

$$HSO_3^-(aq) + H_2O(l) \rightleftharpoons H_3O^+(aq) + SO_3^{2-}(aq) \quad pK_a = 7.00$$

Anhydrous sodium sulfite, Na_2SO_3, is used as a bleach in the pulp and paper industry, as a reducing agent in photography, and as an oxygen scavenger in boiler treatments.

(d) Thiosulfuric acid

KEY POINTS Thiosulfuric acid decomposes but the salts are stable; thiosulfate ion is a moderately strong reducing agent.

Aqueous thiosulfuric acid, $H_2S_2O_3$ (13), decomposes rapidly in a complex process that produces a number of products, such as S, SO_2, H_2S, and H_2SO_4. The anhydrous acid is more stable and decomposes slowly to H_2S and SO_3. In contrast to the acid, the thiosulfate salts are stable and can be prepared by boiling the sulfites or hydrogensulfites with elemental sulfur or by oxidation of polysulfides:

$$8K_2SO_3(aq) + S_8(s) \rightarrow 8K_2S_2O_3(aq)$$

$$2CaS_2(s) + 3O_2(g) \rightarrow 2CaS_2O_3(s)$$

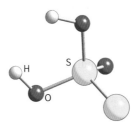

13 Thiosulfuric acid, $H_2S_2O_3$

The thiosulfate ion, $S_2O_3^{2-}$, is a moderately strong reducing agent:

$$\tfrac{1}{2}S_4O_6^{2-}(aq) + e^- \rightarrow S_2O_3^{2-}(aq) \quad E^\ominus = +0.09\,V$$

The reaction with iodine is the basis of iodometric titrations in analytical chemistry:

$$\tfrac{1}{2}I_2(aq) + e^- \rightarrow I^-(aq) \quad E^\ominus = +0.54\,V$$

and thus

$$2S_2O_3^{2-}(aq) + I_2(aq) \rightarrow S_4O_6^{2-}(aq) + 2I^-(aq)$$

The tetrathionate anion, $S_4O_6^{2-}$ (Table 17.8), has three S–S bonds, which accounts for its stability. Stronger oxidizing agents, such as chlorine, oxidize thiosulfate to sulfate, which has led to the use of thiosulfate to remove excess chlorine in the bleaching industries.

(e) Peroxosulfuric acids

KEY POINT Peroxodisulfate salts are strong oxidizing agents.

Peroxodisulfuric acid, $H_2S_2O_8$ (14), is a crystalline solid. Its ammonium and potassium salts are prepared on an industrial scale by the oxidation of their respective sulfates. They are strong oxidizing and bleaching agents.

$$\tfrac{1}{2}S_2O_8^{2-}(aq) + H^+(aq) + e^- \rightarrow HSO_4^-(aq) \quad E^\ominus = +2.12\,V$$

When $K_2S_2O_8$ is heated, ozone and oxygen are evolved.

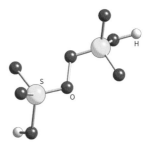

14 Peroxodisulfuric acid, $H_2S_2O_8$

(f) Dithionous and dithionic acids

KEY POINT Dithionite and dithionate are salts of S(III): they contain an S–S bond and are prone to disproportionation. Sodium dithionite is a useful reducing agent.

As mentioned in Section 10.1, radicals of main group elements are rare beyond Period 3, and unpaired electrons tend to result in E–E bonds. Neither anhydrous dithionous acid, $H_2S_2O_4$ (15), nor dithionic acid, $H_2S_2O_6$ (16), can be isolated, but stable crystalline solids are formed with dithionite and dithionate ions, featuring S in oxidation states +3 and +5 respectively.

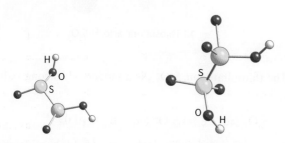

15 Dithionous acid, $H_2S_2O_4$ **16** Dithionic acid, $H_2S_2O_6$

Dithionites, $S_2O_4^{2-}$, can be prepared by the reduction of sulfites with zinc dust or sodium amalgam. Sodium dithionite is an important reducing agent in biochemistry. Neutral and acidic solutions of dithionite disproportionate into HSO_3^- and $S_2O_3^{2-}$:

$$2\,S_2O_4^{2-}(aq) + H_2O(l) \rightarrow 2\,HSO_3^-(aq) + S_2O_3^{2-}(aq)$$

The dithionite ion dissociates into SO_2^- radicals which are stronger reducing agents.

$$S_2O_4^{2-}(aq) \rightleftharpoons 2\,SO_2^-(aq)$$

Dithionates, $S_2O_6^{2-}$, are prepared by oxidation of the corresponding sulfite. Strong oxidizing agents such as MnO_4^- oxidize dithionate to sulfate:

$$SO_4^{2-}(aq) + 2\,H^+(aq) + e^- \rightarrow \tfrac{1}{2}S_2O_6^{2-}(aq) + H_2O(l)$$
$$E^\ominus = -0.25\,V$$

Strong reducing agents reduce it to SO_3^{2-}:

$$\tfrac{1}{2}S_2O_6^{2-}(aq) + 2\,H^+(aq) + e^- \rightarrow H_2SO_3(aq) \quad E^\ominus = +0.57\,V$$

Neutral and acidic solutions of dithionate slowly decompose to SO_2 and SO_4^{2-}:

$$S_2O_6^{2-}(aq) \rightarrow SO_2(aq) + SO_4^{2-}(aq)$$

17.12 Polyanions of sulfur, selenium, and tellurium

KEY POINTS Sulfur forms polyanions with up to six catenated sulfur atoms; polyselenides form chains and rings, and polytellurides form chains and bicyclic structures.

The tendency for catenation, so prevalent in the structures of elemental sulfur, selenium, and tellurium, extends to anions and cations. Here we ouline some properties of the polyanions.

The mineral iron pyrites, also known as 'fools' gold', has the formula FeS_2 and consists of Fe^{2+} and discrete S_2^{2-} anion in a rock-salt structure (Fig. 17.5). The S_3^- radical anion exists in the blue pigment ultramarine, an aluminosilicate that contains S_3^- anions and sodium cations embedded in pores in its structure. The colour can be changed by replacing the sodium with different cations; for example, silver-ultramarine is green.

Many polysulfide salts of electropositive elements have been characterized. They all contain catenated S_n^{2-} anions ranging from disulfide to trisulfide (17) and extended chains such as (18). Typical examples are Na_2S_2, Na_2S_3, Na_4S_4, K_2S_4, and Cs_2S_6. They can be prepared by heating stoichiometric amounts of S and the metal in a sealed tube.

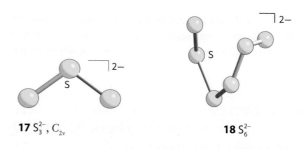

17 S_3^{2-}, C_{2v} **18** S_6^{2-}

The smaller polyselenides and polytellurides resemble the polysulfides. The structures of the larger ones are more complex and depend to some extent on the nature of the cation. The polyselenides up to Se_9^{2-} are chains but larger molecules form rings such as Se_{11}^{2-}, which has a Se atom at the centre of two six-membered rings (19). The polytellurides may also be bicyclic, as in Te_7^{2-} (20).

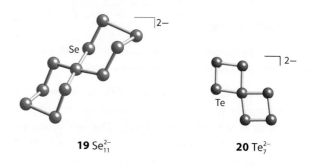

19 Se_{11}^{2-} **20** Te_7^{2-}

Polysulfides, polyselenides and polytellurides act as bidentate ligands, electron density being concentrated at each end of the E_n^{2-} chain. Complexes include $[WS(S_4)_2]^{2-}$ (21) which also contains a terminal sulfido ligand, and $[Ti(Cp)_2Se_5]$ (22).

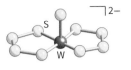

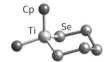

21 $[WS(S_4)_2]^{2-}$

22 $[Ti(Cp)_2Se_5], Cp = C_5H_5^-$

Redox interconversions between sulfur, polysulfides, and ultimately sulfide, in which successive electron transfers cause the S_8 ring to be opened and divided up, are exploited in metal sulfur batteries, an important contributor to renewable energy storage (Box 12.7 and Chapter 30). Reduction at the sulfur cathode produces a range of polysulfides, which may be intermediates or final products, depending on temperature and electrolyte (originally β-alumina):

$$16\,M + S_8 \rightleftharpoons 8\,M_2[S_8^{2-}, S_4^{2-}, S_3^{2-}, S_2^{2-}] \rightleftharpoons 8\,M_2S$$

Sodium-sulfur batteries normally operate at high temperature, yielding a cell potential just over 2 V, the final product formed on the cathode normally being liquid Na_2S_4. There is considerable interest in developing batteries that can operate efficiently at ambient temperatures and yield metal sulfide (M_2S) as the final discharged product.

17.13 Polycations of sulfur, selenium, and tellurium

KEY POINT Polyatomic cations of S, Se, and Te can be produced by the action of mild oxidizing agents on the elements in strong acid media.

Many cationic chain, ring, and cluster compounds of S, Se, and Te have been prepared. The cations are strong oxidizing agents and Lewis acids, thus requiring preparative conditions quite different from those used to synthesize the highly reducing polyanions. For example, S_8 is oxidized by AsF_5 in liquid sulfur dioxide to yield the S_8^{2+} cation:

$$S_8 + 3\,AsF_5 \xrightarrow{SO_2} [S_8][AsF_6]_2 + AsF_3$$

A solvent is used that is more acidic than the polycations, such as fluorosulfuric acid. S, Se, and Te each form ions of the type E_4^{2+}. For example, Se_4^{2+} is formed from elemental Se by reaction with the strongly oxidizing peroxide compound FO_2SOOSO_2F:

$$4\,Se + S_2O_6F_2 \xrightarrow{HSO_3F} [Se_4][SO_3F]_2$$

The E_4^{2+} cations have square-planar (D_{4h}) structures (23). The particular stability of these cations is explained by considering the molecular orbitals. Each E atom has six valence electrons, giving $24 - 2 = 22$ electrons in all, and there are two lone pairs on each E atom. Six electrons fill the a_{2u} and e_g π orbitals, leaving the higher-energy antibonding b_{2u} π orbital vacant.

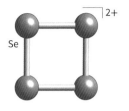

23 Se_4^{2+}

In contrast, most of the larger ring systems can be understood in terms of localized $2c,2e$ bonds. For these larger rings, the removal of two electrons brings about the formation of an additional $2c,2e$ bond, thereby preserving the local electron count on each element. This change is readily seen for the oxidation of S_8 to S_8^{2+} (24). An X-ray single-crystal structure determination shows that the transannular bond in Se_8^{2+} is longer than the other bonds. Long transannular bonds are common in these types of compounds.

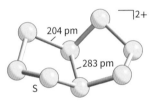

24 S_8^{2+}

17.14 Sulfur–nitrogen compounds

KEY POINTS Neutral heteroatomic ring and cluster compounds of the p-block elements include P_4S_{10} and cyclic S_4N_4. Disulfur dinitride transforms into a polymer that is superconducting at very low temperatures.

Sulfur–nitrogen compounds have structures that can be related to the polycations discussed above. The oldest known, and easiest to prepare, is the pale yellow-orange tetrasulfur tetranitride, S_4N_4 (25), which is made by the reaction of S_2Cl_2 with ammonium chloride:

$$6\,S_2Cl_2(l) + 16\,NH_4Cl(sol) \rightarrow S_4N_4(s) + S_8(s) + 16\,HCl(g)$$

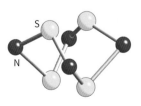

25 S_4N_4

BOX 17.6 How are polymeric sulfur nitrides used for fingerprint detection?

Latent fingerprints are accidental prints left on any surface which are invisible to the naked eye. Fingerprints usually need to be visualized and the traditional method has been to use a soft brush and aluminium powder. However, polysulfide polymers, $(SN)_n$, have been found to be very effective for visualization. This serendipitous discovery was made when zeolites were being exposed to gaseous S_2N_2 and the researchers noticed that fingerprints became visible on glassware and other surfaces. When the S_2N_2 comes into contact with the fingerprint, polymerization to blue-black $(SN)_n$ is induced and the print is exposed. The $(SN)_n$ polythiazyl (polymeric sulfur nitride) produced in this way is even more stable than the bulk material so the prints last for several days under normal conditions, and indefinitely in an inert atmosphere. The S_2N_2 also reacts with traces of inkjet printer ink and can be used to develop print from envelopes that have been in contact with printed materials.

The use of S_2N_2 provides an inexpensive, nondestructive, solvent-free imaging technique for latent fingerprint detection. There are limitations to its use because of its instability; it has to be generated *in situ*, which limits portability. But its versatility and short development times make it viable alongside some metal vacuum deposition techniques in use which visualize fingerprints by deposition of gold and zinc under high vacuum.

Tetrasulfur tetranitride is endergonic ($\Delta_f G^\ominus = +536\,\text{kJ}\,\text{mol}^{-1}$) and may decompose explosively. The 'cradle-like' molecule is an eight-membered ring with the four N atoms in a plane and bridged by S atoms that project above and below the plane. The short S–S distance (258 pm) suggests that there is a weak interaction between pairs of S atoms.

Disulfur dinitride, S_2N_2 (**26**), is formed (together with Ag_2S and N_2) when S_4N_4 vapour is passed over hot silver wool. It is even more sensitive than its precursor and explodes above room temperature. When allowed to stand at 0°C for several days, S_2N_2 transforms into a bronze-coloured zigzag polymer of composition $(SN)_n$ (**27**), which is much more stable than its precursor, not exploding below 240°C. The compound exhibits metallic conductivity along the chain axis and becomes superconducting below 0.3 K.

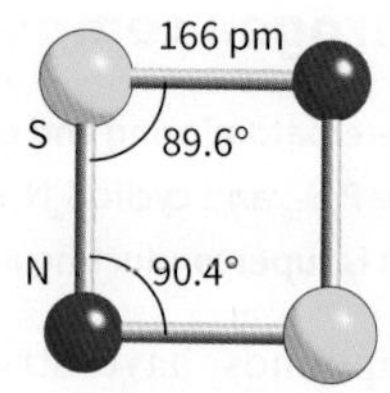

The discovery of this superconductivity was important because it was the first example of a superconductor that had no metal constituents. Halogenated derivatives have been synthesized that have even higher conductivity. For example, partial bromination of $(SN)_n$ produces blue-black single crystals of $(SNBr_{0.4})_n$, which has room-temperature conductivity an order of magnitude greater than $(SN)_n$. Treatment of S_4N_4 with ICl, IBr, and I_2 produces highly conducting, nonstoichiometric polymers with conductivities greater by 16 orders of magnitude than $(SN)_n$.

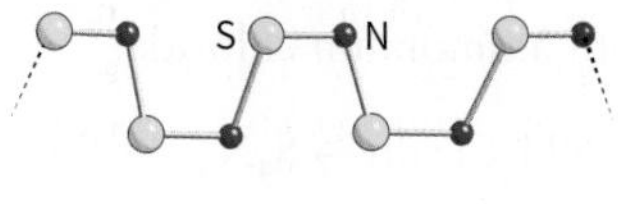

EXAMPLE 17.3 Predicting properties of a sulfur–nitrogen compound

Sulfur–nitrogen compounds are sometimes described as exhibiting inorganic aromaticity; that is, an inorganic compound with $2n + 2$ electrons available for π bonding. Assuming that each sulfur and nitrogen atom carries a lone pair, predict whether S_2N_2 could be described as aromatic.

Answer The sulfur atoms each have six valence electrons. If they have one lone pair of electrons and form two bonds to two nitrogen atoms, then they will each have two electrons left for π bonding. The nitrogen atoms each have five valence electrons: if they each have one lone pair of electrons and use two electrons in forming bonds to two sulfur atoms, then they will have one electron for π bonding. Thus, there are six electrons available for π bonding and S_2N_2 can be considered to be aromatic.

Self-test 17.3 Predict whether S_4N_4 is aromatic.

26 S_2N_2

166 pm, 89.6°, 90.4°

27 $(SN)_n$

FURTHER READING

N. Saunders, *Oxygen and the Elements of Group 17*. Heinemann (2003).

P. Ball, *H₂O: A Biography of Water*. Phoenix (2004). An entertaining look at the chemistry and physics of water.

R. Steudel, *Elemental Sulfur and Sulfur-Rich Compounds*. Springer-Verlag (2003).

C. Benson, *The Periodic Table of the Elements and Their Chemical Properties*. Kindle edition. MindMelder.com (2009).

P. R. Ogilvy, Singlet oxygen: There is indeed something new under the sun. *Chem. Soc. Rev.*, 2010, **39**, 3181.

A. E. Thorarinsdottir, S. S. Veroneau, and D. G. Nocera, Self-healing oxygen evolution catalysts. *Nature Communications*, 2022, **13**, 1243.

P. Chen, C. Wang, and T. Wang, Review and prospects for room-temperature sodium–sulfur batteries. *Materials Research Letters*, 2022, **10**, 691–719. An account of the principles underlying an energy-storage device that uses Earth-abundant resources.

EXERCISES

17.1 State whether the following oxides are acidic, basic, neutral, or amphoteric: CO_2, P_2O_5, SO_3, MgO, K_2O, Al_2O_3, CO.

17.2 The bond lengths in O_2, O_2^+, and O_2^{2-} are 121, 112, and 149 pm, respectively. Describe the bonding in these molecules in terms of molecular orbital theory and use this description to rationalize the differences in bond lengths.

17.3 Correct any inaccuracies in the following statements and, after correction, provide examples to illustrate each statement:

(a) Elements in the middle of Group 16 are easier to oxidize to the group oxidation number than are the lightest and heaviest members.

(b) In its ground state, O_2 is a triplet and it undergoes Diels–Alder electrophilic attack on dienes.

(c) The diffusion of ozone from the stratosphere into the troposphere poses a major environmental problem.

17.4 Explain why SF_6 is a very stable molecule whereas OF_6 is unknown. Make use of the bond enthalpy data given in Table 2.8.

17.5 Which of the solvents, ethylenediamine (which is basic and reducing) or SO_2 (which is acidic and oxidizing), might not react with (a) Na_2S_4, (b) K_2Te_3?

17.6 Rank the following species from the strongest reducing agent to the strongest oxidizing agent: SO_4^{2-}, SO_3^{2-}, $O_3SO_2SO_3^{2-}$.

17.7 Predict which oxidation states of Mn will be reduced by sulfite ions in basic conditions.

17.8 (a) Give the formula for Te(VI) in acidic aqueous solution and contrast it with the formula for S(VI). (b) Offer a plausible explanation for this difference.

17.9 Use the standard potential data in Resource section 3 to predict which oxoanions of sulfur will disproportionate in acidic conditions.

17.10 Use the standard potential data in Resource section 3 to predict whether SeO_3^{2-} is more stable in acidic or basic solution.

17.11 Predict whether any of the following will be reduced by thiosulfate ions, SeO_3^{2-}, in acidic conditions: VO^{2+}, Fe^{3+}, Cu^+, Co^{3+}.

17.12 Discuss the trends in stabilities of oxidation states for Group 16 elements at pH = 0, as represented by the Frost diagram in Fig. 17.2, and state the likely nature of the species prevailing for each entry.

17.13 Tetramethylammonium fluoride (0.70 g) reacts with SF_4 (0.81 g) to form an ionic product. (a) Write a balanced equation for the reaction and (b) sketch the structure of the anion. (c) How many lines would be observed in the ^{19}F-NMR spectrum of the anion?

17.14 Identify the sulfur-containing compounds A–I.

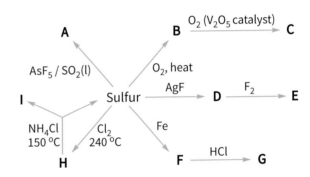

A contains a square cation. It is formed with other cationic species.
B is an acidic gas (b.p. −10°C).
C is a corrosive liquid (b.p. 45°C) that reacts violently with water.
D is a reactive gas containing fluorine atoms in two different environments.
E is a dense, unreactive gas.
F is a semiconducting solid having the NiAs structure.
G is a weakly acidic gas.
H is a yellow liquid containing a S–S bond.
I is a yellow-orange solid formed along with elemental sulfur and HCl: the cradle-like molecule has an equal number of S and N atoms.

17.15 Predict whether the following species will exhibit inorganic aromaticity: (a) $S_3N_3^-$, (b) $S_4N_3^+$, (c) S_5N_5.

17.16 Write a comparative account of the properties of sulfuric, selenic, and telluric acids.

17.17 Use group theory to construct SALCs of the four Se p_z orbitals in the $[Se_4]^{2+}$ ion, that explain why it is square-planar rather than tetrahedral like P_4.

TUTORIAL PROBLEMS

17.1 The isolation and characterization of a Se–Se species with a three-electron sigma bond was described in a paper by Zhang and co-workers (*Nat. Commun.*, 2014, **5**, 4127). (a) Who first proposed the possibility of a three-electron sigma bond? (b) Sketch the orbital interaction diagram for the formation of the bond. (c) What was the Se–Se bond length? Discuss how this compares to a single Se–Se bond and explain the difference between them. (d) Which organic species was used to stabilize the Se–Se bond?

17.2 In their paper 'Spiral chain O_4 form of dense oxygen' (*Proc. Natl. Acad. Sci. U.S.A.*, 2012, **109**, 3, 751), L. Zhu and co-workers describe their predicted chain structure for oxygen. Under what conditions is this structure predicted to exist? Which other Group 16 elements form this structure? How do the predicted properties of this form of oxygen differ from those of other forms of the element?

17.3 The article 'Oxygen, sulfur, selenium, tellurium and polonium' by L. Myongwon Lee and I. Vargas-Baca (*Ann. Rep. Prog. Chem., Sect. A: Inorg. Chem.*, 2012, **108**, 113) summarizes highlights from the 2011 literature on the chemistry of the Group 16 elements. Use the references in this paper to (a) outline the synthetic methods use to prepare $[Te_5Mo_{15}O_{57}]^{8-}$, (b) sketch the structure of aryl tellurium monohalides, (c) sketch the structures of AsSCl and AsS_2, and (d) describe the products of the reaction of ClP_4S_3 with sulfur.

17.4 In their paper 'Formation of tellurium nanotubes through concentration depletion at the surfaces of seeds' (*Adv. Mater.*, 2002, **14**, 279), B. Mayers and Y. Xia describe the synthesis of tellurium nanotubes. Describe how their method differs from those used to produce carbon nanotubes. What were the dimensions of tellurium nanotubes produced? What applications do the authors envisage for tellurium nanotubes?

17.5 In November 2006 the former KGB agent Alexander Litvinenko was found to have been poisoned by radioactive polonium-210. Write a review of the chemical and radiological properties of Po and discuss its toxicity.

17.6 Tetramethyltellurium, $Te(CH_3)_4$, was prepared in 1989 (R. W. Gedrige, D. C. Harris, K. R. Higa, and R. A. Nissan, *Organometallics*, 1989, **8**, 2817), and its synthesis was soon followed by the preparation of the hexamethyl compound (L. Ahmed and J.A. Morrison, *J. Am. Chem. Soc.*, 1990, **112**, 7411). Explain why these compounds are so unusual, give equations for their syntheses, and speculate on why these synthetic procedures are successful. In relation to the last point, speculate on why reaction of TeF_4 with methyllithium does not yield tetramethyltellurium.

17.7 The nature of the sulfur cycle in ancient times has been investigated (J. Farquhar, H. Bao, and M. Thiemen, *Science*, 2000, **289**, 756). What three factors influence the modern-day cycle? When did the authors establish that a significant change in the cycle had occurred, and how were the differences between the ancient and modern cycles explained?

18 The Group 17 elements

Part A: The essentials
18.1 The elements
18.2 Simple compounds
18.3 The interhalogens

Part B: The detail
18.4 Occurrence, recovery, and uses
18.5 Molecular structure and properties
18.6 Reactivity trends
18.7 Pseudohalogens
18.8 Special properties of fluorine compounds
18.9 Structural features
18.10 The interhalogens
18.11 Halogen oxides
18.12 Oxoacids and oxoanions
18.13 Thermodynamic aspects of oxoanion redox reactions
18.14 Trends in rates of oxoanion redox reactions
18.15 Redox properties of individual oxidation states
18.16 Fluorocarbons

Further reading

Exercises

Tutorial problems

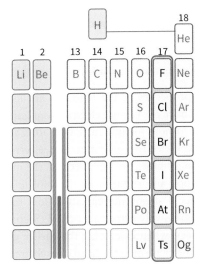

All the Group 17 elements are nonmetals which, apart from fluorine, exhibit a wide range of oxidation states. The elemental forms are reactive diatomic molecules, with activities decreasing down the group. As with the elements in Groups 15 and 16, we shall see that the oxoanions of the halogens are oxidizing agents that often react by atom transfer. There is a useful correlation between the oxidation number of the central atom and the rates of redox reactions.

The Group 17 elements—fluorine, chlorine, bromine, iodine, astatine, and tennessine—are known as the **halogens** from the Greek for 'salt giver'. Fluorine and chlorine are poisonous gases, bromine is a toxic, volatile liquid, and iodine is a sublimable solid. Astatine is highly radioactive with the most stable isotope, ^{210}At, having a half-life of 8.1 hours. Element 117, tennessine, is a very short-lived synthetic element, of which only a few atoms have ever been made. The halogens are among the most reactive nonmetallic elements. The chemical properties of the halogens are extensive, and their compounds have been mentioned many times already and were summarized in Section 9.8. Therefore, in this chapter we highlight their systematic features and their compounds with oxygen, and discuss the interhalogens.

PART A: The essentials

As we discuss the elements of the penultimate group in the p block, we shall see that many of the systematic themes that were helpful for discussing preceding groups again prove useful. For instance, the VSEPR model can be used to predict the shapes of the wide range of molecules that the halogens form among themselves, with oxygen, and with xenon.

18.1 The elements

KEY POINTS Except for fluorine and the highly radioactive astatine, the halogens exist with oxidation numbers ranging from −1 to +7; the small and highly electronegative fluorine atom is effective in oxidizing many elements to high oxidation states.

The atomic properties of the halogens are listed in Table 18.1; they all have the valence electron configuration ns^2np^5. The features to note include their high ionization energies and their high electronegativities and electron affinities. Their electron affinities are high because the incoming electron can occupy an orbital of an incomplete valence shell and experience a strong nuclear attraction: recall that Z_{eff} increases progressively across the period (Section 1.7).

We have seen when discussing the earlier groups in the p block that the element at the head of each group has properties that are distinct from those of its heavier congeners. The anomalies are much less striking for the halogens, and the most notable difference is that F has a lower electron affinity than Cl. Intuitively, this feature seems to be at odds with the high electronegativity of F, but it stems from the larger electron–electron repulsion in the compact F atom as compared with the larger Cl atom. This electron–electron repulsion is also responsible for the weakness of the F–F bond in F_2. Despite this difference in electron affinity, the enthalpies of formation of metal fluorides are generally much greater than those of metal chlorides because the low electron affinity of F is more than offset by the high lattice enthalpies of ionic compounds containing the small F^- ion (Fig. 18.1) and the strengths of bonds in covalent species (for example, the fluorides of metals in high oxidation states).

Fluorine is a pale yellow gas that reacts with most inorganic and organic molecules and the noble gases Kr, Xe, and Rn (Section 19.5). Consequently, it is very difficult to handle, but it can be stored in steel or monel metal (a nickel/copper alloy) as these alloys form a passivating metal fluoride surface film. Chlorine is a green–yellow toxic gas. Bromine is the only liquid nonmetallic element at room temperature and pressure, and is dark red, toxic, and volatile. Iodine is a purple-grey solid that sublimes to a violet vapour. The violet colour persists when it is dissolved in nonpolar solvents such as CCl_4. However, in polar solvents it dissolves to produce red-brown solutions, indicating the presence of polyiodide ions such as I_3^- (Section 18.10(c)).

The halogens are so reactive that they are found naturally only as compounds. They occur mainly as halides, but the most easily oxidized element, I, is also found as sodium

TABLE 18.1 Selected properties of the elements

	F	Cl	Br	I	At
Covalent radius/pm	57	102	120	133	150
Ionic radius/pm	131	181	196	220	
First ionization energy/kJ mol^{-1}	1681	1251	1139	1008	926
Melting point/°C	−220	−101	−7.2	114	302
Boiling point/°C	−188	−34.7	58.8	184	
Pauling electronegativity	4.0	3.2	3.0	2.7	
Electron affinity/ kJ mol^{-1}	328	349	325	295	270
$E^{\ominus}(X_2,X^-)/V$	+3.05	+1.36	+1.09	+0.54	

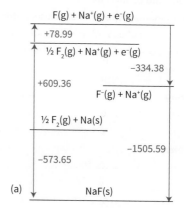
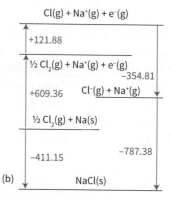

FIGURE 18.1 Thermochemical cycles for (a) sodium fluoride and (b) sodium chloride. All values are in kilojoules per mole (kJ mol^{-1}).

or potassium iodate, KIO_3, in alkali metal nitrate deposits. Because many chlorides, bromides, and iodides are soluble, these anions occur in the oceans and in brines. The primary source of F is calcium fluoride, which has low solubility in water and is often found in sedimentary deposits (as fluorite, CaF_2). Chlorine occurs as sodium chloride in rock salt. Bromine is obtained by the displacement of the element from brine by chlorine. Iodine accumulates in seaweed, from which it was originally extracted. Iodine is now obtained from salt deposits, including those associated with gas and oil fields.

18.2 Simple compounds

KEY POINTS All the halogens form hydrogen halides; HF is a liquid and HCl, HBr, and HI are gases. All the Group 17 elements form oxo compounds and oxoanions.

Because fluorine is the most electronegative of all elements, it is never found in a positive oxidation state (except in the transient gas-phase species F_2^+). With the possible exception of At, the other halogens occur with oxidation numbers ranging from −1 to +7. Compounds of Br(VII) are very unstable compared to those of Cl and I, this being yet another example of the alternation effect (Section 10.4). The dearth of chemical information on At stems from its lack of any stable isotopes and the relatively short half-life (8.1 hours) of the most long-lived of its 33 known isotopes. Astatine solutions are intensely radioactive and can be studied only in high dilution. Astatine appears to exist as the anion At⁻ and as At(I) and At(III) oxoanions; no evidence for At(VII) has yet been obtained.

The high electronegativity of F leads to enhanced Brønsted acidity in compounds containing the element compared to nonfluorine analogues. A combination of high electronegativity and small atomic radius, which means that many atoms can pack round a central atom, results in F being able to stabilize high oxidation states of most elements; for example, UF_6 and IF_7. (Oxygen, however, can often achieve higher because its −2 charge means that fewer oxygen atoms have to pack around the central atom.) The metal fluorides in low oxidation states tend to be predominantly ionic, whereas extensive covalent character is found in metal chlorides, bromides, and iodides.

The synthesis of fluorocarbon compounds, such as polytetrafluoroethene (PTFE), is of great technological importance because these compounds are useful in applications ranging from coatings for nonstick cookware and halogen-resistant laboratory vessels to the volatile fluorocarbons used as refrigerants in air conditioners and refrigerators. Fluorocarbon derivatives have also been the topic of considerable exploratory synthetic research because their derivatives often have unusual properties. Hydrofluorocarbons (HFCs) are used as replacements for chlorofluorocarbons (CFCs) as refrigerants and propellants. They are also used as anaesthetics, where they are less flammable than their nonfluorinated counterparts. Tetrafluoroethane, CHF_2CHF_2, is an important solvent for the extraction of natural products such as vanilla and taxol, which is used in chemotherapy.

All the Group 17 elements form protic molecular hydrides: the hydrogen halides. Largely on account of its ability to participate in extensive hydrogen bonding, the properties of HF contrast starkly with those of the other hydrogen halides (Table 18.2). Because hydrogen bonding (Section 5.9) is extensive, HF is a volatile liquid whereas HCl, HBr, and HI are gases at room temperature. Hydrogen fluoride has a wide liquid range, a high relative permittivity, and high conductivity. All the hydrogen halides are Brønsted acids: aqueous HF is a weak acid ('hydrofluoric acid'), whereas HCl, HBr, and HI are all essentially fully deprotonated in water (Table 18.2).

Although hydrofluoric acid is a weak acid, it is one of the most toxic and corrosive substances known, being able to attack glass, metals, concrete, and organic matter. It is much more hazardous to handle than other acids because it is very readily absorbed through the skin, and even brief contact can lead to severe burning and necrosis of the skin and deep tissue, and damage to bone by decalcification through formation of CaF_2 from calcium phosphate.

Many binary compounds of the halogens and oxygen are known, but most are unstable and not commonly encountered in the laboratory. We shall mention only a few of the most important.

TABLE 18.2 Selected properties of the hydrogen halides

	HF	HCl	HBr	HI
Melting point/°C	−84	−114	−89	−51
Boiling point/°C	20	−85	−67	−34
Relative permittivity	83.6 (at 0°C)	9.3 (at −95°C)	7.0 (at −85°C)	3.4 (at −50°C)
Electrical conductivity/S cm⁻¹	$c.10^{-6}$ (at 0°C)	$c.10^{-9}$ (at −85°C)	$c.10^{-9}$ (at −85°C)	$c.10^{-10}$ (at −50°C)
$\Delta_f G^\ominus$/kJ mol⁻¹	−273.2	−95.3	−54.4	+1.72
Bond dissociation energy/kJ mol⁻¹	567	431	366	298
pK_a	3.45	$c.$ −7	$c.$ −9	$c.$ −11

TABLE 18.3 Selected oxides of chlorine

Oxidation number	+1	+3	+4	+4	+6	+7
Formula	Cl_2O	Cl_2O_3	ClO_2	Cl_2O_4	Cl_2O_6	Cl_2O_7
Colour	brown-yellow	dark brown	yellow	pale yellow	dark red	colourless
State	gas	solid	gas	liquid	liquid	liquid

Oxygen difluoride, OF_2, is the most stable oxygen compound of F and decomposes above 200°C. Chlorine occurs with many different oxidation numbers in its oxides (Table 18.3). Some of these oxides are odd-electron species, including ClO_2, in which Cl has the unusual oxidation number +4, and Cl_2O_6, which exists as a mixed-oxidation-state ionic solid, $[ClO_2^+][ClO_4^-]$. All the chlorine oxides are endergonic ($\Delta_f G^\ominus > 0$) and unstable, and explode when heated. There are fewer oxides of Br than Cl. The best characterized compounds are Br_2O, Br_2O_3, and BrO_2. The oxides of I are the most stable halogen oxides. The most important of these is I_2O_5. Both BrO and IO (which are odd-electron species) have been implicated in ozone depletion, and both are produced naturally by volcanic activity.

All the Group 17 elements apart from fluorine form oxoanions and oxoacids. The wide range of oxoanions and oxoacids of the halogens presents a challenge to those who devise systems of nomenclature. We shall use the common names, such as chlorate for ClO_3^-, rather than the systematic names, such as trioxidochlorate(V). Table 18.4 lists the oxoanions of Cl with common and systematic nomenclature. The strength of the acids can be predicted by using Pauling's rules (Section 5.3(b)). All the oxoanions are strong oxidizing agents.

> **EXAMPLE 18.1** Predicting the strengths of halogen oxoacids
>
> Use Pauling's rules to predict the acid strengths of $HClO_3$ and $HClO_2$.
>
> **Answer** We write $HClO_3$ as $O_2Cl(OH)$. The rules then predict that $pK_a = 8 - 5p$, with $p = 2$. Thus the pK_a is predicted to be approximately −2. Similarly, we write $HClO_2$ as $OCl(OH)$ and $pK_a = 8 - 5p$, with $p = 1$. Thus the pK_a is predicted to be approximately 3. Hence $HClO_3$ is a stronger acid than $HClO_2$. These are theoretical values, but the order is confirmed by the data shown in Table 18.12.
>
> **Self-test 18.1** Sodium chlorite, $NaClO_2$, is a strong oxidizing agent and is used as a disinfectant. Give the equation for the disproportionation of the chlorite ion.

18.3 The interhalogens

KEY POINT All the halogens form compounds with other members of the group.

An interesting class of compounds is the **interhalogens** which are formed between Group 17 elements in a way not observed in other groups. The binary interhalogens are molecular compounds with formulas XY, XY_3, XY_5, and XY_7, where the heavier, less electronegative halogen X is the central atom. They also form ternary interhalogens of the type XY_2Z and XYZ_2, where Z is also a halogen atom. The interhalogens are of special importance as highly reactive intermediates and for providing useful insights into bonding.

The diatomic interhalogens, XY, have been made for all combinations of the elements, but many of them do not survive for long. All the F interhalogen compounds are exergonic ($\Delta_f G^\ominus < 0$). The least labile interhalogen is ClF, but ICl and IBr can also be obtained in pure crystalline form. Their physical properties are intermediate between those of their component elements. For example, the deep red α-ICl (m.p. 27°C, b.p. 97°C) is intermediate between yellowish-green Cl_2 (m.p. −101°C, b.p. −35°C) and dark purple I_2 (m.p. 114°C, b.p. 184°C). Photoelectron spectra indicate that the molecular orbital energy levels in the mixed dihalogen molecules lie in the order $3\sigma^2 < 1\pi^4 < 2\pi^4$, which is the same as in the homonuclear dihalogen molecules (Fig. 18.2).

TABLE 18.4 Chlorine oxoanions

Oxidation number	Formula	Name*	Point group	Shape	Remarks
+1	ClO^-	Hypochlorite [monoxidochlorate(I)]	$C_{\infty v}$	Linear	Good oxidizing agent
+3	ClO_2^-	Chlorite [dioxidochlorate(III)]	C_{2v}	Angular	Strong oxidizing agent, disproportionates
+5	ClO_3^-	Chlorate [trioxidochlorate(V)]	C_{3v}	Pyramidal	Oxidizing agent
+7	ClO_4^-	Perchlorate [tetraoxidochlorate(VII)]	T_d	Tetrahedral	Oxidizing agent, very weak ligand

* IUPAC names are in square brackets.

axial pairs and one of the three equatorial pairs, and then the two axial bonding pairs move away from the two equatorial lone pairs. As a result, XY_3 molecules have a C_{2v} bent T shape. There are some discrepancies: for example, ICl_3 is a Cl-bridged dimer.

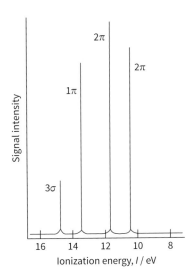

FIGURE 18.2 Photoelectron spectrum of ICl. The 2π levels give rise to two peaks because of spin-orbit interaction in the positive ion.

TABLE 18.5 Representative interhalogens

XY	XY_3	XY_5	XY_7
ClF	ClF_3	ClF_5	
BrF*	BrF_3	BrF_5	
IF	$(IF_3)_n$	IF_5	IF_7
BrCl			
ICl	I_2Cl_6		
IBr			

* Very unstable.

An interesting historical note is that ICl was discovered before Br_2 in the early nineteenth century, and later, when the first samples of the dark red–brown Br_2 (m.p. −7°C, b.p. 59°C) were prepared, they were mistaken for ICl.

Most of the higher interhalogens are fluorides (Table 18.5). The only neutral interhalogen with the central atom in a +7 oxidation state is IF_7, but the cation ClF_6^+, a compound of Cl(VII), is known. The absence of a neutral ClF_7 reflects the destabilizing effect of nonbonding electron repulsions between F atoms (indeed, coordination numbers greater than 6 are not observed for other p-block central atoms in Period 3). The lack of BrF_7 might be rationalized in a similar way, but in addition we shall see later that it is difficult for bromine to achieve its maximum oxidation state. This is another manifestation of the alternation effect (Section 10.4). In this respect, it resembles some other Period 4 p-block elements, notably arsenic and selenium.

The shapes of interhalogen molecules (1), (2), and (3) are largely in accord with the VSEPR model (Section 2.3). For example, the XY_3 compounds (such as ClF_3) have five valence electron pairs around the X atom in a trigonal-bipyramidal arrangement. The Y atoms attach to the two

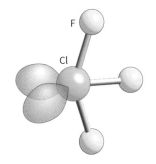

1 ClF_3, C_{2v}

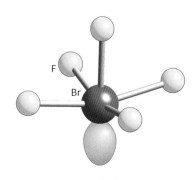

2 BrF_5, C_{4v}

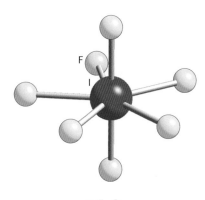

3 IF_7, D_{5h}

The Lewis structure of XF_5 has five bonding pairs and one lone pair on the central X atom and, as expected from the VSEPR model, XF_5 molecules are square-pyramidal. As already mentioned, the only known XY_7 compound is IF_7, which is predicted to be pentagonal-bipyramidal. The experimental evidence for its actual structure is inconclusive. The bonding in IF_7 can be explained without invoking d-orbital participation by adopting a molecular orbital model in which bonding and nonbonding orbitals are occupied but antibonding orbitals are not.

Polymeric interhalogens may also be formed and may be cationic or anionic. Examples of cationic polyhalides are I_3^+ (**4**) and I_5^+ (**5**). Anionic polyhalides are most numerous for iodine. The I_3^- ion is the most stable but others with the general formula $[(I_2)_nI^-]$ are formed. Other anionic polyhalides include Cl_3^- and BrF_4^-.

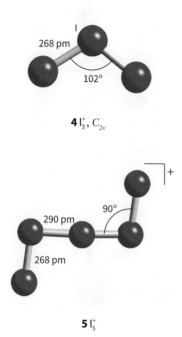

4 I_3^+, C_{2v}

5 I_5^+

> **EXAMPLE 18.2** **Predicting the shapes of interhalogen compounds**
>
> Use the VSEPR model to predict the shape of the cationic interhalogen $[ClF_6]^+$.
>
> **Answer** The central Cl in $[ClF_6]^+$ has a total of 12 electrons which will be arranged in the six bonding pairs; there are no lone pairs around the central Cl atom. According to the VSEPR model this will result in an octahedral ion (**6**).
>
>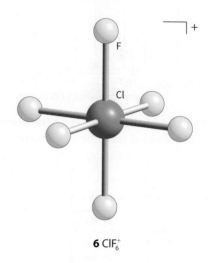
>
> **6** ClF_6^+
>
> **Self-test 18.2** Predict the shape of the Cl_3^- ion.

PART B: The detail

In this section we look in some detail at the chemistry of the halogens. Most elements form halides with the halogens, and these have been dealt with in individual group chapters, so our focus here is particularly on the interhalogens and the halogen oxides.

18.4 Occurrence, recovery, and uses

KEY POINTS Fluorine, chlorine, and bromine are prepared by electrochemical oxidation of halide salts; chlorine is used to oxidize Br⁻ and I⁻ to the corresponding dihalogen.

All the dihalogens (except the radioactive At_2) are produced commercially on a large scale, with chlorine production by far the greatest, followed by fluorine. The principal method of production of the elements is by electrolysis of the halides (Section 6.19). The strongly positive standard potentials $E^\ominus$ (F_2, F^-) = +2.87 V and $E^\ominus$ (Cl_2, Cl^-) = +1.36 V indicate that the oxidation of F^- and Cl^- ions requires a strong oxidizing agent, and only electrolytic oxidation is commercially feasible. An aqueous electrolyte cannot be used for fluorine production because water is oxidized at a much lower potential (+1.23 V), and any fluorine produced would react rapidly with water. The isolation of elemental fluorine is achieved by electrolysis of a 1:2 mixture of molten KF and HF in a cell like that shown in Fig. 18.3. It is important to keep fluorine and the by-product, hydrogen, separate because they react violently.

Most commercial chlorine is produced by the electrolysis of aqueous sodium chloride solution in a **chloralkali cell** (Fig. 18.4). The half-reactions are

Anode: $2Cl^-(aq) \rightarrow Cl_2(g) + 2e^-$
Cathode: $2H_2O(l) + 2e^- \rightarrow 2OH^-(aq) + H_2(g)$

The oxidation of water at the anode is suppressed by using an electrode material that has a higher overpotential for O_2 evolution than for Cl_2 evolution (Section 6.19). The best anode material seems to be RuO_2. This process is the basis of the chloralkali industry, which produces sodium hydroxide on a massive scale (see Box 12.5):

$2NaCl(s) + 2H_2O(l) \rightarrow 2NaOH(aq) + H_2(g) + Cl_2(g)$

Bromine is obtained by the chemical oxidation of Br⁻ ions in seawater. A similar process is used to recover iodine from

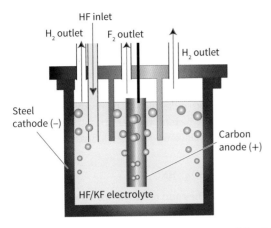

FIGURE 18.3 Schematic diagram of an electrolysis cell for the production of fluorine from KF dissolved in liquid HF.

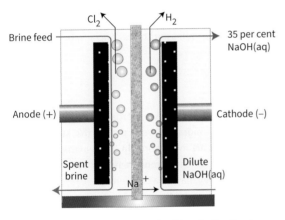

FIGURE 18.4 Schematic diagram of a chloralkali cell using a cation transport membrane, which has high permeability to Na^+ ions and low permeability to OH^- and Cl^- ions.

certain natural brines that are rich in I^-. The more strongly oxidizing halogen, chlorine, is used as the oxidizing agent in both processes, and the resulting Br_2 and I_2 are driven from the solution in a stream of air:

$$Cl_2(g) + 2X^-(aq) \xrightarrow{air} 2Cl^-(aq) + X_2(g) \quad (X = Br \text{ or } I)$$

EXAMPLE 18.3 Analysing the recovery of Br_2 from brine

Show that from a thermodynamic standpoint bromide ions can be oxidized to Br_2 by Cl_2 and by O_2, and suggest a reason why O_2 is not used for this purpose.

Answer We need to consider the relevant standard potentials and recall that a redox couple can be driven in the direction of oxidation by a couple with a more positive standard potential. The two half-reactions we need to consider for oxidation by chlorine are

$$Cl_2(g) + 2e^- \rightarrow 2Cl^-(aq) \quad E^\ominus = +1.358\,V$$
$$Br_2(g) + 2e^- \rightarrow 2Br^-(aq) \quad E^\ominus = +1.087\,V$$

Because $E^\ominus(Cl_2,Cl^-) > E^\ominus(Br_2,Br^-)$, chlorine can be used to oxidize Br^- in the reaction

$$Cl_2(g) + 2Br^-(aq) \rightarrow 2Cl^-(aq) + Br_2(g) \quad E^\ominus = +0.271\,V$$

To encourage the formation of bromine, the Br_2 is removed in a steam–air mixture. Oxygen would be thermodynamically capable of carrying out this reaction in acidic solution:

$$O_2(g) + 4H^+(aq) + 4e^- \rightarrow 2H_2O(l) \quad E^\ominus = +1.229\,V$$
$$Br_2(l) + 2e^- \rightarrow 2Br^- \quad E^\ominus = +1.087\,V$$

resulting in

$$O_2(g) + 4Br^-(aq) + 4H^+(aq) \rightarrow 2H_2O(l) + Br_2(l) \quad E^\ominus_{cell} = +0.142\,V$$

But the reaction is not favourable at pH = 7, when $E^\ominus_{cell} = -0.15\,V$. Even though the reaction is thermodynamically favourable in acidic solution, it is doubtful that the rate would be adequate because an overpotential of about 0.6 V is associated with the reactions of O_2 (Section 6.16 and Box 17.2). Even if the oxidation by O_2 in acidic solution were kinetically favourable, the process would be unattractive because of the cost of acidifying large quantities of brine and then neutralizing the effluent.

Self-test 18.3 One natural source of iodine is sodium iodate, $NaIO_3$. Which of the reducing agents, $SO_2(aq)$ or $Sn^{2+}(aq)$, would seem practical from the standpoints of thermodynamic feasibility and plausible judgements about cost? Standard potentials are given in Resource section 3.

All the Group 17 elements undergo thermal or photochemical dissociation in the gas phase to form radicals. These radicals take part in chain reactions, such as:

$$X_2 \xrightarrow{\Delta/h\nu} X^\bullet + X^\bullet$$
$$H_2 + X^\bullet \rightarrow HX + H^\bullet$$
$$H^\bullet + X_2 \rightarrow HX + X^\bullet$$

A reaction of this type between chlorine and methane is used in the industrial synthesis of chloroform, $CHCl_3$, and dichloromethane, CH_2Cl_2.

Compounds of fluorine are used throughout industry. Fluorine as F^- ions is added to some domestic water supplies and toothpaste to prevent tooth decay; it converts the acid soluble hydroxyapatite, $Ca_5(PO_4)_3OH$, found in tooth enamel to the less soluble fluorapatite, $Ca_5(PO_4)_3F$. It is used to make UF_6 in the nuclear power industry for the separation of the isotopes of uranium (Section 21.10). Hydrogen fluoride is used to etch glass and as a nonaqueous solvent. Chlorine is widely used in industry to make chlorinated hydrocarbons and in applications in which a strong oxidizing agent is needed, including disinfectants and bleaches. These applications are in decline, however, because some organic Cl compounds are carcinogenic and CFCs are implicated in the destruction of ozone in the stratosphere (Box 18.1). HFCs are now replacing CFCs in applications such as refrigeration and air conditioning. Organobromine compounds are used in synthetic organic chemistry: the C–Br bond is not as strong as the C–Cl bond and Br can

BOX 18.1 How did the use of chlorofluorocarbons lead to the ozone hole?

The ozone layer extends from 10 km to 50 km above the Earth's surface and plays a crucial role in protecting us from the harmful effects of the Sun's ultraviolet rays by absorbing radiation at wavelengths below 300 nm, thereby attenuating the spectrum of sunlight at ground level.

Ozone is produced naturally by the action of UV radiation on O_2 in the upper atmosphere:

$$O_2 \xrightarrow{h\nu} O + O$$
$$O + O_2 \rightarrow O_3$$

When ozone absorbs an ultraviolet photon, it dissociates:

$$O_3 \xrightarrow{h\nu} O_2 + O$$

The resulting O atom can then remove ozone in the reaction

$$O_3 + O \rightarrow O_2 + O_2$$

These reactions constitute the principal steps of the **oxygen-ozone cycle** that maintains an equilibrium (but seasonally varying) concentration of ozone. If all the atmospheric O_3 were condensed into a single layer at 1 atm and 25°C, it would cover the Earth to a depth of about 3 mm.

The stratosphere also contains naturally occurring species, such as the hydroxyl radical and nitric oxide (X = OH, NO), that catalyse the destruction of ozone by reactions such as

$$X + O_3 \rightarrow XO + O_2$$
$$XO + O \rightarrow X + O_2$$

However, the main concern about the loss of ozone centres on Cl and Br atoms introduced artificially by industrial activities, which catalyse O_3 destruction very efficiently. Chlorine and bromine are carried into the stratosphere as part of organohalogen molecules, RHal, which release the halogen atoms when the C–Hal bond is fragmented by far-UV photons. The ozone-destroying potential of these molecules was pointed out in 1974 by Mario Molina and Sherwood Rowland, who (together with Paul Crutzen) won the 1995 Nobel Prize in Chemistry for their work.

International action followed thirteen years later (in the form of the 1987 Montreal Protocol), and was given added impetus by the discovery of the 'ozone hole' over Antarctica, which provided dramatic evidence of the vulnerability of atmospheric ozone.

This hole surprised even the scientists working on the problem; its explanation required additional chemistry, involving the polar stratospheric clouds that form in winter. The ice crystals in these clouds adsorb molecules of chlorine or bromine nitrate, $ClONO_2$ or $BrONO_2$, which form when stratospheric ClO or BrO combine with NO_2. Once on the ice surface, these molecules react with water:

$$H_2O + XONO_2 \rightarrow HOX + HNO_3$$

where X = Cl or Br. They also react with co-adsorbed HCl or HBr (formed by attack of Cl$^-$ and Br$^-$ on the methane escaping from the troposphere):

$$HX + XONO_2 \rightarrow X_2 + HNO_3$$

The nitric acid, being very hygroscopic, enters the ice crystals and the HOX or X_2 molecules are released during the dark polar winter. When the sunlight strengthens in the spring, these molecules photolyse, releasing high concentrations of ozone-destroying radicals:

$$HOX \xrightarrow{h\nu} HO^{\cdot} + X^{\cdot}$$
$$X_2 \xrightarrow{h\nu} 2X^{\cdot}$$

To threaten the ozone layer, organohalogen molecules must survive their migration from the Earth's surface. Those containing H atoms are mostly broken down in the troposphere (the lowest region of the atmosphere) by reaction with HO radicals. Even so, they may be a problem if released in sufficient amounts. There is currently a major controversy over the use of bromomethane, CH_3Br, as an agricultural fumigant. However, the greatest potential for ozone destruction rests with molecules that lack H atoms, the chlorofluorocarbons (CFCs), which have been used in many industrial applications, and their brominated analogues (the halons), which are used to extinguish fires. These compounds have no tropospheric sink and eventually reach the stratosphere unaltered. They are the main focus of the international regulatory regime worked out in 1987 (and amended in 1990 and 1992). Most CFCs and halons have been phased out of production, and their atmospheric concentrations are beginning to decline. The CFCs present an additional problem as they are also potent greenhouse gases.

---

be more readily displaced (and recycled). Organobromine compounds are the most widely used chemical flame retardants and are used in electronics, clothes, and furniture.

Iodine is an essential element and iodine deficiency is a cause of goitre, the enlargement of the thyroid gland. For this reason, small amounts of potassium iodide are added to table salt (Box 18.2), and ethylenediammonium diiodide is a widely used supplement in animal feeds. Iodine is used as a co-catalyst in the production of ethanoic acid from methanol (Section 31.7) and AgI is used in cloud seeding.

18.5 Molecular structure and properties

KEY POINTS The F–F bond is weak relative to the Cl–Cl bond; bond strengths decrease down the group from chlorine.

Among the most striking physical properties of the halogens are their colours. In the vapour they range from the almost colourless F_2, through yellow-green Cl_2 and red-brown Br_2, to purple I_2. The progression of the maximum absorption to longer wavelengths reflects the decrease in the HOMO–LUMO gap on descending the group. In each case, the optical absorption spectrum arises primarily from transitions in

BOX 18.2 How is iodine essential to health?

Iodine is an essential element for the production of the iodine-containing hormones thyroxine and triiodothyronine produced in the thyroid gland (Fig. B18.1). These hormones are essential for normal growth, development, and brain function, functioning of the nervous system, and metabolic processes.

Iodine deficiency results in both physical and mental problems. An iodine-deficient person may develop a goitre (a swelling of the thyroid gland in front of the neck), hypothyroidism, and reduced mental function. Hypothyroidism refers to reduced thyroid hormone production, and the symptoms include fatigue, depression, weight gain, coarse hair, dry skin, muscle cramps, and decreased concentration. Iodine-deficient women may give birth to babies with severe birth defects, and iodine deficiency during childhood causes slow mental and physical development.

All iodine has to be ingested from our food. The recommended daily dose is 150 µg per day. Iodine in the form of iodide is absorbed from the bloodstream by a process called iodide trapping. In this process, sodium is transported with iodide into the cell and then concentrated in the thyroid follicles to about thirty times its concentration in the blood.

Foods rich in iodine include seaweed, fish, dairy produce, and vegetables grown in iodine-rich soil. Iodine cannot be stored in the body so intake of these foods has to be regular. Iodine, in the form of KI or NaI, is routinely added to table salt in many countries, and this practice has helped make iodine deficiency relatively uncommon. However, there is recent evidence that a decrease in milk consumption, and the emphasis on the health benefits of reduced salt intake may be leading to the re-emergence of iodine deficiency.

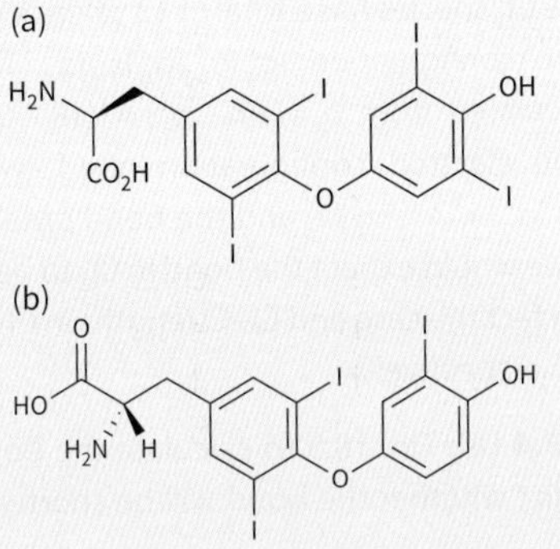

FIGURE B18.1 The structures of (a) thyroxine and (b) triiodothyronine.

---

which an electron is promoted from the highest filled $2\sigma_g$ and $1\pi_g$ orbitals into the vacant antibonding $2\sigma_u$ orbital (Fig. 18.5).

Except for F_2, the analysis of the UV absorption spectra gives precise values for the dihalogen bond dissociation energies (Fig. 18.6). It is found that bond strengths decrease down the group from Cl_2.

The UV spectrum of F_2, however, is a broad continuum that lacks structure because absorption is accompanied by dissociation of the F_2 molecule. The lack of discrete absorption bands makes it difficult to estimate the dissociation energy spectroscopically, and thermochemical methods are complicated by the highly corrosive nature of this reactive halogen. When these problems were solved, the F–F bond enthalpy was found to be less than that of Br_2 and thus out of line with the trend in the group. However, the low F–F bond enthalpy is consistent with the low single-bond enthalpies of N–N, O–O, and various combinations of N, F, and O

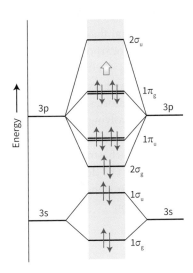

FIGURE 18.5 Schematic molecular orbital energy level diagram for F_2. For Cl_2, Br_2, and I_2 the order of the π_u and upper σ_g orbitals is reversed. The broad vertical arrow indicates how the destabilizing $1\pi_g$ orbital becomes increasingly influential upon ascending the group.

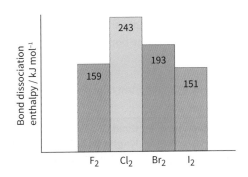

FIGURE 18.6 Bond dissociation enthalpies of the halogens in kilojoules per mole (kJ mol^{-1}).

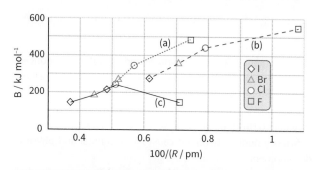

FIGURE 18.7 Dissociation enthalpies of (a) carbon–halogen, (b) hydrogen–halogen, and (c) halogen–halogen bonds plotted against the reciprocal of the bond length.

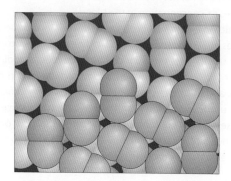

FIGURE 18.8 Solid chlorine, bromine, and iodine have similar structures. The closest nonbonded interactions are relatively less compressed in Cl_2 and Br_2 than in I_2.

(Fig. 18.7). The simplest explanation (like the explanation of the low electron affinity of fluorine) is that the bond is weakened by the strong repulsions between nonbonding electrons in the small F_2 molecule. In molecular-orbital terms, the molecule has numerous electrons in strongly antibonding orbitals.

All the halogens form **halogen bonds** which are similar to hydrogen bonds, but the halogen atoms are the electron acceptors. The order of halogen bond strength for the dihalogens is $F_2 < Cl_2 < Br_2 < I_2$. Chlorine, bromine, and iodine all crystallize in lattices of the same symmetry (Fig. 18.8), so it is possible to make a detailed comparison of distances between bonded and nonbonded adjacent atoms (Table 18.6).

The important conclusion is that nonbonded distances do not increase as rapidly as the bond lengths. This observation

TABLE 18.6 Bonding and shortest nonbonding distances for solid dihalogens

Element	Temperature/°C	Bond length/pm	Nonbonding distance/pm	Ratio
Cl_2	−160	198	332	1.68
Br_2	−106	227	332	1.46
I_2	−163	272	350	1.29

suggests the presence of weak intermolecular bonding interactions that strengthen on going from Cl_2 to I_2. Solid iodine is a semiconductor and under high pressure exhibits metallic conductivity.

EXAMPLE 18.4 Estimating bond lengths

Use Fig. 18.5 to predict the bond order of the Cl_2^+ molecular ion. Predict whether the Cl–Cl bond in Cl_2^+ will be longer or shorter than that in Cl_2.

Answer The Cl_2^+ species has a total of 13 valence electrons giving the electron configuration $1\sigma_g^2 1\sigma_u^2 2\sigma_g^2 1\pi_u^4 1\pi_g^3$. The bond order can be calculated from $b = \frac{1}{2}(n - n^*) = \frac{1}{2}(8 - 5) = 1.5$ (Section 2.10(a)). The electron configuration of Cl_2 with 14 valence electrons is $1\sigma_g^2 1\sigma_u^2 2\sigma_g^2 1\pi_u^4 1\pi_g^4$ and the bond order is $\frac{1}{2}(8 - 6) = 1$. Therefore, we would expect the bond in Cl_2^+ to be shorter than in Cl_2. This is in fact the case and Cl–Cl lengths in Cl_2 and Cl_2^+ are 198 and 189 pm respectively.

Self-test 18.4 Use Fig. 18.5 to calculate the bond order of the F_2^- ion. Predict whether the bond will be shorter or longer than that of F_2.

18.6 Reactivity trends

KEY POINTS Fluorine is the most oxidizing halogen; the oxidizing power of the halogens decreases down the group.

Fluorine, F_2, is the most reactive nonmetal and is the strongest oxidizing agent among the halogens. The rapidity of many of its reactions with other elements may in part be due to a low kinetic barrier associated with the weak F–F bond. Despite the thermodynamic stability of most metal fluorides, fluorine can be handled in containers made from some metals, such as Ni, because a substantial number of them form a passive metal fluoride surface film on contact with fluorine gas. Fluorocarbon polymers, such as PTFE, are also useful materials for the construction of apparatus to contain fluorine and oxidizing fluorine compounds (Fig. 18.9). Few laboratories have the equipment and expertise for research involving elemental F_2.

The standard potentials for the halogens (Table 18.1) indicate that F_2 is a much stronger oxidizing agent than Cl_2 in aqueous solution. The decrease in oxidizing strength continues in more modest steps from Cl_2 through Br_2 to I_2. Although the half-reaction

$$\tfrac{1}{2}X_2(g) + e^- \rightarrow X^-(aq)$$

is favoured by a high electron affinity (which suggests that F should have a lower standard potential than Cl), the process is favoured by the low bond enthalpy of F_2 and by the highly exothermic hydration of the small F^- ion (Fig. 18.10). The net outcome of these three competing effects of size is that F is the most strongly oxidizing element of the group.

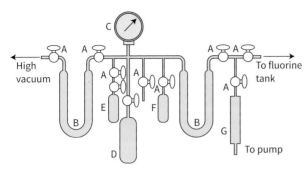

FIGURE 18.9 A typical metal vacuum system for handling fluorine and reactive fluorides. Nickel tubing is used throughout. (A) Monel valves, (B) nickel U-traps, (C) monel pressure gauge, (D) nickel container, (E) PTFE reaction tube, (F) nickel reaction vessel, and (G) nickel canister filled with soda lime to neutralize HF and react with F_2 and fluorine compounds.

TABLE 18.7 Pseudohalides, pseudohalogens, and corresponding acids

Pseudohalide	Pseudohalogen	$E^\ominus$/V	Acid	pK_a
CN^-	NCCN	+0.27	HCN	9.2
Cyanide	Cyanogen		Hydrogen cyanide	
NCS^-	NCSSCN	+0.77	HNCS	−1.9
Thiocyanate	Dithiocyanogen		Hydrogen thiocyanate	
NCO^-			HNCO	3.5
Cyanate			Isocyanic acid	
CNO^-			HCNO	3.66
Fulminate			Fulminic acid	
NNN^-			HNNN	4.92
Azide			Hydrazoic acid	

FIGURE 18.10 Thermochemical cycles for the enthalpy of formation of (a) aqueous sodium fluoride and (b) aqueous sodium chloride. The hydration is much more exothermic for F^- than for Cl^-. All values are in kilojoules per mole (kJ mol^{-1}).

18.7 Pseudohalogens

KEY POINTS Pseudohalogens and pseudohalides mimic halogens and halides, respectively; the pseudohalogens exist as dimers and form molecular compounds with nonmetals and ionic compounds with alkali metals.

A number of compounds have properties so similar to those of the halogens that they are called **pseudohalogens** (Table 18.7). For example, like the dihalogens, cyanogen, $(CN)_2$, undergoes thermal and photochemical dissociation in the gas phase; the resulting CN radicals are isolobal with halogen atoms and undergo similar reactions, such as a chain reaction with hydrogen:

$$NC-CN \xrightarrow{heat\ or\ light} 2CN^\bullet$$
$$H_2 + CN^\bullet \rightarrow HCN + H^\bullet$$
$$H^\bullet + NC-CN \rightarrow HCN + CN^\bullet$$
overall: $H_2 + C_2N_2 \rightarrow 2HCN$

Another similarity is the reduction of a pseudohalogen:

$$\tfrac{1}{2}(CN)_2(g) + e^- \rightarrow CN^-(aq)$$

The anion formally derived from a pseudohalogen is called a pseudohalide ion (Section 15.12). An example is the cyanide anion, CN^-. Covalent pseudohalides similar to the covalent halides of the p-block elements are also common. They are often structurally similar to the corresponding covalent halides (compare (7) and (8)) and undergo similar metathesis reactions.

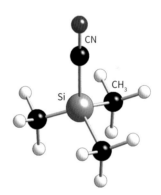

7 $(CH_3)_3SiCN$

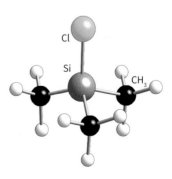

8 $(CH_3)_3SiCl$

As with all analogies, the concepts of pseudohalogen and pseudohalide have many limitations. For example,

TABLE 18.8 Normal boiling points (in °C) of compounds of fluorine and their analogues

F_2	−188.2	H_2	−252.8	Cl_2	−34.0
CF_4	−127.9	CH_4	−161.5	CCl_4	76.7
PF_3	−101.5	PH_3	−87.7	PCl_3	75.5

pseudohalogen ions are not spherical, so the structures of their ionic compounds often differ: NaCl is fcc but NaCN is similar to CaC_2 (Fig. 15.15). The pseudohalogens are generally less electronegative than the lighter halogens, and some pseudohalides have more versatile donor properties. The thiocyanate ion, SCN^-, for instance, acts as an ambidentate ligand with a soft base site, S, and a hard base site, N (Sections 5.10 and 7.1).

18.8 Special properties of fluorine compounds

KEY POINTS Fluorine substituents promote volatility, increase the strengths of Lewis and Brønsted acids, and stabilize high oxidation states.

The boiling points in Table 18.8 demonstrate that molecular compounds of F tend to be highly volatile, in some cases even more volatile than the corresponding hydrogen compounds (compare, for example, PF_3, b.p. −101.5°C, and PH_3, b.p. −87.7°C), and in all cases much more volatile than the Cl analogues. The volatilities of the compounds are a result of variations in the strength of the dispersion interaction (the interaction between instantaneous transient electric dipole moments), which is strongest for highly polarizable molecules. The electrons in the small F atoms are gripped tightly by the nuclei, and consequently F compounds have low polarizabilities and hence weak dispersion interactions.

There are some opposite effects on volatility that can be traced to hydrogen bonding (Section 5.9). The structure of solid HF is a planar zigzag chain polymer of F–H···F units. Although liquid HF has a lower density and viscosity than water, which suggests the absence of an extensive three-dimensional network of H bonds, in the gas phase HF forms H-bonded oligomers, $(HF)_n$, with n up to 5 or 6. As with H_2O and NH_3, the properties of HF, such as a wide liquid range, make it an excellent solvent.

Hydrogen fluoride undergoes autoprotolysis:

$$2HF(l) \rightleftharpoons H_2F^+(sol) + F^-(sol) \quad pK_{auto} = 12.3$$

It is a much weaker acid (pK_a = 3.45 in water) than the other hydrogen halides. Although this difference is sometimes attributed to the formation of an ion pair ($H_3O^+F^-$), theoretical considerations show that its poor proton-donor properties are a direct result of the very strong H–F bond.

Carboxylic acids act as bases in anhydrous HF and are protonated:

$$HCOOH(l) + 2HF(l) \rightarrow HC(OH)_2^+(sol) + HF_2^-(sol)$$

An important characteristic is the ability of an F atom in a compound to withdraw electrons from the other atoms present and, if the compound is a Brønsted acid, to enhance its acidity. An example of this effect is the increase by three orders of magnitude in the acidity of trifluoromethanesulfonic acid, $HOSO_2CF_3$ (pK_a = 3.0 in nitromethane), over that of methanesulfonic acid, $HOSO_2CH_3$ (pK_a = 6.0 in nitromethane). The presence of F atoms in a molecule also results, for the same reason, in an enhanced Lewis acidity. For example, we saw in Sections 5.8 and 6.11 that SbF_5 is one of the strongest Lewis acids of its type and much stronger than $SbCl_5$.

Some examples of high-oxidation-state compounds of F are IF_7, PtF_6, BiF_5, $KAgF_4$, UF_6, and ReF_7. Rhenium(VII) heptafluoride is the only example of a thermally stable metal heptafluoride, and uranium(VI) hexafluoride is important in the separation of U isotopes in the preprocessing of nuclear fuels. All these compounds are examples of the highest oxidation state attainable for these elements, perhaps the most notable being the rare oxidation state Hg(IV): the square-planar d^8 molecule HgF_4 has been observed by matrix isolation in solid neon at 4 K. Another example is the stability of PbF_4, compared to all other Pb(IV) halides.

A related phenomenon is the tendency of fluorine to disfavour low oxidation states. Thus, solid copper(I) fluoride, CuF, is unstable but CuCl, CuBr, and CuI are stable with respect to disproportionation. Similar trends were discussed in Section 4.11 in terms of a simple ionic model in which the small size of the F^- ion in combination with a small, highly charged cation results in a high lattice enthalpy. As a result, there is a thermodynamic tendency for CuF to disproportionate and form copper metal and CuF_2 (because Cu^{2+} is doubly charged, its ionic radius is smaller than that of Cu^+, and these factors lead to a greater lattice enthalpy).

Compounds that accept F^- ions are Lewis acids and compounds that donate F^- are Lewis bases:

$$SbF_5(s) + HF(l) \rightarrow SbF_6^-(sol) + H^+(sol)$$
$$XeF_6(s) + HF(l) \rightarrow XeF_5^+(sol) + HF_2^-(sol)$$

Ionic fluorides dissolve in HF to give highly conducting solutions. The fact that chlorides, bromides, and iodides react with HF to give the corresponding fluoride and HX provides a preparative route to anhydrous fluorides:

$$TiCl_4(l) + 4HF(l) \rightarrow TiF_4(s) + 4HCl(g)$$

18.9 Structural features

Metal difluorides, MF_2, where M is a Group 2 or d metal, generally adopt the CaF_2 or rutile structure and are described well by the ionic model. In contrast, whereas the Group 2 dichlorides, dibromides, and diiodides may be described by the ionic model, the d-metal analogues adopt the CdI_2 or $CdCl_2$ layer structure and their bonding is not described well by either the ionic or covalent models. Many metal trifluorides have three-dimensional ionic structures, but the trichlorides, tribromides, and triiodides have layered structures. The compounds NbF_3 and FeF_3 (at high temperature) adopt the ReO_3 structure type (Section 4.6 and Fig. 25.34), and many other metal trifluorides (including AlF_3, ScF_3, and CoF_3) have a slightly distorted variant of this structure type.

As the oxidation number of the metal atom increases, the halides become more covalent. Thus, all metal hexahalides, such as MoF_6 and WCl_6, are molecular covalent compounds. For intermediate oxidation states (such as MF_4 and MF_5) the structures normally consist of linked MF_6 polyhedra. Titanium tetrafluoride has a structure based on columns of triangular Ti_3F_{15} units formed from three TiF_6 octahedra (**9**), whereas NbF_5 is built from four NbF_6 octahedra forming a square unit of composition Nb_4F_{20} (**10**).

Rb, and Cs and M a dipositive d-metal ion, adopt the perovskite structure (Section 4.9). One example is $KMnF_3$, which precipitates when potassium fluoride is added to Mn(II) solutions. The hybrid organic–inorganic perovskite compound $(CH_3NH_3)PbI_{3-x-y}Cl_xBr_y$ is used in solar cells for its light absorbance and high-power conversion efficiency (Section 30.2). Molten cryolite, Na_3AlF_6, is used to dissolve aluminium oxide in electrochemical extraction of aluminium. Its structure is related to that of perovskite (ABO_3) with Na in the A sites and a mixture of Na and Al in the B sites: the formula is $Na(Al_{1/2}Na_{1/2})F_3$, which is equivalent to Na_3AlF_6. Mixed-anion compounds containing halides are also well characterized and include the superconducting cuprate $Sr_{2-x}Na_xCuO_2F_2$.

18.10 The interhalogens

The halogens form many compounds between themselves, with formulas ranging from XY to XY_7 (Table 18.9). Their structures can usually be predicted accurately by the VSEPR rules and verified by techniques such as ^{19}F-NMR.

EXAMPLE 18.5 Verifying the shape of an interhalogen molecule

BrF_5 is a fluxional molecule that rapidly interconverts between a square-pyramidal (**11**) and a trigonal-bipyramidal structure (**12**). If these two structures could be isolated, explain how ^{19}F-NMR could be used to differentiate between them.

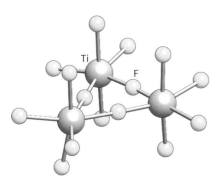

9 Ti_3F_{15}

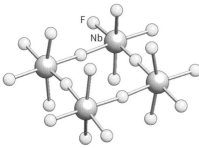

10 Nb_4F_{20}

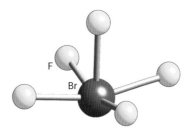

11 BrF_5, C_{4v}

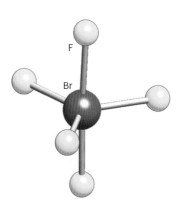

12 BrF_5, D_{3h}

Although not as important in applications as complex oxides, complex solid fluorides and chlorides, such as the ternary phases $MM'F_n$ and $MM'Cl_n$, and the quaternary compounds $MM'M''F_n$ have structures similar to their oxide counterparts. As F^- has an oxidation number of −1 compared to −2 for O^{2-}, the compositionally equivalent fluorides or chlorides generally contain d metals in lower oxidation states than the equivalent oxide. Thus, ternary fluorides of stoichiometry ABF_3 with, for example, A = K,

Answer We need to identify the number of different ^{19}F environments in each case and consider how each is coupled to other F environments. In the case of the square pyramid there are two different F environments of four and one F atoms, respectively. The ^{19}F-NMR would show a signal equivalent to four F atoms split into a doublet by coupling to the other one F atom. A second signal equivalent to one F atom would be split into a quintuplet by coupling to four F atoms. The trigonal-bipyramidal structure has two F environments, equivalent to three and two F atoms, respectively. Their resonances would be split into respectively a triplet and quartet by spin–spin coupling.

Self-test 18.5 Predict the ^{19}F-NMR pattern for IF$_7$.

(a) Chemical properties

KEY POINT Fluorine-containing interhalogens are typically Lewis acids and strong oxidizing agents.

All the interhalogens are oxidizing agents. In general, the rates of oxidation of interhalogens do not bear a simple relation to their thermodynamic stabilities. As with all the known interhalogen fluorides, ClF$_3$ is an exergonic compound, and thermodynamically it is a weaker fluorinating agent than F$_2$ itself. However, the rate at which it fluorinates substances generally exceeds that of fluorine, so it is in fact an aggressive fluorinating agent towards many elements and compounds. The fluorides ClF$_3$ and BrF$_3$ are much more aggressive fluorinating agents than BrF$_5$, IF$_5$, and IF$_7$; iodine pentafluoride, for instance, is a convenient mild fluorinating agent that can be handled in glass apparatus. One use of ClF$_3$ as a fluorinating agent is in the formation of a passivating metal fluoride film on the inside of the nickel apparatus used in fluorine chemistry.

Both ClF$_3$ and BrF$_3$ react vigorously (often explosively) with organic matter, water, ammonia, and asbestos, and expel oxygen from many metal oxides:

$$2Co_3O_4(s) + 6ClF_3(g) \rightarrow 6CoF_3(s) + 3Cl_2(g) + 4O_2(g)$$

Bromine trifluoride autoionizes in the liquid state:

$$2BrF_3(l) \rightleftharpoons BrF_2^+(sol) + BrF_4^-(sol)$$

This Lewis acid–base behaviour is shown by its ability to dissolve a number of halide salts:

$$CsF(s) + BrF_3 \rightarrow Cs^+(sol) + BrF_4^-(sol)$$

Bromine trifluoride is a useful solvent for ionic reactions that must be carried out under highly oxidizing conditions. The Lewis-acid character of BrF$_3$ is shared by other interhalogens, which react with alkali metal fluorides to produce anionic fluoride complexes.

(b) Cationic interhalogens

KEY POINT Cationic interhalogen compounds have structures in accord with the VSEPR model.

Under special strongly oxidizing conditions, such as in fuming sulfuric acid, I$_2$ is oxidized to the blue paramagnetic diiodinium cation, I$_2^+$. The dibrominium cation, Br$_2^+$, is also known. The bonds of these cations are shorter than those of the corresponding neutral dihalogens, which is the expected result for loss of an electron from a π^* orbital and the accompanying increase in bond order from 1 to 1.5 (see Fig. 18.5). Three higher polyhalogen cations, Br$_5^+$, I$_3^+$, and I$_5^+$, are known, and X-ray diffraction studies of the iodine species have established the structures shown in (**4**) and (**5**). The angular shape of I$_3^+$ is in line with the VSEPR model because the central I atom has two lone pairs of electrons.

Another class of polyhalogen cations of formula XF$_n^+$ is obtained when a strong Lewis acid, such as SbF$_5$, abstracts F$^-$ from interhalogen fluorides:

$$ClF_3 + SbF_5 \rightarrow [ClF_2]^+[SbF_6]^-$$

This formulation is idealized because X-ray diffraction of solid compounds that contain these cations indicates that

TABLE 18.9 Properties of interhalogens

XY	XY$_3$	XY$_5$	XY$_7$
ClF	ClF$_3$	ClF$_5$	
Colourless	Colourless	Colourless	
m.p. −156°C	m.p. −76°C	m.p. −103°C	
b.p. −100°C	b.p. 12°C	b.p. −13°C	
BrF*	BrF$_3$	BrF$_5$	
Light brown	Yellow	Colourless	
m.p. ≈ −33°C	m.p. 9°C	m.p. −61°C	
b.p. −20°C	b.p. 126°C	b.p. 41°C	
IF*	(IF$_3$)$_n$	IF$_5$	IF$_7$
	Yellow	Colourless	Colourless
	Dec.† −28°C	m.p. 9°C b.p. 105°C	m.p. 6.5°C (triple point) Subl.† 5°C
BrCl*			
Red-brown			
m.p. ≈ −66°C			
b.p. 5°C			
α-ICl, β-ICl	I$_2$Cl$_6$		
Ruby-red solid, black liquid	Bright yellow		
m.p. 27°C, m.p. 14°C	m.p. 101°C (16 atm)		
b.p. 97–100°C			
IBr			
Black solid			
m.p. 41°C			
b.p. ≈ 116°C			

* Very unstable.
† Dec.: decomposes; Subl.: sublimes.

the F⁻ abstraction from the cations is incomplete and that the anions remain weakly associated with the cations by fluorine bridges (**14**). Table 18.10 lists a variety of interhalogen cations that are prepared in a similar manner.

14 (ClF$_2$)(SbF$_6$)$_2$

EXAMPLE 18.6 Predicting the shapes of interhalogen compounds

Predict the shapes of the reactant and products from the reaction discussed above:

2BrF$_3$ (l) ⇌ BrF$_2^+$ (sol) + BrF$_4^-$ (sol)

Answer BrF$_3$ (and all XY$_3$ interhalogens) have five electron pairs around the central atom, giving a trigonal-bipyramidal shape. The two lone pairs take up equatorial positions, resulting in a distorted T-shaped molecule as in (**1**). BrF$_2^+$ has four electron pairs around the central atom, two of which are lone pairs. This results in an angular, or bent, molecule as in (**4**). BrF$_4^-$ has six electron pairs around the central atom. The two lone pairs are *trans* to each other, giving a square-planar molecule (**13**).

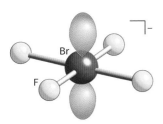

13 BrF$_4^-$

Self-test 18.6 Predict the shapes of the reactants and the two ions in the product for the following reaction: ClF$_3$ + SbF$_5$ → [ClF$_2$][SbF$_6$].

(c) Polyhalides

KEY POINTS Polyiodides, such as I$_3^-$, are formed by adding I$_2$ to I⁻; they are stabilized by large cations. Some of the most stable polyhalogen anions contain fluorine as the substituent; their structures usually conform to the VSEPR model.

TABLE 18.10 Representative interhalogen cations

Compound	Shape
ClF$_2^+$, BrF$_2^+$, ICl$_2^+$	Bent, C_{2v}
ClF$_4^+$, BrF$_4^+$, IF$_4^+$	See-saw, C_{2v}
ClF$_6^+$, BrF$_6^+$, IF$_6^+$	Octahedral, O_h

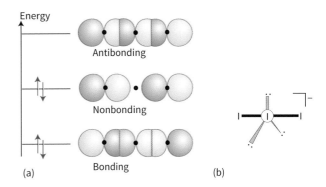

FIGURE 18.11 Some representations of the I$_3^-$ polyiodide ion. (a) The σ interaction. (b) Lewis and VSEPR rationalization of the linear structure, where the five electron pairs are arranged around the central atom in a trigonal-bipyramidal array.

A deep brown colour develops when I$_2$ is added to a solution of I⁻ ions. This colour is characteristic of the polyiodides, which include triiodide ions, I$_3^-$, and pentaiodide ions, I$_5^-$. These polyiodides are Lewis acid–base complexes in which I⁻ and I$_3^-$ act as the bases and I$_2$ acts as the acid (Fig. 18.11). The Lewis structure of I$_3^-$ has three equatorial lone pairs on the central I atom and two axial bonding pairs in a trigonal-bipyramidal arrangement. This hypervalent Lewis structure is consistent with the observed linear structure of I$_3^-$, which is described in more detail below.

An I$_3^-$ ion can interact with other I$_2$ molecules to yield larger mononegative polyiodides of composition [(I$_2$)$_n$I⁻]. The I$_3^-$ ion is the most stable member of this series. In combination with a large cation, such as [N(CH$_3$)$_4$]⁺, it is symmetrical and linear, with a longer I–I bond than in I$_2$. However, the structure of the triiodide ion, like that of the polyiodides in general, is highly sensitive to the identity of the counter-ion. For example, Cs⁺, which is smaller than the tetramethylammonium ion, distorts the I$_3^-$ ion and produces one long and one short I–I bond (**15**). The ease with which the ion responds to its environment is a reflection of the weakness of bonds that just manage to hold the atoms together. An example of sensitivity to the cation is provided by NaI$_3$, which can be formed in aqueous solution but decomposes when the water is evaporated:

Na⁺(aq) + I$_3^-$(aq) $\xrightarrow{\text{remove water}}$ NaI(s) + I$_2$(s)

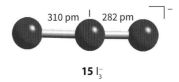

15 I$_3^-$

A more extreme example is NI$_3$·NH$_3$, which is a black powder formed when iodine crystals are added to concentrated ammonia solution. The free NI_3 can be prepared by reacting iodine monofluoride with boron nitride:

$$3IF(g) + BN(s) \rightarrow NI_3(s) + BF_3(g)$$

Nitrogen triiodide and the ammoniate are extremely unstable and detonate at the slightest touch or vibration:

$$2NI_3 \cdot NH_3(s) \rightarrow N_2(g) + 3I_2(s) + 2NH_3(g)$$

Although the formula of nitrogen triiodide is usually written NI_3, it might be more accurate to write I_3N, as the compound is thought to consist of I^+ and N^{3-} ions and its sensitivity to shock is due to the redox instability of these ions. This behaviour is also another example of the instability of large anions in combination with small cations, which, as we saw in Section 4.15, can be rationalized by the ionic model.

A vast number of polyiodide ions exist with the number of iodine atoms varying up to around 30 iodine atoms. Several series of polyiodides are formed with the general formulas I_{2n+1}^-, I_{2n+2}^{2-}, I_{2n+3}^{3-}, and I_{2n+4}^{4-}. The existence and structures of the higher polyiodides are sensitive to the counter-ion for similar reasons, and large cations are necessary to stabilize them in the solid state. In fact, entirely different shapes are observed for polyiodide ions in combination with various large cations, as the structure of the anion is determined in large measure by the manner in which the ions pack together in the crystal. The bond lengths in a polyiodide ion often suggest that it can be regarded as a chain of associated I^-, I_2, I_3^-, and sometimes I_4^{2-} units (Fig. 18.12).

Solids containing polyiodides exhibit electrical conductivity, which may arise either from the hopping of electrons (or holes) or by an ion relay along the polyiodide chain (Fig. 18.13).

Although polyhalide formation is most pronounced for iodine, other polyhalides are also known. New synthetic techniques such as the use of ionic liquids have enabled the synthesis of higher polybromide monoanions in the series

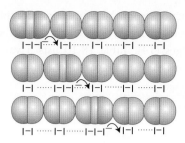

FIGURE 18.13 One possible mode of charge transport along a polyiodide chain is the shift of long and short bonds, resulting in the effective migration of an I^- ion along a chain. Three successive stages in the migration are shown. Note that the iodide ion from the I_3^- on the left is not the same one emerging on the right.

Br_{2n+1}^- and Br_{2n+2}^{2-} up to twenty Br atoms. Polybromides have high electrical conductivity and are used as bromination reagents and electrolytes in batteries or in dye-sensitized solar cells.

Other polyhalides include Cl_3^-, Br_3^-, and BrI_2^-, which are known in solution and (in partnership with large cations) as solids. Ionic liquids of the form $R^+IBr_2^-$ or $R^+BrI_2^-$, where R^+ is 1,3-dialkylimidazolium cation, are used as solvents in dye-sensitized solar cells. Even F_3^- has been detected spectroscopically at low temperatures in an inert matrix. This technique, known as matrix isolation (Box 19.3), makes use of the co-deposition of the reactants with a large excess of noble gas at very low temperatures (in the region of 4–14 K). The solid noble gas forms an inert matrix within which the F_3^- ion can sit in chemical isolation.

In addition to complex formation between dihalogens and halide ions, some interhalogens can act as Lewis acids towards halide ions. The reaction results in the formation of polyhalides that, in contrast to the chain-like polyiodides, are assembled around a central halogen acceptor atom in a high oxidation state. As mentioned earlier, for instance, BrF_3 reacts with CsF to form $CsBrF_4$, which contains the square-planar BrF_4^- anion (**13**). Many of these interhalogen anions have been synthesized (Table 18.11). Their shapes generally agree with the VSEPR model, but there are some interesting exceptions. Two such exceptions are ClF_6^- and BrF_6^-, in which the central halogen has a lone pair of

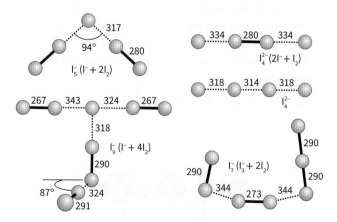

FIGURE 18.12 Some representative polyiodide structures and their approximate description in terms of I^-, I_3^-, and I_2 building blocks. Bond lengths and angles vary with the identity of the cation.

TABLE 18.11 Representative interhalogen anions

Compound	Shape
ClF_2^+, IF_2^-, ICl_2^-, IBr_2^-	Linear
ClF_4^-, BrF_4^-, IF_4^-, ICl_4^-	Square-planar
ClF_6^-, BrF_6^-	Distorted octahedral
IF_6^-	Trigonally distorted octahedron
IF_8^-	Square antiprism

electrons, and the apparent structure is distorted from octahedral. The ion IF_6^- forms an extended array through I–F···I interactions.

EXAMPLE 18.7 Proposing a bonding model for I⁺ complexes

In some cases the interaction of I_2 with strong donor ligands leads to the formation of cationic complexes such as bis(pyridine) iodine(+1), [py–I–py]⁺. Propose a bonding model for this linear complex from (a) the standpoint of the VSEPR model and (b) simple molecular-orbital considerations.

Answer (a) The Lewis electron structure places 10 electrons around the central I⁺ in [py–I–py]⁺, 6 from the iodine cation and 4 from the lone pairs on the two pyridine ligands. According to the VSEPR model, these pairs should form a trigonal bipyramid. The lone pairs will occupy the equatorial positions, and consequently the complex should be linear. (b) From a molecular-orbital perspective, the orbitals of the N–I–N array can be pictured as being formed from an iodine 5p orbital and an orbital of σ symmetry from each of the two ligand atoms. Three orbitals can be constructed: 1σ (bonding), 2σ (nearly nonbonding), and 3σ (antibonding). There are four electrons to accommodate (two from each ligand atom; the iodine 5p orbital is empty). The resulting configuration is $1\sigma^2 2\sigma^2$, which is net bonding.

Self-test 18.7 From the perspective of structure, indicate several polyhalides that are analogous to [py–I–py]⁺, and describe their bonding.

18.11 Halogen oxides

KEY POINTS The only fluorine oxygen compounds are OF_2 and O_2F_2; chlorine oxides are known for Cl oxidation numbers of +1, +4, +6, and +7; the strong and facile oxidizing agent ClO_2 is the most commonly used halogen oxide.

Oxygen difluoride (FOF; m.p. −224°C, b.p. −145°C), the most stable binary compound of O and F, is prepared by passing fluorine through dilute aqueous hydroxide solution:

$$2F_2(g) + 2OH^-(aq) \rightarrow OF_2(g) + 2F^-(aq) + H_2O(l)$$

The pure difluoride survives in the gas phase above room temperature and does not react with glass. It is a strong fluorinating agent, but less so than fluorine itself. As suggested by the VSEPR model, the OF_2 molecule is angular.

Dioxygen difluoride (FOOF; m.p. −154°C, b.p. −57°C) can be synthesized by photolysis of a liquid mixture of the two elements. It is unstable in the liquid state and decomposes rapidly above 100°C, but can be transferred (with some decomposition) as a low-pressure gas in a metal vacuum line. Dioxygen difluoride is an even more aggressive fluorinating agent than ClF_3. For example, it oxidizes plutonium metal and its compounds to PuF_6, which is an intermediate in the reprocessing of nuclear fuels, in a reaction that ClF_3 cannot accomplish:

$$Pu(s) + 3O_2F_2(g) \rightarrow PuF_6(g) + 3O_2(g)$$

Chlorine dioxide is the only halogen oxide produced on a large scale. The reaction used is the reduction of ClO_3^- with HCl or SO_2 in strongly acidic solution:

$$2ClO_3^-(aq) + SO_2(g) \xrightarrow{acid} 2ClO_2(g) + SO_4^{2-}(aq)$$

Because chlorine dioxide is a strongly endergonic compound ($\Delta G^\ominus = +121\,kJ\,mol^{-1}$), it must be kept dilute to avoid explosive decomposition and is therefore used at the site of production. Its major uses are to bleach paper pulp and to disinfect sewage and drinking water. Some controversy surrounds these applications because the action of chlorine (or its product of hydrolysis, HClO) and chlorine dioxide on organic matter produces low concentrations of chlorocarbon compounds, some of which are potential carcinogens. However, the disinfection of water undoubtedly saves many more lives than the carcinogenic by-products may take. Chlorine bleaches are being replaced by oxygen-based bleaches such as hydrogen peroxide (Box 17.3).

The most well-known oxides of bromine are given below:

Oxidation number	+1	+3	+4
Formula	Br_2O	Br_2O_3	BrO_2
Colour	Dark brown	Orange	Pale yellow
State	Solid	Solid	Solid

The structure of BrO_2 has been found to be a mixed Br(I)/Br(VII) oxide, $BrOBrO_3$. All the bromine oxides are thermally unstable above −40°C and explode on heating.

The most stable halogen oxides are those formed by iodine. The most important of these is I_2O_5 (16), which is used to oxidize carbon monoxide quantitatively to carbon dioxide in the analysis of CO in blood and in air. The compound is a white, hygroscopic solid. It dissolves in water to give iodic acid, HIO_3. The less stable iodine oxides I_2O_4 and I_4O_9 are both yellow solids that decompose on heating to give I_2O_5:

$$5I_2O_4(s) \rightarrow 4I_2O_5(s) + I_2(g)$$
$$4I_4O_9(s) \rightarrow 6I_2O_5(s) + 2I_2(g) + 3O_2(g)$$

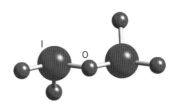

16 I_2O_5

TABLE 18.12 Acidities of chlorine oxoacids

Acid	p/q	pK$_a$
HOCl	0	7.53 (weak)
HOClO	1	2.00
HOClO$_2$	2	−1.2
HOClO$_3$	3	−10 (strong)

18.12 Oxoacids and oxoanions

KEY POINTS The halogen oxoanions are thermodynamically strong oxidizing agents; perchlorates of oxidizable cations are unstable.

The strengths of the oxoacids vary systematically with the number of O atoms on the central atom (Table 18.12; see Pauling's rules in Section 5.9(b)). Periodic acid, H$_5$IO$_6$, is the I(VII) analogue of perchloric acid. It is a weak acid (pK$_{a1}$ = 3.29), which can be explained as soon as we note that its formula is (HO)$_5$IO and that there is only one I=O group. The O atoms in the conjugate base H$_4$IO$_6^-$ are very labile on account of the rapid equilibration:

$$H_4IO_6^-(aq) \rightleftharpoons IO_4^-(aq) + 2H_2O(l) \quad K = 40$$

In basic solution, IO$_4^-$ is the dominant ion. The tendency to have an expanded coordination shell is shared by the oxoacids of the neighbouring Group 16 element tellurium, which in its maximum oxidation state forms the weak acid Te(OH)$_6$.

The halogen oxoanions, like many oxoanions, form metal complexes, including the metal perchlorates and periodates discussed here. In this connection we note that because HClO$_4$ is a very strong acid and H$_5$IO$_6$ is a weak acid, it follows that ClO$_4^-$ is a very weak base and H$_4$IO$_6^-$ is a relatively strong base.

In view of the low Brønsted basicity and single negative charge of the perchlorate ion, ClO$_4^-$, it is not surprising that it is such a weak Lewis base, with little tendency to form complexes with cations in aqueous solution. Therefore, metal perchlorates are often used to study the properties of hexaaqua ions in solution. The ClO$_4^-$ ion is used as a weakly coordinating ion that can readily be displaced from a complex by other ligands, or as a medium-sized anion that might stabilize solid salts containing large cationic complexes with easily displaced ligands.

The ClO$_4^-$ ion is a powerful oxidizing agent so that solid compounds of perchlorate should be avoided whenever there are oxidizable ligands or ions present (which is commonly the case). Conversely, reactions of ClO$_4^-$ are generally slow and it is possible to prepare many metastable perchlorate complexes or salts that may be handled with deceptive ease. However, once reaction has been initiated by mechanical action, heat, or static electricity, these compounds can detonate with disastrous consequences. Such explosions have injured chemists who may have handled a compound many times before it unexpectedly exploded. Some readily available and more docile weakly basic anions may be used in place of ClO$_4^-$; they include trifluoridomethanesulfonate, [SO$_3$CF$_3$]$^-$, tetrafluoridoborate, BF$_4^-$, and hexafluoridophosphate, [PF$_6$]$^-$.

For many years it was believed that perbromate did not exist. However, in 1968 it was prepared by a radiochemical route based on the β decay of ^{83}Se, and chemical syntheses have now been devised. The perbromate ion is more oxidizing than any other oxohalide. The instability of perbromate compared to perchlorate and periodate is an example of the reluctance of post-3d elements to achieve their highest possible oxidation state, and is a manifestation of the alternation effect (Section 10.4).

In contrast to perchlorate, periodate is a rapid oxidizing agent and a stronger Lewis base. These properties lead to the use of periodate as an oxidizing agent in organic chemistry, and stabilizing ligand for metal ions in high oxidation states. Some of the high oxidation states it can be used to form are very unusual: they include Cu(III) in a salt containing the [Cu(HIO$_6$)$_2$]$^{5-}$ complex and Ni(IV) in an extended complex containing the [Ni(IO$_6$)]$^-$ unit. The periodate ligand is bidentate in these complexes, and in the last example it forms a bridge between Ni(IV) ions.

18.13 Thermodynamic aspects of oxoanion redox reactions

KEY POINT The oxoanions of halogens are strong oxidizing agents, especially in acidic solution.

The thermodynamic tendencies of the halogen oxoanions and oxoacids to participate in redox reactions have been extensively studied. As we shall see, we can summarize their behaviour with a Frost diagram that is quite easy to rationalize. It is a very different story with the rates of the reactions, which vary widely. Their mechanisms are only partly understood, despite many years of investigation. Recent progress in the understanding of some of these mechanisms stems from advances in techniques for fast reactions and interest in oscillating reactions (Box 18.3).

We saw in Section 6.13 that, if in a Frost diagram a species lies above the line joining its two neighbours of higher and lower oxidation numbers, then it is unstable with respect to disproportionation into them. From the Frost diagram for the halogen oxoanions and oxoacids in Fig. 18.14 we can see that many of the oxoanions in intermediate oxidation states are susceptible to disproportionation. Chlorous acid, HClO$_2$, for instance, lies above the line joining its two neighbours, and is liable to disproportionation:

$$2HClO_2(aq) \rightarrow ClO_3^-(aq) + HClO(aq) + H^+(aq)$$

$$E_{cell}^{\ominus} = +0.52\text{ V}$$

BOX 18.3 What are oscillating reactions?

Clock reactions and oscillating reactions are an active topic of research and provide fascinating lecture demonstrations. Most oscillating reactions are based on the reactions of halogen oxoanions, apparently because of the variety of oxidation states and their sensitivity to changes in pH.

In 1895, H. Landot discovered that a mixture of sulfite, iodate, and starch in acidic aqueous solution remains nearly colourless for an initial period and then suddenly switches to the dark purple of the I_2-starch complex. When the concentrations are properly adjusted, the reaction oscillates between nearly colourless and opaque blue. The reactions leading to this oscillation are the reduction of iodate to iodide by sulfite in the presence of hexacyanidoferrate(II):

$$IO_3^-(aq) + 3SO_3^{2-}(aq) \rightarrow I^-(aq) + 3SO_4^{2-}(aq)$$

A comproportionation reaction between I^- and IO_3^- then produces I_2, which forms an intensely coloured complex with starch:

$$IO_3^-(aq) + 6H^+(aq) + 5I^-(aq) \rightarrow 3H_2O(l) + 3I_2(starch)$$

Under some conditions the I_2-starch complex is the final state, but adjustment of concentrations may lead to bleaching of the complex by sulfite reduction of iodine to the colourless I^-(aq) ion:

$$3I_2(starch) + 3SO_3^{2-}(aq) + 3H_2O(l) \rightarrow 6I^-(aq) + 6H^+(aq) + 3SO_4^{2-}(aq)$$

The reaction may then oscillate between colourless and blue as the I_2/I^- ratio changes. The reaction of the $[Fe(CN)_6]^{4-}$ causes the oscillation by competing with the sulfite in reactions with IO_3^- and I_2, which are considerably slower than those with sulfite:

$$IO_3^-(aq) + 6[Fe(CN)_6]^{4-}(aq) + 6H^+(aq)$$
$$\rightarrow I^-(aq) + 6[Fe(CN)_6]^{3-}(aq) + 3H_2O(l)$$

$$I_2(starch) + 2[Fe(CN)_6]^{4-}(aq) \rightarrow 2I^-(aq) + 2[Fe(CN)_6]^{3-}(aq)$$

The oscillation begins after all the $[Fe(CN)_6]^{4-}$ has been removed. It is regenerated by reaction with SO_3^{2-}:

$$2[Fe(CN)_6]^{3-} + SO_3^{2-} + H_2O \rightarrow 2[Fe(CN)_6]^{4-} + SO_4^{2-} + 2H^+$$

The Belousov–Zhabotinsky reaction is another oscillating reaction and is based on the dual role of H_2O_2 as an oxidizing and a reducing agent. In this reaction, H_2O_2, KIO_3, H_2SO_4, and starch are mixed together, and hydrogen peroxide reduces iodate to iodine and is itself oxidized to oxygen gas:

$$5H_2O_2(aq) + 2IO_3^-(aq) + 2H^+(aq) \rightarrow I_2(starch) + 5O_2(g) + 6H_2O(l)$$

The hydrogen peroxide also oxidizes iodine to iodate:

$$5H_2O_2(aq) + I_2(starch) \rightarrow 2IO_3^-(aq) + 2H^+(aq) + 4H_2O(l)$$

The net result of these reactions is the iodate catalysis of the disproportionation of hydrogen peroxide, and the reaction oscillates between colourless and blue.

The detailed analysis of the kinetic conditions for oscillating reactions is pursued by chemists and chemical engineers. In the former case, the challenge is to use kinetic data determined separately for the individual steps to model the observed oscillations with a view to testing the validity of the overall scheme. As oscillating reactions have been observed in commercial catalytic processes, the concern of the chemical engineer is to avoid large fluctuations or even chaotic reactions that might degrade the process. Oscillating reactions are of more than industrial interest, for they also maintain the rhythm of the heartbeat and their interruption can result in fibrillation and death.

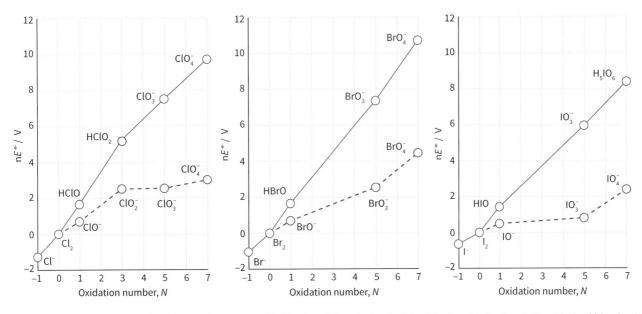

FIGURE 18.14 Frost diagrams for chlorine, bromine, and iodine in acidic solution (solid red line) and in basic solution (dashed blue line).

Although BrO_2^- has been observed, the corresponding I(III) species is so unstable that it does not exist in solution, except perhaps as a transient intermediate.

We also saw in Section 6.13 that the more positive the slope for the line from a lower to higher oxidation state species in a Frost diagram, the stronger the oxidizing power of the couple. A glance at Fig. 18.14 shows that all three Frost diagrams have steep, positively sloping lines, which immediately shows that all the oxidation states except the lowest (Cl^-, Br^-, and I^-) are strongly oxidizing.

Finally, basic conditions decrease reduction potentials for oxoanions as compared with their conjugate acids (Sections 6.6 and 6.14). This decrease is evident in the less steep slopes of the lines in the Frost diagrams for the oxoanions in basic solution. The numerical comparison for ClO_4^- ions in 1 M acid compared with 1 M base makes this clear:

At pH = 0, $ClO_4^-(aq) + 2H^+(aq) + 2e^- \rightarrow ClO_3^-(aq) + H_2O(l)$
$$E^\ominus = +1.20\,V$$

At pH = 14, $ClO_4^-(aq) + H_2O(l) + 2e^- \rightarrow ClO_3^-(aq)$
$$+ 2OH^-(aq) \quad E^\ominus = +0.37\,V$$

The reduction potentials show that perchlorate is thermodynamically a much weaker oxidizing agent in basic solution than in acidic solution.

EXAMPLE 18.8 Predicting the disproportionation of oxoanions

Use Fig. 18.14 to predict which species of chlorine will disproportionate in basic conditions, and give balanced equations for the reactions.

Answer We need to identify which species lie above the line that joins species with neighbouring oxidation numbers (Section 6.13). The species Cl_2 and ClO_2^- lie above a line joining adjacent species and will therefore disproportionate. The equations for the reactions are

$Cl_2 + 2OH^- \rightarrow ClO^- + Cl^- + H_2O$
$2ClO_2^- \rightarrow ClO_3^- + ClO^-$

Self-test 18.8 Predict which species of (a) bromine and (b) iodine will disproportionate in basic conditions, and give balanced equations for the reactions.

18.14 Trends in rates of oxoanion redox reactions

KEY POINTS Oxidation by halogen oxoanions is faster for the lower oxidation states; rates and thermodynamics of oxidation are both enhanced by an acidic medium.

Mechanistic studies show that the redox reactions of halogen oxoanions are complex. Nevertheless, despite this complexity, a few discernible patterns help to correlate the trends in rates of reaction. These correlations have practical value and give some clues about the mechanisms that may be involved.

The oxidation of many molecules and ions by halogen oxoanions becomes progressively faster as the oxidation number of the halogen decreases. Thus the rates observed are often in the order

$ClO_4^- < ClO_3^- < ClO_2^- \approx ClO \approx Cl_2$
$BrO_4^- < BrO_3^- \approx BrO^- \approx Br_2$
$IO_4^- < IO_3^- < I_2$

For example, aqueous solutions containing Fe^{2+} and ClO_4^- are stable for many months in the absence of dissolved oxygen, but an equilibrium mixture of aqueous HClO and Cl_2 rapidly oxidizes Fe^{2+}.

Oxoanions of the heavier halogens tend to react most rapidly, particularly for the elements in their highest oxidation states:

$ClO_4^- < BrO_4^- < IO_4^-$

As we have remarked, perchlorates in dilute aqueous solution are usually unreactive, but periodate oxidations are fast enough to be used for titrations. The mechanistic details are often complex, but the existence of both four- and six-coordinate periodate ions shows that the I atom in periodate is accessible to nucleophiles.

We have already seen that the thermodynamic tendency of oxoanions to act as oxidizing agents increases as the pH is lowered. It is found that their rates are increased too. Thus, kinetics and equilibria unite to bring about otherwise difficult oxidations. The oxidation of halides by BrO_3^- ions, for instance, is second order in H^+:

$$Rate = k_r[BrO_3^-][X^-][H^+]^2$$

and so the rate increases as the pH is decreased. The acid is thought to protonate the oxo group in the oxoanion, so aiding oxygen–halogen bond scission. Another role of protonation is to increase the electrophilicity of the halogen. An example is HClO, where, as described below, the Cl atom may be viewed as an electrophile towards an incoming reducing agent (**17**). An illustration of the effect of acidity on rate is the use of a mixture of H_2SO_4 and $HClO_4$ in the final stages of the oxidation of organic matter in certain analytical procedures.

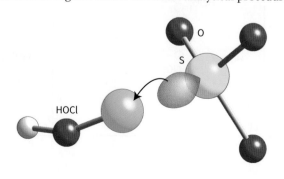

17

18.15 Redox properties of individual oxidation states

KEY POINTS Dihalogen molecules disproportionate in aqueous solution. Hypochlorite is a facile oxidizing agent; hypohalite and halite ions undergo disproportionation. Chlorate ions undergo disproportionation in solution but bromates and iodates do not.

With the general redox properties of the halogens now outlined, we can consider the characteristic properties and reactions of specific oxidation states. Although we are dealing here with halogen oxides, it is convenient to mention, for the sake of completeness, the redox properties of halogen(0) species. Fig. 18.15 summarizes some of the reactions that interconvert the oxoanions and oxoacids of chlorine in its various oxidation states. One point to note is the major role of disproportionation and electrochemical reactions in the scheme. For example, the figure includes the production of Cl_2 by the electrochemical oxidation of Cl^-, which was discussed in Section 18.4.

Disproportionation is thermodynamically favourable for basic solutions of Cl_2, Br_2, and I_2. The equilibria in basic aqueous solution are

$$X_2(aq) + 2OH^-(aq) \rightleftharpoons XO^-(aq) + X^-(aq) + H_2O(l)$$

$$K = \frac{[XO^-][X^-]}{[X_2][OH^-]^2}$$

with $K = 7.5 \times 10^{15}$ for X = Cl, 2×10^8 for Br, and 30 for I.

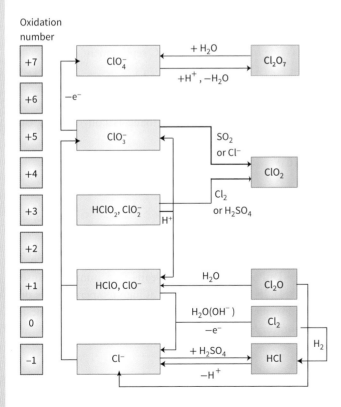

FIGURE 18.15 The interconversion of oxidation states of some important chlorine species.

A NOTE ON GOOD PRACTICE

When writing the associated equilibrium expressions in aqueous solution, we conform to the usual convention that the activity of water is 1, so H_2O does not appear in the equilibrium constant.

Disproportionation is much less favourable in acidic solution, as would be expected from the fact that H^+ is a product of the reaction

$$Cl_2(aq) + H_2O(l) \rightleftharpoons HClO(aq) + H^+(aq) + Cl^-(aq)$$

$$K = 3.9 \times 10^{-4}$$

Because the redox reactions of Cl_2 are often fast, Cl_2 in water is widely used as an inexpensive and powerful oxidizing agent. The equilibrium constants for the hydrolysis of Br_2 and I_2 in acid solution are smaller than for Cl_2, and both elements are unchanged when dissolved in slightly acidified water. Because F_2 is a much stronger oxidizing agent than the other halogens, it produces mainly O_2 and H_2O_2 when in contact with water. As a result, hypofluorous acid, HOF, was discovered long after the other hypohalous acids.

The aqueous Cl(I) species hypochlorous acid, HOC (the angular molecular species H–O–Cl), and hypochlorite ions, ClO^-, are facile oxidizing agents that are used as household bleach and disinfectant, and as laboratory oxidizing agents. The ready access to the unobstructed, electrophilic Cl atom in HOC appears to be one feature that leads to the very fast redox reactions of this compound. These rates contrast with the much slower redox reactions of perchlorate ions, in which access to the Cl atom is blocked by the surrounding tetrahedron of O atoms.

A NOTE ON GOOD PRACTICE

Hypohalous acids are widely denoted HOX to emphasize their structure. We have adopted the formula HXO to emphasize their relationship to the other oxoacids, HXO_n.

Hypohalite ions undergo disproportionation. For instance, ClO^- disproportionates into Cl^- and ClO_3^-:

$$3ClO^-(aq) \rightleftharpoons 2Cl^-(aq) + ClO_3^-(aq) \quad K = 1.5 \times 10^{27}$$

This reaction (which is used for the commercial production of chlorates) is slow at or below room temperature for ClO^- but is much faster for BrO^-. It is so fast for IO^- that this ion has been detected only as a reaction intermediate.

Chlorite ions, ClO_2^-, and bromite ions, BrO_2^-, are both susceptible to disproportionation. However, the rate is strongly dependent on pH, and ClO_2^- (and to a lesser extent BrO_2^-) can be handled in basic solution with only slow decomposition.

By contrast, chlorous acid, $HClO_2$, and bromous acid, $HBrO_2$, both disproportionate rapidly. Iodine(III) is even more elusive, and HIO_2 has been identified only as a transient species in aqueous solution.

The Frost diagram for Cl shown in Fig. 18.14 indicates that chlorate ions, ClO_3^-, are marginally unstable with respect to disproportionation in both acidic and basic solution:

$$4ClO_3^-(aq) \rightleftharpoons 3ClO_4^-(aq) + Cl^-(aq)$$
$$\Delta_r G^\circ = 24 \text{ kJ mol}^{-1} \quad K = 1.5 \times 10^4$$

Because $HClO_3$ is a strong acid, and this reaction is slow at both low and high pH, ClO_3^- ions can be handled readily in aqueous solution. Bromates and iodates are thermodynamically stable with respect to disproportionation.

Of the three XO_4^- ions, BrO_4^- is the strongest oxidizing agent. That perbromate is out of line with its adjacent halogen congeners fits a general pattern for anomalies in the chemistry of p-block elements of Period 4 (Section 10.4). However, the reduction of periodate in dilute acid is faster than that of perchlorate or perbromate, and periodates are therefore used in analytical chemistry as oxidizing titrants and also in syntheses, such as the oxidative cleavage of diols:

A BRIEF ILLUSTRATION

To confirm that perbromate is the most oxidizing perhalate ion we need to consider the slope of the lines joining the perhalate ions to their neighbours in the Frost diagram. The more positive the slope of the line, the stronger the oxidizing power of the couple. Inspection of the diagram reveals that the line joining the BrO_4^-/BrO_3^- couple is the most positive. In fact, the E values for the ClO_4^-/ClO_3^-, BrO_4^-/BrO_3^-, and IO_4^-/IO_3^- couples in acidic conditions are 1.20, 1.85, and 1.60 V, respectively, confirming that perbromate is the strongest oxidizing agent.

18.16 Fluorocarbons

KEY POINT Fluorocarbon molecules and polymers are resistant to chemical reaction.

Fluorocarbons find many useful applications, most of which arise as a result of the inertness of the very strong C–F bond (456 kJ mol^{-1}). The direct reaction of an aliphatic hydrocarbon with an oxidizing metal fluoride leads to the formation of strong C–F bonds and produces HF as a by-product:

$$RH(l) + 2CoF_3(s) \rightarrow RF(sol) + 2CoF_2(s) + HF(sol)$$
$$R = \text{alkyl or aryl}$$

When R is aryl, CoF_3 yields the cyclic saturated fluoride:

$$C_6H_6(l) + 18CoF_3(s) \rightarrow C_6F_{12}(l) + 18CoF_2(s) + 6HF(l)$$

The strongly oxidizing fluorinating agent used in these reactions, CoF_3, is regenerated by the reaction of CoF_2 with fluorine:

$$2CoF_2(s) + F_2(g) \rightarrow 2CoF_3(s)$$

Another important method of C–F bond formation is halogen exchange by the reaction of a nonoxidizing fluoride, such as HF, with a chlorocarbon in the presence of a catalyst, such as SbF_3:

$$CCl_4(l) + HF(l) \rightarrow CCl_3F(l) + HCl(g)$$
$$CHCl_3(l) + 2HF(l) \rightarrow CHClF_2(l) + 2HCl(g)$$

These processes used to be performed on a large scale to produce the CFCs and hydrochlorofluorocarbons (HCFCs) that were used as refrigerant fluids, the propellant in spray cans, and in the blowing agent in plastic foam products. These applications have been banned in most countries and are being phased out worldwide because of the role of CFCs and HCFCs in ozone depletion. Their replacement by HFCs requires significant investment by the chemical industry because, in contrast to the simple one-step synthesis of CFCs and HCFCs, HFC production is a complex, multi-stage process. For example, the preferred route to CF_3CH_2F, which is one of the preferred CFC replacements, is

$$CCl_2=CCl_2 \xrightarrow{HF+Cl_2} CClF_2CCl_2F \xrightarrow{\text{isomerize}} CF_3CCl_3$$
$$\xrightarrow{HF} CF_3CCl_2F \xrightarrow{H_2} CF_3CH_2F$$

When heated, chlorodifluoromethane is converted to the useful monomer C_2F_4:

$$2CHClF_2 \xrightarrow{600-800^\circ C} C_2F_4 + 2HCl$$

The polymerization of tetrafluoroethene is carried out with a radical initiator:

$$nC_2F_4 \xrightarrow{ROO\cdot} (-CF_2-CF_2-)_n$$

PTFE is sold under many trade names, one of which is Teflon® (DuPont); its inertness leads to a number of uses (Box 18.4). Its depolymerization at high temperatures is the most convenient method of preparing tetrafluoroethene in the laboratory:

$$(-CF_2-CF_2-)_n \xrightarrow{600^\circ C} n\, C_2F_4$$

Although tetrafluoroethene is not highly toxic, a by-product, 1,1,3,3,3-pentafluoro-2-trifluoromethyl-1-propene, is toxic, and its presence dictates care in handling crude tetrafluoroethene.

BOX 18.4 Why is PTFE so inert?

Polytetrafluoroethene, PTFE, is a unique product in the plastics industry. It is chemically inert, is thermally stable over a wide temperature range (−196°C to 260°C), is an excellent electrical insulator, and has a low coefficient of friction. It is a white solid that is manufactured by the polymerization of tetrafluoroethene:

$$n\,CF_2=CF_2 \rightarrow (CF_2CF_2)_n$$

PTFE is an expensive polymer because of the cost of synthesizing and purifying the monomer by a multistage process:

$$CH_4(g) + 3Cl_2(g) \rightarrow CHCl_3(g) + 3HCl(g)$$
$$CHCl_3(g) + 2HF(g) \rightarrow CHClF_2(g) + 2HCl(g)$$
$$2CHClF_2(g) \xrightarrow{\Delta} CF_2=CF_2(g) + 2HCl(g)$$

The hydrogen fluoride is generated by the action of sulfuric acid on fluorite:

$$CaF_2(s) + H_2SO_4(l) \rightarrow CaSO_4(s) + 2HF(l)$$

As the process employs HF and HCl, the reactors have to be lined with platinum. Many by-products are produced, which leads to complex purification of the final product.

The tetrafluoroethene is polymerized using a radical initiator in two ways: solution polymerization with vigorous agitation produces a resin known as granular PTFE; emulsion polymerization with a dispersing agent and gentle agitation produces small particles known as dispersed PTFE. The molten polymer does not flow, so the usual methods of processing cannot be used. Instead, processes similar to those used for metals are applied. For example, the dispersed solid form can be cold-extruded (which is a method used for processing lead).

The remarkable properties of PTFE arise from the protective sheath that the F atoms form around the carbon-polymer backbone. The F atoms are just the right size to completely fill the space around the carbon backbone and hence protect it from chemical attack. The sheath also causes a disruption to the intermolecular forces at the surface, leading to a low coefficient of friction and the familiar nonstick properties. The polymer is used in a wide range of applications. Its low electrical conductivity leads to its use in electrical tapes, wires, and coaxial cable. Its mechanical properties make it an ideal material for seals, piston rings, and bearings. It is used as a packaging material, in hose lines, and as thread-sealant tape. Familiar applications are as the nonstick coating on cookware, and as the porous fabric Gore-Tex®.

FURTHER READING

M. Schnürch, M. Spina, A. F. Khan, M. D. Mihovilovic, and P. Stanetty, Halogen dance reactions: a review. *Chem. Soc. Rev.*, 2007, **36**, 1046.

V. Gouverneur, Fluorine chemistry: An outlook to the future. *Adv. Synth. Catal.*, 2024, **366**, 3404.

P. Enghag, *Encyclopedia of the Elements*. John Wiley & Sons (2004).

D. M. P. Mingos, *Essential Trends in Inorganic Chemistry*. Oxford University Press (2004). An overview of inorganic chemistry from the perspective of structure and bonding.

R. B. King (ed.), *Encyclopedia of Inorganic Chemistry*. John Wiley & Sons (2005).

EXERCISES

18.1 Preferably without consulting reference material, write out the halogens as they appear in the periodic table, and indicate the trends in (a) physical state (s, l, or g) at room temperature and pressure, (b) electronegativity, (c) hardness of the halide ion, (d) colour.

18.2 Describe how the halogens are recovered from the naturally occurring halides and rationalize the approach in terms of standard potentials. Give balanced chemical equations and conditions where appropriate.

18.3 Use the molecular orbital diagram in Fig. 18.5 to determine the bond order of the Br_2^+ ion. Will the Br–Br bond be longer or shorter than that in the Br_2 molecule?

18.4 Sketch a chloralkali cell. Show the half-cell reactions and indicate the direction of diffusion of the ions. Give the chemical equation for the unwanted reaction that would occur if OH^- migrated through the membrane and into the anode compartment.

18.5 Sketch the form of the vacant σ^* orbital of a dihalogen molecule and describe its role in the Lewis acidity of the dihalogens.

18.6 Which dihalogens are thermodynamically capable of oxidizing H_2O to O_2?

18.7 Nitrogen trifluoride, NF_3, boils at −129°C and is a very weak Lewis base. By contrast, the lower molar mass compound NH_3 boils at −33°C and is well known as a Lewis base.

(a) Describe the origins of this very large difference in volatility.

(b) Describe the probable origins of the difference in basicity.

18.8 Based on the analogy between halogens and pseudohalogens, write (a) the balanced equation for the probable reaction of cyanogen, $(CN)_2$, with aqueous sodium hydroxide, (b) the equation for the probable reaction of excess thiocyanate with the oxidizing agent $MnO_2(s)$ in acidic aqueous solution, and (c) a plausible structure for trimethylsilyl cyanide.

18.9 Given that 1.84 g of IF_3 reacts with 0.93 g of $[(CH_3)_4N]F$ to form a product X, (a) identify X, (b) use the VSEPR model to predict the shapes of IF_3 and the cation and anion in X, and (c) predict how many ^{19}F-NMR signals would be observed in IF_3 and X.

18.10 Treatment of Br_2 with ozone in $CFCl_3$ at −50°C yields yellow crystals of an unstable diamagnetic compound. Identify this compound and predict the products obtained when it reacts with one mole equivalent of NaOH.

18.11 Indicate the product of the reaction between ClF_5 and SbF_5. Predict the shapes of the reactants and products.

18.12 Predict whether each of the following solutes is likely to make liquid BrF_3 act as a Lewis acid or a Lewis base: (a) SbF_5, (b) SF_6, (c) CsF.

18.13 (a) Use the VSEPR model to predict the probable shapes of $[IF_6]^+$ and IF_7. (b) Give a plausible chemical equation for the preparation of $[IF_6][SbF_6]$.

18.14 Predict the shape of the doubly chlorine-bridged I_2Cl_6 molecule by using the VSEPR model, and assign the point group.

18.15 Predict the structures and identify the point groups of ClO_2F and ClO_3F.

18.16 Given the bond lengths and angles in I_5^- (5), describe the bonding in terms of two-centre and three-centre bonds. Can the structure be accounted for in terms of the VSEPR model?

18.17 Predict whether each of the following compounds is likely to be dangerously explosive in contact with BrF_3, and explain your answer: (a) SbF_5, (b) CH_3OH, (c) F_2, (d) S_2Cl_2.

18.18 The formation of Br_3^- from a tetraalkylammonium bromide and Br_2 is only slightly exergonic. Write an equation for the interaction of $[NR_4][Br_3]$ with I_2 in CH_2Cl_2 solution, and give your reasoning.

18.19 Explain why $CsI_3(s)$ is stable with respect to decomposition but $NaI_3(s)$ is not.

18.20 (a) Give the formulas and the probable relative acidities of perbromic acid and periodic acid. (b) Which is the more stable to decomposition liberating oxygen?

18.21 (a) Describe the expected trend in the standard potential of an oxoanion in a solution with decreasing pH. (b) Demonstrate this phenomenon by calculating the reduction potential of ClO_4^- at pH = 7 and comparing it with the tabulated value at pH = 0.

18.22 With regard to the general influence of pH on the standard potentials of oxoanions, explain why the disproportionation of an oxoanion is often promoted by low pH.

18.23 Use the Latimer diagrams in Resource section 3 to calculate the standard potentials for the following couples in basic solution: (a) ClO_4^-/ClO^-, (b) BrO_4^-/BrO^-, (c) IO_4^-/IO^-. Comment on the relative feasibilities of the reduction reactions.

18.24 (a) For which of the following anions is disproportionation thermodynamically favourable in acidic solution: ClO^-, ClO_2^-, ClO_3^-, and ClO_4^-? (If you do not know the properties of these ions, determine them from a table of standard potentials as in Resource section 3.) (b) For which of the favourable cases is the reaction very slow at room temperature?

18.25 Which of the following compounds present an explosion hazard? (a) NH_4ClO_4, (b) $Mg(ClO_4)_2$, (c) $NaClO_4$, (d) $[Fe(OH_2)_6][ClO_4]_2$. Explain your reasoning.

18.26 Use standard potentials to predict which of the following will be oxidized by ClO^- ions in acidic conditions: (a) Cr^{3+}, (b) V^{3+}, (c) Fe^{2+}, (d) Co^{2+}.

18.27 Identify all the compounds **A** to **G**.

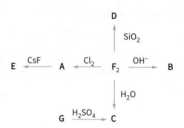

18.28 Many of the acids and salts corresponding to the positive oxidation numbers of the halogens are not listed in the catalogue of a major international chemical supplier: (a) $KClO_4$ and KIO_4 are available but $KBrO_4$ is not, (b) $KClO_3$, $KBrO_3$, and KIO_3 are all available, (c) $NaClO_2$ and $NaBrO_2 \cdot 3H_2O$ are available but salts of IO_2^- are not, (d) only ClO^- salts are available but the bromine and iodine analogues are not. Describe the probable reason for the unavailability of those salts of the oxoanions.

18.33 Identify the incorrect statements among the following descriptions and provide correct statements:

(a) Oxidation of the halides is the only commercial method of preparing the halogens from F_2 to I_2.

(b) ClF_4^- and I_5^- are isolobal and isostructural.

(c) Atom-transfer processes are common in the mechanisms of oxidations by the halogen oxoanions, and an example is the O atom transfer in the oxidation of SO_3^{2-} by ClO^-.

(d) Periodate appears to be a more facile oxidizing agent than perchlorate because the former can coordinate to the reducing agent at the I(VII) centre, whereas the Cl(VII) centre in perchlorate is inaccessible to reducing agents.

TUTORIAL PROBLEMS

18.1 The paper 'The consequences of excess iodine' by A. Leung and E. Braverman (*Nat. Rev.*, 2014, **10**, 136) reviews the health implications of excess iodine ingestion. (a) What is the recommended daily intake of iodine? (b) What are the main dietary sources of iodine? (c) What is usually the source of high levels of iodine? (d) What are the symptoms of ingestion of high levels of iodine? (d) What were the health benefits of a high iodine diet in Japan in the period after 2011? (e) How is sufficient dietary iodine ensured in most countries?

18.2 The phenomenon of the halogen bond has been known for more than a century. In the paper 'The halogen bond in solution' (*Chem. Soc. Rev.*, 2012, **41**, 3547), M. Erdelyi reviews the current state of knowledge of the nature of the halogen bond. Describe what is meant by halogen bonding. Work on the halogen bond received the 1969 Nobel Prize in Chemistry. Who were the recipients of this prize? Describe the relevance of halogen bonding in biology. List the techniques that have been used to probe the nature of halogen bonded interactions, and outline the molecular orbital description of such a bond. Explain what is measured by the diiodide basicity scale, and give the assumptions on which the scale is based. Give the reference in which this scale was first mentioned.

18.3 In their paper 'Recent discoveries of polyhalogen anions—from bromine to fluorine' (*Z. Anorg. Allg. Chem.*, 2014, **640**, 7, 1281), H. Haller and S. Riedel describe the synthesis and structure elucidation of several new polyhalide ions. (a) Outline the synthetic strategies that enabled these polyhalide ions to be synthesized. (b) Sketch the structures of the Br_5^+, Br_{13}^{3-}, and Cl_7^- ions. (c) Referring to earlier work on polyiodides, discuss how I_3^- units coordinate to build up polyiodides.

18.4 The potential of organofluoro compounds in materials chemistry is discussed in a paper by R. Berger and co-workers (*Chem. Soc. Rev.*, 2011, **40**, 3496). One group of compounds discussed is the fluorinated fullerenes. Give an equation for the reaction of the most commonly used fluorinated fullerene with an organic species to form [18]trannulene, and sketch the product. The use of fluorine in pharmaceuticals is also described in the paper. Summarize the reasons why fluorine finds wide application in pharmaceutical compounds.

18.5 The reaction of I^- ions is often used to titrate ClO^-, giving deeply coloured I_3^- ions, along with Cl^- and H_2O. Although never proved, it was once thought that the initial reaction proceeds by O atom transfer from Cl to I. However, it is now believed that the reaction proceeds by Cl atom transfer to give ICl as the intermediate (K. Kumar, R. A. Day, and D. W. Margerum, *Inorg. Chem.*, 1986, **25**, 4344). Summarize the evidence for Cl atom transfer.

18.6 Until the work of K. O. Christe (*Inorg. Chem.*, 1986, **25**, 3721), F_2 could be prepared only electrochemically. Give chemical equations for Christe's preparation, and summarize the reasoning behind it.

18.7 The use of templates to synthesize long-chain polyiodide ions has been described (A. J. Blake et al., *Chem. Soc. Rev.*, 1998, **27**, 195). (a) According to the authors, what is the longest polyiodide that has been characterized? (b) How does the nature of the cation influence the structure of the polyanion? (c) What was the templating agent used for the synthesis of I_7^- and I_{12}^-? (c) Which spectroscopic method was used for the characterization of the polyanions in this study?

18.8 Write a review of the environmental problems associated with the use of chlorine-based bleaches in industry, and suggest possible solutions.

19 The Group 18 elements

Part A: The essentials
19.1 The elements
19.2 Simple compounds

Part B: The detail
19.3 Occurrence and recovery
19.4 Uses
19.5 Synthesis and structure of xenon fluorides
19.6 Reactions of xenon fluorides
19.7 Xenon-oxygen compounds
19.8 Insertion compounds
19.9 Organoxenon compounds
19.10 Coordination compounds
19.11 Other compounds of noble gases

Further reading

Exercises

Tutorial problems

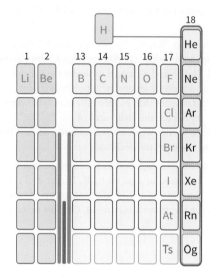

The final group in the p block contains seven elements that are so unreactive that they form only a very limited range of compounds. The existence of the Group 18 elements was not suspected until late in the nineteenth century, and their discovery led to the redrawing of the periodic table and played a key role in the development of bonding theories.

The Group 18 elements—helium, neon, argon, krypton, xenon, radon, and oganesson—are all monatomic gases. They are the least reactive elements and have been given, and then lost, various collective names over the years as different aspects of their properties have been identified and then disproved. Thus they have been called the *rare gases* and the *inert gases*, and are currently called the **noble gases**. The first name is inappropriate, because argon is by no means rare (it is substantially more abundant than CO_2 in the atmosphere). The second has been inappropriate since the discovery of compounds of xenon. The name 'noble gases' is now accepted because it gives the sense of low but significant reactivity.

PART A: The essentials

In this section we survey the limited chemistry of the noble gases and concentrate in particular on the well-characterized compounds of xenon.

19.1 The elements

KEY POINT Of the noble gases, only xenon forms a significant range of compounds with fluorine and oxygen.

All the Group 18 elements are very unreactive. Their unreactivity can be understood in terms of their atomic properties (Table 19.1) and in particular their ground-state valence electron configurations, ns^2np^6. The features to note include their high ionization energies and negative electron affinities. The first ionization energy is high because the effective nuclear charge is high at the far right of the period. The electron affinities are negative because an incoming electron needs to occupy an orbital belonging to a new shell.

By mass, helium makes up 23% of the universe and the Sun, and is the second most abundant element after hydrogen; it is rare in the atmosphere because its atoms travel fast enough to escape from the Earth. All the other noble gases, except the synthetic oganesson, occur in the atmosphere. The abundances of argon (0.94% by volume) and neon (1.5×10^{-3} %) make these two elements more plentiful than many familiar elements, such as arsenic and bismuth, in the Earth's crust (Fig. 19.1). Xenon and radon are the rarest naturally occurring elements of the group. Radon is a product of radioactive decay and, as its atomic number exceeds that of lead, is itself unstable; it accounts for around 50% of background radiation. Oganesson is completely synthetic and probably only three atoms of it have ever been made. Claims for its existence were first made in 2006, but it was only officially added to the periodic table in 2015.

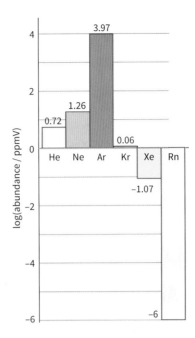

FIGURE 19.1 Abundances of the noble gases in the Earth's crust. The values are log (atmospheric parts per million by volume).

19.2 Simple compounds

KEY POINT Xenon forms fluorides, oxides, and oxofluorides.

The most important oxidation numbers of Xe in its compounds are +2, +4, and +6. Compounds with Xe–F, Xe–O, Xe–N, Xe–H, Xe–C, and Xe–metal bonds are known, and Xe can behave as a ligand. The chemical properties of xenon's lighter congener krypton are much more limited. The study of radon chemistry, like that of astatine, is inhibited by the high radioactivity of the element.

Xenon reacts directly with fluorine to produce XeF_2 (**1**), XeF_4 (**2**), and XeF_6 (**3**). Solid XeF_6 is more complex than its monomeric gas-phase structure and contains fluoride ion-bridged XeF_5^+ cations. The xenon fluorides are strong oxidizing agents and form complexes with F^- ions, such as XeF_5^-. XeF_2 is used as a fluorinating agent in the laboratory as it is easier to handle than fluorine gas.

TABLE 19.1 Selected properties of the elements

	He	Ne	Ar	Kr	Xe	Rn
Atomic radius/pm	28	58	106	116	140	150
Melting point/°C	−272	−249	−189	−157	−112	−71
Boiling point/°C	−269	−246	−186	−152	−108	−62
Electron affinity/ kJ mol⁻¹	−48.2	−115.8	−96.5	−96.5	−77.2	−68.0
First ionization energy/kJ mol⁻¹	2373	2080	1520	1350	1170	1036

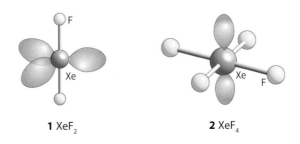

1 XeF_2 **2** XeF_4

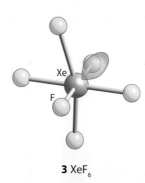

3 XeF_6

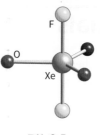

7 XeO_3F_2

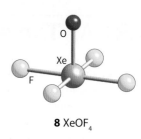

8 $XeOF_4$

Xenon forms xenon trioxide, XeO_3 (4), and xenon tetroxide, XeO_4 (5), which both decompose explosively. The white crystalline perxenates of several alkali metals have been prepared, and contain the XeO_6^{4-} ion.

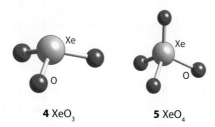

4 XeO_3 **5** XeO_4

Xenon also forms several oxofluorides. The oxofluoride $XeOF_2$ has a T-shape geometry (6) and XeO_3F_2 is a trigonal bipyramid (7). The square-pyramidal molecule $XeOF_4$ (8) is remarkably similar to IF_5 in its physical and chemical properties.

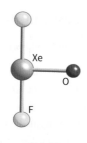

6 $XeOF_2$

Xenon forms a number of hydrides in noble gas matrices, such as HXeH, HXeOH, and HXeOXeH. Xenon, argon, and krypton form clathrates when frozen with water at high pressure (see Box 15.3 and Fig. 5.15). The noble-gas atoms are guests within the three-dimensional ice structure and have the composition $E \cdot 6H_2O$. Clathrates provide a convenient way of handling the radioactive isotopes of krypton and xenon.

EXAMPLE 19.1 Calculating bond orders of noble gas ions

Use the molecular orbital diagram in Fig. 2.12 to predict the bond order of the Xe_2^+ ion.

Answer We will assume that the order of the molecular orbitals for xenon species is similar to those for fluorine. The Xe_2^+ ion will have 15 valence electrons. Consequently, the electron configuration of Xe_2^+ will be $1\sigma_g^2 1\sigma_u^2 2\sigma_g^2 1\pi_u^4 1\pi_g^4 2\sigma_u^1$. The bond order $b = \frac{1}{2}(n - n^*) = \frac{1}{2}(8 - 7) = 0.5$.

Self-test 19.1 Predict the bond order of the XeF^+ ion.

PART B: The detail

In this section we describe the detailed chemistry of the Group 18 elements. Although they are the least reactive of all elements, the noble gases, especially Xe, form a surprisingly wide range of compounds with hydrogen, oxygen, and the halogens.

19.3 Occurrence and recovery

KEY POINTS The noble gases are monatomic; radon and oganesson are radioactive.

Because they are so unreactive and are rarely concentrated in nature, the noble gases eluded recognition until the end of the nineteenth century. Indeed, Mendeleev did not make a place for them in his periodic table because the chemical regularities of the other elements, on which his table was based, did not suggest their existence. However, in 1868 a new spectral line observed in the spectrum of the Sun did not correspond to a known element. This line was eventually attributed to helium, and in due course the element itself and its congeners were found on Earth.

The noble gases were given names that reflect their curious nature; helium from the Greek *helios* for 'sun', neon from the Greek *neos* for 'new', argon from *argos* meaning 'inactive', krypton from *kryptos* meaning 'hidden', and xenon from *xenos* meaning 'strange'. Radon was named after radium as it is a product of the radioactive decay of

> **BOX 19.1   Helium: can we meet increasing demand for a rare gas?**
>
> Helium is separated from natural gases in three steps. First, impurities such as H_2O, CO_2, and H_2S are removed from the gas by passing through a molecular sieve (Section 29.2). This is followed by removal of any high-molecular-weight hydrocarbons by liquid extraction and, finally, low-temperature distillation removes the methane that makes up the bulk of the natural gas. The product is crude helium containing nitrogen along with smaller amounts of argon, neon, and hydrogen. Final purification is carried out using activated charcoal at liquid-nitrogen temperatures and high pressure, or by pressure-swing adsorption.
>
> Demand for helium has been outstripping supply for some years, and some scientists have been predicting that all helium will run out in 20–30 years. The United States used to hold massive underground reserves of helium, but in 1996 legislation required the depletion of this reserve, and as a consequence relatively cheap helium flooded the market, leading to increased use and less recycling. The price of helium has subsequently risen over 500% in 15 years, and users have been encouraged to take a more responsible approach to recycling, with calls for bans on trivial uses such as helium party-balloons. In June 2016 massive reserves of helium were found in Tanzania's Rift Valley. It is believed that volcanic action generated sufficient heat to release the gas from helium-bearing rocks. The discovery of this helium reserve has been hailed as a 'game changer'. The world's current helium use is around 0.2 billion cubic metres per year. The Tanzanian reserves hold around 1.6 billion cubic metres of helium, clearly sufficient to allay any immediate concerns. However, the only long-term solution is recycling and newer, highly efficient, systems are increasingly being used.

that element. Oganesson is named after Yuri Oganessian, a nuclear physicist who played a leading role in the discovery of the heaviest elements in the periodic table.

Helium atoms are too light to be retained by the Earth's gravitational field, so most of the He on Earth (5 parts per million by volume) is the product of the continued α emission in the decay of radioactive elements. High He concentrations of up to 7% by mass are found in certain natural gas deposits (mainly in the United States and eastern Europe), from which it can be recovered by low-temperature distillation (Box 19.1). Some He arrives from the Sun as a solar wind of α particles. Neon, Ar, Kr, and Xe are extracted from liquid air by low-temperature distillation.

The elements are all monatomic gases at room temperature. In the liquid phase they form low concentrations of dimers that are held together by dispersion forces. The low boiling points of the lighter noble gases (Table 19.1) follow from the weakness of these forces between the atoms and the absence of other forces. When helium (specifically ^{4}He, not the rarer isotope ^{3}He) is cooled below 2.178 K it undergoes a transformation into a second liquid phase known as **helium-II**. This phase is classed as a superfluid, as it flows without viscosity. Solid He is formed only under pressure.

19.4 Uses

KEY POINTS Helium is used as an inert gas and as a light source in lasers and electric discharge lamps; liquid helium is a very-low-temperature refrigerant.

On account of its low density and nonflammability, He is used in balloons and lighter-than-air craft. Its very low boiling point leads to its wide use in cryogenics and as a very-low-temperature refrigerant; it is used to cool nuclear reactors, infrared detectors, and superconducting magnets that are used for magnetic resonance imaging (MRI) and NMR. The Large Hadron Collider at CERN uses 96 t of liquid helium to maintain the low temperature of its magnets. The space industry uses helium to purge fuel from its rockets and in satellites. Helium is also used as an inert atmosphere for growing crystals of semiconducting materials, such as Si. It is mixed with O_2 in a 4:1 ratio to provide an artificial atmosphere for divers, where its lower solubility than nitrogen minimizes the danger of causing the 'bends', or decompression sickness. This same gas mixture is used to treat acute asthma attacks, as its density presents less resistance in the lungs compared to oxygen–air mixtures. The helium isotope helium-3, ^{3}He, is important for applications such as neutron detectors and lung imaging. It also shows potential uses in nuclear fusion reactors.

The most widespread use of Ar is to provide an inert atmosphere for the production of air-sensitive compounds and as an inert gas blanket to suppress the oxidation of metals when they are welded. Argon is also used as a cryogenic refrigerant and to fill the gap between the panes in sealed double-glazed windows, as its low thermal conductivity reduces heat loss.

Xenon has anaesthetic properties but is not very widely used because it is approximately 2000 times more expensive than N_2O. An isotope of Xe is also used in medical imaging (Box 19.2).

Radon, which is a product of nuclear power plants and of the radioactive decay of naturally occurring Th and U, is a health hazard because of the ionizing nuclear radiation it produces. It is a contributor to the background radiation arising from cosmic rays and terrestrial sources. However, in regions where the soil, underlying rocks, or building materials contain significant concentrations of uranium

BOX 19.2 How is ^{129}Xe-NMR used in medicine and materials?

Magnetic resonance imaging (MRI) is widely used in medicine to produce high-quality images of soft matter within the body. The technique relies on the protons of water molecules in tissue to provide the NMR signal, but protons are difficult to image in some parts of the body, particularly the lungs and the brain. However, other NMR active nuclei can also be used for MRI, most notably ^{129}Xe, which is 26.4% abundant with $I = \frac{1}{2}$.

Because the extent of polarization of nuclear spin in ^{129}Xe is too low to give a good signal, it is enhanced by using low temperatures or increasing the applied magnetic field. Alternatively, the polarization can be enhanced by a factor of about 10^5 by spin exchange with a polarized alkali metal. MRI images of the lungs can be obtained by inhaling this hyperpolarized ^{129}Xe into the lungs. The xenon is then transferred from the lungs to the blood and then into other tissues. Consequently, images of the circulatory system, the brain, and other vital organs can be obtained by imaging blood vessels.

One challenge of this technique is the short relaxation time of hyperpolarized ^{129}Xe of less than five seconds, meaning that there is only a short window in which to acquire the image. Hyperpolarized ^{129}Xe-NMR is also used in materials chemistry, where it is applied to examine the structures of mesoporous materials such as zeolites and ceramics, and soft matter such as polymer melts and elastomers.

The 26.4% abundance of ^{129}Xe means that it is feasible to record conventional NMR for spectroscopic characterization of compounds of xenon. This produces satellite peaks with relative intensities 13:74:13 in the spectra of other NMR nuclei (Section 8.6).

---

and thorium, excessive amounts of the gas have been found in buildings. Uranium and thorium have half-lives of several billion years. They decay to produce radium which then decays to produce radon, so these very long half-lives ensure that radon will be constantly replenished for billions of years. Radon itself is a gas and can be inhaled. Its decay products, such as isotopes of lead and bismuth, are solids and can stick to surfaces and dust particles, thus persisting in indoor environments.

Because of their lack of chemical reactivity, the noble gases are extensively used in various light sources, including conventional sources such as neon signs and fluorescent lamps, where neon produces red light and helium yellow light. Xenon is used in xenon flash-lamps, which produce short bursts of visible and ultraviolet light, and high-intensity discharge lights (also known as arc lights) which produce a continuous light. They are also used in lasers such as helium–neon, argon-ion, and krypton-ion lasers. In each case, an electric discharge through the gas ionizes some of the atoms and promotes both ions and neutral atoms into excited states that then emit electromagnetic radiation on return to a lower state. Argon was used as the inert atmosphere in incandescent light bulbs, where it reduced the burning of the filament.

19.5 Synthesis and structure of xenon fluorides

KEY POINT Xenon reacts with fluorine to form XeF_2, XeF_4, and XeF_6.

The reactivity of the noble gases has been investigated sporadically ever since their discovery, but all early attempts to coerce them into compound formation were unsuccessful. Until the 1960s the only known complex species were the unstable diatomic species such as He_2^+ and Ar_2^+, which were detected spectroscopically. However, in March 1962, Neil Bartlett, then at the University of British Columbia, observed the reaction of a noble gas. Bartlett's report, and another from Rudolf Hoppe's group in the University of Münster a few weeks later, set off a flurry of activity throughout the world. Within a year, a series of xenon fluorides and oxo compounds had been synthesized and characterized. The field is somewhat limited, but compounds with bonds to nitrogen, carbon, and metals have been prepared.

Bartlett's motivation for studying xenon was based on the observations that PtF_6 can oxidize O_2, to give the solid O_2PtF_6, and that the ionization energy of xenon is similar to that of molecular oxygen. Indeed, reaction of xenon with PtF_6 did give a solid, but the reaction is complex and the products are a mixture of $[XeF]^+[PtF_5]^-$, $[XeF]^+[Pt_2F_{11}]^-$, and $[Xe_2F_3]^+[PtF_6]^-$. The direct reaction of xenon and fluorine leads to a series of compounds with oxidation numbers +2 (XeF_2), +4 (XeF_4), and +6 (XeF_6).

The structures of XeF_2 and XeF_4 are well established from diffraction and spectroscopic methods. Similar measurements on XeF_6 in the gas phase, however, led to the conclusion that this molecule is fluxional. Infrared spectra and electron diffraction on XeF_6 show that a distortion occurs about a three-fold axis, suggesting that a triangular face of F atoms opens up to accommodate a lone pair of electrons, as in (3). One interpretation is that the fluxional process arises from the migration of the lone pair from one triangular face to another. Solid XeF_6 consists of F^--bridged XeF_5^+ units, and in solution it forms Xe_4F_{24} tetramers. The gaseous and solid structures contain units which bear a molecular and electronic structural resemblance to the isoelectronic polyhalide anions I_3^- and ClF_4^- (Section 18.10).

The xenon fluorides are synthesized by direct reaction of the elements, usually in a nickel reaction vessel that has been passivated by exposure to F_2 to form a thin protective NiF_2 coating. This treatment also removes surface oxide, which would react with the xenon fluorides. The synthetic conditions indicated in the following equations show that formation of the higher halides is favoured by a higher proportion of fluorine and higher total pressure:

$$Xe(g) + F_2(g) \xrightarrow{400°C, 1\,atm} XeF_2(g) \quad \text{(Xe in excess)}$$
$$Xe(g) + 2F_2(g) \xrightarrow{600°C, 6\,atm} XeF_4(g) \quad (Xe : F_2 = 1 : 1.5)$$
$$Xe(g) + 3F_2(g) \xrightarrow{300°C, 60\,atm} XeF_6(g) \quad (Xe : F_2 = 1 : 2.0)$$

A simple 'windowsill' synthesis is also possible. Xenon and fluorine are sealed in a glass bulb (rigorously dried to prevent the formation of HF and the attendant etching of the glass) and the bulb is exposed to sunlight, whereupon beautiful crystals of XeF_2 slowly form in the bulb. It will be recalled that F_2 undergoes photodissociation (Section 18.5), and in this synthesis the photochemically generated F atoms react with Xe atoms.

19.6 Reactions of xenon fluorides

KEY POINTS Xenon fluorides are strong oxidizing agents and form complexes with F^- such as XeF_5^-, XeF_7^-, and XeF_8^{2-}; they are used in the preparation of compounds containing Xe–O and Xe–N bonds.

The reactions of the xenon fluorides are similar to those of the high-oxidation-state interhalogens (Section 18.10), and redox and metathesis reactions dominate. One important reaction of XeF_6 is metathesis with oxides:

$$XeF_6(s) + 3H_2O(l) \rightarrow XeO_3(aq) + 6HF(g)$$
$$2XeF_6(s) + 3SiO_2(s) \rightarrow 2XeO_3(s) + 3SiF_4(g)$$

Another striking chemical property of the xenon fluorides is their strong oxidizing power:

$$2XeF_2(s) + 2H_2O(l) \rightarrow 2Xe(g) + 4HF(g) + O_2(g)$$
$$XeF_4(s) + Pt(s) \rightarrow Xe(g) + PtF_4(s)$$

As with the interhalogens, the xenon fluorides react with strong Lewis acids to form xenon fluoride cations:

$$XeF_2(s) + SbF_5(l) \rightarrow [XeF]^+[SbF_6]^-(s)$$

These cations are associated with the counter-ion by F^- bridges.

Another similarity with the interhalogens is the reaction of XeF_4 with the Lewis base F^- in acetonitrile (cyanomethane, CH_3CN) solution to produce the XeF_5^- ion:

$$XeF_4 + [N(CH_3)_4]F \rightarrow [N(CH_3)_4]^+[XeF_5]^-$$

The XeF_5^- ion is pentagonal-planar (**9**), and in the VSEPR model the two electron pairs on Xe occupy axial positions on opposite sides of the plane.

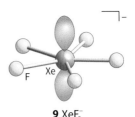

9 XeF_5^-

Similarly, it has been known for many years that reaction of XeF_6 with an F^- source produces the XeF_7^- or XeF_8^{2-} ions, depending on the proportion of fluoride. Only the shape of XeF_8^{2-} is known: it is a square antiprism (**10**), which is difficult to reconcile with the simple VSEPR model because this shape does not provide a site for the lone pair on Xe. However, if the lone pair is considered an 'inert pair', Section 10.7, this structure is more readily understood.

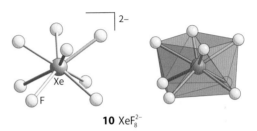

10 XeF_8^{2-}

The xenon fluorides are the gateway to the preparation of compounds of the noble gases with elements other than F and O. The reaction of nucleophiles with a xenon fluoride is one useful strategy for the synthesis of such bonds. For instance, the reaction

$$XeF_2 + HN(SO_2F)_2 \rightarrow FXeN(SO_2F)_2 + HF$$

is driven forward by the stability of the product HF and the energy of formation of the Xe–N bond (**11**).

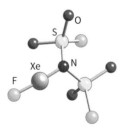

11 $FXeN(SO_2F)_2$

A strong Lewis acid such as AsF_5 can extract F^- from the product of this reaction to yield the cation $[XeN(SO_2F)_2]^+$. Another route to XeN bonds is the reaction of one of the fluorides with a strong Lewis acid:

$$XeF_2 + AsF_5 \rightarrow [XeF]^+[AsF_6]^-$$

followed by the introduction of a Lewis base, such as CH_3CN, to yield $[CH_3CNXe]^+[AsF_6]^-$.

> **EXAMPLE 19.2** Estimating bond enthalpies for xenon compounds
>
> Given that the enthalpy of formation for XeF_4 is -267 kJ mol^{-1} and the F–F bond enthalpy is 159 kJ mol^{-1}, calculate the average Xe–F bond enthalpy in XeF_4.
>
> **Answer** We make use of the fact that the enthalpy of a reaction is equal to the difference between the sum of the bond enthalpies for broken bonds and the sum of the bond enthalpies for the bonds that are formed. The reaction is
>
> $Xe(g) + 2F_2(g) \rightarrow XeF_4(g)$
>
> In this reaction, 2 mol F–F bonds must be broken corresponding to an enthalpy change of $(2 \times 159\,\text{kJ}) = 318$ kJ. So $\Delta_f H^\ominus = -267 = 318 - 4B(\text{Xe–F})$. Rearranging this expression gives $4B(\text{Xe–F}) = 585$ kJ. Therefore, the Xe–F bond enthalpy is 146 kJ mol^{-1}.
>
> **Self-test 19.2** Use the value for the Xe–F bond enthalpy calculated above to determine the enthalpy of formation for XeF_6.

19.7 Xenon–oxygen compounds

KEY POINT The xenon oxides are unstable and highly explosive.

Xenon oxides are all endergonic ($\Delta_f G^\ominus > 0$) and cannot be prepared by direct interaction of the elements. The oxides and oxofluorides are prepared by the hydrolysis of xenon fluorides:

$XeF_6(s) + 3H_2O(l) \rightarrow XeO_3(s) + 6HF(aq)$

$3XeF_4(s) + 6H_2O(l) \rightarrow XeO_3(s) + 2Xe(g) + \tfrac{3}{2}O_2(g) + 12HF(aq)$

$XeF_6(s) + H_2O(l) \rightarrow XeOF_4(s) + 2HF(aq)$

The pyramidal xenon trioxide, XeO_3 (4), presents a serious hazard because this endergonic compound is highly explosive. It is a very strong oxidizing agent in acidic solution, with $E^\ominus(XeO_3, Xe) = +2.10$ V. In basic aqueous solution the Xe(VI) oxoanion $HXeO_4^-$ slowly decomposes in a coupled disproportionation and water oxidation to yield a Xe(VIII) perxenate ion, XeO_6^{4-}, and xenon:

$2HXeO_4^-(aq) + 2OH^-(aq) \rightarrow XeO_6^{4-}(aq) + O_2(g) + 2H_2O(l)$

The perxenates of several alkali metal ions have been prepared by treatment of XeO_3 with ozone in basic conditions. These compounds are white, crystalline solids with octahedral XeO_6^{4-} units (12). They are powerful oxidizing agents in acidic aqueous solution:

$XeO_6^{4-}(aq) + 3H^+(aq) \rightarrow HXeO_4^-(aq) + \tfrac{1}{2}O_2(g) + H_2O(l)$

Treating Ba_2XeO_6 with concentrated sulfuric acid produces the only other known oxide of xenon, XeO_4 (5), which is an explosively unstable gas.

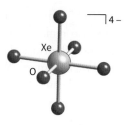

12 XeO_6^{4-}

$[XeOXeOXe]^{2+}$ has been described as the missing Xe(II) oxygen compound. It has been isolated as an adduct with $F(ReO_2F_3)_2$ in the compound $[XeOXeOXe][\mu\text{-}F(ReO_2F_3)_2]_2$, which consists of a zigzag $[XeOXeOXe]^{2+}$ cation with fluorine bridges to the anions (13).

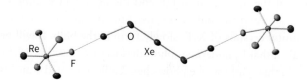

13 $[XeOXeOXe][\mu\text{-}F(ReO_2F_3)_2]_2$

Xenon forms the oxofluorides $XeOF_2$ (6), XeO_3F_2 (7), and $XeOF_4$ (8). When alkali metal fluorides are dissolved in $XeOF_4$, solvated fluoride ions are formed of composition $F^-\cdot3XeOF_4$. Attempted removal of $XeOF_4$ from the solvate yields $XeOF_5^-$ (14), which is a pentagonal pyramid.

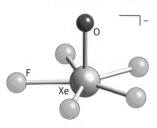

14 $XeOF_5^-$

The only known xenon(II) oxofluoride is the $[FXeOXe\text{–}FXeF][AsF_6]$ salt, the cation of which can be thought of as the $[XeOXe]^{2+}$ cation with one Xe atom bonded to a F atom and the other weakly coordinated to XeF_2 (15).

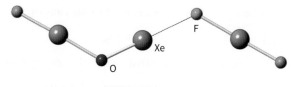

15 $[FXeOXe\text{–}FXeF]^+$

A BRIEF ILLUSTRATION

The structures of compounds of xenon can be probed using ^{129}Xe-NMR spectroscopy. For example, the ^{129}Xe-NMR spectrum of $XeOF_4$ (8) consists of a quintuplet of peaks. These peaks correspond to the single Xe environment, which is coupled to four equivalent ^{19}F atoms.

EXAMPLE 19.3 Describing the synthesis of a noble-gas compound

Describe a procedure for the synthesis of potassium perxenate, starting with xenon and other reagents of your choice.

Answer We know that Xe–O compounds are endergonic, so they cannot be prepared by direct reaction of xenon and oxygen, so we need to look for an indirect method. As was described in the text, the hydrolysis of XeF_6 yields XeO_3, which undergoes disproportionation in basic solution to yield perxenate, XeO_6^{4-}. Thus, XeF_6 could be synthesized by the reaction of xenon and excess F_2 at 300°C and 6 MPa in a stout nickel container. The resulting XeF_6 might then be converted to the perxenate in one step by exposing it to aqueous KOH solution. The resulting potassium perxenate (which turns out to be a hydrate) could then be crystallized.

Self-test 19.3 Write a balanced equation for the decomposition of xenate ions in basic solution for the production of perxenate ions, xenon, and oxygen.

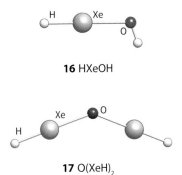

16 HXeOH

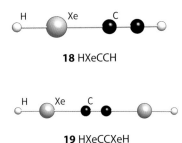

17 O(XeH)$_2$

Xenon atoms have been successfully inserted into H–C bonds of hydrocarbons by photolysis and annealing of a solid mixture of C_2H_2 and Xe. The noble-gas hydrides HXeC≡CH (**18**) and HXeC≡CXeH (**19**) have been prepared in this way.

18 HXeCCH

19 HXeCCXeH

These insertion reactions may hold the key to explaining the so-called 'missing xenon' phenomenon. Relative to the other noble gases the amount of xenon in the atmosphere is depleted by a factor of 20. One theory is that xenon in the Earth's interior can form stable compounds. The formation of these insertion compounds indicates that xenon compound formation under extreme conditions may indeed be possible. In fact, recent studies have indicated that Xe will exchange with Si in silicate minerals under high temperature and pressure.

19.8 Insertion compounds

KEY POINT Xenon and krypton can insert into H–Y bonds.

A number of noble-gas hydrides with the general formula H–E–Y, where E is the Group 18 element and Y is an electronegative element or fragment, have been isolated by matrix isolation at low temperatures. The first species identified were HXeCl, HXeBr, HXeI, and HKrCl. More recently characterized species include HKrC≡N and HXeC≡C–C≡N. They are all prepared by UV photolysis of the HY precursors in the solid noble gas at low temperatures (Box 19.3). When xenon reacts in this way with water, both HXeOH (**16**) and the HXeO radical are produced. The latter can react further with another Xe atom and hydrogen to form HXeOXeH (**17**), which is metastable with respect to H_2O and 2 Xe and more stable than HXeOH, HXeH, and HXeBr.

BOX 19.3 How can noble gases be used to trap reactive species?

The term 'matrix isolation' is used to describe the trapping of reactive species such as radical ions and reaction intermediates in any inert matrix, such as polymers or resins, but is more usually used to refer to the trapping in a noble gas or nitrogen. Noble gases are used because of their low reactivity and broad optical transparency in the solid state.

Matrix isolation allows the study of very reactive or unstable species by various spectroscopic techniques. The reactive species are trapped in a large volume of an inert matrix at low temperatures and under high vacuum. The low temperatures ensure the rigidity of the matrix. The matrix with the sample is condensed onto a surface or optical cell for subsequent investigation. The host matrix is a solid in which guest species are embedded and dilute enough to be effectively isolated from each other. For example, solid krypton can be used as a matrix for isolated F_3^- ions and solid argon has been used to study intermediates in reactions between ozone and alkenes. Matrix isolation in noble gases is also used for the study of reactive noble-gas species. For example, HXeI has been studied in Xe, and HKrCN has been studied in Kr.

19.9 Organoxenon compounds

KEY POINT Organoxenon compounds can be prepared by xenodeborylation of an organoboron compound.

The first compound containing Xe–C bonds was reported in 1989. Since then a wide variety of organoxenon compounds have been prepared. The most useful routes to organoxenon compounds are through the fluorides XeF_2 and XeF_4.

Organoxenon(II) salts can be prepared from organoboranes by **xenodeborylation**, the substitution of B by Xe. For example, tris(pentafluorophenyl)borane reacts with XeF_2 in dichloromethane to produce arylxenon(II) fluoroborates (**20**):

$$(C_6F_5)_3B + 3\,XeF_2 \xrightarrow{CH_2Cl_2} [C_6F_5Xe]^+ + [(C_6F_5)_nBF_{4-n}]^-$$
$$n = 1, 2$$

20 $C_6F_5Xe^+$

When this reaction is carried out in anhydrous HF, all the C_6F_5 groups are transferred to the Xe:

$$(C_6F_5)_3B + 3\,XeF_2 \xrightarrow{HF} 3[C_6F_5Xe]^+ + [BF_4]^- + 2[F(HF)_n]^-$$

A general route that can be used to introduce other organic groups uses organodifluoroboranes, RBF_2:

$$RBF_2 + XeF_2 \xrightarrow{CH_2Cl_2} [RXe]^+ + [BF_4]^-$$

In addition to xenodeborylation, organoxenon(II) compounds (**21**) can be prepared from $C_6F_5SiMe_3$:

$$3\,C_6F_5SiMe_3 + 2\,XeF_2 \xrightarrow{CH_2Cl_2} Xe(C_6F_5)_2 + C_6F_5XeF + 3\,Me_3SiF$$

21 $Xe(C_6H_5)_2$

Organoxenon(II) compounds are thermally unstable and decompose above –40°C.

The first organoxenon(IV) compound was prepared by the reaction between XeF_4 and $C_6F_5BF_2$ in CH_2Cl_2:

$$C_6H_5BF_2 + XeF_4 \xrightarrow{CH_2Cl_2} [C_6F_5XeF_2][BF_4]$$

Organoxenon(IV) compounds are less thermally stable than the analogous Xe(II) compounds. In all Xe(II) and Xe(IV) salts, the Xe atom is bonded to a C atom that is part of a π system. Extended π systems, such as in aryl groups, increase the stability of the Xe–C bond. This stability is further favoured by the presence of electron-withdrawing substituents (such as fluorine) on the aryl group.

19.10 Coordination compounds

KEY POINTS Argon, xenon, and krypton form coordination compounds that are usually studied by matrix isolation; the stability of the complexes decreases in the order Xe > Kr > Ar.

Coordination compounds of noble gases have been known since the mid-1970s. The first stable noble-gas coordination compound to be synthesized was $[AuXe_4]^{2+}[Sb_2F_{11}]^{2-}$, which contains an unusual square-planar $[AuXe_4]^{2+}$ cation (**22**).

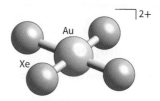

22 $[AuXe_4]^{2+}$

The compound is made by reduction of AuF_3 by HF/SbF_5 in elemental xenon to yield dark red crystals that are stable up to –78°C. Alternatively, addition of Xe to a solution of AuF_3 in HF/SbF_5 gives a dark red solution that is stable up to –40°C. This solution is stable at room temperature under a xenon pressure of 1 MPa (about 10 atm). During the reduction of Au^{3+} to Au^{2+} the extreme Brønsted acidity of HF/SbF_5 (Sections 5.17 and 16.11) is essential, and the overall reaction indicates the role of protons:

$$AuF_3 + 6\,Xe + 3\,H^+ \xrightarrow{HF/SbF_5} [AuXe_4]^{2+} + Xe_2^+ + 3\,HF$$

Green crystals of $[Xe_2]^+[Sb_4F_{21}]^-$ are also produced at –60°C. The Xe–Xe bond length is 309 pm. The linear blue Xe_4^+ cation has also been observed and has bond lengths of 353 and 319 pm (**23**), the longest homonuclear bonds known for a main-group element. $[HXeXe]^+F^-$, $[HArAr]^+F^-$, and $[HKrKr]^+F^-$ have been studied using matrix-isolation techniques.

353 pm 319 pm 353 pm
Xe

23 Xe_4^+

Many complexes of the noble gases are transient species that have been characterized by matrix isolation. The complex $[Fe(CO)_4Xe]$ is formed when $[Fe(CO)_5]$ is photolysed in solid xenon at 12 K. Similarly, [M(CO)$_5$E] is formed when [M(CO)$_6$] (M = Cr, Mo, or W) is photolysed in solid argon, krypton, or xenon at 20 K, where E = Ar, Kr, or Xe, respectively.

An alternative method of synthesizing these complexes is to generate them in an argon, krypton, or xenon atmosphere. The Xe complex has also been isolated in liquid xenon. The stabilities of the complexes decrease in the order W > Mo ≈ Cr and Xe > Kr > Ar. The complexes are octahedral (24) and the bonding is thought to involve interactions between the p orbitals on the noble gas and orbitals on the equatorial CO groups. Thus, noble gases can be thought of as potential ligands, and indeed have been fully characterized as such (including by NMR; Section 24.16) in organometallic compounds such as [Re(η^5-Cp)(CO)(PF$_3$)Xe]. A number of actinide complexes such as [UO$_2$E$_n$]$^+$ have been reported.

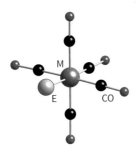

24 [M(C)O)$_5$E], E = Ar, Kr, Xe

19.11 Other compounds of noble gases

KEY POINT Krypton and radon fluorides are known, but their chemical properties are much less extensive than those of xenon.

Radon has a lower ionization energy than Xe, so it can be expected to form compounds even more readily. Evidence exists for the formation of RnF$_2$ and cationic compounds, such as [RnF]$^+$[SbF$_6$]$^-$, but detailed characterization is frustrated by their radioactivity. Krypton has a much higher ionization energy than Xe (Table 19.1), and its ability to form compounds is more limited. Krypton difluoride, KrF$_2$, is prepared by passing an electric discharge or ionizing radiation through a fluorine–krypton mixture at low temperatures (−196°C). As with XeF$_2$, the krypton compound is a colourless volatile solid and the molecule is linear. It has a highly endergonic energy of formation and is a highly reactive compound that must be stored at low temperatures.

When monomeric HF is photolysed in solid argon and annealed to 18 K, HArF is formed. This compound is stable up to 27 K and contains the HAr$^+$ and F$^-$ ions. The related molecular ions HHe$^+$, HNe$^+$, HKr$^+$, and HXe$^+$ have been observed by spectroscopy. The dications ArCF$_2^{2+}$ and ArSiF$_2^{2+}$ have been observed following electron bombardment of CF$_3^{2+}$ or SiF$_3^{2+}$ with argon in a mass spectrometer.

The heavier noble gases form clathrates. Argon, Kr, and Xe form clathrates with quinol (1,4-C$_6$H$_4$(OH)$_2$) with one gas atom to three quinol molecules. They also form clathrate hydrates with water in a ratio of one gas atom to 46 H$_2$O molecules (Section 5.9). Helium and Ne are too small to form stable clathrates. Titan, Saturn's moon, has a dense atmosphere in which levels of Kr and Xe are depleted in comparison to Ar. It is believed that the Kr and Xe are trapped in clathrates, whereas the smaller Ar atoms are trapped less effectively.

Endohedral fullerene complexes, in which the 'guest' atom or ion sits inside a fullerene cage (Section 15.6), have been observed for C$_{60}^{n+}$ and C$_{70}^{n+}$ (n = 1, 2, or 3) with He, and C$_{60}^+$ with Ne. Molecular orbital calculations indicate that Ar should be able to penetrate the C$_{60}^+$ cage, although this complex has not yet been observed.

Other than the fullerene complexes, those identified transiently in high-energy molecular beams, or van der Waals complexes in the gas phase, there are no known compounds of He. However, theoretical calculations predict HeBeO to be exergonic, and He is found as a molecule of crystallization at very high pressure (Section 12.2).

FURTHER READING

W. Grochala, Atypical compounds of gases which have been called 'noble'. *Chem. Soc. Rev.*, 2007, **36**, 1632.

Z. Mazej, Noble-gas chemistry more than half a century after the first report of the noble-gas compound. *Molecules*, 2020, **25**, 3014.

M. S. Albert, G. D. Cates, B. Driehuys, W. Happer, B. Saam, C. S. Springer, and A. Wishnia, Biological magnetic resonance imaging using laser-polarized ^{129}Xe. *Nature*, 1994, **370**, 199.

R. B. King (ed.), *Encyclopedia of Inorganic Chemistry*. John Wiley & Sons (2005).

M. Ozima and F.A. Podosec, *Noble Gas Geochemistry*. Cambridge University Press (2006).

G. J. Schrobilgen, Chemistry at the edge of the periodic table: The importance of periodic trends: on the discovery of the noble gases and the development of noble-gas chemistry. *Structure and Bonding*, 2019, **181**, 157–196. An enjoyable account of the early failure and final success in the quest for compounds of the noble gases.

H. Frohn and V. V. Bardin. Preparation and reactivity of compounds containing a carbon–xenon bond. *Organometallics*, 2001, **20**, 4750. A readable review of the organo compounds of the noble gases.

EXERCISES

19.1 Explain why helium is present in low concentration in the atmosphere even though it is the second most abundant element in the universe.

19.2 Which of the noble gases would you choose as (a) the lowest-temperature liquid refrigerant, (b) an electric discharge light source requiring a safe gas with the lowest ionization energy, (c) the least expensive inert atmosphere?

19.3 By means of balanced chemical equations and a statement of conditions, describe a suitable synthesis of (a) xenon difluoride, (b) xenon hexafluoride, (c) xenon trioxide.

19.4 Given that the bond enthalpies of the Xe–F and the F–F bonds are 144 and 155 kJ mol^{-1} respectively, calculate the enthalpy of formation for XeF_2.

19.5 Draw the Lewis structures of (a) $XeOF_4$, (b) XeO_2F_2, and (c) XeO_6^{2-}.

19.6 Give the formula and describe the structure of a noble-gas species that is isostructural with (a) ICl_4^-, (b) IBr_2^-, (c) BrO_3^-, (d) ClF.

19.7 (a) Give a Lewis structure for XeF_7^-. (b) Speculate on its possible structures by using the VSEPR model and analogy with other xenon fluoride anions.

19.8 When [XeF][RuF$_6$] reacts with excess F$_2$ at elevated temperature the compound [XeF$_5$][RuF$_6$] is formed. Predict the shape of the cation and anion.

19.9 Use molecular orbital theory to calculate the bond order of the diatomic species E_2^+ with E = He and Ne.

19.10 Use VSEPR to predict the structures of (a) XeF_3^+, (b) XeF_3^-, (c) XeF_5^+, and (d) XeF_5^-.

19.11 Identify the xenon compounds A, B, C, D, and E.

$$D \xleftarrow{H_2O} C \xleftarrow{xs\ F_2} Xe \xrightarrow{F_2} A \xrightarrow{MeBF_2} B$$
$$\downarrow 2F_2$$
$$E$$

19.12 Predict the appearance of the ^{19}F-NMR spectrum of XeF_4.

19.13 Predict the appearance of the ^{129}Xe-NMR spectrum of $XeOF_3^+$.

19.14 Predict the appearance of the ^{19}F-NMR spectrum of $XeOF_4$.

TUTORIAL PROBLEMS

19.1 In December 2015 the International Union of Pure and Applied Chemistry (IUPAC) and the International Union of Pure and Applied Physics (IUPAP) jointly announced their acceptance of the evidence that demonstrated the existence of element 118. What was the name and symbol of the element before it was given its new name of oganesson? Which two institutions collaborated to synthesize the new element? Outline the experiments that led to the synthesis of oganesson. Outline the achievements of the scientist after whom the element is named.

19.2 Argon has been found to react with CF_3^{2+} to form $ArCF_2^{2+}$ (*J. Phys. Chem. Lett.* 2010, **1**, 358). (a) What other Ar–C species are formed during the reaction? (b) What are the Ar–C and C–F bond lengths in the product? Give the point group of the dication. (c) Why do He and Ne not form dications of this type with CF_3^{2+}? (d) Why do Kr and Xe not form dications of this type?

19.3 The paper 'Exploring the effects on lipid bilayer induced by noble gases via molecular dynamics simulations' (*Nature*, 2015, **5**, 17, 235) discusses the anaesthetic properties of noble gases. Which noble gases cause anaesthesia? What is the order of their potency? What is the mode of action of noble gas anaesthetics? Why is xenon the most potent noble gas anaesthetic?

19.4 In the paper 'Predicted chemical bonds between rare gases and Au' (*J. Am. Chem. Soc.*, 1995, **117**, 2067), P. Pyykkii used a computational study of the species RgAuRg$^+$ and AuRg$^+$ (where Rg refers to a 'rare' gas) to predict the bond energies and bond lengths of the Au–Rg bond. Why did the author draw parallels between the Au–Rg bonds and H–Rg bonds? Give the values of bond energy and bond length for the series of Au–Rg bonds. How would these values be expected to differ for Cu–Rg$^+$ species? List the values of the analogous Cu–Rg$^+$ bonds and rationalize the differences.

19.5 The paper 'Atypical compounds of gases which have been called "noble"' (*Chem. Soc. Rev.*, 2007, **36**, 1632) provides a thorough account of the range of compounds formed by Group 18 elements. Among the compounds described are XeF_5Cl, HXeOOXeH, and $ClXeFXeCl^+$. Sketch the structures of these molecules. Outline the author's rationale for classing noble-gas atoms as Lewis bases, and describe why XeF_2 acts as a ligand with metal cations. Only one compound of xenon with mercury was described. Give the formula of this compound, outline its synthesis, and give the Hg–Xe bond energy and bond length.

19.6 The first compound containing an Xe–N bond was reported by R. D. LeBlond and K. K. DesMarteau (*J. Chem. Soc., Chem. Commun.*, 1974, **14**, 554). Summarize the method of synthesis and characterization. (The proposed structure was later confirmed in an X-ray crystal structure determination.)

19.7 (a) Use the references in the paper by O. S. Jina, X. Z. Sun, and M. W. George (*J. Chem. Soc., Dalton Trans.*, 2003, 1773) to produce a review of the use of matrix isolation in the characterization of organometallic noble-gas complexes. (b) How did the methods used by these authors differ from the usual techniques of matrix isolation? (c) Place the complexes [MnCp(CO)$_2$Xe], [RhCp(CO)Xe], [MnCp(CO)$_2$Kr], [Mo(CO)$_5$Kr], and [W(CO)$_5$Kr] in order of increasing stability towards CO substitution.

19.8 In their paper 'Xenon as a complex ligand: The tetra-xenon gold(II) cation in AuXe$_4^{2+}$(Sb$_2$F$_{11}^-$)$_2$', S. Seidel and K. Seppelt (*Science*, 2000, **290**, 117) describe the first synthesis of a stable noble-gas coordination compound. Give details of the synthesis and characterization of the compound.

19.9 The synthesis and characterization of the XeOF$_5^-$ anion has been described by A. Ellern and K. Seppelt (*Angew. Chem., Int. Ed. Engl.*, 1995, **34**, 1586). (a) Summarize the similarities between XeOF$_4$ and IF$_5$. (b) Give possible reasons for the differences in structure between XeOF$_5^-$ and IF$_6^-$. (c) Summarize how XeOF$_5^-$ was prepared.

19.10 In their paper 'Helium chemistry: Theoretical predictions and experimental challenge' (*J. Am. Chem. Soc.*, 1987, **109**, 5917), W. Koch and co-workers used quantum-mechanical calculations to demonstrate that helium can form strong bonds with carbon in cations. (a) Give the range of bond lengths calculated for these He–C cations. (b) What are the requirements for an element to form a strong bond with He? (c) To which branch of science do the authors suggest that this work would be particularly relevant?

19.11 In their paper 'Observation of superflow in solid helium' (*Science*, 2004, **305**, 1941), E. Kim and M. Chan described the observed superfluidity of solid helium. Define superfluidity, describe the experiment carried out by the authors to demonstrate the property, and summarize their explanation of superfluidity.

20 The d-block elements

Part A: The essentials

20.1 **Occurrence and recovery**

20.2 **Chemical and physical properties**

Part B: The detail

20.3 **Group 3: scandium, yttrium, and lanthanum**

20.4 **Group 4: titanium, zirconium, and hafnium**

20.5 **Group 5: vanadium, niobium, and tantalum**

20.6 **Group 6: chromium, molybdenum, and tungsten**

20.7 **Group 7: manganese, technetium, and rhenium**

20.8 **Group 8: iron, ruthenium, and osmium**

20.9 **Group 9: cobalt, rhodium, and iridium**

20.10 **Group 10: nickel, palladium, and platinum**

20.11 **Group 11: copper, silver, and gold**

20.12 **Group 12: zinc, cadmium, and mercury**

20.13 **The 6d elements**

Further reading

Exercises

Tutorial problems

The d-block elements are all metallic and their chemical properties are central to biology, industry, and many aspects of contemporary research. In Chapters 9 and 10 we have already described many systematic trends across the block and observed trends in the chemical properties within each group. We now look at the properties of the individual metals and their compounds in more detail.

The two terms **d-block metal** and **transition metal** are often used interchangeably; however, they do not mean the same thing. The name 'transition metal' originally derived from the fact that their chemical properties were transitional between those of the s and p blocks. Now, however, the IUPAC definition of a **transition element** is an element that has an incomplete d subshell in either the neutral atom or its ions. Thus, two of the Group 12 elements (Zn, Cd) are members of the d block but are not transition elements as they do not have any compounds with an incomplete d subshell. The situation for the third Group 12 element, mercury, is different: the report of a mercury(IV) compound (HgF_4), which has the d^8 electron configuration, qualifies mercury as a transition metal.

In the following discussion, it will be convenient to refer to each row of the d block as a series, with the 3d series the first row of the block (Period 4), the 4d series the second row (Period 5), and so on. It will prove important to note the intrusion of the f block, the **inner transition elements**, the lanthanoids, before the 5d series.

Elements towards the left of the d block are often referred to as *early* and those towards the right are referred to as *late*.[1] Although the 6d series (4th row, actinium–copernicium) is complete, the problems associated with synthesizing the elements and their short half-lives once formed (with the exception of actinium, which is discussed as part of the actinoids in Chapter 21) are such that little experimental chemistry has yet been performed, so the discussion here will be limited.

The chemical properties of the d-block elements (or d metals, as we shall commonly call them) will be covered in this chapter and Chapters 22–24. It is appropriate to begin with a survey of their occurrence and properties. The trends

---

[1] When referring to a specific row, the terms *light* and *heavy* are often used in place of *early* and *late*.

in their properties, the correlation with their electronic structures, and therefore their location in the periodic table, should be kept in mind throughout these chapters. Chapters 9 and 10 dealt with many of the trends that occur in the d block, and we concentrate on specific properties of the individual elements and some of their important compounds here.

PART A: The essentials

20.1 Occurrence and recovery

KEY POINTS Chemically soft members of the d block occur as sulfide minerals, and some can be roasted in air to obtain the metal; the more electropositive 'hard' metals occur as oxides and are extracted by reduction.

The elements on the left of the 3d series occur in nature primarily as metal oxides or as metal cations in combination with oxoanions (Table 20.1). The metals are obtained by chemical reduction (Section 6.17). Titanium ores are the most difficult to reduce, and the element is normally produced by heating TiO_2 with chlorine and carbon to produce $TiCl_4$, which is then reduced by molten magnesium at about 1000°C in an inert-gas atmosphere. The oxides of Cr, Mn, and Fe are conventionally reduced with carbon, although mounting efforts are underway to phase out this environmentally damaging approach. To the right of Fe in the 3d series, Co, Ni, Cu, and Zn occur mainly as sulfides and arsenides, which is consistent with the increasingly soft Lewis acid character of their dipositive ions. Sulfide ores are usually roasted in air either to the metal directly (for example, Ni), or to an oxide that is subsequently reduced (for example, Zn). Copper is used in large quantities for electrical applications; electrolysis is used to refine crude copper to achieve the high purity needed for high electrical conductivity.

In the second and third rows of the transition metals the switch to sulfide in ores happens earlier, with only Groups 3 and 4 metals occurring predominantly in the Earth's crust in oxo compounds. The difficulty of reducing the early 4d and 5d metals Mo and W is apparent from Table 20.1, and reflects the tendency of these elements to have stable high oxidation states, as discussed earlier (Section 9.4). The platinum metals (Ru and Os, Rh and Ir, and Pd and Pt), which are found at the lower right of the d block, occur as sulfide and arsenide ores, usually in association with larger quantities of Cu, Ni, and Co. They are collected from the sludge that forms during the electrolytic refinement of copper and nickel. Gold (and to some extent silver) is found in its elemental form.

20.2 Chemical and physical properties

KEY POINTS The chemical and physical properties of the 3d metals are often substantially different from those of the 4d and 5d metals, which show great similarity to each other. The presence of d electrons in the valence shell is responsible for the colour, electronic conduction, magnetism, and rich organometallic chemistry of the d metals and their compounds.

TABLE 20.1 Mineral sources and methods of recovery of some commercially important d metals

Metal	Principal minerals	Method of recovery	Note
Titanium	Ilmenite, $FeTiO_3$ Rutile, TiO_2	$TiO_2 + 2C + Cl_2 \rightarrow TiCl_4 + 2CO$ followed by reduction of $TiCl_4$ with Na or Mg	
Chromium	Chromite, $FeCr_2O_4$	$FeCr_2O_4 + 4C \rightarrow Fe + 2Cr + 4CO$	(a)
Molybdenum	Molybdenite, MoS_2	$2MoS_2 + 7O_2 \rightarrow 2MoO_3 + 4SO_2$ followed by either $MoO_3 + 2Fe \rightarrow Mo + Fe_2O_3$ or $MoO_3 + 3H_2 \rightarrow Mo + 3H_2O$	
Tungsten	Scheelite, $CaWO_4$ Wolframite, $FeMn(WO_4)_2$	$CaWO_4 + 2HCl \rightarrow WO_3 + CaCl_2 + H_2O$ followed by $WO_3 + 3H_2 \rightarrow W + 3H_2O$	
Manganese	Pyrolusite, MnO_2	$MnO_2 + 2C \rightarrow Mn + 2CO$	(b)
Iron	Haematite, Fe_2O_3 Magnetite, Fe_3O_4 Limonite, $FeO(OH)$	$Fe_2O_3 + 3CO \rightarrow 2Fe + 3CO_2$	
Cobalt	CoAsS Smaltite, $CoAs_2$ Linnaeite, Co_3S_4	By-product of copper and nickel production	
Nickel	Pentlandite, $(Fe,Ni)_9S_8$	$NiS + O_2 \rightarrow Ni + SO_2$	(c)
Copper	Chalcopyrite, $CuFeS_2$ Chalcocite, Cu_2S	$2CuFeS_2 + 2SiO_2 + 5O_2 \rightarrow 2Cu + 2FeSiO_3 + 4SO_2$	

(a) The iron–chromium alloy is used directly for stainless steel.
(b) The reaction is carried out in a blast furnace with Fe_2O_3 to produce alloys.
(c) NiS is formed by melting the mineral and separated by physical processes. NiO is used in a blast furnace with iron oxides to produce steel. Nickel is purified by electrolysis or the Mond process via $[Ni(CO)_4]$, Box 24.1.

TABLE 20.2 The sizes and electronegativities of the d metals

Group	3	4	5	6	7	8	9	10	11	12
Metal	Sc	Ti	V	Cr	Mn	Fe	Co	Ni	Cu	Zn
Pauling electronegativity	1.3	1.5	1.6	1.6	1.5	1.9	1.9	1.9	1.9	1.6
metallic radius/pm	163	142	134	130	132	125	125	126	126	138
covalent radius/pm	170	160	153	139	139 (ls) 161 (hs)	132 (ls) 161 (hs)	126 (ls) 150 (hs)	124	132	122
Metal	Y	Zr	Nb	Mo	Tc	Ru	Rh	Pd	Ag	Cd
Pauling electronegativity	1.2	1.4	1.6	1.8	1.9	2.2	2.2	2.2	1.9	1.7
metallic radius/pm	179	158	143	139	136	134	132	142	144	157
covalent radius/pm	190	175	164	154	147	146	142	139	145	144
Metal	La	Hf	Ta	W	Re	Os	Ir	Pt	Au	Hg
Pauling electronegativity	1.0	1.3	1.5	1.7	1.9	2.2	2.2	2.2	2.4	1.9
metallic radius/pm	182	158	145	135	137	135	136	139	144	152
covalent radius/pm	207	175	170	162	151	144	141	136	136	132

Table 20.2 lists the electronegativities and atomic radii of all the d metals. We can see that the 3d metals are significantly smaller than their 4d and 5d congeners; the similarity in size of the 4d and 5d metals is a manifestation of the lanthanoid contraction (Section 10.5). Coordination numbers are typically greater for the larger 4d and 5d metals, with less common geometries, such as square antiprism, being found. Electronegativities generally increase across the d block, resulting in the increasing soft Lewis acid behaviour noted above.

Table 20.3 shows both the common, and maximum, oxidation states of nonorganometallic complexes of the

TABLE 20.3 Oxidation states of the d metals in non organometallic compounds, parentheses denote less common oxidation states

Group	3	4	5	6	7	8	9	10	11	12
	Sc 3	Ti (2) (3) 4	V (2) (3) 4 5	Cr 2 3 (4) (5) 6	Mn 2 3 4 (5) (6) 7	Fe 2 3 (4) (5) (6)	Co 2 3 (4)	Ni 2 3 (4)	Cu 1 2 (3) (4)	Zn 2
	Y 3	Zr (2) (3) 4	Nb (3) 4 5	Mo (2) (3) 4 5 6	Tc (3) 4 5 6 7	Ru 2 3 4 5 (6) (7) 8	Rh 1 (2) 3 (4) (5) (6)	Pd 2 (3) 4	Ag 1 (2) (3)	Cd 2
	La 3	Hf (2) (3) 4	Ta (3) 4 5	W (2) (3) 4 5 6	Re (3) 4 5 6 7	Os (2) 3 4 5 (6) (7) 8	Ir 1 (2) 3 (4) (5) (6)	Pt 2 (3) 4	Au 1 (2) 3 (5)	Hg 1 2 (4)

d metals, and we can easily see the trends of increasing stability of high oxidation states down a group and the maximum state peaking in the middle of the series (Section 9.4). The chemistries of the d metals also reflect their hard/soft nature. The harder first-row transition metals and the early 4d and 5d elements have an extensive chemistry in combination with oxygen, giving simple and complex oxides which form many functional solid-state materials such as heterogeneous catalysts and electronic and optical materials.

The 'oxo wall', lying roughly midway across the d block, separates the earlier elements able to form a multiple bond to an oxido ligand (M=O or M≡O) from the later ones that cannot (Section 10.8). Oxido species formed with metals very close to the wall are particularly reactive and are often encountered as important catalytic intermediates. The later elements have more extensive chemistries with soft ligands such as sulfide. Relativistic effects (Section 10.6), which affect the energies of the outer electrons, are significant for gold and mercury and have considerable bearing on their chemistry.

We can contrast the chemical properties of the d metals with those of the s (Chapters 12 and 13) and p blocks (Chapters 14–18), but the differences all ultimately come down to the presence of d electrons. Chapter 22 deals in detail with the electronic structure of d-metal complexes, but we need to note here that, in complexes, the d orbital energies are significantly split by the ligand field, as described in Chapter 7. Electronic transitions are possible between the orbitals, with the energy required normally corresponding to visible light, resulting in coloured complexes for those with configurations d^1–d^9; Table 20.4

gives characteristic colours for 3d metal ions in aqueous solution. Unpaired electrons in these d orbitals are responsible for electrical conductivity of some compounds and also magnetism; there are configurations where the number of unpaired electrons varies (Section 7.16).

High- and low-spin complexes are possible, though with the increased size of the second- and third-row d metals we see a preponderance for low-spin complexes of these metals. Chapter 24 deals in detail with the organometallic chemistry of the d metals, but we should note that it is the availability of d orbitals that allows species such as alkenes, arenes, and carbonyls to bind to the metals. The organometallic chemistry of the d metals is thus substantially richer than any other metals.

The physical properties of the elemental d metals are, once again, intimately related to the presence and number of d electrons. Metallic bonding becomes stronger across the rows, peaking in the middle (Section 9.3), with a concomitant increase in melting point, density, and enthalpy of atomization.

The use of the metals throughout history relates to a rather eclectic set of physical properties: availability of ores (or even the natural metal), ease of reducing the ores, and ease of working the metal. Thus gold, which occurs as the pure metal and is easily malleable, has been used for many thousands of years; its lack of any real strength restricted its uses to decorative ones and as currency. Copper, readily available and easily reduced, has been used for at least 5000 years: the metal itself is strong enough to be used structurally. However, it was the discovery that mixing copper with tin (roughly 2:1) gives the much stronger bronze, that allowed the manufacture of the first metallic tools (and

TABLE 20.4 Typical colours of the 3d metal ions in aqueous solution

Ion	Typical octahedral example	Colour	Typical tetrahedral example	Colour
Ti^{3+}	$[Ti(H_2O)_6]^{3+}$	purple, blue		
V^{4+}	$[VO(H_2O)_n]^{2+}$	blue		
V^{3+}	$[V(H_2O)_6]^{3+}$	green		
V^{2+}	$[V(H_2O)_6]^{2+}$	purple/lilac		
Cr^{2+}	$[Cr(H_2O)_6]^{2+}$	blue		
Cr^{3+}	$[Cr(H_2O)_6]^{3+}$	green		
Mn^{2+}	$[Mn(H_2O)_6]^{2+}$	very pale pink	$[MnCl_4]^{2-}$	yellow-brown
Mn^{3+}	$[MnF_6]^{3-}$	green, purple		
Fe^{2+}	$[Fe(H_2O)_6]^{2+}$	pale green–blue	$[FeCl_4]^{2-}$	yellow-brown
Fe^{3+}	$[FeF_6]^{3-}$	pale violet		
Co^{2+}	$[Co(H_2O)_6]^{2+}$	pink–red	$[CoCl_4]^{2-}$	intense blue
Co^{3+}	$[Co(NH_3)_6]^{3+}$	orange–brown		
Ni^{2+}	$[Ni(H_2O)_6]^{2+}$	green	$[NiCl_4]^{2-}$	yellow-green
Ni^{2+}	$[Ni(en)_3]^{2+}$	violet		
Cu^{2+}	$[Cu(H_2O)_6]^{2+}$	blue	$[CuCl_4]^{2-}$	yellow

weapons). The development of bronze allowed cultures to progress beyond the Stone Age and closer to civilization as we know it today.

The subsequent development of the much stronger iron (or more precisely, steel) only became possible with more advanced smelting techniques. The harnessing of the strength of iron allowed substantial advances in structural uses of metal and tool making to be developed, giving rise to modern civilization. As a metal, iron still reigns supreme in its use and versatility with, as we note later, more than 90% of all metal refining currently being iron. The development of specialist alloys (e.g. lightweight materials used in the aerospace industry) and composites was a twentieth-century phenomenon, and further technological advances dependent on metals have mostly been incremental rather than step changes. Nowadays, probably every metal in the d block has at least a small-scale specialist use, whether that be in performance alloys, electronics, or as a component in high-temperature superconductors.

Many of the d-block elements are found at the active sites of enzymes, catalysing a wide range of reactions (Chapter 26) and, increasingly, d-block complexes are used as medicines (Chapter 32).

PART B: The detail

It is impossible to do justice to the sheer variety and scope of the chemistry of 30 elements in one chapter, but the next sections provide an introduction to all the key features of the chemistry of the d metals. The next four chapters develop their chemistry in more detail, and additional reading is presented at the end of this chapter.

20.3 Group 3: scandium, yttrium, and lanthanum

(a) Occurrence and uses

KEY POINT The Group 3 metals are all found in lanthanoid ores and are used in small amounts in specialist alloys and in host materials for optical applications.

Scandium is a rare silvery-white metal with chemical properties very like those of the lanthanoids. It is present at low levels in many lanthanoid ores and is extracted in small quantities from these (Section 21.2). Commercial uses of scandium are largely restricted—in part because of the expense of extraction—to alloying it, at low levels, with aluminium in aerospace industry components; current production is of the order of 2 tonnes a year. A small amount of scandium iodide is used in some high-intensity mercury vapour discharge lamps as it produces an output similar to sunlight. Scandium has no known biological role, and is considered nontoxic.

Yttrium is also found in most lanthanoid ores and is separated from them, as are all lanthanoids (Section 21.2). Annual production is of the order of 600 tonnes a year, as yttrium compounds have many uses. Common uses include as host compounds which can be doped with lanthanoid ions for optical applications. Doped complex oxides such as Eu:YVO$_4$ are used as phosphors in display devices and fluorescent and LED lighting (Section 30.1(d)), yttrium aluminium garnet (YAG, Y$_3$Al$_5$O$_{12}$) is used as a component of many lasers, and yttrium barium copper oxide (YBa$_2$Cu$_3$O$_7$) is used as a high-temperature superconductor (Section 30.7). Yttrium has no identified biological role, but it does concentrate in the liver and bone, and soluble salts are regarded as mildly toxic.

Lanthanum is present in most lanthanoid ores, with monazite (Ce, La, Th, Nd, Y)PO$_4$, and bastnäsite (Ce, La, Y)CO$_3$F, providing the main commercial sources. Uses of lanthanum have changed with time: initial use as a component of gas mantles (where it was present at ~20% as La$_2$O$_3$) has largely been superseded by use in scintillation counters. The metal itself is used as a component of *mischmetal* (Section 21.3) and a component in the anode of nickel metal hydride rechargeable batteries. Lanthanum has no well-established biological role, but lanthanum carbonate can be given to people suffering from chronic kidney dysfunction to help prevent phosphate levels in the blood getting too high; LaPO$_4$ is extremely insoluble, with a solubility product constant of 3.7×10^{-23}.

(b) Binary compounds

KEY POINT The Group 3 metals all form compounds in the +3 state, and some subhalides.

Scandium, yttrium, and lanthanum are all electropositive metals with compounds almost exclusively in the +3 oxidation state. Chemically, scandium resembles aluminium and indium more than the other d metals, unlike yttrium and lanthanum which resemble the lanthanoids. Oxides are white solids of the form M$_2$O$_3$, insoluble in water, but soluble in dilute acid. Halides, MX$_3$, can all be considered ionic, though some of the iodides are prone to hydrolysis, precipitating MO(OH). A number of subhalides exist, such as Sc$_7$Cl$_{10}$ and Sc$_7$Cl$_{12}$, which contain multiple metal–metal bonds within chains (Fig. 20.1). Scandium and yttrium form nitrides of the formula MN when the metals are reacted at high temperatures with N$_2$.

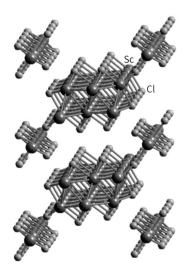

FIGURE 20.1 The structure of Sc_7Cl_{10}.

(c) Complex oxides and halides

KEY POINT Yttrium oxide-based materials doped with lanthanoid ions have important uses in optical applications.

The ionic radius of Y^{3+}, at 102 pm in eight-fold coordination, is similar to that of many trivalent lanthanoid cations, Ln^{3+} (Table 21.4). Unlike the lanthanoids, Y^{3+} has no unpaired f electrons, and this means that complex yttrium oxides and halides are excellent host materials for doping with low levels (a few percent) of many of the Ln^{3+} cations. These lanthanoid-doped yttrium compounds have important applications in optical devices, and their properties rely on the presence of the f electrons of the Ln^{3+} being widely separated in a host material with no other unpaired electrons.

One such class of compounds are the lanthanoid-doped yttrium aluminium garnets used in high-intensity lasers (Section 22.6). Yttrium aluminium garnet (YAG), $Y_3Al_5O_{12}$, has the structure shown in Fig. 20.2, in which the yttrium site has cubic, eight-fold coordination to oxide ions.

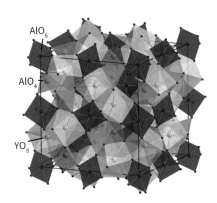

FIGURE 20.2 Yttrium aluminium garnet, YAG $Y_3Al_5O_{12}$.

Replacement of 1% of yttrium by neodymium in the structure, normally given the terminology Nd:YAG, produces a material which, when exposed to high-intensity light from a flash-lamp, produces an intense laser emission at 1064 nm. Doping with other lanthanoids yields laser emission at different wavelengths: Er:YAG produces an emission at 2940 nm, of use in dentistry and medical applications. The complex yttrium fluoride, $LiYF_4$, is also used as a lanthanoid ion host for laser applications; in $LiYF_4$, yttrium adopts a distorted cubic coordination with four fluoride ions at 222 pm and four at 230 pm.

Complex yttrium oxides are also used as lanthanoid ion hosts for phosphor applications where UV light is converted to visible wavelengths. Erbium-doped YVO_4, with the zircon structure of $ZrSiO_4$, produces a red light used in displays and fluorescent lighting. Yttrium iron garnet (YIG), $Y_3Fe_5O_{12}$, with the same structure as YAG but with Fe^{3+} replacing Al^{3+}, is a ferrimagnetic material with a Curie temperature of 500 K (Section 30.6), and is used in a number of magneto-optic applications. Varying the magnetic field applied to a sample of YIG changes the frequency of microwaves that can pass through it and is of importance in mobile phone technology.

Doping ZrO_2 with small levels of Y_2O_3 introduces vacancies in the oxide sublattice, producing a material with the fluorite structure (Section 4.9) known as yttria-stabilized zirconia. Because of the oxide-ion vacancies introduced as a result of replacing Zr^{4+} by Y^{3+}, on heating $(Zr,Y)O_{2-x}$ ($0 \leq x \leq 0.08$) above 800°C, the oxide ions migrate rapidly through the solid structure, leading to applications in oxygen gas sensors and solid oxide fuel cells (Section 25.8 and Box 25.6).

(d) Coordination complexes

KEY POINT The Group 3 metals form complexes of hard ligands with coordination numbers of 6 and more.

The coordination complexes of scandium are mostly octahedral, six-coordinate complexes of hard ligands like $[ScF_6]^{3-}$ or $[ScCl_3(OH_2)_3]$. Higher coordination numbers are possible and complexes in which the metal is present as $[Sc(H_2O)_9]^{3+}$ have been characterized. The tendency to adopt higher coordination numbers is greater for yttrium and lanthanum; lanthanum and the lanthanoids are rarely six-coordinate (Sections 9.6 and 21.7).

(e) Organometallic compounds

KEY POINT The organometallic chemistry of the Group 3 metals is rather limited and very similar to that of the lanthanoids.

The organometallic chemistry of the Group 3 metals is essentially identical to that of the lanthanoids, and is described in

Section 21.8: in the +3 oxidation state the lack of any d electrons that can backbond to organic fragments restricts the number of bonding modes that are available. The strongly electropositive nature of scandium, yttrium, and lanthanum means that they need good donor, not acceptor, ligands. Thus alkoxide, amide, and halide ligands, which are both σ and π donors, are common, whereas CO and phosphine ligands, which are σ donors and π acceptors, are rarely seen. Group 3 organometallic compounds are extremely air- and moisture-sensitive.

Yttrium has only one naturally occurring isotope (^{89}Y): this has spin ½, making it useful for NMR studies of organometallic compounds. Yttrium complexes can also be used as models for lanthanoid complexes, and the NMR spectra of these models can give insight into reactivity and structure of lanthanoid complexes that might otherwise be difficult to obtain because of the paramagnetism of the lanthanoid ions.

20.4 Group 4: titanium, zirconium, and hafnium

(a) Occurrence and uses

KEY POINTS Titanium is widely dispersed and, though it is expensive to purify, has many applications; zirconium and hafnium both find uses in nuclear power plants.

Titanium is the second most abundant d metal, after iron, in the Earth's crust. The principal ores are rutile (TiO_2) and ilmenite ($FeTiO_3$), both of which are widely distributed. Titanium has many desirable properties: it is as strong as steel but only half as dense, it is corrosion-resistant and high-melting. It has many uses in applications where weight is at a premium, such as in the aerospace industry. In spite of the favourable properties of titanium, it is rather costly to extract and refine and so is only used sparing. The most widely used purification process, the Kroll process (Section 6.17), involves heating crude TiO_2 ores with chlorine gas in the presence of carbon to give $TiCl_4$. Fractional distillation of the $TiCl_4$ gives a pure liquid which can then be reduced with molten magnesium under an argon atmosphere. Titanium dioxide is widely used as a white pigment, in paint, sunscreen, and even as a food colourant. Titanium has no known biological role, and is considered nontoxic; the metal itself is used during surgery for pinning bones, hip and knee replacements, cranial plates, and teeth (see Box 32.1).

Zirconium is principally found as the silicate mineral zircon ($ZrSiO_4$), with 80% of zircon mining occurring in Australia and South Africa. Most zircon is used directly for decorative ceramics, where doping with V^{4+}, Pr^{3+}, and Fe^{3+} gives rise to blue, yellow, and orange compounds used as glazes. Small amounts of zircon are converted to the metal via the Kroll process. Zirconium is highly transparent to neutrons and is used as a cladding material for fuel rods in nuclear power plants. Though zirconium currently has no known biological role, some compounds are used medicinally; for example, in the treatment of hyperkalemia (a serious condition associated with elevated potassium levels). In dentistry, ZrO_2 is used in the construction of crowns. Zircon (and zirconium) contain a few percent hafnium, and this is the main source of hafnium, which is obtained via liquid–liquid extraction from solution. Hafnium's major use is in incandescent light bulbs (where it helps protect the tungsten filament) and as a highly effective neutron absorber in control rods in nuclear power plants. Hafnium has no known biological role, and is considered nontoxic.

(b) Binary compounds

KEY POINT The Group 4 metals are electropositive and their chemistry is dominated by the +4 state.

Titanium, zirconium, and hafnium are all electropositive and are normally found in the +4 oxidation state. The MO_2 compounds are high-melting solids, and the MF_4 compounds and MCl_4 (M = Zr, Hf) are solids at room temperature that exhibit complex structures with halides bridging the M ions. By contrast, the MX_4 (X = Cl, Br, I) compounds, with the exception of $ZrCl_4$ and $HfCl_4$ are volatile covalent liquids or low-melting solids which have tetrahedral geometries. These MX_4 (X = Cl, Br, I) compounds are strong Lewis acids and are rapidly hydrolysed by water. Lower-oxidation-state compounds (e.g. MCl_3, M_2O_3, and MO) can be prepared by careful reduction of the normal halide or oxide. Zirconium forms a monohalide when zirconium tetrachloride is reduced with zirconium metal in a sealed tube at high temperatures:

$$3Zr(s) + ZrCl_4(g) \rightarrow 4ZrCl(s)$$

The structure of ZrCl (Fig. 20.3) has Zr ions within bonding distance in two adjacent layers of metal atoms sandwiched between Cl⁻ layers.

Whereas titanium forms a single hydride, TiH_2, zirconium forms a number of hydrides, ZrH_x (1 < x < 4), which exhibit

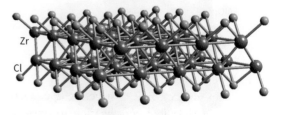

FIGURE 20.3 The structure of ZrCl consists of layers of metal atoms in graphite-like hexagonal nets.

a number of discrete phases and structures, according to composition. All the known hydrides react violently with water. A borohydride complex of Zr(IV) has 12 hydrogens coordinating to a single Zr atom (**1**).

1 [Zr(BH$_4$)$_4$]

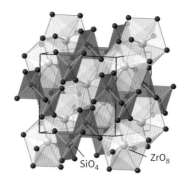

FIGURE 20.4 The zircon, ZrSiO$_4$ structure.

(c) Complex oxides and halides

KEY POINTS Perovskite-based complexes of the Group 4 metals have several technological uses; oxygen vacancies may be incorporated into solid zirconia-based materials, producing oxide ion mobility.

The Group 4 metals form a wide range of complex oxides and, in combination with the larger divalent cations from Group 2 (Ca^{2+}, Sr^{2+}, and Ba^{2+}) and Group 14 (Pb^{2+}), form technologically important phases with the perovskite structure (Section 4.9(b)). In the perovskite stoichiometry, ABO$_3$, the Group 4 cations adopt the B-cation sites with an idealized six-fold octahedral coordination. In BaTiO$_3$ the perovskite structure is distorted from the normal cubic arrangement to a tetragonal phase where the environment around the Ti^{4+} ion has one short (Ti–O, 199.8 pm) and one long (Ti–O, 211 pm) distance. An asymmetry of this type in the coordination geometries of early d-block elements in high oxidation states is a common feature of their structural chemistries and leads to BaTiO$_3$ having a very high dielectric constant, around 2000 times that of air, and thus its use in charge-storage capacitors. Lead zirconate titanate, PZT, (Pb(Zr$_x$Ti$_{1-x}$)O$_3$ with $0 \leq x \leq 1$, shows a strong piezoelectric effect and is used in transducers for converting sound into a voltage or vice versa.

ZrO$_2$ adopts a distorted form of the fluorite structure, but doping Zr^{4+} with a few percent of Y^{3+} leads to a material known as yttria-stabilized zirconia, YSZ, with the ideal cubic fluorite structure. This material, Zr$_{1-x}$Y$_x$O$_{2-x/2}$, contains vacancies in the oxide sublattice which, above 750°C, allows oxide ions to diffuse rapidly through the structure (Chapter 25). This oxide ion mobility leads to application as oxygen gas sensors used in automotive vehicles and in fuel cells.

The zirconium silicate (ZrSiO$_4$) structure (Fig. 20.4) contains Zr^{4+} ions in eight-fold coordination to oxygen. This site may be doped with a large variety of metal cations, giving rise to naturally coloured zircon gemstones (Box 25.1) or brightly coloured glazes for use in ceramics. Octahedral [MX$_6$]$^{2-}$ ions are known in compounds such as BaTiF$_6$ and K$_2$HfF$_6$; in one polymorph, the latter has an antifluorite arrangement of ions.

(d) Coordination complexes

KEY POINTS Complexes of the Group 4 metals are octahedral and the hexaaqua ions are highly acidic.

Complexes of the M^{4+} ion dominate and these are invariably octahedral, with hard donor ligands. The simple hexaaqua ion [M(H$_2$O)$_6$]$^{4+}$ is unknown, as the high charge-to-size ratio results in it being highly acidic; complexes such as [M(H$_2$O)$_4$(OH)$_2$]$^{2+}$ and [M(H$_2$O)$_3$(OH)$_3$]$^+$ dominate in aqueous solution. The metal fluorides are not prone to hydrolysis, and can form the [MF$_6$]$^{2-}$ ion in the presence of extra fluoride; otherwise, species such as [MF$_4$(OH)$_2$]$^{2-}$ or [MF$_4$(H$_2$O)(OH)]$^-$ form. The controlled hydrolysis of the metal alkoxides (in particular titanium), [M(OR)$_4$], provides a sol–gel route (Section 25.9) into composites and nanoparticles of the metal oxide.

(e) Organometallic compounds

KEY POINT Most organometallic compounds of the Group 4 metals have fewer than 18 electrons and are air- and moisture-sensitive.

As befits early, electropositive d metals, the organometallic compounds of Ti, Zr, and Hf are normally electron-deficient: the homoleptic carbonyls, [M(CO)$_6$], would be 16-electron species, and only the Ti complex is known. Relatively unstable [Ti(CO)$_6$] is readily reduced to the 18-electron [M(CO)$_6$]$^{2-}$ anion, and the equivalent Zr anion can also be isolated. Bis(cyclopentadienyl) complexes are normally of the form [M(Cp)$_2$XY] (X, Y = H, Cl, R) (**2**) with nonparallel rings; they still only have 16 valence electrons.

2

Compounds of type (2) are active alkene polymerization catalysts and are used to synthesize highly stereoregular polymers (Section 31.14). Other hydrocarbon complexes include the 16-electron cyclopentadienyl, cycloheptatrienyl complex (3) of Zr which has parallel rings, and the bis(arene) complexes (4) of all three metals, with the sterically very demanding tris(^{t}Bu)benzene ligand. Most organometallic compounds of the Group 4 metals are air- and moisture-sensitive, but this has not precluded significant development of their chemistry.

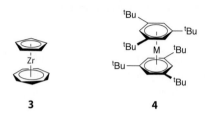

3   4

20.5 Group 5: vanadium, niobium, and tantalum

(a) Occurrence and uses

KEY POINT Vanadium occurs widely and is used to harden steels; niobium and tantalum are much less prevalent and have only specialist uses.

Vanadium occurs naturally in more than 60 different minerals and in fossil-fuel deposits. In some countries it is produced from the slag that results from steel smelting, while in others it is produced from the flue dust that forms from the burning of heavy oils, or as a by-product of uranium mining. It is mainly used to harden steel and to produce alloys such as those used in high-speed tools. The most important industrial vanadium compound, vanadium pentoxide, is used as a catalyst for the production of sulfuric acid. Vanadium is the only element in Group 5 with established biological roles (Chapter 26). It is found at the active site of haloperoxidases, enzymes responsible for the production of halogenated natural products by marine algae, and in place of molybdenum in an alternative class of nitrogenase, the microbial enzyme responsible for ammonia biosynthesis (Section 25.14).

Niobium and tantalum have very similar chemical and physical properties and are difficult to distinguish and separate. A new element, named columbium, was reported in 1801, and eight years later it was erroneously concluded that it was identical to tantalum. In 1846 another 'new' element, named niobium, was discovered; experimentation in 1865 proved that niobium and columbium were the same element (and not tantalum). Niobium has only been the official name for element 41 since 1949, columbium and niobium being used interchangeably up to then. Niobium and tantalum are both extracted from the ores tantalite and columbite, with Brazil supplying 75% of the world's usage. Separation of niobium from tantalum relies on the difference in solubilities of their complex fluorides in water. Niobium has no known biological role, though it is considered toxic.

Niobium is mostly used at low levels in specialist steel alloys, where it imparts increased hardness and strength. Other alloys, where it is present at much higher levels, are superconducting at liquid helium temperature (4.2 K) (the pure metal is also superconducting below 10 K); these alloys, such as niobium–tin (Nb_3Sn), are widely used in the superconducting magnets of NMR spectrometers and MRI scanners. Tantalum also finds use at a low level in steel, enhancing the hardness and corrosion resistance of the metal. Tantalum has no known biological role, and is considered nontoxic, with tantalum plates, bolts, and wires used in surgery.

(b) Binary compounds

KEY POINTS Vanadium forms binary compounds in the +3, +4, and +5 states, with the +5 state being oxidizing; binary compounds of niobium and tantalum are most stable in the +5 state.

While vanadium forms stable compounds in a number of oxidation states, the chemistry of niobium and tantalum is dominated by the +5 state. Vanadium(V) compounds are typically oxidizing, whereas those of niobium and tantalum are not. Oxides, M_2O_5, are known for all three metals, and are formed on burning the metals in air. The structure of V_2O_5 consists of sheets of $V(=O)O_4$ square-based pyramids sharing the four oxygens in the pyramid base (Fig. 20.5).

The oxides are high-melting solids, only sparingly soluble in neutral water. Depending on pH, the form of the oxide in solution can vary from $[MO_4]^{3-}$ through to $[MO_2]^+$, with the formation of a large number of polyoxometallates in between (Box 20.1). The lower oxides, MO_2, are known for all the Group 5 metals, adopting distorted rutile structures, but only vanadium forms a stable M_2O_3 compound. VO_2 undergoes a transition from the semiconducting phase that exists at room temperature, to a metallic phase above 68°C; the room-temperature structure contains weak V–V bonds that localize the electrons but, on heating, these bonds are broken, allowing high electronic conductivity.

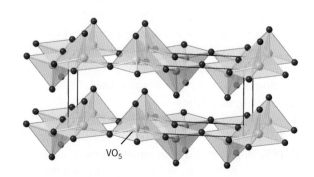

FIGURE 20.5 The structure of V_2O_5.

BOX 20.1 What are polyoxometallates and why are they important?

A **polyoxometallate** is an oxoanion containing more than one metal atom. The H_2O ligand created by protonation of an oxido ligand at low pH can be eliminated from the central metal atom and thus lead to the condensation of mononuclear oxometallates (Section 5.5). A familiar example is the reaction of a basic chromate solution, which is yellow, with excess acid to form the oxido-bridged dichromate ion, which is orange:

$$2H^+ + 2\ [CrO_4]^{2-} \longrightarrow [Cr_2O_7]^{2-} + H_2O$$

$$2[CrO_4]^{2-}(aq) + 2H^+(aq) \rightarrow [Cr_2O_7]^{2-}(aq) + H_2O(l)$$

In highly acidic solution, oxido-bridged Cr(VI) species with longer chains are formed. The tendency for Cr(VI) to form polyoxo species is limited by the fact that the O tetrahedra link only through vertices: edge and face bridging would result in too close an approach of the metal atoms. By contrast, it is found that five- and six-coordinate metal oxido complexes, which are common with the larger 4d- and 5d-series metal atoms, can share oxido ligands between either vertices or edges. These structural possibilities lead to a richer variety of polyoxometallates than are found with the 3d-series metals.

Chromium's neighbours in Groups 5 and 6 form six-coordinate polyoxo complexes (Fig. B20.1). In Group 5, the polyoxometallates are most numerous for vanadium, which forms many V(V) complexes and a few V(IV) or mixed oxidation state V(IV)–V(V) polyoxido complexes. Polyoxometallate formation is most pronounced in Groups 5 and 6 for V(V), Mo(VI), and W(VI).

It is often convenient to represent the structures of the polyoxometallate ions by polyhedra, with the metal atom understood to be in the centre and O atoms at the vertices. For example, the sharing of O atom vertices in the dichromate ion, $[Cr_2O_7]^{2-}$, may be depicted in either the traditional way (**B1**) or in the polyhedral representation (**B2**).

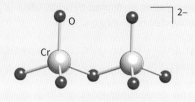

B1 $[Cr_2O_7]^{2-}$

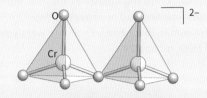

B2 Tetrahedral $[Cr_2O_7]^{2-}$

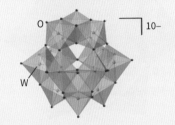

B3 $[W_{12}O_{40}(OH)_2]^{10-}$

Similarly, the important M_6O_{19} structure of $[Nb_6O_{19}]^{8-}$, $[Ta_6O_{19}]^{8-}$, $[Mo_6O_{19}]^{2-}$, and $[W_6O_{19}]^{2-}$ is depicted by the conventional or polyhedral structures shown in Fig. B20.2. The structures for this series of polyoxometallates contain terminal O atoms (those projecting out from a single metal atom) and two types of bridging O atoms: two-metal bridges, M–O–M, and one hypercoordinated O atom in the centre of the structure that is common to all six metal atoms. The structure consists of six MO_6 octahedra, each sharing an edge with four neighbours. The overall symmetry of the M_6O_{19} array is O_h. Another example of a polyoxometallate is $[W_{12}O_{40}(OH)_2]^{10-}$ (**B3**). As shown by this formula, the polyoxoanion is partially protonated.

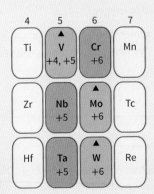

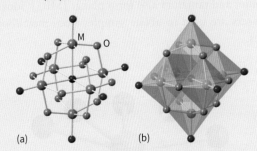

FIGURE B20.1 Elements of the early d block that form polyoxometallates have their symbols in bold. The elements in red boxes with the ▲ symbol form the greatest variety of polyoxometallates.

FIGURE B20.2 (a) Conventional and (b) polyhedral representation of the six edge-shared octahedra as found in $[M_6O_{19}]^{2-}$.

Proton-transfer equilibria are common for the polyoxometallates, and may occur together with condensation and fragmentation reactions. Polyoxometallate anions can be prepared by carefully adjusting pH and concentrations: for example, polyoxomolybdates and polyoxotungstates are formed by acidification of solutions of the simple molybdate or tungstate:

$$6[MoO_4]^{2-}(aq) + 10H^+(aq) \rightarrow [Mo_6O_{19}]^{2-}(aq) + 5H_2O(l)$$

$$8[MoO_4]^{2-}(aq) + 12H^+(aq) \rightarrow [Mo_8O_{26}]^{4-}(aq) + 6H_2O(l)$$

Mixed-metal polyoxometallates are also common, as in $[MoV_9O_{28}]^{5-}$, and there is a large class of heteropolyoxometallates, such as the molybdates and tungstates, that also incorporate P, As, and other heteroatoms. For example, $[PMo_{12}O_{40}]^{3-}$ contains a PO_4^{3-} tetrahedron that shares O atoms with surrounding octahedral MoO_6 groups (**B4**).

Many different heteroatoms can be incorporated into this structure, and the general formulation is $[X(+N)Mo_{12}O_{40}]^{(8-N)-}$ where X(+N) represents the oxidation state of the heteroatom X, which may be As(V), Si(IV), Ge(IV), or Ti(IV). An even broader range of heteroatoms is observed with the analogous tungsten heteropolyoxoanions. Heteropolyoxomolybdates and tungstates can undergo one-electron reduction with no change in structure but with the formation of a deep blue colour. The colour seems to arise from the excitation of the added electron from Mo(V) or W(V) to an adjacent Mo(VI) or W(VI) site.

Polyoxometallates have a number of uses: some can be reversibly reduced and are used as catalysts for the splitting of water to hydrogen and oxygen, as well as for a range of organic reactions; some exhibit luminescence, some with unpaired electrons have unusual magnetic properties and are being investigated as possible nanocomputer storage devices, and finally, many potential medicinal applications have been suggested, such as antitumoral and antiviral treatments.

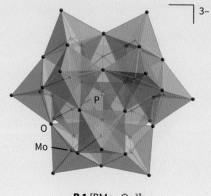

B4 $[PMo_{12}O_{40}]^{3-}$

Of the halogens, only fluorine brings out the +5 state for vanadium in VX_5, whereas MX_5 compounds are also known for Cl, Br, and I with niobium and tantalum. All the MX_5 compounds are volatile solids, prone to hydrolysis and sensitive to oxygen. The lower halides, MX_4 and MX_3, are known for all three members of Group 5; NbF_3 adopts the ReO_3 structure type, but only vanadium forms MX_2 compounds.

Niobium and tantalum also form planar $[MB_{10}]^-$ ions (**5**) when lasers are used to ablate samples from metal/boron mixtures.

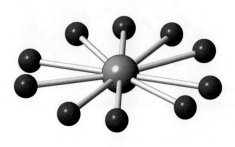

5

EXAMPLE 20.1 Assessing the redox stability of vanadium species in solution

What is the formula and oxidation state of the species that is thermodynamically favoured when an acidic aqueous solution of V^{2+} is exposed to oxygen?

Answer The Latimer diagram for vanadium in aqueous acid is shown below (see Resource section 3; reduction potentials are given in volts).

$$\underset{+5}{VO_2^+} \xrightarrow{+1.000} \underset{+4}{VO^{2+}} \xrightarrow{+0.337} \underset{+3}{V^{3+}} \xrightarrow{-0.255} \underset{+2}{V^{2+}}$$

$$VO^{2+} \xrightarrow{+0.668} V^{3+}$$

Since the reduction potential for the O_2/H_2O couple is 1.229 V, V^{2+} will be readily oxidized to V^{3+} and then all the way to VO_2^+ (+5 oxidation state) as long as sufficient O_2 is present. The net reaction is

$$4V^{2+} + 3O_2 + 2H_2O \rightarrow 4VO_2^+ + 4H^+$$

Self-test 20.1 Use the appropriate Latimer diagram to assess the stability of V^{2+} in alkaline solution.

(c) Complex oxides and halides

KEY POINT Complex oxides of the Group 5 metals have uses in technological applications.

Large crystals of $LiNbO_3$ and $LiTaO_3$ can be grown, and the compounds exhibit ferroelectric and piezoelectric effects and a nonlinear optical polarizability. The structure consists of $Nb(Ta)O_6$ and LiO_6 octahedra. $LiNbO_3$ is used in optical waveguides and surface acoustic wave (SAW) filters in mobile phones. Complex lithium vanadium oxides, such as Li_xVO_2 and $Li_xV_6O_{13}$, are currently being investigated as rechargeable battery materials, as the large range of oxidation states accessible for vanadium offers the potential for high energy-storage capacities. YVO_4 is used as a host material for lanthanoid ions in phosphors.

(d) Coordination complexes

KEY POINTS Vanadium forms many complexes of the vanadyl ion; niobium and tantalum form complexes with large coordination numbers.

The coordination chemistry of vanadium is dominated by the +4 oxidation state vanadyl ion, VO^{2+}, in which V(IV) is coordinated by the strong π-donor oxido ligand (to give a triple bond, V≡O) and a further four ligands to form square-pyramidal complexes. Aqueous solutions of $[V(H_2O)_6]^{3+}$ can be used as precursors to a large number of other octahedral complexes. Though vanadium also forms the $[V(H_2O)_6]^{2+}$ ion in solution from the reduction of higher oxidation state complexes, solutions of it are strongly reducing, and air must be excluded. Niobium and tantalum form many complexes with hard donors, and coordination numbers of 6, 7 (pentagonal-bipyramidal; **6**), and 8 (dodecahedral; **7**) are common.

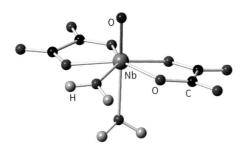

6 $[Nb(OH_2)_2(O)(ox)_2]^-$

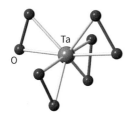

7 $[Ta(\eta^2-O_2)_4]^{3-}$

(e) Organometallic compounds

KEY POINTS The Group 5 metals form complexes with most organometallic ligands, but the paramagnetism of many of the complexes has hindered their characterization.

The organometallic chemistry of the Group 5 metals is generally underdeveloped, due in large part to the paramagnetic nature of most complexes, which makes characterization by NMR very difficult; the chemistry is challenging, but frequently surprising and rewarding. The binary carbonyls $[M(CO)_6]$ are 17-electron species and have only been isolated for vanadium. The 18-electron $[M(CO)_6]^-$ anions are, however, known for all the Group 5 metals: they are the product of reductive carbonylation of metal salts (Section 24.18). Indeed, the synthesis of the neutral $[V(CO)_6]$ can only be achieved from the $[V(CO)_6]^-$ anion. The simple 15-electron $[M(Cp)_2]$ can only be isolated for M = V. Mixed ligand 18-electron $[CpM(CO)_4]$ complexes are readily prepared, and provide a route into many other complexes of the type $[CpM(CO)_3L]$ or $[CpM(CO)_2LL']$. Complexes of almost every organometallic ligand are known, and some Nb and Ta complexes are utilized in Fischer–Tropsch processes that convert CO and H_2 to hydrocarbons (Section 31.11). Tantalum is unusual amongst d metals in that it forms η^8-COT complexes (Section 24.11) such as (**8**), presumably as a result of its large size; even so, these COT ligands can interconvert to the pentalene ligand found in complex (**9**).

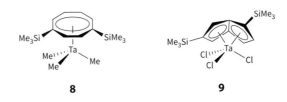

8 **9**

20.6 Group 6: chromium, molybdenum, and tungsten

(a) Occurrence and uses

KEY POINTS All three of the Group 6 metals are widely used in steels, imparting hardness and corrosion resistance; all three metals have biological roles.

Chromium is widely dispersed throughout the Earth's crust, mainly occurring as oxide minerals such as chromite, $FeCr_2O_4$, though occasionally it is found as the metal. Currently South Africa produces 40% of the world's supply via chromite reduction. Chromium is hard and extremely resistant to corrosion, and these properties account for its major use: as a component of steels, especially stainless ones, where it is present at between 10% and 40%. Chromium ('chrome') plating is also used to protect steels and to give a shiny surface. Reduction of chromite ores via electric arc melting produces ferrochrome, which is used directly in steel production. If pure chromium is required, chromite is roasted in air and then soluble chromium(VI) salts are separated from insoluble Fe_2O_3; subsequent reduction is achieved with carbon and aluminium. A number of chromium salts are used as dyes, pigments, and preservatives; chromium compounds such as supported CrO_3 (which reduces *in situ*) or chromocene are used as alkene polymerization catalysts. Chromium is essential to humans, where it is normally utilized in the +3 oxidation state: it is involved in the regulation of glucose levels and is often taken as a dietary supplement, although the mechanism of its action is unclear; in excess, or in the +6 state, chromium is toxic and carcinogenic.

Molybdenum and tungsten are both extremely hard metals and are widely used as components of steels. The use in steels accounts for more than 80% of molybdenum's production, with some steels containing up to 10% molybdenum. Molybdenite (MoS_2) is the principal ore from which the metal is extracted, with the initial step being a roasting in air to give an oxide; the oxide is then reduced with iron for use in steels, or with hydrogen for other uses. Molybdenum is essential to all domains of life, with at least 20 enzymes known in plants and animals, where its main role is as the active site of enzymes that catalyse the transfer of oxygen atoms, thus exploiting the properties of its higher oxidation states (Section 26.12). Molybdenum is also found in the active site (the FeMo cofactor) of nitrogenase, the bacterial enzyme responsible for producing ammonia from atmospheric N_2 (Section 26.14).

Tungsten has the highest melting point of any metal and for this reason found use as the filaments of incandescent lightbulbs. Along with its use in steels, tungsten is extensively used in the form of tungsten carbide, an extremely hard material used to tip drill bits and saw blades. Tungsten is extracted from ores such as wolframite ($(Fe,Mn)WO_4$) or scheelite ($CaWO_4$) by roasting ores to give oxides that are reduced by carbon or hydrogen; China currently produces more than three-quarters of the world's supply. Tungsten is the only third-row d metal to have a biological role: in some microbes it is found at the active site of enzymes that catalyse reactions requiring more demanding reducing conditions than their molybdenum counterparts; the tungsten-containing formate dehydrogenase catalyses the conversion of CO_2 to formic acid.

(b) Binary compounds

KEY POINTS Chromium forms highly oxidizing compounds in the +6 oxidation state, whereas the equivalent compounds of molybdenum and tungsten are not appreciably oxidizing.

All three of the Group 6 metals form a trioxide, MO_3. Whereas CrO_3 (and its aqueous counterparts $[CrO_4]^{2-}$ and $[Cr_2O_7]^{2-}$) is strongly oxidizing, MoO_3 and WO_3 are not oxidizing. This behaviour is partly a reflection of this increased stability of high oxidation states of the second- and third-row metals, compared with first-row metals, and partly a result of the structures of the trioxides. CrO_3 consists of chains of vertex-sharing tetrahedra, while the structures of MoO_3 and WO_3 consist of linked MO_6 octahedra; WO_3 adopts a distorted form of the ReO_3 structure (Section 25.11). Solid WO_3 undergoes reductive insertion reactions where small cations such as Li^+ or a proton become inserted into the structure during reduction of W(VI) to W(V):

$$2n\text{-BuLi(sol)} + 2WO_3(s) \rightarrow 2LiWO_3(s) + (n\text{-Bu})_2(\text{sol})$$

Lower oxides of Mo and W such as MO_2 can be synthesized by careful reduction of MO_3 with hydrogen, whereas chromium trioxide gives the stable +3 oxidation state (either as Cr_2O_3 or Cr^{3+}) with mild reducing agents. CrO_2 may be prepared as long rod-shaped crystals by the reduction of CrO_3 in water at high temperature and pressure; it adopts the rutile structure type. CrO_2 exhibits excellent ferromagnetic properties, which led to its widespread use in high-quality (low-noise) audio and video tapes in the 1970s and 1980s. It is still used in some data-storage and sound-recording applications. Cr_2O_3 is a widely used green pigment, originally called viridian, because of its inertness and stability in sunlight. Chromium also forms a reducing black oxide, CrO.

Sulfides Cr_2S_3 and MS_2 (M = Mo and W) are known, with the MS_2 compounds adopting layered structures (Fig. 20.6) which are responsible for their use as solid-state lubricants.

Hexahalides, MX_6, are known for all metals with X = F, but only Mo and W form hexachlorides and bromides; all the hexahalides are volatile liquids, that are rapidly and violently hydrolysed by water. Lower ionic halides such as CrX_3 are easily prepared, and the strongly reducing CrX_2 salts are made by careful reaction of Cr with HX. Molybdenum and tungsten also form solid MX_3 salts with all the halogens except fluorine, but these are frequently layered structures, or clusters, as in W_6Cl_{18} (**10**).

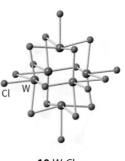

10 W_6Cl_{18}

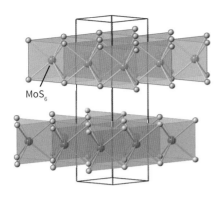

FIGURE 20.6 The layered structure of MoS_2.

(c) Complex oxides

KEY POINTS Tungsten and molybdenum form electrically conducting complex oxides with other metals; these oxides are known as bronzes.

Tungsten and molybdenum bronzes, first discovered in 1824 by Wöhler, are well-defined nonstoichiometric compounds of general formula M_xWO_3 or M_xMoO_3, where M is most commonly an alkali metal and $0 < x \leq 1$. Analogous compounds of vanadium, niobium, and titanium have been prepared and found to have similar properties. The term 'bronze' is now applied to a ternary metal oxide of general formula $M'_xM''_yO_z$, where (i) M″ is a transition metal, (ii) M''_yO_z is its highest binary oxide, (iii) M′ is a second, usually electropositive, metal, and (iv) x is a variable falling in the range $0 < x < 1$.

These chemically inert tungsten and molybdenum bronzes show several characteristic properties which derive from their electronic structure. While WO_3 and MoO_3 are insulators, with d^0 electronic configurations, the presence of variable amounts of alkali metal in the bronzes reduces the transition metal and introduces electrons into a conduction band (partially filling it with x electrons from the M_x atoms). This gives rise to metallic properties including high electrical conductivity and, in crystalline form, lustre, leading to the generic name 'bronze'. Tungsten bronzes can be prepared by an electrolytic reaction in which molten mixtures of an alkali metal tungstate and WO_3 are reduced at a platinum electrode. $NaWO_3$ (a bronze with $x = 1$) prepared in this manner grows as large (up to 1 cm³), golden-brown cubic crystals. The compound adopts the perovskite structure (Section 4.9(b)) and is electronically analogous to the metallic ReO_3 (Section 25.11).

(d) Coordination complexes

KEY POINTS Group 6 metals form a large number of polyoxometallates; they also form dimers with quadruple M–M bonds.

Oxido complexes of the Group 6 metals are typically based upon tetrahedral MO_4^{2-} units but they form a wide variety of polyoxometallates (Box 20.1) which can be considered as consisting of edge-sharing MO_6 octahedra. Otherwise, aqueous solutions of lower-oxidation-state metal ions are octahedral. The d^3 complexes of the Cr^{3+} cations generally have octahedral geometry that is favoured by a large ligand-field stabilization energy (LFSE; Section 7.16) and are normally inert to substitution; those of the high-spin d^4 Cr^{2+} ion are Jahn–Teller distorted (Section 7.16) and usually very labile (Section 23.1).

All three metals form dimers in the +2 oxidation state with quadruple metal–metal bonds. The bonding in such complexes is exemplified by the carboxylates (**11**) and consists of one σ, two π, and one δ bond (Box 20.2).

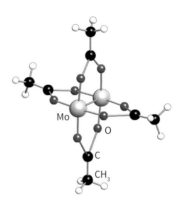

11 $[Mo_2(\mu\text{-}CO_2CH_3)_4]$

(e) Organometallic compounds

KEY POINTS The organometallic chemistry of the Group 6 metals is dominated by 18-electron complexes; neutral $[M(CO)_6]$ complexes are air-stable and neutral bis(arene) complexes are well known.

The organometallic chemistry of the Group 6 metals is vast, and is completely dominated by the readily available 18-electron configuration. Thus the 18-electron neutral hexacarbonyls, $[M(CO)_6]$, are all extremely stable white solids that can be handled in air and in the presence of moisture with no problems. Substitution of carbonyls provides access to complexes containing many different ligands: alkenes, alkynes, NHCs, phosphines, dienes, trienes, and arenes. 18-electron bis(arene) complexes (**12**) are known for Cr and Mo. Cyclopentadienyl derivatives are typically of the type $[(Cp)_2MH_2]$ (**13**), so-called ansa (loop-like) complexes (**14**), or half-sandwich complexes such as $[CpML_3Cl]$ (**15**). The 16-electron compound $[(Cp)_2Cr]$, known as chromocene, is known, but is strongly reducing.

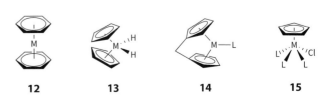

12　　**13**　　**14**　　**15**

BOX 20.2 How do metal–metal bonds arise?

The first d-block metal–metal bonded species to be identified was the $[Hg_2]^{2+}$ ion of mercury(I) compounds, as occurs in Hg_2Cl_2, and examples of metal–metal bonded compounds and clusters are now known for most of the d metals. Some of their common structural motifs are an ethane-like structure (**B5**), an edge-shared bioctahedron (**B6**), a face-shared bioctahedron (**B7**), and the tetragonal prism of $[Re_2Cl_8]^{2-}$ (**B8**).

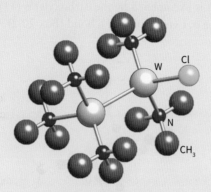

B5 [(Me$_3$N)$_3$W–WCl(NMe$_3$)$_2$]

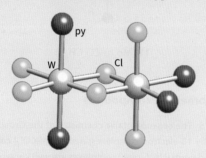

B6 [(py)$_2$Cl$_2$W(μ-Cl)$_2$WCl$_2$(py)$_2$]

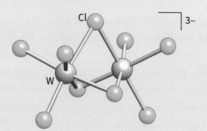

B7 [Cl$_3$W(μ-Cl)$_3$WCl$_3$]$^{3-}$

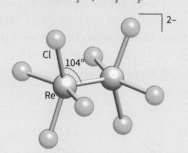

B8 [Re$_2$Cl$_8$]$^{2-}$

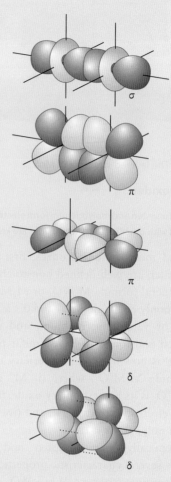

FIGURE B20.3 The origin of σ, π, and δ interactions between the d orbitals of two d-metal atoms situated along the z-axis. Only bonding combinations are shown.

If we consider the possible overlap between d orbitals on adjacent metal atoms, we can see that (Fig. B20.3)

- a σ bond between two metal atoms can arise from the overlap of a d_{z^2} orbital from each atom
- two π bonds can arise from the overlap of d_{zx} or d_{yz} orbitals
- two δ bonds can be formed from the overlap of two face-to-face d_{xy} or $d_{x^2-y^2}$ orbitals

Thus a quintuple bond could result if all the bonding orbitals are occupied to give the electron configuration $\sigma^2\pi^4\delta^4$ (Fig. B20.4).

Five d electrons would be needed from each metal for a quintuple bond, and this is exactly the case for the d^5 Cr(I) centres in (**18**). In this molecule, the two Cr atoms are separated by a very short distance of 183.5 pm, unsupported by any additional bridging ligand interactions; compare this distance with the separation of Cr atoms in the bulk metal, which is 258 pm. The Re compound (**B8**) also lacks bridging

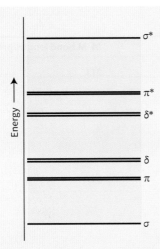

FIGURE B20.4 Approximate molecular orbital energy level scheme for the M–M interactions

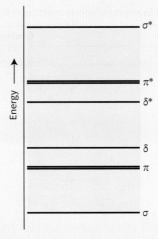

FIGURE B20.5 Approximate molecular orbital energy level scheme for the M–M interactions in a quadruply bonded system, where only the $d_{x^2-y^2}$ is utilized in bonding to the ligands.

ligand interactions, but here the two d^4 Re(III) centres can form only a quadruple Re–Re bond. The four d electrons from each Re atom in (**B8**) result in the configuration $\sigma^2\pi^4\delta^2$, with the $d_{x^2-y^2}$ orbital involved in bonding to the Cl⁻ ligands. Evidence for quadruple bonding comes from the observation that $[Re_2Cl_8]^{2-}$ has an eclipsed array of Cl ligands, which is sterically unfavourable. It is argued that the δ bond, which is formed only when the d_{xy} orbitals are confacial, locks the complex in the eclipsed conformation.

Many other species with multiple metal–metal bonds, where the $d_{x^2-y^2}$ orbital is involved in bonding to ligand species, are known. In all these complexes, the molecular orbital diagram shown in Fig. B20.5 becomes appropriate and the maximum possible bond order is 4. A well-known example is the quadruply bonded compound molybdenum(II) acetate (**11**), which is prepared by heating $[Mo(CO)_6]$ with acetic acid:

$$2[Mo(CO)_6] + 4CH_3COOH \rightarrow [Mo_2(O_2CCH_3)_4] + 2H_2 + 12CO$$

The dimolybdenum complex is an excellent starting material for the preparation of other Mo–Mo bonded compounds. For example, the quadruply bonded chlorido complex is obtained when the acetato complex is treated with concentrated hydrochloric acid below room temperature:

$$[Mo_2(O_2CCH_3)_4](aq) + 4H^+(aq) + 8Cl^-(aq) \rightarrow [Mo_2Cl_8]^{4-}(aq) + 4CH_3COOH(aq)$$

As shown in Table B20.1, incomplete occupation of the bonding orbitals can result in a reduction of the formal bond order to 3.5 or to the triply bonded M≡M systems. These complexes are more numerous than the quadruply bonded complexes and, because δ bonds are weak, M≡M bond lengths are often similar to those of quadruply bonded systems. A decrease of bond order can also stem from the occupation of both the δ^* orbitals and, once these are fully occupied, successive occupation of the two higher-lying π^* orbitals leads to further decrease in the bond order from 2.5 to 1.

As with carbon–carbon multiple bonds, metal–metal multiple bonds are centres of reaction. However, the variety of structures resulting from the reactions of metal–metal multiple bonded compounds is more diverse than for organic compounds. For example:

Cp(CO)₂Mo≡Mo(CO)₂Cp + HI ⟶ Cp(CO)₂Mo(H)(I)Mo(CO)₂Cp

In this reaction, HI adds across a triple bond but both the H and I bridge the metal atoms; the outcome is quite unlike the addition of HX to an alkyne, which results in a substituted alkene. The reaction product can be regarded as containing a 3c,2e MHM bridge and an iodide anion bonding by two conventional 2c,2e bonds, one to each Mo atom.

Larger metal clusters can be synthesized by addition to a metal–metal multiple bond. For example, $[Pt(PPh_3)_4]$ loses two triphenylphosphine ligands when it adds to the Mo–Mo triple bond, resulting in a three-metal cluster:

Cp(CO)₂Mo≡Mo(CO)₂Cp + [Pt(PPh₃)₄] ⟶ Pt(PPh₃)₂ bridging Cp(CO)₂Mo=Mo(CO)₂Cp

20 The d-block elements

TABLE 20.B1 Examples of metal–metal bonded tetragonal-prismatic complexes[†]

Complex	Configuration	Bond order	M–M bond length/pm
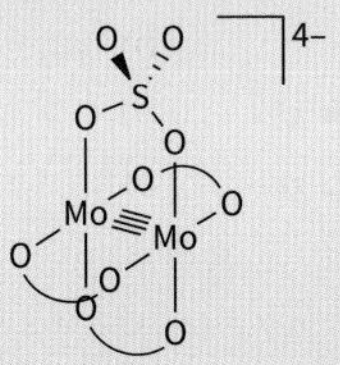 [Mo₂(SO₄)₄]⁴⁻	$\sigma^2 \pi^4 \delta^2$	4	211
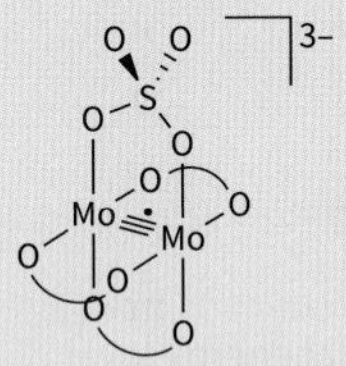 [Mo₂(SO₄)₄]³⁻	$\sigma^2 \pi^4 \delta^1$	3.5	217
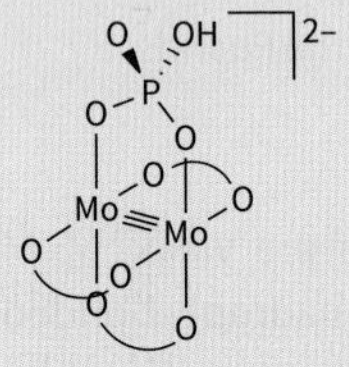 [Mo₂(HPO₄)₄]²⁻	$\sigma^2 \pi^4$	3	222
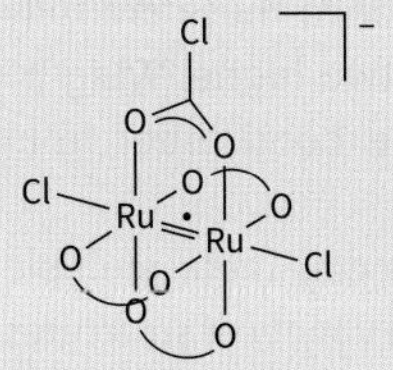 [Ru₂(O₂CR)₄Cl₂]⁻	$\sigma^2 \pi^4 \delta^2 \delta^{*1} \pi^{*2}$	2.5	227
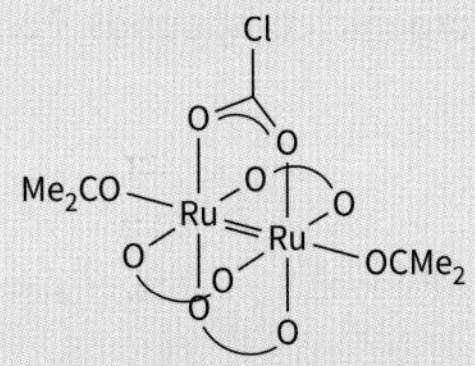 Ru₂(O₂CR)₄Cl(OCMe₂)(Me₂CO)	$\sigma^2 \pi^4 \delta^2 \delta^{*2} \pi^{*2}$	2	238
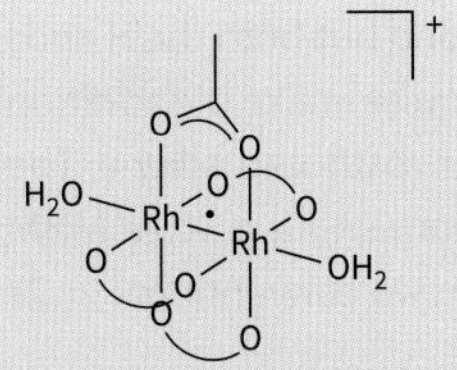 [Rh₂(O₂CMe)₄(OH₂)₂]⁺	$\sigma^2 \pi^4 \delta^2 \delta^{*1} \pi^{*4}$	1.5	232
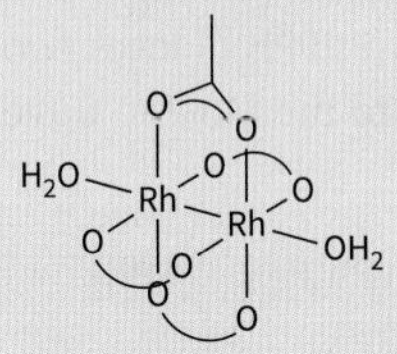 Rh₂(O₂CMe)₄(OH₂)₂	$\sigma^2 \pi^4 \delta^2 \delta^{*2} \pi^{*4}$	1	239

[†] When multiple bridging ligands are present, only one is shown in detail.

Tungsten in particular forms a number of alkylidene and alkylidyne complexes, e.g. (**16**) and (**17**), where formal metal–carbon double and triple bonds, respectively, are present. Multiple bonds between metal atoms are also possible (Box 20.2), and the isolation of (**18**) was the first definitive isolation of a quintuple bond in a stable complex.

20.7 Group 7: manganese, technetium, and rhenium

(a) Occurrence and uses

KEY POINTS Manganese is widely dispersed on Earth, and its compounds have many uses; although technetium has no stable isotopes, it has a number of uses. Rhenium is very rare.

Manganese is distributed widely across the planet, though about 80% of the known reserves are found in South Africa as pyrolusite (MnO_2). It is estimated that about 500 billion tonnes of manganese nodules exist (mainly as MnO_2) on the ocean floor, and economically viable (and, hopefully, environmentally benign) routes to extracting these are being developed. Manganese metal itself is rather brittle, but it is used in steel at levels between 1 and 13% where it improves strength; manganese is also used in aluminium alloys, such as those used in drinks cans, at a level of ~1% to improve corrosion resistance. MnO_2 is a major component of alkaline batteries: MnO_2 is reduced to Mn_2O_3 and zinc metal is oxidized to ZnO, producing a voltage of 1.5 V. Other uses for manganese salts include pigments in ceramics and glasses. Manganese is a very important element in biology and is present as the active site metal in numerous enzymes across all forms of life: by far its most notable role is at the active site of the enzyme uniquely responsible for photosynthetic O_2 evolution (Section 26.10). Manganese(II) compounds are nontoxic, but manganese(VII) compounds are highly oxidizing and poisonous.

Technetium was the first unstable element to be synthesized: there are no stable isotopes and only minute quantities are found in nature. It is a fission by-product of uranium, and is extracted in bulk from spent nuclear fuel rods. Technetium is also formed when molybdenum is bombarded by neutrons, and this route was used to isolate the first samples in 1936. A metastable isotope, ^{99m}Tc, is used medically as a radioactive tracer; it decays via β emission with a half-life of 6 hours, and compounds such as Cardiolyte®, which is used to image the heart, are in widespread use in hospitals (Section 32.9). The two longest-lived isotopes, ^{98}Tc and ^{97}Tc, have half-lives of over 2 million years, and their detection in red giants proves that heavy-atom nucleosynthesis takes place in such stars. The development of technetium chemistry has been retarded by its radioactivity, but what there is suggests a great similarity with that of Re.

Rhenium was the last stable nonradioactive element to be discovered, and was first found in gadolinite (a lanthanoid ore) at a level of only 10 ppm in 1925. It is also found in molybdenum ores and is currently extracted from the flue dusts of molybdenum smelters; it is one of the rarest metals in the Earth's crust. Rhenium is used in high-temperature alloys for jet engines and, together with platinum, as a catalyst to reform alkanes. It has no known biological role, and is considered nontoxic.

(b) Binary compounds

KEY POINTS Manganese forms stable compounds in a large number of oxidation states; by contrast, the chemistries of technetium and rhenium are dominated by higher-oxidation-state compounds.

Manganese forms oxides MnO, Mn_2O_3, MnO_2, and Mn_2O_7. Technetium and rhenium, by contrast, only form MO_2, MO_3, and M_2O_7. The cubic ReO_3 structure has been described in Section 4.9, and ReO_2 adopts the rutile structure type. The +7 oxidation state M_2O_7 oxides are all volatile and dissolve in water to give the $[MO_4]^-$ ions; the Mn complex is the familiar purple 'permanganate' ion, a very powerful oxidizer. The Tc and Re complexes are less oxidizing.

The difficulty of placing seven ligands around a single metal ion means that binary MX_7 halides are only represented by ReF_7, whereas Re and Tc both form MF_6, MF_5, and all four MX_4 halides. Rhenium forms MX_3 with Cl, Br, and I (actually trimers of M_3X_9; **19**); manganese forms MnF_4, MnF_3, and all four MX_2 halides. The equilibrium between MnF_3 and MnF_4 is used to purify fluorine: MnF_3 is reacted with impure $F_2(g)$ to produce $MnF_4(s)$, which is then heated to above 400°C to liberate pure fluorine gas.

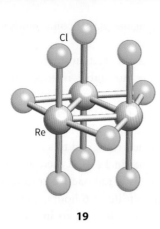

19

(c) Complex oxides and halides

KEY POINT Manganese forms a number of complex oxides and halides with useful magnetic properties.

Complex manganese oxides of the formula $(Ln_{1-x}Sr_x)MnO_3$, where Ln is a trivalent lanthanoid cation, typically Pr^{3+}, show **colossal magnetoresistance**: their electrical resistance changes dramatically (several orders of magnitude) when they are placed in a magnetic field (Section 30.6). These materials adopt the perovskite structure and contain a mixture of Mn^{3+} and Mn^{4+} ions on the B-cation site. Manganese violet is a pigment described as an ammonium manganese pyrophosphate and contains the Mn^{3+} ion. $LiMn_2O_4$, with the spinel structure, is used as the positive electrode in some rechargeable batteries. Mixed manganese–zinc–iron spinels, $(Mn_{1-x}Zn_x)Fe_2O_4$, are known as soft ferrites and are used in transformer cores.

$KMnF_3$, prepared by the addition of KF to Mn^{2+} solutions, adopts the perovskite structure (Section 25.13) with Mn^{2+} octahedrally coordinated to F^- on the B-cation site and K^+ on the A-cation site; it acts as host for Ln^{3+} and has some applications in imaging. Fluorination of MnF_2 in the presence of an alkali metal fluoride leads to the complex Mn(IV) fluorides of the composition M_2MnF_6 (M = K, Rb, Cs).

(d) Coordination complexes

KEY POINTS Manganese(II) is the most stable form of Mn in aqueous solution and has a high-spin d^5 configuration; the solution chemistry of technetium and rhenium is dominated by higher oxidation states.

Manganese (VII) and (VI) complexes (formally d^0 and d^1, respectively) are based around the tetrahedral MO_4 unit (some have a halogen in place of an oxygen) and are strongly oxidizing in solution. The only other oxidation states that have a significant solution chemistry are the +3 state (typically high-spin, Jahn–Teller distorted d^4 octahedral complexes) and the +2 state. Of these two states, it is the +2 state that dominates: complexes are invariably high-spin (even the cyanide complex) with the half-filled d^5 configuration being very considerably stabilized by the quantum-mechanical exchange energy (Section 1.5). Mn(II) complexes have no LFSE and show little preference for any specific coordination geometry; in combination with small coordination species such as OH_2, O^{2-}, and F^-, the size of Mn^{2+} (ionic radius 65 pm) normally produces octahedral complexes. The absence of any allowed electronic transitions means complexes of Mn(II) are essentially colourless.

By contrast, technetium and rhenium show no solution chemistry in the +2 state, and their coordination chemistry is dominated by high-oxidation-state complexes with oxido and nitrido ligands. These complexes show a variety of geometries, with octahedral and square-based pyramidal (with an oxido or nitrido in the apical position) being the most common. In contrast to the intense colour of the permanganate anion $[MnO_4]^-$, the equivalent Tc and Re anions, pertechnate and perrhenate, are colourless as the energy required for the charge-transfer process corresponds to absorption in the UV, illustrating the less oxidizing nature of Re(VII) and Tc(VII) compared to Mn(VII).

(e) Organometallic compounds

KEY POINT The majority of the organometallic compounds of the Group 7 metals are 18-electron species containing carbonyl ligands.

Carbonyl containing compounds make up the bulk of all organometallic compounds known for the Group 7 metals. All three metals form neutral 18-electron bimetallic carbonyl complexes $[M_2(CO)_{10}]$ and the metal–metal bond in these complexes is easily cleaved, either oxidatively (e.g. with Br_2) or reductively (e.g. with Na) to give monomeric octahedral complexes such as $[BrMn(CO)_5]$ or $[MeMn(CO)_5]$. Substitution of these complexes can be used to give monocyclopentadienyl complexes such as $[CpM(CO)_3]$, and complexes of arenes and dihydrogen (**20**) are known.

20

Bis(cyclopentadienyl) complexes would be 17-electron, but only manganese is known to form monomeric species. The properties of the simple $[(Cp)_2Mn]$ are complicated by the strong preference of the formally Mn^{2+} ion to remain high-spin ($S = \tfrac{5}{2}$) (Section 7.16), and it is only with more sterically demanding cyclopentadienyls that the predicted ($S = \tfrac{1}{2}$) structures are found: these compounds are easily reduced to 18-electron species. Technetium forms a dimeric compound of formula $[(Cp)_4Tc_2]$, but the structure is unknown. Photogenerated 16-electron $[(Cp)_2Re]^+$ (**21**) can activate benzene via oxidative addition to give an 18-electron aryl hydride (**22**).

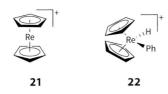

21   22

20.8 Group 8: iron, ruthenium, and osmium

(a) Occurrence and uses

KEY POINTS The very abundant iron is both the most widely used metal on the planet, and the most exploited d-block element in biology; ruthenium and osmium are very rare, and have only specialist uses and no biological roles.

Iron is the most common element (by mass) on Earth as a whole, forming much of Earth's outer and inner core (Box 4.1), and is the fourth most common element in the Earth's crust. The ^{56}Fe nucleus is the most stable of all isotopes, and the ultimate result of nuclear fusion. Iron is the single most important metal in human civilization: it is refined and used on the scale of more than one billion tonnes a year; more than 90% of all metal refined is iron. Iron ores are widely distributed, and most iron is reduced in a blast furnace with coke and limestone. Iron has been used by humans for at least 3000 years, and the alloying of iron—which is actually rather soft when completely pure—with other metals (V, Cr, Mo, W, Mn, Co, Ni) and carbon produces steels of a wide spectrum of properties. Iron's use as a catalyst in the Haber process to make ammonia from dinitrogen and hydrogen is critical to the industrial manufacture of fertilizers.

Iron is the most abundant transition metal in biology, with the average human containing over 4 g of Fe. The numerous roles of iron include O_2 transport (haemoglobin), electron transfer, acid–base, radical, and redox catalysis in a diverse range of enzymes, regulation of gene expression, and even sensing the Earth's magnetic field (Chapter 26).

Ruthenium is extremely rare, and is obtained commercially as a by-product from nickel and copper refining as well as by the processing of platinum group metal ores. It is used in platinum and palladium alloys, which it hardens, to make wear-resistant electrical contacts; ruthenium dioxide and lead and bismuth ruthenates are used in chip resistors. These two electronic applications account for more than 50% of the ruthenium consumption, with the remainder used in a variety of chemical processes such as anodes for chlorine production.

Osmium is the densest of all elements (22.6 g cm^{-3}), roughly twice as dense as lead. Osmium is the least abundant of the naturally occurring elements and is found in nature as the metal (together with iridium), but it is more economically recovered as a by-product of nickel refining; annual production is of the order of 100 kg. Osmium has a very high melting point (3054°C) and is exceptionally hard; consequently, it is very difficult to machine or work. Uses rely on hardness and wear resistance: older uses include nibs of fountain pens, bearings in clocks and compasses, gramophone needles; more modern uses include in electrical contacts and as the tetroxide in organic synthesis. Ruthenium and osmium have no known biological role; while most of their salts are considered nonpoisonous, the volatile tetraoxides, MO_4, are highly toxic.

(b) Binary compounds

KEY POINTS The highest oxidation state observed for iron is lower than that of ruthenium and osmium; low-oxidation-state (+2 and +3) iron compounds are the most stable.

Iron forms a number of oxides: $Fe_{1-x}O$, $x \sim 0.04$ (oxidation state mainly +2), Fe_2O_3 (oxidation state +3), and Fe_3O_4 (a mixed-oxidation-state complex: one-third of the iron is Fe(II) and two-thirds Fe(III)) with the spinel structure. Fe_3O_4 occurs naturally as the mineral magnetite and may be permanently magnetized (whereupon it is known as lodestone); it has been used in navigational compasses in Europe for at least 900 years, and is thought to have been used by the Olmecs in what is now Mexico some 3000 years ago. Though iron is present as Fe(VI) in compounds of the $[FeO_4]^{2-}$ anion, the equivalent oxide, FeO_3, is unknown. Similarly, Fe(IV) complexes of the fluoride, such as $[FeF_6]^{2-}$, are known but the simple FeF_4 is not. Binary halides, FeX_3 and FeX_2, are known for all halogens. Iron forms both a sulfide, FeS, and a disulfide, FeS_2, found naturally as iron pyrites and known as 'fool's gold', a complex of the disulfide ion S_2^{2-} (Section 17.12).

Ruthenium metal is oxidized by air at room temperature to give a passivating layer of RuO_2, and requires higher temperatures to completely oxidize the metal to RuO_2. Osmium, by contrast, is very slowly oxidized by air to give volatile and highly toxic OsO_4, which is responsible for the smell of osmium metal and hence its name (from the Greek *osme*, meaning smell). Osmium dioxide, OsO_2, is formed if the oxidant is not in excess, and is a black crystalline powder. Both Ru and Os dioxides adopt the rutile structure. RuO_4 can be made with the use of strong oxidants, though it is unstable with respect to decomposition to RuO_2 and O_2; both RuO_4 and OsO_4 are yellow, volatile, and extremely toxic. RuO_2 has some uses as an electrocatalyst for the oxygen evolution reaction in fuel cells (Sections 6.16 and 25.16).

Hexa-, penta-, and tetrafluorides are known for both Ru and Os, with tetrachlorides and tetrabromides also known for Os. In the +3 oxidation state all Ru halides are known, together with Os bromide and iodides.

(c) Complex pnictides, oxides, and halides

KEY POINTS The Group 8 metals form a wide variety of complex solids; those of iron have important electronic and magnetic properties.

Complex ferrite oxides with the spinel structure type AB_2O_4, where B is typically Fe^{3+}, are formed with a large variety of divalent cations (A). These include $ZnFe_2O_4$ and mixed A-cation site materials such as $Mn_{1-x}Zn_xFe_2O_4$ (manganese–zinc ferrite) and $Ni_xZn_{1-x}Fe_2O_4$ (nickel iron ferrite). These materials are known as soft ferrites and, because of their ferromagnetic properties and easily reversible magnetization, they have applications in transformer cores, and are also used as brown and black pigments. Hard ferrites, such as the strontium and barium ferrites with the compositions $AFe_{12}O_{19}$, are excellent permanent magnets. Yttrium iron garnet (YIG), $Y_3Fe_5O_{12}$, has applications in various magneto-optical applications (Section 30.1).

Iron-based superconductors include Ln(O,F)FeAs, where Ln is a lanthanoid (La, Ce, Sm, Nd, Gd), with superconducting critical temperatures (T_c) of up to 53 K; and the arsenides LiFeAs and NaFeAs (T_c of 18 and 25 K, respectively); in these pnictides and oxopnictides the iron is coordinated to arsenic in layers, while the more electropositive cations and oxide, where present, form interleaving layers.

In their intermediate oxidation states, osmium and ruthenium form complex oxides such as $Pb_2Os_2O_7$ (oxidation state of Os +5), with the pyrochlore structure (Section 25.13). In their higher oxidation states, all the Group 8 metals form complex oxide compounds with discrete oxoanions such as $BaFeO_4$ (tetrahedral FeO_4^{2-}), Na_4FeO_4, (Fe^{4+} d^4, Jahn–Teller distorted away from the tetrahedral ideal towards a flattened form of FeO_4^{4-}), and K_2OsO_5 (trigonal-bipyramidal OsO_5^{2-}).

(d) Coordination complexes

KEY POINTS Octahedral complexes are common for all the Group 8 metals; in aqueous solution, the balance between high- and low-spin complexes of iron is rather fine.

Though high-oxidation-state complexes of iron derived from the FeO_4^{2-} and FeO_4^{4-} anions exist, Fe(II) and Fe(III) are the more usual ions present in solution. Iron(III) complexes are typically octahedral and can be oxidizing. The balance between high- and low-spin complexes is at its most delicate here, and complexes of low-field ligands (e.g. water, halides) are high-spin, while complexes of high-field ligands (e.g. cyanide, bpy) are low-spin. It is possible to make complexes of Fe(III) that change from low to high spin with temperature, pressure, or solvent ('spin-crossover complexes'). Iron(III) is relatively hard and favours oxygen donor ligands; its polarizing power means that solutions of $[Fe(H_2O)_6]^{3+}$ are rather acidic, and species such as $[Fe(H_2O)_5(OH)]^{2+}$ and $[(H_2O)_5Fe(\mu\text{-}O)Fe(H_2O)_5]^{4+}$ are present in solution. Iron(II) complexes are normally octahedral and reducing: thus $[Fe(H_2O)_6]^{2+}$ is oxidized by air to give the Fe(III) compound. When Fe(II) is complexed with high-field ligands such as cyanide or phenanthroline the complexes become low-spin and, with a d^6 electron configuration, substantially stabilized and rather inert. Complexes of Fe(IV) contain the ferryl group (Fe=O): these are highly reactive (Fe sits on the 'oxo wall', Section 10.8) and generally exist as catalytic intermediates in oxygenation reactions, most notably those catalysed by numerous enzymes (Section 26.10)

In their higher-oxidation-state coordination chemistry, ruthenium and osmium both form octahedral $[MX_6]^{2-}$ ions with halides, and some complicated oxo anions such as $[Ru_4O_6(H_2O)_{12}]^{4+}$ (**23**). In their lower oxidation states their chemistry is dominated by low-spin octahedral complexes, with Ru showing a surprisingly large number of stable complexes in the +2 oxidation state. One complex, $[Ru(bpy)_3]^{2+}$, is typically used as a photosensitizer: oxidation to the +3 state is achieved when visible light excites an electron from the Ru centre to an antibonding orbital of the bpy ligand (Section 23.14).

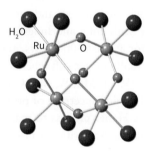

23 $[Ru_4O_6(H_2O)_{12}]^{4+}$

(e) Organometallic compounds

KEY POINT The Group 8 metals form stable bis(cyclopentadienyl) complexes and a wide variety of carbonyl clusters.

Organometallic compounds of the Group 8 metals include the most iconic of all organometallic complexes: ferrocene (**24**).

24 Ferrocene

It was the discovery of, and subsequent elucidation of the bonding of, the 18-electron ferrocene in the early 1950s that inspired the whole of modern organometallic chemistry (Chapter 24). Ruthenium and osmium form similar sandwich complexes known as ruthenocene and osmocene; all three compounds are very stable and can be handled, and indeed sublimed, in air with no precautions. Iron forms a simple 18-electron, trigonal-bipyramidal carbonyl, [Fe(CO)$_5$] (**25**), when finely divided iron metal is treated with carbon monoxide. Loss of carbon monoxide from [Fe(CO)$_5$] is possible to give clusters such as [Fe$_2$(CO)$_9$] (**26**) and [Fe$_3$(CO)$_{12}$] (Section 24.20); the simplest stable carbonyl that Ru and Os form is in fact the [M$_3$(CO)$_{12}$] cluster (**27**).

25 **26** **27**

Organometallic complexes of the Group 8 metals containing combinations of carbonyl, cyclopentadienyl, and many other ligands are easily accessible. There are a vast number of cluster compounds containing Group 8 metals.

Ruthenium carbene complexes, such as (**28**), are active in the alkene metathesis reaction that resulted in the award of the 2005 Nobel Prize to Grubbs, Chauvin, and Schrock (Section 31.1).

28

20.9 Group 9: cobalt, rhodium, and iridium

(a) Occurrence and uses

KEY POINTS Cobalt is used as an important component of many steels, and its salts have been used for millennia as pigments; rhodium and iridium are very rare but have important uses in catalysis.

Cobalt is found at low levels in most rocks and soils and was isolated from a meteorite in 1820. It is of considerable economic importance, with the main source being as a by-product of copper and nickel mining. Alongside its uses in steel, where it imparts hardness, it is extensively used in batteries and magnets. Cobalt compounds have been used for millennia to impart a rich blue colour to glass, glazes, and ceramics: cobalt has been detected in Egyptian sculptures, Persian jewellery from the third century BC, and the ruins of Pompeii (destroyed in AD 79). Most cobalt salts are considered nontoxic. The active site of cobalt-containing enzymes is a macrocyclic complex known as cobalamin, commonly encountered as Vitamin B$_{12}$, which forms a Co-alkyl bond. One of the first complex biological molecules for which the structure was determined by X-rays, cobalamin is synthesized exclusively by bacteria yet is essential, in trace amounts, for all animal life (Section 26.11). Cobalamin-containing enzymes catalyse radical rearrangements and methyl transfer reactions.

Rhodium is one of the least abundant of the naturally occurring elements on Earth, and is found in very small amounts as the free metal. The normal source of the element is as a by-product of copper and nickel refining, and only around 20 tonnes are isolated each year. Rhodium metal is highly resistant to oxidation and is very reflective: it is used, as a thin film, to coat optical fibres, certain mirrors, and reflectors in headlights. The major use, accounting for 80% of all rhodium, is as a component of catalytic converters in car exhaust systems. Rhodium compounds are also used in the Monsanto acetic acid process (Section 31.7), where they catalyse the carbonylation of methanol to give acetic (ethanoic) acid.

Iridium is the most corrosion-resistant metal known, but so rare that only around 3 tonnes per year are extracted and used. Most iridium found naturally is as an alloy with osmium (osmiridium), though the major commercial source is the anode sludge that forms from the electrorefining of copper. Uses of iridium rely on its hardness and corrosion resistance: it is used as alloys with osmium in bearings, fountain pen nibs, and as reflectors in X-ray telescopes. Iridium salts are used in the Cativa process for methanol carbonylation that is superseding the rhodium-based Monsanto process (Section 31.7). Neither rhodium nor iridium has any known biological role, but salts of both are mildly toxic if ingested.

(b) Binary compounds

KEY POINTS The highest oxidation state found for the Group 9 metals is in the fluorides and is +6 for rhodium and iridium and +4 for cobalt; the most stable oxidation state is +2 for cobalt and +3 for rhodium and iridium.

High-oxidation-state compounds of the Group 9 metals are restricted to the fluorides RhF$_6$, IrF$_6$, RhF$_5$, IrF$_5$, and MF$_4$, and the oxides RhO$_2$ and IrO$_2$ (rutile structure type); the fluorides are strongly oxidizing and often unstable. While IrO$_2$ is the stable oxide of iridium, rhodium forms the more common Rh$_2$O$_3$. Cobalt is normally found in the +2 oxidation state, with all four halides and the oxide well known. Cobalt forms relatively few binary compounds in the +3 oxidation state (the fluoride) and partly in the oxide Co$_3$O$_4$, but does form a wealth of coordination complexes. For rhodium and iridium, it is the +3 oxidation state that is the most stable and well-developed: all halides are known, as are oxides.

(c) Complex oxides and halides

KEY POINT Intensely coloured cobalt pigments have Co^{2+} in tetrahedral coordination.

LiCoO$_2$ has a layer structure consisting of linked Co(III) O$_6$ octahedra separated by lithium ions; this lithium may be partly displaced electrochemically, leading to the use of this material in rechargeable battery systems

(Section 30.9). CoAl$_2$O$_4$ has Co(II) in a tetrahedral site, producing a deep royal blue colour for this compound with the spinel structure; it is widely used as a pigment. Tetrahedrally coordinated cobalt(II) is found in other complex oxides and compounds, including the bright blue cobalt glasses formed when Co^{2+} compounds are added to silicates.

(d) Coordination complexes

KEY POINTS Cobalt forms more tetrahedral complexes than any other d metal, whereas rhodium and iridium mostly form octahedral complexes.

The aqueous coordination chemistry of cobalt is dominated by Co(II) and Co(III), with the balance between the stability of the two oxidation states being ligand-dependent: [Co(H$_2$O)$_6$]$^{3+}$ will oxidize water to liberate oxygen and [Co(H$_2$O)$_6$]$^{2+}$, whereas [Co(NH$_3$)$_6$]$^{2+}$ is oxidized by air to give [Co(NH$_3$)$_6$]$^{3+}$. Cobalt(II) forms both octahedral and tetrahedral complexes, with cobalt(II) forming more tetrahedral complexes than all other d metals; this is because for the d^7 configuration the difference in LFSEs of octahedral and tetrahedral geometries is at a minimum (Section 7.16). All of the tetrahedral complexes are high-spin, as are most of the octahedral ones; in general the octahedral complexes are pink or red and the tetrahedral ones are intensely blue. Cobalt(III) complexes are normally octahedral, and the vast majority of those are low-spin d^6 and inert to ligand substitution (Section 23.6).

The coordination chemistry of rhodium and iridium is totally dominated by octahedral low-spin d^6 complexes of the M^{3+} cation. Dissolution of MCl$_3$ in HCl(aq) results in, depending on concentration of chloride, all complexes possible between [M(H$_2$O)$_6$]$^{3+}$ and [MCl$_6$]$^{3-}$. Many other complexes with hard σ donors such as ammonia are known.

(e) Organometallic compounds

KEY POINTS Group 9 metals form 18-electron complexes as well as catalytically active 16-electron square-planar complexes.

The smallest neutral carbonyls known for the Group 9 metals are the dimeric [Co$_2$(CO)$_8$] (29), and the tetrameric [Rh$_4$(CO)$_{12}$] and [Ir$_4$(CO)$_{12}$] compounds (30), with other higher clusters known too. These carbonyl complexes are useful precursors that lead into the catalytically active complexes used in hydroformylation (Section 31.3) and carbonylation (Section 31.7) reactions.

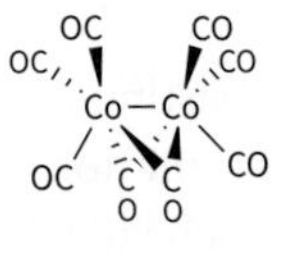

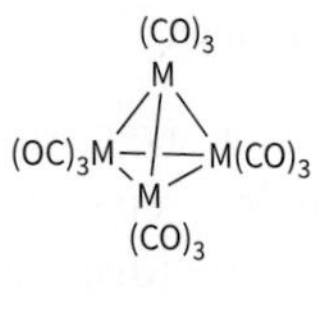

Neutral bis(cyclopentadienyl) complexes would be 19-electron species, and only cobalt forms a simple monomeric species. Rhodium is thought to form a monomeric species in the gas phase and at temperatures below −196°C, but exists as a dimer (where the formal electron count is 18 at each rhodium) under more normal conditions:

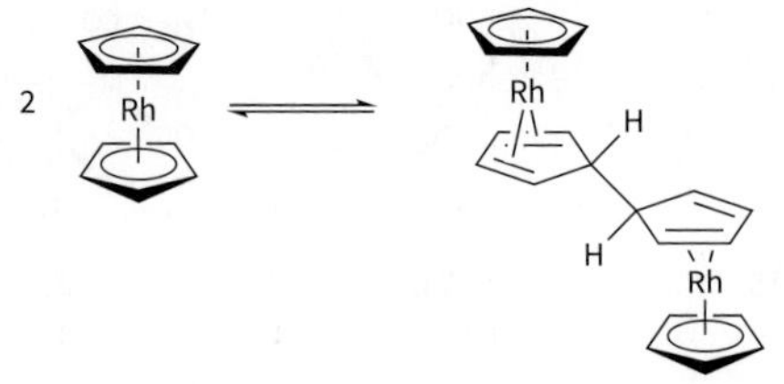

All three metals, however, form the 18-electron [(Cp)$_2$M]$^+$ cations, which show great stability. Mono(cyclopentadienyl) species such as (31) and (32) have a rich chemistry, including in the area of alkane activation.

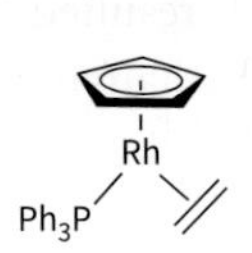

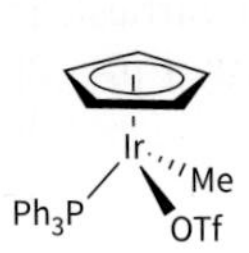

It is in the formally +1 oxidation state that we see 16-electron square-planar Rh(I) and Ir(I) complexes. Complexes of this type, such as the historic Vaska's complex (33), are ideally set up for an oxidative addition reaction to give 18-electron octahedral complexes of M(III). Examples of catalytic reactions that use the ready accessibility of oxidative additions include homogeneous hydrogenation catalysts (Wilkinson's catalyst, Section 31.2) and the synthesis of ethanoic acid via the carbonylation of methanol (Section 31.7). It is interesting to note that the very first industrial processes for methanol carbonylation used a cobalt catalyst, the second generation used rhodium catalysts, and the current generation uses iridium catalysts.

20.10 Group 10: nickel, palladium, and platinum

(a) Occurrence and uses

KEY POINTS All three Group 10 metals have important uses: most nickel is used in steel, and both palladium and platinum are used in catalysis.

Nickel occurs widely throughout the Earth's crust and is thought to make up about 10% of its core. Normally, nickel is found in conjunction with iron in laterite minerals such as nickeliferous limonite (Fe, Ni)O(OH) or magmatic sulfides such as pentlandite (Ni, Fe)$_9$S$_8$, though the huge reserves found near Sudbury, Ontario, in Canada, which are thought to have arrived via the impact of a gigantic meteorite, are millerite, NiS.

Conventional roasting in air and reduction with coke produces nickel of ~75% purity, which is suitable for use in most alloys. Very pure nickel is produced via electrolysis or via the Mond process. Commercialization of the Mond process (1890) relied on the volatility of [Ni(CO)$_4$], a molecule whose formula had been established a couple of years previously, but whose bonding was still unexplained (Section 24.5). Briefly, carbon monoxide is passed over finely divided nickel, forming the carbonyl, which is swept along to a higher-temperature region where it decomposes back to nickel and carbon monoxide (which is then recycled). This process can be used to nickel-plate items: hot items are placed in a flow of [Ni(CO)$_4$] which then decomposes on the surface, depositing nickel.

Some 60% of nickel is used in corrosion-resistant steel alloys, and the bulk of the rest in other alloys, or for plating. Nickel features rarely in higher life forms, but is an important element in the microbial world, where it is found in enzymes that catalyse methane biosynthesis, H$_2$ oxidation, H$_2$ production, and CO$_2$ reduction—reactions of great interest for renewable energy.

A particularly infamous use of Ni in biology is as the active catalytic metal in urease, an enzyme produced by the stomach pathogen *Helicobacter pylori*, which is responsible for gastric ulcers and is implicated in stomach cancer. Some people develop dermatitis when exposed to metallic nickel, though most salts are nontoxic; [Ni(CO)$_4$] is extremely toxic, even at very low doses.

Palladium is rare, and occurs in combination with gold and platinum in nature; however, most palladium is recovered as a by-product of nickel refining. The majority of palladium refined today is used in catalytic convertors in car exhausts, where it helps oxidize partially burnt hydrocarbons. Other uses include dentistry, jewellery, and in the fabrication of microcapacitors. Synthetic chemists make significant use of palladium, either as a hydrogenation catalyst (normally suspended on carbon) or in one of the many palladium-catalysed carbon–carbon bond-forming reactions for which the Nobel Prize in Chemistry was awarded in 2010 (Section 31.6). Palladium has no known biological role, but PdCl$_2$ was once prescribed for tuberculosis, without any harmful—or indeed beneficial—effect.

Although rather rare, platinum is found as the native metal across the planet. Several ancient civilizations are known to have worked the metal, though it was not recognized as a new element until around 1750. Currently it is mined in South Africa, Russia, and Canada, but is normally recovered from nickel and copper refining. As a pure metal, platinum is silvery-white, lustrous, ductile, and malleable (it is the most ductile of all pure metals). It does not oxidize in air at any temperature, is resistant to most chemicals, and has a very high melting point (1768°C). Thus in the laboratory, platinum ware (crucibles, electrodes, etc.) are commonly used when conditions are demanding. These properties also account for its second major use: jewellery. Such is the cachet associated with platinum, it is frequently more highly regarded than gold. Currently the major industrial use of platinum is in catalytic convertors, though this use is waning. While platinum has no known biological role, compounds derived from *cis*-[PtCl$_2$(NH$_3$)$_2$] (cisplatin) (Section 32.1) are widely used as anticancer treatments; it is known that the platinum binds to the DNA of cells and prevents their replication.

(b) Binary compounds

KEY POINT The most commonly observed oxidation state for the Group 10 metals is +2.

All three Group 10 metals give compounds in the +2 oxidation state, and these are very stable under normal conditions. Further oxidation is relatively easy for platinum to the +4 state, and even to the +6 oxidation state. Thus the compounds MX$_2$ are known for all metals and all halogens, PtX$_4$ are known for all halogens, but only MF$_4$ is known for Ni and Pd. The very strongly oxidizing PtF$_6$ can oxidize both dioxygen and xenon. Oxides MO and sulfides MS are known for all metals, and the dioxide PtO$_2$ and mixed-valence Pt$_3$O$_4$ can also be isolated. NiO has the rock-salt structure (octahedral coordination of both Ni^{2+} and O^{2-}), though PdO and PtO adopt a structure with square-planar metal coordination.

Palladium readily absorbs hydrogen at room temperature to form an interstitial hydride with a greater density of hydrogen than solid hydrogen itself. A porous form of nickel, known as Raney nickel, can be made, and this form is particularly able to absorb hydrogen which can then be delivered to organic substrates.

(c) Hydroxides, complex hydrides, and oxides

KEY POINTS Nickel hydroxides are used in batteries and the oxides in solid fuel cells.

Nickel metal hydride rechargeable batteries (Box 11.6) use the transformation between Ni(OH)$_2$ and NiO(OH) as the positive electrode. The α-form of Ni(OH)$_2$, used in these batteries, has a layer structure formed from face-sharing Ni(OH)$_6$ octahedra.

Reaction of the alloy LaNi$_5$ with hydrogen gas leads to the complex LaNi$_5$H$_6$, which contains more hydrogen per unit volume than liquid hydrogen. Heating this material under reduced pressures evolves hydrogen, and this material is of interest as a hydrogen-storage material where weight is not of critical importance (see also Box 12.5).

Composites of NiO and yttrium-stabilized zirconia are used as highly catalytically active anodes in solid oxide fuel cells, and lanthanum nickelate, $La_2NiO_{4+\delta}$, ($0 \leq \delta \leq 0.2$), is used as a cathode. Titanium dioxide doped with nickel and antimony oxides, $(Ti_{0.85}Ni_{0.05}Sb_{0.10})O_2$, is used a yellow pigment for plastics and ceramic glazes (Box 25.12).

(d) Coordination complexes

KEY POINTS Nickel complexes are known with many geometries; Pd(II) and Pt(II) complexes are square-planar.

Complexes of Ni^{2+} in solution show octahedral (e.g. $[Ni(H_2O)_6]^{2+}$), trigonal-bipyramidal (e.g. $[Ni(CN)_5]^{3-}$ with some cations), square-based-pyramidal (e.g. $[Ni(CN)_5]^{3-}$ with some cations), tetrahedral (e.g. $[NiCl_4]^{2-}$), and square-planar (e.g. $[Ni(CN)_4]^{2-}$) geometries. By contrast, those of Pd^{2+} and Pt^{2+} are almost invariably square-planar, such a geometry being particularly favoured for a d^8 configuration with large d orbitals and strong ligand field (Sections 7.7 and 7.16). Platinum (and, to a lesser extent, palladium) exhibits a reasonably diverse set of complexes in the +4 oxidation state: these complexes are all low-spin octahedral d^6, and are typically very inert.

(e) Organometallic compounds

KEY POINTS Nickel forms the homoleptic carbonyl $[Ni(CO)_4]$, but the equivalent palladium and platinum complexes are unknown; square-planar organometallic complexes of palladium are widely used in catalysis.

Organometallic compounds of the Group 10 metals include the two historic compounds, nickel tetracarbonyl, $[Ni(CO)_4]$, and $K[(CH_2=CH_2)PtCl_3]$, known as Zeise's salt (Section 24.9). $[Ni(CO)_4]$ is a tetrahedral 18-electron species which has been used as a precursor to many other formally nickel(0) complexes, but the equivalent Pd and Pt complexes are unknown under normal conditions. It is in the +2 oxidation state that the Group 10 metals show the greatest number of complexes: these are almost all 16-electron square-planar compounds, like Zeise's salt.

Oxidative addition (Section 24.22) to these complexes is often facile, giving 18-electron octahedral complexes (often followed by reductive elimination to generate another M(II) species). Similarly, oxidative addition is possible at formally M(0) complexes, and there are many examples of catalytic reactions of palladium that utilize this type of pathway, such as hydrogenation reactions (Section 31.2), the Wacker process (Section 31.4), and palladium-mediated methods of carbon–carbon bond formation (Section 31.6).

There is considerable interest in the activation of simple alkanes by platinum, as it appears that organometallic platinum complexes possess the best balance of properties to allow a pathway to functionalizing hydrocarbons to be realized. Organometallic nickel complexes find uses as hydrogenation and polymerization catalysts and in the cyclotrimerization of alkynes to give arenes.

20.11 Group 11: copper, silver, and gold

(a) Occurrence and uses

KEY POINTS The Group 11 metals have been used for at least 5000 years; all three are extremely malleable.

All three of the Group 11 metals have been known and worked for at least 5000 years, their value reflected in their use as currency over the ages.

Copper is a soft, reddish metal that is easily worked and can be made into flexible wires; it has very high electrical and thermal conductivity. While its surface does tarnish and weather to a soft green hue (verdigris, a basic copper carbonate), it does not corrode under normal conditions. Copper is extracted at about 15 million tonnes a year, and current economically viable reserves are predicted to only last for a further 35–40 years. Sulfide ores are roasted, treated with limestone, and decomposed by heat to give impure copper. Electrolysis of this material gives copper of 99.99% purity; a sludge of other metals that were present in the crude copper forms beneath the anode and it is from this that many of these elements (Ag, Au, Ir, Os, Pd, Pt, Rh, Ru) are extracted.

The majority of copper is used in electrical equipment and plumbing (including heat exchangers), with around 5% being used in alloys such as brass. Copper is very important in biology: there are numerous copper-dependent enzymes, with cytochrome *c* oxidase (Section 26.10) being required by all higher life forms to produce energy. Copper is also found in haemocyanin, the blue-coloured O_2-transport protein found in arthropods and molluscs, and in enzymes responsible for metabolizing methane (Section 26.10). Copper salts find some use as antifungal agents, and large quantities are toxic by ingestion.

Silver is widely dispersed and mined as the sulfide ore argentite (Ag_2S), chlorargyrite (AgCl), and pyrargyrite (Ag_3SbS_3), though it is also produced as a by-product of the electrolytic refining of copper. Silver is a very ductile, malleable metal, with a brilliant white metallic lustre that can take a high degree of polish, though it does tarnish in the atmosphere. It has the highest electrical and thermal conductivity of all metals. Alongside its familiar use in jewellery, silver is used in mirrors and in the electrical industry (utilizing its excellent conductivity). The use of silver salts in traditional photographic films has almost disappeared with the advent of digital photography. Silver has no known biological role, but the Ag^+ cation is deadly to both bacteria and viruses, and some medical dressings and preparations contain silver. A more mundane use of this property is the incorporation of silver metal at low levels into clothing (e.g. socks) to prevent the growth of bacteria that are responsible for offensive odours.

Gold was traditionally the most highly prized of all metals, though nowadays it is not always the most expensive. It occurs as the metal in many locales and is extracted at around 3000 tonnes a year. Gold is familiar as a bright-yellow metal, with the colour deriving from relativistic stabilization of the 6s orbital relative to the 5d (Section 10.6). Gold does not tarnish and it is the most malleable of all metals: 1 g of it—about the size of a grain of rice—can be beaten into a sheet more than 1 m² in area. Traditional uses in jewellery account for around 75% of current gold production and the bulk of the remainder is used as investments, with significant amounts used as electrical contacts. Colloidal gold nanoparticles vary in colour from red to purple, depending on size: they have been used as pigments for glass and porcelain for many centuries. Although gold has no known biological role, complexes are used to treat rheumatoid arthritis (Section 32.2); the metal itself has been used in dentistry for at least 2500 years.

(b) Binary compounds

KEY POINTS The +2 oxidation state is the most stable for copper; silver and gold form most compounds in the +1 and +3 states.

The Group 11 metals all show binary compounds in oxidation states +1 and +2, with gold and silver also showing compounds in the +3 state, and gold also able to form AuF_5. Copper forms no binary compounds in the +3 state, though some complex fluorides and oxides are known; the most stable state for copper is +2, and all the Cu(II) halides except the iodide are known, together with the oxide and sulfide. Copper (I) compounds are normally unstable with respect to disproportionation in solution, but the equilibrium position of the reaction is affected by the presence of certain ligands. Cu_2O (Fig. 20.7) is a stable, red solid formed by the reduction of Cu^{2+} solutions and contains Cu^+ ions in linear coordination to oxygen. It is used as a pigment and component in antifouling paint for ships. CuO is a dark brown solid that contains square-planar Cu^{2+}.

By contrast, the +2 oxidation state is not common for silver and very rare for gold. Ag(II) is strongly oxidizing but is known in the compound AgF_2. Silver and gold halides are known in the +3 oxidation state as AgF_3, AuF_3, $AuCl_3$, and $AuBr_3$. Gold's stable oxide is Au_2O_3, whereas that of silver is Ag_2O. Ag_2O is used in the silver oxide battery, where it is reduced along with zinc. The sulfides are also known and are small band-gap semiconductors.

(c) Complex chalcogenides and halides

KEY POINTS High-oxidation-state complexes of the Group 11 metals can be stabilized with fluoride and oxide anions; complex copper oxides form high-temperature superconductors.

The Cu(III) oxidation state may be stabilized in complex oxides and fluorides such as $LiCuO_2$ and $CsCuF_4$, and the highly oxidizing Cu(IV) is known in Cs_2CuF_6. High-oxidation-state complex silver and gold fluorides may be synthesized, as in $KAgF_4$, $La(AuF_4)_3$, and $KAuF_6$.

The high-temperature superconductors are a family of complex copper oxides. Following the discovery of $La_{2-x}Ba_xCuO_4$ by Bednorz and Muller in 1986, more than 50 different complex copper oxide compositions and structures have been discovered. These include the widely studied $YBa_2Cu_3O_{7-dx}$ (YBCO) and $Bi_2Sr_2CaCu_2O_8$ (BISCO) phases. These compounds, whose structures are all related to that of perovskite, contain copper in an average oxidation state between +2.15 and +2.35. A key structural feature of these superconductors is a sheet of CuO_4 square planes linked through all vertices, with overall composition CuO_2 (Fig. 20.8).

Copper indium diselenide (CIS, $CuInSe_2$) and the gallium-doped form, copper indium gallium selenide (CIGS, $CuIn_{1-x}Ga_xSe_2$), are semiconductors of importance in photovoltaic cells because of their high absorption coefficient for photons with energies above 1.5 eV, leading to solar cell efficiencies near 20%.

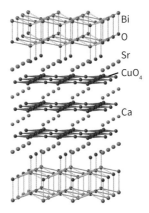

FIGURE 20.8 The structure of BISCO, showing the linked CuO_4 square planes.

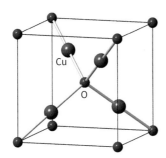

FIGURE 20.7 The structure of Cu_2O.

(d) Coordination complexes

KEY POINTS Copper(II) complexes are Jahn–Teller distorted; silver complexes exhibit little coordination geometry preference; Au(III) complexes are square-planar and Au(I) complexes are linear.

The coordination chemistry of copper is dominated by that of the Jahn–Teller distorted (Section 7.16) d^9 Cu^{2+} ion: notionally octahedral complexes with two opposite ligands either closer in than, or further away from, the other four. The hexaaqua ion is blue, and substitution of the water by amine ligands results in intense blue colours. Copper(I) salts in solution are unstable with respect to disproportionation and rapidly convert to metallic Cu and Cu(II). Silver's coordination chemistry is largely that of the d^{10} Ag^+ complexes, where little preference for coordination geometry is exhibited: for instance, linear $[Ag(NH_3)_2]^+$, trigonal-planar $[Ag(NH_3)_3]^+$, and tetrahedral $[Ag(NH_3)_4]^+$ geometries are all known. Au(III) d^8 complexes are invariably square-planar, as in $[AuCl_4]^-$, and Au(I) complexes are normally linear.

(e) Organometallic compounds

KEY POINT With the exception of gold, the organometallic chemistry of the Group 11 metals is poorly represented.

The Group 11 metals show a rather restrictive organometallic chemistry, largely due to the predominance of the +1 oxidation state with its d^{10} configuration in their organometallic complexes.

Copper organometallics are essentially restricted to simple η^1-alkyl and -aryl complexes of formally Cu(I): a small number of unstable carbonyl compounds and a few examples of coordinated alkenes and arenes have been reported.

Silver organometallics are likewise mostly η^1-alkyl and -aryl complexes, though a limited chemistry of square-planar d^8 complexes of Ag(III) has been reported: these complexes are all rather unstable and oxidizing. The recent popular use of silver oxide as a base to deprotonate imidazolium salts to give NHC ligands (Section 24.15) bound to silver via a carbon has resulted in a large number of new, formally organometallic complexes, but in reality the role of the silver is simply one of a convenient conduit for delivery of the NHC ligand.

Gold shows the most developed organometallic chemistry of all the Group 11 metals, with the well-established Au(I) chemistry supplemented by a number of reactive (but not prohibitively so) and catalytically active square-planar complexes of Au(III). η^2-Alkene complexes are known, but no complexes with a higher hapticity have yet been reported. The only carbonyl complexes with appreciable stability are [Au(CO)Cl] and [Au(CO)Br].

20.12 Group 12: zinc, cadmium, and mercury

(a) Occurrence and uses

KEY POINTS Zinc is important both industrially and biologically; cadmium has uses in batteries; uses of mercury are being phased out because of its toxicity.

Though not initially recognized as a separate element, zinc would appear to have been used for at least 2300 years: certainly the Romans were aware of brass, an alloy of copper and zinc. Zinc is currently the fourth most widely used metal (after Fe, Al, and Cu) on Earth, with some 10 million tonnes or so produced each year.

Zinc's principal ores are sphalerite and wurtzite (both ZnS), and these are roasted to give the oxide which can then be reduced with coke. More than half the zinc produced is used to protect steel in the form of zinc plating (galvanizing). Most of the remainder is either used as the pure metal or in alloys such as brass, or as the oxide.

Zinc oxide, ZnO, is used in the vulcanization of rubber, in a variety of medical applications such as calamine lotion and antibacterial creams, as a white pigment—for example, in coatings on paper—and as a UV-blocking component of plastics.

Zinc sulfide, ZnS, is used as a host and activator in many phosphors. When doped with manganese, it emits an orange light when excited with UV or X-ray radiation; with silver doping, the colour produced is blue. Copper-doped ZnS is a phosphorescent material used in glow-in-the-dark paints.

Zinc is the second most abundant d metal in biology, and occurs in more than 200 different enzymes. Its main roles are as the active site metal in a very large number of enzymes that perform acid–base catalysis, such as carbonic anhydrase (Section 26.9), and as the structure-forming metal in 'Zn-finger' transcription factors, proteins that recognize particular sequences of DNA and thus process the genetic code. Zinc is also being discovered to play a vital role in neurochemistry; zinc salts are largely nontoxic, and zinc oxide is used in sunblock and to treat skin infections.

Cadmium is a soft, silvery bluish metal that tarnishes in air. It is present as an impurity in most zinc ores, and recovery of cadmium from these sources provides more than enough for current consumption. Currently around 90% of cadmium is used in rechargeable batteries, with the bulk of the remainder being used as coatings in specialist steels. Cadmium in the soil is absorbed by many plants, but is not known to have any beneficial roles except in one marine diatom (where it is found in a carbonic anhydrase); it is an accumulative poison in most animals, most likely because it can nonreversibly replace zinc in certain proteins, and destroy their function.

Mercury is unique amongst all the metallic elements in that it is a liquid at room temperature, a property largely attributable to the relativistic stabilization of the $6s^2$ shell (Section 10.6). Instantly recognizable, mercury is a fascinating liquid—an unmistakably metallic fluid that it is possible to float lead on. The most important ore of mercury is cinnabar (HgS), which is roasted at 500°C, whereupon metallic mercury distils out. As a metal, mercury finds use in thermometers, in older fluorescent lights (as the vapour), and as various alloys ('amalgams'), including some used in dentistry. Cinnabar, also known as vermilion, has long been used as a red pigment: the Palaeolithic paintings from around 30 000 years ago in caves in Spain and France were made with vermilion. Because of increasing environmental concerns, use of mercury is being phased out in most applications. Mercury has no known biological roles, and is intensely poisonous (Box 20.3).

BOX 20.3 Why is mercury toxic?

Mercury salts are actually relatively harmless because membranes present a barrier to small ionic species including Hg^{2+}. Likewise, metallic mercury is not absorbed through the gut, so it is not toxic when swallowed. However, mercury vapour is highly toxic (that is why all mercury spills must be rigorously cleaned up) because the neutral atoms readily pass through the lungs' membranes and also across the blood–brain barrier. Once in the brain, the Hg(0) is oxidized to Hg(II) by the vigorous oxidizing activity of brain-cell mitochondria, and the Hg(II) binds tightly to crucial thiolate groups of neuronal proteins. Mercury is a powerful neurotoxin, but only if it gets inside nerve cells. Even more hazardous than mercury vapour are organomercury compounds, particularly methylmercury. Thus, CH_3Hg^+ is taken up through the gut because it is complexed by the stomach's chloride, forming CH_3HgCl, which, being electrically neutral, can pass through a membrane. Once inside cells, CH_3Hg^+ binds to thiolate groups and accumulates.

Nowadays the environmental toxicity of mercury is associated almost entirely with eating fish. Methylmercury is produced by the action of sulfate-reducing bacteria on Hg^{2+} in sediments, and accumulates as little fish are eaten by bigger fish further up the aquatic food chain. Fish everywhere have some level of mercury present. Mercury levels can increase markedly if sediments are contaminated by additional mercury. The worst known case of environmental mercury poisoning occurred in the 1950s in the Japanese fishing village of Minamata. A polyvinyl chloride plant, using Hg^{2+} as a catalyst, discharged mercury-laden residues into the bay, where fish accumulated methylmercury to levels approaching 100 ppm. Thousands of people were poisoned by eating the fish, and a number of infants suffered mental disabilities and motor disturbance from exposure *in utero*. This disaster led to stricter standards for fish consumption. Limited consumption is advised for fish at the top of the food chain, such as pike and bass in fresh waters, and swordfish and tuna in the oceans.

Regulatory action has been aimed at reducing mercury discharges and emissions, and industrial point sources have been largely controlled. Chloralkali plants, producing the large-volume industrial chemicals Cl_2 and NaOH by electrolysis of NaCl, were a major source of the problem, as a mercury-pool electrode was used to transfer metallic sodium to a separate hydroxide-generating compartment. The process has now been modified by separating the two electrode compartments with a cation-exchange membrane, which prevents migration of the anions. Combustion can vent mercury to the atmosphere if the fuel contains mercury compounds. Whereas municipal waste and hospital incinerators have been equipped with filters to reduce mercury emissions to the air, coal contains small amounts of mercury minerals. Because of the huge quantities burned, coal is a major contributor to environmental mercury.

Mercury poses a global environmental problem, because mercury vapour and volatile organomercurial compounds can travel long distances in the atmosphere. Eventually, elemental mercury is oxidized and organomercurials are decomposed, both to Hg^{2+}, by reaction with ozone or by atmospheric hydroxyl or halogen radicals. The Hg^{2+} ions are solvated by water molecules and deposited in rainfall. Thus mercury deposition can occur far from the emission source, and is distributed fairly uniformly around the globe. For example, it is estimated that only a third of North American mercury emissions are in fact deposited in the United States, and this accounts for only half of the mercury deposition in that country. Even the gold fields of Brazil contribute, because miners use mercury to extract the gold, which is recovered by heating the resultant amalgam to drive off the mercury. This practice is estimated to account for 2% of global mercury emissions (half of South American emissions).

The picture is further complicated by the fact that much of the deposited mercury is recirculated through processes that produce volatile compounds or mercury vapour. For example, much of the biomethylation activity of sulfate-reducing bacteria produces dimethylmercury, $(CH_3)_2Hg$, which, being volatile, is vented to the atmosphere. Other bacteria have an enzyme (methylmercury lyase) that breaks the methyl–mercury bond of CH_3Hg^+, and another enzyme (methylmercury reductase) that reduces the resulting Hg(II) to Hg(0); this is a protective mechanism for the microorganisms, ridding them of mercury as volatile Hg(0).

(b) Binary compounds

KEY POINTS The chemistry of zinc and cadmium is dominated by the +2 oxidation state; mercury also forms complexes of the $[Hg_2]^{2+}$ cation, which contains an Hg–Hg single bond.

The chemistries of zinc and cadmium are quite similar to each other and resemble in some ways that of the Group 2 metals. Most of the differences between Zn and Cd can be attributed to the increased size of cadmium; their chemistry is almost exclusively that of the M^{2+} d^{10} state, with oxides, sulfides, and halides all known. Zinc forms a stable hydride, ZnH_2. Mercury forms a large number of Hg^{2+} compounds but also forms compounds of the $[Hg_2]^{2+}$ cation in which two Hg ions are singly bonded to each other. Recent reports have identified HgF_4 in low-temperature matrices (which, with its d^8 configuration, would imply mercury should be considered to be a transition metal), but it is still some way away from being an isolated compound.

Zinc oxide exists in two polymorphs adopting the wurtzite and zinc-blende structures, with the former being the thermodynamically more stable form. Under high pressures, >10 GPa, these 4:4 coordination structures transform to the rock-salt structure. White ZnO reversibly loses a small amount of oxygen on heating, forming the bright yellow $Zn_{1+x}O$, with interstitial zinc as Frenkel defects (Section 25.1). The larger Cd^{2+} ion in CdO yields the rock-salt structure. HgO adopts a structure which contains linear O–Hg–O units linked into chains (Fig. 20.9). In the polycrystalline form it is a red solid, but in the small-particle form produced by rapid precipitation from solution it is yellow. On heating it decomposes to mercury metal and oxygen gas, and it was by this route that Joseph Priestley first produced pure oxygen in 1774.

Like ZnO, ZnS is polymorphic, and the two structure types hexagonal wurtzite and cubic zinc blende (sphalerite) are named after these naturally occurring mineral forms. CdS shows the same polymorphic behaviour as ZnS, forming 4:4 coordination structures. It is yellow and is used as a pigment (cadmium yellow). CdSe is polymorphic too, and the red wurtzite structure is also used as a pigment (cadmium red). CdTe is a small-band-gap semiconductor and is used in photovoltaic cells. Red HgS has a structure consisting of helices formed from linear S–Hg–S units. The semiconductor properties and uses of the Group 12 chalcogenides are discussed in more detail in Box 20.4.

EXAMPLE 20.2 Metal–metal bonding and clusters

Suggest which interactions might be responsible for, and thus the bond order of, the metal–metal bond in the $[Hg_2]^{2+}$ ion.

Answer We need to judge the types of bonds that can form from the available atomic orbitals on each metal atom and the bond order that results from their occupation. The oxidation state of mercury in $[Hg_2]^{2+}$ is Hg(I), and therefore its electron configuration is $d^{10}s^1$. Although overlap of the d orbitals in the manner depicted in Fig. B20.3 is possible, the 20 d electrons from the two Hg ions would result in complete filling of both the bonding and the antibonding orbitals, with no effective bonding. Therefore the bonding must come from the overlap of s orbitals on each ion and the two remaining s electrons: σ bonding and antibonding orbitals can be constructed and, as only the former is occupied, it results in a single Hg–Hg bond. Even if a degree of sd hybridization is invoked, the description still demands the involvement of s orbitals in the bonding, and the bond order remains 1.

Self-test 20.2 Describe the probable structure of the compound formed when Re_3Cl_9 is dissolved in a solvent containing PPh_3.

Zinc halides, ZnX_2, are all solids at room temperature, though in the gas phase they are linear X–Zn–X molecules. Zinc fluoride, ZnF_2 is a very high-melting point solid with the rutile structure, whereas the other three have layered structures, lower melting points, and high solubility in water. Cadmium likewise forms a high-melting fluoride (fluorite structure) and layered structures for the chloride, bromide, and iodide. Mercury forms both HgX_2 and Hg_2X_2, though with the exception of Hg_2I_2 the Hg_2X_2 compounds easily disproportionate to Hg and HgX_2. The compound Hg_2I_2, also known as protoiodide, was used as a medicine in the nineteenth century to treat everything from acne to kidney disease, and, in particular, syphilis; the side effects of protoiodide are so bad that the 'cure' was often feared more than the disease.

(c) Complex oxides and halides

KEY POINT Tetrahedrally coordinated zinc ions form porous structures analogous to zeolites.

Zinc oxide is amphoteric, and under basic conditions, solutions containing tetrahedral $[Zn(OH)_4]^{2-}$ ions are formed. These may be condensed in combination with other tetrahedral species such as phosphate to form porous framework structures—analogous to the aluminosilicate zeolites (Section 25.16)—known as zincophosphates. Zinc

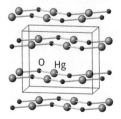

FIGURE 20.9 The structure of HgO

BOX 20.4 What are the uses of the Group 12 chalcogenides?

The compounds between the Group 12 and 16 elements have a number of important applications derived from their semiconducting and associated optical properties. In these metal chalcogenides, MX, the valence band is derived mainly from X^{2-} orbitals and the conduction band from M^{2+} orbitals, giving rise to the simple band structure shown in Fig. B20.6.

The band gap in this system depends upon the relative energies of the orbitals from M and X contributing to the bands, as this controls both the bandwidth and the separation between M^{2+} and X^{2-} levels. For a particular chalcogenide, say Se, as the group Zn, Cd, Hg is descended the orbitals become more closely matched in energy and the band gap becomes smaller, while for the series $MO \rightarrow MS \rightarrow MSe \rightarrow MTe$ the rising chalcogenide orbital energies also decrease the band gap. The observed band gaps for the Group 12 MX compounds are summarized in Table B20.2.

The interaction of light with these materials of varying band gap leads to several important applications. Visible light covers the range 1.76 eV (red light) to 3.1 eV (blue light), and light of sufficient energy will promote an electron in a compound MX from the valence band to the conduction band. A large-band-gap semiconductor such as ZnO (band gap greater than 3.1 eV) will only absorb light in the ultraviolet, and this leads to its use in sunscreen creams; ZnO and ZnS are also used as white pigments, with no absorption in the visible region.

As the band gap diminishes in materials such as CdS and CdSe, the absorption moves into the visible region; CdS absorbs blue light and CdSe absorbs everything except red light. Thus CdS is a bright yellow solid (the complementary colour of blue; Fig. 8.14) and CdSe is red, and this leads to the application of these materials as pigments, the artist's cadmium yellow and cadmium red. Intermediate shades such as orange can be obtained with the solid solution $CdS_{1-x}Se_x$.

With smaller-band-gap materials such as CdTe and mixed Group 12 metal systems, (Cd,Hg)Te, the band gap becomes very small and the materials absorb across the entire ultraviolet and visible spectrum and into the near infrared. This is exploited in solar cells based on CdTe and infrared radiation detectors using cadmium mercury telluride.

TABLE B20.2 Band gaps (eV) at 300 K

	O	S	Se	Te
Zn	3.37	3.54/3.91*	2.7	2.25
Cd	2.37	2.42	1.84	1.49
Hg/Cd	2.15	2.1(α)	0.8	~0.1

* Depends on polymorph: sphalerite/wurtzite.

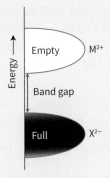

FIGURE B20.6 The simple band structure of MX salts.

phosphate, $Zn_3(PO_4)_2$, is used as a corrosion-resistant coating on metal surfaces and in dental cements.

Relatively few complex mercury oxides have been synthesized, though of note is $HgBa_2Ca_2Cu_3O_{8-x}$ ($0 \leq x \leq 0.35$) with a very high superconducting critical temperature of 133 K.

(d) Coordination complexes

KEY POINTS Tetrahedral and octahedral complexes of the M^{2+} cations are found; those of the $[Hg_2]^{2+}$ cation are linear.

Both tetrahedral and octahedral complexes of the Group 12 metals are known for the M^{2+} cations, and the lack of any ligand-field stabilization energy for the d^{10} configuration means there is no strong preference for either geometry. For instance, cadmium forms the tetrahedral $[Cd(NH_3)_4]^{2+}$ ion in dilute ammonia solutions, but the octahedral $[Cd(NH_3)_6]^{2+}$ in more concentrated solutions; mercury even forms trigonal complexes such as $[HgI_3]^-$. Zn^{2+} is borderline between hard and soft and readily forms complexes with both types of donor, whereas Cd^{2+} and Hg^{2+} cations are distinctly soft. Complexes of the $[Hg_2]^{2+}$ cation are normally linear of the type X–Hg–Hg–X.

(e) Organometallic compounds

KEY POINT The Group 12 organometallic complexes are limited in scope, but have significant uses.

Diethyl zinc was first isolated in 1848, and synthetic uses of dialkyl and diaryl zinc reagents, R_2Zn, and their derivatives, as alternatives to Grignard or oganolithium reagents, are well established. However, zinc actually exhibits a very limited organometallic chemistry: zinc organometallic complexes are never more than four-coordinate and never show π interactions. Thus η^2-alkene and η^5-cyclopentadienyl compounds are inaccessible; likewise carbonyls are unknown. Similarly, cadmium and mercury organometallics are restricted to σ-bonded alkyls and aryls. Organocadmium complexes have few interesting features or practical uses. Organomercury compounds such as dialkyl- and diarylmercury, unlike the equivalent lithium, magnesium, and zinc reagents, show remarkable water and air stability, and thus, in spite of their toxicity, they find many practical uses in the small-scale synthetic laboratory.

20.13 The 6d elements

The fourth row of the d-block comprises superheavy elements extending in nuclear mass from actinium (discussed further in Chapter 21) to copernicium. All the 6d elements have unstable nuclei that are subject to radioactive decay; with the exception of actinium, they are all synthetic and have very short half-lives. The 6d elements represent the frontier of the periodic table and their investigation is inspired not by practical application but by curiosity, as well as the desire to overcome technical challenges.

It is not obvious whether the chemical properties of these elements will agree with those predicted from periodic trends. If their properties do follow normal trends, the compounds formed should be analogues of the lighter members of the same group, but with important modifications. Relativistic effects are predicted to play a major role: destabilization and expansion of the 6d orbitals is expected to lead to species in Group 11 such as RgF_7 where the oxidation state is +7, compared with gold (directly above it in the group) that achieves a maximum oxidation state of only +5 in Au_2F_{10}. Relativistic stabilization of the 7s orbital could render elemental copernicium (below mercury in Group 12) a heavy inert gas. Almost all our chemical insight so far has stemmed from theoretical calculations, as the nuclear instability makes experimental studies extremely challenging.

Elements heavier than fermium cannot be produced by neutron capture but must be synthesized by bombarding the nucleus of a suitable precursor (the target) with a light atom (the projectile). A strategy for obtaining a brief but accurate glimpse of the compounds formed by the new element is to pass a jet of the metal atoms produced by the nuclear fusion process (and attached to aerosol particles) through a gas mixture with which reactions occur to produce the desired compound. The compound, which must be volatile (and hence electrically neutral), collides with a variable temperature detector with which it interacts, enabling its enthalpy of adsorption to be determined rapidly and compared with values for the lighter analogues from the same group, Fig. 20.10. This procedure was first adopted to produce $Sg(CO)_6$, from ^{265}Sg (half-life 8.5 s) derived from the fusion of ^{248}Cm with ^{22}Ne nuclei. Although ^{265}Sg has a shorter half-life than other nuclides (that of ^{260}Sg is 18 min) the ability to make a chemical compound and characterize it depends also on how rapidly and efficiently it is produced.

$$^{248}Cm_2O_3(s) + {}^{22}Ne \xrightarrow{fusion} {}^{265}Sg(g) + 5\,{}^1n$$

$$Sg(g) \xrightarrow{CO(g)} Sg(CO)_6(g)$$

The adsorption characteristics of the product matched those expected by extrapolation from the stable hexacarbonyl analogues of Cr, Mo, and W, thus identifying it as $Sg(CO)_6$ and suggesting that the 18-electron rule (Section 24.2) holds for the fourth row.

A similar strategy was used to prepare a volatile compound of dubnium (^{262}Db, half-life 34 s), in this case the oxychloride $DbOCl_3$ (34), an analogue of compounds of the group 5 metals Nb and Ta.

$$^{248}Cm_2O_3(s) + {}^{19}F \xrightarrow{fusion} {}^{262}Db(g) + 5\,{}^1n$$

$$Db(g) \xrightarrow{SOCl_2(g)} DbOCl_3(g)$$

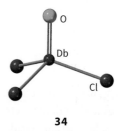

34

While the adsorption enthalpy of the product continued the expected trend for $-\Delta H_{ads}$ values, i.e. $NbOCl_3 < TaOCl_3 < DbOCl_3$, the slightly lower-than-expected binding strength of the Db compound suggests a smaller molecular dipole and thus greater covalence—a likely consequence of the larger 6d orbitals.

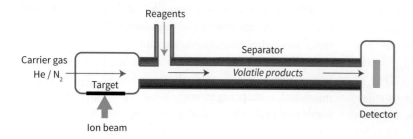

FIGURE 20.10 The principle of gas-phase production and detection of volatile compounds of the superheavy elements.

FURTHER READING

J. Emsley, *Nature's Building Blocks*. Oxford University Press (2011). An A–Z guide to the elements.

J. A. McCleverty and T. J. Meyer (eds.), *Comprehensive Coordination Chemistry II*. Elsevier (2004).

R. H. Crabtree, *The Organometallic Chemistry of the Transition Metals*. John Wiley & Sons (2009).

C. Elschenbroich, *Organometallics*. Wiley-VCH (2006).

R. J. P. Williams and R. E. M. Rickaby, *Evolution's Destiny*. RSC Publishing (2012). A stimulating book that looks at how life and the environment evolved together.

F. Neve, Chemistry of superheavy transition elements. *J. Coord. Chem.* 2022, **75**, 2287. A brief review of the pioneering research being carried out to determine the chemical properties of the fourth-row transition elements.

N. M. Chiera et al., Chemical characterization of a volatile dubnium compound, $DbOCl_3$. *Angewandte Chem. Int. Ed.* 2021, **60**, 17871. This article contains interesting insight into how the chemistry of a very unstable element is determined and compared with lighter members of the group. It complements an earlier article on the synthesis and characterization of $Sg(CO)_6$, by J. Even et al., *Science*, 2014, **345**, 1491.

EXERCISES

20.1 What is the highest group oxidation state observed for the first-row transition metals? Give an example of a group oxidation state oxo species of this metal ion. What is the highest group oxidation state observed in the second and the third row? Contrast the stability down the group.

20.2 Use the information in Resource section 3 to construct Frost diagrams for the Group 6 elements Cr, Mo, and W in acidic conditions. Use these diagrams to predict (a) which oxidation state of each element is most oxidizing and (b) whether any of the oxidation states are susceptible to disproportionation.

20.3 Sketch the following ions: (a) dichromate(VI), (b) vanadyl, (c) orthovanadate, (d) manganate(VI).

20.4 Explain why TiO_2, V_2O_5, and CrO_3 are well-known compounds, but FeO_4 and Co_2O_9 have not been prepared.

20.5 Which of the following is *not* true:

(a) The chemistry of molybdenum is more similar to that of tungsten than that of chromium.

(b) Higher oxidation states are most prevalent in the second and third rows of transition elements.

(c) Transition metals have high atomic volumes and are therefore not very dense.

(d) Enthalpies of atomization reach a maximum in the middle of a row.

20.6 Look up the electron gain enthalpies for Cu, Ag, and Au and the ionization energies of the Group 1 metals. Discuss the likely stability of compounds $M^+M'^-$, where M = Group 1 metal, M' = Group 11 metal.

20.7 Explain why isostructural HfO_2 and ZrO_2 have densities of 9.68 g cm^{-3} and 5.73 g cm^{-3}, respectively.

20.8 Use the information in Resource section 3 to construct a Frost diagram for mercury in acidic solution. Comment on the tendency for $[Hg_2]^{2+}$ to disproportionate.

20.9 Many d-metal compounds are used as pigments. Apart from colour, what properties must a compound possess to be useful as a pigment?

TUTORIAL PROBLEMS

20.1 Describe and account for the trends in the ionic radii and the stability of high oxidation states upon moving from the second to the third row of the transition series.

20.2 Discuss the benefits and the costs of using lightweight titanium alloys over the more conventional steel in (a) cars and (b) aeroplanes.

20.3 TiO_2 can be manufactured via the chloride process or the sulfate process. Outline the advantages and disadvantages of each of these processes in terms of the raw material required, the nature of the pigment produced, and the environmental impact of each process.

20.4 Iron is essential to all life forms. Assess the solubility of Fe in +2 and +3 oxidation states, and combine this with likely stable oxidation states under normal conditions to assess the bioavailability of iron. Consider how the atmosphere changed as oxygenic photosynthesis took over (Box 7.1) and comment on how this might have affected the bioavailability of iron.

20.5 Silica-supported compounds of chromium have been used as catalysts for olefin polymerization for many years. Write a review of this application of chromium chemistry. Include discussion of the Phillips and Union Carbide systems and include discussion of one published paper on the topic from the past two years.

20.6 Gold, platinum, and palladium are all known as precious metals. Review their uses in technology and other areas, and discuss why the prices of gold, platinum, and palladium have varied over time.

20.7 Discuss the historical production and use of gold nanoparticles. Account for the colour of the nanoparticles.

20.8 In their paper 'Shape control in gold nanoparticle synthesis' (*Chem. Soc. Rev.*, 2008, **37**, 1783), M. Grzelczak and co-workers discuss the synthesis of gold nanoparticles with a range of morphologies. Describe the different morphologies summarized in this paper, and outline the proposed growth mechanisms for the nanoparticles when synthesized with and without the presence of silver ions.

20.9 Discuss the side effects of protiodide, Hg_2I_2, when ingested, and the risks of overdose.

20.10 A paper was published in 2014 with the title 'Identification of an iridium-containing compound with a formal oxidation state of IX' (*Nature*, 2014, **514**, 475), and another was published in 2016 with the title 'Oxidation State 10 Exists' (*Angew. Chem. Int. Ed.*, 2016, **55**, 9004). Review these papers, and comment on the veracity of their titles.

20.11 Discuss the historical role of mercury in the production of felt, and the likely etymology of the expression 'mad as a hatter'.

20.12 Consider the 6d elements Rf–Cn. Make predictions of the key features of their chemistry were sufficient quantities ever to be isolated.

The f-block elements

21

The elements

21.1 The valence orbitals
21.2 Occurrence and recovery
21.3 Physical properties and applications

Lanthanoid chemistry

21.4 General trends
21.5 Binary ionic compounds
21.6 Ternary and complex oxides
21.7 Coordination compounds
21.8 Organometallic compounds

Actinoid chemistry

21.9 General trends
21.10 Thorium and uranium
21.11 Neptunium, plutonium, and americium
21.12 The heavy actinoids

Further reading

Exercises

Tutorial problems

The f block is a distinctive section of the periodic table comprising elements that collectively highlight many of the most important rules of atomic structure and bonding, yet individually demonstrate how these rules can be challenged and exploited. The characteristic uniformity of the lanthanoids (Ln) reflects the fact that the 4f orbitals are core-like and overlap little with orbitals on donor atoms. The lanthanoids are electropositive metals that behave in many respects like Group 2 elements, as ionization of the $6s^2$ and $5d^1$ electrons gives rise to a stable cation (3+) that interacts with other species in an almost entirely ionic manner. This striking insensitivity to the surrounding environment means that the chemistry and physics of Ln^{3+} compounds can be understood largely on the basis of inherent atomic structures and properties, as introduced in Chapter 1. Compounds of Ln(II) and Ln(IV) can be synthesized whenever the relatively favourable energy of a particular $4f^n/5d^1$ electron configuration presents an opportunity, and these atypical species are often highly reactive. The lanthanoids are widely exploited in everyday technologies, including light-emitting diodes (LEDs), medical diagnostics, and powerful magnets used in devices ranging from computers to wind turbines, an economically crucial and geopolitically sensitive resource. The 5f elements (the actinoids, An) are divided into two groups. The light (early) elements (Th–Am) resemble many d metals in displaying a rich variety of redox and coordination chemistry. The stable, collinear AnO_2^{n+} unit ($n = 1,2$) is an important feature in the chemistry of the naturally occurring U as well as the synthetic elements Np, Pu, and Am. These elements are highly radioactive and require special handling. Those having long half-lives pose long-term environmental hazards. The heavy (later) actinoids more closely resemble the lanthanoids but they have little nuclear stability and their chemistry is very difficult to study.

The two series of elements in the f block derive from the filling of the seven 4f and 5f orbitals, respectively. This occupation of f orbitals from f^1 to f^{14} corresponds to the elements cerium (Ce) to lutetium (Lu) in Period 6 and from thorium (Th) to lawrencium (Lr) in Period 7; however, given the similarity of their chemical properties, the elements lanthanum (La) and actinium (Ac) are normally included in discussion of the f block, as we do here.

The 4f elements are collectively the **lanthanoids** (formerly and still commonly 'the lanthanides') and the 5f elements are the **actinoids** (the 'actinides'). The lanthanoids are commonly referred to as the 'rare earth elements'; however, they are not particularly rare, except for promethium, which has no stable isotope. A general lanthanoid is represented by the symbol Ln and an actinoid by An. Within each row, the elements are further classified as 'light' or 'heavy' based on their atomic number. The light lanthanoids range from La (57) to promethium (61), and when referred to as rare earths they also include scandium and yttrium—two lighter elements having lanthanoid-like properties. The heavy lanthanoids are those ranging from samarium (62) to lutetium (71). The light actinides range from actinium (89) to plutonium (94), while the heavy actinoids are considered to begin with berkelium (95).

The chemical properties of the lanthanoids are quite different from those of the actinoids, and after some general introductions we discuss them separately. As we shall see, there is a striking uniformity in the properties of the lanthanoids, punctuated by interesting exceptions, whereas the actinoids display greater diversity and many form compounds akin to those of the d block.

The elements

We begin our account of the f-block elements by considering their general properties and methods for their extraction.

21.1 The valence orbitals

KEY POINTS The 4f orbitals make very little contribution to bonding: their radial distribution functions lie within the 6s and 5d orbitals from which electrons are easily removed to form the 3+ ion. Lanthanoids thus display predominantly ionic bonding. The 5f orbitals are more diffuse, and the actinoids have a richer chemistry that includes covalent bonding and a variety of oxidation states for the lighter members of the series.

To compare and contrast the properties of the lanthanoids and actinoids it is first necessary to consider how the 4f and 5f orbitals project out from the core and into their respective outer orbital regions that include s, p, and d orbitals. As we saw in Chapter 1 (Figure 1.16), the angular wavefunctions of f orbitals and hence the covalent bonds that result from their involvement should be highly directional. It is found, however, that apart from the light actinoids (Th–Pu), f orbitals make relatively little contribution to covalent bonding, and to understand this we need to consider their radial projections beyond the core. The f and d orbitals have been compared to flower petals, with the f orbital lobes likened to daisy petals, while the d orbitals are like petals of the giant poppy! The 4f orbitals of the lanthanoids lack a radial node (Sections 1.3 and 10.1) and hence are poorly shielded from the nuclear charge: they lie buried below the 5d and 6s orbitals and contract sharply and become core-like in response

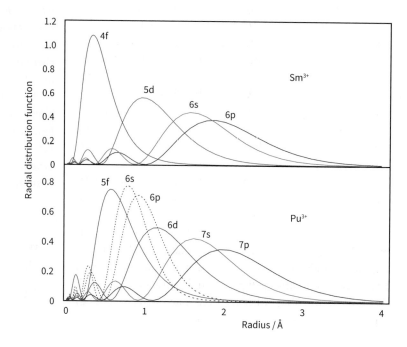

FIGURE 21.1 Comparison of the radial distribution functions for valence orbitals of Sm^{3+} and Pu^{3+}. The graph for Pu^{3+} includes the 6s and 6p orbitals that are considered to lie just within the core.

to even the smallest increase in effective nuclear charge. Fig. 21.1 shows the outer orbital radial distribution functions for a typical lanthanoid ion (Sm^{3+}) and its actinoid counterpart (Pu^{3+}). In both cases the f orbitals are much more contracted than the d orbitals.

The 5f orbitals of the actinoids have a radial node so they are more penetrating and provide better shielding from the nuclear charge (Section 10.1). Relativistic effects—a consequence of the high nuclear charge (Sections 1.7 and 10.6)—also have a strong influence. The 5f orbitals are therefore more diffuse than the 4f orbitals and more likely to engage in effective overlap with ligand orbitals. Electron–electron repulsion, which is a key feature of 4f orbitals, is less important for 5f orbitals. The 6d orbitals extend slightly further out than their 5d counterparts and, continuing the 3d–5d trend we saw in Chapter 20, we expect 6d orbitals to be even more effective in covalent bond formation. In fact, we will see that the early actinoids resemble the d metals in exhibiting a rich variety of complexes and oxidation states. In contrast, the lanthanoids behave more like Group 2 metals, in that the chemistry of the dominant oxidation state is concerned almost exclusively with nondirectional electrostatic bonding.

21.2 Occurrence and recovery

KEY POINTS The principal sources of the lanthanoids are carbonate–fluoride and phosphate minerals; the most important actinoid, uranium, is recovered from its oxide.

Other than promethium (Pm), the lanthanoids are quite common in the Earth's crust; indeed, even the 'rarest' lanthanoid, thulium, has a crustal abundance that is similar to that of iodine (Table 21.1). The uniformity between the elements is reflected in the way that they tend to occur together in geological environments, with just a blunt distinction due to size as the light Ln^{3+} have larger radii. The principal mineral source for the light lanthanoids is bastnaesite ($LnCO_3F$), while xenotime ($LnPO_4$) is the principal source of the heavy lanthanoids. Another phosphate mineral, monazite $(Ln,Th)PO_4$, is also an important source of light lanthanoids, although the need to deal with radioactive thorium makes it less attractive. Fig. 21.2 shows simplified schemes for lanthanoid extraction from these ores. The uniformity and dominance of the 3+ oxidation state for all the lanthanoids makes separation difficult, although cerium, which can be oxidized to Ce(IV), and europium, which can be reduced to Eu(II), are separable from the other lanthanoids by exploiting their redox chemistry. Isolation of the remaining lanthanoids, as their Ln^{3+} ions, is accomplished on a large scale by multistep liquid–liquid extraction in which the ions are distributed between an aqueous phase and an organic phase containing complexing agents. Ion-exchange chromatography is used to separate the individual lanthanoid ions when high purity is required. Pure and mixed lanthanoid metals are prepared by the electrolysis of molten lanthanoid halides.

Beyond lead ($Z = 82$), no element has a stable isotope, but two of the actinoids, thorium (Th, $Z = 90$) and uranium (U, $Z = 92$), have isotopes that are sufficiently long-lived that large quantities, comparable to those of Sn and I, have persisted from their formation in the supernova that preceded the formation of the Sun (Chapter 1). Uranium is among the more common elements in the Earth's crust, and is extracted in large quantities for the nuclear fuels industry. The main sources are oxide minerals, particularly uraninite UO_2 (also known as pitchblende).

The transuranium elements up to fermium (^{257}Fm) are synthesized by neutron bombardment of light actinoids:

$$^{283}_{92}U + ^{1}_{0}n \rightarrow ^{239}_{92}U \rightarrow ^{239}_{93}Np + e^- + \nu$$

$$^{239}_{93}Np \rightarrow ^{239}_{94}Pu + e^- + \nu$$

TABLE 21.1 Crustal abundance of lanthanoids (bold) compared to some other widely used metallic elements

Element	Abundance/ppm
Fe	43 200
Cr	126
Ce	**60**
Ni	56
La	**30**
Nd	**27**
Co	24
Pb	15
Pr	**6.7**
Sm	**5.3**
Gd	**4.0**
Dy	**4.0**
Er	**2.1**
Yb	**2.0**
Eu	**1.3**
Mo	1.1
W	1.0
Ho	**0.8**
Tb	**0.7**
Lu	**0.35**
Tm	**0.30**
Ag	0.07
Hg	0.04
Au	0.0025
Pt	0.0004
Rh	0.00006

(a)

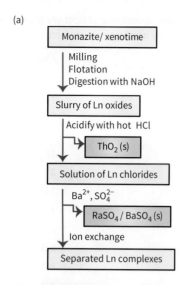

(b)

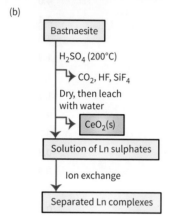

FIGURE 21.2 Summary of the steps employed in isolating the lanthanoids from their ores.

TABLE 21.2 Half-lives of the most stable actinoid isotopes

Z	Name	Symbol	Mass number	Half-life
89	actinium	Ac	227	21.8 y
90	thorium	Th	232	1.41×10^{10} y
91	protactinium	Pa	231	3.28×10^4 y
92	uranium	U	238	4.47×10^9 y
93	neptunium	Np	237	2.14×10^6 y
94	plutonium	Pu	244	8.1×10^7 y
95	americium	Am	243	7.38×10^3 y
96	curium	Cm	247	1.6×10^7 y
97	berkelium	Bk	247	1.38×10^3 y
98	californium	Cf	251	900 y
99	einsteinium	Es	252	460 d
100	fermium	Fm	257	100 d
101	mendelevium	Md	258	55 d
102	nobelium	No	259	1.0 h
103	lawrencium	Lr	260	3 min

By contrast, heavier actinoids must be produced by nuclear fusion—light-atom bombardment of light actinoids producing tiny quantities of product (often only a few very unstable atoms):

$$^{249}_{97}\text{Bk} + ^{18}_{8}\text{O} \rightarrow ^{260}_{103}\text{Lr} + ^{4}_{2}\text{He} + 3\,^{1}_{0}\text{n}$$

Characterizing these exotic elements and their compounds has been a triumph for chemical ingenuity (see also Section 20.16).

Table 21.2 gives the half-lives of the most stable actinoid isotopes. Californium, traces of which appear in uranium-rich deposits as a product of successive neutron capture and beta-decay processes, is notable as being the heaviest naturally occurring element.

21.3 Physical properties and applications

KEY POINTS Lanthanoids are reactive metals: their compounds are being increasingly applied in a wide range of technologies. The primary uses of actinoids are restricted by their radioactivity.

The lanthanoids are soft white metals having densities comparable to those of the 3d metals (6–10 g cm^{-3}). For metals, they are relatively poor conductors of both heat and electricity, with thermal and electrical conductivities 25 and 50 times less than that of copper, respectively. The metals react with steam and dilute acids but are somewhat passivated by an oxide coating. The majority of the metals adopt the hexagonal cubic close-packed structure type, though cubic close-packed forms are also known for most of the elements, particularly under high pressure.

The lanthanoids and their compounds are important in a wide range of applications, many of which lie at the 'cutting edge' of technologies and depend on their optical and magnetic properties (Sections 22.6 and 22.10).

Mixtures of light lanthanoids, particularly rich in lanthanum and cerium, are referred to as *mischmetal* and are used in steel-making to remove impurities such as oxygen, hydrogen, sulfur, and arsenic, which decrease the mechanical strength and ductility of steel. Lanthanoid-based magnetic materials underpin numerous technologies (Section 30.6). Neodymium iron boride, Nd$_2$Fe$_{14}$B, is widely used as a permanent magnet, its magnetic strength being more than 10 times greater than Fe-based magnets. Samarium cobalt magnets SmCo$_5$ and Sm$_2$Co$_{17}$ have comparable performance to Nd$_2$Fe$_{14}$B but with improved resistance to corrosion. Applications of these high-strength magnetic materials include headphones, microphones, magnetic switches, wind turbines, car energy generation, and components for the guidance of particle beams.

The phosphors used for displays and lighting are based on europium and terbium ions that emit sharply defined colours: red from Eu^{3+}, blue from Eu^{2+}, and green from Tb^{3+} (Section 22.6).

Neodymium (as Nd^{3+}), samarium (as Sm^{3+}), and holmium (as Ho^{3+}) are used in solid-state lasers. Complexes of Gd^{3+} with elaborate and often ingeniously designed organic ligands are used in the medical sciences as tissue-selective magnetic resonance imaging (MRI) agents (Section 32.9).

Numerous exploitations of lanthanoid luminescence (Section 22.6) make use of the fact that emissive intensity is greatly increased and controlled if the ligand incorporates a functionality (an 'antenna') that is a strong absorber of light or can be excited electrochemically. Lanthanoid complexes are also finding important applications as catalysts for a range of organic reactions, and a Ln^{3+} ion has even been identified at the active site of an enzyme catalysing the oxidation of alcohols.

The density of the actinoids increases from $10.1\,g\,cm^{-3}$ for actinium to $20.4\,g\,cm^{-3}$ for neptunium, before decreasing for the remainder of the series. Plutonium metal has at least six phases at atmospheric pressure with density differences of more than 20%. Many of the physical and chemical properties of the actinoids are unknown because only minuscule quantities have ever been isolated: in addition to being radioactive, many actinoids are known to be chemically toxic and they are all treated as hazardous. Their primary peaceful use is in nuclear reactors for electricity generation, but trace amounts of actinoids have long been applied in various everyday technologies, examples being the use of uranium in certain glasses and americium in smoke detectors. They are also used as neutron sources and as precursors for the synthesis (by fusion with light atoms) of exotic, superheavy transuranium and 6d elements (Section 20.13).

Lanthanoid chemistry

The lanthanoids are all electropositive metals with a remarkable uniformity of chemical properties. The significant difference between two lanthanoids is often only their size, and the ability to choose a lanthanoid of a particular size allows the 'tuning' of the properties of their compounds. For example, the magnetic and electronic properties of a material often depend on the exact separation of the atoms present and degree of overlap of various atomic orbitals. By choosing a lanthanoid of an appropriate size this separation can be controlled, with consequences for its electrical conductivity or magnetic ordering temperature.

21.4 General trends

KEY POINTS The lanthanoids are highly electropositive metals and most commonly occur in their compounds as Ln(III). Other oxidation states are stabilized mainly when an empty, half-filled, or full f subshell is produced.

Lanthanoids favour the oxidation state Ln(III) with a uniformity that is unprecedented in the periodic table, and in many other ways their chemistry resembles that of the Group 2 elements. All the metals evolve H_2 from dilute acid solutions. A Ln^{3+} ion is a hard Lewis acid, as indicated by its strong preference for F^- or ligands having O atoms as donors, and its common occurrence with carbonate or phosphate in minerals.

The prevalence of the 3+ oxidation state can be traced to the core-like nature of the 4f orbitals which are relatively deeply buried inside the atom and interact only weakly with the orbitals of neighbouring atoms. Upon crossing the row, increases in atomic number are largely just a matter of increasingly populating the stable 4f subshell, thus producing little change in chemical properties. Our starting point for understanding the chemistry of lanthanoids is a brief description of the free atom or ion, from which we will see that the subtle differences between each lanthanoid represent an interesting balance between the energies of ionization, atomization, and bond formation, the latter being little influenced by ligand-field effects except in special cases. We will now see how these properties arise.

(a) Electronic structures and ionization energies

Electron configurations of lanthanoid atoms and corresponding Ln^{3+} ions are given in Table 21.3. In the atomic

TABLE 21.3 Atomic properties of lanthanoids

Z	Name	Symbol	Electron configuration	
			M	M^{3+}
57	lanthanum	La	$[Xe]5d^16s^2$	$[Xe]$
58	cerium	Ce	$[Xe]4f^15d^16s^2$	$[Xe]4f^1$
59	praseodymium	Pr	$[Xe]4f^36s^2$	$[Xe]4f^2$
60	neodymium	Nd	$[Xe]4f^46s^2$	$[Xe]4f^3$
61	promethium	Pm	$[Xe]4f^56s^2$	$[Xe]4f^4$
62	samarium	Sm	$[Xe]4f^66s^2$	$[Xe]4f^5$
63	europium	Eu	$[Xe]4f^76s^2$	$[Xe]4f^6$
64	gadolinium	Gd	$[Xe]4f^75d^16s^2$	$[Xe]4f^7$
65	terbium	Tb	$[Xe]4f^96s^2$	$[Xe]4f^8$
66	dysprosium	Dy	$[Xe]4f^{10}6s^2$	$[Xe]4f^9$
67	holmium	Ho	$[Xe]4f^{11}6s^2$	$[Xe]4f^{10}$
68	erbium	Er	$[Xe]4f^{12}6s^2$	$[Xe]4f^{11}$
69	thulium	Tm	$[Xe]4f^{13}6s^2$	$[Xe]4f^{12}$
70	ytterbium	Yb	$[Xe]4f^{14}6s^2$	$[Xe]4f^{13}$
71	lutetium	Lu	$[Xe]4f^{14}5d^16s^2$	$[Xe]4f^{14}$

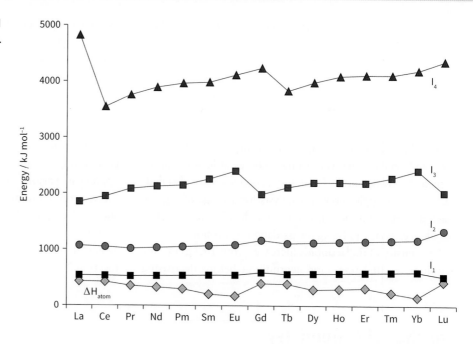

FIGURE 21.3 Variation of ionization and atomization energies for the lanthanoids.

state, all lanthanoids have the configuration [Xe]4f^{n}6s^2 except Ce, Gd, and Lu where a 5d orbital is singly occupied. As we saw in Fig. 21.1, the 4f orbitals have no radial node, consequences of which are that 4f electrons are poorly screened from the nuclear charge and electron–electron repulsions are significant (Section 10.1). Once the two valence 6s electrons and one further electron (from either the 4f or 5d orbital) have been removed, the remaining 4f electrons are held tightly by the nucleus and do not extend beyond the xenon-like core.

Fig. 21.3 shows how ionization energies vary across the lanthanoids. As a rough guide, I_4 is approximately the same as the sum of the first three ionization energies ($I_1 + I_2 + I_3$), largely explaining the scarcity of compounds of Ln(IV). The increased nuclear attraction across the series leads to a very gradual increase in ionization energies that is modulated by changes in electronic structure and total orbital angular momentum L (we discuss ground-state terms in Section 22.6). The trend in I_3 is interrupted by four discontinuities, the two largest of which occur between Eu and Gd and between Yb and Lu. Ionization of Eu^{2+} ([Xe]4f^7) incurs a large loss of exchange energy (Section 1.7), whereas ionization of Gd^{2+} ([Xe]4f^{7}5d^1) involves a less tightly bound d-electron. An analogous argument applies for Yb^{2+} ([Xe]4f^{14}) compared to Lu^{2+} ([Xe]4f^{14}5d^1). The two other discontinuities are less marked and occur across the triads Pr–Nd–Pm (the quarter-shell effect) and Dy–Ho–Er (the three-quarter shell effect). The 'flattening off' effects are due to respective changes (loss, no change, gain) in orbital angular momentum (L) upon ionization, which will be outlined later in Section 22.6. The trend for I_4 resembles I_3 except for the shift to the next highest atomic number.

(b) Atomization energies

The variation of atomization enthalpies $\Delta_aH^\ominus$ across the lanthanoids is almost the mirror image of I_3. Metallic bonding barely involves the core-like 4f electrons but varies with occupancy of the 5d orbitals that overlap to form a band containing delocalized ('itinerant') electrons (Section 2.11). When atoms condense to form the metallic state it is energetically favourable, in most cases, to promote a 4f electron into the 5d orbital. Such promotion is not required for Gd ([Xe]4f^{7}5d^{1}6s^2) but incurs a greater energy cost for Eu ([Xe]4f^{7}6s^2)—each atom containing a stable, half-filled 4f shell. As a consequence, we see an increase in $\Delta_aH^\ominus$ between Eu and Gd.

(c) Radii

The metallic radii (Fig. 21.4) decrease smoothly from La to Lu apart from two exceptions, at Eu and Yb. These exceptions are due to those metals retaining their 4f^7 and 4f^{14} electron configurations (Table 21.3) in preference to promoting an electron into the band-forming 5d orbital which (as just mentioned) would achieve stronger metallic bonding and thus a smaller M–M distance. The metallic radii thus correspond to electron configurations [Xe]4f^7 and [Xe]4f^{14} (corresponding to Ln^{2+} (e$^-$)$_2$) for Eu and Yb, respectively, and [Xe]4f^{n}5d^1 (Ln^{3+} (e$^-$)$_3$) for the other lanthanoids, where the subscript 'n' in (e$^-$)$_n$ is the number of itinerant s and d electrons formally occupying the band (but normally omitted where $n = 1$).

All Ln^{3+} ions have the electron configuration [Xe]4f^n and their radii (based on typical eight-fold coordination) contract steadily from 116 pm for La^{3+} to 98 pm for Lu^{3+}. As we mentioned earlier (Section 10.5), the lanthanoid contraction has important implications for the early 4d and 5d metals as it results in their radii and chemical properties being very

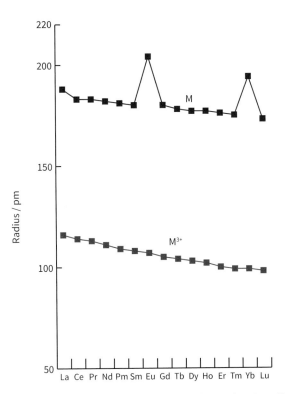

FIGURE 21.4 Variation of the metallic radii of Ln and ionic radii of Ln^{3+}.

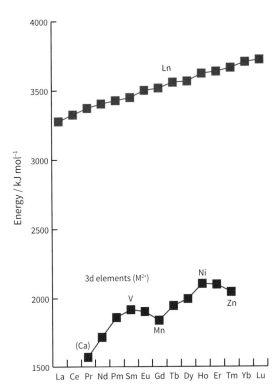

FIGURE 21.5 Enthalpies of hydration of Ln^{3+} ions and comparison with M^{2+} cations of the 3d metals.

similar despite an increase in principal quantum number; consequently, the pairs of elements Zr, Hf, and Nb, Ta, show almost identical properties. The smooth decrease in ionic radius is not observed elsewhere in the periodic table: it is attributed mainly to the increase in Z_{eff} as electrons are added to the poorly shielding 4f subshell, although detailed calculations indicate that relativistic effects also make a contractive contribution. Unlike the d-block elements, ligand-field effects have very little influence, as we discuss next.

(d) Ligand-field effects

A consequence of the f orbitals being buried is that a Ln^{3+} ion has no frontier orbitals with directional preference; consequently, ligand-field effects are sufficiently small as to ensure thermal population of all states arising from the free ion term (see Section 22.6) and have little significant influence on the chemistry. The weakness of ligand-field effects is very evident from Fig. 21.5, which shows that the enthalpies of hydration for Ln^{3+} ions display a smooth increase compared to the 'double hump' plot displayed by the 3d M^{2+} ions.

Size alone controls the coordination numbers for $[Ln(OH_2)_n]^{3+}$ in aqueous solution: these start at 9 for the early lanthanoids, which adopt a tricapped trigonal-prismatic geometry with idealized D_{3h} symmetry (**1**), and decrease to 7–8 for the later, smaller members of the series by successive loss of capping H_2O (**2**, **3**) or reorganization into a square antiprism (**4**). The aqua ions are highly labile (Fig. 23.1), as expected for large cations lacking much ligand-field stabilization, and weakly acidic, with pK values ranging from 8.5 for La^{3+}(aq) to 7.6 for Lu^{3+}(aq).

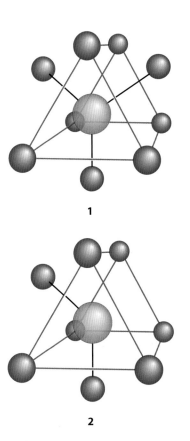

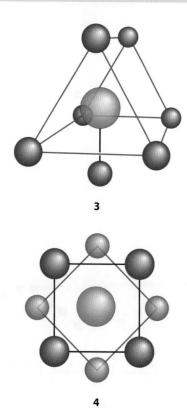

3

4

TABLE 21.4 Standard potentials, ionic radii, and oxidation numbers (O.N.) for the lanthanoids

	E (Ln^{3+}/Ln)	r (Ln^{3+})/pm	O.N.†
La	−2.38	116	**3**,
Ce	−2.34	114	**3**, 4
Pr	−2.35	113	**3**, 4
Nd	−2.32	111	2, **3**
Pm	−2.29	109	**3**
Sm	−2.30	108	2, **3**
Eu	−1.99	107	2, **3**
Gd	−2.28	105	**3**
Tb	−2.31	104	**3**, 4
Dy	−2.29	103	2, **3**
Ho	−2.33	102	**3**
Er	−2.32	100	**3**
Tm	−2.32	99	2, **3**
Yb	−2.22	99	2, **3**
Lu	−2.30	98	**3**

† Principal oxidation numbers in bold.

(e) Standard potentials and redox chemistry

The dominance of oxidation state Ln(III) throughout the 4f series is mainly due to the comparative ease with which the cumulative ionization energies ($I_1 + I_2 + I_3$) to form Ln^{3+} (Fig. 21.3) are compensated by strong hydration or lattice energies. Consequently, Ln^{3+} is favoured over Ln^{2+}, whereas I_4 usually presents an uncompensatable cost for forming Ln^{4+}. The 18% decrease in ionic radius from La^{3+} to Lu^{3+} leads to an increase in the hydration enthalpy across the series which counterbalances the increase in ($I_1 + I_2 + I_3$) and mirrored trend in atomization energy. As a result, the standard potentials (Ln^{3+}/Ln) of the lanthanoids are all very similar and are close to that of Mg^{2+}/Mg (Table 21.4), with $E^{\ominus}$(La^{3+}/La) = −2.38 V being almost identical to $E^{\ominus}$(Lu^{3+}/Lu) = −2.30 V at the other end of the series.

Oxidation states apart from Ln(III) pose important and interesting challenges in lanthanoid chemistry. A guide to the trends in frontier f- and d-orbital energies across the lanthanoid series is shown in Fig. 21.6.

The energies are derived from spectroscopic measurements performed on lanthanoid ions doped into YPO$_4$: the higher the energy, the greater the ease of ionization (or promotion) from that state; the lower the energy, the more stable that oxidation state should be. Focusing only on the 4f energies, Ce^{3+} and Tb^{3+} are most susceptible to oxidation to the 4+ state, whereas Eu, Sm, and Yb are more likely to form compounds in the 2+ state. The peak (2+) and trough (3+) for Gd^{3+} show that it is least susceptible to either oxidation or reduction. The atypical oxidation states are associated with empty (f^0), half-filled (f^7), and full (f^{14}) subshells: this is a consequence of increasing Z_{eff} across the series (hence I_4 is lowest for Ce) along with the exchange energy that is maximized at f^7 and again at f^{14} (Fig. 21.3).

In terms of their aqueous chemistry, the elements Ce and Eu exhibit by far the most important and useful redox properties. Thus, Ce^{3+} (f^1) can be oxidized to Ce^{4+} (f^0), which is a strong and useful oxidizing agent, and aqueous solutions of the Eu^{2+} ion are sufficiently stable that it is a convenient one-electron reducing agent. Metallic Eu and Yb react with liquid NH$_3$ to give blue solutions containing the solvated electron along with Eu^{2+} and Yb^{2+}, analogous to the chemistry of Group 2 elements. The aqueous ions Sm^{2+} and Yb^{2+} rapidly reduce water to hydrogen.

An increasing number of molecular compounds are being synthesized that contain Ln(IV) and particularly Ln(II). The former is largely limited to Ce, whereas examples of Ln(II) now include, in addition to Eu^{2+} and Yb^{2+}, many complexes of Sm^{2+}, Tm^{2+}, Dy^{2+}, and Nd^{2+} that can be prepared in nonreducible solvents such as ethers. The field of Ln(II) complexes has expanded rapidly in recent years, particularly in regard to organometallic chemistry.

Many more examples of '2+' and '4+' lanthanoid ions occur in the solid state. The ease of forming Ln(IV) compounds is related to I_4. Apart from Ce(IV), notable examples include Pr(IV) and Tb(IV), and the oxides of these elements formed in air, Pr$_6$O$_{11}$ and Tb$_4$O$_7$, contain mixtures of Ln(III) and Ln(IV). Under very strongly oxidizing conditions Dy(IV) and Nd(IV) can also be obtained. Many Ln(II) binary compounds, including the sulfides, are electronic conductors because a 5d electron is released into the conduction band.

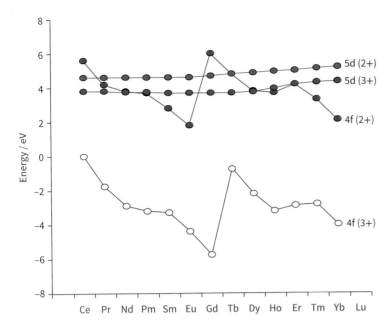

FIGURE 21.6 Relative energies (vs Ce^{3+}) of the 4f and 5d orbitals of a lanthanoid ion doped into YPO_4.

21.5 Binary ionic compounds

KEY POINTS The structures of binary lanthanoid(III) compounds are determined by the size of the Ln(III) cation: oxides, halides, hydrides, and nitrides are all known. Lanthanoid(IV) compounds are restricted to oxides and fluorides of Ce, Pr, and Tb. The properties of lanthanoid(II) solids LnO and LnX_2 (X = H, Cl, Br, I) across the row reflect the difference in energies between 4f and 5d orbitals, with increasing stabilization of 4f vs 5d orbitals favouring salt-like rather than metallic behaviour.

Lanthanoid(III) ions have radii that vary between 116 and 98 pm (Table 21.4); by comparison, Fe^{3+} has an ionic radius of 64 pm. Thus, the volume occupied by a Ln^{3+} ion is typically four to five times that occupied by a typical 3d-metal ion. Unlike the 3d metals, which rarely exceed a coordination number of 6 (with 4 being common too), compounds of lanthanoids often have high coordination numbers, typically between 6 and 12, and a wide variety of coordination environments.

All the lanthanoids react with O_2 at high temperatures to produce oxides: in most cases these are the sesquioxides Ln_2O_3 but lanthanoids having reasonably low values of I_4 form higher oxides, specifically the dioxide CeO_2 and non-stoichiometric oxides Pr_6O_{11} and Tb_4O_7. The latter react further with O_2 at high pressure to produce PrO_2 and TbO_2.

$$2\,Ln(s) + 3\,O_2(g) \rightarrow Ln_2O_3(s)$$
$$Ce(s) + O_2(g) \rightarrow CeO_2(s)$$

Cerium dioxide is widely used in industry as a catalyst and catalyst support: it also has promising applications in solar hydrogen production by a chemical loop process (Section 6.17), since it loses O_2 at very high temperature (as generated in a solar furnace) and reacts with water upon cooling, whereupon H_2 is produced and CeO_2 is regenerated (Box 11.4). All the dioxides adopt the fluorite structure, as expected from radius-ratio rules (Section 4.10), whereas the sesquioxides have more complex structures in which the average coordination number of the Ln^{3+} ions is typically 7.

Three main structure types termed A-, B-, C-Ln_2O_3 are known, and many of the oxides are polymorphic with transitions between the structures occurring as the temperature is changed. The coordination geometries are determined by the radius of the lanthanoid ion, with the average cation coordination number in the structures decreasing with decreasing ionic radius: for example, the La^{3+} ion in La_2O_3 has coordination number 7, whereas the Lu^{3+} ion in Lu_2O_3 has coordination number 6.

Phases of composition $Ln_2S_nO_{3-n}$ can be obtained by subjecting mixtures of Ln_2O_3 and Ln_2S_3 to intense heating (> 1500°C). Due to their intense red–orange–yellow colours, they have been studied as possible pigments to replace the toxic CdS and CdSe.

All the lanthanoids form nitrides of composition LnN that adopt the rock-salt structure with alternating Ln^{3+} and N^{3-} ions. These compounds have important technological applications due to their electronic, optical, and magnetic properties. Bulk nitrides are synthesized by reaction of anhydrous trihalides with lithium nitride under vacuum at 400°C.

$$LnCl_3(s) + Li_3N(s) \rightarrow LnN(s) + 3\,LiCl(s)$$

Lanthanide nitrides deposited as thin wafer-like films on a substrate are important in optoelectronics. Thin films of GdN on Si are formed by decomposition of a nitrogen-rich tris(guanidinate)Gd complex using a technique called metal–organic chemical vapour deposition (MOCVD), described in Section 28.4. Nanostructured LnN foams are formed by a similar procedure.

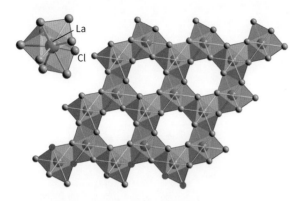

FIGURE 21.7 The structure of $LaCl_3$ shown as vertex-linked $LaCl_9$-capped antiprisms: a single unit is shown in the inset.

The lanthanoids generally react directly with halogens to form trihalides, LnX_3, which have complex structural characteristics as a result of the high coordination numbers for these large ions. For example, in LaF_3 the La^{3+} ion is in an irregular 11-coordinate environment. and in $LaCl_3$ it is in a 9-coordinate tricapped trigonal-prismatic environment (Fig. 21.7).

The trihalides of the smaller lanthanoids at the end of the series have different structure types, with distorted tricapped trigonal prisms for LnF_3 (5) and layer structures based on 6-coordinate cubic close packing for $LnCl_3$. Apart from oxygen, fluoride is the only element able to stabilize Ln(IV) compounds. Cerium reacts with F_2 at room temperature to produce CeF_4, which crystallizes with a structure formed from vertex-sharing CeF_8 polyhedra (Fig. 21.8). In contrast, PrF_4 and TbF_4 are highly reactive and are synthesized only under extreme conditions, by reacting Pr_6O_{11} or Tb_4O_7 with F_2 under UV irradiation in liquid HF for several days.

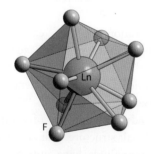

5 LnF_3

Binary compounds of Ln(II) are known for O, S, H, and the halides I, Br, and Cl. The properties of these compounds depend on how readily an electron is promoted from the 4f orbital to the diffuse 5d orbital that is able to form a broad conduction band (Section 2.11). The compounds are commonly classified as being 'metallic', in which the cation is usually formulated as (Ln^{3+}, e^-) as explained in Section 21.4(c), or 'salt-like', in which the cation is formulated as Ln^{2+}. Metallic Ln(II) compounds are electronically conducting and dark in colour, whereas salt-like Ln(II) compounds are insulators and pale-coloured.

Monoxides, LnO, having the rock-salt structure are formed by comproportionation of the respective Ln_2O_3 with a stoichiometric amount of powdered Ln metal, under a pressurized inert-gas atmosphere:

$$Ln_2O_3(s) + Ln(s) \rightarrow 3LnO(s)$$

Among the best-known examples, PrO and NdO are metallic oxides, that is, $Ln^{3+}(O^{2-})(e^-)$, whereas EuO and YbO are pale, insulating solids (Ln^{2+}, O^{2-}).

EXAMPLE 21.1 Explaining why some Ln(II) oxides are metallic whereas others are insulators

Explain how metallic or salt-like properties may be predicted for various oxides LnO, using PrO, NdO, EuO, and YbO as examples.

Answer Metallic character arises because an itinerant electron is present in a band that is derived from overlap of the diffuse 5d orbitals (Section 2.11): electrons in the contracted 4f orbitals remain localized. Whether the 5d orbital is easily occupied or otherwise depends on its energy relative to the 4f orbital in Ln atoms that are formally in the 2+ oxidation state. Using Fig. 21.6, which refers to Ln^{2+} or Ln^{3+} ions doped into YPO_4, we see that the 4f energies for isolated Eu^{2+} and Yb^{2+} sites are substantially lower than the corresponding 5d energies, so EuO and YbO should be salt-like. To be metallic, the energy of the 5d orbital in the 2+ oxidation state should not be too high relative to that of the 4f orbital, so that 5d band occupancy is favourable: PrO and NdO are predicted to be metallic.

Self-test 21.1 Predict the stability and properties of GdO.

Sulfides of stoichiometry LnS having the rock-salt structure may be obtained by direct reaction of the elements at 1000°C. Except for SmS, EuS, and TmS, the sulfides are electronically conducting and formulated as $Ln^{3+}(S^{2-})(e^-)$: similar compounds are formed with selenium and tellurium.

Dihalides exist for most of the lanthanoids and are generally formed by comproportionation of LnX_3 and Ln powder at high temperatures, in an inert atmosphere and under pressure:

$$2LnX_3(s) + Ln(s) \rightarrow 3LnX_2(s)$$

The dihalides of Eu, Yb, Sm, Nd, Tm, and Dy are salt-like solids, whereas those of La, Ce, Pr, Gd, and Y are dark metallic solids formulated as $Ln^{3+}(X^-)_2(e^-)$. The diiodides are useful starting materials for Ln(II) complexes, particularly

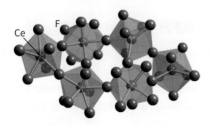

FIGURE 21.8 The structure of CeF_4 contains vertex-sharing CeF_8 antiprisms.

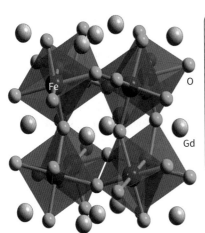
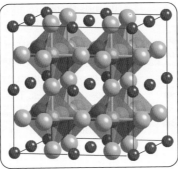

FIGURE 21.9 The GdFeO$_3$ structure type, with the FeO$_6$ octahedra outlined. Inset shows how the GdFeO$_3$ structure is related to the perovskite structure.

SmI$_2$, which is used as a reagent in organic synthesis: SmI$_2$ is conveniently prepared on the laboratory scale by reacting Sm powder with diiodoethane.

$$Sm(s) + ICH_2CH_2I(l) \rightarrow SmI_2(s) + C_2H_4(g)$$

All the metals react with H$_2$ to give binary hydrides that vary in stoichiometry between LnH$_2$ and LnH$_3$. The dihydrides adopt the fluorite structure (Section 4.9) based on cubic close-packed Ln ions with hydride ions in all the tetrahedral holes: all are dark, metallic materials apart from EuH$_2$ and YbH$_2$, which are salt-like. Further hydrogenation produces nonstoichiometric compounds LnH$_{3-x}$, although the smaller lanthanoids (for instance, Dy, Yb, and Lu) form stoichiometric trihydrides, LnH$_3$. Complex metal hydrides such as LaNi$_5$H$_6$ have been studied intensively as reversible hydrogen-storage materials (Box 11.5 and Section 30.8). though they are too heavy for transport purposes.

Three different lanthanoid carbide stoichiometries are known: M$_3$C, M$_2$C$_3$, and MC$_2$. The M$_3$C phases that are formed with the heavier lanthanoids contain isolated, interstitial C atoms and are hydrolysed by water to produce methane. The M$_2$C$_3$ phases that form for the lighter lanthanoids La–Ho contain the dicarbide anion C$_2^{2-}$, as is found in CaC$_2$ (Section 4.9). Some MC$_2$ phases are metallic and can be expressed as Ln^{3+}(C$_2^{2-}$, e$^-$) with the d-electron entering the conduction band; they react with water to form ethyne. The lanthanoid nickel borocarbides, LnNi$_2$B$_2$C, have structures containing alternating layers of the stoichiometries LnC and Ni$_2$B$_2$. These borocarbides are superconductors at low temperature: the transition temperature for LuNi$_2$B$_2$C, for instance, is 16 K.

21.6 Ternary and complex oxides

KEY POINT Lanthanoid ions are often found in perovskites and garnets, where the ability to change the size of the ion allows the properties of the materials to be modified.

The lanthanoids are a good source of large, stable tripositive cations having a reasonable range of ionic radii without demanding ligand-field preferences. As a result, they can occupy one or more of the cation positions in ternary and more complex oxides. For example, perovskites of the type ABO$_3$ (Section 4.9) can readily be prepared with La on the larger A cation site; an example is LaFeO$_3$. Indeed, some distorted structure types are named after lanthanoids; an example is the structural type GdFeO$_3$ (Fig. 21.9), which has vertex-linked FeO$_6$ octahedra around the Gd^{3+} ion (as in the parent perovskite structure, Fig. 4.43) but with the octahedra being tilted relative to each other. This tilting allows better coordination to the central Gd^{3+} ion.

The ability to change the size of the B^{3+} ion in a series of compounds LnBO$_3$ allows the physical properties of the complex oxide to be modified in a controlled manner. For example, in the series of compounds LnNiO$_3$ for Ln = Pr to Eu, the insulator–metallic transition temperature T_{IM} increases with decreasing lanthanoid ionic radius (Table 21.5).

The perovskite unit cell is a structural building block often found in more complex oxide structures, and lanthanoids are frequently used in such materials. Well-known examples are the original high-temperature superconducting cuprate, La$_{1.8}$Ba$_{0.2}$CuO$_4$ and the family of '123' complex oxides, LnBa$_2$Cu$_3$O$_7$, which become superconducting below 93 K (Chapter 25). The best-known of these high-temperature superconductors is the d-block (yttrium) compound YBa$_2$Cu$_3$O$_7$, but they are also found for all the lanthanoids. Other materials where the choice of lanthanoid is crucial in obtaining the required property include the complex

TABLE 21.5 Physical properties of some ternary oxides (LnNiO$_3$)

	PrNiO$_3$	NdNiO$_3$	EuNiO$_3$
r(Ln^{3+})/pm	113	111	107
T_{IM}/K	135	200	480

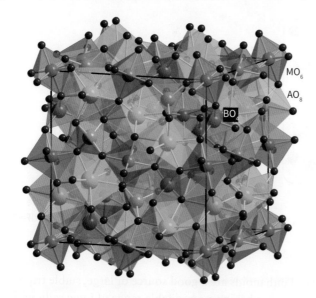

FIGURE 21.10 The garnet structure, shown as linked AO_8, BO_4, and MO_6 polyhedra. The 8-coordinate A sites often occupied by yttrium can be occupied by other lanthanoids.

manganites $Ln_{1-x}Sr_xMnO_3$ which exhibit resistance effects that depend strongly on the applied magnetic field and temperature; the optimized properties are found when Ln = Pr.

The spinel structure (Fig. 4.45) has only small tetrahedral and octahedral holes in the close-packed O^{2-} ion array and thus cannot accommodate bulky lanthanoid ions. However, the garnet structure adopted by materials of stoichiometry $M_3M'_2(XO_4)_3$, where M and M' are normally di- and tripositive cations and X includes Si, Al, Ga, and Ge, has 8-coordinate sites that can be occupied by lanthanoid ions (Fig. 21.10). Aside from yttrium aluminium garnet, which is the host material for neodymium ions in the laser material Nd:YAG, the material known as yttrium iron garnet (YIG) is an important ferrimagnet used in microwave and optical communication devices.

21.7 Coordination compounds

KEY POINTS An abundance of lanthanoid (III) complexes are formed with anionic polydentate ligands containing oxygen-atom donors suited for bonding that is electrostatic in nature. Coordination numbers usually exceed six and the ligands adopt geometries that minimize inter-ligand repulsions. Complexes of Ln(II) are highly reducing. Lanthanoid(IV) complexes are represented by a number of complexes of cerium.

(a) Complexes of Ln(III)

Without the means for strong overlap of orbitals, the bonds formed between Ln^{3+} ions and ligands are electrostatic in nature, and stable complexes are only achieved with polydentate anionic chelating ligands. The spatially buried f electrons have no significant stereochemical influence, and ligands therefore adopt positions that minimize inter-ligand repulsions. Polydentate ligands must satisfy their own stereochemical constraints, much as for the s-block ions and Al^{3+} complexes. As with the aquo ions $[Ln(OH_2)_n]^{3+}$ (**1–4**) the coordination numbers and structures of complexes vary across the series. For example, the small ytterbium cation, Yb^{3+}, forms the 7-coordinate complex $[Yb(acac)_3(OH_2)]$, and the larger La^{3+} is 8-coordinate in $[La(acac)_3(OH_2)_2]$. The structures of these two complexes are (approximately) a monocapped trigonal prism (**6**) and a square antiprism (**7**), respectively. A square antiprism imposes less ligand-ligand repulsion than cubic geometry.

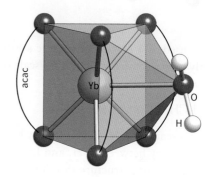

6 Monocapped trigonal prism (Yb^{3+}, a small cation)

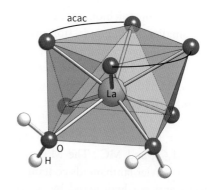

7 Square antiprism (Ln^{3+}, large cations)

Many lanthanoid complexes are formed with crown ether and β-diketonate ligands. The partially fluorinated β-diketonate ligand $[CF_3COCHCOCF_3]^-$, nicknamed 'fod', produces complexes with Ln^{3+} that are volatile and soluble in organic solvents. On account of their volatility, these complexes are used as precursors for the synthesis of lanthanoid-containing superconductors by vapour deposition (Section 28.4).

Charged ligands generally have the highest affinity for the smallest Ln^{3+} ions, and the resulting increase in formation constants from large, lighter Ln^{3+} (on the left of the series) to small, heavier Ln^{3+} (on the right of the series) provides a convenient method for the chromatographic purification of these ions (Fig. 21.11). In the early days of lanthanoid chemistry, before ion-exchange chromatography was

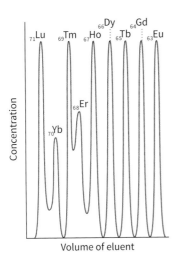

FIGURE 21.11 Elution of heavy Ln(III) ions from a cation-exchange column using ammonium 2-hydroxyisobutyrate as the eluent. Note that the higher atomic number lanthanoids elute first because they have smaller radii and are more strongly complexed by the eluent.

developed, tedious repetitive crystallizations were required to separate the elements.

The coordination chemistry of lanthanoids is a flourishing field of research that owes a great deal to their applications in bioimaging, either as optical or magnetic tracers. The trick here is to attach a target functionality, such as an antenna or bioreceptive group, to a polydentate/macrocyclic ligand that forms a strong, nonlabile complex with the lanthanoid. Many robust complexes are based upon a macrocyclic scaffold, an example being one in which fluorescein is attached to cyclen (8). Further examples will be mentioned in Section 32.9 where we deal with bioimaging in more detail. Unlike aqua ions which exchange water molecules on the ns timescale, many macrocyclic complexes have exchange half-lives that are measured in years.

8 Cyclen

Lanthanoid 3+ ions can be used to construct three-dimensional metal–organic frameworks, abbreviated MOFs

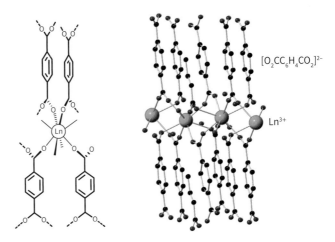

FIGURE 21.12 Part of a metal–organic framework based upon eight-fold coordination of Ln^{3+} by terephthalate (benzenedicarboxylate anion (bdc)).

(Section 29.4). The local structure of one such example, formed with terephthalate (benzene-1,4-dicarboxylic acid, bdc), has eight-fold coordination, with each Ln^{3+} sharing a carboxylate group with another Ln^{3+} (Fig. 21.12). The use of Ln^{3+} ions gives rise to printable materials (inks) with characteristic luminescent properties that have potential future applications for displaying encrypted information.

(b) Complexes of Ln(II) and Ln(IV)

The standard potentials for the Ln^{3+}/Ln^{2+} redox couple given in Table 21.6 provide a good practical guide for predicting the stability of Ln(II) complexes. Europium is the only lanthanoid showing an extensive aqueous chemistry in the 2+ oxidation state—a property that originates in its relatively high I_3 value (Fig. 21.3) attributed to the relative stability of the $4f^7$ configuration (Fig. 21.6). Indeed, many Eu(II) complexes can be conveniently prepared in water by adding a polyanionic ligand to a solution of Eu^{2+} or by electrochemical reduction of the Eu(III) complex. Complexes of Eu(II) with EGTA (etheneglycol-bis(2-aminoethyl)-N,N,N',N'-tetraacetate(4^-)) or DTPA (diethylenetriamine-N,N,N',N''-pentaacetate(5^-)) are powerful one-electron reducing agents for use in aqueous solution; they have much more negative reduction potentials than $Eu^{2+}(aq)$ due to the preference of the polyanionic ligands for the 3+ oxidation state, but they are slow to evolve H_2 and persist for long periods.

TABLE 21.6 Standard potentials/V for the Ln^{3+}/Ln^{2+} couple in H_2O

Eu	−0.35
Yb	−1.15
Sm	−1.55
Tm	−2.3
Dy	−2.5
Nd	−2.6

Aside from organometallic compounds, the most important coordination complexes of other Ln(II) are those synthesized with nonreducible chelating ethers as solvent. Starting from LnI_2 a variety of highly reactive Ln(II) ethoxy complexes are formed, the stability order Sm > Tm > Dy tracking the trend in standard potentials given in Table 21.6. Samarium diiodide, SmI_2, is commercially available as a deep blue tetrahydrofuran solution containing species formulated as $[SmI_2(thf)_n]$ (**9**) in which the iodide ligands occupy axial positions. Dark green crystals of $[TmI_2(dme)_3]$ (**10**) are produced by comproportionation of TmI_3 and powdered Tm in dimethoxyethane (dme):

$$2\,TmI_3 + Tm + 9\,dme \rightarrow 3[TmI_2(dme)_3]$$

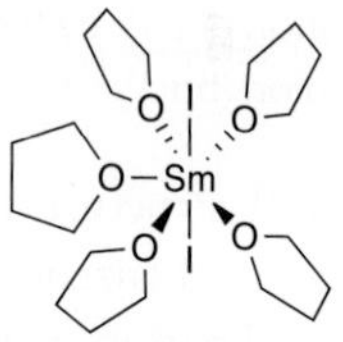

9 $[SmI_2(thf)_n]$

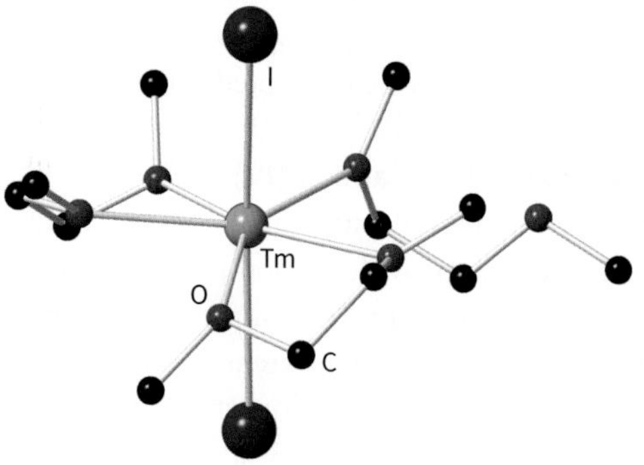

10 $[TmI_2(dme)_3]$

These complexes are powerful one-electron reductants in nonaqueous etheric solvents and useful reagents for reductive C–C coupling reactions, an example being the formation of tertiary alcohols from alkyl halides and ketones.

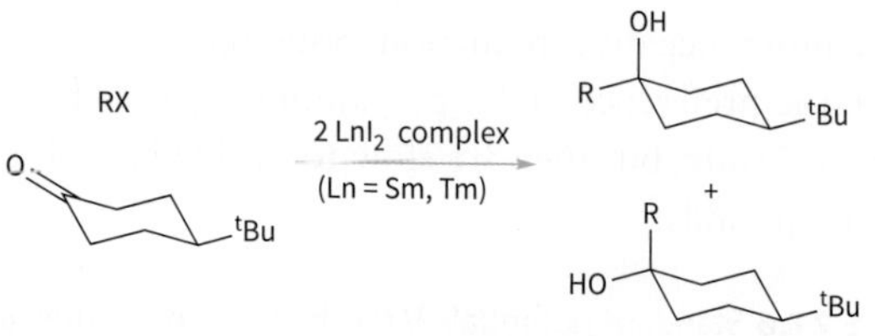

Carbon–carbon coupling reactions can also occur between ligands. Reaction of TmI_2 with pyridine results in formation of the complex $[((py)_4TmI_2)_2(\mu\text{-}C_{10}H_{10}N_2)]$ (**11**) in which the product, 1,1-dihydro-4,4′-bipyridyl, bridges the two Tm(III). The Tm=N bonds are drawn only to stress that the N atom is deprotonated.

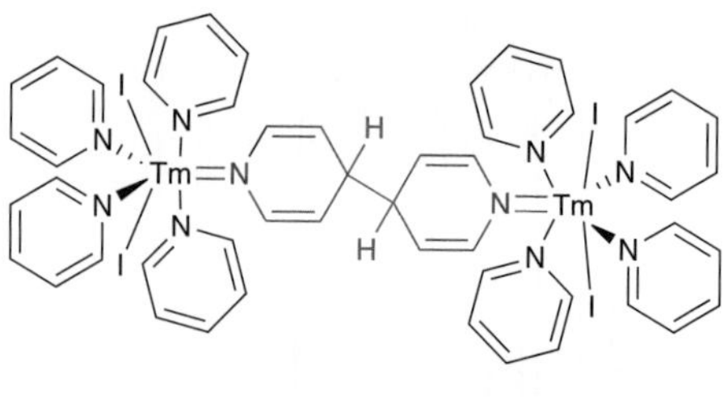

11 $[((py)_4TmI_2)_2(\mu\text{-}C_{10}H_{10}N_2)]$

An extensive range of coordination chemistry exists for Ce(IV). Apart from $[Ce(NO_3)_6]^{2-}$, most Ce(IV) complexes are based on alkoxide or carboxylate ligands. A binuclear Ce(IV) complex with bridging oxido ligands is formed by reacting a tris(amido)Ce(III) precursor with O_2.

$$Ce(NR_2)_3 \xrightarrow[\text{hexane}]{O_2,\ -27°C} \text{(binuclear Ce complex)}$$

21.8 Organometallic compounds

KEY POINTS The organometallic chemistry of the lanthanoids is dominated by Ln(III) compounds with bonding that is predominantly ionic or stabilized by strong donor ligands. The 18-electron rule that governs much of the organometallic chemistry of the d-block elements does not hold for lanthanoid organometallic compounds, which tend to resemble those of the early, more electropositive d metals. Steric effects are far more important than electronic effects and some cyclopentadienyl Ln(III) compounds have important applications in stereoselective catalysis. Organometallic compounds of Ln(II) are powerful reducing agents and some of these have $5d^1$ ground states.

The organometallic chemistry of the lanthanoids is much less developed than for the d-block elements, which is described in detail in Chapter 24. The majority of lanthanoid organometallic compounds formally contain Ln(III) for which the lack of any orbitals that can engage in π back bonding to organic fragments (the 5d orbitals are empty and the 4f orbitals are too buried) restricts the number of covalent bonding modes that are available.

The first η² alkene complex of a lanthanoid, $[(Cp^*)_2Yb(C_2H_4)Pt(PPh_3)_2]$ (**12**), was characterized only in 1987, a century and a half after the first d-metal alkene complex, Zeise's salt, was isolated. However, the alkene ligand in this complex is particularly electron-rich as it is already bound to a π-donating Pt(0) centre, and back-bonding from the Yb atom is therefore unnecessary for stability. The strongly electropositive nature of the lanthanoids means that they need good donor, not good acceptor, ligands. Consequently, CO, phosphine, and alkene ligands are rarely seen in lanthanoid chemistry, whereas strong donors such as N-heterocyclic carbenes are found in a wide variety of complexes, one example being the biscarbene, tris-alkyl Er complex (**13**) with five metal–carbon bonds.

12 [(Cp*)$_2$Yb(C$_2$H$_4$)Pt(PPh$_3$)$_2$]

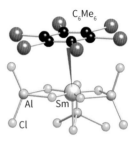

13 The biscarbene, tris-alkyl Er complex

The 18-electron rule, which we will see in Chapter 24 governs much of the organometallic chemistry of the d-block elements, is not applicable for organometallic lanthanoid compounds. In general, organometallic lanthanoid compounds depend much more upon ionic bonding than the d-block organometallics, and they are very sensitive to air and moisture. Instead of being controlled by electron configuration, the chemistry is influenced by steric restraints that become more demanding as the radii decrease across the series La–Lu.

By far the most organometallic compounds are formed with cyclopentadienyl ligands, and in most cases it is appropriate to consider these compounds as containing Cp$^-$ groups electrostatically bound to a central Ln^{3+} cation. The larger lanthanoid ions can easily accommodate three cyclopentadienyl ligands, and they even tend to oligomerize, indicating that there is yet more space for additional ligands. Control over the reactivity of the complexes is greatly improved by using the sterically demanding pentamethylcyclopentadienyl (Cp*) ligand, although it was nearly forty years after the original [Ln(Cp)$_3$] compounds were isolated that [Ln(Cp*)$_3$] compounds were obtained, and even then they exist in equilibrium between η^5 and η^1 forms.

Current research into lanthanoid organometallic compounds typically involves compounds of the type [(Cp)$_2$LnR]$_2$, [(Cp)$_2$LnR(sol)], [(Cp*)LnRX(sol)], and [(Cp*)LnR$_2$(sol)]. σ-Bonded alkyl groups are common, with compounds containing cyclopentadienyl ligands tending to dominate. Compounds containing η^8-cyclooctatetraene ligands are known, such as [Ce(C$_8$H$_8$)$_2$] (**14**), and as noted in Section 24.11, it is best to consider these to be complexes of the electron-rich C$_8$H$_8^{2-}$ ligand. A number of arene complexes are also known, such as [(C$_6$Me$_6$)Sm(AlCl$_4$)$_3$] (**15**), where it is thought that the bonding is largely the result of an electrostatically induced dipole between the Sm^{3+} ion and the electron-rich ring.

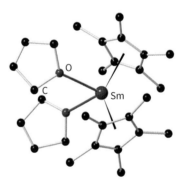

14 [Ce(C$_8$H$_8$)$_2$]

15 [(C$_6$Me$_6$)Sm(AlCl$_4$)$_3$]

Most of the Ln(II) molecular compounds that have been synthesized contain Cp* ligands, a typical example being [Sm(Cp*)$_2$(thf)$_2$] (**16**) which is isolated from tetrahydrofuran (thf) solution.

16 [Sm(Cp*)$_2$(thf)$_2$]

As expected, the occurrence and reactivities of the Ln(II) metallocenes correlate quite well with having low values for I_3, so the compounds of Eu(II), Yb(II), and Sm(II) are most easily accessed and those of Tm(II), Dy(II) less so; but even Er(II) and Ho(II) metallocenes are known. Trigonal Ln(II) complexes containing three C$_5$H$_4$Si(Me)$_3$ ligands have been synthesized for which the ground state is 5d^1 instead of 4f^n (Box 21.1).

From a practical angle, Ln(II) metallocenes are useful one-electron reducing agents covering a range of reactivity. Samarocene, Sm(Cp*)$_2$, reacts with N$_2$ to give dark red crystals of [(Sm(Cp*))$_2$N$_2$] (**18**), the first f-block compound

BOX 21.1 How do d orbitals participate in the bonding in Ln(II) complexes?

Potassium salts of the anion (**17**), containing a Ln(II) cation and three sterically demanding $C_5H_4Si(Me)_3$ (Cp') ligands, can be synthesized for all lanthanoids.

17

The salts are prepared from the appropriate neutral Ln(III) complex under anaerobic conditions using potassium graphite as the reducing agent in a solvent such as THF:

$$Cp'_3Ln \xrightarrow{KC_8, 2,2,2\text{-cryptand}} [K(crypt)]^+ [Cp'_3Ln]^- + graphite$$

The Ln(II) complexes may be divided into two classes, 'traditional' and 'new', based upon spectral and structural data. The 'new' complexes display particularly intense UV-visible absorptions ($\varepsilon_{max} > 3000$ mol^{-1} dm^3) and show little expansion in Ln–(ring centroid) distance (d) compared to the starting Ln(III) complex. By contrast, the 'traditional' complexes do not have the intense absorption but do show a large expansion in the Ln–(ring centroid) distance (Table B21.1).

Comparison with the Ln(II) reference complexes formed from $Cp'_3Lu(f^{14})$, which must have a $f^{14}d^1$ configuration, and that formed from Cp'_3Y (f^0), which would have a $4d^1$ configuration, shows that the 'new' complexes also have a $5d^1$ ground state rather than $4f^n$. The low sensitivity of bond distance to oxidation state is expected for bonding that is predominantly covalent. The intense absorption bands are due to metal–ligand charge transfers, rather than the $4f^{(n-1)}5d^1 \leftarrow 4f^n$ seen with 'traditional' Ln(II) species.

Fig. 21.6 helps to provide an explanation for why Sm(II), Eu(II), Tm(II), and Yb(II) have the greatest tendency to adopt a $4f^n$ ground state: their f orbitals are substantially lower in energy than the corresponding d orbitals in the 2+ state. For the other Ln(II) ions (except for Ce and Gd, where the d orbitals are substantially lower in energy and a d^1 configuration would

be expected anyway), the f orbitals are only slightly lower in energy than the d orbitals, and we might expect the f orbitals to be preferentially populated. The important point that we have not yet taken into consideration is the ligand-field effects that become influential when d orbitals are used. The stability of the 'new' Ln(II) complexes can be rationalized in terms of the additional stabilization of the d_{z^2} orbital that occurs with a trigonal-planar ligand field, as shown in Fig. B21.1. The $5d_{z^2}$ orbital becomes sufficiently low in energy that the added electron can reside there rather than in a 4f orbital.

Use of the 5d orbital and enhanced covalence that occurs with these complexes resembles the behaviour observed with solid-state LnO and LnX_2 compounds (X = Cl, Br, I, H) which can be either 'salt-like' or 'metallic' (Section 21.5). Although metallic divalent compounds are commonly formulated as (Ln^{3+}, e^-), the electron is itinerant because it enters a band formed by overlap of 5d orbitals.

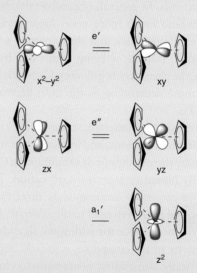

FIGURE B21.1 Ordering of the d-orbital energies in D_{3h} symmetry and stabilization of the $5d^1$ state in trigonal $[(Cp')_3Ln(II)]^-$ complexes.

TABLE B21.1

Ln	La	Ce	Pr	Nd	Sm	Eu	Gd	Tb	Dy	Ho	Er	Tm	Yb	Lu	Y
Δd/pm	2.7	2.9	2.7	3.0	14.9	15.6	3.1	3.1	3.6	3.2	3.0	12.3	14.3	3.1	3.1
Intense UV/Vis	√	√	√	√	X	X	√	√	√	√	√	X	X	√	√
T/N/R*	N	N	N	N	T	T	N	N	N	N	N	T	T	R/N	R

*T = Traditional, N = New, R = Reference.

discovered to contain coordinated molecular N₂: in contrast, the more powerfully reducing thulium(II) analogue, Tm(Cp*)₂, reacts reductively with N₂ to give a structurally similar, pale-coloured binuclear Tm(III) complex with a bridging diazenide $(N_2)^{2-}$ ligand.

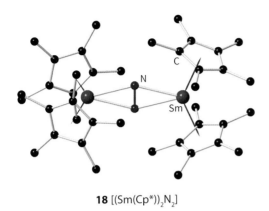

18 [(Sm(Cp*))₂N₂]

Lanthanoid organometallic compounds are important homogeneous catalysts. Unlike the d-block elements, they are unable to carry out oxidative addition or reductive elimination reactions because no f-block element has a pair of accessible oxidation states differing by 2. But Ln(III) organometallics, specifically Ln(Cp*)X complexes, are highly active catalysts for carbon–carbon insertion reactions and σ-bond metathesis. Unlike most d-block compounds, they are not poisoned by CO or sulfides. A familiar class of carbon–carbon insertion reaction is the Ziegler–Natta polymerization of alkenes, described in more detail in Section 31.14.

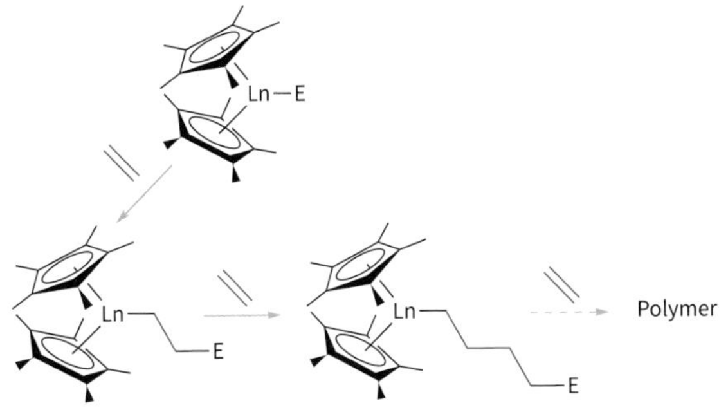

Another reaction, σ-bond metathesis, is represented as follows:

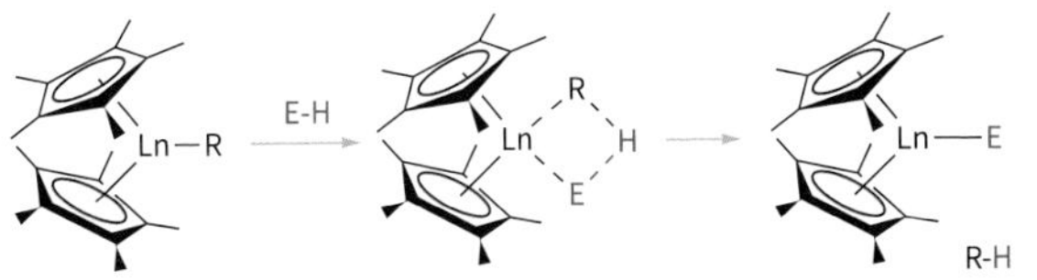

An important event that focused serious attention on lanthanoid organometallic chemistry was the discovery that they could activate the C–H bond in methane. This discovery was based on the observation that $^{13}CH_4$ exchanges ^{13}C with the CH₃ group attached to Lu:

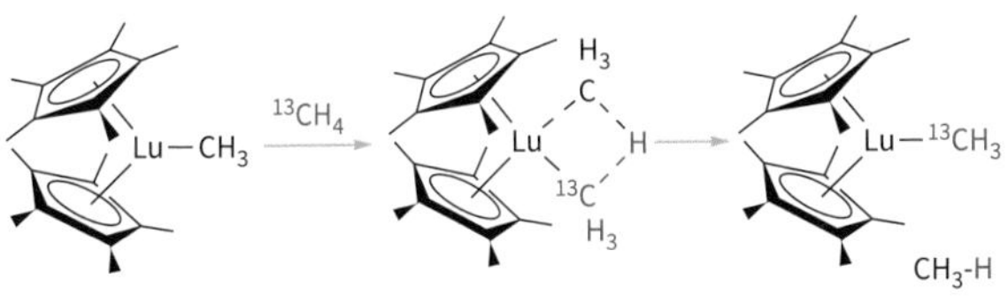

EXAMPLE 21.2 **Accounting for the organometallic reactivity of a lanthanoid**

Suggest a likely reaction pathway for the following transformation:

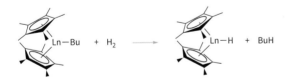

Answer The reaction with dihydrogen cannot proceed through formation of a dihydrogen complex, oxidative addition to form a dihydride, and then reductive elimination of butane, because the lanthanoid is incapable of oxidative addition reactions. A likely intermediate is therefore

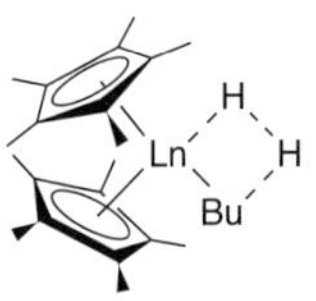

Self-test 21.2 The product of the reaction above is in fact a hydrido bridged dimer. Suggest a strategy to ensure that the hydride is monomeric.

Actinoid chemistry

The chemical properties of the actinoids show much less uniformity across the series than those of the lanthanoids, and the light (early) members (Ac–Am) resemble the early d metals in exhibiting a wide range of oxidation states. However, the nuclear instability associated with most of the actinoids has hindered their study. On the one hand,

BOX 21.2 What are the pros and cons of nuclear fission?

The fission of heavy elements, such as ^{235}U, can be induced by bombardment by neutrons. Thermal neutrons (neutrons with low velocities) bring about the fission of ^{235}U to produce two nuclides of medium mass, and a large amount of energy is released because the binding energy per nucleon decreases steadily for atomic numbers beyond about 26 (Fe; see Fig. 1.1). That unsymmetrical fission of the uranium nucleus occurs is shown by the double-humped distribution of fission products (see Fig. B21.2), with maxima close to mass numbers 95 (isotopes of Mo) and 135 (isotopes of Ba). Almost all the fission products are unstable nuclides. The most troublesome are those with half-lives in the range of years to centuries: these nuclides decay fast enough to be highly radioactive but not sufficiently fast to disappear in a convenient time.

The heat produced by nuclear fission is used to produce steam, which can drive turbines in much the same way as conventional power plants use the burning of fuels to produce heat. However, the energy produced by the fission of a heavy element is huge in comparison with the burning of conventional fuels: for instance, the complete combustion of 1 kg of octane produces approximately 50 MJ, whereas the energy liberated by the fission of 1 kg of ^{235}U is approximately 2 TJ (1 TJ = 10^{12} J), 40 000 times as much. Later designs of nuclear power plants have also used plutonium, normally mixed with uranium. Although nuclear power is carbon-neutral during operation and offers the potential of enormous quantities of energy at low cost, no satisfactory method of disposal of the radioactive waste it produces has yet been found.

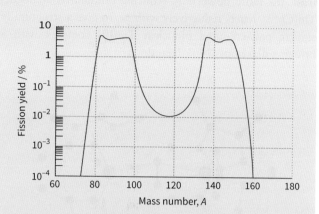

FIGURE B21.2 The double-humped distribution of fission products of uranium.

---

long-lived but intensely radioactive isotopes present severe hazards for handling and disposal; on the other hand, very short-lived isotopes present challenges for synthesis and characterization, difficulties that increase with atomic number as we have already seen for the 6d elements (Section 20.13).

Because the heavy actinoids are available in such tiny amounts, little is known about their reactions, and most of the chemical properties of the transamericium elements (the elements following americium, Z = 95) have been established by experiments carried out on a microgram scale or even on just a few hundred atoms. For example, the actinoid ion complexes have been adsorbed on and eluted from a single bead of ion-exchange material of diameter 0.2 mm. Establishing the chemical properties of the heavy actinoids depends on rapid characterization methods and upon finding an isotope that not only has a sufficiently long half-life, but can be produced easily and rapidly, the balance between formation and decay being crucial. One example is a complex of einsteinium, described later. The early actinoids, particularly U and Pu, are of great importance in the generation of power through nuclear fission (Box 21.2).

21.9 General trends

KEY POINTS The light actinoids (Th–Pu) do not exhibit the chemical uniformity of the lanthanoids, but behave more like d-block elements. A prevailing motif is the collinear O–An–O unit that is stabilized by strong σ- and π-donation into metal 6d and 5f orbitals. As the 5f block is traversed the 3+ oxidation state becomes increasingly dominant, and beyond americium the elements more closely resemble the lanthanoids.

The 15 elements from actinium (Ac, Z = 89) to lawrencium (Lr, Z = 103) involve the progressive completion of the 5f subshell, and in this sense are analogues of the lanthanoids. Although many actinoids do occur as An(III) analogous to the lanthanoids, the light actinoids also occur in a rich variety of other oxidation states. The basis for this difference is found in the differing penetration of the 4f and 5f orbitals. As we saw in Fig. 21.1, the 5f orbital contains a radial node and has an inner region that screens the outer orbitals from the nuclear charge: the 5f orbitals are therefore much less core-like than the 4f, at least until we reach Pu. The 6d orbitals are also more diffuse than the 5d orbitals.

Table 21.7 lists the electronic configurations and the oxidation states (principal ones highlighted) that are encountered for each actinoid. Compared with the lanthanoids (Tables 21.3 and 21.4) we immediately see how the actinoids make greater use of the d orbitals and are much more versatile in regard to their oxidation states, but we see also that these features are reserved for the light members. For americium (Am, Z = 95) and beyond, the properties of the actinoids begin to converge with those of the lanthanoids. With increasing atomic number, the 3+ oxidation state becomes progressively more stable relative to higher

TABLE 21.7 Electron configurations and oxidation numbers (O.N.) of the actinoids

Z	Name	Symbol	Electron configuration of metal	O.N.*
89	actinium	Ac	[Rn]6d^{1}7s^2	**3**
90	thorium	Th	[Rn]6d^{2}7s^2	**4**
91	protactinium	Pa	[Rn]5f^{2}6d^{1}7s^2	3, **4**, **5**
92	uranium	U	[Rn]5f^{3}6d^{1}7s^2	3, **4**, 5, **6**
93	neptunium	Np	[Rn] 5f^{4}6d^{1}7s^2	3, 4, **5**, 6, 7
94	plutonium	Pu	[Rn] 5f^{6}7s^2	2, 3, **4**, **5**, 6, 7
95	americium	Am	[Rn] 5f^{7}7s^2	2, **3**, 4, 5, 6
96	curium	Cm	[Rn] 5f^{7}6d^{1}7s^2	**3**, 4
97	berkelium	Bk	[Rn] 5f^{9}7s^2	**3**, 4
98	californium	Cf	[Rn] 5f^{10}7s^2	2, **3**, 4
99	einsteinium	Es	[Rn]5f^{11}7s^2	2, **3**
100	fermium	Fm	[Rn]5f^{12}7s^2	2, **3**
101	mendelevium	Md	[Rn]5f^{13}7s^2	2, **3**
102	nobelium	No	[Rn]5f^{14}7s^2	**2**, 3
103	lawrencium	Lr	[Rn]5f^{14}6d^{1}7s^2	2, **3**

* Principal oxidation numbers are highlighted.

oxidation states and is dominant for Cm, Bk, Cf, and Es; the heavy actinoids therefore resemble the lanthanoids. The 2+ oxidation state makes an initial appearance at Am, reflecting the special stability of the half-filled shell (5f^7), then appears consistently from Cf onwards.

The striking differences between the chemical properties of the lanthanoids and early actinoids led to controversy about the most appropriate placement of the actinoids in the periodic table. Before 1945, periodic tables usually showed U below W because both elements have a maximum oxidation number of +6. The emergence of oxidation state An(III) for the heavy actinoids was a key point in determining their current placement. The similarity of the heavy actinoids and the lanthanoids is illustrated by their similar elution behaviour in ion-exchange separation (compare Figs. 21.11 and 21.13).

The Frost diagrams in Fig. 21.14 show how the stabilities of different oxidation states of aqua species vary across the series.

EXAMPLE 21.3 Assessing the redox stability of actinoid ions

Use the Frost diagram for thorium (Fig. 21.14) to describe the relative stabilities of Th(II), Th(III), and Th(IV) in aqueous solution.

Answer We need to use the interpretation of Frost diagrams described in Section 6.13. The initial slope in the Frost diagrams indicates that the Th^{2+} ion might be readily attained with a mild oxidant. However, Th^{2+} lies far above the lines connecting Th(0) with the higher oxidation states, so it is highly susceptible to disproportionation. In turn, Th(III) is readily oxidized to Th(IV) and the steep negative slope indicates that it is susceptible to oxidation by water:

$$Th^{3+}(aq) + H^+(aq) \rightarrow Th^{4+}(aq) + \tfrac{1}{2}H_2(g)$$

We can confirm from Resource section 2 that because $E^\ominus = -3.8\,V$, this reaction is highly favoured. Thus, Th(IV) will be the exclusive oxidation state in aqueous solution.

Self-test 21.3 Use the Frost diagrams and data in Resource section 2 to determine the most stable oxidation number for uranium ions in acid aqueous solution in the presence of air.

Linear or nearly linear dioxido units (AnO$_2^+$ and AnO$_2^{2+}$) dominate the chemistry for oxidation numbers +5 and +6 of the early actinoids (U, Np, Pu, and Am). The remaining ligands occupy equatorial or near-equatorial positions. The collinear An–O bonds are very strong: gas phase dissociation energies (AnO$_2^{2+}$) are 618, 514, and 421 kJ mol^{-1} for An = U, Np, Pu, respectively and oxygen atom exchange is extremely slow (the half-life for UO$_2^{2+}$ in aqueous acid is of the order of 10^9 s). The dominance of the linear dioxido unit in early actinoid chemistry is strong evidence for the covalent nature of the bonding that involves the 5f and 6d orbitals. This property is completely different from the nondirectional, electrostatic bonding displayed by the lanthanoids and to understand the bonding in the AnO$_2^{2+}$ unit we look at the molecular orbitals shown in Fig. 21.15. We consider that the AnO$_2^{2+}$ unit has $D_{\infty h}$ symmetry.

The *trans* geometry maximizes σ-bonding by combining the O2p σ orbitals with an actinoid 6d$_{z^2}$ (g-symmetry) and a hybrid formed by mixing 5f$_{z^3}$ with the semi-core 6p$_z$ orbital (u-symmetry). The bonding thus uses both symmetric (g)

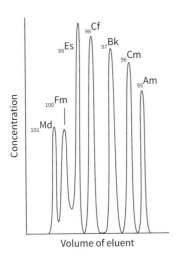

FIGURE 21.13 Elution of heavy actinoid ions from a cation exchange column using ammonium 2-hydroxyisobutyrate as the eluent. Note the similarity in elution sequence to Fig. 21.11: the heavy (smaller) An^{3+} ions elute first.

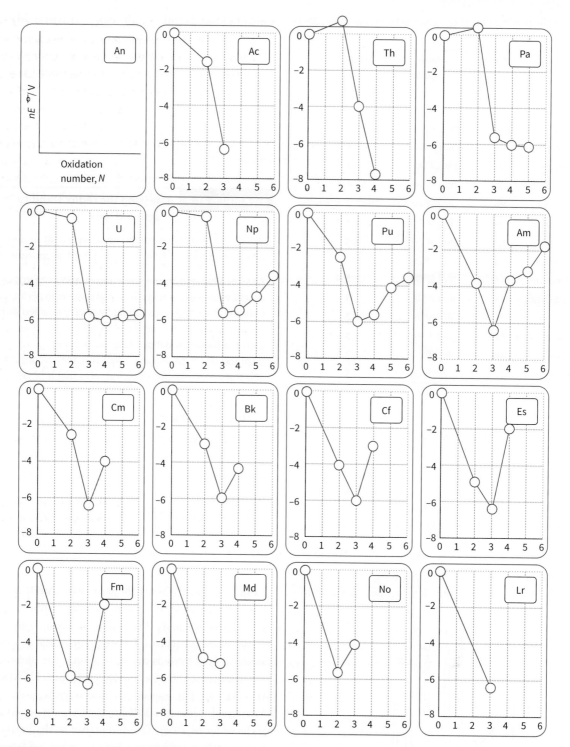

FIGURE 21.14 Frost diagrams of the actinoids in acidic solution. A labelled generic diagram is shown in the upper left corner. (Based on J. J. Katz, G. T. Seaborg, and L. Morss, *Chemistry of the Actinide Elements*. Chapman and Hall (1986).)

and antisymmetric (u) O σ SALCs. Four An–O π bonds are formed by combining the O2p π orbitals with two 6dπ (xz, yz) and two 5f orbitals of π symmetry on the actinoid. The AnO_2^{2+} unit may be described as O≡An≡O, i.e. with each collinear An–O fragment having an overall bond order of 3. Such a scheme holds for all the lighter actinoids because the 5fδ and 5fφ sets are nonbonding.

Compared to the core-like nature of the 4f orbitals of the lanthanoids, the 5f orbitals of the light actinoids are more diffuse (Fig. 21.1)—a consequence being that the spectra of actinoid complexes are more strongly affected by ligands. A convincing demonstration of the extension of 5f orbitals into the bonding region is provided by the EPR spectrum of Pu^{3+} doped into CaF_2, in which Pu^{3+} occupies a fraction of

Actinoid chemistry

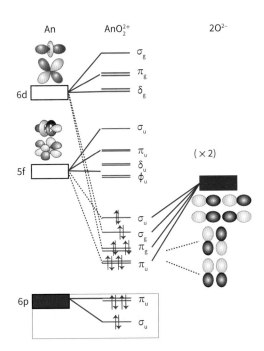

FIGURE 21.15 Molecular orbital diagram for the AnO_2^{2+} unit showing how the An 6d and 5f orbitals interact with the O 2p atomic orbitals of s and p symmetry.

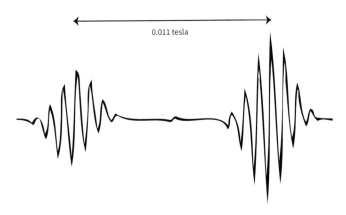

FIGURE 21.16 The EPR spectrum of Pu^{3+} doped into CaF_2 showing the doublet due to hyperfine coupling with ^{239}Pu ($I = ½$) and the superhyperfine coupling with eight equivalent F^- ($I = ½$). The outer lines are too faint to see in this spectrum.

the cubic sites (Fig. 21.16). Each component of the hyperfine doublet that arises from the unpaired electron interacting with the ^{239}Pu nucleus ($I = ½$) is further split into nine lines by the electron's further interaction with ligand nuclei (known as superhyperfine coupling) with eight equivalent ^{19}F ($I = ½$).

There is an actinoid contraction, but unlike for the lanthanoids this does not have consequences of any practical nature for the d-block elements, the chemistry of the 6d elements still being largely unexplored (Section 20.13). The actinoids have large atomic and ionic radii (the radius of an An^{3+} ion is typically about 5 pm larger than its Ln^{3+} congener) and high coordination numbers are expected. Like the larger lanthanoids, the 3+ aqua-cations are 9-coordinate. Uranium in solid UCl_4 is 8-coordinate and in solid UBr_4 it is 7-coordinate in a pentagonal-bipyramidal array, but solid-state structures with coordination numbers up to 12 have also been observed.

21.10 Thorium and uranium

KEY POINTS The common nuclides of thorium and uranium exhibit only low levels of radioactivity, so their chemical properties have been extensively developed. The uranyl cation is found in complexes with many different ligand donor atoms, and the organometallic compounds of the elements are dominated by pentamethylcyclopentadienyl complexes.

Because of their ready availability and relatively low level of radioactivity, the chemical manipulation of Th and U can be carried out with ordinary laboratory techniques. As indicated in Fig. 21.14, the most stable oxidation state of thorium in aqueous solution is Th(IV). This oxidation state also dominates the solid-state chemistry of the element. Eight-coordination is common in simple Th(IV) compounds. For example, ThO_2 has the fluorite structure (in which a Th atom is surrounded by a cubic array of O^{2-} ions) and in $ThCl_4$ and ThF_4 the coordination numbers are also 8 with dodecahedral and square antiprism symmetry respectively. The coordination number of Th in $[Th(NO_3)_4(OPPh_3)_2]$ (**19a**) is 10, with the NO_3^- ions and triphenylphosphine oxide groups arranged in a capped cubic array around the Th atom. As already mentioned in Section 9.6, the actinoids may exhibit very high coordination numbers, an example being the 11-coordinate Th existing in its hydrated nitrato complex, $Th(NO_3)_4 \cdot 3H_2O$, in which the Th^{4+} ion is bonded to four NO_3^- ions in a bidentate fashion and to three H_2O molecules (**19b**).

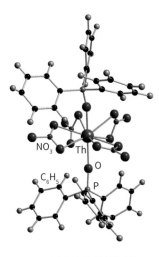

19a $[Th(NO_3)_4(OPPh_3)_2]$

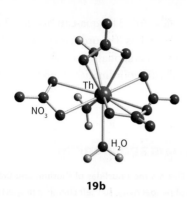

19b

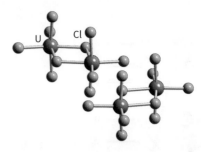

FIGURE 21.17 The crystal structure of U_2Cl_{10} consists of discrete molecules formed from pairs of edge-sharing UCl_6 octahedra.

The chemical properties of U are more varied than those of Th because the element has access to oxidation states from U(III) to U(VI), with U(IV) and U(VI) the most common. As suggested from the Frost diagram, the U^{3+} aqua ion is fairly strongly reducing whereas U(V) (as the UO_2^+ ion) disproportionates. Uranium metal does not form a passivating oxide coating, so it becomes corroded upon prolonged exposure to air to give a complex mixture of oxides including UO_2, U_3O_8, and several polymorphs of the stoichiometry UO_3. The dioxide UO_2 adopts the fluorite structure but also takes up interstitial O atoms to form the nonstoichiometric series UO_{2+x}, $0 < x < 0.25$. The most important oxide is UO_3, one form of which, $\delta\text{-}UO_3$, adopts the ReO_3 structure type (Section 25.4). The very stable UO_2^{2+} cation, obtained by dissolving UO_3 in acid, forms complexes with many anions, such as NO_3^- and SO_4^{2-} which occupy equatorial positions. The linear UO_2^{2+} unit persists in the solid state; for instance UO_2F_2 has a slightly puckered ring of six F^- ions around the UO_2^{2+} unit.

Uranium halides are known for the full range of oxidation states U(III) to U(VI), with a trend towards decreasing coordination number with increasing oxidation number. The most important fluoride is UF_6 which is synthesized on a large scale from UO_2.

$$UO_2 + 4HF \rightarrow UF_4 + 2H_2O$$
$$3UF_4 + 2ClF_3 \rightarrow 3UF_6 + Cl_2$$

The high volatility of UF_6 (it sublimes at 57°C) together with the occurrence of fluorine in a single isotopic form account for the use of this compound in the separation of the uranium isotopes by gaseous diffusion or centrifugation. The tetrachloride UCl_4—a useful starting material for synthesizing many U(IV) compounds—is made by reacting UO_3 with hexachloropropene. The U atom is 9-coordinate in solid UCl_3, 8-coordinate in UCl_4, and 6-coordinate for the U(V) and U(VI) chlorides U_2Cl_{10} (Fig. 21.17) and UCl_6, both of which are molecular compounds.

The separation of uranium from most other metals is accomplished by extraction of the neutral uranyl nitrato complex $[UO_2(NO_3)_2(OH_2)_4]$ from the aqueous phase into a polar organic phase, such as a solution of tributyl phosphate dissolved in a hydrocarbon solvent. This kind of solvent extraction process is used to separate actinoids from other fission products in spent nuclear fuel.

The organometallic chemistry of U and Th is reasonably well developed and shows many similarities to that of the lanthanoids, except that Th and U occur in a number of oxidation states and are larger than the typical Ln ion. Thus, compounds are dominated by those containing good donor ligands, such as σ-bonded alkyl and cyclopentadienyl groups as well as N-heterocyclic carbenes. The increased size of Th and U compared with typical lanthanoids means that the tetrahedral species $[Th(Cp)_4]$ and $[U(Cp)_4]$ (**20**) can be isolated as monomers, and not only can $[U(Cp^*)_3]$ be isolated but so too can $[U(Cp^*)_3Cl]$ (**21**).

20 $[Th(Cp)_4]$ and $[U(Cp)_4]$

21 $[U(Cp^*)_3Cl]$

Compounds with fullerenes are also known in which U and Th are trapped inside their nanocontainer. As with lanthanoid organometallic compounds, actinoid organometallics do not obey the 18-electron rule (Section 24.2). But unlike the lanthanoids, metal–metal binding makes an appearance, as shown with the trithorium cluster anion $[\{Th(\eta^8\text{-}C_8H_8)(\mu_3\text{-}Cl)_2\}_3]^{2-}$ (**22**). The 6d and 5f orbitals contributed by one Th(IV) and two Th(III) combine to form a delocalized three-centre two-electron metal–metal bond.

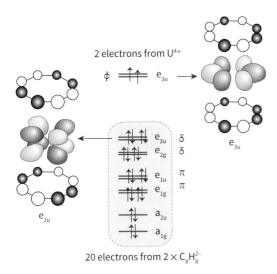

FIGURE 21.18 Partial molecular orbital diagram for actinoid η^8-cyclooctatetraene sandwich complexes showing how the 6d and 5f atomic orbitals interact with the C p_z orbitals (phases pointing at the central An are represented schematically; only one lobe is shown).

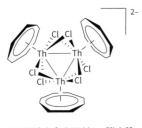

22 $[\{Th(\eta^8\text{-}C_8H_8)(\mu_3\text{-}Cl)_2\}_3]^{2-}$

Sandwich compounds are possible with the η^8-cyclooctatetraene ligand, representative examples being thorocene, $[Th(C_8H_8)_2]$, and uranocene, $[U(C_8H_8)_2]$ (**23**), both of which have D_{8h} symmetry with eclipsed rings. The bonding orbitals are shown in Fig. 21.18. Twenty electrons from the two $C_8H_8^{2-}$ fill the bonding orbitals, leaving a weakly bonding e_{3u} (fϕ) orbital having actinoid character. Uranocene (e_{3u}^2) has a triplet ground state. So far, actinoid cyclooctatetraene compounds are the only 'real' compounds (as opposed to metal dimers in the gas phase) that might contain any contribution from ϕ bonding.

23 $[U(C_8H_8)_2]$

21.11 Neptunium, plutonium, and americium

KEY POINTS Like uranium, neptunium and plutonium exhibit several stable oxidation states dominated by complexes with oxido ligands. Americium marks the beginning of the heavy actinoids, in which An(III) dominates.

The intensely radioactive elements Np, Pu, and Am form compounds containing similar species, though there are significant differences in the stabilities of the main oxidation states. The Frost diagrams in Fig. 21.14 summarize their behaviour. Neptunium dissolves in dilute acids to produce Np^{3+} which is oxidized by air to produce Np^{4+}. Increasingly strong oxidizing agents produce NpO_2^+ (Np(V)) and NpO_2^{2+} (Np(VI)). The four common oxidation states of plutonium, Pu(III), Pu(IV), Pu(V), and Pu(VI), are separated from each other by less than 1 V, and solutions of Pu often contain a mixture of the species Pu^{3+}, Pu^{4+}, and PuO_2^{2+} (PuO_2^+ has a tendency to disproportionate to Pu^{4+} and PuO_2^{2+}). The Pu(VIII) species $[PuO_4(OH)_2]^{2-}$ formed by oxidation under alkaline conditions (>1 M NaOH) has four strongly bound oxygen atoms in the equatorial plane (**24**). The Am^{3+} ion is the most stable americium species in aqueous solution, reflecting the tendency for An(III) to dominate actinoid chemistry for the high atomic number elements. Under strong oxidizing conditions AmO_2^+ and AmO_2^{2+} can be formed; Am(IV) disproportionates in acidic solutions.

24

The An(IV) oxides NpO_2, PuO_2, and AmO_2, which are formed by heating the elements or their salts in air, all adopt the fluorite structure (Section 4.9). Lower oxides include Np_3O_8, Pu_2O_3, and Am_2O_3. The trichlorides, $AnCl_3$, can be obtained by direct reaction of the elements at 450°C and have structures analogous to that of $LnCl_3$ with a 9-coordinate An atom. Tetrafluorides are known for all three actinoids, though only Np and Pu form tetrachlorides, further demonstrating the difficulty in oxidizing americium to Am(IV). Both Np and Pu form hexafluorides which, like UF_6, are volatile solids.

All three metals exhibit chemical properties analogous to that of the uranyl ion, forming NpO_2^{2+}, PuO_2^{2+}, and AmO_2^{2+} which can be extracted from aqueous solution by tributyl phosphate as $AnO_2(NO_3)_2\{OP(OBu)_3\}_2$. The tetrahalides are Lewis acids and form adducts with electron-pair donors such as DMSO, as in $AnCl_4(Me_2SO)_7$. Neptunium forms a number of organometallic compounds that are analogues of those of uranium, such as $Np(Cp)_4$. In a landmark example of nuclear chemistry, a sample of the complex $[Np(Cp)_3X]$ was prepared by neutron bombardment of its uranium analogue (Fig. 21.19).

Plutonium is a potentially dangerous environmental hazard in regions near to nuclear processing plants, where there is always the possibility that it may leach into groundwater sources and contaminate soils. There is therefore much

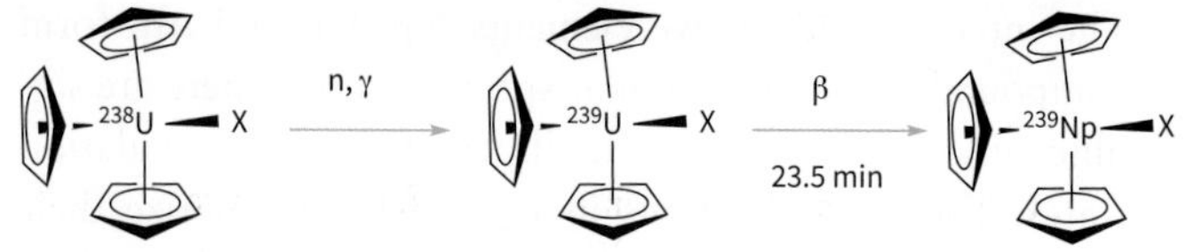

^{238}U—X $\xrightarrow{n, \gamma}$ ^{239}U—X $\xrightarrow{\beta}_{23.5 \text{ min}}$ ^{239}Np—X

FIGURE 21.19 Synthesis of a Np(IV) complex using a nuclear reaction (X = F, Cl). Bombardment of the original ^{238}U complex with neutrons produces the unstable ^{239}U intermediate which undergoes β decay to produce ^{239}Np (half-life 2.4 d) with retention of the original coordination sphere.

interest in the nature and properties of the complexes formed by Pu in different oxidation states with ubiquitous ligands such as chloride, nitrate, carbonate, or phosphate. Carbonate stabilizes Pu(VI) through formation of the hexagonal-bipyramidal $[PuO_2(CO_3)_3]^{4-}$ anion (**25**). As we will see in Section 32.8, specific complexing agents are also available for sequestering plutonium that has found its way into the human body.

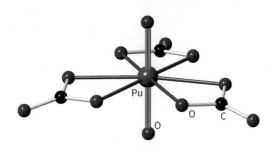

25 $[PuO_2(CO_3)_3]^{4-}$ anion

21.12 The heavy actinoids

From curium to lawrencium, the elements behave more like the lanthanoids in terms of preference for high coordination numbers, decreasing covalence, and lower oxidation numbers. Moreover, as can be seen from Fig. 21.14, the 2+ oxidation state becomes increasingly stable beyond californium.

The challenges facing investigations and applications of these elements stem not so much from hazards associated with their intense radioactivity and difficulty of disposal than their short half-lives; the heaviest elements, nobelium and lawrencium, have only a transient existence. Where producible in visible quantity, they have a metallic appearance and may glow in the dark.

Beyond fermium, elements cannot be synthesized by neutron capture and nuclear fusion is required, yields from the latter being too low to allow their chemical properties to be determined by conventional methods. An isotope having a shorter half-life may be preferred over one that

is much more stable if it can be produced more easily. For instance, although ^{259}No is the longest lived isotope of nobelium, the less stable ^{255}No (half-life 190 s) is preferred for chemical studies because it is produced more easily at sufficient scale.

Characterization of the structures and chemical properties of compounds formed by these elements requires spectroscopic techniques able to acquire rapid and precise information on less than a microgram of sample, synchrotron-based X-ray methods (Chapter 4) being appropriate. The rapid formation of a complex of einsteinium (^{254}Es, half-life 276 d), characterized by XAS, EXAFS, and luminescence spectroscopy is shown in Fig. 21.20. The eight-coordinate Es^{3+} complex resembles those formed by Ln^{3+} ions.

Despite such limitations, an increasing number of peaceful applications are being found for some heavy actinoids. As a strong neutron emitter, ^{252}Cf is used as the source in many neutron-based technologies, including spectroscopy, diffraction, medical radiography, and detection of faults in materials, as well as for priming nuclear reactors. Californium and curium have been used as targets to synthesize super-heavy elements by nuclear fusion: these include transactinides, members of the 6d row (Section 20.13).

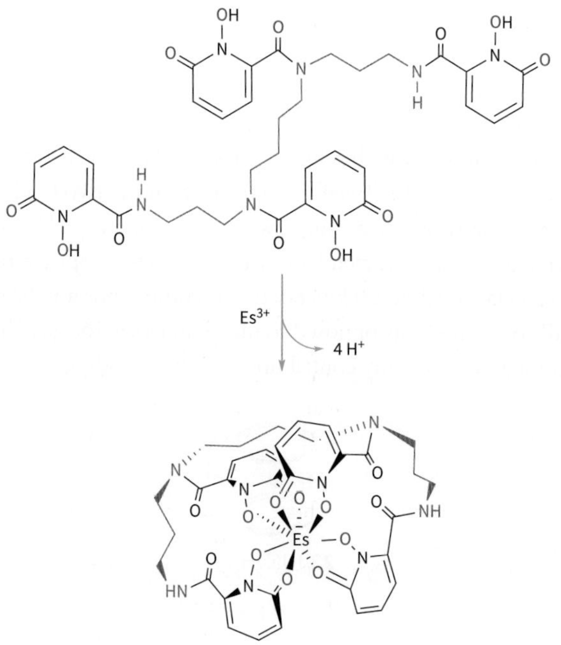

FIGURE 21.20 Rapid complexation of Es^{3+} with an octadentate hydroxypyridinone (HOPO) ligand.

FURTHER READING

D. L. Clarke, The chemical complexities of plutonium. *Los Alamos Sci.*, 2000, **26**, 364. A valuable guide to Pu coordination chemistry.

D. M. P. Mingos and R. H. Crabtree (eds.), *Comprehensive Organometallic Chemistry III*. Elsevier (2006). Volume 4 (ed. M. Bochmann) deals with Groups 3 and 4 and the lanthanoids and actinoids.

W. J. Evans, The importance of questioning scientific assumptions: Some lessons from f element chemistry. *Inorg. Chem.*, 2007, **46**, 3435. A substantial tutorial paper explaining the discovery, characterization, and significance of interesting Ln^{2+} complexes.

P. L. Arnold and I. J. Casely, f-Block N-heterocyclic carbene complexes. *Chem. Rev.*, 2009, **109**, 3599. A comprehensive review of f-block organometallic compounds, focusing on those featuring strong donor ligands.

M. L. Neidig, D. L. Clarke, and R. L. Martin, Covalency in f-element complexes. *Coord. Chem. Rev.*, 2013, **257**, 394. An excellent article explaining the role of covalent bonding in actinoid complexes.

R. Sessoli and A. K. Powell, Strategies towards single molecule magnets based on lanthanide ions. *Coord. Chem. Rev.*, 2009, **253**, 2328. This article gives a good account of the magnetic properties of molecules containing several lanthanoid ions.

G. J. Stasiuk, S. Faulkner, and N.J. Long, Novel imaging chelates for drug discovery. *Curr. Opin. Pharmacol.*, 2012, **12**, 576. An interesting article on the design of ligands for applications in Ln^{3+} coordination chemistry.

J.-C. G. Bunzli, in *Lanthanide Luminescence: Photophysical, Analytical and Biological Aspects* (eds. P. Hanninen and H. Harma), Volume 7 in the Springer Series on Fluorescence, 2011. A good account of the optical properties of lanthanoids.

J.-C. G. Bunzli, Lanthanide luminescence for biomedical analyses and imaging. *Chem. Rev.*, 2010, **110**, 2729. A comprehensive review of the biomedical applications of Ln compounds.

X. Huang, S. Han, W. Huang, and X. Liu, Enhancing solar cell efficiency: The search for luminescent materials as spectral converters. *Chem. Soc. Rev.*, 2013, **42**, 173. An article describing 'quantum cutting' and other photophysical properties of Ln materials.

W. J. Evans, Tutorial on the role of cyclopentadienyl ligands in the discovery of molecular complexes of the rare-earth and actinide metals in new oxidation states. *Organometallics*, 2016, **35**, 3088. An account of the role of d orbitals in the chemistry of Ln(II) and stabilization of U(II).

P. L. Arnold, M. W. McMullor, J. Rieb, and F. E. Kuehn, C–H bond activation by f-block complexes. *Angew. Chem. Int. Ed.*, 2015, **54**, 82. A review of the progress made in developing f-block chemistry in homogeneous catalysis.

O. Guillou, C. Daiguebonne, G. Calvez, and K. Bernot, A long journey in lanthanide chemistry: from fundamental crystallogenesis studies to commercial anticounterfeiting taggants. *Acc. Chem. Res.*, 2016, **49**, 844. An article that highlights the future potential for lanthanoid chemistry in high technology.

J. T. Boronski et al., A crystalline tri-thorium cluster with σ-aromatic metal–metal bonding. *Nature* 2021, **598**, 72–8. An important report showing that actinoids can form cluster complexes containing metal–metal bonds.

K. P. Carter et al., Structural and spectroscopic characterization of an einsteinium complex. *Nature* 2021, **590**, 85–90. An article describing the characterization of a lanthanoid-like complex of a highly unstable heavy actinide.

EXERCISES

21.1 (a) Give a balanced equation for the reaction of any of the lanthanoids with aqueous acid. (b) Justify your answer with reduction potentials and with a generalization on the most stable positive oxidation states for the lanthanoids. (c) Name two lanthanoids that have the greatest tendency to deviate from the usual oxidation state, and correlate this deviation with electronic structure.

21.2 Explain the variation in ionic radii between La^{3+} and Lu^{3+}.

21.3 From a knowledge of their chemical properties, speculate on why cerium and europium were the easiest lanthanoids to isolate before the development of ion-exchange chromatography.

21.4 Explain why $Nd(C_5H_5)_3$ rapidly exchanges its cyclopentadienyl ligands with Cl^- and many other anions, but the cyclopentadienyl ligands in $U(C_5H_5)_3$ are not labile.

21.5 Discuss the assertion that the chemistry of the actinoids is closer to that of the d-block elements than to that of the lanthanoids.

21.6 Explain why UF_3 and UF_4 are high melting point solids, whereas UF_6 sublimes at 57°C.

21.7 Predict what species are formed when Pu metal is dissolved in dilute HCl and the nature of the solid product that forms when HF is subsequently added.

21.8 Account for the strength of the An–O bonds in the following AnO_2^{2+} (aq) cations: UO_2^{2+}, NpO_2^{2+}, PuO_2^{2+}, AmO_2^{2+} and explain why a linear arrangement is found in all cases.

21.9 Compare and contrast the types of ligands that are able to form stable complexes with lanthanoids and actinoids, distinguishing light and heavy elements where appropriate.

21.10 Suggest reasons for the appearance of the 2+ oxidation state in compounds of f-block elements.

21.11 A lanthanoid compound is reported in which Ln appears to be in the 5+ oxidation state. Explain why such a discovery (fictitious in this case) would be highly significant, and predict the most likely identity of the lanthanoid.

21.12 The elements Eu and Gd display certain similarities with Mn and Fe. Discuss this statement.

21.13 Explain why stable and readily isolable carbonyl complexes are unknown for the lanthanoids.

21.14 Suggest a synthesis of neptunocene from $NpCl_4$.

21.15 Account for the similar electronic spectra of Eu^{3+} complexes with various ligands, compared to the variation of the electronic spectra of Am^{3+} complexes as the ligand is varied.

21.16 Predict a structure type for BkN based on the ionic radii $r(Bk^{3+}) = 96$ pm and $r(N^{3-}) = 146$ pm.

21.17 Explain why the U–O bond length in the uranyl ion $[UO_2]^{2+}$ typically has a value around 180 pm, just slightly more than the Os–O bond length in $[OsO_2]^{2+}$ (174 pm), whereas the conventional ionic radius of U^{6+} (73 pm) is much larger than that of Os^{6+} (54 pm).

21.18 Comment on the finding that atomic lawrencium has the ground state s^2p, whereas atomic lutetium has the ground state s^2d.

TUTORIAL PROBLEMS

21.1 Lanthanoid coordination compounds rarely exhibit isomerism in solution. Suggest two factors that might cause this phenomenon, explaining your reasoning. (See D. Parker, R. S. Dickins, H. Puschmann, C. Crossland, and J. A. K. Howard, *Chem. Rev.*, 2002, **102**, 1977.)

21.2 Neither lanthanoid nor actinoid organometallic compounds obey the 18-electron rule. Discuss the reasons, using the structures of the tris(Cp) and tris(Cp*) Ln and An complexes as examples. (See W. J. Evans and B. L. Davis, *Chem. Rev.*, 2002, **102**, 2119.)

21.3 The catalytic activity of many organolanthanoids is controlled by the ionic radius of Ln^{3+}. In general, $Ln(Cp^*)_2X$ complexes that are catalysts for olefin polymerization show decreasing reactivity as La is replaced by Lu. In contrast, some reactions proceed more rapidly as the catalyst is changed from La to Lu. Discuss the principles involved in tuning the catalytic activities of organolanthanoid catalysts in terms of rates and selectivity. (See C. J. Weiss and T. J. Marks, *Dalton Trans.*, 2010, **39**, 6576.)

21.4 The existence of a maximum oxidation number of +6 for both uranium (Z = 92) and tungsten (Z = 74) prompted the placement of U under W in early periodic tables. When the element after uranium, neptunium (Z = 93), was discovered in 1940, its properties did not correspond to those of rhenium (Z = 75), and this cast doubt on the original placement of uranium. (See G. T. Seaborg and W. D. Loveland, *The Elements Beyond Uranium*. Wiley-Interscience (1990), p. 9 *et seq*.) Using standard potential data from Resource section 3, discuss the differences in oxidation state stability between Np and Re.

21.5 In their review on new commercial applications of Ln chemistry (*Acc. Chem. Res.*, 2016, **49**, 844), Guillou and co-workers explain how the coordination of different Ln(III) ions within metal–organic frameworks can be exploited to produce unique luminescent inks for anticounterfeiting signatures. Discuss the principles of this technology.

21.6 The processing of spent nuclear fuel and separation of the lighter actinoid elements, U, Np, and Pu, is an important industrial process. Discuss the chemistry involved in the various methods used to extract and separate these elements.

21.7 Based on their article reporting the synthesis of new complexes of Np(III) and Pu(III) (*Chemical Communications*, 2022, **58**, 997), Evans and co-workers propose that cryptand ligands offer important new opportunities to develop transuranium chemistry. Outline the advances that such an approach could provide.

21.8 The 24-membered U ring complex shown below is very unusual, as it is rigid and contains both bridging azido and bridging nitrido ligands. Describe how this complex was synthesized, and the significance of the bonding that is observed. (See W. J. Evans, S. A. Kozimor, and J. W. Ziller, *Science*, 2005, **309**, 1835.)

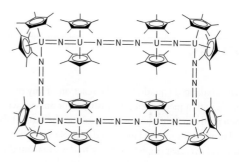

d- and f-Metal complexes: electronic and magnetic properties

22

Electronic spectra

22.1 Electronic spectra of atoms
22.2 Electronic spectra of d-metal complexes
22.3 Charge-transfer bands
22.4 Selection rules and intensities
22.5 Luminescence
22.6 Electronic spectra of the lanthanoids
22.7 Electronic spectra of the actinoids

Magnetism

22.8 Cooperative magnetism
22.9 Spin-crossover complexes
22.10 Magnetic properties of the lanthanoids

Further reading

Exercises

Tutorial problems

The electronic and magnetic properties of the d- and f-metal complexes play an important role in their uses. In this chapter we examine the electronic spectra of complexes and see how ligand-field theory allows us to interpret the energies and intensities of electronic transitions. Magnetic properties are also discussed.

Electronic spectra

In Chapter 7 we introduced crystal-field and ligand-field theory, and saw how they can simplistically account for the colour exhibited by many d-metal complexes. However, the basic idea of nondegenerate d-orbitals and electronic transitions between these levels is insufficient to account for the full complexity of the electronic spectra. The magnitudes of ligand-field splittings are such that the energy of electronic transitions often corresponds to the absorption of ultraviolet and visble light and hence produces colour. However, the presence of electron–electron repulsions within and between the metal orbitals means that the absorption frequencies are not in general a direct portrayal of the ligand-field splitting. The role of electron–electron repulsion was originally determined by the analysis of spherically symmetrical atoms and ions in the gas phase, and much of that information can be used in the analysis of the spectra of metal complexes provided we take into account the lower symmetry of a complex.

Throughout this discussion, keep in mind that the purpose is to find a way to understand and quantify the values of the ligand-field splitting parameters from the electronic absorption spectrum of a complex with more than one d electron, when electron–electron repulsions are important. First, we discuss the spectra of free atoms to understand how to take electron–electron repulsions into account. Then we see what energy states atoms adopt when they are embedded in the commonly adopted octahedral ligand field. Finally, we see how to represent the energies of these states for various field strengths and electron–electron repulsion energies (in the Tanabe–Sugano diagrams of Section 22.2), and how to use these diagrams to extract the value of the ligand-field splitting parameter.

22.1 Electronic spectra of atoms

KEY POINT Electron–electron repulsions result in multiple absorptions in the electronic spectrum.

Fig. 22.1 sets the stage for our discussion by showing the electronic absorption spectrum of the d^3 complex $[Cr(NH_3)_6]^{3+}$ in aqueous solution. The band at lowest energy (longest wavelength) is very weak; later we shall see that it is an example of a 'spin-forbidden' transition. Next are two bands with intermediate intensities; these are 'spin-allowed' transitions between the t_{2g} and e_g orbitals of the complex, which are mainly derived from the metal d orbitals. The third feature in the spectrum is an intense charge-transfer band at short wavelength (labelled CT, denoting 'charge transfer'), of which only the low-energy tail is evident in the illustration.

One problem that immediately confronts us is why two absorptions can be ascribed to the apparently single transition $t_{2g}^2 e_g^1 \leftarrow t_{2g}^3$. This splitting of a single transition into two bands is in fact an outcome of the electron–electron repulsions mentioned above. To understand how it arises, and to extract the information the spectrum contains, we first need to consider the spectra of free atoms and ions.

(a) Spectroscopic terms

KEY POINTS Different microstates exist for the same electronic configuration; for light atoms, Russell–Saunders coupling is used to describe the terms, which are specified by symbols in which the value of L is indicated by one of the letters S, P, D, ..., and the value of $2S + 1$ is given as a left superscript.

In Chapter 1 we expressed the electronic structures of atoms by giving their electronic configurations, the designation of the number of electrons in each orbital (as in $1s^2 2s^1$ for Li). However, a configuration is an incomplete description of the arrangement of electrons in atoms. In the configuration $2p^2$, for instance, the two electrons might occupy orbitals with different orientations of their orbital angular momenta (that is, with different values of m_l from among the possibilities +1, 0, and −1 that are available when $l = 1$). Similarly, the designation $2p^2$ tells us nothing about the spin orientations of the two electrons, for each electron m_s might be either $+\frac{1}{2}$ or $-\frac{1}{2}$. The atom may in fact have several different states of total orbital and spin angular momenta, each one corresponding to the occupation of orbitals with different values of m_l by electrons with different values of m_s. The different ways in which the electrons can occupy the orbitals specified in the configuration are called the **microstates** of the configuration. For example, one microstate of a $2p^2$ configuration is $(1^+,1^-)$; this notation signifies that both electrons occupy an orbital with $m_l = +1$ but do so with opposite spins, the superscript + indicating $m_s = +\frac{1}{2}$ and − indicating $m_s = -\frac{1}{2}$. Another microstate of the same configuration is $(-1^+,0^+)$. In this microstate, both electrons have $m_s = +\frac{1}{2}$, but one occupies the 2p orbital with $m_l = -1$ and the other occupies the orbital with $m_l = 0$.

The microstates of a given configuration have the same energy only if electron–electron repulsions on the atom are negligible. However, because atoms and most molecules are compact, interelectronic repulsions are strong and cannot always be ignored. As a result, microstates that correspond to different relative spatial distributions of electrons have different energies. If we group together the microstates that have the same energy when electron–electron repulsions are taken into account, we obtain the spectroscopically distinguishable energy levels called **terms**.

For light atoms and the 3d and 4f series, it turns out that the most important property of a microstate that helps us decide its energy is the relative orientation of the spins of the electrons. Next in importance is the relative orientation of the orbital angular momenta of the electrons. It follows that we can identify the terms of light atoms and put them in

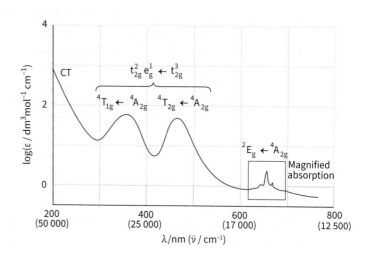

FIGURE 22.1 The spectrum of the d^3 complex $[Cr(NH_3)_6]^{3+}$, which illustrates the features studied in this section, and the assignments of the transitions as explained in the text.

order of increasing energy by sorting the microstates according to their total spin quantum number S (which is determined by the relative orientation of the individual spins) and subsequently (within values of S) according to their total orbital angular momentum quantum number L (which is determined by the relative orientation of the individual orbital angular momenta of the electrons). The process of combining electron angular momenta by summing first the spins, then the orbital momenta, and finally combining the two resultants is called **Russell–Saunders coupling**.

For heavy atoms, such as those of the 4d and 5d series, the relative orientations of orbital momenta or of spin momenta are weaker effects and, therefore, less important in initially determining relative energies. In these atoms the spin and orbital angular momenta of individual electrons are strongly coupled together by **spin–orbit coupling**, so the relative orientation of the spin and orbital angular momenta of each electron is the most important feature for determining the energy. The terms of heavy atoms are therefore sorted on the basis of the values of the total angular momentum quantum number j for an electron in each microstate. This scheme is called *jj-coupling*, but we shall not consider it further in this section.

Returning to the application of Russell–Saunders coupling to 3d metals, our first task is to identify the values of L and S that can arise from combining the orbital and spin angular momenta of individual electrons. Suppose we have two electrons with quantum numbers l_1, s_1 and l_2, s_2. Then, according to the **Clebsch–Gordan series**, the possible values of L and S are

$$L = l_1 + l_2, l_1 + l_2 - 1, \ldots, |l_1 - l_2|$$
$$S = s_1 + s_2, s_1 + s_2 - 1, \ldots, |s_1 - s_2| \quad (22.1)$$

(The modulus signs appear because neither L nor S can, by definition, be negative.) For example, an atom with configuration d² ($l_1 = 2$, $l_2 = 2$) can have the following values of L:

$$L = 2 + 2, 2 + 2 - 1, \ldots, |2 - 2| = 4, 3, 2, 1, 0$$

The total spin (because $s_1 = \frac{1}{2}$, $s_2 = \frac{1}{2}$) can be

$$S = \tfrac{1}{2} + \tfrac{1}{2}, \tfrac{1}{2} + \tfrac{1}{2} - 1, \ldots, |\tfrac{1}{2} - \tfrac{1}{2}| = 1, 0$$

To find the values of L and S for atoms with three electrons, we continue the process by combining l_3 with the value of L just obtained, and likewise for s_3.

Once L and S have been found, we can write down the allowed values of the quantum numbers M_L and M_S:

$$M_L = L, L-1, \ldots, -L \qquad M_S = S, S-1, \ldots, -S$$

These quantum numbers give the orientation of the angular momentum relative to an arbitrary axis: there are $2L + 1$ values of M_L for a given value of L and $2S + 1$ values of M_S for a given value of S. The values of M_L and M_S for a given microstate can be found very easily by adding together the values of m_l or m_s for the individual electrons. Therefore, if one electron has the quantum number m_{l1} and the other has m_{l2}, then

$$M_L = m_{l1} + m_{l2}$$

A similar expression applies to the total spin:

$$M_S = m_{s1} + m_{s2}$$

The microstate $(0^+, -1^-)$ is one microstate with $M_L = 0 - 1 = -1$ and $M_S = \tfrac{1}{2} + (-\tfrac{1}{2}) = 0$. Other microstates with the same energy and same overall M_L and M_S values include $(+1^-,-2^+)$ $(+1^+,-2^-)$ and $(0^-,-1^+)$.

By analogy with the notation s, p, d, ... for orbitals with $l = 0, 1, 2, \ldots$, the total orbital angular momentum of an atomic term is denoted by the equivalent uppercase letter:

$L =$	0	1	2	3	4	
	S	P	D	F	G	then alphabetical (omitting J)

The total spin is normally reported as the value of $2S + 1$, which is called the **multiplicity** of the term:

$S =$	0	½	1	3/2	2
$2S + 1 =$	1	2	3	4	5

The multiplicity is written as a left superscript on the letter representing the value of L, and the entire label of a term is called a **term symbol**. Thus, the term symbol ³P denotes a term (a collection of nearly degenerate states) with $L = 1$ and $S = 1$, and is called a *triplet term* (from the value of the multiplicity).

EXAMPLE 22.1 Deriving term symbols

Give the term symbols for an atom with the configurations (a) s¹, (b) p¹, and (c) s¹p¹.

Answer We need to use the Clebsch–Gordan series to couple any angular momenta, identify the letter for the term symbol from the table above, and then attach the multiplicity as a left superscript. (a) The single s electron has $l = 0$ and $s = ½$. Because there is only one electron, $L = 0$ (an S term), $S = s = ½$ and $2S + 1 = 2$ (a doublet term). The term symbol is therefore ²S. (b) For a single p electron, $l = 1$, so $L = 1$ and the term is ²P. (These terms arise in the spectrum of an alkali metal atom, such as Na.) (c) With one s and one p electron, $L = 0 + 1 = 1$, a P term. The electrons may be paired ($S = 0$) or parallel ($S = 1$). Hence both ¹P and ³P terms are possible.

Self-test 22.1 What terms arise from a p¹d¹ configuration?

(b) The classification of microstates

KEY POINT The allowed terms of a configuration are found by identifying the values of L and S to which the microstates of an atom can contribute.

The Pauli principle restricts the microstates that can occur in a configuration and consequently affects the terms that can occur. For example, two electrons cannot both have the same spin and be in a d orbital with $m_l = +2$. Therefore, the microstate $(2^+, 2^+)$ is forbidden, as is a term symbol associated with this microstate. We shall illustrate how to determine what terms are allowed by considering a d^2 configuration, as the outcome will be useful in the discussion of the complexes encountered later in the chapter. An example of a species with a d^2 configuration is a Ti^{2+} ion.

We start the analysis by setting up a table of microstates of the d^2 configuration (Table 22.1); only the microstates allowed by the Pauli principle have been included. We then use a process of elimination to classify all the microstates. First, we note the largest value of M_L, which for a d^2 configuration is +4. This state must belong to a term with $L = 4$ (a G term). Table 22.1 shows that the only value of M_S that occurs for this term is $M_S = 0$, so the G term must be a singlet. Moreover, as there are nine possible values of M_L when $L = 4$ (4, 3, 2, 1, 0, −1, −2, −3, −4), one of the microstates in each of the boxes in the column below $(2^+, 2^-)$ must belong to this term.[1] We can therefore strike out one microstate from each row in the central column of Table 22.1 (including one from each of $M_L = -1$ to -4), which leaves 36 microstates.

The next-largest value is $M_L = +3$, which must stem from $L = 3$ and hence belong to an F term. That row contains one microstate in each column (that is, each box contains one unassigned combination for $M_S = -1, 0$, and $+1$), which signifies $S = 1$ and therefore a triplet term. Hence the microstates belong to ^3F. The same is true for one microstate in each of the rows and each of the columns down to $M_L = -3$; altogether the seven values of M_L each with 3 possible M_S values account for a total of $3 \times 7 = 21$ microstates. If we strike out one state in each of these 21 boxes, we are left with 15 to be assigned.

There is one unassigned microstate in the row with $M_L = +2$ (which must arise from $L = 2$) and the column under $M_S = 0$ ($S = 0$), which must therefore belong to a ^1D term. This term has five possible values of M_L, which removes one microstate from each row in the column headed $M_S = 0$ down to $M_L = -2$, leaving 10 microstates now unassigned. Because these unassigned microstates include one with $M_L = +1$ and $M_S = +1$, nine (3 M_L values each with 3 M_S values) of these microstates must belong to a ^3P term. There now remains only one microstate in the central box of the table, with $M_L = 0$ and $M_S = 0$. This microstate must be the one and only state of a ^1S term (which has $L = 0$ and $S = 0$).

At this point we can conclude that the terms of a $3d^2$ configuration are ^1G, ^3F, ^1D, ^3P, and ^1S. These terms account for all 45 permitted states:

Term	Number of states
^1G	$9 \times 1 = 9$
^3F	$7 \times 3 = 21$
^1D	$5 \times 1 = 5$
^3P	$3 \times 3 = 9$
^1S	$1 \times 1 = 1$
Total:	45

(c) The energies of the terms

KEY POINT Hund's rules indicate the ground term of a gas-phase atom or ion.

Once the values of L and S that can arise from a given configuration are known, it is possible to identify the term of lowest energy by using Hund's rules. The first of these empirical rules was introduced in Section 1.5, where it was expressed as 'the lowest energy configuration is achieved if the electron spins are parallel'. Because a high value of S stems from parallel electron spins, an alternative statement is:

1. For a given configuration, the term with the greatest multiplicity lies lowest in energy.

The rule implies that a triplet term of a configuration (if one is permitted) has a lower energy than a singlet term of the same configuration. For the d^2 configuration, this rule predicts that the ground state will be either ^3F or ^3P.

TABLE 22.1 Microstates of the d^2 configuration

M_L	M_S		
	−1	0	+1
+4		$(2^+,2^-)$	
+3	$(2^-,1^-)$	$(2^+,1^-)(2^-,1^+)$	$(2^+,1^+)$
+2	$(2^-,0^-)$	$(2^+,0^-)(2^-,0^+)(1^+,1^-)$	$(2^+,0^+)$
+1	$(2^-,-1^-)(1^-,0^-)$	$(2^+,-1^-)(2^-,-1^+)(1^+,0^-)$ $(1^-,0^+)$	$(2^+,-1^+)(1^+,0^+)$
0	$(1^-,-1^-)(2^-,-2^-)$	$(1^+,-1^-)(1^-,-1^+)(2^+,-2^-)$ $(2^-,-2^+)(0^+,0^-)$	$(1^+,-1^+)(2^+,-2^+)$
−1 to −4*			

*The lower half of the diagram is a reflection of the upper half.

---

[1] In fact, it is unlikely that one of the microstates itself will correspond to one of these states: in general, a state is a linear combination of microstates. However, as N linear combinations can be formed from N microstates, each time we cross off one microstate we are taking one linear combination into account, so the bookkeeping is correct even though the detail may be wrong.

By inspecting spectroscopic data, Hund also identified a second rule for the relative energies of the terms of a given multiplicity:

2. For terms of given multiplicity, the term with the greatest value of L lies lowest in energy.

The physical justification for this rule is that when L is high, the electrons are orbiting in the same direction and less likely to meet one another and hence experience a lower repulsion. If L is low, the electrons are more likely to be closer to each other, and hence repel one another more strongly.[2] The second rule implies that, of the two triplet terms of a d^2 configuration, the ^3F term is lower in energy than the ^3P term. It follows that the ground term of a d^2 species such as Ti^{2+} is expected to be ^3F.

The spin multiplicity rule is fairly reliable for predicting the ordering of terms, but the 'greatest L' rule is reliable only for predicting the ground term, the term of lowest energy; as a result of more complex interelectron repulsions there is generally little correlation of L with the order of the higher terms. Thus, for d^2 the rules predict the order

$$^3F < {}^3P < {}^1G < {}^1D < {}^1S$$

but the order observed for Ti^{2+} from spectroscopy is

$$^3F < {}^1D < {}^3P < {}^1G < {}^1S$$

The reasons for this difference are explored in the next section.

Normally, all we want to know is the identity of the ground term of an atom or ion. The procedure may then be simplified and summarized as follows:

1. Identify the microstate that has the highest value of M_S.

This step tells us the highest multiplicity of the configuration.

2. Identify the highest permitted value of M_L for that multiplicity.

This step tells us the highest value of L consistent with the highest multiplicity.

EXAMPLE 22.2 **Identifying the ground term of a configuration**

What is the ground term of the configurations (a) $3d^5$ of Mn^{2+} and (b) $3d^3$ of Cr^{3+}?

Answer First we need to identify the term with maximum multiplicity, as this will be the ground term. Then we need to identify the L value for any terms that have the maximum multiplicity, for the term with the highest L value will be the ground term.

---

[2] On an atomic level we can see that when we have two electrons in a d shell they are less likely to interact or meet if their orbital angular momenta are in the same direction with $m_l = +2$ and $m_l = +1$ than in opposite directions with $m_l = +2$ and $m_l = -2$.

(a) Because the d^5 configuration permits occupation of each d orbital singly with parallel spins, the maximum value of S is $5/2$, giving a multiplicity of $2 \times 5/2 + 1 = 6$, a sextet term. If each of the electrons is to have the same spin quantum number, all must occupy different orbitals and hence have different M_L values. Thus, the M_L values of the occupied orbitals will be $+2, +1, 0, -1$, and -2. This configuration is the only one possible for a sextet term. Because the sum of the M_L is 0, it follows that $L = 0$ and the term is ^6S.
(b) For the configuration d^3, the maximum multiplicity corresponds to all three electrons having the same spin quantum number, so $S = 3/2$. The multiplicity is therefore $2 \times 3/2 + 1 = 4$, a quartet. Again, the three M_L values must be different if the electrons are all parallel. There are several possible arrangements that give quartet terms, but the one that gives a maximum value of M_L has the three electrons with $M_L = +2, +1$, and 0, giving a total of $+3$, which must arise from a term with $L = 3$, an F term. Hence, the ground term of d^3 is ^4F.

Self-test 22.2 Identify the ground terms of (a) $2p^2$ and (b) $3d^9$. (Hint: Because d^9 is one electron short of a closed shell with $L = 0$ and $S = 0$, treat it on the same footing as a d^1 configuration.)

Fig. 22.2 shows the relative energies of the terms for the d^2 and d^3 configurations of free atoms. Later, we shall see how to extend these diagrams to include the effect of a ligand field (Section 22.2).

(d) Racah parameters

KEY POINTS The Racah parameters summarize the effects of electron–electron repulsion on the energies of the terms that arise from a single configuration; the parameters are the quantitative expression of the ideas underlying Hund's rules and account for deviations from them.

Different terms of a configuration have different energies on account of the repulsion between electrons. To calculate the energies of the terms we must evaluate these electron–electron repulsion energies as complicated integrals over

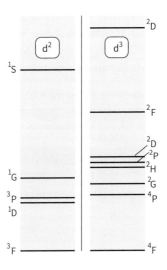

FIGURE 22.2 The relative energies of the terms arising from d^2 (left) and d^3 (right) configurations of a free atom.

the orbitals occupied by the electrons. Mercifully, however, all the integrals for a given configuration can be collected together in three specific combinations, and the repulsion energy of any term of a configuration can be expressed as a sum of these three quantities. The three combinations of integrals are called the **Racah parameters** and denoted A, B, and C. The parameter A corresponds to an average of the total interelectron repulsion, and B and C relate to the repulsion energies between individual d electrons. We do not even need to know the theoretical values of the parameters or the theoretical expressions for them, because it is more reliable to use A, B, and C as empirical quantities obtained from gas-phase atomic spectroscopy.

Each term stemming from a given configuration has an energy that may be expressed as a linear combination of all three Racah parameters. For a d^2 configuration a detailed analysis shows that

$$E(^1S) = A + 14B + 7C \qquad E(^1G) = A + 4B + 2C$$
$$E(^1D) = A - 3B + 2C \qquad E(^3P) = A + 7B$$
$$E(^3F) = A - 8B$$

The values of A, B, and C can be determined by fitting these expressions to the observed energies of the terms. Note that A is common to all the terms (as remarked above, it is the average of the total interelectron repulsion energy); therefore, if we are interested only in their relative energies, we do not need to know its value. All three Racah parameters are positive as they represent electron–electron repulsions. Therefore, provided $C > 5B$, the energies of the terms of the d^2 configuration lie in the order

$$^3F < ^3P < ^1D < ^1G < ^1S$$

This order is nearly the same as obtained by using Hund's rules. However, if $C < 5B$, the advantage of having an occupation of orbitals that corresponds to a high orbital angular momentum is greater than the advantage of having a high multiplicity, and the ^3P term lies above ^1D (as is in fact the case for Ti^{2+}). Table 22.2 shows some experimental values of B and C. The values in parentheses indicate that $C \approx 4B$, so

TABLE 22.2 Racah parameters for some d-block ions*

	1+	2+	3+	4+
Ti		720 (3.7)		
V		765 (3.9)	860 (4.8)	
Cr		830 (4.1)	1030 (3.7)	1040 (4.1)
Mn		960 (3.5)	1130 (3.2)	
Fe		1060 (4.1)		
Co		1120 (3.9)		
Ni		1080 (4.5)		
Cu	1220 (4.0)	1240 (3.8)		

*The table gives the B parameter in cm^{-1} with the value of C/B in parentheses.

the ions listed there are in the region where Hund's rules are not reliable for predicting anything more than the ground term of a configuration.

The parameter C appears only in the expressions for the energies of states that differ in multiplicity from the ground state. Hence if, as is usual, we are interested only in the relative energies of terms of the same multiplicity as the ground state (that is, excitation without a change in spin state), we do not need to know the value of C. The parameter B is of the most interest, and we return to factors that affect its value in Section 22.2(f).

22.2 Electronic spectra of d-metal complexes

The preceding discussion related only to free atoms, but these are not generally encountered in normal chemistry, so we will now expand our discussion to encompass complex ions, where the energy levels of the orbitals are affected by the ligand field and electron–electron repulsion.

Consider the spectrum of $[Cr(NH_3)_6]^{3+}$ in Fig. 22.1: there are two central bands with intermediate intensities. Because both the transitions are between orbitals that are predominantly metal d orbital in character, with a separation characterized by the strength of the ligand-field splitting parameter Δ_O, these two transitions are called **d–d transitions** or **ligand-field transitions**.

To understand experimental ligand-field spectra, such as that in Fig. 22.1, we will initially consider the type of transition that can occur where d-orbitals are nondegenerate and then apply similar arguments to those in Section 22.1 relating to electron–electron repulsion.

(a) Ligand-field transitions

KEY POINT Electron–electron repulsion splits ligand-field transitions into components with different energies.

According to the discussion in Section 7.16, in an octahedral ligand field the five d orbitals are split into two sets of orbitals with different energies, the t_{2g} and e_g sets. We expect an octahedral d^3 complex $[Cr(NH_3)_6]^{3+}$ to have the ground-state configuration t_{2g}^3. The absorption near 25 000 cm^{-1} in the spectrum of this complex (Fig. 22.1) can be identified as arising from the excitation $t_{2g}^2 e_g^1 \leftarrow t_{2g}^3$ because the corresponding energy is typical of ligand-field splittings in complexes.

Before we embark on a full analysis of the energy levels involved in the transition, it will be helpful to see qualitatively from the viewpoint of molecular orbital theory why the transition gives rise to two bands. First, note that a $d_{z^2} \leftarrow d_{xy}$ transition, which is one way of achieving $e_g \leftarrow t_{2g}$, promotes an electron from an orbital in the xy-plane into the already electron-rich z-direction: that axis is electron-rich because both d_{yz} and d_{zx} are occupied (Fig. 22.3). However,

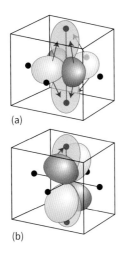

(a)

(b)

FIGURE 22.3 The shifts in electron density that accompany the two transitions discussed in the text. There is a considerable relocation of electron density towards the ligands on the z-axis in (a), but a much less substantial relocation in (b).

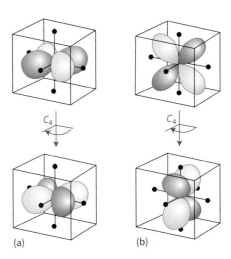

(a)   (b)

FIGURE 22.4 The changes in sign that occur under C_4 rotations about the z-axis: (a) a d_{xy} orbital is rotated into the negative of itself; (b) a d_{yz} orbital is rotated into a d_{zx} orbital.

a $d_{z^2} \leftarrow d_{zx}$ transition, which is another way of achieving $e_g \leftarrow t_{2g}$, merely relocates an electron that is already largely concentrated along the z-axis. In the former case, but not in the latter, there is a distinct increase in electron repulsion and, as a result, the two $e_g \leftarrow t_{2g}$ transitions lie at different energies. There are six possible $t_{2g}^2 e_g^1 \leftarrow t_{2g}^3$ transitions, and all resemble one or other of these two cases: three of them fall into one group and the other three fall into the second group, thus giving rise to the two bands observed in Fig. 22.1.

(b) The spectroscopic terms

KEY POINTS The terms of an octahedral complex are labelled by the symmetry species of the overall orbital state; a superscript prefix shows the multiplicity of the term.

The two bands we are discussing in Fig. 22.1 are labelled $^4T_{2g} \leftarrow {}^4A_{2g}$ (at 21 550 cm^{-1}) and $^4T_{1g} \leftarrow {}^4A_{2g}$ (at 28 500 cm^{-1}). The labels are **molecular term symbols** and serve a purpose similar to that of atomic term symbols introduced earlier. The left superscript denotes the multiplicity, so the superscript 4 denotes a quartet state with $S = \frac{3}{2}$, as expected when there are three unpaired electrons. The rest of the term symbol is the symmetry label of the overall electronic orbital state of the complex. For example, the nearly totally symmetric ground state of a d^3 complex (with an electron in each of the three t_{2g} orbitals) is denoted A_{2g}. We say *nearly* totally symmetric because close inspection of the behaviour of the three occupied t_{2g} orbitals shows that the C_3 rotation of the O_h point group transforms the product $t_{2g} \times t_{2g} \times t_{2g}$ into itself, which identifies the complex as an A symmetry species (see the character table in Resource section 4). Moreover, because each orbital has even parity (g), the overall parity is also g. However, each C_4 rotation transforms one t_{2g} orbital into the negative of itself and the other two t_{2g} orbitals into each other (Fig. 22.4), so overall there is a change of sign under this operation and its character is −1. The term is therefore A_{2g} rather than the totally symmetric A_{1g} of a closed shell.

It is more difficult to establish that the term symbols that can arise from the quartet $t_{2g}^2 e_g^1$ excited configuration are $^4T_{2g}$ and $^4T_{1g}$; they are derived by direct product analysis and we shall not consider this aspect here. The superscript 4 implies that the upper configuration continues to have the same number of unpaired spins as in the ground state, and the subscript g stems from the even parity of all the contributing orbitals.

(c) Correlating the terms

KEY POINT In the ligand field of an octahedral complex, the free atom terms split and are then labelled by their symmetry species, as enumerated in Table 22.3.

In a free atom, where all five d orbitals in a shell are degenerate, we needed to consider only the electron–electron repulsions to arrive at the relative ordering of the terms of a given d^n configuration. In a complex, the d orbitals are not all degenerate and it is necessary to take into account the difference in energy between the t_{2g} and e_g orbitals as well as the electron–electron repulsions.

Consider the simplest case of an atom or ion with a single valence electron. Because a totally symmetric orbital

TABLE 22.3 The correlation of spectroscopic terms for d electrons in O_h complexes

Atomic term	Number of states	Terms in O_h symmetry
S	1	A_{1g}
P	3	T_{1g}
D	5	$T_{2g} + E_g$
F	7	$T_{1g} + T_{2g} + A_{2g}$
G	9	$A_{1g} + E_g + T_{1g} + T_{2g}$

in one environment becomes a totally symmetric orbital in another environment, an s orbital in a free atom becomes an a_{1g} orbital in an octahedral field. We express the change by saying that the s orbital of the atom 'correlates' with the a_{1g} orbital of the complex. Similarly, the five d orbitals of a free atom correlate with the triply degenerate t_{2g} and doubly degenerate e_g sets in an octahedral complex.

We can now consider a many-electron atom. In exactly the same way as for a single electron, the totally symmetric overall S term of a many-electron atom correlates with the totally symmetric A_{1g} term of an octahedral complex. Likewise, an atomic D term splits into a T_{2g} term and an E_g term in O_h symmetry. The same kind of analysis can be applied to other states, and Table 22.3 summarizes the correlations between the free atom terms and the terms in an octahedral complex.

EXAMPLE 22.3 Identifying correlations between terms

What terms in a complex with O_h symmetry correlate with the ^3P term of a free atom with a d^2 configuration?

Answer We argue by analogy: if we know how p orbitals correlate with orbitals in a complex, then we can use that information to express how the overall states correlate, simply by changing to uppercase letters. The three p orbitals of a free atom become the triply degenerate t_{1u} orbitals of an octahedral complex. Therefore, if we disregard parity for the moment, a P term of a many-electron atom becomes a T_1 term in the point group O_h. Because d orbitals have even parity, the term overall must be g, and specifically T_{1g}. The multiplicity is unchanged in the correlation, so the ^3P term becomes a $^3T_{1g}$ term.

Self-test 22.3 What terms in a d^2 complex of O_h symmetry correlate with the ^3F and ^1D terms of a free atom?

(d) The energies of the terms: weak- and strong-field limits

KEY POINTS For a given metal ion, the energies of the individual terms respond differently to ligands of increasing field strength, and the correlation between free atom terms and terms of a complex can be displayed on an Orgel diagram.

As we have seen for free ions, electron–electron repulsions are difficult to take into account using simple theories but the discussion is simplified by considering two extreme cases. In the weak-field limit the ligand field, as measured by the size of Δ_O, is so weak that only electron–electron repulsions are important. This analysis is analogous to that used for the free ion and, as the Racah parameters B and C fully describe the interelectron repulsions, these are the only parameters we need at this limit. In the strong-field limit the ligand field is so strong that electron–electron repulsions can be ignored and the energies of the terms can be expressed solely in terms of Δ_O. Then, with the two extremes established, we can consider intermediate cases by drawing a correlation diagram between the two. We shall illustrate what is involved by considering two simple cases; namely, d^1 and d^2. Then we show how the same ideas are used to treat more complicated cases.

The only term arising from the d^1 configuration of a free atom is ^2D. In an octahedral complex the configuration is either t_{2g}^1, which gives rise to a $^2T_{2g}$ term, or e_g^1, which gives rise to a ^2E_g term. Because there is only one electron, there is no electron–electron repulsion to include, and the separation of the $^2T_{2g}$ and ^2E_g terms is the same as the separation of the t_{2g} and e_g orbitals, which is Δ_O. The correlation diagram for the d^1 configuration will therefore resemble that shown in Fig. 22.5.

We saw earlier that for a d^2 configuration the lowest energy term in the free atom is the triplet ^3F. We need consider only electronic transitions that start from the ground state, and in this section we will discuss only those in which there is no change in spin; that is, triplet states. There is one additional triplet term (^3P); relative to the lower term (^3F), the energies of the terms in the free ion are $E(^3F) = 0$ and $E(^3P) = 15B$; see Section 22.1. These two energies are marked on the left of Fig. 22.6.

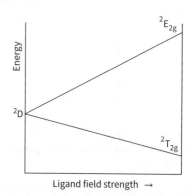

FIGURE 22.5 Correlation diagram for a free ion (left) and the strong-field terms (right) of a d^1 configuration.

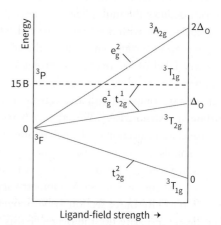

FIGURE 22.6 Correlation diagram for a free ion (left) and the strong-field terms (right) of a d^2 configuration.

Now consider the very-strong-field limit. A d^2 atom has the configurations

$$t_{2g}^2 < t_{2g}^1 e_g^1 < e_g^2$$

In an octahedral field, these configurations have different energies; that is, as we noted earlier, the ^3F term splits into three terms. From the analysis in Chapter 7 (Fig. 7.9), we can write their energies as

$$E(t_{2g}^2) = 2(-\tfrac{2}{5}\Delta_O) = -0.8\Delta_O$$
$$E(t_{2g}^1 e_g^1) = (-\tfrac{2}{5} + \tfrac{3}{5})\Delta_O = +0.2\Delta_O$$
$$E(e_g^2) = 2(\tfrac{3}{5})\Delta_O = +1.2\Delta_O$$

Therefore, relative to the energy of the lowest term, their energies are

$$E(t_{2g}^2, T_{1g}) = 0 \qquad E(t_{2g}^1 e_g^1, T_{2g}) = \Delta_O \qquad E(e_g^2, A_{2g}) = 2\Delta_O$$

These energies are marked on the right in Fig. 22.6.

Our problem now is to account for the energies when neither the ligand-field nor the electron repulsion term is dominant. To do so, we correlate the terms in the two extreme cases. The triplet t_{2g}^2 configuration gives rise to a $^3T_{1g}$ term, and this correlates with the ^3F term of the free atom. The remaining correlations can be established similarly, and we see that the $t_{2g}^1 e_g^1$ configuration gives rise to a $^3T_{2g}$ term and that the e_g^2 configuration gives rise to a $^3A_{2g}$ term; both terms correlate with the ^3F term of the free atom. Note that some terms, such as the $^3T_{1g}$ term that correlates with ^3P, are independent of the ligand-field strength, as would be expected with one electron in a t_{2g} orbital and one in an e_g orbital.

All the correlations are shown in Fig. 22.6, which is a simplified version of an **Orgel diagram**. An Orgel diagram is a plot of the energies of the various electron configurations as a function of the ligand-field strength and can be constructed for any d-electron configuration; several electronic configurations can be combined on the same diagram. Orgel diagrams are of considerable value for simple discussions of the electronic spectra of complexes; however, they consider only some of the possible transitions (the spin-allowed transitions, which is why we considered only the triplet terms). Orgel diagrams are qualitative and cannot be used in association with experimental spectra to extract a value for the ligand-field splitting parameter Δ_O.

(e) Tanabe–Sugano diagrams

KEY POINT Tanabe–Sugano diagrams are correlation diagrams that depict the energies of electronic states of complexes as a function of the strength of the ligand field.

More detailed theory can be used to construct complete diagrams showing the correlation of all possible terms for any electron configuration with strength of ligand field. The most widely used versions are called **Tanabe–Sugano diagrams**,

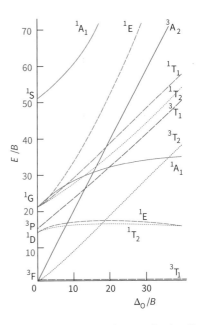

FIGURE 22.7 The Tanabe–Sugano diagram for the d^2 configuration. Note that the lefthand axis corresponds to Fig. 22.2 (left). A complete collection of diagrams for d^n configurations is given in Resource section 6. The parity subscript g has been omitted from the term symbols for clarity.

after the scientists who devised them. Fig. 22.7 shows the diagram for d^2, and we can see all the splittings that occur; thus the ^3F splits into three, the ^1D into two, and the ^1G into four. In these diagrams the term energies, E, are expressed as E/B and plotted against Δ_O/B, where B is the Racah parameter.

The relative energies of the terms arising from a given configuration are independent of A, and by choosing a value of C (typically setting $C \approx 4B$), terms of all energies can be plotted on the same diagrams. Some lines in Tanabe–Sugano diagrams are curved because of the mixing of terms of the same symmetry type. Terms of the same symmetry obey the **noncrossing rule**, which states that, if the increasing ligand field causes two weak-field terms of the same symmetry to approach, then they do not cross but bend apart from each other (Fig. 22.8). The effect of the noncrossing rule can be seen for the two ^1E terms, the two $^1T_2$ terms, and the two $^1A_1$ terms in Fig. 22.7.

Tanabe–Sugano diagrams for O_h complexes with configurations d^2 to d^8 are given in Resource section 6. The zero of energy in a Tanabe–Sugano diagram is always taken as that of the lowest term. Hence the lines in the diagrams have abrupt changes of slope when there is a change in the identity of the ground term brought about by the change from high-spin to low-spin with increasing field strength (see the diagram for d^4, for instance).

The purpose of the preceding discussion has been to find a way to extract the value of the ligand-field splitting parameter from the electronic absorption spectrum of a complex with more than one d electron when electron–electron repulsions are important, such as that in Fig. 22.1. The strategy involves fitting the observed transitions to the correlation

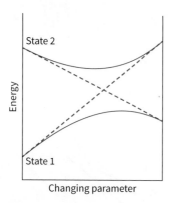

FIGURE 22.8 The noncrossing rule states that if two states of the same symmetry are likely to cross as a parameter is changed (as shown by the dashed blue lines), they will in fact mix together and avoid the crossing (as shown by the solid red lines).

lines in a Tanabe–Sugano diagram and identifying the values of Δ_O and B where the observed transition energies match the pattern. This procedure is illustrated in Example 22.4.

As we shall see in Section 22.4, certain transitions are allowed and certain transitions are forbidden. In particular, those that correspond to a change of spin state are forbidden, whereas those that do not are allowed. In general, spin-allowed transitions are found to dominate the UV–visible absorption spectrum and show easily observed absorption bands; thus, we might only expect to see three transitions for a d^2 ion, specifically $^3T_{2g} \leftarrow {}^3T_{1g}$, $^3T_{1g} \leftarrow {}^3T_{1g}$, and $^3A_{2g} \leftarrow {}^3T_{1g}$. However, some complexes (for instance, high-spin d^5 ions such as Mn^{2+}) do not have any spin-allowed transitions, all of the absorptions are weak, and none of the 11 possible transitions dominates.

EXAMPLE 22.4 Calculating Δ_O and B using a Tanabe–Sugano diagram

Deduce the values of Δ_O and B for $[Cr(NH_3)_6]^{3+}$ from the spectrum in Fig. 22.1 and a Tanabe–Sugano diagram.

Answer We need to identify the relevant Tanabe–Sugano diagram and then locate the position on the diagram where the observed ratio of transition energies (as wavenumbers) matches the theoretical ratio. The relevant diagram (for d^3) is shown in Fig. 22.9. We need concern ourselves only with the spin-allowed transitions, of which there are three for a d^3 ion (a $^4T_{2g} \leftarrow {}^4A_{2g}$ and two $^4T_{1g} \leftarrow {}^4A_{2g}$ transitions). We have seen that the spectrum in Fig. 22.1 exhibits two low-energy ligand-field transitions at 21 550 and 28 500 cm^{-1}, which correspond to the two lowest-energy transitions ($^4T_{2g} \leftarrow {}^4A_{2g}$ and $^4T_{1g} \leftarrow {}^4A_{2g}$). The ratio of the energies of the transitions is 1.32, and the only point in Fig. 22.9 where this energy ratio is satisfied is on the far right. Hence, we can read off the value of $\Delta_O/B = 33.0$ from the location of this point. The tip of the arrow representing the lower energy transition lies at $32.8B$ vertically, so equating $32.8B$ and 21 550 cm^{-1} gives $B = 657$ cm^{-1} and therefore $\Delta_O = 21\ 700$ cm^{-1}.

Self-test 22.4 Use the same Tanabe–Sugano diagram to predict the energies of the first two spin-allowed quartet bands in the spectrum of $[Cr(OH_2)_6]^{3+}$ for which $\Delta_O = 17\ 600$ cm^{-1} and $B = 700$ cm^{-1}.

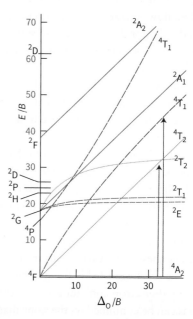

FIGURE 22.9 The Tanabe–Sugano diagram for the d^3 configuration. Note that the left-hand axis corresponds to Fig. 22.2 (right). A complete collection of diagrams for d^n configurations is given in Resource section 6. The parity subscript g has been omitted from the term symbols for clarity.

A Tanabe–Sugano diagram also provides some understanding of the widths of some absorption lines. Consider Fig. 22.7 and the $^3T_{2g} \leftarrow {}^3T_{1g}$ and $^3A_{2g} \leftarrow {}^3T_{1g}$ transitions. The Franck–Condon principle states that electronic transitions are so fast that they occur without nuclear motion (vibration); conversely, in a molecule with a variety of vibrational states, any electronic transition will sample all these states instantaneously, resulting in a broad electronic transition if the vibration affects the energy of that transition. Thus, because the line representing $^3T_{2g}$ is not parallel to that representing the ground state $^3T_{1g}$, any variation in the value of Δ_O (such as those caused by molecular vibration) will result in a change in the energy of the electronic transition and hence a broadening of the absorption band. The line representing $^3A_{2g}$ is even less parallel to the line representing $^3T_{1g}$; the energy of this transition will be affected even more by variations in Δ_O and will thus be even broader. By contrast, the line representing the lower $^1T_{2g}$ term is almost parallel to that representing $^3T_{1g}$ and thus the energy of this transition is largely unaffected by variations in Δ_O, and the absorption is consequently very sharp (albeit weak, because it is spin forbidden).

(f) The nephelauxetic series

KEY POINTS Electron–electron repulsions are lower in complexes than in free ions because of electron delocalization; the nephelauxetic parameter is a measure of the extent of d-electron delocalization on to the ligands of a complex; the softer the ligand, the smaller the nephelauxetic parameter.

In Example 22.4 we found that $B = 657$ cm^{-1} for $[Cr(NH_3)_6]^{3+}$, which is only 64% of the value for a Cr^{3+} ion in the gas phase. This reduction is a general observation and indicates that electron repulsions are weaker in complexes than in the free atoms and ions. The weakening occurs because the occupied molecular orbitals are delocalized over the ligands and away from the metal. The delocalization increases the average separation of the electrons and hence reduces their mutual repulsion.

The reduction of B from its free ion value is normally reported in terms of the **nephelauxetic parameter**, β:[3]

$$\beta = B(\text{complex})/B(\text{free ion}) \quad (22.2)$$

The values of β depend on the identity of the metal ion and the ligand, and a list of ligands ordered by the value of β gives the **nephelauxetic series**:

$$Br^- < CN^- < Cl^- < NH_3 < H_2O < F^-$$

A small value of β indicates a large measure of d-electron delocalization on to the ligands and hence a significant covalent character in the complex. Thus the series shows that a Br^- ligand results in a greater reduction in electron repulsions in the ion than an F^- ligand, which is consistent with a greater covalent character in bromido complexes than in analogous fluorido complexes. As an example, a free Ni^{2+} ion has $B = 1041$ cm^{-1}, whereas $[NiF_6]^{4-}$ has $B = 843$ cm^{-1} ($\beta = 0.81$), and $[NiBr_4]^{2-}$ has $B = 600$ cm^{-1} ($\beta = 0.58$). Another way of expressing the trend represented by the nephelauxetic series is: *the softer the ligand, the smaller the nephelauxetic parameter*. In addition, the nephelauxetic parameter depends on the charge on the metal ion so, for example β Mn(II) < β Mn(III) < β Mn(IV), again following increased covalency in bonding.

22.3 Charge-transfer bands

KEY POINTS Charge-transfer bands arise from the movement of electrons between orbitals that are predominantly ligand in character and orbitals that are predominantly metal in character; such transitions are identified by their high intensity and the sensitivity of their energies to solvent polarity.

[3] The name is from the Greek words for 'cloud-expanding'.

Another feature in the spectrum of $[Cr(NH_3)_6]^{3+}$ in Fig. 22.1 that remains to be explained is the very intense shoulder of an absorption that appears to have a maximum at well above 50 000 cm^{-1}. The high intensity suggests that this transition is not a simple ligand-field transition, but is consistent with a **charge-transfer transition** (CT transition).

In a CT transition, an electron migrates between orbitals that are predominantly ligand in character and orbitals that are predominantly metal in character. The transition is classified as a **ligand-to-metal charge-transfer transition** (LMCT transition) if the migration of the electron is from the ligand to the metal, and as a **metal-to-ligand charge-transfer transition** (MLCT transition) if the charge migration occurs in the opposite direction.

An example of an MLCT transition is the one responsible for the red colour of tris(bipyridyl)iron(II), the complex used for the colourimetric analysis of Fe(II). In this case, an electron makes a transition from a d orbital of the central metal into a π^* orbital of the ligand. Fig. 22.10 summarizes the transitions we classify as charge-transfer.

Several lines of evidence are used to identify a band as due to a CT transition. The high intensity of the band, which is evident in Fig. 22.1, is one strong indication, as CT transitions are fully allowed (Section 22.4). Another indication is if such a band appears following the replacement of one ligand with another, as this implies that the band is strongly dependent on the ligand. The CT character is most often identified (and distinguished from $\pi^* \leftarrow \pi$ transitions on ligands) by demonstrating **solvatochromism**, the variation of the transition frequency with changes in solvent permittivity. Solvent permittivity affects the stabilization of any charges associated with an ion, and will only have an effect on transition frequency if there is a large shift in electron density as a result of the transition, which is more consistent

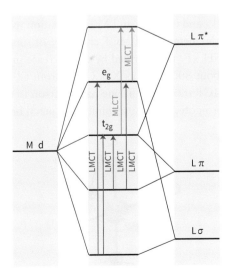

FIGURE 22.10 A summary of the charge transfer transitions in an octahedral complex.

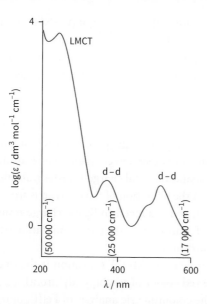

FIGURE 22.11 The absorption spectrum of $[CrCl(NH_3)_5]^{2+}$ in water in the visible and ultraviolet regions. The peak corresponding to the transition $2E \leftarrow 4A$ is not visible on this magnification.

with a metal–ligand transition than a ligand–ligand or metal–metal transition.

Fig. 22.11 shows another example of a CT transition in the visible and UV spectrum of $[CrCl(NH_3)_5]^{2+}$ (**1**). If we compare this spectrum with that of $[Cr(NH_3)_6]^{3+}$ in Fig. 22.1, we can recognize the two ligand-field bands in the visible region. The replacement of one NH_3 ligand by a weaker-field Cl^- ligand moves the lowest energy ligand-field bands to lower energy than those of $[Cr(NH_3)_6]^{3+}$. Also, a shoulder appears on the high-energy side of one of the ligand-field bands, indicating an additional transition that is the result of the reduction in symmetry from O_h to C_{4v}. The major new feature in the spectrum is the strong absorption maximum in the ultraviolet, near $42\,000\,\text{cm}^{-1}$. This band is at lower energy than the corresponding band in the spectrum of $[Cr(NH_3)_6]^{3+}$ and is due to an LMCT transition from the Cl^- ligand to the metal. The LMCT character of similar bands in $[CoX(NH_3)_5]^{2+}$ is confirmed by the decrease in energy in steps of about $8000\,\text{cm}^{-1}$ as X is varied from Cl to Br to I. In this LMCT transition, a lone-pair electron of the halide ligand is promoted into a predominantly metal orbital.

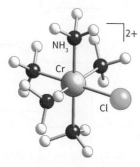

1 $[CrCl(NH_3)_5]^{2+}$

(a) LMCT transitions

KEY POINTS Ligand-to-metal charge-transfer transitions are observed in the visible region of the spectrum when the metal is in a high oxidation state and ligands contain nonbonding electrons; the variation in the position of LMCT bands can be correlated with the order of the electrochemical series.

Charge-transfer bands in the visible region of the spectrum (and hence contributing to the intense colours of many complexes) may occur if the ligands have lone pairs of relatively high energy (as in sulfur and selenium) or if the metal atom has low-lying empty orbitals.

The tetraoxidoanions of metals with high oxidation numbers (such as $[MnO_4]^-$) provide what are probably the most familiar examples of LMCT bands. In these, an O lone-pair electron is promoted into a low-lying empty metal e orbital. High metal oxidation numbers correspond to a low d-orbital population (many are formally d^0), so the acceptor level is available and low in energy. The trend in LMCT energies is:

Oxidation number	
+7	$[MnO_4]^- < [TcO_4]^- < [ReO_4]^-$
+6	$[CrO_4]^{2-} < [MoO_4]^{2-} < [WO_4]^{2-}$
+5	$[VO_4]^{3-} < [NbO_4]^{3-} < [TaO_4]^{3-}$

The UV–visible spectra of the tetraoxido anions of the Group 6 metals, $[CrO_4]^{2-}$, $[MoO_4]^{2-}$, and $[WO_4]^{2-}$ are shown in Fig. 22.12. The energies of the transitions correlate with the order of the electrochemical series (Section 6.4), with the lowest-energy transitions taking place to the most easily reduced metal ions. This correlation is consistent with the transition being the transfer of an electron from the ligands to the metal ion, corresponding, in effect, to the transient reduction of the metal ion by the ligands. Polymeric and

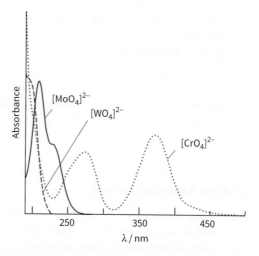

FIGURE 22.12 Optical absorption spectra of the ions $[CrO_4]^{2-}$, $[WO_4]^{2-}$, and $[MoO_4]^{2-}$. On descending the group, the absorption maximum moves to shorter wavelengths, indicating an increase in the energy of the LMCT band.

monomeric oxidoanions follow the same trends, with the oxidation state of the metal the determining factor. The similarity suggests that these LMCT transitions are localized processes that take place on discrete molecular fragments.

The correlation of the energy of the LMCT with the ease of reduction of the metal centre is often reflected in the colour of these tetraoxido complexes. For example, consider the colours of $[VO_4]^{3-}$, $[CrO_4]^{2-}$, and $[MnO_4]^-$ in aqueous solution: colourless/very pale yellow, yellow-orange, and purple, respectively. As the charge on the transition metal centre increases the energy of the LMCT decreases and the absorption band shifts from being almost totally in the UV region (λ_{max} = 280 nm, $[VO_4]^{3-}$), at the border of the UV-visible region (λ_{max} = 373 nm $[CrO_4]^{2-}$), and fully in the visible region (λ_{max} = 520 nm $[MnO_4]^-$. Similarly, as Mo(VI) is harder to reduce than Cr(VI) the $[MoO_4]^{2-}$ LMCT band lies in the UV region, and this anion is almost colourless. LMCT absorptions observed in metal complexes in solution have analogous transitions in many solids where an electron can be promoted from the valence band (filled and usually formed from the overlap of anion p orbitals) to the conduction band (usually empty or partially empty and formed from overlap of orbitals centred on the metal), Section 25.5.

(b) MLCT transitions

KEY POINT Charge-transfer transitions from metal to ligand are observed when the metal is in a low oxidation state and the ligands have low-lying acceptor orbitals.

Charge-transfer transitions from metal to ligand are most commonly observed in complexes with ligands that have low-lying π* orbitals, especially aromatic ligands. The transition occurs at low energy and appears in the visible spectrum if the metal ion is in a low oxidation state, as its d orbitals are then relatively close in energy to the empty ligand orbitals.

The family of ligands most commonly involved in MLCT transitions are the diimines, which have two N donor atoms: two important examples are 2,2′-bipyridine (bpy, **2**) and 1,10-phenanthroline (phen, **3**).

2 2,2′-Bipyridine (bpy)

3 1,10-Phenanthroline (phen)

Complexes of diimines with strong MLCT bands include tris(diimine) species such as tris(2,2′-bipyridyl)ruthenium(II) (**4**), which is orange. A diimine ligand may also be easily substituted into a complex with other ligands that favour a low oxidation state. Two examples are $[W(CO)_4(phen)]$ and $[Fe(CO)_3(bpy)]$. However, the occurrence of MLCT transitions is by no means limited to diimine ligands. Another important ligand type that shows typical MLCT transitions is dithiolene, $S_2C_2R_2^{2-}$ (**5**). Resonance Raman spectroscopy (Section 8.5) is a powerful technique for the study of MLCT transitions.

4 Tris(2,2′-bipyridyl)ruthenium(III)

5 Dithiolate

The MLCT excitation of tris(2,2′-bipyridyl)ruthenium(II) has been the subject of intense research efforts because the excited state that results from the charge transfer has a lifetime of microseconds, and the complex is a versatile photochemical redox reagent. The photochemical behaviour of a number of related complexes has also been studied on account of their relatively long excited-state lifetimes.

22.4 Selection rules and intensities

KEY POINT The strength of an electronic transition is determined by the transition dipole moment.

The contrast in intensity between typical charge-transfer bands and typical ligand-field bands raises the question of the factors that control the intensities of absorption bands. In an octahedral, nearly octahedral, or square-planar complex, the maximum molar absorption coefficient ε_{max} (which measures the strength of the absorption)[4] is typically less than or close to 100 dm³ mol⁻¹ cm⁻¹ for ligand-field transitions. In tetrahedral complexes, which have no centre of

---

[4] The molar absorption coefficient is the constant in the Beer–Lambert law for the transmittance $T = I_t/I_i$ when light passes through a length L of solution of molar concentration $[X]$ and is attenuated from an intensity I_i to an intensity I_t: $\log T = -\varepsilon[X]L$ = Absorbance (the logarithm is a common logarithm, to the base 10). Its older but still widely used name is the 'extinction coefficient'.

symmetry, ε_{max} for ligand-field transitions might exceed 250 dm³ mol⁻¹ cm⁻¹. By contrast, charge-transfer bands usually have an ε_{max} in the range 1000–50 000 dm³ mol⁻¹ cm⁻¹.

To understand the intensities of transitions in complexes, we have to explore the strength with which the complex couples with the electromagnetic field. Intense transitions indicate strong coupling; weak transitions indicate feeble coupling. The strength of coupling when an electron makes a transition from a state with wavefunction ψ_i to one with wavefunction ψ_f is measured by the **transition dipole moment**, which is defined as the integral

$$\mu_{fi} = \int \psi_f^* \mu \psi_i d\tau \qquad (22.3)$$

where μ is the electric dipole moment operator, $-er$. The transition dipole moment can be regarded as a measure of the impulse that a transition imparts to the electromagnetic field: a large impulse corresponds to an intense transition; zero impulse corresponds to a forbidden transition. The intensity of a transition is proportional to the square of its transition dipole moment.

A spectroscopic **selection rule** is a statement about which transitions are allowed and which are forbidden. An **allowed transition** is a transition with a nonzero transition dipole moment, and hence nonzero intensity. A **forbidden transition** is a transition for which the transition dipole moment is calculated as zero. Formally forbidden transitions may occur in a spectrum if the assumptions on which the transition dipole moments were calculated are invalid, such as the complex having a lower symmetry than assumed. Charge-transfer transitions are fully allowed and are thus associated with intense absorptions.

(a) Spin selection rules

KEY POINTS Electronic transitions with a change of multiplicity are forbidden; intensities of spin-forbidden transitions are greater for 4d- or 5d-series metal complexes than for comparable 3d-series complexes.

The electromagnetic field of the incident radiation cannot change the relative orientations of the spins of the electrons in a complex. For example, an initially antiparallel pair of electrons cannot be converted to a parallel pair, so a singlet ($S = 0$) cannot undergo a transition to a triplet ($S = 1$). This restriction is summarized by the rule $\Delta S = 0$ for **spin-allowed transitions** (Fig. 22.13).

The coupling of spin and orbital angular momenta can relax the spin selection rule, but such **spin-forbidden** ($\Delta S \neq 0$) transitions are generally much weaker than spin-allowed transitions. The intensity of spin-forbidden bands increases as the atomic number increases, because the strength of the spin–orbit coupling is greater for heavy atoms than for light atoms. The breakdown of the spin selection rule by spin–orbit coupling is often called the **heavy-atom effect**. In the 3d series, in which spin–orbit coupling is weak, spin-forbidden

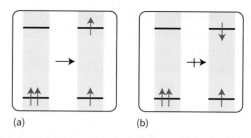

FIGURE 22.13 (a) A spin-allowed transition does not change the multiplicity. (b) A spin-forbidden transition results in a change in the multiplicity.

bands have ε_{max} less than about 1 dm³ mol⁻¹ cm⁻¹; however, spin-forbidden bands are a significant feature in the spectra of heavy d-metal complexes.

The very weak transition labelled $^2E_g \leftarrow {}^4A_{2g}$ in Fig. 22.1 is an example of a spin-forbidden transition. Some metal ions, such as the high-spin d⁵ Mn²⁺ ion, have no spin-allowed transitions, and hence are only weakly coloured.

(b) The Laporte selection rule

KEY POINTS Transitions between d orbitals are forbidden in octahedral complexes; asymmetrical vibrations relax this restriction.

The **Laporte selection rule** states that *in a centrosymmetric molecule or ion, the only allowed transitions are those accompanied by a change in parity*. That is, transitions between g and u terms are permitted, but a g term cannot undergo a transition to another g term and a u term cannot undergo a transition to another u term:

$$g \leftrightarrow u \quad g \not\leftrightarrow u \quad u \not\leftrightarrow u$$

In many cases it is enough to note that in a centrosymmetric complex, if there is no change in quantum number l, then there can be no change in parity. Thus, s–s, p–p, d–d, and f–f transitions are forbidden. Since s and d orbitals are g, whereas p and f orbitals are u, it follows that s–p, p–d, and d–f transitions are allowed, whereas s–d and p–f transitions are forbidden.

A more formal treatment of the Laporte selection rule is based on the properties of the transition dipole moment, which is proportional to r. Because r changes sign under inversion (and is therefore u), the entire integral in eqn. 22.3 also changes sign under inversion if ψ_i and ψ_f have the same parity, because g × u × g = u and u × u × u = u. Therefore, because the value of an integral cannot depend on the choice of coordinates used to evaluate it,⁵ it vanishes if ψ_i and ψ_f have the same parity. However, if they have opposite parity, the integral does not change sign under inversion of the coordinates, because g × u × u = g and therefore need not vanish.

---

⁵ An integral is an area, and areas are independent of the coordinates used for their evaluation. An integral vanishes unless it contains the totally symmetric function (A_1).

In a centrosymmetric complex, d–d ligand-field transitions are g ↔ g and are therefore forbidden. Their forbidden character accounts for the relative weakness of these transitions in octahedral complexes (which are centrosymmetric) compared with those in tetrahedral complexes, for which the Laporte rule does not apply (they are noncentrosymmetric, and their orbitals have no g or u as a subscript).

The question remains why d–d ligand-field transitions in octahedral complexes occur at all, even weakly. The Laporte selection rule may be relaxed in two ways. First, a complex may depart slightly from perfect centrosymmetry in its ground state, perhaps on account of the intrinsic asymmetry in the structure of polyatomic ligands or a distortion imposed by the environment of a complex packed into a crystal. Alternatively, the complex might undergo an asymmetrical vibration, which also destroys its centre of inversion. In either case, a Laporte-forbidden d–d ligand-field band tends to be much more intense than a spin-forbidden transition.

Table 22.4 summarizes typical intensities of electronic transitions of complexes of the 3d-series elements. The width of spectroscopic absorption bands is due principally to the simultaneous excitation of vibration when the electron is promoted from one distribution to another. According to the **Franck–Condon principle**, the electronic transition takes place within a stationary nuclear framework. As a result, after the transition has occurred, the nuclei experience a new force field and the molecule begins to vibrate anew.

TABLE 22.4 Intensities of spectroscopic bands in 3d complexes

Band type	$\varepsilon_{max}/(dm^3\,mol^{-1}\,cm^{-1})$
Spin-forbidden	< 1
Laporte-forbidden d–d	20–100
Laporte-allowed d–d	c. 250
Symmetry-allowed (e.g. CT)	1000–50 000

EXAMPLE 22.5 Assigning a spectrum using selection rules

Assign the bands in the spectrum in Fig. 22.11 by considering their intensities.

Answer If we assume that the complex is approximately octahedral, examination of the Tanabe–Sugano diagram for a d^3 ion reveals that the ground term is $^4A_{2g}$. Transitions to the higher terms ^2E_g, $^2T_{1g}$, and $^2T_{2g}$ are spin-forbidden and will have ε_{max} < 1 dm^3 mol^{-1} cm^{-1}. Thus, very weak bands for these transitions are predicted and will be difficult to distinguish. The next two higher terms of the same multiplicity are $^4T_{2g}$ and $^4T_{1g}$. These terms are reached by spin-allowed but Laporte-forbidden ligand-field transitions, and have $\varepsilon_{max} \approx 100$ dm^3 mol^{-1} cm^{-1}: these are

the two bands at 360 and 510 nm. In the near UV, the band with $\varepsilon_{max} \approx 10\,000$ dm^3 mol^{-1} cm^{-1} corresponds to the LMCT transitions in which an electron from a chlorine π lone pair is promoted into a molecular orbital that is principally metal d orbital in character.

Self-test 22.5 The spectrum of $[Cr(NCS)_6]^{3-}$ has a very weak band near 16 000 cm^{-1}, a band at 17 700 cm^{-1} with ε_{max} = 160 dm^3 mol^{-1} cm^{-1}, a band at 23 800 cm^{-1} with ε_{max} = 130 dm^3 mol^{-1} cm^{-1}, and a very strong band at 32 400 cm^{-1}. Assign these transitions using the d^3 Tanabe–Sugano diagram and selection rule considerations. (*Hint*: NCS$^-$ has low-lying π* orbitals.)

22.5 Luminescence

KEY POINTS A luminescent complex is one that re-emits radiation after it has been electronically excited. Fluorescence occurs when there is no change in multiplicity, whereas phosphorescence occurs when an excited state undergoes intersystem crossing to a state of different multiplicity and then undergoes radiative decay.

A complex is **luminescent** if it emits radiation after it has been electronically excited by the absorption of radiation. Luminescence competes with nonradiative decay by thermal degradation of energy to the surroundings. Relatively fast radiative decay is not especially common at room temperature for d-metal complexes, so strongly luminescent systems are comparatively rare. Nevertheless, they do occur, and we can distinguish two types of process. Traditionally, rapidly decaying luminescence was called 'fluorescence', and luminescence that persists after the exciting illumination is extinguished was called 'phosphorescence'. However, because the lifetime criterion is not reliable, the modern definitions of the two kinds of luminescence are based on the distinctive mechanisms of the processes. **Fluorescence** is radiative decay from an excited state of the same multiplicity as the ground state. The transition is spin-allowed and is fast; fluorescence half-lives are a matter of nanoseconds. **Phosphorescence** is radiative decay from a state of different multiplicity from the ground state. It is a spin-forbidden process, and hence is often slow. Phosphorescence is exploited in phosphors (Box 30.1).

The initial excitation of a phosphorescent complex usually populates a state by a spin-allowed transition, so the mechanism of phosphorescence involves **intersystem crossing**, the nonradiative conversion of the initial excited state into another excited state of different multiplicity. This second state acts as an energy reservoir because radiative decay to the ground state is spin-forbidden. However, just as spin–orbit coupling allows the intersystem crossing to occur, it also breaks down the spin selection rule, so the radiative decay can occur. Radiative decay back to the ground state is slow, so a phosphorescent state of a d-metal complex may survive for microseconds or even longer.

An important example of phosphorescence is provided by ruby, which consists of a low concentration of Cr^{3+} ions in place of Al^{3+} in alumina. Each Cr^{3+} ion is surrounded octahedrally by six O^{2-} ions, and the initial excitations are the spin-allowed processes

$$t_{2g}^2 e_g^1 \leftarrow t_{2g}^3 : \ {}^4T_{2g} \leftarrow {}^4A_{2g} \text{ and } {}^4T_{1g} \leftarrow {}^4A_{2g}$$

These absorptions occur in the green and violet regions of the spectrum and are responsible for the red colour of the gem, as shown in Fig. 22.14, a Jablonski diagram. Intersystem crossing to a ${}^2E$ term of the t_{2g}^3 configuration occurs in a few picoseconds or less, and red 627 nm phosphorescence occurs as this doublet decays back into the quartet ground state. This red emission adds to the red perceived by the subtraction of green and violet light from white light, and adds brilliance to the gem's appearance. This effect was utilized in the first laser to be constructed (in 1960).

A similar ${}^2E \rightarrow {}^4A$ phosphorescence can also be observed from a number of Cr(III) complexes in solution. The ${}^2E$ term arises from the t_{2g}^3 configuration, which is the same as the ground state, and thus the strength of the ligand field is not important and the band is very narrow. The emission is always in the red (and close to the wavelength of ruby emission). If the ligands are rigid, as in $[Cr(bpy)_3]^{3+}$, the ${}^2E$ term may live for several microseconds in solution.

Another interesting example of a phosphorescent state is found in $[Ru(bpy)_3]^{2+}$. The excited singlet term produced by a spin-allowed MLCT transition of this d^6 complex

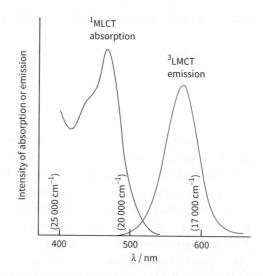

FIGURE 22.15 The absorption and phosphorescence spectra of $[Ru(bpy)_3]^{2+}$.

undergoes intersystem crossing to the lower-energy triplet term of the same configuration, $t_{2g}^5 \pi^{*1}$. Bright orange emission, as the excited electron returns to the metal (an LMCT transition), then occurs with a lifetime of about 1 ms (Fig. 22.15). The effects of other molecules (quenchers) on the lifetime of the emission may be used to monitor the rate of electron transfer from the excited state.

22.6 Electronic spectra of the lanthanoids

Much of the commercial and technological value of the lanthanoids stems from their optical and magnetic properties.

(a) Electronic absorption spectra

KEY POINTS Lanthanoid 3+ ions typically display weak but sharp absorption spectra because the f orbitals are shielded from the ligands: more intense, broader Laporte-allowed f–d transitions are restricted to the high-energy UV region of the spectrum. In contrast, lanthanoid 2+ ions are often highly coloured due to allowed transitions (f–d or charge transfer) that occur in the visible region.

Lanthanoid(III) ions are weakly coloured with absorptions that are associated with f–f transitions (Table 22.5). The spectra of their complexes generally show much narrower and more distinct absorption bands than those of d-metal complexes. Both the narrowness of the spectral features and their insensitivity to the nature of coordinated ligands are a consequence of the 4f orbitals having a smaller radial extension than the filled 5s and 5p orbitals. In contrast to the pale colours of Ln(III) complexes, many complexes of Ln(II) and Ln(IV) are intensely coloured, due to f–d or charge-transfer transitions in the visible region of the spectrum.

An analysis of f–f electronic transitions can be very complex; for instance, there are 91 microstates of an f^2 configuration. However, the discussion is simplified by the fact

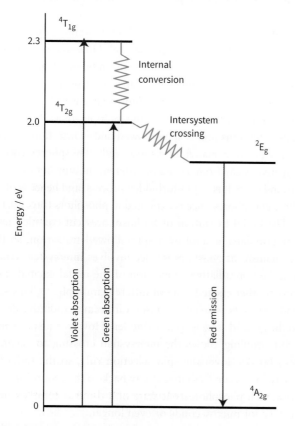

FIGURE 22.14 A Jablonski diagram that shows the transitions responsible for the absorption and luminescence of Cr^{3+} ions in ruby.

TABLE 22.5 Colours, term symbols, and magnetic moments of Ln^{3+} ions

	Colour in aqueous solution	Ground State	μ/μ_B	
			Theory	Observed[§]
La^{3+}	colourless	$^1S_0$	0	0
Ce^{3+}	colourless	$^2F_{5/2}$	2.54	2.46
Pr^{3+}	green-yellow	$^3H_4$	3.62	3.47–3.61
Nd^{3+}	violet	$^4I_{9/2}$	3.68	3.44–3.65
Pm^{3+}	–	$^5I_4$	2.83	–
Sm^{3+}	yellow	$^6H_{5/2}$	0.84 (1.55–1.65)[¶]	1.54–1.65
Eu^{3+}	pink	$^7F_0$	0 (3.40–3.51)[¶]	3.32–3.54
Gd^{3+}	colourless	$^8S_{7/2}$	7.94	7.9–8.0
Tb^{3+}	pink	$^7F_6$	9.72	9.69–9.81
Dy^{3+}	yellow-green	$^6H_{15/2}$	10.63	10.0–10.6
Ho^{3+}	yellow	$^5I_8$	10.60	10.4–10.7
Er^{3+}	lilac	$^4I_{15/2}$	9.59	9.4–9.5
Tm^{3+}	green	$^3H_6$	7.57	7.0–7.5
Yb^{3+}	colourless	$^2F_{7/2}$	4.54	4.0–4.5
Lu^{3+}	colourless	$^1S_0$	0	0

[¶] The values in parentheses include expected contributions from terms other than the ground state.
[§] From compounds such as $Ln_2(SO_4)_3 \cdot 8H_2O$ and $Ln(Cp)_3$.

that f orbitals lie relatively deep inside the atom and overlap only weakly with ligand orbitals. Hence, as a first approximation, their electronic states (and therefore electronic spectra) can be discussed in the free-ion limit with no splitting of the terms (Section 22.1).

Despite the elements having high atomic numbers the Russell–Saunders coupling scheme is still a good approximation and the best place to start. Thus the multiplicity and the orbital angular quantum momentum are used to derive an initial term symbol in an identical manner to that used for d-metals (Section 22.1(a)). However, for lanthanoid ions the spin-orbit coupling is sufficiently large that they have a single well-defined value of the total angular momentum, J, in the ground state. J can take values from $L + S$ to $|L−S|$. According to Hund's rules, for a less than half-filled shell, the lowest value of J lies lowest in energy, i.e $J = |L−S|$, whereas for a more than half-filled shell, it is the highest value of J that lies lowest, i.e $J = L + S$. The total angular momentum, J, is normally indicated as a subscript after the main letter of the term symbol.

EXAMPLE 22.6 Deriving the ground state term symbol of a lanthanoid ion

What is the ground state term symbol of Pr^{3+} (f^2)?

Answer The procedure for deriving the ground state term symbols, which have the general form $^{2S+1}\{L\}_J$, for d-block elements was summarized in Section 22.1(a), and we can proceed in a similar way. According to Hund's rules, the ground state will have the two electrons in different f orbitals, each with $l = 3$, so the maximum value of $M_L = m_{l_1} + m_{l_2}$ will be $M_L = (+3) + (+2) = +5$, which must stem from a state with $L = 5$, an H term. The lower energy spin arrangement of two electrons in different orbitals is a triplet with $S = 1$, so the term will be ^3H. The total angular momentum of a term with $L = 5$ and $S = 1$ will be $J = 6, 5,$ or 4. According to Hund's rules, for a less than half-full shell the level with the lowest value of J lies lowest (in this case, $J = 4$), so we can expect its term symbol to be $^3H_4$.

Self-test 22.6 What is the ground-state term symbol of (i) Gd^{3+} (f^7) and (ii) Er^{3+} (f^{11})?

The large number of microstates for each electronic configuration means a correspondingly large number of terms and hence of possible transitions between them. As the terms are derived almost purely from f orbitals, and there is little d–f orbital mixing or mixing with ligand orbitals, the transitions are Laporte-forbidden (Section 22.4). In addition, electrons in 4f orbitals interact only weakly with the ligands so there is little coupling of the electronic transitions with molecular vibrations, with the consequence that the bands are narrow and gain little intensity from vibronic coupling. Hence, in contrast to the d metals which normally show one or two broad bands of moderate intensity, the visible spectra of lanthanoids usually consist of a large number of sharp, low intensity peaks that are barely affected by changing the coordination environment of the lanthanoid. Molar absorption coefficients (ε) are typically 1–$10\,dm^3\,mol^{-1}\,cm^{-1}$ compared with octahedral d-metal complexes that are enhanced by vibronic coupling (close to $100\,dm^3\,mol^{-1}\,cm^{-1}$).

Ground state terms for all the Ln^{3+} ions are given in Table 22.5, and Fig. 22.16 shows a simplified diagram of the energy levels of excited states. Absorption bands arise due to excitation from the ground state into excited states, and Fig. 22.17 shows the experimental absorption spectrum of Pr^{3+}(aq) from the near-infrared to the ultraviolet region. The absorptions (upward arrows in Fig. 22.16) occur mainly between 450–500 nm (blue) and at 580 nm (yellow), so the residual light that reaches the eye after reflection from a Pr^{3+} compound is mainly green, giving this ion its characteristic pale green colour. The ion Gd^{3+} is colourless, as expected from the large energy gap. The Nd^{3+} ion has a sharp absorption at 580 nm due to the transition $^4I_{9/2} \leftarrow {}^4F_{3/2}$ giving compounds of Nd a pale violet colour. This wavelength corresponds almost exactly to the main yellow emission from excited sodium atoms (Section 12.1). As a result, neodymium ions are incorporated into goggles used by glassblowers, to reduce the glare produced by hot sodium silicate glasses.

Transitions between 4f and 5d orbitals for Ln^{3+} ions generally occur in the higher energy UV region of the spectrum. Since the Laporte rule is relaxed and the 5d orbitals

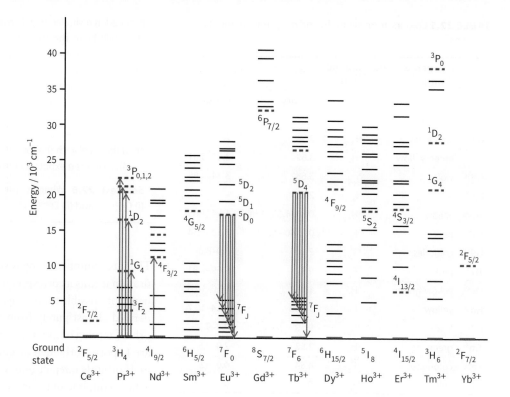

FIGURE 22.16 Simplified energy level diagram for the lanthanoid trivalent cations. Dashed red lines represent the major excited state luminescent levels. Upward arrows for Pr^{3+} and Nd^{3+} represent principal absorption lines for those ions. Downward arrows for Eu^{3+} and Tb^{3+} represent the principal emission lines for those ions.

have greater interaction with the ligand field, these transitions give rise to stronger and broader bands that are more sensitive to the environment. A guide to the trends in f- and d-orbital energies across the lanthanoid series was given in Fig. 21.6. From this graph we can see that the lowest energy bands are observed with Ce^{3+} for which $5d^1 \leftarrow 4f^1$ transitions start at about 250 nm. In contrast to Ln^{3+}, many compounds of Ln^{2+} are intensely coloured because the $4f^{n-1}5d^1 \leftarrow 4f^n$ transitions occur in the near-IR and visible regions of the spectrum.

(b) Luminescence

KEY POINT Lanthanoid ions show useful emission spectra with applications in phosphors, lasers, and imaging.

Some of the most important applications of the lanthanoids derive from their emission spectra produced after excitation of the f electrons. The emission spectra show similarities to the absorption spectra in that they consist of sharply defined frequencies characteristic of the lanthanoid cation and are mainly independent of the ligand.

Natural emission is limited by the low probability of absorption, which means that excited states are not easily populated. Even so, all the lanthanoid ions except La^{3+} (f^0) and Lu^{3+} (f^{14}) show inherent luminescence, and the strongest and most useful emissions are associated with Eu^{3+} ($^7F_{0-6} \leftarrow {}^5D_0$, giving red) and Tb^{3+} ($^7F_{6-0} \leftarrow {}^5D_4$, giving green), as indicated by the downward arrows in Fig. 22.16. In part, the relatively strong luminescence of Eu^{3+} and Tb^{3+} is due to the large number of excited states that exist, which increases the probability of intersystem crossing into excited states of different spin multiplicity from the ground state, thus giving rise to phosphorescence. More generally, the strength of the emission also stems from the excited electron interacting only weakly with its environment (due to the contracted nature of f orbitals) and thus having a long radiative lifetime (milliseconds to nanoseconds).

Whereas natural emission is limited by the fact that the excitation by absorption is only weakly allowed, luminescence can be greatly enhanced by placing a light-absorbing

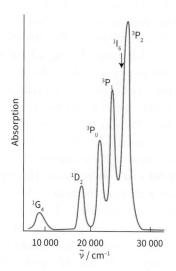

FIGURE 22.17 Absorption spectrum of Pr^{3+}. The energies are also indicated in Fig. 22.16.

> **BOX 22.1** What are lanthanoid-based phosphors and lasers?

Phosphors are fluorescent materials that normally convert high-energy photons, typically in the UV region of the electromagnetic spectrum, into lower-energy visible wavelengths. Whereas a number of fluorescent materials are based on d-metal systems (e.g. Cu and Mn doped into zinc sulfide) for some emitted energies lanthanoid-containing materials offer the best performance. This is particularly so of the red phosphors needed in displays, such as LED screens, and fluorescent lighting where the emission spectrum of Eu^{3+} occurs in a series of lines between 580 nm (orange) and 700 nm (red). Terbium, as Tb^{3+}, is also widely used in similar phosphors, producing emissions between 480 and 580 nm (green). The phosphor compounds, such as Eu^{3+} doped into YVO_4, are coated on the interiors of fluorescent light tubes and convert the UV radiation generated from the mercury discharge into visible light. By using a combination of phosphors fluorescing in different regions of the visible spectrum, white light is emitted by the tube (see Fig. B22.1). Similar phosphors are used in LED lighting (Section 30.1).

Neodymium can be doped into the garnet structure of yttrium aluminium oxide (YAG), $Y_3Al_5O_{12}$, at levels of 1% Nd for Y, to produce the material that is used in Nd:YAG lasers. The Nd^{3+} ions are strongly absorbing at wavelengths between 730–760 nm and 790–820 nm, such as the high-intensity light produced by krypton-containing flash lamps. The ions are excited from their ground $^4I_{9/2}$ state to a number of excited states which then transfer into a relatively long-lived excited $^4F_{3/2}$ state. Radiative decay of this state to $^4I_{11/2}$ (which lies just above the ground state $^4I_{9/2}$) is stimulated by a photon of the same frequency and results in laser radiation. So, once a large number of ions exist in an excited state, they can all be stimulated to emit light simultaneously, so producing very intense radiation.

A Nd:YAG laser typically emits at 1.064 μm, in the near-infrared region of the electromagnetic spectrum in pulsed and continuous modes. There are weaker transitions near 0.940,

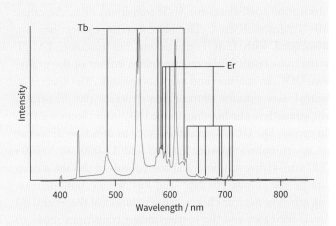

FIGURE B22.1 Spectrum of a fluorescent lamp showing lines due to lanthanoid ions in the phosphor.

1.120, 1.320, and 1.440 μm. The high-intensity pulses may be frequency-doubled to generate laser light at 532 nm (in the visible region) or even higher harmonics at 355 and 266 nm. Applications of Nd:YAG lasers include removal of cataracts and unwanted hair, range finders, and for marking plastics and glass. In the latter application, a white pigment that absorbs strongly in the infrared region, near the main Nd:YAG emission line of 1.064 μm, is added to a transparent plastic. When the laser is shone on the plastic, the pigment absorbs the energy and heats up sufficiently to burn the plastic at the point exposed to the laser, which becomes indelibly marked.

There are several YAG gain media with other lanthanoid dopants; for example, Yb:YAG emits at either 1.030 μm (strongest line) or 1.050 μm and is often used where a thin disk of lasing material is needed, and Er:YAG lasers emit at 2.94 μm and are used in dentistry and for skin resurfacing.

group (an 'antenna') on the ligand in a complex or nearby in a solid material. The antenna is excited by allowed transitions and the energy is transferred either directly or indirectly (via intersystem crossing) into excited states of the lanthanoid. The processes are usually represented by a Jablonski diagram like that shown previously in Fig. 22.14; here we can imagine the excitation of the antenna being equivalent to the absorptions shown as violet and green, with the energy transfer being equivalent to the intersystem crossing.

In electro-luminescence, lanthanoid compounds are excited electrochemically, using electrodes to inject electrons into the higher orbitals of suitable organic groups on the ligand and remove electrons from the occupied orbitals. Electroluminescence devices based on lanthanoids have been proposed for use in organic light-emitting diodes (OLEDs).

Exploiting the luminescence properties of lanthanoids and their compounds is a flourishing area of high technology, in which applications range across medical imaging, electronic displays, and encrypted signatures known as taggants. By stimulating the emission from the excited state, high intensity laser radiation can be obtained, as in neodymium–yttrium aluminium garnet (Nd:YAG) lasers (Box 22.1).

22.7 Electronic spectra of the actinoids

KEY POINTS The electronic spectra of the early actinoids have contributions from ligand-to-metal charge transfer, $5f \rightarrow 6d$, and $5f \rightarrow 5f$ transitions. The uranyl ion fluoresces strongly.

Transitions between electronic states involving only the f orbitals, the 5f and 6d orbitals, and LMCT are all possible for the actinoid ions; they may have very similar energies, and hence overlap, complicating the picture. The f–f transitions are broader and more intense than for the lanthanoids because the 5f orbitals extend beyond the core and interact more strongly with the ligands. Typical molar absorption coefficients for 5f–5f transitions lie in the range 10–100 dm^3 mol^{-1} cm^{-1}. The most intense absorptions are associated with LMCT transitions. For instance, LMCT transitions result in the intense yellow colour of the uranyl ion, UO_2^{2+}, in solution and in its compounds. The spectra of uranyl ions contain vibronic fine structure that illustrates the strength of the bonding in the UO_2^{2+} unit (Section 21.9). In species like U^{3+} (f^3) transitions such as $5f^2 6d^1 \leftarrow 5f^3$ occur at wavenumbers between 20 000 and 33 000 cm^{-1} (500–300 nm), giving solutions and compounds of this ion a deep orange-red colour. For Np^{3+} and Pu^{3+} which show increasing effective nuclear charge, the separation of the 5f and 6d levels increases and the corresponding transitions move to higher energies, into the UV region of the spectrum: Np^{3+} solutions are violet, and those of Pu^{3+} are light violet–blue due mainly to f–f transitions.

The uranyl ion, UO_2^{2+}, is also strongly fluorescent, with a strong emission between 500 and 550 nm when excited with UV radiation (Fig. 22.18). This property has been used for colouring glass, where the addition of 0.5–2% of uranyl salts produces a bright golden yellow colour due to the charge-transfer absorption discussed previously. The glass has a bright, green–yellow fluorescence in sunlight which adds to its attractiveness; when viewed under UV radiation it glows an intense green. However, as this glass is radioactive its commercial production has been largely phased out in recent years.

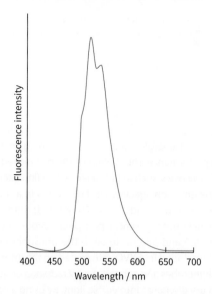

FIGURE 22.18 Emission spectrum of the uranyl ion, excited with UV radiation.

Magnetism

The diamagnetic and paramagnetic properties of d-metal complexes were introduced in Section 7.16, but the discussion was restricted to magnetically dilute species, where the individual paramagnetic centres—the atoms with unpaired d-electrons—are separate from each other. We now consider two further aspects of magnetism—one where magnetic centres can interact with one another, and one where the spin state may change. Materials that exhibit cooperative magnetism are discussed in detail in Section 30.6.

22.8 Cooperative magnetism

KEY POINT In solids, the spins on neighbouring metal centres may interact to produce magnetic behaviour, such as ferromagnetism and antiferromagnetism, that are representative of the whole solid.

In the solid state the individual magnetic centres are often close together and separated by only a single atom, typically O. In such arrays, cooperative properties can arise from interactions between electron spins on different atoms.

The **magnetic susceptibility**, χ, of a material is a measure of how easy it is to align electron spins with the applied magnetic field in the sense that the induced magnetic moment is proportional to the applied field, with χ the constant of proportionality. A paramagnetic material has a positive susceptibility, and a diamagnetic material has a negative susceptibility. Magnetic effects arising from cooperative phenomena can be very much larger than those arising from individual atoms and ions. The susceptibility and its variation with temperature are different for different types of magnetic materials, and are summarized in Table 22.6 and Fig. 22.19.

The application of a magnetic field to a paramagnetic material results in the partial alignment of the spins parallel to the field. As a paramagnetic material is cooled, the disordering effect of thermal motion is reduced, more spins become aligned, and the magnetic susceptibility increases. In a **ferromagnetic substance**, which is one example of a cooperative magnetic property, the spins on different metal centres are coupled into a parallel alignment that is sustained over thousands of atoms to form a **magnetic domain** (Fig. 22.20).

The net magnetic moment, and hence the magnetic susceptibility, may be very large because the magnetic moments of individual spins add to each other. Moreover, once established and with the temperature maintained below the **Curie temperature** (T_C), the magnetization persists after the applied field is removed because the spins are locked together. Ferromagnetism is exhibited by materials

TABLE 22.6 Magnetic behaviour of materials

Magnetic behaviour	Typical value of χ	Variation of χ with increasing temperature	Field dependence
Diamagnetism (no unpaired spins)	-8×10^{-6} for Cu	None	No
Paramagnetism	$+4 \times 10^{-3}$ for FeSO$_4$	Decreases	No
Ferromagnetism	$+5 \times 10^3$ for Fe	Decreases	Yes
Antiferromagnetism	$0{-}10^{-2}$	Increases	(Yes)

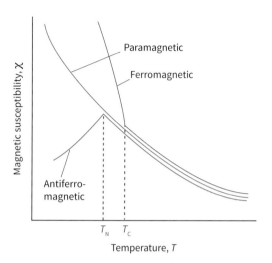

FIGURE 22.19 The temperature dependence of the susceptibilities of paramagnetic, ferromagnetic, and antiferromagnetic substances.

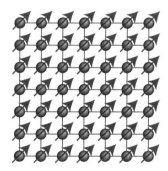

FIGURE 22.20 The parallel alignment of individual magnetic moments in a ferromagnetic material.

containing unpaired electrons in d or, more rarely, f orbitals that couple with unpaired electrons in similar orbitals on surrounding atoms. The key feature is that this interaction is strong enough to align spins, but not so strong as to form covalent bonds, in which the electrons would be paired. At temperatures above T_C the disordering effect of thermal motion overcomes the ordering effect of the interaction and the material becomes paramagnetic (Fig. 22.19).

The magnetization, M, of a ferromagnet—its bulk magnetic moment—is not proportional to the applied field strength H. Instead, a 'hysteresis loop' is observed like that shown in Fig. 22.21. For **hard ferromagnets** the loop is broad and M remains large when the applied field has been reduced to zero. Hard ferromagnets are used for permanent magnets where the direction of the magnetization does not need to be reversed. A **soft ferromagnet** has a narrower hysteresis loop and is therefore much more responsive to the applied field. Soft ferromagnets are used in transformers, where they must respond to a rapidly oscillating field.

In an **antiferromagnetic material**, neighbouring spins are locked into an antiparallel alignment (Fig. 22.22). As a result, the collection of individual magnetic moments cancel and the sample has a low magnetic moment and magnetic susceptibility (tending, in fact, to zero). Antiferromagnetism is often observed when a paramagnetic material is cooled to a low temperature, and is indicated by a sharp decrease in magnetic susceptibility at the **Néel temperature**, T_N (Fig. 22.19). Above T_N the magnetic susceptibility is that of a paramagnetic material, and decreases as the temperature is raised.

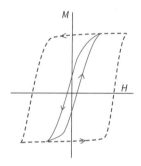

FIGURE 22.21 Magnetization curves for ferromagnetic materials. A hysteresis loop results because the magnetization of the sample with increasing field ($\rightarrow$) is not retraced as the field is decreased ($\leftarrow$). Dashed blue line: hard ferromagnet; solid red line: soft ferromagnet.

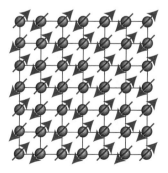

FIGURE 22.22 The antiparallel arrangement of individual magnetic moments in an antiferromagnetic material.

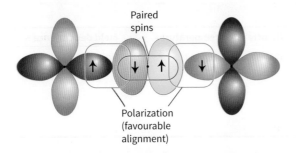

FIGURE 22.23 Antiferromagnetic coupling between two metal centres created by spin polarization of a bridging ligand.

The spin coupling responsible for antiferromagnetism generally occurs through intervening ligands by a mechanism called **superexchange**. As indicated in Fig. 22.23, the spin on one metal atom induces a small spin polarization on an occupied orbital of a ligand, and this spin polarization results in an antiparallel alignment of the spin on the adjacent metal atom. This alternating . . . ↑ ↓↑ ↓ . . . alignment of spins then propagates throughout the material. Many d-metal oxides exhibit antiferromagnetic behaviour that can be ascribed to a superexchange mechanism involving O atoms; for example, MnO is antiferromagnetic below 122 K and Cr_2O_3 is antiferromagnetic below 310 K. Coupling of spins through intervening ligands is frequently observed in molecular complexes containing two ligand-bridged metal ions, but it is weaker than with a simple O^{2-} link between metal sites, and as a result the ordering temperatures are much lower, typically below 100 K.

In **ferrimagnetism**, a net magnetic ordering of ions with different individual magnetic moments is observed below the Curie temperature. These ions can order with opposed spins, as in antiferromagnetism, but because the individual spin moments are different, there is incomplete cancellation and the sample has a net overall moment. As with antiferromagnetism, these interactions are generally transmitted through the ligands; an example is magnetite, Fe_3O_4.

There are a large number of molecular systems where magnetic coupling is observed. Typical systems have two or more metal atoms bridged by ligands that mediate the coupling. Simple examples include copper acetate (**6**), which exists as a dimer with antiferromagnetic coupling between the two d^9 centres. Many metalloenzymes (Sections 26.8 to 26.15) have multiple metal centres that show magnetic coupling.

22.9 Spin-crossover complexes

KEY POINT When the factors that determine the spin state of a d-metal centre are closely matched, complexes that change spin state in response to external stimuli are possible.

We have seen how a number of factors, such as oxidation state and ligand type, determine whether a d-metal complex is high- or low-spin. With some complexes, normally of the 3d-series metals, there is only a very small energy difference between the two states, leading to the possibility of **spin-crossover** complexes. Such complexes change their spin state in response to an external stimulus (such as heat or pressure), which in turn leads to a change in their bulk magnetic properties. An example is the d^6 iron complex of two diphenylterpyridine ligands (**7**), which is low-spin ($S = 0$) below 300 K, but high-spin ($S = 2$) above 323 K.

The transition from one spin state to another can be abrupt, gradual, or even stepped (Fig. 22.24). In the solid state, a further feature of spin-crossover complexes is the existence of cooperativity between the magnetic centres, which can lead to hysteresis like that shown in Fig. 22.25.

A preference normally exists for the low-spin state under high pressure and low temperature. This preference can be understood on the basis that the e_g orbitals, which are more extensively occupied in the high-spin state, have significant metal–ligand antibonding character. Thus, the high-spin form occupies a larger volume, which is favoured by low pressure or high temperature.

Spin-crossover complexes occur in many geological systems, and have the potential to be exploited in both practical

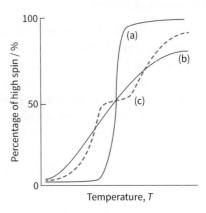

FIGURE 22.24 The change to high-spin might be (a) abrupt, (b) gradual, or (c) stepped.

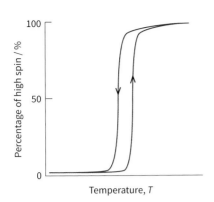

FIGURE 22.25 A hysteresis loop that can occur with some spin-crossover systems.

magnetic information storage and pressure-sensitive devices (Section 30.6).

22.10 Magnetic properties of the lanthanoids

KEY POINT The core-like nature of the unpaired 4f electrons in lanthanoid compounds leads them to exhibit magnetic moments in close agreement with values predicted for the free ions based upon Russell–Saunders coupling.

The magnetic properties of lanthanoid complexes differ greatly from those of the d block and are relatively easy to predict and explain. As we have seen, the spins of unpaired electrons in the 4f orbitals couple strongly with the orbital angular momentum but have little interaction with the ligand environment; as a result, the magnetic moment for a given $4f^n$ configuration, regardless of its chemical complexity, is close to the value calculated for the free ion. The magnetic moment μ is expressed in terms of the total angular momentum quantum number J:

$$\mu = g_J \{J(J+1)\}^{1/2} \mu_B$$

where the Landé g-factor is

$$g_J = 1 + \frac{S(S+1) - L(L+1) + J(J+1)}{2J(J+1)}$$

and μ_B is the Bohr magneton. Theoretical values of the magnetic moment of the ground states of the Ln^{3+} ions are summarized in Table 22.5; in general, these values agree well with experimental data.

A BRIEF ILLUSTRATION

As we have seen, the ground state term symbol for Pr^{3+} (f^2) is $^3H_4$ with $L = 5$, $S = 1$, and $J = 4$. It follows that

$$g_J = 1 + \frac{1(1+1) - 5(5+1) + 4(4+1)}{2 \times 4(4+1)} = 1 + \frac{2 - 30 + 20}{40} = \frac{4}{5}$$

Therefore

$$\mu = g_J \{J(J+1)\}^{1/2} \mu_B = \tfrac{4}{5}\{4(4+1)\}^{1/2} \mu_B = 3.58 \ \mu_B$$

The analysis in A Brief Illustration assumes that only one $^{2S+1}\{L\}_J$ level is occupied at the temperature of the experiment, and this is a good assumption for most lanthanoid ions. For example, the first excited state of Ce^{3+} ($^2F_{7/2}$) is approximately 2500 cm^{-1} above the ground state ($^2F_{5/2}$), and nearly unpopulated at room temperature when $kT \gg 200$ cm^{-1}. Small contributions from the higher energy terms result in minor deviations of observed values from those based on the population of a single term. For Eu^{3+}, and to a lesser extent Sm^{3+}, the first excited state lies close to the ground state (for Eu^{3+}, $^7F_1$ lies only 300 cm^{-1} above the $^7F_0$ ground state) and is partly populated even at room temperature. Although the value of μ based on occupation of only the ground state is zero (because $J = 0$), the experimentally observed value is nonzero and varies with temperature according to the Boltzmann population of the upper state.

Long-range magnetic ordering effects, ferromagnetism and antiferromagnetism (Section 22.8), are observed for many lanthanoid compounds in the solid state, though in general the coupling between metal atoms is much weaker than in the d-block compounds due to the contracted nature of the 4f orbitals that house the electrons. As a result of this weaker coupling the magnetic ordering temperatures for lanthanoid compounds are much lower. In $BaTbO_3$ the magnetic moments of the Tb^{3+} ion become antiferromagnetically ordered below 36 K. Lanthanoid complexes lie at the centre of research into **single molecule magnets** that offer important new possibilities for storing information at the molecular level (Section 25.19), one particularly special example being the triangular Dy(III) complex mentioned in Tutorial Problems.

FURTHER READING

E. U. Condon and G. H. Shortley, *The Theory of Atomic Spectra*. Cambridge University Press (1935). Revised as E. U. Condon and H. Odabasi, *Atomic Structure*. Cambridge University Press (1980). The standard reference text on atomic spectra.

E. I. Solomon and A. B. P. Lever, *Inorganic Electronic Structure and Spectroscopy*. John Wiley & Sons (2006). A thorough account of the material covered in this chapter, including a useful discussion of solvatochromism.

S. F. A. Kettle, *Physical Inorganic Chemistry: A Co-ordination Chemistry Approach*. Oxford University Press (1998).

B. N. Figgis and M. A. Hitchman, *Ligand Field Theory and Its Applications*. John Wiley & Sons (2000).

M. T. Weller and N. A. Young, *Characterisation Methods in Inorganic Chemistry*. Oxford University Press (2017). Contains a more detailed analysis of electronic spectroscopy of transition metals and lanthanoids.

K. Sridharan, *Spectral Methods in Transition Metal Complexes*, 1st ed. Elsevier (2016). A conceptual understanding on how to interpret the optical UV-vis spectroscopy of transition metal complexes.

J.-C. G. Bunzli, in *Lanthanide Luminescence: Photophysical, Analytical and Biological Aspects* (eds. P. Hanninen and H. Harma), Volume 7 in the Springer Series on Fluorescence, 2011. A good account of the optical properties of lanthanoids.

V. W. W. Yam, V. K. M. Au, and S. Y. L. Leung, Light-emitting self-assembled materials based on d^8 and d^{10} transition metal complexes. *Chem. Rev.*, 2015, **115**, 7589–7728.

A. F. Orchard, *Magnetochemistry*. Oxford University Press (2003). This book provides detailed, modern explanations, based on ligand-field theory, of the origins and interpretations of magnetic effects in complexes and materials.

X. Huang, S. Han, W. Huang, and X. Liu, Enhancing solar cell efficiency: The search for luminescent materials as spectral converters. *Chem. Soc. Rev.*, 2013, **42**, 173. An article describing 'quantum cutting' and other photophysical properties of Ln materials.

EXERCISES

22.1 Write the Russell–Saunders term symbols for states with the angular momentum quantum numbers (L,S): (a) $(0, 5/2)$, (b) $(3, 3/2)$, (c) $(2, 1/2)$, (d) $(1,1)$.

22.2 Identify the ground term from each set of terms: (a) ^1P, ^3P, ^3F, ^1G; (b) ^3P, ^5D, ^3H, ^1I, ^1G; (c) ^6S, ^4P, ^4G, ^2I.

22.3 Give the Russell–Saunders terms of the configurations (a) $4s^1$, (b) $3p^2$. Identify the ground term.

22.4 The gas-phase ion V^{3+} has a ^3F ground term. The ^1D and ^3P terms lie, respectively, 10 642 and 12 920 cm^{-1} above it. The energies of the terms are given in terms of Racah parameters as $E(^3F) = A - 8B$, $E(^3P) = A + 7B$, $E(^1D) = A - 3B + 2C$. Calculate the values of B and C for V^{3+}.

22.5 Write the d-orbital configurations and use the Tanabe–Sugano diagrams (Resource section 6) to identify the ground term of (a) low-spin $[Rh(NH_3)_6]^{3+}$, (b) $[Ti(OH_2)_6]^{3+}$, (c) high-spin $[Fe(OH_2)_6]^{3+}$.

22.6 Using the Tanabe–Sugano diagrams in Resource section 6, estimate Δ_O and B for (a) $[Ni(OH_2)_6]^{2+}$ (absorptions at 8500, 15 400, and 26 000 cm^{-1}) and (b) $[Ni(NH_3)_6]^{2+}$ (absorptions at 10 750, 17 500, and 28 200 cm^{-1}).

22.7 Match each of the extinction coefficients, ε_{max}, <1, 10, 100, >10^4 (in dm^3 mol^{-1} cm^{-1}), with one of the species: $[MnO_4]^-$ in H_2O; Fe^{3+} (aq) in 1.0 M $HClO_4$; Co^{2+} (aq) in 1.0 M $HClO_4$; $[CoCl_4]^{2-}$ in 10 M HCl.

22.8 The spectrum of $[Co(NH_3)_6]^{3+}$ has a very weak band in the red and two moderate intensity bands in the visible to near-UV. How should these transitions be assigned?

22.9 Explain why $[FeF_6]^{3-}$ is almost colourless whereas $[CoF_6]^{3-}$ is coloured and exhibits only a single band in the visible region of the spectrum.

22.10 The Racah parameter B is 460 cm^{-1} in $[Co(CN)_6]^{3-}$ and 615 cm^{-1} in $[Co(NH_3)_6]^{3+}$. Consider the nature of bonding with the two ligands and explain the difference in nephelauxetic effect.

22.11 An approximately 'octahedral' complex of Co(III) with ammine and chlorido ligands gives two bands with ε_{max} between 60 and 80 dm^3 mol^{-1} cm^{-1}, one weak peak with $\varepsilon_{max} = 2$ dm^3 mol^{-1} cm^{-1}, and a strong band at higher energy with $\varepsilon_{max} = 2 \times 10^4$ dm^3 mol^{-1} cm^{-1}. What do you suggest for the origins of these transitions?

22.12 Ordinary bottle glass appears nearly colourless when viewed through the wall of the bottle, but visibly coloured (green) when viewed from the end, so that the light has a long path through the glass. The colour is associated with the presence of Fe^{3+} in the silicate matrix. Explain this observation.

22.13 Solutions of $[Cr(OH_2)_6]^{3+}$ ions are pale blue-green but the chromate ion, $[CrO_4]^{2-}$, is an intense yellow. Characterize the origins of the transitions and explain the relative intensities.

22.14 Classify the symmetry type of the d orbitals in a tetragonal C_{4v} symmetry complex, such as $[CoCl(NH_3)_5]^{2+}$, where the Cl lies on the z-axis. (a) Which orbitals will be displaced from their position in the octahedral molecular orbital diagram by π interactions with the lone pairs of the Cl^- ligand? (b) Which orbital will move because the Cl^- ligand is not as strong a σ base as NH_3? (c) Sketch the qualitative molecular orbital diagram for the C_{4v} complex.

22.15 Consider the molecular orbital diagram for a tetrahedral complex (based on Fig. 7.15) and the relevant d-orbital configuration, and show that the purple colour of $[MnO_4]^-$ ions cannot arise from a ligand-field transition. Given that the wavenumbers of the two transitions in $[MnO_4]^-$ are 18 500 and 32 200 cm^{-1}, explain how to estimate Δ_T from an assignment of the two charge-transfer transitions, even though Δ_T cannot be observed directly.

22.16 Derive the ground state term symbol for the following ions: Tb^{3+}, Nd^{3+}, Ho^{3+}, Er^{3+}, Lu^{3+}.

22.17 Account for the similar electronic spectra of Eu^{3+} complexes with various ligands, compared to the variation of the electronic spectra of Am^{3+} complexes as the ligand is varied.

TUTORIAL PROBLEMS

22.1 Vanadium(IV) species that have the V=O group have quite distinct spectra. What is the d-electron configuration of V(IV)? The most symmetrical of such complexes are [VOL$_5$] with C_{4v} symmetry, with the O atom on the z-axis. What are the symmetry species of the five d orbitals in [VOL$_5$] complexes? How many d–d bands are expected in the spectra of these complexes? A band near 24 000 cm^{-1} in these complexes shows vibrational progressions of the V=O vibration, implicating an orbital involving V=O bonding. Which d–d transition is a candidate? (See C. J. Ballhausen and H. B. Gray, *Inorg. Chem.*, 1962, **1**, 111.)

22.2 MLCT bands can be recognized by the fact that the energy is a sensitive function of the polarity of the solvent (because the excited state is more polar than the ground state). Two simplified molecular orbital diagrams are shown in Fig. 22.26. In (a) is a case with a ligand π level higher than the metal d orbital. In (b) is a case in which the metal d orbital and the ligand level are at the same energy. Which of the two MLCT bands should be more solvent-sensitive? These two cases are realized by [W(CO)$_4$(phen)] and [W(CO)$_4$(^{i}Pr-DAB)], respectively, where DAB = 1,4-diaza-1,3-butadiene. (See P. C. Servas, H. K. van Dijk, T. L. Snoeck, D. J. Stufkens, and A. Oskam, *Inorg. Chem.*, 1985, **24**, 4494.) Comment on the CT character of the transition as a function of the extent of backdonation by the metal atom.

22.3 Determine the ground-state term symbols for the initial and final states for ionization of Pr^{2+}, Nd^{2+}, Pm^{2+}, Dy^{2+}, Ho^{2+}, and Er^{2+}, to Ln^{3+}, and hence comment on the influence of changes in orbital angular momentum (L) on corresponding I_3 values (the so-called quarter-/three-quarter shell effects).

22.4 Consider spin-crossover complexes and identify the features that a complex would need for it to be used in (a) a practical pressure sensor and (b) a practical information-storage device (see P. Gütlich, Y. Garcia, and H. A. Goodwin, *Chem. Soc. Rev.*, 2000, **29**, 419).

22.5 'Quantum cutting' is a process by which absorption of a single high-energy photon, corresponding to incident light in the UV spectral region, results in the emission of two photons of lower energy, in the visible region. The process is highly efficient and has important applications for improving the spectral range of solar photovoltaic cells, as well as providing more efficient domestic lighting. With reference to articles by Huang et al. (*Chem. Soc. Rev.* 2013, **42**, 173) and Lorbeer et al. (*Chem. Commun.*, 2010, **46**, 571), explain why Gd, in conjunction with other lanthanoids, is important for developing this technology.

22.6 Consider the octahedral [NiF$_6$]$^{3-}$ ion. What spin states and magnetic moments could it have? How would the Jahn–Teller effect affect the two spin states, and how might this complicate an interpretation of the UV/vis absorption spectrum? How has evidence from the magnetic moment and absorption spectrum allowed a definitive assignment of the spin state? (See T. I. Court and M. F. A. Dove, *J. Chem. Soc., Dalton*, 1973, p 1995.)

22.7 'Single molecule magnets' (abbreviated SMMs) offer intriguing new possibilities for storing and processing data: lanthanoids, due to their large number of unpaired electrons, are attracting a great deal of interest. The triangular complex [Dy$_3$(OH)$_2$(o-van)$_3$(H$_2$O)$_5$Cl]$^{3+}$ (**8**) consists of three Dy^{3+}, each coordinated by a molecule of vanillin and interlinked by hydroxide ions. With reference to the article by R. Sessoli and A. K. Powell (*Coord. Chem. Rev.*, 2009, **253**, 2328), explain why Dy(III) triangles are particularly interesting to those working in the SMM field.

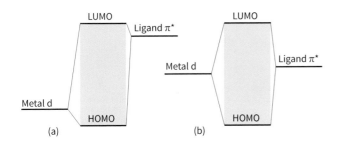

FIGURE 22.26 Representation of the orbitals involved in MLCT transitions for cases in which the energy of the ligand π* orbital varies with respect to the energy of the metal d orbital. See Tutorial Problem 22.2.

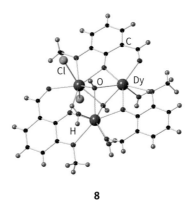

8

23 Coordination chemistry: reactions of complexes

Ligand substitution reactions
23.1 Rates of ligand substitution
23.2 The classification of mechanisms

Ligand substitution in square-planar complexes
23.3 The nucleophilicity of the entering group
23.4 The shape of the transition state

Ligand substitution in octahedral complexes
23.5 Rate laws and their interpretation
23.6 The activation of octahedral complexes
23.7 Base hydrolysis
23.8 Stereochemistry
23.9 Isomerization reactions

Redox reactions
23.10 The classification of redox reactions
23.11 The inner-sphere mechanism
23.12 The outer-sphere mechanism

Photochemical reactions
23.13 Prompt and delayed reactions
23.14 d–d and charge-transfer reactions
23.15 Transitions in metal–metal bonded systems

Further reading

Exercises

Tutorial problems

In this chapter we look at the evidence and experiments that are used in the analysis of the reaction pathways of metal complexes and so develop a deeper understanding of their mechanisms. Because a mechanism is rarely known definitively, the nature of the evidence for it should always be kept in mind in order to recognize that there might be other consistent possibilities. In the first part of this chapter we consider ligand-exchange reactions and describe how reaction mechanisms are classified. We consider the steps by which the reactions take place and the details of the formation of the transition state. These concepts are then used to describe the mechanisms of the redox reactions of complexes.

This chapter builds on the introduction to coordination chemistry in Chapter 7, together with the descriptive chemistry of the s-, d- and f-metals (Chapters 12, 13, 20, and 21). Most of the material here applies to all metals, regardless of the block to which they belong. However, there are special features of each block, and we shall point them out when they arise.

Ligand substitution reactions

The most fundamental reaction a complex can undergo is **ligand substitution**, a reaction in which one Lewis base displaces another from a Lewis acid:

$$Y + M-X \rightarrow M-Y + X$$

This class of reaction includes complex formation reactions, in which the **leaving group**, the displaced base X, is a solvent molecule and the **entering group**, the displacing base Y, is some other ligand. An example is the replacement of a water ligand by Cl^-:

$$[Co(OH_2)_6]^{2+}(aq) + Cl^-(aq) \rightarrow [CoCl(OH_2)_5]^+(aq) + H_2O(l)$$

The thermodynamic aspects of complex formation are discussed in Sections 7.12–7.15.

23.1 Rates of ligand substitution

KEY POINTS The rates of substitution reactions span a very wide range and correlate with the structures of the complexes; complexes that react quickly are called labile; those that react slowly are called inert or nonlabile.

Rates of reaction are as important as equilibria in coordination chemistry. The numerous isomers of the ammines of Co(III) and Pt(II), which were so important to the development of the subject, could not have been isolated if ligand substitutions and interconversion of the isomers had been fast. But what determines whether one complex will survive for long periods whereas another will undergo rapid reaction?

The rate at which one complex converts into another is governed by the height of the activation energy barrier that lies between them. Thermodynamically unstable complexes that survive for long periods (by convention, at least a minute) are commonly called 'inert', but **nonlabile** is perhaps more appropriate and is the term we shall use. Complexes that undergo more rapid equilibration are called **labile**. An example of each type is the labile complex $[Mn(OH_2)_6]^{2+}$, which has a half-life of the order of microseconds before the H_2O is replaced by another H_2O or a stronger base, and the nonlabile complex $[Co(NH_3)_5(OH_2)]^{3+}$, in which H_2O survives for many hours as a ligand before it is replaced by a stronger base.

Fig. 23.1 shows the characteristic lifetimes of the important aqua metal ion complexes. We see a range of lifetimes starting at about 1 ns, which is approximately the time it takes for a molecule to diffuse one molecular diameter in solution. At the other end of the scale are lifetimes in years. Even so, the illustration does not show the longest times that could be considered, which are comparable to geological eras.

We shall examine the lability of complexes in greater detail when we discuss the mechanism of reactions later in this section, but we can make two broad generalizations now. The first is that complexes of metals that have no additional factor to provide extra stability—for instance, the ligand field stabilization energy (LFSE) and chelate effects—are among the most labile. Any additional stability of a complex results in an increase in activation energy for a ligand replacement reaction and hence decreases the lability of the complex. A second generalization is that very small ions are often less labile because they have greater M–L bond strengths and it is sterically very difficult for incoming ligands to approach the metal atom closely.

Some further generalizations are as follows:

- All complexes of s-block ions except the smallest (Be^{2+} and Mg^{2+}) are extremely labile.
- Complexes of the M(III) ions of the f block are all very labile.

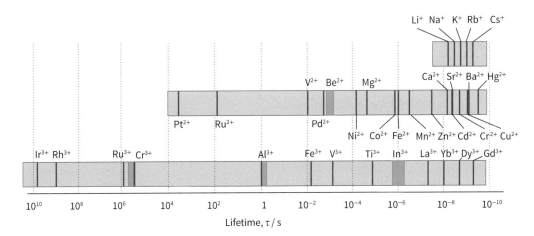

FIGURE 23.1 Characteristic lifetimes for exchange of water molecules in aqua complexes.

- Complexes of the d¹⁰ ions (Zn^{2+}, Cd^{2+}, and Hg^{2+}) are normally very labile.
- Across the 3d series, complexes of d-block M(II) ions are generally moderately labile, with distorted Cu(II) complexes among the most labile.
- Complexes of d-block M(III) ions are distinctly less labile than d-block M(II) ions.
- d-Metal complexes with d^3 and low-spin d^6 configurations (for example Cr(III), Fe(II), and Co(III)) are generally nonlabile as they have large LFSEs. Chelate complexes with the same configuration, such as $[Fe(phen)_3]^{2+}$, are particularly nonlabile.
- Nonlability is common among the complexes of the 4d and 5d series, which reflects the high LFSE and strength of the metal–ligand bonding.

Table 23.1 illustrates the range of timescales for a number of reactions.

The natures of the ligands in the complex also affect the rates of reactions. The identity of the incoming ligand has the greatest effect, and equilibrium constants of displacement reactions can be used to rank ligands in order of their strength as Lewis bases. However, a different order may be found if bases are ranked according to the rates at which they displace a ligand from the central metal ion. Therefore, for kinetic considerations, we replace the equilibrium concept of basicity by the kinetic concept of **nucleophilicity**, the rate of attack on a complex by a given Lewis base relative to the rate of attack by a reference Lewis base. The shift from equilibrium to kinetic considerations is emphasized by referring to ligand displacement as **nucleophilic substitution**.

Ligands other than the entering and leaving groups may play a significant role in controlling the rates of reactions; these ligands are referred to as **spectator ligands**. For instance, it is observed for square-planar complexes that the ligand *trans* to the leaving group X has a great effect on the rate of substitution of X by the entering group Y.

23.2 The classification of mechanisms

The **mechanism** of a reaction is the sequence of elementary steps by which the reaction takes place. Once a suitable mechanism has been identified, attention turns to the details of the activation process of the rate-determining step. In some cases the overall mechanism is poorly resolved, and the only information available is the rate-determining step.

(a) Association, dissociation, and interchange

KEY POINTS The mechanism of a nucleophilic substitution reaction is the sequence of elementary steps by which the reaction takes place and is classified as associative, dissociative, or interchange; an associative mechanism is distinguished from an interchange mechanism by demonstrating that the intermediate has a relatively long life.

The first stage in the kinetic analysis of a reaction is to study how its rate changes as the concentrations of reactants are varied. This type of investigation leads to the identification of **rate laws**, the differential equations governing the rate of change of the concentrations of reactants and products. For example, the observation that the rate of formation of $[Ni(NH_3)(OH_2)_5]^{2+}$ from $[Ni(OH_2)_6]^{2+}$ is proportional to the concentration of both NH_3 and $[Ni(OH_2)_6]^{2+}$ implies that the reaction is first order in each of these two reactants, and that the overall rate law is

$$\text{rate} = k[Ni(OH_2)_6^{2+}][NH_3] \quad (23.1)$$

> **A NOTE ON GOOD PRACTICE**
>
> In rate equations, as in expressions for equilibrium constants, we omit the brackets that are part of the chemical formula of the complex; the surviving brackets denote molar concentration. We denote rate constants by k, and there is usually no confusion with Boltzmann's constant as it normally occurs in the denominator of exponents; where there is potential for confusion, Boltzmann's constant is written k_B.

TABLE 23.1 Representative timescales of chemical and physical processes

Timescale*	Process	Example
10^8 s	Ligand exchange (inert complex)	$[Cr(OH_2)_6]^{3+}$ – H_2O (c. 32 years)
60 s	Ligand exchange (nonlabile complex)	$[V(OH_2)_6]^{3+}$ – H_2O (50 s)
1 ms	Ligand exchange (labile complex)	$[Mn(OH_2)_6]^{2+}$ – H_2O (0.2 ms)
1 μs	Intervalence charge transfer	$(H_3N)_5Ru^{II}$—N⟨⟩N—$Ru^{III}(NH_3)_5$ (0.5 μs)
1 ns	Hydrogen bond rearrangement	C_4H_9OH (1 ns)
10 ps	Ligand association	$Cr(CO)_5$ + THF (10 ps)
1 ps	Rotation time in liquid	CH_3CN (1 ps)
1 fs	Molecular vibration	Sn–Cl stretch (300 fs)

* Approximate time at room temperature.

In simple sequential reaction schemes, the slowest elementary step of the reaction dominates the overall reaction rate and the overall rate law, and is called the **rate-determining step**. However, in general, all the steps in the reaction may contribute to the rate law and affect its rate. Therefore, in conjunction with stereochemical and isotopic labelling studies, the determination of the rate law is the route to the elucidation of the mechanism of the reaction.

Three main classes of reaction mechanism have been identified. A **dissociative mechanism,** denoted D, is a reaction sequence in which an *intermediate* of reduced coordination number is formed by the departure of the leaving group:

$$ML_nX \rightarrow ML_n + X$$
$$ML_n + Y \rightarrow ML_nY$$

Here ML_n (the metal atom and any spectator ligands) is a true intermediate that can, in principle, be detected (or even isolated). The typical form of the corresponding reaction profile is shown in Fig. 23.2.

A BRIEF ILLUSTRATION

The substitution of hexacarbonyltungsten (0) by phosphine takes place by dissociation of CO from the complex

$$[W(CO)_6] \rightarrow [W(CO)_5] + CO$$

followed by coordination of phosphine:

$$[W(CO)_5] + PPh_3 \rightarrow [W(CO)_5(PPh_3)]$$

Under the conditions in which this reaction is usually performed in the laboratory, the intermediate $[W(CO)_5]$ is rapidly captured by the solvent, such as tetrahydrofuran, to form $[W(CO)_5(THF)]$. This complex in turn is converted to the phosphine product, presumably by a second dissociative process.

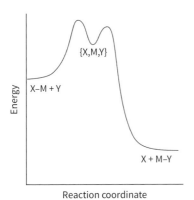

FIGURE 23.2 The typical form of the reaction profile of a reaction with a dissociative mechanism

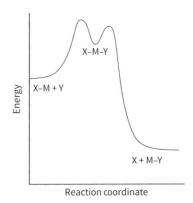

FIGURE 23.3 The typical form of the reaction profile of a reaction with an associative mechanism.

An **associative mechanism,** denoted A, involves a step in which an *intermediate* is formed with a higher coordination number than the original complex:

$$ML_nX + Y \rightarrow ML_nXY$$
$$ML_nXY \rightarrow ML_nY + X$$

Once again, the intermediate ML_nXY can, in principle at least, be detected. This mechanism plays a role in many reactions of square-planar Au(III), Pt(II), Pd(II), Ni(II), and Ir(I) d^8 complexes. The typical form of the reaction profile is similar to that of the dissociative mechanism, and is shown in Fig. 23.3.

A BRIEF ILLUSTRATION

The first step in the exchange of $^{14}CN^-$ with the ligands in the square-planar complex $[Ni(CN)_4]^{2-}$ is the coordination of a ligand to the complex:

$$[Ni(CN)_4]^{2-} + {}^{14}CN^- \rightarrow [Ni(CN)_4({}^{14}CN)]^{3-}$$

A ligand is then discarded:

$$[Ni(CN)_4({}^{14}CN)]^{3-} \rightarrow [Ni(CN)_3({}^{14}CN)]^{2-} + CN^-$$

The radioactivity of carbon-14 provides a means of monitoring this reaction, and the intermediate $[Ni(CN)_5]^{3-}$ has been detected and isolated.

An **interchange mechanism,** denoted I, takes place in one step:

$$ML_nX + Y \rightarrow X \cdots ML_n \cdots Y \rightarrow ML_nY + X$$

The leaving and entering groups exchange in a single step by forming a transition state but not a true intermediate. The interchange mechanism is common for many reactions of six-coordinate complexes. The typical form of the reaction profile is shown in Fig. 23.4.

The distinction between the A and I mechanisms hinges on whether or not the *intermediate* persists long enough to be detectable. One type of evidence is the isolation of an

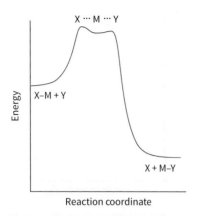

FIGURE 23.4 The typical form of the reaction profile of a reaction with an interchange mechanism.

intermediate in another related reaction or under different conditions. If an argument by extrapolation to the actual reaction conditions suggests that a moderately long-lived intermediate might exist during the reaction in question, then the A path is indicated. For example, the synthesis of the trigonal-bipyramidal Pt(II) complex, $[Pt(SnCl_3)_5]^{3-}$, indicates that a five-coordinate platinum complex may be plausible in substitution reactions of square-planar Pt(II) ammine complexes. Similarly, the fact that $[Ni(CN)_5]^{3-}$ is observed spectroscopically in solution, and that it has been isolated in the crystalline state, provides support for the view that it is involved when CN^- exchanges with the square-planar tetracyanidonickelate(II) ion.

A second indication of the persistence of an intermediate is the observation of a stereochemical change, which implies that the intermediate has lived long enough to undergo rearrangement. *Cis*-to-*trans* isomerization is observed in the substitution reactions of certain square-planar phosphine Pt(II) complexes, which is in contrast to the retention of configuration usually observed. This difference implies that the trigonal-bipyramidal intermediate lives long enough for an exchange between the axial and equatorial ligand positions to occur.

Direct spectroscopic detection of the intermediate, and hence an indication of A rather than I, may be possible if a sufficient amount accumulates. Such direct evidence, however, requires an unusually stable intermediate with favourable spectroscopic characteristics.

(b) The rate-determining step

KEY POINT The rate-determining step is classified as associative or dissociative according to the dependence of its rate on the identity of the entering group.

Now we consider the rate-determining step of a reaction and the details of its formation. The step is called **associative** and denoted *a* if its rate depends strongly on the identity of the incoming group. Examples are found among reactions of the d^8 square-planar complexes of Pt(II), Pd(II), and Au(III), including

$$[PtCl(dien)]^+(aq) + I^-(aq) \rightarrow [PtI(dien)]^+(aq) + Cl^-(aq)$$

where dien is diethylenetriamine ($NH_2CH_2CH_2NHCH_2CH_2NH_2$). It is found, for instance, that use of I^- instead of Br^- increases the rate constant by an order of magnitude. Experimental observations on the substitution reactions of square-planar complexes support the view that the rate-determining step is associative.

The strong dependence of the rate-determining step on entering group Y indicates that the *transition state* must involve significant bonding to Y. A reaction with an associative mechanism (A) will be associatively activated (*a*) if the attachment of Y to the initial reactant ML_nX is the rate-determining step; such a reaction is designated A_a, and in this case the intermediate ML_nXY would not be detected. A reaction with a dissociative mechanism (D) is associatively activated (*a*) if the attachment of Y to the intermediate ML_n is the rate-determining step; such a reaction is designated D_a. Fig. 23.5 shows the reaction profiles for associatively activated A and D mechanisms. For the reactions to proceed, it is necessary to have established a population of an encounter complex $\{X-M, Y\}$ in a pre-equilibrium step.

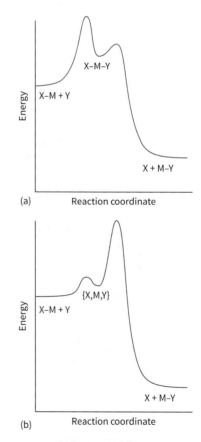

FIGURE 23.5 The typical form of the reaction profile of reactions with associatively activated steps: (a) associative mechanism, A_a; (b) a dissociative mechanism, D_a.

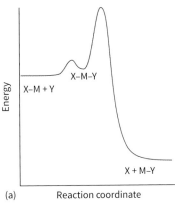

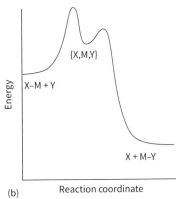

FIGURE 23.6 The typical form of the reaction profile of reactions with a dissociatively activated step: (a) an associative mechanism, A_d; (b) a dissociative mechanism, D_d.

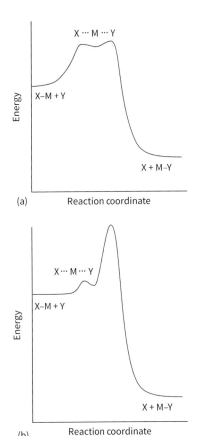

FIGURE 23.7 The typical form of the reaction profile of reactions with an interchange mechanism: (a) associatively activated, I_a; (b) dissociatively activated, I_d.

The rate-determining step is called **dissociative** and denoted d if its rate is largely independent of the identity of Y. This category includes some of the classic examples of ligand substitution in octahedral d-metal complexes, including

$$[Ni(OH_2)_6]^{2+}(aq) + NH_3(aq) \rightarrow [Ni(NH_3)(OH_2)_5]^{2+}(aq) + H_2O(l)$$

where it is found that the rate changes only by a few percent at most when pyridine is used as the incoming ligand instead of NH_3.

The weak dependence on Y of a dissociatively activated process indicates that the rate of formation of the transition state is determined largely by the rate at which the bond to the leaving group X can break. A reaction with an associative mechanism (A) will be dissociatively activated (d) provided the loss of X from the intermediate YML_nX is the rate-determining step; such a reaction is designated A_d. A reaction with a dissociative mechanism (D) is dissociatively activated (d) if the initial loss of X from the reactant ML_nX is the rate-determining step; such a reaction is designated D_d. In the latter case, the intermediate ML_n would not be detected. Figure 23.6 shows the reaction profiles for dissociatively activated A and D mechanisms.

A reaction that has an interchange mechanism (I) can be either associatively or dissociatively activated, and is designated either I_a or I_d, respectively. In an I_a mechanism, the rate of reaction depends on the rate at which the M...Y bond forms, whereas in an I_d reaction the rate of reaction depends on the rate at which the M...X bond breaks (Fig. 23.7).

The distinction between these possibilities may be summarized as follows, where ML_nX denotes the initial complex:

Mechanism:	A		I		D	
Activation:	a	d	a	d	a	d
Rate-determining step	Y attaching to ML_nX	Loss of X from YML_nX	Y attaching to ML_nX	Loss of X from YML_nX	Y attaching to ML_n	Loss of X from ML_nX
Detect intermediate?	no	ML_nXY detectable	no	no	ML_n detectable	no

Ligand substitution in square-planar complexes

The mechanism of ligand exchange in square-planar Pt complexes has been studied extensively, largely because the reactions occur on a timescale that is very amenable to investigation. We might expect an associative mechanism for ligand exchange because square-planar complexes are sterically uncrowded—they can be considered as octahedral complexes with two ligands missing—but it is rarely that simple. The elucidation of the mechanism of the substitution of square-planar complexes is often complicated by the occurrence of alternative pathways. For instance, if a reaction such as

$$[PtCl(dien)]^+(aq) + I^-(aq) \rightarrow [PtI(dien)]^+(aq) + Cl^-(aq)$$

is first-order in the complex and independent of the concentration of I^-, then the rate of reaction will be equal to $k_1[PtCl(dien)^+]$. However, if there is a pathway in which the rate law is first order in the complex *and* first order in the incoming group (that is, overall second order) then the rate would be given by $k_2[PtCl(dien)^+][I^-]$. If both reaction pathways occur at comparable rates, the rate law has the form

$$\text{rate} = (k_1 + k_2[I^-])[PtCl(dien)^+] \tag{23.2}$$

A reaction like this is usually studied under the conditions $[I^-] \gg [\text{complex}]$ so that $[I^-]$ does not change significantly during the reaction. This simplifies the treatment of the data as $k_1 + k_2[I^-]$ is effectively constant and the rate law is now pseudo-first order:

$$\text{rate} = k_{obs}[PtCl(dien)^+] \quad k_{obs} = k_1 + k_2[I^-] \tag{23.3}$$

A plot of the observed pseudo-first-order rate constant against $[I^-]$ gives k_2 as the slope and k_1 as the intercept, an example is shown in Fig. 23.24.

In the following sections we examine the factors that affect the second-order reaction and then consider the first-order process.

23.3 The nucleophilicity of the entering group

KEY POINTS The nucleophilicity of an entering group is expressed in terms of the nucleophilicity parameter defined in terms of the substitution reactions of a specific square-planar platinum complex; the sensitivity of other platinum complexes to changes in the entering group is expressed in terms of the nucleophilic discrimination factor.

We start by considering the variation of the rate of the reaction as the entering group Y is varied. The reactivity of Y (for instance, I^- in the reaction above) can be expressed in terms of a **nucleophilicity parameter**, n_{Pt}:

$$n_{Pt} = \log \frac{k_2(Y)}{k_2^\circ} \tag{23.4}$$

where $k_2(Y)$ is the second-order rate constant for the reaction

$$trans\text{-}[PtCl_2(py)_2] + Y \rightarrow trans\text{-}[PtClY(py)_2]^+ + Cl^-$$

and k_2° is the rate constant for the same reaction with the reference nucleophile methanol. The entering group is highly nucleophilic, or has a high nucleophilicity, if n_{Pt} is large.

Table 23.2 gives some values of n_{Pt}. One striking feature of the data is that, although the entering groups in the table are all quite simple, the rate constants span nearly nine orders of magnitude. Another feature is that the nucleophilicity of the entering group towards Pt appears to correlate with soft Lewis basicity (Section 4.9), with $Cl^- < I^-$, $O < S$, and $NH_3 < PR_3$.

The nucleophilicity parameter is defined in terms of the reaction rates of a specific platinum complex. When the complex itself is varied we find that the reaction rates show a range of different sensitivities towards changes in the entering group. To express this range of sensitivities we rearrange eqn. 23.4 into

$$\log k_2(Y) = n_{Pt}(Y) + C \tag{23.5}$$

where $C = \log k_2^\circ$. Now consider the analogous substitution reactions for the general complex $[PtL_3X]$:

$$[PtL_3X] + Y \rightarrow [PtL_3Y] + X$$

The relative rates of these reactions can be expressed in terms of the same nucleophilicity parameter n_{Pt}, provided we replace eqn. 23.5 by

$$\log k_2(Y) = S n_{Pt}(Y) + C \tag{23.6}$$

The parameter S, which characterizes the sensitivity of the rate constant to the nucleophilicity parameter, is called the **nucleophilic discrimination factor**. We see that the straight

TABLE 23.2 A selection of n_{Pt} values for a range of nucleophiles

Nucleophile	Donor atom	n_{Pt}
CH_3OH	O	0
Cl^-	Cl	3.04
Br^-	Br	4.18
I^-	I	5.42
CN^-	C	7.14
SCN^-	S	5.75
N_3^-	N	3.58
C_6H_5SH	S	4.15
NH_3	N	3.07
$(C_6H_5)_3P$	P	8.93

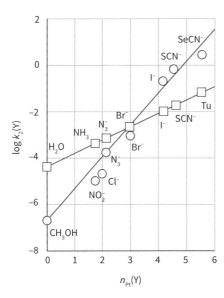

FIGURE 23.8 The slope of the straight line obtained by plotting log $k_2(Y)$ against the nucleophilicity parameter $n_{Pt}(Y)$ for a series of ligands is a measure of the responsiveness of the complex to the nucleophilicity of the entering group. (Tu = thiourea)

line obtained by plotting log $k_2(Y)$ against n_{Pt} for reactions of Y with trans-[PtCl$_2$(PEt$_3$)$_2$] (the red circles in Fig. 23.8) is steeper than that for reactions with cis-[PtCl$_2$(en)] (the blue squares in Fig. 23.8). Hence, S is larger for the former reaction, which indicates that the rate of the reaction is more sensitive to changes in the nucleophilicity of the entering group.

Some values of S are given in Table 23.3. Note that S is close to 1 in all cases, so all the complexes are quite sensitive to n_{Pt}. This sensitivity is what we expect for associatively activated reactions. Another feature to note is that larger values of S are found for complexes of platinum with softer base ligands.

TABLE 23.3 Nucleophilic discrimination factors

	S
trans-[PtCl$_2$(PEt$_3$)$_2$]	1.43
trans-[PtCl$_2$(py)$_2$]	1.00
[PtCl$_2$(en)]	0.64
[PtCl(dien)]$^+$	0.65

EXAMPLE 23.1 Using the nucleophilicity parameter

The second-order rate constant for the reaction of I$^-$ with trans-[Pt(CH$_3$)Cl(PEt$_3$)$_2$] in methanol at 30°C is 40 dm^3 mol^{-1} s^{-1}. The corresponding reaction with azide, N$_3^-$, has k_2 = 7.0 dm^3 mol^{-1} s^{-1}. Estimate S and C for the reaction, given n_{Pt} values of 5.42 and 3.58, respectively, for the two nucleophiles.

Answer To determine S and C, we need to use the two pieces of information to set up and solve two simultaneous equations based on eqn. 23.6. Substituting the two values of n_{Pt} into eqn. 23.6 gives

$$1.60 = 5.42S + C \text{ (for I}^-\text{)}$$
$$0.85 = 3.58S + C \text{ (for N}_3^-\text{)}$$

Solving these two simultaneous equations gives S = 0.41 and C = 20.62. The value of S is fairly small, showing that the discrimination of this complex among different nucleophiles is not great. This lack of sensitivity is related to the fairly large value of C, which corresponds to the rate constant being large and hence to the complex being reactive. It is commonly found that high reactivity correlates with low selectivity.

Self-test 23.1 Calculate the second-order rate constant for the reaction of the same complex with NO$_2^-$ for which n_{Pt} = 3.22.

23.4 The shape of the transition state

Careful studies of the variation of the reaction rates of square-planar complexes with changes in the composition of the reactant complex and the conditions of the reaction shed light on the general shape of the transition state. They also confirm that substitution almost invariably has an associative rate-determining stage; hence intermediates are rarely detected.

(a) The trans effect

KEY POINT A strong σ-donor ligand or π-acceptor ligand greatly accelerates substitution of a ligand that lies in the trans position in square-planar complexes.

The spectator ligands T that are trans to the leaving group in square-planar complexes influence the rate of substitution. This phenomenon is called the **trans effect**. It is generally accepted that the trans effect arises from two separate influences—one arising in the ground state and the other in the transition state itself.

The **trans influence** is the extent to which the ligand T weakens the bond trans to itself in the ground state of the complex. The trans influence correlates with the σ-donor ability of the ligand T because, broadly speaking, ligands trans to each other use the same orbitals on the metal for bonding. Thus if one ligand is a strong σ donor, then the ligand trans to it cannot donate electrons to the metal so well, and thus has a weaker interaction with the metal. The trans influence is assessed quantitatively by measuring bond lengths, stretching frequencies, and metal-to-ligand NMR coupling constants (Section 8.6). The **transition-state effect** correlates with the π-acceptor ability of the ligand. Its origin is thought to be related to the increase in electron density on the metal atom arising from the incoming ligand: any ligand that can accept this increased electron density will stabilize the transition state (**1**).

TABLE 23.4 The effect of the *trans* ligand in reactions of *trans*-[PtCl(PEt$_3$)$_2$L]

L	k_1/s^{-1}	k_2/(dm^{3}mol^{-1}s^{-1})
CH$_3^-$	1.7 × 10^{-4}	6.7 × 10^{-2}
C$_6$H$_5^-$	3.3 × 10^{-5}	1.6 × 10^{-2}
Cl$^-$	1.0 × 10^{-6}	4.0 × 10^{-4}
H$^-$	1.8 × 10^{-2}	4.2
PEt$_3$	1.7 × 10^{-2}	3.8

k_1 and k_2 are defined in eqn. 23.2.

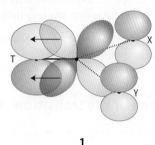

1

The *trans* effect is the combination of both effects; it should be noted that the same factors contribute to a large ligand-field splitting in d-metal complexes. *Trans* effects are listed in Table 23.4 and follow the order:

For a *trans* σ-donor:
OH$^-$ < NH$_3$ < Cl$^-$ < Br$^-$ < CN$^-$, CH$_3^-$ < I$^-$ < SCN$^-$ < PR$_3$, H$^-$

For a *trans* π-acceptor:
Br$^-$ < I$^-$ < NCS$^-$ < NO$_2^-$ < CN$^-$ < CO, < C$_2$H$_4$

EXAMPLE 23.2 Using the *trans* effect synthetically

Use the *trans* effect series to suggest synthetic routes to *cis*- and *trans*-[PtCl$_2$(NH$_3$)$_2$] from [Pt(NH$_3$)$_4$]$^{2+}$ and [PtCl$_4$]$^{2-}$.

Answer If we consider the reaction of [Pt(NH$_3$)$_4$]$^{2+}$ with HCl we can see it initially leads to [PtCl(NH$_3$)$_3$]$^+$. Now, because the *trans* effect of Cl$^-$ is greater than that of NH$_3$, substitution reactions will occur preferentially *trans* to Cl$^-$, and further action of HCl gives *trans*-[PtCl$_2$(NH$_3$)$_2$]:

[Pt(NH$_3$)$_4$]$^{2+}$ + Cl$^-$ → [PtCl(NH$_3$)$_3$]$^+$ → *trans*-[PtCl$_2$(NH$_3$)$_2$]

However, when the starting complex is [PtCl$_4$]$^{2-}$, reaction with NH$_3$ leads first to [PtCl$_3$(NH$_3$)]$^-$. A second step should substitute one of the two mutually *trans* Cl$^-$ ligands with NH$_3$ to give *cis*-[PtCl$_2$(NH$_3$)$_2$]:

[PtCl$_4$]$^{2-}$ + NH$_3$ → [PtCl$_3$(NH$_3$)]$^-$ → *cis*-[PtCl$_2$(NH$_3$)$_2$]

Self-test 23.2 Given the reactants PPh$_3$, NH$_3$, and [PtCl$_4$]$^{2-}$, propose efficient routes to both *cis*- and *trans*-[PtCl$_2$(NH$_3$)(PPh$_3$)].

(b) Steric effects

KEY POINT Steric crowding at the reaction centre usually inhibits associative reactions and facilitates dissociative reactions.

Steric crowding at the reaction centre by bulky groups that can block the approach of attacking nucleophiles will inhibit associative reactions. The rate constants for the replacement of Cl$^-$ by H$_2$O in *cis*-[PtClL(PEt$_3$)$_2$]$^+$ complexes at 25°C illustrate the point:

L =	pyridine	2-methylpyridine	2,6-dimethylpyridine
k/s^{-1}	8 × 10^{-2}	2.0 × 10^{-4}	1.0 × 10^{-6}

The methyl groups adjacent to the N donor atom greatly decrease the rate. In the 2-methylpyridine complex they block positions either above or below the plane. In the 2,6-dimethylpyridine complex they block positions both above and below the plane (**2**). Thus, along the series, the methyl groups increasingly hinder attack by H$_2$O.

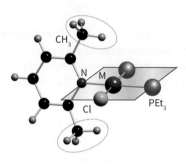

2

The effect is smaller if L is *trans* to Cl$^-$. This difference is explained by the methyl groups then being further from the entering and leaving groups in the trigonal-bipyramidal transition state if the pyridine ligand is in the trigonal plane (**3**). Conversely, the decrease in coordination number that occurs in a dissociative reaction can relieve the steric overcrowding and thus increase the rate of the dissociative reaction.

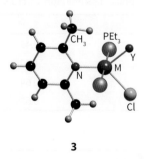

3

(c) Stereochemistry

KEY POINT Substitution of a square-planar complex preserves the original geometry, which suggests a trigonal-pyramidal transition state.

Further insight into the nature of the transition state is obtained from the observation that substitution of a

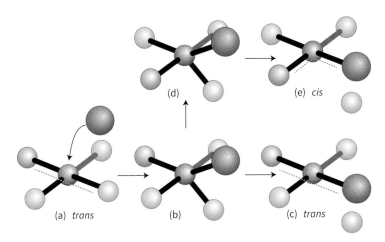

FIGURE 23.9 The stereochemistry of substitution in a square-planar complex. The normal path (resulting in retention) is from (a) to (c). However, if intermediate (b) is sufficiently long-lived, it can undergo pseudorotation to (d), which leads to isomer (e).

square-planar d-metal complex usually preserves the original geometry. That is, a *cis* complex gives a *cis* product, and a *trans* complex gives a *trans* product. This behaviour is explained by the formation of an approximately trigonal-bipyramidal transition state with the entering, leaving, and *trans* groups in the trigonal plane (**4**).[1] Trigonal-bipyramidal intermediates of this type account for the relatively small influence that the two *cis* spectator ligands have on the rate of substitution, as their bonding orbitals will be largely unaffected by the course of the reaction.

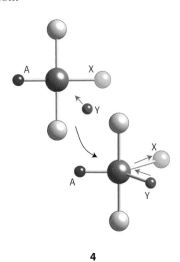

4

The steric course of the reaction is shown in Fig. 23.9. We can expect a *cis* ligand to exchange places with the T ligand in the trigonal plane only if the intermediate lives long enough to be stereomobile. That is, it must be a long-lived associative (*A*) intermediate, with release of the ligand from the five-coordinate intermediate being the rate-determining step.

[1] Note that the stereochemistry is quite different from that of p-block central atoms, such as Si(IV) and P(V), where the leaving group departs from the more crowded axial position.

(d) Temperature and pressure dependence

KEY POINT Negative volumes and entropies of activation support the view that the rate-determining step of square-planar Pt(II) complexes is associative.

Another clue to the nature of the transition state comes from the entropies and volumes of activation for reactions of Pt(II) and Au(III) complexes (Table 23.5). The entropy of activation is obtained from the temperature dependence of the rate constant, and indicates the change in disorder (of reactants and solvent) when the transition state forms. Likewise, the volume of activation, which is obtained (using special equipment) from the pressure dependence of the rate constant, is the change in volume that occurs on formation of the transition state. The limiting cases for the volume of activation in ligand substitution reactions correspond to the increase in molar volume of the outgoing ligand (for a dissociative reaction) and the decrease in molar volume of the incoming ligand (for an associative reaction). As an example, the molar volume of H_2O is $18\,cm^3$ ($0.018\,dm^3$) so a reaction in which the transition state was reached by complete dissociation of an H_2O ligand would have a volume of activation approaching $0.018\,dm^3$.

TABLE 23.5 Activation parameters for substitution in square-planar complexes (in methanol)

Reaction	k_1			k_2		
	$\Delta^\ddagger H$	$\Delta^\ddagger S$	$\Delta^\ddagger V$	$\Delta^\ddagger H$	$\Delta^\ddagger S$	$\Delta^\ddagger V$
trans-[PtCl(NO_2)(py)$_2$] + py				50	−100	−38
trans-[PtBrP$_2$(mes)] + SC(NH_2)$_2$	71	−84	−46	46	−138	−54
cis-[PtBrP$_2$(mes)] + I$^-$	84	−59	−67	63	−121	−63
cis-[PtBrP$_2$(mes)] + SC(NH_2)$_2$	79	−71	−71	59	−121	−54
[AuCl(dien)]$^{2+}$ + Br$^-$				54	−17	

[PtBrP$_2$(mes)] is [PtBr(PEt$_3$)$_2$(2,4,6-Me$_3$C$_6$H$_2$)].

Enthalpy in kJ mol^{-1}; entropy in J K^{-1} mol^{-1}; volume in cm^3 mol^{-1}.

The two striking aspects of the data in the table are the consistently strongly negative values of both quantities. The simplest explanation of the decrease in disorder and the decrease in volume is that the entering ligand is being incorporated into the transition state without release of the leaving group. That is, we can conclude that the rate-determining step is associative.

(e) The first-order pathway

KEY POINT The first-order contribution to the rate law is a pseudo-first-order process in which the solvent participates.

Having considered factors that affect the second-order pathway, we can now consider the first-order pathway for the substitution of square-planar complexes. The first issue we must address is the first-order pathway in the rate equation, and decide whether k_1 in the rate law in eqn. 23.2 and its generalization

$$\text{rate} = (k_1 + k_2[Y])[PtL_4] \qquad (23.7)$$

does indeed represent the operation of an entirely different reaction mechanism. It turns out that it does not, and k_1 represents an associative reaction involving the solvent. In this pathway the substitution of Cl^- by pyridine in methanol as solvent proceeds in two steps, with the first rate-determining:

$$[PtCl(dien)]^+ + CH_3OH \rightarrow [Pt(CH_3OH)(dien)]^{2+} + Cl^- \quad (\text{slow})$$

$$[Pt(CH_3OH)(dien)]^{2+} + py \rightarrow [Pt(py)(dien)]^{2+} + CH_3OH \quad (\text{fast})$$

The evidence for this two-step mechanism comes from a correlation of the rates of these reactions with the nucleophilicity parameters of the solvent molecules and the observation that reactions of entering groups with solvent complexes are rapid when compared to the step in which the solvent displaces a ligand. Thus, the substitution of a ligand at a square-planar platinum complex results from two competing associative reactions.

Ligand substitution in octahedral complexes

Octahedral complexes occur for different metals in a wide range of oxidation states and with a great diversity of bonding modes. We might therefore expect a huge variety of mechanisms of substitution; however, almost all octahedral complexes react by the interchange (*I*) mechanism. The only real question is whether the rate-determining step is associative or dissociative. The analysis of rate laws for reactions that take place by such a mechanism helps to formulate the precise conditions for distinguishing these two possibilities and identifying the substitution as I_a (interchange with an associative rate-determining stage) or I_d (interchange with a dissociative rate-determining stage). The difference between the two classes of reaction hinges on whether the rate-determining step is the formation of the new Y···M bond or the breaking of the old M···X bond.

23.5 Rate laws and their interpretation

Rate laws provide an insight into the detailed mechanism of a reaction in the sense that any proposed mechanism must be consistent with the observed rate law. In the following section we see how the rate laws found experimentally for ligand substitution are interpreted.

(a) The Eigen–Wilkins mechanism

KEY POINTS In the Eigen–Wilkins mechanism, an encounter complex is formed in a pre-equilibrium step and the encounter complex forms products in a subsequent rate-determining step.

As an example of a ligand substitution reaction, we consider

$$[Ni(OH_2)_6]^{2+} + NH_3 \rightarrow [Ni(NH_3)(OH_2)_5]^{2+} + H_2O$$

The first step in the **Eigen–Wilkins mechanism** is an encounter in which the complex ML_6, in this case $[Ni(OH_2)_6]^{2+}$, and the entering group Y, in this case NH_3, diffuse together and come into contact:

$$[Ni(OH_2)_6]^{2+} + NH_3 \rightarrow \{[Ni(OH_2)_6]^{2+}, NH_3\}$$

The two components of the encounter pair, the entity {A,B}, may also separate at a rate governed by their ability to migrate by diffusion through the solvent:

$$\{[Ni(OH_2)_6]^{2+}, NH_3\} \rightarrow [Ni(OH_2)_6]^{2+} + NH_3$$

Because in aqueous solution the lifetime of an encounter pair is approximately 1 ns, the formation of the pair can be treated as a pre-equilibrium in all reactions that take longer than a few nanoseconds. Consequently, we can express the concentrations in terms of a pre-equilibrium constant K_E:

$$ML_6 + Y \rightleftharpoons \{ML_6, Y\} \qquad K_E = \frac{[\{ML_6, Y\}]}{[ML_6][Y]}$$

The second step in the mechanism is the rate-determining reaction of the encounter complex to give products:

$$\{[Ni(OH_2)_6]^{2+}, NH_3\} \rightarrow [Ni(NH_3)(OH_2)_5]^{2+} + H_2O$$

and in general

$$\{ML_6, Y\} \rightarrow ML_5Y + L \qquad \text{rate} = k[\{ML_6, Y\}]$$

TABLE 23.6 Complex formation by the $[Ni(OH_2)_6]^{2+}$ ion

Ligand	k_{obs}/(dm^3 mol^{-1} s^{-1})	K_E/(dm^3 mol^{-1})	(k_{obs}/K_E)/s^{-1}
$CH_3CO_2^-$	1×10^5	3	3×10^4
F^-	8×10^5	1	8×10^3
HF	3×10^3	0.15	2×10^4
H_2O^*			3×10^3
NH_3	5×10^3	0.15	3×10^4
$[NH_2(CH_2)_2NH_3]^+$	4×10^2	0.02	2×10^4
SCN^-	6×10^3	1	6×10^3

* The solvent is always in encounter with the ion so that K_E is undefined and all rates are inherently first-order.

We cannot simply substitute $[\{ML_6,Y\}] = K_E[ML_6][Y]$ into this expression because the concentration of ML_6 must take into account the fact that some of it is present as the encounter pair; that is, $[M]_{tot} = [\{ML_6,Y\}] + [ML_6]$, the total concentration of the complex. It follows that

$$\text{rate} = \frac{kK_E[M]_{tot}[Y]}{1+K_E[Y]} \qquad (23.8)$$

It is rarely possible to conduct experiments over a range of concentrations wide enough to test eqn. 23.8 exhaustively. However, at such low concentrations of the entering group that $K_E[Y] \ll 1$, the rate law reduces to

$$\text{rate} = k_{obs}[M]_{tot}[Y] \quad k_{obs} = kK_E \qquad (23.9)$$

Because k_{obs} can be measured and K_E can be either measured or estimated as we describe below, the rate constant k can be found as it is equal to k_{obs}/K_E. The results for reactions of Ni(II) hexaaqua complexes with various nucleophiles are shown in Table 23.6. The very small variation in k indicates a model I_d reaction with very slight sensitivity to the nucleophilicity of the entering group.

When Y is a solvent molecule the encounter equilibrium is 'saturated' in the sense that, because the complex is always surrounded by solvent, a solvent molecule is always available to take the place of one that leaves the complex. In such a case $K_E[Y] \gg 1$ and $k_{obs} = k$. Thus, reactions with the solvent can be directly compared to reactions with other entering ligands without needing to estimate the value of K_E.

(b) The Fuoss–Eigen equation

KEY POINT The Fuoss–Eigen equation provides an estimate of the pre-equilibrium constant based on the strength of the Coulombic interaction between the reactants and their distance of closest approach.

The equilibrium constant K_E for the encounter pair can be estimated by using a simple equation proposed independently by R. M. Fuoss and M. Eigen. Both sought to take the complex size and charge into account, expecting larger, oppositely charged ions to meet more frequently than small ions of the same charge. Fuoss used an approach based on statistical thermodynamics and Eigen used one based on kinetics. Their result, which is called the **Fuoss–Eigen equation**, is

$$K_E = \tfrac{4}{3}\pi a^3 N_A e^{-V/k_BT} \qquad (23.10)$$

In this expression a is the distance of closest approach of ions of charge numbers z_1 and z_2 in a medium of permittivity ε, V is the Coulombic potential energy $(z_1 z_2 e^2/4\pi\varepsilon a)$ of the ions at that distance, and N_A is Avogadro's constant. Although the value predicted by this equation depends strongly on the details of the charges and radii of the ions, typically it clearly favours the encounter if the reactants are large (so a is large) or oppositely charged (V negative).

A BRIEF ILLUSTRATION

If one of the reactants is uncharged (as in the case of substitution by NH_3), then $V = 0$ and $K_E = \tfrac{4}{3}\pi a^3 N_A$. For an encounter distance of 200 pm for neutral species, we find

$$K_E = \frac{4\pi}{3} \times (2.00 \times 10^{-10}\,\text{m})^3 \times (6.022 \times 10^{23}\,\text{mol}^{-1}) = 2.02 \times 10^{-5}\,\text{m}^3\,\text{mol}^{-1}$$

or, in the units normally used, 2.02×10^{-2} dm^3 mol^{-1}. For two singly charged ions of opposite charge in water at 298 K (when $\varepsilon_r = 78$), other factors being equal, the value of K_E is increased by a factor of

$$e^{-V/k_BT} = e^{e^2/4\pi\varepsilon a k_BT} = 36$$

23.6 The activation of octahedral complexes

Many studies of substitution in octahedral complexes support the view that the rate-determining step is dissociative, and we summarize these studies first. However, the reactions of octahedral complexes can acquire a distinct associative character in the case of large central ions (as in the 4d and 5d series) or where the d-electron population at the metal is low (the early members of the d block). More room for attack or lower π^* electron density appears to facilitate nucleophilic attack and hence permit association.

(a) Leaving-group effects

KEY POINTS A large effect of the leaving group X is expected in I_d reactions; a linear relation is found between the logarithms of the rate constants and equilibrium constants.

We can expect the identity of the leaving group X to have a large effect in dissociatively activated reactions because their rates depend on the scission of the M...X bond. When X is the only variable, as in the reaction

$$[CoX(NH_3)_5]^{2+} + H_2O \rightarrow [Co(NH_3)_5(OH_2)]^{3+} + X^-$$

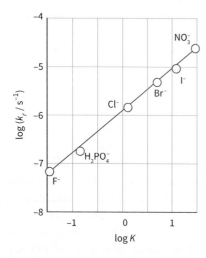

FIGURE 23.10 The straight line obtained when the logarithm of a rate constant is plotted against the logarithm of an equilibrium constant shows the existence of a linear free energy relation. The graph is for the reaction $[Co(NH_3)_5X]^{2+} + H_2O \rightarrow [Co(NH_3)_5(OH_2)]^{3+} + X^-$ with different leaving groups X.

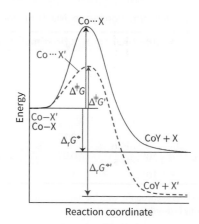

FIGURE 23.11 The existence of a linear free energy relation with unit slope shows that changing X has the same effect on $\Delta^{\ddagger}G$ for the conversion of M–X to the transition state as it has on $\Delta_r G^{\ominus}$ for the complete elimination of X^-. The reaction profile shows the effect of changing the leaving group from X to X′.

it is found that the rate constant and equilibrium constant of the reaction are related by

$$\ln k = \ln K + c \tag{23.11}$$

This correlation is illustrated in Fig. 23.10. Because both logarithms are proportional to Gibbs energies ($\ln k$ is approximately proportional to the activation Gibbs energy, $\Delta^{\ddagger}G$, and $\ln K$ is proportional to the standard reaction Gibbs energy, $\Delta_r G^{\ominus}$), we can write the following **linear free-energy relation (LFER)**:

$$\Delta^{\ddagger}G = p\Delta_r G^{\ominus} + b \tag{23.12}$$

with p and b constants (and $p \approx 1$).

The existence of an LFER of unit slope, as for the reaction of $[CoX(NH_3)_5]^{2+}$, shows that changing X has the same effect on $\Delta^{\ddagger}G$ for the conversion of Co–X to the transition state as it has on $\Delta_r G^{\ominus}$ for the complete elimination of X^- (Fig. 23.11). This observation in turn suggests that in a reaction with an interchange mechanism and a dissociative rate-determining step (I_d), the leaving group (an anionic ligand) has already become a solvated ion in the transition state. An LFER with a slope of less than 1, indicating some associative character, is observed for the corresponding complexes of Rh(III). For Co(III), the reaction rates are in the order $I^- > Br^- > Cl^-$, whereas for Rh(III) they are reversed, and are in the order $I^- < Br^- < Cl^-$. This difference should be expected as the softer Rh(III) centre forms more stable complexes with I^-, compared with Br^- and Cl^-, whereas the harder Co(III) centre forms more stable complexes with Cl^-.

(b) The effects of spectator ligands

KEY POINTS In octahedral complexes, spectator ligands affect rates of substitution; the effect is related to the strength of the metal–ligand interaction, with stronger donor ligands increasing the reaction rate by stabilizing the transition state.

In Co(III), Cr(III), and related octahedral complexes, both *cis* and *trans* ligands affect rates of substitution in proportion to the strength of the bonds they form with the metal atom. For instance, hydrolysis reactions such as

$$[NiXL_5]^+ + H_2O \rightarrow [NiL_5(OH_2)]^{2+} + X^-$$

are much faster when L is NH_3 than when it is H_2O. This difference can be explained on the grounds that, as NH_3 is a stronger σ donor than H_2O, it increases the electron density at the metal atom and hence facilitates the scission of the M–X bond and the formation of X^-. In the transition state, the stronger donor stabilizes the reduced coordination number.

(c) Steric effects

KEY POINT Steric crowding favours dissociative activation because formation of the transition state can relieve strain.

Steric effects on reactions with dissociative rate-determining steps can be illustrated by considering the rate of hydrolysis of the first Cl^- ligand in two complexes of the type $[CoCl_2(bn)_2]^+$:

$$[CoCl_2(bn)_2]^+ + H_2O \rightarrow [CoCl(OH_2)(bn)_2]^{2+} + Cl^-$$

The ligand bn is 2,3-butanediamine, and may be coordinated in either a chiral (5) or achiral (6) fashion. The important observation is that the complex formed with the chiral form of the ligand hydrolyses 30 times more slowly than the complex of the achiral form. The two ligands have very similar electronic effects, but the CH_3 groups are on the opposite sides of the chelate ring in (5) but adjacent and crowded in (6). The latter arrangement is more reactive because the

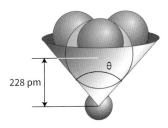

FIGURE 23.12 The determination of the ligand cone angles from space-filling molecular models of the ligand and an assumed M–P bond length of 228 pm.

TABLE 23.7 Tolman cone angles for various ligands

Ligand	θ/°	Ligand	θ/°
CH_3	90	$P(OC_6H_5)_3$	127
CO	95	PBu_3	130
Cl, Et	102	PEt_3	132
PF_3	104	$\eta^5\text{-}C_5H_5$ (Cp)	136
Br, Ph	105	PPh_3	145
I, $P(OCH_3)_3$	107	$\eta^5\text{-}C_5Me_5$ (Cp*)	165
PMe_3	118	$2,4\text{-}Me_2C_5H_3$	180
t-Butyl	126	$P(t\text{-}Bu)_3$	182

strain is relieved in the dissociative transition state with its lowered coordination number. In general, steric crowding favours an I_d process because the five-coordinate transition state can relieve strain.

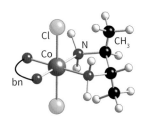

5 Chiral $[Co(Cl)_2(bn)_2]^+$

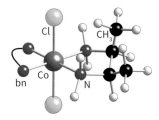

6 Achiral $[Co(Cl)_2(bn)_2]^+$

Quantitative treatments of the steric effects of ligands have been developed using molecular modelling computer software that takes van der Waals interactions into account. However, a more pictorial semiquantitative approach was introduced by C. A. Tolman. In this approach, the extent to which various ligands (especially phosphines) crowd each other is assessed by approximating the ligand by a cone with an angle determined from a space-filling model and, for phosphine ligands, an M–P bond length of 228 pm (Fig. 23.12 and Table 23.7).[2] The ligand CO is small in the sense of having a small cone angle; P(^{t}Bu)$_3$ is regarded as bulky because it has a large cone angle. Bulky ligands have considerable steric repulsion with each other when packed around a metal centre. They favour dissociative activation and inhibit associative activation.

---

[2] Tolman's studies were on nickel complexes, so strictly speaking we are talking about a Ni–P distance of 228 pm.

As an illustration, the rate of the reaction of $[Ru(CO)_3(PR_3)(SiCl_3)_2]$ (7) with Y to give $[Ru(CO)_2Y(PR_3)(SiCl_3)_2]$ is independent of the identity of Y, which suggests that the rate-determining step is dissociative. Furthermore, it has been found that there is only a small variation in rate for the substituents Y with similar cone angles but significantly different values of pK_a. This observation supports the assignment of the rate changes to steric effects because changes in pK_a should correlate with changes in electron distributions in the ligands.

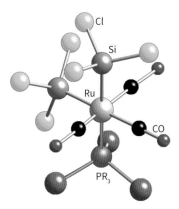

7 $[Ru(CO)_3(PR_3)(SiCl_3)_2]^+$

(d) Activation energetics

KEY POINT A significant loss of LFSE on going from the starting complex to the transition state results in nonlabile complexes.

One factor that has a strong bearing on the activation of complexes is the difference between the ligand-field stabilization energy (LFSE, Section 7.16) of the reacting complex and that of the transition state (the LFSE‡). This difference is known as the **ligand-field activation energy** (LFAE):

$$\text{LFAE} = \text{LFSE} - \text{LFSE}^‡ \qquad (23.13)$$

Table 23.8 gives calculated values of LFAE for the replacement of a ligand at an octahedral complex, assuming a square-pyramidal transition state (that is, a dissociatively activated reaction). If we now revisit Fig. 23.1 we can see

TABLE 23.8 Ligand-field activation energies for octahedral complexes with different d electron configurations

Config	Example	LFSE/Δ_o*	LFSE‡/Δ_o†	LFAE/Δ_o
d^0	Ca^{2+}, Sc^{3+}, La^{3+}	0	0	0
d^1	Ti^{3+}	−0.4	−0.46	−0.06
d^2	Ti^{2+}	−0.8	−0.91	−0.11
d^3	V^{2+}, Cr^{3+}	−1.2	−1	0.2
d^4, hs	Cr^{2+}	−0.6	−0.91	−0.31
d^4, ls	Re^{3+}	−1.6	−1.46	0.13
d^5, hs	Mn^{2+}	0	0	0
d^5, ls	Fe^{3+}	−2	−1.91	0.09
d^6, hs	Fe^{2+}	−0.4	−0.46	−0.06
d^6, ls	Ru^{2+}, Ir^{3+}, Pt^{4+}	−2.4	−2	0.4
d^7, hs	Co^{2+}	−0.8	−0.91	−0.11
d^7, ls	Rh^{2+}	−1.8	−1.91	−0.11
d^8	Ni^{2+}	−1.2	−1	0.2
d^9	Cu^{2+}	−0.6	−0.91	−0.31
d^{10}	Zn^{2+}	0	0	0

* Octahedral.
† Square-pyramidal.
hs, high spin; ls, low spin.

that a large LFAE can be used to account for the nonlabile nature of V^{2+}, Cr^{3+}, Ru^{2+}, Ir^{3+}, and Ni^{2+} complexes: in these examples there is a significant loss of LFSE on moving from an octahedral complex to a transition state.

The corollary to this argument, that complexes with a zero or negative LFAE are particularly labile, is borne out when we look at exchange lifetimes for the aqua complexes of Ca^{2+}, La^{3+}, Cr^{2+}, Mn^{2+}, Cu^{2+}, and Zn^{2+}. Some measured $\Delta^{\ddagger}H$ values are given in Table 23.9 where we can see that the value for V^{2+} and Ni^{2+} (LFAE = 0.2 Δ_o) is substantially larger than that of Mn^{2+} (LFAE = 0) and Fe^{2+} (LFAE = −0.06 Δ_o).

(e) Associative activation

KEY POINT A negative volume of activation indicates association of the entering group into the transition state.

As we have seen, the activation volume reflects the change in compactness (including that of the surrounding solvent) when the transition state forms from the reactants. Table 23.9 gives $\Delta^{\ddagger}V$ for some H_2O ligand self-exchange reactions.

TABLE 23.9 Activation parameters for the H_2O exchange reactions $[M(OH_2)_6]^{2+} + H_2^{17}O \rightarrow [M(OH_2)_5(^{17}OH_2)]^{2+} + H_2O$

	$\Delta^{\ddagger}H$/(kJ mol^{-1})	$\Delta^{\ddagger}V$/(cm^3 mol^{-1})
$V^{2+}(d^3)$	61.8	−4.1
$Mn^{2+}(d^5$, hs$)$	32.9	−5.4
$Fe^{2+}(d^6$, hs$)$	41.4	+3.8
$Co^{2+}(d^7$, hs$)$	46.9	+6.1
$Ni^{2+}(d^8)$	56.9	+7.2

TABLE 23.10 Kinetic parameters for anion attack on Cr(III)

X*	L = H_2O			L = NH_3
	k/ (10^{-8} mol^{-1} s^{-1})	$\Delta^{\ddagger}H$/ (kJ mol^{-1})	$\Delta^{\ddagger}S$/ (J K^{-1} mol^{-1})	k/ (10^{-4} dm^3 mol^{-1} s^{-1})
Br^-	0.46	122	8	3.7
Cl^-	1.15	126	38	0.7
NCS^-	48.7	105	4	4.2

* Reaction is $[CrL_5(OH_2)]^{3+} + X^- \rightarrow [CrL_5X]^{2+} + H_2O$.

A negative volume of activation can be interpreted as the result of shrinkage when an H_2O molecule becomes part of the transition state (implying significant associative character), whereas a positive volume of activation can be interpreted as the result of expansion when an H_2O molecule leaves to form the transition state (implying significant dissociative character).[3] We see that $\Delta^{\ddagger}V$ becomes more positive, from −4.1 cm^3 mol^{-1} for V^{2+} to +7.2 cm^3 mol^{-1} for Ni^{2+}, corresponding to a decrease in associative character across the 3d series. In part, this arises because of the decrease in ionic radius as the first row is traversed and in part from the increase in the number of nonbonding d electrons from d^3 to d^8 across the 3d series: associative activation needs the metal centre to be accessible to nucleophilic attack, either by being large or having a low (nonbonding or π*) d-electron population (so that incoming lone pairs can be donated into these orbitals). Negative volumes of activation are also observed for the larger 4d and 5d ions, such as for Rh(III), and indicate an associative interaction of the entering group in the transition state of the reaction.

Table 23.10 shows some data for the formation of Br^-, Cl^-, and NCS^- complexes from $[Cr(OH_2)_6]^{3+}$ and $[Cr(NH_3)_5(OH_2)]^{3+}$. In contrast to the strong dependence of the hexaaqua complex, the pentaammine complex shows only a weak dependence on the identity of the nucleophile, suggesting a transition from I_a to I_d. In addition, the rate constants for the replacement of H_2O in $[Cr(OH_2)_6]^{3+}$ by Cl^-, Br^-, or NCS^- are smaller by a factor of about 10^4 than those for the analogous reactions of $[Cr(NH_3)_5(OH_2)]^{3+}$. This difference suggests that the NH_3 ligands, which are stronger σ donors than H_2O, promote dissociation of the sixth ligand more effectively. As we saw above, this behaviour is to be expected in dissociatively activated reactions.

EXAMPLE 23.3 Interpreting kinetic data in terms of a mechanism

The second-order rate constants for formation of $[VX(OH_2)_5]^+$ from $[V(OH_2)_6]^{2+}$ and X^- for X^- = Cl^-, NCS^-, and N_3^- are in the ratio 1:2:10. What do the data suggest about the rate-determining step for the substitution reaction?

---

[3] Recall that the limiting value for $\Delta^{\ddagger}V$ is approximately ±18 cm^3 mol^{-1}, the molar volume of water, with A reactions having negative and D reactions positive values.

Answer We need to consider the factors that might affect the rate of the reaction. All three ligands are singly charged anions of similar size, so we can expect the encounter equilibrium constants to be similar. Therefore, the second-order rate constants are proportional to first-order rate constants for substitution in the encounter complex. The second-order rate constant is equal to $k_2 K_E$, where K_E is the pre-equilibrium constant and k_2 is the first-order rate constant for substitution of the encounter complex. The greater rate constants for NCS^- than for Cl^-, and especially the five-fold difference of NCS^- from its close structural analogue N_3^- suggest some contribution from nucleophilic attack and an associative reaction. By contrast, there is no such systematic pattern for the same anions reacting with Ni(II), for which the reaction is believed to be dissociative.

Self-test 23.3 Use the data in Table 23.6 to estimate an appropriate value for K_E and calculate k_2 for the reactions of V(II) with Cl^- if the observed second-order rate constant is 1.2×10^2 $dm^3\ mol^{-1}\ s^{-1}$.

23.7 Base hydrolysis

KEY POINT Octahedral substitution can be greatly accelerated by OH^- ions, when ligands with acidic hydrogens are present, as a result of the decrease in charge of the reactive species and the increased ability of the deprotonated ligand to stabilize the transition state.

Consider a substitution reaction in which the ligands possess acidic protons, such as

$$[CoCl(NH_3)_5]^{2+} + OH^- \rightarrow [Co(OH)(NH_3)_5]^{2+} + Cl^-$$

An extended series of studies has shown that though the rate law is overall second order, with rate = $k[CoCl(NH_3)_5^{2+}][OH^-]$ the mechanism is not a simple bimolecular attack by OH^- on the complex. For instance, whereas the replacement of Cl^- by OH^- is fast, the replacement of Cl^- by F^- is slow, even though F^- resembles OH^- more closely in terms of size and nucleophilicity. There is a considerable body of indirect evidence relating to the problem, but one elegant experiment makes the essential point. This conclusive evidence comes from a study of the $^{18}O/^{16}O$ isotope distribution in the product $[Co(OH)(NH_3)_5]^{2+}$. It is known that the $^{18}O/^{16}O$ ratio differs between H_2O and OH^- at equilibrium, and this fact can be used to establish whether the incoming group is H_2O or OH^-. The $^{18}O/^{16}O$ isotope ratio in the cobalt product matches that for H_2O, not that for the OH^- ions, proving that it is an H_2O molecule that is the entering group.

The mechanism that takes these observations into account supposes that the role of OH^- is to act as a Brønsted base, not an entering group:

$$[CoCl(NH_3)_5]^{2+} + OH^- \rightleftharpoons [CoCl(NH_2)(NH_3)_4]^+ + H_2O$$
$$[CoCl(NH_2)(NH_3)_4]^+ \rightarrow [Co(NH_2)(NH_3)_4]^{2+} + Cl^-$$
$$\text{(slow, rate-determining step)}$$
$$[Co(NH_2)(NH_3)_4]^{2+} + H_2O \rightarrow [Co(OH)(NH_3)_5]^{2+} \text{ (fast)}$$

In the first step, a coordinated NH_3 ligand acts as a Brønsted acid and a rapid equilibrium is established between the starting complex and its conjugate base which contains an amido (NH_2^-) ligand. The deprotonated form of the complex has a lower charge and it will be able to lose a Cl^- ion more readily than the protonated form, thus accelerating the reaction. In addition, the amido ligand is both a stronger σ donor than NH_3 and a good π donor. This strong donation by NH_2^- labilizes the Cl^- ligand occupying the *trans* position, *and* stabilizes the five-coordinate transition state (see the next Brief Illustration for a discussion of the stereochemical consequences). The final steps are the rapid attachment of the incoming water and a proton transfer to the amide.

23.8 Stereochemistry

KEY POINT Reaction through a square-pyramidal intermediate results in retention of the original geometry, but reaction through a trigonal-bipyramidal intermediate can lead to isomerization.

Classic examples of octahedral substitution stereochemistry are provided by Co(III) complexes. Table 23.11 shows some data for the hydrolysis of *cis-* and *trans-*$[CoAX(en)_2]^+$ (8) and (9), respectively, where X is the leaving group (either Cl^- or Br^-) and A is OH^-, NO_2^-, NCS^-, or Cl^-. The stereochemical consequences of substitution of octahedral complexes are much more intricate than those of square-planar complexes. The *cis* complexes do not undergo isomerization when substitution occurs, whereas the *trans* forms show a tendency to isomerize in the order A = $NO_2^- < Cl^- < NCS^- < OH^-$.

TABLE 23.11 Stereochemical course of hydrolysis reactions of $[CoAX(en)_2]^+$

	A	X	Percentage *cis* in product
cis	OH^-	Cl^-	100
	Cl^-	Cl^-	100
	NCS^-	Cl^-	100
	Cl^-	Br^-	100
trans	NO_2^-	Cl^-	0
	NCS^-	Cl^-	50–70
	Cl^-	Cl^-	35
	OH^-	Cl^-	75

X is the leaving group.

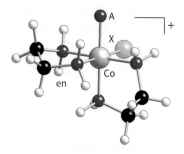

8 cis-[CoAX(en)$_2$]$^+$

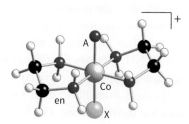

9 trans-[CoAX(en)$_2$]$^+$

The data can be understood in terms of an I_d mechanism and by recognizing that the five-coordinate metal centre in the transition state may resemble either of the two stable geometries for five-coordination; namely, square-pyramidal or trigonal-bipyramidal. As can be seen from Fig. 23.13, reaction through the square-pyramidal complex results in retention of the original geometry, but reaction through the trigonal-bipyramidal complex can lead to isomerization. The cis complex gives rise to a square-pyramidal intermediate, but the trans isomer gives a trigonal-bipyramidal intermediate. For d-metals, trigonal-bipyramidal complexes are favoured when the ligands in the equatorial positions are good π donors, and a good π-donor ligand trans to the leaving group Cl$^-$ favours isomerization (**10**).

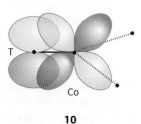

10

A BRIEF ILLUSTRATION

Substitution of Co(III) complexes of the type [CoAX(en)$_2$]$^+$ results in trans-to-cis isomerization, but only when the reaction is catalysed by a base. In a base hydrolysis reaction, one of the NH$_2$R groups of the en ligands loses a proton and becomes its conjugate base, :NHR$^-$. The :NHR$^-$ ligand group is a strong π-donor ligand and favours a trigonal bipyramid of the type shown in Fig. 23.13, and it may be attacked in the different ways shown there. If the direction of attack of the incoming ligands were random, we would expect 33% trans and 67% cis product.

23.9 Isomerization reactions

KEY POINTS Isomerization of a complex can take place by mechanisms that involve substitution, bond cleavage, and reformation, or twisting.

Isomerization reactions are closely related to substitution reactions; indeed, a major pathway for isomerization is often via substitution. The square-planar Pt(II) and

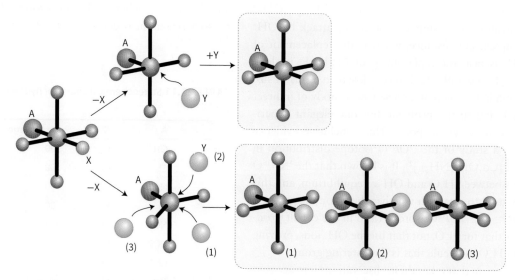

FIGURE 23.13 Reaction through a square-pyramidal complex (top path) results in retention of the original geometry, but reaction through a trigonal-bipyramidal complex (bottom path) can lead to isomerization.

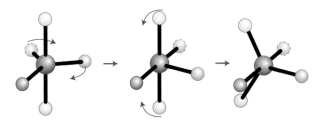

FIGURE 23.14 The exchange of axial and equatorial ligands by a twist through a square-pyramidal conformation of the complex.

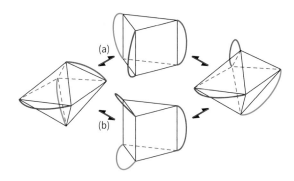

FIGURE 23.15 (a) The Bailar twist and (b) the Ray–Dutt twist by which an octahedral complex can undergo isomerization without losing a ligand or breaking a bond.

octahedral Co(III) complexes we have discussed can form five-coordinate trigonal-bipyramidal transition states. The interchange of the axial and equatorial ligands in a trigonal-bipyramidal complex can be pictured as occurring by a Berry pseudorotation through a square-pyramidal conformation (Section 7.9 and Fig. 23.14). As we have seen, when a trigonal-bipyramidal complex adds a ligand to produce a six-coordinate complex, a new direction of attack of the entering group can result in isomerization.

If a chelate ligand is present, isomerization can occur as a consequence of metal–ligand bond breaking, and substitution need not occur. An example is the exchange of the 'outer' CD_3 group with the 'inner' CH_3 group during the isomerization of a substituted tris(acetylacetonato)cobalt(III) complex, (11) → (12).

An octahedral complex can also undergo isomerization by an intramolecular twist without loss of a ligand or breaking of a bond. There is evidence, for example, that

11

12

racemization of $[Ni(en)_3]^{2+}$ occurs by such an internal twist. Two possible paths are the **Bailar twist** and the **Ray–Dutt twist** (Fig. 23.15).

Redox reactions

As remarked in Chapter 6, redox reactions can occur by the direct transfer of electrons (as in some electrochemical cells and in many solution reactions) or by the transfer of atoms and ions (as in the transfer of O atoms in reactions of oxidoanions). Because redox reactions in solution involve both an oxidizing and a reducing agent, they are usually bimolecular in character. The exceptions are reactions in which one molecule has both oxidizing and reducing centres.

23.10 The classification of redox reactions

KEY POINTS In an inner-sphere redox reaction a ligand is shared to form a transition state; in an outer-sphere redox reaction there is no bridging ligand between the reacting species.

In the 1950s, Henry Taube identified two mechanisms of redox reactions for metal complexes. One is the **inner-sphere mechanism**, which includes atom-transfer processes. In an inner-sphere mechanism, the coordination spheres of the reactants share a ligand transitorily and form a bridged transition state. The other is an **outer-sphere mechanism**, which includes many simple electron transfers. In an outer-sphere mechanism, the complexes come into contact without sharing a bridging ligand and the electron tunnels from one metal atom to the other.

The mechanisms of some redox reactions have been definitively assigned as inner- or outer-sphere. However, the mechanisms of a vast number of reactions are unknown because it is difficult to make unambiguous assignments when complexes are labile. Much of the study of well-defined examples is directed towards the identification of the parameters that differentiate the two paths, with the aim of being able to make correct assignments in more difficult cases.

23.11 The inner-sphere mechanism

KEY POINT The rate-determining step of an inner-sphere redox reaction may be any one of the component processes, but a common one is electron transfer.

The inner-sphere mechanism was first confirmed for the reduction of the nonlabile complex [CoCl(NH$_3$)$_5$]$^{2+}$ by labile Cr^{2+}(aq). The products of the reaction included both labile Co^{2+}(aq) and nonlabile [CrCl(OH$_2$)$_5$]$^{2+}$, and addition of ^{36}Cl$^-$ to the solution did not lead to the incorporation of any of the isotope into the Cr(III) product. The electron-transfer step is much faster than reactions that remove Cl$^-$ from nonlabile Co(III) or introduce Cl$^-$ into the nonlabile [Cr(OH$_2$)$_6$]$^{3+}$ complex. These observations suggested that Cl moves directly from the coordination sphere of one complex to that of the other during the reaction. Since the Cl$^-$ attached to nonlabile Co(III) can easily enter into the labile coordination sphere of [Cr(OH$_2$)$_6$]$^{2+}$ to produce the bridged complex (**13**), it was suggested that this type of complex was an intermediate in the reaction.

TABLE 23.12 Second-order rate constants for selected inner-sphere reactions with variable bridging ligands

Oxidant	Reductant	Bridging ligand	$k/(\text{dm}^3\,\text{mol}^{-1}\text{s}^{-1})$
[Co(NH$_3$)$_6$]$^{3+}$	[Cr(OH$_2$)$_6$]$^{2+}$		8×10^{-5}
[CoF(NH$_3$)$_5$]$^{2+}$	[Cr(OH$_2$)$_6$]$^{2+}$	F$^-$	2.5×10^5
[CoI(NH$_3$)$_5$]$^{2+}$	[Cr(OH$_2$)$_6$]$^{2+}$	Cl$^-$	6.0×10^5
[CoI(NH$_3$)$_5$]$^{2+}$	[Cr(OH$_2$)$_6$]$^{2+}$	I$^-$	3.0×10^6
[Co(NCS)(NH$_3$)$_5$]$^{2+}$	[Cr(OH$_2$)$_6$]$^{2+}$	NCS$^-$	1.9×10^1
[Co(SCN)(NH$_3$)$_5$]$^{2+}$	[Cr(OH$_2$)$_6$]$^{2+}$	SCN$^-$	1.9×10^5
[Co(OH$_2$)(NH$_3$)$_5$]$^{2+}$	[Cr(OH$_2$)$_6$]$^{2+}$	H$_2$O	1.0×10^{-1}
[CrF(OH$_2$)$_5$]$^{2+}$	[Cr(OH$_2$)$_6$]$^{2+}$	F$^-$	7.4×10^{-3}

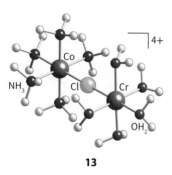

13

Inner-sphere reactions, though involving more steps than outer-sphere reactions, can be fast. Fig. 23.16 summarizes the steps necessary for such a reaction to occur. The first two steps of an inner-sphere reaction are the formation of a precursor complex and the formation of the bridged binuclear intermediate. The first step is identical to the first step in the Eigen–Wilkins mechanism (Section 23.5). The final steps are electron transfer through the bridging ligand to give the successor complex, followed by dissociation to give the products.

The rate-determining step of the overall reaction may be any one of these processes, but the most common one is the electron-transfer step. However, if both metal ions have a nonlabile electron configuration after electron transfer, then the break-up of the bridged complex is rate-determining. An example is the reduction of [RuCl(NH$_3$)$_5$]$^{2+}$ by [Cr(OH$_2$)$_6$]$^{2+}$, in which the rate-determining step is the dissociation of the Cl-bridged complex [Ru^{II}(NH$_3$)$_5$(μ-Cl)Cr^{III}(OH$_2$)$_5$]$^{4+}$. Reactions in which the formation of the bridged complex is rate-determining tend to have similar rate constants for a series of partners of a given species. For example, the oxidation of V^{2+}(aq) has similar rate constants for a long series of Co(III) oxidants with different bridging ligands. The explanation is that the rate-determining step is the substitution of an H$_2$O molecule from the coordination sphere of V(II), which is quite slow (Table 23.9).

The numerous reactions in which electron transfer is rate-determining do not display such simple regularities. Rates vary over a wide range, as metal ions and bridging ligands are varied.[4] The data in Table 23.12 show some typical variations as bridging ligand, oxidizing metal, and reducing metal are changed.

All the reactions in Table 23.12 result in a change of oxidation number by ±1. Such reactions are still often called **one-equivalent processes**, the name reflecting the largely outmoded term 'chemical equivalent'. Similarly, reactions that result in the change of oxidation number by ±2 are often called **two-equivalent processes** and may resemble nucleophilic substitutions. This resemblance can be seen by considering the reaction between Pt(II) and Pt(IV) chlorido complexes which occurs through a Cl$^-$ bridge (**14**). The reaction depends on the transfer of a Cl$^-$ ion in the break-up of the successor complex.

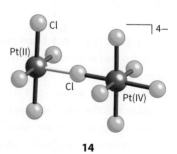

14

There is no difficulty in assigning an inner-sphere mechanism when the reaction involves ligand transfer from an initially nonlabile reactant to a nonlabile product. With more labile complexes, an inner-sphere mechanism should always be considered likely when ligand transfer occurs as well as electron transfer, and if good bridging groups such as Cl$^-$,

---

[4] Some bridged intermediates have been isolated with the electron clearly located on the bridge, but we shall not consider these here.

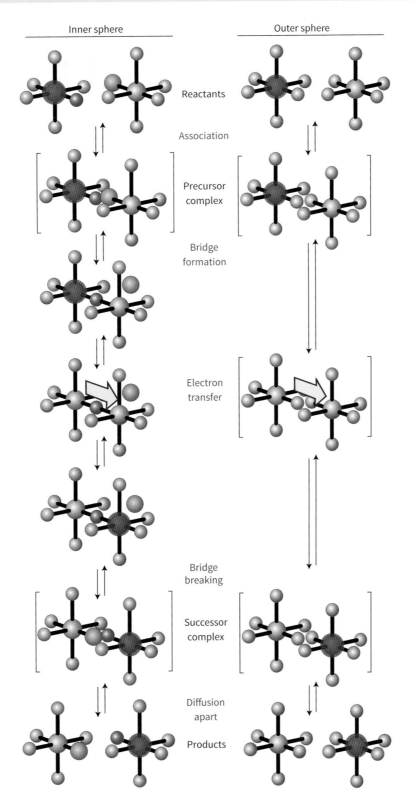

FIGURE 23.16 The different pathways followed by inner- and outer-sphere mechanisms.

Br$^-$, I$^-$, N$_3^-$, CN$^-$, SCN$^-$ pyrazine (**15**), 4,4'-bipyridine (**16**), and 4-dimethylaminopyridine (**17**) are present. Although all these ligands have lone pairs to form the bridge, this may not be an essential requirement. For instance, just as the carbon atom of a methyl group can act as a bridge between OH$^-$ and I$^-$ in the hydrolysis of iodomethane, so it can act as a bridge between Cr(II) and Co(III) in the reduction of methylcobalt species by Cr(II).

15 Pyrazine

16 4,4′-Bipyridine

17 Dimethylaminopyridine

The oxidation of a metal centre by oxidoanions is also an example of an inner-sphere process, and is important in some enzymes (Section 26.12). For example, in the oxidation of Mo(IV) by NO_3^- ions, an O atom of the nitrate ion binds to the Mo atom, facilitating the electron transfer from Mo to N, and then remains bound to the Mo(VI) product:

$$Mo(IV) + NO_3^- \rightarrow Mo-O-NO_2^- \rightarrow Mo=O + NO_2^-$$

A BRIEF ILLUSTRATION

The rate constant for the oxidation of the Ru(II) centre by the Co(III) centre in the bimetallic complex (**18**) is 1.0×10^2 dm^3 mol^{-1} s^{-1}, whereas the rate constant for complex (**19**) is 1.6×10^{-2} dm^3 mol^{-1} s^{-1}. In both complexes there is a pyridine carboxylic acid group bridging the two metal centres. These groups are bound to both metal atoms and could facilitate an electron-transfer process through the bridge, suggesting an inner-sphere process. The fact that the rate constant changes between the two complexes, when the only substantive difference between them is in the substitution pattern of the pyridine ring, confirms that the bridge must be playing a role in the electron-transfer process.

18

19

23.12 The outer-sphere mechanism

KEY POINTS An outer-sphere redox reaction involves electron tunnelling between two reactants without any major disturbance of their covalent bonding or inner coordination spheres; the rate constant depends on the electronic and geometrical structures of the reacting species and on the Gibbs energy of reaction.

A conceptual starting point for understanding the principles of outer-sphere electron transfer is the deceptively simple reaction called **electron self-exchange**. A typical example is the exchange of an electron between $[Fe(OH_2)_6]^{3+}$ and $[Fe(OH_2)_6]^{2+}$ ions in water:

$$[Fe(OH_2)_6]^{3+} + [Fe(OH_2)_6]^{2+} \rightarrow [Fe(OH_2)_6]^{2+} + [Fe(OH_2)_6]^{3+}$$

Self-exchange reactions can be studied over a wide dynamic range with techniques ranging from isotopic labelling to NMR, with EPR being useful for even faster reactions. The rate constant of the Fe^{3+}/Fe^{2+} reaction is about 1 dm^3 mol^{-1} s^{-1} at 25°C.

To set up a scheme for the mechanism, we suppose that Fe^{3+} and Fe^{2+} come together to form a weak outer-sphere complex (Fig. 23.17). Note that the Fe–OH$_2$ bond lengths are different in the two different oxidation state complexes. We need to consider, by assuming that the overlap of their respective acceptor and donor orbitals is sufficient to give a reasonable tunnelling probability,[5] how rapidly an electron transfers between the two metal ions. To explore this problem, we invoke the Franck–Condon principle, introduced originally to account for the vibrational structure of electronic transitions in spectroscopy, which states that

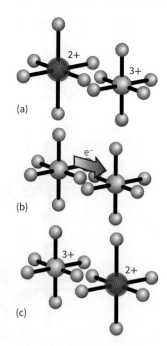

FIGURE 23.17 Electron transfer between two metal ions in a precursor complex is not productive until their coordination shells have reorganized to be of equal size. (a) Reactants, (b) reactant complexes having distorted into the same geometry, (c) products.

---

[5] Tunnelling refers to a process where, according to classical physics, the electrons do not have sufficient energy to overcome the barrier but penetrate into or through it.

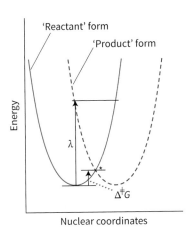

FIGURE 23.18 The energy curves for electron self-exchange. The nuclear motions of both the oxidized and reduced species (shown displaced along the reaction coordinate) and the surrounding solvent are represented by potential wells. Electron transfer to oxidized metal ion (left) occurs once fluctuations of its inner and outer coordination shell bring it to a point (denoted *) on its energy surface that coincides with the energy surface of its reduced state (right). This point is at the intersection of the two curves. The activation energy depends on the horizontal displacement of the two curves (representing the difference in sizes of the oxidized and reduced forms).

electronic transitions are so fast that they take place in a stationary nuclear framework.

In Fig. 23.18 the nuclear motions associated with the 'reactant' Fe^{3+} and its 'conjugate product' Fe^{2+} are represented as displacements along a reaction coordinate. If $[Fe(OH_2)_6]^{3+}$ lies at its energy minimum, then an instantaneous electron transfer would give a compressed state of $[Fe(OH_2)_6]^{2+}$ Likewise, the removal of an electron from Fe^{2+} at its energy minimum would give an expanded state of $[Fe(OH_2)_6]^{3+}$ The only instant at which the electron can transfer within the precursor complex is when both $[Fe(OH_2)_6]^{3+}$ and $[Fe(OH_2)_6]^{2+}$ have achieved the same nuclear configuration by thermally induced fluctuations. That configuration corresponds to the point of intersection of the two curves, and the energy required to reach this position is the Gibbs energy of activation, $\Delta^{\ddagger}G$. If $[Fe(OH_2)_6]^{3+}$ and $[Fe(OH_2)_6]^{2+}$ differ in their nuclear configurations, $\Delta^{\ddagger}G$ is larger and electron exchange is slower. The rate of electron transfer across the encounter complex is expressed quantitatively by the equation

$$k_{ET} = \nu_N \kappa_e e^{-\Delta^{\ddagger}G/RT} \quad (23.14)$$

in which k_{ET} is the rate constant for electron transfer and $\Delta^{\ddagger}G$ is given by the **Marcus equation**,[6]

$$\Delta^{\ddagger}G = \tfrac{1}{4}\lambda\left(1 + \frac{\Delta_r G^{\ominus}}{\lambda}\right)^2 \quad (23.15)$$

[6] Rudolph A. Marcus was awarded the 1992 Nobel Prize in Chemistry for developing a theory explaining the rates of electron transfer between molecules in solution.

where $\Delta_r G^{\ominus}$ is the standard reaction Gibbs energy (obtained from the difference in standard potentials of the redox partners), and λ is the **reorganization energy**—the energy required to move all the nuclei associated with the reactant to the positions they adopt in the product but without transferring the electron. This energy depends on the changes in metal–ligand bond lengths (the so-called *inner-sphere reorganization energy*) and alterations in solvent polarization, principally the orientation of the solvent molecules around the complex (the *outer-sphere reorganization energy*).

The pre-exponential factor in eqn. 23.14 has two components, the **nuclear frequency factor** ν_N and the **electronic factor** κ_e. The former is the frequency at which the two complexes, having already encountered each other in the solution, attain the transition state. The electronic factor gives the probability on a scale from 0 to 1 that an electron will transfer when the transition state is reached; its precise value depends on the extent of overlap of the donor and acceptor orbitals, and it increases as orbital overlap improves.

A small reorganization energy and a value of κ_e close to 1 correspond to a redox couple capable of fast electron self-exchange. The first requirement is achieved if the transferred electron is removed from or added to a nonbonding orbital, as the change in metal–ligand bond length is then minimized. It is also likely if the metal ion is shielded from the solvent, in the sense of it being sterically difficult for solvent molecules to approach close to the metal ion, because the polarization of the solvent is also a component of the reorganization energy. Simple metal ions such as aqua species typically have λ well in excess of 1 eV, whereas buried redox centres in enzymes, which are very well shielded from the solvent, can have values as low as 0.25 eV.

For a self-exchange reaction, $\Delta_r G^{\ominus} = 0$, and therefore, from eqn. 23.15, $\Delta^{\ddagger}G = \tfrac{1}{4}\lambda$, and the rate of electron transfer is controlled by the reorganization energy (Fig. 23.19(a)).

To a considerable extent, the rates of self-exchange can be interpreted in terms of the types of orbitals involved in the transfer (Table 23.13). In the $[Cr(OH_2)_6]^{3+/2+}$ self-exchange reaction, an electron is transferred between antibonding σ^* orbitals, and the consequent extensive change in metal–ligand bond lengths results in a large inner-sphere reorganization energy and therefore a slow reaction. The $[Co(NH_3)_6]^{3+/2+}$ couple has an even greater reorganization energy because two electrons are moved into the σ^* orbital as rearrangement occurs and the reaction is even slower. With the other hexaaqua and hexaammine complexes in the table, the electron is transferred between weakly antibonding or nonbonding π orbitals, the inner-sphere reorganization is less extensive, and the reactions are faster. The bulky, hydrophobic chelating ligand bipyridyl acts as a solvent shield, thus decreasing the outer-sphere reorganization energy.

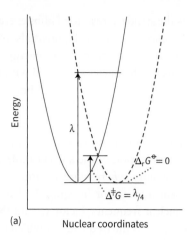
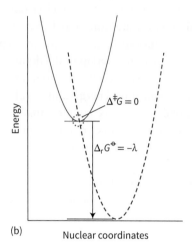
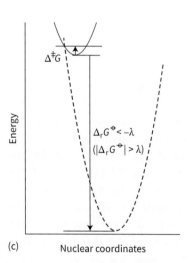

FIGURE 23.19 Variation of the activation Gibbs energy ($\Delta^\ddagger G$) with reaction Gibbs energy ($\Delta_r G^\ominus$). (a) In a self exchange reaction, $\Delta_r G^\ominus = 0$ and $\Delta^\ddagger G = \lambda/4$. (b) A reaction is 'activationless' when $\Delta_r G^\ominus = -\lambda$. (c) $\Delta^\ddagger G$ increases (the rate diminishes) as $\Delta_r G^\ominus$ becomes more negative beyond $\Delta_r G^\ominus = -\lambda$.

TABLE 23.13 Correlations between rate constants and electron configurations for electron self-exchange reactions

Reaction	Electron Configuration	Δd/pm*	k_{11}/(dm^3 mol^{-1}s^{-1})
[Cr(OH$_2$)$_6$]$^{3+/2+}$	$t_{2g}^3/t_{2g}^3 e_g$	20	1×10^{-5}
[V(OH$_2$)$_6$]$^{3+/2+}$	t_{2g}^2/t_{2g}^3	13	1×10^{-5}
[Fe(OH$_2$)$_6$]$^{3+/2+}$	$t_{2g}^3 e_g^2/t_{2g}^4 e_g^2$	13	1.1
[Ru(OH$_2$)$_6$]$^{3+/2+}$	t_{2g}^5/t_{2g}^6	9	20
[Ru(NH$_3$)$_6$]$^{3+/2+}$	t_{2g}^5/t_{2g}^6	4	6.6×10^3
[Co(NH$_3$)$_6$]$^{3+/2+}$	$t_{2g}^6/t_{2g}^5 e_g^2$	22	2×10^{-8}
[Fe(bpy)$_3$]$^{3+/2+}$	t_{2g}^5/t_{2g}^6	0	3×10^8
[Ru(bpy)$_3$]$^{3+/2+}$	t_{2g}^5/t_{2g}^6	0	4×10^8
[Ni(bpy)$_3$]$^{3+/2+}$	$t_{2g}^6 e_g/t_{2g}^6 e_g^2$	12	1.5×10^3

* Δd is the change in mean M–L bond length

the reorganization energy when the electron is transferred between π orbitals, as occurs with Fe and Ru, where the electron transfer is between t_{2g} orbitals (which, as explained in Section 7.17, can participate in π bonding), but not with Ni, where the electron transfer is between e_g orbitals. Delocalization can also increase the electronic factor.

Self-exchange reactions are helpful for pointing out the concepts that are involved in electron transfer, but chemically useful redox reactions occur between different species and involve net electron transfer. For the latter reactions, $\Delta_r G^\ominus$ is nonzero and contributes to the rate through eqns. 23.14 and 23.15. Provided $|\Delta_r G^\ominus| \ll |\lambda|$, eqn. 23.15 becomes

$$\Delta^\ddagger G = \tfrac{1}{4}\lambda\left(1 + \frac{\Delta_r G^\ominus}{\lambda}\right)^2 \approx \tfrac{1}{4}\lambda\left(1 + \frac{2\Delta_r G^\ominus}{\lambda}\right) = \tfrac{1}{4}(\lambda + 2\Delta_r G^\ominus)$$

and then, according to eqn. 23.14,

$$k_{ET} \approx \nu_N \kappa_e e^{-(\lambda + 2\Delta_r G^\ominus)/4RT}$$

Because $\lambda > 0$ and $\Delta_r G^\ominus < 0$ for thermodynamically feasible reactions, provided $|\Delta_r G^\ominus| \ll |\lambda|$, the rate constant increases exponentially as $\Delta_r G^\ominus$ becomes increasingly favourable (that is, more negative). However, as $|\Delta_r G^\ominus|$ becomes comparable to $|\lambda|$, this equation breaks down and we see that the reaction rate peaks before declining as $|\Delta_r G^\ominus| > |\lambda|$.

Eqn. 23.15 shows that $\Delta^\ddagger G = 0$ when $\Delta_r G^\ominus = -\lambda$. That is, the reaction becomes 'activationless' when the standard reaction Gibbs energy and the reorganization energy cancel (Fig. 23.19(b)). The activation energy now *increases* as $\Delta_r G^\ominus$ becomes more negative and the reaction rate decreases. This slowing of the reaction as the standard Gibbs energy of the reaction becomes more exergonic is called **inverted behaviour** (Fig. 23.19(c)).

Inverted behaviour has important practical consequences, a notable one relating to the long-range electron transfer involved in photosynthesis. Photosystems are complex proteins containing light-excitable pigments such as chlorophyll and a

Bipyridyl and other π-acceptor ligands allow electrons in an orbital with π symmetry on the metal ion to delocalize on to the ligand. This delocalization effectively lowers

chain of redox centres having low reorganization energies. In this chain, one highly exergonic recombination of the photoelectron with oxidized chlorophyll is sufficiently retarded (to 30 ns) as to allow the electron to escape (in 200 ps) and proceed down the photosynthetic electron-transport chain, ultimately to produce reduced carbon compounds (Section 26.8).

The theoretical dependence of the reaction rate of a reaction on the standard Gibbs energy is plotted in Fig. 23.20, and Fig. 23.21 shows the observed variation of reaction rate with $\Delta_r G^\ominus$ for the iridium complex (**20**). The results plotted in Fig. 23.21 represent the first unambiguous experimental observation of an inverted region for a synthetic complex. Practically, intramolecular electron-transfer reactions are more straightforward to study because the rate will not be limited by diffusion; in these experiments the rates of electron transfer between the Ir(I) dimer and various R-substituted pyridinium groups were measured following photoexcitation of Ir(I), Section 23.13.

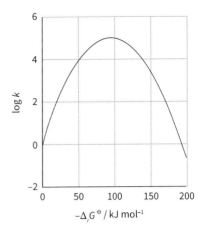

FIGURE 23.20 The theoretical dependence of the $\log_{10}$ of the reaction rate (in arbitrary units) on $\Delta_r G^\ominus$ for a redox reaction with $\lambda = 1.0$ eV (96.4 kJ mol^{-1}).

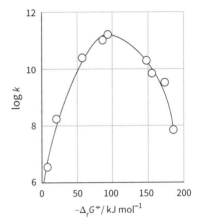

FIGURE 23.21 Plot of $\log_{10} k$ against $-\Delta_r G^\ominus$ for the iridium complex (**20**) with various R substituents, measured in acetonitrile solution at room temperature.

20

The Marcus equation can be used to predict the rate constants for outer-sphere electron-transfer reactions between different species. Consider the electron-transfer reaction between an oxidant Ox$_1$ and reductant Red$_2$:

$$\text{Ox}_1 + \text{Red}_2 \rightarrow \text{Red}_1 + \text{Ox}_2$$

If we suppose that the reorganization energy for this reaction is the average of the values for the two self-exchange processes, we can write $\lambda_{12} = \frac{1}{2}(\lambda_{11} + \lambda_{22})$, and then manipulation of eqns. 23.14 and 23.15 gives the **Marcus cross-relation**,

$$k_{12} = (k_{11} k_{22} K_{12} f_{12})^{1/2} \tag{23.16}$$

in which k_{12} is the rate constant, K_{12} is the equilibrium constant obtained from $\Delta_r G^\ominus$, and k_{11} and k_{22} are the respective self-exchange rate constants for the two reaction partners. For reactions between simple ions in solution, when the standard Gibbs reaction energy is not too large, an LFER of the kind expressed by eqn. 23.12 exists between $\Delta^\ddagger G$ and $\Delta_r G^\ominus$ and f_{12} can normally be set to 1. However, for reactions that are highly favourable thermodynamically (that is, $\Delta_r G^\ominus$ large and negative), the LFER breaks down. The term f_{12} takes into consideration the nonlinearity of the relationship between $\Delta^\ddagger G$ and $\Delta_r G^\ominus$ and is given by

$$\log f_{12} = \frac{(\log K_{12})^2}{4 \log(k_{11} k_{22}/Z)} \tag{23.17}$$

where Z is the constant of proportionality between the encounter density in solution (in moles of encounters per cubic decimetre per second) and the molar concentrations of the reactants; it is often taken to be 10^{11} mol^{-1} dm^3 s^{-1}.

A BRIEF ILLUSTRATION

The rate constants for

$$[\text{Co(bpy)}_3]^{2+} + [\text{Co(bpy)}_3]^{3+} \xrightarrow{k_{11}} [\text{Co(bpy)}_3]^{3+} + [\text{Co(bpy)}_3]^{2+}$$
$$[\text{Co(terpy)}_2]^{2+} + [\text{Co(terpy)}_2]^{3+} \xrightarrow{k_{22}} [\text{Co(terpy)}_2]^{3+} + [\text{Co(terpy)}_2]^{2+}$$

(where bpy is bipyridyl and terpy is tripyridyl) are $k_{11} = 9.0$ dm^3 mol^{-1} s^{-1} and $k_{22} = 48$ dm^3 mol^{-1} s^{-1}, and $K_{12} = 3.57$. Then for the outer-sphere reduction of $[\text{Co(bpy)}_3]^{3+}$ by $[\text{Co(terpy)}_2]^{2+}$, eqn. 23.16 with $f_{12} = 1$ (as noted above) gives

$$k_{12} = (9.0 \times 48 \times 3.57)^{\frac{1}{2}} \text{ dm}^3 \text{ mol}^{-1} \text{ s}^{-1} = 39 \text{ dm}^3 \text{ mol}^{-1} \text{ s}^{-1}$$

This result compares reasonably well with the experimental value, which is 64 dm^3 mol^{-1} s^{-1}. Note how electron-transfer rates can vary by many orders of magnitude.

Photochemical reactions

The absorption of a photon of ultraviolet radiation or visible light increases the energy of a complex by between 170 and 600 kJ mol^{-1}. Because these energies are larger than typical activation energies, it should not be surprising that new reaction channels are opened. However, when the high energy of a photon is used to provide the energy of the primary forward reaction, the back reaction is almost always very favourable, and much of the design of efficient photochemical systems lies in trying to avoid the back reaction.

23.13 Prompt and delayed reactions

KEY POINT Reactions of electronically excited species are classified as prompt or delayed.

In some cases, the excited state formed after absorption of a photon dissociates almost immediately after it is formed. Examples include formation of the pentacarbonyl intermediates that initiate ligand substitution in metal carbonyl compounds:

$$[Cr(CO)_6] \xrightarrow{h\nu} [Cr(CO)_5] + CO$$

and the scission of Co-Cl bonds:

$$[Co^{III}Cl(H_2O)_5]^{2+} \xrightarrow{h\nu(\lambda=350\text{ nm})} [Co^{II}(H_2O)_5]^{2+} + Cl^-$$

Both processes occur in less than 10 ps and hence are called **prompt reactions**.

In the second reaction, the **quantum yield**, the amount of reaction per mole of photons absorbed, increases as the wavelength of the radiation is decreased (and the photon energy correspondingly increased, $E_{photon} = hc/\lambda$). The energy in excess of the bond energy is available to the newly formed fragments and increases the probability that they will escape from each other through the solution before they have an opportunity to recombine.

Some excited states have long lifetimes. They may be regarded as energetic isomers of the ground state that can participate in **delayed reactions**. The excited state of $[Ru^{II}(bpy)_3]^{2+}$ created by photon absorption in the metal-to-ligand charge-transfer band (Section 22.3) may be regarded as a Ru(III) cation complexed to a radical anion of the ligand. Its redox reactions can be explained by adding the excitation energy (expressed as a potential by using $-FE = \Delta_r G$ and equating $\Delta_r G$ to the molar excitation energy) to the ground-state reduction potential (Fig. 23.22).

23.14 d–d and charge-transfer reactions

KEY POINT A useful first approximation is to associate photosubstitution and photoisomerization with d–d transitions and photoredox reactions with charge-transfer transitions, but the rule is not absolute.

There are two main types of spectroscopically observable electron promotion in d-metal complexes; namely, d–d transitions and charge-transfer transitions (Sections 22.2 and 22.3). A d–d transition corresponds to the essentially angular redistribution of electrons within a d shell. In octahedral complexes, this redistribution often corresponds to the occupation of M–L antibonding e_g orbitals. An example is the $^4T_{1g} \leftarrow {}^4A_{2g}(t_{2g}^2 e_g^1 \leftarrow t_{2g}^3)$ transition in $[Cr(NH_3)_6]^{3+}$. The occupation of the antibonding e_g orbital results in a quantum yield close to 1 (specifically 0.6) for the photosubstitution

$$[Cr(NH_3)_6]^{3+} + H_2O \xrightarrow{h\nu} [Cr(NH_3)_5(OH_2)]^{3+} + NH_3$$

This is a prompt reaction, occurring in less than 5 ps.

Charge-transfer transitions correspond to the *radial* redistribution of electron density. They correspond to the promotion of electrons into predominantly ligand orbitals if the transition is metal-to-ligand, or into orbitals of predominantly metal character if the transition is ligand-to-metal. The former process corresponds to oxidation of the metal centre and the latter to its reduction. These excitations commonly initiate photoredox reactions of the kind already mentioned in connection with Co(III) and Ru(II).

Although a useful first approximation is to associate photosubstitution and photoisomerization with d–d transitions and photoredox with charge-transfer transitions, the rule is not absolute. For example, it is not uncommon for a charge-transfer transition to result in photosubstitution by an indirect path:

$$[Co^{III}Cl(NH_3)_5]^{2+} + H_2O \xrightarrow{h\nu} [Co^{II}(NH_3)_5(OH_2)]^{2+} + Cl^-$$
$$[Co^{II}(NH_3)_5(OH_2)]^{2+} + Cl^- \rightarrow [Co^{III}(NH_3)_5(OH_2)]^{3+} + Cl^-$$

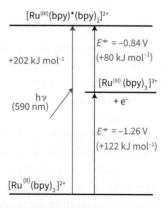

FIGURE 23.22 The photoexcitation of $[Ru^{II}(bpy)_3]^{2+}$ can be treated as if the excited state is a Ru(III) cation complexed to a radical anion of the ligand.

BOX 23.1 How are metal complexes used in solar cells?

Commercial solar cells that convert sunlight to electricity are mainly based on silicon, though systems based on semiconducting oxides such as TiO_2 are also being developed. The harvesting of sunlight generally involves the excitation of an electron in the semiconductor from a valence band to a conduction band; the energy difference between these two bands controls the wavelengths of sunlight that can be converted to electricity. In pure TiO_2 the band gap is large (>3eV), so only UV light can be harvested directly in this material, leading to low conversion efficiencies of only a few percent with natural sunlight. However, through the use of dyes that absorb visible light, the proportion of sunlight that can be harvested increases significantly. Cells which combine TiO_2 with dyes, known as Grätzel cells or dye solar cells (DSCs), have demonstrated efficiencies for light-to-electricity conversion of around 11%. The key to this increased efficiency is the dye, and those most commonly used are based on ruthenium(II).

Complexes such as compound **S1** (known as an N-3 dye) exhibit ligand-centred charge transfer (LCCT) transitions (π–π*) as well as metal-to-ligand charge transfer (MLCT) transitions (4d–π*). These transitions give rise to strong light absorption between 400 and 600 nm (Fig. B23.1). In the Grätzel cell, sunlight is absorbed by a monomolecular layer of a dye coated onto a thin film of nanocrystalline TiO_2. This photon promotes an electron from the Ru^{2+}-based ground state to an excited state $(Ru^{2+})^*$. The excited electron is then transferred, within a picosecond, into the conduction band of TiO_2. This leads to an effective charge separation, with the electron in the TiO_2 and a positive charge on surface-adsorbed Ru^{3+} dye molecule. The Ru^{3+} species is then reduced within nanoseconds by iodide (I^-) present in the electrolyte system of the cell. The electron injected into TiO_2 diffuses to a conductive surface for current collection and the generation of electricity.

The effectiveness of the light conversion in DSCs depends on the kinetics of the various processes. The electron transfer to the TiO_2 conduction band from the excited state of the ruthenium dye is much faster than electronic relaxation processes back to the ground state or chemical side reactions. Moreover, the reduction of the oxidized dye (Ru^{3+}) by I^- is significantly faster than the direct recombination reaction between the injected electron and Ru^{3+}.

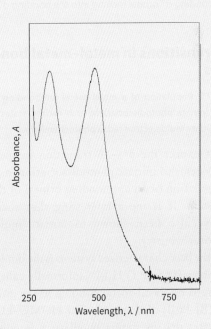

FIGURE B23.1 The absorption spectrum of **S1**.

S1

In this case, the aqua complex formed after the homolytic fission of the Co–Cl bond is reoxidized by the Cl atom. The net result leaves the Co substituted. Conversely, some excited states show no differences in substitutional reactivity compared with the ground state: the long-lived excited ^2E state of $[Cr(bpy)_3]^{3+}$ results from a pure d–d transition and its lifetime of several microseconds allows the excess energy to enhance its redox reactions. The standard potential (+1.3V), calculated by adding the excitation energy to the ground-state value, accounts for its function as a good oxidizing agent, in which it undergoes reduction to $[Cr(bpy)_3]^{2+}$.

The use of transition-metal ions in solar cells is described in Box 23.1.

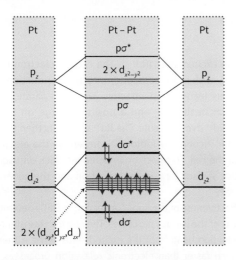

FIGURE 23.23 The dinuclear complex $[Pt_2(\mu-P_2O_5H_2)_4]^{4-}$ (**21**) consists of two face-to-face square-planar complexes held together by a bridging pyrophosphito ligand. The metal p_z and d_{z^2} orbitals interact along the Pt–Pt axis. The other p and d orbitals are considered to be nonbonding. Photoexcitation results in an electron in the antibonding σ* orbital moving into the bonding σ orbital.

23.15 Transitions in metal–metal bonded systems

KEY POINTS Population of a metal–metal antibonding orbital can sometimes initiate photodissociation; such excited states have been shown to initiate multielectron redox photochemistry.

We might expect the δ*← δ transition in metal–metal bonded systems to initiate photodissociation, as it results in the population of an antibonding orbital of the metal–metal system. It is more interesting that such excited states have also been shown to initiate multielectron redox photochemistry.

One of the best-characterized systems is the dinuclear platinum complex $[Pt_2(\mu-P_2O_5H_2)_4]^{4-}$, informally called 'PtPOP' (**21**). There is no metal–metal bonding in the ground state of this Pt(II)–Pt(II) d^8–d^8 species. The HOMO–LUMO pattern indicates that excitation populates a bonding orbital between the two metal atoms (Fig. 23.23). The lowest-lying excited state has a lifetime of 9 μs and is a powerful reducing agent, reacting by both electron and halogen-atom transfer. The most interesting oxidation products contain single Pt(III)–Pt(III) bonds with X⁻ ligands at both ends (the X⁻ is a halide or pseudohalide and was previously present in solution). Irradiation in the presence of (Bu)₃SnH gives a dihydrido product that can eliminate H_2.

21 $[Pt_2(\mu-P_2O_5H_2)_4]^{4-}$, PtPOP

Irradiation of the quadruply bonded dinuclear cluster $[Mo_2(O_2P(OC_6H_5)_2)_4]$ (**22**) at 500 nm in the presence of $ClCH_2CH_2Cl$ results in production of ethene and the addition of two Cl atoms to the two Mo atoms, with a two-electron oxidation. The reaction proceeds in one-electron steps, and requires a complex with the metal atoms shielded by sterically crowding ligands. If smaller ligands are present, the reaction that occurs instead is a photochemical oxidative addition of the organic molecule.

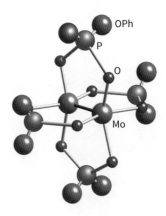

22 $[Mo_2(O_2P(OPh)_2)_4]$

FURTHER READING

G. J. Leigh and N. Winterbottom (eds.), *Modern Coordination Chemistry: The Legacy of Joseph Chatt*. Royal Society of Chemistry (2002). A readable historical discussion of this area.

M. L. Tobe and J. Burgess, *Inorganic Reaction Mechanisms*. Longman (1999).

R. G. Wilkins, *Kinetics and Mechanism of Reactions of Transition Metal Complexes*. VCH (1991).

A special issue of *Coordination Chemistry Reviews* has been dedicated to the work of Henry Taube. See *Coord. Chem. Rev.*, 2005, **249**.

Two readable discussions of redox processes are to be found in Taube's 1983 Nobel Prize lecture, reprinted in *Science*, 1984, 226, 1028, and in Marcus's 1992 Nobel Prize lecture, published in *Nobel Lectures: Chemistry 1991–1995*, World Scientific (1997).

EXERCISES

23.1 The rate constants for the formation of $[CoX(NH_3)_5]^{2+}$ from $[Co(NH_3)_5OH_2]^{3+}$ for $X = Cl^-, Br^-, N_3^-$, and SCN^- differ by no more than a factor of 2. What is the mechanism of the substitution?

23.2 If a substitution process is associative, why may it be difficult to characterize an aqua ion as labile or inert?

23.3 The reactions of $[Ni(CO)_4]$ in which phosphines or phosphites replace CO to give $[Ni(CO)_3L]$ all occur at the same rate regardless of which phosphine or phosphite is being used. Is the reaction *d* or *a*?

23.4 Write the rate law for formation of $[MnX(OH_2)_5]^+$ from the aqua ion and X^-. How would you determine if the reaction is *d* or *a*?

23.5 Octahedral complexes of metal centres with high oxidation numbers or of d metals of the second and third series are less labile than those of low oxidation number and d metals of the first series of the block. Account for this observation on the basis of a dissociative rate-determining step.

23.6 A Pt(II) complex of tetramethyldiethylenetriamine is attacked by Cl^- 10^5 times less rapidly than the diethylenetriamine analogue. Explain this observation in terms of an associative rate-determining step.

23.7 The rate of loss of chlorobenzene, PhCl, from $[W(CO)_4L(PhCl)]$ increases with increase in the cone angle of L. What does this observation suggest about the mechanism?

23.8 The pressure dependence of the replacement of chlorobenzene (PhCl) by piperidine in the complex $[W(CO)_4(PPh_3)(PhCl)]$ has been studied. The volume of activation is found to be $+11.3\ cm^3\ mol^{-1}$. What does this value suggest about the mechanism?

23.9 Does the fact that $[Ni(CN)_5]^{3-}$ can be isolated help to explain why substitution reactions of $[Ni(CN)_4]^{2-}$ are very rapid?

23.10 Reactions of $[Pt(Ph)_2(SMe_2)_2]$ with the bidentate ligand 1,10-phenanthroline (phen) give $[Pt(Ph)_2phen]$. There is a kinetic pathway with activation parameters $\Delta^{\ddagger}H = +101\ kJ\ mol^{-1}$ and $\Delta^{\ddagger}S = +42\ J\ K^{-1}\ mol^{-1}$. Propose a mechanism.

23.11 Design two-step syntheses of *cis*- and *trans*-$[PtCl_2(NO_2)(NH_3)]^-$ starting from $[PtCl_4]^{2-}$.

23.12 How does each of the following modifications affect the rate of a square-planar complex substitution reaction? (a) Changing a *trans* ligand from H^- to Cl^-; (b) changing the leaving group from Cl^- to I^-; (c) adding a bulky substituent to a *cis* ligand; (d) increasing the positive charge on the complex.

23.13 The rate of attack on $[Co(OH_2)_6]^{3+}$ by an entering group Y is nearly independent of Y, with the spectacular exception of the rapid reaction with OH^-. Explain the anomaly. What is the implication of your explanation for the behaviour of a complex lacking Brønsted acidity on the ligands?

23.14 Predict the products of the following reactions:
(a) $[Pt(PR_3)_4]^{2+} + 2Cl^-$ (b) $[PtCl_4]^{2-} + 2PR_3$ (c) *cis*-$[Pt(NH_3)_2(py)_2]^{2+} + 2Cl^-$

23.15 Put in order of increasing rate of substitution by H_2O the complexes (a) $[Co(NH_3)_6]^{3+}$, (b) $[Rh(NH_3)_6]^{3+}$, (c) $[Ir(NH_3)_6]^{3+}$, (d) $[Mn(OH_2)_6]^{2+}$, (e) $[Ni(OH_2)_6]^{2+}$.

23.16 State the effect on the rate of dissociatively activated reactions of Rh(III) complexes of (a) an increase in the overall charge on the complex, (b) changing the leaving group from NO_3^- to Cl^-, (c) changing the entering group from Cl^- to I^-, (d) changing the *cis* ligands from NH_3 to H_2O.

23.17 Write out the inner- and outer-sphere pathways for reduction of azidopentaamminecobalt(III) ion with $V^{2+}(aq)$. What experimental data might be used to distinguish between the two pathways?

23.18 The compound $[Fe(SCN)(OH_2)_5]^{2+}$ can be detected in the reaction of $[Co(NCS)(NH_3)_5]^{2+}$ with $Fe^{2+}(aq)$ to give $Fe^{3+}(aq)$ and $Co^{2+}(aq)$. What does this observation suggest about the mechanism?

23.19 The rate of reduction of $[Co(NH_3)_5(OH_2)]^{3+}$ by Cr(II) is seven orders of magnitude slower than reduction of its conjugate base, $[Co(NH_3)_5(OH)]^{2+}$, by Cr(II). For the corresponding reductions with $[Ru(NH_3)_6]^{2+}$, the two differ by less than a factor of 10. What do these observations suggest about mechanisms?

23.20 Predict the rate constants for electron transfer in the oxidation of $[V(OH_2)_6]^{2+}$ ($E^{\ominus}(V^{3+}/V^{2+}) = -0.255\ V$) by the oxidants (a) $[Ru(NH_3)_6]^{3+}$ ($E^{\ominus}(Ru^{3+}/Ru^{2+}) = +0.07\ V$), (b) $[Co(NH_3)_6]^{3+}$ ($E^{\ominus}(Co^{3+}/Co^{2+}) = +0.10\ V$). Comment on the relative sizes of the rate constants.

23.21 Predict the rate constants for electron transfer in the oxidation of $[Cr(OH_2)_6]^{2+}$ ($E^{\ominus}(Cr^{3+}/Cr^{2+}) = -0.41\ V$) and each of the oxidants $[Ru(NH_3)_6]^{3+}$ ($E^{\ominus}(Ru^{3+}/Ru^{2+}) = +0.07\ V$), $[Fe(OH_2)_6]^{3+}$ ($E^{\ominus}(Fe^{3+}/Fe^{2+}) = +0.77\ V$) and $[Ru(bpy)_3]^{3+}$ ($E^{\ominus}(Ru^{3+}/Ru^{2+}) = +1.26\ V$). Comment on the relative sizes of the rate constants.

23.22 The photochemical substitution of $[W(CO)_5(py)]$ (py = pyridine) with triphenylphosphine gives $[W(CO)_5(P(C_6H_5)_3)]$. In the presence of excess phosphine, the quantum yield is approximately 0.4. A flash photolysis study reveals a spectrum that can be assigned to the intermediate $[W(CO)_5]$. What product and quantum yield do you predict for substitution of $[W(CO)_5(py)]$ in the presence of excess triethylamine? Is this reaction expected to be initiated from the ligand field or MLCT excited state of the complex?

23.23 From the spectrum of $[CrCl(NH_3)_5]^{2+}$ shown in Fig. 22.11, propose a wavelength for photoinitiation of reduction of Cr(III) to Cr(II) accompanied by oxidation of a ligand.

TUTORIAL PROBLEMS

23.1 Given the following mechanism for the formation of a chelate complex,

$[Ni(OH_2)_6]^{2+} + L-L \rightleftharpoons [Ni(OH_2)_6]^{2+}, L-L \quad K_E, \text{rapid}$

$[Ni(OH_2)_6]^{2+}, L-L \rightleftharpoons [Ni(OH_2)_5 L-L]^{2+} + H_2O \quad k_a, k_a'$

$[Ni(OH_2)_5 L-L]^{2+} \rightleftharpoons [Ni(OH_2)_4 L-L]^{2+} + H_2O \quad k_b, k_b'$

derive the rate law for the formation of the chelate. Discuss the step that is different from that for two monodentate ligands. The formation of chelates with strongly bound ligands occurs at the rate of formation of the analogous monodentate complex, but the formation of chelates of weakly bound ligands is often significantly slower. Assuming an I_d mechanism, explain this observation. (See R. G. Wilkins, *Acc. Chem. Res.*, 1970, **3**, 408.)

23.2 Discuss the following set of rate constants (k) and activation parameters for water exchange reactions of metal aqua ions.

Species	k/s^{-1}	$\Delta H^{\ddagger}/kJ\ mol^{-1}$	$\Delta V^{\ddagger}/cm^3\ mol^{-1}$
$[Ti(OH_2)_6]^{3+}$	1.8×10^5	43.4	−12.1
$[V(OH_2)_6]^{2+}$	87	61.8	−4.1
$[V(OH_2)_6]^{3+}$	500	49.4	−12.1
$[Cr(OH_2)_6]^{3+}$	2.4×10^{-6}	109	−9.6
$[Cr(OH_2)_5(OH)]^{2+}$	1.8×10^{-4}	111	+2.7
$[Mn(OH_2)_6]^{2+}$	2.1×10^7	32.9	−5.4
$[Fe(OH_2)_6]^{3+}$	1.6×10^2	63.9	−5.4
$[Co(OH_2)_6]^{2+}$	3.2×10^6	46.9	+6.1
$[Ni(OH_2)_6]^{2+}$	3.2×10^4	56.9	+7.2
$[Cu(OH_2)_6]^{2+}$	4.4×10^9	11.5	+2.0

23.3 The complex $[PtH(PEt_3)_3]^+$ was studied in deuterated acetone in the presence of excess PEt_3. In the absence of excess ligand the ^1H-NMR spectrum in the hydride region exhibits a doublet of triplets. As excess PEt_3 ligand is added the hydride signal begins to change, the line shape depending on the ligand concentration. Suggest a mechanism to account for the effects of excess PEt_3.

23.4 Solutions of $[PtH_2(PMe_3)_2]$ exist as a mixture of *cis* and *trans* isomers. Addition of excess PMe_3 led to formation of $[PtH_2(PMe_3)_3]$ at a concentration that could be detected using NMR. This complex exchanged phosphine ligands rapidly with the *trans* isomer but not the *cis*. Propose a pathway. What are the implications for the *trans* effect of H versus PMe_3? (See D. L. Packett and W. G. Trogler, *Inorg. Chem.*, 1988, **27**, 1768.)

23.5 Fig. 23.24 (which is based on J. B. Goddard and F. Basolo, *Inorg. Chem.*, 1968, **7**, 936) shows the observed first-order rate constants for the reaction of $[PdBrL]^+$ with various Y^- to give

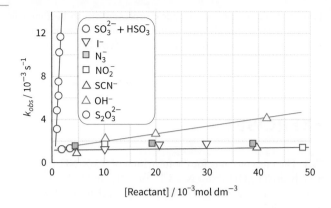

FIGURE 23.24 The data required for Problem 23.5.

$[PdYL]^+$, where L is $Et_2NCH_2CH_2NHCH_2CH_2NEt_2$. Note the large slope for $S_2O_3^{2-}$ and zero slopes for $Y^- = N_3^-, I^-, NO_2^-$, and SCN^-. Propose a mechanism.

23.6 The activation enthalpy for the reduction of *cis*-$[CoCl_2(en)_2]^+$ by Cr^{2+} (aq) is −24 kJ mol^{-1}. Explain the negative value. (See R. C. Patel, R. E. Ball, J. F. Endicott, and R. G. Hughes, *Inorg. Chem.*, 1970, **9**, 23.)

23.7 Consider the complexes (**18**) and (**19**) discussed in the Brief Illustration in Section 23.11. Think about the potential routes of the electron transfer in the two complexes, and suggest why there is such a difference in the rates of electron transfer between the two complexes.

23.8 Predict the rate constants for outer-sphere reactions from the following data. Compare your results to the measured values in the last column.

Reaction	$k_{11}/dm^3\ mol^{-1}\ s^{-1}$	$k_{22}/dm^3\ mol^{-1}\ s^{-1}$	$E^{\ominus}/V$	$k_{obs}/dm^3\ mol^{-1}\ s^{-1}$
$Cr^{2+} + Fe^{2+}$	2×10^{-5}	4.0	+1.18	2.3×10^3
$[W(CN)_8]^{4-} + Ce(IV)$	$>4 \times 10^4$	4.4	+0.54	$>10^8$
$[Fe(CN)_6]^{4-} + [MnO_4]^-$	7.4×10^2	3×10^3	+1.30	1.7×10^5
$[Fe(phen)_3]^{2+} + Ce(IV)$	$>3 \times 10^7$	4.4	+0.66	1.4×10^5

23.9 In the presence of catalytic amounts of $[Pt(P_2O_5H_2)_4]^{4-}$ (**21**) and light, 2-propanol produces H_2 and acetone (E. L. Harley, A. E. Stiegman, A. Vlcek Jr, and H. B. Gray, *J. Am. Chem. Soc.*, 1987, **109**, 5233; D. C. Smith and H. B. Gray, *Coord. Chem. Rev.*, 1990, **100**, 169). (a) Give the equation for the overall reaction. (b) Give a plausible molecular orbital scheme for the metal–metal bonding in this tetragonal-prismatic complex, and indicate the nature of the excited state that is thought to be responsible for the photochemistry. (c) Indicate the metal complex intermediates and the evidence for their existence.

d-Metal organometallic chemistry

24

Bonding

24.1 Stable electron configurations
24.2 Electron-count preference
24.3 Electron counting and oxidation states
24.4 Nomenclature

Ligands

24.5 Carbon monoxide
24.6 Phosphines
24.7 Hydrides and dihydrogen complexes
24.8 η^1-Alkyl, -alkenyl, -alkynyl, and -aryl ligands
24.9 η^2-Alkene and -alkyne ligands
24.10 Nonconjugated diene and polyene ligands
24.11 Butadiene, cyclobutadiene, and cyclooctatetraene
24.12 Benzene and other arenes
24.13 The allyl ligand
24.14 Cyclopentadiene and cycloheptatriene
24.15 Carbenes
24.16 Alkanes, agostic hydrogens, and noble gases
24.17 Dinitrogen and nitrogen monoxide

Compounds

24.18 d-Block carbonyls
24.19 Metallocenes
24.20 Metal–metal bonding and metal clusters

Reactions

24.21 Ligand substitution
24.22 Oxidative addition and reductive elimination
24.23 σ-Bond metathesis
24.24 1,1-Migratory insertion reactions
24.25 1,2-Insertions and β-hydride elimination
24.26 α-, γ-, and δ-Hydride eliminations and cyclometallations

Further reading

Exercises

Tutorial problems

Organometallic chemistry is the chemistry of compounds containing metal–carbon bonds. Much of the basic organometallic chemistry of the s- and p-block metals was understood by the early part of the twentieth century, and has been discussed in Chapters 9–19. The organometallic chemistry of the d and f blocks has been developed much more recently. Since the mid-1950s this field has grown into a thriving area that spans new types of reactions, unusual structures, and practical applications in organic synthesis and industrial catalysis.

A few d-block organometallic compounds were synthesized and partially characterized in the nineteenth century. The first of them (**1**), an ethene complex of platinum(II), was prepared by W. C. Zeise in 1827, with the first metal carbonyls, [PtCl$_2$(CO)$_2$] and [PtCl$_2$(CO)]$_2$, being reported by P. Schützenberger in 1868.

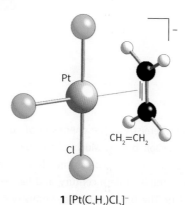

1 [Pt(C$_2$H$_4$)Cl$_3$]$^-$

The next major discovery was tetracarbonylnickel (**2**), which was synthesized by L. Mond, C. Langer, and F. Quinke in 1890. Beginning in the 1930s, W. Hieber synthesized a wide variety of metal carbonyl cluster compounds, many of which are anionic, including [Fe$_4$(CO)$_{13}$]$^{2-}$ (**3**).

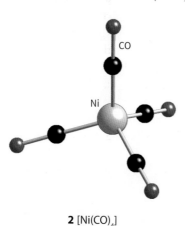

2 [Ni(CO)$_4$]

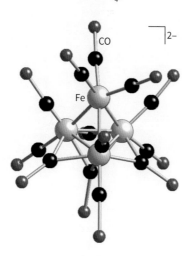

3 [Fe$_4$(CO)$_{13}$]$^{2-}$

It was clear from this work that metal carbonyl chemistry was potentially a very rich field. However, as the structures of these and other d- and f-block organometallic compounds are difficult or impossible to deduce by chemical means alone, fundamental advances had to await the development of X-ray diffraction for precise structural data on solid samples, and of IR and NMR spectroscopy for structural information in solution. The discovery of the remarkably stable organometallic compound ferrocene, [Fe(C$_5$H$_5$)$_2$] (**4**), occurred at a time (in 1951) when these techniques were becoming widely available. The 'sandwich' structure of ferrocene was soon correctly inferred from its IR spectrum and then determined in detail by X-ray crystallography.

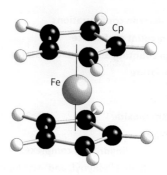

4 [FeCp$_2$], Cp = C$_5$H$_5$

The stability, structure, and bonding of ferrocene defied the classical Lewis description and therefore captured the imagination of chemists. This puzzle in turn set off a train of synthesizing, characterizing, and theorizing that led to the rapid development of d-block organometallic chemistry. Two highly productive research workers in the formative stage of the subject—E. O. Fischer in Munich and G. Wilkinson in London—were awarded the Nobel Prize in 1973 for their contributions. Similarly, f-block organometallic chemistry blossomed soon after the discovery in the late 1970s that the pentamethylcyclopentadienyl ligand, C$_5$Me$_5^-$, forms stable f-block compounds (**5**).

5 [Th(Cp*)$_2$(H)(OR)]

We adhere to the convention that an organometallic compound contains at least one metal–carbon (M–C) bond. Thus, compounds (**1**) to (**5**) clearly qualify as organometallic, whereas a complex such as [Co(en)$_3$]$^{3+}$, which contains carbon but has no M–C bonds, does not. Cyano complexes, such as hexacyanidoferrate(II) ions, do have M–C bonds, but as their properties are more akin to those of conventional coordination complexes they are generally not considered as organometallic. In contrast, complexes of the isoelectronic ligand CO are considered to be organometallic. The justification for this somewhat arbitrary distinction is that many metal carbonyls are significantly different from coordination complexes, both chemically and physically.

In general, the distinctions between the two classes of compounds are clear: coordination complexes normally are charged, with variable d-electron count, and are soluble in water; organometallic compounds are often neutral, with fixed d-electron count, and are soluble in organic solvents such as tetrahydrofuran. Most organometallic compounds have properties that are much closer to organic compounds than inorganic salts, with many of them having low melting points (some are liquids or gases at room temperature).

Bonding

Although there are many organometallic compounds of the s and p blocks, the bonding in these compounds is often relatively simple and normally adequately described solely by σ bonds. The d metals, in contrast, form a large number of organometallic compounds with many different bonding modes. For instance, to describe fully the bonding of a cyclopentadienyl group to iron in ferrocene (and in general to any d metal), we need to invoke σ, π, and δ bonds.

Unlike coordination compounds, d-metal organometallic compounds normally have relatively few stable electron configurations and often have a total of 16 or 18 valence electrons around the metal atom. This restriction to a limited number of electronic configurations is due to the strength of the π (and δ, where appropriate) bonding interactions between the metal atom and the carbon-containing ligands.

24.1 Stable electron configurations

We start by examining the bonding patterns so that we can appreciate the importance of π bonds and understand the origin of the restriction of the d-metal organometallic compounds to certain electron configurations.

(a) 18-Electron compounds

KEY POINTS Six σ-bonding interactions are possible in an octahedral complex and, when π-acceptor ligands are present, bonding combinations can be made with the three orbitals of the t_{2g} set, leading to nine bonding MOs, and space for a total of 18 electrons.

In the 1920s, N. V. Sidgwick recognized that the metal atom in a simple metal carbonyl, such as [Ni(CO)$_4$] (**2**), has the same valence electron count (18) as the noble gas that terminates the long period to which the metal belongs. Sidgwick coined the term 'inert gas rule' for this indication of stability, but it is now usually referred to as the **18-electron rule**.[1]

It becomes readily apparent, however, that the 18-electron rule is not as uniformly obeyed for d-block organometallic compounds, as the octet rule is obeyed for compounds of Period 2 elements, and we need to look more closely at the bonding to establish the reasons for the stability of both the compounds that have the 18-electron configurations and those that do not.

Fig. 24.1 shows the energy levels that arise when a strong-field ligand such as carbon monoxide bonds to a d-metal atom (Section 7.17). Carbon monoxide is a strong-field ligand, even though it is a poor σ donor, because it can use its empty π* orbitals to act as a good π acceptor. In this picture of the bonding, the t_{2g} orbitals of the metal atom are no longer nonbonding, as they would be in the absence of π interactions, but are bonding. The energy-level diagram shows six bonding MOs that result from the ligand–metal σ interactions, and three bonding MOs that result from π interactions. Thus, up to 18 electrons can be accommodated in the nine bonding MOs. Compounds that have this configuration are remarkably stable; for instance, the 18-electron [Cr(CO)$_6$] is a colourless, air-stable compound.

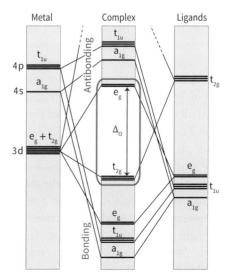

FIGURE 24.1 The energy levels of the molecular orbitals of an octahedral complex with strong-field ligands.

---

[1] The 18-electron rule is sometimes referred to as the *effective atomic number* or *EAN* rule.

An indication of the size of the HOMO–LUMO gap (Δ_O) can be gained from a consideration of its lack of colour, which results from a lack of any electronic transitions in the visible region of the spectrum; that is, Δ_O is so large that such transitions are shifted to the UV.

The only way to accommodate more than 18 valence electrons in an octahedral complex with strong-field ligands is to populate an antibonding orbital. As a result, such complexes are unstable, being particularly prone to electron loss and act as reducing agents. Compounds with fewer than 18 electrons will not necessarily be very unstable, but such complexes will find it energetically favourable to acquire extra electrons by reaction and so populate their bonding MOs fully. As we shall see later, compounds with fewer than 18 electrons often occur as intermediates in reaction pathways.

The bonding characteristic of the carbonyl ligand is replicated with other ligands, which are often poor σ donors but good π acceptors. Hence, octahedral organometallic compounds are most stable when they have a total of 18 valence electrons around their central metal ion.

Similar arguments can be used to rationalize the stability of the 18-electron configuration for other geometries, such as tetrahedral and trigonal-bipyramidal. The steric requirements of most ligands normally preclude coordination numbers greater than six for d-metal organometallic compounds.

(b) 16-Electron square-planar compounds

KEY POINT With strong-field ligands, a square-planar complex has only eight bonding MOs, so a 16-electron configuration is the most energetically favourable configuration.

One other geometry already discussed in the context of coordination chemistry is the square-planar arrangement of four ligands (Section 7.16), where we noted that it occurred generally for strong-field ligands and a d^8 metal ion. Because organometallic ligands often produce a strong field, many square-planar organometallic compounds exist. Stable square-planar complexes are normally found with a total of 16 valence electrons, which results in the population of all the bonding and none of the antibonding MOs (Fig. 24.2).

The ligands in square-planar complexes can normally provide only two electrons each, for a total of eight electrons. Therefore, to reach 16 electrons, the metal ion must provide an additional eight electrons. As a result, organometallic compounds with 16 valence electrons are common only on the right of the d block, particularly in Groups 9 and 10 (Table 24.1). Examples of such complexes include [IrCl(CO)(PPh$_3$)$_2$] (**6**) and the anion of Zeise's salt, [Pt(C$_2$H$_4$)Cl$_3$]$^-$ (**1**). Square-planar 16-electron complexes are particularly common for the heavier elements in Groups 9 and 10,

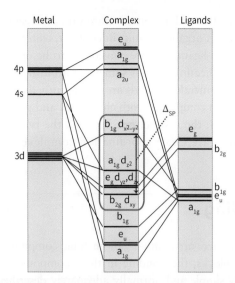

FIGURE 24.2 The energy levels of the molecular orbitals of a square-planar complex with strong-field ligands. In the green box, the four lowest MOs correspond to bonding interactions, and the highest corresponds to an antibonding interaction; the MOs are labelled with the d-orbitals from which they are derived.

TABLE 24.1 Validity of the 16/18-electron rule for d-metal organometallic compounds

Usually less than 18 electrons			Usually 18 electrons			16 or 18 electrons	
Sc	Ti	V	Cr	Mn	Fe	Co	Ni
Y	Zr	Nb	Mo	Tc	Ru	Rh	Pd
La	Hf	Ta	W	Re	Os	Ir	Pt

especially for Rh(I), Ir(I), Pd(II), and Pt(II), because the ligand-field splitting is large and the ligand-field stabilization energy of these complexes favours the square-planar configuration.

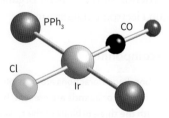

6 *trans*-[IrCl(CO)(PPh$_3$)$_2$]

24.2 Electron-count preference

KEY POINT On the left of the d block, steric requirements can mean that it is not possible to assemble enough ligands around a metal atom to achieve a total of 16 or 18 valence electrons.

The preference of a metal atom for one particular geometry and electron count is not normally so strong that other geometries do not occur. For instance, although the chemistry of both Pd(II) and Rh(I) is dominated by 16-electron square-planar complexes, the cyclopentadienyl palladium(II) complex (**7**) and the rhodium(I) complex (**8**) are both 18-electron compounds.

7

8 [RhMe(PMe$_3$)$_4$]

Steric factors may restrict the number of ligands that can bond to a metal atom, and stabilize compounds with lower electron counts than might have been expected. For instance, the tricyclohexylphosphine ligands (**9**) in the trigonal-planar Pt(0) compound [Pt(PCy$_3$)$_3$] are so large that only three can be fitted around the metal atom, which consequently has only 16 valence electrons. Steric stabilization of the metal centre is partially a kinetic effect, in which the large groups protect the metal centre from reaction. Many complexes readily lose or gain ligands, forming other configurations transitorily during reactions; indeed, the accessibility of these other configurations is precisely why the organometallic chemistry of the d metals is so interesting.

9 PCy$_3$, Cy = *cyclo*-C$_6$H$_{11}$

Unusual electron configurations are common on the left of the d block, where the metal atoms have fewer electrons, and it is often not possible to crowd enough ligands around the atom to bring the electron count up to 16 or 18. For example, the simplest carbonyl in Group 5, [V(CO)$_6$], is a 17-electron complex. Other examples include [W(CH$_3$)$_6$], which has 12 valence electrons, and [Cr(Cp)(CO)$_2$(PPh$_3$)] (**10**), with 17. The latter compound provides another good example of the role of steric crowding. When the compact CO ligand is present in place of bulky triphenylphosphine, a dimeric compound, (**11**), with a long but definite Cr–Cr bond is observed in the solid state and in solution. The formation of the Cr–Cr bond in [Cr(Cp)(CO)$_3$]$_2$ raises the electron count on each metal to 18, as two electrons are shared in that bond.

10 [(C$_5$H$_5$)Cr(CO)$_2$(PPh$_3$)]   **11** [(Cp)(CO)$_3$Cr–Cr(CO)$_3$(Cp)]

24.3 Electron counting and oxidation states

The dominance of 16- and 18-electron configurations in organometallic chemistry makes it imperative to be able to count the number of valence electrons on a central metal atom because knowing that number allows us to predict the stabilities of compounds and to suggest patterns of reactivity. Although the concept of 'oxidation state' for organometallic compounds is regarded by many as tenuous at best, the vast majority of the research community uses it as a convenient shorthand for describing electron configurations. Oxidation states (and the corresponding oxidation number) help to systematize reactions such as oxidative addition (Section 24.22), and also bring out analogies between the chemical properties of organometallic and coordination complexes. Fortunately, the business of counting electrons and assigning oxidation numbers can be combined.

Two models are routinely used to count electrons, the so-called **neutral-ligand method** (sometimes called the *covalent method*) and the **donor-pair method** (sometimes known as the *ionic method*). We introduce both briefly—they produce identical results for electron counting—but in the remainder of the chapter we use the donor-pair method as it may also be used to assign oxidation numbers. A recent article, entitled 'In defence of oxidation states' (see Further Reading) goes into some detail as to the merits of both models.

(a) Neutral-ligand method

KEY POINT All ligands are treated as neutral and are categorized according to how many electrons they are considered to donate.

For the sake of counting electrons, each metal atom and ligand is treated as neutral. We include in the count all valence electrons of the metal atom and all the electrons donated by the ligands. If the complex is charged, we simply add or subtract the appropriate number of electrons to the total. Ligands are defined as **L type** if they are neutral two-electron donors (like CO, PMe$_3$) and **X type** if, when they are considered to be neutral, they are one-electron radical donors (like halogen atoms, H, CH$_3$). For example,

[Fe(CO)$_5$] acquires 18 electrons from the 8 valence electrons on the Fe atom and the 10 electrons donated by the five CO ligands. Some ligands are considered combinations of these types; for instance, cyclopentadienyl is considered as a five-electron L$_2$X donor; see Table 24.2.

The advantage of the neutral-ligand method is that, given the information in Table 24.2, it is trivial to establish the electron count. The disadvantage, however, is that the method overestimates the degree of covalence and thus underestimates the charge at the metal. Moreover, it becomes confusing to assign an oxidation number to a metal, and meaningful information on some ligands is lost.

TABLE 24.2 Typical ligands and their electron counts

(a) Neutral-ligand method			
Ligand	**Formula**	**Designation**	**Electrons donated**
Carbonyl	CO	L	2
Phosphine	PR$_3$	L	2
Hydride	H	X	1
Chloride	Cl	X	1
Dihydrogen	H$_2$	L	2
η^1-Alkyl, -alkenyl, -alkynyl, and -aryl groups	R	X	1
η^2-Alkene	CH$_2$=CH$_2$	L	2
η^2-Alkyne	RCCR	L	2
Dinitrogen	N$_2$	L	2
Butadiene	CH$_2$=CH–CH=CH$_2$	L$_2$	4
Benzene	C$_6$H$_6$	L$_3$	6
η^3-Allyl	CH$_2$CHCH$_2$	LX	3
η^5-Cyclopentadienyl	C$_5$H$_5$	L$_2$X	5

(b) Donor-pair method*		
Ligand	**Formula**	**Electrons donated**
Carbonyl	CO	2
Phosphine	PR$_3$	2
Hydride	H$^-$	2
Chloride	Cl$^-$	2
Dihydrogen	H$_2$	2
η^1-Alkyl, -alkenyl, -alkynyl, and -aryl groups	R$^-$	2
η^2-Alkene	CH$_2$=CH$_2$	2
η^2-Alkyne	RCCR	2
Dinitrogen	N$_2$	2
Butadiene	CH$_2$=CH–CH=CH$_2$	4
Benzene	C$_6$H$_6$	6
η^3-Allyl	CH$_2$CHCH$_2^-$	4
η^5-Cyclopentadienyl	C$_5$H$_5^-$	6

* We use this method throughout this book.

EXAMPLE 24.1 Counting electrons using the neutral-ligand method

Do (a) [IrBr$_2$(CH$_3$)(CO)(PPh$_3$)$_2$] and (b) [Cr(η^5-C$_5$H$_5$)(η^6-C$_6$H$_6$)] obey the 18-electron rule?

Answer (a) We start with the Ir atom (Group 9), which has nine valence electrons, then add in the electrons from the two Br atoms and the CH$_3$ group (each is a one-electron X donor) and finally add in the electrons from the CO and PPh$_3$ (both are two-electron L donors). Thus, the number of valence electrons on the metal atom is $9 + (3 \times 1) + (3 \times 2) = 18$.

(b) In a similar fashion, the Cr atom (Group 6) has six valence electrons, the η^5-C$_5$H$_5$ ligand donates five electrons, and the η^6-C$_6$H$_6$ ligand donates six, so the number of metal valence electrons is $6 + 5 + 6 = 17$. This complex does not obey the 18-electron rule and is not stable. A related but stable 18-electron compound is [Cr(η^6-C$_6$H$_6$)$_2$].

Self-test 24.1 Is [Mo(CO)$_7$] likely to be stable?

(b) Donor-pair method

KEY POINT Ligands are considered to donate electrons in pairs, resulting in the need to treat some ligands as neutral and others as charged.

The donor-pair method requires a calculation of the oxidation number. The rules for calculating the oxidation number of an element in an organometallic compound are the same as for conventional coordination compounds. Neutral ligands, such as CO and phosphine, are considered to be two-electron donors and are formally assigned an oxidation number of 0. Ligands such as halogens, H, and CH$_3$ are formally considered to take an electron from the metal atom, and are treated as halides (e.g. Cl$^-$), H$^-$, and CH$_3^-$ (and hence are assigned oxidation number -1); in this anionic state they are considered to be two-electron donors. The cyclopentadienyl ligand, C$_5$H$_5$ (Cp), is treated as C$_5$H$_5^-$ (it is assigned an oxidation number of -1); in this anionic state it is considered to be a six-electron donor. then:

The *oxidation number* of the metal atom is the total charge of the complex minus the charges of any ligands.

The *number of electrons* the metal provides is its group number minus its oxidation number.

The *total electron count* is the sum of the number of electrons on the metal atom and the number of electrons provided by the ligands.

The main advantage of this method is that with a little practice both the electron count and the oxidation number may be determined in a straightforward manner. The main disadvantage is that it overestimates the charge on the metal atom and can suggest reactivity that might be incorrect (see Section 24.7 on hydrides). Table 24.2 lists the maximum number of electrons available for donation to a metal for most common ligands.

EXAMPLE 24.2 Assigning oxidation numbers and counting the electrons using the donor-pair method

Assign the oxidation number and count the valence electrons on the metal atom in (a) [IrBr$_2$(CH$_3$)(CO)(PPh$_3$)$_2$], (b) [Cr(η^5-C$_5$H$_5$)(η^6-C$_6$H$_6$)], and (c) [Mn(CO)$_5$]$^-$.

Answer (a) We treat the two Br groups and the CH$_3$ as three singly negatively charged two-electron donors and the CO and the two PPh$_3$ ligands as three two-electron donors, providing 12 electrons in all. Because the complex is neutral overall, the Group 9 Ir atom must have a charge of +3 (that is, have oxidation number +3) to balance the charge of the three anionic ligands, and thus contributes 9 − 3 = 6 electrons. This analysis gives a total of 18 electrons for the Ir(III) complex.

(b) We treat the η^5-C$_5$H$_5$ ligand as C$_5$H$_5^-$ and thus it donates six electrons, with the η^6-C$_6$H$_6$ ligand donating a further six. To maintain neutrality, the Group 6 Cr atom must have a charge of +1 (and an oxidation number of +1) and contributes 6 − 1 = 5 electrons. The total number of metal electrons is 12 + 5 = 17 for a Cr(I) complex. As noted before, this complex does not obey the 18-electron rule and is unlikely to be stable.

(c) We treat each CO ligand as neutral and contributing two electrons, giving 10 electrons. The overall charge of the complex is −1; because all the ligands are neutral, we consider this charge to reside formally on the metal atom, giving it an oxidation number of −1. The Group 7 Mn atom thus contributes 7 + 1 electrons, giving a total of 18 for a Mn(−1) complex.

Self-test 24.2 What is the electron count for, and oxidation number of, platinum in the anion of Zeise's salt, [Pt(CH$_2$=CH$_2$)Cl$_3$]$^-$? Treat CH$_2$=CH$_2$ as a neutral two-electron donor.

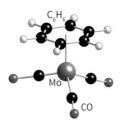

12 [Mo(η^6-C$_6$H$_6$)(CO)$_3$]

The IUPAC recommendation for the formula of an organometallic compound is to write it in the same form as for a coordination complex: the symbol for the metal is written first, followed by the ligands, listed in alphabetical order based on their chemical symbol. We shall follow these conventions unless a different order of ligands helps to clarify a particular point.

Often a ligand with carbon-donor atoms can exhibit multiple bonding modes—for instance, the cyclopentadienyl group can commonly bond to a d-metal atom in three different ways—so we need some additional nomenclature. Without going into the intimate details of the bonding of the various ligands (we do that later in this chapter), the extra information we need to describe a bonding mode is the number of points of attachment. This procedure gives rise to the notion of **hapticity**, the number of ligand atoms that are considered formally to be bonded to the metal atom. The hapticity is denoted η^n, where n is the number of atoms (η is the Greek letter eta). For example, a CH$_3$ group attached by a single M–C bond is monohapto, η^1, and, if the two C atoms of an ethene ligand are both within bonding distance of the metal, the ligand is dihapto, η^2. Thus, three cyclopentadienyl complexes might be described as having η^1 (**13**), η^3 (**14**), or η^5 (**15**) cyclopentadienyl groups.

13 η^1-Cyclopentadienyl

14 η^3-Cyclopentadienyl

15 η^5-Cyclopentadienyl

Some ligands (including the simplest of them all, the hydride ligand, H$^-$) can bond to more than one metal atom in the same complex, and are then referred to as **bridging ligands**. We do not require any new concepts to understand bridging ligands other than those introduced in Section 3.11. Recall from Section 7.2 that the Greek letter μ (mu) is used to indicate how many atoms the ligand bridges. Thus, a μ$_2$-CO is a carbonyl group that bridges two metal atoms and a μ$_3$-CO bridges three.

24.4 Nomenclature

KEY POINT The naming of organometallic compounds is similar to the naming of coordination compounds, but certain ligands have multiple bonding modes, which are described using η and μ.

According to the recommended convention, we use the same system of nomenclature for organometallic compounds as set out for coordination complexes in Section 7.2. Thus, ligands are listed in alphabetical order followed by the name of the metal, all of which is written as one word. The name of the metal should be followed by its oxidation number in parentheses. The nomenclature used in research journals, however, does not always obey these rules, and it is common to find the name of the metal buried in the middle of the name of the compound and the oxidation number omitted. For example, (**12**) is often referred to as benzene-molybdenum-tricarbonyl, rather than the preferred name benzene(tricarbonyl)molybdenum(0).

EXAMPLE 24.3 Naming organometallic compounds

Give the formal names of (a) ferrocene (**4**) and (b) [RhMe(PMe$_3$)$_4$] (**8**).

Answer (a) Ferrocene contains two cyclopentadienyl groups that are both bound to the metal atom through all five carbon atoms; thus both groups are designated η^5. The full name for ferrocene is thus bis(η^5-cyclopentadienyl)iron(II). (b) The rhodium compound contains one formally anionic methyl group and four neutral trimethylphosphine ligands; therefore the formal name is methyltetrakis-(trimethylphosphine)rhodium(I).

Self-test 24.3 What is the formal name of [Ir(Br)$_2$(CH$_3$)(CO)(PPh$_3$)$_2$]?

Ligands

A large number of ligands are found in organometallic complexes, with many different bonding modes. Because the reactivity of the metal atom and the ligands is affected by the M–L bonding, it is important to look at each ligand in some detail.

24.5 Carbon monoxide

KEY POINT The 3σ orbital of CO serves as a very weak donor and the π* orbitals act as acceptors.

Carbon monoxide is a very common ligand in organometallic chemistry, where it is known as the *carbonyl group*. Carbon monoxide is particularly good at stabilizing very low oxidation states, with many compounds (such as [Fe(CO)$_5$]) having the metal in its zero oxidation state. We described the molecular orbital structure of CO in Section 2.9, and it would be sensible to review that section.

A simple picture of the bonding of CO to a metal atom is to treat the lone pair on the carbon atom as a Lewis σ base (an electron-pair donor) and the empty CO antibonding orbital as a Lewis π acid (an electron-pair acceptor), which accepts π-electron density from the filled d orbitals on the metal atom. In this picture, the bonding can be considered to be made up of two parts: a σ bond from the ligand to the metal atom (**16**) and a π bond from the metal atom to the ligand (**17**). This type of π bonding is sometimes referred to as π **backbonding**.

Carbon monoxide is not appreciably nucleophilic, which suggests that σ bonding to a d-metal atom is weak. As many d-metal carbonyl compounds are very stable, we can infer that π backbonding is strong and the stability of carbonyl complexes arises mainly from the π-acceptor properties of CO. Further evidence for this view comes from the observation that stable carbonyl complexes exist only for metals that have filled d orbitals of an energy suitable for donation to the CO antibonding orbital. For instance, elements in the s and p blocks do not form stable carbonyl complexes. However, the bonding of CO to a d-metal atom is best regarded as a synergistic (that is, mutually enhancing) outcome of both σ and π bonding: the π backbonding from the metal to the CO increases the electron density on the CO, which in turn increases the ability of the CO to form a σ bond to the metal atom.

A more formal description of the bonding can be derived from the molecular orbital scheme for CO (Fig. 24.3), which shows that the HOMO has σ symmetry and is essentially a lobe that projects away from the C atom. When CO acts as a ligand, this 3σ orbital serves as a very weak donor to a metal atom, and forms a σ bond with the central metal

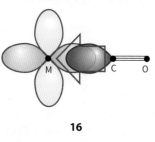

16

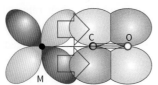

17

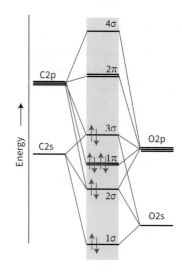

FIGURE 24.3 The molecular orbital scheme for CO shows that the HOMO has σ symmetry and is essentially a lobe that projects away from the C atom. The LUMO has π symmetry.

atom. The LUMOs of CO are the antibonding 2π orbitals. These two orbitals play a crucial role because they can overlap with metal d orbitals that have local π symmetry (such as the t_{2g} orbitals in an O_h complex). The π interaction leads to the delocalization of electrons from filled d orbitals on the metal atom into the empty π^* orbitals on the CO ligands, so the ligand also acts as a π acceptor.

One important consequence of this bonding scheme is the effect on the strength of the CO triple bond: the stronger the metal–carbon bond becomes through pushing electron density from the metal atom into the 2π orbitals, the weaker the CO bond becomes, as this electron density enters a CO antibonding orbital. In the extreme case, when two electrons are fully donated by the metal atom, a formal metal–carbon double bond is formed; because the two electrons occupy a CO antibonding orbital, this donation results in a decrease in the bond order of the CO to 2.

In practice, the bonding is somewhere between M–C≡O, with no backbonding, and M=C=O, with complete backbonding. Infrared spectroscopy is a very convenient method of assessing the extent of π bonding; the CO stretch is clearly identifiable as it is both strong and (normally) clear of all other absorptions. In CO gas, the absorption for the triple bond is at 2143 cm^{-1}, whereas a typical metal carbonyl complex has a stretching mode in the range 2100–1700 cm^{-1} (Table 24.3). The number of IR absorptions that occur in a particular carbonyl compound is discussed in Section 24.18(g).

Carbonyl stretching frequencies are often used to determine the order of acceptor or donor strengths for the other ligands present in a complex. The basis of the approach is that the CO stretching frequency is decreased when it serves as a π acceptor. However, as other π acceptors in the same complex compete for the d electrons of the metal atom, they cause the CO frequency to increase. This behaviour is opposite to that observed with donor ligands, which cause the CO stretching frequency to decrease as they supply electrons to the metal atom and hence, indirectly, to the CO π^* orbitals. Thus, strong σ-donor ligands attached to a metal carbonyl and a formal negative charge on a metal carbonyl anion both result in slightly greater CO bond lengths and significantly lower CO stretching frequencies.

TABLE 24.3 The influence of coordination and charge on CO stretching bands

Compound	$\tilde{\nu}$/cm^{-1}
CO	2143
[Mn(CO)$_6$]$^+$	2090
Cr(CO)$_6$	2000
[V(CO)$_6$]$^-$	1860
[Ti(CO)$_6$]$^{2-}$	1750

Carbon monoxide is versatile as a ligand because, as well as the bonding mode we have described so far (often referred to as 'terminal'), it can bridge two (**18**) or three (**19**) metal atoms.

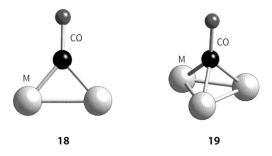

Although the description of the bonding is now more complicated, the concepts of σ-donor and π-acceptor ligands remain useful. The CO stretching frequencies generally follow the order MCO > M$_2$CO > M$_3$CO, which suggests an increasing occupation of the π^* orbital as the CO molecule bonds to more metal atoms and more electron density from the metals enters the CO π^* orbitals. As a rule of thumb, carbonyls bridging two metal atoms typically have stretching bands in the range 1900–1750 cm^{-1}, and those that bridge three atoms have stretching bands in the range 1800–1600 cm^{-1} (Fig. 24.4). We consider carbon monoxide to be a two-electron neutral ligand when terminal or bridging (thus a carbonyl bridging two metals can be considered to give one electron to each).

A further bonding mode for CO that is sometimes observed is when the CO is terminally bound to one metal atom and the CO triple bond binds side-on to another metal (**20**). This description is best considered as two separate bonding interactions, with the terminal interaction being the same as that described above and the side-on bonding being essentially identical to that of other side-on π donors such as alkynes and N$_2$, which are discussed later.

The synthesis, properties, and reactivities of compounds containing the carbonyl ligand are discussed in more detail in Section 24.18.

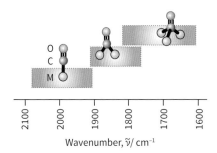

FIGURE 24.4 Approximate ranges for CO stretching bands in neutral metal carbonyls. Note that high wavenumbers (and hence high frequencies) are on the left, in keeping with the way infrared spectra are generally plotted.

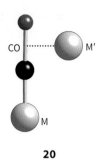

20

24.6 Phosphines

KEY POINT Phosphines bond to metals by a combination of σ donation from the P atom and π backbonding from the metal atom.

Although phosphine complexes are not formally organometallic because the ligands do not bond to the metal atom through a carbon atom, they are best discussed here as their bonding has many similarities to that of carbon monoxide.

Phosphine, PH_3 (formally, phosphane), is a reactive, noxious, poisonous, and flammable gas (Section 16.10). Like ammonia, phosphine can behave as a Lewis base and use its lone pair to donate electron density to a Lewis acid, and so act as a ligand. However, given the problems associated with handling phosphine, it is rarely used as a ligand. Substituted phosphines, on the other hand—such as trialkylphosphines (for example, PMe_3, PEt_3), triarylphosphines (for example, PPh_3), or trialkyl- or triarylphosphites (for example, $P(OMe)_3$, $P(OPh)_3$, **21**), and a whole host of bridged multidentate di- and triphosphines (for example, $Ph_2PCH_2CH_2PPh_2$ = dppe, **22**)—are easy to handle (indeed some are air-stable, odourless solids with no appreciable toxicity) and are widely used as ligands; all are colloquially referred to as 'phosphines'.

Phosphines have a lone pair on the P atom that is appreciably basic and nucleophilic, and can serve as a σ donor. Phosphines also have empty orbitals on the P atom that can overlap with filled d orbitals on 3d-metal ions and behave as π acceptors (**23**). The bonding of phosphines to a d-metal atom, made up of a σ bond from the ligand to the metal and a π bond from the metal back to the ligand, is completely analogous to the bonding of CO to a d-metal atom. Quite which orbitals on the P atom behave as π acceptors has been the subject of considerable debate in the past, with some groups claiming a role for unoccupied 3d orbitals on P and some claiming a role for the P–R σ* orbitals; current consensus favours the σ* orbitals. In any case, each phosphine provides an additional two electrons to the valence electron count.

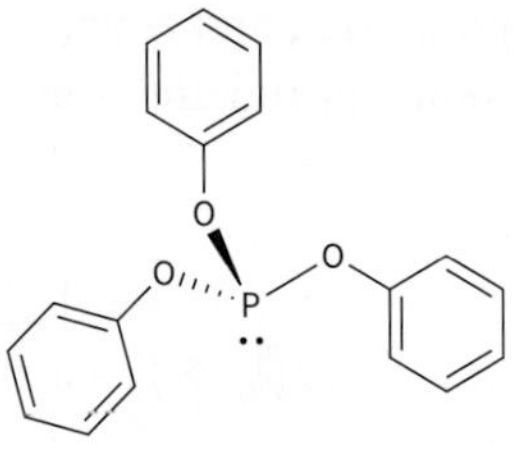

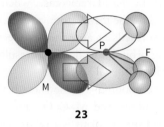

23

As we have remarked, a huge variety of phosphines are both possible and widely available, including chiral systems such as 2,2′-bis(diphenylphosphino)-1,1′-binaphthyl (BINAP, **24**), in which steric constraints result in compounds that can be resolved into diastereomers. Generally there are two properties of phosphine ligands that are considered important in discussions of the reactivity of their complexes: their steric bulk and their electron-donating (and -accepting) ability.

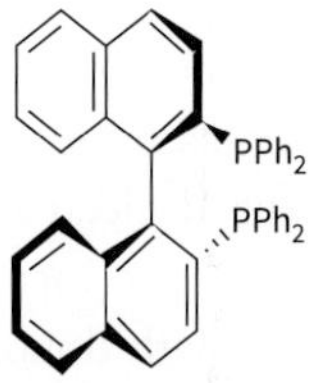

24 2,2′-bis(diphenylphosphino)-1,1′-binaphthyl, BINAP

We described in Section 23.6 how the steric bulk of phosphines can be expressed in terms of the notional cone occupied by the bonded ligand, and Table 24.4 lists some of the derived cone angles.

The bonding of phosphines to d-metal atoms is, as we have seen, a composite of σ bonding from the ligand to the metal atom and π backbonding from the metal atom to the ligand. The σ-donating ability and π-acceptor ability of phosphines are inversely correlated in the sense that electron-rich phosphines, such as PMe_3, are good σ donors and poor π acceptors, whereas electron-poor phosphines, such as PF_3, are poor σ donors and good π acceptors. Thus Lewis basicity can normally be used as a single scale to

21 $P(OPh)_3$

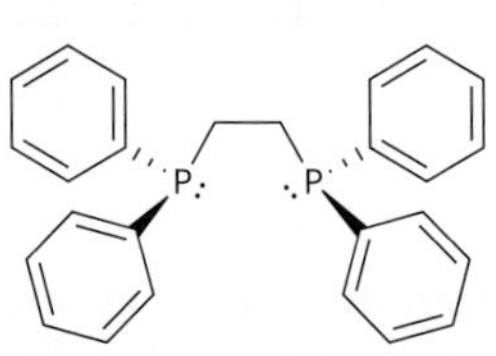

22 $Ph_2PCH_2CH_2PPh_2$, dppe

indicate their donor/acceptor ability. The generally accepted order of basicity of phosphines is

$$PCy_3 > PEt_3 > PMe_3 > PPh_3 > P(OMe)_3 > P(OPh)_3 > PCl_3 > PF_3$$

and is easily understood in terms of the electronegativity of the substituents on the P atom. The basicity of a phosphine is not simply related to the strength of the M–P bond in a complex; for instance, an electron-poor metal atom forms a stronger bond with an electron-rich (basic) phosphine, whereas an electron-rich metal atom will form a stronger bond with an electron-poor phosphine.

If there are carbonyl ligands present in a metal–phosphine complex, then the carbonyl stretching frequency can be used to assess the basicity of the phosphine ligand: this method allows us to conclude that PF_3 is a π acceptor comparable to CO.

The vast range of phosphines that are commonly used in organometallic chemistry is a testament to their versatility as ligands: judicious choice allows control over both steric and electronic properties of the metal atom in a complex. Phosphorus-31 (which occurs in 100% natural abundance, and has I = ½) is easy to observe by NMR, and both the ^{31}P chemical shift and the coupling constant to the metal atom (where appropriate) give considerable insight into the bonding and reactivity of a complex. Like carbonyls, phosphines can bridge either two or three metal atoms, providing additional variety in bonding modes.

EXAMPLE 24.4 Interpreting carbonyl stretching frequencies and phosphine complexes

(a) Which of the two isoelectronic compounds $[Cr(CO)_6]$ and $[V(CO)_6]^-$ will have the higher CO stretching frequency? (b) Which of the two chromium compounds $[Cr(CO)_5(PEt_3)]$ and $[Cr(CO)_5(PPh_3)]$ will have the lower CO stretching frequency? Which will have the shorter M–C bond?

Answer We need to think about whether backbonding to the CO ligands is enhanced or diminished: more backbonding results in a weaker carbon–oxygen bond. (a) The negative charge on the V complex will result in greater π backbonding to the CO π* orbitals, compared to the Cr complex. This backbonding results in a weakening of the CO bond, with a corresponding decrease in stretching frequency. Thus, the Cr complex has the higher CO stretching frequency. (b) PEt_3 is more basic than PPh_3 and thus the PEt_3 complex will have greater electron density on the metal atom than the PPh_3 complex. The greater electron density will result in greater backbonding to the carbonyl ligand and thus both a lower CO stretching frequency and a shorter M–C bond.

Self-test 24.4 Which of the two iron compounds $[Fe(CO)_5]$ and $[Fe(CO)_4(PEt_3)]$ will have the higher CO stretching frequency? Which will have the longer M–C bond?

TABLE 24.4 Tolman cone angles (in degrees) for selected phosphines

PF_3	104
$P(OMe)_3$	107
PMe_3	118
PCl_3	125
$P(OPh)_3$	127
PEt_3	132
PPh_3	145
PCy_3	169
P^tBu_3	182
$P(o\text{-tolyl})_3$	193

24.7 Hydrides and dihydrogen complexes

KEY POINTS The bonding of a hydrogen atom to a metal atom is a σ interaction, whereas the bonding of a dihydrogen ligand involves π backbonding.

A hydrogen atom directly bonded to a metal is commonly found in organometallic complexes and is referred to as a **hydrido ligand**, or a **hydride**. The name 'hydride' can sometimes be misleading as it implies a H^- ligand. Although the formulation H^- might be appropriate for most hydrides, such as $[CoH(PMe_3)_4]$, some hydrides are appreciably acidic and behave as though they contain H^+; for example, $[CoH(CO)_4]$ is an acid with pK_a = 8.3 (in acetonitrile). The acidity of organometallic carbonyls is described in Section 24.18. In the donor-pair method of electron counting, we consider the hydride ligand to contribute two electrons and to have a single negative charge (that is, to be H^-).

The bonding of a hydrogen atom to a metal atom is simple because the only orbital of appropriate energy for bonding on the hydrogen is H1s and the M–H bond can be considered as a σ interaction between the two atoms. Hydrides are readily identified by NMR spectroscopy, as their chemical shift is rather unusual, typically occurring in the range $-50 < \delta < 0$. Infrared spectroscopy can also be useful in identifying metal hydrides, as they normally have a sharp stretching band in the range 2250–1650 cm^{-1}. X-ray diffraction—normally so valuable for identifying the structure of crystalline materials—might not always identify hydrides, as the diffraction is related to electron density and the hydride ligand will have at most two electrons around it, compared with, for instance, 78 for a platinum. Neutron diffraction is of more use in locating hydride ligands, especially if the hydrogen atom is replaced by a deuterium atom, because deuterium has a large neutron scattering cross-section (Section 8.1).

An M–H bond can sometimes be produced by protonation of an organometallic compound, such as neutral and anionic metal carbonyls (Section 24.18(e)). For example, ferrocene can be protonated in strong acid to produce an Fe–H bond:

[Cp$_2$Fe] + HBF$_4$ → [Cp$_2$FeH]$^+$[BF$_4$]$^-$

Bridging hydrides exist, where an H atom bridges either two or three metal atoms: here the bonding can be treated in exactly the same way we considered bridging hydrides in diborane, B$_2$H$_6$ (Section 3.11).

Although the first organometallic metal hydride was reported in 1931, complexes of hydrogen gas, H$_2$, were identified some 50 years later, in 1984. In such compounds, the dihydrogen molecule, H$_2$, bonds side-on to the metal atom (in the older literature, such compounds were sometimes called *nonclassical hydrides*). The bonding of dihydrogen to the metal atom is considered to be made up of two components: a σ donation of the two electrons in the H$_2$ bond to the metal atom (**25**) and a π backdonation from the metal to the σ* antibonding orbital of H$_2$ (**26**). This picture of the bonding raises a number of interesting issues. In particular, as the π backbonding from the metal atom increases, the strength of the H–H bond decreases and the structure tends to that of a dihydride:

no longer available to the metal atom for further bonding. This transformation of the dihydrogen molecule to a dihydride is an example of *oxidative addition*, and is discussed more fully later in this chapter (Section 24.22).

It is now recognized that complexes exist with structures at all points between these two extremes, and in some cases an equilibrium can be identified between the two. Work by G. Kubas on tungsten complexes used the H–D coupling constant to show that it is possible to detect both the dihydrogen complex (**27**, $^1J_{HD}$ = 34 Hz) and the dihydride (**28**, $^2J_{HD}$ < 2 Hz), and to follow the conversion from one to the other. Certain microbes contain enzymes known as *hydrogenases* that use Fe and Ni at their catalytic centres to catalyse the rapid oxidation of H$_2$ and reduction of H$^+$, via intermediate metal dihydrogen and hydride species (Section 26.13).

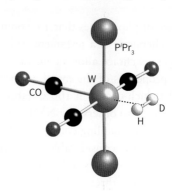

27 [W(HD)(P^{i}Pr$_3$)$_2$(CO)$_3$]

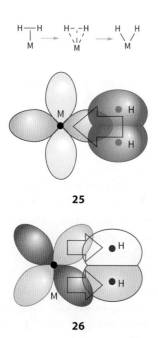

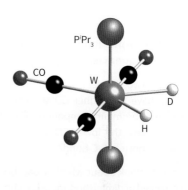

28 [W(H)(D)(P^{i}Pr$_3$)$_2$(CO)$_3$]

A dihydrogen molecule is treated as a neutral two-electron donor. Thus the transformation of the dihydrogen into two hydrides (each considered to have a single negative charge and to contribute two electrons) requires the formal charge on the metal atom to increase by two. That is, the metal is oxidized by two units and the dihydrogen is reduced. Although it might seem that this oxidation of the metal is just an anomaly thrown up by our method of counting electrons, two of the electrons on the metal atom have been used to backbond to the dihydrogen, and these two electrons are

24.8 η^1-Alkyl, -alkenyl, -alkynyl, and -aryl ligands

KEY POINT The metal–ligand bonding of η^1-hydrocarbon ligands is a σ interaction.

Alkyl groups are often found as ligands in d-metal organometallic chemistry, and their bonding presents no new features: it is best considered a simple covalent σ interaction between the metal atom and the carbon atom of the organic fragment. Alkyl groups with a hydrogen atom on a carbon atom adjacent to the one that bonds to the metal are prone to decompose by a process known as *β-hydrogen elimination* (Section 24.25), and hence those alkyl groups that cannot react in this fashion, such as methyl, benzyl ($CH_2C_6H_5$), neopentyl (CH_2CMe_3), and trimethylsilylmethyl (CH_2SiMe_3), are more stable than those that can, such as ethyl.

Alkenyl (**29**), alkynyl (**30**), and aryl (**31**) groups can bond to a metal atom in a similar fashion, binding to the metal atom through a single carbon atom, and hence are described as monohapto (η^1). Although there is potential for each of these three groups to accept π electron density into antibonding orbitals, there is little evidence that this happens. For instance, even though an η^1-alkynyl group might be considered analogous to a CO group, the stretching frequency of the triple bond in alkynyl complexes changes little on attachment to a metal. Bridging alkyl and aryl groups also exist, and the bonding can be considered in the same way as we have considered other bridging ligands, with $3c,2e$ bonds.

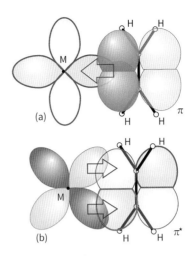

FIGURE 24.5 The interaction of ethene with a metal atom. (a) Donation of electron density from the filled π molecular orbital of ethene to a vacant metal σ orbital. (b) Acceptance of electron density from a filled dπ orbital into the vacant π* orbital of ethene.

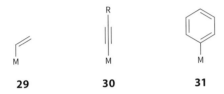

Alkyl, alkenyl, alkynyl, and aryl groups are commonly introduced into organometallic complexes by the displacement of a halide at a metal centre with a lithium or Grignard reagent; for example:

$$Cl_{//}Pd_{\cdots}PPh_3 / Ph_3P\,Cl \xrightarrow{2\,PhLi} Ph_{//}Pd_{\cdots}PPh_3 / Ph_3P\,Ph + 2\,LiCl$$

We consider alkyl, alkenyl, alkynyl, and aryl ligands to be two-electron donors with a single negative charge (for example, Me^-, Ph^-) in the donor-pair scheme of electron counting.

24.9 η²-Alkene and -alkyne ligands

KEY POINT The bonding of an alkene or an alkyne to a metal atom is best described as a σ interaction from the multiple bond to the metal atom, with a π backbonding interaction from the metal atom to the π* antibonding orbital on the alkene or alkyne.

Alkenes are routinely found bound to metal centres: the first organometallic compound isolated, Zeise's salt (**1**), was a complex of ethene. Alkenes normally bond side-on to a metal atom with both carbon atoms of the double bond equidistant from the metal with the other groups on the alkene approximately perpendicular to the plane of the metal atom and the two carbon atoms (**32**). In this arrangement, the electron density of the C=C π bond can be donated to an empty orbital on the metal atom to form a σ bond. In parallel with this interaction, a filled metal d orbital can donate electron density back to the empty π* orbital of the alkene to form a π bond. This description is called the **Dewar–Chatt–Duncanson model** (Fig. 24.5), and η²-alkenes are considered to be two-electron neutral ligands.

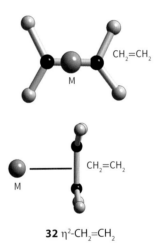

32 η²-CH_2=CH_2

Electron-donor and -acceptor character appear to be fairly evenly balanced in most ethene complexes of the d metals, but the degree of donation and backdonation can be altered by substituents on the metal atom and on the alkene. When the π backbonding from the metal atom increases, the strength of the C=C bond decreases as the electron density is located in the C=C antibonding orbital, and the structure tends to that of a C–C singly bonded structure, a metallocyclopropane:

Dihaptoalkenes with only a small degree of electron donation from the metal have their substituents bent slightly away from the metal atom, and the C=C bond length is only slightly greater than in the free alkene (134 pm). When the degree of backdonation is greater, substituents on the alkene are bent away more from the metal atom, and the C=C bond length approaches that characteristic of a single bond. Steric constraints can also force the other groups on the alkene to bend away from the metal atom.

Alkynes have two orthogonal π bonds and hence the potential to be four-electron donors, though not to the same metal atom. It is best to consider the two π bonds separately: when side-on to a single metal atom, only one of the two π bonds can overlap with metal orbitals. Thus, a C≡C bond is best considered as forming a two-electron, η^2-interaction, with the π* orbitals accepting electron density from a metal atom, like an alkene. When strongly electron-withdrawing groups are attached to an alkyne, the ligand can become an excellent π acceptor and displace other ligands such as phosphines; the compound commonly known as dimethylacetylenedicarboxylate, $CH_3OCOC\equiv CCOOCH_3$, is a good example.

Substituted alkynes can form very stable polymetallic complexes in which the alkyne can be regarded as a four-electron donor. An example is η^2-diphenylethyne-(hexacarbonyl)dicobalt(0), in which we can view one π bond as donating to one of the Co atoms and the second π bond as overlapping with the other Co atom (33). In this example, the alkyl or aryl groups present on the alkyne impart stability by lowering the tendency towards secondary reactions of the coordinated ethyne, such as loss of the slightly acidic ethynic H atom to the metal atom.

33 $[Co_2(PhC\equiv CPh)(CO)_6]$

24.10 Nonconjugated diene and polyene ligands

KEY POINT The bonding of nonconjugated alkenes to a metal atom is best described as independent multiple alkenes bonding to a metal centre.

Nonconjugated diene (–C=C–X–C=C–) and polyene ligands can also bond to metal atoms. It is simplest to consider them as linked alkenes, and hence they present no new bonding concepts. As with the chelate effect in coordination complexes (Section 7.14), the resulting polyene complexes are usually more stable than the equivalent complex with individual ligands because the entropy of dissociation of the complex is much smaller than when the liberated ligands can move independently. For example, bis(η^4-cycloocta-1,5-diene)nickel(0) (**34**) is more stable than the corresponding complex containing four ethene ligands. Cycloocta-1,5-diene (**35**) is a fairly common ligand in organometallic chemistry, where it is referred to engagingly as 'cod', and is normally introduced into the metal coordination sphere by simple ligand displacement reactions. An example is:

34 $[Ni(cod)_2]$

35 Cycloocta-1,5-diene, cod

Metal–cod complexes are commonly used as starting materials because they often have intermediate stability. Many of them are sufficiently stable to be isolated and handled, but cod can be displaced by many other ligands. For example, if the highly toxic $[Ni(CO)_4]$ molecule is needed in a reaction, then it may be generated from $[Ni(cod)_2]$ directly in the reaction flask:

$$[Ni(cod)_2](sol) + 4\,CO(g) \rightarrow [Ni(CO)_4](sol) + 2\,cod(sol)$$

24.11 Butadiene, cyclobutadiene, and cyclooctatetraene

KEY POINTS Some insight into the bonding of butadiene and cyclobutadiene can be gained by treating them as containing two alkene units, but a full understanding needs a consideration of the molecular orbitals. Cyclooctatetraene bonds in a number of different ways; the most common mode in d-metal chemistry is as an η^4 donor, analogous to butadiene.

The temptation with both butadiene and cyclobutadiene is to treat them like two isolated double bonds. However, a proper molecular orbital approach is necessary to understand the bonding fully because the ligand–metal atom interactions are different in the two cases.

Fig. 24.6 shows the molecular orbitals for the π system in butadiene. The two occupied lower energy MOs can behave as donors to the metal—the lowest a σ donor and the next a π donor. The next-higher unoccupied MO, the LUMO, can act as a π acceptor from the metal atom. Thus, attachment of a butadiene molecule to a metal atom results in population of an MO that is bonding between the two central C atoms (which are already nominally singly bonded) and antibonding between the nominally doubly bonded

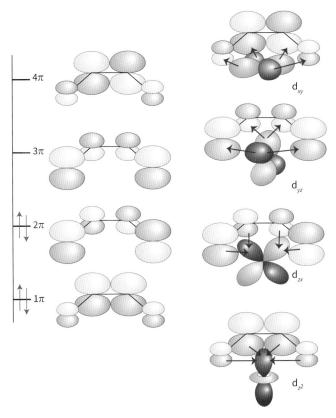

FIGURE 24.6 The molecular orbitals of the π system in butadiene; also shown are metal d orbitals of appropriate symmetry to form bonding interactions.

C atoms. The resulting modification of electron density results in a shortening of the central C–C bond and a lengthening of the double C–C bonds; in some complexes the central C–C bond is found to be even shorter than the other two C–C bonds. In theory, a δ-bonding interaction is possible between the d_{xy} orbital of the metal atom and the most antibonding of the butadiene MOs, but there is no definitive evidence that it occurs. Butadiene is therefore considered to be a four-electron neutral ligand in electron-counting schemes.

Cyclobutadiene is rectangular (D_{2h}) and unstable as a free molecule, as a result of bond angle constraints and its four-electron anti-aromatic configuration. However, stable complexes are known, including [Ru(η^4-C$_4$H$_4$)(CO)$_3$] (36). This species is one of many in which coordination to a metal atom stabilizes an otherwise unstable molecule.

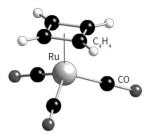

36 [Ru(C$_4$H$_4$)(CO)$_3$]

As a result of a distortion from the square arrangement,[2] cyclobutadiene has an MO diagram similar to that of butadiene (Fig. 24.7). Population of the LUMO by backbonding now leads to greater bonding on the long sides of the rectangular cyclobutadiene molecule, tending towards a square (D_{4h}) arrangement. If the cyclobutadiene formally accepts two electrons from the metal atom, it will have six π electrons with three MOs occupied and no incentive to distort from square; two of the MOs are degenerate in this configuration. All cyclobutadiene–metal complexes are found to be square, suggesting that a six-electron aromatic configuration more accurately describes the bonding within the carbon ring than a four-electron configuration does. This view of the bonding has led some to regard cyclobutadiene complexes as complexes of the R$_4$C$_4^{2-}$ dianion (a six-electron donor), although for convenience most treat cyclobutadiene as a four-electron neutral ligand. Once again, a δ-bonding interaction is possible between the d_{xy} orbital of the metal atom and the most antibonding of the butadiene MOs, but again there seems to be no definitive evidence that it occurs.

Because cyclobutadiene is unstable, the ligand must be generated in the presence of the metal to which it is to be coordinated. This synthesis can be accomplished in a variety of ways. One method is the dehalogenation of a halogenated cyclobutene:

Another procedure is the dimerization of a substituted ethyne:

Cyclooctatetraene (37) is a large ligand that is found in a wide variety of bonding arrangements. Like cyclobutadiene, it is anti-aromatic as the free molecule. Cyclooctatetraene can bond to a metal atom in an octahapto arrangement, in which it is planar and all C–C bond lengths are equal. In a similar fashion to cyclobutadiene, in this arrangement cyclooctatetraene is considered to extract two electrons and to become (formally) the aromatic dinegative ligand [C$_8$H$_8$]$^{2-}$ (that is, a 10-electron donor). Cyclooctatetraenes bound in

---

[2] We can regard this distortion as an organic example of the Jahn–Teller effect (Section 7.16(g)).

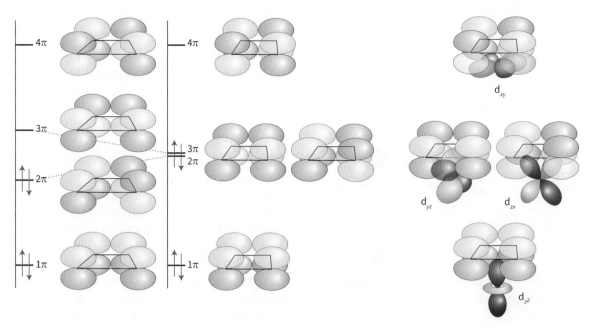

FIGURE 24.7 The molecular orbitals of the π system in cyclobutadiene; also shown are metal d orbitals of appropriate symmetry to form bonding interactions.

this fashion are rarely found in d-metal compounds and are normally found only with lanthanoids and actinoids (Sections 21.8 and 21.10). For example, two such dinegative ligands are found in bis(η^8-cyclooctatetraenyl)-uranium(IV), [U(η^8-C$_8$H$_8$)$_2$], commonly referred to as uranocene.

37 Cyclooctatetraene

The more usual bonding modes for cyclooctatetraenes in d-metal compounds are as puckered η^4-C$_8$H$_8$ ligands such as (**38**), where the bonding part of the ligand can be treated as a butadiene. Bridging modes, (**39**) and (**40**), are also possible.

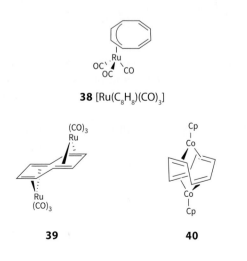

38 [Ru(C$_8$H$_8$)(CO)$_3$]

39   **40**

24.12 Benzene and other arenes

KEY POINT A consideration of the MOs of benzene leads to a picture of bonding of benzene to a metal atom that includes a significant δ backbonding interaction.

If benzene is considered to have three localized double bonds, each double bond could behave as a ligand and the molecule could behave as a tridentate η^6-ligand. A compound such as bis(η^6-benzene)chromium(0) (**41**) could then be considered to be made up of six coordinated double bonds, each donating two electrons, bound to a d^6 metal atom, giving a total of 18 valence electrons for the octahedral complex. Bis(η^6-benzene)chromium(0) does exist and is remarkably stable: it can be handled in air and sublimes with no decomposition. Although this description of the bonding is a first step towards understanding its structure, the true picture needs a deeper consideration of the molecular orbitals involved.

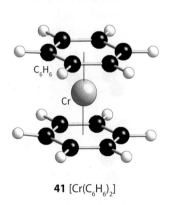

41 [Cr(C$_6$H$_6$)$_2$]

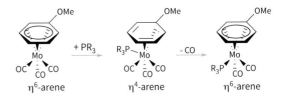

In fact, some η^4-arene complexes, such as (**42**), have been isolated and crystallographically characterized.

42

η^2-Arenes are also known and are analogous to η^2-alkenes; they have an important role in the activation of arenes by metal complexes; once again, some examples, such as (**43**), have been isolated.

43

24.13 The allyl ligand

KEY POINTS A consideration of the MOs of η^3-allyl complexes leads to a picture that gives two identical C–C bond lengths; because the type of bonding of the allyl ligand is so variable, η^3-allyl complexes are often highly reactive.

The allyl ligand, $CH_2 = CH - CH_2^-$, can bind to a metal atom in either of two configurations. As an η^1-ligand (**44**) it should be considered just like an η^1-alkyl group (that is, as a two-electron donor with a single negative charge). However, the allyl ligand can also use its double bond as an additional two-electron donor and act as an η^3-ligand (**45**); in this arrangement it acts as a four-electron donor with a single negative charge. The η^3-allyl ligand can be thought of as a resonance between two forms (**46**), and because all evidence points towards a symmetrical structure, it is often depicted with a curved line representing all the bonding electrons (**47**).

44 η^1-($CH_2CH=CH_2$)

45 **46** **47** η^3-(CH_2CHCH_2)

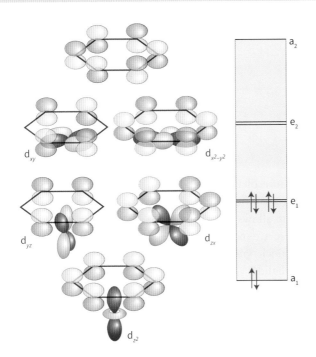

FIGURE 24.8 The molecular orbitals of the π system in benzene; also shown are metal d orbitals of appropriate symmetry to form bonding interactions.

In the molecular-orbital picture of the π bonding in benzene there are three bonding and three antibonding orbitals. If we consider a single benzene molecule bonding to a single metal, and consider only the d orbitals, the strongest interaction is a σ interaction between the most strongly bonding a_2 benzene MO and the d_{z^2} orbital of the metal atom. π bonds are possible between the two other bonding benzene MOs and the d_{zx} and d_{yz} orbitals. Backbonding from the metal atom to the benzene is possible as a δ interaction between the $d_{x^2-y^2}$ and d_{xy} orbitals and the empty antibonding e_2 orbitals of benzene (Fig. 24.8). η^6-Arenes are considered to be neutral ligands that donate six electrons and are normally considered to take up three coordination sites at a metal.

Hexahapto (η^6) arene complexes are very easy to make, often simply by dissolving a compound that has three replaceable ligands in the arene and refluxing the solution:

Mo(CO)$_6$ $\xrightarrow{\text{PhOMe}}$ [η⁶-arene Mo(CO)₃] + 3 CO

One commonly invoked reaction intermediate of η^6-arene complexes is a 'slipping' to an η^4 complex, which donates only four electrons to the metal, and therefore allows a substitution reaction to proceed without an initial ligand loss:

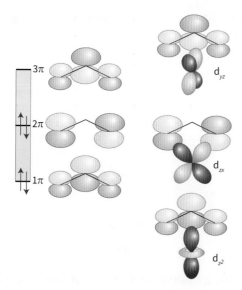

FIGURE 24.9 The molecular orbitals of the π system of the allyl group; also shown are metal d orbitals of appropriate symmetry to form bonding interactions.

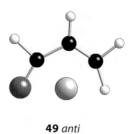

49 *anti*

As with benzene, a more detailed understanding of the bonding of an allyl group needs a consideration of the molecular orbitals of the organic fragment (Fig. 24.9), whereupon it becomes apparent why a symmetrical arrangement is the correct description of the η^3-bonding mode. The filled 1π orbital on the allyl group behaves as a σ donor (into the d_{z^2} orbital), the 2π orbital behaves as a π donor (into the d_{zx} orbital), and the 3π orbital behaves as a π acceptor (from the d_{yz} orbital). Thus, the interactions of the metal atom with each of the terminal carbon atoms are identical and a symmetrical arrangement results.

The terminal substituents of an η^3-allyl group are bent slightly out of the plane of the three-carbon backbone and are either *syn* (**48**) or *anti* (**49**) relative to the central hydrogen. It is common to observe *anti* and *syn* group exchange, which in some cases is fast on an NMR timescale. A mechanism that involves the transformation η^3 to η^1 to η^3 is often invoked to explain this exchange.

Because of this flexibility in the bonding, η^3-allyl complexes are often highly reactive as transformation to the η^1-form allows them to bind readily to another ligand.

There are many synthetic routes to allyl complexes. One is the nucleophilic attack of an allyl Grignard reagent on a metal halide:

$$2\,C_3H_5MgBr + NiCl_2 \rightarrow [Ni(\eta^3{-}C_3H_5)_2] + 2\,MgBrCl$$

Nucleophilic attack on a haloalkane by a metal atom in a low oxidation state also yields allyl complexes:

In complexes where the metal centre is not protonated directly, the protonation of a butadiene ligand can lead to an η^3-allyl complex:

24.14 Cyclopentadiene and cycloheptatriene

KEY POINTS The common η^5-bonding mode of a cyclopentadienyl ligand can be understood on the basis of both σ and π donation from the organic fragment to the metal, in conjunction with δ backbonding; cycloheptatriene commonly forms either neutral η^6-complexes or η^7-complexes of the aromatic cycloheptatrienyl cation $(C_7H_7)^+$.

Cyclopentadiene, C_5H_6, is a mildly acidic hydrocarbon that can be deprotonated to form the cyclopentadienyl anion, $C_5H_5^-$. The stability of the cyclopentadienyl anion can be understood when it is realized that the six electrons in its π system make it aromatic. The delocalization of these six electrons results in a ring structure with five equal bond lengths. As a ligand, the cyclopentadienyl group has played a major role in the development of organometallic chemistry and continues to be the archetype of cyclic polyene ligands.

We have already alluded to the role ferrocene (**4**) had in the development of organometallic chemistry. A huge

48 *syn*

number of metal cyclopentadienyl and substituted cyclopentadienyl compounds are known. Some compounds have $C_5H_5^-$ as a monohapto ligand (13), in which case it is treated like an η^1-alkyl group; others contain $C_5H_5^-$ as a trihapto ligand (14), in which case it is treated like an η^3-allyl group. Usually, though, $C_5H_5^-$ is present as a pentahapto ligand (15), bound through all five carbons of its ring. Examples of complexes containing η^1- and and η^5-cyclopentadienyl groups include (50); (51) is an example of a complex containing η^3- and η^5-cyclopentadienyl groups.

50   **51**

We treat the η^5–$C_5H_5^-$ group as a six-electron donor. Formally, the electron donation to the metal now comes from the filled 1π (σ bonding) and 2π (π bonding) MOs (Fig. 24.10) with δ backbonding from the d_{xy} and $d_{x^2-y^2}$ orbitals on the metal atom. As we shall see in Section 24.19, coordinated Cp ligands behave as though they maintain their six-electron aromatic structure.

The electronic and steric properties of cyclopentadiene can easily be tuned: electron withdrawing and donating groups can be attached to the five-membered ring, and steric bulk can be enhanced by additional substitution. The pentamethylcyclopentadienyl group (Cp*) is commonly used to provide greater electron density on, and greater steric protection of, a metal atom. Chiral groups are often added to Cp groups so that complexes can be used in stereoselective reactions: the *neo*-menthyl group (52) is commonly used. The synthesis, properties, and reactivities of compounds containing the cyclopentadienyl ligand are discussed in more detail in Section 24.19.

52 *neo*-Methylcyclopentadienyl

Cycloheptatriene, C_7H_8 (53), can form η^6 complexes such as (54), which may be treated as having three η^2-alkene molecules bound to the metal atom. Hydride abstraction from these complexes results in formation of η^7 complexes of the six-electron aromatic cycloheptatrienyl cation $C_7H_7^+$ (55), formerly known as the tropylium cation. In η^7-cycloheptatrienyl complexes, for example (56), all carbon–carbon bond lengths are equal; bonding to the metal atom and backbonding from the metal are similar to those found in arene and cyclopentadienyl complexes.

53 Cycloheptatriene

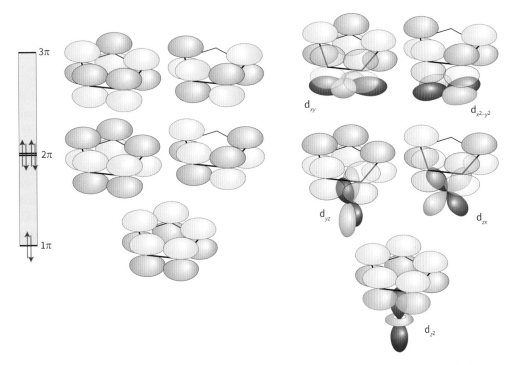

FIGURE 24.10 The molecular orbitals for the π systems of the cyclopentadienyl⁻ group; also shown are metal d orbitals of appropriate symmetry to form bonding interactions.

54   **55**   **56**

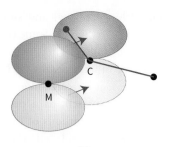

59

24.15 Carbenes

KEY POINTS Fischer- and Schrock-type carbene complexes are considered to have a metal–carbon double bond; *N*-heterocyclic carbenes are considered to have a metal–carbon single bond, together with π backbonding.

Carbene, CH_2, has only six electrons around its C atom and is consequently highly reactive. Other substituted carbenes exist and are substantially less reactive and can behave as ligands towards metals.

In principle, carbenes can exist in one of two limiting electronic configurations: with a linear arrangement of the two groups bound to the carbon atom and the two remaining electrons unpaired in two p orbitals (**57**), or with the two groups bent, the two remaining electrons paired, and an empty p orbital (**58**).

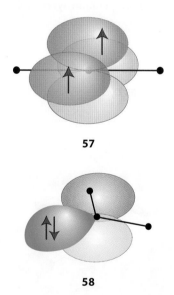

57

58

Carbenes with a linear arrangement of groups are referred to as 'triplet carbenes' (because the two electrons are unpaired and $S = 1$) and are favoured when sterically very bulky groups are attached to the carbene carbon. Carbenes with the bent arrangement are known as 'singlet carbenes' (the two electrons are paired and $S = 0$) and are the normal form for carbenes. The electron pair on the carbon atom of a singlet carbene is suitable to bond to a metal atom, resulting in a ligand-to-metal bond. The empty p orbital on the C atom can then accept electron density from the metal atom, thus stabilizing the electron-poor carbon atom (**59**). For historical reasons, carbenes bonded to metal atoms in this fashion are known as **Fischer carbenes** and are represented by a metal–carbon double bond. Fischer carbenes are electron-deficient at the C atom and consequently are easily attacked by nucleophiles. When the backbonding to the C atom is very strong, the carbene can become electron-rich and is thus prone to attack by electrophiles. Carbenes of this type are known as **Schrock carbenes**, after their discoverer. The term **alkylidene** technically refers only to carbenes with alkyl substituents (CR_2), but is sometimes used to mean both Fischer and Schrock carbenes.

More recently, a large number of derivatives of what are known as *N*-heterocyclic carbenes (NHCs) have been used as ligands. In most NHCs, two nitrogen atoms are adjacent to the carbene carbon atom, and if the lone pair on the nitrogen is considered to be largely p-orbital based, then strong π-donor interactions from the two nitrogen atoms can help to stabilize the carbene (**60**).

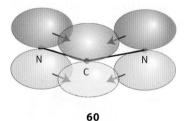

60

Constraining the carbene C atom and the two N atoms into a ring helps to stabilize the carbene, and five-membered rings are common (**61**). Additional stability can be achieved with a double bond in the ring, which provides an additional two electrons that may be considered part of a six-electron aromatic resonance structure (**62**). NHC ligands are considered to be two-electron σ donors and initial descriptions of the bonding suggested only minimal π backbonding from the metal atom. However, the current view is that there is actually significant π backbonding from the metal atom to the NHC.

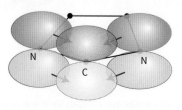

61

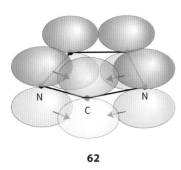

62

24.16 Alkanes, agostic hydrogens, and noble gases

KEY POINT Alkanes can donate the electron density from C–H single bonds to a metal atom and, in the absence of other donors, even the electron density of a noble gas atom can enable it to behave as a ligand.

Highly reactive metal intermediates can be generated by photolysis and, in the absence of any other ligands, alkanes and noble gases have been observed to coordinate to the metal atom. Such species were first identified in the 1970s in solid methane and noble-gas matrices, and were initially regarded as mere curiosities. However, both species have recently been fully characterized in solution and are now accepted as important intermediates in some reactions.

Alkanes are considered to donate electron density from a C–H σ bond to the metal atom (**63**), and accept π-electron density back from the metal atom into the corresponding σ* orbital (**64**), just like dihydrogen (Section 24.7).

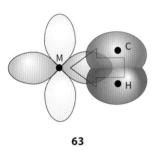

63

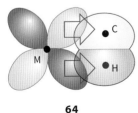

64

Although most alkane complexes are short-lived, with the alkane being readily displaced, in 1998 the cyclopentane complex (**65**) became the first alkane complex to be unambiguously identified in solution by NMR; in 2009 a complex of the simplest hydrocarbon, methane, was characterized by NMR (**66**), and in 2012 a crystal structure of an alkane complex was reported.

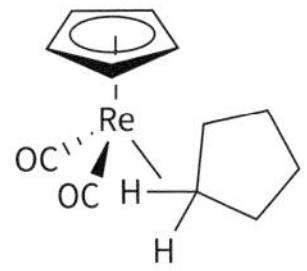

65

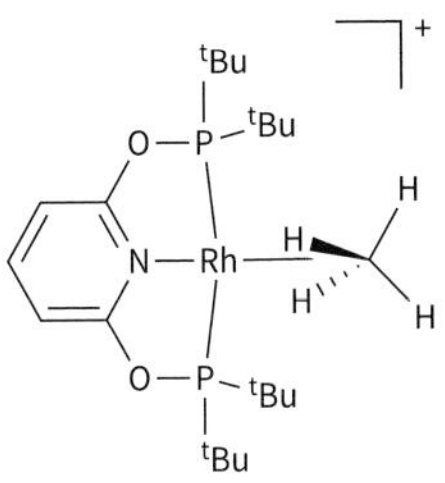

66

Interactions between the C–H bond of an already coordinated ligand and the metal atom have also been observed. These species are referred to as having **agostic** C–H interactions, from the Greek for 'to clasp or to hold to oneself'—like a shield would be—and are thought to have additional stability because of the chelate effect (Section 7.14). Many examples of compounds with agostic interactions are now known; an example is (**67**). Though weakly bonding, each C–H to metal–atom interaction, whether it be agostic or not, is considered formally to donate two electrons to the metal.

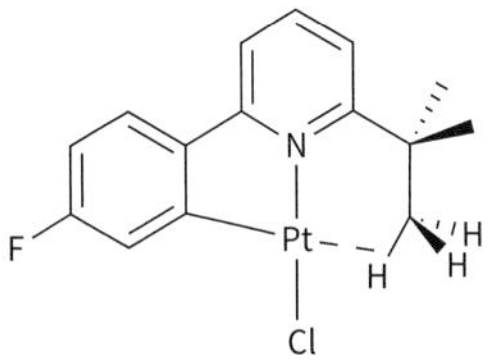

67

Unlikely as it might seem, noble gas atoms can behave as ligands towards metal centres, and a number of complexes of Kr and Xe have been identified by IR spectroscopy, with the relatively long-lived Xe complex (**68**) characterized in solution by NMR in 2005. These complexes are stable only in the absence of better ligands (such as alkanes). The noble gas is formally considered to be neutral and to donate two electrons.

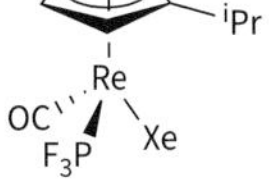

68

EXAMPLE 24.5 Counting electrons in complexes

Which of the following compounds have 18 valence electrons: (a) the agostic Pt compound shown as (**67**), (b) [Re(ⁱPr-Cp)(CO)(PF₃)Xe] (**68**)?

Answer (a) We consider both the η¹-aryl and the chloride ligands as two-electron singly negatively charged donors, with the pyridine also being considered as a two-electron neutral donor. Thus the compound is a complex of Pt(II), which provides eight further electrons. The total number of electrons, before we consider the agostic interaction, is therefore (3 × 2) + 8 = 14.

The agostic interaction can be considered to donate a further two electrons, leading to a 16-electron compound. In fact, the crystal structure of (**67**) indicates that two of the hydrogens of one methyl group interact, implying a formally 18-electron species. (b) The ^{i}Pr–Cp ligand is considered a six-electron donor with a single negative charge and the CO, PF$_3$, and Xe ligands are each considered to be two-electron neutral donors, implying that the Re must have oxidation number +1, so providing a further six electrons. The total electron count is therefore $6 + 2 + 2 + 2 + 6 = 18$.

Self-test 24.5 Show that both (a) [Mo(η^6-C$_7$H$_8$)(CO)$_3$] (**54**) and (b) [Mo(η^7-C$_7$H$_7$)(CO)$_3$]$^+$ (**56**) are 18-electron species.

24.17 Dinitrogen and nitrogen monoxide

KEY POINTS The bonding of dinitrogen to a metal atom is weak but contains both a σ-donating and a π-accepting component; nitrogen monoxide can bind to a metal in two different ways, either bent or straight.

Neither of dinitrogen, N$_2$, nor nitrogen monoxide, NO, is strictly an organometallic ligand, although they are sometimes found in organometallic compounds. Dinitrogen is a much-sought-after ligand, as complexes can potentially take part in a catalytic reduction of nitrogen to more useful species. Dinitrogen can bind to metals in a number of different ways. The majority of complexes have a terminal monohapto link, η^1-N$_2$, in which the bonding can be considered to be like that of the isoelectronic CO ligand (**69**). Dinitrogen is both a weaker σ donor and a weaker π acceptor than CO, and hence is bound less strongly; in fact only good π-donor metal atoms bind N$_2$. Like CO, the N$_2$ ligand has a distinctive IR stretching band lying in the range 2150–1900 cm^{-1}.

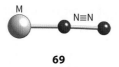

69

A dinitrogen molecule can participate in two bonding interactions and bridge two metal atoms (**70**). If the backdonation to the nitrogen is extensive in this kind of complex, it can formally be considered to have been reduced to a hydrazine (**71**). Occasionally, dinitrogen ligands are found bound in a dihapto (η^2) side-on fashion (**72**). In these complexes the ligand is best considered analogous to an η^2-alkyne. The side-on bonding mode seems to be particularly common in complexes of the f metals (Chapter 21).

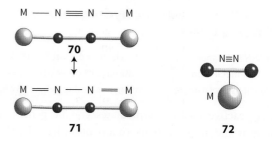

Nitrogen monoxide (nitric oxide) is a radical with 11 valence electrons. When bound, NO is referred to as the nitrosyl ligand and can bond in one of two modes to d-metal atoms, in either a bent or a linear fashion. In the linear arrangement (**73**), the ligand is considered to be the NO$^+$ cation. The NO$^+$ cation is isoelectronic with CO, and the bonding can be considered in a similar fashion (a two-electron σ donor, with strong π-acceptor ability). In the bent arrangement (**74**), NO is considered to behave as NO$^-$, again donating two electrons. In many complexes, NO can change its coordination mode; in effect, moving from the linear to the bent mode reduces the number of electrons on the metal by two.

M—N≡O   M—N⃛
                   ‖
                   O

73   **74**

Compounds

The preceding discussion of ligands and their bonding modes suggests that there are likely to be a large number of organometallic compounds that have either 16 or 18 valence electrons. A detailed discussion of all these compounds is well beyond the scope of this book, see Further Reading. However, we shall examine a number of different classes of compounds because they provide insight into the structures and properties of many other compounds that can be derived from them. Thus, we shall now consider the structures, bonding, and reactions of metal carbonyls, which historically formed the foundation of much of d-block organometallic chemistry. Then we consider some sandwich compounds, before describing the structures and reactions of metal cluster compounds.

24.18 d-Block carbonyls

d-Block carbonyls have been studied extensively since the discovery of tetracarbonylnickel in 1890. Interest in carbonyl compounds has not waned, with many important industrial processes relying on carbonyl intermediates.

(a) Homoleptic carbonyls

KEY POINTS The carbonyls of the Period 4 elements of Groups 6 to 10 obey the 18-electron rule; they have alternately one and two metal atoms and a decreasing number of CO ligands.

A **homoleptic complex** is a complex with only one kind of ligand. Simple homoleptic metal carbonyls can be prepared for most of the d metals, but those of Pd and Pt are so unstable that they exist only at low temperatures. No simple neutral metal carbonyls are known for Cu, Ag, or Au, or for the members of Group 12. The metal carbonyls are useful synthetic precursors for other organometallic compounds and are used in organic syntheses and as industrial catalysts.

The 18-electron rule helps to systematize the formulas of metal carbonyls. As shown in Table 24.5, the neutral carbonyls of the Period 4 elements of Groups 6 to 10 have alternately one and two metal atoms and a decreasing number of CO ligands. The binuclear carbonyls are formed by elements of the odd-numbered groups, which have an odd number of valence electrons and therefore dimerize by forming metal–metal (M–M) bonds (each M–M bond effectively increases the electron count on the metal by 1). The decrease in the number of CO ligands from left to right across a period matches the need for fewer CO ligands to achieve 18 valence electrons. The simple vanadium carbonyl $[V(CO)_6]$ is an exception as it only has 17 valence electrons and is too sterically crowded to dimerize; it is, however, readily reduced to the 18-electron $[V(CO)_6]^-$ anion.

Simple metal carbonyl molecules often have well-defined, simple, symmetrical shapes that correspond to the CO ligands taking up the most distant locations, like regions of enhanced electron density in the VSEPR model. Thus, the Group 6 hexacarbonyls are octahedral, pentacarbonyliron(0) is trigonal-bipyramidal, tetracarbonylnickel(0) is tetrahedral, and decacarbonyldimanganese(0) consists of two square-pyramidal $Mn(CO)_5$ groups joined by a metal–metal bond. Bridging carbonyls are also found; for example, one isomer of octacarbonyldicobalt(0) has its metal–metal bond bridged by two CO ligands.

(b) Synthesis of homoleptic carbonyls

KEY POINTS Some metal carbonyls are formed by direct reaction, but of those that can be formed in this way most require high pressures and temperatures; metal carbonyls are commonly formed by reductive carbonylation.

The two principal methods for the synthesis of monometallic metal carbonyls are direct combination of carbon monoxide with a finely divided metal and the reduction of a metal salt in the presence of carbon monoxide under pressure. Many polymetallic carbonyls are synthesized from monometallic carbonyls.

TABLE 24.5 Formulas and electron count for some 3d-series carbonyls

Group	Formula	Valence electrons		Structure
6	$[Cr(CO)_6]$	Cr	6	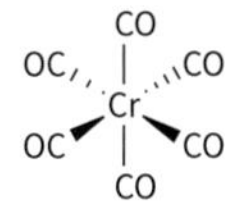
		6(CO)	12	
		Total	18	
7	$[Mn_2(CO)_{10}]$	Mn	7	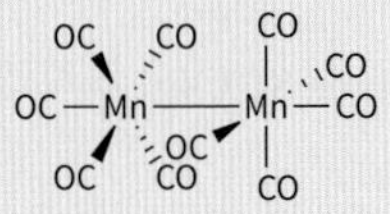
		5(CO)	10	
		M–M	1	
		Total	18	
8	$[Fe(CO)_5]$	Fe	8	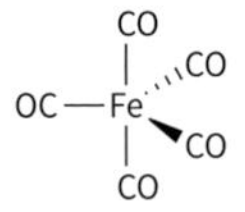
		5(CO)	10	
		Total	18	
9	$[Co_2(CO)_8]$	Co	9	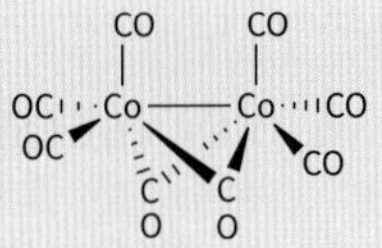
		4(CO)	8	
		M–M	1	
		Total	18	
10	$[Ni(CO)_4]$	Ni	10	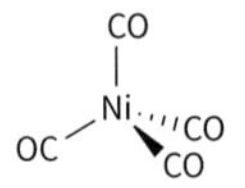
		4(CO)	8	
		Total	18	

BOX 24.1 The Mond process

Ludwig Mond, Carl Langer, and Friedrich Quinke discovered [Ni(CO)$_4$] in the course of studying the corrosion of nickel valves in process gas containing CO in 1890. They were not able to fully characterize the new compound (calling it 'nickel-carbon-oxide'), commenting: 'We have at present no suggestion to offer as to the constitution of this remarkable compound.' They were, however, able to assign the formula 'Ni(CO)$_4$' to their new compound and were quick to apply their discovery to develop a new industrial process (the Mond process) for the purification of nickel. This process was such a success that it brought nickel all the way from Canada to Mond's factory in South Wales.

The Mond process relies on the ease of synthesis of [Ni(CO)$_4$]: at a pressure of a single atmosphere of carbon monoxide, nickel metal will react at around 50°C to give [Ni(CO)$_4$]:

$$\text{Ni} + 4\text{CO} \rightarrow [\text{Ni(CO)}_4]$$

At this temperature, [Ni(CO)$_4$] is a gas (b.p. 34°C) and is easily separated from the residues of the impure nickel. Nickel tetracarbonyl decomposes to give pure nickel at about 220°C, liberating carbon monoxide, which can then be reused. Typically, the impure nickel is generated from the reduction of nickel oxide ores with a mixture of hydrogen and carbon monoxide.

Mond, Langer, and Quinke also attempted the synthesis of analogous compounds with other metals, but were unable to isolate anything new. They did, however, succeed in extracting nickel contaminants from samples of cobalt, suggesting a way to purify cobalt too.

The centennial of the discovery of nickel tetracarbonyl is celebrated in a special volume of *J. Organomet. Chem.*, 1990, **383**, which is devoted to metal carbonyl chemistry.

In 1890, Mond, Langer, and Quinke discovered that the direct combination of nickel and carbon monoxide produced tetracarbonylnickel(0), [Ni(CO)$_4$]—a reaction that is used in the Mond process for purifying nickel (Box 24.1):

$$\text{Ni(s)} + 4\,\text{CO(g)} \xrightarrow{50°C,\,1\,\text{atm CO}} [\text{Ni(CO)}_4](\text{g})$$

Tetracarbonylnickel(0) is in fact the metal carbonyl that is most readily synthesized in this way, with other metal carbonyls, such as [Fe(CO)$_5$], being formed more slowly. They are therefore synthesized at high pressures and temperatures (Fig. 24.11):

$$\text{Fe(s)} + 5\,\text{CO(g)} \xrightarrow{200°C,\,200\,\text{atm}} [\text{Fe(CO)}_5](\text{l})$$
$$2\,\text{Co(s)} + 8\,\text{CO(g)} \xrightarrow{150°C,\,35\,\text{atm}} [\text{Co}_2(\text{CO})_8](\text{s})$$

Direct reaction is impractical for most of the remaining d metals, and **reductive carbonylation**, the reduction of a salt or metal complex in the presence of CO, is normally employed instead. Reducing agents vary from active metals such as aluminium and sodium to alkyl-aluminium compounds, H$_2$, and CO itself:

$$\text{CrCl}_3(\text{s}) + \text{Al(s)} + 6\,\text{CO(g)} \xrightarrow{\text{AlCl}_3\,\text{benzene}} \text{AlCl}_3(\text{sol}) + [\text{Cr(CO)}_6](\text{sol})$$

$$3\,\text{Ru(acac)}_3(\text{sol}) + \text{H}_2(\text{g}) + 12\,\text{CO(g)} \xrightarrow{150°C,\,200\,\text{atm},\,\text{CH}_3\text{OH}} [\text{Ru}_3(\text{CO})_{12}](\text{sol}) + \cdots$$

$$\text{Re}_2\text{O}_7(\text{s}) + 17\,\text{CO(g)} \xrightarrow{250°C,\,350\,\text{atm}} [\text{Re}_2(\text{CO})_{10}](\text{s}) + 7\,\text{CO}_2(\text{g})$$

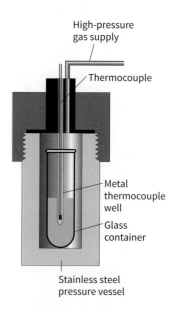

FIGURE 24.11 A high pressure reaction vessel. The reaction mixture is in a glass container.

(c) Properties of homoleptic carbonyls

KEY POINTS All the mononuclear carbonyls are volatile; all the mononuclear and many of the polynuclear carbonyls are soluble in hydrocarbon solvents; polynuclear carbonyls are coloured.

Iron and nickel carbonyls are liquids at room temperature and pressure, but all other common carbonyls are solids. All the mononuclear carbonyls are volatile; their vapour pressures at room temperature range from approximately 50 kPa for tetracarbonylnickel(0) to approximately 10 Pa for hexacarbonyltungsten(0). The high volatility of [Ni(CO)$_4$], coupled with its extremely high toxicity, means that unusual care is required in its handling. Although the other carbonyls appear to be less toxic, they too must not be inhaled or allowed to touch the skin.

Because they are nonpolar, all the mononuclear and many of the polynuclear carbonyls are soluble in hydrocarbon solvents. The most striking exception among the common carbonyls is nonacarbonyldiiron(0), [Fe$_2$(CO)$_9$], which has a very low vapour pressure and is insoluble in solvents with which it does not react.

Most of the mononuclear carbonyls are colourless or lightly coloured. Polynuclear carbonyls are coloured, the intensity of the colour increasing with the number of metal atoms. For example, pentacarbonyliron(0) is a light straw-coloured liquid, nonacarbonyldiiron(0) forms golden-yellow flakes, and dodecacarbonyltriiron(0) is a deep green compound that looks black in the solid state. The colours of polynuclear carbonyls arise from electronic transitions between orbitals that are largely localized on the metal framework.

The principal reactions of the metal centre of simple metal carbonyls are substitution (Section 24.21), oxidation, reduction, and condensation into clusters (Section 24.20). In certain cases, the CO ligand itself is also subject to attack by nucleophiles or electrophiles.

(d) Oxidation and reduction of carbonyls

KEY POINTS Most metal carbonyls can be reduced to metal carbonylates; some metal carbonyls disproportionate in the presence of a strongly basic ligand, producing the ligated cation and a carbonylate anion; metal carbonyls are susceptible to oxidation by air; metal–metal bonds undergo oxidative cleavage.

Most neutral metal carbonyl complexes can be reduced to an anionic form known as a **metal carbonylate**. In monometallic carbonyls, two-electron reduction is generally accompanied by loss of the two-electron donor CO ligand, thus preserving the electron count at 18:

$$2\,Na + [Fe(CO)_5] \xrightarrow{THF} (Na^+)_2[Fe(CO)_4]^{2-} + CO$$

The metal carbonylate contains Fe with oxidation number −2, and it is rapidly oxidized by air. That much of the negative charge is delocalized over the CO ligands is confirmed by the observation of a low CO stretching band in the IR spectrum at about 1785 cm^{-1}. Polynuclear carbonyls, which obey the 18-electron rule through the formation of M–M bonds, are generally cleaved by strong reducing agents. The 18-electron rule is obeyed in the product, and a mononegative mononuclear carbonylate results:

$$2\,Na + [(OC)_5Mn-Mn(CO)_5] \xrightarrow{THF} 2\,Na[Mn(CO)_5]$$

Some metal carbonyls disproportionate in the presence of a strongly basic ligand, producing the ligated cation and a carbonylate. Much of the driving force for this reaction is the stability of the metal cation when it is surrounded by strongly basic ligands. Octacarbonyldicobalt(0) is highly susceptible to this type of reaction when exposed to a good Lewis base such as pyridine (py):

$$3[Co_2^{(0)}(CO)_8] + 12\,py \rightarrow 2[Co^{(+2)}(py)_6][Co^{(-1)}(CO)_4]_2 + 8\,CO$$

It is also possible for the CO ligand to be oxidized in the presence of the strongly basic ligand OH$^-$, the net outcome being the reduction of a metal centre:

$$3[Fe^{(0)}(CO)_5] + 4\,OH^- \rightarrow [Fe_3^{(-2/3)}(CO)_{11}]^{2-} + CO_3^{2-} + 2\,H_2O + 3\,CO$$

Carbonyl compounds that have only 17 electrons are particularly prone to reduction to give 18-electron carbonylates.

Metal carbonyls are susceptible to oxidation by air. Although uncontrolled oxidation produces the metal oxide and CO or CO$_2$, of more interest in organometallic chemistry are the controlled reactions that give rise to organometallic halides. One of the simplest of these is the oxidative cleavage of an M–M bond:

$$[(OC)_5Mn^{(0)}-Mn^{(0)}(CO)_5] + Br_2 \rightarrow 2[Mn^{(+1)}Br(CO)_5]$$

In keeping with the loss of electron density from the metal when a halogen atom is attached, the CO stretching frequencies of the product are significantly higher than those of [Mn$_2$(CO)$_{10}$].

(e) Metal carbonyl basicity

KEY POINTS Most organometallic carbonyl compounds can be protonated at the metal centre; the acidity of the protonated form depends on the other ligands on the metal.

Many organometallic compounds can be protonated at the metal centre. Metal carbonylates provide many examples of this basicity:

$$[Mn(CO)_5]^- + H^+ \rightarrow [MnH(CO)_5]$$

The affinity of metal carbonylates for the proton varies widely (Table 24.6). It is observed that the greater the electron density on the metal centre of the anion, the higher its Brønsted basicity and hence the lower the acidity of its conjugate acid (the metal carbonyl hydride).

As we noted in Section 24.7, d-block M–H complexes are commonly referred to as 'hydrides', which reflects the assignment of oxidation number −1 to an H atom attached to a metal atom (Section 1.6(d)). Nevertheless, most of the carbonyl hydrides of metals to the right of the d block are Brønsted acids. The Brønsted acidity of a metal carbonyl hydride is a reflection of the π-acceptor strength of a CO ligand, which stabilizes the conjugate base. Thus, [CoH(CO)$_4$] is acidic whereas [CoH(PMe$_3$)$_4$] is strongly hydridic. In striking contrast to p-block hydrogen compounds, the Brønsted acidity of d-block M–H compounds decreases on descending a group.

TABLE 24.6 Acidity constants of d-metal hydrides in acetonitrile at 25°C

Hydride	pK_a
$[CoH(CO)_4]$	8.3
$[CoH(CO)_3P(OPh)_3]$	11.3
$[Fe(H)_2(CO)_4]$	11.4
$[CrH(Cp)(CO)_3]$	13.3
$[MoH(Cp)(CO)_3]$	13.9
$[MnH(CO)_5]$	15.1
$[CoH(CO)_3PPh_3]$	15.4
$[WH(Cp)(CO)_3]$	16.1
$[MoH(Cp^*)(CO)_3]$	17.1
$[Ru(H)_2(CO)_4]$	18.7
$[FeH(Cp)(CO)_2]$	19.4
$[RuH(Cp)(CO)_2]$	20.2
$[Os(H)_2(CO)_4]$	20.8
$[ReH(CO)_5]$	21.1
$[FeH(Cp^*)(CO)_2]$	26.3
$[WH(Cp)(CO)_2PMe_3]$	26.6

Neutral metal carbonyls (such as pentacarbonyliron, $[Fe(CO)_5]$) can be protonated in air-free concentrated acid; the Brønsted basicity of a metal atom with oxidation number zero is associated with the presence of nonbonding d electrons. Compounds having metal–metal bonds, such as clusters (Section 24.20), are even more easily protonated; here the Brønsted basicity is associated with the ready protonation of M–M bonds to produce a formal 3c,2e bond like that in diborane:

$$[3Fe(CO)_{11}]^{2-} + H^+ \rightarrow [Fe_3H(CO)_{11}]^-$$

The M–H–M bridge is by far the most common bonding mode of hydrogen in clusters.

Metal basicity is turned to good use in the synthesis of a wide variety of organometallic compounds. For example, alkyl and acyl groups can be attached to metal atoms by the reaction of an alkyl or acyl halide with an anionic metal carbonyl:

$$[Mn(CO)_5]^- + CH_3I \rightarrow [Mn(CH_3)(CO)_5] + I^-$$
$$[Co(CO)_4]^- + CH_3COI \rightarrow [Co(COCH_3)(CO)_4] + I^-$$

A similar reaction with organometallic halides may be used to form M–M bonds:

$$[Mn(CO)_5]^- + [ReBr(CO)_5] \rightarrow [(OC)_5Mn-Re(CO)_5] + Br^-$$

(f) Reactions of the CO ligand

KEY POINTS The C atom of CO is susceptible to attack by nucleophiles if it is attached to a metal atom that is electron-poor; the O atom of CO is susceptible to attack by electrophiles in electron-rich carbonyls.

The C atom of CO is susceptible to attack by nucleophiles if it is attached to a metal atom that is not electron-rich. Thus, terminal carbonyls with high CO stretching frequencies are liable to attack by nucleophiles. The d electrons in these neutral or cationic metal carbonyls are not extensively delocalized on to the carbonyl C atom and so that atom can be attacked by electron-rich reagents. For example, strong nucleophiles (such as methyllithium; Section 12.17) attack the CO in many neutral metal carbonyl compounds:

$$\tfrac{1}{4}[Li_4(CH_3)_4] + [Mo(CO)_6] \rightarrow Li[Mo(COCH_3)(CO)_5]$$

The resulting anionic acyl compound reacts with carbocation reagents to produce a stable and easily handled neutral product:

The product of this reaction, with a direct M=C bond, is a Fischer carbene (Section 24.15). The attack of a nucleophile on the C atom is also important for the mechanism of the hydroxide-induced dissociation of metal carbonyls:

$$[(OC)_{n-1}M(CO)] + OH^- \rightarrow [(OC)_{n-1}M(COOH)]^-$$
$$[(OC)_{n-1}M(COOH)]^- + 3OH^- \rightarrow [M(CO)_{n-1}]^{2-} + CO_3^{2-} + 2H_2O$$

In electron-rich metal carbonyls, considerable electron density is delocalized on the CO ligand. As a result, in some cases the O atom of a CO ligand is susceptible to attack by electrophiles. Once again, IR data provide an indication of when this type of reaction should be expected, as a low CO stretching frequency indicates significant backdonation to the CO ligand and hence appreciable electron density on the O atom. Thus, a bridging carbonyl is particularly susceptible to attack at the O atom:

The attachment of an electrophile to the oxygen of a CO ligand, as in the structure on the right of this equation, promotes migratory insertion reactions (Section 24.24) and C–O cleavage reactions.

The ability of some alkyl-substituted metal carbonyls to undergo a migratory insertion reaction to give acyl ligands, $-(CO)R$, is discussed in detail in Section 24.24.

EXAMPLE 24.6   Converting CO to carbene and acyl ligands

Propose a set of reactions for the formation of [W(C(OCH$_3$)Ph)(CO)$_5$] starting with hexacarbonyltungsten(0) and other reagents of your choice.

Answer We know that CO ligands in hexacarbonyltungsten(0) are susceptible to attack by nucleophiles, and therefore that the reaction with phenyllithium should give a C-phenyl intermediate:

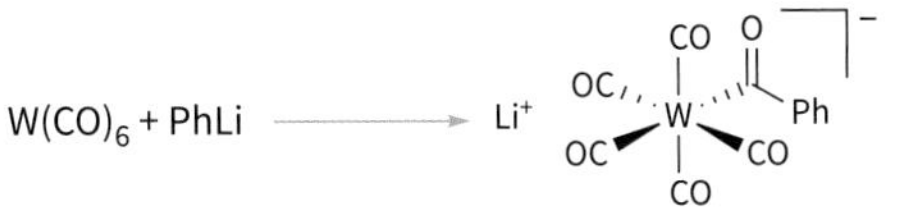

W(CO)$_6$ + PhLi ⟶ Li$^+$ [(OC)$_5$W–C(O)Ph]$^-$

This anion can then react with a carbon electrophile to attach an alkyl group to the O atom of the CO ligand:

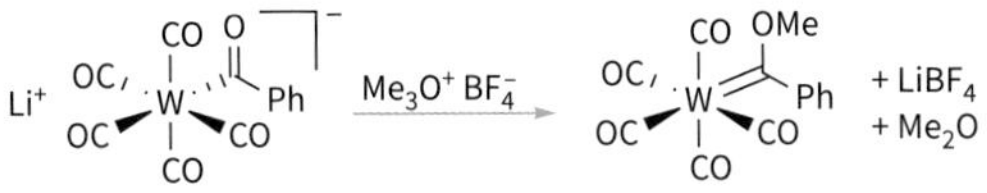

Li$^+$ [(OC)$_5$W–C(O)Ph]$^-$ + Me$_3$O$^+$BF$_4^-$ ⟶ (OC)$_5$W=C(OMe)Ph + LiBF$_4$ + Me$_2$O

Self-test 24.6 Propose a synthesis for [Mn(COCH$_3$)(CO)$_4$(PPh$_3$)] starting with [Mn$_2$(CO)$_{10}$], PPh$_3$, Na, and CH$_3$I.

(g) Spectroscopic properties of carbonyl compounds

KEY POINTS The CO stretching frequency is decreased when it serves as a π acceptor; donor ligands cause the CO stretching frequency to decrease as they supply electrons to the metal; ^{13}C-NMR is of less use as many carbonyl compounds are fluxional on the NMR timescale.

Infrared and ^{13}C-NMR spectroscopy are widely used to determine the arrangement of atoms in metal carbonyl compounds, as separate signals are observed for inequivalent CO ligands. NMR spectra generally contain more detailed structural information than IR spectra, provided the molecule is not fluxional (the timescales of the IR and NMR transition are different; Sections 8.5 and 8.6). However, IR spectra are often simpler to obtain, and are particularly useful for following reactions. Most CO stretching bands occur in the range 2100–1700 cm^{-1}, a region that is generally free of bands arising from organic groups. Both the range of CO stretching frequencies (see Fig. 24.4) and the number of CO bands (Table 24.7) are important for making structural inferences.

Group theory allows us to predict the number of active CO stretches in both the IR and Raman spectra (Section 8.5). If the CO ligands are not related by a centre of inversion or a three-fold or higher axis of symmetry, a molecule with N CO ligands will have N CO stretching absorption bands. Thus, a bent OC–M–CO group (with only a two-fold symmetry axis) will have two infrared absorptions because both the symmetric (**75**) and antisymmetric (**76**) stretches cause the electric dipole moment to change and are IR active.

TABLE 24.7 Relation between the structure of a carbonyl complex and the number of CO stretching bands in its IR spectrum

Complex	Isomer	Structure	Point group	Number of bands*
[M(CO)$_6$]		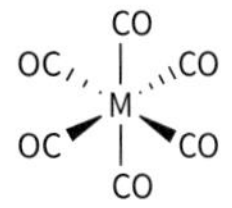 octahedral	O_h	1
[M(CO)$_5$L]		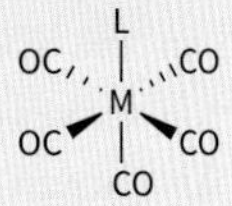	C_{4v}	3†
[M(CO)$_4$L$_2$]	trans	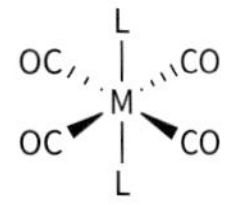	D_{4h}	1
[M(CO)$_4$L$_2$]	cis	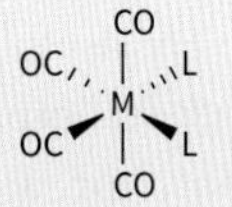	C_{2v}	4‡
[M(CO)$_3$L$_3$]	mer	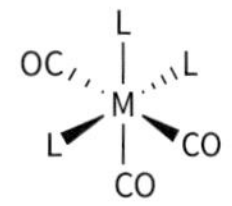	C_{2v}	3‡
[M(CO)$_3$L$_3$]	fac	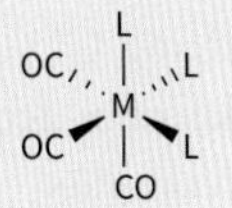	C_{3v}	2
[M(CO)$_5$]		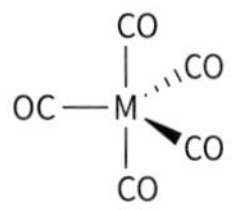 trigonal bipyramidal	D_{3h}	2
[M(CO)$_4$L]	ax	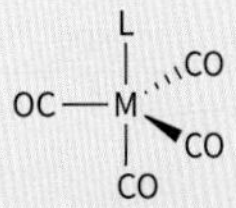	C_{3v}	3§
[M(CO)$_4$L]	eq	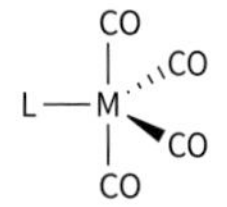	C_{2v}	4
[M(CO)$_3$L$_2$]	trans	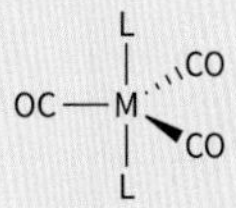	D_{3h}	1
[M(CO)$_3$L$_2$]	cis	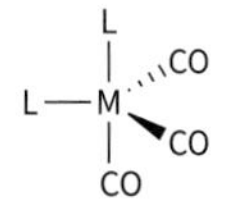	C_s	3
[M(CO)$_4$]		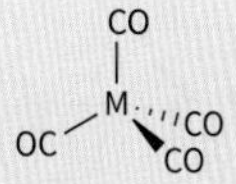 tetrahedral	T_d	1

*The number of IR bands expected in the CO stretching region is based on formal selection rules, and in some cases fewer bands are observed, as explained below.
† If the fourfold array of CO ligands lies in the same plane as the metal atom, two bands will be observed.
‡ If the *trans* CO ligands are nearly collinear, one fewer band will be observed.
§ If the threefold array of CO ligands is nearly planar, only two bands will be observed.

75   76

Highly symmetrical molecules have fewer bands than CO ligands. Thus, in a linear OC–M–CO group, only one IR band (corresponding to the out-of-phase stretching of the two CO ligands) is observed in the CO stretching region because the symmetrical stretch leaves the overall electric dipole moment unchanged. As shown in Fig. 24.12, the positions of the CO ligands in a metal carbonyl may be more symmetrical than the point group of the whole compound suggests, and then fewer bands will be observed than are predicted on the basis of the overall point group. Raman spectroscopy can be very useful in assigning structures because the selection rules complement those of IR (Sections 22.4 and 8.5). Thus, for a linear OC–M–CO group, the symmetric stretching of the two CO ligands is observed in the Raman spectrum.

As we noted in Section 24.5, infrared spectroscopy is also useful for distinguishing terminal CO (M–CO) from bridging CO (μ_2-CO) and face-bridging CO (μ_3-CO). It can also be used to determine the order of π-acceptor strengths for the other ligands present in a complex.

A motionally averaged NMR signal is observed when a molecule undergoes changes in structure more rapidly than the technique can resolve (Section 8.6). Although this phenomenon is quite common in the NMR spectra of organometallic compounds, it is not normally observed in their IR or Raman spectra. An example of this difference is [Fe(CO)$_5$], for which the ^{13}C-NMR spectrum shows a single signal at $\delta = 210$, whereas IR and Raman spectra are consistent with a trigonal-bipyramidal structure.

EXAMPLE 24.7 Determining the structure of a carbonyl from IR data

The complex [Cr(CO)$_4$(PPh$_3$)$_2$] has one very strong IR absorption band at 1889 cm^{-1} and two other very weak bands in the CO stretching region. What is the probable structure of this compound? (The CO stretching frequencies are lower than in the corresponding hexacarbonyl because the phosphine ligands are better σ donors and poorer π acceptors than CO.)

Answer If we consider the possible isomers, we can see that a disubstituted hexacarbonyl may exist with either *cis* or *trans* configuration. In the *cis* isomer, the four CO ligands are in a low-symmetry (C_{2v}) environment and therefore four IR bands should be observed, as indicated in Table 24.7. The *trans* isomer has a square-planar array of four CO ligands (D_{4h}), for which only one band in the CO stretching region is expected (Table 24.7). The *trans* CO arrangement is indicated by the data because it is reasonable to assume that the weak bands reflect a small departure from D_{4h} symmetry imposed by the PPh$_3$ ligands.

Self-test 24.7 The IR spectrum of [Ni$_2$(η^5-Cp)$_2$(CO)$_2$] has a pair of CO stretching bands at 1857 cm^{-1} (strong) and 1897 cm^{-1} (weak). Does this complex contain bridging or terminal CO ligands, or both? (Substitution of η^5-C$_5$H$_5$ ligands for CO ligands leads to small shifts in the CO stretching frequencies for a terminal CO ligand.)

24.19 Metallocenes

As we have already noted, cyclopentadienyl compounds are often remarkably stable, and the discovery of ferrocene, [(Cp)$_2$Fe], in 1951 sparked renewed interest in the whole field of d-block organometallic compounds. Many Cp complexes have two ring systems, with the metal sandwiched between the two rings, and it was for work on so-called 'sandwich compounds', formally **metallocenes**, that Wilkinson and Fischer were awarded the Nobel Prize in 1972.

In keeping with the picture of a metallocene as a metal sandwiched between two planar carbon rings, we can consider compounds of η^4-cyclobutadiene, η^5-cyclopentadienyl, η^6-arenes, η^7-cycloheptatrienyl C$_7$H$_7^+$, η^8-cyclooctatetraene, and even η^3-cyclopropenium (C$_3$H$_3^+$) to be metallocenes. As all the carbon–carbon bond lengths in each of these bound ligands are identical, it makes sense to treat each ligand as having an aromatic configuration; that is,

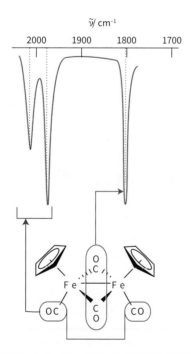

FIGURE 24.12 The infrared spectrum of [Fe$_2$(Cp)$_2$(CO)$_4$]. Note the two high-frequency terminal CO stretches and the lower frequency absorption of the bridging CO ligands. Although two bridging CO bands would be expected on account of the low symmetry of the complex, a single band is observed because the two bridging CO groups are nearly collinear.

π electrons:	2	6	10
	η^3-Cyclopropenium$^+$	η^4-Cyclobutadiene^{2-}	η^8-Cyclooctatetraenyl^{2-}
		η^5-Cyclopentadienyl$^-$	η^9-Cyclononatetraenyl$^-$
		η^6-Arenes	
		η^7-Cycloheptatrienyl$^+$	

This picture of the structure of metallocenes is not wholly in keeping with either system of electron counting, but is closer to the donor-pair method.

We touched on the structures and reactivities of some metallocenes when we discussed the mode of bonding of ligands such as the cyclopentadienyl group, and here we look at some further aspects of their bonding and reactivity.

(a) Synthesis and reactivity of cyclopentadienyl compounds

KEY POINTS Deprotonation of cyclopentadiene gives a convenient precursor to many metal cyclopentadienyl compounds; bound cyclopentadienyl rings behave as aromatic compounds and will undergo Friedel–Crafts electrophilic reactions.

Sodium cyclopentadienide, NaCp, is a common starting material for the preparation of cyclopentadienyl compounds. It can conveniently be prepared by the action of metallic sodium on cyclopentadiene in tetrahydrofuran solution:

$$2\,\text{Na} + 2\,\text{C}_5\text{H}_6 \xrightarrow{\text{THF}} 2\,\text{Na}[\text{C}_5\text{H}_5] + \text{H}_2$$

Sodium cyclopentadienide can then be used to react with d-metal halides to produce metallocenes. Cyclopentadiene itself is acidic enough that potassium hydroxide will deprotonate it in solution and, for example, ferrocene can be prepared with a minimum of fuss:

$$2\,\text{KOH} + 2\,\text{C}_5\text{H}_6 + \text{FeCl}_2 \xrightarrow{\text{DMSO}} [\text{Fe}(\text{C}_5\text{H}_5)_2] + 2\,\text{H}_2\text{O} + 2\,\text{KCl}$$

Because of their great stability, the 18-electron Group 8 compounds ferrocene, ruthenocene, and osmocene maintain their ligand–metal bonds under rather harsh conditions, and it is possible to carry out a variety of transformations on the cyclopentadienyl ligands. For example, they undergo reactions similar to those of simple aromatic hydrocarbons, such as Friedel–Crafts acylation:

It is also possible to replace H on a C_5H_5 ring by Li:

As might be imagined, the lithiated product is an excellent starting material for the synthesis of a wide variety of ring-substituted products, and in this respect resembles simple organolithium compounds (Section 12.17). Most Cp complexes of other metals undergo reactions similar to these two types, where the five-membered ring behaves as an aromatic system.

(b) Bonding in bis(cyclopentadienyl)metal complexes

KEY POINTS The MO picture of bonding in bis(Cp) metal complexes shows that the frontier orbitals are neither strongly bonding nor strongly antibonding; thus, complexes that do not obey the 18-electron rule are possible.

We start by looking at ferrocene, where the molecular orbital energy-level diagram shown in Fig. 24.13 accounts for a number of experimental observations. This diagram refers to the eclipsed (D_{5h}) form of the complex which, in the gas phase, is about $4\,\text{kJ}\,\text{mol}^{-1}$ lower in energy than the staggered conformation. We shall focus our attention on the frontier orbitals. As shown in Fig. 24.13, the e''_1 symmetry-adapted linear combinations of ligand orbitals have the same symmetry as the d_{zx} and d_{yz} orbitals of the metal atom. The lower-energy frontier orbital (a'_1) is composed of the d_{z^2} and the corresponding SALC of ligand orbitals. However, there is little interaction between the ligands and the metal orbitals because the ligand π orbitals happen to lie in the conical nodal surface of the d_{z^2} orbital of the metal atom. In ferrocene and the other 18-electron bis(cyclopentadienyl) complexes, the a'_1 frontier orbital and all lower orbitals are full, but the e''_1 frontier orbital and all higher orbitals are empty.

The frontier orbitals are neither strongly bonding nor strongly antibonding. This characteristic permits the possibility of the existence of bis(cyclopentadienyl) complexes that diverge from the 18-electron rule. Thus the easy oxidation of ferrocene to the 17-electron complex $[\text{Fe}(\eta^5\text{-Cp})_2]^+$ corresponds to the removal of an electron from the nonbonding a'_1 orbital. Population of the e''_1 orbitals is seen in the 19-electron complex $[\text{Co}(\eta^5\text{-Cp})_2]$ and the 20-electron complex $[\text{Ni}(\eta^5\text{-Cp})_2]$. Deviations from the 18-electron rule, however, do lead to significant changes in M–C bond lengths that correlate fairly well with the molecular orbital scheme (Table 24.8).

A useful comparison can be made with octahedral coordination complexes. The e''_1 frontier orbital of a metallocene is the analogue of the e_g orbital in an octahedral

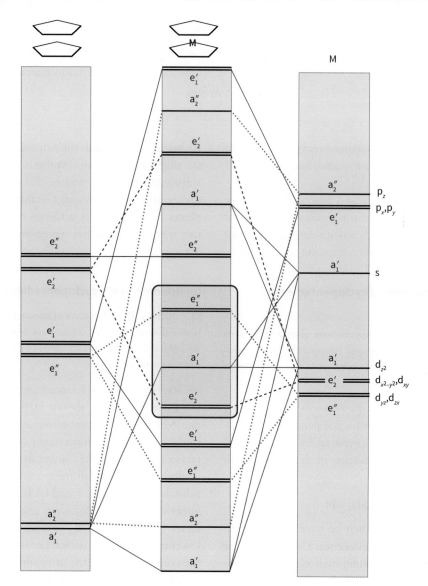

FIGURE 24.13 Molecular orbital energy diagram for a [M(Cp)$_2$] with D_{5h} symmetry. The energies of the symmetry-adapted π orbitals of the C$_5$H$_5$ ligands are shown on the left, relevant d orbitals of the metal are on the right, and the resulting molecular orbital energies are in the centre. Eighteen electrons can be accommodated by filling the molecular orbitals up to and including the a'_1 orbital in the box. The box denotes the orbitals typically regarded as frontier orbitals in these molecules.

TABLE 24.8 Electronic configuration and M–C bond length in [M(η^5-Cp)$_2$] complexes

Complex	Valence electrons	Electron configuration	M–C bond length/pm
[V(η^5-Cp)$_2$]	15	$e'^2_2 a'^1_1$	228
[Cr(η^5-Cp)$_2$]	16	$e'^3_2 a'^1_1$	217
[Mn(η^5-Me-C$_5$H$_4$)$_2$]*	17	$e'^3_2 a'^2_1$	211
[Fe(η^5-Cp)$_2$]	18	$e'^4_2 a'^2_1$	206
[Co(η^5-Cp)$_2$]	19	$e'^4_2 a'^2_1 e''^1_1$	212
[Ni(η^5-Cp)$_2$]	20	$e'^4_2 a'^2_1 e''^2_1$	220

* Data are quoted for this complex because [Mn(η^5-Cp)$_2$] has a high-spin configuration and hence an anomalously long M–C bond (238 pm).

complex, and the a'_1 orbital plus the e'_2 pair of orbitals are analogous to the t_{2g} orbitals of an octahedral complex. This formal similarity extends to the existence of high- and low-spin bis(cyclopentadienyl) complexes.

EXAMPLE 24.8 Identifying metallocene electronic structure and stability

Refer to Fig. 24.13. Discuss the occupancy and nature of the HOMO in [Co(η^5-Cp)$_2$]$^+$ and the change in metal–ligand bonding relative to neutral cobaltocene.

Answer If we consider the [Co(η^5-Cp)$_2$]$^+$ ion we can see that it contains 18 valence electrons (six from Co(III), 12 from the two Cp⁻ ligands). If we then assume that the molecular orbital energy-level diagram for ferrocene is applicable, the 18-electron count leads to double occupancy of the orbitals up to a_1'. The 19-electron cobaltocene molecule has an additional electron in the e_1'' orbital, which is antibonding with respect to the metal and ligands, and is very easy to remove (cobaltocene is oxidized much more readily than ferrocene.) Therefore, the metal–ligand bonds should be stronger and shorter in $[Co(\eta^5\text{-}Cp)_2]^+$ than in $[Co(\eta^5\text{-}Cp)_2]$. This conclusion is borne out by structural data.

Self-test 24.8 By using the same molecular orbital diagram, comment on whether the removal of an electron from $[Fe(\eta^5\text{-}Cp)_2]$ to produce $[Fe(\eta^5\text{-}Cp)_2]^+$ should produce a substantial change in M–C bond length relative to neutral ferrocene.

(c) Fluxional behaviour of metallocenes

KEY POINT Many metallocenes exhibit fluxionality and undergo internal rotation because the barrier to the interconversion of the various forms is low.

One of the most remarkable aspects of many cyclic polyene complexes is their stereochemical nonrigidity (fluxionality). For example, at room temperature the two rings in ferrocene rotate rapidly relative to each other as there is only a low staggered/eclipsed conversion barrier. This type of fluxional process is called **internal rotation**, and is similar to the process by which the two CH_3 groups rotate relative to each other in ethane. We have already noted how, in the gas phase, the eclipsed conformation of ferrocene is slightly more stable than the staggered one; this difference arises from the improved overlap of the metal d orbitals with those of the Cp rings in the eclipsed conformation. However, the steric bulk of substituents on the metallocene rings can destabilize the eclipsed conformation and can make the sterically less congested staggered form the preferred conformation. The rings of metallocenes are often drawn in a staggered conformation simply because there is then a little more space to illustrate substitutions.

Of greater interest is the stereochemical nonrigidity that is often seen when a conjugated cyclic polyene is attached to a metal atom through some, but not all, of its C atoms.

In such complexes the metal–ligand bonding may hop around the ring—a process known informally as 'ring whizzing'. A simple example is found in $[Ge(\eta^1\text{-}Cp)(CH_3)_3]$, in which the single site of attachment of the Ge atom to the cyclopentadiene ring hops around the ring in a series of **1,2-shifts**—a motion in which a C–M bond is replaced by a C–M bond to the next C atom around the ring; this motion is known as a 1,2-shift, because the bond starts on atom 1 and ends up on the adjacent atom 2 (Fig. 24.14). The great majority of fluxional conjugated polyene complexes that have been investigated migrate by 1,2-shifts, but it is not known whether these shifts are controlled by a principle of least motion or by some aspect of orbital symmetry.

Nuclear magnetic resonance provides the primary evidence for the existence and mechanism of these fluxional processes, as they occur on a timescale of 10^{-2} to 10^{-4} s, and can be studied by ¹H- and ¹³C-NMR. The compound $[Ru(\eta^4\text{-}C_8H_8)(CO)_3]$ (77) provides a good illustration of the approach. At room temperature, its ¹H-NMR spectrum consists of a single, sharp line that could be interpreted as arising from a symmetrical $\eta^8\text{-}C_8H_8$ ligand. However, X-ray diffraction studies of single crystals show unambiguously that the ligand is tetrahapto. This conflict is resolved by ¹H-NMR spectra at lower temperatures, because as the sample is cooled, the signal broadens and then separates into four peaks. These peaks are expected for the four pairs of protons of an $\eta^4\text{-}C_8H_8$ ligand. The interpretation is that at room temperature the ring is 'whizzing' around the metal atom rapidly compared with the timescale of the NMR experiment (Section 8.6), so an averaged signal is observed. At lower temperatures the motion of the ring is slower, and the distinct conformations exist long enough to be resolved. A detailed analysis of the line shape of the NMR spectra can be used to measure the activation energy of the migration.

77

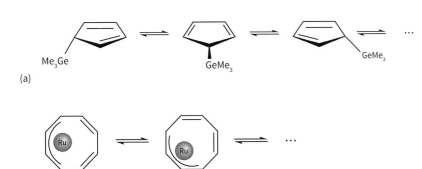

FIGURE 24.14 (a) The fluxional process in $[Ge(\eta^1\text{-}Cp)(Me)_3]$ occurs by a series of 1,2-shifts. (b) The fluxionality of $[Ru(\eta^4\text{-}C_8H_8)(CO)_3]$ can be described similarly. We need to imagine that the Ru atom is out of the plane of the page with the CO ligands omitted for clarity.

(d) Bent metallocene complexes

KEY POINT The structures of bent-sandwich compounds can be systematized in terms of a model in which three metal atom orbitals project towards the open face of the bent Cp_2M fragment.

In addition to the simple bis(cyclopentadienyl) and bis(arene) complexes with parallel rings, there are many related structures. In the jargon of this area, these species are referred to as 'bent-sandwich compounds' (**78**), 'half-sandwich' or 'piano-stool' compounds (**79**), and, inevitably, 'triple deckers' (**80**). Bent-sandwich compounds play a major role in the organometallic chemistry of the early and middle d-block elements, and examples include [Ti(η^5-Cp)$_2$Cl$_2$], [Re(η^5-Cp)$_2$Cl], [W(η^5-Cp)$_2$(H)$_2$], and [Nb(η^5-Cp)$_2$Cl$_3$].

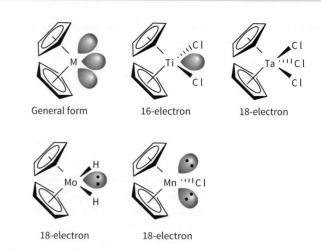

FIGURE 24.15 Bent sandwich compounds with their electron counts.

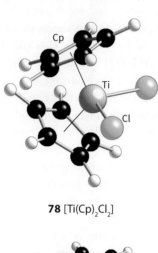

78 [Ti(Cp)$_2$Cl$_2$]

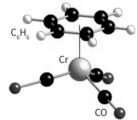

79 [Cr(η^6-C$_6$H$_6$)(CO)$_3$]

80 [Ni$_2$(Cp)$_3$]$^+$

As shown in Fig. 24.15, bent-sandwich compounds occur with a variety of electron counts and stereochemistries. Their structures can be systematized in terms of a model in which three metal atom orbitals project out of the open face of the bent M(Cp)$_2$ fragment. According to this model, the metal atom often satisfies its electron deficiency when the electron count is less than 18 by interaction with lone pairs or agostic C–H groups on the ligands.

24.20 Metal–metal bonding and metal clusters

Organic chemists must go to extravagant efforts to synthesize their cage-like molecules, such as cubane (**81**). By contrast, one of the distinctive characteristics of inorganic chemistry is the large number of closed polyhedral molecules, such as the tetrahedral P$_4$ molecule (Section 16.1), the octahedral halide-bridged early d-block clusters (Box 20.2), the polyhedral carboranes (Section 14.12), and the organometallic cluster compounds that we discuss here. The structures of clusters often resemble the close-packed structures of the metal itself, and this similarity provides the main rationale for studying them—the idea that the chemical properties of the ligands of a cluster reflect the behaviour on a metal surface. In recent years, clusters have inspired research for different reasons: materials such as quantum dots and nanoparticles (Section 28.7) have electronic properties that depend on how large the cluster is.

81 Cubane, C$_8$H$_8$

(a) Structure of clusters

KEY POINT A cluster includes all compounds with metal–metal bonds that form triangular or larger cyclic structures.

A rigorous definition of **metal clusters** restricts them to molecular complexes with metal–metal bonds that form triangular or larger cyclic structures. This definition excludes linear M–M compounds and cage compounds, in which several metal atoms are held together exclusively by ligand bridges. However, this rigorous definition is normally relaxed, and we shall consider any M–M bonded system as a cluster.

The distinction between cage and cluster compounds can seem arbitrary, as the presence of bridging ligands in a cluster such as (82) raises the possibility that the atoms are held together by M–L–M interactions rather than M–M bonds. Bond lengths are of some help in resolving this issue. If the M–M distance is much greater than twice the metallic radius, then it is reasonable to conclude that the M–M bond is either very weak or absent. However, if the metal atoms are within a reasonable bonding distance, the proportion of the bonding that is attributable to direct M–M interaction is ambiguous.

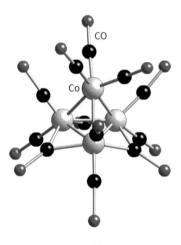

82 $[Co_4(CO)_{12}]$

For example, following the isolation of $[Fe_2(CO)_9]$ (83) in 1938, there was much debate about the extent of Fe–Fe bonding. Initial suggestions of no Fe–Fe interaction were superseded by the general acceptance of an M–M bond some 20 years later, though some current bonding theories can account for the structure without invoking any M–M bonding (see Tutorial Problem 24.13).

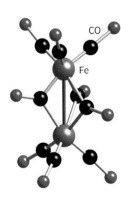

83 $[Fe_2(CO)_9]$

Metal–metal bond strengths in metal complexes cannot be determined with great precision, but a variety of pieces of evidence—such as the stability of compounds and M–M force constants—indicate that there is an increase in M–M bond strengths down a group in the d block. This trend contrasts with that in the p block, where element–element bonds are usually weaker for the heavier members of a group. As a consequence of this trend, metal–metal bonded systems are most numerous for the 4d- and 5d-series metals.

(b) Electron counting in clusters

KEY POINTS The 18-electron rule is suitable for identifying the correct number of electrons for clusters with fewer than six metal atoms; the Wade-Mingos-Lauher rules identify a correlation between the valence electron count and the structures of larger organometallic complexes.

Organometallic cluster compounds are rare for the early d metals and almost unknown for the f metals (Section 21.10), but a large number of metal carbonyl clusters exist for the elements of Groups 6 to 10. The bonding in the smaller clusters can be readily explained in terms of local M–M and M–L electron-pair bonding and the 18-electron rule.

If we take $[Mn_2(CO)_{10}]$ and $[Os_3(CO)_{12}]$ as examples we can arrive at a simple, yet illuminating, picture. In $[Mn_2(CO)_{10}]$ (84), each Mn atom is coordinated by five terminal carbonyls and is considered to have 17 electrons (7 from Mn and 10 from the five CO ligands) before we take into account the Mn–Mn bond. This Mn–Mn bond consists of two electrons shared between the two metal atoms, and hence raises the electron count of each by 1, resulting in two 18-electron metal atoms, but with a total electron count of only 34, not 36.

$$\begin{array}{c} OC \quad CO \quad\quad CO \\ \ \ \ \backslash \ | \ \ \ \ \ \ \ \ \ |\ \ / \\ OC-Mn-Mn-CO \\ \ \ / \ |\ \ \ \ \ \ \ \ \ | \ \backslash \\ OC \ \ CO \ \ \ \ CO \end{array}$$

84 $[Mn_2(CO)_{10}]$

In $[Os_3(CO)_{12}]$ (85), each Os atom is coordinated by four carbonyls and, as an $[Os(CO)_4]$ fragment, has 16 electrons (eight from the metal and eight from four carbonyls) before metal–metal bonding is taken into consideration. Each metal is bonded to two others to give a triangular arrangement of metals (three M–M bonds in the cluster) with the two bonds to each metal increasing the number of electrons around the metals to 18, but with the total electron count of only 48, not 54, electrons.

$$\begin{array}{c} (CO)_4 \\ Os \\ / \ \ \backslash \\ (OC)_4Os-Os(CO)_4 \end{array}$$

85 $[Os_3(CO)_{12}]$

TABLE 24.9 Correlation of cluster valence electron (CVE) count and structure

Number of metal atoms	Structure of metal framework		CVE count	Example
1	Single atom	M	18	[Ni(CO)$_4$] (**2**)
2	Linear	M—M	34	[Mn$_2$(CO)$_{10}$] (**84**)
3	Closed triangle	M—M \ M	48	[Os$_3$(CO)$_{12}$] (**85**)
4	Tetrahedron	M/M\M / M	60	[Co$_4$(CO)$_{12}$] (**82**)
	Butterfly	M\M/M—M	62	[Fe$_4$(CO)$_{12}$C]$^{2-}$
	Square	M—M / M—M	64	[Os$_4$(CO)$_{16}$]
5	Trigonal bipyramid	M-M\M/M-M	72	[Os$_5$(CO)$_{16}$]
	Square pyramid	M/M\M—M	74	[Fe$_5$C(CO)$_{15}$]
6	Octahedron	M/M\M-M/M\M	86	[Ru$_6$C(CO)$_{17}$]
	Trigonal prism	M—M, M-M, M—M	90	[Rh$_6$C(CO)$_{15}$]$^{2-}$

The bonding electrons we are dealing with are referred to as the **cluster valence electrons** (CVEs), and it quickly becomes apparent that a cluster of x metal atoms with y metal–metal bonds needs $18x – 2y$ electrons. Octahedral M$_6$ and larger clusters do not always conform to this pattern, and the polyhedral skeletal electron-pair rules, known as Wade's rules (Section 14.11), have been refined by D. M. P. Mingos and J. Lauher to apply to metal clusters. The structures that the **Wade–Mingos–Lauher rules** suggest are summarized in Table 24.9; the rules apply most reliably to metal clusters in Groups 6–9. In general, and as in the boron hydrides where similar considerations apply, more open structures (which have fewer metal–metal bonds) occur when there is a higher CVE count.

EXAMPLE 24.9 Correlating spectroscopic data, cluster valence electron count, and structure

The reaction of chloroform (trichloromethane) with [Co$_2$(CO)$_8$] yields a compound of formula [Co$_3$(CH)(CO)$_9$]. Both NMR and IR data indicate the presence of only terminal CO ligands and the presence of a CH group. Propose a structure consistent with the spectra and the correlation of CVE with structure.

Answer We assume that the CH ligand is simply C-bonded: one C electron is used for the C–H bond, so three are available for bonding in the cluster. Electrons available for the cluster are then 27 from three Co atoms, 18 from nine CO ligands, and 3 from the CH. The resulting total CVE of 48 indicates a triangular cluster of metals (see Table 24.9). A structure consistent with this conclusion and the presence of only terminal CO ligands and a capping CH ligand is (**86**).

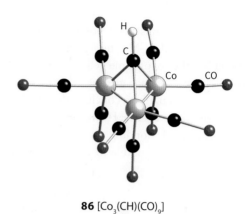

86 [Co$_3$(CH)(CO)$_9$]

Self-test 24.9 The compound [Fe$_4$(Cp)$_4$(CO)$_4$] is a dark-green solid. Its IR spectrum shows a single CO stretch at 1640 cm^{-1}. The ^{1}H-NMR spectrum is a single line, even at low temperatures. From this spectroscopic information and the CVE, propose a structure for [Fe$_4$(Cp)$_4$(CO)$_4$].

(c) Isolobal analogies

KEY POINTS Structurally analogous fragments of molecules are described as isolobal; groups of isolobal molecular fragments are used to suggest patterns of bonding between seemingly unrelated fragments, allowing the rationalization of diverse structures.

We can identify analogies in the structures of apparently unrelated molecules. Thus, we may view N(CH$_3$)$_3$ as derived from NH$_3$ by substitution of a CH$_3$ fragment for each H atom. In current terminology, the structurally analogous fragments are said to be **isolobal**, and the relationship is expressed by the symbol ⟝⟞. The origin of the name is the lobe-like shape of a hybrid orbital in a molecular fragment. Two fragments are isolobal if their highest-energy orbitals have the same symmetry (such as the σ symmetry of the H1s and a Csp^3 hybrid orbital), similar energies, and the same electron occupation (one in each case in H1s and Csp^3). Table 24.10 lists some selected isolobal fragments, and the first line shows isolobal fragments with a single frontier orbital. The recognition of this family permits us to anticipate by analogy with H–H that molecules such as H$_3$C–CH$_3$ and (OC)$_5$Mn–CH$_3$ can be formed. The second line of Table 24.10 lists some isolobal fragments with two frontier orbitals, and the third line lists some with three.

Isolobal analogies provide a good way to picture the incorporation of hetero atoms into a metal cluster. These analogies allow us to draw a parallel between [Co$_3$(CH)(CO)$_9$] (**86**) and [Co$_4$(CO)$_{12}$] (**82**), both of which can be regarded as triangular Co$_3$(CO)$_9$ fragments capped on one side, either

TABLE 24.10 Selected isolobal fragments

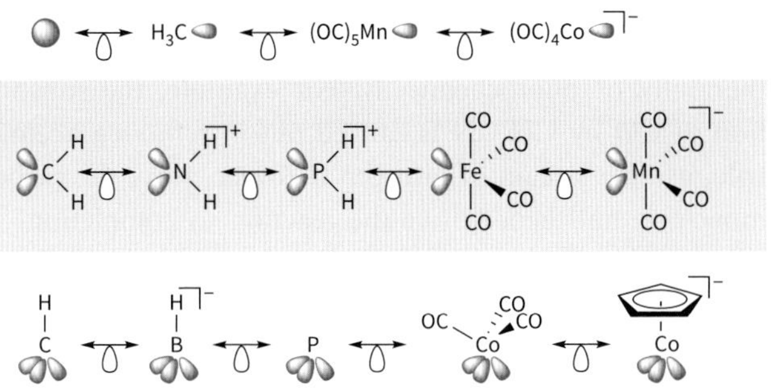

Note that electrons can be added to or subtracted from each member of the isolobal group and still maintain isolobality. For example, $CH_3^+ \longleftrightarrow Mn(CO)_5^+ \longleftrightarrow Co(CO)_4$.

by $Co(CO)_3$ or by CH. A minor complication in this comparison is the occurrence of $Co(CO)_2$ groups together with bridging CO ligands in $[Co_4(CO)_{12}]$, because bridging and terminal ligands often have similar energies. Further inspection of the isolobal fragments given in Table 24.10 shows that a P atom is isolobal with CH; accordingly, a cluster similar to $[Co_3(CH)(CO)_9]$ (86) is known, but with a capping P atom. Similarly, the ligands CR_2 and $Fe(CO)_4$ are both capable of bonding to two metal atoms in a cluster; CH_3 and $Mn(CO)_5$ can bond to one metal atom.

As a final example of isolobal analogies, consider the mixed manganese and platinum complex (87): the three-membered {Mn,Pt,P} rings can be considered as the metallocyclopropane form of a coordinated double bond (88). Both halves of the Mn=P fragments are isolobal with CH_2 if we treat them as PR_2^+ and $Mn(CO)_4^-$, and the complete fragment can then be treated as analogous to an ethene molecule. This treatment means that (87) can be considered analogous to bis-(ethene)carbonylplatinum(0) (89), a known simple 16-electron organometallic compound.

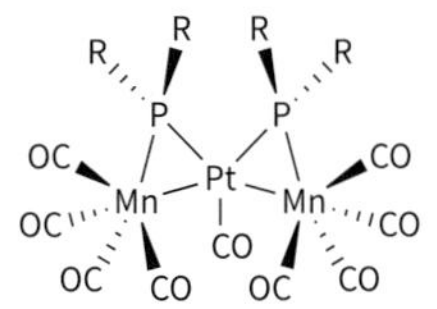

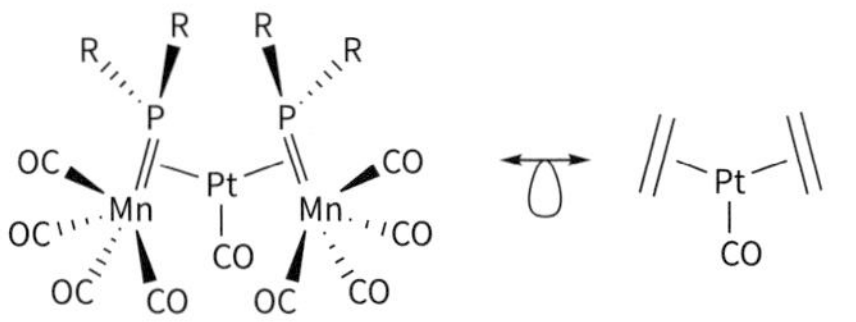

(d) Synthesis of clusters

KEY POINTS Three methods are commonly used to prepare metal clusters: thermal expulsion of CO from a metal carbonyl, the condensation of a carbonyl anion and a neutral organometallic complex, and the condensation of an organometallic complex with an unsaturated organometallic compound.

One of the oldest methods for the synthesis of metal clusters is the thermal expulsion of CO from a metal carbonyl. The pyrolytic formation of metal-cluster compounds can be viewed from the standpoint of electron count: a decrease in valence electrons around the metal resulting from loss of CO is compensated by the formation of M–M bonds. One example is the synthesis of $[Co_4(CO)_{12}]$ by heating $[Co_2(CO)_8]$:

$$2[Co_2(CO)_8] \rightarrow [Co_4(CO)_{12}] + 4CO$$

This reaction proceeds slowly at room temperature, so samples of octacarbonyldicobalt(0) are usually contaminated with dodecacarbonyltetracobalt(0).

A widely used and more controllable reaction is based on the condensation of a carbonyl anion and a neutral organometallic complex:

$$[Ni_5(CO)_{12}]^{2-} + [Ni(CO)_4] \rightarrow [Ni_6(CO)_{12}]^{2-} + 4CO$$

The Ni_5 complex has a CVE of 76, whereas the Ni_6 complex has a count of 86. The descriptive name **redox condensation** is often given to reactions of this type, which are very useful for the preparation of anionic metal carbonyl clusters. In this example, a trigonal-bipyramidal cluster containing Ni with formal oxidation number $-\frac{2}{5}$ and $[Ni(CO)_4]$ containing Ni(0) is converted into an octahedral cluster having Ni with oxidation number $-\frac{1}{3}$. The $[Ni_5(CO)_{12}]^{2-}$ cluster, which has four electrons in excess of the 72 expected for a trigonal bipyramid, illustrates a fairly common tendency for the Group 10 metal clusters to have an electron count in excess of that expected from the Wade–Mingos–Lauher rules.

A third method, pioneered by F. G. A. Stone, is based on the condensation of an organometallic complex containing displaceable ligands with an unsaturated organometallic compound. The unsaturated complex may be a metal alkylidene, $L_nM=CR_2$, a metal alkylidyne, $L_nM\equiv CR$, or a compound with multiple metal–metal bonds:

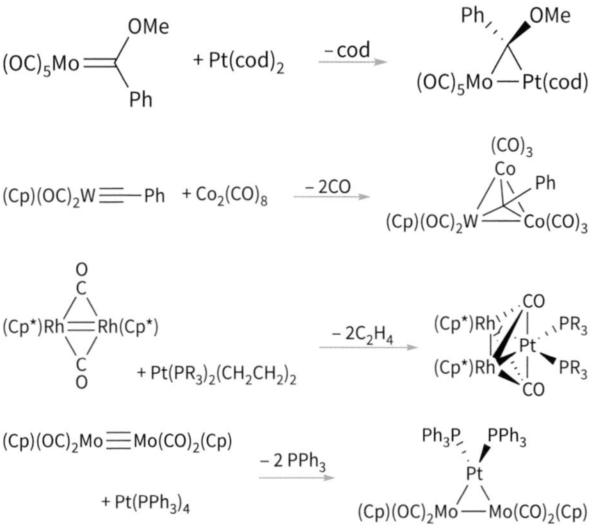

Reactions

The fact that most organometallic compounds can be made to react in a variety of ways is responsible for their use as catalysts. In the previous sections we looked at ligands and how to introduce them into a metal centre, and in this section we look at how ligands might react further or with each other. Implicit in the following discussion is the fact that coordinatively saturated complexes are less reactive than unsaturated ones.

24.21 Ligand substitution

KEY POINTS The substitution of ligands in organometallic complexes is very similar to the substitution of ligands in coordination complexes, with the additional constraint that the valence electron count at the metal atom does not increase above 18; steric crowding of ligands increases the rate of dissociative processes and decreases the rate of associative processes.

Extensive studies of CO substitution reactions of simple carbonyl complexes have revealed systematic trends in mechanisms and rates, and much that has been established for these compounds is applicable to all organometallic complexes. The simple replacement of one ligand by another in organometallic complexes is very similar to that observed with coordination compounds, where reactions sometimes go by an associative, a dissociative, or an interchange pathway, with the reaction being either associatively or dissociatively activated (Section 23.2).

The simplest examples of substitution reactions involve the replacement of CO by another electron-pair donor, such as a phosphine. Studies of the rates at which trialkylphosphines and other ligands replace CO in $[Ni(CO)_4]$, $[Fe(CO)_5]$, and the hexacarbonyls of the chromium group show that they are relatively insensitive to the incoming group, indicating that a dissociatively activated mechanism is in operation. In some cases, a solvated intermediate such as $[Cr(CO)_5(THF)]$ has been detected. This intermediate then combines with the entering group in a bimolecular process:

$$[Cr(CO)_6] + sol \rightarrow [Cr(CO)_5](sol) + CO$$
$$[Cr(CO)_5](sol) + L \rightarrow [Cr(CO)_5L] + sol$$

A dissociatively activated substitution reaction would be expected with metal carbonyl complexes as associative activation would require reaction intermediates with more than 18 valence electrons, formation of which corresponds to populating high-energy antibonding MOs.

Whereas the loss of the first CO group from $[Ni(CO)_4]$ occurs easily, and substitution is fast at room temperature, the CO ligands are much more tightly bound in the Group 6 carbonyls, and loss of CO often needs to be promoted thermally or photochemically. For example, the substitution of CO by CH_3CN is carried out in refluxing acetonitrile, using a stream of nitrogen to sweep away the carbon monoxide and hence drive the reaction to completion. To achieve photolysis, mononuclear carbonyls (which do not absorb strongly in the visible region) are exposed to near-UV radiation in an apparatus like that shown in Fig. 24.16. As with the thermal process, there is strong evidence that the photoassisted substitution reaction leads to the formation of a labile intermediate complex with the solvent, which is then displaced by the entering group. Solvated intermediates in the photolysis of metal carbonyls have been detected, not only in polar solvents such as THF but also in every solvent that has been tried, even in alkanes and noble gases.

The rates of substitution of ligands in 16-electron complexes are sensitive to the identity and concentration of the entering group, which indicates associative activation. For example, the reactions of $[Ir(CO)Cl(PPh_3)_2]$ with triethylphosphine are associatively activated:

$$[Ir(CO)Cl(PPh_3)_2] + PEt_3 \rightarrow [Ir(CO)Cl(PPh_3)_2(PEt_3)] \rightarrow$$
$$[Ir(CO)Cl(PPh_3)(PEt_3)] + PPh_3$$

Sixteen-electron organometallic compounds appear to undergo associatively activated substitution reactions because the 18-electron activated complex is energetically more favourable than the 14-electron activated complex that would occur in dissociative activation.

As in the reactions of coordination complexes, we can expect steric crowding between ligands to accelerate dissociative processes and to decrease the rates of associative processes (Section 23.6). The extent to which various ligands crowd each other is approximated by the Tolman cone angle (Table 23.7), and we can see how it influences

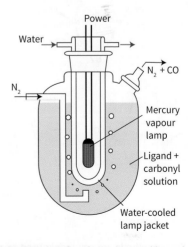

FIGURE 24.16 Apparatus for photochemical ligand substitution of metal carbonyls.

TABLE 24.11 Cone angles and dissociation constants for some Ni complexes

L	θ/°	K_d
PMe$_3$	118	<10^{-9}
PEt$_3$	137	1.2 × 10^{-5}
PMePh$_2$	136	5.0 × 10^{-2}
PPh$_3$	145	Large
P^{t}Bu$_3$	182	Large

Data are for NiL$_4$ ⇌ NiL$_3$ + L in benzene at 25°C.

the equilibrium constant for ligand binding by examining the dissociation constants of [Ni(PR$_3$)$_4$] complexes (Table 24.11). These complexes are slightly dissociated in solution if the phosphine ligands are compact, such as PMe$_3$, with a cone angle of 118°. However, a complex such as [Ni(P^{t}Bu$_3$)$_4$], where the cone angle is huge (182°), is highly dissociated.

The rate of CO substitution in six-coordinate metal carbonyls often decreases as more strongly basic ligands replace CO, and two or three alkylphosphine ligands often represent the limit of substitution. With bulky phosphine ligands, further substitution may be thermodynamically unfavourable on account of ligand crowding, but increased electron density on the metal centre, which arises when a π-acceptor ligand is replaced by a net donor ligand, appears to bind the remaining CO ligands more tightly and therefore reduce the rate of CO dissociative substitution. The explanation of the influence of σ-donor ligands on CO bonding is that the increased electron density contributed by the phosphine leads to stronger π backbonding to the remaining CO ligands and therefore strengthens the M–CO bonds. This stronger M–C bond decreases the tendency of CO to leave the metal atom and therefore decreases the rate of dissociative substitution. It is also observed that the second carbonyl that is replaced is normally *cis* to the site of the first, and that replacement of a third carbonyl results in a *fac* complex. The reason for this regiochemistry is that CO ligands have very high *trans* effects (Section 23.4).

EXAMPLE 24.10 Preparing substituted metal carbonyls

Starting with MoO$_3$ as a source of Mo, and CO and PPh$_3$ as the ligand sources, plus other reagents of your choice, give equations and conditions for the synthesis of [Mo(CO)$_5$PPh$_3$].

Answer Considering the materials available to us, a sensible procedure might be to synthesize [Mo(CO)$_6$] first and then carry out a ligand substitution. Reductive carbonylation of MoO$_3$ can be performed using Al(CH$_2$CH$_3$)$_3$ as a reducing agent in the presence of carbon monoxide under pressure. The temperature and pressure required for this reaction are less than those for the direct combination of molybdenum and carbon monoxide:

$$\text{MoO}_3 + \text{Al(CH}_2\text{CH}_3)_3 + 6\,\text{CO} \xrightarrow{50\,\text{atm, 150°C, heptane}} [\text{Mo(CO)}_6]$$
$$+ \text{oxidation products of Al(CH}_2\text{CH}_3)_3$$

The subsequent substitution could be carried out photochemically by using the apparatus illustrated in Fig. 24.16:

$$[\text{Mo(CO)}_6] + \text{PPh}_3 \rightarrow [\text{Mo(CO)}_5\text{PPh}_3] + \text{CO}$$

The progress of the reaction can be followed by IR spectroscopy in the CO stretching region using small samples that are removed periodically from the reaction vessel.

Self-test 24.10 If the highly substituted complex [Mo(CO)$_3$L$_3$] is desired, which of the ligands PMe$_3$ or P(^{t}Bu)$_3$ would be preferred? Give reasons for your choice.

Although the generalizations above apply to a wide range of reactions, some exceptions are observed, especially if cyclopentadienyl or nitrosyl ligands are present. In these cases it is common to find evidence of associatively activated substitution even for 18-electron complexes. The common explanation is that NO may switch from being linear (as in **73**) to being angular (as in **74**), whereupon it donates two fewer electrons (Section 24.17). Similarly, the η^5-Cp$^-$ six-electron donor can slip relative to the metal and become an η^3-Cp$^-$ four-electron donor. In this case, the C$_5$H$_5^-$ ligand is regarded as having a three-carbon interaction with the metal, while the remaining two electrons form a simple C=C bond that is not engaged with the metal (**90**), and the relatively electron-depleted central metal atom becomes susceptible to substitution:

$$[\text{V(CO)}_5(\text{NO})] + \text{PPh}_3 \rightarrow [\text{V(CO)}_4(\text{NO})(\text{PPh}_3)] + \text{CO}$$
$$[\text{Re}(\eta^5\text{–Cp})(\text{CO})_3] + \text{PPh}_3 \rightarrow [\text{Re}(\eta^3\text{–Cp})(\text{CO})_2(\text{PPh}_3)] + \text{CO}$$

90

It has been found that, for some metal carbonyls, the displacement of CO can be catalysed by electron-transfer processes that create anion or cation radicals. These radicals do not have 18 electrons, and a typical process of this type is illustrated in Fig. 24.17. As can be seen, the key feature is the lability of CO in the 19-electron anion radical compared to the metal carbonyl starting material. Similarly, the less common 19- and 17-electron metal compounds are labile with respect to substitution.

The substitution of ligands at a cluster is often not a straightforward process, because fragmentation is common. Fragmentation occurs because the M–M bonds in a cluster are generally comparable in strength to the M–L bonds, and

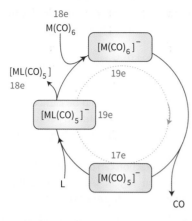

FIGURE 24.17 Schematic diagram of an electron transfer catalysed CO substitution. After addition of a small amount of a reducing initiator, the cycle continues until the limiting reagent $[M(CO)_6]$ or L has been consumed.

so the breaking of the M–M bonds provides a reaction pathway with a low activation energy. For example, dodecacarbonyltriiron(0) reacts with triphenylphosphine under mild conditions to yield simple mono- and disubstituted products as well as some cluster fragmentation products:

$$3[Fe_3(CO)_{12}] + 6\,PPh_3 \rightarrow [Fe_3(CO)_{11}PPh_3] + [Fe_3(CO)_{10}(PPh_3)_2]$$
$$+ [Fe(CO)_5] + [Fe(CO)_4PPh_3] + [Fe(CO)_3(PPh_3)_2] + 3\,CO$$

However, for somewhat longer reaction times or elevated temperatures, only products containing single iron atoms are obtained. Because the strength of M–M bonds increases down a group, substitution products of the heavier clusters, such as $[Ru_3(CO)_{10}(PPh_3)_2]$ or $[Os_3(CO)_{10}(PPh_3)_2]$, can be prepared without significant fragmentation into mononuclear complexes.

EXAMPLE 24.11 Assessing substitutional reactivity

Which of the compounds (**91**) or (**92**) will undergo substitution of a CO ligand for a phosphine more readily?

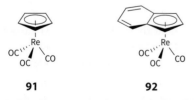

Answer If we consider the ligands present on the compounds, we can see that the 18-electron compound (**92**) contains the indenyl ligand, which can ring-slip to a 16-electron η^3 bound form (**93**) more readily than the normal Cp compound (**91**), as the double bond that forms becomes part of a six-membered aromatic ring. This ring-slipping provides a low-energy route to coordinative unsaturation and thus the indenyl compound reacts with an incoming ligand much more rapidly than the Cp compound.

93

Self-test 24.11 Assess the relative substitutional reactivities of indenyl and fluorenyl (**94**) compounds.

94

24.22 Oxidative addition and reductive elimination

KEY POINTS Oxidative addition occurs when a molecule X–Y adds to a metal atom to form new M–X and M–Y bonds with cleavage of the X–Y bond; oxidative addition results in an increase in the coordination number of the metal atom by 2 and an increase in the oxidation number by 2; reductive elimination is the reverse of oxidative addition.

When we discussed the bonding of dihydrogen to a metal atom in Section 24.7, we noted that the oxidation number on the metal atom increased by 2 when the dihydrogen reacted to give a dihydride:

$$M(N_{ox}) + H_2 \rightarrow [M(N_{ox}+2)(H)_2]$$

The increase in oxidation number of the metal by 2 arises because dihydrogen is treated as a neutral ligand, whereas the hydride ligands are treated as H⁻: thus the formation of two M–H bonds from an H_2 molecule corresponds to a formal increase in the charge on the metal by 2. As we previously noted, it might seem that this oxidation of the metal is just a function of our method of counting electrons. However, it is worth emphasizing that two of the electrons on the metal atom have been used to backbond to the dihydrogen, and these two electrons are no longer available to the metal for further bonding, i.e. an oxidation. This type of reaction is quite general and is known as **oxidative addition**. A large number of molecules add oxidatively to a metal atom, including the alkyl and aryl halides, dihydrogen, and simple hydrocarbons. In general, the addition of any molecule X–Y to a metal atom to give M(X)(Y) can be classed as oxidative addition. Thus the reaction of a metal complex, $[ML_n]$, with an acid such as HCl to give $[ML_n(H)(Cl)]$ is an oxidative addition reaction. Oxidative addition reactions are not restricted to d-block metals: the reaction of magnesium to form Grignard reagents (Section 13.13) is an oxidative addition reaction.

Oxidative addition reactions result in two more ligands bound to the metal with an increase in the total electron count at the metal of 2. Thus, oxidative addition reactions normally require a coordinatively unsaturated metal centre, and are particularly common for 16-electron square-planar metal complexes:

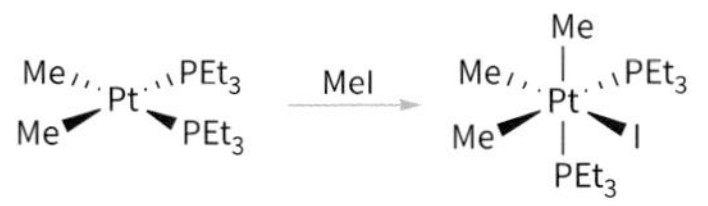

$$\text{Me}_{\prime\prime\prime}\text{Pt}^{\backslash\backslash\backslash\text{PEt}_3}_{\text{PEt}_3}\text{Me} \quad \xrightarrow{\text{MeI}} \quad \text{Me}_{\prime\prime\prime}\overset{\overset{\text{Me}}{|}}{\underset{\underset{\text{PEt}_3}{|}}{\text{Pt}}}\text{I}^{\backslash\backslash\backslash\text{PEt}_3}\text{Me}$$

16e Pt(II) → 18e Pt(IV)

The oxidative addition of hydrogen is a concerted reaction: dihydrogen coordinates to form a σ-bonded H$_2$ ligand, and then backbonding from the metal results in cleavage of the H–H bond and the formation of *cis* dihydrides:

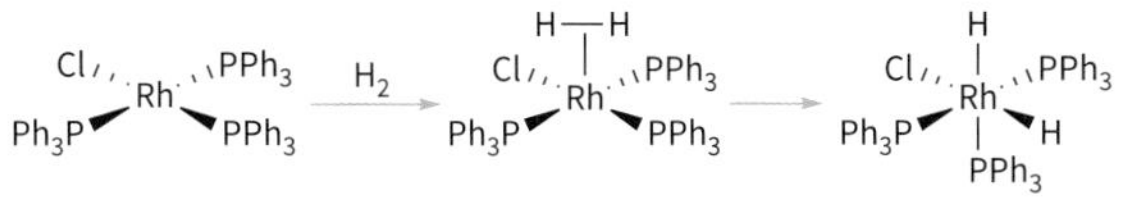

Four coordinate 16e Rh(I) → Five coordinate 18e Rh(I) → Six coordinate 18e Rh(III)

Other molecules, such as alkanes and aryl halides, are known to react in a concerted fashion, and in all these cases the two incoming ligands end up *cis* to each other.

Some oxidative addition reactions are not concerted, and either go through radical intermediates or are best thought of as S_N2 displacement reactions. Radical oxidative addition reactions are rare, and will not be discussed further here. In an S_N2 oxidative addition reaction, a lone pair on the metal attacks the X–Y molecule displacing Y$^-$, which subsequently bonds to the metal:

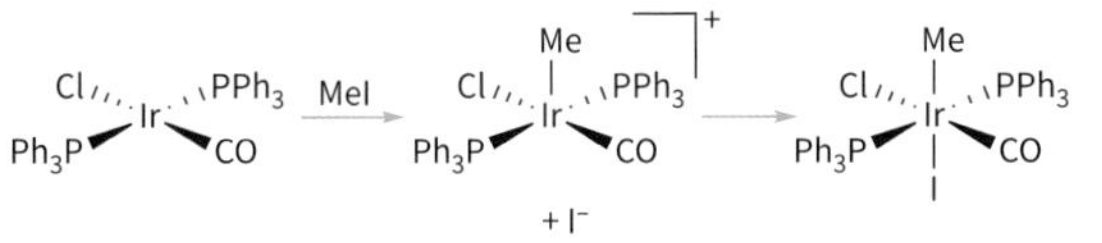

There are two stereochemical consequences of this reaction. Firstly, the two incoming ligands need not end up *cis* to each other; indeed, normally they do not. Secondly, unlike the concerted reaction, any chirality at the X group is inverted. An S_N2-type oxidative addition is common for polar molecules such as alkyl halides.

The opposite of oxidative addition, where two ligands couple and eliminate from a metal centre, is known as **reductive elimination**:

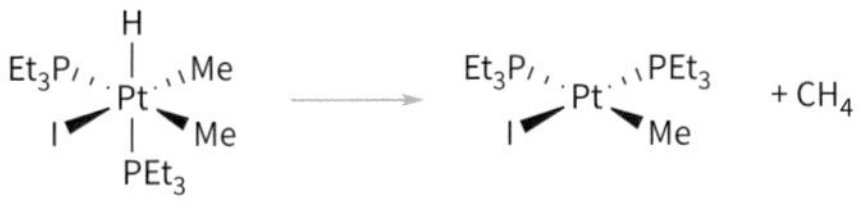

18e Pt(IV) → 16e Pt(II) + CH$_4$

Reductive elimination reactions require both eliminating fragments to be *cis* to each other, and are best thought of as the reverse of the concerted form of oxidative addition.

Oxidative addition and reductive elimination reactions are, in principle, reversible. However, in practice, one direction is normally thermodynamically favoured over the other. Oxidative addition and reductive elimination reactions play a major role in many catalytic processes (Sections 31.1–31.7).

EXAMPLE 24.12 Identifying oxidative addition and reductive elimination

Show that the reaction

is an example of an oxidative addition reaction.

Answer In order to identify an oxidative addition reaction, we need to establish the valence electron counts and oxidation states of both the starting material and the product. The four-coordinate square-planar Rh starting material contains an η^1-alkynyl ligand, as well as three neutral phosphine ligands; it is therefore a 16-electron Rh(I) species. The six-coordinate octahedral product contains two η^1-alkynyl ligands, a hydride ligand, and three neutral phosphine ligands; it is therefore an 18-electron Rh(III) species. The increase in both coordination number and oxidation number by 2 identifies it as an oxidative addition.

Self-test 24.12 Show that the reaction:

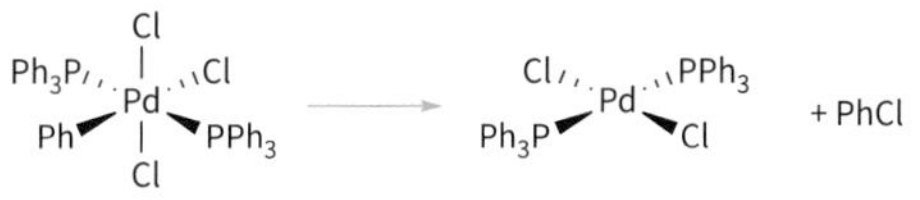

is an example of reductive elimination.

24.23 σ-Bond metathesis

KEY POINT A σ-bond metathesis reaction is a concerted process that sometimes occurs when oxidative addition cannot take place.

A reaction sequence that appears to be an oxidative addition followed by a reductive elimination may in fact be the exchange of two species by a process known as **σ-bond metathesis**. σ-Bond metathesis reactions are common for early d-metal complexes where there are not enough electrons on the metal atom for it to participate in oxidative addition. For instance, the 16-electron compound [(Cp)$_2$ZrHMe] cannot react with H$_2$ to give a trihydride, as all its electrons are involved in bonding to the existing ligands. A four-membered transition state is proposed in such cases, and a concerted bond-making and bond-breaking step results in elimination of methane:

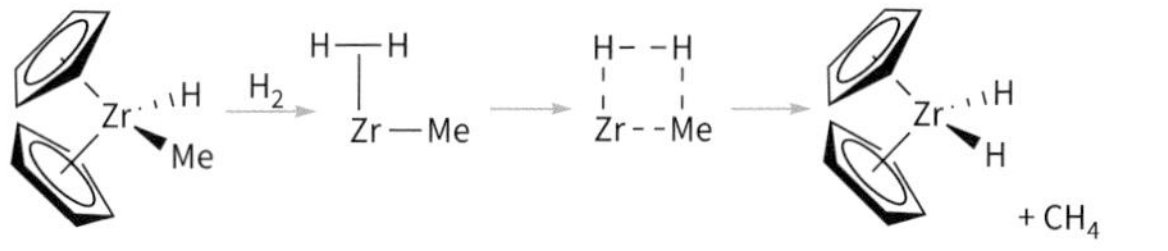

24.24 1,1-Migratory insertion reactions

KEY POINT 1,1-Migratory insertion reactions result from the migration of a species such as a hydride or alkyl group to an adjacent ligand such as carbonyl, to give a metal complex with two fewer electrons on the metal atom.

A **1,1-migratory insertion reaction** is exemplified by reactions of the η^1-CO ligand, where the following change can take place:

The reaction is called a '1,1-reaction' because the X group that was one bond away from the metal atom ends up on an atom that is one bond away from the metal atom. Typically, the X group is an alkyl or aryl species, and then the product contains an acyl group. In principle, the reaction could proceed by a migration of the X group, or an insertion of the CO into the M–X bond. An uncertainty about the actual mechanism has led to the apparently contradictory name **migratory insertion**. Colloquially, however, the terms 'migratory insertion', 'migration', and 'insertion' are used interchangeably. The overall reaction results in a decrease in the number of electrons on the metal atom by 2, with no change in the oxidation state. It is therefore possible to induce 1,1-migratory insertion reactions by the addition of another species that can act as a ligand:

$$[Mn(Me)(CO)_5] + PPh_3 \rightarrow [Mn(MeCO)(CO)_4PPh_3]$$

The classic study of the migratory insertion of CO with $[Mn(Me)(CO)_5]$ illustrates a number of key features of reactions of this type.[3] First, in the reaction

$$[Mn(Me)(CO)_5] + {}^{13}CO \rightarrow [Mn(MeCO)(CO)_4({}^{13}CO)]$$

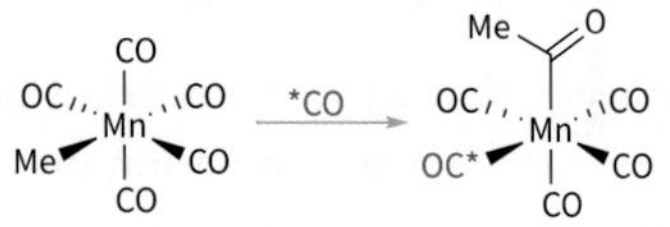

the product has only one labelled CO, and that group is *cis* to the newly formed acyl group. This stereochemistry demonstrates that the incoming CO group does not insert into the Mn–Me bond, and that either the methyl group migrates to an adjacent CO ligand, or a CO ligand adjacent to the methyl group inserts into the Mn–Me bond. Second, in the reverse reaction

$$cis-[Mn(MeCO)(CO)_4({}^{13}CO)] \rightarrow [Mn(Me)(CO)_5] + CO$$

it is possible to distinguish between the migration of the methyl group and the insertion of the CO ligand. For the reaction to proceed, cis-$[Mn(MeCO)(CO)_4({}^{13}CO)]$ must first lose one of the four CO ligands *cis* to the acyl group. There is no preference for which of the four leave: each is as statistically likely to leave as the other three. The scheme below summarizes the potential reaction pathways.

In one-quarter of the instances, the ligand lost will be the labelled CO and there will be no significant information gained. In half the instances, a nonlabelled CO will be lost, leaving a vacant site *cis* to both the labelled CO ligand and the acyl group. In this case, either (a) migration of the methyl group back to the metal atom or (b) extrusion of CO will lead to the methyl group and the ^{13}CO ligand being *cis* to each other, and no information is gained. However, in the remaining one-quarter of the instances, the CO that is *trans* to the labelled CO ligand will be lost, and in this case it is possible to distinguish (c) CO ligand extrusion from (d) methyl group migration. If the methyl group migrates, it ends up *trans* to the labelled CO, whereas if the CO is extruded, the methyl group ends up *cis* to the labelled CO.

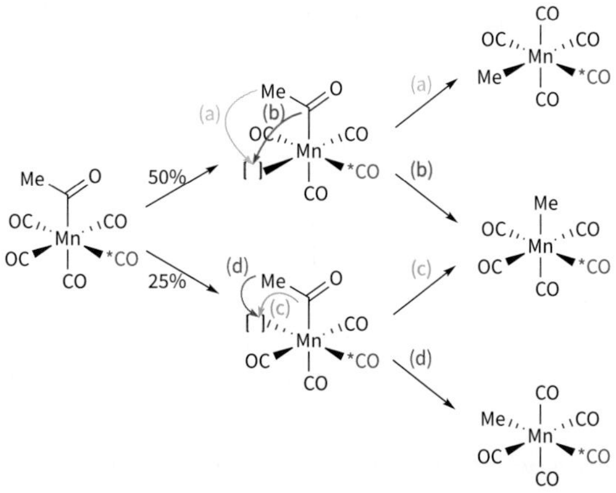

Because the product with Me and ^{13}CO *trans* to each other constitutes about 25% of the product, we can conclude that the Me group does indeed migrate. Application of the principle of microscopic reversibility[4] allows us to conclude that the forward reaction proceeds by methyl-group migration. All 1,1-migratory insertions are now thought to proceed by migration of the X group. An important consequence of this pathway is that the relative positions of the other groups on the migrating atom are left unchanged, so the stereochemistry at the X group is preserved.

---

[3] The Mn species are not fluxional and do not rearrange; if they did, we would not be able to draw the conclusions made here.

[4] The principle of microscopic reversibility states that both forward and reverse reactions proceed by the same mechanism.

24.25 1,2-Insertions and β-hydride elimination

KEY POINTS 1,2-Insertion reactions are observed with η^2-ligands such as alkenes and result in the formation of an η^1-ligand with no change in oxidation state of the metal; β-hydride elimination is the reverse of 1,2-insertion.

1,2-Insertion reactions are commonly observed with η^2-ligands, such as alkenes and alkynes, and are exemplified by the reaction:

The reaction is a **1,2-insertion** because the X group that was one bond away from the metal atom ends up on an atom that is two bonds away from the metal. Typically, the X group is a hydride, alkyl, or aryl species, in which case the product contains a (substituted) alkyl group. Like 1,1-insertion reactions, the overall reaction results in a decrease in the number of electrons on the metal atom by 2, with no change in the oxidation state.

If, in the above reaction with X = H, another ethene molecule were to coordinate, the resultant ethyl group could migrate to give a butyl group:

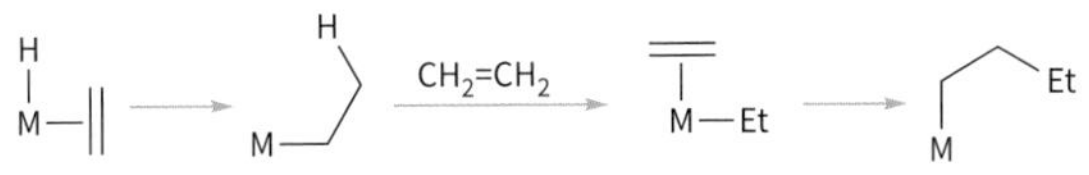

Repetition of this process gives polyethene. Catalytic reactions of this kind are of considerable industrial importance, and are discussed in Section 31.14.

The reverse of 1,2-insertions can occur, but this is rare except when X = H, when the reaction is known as β-**hydride elimination**:[5]

The experimental evidence shows that both 1,2-insertion and β-hydride elimination proceed through a syn intermediate:

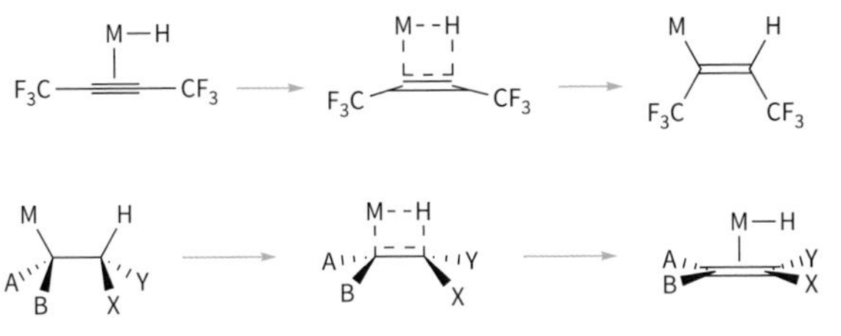

---
[5] The reaction is known as 'β-hydride elimination' because the H atom that is eliminated is on the second carbon atom from the metal atom (the carbon atom bonded to the metal atom is the α carbon, the third one the γ carbon, etc.).

As noted in Section 24.8, a β-hydride elimination reaction can provide a facile route for decomposition of alkyl-containing compounds. The 1,2-insertion reaction, coupled with the β-hydride elimination, can also provide a low-energy route to alkene isomerization:

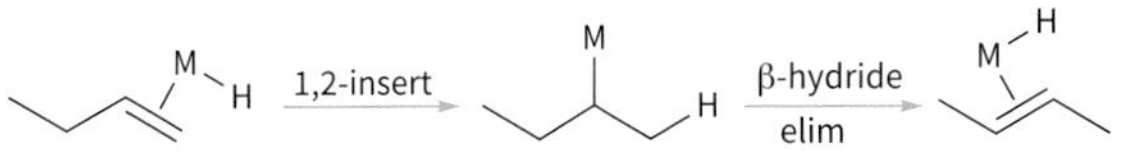

24.26 α-, γ-, and δ-Hydride eliminations and cyclometallations

KEY POINT Cyclometallation reactions, in which a metal inserts into a remote C–H bond, are equivalent to hydride-elimination reactions.

α-Hydride eliminations are occasionally found for complexes that have no β hydrogens, and the reaction gives rise to a carbene that is often highly reactive:

γ-Hydride and δ-hydride eliminations are more commonly observed. Because the product contains a **metallocycle**, a cyclic structure incorporating a metal atom, these reactions are normally described as **cyclometallation** reactions:

A cyclometallation reaction is often also thought of as the oxidative addition of a remote C–H bond. Both α- and β-hydride eliminations can also be considered as cyclometallation reactions. This identification is more obvious for the β-hydride elimination if we consider an alkene in its metallacyclopropane form:

EXAMPLE 24.13 Predicting the outcomes of insertion and elimination reactions

What product, including its stereochemistry, would you expect from the reaction between [Mn(Me)(CO)$_5$] and PPh$_3$?

Answer If we consider the reaction between [Mn(Me)(CO)$_5$] and PPh$_3$ we can see that it is unlikely to be the simple replacement of a carbonyl ligand by the phosphine, as this reaction would require a strongly bound carbonyl ligand to dissociate. A more likely reaction would be the migration of the methyl group on to an adjacent CO ligand to give an acyl group, with the phosphine filling the vacated coordination site. This reaction has a low activation barrier, and the product would therefore be expected to be cis-[Mn(MeCO)(PPh$_3$)(CO)$_4$].

Self-test 24.13 Explain why [Pt(Et)(Cl)(PEt$_3$)$_2$] readily decomposes, whereas [Pt(Me)(Cl)(PEt$_3$)$_2$] does not.

FURTHER READING

J. F. Hartwig, *Organotransition Metal Chemistry: From Bonding to Catalysis*. University Science Books (2010). The best single-volume book on the subject.

R. H. Crabtree, *The Organometallic Chemistry of the Transition Metals*, 7th ed. John Wiley & Sons (2019).

C. Elschenbroich, *Organometallics*. Wiley-VCH (2006).

R. H. Crabtree and D. M. P. Mingos (eds.), *Comprehensive Organometallic Chemistry III*. Elsevier (2006). The definitive reference work that builds on the two earlier editions.

N. C. Norman and P. G. Pringle, In defence of oxidation states. *Dalton Trans.*, 2022, **51**, 400. An article that looks into the merits of the two main methods of electron counting. See also G. Parkin, *Dalton Trans.*, 2022, **51**, 411 for additional arguments.

N. C. Norman and P. G. Pringle, The d^n number in transition metal chemistry: Its utility and limitations. *Chemistry*, 2023, **5**, 2630–56. A more in depth look at electron counting.

G. Wilkinson, The iron sandwich. A recollection of the first four months. *J. Organomet. Chem.*, 1975, **100**, 273. Wilkinson's personal account of the development of metallocene chemistry.

G. J. Kubas, Fundamentals of H_2 binding and reactivity on transition metals underlying hydrogenase function and H_2 production and storage. *Chem. Rev.*, 2007, **107**, 4152. A historical perspective and a full account of the discovery of dihydrogen complexes.

D. M. P. Mingos and D. J. Wales, *Introduction to Cluster Chemistry*. Prentice Hall (1990).

J. W. Lauher, The bonding capabilities of transition metal clusters. *J. Am. Chem. Soc.*, 1978, **100**, 5305. Descriptions of some of the concepts of cluster bonding.

R. Hoffmann, Building bridges between inorganic and organic chemistry. *Angew. Chem., Int. Ed. Engl.*, 1982, **21**, 711. The application of isolobal analogies to metal cluster compounds (Hoffmann's Nobel Prize lecture).

EXERCISES

24.1 Name the species, draw the structures of, and give valence electron counts to the metal atoms in: (a) [Fe(CO)$_5$], (b) [Mn$_2$(CO)$_{10}$], (c) [V(CO)$_6$], (d) [Fe(CO)$_4$]$^{2-}$, (e) [La(η^5-Cp*)$_3$], (f) [Fe(η^3-allyl)(CO)$_3$Cl], (g) [Fe(CO)$_4$(PEt$_3$)], (h) [Rh(Me)(CO)$_2$(PPh$_3$)$_2$], (i) [Pd(Me)(Cl)(PPh$_3$)$_2$], (j) [Co(η^5-C$_5$H$_5$)(η^4-C$_4$Ph$_4$)], (k) [Fe(η^5-C$_5$H$_5$)(CO)$_2$]$^-$, (l) [Cr(η^6-C$_6$H$_6$)(η^6-C$_7$H$_8$)], (m) [Ta(η^5-C$_5$H$_5$)$_2$Cl$_3$], (n) [Ni(η^5-C$_5$H$_5$)NO]. Do any of the complexes deviate from the 18-electron rule? If so, how is this reflected in their structure or chemical properties?

24.2 (a) Sketch an η^2 interaction of 1,3-butadiene with a metal atom and (b) do the same for an η^4 interaction.

24.3 What hapticities are possible for the interaction of each of the following ligands with a single d-block metal atom such as cobalt? (a) C_2H_4, (b) cyclopentadienyl, (c) C_6H_6, (d) cyclooctadiene, (e) cyclooctatetraene.

24.4 Draw plausible structures and give the electron count of (a) [Ni(η^3-C$_3$H$_5$)$_2$], (b) η^4-cyclobutadiene-η^5-cyclopentadienylcobalt, (c) [Co(η^3-C$_3$H$_5$)(CO)$_2$]. If the electron count deviates from 18, is the deviation explicable in terms of periodic trends?

24.5 State the two common methods for the preparation of simple metal carbonyls and illustrate your answer with chemical equations. Is the selection of method based on thermodynamic or kinetic considerations?

24.6 Suppose that you are given a series of metal tricarbonyl compounds having the respective symmetries C_{2v}, D_{3h}, and C_s. Without consulting reference material, which of these should display the greatest number of CO stretching bands in the IR spectrum? Check your answer and give the number of expected bands for each by consulting Table 24.7.

24.7 Provide plausible reasons for the differences in IR stretching frequencies between each of the following pairs: (a) [Mo(CO)$_3$(PF$_3$)$_3$] 2040, 1991 cm^{-1} versus [Mo(CO)$_3$(PMe$_3$)$_3$] 1945, 1851 cm^{-1}, (b) [Mn(Cp)(CO)$_3$] 2023, 1939 cm^{-1} versus [Mn(Cp*)(CO)$_3$] 2017, 1928 cm^{-1}.

24.8 The compound [Ni$_3$(C$_5$H$_5$)$_3$(CO)$_2$] has a single CO stretching absorption at 1761 cm^{-1}. The IR data indicate that all C$_5$H$_5$ ligands are pentahapto and probably in identical environments. (a) On the basis of these data, propose a structure. (b) Does the electron count for each metal in your structure agree with the 18-electron rule? If not, is nickel in a region of the periodic table where deviations from the 18-electron rule are common?

24.9 Decide which of the two complexes (a) [W(CO)$_6$] or (b) [Ir(CO)Cl(PPh$_3$)$_2$] should undergo the fastest exchange with ^{13}CO. Justify your answer.

24.10 Which metal carbonyl in each of (a) [Fe(CO)$_4$]$^{2-}$ or [Co(CO)$_4$]$^-$, (b) [Mn(CO)$_5$]$^-$ or [Re(CO)$_5$]$^-$ should be the most basic towards a proton? Explain your answer.

24.11 Using the 18-electron rule as a guide, indicate the probable number of carbonyl ligands in (a) [W(η^6-C$_6$H$_6$)(CO)$_n$], (b) [Rh(η^5-C$_5$H$_5$)(CO)$_n$], and (c) [Ru$_3$(CO)$_n$].

24.12 Propose two syntheses for [MnMe(CO)$_5$], both starting with [Mn$_2$(CO)$_{10}$], with one using Na and one using Br$_2$. You may use other reagents of your choice.

24.13 Give the probable structure of the product obtained when [Mo(CO)$_6$] is allowed to react first with LiPh and then with the strong carbocation reagent, CH$_3$OSO$_2$CF$_3$.

24.14 Na[W(η^5-C$_5$H$_5$)(CO)$_3$] reacts with 3-chloroprop-1-ene to give a solid, **A**, which has the molecular formula [W(C$_3$H$_5$)(C$_5$H$_5$)(CO)$_3$]. Compound **A** loses carbon monoxide on exposure to light and forms compound **B**, which has the formula [W(C$_3$H$_5$)(C$_5$H$_5$)(CO)$_2$]. Treating compound **A** with hydrogen chloride and then potassium hexafluorophosphate, K$^+$PF$_6^-$, results in the formation of a salt, **C**. Compound **C** has the molecular formula [W(C$_3$H$_6$)(C$_5$H$_5$)(CO)$_3$]PF$_6$. Use this information and the 18-electron rule to identify the compounds **A**, **B**, and **C**. Sketch a structure for each, paying particular attention to the hapticity of the hydrocarbons.

24.15 Suggest syntheses of (a) [Mo(η^7-C$_7$H$_7$)(CO)$_3$]BF$_4$ from [Mo(CO)$_6$] and (b) [Ir(COMe)(CO)(Cl)$_2$(PPh$_3$)$_2$] from [Ir(CO)Cl(PPh$_3$)$_2$].

24.16 When [Fe(CO)$_5$] is refluxed with cyclopentadiene, compound **A** is formed which has the empirical formula C$_8$H$_6$O$_3$Fe and a complicated ^{1}H-NMR spectrum. Compound **A** readily loses CO to give compound **B** with two ^{1}H-NMR resonances—one at negative chemical shift (relative intensity 1) and one at around 5 ppm (relative intensity 5). Subsequent heating of **B** results in the loss of H$_2$ and the formation of compound **C**. Compound **C** has a single ^{1}H-NMR resonance and the empirical formula C$_7$H$_5$O$_2$Fe. Compounds **A**, **B**, and **C** all have 18 valence electrons: identify them, and explain the observed spectroscopic data.

24.17 Treatment of TiCl$_4$ at low temperature with EtMgBr gives an organometallic compound that is unstable above −70°C. However, treatment of TiCl$_4$ at low temperature with MeLi or LiCH$_2$SiMe$_3$ gives organometallic compounds that are stable at room temperature. Rationalize these observations.

24.18 Treatment of TiCl$_4$ with 4 equivalents of NaCp gives a single organometallic compound (together with NaCl by-product). At room temperature, the ^{1}H-NMR spectrum shows a single sharp singlet; on cooling to −40°C this singlet separates into two singlets of equal intensity; further cooling results in one of the singlets separating into three signals with intensity ratios 1:2:2. Explain these results.

24.19 Give the equations for workable reactions that will convert [Fe(η^5-C$_5$H$_5$)$_2$] into (a) [Fe(η^5-C$_5$H$_5$)(η^5-C$_5$H$_4$COCH$_3$)] and (b) [Fe(η^5-C$_5$H$_5$)(η^5-C$_5$H$_4$CO$_2$H)].

24.20 Sketch the a_1' symmetry-adapted orbitals for the two eclipsed C$_5$H$_5$ ligands stacked together with D_{5h} symmetry. Identify the s, p, and d orbitals of a metal atom lying between the rings that may have nonzero overlap, and state how many a_1' molecular orbitals may be formed.

24.21 The compound [Ni(η^5-C$_5$H$_5$)$_2$] readily adds one molecule of HF to yield [Ni(η^5-C$_5$H$_5$)(η^4-C$_5$H$_6$)]$^+$, whereas [Fe(η^5-C$_5$H$_5$)$_2$] reacts with strong acid to yield [Fe(η^5-C$_5$H$_5$)$_2$H]$^+$. In the latter compound the H atom is attached to the Fe atom. Provide a reasonable explanation for this difference.

24.22 Write a plausible mechanism, giving your reasoning, for the reactions

(a) [Mn(CO)$_5$(CF$_2$)]$^+$ + H$_2$O → [Mn(CO)$_6$]$^+$ + 2 HF

(b) [Rh(CO)(C$_2$H$_5$)(PR$_3$)$_2$] → [RhH(CO)(PR$_3$)$_2$] + C$_2$H$_4$

24.23 Suggest two plausible routes by which a carbonyl ligand in [Mo(Cp)(CO)$_3$Me] might exchange for a phosphine. Neither route should invoke the initial dissociation of a CO.

24.24 (a) What cluster valence electron (CVE) count is characteristic of octahedral and trigonal-prismatic complexes? (b) Can these CVE values be derived from the 18-electron rule? (c) Determine the probable geometry (octahedral or trigonal-prismatic) of [Fe$_6$(C)(CO)$_{16}$]$^{2-}$ and [Co$_6$(C)(CO)$_{16}$]$^{2-}$. (In both cases, the C atom resides in the centre of the cluster and can be considered to be a four-electron donor.)

24.25 Based on isolobal analogies, choose the groups that might replace the group in bold font in

(a) [Co$_2$(CO)$_9$**CH**]: OCH$_3$, N(CH$_3$)$_2$, or SiCH$_3$

(b) [(OC)$_5$Mn**Mn**(CO)$_5$]: I, CH$_2$, or CCH$_3$

24.26 Ligand substitution reactions on metal clusters are often found to occur by associative mechanisms, and it is postulated that these occur by initial breaking of an M–M bond, thereby providing an open coordination site for the incoming ligand. If the proposed mechanism is applicable, which would you expect to undergo the fastest exchange with added ^{13}CO, [Co$_4$(CO)$_{12}$] or [Ir$_4$(CO)$_{12}$]? Suggest an explanation.

TUTORIAL PROBLEMS

24.1 Propose the structure of the product obtained by the reaction of [Re(CO)(η^5-C$_5$H$_5$)(NO)(PPh$_3$)]$^+$ with Li[HBEt$_3$]. The latter contains a strongly nucleophilic hydride. (For full details, see W. Tam, G. Y. Lin, W. K. Wong, W. A. Kiel, V. Wong, and J. A. Gladysz, *J. Am. Chem. Soc.*, 1982, **104**, 141.)

24.2 When several CO ligands are present in a metal carbonyl, an indication of the individual bond strengths can be obtained by means of force constants derived from the experimental IR frequencies. In [Cr(CO)$_5$(PPh$_3$)] the *cis*-CO ligands have the higher force constants, whereas in [Ph$_3$SnCo(CO)$_4$] the force constants are higher for the *trans*-CO. Suggest why, and explain which carbonyl C atoms should be susceptible to nucleophilic attack in these two cases. (For details, see D. J. Darensbourg and M. Y. Darensbourg, *Inorg. Chem.*, 1970, **9**, 1691.)

24.3 Discuss the following data for M–M and M–CO bond enthalpies in a series of metal carbonyl cluster compounds. The enthalpies given in each case are average values for that compound.

	B_{M-M}/kJ mol^{-1}	B_{M-C}/kJ mol^{-1}
Fe(CO)$_5$		117
Fe$_3$(CO)$_{12}$	65	126
Ru$_3$(CO)$_{12}$	78	182
Os$_3$(CO)$_{12}$	94	201
Co$_4$(CO)$_{12}$	74	140
Rh$_4$(CO)$_{12}$	86	178
Ir$_4$(CO)$_{12}$	117	196

24.4 It is often possible to assign different resonance structures to an organometallic compound; for example, there can be competition between carbene and zwitterionic forms. Suggest ways of distinguishing the two forms, and describe conditions that might favour one form over the other. (See N. Ashkenazi, A. Vigalok, S. Parthiban, Y. Ben-David, L. J. W. Shimon, J. M. L. Martin, and D. Milstein, *J. Am. Chem. Soc.*, 2000, **122**, 8797;

and C. P. Newman, G. J. Clarkson, N. W. Alcock, and J. P. Rourke, *Dalton Trans.*, 2006, 3321.)

24.5 Agostic interactions are often thought of as weak. Discuss examples where agostic interactions have been shown to displace other ligand interactions. (See B. L. Conley and T. J. Williams, *J. Am. Chem. Soc.*, 2010, **132**, 1764; and S. H. Crosby, G. J. Clarkson, R. J. Deeth, and J. P. Rourke, *Dalton Trans.*, 2011, **40**, 1227.)

24.6 Describe how alkane complexes of d metals have been unambiguously identified by NMR. What additional insights can be gained from X-ray diffraction? (See S. Geftakis and G. E. Ball, *J. Am. Chem. Soc.*, 1998, **120**, 9953; W. H. Bernskoetter, C. K. Schauer, K. I. Goldberg, and M. Brookhart, *Science*, 2009, **326**, 553; and S. D. Pike, A. L. Thompson, A. G. Algarra, D. C. Apperley, S. A. Macgregor, and A. S. Weller, *Science*, 2012, **337**, 1648.)

24.7 It is possible to distinguish two types of notionally agostic interactions: agostic and anagostic. Describe the differences between the two types. (See M. Brookhart, M. L. H. Green, and G. Parkin, *Proc. Nat. Acad. Sci. U.S.A.*, 2007, **104**, 6908.)

24.8 The dinitrogen complex $[Zr_2(\eta^5\text{-}Cp^*)_4(N_2)_3]$ has been isolated and its structure determined by single-crystal X-ray diffraction. Each Zr atom is bonded to two Cp* and one terminal N_2. The third N_2 bridges between the Zr atoms in a nearly linear ZrNNZr array. Before consulting the reference, write a plausible structure for this compound that accounts for the ^{1}H-NMR spectrum obtained on a sample held at 27°C. This spectrum shows two singlets, indicating that the Cp* rings are in two different environments. At somewhat above room temperature these rings become equivalent on the NMR timescale, and ^{15}N-NMR indicates that N_2 exchange between the terminal ligands and dissolved N_2 is correlated with the process that interconverts the Cp* ligand sites. Propose a way in which this equilibration could interconvert the sites of the Cp* ligands. (For further details, see J. M. Manriquez, D. R. McAlister, E. Rosenberg, H. M. Shiller, K. L. Willamson, S. I. Chan, and J. E. Bercaw, *J. Am. Chem. Soc.*, 1978, **100**, 3078.)

24.9 How might you unambiguously identify, in solution, an organometallic complex containing a noble gas atom as a ligand? (See G. E. Ball, T. A. Darwish, S. Geftakis, M. W. George, D. J. Lawes, P. Portius, and J. P. Rourke, *Proc. Natl Acad. Sci.*, 2005, **102**, 1853.)

24.10 What conclusions can you draw from the bonding and reactivity of dihydrogen bound to a low-oxidation-state d-block metal that might be applicable to the bonding of an alkane to a metal? What implications might this have for the oxidative addition of a hydrocarbon to a metal atom? (See, for example, R. H. Crabtree, *J. Organomet. Chem.*, 2004, **689**, 4083.)

24.11 Compare and contrast Fischer and Schrock carbenes. (See E. O. Fischer, *Adv. Organomet. Chem.*, 1976, **14**, 1; and R. R. Schrock, *Acc. Chem. Res.*, 1984, **12**, 98.)

24.12 Rearrangements of ligands so that differing numbers of carbons interact with the central metal atom are known as *haptotropic rearrangements*. Consider the (fluorenyl)(cyclopentadienyl)iron complex. What haptotropic isomers are possible? (See E. Kirillov, S. Kahlal, T. Roisnel, T. Georgelin, J. Saillard, and J. Carpentier, *Organometallics*, 2008, **27**, 387.)

24.13 It is possible to consider bridging carbonyl ligands in several different ways, with differing numbers of electrons donated to metals. Describe how it is possible to account for the structure of $[Fe_2(CO)_9]$ as having an 18-electron count at both metals, but without invoking the presence of an Fe–Fe bond. (See J. C. Green, M. L. H. Green, and G. Parkin, *Chem. Comm.*, 2012, **48**, 11481.)

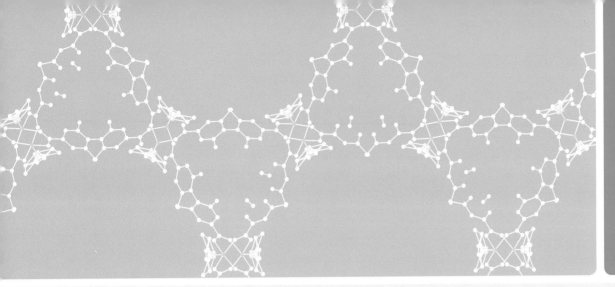

PART 3

Inorganic chemistry of complex solids and biological systems

The fundamental concepts and theories described in Part 1 of this text need to be extended for two further important areas of inorganic chemistry: materials chemistry and biological inorganic chemistry. While Chapters 1–7 and 9–10 introduced many of underlying principles needed to understand the chemistry of simple molecular and solid systems described throughout Part 2, the complexity of many materials and bioinorganic species requires a deeper understanding of additional concepts and structural chemistry. This more detailed description of complex solid materials and of the inorganic chemistry thriving within living cells is central to many important advances and applications across inorganic chemistry discussed in Part 4.

Chapter 25 introduces advanced solid state chemistry, focusing initially on the important roles of defects and dopants that occur in solid compounds with extended structures. We develop band theory to discuss doped materials and describe the role of defects in ion motion. The synthesis and structures of multicomponent materials are described—particularly complex oxides, but also halides, nitrides, and sulfides. Complex framework structures formed from linked polyhedral units are also discussed, as are molecular solid materials, including fullerides. Some of the useful electronic, magnetic, and optical properties of complex solids are described as introductory material for Part 4.

Chapter 26 discusses the natural roles of different elements in biological systems and the various and extraordinarily subtle ways in which each one is exploited. Evolution has selected the most suitable element for each application, be it transport, signalling, sensing, catalysis, or solid-state systems, and produced molecules and materials having exquisite structures and unique properties. Metalloenzymes are highly active and selective catalysts. and it is important to establish their modes of action and gain insight to help design new catalysts for technology.

Advanced solid-state chemistry: structures, compounds, and electronic properties

25

Defects and nonstoichiometry
- 25.1 **The origins and types of defects**
- 25.2 **Nonstoichiometric compounds and solid solutions**
- 25.3 **Complex alloys and interstitials**
- 25.4 **Extended defects**

The electronic structures of solids
- 25.5 **The conductivities of inorganic solids**
- 25.6 **Semiconductors: Si analogues and extrinsic semiconductors**

Ion transport in solids
- 25.7 **Atom and ion diffusion and conduction**
- 25.8 **Solid electrolytes and electrodes**

Synthesis of materials
- 25.9 **The formation of bulk materials**

Metal oxides, nitrides, and fluorides
- 25.10 **Monoxides of the 3d metals**
- 25.11 **Higher oxides and complex oxides**
- 25.12 **Oxide glasses**
- 25.13 **Nitrides, fluorides, and mixed-anion phases**

Sulfides, intercalation compounds, and metal-rich phases
- 25.14 **Metal sulfides, selenides, and tellurides**
- 25.15 **Metal hydrides and solids for hydrogen storage**

Framework structures
- 25.16 **Structures based on tetrahedral oxoanions**
- 25.17 **Structures based on linked polyhedral metal centres**

Molecular inorganic materials and fullerides
- 25.18 **Fullerides**
- 25.19 **Molecular inorganic materials chemistry**

Further Reading

Exercises

Tutorial Problems

An introduction to the inorganic chemistry of solid-state materials was presented in Chapters 2 and 4; it included the ionic model, simple band theory, and the structures of binary compounds. In this chapter we extend those fundamental concepts by describing the structures and synthesis of more complex solids and introducing defect chemistry and nonstoichiometry. The thermodynamics and structural chemistry behind the origins and roles of defects, present in all solids, are discussed, together with a description of the main types of point defect. Where high levels of defects and dopants are present in a solid these result in the formation of solid solutions or nonstoichiometric materials that exhibit variable composition. The role of defects in allowing ion motion through solids is also described. We discuss the types of solids that exhibit high ion conductivity and their resultant useful properties. The electronic structures and properties of materials are discussed further, extending band theory to cover the role of defects in controlling electronic properties. We describe how inorganic materials are synthesized as bulk solids, either through high-temperature reaction or deposition from solutions. We then focus on describing the key classes of solids, metal oxides and complex oxides, glasses, metal nitrides, and halides, and metal chalcogenides with references to their structures and properties. The compositional and structural chemistry of framework solids, formed from linked polyhedral species and covering zeolites and metal–organic frameworks, is introduced. Materials formed from interacting molecules and molecular ions are also described. This advanced solid-state chemistry provides further foundations for understanding the exciting research and developments currently taking place in materials chemistry, which we discuss in Part 4.

Defects and nonstoichiometry

KEY POINTS Defects, vacant sites, and misplaced atoms are a feature of all solids as their formation is thermodynamically favourable.

All solids contain **defects**, or imperfections of structure or composition. Defects, which occur naturally when any solid is formed during synthesis, Section 25.9, or through the deliberate introduction of dopants, are important because they influence properties such as mechanical strength, electrical conductivity, optical properties, and chemical reactivity. We need to consider both **intrinsic defects**, which are defects that occur in the pure substance, and **extrinsic defects**, which stem from the presence of impurities or dopants. It is also common to distinguish **point defects**, which occur at single sites, from **extended defects**, which are ordered in one, two, and three dimensions. Point defects are random errors in a periodic lattice, such as the absence of an atom at its usual site or the presence of an atom at a site that is not normally occupied. Extended defects involve the formation of various irregularities in the stacking of the planes of atoms.

25.1 The origins and types of defects

Solids, when synthesized, naturally contain a level of defects because they introduce disorder into an otherwise perfect structure and hence increase its entropy. The Gibbs energy, $G = H - TS$, of a solid with defects has contributions from the enthalpy and the entropy of the sample. The creation of defects is normally endothermic because the lattice is disrupted so the enthalpy of the solid increases. However, the term $-TS$ becomes more negative as defects are formed because they introduce disorder into the lattice and the entropy rises. Therefore, provided $T > 0$, the Gibbs energy will have a minimum at a nonzero concentration of defects and their formation will be spontaneous (Fig. 25.1(a)). Moreover, as the temperature of the solid is raised, the minimum in G shifts to higher defect concentrations (Fig. 25.1(b)); this means that solids have a greater number of defects at high temperatures and as their melting points are approached.

(a) Intrinsic point defects

KEY POINTS Schottky defects are site vacancies, formed in cation/anion pairs, and Frenkel defects are displaced, interstitial atoms; the structure of a solid influences the type of defect that occurs, with Frenkel defects forming in solids with lower coordination numbers and more covalency and Schottky defects forming in more ionic materials.

The solid-state physicists W. Schottky and J. Frenkel identified two specific types of point defect. A **Schottky defect** (Fig. 25.2) is a vacancy in an otherwise perfect arrangement of atoms or ions in a structure. That is, it is a point defect in which an atom or ion is missing from its normal

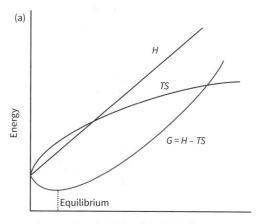

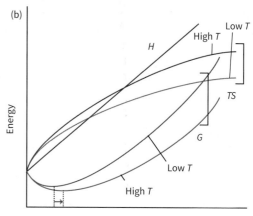

FIGURE 25.1 (a) The variation of the enthalpy and entropy of a crystal as the number of defects increases. The resulting Gibbs energy, $G = H - TS$, has a minimum at a nonzero concentration, and hence defect formation is spontaneous. (b) As the temperature is increased, the minimum in the Gibbs energy moves to higher defect concentrations, so more defects are present at equilibrium at higher temperatures than at low temperatures.

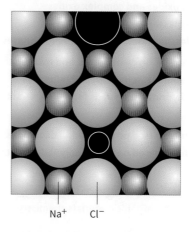

FIGURE 25.2 A Schottky defect is the absence of ions on normally occupied sites; for charge neutrality there must be equal numbers of cation and anion vacancies in a 1:1 compound.

site in the structure. The overall stoichiometry of a solid is not affected by the presence of Schottky defects because, to ensure charge balance, the defects occur in pairs; in a compound of stoichiometry MX there are equal numbers of vacancies at cation and anion sites. In solids of different composition, for example MX_2, the defects must occur with balanced charges, so two anion vacancies must be created for each cation lost. Schottky defects occur at low concentrations in purely ionic solids, such as NaCl; they occur most commonly in structures with high coordination numbers, such as close-packed ions and metals, where the enthalpy penalty of reducing the average coordination number of the remaining atoms (from 12 to 11, for instance) is relatively low.

A **Frenkel defect** (Fig. 25.3) is a point defect in which an atom or ion has been displaced onto an interstitial site. For example, in silver chloride, which has the rock-salt structure, a small number of Ag^+ ions reside in tetrahedral sites (**1**), leaving vacancies elsewhere on octahedral sites normally occupied. The stoichiometry of the compound is unchanged when a Frenkel defect forms, and it is possible to have Frenkel defects involving either one (M or X displaced) or both (some M and some X interstitials) of the ion types in a binary compound, MX. Thus, the Frenkel defects that occur in, for example, PbF_2, involve the displacement of a small number of F^- ions from their normal sites in the fluorite structure, i.e. the tetrahedral holes in the close-packed Pb^{2+} ion array, to sites that correspond to the octahedral holes. A useful generalization is that Frenkel defects are most often encountered in structures such as wurtzite and sphalerite in which coordination numbers are low (6 or less, 4:4 for these two structures) and the more open structure provides sites that can accommodate the interstitial atoms. This is not to say that Frenkel defects are exclusive to such structures; as we have seen, the 8:4-coordination fluorite structure can accommodate such interstitials although some local repositioning of adjacent anions is required to allow for the presence of the displaced anion.

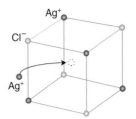

1 Interstitial Ag^+

The concentration of Schottky defects varies considerably from one type of compound to the next. The concentration of vacancies is very low in the alkali metal halides, being of the order of $10^6\,cm^{-3}$ at 130°C, corresponding to about one defect per 10^{14} formula units. Conversely, some d-metal oxides, sulfides, and hydrides have very high concentrations of vacancies. An extreme example is the high-temperature form of TiO, which has vacancies on both the cation and anion sites at a concentration corresponding to about one defect per seven formula units.

Defects, when present in large numbers, may affect the density of a solid. Significant numbers of Schottky defects, as vacancies, will lead to a decrease in density. For example, TiO, with 14% of both the anion and cation sites vacant, has a measured density of $4.96\,g\,cm^{-1}$, much less than that expected for a perfect TiO structure, $5.81\,g\,cm^{-1}$. Frenkel defects have little effect on density as they involve displaced atoms or ions, leaving the number of species in the unit cell unchanged.

EXAMPLE 25.1 Predicting defect types

What type of intrinsic defect would you expect to find in (a) MgO and (b) CdTe?

Answer The type of defect that is formed depends on factors such as the coordination numbers and the level of covalency in the bonding, with high coordination numbers and ionic bonding favouring Schottky defects and low coordination numbers, with partial covalency in the bonding, favouring Frenkel defects. (a) MgO has the rock-salt structure and the ionic bonding in this compound generally favours Schottky defects. (b) CdTe adopts the wurtzite structure with 4:4 coordination, favouring Frenkel defects.

Self-test 25.1 Predict the most likely type of intrinsic defects for (a) HgS and (b) CsF.

Schottky and Frenkel defects are only two of the many possible types of defect. Another type is an **atom-interchange** or **anti-site defect**, which consists of an interchanged pair of atoms. This type of defect is common in

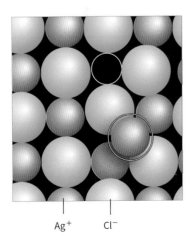

FIGURE 25.3 A Frenkel defect forms when an ion moves to an interstitial site.

metal alloys with exchange of neutral atoms. It is expected to be very unfavourable for binary ionic compounds on account of the introduction of strongly repulsive interactions between neighbouring similarly charged ions. For example, a copper/gold alloy of exact overall composition CuAu has extensive disorder at high temperatures, with a significant fraction of Cu and Au atoms interchanged (Fig. 25.4). The interchange of similarly charged species on different sites in ternary and compositionally more complex compounds is common; thus in spinels (Section 25.11(c)) the partial swapping of the metal ions between tetrahedral and octahedral sites is often observed.

(b) Extrinsic point defects

KEY POINT Extrinsic defects are defects introduced into a solid as a result of doping with an impurity atom.

Extrinsic defects—those resulting from the presence of impurities—are inevitable because perfect purity is unattainable in practice in crystals of any significant size. This behaviour is commonly seen in naturally occurring minerals. The incorporation of low levels of Cr into the Al_2O_3 structure produces the gemstone ruby, whereas replacement of some Al by both Fe and Ti results in the blue gemstone

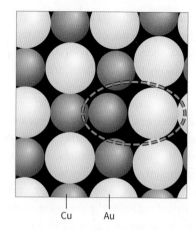

FIGURE 25.4 Atom exchange can give rise to a point defect as in CuAu.

sapphire (Box 25.1). The substituting species normally has a similar atomic or ionic radius to the species which it replaces; Cr^{3+} substituting in ruby has a similar ionic radius to Al^{3+}. Impurities can also be introduced intentionally by doping one material with another. A **dopant** is a small level, typically 0.1–5%, of an element that replaces another in a structure; an example is the introduction of As into Si to modify the latter's semiconducting properties. Synthetic

---

BOX 25.1 Why do defects and dopants give colour to gemstones?

Defects and dopant ions are responsible for the colours of many gemstones. Whereas aluminium oxide (Al_2O_3), silica (SiO_2), and fluorite (CaF_2) in their pure forms are colourless, brightly coloured materials may be produced by substituting in small levels of dopant ions or producing vacant sites that trap electrons. The impurities and defects are often produced in naturally occurring minerals on account of the geological and environmental conditions under which they are formed. For example, d-metal ions were often present in the solutions from which the gemstones grew, and the presence of ionizing radiation from radioactive species in the natural environment generated electrons that became trapped in their structure.

The most common origin of colour in a gemstone is a d-metal ion dopant (see Table B25.1). Thus, ruby is Al_2O_3 containing around 0.2–1 atom per cent Cr^{3+} ions in place of the Al^{3+} ions, and its red colour results from the absorption of green light in the visible spectrum due to the excitation of Cr 3d electrons (Section 22.2). The same ion is responsible for the green of emeralds; the different colour reflects a different local coordination environment of the dopant. The host structure is

TABLE B25.1 Gemstones and the origin of their colours

Mineral or gemstone	Colour	Parent formula	Dopant or defect responsible for the colour
Ruby	Red	Al_2O_3	Cr^{3+} replacing Al^{3+} in octahedral sites
Emerald	Green	$Be_3Al_2(SiO_3)_6$	Cr^{3+} replacing Al^{3+} in octahedral sites
Tourmaline	Green or pink	$Na_3Li_3Al_6(BO_3)_3(SiO_3)_6F_4$	Cr^{3+} or Mn^{2+} replacing Li^+ and Al^{3+} in octahedral sites, respectively
Garnet	Red	$Mg_3Al_2(SiO_4)_3$	Fe^{2+} replacing Mg^{2+} in 8-coordination sites
Peridot	Yellow-green	Mg_2SiO_4	Fe^{2+} replacing Mg^{2+} in 6-coordination sites
Sapphire	Blue	Al_2O_3	Electron transfer between Fe^{2+} and Ti^{4+} replacing Al^{3+} in adjacent octahedral sites
Diamond	Colourless, pale blue, or yellow	C	Colour centres from N
Amethyst	Purple	SiO_2	Colour centre based on Fe^{3+}/Fe^{4+}
Fluorite	Purple	CaF_2	Colour centre based on trapped electron

beryl (beryllium aluminium silicate, $Be_3Al_2(SiO_3)_6$), and the Cr^{3+} ion is surrounded by six silicate ions, rather than the six O^{2-} ions in ruby, producing absorption at a different energy compared with ruby. Other d-metal ions are responsible for the colours of other gemstones. Iron(II) produces the red of garnets and the yellow-green of peridots. Manganese(II) is responsible for the pink colour of some tourmalines.

In ruby and emerald the colour is caused by excitation of electrons on a single dopant d-metal ion, Cr^{3+}, Section 22.2. When more than one dopant species, which may be of different type or oxidation state, is present, it is possible to transfer an electron between them. One example of this behaviour is sapphire. Sapphire, like ruby, is alumina but in this gemstone some adjacent pairs of Al^{3+} ions are replaced by neighbouring Fe^{2+} and Ti^{4+}. This material absorbs visible radiation of a wavelength corresponding to yellow light as an electron is transferred from Fe^{2+} to Ti^{4+}, so producing a brilliant blue colour (the complementary colour of yellow).

In other gemstones and minerals, their colour is a result of doping a host structure with a species that has a different charge from the ion that it replaces or by the presence of a vacancy (Schottky-type defect). In both cases, a colour-centre or F-centre (F from the German word *farbe* for colour) can be formed. As the charge at an F-centre (a dopant or a vacancy) is different from that of a normally occupied site in the same structure, it can supply an electron to, or receive an electron from, another ion through a low-energy process. This electron can then be excited by absorbing visible light, so producing colour. For instance, in purple fluorite, CaF_2, an F-centre is formed from a vacancy (Schottky-type defect) on a normally occupied F^- ion site. This site can then trap an electron—often generated by exposure of the mineral to ionizing radiation in the natural environment. Excitation of the trapped electron, which acts like a particle in a box, absorbs visible light in the wavelength range 530–600 nm, producing the common violet/purple colour of this mineral.

In amethyst, the purple derivative of quartz, SiO_2, some Si^{4+} ions are substituted by Fe^{3+} ions. This replacement leaves a hole (one missing electron), and excitation of this hole, by ionizing radiation for instance, traps it by forming Fe^{4+} or O^- in the quartz structure. Further excitation of the electrons associated with Fe^{4+} or O^- in this material now occurs by the absorption of visible light at 540 nm, producing the observed purple colour. If an amethyst crystal is heated to 450°C, the hole is freed from its trap. The colour of the crystal reverts to that typical of iron-doped silica and is a characteristic of the yellow semiprecious gemstone citrine. If citrine is irradiated the trapped-hole is regenerated and the original amethyst colour restored.

Colour centres can also be produced by nuclear transformations. An example of such a transformation is the β-decay of ^{14}C in diamond. This decay produces a ^{14}N atom, with an additional valence electron, embedded in the diamond structure. The electron energy levels associated with these N atoms allow absorption in the visible region of the spectrum and produce the colouration of blue and yellow diamonds.

equivalents of ruby and sapphire can also be synthesized easily in the laboratory by doping small levels of Cr, or Fe and Ti, into the Al_2O_3 structure, replacing aluminium.

When the dopant species is introduced into the host the latter's structure remains essentially unchanged. If attempts are made to introduce high levels of the dopant species, a new structure often forms or the dopant species is not incorporated. This behaviour usually limits the level of extrinsic point defects to low levels. The composition of ruby is typically $(Al_{0.998}Cr_{0.002})_2O_3$, with 0.2% of metal sites as extrinsic Cr^{3+} dopant ions. Some solids may tolerate much higher levels of defects (Section 25.1(a)). Dopants often modify the electronic structure of the solid. Thus, when an As atom replaces an Si atom, the additional electron from each As atom can be thermally promoted into the conduction band, improving the overall conductivity of the extrinsic semiconductor, Section 25.6(b). In the more ionic substance ZrO_2, the introduction of Ca^{2+} dopant ions in place of Zr^{4+} ions is accompanied by the formation of an O^{2-} ion vacancy to maintain charge neutrality (Fig. 25.5). The induced vacancies allow oxide ions to migrate through the structure, increasing the ionic conductivity of the solid.

Another example of an extrinsic point defect is a **colour centre or F-centre**, Fig. 25.6 and Box 25.1. One type of colour centre can be produced by heating an alkali metal

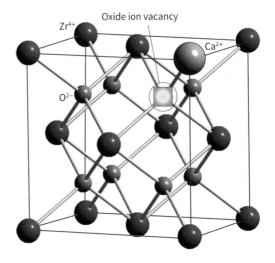

FIGURE 25.5 Introduction of a Ca^{2+} ion into the ZrO_2 lattice produces a vacancy on the O^{2-} sublattice. This substitution helps to stabilize the cubic fluorite structure for ZrO_2.

halide crystal in the vapour of the alkali metal; this gives a material with a colour characteristic of the system: NaCl becomes orange, KCl violet, and KBr blue-green. The process results in the introduction of an alkali metal cation at a normal cation site and the associated electron from the metal atom occupies a halide ion vacancy, generating the colour centre, Fig. 25.6. The characteristic colour results

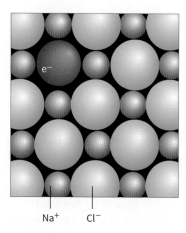

FIGURE 25.6 An F-centre is an electron that occupies an anion vacancy. The energy levels of the electron resemble those of a particle in a three-dimensional square well.

from the excitation of the electron in the localized environment of its surrounding ions. As the size of the cations and anions changes in the series NaCl, KCl, and KBr, the dimensions of the vacancy in which the trapped electron resides increase and the energy required to excite the electron decreases producing the different colours observed.

An alternative method of producing F-centres involves exposing a material to an X-ray beam that ionizes electrons into anion vacancies. F-centres and extrinsic defects are important in producing colour in gemstones (Box 25.1).

EXAMPLE 25.2 Predicting possible dopant ions

What transition metal ions might substitute for Al^{3+} in beryl, $Be_3Al_2(SiO_3)_6$, forming extrinsic defects?

Answer We need to identify ions of similar charge and size. Ionic radii are listed in Resource section 1. Triply charged cations with ionic radii similar to Al^{3+} ($r = 53\,pm$) should prove to be suitable dopant ions. Candidates could be Fe^{3+} ($r = 55\,pm$), Mn^{3+} ($r = 65\,pm$), and Cr^{3+} ($r = 62\,pm$). Indeed, when the extrinsic defect is Cr^{3+} the material is a bright green beryl, the gemstone emerald. For Mn^{3+} the material is a red or pink beryl, sometimes also known as morganite, and for Fe^{3+} it is the yellow beryl heliodor.

Self-test 25.2 What elements other than As might be used to form extrinsic defects in silicon?

25.2 Nonstoichiometric compounds and solid solutions

The statement that the stoichiometry of a compound is fixed by its chemical formula is not always true for solids, as differences in the content of unit cells can occur throughout a solid. Changes in unit cell composition can arise through vacancies at one or more atom sites, the presence of interstitial atoms, or the substitution of one atom type by another.

(a) Nonstoichiometry

KEY POINT Deviations from ideal stoichiometry are common in the solid-state compounds of the d-, f-, and heavier p-block elements.

A **nonstoichiometric compound** is a substance that exhibits variable composition but retains the same structure type. For example, at 1000°C the composition of 'iron monoxide', which is sometimes referred to as wüstite, $Fe_{1-x}O$, varies from $Fe_{0.89}O$ to $Fe_{0.96}O$. Gradual changes in the size of the unit cell occur as the composition is varied, but all the features of the rock-salt structure are retained throughout this composition range. The fact that the lattice parameter of the compound varies smoothly with composition is a defining criterion of a nonstoichiometric compound because a discontinuity in the value of the lattice parameter indicates the formation of a new crystal phase. Moreover, the thermodynamic properties of nonstoichiometric compounds also vary continuously as the composition changes. For example, as the partial pressure of oxygen above a metal oxide is varied, both the lattice parameter and the equilibrium composition of the oxide change continuously (Figs. 25.7 and 25.8). The gradual change in the lattice parameter of a solid as a function of its composition is known as **Vegard's rule**.

Table 25.1 lists some representative nonstoichiometric hydrides, oxides, and sulfides. Note that as the formation of a nonstoichiometric compound requires overall changes in composition, it also requires at least one element to exist in more than one oxidation state. Thus, in wüstite, $Fe_{1-x}O$, as x increases some iron(II) must be oxidized to iron(III) in the structure. Hence deviations from stoichiometry are usual only for d- and f-block elements, which commonly adopt

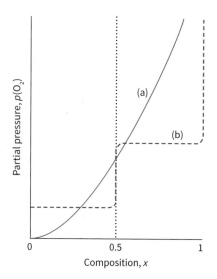

FIGURE 25.7 Schematic representation of the variation of the partial pressure of oxygen with composition at constant pressure for (a) a nonstoichiometric oxide MO_{1+x} and (b) a stoichiometric pair of metal oxides MO and MO_2. The x-axis is the atom ratio in MO_{1+x}.

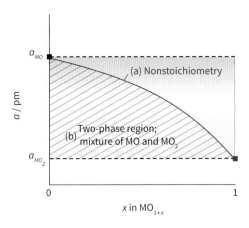

FIGURE 25.8 Schematic representation of the variation of one lattice parameter with composition for (a) a nonstoichiometric oxide MO_{1+x}, and (b) a stoichiometric pair of metal oxides MO and MO_2 with no intermediate stoichiometric phases (which would produce a two-phase mixture for $0 < x < 1$, each phase in the mixture having the lattice parameter of the end member).

two or more oxidation states, and for some heavy p-block metals that have two accessible oxidation states through the inert pair effect, Section 10.7.

(b) Solid solutions in compounds

KEY POINT *A solid solution occurs where there is a continuous variation in compound stoichiometry without a change in structural type. This behaviour can occur in many ionic solids such as metal oxides.*

Because many substances adopt the same structural type, it is often energetically feasible to replace one type of atom or ion with another. This behaviour is seen in many simple metal alloys such as those discussed in detail in Section 25.3. Thus zinc/copper brasses exist for the complete range of compositions $Cu_{1-x}Zn_x$ with $0 < x < 0.38$, where Cu atoms in the structure are gradually replaced by Zn atoms. This replacement occurs randomly throughout the solid, and individual unit cells contain an arbitrary number of Cu and Zn atoms (but such that the sum of their contents gives the overall brass stoichiometry). The continuous variation in composition without a change of structure type is termed a **solid solution**, as one element is effectively 'dissolved' into the structure of the other.

As well as some alloys, solid solutions are also found for compounds, including oxides, sulfides, and halides. Good examples are found for the perovskite structure adopted by many compounds of stoichiometry ABX_3 (Section 25.11(d)), composed of the ions A^{n+}, B^{m+}, and X^{x-}, in which the composition can be varied continuously by varying the ions that occupy some or all of the A, B, and X sites. For instance, both $LaFe(III)O_3$ and $SrFe(IV)O_3$ adopt the perovskite structure, and we can consider a perovskite crystal that has, randomly distributed, half $SrFeO_3$ unit cells (with Sr^{2+} on the A-type cation site) and half $LaFeO_3$ unit cells (with La^{3+} on the A-site). The overall compound stoichiometry is $LaFeO_3 + SrFeO_3 = LaSrFe_2O_6$, which is better written $(La_{0.5}Sr_{0.5})FeO_3$, to reflect the normal ABO_3 perovskite stoichiometry.

Other proportions of these unit cells are possible, and the series of compounds $La_{1-x}Sr_xFeO_3$ for $0 \leq x \leq 1$ can be prepared. This system is also called a solid solution because all the phases formed as x is varied have the same perovskite structure. In a solid solution, all the sites in the structure remain fully occupied, the overall compound stoichiometry remains constant (albeit with different proportions of atom types on some sites), and there is a smooth variation in lattice parameter over its composition range.

Solid solutions occur most frequently for d-metal compounds because the change in one component might require a change in the oxidation state of another component

TABLE 25.1 Representative composition ranges* of nonstoichiometric binary hydrides, oxides, and sulfides

d Block				f Block	
Hydrides					
				Fluorite type	Hexagonal
TiH_x	1–2				
ZrH_x	1.5–1.6	GdH_x		1.9–2.3	2.85–3.0
HfH_x	1.7–1.8	ErH_x		1.95–2.31	2.82–3.0
NbH_x	0.64–1.0	LuH_x		1.85–2.23	1.74–3.0
Oxides					
	Rock-salt type	Rutile type			
TiO_x	0.7–1.25	1.9–2.0			
VO_x	0.9–1.20	1.8–2.0			
NbO_x	0.9–1.04				
Sulfides					
ZrS_x	0.9–1.0				
YS_x	0.9–1.0				

* Expressed as the range of values x may take.

to preserve the charge balance. Thus, as x increases in $La_{1-x}Sr_xFeO_3$ and La(III) is replaced by Sr(II), the oxidation state of iron must change from Fe(III) to Fe(IV). This change can occur through a gradual replacement of one exact oxidation state, here Fe(III), by another, Fe(IV), on a proportion of the cation sites within the structure. Some other solid solutions include the high-temperature superconductors of composition $La_{2-x}Ba_xCuO_4$ ($0 \leq x \leq 0.4$), which are superconducting for $0.12 \leq x \leq 0.25$, and the spinels $Mn_{1-x}Fe_{2+x}O_4$ ($0 \leq x \leq 1$) and $Li_{1+x}Mn_{2-x}O_4$ ($0 \leq x \leq 0.10$), Section 25.11(c). It is also possible to combine solid-solution behaviour on a cation site with nonstoichiometry caused by defects on a different ion site. An example is the system $La_{1-x}Sr_xFeO_{3-y}$, with $0 \leq x \leq 1.0$ and $0.0 \leq y \leq 0.5$, which has vacancies on the O^{2-} ion sites coupled with a solid solution from the variation in the occupancy of the La/Sr site.

25.3 Complex alloys and interstitials

The chemistry and structures of simple alloys such as brass have been introduced in Section 4.8 together with a basic discussion of interstitial solid solutions and carbon steels. In this Section we extend this discussion of **substitutional alloys** and **interstitial solid solutions**. Fig. 25.9 illustrates the arrangement of atoms in these types of structure.

(a) Substitutional alloys

KEY POINT A substitutional solid solution or alloy involves the replacement of one type of metal atom in a structure by another.

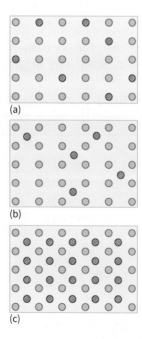

FIGURE 25.9 (a) Substitutional and (b) interstitial alloys. (c) A regular arrangement of interstitial atoms can lead to a new structure.

Substitutional solid solutions are generally formed if three criteria are fulfilled:

- The metallic radii (Resource section 2) of the elements are within about 15% of each other.
- The crystal structures of the two pure metals are the same; this similarity indicates that the directional forces between the two types of atom are compatible with each other.
- The electropositive characters of the two components are similar; otherwise compound formation, where electrons are transferred between species, would be more likely.

Thus, although sodium and potassium are chemically similar and have bcc structures, the metallic radius of Na (180 pm) is 22% smaller than that of K (230 pm), and the two metals do not form a solid solution. Copper and nickel, however, two neighbours late in the d block, have similar electropositive character, similar crystal structures (both ccp), and identical metallic radii (Ni 126 pm, Cu 126 pm, only 2.3% different), and form a continuous series of solid solutions, ranging from pure nickel to pure copper. Zinc, copper's other neighbour in Period 4, has a similar metallic radius (138 pm, 9% larger), but it is hcp, not ccp. In this instance, zinc forms a solid solution with copper for low levels of zinc, forming phases known as 'α-brasses' with the composition $Cu_{1-x}Zn_x$, $0 \leq x \leq 0.38$, and the same structural type as pure copper (ccp). Where alloys adopt a different crystal structure from that of the pure metal, these materials are usually referred to as intermetallics (Section 25.3(c)).

(b) Interstitial atoms in metals

KEY POINT In an interstitial solid solution, additional small atoms occupy holes within the lattice of the original metal structure.

As discussed in Section 4.8, interstitial solid solutions are often formed between metals and small atoms (such as boron, carbon, or nitrogen) that can inhabit the interstices, normally the octahedral or tetrahedral holes, in the structure. The small atoms enter the host solid with preservation of the crystal structure of the original metal and without the transfer of electrons and formation of ionic species. There is either a simple whole-number ratio of metal and interstitial atoms (as in tungsten carbide, WC) or the small atoms are distributed randomly in the available spaces or holes in the structure between the packed atoms. The former substances are true compounds and the latter can be considered as interstitial solid solutions or, on account of the variation in the atomic ratio of the two elements, so-called nonstoichiometric compounds (Section 25.2(a)).

Considerations of size can help to decide where the formation of an interstitial solid solution is likely to occur.

Thus, the largest solute atom that can enter a close-packed solid without distorting the structure appreciably is one that just fits an octahedral hole, which, as we have seen, has a radius of $0.414r$, where r is the radius of the close-packed atom. For small atoms, such as B, C, or N, the atomic radii of the possible host metal atom structures include those of the d metals such as Fe, Co, and Ni, and the carbon steels (Box 4.2).

(c) Intermetallic compounds and Zintl phases

KEY POINT Intermetallic compounds are alloys in which the structure adopted is different from the structures of either component metal. Zintl phases have characteristics of metal alloys and ionic compounds.

When some liquid mixtures of metals are cooled, they form phases with definite structures that are often unrelated to the parent structure. These phases are called **intermetallic compounds**. Examples include β-brass (CuZn) and compounds of the compositions $MgZn_2$, Cu_3Au, $NaTl$, and Na_5Zn_{21}. β-Brass with the composition $Cu_{0.52}Zn_{0.48}$ adopts a bcc structure type at high temperature and a hcp structure at lower temperatures.

Particular intermetallic alloy structures are found for specific values of the electron-per-atom ratio (e/a) and a series of rules, proposed by **Hume-Rothery**, allow the prediction of the most stable structure (normally bcc, hcp, or fcc) for a chosen alloy composition; where e/a is the number of electrons contributed to the conduction band per atom. These rules, which require a detailed understanding of band structure (Section 2.11), predict that a Cu:Zn alloy with e/a < 1.4 (α-brass, $Cu_{1-x}Zn_x$ with $0 \leq x \leq 0.38$) should have a ccp lattice, and an alloy with e/a of 1.5 (CuZn, β-brass) a bcc lattice. These rules count pure copper (with the configuration $3d^{10}4s^1$) as contributing one electron and zinc ($3d^{10}4s^2$) as contributing two electrons, so the alloys between pure copper and pure zinc span e/a values from 1 to 2. On this

---

BOX 25.2 What are quasicrystals?

Most crystalline solids have structures with long-range periodic order that can be described using a unit cell that repeats at regular intervals (Section 4.1). For unit cells to pack and fill three-dimensional space fully there are restrictions on the symmetry of the unit cell: two-, three-, four-, and six-fold rotational symmetry is allowed, but five-, seven-, and all higher-fold symmetry axes are forbidden. (The essential symmetries for the seven different crystal systems were given in Table 4.1.) The reasons for these symmetry restrictions can be readily visualized: tiling two-dimensional space, so that there is no space between tiles, is possible with square and hexagonal tiles (four- and six-fold rotational symmetry, respectively) but impossible with pentagonal tiles (five-fold symmetry).

Quasicrystals (quasiperiodic crystals) have a more complex long-range order. In a quasiperiodic structure, the atomic positions along each direction in the crystal are repeated at an irrational number of units (rather than integer values as in the repeat of unit cells of crystalline materials). This difference exempts quasicrystals from the crystallographic restrictions, and they can exhibit rotational symmetries forbidden to crystals, including five-fold symmetry. Unlike with normal crystals, the pattern never repeats itself exactly; one example is shown in Fig. B25.2. Despite this, quasicrystals are well-ordered materials, often intermetallics, that produce sharp diffraction patterns (Section 8.1). The concept of quasicrystals was introduced in 1984 following the study of a rapidly quenched alloy of 86% Al and 14% Mn that showed icosahedral symmetry; Dan Shechtman was awarded the Nobel Prize in Chemistry in 2011 for this work. In the last 35 years, hundreds of different quasicrystalline systems have been identified, including

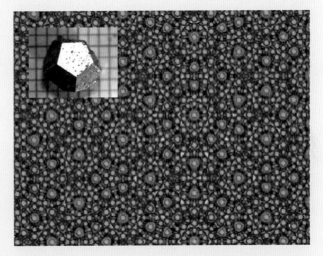

FIGURE B25.2 Computer-generated representation of a quasicrystal showing five-fold symmetry. Inset: a photograph of a dodecahedral quasicrystal.

complex compositions and structures such as the decagonal $Al_{70.5}Mn_{16.5}Pd_{13}$.

Quasicrystal materials tend to be hard but are relatively poor conductors of heat (as regular periodic arrays of atoms, such as occurs in diamond and graphene, produce the highest rates of thermal energy conduction) and electricity. These properties make them useful as coatings for frying pans and as insulating materials for electrical wires. They are also used in the most durable steels, in razor blades, and in ultrafine needles for eye surgery.

basis the β-brass composition close to $Cu_{0.5}Zn_{0.5}$ (e/a = 1.5) is predicted to have a bcc structure, which, as noted above, it does adopt at high temperatures.

Some alloys, including γ-brass in the Cu/Zn system of the stoichiometry $Cu_{0.39}Zn_{0.61}$, form very complex structures due to the arrangements of copper and zinc atoms and some vacant sites; the unit cell of γ-brass is 27 times the size of that of β-brass. Other alloys such as $Al_{0.88}Mn_{0.12}$ form structures which include elements of five-fold symmetry and therefore do not repeat perfectly when translated as a unit cell. These materials are known as quasicrystals (Box 25.2).

Intermetallic compounds are normally high-melting, hard, and more brittle than most metals and alloys. Examples include alnico, A_3B niobium–tin and niobium–germanium superconducting compounds, 'A15 systems' such as $LaNi_5$ (see Box 25.3), which can be used as hydrogen storage materials, NiAl and Ni_3Al superalloys, and titanium nickel shape-memory alloys. For more information,

BOX 25.3 How creating intermetallics leads to useful properties

Intermetallics with the acronym '**alnico**' consist mainly of iron with Al, Ni, and Co, and sometimes small levels of C and Ti. Alnico alloys are ferromagnets (Section 22.8) that have a very good resistance to loss of magnetism (a high coercivity), even at high temperatures, and as a result they have widespread applications as permanent magnets. Before the development of rare-earth magnets in the 1970s (Section 30.6), they were the strongest available magnets. One such intermetallic, alnico-500, contains 50% Fe with the remainder as 24% cobalt, 14% nickel, 8% aluminium, 3% copper, and 0.45% niobium. The structures of these alnico alloys are essentially the simple bcc unit cell with a random distribution of the various metal atom components (Section 4.8), but at the microscopic level the structure is more complex, with small portions of the crystals (domains) being richer in one component and poorer in another.

The so-called A15 phases are a series of intermetallic compounds of the composition A_3B (where A is a transition metal and B can be a transition metal or a Group 13 or 14 element). The structure is that shown in Fig. B25.3 and consists of a cube with a B-type atom at the corners and at the body centre, and two A atoms in each face (giving the overall unit cell stoichiometry $2 \times B + 6 \times 2 \times ½ \times A = A_6B_2$ or A_3B). Intermetallics from this family include Nb_3Ge, which is superconducting below 23.2 K—the highest value known until the discovery of the cuprate superconductors in 1986 (Section 30.7).

Intermetallic compounds of the composition AB_5 (A = lanthanoid, alkaline earth, or transition element; B = d- or p-block element)—in particular, those crystallizing with a hexagonal unit cell—are researched for various technological applications. $LaNi_5$ absorbs up to six hydrogen atoms per formula unit under 1.5 atm of $H_2(g)$ at 300 K, making this phase of interest for hydrogen-storage applications (Section 30.8). The hydrogen is released on heating to 350 K. The relatively low proportion of hydrogen in $LaNi_5H_6$, 1.4 weight per cent, probably makes this material unsuitable for transport applications, but further alloying with magnesium improves this value.

Several intermetallic phases show the properties of a **shape-memory alloy (SMA)**, of which Cu–Al–Ni (Cu_3Al with a small level of nickel) and **Nitinol** (**Ni**ckel **Ti**tanium **Na**val **O**rdnance **L**aboratory, NiTi) are perhaps the most important. Nitinol can undergo a phase change between two forms known as martensite (with a tetragonal unit cell) and austenite (face-centred cubic); see also Box 4.2.

In the martensite form Nitinol can be bent into various shapes, but when this form is heated it converts to a rigid austenite phase. On cooling the austenite phase, the SMA converts back to the martensite form but with a 'memory' of the shape of the high-temperature austenite form. It will convert back to this martensite form if heated above a specific temperature, the transition temperature, M_s. By varying the Ni:Ti ratio, M_s can be adjusted between −100°C and +150°C. Thus, if a straight wire of Nitinol, with M_s = 50°C and previously heated to 500°C, is bent into a complex shape at room temperature, it will retain that shape indefinitely under ambient conditions; however, on heating to over 50°C it will spring back to its original straight form. This cycle can be repeated millions of times.

SMAs are used in some actuators, where a material that changes shape, stiffness, or position in response to temperature is needed. Applications have included variable-geometry parts in aircraft engines to reduce noise as their temperature increases, brackets and wires in dental surgery, resilient frames for glasses, and expandable coronary stents. A collapsible stent can be inserted into a vein, and when heated it will return to its original expanded shape, thus improving the blood flow.

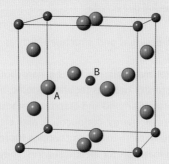

FIGURE B25.3 The structure of the A15 intermetallic phases.

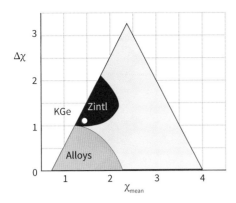

FIGURE 25.10 The approximate locations of Zintl phases in a Ketelaar triangle. The point marks the location of one exemplar, KGe.

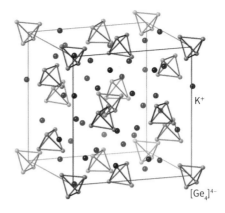

FIGURE 25.11 The structure of the Zintl phase KGe showing the $[Ge_4]^{4-}$ tetrahedral units and interspersed K ions.

see Box 25.3. Some of these intermetallic compounds contain a very electropositive metal (e.g. K or Ba) in combination with a less electropositive metal or metalloid (e.g. Ge or Zn), and in a Ketelaar triangle they lie above the true alloys (Fig. 25.10). Such combinations are called **Zintl phases** and are stoichiometric materials.

Zintl phases are not fully ionic, though they are often brittle. They exhibit some metallic properties, including lustre, though their electrical conductivity is poor. They are often composed of the combination of an electropositive Group 1 or 2 metal, such as Cs^+ or Ba^{2+}, and a complex metal anion formed from a post-transition metal from Groups 13–16; for example, $[Tl_4]^{8-}$. A classic example of a Zintl phase is KGe (K_4Ge_4), with the structure shown in Fig. 25.11, consisting of K^+ cations and $[Ge_4]^{4-}$ tetrahedral anions.

The complex metal anion in a Zintl phase can exist as discrete anions, through one- and two-dimensional structural units to a negatively charged three-dimensional framework. Thus, $BaSi_2$ contains discrete tetrahedral $[Si_4]^{4-}$ units, $CaSi_2$ has infinite 2D puckered sheets of the stoichiometry $[Si_2]^{2-}$, and $SrSi_2$ consists of a 3D network, also of stoichiometry $[Si_2]^{2-}$, in which Sr^{2+} cations reside; in all these structures

each Si atom is bonded to three other Si atoms, Fig. 25.12. Other compounds of this class include Ba_3Si_4, KTl, and $Ca_{14}Si_{25}$. Zintl phases are being actively researched due to the excellent thermoelectric properties exhibited by some materials of this type, Section 30.4.

25.4 Extended defects

KEY POINT Wadsley defects are shear planes that collect defects along certain crystallographic directions.

The incorporation of point defects into a solid entails a significant local distortion of the structure and in some instances localized charge imbalances too; this leads to high enthalpies of formation of the defects, Fig. 25.1. Therefore, it should not be surprising that defects may cluster together significantly, reducing their average enthalpy of formation. Frequently high levels of defects group together to form lines and planes known as **extended defects**.

Tungsten oxides illustrate the formation of extended defects as planes. As illustrated in Fig. 25.13, the idealized structure of WO_3 (which is usually referred to as the 'ReO$_3$

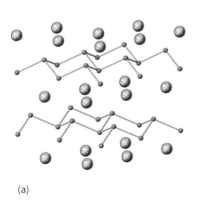
(a)

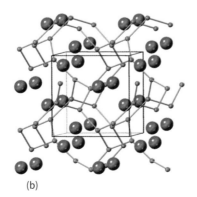
(b)

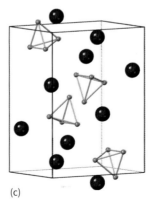
(c)

FIGURE 25.12 The structures of (a) $CaSi_2$, (b) $SrSi_2$, and (c) $BaSi_2$; silicon is shown as a small orange sphere and the metal cation as a larger grey sphere.

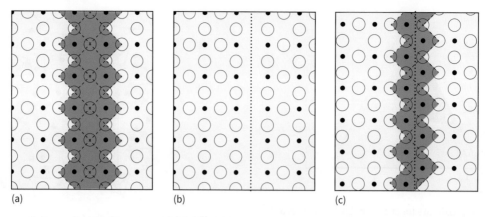

FIGURE 25.13 The concept of a crystallographic shear plane, illustrated by the (100) plane of the ReO_3 structure. (a) A plane of metal (Re = dots) and oxygen (O) atoms. The octahedron around each metal atom is completed by a plane of oxygen atoms above and below the plane illustrated here. Some of the octahedra are shaded to clarify the processes that follow. (b) Oxygen atoms in the plane perpendicular to the page are removed, leaving two planes of metal atoms that lack their sixth oxygen ligand. (c) The octahedral coordination of the two planes of metal atoms is restored by translating the right slab as shown. This creates a plane (labelled a shear plane), vertical on the page, in which the MO_6 octahedra share edges.
Adapted from S. A. Barnett and A. Madan, Superhard superlattices. *Physics World*, 1998, **11**(1), 45–8.

structure'; see Section 25.11(b)) consists of WO_6 octahedra sharing all vertices. To picture the formation of the defect plane, we imagine the removal of shared O atoms along a diagonal. Then, adjacent slabs slip past each other in a motion that results in the completion of the vacant coordination sites around each W atom. This shearing motion creates edge-shared octahedra along a diagonal, replacing the purely corner-sharing octahedra in the ReO_3 structure and increasing the oxygen coordination number to compensate for the lower overall oxygen stoichiometry. The resulting structure was named a **crystallographic shear plane** by A. D. Wadsley, who first devised this way of describing extended planar defects.

Crystallographic shear planes randomly distributed in the solid are called **Wadsley defects**. Such defects lead to a continuous range of compositions, as in tungsten oxide (made by heating and reducing WO_3 with tungsten metal), which ranges from WO_3 to $WO_{2.93}$. If, however, the crystallographic shear planes are distributed in a nonrandom, periodic manner, so giving rise to a new unit cell, then we should regard the material as a new stoichiometric phase. Thus, when even more O^{2-} ions are removed from tungsten oxide, a series of discrete phases having ordered crystallographic shear planes and compositions W_nO_{3n-2} (n = 20, 24, 25, and 40) are observed. Compounds with closely spaced compositions that contain shear planes are known for oxides of W, Mo, Ti, V, and some of their complex oxides; for example, the tungsten bronzes $M_8W_9O_{47}$ (M = Nb, Ta), and the 'Magnéli phases', V_nO_{2n-1} (n = 3–9). Electron microscopy (Section 8.17) provides an excellent method of observing these extended defects experimentally, because it reveals both ordered and random arrays of shear planes (Fig. 25.14).

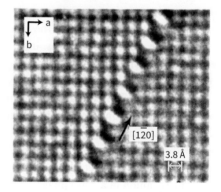

FIGURE 25.14 (a) High-resolution electron micrograph lattice image of a crystallographic shear plane in WO_{3-x}. (b) The oxygen octahedral polyhedra that surround the W atoms imaged in the electron micrograph; note the edge-shared octahedra along the crystallographic shear plane. (Reproduced by permission from S. Iijima et al., *J. Solid State Chem.*, 1975, **14**, 52.)

The electronic structures of solids

Chapter 4 and the previous sections in this chapter have introduced concepts associated with the structures, energetics, and defects of ionic solids where it was necessary to consider almost infinite arrays of ions and the interactions between them. Similarly, an understanding of the electronic structures of solids, and the derived properties, including electronic conductivity, magnetism, and many optical effects, needs to consider the interactions of electrons with each other and extended arrays of atoms or ions. Chapter 2 introduced a simple theoretical approach to the electronic properties of solids, band theory, extending the ideas of molecular orbital (MO) theory, and treating solids as a single huge molecule with very large numbers of orbitals. Band theory of simple compounds is extended in this section to cover the electronic structures of compositionally more complex solids and those containing dopants. Similar concepts are used later in this chapter and elsewhere to understand other key properties of large three-dimensional arrays of electronically interacting centres including ferromagnetism (Section 22.8), superconductivity (Section 30.7), and the colours of solids (Section 30.1).

25.5 The conductivities of inorganic solids

KEY POINTS A metallic conductor is a substance with an electric conductivity that decreases with increasing temperature; a semiconductor is a substance with an electric conductivity that increases with increasing temperature.

In Chapter 2, Section 2.11, the MO theory of small molecules was extended to account for the properties of solids, which are aggregations of an almost infinite number of atoms, by considering the energy levels available to electrons where many atomic orbitals overlap. This approach is strikingly successful for the description of metals; it can be used to explain their characteristic lustre, their good electrical and thermal conductivity, and their malleability. All these properties arise from the ability of the atoms to contribute electrons to an almost infinite array of energy levels, sometimes referred to as a 'sea of electrons'. The lustre and electrical conductivity stem from the mobility of these electrons in response to either the oscillating electric field of an incident ray of light, or to a potential difference. The high thermal conductivity is also a consequence of electron mobility because an electron can collide with a vibrating atom, pick up its energy, and transfer it to another atom elsewhere in the solid. The ease with which metals can be mechanically deformed is another aspect of electron mobility, because the electron 'sea' can quickly readjust to a deformation of the solid and continue to bind the atoms together.

Electronic conduction is also a characteristic of semiconductors. The criterion for distinguishing between a metallic conductor and a semiconductor is the temperature dependence of the electric conductivity (Fig. 25.15):

- A **metallic conductor** is a substance with a high electric conductivity, $10^4 – 10^8\,\text{S m}^{-1}$, that *decreases* with increasing temperature.
- A **semiconductor** is a substance with a relatively low electric conductivity, $10^{-6} – 10^2\,\text{S m}^{-1}$, that *increases* with increasing temperature.

It is also generally the case (but not the criterion for distinguishing them) that the conductivities of metals at room temperature are higher than those of semiconductors. Typical values are given in Fig. 25.15. A solid **insulator** is a substance with a very low electrical conductivity. However, when that conductivity can be measured, it is found to increase with temperature, like that of a semiconductor. For some purposes, therefore, it is possible to disregard the classification 'insulator' and to treat all solids as either metals or semiconductors, with an insulator having a very large band gap that exhibits extremely low electrical conductivity at room temperature, $>10^{-10}\,\text{S m}^{-1}$. **Superconductors** are a special class of materials that have zero electrical resistance below a critical temperature, and these materials are discussed in detail in Section 30.7.

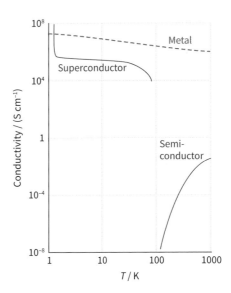

FIGURE 25.15 The variation of the electrical conductivity of a substance with temperature is the basis of the classification of the substance as a metallic conductor, a semiconductor, or a superconductor.

25.6 Semiconductors: Si analogues and extrinsic semiconductors

Intrinsic semiconductivity, where a material has a small band gap allowing promotion of an electron from a valence band to a conduction band, was described in Section 2.11(e). Example intrinsic semiconductor materials include the elements Si and Ge and the compounds GaAs and SiC. The properties of these intrinsic semiconducting solids may be controlled by changing the electronegativity of the elements present. For example, band gaps (Fig. 2.39) can be increased by using combinations of more electropositive and electronegative elements; the band gap in GaAs (χ(Ga) = 1.81 and χ(As) = 2.18) is 1.35 eV while that in InP (χ(In) = 1.78 and χ(P) = 2.19) is 1.42 eV. Introducing dopants and defects into solids (Section 25.1) also modifies the band structures and band gaps of solids, and the electronic properties of these **extrinsic semiconductor** materials are described in this Section.

(a) Semiconductor systems isoelectronic with silicon

KEY POINTS Semiconductors formed from equal amounts of Group 13/15 or Group 12/16 elements are isoelectronic with silicon and can have enhanced properties based on changes in the electronic structure and electron motion.

Gallium arsenide, GaAs, is one of a number of so-called Group 13/15 (or, still more commonly, III/V) semiconductors, which also include GaP, InP, AlAs, and GaN, formed by combination of equal amounts of a Group 13 and a Group 15 element. Ternary and quaternary Group 13/15 compounds, such as $Al_xGa_{1-x}As$, $InAs_{1-y}P_y$, and $In_xGa_{1-x}As_{1-y}P_y$, can also be formed, and many of them also have valuable semiconducting properties. Note that these compositions are isoelectronic with pure Group 14 elements, but the changes in the element electronegativity and thus bonding type (for instance, pure Si can be considered as having purely covalent bonding whereas GaAs has a small degree of ionic character due to the difference in the electronegativities of Ga and As) leads to changes in the band structures and fundamental properties associated with electron motion through the structures.

One of the advantageous properties of GaAs is that semiconductor devices based on it respond more rapidly to electrical signals than those based on silicon, as the electron mobility through the GaAs structure is around six times higher than that in Si. This responsiveness makes GaAs better than silicon for a number of tasks, such as amplifying the high-frequency (1–10 GHz) signals of satellite TV. Gallium arsenide can be used with signal frequencies up to about 100 GHz. At even higher frequencies, materials such as indium phosphide (InP) may be used. At present, frequencies above about 50 GHz are rarely used commercially, so most of the electronics in the world tend to be based on silicon, with some GaAs and only a few InP devices. Gallium arsenide is also far more expensive than Si, in terms of both the cost of raw materials and the chemical processes required to produce the pure material.

In some Group 13/15 semiconductors, such as GaN, the cubic, diamond-like sphalerite structure is only metastable, the stable polymorph being the hexagonal, wurtzite structure. Both structures can be grown by altering the synthetic routes and conditions. Because of their large and direct energy band gaps (Section 30.2), these semiconductors allow the fabrication of luminescent devices that produce blue light at high intensity (Box 25.4), and their stabilities to high temperatures and good thermal conductivities also make them valuable for the fabrication of high-power transistors.

The Group 12/16 (II/VI) semiconductors comprise the compounds containing Zn, Cd, and Hg as cations, and O, S, Se, and Te as anions. These semiconductor materials can crystallize in either the cubic sphalerite phase or the hexagonal wurtzite phase, and the form synthesized has characteristic semiconducting properties. For example, the band gap is 3.64 eV in cubic ZnS but 3.74 eV in hexagonal ZnS. These Group 12/16 compounds are more ionic in nature than the Group 13/15 semiconductors and Group 14 elements, particularly for the lighter elements, and the band gap is around 3–4 eV for ZnO and ZnS but 1.475 eV for CdTe. Although amorphous Si is the leading thin-film photovoltaic (PV) material, cadmium telluride (CdTe) and other related materials are being studied for this application, Section 30.2.

(b) Extrinsic semiconductors

KEY POINTS p-Type semiconductors are solids doped with atoms that remove electrons from the valence band; n-type semiconductors are solids doped with atoms that supply electrons to the conduction band.

An extrinsic semiconductor is a substance that is a semiconductor on account of the presence of intentionally added impurities. The number of electron carriers can be increased if atoms with more electrons than the parent element can be introduced by doping. Remarkably low levels of dopant concentration are needed—only about one atom per 10^9 of the host material—so it is essential to achieve very high purity of the parent element initially.

If arsenic atoms ([Ar]$4s^2 4p^3$) are introduced into a silicon crystal ([Ne]$3s^2 3p^2$), one additional electron will be available for each dopant atom that is substituted. Note that the doping is *substitutional* in the sense that the dopant atom takes the place of an Si atom in the silicon structure. If the

BOX 25.4 What is a p–n junction, and how are they used in LEDs?

A p–n junction can be constructed from two pieces of silicon, one of which is n-type and the other p-type (Fig. B25.4), where the illustration shows the band structure of the junction. The Fermi levels in the differently doped materials are different, but when they are placed in contact electrons will flow from the n-type (high potential) to the p-type (low potential) region across the junction so as to reach an equilibrium distribution in which the Fermi levels are equal. If a potential difference is applied in the right direction across the junction, this process will continue with electrons able to move from n-type to p-type, and a current flows in this direction. However, current can flow only in this direction, from n-type to p-type. The p–n junction thus forms the basis of a rectifier, allowing current to pass in only one direction.

A light-emitting diode (LED) is a p–n junction semiconductor diode that emits light when current is passed through it. The transfer of the electron from the conduction band of the n-type material to the valence band of the p-type semiconductor is accompanied by the emission of light. LEDs are highly monochromatic, emitting a pure colour in a narrow frequency range. The colour is controlled by the band gap, with small band gaps producing radiation in the infrared and red regions of the electromagnetic spectrum and larger band gaps resulting in emission in the blue and ultraviolet regions:

LED colour	Chip material	
	Low brightness	High brightness
Red	GaAsP/GaP	AlInGaP
Orange	GaAsP/GaP	AlInGaP
Amber	GaAsP/GaP	AlInGaP
Yellow	GaP	–
Green	GaP	GaN
Turquoise	–	GaN
Blue	–	GaN

It is possible to produce white light with a single LED by using a phosphor layer (Ce^{3+} doped yttrium aluminium garnet, Box 22.1) on the surface of a blue (gallium nitride) LED. These white-light LEDs are highly efficient at converting electricity into light—much more so than incandescent lamps, and even 'low-energy' fluorescent lights.

Another advantage of being able to produce blue LEDs is that they can be used as lasers in high-capacity optoelectronic storage devices such as Blu-ray® disks. The wavelength of blue light produced by these LEDs (405 nm) is shorter than that used in DVD-format devices (red light, 650 nm), which allows data to be written in smaller bits on an optical disk.

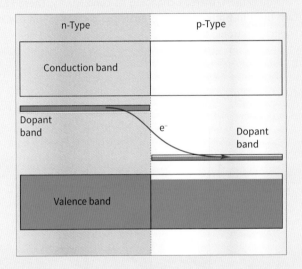

FIGURE B25.4 The structure of a p–n junction.

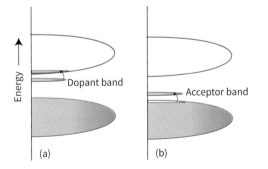

FIGURE 25.16 The band structure in (a) an n-type semiconductor and (b) a p-type semiconductor.

donor atoms, the As atoms, are far apart from each other, their electrons will be localized and the donor band will be very narrow (Fig. 25.16(a)). Moreover, the foreign atom levels will lie at higher energy than the valence electrons of the host structure, and the filled dopant band is commonly near the empty conduction band. For $T > 0$, some of its electrons will be thermally promoted into the empty conduction band. In other words, thermal excitation will lead to the transfer of an electron from an As atom into the empty orbitals on a neighbouring Si atom. From there it will be able to migrate through the structure in the band formed by Si–Si overlap. This process gives rise to **n-type semiconductivity**, the 'n' indicating that the charge carriers are negatively charged (i.e. electrons).

An alternative substitutional procedure is to dope the silicon with atoms of an element with fewer valence electrons on each atom, such as gallium ($[Ar]4s^2 4p^1$). A dopant atom of this kind effectively introduces holes into the solid. More formally, the dopant atoms form a very narrow, empty **acceptor band** that lies above the full Si band

BOX 25.5 How does the doping of semiconductors lead to applications in electronics?

Semiconductors have many applications because their properties can be easily modified by the addition of impurities to produce, for example, n- and p-type semiconductors. Furthermore, their electrical conductivities can be controlled by application of an electric field, by exposure to light, by pressure, or by heat, and as a result they can be used in many sensor devices.

When the junction of a p-type and an n-type semiconductor is under 'reverse bias' (i.e. with the p-side at a lower electric potential), the flow of current is very small, but it is high when the junction is under 'forward bias' (with the p-side at a higher electric potential). The exposure of a semiconductor to light can generate electron–hole pairs, which increases its conductivity through the increased number of free carriers (electrons or holes). Diodes that use this phenomenon are known as photodiodes. Compound semiconductor diodes can also be used to generate light, as in light-emitting diodes and laser diodes.

Bipolar junction transistors (BJT) are formed from two p-n junctions, in either an npn or a pnp configuration, with a narrow central region termed the *base*. The other regions, and their associated terminals, are known as the *emitter* and the *collector*. A small potential difference applied across the base and the emitter junction changes the properties of the base–collector junction so that it can conduct current even though it is reverse-biased. Thus, a transistor allows a current to be controlled by a small change in potential difference and is consequently used in amplifiers. Because the current flowing through a BJT is dependent on temperature, it can be used as a temperature sensor. Another type of transistor, the field-effect transistor (FET), operates on the principle that semiconductor conductivity can be increased or decreased by the presence of an electric field. The electric field increases the number of charge carriers, thereby changing its conductivity. These FETs are used in both digital and analogue circuits to amplify or switch electronic signals.

(Fig. 25.16(b)). At $T = 0$ the acceptor band is empty, but at higher temperatures it can accept thermally excited electrons from the Si valence band. By doing so, it introduces holes into the latter and hence allows the remaining electrons in the band to be mobile. Because the charge carriers are now effectively positive holes in the lower band, this type of semiconductivity is called **p-type semiconductivity**. Semiconductor materials are essential components of all modern electronic circuits, and some devices based on them are described in Boxes 25.4 and 25.5.

Several d-metal oxides, including ZnO and Fe_2O_3, are n-type semiconductors. In their case, the property is due to small variations in stoichiometry and a small deficit of O atoms. The electrons that should be in localized O atomic orbitals (giving a very narrow oxide band, essentially localized individual O^{2-} ions) occupy a previously empty conduction band formed by the metal orbitals (Fig. 25.17). The electrical conductivity decreases after the solids have been heated in oxygen and cooled slowly back to room temperature, because the deficit of O atoms is partly replaced and, as the atoms are added, electrons are withdrawn from the conduction band to form oxide ions. However, when measured at high temperatures the conductivity of ZnO increases as further oxygen is lost from the structure, so increasing the number of electrons in the conduction band.

p-Type semiconduction is observed for some low oxidation number d-metal chalcogenides and halides, including Cu_2O, FeO, FeS, and CuI. In these compounds, the loss of

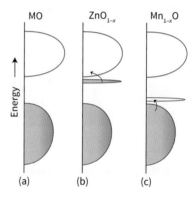

FIGURE 25.17 The band structure in (a) a stoichiometric oxide, (b) an anion-deficient oxide, and (c) an anion-excess oxide.

electrons can occur through a process equivalent to the oxidation of some of the metal atoms, with the result that holes appear in the predominantly metal band. The conductivity increases when these compounds are heated in oxygen (or sulfur and halogen sources for FeS and CuI, respectively) because more holes are formed in the metal band as oxidation progresses. n-Type semiconductivity, however, tends to occur for oxides of metals in higher oxidation states, as the metal can be reduced to a lower oxidation state by occupation of a conduction band formed from the metal orbitals. Thus, typical n-type semiconductors include Fe_2O_3, MnO_2, and CuO. By contrast, p-type semiconductivity occurs when the metal is in a low oxidation state, such as MnO and Cr_2O_3.

EXAMPLE 25.3 Predicting extrinsic semiconducting properties

Which of the oxides WO_3, MgO, and CdO are likely to show p- or n-type extrinsic semiconductivity?

Answer The type of semiconductivity depends on the defect levels that are likely to be introduced, which is, in turn, determined by whether the metal present can easily be oxidized or reduced. If the metal can easily be oxidized (which may be the case if it has a low oxidation number), then p-type semiconductivity is expected. On the other hand, if the metal can easily be reduced (which may be the case if it has a high oxidation number), then n-type semiconductivity is expected. Thus, WO_3, with tungsten present in the high oxidation state, W(VI), is readily reduced and accepts electrons from the O^{2-} ions, which escape as elemental oxygen. The excess electrons enter a band formed from the W d orbitals, resulting in n-type semiconductivity. Similarly, CdO, like ZnO, readily loses oxygen and is predicted to be an n-type semiconductor. In contrast, Mg^{2+} ions are neither easily oxidized nor reduced, so MgO does not lose or gain even small quantities of oxygen and is an insulator.

Self-test 25.3 Predict p- or n-type extrinsic semiconductivity for V_2O_5 and CoO.

Ion transport in solids

In an ionic solid, which has an extended repeat structure, the potential exists for both anions and cations to move through the crystals giving rise to **ionic conductivity**. As discussed in Section 25.1, all solids contain a level of defects—imperfections of structure or composition—and it is also possible to introduce defects (extrinsic defects) through doping. These defects, which are mainly interstitial (Frenkel-type) or vacancies (Schottky-type), are important because they provide pathways for ions to move through a solid-state structure, and ionic conductivity is usually enhanced by the presence of defects. Ion transport through solids is also necessary for solids to react with each other; Section 25.9.

Materials with high ionic conductivity have important applications in sensors, rechargeable Li-ion batteries, and fuel cells. This section covers the methods by which ions travel through solids, and how this property can be applied in devices such as an oxygen sensor. More detailed discussion of specific material types; for example, solids exhibiting high lithium ion conductivity for rechargeable batteries are covered later in this chapter.

25.7 Atom and ion diffusion and conduction

KEY POINT The diffusion of ions in solids is strongly dependent on the presence of defects.

Diffusion of atoms or ions in solids at room temperature is very much slower than the diffusion processes that occur in gases and liquid. This is why most solid-state reactions are undertaken at very high temperatures (Section 25.9), where ion mobility increases significantly. However, there are some striking exceptions to this generalization and some solids demonstrate ionic conductivity as high as $10^{-4}\,Sm^{-1}$ at room temperature. Diffusion of atoms or ions in solids is in fact very important in many areas of solid-state technology, such as semiconductor manufacture, the synthesis of new solids, fuel cells, sensors, metallurgy, and heterogeneous catalysis.

The rates at which ions move through a solid can be understood in terms of the mechanism for their migration, and the activation barriers the ions encounter as they move. The lowest-energy pathway generally involves defect sites with the roles summarized in Fig. 25.18.

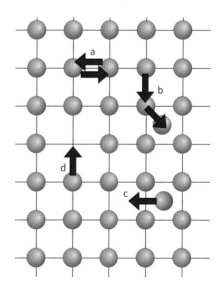

FIGURE 25.18 Possible diffusion mechanisms for ions or atoms in a solid: (a) two atoms or ions exchange positions; (b) an ion hops from a normally occupied site in the structure to an interstitial site, which produces a vacancy which can then be filled by movement of an ion from another site; (c) an ion hops between two different interstitial sites; (d) an ion or atom moves from a normally occupied site to a vacancy, so producing a new vacant site.

Materials that show high rates of diffusion at moderate temperatures have the following characteristics:

- Low-energy barriers: so temperatures at (or a little above) 300 K are sufficient to permit ions to jump from site to site.
- Low charges and small radii: so, for example, the most mobile cation (other than the proton) and anion are Li^+ and F^-, respectively. Reasonable mobilities are also found for Na^+ and O^{2-}. More highly charged ions develop stronger electrostatic interactions and are less mobile.
- High concentrations of intrinsic or extrinsic defects: defects typically provide a low-energy pathway for diffusion through a structure that does not involve the large energy penalties associated with continuously displacing ions from normal, favourable ion sites. These defects should not be ordered, as for crystallographic shear planes (Section 25.4), because such ordering removes the low energy diffusion pathway.
- Mobile ions are present as a significant proportion of the total number of ions.

Fig. 25.19 shows the temperature dependence of the diffusion coefficients, which are a measure of mobility, for specified ions in a selection of solids at high temperatures. The slopes of the lines are proportional to the activation energy for migration. Thus Na^+ is highly mobile and has a low activation energy for motion through β-alumina, whereas Ca^{2+} in CaO is much less mobile and has a high activation energy for hopping through the rock-salt structure.

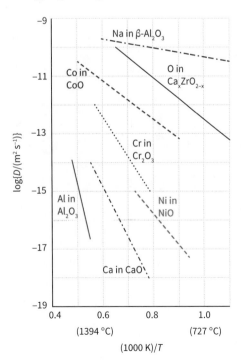

FIGURE 25.19 Diffusion coefficients (on a logarithmic scale) as a function of inverse temperature, for mobile ions in selected solids.

25.8 Solid electrolytes and electrodes

Any electrochemical cell, such as a battery, fuel cell, electrochromic display, or electrochemical sensor, requires an electrolyte to transport charge. In many applications, an ionic solution (e.g. dilute sulfuric acid in a lead–acid battery) is an acceptable electrolyte, but because it is often desirable to avoid a liquid phase because of the possibility of spillage, there is considerable interest in the development of solid electrolytes. Historically, two important and thoroughly studied solid electrolytes with mobile cations are silver tetraiodidomercurate(II), Ag_2HgI_4, and sodium β-alumina with the composition $Na_{1+x}Al_{11}O_{17+x/2}$. Other important fast-cationic conductors include NASICON (a name formed from the letters in '**Na** **s**uperionic **con**ductor') of composition $Na_{1+x}Zr_2P_{3-x}Si_xO_{12}$, lithium garnets such as $Li_{7-x}La_3Zr_{2-x}Ta_xO_{12}$, and a number of proton conductors that operate at, or moderately above, room temperature, such as $CsHSO_4$ above 160°C.

Solids exhibiting high anion mobility are rarer than cationic conductors and generally show high conductivity only at elevated temperatures: anions are typically larger than cations, and so the energy barrier for diffusion through the solid is high. As a consequence, fast anion conduction in solids is limited to F^- and O^{2-} (with ionic radii 133 pm and 140 pm, respectively). Despite these limitations, anionic conductors play an important role in sensors and fuel cells, where a typical material is yttrium-stabilized zirconia (YSZ), of composition $Y_xZr_{1-x}O_{2-x/2}$. Table 25.2 summarizes some typical ionic conductivity values of solid electrolytes and other ionically conducting media.

Solid electrolytes transport ions without any change in composition, oxidation state, or structure. Ions may be

TABLE 25.2 Comparative values of ionic and electronic conductivity

Material	Conductivity/S m^{-1}
Ionic conductors	
Ionic crystals	10^{-16}–10^{-2}
Example: LiI at 298°C	10^{-4}
Example: CaO at 800°C	10^{-6}
Solid electrolytes	10^{-1}–10^3
Example: YSZ at 600°C	1
Example: AgI at 500°C	10^2
Strong (liquid) electrolytes	10^{-1}–10^3
Example: 1 M NaCl(aq)	10^2
Electronic conductors	
Metals	10^3–10^6
Semiconductors	10^{-3}–10^2
Insulators	$>10^{-7}$

inserted, or removed, from a solid, and as these are charged species there will be an associated transfer of electrons and change in the oxidation state. There can also be changes in the structure of the solid to accommodate additional or fewer ions. Solid materials that can transport ions in and out of their structures have important applications, as both cathodes and anode in rechargeable batteries, where they are termed **solid electrodes**.

(a) Solid cationic electrolytes

KEY POINTS Solid inorganic electrolytes often have a low-temperature form in which the ions are ordered on a subset of sites in the structure; at higher temperatures the ions become disordered over the sites and the ionic conductivity increases.

While there is a rapidly growing number of different compositions and structures that exhibit high cation mobility near or just above room temperature, the basis for this property is readily illustrated through consideration of two well-studied materials: Ag_2HgI_4 and sodium β-alumina.

Below 50°C, Ag_2HgI_4 has an ordered crystal structure in which Ag^+ and Hg^{2+} ions are tetrahedrally coordinated by I^- ions and there are unoccupied tetrahedral holes (Fig. 25.20(a)), and its ionic conductivity is low. Above 50°C, however, the Ag^+ and Hg^{2+} ions are randomly distributed over the tetrahedral sites (Fig. 25.20(b)), and as a result there are many more sites that Ag^+ ions can occupy within the structure than there are Ag^+ ions present. At this temperature the material is a good ionic conductor, largely on account of the mobility of the Ag^+ ions between the different sites available for them.

The close-packed array of polarizable I^- ions is easily deformed and results in a low activation energy for the migration of an Ag^+ ion from one ion site to the next vacant one. There are many related solid electrolytes having similar structures containing soft anions, such as AgI and $RbAg_4I_5$, both of which have highly mobile Ag^+ ions, so that the conductivity of $RbAg_4I_5$ at room temperature is greater than that of aqueous sodium chloride.

Sodium β′-alumina is an example of a mechanically hard material that is a good ionic conductor. In this case, the rigid and dense Al_2O_3 slabs are bridged by a sparse array of O^{2-} ions (Fig. 25.21). The plane containing these bridging oxide ions also contains Na^+ ions, which can move from site to site because there are no major bottlenecks to hinder their motion. Many similar rigid materials having planes or channels through which ions can move are known; they are called **framework electrolytes**.

Another closely related material, sodium β′-alumina, has even less restricted motion of ions than β-alumina, and it has been found possible to substitute doubly charged cations such as Mg^{2+} or Ni^{2+} for Na^+. Even the large lanthanoid cation Eu^{2+} can be introduced into β′-alumina, although the

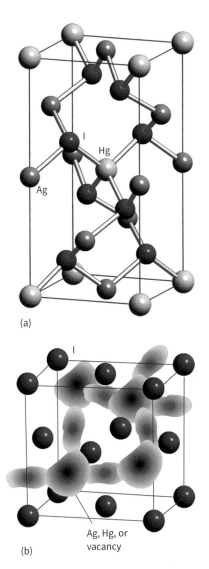

FIGURE 25.20 (a) Low-temperature ordered structure of Ag_2HgI_4; a tetrahedral hole is surrounded by four iodide anions. (b) High-temperature disordered structure showing the cation disorder. Ag_2HgI_4 is an Ag^+ ion conductor in the high-temperature form.

diffusion of such ions is slower than that of their smaller counterparts. The material NASICON, mentioned earlier, is a nonstoichiometric, solid-solution system with a framework constructed from ZrO_6 octahedra and PO_4 tetrahedra, corresponding to the parent phase of composition $NaZr_2P_3O_{12}$ (Fig. 25.22). A solid solution can be obtained by partially replacing P by Si to give $Na_{1+x}Zr_2P_{3-x}Si_xO_{12}$, with an increase in the number of Na^+ ions for charge balance. In this material, the full set of possible Na^+ sites is only partially filled, and these sites lie within a three-dimensional network of channels that allow rapid migration of the remaining Na^+ ions.

Other classes of materials currently being investigated as fast cation conductors include Li_4GeO_4 doped with V on the Ge sites, $Li_{4-x}(Ge_{1-x}V_x)O_4$, a lithium-ion conductor with vacancies on the Li^+ ion sublattice; the perovskite

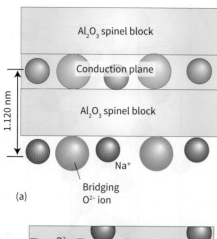

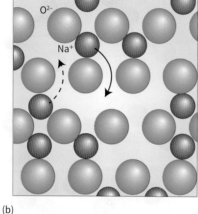

FIGURE 25.21 (a) Schematic side view of β-alumina, showing the Na$_2$O conduction planes between Al$_2$O$_3$ slabs. The O atoms in these planes bridge the two slabs. (b) A view of the conduction plane; note the abundance of mobile ions and vacancies in which they can move.

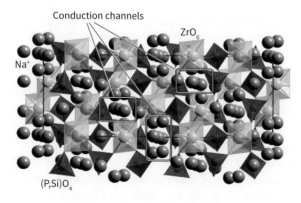

FIGURE 25.22 The Na$_{1+x}$Zr$_2$P$_{3-x}$Si$_x$O$_{12}$ (NASICON) structure shown as linked (P,Si)O$_4$ tetrahedra and ZrO$_6$ octahedra.

La$_{0.6}$Li$_{0.2}$TiO$_3$; and sodium yttrium silicate, Na$_5$YSi$_4$O$_{12}$, a sodium-ion conductor. One of the best lithium-ion conductivities at room temperature, 1.1×10^{-3} S cm^{-1}, has been reported for Li$_{6.4}$La$_3$Zr$_{1.4}$Ta$_{0.6}$O$_{12}$. Solids with high lithium-ion conductivity are of interest for use as electrolytes in all solid state Li-ion batteries. Most commercial rechargeable Li-ion batteries use a liquid or gel-like electrolyte, such as LiPF$_6$ in an organic solvent, with associated safety problems and battery design compromises (liquid electrolytes cannot be used with a lithium metal anode). Recent advances on solid electrolytes for use in rechargeable batteries are discussed further in Section 30.9.

EXAMPLE 25.4 Correlating conductivity and ion size in a framework electrolyte

Conductivity data on β-alumina containing monopositive ions of various radii show that Ag$^+$ and Na$^+$ ions, both of which have radii close to 100 pm, have activation energies for conductivity close to 17 kJ mol^{-1}, whereas that for Tl$^+$ (radius 149 pm) is about 35 kJ mol^{-1}. Suggest an explanation of this difference.

Answer One way of approaching this problem is to think of the constrictions to migration that are related to the sizes of the ions. In sodium β-alumina and related β-aluminas, a fairly rigid framework provides a two-dimensional network of passages that permit ion migration. Judging from the experimental results, the bottlenecks for ion motion appear to be large enough to allow Na$^+$ or Ag$^+$ (ionic radii close to 100 pm) to pass quite readily (with a low activation energy) but too small to let the larger Tl$^+$ (ionic radius 149 pm) pass through as readily.

Self-test 25.4 Why does increased pressure reduce the conductivity of K$^+$ in β-alumina more than that of Na$^+$ in β-alumina?

(b) Solid anionic electrolytes

KEY POINT Anion mobility can occur at high temperatures in certain structures that contain high levels of anion vacancies.

In 1834, Michael Faraday reported that red-hot solid PbF$_2$ (near 800°C) is a good conductor of electricity. Much later it was recognized that the conductivity arises from the mobility of F$^-$ ions through the solid. The property of anion conductivity is shared by other crystals having the fluorite structure. Ion transport in these solids is thought to be by an **interstitial mechanism** in which an F$^-$ ion first migrates from its normal position into an interstitial site (a Frenkel-type defect, Section 25.1) and then moves to a vacant F$^-$ site.

Structures that have large numbers of vacant sites generally show the highest ionic conductivities because they provide a path for ion motion (although at very high levels of defects, clustering of the defects or the vacancies can lower the conductivity). These vacancies, which are equivalent to extrinsic defects, can be introduced in fairly high numbers into many simple oxides and fluorides by doping with appropriately chosen metal ions in different oxidation states. Zirconia, ZrO$_2$, has a fluorite structure at high temperatures, but on cooling the pure material to room temperature it distorts to a monoclinic polymorph. The cubic fluorite structure may be stabilized at room temperature by replacing some Zr^{4+} with other ions, such as the similarly

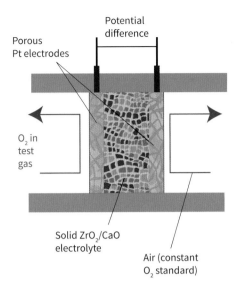

FIGURE 25.23 An oxygen sensor based on the solid electrolyte $Zr_{1-x}Ca_xO_{2-x}$.

The cell potential is related to the two oxygen partial pressures (p_1 and p_2) by the Nernst equation (Section 6.5), for the half-cell reaction $O_2 + 4e^- \rightarrow 2O^{2-}$, which occurs at both electrodes:

$$E_{cell} = \frac{RT}{4F} \ln \frac{p_1}{p_2} \qquad (25.1)$$

so a simple measurement of the potential difference provides a measure of the oxygen partial pressure in the exhaust gases.

A BRIEF ILLUSTRATION

According to Eqn. 25.1, the potential difference produced by an oxygen sensor operating at 1000 K in an exhaust system, with air on one side ($p(O_2) = 0.2$ atm) and a burnt fuel/air mixture ($p(O_2) = 0.001$ atm) on the other, is about 0.1 V.

(c) Mixed ionic–electronic conductors

KEY POINT Solid materials can exhibit both ionic and electronic conductivity.

Most ionic conductors, such as sodium β′-alumina and YSZ, have low electronic conductivity (i.e. conduction by electron rather than ion motion). Their application as solid electrolytes, in sensors for instance, requires this feature to avoid shorting out the cell. In some cases a combination of electronic and ionic conductivity is desirable, and this type of behaviour can be found in some d-metal compounds where defects allow O^{2-} conduction and the metal d orbitals provide an electronic conduction band. Many such materials are perovskite-based structures with mixed oxidation states at the B cation sites (Sections 4.9 and 25.11(d)). Two examples are $La_{1-x}Sr_xCoO_{3-y}$ and $La_{1-x}Sr_xFeO_{3-y}$. These oxide systems are good electronic conductors with partially filled bands as a result of the nonintegral d-metal oxidation number, and can conduct by O^{2-} migration through the perovskite O^{2-} ion sites. This type of material is of use in solid oxide fuel cells (Box 25.6), one of the fuel cell types mentioned in Box 6.1, in which one electrode has to allow diffusion of ions through an electrically conducting electrode.

(d) Solid electrodes and lithium ion batteries

KEY POINT The extraction and insertion of metal ions from and into solid structures is exploited in the electrodes of rechargeable batteries.

The transport of ions through a solid electrolyte involves no change in stoichiometry. Ions may also be inserted into or removed from a solid, and because they carry charge this involves simultaneous oxidation or reduction of the host solid. Structures that allow the rapid transport of ions in or

sized Ca^{2+} and Y^{3+} ions. Doping with these ions of lower oxidation state results in the introduction of vacancies on the anion sites to preserve the charge neutrality of the material, and produces, for example, $Y_xZr_{1-x}O_{2-x/2}$, the material mentioned previously as yttrium-stabilized zirconia (YSZ). This material has completely occupied cation sites in the fluorite structure but high levels of anion vacancies, with $0 \leq x \leq 0.15$. These vacant sites provide a path for oxide-ion diffusion through the structure so that a typical electrical conductivity in, for example, $Ca_{0.15}Zr_{0.85}O_{1.85}$ is $5\,S\,cm^{-1}$ at 1000°C; note that this conductivity is much lower than typical solid-state cation conductivities, even at these very high temperatures, because of the large anion size.

The high oxide-ion conductivity of calcium-oxide-doped zirconia is exploited in a solid-state electrochemical sensor for measuring the partial pressure of oxygen in automobile exhaust systems (Fig. 25.23).[1] The platinum electrodes in this cell adsorb O atoms and, if the partial pressures of oxygen are different between the sample and reference sides, there is a thermodynamic tendency for oxygen to migrate through the electrolyte as the O^{2-} ion. The thermodynamically favoured processes are:

High-$p(O_2)$ side:

$$\tfrac{1}{2}O_2(g) + Pt(g) \rightarrow O(Pt, surface)$$
$$O(Pt, surface) + 2e^- \rightarrow O^{2-}(ZrO_2)$$

Low-$p(O_2)$ side:

$$O^{2-}(ZrO_2) \rightarrow O(Pt, surface) + 2e^-$$
$$O(Pt, surface) \rightarrow \tfrac{1}{2}O_2(g) + Pt(s)$$

---

[1] The signal from this sensor is used to adjust the air/fuel ratio and thereby the composition of the exhaust gas being fed to the catalytic converter.

BOX 25.6 How does a solid oxide fuel cell work?

A fuel cell consists of an electrolyte sandwiched between two electrodes; oxygen passes over one electrode and the fuel over the other, generating electricity, water, and heat. The general operation and construction of a fuel cell that converts a fuel, such as hydrogen, methane, or methanol, by reaction with oxygen into electrical energy (and combustion products H_2O and CO_2) was described in Box 6.1. A variety of materials can be used as the electrolyte in such cells, including phosphoric acid, proton-exchange membranes, or, in *solid oxide fuel cells* (SOFCs), oxide-ion conductors.

SOFCs operate at high temperatures and use an oxide-ion conductor as the electrolyte. The design of a typical SOFC is shown in Fig. B25.5. Each cell generates a limited potential difference but, as with the cells of a battery, a connected stack may be constructed in series to increase the potential difference and power supplied. Cells are connected electrically through an 'interconnect' that can also be used to isolate the fuel and air supplies for each cell.

The attraction of SOFCs is based on several aspects, including the clean conversion of fuel to electricity, low levels of noise pollution, the ability to cope with different fuels, and, most significantly, a high efficiency. Their high efficiency is a result of their high operating temperatures, typically between 500°C and 1000°C. In high-temperature SOFCs the interconnect may be a ceramic such as lanthanum chromite (the perovskite $LaCrO_3$); if the temperature is below 1000°C, an alloy such as Y/Cr may be used. The oxide-ion conductor used as the electrolyte in these very high-temperature SOFCs is normally YSZ. The development of new materials for use in SOFCs is discussed further in Section 30.5.

Intermediate-temperature SOFCs, which typically operate between 500°C and 700°C, have a number of advantages over very high-temperature devices in that there is reduced corrosion, a simpler design, and a much-reduced time to heat the system to the operating temperature. However, such devices require a material with an excellent oxide-ion conductivity at lower temperatures to act as the electrolyte; Section 30.5. One of the main advantages of SOFCs over other fuel-cell types is their ability to handle more convenient hydrocarbon fuels, while other types of fuel cell have to rely on a clean supply of hydrogen for their operation. Because SOFCs operate at high temperature there is the opportunity to convert hydrocarbons catalytically to hydrogen and carbon oxides within the system. Because of their size. and the requirement to be heated to, and operate at, high temperatures, the applications of SOFCs focus on medium- to large-scale static systems, including small residential systems producing about 2 kW.

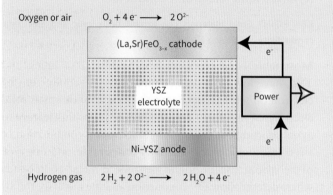

FIGURE B25.5 The structure of a solid oxide fuel cell.

out of their structures often consist of layers, similar to the β-aluminas, Section 25.8(a), and the inserted or deinserted ions migrate rapidly between the layers.

The demonstration that high ionic conductivity in some solids is associated with the ability to vary the oxidation state of a d-metal ion has led to the development of materials for use as the solid electrode in rechargeable batteries (see Box 12.3). Example cathode materials include $LiCoO_2$, with a layer-type structure based on sheets of edge-linked CoO_6 octahedra separated by Li^+ ions (Fig. 25.24), and various lithium manganese spinels, such as $LiMn_2O_4$. In each of these compounds, the battery is charged by removing the mobile Li^+ ions from the complex metal oxide, as in

$$LiCoO_2 \rightarrow CoO_2 + Li^+ + e^-$$

The battery is discharged through the reverse electrochemical reaction, Fig. 25.25.

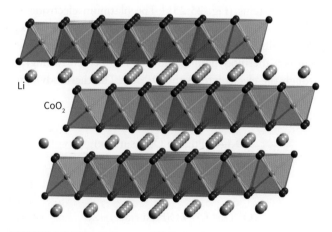

FIGURE 25.24 The structure of $LiCoO_2$ shown as layers of linked CoO_6, overall composition CoO_2, octahedra separated by Li^+ cations.

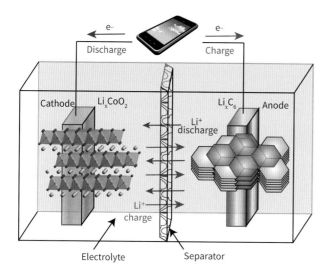

FIGURE 25.25 Schematic of the processes taking place in a rechargeable Li-ion battery, using $LiCoO_2$ as the cathode and graphite as the anode, during charge and discharge.

The other solid electrode, the anode, in a rechargeable Li-ion battery can simply be Li metal, which completes the overall cell reaction through the process $Li(s) \rightarrow Li^+ + e^-$. However, the use of Li metal has a number of problems associated with its reactivity and volume changes that occur in the cell. Therefore, an alternative anode material that is frequently used in rechargeable batteries is graphitic carbon, which is light and can electrochemically intercalate large quantities of Li to form LiC_6 (Section 15.5). As the cell is discharged, Li is transferred from between the carbon layers at the anode and intercalated into the metal oxide at the cathode (and vice versa on charging), with the following overall processes (Fig. 25.25):

$$Li_yC_6 + Li_{1-x}CoO_2 \underset{\text{charge}}{\overset{\text{discharge}}{\rightleftharpoons}} C_6 + Li_{1-x+y}CoO_2$$

Recent advances in lithium-ion battery materials (cathode, anode, and electrolyte) are discussed in detail in Section 30.9.

Synthesis of materials

Much of synthetic inorganic chemistry, including the coordination chemistry of the metals and organometallic chemistry, involves the conversion of molecules by replacing one ligand by another in a solution-based reaction. Processes of this type generally have relatively small activation energies (typically $<200\,kJ\,mol^{-1}$) and can be undertaken at low temperatures, typically between 0°C and 150°C, and in solvents that permit the migration of reacting species. Rapid molecular migration in solvents also results in fairly short reaction times of minutes or hours.

The formation of a new solid material by the reaction of two precursor solids, however, involves rather different processes as the high lattice energies of their extended structures, often greater than $2000\,kJ\,mol^{-1}$, need to be overcome. Furthermore, ion migration in the solid state is normally slow compared to diffusion processes in solution, except at very elevated temperatures; Section 25.9. Some inorganic materials can be prepared from solutions at much lower temperatures, where the building blocks are condensed together to form the extended structure.

This section focuses on the synthesis of a bulk single-phase material rather than the control of crystallite size, particle morphology, and the production of thin films; these additional aspects of synthesis in solid-state chemistry, of great relevance to nanomaterial chemistry, are covered in Chapter 29.

25.9 The formation of bulk materials

New materials can be obtained by two main methods. One is the direct reaction of two or more solids which involves breaking down their lattices and reforming the new structure; the other is the linking of polyhedral units from solution and deposition of the newly formed solid.

(a) Direct synthesis at high temperature

KEY POINTS Many complex solids can be obtained by direct reaction of the components at high temperatures, typically with a time scale of days. To improve reaction rates components may be mixed initially on the atomic scale via solution and sol–gel processes.

The most widely used method for the synthesis of bulk inorganic solids involves heating the solid reactants together at a high temperature, normally between 500°C and 1500°C, for an extended period. A complex oxide may be obtained by heating a mixture of all the oxides of the various metals present; alternatively, simple compounds that decompose to produce the oxides may be used instead of the oxide itself. Thus, ternary oxides, such as $BaTiO_3$, and quaternary oxides, such as $YBa_2Cu_3O_7$, are synthesized by heating together the following mixtures for several days:

$$BaCO_3(s) + TiO_2(s) \xrightarrow{1000°C} BaTiO_3(s) + CO_2(g)$$

$$\tfrac{1}{2}Y_2O_3(s) + 2\,BaCO_3(s) + 3\,CuO(s) + \tfrac{1}{4}O_2(g)$$
$$\xrightarrow{930°C/\text{air and } 450°C/O_2} YBa_2Cu_3O_7(s) + 2\,CO_2(g)$$

High temperatures are used in these syntheses to accelerate the slow diffusion of ions in solids and to overcome the high Coulombic attractions between the ions. Reactants are normally in the form of powders, with small particle sizes of typically <10 μm, that are thoroughly ground together before heating, with the aim of reducing the ion diffusion path-lengths. The direct method is applicable to many other inorganic material types, such as the syntheses of complex chlorides and dense, anhydrous metal aluminosilicates:

$$3\,CsCl(s) + 2\,ScCl_3(s) \rightarrow Cs_3Sc_2Cl_9(s)$$
$$NaAlO_2(s) + SiO_2(s) \rightarrow NaAlSiO_4(s)$$

Most simple binary oxides are available commercially as pure, polycrystalline powders with typical particle dimensions of a few micrometres. Alternatively, the decomposition of a simple metal salt precursor, either prior to or during reaction, leads to an oxide with small particle sizes (<1 μm). Example precursors include metal carbonates, hydroxides, oxalates, and nitrates, which all decompose to the metal oxide and normally at temperatures well below 1000°C. An additional advantage of precursors is that they are often stable in air, whereas many oxides are hygroscopic or react with carbon dioxide in the air. Thus, in the synthesis of $BaTiO_3$, barium carbonate, $BaCO_3$, which starts to decompose to BaO above 900°C, would be ground together with TiO_2 in the correct stoichiometric proportions, using a pestle and mortar or ball mill (where the solid is shaken vigorously with small hard spheres inside a canister). The mixture is then transferred to a crucible, normally constructed of an inert material such as vitreous silica, recrystallized alumina, or platinum, and placed in a furnace. Even at high temperatures, 1100°C, the reaction is slow and typically takes several days.

A variety of methods can be used to improve reaction rates, including pelletizing the reaction mixture under high pressure to increase the contact between the reactant particles, regrinding the mixture periodically to introduce virgin reactant interfaces, and the use of 'fluxes'—low-melting solids that aid the ion-diffusion processes. The size of the reactant particles is a major factor in controlling the time it takes a reaction to proceed to completion. The larger the particles, the lower their total surface area, and therefore the smaller the area at which the reaction can take place. Furthermore, the distances over which the diffusion of the reactant ions must occur, to reach the reaction surface, are much greater for larger particles, which can be micrometres for a commercial polycrystalline material. In order to increase the rate of reaction and allow solid-state reactions to occur at lower temperatures, reactants having small particle sizes, between 10 nm and 1 μm, and large surface areas are often deliberately employed; reactants with small particle sizes may be synthesized initially before being used in the main target-phase forming reaction.

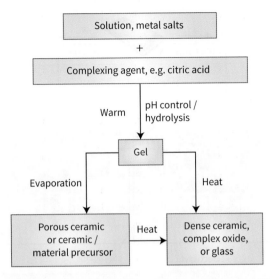

FIGURE 25.26 Schematic diagram of the sol–gel process. When the gel is dried at high temperatures, dense ceramics or glasses are formed. Drying at low temperatures above the critical pressure of water produces porous solids known as xerogels or aerogels.

Improved mixing of the reactant species can also be achieved by using solutions at an early stage of the reaction process. These methods are embodied in **sol–gel processes** which are described schematically in Fig. 25.26. Sol–gel processes can be used to produce crystalline complex metal oxides, ceramics, nanoparticles (Chapter 28), and high-surface-area compounds for catalysis, such as silica gel and alumina (Section 31.8).

The advantages of starting with solutions is that the reactants are mixed as solvated species at the atomic level, so overcoming the problems associated with the direct reaction of two or more solid phases consisting of micrometre-sized particles. In the simplest reaction of this type, a solution of metal ions (for example, metal nitrates) is converted to a solid through a variety of methods, including evaporation of the solvent or precipitation as a simple mixed metal salt. This solid is then heated to produce the target material. Two simple examples, as routes to producing the complex oxides La_2CuO_4 and $ZnFe_2O_4$, respectively, are

$$2\,La^{3+}(aq) + Cu^{2+}(aq) \xrightarrow{5OH^-(aq)} 2\,La(OH)_3 \cdot Cu(OH)_2(s)$$
$$\xrightarrow{600°C} La_2CuO_4(s) + 4\,H_2O(g)$$

$$Zn^{2+}(aq) + 2\,Fe^{2+}(aq) + 3\,C_2O_4^{2-}(aq) \rightarrow ZnFe_2(C_2O_4)_3(s)$$
$$\xrightarrow{700°C} ZnFe_2O_4(s) + 4\,CO(g) + 2\,CO_2(g)$$

A typical sol–gel process involves producing an aqueous solution of the various metal salts, adding complexing agents such as carboxylic acids or alcohols, and slow evaporation of the water leading to a viscous solution or gel. Alternatively, metal alkoxide precursors can be dissolved in an alcohol, and the addition of water then causes hydrolysis to produce a thick gel. Further drying of these gels (which consist of the metal species linked through organic alkoxides and carboxylates) at high temperatures leads to solids which can be decomposed at relatively low temperatures (300–600°C) to produce the desired complex oxide product.

One specific method of this type is the 'Pechini method' in which metal salts or alkoxides are added to a solution of citric acid and ethylene glycol. Heating this solution to above 100°C results in a condensation reaction to produce a polymeric mixed metal citrate gel, which then decomposes on heating above 400°C to form a complex oxide. As well as the advantages of reduced reaction times, a result of the intimate mixing of the reactants, the temperature of the final decomposition reaction (evolving water and carbon oxides) is somewhat lower than that needed for the direct reaction of the oxides. The use of a lower temperature can also have the effect of reducing the size of the particles formed as the product in the final reaction stage. Further applications of this method for the synthesis of nanoparticles are covered in Chapter 28.

The reaction environment may need to be controlled if a particular oxidation state is required or one of the reactants is volatile. Solid-state reactions can be carried out in a controlled atmosphere, using a tube furnace in which a gas can be passed over the reaction mixture while it is being heated. An example of a reaction of this type is the use of an inert gas to prevent oxidation, as in the preparation of $TlTaO_3$:

$$Tl_2O(s) + Ta_2O_5(s) \xrightarrow{N_2/600°C} 2\,TlTaO_3(s)$$

where the oxidation of Tl(I) to Tl(III), which would occur in air, is avoided.

High gas pressures may also be used to control the composition of the product. For example, Fe(III) is normally obtained in complex iron oxides synthesized in air or oxygen at or near normal pressures, but Fe(IV) may be formed, as in the production of Sr_2FeO_4 from mixtures of SrO and Fe_2O_3, under several hundred atmospheres of oxygen. For volatile reactants, the reaction mixture is normally sealed in a glass tube, under vacuum, prior to heating. Examples of such reactions are

$$Ta(s) + S_2(l) \xrightarrow{500°C/\text{sealed silica tube}} TaS_2(s)$$

$$Tl_2O_3(l) + 2\,BaO(s) + 3\,CaO(s) + 4\,CuO(s) \xrightarrow{860°C/\text{sealed gold capsule}} Tl_2Ba_2Ca_3Cu_4O_{12}(s)$$

Sulfur and thallium(III) oxide are volatile at the respective reaction temperatures and would be lost from the reaction mixture in an open vessel, leading to products of the incorrect stoichiometry.

High pressures can also be used to affect the outcome of a solid-state chemical reaction. Specialized apparatus, typically based on large presses, allows reactions between solids to take place at pressures of up to about 100 GPa (1 Mbar) at temperatures close to 1500°C. Reactions carried out under these conditions promote the formation of dense, higher-coordination-number structures. An example is the production of $MgSiO_3$, with a perovskite-like structure (Sections 4.9(b) and 25.11(d)) and a six-coordinate Si in an octahedral SiO_6 unit, rather than the normal tetrahedral SiO_4 unit found in most silicates. This type of apparatus can also be used to make diamonds from graphite (Box 15.2). Small-scale reactions can be undertaken at very high pressures in 'diamond anvil cells', in which the faces of two opposed diamonds are pushed together in a vice-like apparatus to generate pressures of up to 100 GPa.

(b) Chemical vapour transport

KEY POINT Sealed tube chemical vapour transport can be used to purify solids including metal disulfides synthesized by the direct reaction of the elements.

The synthesis of solids by direct reaction of the elements often leads to impure products, especially if the reaction is slow and incomplete. For example, compounds of the chalcogens with d-metals are usually prepared by heating mixtures in a sealed tube; however, the products obtained in this manner can have a variety of compositions and stoichiometries. The preparation of single-phase, crystalline dichalcogenides suitable for chemical and structural studies is often performed by **chemical vapour transport** (CVT). It is possible in some cases simply to sublime a compound, but the CVT crystal growth technique can also be applied to a wide variety of nonvolatile compounds in solid-state chemistry. One oxide example is ReO_3 which can be grown as very pure, single crystals using this method.

In a typical procedure, the impure material is loaded into one end of a borosilicate or fused-quartz tube. After evacuation, a small amount of a CVT agent is introduced and the tube is sealed and placed in a furnace with a temperature gradient. The polycrystalline and possibly impure metal chalcogenide is vaporized at one end and redeposited as pure crystals at the other (Fig. 25.27). The technique is called chemical vapour *transport* rather than *sublimation*, because the CVT agent, which is often a halogen, produces an intermediate volatile species, such as a metal halide. Generally, only a small amount of transport agent is needed because on crystal formation it is released and diffuses back to pick up

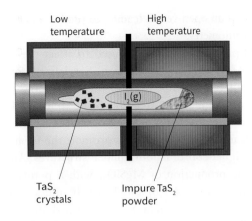

FIGURE 25.27 Vapour transport crystal growth and purification of TaS_2. A small quantity of I_2 is present to serve as a transport agent.

more reactant. For example, TaS_2 can be transported with I_2 in a temperature gradient. The reaction with I_2 to produce gaseous products,

$$TaS_2(s) + 2I_2(g) \rightarrow TaI_4(g) + S_2(g)$$

is endothermic, so the equilibrium lies further to the right at 850°C than it does at 750°C. Consequently, although TaI_4 is formed at 850°C, at 750°C the mixture deposits TaS_2. If, as occasionally is the case, the transport reaction is exothermic, the solid is carried from the cooler to the hotter end of the tube. Purification of ReO_3 also uses iodine as the CVT agent with the intermediate volatile species as ReO_2I_2; other solids which are purified by CVT (CVT-agent given in parentheses) are WO_2 (I_2), $CrSi_2$ ($SiCl_4$) and VCl_3 (Br_2).

(c) Solution methods

KEY POINT Frameworks formed from linked polyhedral species can be obtained by condensation reactions in solution.

Many inorganic materials, especially framework structures formed from linked tetrahedral or octahedra units, can be synthesized by crystallization from solution. Although the methods used are very diverse, the following are typical reactions that occur in water and lead to materials with extended structures consisting of metal centres linked through anions:

$$ZrO_2(s) + 2\,H_3PO_4(l) \rightarrow Zr(HPO_4)_2 \cdot H_2O(s) + H_2O(l)$$

$$3\,KF(aq) + MnBr_2(aq) \rightarrow KMnF_3(s) + 2\,KBr(aq)$$

$KMnF_3$ adopts the perovskite structure type (Sections 4.9(b) and 25.11(d)).

Room-temperature solution methods can be extended by using **hydrothermal techniques**, in which the reacting solution is heated above its normal boiling point in a sealed vessel. Hydrothermal reactions are important for the synthesis of open-structure aluminosilicates (**zeolites**), Sections 15.10 and 29.1, analogous porous structures based on linked oxopolyhedra, and related **metal–organic frameworks** (MOFs), in which metal ions are linked by coordinating organic species, most frequently as carboxylates and amines (Sections 25.17(c) and 29.3); the MOF $[Cu_2(tpt)_2(SO_4)_2(H_2O)_2]\cdot 4H_2O$ is made by the reaction of a copper sulfate solution in water with 2,4,6-tris(4-pyridyl)-1,3,5-triazine (tpt) at 180°C.

While some zeolites can be made below the boiling point of water, as with the synthesis of zeolite **LTA**,[2]

$$12\,NaAlO_2(s) + 12\,Na_2SiO_3(s) + (12+n)H_2O \xrightarrow{90°C}$$
$$Na_{12}[Si_{12}Al_{12}O_{48}]\cdot nH_2O(\text{zeolite LTA})(s) + 24\,NaOH(aq)$$

others require higher temperatures and the addition of a **structure-directing agent** (SDA) which controls the topology of the framework produced. Thus, the synthesis of zeolite **BEA**,[2] with the composition $Na_{0.92}K_{0.62}(TEA)_{7.6}[Al_{4.53}Si_{59.47}O_{128}]$ (TEA = tetraethylammonium cation), involves the reaction of sodium aluminate, silica, NaCl, KCl, and the SDA tetraethylammonium hydroxide. These porous structures are often thermodynamically metastable with respect to conversion to denser structure types, so they cannot be made by direct high-temperature reactions, which would produce the thermodynamically most stable phase. For example, the sodium aluminosilicate zeolite LTA, $Na_{12}[Si_{12}Al_{12}O_{48}]\cdot nH_2O$, formed in solution converts on heating above 800°C to the dense aluminosilicate $NaSiAlO_4$ ($\Delta G \sim -15\,\text{kJ mol}^{-1}$).

Recently, solvents such as liquid ammonia, supercritical CO_2, and organic amines have been used in so-called **solvothermal reactions**. Reactions in ionic liquids, often salts of organic cationic species with low melting points (below or near room temperature, Section 5.15(e)), can also be undertaken to produce zeolites and are known as **ionothermal** reactions.

Although high-temperature direct-combination methods and solvothermal techniques are the most commonly used methods in materials chemistry, some reactions involving solids can occur at low temperatures if there is no major change in structure. These so-called 'intercalation reactions' are discussed in Section 25.14(c).

---

[2] **LTA** and **BEA** are examples of the three-letter codes used by the International Zeolite Association to identify different zeolite (aluminosilicate) structures (Section 29.1).

EXAMPLE 25.5 How would you synthesize a complex oxide?

Describe a synthetic route to the high-temperature superconducting complex oxide $ErBa_2Cu_3O_{7-x}$.

Answer We need to think of an analogous compound and adapt its preparation to this compound. The same method as used for preparing $YBa_2Cu_3O_7$ can be used but with the appropriate lanthanoid oxide. That is, use the reaction of erbium oxide (Er_2O_3), barium carbonate (which decomposes to yield BaO), and copper(II) oxide at 940°C, for a period of days, followed by annealing under pure oxygen at 450°C. Alternatively an initial solution based sol–gel route could be used where $Eu(NO_3)_3 \cdot nH_2O$, $Ba(CH_3COO)_2 \cdot nH_2O$ and $Cu(CH_3COO)_2 \cdot H_2O$ are dissolved in water and tartaric acid added and the solution concentrated by evaporation. Heating the gel so-formed at 800°C under oxygen for 12 hours would yield $ErBa_2Cu_3O_{7-x}$.

Self-test 25.5 How could you prepare samples of (a) $SrTiO_3$, (b) Sr_2VO_4, and (c) a porous aluminophosphate $AlPO_4$?

Metal oxides, nitrides, and fluorides

In this section we explore the binary compounds of O, N, and F with metals. These compounds, particularly oxides, are central to much solid-state chemistry on account of their stability, ease of synthesis, and variety in composition and structure. These attributes lead to the vast number of compounds that have been synthesized, and the ability to tune the properties of a compound for a specific application based on their electronic, magnetic, or optical characteristics. As we shall see, a discussion of the chemical properties of these compounds also provides insight into defects, nonstoichiometry, ion diffusion, and the influence of these characteristics on physical properties.

The chemical properties of metal fluorides parallel much of that of metal oxides, but the lower charge of the F^- ion means that equivalent stoichiometries are produced with cations having lower charges, as in $KMn(II)F_3$ compared with $SrMn(IV)O_3$. Compounds containing the nitride ion, N^{3-}, in combination with one or more metal ions have only been developed to a significant extent in the last 25 years. The area of mixed-anion solid-state chemistry, where compounds containing transition metals in combination with more than one anion, such as oxide-nitrides and oxide-fluorides, is one where rapid advances are being made in terms of new materials and structure types.

25.10 Monoxides of the 3d metals

The monoxides of most of the 3d metals adopt the rock-salt structure (Table 25.3), though these seemingly simple compounds are usually obtained with significant deviations from the nominal MO stoichiometry.

(a) Defects and nonstoichiometry

KEY POINT The nonstoichiometry of $Fe_{1-x}O$ arises from the creation of vacancies on the Fe^{2+} octahedral sites, with each vacancy charge-compensated by the conversion of two Fe^{2+} ions to two Fe^{3+} ions.

TABLE 25.3 Monoxides of the 3d-series metals

Compound as MO_x	Structure	Composition, x	Electrical character
CaO_x	Rock-salt	1	Insulator
TiO_x	Rock-salt	0.65–1.25	Metallic
VO_x	Rock-salt	0.79–1.29	Metallic
MnO_x	Rock-salt	1–1.15	Semiconductor
FeO_x	Rock-salt	1.04–1.17	Semiconductor
CoO_x	Rock-salt	1–1.01	Semiconductor
NiO_x	Rock-salt	1–1.001	Insulator
CuO_x	PtS (linked CuO_4 square planes)	1	Semiconductor
ZnO_x	Wurtzite	Slight Zn excess $x\sim0.98$ at 1400 K	Wide band gap n-type semiconductor

The origin of nonstoichiometry in 'FeO' has been studied in more detail than that in most other MO compounds. In fact, it is found experimentally that stoichiometric FeO does not exist, but rather a range of iron-deficient compounds $Fe_{1-x}O$, $0.13 > x > 0.04$ can be obtained by quenching (cooling very rapidly) iron(II) oxide from high temperatures. The compound $Fe_{1-x}O$ is in fact **metastable** at room temperature, meaning that it is thermodynamically unstable with respect to disproportionation into iron metal and Fe_3O_4 but does not convert for kinetic reasons. The general consensus is that the structure of $Fe_{1-x}O$ is derived from the rock-salt structure of an idealized 'FeO' by the presence of vacancies on the Fe^{2+} octahedral sites, and that each vacancy is charge-compensated by the conversion of two adjacent Fe^{2+} ions to two Fe^{3+} ions. The relative ease of oxidizing Fe(II) to Fe(III) accounts for the fairly broad range of compositions of $Fe_{1-x}O$. At high temperatures, the interstitial Fe^{3+} ions associate with the Fe^{2+} vacancies (or defects) to form clusters distributed throughout the structure (Fig. 25.28).

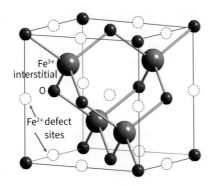

FIGURE 25.28 Defect sites proposed for $Fe_{1-x}O$. Note that the tetrahedral Fe^{3+} interstitials (large grey spheres) and octahedral Fe^{2+} vacancies (dotted circles) are clustered together.

Similar defects and the clustering of defects appear to occur with all other 3d-metal monoxides, with the possible exceptions of CoO and NiO. The range of nonstoichiometry in $Ni_{1-x}O$ is extremely narrow, but conductivity and the rate of ion diffusion vary with oxygen partial pressure in a manner that suggests the presence of isolated point defects. As indicated by standard potentials in aqueous solution, Fe(II) is more easily oxidized than either Co(II) or Ni(II); this solution redox chemistry correlates well with the much smaller range of oxygen deficiency in NiO and CoO. Chromium(II) oxide, like $Fe_{1-x}O$, spontaneously disproportionates at high temperature:

$$3\,Cr(II)O(s) \rightarrow Cr(III)_2O_3(s) + Cr(0)(s)$$

However, the material can be stabilized by crystallization in a copper(II) oxide matrix.

Both CrO and TiO have structures that show high levels of defects on both the cation and anion sites, forming metal-rich or metal-deficient stoichiometries ($Ti_{1-x}O$ and TiO_{1-x}). In fact, TiO has large numbers of vacancies, in equal amounts, on both cation and anion sublattices.

(b) Electronic properties

KEY POINTS The 3d-metal monoxides MnO, FeO, CoO, and NiO are semiconductors; TiO and VO are metallic conductors.

The 3d-metal monoxides MnO, $Fe_{1-x}O$, CoO, and NiO have low electrical conductivities that increase with temperature (corresponding to semiconducting behaviour) or have such large band gaps that they are insulators. The electron or hole migration in these oxide semiconductors is attributed to a hopping mechanism. In this model, the electron hops from one localized metal atom site to the next. When it lands on a new site it causes the surrounding ions to adjust their locations and the electron or hole is trapped temporarily in the potential well produced by this distortion. The electron resides at its new site until it is thermally activated to migrate into another nearby site. Another aspect of this charge-hopping mechanism is that the electron or hole tends to associate with local defects, so the activation energy for charge transport may also include the energy of freeing the hole from its position next to a defect.

Hopping contrasts with the band model for semiconductivity, discussed in Section 2.11, where the conduction and valence electrons occupy orbitals that spread through the whole crystal. The difference stems from the less-diffuse d orbitals in the monoxides of the mid-to-late 3d metals, which are too compact to form the broad bands necessary for metallic conduction. When NiO is doped with Li_2O in an O_2 atmosphere a solid solution $Li_x(Ni^{2+})_{1-2x}(Ni^{3+})_xO$ is obtained, which has greatly increased conductivity for reasons similar to the increase in conductivity of Si when doped with In (Section 25.6). The characteristic pronounced increase in electronic conductivity with increasing temperature of metal oxide semiconductors is used in 'thermistors' to measure temperature.

In contrast to the semiconductivity of the monoxides in the centre and right of the 3d series, TiO and VO have high electronic conductivities that decrease with increasing temperature. This metallic conductivity persists over a broad composition range from highly oxygen-rich $Ti_{1-x}O$ to metal-rich TiO_{1-x}. In these compounds, a conduction band is formed by the overlap of the t_{2g} orbitals of metal ions in neighbouring octahedral sites that are oriented towards each other (Fig. 25.29).

The radial extension of the d orbitals of these early d-block elements is greater than for elements later in the period, and a band results from their overlap (Fig. 25.30); this band is only partly filled. The widely varying composition of TiO appears to be associated with the electronic delocalization: the conduction band serves as a rapidly accessible source and sink of electrons that can readily compensate for the formation of vacancies.

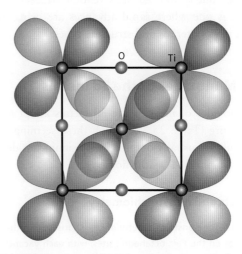

FIGURE 25.29 One face of the TiO rock-salt structure showing how orbital overlap can occur for the d_{xy}, d_{yz}, and d_{zx} orbitals.

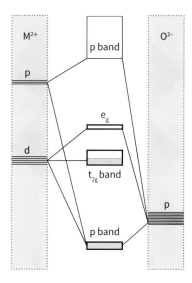

FIGURE 25.30 Band structure energy-level diagram for early d-metal monoxides. The t_{2g} band is partly filled for some early transition metals and metallic conduction results.

(c) Magnetic properties

KEY POINT The 3d-metal monoxides MnO, FeO, CoO, and NiO order antiferromagnetically with Néel temperatures that increase from Mn to Ni.

As well as having electronic properties that are the result of interactions between the d electrons, d-metal monoxides have magnetic properties that derive from cooperative interaction of the individual atomic magnetic moments (Section 22.8). The overall magnetic structure of MnO and the other 3d-series metal monoxides is shown in Fig. 25.31. The Néel temperatures (T_N, the temperature of the paramagnetic to antiferromagnetic transition; Section 22.8) of the series of d-metal oxides are as follows:

MnO	FeO	CoO	NiO
122 K	258 K	271 K	523 K

These values reflect the strength of the superexchange spin interactions (Section 22.8) along the M–O–M directions, which in the rock-salt type structure propagate in all three unit-cell directions. As the size of the M^{2+} ion decreases from Mn to Ni, the superexchange mechanism becomes stronger because of the increased metal–oxygen orbital overlap, and T_N increases.

25.11 Higher oxides and complex oxides

Binary metal oxides that do not have a 1:1 metal:oxygen ratio are known as **higher oxides**. Compounds containing ions of more than one metal are often termed **complex oxides** or **mixed oxides**, and include compounds containing three elements (ternary oxides; for instance, $LaFeO_3$), four (quaternary; for instance, $YBa_2Cu_3O_7$), or more elements. This section describes the structures and properties of some of the more important complex, mixed-metal oxides.

(a) The M_2O_3 corundum structure

KEY POINT The corundum structure is adopted by many oxides of the stoichiometry M_2O_3, including Cr-doped aluminium oxide (ruby).

α-Aluminium oxide (the mineral *corundum*) adopts a structure that can be modelled as a hexagonal close-packed array of O^{2-} ions with the cations in two-thirds of the octahedral holes (Fig. 25.32).

The corundum structure is also adopted by the oxides of Ti, V, Cr, Rh, Fe, and Ga in their +3 oxidation states. Two of these oxides, Ti_2O_3 and V_2O_3, exhibit metallic-to-semiconducting transitions below 410 and 150 K, respectively

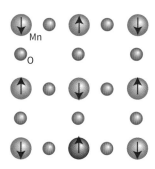

FIGURE 25.31 The overall magnetic structure of MnO and the other 3d-series metal monoxides.

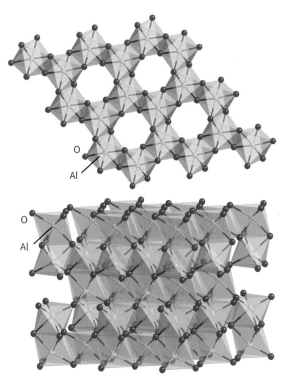

FIGURE 25.32 The corundum structure, as adopted by Al_2O_3, with cations occupying two-thirds of the octahedral holes between layers of close-packed oxide ions.

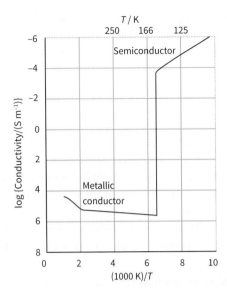

FIGURE 25.33 The temperature dependence of the electrical conductivity of V_2O_3, showing the metal-to-semiconductor transition.

(Fig. 25.33). In V_2O_3, the transition is accompanied by antiferromagnetic ordering of the spins. The two insulators Cr_2O_3 and Fe_2O_3 also display antiferromagnetic ordering.

Another interesting aspect of the M_2O_3 compounds is the formation of solid solutions of dark-green Cr_2O_3 and colourless Al_2O_3 to form brilliant red ruby (Section 22.5). The blue shift in the ligand-field transitions of Cr^{3+} stems from the compression of the O^{2-} ions around Cr^{3+} in the Al_2O_3 host structure. (In Al_2O_3, the lattice constants are $a = 475$ pm and $c = 1300$ pm, while in Cr_2O_3 the corresponding values are larger at 493 pm and 1356 pm, respectively.) The compression shifts the absorption towards the blue as the strength of the ligand field increases, and the solid appears red in white light. The responsiveness of the absorption (and fluorescence) spectrum of Cr^{3+} ions to compression is sometimes used to measure pressure in high-pressure experiments. In this application, a tiny crystal of ruby in one segment of the sample can be interrogated by visible light, and the shift in its fluorescence spectrum provides an indication of the pressure inside the cell.

(b) Rhenium trioxide structure

KEY POINT The rhenium trioxide structure can be constructed from ReO_6 octahedra sharing all vertices in three dimensions.

The rhenium trioxide structure type is very simple, consisting of a primitive cubic unit cell with Re atoms at the corners and O atoms at the midpoint of each edge (Fig. 25.34). Alternatively, the structure can be considered to be derived from ReO_6 octahedra sharing all vertices. The structure is also closely related to a perovskite structure (Section 25.11(d)) in which the A-type cation has been removed from the unit cell. Materials adopting the rhenium trioxide structure are

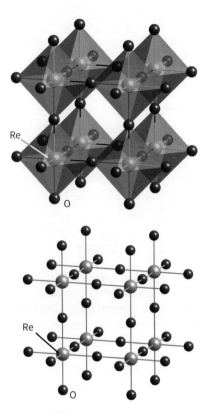

FIGURE 25.34 The ReO_3 structure shown as the unit cell, and the ReO_6 octahedra forming the unit cell.

relatively rare. This rarity is in part due to the requirement of the oxidation state M(VI) when M is in combination with oxygen. Rhenium(VI) oxide, ReO_3, itself and one form of UO_3 (δ-UO_3) have this structure type as do some metal trifluorides such as ScF_3. At room temperature, WO_3 exists in a modified version of the ReO_3 structure where the WO_6 octahedra are slightly distorted and tilted relative to each other so that the W–O–W bond angle is not 180°.

Rhenium trioxide itself is a bright-red, lustrous solid. Its electrical conductivity at room temperature is similar to that of copper metal. The band structure for this compound contains a band derived from the Re t_{2g} orbitals and the O 2p orbitals (Fig. 25.35). This band can contain up to six electrons per Re atom but is only partially filled for the Re^{6+} d^1 configuration, so producing the observed metallic properties.

(c) Spinels

KEY POINT The observation that many d-metal spinels do not have the normal spinel structure is related to the effect of ligand-field stabilization energies on the site preferences of the ions.

The d-block higher oxides Fe_3O_4, Co_3O_4, and Mn_3O_4, and many related mixed-metal compounds, such as $ZnFe_2O_4$, have very useful magnetic properties. They all adopt the structural type of the mineral spinel, $MgAl_2O_4$, and have the general formula AB_2O_4. Most oxide spinels are formed with

Metal oxides, nitrides, and fluorides

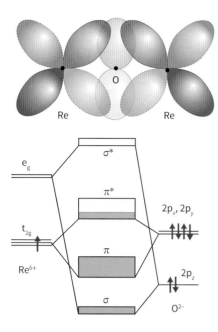

FIGURE 25.35 The band structure of ReO$_3$.

TABLE 25.4 Occupation factor, λ, in some spinels

	A*	Mg^{2+}	Mn^{2+}	Fe^{2+}	Co^{2+}	Ni^{2+}	Cu^{2+}	Zn^{2+}
	B	d^0	d^5	d^6	d^7	d^8	d^9	d^{10}
Al^{3+}	d^0	0		0	0	0	0.38	0
Cr^{3+}	d^3	0	0	0	0	0	0	0
Mn^{3+}	d^4	0						0
Fe^{3+}	d^5	0.45	0.1	0.5	0.5	0.5	0.5	0
Co^{3+}	d^6	0.5			0	0.5	0.25	0

* $\lambda = 0$ corresponds to a normal spinel; $\lambda = 0.5$ corresponds to an inverse spinel.

The observation that many d-metal spinels do not conform to this expectation has been traced to the effect of ligand-field stabilization energies on the site preferences of the ions.

The **occupation factor**, λ, of a spinel is the fraction of B atoms in the tetrahedral sites: $\lambda = 0$ for a normal spinel and $\lambda = 0.5$ for an inverse spinel, B[AB]O$_4$. Intermediate λ values indicate a level of disorder in the distribution, where B-type cations occupy that portion of the tetrahedral sites. The distribution of cations in (A^{2+},B^{3+}) spinels (Table 25.4) illustrates that, for d^0 A and B ions, the normal structure is preferred, as predicted by electrostatic considerations. Table 25.4 shows that when A^{2+} is a d^6, d^7, d^8, or d^9 ion and B^{3+} is Fe^{3+}, the inverse structure is generally favoured. This arrangement can be traced to the lack of ligand-field stabilization (Section 7.16) of the high-spin d^5 Fe^{3+} ion in either the octahedral or the tetrahedral site, and the ligand-field stabilization of the other d^n ions in the octahedral site.

For other combinations of d-metal ions on the A and B sites, the relative ligand-field stabilization energies of the different arrangements of the two ions on the octahedral and tetrahedral sites need to be calculated. It is also important to note that simple ligand-field stabilization appears to work over this limited range of cations. More detailed analysis is necessary when cations of different radii are present, or if any ions that are present do not adopt the high-spin configuration typical of most metals in spinels (for instance, Co^{3+} in Co$_3$O$_4$, which is low-spin d^6). Moreover, because λ is often found to depend on the temperature, care has to be taken in the synthesis of a spinel with a specific distribution of cations because slow cooling or quenching of a sample from a high reaction temperature can produce quite different cation distributions.

a combination of A^{2+} and B^{3+} cations (i.e. as A^{2+}B$_2^{3+}$O$_4$ in Mg^{2+}[Al^{3+}]$_2$O$_4$), although there are a number of spinels that can be formulated with A^{4+} and B^{2+} cations (as A^{4+}B$_2^{2+}$O$_4$, as in Ge^{4+}[Co^{2+}]$_2$O$_4$). The spinel structure was described briefly in Section 4.9, where we saw that it consists of an fcc array of O^{2-} ions in which the A ions reside in one eighth of the tetrahedral holes and the B ions inhabit half the octahedral holes (Figs. 4.45 and 25.36); this structure is commonly denoted A[B$_2$]O$_4$, where the atom type in the square bracket represents that occupying the octahedral sites. In the inverse spinel structure, the cation distribution is B[AB]O$_4$, with the more abundant B-type cation distributed over both coordination geometries. Lattice-enthalpy calculations based on a simple ionic model indicate that, for A^{2+} and B^{3+}, the normal spinel structure, A[B$_2$]O$_4$, should be the more stable.

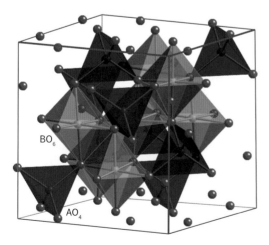

FIGURE 25.36 A segment of the spinel (AB$_2$O$_4$) unit cell, showing the tetrahedral environment of A ions and the octahedral environments of B ions (see also Fig. 4.45).

EXAMPLE 25.6 Predicting the structures of spinel compounds

Is MnCr$_2$O$_4$ likely to have a normal or inverse spinel structure?

Answer We need to consider whether there is a ligand-field stabilization. Because Cr^{3+} (d^3) has a large ligand-field stabilization energy (LFSE, 1.2Δ_O from Table 7.6) in the octahedral site (but a much smaller one in a tetrahedral field), whereas the high-spin d^5 Mn^{2+} ion does not have any LFSE, a normal spinel structure is expected. Table 25.4 shows that this prediction is verified experimentally.

Self-test 25.6 Table 25.4 indicates that $FeCr_2O_4$ is a normal spinel. Rationalize this observation.

The inverse spinels of formula AFe_2O_4 are sometimes classified as **ferrites** (the same term also applies in different circumstances to other iron oxides), and the magnetic properties and applications of these materials are discussed further in Section 30.6.

The compound $CoAl_2O_4$ is among the normal spinels in Table 25.4 with $\lambda = 0$, and thus has the Co^{2+} ions at the tetrahedral sites. The colour of $CoAl_2O_4$ (an intense blue) is that expected of tetrahedral Co^{2+}. This property, coupled with the ease of synthesis and stability of the spinel structure, has led to cobalt aluminate being used as a pigment ('cobalt blue'). Other mixed d-metal spinels that exhibit strong colours—for example, $CoCr_2O_4$ (green), $CuCr_2O_4$ (black), and $(Zn,Fe)Fe_2O_4$ (orange-brown)—are also used as pigments (Section 30.1), with applications that include colouring various construction materials, such as concrete.

Spinels which formally have the A-type cation as lithium and the B type cation as a first row transition metal or mixture thereof, i.e. LiM_2O_4, M = Ni, Co, Fe, Mn, are of interest as lithium-ion battery cathode materials (Section 30.7). Example phases include $LiMn_2O_4$ and $LiNi_{0.4}Mn_{1.6}O_4$. During charging lithium may be extracted from the tetrahedral sites and the transition metal is oxidized on the octahedral site to compensate.

(d) Perovskites and related phases

KEY POINTS The perovskites have the general formula ABX_3, in which the 12-coordinate hole of a ReO_3-type BX_3 structure is occupied by a large A ion; the perovskite barium titanate, $BaTiO_3$, exhibits ferroelectric and piezoelectric properties associated with cooperative displacements of the ions.

The perovskites have the general formula ABX_3, in which the 12-coordinate hole of BX_3 (as in ReO_3) is occupied by a large A ion (Fig. 25.37; different views of this structure are given in Figs. 4.43 and 4.44). The X ion is most frequently O^{2-} or F^- (as in $NaFeF_3$), although nitride- and hydride-containing perovskites can also be synthesized, as in $LiSrH_3$. The heavier larger halides, where X = Cl^-, Br^-, or I^-, can also form perovskites in combination with larger cations as A and B, and the semiconducting properties of some of these phases, such as $CsPbI_3$, are currently of great research interest for photovoltaic applications (Box 15.4, Section 30.2). 'Antiperovskites', with the formula $XX'A_3$, where X and X' are two different sized anions and A is a cation, are also known; one example is $OBrLi_3$.

The name 'perovskite' comes from the naturally occurring oxide mineral $CaTiO_3$, and the largest class of perovskites

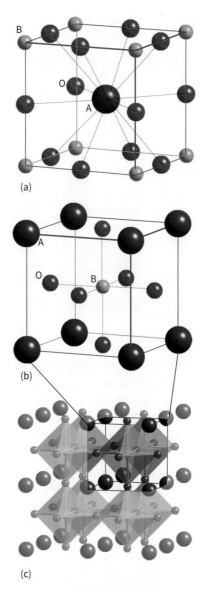

FIGURE 25.37 Views of the perovskite (ABO_3) structure (a) emphasizing the 12-fold coordination of the larger A cation and showing the relationship with the ReO_3 structure of Fig. 25.34, (b) highlighting the octahedral coordination of the B cation. (c) A polyhedral representation accentuating the BO_6 octahedra.

are those with the anion as oxide. This breadth of perovskites is widened by the observation that solid solutions and nonstoichiometry are also common features of the perovskite structure, as in $Ba_{1-x}Sr_xTiO_3$ and $SrFeO_{3-y}$. Some metal-rich materials adopt the perovskite structure with the normal distribution of cations and anions partially inverted; for instance, $SnNCo_3$. The use of perovskites in SOFCs has been discussed in Box 25.6.

The perovskite structure is often observed to be distorted such that the unit cell is no longer centrosymmetric, and a portion of the crystal can acquire an overall permanent electric polarization as a result of aligning the directions of ion displacements within that part of the crystal. Some polar crystals are **ferroelectric** in the sense that they resemble ferromagnets, but instead of the electron spins being aligned over a region of the crystal (often termed a *domain*), the electric dipole moments of many unit cells are aligned. As a result, the relative permittivity, which reflects the polarity of a compound, for a ferroelectric material often exceeds 1×10^3 and can be as high as 1.5×10^4; for comparison, the relative permittivity of liquid water is about 80 at room temperature.

Barium titanate, $BaTiO_3$, is the most extensively studied example of such a material. At temperatures above 120°C this compound has the perfect cubic perovskite structure. At room temperature it adopts a lower-symmetry tetragonal unit cell in which the various ions can be considered as having been displaced from their normal high-symmetry sites (Fig. 25.38). This displacement results in a spontaneous polarization of the unit cell and formation of an electric dipole; coupling between these ion displacements, and therefore the induced dipoles, is very weak. Application of an external electric field aligns these dipoles throughout the material, resulting in a bulk polarization in a particular direction, which can persist after removal of the electric field. The temperature below which this spontaneous polarization can occur and the material behaves as a ferroelectric is called the **Curie temperature** (T_C; Section 22.8). For $BaTiO_3$, $T_C = 120°C$.

The high relative permittivity of barium titanate at room temperature leads to its use in capacitors, where its presence allows up to 1000 times the charge to be stored in comparison with a capacitor with air between the plates. The introduction of dopants into the barium titanate structure, forming solid solutions, allows various properties of the compound to be tuned. For example, the replacement of Ba by Sr or of Ti by Zr causes a sharp lowering of T_C.

Another characteristic of many crystals, including a number of perovskites that lack a centre of symmetry, is **piezoelectricity**, the generation of an electrical field when the crystal is under stress or the change in dimensions of the crystal when an electrical field is applied. Piezoelectric materials are used for a variety of applications, such as pressure transducers (e.g. in mechanical ignition devices for gas cookers and fires), ultramicromanipulators (where very small movements can be controlled), sound detectors, energy harvesting (Section 30.4), and as the probe support in scanning tunnelling microscopy (Section 8.16). Some important examples are $BaTiO_3$, $NaNbO_3$, $NaTaO_3$, and $KTaO_3$.

Another common structure type, that adopted by potassium tetrafluoridonickelate(II), K_2NiF_4 (Fig. 25.39), is related to perovskite. The compound can be thought of as containing individual slices from the perovskite structure that share the four F atoms from the octahedra within the layer, but have terminal F atoms above and below the layer. These layers are displaced relative to each other and separated by the K^+ ions (which are nine-coordinate, to eight F^- ions of one layer and one terminal F^- ion from

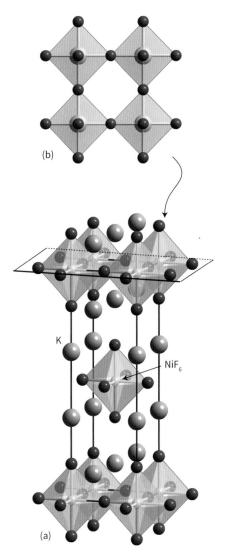

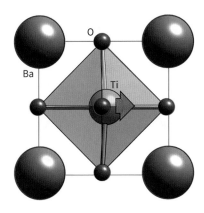

FIGURE 25.38 The tetragonal $BaTiO_3$ structure, showing the local Ti^{4+} ion displacement that leads to the ferroelectric behaviour of this material.

FIGURE 25.39 The K_2NiF_4 structure: (a) the displaced layers of NiF_6 octahedra interspersed with K^+ ions, and (b) a view of one layer of composition NiF_4 showing the corner-sharing octahedra linked through F.

the next). Compounds with the K_2NiF_4 structure type have come under renewed investigation because some high-temperature superconductors, such as $La_{1.85}Sr_{0.15}CuO_4$, crystallize with this structure. Apart from their importance in superconductivity, compounds with the K_2NiF_4 structure also provide an opportunity to investigate two-dimensional magnetic domains, as coupling between electron spins is much stronger within the layers of linked octahedra than between the layers.

The K_2NiF_4 structure has been introduced as being derived from a single slice of the perovskite structure; other related structures are possible where two or more perovskite layers are displaced horizontally relative to each other. Structures with K_2NiF_4 at one end of the range (a single perovskite layer) and perovskite itself at the other (an infinite number of such layers) are known as **Ruddlesden–Popper phases**. They include $Sr_3Fe_2O_7$ with double layers and $Ca_4Mn_3O_{10}$ with triple layers (Fig. 25.40).

The potentially important photovoltaic applications of lead halide perovskites have been discussed in Box 15.4. Materials such as methylammonium lead iodide, $[CH_3NH_3]^+[PbI_3]^-$, are often referred to as 'hybrid materials' in that they contain both inorganic and organic species in their compositions and structures. This area, at the frontiers of inorganic chemistry, is exciting for the development of new functional materials whose properties, in this case optical and semiconducting, can be designed and controlled through their chemistry. Current research is this area is focusing on synthesizing new hybrid perovskite materials based on less toxic, post-transition metals, such as tin and bismuth, in combination with chalcogenides and halides.

(e) Pyrochlores

KEY POINT The pyrochlore structure, found for phases of the composition $A_2B_2O_7$, has a structure related to that of fluorite.

The pyrochlore structure is adopted by many materials of the composition $A_2B_2X_7$ where A and B are different-sized cations and X is an anion (usually O^{2-} or F^-) or mixture of anions (typically O^{2-}, F^-, or sometimes N^{3-}). The A cation is often a Group 1, Group 2, Group 3, lanthanoid, or post-transition metal cation, such as Cs, Ca, La, or Cd, and B is a heavy d-block metal, such as Nb, W, Zr, Ru. Example phases include $La_2Zr_2O_7$, $(Ca,Na)_2Nb_2O_6F$, and $Bi_2Ru_2O_7$. The pyrochlore structure can be considered to be related to the fluorite structure, where the A and B cations (replacing the Ca^{2+} sites of the fluorite, CaF_2, structure) are ordered in one direction and an ordered anion vacancy exists on tetrahedral sites between adjacent B-site cations, i.e. one eighth of the anion sites (formally occupying all tetrahedral holes in fluorite) are vacant, Fig. 25.41. Defect pyrochlores, $A_2B_2O_{7-x}$ ($0 < x < 1$) are also known; these have additional vacancies on the anion sub-lattice. In the $A_2B_2O_7$ composition, the pyrochlore structure is found to be stable when the ionic radius ratio r_A/r_B is greater than 1.26. At lower values the fluorite structure with disordered A and B cations is adopted.

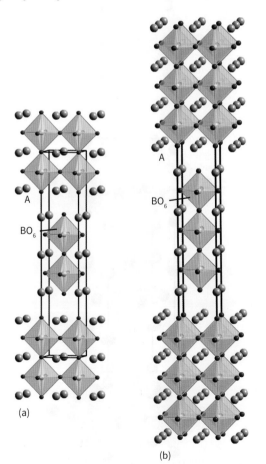

FIGURE 25.40 The Ruddlesden–Popper phases of stoichiometry (a) $A_3B_2O_7$ and (b) $A_4B_3O_{10}$, formed respectively from two and three perovskite layers of linked BO_6 octahedra separated by A-type cations.

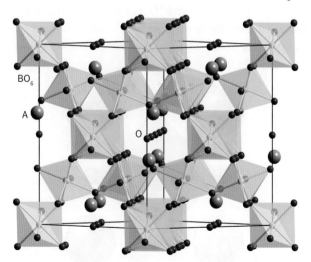

FIGURE 25.41 The pyrochlore structure adopted by many compounds of stoichiometry, $A_2B_2O_7$. The BO_6 octahedra are shown which form channels containing the A-type cations and oxide ions.

Lanthanum zirconate, $La_2Zr_2O_7$, exhibits a low thermal conductivity and high-temperature stability leading to applications in thermal barrier coatings and as a reinforcement material. Some pyrochlore and defect pyrochlore phases show superconductivity; for example, $Cd_2Re_2O_7$ is superconducting below 1.4 K, and KOs_2O_6 is superconducting below 10 K. $Bi_2Ru_2O_7$ is an excellent electrocatalyst for the oxygen evolution reaction in water-splitting; Section 30.3.

25.12 Oxide glasses

The term **ceramic** is often applied to all inorganic nonmetallic, nonmolecular materials, including both amorphous and crystalline materials, but the term is more normally reserved for compounds or mixtures that have undergone heat treatment and sintered to form dense complex oxides. The term **glass** is used in a variety of contexts, but for our present purposes it implies an amorphous ceramic with a viscosity so high that it can be considered to be rigid. A substance in its glassy form is said to be in its **vitreous** state. Although ceramics and glasses have been utilized since antiquity, their development is currently an area of rapid scientific and technological progress. This enthusiasm stems from interest in the scientific basis of their properties and the development of novel synthetic routes to new high-performance materials. We confine our attention here to glasses. The most familiar glasses are alkali-metal or alkaline-earth-metal silicates and borosilicates.

(a) Glass formation

KEY POINTS Silicon dioxide readily forms a glass because the three-dimensional network of strong covalent Si–O bonds in the melt does not easily break and reform on cooling; the Zachariasen rules summarize the properties likely to lead to glass formation.

A glass is prepared by cooling a melt more quickly than it can crystallize. Cooling molten silica, for instance, produces vitreous quartz. Under these conditions the solid has no long-range order as judged by the lack of X-ray diffraction peaks, but spectroscopic and other data indicate that each Si atom is surrounded by a tetrahedral array of O atoms.

The lack of long-range order results from variations of the Si–O–Si angles. Fig. 25.42(a) illustrates in two dimensions how a local coordination environment can be preserved but long-range order lost by variation of the bond angles around O. This loss of long-range order is readily apparent when X-rays are scattered from a glass (Fig. 25.42(b)): in contrast to a long-range, periodically ordered crystalline material where diffraction gives rise to a series of diffraction maxima (Section 8.1), the X-ray diffraction pattern obtained from a glass shows only broad features as the long-range order is lost.

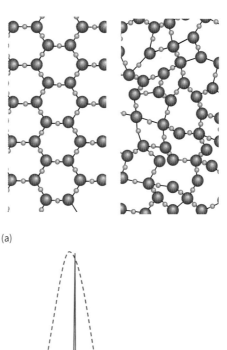

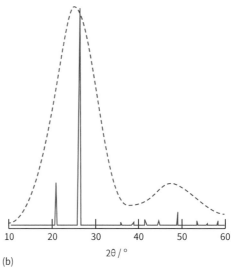

FIGURE 25.42 (a) Schematic representation of a two-dimensional crystal (left), compared with a two-dimensional glass (right). (b) The powder X-ray diffraction pattern of a glass (SiO_2, dashed red line) contrasted with that of a crystalline solid (quartz SiO_2, solid blue line). The long-range order, which gives rise to the sharp diffraction maxima in quartz, is no longer present in glassy SiO_2 and only broad features are seen.

Silicon dioxide readily forms a glass because the three-dimensional network of strong covalent Si–O bonds in the melt does not readily break and reform on cooling. The lack of strong directional bonds in metals and simple ionic substances makes it much more difficult to form glasses from these materials. Recently, however, techniques have been developed for ultrafast cooling and, as a result, a wide variety of metals and simple inorganic materials can now be frozen into a vitreous state.

The concept that the local coordination sphere of the glass-forming element is preserved, but that bond angles around the linking oxygen atom are variable was originally proposed by W. H. Zachariasen in 1932. He reasoned that these conditions would lead to similar molar Gibbs energies and molar volumes for the glass and its crystalline counterpart. Zachariasen also proposed that the vitreous state is

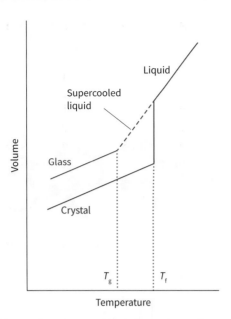

FIGURE 25.43 Comparison of the volume change for supercooled liquids and glasses with that for a crystalline material. The glass transition temperature is T_g and T_f the melting point.

(b) Glass composition, production, and application

KEY POINTS Low-valence metal oxides, such as Na_2O and CaO, are often added to silica to reduce its softening temperature by disrupting the silicon–oxygen framework. Other cations may be incorporated into glasses, giving applications as diverse as lasers and nuclear-waste containment.

Although vitreous silica is a strong glass that can withstand rapid cooling or heating without cracking, it has a high glass transition temperature and must be worked at inconveniently high temperatures. Therefore, a **modifier**, such as Na_2O or CaO, is commonly added to SiO_2. A modifier disrupts some of the Si–O–Si linkages and replaces them with terminal Si–O$^-$ links that associate with the cation (Fig. 25.44). The consequent partial disruption of the Si–O network leads to glasses that have lower softening points. The common glass used in bottles and windows is called 'sodalime glass' and contains Na_2O and CaO as modifiers. When B_2O_3 is used as a modifier, the resulting 'borosilicate glasses' have lower thermal expansion coefficients than sodalime glass and are less likely to crack when heated. Borosilicate glass (such as Pyrex®) is therefore widely used for ovenware and laboratory glassware.

Glass formation is a property of many oxides, and practical glasses have been made from sulfides, fluorides, and other anionic constituents. Some of the best glass formers are the oxides of elements near silicon in the periodic table (B_2O_3, GeO_2, and P_2O_5), but the solubility in water of most borate and phosphate glasses and the high cost of germanium limit their usefulness.

favoured by polyhedral corner-sharing O atoms, rather than edge- or face-shared, which would enforce greater order. These and other **Zachariasen rules** hold for common glass-forming oxides, but exceptions are known.

An instructive comparison between vitreous and crystalline materials is seen in their change in volume with temperature (Fig. 25.43). When a molten material crystallizes, an abrupt change in volume (usually a decrease) occurs, representing the more efficient and closer packing of atoms, ions, or molecules in the solid phase. This is known as a **first-order phase transition** accompanied by a change in enthalpy. By contrast, a glass-forming material that is cooled sufficiently rapidly persists as a metastable supercooled liquid. When cooled below the **glass transition temperature**, T_g, the supercooled liquid becomes rigid, and this change is accompanied by only an inflection in the cooling curve rather than an abrupt change in slope; this observation is consistent with no change in enthalpy but a change in heat capacity of the system, and indicates that the structure of the glassy, disordered solid is similar to that of the liquid. The rates of crystallization are very slow for many complex metal silicates, phosphates, and borates, and it is these compounds that often form glasses.

Glasses or ceramics containing TiO_2 and Al_2O_3 can be prepared using sol–gel methods (Section 25.9) at much lower temperatures than required to produce the ceramic from the simple oxides. Often a special shape can be fashioned at the gel stage. Thus, the gel may be shaped into a fibre and then heated to expel water; this process produces a glass or ceramic fibre at much lower temperatures than would be necessary if the fibre were made from the melt of the components.

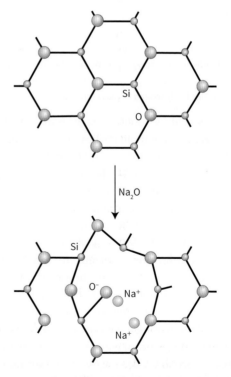

FIGURE 25.44 The role of a modifier is to introduce terminal O$^-$ units and cations that disrupt the lattice.

The development of transparent crystalline and vitreous materials for light transmission and processing has led to a revolution in signal transmission. For example, optical fibres are currently being produced with a composition gradient from the interior to the surface. This composition gradient modifies the refractive index, and thereby decreases loss of light from the sides of the glass fibre. Fluoride glasses are also being investigated as possible substitutes for oxide glasses, because oxide glasses contain small numbers of OH groups which absorb near-infrared radiation and attenuate the signal. Doping lanthanoid ions into the glass fibres produces materials that can be used to amplify the signals by using lasing effects (Box 22.1). Optical circuit elements are being developed that may eventually replace all the components in an electronic integrated circuit and lead to very fast optical computers.

As modifier cations effectively become trapped in a glass, which is chemically inert and thermodynamically very stable with respect to transformation to a soluble crystalline phase, such glassy materials offer a potential method for containing and storing nuclear waste as 'Synroc' ('synthetic rock'). Thus, vitrification of metal oxides containing a radioactive species with glass-forming oxides produces a stable glass that can be stored for long periods, allowing the radionuclides to decay.

The incorporation of functional inorganic compounds into glasses has led to the development of so-called 'smart glasses' that have switchable properties, including **electrochromism**, the ability to change colour or light transmission properties in response to application of a potential difference, and reversible **photochromism**, the ability to change colour under certain light conditions.

An electrochromic glass generally consists of a glass coated with colourless tungsten trioxide, WO_3, or a sandwich-like layer structure consisting of these two components. When a potential difference is applied to the WO_3 layer, cations are inserted and the W is partially reduced to form $M_xW(VI,V)O_3$, which is dark blue. The coating maintains its dark colour until the potential difference is reversed, when the glass becomes colourless.

Applications of electrochromic glasses include privacy glass, auto-dimming rear-view mirrors in vehicles, and aircraft windows. A similar coating is used in self-cleaning glasses, where a metal oxide coating on the glass acts as a photocatalyst for the breakdown of organic dirt on their surfaces (Section 30.3(b) and Fig. 30.11). A photochromic glass incorporates a small amount of a colourless silver halide, normally AgCl, within the glass. When exposed to UV radiation, a portion of sunlight, the AgCl dissociates to form small clusters of Ag atoms, which absorb light across the visible region of the spectrum, imparting a grey colour to the glass. When the UV radiation is removed, the AgCl reforms and the glass returns to its optically transparent state. Photochromic glasses are widely used in the lenses of spectacles and sunglasses.

25.13 Nitrides, fluorides, and mixed-anion phases

The solid-state chemistry of metals in combination with anions other than O^{2-} is not as highly developed or extensive as that of the complex oxides described so far. However, the solid-state chemistries of complex nitrides and fluorides and of mixed-anion compounds (oxide-nitrides, oxide-sulfides, and oxide-fluorides) are of growing importance. The chemistry of carbides was described in Section 15.13.

(a) Nitrides

KEY POINTS Complex metal nitrides and oxide-nitrides are materials containing the N^{3-} anion; many new compounds of this type have recently been synthesized.

Simple metal nitrides of main-group elements, such as AlN, GaN, and Li_3N, have been known for decades. Many of the recent advances in nitride chemistry have centred on d-metal compounds and complex nitrides. That nitrides are less common than oxides stems, in part, from the high enthalpy of formation of N^{3-} compared with that of O^{2-}. Furthermore, because many nitrides are sensitive to oxygen and water, their synthesis and handling are problematic. Some simple metal nitrides can be obtained by the direct reaction of the elements; for example, Li_3N is obtained by heating lithium in a stream of nitrogen at 400°C. The instability of sodium nitride allows sodium azide to be used as a nitriding agent:

$$2\,NaN_3(s) + 9\,Sr(s) + 6\,Ge(s) \xrightarrow{750°C,\ sealed\ Nb\ tube} 3\,Sr_3Ge_2N_2(s) + 2\,Na(g)$$

The ammonolysis of oxides (the dehydrogenation of NH_3 by oxides with the formation of water as a by-product) provides a convenient route to some nitrides. For instance, tantalum nitride can be obtained by heating tantalum pentoxide in a fast-flowing stream of ammonia:

$$3\,Ta_2O_5(s) + 10\,NH_3(l) \xrightarrow{700°C} 2\,Ta_3N_5(s) + 15\,H_2O(g)$$

In such reactions, the equilibrium is driven towards the products by removal of the steam in the gas flow. Similar reactions may be used for the preparation of complex nitrides from complex oxides, although competing reactions involving partial reduction of the metal oxide by ammonia can also occur. There is also a tendency with nitrides for the formation of compounds in which the metallic element is in a lower oxidation state because nitrogen, on account of its high bond energy, is not as potent an oxidant as oxygen or fluorine. Thus, whereas heating titanium in oxygen readily produces TiO_2, and Ti_2N and TiN are known, Ti_3N_4 is difficult to prepare and poorly characterized. Likewise, V_3N_5 is unknown, whereas V_2O_5 is readily obtained from the decomposition of many vanadium salts in air.

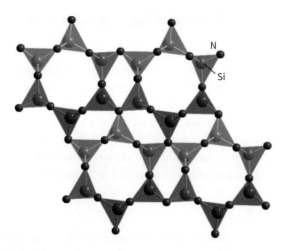

FIGURE 25.45 The structure of Si_3N_4, shown as linked SiN_4 tetrahedra.

Many of the early d-metal nitrides are interstitial compounds and are used as high-temperature refractory ceramics. Similarly, the nitrides of Si and Al, including Si_3N_4 (Fig. 25.45), are stable at very high temperatures, particularly under nonoxidizing conditions, and are used for crucibles and furnace elements.

Recently, GaN, which can exist in both wurtzite and sphalerite structural types, has been the focus of considerable research on account of its semiconducting properties. The nitride Li_3N has an unusual structure based on hexagonal Li_2N^- layers separated by Li^+ ions (Section 12.12 and Fig. 25.46). These Li^+ ions are highly mobile, as is expected from the existence of free space between the layers, and this compound and other structurally related materials are being studied for possible use in rechargeable batteries. Among the many complex nitrides that have been synthesized are materials of stoichiometry AMN_2, such as $SrZrN_2$ and $CaTaN_2$, with structures based on sheets formed from MN_6 octahedra sharing edges (see the discussion of $LiCoO_2$ in Section 25.8(d)) and A_2MN_3.

(b) Oxide-nitrides

KEY POINT Partially replacing oxide by nitride in solids allows control of their band gaps and leads to their application as pigments and photocatalysts.

As noted above, heating oxides under ammonia can lead to the complete or partial displacement of oxide by nitride. In general, complete elimination of O^{2-} ions from the product can be difficult, requiring large quantities of ammonia to drive the equilibrium over to the formation of the nitride, so the reactions often give products that contain both the oxide and nitride ions:

$$Ca_2Ta_2O_7(s) + 2\,NH_3(g) \xrightarrow{800°C} 2\,CaTaO_2N(s) + 3\,H_2O(g)$$

In comparison with the O^{2-} ion, the higher charge of the N^{3-} ion results in a greater degree of covalency in its bonding and therefore nitrides—particularly those of less electropositive elements such as d metals—should not be described in purely ionic terms. In terms of their band structures, the higher energy of a nitride valence band causes a narrowing of the band gap in oxide-nitrides in comparison with pure oxides. This ability to reduce the band gap is of importance in designing materials that absorb visible light (effectively a valence-to-conduction-band or nitride-to-metal charge-transfer electron transition), as opposed to many oxides (such as TiO_2 and Ta_2O_5) which only absorb in the UV region because of their large band gaps. Thus, titanium and tantalum oxide-nitrides, such as $CaTaO_2N$ synthesized as above, are intensely coloured—typically yellow, orange, and red—and have been used as pigments. Coloured oxide-nitrides are also being considered for applications as photocatalysts (Section 30.3), where again, absorption of light across the visible region is beneficial.

(c) Fluorides and other halides

KEY POINT Because fluorine and oxygen have similar ionic radii, fluoride solid-state chemistry parallels much of oxide chemistry.

The ionic radii of F^- and O^{2-} are very similar (at between 130 and 140 pm), and as a result, metal fluorides show many stoichiometric and structural analogies with the complex oxides but with lower charge on the metal ion to reflect the lower charge on the F^- ion. Many binary metal fluorides adopt the simple structural types expected on the basis of the radius ratio rule (Section 4.10). For example, FeF_2 and PdF_2 have a rutile structure and AgF has a rocksalt structure; similarly, NbF_3 adopts the ReO_3 structure. For complex fluorides, analogues of typical oxide structural types are well known, including perovskites (such as $KMnF_3$), Ruddlesden–Popper phases (e.g. $K_3Co_2F_7$), and

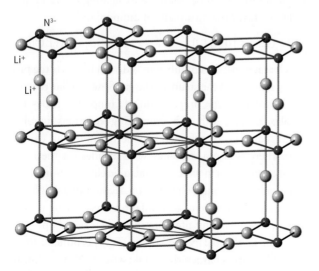

FIGURE 25.46 The Li_3N structure, showing hexagonal layers of composition $[Li_2N]^-$ separated by Li^+ cations.

spinels (Li_2NiF_4) (Sections 25.11(c) and (d)). Synthetic routes to complex fluorides also parallel those for oxides. For instance, the direct reaction of two metal fluorides yields the complex fluoride, as in

$$2\,LiF(s) + NiF_2(s) \rightarrow Li_2NiF_4(s)$$

Like some complex oxides, some complex fluorides may be precipitated from solution:

$$MnBr_2(aq) + 3\,KF(aq) \rightarrow KMnF_3(s) + 2\,KBr(aq)$$

As with the ammonolysis of oxides to produce nitrides and oxide-nitrides, the formation of oxide-fluorides is possible by the appropriate treatment of a complex oxide, as in

$$Sr_2CuO_3(s) \xrightarrow{F_2,\,200°C} Sr_2CuO_2F_{2+x}(s)$$

$Sr_2CuO_2F_{2+x}$ is a superconductor with $T_c = 45\,K$.

Fluoride analogues of the silicate glasses, which are based on linked SiO_4 tetrahedra, exist for small cations that form tetrahedral units in combination with F^-; an example is $LiBF_4$, which contains linked BF_4 tetrahedra. Lithium borofluoride glasses are used to contain samples for X-ray work because they are highly transparent to X-rays on account of their low electron densities. Framework and layer structures based on linked MF_4 (M = Li, Be) tetrahedra have also been described. Some metal fluorides are used as fluorination agents in organic chemistry. However, few of the solid complex metal fluorides are technologically important in comparison with the wealth of applications associated with analogous complex oxides.

Metal chloride structures reflect the greater covalence associated with bonding to chloride in comparison with fluoride: the chlorides are less ionic and have structures with lower coordination numbers than the corresponding fluorides. Thus, simple metal chlorides, bromides, and iodides normally adopt the cadmium-chloride or cadmium-iodide structures based on sheets formed from edge-sharing MX_6 octahedra. Complex chlorides often contain the same structural unit; for example, $CsNiCl_3$ has chains of edge-sharing $NiCl_6$ octahedra separated by Cs^+ ions. Many analogues of oxide structures also occur among the complex chlorides, such as $KMnCl_3$, K_2MnCl_4, and Li_2MnCl_4, which have the perovskite, K_2NiF_4, and spinel structures, respectively.

Sulfides, intercalation compounds, and metal-rich phases

25.14 Metal sulfides, selenides, and tellurides

The soft chalcogens S, Se, and Te form binary compounds with metals that can have quite different structures from the corresponding oxides, nitrides, and fluorides. As we saw in Sections 4.9 and 17.9 and Chapter 9, this difference is consistent with the greater covalence of the compounds of sulfur and its heavier congeners. For example, we noted there that MO compounds generally adopt the rock-salt structure, whereas ZnS and CdS can crystallize with either of the sphalerite or the wurtzite structures, in which the lower coordination numbers indicate the presence of directional bonding. Similarly, the d-block monosulfides generally adopt the more characteristically covalent nickel-arsenide structure rather than the rock-salt structure of alkaline-earth oxides such as MgO. Even more striking are the layered MS_2 compounds formed by many d-block elements, in contrast to the fluorite or rutile structures of many d-block dioxides.

(a) Main group metal chalcogenides

KEY POINT Electropositive s-block metal chalcogenides adopt structures derived from the close-packing of the anions.

The s-block metals form simple sulfides of the expected compositions M_2S and MS. The alkali metal sulfides (M_2S, M = Li, Na, K, Rb, Cs) adopt the antifluorite structure (Section 4.9). Of the Group II sulfides, MS, BeS adopts the cubic sphalerite structure with tetrahedral coordination for Be, while the remainder of the MS phases (M = Mg, Ca, Sr, Ba, Ra) adopt the rock-salt structure type with 6:6 coordination. Selenides M_2Se and MSe adopt analogous structures.

Sulfides are known for the Group 13 heavy metals in their +1 and +3 oxidation states; for example, In_2S_3 and Tl_2S. Phases of the composition In_5S_6 and In_6S_7 are known, and In_5S_7 contains close packed S_2^- anions with a mixture of In^+ and In^{3+} cations in octahedral holes.

In Group 14, SnS_2 adopts the CdI_2 structure type, while GeS and SnS adopt layer-like structures with five-coordinate Sn^{2+}/Ge^{2+} and the lone pair directed into the interlayer space. PbS, PbSe, and PbTe all adopt the rock-salt structure type.

(b) Transition metal chalcogenide chemistry

KEY POINTS Transition metal chalcogenides often exhibit a degree of covalency and weak metal–metal bonding. Dichalcogenide anions, $[X_2]^{2-}$, are often found in these materials.

Transition metal chalcogenides are known for a large range of stoichiometries, though the most common are the chalcogenides with the stoichiometries MX and MX_2. Metal-rich compositions, such as M_2S, often exhibit extensive metal–metal bonding, and chalcogenide-rich materials, such as Re_2S_7, contain chalcogen–chalcogen bonding. The extensive chemistry of layered MX_2 phases is described in Section 25.14(c) and Chevrel phases, which contain metal-chalcogenide clusters, are discussed in Section 25.14(d).

Phases of the composition MX are known for many divalent transition metals—particularly the later first-row transition metals. They typically adopt one of three main structure types described in Chapter 4, with a degree of covalent bonding cubic ZnS (sphalerite), hexagonal ZnS (wurtzite), or NiAs. Metals from Groups 9, 10, 11, and 12 are often found as sulfides in Nature; for example, the ores millerite (NiS), greenockite (CdS), and cinnabar (HgS). The Group 11 metals are also often found in Nature as the sulfides, with the overall stoichiometry M_2S; Cu_2S (chalcocite), Ag_2S (acanthite), and $Ag_{1.5}Au_{0.5}Te$ (petzite). Industrially important monochalcogenides include the pigments CdS (yellow) and CdSe (red) (Section 30.1).

The persulfido anion $[S_2]^{2-}$ occurs in iron pyrites, FeS_2, and other mid- and late- transition metal disulfides (Mn, Co, Ni) adopt the pyrite or the closely related marcasite structure (adopting an orthorhombic unit cell). Early transition metals such as Ti, V, Mo, and W, which adopt a tetravalent oxidation state, form compounds with two $[S_2]^{2-}$ anions, $[M^{4+}][S_2^{2-}]_2$. The mineral patrónite, VS_4, is of this type. Several early transition metals (Ti, V, Cr, Mn groups) also form trichalcogenides (X = S, Se) with the overall stoichiometry 'MX_3'; these contain a mixture of $[X_2]^{2-}$ and X^{2-} anions as $[M]^{4+}[[X_2]^{2-}[X]^{2-}$; an example is NbS_3.

Thiospinels of the composition AB_2S_4 adopt the spinel structure with a close-packed array of S^{2-} anions (Section 25.11(c)), and include the compounds $ZnCr_2S_4$, $FeNi_2S_4$ (Fe^{2+} and Ni^{3+}), and Co_3S_4 ($Co^{2+}[Co^{3+}]_2S_4$). The Cr_3S_4 structure-type is derived from the CdI_2 structure-type (Section 4.9) with close-packed S^{2-} anions. In Cr_3S_4, Cr^{3+} ions occupy the octahedral holes between alternate close-packed layers (as Cd does in CdI_2), and half the octahedral sites in the other inter-sulfide layer gap are also occupied (by Cr^{2+}) in an ordered manner.

(c) Layered MX_2 (X = S, Se) compounds and intercalation

The layered metal sulfides and their intercalation compounds were introduced in Section 17.9.

In TaS_2 and many other layered disulfides, the d-metal ions are located in octahedral holes between close-packed AB layers (Fig. 25.47(a)). The Ta ions form a close-packed layer denoted X, so the metal and adjoining sulfide layers can be portrayed as an AXB sandwich. These sandwich-like slabs form a three-dimensional crystal by stacking in sequences such as ...AXBAXBAXB..., where the strongly bound AXB slabs are held to their neighbours by weak dispersion forces.

An alternative view of these MS_2 structures, which have the metal ions in octahedral holes, is as MS_6 octahedra sharing edges (Fig. 25.48(a)). This reinforces the idea of the greater degree of covalent bonding that occurs in these materials than in, for instance, Li_2S, which has the antifluorite structure.

The Nb atoms in NbS_2 reside in the trigonal-prismatic holes between sulfide layers that are in register with one another (AA, Figs. 25.47(b) and 25.48(b)). The Nb atoms, which are strongly bonded to the adjacent sulfide layers,

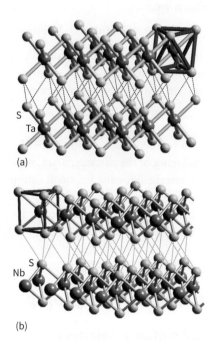

FIGURE 25.47 (a) The structure of TaS_2 (CdI_2-type); the Ta atoms reside in octahedral sites between the AB layers of S atoms. (b) The NbS_2 structure; the Nb atoms reside in trigonal-prismatic sites, between the sulfide layers.

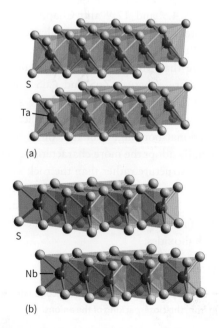

FIGURE 25.48 The metal-disulfide structures of Fig. 25.47 drawn as layers of MS_6 polyhedra sharing edges: (a) octahedra in TaS_2, (b) trigonal prisms in NbS_2.

form a close-packed array denoted m, so we can represent each slab as AmA or CmC. These slabs form a three-dimensional crystal by stacking in a pattern such as ...AmACmCAmACmC.... Weak dispersion forces also contribute to holding these AmA and CmC slabs together. Polytypes (versions that differ only in the stacking arrangement along a direction perpendicular to the plane of the slabs) can occur; thus, NbS$_2$ and MoS$_2$ form several polytypes, including one with the sequence ...CmCAmABmB.... Molybdenum sulfide, MoS$_2$, is used as a high-performance lubricant in, for example, racing cars and machining, as it can be used at much higher temperatures and pressures than can oils. Lubrication occurs with a dry coating of the material, as the MoS$_2$ layers are able to slip over each other easily because of the weak interlayer interactions.

(d) Metal-rich chalcogenides and Chevrel phases

In most of their chalcogenides, transition metals adopt oxidation states of II or greater. Nonetheless, several examples exist where the metallic atoms far outnumber the chalcogens. Such compounds typically have extensive metal–metal bonding, and one example is Nb$_{21}$S$_8$, shown in Fig. 25.49.

Chevrel phases form an important class of metal-rich ternary chalcogenide compounds first reported by R. Chevrel in 1971. These compounds, which illustrate three-dimensional intercalation, have formulas such as Mo$_6$X$_8$ and A$_x$Mo$_6$S$_8$; Se or Te may take the place of S and the intercalated A atom may be a variety of metals including Li, Mn, Fe, Cd, or Pb. The parent compounds, Mo$_6$Se$_8$ and Mo$_6$Te$_8$, are prepared by heating the elements at about 1000°C. A structural unit common to this series is M$_6$S$_8$, which may be viewed as an octahedron of M atoms face-bridged by S atoms, or alternatively as an octahedron of M atoms in a cube of S atoms (Fig. 25.50(a)). This type of cluster is also observed for some halides of the Period 4 and 5 early d-block elements, including the [M$_6$X$_8$]$^{4+}$ cluster found in Mo and W dichlorides, bromides, and iodides.

Figure 25.50(b) shows that in the three-dimensional solid the Mo$_6$S$_8$ clusters are tilted relative to each other and relative to the sites occupied by intercalated ions. This tilting allows a secondary donor–acceptor interaction between vacant Mo4d$_{z^2}$ orbitals (which project outward from the faces of the Mo$_6$S$_8$ cube) and a filled donor orbital on the S atoms of adjacent clusters. One of the physical properties that has drawn attention to the Chevrel phases is their superconductivity. Superconductivity exists up to 14 K in PbMo$_6$S$_8$, and it also persists to very high magnetic fields, which is of considerable practical interest because many applications involve high fields (over 25 T)—including NMR instruments and medical MRI (magnetic resonance imaging) scanners. In this respect the Chevrel phases appear to be significantly superior to the newer oxocuprate high-temperature superconductors (Section 30.7).

25.15 Metal hydrides and solids for hydrogen storage

Hydrogen forms metal hydrides by reaction with many metals and metal alloys (Sections 11.6(b) and (c)). In many cases this process can be reversed, liberating H$_2$ and regenerating the metal or alloy, by heating the (complex) metal hydride. These metal hydrides have an important safety advantage over pressurized or liquefied hydrogen (and many physisorbed hydrogen systems) that are also being considered for hydrogen storage, where rapid uncontrolled release

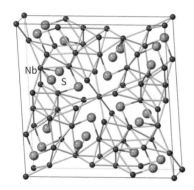

FIGURE 25.49 The structure of the metal-rich sulfide Nb$_{21}$S$_8$, showing the Nb-Nb bonding.

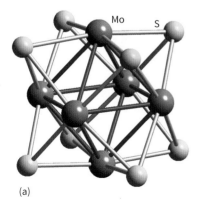

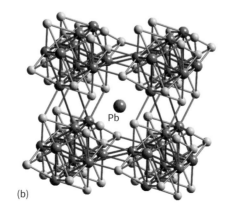

FIGURE 25.50 (a) The Mo$_6$S$_8$ unit present in a Chevrel-phase Pb$_x$Mo$_6$S$_8$. (b) The arrangement of the [M$_6$X$_8$]$^{4-}$ clusters in Chevrel phases.

of hydrogen could be a concern (Boxes 11.5 and 13.4). For hydrogen-storage applications, compounds that contain hydrogen in combination with the light elements (such as Li, Be, Na, Mg, B, and Al) offer some of the most promising materials, particularly as hydrogen to metal ratios can reach 2 or more. Simple metal-hydride systems have been dealt with in the appropriate sections of Chapters 11, 12, and 13. Here we expand on the solid-state chemistry of these compounds and the related complex metal hydrides. Recent research on metal hydrides and their applications is described in Section 30.8.

(a) Magnesium-based metal hydrides

KEY POINT Magnesium hydride contains a high weight per cent of hydrogen and is a potential hydrogen-storage material if the hydrogen release temperature can be reduced.

Magnesium hydride, MgH_2, adopts the rutile structure type, Fig. 4.40, and potentially offers a high capacity of 7.7% hydrogen by mass combined with the low cost of magnesium and good reversibility for hydrogen uptake and evolution at high temperatures. However, its decomposition temperature of 300°C under 1 atm of hydrogen gas is uneconomically high, and considerable effort has been expended to develop new related materials that have lower hydrogen-release temperatures. Doping of MgH_2 with a variety of metals has been undertaken to produce the solid solutions $Mg_{1-x}M_xH_{2\pm y}$ for M = Al, V, Ni, Co, Ti, Ge, and La/Ni mixtures, some of which have slightly lower decomposition temperatures. More promising are changes in the microstructure of MgH_2 which are caused by ball-milling (Sections 25.9, 28.2, and 30.8), particularly when done in combination with other materials, such as metals (Ni, Pd) and metal oxides (V_2O_5, Cr_2O_3). Ball-milling of MgH_2 increases the surface area of the solid (by a factor of about 10) and the number of defects in the crystallites, promoting both adsorption and deadsorption of hydrogen.

(b) Complex hydrides

KEY POINT Complex metal hydrides such as hydridoaluminates (alanates), amides, and tetrahydridoborates release H_2 when heated.

Complex hydrides include the tetrahydridoaluminates (containing AlH_4^-), amides (NH_2^-), and boranates (Box 14.1, including tetrahydridoborates containing BH_4^-), which can contain very high levels of hydrogen; for example, $LiBH_4$ is 18% hydrogen by mass. For these systems, their decomposition to liberate hydrogen is a problem: tetrahydridoborates, for instance, decompose only near 500°C. For some applications that do not require reversibility—the so-called 'one-pass hydrogen-storage systems'—the hydrogen can be liberated by treatment with water.

A BRIEF ILLUSTRATION

The molar mass of Li_2BeH_4 is $(2 \times 6.94) + 9.01 + (4 \times 1.01)$ g mol^{-1} = 26.93 g mol^{-1}, of which 4.04 g mol^{-1} is H, so the hydrogen content is $[(4.04 \text{ g mol}^{-1})/(26.93 \text{ g mol}^{-1})] \times 100\% = 15.0\%$.

Both $NaAlH_4$ and Na_3AlH_6 have good theoretical hydrogen-storage capacities, totalling 7.4 and 5.9% by mass, respectively, and low cost. Their structures are based on the tetrahedral AlH_4^- and octahedral AlH_6^{3-} complex anions (Fig. 25.51). Other complex hydrides include the lithium hydridoaluminates $LiAlH_4$ and Li_3AlH_6, which contain the complex anions $[AlH_4]^-$ and $[AlH_6]^{3-}$, have higher mass percentages of hydrogen than the corresponding sodium compounds, and would be attractive for hydrogen storage except for their chemical instability. The development of complex hydrides for hydrogen storage applications are discussed further in Section 30.8.

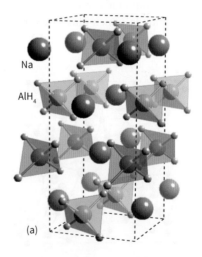

(a)

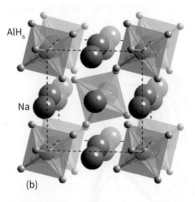

(b)

FIGURE 25.51 The structures of (a) $NaAlH_4$ and (b) Na_3AlH_6, depicting the hydride coordination around aluminium.

(c) Intermetallic compounds

KEY POINT Intermetallic phases based on first-row transition metals react with hydrogen reversibly.

Several types of intermetallic compound are being investigated as potential hydrogen-storage systems as, in general, they have excellent reversible hydrogen uptake at low pressures (1–20 bar) at just above room temperature (Table 25.5). However, the mass percentage of hydrogen that can be adsorbed by these materials is relatively low because of the high molar masses of the metals. One system, the AB_5-type phases, is typified by $LaNi_5$, which absorbs hydrogen to give $LaNi_5H_6$, but this corresponds to only 1.5% hydrogen by mass.

A series of alloys termed **Laves phases** are adopted by some intermetallics of the stoichiometry AB_2, where A = Ti, Zr, or Ln and B is a 3d metal such as V, Cr, Mn, or Fe. These materials have high capacities and good kinetics for hydrogen adsorption, forming compounds such as $ZrFe_2H_{3.5}$ and $ErFe_2H_5$ with up to 2% hydrogen by mass. However, the hydrides of these Laves phases are thermodynamically very stable at room temperature, restricting the reverse desorption of hydrogen.

The compound Mg_2NiH_4, which contains 3.6% hydrogen by mass, is formed by heating the alloy Mg_2Ni to 300°C under 25 kbar of hydrogen:

$$Mg_2Ni(s) + 2H_2(g) \rightarrow Mg_2NiH_4(s)$$

The structure of Mg_2NiH_4 consists, at low temperature, of an ordered array of Mg, Ni, and H, but transforms, at high temperature or with extended ball-milling, to a cubic phase with H^- ions randomly distributed throughout the arrangement of Mg and Ni atoms (Fig. 25.52).

A third class of intermetallic compounds being studied is the so-called 'Ti-based bcc alloys'; these are exemplified by FeTi and alloys of similar composition containing various quantities of other d-metals, including V, Cr, or Mn, such as $TiV_{0.28}Cr_{0.72}$. Hydrogen capacities of nearly 2.5% by mass can be achieved for these alloys, but high temperatures and pressures are needed to reach these values.

TABLE 25.5 Intermetallic structure types for hydrogen storage

Type	Metals	Typical hydride composition	Mass per cent H_2	p_{eq}, T^*
Element	Pd	$PdH_{0.6}$	0.56	0.020 bar, 298 K
AB_5	$LaNi_5$	$LaNi_5H_6$	1.5	2 bar, 298 K
AB_2 (Laves)	ZrV_2	$ZrV_2H_{5.5}$	3.0	10^{-8} bar, 323 K
AB	FeTi	$FeTiH_2$	1.9	5 bar, 303 K
A_2B	Mg_2Ni	Mg_2NiH_4	3.6	1 bar, 555 K
BCC	TiV_2	TiV_2H_4	2.6	10 bar, 313 K

* The values of p_{eq} and T are the pressure and temperature conditions for phase formation/decomposition.

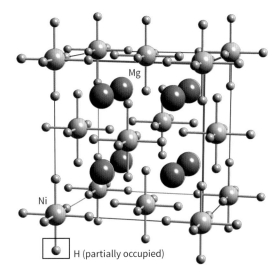

FIGURE 25.52 The idealized structure of cubic Mg_2NiH_4; hydrogen sites are only partially occupied.

Framework structures

The majority of the descriptive structural chemistry of metals (transition, s-block, and heavy p-block) in combination with oxides, sulfide, nitride, and halide anions covered so far in this chapter, has focused on compounds adopting structure types based on close-packed anions (Chapter 4). For the light p-block metals and metalloids, such as Si, Al, and P, tetrahedral coordination to anions, in particular oxide, is most prevalent, and structure descriptions based on linking together these tetrahedral units become a more useful concept. This structural chemistry, using building blocks of linked tetrahedra, is also usefully extended to other common polyhedral units. Octahedral units, MX_6, are most prevalent, where M is often a transition metal and X can be a simple anion—for example, F^- or O^{2-}—or more complex anions such as CN^- or an organic carboxylate species.

Linking these polyhedral units together through sharing of the anions at the polyhedra vertices gives rise to **framework structures**, Fig. 25.53. These frameworks may contain other species such as cations or small molecules which may only bond weakly with the framework. Removal of the small molecules, such as water, from within the framework gives rise to solids that contain small pores with dimensions in the range 0.3–20 nm, known as **microporous solids**.

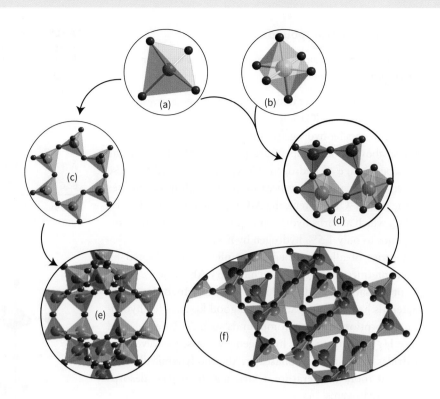

FIGURE 25.53 Tetrahedral (a) and octahedral (b) units are linked together through their vertices to form larger units, called secondary building units such as (c) and (d), which in turn bridge through their vertices to form framework structures such as (e) and (f).

25.16 Structures based on tetrahedral oxoanions

As remarked above, the elements able to form very stable tetrahedral MO_4 species that can link together into framework structures are lighter p-block metals and metalloids, though some of the later 3d transitional metals also form stable tetrahedral MO_4 units. These ions are so small that they coordinate strongly to four O atoms in preference to higher coordination numbers; the principal examples are SiO_4, AlO_4, and PO_4, although GaO_4, GeO_4, AsO_4, BO_4, BeO_4, $Co(II)O_4$, and ZnO_4 are all well known in these structure-types. Other tetrahedral units that have been found only rarely in framework structures include $Ni(II)O_4$, $Cu(II)O_4$, and InO_4. In this chapter we concentrate on the structures derived from the most commonly found units in the zeolites (framework aluminosilicates with large non-framework pores within their structures), aluminophosphates, and phosphates.

(a) Aluminosilicate zeolites

KEY POINTS Zeolites, with three-dimensional frameworks formed from linked SiO_4 and AlO_4 tetrahedra only, have the general formula $(M^{n+})_{x/n}[(AlO_2)_x(SiO_2)_{2-x}] \cdot mH_2O$, where M is a cation, typically Na^+, K^+, or Ca^{2+}; the cations and water molecules occupy sites within pores or channels of the fully connected framework.

In zeolites synthesized from solution (Sections 25.9 and 29.1), where all the vertices of the SiO_4 and AlO_4 building units are shared between two tetrahedra, **Lowenstein's rule** states that no oxygen atom is shared between two AlO_4 tetrahedra. Lowenstein's rule applies only to frameworks synthesized from solution, as directly linked AlO_4 tetrahedra are readily obtained in compounds generated from reactions at high temperatures. This indicates that in aqueous solutions the bridging oxygen in a $[O_3Al–O–SiO_3]$ unit is stabilized relative to one in a $[O_3Al–O–AlO_3]$ unit, presumably because of the higher formal charge on Si.

A BRIEF ILLUSTRATION

To determine the maximum Al:Si ratio in a zeolite we note that the maximum Al content is reached when alternate tetrahedra in the structure are those of AlO_4 and each vertex is linked to four SiO_4 units (and, similarly, each SiO_4 tetrahedron is surrounded by four AlO_4 units). Thus the highest ratio achievable is 1:1, equivalent to $x = 1.0$ in the general formula of a zeolite given above, which for M = Na^+ becomes $(Na)[(AlO_2)(SiO_2)] \cdot mH_2O \equiv NaAlSiO_4 \cdot mH_2O$.

At the other extreme, where $x = 0$, a zeolite is constructed purely of SiO_4 tetrahedra and no cations are needed to balance a charge on the framework. These materials are often hydrophobic, and the overall zeolite formula becomes simply 'SiO_2'; that is, (porous) polymorphs of silica.

Tetrahedral units, SiO_4 and AlO_4, may be linked together through their vertices in many different ways, and this structural flexibility gives rise to an enormous variety of structures. Over 200 different zeolite structures are known (more than 50 of which occur naturally as minerals), and Fig. 25.54 shows two such frameworks The various different zeolite

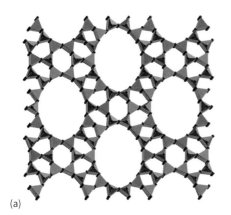
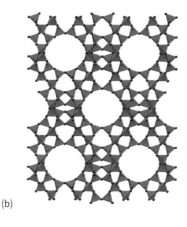

FIGURE 25.54 Representations as linked tetrahedra of synthesized zeolites showing the main channels in two such materials (a) UTD-1 (b) FAU.

structures can be synthesized by modifying the reaction conditions (for example, temperature, pressure) and the use of structure-directing agents or **templates**. The role of structure-directing agents in the synthesis of zeolitic materials was introduced in Section 15.15 and is discussed further in Chapter 29. Heterogeneous catalysis using zeolites is described in detail in Section 31.13, and two additional major applications of zeolites, as absorbents for small molecules and as ion exchangers, are described in Section 29.1.

(b) Aluminophosphates

KEY POINTS Structures built from repeating AlO_4PO_4 tetrahedral units are known as AlPOs.

The structural and electronic equivalence of two silicate tetrahedra, $(SiO_4)_2$, and the aluminophosphate unit, AlO_4PO_4, can be recognized in the simple compounds SiO_2 and $AlPO_4$, both of which adopt a similar range of dense polymorphs, including the quartz structure. The development of zeolites with very high Si content, which are effectively silica polymorphs, in turn led to the discovery of the aluminophosphate (ALPO) framework structures based on a 1:1 mixture of AlO_4 and PO_4 tetrahedra; $AlPO_4$ itself has the same structure as quartz (SiO_2) but with an arrangement of alternating Al and P atoms at the centres of the tetrahedra. A wide range of ALPOs has been developed and these materials are discussed further in Section 29.1(d). Organic template molecules have been designed and used to prepare many different ALPO structures, such as those with the structure codes VPI (with a large channel formed from 18 AlO_4/PO_4 tetrahedra), DAF, BOZ, and CIT-5 (Fig. 25.55).

(c) Phosphates

KEY POINT Calcium hydrogenphosphates are inorganic materials used in bone formation.

Another tetrahedral oxoanion that is frequently incorporated into framework materials is the phosphate group, PO_4^{3-}, although, as we shall see in the next section, many other tetrahedral units also form such structures.

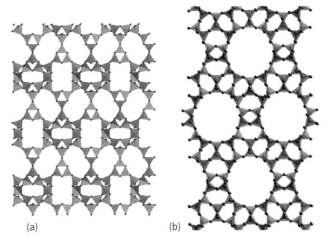

FIGURE 25.55 The frameworks, formed from linked oxotetrahedra, of (a) BOZ and (b) CIT-5. In each case the main channels present are emphasized by viewing the structure along them.

Simple phosphate structures, described in Sections 16.15 and 16.16, are generally formed from linked PO_4 tetrahedra in chains, cross-linked chains, and cyclic units. Here we consider in more detail just one metal phosphate material—namely, calcium hydrogenphosphate—and closely related materials. The principal mineral present in bone and teeth is hydroxyapatite, $Ca_5(OH)(PO_4)_3$, the structure of which consists of Ca^{2+} ions coordinated by PO_4^{3-} and OH^- groups to produce a rigid three-dimensional structure (Fig. 25.56). The mineral apatite is the partially fluoride-substituted $Ca_5(OH,F)(PO_4)_3$. Related biominerals are $Ca_8H_2(PO_4)_6$ and amorphous forms of calcium phosphate itself. Biominerals are discussed in further detail in Section 26.18.

25.17 Structures based on linked polyhedral metal centres

Many metals adopt MO_6 octahedral coordination in their oxo compounds. This polyhedral structural unit can be regarded as being formed by locating metal ions in the octahedral holes of a close-packed O^{2-} ion array in, for example,

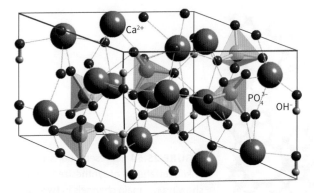

FIGURE 25.56 The structure of hydroxyapatite, $Ca_5(OH)(PO_4)_3$, shown as Ca^{2+} ions coordinated by phosphate and OH^- ions into a strong three-dimensional structure. The isolated O atoms are OH^- ions with the H lying outside the unit cell.

MgO with the rock-salt structure. The MO_6 polyhedral building unit may also be incorporated into framework-type structures, often in combination with tetrahedral oxo species.

(a) Clays, pillared clays, and layered double hydroxides

KEY POINT Sheet-like structures, found in many metal hydroxides and clays, can be constructed from linked metal oxotetrahedra and oxooctahedra.

The diameters of the largest pores found in synthetic zeolites are of the order of 1.2 nm. In an attempt to increase this diameter and allow larger molecules to be absorbed into inorganic structures, chemists have turned to MOFs (Sections 25.17(c) and 29.4), mesoporous materials (Section 28.10), and structures produced by 'pillaring' (that is, stacking and connecting together two-dimensional materials). The two-dimensional nature of many d-metal disulfides and some of their intercalation compounds are discussed in Sections 17.9 and 25.14. Similar intercalation reactions, when applied to aluminosilicates from the clay family, allow the synthesis of large-pore materials.

Naturally occurring clay minerals, such as kaolinite, $Al_2Si_2O_5(OH)_4$, hectorite, and montmorillonite have layer structures like those shown in Fig. 25.57. The layers are constructed from vertex- and edge-sharing octahedra (MO_6) and tetrahedra (TO_4), and exist as double-layer systems (two layers, one formed from octahedra and one from tetrahedra; Fig. 25.57), as in kaolinite, and triple-layer (a central layer based on octahedra sandwiched between two tetrahedra-based layers) systems, as in bentonite. The atoms M and T contained within the layers, which have an overall negative charge, are typically Si (on tetrahedral sites) and Al (occupying octahedral and tetrahedral sites). Small singly and doubly charged ions, such as Li^+ and Mg^{2+}, occupy sites between the layers. These interlayer cations are often hydrated and can readily be replaced by ion exchange. Other materials with similar structures are the

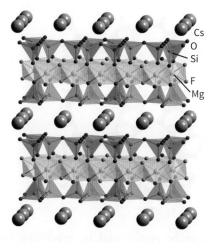

FIGURE 25.57 Sheet-like structures of the clay hectorite, which consist of layers of linked octahedra and tetrahedra typically centred on Al, Si, or Mg and separated by cations such as K^+ or Cs^+.

layered double hydroxides with structures similar to that of $Mg(OH)_2$, the naturally occurring mineral brucite.

Clay minerals have been used for millennia in the fabrication of pottery, tiles, and other structural ceramic materials. Natural clays, containing a variety of clay minerals, consist of plate-like crystallites separated by water molecules that hydrogen bond between the particles. This leads to a very deformable and malleable material, as the plate-like crystals can slip and move easily past each other, but also a material that will hold its shape once it has been moulded. When clays are dried in air, the loss of water between the crystallites leaves a hardened material with a more rigid structure, but one that can be wetted and reformed. Heating a clay to high temperatures (>900°C) in a kiln results in complete dehydration, removing the hydroxide ions and water molecules between the layers in the structure and on the particle surfaces. This dehydration reaction also sinters the individual clay particles together, forming a hard monolithic material: the fired clay or pottery. For kaolinite, $Al_2Si_2O_5(OH)_4$, the following reactions occur:

$$6\,Al_2Si_2O_5(OH)_4 \rightarrow 6\,Al_2Si_2O_7 \rightarrow 3\,Si_3Al_4O_{12} + 3\,SiO_2 \rightarrow 2(3\,Al_2O_3.2\,SiO_2) + 8\,SiO_2$$

The initial reaction product on heating to around 600°C is metakaolin, $Al_2Si_2O_7$. The final product, formed above 1000°C, consists of a mixture of mullite, $3\,Al_2O_3.2\,SiO_2$, and silica as a hard, brittle material that does not rehydrate.

In the pillaring of clays, the species exchanged into the interlayer region is selected for size. Large cations, including alkylammonium ions and polynuclear hydroxometal ions, may replace the alkali metal, as shown schematically in Fig. 25.58. The most widely used pillaring species are of the polynuclear hydroxide type and include $[Al_{13}O_4(OH)_{28}]^{3+}$, $[Zr_4(OH)_{16-n}]^{n+}$, and $Si_8O_{12}(OH)_8$; the first consists of a central AlO_4 tetrahedron surrounded by octahedrally coordinated Al^{3+} ions as $Al(O,OH)_6$ species. The pillaring process

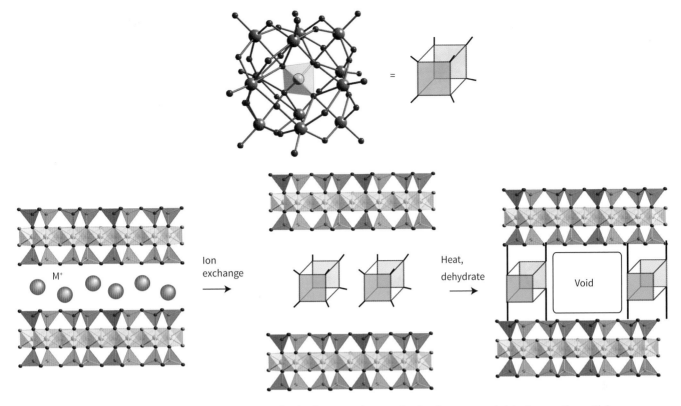

FIGURE 25.58 Schematic representation of pillaring of a clay by ion exchange of a simple monatomic interlayer cation with large polynuclear hydroxometallate, followed by dehydration and cross-linking of the layers to form the cavities. Spheres represent M⁺.

can be followed by powder X-ray diffraction because it leads to expansion of the interlayer spacing, corresponding to an increase in the c lattice parameter.

Once an ion such as $[Al_{13}O_4(OH)_{28}]^{3+}$ has been incorporated between the layers, heating the modified clay results in its dehydration (similar to the decomposition of kaolinite to metakaolin described earlier) and the formation of directly bonded layers (Fig. 25.58). The resulting product is a pillared clay with excellent thermal stability to at least 500°C. The expanded interlayer region can now absorb large molecules in the same way as zeolites. However, because the distribution of pillaring ions between the layers is difficult to control, the pillared clay structures are less regular than zeolites. Despite this lack of uniformity, pillared clays have been widely studied for their potential as heterogeneous catalysts (Section 31.13) because they act in a similar way to zeolites, as acid catalysts promoting isomerization and dehydration.

(b) Oxometal polyhedral units in inorganic framework chemistry

KEY POINT Enormous structural diversity can be obtained from linking various metal-centred polyhedra into frameworks.

The extensive development of aluminosilicates and zeolites has motivated synthetic inorganic chemists to seek similar structural types built from other tetrahedral and octahedral polyhedra for use in ion exchange, absorption, and catalysis. The use of different and larger polyhedra provides more flexibility in the framework topologies, and there is also the potential to incorporate d-metal ions with their associated properties of colour, redox properties, and magnetism. Tetrahedral species that have been incorporated into such frameworks include ZnO_4, AsO_4, CoO_4, GaO_4, and GeO_4, as well as the aluminate, silicate, and phosphate groups mentioned previously. Octahedral units are mainly based on d metals and the heavier and larger metals from Groups 13 and 14. Other polyhedral units such as five-coordinate square pyramids also occur, but less commonly.

Pore sizes in these inorganic materials are generally small and inaccessible to larger organic molecules but frequently incorporate small cations (Group 1 or Group 2, protonated small organic amines) that charge balance the negatively charged framework. Applications of these materials are therefore focused on ion exchange rather than catalysis. Compounds of this type and their ion exchange applications are discussed in detail in Section 29.3.

(c) Metal–organic frameworks

KEY POINT The use of organic linkers, such as carboxylates, between metal centres produces hybrid inorganic–organic materials which can exhibit high porosity.

Metal–organic frameworks (MOFs), sometimes also called coordination polymers, have structures based on bidentate or

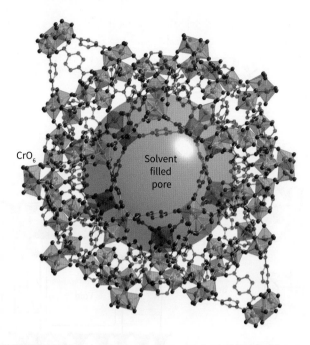

FIGURE 25.59 A metal–organic framework (MOF) formed from chromium terephthalate units showing CrO$_6$ octahedra (gold) linked by the organic anions. The large central sphere shows the extent of the pore in this material, which can be filled with solvent or gas molecules.

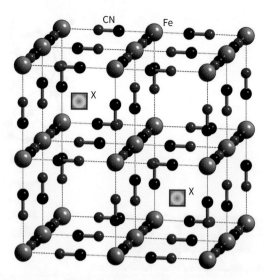

FIGURE 25.60 Prussian blue, [Fe(III)]$_4$[Fe(II)(CN)$_6$]$_3$. The iron ions are linked through cyanide, forming a cubic unit cell. Shaded squares (X) show regions in the structure where cations or water molecules may reside.

polydentate organic ligands lying between the metal atoms. Examples of simple ligands used to build these, often porous, frameworks are CN$^-$, nitriles, amines, imidazoles, and, most commonly, carboxylates. Many thousands of these compounds have now been made, and include positively charged frameworks with balancing anions in the cavities (for instance, Ag(4,4′-bpy)NO$_3$) and electrically neutral frameworks, such as Zn$_2$(1,3,5-benzenetricarboxylate)NO$_3$·H$_2$O·C$_2$H$_5$OH, from which the H$_2$O molecules may be removed reversibly.

MOFs have properties analogous to those of zeolites, with high porosities leading to potential applications in gas adsorption, separation, and storage, and the potential for similar catalytic reactions to those that occur inside zeolite pores and channels (Section 31.13), so they are being very actively researched for these applications, Section 29.2. Potential advantages of MOFs over zeolites include the much greater structural diversity available with large numbers of linking organic groups and, through the use of long linking groups based on aromatic carboxylates, very highly porous structures. One MOF of note is the framework of a chromium terephthalate (1,4-benzenedicarboxylate), which has pore diameters of about 3 nm and a specific internal surface area for N$_2$ of over 5000 m^2 g^{-1}, Fig. 25.59.

(d) Prussian blue and analogues

KEY POINT Prussian blues are a family of isostructural materials, based on transition metals linked through cyanide, that exhibit useful optical and magnetic properties.

The ability to link transition metal ions into framework structures using multidentate organic ligands has been illustrated in Section 25.17(c) with a discussion of metal–organic frameworks. Simple inorganic cations such as cyanide, CN$^-$, which can coordinate through both C and N ends of the anion, can also link metal cations. Prussian blue, [Fe(III)]$_4$[Fe(II)(CN)$_6$]$_3$ (Fig. 25.60), is one such compound, and its dark blue colour has resulted in its widespread use as a pigment and in inks. This intense colour results from an allowed transition between the Fe(II) and Fe(III) centres through the cyanide ligand.

Prussian blue also demonstrates the ability to take up large cations into its cubic cavity (Fig. 25.60) and so trap or sequester them. This has also led to its use to treat poisoning by thallium (as Tl$^+$) or radioactive ^{137}Cs$^+$.

Current interest in Prussian blue, and related phases with different compositions but similar structure type, extends beyond its pigment properties, as this material also shows ferromagnetic ordering (Section 22.8) of the iron centres below 5.6 K. Substitution of the iron ions may take place at both the trivalent and divalent ion sites of [Fe(III)]$_4$[Fe(II)(CN)$_6$]$_3$·xH$_2$O; cations may be introduced into cavities along with the water molecules. and various levels of vacancies may be present on the transition metal sites. This produces a family of materials of general composition A$_y^{n+}$(M′, M″...)$^{2+}$[M^{3+}(CN)$_6$]·xH$_2$O known as **Prussian blue analogues**; for example, the series of compounds M$_{1.5}^{2+}$[Cr^{3+}(CN)$_6$]·xH$_2$O (M = V, Cr, Mn, Ni, Cu) has been synthesized, as has KV(II)[Cr(III)(CN)$_6$]. Ferromagnetic ordering temperatures (the Curie temperature, T_C) in these compounds can be much higher than in Prussian blue because of stronger interactions between the metal centres, so that a material with the composition V(II)[Cr(III)(CN)]$_{0.86}$·2.8H$_2$O exhibits T_C = 315 K and KV(II)[Cr(III)(CN)$_6$] has T_C = 385 K; that is, these materials are ferromagnetic at room temperature.

Magneto-optical effects are based on the interaction between light and a magnetic material. For example, the **Faraday effect** is observed when linearly polarized light propagates through a magnetically ordered material. The polarized light undergoes a rotation of the plane of polarization that is directly (linearly) proportional to the component of the magnetic field in the direction of propagation. In Prussian blue analogues, which are often intensely coloured because of the presence of transition-metal ions and intervalence electron transfer between the metal centres, the Faraday effect is wavelength-dependent. This behaviour has potential use in optomagnetic data storage devices, in which light of a specific wavelength may be used to write or read information that is stored in the form of local magnetic-moment orientations in the material.

The rigid three-dimensional frameworks (and the associated windows allowing ion transport between neighbouring cages) of Prussian blue analogues coupled with the redox properties of the transition metals has led to their being investigated as robust rechargeable battery materials. Cations, such as Na$^+$ and K$^+$, can be extracted and inserted electrochemically with minimal change in material volume, leading to high battery stability and longevity when cycled.

Molecular inorganic materials and fullerides

The majority of compounds discussed so far in this chapter have been materials with extended structures in which ionic or covalent interactions link all the atoms and ions together into a three-dimensional structure. Examples include infinite structures based on ionic, as in NaCl, or covalent interactions, as in SiO$_2$. These materials are widely used in applications such as heterogeneous catalysis, rechargeable batteries, and electronic devices because of the chemical and thermal stability that derives from their linked structures.

It is often possible to tune the properties of many solids exactly; for example, by doping, introducing defects, or forming solid solutions. However, control of the arrangement of atoms into a particular structure cannot be achieved to the same degree for solids as for molecular systems. Thus, a chemist working with coordination compounds or organometallics can modify a complex or molecule by introducing a wide range of ligands, often through simple substitution reactions.

The desire to combine the synthetic and chemical flexibility of molecular chemistry with the properties of classical solid-state materials has led to the rapid emergence of the area of **molecular materials chemistry**, where functional solids are produced from linked and interacting molecules or molecular ions.

25.18 Fullerides

KEY POINTS Solid C$_{60}$ can be considered as a close-packed array of fullerene molecules interacting only weakly through van der Waals forces; holes in arrays of C$_{60}$ molecules may be filled by simple and solvated cations and small inorganic molecules.

The chemical properties of C$_{60}$ span many of the conventional borders of chemistry and include the chemistry of C$_{60}$ as a ligand (Section 15.6). In this section we describe the solid-state chemistry of solid fullerene, C$_{60}$(s), and the M$_n$C$_{60}$ fulleride derivatives that contain discrete C$_{60}^{n-}$ molecular anions. The synthesis and chemistry of the more complex carbon nanotubes are discussed in Section 28.8.

Crystals of C$_{60}$ grown from solution may contain included solvent molecules, but with the correct crystallization and purification methods; for example, by using sublimation to eliminate the solvent molecules, pure C$_{60}$ crystals may be grown. The solid structure has a face-centred cubic array of C$_{60}$ molecules, as shown in Fig. 25.61, as would be expected on the basis of efficient packing of these almost spherical molecules. At room temperature the molecules can rotate freely in their lattice positions, and powder X-ray diffraction data collected from crystalline C$_{60}$ are typical of an fcc lattice with a lattice parameter of 1417 pm. The molecules are separated by a distance of 296 pm, which is similar to the value found for the interlayer separation in graphite (335 pm). On cooling of the solid, the rotation halts and adjacent molecules align relative to each other such that

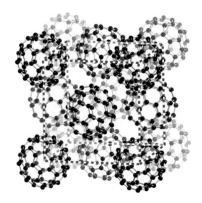

FIGURE 25.61 The arrangement of C$_{60}$ molecules in a face-centred cubic lattice in the crystalline material.

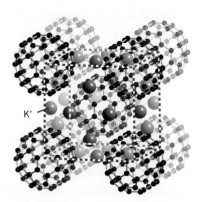

FIGURE 25.62 The structure of K_6C_{60}, with a body-centred cubic unit cell with C_{60}^{6-} molecular ions at the cell corners and body centre, and K^+ ions occupying half of the sites in the faces of the cell which have approximate tetrahedral coordination to four C_{60} molecular ions.

an electron-rich region of one C_{60} molecule is close to an electron-poor region in its neighbour.

Exposure of solid C_{60} to alkali metal vapour results in the formation of a series of compounds of formula M_xC_{60}, the precise stoichiometry of the product depending on the composition of the reactant mixture. With excess alkali metal, compounds of composition M_6C_{60} (M = Li, Na, K, Rb, Cs) are formed. The structure of K_6C_{60} is body-centred cubic; C_{60}^{6-} molecular ions occupy sites at the cell corners and body centre and the K^+ ions fill a portion of the sites, with approximately tetrahedral coordination to four C_{60} molecular ions, near the centre of each of the faces (Fig. 25.62).

Of most interest are the compounds with the stoichiometries M_3C_{60}, which become superconducting in the temperature range 10–40 K depending on the type of metal. The stoichiometry K_3C_{60} is obtained by filling all the tetrahedral and all the octahedral holes in the cubic close-packed arrangement of C_{60}^{3-} (Fig. 25.63). K_3C_{60} becomes superconducting on being cooled to 18 K, although gradual replacement of K by the larger alkali metal ions raises T_c, so that for Rb_3C_{60}, $T_c = 29$ K, and for $CsRb_2C_{60}$, $T_c = 33$ K. Note that Cs_3C_{60} does not form the same fcc structure as the other M_3C_{60} phases (in fact it has a structure based on a body-centred arrangement C_{60}^{3-} anions) and is not superconducting at normal pressures; however, it can be made superconducting, with a critical temperature of 40 K at 12 kbar.

Other more complex species can be incorporated into a matrix of C_{60} units. Molecular species such as iodine (I_2) or phosphorus molecules (P_4 tetrahedra) can fill spaces between the C_{60} molecules in close-packed arrays. Solvated cations can also occupy the tetrahedral and octahedral holes in a similar way to the simple alkali metal cations. Thus, $Na(NH_3)_4CsNaC_{60}$, which is obtained by the reaction of Na_2CsC_{60} with ammonia, contains Na^+ ions solvated with ammonia molecules on the octahedral site and nonsolvated Cs^+ and Na^+ ions on the (twice as abundant) tetrahedral sites in an fcc arrangement of $[C_{60}]^{3-}$ molecular ions.

25.19 Molecular inorganic materials chemistry

The ability to modify the shapes, and thus the packing and arrangements of inorganic molecules in the solid state, is one valuable aspect of molecular materials chemistry. This capability, when associated with some of the specific properties of inorganic compounds, such as the unpaired electrons of d metals, can allow control of magnetic and electronic properties. This section considers a number of such inorganic molecular materials being developed at the frontiers of the subject.

(a) One-dimensional metals

KEY POINTS A stack of molecules that interact with each other along one dimension, as occurs in a number of crystalline platinum complexes, can show conductivity in that direction; a Peierls distortion ensures that no one-dimensional solid is a metallic conductor below a critical temperature.

A one-dimensional metal is a material that exhibits metallic properties along one direction in the crystal and nonmetallic properties orthogonal to that direction. This should be contrasted to one-dimensional structures and nanomaterials, including those discussed in Chapter 28, which have a structural or physical component that is one-dimensional in nature. One-dimensional metallic properties arise when the orbital overlap occurs along a single direction in the crystal (as in VO_2, Section 25.11). Several classes of one-dimensional metals are known and include $(SN)_x$ and organic polymers such as doped polyacetylenes $[(CH)I_{0.25}]_n$, but this section is concerned specifically with chains of interacting d metals, particularly Pt.

The structural requirements for a one-dimensional metal are satisfied in some square-planar transition metal complexes that stack one above another in the solid state (Fig. 25.64).

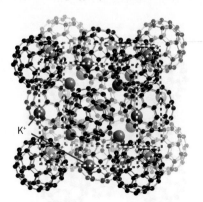

FIGURE 25.63 The structure of K_3C_{60} is obtained by filling all the tetrahedral and all the octahedral holes in the close-packed C_{60}^{3-} lattice with K^+ ions.

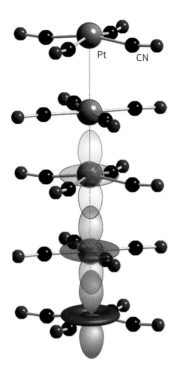

FIGURE 25.64 A representation of the infinite-chain structure of KCP (K$_2$Pt(CN)$_4$Br$_{0.3}$·3H$_2$O) and a schematic illustration of its conduction band formed by d-orbital overlap.

The ligands surrounding the metal atom ensure large interchain separations, of at least 900 pm, while the average intrachain metal–metal distance is less than 300 pm. Square-planar complexes are commonly found for metal ions with d^8 configurations, and the overlap of orbitals between d^8 species is greatest for the heavy d metals of Period 6 (which use 5d orbitals). Hence the compounds of interest are mainly associated with Pt(II) and Ir(I), where a band is formed from overlapping d$_{z^2}$ and p$_z$ orbitals. The d$_{z^2}$ is full for Pt(II), and a partially full level is achieved by oxidation of the platinum. Many d^8 tetracyanidoplatinate(II) complexes are semiconductors with $d_{Pt-Pt} \sim 310$ pm, and the partial oxidation of these salts results in Pt–Pt distances of less than 290 pm and metallic behaviour. Typical means of oxidizing the chain are the incorporation of extra anions into the structure or the removal of cations. The first one-dimensional metal Pt complex was made as early as 1846 by oxidation of a solution of K$_2$Pt(CN)$_4$·3H$_2$O with bromine, which on evaporation gave crystals of K$_2$Pt(CN)$_4$Br$_{0.3}$·3H$_2$O, also known as KCP (as in Fig. 25.64).

The electronic properties of one-dimensional metals are not quite as simple as has been implied by the discussion so far, as a theorem due to Rudolph Peierls states that, at $T = 0$, no one-dimensional solid is a metal! The origin of Peierls' theorem can be traced to a hidden assumption in the discussion so far: we have supposed that the atoms lie in a line with a regular separation. However, the actual spacing in a one-dimensional solid (and any solid) is determined by the distribution of the electrons, not vice versa, and there is no guarantee that the state of lowest energy is a solid with a regular lattice spacing. In fact, in a one-dimensional solid at $T = 0$, there always exists a distortion, a **Peierls distortion**, which leads to a lower energy than in the perfectly regular solid.

An idea of the origin and effect of a Peierls distortion can be obtained by considering a one-dimensional solid of N atoms and N valence electrons (Fig. 25.65). Such a line of atoms distorts to one that has alternating long and short bonds. Although the longer bond is energetically unfavourable, the strength of the short bond more than compensates for the weakness of the long bond and the net effect is a lowering of energy below that of the regular solid. Now, instead of the electrons near the Fermi surface being free to move through the solid, they are trapped between the longer-bonded atoms (these electrons have antibonding character, and so are found outside the internuclear region between strongly bonded atoms). The Peierls distortion introduces a band gap in the centre of the original conduction band, and the filled orbitals are separated from the empty orbitals. Hence, the distortion results in a semiconductor or insulator, not a metallic conductor.

The conduction band in KCP is a d band formed principally by overlap of Pt 5d$_{z^2}$ orbitals. The small proportion of Br in the compound, which is present as Br$^-$, removes a small number of electrons from this otherwise full d band, so turning it into a conduction band. Indeed, at room temperature, doped KCP is a lustrous bronze colour with its highest conductivity along the axis of the Pt chain. However, below 150 K the conductivity drops sharply on account of the onset of a Peierls distortion. At higher temperatures the motion of the atoms averages the distortion to zero, the separation is regular (on average), the gap is absent, and the solid is a one-dimensional metal. Mixed-valence platinum chain complexes, such as [Pt(en)$_2$][PtCl$_2$(en)$_2$](ClO$_4$)$_4$, may be isolated as insulated, one-dimensional wires by surrounding the Pt chain with a sheath of electronically inert molecules such as anionic lipids. Nanowires are discussed more fully in Section 28.8.

The observation of metallic behaviour in one-dimensional solids formed from interacting molecules has in turn led to a

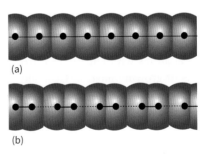

FIGURE 25.65 The formation of a Peierls distortion. The energy of the line of atoms with alternating bond lengths (b) is lower than that of the uniformly spaced atoms (a).

search for superconductivity in this class of materials. One type of material in which superconductivity has been found is a series of metal complexes derived from the organic metal and superconductors based on stacked molecules with interacting π systems. Thus salts of TCNQ (7,7,8,8-tetracyano-*p*-quinodimethane (**2**) with TTF (tetrathiafulvalene, **3**) show some metallic properties, in that they are conducting and absorb a range of wavelengths of light.

2 TCNQ (7,7,8,8-tetracyano-*p*-quinodimethane)

3 TTF (tetrathiafulvalene)

Salts of tetramethyltetraselenafulvalene (TMTSF, **4**) such as $(TMTSF)_2ClO_4$ show superconductivity, albeit at less than 10 K and often only under pressure. Molecular metal complexes involving types of sulfur-containing ligand, such as dmit (**5**) ($dmit^{2-}$ = 1,3-dithiol-2-thione-4,5-dithiolato, the anion derived from 4,5-dimercapto-1,3-dithiole-2-thione), can also show superconductivity. The compound $[TTF][Ni(dmit)_2]_2$, which consists of stacks of both TTF and $[Ni(dmit)_2]$, shows superconductivity at 10 kbar and below 2 K.

4 Tetramethyltetraselenafulvalene (TMTSF)

5 $dmit^{2-}$

(b) Molecular magnets

KEY POINT Molecular solids containing individual molecules, clusters, or linked chains of molecules can show bulk magnetic effects such as ferromagnetism.

Molecular inorganic magnetic materials, in which individual molecules, or units constructed from such molecules, contain d-metal atoms with unpaired electrons, are a class of compounds of growing interest. Generally the phenomena associated with long-range interaction of electron spins, such as ferromagnetism and antiferromagnetism, are much weaker in these complexes as the short, superexchange-type pathways that are found in metal oxides do not exist. However, as with all molecular systems the opportunity exists to tune interactions between metal centres by tailoring the ligand properties.

Examples of ferromagnetic molecular inorganic compounds are decamethylferrocene tetracyanoethenide (TCNE), $[Fe(\eta^5-Cp^*)_2](C_2(CN)_4)$ (**6**; $Cp^* = C_5Me_5$), and the analogous manganese compound. These materials, which have structures based on chains of alternating $[M(\eta^5-Cp^*)_2]^+$ and $TCNE^-$ ions, show ferromagnetism along the chain direction below $T_C = 4.8$ K (for M = Fe) or 6.2 K (for M = Mn). An alternative approach to molecular-based compounds exhibiting magnetic ordering consists of assembling chains of magnetically interacting centres. For example, MnCu(2-hydroxy-1,3-propenebisoxamato)·$3H_2O$ (**7**) consists of chains of alternating Mn(II) and Cu(II) ions bridged by the $[HOCH(CH_2NC(O)COO)_2]^{4-}$ ligand. The magnetic moments on the metal ions in the chains order ferromagnetically below 115 K. This ferromagnetic ordering occurs initially only along the chains—that is, in one dimension—because of the stronger interactions through the ligands and shorter Mn–Cu distances, but this material orders fully in three dimensions at 4.6 K once the individual chains interact with each other magnetically.

6 Decamethylferrocene tetracyanoethenide (TCNE), $[Fe(\eta^5-Cp^*)_2(C_2(CN)_4)]$

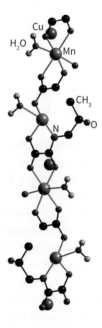

7 MnCu(2-hydroxy-1,3-propenebisoxamato)·$3H_2O$

The incorporation of several d-metal ions into a single complex provides an opportunity to produce a molecule that acts as a tiny magnet. Such compounds have been termed **single-molecule magnets** (SMMs). One example is the complex manganese

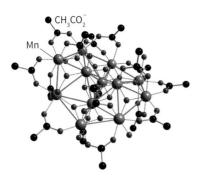

FIGURE 25.66 The core of the single-molecule magnetic compound $Mn_{12}O_{12}(O_2CMe)_{16}(H_2O)_4 \cdot 2\,MeCO_2H \cdot 4\,H_2O$ that contains 12 Mn(III) and Mn(IV) centres.

acetate $[Mn_{12}O_{12}](O_2CMe)_{16}(H_2O)_4 \cdot 2\,MeCO_2H \cdot 4\,H_2O$, which contains a cluster of 12 Mn(III) and Mn(IV) ions linked through O atoms with the metal oxide unit terminated by the acetate groups (Fig. 25.66). Another is $[Mn_{84}O_{72}(O_2CMe)_{78}(OMe)_{24}(MeOH)_{12}(H_2O)_{42}(OH)_6] \cdot xH_2O \cdot yCHCl_3$, which contains 84 Mn(III) ions in a large doughnut-shaped molecule 4 nm in diameter. The ability to magnetize such individual SMMs potentially provides a route to storing information at extremely high densities, because their dimensions—a few nanometres—are much less than that of the typical domain of magnetic material used in a conventional magnetic data-storage medium.

(c) Inorganic liquid crystals

KEY POINT Inorganic metal complexes with disc- or rod-like geometries can show liquid crystalline properties.

Liquid crystalline, or **mesogenic**, compounds possess properties that lie between those of solids and liquids and include both. For instance, they are fluid, but with positional order in at least one dimension. These materials have become widely used in displays. The molecules that form liquid crystalline materials are generally **calamitic** (rod-like) or **discotic** (disc-like), and these shapes lead to the ordered liquid-type structures in which the molecules align in a particular direction (Fig. 25.67). Although most liquid crystalline materials are totally organic, there are a growing number of inorganic liquid crystals based on the coordination compounds of metals and on organometallic compounds. These metal-containing liquid crystals show similar properties to the purely organic systems, but offer additional properties associated with a d-metal centre, including as redox and magnetic effects.

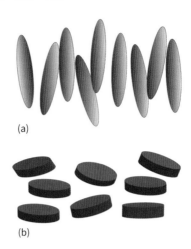

FIGURE 25.67 Schematic diagram of liquid crystalline materials based on (a) calamitic (rod-like) and (b) discotic (disc-like) molecules.

As the requirement for liquid crystalline behaviour is a rod- or disc-shaped molecule, many of the metal-containing systems are based around the low-coordination geometries of the later d metals, particularly the square-planar complexes found for Groups 10 and 11. Thus the β-diketone complex (8) has a square-planar Cu^{2+} ion coordinated to four O atoms from two β-diketones with long, pendant alkyl groups. This copper(II) material is paramagnetic, but also forms a **nematic phase** in which the rod-shaped molecules align predominantly in one direction (as in Fig. 25.67).

8 β-diketone complex of copper

FURTHER READING

A. F. Wells, *Structural Inorganic Chemistry*. Oxford University Press (1985). The standard reference book, which surveys the structures of a huge number of inorganic solids.

J. K. Burdett, *Chemical Bonding in Solids*. Oxford University Press (1995). Further details of the electronic structures of solids.

U. Müller, *Inorganic Structural Chemistry*. John Wiley & Sons (1993).

A. R. West, *Solid State Chemistry and its Applications, Student Edition*, 2nd ed. John Wiley & Sons (2014).

S. E. Dann, *Reactions and Characterization of Solids*. Royal Society of Chemistry (2000).

L. E. Smart and E. A. Moore, *Solid State Chemistry: An Introduction*, 5th ed. CRC Press (2020).

P. A. Cox, *The Electronic Structure and Chemistry of Solids*. Oxford University Press (1987).

D. K. Chakrabarty, *Solid State Chemistry*, 2nd ed. New Age Science Ltd (2010).

P. M. Woodward, P. Karen, J. S. O. Evans, and T. Vogt, *Solid State Materials Chemistry*. Cambridge University Press (2021).

A. K. Cheetham and P. Day (eds.), *Solid State Chemistry: Compounds*. Oxford University Press (1992). A useful collection of chapters covering the key compound types in materials chemistry.

R. C. Mehrotra, Present status and future potential of the sol–gel process. *Struct. Bonding*, 1992, 77, 1. A good review of sol–gel chemistry.

A. K. Cheetham, G. Férey, and T. Loiseau, Open-framework inorganic materials. *Angew. Chem., Int. Ed.*, 1999, **38**, 3268. An excellent review of progress in, and the structural chemistry of, framework solids.

D. W. Bruce, D. O'Hare, and R. I. Walton, *Inorganic Materials Series*. John Wiley & Sons (2011). A five-volume set covering (i) functional oxides, (ii) molecular materials, (iii) low-dimensional solids, (iv) porous materials, and (v) energy materials. Individual volumes are available separately.

M. T. Weller, *Inorganic Materials Chemistry*. Oxford Chemistry Primers 23. Oxford University Press (1994). Introductory text covering some aspects of solid-state chemistry and materials characterization.

L. V. Interrante, L. A. Casper, and A. B. Ellis (eds.), *Materials Chemistry: An Emerging Discipline*, Advances in Chemistry Series, no. 245. American Chemical Society (1995). A series of chapters on a broad range of inorganic and organic solids.

B. D. Fahlman, *Materials Chemistry*, 2nd edn. Springer (2014). Good coverage of semiconductors, metals and alloys, and characterization methods.

P. Day, *Molecules Into Materials; Case Studies in Materials Chemistry: Mixed Valency, Magnetism and Superconductivity*. World Scientific Publishing (2007). A collection of papers demonstrating how this important area of materials chemistry developed.

P. Ball, *Made to Measure: New Materials for the 21st Century*. Princeton University Press (1997). A very readable overview of materials from the perspective of applications, covering fuel cells, ultrahard materials, and smart materials.

J. N. Lalena and D. A. Cleary, *Principles of Inorganic Materials Design*. John Wiley & Sons (2005). A good overview of inorganic materials with a strong theoretical perspective.

EXERCISES

25.1 Predict what type of intrinsic defect is most likely to occur in (a) Ca_3N_2, (b) HgS.

25.2 By considering which dopant ions produce the blue of sapphires, provide an explanation for the origin of the colour in the blue form of beryl known as aquamarine.

25.3 For which of the following compounds might significant levels of nonstoichiometry be found: aluminium oxide, vanadium carbide, manganese sulfide?

25.4 Explain why higher levels of defects are found in solids at high temperatures and close to their melting points. How would pressure affect the equilibrium number of defects in a solid?

25.5 By considering the effect on the lattice energies of incorporating large numbers of defects and the resultant changes in oxidation numbers of the ions making up the structure, predict which of the following systems should show nonstoichiometry over a large range of x: $Zn_{1+x}O$, $Fe_{1-x}O$, UO_{2+x}.

25.6 When NiO is doped with small quantities of Li_2O, the electronic conductivity of the solid increases. Provide a plausible chemical explanation for this observation. (Hint: Li^+ occurs on Ni^{2+} sites.)

25.7 The compound Fe_xO generally has $x < 1$. Describe the probable metal ion defect that leads to x being less than 1.

25.8 Consider the possible binary combinations that could be made from the four elements Na, Cs, Si, and Bi; for example Na_nBi_m. Which combinations are likely to be (a) alloys or (b) Zintl phases?

25.9 How might you distinguish experimentally the existence of a solid solution from a series of crystallographic shear-plane structures for a material that appears to have variable composition?

25.10 Describe the difference between a semiconductor and a semimetal. How do dopants change the electronic properties of (a) a semiconductor and (b) a semimetal?

25.11 Classify the following as to whether they are likely to show n- or p-type semiconductivity: Ag_2S, VO_2, CuBr.

25.12 Graphite is a semimetal with a band structure of the type shown in Fig. 2.37. Reaction of graphite with potassium produces C_8K while reaction with bromine yields C_8Br. Assuming the graphite sheets remain intact and potassium and bromine enter the graphite structure as K^+ and Br^- ions, respectively, discuss whether you would expect the compounds C_8K and C_8Br to exhibit metallic, semimetallic, semiconducting, or insulating properties.

25.13 Classify the following as n- or p-doped semiconductors: (a) Ga-doped Ge, (b) As-doped Si, (c) $In_{0.49}As_{0.51}$.

25.14 Which of VO or NiO would be expected to show metallic properties?

25.15 Describe the main structural features of a solid that has high ionic conductivity.

25.16 Use ionic radii data (Resource section 1) to suggest possible dopants to increase anion conductivity in (a) PbF_2 and (b) Bi_2O_3 (six-coordinate Bi^{3+}).

25.17 Outline how you could prepare samples of (a) $MgCr_2O_4$, (b) $LaFeO_3$, (c) Ta_3N_5, (d) $LiMgH_3$, (e) $KCuF_3$, (f) the zeolite A analogue with Ga replacing Al, $Na_{12}[Si_{12}Ga_{12}O_{48}]\cdot nH_2O$.

25.18 Identify the likely products of the reactions

(a) $Li_2CO_3 + CoO \xrightarrow{800°C, O_2}$

(b) $2Sr(OH)_2 + WO_3 + MnO \xrightarrow{900°C, O_2}$

25.19 Draw one unit cell in the ReO_3 structure, showing the M and O atoms. Does this structure appear to be sufficiently open to undergo Na^+ ion intercalation? If so, where might the Na^+ ions reside? Describe the likely structure type for a material where the Na:Re ratio is 1:1.

25.20 'Double perovskites' of the composition $AA'BB'O_6$ have been widely studied in respect of their magnetic properties. For A = Rb and A' = Sr, suggest some possible ions for the B and B' sites in a double perovskite.

25.21 The iodostannate perovskites, $ASnI_3$, have been studied as potential semiconductors for applications as photovoltaics. By reference to the known $APbI_3$ phases that adopt the perovskite structure, suggest which cations, A^+, might be of a suitable size to form a perovskite in the $ASnI_3$ system.

25.22 Predict the Néel temperature for CrO.

25.23 Write possible formulas (using elements from the periodic table) for a sulfide and a fluoride that might adopt the spinel structure.

25.24 State Zachariasen's two generalizations that favour glass formation and apply them to the observation that cooling molten CaF_2 leads to a crystalline solid whereas cooling molten SiO_2 at a similar rate produces a glass.

25.25 Classify the oxides (a) BeO, (b) TiO_2, (c) La_2O_3, (d) B_2O_3, (e) GeO_2 into glass-forming and non-glass-forming.

25.26 Which metal fluorides might be glass-forming?

25.27 Describe the interactions between Mo_6S_8 units in a Chevrel phase.

25.28 Describe Lowenstein's rule and the limits it imposes on the values of x for a zeolite of the general formula $(M^{n+})_{x/n}[(AlO_2)_x(SiO_2)_{2-x}]$ synthesized under hydrothermal conditions.

25.29 Discuss briefly the structural chemistry of clay minerals, including the differences between the structures of kaolinite and bentonite.

25.30 Which elements in the periodic table from stable oxotetrahedral species are known in framework structures?

25.31 Summarize the key structural and compositional elements of an MOF.

25.32 Calculate the mass percentage of hydrogen in $NaBH_4$, and state whether or not this material might be suitable for hydrogen storage.

25.33 Why is BeH_2 not considered to be a suitable hydrogen-storage material?

25.34 Describe the structures of Na_2C_{60} and Na_3C_{60} in terms of hole-filling in a close-packed array of fulleride molecular ions.

TUTORIAL PROBLEMS

25.1 The International Gem Society has published a guide to understanding synthetic gemstones. Discuss the history and chemistry of synthetic gemstones.

25.2 Describe the inorganic chemistry involved in the operation of a lambda sensor (an exhaust-gas oxygen sensor) in a vehicle engine.

25.3 To obtain high oxidation states for first-row d-metals in complex oxides—for example, $Sr_2Fe(IV)O_4$—compounds are normally prepared at as low a temperature as possible commensurate with the reaction. Discuss the thermodynamic reasons for the use of these conditions, and explain why the optimum temperature for producing YBCO involves a final annealing stage at about 450°C under pure oxygen.

25.4 Heating a complex oxide under ammonia can lead to nitridation (with nitrogen replacing oxygen) or reduction (with the production of nitrogen and water). Describe possible products that might be formed when $SrWO_4$ is heated in ammonia. By consideration of the relative entropy changes for the two reactions, explain how the reaction temperature might affect the outcome.

25.5 What are the advantages of sol–gel routes, in comparison with direct high-temperature reaction methods, in the synthesis of complex metal oxides? Why are sol–gel routes often used to prepare finely divided or nanoparticulate material?

25.6 Describe the delafossite structure type adopted by some compounds of the stoichiometry $MM'X_2$. Which elemental combinations are known to form this structure type? Discuss their electronic properties when X = O, and potential applications as photocatalysts.

25.7 Compare and contrast the chemistries of graphite and C_{60} with respect to their compounds in association with the alkali metals.

25.8 'The increased reactivity that allows the design and synthesis of materials based on molecular units also means that these compounds are unsuitable for many applications that currently use inorganic materials.' Discuss this remark.

26 Biological inorganic chemistry

The organization of cells
26.1 The physical structure of cells
26.2 The inorganic composition of living organisms
26.3 Biological metal-coordination sites

Metal ions in transport and communication
26.4 Sodium and potassium transport
26.5 Calcium signalling proteins
26.6 Selective transport and storage of iron
26.7 Oxygen transport and storage
26.8 Electron transfer

Catalytic processes
26.9 Acid–base catalysis
26.10 Enzymes dealing with H_2O_2 and O_2
26.11 Enzymes dealing with radical rearrangements and alkyl group transfer
26.12 Oxygen atom transfer by molybdenum and tungsten enzymes
26.13 Hydrogenases, enzymes that activate H_2
26.14 The nitrogen cycle

Metals in gene regulation
26.15 Transcription factors and the role of Zn
26.16 Iron proteins as sensors
26.17 Proteins that sense Cu and Zn levels
26.18 Biomineralization

Perspectives
26.19 The contributions of individual elements
26.20 Future directions

Further reading

Exercises

Tutorial problems

Organisms have exploited the chemical properties of the elements in remarkable ways, providing examples of coordination specificities that are far higher than observed in simple compounds. This chapter describes how elements are taken up selectively by different cells and intracellular compartments, and the various ways they are exploited. We discuss the structures and functions of complexes and materials that are formed in the biological environment in the context of the chemistry covered earlier in the text.

Biological inorganic chemistry ('bioinorganic chemistry') is the study of the 'inorganic' elements as they are utilized in biology. The main focus is on metal ions, where we are interested in their interaction with biological ligands and the important chemical properties they are able to exhibit and impart to an organism. These properties and the roles they underpin include electron transfer, catalysis, signalling, regulation, sensing, defence, and structural support.

The organization of cells

To appreciate the role of the elements (other than C, H, O, and N) in the structure and function of organisms we need to know a little about the organization of the 'atom' of biology, the living cell, and its 'fundamental particles', the cell's constituent organelles.

26.1 The physical structure of cells

KEY POINTS Living cells and organelles are enclosed by membranes; the concentrations of specific elements may vary greatly between different compartments due to the actions of ion pumps and gated channels.

Cells, the basic units of any living organism, range in complexity from the simplest types found in prokaryotes (bacteria and bacteria-like organisms classified as archaea) to the much larger and more complex examples found in eukaryotes (which include animals and plants). The main features of these cells are illustrated in the generic model shown in Fig. 26.1. Crucial to all cells are membranes, which act as barriers to water and ions and make possible the management of all mobile species and of electrical currents. Membranes are lipid bilayers, approximately 4 nm thick, in which are embedded protein molecules and other components. Bilayer membranes have great lateral strength but they are easy to bend. The long hydrocarbon chains of lipids make the membrane interior very hydrophobic and impermeable to ions, which must instead travel through specific channels, pumps, and other receptors provided by special membrane proteins. The structure of a cell also depends on osmotic pressure, which is maintained by high concentrations of solutes, including ions, imported during active transport by pumps.

Prokaryotic cells consist of an enclosed aqueous phase, the **cytoplasm**, which contains the DNA and most of the materials used and transformed in the biochemical reactions: these are often confined within enclosures called 'microcompartments' specializing in particular tasks. Bacteria are classified according to whether they are enclosed by a single membrane or have an additional intermediate aqueous space, the **periplasm**, between the outer membrane and the cytoplasmic membrane, and are known as 'Gram-positive' or 'Gram-negative', respectively, depending on their response to a staining test with the dye crystal violet.

The much more extensive cytoplasm of eukaryotic cells contains subcompartments (also enclosed within lipid bilayers) known as **organelles**, which have highly specialized functions. Organelles include the **nucleus** (which houses DNA), **mitochondria** (the 'fuel cells' that carry out respiration), **chloroplasts** (the 'photocells' that harness light energy and 'fix' CO_2), the **endoplasmic reticulum** (for protein synthesis), **Golgi** (vesicles containing proteins for export), **lysosomes** (which contain degradative enzymes and help rid the cell of waste), **peroxisomes** (which remove harmful hydrogen peroxide), and other specialized processing zones.

26.2 The inorganic composition of living organisms

KEY POINTS The major biological elements are oxygen, hydrogen, carbon, nitrogen, phosphorus, sulfur, sodium, magnesium, calcium, and potassium. The trace elements include many d metals, as well as selenium, iodine, silicon, and boron. The pool of metal-containing species and nonmetal trace elements in living organisms is known as the metallome. Different elements are strongly segregated inside and outside a cell and among different internal compartments.

The periodic table in Fig. 26.2 highlights the elements known to be used naturally by living organisms. All the second- and third-period elements except Li, Be, Al, and the noble gases are used, as are most of the 3d elements, whereas Cd, Br, I, Mo, and W are the only heavier elements known to have an established biological function (although a role has recently been confirmed for lanthanoids, Ln^{3+}). Several others, such as Li, Ga, Tc, Ru, Gd, Pt, and Au, have important and increasingly well-understood applications in medicine and we will review their roles in Chapter 32.

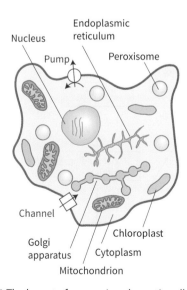

FIGURE 26.1 The layout of a generic eukaryotic cell showing the cell membrane, various kinds of compartments (organelles) that may be present in different cases, and the membrane-bound pumps and channels that control the flow of ions between compartments.

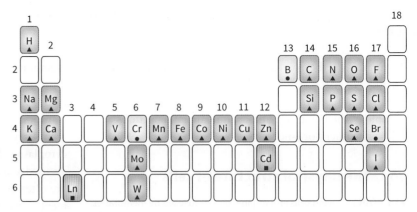

FIGURE 26.2 Periodic table showing the elements used by biology. Green (▲) indicates large-scale and widespread use, blue (●) indicates elements that are only required (or suspected to be so) in ultratrace amounts, orange (■) indicates elements that are used extremely rarely by certain organisms.

(a) Exploitation of different elements by biology

The biologically essential elements can be classified as either 'major' or 'trace', although the levels vary considerably among organisms and different components of organisms. For example, Ca has little role in microorganisms but is abundant in higher life forms, whereas the use of Co by higher organisms depends upon it being incorporated into a special macrocyclic ligand (known as cobalamin) by microorganisms. Cobalamin is one of many metal-containing cofactors that are responsible for the catalytic activity of the enzyme in which it is housed. There is probably a universal requirement for K, Mg, Fe, and Mo. Vanadium is used by lower animals and plants as well as some bacteria. Nickel is essential for most microorganisms, including pathogens such as *Helicobacter pylori* (Section 32.3), and is used by plants, but there is no evidence for any direct role in animals.

Nature's use of different elements is largely based on their availability. For example, Zn has widespread use (and, together with Fe, ranks among the most abundant biological trace elements), whereas Co (a comparatively rare element) is essentially restricted to cobalamin. The early atmosphere (over 2.3 billion years ago[1]), being highly reducing, enabled Fe to be freely available as soluble Fe(II) salts, whereas Cu was trapped as insoluble sulfides (as was Zn). Indeed, Cu is not found in the archaea (which are believed to have evolved before the Great Oxygenation Event, Box 17.1) which include hyperthermophiles—organisms that are able to survive at temperatures in excess of 100°C. These organisms are found in deep-sea hydrothermal vents and terrestrial hot springs, and are good sources of enzymes that contain W, the heaviest element known to be essential to life. The finding that cofactors containing W, Co, and for the most part Ni, are produced only by more primitive life forms probably reflects their special role in the early stages of evolution.

(b) Metallomics and compartmentalization

The term **metallomics** refers to the systematic study of the metallome—the pool of metal-containing species and some nonmetal trace elements in living organisms. A variety of analytical techniques are used to determine the organization, speciation, distribution, storage, regulation, dynamic properties, and pathogenicity of metal ions. A flow chart is shown in Fig. 26.3.

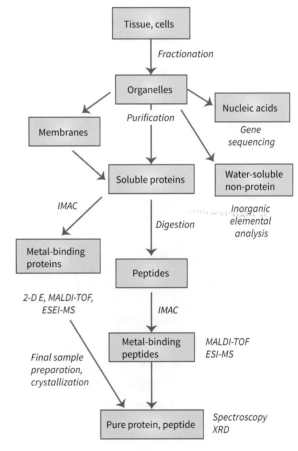

FIGURE 26.3 Flow chart for metallomic investigations, outlining how metal-binding proteins are identified in biological material using a systematic sequence of separation and analytical procedures. IMAC = immobilized metal affinity chromatography; 2-D E = 2-D electrophoresis; XRD = X-ray diffraction. Other terms are defined in Chapter 8.

---

[1] Current geological and geochemical evidence dates the advent of atmospheric O_2 at between 2.2 and 2.4 billion years ago: the Great Oxygenation Event (Box 17.1). It is likely that O_2 arose by the earliest catalytic actions of the photosynthetic Mn cluster described in Section 26.10.

TABLE 26.1 The approximate concentrations, (/mol dm^{-3}), where known, of elements (apart from C, H, O, N, P, S, Se) in different biological zones

Element	External fluids (sea water)	Free ions in external fluids (blood plasma)	Cytoplasm (free ions)	Comments on status in cell
Na	>10^{-1}	10^{-1}	<10^{-2}	Not bound
K	10^{-2}	4×10^{-3}	<3×10^{-1}	Not bound
Mg	>10^{-2}	10^{-3}	c.10^{-3}	Weakly bound as ATP complex
Ca	>10^{-3}	10^{-3}	c.10^{-7}	Concentrated in some vesicles
Cl	10^{-1}	10^{-1}	10^{-2}	Not bound
Fe	10^{-10} (variable) (Fe(III))	10^{-16} (Fe(III))	<10^{-7} (Fe(II))	Too much unbound Fe is toxic (Fenton chemistry) in and out of cells
Zn	<10^{-8}	10^{-9}	<10^{-11}	Totally bound, but may be exchangeable (released transiently)
Cu	<10^{-10} (Cu(II))	10^{-12}	<10^{-15} (Cu(I))	Totally bound, not mobile. Mostly outside cytoplasm
Mn	10^{-9}		c.10^{-6}	Higher in chloroplasts and vesicles
Co	10^{-11}		<10^{-9}	Totally bound (cobalamin)
Ni	10^{-9}		<10^{-10}	Totally bound
Mo	10^{-7}		<10^{-7}	Mostly bound

The principal goal of metallomics research is to identify *where* a particular metal ion or other trace element binds; that is, which cells, which organelles, which molecules, and which group of ligands. A further aim is to establish the dynamics (mobility) of each metal ion. Table 26.1 lists the metals (elements apart from C, N, O, S, P, and Se) and their abundances in different biological zones. An investigation in metallomics begins with organ/tissue sampling, cell disruption, and separation of different components. Instrumental techniques are then applied—such as mass spectrometry and ultrasensitive absorption and emission spectroscopic methods, high-resolution imaging, and sequencing of proteins and nucleic acids. These experimental procedures are complemented with computational methods. Information derived from metallomics plays an important role in drug development, as we will mention in Chapter 32.

Compartmentalization is the distribution of elements inside and outside a cell and between different internal compartments. The maintenance of constant ion levels in different biological zones is an example of 'homeostasis', and is achieved as a result of membranes being barriers to ion flow. An example is the large difference in concentration of K$^+$ and Na$^+$ ions across cell membranes. In the cytoplasm, the K$^+$ concentration may be as high as 0.3 M whereas outside it is usually less than 5×10^{-3} M. By contrast, Na$^+$ is abundant outside a cell but scarce inside; indeed, the low intracellular concentration of Na$^+$, which has characteristically weak binding to ligands, means that it has few specific roles in biochemistry. Another important example is Ca^{2+}, which is almost absent from the cytoplasm (its concentration is below 1×10^{-7} M) yet is a common cation in the extracellular environment, and is concentrated in certain organelles, such as mitochondria. That pH may also vary greatly between different compartments has particularly important implications, because sustaining a transmembrane proton gradient is a key feature in photosynthesis and respiration.

The distributions of Cu and Fe provide another example: Cu enzymes are often **extracellular**; that is, they are synthesized in the cell and then secreted outside the cell, where they catalyse reactions involving O$_2$. By contrast, Fe enzymes are contained inside the cell. This difference can be rationalized on the basis that the inactive trapped states of these elements are Fe(III) and Cu(I) (or even metallic Cu) and organisms have stumbled upon the expediency of keeping Fe in a relatively reducing environment and Cu in a relatively oxidizing environment.

The selective uptake of metal ions has nutritional and other applications, for many organisms and organs are known to concentrate particular elements. Thus, liver cells are a good source of cobalamin[2] (Co), and milk is rich in Ca. Certain bacteria accumulate Au and thus provide an unusual way for procuring this precious metal. Compartmentalization is an important factor in the design of metal complexes that are used in medicine (Chapter 32).

The very small size of bacteria and organelles raises an interesting point about scale, as species present at very low concentrations in tiny volumes may be represented by only a few individual atoms or molecules. For example, the cytoplasm in a bacterial cell of volume 10^{-15} dm^3 at pH = 6 will

---

[2] In nutrition, the common complexes of cobalamin that are ingested are known as vitamin B$_{12}$.

contain less than 1000 'free' H^+ ions. Indeed, any element nominally present at less than $1\,\text{nmol dm}^{-3}$ (abbreviated 1 nM) may be completely absent in individual cases. The word 'free' is significant, particularly for metal ions such as Zn^{2+} that are high in the Irving–Williams series (Section 7.16(i)); even a eukaryotic cell with a total Zn concentration of $0.1\,\text{mmol dm}^{-3}$ (0.1 mM) may contain very few uncomplexed Zn^{2+} ions.

Two important issues arise in the context of compartmentalization. First, the process requires energy because ions must be pumped against an adverse gradient of chemical potential. However, once a concentration difference has been established there is a difference in electrical potential across the membrane dividing the two regions. For instance, if the concentrations of K^+ ions on either side of a membrane are $[K^+]_{in}$ and $[K^+]_{out}$, then the contribution to the potential difference across the membrane is

$$\Delta\phi = \frac{RT}{F}\ln\frac{[K^+]_{in}}{[K^+]_{out}} \qquad (26.1)$$

This difference in electrical potential is a way of storing energy, which is released when the ions flood back to their natural concentrations. Second, the selective transport of ions must occur through **ion channels** built from membrane-spanning proteins, some of which release ions upon receipt of an electrical or chemical signal, whereas others, the **transporters** and **pumps**, transfer ions against the concentration gradient by using energy provided by adenosine triphosphate (ATP) hydrolysis. The selectivity of these channels is exemplified by the highly discriminatory transport of K^+ as distinct from Na^+ (Section 26.4).

Proteins—the most important sites for metal ion coordination—are not permanent species but are ceaselessly degraded by enzymes (proteases), releasing both amino acids and metal ions to provide materials for new molecules.

EXAMPLE 26.1 Assessing the role of phosphate ions

Phosphate is the most abundant small anion in the cytoplasm. What implications does this abundance have for the biochemistry of Ca^{2+}?

Answer We can approach this problem by considering how Ca^{2+} is compartmentalized. In a eukaryotic cell, Ca^{2+} is pumped out of the cytoplasm (to the exterior or into organelles such as mitochondria) using energy derived from ATP hydrolysis. Spontaneous influx of Ca^{2+} occurs under the action of special channels or if the cell boundary is damaged. The solubility product of $Ca_3(PO_4)_2$ is very low, and it could precipitate inside the cytoplasm if the Ca^{2+} concentration rises above a critical value.

Self-test 26.1 Is Fe(II) expected to exist as the simple $[Fe(H_2O)_6]^{2+}$ cation in a living cell?

26.3 Biological metal-coordination sites

KEY POINTS The major binding sites for metal ions are provided by the amino acids that make up protein molecules; the ligands range from backbone peptide carbonyls to the side chains that provide more specific complexation; nucleic acids and lipid head groups are usually coordinated to major metal ions. Special ligands such as macrocycles are used to impart particular catalytic properties to metal cofactors in proteins.

Metal ions coordinate to proteins, nucleic acids, lipids, and a variety of other molecules. For instance, ATP (Section 5.5) is a tetraprotic acid and is always found as its Mg^{2+} complex (**1**); DNA is stabilized by weak coordination of K^+ and Mg^{2+} to its phosphate groups but destabilized by binding of soft metal ions such as Cu(I) to the secluded bases. Ribozymes may represent an important stage in the early evolution of life forms and are catalytic molecules composed of RNA and Mg^{2+}. The binding of Mg^{2+} to phospholipid head groups is important for stabilizing membranes. There are a number of important small ligands, apart from water and free amino acids, which include sulfide, sulfate, carbonate, cyanide, carbon monoxide, and nitrogen monoxide, as well as organic acids such as citrate that form reasonably strong polydentate complexes with Fe(III).

1 Mg-ATP complex

As will be familiar from introductory chemistry, a protein is a polymer with a specific sequence of amino acids linked by peptide bonds (**2**). The units are often referred to as 'residues'. In macromolecular chemistry, it is customary to refer to molecular mass in terms of kilodaltons (kDa), where $1\,\text{kDa} = 1\,\text{kg mol}^{-1}$. A 'small' protein is generally regarded as one with molar mass below 20 kDa, whereas a 'large' protein is one having a molar mass above 100 kDa. Thanks to massive advances in X-ray diffraction (particularly using a synchrotron), NMR spectroscopy, and most recently cryo-electron microscopy (cryo-EM), the structures of over 50,000 proteins are now known.

2 Peptide bond

The principal amino acids are listed in Table 26.2. Proteins are synthesized—a process called **translation** (of the genetic code carried by DNA)—on a special assembly called a *ribosome*. In referring to such a process, we say that a gene is *expressed* to produce a particular protein. The protein may be processed further by **post-translational modification**, a change made to the protein structure, which includes the binding of **cofactors** such as metal ions. The total number of different proteins expressed by an organism is known as the *proteome*.

Metalloproteins—proteins containing one or more metal ions—perform a wide range of specific functions. These functions include catalysis of oxidation and reduction reactions (for which the most important elements are Fe,

TABLE 26.2 The amino acids and their codes

Amino acid	Structure in peptide chain (side chain shown in bold blue)	Three-letter abbreviation	One-letter abbreviation	Amino acid	Structure in peptide chain (side chain shown in bold blue)	Three-letter abbreviation	One-letter abbreviation
Alanine	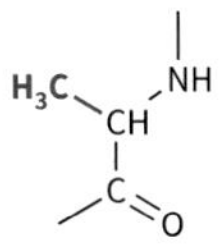	Ala	A	Leucine	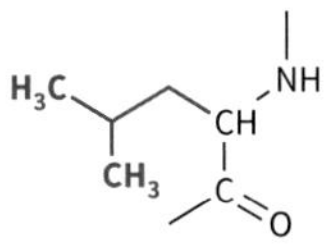	Leu	L
Arginine	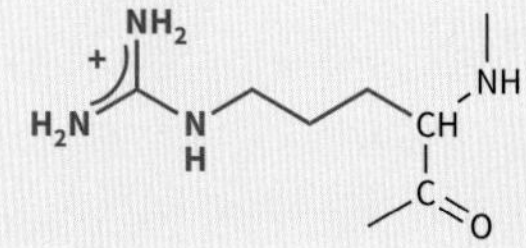	Arg	R	Lysine	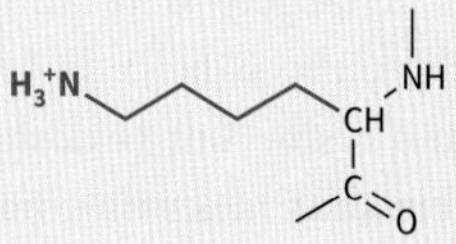	Lys	K
Asparagine	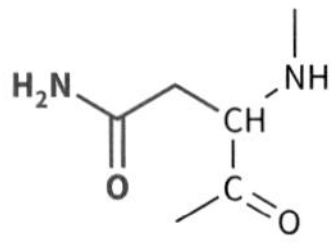	Asn	N	Methionine	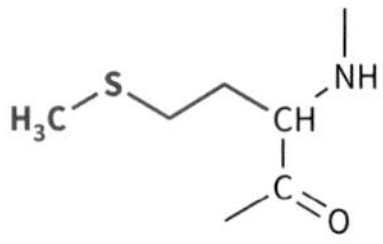	Met	M
Aspartic acid	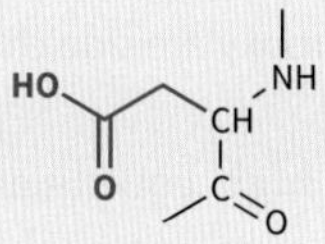	Asp	D	Phenylalanine	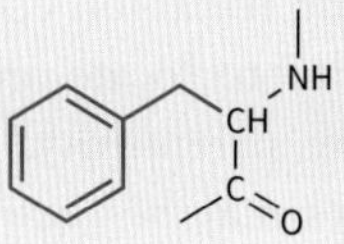	Phe	F
Cysteine	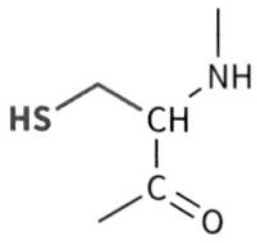	Cys	C	Proline	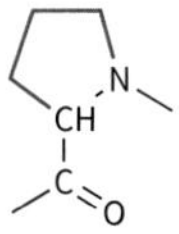	Pro	P
Glutamic acid	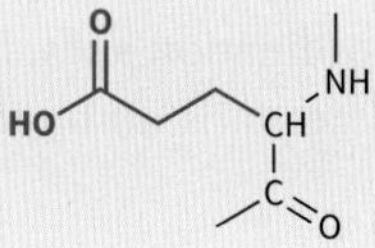	Glu	E	Serine	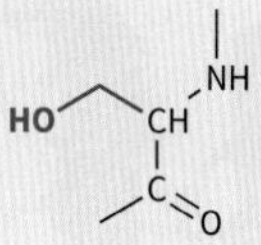	Ser	S
Glutamine	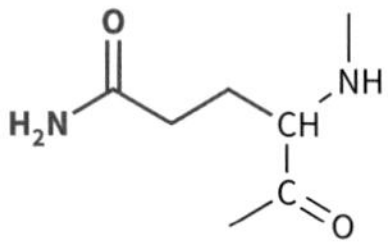	Gln	Q	Threonine	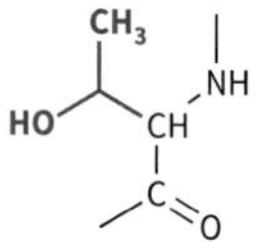	Thr	T
Glycine	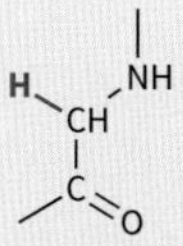	Gly	G	Tryptophan	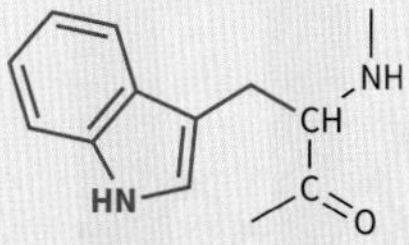	Trp	W
Histidine	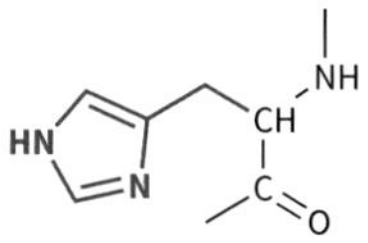	His	H	Tyrosine	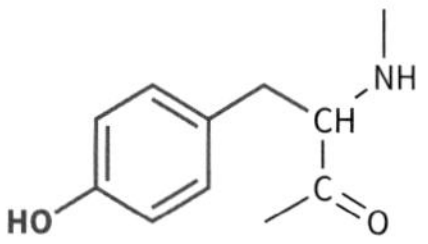	Tyr	Y
Isoleucine	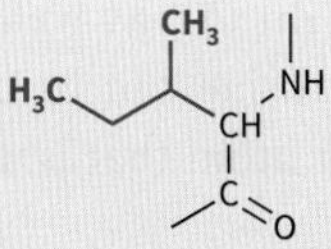	Ile	I	Valine	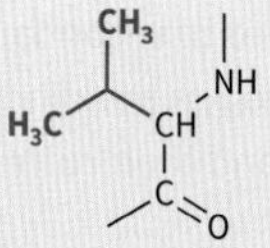	Val	V

Mn, Cu, and Mo), radical-based rearrangement reactions, methyl-group transfer and dehalogenation (Co, Fe), hydrolysis (Zn, Fe, Mg, Mn, and Ni), and DNA decryption (Zn). Special proteins are required for transporting and storing different metal atoms. The action of Ca^{2+} is to alter the conformation (shape) of a protein as a step in cell signalling (a term used to describe the transfer of information between and within cells). Such proteins are often known as **metal ion-activated proteins**.

Hydrogen bonding between main-chain –NH and CO groups of different amino acids results in **secondary structure** (Fig. 26.4). The **α-helix** regions of a polypeptide provide flexible mobility (like springs) and are important in converting processes that occur at the metal site into conformational changes; by contrast, a **β-sheet** region confers rigidity to support a preorganized coordination sphere suited to a particular metal ion (Sections 7.14 and 12.16).

The secondary structure is largely determined by the sequence of amino acids: thus, the α-helix is favoured by chains containing alanine and lysine but is destabilized by glycine and proline. A protein that lacks its cofactor (such as the metal ion itself or a special metal complex required for normal activity) is called an **apoprotein**; an enzyme with a complete complement of cofactors is known as a **holoenzyme**.

An important factor influencing metal-ion coordination in proteins is the energy required to locate an electrical charge inside a medium of low permittivity. To a first approximation, protein molecules may be regarded as oil droplets in which the interior has a much lower relative permittivity (about 4) than water (about 78). This difference leads to a strong tendency to preserve electrical neutrality at the metal site, and hence influence the redox chemistry and Brønsted acidity of its ligands. We expect metal ions with higher charge to be coordinated by anionic ligands; that is, in their deprotonated forms (carboxylates, thiolates, as well as hydroxide/oxide).

(a) Amino acids as ligands

KEY POINT All amino acid residues can use their peptide carbonyl as a donor group, and this ligand is most important in coordinating Na^+ and K^+, but it is the different side chain functionalities that provide the rich variety of ligands for selective coordination of other metal ions.

By referring to Table 26.2 and from the discussion in Section 5.10, we can recognize donor groups that are chemically either hard or soft and which therefore confer a particular affinity for specific metal ions. Aspartate and glutamate each provide a hard carboxylate group, and may use one or both O atoms as donors (**3**).

3 Ca^{2+} coordination by carboxylate side chains

The ability of Ca^{2+} to have a high coordination number and its preference for hard donors are such that certain Ca^{2+}-binding proteins also contain the unusual amino acids γ-carboxyglutamate and hydroxyaspartate (generated by post-translational modification), which provide additional functionalities to enhance binding. Histidine, which has an imidazole group with two coordination sites, the ε-N atom (more common) and the δ-N atom, is an important ligand for Fe, Cu, and Zn (**4**).

4 Cu–imidazole coordination

Cysteine has a thiol S atom that is expected to be deprotonated (thiolate) when involved in metal coordination. It is a good ligand for Fe, Cu, and Zn (**5**), as well as for toxic metals such as Cd and Hg. Tyrosine can be deprotonated to provide a phenolate O donor atom that is a good ligand for Fe(III) (**6**).

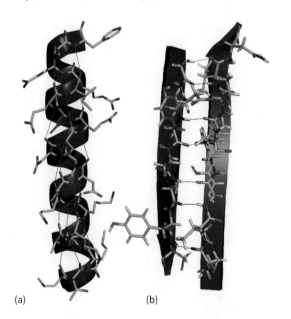

FIGURE 26.4 The most important regions of secondary structure, (a) α-helix and (b) β-sheet, showing hydrogen bonding between main-chain amide and carbonyl groups and their corresponding representations.

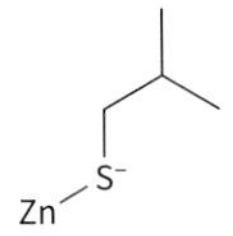

5 Zn–cysteine coordination

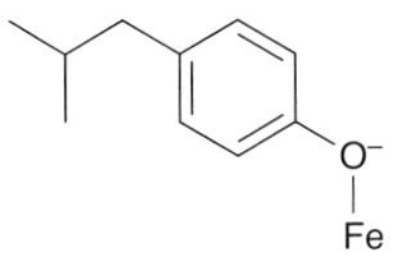

6 Fe–tyrosine coordination

Selenocysteine (a specially coded amino acid in which Se replaces S) has also been identified as a ligand; for example, it is found as a ligand to Ni in some hydrogenases (Section 26.13). A modified form of lysine, in which the side-chain $-NH_2$ has reacted with CO_2 to form a carbamate, is found as a ligand to Mg in the crucial photosynthetic enzyme known as Rubisco (Section 26.9) and in other enzymes such as urease, where it is a ligand for Ni(II).

Proteins can enforce unusual metal coordination geometries and activities that are rarely encountered in small complexes. The two convenient representations of peptide structure shown in Fig. 26.4 do not tell the full story because they ignore the side chains that make up the bulk of the protein; indeed, once the various side chains are included it usually becomes very difficult to 'see' the metal atom even if it is just below the protein surface. The protein provides very specific steric hindrance in the metal coordination sphere that is difficult to mimic with organic ligands: accordingly, chemists attempting to model the active sites of metalloenzymes make extensive use of bulky ligands in order to protect a coordinatively unsaturated site. Protein-induced strain is another important possibility; for example, the protein may impose a coordination geometry on the metal ion that resembles the transition state for the particular process being executed.

Many of the structures of active sites shown in this chapter are direct representations based on X-ray diffraction data and constructed using commercially available software known as Pymol®. Instead of the aesthetically pleasing structures that are familiar to chemists working with small complexes, we will frequently see severe distortions—twisted angles and unusual bond lengths—that are imposed by the surrounding protein.

(b) Special ligands

KEY POINT Metal ions may be coordinated in proteins by special organic ligands such as porphyrins and pterin-dithiolenes.

The porphyrin group (**7**) was first identified in haemoglobin (Fe) and a similar macrocycle is found in chlorophyll (Mg). There are several classes of this hydrophobic macrocycle, each differing in the nature of the side chains. Complexed to Fe, it is commonly referred to as haem (alternatively, heme).

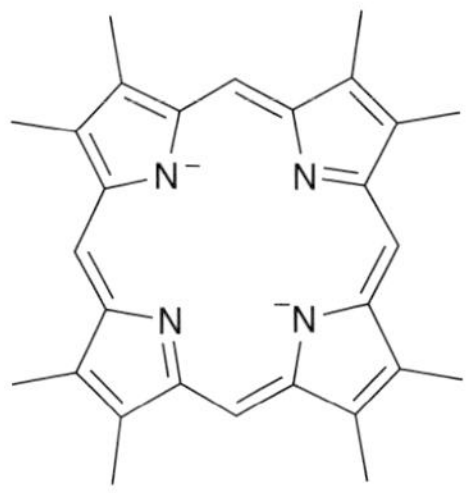

7 The porphyrin macrocyclic ligand

The corrin ligand (**8**) has a slightly smaller ring size and coordinates Co in cobalamin (Section 26.11).

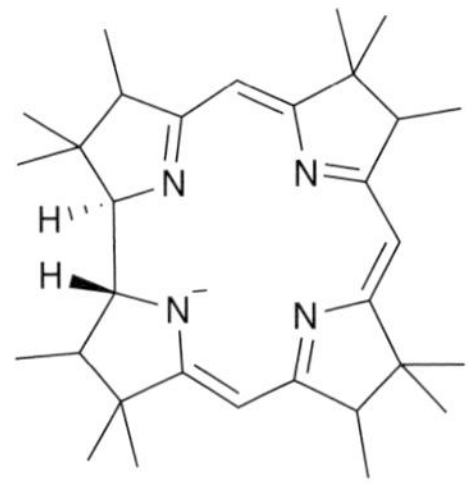

8 The corrin macrocyclic ligand

Rather than show these macrocycles in full, we shall use shorthand symbols such as (**9**) to show the complexes they form with metals.

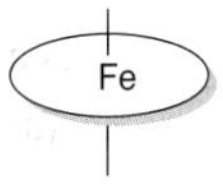

9 Shorthand representation of Fe porphyrin

Almost all Mo and W enzymes have the metal coordinated by a special ligand known as molybdopterin (**10**). The donors to the metal are a pair of S atoms from a dithiolene group that is covalently attached to a pterin. The phosphate group is often joined to a nucleoside base X, such as guanosine 5′-phosphate (GMP) resulting in the formation of a diphosphate bond. Why Mo and W are coordinated by this complex ligand is unknown, but the pterin group could provide a good electron conduit and facilitate redox reactions.

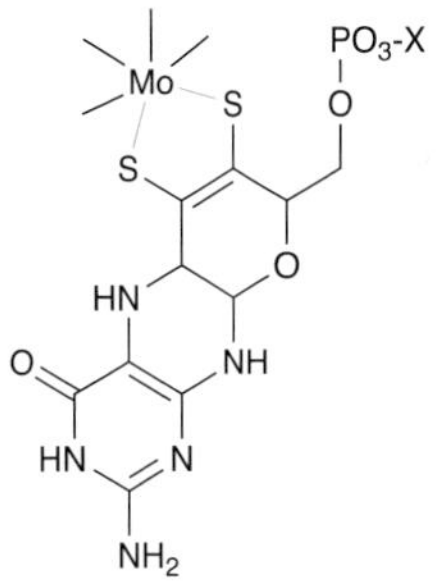

10 Molybdopterin as a ligand, coordinating Mo

(c) The structures of inner and outer coordination shells

KEY POINTS The likelihood that a protein will coordinate a particular kind of metal centre can be inferred from the amino acid sequence and ultimately from the gene itself. Spectroscopic methods, where applicable, are powerful tools for investigating the inner coordination shell. The outer coordination shell, which is crucial for selective binding of exogenous reagents and catalytic activity, can be systematically altered by genetic engineering.

The structures of metal coordination sites have been determined mainly by X-ray diffraction (now mostly by using a synchrotron, Section 8.1) and sometimes by nuclear magnetic resonance (NMR) spectroscopy (Section 8.6).[3] Metalloenzymes are important subjects in the development of time-resolved diffraction using an X-ray free-electron laser (XFEL)—a technology able to reveal the structures of catalytic intermediates on the millisecond timescale.

The basic structure of the protein can be determined even if the resolution is too low to reveal details of the coordination at the metal site. The packing of amino acids in a protein is far denser than is commonly conveyed by simple representations, as may be seen by comparing two views of the structure of the K^+ channel in Fig. 26.5. Thus, even the substitution of an amino acid that is far from a metal centre may result in significant structural changes to its coordination shell and properties.

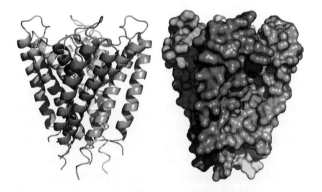

FIGURE 26.5 Illustrations of how protein structures are represented to reveal either (left) secondary structure or (right) the filling of space by nonhydrogen atoms. The example shows the four subunits of the K^+ channel, which is found mainly embedded in the cell membrane.

---

[3] The atomic coordinates of proteins and other large biological molecules are stored in a public repository known as the Protein Data Bank located at http://www.rcsb.org/pdb/home/home.do. Each set of coordinates corresponding to a particular structure determination is identified by its 'pdb code'. A variety of software packages are available to construct and examine protein structures generated from these coordinates.

Of special interest are tunnels or clefts that allow a substrate (the reactant or product) selective access to the active site, pathways for long-range electron transfer (metal centres positioned less than 1.5 nm apart), pathways for long-range proton transfer (comprising chains of Brønsted acid–base groups, such as carboxylates and water molecules, in close proximity, usually less than 0.3 nm apart), and channels for small gaseous molecules (which can be revealed in X-ray diffraction by placing the crystal under Xe, an electron-rich gas).

EXAMPLE 26.2 Interpreting the coordination environments of metal ions

Simple Cu(II) complexes have four to six ligands with trigonal-bipyramidal or tetragonal geometries, whereas simple complexes of Cu(I) have four or fewer ligands, and geometries that range between tetrahedral and linear. With reference to Section 23.12, predict how a Cu-binding protein will have evolved so that the Cu can act as an efficient electron-transfer site.

Answer An efficient electron-transfer reaction is one that is fast despite having a small driving force. The Marcus equation tells us that an efficient electron-transfer site is one for which the reorganization energy is small. The protein enforces on the Cu atom a coordination sphere that should be a compromise between Cu(II) and Cu(I) states and probably resembles the transition state for electron transfer (see Section 26.8).

Self-test 26.2 In certain chlorophyll cofactors the Mg is axially coordinated by a methionine-S ligand. Suggest how this unexpected choice of ligand is achieved in a protein, whereas it would be extremely unusual in simple complexes.

Other physical methods described in Chapter 8 provide less information on the overall structure but are useful for identifying ligands. Thus, EPR spectroscopy is very important for studying d-block metals, especially those engaged in redox chemistry, since at least one oxidation state usually has an unpaired electron. The use of NMR is usually restricted to proteins smaller than 20–30 kDa because tumbling rates for larger proteins are too slow and ^1H resonances are too broad to observe unless shifted away from the normal region ($\delta \approx 1$–10) by a paramagnetic metal centre. Extended X-ray absorption fine-structure spectroscopy (EXAFS, Section 8.10) can provide structural information on metal sites in amorphous solid samples, including frozen solutions. Vibrational spectroscopy (Section 8.5) is now widely used: IR spectroscopy is particularly useful for ligands such as CO and CN^-, and resonance Raman spectroscopy is very helpful when the metal centre has strong electronic transitions, such as occur with Fe porphyrins. Mössbauer spectroscopy (Section 8.8) plays a special role in studies of Fe sites. Perhaps the greatest challenge is presented by Zn^{2+}, which has a d^{10} configuration that provides no useful magnetic or electronic signatures (see Tutorial Problem 26.1 at the end of this chapter).

Metal ion binding sites can often be predicted from a gene sequence. **Bioinformatics**, the development and use of software to analyse and compare DNA sequences, is a powerful tool because many proteins that bind metal ions or have a metal-containing cofactor occur at cellular levels below that normally detectable directly by analysis and isolation.

A particularly common sequence of the human genome encodes the so-called **Zn finger domain**, thereby identifying proteins that are involved in DNA binding (Section 26.15). Likewise, it can be predicted whether the protein that is encoded is likely to bind Cu, Ca, Fe-porphyrin, or different types of Fe–S clusters. The gene can be cloned, and the protein for which it encodes can be produced in sufficiently large quantities by 'overexpression' in suitable hosts, such as the common gut bacterium *Escherichia coli* or yeast, to enable it to be characterized.

Furthermore, the use of genetic engineering to alter the amino acids in a protein, the technique of **site-directed mutagenesis**, is a powerful principle in biological inorganic chemistry. This technique often permits identification of the ligands that coordinate particular metal ions and the participation of residues in the outer shell that are equally essential for biological activity. The outer coordination shell includes functionalities provided by the protein that control the access of reagents, including water and protons, to the inner shell complex and influence their binding by steric or polar interactions.

Although structural and spectroscopic studies give a good idea of the basic coordination environment of a metal centre, it is by no means certain that the same structure is retained in key stages of a catalytic cycle, in which unstable states are formed as intermediates. The most stable state of an enzyme, in which form it is usually isolated, is called the 'resting state'. Many enzymes are catalytically inactive upon isolation and must be subjected to an activation procedure that may involve reinsertion of a metal ion or other cofactor or removal of an inhibitory ligand.

Intense efforts have been made to model the active sites of metalloproteins by synthesizing analogues. The models may be divided into two classes: those designed to mimic the structure and spectroscopic properties of the real site, and those synthesized with the intention of mimicking a functional activity, most obviously catalysis. Synthetic models not only illuminate the chemical principles underlying biological activity but also generate new directions for coordination chemistry. As we shall see throughout this chapter, the difficulty for chemists is that an enzyme not only imposes strain on the inner coordination sphere of a metal atom (even a porphyrin ring is puckered in most cases), but also provides, at fixed distances, the outer-sphere (2nd, 3rd ... nth shell) functional groups that are often equally important. The active site of a metalloenzyme is the ultimate example of supramolecular chemistry.

Metal ions in transport and communication

In this large section we turn to the transport of ions such as Na^+ and K^+ through membranes, the role of Ca^{2+} as a messenger, and the transport and storage of two key elements—Fe and O_2—through tissue and cells.

26.4 Sodium and potassium transport

KEY POINTS Transport across a membrane is active (energized) or passive (spontaneous); the flow of ions is achieved by proteins known as ion pumps (active) and channels (passive).

In Section 12.16 we saw that differentiating between Na^+ and K^+—two ions that are very similar except for their radii (102 pm and 138 pm, respectively)—is achieved through their selective complexation by special ligands, such as crown ethers and cryptands, with dimensions appropriate for coordination to one particular kind of ion. Organisms use this principle in the molecules known as **ionophores**, which have hydrophobic exteriors that render them soluble in lipids. The antibiotic valinomycin (Section 12.16) is an ionophore with a high selectivity for K^+, which is coordinated by six carbonyl groups. Valinomycin enables K^+ to pass through a bacterial cell membrane and thereby dissipate the electrical potential difference, so causing the bacterium's death. Transport driven by a concentration gradient is known as passive transport.

Ion channels are large membrane-spanning proteins that allow selective, passive transport of K^+ and Na^+ (as well as Ca^{2+} and Cl^-) and are responsible for electrical conduction in nervous systems as well as in coupled transport of solutes.[4] Fig. 26.6 shows different structural aspects of the potential-gated K^+ channel. Moving from the inside surface of the membrane, the enzyme has a pore (which can open and close on receipt of a signal) leading into a central cavity about 1 nm in diameter; up to this stage, K^+ ions can remain hydrated. Polypeptide helices pointing at this cavity have their partial charges directed in such a way as to favour

---

[4] Roderick MacKinnon shared the 2003 Nobel Prize for Chemistry for his elucidation of the structures and mechanisms of ion channels.

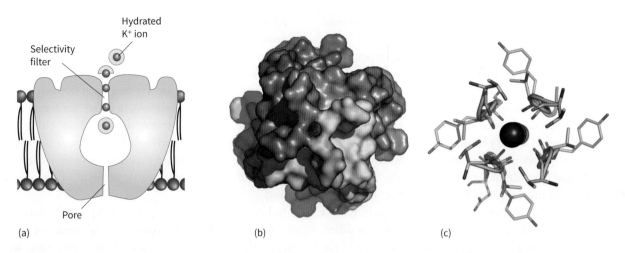

FIGURE 26.6 (a) Schematic structure of the K⁺ channel showing the different components and the transport of K⁺ ions: the blue halo represents hydration. (b) View of the enzyme from inside the cell showing the entrance pore that admits hydrated ions. (c) View looking up the selectivity filter showing how mobile dehydrated K⁺ ions are coordinated by peptide carbonyl O atoms provided by each of the four subunits. Note the almost four-fold symmetry axis.

population by cations, resulting in a local K⁺ concentration of approximately 2 M.

Above the central cavity the tunnel contracts into a **selectivity filter** consisting of helical ladders of closely spaced peptide carbonyl-O donors that form a sequence of four cubic eight-fold coordination sites. During operation of the channel these sites are occupied, at any one time, by an alternating sequence of two K⁺ ions and two H₂O molecules; that is, $\cdots K^+\cdots H_2O\cdots K^+\cdots H_2O\cdots$. The rate of passage of K⁺ ions through the selectivity filter is close to the limit for diffusion control. A plausible mechanism for selective K⁺ transport (Fig. 26.7) involves concerted displacement of K⁺ ions between adjacent cubic, eight-fold carbonyl-O sites through intermediate, unstable octahedral states in which the K⁺ ions are coordinated equatorially by four carbonyl-O donors and axially by the two intervening H₂O molecules.

This mechanism is not effective for Na⁺ because the cavity is too large, which accounts for the 10^4-fold selectivity of the channel for K⁺ over Na⁺. The binding is weak and fast because it is important to convey K⁺ but not trap it.

Overall, the K⁺ channel works by the principle illustrated in Fig. 26.8. Charged groups on the molecule move in response to a change in the membrane potential and cause the intracellular pore to open, so allowing entry of hydrated K⁺ ions. Selective binding of dehydrated K⁺ ions occurs in the filter region, a potential drop across the membrane is sensed, and the pore closes. At this point the filter opens up to the external surface, where the K⁺ concentration is low and the K⁺ ions are hydrated and released. This release causes the protein to switch back to the original conformation, and K⁺ ions again enter the filter.

The Na pump (Na⁺/K⁺-ATPase), the enzyme that maintains the concentration differential of Na⁺ and K⁺ inside and outside a cell, is another example of the high discrimination

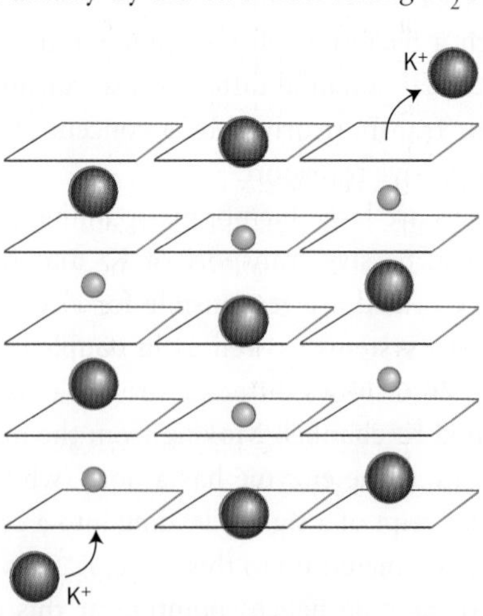

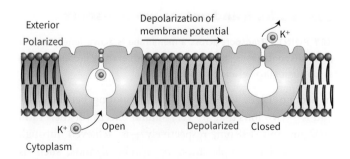

FIGURE 26.7 Mechanism of transport of K⁺ ions through the selectivity filter of the K⁺ channel. Small green spheres represent water molecules. The middle column shows the intermediate state in which K⁺ has a lower coordination number.

FIGURE 26.8 General principle of the action of the K⁺ channel. The potential difference across the membrane is sensed by the protein, which causes the pore to open, allowing hydrated ions to enter the cavity. After shedding their hydration sphere, K⁺ ions pass up the selectivity filter at rates close to diffusion control.

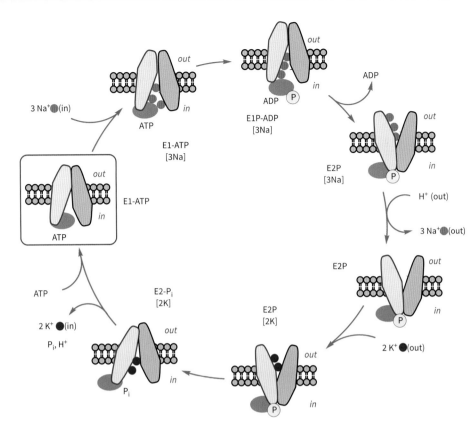

FIGURE 26.9 General principle of the Na⁺/K⁺-ATPase (the Na pump). Release of two K⁺ ions into the cytoplasm is accompanied by binding of ATP (from the cytoplasm) and conversion of the enzyme into state E1, which binds three Na⁺ ions from the cytoplasm. A phosphate group (P) is transferred from ATP to the enzyme, which then opens to the external side (state E2), expels three Na⁺ ions, then binds two K⁺ ions. Hydrolysis of the phosphate ester to produce phosphate ion (P_i) opens the pump to the internal side, K⁺ is released into the cytoplasm and the cycle begins again.

between alkali metal ions that has evolved with biological ligands. Here the transport is active: the ions are pumped against their concentration gradients by coupling the process to ATP hydrolysis. The mechanism of the catalytic cycle, outlined in Fig. 26.9, involves conformational changes induced by ATP-driven protein phosphorylation. Three Na⁺ ions are pumped out of the cytoplasm in exchange for two K⁺ (net transfer of an H⁺ maintains transmembrane ionic balance). At the detailed level, X-ray diffraction studies reveal that the three Na⁺ ions that are taken up by the E1 state are bound between two α-helices that move closely together (Fig. 26.10).

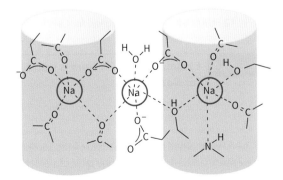

FIGURE 26.10 Representation of the binding site for three Na⁺ ions located across two α-helices of the Na⁺/K⁺-ATPase.

EXAMPLE 26.3 Assessing the role of ions in active and passive transport

The toxic species Tl⁺ (radius 150 pm) is used as an NMR probe for K⁺ binding in proteins. Explain why Tl⁺ is suited for this purpose and account for its high toxicity.

Answer To address this question we need to recall from Chapter 14 that Tl, in common with other heavy, post-d-block elements, displays the inert-pair effect (Section 10.7), a preference for forming compounds in which its oxidation number is 2 less than the group oxidation number. Thallium (Group 13) thus resembles the heavy Group 1 elements (in fact, TlOH is a strong base) and Tl⁺ can replace K⁺ in complexes, with the advantage that it can be studied by NMR spectroscopy (²⁰³Tl and ²⁰⁵Tl have $I = \frac{1}{2}$). The similarity with K⁺ allows Tl⁺, a toxic element, free entry into a cell because it is 'recognized' by the Na⁺/K⁺-ATPase. But once inside, more subtle differences in chemical properties, such as the tendency of Tl to form more stable complexes with soft ligands, are manifested and become lethal.

Self-test 26.3 Explain why the intravenous fluid used in hospital procedures contains NaCl rather than KCl.

26.5 Calcium signalling proteins

KEY POINT Calcium ions are suitable for signalling because they exhibit fast ligand exchange and a large, flexible coordination geometry.

Calcium ions play a crucial role in higher organisms as a cell (intercompartmental) messenger. Fluxes of Ca^{2+} trigger enzyme action in cells in response to receiving a hormonal or electrical signal from elsewhere in the organism. Calcium is especially suited for signalling because it has high ligand exchange rates, intermediate binding constants, and a large, flexible coordination sphere.

Calcium signalling proteins are small proteins that change their conformation depending upon the binding of Ca^{2+} at one or more sites; they are thus examples of metal ion-activated proteins mentioned earlier. Every muscle movement we make is stimulated by Ca^{2+} binding to a protein known as troponin C. The best studied Ca^{2+}-regulatory protein is calmodulin (17kDa, Fig. 26.11): its roles include activating protein kinases that catalyse phosphorylation of proteins and activation of NO synthase, an Fe-containing enzyme responsible for generating the intercellular signalling molecule nitric oxide. Calmodulin has four Ca^{2+}-binding sites (one is shown as **11**) with dissociation constants lying in the micromolar range.[5] The binding of Ca^{2+} to the four sites alters the protein conformation, and it is then recognized by a target enzyme.

Calcium signalling requires special Ca^{2+} pumps, which are large, membrane-spanning enzymes that pump Ca^{2+} out of the cytoplasm, either out of the cell altogether or into Ca-storing organelles such as the endoplasmic reticulum or the mitochondria. As with Na^+/K^+-ATPases, the energy for Ca^{2+} pumping comes from ATP phosphate transfer and hydrolysis. Hormones or electrical stimuli open specific channels (analogous to K^+ channels) that release Ca^{2+} into the cell. Because the level in the cytoplasm before the pulse is low, the influx easily raises the Ca^{2+} concentration above that needed for Ca^{2+}-binding proteins such as calmodulin (or troponin C in muscle). The action is short-lived, as after a pulse of Ca^{2+} the cell is quickly evacuated by the calcium pumps.

Although Ca^{2+} is 'invisible' to most spectroscopic methods, some Ca proteins, such as calmodulin or troponin C, are small enough to be studied by NMR. Because of their preference for forming complexes with large multicarboxylate ligands, the lanthanoid ions (Chapter 21) have been used as probes for Ca binding, exploiting their properties of paramagnetism (as chemical shift reagents in NMR spectroscopy) and fluorescence (Section 22.5). Intracellular concentrations of Ca are monitored by using special fluorescent polycarboxylate ligands (**12**) that are introduced to the cell as their esters, which are hydrophobic and able to cross the membrane lipid barrier. Once in the cell, enzymes known as esterases hydrolyse the esters and release the ligands, which respond to changes in Ca^{2+} concentration in the range 10^{-7}–10^{-9} M.

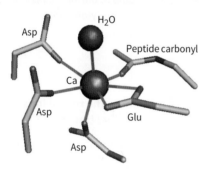

11 A Ca^{2+} binding site in calmodulin

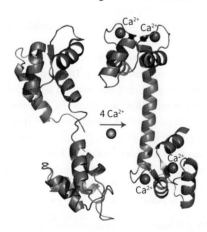

FIGURE 26.11 The binding of four Ca^{2+} to apocalmodulin causes a change in the protein conformation, converting it to a form that is recognized by many enzymes. The high proportion of α-helix is typical of proteins that are activated by metal ion binding.

[5] In simple terms, the occupation of a binding site is 50% when the concentration (in $mol\,dm^{-3}$) of species is equal to the dissociation constant (the reciprocal of the association constant).

12 FURA-2, a fluorescent ligand for Ca^{2+}

EXAMPLE 26.4 Explaining why calcium is suitable for signalling

Why is Ca^{2+} more suitable than Mg^{2+} for fast signalling processes in cells?

Answer To answer this question we need to refer to Section 23.1, in which we saw that the ligand-exchange rates of s-block metal ions increase down each group. The exchange of coordinated H_2O molecules is 10^3–10^4 times faster for Ca^{2+} than for Mg^{2+}. The speed at which Ca^{2+} can shed its ligands and bind to a target protein is crucial for ensuring fast signalling. As an example, rapid muscle contractions that may protect an organism against sudden attack are initiated by Ca^{2+} binding to troponin C.

Self-test 26.4 The Ca^{2+} pump is activated by calmodulin. Explain the significance of this observation. Hint: Consider how a feedback mechanism could control Ca^{2+} levels in the cytoplasm.

26.6 Selective transport and storage of iron

KEY POINTS The uptake of Fe into organisms involves special ligands known as siderophores; transport in the circulating fluids of higher organisms requires a protein called transferrin; Fe is stored as ferritin.

Iron is essential for almost all life forms; however, Fe is also difficult to obtain, yet any excess presents a serious toxic risk. Nature has at least two problems in dealing with this element. The first is the insolubility of Fe(III), which is the stable oxidation state found in most minerals. As the pH increases, hydrolysis, polymerization, and precipitation of hydrated forms of the oxide occur. Polymeric oxide-bridged Fe(III) is the thermodynamic sink of aerobic Fe chemistry (as seen in a Pourbaix diagram, Section 6.14). The obvious insolubility of rust renders the straightforward uptake by a cell very difficult. The second problem is the toxicity of 'simple Fe' species, particularly through the generation of ·OH radicals. To prevent Fe^{2+} or Fe^{3+} from reacting with oxygen species in an uncontrolled manner, a protective coordination environment is required. Nature has evolved sophisticated chemical systems to execute and regulate all aspects, from the primary acquisition of Fe, to its subsequent transport, storage, and utilization in tissue. The 'Fe cycle' as it affects a human is summarized in Fig. 26.12.

(a) Siderophores

Siderophores are small polydentate ligands which have a very high affinity for Fe(III). They are secreted from many bacterial cells into the external medium, where they sequester Fe to give a soluble complex that re-enters the organism at a specific receptor. Once inside the cell, the Fe is released.

Aside from citrate (the Fe(III) citrate complex is probably the simplest Fe transport species in biology) there are two main types of siderophore. The first type is based on phenolate or catecholate ligands, and is exemplified by enterobactin (**13**) for which the value of the association constant for Fe(III) is 10^{52}—an affinity so great that enterobactin enables bacteria to erode steel bridges! The second type of siderophore is based on hydroxamate ligands and is exemplified by ferrichrome (**14**)—a cyclic hexapeptide consisting of three glycine and three N-hydroxyl-L-ornithines.

13 Enterobactin

14 Ferrichrome

All Fe(III) siderophore complexes are octahedral and high spin. Because the donor atoms are hard O or N atoms and negatively charged, they have a relatively low affinity for Fe(II). Synthetic siderophores are proving to be very useful agents for the control of 'iron overload'—a serious condition affecting large populations of the world, particularly in South-East Asia (Section 32.8).

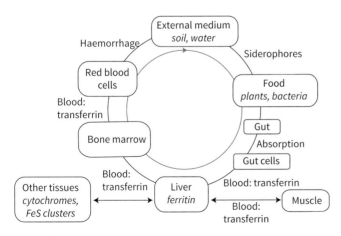

FIGURE 26.12 The biological Fe cycle showing how Fe is taken up from the external medium and guarded carefully in its travels through organisms.

(b) Iron-transport proteins in higher organisms

There are several important, structurally similar Fe-transport proteins known collectively as **transferrins**. The best-characterized examples are serum transferrin (in blood plasma), ovotransferrin (in egg-white), and lactoferrin (in milk). The apoproteins are potent antibacterial agents as they deprive microbes of their iron. Transferrins are also present in tears, serving to cleanse eyes after irritation. All these transferrins are glycoproteins (protein molecules modified by covalently bound carbohydrate) with molar masses of about 80 kDa and containing two separated and equivalent binding sites for Fe. Complexation of Fe(III) at each site involves simultaneous binding of HCO_3^- or CO_3^{2-} and release of H^+:

$$\text{apo-TF} + \text{Fe(III)} + HCO_3^- \rightarrow \text{TF–Fe(III)–}CO_3^{2-} + H^+$$

where TF denotes transferrin. For each site, the association constant under physiological conditions (pH = 7) lies in the range 10^{22}–10^{26}. However, its value depends strongly on the pH, and this dependence is the main factor controlling Fe uptake and release.

Transferrin consists of two very similar parts, termed the **N-lobe** and the **C-lobe** (Fig. 26.13). The protein is a product of gene duplication, and the structure of the first half of the molecule can almost be overlaid on the second half. Each half consists of two domains, 1 and 2, which together form a cleft with a binding site for Fe(III). There is a considerable proportion of α-helix, resulting in flexibility. Complexation with Fe(III) causes a conformational change consisting of a hinge motion involving domains 1 and 2 at each lobe. Binding of Fe(III) at each half causes the domains to come together.

In each site (**15**), a single Fe atom is coordinated by widely dispersed amino acid side chains from both domains and the connecting region, hence the change in conformation that occurs. The protein ligands are carboxylate-O (Asp), two phenolate-O (Tyr), and an imidazole-N (His). Only one of the aspartate carboxylate-O atoms is coordinated to the Fe. The protein ligands form part of a distorted octahedral coordination sphere. The coordination is completed by bidentate binding to an exogenous carbonate, although in certain cases phosphate is bound instead. As expected from the predominantly anionic ligand set, Fe(III) binds much more tightly than Fe(II). However, ions similar to Fe(III), particularly Ga(III) and Al(III), also bind tightly, enabling them to use the same transport system to gain access to tissues.

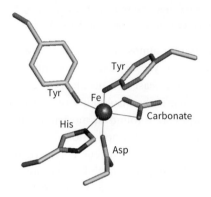

15 An Fe-binding site in transferrin

(c) Release of iron from transferrin

Cells in need of Fe produce large amounts of a protein called the **transferrin receptor** (180 kDa), which is incorporated within their plasma membrane. This protein binds Fe-loaded transferrin. The favoured mechanism for Fe uptake involves the Fe-loaded transferrin receptor complex entering the cell by a process known as *endocytosis*, in which a section of the cell membrane is engulfed by the wall along with its component membrane-bound proteins, to form a vesicle. The pH within this vesicle is then lowered by a membrane-bound H^+-pumping enzyme that is also swallowed by the cell. The subsequent release of Fe(III) is probably linked to the coordination of carbonate, which is **synergistic** in the sense that it is necessary for the binding of Fe, but unstable at low pH since carbonate becomes protonated (Sections 5.8 and 15.9). Indeed, from *in vitro* studies it is known that Fe is released by lowering the pH to about 5 for serum transferrin and to 2–3 for lactoferrin. The vesicle then splits and the TF-receptor complex is returned to the plasma membrane by **exocytosis**, while Fe(III), probably now complexed by citrate, is released to the cytoplasm.

(d) Ferritin, the cellular Fe store

Ferritin is the principal store of non-haem Fe in animals (most Fe is occupied in haemoglobin and myoglobin) and, when fully loaded, contains 20% Fe by mass! It occurs in all types of organism, from mammals to prokaryotes. In mammals, it is found particularly in the spleen and in blood. Ferritins have two components: a 'mineral' core that

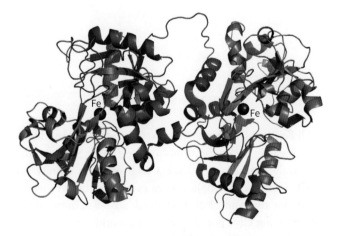

FIGURE 26.13 Structure of the Fe-transport protein transferrin: the identical halves of the molecule each coordinate a single Fe(III) atom (shown as black spheres) between two lobes. Binding of two Fe(III) causes a conformational change that allows transferrin to be recognized by the transferrin receptor.

contains up to 4500 Fe atoms (mammalian ferritin), and a protein shell. Apoferritin (the protein shell devoid of Fe) can be prepared by treatment of ferritin with reducing agents and an Fe(II) chelating ligand (such as 1,10-phenanthroline or 2,2′-bipyridyl). Dialysis then yields the intact shell.

Apoferritins have average molar masses in the range 460–550 kDa. The protein shell (Fig. 26.14) consists of 24 interlinked subunits that form a hollow sphere with two-fold, three-fold (which is shown in the illustration), and four-fold symmetry axes. Each subunit consists of a bundle of four long and one short α-helices, with a loop that forms a section of β-sheet with a neighbouring subunit.

The mineral core is composed of hydrated Fe(III) oxide with varying amounts of phosphate which helps anchor it to the internal surface. The structure of the core as revealed by X-ray or electron diffraction resembles that of ferrihydrite, $5Fe_2O_3 \cdot 9H_2O$, a mineral based on an hcp array of O^{2-} and OH^- ions, with Fe(III) layered in both octahedral and tetrahedral sites (16).

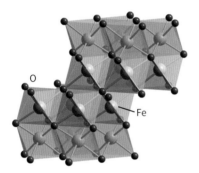

16 Ferrihydrite

The three- and four-fold symmetry axes of apoferritin are, respectively, hydrophilic, and hydrophobic pores. The three-fold-axis pores are suited for the passage of ions. However, the ferrihydrite core is insoluble and Fe must be mobilized. The most feasible mechanism so far proposed for the reversible incorporation of Fe in ferritin involves its transport in and out as Fe(II), possibly as the Fe^{2+} aqua ion, which is soluble at neutral pH, but more likely as some type of 'chaperone' complex. Oxidation to Fe(III) is thought to occur at specific di-iron binding sites known as **ferroxidase centres**, present in each of the subunits. Oxidation to Fe(III) involves the coordination of O_2 and inner-sphere electron transfer:

$$2\,Fe(II) + O_2 + 2H^+ \rightarrow 2\,Fe(III) + H_2O_2$$

The mechanism by which Fe is released almost certainly involves its reduction back to the more mobile Fe(II).

26.7 Oxygen transport and storage

Dioxygen, O_2, is a special molecule that has not always been available to biology; in fact, to many life forms it is highly toxic. The early atmosphere on Earth had no O_2: indeed, as the waste product of oxygenic photosynthesis that began with cyanobacteria more than 2 billion years ago, O_2 is a biogenic substance (Box 17.1, Section 17.3), owing its existence to solar energy capture by living organisms. As we shall see in Section 26.10, the great thermodynamic advantage of having such a powerful oxidant available undoubtedly led to the evolution of higher organisms that now dominate Earth. Indeed, the requirement for O_2 became so important as to necessitate special systems for transporting and storing it. Apart from the difficulty in supplying O_2 to buried tissue, there is the problem of achieving a sufficiently high concentration in aqueous environments. This problem is overcome by special metalloproteins known as O_2 **carriers**. In mammals and most other animals and plants, these special proteins (myoglobin and haemoglobin) contain an Fe porphyrin cofactor. Animals such as molluscs and arthropods use a Cu protein called haemocyanin, and some lower invertebrates use an alternative type of Fe protein, haemerythrin, which contains a dinuclear Fe site.

(a) Myoglobin

KEY POINTS The deoxy form of myoglobin which contains high-spin five-coordinate Fe(II) reacts rapidly and reversibly with O_2 to produce low-spin six-coordinate Fe(II); a slow autoxidation reaction releases superoxide and produces Fe(III), which is inactive in binding O_2.

Myoglobin[6] is an Fe protein (17 kDa, Fig. 26.15) which coordinates O_2 reversibly and controls its concentration in tissue. The molecule contains several regions of α-helix, implying mobility, with the single Fe porphyrin group located in a cleft between helices labelled E and F. Two propionate substituents on the porphyrin interact with solvent

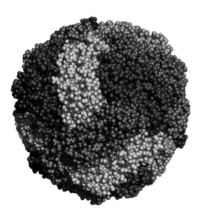

FIGURE 26.14 The structure of ferritin, showing the arrangement of subunits that make up the protein shell.

---

[6] Myoglobin was the first protein for which the three-dimensional structure was determined by X-ray diffraction. For this achievement, John Kendrew shared the 1962 Nobel Prize for Chemistry with Max Perutz, who solved the structure of haemoglobin.

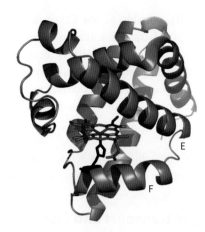

FIGURE 26.15 Structure of myoglobin, showing the Fe porphyrin group located between helices E and F.

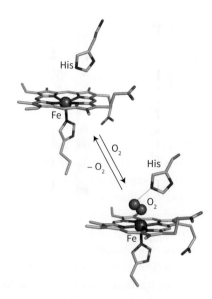

FIGURE 26.16 Reversible binding of O_2 to myoglobin: coordination by O_2 causes the Fe to become low-spin and move into the plane of the porphyrin ring.

H_2O molecules on the protein surface. The fifth ligand to the Fe is provided by a histidine-N from helix F, and the sixth position is the site at which O_2 is coordinated. In common terminology, the side of the haem plane at which exchangeable ligands are bound is known as the **distal region**, while that below the haem plane is known as the **proximal region**. The proximal histidine on helix F is conserved in myoglobins of all species. Such 'highly conserved' amino acids are a strong indication that evolution has determined that they are essential for function. The other conserved histidine is located on helix E.

Deoxymyoglobin (Mb) is bluish red and contains Fe(II)—the oxidation state that binds O_2 reversibly to give the familiar bright red oxymyoglobin (oxyMb). Metmyoglobin (metMb), a brownish-coloured Fe(III) product that is unable to bind O_2, is formed by one-electron oxidation reactions, including a ligand substitution-induced redox reaction in which Cl^- ions displace bound O_2 as superoxide:

$$Fe(II)O_2 + Cl^- \rightarrow Fe(III)Cl + O_2^-$$

In healthy tissue, an enzyme (methaemoglobin reductase) is available to catalyse the reduction of the met form back to the Fe(II) form.

The Fe in deoxymyoglobin is five-coordinate and high-spin. When O_2 binds it is coordinated in end-on fashion to the Fe atom, the electronic structure of which is tuned by the proximal histidine ligand on helix F (Fig. 26.16). The unbound end of the O_2 molecule is fastened by a hydrogen bond to the imidazole-NH of the distal histidine in helix E. Coordination of O_2 (a strong-field π-acceptor ligand) causes the Fe(II) to switch from high-spin (equivalent to $t_{2g}^4 e_g^2$) to low-spin (t_{2g}^6), and with no d electrons in antibonding orbitals, to shrink slightly and move into the plane of the ring.

The bonding is often expressed in terms of Fe(II) coordination by singlet O_2 ($^1\Delta_g$, see Section 17.3), where both electrons occupy one of the $2\pi_g$ orbitals. The doubly occupied antibonding $2\pi_g$ orbital acts as a σ donor and the empty $2\pi_g$

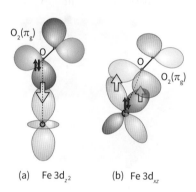

FIGURE 26.17 The orbitals used to form the $Fe-O_2$ adduct of myoglobin and haemoglobin. This model considers the O_2 ligand to be in a singlet state, in which the full $2\pi_g$ orbital donates an electron pair and the empty $2\pi_g$ orbital acts as a π-electron pair acceptor.

orbital accepts an electron pair from the Fe (Fig. 26.17). Alternative descriptions are often considered, including one in which the bonding is expressed in terms of low-spin Fe(III) coordinated by superoxide, O_2^-. With the latter model, the formation of metmyoglobin by reaction with anions is a simple ligand displacement.

(b) Haemoglobin

KEY POINT Haemoglobin consists of a tetramer of myoglobin-like subunits, with four Fe sites that bind O_2 cooperatively.

Haemoglobin (Hb, 68 kDa, Fig. 26.18) is the familiar O_2 transport protein found in special cells known as *erythrocytes* (red blood cells), where it is highly concentrated (0.005 M, implying that the macromolecules are almost in contact with each other). A litre of human blood contains

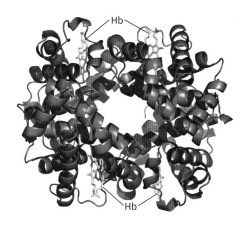

FIGURE 26.18 Haemoglobin is an $\alpha_2\beta_2$ tetramer. Its α and β subunits are very similar to myoglobin. The four haem groups are shown at the top and bottom and are labelled 'Hb'.

about 150 g of Hb. Simplistically, Hb can be thought of as a tetramer of myoglobin-like units with a cavity in the middle. There are in fact two types of Mb-like subunits, which differ slightly in their structures, and Hb is referred to as an $\alpha_2\beta_2$ **tetramer**.

The O_2 binding curves for Mb and Hb are compared in Fig. 26.19: the sigmoidal curve for Hb indicates that uptake and release of successive O_2 molecules is cooperative. At low O_2 partial pressure and greater acidity (as in venous blood and muscle tissue following aggressive exercise) Hb has a low affinity, which enables it to transfer its O_2 to Mb. As the pressure increases, so does the affinity of Hb for O_2 and as a result Hb can pick up O_2 in the lungs. This change in affinity is attributed to there being two conformations. The **tensed**

state (T) has a low affinity, and the **relaxed state** (R) has a high affinity. Deoxy-Hb is T and fully loaded oxy-Hb is R.

A model for the molecular basis of cooperativity arises by considering that binding of the first O_2 molecule to the T-state molecule is weak, but the resulting decrease in the size of the Fe allows it to move into the plane of the porphyrin ring (Fig. 26.16). This motion is particularly important for Hb because it pulls on the proximal histidine ligand and helix F moves. This movement is transmitted to the other O_2 binding sites, the effect being to push the other Fe atoms closer to their respective ring planes and thereby convert the protein into the R state. The way is thereby opened for them to bind O_2, which they now do with greater ease, although the statistical probability decreases as saturation is approached.

(c) Other oxygen transport systems

KEY POINT Arthropods and molluscs use haemocyanin, and certain marine worms use haemerythrin.

In many organisms, such as arthropods and molluscs, O_2 is transported by the Cu protein haemocyanin, which, unlike haemoglobin, is extracellular, as is common for Cu proteins. Haemocyanin is oligomeric, with each monomer containing a pair of Cu atoms in close proximity. Deoxyhaemocyanin (Cu(I)) is colourless, but becomes bright blue when O_2 binds.

The active site is shown in Fig. 26.20. In the deoxy state, each Cu atom is three-coordinate and bound in a pyramidal array by three histidine residues. The two Cu atoms are so far apart (460 pm) that there is no direct interaction between them. The low coordination number is typical of Cu(I), which is normally two- to four-coordinate. Rapid and reversible coordination of O_2 occurs between the two Cu atoms in a bridging dihapto manner (μ-$\eta^2\eta^2$), and the low

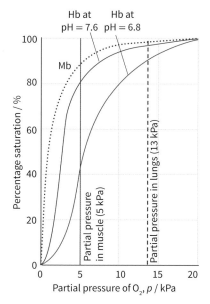

FIGURE 26.19 Oxygen binding curves for myoglobin and haemoglobin showing how cooperativity between the four sites in haemoglobin gives rise to a sigmoidal curve. The binding of the first O_2 molecule to haemoglobin is unfavourable, but it results in a greatly enhanced affinity for subsequent O_2 molecules.

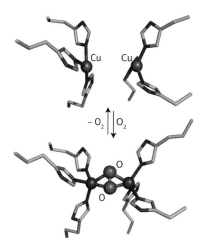

FIGURE 26.20 Binding of O_2 at the active site of haemocyanin causes the two Cu atoms to be brought closer together. The O_2 complex is regarded as a binuclear Cu(II) centre in which the two Cu atoms are bridged by an η^2,η^2-peroxide.

vibrational wavenumber of the coordinated O_2 molecule (750 cm^{-1}) shows it has been reduced to peroxide O_2^{2-}, with an accompanying lowering of the bond order from 2 to 1. To accommodate the binding of O_2, the protein adjusts its conformation to bring the two Cu atoms closer together. The Cu sites become five-coordinate, which is typical for Cu(II).

Haemerythrin—an O_2 carrier protein found in some marine invertebrates—contains a dinuclear Fe centre that is encountered in a number of proteins with diverse functions, such as methane monooxygenase (Section 26.10(c)), and some ribonucleotide reductases and acid phosphatases (Section 26.9(c)). The two Fe atoms in the active site of haemerythrin (**17**) are individually coordinated by amino acid side chains, but are also linked by two bridging carboxylate groups and a small ligand. In the reduced (Fe(II)) form, which binds O_2 reversibly, this small ligand is an OH$^-$ ion. Coordination of O_2 occurs at only one of the Fe atoms, and the distal O atom forms a hydrogen bond to the H atom of the bridging hydroxide.

designed to protect the coordinated O–O ligand, preventing it from reacting further. Sterically hindered Fe(II) complexes such as the 'basket' porphyrin (**18**) achieve this protection by preventing a second Fe(II) complex from attacking the distal O atom of the superoxo species to form a bridged peroxo intermediate. As we shall see in Section 26.10, peroxo complexes of Fe(III) tend to undergo rapid O–O bond proteolysis, resulting in formation of H_2O and Fe(IV)=O along with a cation radical. Controlling the access of electrons and H$^+$ to the active site plays a crucial role in determining the fate of the O_2 molecule.

18 'Basket' Fe-porphyrin with O_2 bound

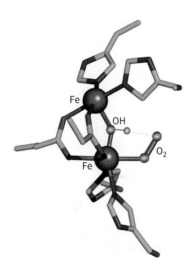

17 The active site of haemerythrin

Simple Cu complexes that can coordinate O_2 reversibly are also rare, but studies have revealed interesting chemistry that is particularly relevant for developing catalysts for oxygenation reactions. Analogues of the dinuclear Cu(I) centre of haemocyanin react with O_2 but have a strong tendency to undergo further reactions that involve cleavage of the O–O bond, an example being the rapid equilibrium between μ-η^2,η^2 peroxo-dicopper(II) and bis(μ-oxo)copper(III) complexes shown in Fig. 26.21.

(d) Reversible O_2 binding by small-molecule analogues

KEY POINTS Proteins binding O_2 reversibly do so by preventing its reduction and eventual O–O bond cleavage. This protection is difficult to achieve with small molecules. Certain elaborate macrocyclic Fe(II) complexes exhibit reversible O_2 binding by providing steric hindrance to attack on the coordinated O_2.

Much effort has been spent on synthesizing simple complexes that coordinate O_2 reversibly and could be used as blood substitutes in special circumstances, such as emergency surgery. The problem is that although O_2 reacts with d-block metal ions to form complexes in which the O–O bond is retained (as in superoxo and peroxo species), these products tend to undergo irreversible decomposition involving rapid O–O bond cleavage and formation of water or oxides. Overcoming this problem requires complexes

FIGURE 26.21 Rapid equilibrium between μ-η^2,η^2 peroxodicopper(II) and bis(μ-oxo)copper(III) in a model complex for the active site of haemocyanin.

EXAMPLE 26.5 Identifying how biology compensates for strong competition by CO

Carbon monoxide is well known to be a strong inhibitor of O_2 binding by myoglobin and haemoglobin, yet relative to O_2 its binding is much weaker in the protein compared to a simple Fe-porphyrin complex. This suppression of CO binding is important, as even trace levels of CO would otherwise have serious consequences for aerobes. Suggest an explanation.

Answer We need to consider how CO and O_2, both of which are π-acceptor ligands, differ in terms of the orbitals they use for bonding to a metal atom. The binding of O_2 is nonlinear (see Figs. 26.16 and 26.17), and the distal O atom is well positioned to form a hydrogen bond to the distal imidazole. By contrast, CO adopts a linear Fe–C–O bonding arrangement (Section 24.5) and does not participate in the additional H-bonding.

Self-test 26.5 Suggest a reaction sequence accounting for why simple Fe-porphyrin complexes are unable to bind O_2 reversibly, but instead give products that include oxobridged dinuclear Fe(III) porphyrin species.

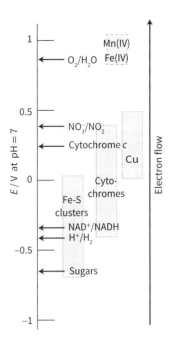

FIGURE 26.22 The 'redox spectrum' of life.

26.8 Electron transfer

For all but some unusual organisms,[7] the energy for life stems ultimately from the Sun, either directly in photosynthesis or indirectly by acquiring energy-rich compounds (fuel) from photosynthesizing organisms. Energy acquired as a flow of electrons from fuel to oxidant is tapped off to drive all kinds of reactions, including ion pumping and the synthesis of ATP, itself a 'fuel', as its close coupling to reactions that are otherwise thermodynamically uphill delivers the energy to drive that reaction. Important primary fuels include fats, sugars, and H_2, and important biological oxidants include O_2, nitrate, and even H^+ which can be reduced to H_2 (Section 26.13). As seen from Fig. 26.22, oxidation of sugars by O_2 provides a great deal of energy (over 4 eV per O_2 molecule), and is the reason for the great success of aerobic organisms over the anaerobic ones that once dominated the Earth. Evolution has perfected electron-transfer (ET) centres to enable these processes to occur with maximum speed and efficiency, almost as if the different catalysts were 'wired' together.

(a) General considerations

KEY POINTS Electron flow along electron-transport chains is coupled to chemical processes such as the uphill transfer of ions (particularly H^+) across membranes. The simplest electron-transfer centres have evolved to optimize fast electron transfer.

In organisms, electrons are abstracted from food (fuel) and flow to an oxidant, down the potential gradient formed by the sequence of ET acceptors and donors known as a **respiratory chain** (Fig. 26.23).[8] Apart from flavins, quinones, and certain amino acids (such as tyrosine or tryptophan in special cases) which are redox-active organic cofactors, these acceptors and donors are protein-bound metal complexes, which fall into three main classes; namely Fe-S clusters, cytochromes, and Cu centres. The enzymes of the respiratory chain are generally bound in a membrane, across which the energy from ET is used to sustain a transmembrane proton gradient: this is the basis of the **chemiosmotic theory**. The counterflow of H^+ through a rotating enzyme known as ATP synthase, drives the phosphorylation of ADP to ATP. Many membrane-bound redox enzymes are **electrogenic proton pumps**, which means they directly couple long-range ET to transmembrane proton transfer through specific internal channels.

We shall examine the properties of the three main types of ET centre. The same rules concerning outer-sphere electron transfer that were discussed in Section 23.12 apply to metal centres in proteins, and we should note that organisms have optimized the structures and properties of these centres to achieve efficient long-range electron transfer.

In the following discussion it will be useful to keep in mind that reduction potentials depend on several factors (Chapter 6). Besides ionization energy and ligand environment (strong anionic donors stabilize high oxidation states and lower the reduction potential; weak donors, π acceptors, and protons stabilize low oxidation states and raise

---

[7] Lithotrophs are primitive organisms possessing the ability to derive their energy from minerals.

[8] An overall current of about 80 A flows through the mitochondrial respiratory chains in an average human!

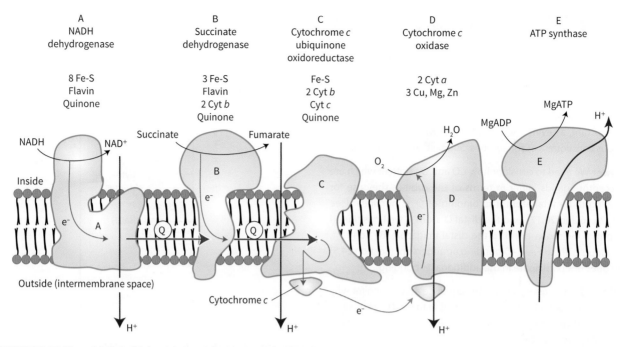

FIGURE 26.23 The mitochondrial respiratory electron-transfer (ET) chain consists of several metalloenzyme molecules that use the energy of electron transport (green arrows) to transport protons across a membrane. The proton gradient is used to drive ATP synthesis. Stoichiometries are not shown.

the reduction potential), an active site in a protein is also influenced by the relative permittivity (which stabilizes centres with low overall charge), the presence of neighbouring charges, including those provided by other bound metal ions, and the availability of hydrogen-bonding interactions that will also stabilize reduced states.

When we consider the kinetics of electron transfer, it will similarly be useful to keep in mind that 'efficiency' means that electron transfer is fast even when the reaction Gibbs energy is low, and therefore the reorganization energy λ of Marcus theory (Section 23.12) should be low. This requirement is met by providing a ligand environment that does not alter significantly when an electron is added and by burying the site so that water molecules are excluded. Intersite distances are generally less than 1.4 nm in order to facilitate electron tunnelling, although it is still debated whether electron transfer in proteins depends mainly on distance alone or whether the protein can provide special pathways.

(b) Cytochromes

KEY POINTS Cytochromes operate in the potential region −0.3 to +0.4 V vs SHE under biological pH conditions; they have a combination of low reorganization energy and extended electron coupling through delocalized orbitals.

Cytochromes were identified many years ago as cell pigments (hence the name). They contain an Fe porphyrin group, and the term 'cytochrome' can refer to both an individual protein and a subunit of a larger enzyme that contains the cofactor. Electron-transferring cytochromes use the Fe^{3+}/Fe^{2+} couple and are generally six-coordinate, with two stable axial bonds to amino acid donors, and the Fe is usually low spin in both oxidation states. This contrasts with ligand-binding Fe-porphyrin proteins such as haemoglobin, for which the sixth coordination site is either empty or is occupied by an H_2O molecule.

A good way to consider the capability of cytochromes for fast electron transfer is to treat the d orbitals of Fe(III) and Fe(II) in terms of an octahedral ligand field and to consider the overlap between the electron-rich but nearly nonbonding t_{2g} orbitals (the configurations are t_{2g}^5 and t_{2g}^6 in Fe(III) and Fe(II), respectively) and the orbitals of the porphyrin. The electron enters or leaves an orbital having π overlap with a π* antibonding molecular orbital on the ring system. This arrangement provides enhanced electron transfer because the d orbitals of the Fe atom are effectively extended out to the edge of the porphyrin ring, so decreasing the distance over which an electron must transfer between redox partners (Fig. 26.24).

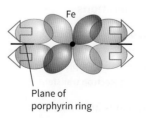

FIGURE 26.24 Overlap between the t_{2g} orbitals of the Fe and low-lying empty π* orbitals on the porphyrin effectively extends the Fe orbitals out to the periphery of the ring.

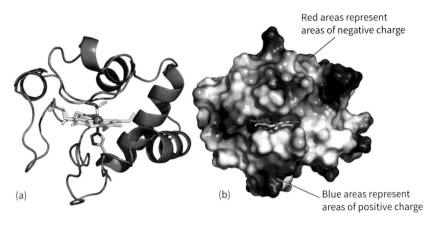

FIGURE 26.25 Different views (but from the same viewpoint) of mitochondrial cytochrome c. (a) The secondary structure and the position of the haem cofactor. (b) The surface charge distribution that guides the docking with its natural redox partners (red and blue areas represent patches of negative and positive charge, respectively).

The paradigm of cytochromes is **mitochondrial cytochrome c** (12 kDa, Fig. 26.25). This water-soluble protein is found in the mitochondrial intermembrane space, where it supplies electrons to cytochrome c oxidase, the enzyme responsible for reducing O_2 to H_2O at the end of the energy-transducing respiratory chain (Section 26.10).

The fifth and sixth ligands to Fe in cytochrome c are histidine (imidazole-N) and methionine (thioether-S) (**19**). Methionine is not a common ligand in metalloproteins, but because it is a neutral, soft donor it is expected to stabilize Fe(II) rather than Fe(III). The reduction potential of cytochrome c is +0.26 V vs SHE, at the higher end of values for cytochromes in general. Cytochromes vary in the identity of the axial ligands as well as the structure of the porphyrin ligand (the notation a, b, c, d, ... historically defines positions of absorption maxima in the visible region, but also refers to variations in the substituents on the porphyrin ring). Many cytochromes, in particular those sandwiched between membrane-spanning helices, have bis(histidine) axial ligation (**20**).

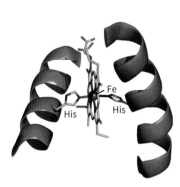

20 Fe-porphyrin having bis(histidine) axial ligation

In cytochrome c, the edge of the porphyrin ring is exposed to solvent and is the most likely site for electrons to enter or leave. Specific protein–protein interactions are important for obtaining efficient electron transfer, and the region around the exposed edge of the porphyrin ring in cytochrome c provides a pattern of charges that are recognized by cytochrome c oxidase and other redox partners. One example in particular has been well studied, that of cytochrome c with yeast cytochrome c peroxidase (Section 26.10(a)). The driving force for electron transfer from either of the two catalytic intermediates of peroxidase to each reduced cytochrome c is approximately 0.5 V. Fig. 26.26 shows the structure of a bimolecular complex formed between cytochrome c and cytochrome c peroxidase. Electrostatic interactions guide the two proteins together within a distance that is favourable for fast electron tunnelling between cytochrome c and two redox centres on the peroxidase, the haem cofactor and tryptophan-191.

(c) Iron–sulfur clusters

KEY POINTS Iron–sulfur clusters generally operate at more negative potentials than cytochromes: examples of mixed valent species, they contain high-spin Fe(III) and Fe(II) in various ratios connected by sulfido ligands, in a mainly tetrahedral environment.

Widespread throughout biology, Fe-S centres are represented by square brackets representing the number of Fe and nonprotein S atoms present in the 'inorganic core', as in [2Fe-2S] (**21**), [4Fe-4S] (**22**), and [3Fe-4S] (**23**).

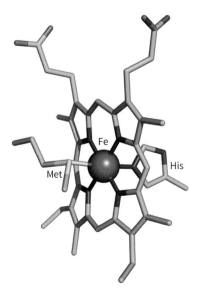

19 Active site of cytochrome c

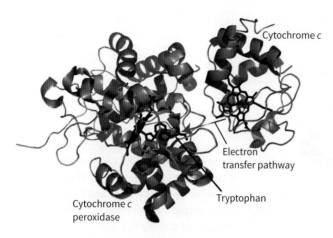

FIGURE 26.26 The bimolecular ET complex between cytochrome *c* and cytochrome *c* peroxidase produced by co-crystallization of cytochrome *c* with the Zn derivative of cytochrome *c* peroxidase (Zn substitution renders the peroxidase inactive, simplifying the procedure). The orientation suggests an electron-transfer pathway between the haem groups of cytochrome *c* and cytochrome *c* peroxidase that includes a tryptophan (Table 26.2).

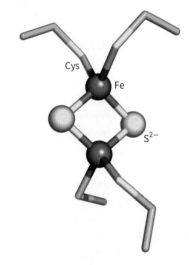

21 [2Fe-2S]

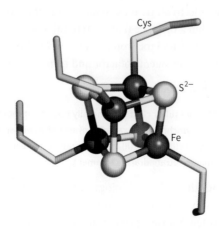

22 [4Fe-4S]

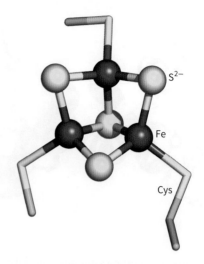

23 [3Fe-4S]

The efficacy of Fe-S clusters as fast ET centres is largely due to the covalent nature of the bonding and being able to delocalize the added electron to varying degrees, which minimizes bond length changes and decreases the reorganization energy. The presence of sulfur ligands to provide good electron lead-in groups is also important. Small, mobile electron-carrier proteins containing Fe-S clusters are known as **ferredoxins**. In many large enzymes, Fe-S clusters are arranged in a relay, less than 1.5 nm apart, to link remote redox sites in the same molecule. The relay concept is illustrated in Fig. 26.27 which shows the membrane-extrinsic (protruding into the aqueous-phase) domain of NADH dehydrogenase (Complex I of mitochondria; see also Fig. 26.23).

In nearly all cases, the Fe atoms are tetrahedrally coordinated by sulfido bridging ligands using cysteine thiolate (RS$^-$) groups as the protein ligands. The overall assembly including the protein ligands is known as an 'Fe-S centre'. Examples are known in which one or more of the Fe atoms are coordinated by nonthiolate amino acid ligands, such as carboxylate, imidazole, and alkoxyl (serine), or by an exogenous ligand such as H$_2$O or OH$^-$, and the coordination number about the Fe subsite may be increased to six. The cubane [4Fe-4S] (**22**) and cuboidal [3Fe-4S] (**23**) clusters are obviously closely related, and may even interconvert within a protein by the addition or removal of Fe from one subsite. Larger clusters also occur, such as the 'super clusters' [8Fe-7S] and [Mo7Fe-9S-C] found in nitrogenase (Section 26.14).

Iron–sulfur clusters are excellent examples of mixed-valence systems. The redox state of a cluster is represented by summing the charges due to Fe (3+ or 2+, respectively) and S atoms (2−) and the resultant overall charge, which is referred to as the **oxidation level**, is written as a superscript. Despite the presence of more than one Fe atom, Fe-S clusters are generally restricted to single electron transfers.

because they were originally discovered in a protein called *high-potential iron protein* for which the reduction potential is 0.35 V) and some members of a group called **Rieske centres**, which are [2Fe-2S] clusters having one Fe subsite coordinated by two neutral imidazole ligands rather than cysteine (24).

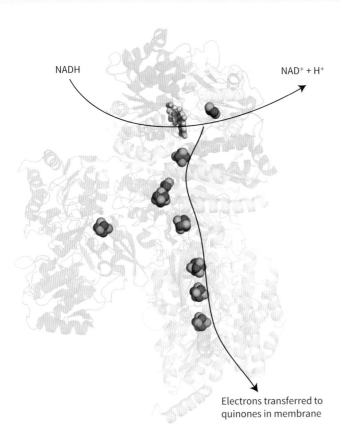

Electrons transferred to quinones in membrane

FIGURE 26.27 A series of Fe-S clusters leading from the active site organic (flavin) cofactor provides a long-range electron-transfer relay in the membrane-extrinsic domain of mitochondrial Complex I.

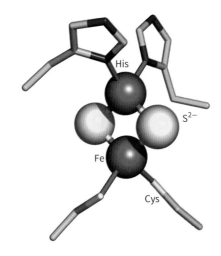

24 [2Fe-2S] cluster with two imidazole (histidine) ligands

The half-reactions have been written to include the spin states of Fe-S clusters: individual Fe atoms are high spin, as expected for tetrahedral coordination by S^{2-}, and different magnetic states arise from ferromagnetic and antiferromagnetic coupling (Section 22.8). These magnetic properties are very important, as they allow the centres to be investigated by EPR spectroscopy (Section 8.7).

[2Fe-2S]$^{2+}$ + e$^-$ → [2Fe-2S]$^+$ $E^\ominus$ = 0 to −0.4 V
2 Fe(III)   {Fe(III):Fe(II)}
$S = 0$ $S = \frac{1}{2}$

[3Fe-4S]$^+$ + e$^-$ → [3Fe-4S]^0 $E^\ominus$ = +0.1 to −0.4 V
3 Fe(III)   {2 Fe(III):Fe(II)}
$S = \frac{1}{2}$ $S = 2$

[4Fe-4S]$^{2+}$ + e$^-$ → [4Fe-4S]$^+$ $E^\ominus$ = −0.2 to −0.7 V
{2 Fe(III):2Fe(II)}   {Fe(III):3Fe(II)}
$S = 0$ $S = \frac{1}{2}$

Most Fe-S centres have negative reduction potentials (usually more negative than 0 V vs SHE at biological pH values) so the reduced forms are good reducing agents: exceptions are [4Fe-4S] clusters that operate instead between the +3 and +2 oxidation states (these are called 'HiPIP' centres

A major question is how Fe-S centres are synthesized and inserted into the protein. This process has been studied mostly in prokaryotes, from which it is known that specific proteins are involved in the supply and transport of Fe and S atoms, their assembly into clusters, and their transfer to target proteins. Free sulfide (H_2S, HS^-, or S^{2-}) in a cell is highly poisonous, so it is produced only when required, usually by an enzyme called cysteine desulfurase, which breaks down cysteine to yield S^{2-} ions and alanine.

(d) Copper electron-transfer centres

KEY POINT The protein overcomes the large inherent difference in preferred geometries for Cu(II) and Cu(I) by constraining Cu in a co-ordination environment that does not change significantly upon electron transfer.

The so-called 'blue Cu' centre is the active site of a number of small electron-transfer proteins as well as larger enzymes (the blue Cu oxidases) that also contain other Cu sites. Blue Cu centres have reduction potentials for the Cu(II)/Cu(I) redox couple that lie in the range 0.15–0.7 V, and so they are generally more oxidizing than cytochromes. The name stems from the intense blue colour of pure samples in the oxidized state, which arises from ligand(thiolate)-to-metal charge transfer. In all cases, the Cu is shielded from solvent water and coordinated by a minimum of two imidazole-N and one cysteine-S in a nearly trigonal-planar manner, with one or two longer

EXAMPLE 26.6 Facilitating sequential two-electron transfers

An interesting type of Fe-S cluster that has recently been discovered has an unusual [4Fe-3S] core that is coordinated by six cysteines rather than four. The sulfur atom of one of the extra cysteines bridges two Fe atoms, replacing one of the μ_3 sulfido atoms that would normally be present. The resulting cluster is more pliable, and in addition to the normal redox couple, a second, fully reversible electron transfer occurs at just a slightly higher potential. The 'superoxidized' form of the cluster (**25**) contains a localized Fe(III) that is coordinated by an adjacent peptide N that has become deprotonated during the oxidation process. Offer an explanation as to why this second redox couple is allowed in this instance when normally it is not.

Answer The question may be addressed most simply by considering the severe Coulombic restriction that is placed on a redox-active centre inside a protein molecule. Whereas altering the charge by one unit is quite easily accommodated, altering the charge by two units within such a low dielectric medium is prohibitive. The [4Fe-3S] cluster circumvents this problem because removal of the second electron is electrostatically compensated by removal of a proton from the same locality. The deprotonated peptide-N atom is an excellent donor ligand that stabilizes Fe(III).

$$HN \xrightarrow{e^-} HN \xrightarrow{e^-, H^+} N^-$$

$$[4Fe\text{-}3S]^{3+} \quad\quad [4Fe\text{-}3S]^{4+} \quad\quad [4Fe\text{-}3S]^{5+}$$

Self-test 26.6 Which of the following factors should raise the reduction potential of an Fe-S cluster in a protein: hydrogen bonding between adjacent side chains and S-atoms of the cluster; the presence of nearby negatively-charged side chains; replacing one cysteine by a histidine?

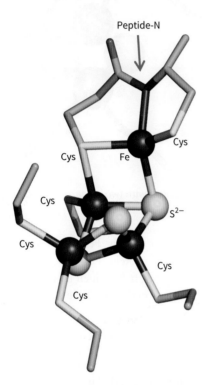

25 Superoxidized [4Fe-3S] cluster

bonds to axial ligands. The most studied examples are plastocyanin (Fig. 26.28)—a small electron-carrier protein in chloroplasts—and azurin, a bacterial electron carrier. These small proteins have a 'β-barrel' structure, in which a barrel of β-sheet (Fig. 26.4) holds the Cu coordination sphere in a rigid geometry.

Indeed, the crystal structures of oxidized, reduced (**26**), and apo forms reveal that the ligands remain in essentially the same position in all cases. As a result, the blue Cu centre is well suited to undergo fast and efficient electron transfer because the reorganization energy is small.

FIGURE 26.28 The plastocyanin molecule.

26 Cu coordination in oxidized and reduced forms of plastocyanin

The dinuclear Cu centre known as Cu_A is present in cytochrome c oxidase (Section 26.10(b)) and N_2O reductase (Section 26.14). The two Cu atoms (**27**) are each coordinated by two imidazole groups, and a pair of cysteine thiolate ligands act as bridging ligands. In the reduced form, both Cu atoms are Cu(I). This form undergoes one-electron oxidation to give a purple, paramagnetic species in which the unpaired electron is shared between the two Cu atoms. Once again, we see how delocalization assists electron transfer because the reorganization energy is lowered.

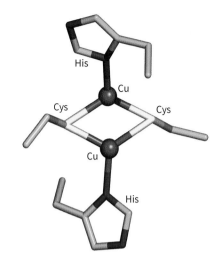

27 Cu_A, a dinuclear Cu centre used for electron transfer

EXAMPLE 26.7 **How ET centres are equipped for their function**

The reduction potential of Rieske Fe-S centres is very pH dependent, unlike the standard Fe-S centres that have only thiolate ligation. Suggest an explanation.

Answer We need to refer back to Sections 6.6 and 6.14 to see how protonation equilibria influence reduction potentials. Each of the two imidazole ligands that coordinate one of the Fe atoms in the Rieske [2Fe-2S] cluster is electrically neutral at pH = 7, and the proton located on the noncoordinating N atom is easily removed. The pK_a depends on the oxidation level of the cluster, and there is a large region of pH in which the imidazole ligands are protonated in the reduced form but not in the oxidized form. As a result, the reduction potential depends on pH.

Self-test 26.7 Simple Cu(II) compounds show a large EPR hyperfine coupling to the Cu nucleus ($I = \frac{3}{2}$ for ^{65}Cu and ^{63}Cu), whereas the EPR spectra of blue Cu proteins show a much smaller hyperfine coupling. What does this suggest about the nature of the ligand coordination at blue Cu centres?

Catalytic processes

The classic role of enzymes is as highly selective catalysts for the myriad chemical reactions that take place in organisms and sustain the activities of life. In this section we view some of the most important examples in terms of the suitability of certain elements for their roles.

26.9 Acid–base catalysis

Biological systems rarely have the extreme pH conditions under which catalysis by 'free' solvated H^+ or OH^- can occur; indeed, the result would be indiscriminate because all hydrolysable bonds would be targets. One way that organisms have solved this problem has been to harness properties of certain metal ions and build them into protein structures designed to accomplish specific (Brønsted) acid–base reactions (Section 5.1).

Organisms make extensive use of Zn for achieving acid–base catalysis, but not to the exclusion of other metals. For example, in addition to the numerous enzymes that feature Fe(II) and Fe(III), Mg(II) serves as the catalyst in pyruvate kinase (phosphate ester hydrolysis) and ribulose bisphosphate carboxylase (CO_2 incorporation into organic molecules), Mn is the catalyst in arginase (for the hydrolysis of arginine, yielding urea and L-ornithine), and Ni(II) is the active metal in urease (for the hydrolysis of urea, yielding ammonia and, ultimately, carbon dioxide), an enzyme that is crucial for the virulence of *Helicobacter pylori*, a notorious human pathogen (Section 32.3). Because of their importance to industry and medicine, many of these enzymes have been studied in great detail, and model systems have been synthesized in efforts to reproduce catalytic properties and understand the mode of action of inhibitors. Many of these sites (including arginase-[Mn,Mn] and urease-[Ni,Ni]) contain two or more metal ions in an arrangement unique to the protein and difficult to model with simple ligands.

(a) Zinc enzymes

KEY POINT Zinc is well suited for catalysing acid–base reactions, as it is abundant, redox inactive, forms strong bonds to donor groups of amino acid residues, and exogenous ligands such as H_2O are exchanged rapidly.

The biological roles of Zn are either catalytic, or structural and regulatory. Unlike Ca and Mg, Zn forms more stable complexes with softer donors, so it is not surprising that it is usually found coordinated in proteins through histidine and cysteine residues. Typical catalytic sites (**28**) commonly have three permanent protein ligands and an exchangeable ligand (H_2O), whereas structure-stabilizing Zn sites (**29**) are coordinated by four 'permanent' protein ligands.

28 A catalytic Zn site 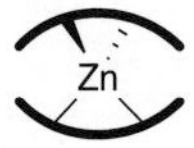   **29** A structural Zn site

A Zn^{2+} ion has high rates of ligand exchange and its polarizing power means that the pK_a of a coordinated H_2O molecule is quite low. Combined with strong binding to protein ligands, rapid ligand exchange (coordinated H_2O or substrate molecules), a reasonably high electron affinity, flexibility of coordination geometry, and no complicating redox chemistry, Zn is well suited to its role in catalysing specific acid–base reactions. The large family of Zn enzymes include carbonic anhydrase, carboxypeptidases, alkaline phosphatase, β-lactamase (responsible for penicillin resistance in bacteria, Fig. 32.9), and alcohol dehydrogenase.

The lack of good spectroscopic probes for Zn, however, has meant that even though it is tightly bound in a protein, it has been difficult to confirm its binding or deduce its coordination geometry in the absence of direct structural information from X-ray diffraction or NMR. However, some elegant measurements have exploited the ability of Co^{2+}, which is coloured and paramagnetic, or Cd^{2+}, which has useful NMR properties, to substitute for and report on the Zn site. These substitutions depend on strong similarities between the metal ions: like Zn^{2+}, Co^{2+} readily forms tetrahedral complexes, whereas Cd lies directly below Zn in the periodic table.

The mechanisms of Zn enzymes are normally discussed in terms of two limiting cases, originally noted for ester and peptide hydrolysis, that are relevant for other metal ion catalysts. In the **Zn-hydroxide mechanism**, the Zn functions by promoting deprotonation of a bound water molecule, so creating an optimally positioned OH^- nucleophile that can go on to attack the carbonyl C atom:

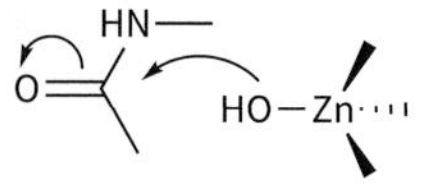

In the **Zn-carbonyl mechanism**, the Zn ion acts directly as a Lewis acid to accept an electron pair from the carbonyl O atom, and its role is therefore analogous to H^+ in acid catalysis:

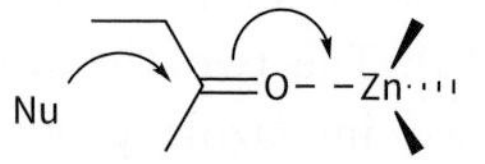

Similar reactions occur with other X=O groups, particularly the P=O of phosphate esters. There is an obvious advantage of achieving such 'acid' catalysis at a metal site that is anchored within a stereoselective outer-sphere environment.

The formation and transport of CO_2 is a fundamental process in biology. The solubility of CO_2 in water depends upon its hydration and deprotonation to form HCO_3^-. However, the uncatalysed reaction at pH = 7 is very slow, the forward process occurring with a rate constant of less than $10^{-3}\,s^{-1}$. Because turnover of CO_2 by biological systems is very high, such a rate is far too slow to sustain vigorous aerobic life in a complex organism. In photosynthesis, only CO_2 can be used by the enzyme known as 'Rubisco' (Section 26.9(b)) so rapid dehydration of HCO_3^- is essential and is one of the first steps in the production of biomass. The CO_2/HCO_3^- equilibrium is also important (in addition to its role in CO_2 transport) because it provides a way of regulating tissue pH.

Carbonic anhydrase (CA, or carbon dioxide dehydratase) catalyses this reaction with a rate enhancement well above 10^6. There are several forms of CA, all of which are monomers with a molar mass close to 30 kDa containing one Zn atom. The best studied enzyme is CA II from red blood cells, which has a turnover frequency for CO_2 hydration of about $10^6\,s^{-1}$, making it one of the most active of all enzymes. The crystal structure of human CA II shows that the Zn atom is located in a conical cavity about 1.6 nm deep, which is lined with several histidine residues. The Zn is coordinated by three His-N ligands and one H_2O molecule in a tetrahedral arrangement (Fig. 26.29), although some CAs from higher plants have two cysteines and one histidine. The pK_a of the bound H_2O is lowered by about 3 units compared to the aqua ion, thus creating an extremely high local concentration of OH^- as attacking nucleophile, even at pH 7. Other groups in the active site pocket, including noncoordinating histidines and ordered water molecules, are important for mediating proton transfer (the rate-limiting factor) and for binding the CO_2 substrate (which does not coordinate to Zn).

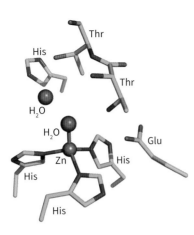

FIGURE 26.29 The active site of carbonic anhydrase.

The mechanism of action of CA (Fig. 26.30) is best described in terms of a Zn-hydroxide mechanism. Rapid proton transfers are aided by a hydrogen bonding network that extends from the protein surface to the active site. The key feature is the acidity of the H_2O molecule coordinated to Zn, as the coordinated HO^- ion that is produced after deprotonation is sufficiently nucleophilic to attack a nearby CO_2 molecule bound noncovalently in an adjacent position. This attack results in a coordinated HCO_3^- ion, which is then released.

Small analogues of CA that have been studied, such as (30), reproduce the substrate binding and acid–base properties of the enzyme, but their catalytic activity is orders of magnitude lower.

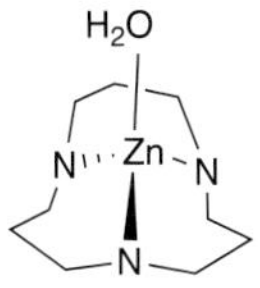

Carboxypeptidase (CPD, approx. 35 kDa) is an exopeptidase—an enzyme that catalyses the hydrolytic cleavage of C-terminal amino acids containing an aromatic or bulky aliphatic side chain. There are two types of Zn-containing enzyme, and both are synthesized as inactive precursors in the pancreas for secretion into the digestive tract. The better studied is CPD A, which acts upon terminal aromatic residues; whereas CPD B acts upon terminal basic residues. The X-ray structure shows that the Zn is located to one side of a groove in which the substrate is bound and is coordinated by two histidine-N ligands, one glutamate-CO_2^- (bidentate), and one weakly bound H_2O molecule (Fig. 26.31). The structure obtained in the presence of a small inhibitor (glycyl tyrosine) shows that the H_2O molecule has moved away from the Zn atom, which has become coordinated instead to the carbonyl-O of the glycine, suggesting a Zn-carbonyl mechanism. The guanidinium group of a nearby arginine binds the terminal carboxylate, while the tyrosine provides aromatic/hydrophobic π-stacking recognition.

Alkaline phosphatase (AP) introduces us to catalytic Zn centres that contain more than one metal atom: it occurs in tissues as diverse as the intestine and bone, where it is found

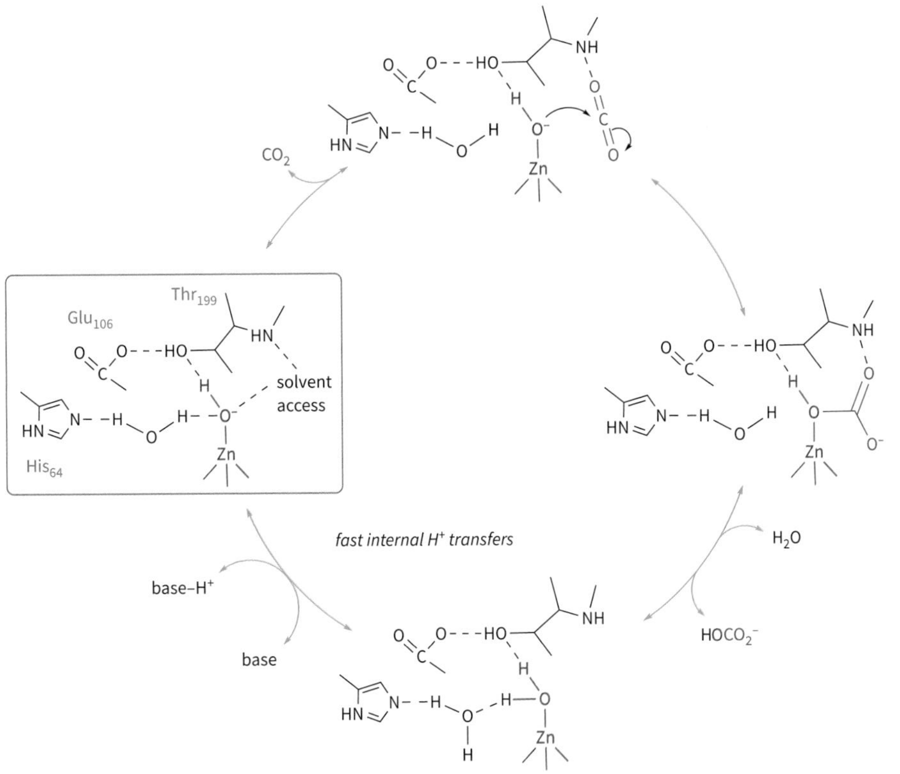

FIGURE 26.30 The mechanism of action of carbonic anhydrase indicating the importance of internal proton transfer steps in this very fast reaction.

FIGURE 26.31 Structure of the active site of carboxypeptidase with a peptide inhibitor (bold red) bound to it.

in the membranes of osteoblasts, cells that form the sites of nucleation of hydroxyapatite crystals. Alkaline phosphatase catalyses the general breakdown of organic phosphates, including ATP, to provide the phosphate required for bone growth.

$$R - OPO_3^{2-} + H_2O \rightarrow ROH + HOPO_3^{2-}$$

As its name implies, its optimum pH is in the mild alkaline region. The active site of AP contains two Zn atoms located only about 0.4 nm apart with a Mg ion nearby. The crystal structure of the enzyme–phosphate complex (**31**) reveals that the phosphate ion (the product of the normal reaction) bridges the two Zn atoms.

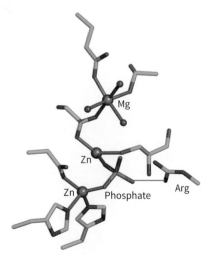

31 Active site of alkaline phosphatase

Alcohol dehydrogenase (ADH) is discussed here even though it is classed as a redox enzyme, because the role of Zn once again is as a Lewis acid. The reaction catalysed is the reduction of NAD$^+$ by alcohol:

The Zn activates the C–OH group towards transfer of the H as a hydride entity to a molecule of NAD$^+$. It is easy to visualize this reaction in the direction of aldehyde oxidation, in which the Zn atom polarizes the carbonyl group and induces attack by the nucleophilic hydridic H atom from NADH. Alcohol dehydrogenase is an α_2 dimer that contains both catalytic and structural Zn sites.

Cadmium—the element below Zn in Group 12 and normally regarded as highly poisonous—is now recognized as being an essential nutrient for certain organisms. In 2005, a carbonic anhydrase isolated from the photosynthetic marine microalga *Thalassiosira weissflogii* was discovered to contain Cd at its active site. In contrast to cases in which Cd is simply able to substitute for Zn, this enzyme is specific for Cd^{2+}. The surface waters in which the Cd-using *T. weissflogii* are found are extremely low in Zn^{2+}, and its growth in the laboratory is stimulated by adding Cd^{2+}. As we will see, it is CO$_2$, not HCO$_3^-$, that is the carbon source for photosynthesis, and organisms must adapt to survive even if this means using a metal ion that is otherwise lethal.

(b) Magnesium enzymes

KEY POINTS Compared to Zn^{2+}, Mg^{2+} forms much weaker complexes with biological ligands, but the differential is largely offset by the relatively high concentration of Mg^{2+} in cells. Aside from its incorporation into chlorophyll, Mg^{2+} has important roles in acid–base catalysis—in enzymes called endonucleases that 'cut' DNA at specific points and as the catalytic centre of ribulose bisphosphate carboxylase (Rubisco), the photosynthetic enzyme responsible for incorporating CO$_2$ into organic molecules.

The Mg^{2+} cation confers less polarization of coordinated ligands than Zn^{2+} (we often refer to Mg^{2+} being a 'weaker acid' than Zn^{2+}); however, compared to Zn^{2+} it is much more mobile, and cells contain high concentrations of uncomplexed Mg^{2+} ions. Its wide yet indirect role in enzyme catalysis is as the Mg–ATP complex (**1**), which is the substrate for **kinases**, the enzymes that selectively transfer phosphate groups to a target (a process known as phosphorylation), thereby activating it for reaction or (for a protein) causing it to change its conformation. Phosphate esters are crucial in biology because tightly coupling their hydrolysis to the catalysis of a specific reaction renders that reaction much less reversible: the negative shift in ΔG is equivalent to providing energy (Section 5.5). Kinases are controlled by calmodulin (Section 26.5) and other proteins, so they are part of the internal signalling capability of higher organisms.

Endonucleases are phosphatases that hydrolyse phosphate ester linkages in DNA, causing cleavage at a specific point. These enzymes, which bind at least two Mg^{2+} ions per catalytic monomer, have great importance in biotechnology, particularly in the rapidly advancing field of gene therapy that incorporates them in DNA-sequence recognition systems such as 'CRISPR' (cluster-regularly-interspaced-short-palindromic-repeats) and zinc fingers (Section 26.15). The Mg^{2+} ions are not bound tightly, but one is essential for activity. Fig. 26.32 shows the structure of a dimeric endonuclease bound to DNA, along with a close-up view of one of the active sites. To obtain an inactive enzyme for crystal structure determination, Mg^{2+} was replaced by Mn^{2+}, but the coordination to protein carboxylate side chains and formation of the bond to a DNA phosphate group that polarizes it for hydrolysis, are expected to be similar. The metal ion binds to the O=P of a juxtaposed phosphate, polarizing the bond and activating the phosphate group to attack by H_2O.

An Mg-containing enzyme that lies at the very root of the carbon cycle and biosynthesis of organic compounds is ribulose 1,5-bisphosphate carboxylase, commonly known as **Rubisco**. This enzyme is responsible for the production of biomass by oxygenic photosynthetic organisms and removal of CO_2 from the atmosphere (to the extent globally, of over 10^{11} t of CO_2 per year). Rubisco is an enzyme of the **Calvin cycle**, the stages of photosynthesis that can occur in the dark, in which it catalyses the incorporation of CO_2 into a molecule of ribulose 1,5-bisphosphate (Fig. 26.33).

Rubisco is believed to be the most abundant enzyme on Earth, but its turnover frequency is low, and efforts are underway to improve its efficiency. The Mg^{2+} ion is octahedrally coordinated by carboxylate groups from glutamate and aspartate residues, three coordinated H_2O molecules, and a carbamate derived from a lysine residue. The carbamate is formed by a reaction between CO_2 and the terminal $-NH_2$, in an activation process that is also necessary in order for Mg^{2+} to bind. In the catalytic cycle, the binding of ribulose 1,5-bisphosphate displaces two H_2O molecules, and proton abstraction assisted by the carbamate results in a coordinated enolate. This nucleophilic intermediate reacts with CO_2, forming a new C–C bond, then the product is cleaved to yield two new three-carbon species and the cycle continues. The C3 product, 3-phospho-D-glycerate, is reduced during further steps of the Calvin cycle. Thus, in contrast to the direct electrocatalytic reduction of CO_2 which is of interest to industry, biology 'fixes' CO_2 by first incorporating it into an activated organic acceptor.

The reactive enolate will also react with O_2, in which case the result is an oxidative degradation of substrate: for this reason the enzyme is often called ribulose 1,5-bisphosphate carboxylase-oxygenase. We note the contrast with Zn^{2+}, which would favour ligation by softer ligands and a lower coordination number. Rubisco requires a metal ion that combines good Lewis acidity with weak binding and high abundance.

In contrast to many industrial processes for converting CO_2 into reduced products, which involve use of H_2 or electrochemical reduction to form CO, formate, or other C_1 products, each turn of the Calvin cycle introduces CO_2 into an activated recipient, the C_5 sugar that has been phosphorylated by ATP (which renders it more reactive) and no direct reduction of CO_2 is involved.

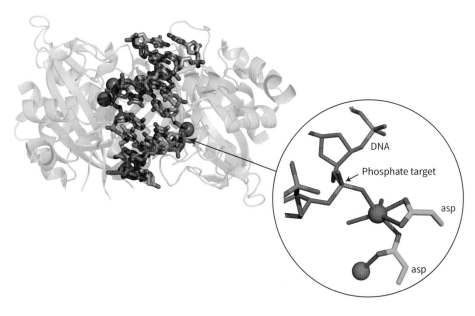

FIGURE 26.32 Structure of endonuclease II (in the presence of Mn^{2+}, shown as spheres) complexed with DNA. The enlarged inset indicates the way that Mg^{2+} can bind to a phosphate group and polarize it for hydrolysis.

FIGURE 26.33 Mechanism of action of ribulose 1,5-bisphosphate carboxylase, the enzyme responsible for removing CO_2 from the atmosphere and 'fixing' it in organic molecules in plants.

(c) Hydrolytic iron enzymes

KEY POINTS Acid phosphatases contain a dinuclear metal site containing Fe(III) in conjunction with Fe, Zn, or Mn; aconitase contains a [4Fe-4S] cluster, one subsite of which is modified to manipulate the substrates.

Acid phosphatases, sometimes known as 'purple' acid phosphatases (PAPs) on account of their intense colour, occur in various mammalian organs, particularly the bovine spleen and porcine uterus. Acid phosphatases catalyse hydrolysis of phosphate esters, with optimal activity under mild acid conditions. They are involved in bone maintenance and hydrolysis of phosphorylated proteins (they are therefore important in signalling). They may also have other functions such as Fe transport. The pink or purple colours of acid phosphatases are due to a tyrosinate → Fe(III) charge-transfer transition at 510–550 nm ($\varepsilon = 4000\,dm^3\,mol^{-1}\,cm^{-1}$). The active site contains two Fe atoms linked by ligands, similar to haemerythrin (**17**). Acid phosphatases are inactive in the oxidized {Fe(III)Fe(III)} state in which they are often isolated. In the active state, one Fe is reduced to Fe(II). Both Fe atoms are high spin, and remain so throughout the various stages of reactions.

Acid phosphatases also occur in plants, and in these enzymes the reducible Fe is replaced by Zn or Mn. The active site of an acid phosphatase from sweet potato (**32**) shows how phosphate becomes coordinated to both Fe(III) and Mn(II) ions. In the mechanism shown in Fig. 26.34, rapid binding of the phosphate group of the ester occurs to the M(II) subsite, then the P atom is attacked by an OH^- ion that is formed at the more acidic Fe(III) subsite.

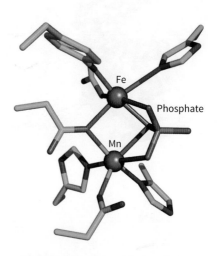

32 Active site of an acid phosphatase

FIGURE 26.34 Proposed mechanism of action of acid phosphatase. The second metal site M(II) is occupied by Fe (most common in animals) or by Mn or Zn (in plants).

Aconitase is an essential enzyme of the **tricarboxylic acid cycle** (also known as the Krebs cycle, or citric acid cycle), the main source of energy production in higher organisms. Aconitase catalyses the interconversion of citrate and isocitrate in a reaction that formally involves dehydration and rehydration, and proceeds through an intermediate, aconitate, which is released in small amounts:

citrate → aconitate → isocitrate

The active form of the enzyme contains a [4Fe-4S] cluster, which degrades to [3Fe-4S] when the enzyme is exposed to air. The specific site of catalysis is the Fe atom that is lost upon oxidation. This unique subsite is not coordinated by a protein ligand but by an H_2O molecule, which helps explain why this Fe is more readily removed.

A plausible mechanism for the action of aconitase, based on structural, kinetic, and spectroscopic evidence, involves the binding of citrate to the active Fe subsite, which increases its coordination number to 6. An intermediate in the catalytic cycle is 'captured' for X-ray diffraction investigation, using a site-directed mutant that can bind citrate but cannot complete the reaction (**33**). The Fe atom polarizes a C–O bond and OH is abstracted, while a nearby base accepts a proton. The substrate swings round, and the OH and H are reinserted onto a different position. A form of aconitase found in cytoplasm has another intriguing role, that of an Fe sensor (Section 26.16).

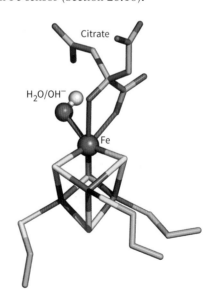

33 Active site of aconitase with citrate bound

26.10 Enzymes dealing with H₂O₂ and O₂

In Section 26.7 we saw how organisms have evolved systems that transport O_2 reversibly and deliver it unchanged to where it is required. In this section we describe how O_2 is reduced catalytically, either for production of energy or synthesis of oxygenated organic molecules. We start by considering a simpler case: that of the reduction of hydrogen peroxide, as this discussion introduces Fe(IV) as a key intermediate in so many biological processes. We end with the reaction that closes the remarkable 'oxygen cycle', the production of all atmospheric O_2 from H_2O, catalysed by a unique Mn/Ca-containing cofactor contained in a giant photosynthetic enzyme present in all green plants.

(a) Peroxidases

KEY POINTS Peroxidases catalyse reduction of hydrogen peroxide; they provide important examples of Fe(IV) intermediates that can be isolated and characterized.

Haem-containing peroxidases, as exemplified by horseradish peroxidase (HRP) and cytochrome c peroxidase (CcP), catalyse the reduction of harmful hydrogen peroxide:

$$H_2O_2(aq) + 2e^- + 2H^+(aq) \rightarrow 2H_2O(l)$$

Part of the intense chemical interest in these enzymes lies in the fact that they provide the best structurally defined examples of Fe(IV) complexes in chemistry. Iron(IV) is an important catalytic intermediate in numerous biological processes involving oxygen (Section 10.8). Catalase, which catalyses the thermodynamically favourable disproportionation of H_2O_2 and is one of the most active enzymes known, is also a peroxidase. The active site of yeast cytochrome c peroxidase shown in Fig. 26.35 suggests how the substrate is manipulated during the catalytic cycle. The proximal ligand is the imidazole side chain of a histidine and the distal pocket, like myoglobin, also contains an imidazole side chain, but there is also a guanidinium group from arginine.

The catalytic cycle shown in Fig. 26.36 starts from the Fe(III) form. A molecule of H_2O_2 coordinates to Fe(III) and a proton is transferred to the distal histidine, resulting in an Fe(III)-OOH intermediate. Upon arrival of another proton from an external donor the O–O bond is cleaved: one half leaves as H_2O and the other remains bound to the Fe atom as a reactive Fe–O species. Although it is instructive to regard this intermediate as a trapped O atom (or an O^{2-} ion bound to Fe(V)), detailed measurements by EPR and Mössbauer spectroscopy show the highly oxidizing state (known historically as 'Compound I') to consist of an Fe(IV)=O entity (referred to as oxido-Fe(IV) or ferryl) and an oxidized organic cation radical. In HRP the radical is located on the porphyrin ring whereas in CcP it is located on nearby peptide residue tryptophan-191. The O^{2-} (oxido) ligand is a very strong π-donor and is well suited to stabilizing high oxidation states, and we will see later that Fe(IV)=O plays a major role in the enzymatic reactions of Fe with O_2. Compound I is reduced back to the resting Fe(III) state by two one-electron transfers either from organic substrates or cytochrome c. The first electron reduces the organic radical to produce Compound II, which retains the ferryl entity: it is shown as Fe(IV)=O, but possibly exists in a protonated form Fe(IV)-OH.

Hydrogen peroxide is used in biology to synthesize halogenated compounds. This process, which is particularly important in macroalgae (seaweed), is catalysed by a class of peroxidases that do not contain Fe but instead use vanadium in a way that exploits the Lewis acidity of V(V). The active site of bromoperoxidase (**34**) consists of an oxo-V^V unit coordinated by a single histidine-N. The catalytic mechanism involves the activation of η^2-coordinated peroxide, whereby an oxygen atom is transferred to an incoming Br^- ion resulting in a reactive BrO^- species that attacks the organic substrate.

34 Active site of bromoperoxidase

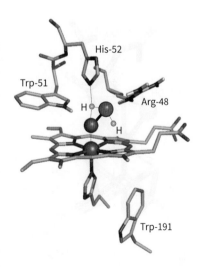

FIGURE 26.35 The active site of yeast cytochrome c peroxidase showing amino acids essential for activity and indicating how peroxide is cleaved in the distal pocket.

(b) Oxidases that use O₂ as a substrate

KEY POINTS Oxidases catalyse the reduction of O_2 to water or hydrogen peroxide without incorporation of O atoms into the oxidizable substrate; they include cytochrome c oxidase, the enzyme that is a basis for all higher life forms.

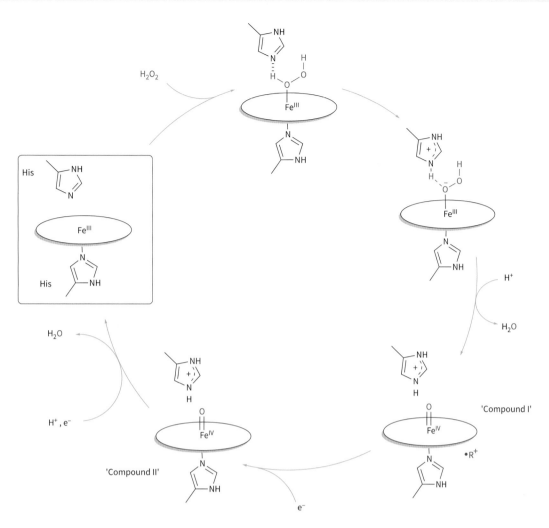

FIGURE 26.36 The catalytic cycle of haem-containing peroxidases (·R$^+$ is an oxidized cation radical, formed on the porphyrin ring or a nearby amino acid). Likely intermediates and products (excluding solvent-exchanged H) are shown in red.

Cytochrome *c* oxidase is a membrane-bound enzyme that catalyses the four-electron reduction of O_2 to water, using cytochrome *c* as the electron donor. The potential difference between the two half-cell reactions is over 0.5 V, but this value does not reflect the true thermodynamics because the actual reaction catalysed by cytochrome *c* oxidase is:

$$O_2 + 4e^- + 8H^+(\text{inside}) \rightarrow 2H_2O + 4H^+(\text{outside})$$

This reaction includes four H^+ that are not consumed chemically but are 'pumped' across the membrane (from 'inside' to 'outside') against a concentration gradient. Such an enzyme is an electrogenic ion pump (or proton pump). In eukaryotes, cytochrome oxidase is located in the inner membrane of mitochondria and has many subunits (although a simpler enzyme is produced by some bacteria). The enzyme (Fig. 26.37) contains three Cu atoms and two haem-Fe atoms, as well as an Mg atom and a Zn atom that may have structural importance.

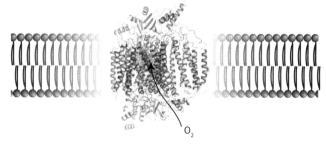

FIGURE 26.37 The structure of cytochrome *c* oxidase as it occurs in the membrane, showing the locations of the redox centres and the sites for reaction with O_2 and cytochrome *c*. See (**35**) for a more detailed view of the active site.

The Cu and Fe atoms are arranged in three main sites. The active site for O_2 reduction consists of a myoglobin-like Fe-porphyrin (haem-a$_3$) that is adjacent to a 'semi-haemocyanin-like' Cu (known as Cu$_B$) coordinated by three histidine ligands (**35**).

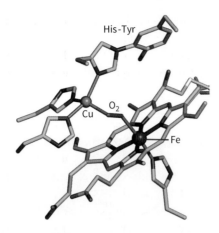

35 Active site of cytochrome *c* oxidase showing how O_2 may be bound in an intermediate state

One of the histidine imidazole ligands to the Cu is modified by formation of a covalent bond to an adjacent tyrosine. Electrons are supplied to the dinuclear site by a second Fe porphyrin (haem-a) that is six-coordinate, as expected for an electron-transfer centre. All these centres are located in subunit 1. The electron arriving from cytochrome *c* is accepted initially by the dinuclear Cu_A centre (**27**, Section 26.8(d)) located in subunit 2. The electron transfer sequence is therefore

Cytochrome *c* → Cu_A → haem *a* → binuclear site

Cytochrome *c* oxidase contains two proton-transfer channels, one of which is used to supply the protons needed for H_2O production, while the other is used for protons that are being pumped across the membrane. Fig. 26.38 shows the proposed catalytic cycle. Starting from state O, two electrons are added to produce the reduced (R) state in which the active site is Fe(II)–Cu(I), then O_2 binds to give an intermediate 'oxy' form, (A), that resembles oxymyoglobin. However, unlike oxymyoglobin, this state takes up the other electron that is immediately available from Cu(I), producing a peroxy species that rapidly breaks down to give a state known as P. Species P has been trapped and studied by

FIGURE 26.38 The catalytic cycle of cytochrome *c* oxidase. The intermediates are labelled according to current convention. Electrons are provided from the other haem centre and Cu_A. During the cycle, an additional four H^+ are pumped across the membrane.

optical and EPR spectroscopy, which show that it contains Fe(IV)=O and an organic radical formed on the unusual His–Tyr pair. In the next stage, the radical is reduced leaving the ferryl intermediate F, which is then reduced back to O.

The role of a cation radical in species P is notable: without such a radical, the Fe would have to be assigned as Fe(V). Use of tyrosine as a redox-active site is common in enzymes that catalyse reactions involving strong oxidants: formation of the tyrosyl radical (usually abbreviated Y·) from tyrosine is an example of proton-coupled electron transfer (Section 6.14):

$$\text{—O}^\bullet + H^+ + e^- \rightleftharpoons \text{—OH}$$

It is vital that intermediates such as peroxide are not released during conversion of O_2 to water. Studies with an elaborate model complex (**36**) which can be attached to an electrode, show that the presence of the phenol (acting in place of the naturally-occurring tyrosine) is crucial, because it allows all four electrons necessary for the reduction of the O_2 to be provided rapidly without relying on long-range electron transfer, which is slow through the long-chain aliphatic linker. If the phenolic –OH group is replaced by –OCH$_3$, hydrogen peroxide is released during O_2 reduction because the methoxy derivative is unable to form an oxidized radical.

Blue Cu oxidases contain a blue Cu centre that removes an electron from a substrate and passes it to a trinuclear Cu site that catalyses the reduction of O_2 to H_2O. Two examples—ascorbate oxidase and a larger class known as laccases—are well characterized, whereas another protein, ceruloplasmin, occurs in mammalian tissue and is the least well understood. Ascorbate oxidase occurs in the skins of fruit such as cucumbers and pumpkins. Its role may be two-fold: to protect the flesh of the fruit from O_2 and to oxidize phenolic substrates to intermediates that will form the skin of the fruit. Laccases are widely distributed, particularly in plants and fungi from which they are secreted to catalyse the oxidation of phenolic substrates. The active site at which O_2 is reduced (**37**) contains a pair of Cu atoms linked in the oxidized form by a bridging O atom, with a third Cu atom situated very close by, completing an almost triangular arrangement.

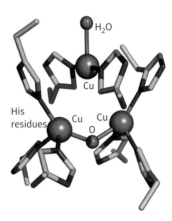

37 The trinuclear active site of a blue Cu oxidase

Amine oxidases catalyse the oxidation of amines to aldehydes by using just a single Cu atom that shuttles between Cu(II) and Cu(I), yet the enzyme carries out the two-electron reduction of O_2, producing a molecule of H_2O_2. The problem is overcome because, like cytochrome *c* peroxidase and cytochrome *c* oxidase, amine oxidases have an additional oxidizing source located near to the metal, in this case a special organic cofactor called topaquinone (TPQ) that is formed by post-translational oxidation of tyrosine (**38**).

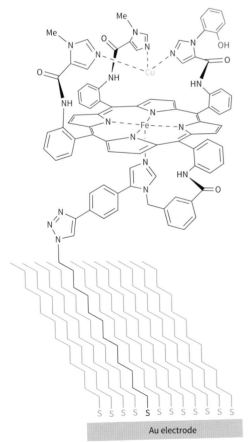

36 A model for the active site of cytochrome *c* oxidase, tethered to an electrode

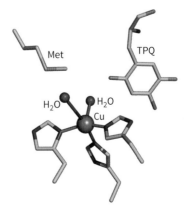

38 The active site of an amine oxidase

EXAMPLE 26.8 Interpreting reduction potentials

The four-electron reduction potential for O_2 is +0.82 V at pH = 7. Cytochrome c, the electron donor to cytochrome c oxidase, has a reduction potential of +0.26 V, whereas the organic substrates of fungal laccases often have values as high as +0.7 V. What is the significance of these data in terms of energy conservation?

Answer Although cytochrome c oxidase and laccase both catalyse the efficient four-electron reduction of O_2 we need to consider their different biological functions. Laccases are efficient catalysts of phenol oxidation, the driving force being small. Cytochrome oxidase is a proton pump, and approximately 2 eV (4 × 0.56 eV) of Gibbs energy is available from oxidation of cytochrome c to drive proton transfer across the mitochondrial inner membrane.

Self-test 26.8 Before the discovery of the unusual active site structures in amine oxidase and another Cu enzyme called galactose oxidase, Cu(III) was proposed as a catalytic intermediate. What properties would be expected of this state?

(c) Oxygenases

KEY POINT Oxygenases normally catalyse the insertion of one or both O atoms derived from O_2 into an organic substrate.

Oxygenases catalyse the *insertion* of one or both O atoms of O_2 into an organic substrate: they are often referred to as *hydroxylases* when an O atom is inserted into a C–H bond. This process is remarkable as it amounts to the 'taming' of a highly reactive O atom through its attachment to a catalyst. Most oxygenases contain Fe, the rest contain Cu or the organic cofactor flavin. There are many variations. One or both O atoms may be incorporated (represented here as [O]) into organic molecules.

Monooxygenases catalyse reactions of the type

$$R + O_2 + 2H^+ + 2e^- \rightarrow R-[O] + H_2O$$

Only one O atom is incorporated into an organic compound, the other forming a water molecule. The electrons are supplied by an electron donor such as an Fe-S protein.

Dioxygenases catalyse the insertion of both atoms of O_2 into one or two organic substrates (which may lead to further reactions such as C–C bond cleavage) and no additional electron donor is required.

$$R + R' + O_2 \rightarrow R-[O] + R'-[O]$$
$$R + O_2 \rightarrow [O]-R-[O]$$

The Fe enzymes are divided into two main classes: haem and non-haem. We discuss the haem enzymes first, the most important type being cytochrome P450.

Cytochrome P450 (or just 'P450') refers to an important and widely distributed group of haem-containing monooxygenases. In eukaryotes they are localized particularly in mitochondria, and in higher animals they are concentrated in liver tissue. The enzymes play an essential role in detoxification as well as in biosynthesis (for example steroid transformations), such as the production of progesterone. The designation 'P450' arises from the intense absorption band that appears at 450 nm when solutions containing the enzyme or even crude tissue extracts are treated with a reducing agent and carbon monoxide, which produces the Fe(II)–CO complex. Most P450s are complex membrane-bound enzymes that are difficult to isolate. Much of what we know about them stems from studies carried out with an enzyme known as P450$_{cam}$, which is isolated from the bacterium *Pseudomonas putida*. This organism uses camphor as its sole source of carbon, and the first stage is oxygenation of the 5-position:

$$\text{camphor} \xrightarrow[\text{Cytochrome P-450}]{O_2,\ 2H^+,\ 2e^-} \text{5-hydroxycamphor} + H_2O$$

The catalytic cycle (Fig. 26.39) has been studied by using a combination of kinetic, spectroscopic, and site-directed mutagenesis methods. Starting from the enzyme in the Fe(III) form (A), the binding of the substrate in the active site pocket (stage 1) occurs with a change in spin state from low spin ($S = \frac{1}{2}$) to high spin ($S = \frac{5}{2}$), and the reduction potential increases, causing an electron to be transferred (stage 2) from a small [2Fe-2S]-containing ferredoxin. The five-coordinate Fe(II) that is formed resembles deoxymyoglobin and binds O_2 (stage 3). It is helpful to regard the product as an Fe(III)-superoxido species. Unlike in myoglobin, addition of a second electron is both thermodynamically and kinetically favourable. The subsequent reactions (stages 4–6) are very fast, but it is thought that an Fe(III) peroxide intermediate

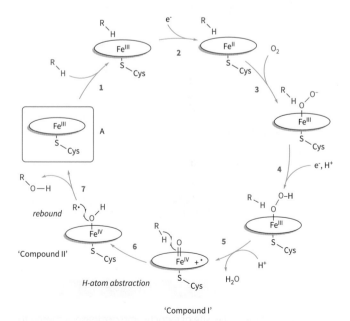

FIGURE 26.39 The catalytic cycle of cytochrome P450.

is formed that undergoes rapid heterolytic O–O cleavage to produce a ferryl species accompanied by a cation radical (in this case on the porphyrin ring), that is similar to Compound I of peroxidases. In what is known as the **oxygen rebound mechanism**, the Fe(IV)=O group abstracts an H atom from the substrate and then inserts it back as an OH radical.

It is useful to reflect on the fact that Fe lies at the 'oxo wall' (Section 10.8)—the boundary within the d block beyond which multiply bonded octahedral (or tetragonal) oxido complexes cannot exist because the metal is unable to attain an oxidation number high enough to leave a vacancy in the π-symmetry (t_{2g}) orbitals. The ferryl group (Fe(IV)=O) is a powerful hydrogen atom abstractor, and its potency is increased further by the porphyrin-based cation radical.

Other P450s are thought to operate by similar mechanisms but differ in the architecture of the active site pocket. The P450 active site, unlike that in peroxidases, is predominantly hydrophobic, with specific polar groups present to orient the organic substrate, so the correct R–H bond is brought close to the Fe(IV)=O entity.

Non-haem Fe oxygenases are distributed across all forms of life where they perform a wide range of functions, ranging from degradation of metabolites to the synthesis of antibiotics and sensing of O_2 in tissue. Two major classes, having very different modes of catalytic action, may be distinguished.

For the oxygenases in the first category, the Fe remains throughout the catalytic cycle as high-spin Fe(III), its role being to function as a Lewis acid and activate an organic substrate to attack by a noncoordinating O_2 molecule in the manner shown for protocatechuate 3,4-dioxygenase.

In most cases the Fe is coordinated by two histidine-N and a carboxylate-O (from aspartate or glutamate) at three facial sites of an octahedron, leaving three sites remaining to bind O_2 and other ligands (Fig. 26.40).

The fact that Fe(II) is low in the Irving–Williams series (Section 7.16) combined with the small number of protein ligands renders the metal ion rather labile. This instability, the high reactivity with O_2, and shortage of convenient spectroscopic characteristics, have complicated investigations of the Fe(II) resting state and reaction intermediates—problems that have been largely overcome by ingenuity—substituting Mn(II) for Fe(II) and most recently by using time-resolved spectroscopy and X-ray diffraction to investigate intermediates. The Fe(II)-oxygenases in particular are responsible for catalysing a host of important reactions, some of which are listed in Table 26.3.

An important class of Fe(II) oxygenases uses the central metabolite 2-oxoglutarate (2OG) as a co-substrate. The principle of **2OG-dependent oxygenases** is that the transfer of one O atom of O_2 to coordinated 2OG induces its breakdown to succinate (which remains temporarily coordinated) and CO_2 (which is released)—the irreversible reaction driving insertion of the other O atom into the primary substrate.

The Fe is tightly coordinated by a set of protein ligands that includes two His-N and two Tyr-O, the latter hard donors being particularly suitable for stabilizing Fe(III) relative to Fe(II). The Fe(III) enzymes are deep red in colour, due to an intense tyrosinate-to-Fe(III) charge transfer transition.

In the second category, the Fe is coordinated as Fe(II) and reacts directly with O_2 to generate a highly reactive intermediate, a ferryl (Fe^{IV}=O) or ferric peroxide (Fe^{III}-O-O–) species, that attacks the substrate in the manner shown for catechol 2,3-dioxygenase.[9]

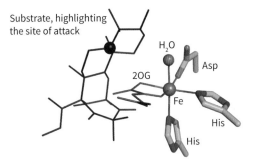

FIGURE 26.40 Active site of a 2OG-dependent oxygenase showing how the Fe is coordinated by a His, His, Asp *facial* triad. Also shown is a bound water, 2OG, and a nearby substrate poised for O-atom insertion into a C–H bond.

[9] The Fe(III)-based oxygenases were long referred to as *intradiol* oxygenases, while the Fe(II) enzymes were known as *extradiol* oxygenases. This historic classification followed discovery of their roles in the biodegradation of hydroxylated aromatic compounds and the observation of whether ring cleavage induced by oxygen attack occurred within or outside two adjacent OH groups.

TABLE 26.3 Some reactions catalysed by Fe(II)-oxygenases

Enzyme	Primary reaction	Co-substrate/electron donor	Biological application
Catechol dioxygenase	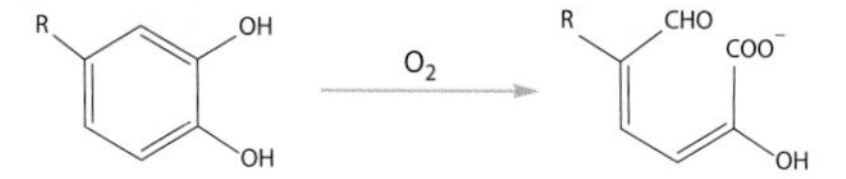 R-catechol + O₂ → R-muconate (CHO, COO⁻, OH)	None	Degradation of small molecules
Naphthalene 1,2-dioxygenase	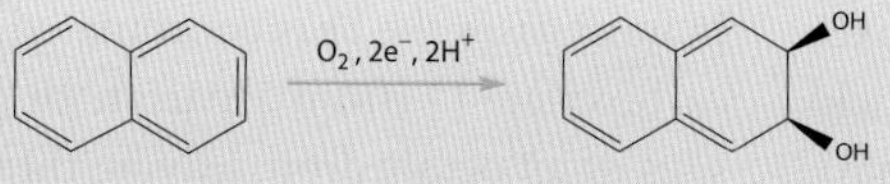 naphthalene + O₂, 2e⁻, 2H⁺ → cis-dihydrodiol	Fe-S	Detoxification
Demethylases (e.g. lysine)	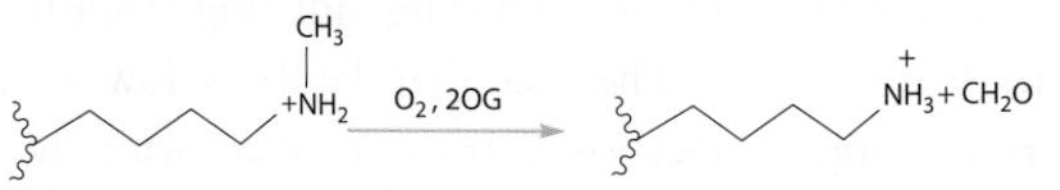 R-NH-CH₃ + O₂, 2OG → R-NH₂ + CH₂O (shown as +NH₂CH₃ → +NH₃ + CH₂O)	2OG	Regulation of protein/DNA interactions
Proline (prolyl) hydroxylase	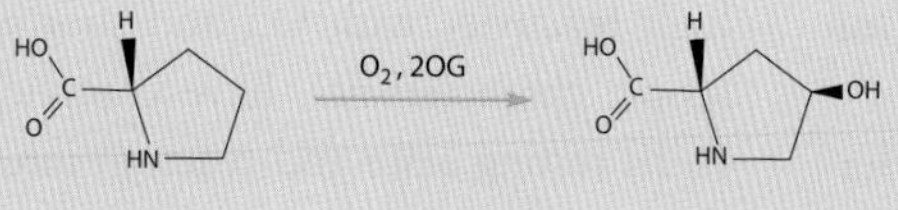 proline + O₂, 2OG → hydroxyproline	2OG	Production of connective tissue (collagen); O₂ sensing by hypoxia-inducible factor (Section 26.16)
Halogenases	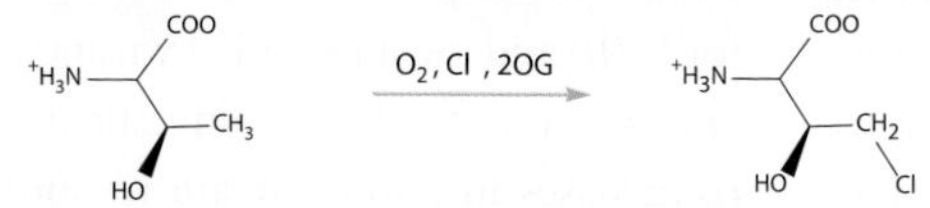 threonine + O₂, Cl⁻, 2OG → chloro-product	2OG	Biosynthesis

A general mechanism for 2OG-dependent oxygenases is shown in Fig. 26.41. Roles of 2OG-dependent oxygenases include cell signalling by modifying an amino acid in certain transcription factors (Section 26.16) and in the biosynthesis of antibiotics containing a β-lactam ring (Box 26.1).

Halogenases—a subclass of 2OG-dependent oxygenases—catalyse the replacement of a hydrogen atom by a halogen atom derived from a halide anion (usually Cl⁻). The active sites of these enzymes differ from typical Fe(II) oxygenases in that the halide replaces the carboxylate ligand derived from an aspartate or glutamate residue of the protein (**39**).

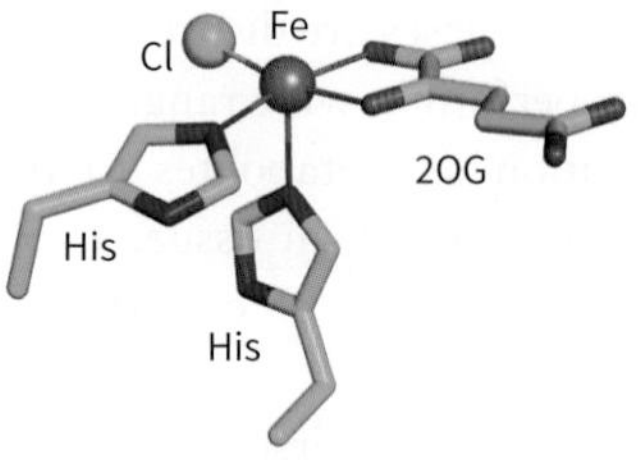

39

The common factor connecting the activities of oxygenase and halogenase is the nature of the 'rebound' intermediate. The halide ion is positioned with respect to the substrate in such a way that it is favoured to attack (as a halogen atom) in place of the ·OH radical.

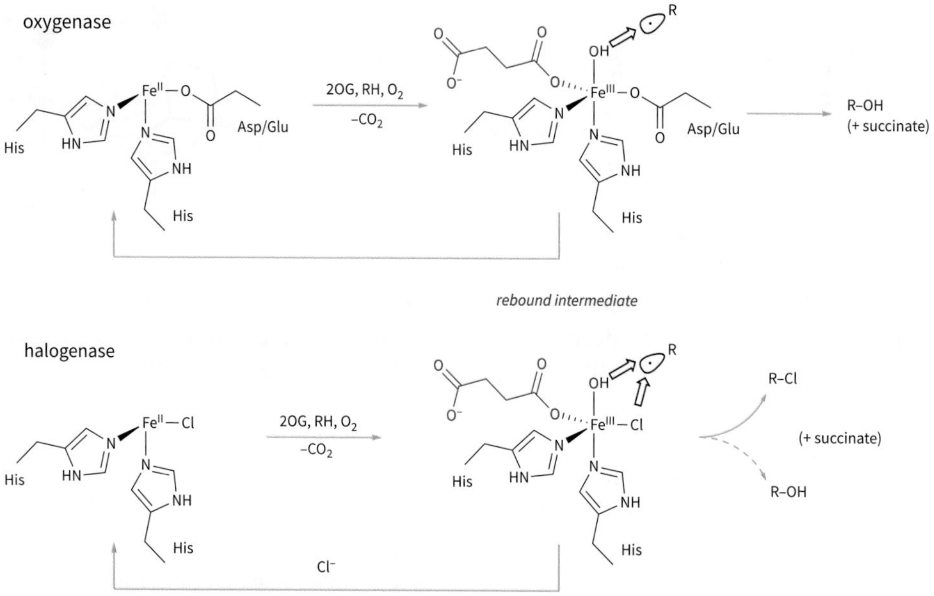

oxygenase:
$\text{Fe}^{II}\text{—O—Asp/Glu}$ (with two His, NH ligands) $\xrightarrow{\text{2OG, RH, O}_2,\ -\text{CO}_2}$ rebound intermediate $\text{Fe}^{III}\text{—O—Asp/Glu}$ with OH and R → R-OH (+ succinate)

halogenase:
$\text{Fe}^{II}\text{—Cl}$ (with two His, NH ligands) $\xrightarrow{\text{2OG, RH, O}_2,\ -\text{CO}_2}$ $\text{Fe}^{III}\text{—Cl}$ rebound intermediate → R-Cl (+ succinate), R-OH

Cl⁻

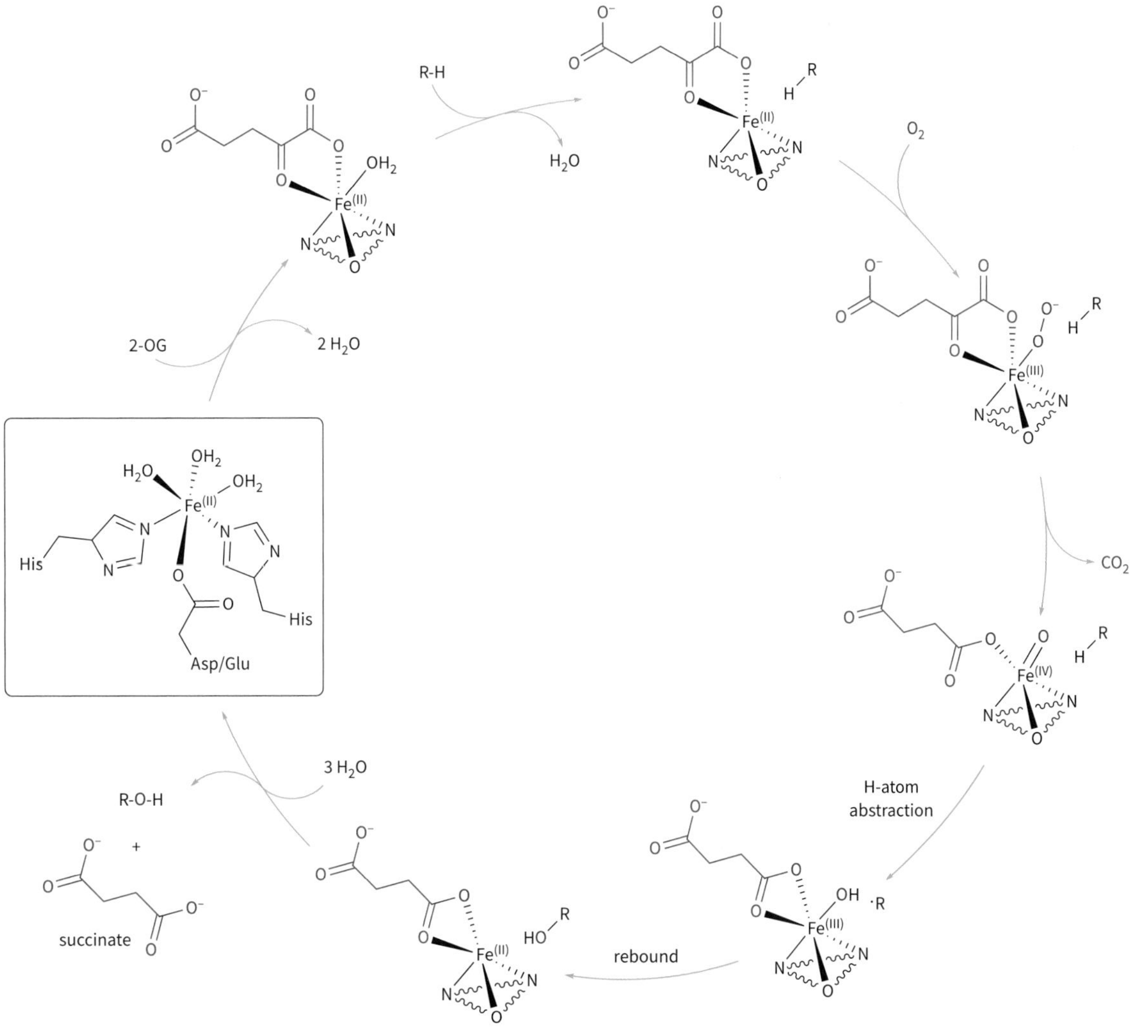

FIGURE 26.41. A general mechanism for catalysis by 2OG-dependent oxygenases.

BOX 26.1 How bacteria make penicillin

Penicillin was first discovered by Alexander Fleming in 1939 before its chemistry was uncovered by Howard Florey and Ernst Chain: the three scientists shared the Nobel Prize in Physiology or Medicine in 1945. The biosynthesis of isopenicillin N (IPN), the immediate precursor of penicillin N (into which it is converted by an epimerase), is one of the most remarkable catalytic reactions known.

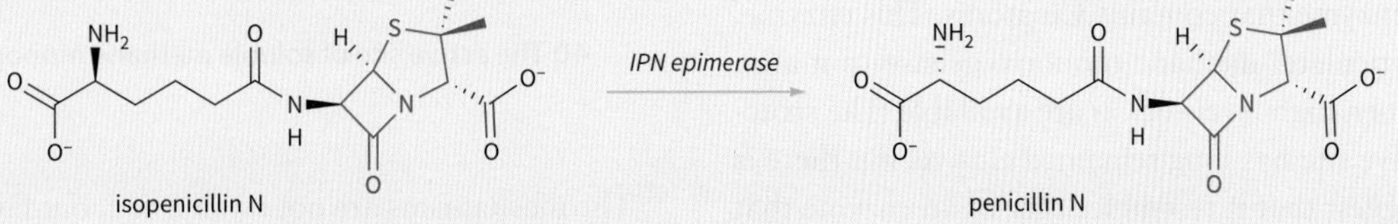

isopenicillin N → (IPN epimerase) → penicillin N

The enzyme responsible for catalysing the formation of isopenicillin from a peptide derivative is known as isopenicillin N synthase (IPNS). It is related to the 2OG dependent oxygenases, but does not use 2OG as a cosubstrate, instead both O-atoms of O_2 are used to make H_2O, the free energy being used to drive two sequential ring-closure reactions (Fig. B26.1). The two demanding steps are achieved while the substrate remains bound to the Fe.

FIGURE B26.1 Biosynthesis of β-lactam ring

Oxygenases play a crucial role in the metabolism of methane, a greenhouse gas. Of all hydrocarbons, methane contains the strongest C–H bonds (see Figure 16.3) and is the most difficult to activate. Methane-metabolizing bacteria produce two types of enzyme that catalyse the conversion of methane to methanol (a more useful chemical and fuel) and thus attract much industrial interest. One is a membrane-bound enzyme that contains Cu atoms. This enzyme, known as 'particulate' methane monooxygenase (p-mmo), is expressed when high levels of Cu are available. The structure of the active site has long remained elusive, but there is evidence to suggest that it resembles that of an enzyme that degrades cellulose (Box 26.2). The other enzyme, soluble methane monooxygenase (s-mmo), contains a dinuclear Fe active site (**40**) that is related to haemerythrin (**17**) and acid phosphatase (**32**).

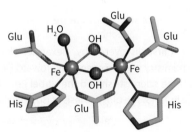

40 The active site of soluble methane monooxygenase

The mechanisms are not established, but Fig. 26.42 shows a plausible catalytic cycle for s-mmo. The intermediate Fe(IV) species that is proposed differs from those we have encountered up to now, as the O_2-derived oxido ligands occupy bridging positions.

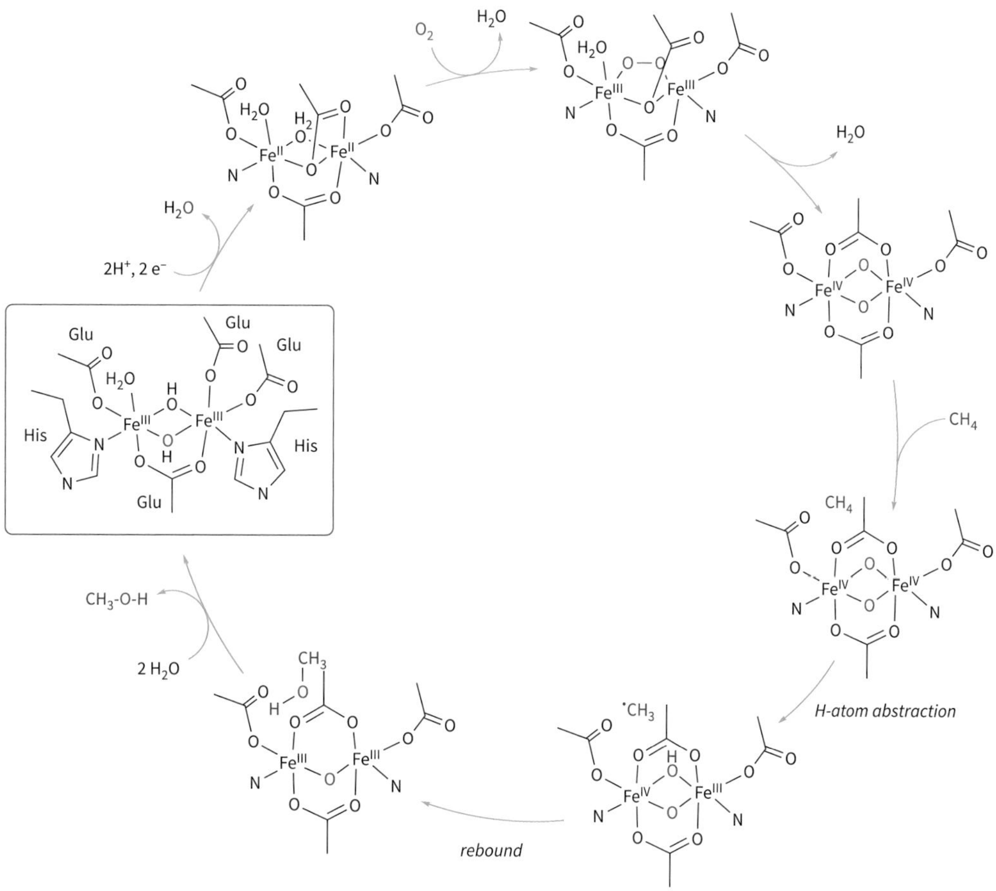

FIGURE 26.42 A plausible catalytic cycle for soluble methane monooxygenase.

Despite the importance of Fe(IV) as an enzyme intermediate, small Fe(IV) complexes that could provide important models for understanding the enzymes have been elusive, not least because they are powerfully oxidising. The easiest compounds to synthesize are haem analogues in which Fe is equatorially ligated by porphyrin: these can be prepared by reacting the Fe(II) or Fe(III) forms with a peroxo acid, and they show catalytic activity. Small molecule analogues of non-haem Fe oxygenases have also been synthesized, and those Fe(IV) compounds having sufficient stability to be crystallized display short Fe–O bonds, ranging between 164 and 168 pm. However, functional analogues—those having significant catalytic activity—have been more elusive. Despite the difficulty in obtaining a structure of the reactive Fe(IV) intermediate in enzymes, an important clue lay in its spin state, established using Mössbauer spectroscopy to characterize samples trapped during catalytic turnover by rapid freeze-quenching. In all cases the reactive intermediate has an $S = 2$ spin state, the implication being that the ligand field is trigonal or pseudo octahedral but with a weakened equatorial field.

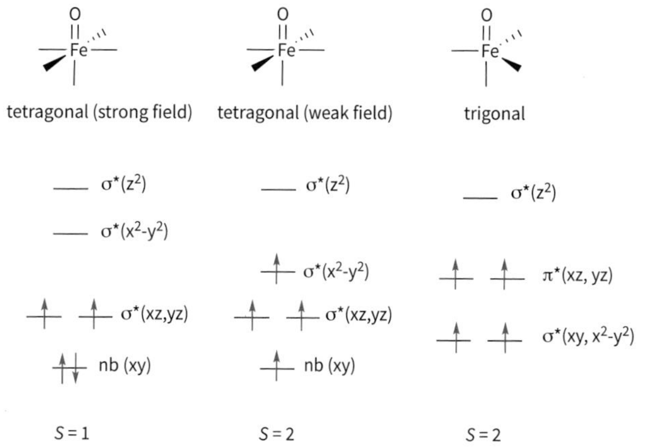

An active-site analogue (**41**) having similar magnetic properties is synthesized using a bulky ligand that provides three equatorial and one apical N donor (an acetonitrile is also bound). The compound shows high catalytic activity for converting cyclohexane to cyclohexanol at −40 °C.

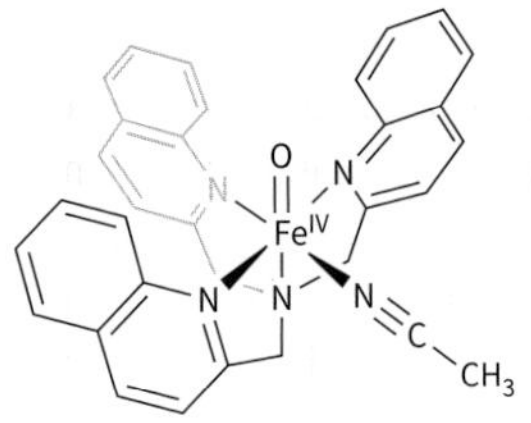

41

Tyrosinase and catechol oxidase—two enzymes responsible for producing melanin-type pigments—each contain a strongly coupled dinuclear Cu centre that coordinates O_2 in a manner similar to haemocyanin. However, unlike in haemocyanin, the coordinated O_2 is activated for electrophilic attack at a phenolic ring of the substrate. The structure of the active site of catechol oxidase complexed with the inhibitor phenolthiourea (**42**) shows how the phenol ring of the substrate can be oriented in close proximity to a bridging O_2. Copper enzymes are also responsible for the production of important neurotransmitters and hormones, such as dopamine and noradrenaline. These enzymes contain two Cu atoms that are well separated in space and uncoupled magnetically. An oxygenase having a single Cu at its active site catalyses the oxidative breakdown of solid biomass, a reaction of great industrial importance. The properties of this enzyme are described in Box 26.2.

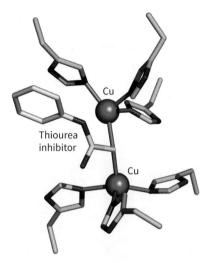

42 The active site of catechol oxidase with an inhibitor bound

(d) Photosynthetic O_2 production

KEY POINTS Biological solar energy capture by photoactive centres results in the generation of species with sufficiently negative reduction potentials to generate NADPH (the biological equivalent of BH_4^-) from NADP$^+$. With the help of ATP that is also produced during photosynthetic ET, atmospheric CO_2 is incorporated into organic molecules in the Calvin cycle that involves Rubisco. In all higher plants and cyanobacteria, the electrons are derived from water, which is converted to O_2 by a unique catalytic centre containing four Mn atoms and one Ca atom. The O_2 that is so central to life on Earth as we know it arises as a waste product of photosynthesis.

BOX 26.2 How does a copper enzyme degrade cellulose?

Most of the organic material that is produced by photosynthesis is unavailable for use by industry or as fuels. Biomass largely consists of polymeric carbohydrates—polysaccharides such as cellulose and lignin, that are very difficult to break down to simpler sugars as they are resistant to hydrolysis. However, a breakthrough has occurred with the discovery that certain fungi and bacteria produce enzymes that are able to degrade polysaccharides to simpler sugars by initiating the oxidation reaction. The enzymes responsible are known as polysaccharide monooxygenases (more commonly as LPMOs, the 'L' representing 'lytic') and contain Cu. Whether by use of the organisms or the enzymes themselves, industry has a new tool by which to increase the yield of biofuels.

The following scheme shows the principle by which cellulose is broken down.

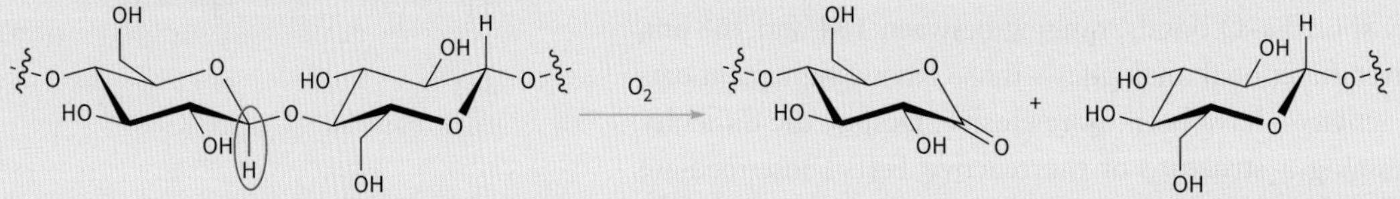

Cleavage of the glycosidic bond is initiated by insertion of an O atom into the C–H bond indicated. The resulting arrangement is unstable and (HO)C–O–C scission leads to the ketone product.

The crystal structure of an LPMO has been determined in the presence of a cellotriose, a small water-soluble substrate (Fig. B26.2). The Cu is located close to the surface of the enzyme: it is coordinated by an *N*-methylhistidine that lies at the N-terminus, along with its peptide-N, a second histidine, and a tyrosine. The coordination is completed by a chloride ion which indicates the position at which an O_2 molecule is likely to be

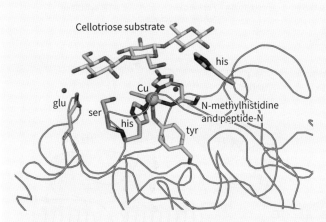

FIGURE B26.2 The active site of a 'lytic' polysaccharide monooxygenase.

coordinated during the catalytic cycle. The cellobiose molecule is tethered by several polar contacts with amino acid side chains and structured water molecules that position the target C–H bond close to the position occupied by the Cl ligand.

How the reaction occurs is currently under debate. The popular description involves LPMO acting as a monooxgenase in a catalytic cycle that requires an electron donor such as ascorbate. Upon reaction with O_2, an initial Cu^I–O_2 adduct is converted into a powerful oxidant formulated as $Cu(III)$-OH or an oxyl radical species $Cu(II)$-O· that can abstract the target H atom and replace it by –OH in a rebound mechanism. However, this description needs to be modified based on observations that LPMOs can use H_2O_2 as the oxidant. Given the likelihood that some H_2O_2 is released during an oxygenase cycle, there would indeed be an advantage in securing its recapture.

Photosynthesis is the production of organic molecules using solar energy. It is conveniently divided (although erroneously so) into the **light reactions** (the processes by which electromagnetic energy is trapped) and the non-light-dependent **dark reactions** (in which the energy acquired in the light reactions is used to convert CO_2 and H_2O into carbohydrates). We have already mentioned the most important of the dark reactions (which actually occur during both day and night), the incorporation of CO_2 into organic molecules, which is catalysed by Rubisco (Fig. 26.33). In this section we describe some of the roles that metals play in the light reactions.

The basic principle of photochemical energy capture, applied in a technology to produce H_2 from water, is described elsewhere (Sections 11.4 and 30.3). We can view photosynthesis in an analogous way, in that H_2 is 'stored' by reaction with CO_2. In biology, photons from the Sun excite pigments present in giant membrane-bound proteins known as **photosystems**. The most important pigment, the familiar green-coloured chlorophyll, is an Mg complex that is very similar to a porphyrin (7). Most chlorophyll is located in giant proteins known as **light-harvesting antennae**, the name perfectly describing their function, which is to collect photons and funnel their energy to chlorophyll complexes in enzymes, producing an excited state in each case. Each electron released by excited chlorophyll travels rapidly down a sequence of protein-bound acceptors, including Fe-S clusters, and (through the agency of ferredoxin and other enzymes) is eventually used to incorporate CO_2 into carbohydrate. Immediately after releasing a 'hot' electron, the chlorophyll cation, a powerful oxidant, must be rapidly reduced by using a 'cold' electron from another site to avoid wasting the energy by recombination (simple reversal of electron flow).

In green plants, photosynthesis occurs in special organelles known as *chloroplasts*. Plant chloroplasts have two photosystems, I and II, operating in series, that allow low-energy light (approximately 680–700 nm, >1 eV) to span the large potential range (>1 V) within which water is stable. Biology thus adopted the principle of the tandem two-photon Z-scheme discussed in Section 30.3. The arrangement of proteins is depicted in Fig. 26.43. Some of the energy of the photosynthetic electron-transfer chain is used to generate a transmembrane proton gradient which in turn drives the synthesis of ATP, as in mitochondria. Photosystem I lies at the low-potential end; its electron donor is the blue Cu protein plastocyanin (Fig. 26.28) that has been reduced using the electrons generated by photosystem II; in turn, the electron donor to photosystem II is H_2O. Thus green plants dispose of the oxidizing power by converting H_2O into O_2. This four-electron reaction is remarkable because no intermediates are released. The catalyst, called the 'oxygen evolving centre' (OEC), also has a special significance because its action, commencing over 2 billion years ago, has essentially provided all the O_2 we have in the atmosphere. The OEC is the only enzyme active site known to produce an O–O bond from two H_2O molecules, and there is much interest in producing functional models of this catalyst for photochemical water splitting (Box 11.4).

The OEC is a metal oxide cluster containing four Mn atoms and one Ca atom that is located in subunit D1 of photosystem II. Subunit D1 has long attracted interest because the living cell replaces it at frequent intervals (every 20–30 minutes) as it quickly becomes worn out by oxidative damage. Synchrotron

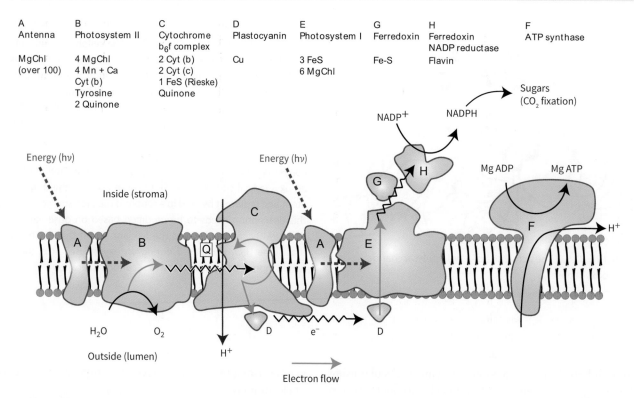

FIGURE 26.43 The arrangement of proteins in the photosynthetic electron-transport chain (the Mg-chlorophyll complex is represented 'MgChl'): **A.** Antenna ('light harvesting') complex. **B.** Photosystem II. **C.** The 'cytochrome b_6f complex' (this is similar to complex III in the mitochondrial ET chain). **D.** Plastocyanin (soluble). **E.** Photosystem I. **F.** ATP synthase. **G.** Ferredoxin (Fe-S). **H.** Ferredoxin-NADP$^+$ reductase (flavin). Blue dashed arrows show inputs of solar energy. Electron transfers are represented as green arrows within protein complexes, and jagged arrows between proteins. Note how the overall transfer of electrons is from Mn (high potential) to Fe-S (low potential): this apparently 'uphill' flow reflects the crucial input of energy at each photosystem. Stoichiometries are not shown.

X-ray diffraction data show that the metal atoms are arranged as a [3MnCa-4O] cubane linked to a fourth Mn to give a [4MnCa-5O] 'chair-like' cluster (Fig. 26.44).

The Mn atoms are in high oxidation states and easily reducible by photoelectrons, so short X-ray exposure times have been crucial. Indeed, elucidating the structure of the OEC in its different oxidation states has done much to advance the development of *time-resolved* X-ray diffraction using free-electron lasers (XFEL). The OEC exploits the oxidizing abilities of Mn(IV) and Mn(V) (or Mn(IV)-oxyl radical), coupled with that of a nearby tyrosine residue, to oxidize H_2O to O_2.

Successive photons received by photosystem II result in the OEC being progressively oxidized (the acceptor, an oxidized chlorophyll known as P680$^+$, has a reduction potential of approximately 1.3 V at pH 7) through a series of states designated S0 to S4, as shown in Fig. 26.45. Apart from S4, which breaks down very quickly to release O_2,

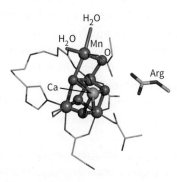

FIGURE 26.44 The [4MnCa-5O] active site of photosynthetic O_2 production.

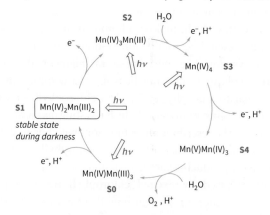

FIGURE 26.45 The 'S-cycle' for evolution of O_2 by successive one-electron oxidation of the [4MnCa-5O] cluster of photosystem II. Progression around the cycle is represented by equivalent increases in Mn oxidation level, but H$^+$ transfers are omitted. In chloroplasts that have become adapted to dark conditions, the cycle 'rests' in the S1 state.

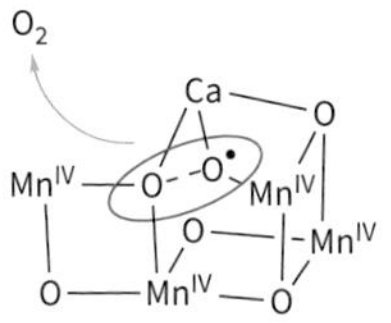

FIGURE 26.46 A possible transition state leading to release of O_2 from the S4 state of the [4MnCa-5O] active site.

the states of the 'S-cycle' are identifiable in kinetic studies by their characteristic spectroscopic properties; for example, S2 shows a complex EPR spectrum that is the result of different couplings within a cluster composed of three Mn(IV) and one Mn(III). Note that the Mn ligands are hard O-atom donors, and Mn(III) (d^4) and Mn(IV) (d^3) are hard Lewis acids.

Based on the available structural evidence, different models have been proposed for the mechanism of O_2 evolution from two H_2O molecules. Intuitively, as the Mn sites are progressively oxidized, coordinated H_2O molecules become increasingly acidic and lose protons (Section 6.14), progressing from H_2O through OH^- to O^{2-}. The high reactivity of the Mn(V) formally generated in the highest S-state is explained by the oxo wall concept (Section 10.8). The O-entity in Mn(V)≡O is a powerful electrophile that is able to attack a second O^{2-}, or, alternatively, it can take on oxyl radical (O^-) character. Ultimately, the overriding barrier to O_2 formation is probably formation of the weak peroxidic O–O bond, following which, the formation of O=O is energetically downhill (Section 17.3 and Resource section 3). The nature of the transition state cannot be established experimentally, but density functional theory, based upon increasingly refined structures determined using XFEL, has provided crucial insight into the mechanism. One possible transition state arising from S4 involves the attack of an oxyl radical on an oxido-O (Fig. 26.46).

26.11 Enzymes dealing with radical rearrangements and alkyl group transfer

KEY POINTS Nature uses cobalt in the form of complexes with a macrocyclic ligand, known as corrin. Complexes in which the fifth ligand is a benzimidazole that is covalently linked to the corrin ring are known as cobalamins. Cobalamin enzymes catalyse methyl transfer and dehalogenation. In coenzyme B_{12} the sixth ligand is deoxyadenosine, which is coordinated through a Co–C bond; enzymes containing coenzyme B_{12} catalyse radical-based rearrangements.

(a) Reactions of enzymes containing cobalamin (B_{12})

Cobalt macrocycle complexes are cofactors in enzymes that catalyse methyl transfer, radical-based rearrangements (e.g. isomerizations), and the reductive dehalogenation of organic molecules. The macrocycle is a corrin ring (**8**), which is similar to porphyrin (**7**), except there is less conjugation and it has a smaller ring (15-membered instead of 16-membered). Like Fe porphyrins, the Co corrins are enzyme cofactors and exert their activities when bound within a protein. The five-coordinate complex known as **cobalamin** or **coenzyme B_{12}** (**43**) includes a fifth nitrogen donor in one of the axial positions. This ligand is usually a dimethylbenzimidazole that is covalently linked to the corrin ring through a nucleotide so it cannot escape; but it is often found hanging free (in the 'base-off' mode), and the axial coordination site is occupied instead by an amino acid, particularly histidine.

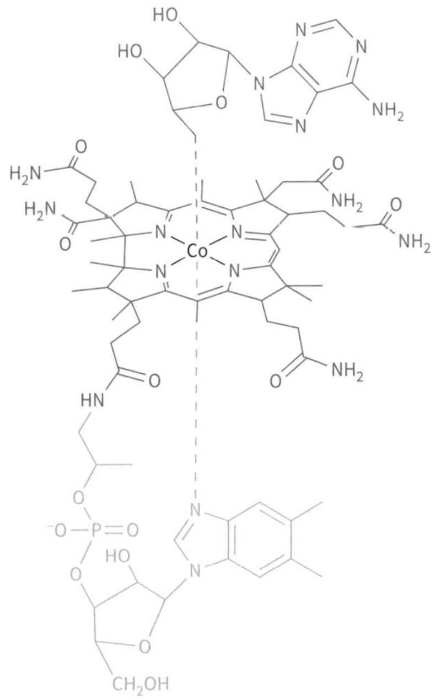

43

The sixth ligand is exchangeable and the complex is ingested in the form of species such as aquacobalamin, hydroxocobalamin, or cyanocobalamin, known collectively as vitamin B_{12}. In enzymes that catalyse radical rearrangements, the sixth ligand is normally 5′-deoxyadenosine, bonded to the Co atom through the –CH_2– group, making coenzyme B_{12} a rare example of a naturally occurring organometallic compound.[10] Cobalamin is essential for higher organisms (the human requirement is just a few milligrams per day) but is synthesized only by microorganisms.

The Co atom can exist in three oxidation states under physiological conditions, Co(III), Co(II), and Co(I), all of which are low spin. The electronic structure of Co is crucial to its biological activity. As expected, the Co(III) form (d^6) is an 18-electron, six-coordinate species (**44**). The Co(II) form

---

[10] Coenzyme B_{12} was one of the earliest biologically important molecules to be structurally characterized by X-ray diffraction methods (Dorothy Crowfoot Hodgkin, Nobel Prize for Chemistry, 1964). In 1973, Robert Woodward and Albert Eschenmoser published the total synthesis of B_{12}, the most complex natural product to be synthesized at that time, involving almost one hundred steps.

(45) is 17-electron, five-coordinate and has its unpaired electron in the d_{z^2} orbital. The Co(I) form (46) is a classic 16-electron, four-coordinate square-planar species, due to dissociation of both axial ligands.

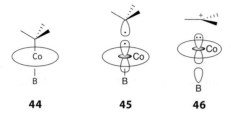

Methyl transfer reactions of cobalamins exploit the high nucleophilicity of square-planar Co(I). A particularly important example is methionine synthase, which is responsible for the biosynthesis of methionine. Methionine is produced by transferring a CH_3 group, derived from the methyl carrier methyl hydrofolate, to homocysteine. Not only is methionine an essential amino acid, but also accumulation of homocysteine (which occurs if activity is impaired) is associated with serious medical problems. The mechanism involves a cycle (analogous to oxidative addition/reductive elimination) in which Co(I) abstracts an electrophilic –CH_3 group (effectively CH_3^+) from a quaternary N atom on N^5-tetrahydrofolate to produce methylcobalamin, which then transfers –CH_3 to homocysteine (Fig. 26.47).

The structure of the Co(III) intermediate (47) shows the cobalamin is bound to the enzyme in 'base-off' mode, whereby the nucleotide and benzimidazole dangle free, leaving a histidine residue to coordinate at the 5th position. Methylcobalamin is the methyl-transferring cofactor for a wide variety of biosynthetic pathways, including the production of antibiotics. In anaerobic microbes, methylcobalamin is involved in the synthesis of acetyl coenzyme A, an essential metabolite, and in production of methane by methanogens.

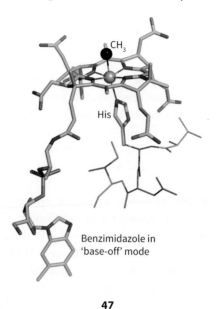

47

EXAMPLE 26.9 **Identifying the significance of the d-electron configuration of cobalamin**

Why is a Co-based macrocyclic complex (rather than an Fe complex like haem) well suited for radical-based rearrangements?

Answer To answer this question, we need to consider the electron configuration of Co(II) complexes in which there is a strong equatorial ligand field. Radical-based rearrangements depend on homolytic cleavage of the Co–C bond, which generates the adenosine radical and leaves an electron in the d_{z^2} orbital of the Co. This is the stable configuration for a low-spin Co(II) (d^7) complex, but for Fe, this configuration would require the oxidation state Fe(I), which is unstable without special ligands such as π-acceptors.

Self-test 26.9 Provide an explanation for why the toxicity of mercury is greatly increased by the action of enzymes containing cobalamin.

Radical-based rearrangements catalysed by coenzyme B_{12} include isomerizations (mutases) and dehydration or deamination (lyases). The generic reaction is

$$R_1\underset{R_2}{\overset{X\quad H}{-}}\underset{R_3}{\overset{}{-}}R_4 \longrightarrow R_1\underset{R_2}{\overset{H\quad X}{-}}\underset{R_3}{\overset{}{-}}R_4$$

Dehydration and deamination occur after two –OH or –OH and –NH_2 become placed on the same carbon atom, and hence are triggered by isomerization:

FIGURE 26.47 General principle of the mechanism of methionine synthase. Co(I) is a strong nucleophile and attacks the electrophilic quaternary-CH_3 group on the methyl carrier tetrahydrofolate. The resulting Co(III) methyl complex transfers CH_3^+ to homocysteine.

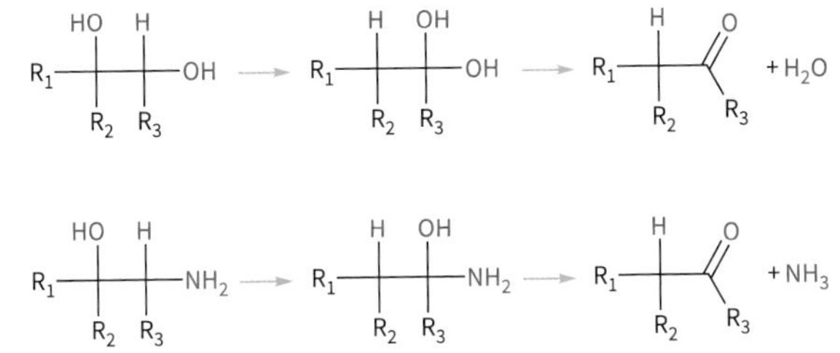

Radical-based rearrangements occur by a mechanism involving initiation of radical formation that begins with enzyme-induced weakening of the Co–C (adenosine) bond. In the free state, the Co–C bond dissociation energy is about 130 kJ mol^{-1}, but when bound in the enzyme the bond is substantially weakened, resulting in homolytic cleavage of the Co–CH$_2$R bond. This step generates five-coordinate low-spin Co(II) and a ·CH$_2$R radical, which gives rise to controlled radical chemistry in the enzyme active site pocket (Fig. 26.48). Important examples include methylmalonyl CoA mutase and diol dehydratases.

Cobalamin-containing enzymes are also important catalysts for the dehalogenation of organic molecules by microorganisms—a process of considerable environmental importance.

(b) Radical SAM enzymes: Fe-S clusters in radical-based rearrangements

The discovery in 1970 of an enzyme, lysine 2,3-aminomutase, which catalyses the radical-based rearrangement of an amino acid without any involvement of coenzyme B$_{12}$, initiated development of a whole new area of biochemistry. Lysine 2,3-aminomutase belongs to a large class of Fe-S enzymes now known as the radical S-adenosylmethionine (SAM) superfamily, that (unlike cobalamin) is produced by all forms of living organisms. Radical SAM enzymes include those responsible for synthesis of essential vitamins, such as vitamin H (biotin), vitamin B$_1$ (thiamine), haem, and molybdopterin (Section 26.12), as well as those that undertake routine repair of DNA. The enzymes play an important role in the attack of organisms by viruses.

Interconversion of L-lysine and L-β-lysine involves migration of the α-amino group to the β-carbon atom. L-β-Lysine is required by certain bacteria for antibiotic synthesis.

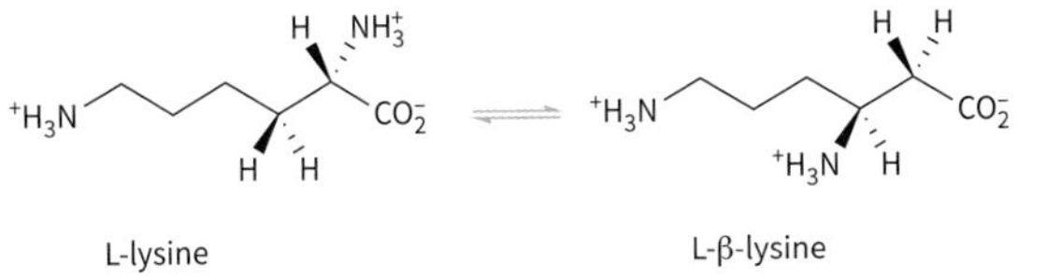

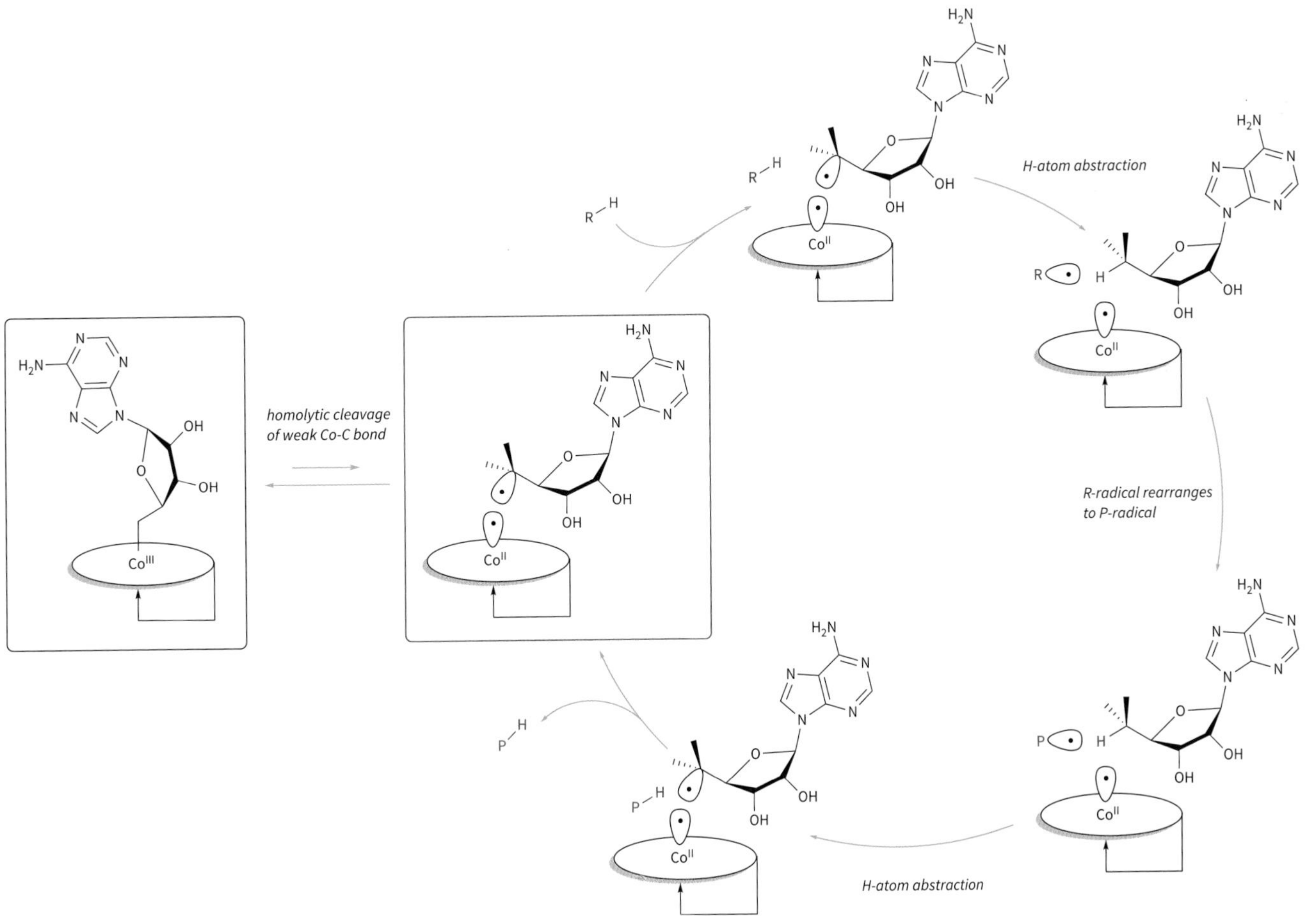

FIGURE 26.48 The principle of radical-based rearrangements by coenzyme B$_{12}$. Homolytic cleavage of the Co–C bond results in low-spin Co(II) (d$^1_{z^2}$) and a carbon radical that abstracts an H atom from the substrate RH. The substrate radical is retained in the active site and undergoes rearrangement before the hydrogen atom transfers back to form the product PH.

As with B$_{12}$ enzymes, this reaction involves a 5′-deoxyadenosyl radical, but in radical SAM enzymes the radical is generated by reductive cleavage of the SAM cation using a special [4Fe-4S] cluster. The reaction sequence begins with reduction of the [4Fe-4S] cluster: [4Fe-4S]$^+$ is a powerful reductant and SAM is reductively cleaved at the tertiary S$^+$ to produce methionine, which remains coordinated to the special Fe of the cluster, and the deoxyadenosyl radical which now abstracts a hydrogen atom from lysine and induces rearrangement. The X-ray structure of the precursor (Fig. 26.49) reveals how the different groups are arranged in space.

Based on various lines of spectroscopic evidence, likely mechanisms for forming a 5′-deoxyadenosyl radical (Fig. 26.50) involve the tertiary S attacking the unique Fe subsite (a) or a cluster S atom (b). In either case this attack is accompanied by rapid electron transfer and bond cleavage.

More than 70 different biochemical processes are now known to include a step catalysed by a radical SAM enzyme. The amino acid part sequence –C*xxx*C*xx*C– is the characteristic coordination motif for the [4Fe-4S] cluster in a radical SAM protein, and more than 100,000 different proteins have been identified by searching for the equivalent base sequence occurring in genes across all living organisms.

(c) How methane is produced by an enzyme containing a Ni-macrocycle

Methane, a potent greenhouse gas, is released in enormous amounts (in the region of 10^9 tonnes/year, and 90% of the total global production) by methanogenic archaea. The final stage of methanogenesis involves the oxidative coupling of a thiol (coenzyme B) and a methyl thioether (coenzyme M).

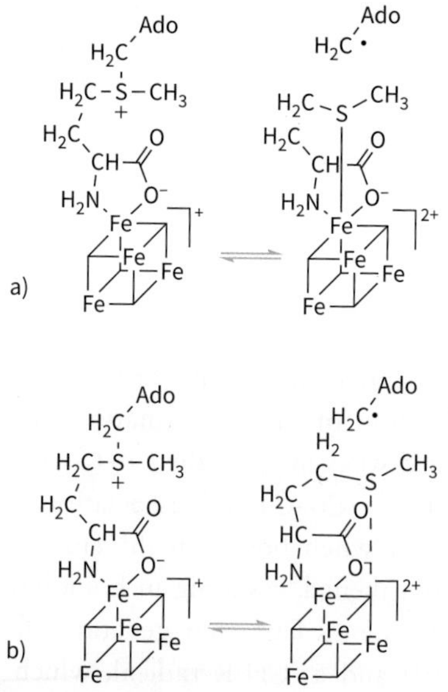

FIGURE 26.50 Two possible mechanisms by which the reactive 5′-deoxyadenosyl radical can be generated by the reduced [4Fe-4S]$^+$ cluster in radical SAM enzymes.

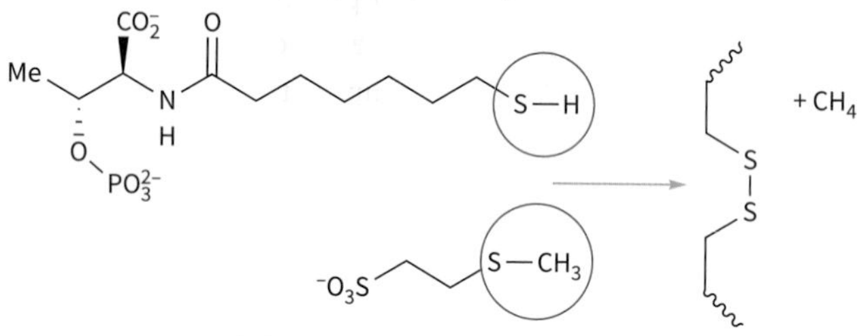

The reaction is catalysed by an enzyme known as methyl-coenzyme M reductase, which contains an elaborate macrocyclic Ni complex known as F430 (**48**) as the active site. Compared to porphyrin (**7**) and corrin (**8**) cofactor, F430 lacks extensive conjugation, and the coordination allows a more flexible geometry for the Ni. An amide carbonyl is bound at the fifth position.

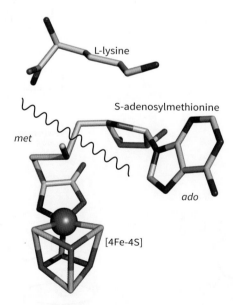

FIGURE 26.49 Structure of the active site of a radical S-adenosylmethionine (SAM) enzyme. In the precursor state shown, the SAM is coordinated to the [4Fe-4S] cluster by the main-chain N and O atoms of the methionine. The substrate, lysine, is held close by. The cleavage point, yielding *met* and *ado*, is represented by the wavy line.

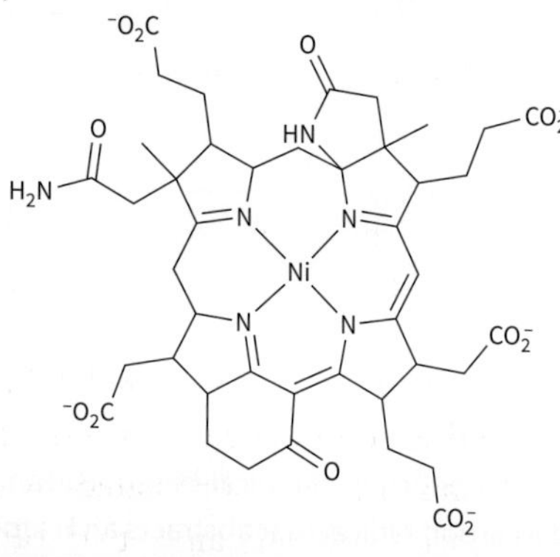

48 The Ni-macrocyclic complex F430

FIGURE 26.51 The methane evolving cycle of methyl coenzyme M reductase.

In a proposed mechanism shown in Fig. 26.51, the methane molecule is formed in a concerted process involving simultaneous transfer of the unpaired electron on Ni(I) to the methyl group of the thioether, coupled to H-atom abstraction from coenzyme B. The cycle is closed by disulfide bond formation and regeneration of Ni(I).

26.12 Oxygen atom transfer by molybdenum and tungsten enzymes

KEY POINTS Mo is used to catalyse O atom transfer in which the O atom is provided by a water molecule; related chemistry, but in more reducing environments, is displayed by W.

Molybdenum and tungsten are the only heavier elements known so far to have major functions in biology. Molybdenum is widespread across all life forms, and this section deals with its presence in enzymes other than nitrogenase (Section 26.14). By contrast, tungsten has so far only been found in prokaryotes. Molybdenum enzymes catalyse the oxidation and reduction of small molecules, particularly inorganic species. Reactions include oxidation of sulfite, arsenite, xanthine, aldehydes, and carbon monoxide, and reduction of nitrate and dimethyl sulfoxide (DMSO).

Both Mo and W are found in combination with an unusual class of cofactor known as molybdopterin (**10**) at which the metal is coordinated by a dithiolene group. In higher organisms, Mo is coordinated by one pterin dithiolene along with other ligands that often include cysteine. The structure is illustrated by the active site of sulfite oxidase, where we see also how the pterin ligand is twisted (**49**). In prokaryotes, Mo enzymes have two molybdopterin cofactors coordinated to the metal atom and, although the role of such elaborate ligands is not entirely clear, they are potentially redox active and may mediate long-range electron transfer. In humans and other mammals, an inability to synthesize molybdopterin has serious consequences. For example, sulfite oxidase deficiency is a rare, inherited defect (sulfite ions are very toxic) and is often fatal.

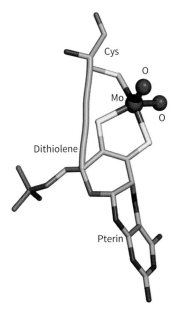

49 The active site of sulfite oxidase

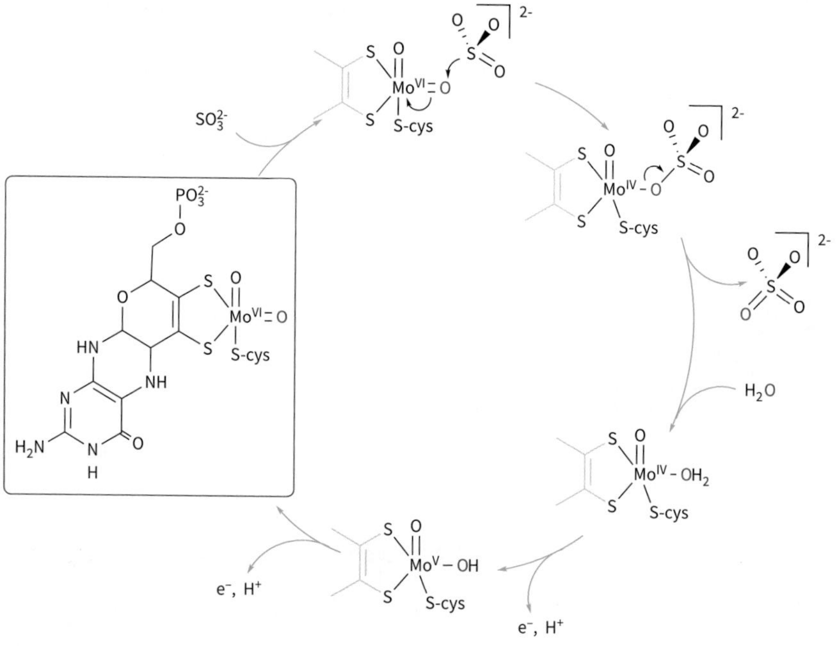

FIGURE 26.52 Catalytic cycle for the oxidation of sulfite to sulfate by sulfite oxidase, illustrating the direct O-atom transfer mechanism for Mo enzymes.

Coordination of the Mo is usually completed by ligands derived from H_2O, specifically H_2O itself, OH^-, and O^{2-}. Molybdenum is suited for its role because it provides a series of three stable oxidation states, Mo(IV), Mo(V), and Mo(VI), related by one-electron transfers that are coupled to proton transfer. Typically, Mo(IV) and Mo(VI) differ in the number of oxido ligands they contain, and Mo enzymes are commonly considered to couple one-electron transfer reactions with O-atom transfer.

A mechanism often considered for Mo enzymes is direct O atom transfer, which is illustrated in Fig. 26.52 for sulfite oxidase. The S atom of the sulfite ion attacks an electron-deficient O atom coordinated to Mo(VI), leading to Mo–O bond cleavage, formation of Mo(IV), and dissociation of SO_4^{2-}. Reoxidation back to Mo(VI), during which a transferable O atom is regained, occurs by two one-electron transfers from an Fe-porphyrin that is located on a mobile 'cytochrome' domain of the enzyme (Fig. 26.53). The intermediate state containing Mo(V) (d^1) is detectable by EPR spectroscopy.

This kind of oxygenation reaction can be distinguished from that of the Fe and Cu enzymes described previously, because with Mo enzymes the oxo group that is transferred is not derived from molecular O_2 but from water. The Mo(VI)=O unit can transfer an O atom, either directly (inner sphere) or indirectly to reducing (oxophilic) substrates, such as SO_3^{2-}, but cannot oxygenate C–H bonds. Fig. 26.54 shows the reaction enthalpies for O-atom transfer: we see that the highly oxidizing Fe species formed by reaction with O_2 are able to oxygenate all substrates, whereas Mo(VI) oxo species are limited to more reducing substrates and Mo(IV) is able to extract an O atom from nitrate.

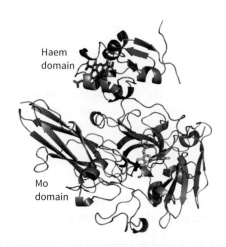

FIGURE 26.53 Structure of sulfite oxidase showing the Mo and haem domains. The region of polypeptide linking the two domains is highly mobile and is not resolved by crystallography.

As expected from its position below Mo in Group 6, the lower oxidation states of W are less stable than those of Mo, so W(IV) species are usually potent reducing agents. This potency is illustrated by the W-containing formate dehydrogenases present in certain primitive organisms, which catalyse the reduction of CO_2 to formate, the first stage in nonphotosynthetic carbon assimilation. This reaction does not involve O-atom insertion but rather the formation of a C–H bond. One mechanism that has been proposed, offering the equivalent of a hydride transfer, is

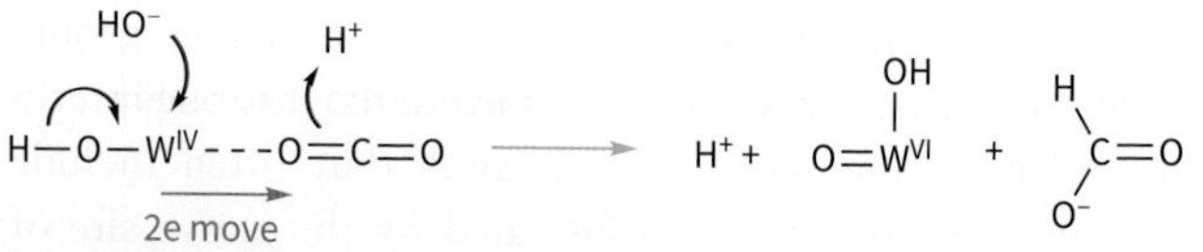

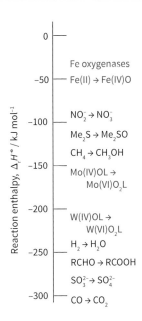

FIGURE 26.54 Scale showing relative enthalpies for oxygen-atom transfer. Fe(IV) oxy species are O-atom donors, whereas Mo(IV) and W(IV) are good O-atom acceptors.

Microbial oxidation of CO to CO_2 is important for removing more than 100 Mt of this toxic gas from the atmosphere every year. As shown in Box 26.3, this reaction is carried out by enzymes that contain either an Mo-pterin/Cu cofactor or an air-sensitive [Ni-4Fe-4S] cluster.

26.13 Hydrogenases, enzymes that activate H_2

KEY POINT The active sites of hydrogenases contain Fe or Ni, along with CO and CN ligands.

It has been estimated that 99% of all organisms utilize H_2. Even if these species are almost entirely microbes, the fact remains that almost all bacteria and archaea possess extremely active metalloenzymes, known as hydrogenases, that catalyse the interconversion of H_2 and H^+ (as water). The elusive molecule H_2 is produced by some organisms (it is a waste product) and used by others as a fuel, helping to explain why so little H_2 is in fact detected in the atmosphere (Box 11.1). Human breath contains measurable amounts of

BOX 26.3 How do organisms live with carbon monoxide?

Contrary to what is expected from its well-known toxicity, carbon monoxide is one of Nature's most essential small molecules. Even at atmospheric levels (0.05–0.35 ppm) CO is scavenged by a diverse range of microbial organisms for which it provides a source of carbon for growth and a 'fuel' for energy (CO is a stronger reducing agent than H_2). Two unusual enzymes, known as carbon monoxide dehydrogenases, catalyse the rapid oxidation of CO to CO_2. Aerobes use an enzyme containing an unusual Mo-pterin group (Fig. B26.3). A sulfido ligand on the Mo atom is shared with a Cu atom that is coordinated to one other ligand, a cysteine-S, completing the linear arrangement that is so common for Cu(I). A possible mechanism for CO_2 formation involves one of the oxogroups of Mo(VI) attacking the C-atom of a CO that is coordinated to Cu(I).

In contrast, certain anaerobes with the unique ability to live on CO as sole energy and carbon source use an enzyme that contains an unusual [Ni-4Fe-4S] cluster. X-ray diffraction studies on crystals of the Ni enzyme incubated in the presence of HCO_3^- at different potentials have revealed the structure of an intermediate showing CO_2 coordinated at the carbon atom by Ni and Fe (Fig. B26.4). This intermediate supports a mechanism in which, during the conversion of CO to CO_2, CO binds to the Ni site (which is Ni(II) square-planar) and the C-atom is attacked by a OH^- ion that was coordinated to the pendant Fe atom.

Carbon monoxide is carried in mammals by haemoglobin, and it is estimated that about 0.6% of the total haemoglobin of an average healthy human is in the carbonylated form. Free CO is

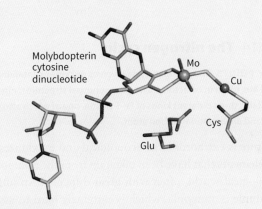

FIGURE B26.3 Active site of an Mo-Cu carbon monoxide dehydrogenase.

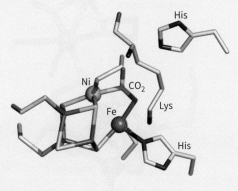

FIGURE B26.4 Active site of a [Ni-4Fe-4S] carbon monoxide dehydrogenase showing a catalytic intermediate containing an Ni-C bond.

produced by the action of haem oxygenase, a P450-type enzyme that catalyses the first step in haem degradation. In addition to releasing Fe and CO, the breakdown of haem produces biliverdin and bilirubin, the familiar green and yellow pigments responsible for the appearance of a bruise. Like NO, CO appears to be a cell signalling agent, and in low amounts it has important therapeutic effects, including suppression of hypertension (high blood pressure) and protection against tissue rejection following an organ transplant. There is therefore considerable interest in developing pharmaceutical agents, such as the water-soluble complex [Ru(CO)$_3$Cl(glycinate)], to release CO slowly and supplement the action of haem oxygenase (Section 32.7).

H$_2$ due to the action of bacteria in the gut. Hydrogenases are very active enzymes, with turnover frequencies (molecules of substrate transformed per second per molecule of enzyme) exceeding 10,000 s^{-1}. They are therefore attracting much attention for the insight they can provide regarding clean production of H$_2$ (Section 11.4) and oxidation of H$_2$ in low-temperature fuel cells—technology that otherwise depends greatly on Pt as a catalyst (Box 6.1).

There are three classes of hydrogenase, based on the structure of the active site: all contain Fe and some also contain Ni. The two best-characterized active sites are known as [NiFe] (**50**) and [FeFe] (**51**), and they all contain at least one CO ligand, with further ligation being provided by CN$^-$, cysteine (and sometimes selenocysteine) ligands.

The active site of [FeFe]-hydrogenases (the site is historically known as the 'H-cluster') contains an unusual bidentate ligand assigned as azadithiolate (adt, (SCH$_2$)$_2$NH) that forms a bridge between the Fe atoms, and a [4Fe-4S] cluster that is linked to one Fe by a bridging cysteine thiolate ligand. These fragile active sites, requiring several enzyme-catalysed steps for their biosynthesis, are buried deeply within the enzyme, thus necessitating special intramolecular pathways to convey H$_2$ and H$^+$, and a relay of Fe-S clusters for long-range electron transfer (analogous to that shown in Fig. 26.27).

A plausible mechanism of catalysis for the [FeFe]-hydrogenases involves the participation of lower oxidation states of Fe and Fe–H (hydrido) species, as shown in Fig. 26.55. Viewed in the direction of oxidation, H$_2$ might first bind to the 'distal' Fe (the one furthest from the [4Fe-4S] cluster) in an analogous manner to η^2-dihydrogen complexes (Section 24.7); this attack is expected because the distal Fe is formally low-spin Fe(II) with a vacant coordination site and 16 valence electrons. Heterolytic cleavage is then mediated by a 'frustrated Lewis pair' (Section 5.10), the H$^+$ being abstracted by the optimally positioned 'bridgehead'-N base on the azadithiolate ligand, leaving an Fe(II)-hydrido complex. Deprotonation of the bridgehead-N and its exit to solvent is followed by internal proton-coupled electron transfers that reprotonate the bridgehead-N and generate Fe(I) and reduced [4Fe-4S]$^+$ cluster. The initial state is restored by a second deprotonation and external electron transfers (usually to a ferredoxin). Similar principles may apply for [NiFe] enzymes.

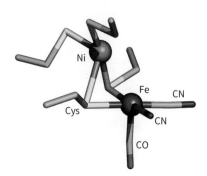

50 Active site of [NiFe]-hydrogenase

26.14 The nitrogen cycle

KEY POINTS The nitrogen cycle involves enzymes containing Fe, Cu, and Mo, often in cofactors having very unusual structures; nitrogenase contains three different kinds of Fe-S cluster, one of which also contains Mo and an interstitial carbon atom.

Nature is extraordinarily economical and maximizes its use of elements that have been taken up from the nonbiological, geological world, often with great difficulty. An important example is ammonia, which is extremely hard to assimilate from its unreactive gaseous source, N$_2$. The global biological nitrogen cycle involves organisms of all types and a diverse variety of metalloenzymes (Fig. 26.56).

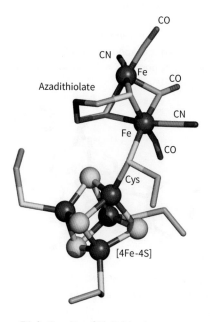

51 Active site of [FeFe]-hydrogenase

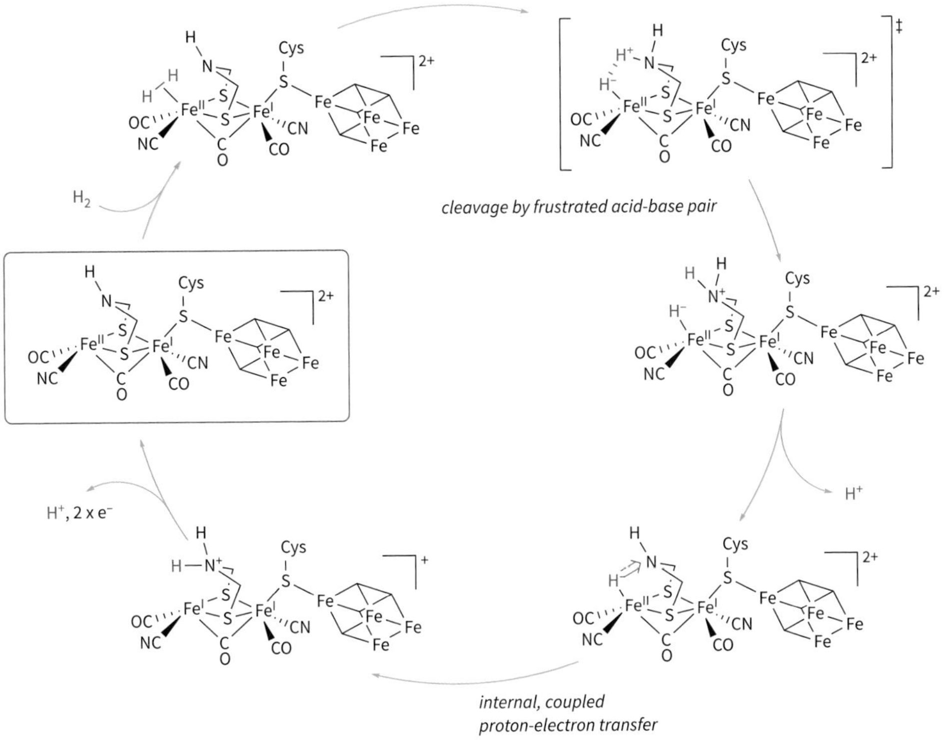

FIGURE 26.55 Proposed catalytic cycle for a [FeFe]-hydrogenase, showing a possible transition state in which the pendant base induces heterolytic cleavage of the H–H bond. The charge indicated as 2+ or + is localized on the [4Fe-4S] cluster component.

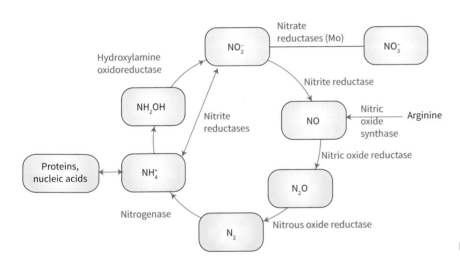

FIGURE 26.56 The biological nitrogen cycle.

The cycle can be divided into uptake of usable nitrogen (assimilation) from nitrate or N_2, and denitrification (dissimilation). The nitrogen cycle involves many different organisms and a variety of metal-containing enzymes. Many of the compounds are toxic or environmentally challenging. Ammonia is crucial for the biosynthesis of amino acids and NO_3^- is used as an oxidant. Molecules such as NO are produced in small amounts to serve as cell signalling agents that play a crucial role in physiology and health. Nitrous oxide, which is isoelectronic with CO_2, is a greenhouse gas: its release to the atmosphere depends on the balance between activities and abundances of NO reductase and N_2O reductase across the biological world.

The so-called 'nitrogen-fixing' bacteria found in soil and root nodules of certain plants contain an enzyme called nitrogenase which catalyses the reduction of N_2 to ammonia in a reaction that is coupled to the hydrolysis of 16 molecules of ATP and the production of H_2:

$$N_2 + 8\,H^+ + 8\,e^- + 16\,ATP \rightarrow 2\,NH_3 + H_2 + 16\,ADP + 16\,HPO_4^{2-}$$

'Fixed' nitrogen is essential for the synthesis of amino acids and nucleic acids, so it is central to agricultural production. Industrial production of ammonia by the Haber–Bosch process (Sections 16.6 and 31.9) involves reaction of N_2 and H_2 at high pressures and high temperatures; by contrast, nitrogenase produces NH_3 under normal conditions, and it is

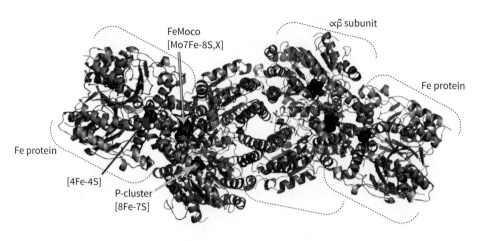

FIGURE 26.57 The structure of nitrogenase showing the Fe protein and the MoFe protein complexed with each other. Positions of the metal centres (black) are indicated. The MoFe protein is an $\alpha_2\beta_2$ tetramer of different subunits (red and blue) and it contains two 'P-clusters' and two MoFe cofactors. The Fe protein (green) has a [4Fe-4S] cluster and it is also the site of binding and hydrolysis of Mg-ATP.

small wonder that it has attracted so much attention. Indeed, the mechanism of activation of the N_2 molecule by nitrogenase has inspired coordination chemists for several decades. Regardless of whether it is biological or nonbiological, the process is very costly in terms of energy for a reaction that is not very unfavourable thermodynamically. However, as we saw in Chapter 16, N_2 is an unreactive molecule and energy is required to overcome the high activation barrier for its reduction.

Nitrogenase is a complex enzyme that consists of two types of protein (Fig. 26.57): the larger of the two is called the 'MoFe-protein', and the smaller is the 'Fe-protein'. The Fe-protein contains a single [4Fe-4S] cluster that is coordinated by two cysteine residues from two subunits. The role of the Fe-protein is to transfer electrons to the MoFe-protein in a reaction that is far from understood: in particular, it is unclear why each electron transfer is accompanied by the hydrolysis of two ATP molecules which are bound to the Fe-protein, although adding energy to supplement the limited electrochemical driving force may be important for activating such an inert molecule.

The MoFe protein is a $\alpha_2\beta_2$ 'tetramer', each $\alpha\beta$ pair of which contains two types of 'super cluster'. The [8Fe-7S] cluster (**52**) is known as the 'P-cluster' and is thought to be an electron-transfer centre, whereas the other cluster (**53**), formulated as [Mo7Fe-9S,C] and known as 'FeMoco' (FeMo cofactor), is the site at which N_2 is reduced to NH_3. The Mo is also coordinated by an imidazole-N from histidine and two O atoms from an exogenous molecule of *R*-homocitrate. A methide carbon atom lies at the centre of the cage formed by the six central Fe atoms and the belt of three central bridging sulfido ligands.

The framework of the catalytic cycle, Fig. 26.58, proposed in the 1970s and known as the L-T cycle after its originators David Lowe and Roger Thorneley, has stood

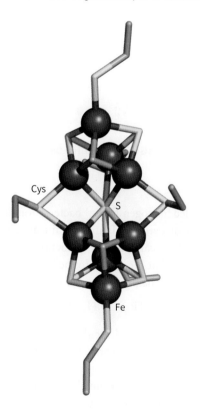

52 Nitrogenase 'P-cluster' [8Fe-7S]

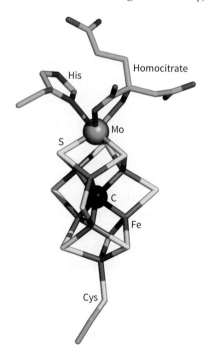

53 Nitrogenase 'FeMoco' [Mo7Fe-9S,C]

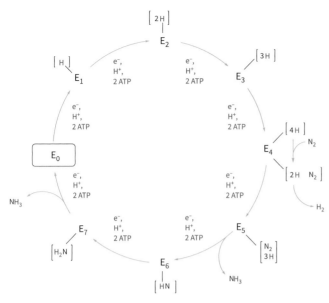

FIGURE 26.58 The Lowe–Thorneley catalytic cycle for biological nitrogen fixation by nitrogenase. Square brackets indicate the species likely to be accumulating on FeMoco.

V-nitrogenases catalyse the reduction of CO to ethene and ethane.

Nitrate reductase is another example of an Mo enzyme involved in the transfer of an O atom, in this case catalysing a reduction reaction (the formal potential for the NO_3^-/NO_2^- couple at pH = 7 is +0.4 V; thus NO_3^- is quite strongly oxidizing). The other enzymes in the nitrogen cycle contain either haem or Cu as their active sites. There are two distinct classes of nitrite reductase. One is a multi-haem enzyme that can reduce nitrite all the way to NH_3. The other class contains Cu and carries out one-electron transfer, producing NO: it is a trimer of identical subunits, each of which contains one 'blue' Cu centre (mediating long-range electron transfer to the electron donor, usually a small 'blue' Cu protein) and a Cu centre with more conventional tetragonal geometry that is thought to be the site of nitrite binding.

The nitrogen cycle is notable for using some of the most unusual redox centres yet encountered, as well as some of the strangest reactions. Another intriguing cofactor is a [4Cu-2S] cluster, named Cu_Z, that is found in N_2O reductase. It is quite a challenge to bind and activate N_2O, which is a poor ligand, so it is interesting that a crystal structure of the enzyme obtained under 15 bar N_2O reveals an N_2O molecule occupying a noncoordinating site just over 3 Å from Cu_Z (**54**). The four Cu atoms of the [4Cu-2S] cluster are coordinated to the protein by imidazole ligands and linked together by two inorganic sulfide ions. Long-range electron transfer in N_2O reductase is carried out by a Cu_A centre (**27**), the same centre as found in cytochrome *c* oxidase.

the test of time. Eight steps are involved, each requiring the addition of one electron and hydrolysis of two ATP—the fixed energy package resembling the constant supply of an electrocatalytic overpotential (Section 6.16).

However, the actual mechanism of N_2 reduction remains largely unresolved at the atomic level, the major questions being the role of reactive metal hydrido species that are generated as intermediates, and where and how N_2 is bound. In the former case it is known that the binding of N_2, which occurs after four $1e^-/2ATP$ packages have been delivered, coincides with the release of an H_2 molecule. Regarding how N_2 is bound, it is significant that nitrogenases are known in which the Mo atom is replaced by a V or Fe atom, arguing against a specific role for Mo. Instead, reactive hydrido (Fe-H) intermediates may play an important role. Carbon monoxide, long known to be an inhibitor of nitrogenase, has been found to be a substrate, being converted by V-nitrogenases to C_2 hydrocarbons in a manner that may resemble that of the Fischer–Tropsch reaction (Section 31.11). Crystal structures of reaction intermediates reveal that the first CO displaces one of the central belt sulfido ligands of FeVco (the sulfide is moved to a nearby 'holding position'). At higher pressure, a second CO is coordinated as a terminal ligand in an adjacent position. The side-by-side arrangement helps to explain why

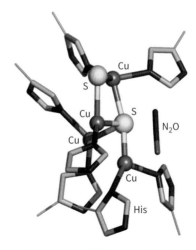

54 Cu_Z showing the position of the N_2O molecule that is reduced

Of particular importance for humans are two enzymes that manipulate NO. One of these, NO synthase, is a haem enzyme responsible for producing NO, by oxidation of L-arginine, upon receipt of a signal. Its activity is controlled by calmodulin (Section 26.5). The other enzyme is guanylyl cyclase which catalyses the formation of the important regulator cyclic guanosine monophosphate (cGMP) from guanosine triphosphate. Nitric oxide binds to the haem Fe of guanylyl cyclase, displacing a histidine ligand and activating the enzyme. A mischievous NO-binding protein is nitrophorin, which is found in some bloodsucking parasites, notably 'kissing bugs' (predacious bugs of the family *Reduviidae*). Nitrophorin binds NO tightly until it is injected into a victim, where a change in pH causes its release. The free NO causes dilation of the surrounding blood vessels, rendering the victim a more effective blood donor.

Metals in gene regulation

Reading the code encrypted in DNA requires proteins that can recognize specific sequences by exploiting the fidelity that stems from binding at multiple contact points. Proteins known as transcription factors, which locate the position at which to start 'reading life's blueprint' and thus 'transcribe' DNA-based information into RNA-based information, make extensive use of Zn to create structural folds that project 'feelers' into the DNA. The so-called 'Zn-finger domains' are exploited technologically in engineered enzymes known as Zn-finger nucleases, which are used in gene therapy. A number of DNA-binding metalloproteins are used to detect and quantify the presence of small molecules, particularly O_2, NO, and CO. These proteins therefore act as sensors, alerting an organism to an excess or deficit of particular species, and triggering some kind of remedial action. Special proteins are used to sense the levels of metals such as Cu and Zn that are otherwise strongly complexed in a cell. Cellular levels of Fe are sensed by an iron–sulfur protein that can bind to RNA.

26.15 Transcription factors and the role of Zn

KEY POINTS Zinc fingers are protein structural features produced by coordination of Zn to specific histidine and cysteine residues; a sequence of these fingers enables the protein to recognize and bind to precise sequences of DNA base pairs and plays a crucial role in transferring information from the gene.

As well as being an excellent catalyst, which we outlined in Section 26.9(a), Zn^{2+} has a unique structural role (**29**) as it is particularly suited for binding to proteins to retain them in a particular conformation: Zn^{2+} is high in the Irving–Williams series (Section 7.16) and thus forms stable complexes, particularly to S and N donors. Zinc is also redox inactive, which is an important factor because it is crucial to avoid oxidative damage to DNA. Other examples of structural zinc include insulin and alcohol dehydrogenase.

Transcription factors are proteins that recognize certain regions of DNA and control how the genetic code is interpreted as RNA. Many DNA-binding proteins contain repeating domains that are folded in place by the binding of Zn^{2+} and form characteristic folds known as 'zinc fingers' (Fig. 26.59). In a typical case, one side of the finger provides two cysteine-S donors and the other side provides two histidine-N donors and folds as an α-helix. Each 'fingertip' makes recognitory contacts with specific DNA bases.

As shown in Fig. 26.60, the zinc fingers wrap around sequences of DNA that they are able to recognize by acting collectively. The high fidelity of transcription factors is the result of a number of such contacts being made along the DNA chain at the beginning of the sequence that is transcribed.

The characteristic residue sequence for a Zn-finger motif is

$$-(\text{Tyr}, \text{Phe})-X-\text{Cys}-X_{2-4}-\text{Cys}-X_3-\text{Phe}-X_5-\text{Leu}-X_2-\text{His}-X_{3-5}-\text{His}-$$

where the amino acid X is variable. Aside from the 'classical' $(\text{Cys})_2(\text{His})_2$ zinc finger, others have been discovered

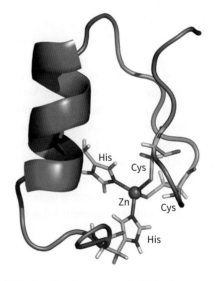

FIGURE 26.59 Zinc fingers are protein folds that expose and project a series of amino acids able to recognize and interact with specific base sequences on DNA. A typical finger is formed by the coordination of Zn(II) to two pairs of amino acid side chains located either side of the 'fingertip'.

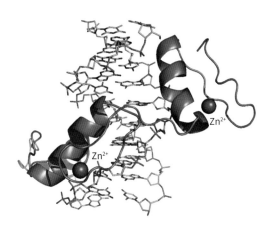

FIGURE 26.60 A pair of zinc fingers interacting with a section of DNA.

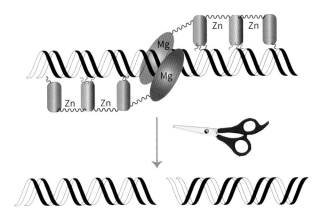

FIGURE 26.61 Principle of a zinc-finger nuclease used to cut DNA at specific sites. The Zn fingers locate the cleavage site and the Mg catalyst carries out hydrolysis.

that have (Cys)$_3$His or (Cys)$_4$ coordination, together with more elaborate examples having 'Zn-thiolate clusters', such as the so-called GAL4 transcription factor in which two Zn atoms are linked by bridging cysteine-S ligands (55). Various protein folds are produced, with faintly jocular names, such as 'zinc knuckles', joining an increasingly large family. Higher-order Zn-thiolate clusters are found in proteins known as metallothioneins, and some Zn-sensor proteins (see Section 26.17).

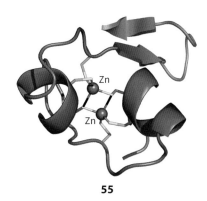

55

Zinc-finger nucleases (ZFNs) are synthetic enzymes formed by fusing an endonuclease (Section 26.9(b) and Fig. 26.32) with a Zn-finger protein. By appropriate selection of Zn finger and endonuclease, it is possible to excise particular genes. Together with other systems, particularly CRISPR-associated nucleases, ZFNs have important applications in biotechnology including gene therapy. The principle is shown in Fig. 26.61.

26.16 Iron proteins as sensors

KEY POINT Organisms use sophisticated regulatory systems based on Fe-containing proteins to adapt quickly to changes in cellular concentrations of Fe and O_2.

We have already seen how Fe S clusters are used in electron transfer and many types of catalysis. The coordination of an Fe-S cluster ties together different parts of a protein and thus controls its tertiary structure. The sensitivity of the cluster to O_2, electrochemical potential, or Fe and S concentrations makes it able to be an important sensory device. In the presence of O_2 or other potent oxidizing agents, [4Fe-4S] clusters have a tendency (controlled by the protein) to degrade, producing [3Fe-4S] and [2Fe-2S] species. The cluster may be removed completely under some conditions (Fig. 26.62). The principle behind an organism's exploitation of Fe-S clusters as sensors is that the presence or absence of a particular cluster (the structure of which is very sensitive

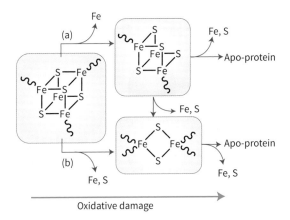

FIGURE 26.62 The degradation of Fe-S clusters forms the basis for a sensory system. The [4Fe-4S] cluster cannot support a state in which all Fe are Fe(III); thus severe oxidizing conditions, including exposure to O_2, causes their breakdown to [3Fe-4S] or [2Fe-2S] and eventually complete destruction. Degradation to [3Fe-4S] (a) requires only removal of an Fe subsite whereas degradation to [2Fe-2S] (b) may require rearrangement of the ligands (cysteine) and produce a significant protein conformational change. These processes link cluster status to availability of Fe as well as O_2 and other oxidants, and provide the basis for sensors and feedback control.

to Fe or oxygen) alters the conformation of the protein and determines its ability to bind to nucleic acids.

In higher organisms, the protein responsible for regulating Fe uptake (transferrin) and storage (ferritin) is an Fe-S protein known as the **iron regulatory protein** (IRP), which is closely related to aconitase (Section 26.9(c)) but is found in the cytoplasm rather than in mitochondria. It acts by binding to specific regions of messenger RNA (mRNA) that carry the genetic command (transcribed from DNA) to synthesize transferrin receptor or ferritin. A specific interaction region on the RNA is known as the **iron-responsive element** (IRE). The principle is outlined in Fig. 26.63. When Fe levels are high, a [4Fe-4S] cluster is present, and the protein does not bind to the IRE that controls translation of ferritin. In this case, binding would be a 'stop' command: instead, the cell now responds by synthesizing ferritin. Simultaneously, only the protein lacking a [4Fe-4S] cluster can bind to the transferrin receptor IRE, which stabilizes it so transferrin receptor is made. Overall, when Fe levels are high, the opposite actions occur: a [4Fe-4S] cluster is formed, ferritin synthesis is activated, and transferrin receptor synthesis is switched off (repressed).

The common gut bacterium *E. coli* derives energy either by aerobic respiration (using a terminal oxidase related to cytochrome *c* oxidase, Section 26.10(b)) or by anaerobic respiration with an oxidant such as fumaric acid, or nitrate, using the Mo enzyme nitrate reductase (Section 26.12). The problem that the organism faces is how to sense whether O_2 is present at a sufficiently low level to warrant inactivating the genes functioning in aerobic respiration and instead activate the genes producing enzymes necessary for the less efficient anaerobic respiration. This detection is achieved

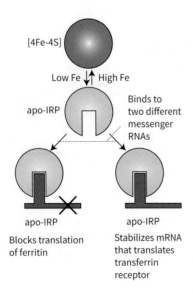

FIGURE 26.63 Interactions of iron-regulatory protein with iron-responsive elements on the RNAs responsible for synthesizing ferritin or transferrin receptor depend on whether an Fe-S cluster is present and form the basis for regulation of cellular Fe levels.

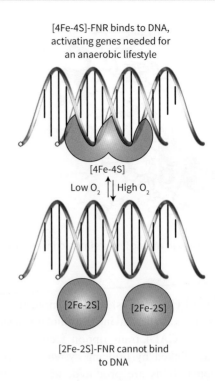

FIGURE 26.64 The principle of operation of the fumarate–nitrate regulatory system that controls aerobic vs anaerobic respiration in bacteria.

by a transcription factor called fumarate nitrate regulator (FNR) and the principle is outlined in Fig. 26.64. In the absence of O_2, FNR is a dimeric protein with one [4Fe-4S] cluster per subunit. In this form it binds to specific regions of DNA, repressing transcription of the aerobic enzymes and activating transcription of enzymes such as nitrate reductase. When O_2 is present, the [4Fe-4S] cluster is degraded to a [2Fe-2S] cluster and the dimer breaks up so that it cannot bind to DNA. The genes encoding aerobic respiratory enzymes are thus able to be transcribed, whereas those for anaerobic respiration are repressed.

In higher animals, the system that regulates the ability of cells to cope with O_2 shortage involves an Fe oxygenase. In Section 26.10(c) we mentioned prolyl hydroxylases, which catalyse the hydroxylation of specific proline residues in proteins, thus altering their properties. In higher animals, one such target protein is the transcription factor called hypoxia inducible factor (HIF, there are various forms, 1, 2, and so on). Comprised of two subunits, α and β, HIF mediates the expression of genes responsible for adapting cells to low-O_2 conditions (hypoxia). We should bear in mind here that the internal environment of cells and cell compartments is usually quite reducing, equivalent to an electrode potential below −0.2 V, and even though we regard O_2 as essential for higher organisms, its actual levels may be fairly low. When O_2 levels are above a safe threshold, prolyl hydroxylases catalyse oxygenation of conserved proline residues on the α subunit (Fig. 26.65), causing it to be recognized

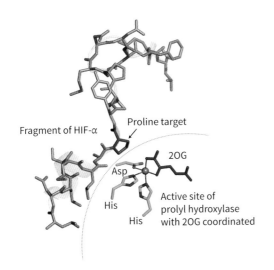

FIGURE 26.65 In response to sufficiently high O_2 levels in tissue, a prolyl oxygenase catalyses the hydroxylation of proline residues on the surface of a transcription factor (HIFα).

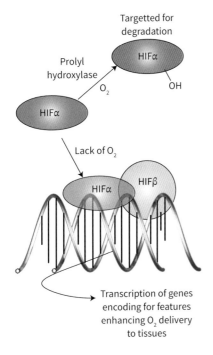

FIGURE 26.66 The principle of O_2 sensing by prolyl oxygenases.

by a protein that induces its degradation by proteases (Fig. 26.66), thus switching off the signal. Hence, genes such as those ultimately responsible for producing more red blood cells (which help an individual to cope better under lower O_2 levels) are not activated.

Although essential for higher organisms, O_2 requires stringent control of its four-electron reduction, and increasing amounts of research are being carried out to prevent and cure malfunctions of normal O_2 consumption. The term 'oxidative stress' is used to describe conditions in which the normal function of an organism is threatened by production of partially reduced O_2 intermediates, such as superoxides, peroxides, and hydroxyl radicals, known collectively as reactive oxygen species (ROS). Prolonged exposure to ROS may cause premature aging and certain cancers. To avoid or minimize oxidative stress, cells must first sense ROS and then produce agents to destroy them. Both sensory and attack agents are proteins with active groups such as metal ions (particularly Fe, Cu) and exposed redox-active cysteine thiols.

An underlying principle of haem-based sensors is that the small molecules being sensed are π acceptors that can bind strongly to the Fe and displace an indigenous ligand. This binding results in a change in conformation that alters catalytic activity or the ability of the protein to bind to DNA. Of the indigenous ligands, two classic examples are NO and CO; although we have long thought of these molecules as toxic to higher life forms, they are becoming well established as hormones (Section 32.7). Indeed, there is evidence that the sensing of trace levels of CO is important for controlling circadian rhythms in mammals.

The enzyme guanyl cyclase senses NO, a molecule that is now well established as a hormone that delivers messages between cells. Guanyl cyclase catalyses the conversion of guanidine monophosphate (GMP) to cyclic GMP (cGMP), which is important for activating many cellular processes. The catalytic activity of guanyl cyclase increases greatly (by a factor of 200) when NO binds to the haem, but (as is obviously important) by a factor of only 4 when CO is bound.

An excellent atomically-defined example of CO sensing is provided by a transcription factor known as CooA. This haem-containing protein is found in some bacteria that are able to grow on CO as their sole energy source under anaerobic conditions. Whether growth on CO takes place depends on the ambient CO level, as an organism will not waste its resources synthesizing the necessary enzymes when the essential substrate is not present.

CooA is a dimer, each subunit of which contains a single *b*-type cytochrome (the sensor) and a 'helix-turn-helix' protein fold that binds to DNA (Fig. 26.67). In the absence of CO, each Fe(II) is six-coordinate, and both its axial ligands are amino acids of the protein, a histidine imidazole, and, unusually, the main-chain $-NH_2$ group of a proline that is also the N-terminal residue of the other subunit. In this form, CooA cannot bind to the specific DNA sequence to transcribe the genes for synthesizing the CO-oxidizing enzymes necessary for existing on CO. When CO is present it binds to the Fe, displacing the distal proline residue and causing CooA to adopt a conformation that will bind to the DNA. The likelihood of NO binding in place of CO to cause a false transcriptional response is prevented because NO not only displaces proline but also results in dissociation of the proximal histidine; the NO complex is thus not recognized.

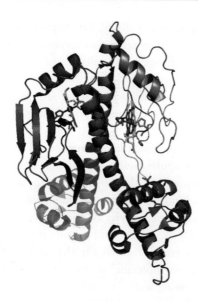

FIGURE 26.67 The structure of CooA, a bacterial CO sensor and transcription factor. The molecule, a dimer of two identical subunits, has two haem binding domains and two 'helix–turn–helix' domains that recognize a section of DNA. The protein ligands to the Fe atoms are a histidine from one subunit and an N-terminal proline from the other. The binding of CO and displacement of proline disrupts the assembly and allows CooA to bind to DNA.

26.17 Proteins that sense Cu and Zn levels

KEY POINT Cu and Zn are sensed by proteins with binding sites specially tailored to meet the specific coordination preferences of each metal atom.

The levels of Cu in cells are so strictly controlled that almost no uncomplexed Cu is present. An imbalance in Cu levels is associated with serious health problems such as Menkes disease (Cu deficiency) and Wilson's disease (Cu accumulation). Most of what we know about how Cu levels are sensed and converted to cell signals stems from studies on the *E. coli* system, which involves a transcription factor called CueR (Fig. 26.68). This protein binds Cu(I) with high selectivity, although it also binds Ag(I) and Au(I).

Metal coordination causes a conformational change that enables CueR to bind to DNA at a receptor site that controls transcription of an enzyme known as CopA, which is an ATP-driven Cu pump. CopA is located in the cytoplasmic membrane and exports Cu into the periplasm. In CueR, the Cu(I) is coordinated by two cysteine-S atoms arranged in a linear coordination geometry. Titrations using CN^- as a buffer show that Cu^+ is bound with a dissociation constant of approximately 10^{-21}. As may be understood by reference to (Section 7.3), this ligand environment leads to remarkably selective binding for d^{10} ions, and measurements with Ag and Au show that these ions are taken up with similar affinities.

Most of our current insight about Zn sensing, as for Cu, is provided by studies of bacterial systems. The major difference

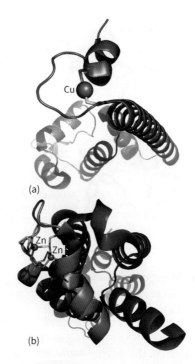

FIGURE 26.68 Comparison of (a) the Cu and (b) the Zn binding sites in the respective transcription factors CueR and ZntR. Note how Cu(I) is recognized by a linear binding site (to two cysteines), whereas Zn is recognized by an arrangement of Cys and His ligands that binds two Zn(II) atoms together with a bridging phosphate group.

with respect to Cu is that although Zn is also coordinated (mainly) by cysteine thiolates, the geometry is tetrahedral rather than linear. *E. coli* contains a Zn^{2+}-sensing transcription factor known as ZntR that is closely related to CueR. The factor ZntR contains two Zn binding domains, each of which coordinates a pair of Zn using cysteine and histidine ligands. The surrounding protein fold is shown in Fig. 26.68, for comparison with CueR. The extent to which these dynamic Zn binding sites can be identified with zinc fingers remains unclear.

EXAMPLE 26.10 Identifying links between redox chemistry and metal ion sensing

Suggest a way in which the binding of Cu or Zn to their respective sensor proteins might be linked to the level of cellular O_2.

Answer To address this problem we recall from Chapter 17 that strong S–S bonds arise from the combination of S atoms or radicals. A pair of cysteines that are coordinating a metal ion or are able to approach each other at close range can undergo oxidation by O_2 or other oxidants, resulting in the formation of a disulfide bond (cystine). This reaction prevents the cysteine-S atoms from acting as ligands, and provides a way in which even redox inactive metals such as Zn may be involved in sensing O_2.

Self-test 26.10 Why might Cu sensors be suited to bind Cu(I) rather than Cu(II)?

26.18 Biomineralization

KEY POINTS Calcium compounds are used in exoskeletons, bones, teeth, and other devices; some organisms use crystals of magnetite, Fe_3O_4, as a compass; plants produce silica-based protective devices.

Biominerals can be either infinite covalent networks or ionic solids. The former include the silicates, which occur extensively in the plant world. Leaves, even whole plants, are often covered with silica hairs or spines that offer protection against predatory herbivores. Ionic biominerals are mainly based on calcium salts, and exploit the high lattice energy and low solubilities of these compounds. Calcium carbonate (calcite or aragonite, Box 13.7) is the material present in seashells and eggshells. These minerals persist long after the organism has died; indeed, chalk is a biogenic mineral, a result of the process of **calciferation** of prehistoric organisms. Calcium phosphate (hydroxyapatite) is the mineral component of bones and teeth, which are particularly good examples of how organisms fabricate 'living' composite materials. Indeed, the different properties of bone found among species (such as stiffness) are produced by varying the amount of organic component, mostly the fibrous protein collagen, with which hydroxyapatite is associated. High hydroxyapatite/collagen ratios are found for large marine animals, whereas low ratios are found for animals requiring agility and elasticity.

Biomineralization is crystallization under biological control. There are some striking examples that have no counterparts in the laboratory. Perhaps the most familiar example is the exoskeleton of the sea urchin, which comprises large sponge-like plates containing continuous macropores 15 μm in diameter; each of these plates is a single crystal of Mg-rich calcite. Large single crystals support other organisms, such as diatoms and radiolarians, with their skeletons of silica cages (Fig. 26.69).

Biominerals produce some intriguing gadgetry. Fig. 26.70 shows crystals of calcite that are part of the gravity sensor device in the inner ear. These crystals are located on a membrane above sensory cells, and any acceleration or change

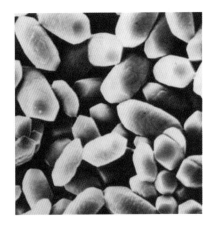

FIGURE 26.70 Our gravity sensor. Crystals of biologically formed calcite that are found in the inner ear. (Photograph supplied by Professor S. Mann, University of Bristol.)

of posture that causes them to move results in an electrical signal to the brain. The crystals are uniform in size and spindle-shaped, so that they move evenly without becoming hooked together.

The same property is seen for crystals of magnetite (Fe_3O_4) that are found in a variety of magnetotactic bacteria (Fig. 26.71). There are considerable variations in size and shape across different species, but within any one species, the crystals, formed in magnetosome vesicles, are uniform. Magnetotactic bacteria inhabit fluid sediment suspensions in marine and freshwater environments, and it is thought that their microcompasses allow them to always swim in a downwards direction to maintain their chemical environment during turbulent conditions.

There is great interest in how biomaterials are formed, not least because of the inspiration they provide for nanotechnology. The formation of biominerals involves the following hierarchy of control mechanisms:

1. Chemical control (solubility, supersaturation, nucleation on peptide or sugar templates).
2. Spatial control (confinement of crystal growth by boundaries such as cells, subcompartments, and even proteins in the case of ferritin, Section 26.6).

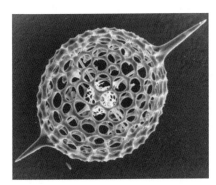

FIGURE 26.69 The porous silica structure of a radiolarian microskeleton showing the large radial spines. (Photograph supplied by Professor S. Mann, University of Bristol.)

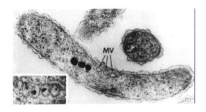

FIGURE 26.71 Magnetite crystals in magnetotactic bacteria. These are tiny compasses that guide these organisms to move vertically in river bed sludge. MV indicates empty vesicles. (Photograph supplied by Professor S. Mann, University of Bristol.)

3. Structural control (growth is favoured on a specific crystal face).
4. Morphological control (growth of the crystal is limited by boundaries imposed by organic material that grows with time).
5. Constructional control (interweaving inorganic and organic materials to form a higher-order structure, such as bone).

Bone is continually being dissolved and reformed; indeed, it functions not only as a structural support but also as the central Ca store. Thus, during pregnancy, bones tend to be raided for their Ca in a process called **demineralization**, which occurs in special cells called *osteoclasts*. Depleted or damaged bones are restored by mineralization, which occurs in cells called *osteoblasts*. These processes involve phosphatases (Section 26.9).

Perspectives

In this final section we stand back from the material in the chapter and review it from a variety of different perspectives, from the point of view of individual elements, and from the point of view of the contribution of bioinorganic chemistry to urgent social problems.

26.19 The contributions of individual elements

KEY POINT Elements are selected for their inherent useful properties and their availability.

In this section we summarize the major roles of each element and correlate what we have discussed, with emphasis on the element rather than the type of reaction that is involved.

Na and K The ions of these elements are characterized by weak binding to hard ligands, and their specificity is based on size and hydrophobicity that arises from a lower charge density. Both ions are highly mobile: compared to Na^+, which is normally 6-coordinate, K^+ is more easily dehydrated and likely to be found coordinated within a protein, often being 8-coordinate. Both Na^+ and K^+ are important agents in controlling cell structure through osmotic pressure, but whereas Na^+ is ejected from cells, K^+ is accumulated, contributing to a sizeable potential difference across the cell membrane. This differential is maintained by ion pumps; in particular, the Na^+/K^+-ATPase, also known as the Na pump (Fig. 26.9). The electrical energy is released by specific gated ion channels, of which the K^+ channel (Section 26.4) has been studied the most.

Mg The magnesium ion is the dominant 2+ ion in cytoplasm and the only one to occur above millimolar levels as a free, uncomplexed aqua species. The energy currency for enzyme catalysis, ATP, is always present as its Mg^{2+} complex. Magnesium has a passive but special role in the light-harvesting molecule chlorophyll, because it is a small 2+ cation that is able to adopt octahedral geometry and can stabilize a structure without promoting energy loss by fluorescence. The mobile Mg^{2+} ion is a weak acid catalyst and is the active metal ion in endonucleases—enzymes that 'cut' DNA at a specific location, and Rubisco, the highly abundant enzyme responsible for removing from the atmosphere some 100 Gt of CO_2 per year.

Ca Calcium ions are important only in eukaryotes. The bulk of biological Ca is used for structural support and devices such as teeth. The selection of Ca for this function is due to the insolubility of Ca carbonate and phosphate salts. Crucially, a relatively small amount of Ca forms the basis of a sophisticated intracellular (intercompartmental) signalling system. The principle of this process is that Ca is suited for rapid coordination to hard acid ligands, especially carboxylates from protein side chains, but has no preference for any particular coordination geometry. Binding constants are such that a particular system that is inactive under very low Ca^{2+} levels (maintained by Ca^{2+} pumps) will be activated when Ca^{2+} is allowed to enter the compartment.

Cr A definite biological role for chromium has long remained elusive, despite evidence dating back to the 1950s which implicated its importance in regulating metabolism and reducing high blood sugar levels (hyperglycemia). Despite uncertainty about its role, chromium is promoted commercially as a nutritional supplement, usually as the tris(pyridine-2-carboxylato) and tris(pyridine-3-carboxylato) complexes. The main obstacle hindering firm scientific investigations of chromium's role is its ultra-low requirement, which poses a huge challenge for obtaining sufficient material to examine. Of the three oxidation states having aqueous chemistry, only Cr(III) is likely to be encountered coordinated to a protein or other biological ligand. (Cr(II) is labile and strongly reducing, whereas Cr(VI) compounds like CrO_4^{2-} are powerfully oxidizing and carcinogenic.)

In 2023, several mitochondrial proteins directly involved in controlling glucose metabolism were identified to bind Cr(III) selectively, one being ATP synthase, the enzyme catalysing the recycling of ATP from ADP. This discovery, made using a sensitive and selective fluorescent probe to map the Cr(III) proteome in live cells, opens the way to a greater understanding of biological chromium (see Further Reading).

Mn Manganese has several oxidation states, most of which are very oxidizing. It is well suited as a redox catalyst for reactions involving positive reduction potentials. One reaction in particular, in which H_2O is used as the electron donor in photosynthesis, is responsible for producing the O_2 in the Earth's atmosphere. This reaction involves a special [4MnCa-5O] cluster. Compounds of Mn(II) are also used as weak acid–base catalysts in some enzymes. Spectroscopic detectability varies depending on oxidation state: EPR is useful for Mn(II) and for particular states of the [4MnCa-5O] cluster.

Fe Versatile Fe is essential to almost all organisms and was certainly a very early element in biology. Three oxidation states are important; namely, Fe(II), Fe(III), and Fe(IV). Active sites based on Fe catalyse a great variety of redox reactions ranging from electron transfer to oxygenation, as well as acid–base reactions that include reversible O_2 binding, dehydration/hydration, and ester hydrolysis. Iron-containing active sites feature ligands ranging from soft donors such as sulfide (as in Fe-S clusters) to hard donors such as carboxylate. The porphyrin macrocycle is particularly important as a ligand.

Iron(II) in various coordination environments is used to bind O_2, either reversibly or as a prerequisite for activation. Iron(III) is a good Lewis acid, whereas the Fe(IV)=O (ferryl) group may be considered as Nature's way of managing a reactive O atom for insertion into C–H bonds. Cells contain very little uncomplexed Fe(II) and extremely low levels of Fe(III). These ions are toxic, particularly in terms of their reaction with peroxides, which generates the hydroxyl radical. Primary uptake into organisms from minerals poses problems, because Fe is found predominantly as Fe(III), salts of which are insoluble at neutral pH (see Fig. 6.15).

Iron uptake, delivery, and storage are controlled by sophisticated transport systems, including a special storage protein known as ferritin. Iron porphyrins (as found in cytochromes) show intense UV–visible absorption bands, and most active sites with unpaired electrons give rise to characteristic EPR spectra.

Co Cobalt is generally processed only by microorganisms, and higher organisms must ingest it as 'vitamin B_{12}' in which Co is complexed by a macrocycle called corrin. Complexes, typically those in which the fifth ligand is a benzimidazole that is covalently linked to the corrin ring, are known as cobalamins. Cobalamins are cofactors in enzymes that catalyse alkyl transfer reactions, many radical-based rearrangements, and reductive dehalogenation. Alkyl transfer reactions exploit the high nucleophilicity of Co(I). In the special cofactor known as coenzyme B_{12}, the sixth ligand to Co(III) is a carbanion donor atom from deoxyadenosine. Radical-based rearrangements involve the ability of coenzyme B_{12} to undergo facile homolytic cleavage of the Co–C bond, producing stable low-spin Co(II) and a carbon radical that can abstract a hydrogen atom from substrates. Cobalamin-containing enzymes show strong UV–visible absorption bands; EPR spectra are observed for Co(II). Rare examples of Co proteins that do not contain the corrin macrocycle include an enzyme known as nitrile dehydratase.

Ni Nickel is important in prokaryotes, particularly as the active site of hydrogenases which are widely distributed throughout microbiology. As the active metal in some remarkable enzymes, Ni displays unexpectedly rich redox chemistry, utilizing the +3 and +1 oxidation states that are rare in conventional coordination chemistry. Coenzyme A synthase uses Ni to produce CO (in an attached form of CO dehydrogenase) and then react it with CH_3^- (provided by a cobalamin enzyme) to produce a C–C bond in the form of an acetyl ester. Methyl coenzyme M reductase, which contains a complex, flexible Ni-macrocyclic complex as the active site, is responsible for generating most of the methane on Earth. Nickel is also found in plants as the active site of urease, a hydrolytic enzyme. Urease was the first enzyme to be crystallized (in 1926), yet it was not until 1976 that it was discovered to contain Ni.

Cu Unlike Fe, copper probably became important only after O_2 had become established in the Earth's atmosphere, and it became available as soluble Cu(II) salts rather than insoluble sulfides (Cu_2S). The main role of Cu is in electron transfer reactions at the higher end of the potential scale and catalysis of redox reactions involving O_2 and N_2O. It is also used for reversible O_2 binding. Both Cu(II) and Cu(I) are strongly bound to biological ligands, particularly soft bases. Free Cu ions (complexed only by water) are highly toxic and virtually absent from cells, as they are sequestered by metallothionein (see below), better known as a Zn^{2+} storage protein.

Zn Zinc is an excellent Lewis acid, forming stable complexes with ligands such as N and S donors and catalysing reactions such as ester and peptide hydrolysis. The biological importance of Zn stems largely from its lack of interfering redox chemistry, although its common adoption of di-, tri-, and tetrathiolate ligation provides a link to the redox chemistry of cysteine/cystine interconversions. Zinc is used as a structure former in enzymes and proteins that bind to DNA at specific locations. A major problem has been the shortage of spectroscopic methods for studying this d^{10} ion. In some cases, Zn enzymes have been studied by EPR, after substituting the Zn by Co(II). In the cell, Zn^{2+} concentrations are buffered by cysteine-rich proteins known as metallothioneins.

Mo and W Molybdenum is an abundant element that is probably used by all organisms as a redox catalyst for the transfer of O atoms derived from H_2O. In these oxotransfer enzymes the Mo is always part of a larger pterin-containing cofactor in which it is coordinated by a special dithiolene ligand. Interconversion between Mo(IV) and Mo(VI) usually results in a change in the number of terminal oxido (O^{2-}) ligands, and recovery of the starting material occurs by single-electron transfer reactions with Mo(V) as an intermediate. Aside from oxo-transfer and related reactions, Mo has another intriguing role, that of nitrogen fixation, in which it is part of a special Fe-S 'super cluster'. Use of W is confined to prokaryotes, where it is also used as a redox catalyst but in reactions where a stronger reducing agent is required.

Si Silicon is often neglected among biological elements, yet its turnover in some organisms is comparable to that of carbon. Silica is an important material for the fabrication of the exoskeleton and of prickly defensive armour in plants.

Lanthanoids Recently discovered in an enzyme that oxidizes alcohols, and possibly more widespread in organisms inhabiting niche environments.

26.20 Future directions

KEY POINTS Biological metals and metalloproteins have important futures in medicine, energy production, green synthesis, and nanotechnology.

The Group 1 and 2 elements, so important to life, are often ignored by chemists, though this may be due in part to the difficulty of tracking the movements of labile and spectroscopically challenging metal ions. The pioneering studies of the structures and mechanism of ion channels mentioned in Section 26.4 are providing important new leads in neurophysiology, including the rational design of drugs that can block or modify their action in some way. New functions for Ca are continually emerging, and one intriguing aspect is its role in determining the left–right asymmetry of higher organisms, a prime example being the specific placements of heart and liver in the body cavity. The so-called **Notch signalling pathway** in embryonic cells depends on transient extracellular bursts of Ca^{2+} that depend on the activity of an H^+/K^+-ATPase.

There is a growing interest in the role of Zn and Zn transport proteins in control of cellular activity, and also of neural transmission. Indeed, the term **metalloneurochemistry** has been coined to describe the study of metal-ion function in the brain and nervous system at the molecular level. It is now becoming clear that transient, mobile Zn^{2+} ions, ejected from cells following stimuli or stress, are involved in cell-to-cell communication in various tissues, including the brain. An important challenge is to map out the distribution and flow of Zn in tissue such as brain, and advances are being made in the design of fluorescent ligands that will bind Zn selectively at cellular levels and report on its transport across different zones; for example, the synaptic junctions. Metal ions are involved in protein folding, and it is believed that Cu in particular may have an important role in fatal neurodegenerative disorders. These roles include controlling the behaviour of proteins called prions that are involved in transmittable diseases such as spongiform encephalopathy (Creutzfeldt–Jakob disease, the human form of 'mad cow' disease), as well as amyloid peptides that are implicated in Alzheimer's disease.

In many regions of the world, rice is the staple food, but it is not a good source of Fe. Thus, transgenic techniques are being used to improve Fe content. The object is to produce better plant siderophores and improve Fe storage (by enhanced expression of the ferritin gene).

Enzymes tend to show much higher catalytic rates and far higher selectivity than synthetic catalysts, leading naturally to greater efficiency. Directed evolution (Box 26.4) can generate *de novo* enzymes able to catalyse important reactions previously unencountered in biology. The principal disadvantages of using isolated enzymes as industrial catalysts are their lower thermal stability, limitations on solvent and pH conditions, and a large mass per active unit; however, these must be balanced against the greater scope for producing complex, high-value chemical products. There is much interest in achieving enzyme-like catalytic performance with small synthetic molecules—a concept that is popularly known as 'bioinspired catalysis'. The idea is to reproduce, using all the tools of synthetic chemistry, the properties of an enzyme trimmed down to its smallest fully functional component. It is not necessary to use the same building blocks, as chemists have at their disposal a much wider repertoire than naturally occurring peptides and macrocycles; instead, the aim is to mimic *function*. Examples of bioinspired catalysts have already been described: areas of particular interest for industrial production are the conversion of methane to methanol (Section 26.10), activation of N_2 to produce cheap fertilizers, and production of hydrogen.

How can chemists achieve the complexity that appears so important in a metalloenzyme? As an example, sulfite reductase and nitrite reductase each catalyse a six-electron reaction and share a similar active site: a modified haem in which the proximal ligand to the Fe, a cysteine-S, is shared with a [4Fe-4S] cluster (56), thereby electronically coupling the two redox centres. Evolution clearly favoured bolting on an additional electron reservoir to facilitate the multi-electron reaction. The transformations of sulfite to sulfide or nitrite to ammonia occur in the distal pocket, which is laced with precisely positioned positively-charged residues (arginine and lysine) to help direct the binding of the anionic reactant (a phosphate is found coordinated in the crystal structure) to the Fe.

BOX 26.4 Directed evolution – coaxing biology to produce enzymes that catalyse new reactions

Biology has produced the most active and selective catalysts known, but these are often fragile (recycling of building blocks being more important than long-term stability), and catalyse only reactions that are necessary for the living cell to flourish. The enzymes discovered and characterized by scientists in less than a century have evolved over billions of years by mutation and natural selection. But just as industries flourish by quickly finding ways to respond to changing market conditions, organisms can be forced to confront new chemical challenges in the 'wink of an eye' by adapting their existing machinery. The strategy, exploited to produce enzymes for new, unusual reactions, is known as Directed Evolution.[11]

The principle is to select or screen for an enzyme that is a good candidate for becoming a proficient catalyst (once a number of changes to its structure have been made) for a reaction that is new to biology, the obvious problem being that no-one can know what changes will work. A cycle of directed evolution is shown in Fig. B26.5. The gene encoding that enzyme is isolated (on a piece of circular DNA known as a plasmid) and subjected to random mutagenesis, using radiation or chemical mutagens, and amplification by PCR (the polymerase chain reaction) to produce a 'library' of mutant genes. The library of genes is then transformed into a suitable microorganism host. Colonies showing a positive result (the production of the desired new

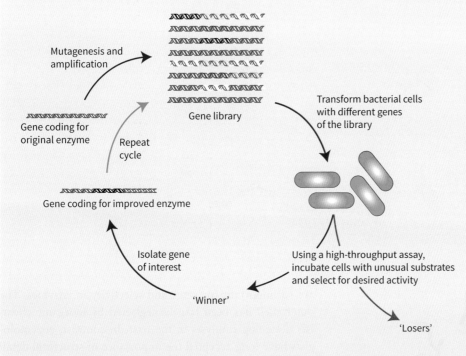

FIGURE B26.5 A cycle of directed evolution.

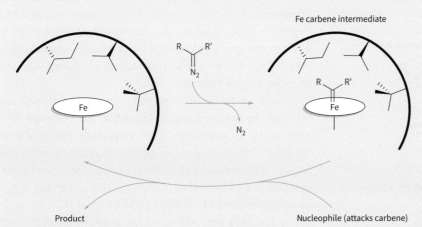

FIGURE B26.6 A carbene-transfer reaction catalysed by an evolved cytochrome P-450.

[11] For her pioneering research on generating new enzymes by Directed Evolution, Frances Arnold shared the 2018 Nobel Prize in Chemistry.

product) are identified, the 'improved' gene extracted, and the cycle restarted. Two or more cycles are carried out until optimization is established.

Exemplary new-to-Nature carbene transfer reactions have used a haem-containing protein as the starting point (Fig. B26.6). The Fe porphyrin reacts first with a diazo precursor to produce a reactive metal-carbene intermediate, not known to be used in biology but well-known in transition metal catalysis (Sections 24.15, 31.1). A nucleophile can attack the carbene, speed and selectivity each depending on properties introduced by the protein environment tuned by directed evolution. As expected, the most important modifications occur in the distal pocket, although beneficial changes may occur far from the active site.

Other reactions catalysed include the syntheses of pure isomers of compounds containing C–B or C–Si bonds, also unknown in biology, or highly strained ring systems.

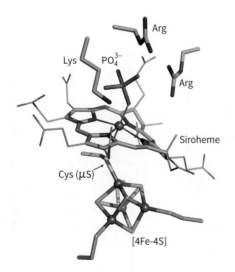

56 The active site of sulfite oxidase, which catalyses a 6-electron reaction

Although it has become commonplace to synthesize elaborate complexes containing two or more different metal atoms coordinated at specific primary coordination sites, a major challenge remains in creating the *outer* shells that are so important for rates and selectivity in enzymes. The outer shell has been largely neglected by inorganic chemists designing catalysts in the past. In contrast, enzymologists have long accepted the importance of structural detail extending far out from the actual site where catalytic bond transformations occur. Artificial metalloenzymes, in which organometallic catalysts are incorporated into proteins, exploit the powerful second/outer-sphere selectivities of metalloenzymes in a new approach to catalysing reactions of technological interest.

As use of fossil fuels becomes increasingly restricted, whether by environmental legislation or resource limitations, H_2 is emerging as an important energy carrier. Hydrogen can be used either directly or indirectly (after conversion into fuels such as alcohol) to store energy and power vehicles of all kinds. One of the scientific challenges is how to obtain efficient electrolytic production of H_2 from water, given that electricity will be widely available from a variety of sources. This process requires demanding conditions of temperature and overpotential (Section 11.4), or catalysts that are based on Pt and other precious metals. However, Nature has already shown us that rapid hydrogen cycling is possible under mild conditions by using just the common metals Fe and Ni. A related challenge is the synthesis of efficient electrocatalysts that can convert water to O_2 without requiring a large overpotential, not because there is a great need for O_2 itself, but because it is an essential by-product of electrolytic or photolytic H_2 production (Sections 6.16 and 11.4). Once again, we can turn to the biosphere for inspiration, because by elucidating the mechanism of the Mn catalyst, we might synthesize new catalysts that are both cheap and durable (Box 17.2).

We have seen the exquisite structures of materials that are produced by organisms. This understanding is now leading to new directions in nanotechnology. For example, sponge-like single crystals of calcite, having intricate morphological features, have been produced on polymer membranes formed by templating the skeletal plates of the sea urchin. Another development is the production of Pd nanoclusters by hydrogen-oxidizing bacteria, the action of hydrogenases making available controlled electron flow to effect the electroplating of Pd onto microscopic sites.

FURTHER READING

S. Mounicou, J. Szpunar, and R. Lobinski, Metallomics: The concept and methodology. *Chem. Soc. Rev.*, 2009, **38**, 1119.

J. J. R. Frausto da Silva and R. J. P. Williams, *The Biological Chemistry of the Elements*. Oxford University Press (2001). A classic text, still highly relevant, that looks at the broader picture of the relationship between elements and life.

E. Gouaux and R. MacKinnon, Principles of selective ion transport in channels and pumps. *Science*, 2005, **310**, 1461. An article linking detailed 3D structural data of giant proteins with physiological function and the chemistry of Group 1 and 2 metal ions and Cl⁻.

R. K. O. Sigel and A. M. Pyle, Alternative roles for metal ions in enzyme catalysis and the implications for ribozyme chemistry. *Chem. Rev.*, 2007, **107**, 97. A review describing the role of metal ions, particularly Mg, as active centres in catalysts based on RNA instead of proteins.

J. L. Lee, D. L. Ross, S. K. Barman, J. W. Ziller, and A. S. Borovik, C–H bond activation by nonheme metal complexes. *Inorganic Chemistry*, 2021, **60**, 13759. An excellent account of the chemical principles underlying catalytic oxygenation by enzymes and bioinspired analogues.

S. C. Wang and P. A. Frey, S-Adenosylmethionine as an oxidant: The radical SAM superfamily. *Trends Biochem. Sci.*, 2007, **32**, 101. An authoritative account of the discovery of radical SAM enzymes. It categorizes the different classes of enzymes and describes the mechanisms by which the [4Fe-4S] cluster cleaves SAM to initiate radical reactions.

S. Mann, *Biomineralization: Principles and Concepts in Bioinorganic Materials Chemistry*. Oxford University Press (2001).

N. Cox, D. A. Pantazis, and W. Lubitz, Current understanding of the mechanism of water oxidation in photosystem II and its relation to XFEL data. *Annu. Rev. Biochem.*, 2020, **89**, 19.1–19.26. A recent account of progress in determining the detailed mechanism by which plants produce O_2.

O. Einsle and D. C. Rees, Structural enzymology of nitrogenase enzymes. *Chemical Reviews*, 2020, **120**, 4969. An account of progress in understanding how the active sites of nitrogenases activate small molecules.

F. H. Arnold, Directed evolution: bringing new chemistry to life. *Angewandte Chemie Int. Ed.*, 2018, **57**, 4143. An essay about adapting enzymes to catalyse new reactions.

R. K. Thauer, Methyl (alkyl)-coenzyme M reductases: Nickel F-430-containing enzymes involved in anaerobic methane formation and in anaerobic oxidation of methane or of short chain alkanes. *Biochemistry*, 2019, **58**, 5198. An excellent review by one of the pioneers of methane enzymology.

A. Pomowski, W. G. Zumft, P. M. H. Kroneck, and O. Einsle, N_2O binding at a [4Cu:2S] copper–sulphur cluster in nitrous oxide reductase. *Nature*, 2011, **477**, 234. This paper gives an excellent example of the detective work being carried out using synchrotron-based crystallography.

E. M. Nolan and S. J. Lippard, Small-molecule fluorescent sensors for investigating zinc metalloneurochemistry. *Acc. Chem. Res.*, 2009, **42**, 193. An introduction to an exciting field, and an account of the development of an elegant technique for detecting zinc in tissue.

R. D. Britt, D. L. M. Suess, and T. A. Stich, An Mn(V)-oxo role in splitting water. *Proc. Natl. Acad. Sci. U.S.A.*, 2015, **112**, 5265. A commentary on the possible role of Mn(V) in catalysing the evolution of atmospheric O_2.

T. R. Ward, Artificial metalloenzymes based on the biotin-avidin technology: Enantiomeric catalysis and beyond. *Acc. Chem. Res.*, 2011, **44**, 47. Describes recent progress in making new catalysts by incorporating organometallic complexes inside proteins.

A. Klug, The discovery of zinc fingers and their applications in gene regulation and genome manipulation. *Annu. Rev. Biochem.*, 2010, **79**, 213. A highly readable account of the history and future of the role of zinc in molecular genetics.

P. H. Walton and G. J. Davies, On the catalytic mechanisms of lytic polysaccharide monooxygenases. *Curr. Opin. Chem. Biol.*, 2016, **31**, 195. Article describing progress in understanding and exploiting a class of Cu enzymes that can digest biomass.

C. Van Stappen, Y. Deng, Y. Liu, H. Heidari, J.-X. Wang, Y. Zhou, A. P. Ledray, and Y. Lu, Designing artificial metalloenzymes by tuning of the environment beyond the primary coordination sphere. *Chem. Rev.*, 2022, **122**, 11974–12045. A review of progress made in developing 'artificial metalloenzymes' by designing and synthesizing complexes having special environments outside their primary coordination sphere.

H. Wang et al., Mitochondrial ATP synthase as a direct molecular target of chromium (III) to ameliorate hyperglycaemia stress. *Nature Communications*, 2023, **14**, 1738. A report about a successful investigation to uncover the role of Cr(III) in living cells.

EXERCISES

26.1 With reference to details that have been discussed in Section 26.3 for the K^+ channel, predict the properties of Na^+, Ca^{2+}, and Cl^- binding sites that would be important for providing selectivity in their respective transmembrane ion transporters.

26.2 Calcium-binding proteins can be studied by using lanthanoid ions (Ln^{3+}). Compare and contrast the coordination preferences of the two types of metal ion, and suggest techniques in which lanthanoid ions would be useful.

26.3 In zinc enzymes, 'spectroscopically silent' Zn(II) can often be replaced by Co(II) with high retention of activity. Explain the principles by which this substitution can be exploited to obtain structural and mechanistic information.

26.4 Compare and contrast the acid–base catalytic activities of Zn(II), Fe(III), and Mg(II).

26.5 Propose physical methods that would allow you to determine whether a reactive intermediate isolated by rapid freeze quenching contains Fe(V).

26.6 Suggest likely intermediates formed during the six-electron reduction of an N_2 molecule to NH_3.

26.7 The structure of the P-cluster in nitrogenase differs significantly between oxidized and reduced states. Comment on this observation in the light of proposals that it participates in long-range electron transfer.

26.8 The MoFe cofactor can be extracted from nitrogenase by using dimethylformamide (DMF), although it is catalytically inactive in this state. Suggest experiments that could establish if the structure of the species in DMF solution is the same as present in the enzyme.

26.9 Microorganisms can synthesize the acetyl group (CH_3CO-) by direct combination of methyl groups with CO. Make some predictions about the metals that are involved.

26.10 Comment on the implications of discovering, by microwave detection, substantial levels of O_2 on a planet in another solar system (an exoplanet).

26.11 Reductive dehalogenases are cobalamin-containing enzymes which catalyse the replacement of a halogen atom by a hydrogen atom in organic molecules (X is typically Cl).

$$R-X + 2e^- + H^+ \rightleftharpoons R-H + X^-$$

With reference to the principles of halogen bonding (Section 5.8) and the properties of cobalamins explained in Section 26.11, propose an outline mechanism for dehalogenation of organic halides starting from Co(I), and suggest how a conserved tyrosine positioned close to the site of reaction may play a key role.

26.12 The demanding cleavage of a C–H bond is achieved in the laboratory by oxidative addition to a metal in a low oxidation state (Section 24.22), whereas biology achieves this by using a high-valent metal. Comment on this statement and its implications.

26.13 Crystal structures revealing the active sites of non-haem Fe oxygenases are often determined using Mn^{2+} in place of Fe^{2+}. Explain why this strategy is adopted.

26.14 A research group has just reported their discovery of an enzyme having chromium as the active site metal. Explain why such a role for chromium should be so surprising.

TUTORIAL PROBLEMS

26.1 In an article on the detection of Zn(II) that is released from neuronal tissue (brain, nerves) following trauma, E. Tomat and S. J. Lippard describe the development of special ligands that are highly selective for Zn and allow it to be imaged by a technique called confocal microscopy (*Curr. Opin. Chem. Biol.*, 2010, **14**, 225). Using your knowledge of the coordination chemistry of Zn, explain the principles underlying this research.

26.2 In justifying research into small molecule catalysts for producing NH_3 from N_2, it is sometimes stated that nitrogenase is an 'efficient' enzyme. How true is this statement? Comment critically on the argument that 'knowing the 3D structure of nitrogenase has not enlightened us as to its mechanism of action', and discuss how this view might be valid more generally for enzymes for which a structure is known.

26.3 'In Mo enzymes, the bond between a terminal oxo anion and Mo(VI) is usually written as a double bond, whereas it is more correctly assigned as a triple bond.' Discuss this statement. Suggest how a terminal oxido ligand influences the reactivity of other coordination sites on the Mo atom, and explain how a terminal sulfido ligand (as occurs in xanthine oxidase) would alter the properties of the active site.

26.4. In a surprising development, Klebensberger and colleagues describe their discovery of an alcohol dehydrogenase in which the active site contains a lanthanoid ion (M. Wehrmann et al., Functional role of lanthanides in enzymatic activity and transcriptional regulation of pyrroloquinoline quinone-dependent alcohol dehydrogenases in *Pseudomonas putida* KT2440. *mBio.*, 2017, **8**, e00570-17). Explain why biology may have a special place for such elements.

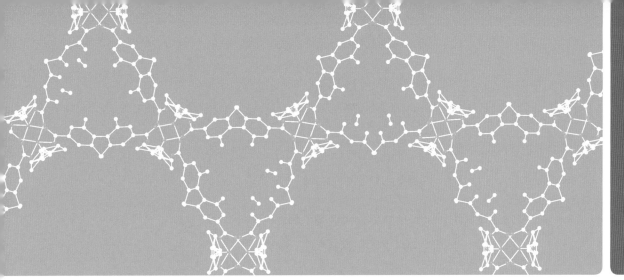

PART 4

Expanding our horizons: advances and applications

Inorganic chemistry is advancing rapidly at its frontiers, especially where research impinges on other disciplines such as the life sciences, condensed-matter physics, materials science, and environmental chemistry. These swiftly developing fields also represent many areas of inorganic chemistry where new types of compounds are utilized in catalysis, electronics, and pharmaceuticals. Building on the introductory, descriptive, and advanced material in Parts 1–3, the aim of this section of the book is to demonstrate the vigorous nature of contemporary inorganic chemistry and the importance of its applications.

In Chapter 27 we highlight the important role that inorganic chemistry is playing in promoting environmentally sustainable technologies in industry. Green chemistry pioneers are striving to exploit the many properties of different elements and their compounds in order to improve energy efficiency, reduce waste, and minimize use of hazardous materials.

In Chapters 28–30 we discuss three key areas of materials chemistry that are advancing rapidly and where numerous modern-day applications involve inorganic solids. In Chapter 28 we describe nanochemistry and how inorganic materials are manipulated at the nanoscale to enhance their electronic, magnetic, and optical properties. Chapter 29 describes the chemistry of porous solids and how controlling the internal structures of framework materials leads to numerous current and developing applications. The quest for sustainable and renewable energy has become a vitally important topic in the 21st century, and Chapter 30 describes how advances in inorganic chemistry are central to this rapidly developing field.

In Chapter 31 we explore how the most important industrial processes depend on catalysts. Several 'state-of-the art' examples are highlighted, each serving to emphasize the need for continued research and development to improve performance, maintain economic competitiveness, and decrease environmental impact.

Finally, Chapter 32 describes how medical science is exploiting inorganic chemistry to detect and treat illness and injury. In pharmacology, advances in drugs and diagnostics are often focused on the 'stranger' elements, such as platinum, bismuth, and synthetic technetium, while replacement surgery depends on inorganic materials that combine strength and biocompatibility.

Green chemistry

Twelve principles

27.1 **Prevention**

27.2 **Atom economy**

27.3 **Less hazardous chemical species**

27.4 **Designing safer chemicals**

27.5 **Safer solvents and auxiliaries**

27.6 **Design for energy efficiency**

27.7 **Use of renewable feedstocks**

27.8 **Minimize derivatives**

27.9 **Catalysis**

27.10 **Design for degradation**

27.11 **Real-time analysis for pollution prevention**

27.12 **Inherently safer chemistry for accident prevention**

Further reading

Exercises

Tutorial problems

Sustainable development has become the cultural, scientific, and technical imperative of the twenty-first century. The mantra of sustainability is meeting the needs of the present without compromising the ability of future generations to meet their own needs. For chemistry, sustainable development is conceptualized as the practice of green chemistry, many aspects of which have been described throughout the earlier chapters of this book, including basic catalysis and the hydrogen energy economy. In this chapter we formalize our treatment of the concepts of green inorganic chemistry by using the 12 guiding principles first proposed by Anastas and Warner in 1998.

The chemical industry has embraced green chemistry in recent decades as it strives to move towards being more environmentally sustainable and economically viable. The principles of green chemistry are also just as applicable to laboratory-scale chemistry, although sometimes less of a consideration in that context. These principles cover issues from efficiency of reactions, waste production, seeking alternative and safer solvents, to reducing hazardous by-products and products and inherently safer processes. We approach the concept of green chemistry specifically from the point of view of inorganic chemistry, and see the important role many inorganic compounds have in facilitating greener processes.

For most of the decades in the twentieth century the chemical industry was seen by many as dirty and polluting. Although the industry has made enormous strides in cleaning up its processes and minimizing its impact on the environment and on natural resources, there is still much to do. The green chemistry movement takes these changes to the next level and designs sustainable practices into chemical processes. The motivations for green chemistry are not entirely altruistic: many innovations arise in response to legislative changes and economic pressures. These external factors help to make greener chemical processes more economically viable than the original processes, with reduced energy costs, less waste treatment, lower energy usage, and lower capital costs.

Twelve principles

Green chemistry is conveniently described by the 12 principles that were proposed by Anastas and Warner: here we will review each in turn and illustrate them through examples from inorganic chemistry.

27.1 Prevention

KEY POINT The first principle of green chemistry is that it is better to prevent the production of waste than to have to treat waste.

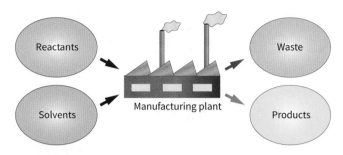

FIGURE 27.1 Waste production in the chemical industry.

Many chemical processes are inherently very wasteful, sometimes producing a much greater mass of waste than of valuable product (Fig. 27.1). The first principle of green chemistry states that avoiding producing waste products is beneficial compared to the effort and cost involved in cleaning up or separating out any waste produced. The extent of this chemical wastage is measured using the E-factor given by the expression below

$$\text{E-factor} = \text{mass waste/mass product} \tag{27.1}$$

The E-factor is easy to measure, as any manufacturing site knows how much material enters and how much leaves as waste and products. E-factors vary enormously, from the oil refining industry, with E-factors around 0.1, to the pharmaceutical industry with values in the range 10–1000 (Table 27.1). The economics of these industries explains these figures to a large extent. Profit margins are small in oil-refining, so much effort has been put into reducing waste and finding alternative uses for by-products. In contrast, profit margins are high in the pharmaceutical industries where the focus is more on producing a high-quality final product than on minimizing waste. E-factors do not take into account recycled solvents or the reuse of catalysts.

An example of waste prevention in inorganic chemistry is the manufacture of titanium dioxide pigment. Titanium dioxide is the most widely used white pigment globally (Sections 20.4 and 30.1), and is manufactured on a massive scale by both the sulfate and the chloride processes. The older sulfate process produces TiO_2 from the cheap ilmenite ore ($FeTiO_3$) or titanium slag as a starting material. One of the key steps in the manufacturing process is the digestion of the ore by concentrated sulfuric acid followed by hydrolysis to produce a hydrated TiO_2:

$$FeTiO_3(s) + 2\,H_2SO_4(l) \rightarrow TiOSO_4(s) + FeSO_4(s) + 2\,H_2O(l)$$
$$TiOSO_4(s) + (n+1)H_2O(l) \rightarrow TiO_2 \cdot nH_2O(s) + H_2SO_4(aq)$$

This process can result in 3–4 tonnes of solid waste (largely $FeSO_4$) and 8 tonnes of dilute sulfuric acid per tonne of pure $TiO_2(s)$ produced. The solid waste can be sold on for use in water treatment and building materials, and the sulfuric acid can be reclaimed and recycled, but the additional steps involved in isolating them are obviously expensive, energy-intensive, and inconvenient. The sulfate process is increasingly being replaced by the chloride process, which produces pure TiO_2 from the more expensive rutile ore (brown impure TiO_2). In this process, the ore is first converted to $TiCl_4$ (gaseous, at the temperature of the reaction) by heating in chlorine gas in the presence of carbon:

$$TiO_2(s) + 2\,Cl_2(g) + C(s) \rightarrow TiCl_4(g) + CO_2(g)$$

This first stage is similar to that of the Kroll process for obtaining pure titanium metal (Section 6.17). In the chloride process, the $TiCl_4$ is purified by distillation and then reacted with oxygen to form the pure white pigment and regenerate Cl_2 gas which is then recycled:

$$TiCl_4(g) + O_2(g) \rightarrow TiO_2(s) + 2\,Cl_2(g)$$

The chloride process was developed in the 1950s and is viewed as having less environmental impact than the sulfate process, as it addresses the huge amounts of waste produced during the older process. However, although the solid waste

TABLE 27.1 Typical E-factors for various industries

Industry	Annual production/tonnes	E-factor	Annual waste produced/tonnes
Oil refining	10^6–10^8	<0.1	<10^5 to 10^7
Bulk chemicals	10^4–10^6	<1–5	10^4 to 5×10^6
Fine chemicals	10^2–10^4	5 to >50	500 to >500,000
Pharmaceutical	10–10^3	25 to >100	250 to >100,000

is eliminated, the chloride process produces approximately 1 tonne of CO_2 for every 2 tonnes of TiO_2. The process was introduced at a time when CO_2 was not considered a pollutant or a greenhouse gas, as it is now. In 2000 the Altair process for TiO_2 manufacture was patented. This process is based on the digestion of the cheap ilmenite ore in concentrated hydrochloric acid followed by solvent extraction and spray hydrolysis to produce a TiO_2 hydrate, which is then calcined to produce the pure white TiO_2 pigment. Waste is produced in the form of iron chlorides and oxides, but all of it can be used in other processes. The process claims to be the greenest of the three processes, though it is energy intensive. However, the process is not currently in widespread use as it can only be operated under licence and is predominantly used for the manufacture of high-value nanoparticulate TiO_2, which is used in high-factor sunscreens and as photocatalysts (Section 30.3(b)).

EXAMPLE 27.1 Calculating E-factors

From the data given in the text, calculate the E-factor for the manufacture of TiO_2 by the sulfate process.

Answer We will assume that 3.5 tonnes of waste $FeSO_4$ and 8 tonnes of dilute sulfuric acid are produced per tonne of TiO_2. Using eqn. 27.1 we have E = 11.5/1 = 11.5. Although the process produces a lot of waste, we can see that it is nowhere near as wasteful as many other industrial processes.

Self-test 27.1 1,3,5-Benzenetriol is an intermediate in the manufacture of pharmaceuticals and explosives. The production of 1 kg of 1,3,5-benzenetriol produces 40 kg of solid waste containing $Cr_2(SO_4)_3$, NH_4Cl, $FeCl_2$, and $KHSO_4$. Does the E-factor for this process suggest that this level of waste might be better tolerated by the bulk chemicals, fine chemicals, or the pharmaceutical industries?

27.2 Atom economy

KEY POINT The second principle of green chemistry states that synthetic methods should be designed to maximize the incorporation of all materials used in the process into the final product.

Atom economy is one of the most important concepts in green chemistry and gives an indication of the theoretical efficiency of a chemical reaction in terms of conversion of starting materials into products. In an ideal reaction, all atoms in the starting materials are present in the product. Atom economy is given by the equation

$$\text{Atom economy}(\%) = \frac{\text{Molecular mass of desired product}}{\text{Molecular mass of all reactants}} \times 100 \quad (27.2)$$

Efficient reactions have high atom economies, produce little waste, and are viewed as environmentally sustainable. Atom economy should not be confused with yield, as percentage yield provides no indication of the quantity of waste or by-products. Atom economy can be poor, even when the yield is high. For example, consider the production of steel in a blast furnace (Section 6.17):

$$Fe_2O_3(s) + 3CO(g) \rightarrow 2Fe(l) + 3CO_2(g)$$

If we assume that we begin with 1 mole of Fe_2O_3 (160 g) and produce 2 moles of iron (112 g), then the yield would be 100%. However, the atom economy would only be $112/(160 + 84) \times 100 = 45.9\%$, which is quite clearly not as efficient as it could be. In fact, reactions of the form A + B → C + D, where C or D is the desired product, usually have poor atom economies. Reactions of the type A + B → C are far more desirable and efficient. For example, the synthesis of ammonia

$$N_2(g) + 3H_2(g) \rightarrow 2NH_3(g)$$

has an atom economy of $34/(28 + 6) \times 100$ or 100%. For a multi-step reaction such as

$$A + B \longrightarrow C + D \xrightarrow{E} F + G$$

where F is the desired product, the atom economy is given by (molecular mass F)/Σ(molecular mass A, B, and E).

Be aware that atom economy calculations do not take into account solvents, the stoichiometry of reagents, reagents that are not incorporated into the product, or catalysts. However, atom economy calculations do provide a very simple indication of the relative green credentials of a process. A high atom economy is always preferable to a low atom economy.

Catalysts play a crucial role in increasing the atom economies of reactions. Catalytic routes typically involve fewer reaction steps, greater selectivity, and regeneration of the catalyst. For example, the now obsolete noncatalytic route to the industrial production of ethylene oxide, or oxirane, $(CH_2CH_2)O$, of which in excess of 15 million tonnes is produced worldwide each year, was a multistep process with an atom economy of 26% and produced 3.5 kg of $CaCl_2$ waste for every 1 kg of ethylene oxide produced.

$$H_2O + Cl_2 \rightarrow HOCl + HCl$$
$$CH_2=CH_2 + HOCl \rightarrow CH_2OHCH_2Cl$$
$$CH_2=CH_2 + Cl_2 \rightarrow CH_2ClCH_2Cl$$
$$2CH_2OHCH_2Cl + Ca(OH)_2 \rightarrow 2(CH_2CH_2)O + CaCl_2 + 2H_2O$$

The catalytic route is now used exclusively. It uses a heterogeneous silver catalyst (Section 31.8) in a one-step addition reaction of oxygen to ethene with an atom economy of 100%:

$$2CH_2=CH_2(g) + O_2(g) \rightarrow 2(CH_2CH_2)O(g)$$

EXAMPLE 27.2 Calculating atom economies

When 2 g of powdered magnesium were reacted with dilute nitric acid the isolated product was 12.10 g of magnesium nitrate. Calculate the yield and the atom economy for this reaction.

Answer The equation for the reaction is $Mg + 2\,HNO_3 \rightarrow Mg(NO_3)_2 + H_2$. 2 g of Mg is equivalent to 0.082 moles. So the theoretical yield of $Mg(NO_3)_2$ is 0.082 moles or 12.16 g. Therefore, the yield of $Mg(NO_3)_2 = \frac{12.10}{12.16} \times 100 = 99.5$ per cent. The atom economy is given by $\frac{148.3}{24.3+2(63)} \times 100 = 98.7$ per cent.

The atom economy is high because the only atoms not to be incorporated into the product are those of low-molecular-mass hydrogen.

Self-test 27.2 Silicon for the electronics industry is extracted from silica, SiO_2, by high-temperature reduction with carbon in an arc furnace: $SiO_2(s) + 2\,C(s) \rightarrow Si(s) + 2\,CO(g)$. Calculate the atom economy for this reaction.

27.3 Less hazardous chemical species

KEY POINT The third principle of green chemistry states that wherever practicable, syntheses should be designed to use and generate substances that cause little damage to human health or the environment.

The eradication of molecules that are hazardous to human health or damaging to the environment is one of the guiding principles of green chemistry. Chemists sometimes need to synthesize toxic or biologically active molecules, including some pharmaceuticals or pesticides, or use toxic or hazardous reactants in the synthesis of a functional compound or material. However, the chemical industry has made great strides in modifying synthetic routes so that less hazardous substances are used or produced. For example, polycarbonates are a family of thermoplastic polymers that are strong and transparent and are used for a variety of applications, including in the building industry. Polycarbonates can be synthesized via carbonyl dichloride (phosgene), $COCl_2$, or dimethyl carbonate, $(CH_3O)_2CO$, precursors. Carbonyl dichloride (Section 15.8) is manufactured from the hazardous carbon monoxide and chlorine gas and is itself toxic (it was used as a chemical weapon in the First World War). The less hazardous dimethyl carbonate is manufactured via a liquid phase reaction of methanol with carbon monoxide and oxygen, and so is a much more benign route. Another route to polycarbonates is based on the copolymerization of carbon dioxide with an epoxide:

This reaction, which was described in Section 15.9, represents the replacement of dimethyl carbonate synthesized from methanol from syngas by a renewable resource, i.e. CO_2. Although the process offers another use for CO_2, the scale of manufacture is unlikely to have an impact on atmospheric CO_2 levels. This catalysed polymerization reaction is carried out at some carbon-capture and storage (CCS) facilities at power stations that burn fossil fuels (Box 15.6). The homogeneous catalysts for this reaction all contain M(III) species and are based on a porphyrin (**1**) or a salen (**2**) structure. The catalysts are remarkably resistant to poisoning by impurities in the captured CO_2, including water, CO, and N_2.

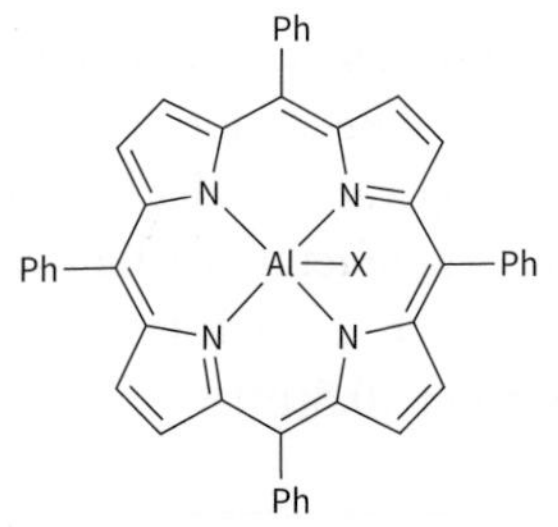

X = Cl, OMe, Me, OR

1

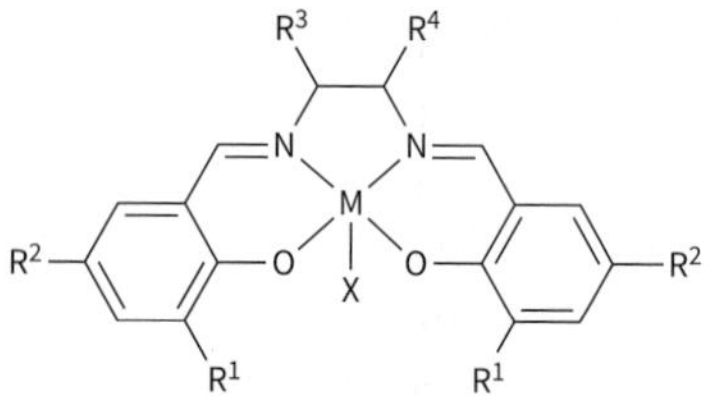

X = co-ligand

2

EXAMPLE 27.3 Capturing atmospheric carbon dioxide

One of the epoxides commonly used in the synthesis of polycarbonate by copolymerization with CO_2 is cyclopropene oxide. Calculate the percentage mass of the resultant polymer that is due to captured CO_2.

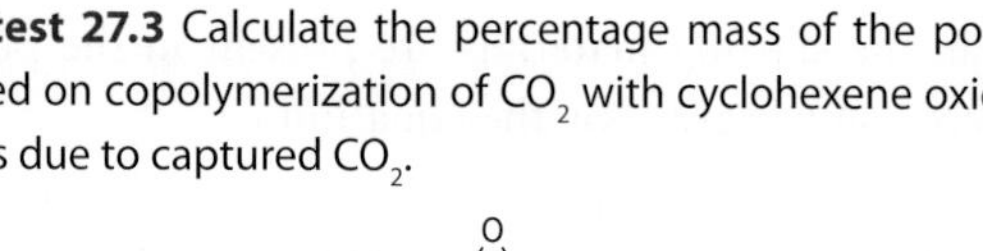

Answer The copolymerization reaction is shown above. The molecular mass of one repeat unit is 102 g mol^{-1}, of which 44 g mol^{-1} derives from the CO_2. Therefore, 43.1 mass per cent of the polymer is synthesized from captured CO_2.

Self-test 27.3 Calculate the percentage mass of the polymer formed on copolymerization of CO_2 with cyclohexene oxide (**3**) that is due to captured CO_2.

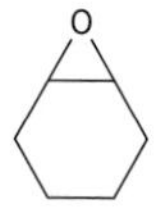

3

The use of carbonyl dichloride has also been eliminated from the synthesis of polyurethane polymers which are used to make rigid foams and durable items such as wheels and tyres. The new route developed by Monsanto replaced the carbonyl dichloride in the synthesis of the isocyanate precursors with carbon dioxide. The additional benefit of the new route is that it produces water rather than hydrochloric acid as a by-product.

$$RNH_2 + COCl_2 \rightarrow RNCO + 2HCl$$
$$RNH_2 + CO_2 \rightarrow RNCO + H_2O$$

Monsanto synthesized disodium iminodiacetate for use in the manufacture of its weedkiller Roundup®, and the original manufacturing route used hydrogen cyanide, hydrochloric acid, formaldehyde, and ammonia:

This route also produced 7 kg of waste for every 1 kg of product and was exothermic, which meant there was the possibility of a runaway reaction. Obviously, all of the reagents involved in this synthesis are hazardous, and hydrogen cyanide is toxic and should be eliminated from the process if at all possible. Monsanto has developed an alternative, greener route based on the copper-catalysed dehydrogenation of diethanolamine:

This new reaction avoids the use of hydrogen cyanide, formaldehyde, hydrochloric acid, and ammonia. It is also endothermic, so even though energy input is required for industrial, large-scale production, it is advantageous as it avoids the risk of a runaway reaction. The reaction also has fewer steps, gives a higher yield, and produces such a high-quality product that further purification is unnecessary.

The replacement of hazardous chemicals with less hazardous ones appears to be a relatively simple way to ensure that a process becomes greener. However, the environmental impact of a chemical process can only be properly evaluated by using life-cycle analysis (LCA). An LCA evaluates the cradle-to-grave impact of manufacturing a product and includes energy requirements, raw materials, the manufacturing process, reactors, solvents, treatment of any waste or by-products, and disposal of the product at the end of its useful lifetime. Only after such an analysis has been undertaken can the true environmental impact of a process or product be judged.

27.4 Designing safer chemicals

KEY POINTS Chemical products should be designed to carry out their desired function whilst minimizing toxicity.

The history of the synthesis of toxic chemicals over the last century did much to taint the reputation of chemistry with the general population. The fallout from the use of infamous examples like thalidomide and DDT did, however, greatly increase the scientific community's knowledge about toxicity to humans and the environment, and has led to a continuous endeavour to produce molecules that demonstrate functionality whilst minimizing harmful effects. One example of a rare occasion when the chemical industry worked in concert to solve an environmental problem was the search for replacements for chlorofluorocarbons (CFCs). Until the 1970s CFCs were widely used as refrigerants and propellants (Section 18.16) until they were found to deplete atmospheric ozone (Box 18.1). The chemical industry worked collaboratively to find less environmentally damaging replacements for CFCs and fairly quickly developed hydrofluorocarbons (HFCs), including the most widely used CF_3CH_2F. However, even though HFCs are not ozone-depleting they are greenhouse gases with a potency many times greater than CO_2. HFCs are very stable and persist in the atmosphere for up to 50 years, therefore posing a long-term threat to the environment. Thus, a development that provided a solution to one environmental issue has unintentionally exacerbated another. For applications such as car air-conditioning systems and industrial refrigeration units HFCs are now being replaced by NH_3 and CO_2.

27.5 Safer solvents and auxiliaries

KEY POINT The use of auxiliary substances in a system should be avoided wherever possible and should be innocuous when used.

Solvents are widely used in chemistry, and many chemical transformations could not happen without them. However, solvents are often toxic, flammable, or volatile, and in such cases finding safer alternatives is imperative. Solvents are also responsible for much of the energy used in a chemical synthesis. They have to be heated, filtered, distilled, refluxed, pumped, and, ideally, recovered and recycled. Therefore, finding synthetic routes that require no or little solvent is very desirable. One approach to the problem of using solvents is to design solventless reactions where the components of the reaction are mixed in a ball mill, Section 28.2, or in which the reactants themselves serve as the solvent or where the reactants are adsorbed onto a solid surface such as a clay material.

Auxiliary substances are agents that are added to reactions, often to facilitate separation of the products from the solvents. These may be solids, liquids, or gases, and may aid complexation, precipitation, or adsorption. These species may themselves be toxic or harmful to the environment and may need to be recovered and recycled. Avoiding the use of these auxiliary substances can minimize safety issues and reduce energy consumption.

One industry that has made major strides in reducing its use of harmful solvents is the paint industry, albeit mostly in response to changes in legislation. Paint for domestic and industrial uses typically comprises an inorganic pigment, a polymer resin to bind the pigment particles and adhere to the painted surface, and a solvent. Until a few decades ago the solvents used in paints were overwhelmingly organic, typically toluene, xylene, turpentine, or methyl ethyl ketone. These volatile solvents evaporated during drying, producing a strong odour and often inducing nausea. The industry has invested heavily in eliminating these volatile organic solvents from many paints, and has replaced them with water. Obviously, formulations had to change to ensure performance was maintained and that the components remained dispersed in the solvent.

Supercritical fluids are replacing organic solvents in many chemical processes and are classed as green solvents even though operating supercritical processes may require more energy. Supercritical CO_2, $scCO_2$, is used as an extraction solvent (Section 5.15(f)). The extracted species is then isolated by removing the $scCO_2$ by reducing the pressure, removing the need for a distillation step and saving energy. The CO_2 released into the atmosphere during this process does not contribute to the net level of greenhouse gas, because it is isolated as a by-product from industry, typically brewing. $scCO_2$ is increasingly being used for small- to medium-scale extractions in the food, fragrance, and flavour industries; for example, decaffeinating coffee and extracting essential oils. It is also used to remove fat from potato crisps to produce low-fat crisps and to extract vegetable oils from soya protein. It is also replacing chlorinated solvents in the dry-cleaning industry.

Supercritical water, scH_2O (Section 5.15(f)), can be used as a reaction solvent as it is fully miscible with many organic materials to form a homogeneous phase, thus reducing the need for organic solvents. It is also used in water and waste treatment as many organic compounds are readily oxidized in scH_2O. High-temperature water oxidation of waste is common, but scH_2O oxidizes a greater range of compounds and makes the treatment much faster; most organic compounds are completely oxidized in scH_2O within two minutes. scH_2O has also been used to decontaminate soil impregnated with industrial waste such as polyaromatic hydrocarbons and polychlorinated biphenyls.

Ionic liquids are ionic materials that are liquid near room temperature because of poor packing of the ions (Section 5.15(e)). Ionic liquids are typically derived from 1,3-dialkylimidazolium cations (4) with counter-ions such as PF_6^-, BF_4^-, and $CF_3SO_3^-$. These systems have melting points of less than (and often much less than) 100°C, and may have a very high viscosity and an effectively zero vapour pressure. Properties such as viscosity and melting point can be tuned by modifying the alkyl chain length or the cation:anion ratio. Density and viscosity vary with the anion. Density varies in the order $BF_4^- < PF_6^- < CF_3SO_3^-$, and viscosity varies in the reverse order. Many ionic liquids are stable up to around 300°C and so are suitable for high-temperature syntheses. They are immiscible with water and organic solvents, so are very useful for liquid–liquid extractions.

In addition to the multiple examples of organic reactions that can be carried out in ionic liquids, there are an increasing number of inorganic reactions that lead to products that are difficult to isolate otherwise. For instance, antimony and selenium will react to give a variety of different heteropolycations of general formula $[Sb_xSe_y]^{n+}$—the exact form of the product depending on the composition of the solvent.

Ionic liquids are also finding applications as solvents in homogeneously catalysed reactions (Sections 31.1 to 31.7), as the catalyst is soluble in the ionic liquid and can be readily isolated and reused. This brings many of the advantages of heterogeneous catalysis to homogeneous systems. If it can be arranged that the catalysts are preferentially soluble in the ionic liquid phase (Section 5.15(e)) (for instance, by making them ionic), immiscible organic solvents can be used to extract organic products. As an example, consider the hydroformylation of alkenes by a Rh catalyst with phosphine ligands (Section 31.3). When the ligand used is triphenylphosphine, the catalyst is extracted from the ionic liquid together with the products. However, when an ionic sulfonated triphenylphosphine is used as the ligand, the catalyst remains in the ionic liquid, and product separation from the catalyst is complete. It should be borne in mind that the ionic liquid phase is not always unreactive and may induce alternative reactions.

Another example of the use of an ionic liquid as a solvent in a metal catalysed reaction is in the synthesis of the nonsteroidal anti-inflammatory drug naproxen. This drug is synthesized by asymmetric hydrogenation catalysed by $[Ru(BINAP)Cl_2]$ (Section 31.2) in $[bmim]BF_4$ (5) which enables the product to be easily separated in high enantiomeric excess:

[Ru(BINAP)Cl$_2$/ H$_2$]
[bmin]BF$_4$/ i-PrOH

(S)-Naproxen

27.6 Design for energy efficiency

KEY POINTS The environmental and economic impacts of the energy requirements of chemical processes should be considered and minimized. Reactions should be conducted at ambient temperature and pressure whenever possible.

Producing chemicals on a large scale in the chemical industry can be very energy intensive, with the need for multiple steps, heating, and high pressures. Even laboratory-scale chemistry can be very energy-intensive. Solutions are heated or refluxed, solvents are distilled off and recycled, and products are isolated. Some reactions rely on energy input to drive the chemical reaction, such as in electrolysis. This principle of green chemistry urges chemists to take responsibility for the impact that the energy requirement of reactions has on the environment and to try to design reactions that minimize it. Chemists should be considering energy implications in just the same way that they try to maximize yield or minimize the number of steps in a synthesis.

Most of the energy used is still generated from fossil or nuclear fuels with all the associated environmental and resource implications. Recent data from the International Renewable Energy Agency indicates that the global chemicals and petrochemical industries use 30–40 × 10^{18} J per year. To put this figure into perspective, the total energy use of the US is 95 × 10^{18} J per year. This figure is so high because of the very high operating temperatures of most of the industrial processes. Certainly, the chemicals industry has taken note of this principle and made real savings in energy consumption over the past thirty years or so (Fig. 27.2), reducing energy use whilst increasing productivity.

This principle is illustrated well by two processes that we have already discussed. The industrial production of titanium dioxide pigment proceeds by the sulfate or the chloride process (Section 27.1). The sulfate process uses five times as much energy as the chloride process, mainly because of the need to evaporate large volumes of aqueous solution in order to retrieve the waste titanium and iron sulfates. This is another reason why the chloride process has become more widely used than the sulfate process. The Haber process for ammonia synthesis (Sections 16.6 and 31.9) uses a catalyst to lower the activation energy of the reaction. The uncatalysed reaction would require temperatures of over 3000°C, but the modern-day production is one of the most energy-efficient large-scale chemical syntheses. However, even the catalysed process required elevated temperatures of 400–500°C and pressure of 200–250 atm.

In Nature, the enzyme nitrogenase converts N$_2$ to ammonia with no need for pressure or heat, but 16 moles of ATP are used for each N$_2$ that is converted to NH$_3$. The biological process is therefore not very efficient either, but does provide a route to NH$_3$ at ambient temperatures. There are intense efforts to understand how the active sites of nitrogenase and other enzymes catalyse important reactions by

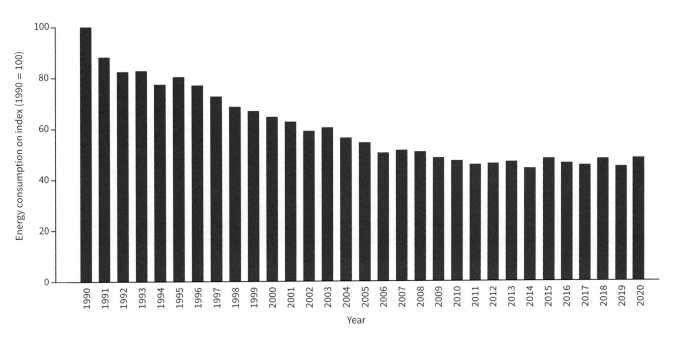

FIGURE 27.2 Energy intensity for European chemical industry 1990–2020. Energy intensity is defined as energy input per unit of chemicals produced. 1990 = 100. (Data from Cefic, 2021 Fact and Figures.)

using abundant metals and mild conditions (Chapter 26). Biomimicry is likely to play a major role in future catalyst development.

Ultrasound and microwaves are increasingly finding applications in synthetic chemistry, Sections 28.9 and 29.4. Molecules undergo excitation in microwave radiation, and heat is generated. When superheating is performed in a sealed vessel under high pressure, reactions which may normally take hours can be completed in minutes. For example, the synthesis of transition metal phthalocyanine dyes is typically performed by heating the dry reactants together in a conventional oven or by refluxing in a solvent for three hours. Using a microwave oven, the reactions are complete within five minutes. Ultrasound causes rapid changes in pressure in a liquid leading to the formation of cavities, or bubbles. The collapse of these cavities causes shock waves in the system and the formation of reactive free radicals, and can lead to the formation of chemical species not attainable under conventional conditions. Ultrasound treatment of $[Mo(CO)_6]$ in the presence of sulfur leads to the formation of nanoparticles of MoS_2, Chapter 28, which are used as a desulfurization catalyst, Section 17.4, to remove sulfur from fossil fuels.

Many chemical processes produce heat that can be utilized rather than be allowed to dissipate into the immediate environment. For example, industrial ammonia synthesis produces heat energy and carbon dioxide, through brown or grey hydrogen production, Box 11.3. Some enlightened producers sell on the released energy to heat glasshouses used by the vegetable- and fruit-growing industries. Some industries produce their own electricity on site, which saves losses through transmission.

Ultimately, the energy efficiency of a chemical process should not be studied in isolation but as part of a wider life cycle analysis (LCA).

of hydrogen and carbon monoxide with traces of carbon dioxide, via steam reforming and thermochemical processes, as well as through heterogeneous catalytic processes involving zeolites (Sections 25.16 and 29.1). Syngas is then a versatile feedstock for the synthesis of almost any other organic compound (Section 11.4), and plays a crucial role in the generation of synthetic fuels through the Fischer–Tropsch process (Section 31.11).

The term renewable feedstocks also applies to feedstocks that are renewable on a human timescale. So, carbon dioxide would be considered renewable as it can be generated from, for example, fermentation. Methane could also be considered to be renewable if it is generated from the anaerobic digestion of biomass similar to the natural process that produces marsh gas. The production of methane as a renewable feedstock has great potential, and a range of biomass sources is useful in this application, including sewage, municipal waste, waste crops, and logging waste. However, some of these potential feedstocks require significant pretreatment to separate out usable biomass. The danger of releasing methane into the atmosphere is also a challenge. One of the most promising biomass sources is seaweed, or kelp, which requires little pretreatment, has high-energy efficiency, and does not require the use of precious farmland for growing it, which many of the other renewable feedstocks do.

Hydrogen can be thought of as a renewable feedstock and fuel as methods of generating it include the electrolysis of water (Section 11.4), and a huge amount of effort is currently being devoted to developing safe methods of storage and release (Box 11.5). When the electrolysis of water utilizes green energy—for instance, from solar power—hydrogen can be considered a fully renewable feedstock, with minimal environmental impact.

27.7 Use of renewable feedstocks

KEY POINT Whenever possible, raw materials and feedstocks for chemical reactions and processes should be renewable.

Humans have made use of fossil fuels and minerals to build and power modern society. These are not renewable resources and are steadily being depleted. Their use is also contributing to the increasing levels of greenhouse gases in the atmosphere. The term 'renewable feedstocks' usually refers to biological feedstocks. Carbon dioxide is removed from the atmosphere through photosynthetic processes to grow biomass: trees, crops, plants, and algae. This biomass is the feedstock that will contribute to powering the future through the synthesis of biofuels, bioethanol, biodiesel, polymers, resins, cellulose, and chemicals, whilst reducing atmospheric levels of CO_2. In fact, almost any organic matter can be converted to syngas, or synthesis gas—a mixture

27.8 Minimize derivatives

KEY POINT Derivatization should be avoided if possible, as it can lead to the use of more chemicals, produce more waste, and use more energy.

This principle of green chemistry states that the number of steps involved in a chemical reaction should be minimized in order to reduce waste produced and energy consumed. Logically, this includes avoiding any derivatization steps such as the use of protecting groups during a synthesis to direct the reaction in a specific direction. Derivatization in organic chemistry reactions can largely be avoided by using enzymes in synthesis because they usually react with just one site in a molecule, leaving the rest of it unchanged. Enzymes are discussed in Chapter 26. That there are very few examples in inorganic chemistry does not diminish the importance of the principle.

27.9 Catalysis

KEY POINT Catalysed reactions are more efficient than uncatalysed reactions.

Catalysis was introduced in Chapter 2 (Section 2.15), where the concepts of lowering the energetic barrier of a reaction, leading to increased reaction rate, was discussed.

Catalysts are green in the sense that they play a crucial role in increasing the atom economies of reactions, and catalytic routes typically involve fewer reaction steps, greater selectivity, and regeneration of the catalyst, thus reducing waste. In industry, homogeneous catalytic processes, in which reactants and catalysts are in the same phase, offer high selectivity and are preferred for exothermic reactions where heat needs to be dissipated effectively. Heterogeneous catalysts, where the catalyst and reactants are in separate phases, tolerate high temperatures and a wide range of operating conditions, producing higher outputs for a given amount of catalyst and reaction time than homogeneous catalysts operating at lower temperatures. Extra steps are not needed to separate product from a heterogeneous catalyst, resulting in more efficient processes.

The field of inorganic catalysis is sufficiently diverse and important that it has frequently been discussed in Part 2 and the whole of Chapter 31 is devoted to it.

27.10 Design for degradation

KEY POINT Once their usefulness is over, chemical products should degrade so that they cause no hazard nor persist in the environment.

Application of the first principle of green chemistry discussed in Section 27.1 leads to the reduction of waste produced during chemical reactions. However, the useful products themselves need to be degradable into products that are harmless to the environment once their useful life is over. Ideally, biodegradability needs to be built into the design of products.

One of the most troublesome classes of nondegradable products is polymers. The vast majority of polymers are manufactured from nonrenewable fossil fuel feedstocks, and persist in the environment for many hundreds of years, littering the land and the seas and filling the limited capacity of landfill sites. They can be dealt with by incineration, but this process produces CO_2, which adds to the greenhouse-gas problem. So, biodegradable polymers are needed and are ideally produced from renewable biomass feedstocks.

The most widely used biodegradable polymer is polylactic acid, obtained from the fermentation of polysaccharides, which is used in packaging. The polymer degrades by first being hydrolysed to shorter-chain-length oligomers. This is followed by bacterial action, which converts the oligomers to water and carbon dioxide. Unfortunately, many of the other properties of polylactic acid, such as flexibility, heat distortion, gas permeability, and the viscosity of the molten polymer (which is important for processing), make the material unsuitable for many applications.

Although polylactic acid is an organic species, its functionality can be enhanced by incorporation of inorganic materials into its structures. Composite materials can be formed from polylactic acid with silicate minerals and clays modified with organic species. For example, a biodegradable polymer can be formed from polylactic acid with montmorillonite clay, a soft silicate mineral, Sections 25.17(a) and 28.10, modified with the octadecylammonium cation, $[CH_3(CH_2)_{17}NH_4]^+$. These modified polymers also exhibit improved biodegradability compared with the unmodified polylactic acid. This improvement is thought to be due to the presence of hydroxyl groups on the exposed silicate layers initiating hydrolysis of the polymer. The use of inorganic MOFs and zeolites as carriers of active molecules are finding increased usage as the carrier degrades harmlessly into the environment, Section 29.2.

The use of metallic copper to sheath the hulls of wooden boats in order to prevent the attack of marine growth, including molluscs, barnacles, and shipworms, was developed by the British Royal Navy in the 18th century. However, this costly solution was discarded in the 20th century after cheaper treatments utilizing tributyltin oxide were developed. Subsequently, it was discovered that tributyltin is very persistent in the environment, is very toxic, and bioaccumulates in shellfish. These issues have been circumvented by the development of alternatives, such as the Sea-Nine® antifouling agent that contains 4,5-dichloro-2-n-octyl-4-isothiazolin-3-one, which degrades rapidly once released. Whereas tributyltin oxide has a half-life of up to nine months in sediment, Sea-Nine has a half-life of just one hour.

Copper plate does not suffer from the issues associated with tributyl tin, and now complexes such as copper pyrithione (6) are finding a role in antifouling treatments, and zinc pyrithione is similarly effective.

6

27.11 Real-time analysis for pollution prevention

KEY POINT Analytical techniques should be used during chemical processes to monitor and control the formation of hazardous substances.

Constant monitoring of chemical reactions is important for checking a reaction's progress, to check the temperature or pH, or to monitor the health of a catalyst, the production of by-products, or the quality of effluent streams. This area of endeavour is known as process analytical chemistry and is increasingly important in the chemical industry, where much analysis is moving out of the laboratory and onto the plant by sampling in real time and *in situ*, which may be within the process or within a reaction stream.

It would seem that there was little role for inorganic chemistry within this aspect of green chemistry, but, as we have seen previously, inorganic chemistry can be a good enabler. Inorganic materials are often incorporated into the sensors that are used in the continuous monitoring of various analytes. Semiconducting metal oxides such as SnO_2, ZnO, WO_3, and In_2O_3 (Section 25.10) are used in gas monitors to detect the presence of small molecules such as H_2, CO, and CO_2. The use of graphene (Box 15.1), and functionalized graphene, is increasingly being explored in the development of ultrasensitive sensors (Section 28.9).

Graphene offers enhanced sensitivity because every atom of the structure is exposed to the analyte. Sensors using single-walled carbon nanotubes are being used to detect nitrogen-containing organic species such as pyridine and aniline. In these sensors a small gas molecule is chemisorbed onto the surface, leading to a change in the concentration of charge carriers. This results in a change in the conductance of the material, and this is measured as the sensor response. A ferrocene-based biosensor has been developed to monitor glucose levels *in situ* in fermentation vessels.

27.12 Inherently safer chemistry for accident prevention

KEY POINT Substances used in a chemical process should be chosen to minimize the risk of chemical accidents, including explosions and fires.

This principle is of paramount importance irrespective of whether chemistry is being carried out on an industrial or a laboratory scale. Safer chemistry and accident prevention is a culture that should pervade all environments where chemistry is carried out. In the past, novel chemistry or economics have taken precedent over safety, whether it be a new laboratory scale synthesis or a large-scale chemical process. Increasingly, chemists consider the safety of a reaction or process during the design of a process or reaction. This principle is known as the safety principle, and in many ways is the culmination of many of the preceding ones.

The safety principle is concerned with controlling known hazards to achieve an acceptable level of risk. The term hazard refers to anything that has potential to do harm to people, property, or the environment. Risk is a measure of how likely it is that the hazard will actually do harm. So properly managing, or reducing, hazards will minimize risk. Therefore, chemists and the chemical industry should work towards reducing the hazards; for instance, by using safer materials, reducing the use of solvents, and reducing the production of waste. Risk can be reduced by strictly controlling the exposure to hazardous materials; for example, by monitoring their production. The place of green chemistry in risk management is shown in Fig. 27.3.

The development of green chemistry practice is undoubtedly beneficial for the environment. Application of many of the principles also leads to economic benefits for the chemical industry. The green chemistry movement is likely to continue into the future, as much of its practice has been stimulated as a response to increasing levels of regulation at a national level in all parts of the industrialized world. This legislation seeks to control environmental and occupational exposure to hazards, and is likely to continue to shape how chemicals are manufactured.

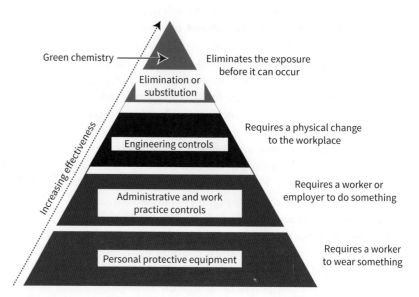

FIGURE 27.3 The green chemistry safety principle. (Source: United States Bureau of Statistics.)

FURTHER READING

P. T. Anastas and J. C. Warner, *Green Chemistry: Theory and Practice*. Oxford University Press (1998). The definitive guide to green chemistry.

M. Lancaster, *Green Chemistry: An Introductory Text*. 3rd ed. Royal Society of Chemistry (2016). A readable text with industrial examples.

A. Matlack, *Introduction to Green Chemistry*. CRC Press (2010). Discussion from an industrial perspective with interesting case studies.

R. A. Sheldon, I. Arends, and U. Hanefeld, *Green Chemistry and Catalysis*. John Wiley & Sons (2007). A focus on the importance of catalysis in green chemistry.

The European Chemical Industry Council, http://www.cefic.org/. This very useful website provides lots of data relating to the European chemical industry, and focuses on sustainability and green chemistry.

O. V. Kharissova, B. I. Kharisov, C. M. Oliva González, Y. P. Méndez, and I. López, Greener Synthesis of Chemical Compounds and Materials. Royal Society Open Science, 2019, **6**, 191378. http://dx.doi.org/10.1098/rsos.191378. A review that focuses on synthetic procedures.

P. T. Anastas and J. B. Zimmerman, The periodic table of the elements of green and sustainable chemistry. *Green Chem.*, 2019, **21**, 6545–6566. https://doi.org/10.1039/C9GC01293A. In this article the authors describe the quest for sustainability in terms of an alternative periodic table.

R. A. Sheldon, The E factor at 30: A passion for pollution prevention. *Green Chem.*, 2023, **25**, 1704–1728. https://DOI:10.1039/D2GC04747K. In this review, a leading pioneer of sustainable chemistry discusses the advances made in improving the efficiency of chemical processes and minimizing waste.

EXERCISES

27.1 The noncatalytic route to manufacture of ethylene oxide produces 3.5 kg of waste for every 1 kg of product. Calculate the E-factor for this process.

27.2 The old Monsanto process for manufacturing disodium iminodiacetate produces 7 kg of waste for every 1 kg of product. Calculate the E-factor for this process.

27.3 Tetramethyllead can be synthesized in the laboratory from lead(II) chloride and a Grignard reagent according to the following reaction:

$$2PbCl_2 + 4CH_3MgBr \rightarrow (CH_3)_4Pb + Pb + 4MgBrCl$$

Assuming that the synthesis was attempted with 4.0 g of $PbCl_2$ and an excess of the Grignard reagent, and that 1.54 g of the product were isolated, calculate the yield and the atom economy. Compare the two values and comment on the efficiency of the reaction.

27.4 Ferrocene can be acetylated by reacting with excess acetic anhydride according to the following reaction:

If the reaction used 1.5 g of ferrocene and produced 1.15 g of acetylferrocene, calculate the yield and the atom economy of the reaction and comment on the values.

27.5 Define the terms 'risk' and 'hazard'.

27.6 What is the green chemistry safety principle?

27.7 Why is real-time in-process monitoring beneficial to green chemical processes?

TUTORIAL PROBLEMS

27.1 DDT was one of the first synthetic pesticides and was used as an effective control measure for malaria, typhoid, and other insect-borne diseases during the Second World War. During the 1950s–1960s it became very widely used for insect control on crops, in animals, and in the home. However, it was banned in the 1970s, and the subsequent furore led to changes in legislation in many countries. Discuss the history of DDT, the problems it caused, and legislation stimulated by the case. Describe in what circumstances DDT is currently recommended for use.

27.2 In their article 'Industrial research: drug companies must adopt green chemistry' (*Nature*, 2016, **534**, 27), J. Tucker and M. Faul describe how they transformed their approach to manufacturing to save their company time and money by making drugs more sustainably. Summarize the approaches they used, and identify which of the 12 principles of green chemistry were applied.

27.3 In the 1960s, the antinausea drug thalidomide caused severe birth defects in babies born to mothers who had taken the drug to relieve morning sickness. Discuss the reasons why this was a serious issue in Europe and yet not in the United States. Describe why the drug is toxic and under what circumstances it is currently prescribed. What changes in legislation were introduced in Europe following these cases?

27.4 Discuss how recycling is different from green chemistry. Evaluate the advantages and disadvantages of each.

27.5 Chlorofluorocarbons (CFCs) were developed as novel, safe refrigerants. What refrigerants were they designed to replace, and what were their hazards? What were the problems associated with CFCs? Review the latest compounds used as refrigerants.

27.6 A. Arbaoui and C. Redshaw (*Polym. Chem.*, 2010, **1**, 801) review catalysts for the synthesis of biodegradable polymers via ring-opening metathesis polymerization. Summarize the need for biodegradable polymers and why new catalysts are required. From the details given, identify which groups of metals give the most active catalysts and with which types of ligands. Illustrate with examples.

27.7 Sensors are increasingly being used *in situ* to monitor reactions and effluent streams in industry. What are the advantages and disadvantages of using *in situ* sensors compared to traditional laboratory analysis?

27.8 Ultimately, the overall 'green' credentials of any process can only be evaluated through life-cycle analysis (LCA). Identify what components would be considered in the LCA of the synthesis of ammonia via the Haber–Bosch process.

27.9 Environmental injustices have, historically, disproportionately affected those who are already disadvantaged. Discuss how green chemistry approaches can address these injustices. See *Nature Sustainability*, 2022, https://doi.org/10.1038/s41893-022-01050-z.

Nanoparticle inorganic chemistry

Introduction to nanomaterials and their synthesis

28.1 **Nanomaterial terminology and history**

28.2 **Solution-based synthesis of nanoparticles**

28.3 **Vapour-phase synthesis of nanoparticles via solutions or solids**

28.4 **Templated synthesis of nanomaterials using frameworks, supports, and substrates**

28.5 **Characterization and formation of nanomaterials using microscopy**

Nanostructures and mesoporous materials: properties and applications

28.6 **Zero-dimensional (0D) control, nanoparticles of carbon, metals, metal oxides, and metal chalcogenides**

28.7 **Special optical properties of 0D nanoparticles and quantum dots**

28.8 **One-dimensional control: carbon nanotubes and inorganic nanowires**

28.9 **Two-dimensional control: graphene, quantum wells, and solid-state superlattices**

28.10 **Three-dimensional control of nanostructural units: mesoporous materials and composites**

Further Reading

Exercises

Tutorial Problems

The properties of an inorganic solid can be controlled by the size and shape of the particles in which it exists and was synthesized. This experimental observation stems from the fact that many of the properties of a solid (for example, electronic conductivity, interaction with visible light, and surface catalysis) are related to the *extended* crystal structure of a material, including its surfaces. For example, electronic bands are formed from overlapping atomic orbitals across the whole solid, so the electronic properties of a solid material depend on the size and dimensionality ('zero'-, one-, two-, or three-dimensional particle shape) of the material particle. The ability to modify the properties of inorganic materials is most apparent at submicron scales (<0.1 μm); that is, the nanometer range, 1–100 nm.

In this chapter we discuss how nanoscale inorganic materials, with one or more dimensions in the range 1–100 nm, can be synthesized and how their properties, in one, two, and three dimensions, differ from those of large crystallites of the same compound. Many applications of these nanomaterials are presented in this chapter, and others are described in the following chapters concerned with porous structures, energy materials, and catalysts, and where nanoparticulate forms of inorganic solids have specific useful properties and applications.

Introduction to nanomaterials and their synthesis

KEY POINTS A nanomaterial is any material that has one dimension on the scale of 1–100 nm; a more exclusive definition is that a nanomaterial is a substance that exhibits properties absent in both the molecular and bulk solid state.

A **nanomaterial** is taken to be a solid material that exists over the scale of 1–100 nm and exhibits novel properties that are related to its scale. Likewise, **nanoscience** is also sometimes restricted to the study of the new effects that arise only in materials that exist on the nanoscale, and **nanotechnology** is similarly restricted to the procedures for creating new functionalities that are possible only by manipulating matter on the nanometre scale.

28.1 Nanomaterial terminology and history

KEY POINTS Nanoparticles are classified in terms of their dimensionality. Top-down fabrication methods carve out or add on nanoscale features to a bulk material by using physical methods; bottom-up fabrication methods assemble atoms or molecules in a controlled manner to build nanomaterials.

Nanomaterials are often described as being zero-, one-, two-, or (sometimes) three-dimensional (0D, 1D, 2D, and 3D), where the dimensionality references the major growth characteristic of the nanoparticles. 1D nanoparticles can be synthesized as chains, filaments, ribbons, and tubes of atoms, and may have one dimension (along the filament direction) greater than 100 nm. 2D nanoparticles form as layers, plates, and films, and can have two dimensions greater than 100 nm. Nanoparticles with all three dimensions less than 100 nm are generally referred to as 0D; if all dimensions are greater than 100 nm, then the material is generally not considered as nanoparticulate, though collections of 0D, 1D, and 2D nanomaterials forming bulk materials (micron and submicron sizes, >100 nm, in all dimensions) are often described as 3D nanomaterials. Similarly, a 3D bulk material may contain nanosized structural elements—for example, embedded nanoparticles or nanopores in mesoporous materials (Section 28.10)—and these are usually termed 3D nanomaterials.

Humans have, in fact, practised nanotechnology for centuries, though it is only recently that the chemistry of nanosized particles has engendered such high levels of interest, research funding, and widespread applications. For example, gold and silver compounds have been used to produce red and yellow stained glass, respectively, for centuries. In stained glass, the gold and silver metal atoms exist as nanoparticles (described previously as 'colloidal particles') with optical properties that depend strongly on their size. Metallic nanopigments have now become a focal point of biomedical nanotechnology because they can be used to tag DNA or deliver a drug to a specific target (Section 32.10). Other historical examples of nanotechnology include the photosensitive nanosized particles in silver halide emulsions used in photography, nanoparticulate TiO_2 pigments in paints and coatings, and the nanosized carbon granules in the 'carbon black' used for reinforcing tyres and in printer's ink (Section 30.1).

There are two basic techniques for the fabrication of nanoscale entities (Fig. 28.1). The first is to take a macroscale (or microscale) object and carve out nanoscale patterns; methods of this sort are called **top-down approaches**. The second technique is to build larger objects by controlling the arrangement of their smaller-scale components; methods of this sort are called **bottom-up approaches**.

Physical interactions are used in top-down fabrication approaches, such as photolithography, electron-beam (e-beam) lithography, and soft lithography; photolithography is used to fabricate very-large-scale integrated circuits having feature dimensions on the 100 nm scale.

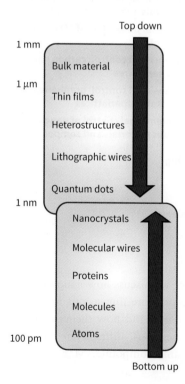

FIGURE 28.1 The two techniques for making nanoscale structures. The top-down technique starts with larger objects that are whittled down into nanoscale objects; the bottom-up technique starts with smaller objects that are combined into nanoscale objects.

The bottom-up approach to nanoscale fabrication is emphasized in this text because it focuses on the chemical interactions of atoms and molecules, and their controlled arrangement, using chemistry, into larger functional structures. The two basic approaches most widely used to prepare nanomaterials using bottom-up methods are solution and vapour-phase methods. As well as forming a nanoparticle of a single uniform phase, these techniques can be used to form structured nanoparticles containing two or more different compounds or materials. One example is a **core-shell** nanoparticle where a second, different phase (shell) is grown around a core of a nanoparticle.

28.2 Solution-based synthesis of nanoparticles

KEY POINTS Solution-based, bottom-up, synthetic methods are mainly used for nanoparticle synthesis because they have atomically mixed and highly mobile reagents; the two stages of crystallization from solution are nucleation and growth.

As discussed in Section 25.9, the two basic techniques used to generate inorganic solids are high-temperature direct-combination methods and solution-based methods. The former does not lend itself well to the synthesis of nanoparticles because the reactants tend to be micrometre-sized particles which react over long periods to reach equilibrium. Moreover, the use of elevated temperatures leads to particle growth during the reaction, resulting in large crystallites, typically of sizes greater than 1 μm. There are, however, some examples of the use of mechanical ball-milling at low temperature to break micrometre-sized powders into nanoparticles; such methods have been used for producing nanosized particles of metal hydrides important as hydrogen storage materials (Section 30.8). Solution-based methods, however, permit excellent control over the crystallization of inorganic materials and are widely used in nanochemistry. By fine-tuning the crystallization process from solution, highly monodispersed, uniformly-shaped nanoparticles of a wide range of compositions can be prepared by appropriate combinations of elements from throughout the periodic table.

Because the reactants in solution-based methods are mixed on an atomic scale and solvated in a liquid medium, diffusion is fast and diffusion distances are typically small. Therefore, reactions can be carried out at low temperature, which minimizes the thermally driven particle growth process that is problematic in direct-combination synthesis methods (Section 25.9). Although the specific features of each reaction differ greatly, the key stages in solution chemistry are:

1. Solvation of reactant species and additives.
2. Formation of stable solid crystallization nuclei from the solution.
3. Controlled growth of the solid nuclei to a specific nanoparticle size through the addition of controlled amounts of reactant species.

The aim in solution synthesis of nanoparticles is to generate, in a controlled manner, the simultaneous formation of large numbers of stable nuclei that undergo little further growth. If growth is to occur, it should occur independently of the nucleation step, because then all particles have a chance to grow to similar sizes. If performed successfully, the particles will be monodispersed; that is, all of a similar size in the nanometre range. The drawback to the solution method is that the particles can undergo **Ostwald ripening**,[1] in which smaller particles in the distribution redissolve and their solvated species subsequently reprecipitate onto larger particles, so increasing the average particle size and decreasing the total particle count. To prevent this unwanted ripening, **stabilizers** (surfactant molecules that help prevent both dissolution and growth of small particles) are added. As there are many methods to synthesize nanoparticles, we limit this discussion to a few well-known and widely used examples.

(a) Gold nanoparticles

KEY POINT Gold nanoparticles can be obtained by the controlled reduction of solutions of $[AuCl_4]^-$ in the presence of stabilizers.

In 1857, Michael Faraday found that reduction of an aqueous solution of $[AuCl_4]^-$ with phosphorus in CS_2 produced a deep-red coloured suspension; this solution contained nanoparticles of gold. Because sulfur ligands form stable complexes with gold (soft–soft interaction; Section 5.10(a)), sulfur-containing species are good stabilizing agents and the most widely used stabilizers for gold nanoparticles contain a thiol group (–SH). An approach has been developed, in the same spirit as Faraday's, to control the size and dispersity of gold nanoparticles obtained from $[AuCl_4]^-$ using thiol stabilizers; this produces air-stable gold nanoparticles with diameters between 1.5 and 5.2 nm. In the so-called **Brust–Schiffrin method**, $[AuCl_4]^-$ is first transferred from a solution in water to methylbenzene (toluene) by using

---

[1] This is a thermodynamically driven process that reduces the surface energy of the particles in the system.

tetraoctylammonium bromide as a phase-transfer agent. The methylbenzene contains dodecanethiol as a stabilizer, and, after transfer, $NaBH_4$ is used as a reducing agent to precipitate Au nanoparticles with dodecanethiol surface groups:

Transfer:

$$[AuCl_4]^-(aq) + N(C_8H_{17})_4^+(sol, aq) \rightarrow N(C_8H_{17})_4^+(sol) + [AuCl_4]^-(sol)$$

Precipitation (of the $[Au_m(C_{12}H_{25}SH)_n]$ nanoparticle):

$$m[AuCl_4]^-(sol) + nC_{12}H_{25}SH(sol) + 3me^- \xrightarrow{NaBH_4} 4mCl^-(sol) + [Au_m(C_{12}H_{25}SH)_n](sol)$$

where 'sol' is methylbenzene. The ratio of stabilizer ($C_{12}H_{25}SH$) to metal (Au) controls the particle size, with higher stabilizer:metal ratios leading to smaller metal particles. By adding the $NaBH_4$ reductant quickly and cooling the system as soon as possible after the reaction terminates, smaller and more uniformly sized (monodispersed) nanoparticles are formed. The rapid addition of reductant increases the probability of simultaneous formation of all nuclei. By cooling the solution quickly, both post-nucleation growth and dissolution of particles are minimized. Similar approaches can be used for other metal nanoparticles, including those of silver, copper, nickel, platinum, and manganese.

Core-shell nanoparticles are typically synthesized by using a two-stage solution method. In the simplest case, nanoparticles of one material are grown and separated and then the reaction conditions are changed to deposit a second phase on the surfaces of the original nanoparticle. Functionalization of the surface of the original core nanoparticle often helps direct the precipitation of the second phase onto the nanoparticle surface rather than it nucleating as new nanoparticles, Fig. 28.2. Core-shell nanoparticles of Fe_3O_4 on gold (termed $Au@Fe_3O_4$) are formed by the production of Au nanoparticles by using the previously described method; these core nanoparticles are then functionalized using a octadecene/oleylamine mixture and then $Fe(CO)_5$ is added. The solution, with its suspension of gold nanoparticles, is heated while air is passed it through to deposit Fe_3O_4 as the shell.

(b) Metal nanoparticles on surfaces

KEY POINT Metal nanoparticles may be grown on metal oxide surfaces for use in heterogeneous catalysis.

Metal nanoparticles supported on inert, high-surface area metal oxides, for example SiO_2, Al_2O_3, and TiO_2 are important industrial heterogeneous catalysts (Sections 31.8 and 31.12). The size and shape (including which metal surface planes are exposed), and distribution in size, of the nanoparticles are all of considerable importance in controlling the catalytic behaviour of the nanoparticle. The formation and controlled growth of metal nanoparticles, including Pt and Pt/Rh, on substrates and templates for catalysis can be achieved using the methods outlined in Section 28.6.

(c) Semiconductor and oxide nanoparticles

KEY POINT Metal oxide and semiconductor nanoparticles may be obtained by controlled precipitation from solution.

Nanoparticles of materials such as GaN, GaP, GaAs, InP, InAs, ZnO, ZnS, ZnSe, CdS, and CdSe have been investigated for their optical properties (Section 28.7, Boxes 20.4 and 28.1) because their interband absorption and fluorescence occur in the visible spectrum. Nanoparticles of materials exhibiting new and useful optical, electronic, and magnetic properties are also known as **quantum dots** (QDs; see Section 28.7). An early description of the preparation of CdSe nanoparticles involved dissolving dimethylcadmium, $Cd(CH_3)_2$, in a solvent mixture of trioctylphosphine (TOP) and trioctylphosphine oxide (TOPO) and adding a solution of Se, also dissolved in TOP or TOPO, at room temperature. The solution mixture was then injected into a reactant vessel containing vigorously stirred hot TOPO, leading to the nucleation of TOPO-stabilized CdSe nanoparticles. Careful cooling and heating of the solution, to control the rates of nucleation and particle growth, led to nanoparticles with narrow size distributions and sizes in the range 2–12 nm. Alternative, less hazardous synthetic methods, avoiding the highly toxic $Cd(CH_3)_2$, have been developed more recently.

Cadmium sulfide can be grown in pH-controlled aqueous solutions of Cd(II) salts with polyphosphate stabilizers by the addition of a sulfur source. For example, at pH = 10.3 the addition of Na_2S causes the precipitation of CdS

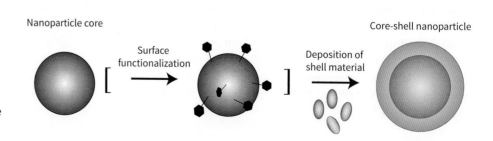

FIGURE 28.2 Schematic showing the synthesis of core-shell nanoparticles.

nanoparticles from aqueous solutions containing $Cd(NO_3)_2$ and sodium polyphosphate. These QDs range in size from 1 to 10 nm, and the size is controllable through the reactant concentrations and the rate of addition of the reactant, Na_2S.

Many applications use nanoparticles of white oxides, most commonly SiO_2 and TiO_2, as pigments and brightening agents in food additives, ink, paints, and coatings, and these can also be grown by solution methods. Efforts to achieve controlled oxide-nanoparticle growth stem from historical studies in traditional ceramic and colloidal applications, where particle sizes from 1 nm to 1 μm are widely used. Silica (SiO_2) and titania (TiO_2) are probably the best-known nanoparticulate oxides grown from solution, and schemes typically involve the controlled hydrolysis of metal alkoxides forming metal hydroxides and hydrated oxides in a process analogous to the initial stages of sol–gel synthesis (Section 25.9). In all cases, strict monitoring of the pH, precursor chemistry, reactant concentration, rate of addition of reactant, and temperature, is required to control the final size and shape of the oxide nanoparticles.

An important example of nanoparticle oxide use is that of TiO_2 in the photoelectrochemical solar cell known as the **Grätzel cell** (Box 23.1). Nucleation of nanoparticulate TiO_2 occurs in the hydrolysis of titanium isopropoxide that is added dropwise into vigorously stirred 0.1 M HNO_3(aq). The TiO_2 nanoparticles are allowed to grow under these synthesis conditions with the size, shape, and state of agglomeration of the nanoparticles controlled by adjusting the pH, temperature, and reaction time in either the nucleation or the growth stage.

28.3 Vapour-phase synthesis of nanoparticles via solutions or solids

KEY POINTS Vapour-phase synthetic methods are alternative techniques for nanoparticle synthesis and can employ solutions or solids as the precursors.

The same fundamentals concerning nucleation and growth that are relevant to solution synthesis apply to vapour-phase synthesis. The vapour phase needs to be supersaturated to the point at which a high density of homogeneous nucleation events produces solid particles in one short burst, and growth must be limited and controlled in a subsequent step, if it is to occur at all. Commercially, vapour-phase synthesis is carried out to produce nanoscale carbon black (graphite particles sizes in the range 10–100 nm) and fumed silica (SiO_2 particle sizes in the range 5–50 nm) in large quantities. Metals, oxides, nitrides, carbides, and chalcogenides can also easily be formed by using vapour-phase techniques.

There are significant differences between vapour-phase and solution-based techniques. In the latter, stabilizers can be added in a straightforward and controllable fashion, and particles remain dispersed and independent of one another.

In vapour-phase techniques, however, surfactants or stabilizers are not easily added, and, without surface stabilizers, nanoparticles tend to agglomerate into larger particles. The size dispersion of nanoparticles is generally better for solution-based techniques than for vapour-phase techniques.

Vapour-phase techniques can be classified by the physical state of the precursor used as a reagent and by the reaction method conditions; for example, **plasma synthesis** or **flame pyrolysis**. In each case, the reagent or reagent solution is converted to a supersaturated or superheated vapour that is allowed to react, decompose, or cool to initiate nucleation. Solid reagents are vaporized and then recondensed in **gas-condensation** (thermal evaporation to produce vapour), **laser-ablation**, **sputtering**, and **spark-discharge** methods. Liquid, solution, or vapour precursors are used in **spray pyrolysis, flame synthesis, laser pyrolysis, plasma synthesis,** and **chemical vapour deposition (CVD)**. In spray pyrolysis a solution is directed onto a hot surface that causes the solvent to evaporate very quickly, leaving the solid nanoparticulate product on the surface. In flame synthesis or pyrolysis, a liquid or solution is directed into a flame and thermal decomposition yields a nanoparticulate product; this is the main method used for the synthesis of fumed silica where $SiCl_4$ is heated in a flame with oxygen.

Laser pyrolysis uses a direct laser beam to rapidly heat a solution, causing evaporation of the solvent and decomposition to the nanoparticulate. In chemical vapour deposition, the vapour is transported to a substrate where reaction and solid nucleation occurs (Fig. 28.3). In a variation of this procedure, gaseous precursors are delivered into a hot-walled reactor and allowed to react homogeneously in the vapour phase to nucleate solids, and particles are collected downstream. Particle

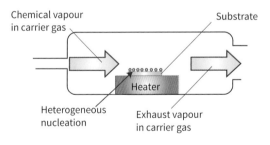

(a)

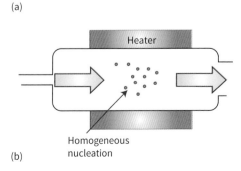

(b)

FIGURE 28.3 Chemical vapour methods to achieve (a) thin film growth and (b) nanoparticle production.

sizes are controlled by the flow rates, precursor chemistry, concentrations, and residence times in the reactor.

Plasma synthesis can be used to synthesize elemental solids, alloys, and oxides, as well as core–shell nanoparticles. In this method, gas or solid particles are fed into a plasma where they vaporize and ionize to highly energetic charged species. Inside a plasma, the temperatures can exceed 10 000 K. On leaving the plasma, the temperature falls rapidly and crystallization occurs under conditions far from equilibrium. The nanoparticles are collected downstream from the plasma zone. Depending on the carrier gas (i.e. the oxidation conditions), elemental solids, core–shell, or compound particles can be formed. Examples of materials synthesized by vapour-phase techniques include metals, oxides, nitrides, and carbides such as SiC, SiO_2, Si_3N_4, $SiC_xO_yN_z$, TiO_2, TiN, ZrO_2, and ZrN.

Both solution and vapour-phase methods can be used to make composite nanoparticles, including core–shell nanoparticles. In both techniques the approach involves growing a second phase on an initial nucleus or nanoparticle. Reaction design to produce core–shell nanoparticles by solution is straightforward provided the solution characteristics of both materials are similar. In practice, however, it is difficult to find materials where the synthesis conditions overlap. Vapour-phase techniques offer another approach to the design of core–shell particles, by injecting a second vapour into a reactor at the growth stage.

28.4 Templated synthesis of nanomaterials using frameworks, supports, and substrates

Heterogeneous nucleation—nucleation that occurs on an existing inert surface—can be used to generate the most important kinds of nanostructures, including zero-, one-, and two-dimensional materials. The methods are similar to those already described: they are either physical or chemical and involve crystallization from either liquids or vapours. The principal difference is that an outside agent is also involved and permits the direct control of the nanoparticle formation. The outside agent can be a framework or a support structure that limits the size of the reaction volume for nanoparticle synthesis by using, for example, inverse micelle frameworks. During two-dimensional thin film growth, the outside agent is a substrate surface.

(a) Nanosized reaction vessels

KEY POINTS By carrying out reactions in nanoscale reaction vessels, the ultimate dimensions of solid products are confined to the vessel size; a reverse micelle has an aqueous core in which reactions can occur.

When the synthesis of a particle is carried out in a 'nanovessel', the ultimate particle size is limited by the vessel size.

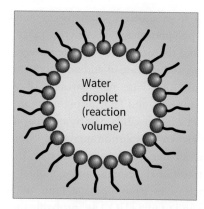

FIGURE 28.4 An inverse micelle. The hydrophilic heads of molecules surround the water droplet and the hydrophobic tails contact the nonpolar solvent, which acts as the dispersed medium. Reactants are solvated in the water droplet and then caused to react in these spatially confined vessels.

One popular route is the **inverse micelle synthesis** approach (Fig. 28.4). An inverse micelle consists of a two-phase dispersion of immiscible liquids; for example, water and a nonpolar oil. By including amphipathic surfactant molecules (molecules having a polar and a nonpolar end), the aqueous phase can be stabilized as dispersed spheres with a size dictated by the water:surfactant ratio. The size of the crystalline particles is limited by the micelle volume, which can be controlled on the nanoscale. Examples of nanoparticles formed by this technique are Cu, Fe, Au, Co, CdS, CdSe, ZrO_2, ferrites, and core–shell particles such as Fe@Au.

(b) Physical vapour deposition

KEY POINTS In physical vapour deposition methods, atoms, ions, or clusters, present as a vapour, adsorb on to the surface and combine with other species to create a solid; molecular beam epitaxy is a technique in which evaporated species from a pure element target are directed as a beam at a substrate where growth occurs.

In **physical vapour deposition** (PVD) methods, vapours are delivered from their source to a solid substrate on which they crystallize. The arriving gaseous species are typically atoms, ions, or clusters of elements. There are several general forms of PVD that are widely used: **molecular beam epitaxy** (MBE), **sputtering**, and **pulsed-laser deposition** (PLD, Fig. 28.5). The gas-phase species can have either relatively low kinetic energies on arrival (as in MBE) or relatively high kinetic energies (as in sputtering and PLD). The most important feature that all PVD methods have in common is the ability to achieve complex film stoichiometries, and because vapour deposition methods allow single atomic layers to be deposited in a controlled fashion on a support or substrate, nanoscale architectures can be built from the bottom up.

Molecular beam epitaxy (MBE) is an ultrahigh vacuum technique for growing thin **epitaxial films**, films that have a definite crystallographic relationship with the underlying

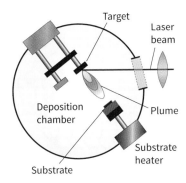

FIGURE 28.5 A pulsed-laser deposition chamber used to produce nanostructured superlattices and artificially layered thin films.

substrate. In MBE, molecular beams are formed by heating elemental sources until atoms evaporate and are transported ballistically at low pressures to the substrate surface. Film stoichiometry and film-growth rates are highly dependent on the beam fluxes, which can be controlled by adjusting the temperatures of the elemental sources. **Homoepitaxy** is the epitaxial growth of a thin film of a material on a substrate of the same material. **Heteroepitaxy** is the epitaxial growth of a thin film of a material on a substrate of a different material. Heteroepitaxy introduces a strain between the growing material and the substrate, caused by crystallographic mismatch of their lattice parameters.

Pulsed-laser deposition (PLD) is a versatile PVD technique that can be used to synthesize a wide variety of high-quality thin films (Fig. 28.5). In PLD, a pulsed laser is used to ablate a target, which releases a plume of atomized and ionized particles from its surface that condenses onto a nearby target. The PLD process usually produces films of composition identical to that of the substrate, which is a major simplification compared with techniques that require fine-tuning or expensive control equipment to achieve a specific stoichiometry. PLD has been used to grow a variety of high-quality superlattices; for example, alternating layers of $SrMnO_3$ and $PrMnO_3$, each only 1 nm thick.

(c) Chemical vapour deposition

KEY POINT In chemical vapour deposition methods, a vapour of molecules chemically interact or decompose at or near the substrate, where they adsorb on the surface and combine with other species to create a solid and residual gaseous products.

The control over complex stoichiometries, the ability to achieve monolayer-by-monolayer growth, and the attainment of high-quality films is not limited to physical vapour techniques. Chemical techniques, such as **metal–organic chemical vapour deposition** (MOCVD) and **atomic layer deposition** (ALD), also provide these levels of control.

In contrast to physical methods, in which species condense directly on to a substrate and react with one another, chemical techniques require that a precursor decomposes chemically on or near the substrate in order to deliver the reactant species to the growing film. Therefore, in chemical vapour methods, the decomposition thermodynamics of the selected precursors must be considered because the vapour often contains elements that must not be incorporated in the growing films. The layout of a typical chemical vapour deposition (CVD) system is shown in Fig. 28.3. These systems normally operate at moderate vacuum or even at atmospheric pressure (in the region of 0.1–100 kPa). Their growth rates can be quite high—more than 10 times that of MBE or PLD.

In CVD techniques, chemical decomposition of the feed molecules proceeds upstream of a substrate surface on which the desired product will be grown. The decomposition of the gaseous reactants is activated by high temperatures, lasers, or plasmas. Large numbers of different materials have been grown using CVD methods. For Group 13/15 (III/V) semiconductors (for example GaAs), the typical sources are organometallic precursors for the Group 13 element (such as $Ga(CH_3)_3$), and hydrides or chlorides for the Group 15 element (such as AsH_3). A typical reaction is

$$Ga(CH_3)_3(g) + AsH_3(g) \xrightarrow{550-650°C} GaAs(s) + 3CH_4(g)$$

carried out in a hydrogen environment.

For complex oxides, as an example the superconducting cuprate $YBa_2Cu_3O_7$, suitable precursors are required for each of the metals; the challenge is to find volatile molecules of the electropositive elements Ba and Y, which normally form ionic compounds. Metal β-diketonates have been found to be useful in this respect, as these compounds often sublime at relatively low temperatures; for example, the 2,2,6,6-tetramethyl-3,5-heptanedione complex of yttrium has a usable vapour pressure at 150°C.

Another approach to molecules that can be used for CVD involves incorporating into a single-molecule precursor more than one of the atom types to be deposited. This procedure has the potential advantage of improving the control of the product stoichiometry. Thus, zinc sulfide can be deposited from a variety of zinc thiocomplexes, such as $Zn(S_2PMe_2)_2$. Ongoing research has the goal of making complex volatile molecules containing two or more different metal atoms in the correct stoichiometric proportions so that they can be deposited simultaneously to make, most commonly, a complex metal oxide or sulfide.

The CVD technique can be fine-tuned to produce very high-quality films. The drawbacks of CVD include the use of toxic chemicals, turbulent flow in the reaction chamber, and the incorporation of unwanted chemical species owing to incomplete decomposition. This technique can be used without a substrate to create nanoparticles by pyrolysis of the vapour species. Furthermore, MOCVD and MBE have been combined to allow for *in situ* monitoring of growth. This technique has been called **chemical beam epitaxy** (CBE).

The final chemical approach described here aims to control the precise chemical interactions that occur at a surface. In the process called **atomic layer deposition** (**ALD**), chemical species are delivered sequentially in controlled amounts to a substrate on which a single monolayer deposits. The excess reactant is then removed. Repetition of this monolayer coverage, subsequent reaction, and removal of excess reactants, allows for precise control over the growth of complex materials layer by layer. In this process, it is necessary to control the chemical species and their interactions to ensure production of monolayer-only coverage. To do so means that the vapour of each reagent must interact in the appropriate manner with any film layer deposited previously. When successful, this technique produces flat, homogeneously coated layers. Some examples of ALD-grown nanomaterials are Al_2O_3, ZrO_2, HfO_2, CuS, SnS, and $BaTiO_3$.

28.5 Characterization and formation of nanomaterials using microscopy

The great advances made in nanoscience and nanotechnology would not have occurred without the ability to characterize the nanoscale structural, chemical, and physical properties of materials. Moreover, the direct observation of nanostructure allows meaningful relationships between processing and properties to be made. The important characterization techniques under the general umbrella of

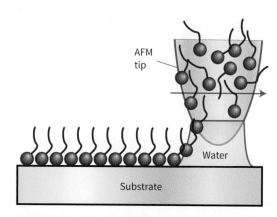

FIGURE 28.6 The process of dip-pen nanolithography. Organothiols move from an AFM tip through a water meniscus, forming a self-assembled monolayer on a gold substrate. (Adapted from C. Mirkin, Nanoscience Boot Camp at Northwestern University, 2001.)

scanning probe microscopy are discussed in Section 8.16. In one related method, **dip-pen nanolithography** (**DPN**, Fig. 28.6), the tip of an atomic force microscope (AFM) is used as an ink pen. By using an ink containing molecular entities, **self-assembled monolayers** (**SAMs**) can be formed through molecular transport of the nanoink to the solid substrate surface. These monolayers often involve specific covalent interactions between the S atoms of organothiols and Au surfaces. Electron microscopy methods (TEM and SEM), described in Section 8.17, have been essential for the visualization of nanoscale structure.

Nanostructures and mesoporous materials: properties and applications

Control over dimensionality in materials can yield unique control over their physical properties; for example, dimensionality has a noted influence on the density of electronic states (Section 2.11) and, therefore, optoelectronic properties. Nanostructured materials also exhibit enhanced physical properties over bulk-phased ones, including strength, toughness, surface area, and mesoporosity. In the next few sections we examine specific examples of how control of dimensionality has been achieved and the resultant useful physical properties that have been reported and exploited in applications. The roles of metal nanoparticles and nanoparticulate metal oxide substrates in heterogeneous catalysts are discussed later in Section 31.12.

28.6 Zero-dimensional (0D) control, nanoparticles of carbon, metals, metal oxides, and metal chalcogenides

Zero-dimensional (0D) nanoparticles have been synthesized for the majority of metals and a wide variety of binary compounds, particularly metal oxides and chalcogenides. In this section, some of the commercially most important 0D inorganic nanomaterials are described and discussed. 0D nanoparticles are synthesized by bottom-up approaches, and a wide variety of shapes—nanospheres, nanorods, nanochains, nanoflowers, nanoreefs, nanoboxes, and nanostars have all been described with reference to the shape of the nanoparticles observed under the electron microscope, Fig. 28.7. Core-shell particles have also been imaged using SEM and TEM, Section 8.17. A key property of a 0D nanoparticle is its high surface area, leading to applications in catalysis (Sections 29.2(c), 31.8, and 31.12) and as adsorbents (Section 29.2(b)).

(a) Carbon black

KEY POINT Carbon black is used as a pigment and reinforcing agent in tyres.

Thermal decomposition or partial combustion of hydrocarbons, mainly oil and gas, leads to the formation of carbon black, which contains a high level of nanoparticles in the

FIGURE 28.7 A variety of shapes observed experimentally for 0D nanoparticles. (a)–(f) show schematics of some of the observed shapes, nanosphere, nanorod, nanocube/nanobox, nanohexagonal prism, nanotrigonal prism, and nanostar/nanoflower: (g)—(j) are experimental SEM/TEM photographs of nanospheres, nanorods, nanocubes, and nanoflowers of silver; (k) shows nanospheres and nanorods of core-shell nanoparticles.

size range 10–100 nm, Fig. 28.8(a). A variety of manufacturing processes are used to produce carbon black commercially. The major ones are: the furnace black process, which involves passing a heavy aromatic crude oil into a furnace at 500°C, the thermal black process in which methane is decomposed at 1200°C in the absence of air, producing hydrogen as a by-product which is then used to heat the furnaces, and the acetylene black process, which thermally decomposes acetylene (ethyne), affording more uniform nanocrystallite sizes. The principal uses of carbon black are as a reinforcing agent in rubber compounds (especially tyres) and as a black pigment in printing inks, surface coatings, paper, and plastics, Section 30.1. Porous carbon blacks are used as absorbents, particularly for gases, and were widely used in gas masks.

(b) Silver and gold nanoparticles

KEY POINT Silver nanoparticles have antibacterial properties and applications.

Silver and gold nanoparticles are typically synthesized using a bottom-up approach involving reduction of their M(I) containing solutions, Fig. 28.7(g–j). Commonly used reducing agents include organic compounds (for example, the reducing sugars maltose and glucose and the citrate anion) and the mild inorganic reductant $[BH_4]^-$. Biological reducing systems, including *Lactobacillus*, may also be used to produce silver nanoparticles. Spherical, triangular, and nanocubic forms have all been synthesized.

While silver (and gold) nanoparticles have demonstrated applications in heterogeneous catalysis (Section 31.12), the main interest is in their biological activity and antimicrobial materials (Section 32.1). Silver nanoparticles are believed to attach to the cell walls of bacteria, and Ag^+ ions can then pass into the cell, causing cell death (through, for example, denaturing ribozomes and interrupting ATP production) or interfering with DNA replication. Commercial applications include refrigerator surfaces, food storage bags, and personal care products. Silver nanoparticles can also be used to carry therapeutic drugs into the human body (Section 32.10). The silver nanoparticles can be coated with a functionalized surface to which a pharmaceutical compound is attached. Once inside the human body, the therapeutic agent can be released from the silver nanoparticles at the specific site of action; for example, a tumour, using a locally focused UV light.

(c) Metal oxide nanoparticles; SiO_2, TiO_2, and ZnO

KEY POINTS Oxide nanoparticles have very high surface areas, leading to applications as absorbents, catalysts, and in sunscreen lotions.

The synthesis of **fumed silica** is typically through flame pyrolysis of $SiCl_4$ with H_2, and the material consists of amorphous nanoparticles of 3–50 nm that aggregate into large units with dimensions 200–500 nm, Fig. 28.8(b). The surfaces of these nanoparticles are covered under ambient conditions with highly reactive Si–OH groups that can also form hydrogen bonds with other molecules. These characteristics lead to the use of fumed silica as an anticaking additive in powders to aid free-flowing, as a reinforcing filler in elastomers, and as a thickener for liquids. Fumed silica can also be used as a mild abrasive in toothpaste. The high surface area of fumed silica, between 50 and 500 m²/g, leads to its application as a catalyst's support in heterogeneous catalysis, Sections 31.8 and 31.15.

Nanoparticulate TiO_2 can be synthesized by rapid precipitation from aqueous solutions formed by dissolving titanium ores in sulfuric acid or by vapour-phase oxidation of $TiCl_4$. TiO_2 pigments produced by these routes have particle sizes in the range 200–300 nm. The ease of its dispersion in oils and the strong UV-absorption properties (wavelength range 280–400 nm) of nanoparticulate (20–50 nm), colourless TiO_2 leads to its widespread use in sunscreen formulations.

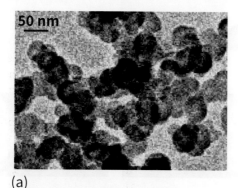

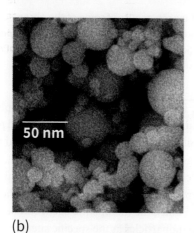

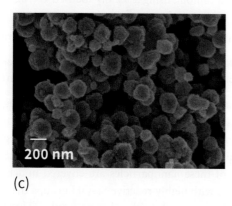

FIGURE 28.8 SEM/TEM micrographs of nanoparticles of commercial (a) carbon black, (b) fumed silica, and (c) TiO_2 pigment nanoparticles.

The pure, bright-white colour of pigment TiO_2 with particle sizes near 225 nm, Fig. 28.8(c), derives from its excellent reflection of visible light, high opacity (poor ability of light to penetrate into the TiO_2 nanoparticles) coupled with almost no absorption. Pigment TiO_2 is widely used in paints, including many coloured paints where its opacity and reflectivity enhances their colour and brightness. TiO_2 pigments are also used in many other consumer products where a bright white appearance is pleasing, including plastic, coatings, paper, and medicines. As TiO_2 is regarded as nontoxic, it is licensed for use in foods including sweets, chewing gum, and cake icing. Nanoparticulate TiO_2, adopting the anatase structure type, is used as a photocatalyst, Section 30.3.

28.7 Special optical properties of 0D nanoparticles and quantum dots

Confinement effects lead to some of the most fundamental manifestations of nanoscale phenomena in materials and are frequently used as a point of departure for the study of nanoscience. Novel optical properties appear in nanoparticles as a result of such effects and are being exploited for information, biological sensing, and energy technologies.

(a) Semiconducting nanoparticles

KEY POINTS The colour of quantum dots is dictated by quantum confinement phenomena and particle localization; quantization of the HOMO–LUMO bands leads to new optical effects involving interband and intraband transitions.

Semiconducting nanoparticles have been investigated intensively for their optical properties. These particles are often called **quantum dots** (QDs) because quantum effects become important in these three-dimensionally confined particles (dots).[2] Two important effects occur in semiconductors when electrons are confined to tiny regions, and the description of the electronic structure reverts partly towards a molecular orbital (MO) model with a large number of electronic energy levels (Section 2.11). Firstly, the band gap increases from the value observed in bulk crystals; secondly, the energy levels of electrons in the conduction band or LUMOs (and holes—the absence of electrons—in the valence band or HOMOs) become quantized (MO-like). Both effects play an important role in determining the optical properties of QDs.

Quantum confinement—the trapping of electrons and holes in tiny regions—thus provides a method of tailoring or engineering the band gap of materials, Section 2.11. The crucial feature is that as the critical nanodimension of a material decreases, the band gap increases. Transitions of electrons between states in the valence band (the so-called HOMO states) and the conduction band (the LUMO states) are called **interband transitions**, and the minimum energy for these transitions is increased in QDs relative to those in bulk semiconductors.

The wavelengths of interband transitions depend on the size of the dots, and it is possible to tailor their luminescence simply by changing their size. A key example of these QD materials is CdSe. By varying the size of the CdSe nanoparticle it is possible to tune luminescent emission over the entire visible spectrum, making them ideal for LED and fluorescent display technologies (Box 28.1).

---

[2] Moungi G. Bawendi, Louis E. Brus, and Aleksey Yekimov were awarded the Nobel Prize in Chemistry in 2023 for the discovery and development of quantum dots.

BOX 28.1 How are CdSe nanocrystals used in LEDs?

Cadmium selenide (CdSe) nanocrystals have found application in a wide variety of commercial venues, including LEDs, solar cells, fluorescent displays, and *in vivo* cellular imaging of cancer cells. Their high photoluminescence efficiencies and emission colour tunability based on nanocrystal size make them attractive for full-colour displays. The most recently developed QD displays also use cadmium-free materials based on indium.

Close-packed QD monolayers have been self-assembled using soft nanolithography contact printing to achieve light-emitting devices (Fig. B28.1). In soft nanolithography, photolithography or electron-beam lithography is used to produce a pattern in a layer of photoresist on the surface of a silicon wafer. A chemical precursor to polydimethylsiloxane (PDMS), a free-flowing liquid, is then poured over the patterned surface and cured into a rubbery solid. After curing, a PDMS stamp that matches the original pattern reproduces features from the master as small as a few nanometres.

Whereas making a master is expensive, copying the pattern on PDMS stamps is inexpensive and easy. One stamp can be used in various inexpensive ways, including microcontact printing and micromoulding to make nanostructures. These techniques can be employed to produce subwavelength optical devices (in which the dimensions of the device element are shorter than the wavelength of the electromagnetic radiation used), waveguides, and optical polarizers used in optical-fibre networks—and eventually, perhaps, in all-optical computers.

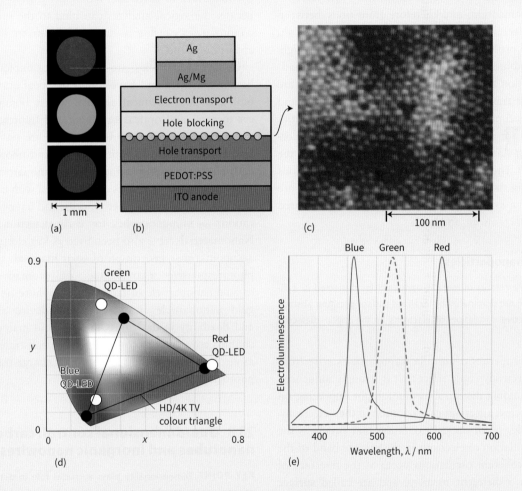

FIGURE B28.1 (a) Electroluminescent (EL) red, green, and blue QD-LED pixels with the device structure shown in (b). (b) Schematic cross-section of a typical QD-LED. (c) High-resolution AFM micrograph showing a close-packed monolayer of QDs deposited on top of the hole-transporting polymer layer, prior to deposition of hole-blocking and electron-transporting layers. (d) Chromaticity diagram showing the positions of red, green, and blue QD-LED colours; an HDTV colour triangle is shown for comparison. (e) Normalized EL spectra of fabricated QD-LEDs corresponding to the colour coordinates in (d). QD-LED images and EL spectra are taken at video brightness (100 cd m^{-2}), which corresponds to the applied current density of 10 mA cm^{-2} for red QD-LEDs, 20 mA cm^{-2} for green QD-LEDs, and 100 mA cm^{-2} for blue QD-LEDs. (Based on *Nano Lett.* 2008, **8**, 12, 4513–17, Copyright 2008, American Chemical Society.)

Another exciting use of QDs is as chromophores for 'biotags', in which dots of different sizes are functionalized to detect different biological analytes. It turns out that although they emit at specific tuneable wavelengths, QDs exhibit broadband absorption for energies above the band gap.

The intriguing point for bioapplications is that it is possible to excite an array of distinct QD chromophores with a single broadband excitation and to simultaneously detect multiple analytes by their distinct optical emissions. These materials have been used to image breast-cancer cells and live nerve cells to track small-molecule transport to specific organelles.

The second manifestation of quantum confinement is that the available quantized energy levels inside a QD have no net linear momentum, and therefore transitions between them do not require any momentum transfer. As a result, the transition probabilities between any two states are high. This lack of momentum dependence also explains the broadband absorption nature of QDs because probabilities are high for most transitions from the occupied valence-band states to unoccupied conduction-band states. Probabilities are also high for **intraband transitions**, transitions of electrons between states in the LUMO band, or of holes in the HOMO band. These relatively intense intraband transitions are typically in the infrared region of the spectrum and are currently being exploited to make devices such as infrared photodetectors, sensors, and lasers.

(b) Metallic nanoparticles

KEY POINT The colours of metallic nanoparticles dispersed in a dielectric medium are dominated by localized surface plasmon absorption, the collective oscillation of electrons at the metal–dielectric interface.

The optical properties of metallic nanoparticles arise from a complex electrodynamic effect that is strongly influenced by the surrounding dielectric medium. Light impinging on metallic particles causes optical excitation of their electrons. The principal type of optical excitation that occurs is the collective oscillation of electrons in the valence band of the metal. Such coherent oscillations occur at the interface of a metal with a dielectric medium, and are called **surface plasmons**.

In bulk particles, the surface plasmons are travelling waves and are characterized by a linear momentum. To excite plasmons using photons in bulk metals, the momenta of the plasmon and the photon must match. This matching is possible only for very specific geometries of the interaction between light and matter, and is a weak contributor to the optical properties of the metal. In nanoparticles, however, the surface plasmons are localized and have no characteristic momentum. As a result, the momenta of the plasmon and the photon do not need to match, and plasmon excitation occurs with a greater intensity. The peak intensity of the surface plasmon absorption for gold and silver occurs in the optical region of the spectrum, and so these metallic nanoparticles are useful as pigments.

The characteristics of plasmon absorption depend strongly on the metal and the dielectric surroundings as well as on the size and shape of the nanoparticle. To control the dielectric surroundings, so-called **core–shell composite nanoparticles** (Fig. 28.2) have been designed in which metallic shells of nanometre-scale thickness encapsulate a dielectric nanoparticle. Metallic nanoparticles and metallic nanoshells are used as dielectric sensors because their optical properties change when they come into contact with different dielectric materials. In particular, biological sensing is of interest because biological analytes can bind to the surface of the nanoparticle, causing a detectable shift in the plasmon absorption band.

Gold nanoparticles are common examples of metallic nanoparticles and have found practical applications as biological and chemical sensors, 'smart bombs' for cancer therapy, and optical switching and fluorescent display materials. Many of these applications are possible owing to the development of techniques to bind photoresponsive chromophores to the surface of the nanoparticles (Section 28.10). Gold nanoparticles can be tagged with biomolecules that permit delivery to specific cells, and have found applications as immunoprobes for early detection of disease. Nanomaterials have also been used as key electron-transfer agents in a new generation of solar materials that couples photoresponsive π-conjugated, polyaromatic molecules to gold nanoparticle surfaces. Chromophore-functionalized gold nanoparticles lead to unique device architectures and flexibility in design, as these hybrid materials can be coupled covalently to conductive glass substrates that serve as electrodes, so leading to improved charge transport and photoefficiencies.

28.8 One-dimensional control: carbon nanotubes and inorganic nanowires

KEY POINT Dimensionality plays a crucial role in determining the properties of materials.

The properties of elongated, one-dimensional nanorods, nanowires, nanofibres, nanowhiskers, nanobelts, and nanotubes have been studied extensively because these one-dimensional systems are the lowest-dimensional structures that can be used for efficient transport of electrons. 1D nanomaterials also demonstrate very high strengths along the length of the nanofibre, -tape, or -tube. The many applications where one-dimensional nanostructures are being exploited include nanoelectronics, very strong and tough composites, functional nanostructured materials, and scanning probe microscopy tips.

A key class of nanomaterials is **carbon nanotubes** (CNTs), introduced in Section 15.6. Carbon nanotubes are perhaps the best example of new nanostructures fabricated through bottom-up chemical synthesis approaches. They have a very simple chemical composition and atomic bonding configuration, but exhibit remarkably diverse structures and unparalleled physical properties. These new nanomaterials have developing applications as chemical sensors, fuel cells, field-effect transistors, electrical interconnects, and mechanical reinforcers. Carbon nanotubes are cylindrical shells formed conceptually by rolling graphene sheets (Section 28.9) into closed tubular nanostructures with diameters matching that of C_{60} (0.5 nm) but exhibiting lengths up to micrometres.

A single-walled nanotube (SWNT) is formed by rolling a sheet of graphene into a cylinder along an (m,n) lattice vector in the graphene plane (Fig. 28.9). The (m,n) indices determine the diameter and chirality of the CNT, which in turn control its physical properties. Most CNTs have closed ends where hemispherical units cap the hollow tubes.

Carbon nanotubes self-assemble into two distinct classes: SWNTs and multiwalled carbon nanotubes (MWNTs). In MWNTs, the tube wall is composed of multiple graphene sheets wrapped concentrically around each other.

CNTs may be synthesized by using a variety of techniques. Laser vaporization methods typically make relatively small amounts of nanocarbons, while specialized CVD techniques were originally developed to synthesize CNTs in quantities in excess of a few milligrams. In the CVD approach, a hydrocarbon gas such as methane is decomposed at elevated temperatures, and C atoms are condensed on to a cooled substrate that may contain various catalysts, such as Fe. This CVD method is attractive because it produces open-ended tubes (which are not produced in the other methods), allows continuous fabrication, and can easily be scaled up to large-scale production. Because the tubes are open, the method also allows for the subsequent use of the nanotube as a templating agent. MWNTs, and to a lesser extent SWNTs, are now manufactured on an industrial scale worldwide, with capacities reaching 100 tonnes per annum in the early 2020s.

In the arc-discharge method, extremely high temperatures are obtained by shorting two carbon rods together, which causes a plasma discharge. Low potential differences and moderately high currents are needed to produce this arc, but these plasmas easily achieve temperatures in excess of where carbon vaporizes (at about 4500 K).

The typical CNT formed by either the arc-discharge method or CVD is multiwalled. To encourage SWNT formation it is necessary to add a metal catalyst—for example, Co, Fe, or Ni—to the carbon source. These metal catalyst

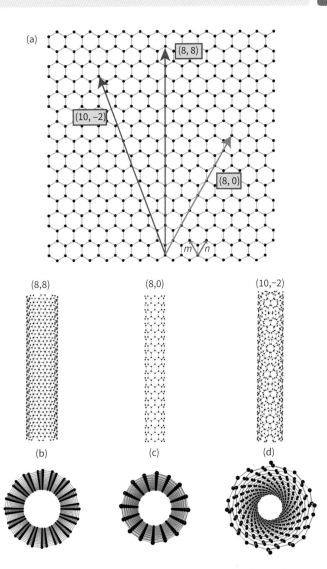

FIGURE 28.9 (a) The honeycomb structure of a graphene sheet. Single-walled carbon nanotubes can be formed by folding the sheet along lattice vectors, two of which are shown as m and n. Folding along the (8,8), (8,0), and (10,−2) vectors leads to armchair (b), zigzag (c), and chiral (d) tubes, respectively. (Based on H. Dai, *Acc. Chem. Rev.*, 2002, **35**, 12, 1035–44. Copyright 2002. Reproduced with permission from the American Chemical Society.)

particles often block the open endcap of each carbon hemisphere and thus promote SWNT growth. In addition, the growth directions of the nanotubes can be controlled by applied electric fields and patterning of the metal catalyst onto different substrates. The patterned-growth approach is feasible with discrete catalytic nanoparticles, and is scalable on large wafers to achieve arrays of nanowires (Fig. 28.10).

The repeating axial hexagonal patterns of CNTs are graphitic structures; however, the electrical properties of nanotubes depend on the relative orientation of the repeating hexagons. The nanotubes can be either semiconductors or metallic conductors. When oriented in the armchair configuration (Fig. 28.9(b)), CNTs exhibit remarkably high electrical

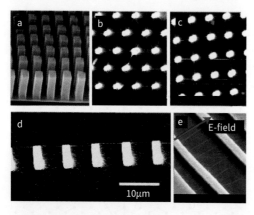

FIGURE 28.10 Ordered carbon nanotube structures obtained by direct chemical vapour deposition synthesis. (a) An SEM image of self-oriented MWNT arrays. Each tower-like structure is formed by many closely packed, multiwalled nanotubes. Nanotubes in each tower are oriented perpendicular to the substrate. (b) SEM top view of a hexagonal network of SWNTs (the line-like structures) suspended on top of silicon posts (the bright dots). (c) SEM top view of a square network of suspended SWNTs. (d) Side view of a suspended SWNT power line on silicon posts (the bright lines). (e) SWNTs suspended by silicon structures (the bright regions). The nanotubes are aligned along the electric field direction. (Based on H. Dai, *Acc. Chem. Rev.*, 2002, **35**, 12, 1035–1044. Copyright 2002. Reproduced with permission from the American Chemical Society.)

conductivity. Electrons can travel through the micrometre-length nanowire with zero scattering and zero heat dissipation; hence CNTs also have very high thermal conductivity, comparable to the best thermal conductors known (diamond, graphite, and graphene). In common with graphene, they are being heralded as the ideal nanomaterial for the development of interconnects for integrated circuits. They may solve two key challenges in the computer industry: heat dissipation, and increased processing speeds.

The main commercial applications of CNTs are derived from their strength and durability and their excellent electrical and thermal conductivity. Production costs are much higher than for carbon black (Section 28.6(a)), which exhibits similar but poorer properties, so applications are limited to those where high performance is required. Current applications include Li-ion battery anodes, where adding a small level to silicon (used for this purpose in some batteries) mitigates against degradation when cycling; the CNTs ensure conductivity between silicon particles as the anode expands and contracts on charge and discharge. Addition of CNTs to polymers, including rubber, silicones, latexes, and thermosetting plastics, improves the strength of the composite and adds electrical conductivity. The latter property leads to applications in antistatic plastic and rubber products.

The strong absorption of visible light found for carbon black, Section 28.6(a), is replicated in bulk assemblies of carbon nanotubes. Growing carbon nanotubes vertically on a substrate leads to a coating which has exceptionally high absorption, with over 99.95% of 663nm light absorbed, and total reflectance across the visible spectrum of below 1.5%. The aligned CNT material can be removed from the substrate surface and added to a solvent for subsequent spray coating. Vantablack® is a commercial black pigment which has applications in removing stray light in optical devices (including cameras and telescopes), and as a specialist artist's pigment, Section 30.1. More recently it has been found that growing CNTs on a cleaned aluminium surface produced a material that absorbs an even higher 99.995% of visible light.

In addition to CNTs, parallel methods have been discovered for making nanotubes out of materials that share bonding characteristics with carbon, including semiconductors and metal oxides. More specifically, BN, ZnO, ZnSe, ZnS, InP, GaAs, InAs, and GaN have all been made into nanotubes. The new electronic properties and small sizes of these nanotubes make them appealing as inorganic nanowires.

Core–sheath nanowires, similar in concept to macroscopic coaxial wires, are also of great interest. For instance, by using a novel nanowire-templating technique based on the layer-by-layer approach and calcining (heating in air to drive off volatile templating agents), ordered Au/TiO_2 core–sheath nanowire arrays have been fabricated. A template-grown gold nanowire array is used as a positive template, and then a cationic polyelectrolyte and an inorganic precursor are assembled on gold nanowires by the layer-by-layer technique. Calcination then converts the inorganic precursor to titanium dioxide (Fig. 28.11).

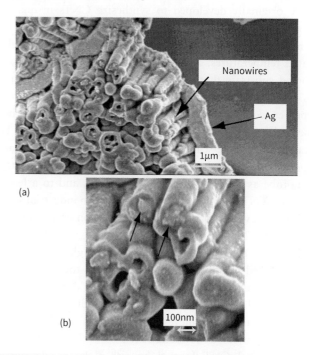

FIGURE 28.11 (a) Low-magnification and (b) high-magnification SEM images of Au/TiO_2 core–sheath nanowire arrays. (Reprinted with permission from Y. G. Guo, et al., *J. Phys. Chem. B*, 2003, **107**, 5441. Copyright 2003, American Chemical Society.)

28.9 Two-dimensional control: graphene, quantum wells, and solid-state superlattices

The best-known two-dimensional structure is graphene: single sheets of carbon atoms forming a hexagonal net as in a single sheet of the graphite structure (Section 15.5). While the interest in graphene centres on its highly unusual and useful physical properties rather than its chemistry, it is worthwhile considering the various routes to its preparation. Other materials which form layered structures, including many transition metal sulfides (Section 25.14), can also be produced in single-sheet form less than a nanometre in thickness.

Several of the processing methods described in Section 28.2 permit the deposition of films only one atomic (or one unit-cell) layer thick, and of many materials. By varying sequentially the types of atomic (or unit-cell) layers being deposited, it is possible to control the material architecture along the growth direction at a subnanometre scale, thereby allowing the bottom-up development of artificially layered nanostructures. A quantum well (QW) is a thin layer of one material sandwiched between two thick layers of another material, and is the two-dimensional equivalent of a zero-dimensional quantum dot (QD, Section 28.7). In a superlattice, two (or more) materials are grown alternately with an artificially induced periodicity along the growth direction. Superlattices often have repeat periods of about 1.5–20 nm or greater, and sublayer thicknesses that range from two unit cells to many tens of unit cells.

Artificial crystal structures, grown layer by layer, usually have repeat distances similar to bulk crystals (about 0.3–2.0 nm) and have sublayer thicknesses that range from an atomic layer to two unit cells (about 1 nm). These structures have found broad commercial application as key electronic device elements in computer chip manufacturing, including hard-disk read heads.

(a) Graphene and other single-layer nanomaterials

KEY POINT Single sheets of graphite, known as graphene, may be obtained by the exfoliation of bulk graphite or by chemical vapour deposition.

Graphene is a single sheet of graphite (Box 15.1) produced by rubbing a lump of graphite; for example, drawing a line with a pencil on a hard surface will produce some graphene sheets. In 2004, physicists at the University of Manchester (UK) and the Institute for Microelectronics Technology in Chernogolovka (Russia) produced graphene from graphite by using adhesive tape. They used the sticky tape (this is sometimes referred to as the Scotch®-tape method) to split or exfoliate graphite into increasingly thinner sheets. Repeated application of this process eventually produces graphite particles less than 0.05 mm thick, including some single layers. The tape can be dissolved in acetone to detach the graphene sheets, which end up suspended in the solvent and can then be deposited by sedimentation or evaporation on to a substrate, e.g. a silicon wafer. In the original experiment, graphene sheets were identified using optical microscopy. Similar exfoliation procedures have been scaled up, and it is now possible to buy multiple-layer graphene in multigram quantities.

This method of producing graphene leads to material with a variety of particle shapes and sizes, and is not ideal for detailed investigation and applications. Other methods of producing graphene as large orientated sheets on substrates mainly involve CVD (Section 28.4). Epitaxial growth via CVD may be undertaken on metal substrates by using various carbon-containing sources, and a typical deposition uses a CH_4/H_2 mixture at $\geq 1000\,°C$. By using a copper foil as the substrate and very low pressures, the growth of graphene automatically stops after the formation of a single graphene layer. Following deposition on a copper foil, the graphene film can be transferred in a roll-to-roll process on to a polymer backing. In this process, a polymer film is pressed on to the top of the graphene-coated copper foil, and the metal is then etched away by acid. In this way, single-layer graphene films with dimensions over 50 cm can be produced—of potential importance for many applications in the electronics industry.

A form of graphene can also be obtained by exfoliation of graphite by using an oxidizing solution of potassium permanganate, $NaNO_3$ and H_2SO_4, though under these conditions a high level of oxygen is incorporated, producing '**graphene oxide**', **GO**. Recently it has been found that microwaves can be used to eliminate much of this oxygen from **GO**, potentially leading to cheap, large-scale production of graphene. However, graphene produced by this route retains a large number of structural defects in the graphene sheets that degrade its mechanical and electrical properties in comparison with graphene produced via exfoliation and deposition techniques.

Thin films produced by the exfoliation of graphite have been known for many decades, but the discovery by Andre Geim and Konstantin Novoselov that graphene has highly unusual but useful physical properties led to the award of the Nobel Prize for Physics in 2010. Graphene is an excellent electron conductor and should theoretically display a resistivity lower than that of the best simple metals, including silver. Graphene also displays very high thermal conductivity ($>5000\,W\,m^{-1}\,K^{-1}$) at room temperature—a value better than those exhibited by carbon nanotubes (Section 28.8), graphite, and diamond. This is potentially important for any future electronic applications based on graphene-containing components. As electronic devices continue to shrink and circuit density increases, low resistive heat losses and high thermal conductivities (that dissipate efficiently any heat generated) should lead to higher device reliability.

Parallel to the layers, graphene is one of the strongest and stiffest materials known, and it is also very lightweight; even so, it can be stretched in the direction of the layers by up to 20% of its initial length. These properties mean that graphene can be added to polymers to make composites (Section 28.10(b)) that have good specific physical properties (e.g. strength per unit mass). As graphene is also electrically conductive, its incorporation into polymers instils a level of conductivity, and such composite plastics do not build up static electrical charges on being rubbed; plastics of this type have uses where static discharges can be detrimental; for example, in the packaging of circuit boards.

When a light source is viewed through graphene it absorbs 2.3% of the light, which makes a single film visible to the naked eye. However, the high overall light transmission, in conjunction with the high electrical conductivity, leads to potential applications in displays—particularly flexible electronic displays and 'smart windows'. In a smart window of this type, a layer of polar liquid-crystal molecules (Section 25.19(c)) is sandwiched between two flexible electrodes comprised of graphene and a transparent polymer. With no voltage applied to the device, the random alignment of liquid crystals scatters light and the smart window is opaque. Application of a voltage across the graphene layers will align the polar molecules, allowing some light to pass through the device, and the smart window becomes transparent.

Similar to the surfaces of most metals and graphite, graphene can adsorb various atoms and molecules; for example, NO_2, NH_3, K, and H_2O/OH. These adsorbate species act as donors or acceptors to the graphene layer and lead to changes in the number of mobile electrons in the film. Measurement of this change in conductivity can be exploited in a sensor for the adsorbed species. Point defects can be introduced into graphene sheets in the form of carbon vacancies, substituted sites (for example, replacing carbon by nitrogen), or atoms added onto its surface. This may lead to applications in spintronics (Section 30.6(c)) based on 'magnetic graphene'. Finally, graphite has important applications as a battery anode material (Section 30.9), and graphene, with its very high surface area, may show enhanced properties for these applications when incorporated into the graphitic anode.

Other recent research on graphene has included applications as a hydrogen storage material, a surface layer to protect glass from corrosion, transparent neural electrodes, quantum LEDs, water purification, high-efficiency photodetectors, and water-impermeable packaging.

Since the discovery that graphite could be exfoliated into, and deposited as, single sheets, chemists have returned to the study of other compounds with layered structures, with a view to investigating their properties when produced in this form. Examples include the metal disulfides MS_2 (M = Ti, Nb, Ta, Mo, and W; Section 25.14) and oxides with layer structures, including V_2O_5. Transition metal carbides, nitrides, and carbonitrides, with formulas that include M_2C, M_2N, M_3C_2, M_4C_3, and M_3CN, can be synthesized as nanosheets and are known as **MXenes**. Boron and germanium may also be grown as one-atom-thick layers, known as **borophene** (various arrangements of boron atoms are known, giving rise to various 'borophenes') and **germanene**, respectively. Borophenes have similar properties to graphene though they can be stronger and more flexible.

(b) Quantum wells

KEY POINTS Quantum wells consist of a thin, small-band-gap material sandwiched between thick layers of large-band-gap materials; multiple quantum wells can enhance the effects of quantum wells when the wells do not interact.

A quantum well is typically composed of two semiconductor materials with different band gaps, such as $Al_{1-x}Ga_xAs$ and GaAs. The smaller-band-gap material (GaAs) is sandwiched between layers of the larger-band-gap material ($Al_{1-x}Ga_xAs$), and the thickness of the layer of the small-band-gap material is confined to the nanometre scale (Fig. 28.12(a)).

The optical properties of quantum wells can be tailored, and both interband (between the valence and conduction band) and intraband (between the quantized sub-bands that are present because of the nanoscale thickness of the material) absorption and emission can be controlled. Quantum-well materials based on both $In_{1-x}Ga_xAs$/GaAs and $Al_{1-x}Ga_xAs$/GaAs have been widely studied, and the optical transition has been observed to move to higher energies compared to bulk GaAs when the nanolayer thickness

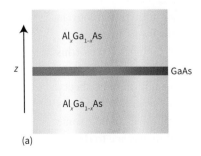
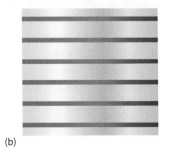

FIGURE 28.12 (a) An $(Al_xGa_{1-x}As)$–(GaAs)–$(Al_xGa_{1-x}As)$ quantum well. The thickness of the GaAs layer is on the nanoscale. (b) A multiple quantum well (MQW) structure formed of the same materials.

drops to below 25 nm. The main use of quantum wells is in semiconductor lasers, where the small-band-gap quantum well is the active layer in the device.

Many of the effects that occur in quantum wells can be enhanced by using **superlattice structures**, the periodic repetition of quantum wells along one direction (Fig. 28.12(b)). In the context of semiconductors these superlattices are called **multiple quantum well** (MQW) structures. If the active layers (the small-band-gap layers) do not interact, then the electrons are confined to a given layer and are unable to tunnel between them. The use of MQW structures in this case increases the absorption or emission from a given device, as there are multiple wells. For example, an MQW laser has much higher power output than the corresponding single-quantum-well laser.

If the large-band-gap layers are thin enough, one QW interacts with the adjacent QW and electrons can tunnel between them. This phenomenon is used in **quantum cascade** (QC) lasers, which operate at high powers in the IR region of the electromagnetic spectrum. The laser characteristics of these materials are fundamentally different from those of semiconductor diodes and MQW lasers in a variety of ways. In a QC laser, only one type of carrier (electrons) is necessary to achieve laser action; in the other two, both electrons and holes are required. Furthermore, in the QC laser the transitions are intraband transitions that arise from the quantization of the valence band. QC lasers are constructed from 13/15 (III/V) semiconducting materials such as GaAs, InAs, and AlAs. All these superlattice systems have been grown by solid-source molecular beam epitaxy on single-crystal substrates.

(c) Solid-state superlattices

KEY POINTS Artificially layered materials have a periodic repeat along the growth direction of a thin film; the periodic repeat is controlled by the number and type of sublayers deposited in sequence.

The ability to grow nanolayers of controlled thickness on substrates, described in Section 28.3, permits the fabrication of artificial nanolayered materials. The periodic repeat along the growth direction of a thin film in a superlattice is controlled by the number and type of sublayers deposited in sequence, whereas the lateral periodicity is determined by the coherence—the matching of lattice characteristics—between the sublayers (Figs. 28.13 and 28.14). The superlattice is built bottom-up, with periods in the nanometre range to produce total thicknesses in the micrometre range.

Superlattice nitrides are ranked among the hardest-known materials. The superlattice period and the chemical composition play important roles in determining the mechanical properties of these nanomaterials, as the interfaces between the two nitride layers are responsible for the enhancement in hardness. Figure 28.15 shows that the maximum hardness of typical superlattice nitrides occurs for periodicities in the range 5–10 nm. These superlattices have been successfully deposited by using sputtering, pulsed laser deposition, and molecular beam epitaxy. Sputtering, by which the source material to be laid down as a thin film on a substrate is bombarded with electrons or ions to produce a gas-phase species, is an economic method for the production of very hard cutting tools.

Perovskite oxides (Section 25.11(d)) are used in numerous applications on account of their ferroelectric, acoustic,

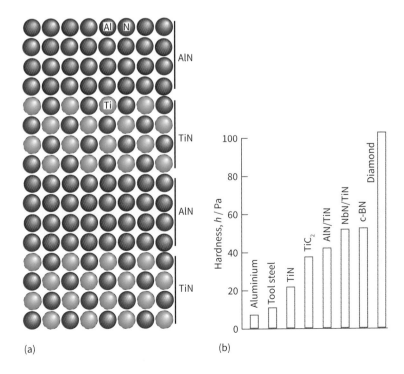

FIGURE 28.13 (a) The structure of an ultrahard AlN/TiN superlattice and (b) the hardness of nitride superlattices and commonly used hard materials. (Adapted from S. A. Barnett and A. Madan, *Phys. World*, 1998, **11**, 45.)

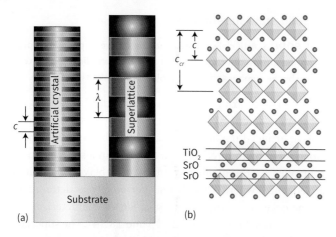

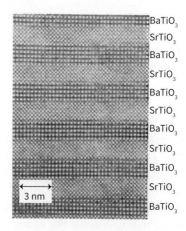

FIGURE 28.14 (a) The structure of an AB artificial crystal and an AB superlattice; c and λ represent the repeat periods during growth of the artificial structure and the superlattice period, respectively. (b) The structure of Sr_2TiO_4 as an artificially layered oxide. The polyhedra represent Ti-centred, corner-sharing TiO_6 octahedra, and the spheres represent Sr^{2+} cations. The lateral coherency between the SrO and TiO_2 sublayers is excellent in this layered structure, and the crystallographic repeat parameter, c_{cr} is twice the growth repeat period, c.

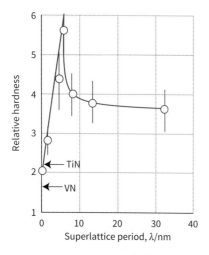

FIGURE 28.15 The dependence of hardness on the superlattice period for $(TiN)_m(VN)_m$ superlattices. The TiN and VN hardnesses are shown for comparison. (Adapted and reprinted from U. Helmersson, et al., Growth of single-crystal TiN/VN strained-layer superlattices with extremely high mechanical hardness. *J. Appl. Phys.* 15 July 1987, **62** (2), 481–4. With the permission of AIP Publishing.)

microwave, electronic, magnetic, and optical properties. The most widely used perovskite is $BaTiO_3$, which is a very important dielectric material (Section 25.11(d)). It has been discovered that nanolayering ferroelectric $BaTiO_3$ with the isostructural perovskite $SrTiO_3$ can lead to enhanced dielectric properties arising from lattice strain. The techniques of PLD, MBE, and CVD have all been used to create $SrTiO_3/BaTiO_3$ superlattice thin films (Fig. 28.16). Although the misfit between the two crystal structures is small enough to allow for epitaxial growth and the formation of coherent interfaces between

FIGURE 28.16 TEM image of an $SrTiO_3/BaTiO_3$ superlattice. (Reprinted from D. G. Schlom, et al., Oxide nanoengineering using MBE. *Mater. Sci. Eng. B*, 2001, **87**, 282. With permission from Elsevier.)

each bilayer, the stress introduced at the interface is sufficient to improve the dielectric response—specifically the remanent polarization (the polarization in the absence of an applied field) of the superlattices, compared to undoped $BaTiO_3$.

Perovskite-based structures can also exhibit interesting magnetic effects. In particular, manganese-based perovskite films of $(La,Sr)MnO_3$ possess useful ferromagnetic and magnetoresistive properties (Section 30.6). The PLD procedure has been used to deposit superlattices of $LaMnO_3/SrMnO_3$; the Mn^{3+} cations are found in the $LaMnO_3$ layers, and the Mn^{4+} are found in the $SrMnO_3$ layers. Thus, the thin-film superlattice technique allows for precise ordering of A-site cations (La, Sr), which in turn causes ordering of the Mn in its different oxidation states. In superlattices of $(LaMnO_3)_m(SrMnO_3)_m$, samples with $m \leq 4$ possess magnetic properties just like solid solutions of $La_{0.5}Sr_{0.5}MnO_3$, whereas superlattices with larger periods have significantly higher resistivities and lower Curie temperatures. Similar effects are observed in superlattices with unequal layer thicknesses, such as $(LaMnO_3)_m(SrMnO_3)_n$ and $(PrMnO_3)_m(SrMnO_3)_m$ systems prepared by laser-MBE (Fig. 28.17).

28.10 Three-dimensional control of nanostructural units: mesoporous materials and composites

The design and synthesis of three-dimensional (3D) supramolecular architectures with tuneable, nanoporous, open-channel structures have attracted considerable attention because of their potential applications as molecular sieves, sensors, size-selective separators, and catalysts (Chapter 30). The ability to fabricate mesoporous materials with pore sizes considerably larger than those achievable with zeolites (Sections 25.16 and 29.1), for which the largest pore size is ~1 nm, is an important advance in the nanomaterials area.

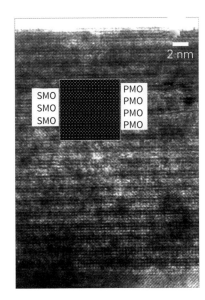

FIGURE 28.17 Cross-sectional TEM image of a 2 × 2 superlattice of SrMnO$_3$ and PrMnO$_3$ prepared using laser-MBE. The inset shows a calculated image confirming the (SrMnO$_3$)$_2$(PrMnO$_3$)$_2$ structure. (Adapted and reprinted from B. Mercey, et al., *In situ* monitoring of the growth and characterization of (PrMnO$_3$)$_n$(SrMnO$_3$) in superlattices. *J. Appl. Phys.* 15 August 2003, **94**, 2816. With the permission of AIP Publishing.)

(a) Mesoporous materials

KEY POINTS Mesoporous materials have ordered pore structures that are defined over the nanoscale; the synthesis of these materials relies upon control of self-assembly; guest species can be incorporated into the inorganic host framework.

An important class of three-dimensionally ordered nanomaterials is **mesostructured nanomaterials**, which includes **mesoporous** materials. Mesoporous materials are exploited in heterogeneous catalysis (Sections 31.8, 31.12, and 31.13) and are of great interest because of the ability to tune the size of the pores present in their structures from 1.5 to 10 nm (Fig. 28.18).

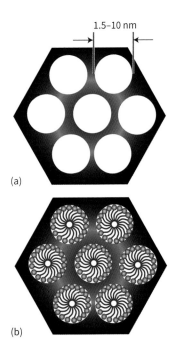

FIGURE 28.18 A hexagonal mesoporous structure with (a) controlled nanoporosity and (b) functionalized pores.

Mesoporous inorganic nanomaterials are synthesized in a multistep process based on the initial self-assembly of surfactant molecules and block copolymers that self-organize into supramolecular structures (which are liquid crystalline assemblies of cylindrical, spherical, or lamellar micelles). These supramolecular frameworks serve as structure-directing templates for the growth of mesostructured inorganic materials (often silica or titania). During a solvothermal reaction step (Section 25.9), oxide particles (silica, Fig. 28.19) form at the surfaces of the hexagonal rods, assembling around the supramolecular structures. The templating agent can be removed through an acid wash or by calcination to yield inorganic materials having hexagonal pores with uniform and controllable dimensions.

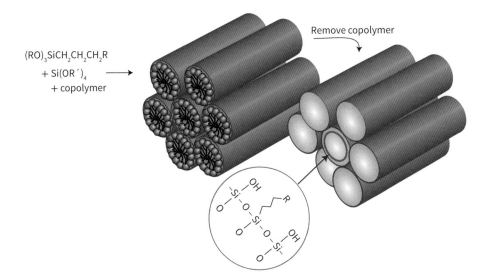

FIGURE 28.19 Block-copolymer structure-directing agents self-assemble into micellar rods forming hexagonal arrays that can be removed to yield nanoporous silica matrices for inclusion chemistry and catalysis. (Adapted with permission from A. P. Wright and M. E. Davis, *Chem. Rev.*, 2002, **102**, 10, 3589–614, Copyright 2002, American Chemical Society.)

A wide range of mesostructured and mesoporous inorganic material structures have been obtained by varying the choice of templating agent and reaction conditions. For instance, the family known as M41S has silica or alumina–silica inorganic phases and various cationic surfactants, leading to three distinct types of structure: hexagonal lamellar (MCM-50), cubic (MCM-48), and hexagonal (MCM-41) (Fig. 28.20). The shape of the surfactant and the water content used during synthesis control the resulting nanoarchitecture, as can be seen in the figure. Surfactants can also be used to tailor the structure; for example, surfactant cetyltrimethylammonium cations (C_{16}TMAC) have been used to fabricate silica nanofibres with hexagonal pores.

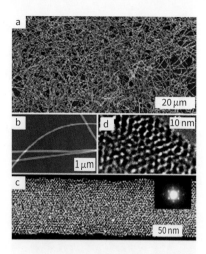

FIGURE 28.21 (a) SEM image of mesoporous silica nanofibres. (b) Low-magnification TEM image of the nanofibres. (c) High-magnification TEM image of one nanofibre; the inset is a selected area electron diffraction pattern of the nanofibre. (d) High-resolution TEM image recorded at the edge of one nanofibre. (From J. Wang, et al., *Chem. Mater.*, 2004, **16**, 5169, Copyright 2004, American Chemical Society.)

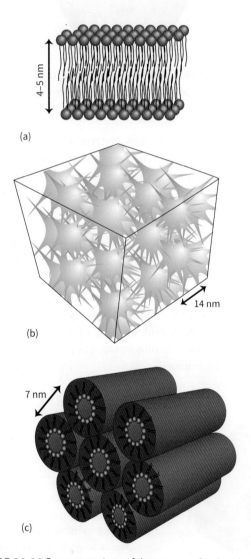

FIGURE 28.20 Representations of three types of ordered mesoporous solids: (a) hexagonal lamellar (layered materials), (b) cubic (complex arrangements), and (c) hexagonal (honeycomb). (Adapted with permission from A. Mueller and D. F. O'Brien, *Chem. Rev.*, 2002, **102**, 3, 727–58, Copyright 2002, American Chemical Society.)

Mesoporous nanomaterials also offer routes to functionalize pores to increase catalytic activity and selectivity. They have received much attention as host materials for the inclusion of guests such as organometallic complexes, polymers, d-metal complexes, macromolecules, and optical laser dyes. The silica nanofibres shown in Fig. 28.21 can be used as hosts to grow nanowires of various other oxide materials.

(b) Inorganic–organic nanocomposites

KEY POINTS Class I inorganic–organic composite nanomaterials have noncovalent interactions, and class II inorganic–organic composite nanomaterials have some covalent interactions; sol-gel and self-assembly methods are key chemical routes to hybrid nanocomposite design and synthesis.

Inorganic–organic nanocomposites are a class of three-dimensional ordered materials whose chemical and physical properties can be tuned by the association of organic and inorganic components at nanoscale. These hybrid materials originated in the paint and polymer industries, where inorganic fillers and pigments were dispersed in organic materials (including solvents, surfactants, and polymers) to fabricate commercial products with improved materials performance. Hybrid nanomaterials offer the materials chemist new routes that use rational materials design to optimize structure–property relationships. Their nanoarchitectures and resulting properties depend on the chemical nature of the components and the synergy between them, producing, for example, enhanced strength and robustness. A key part of the design of these hybrids is the selective tuning of the nature, extent, and accessibility of the interfaces between the inorganic and organic building blocks.

Nanocomposites have already entered the marketplace in sunscreens, fire-retardant fabrics, stain-resistant clothing, thermoplastics, water filters, and automobile parts. Examples include the use of nylon-6/montmorillonite clay nanocomposites for timing-belt covers, television screens that are coated with indigo dyes embedded in a silica/zirconia matrix, organically doped sol–gel glassware, and sol–gel-entrapped enzymes.

Polymer nanocomposites (PNCs) are composed of inorganic nanoparticles dispersed in a polymeric matrix. Early commercial PNCs used two-dimensional ordered or lamellar clays (Section 25.17(a)), such as sodium montmorillonite (Na-MMT), dispersed in the polymer matrix. Such dispersants (or fillers) have sandwich-type structures (with channels between layers), a total thickness of 0.3–1 nm, and a length of 50–100 nm for each layer; these sandwich structures typically form micrometre-sized agglomerates. The dispersion of the inorganic phase within the organic polymeric matrix in a PNC is accomplished by intercalation (inserting the polymer between the layered sheets of the clay). Intercalation involves the expansion of the interlamellar spacing as a consequence of ion exchange, using organic amines or quaternary ammonium salts as described in Section 25.17(b). An alternative route involves exfoliation of the clay into individual nanosheets followed by condensing the two phases, clay and polymer, together; this can be achieved through dissolution methods or by intensive melt mixing of the phases.

The nature of the filler (dispersant) and any cavities (voids) present in a PNC greatly influences the mechanical properties of the composites because these can control the distribution of stress through the composite matrix. Both the yield strength (a measure of the material's ability to resist permanent deformation) and toughness (a measure of the energy absorbed prior to fracture) of nanocomposites are controlled by the nanoparticle size and dispersion, and nanoparticle-to-polymer contact interactions. Therefore, control of the nanoparticle dispersion and alignment of these fillers offers a way to tailor the mechanical properties of the nanocomposites.

One important class of PNCs uses nanocarbons as the dispersants. SWNTs (Section 28.8) have exceptional mechanical properties; their very high Young's modulus (resistance to mechanical stress), low densities, and high length-to-width ratios make them very attractive as fillers to enhance polymer strength. SWNTs have high tensile strength—almost 100 times that of steel—whereas their density is only about one-sixth that of steel. SWNT-reinforced composites have been developed by the physical mixing of SWNTs in solutions of preformed polymers, *in situ* polymerization in

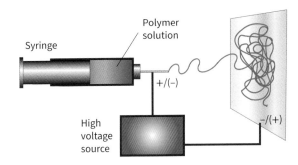

FIGURE 28.22 The electrospinning apparatus. (Adapted with permission from *Nano Lett.*, 2004, **4**, 3, 459–64, Copyright 2004, American Chemical Society.)

the presence of SWNTs, surfactant-assisted processing of SWNT/polymer composites, and chemical modification of the incorporated SWNTs.

Melt processing can be carried out easily by compression moulding at high temperatures and pressures, followed by rapid quenching. So-called **electrospinning methods** use electrostatic forces to distort a droplet of polymer solution into a fine filament, which is then deposited on a substrate (Fig. 28.22).

The nanofibres created from electrospinning can be configured into a variety of forms, including membranes, coatings, and films, and can be deposited on targets of different shapes. Fibres can be prepared with diameters smaller than 3 nm, with high surface-to-volume and length-to-diameter ratios, and with controlled pore sizes. Electrospun polymer nanocomposites have been made with homogeneously dispersed SWNTs and have exhibited significantly improved mechanical strength.

Different organofunctional groups, covalently attached to the nanotubes through oxidation reactions, have been used to improve the chemical compatibility of the nanotubes with specific polymers (Fig. 28.23). The use of SWNTs with different functionalities allows the study of interfacial interactions between the filler and the polymer matrix. This chemical functionalization is an effective approach to improving the processability of the nanocomposite and the chemical

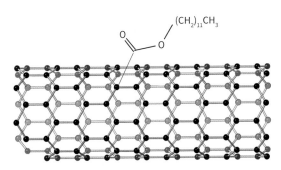

FIGURE 28.23 Ester-functionalized SWNT (SWNT–COO(CH$_2$)$_{11}$CH$_3$). (Based on *Nano Lett.*, 2004, **4**, 3, 459–64, Copyright 2004, American Chemical Society.)

compatibility of the components. Specific functionalization can be used to separate the SWNT bundles and prevent their agglomeration.

Metal-oxide fillers have also been used to optimize polymer strengths and thermal properties. A particularly interesting example involves control over the mechanical properties of alumina/polymethylmethacrylate (PMMA) nanocomposites by engineering the distribution of weak particle-to-polymer interactions. The filler nanoparticles (alumina) are dispersed in the composite matrix in the form of 'net-like' structures as opposed to 'isolated-island' structures formed when microparticles are dispersed in the polymer matrix. In contrast to microparticle/polymer composites, these net-like structures inhibit crazing, the propagation of microscopic cracks during tensional loading. Alumina/PMMA composites also display higher yield strengths and a shift from brittle to ductile behaviour when nanoparticulate alumina is used (Fig. 28.24). Ductility is advantageous, as the composite can then be drawn into thin wires and can withstand sudden impact, making the composite more mechanically resilient. The addition of nanoalumina with the antiagglomerant methacrylic acid lowers the glass transition temperature enough to change the stress–strain curve of the composite and shift from brittle to ductile under tensile loading.

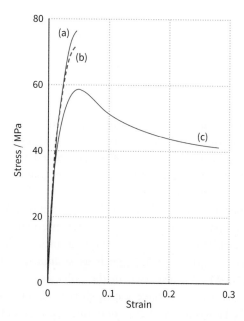

FIGURE 28.24 Typical stress–strain curves for (a) neat PMMA, (b) 2% by mass as-received micrometre-sized alumina-filled PMMA composite, and (c) 2.2% by mass 38 nm (MAA) alumina/PMMA nanocomposite. Although the strength decreases slightly for (c) (related to the decreased stress at the curve maximum), the overall ductility (related to the total strain) and toughness (related to the area under the curve) are greatly improved. (Adapted with permission from B. J. Ash et al., *Macromolecules* 2004, **37**, 1358–69, Copyright 2004, American Chemical Society.)

FURTHER READING

G. A. Ozin and A. C. Arsenault, *Nanochemistry: A Chemical Approach to Nanomaterials*. Royal Society of Chemistry (2008). An invaluable reference book for undergraduate and graduate students, with a discussion of advances in chemical patterning, self-assembly, and nanomaterial synthesis.

C. P. Poole and F. J. Owens, *Introduction to Nanotechnology*. Wiley-Interscience (2003). Key chapters on a variety of nanomaterials systems, including quantum structures, magnetic nanomaterials, nanoelectromechanical systems (NEMS), carbon nanotubes, and nanocomposites, with emphasis on characterization and synthesis strategies.

M. Wilson, K. Kannangara, G. Smith, M. Simmons, and B. Raguse (eds.), *Nanotechnology: Basic Science and Emerging Technologies*. CRC Press (2002). This is another good introductory book on many areas of nanotechnology.

M. Meyyappan, *Inorganic Nanowires: Applications, Properties, and Characterization*. CRC Press (2012). Comprehensive and coherent account of advances in nanofabrication, characterization tools, and research on inorganic nanowires (INWs).

T. K. Sau and A. L. Rogach (eds.), *Complex-Shaped Metal Nanoparticles: Bottom-up Syntheses and Applications*. Wiley-VCH (2012). Covers all important aspects and techniques of preparation and characterization of nanoparticles with controlled morphology and architecture.

W. Choi and J.-W. Lee (eds.), *Graphene: Synthesis and Applications*. CRC Press (2012). Reviews the advancement and future directions of graphene research in the areas of synthesis and properties, and explores applications such as electronics, heat dissipation, field emission, sensors, composites, and energy.

C. Altavilla and E. Ciliberto, *Inorganic Nanoparticles: Synthesis, Applications, and Perspectives*. CRC Press (2012). Presents an overview of these materials and explores the myriad ways in which they are used.

J. Liu, S. Bashir, and X. Wang. *NanoChemistry: From Theory to Application for In-Depth Understanding of Nanomaterials*. De Gruyter (2022).

M. V. Putz (ed.), *New Frontiers in Nanochemistry: Concepts, Theories, and Trends: Volume 1: Structural Nanochemistry; Volume 2: Topological Nanochemistry; Volume 3: Sustainable Nanochemistry*. Apple Academic Press (2020). A comprehensive three-volume set on all aspects of nanochemistry.

EXERCISES

28.1 (a) Compare the surface area for two spherical objects: one has a diameter of 10 nm and the other has a diameter of 1000 nm. (b) Describe whether these two objects are considered nanoparticles by using the size-based definition of a nanomaterial. (c) Using a surface-area-related property, describe what must hold true for either of these objects to be considered nanoparticles, using the size/properties-based definition of a nanomaterial.

28.2 (a) Explain the difference between the top-down and bottom-up methods of fabrication of materials. Be specific, and provide one example of each. (b) Give one advantage and one disadvantage for each synthesis method.

28.3 (a) Describe the three basic steps in nanoparticle formation from solution. (b) Explain why two steps should occur independently to achieve a uniform size distribution. (c) What are stabilizer molecules used for in nanoparticle synthesis?

28.4 Describe the processes occurring during Ostwald ripening.

28.5 (a) Draw a schematic diagram of a core–shell nanoparticle. (b) Briefly describe how core–shell nanoparticles could be made by using either vapour-phase or solution-based techniques.

28.6 Explain whether vapour-phase or solution-based techniques typically lead to (a) larger size distributions in nanoparticle synthesis, (b) agglomerated particles that are strongly bonded to one another in so-called hard agglomerates.

28.7 (a) Discuss the difference between homogeneous and heterogeneous nucleation from the vapour phase. (b) Which type of nucleation is preferred for the growth of a thin film in this process? (c) Which type of nucleation is preferred for the growth of nanoparticles in this process?

28.8 Compare and contrast the band energies for a quantum-dot nanocrystal and a bulk semiconductor.

28.9 Describe the difference between a physical vapour and a chemical vapour with respect to the type and stability of the vapour species.

28.10 Compare the structures and properties of carbon black, graphite, and graphene.

28.11 Which inorganic materials, other than graphite, have structures with strong in-layer bonds but much weaker interactions between layers, and so might be exfoliated into single sheets?

28.12 Describe the different structures and properties of borophene.

28.13 (a) Describe the two classes of inorganic–organic nanocomposites based on their bonding types. (b) Give one example of a nanocomposite in each class.

TUTORIAL PROBLEMS

28.1 Describe the latest advances in the synthesis, properties, and applications of core–shell nanoparticles.

28.2 (a) Give two examples of applications of quantum wells. (b) Describe why quantum wells are used and if either molecular materials or traditional solid-state materials can exhibit similar properties. (c) How are quantum wells made?

28.3 (a) What is the relevance of self-assembly to the fabrication of nanomaterials? (b) What role will it play in nanotechnology?

28.4 Discuss the synthesis, properties, and potential applications of ultrathin two-dimensional nanosheets of layered transition metal dichalcogenides. (See M. Chhowalla et al., *Nature Chemistry*, 2013, **5**, 263; S. Chandrasekaran, *Chem. Soc. Rev.*, 2019, **48**, 4178–4280.)

28.5 The synthesis method described in this chapter for generating CdSe quantum dots involves the use of rather toxic compounds. Explore the chemical literature to find a more recent example that uses less toxic substances. Describe the solvation step, the nucleation step, and the growth step in each case. Comment on the size dispersions in each case. (See the following articles as a start: G. C. Lisensky and E.M. Boatman, *J. Chem. Educ.*, 2005, **82**, 1360; W. William Yu and X.-G. Peng, *Angew. Chem., Int. Ed. Engl.*, 2002, **41**, 2368.)

28.6 The Grätzel cell has been described as a useful photoelectrochemical cell. Describe how the photoelectrochemical cell differs from the photovoltaic cell. Describe why the nanostructure TiO_2 is important for the improved performance. What other inorganic species play an important role in functionalizing the Grätzel cell? (See M. Grätzel, *Nature*, 2001, **414**, 338.) What advances have been made with Grätzel cells since 2010? (See A. B. Muñoz-García, *Chem. Soc. Rev.*, 2021, **50**, 12450–12550.)

28.7 Discuss how the various physical properties of carbon nanotubes and graphene are leading to the incorporation of these materials into current and developing technologies. How do the costs of graphene and carbon nanotubes determine their use in these technologies instead of the much cheaper carbon black?

28.8 Carbon nanotubes have been suggested for use as wires in molecular electronics. Describe the challenges of using nanotubes as wires in terms of connecting two functional electronic devices. Describe a possible technique to overcome some of these problems.

28.9 Nanotubes are widely known for carbon. Find an example of an inorganic nanotube not based on carbon, and describe its synthesis and properties as compared to the corresponding bulk material. Compare its structure to the carbon nanotubes discussed in this chapter.

29 Framework and porous materials

Zeolites

29.1 Advances in zeolite design and synthesis

29.2 Applications of zeolites

Zeotypes and metal–organic frameworks

29.3 Inorganic structures based on linked polyhedral metal centres

29.4 Metal–organic frameworks

Further reading

Exercises

Tutorial problems

Framework structures and microporous solids have been introduced in Section 15.15 and, in particular, Sections 25.16 and 25.17, with the initial descriptions of zeolites, zeotypes, and MOFs. This area of inorganic chemistry research is highly active as a result of these materials' great functionality that allows them to absorb, store, and release small molecules, facilitate catalytic reactions, and undertake cation exchange reactions. These properties lead to widespread applications in molecular separation and storage, ion exchange, and heterogenous catalysis. In this chapter we describe the latest advances in the design and synthesis of materials of this type, focusing on their applications in molecular sieving and separation and ion exchange. Heterogeneous catalysis involving porous materials is discussed in Section 31.13.

Zeolites

29.1 Advances in zeolite design and synthesis

KEY POINT New zeolite framework structures are often synthesized by using complex template molecules.

Zeolites are constructed from TO_4 tetrahedra (T = Si, Al, P, Be...) linked through all vertices forming a three-dimensional framework, Section 15.15. Other elements able to form stable tetrahedral MO_4 species that can link together into framework structures are the later 3d series and the lighter p-block metals and metalloids, and frameworks containing these species are often classed as **zeotypes**. All the aforementioned ions are so small that they coordinate strongly to four O atoms in preference to higher coordination numbers; the principal examples are SiO_4, AlO_4, and PO_4, although GaO_4, GeO_4, AsO_4, BO_4, BeO_4, LiO_4, $Co(II)O_4$, and ZnO_4 are all well known in these structural types. Other tetrahedral units that have been found only rarely in framework structures include $Ni(II)O_4$, $Cu(II)O_4$, and InO_4.

In this section we concentrate on the design and synthesis of structures derived from the most commonly found, very stable, and commercially important tetrahedral units: the aluminosilicate zeolites, $(Si,Al)O_4$ frameworks, and aluminophosphates, $(Al,P)O_4$ frameworks.

The role of structure-directing agents in controlling the synthesis of zeolitic materials with specific framework topology was described in Sections 15.15 and 25.16. Following the discovery of a large number of new microporous structures from the 1950s onwards, many of which were prepared in the laboratory using organic templates, more recent work on zeolites has been directed towards the systematic study of template–framework relationships. These studies can be divided into two major categories. One is to understand the interaction between the framework and template through computer modelling and experiment. The other is to design templates with specific geometries to direct the formation of zeolites with particular pore sizes and connectivity for applications in molecular separation and catalysis.

One area that has become the focus of much attention is the use of bulky and complex organic and organometallic molecules as templates in the quest for new, very large pore structures. These large-pore zeolites have applications in catalysis, where more complex and large organic molecules can enter the zeolite pores and undergo transformations. Another area of research focuses on small-pore zeolites for small-molecule separations and CO_2 sequestration.

(a) Small- and large-pore zeolites and their applications

KEY POINT Different zeolites have applications based on their pore size, which can range between 0.3 and 1.0 nm (3–10 Å).

The 248 different known zeolite structures (classified according to a three-letter code; for example, SOD (sodalite), FER, FAU (faujasite), etc., may be further classified according to the largest pore-window or ring size present, Fig. 29.1. These windows are formed from the linked tetrahedra creating the zeolite framework; zeolites will contain a variety of ring sizes ranging from between 4 and 14 or more tetrahedra. Combinations of 4, 6, and 8 rings are common to the majority of zeolite structures, while larger ring sizes, which are less stable, only occur in a few zeolites. The number of SiO_4 or AlO_4 tetrahedra forming the window or ring delineates its size, and the sizes of the window for some of these configurations are shown in Fig. 29.1.

There is some variation in the diameter of the windows for the same number of tetrahedra, as the T-O-T angle (T= Si, Al) can vary between 110° and 170°, and the framework is flexible to deformations. Furthermore, windows are often not regular in shape, in that a window formed from 8 tetrahedra may appear octagonal but is more likely distorted octagonal or approximately rectangular (derived from alternating T-O-T angles in the ring of near 90° and 180°) in geometry. Windows formed from double rings also occur.

The largest ring size present in different zeolites extends from a 6-membered ring in the sodalite structure (SOD), through 8 in the GOO and EDI structure codes, to 12 or 14 in, for example, DON and CFI. Small-pore zeolites typically have 8 membered rings of dimensions 3.0–4.5 Å; medium-pore zeolites—for example, the FER and MFI structure codes—have 10 rings; large-pore zeolites, which contain 12 rings—for example, FAU—have windows in the range 6–8 Å; and extra-large-pore zeolites have ring sizes greater than 8 Å (for example, DON, 14-ring pore of dimension 8.8 Å).

Since the discovery of large-pore zeolites in the 1960s, most research has concerned their applications in the petrochemical industries and catalysis. These applications involving heterogeneous catalysis using zeolites, including catalytic

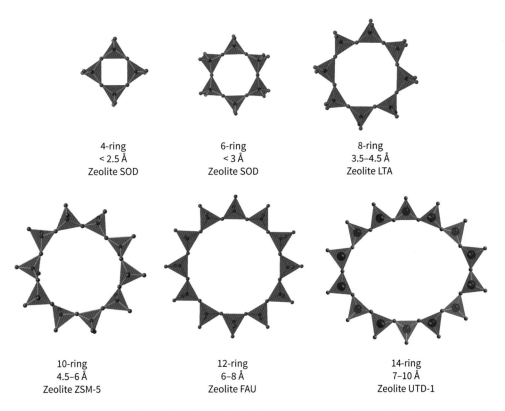

FIGURE 29.1 The ring sizes in zeolites and the number of tetrahedra forming them ranges from 4 to 14; the associated pore or window dimension increases from under 3 Å to 10 Å.

cracking, transalkylation, xylene isomerization, and ethylbenzene synthesis, are covered in more detail in Chapter 31. Recently a considerable level of research on zeolites has focused on small-pore systems for catalysis, particularly the CHA structure type, which are used commercially for NOx reduction and methanol to alkene conversions.

(b) Synthesis of zeolites: classical experimental methods and large pores.

KEY POINT New zeolite framework structures are synthesized using complex template molecules.

The introduction of template molecules or structure directing agents (SDA) into hydrothermal reaction mixtures to control the zeolite framework formed has been described in Section 25.16. This synthesis method has been developed over the last 50 years with the use of ever larger and more complex SDAs. For example, this approach was used to make the first zeolitic materials containing 14-ring channels. Thus, the microporous silica UTD-1 (Fig. 25.54(a)) has been prepared by using the bulky bis(pentamethylcyclopentadienyl) cobalt(II) [Co(Cp*)$_2$], and the siliceous CIT-5 (CFI structure type) has been prepared by using a polycyclic amine and lithium (Fig. 25.54(b)).

Other more complex amines have been synthesized with the aim of using them as templates in zeolite synthesis and to obtain desirable pore geometries for catalysis (Section 31.13). The pure silica zeolite ITE, Fig. 29.2, has been synthesized using a variety of SDAs, including the 1-azoniatricyclo[4.4.4.0^{1,6}]tetradecane cation, the 1,3,3,6,6-pentamethyl-6-azoniabicyclo[3.2.1]octane cation, and [Ni(PEHA)] (PEHA = pentaethylenehexamine, (N'-[2-[2-[2-(2-aminoethylamino)ethylamino]ethylamino]ethyl] ethane-1,2-diamine). All these SDAs are of similar size and template the formation of the main ITE cavity.

The addition of fluorides into the zeolite precursor gel improves reaction rates and acts as a template for some of the smaller cage units; for example, where linked TO$_4$ (T = Si, Al, P, etc.) tetrahedra arranged at the corners of a cube surround a central F$^-$ ion.

There has been a significant increase in the proportion of synthetic zeolites that can be made in (essentially) pure silica form, so that more than 20 structural types of zeolitic silica polymorphs are now known. The use of low H$_2$O/SiO$_2$ ratios in the reaction mixture is a key factor in producing these materials, and the new 'silica' phases so produced are of unusually low density. For example, a purely siliceous framework with the same topology as the naturally occurring mineral chabazite, Ca$_{1.85}$(Al$_{3.7}$Si$_{8.3}$O$_{24}$), is the least dense silica (SiO$_2$) polymorph known, with, according to the normal atomic radii, only 46% of the unit cell volume occupied. These pure silica zeolites are hydrophobic, leading to specific applications in molecular absorption of low-polarity molecules and catalysis (Section 31.13). Zeolite ITE, Fig. 29.2, is another example of a purely siliceous zeolite.

(c) Theoretical design and controlled synthesis of large-pore zeolites

KEY POINT Complex, large molecular ions, sometimes designed through computer modelling, can template zeolites with large pores and channels.

The majority of the 248 known and different zeolite topologies have been obtained using hydrothermal methods and, as discussed previously, trialling a wide variety of different template species (SDAs) experimentally. More recently, research has focused on a designed approach to zeolite synthesis that allows the control of zeolite pore size and interconnectivity, avoids the use of environmentally unfriendly and costly organic templates and solvents, and permits good control of zeolite-particle morphology. For example, the synthesis of some commercially important zeolites uses large and complex organic amines such as (N,N,N-trimethyl-1-adamantammonium hydroxide) for SSZ-13 and (N,N-diethyl-5,8-dimethylazonium bicyclo[3.2.2.]nonane) for SSZ-52. Computational design of templating species that mimicked the pore-structure-directing properties of the complex amines showed that cheaper and more environmentally friendly alternatives (copper tetraethylenepentamine for SSZ-13 and N-ethyl-N-(3,3,5-trimethylcyclohexyl)pyrrolidinium for SSZ-52) could instead be employed successfully.

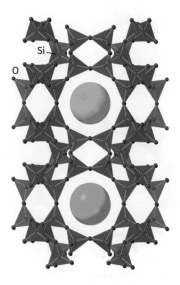

FIGURE 29.2 The zeolite ITE structure showing the cavities, represented by the spheres, templated by the SDAs.

It is also important that the template species promotes the formation of a zeolite crystal of the desired crystal structure (which can then grow in size without the need for further template) rather than be consumed as an initial seed crystal grows, as the latter method wastes expensive template molecules when it is removed from the zeolite through thermal decomposition. This waste and expense is contrary to a number of the 12 principles of green chemistry, Chapter 27.

Almost organotemplate-free syntheses of zeolites, including zeolite beta, silicalite, and ZSM-22, have been obtained by seeding reaction solutions containing no template, particularly in the presence of ethanol, which fills the growing pores. Methods of synthesizing zeolites to reduce the amount of solvent, including waste water, are also being researched, Section 27.5. Nanoparticulate forms of zeolites, Chapter 28, including sheets and rods, can be synthesized, and nanosheets of zeolite FER with a thickness of only 3 nm can be obtained using a template species that promotes growth in just two dimensions.

(d) Synthesis of small-pore zeolites

KEY POINT Zeolites with a maximum 8-ring pore size can be synthesized with controlled Si:Al ratios.

With the increasing importance of small-pore zeolites in catalysis and NOx removal, Section 31.13(c) and Box 31.1, efficient methods of generating zeolites with 8-rings as the maximum window size and controlled aluminium levels are being developed. In common with large-pore zeolites, materials with high silicon content or purely siliceous frameworks are targets; these compositions demonstrate the best small-molecule separation efficiencies (for example, CH_4/CO_2 and propene/propane) and the highest catalytic activities and specificities for processes such as partial methane oxidation and methanol to olefin (MTO) reactions.

High-silicon-content zeolites are also stable to higher temperatures, and decompose more slowly in steam compared to materials with low Si:Al ratios. As examples, the zeolites AEI, KFI, and CHA all have structures based on 8-rings, but while CHA can be made in a purely siliceous form, the highest Si:Al ratios are 25 and near 4 for AEI and KFI respectively. Methods of increasing the silicon content of AEI and KFI zeolites without using expensive SDAs are being sought. A KFI zeolite structure with Si:Al ratio of 3.5 can be templated using a mixture of Na^+ and Cs^+ or Sr^{2+}.

(e) Aluminophosphates, ALPOs, and SAPOs

KEY POINT ALPOs are a large class of zeolites with applications in hydrocracking.

The formula $AlPO_4$ is isoelectronic with $2 \times SiO_2$, and **ALPOs** (aluminium phosphates) represent a large class of inorganic materials that are isostructural with the purely siliceous zeolites. They are normally synthesized by utilizing hydrothermal methods using phosphoric acid and an aluminium source, such as aluminium nitrate or aluminium ethoxide, in the presence of an SDA. Around 50 different ALPO structures are known, of which 20 have zeolite equivalents, including ALPO-42 with the zeolite LTA structure code, ALPO-37 (FAU), and ALPO-34 (CHA). The framework structures have strictly alternating $[PO_4]$ and $[AlO_4]$ tetrahedra, and because overall the framework is uncharged (formally consisting of $[PO_{4/2}]^+$ and $[AlO_{4/2}]^-$ tetrahedra with all oxygen atoms shared) there are no charge-balancing cations in the pores. However, in comparison with siliceous zeolites, ALPOs are hydrophilic. The inclusion of a small level of silicon, yielding **SAPOs** (silicon aluminium phosphates), produces zeolites containing a large number of acidic sites, randomly distributed in the structure, for catalysis, including hydrocracking (Section 31.13(a)).

29.2 Applications of zeolites

Zeolites are used in a wide variety of applications covering ion exchange (in detergents and nuclear waste remediation), catalysis, and molecular absorption: other bulk applications of zeolites include as additives to cements to increase strength and as soil improvers. Around four million tonnes of zeolite are used commercially each year with a market size of $6.5 billion. The major uses of zeolites, as bulk chemicals, absorbents for small molecules, and as ion exchangers, are described in this chapter. Heterogeneous catalysis using zeolites (and related microporous materials including the aluminophosphates) is described in detail in Section 31.13.

(a) Natural zeolites in the construction, agricultural, and animal feed industries

KEY POINT The naturally occurring zeolite clinoptilolite, HEU, is used widely used in cements and fertilizers.

Bulk chemical applications of mined natural zeolites derive from their ability to absorb small molecules, including water and ammonia, and undergo ion exchange reactions that trap or release ions. Around 3 million tonnes of zeolites are open-pit mined annually, mainly clinoptilolite, in China, South Korea, Japan, and Jordan. Naturally occurring zeolites, including clinoptilolite, can be added to cements and concretes to improve their properties. By absorbing and releasing water and other species, including sulfate and chloride ions, during the cement curing process (Box 13.2), the presence of a zeolite modifies the physical properties of the final cement or concrete. These modifications to the reaction chemistry during concrete formation leads to enhancements in engineering properties, including reduced shrinkage, lower water permeability and associated frost resistance, carbonation resistance, and sulfate resistance.

Natural clinoptilolite is widely added to farm and pet animal feeds in doses up to 1g/kg. The zeolite improves digestion of organic matter by slowing *digesta* transit, reduces the level of ammonia, and traps unwanted species including heavy metal ions, 4-methylphenol, and bioamines. Zeolites are classified as nontoxic and safe for human consumption, and have widespread applications in agriculture. A high proportion of chemical fertilizers, containing the key nutrients K^+ and NH_4^+, applied to the soil can be washed off and into groundwaters, causing environmental problems. Slow-release fertilizers use natural clinoptilolite containing potassium and ammonium ions as the extra framework cations, which are then made available more slowly to crops as they leach from the zeolite pores. In similar applications, zeolites surface treated with cationic surfactants can be used to slow the release of PO_4^{3-} into the soil, while herbicides can also be absorbed by zeolites to slow their release on treated soils. Zeolites are also used as soil conditioners, where they help increase moisture retention and control soil pH.

(b) Zeolites as absorbents and for gas separation

KEY POINT Important commercial applications of zeolites involve small molecule separation and absorption.

Zeolites are excellent absorbents for most small molecules, including H_2O, NH_3, H_2S, NO_2, SO_2, and CO_2, linear and branched hydrocarbons, aromatic hydrocarbons, alcohols, and ketones in the gas or liquid phase. Zeolites with different-sized pores may be used to separate mixtures of molecules based on size, and this application has led to their description as **molecular sieves**. Through the correct selection of zeolite pore, it is possible to control the rates of diffusion of various molecules with different effective diameters, leading to separation and purification. Fig. 29.3 illustrates this application schematically.

The sieving properties (and many related catalytic properties, Section 31.13) are controlled by the size of the pores in the zeolite and the ease by which molecules can diffuse through the structure. Of most importance in terms of molecular separations are the *smallest* windows present in the zeolite through which a molecule can diffuse to travel into and between larger pores within the zeolite material. The size of molecule is generally measured by the **kinetic diameter**—a parameter derived from how often molecules collide in the gas phase, Table 29.1.

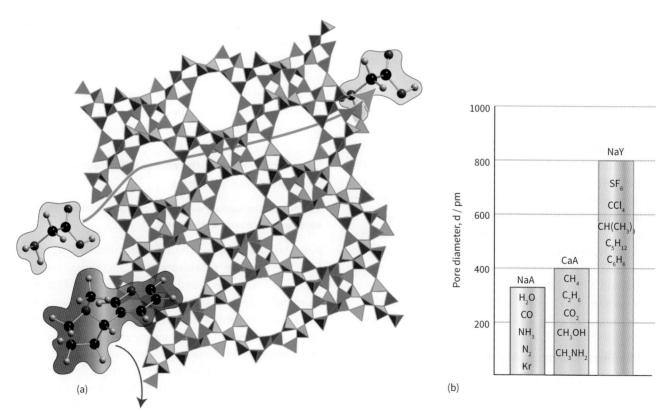

FIGURE 29.3 The use of porous zeolite structures for the separation of molecules of various sizes. (a) Only the smaller molecule can diffuse into and then out of the zeolite pore, so a membrane of this material may be used to separate this mixture. (b) The maximum molecular size that can be absorbed into various zeolite channel diameters. NaA and CaA are the zeolite A, LTA, containing the extra-framework cations of Na^+ and Ca^{2+}; NaY is faujasite FAU, a moderately large-pore zeolite.

TABLE 29.1 Kinetic diameters and key physical constants of small molecules

Molecule	Kinetic diameter (Å)	Polarizability (Å³)	Dipole moment(D)	Quadrupole moment (DÅ)
CO_2	3.30	2.650	0.000	4.30
CH_4	3.80	2.600	0.000	0.02
H_2	2.89	0.80	0.000	0.66
O_2	3.46	1.600	0.000	0.39
CO	3.73	1.95	0.112	2.50
N_2	3.64	1.760	0.000	1.52
C_2H_2 ethyne	3.30			
C_2H_4 ethene	3.80			
C_2H_6 ethane	3.90			
C_3H_6 propene	4.50			
C_3H_8 propane	4.30			
C_6H_6 benzene	5.85			
CCl_4	5.90			
$CH_3CH_2C_6H_5$ ethylbenzene	6.2			

Overall, the permeation of a small gaseous molecule through a zeolite depends on:

(i) The absorption selectivity; a low temperature phenomenon which controls how strongly different molecules initially get adsorbed onto the surface of the zeolite.

(ii) Molecular sieving; where the molecular size (kinetic diameter) determines whether a molecule can enter the zeolite through the pore window while another cannot.

(iii) Diffusion selectivity; where the rates of diffusion of different molecules through a zeolite vary.

Molecular sieving leads to very high selectivities in separating molecules where two molecules have significantly different sizes. One example is in the separation of para-xylene (1,4-dimethylbenzene, PX) from ortho-xylene (1,2-dimethylxylene, OX) and meta-xylene (1,3-dimethylbenzene, MX) where the kinetic diameter of PX (5.9 Å) is much lower than those of OX and MX (6.8 Å) due to its lower cross-section. Zeolite MFI has pore windows of dimension 6 Å which allow PX to diffuse through it but not OX or MX leading to high selectivities (ratio of PX/OX passing through a zeolite membrane) of over 2000 from a dilute hydrocarbon feed. Under industrial process conditions, with xylene partial pressures around 50 kPa, selectivities using zeolite MFI are much lower (~10) but may be improved to near 60 using a mixture of two different zeolites (MFI and MEL) produced as nanosheets (Section 28.9).

Other industrial applications of zeolites for separation and purification include petroleum refining processes, where they are used to remove water, CO_2, and mercury; the desulfurization of natural gas, where the removal of H_2S and other sulfurous compounds protects transmission pipelines and removes the undesirable smell from home supplies; the removal of H_2O and CO_2 from air before liquefaction and separation by cryogenic distillation; and the drying and the removal of odours from pharmaceutical products.

Absorption and diffusion selectivities, which derive from molecular properties that include polarizability, dipole moment, and quadrupolar moment, are much lower but can still usefully separate molecules, including O_2 and N_2, from air without using expensive cryogenic methods. Many dehydrated zeolites adsorb N_2 more strongly into their pores than O_2. This is believed to be due to the fact that N_2 has a higher quadrupole moment than O_2 and interacts more strongly with the resident, extra-framework cations. By passing air over a bed of zeolite at controlled pressure it is possible to produce oxygen of over 95% purity. Zeolites such as NaX (the sodium-exchanged form of an aluminium-rich zeolite of structure type X) and CaA (a calcium-exchanged form of zeolite framework structure type A, also known as LTA) were originally developed for this purpose.

The selectivity of the process can be much improved by exchanging selected cations into the zeolite pores and thus changing the dimensions of the sites onto which the nitrogen and oxygen molecules adsorb. Thus lithium-, calcium-, strontium-, and magnesium-exchanged forms of the faujasite (FAU, type X) and Linde Type A (LTA) structures are now used very effectively in this process, Fig. 29.3.

The separation of alkenes from both alkanes and alkynes is an importance stage in the conversion of crude oil into chemical feedstocks for the production of polymers, where pure alkenes are required. Distillation methods are energy intensive and zeolites offer a low-energy route, particularly for the separation of highly important ethene and propene from alkanes. Because of the very similar kinetic diameters of the alkane and corresponding alkene, Table 29.1, zeolite molecular sieving selectivities for ethene from ethane are typically <10. However, zeolite ITQ-55 has flattened 8-ring windows of dimensions 5.9 × 2.1 Å that are sufficiently flexible to allow ethene absorption into its pores but not the slightly larger ethane; this yields ethene selectivities of around 50 at 110°C.

The molecular sieving properties of zeolites have also been used commercially for desalination of water and removing water from high-temperature gas mixtures. Sea water can be desalinated by using zeolite membranes with small pore windows that allow water molecules (kinetic diameter 2.7 Å) to pass through them, but not the much larger hydrated cations (mainly $Na^+(H_2O)_n$). Efficient systems use very thin zeolite ZSM-5 membranes under 500 nm thick.

The presence of hydrated extra-framework cations in small-pore zeolites, such as Na-LTA (largest window size 8-ring, Fig. 29.1), enhances the rate of water molecule diffusion through the zeolite in comparison with nonpolar and lower polarity molecules that do not coordinate to Na^+ ions. This diffusion mechanism, hopping from being coordinated to one Na^+ cation and on to the next cation, can be used to separate water molecules from other species including H_2, CH_3OH, and CO_2 in gas mixtures. This removal of water can be used to drive the important CO_2 hydrogenation reaction to completion on the right-hand side.

$$CO_2(g) + 3H_2(g) \rightleftharpoons CH_3OH(g) + H_2O(\text{zeolite})$$

(c) Zeolites as ion exchangers

KEY POINTS Zeolites can ion-exchange small cations with very high selectivities and are used for this purpose in remediation of nuclear waste.

The excellent ion-exchange properties of zeolites result from their open structures and ability to trap significant quantities of cations selectively within these pores. High capacities for ion exchange are found for zeolites with a high proportion of Al in the framework, which includes the zeolites with the LTA and gismondine (GIS) topologies that have the highest attainable Al:Si ratio from hydrothermal synthesis of 1:1. Consideration of the generalized zeolite formula $(M^{n+})_{x/n}[(AlO_2)_x(SiO_2)_{2-x}] \cdot mH_2O$, where M is a cation, show that for charge balance a high level of aluminium affords high levels of exchangeable extra-framework cations.

Ion-exchange selectivity has led to a major application of zeolites as a 'builder' in laundry detergents, where they are used to remove the 'hard' ions Ca^{2+} and Mg^{2+} and replace them with 'soft' ions such as Na^+. Phosphates, which may also be used as detergent builders, have been the subject of environmental concerns linked to rapid algal growth and eutrophication of natural waters (Section 6.15). Spherical Na-form LTA zeolite particles, a few micrometres in diameter, are small enough to pass through the openings in the weave of clothing and so may be added to detergents to remove the hard cations in natural waters, and are then washed away harmlessly into the environment as they contain no toxic species.

Another important area where the ion-exchange properties of zeolites are exploited is in the trapping and removal of radionuclides from nuclear waste. Several zeolites, including the widely used clinoptilolite (HEU), have high selectivities for the larger alkali metal and alkaline earth metal cations, which in nuclear waste include ^{137}Cs and ^{90}Sr (Fig. 29.4). The site ion exchange effluent plant (SIXEP) at Sellafield in the UK has used a naturally occurring clinoptilolite (from Mud Hills, California) of the approximate empirical formula $Na_{2.20}Mg_{0.40}Ca_{0.99}Sr_{0.07}Al_6Si_{30}O_{72} \cdot 24H_2O$ for more than 40 years.

The natural clinoptilolite structure contains different cation sites, and the sodium ions mainly occupy sites with 6- to 8-fold coordination to framework oxygen atoms and water molecules present in the pores, Figs. 29.4(a) and 29.4(c). The pore size in clinoptilolite is in the range 3.5–3.9 Å and similar in size to hydrated Cs^+ and Sr^{2+} cations, and in $Cs_6(Al_6Si_{30}O_{72}) \cdot nH_2O$ the Cs^+ ions are coordinated to 10 oxygen atoms in the framework and water molecules, Figs. 29.4(b) and 29.4(d). This idealized coordination leads to a strong thermodynamic preference of the larger Cs^+ and Sr^{2+} cations over Na^+ (which occurs in the naturally formed zeolite). Hence, 1 mol of Sr^{2+} and 20 mol of Cs^+ may be selectively extracted in the presence of vast excesses of Na^+ (7.5×10^5 mol), Mg^{2+} (6.5×10^3 mol), and Ca^{2+} (5×10^3 mol). In the commercial SIXEP process the effluent from the nuclear storage ponds is adjusted to a pH of 7 before being passed over two clinoptilolite beds with a contact time of approximately 8 minutes. This process effectively removes all the ^{137}Cs and ^{90}Sr from the nuclear waste. Once formed, ion-exchanged zeolites may be vitrified by further reaction with glass-forming oxides.

(d) Emerging applications of zeolites

KEY POINTS Zeolites can host nanoparticles and metal clusters, conferring them with optical functionality.

The open porous structures of zeolites are chemically and thermally very stable, and therefore can act as host structures or 'scaffolds' for species other than simple cations and water molecules. Incorporation of guest species, which can include nanoparticles (Section 28.10) and metal complexes, often results in new and enhanced physical properties for the **host-guest compound**. These host-guest species are often described using the nomenclature X@zeolite, where X is the guest species.

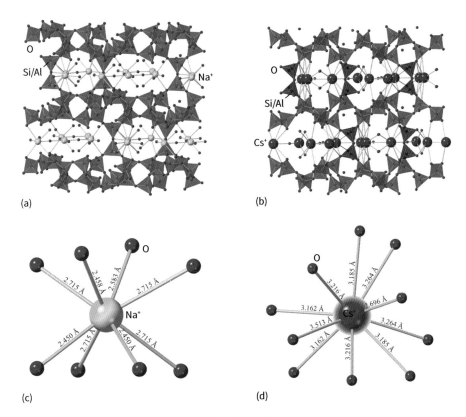

FIGURE 29.4 The clinoptilolite structure of (a) a sodium clinoptilolite and (b) a caesium clinoptilolite $Cs_{7.39}(Al_6Si_{30}O_{72})\cdot 7.39H_2O$; (c) and (d) show the local coordination by oxygen (to both framework and H_2O atoms) to one cation site in the sodium- (8-fold, distances 2.45–2.72 Å) and caesium-forms (10-fold, distances 3.16–3.51 Å).

Silver@zeolite materials—for example, $[Ag_{0.5}]@$[sodium zeolite FAU]—contain $[Ag_4]$ metal clusters that demonstrate excellent fluorescent properties with efficiencies nearing 100% for converting ultraviolet light (300 nm) to visible light (500nm) (Section 30.1). These materials also demonstrate electroluminescent properties where application of an electrical potential yields visible light; by varying the silver content in the zeolite the colour of the electroluminescence can be tuned from blue to red. Carbon dots (CDs) in zeolites (CDs@zeolite) also exhibit fluorescent and phosphorescent properties (Section 30.1).

A Zeolite Encapsulated (transition) Metal Complex (ZEMC) can be synthesized by incorporating the complex during synthesis (sometimes as the SDA) or by formation *in situ* using 'ship-in-a bottle' methods, whereby the central metal ion is ion-exchanged into the zeolite and the ligands are subsequently introduced and coordinate to the metal centre. ZEMC materials can demonstrate enhanced catalytic properties derived from a combination of the homogeneous (metal complex) and heterogeneous (zeolite) functions, Sections 31.13 and 31.15.

Small molecules—for example, ethanol and iodine—introduced into zeolite pores can form ordered or supramolecular networks; this process is driven by the confinement and ordered orientation of the absorbed molecules. In the case of iodine absorbed into the zeolite silicalite, the proximity of the ordered I_2 molecules results in an electrical conductivity many orders of magnitude higher than that of pure iodine.

Zeotypes and metal–organic frameworks

29.3 Inorganic structures based on linked polyhedral metal centres

KEY POINTS Porous materials may be formed from a variety of linked MO_n (n is usually 4 or 6) polyhedra or MS_4 tetrahedra; these framework materials often exhibit useful ion exchange and catalytic properties.

The classification of a material as a zeolite requires that the structure is formed from oxotetrahedra (normally SiO_4, AlO_4, BeO_4, and PO_4) linked through all vertices. Zeolite analogues which are three-dimensional porous structures formed from other oxopolyhedra are often referred to as **zeotypes** or **hypertetrahedral frameworks**. Ring sizes larger than 12 tetrahedrally coordinated ring atoms were first seen in metallophosphate systems and structures with 20 and 24 linked tetrahedra have been made.

Other well-established framework structures include the titanosilicate families (which have four-coordinate Si and five- and six-coordinate Ti sites) and a series of octahedral molecular sieves based on linked MnO_6 units. These compounds are made under the hydrothermal conditions similar to those used for synthesizing many zeolites, but by using a source of Ti, such as $TiCl_4$ or $Ti(OC_2H_5)_4$, that hydrolyses under the basic conditions in the reaction vessel. Templates may also be used in such media and act as structure-directing units, giving rise to particular pore sizes and framework topologies.

In a typical reaction the titano-silicate ETS-10 (Engelhard TitanoSilicate 10) is prepared by the reaction of $TiCl_4$, sodium silicate, sodium hydroxide, and sometimes a template such as tetraethylammonium bromide, in a sealed PTFE-lined autoclave at between 150 and 230°C. Two titanosilicates—ETS-10 and $Na_2Ti_2O_3SiO_4 \cdot 2H_2O$—are worthy of further consideration here as specific examples of this type of material and their applications.

ETS-10, $Na_2[Si_5TiO_{13}] \cdot nH_2O$ is a microporous material built from TiO_6 octahedra and SiO_4 tetrahedra, with the TiO_6 groups linked together in chains (Fig. 29.5(a)). The structure has 12-membered rings (i.e. pores formed from 12-unit polyhedra) in all three directions—a connectivity that gives rise to large octagonal pores containing hydrated Na^+ ions. This compound, in common with several other titanosilicates, including $K_3H(TiO)_4(SiO_4)_3 \cdot 4H_2O$ (a synthetic analogue of a natural mineral pharmacosiderite, Fig. 29.5(b)), shows excellent ion-exchange properties, particularly with large cations. These large ions replace Na^+ with very high selectivity, so that, for example, Cs^+ and Sr^{2+} ions may be extracted into the titanosilicate structure from their dilute solutions. This property has led to the development of these materials for the removal of radionuclides from nuclear waste, Section 29.2(c); the high selectivities for ^{137}Cs and ^{90}Sr exhibited by ETS-10 make this the preferred material for this application.

The aim of some of the recent research on porous framework structures has been to incorporate d-metal ions into the material to engender properties associated with transition metal ions; for example, magnetism, colour, and redox catalysis. Materials that have been synthesized include antiferromagnetic porous iron(III) fluorophosphates with Néel temperatures in the range 10–40 K. These temperatures are relatively high for iron clusters linked by phosphate groups and indicate the presence of moderately strong magnetic interactions. Zeotypic cobalt(II), vanadium(V), titanium(IV), and nickel(II) phosphates have been prepared, including the material known as VSB-1 (Versailles–Santa Barbara), which was the first microporous solid with 24-membered ring tunnels, and is simultaneously porous, magnetic, and an ion-exchange medium (Fig. 29.6).

Tetrahedral coordination is common in simple metal sulfide chemistry, and structures are often formed from

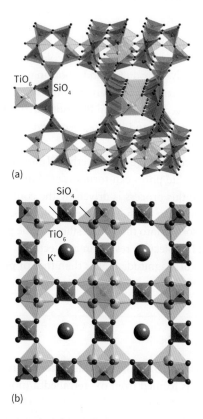

FIGURE 29.5 The structures of (a) the titanosilicate ETS-10, which is built from chains of TiO_6 octahedra linked by SiO_4 tetrahedra, and (b) $K_3H(TiO)_4(SiO_4)_3 \cdot 4H_2O$, a synthetic analogue of a natural mineral, pharmacosiderite.

FIGURE 29.6 The framework structure of VSB-1 consists of linked NiO_6 octahedra and PO_4 tetrahedra. The main channel is bounded by 24 such units and has a diameter such that molecules up to 0.88 nm in diameter may pass through it.

Zeotypes and metal–organic frameworks

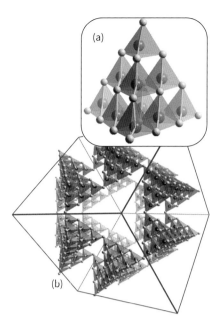

FIGURE 29.7 (a) An isolated supertetrahedral unit of stoichiometry $Zn_{10}S_{20}$ formed from individual ZnS_4 tetrahedra. (b) These supertetrahedral units may be linked together as building blocks to form large, porous, three-dimensional structures.

linked MS_4 tetrahedra, though the larger size of the sulfide ion means that the CN of sulfur is normally greater than 2. In the discussion of ZnS adopting the wurtzite and sphalerite structures (Section 4.9) the CNs of Zn and S are both 4, and some compounds may be considered as containing linked fragments of these structures. These are known as **supertetrahedral clusters**, and an example is $[N(CH_3)_4]_4$ $[Zn_{10}S_4(SPh)_{16}]$, which contains the isolated supertetrahedral unit $Zn_{10}S_{20}$ terminated by phenyl groups (Fig. 29.7).

29.4 Metal–organic frameworks

Metal–organic frameworks (MOFs) or porous coordination polymers (PCPs) were introduced in Section 25.17(c) as three-dimensional structures formed from metal ions linked through bidentate or polydentate ligands. MOFs are also frequently referred to as **hybrid** structures in that they contain both inorganic and organic elements. Examples of the ligands commonly used to build these, often porous, frameworks include CN^-, nitriles, amines, imidazoles, and, most commonly, carboxylates; example linker ligands are summarized in Fig. 29.8.

The nomenclature used for MOFs is less formalized than for zeolites, and new structures are often named initially after the place or chemical laboratory in which they were first synthesized and studied; for example, **HKUST-1** (Hong Kong University of Science and Technology), UiO-66 (Universitet i Oslo), **MIL-88A** (Materials of Institut Lavoisier), and **VSB-5** (Versailles-Santa Barbara). A more systematic nomenclature using the system MOF-*n*, where HKUST-1 is MOF-199, is also used.

(a) MOF structures

KEY POINT The use of organic linkers, such as carboxylates, between metal centres produces hybrid inorganic–organic materials which can exhibit high porosity.

Many thousands of MOFs have been synthesized and various sub-classes of material described, including zeolite-like MOFs (ZMOF-n), ZIF-n (zeolitic imidazolate framework), and MPF-n (metal peptide framework). In 2021 the Cambridge Crystallographic Data Centre reported around

Name	Structural formula	Name	Structural formula
Ethanedioic acid		Benzene-1,2-dicarboxylic acid (o-phthalic acid)	
Propanedioic acid		Benzene-1,3-dicarboxylic acid (m-phthalic acid)	
Butanedioic acid		Benzene-1,4-dicarboxylic acid (p-phthalic acid)	
Pentanedioic acid			

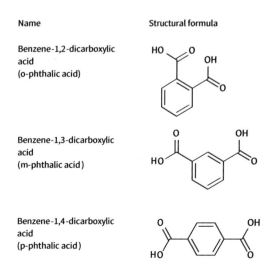

FIGURE 29.8 (*Continued on next page*)

3,4-Dihydroxy-3-cyclobutene-1,2-dione

(Squaric acid)

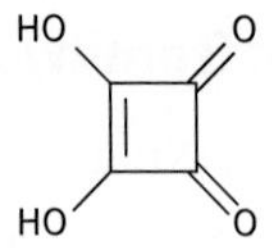

Butynedioic acid

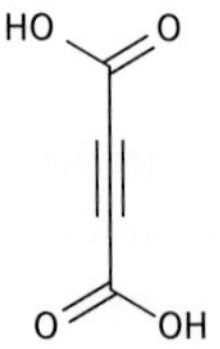

Citric acid

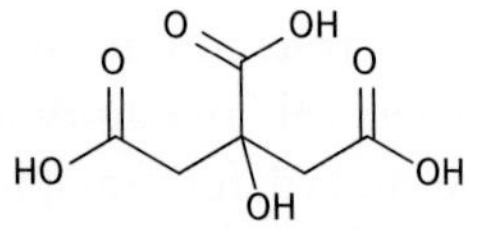

Adamantanetetracarboxylic acid

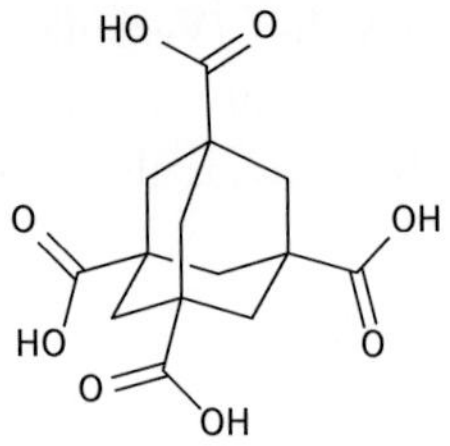

1H-1,2,3-triazole

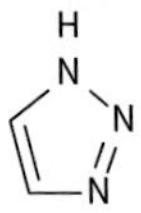

1H-1,2,4-triazole

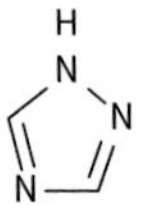

2,5-Dihydroxyterephthalic acid (DOT) and extended forms

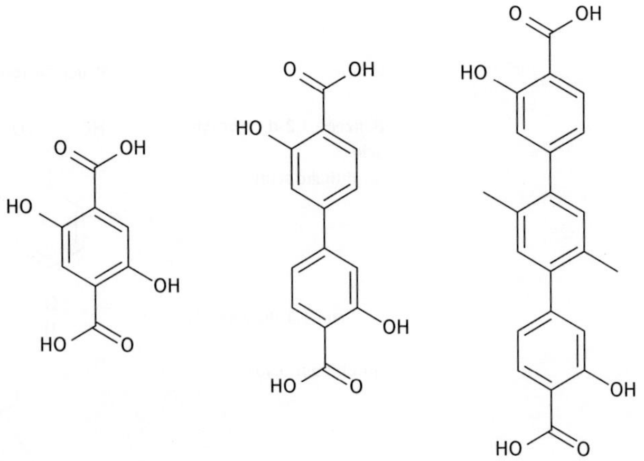

Benzene-1,3,5-tricarboxylic acid

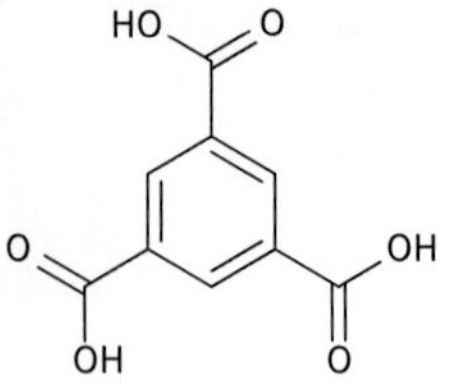

2,6-naphthalenelcarboxylic acid

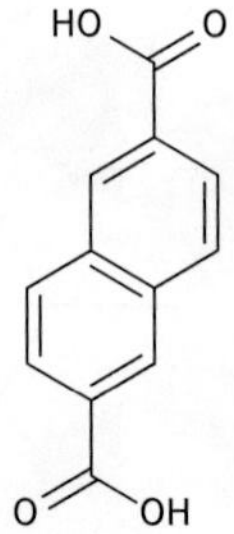

Benzentribenzylic acid

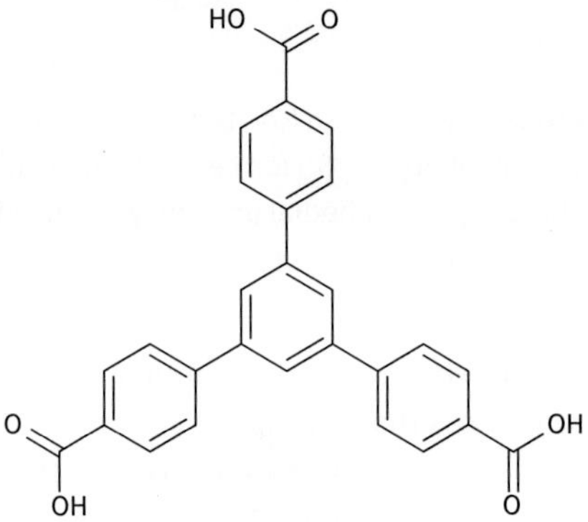

Methanetetrabenzylic acid

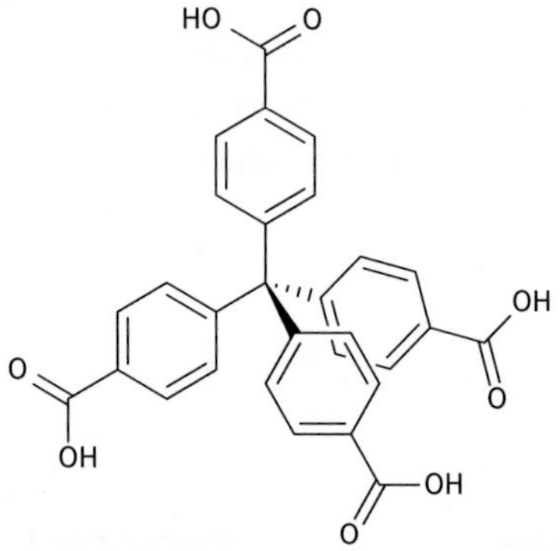

FIGURE 29.8 Common organic linkers for the construction of metal–organic frameworks (MOFs). In an MOF, the carboxylate anion of the corresponding acid is normally the linking species.

100,000 MOF-like structures, of which around 10,000 were porous frameworks.

These MOFs include positively charged frameworks with balancing anions in the cavities (for instance, $Ag(4,4'\text{-bpy})\cdot NO_3$ with free nitrate anions in the pores) and electrically neutral frameworks, such as $Zn_2(1,3,5\text{-benzenetricarboxylate})NO_3 \cdot H_2O \cdot C_2H_5OH$ with nitrate forming part of the framework and from which the H_2O molecules may be removed reversibly. MOF structures of most interest are those which have large pores for gas and molecular absorption, those that strongly absorb specific gases, such as CO_2 and H_2, and those with high thermal stability. Some illustrative and important carboxylate-based MOFs are:

UiO-66 $Zr_6O_4(OH)_4(BDC)_6$, with surface areas up to $1300 m^2/g$ and high thermal stability to over 500°C; Fig. 29.9(a).

MOF-602 $Cu_2(2\text{-MeBPDC})_2(DMA)_2\cdot(DMA)_2(DMF)(Py)(H_2O)_3$; Fig. 29.9(b).

HKUST-1 MOF199 $Cu_3(BTC)_2(H_2O)_3 \cdot xH_2O$; Fig. 29.9(c).

MOF-5 $Zn_4O(BDC)_3$, where BDC = 1,4-benzodicarboxylate; Fig. 29.9(d).

ZIFs are another large class of MOF which have metal centres coordinated by the nitrogen atoms of imidazole rings. ZIFs, including ZIF-8 (zinc bis(2-methylimidazol-1-ide), Fig. 29.10, demonstrate higher thermal and chemical stability than most MOFs and are hydrophobic, leading to much greater stability in wet atmospheres. ZIFs may also be synthesized in a glassy state, retaining some porosity.

(b) Synthesis and modification of MOFs

KEY POINTS MOFs are typically synthesized using solvothermal techniques wherein metal cations are linked using long, bidentate, or polydentate ligands; once the framework is formed it may be chemically modified in subsequent reactions. Lower-temperature synthesis methods include electrochemical, microwave-assisted, and mechanical techniques.

The main synthesis route for MOFs uses solvothermal conditions in which the solvent water used under hydrothermal conditions is replaced by an organic solvent or

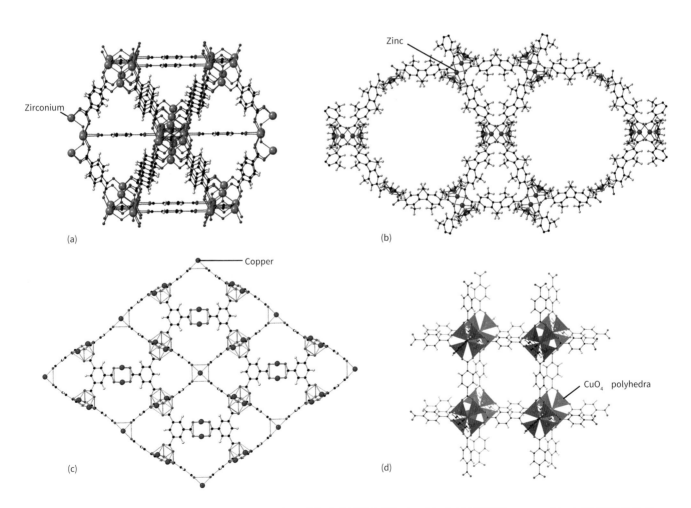

FIGURE 29.9 MOF frameworks; (a) UiO-66, zirconium (grey/blue); (b) MOF-602, zinc (blue); (c) HKUST-1, copper (blue); (d) MOF-5, CuO_4 polyhedra (red).

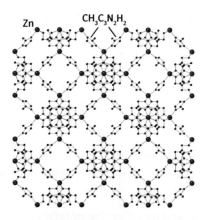

FIGURE 29.10 The framework structure of ZIF-8 with Zn^{2+} linked through bidentate imidazolate anions.

solvent–water mixture. Frequently used solvents with boiling points in the range 50–250°C include methanol, ethanol, acetonitrile, N,N-dimethylformamide (DMF), and N,N-diethylformamide (DEF). In a typical reaction, a metal salt and organic ligands are dissolved in the solvent and placed in a closed pressure vessel or autoclave. The closed pressure vessel is heated in an oven to temperatures in the range 80–220°C, for periods ranging from several hours to weeks, but typically a few days. The solid MOF product is usually isolated by filtration. Variations in reactant concentrations, solvent, or solvent mixtures, and the reaction temperature, all play a major role in the MOF synthesized and its particle morphology.

In a typical synthesis of the well-known MOF HKUST-1, $[Cu_3(BTC)_2(H_2O)_3]_n$ (BTC = benzene 1,3,5-tricarboxylic acid (BTC), also known as trimesic acid (TMA), Fig. 29.8), copper (II) nitrate trihydrate is heated with BTC in a 50:50 H_2O:EtOH solvent mixture at 180°C for 12 hours in a Teflon-lined pressure vessel. This yields turquoise crystals of HKUST-1, Fig. 29.9(c).

The MOF formation may also be altered using reaction additives—for example, surfactants, other metal-coordinating species, and pH modifiers—which often allow the control of MOF particle size and morphology. For example, the addition of ethanoic acid to the reaction mixture in the synthesis of $[Cu_2(\text{naphthalene dicarboxylic acid})_2(\text{dabco})_2]_n$ inhibits three-dimensional crystal growth leading to nanosheets and nanorods, while the addition of dodecanoic acid to the HKUST-1 reaction mixture yields highly regular 2-μm cubic crystallites.

Other more recently developed synthesis methods for MOFs include room-temperature reactions, direct milling of the reactants, and electrochemical, ultrasonic, and microwave-assisted techniques. These lower temperature and mechanical methods allow greater control over crystal growth, including the formation of nanostructured MOFs (Section 28.2) and small particle size distributions. For example, the application of microwave (MW) or ultrasonic (US) irradiation typically results in smaller particle sizes with narrower size distributions compared to hydrothermal methods. Electrochemical methods can be used to deposit thin films of MOFs and alter the particle morphology; MOF-5 can be deposited electrochemically from ionic liquids as flower-like micron-sized particles instead of the usual cubic 10-μm particles obtained hydrothermally.

Once a MOF has been synthesized, the solvent molecules or other species present in the pore may be removed or replaced to change its physical properties; for example, incorporation of amines improves uptake of CO_2 into the pores, while the encapsulation of metal complexes enhances the absorption of dye molecules, Section 28.10(b).

The presence of an organic linker in MOFs allows post-framework formation reactions to be undertaken on the solid MOF which modify the organic chemistry of the linker ligand. Reactions of this type can 'decorate' the organic species, lining the MOF pores to enhance its physical properties in respect of interactions with an absorbed species or to introduce a new functionality. One example is the post-phase formation functionalization of the MOF IRMOF-3, Fig. 29.11, which has benzene 2-amino-1,4-dicarboxylic acid (2-aminoterephthalic acid) as the linking species between zinc centres; this material is the $-NH_2$ functionalized, isoreticular[1] analogue of MOF-5 (Fig. 29.9(d)). Reaction of IRMOF-3 with acetic anhydride leads to an amide functionalized MOF IRMOF-3(AM1), and powder X-ray diffraction data showed that the MOF framework topology was unchanged by this reaction.

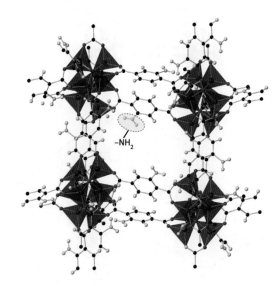

FIGURE 29.11 The structure of IRMOF-3 highlighting the $-NH_2$ functionalization compared to MOF-5; Fig. 29.9(d).

---

[1] Isoreticular: having the same or similar topology so that the framework geometry is the same or similar but has a different linking ligand.

(c) MOF applications

KEY POINT The very high porosity of MOFs is leading to applications in gas absorption, separation, and storage.

MOFs have many properties that are analogous to those of zeolites, with high porosities leading to existing, and likely future, applications based on gas adsorption, separation, and storage. There is also the potential for MOFs to facilitate catalytic reactions analogous to those that occur inside zeolite pores and channels (Section 29.1), though with the larger MOF pore sizes, reactions involving complex, large molecules become possible.

Other potential advantages of MOFs over zeolites include the much greater structural diversity derived from the vast number of different linking organic groups that can be tailored for specific functionality. For example, the use of long linking groups based on lengthy aromatic dicarboxylates (Fig. 29.8) permits the synthesis of highly porous structures that can adsorb or incorporate high volumes of small gaseous molecules, very large species, and even nanoparticles. The ability to chemically modify the organic linker so that it can interact with a specific molecule in the pore also adds to the ability to design MOFs with targeted properties and applications.

Disadvantages of MOFs over zeolites and other purely inorganic porous materials is that they are thermally and chemically much less stable because of the organic component in their structures, though a few MOFs can persist to above 500°C and are stable in acidic or basic conditions.

MOFs have been proposed as materials that can address many of the global challenges currently facing humankind. These advances impact upon energy storage and sustainability (Chapter 30) (for example, as hydrogen storage materials), environmental concerns (as an example, reducing the need for, or eliminating. the use of fertilizers and agrochemicals), and improved human and animal health (for example, targeted pharmaceuticals and cancer treatments).

(i) Small-molecule gas absorption, storage, and separation

A material that has the ability to selectively adsorb and store large quantities of small gaseous molecules will have an enormous impact in various areas of green chemistry. Molecules of particular interest are CO_2 (for the capture of the molecule produced by burning fossil fuels and other industrial processes, including cement manufacture), H_2 (hydrogen economy storage and transport), O_2 and N_2 (in their separation from air), and small alkanes, CH_4 (storage at ambient conditions for transport applications). MOFs have very large specific internal surface areas that can adsorb small molecules. One example of note is the framework of a chromium terephthalate (1,4-benzenedicarboxylate), MIL-101, which has pore diameters of about 3 nm (30 Å)

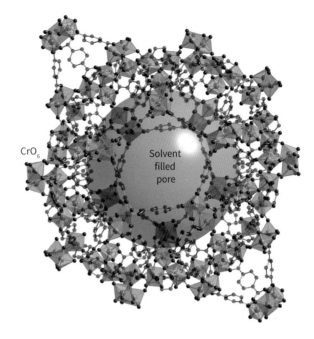

FIGURE 29.12 MIL-101 units showing CrO_6 octahedra linked by terephthalate anions. The large central sphere shows the extent of the pore in this material, which can be filled with solvent or gas molecules.

and a specific internal surface area for N_2 of over 5000 m² g⁻¹ (Fig. 29.12).

Systems that adsorb CO_2 in preference to other molecules, especially N_2 and O_2, are being studied for the sequestration of CO_2 from air and power plant waste gases as part of carbon-capture schemes. The properties required of the CO_2-absorbing MOF for commercial application are: selectivity for CO_2, hydrophobicity to engender stability in wet atmospheres, MOF-regeneration, and ease of synthesis. CO_2 adsorption has been demonstrated for many MOF structures at 273 K. For example, UiO-66 and HKUST-1 absorb around 2–4 mmol of CO_2 per gram of MOF, and a value of almost 10 mmol/g has been reported for a Cu-BTC MOF.

Functionalization of as-synthesized MOFs using basic amines or amino groups on the organic linker (e.g. Fig. 29.11), that coordinate strongly with acidic CO_2, improves adsorption. Organic linkers containing an $-NH_2$ side group that points towards the centre of the MOF pores have been found to coordinate CO_2 molecules more strongly. One further method of improving CO_2 uptake is to incorporate polyethylenimine, $[-CH_2CH_2NH-]_n$, into the MOF pores. Limitations with carbon-capture by MOFs derive from low thermal stability (making cycling of CO_2 capture and release problematic) and the tendency of the materials to adsorb water which reacts with and slowly degrades the framework.

MOFs also have potential applications in gas storage, including H_2 and the light hydrocarbons CH_4 and C_2H_6. Hydrogen incorporation into MOFs is mainly through physisorption onto the internal pore surfaces, and increasing this internal surface area is central to maximizing the weight percent of hydrogen absorbed. Typical experimental weight percents for absorbed hydrogen under cryogenic conditions at 77 K and 50 kbar have been reported in the range 2–12%, with some of the highest values found for MOF-5 (6.5 wt%, 40 kbar) and MIL-101 (6.1, 40kbar). However, under ambient conditions absorbed hydrogen levels are much lower, typically near 1 wt%, reflecting the weak interaction between the H_2 molecule and MOF internal surface at room temperatures.

Various methods of modifying MOFs to improve hydrogen absorption under ambient conditions have been explored; for example, the inclusion of metal cations, nanoparticles, and the formation of composites with graphene oxide. Problems remain with structure collapse of the metal–organic framework when the solvent is fully removed, and degradation of the structure following repeated absorption–desorption cycling.

One commercial use of MOFs has been to reduce the pressure during the cylinder-storage of liquid natural gases (methane, propanes, and butanes), and another is to store the toxic gases AsH_3, PH_3, and BF_3 (all used in the electronics industry) in cylinders at ambient pressure.

Methane storage in gas cylinders containing MOFs has been improved by 30 wt%.

(ii) Active organic compounds

The very large pores of MOFs allow large, complex molecules, such as many pharmaceutically or agrochemically active compounds, to be inserted into them. This aspect mimics the properties of zeolites discussed in Section 29.2(a), though the larger pore size of MOFs means that much larger and complex molecules can be incorporated. Further proposed applications involve using MOFs as highly effective drug-delivery agents (see also Section 32.10). In this technique an active compound—for example, the painkiller ibuprofen—is first absorbed within the MOF pore. When the impregnated MOF is ingested or injected into the human body the active pharmaceutical ingredient can be slowly released and, potentially, just where required; for example, at the site of a tumour, Fig. 29.13. MOFs are also being investigated for applications that allow the slow desorption of ethylene for fruit ripening and in concentrating water as a liquid from low-humidity air in deserts.

Agricultural applications of MOFs are being developed, associated with their ability to absorb and degrade pollutants in natural environmental waters and slowly release

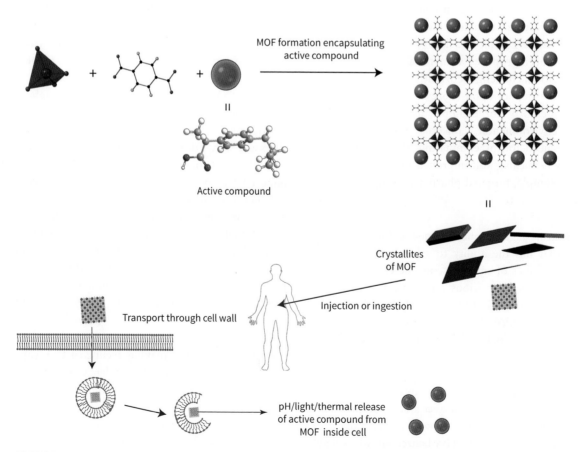

FIGURE 29.13 Schematic of the protection and delivery of an active organic compound to a target cell by using an MOF.

and deliver functional agrochemicals. MOFs show excellent functionality in absorbing herbicides, fungicides, and insecticides; these contaminants are often present at low levels in water run-off from fields. For example, the MOF MIL-53(Cr), [CrOH(BDC)], shows much improved properties over activated carbon for the absorption and removal of the herbicide 2,4-dichlorophenoxyacetic acid from water. Once incorporated within the MOF, the active organic compound can often be degraded and made inactive through subsequent photolysis or photocatalytic reactions (Section 30.3).

One of the most promising applications of MOFs is as slow-release agents for the controlled delivery of agrochemicals, including active herbicides, insecticides, fungicides, plant-growth regulators, and fertilizers. In comparison with zeolites, MOFs can trap much larger active species and degrade more quickly in natural environments, releasing the functional molecule; MOFs, however, are much more expensive to synthesize on a bulk commercial scale. The fumigant cis-1,3-dichlorophropene (1,3-DCPP) can be incorporated into the pores of MOF-1201, [Ca_{14}(L-lactate)$_{20}$(CH_3CO_2)$_8$·solvent], with a loading of 13 wt%. Release of the fumigant from the MOF in air is a hundred times slower than that from the pure liquid, allowing it to be active in the environment for much longer periods. Tebuconazole, a triazide fungicide, can be loaded into the MOF MIL-101(Fe) (constructed of iron and tannate organic linkers). Exposure of the loaded MOF to sunlight, or acidic (pH<5) or alkaline (pH>9) environments, releases the fungicide in a controlled manner, either directly from the pores or by degradation of the MOF.

(iii) Catalysts

Further proposed applications of MOFs are in catalysis, where the ability to control the dimensions and shape of the pore and also to introduce chiral motifs, using a chiral organic linking unit, is a highly exciting area for research, discussed in general terms in Chapter 31. The linker ligand H_2BDC-NH_2 frequently used in MOF formation is easily functionalized at the amine group through a typical Schiff base condensation and subsequent reduction. Attaching a chiral R group, (S-)R, to give H_2BDC-NH(S-)R then allows the synthesis of chiral, isoreticular MOFs that show asymmetric photocatalysis.

(iv) Optical and electronic applications

MOF structures, composed of organic molecules linking metal cation centres, generally show very low electronic conductivities due to poor electron transport along the organic linkers. Methods of improving the electrical conductivity in MOFs through the use of organic linkers with conjugated π-systems have been explored. As one example, nanosheets of [Ni_3(BHT)$_2$], BHT = benzenehexathiol, show strong π- and sulfur 3p conjugation yielding a conductivity of 0.15 Scm^{-1} within the sheet. [(NBu_4)$_2$(Fe^{III})$_2$(dhbq)$_3$] (dhbq = dihydroxybenzoquinone) shows an intrinsic conductivity of over 0.15 Scm^{-1} in three dimensions. Conductivity of MOFs may also be improved by incorporating species such as tetracyanoquinodimethane into the pores. MOFs demonstrating reasonable electronic conductivities have potential applications as thermoelectrics (Section 30.4) and as battery materials (Section 30.9).

The large accessible internal surface areas of MOFs have suggested their use as high-activity photocatalysts (Section 30.3) for waste water treatment and air purification, and as disinfectants. Typical band gaps in MOFs such as MOF-5 and UiO-66(Zr) are between 3 and 4 eV, making them photoactive only with UV radiation, though tuning the band gap through the use of appropriate metals in the framework (e.g. Fe^{3+}) can lower the band gap to between 1.5 and 2.0 eV. MOFs of this type have shown high activity in the photocatalytic degradation of many model organic pollutants, including acid orange 7, rhodamine B, and methylene B, using visible light in the presence of H_2O_2. For many electronic and optical applications of MOFs, problems with long-term stability under working conditions remain.

FURTHER READING

A. K. Cheetham, G. Férey, and T. Loiseau, Open-framework inorganic materials. *Angew. Chem., Int. Ed.*, 1999, 38, 3268. An excellent review of progress in, and the structural chemistry of, framework solids.

D. W. Bruce, D. O'Hare, and R. I. Walton, *Inorganic Materials Series*. John Wiley & Sons (2011). A five-volume set covering (i) functional oxides, (ii) molecular materials, (iii) low-dimensional solids, (iv) porous materials, and (v) energy materials. Individual volumes are available separately.

P. Zarshenas, *Zeolite: Stubborn Network for Chemical Stability!* Lambert Academic Publishing (2022).

S. A. Majid, M. A. Bhat, and M. A. Mir, *Zeolite: Synthesis & Application*. Lambert Academic Publishing (2018).

F. R. Ribeiro, *Zeolites: Science and Technology: 80*. NATO Science Series E, 80 (2012).

F.-S. Xiao and X. Meng, *Zeolites in Sustainable Chemistry: Synthesis, Characterization and Catalytic Applications*. Springer (2016).

M. Król and P. Florek, *Zeolites*. MDPI AG (2022).

L. R. MacGillivray, *Metal–Organic Frameworks: Design and Application*. Wiley; 1st ed. (2010).

O. M. Yaghi, M. J. Kalmutzki, and C. S. Diercks, *Introduction to Reticular Chemistry: Metal–Organic Frameworks and Covalent Organic Frameworks*. Wiley-VCH; 1st ed. (2019).

J. Bedia and C. Belver, *MOFs for Advanced Applications*. MDPI AG (2022).

P. Reeves, *Metal-Organic Frameworks (MOFs): Chemistry, Technologies & Applications*. Nova Science Publishers, Inc. (2016).

S. Rojas, A. R. Rodriguez-Dieguez, and P. Horcajada, Metal–organic frameworks in agriculture. *ACS Applied Materials & Interfaces*, 2022, **14**(15), 16983–17007.

EXERCISES

29.1 Which of the elements Be, Mg, Ga, Zn, P, and Cl might form structures in which the element is incorporated into a framework as an oxotetrahedral species?

29.2 Propose formulas for compounds that could be isomorphous with zeolites having the composition 'SiO$_2$' but using mixtures of Al, P, B, and Zn replacing Si.

29.3 Aluminosilicate surfaces in zeolites act as strong Brønsted acids, whereas silica gel is a very weak acid. (a) Give an explanation for the enhancement of acidity by the presence of Al^{3+} in a silica lattice. (b) Name three other ions that might enhance the acidity of silica.

29.4 Describe what sorts of ligands are needed to produce metal–organic frameworks (MOFs).

29.5 Why are many MOFs with tetrahedrally coordinated metal centres unstable in moist environments?

29.6 Describe methods of functionalizing MOF frameworks, including discussion of pre- and post-framework formation reactions.

TUTORIAL PROBLEMS

29.1 Discuss differences in the physical properties of zeolites, zeotypes (frameworks built from oxotetrahedral species other than AlO$_4$ and SiO$_4$), and metal–organic frameworks (MOFs).

29.2 The book *AI-Guided Design and Property Prediction for Zeolites and Nanoporous Materials* (G. Sastre and F. Daeyaert (eds.), Wiley, ISBN: 978-1-119-81977-6. 2023) discusses how machine-aided learning might in the future drive the search for new nanoporous materials for applications. Discuss how AI might change future research on porous materials, and whether there is still a role for creative human scientific thought.

29.3 Describe recent progress towards the commercialization of MOFs. (See N. Notman, *Chem. World*, 2017, 5, 44; A. M. Wright *et al.*, *Nat. Mater.*, 2024, https://doi.org/10.1038/s41563-024-01947-4.)

30 Energy materials

Energy harvesting, conversion, optical and magnetic materials

30.1 **Pigments, coatings, and phosphors**
30.2 **Photovoltaics and solar cell materials**
30.3 **Photocatalysts and water-splitting**
30.4 **Thermal solar farms; piezoelectric and thermoelectric energy materials**
30.5 **Fuel cell materials**
30.6 **Strong permanent magnets and magnetoresistive compounds**

Energy transmission and storage

30.7 **Superconductors**
30.8 **Hydrogen storage**
30.9 **Rechargeable battery materials**

Further Reading

Exercises

Tutorial Problems

New inorganic materials are central to the development of high efficiency, low cost, and clean-energy technologies, including the harvesting of sunlight, production of hydrogen, and the storage of renewable electrical and chemical energy. The need for these new materials is becoming increasingly important with the depletion of fossil fuels and environmental concerns associated with CO_2 emissions.

Energy harvesting, conversion, optical and magnetic materials

Devices for harvesting renewable energy, including solar cells and wind turbines, are reliant on advanced materials developed through inorganic chemistry research. The intermittent nature of renewable electrical energy requires technologies and materials that convert or store electrical energy; for example, those that generate green hydrogen by electrolysis or employ grid-scale rechargeable batteries.

Inorganic energy materials also have a major role in the efficient use of the energy that is generated, including applications in low-energy lighting, high-performance, rechargeable batteries in electric vehicles, and efficient fuel cells that locally convert green hydrogen to electrical power. Ferromagnetic materials have key applications in renewable energy generation, energy recuperation, motors, and in efficient electronic devices and low-energy data-processing and storage.

In this chapter we discuss the exciting research and developments in inorganic materials chemistry underpinning these advances and applications.

30.1 Pigments, coatings, and phosphors

When light of any wavelength impinges on an inorganic material there are a number of possible outcomes, Fig. 30.1. Initially the light can be transmitted through the material (transmittance), reflected, or scattered (specular and diffuse reflectance), and the photon energy is unaffected; materials that reflect or scatter sunlight can be used as white pigments and have applications in paints and coatings to help keep buildings cool. Alternatively, the solid may absorb the light, usually through exciting electronic or vibrational transitions. These phenomena are usually dependent on the wavelength of light, and materials may transmit light at one wavelength and absorb it at another.

If a photon is absorbed then the energy associated with it can promote a variety of different processes in an inorganic material. Transitions between different electronic energy levels can occur; for example, d–d transitions (Section 22.2) in transition-metal complexes, or the promotion of valence

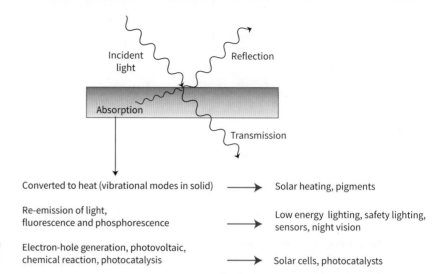

FIGURE 30.1 Pathways of light interacting with an inorganic material and example applications.

band electrons to the conduction band in a semiconductor, Section 2.11.

Relaxation of the excited electron back to the ground state via vibrational modes results in conversion of the photon energy into heat (that is, the vibrations in the solid) and thus the light is just absorbed. Zero or partial vibrational relaxation followed by an electronic transition to a ground or lower electronic level with the emission of light leads to **luminescence**; that is **fluorescence** (immediate emission) or **phosphorescence** (delayed emission), where the transition spin is forbidden. Alternatively, in a solid with a conduction band, the electron (or hole) can migrate through the material producing an electric current at a collector (a photovoltaic material, Section 30.2) or the energized electron can become involved in a chemical reaction (a photocatalyst, Section 30.3). All these pathways are exploited by different inorganic materials in various applications.

Materials that transmit light have applications as windows, optical instruments, smart windows, and transparent conducting layers, Section 30.1(a). Many inorganic solids are intensely coloured due to their absorption of visible light of specific energies, and are used as pigments in colouring inks, plastics, glasses, and glazes. Specialist inorganic pigments use the heat produced from the absorption of light in solar heating devices and thermochromic technologies, Section 30.4. Materials with luminescent properties have applications as inorganic phosphors, in LED lighting, and in lasers, Section 30.1(d) and Box 22.1. The development of stable inorganic materials that can harvest sunlight for the generation of electricity, or for water-splitting and the production of hydrogen, has become a much-researched area. Photovoltaic semiconductor systems are covered in Section 30.2, while in Section 30.3 we consider inorganic photocatalysts used water-splitting and degrading pollutants.

(a) Transparent functional inorganic materials.

KEY POINT Materials that are transparent to visible light but demonstrate electrical conductivity are used in optical devices and touchscreens.

The properties and applications of the transparent conducting material indium tin oxide (ITO, $In_{2-x}Sn_xO_3$) were introduced in Section 14.16. Combining the properties of electrical conductivity and transparency involves doping a material (usually a metal oxide for most applications) sufficiently to produce a large enough number of charge carriers while avoiding strong absorption of light through various electron excitation routes (Section 25.6). Figure 30.2(a) shows the band structure of ITO, $In_{1.8}Sn_{0.2}O_3$, where the introduction of Sn replacing In in In_2O_3 generates an n-type dopant level and additional electrons that enter the conduction band producing a conductivity of almost $10^{-3}\,\Omega\text{cm}$. The absorption spectrum in the visible region is almost unchanged, Fig. 30.2(b), with transmittance remaining very high.

While ITO is employed extensively in many thin-film applications, including touchscreens (Box 14.7) and smart windows, the cost of indium prohibits its use in bulk applications where a colourless or transparent conducting material is required. These applications include antistatic coatings and pigments, where a colourless or white finish is required so that carbon black (Section 30.1(c)) cannot be used.

One application where an antistatic layer is needed is in printing on polymers; for example, packaging labels and banknotes. An antistatic transparent coating is required to stop a charge building up on the polymer sheets during high-speed printing, as an induced static charge can result in the polymer sheets not passing freely through the printing machinery. For these applications, cheaper transparent conducting materials are being developed, including

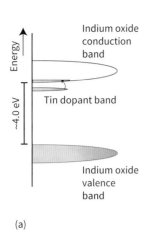
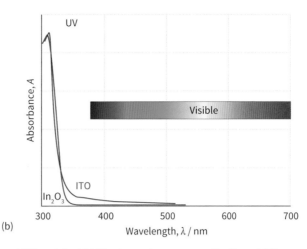

FIGURE 30.2 The band structure of ITO and the UV–Vis absorption spectra of In_2O_3 and ITO.

aluminium-doped zinc oxide (AZO), tin-doped antimony oxide, and fluorine-doped tin oxide (FTO). These n-doped extrinsic semiconducting materials (Section 25.6) demonstrate sufficient electronic conductivity while remaining almost colourless.

(b) Coloured inorganic solids and pigments

KEY POINTS Intense colour in inorganic solids can arise through d–d transitions, charge transfer (and the analogous interband electron transfer), or intervalence charge transfer. Inorganic pigments are widely used in paints, coatings, and colouring plastics.

Inorganic materials often have advantages over organic pigments for applications requiring excellent long-term chemical, light, and thermal stability. Pigments were originally developed from naturally occurring compounds, including hydrated iron oxides or ochres (yellow, orange, and brown), manganese oxides (purple and brown), lead carbonate (white), vermilion (HgS, red), orpiment (As_2S_3, yellow-orange), and copper carbonates (green and blue). These compounds were originally used in prehistoric cave paintings over 60,000 years ago.

Synthetic pigments, which are often analogues of naturally occurring compounds, were developed by some of the earliest chemists and alchemists, and the first true inorganic chemists were probably those involved in making pigments! Thus, the pigment Egyptian blue ($CaCuSi_4O_{10}$) was made from sand, calcium carbonate, and copper ores as long as 3000 years ago. This compound and a structural analogue, Chinese blue ($BaCuSi_4O_{10}$), which was first produced about 2500 years ago, adopt structures containing square-planar copper(II) ions surrounded by Si_4O_{10} groups (Fig. 30.3). Cobalt blue, $CoAl_2O_4$, adopts the spinel structure type (Section 25.11(c)) and was manufactured commercially in France from 1807. Many other transition-metal spinels are

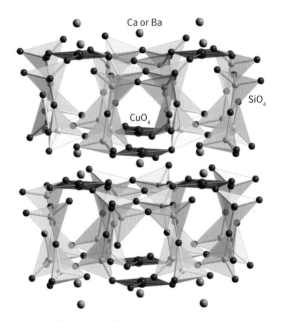

FIGURE 30.3 The square-planar copper/oxygen environment formed by Si_4O_{10} groups in Egyptian blue ($CaCuSi_4O_{10}$) and Chinese blue ($BaCuSi_4O_{10}$).

intensely coloured and used as pigments, including $CuCr_2O_4$ (black), $Zn_xNi_{1-x}Cr_2O_4$ (green), and $CoFe_2O_4$ (brown). Inorganic pigments continue to be important commercial materials for controlling and using light energy in paints, coatings, and optical filters.

Transition-metal oxide pigments are widely used where chemical and high thermal stability are paramount, including in ceramics and glazes, plastics exposed to sunlight for long periods, and coloured cements. The blue colours of $CoAl_2O_4$ and $CaCuSi_4O_{10}$, both used for these applications, stem from the presence of d–d transitions in the visible region of the electromagnetic spectrum (Section 22.2). Other very stable inorganic pigments with colours based on d–d transitions include Ni-doped TiO_2 (yellow), $Cr_2O_3 \cdot nH_2O$ (green), and $YIn_{1-x}Mn_xO_3$ (blue). The intensity of colour in these materials is usually not strong, as d–d transitions are forbidden according to the Laporte selection rule, Section 22.4(b). The characteristic strong colour of cobalt aluminate is a result of the Co^{2+} cation having a noncentrosymmetric tetrahedral site, which removes the constraint of the Laporte selection rule for centrosymmetric environments.

Colour also arises in many inorganic compounds from charge transfer (Section 22.3) or what is electronically often an equivalent process in solids, the promotion of an electron from a valence band (associated with anion orbitals) into a conduction band (associated with the oxidized metal cation orbitals). Charge-transfer pigments include compounds such as lead chromate ($PbCrO_4$), containing the yellow-orange chromate(VI) anion, and $BiVO_4$, with the yellow vanadate(V) anion. The compounds CdS (yellow) and CdSe (red) both adopt the wurtzite structure and their colour arises from transitions from the filled valence band (which is mainly derived from chalcogenide p orbitals) to a conduction band formed from orbitals based mainly on Cd^{2+} (Section 2.11, Fig. 2.38). For a material with a band gap of 2.4 eV (as for CdS at 300 K) these transitions occur as a broad absorption corresponding to wavelengths shorter than 515 nm; thus CdS is bright yellow, as the blue part of the visible spectrum is fully absorbed. For CdSe the band gap is smaller on account of the higher energies of the Se 4p orbitals, and the absorption edge shifts to lower energies; as a result, only red light is *not* absorbed by the material.

In some mixed-valence compounds, electron transfer between differently charged metal centres can also occur in the visible region, and as these transfers are often fully allowed they give rise to intense colour. Prussian blue, $[Fe(III)]_4[Fe(II)(CN)_6]_3$ (Fig. 30.4), is one such compound, and its dark-blue colour has resulted in its widespread use in inks. Intensely coloured, often dark brown, blue, or black Ru compounds, such as the tris(carboxyl)-terpyridine complex $[Ru(2,2',2''-(COOH)_3-terpy)(NCS)_3]$, absorb effectively right across the visible and near-IR regions of the spectrum and are used as photosensitizers in Grätzel-type solar cells (Box 23.1).

Inorganic radicals often have fairly low-energy electronic transitions that can occur in the visible region. Two well-known examples are NO_2 (brown) and ClO_2 (yellow). One inorganic pigment is based on an inorganic radical, but because of the high reactivity normally associated with main-group compounds containing unpaired electrons, this species is trapped inside a small zeolite cage (Section 25.16). Thus, the royal-blue pigment ultramarine, a synthetic analogue of the naturally occurring semiprecious stone lapis lazuli, has the idealized formula $Na_8[SiAlO_4]_6 \cdot (S_3)_2$ and

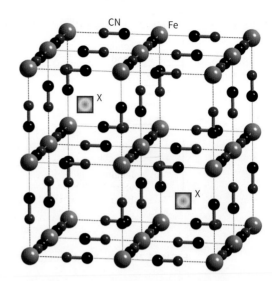

FIGURE 30.4 Prussian blue, $[Fe(III)]_4[Fe(II)(CN)_6]_3$. The iron cations are linked through cyanide anions, forming a cubic unit cell. Shaded squares (X) show regions in the structure where cations or water molecules may reside.

contains the S_3^- polysulfide radical anion occupying a cage formed by the aluminosilicate framework (Fig. 30.5).

Current developments in inorganic pigment chemistry are focused on finding replacements for some of the yellow and red materials that contain heavy metals; for example, toxic Cd and Pb. Although these materials themselves are not toxic, as the compounds are very stable and the metal is difficult to leach into the environment, their synthesis and disposal

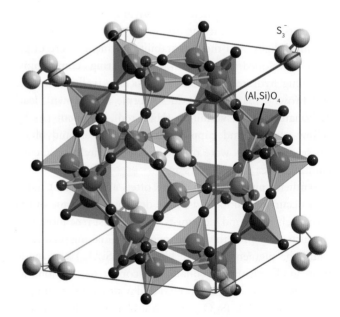

FIGURE 30.5 One of the cages present in ultramarine, $Na_8[SiAlO_4]_6 \cdot (S_3)_2$. The zeolite framework (with the code SOD) consists of linked SiO_4 and AlO_4 tetrahedra surrounding a cavity containing the polysulfide radical ion S_3^- and Na^+ ions (the latter omitted for clarity).

can be problematic. Compounds that have been investigated to replace cadmium chalcogenides and lead-based pigments are the lanthanoid sulfides, such as Ce_2S_3 (red), and early d-metal oxide-nitrides, such as $Ca_{0.5}La_{0.5}Ta(O_{1.5}N_{1.5})$ (orange), Section 25.14. In both cases, the replacement of O^{2-} by S^{2-} or N^{3-} ion has the effect of narrowing the band gap in these solids (compared to the colourless solids CeO_2 and $Ca_2Ta_2O_7$, which have large band gaps and absorb only in the UV region of the spectrum) and bringing the electron excitation energy into the visible. However, neither of these materials has the stability of the cadmium- and lead-based pigments.

Examples of more specialist inorganic pigments are magnetic pigments based on coloured ferromagnetic compounds, including Fe_3O_4 (brownish grey) and CrO_2 (black), and anticorrosive pigments that can be applied to metal surfaces, such as zinc phosphate, $Zn_3(PO_4)_2$. The deposition of inorganic pigments on to surfaces in thin layers can produce additional optical effects beyond light absorption. Thus, deposition of TiO_2 or Fe_3O_4, as layers a few hundred nanometres thick on flakes of mica (Section 28.2) produces lustrous or pearlescent pigments in which interference effects between light scattered from the various surfaces and layers produces shimmering and iridescent colours.

(c) White, black, and interference-effect pigments

KEY POINTS Titanium dioxide is used extensively as a white pigment, and carbon black and CNTs are important black pigments. Special colour, light-absorbing, and interference effects can be induced in inorganic nanomaterials used as pigments.

Some of the most important compounds used to modify the visual characteristics of polymers and paints have visible-region absorption spectra that result in them appearing either white (ideally no absorption in the visible region) or black (complete absorption between 380 and 800 nm). White inorganic materials are classified as pigments, and vast quantities of these compounds are synthesized for applications including the production of white plastics, coatings, and paints. Important commercial compounds of this class that have been used extensively historically are TiO_2, ZnO, ZnS, lead(II) carbonate, and lithopone (a mixture of ZnO and $BaSO_4$); note that none of the metals in these materials has an incomplete d-electron shell that might otherwise induce colour through d–d transitions.

Titanium dioxide, TiO_2, is manufactured in either its rutile or anatase forms (Fig. 30.6(a) and (b)), free from impurities (which is essential for a bright white pigment), and of controlled particle size, including nanoparticles (Section 28.2). The desirable qualities of TiO_2 as a white pigment derive from its large band gap (>3 eV, so no transition absorbing visible light can occur, Fig. 30.6(c)), its excellent light-scattering power (a result of its high refractive index, $n_r = 2.70$), the ability to produce very pure materials of a desired particle size, its good light-fastness and weather resistance, and its nontoxicity (compared with the previously used lead carbonate). Uses of titanium dioxide, which nowadays dominates the white-pigment market, include paints, coatings, printing ink (where it is often used in combination with coloured pigments to increase their brightness and hiding power), plastics, fibres, paper, solar cells, white cements, and even foodstuffs (where it can be added to icing sugar, sweets, processed fish fillets, and flour to improve their brightness).

The nanoparticulate form of TiO_2 (Section 28.2) is widely used in sunscreens due to its excellent absorption and scattering of ultraviolet light at wavelengths below 340 nm, Fig. 30.6(c)). When the nanoparticle size is in the range 20–150 nm, the cosmetic appearance of sunscreens containing TiO_2 is colourless; the nanoparticles are often coated with silica to prevent agglomeration. ZnO is also used widely as a cheaper alternative to TiO_2 and its absorption spectrum, Fig. 30.6(c), lies mainly in the UV region. One large-scale use suggested for white pigments is in building materials and roof paints, tiles, and coatings to reduce the building temperature during the daytime. Painting or tiling a roof area of 100 m² with a material containing an inorganic white pigment enables it to reflect up to 98% of sunlight, lowering the temperature of the roof by almost 5°C.

The most important black pigment is carbon black, Sections 15.6(d) and 28.3. This material has excellent absorption properties right across the visible region of the spectrum, and applications include printing inks, paints, plastics, and rubber. Nanoparticulate forms of carbon as nanotubes, CNTs, or graphene show exceptionally high absorption across the visible light region leading to the blackest materials known, Section 28.6; these materials have specialist optical applications to absorb stray light. Copper(II) chromite, $CuCr_2O_4$, with the spinel structure (Fig. 25.36), is used less frequently as a black pigment.

Carbon black and copper chromite pigments also absorb light outside the visible region, including the infrared, which means that they heat up readily on exposure to sunlight, which contains around 40% of its energy above 800 nm. While this is useful for solar heat harvesting (Section 30.4), this heating can have drawbacks in a number of other applications. There is interest in developing new materials that absorb in the visible region but reflect infrared wavelengths; the complex oxide $Bi_2Mn_4O_{10}$ is one compound that exhibits these properties.

(d) Fluorescence, phosphors, and LEDs

KEY POINT Materials that re-emit absorbed light at different, lower, energies have applications in low-energy lighting.

Luminescence is the emission of light by materials that have absorbed energy in some form. Photoluminescence occurs

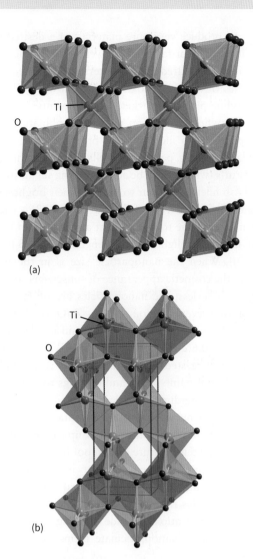
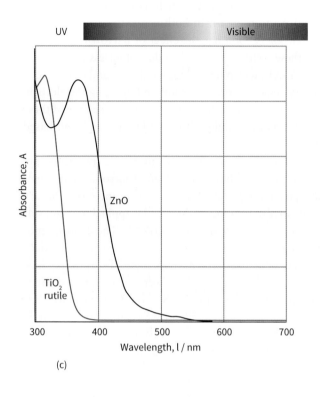

FIGURE 30.6 TiO_2 exists as several polymorphs that can be described in terms of linked TiO_6 octahedra, including the (a) rutile and (b) anatase forms. (c) UV-Vis absorption spectra of TiO_2 and ZnO.

when photons, usually in the ultraviolet region of the electromagnetic spectrum, are the source of energy and the output is usually visible light. Cathodoluminescence uses electron beams as a source of energy, and electroluminescence uses standard electrical energy supplied through a conducting material. Two types of photoluminescence can be distinguished: fluorescence, which has a period of less than 10^{-8} s between photon absorption and emission; and phosphorescence, for which there are much longer delay times (Section 22.5), sometimes of seconds or more. The luminescent properties of lanthanoid ions were introduced in Section 22.6, together with a discussion of the Jablonski diagram, Fig. 22.14.

Photoluminescent materials, frequently referred to as **phosphors**, generally consist of a host structure, such as ZnS (wurtzite structure; Section 4.9), $CaWO_4$ (scheelite structure type with discrete WO_4^{2-} tetrahedra separated by Ca^{2+} ions), or Zn_2SiO_4, into which an **activator ion** is introduced by doping. The host often absorbs the incident UV light before the energy is transferred to the activator ion. The activator ions are usually specific d-metal or lanthanoid ions; for example, Mn^{2+}, Cu^{2+}, Eu^{3+}, and Eu^{2+}, which have the ability to emit light of the desired wavelengths through the electronic process described at the start of this section and in Section 22.6. In some cases, a second dopant is added as a **sensitizer**, which aids the absorption of light of the desired wavelengths; the sensitizer ion then rapidly transfers the absorbed light energy by electron transfer to the activator ion for emission.

Widespread applications of inorganic phosphors include fluorescent and LED lighting, and security marking of banknotes, passports, and valuable goods; in both these applications there is a need for materials that fluoresce with high efficiency in specific regions of the visible spectrum following excitation using UV or blue light.

Many host/activator combinations have been studied as potential phosphor materials, with the aim of producing a material that efficiently converts UV radiation to

TABLE 30.1 Commercial phosphors

Phosphor host	Activator	Colour
Zn_2SiO_4	Mn^{2+}	Green
$CaMg(SiO_3)_2$	diopside Ti	Blue
$CaSiO_3$	Mn	Yellow-orange
$Ca_5(PO_4)_3(F,Cl)$	Mn	Orange
ZnS	Ag^+, Cu^{2+}, Mn^{2+}	Blue, green, yellow
YAG, $Y_3Al_5O_{12}$	Ce^{3+}	Yellow
YVO_4	Eu^{3+}	(Orangey) Red
$SrAl_2O_4$	Eu^{2+} and Dy^{3+}	Green

pure emission of a desirable colour in the visible region, Table 30.1. The modification of the host structure and the nature and environment of the activator ion allows these properties to be tuned.

Older fluorescent tube lamps used a mercury discharge as the source of the UV radiation at wavelengths of 254 and 185 nm. A coating of phosphor on the inside of the tube produced fluorescence at several wavelengths that combined to produce an effective white light. In modern LED lighting—for example, household LED lightbulbs, high intensity torches, and car headlights—the source of the UV radiation is normally a GaN-based LED (Box 25.4) that emits blue ($\lambda = 430$ nm) or long-wavelength ultraviolet light ($\lambda = 365$ nm). This LED is then coated with one or more phosphors which convert the blue or UV light into white light. Combining a blue-emitting LED with the yellow phosphor Ce^{3+}:YAG produces a white light (the human eye sees the combination of the yellow luminescence and some unchanged emitted blue light as white light with a cold blue tinge). The use of a combination of red-, blue-, and green-emitting phosphors with a UV-emitting LED renders a visible white light closer to sunlight, a warmer white. Commercial phosphors, Table 30.1, yield reasonably high-efficiency LED lights, where up to 80% of electrical energy is converted into light via the phosphor.

Current research focuses on materials with even higher conversion efficiencies (over 90%) and purer colours—particularly a clean red emission at wavelengths between 650 and 700 nm. In YVO_4:Eu^{3+} the vanadate group absorbs the incident electron energy and the activator is Eu^{3+}. The emission mechanism involves electron transfer between the vanadate group and Eu^{3+}, and the efficiency of this process depends on the M–O–M bond angle in a similar way to the superexchange process in antiferromagnets (Section 22.8). The nearer this angle is to 180°, the faster and more efficient the electron transfer process. In YVO_4:Eu this angle is 170°, and this material is a highly efficient phosphor.

A perceived white light that mimics sunlight to the human eye can be produced with a combination of yellow and blue light or, alternatively, a mixture of red, green, and blue light (RGB). The observed orange-red fluorescence of Eu^{2+} and Eu^{3+} in different oxide hosts (materials, including YVO_4:Eu, that have been used as red-light phosphors) results from several emissions between 540 and 700 nm. An efficient red phosphor luminescing above 650 nm is required for LED applications, producing a perceived pure white light. More recently discovered red fluorescent materials include $Ca_{1-x}Sr_xS$:Eu^{2+} and $Sr_2Si_5N_8$:Eu^{2+}, where the use of a sulfide or nitride environment around the lanthanoid ion shifts the emission to a brighter, pure red colour compared to Eu^{2+} in oxide environments.

Anti-Stokes phosphors convert two or more photons of lower energy to one of higher energy (such as infrared radiation to visible light). They act by absorbing two or more photons in the excitation process before emitting one photon. The best anti-Stokes phosphors have host structures formed from metal-centred polyhedra, such as YF_3, $NaLa(WO_4)_2$, or $NaYF_4$, doped with Yb^{3+} as a sensitizer ion (to absorb two photons of the IR radiation) and Er^{3+} as an activator (emitting visible light, Box 22.1). Applications include night-vision binoculars.

30.2 Photovoltaics and solar cell materials

KEY POINTS Semiconductors are widely used in harvesting solar energy, and new inorganic materials are being developed to allow thin film photovoltaic devices to be printed.

The semiconducting properties of silicon, doped silicon, and related III/V materials have been introduced in Section 25.6. These semiconductors typically have band gaps between 0.8 and 2.2 eV, and excitation of an electron from the valence band to the conduction band can be achieved using visible light (375 nm visible violet light is equivalent to 3.3 eV and 750 nm red light to 1.65 eV). These properties of semiconductors lead to their use in solar cell devices. Materials that generate an electrical voltage on illumination by light are known as **photovoltaic** (**PV**) materials.

The silicon solar cell, Fig. 30.7(a), consists of two layers; one is an n-type semiconductor, the other is a thin layer of p-type semiconductor, sandwiched between reflective coatings and electrodes. When light impinges at the junction of the n-type and p-type semiconductor layers, an electron–hole pair is created whereby the electron is excited into the conduction band of the n-type semiconductor and the hole forms in the valence band of the p-type semiconductor, Fig. 30.7(b). The electron and hole can then migrate towards electrodes on opposite faces of the semiconductor: in a typical silicon-based solar cell the n-type Si is at the front and electrons are collected here, while holes migrate towards the p-type semiconductor at the rear.

The maximum effective voltage produced by the solar cell, qV, depends on the separation of the Fermi levels of

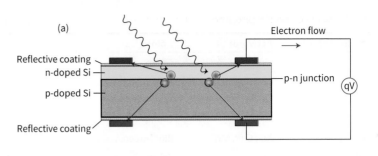

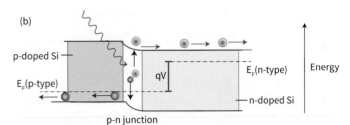

FIGURE 30.7 (a) Schematic of a p-n junction silicon solar cell. (b) Band structure and electron flow in a silicon solar cell on illumination.

the n- and p-type semiconductors, which for a typical silicon solar cell is around 0.6 V, the separation of the dopant energy levels. Electrons generated at the n-type semiconductor can flow in the external circuit, powering a load, and recombine with holes generated at the p-type semiconductor.

The ideal physical characteristics of a semiconductor material used in a solar cell are that it has a band gap of 0.9–1.8 eV, a high optical absorption across the solar spectrum, and demonstrates good electron mobility. The ideal band gap in single p-n junction solar cell materials is derived from a consideration of a number of factors including the photon energy distribution in the Sun's light reaching the Earth, and the recombination of electron–hole pairs before they migrate to the external electrodes. The maximum theoretical solar cell efficiency for a single junction solar cell is 32–34%, a value known as the **Shockley–Queisser (SQ) limit**, which can be achieved theoretically for semiconductors having a band gap between 1.1 and 1.5 eV. Higher solar cell efficiencies are achievable by using multiple semiconductor junctions.

While **crystalline Si, c-Si**, has a near-ideal band gap for solar cell applications (1.12 eV), its absorption of visible light is poor. The reason for this limitation is that light absorption by a semiconductor requires conservation of momentum, and while a photon has zero momentum, the valence band-to-conduction band transition in silicon is **indirect** and requires a change in momentum (which can be obtained from a vibrational mode in the solid). Absorption of light by a solar cell material is improved by using **direct** band gap materials (where there is no change in momentum involved in exciting an electron to the conduction band); direct band gap materials include amorphous silicon, GaAs, CdTe, CIGS, and CZTS (Table 30.2).

In practice, the best commercial solar cells based on c-Si have efficiencies of around 15%, well below the SQ limit. The discovery and characterization of new semiconducting materials for use in solar cells is a major current research topic. As well as optimizing the band gap of the semiconducting material for solar cell performance the research targets include environmental and economic aspects, including toxicity of components, cost, recyclability, manufacturing parameters, and the ability to print or mass produce devices.

Amorphous silicon, a-Si, may be synthesized using plasma-enhanced CVD techniques (Section 28.4(c)) involving the thermal decomposition of SiH_4 between 30°C and 300°C (Section 15.7). The material has a disordered silicon structure, with many silicon atoms having coordination numbers below 4 (as found in c-Si with the diamond structure). As a result, the highly disordered structure with a large number of defects exhibits low electrical conductivity.

Reaction of a-Si with hydrogen gas leads to hydrogenated amorphous silicon, a-Si:H, wherein hydrogen atoms bond to low-coordination Si sites; this decreases the number of defects and markedly improves electron mobility. While a-Si and a-Si:H demonstrate lower efficiencies in solar cells than c-Si, the ability to use low temperatures in the CVD deposition process, and across a large surface area, leads to much lower manufacturing costs. Because of the low deposition temperature, a-Si can also be deposited on plastics and other thermally less stable substrates. Currently, c-Si remains dominant in the silicon solar cell market due to its much higher efficiency.

Table 30.2 summarizes information on various inorganic semiconductors that are being investigated for use in solar cells. The majority of materials investigated as alternatives to c-Si are semiconductors of the III/V and II/VI types.

GaAs has many of the ideal properties required of a photovoltaic material with a direct band gap of 1.42 eV and high electron mobility leading to device efficiencies close to 30%. However, the high cost of gallium and the need to use expensive epitaxial techniques (Section 28.4(b)) to deposit

TABLE 30.2 PV inorganic semiconductor properties

Material	Band gap / eV	Crystal structure	Direct or indirect band gap	Comments (best commercial efficiency)
Crystalline Si, c-Si	1.12	Diamond	Indirect	Cheap, high-temperature manufacturing, (15–20%)
Amorphous Si, a-Si; hydrogenated amorphous silicon, a-Si-H	(0)	Disordered, glassy diamond type	Direct	Cheap, low temperature deposition, low electron mobility, lower efficiency than c-Si, (~7%)
GaAs	1.42	ZnS, zinc blende	Direct	Expensive, toxicity, (25%)
CdTe	1.45	ZnS, wurtzite	Direct	Commercialized, toxicity, Te scarcity, (18%)
$CuIn_{1-x}Ga_xSe_2$, CIGS	1.05–1.6	Zinc blende/chalcopyrite type with ordered cations	Direct	Low efficiency due to electron–hole recombination; costly Ga/In, (20%)
Cu_2ZnSnS_4, CZTS	1.1–1.45	Kesterite, zinc blende/chalcopyrite type with ordered cations	Direct	Low efficiency due to electron–hole recombination, cheap Earth-abundant elements, (11%)
Dye-sensitized TiO_2	~1.5	Rutile with dye layer	—	(12%)
Hybrid lead perovskites	1.57	Perovskite	Nearly direct	Low cost, potential toxicity with lead, (18%)

thin layers of GaAs means that GaAs-based solar cells are only used when optimum performance is required, as in the aerospace industry. The International Space Station and the James Webb Telescope use GaAs-based solar cells that are highly efficient, very lightweight, and flexible.

CdTe has been commercialized for use in solar cells, and in the early 2020s CdTe-devices represent around 5% of the global market. The advantages of CdTe solar cells are high efficiencies (near 20%) and low-cost manufacturing resulting from high-throughput techniques that allow CdTe to be deposited through sublimation.

CIGS, $CuIn_{1-x}Ga_xSe_2$, and CZTS, Cu_2ZnSnS_4, are two semiconducting materials based on an ordered arrangement of cations in the zinc blende structure type (ZnS = MX, equivalent to M_2X_2 and M_4X_4, with X = S or Se plus the different cations), Fig. 30.8. CIGS may be deposited as very thin films for use in solar cells; the direct band gap of this material provides these films with exceptionally high solar absorption. Flexible CIGS solar cells with efficiencies above 20% can be produced by depositing CIGS on to a polymer; the metals are sputtered or evaporated on to the substrate and this is then treated with a selenium vapour or H_2Se. CZTS demonstrates similar properties to CIGS though lower solar cell efficiencies of around 5–11% are found due to difficulties in synthesizing homogeneous thin films. However, CZTS has the advantage of containing only Earth-abundant elements, making devices very cheap to produce and environmentally sustainable.

The structures and properties of hybrid lead iodide perovskite solar cells were introduced in Box 15.4. These materials, $MPbI_3$ (M = methylammonium, formamidinium, or

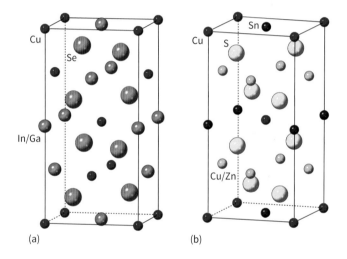

FIGURE 30.8 (a) The structure of CIGS (b) The structure of CZTS. The structures have close-packed S or Se with ordered arrangements of metal cations in half the tetrahedral holes.

caesium and mixtures thereof) offer the potential of high-efficiency solar cells (values of over 20% have been reported for single-junction devices) and low-cost fabrication. The perovskite structure formed for stoichiometries ABX_3, Sections 4.9 and 25.11(d), with B = Pb and X = I, is only stable for the large cations A = $CH_3NH_3^+$ (methylammonium) and $[HC(NH_2)_2]^+$ (formamidinium) and for A = Cs^+ at temperatures above 340°C, Fig. 30.9. These compounds have almost direct band gaps around 1.5–1.6 eV, making them ideal for thin-film solar cell applications, as they may be readily deposited by precipitation from a solution containing PbI_2 and MI and through spin-coating techniques

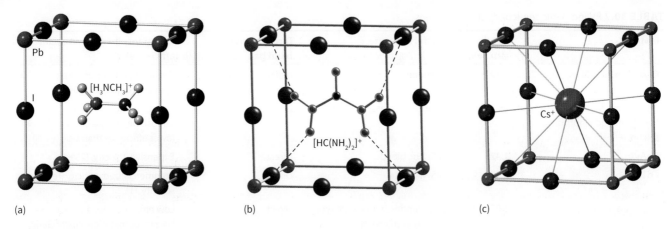

FIGURE 30.9 The structures of the cubic forms of the lead halide perovskites: (a) methylammonium, [CH$_3$NH$_3$]PbI$_3$, (b) formamidinium, [HC(NH$_2$)$_2$]PbI$_3$, and (c) CsPbI$_3$.

(where a substrate disc is rotated and a fluid containing the solid is applied to the disc centre).

Some serious problems remain with lead iodide perovskites because of their limited thermal and chemical stability, though the first commercial solar cells emerged in 2023. Formamidinium lead perovskites readily decompose at room temperature, and all phases of this type are susceptible to reaction with moisture, degrading solar-cell performance over time. Ongoing chemistry research concerns the use of mixed cation phases to improve stability, studies of perovskite phases containing tin and bromide, and the use of additives, including CNTs, to reduce sensitivity to water. Since the discovery and investigation of lead halide perovskites many other compositionally related materials have been studied for use as photovoltaics. Research has focused on complex lead iodides with Ruddlesden–Popper-type structures (Section 25.11(d)), and complex iodides, bromides, selenides, and tellurides of other post-transition metals, particularly Sb, Bi, and Sn.

Dye sensitized solar cells, also known as Grätzel cells, have been described in Box 23.1.

30.3 Photocatalysts and water-splitting

In a photovoltaic material (Box 15.4 and Section 30.2), the light energy excites an electron, and the separation of this electron and the remaining hole, provided they do not recombine immediately, can give rise to electrical current. Alternatively, the photogenerated electron and hole can catalyse a chemical process, such as the splitting of water into H$_2$ and O$_2$, or help facilitate the reaction or decomposition of an organic molecule. Materials that promote these processes are known as **photocatalysts**. The production of hydrogen by using solar energy has been introduced in Box 11.4, where the stages involved in converting light and water into H$_2$ and O$_2$ were described.

(a) Water-splitting photocatalysis

KEY POINT Wide band gap semiconductors, with E_g >3 eV, can use visible light to split water.

The basis of a water-splitting system based on a light-absorbing particle was introduced in Box 11.4. The essential stages involve photon (λ <1000 nm) capture to form an electron–hole pair in the valence and conduction bands, followed by separation and migration of the electron to the cathode to reduce protons to H$_2$(g) (the Hydrogen Evolution Reaction, **HER**) and of the hole to the anode to oxidize water to O$_2$(g) (the Oxygen Evolution Reaction, **OER**), Fig. 30.10(a). The band gap E_g in the photocatalyst

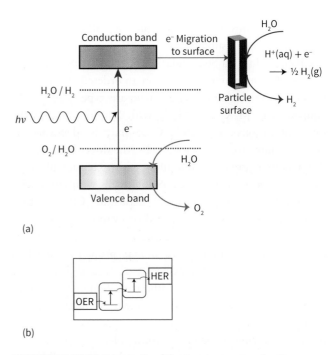

FIGURE 30.10 (a) Schematic of the processes occurring in a semiconductor photocatalyst used for water-splitting. (b) The energy levels in a tandem photocatalytic cell.

must exceed the minimal electrochemical potential required to split water (> 1.23V).

One of the most studied photocatalysts is TiO_2, whose optical properties have been discussed previously, Section 30.1(c). The wide band gap in TiO_2 (anatase polymorph, 3.2 eV; rutile, 3.1 eV) leads to absorption of light with wavelengths below ~390 nm—that is, in the UV region of the electromagnetic spectrum (Fig. 30.6(c)). A TiO_2 surface placed in water photocatalytically produces H_2 and O_2 gases when illuminated by UV light or sunlight (which contains ~5% of its light as UV radiation). However, such processes are very inefficient for harvesting solar energy for many reasons; these include the low efficiency of sunlight absorption with only UV wavelengths absorbed, inefficiencies in the electron–hole separation process, the low surface area of bulk TiO_2 at which the photocatalysis reactions can occur, and the slow O_2 evolution reaction (Section 6.16).

A variety of methods is being investigated to improve the efficiencies of such photocatalytic cells. These include the use of dyes to extend the range of solar light wavelengths absorbed (as in Grätzel cells; Box 23.1), the formation of TiO_2 nanoparticles with high surface areas (Section 28.6(c)), and the development of coatings for TiO_2. Example electrode coatings for TiO_2 include NiO and $(CH_3NH_2)PbI_3$, which promote photocatalysis using visible wavelengths of light (a much larger component of the solar spectrum).

Inorganic materials other than TiO_2 that perform the photocatalytic water-splitting process are also being very actively researched. These materials include alternative transition-metal oxides with smaller band gaps, such as Fe_2O_3, and compounds where the band gap is narrowed through the incorporation of dopants and the formation of solid solutions (Sections 25.1 and 25.6). By tuning the band gap (**band-gap engineering**) it is possible to optimize the efficiency of the photocatalytic material represented by the **apparent quantum efficiency, AQE**. The AQE measures the number of electrons and holes which undergo a reaction to produce hydrogen and oxygen relative to the number of incident photons.

A better practical measure of the efficiency of a photocatalyst for water-splitting using solar radiation is the **solar-to-hydrogen efficiency (STH)**, which is the ratio of the output energy (usually as H_2 gas) to the energy of the incident solar light. The design of new photocatalytic materials through band-gap engineering requires compounds that absorb as much solar light as possible (the band gap being narrowed so that all visible and some near infrared light is absorbed), while having good electron–hole mobility and separation to avoid their recombination. Materials must also have highly efficient surface catalytic properties to facilitate the generation of H_2 and O_2.

Yellow $BiVO_4$, which has a band gap of 2.4 eV, and other bismuth transition-metal oxides with similar band gaps are candidate photocatalytic materials. Complex oxides, including $NaTaO_3$, $K_2Ta_3B_2O_{12}$, and $SnWO_4$, and transition-metal oxide-nitrides are also being studied. One promising area of research are mixed anion systems where replacement of oxide ions by nitride or halide anions leads to a narrowed band gap. Thus, while the band gap in white Ta_2O_5 is 3.9 eV (too high for efficient harvesting of the full solar spectrum), it is 2.1 eV (650 nm) in Ta_3N_5, and for TaON the value is 2.4 eV. Similarly, $MgTa_2O_6$ only absorbs UV light, while nitridation (Section 25.13) yields the red material $MgTa_2O_{6-x}N_x$ with efficient light absorption at wavelengths below 570 nm. These oxide-nitrides show much higher STHs compared to the parent oxide. Introduction of sulfide anions, replacing oxide, also narrows band gaps and the oxysulfide $Y_2Ti_2O_5S_2$ is a promising photocatalyst. Nanoparticulate forms of these materials (Chapter 28) with their very high surface areas also show improved STHs.

An extension of the single material system is a **tandem two-photon device**, in which two semiconductors are joined together in an arrangement known as a **Z-scheme**, Fig. 30.10(b). Instead of the excited electron in a lower semiconductor unit migrating directly to a HER, it transfers across a junction to replace the electron excited from the valence band in the upper semiconductor unit (equivalently, it combines with the hole in this material). The two-stage, two-phase excitation process increases efficiencies by absorbing a greater range of wavelengths from a light source. The use of two semiconductors allows these each to have smaller band gaps for efficient visible-light absorption. Thus, a tandem device can be constructed from WO_3 (E_g = 2.6 eV) and a dye sensitized TiO_2 (E_g = 1.6 eV) and the two-stage light absorption at wavelengths above 500 nm promotes the electron through over 3.5 eV, sufficient for effective water-splitting. Natural photosynthesis (Section 26.10(d)) uses the Z-scheme principle.

(b) Photocatalysts for degrading pollutants

KEY POINT Photocatalysts, including TiO_2, can be used for the oxidation and degradation of organic compounds present on their surfaces.

As well as photocatalysts for water-splitting, the light energy may also be used to catalyse the oxidation and decomposition of organic molecules on the photocatalyst surface. Applications include the destruction of pollutants in the environment and self-cleaning surfaces. The earliest applications of TiO_2 as a photocatalyst were for the removal of organic pollutants in water. A TiO_2 surface in polluted water (containing compounds such as chlorinated biphenyls, phenols, and chlorobenzenes) when exposed to sunlight or UV light generates hydroxyl radicals on the surface which then react with the organic compound, oxidizing and degrading it. Other wide-band-gap semiconductors that absorb UV light, including WO_3 and ZnO, promote similar photocatalytic reactions.

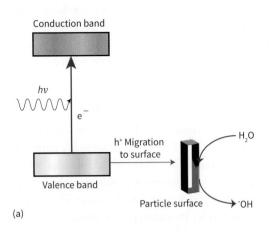

FIGURE 30.11 (a) Generation of hydroxyl radicals on a TiO_2 surface. (b) Windows with (right) and without (left) a photocatalytic self-cleaning layer.

Self-cleaning windows (Fig. 30.11), such as Pilkington Active®, are formed of glass coated with a thin transparent layer of titanium dioxide, in anatase form, which helps clean glass through two distinct properties: photocatalysis and hydrophilicity. When ultraviolet light in sunlight strikes the TiO_2 coating of a self-cleaning window, mobile electrons and holes are generated, Fig. 30.11(a). The mobile holes, h^+, migrate to the TiO_2 surface where they react with adsorbed water molecules to form hydroxyl radicals ($^{\bullet}OH$).

$$h^+(TiO_2) + H_2O(ads) \rightarrow {}^{\bullet}OH(ads) + H^+(aq)$$

The highly reactive hydroxyl radicals (Section 17.6) attack aliphatic sidechains of hydrophobic organic molecules that form greasy deposits on the window surface, reducing their chain length, and producing more hydrophilic, oxygen-containing organic species. When the external surface of the glass is wetted by rain, the hydrophilicity of the TiO_2 surface reduces contact angles for water droplets to very low values, causing the water to form a thin layer and wash the residual, water-soluble, organic species away, Fig. 30.11(b). Since the photocatalytic reactions occur on the surface of the TiO_2 coating they are very effective at loosening the lowest layers of the greasy deposits and all the dirty material gets washed away.

30.4 Thermal solar farms; piezoelectric and thermoelectric energy materials

The main method of harvesting energy directly from the Sun is through solar farms using photovoltaics, Section 30.2. Alternative methods of solar power harvesting include using the thermal energy of the Sun's radiation to drive a chemical reaction, as described in Box 11.4, or to cause a phase change from a solid to a liquid state. As well as using the Sun as an energy source, methods of converting and recycling energy from terrestrial heat and pressure forms also employ inorganic materials.

(a) Thermal solar energy: collection and storage

KEY POINT Sunlight, including infrared radiation, can be harvested through its ability to drive inorganic reactions.

Solar thermal collectors can be classed across low- to high-temperature systems. Low- and medium-temperature collectors are typically flat-plates or tubes coated with a black pigment, such as carbon black (Sections 15.6 and 28.6(a)), and can be used to heat domestic and industrial water supplies and swimming pools to temperatures below 100°C. Solar energy captured in this way can also be used to heat buildings and in commercial drying and cooking processes.

High-temperature collectors can heat water above 100°C and typically concentrate sunlight with mirrors or lenses. Rather than heat water, these high-temperature technologies convert and store solar energy in a chemical reaction or a phase change (using the latent heat for that process); the reverse of this process or reaction liberates heat that can be used to produce steam to drive electric power production by using turbines. Commercial heat-storage systems at thermal solar farms often use a molten salt technology. Many Group 1 metal salts with large monovalent anions have low melting points resulting from the low lattice enthalpies in a salt with low charges (Section 4.12); for example, the melting points of $NaNO_3$ and KNO_3 are 308°C and 220°C respectively, and these can be lowered still further (to 130°C) in mixtures with $Ca(NO_3)_2$. The latent heat (that of KNO_3 is 2.33 kJ mol^{-1}) is thus used to store energy. Daytime solar heat can be used to melt and heat the solid salt or salt mixture, and at night the cooling and solidifying salt can be used to boil water to drive steam turbines.

Other systems use lower-temperature **phase change materials** (**PCMs**), including metal-salt hydrates and hydrated zeolites (Section 29.1), which dehydrate and rehydrate to store thermal energy. Methods of solar thermal heat conversion using chemical reactions to generate green hydrogen have been described in Box 11.4.

(b) Piezoelectric materials for energy conversion and collection

KEY POINT Piezoelectric materials with the perovskite structure can harvest vibrational energy.

The transducer properties of piezoelectric perovskite materials (barium titanate and PZT) exploited for interconverting pressure and vibrational energy into electrical energy, and vice versa, were discussed in Section 25.13(d). The main areas for use of piezoelectric materials have been microphones, loudspeakers, hearing aids, inkjet printer heads, diesel fuel injectors, ultrasonic cleaning, and sensors. A further small-scale application of piezoelectronic materials is in a heart pacemaker, in which body and muscle movements apply pressure to a piezoelectric device that is sufficient to charge a nanobattery to power the pacemaker. On a larger scale, the vibrational energy of vehicles travelling on road surfaces can be harvested by using piezoelectric devices. Lead zirconium titanate, PZT, $Pb[Zr_xTi_{1-x}]O_3$, based transducers buried in road surfaces have been used to power nearby traffic lights.

(c) Thermoelectric materials for thermal energy salvaging

KEY POINT Waste thermal heat can be converted into electrical power by using thermoelectric materials.

Metal chalcogenide compounds and Chevrel phases were introduced in Section 25.14. Many of these materials have applications as thermoelectrics. A **thermoelectric** material generates an electrical potential in response to a temperature difference across it (the **Seebeck effect**). Alternatively, the **Peltier effect** produces a heat flow in a material when an electrical current is applied to it. Materials demonstrating the Seebeck effect can be used to turn low-grade or waste heat energy into electrical power, and those that exhibit the Peltier effect can be used in small-scale cooling devices.

Thermoelectric materials can be classified according to their **figure of merit**, zT, (based on the ratio of electrical to thermal conductivities), which specifies the efficiency of energy conversion from a thermal gradient into electrical energy. For a high zT, a thermoelectric material needs to have a low thermal conductivity but a relatively high electrical conductivity; the highest zT values in materials range between 0.5 and approximately 10. These combined properties are often found in heavy-atom semiconducting or semimetallic compounds. This is because the heavy atoms and clusters in a solid have small vibrational motions at a given temperature, limiting the transmission of heat. Good thermoelectric materials also often contain defects, intrinsic or extrinsic, which disrupt the flow of heat, or low dimensional structures that restrict the thermal motion in one or more directions. Many of the best thermoelectric materials are the heavy chalcogenide (Se, Te) or halide (Br, I) compounds of the post-transition metals (Pb, Sn, Bi).

The most commonly used thermoelectric materials are bismuth telluride, Bi_2Te_3, Fig. 30.12, and selenide, Bi_2Se_3. These compounds, in common with many metal chalcogenides, have layer-like structures similar to those of the MS_2 phases described in Section 25.14(b), with good electrical conductivity within the layers but poor thermal conductivity perpendicular to them, leading to useful thermoelectric efficiencies ($zT \sim 1.0$). Nanostructuring materials with alternating layers of Bi_2Te_3 and Bi_2Se_3 (using routes similar to those described in Section 28.9(c)) disrupt the structures further and lead to a decrease in thermal conductivity compared to the single bulk phases and improvement of zT to 2.5.

Designing new thermoelectric materials requires considerable ingenuity. The combination of structural elements must allow for rapid electron transport (a property associated with crystalline, three-dimensional solids, where there is little electron scattering by the regular placement of atoms) and low thermal conductivity (a property associated with disordered or glassy structural feature, which scatters the vibrational modes responsible for heat transport). In Chevrel phases the ability to incorporate various cations between the M_6X_8 blocks that 'rattle' around on their sites

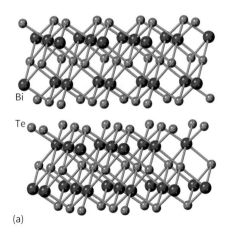

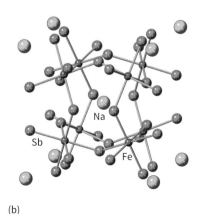

FIGURE 30.12 (a) The structure of Bi_2Te_3. (b) The structure of $Na_{0.25}FeSb_3$.

reduces the thermal conductivity of a material but not at the expense of electronic conductivity. Many other heavy-metal chalcogenide and pnictide phases are being investigated for thermoelectric device applications.

One important family of thermoelectric materials are the so-called 'skutterudites' (named after the mineral skutterudite, $CoAs_3$), having the general formula $M_x(Co,Fe,Ni)(P,As,Sb)_3$ with various inserted cations, M^{n+} including Ln or Na, as in the material $Na_{0.25}FeSb_3$, Fig. 30.12(b). The skutterudite structure is similar to that of ReO_3 (Section 25.11(b)), although the $CoAs_6$ octahedra are tilted to produce some large cavities within the structure, into which the cations may be inserted (as in the perovskite structure with the occupied A-type sites). These cations lower the thermal conductivity of the structure by absorbing heat to 'rattle around' within their cavities, while the high electronic conductivity of the $(Co,Fe,Ni)(P,As,Sb)_3$ three-dimensionally linked network is maintained.

Other inorganic materials of interest as thermoelectrics include the sulfide spinels $CuCr_2S_4$, and Cu_5FeS_4, Mg_2B, and Zintl-type phases (Section 25.3(c)); for example, $Ba_8Ga_{16}Ge_{30}$.

30.5 Fuel cell materials

KEY POINT Solid oxide fuel cells convert the chemical energy in hydrogen or organic fuels into electrical energy.

The functioning of fuel cells, and particularly solid oxide fuel cells (SOFCs), depends on the ionic and electronic conductivity properties of inorganic oxide materials, Section 25.8 and Box 6.1. Oxide-ion conductors are required for the electrolyte and mixed ionic–electronic conductors for the cathode and anode; anodes may also be constructed from mixtures of a nanoparticulate, conducting metal powder and an oxide ion conductor; for example, Ni/YSZ (yttrium-stabilized zirconia). To improve materials towards commercialization and the widespread use of SOFCs, there is a need to reduce their high operating temperatures from above 700°C, as the external energy input needed to heat to these temperatures lowers the overall efficiency of the fuel cell. As discussed in Section 25.8, anionic conductivities in materials normally remain low even at these high temperatures, with the best values in yttrium-stabilized zirconia reaching only $0.2\,S\,cm^{-1}$ at 1000°C. Many complex metal oxides are currently being investigated with the aim of achieving high oxide-ion mobilities at lower temperatures for SOFC applications. The criteria listed in Section 25.8—particularly the need for vacancies on oxide-ion sites to facilitate hopping mechanisms—are central to the design of new oxide-ion conductors.

Compounds that show promising behaviour for the oxide-ion electrolyte include $La_2Mo_2O_9$, barium indium oxide ($Ba_2In_2O_5$), BIMEVOX (a d-metal doped **bi**smuth **v**anadium **ox**ide), the apatite structure of $La_{9.33}Si_6O_{26}$, and strontium- and magnesium-doped lanthanum gallate (Sr,Mg-doped $LaGaO_3$, or LSGM). The structure of $Bi_4V_2O_{11}$ at room temperature is shown in Fig. 30.13(a) and contains layers with the composition Bi_2O_2 interleaved with sheets of the composition V_2O_7 formed from strongly distorted VO_6 octahedra.

At high temperatures, $Bi_4V_2O_{11}$ undergoes a phase transition to a tetragonal unit cell. Doping, in this case replacing V^{5+} by a lower-charged transition-metal ion—for example, Cu^{2+} or Ni^{2+}—is charge-balanced by creating vacancies on oxygen sites in the V_2O_7 layer and yielding materials of the composition $Bi_4(V_{1-x}M_x)_2O_{11-1.5x}$. These materials have a disordered arrangement of oxide ions and vacancies in the $(V_{1-x}M_x)_2O_{7-1.5x}$ layers, providing a facile pathway for oxide-ion hops and diffusion, Fig. 30.13(b).

Materials for the mixed conductor cathode in an SOFC are usually based on the perovskite structure ABO_3 (Sections 4.9 and 25.11(d)), with dopants on the cation sites that are charge balanced with vacancies on the oxide-ion sites. As well as acting as a layer for oxide-ion conduction and electron motion to an external circuit, the anode material also has to catalyse the reaction between hydrogen gas (or an alternative fuel such as methane or methanol) and oxide ions. Perovskite materials that have demonstrated these required properties include $(La,Sr)(Fe,Co)O_{3-x}$, $(La_4Sr_8)(Ti_{11}Mn_{0.5}Ga_{0.5})O_{36-x}$, $La_{0.4}Sr_{0.6}Ti_{1-x}Mn_xO_{3-x}$, $(La_{1-x}Sr_x)Cr_{0.5}Mn_{0.5}O_{3-x}$, and $Sr_{2-x}La_xMg_{1-x}Mn_xMoO_{6-x}$.

One of the best-developed intermediate-temperature (less than about 600°C) SOFCs consists of an anode of Gd-doped CeO_2 (CGO)/Ni, an electrolyte of Gd-doped cerium oxide (CGO), and a cathode of the perovskite LSCF, $(La,Sr)(Fe,Co)O_{3-x}$. The electrolyte material, CGO, possesses a much higher ionic conductivity than YSZ at these lower

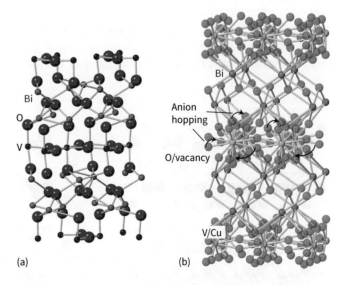

FIGURE 30.13 (a) The structure of $Bi_4V_2O_{11}$ at room temperature; (b) the structure of $Bi_4(V_{1-x}Cu_x)_2O_{11-1.5x}$, showing the disordered oxide ion and vacancy arrangement in the $(V_{1-x}M_x)_2O_{7-1.5x}$ layers.

temperatures. Unfortunately, however, its electronic conductivity (as an n-type semiconductor) is also higher, and the use of a CGO electrolyte can result in reduced efficiency as energy is wasted as a result of electrons flowing through the electrolyte.

30.6 Strong permanent magnets and magnetoresistive compounds

Cooperative magnetism and the ferromagnetic properties of solid materials were introduced in Section 22.8. The ferromagnetic properties of simple metals (Fe) and iron oxides (Fe_3O_4) have been known and utilized since ancient times. Improving the properties of ferromagnetic and ferrimagnetic solids, including the strength of permanent magnets, the ease with which they can be magnetized and demagnetized (**magnetic coercivity**), and controlling their Curie temperatures, requires the design and synthesis of new inorganic materials. Ferromagnetic materials lie at the heart of many clean energy technologies, including the dynamos of wind turbines and the energy recuperation systems in electric vehicles. Magnetic materials also have a role in energy-efficient electric motors and data-storage devices.

(a) Spinels: ferrites

KEY POINT Complex iron oxides adopting the spinel structure have important applications as soft and hard magnets.

Ferrite spinels of the composition AFe_2O_4, with A^{2+} = Zn, Mg, Fe, Co, Ni, Mn or a mixture thereof, and adopting the spinel structure type, were introduced in Section 25.11(c). In these materials the Fe^{3+} ions (high spin d^5) would normally occupy the octahedral sites (B-type sites, normal spinel), as in $ZnFe_2O_4$ ($\lambda = 0$). However, as a d^5 high spin ion Fe^{3+} can be displaced to a tetrahedral site with no loss of CFSE, many ferrite spinels with A = Fe^{3+} adopt the inverse or partially inverse spinel structure type, Table 25.4.

The magnetic moment of the high spin, d^5 Fe^{3+} ion can couple strongly with those on neighbouring B-type cation sites and, therefore, in ferrite spinels there is the potential for cooperative magnetic interactions. The large number of different cations that can be incorporated into the oxide spinel structure allows the control of these interactions through tailoring of the ferrite chemical composition. If J is the energy of interaction of the spins on different ion sites, when $RT > J$, ferrites are paramagnetic, and when $RT < J$, a ferrite may be either ferrimagnetic or antiferromagnetic.

The antiparallel alignment of spins characteristic of antiferromagnetism is illustrated by $ZnFe_2O_4$, which has the cation distribution $Fe[ZnFe]O_4$. In this compound the Fe^{3+} ions (with S = 5/2) in the tetrahedral and octahedral sites are antiferromagnetically coupled, through a superexchange mechanism (Section 22.8) below 9.5 K, to give nearly zero net magnetic moment to the solid as a whole; note that Zn^{2+} as a d^{10} ion makes no contribution to the magnetic moment of the material. Fe_3O_4, $[Fe^{2+}][Fe^{3+}]_2O_4$, magnetite, is an example of a ferrite spinel that has ferrimagnetic ordering at room temperature, due to a large J value.

Ferrimagnetic ferrites are divided into two main classes, depending on their magnetic coercivity: soft and hard ferrites. Both classes have important applications. **Soft ferrites** include the solid-solution nickel-zinc ferrites, $Ni_xZn_{1-x}Fe_2O_4$, and manganese–zinc ferrites, $Mn_xZn_{1-x}Fe_2O_4$. These materials have low magnetic coercivity; that is, they are easily magnetized and demagnetized at room temperature. This property leads to widespread applications in **transformers**. Applying an alternating voltage to a transformer soft ferrite core through one primary copper coil causes the ferrite to be magnetized, demagnetized, and remagnetized in the opposite direction with minimal energy losses. As the ferrite magnetizes and demagnetizes, a second copper coil, with a different number of coil turns to the first and wrapped around the same ferrite core, develops an induced current with a voltage different from that applied to the primary coil.

Hard ferrites, which include $AFe_{12}O_{19}$, A = Sr, Ba, with hexagonal structures closely related to the cubic spinel unit cell, have high coercivity and high **magnetic remanence**, meaning that once magnetized they resist becoming demagnetized. Hard ferrite magnets are used in magnetic recording devices, loudspeakers, and household applications, including fridge magnets and furniture closure devices.

(b) Lanthanoid-based permanent magnetic materials

KEY POINT The strongest permanent magnetic materials are alloys between neodymium and samarium with iron and cobalt.

Strong permanent magnets containing lanthanoids were developed between the 1960s and 1980s. Because many lanthanoids have large numbers of unpaired f-electrons (Chapters 21 and 22) and their compounds often adopt structures with high anisotropy (one direction in the crystal structure is markedly different from the others), some of their alloys and compounds, particularly in combination with iron, cobalt, and nickel, can be strongly magnetized in one direction. These lanthanoid-based materials are the strongest permanent magnets known, with both high coercivity and high remanence; while the ferrite $SrFe_{12}O_{19}$ has a magnetic remanence (B_r) of 0.3 T, values for lanthanoid-based materials reach 1.4 T.

$SmCo_5$, Fig. 30.14(a), with B_r of around 1.0 T, also has a high Curie temperature of 720°C and resists oxidation. These properties and the high cost of samarium lead to specialized applications where magnetic materials need to operate above 300°C; for example, aerospace applications. $Nd_2Fe_{14}B$, Fig. 30.14(b), is a low-cost lanthanoid permanent

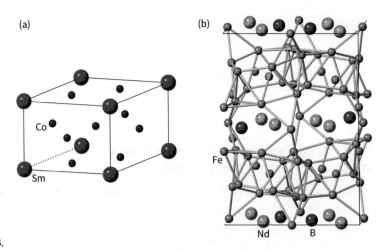

FIGURE 30.14 The structures of (a) $SmCo_5$ and (b) $Nd_2Fe_{14}B$.

magnet with magnetic remanence values of up to 1.6 T and a Curie temperature of 315°C, though it is very susceptible to oxidation and corrosion in moist atmospheres and is often coated with Ni, Cu, or Cr. Partial replacement of neodymium by dysprosium in the solid solution $Nd_{2-x}Dy_xFe_{14}B$, $0 < x < 0.25$, improves the coercivity, and the majority of applications use a composition close to $Nd_{1.8}Dy_{0.2}Fe_{14}B$, despite the much higher cost of Dy. Other lanthanoid-based permanent-magnetic materials that have been researched for applications include $Sm_2Fe_{17}N_3$ ($B_r = 1.54$, $T_C = 477°C$) and $(Sm_{0.8}Zr_{0.2})(Fe_{0.75}Co_{0.25})_{11.5}Ti_{0.5}$ ($B_r = 1.63$, $T_C = 528°C$).

$Nd_2Fe_{14}B$ permanent magnets originally developed widespread applications in speakers and headphones, bicycle dynamos, MRI scanners, and high-performance motors. Major uses of these magnets are now associated with the generation and use of sustainable energy. Modern wind turbines use a permanent magnet synchronous generator to convert the rotational motion of the turbine blades to electricity. A typical modern wind turbine producing 2.5 MW of electricity uses a motor containing around 500 kg of Nd and Dy as $Nd_{1.8}Dy_{0.2}Fe_{14}B$ in the permanent magnet. Permanent magnets are also used in the rotors of many electric vehicles to convert battery power into traction, and full EVs use between 2 and 4 kg of neodymium-based magnetic material.

(c) Giant and colossal magnetoresistance

KEY POINT Perovskites with Mn on the B cation sites can show very large changes in resistance, known as giant or colossal magnetoresistance, on application of a magnetic field.

Manganites are Mn(III) and Mn(IV) complex perovskite structure-type oxides, with the generic solid solution formula $Ln_{1-x}A_xMnO_3$ (A = Ca, Sr, Pb, Ba; Ln = typically La, Pr, or Nd). They order ferromagnetically on cooling below room temperature, with Curie temperatures typically between 100 K and 250 K, and simultaneously transform

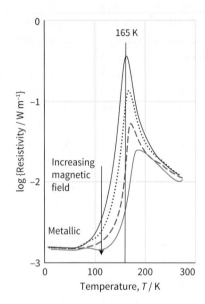

FIGURE 30.15 The resistivity as a function of temperature at various magnetic fields of a material displaying colossal magnetoresistance. At 165 K, application of a magnetic field will cause a change in resistivity of about two orders of magnitude.

from insulators (at the higher temperature) to poor metallic conductors. These materials also exhibit **magnetoresistance**, a marked decrease of their resistance on the application of a magnetic field near and just above their Curie temperatures (Fig. 30.15). Research has shown that for these manganites the decrease in resistance can be as much as 11 orders of magnitude, and for this reason these compounds have been named **giant** or (at the highest levels of change in resistance) **colossal magnetoresistance manganites**.[1]

Giant or colossal magnetoresistance (GMR or CMR) manganites have perovskite-type structures with the A-cation sites occupied by a mixture of Ln^{3+} and A^{2+} cations

---

[1] Albert Fert and Peter Grünberg were awarded the 2007 Nobel Prize in Physics for their discovery of this phenomenon.

> **BOX 30.1** Where are magnetoresistive materials used in hard-disk data storage devices?
>
> Information on computer **hard-disk drives (HDDs)** is encoded using minute magnetic domains with their direction of magnetization representing the logical levels 0 and 1. This information can be read using a simple iron spinel ferrite magnet 'head' and coil, where passing the head over the encoded surface generates small electrical currents in the coil, depending on the direction of the magnetization. However, such devices have a low sensitivity, demanding relatively large regions of magnetic material, and the amount of information that can be stored on such a device is limited. In the 1990s HDDs were developed that used magnetoresistive effects, allowing much greater sensitivity and hence higher data-storage densities.
>
> A giant magnetoresistive (GMR)-based HDD in essence consists of two ferromagnetic layers separated by a spacer layer, which results in a weak coupling between the layers; a layer of an antiferromagnetic material fixes or 'pins' the orientation of one of the ferromagnetic layers. This overall structure is known as a **spin valve**.
>
> When a weak magnetic field, such as that from an encoded bit on a hard disk, passes beneath such a structure, the magnetic orientation of the unpinned magnetic layer is modified relative to that of the pinned layer, and this reorientation generates a significant change in the electrical resistance of the device through the magnetoresitive effect. This can be easily measured electronically as the readout of the encoded surface.
>
> The majority of materials used in HDDs are layers of simple ferromagnetic metals or alloys—for example, FeCr—but the interest in developing even more sensitive devices has led to research on materials exhibiting CMR effects, including the manganites described in the main text.

and the B site occupied by Mn. The oxidation number of Mn in these solid solutions varies between +3 and +4 as the proportion of A^{2+} is changed. Pure $LaMnO_3$ orders antiferromagnetically below its Néel temperature ($T_N = 150\,K$), but with an increase in x in $Ln_{1-x}A_xMnO_3$, corresponding to an increase in Mn^{4+} content, the manganites order ferromagnetically on cooling.

Neither the observation that the transition to conducting electron behaviour and ferromagnetism occurs simultaneously (on cooling the $Ln_{1-x}A_xMnO_3$ manganites) nor the origin of the CMR effect are completely understood; but it is known that they are based on a so-called **double exchange** (**DE**) mechanism between the Mn(III) and Mn(IV) species present in these materials. The basic process of this DE mechanism is the transfer of an electron from $Mn^{3+}\left(t_{2g}^3 e_g^1\right)$ to $Mn^{4+}\left(t_{2g}^3\right)$ through the O atom, so that the Mn(III) and Mn(IV) positions effectively swap places. In manganites at high temperatures, the electrons in these systems become trapped on a specific site, leading to ordering of the Mn(III) and Mn(IV) species; that is, the electron trapping results in **charge ordering**. The charge-ordered state is generally associated with insulating and paramagnetic behaviour, whereas the charge-disordered state, in which the electron can move between sites, is associated with metallic behaviour and ferromagnetism. The high-temperature charge-ordered state can be transformed into a metallic spin-ordered (ferromagnetic) state just by the application of a magnetic field; therefore, application of a magnetic field to a manganite just above its critical temperature causes the transformation of charge ordering into electron delocalization and hence a massive decrease in resistance, Fig. 30.15.

The GMR effect is used in magnetic data-storage devices, such as computer hard drives (Box 30.1), and manganites exhibiting CMR have been studied for such applications. Further work using these compounds is aimed at **spintronics**, where, rather than using electron movements to transmit information, as is the basis of electronics and the functioning of the silicon chip, the movement of spin through materials could be used in a similar way. As spin transfer is much faster than electron motion and does not develop heat through resistive effects, computing devices based on spintronics should have much higher processing powers and not require energy-demanding cooling; the latter is an increasing problem with semiconductor technology as transistors become ever more densely packed on computer processors.

Energy transmission and storage

Once renewable energy has been harvested, in electrical or chemical forms, efficient methods of storing and distributing it are required. Thus, for electrical energy produced from solar cells during the daytime the energy needs to be stored using batteries, converted immediately to another form (e.g. electrolysis to produce green hydrogen), or transmitted through a network for immediate use or remote storage using, for example, physical methods (pumped hydroelectric, pressurized gas, or cryogenic/thermal methods). Similarly, the generation of green hydrogen (either directly from sunlight or through electrolysis of water) then requires its storage and transport. Efficient methods of transmission

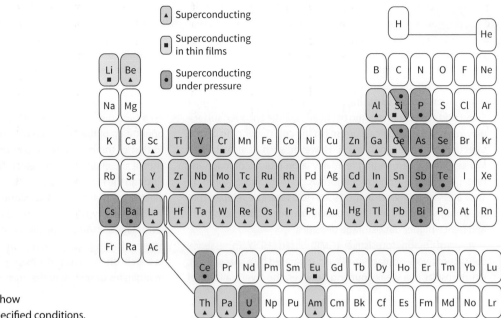

FIGURE 30.16 Elements that show superconductivity under the specified conditions.

and storage of energy from renewable sources are therefore required, and the future development of many of these technologies will be dependent on advances made with inorganic materials.

30.7 Superconductors

Around 6% of generated electricity is lost during distribution due to the resistivity in copper and aluminium cabling and conversion losses in transformers. Further losses to heat also occur in many electrical devices; for example, pumps and fans waste approximately 40% of the energy they use as heat. Minimizing the resistive losses in electricity distribution and use is an important research topic that has focused on discovering and developing new superconducting materials.

(a) Superconducting alloys and high-temperature superconductors

KEY POINTS Many elements and alloys demonstrate superconductivity at temperatures below 77 K. High-temperature cuprate superconductors have structures related to perovskite.

Superconductors have two striking characteristics. Below a critical temperature, T_c (not to be confused with the Curie temperature of a ferroelectric, T_C), they enter the superconducting state and have zero electrical resistance. In this superconducting state they also exhibit the **Meissner effect**, the exclusion of a magnetic field. The Meissner effect is the basis of the common demonstration of superconductivity in which a pellet of superconductor levitates above a magnet. It is also the basis for a number of applications of superconductors that include magnetic levitation, as in 'maglev' trains.

Following the discovery in 1911 that mercury becomes a superconductor when cooled below 4.2 K, physicists and chemists made slow but steady progress in the discovery of superconductors with higher values of T_c; after 75 years, T_c had been edged up to 23 K in Nb_3Ge. Most of these superconducting materials were elements (Fig. 30.16) and metal alloys, although superconductivity had been found in many oxides and sulfides[2] (Table 30.3); magnesium diboride is

TABLE 30.3 Some materials that exhibit superconductivity under ambient pressure below the critical temperature, T_c

Element	T_c/K	Compound	T_c/K
Zn	0.88	Nb_3Ge	23.2
Cd	0.56	Nb_3Sn	18.0
Hg	4.15	$LiTi_2O_4$	13.7
Pb	7.19	$K_{0.4}Na_{0.6}BiO_3$	29.8
Nb	9.50	$YBa_2Cu_3O_7$	93
		$Bi_2Sr_2CaCu_2O_8$	96
		$Tl_2Ba_3Ca_3Cu_4O_{12}$	134
		MgB_2	39
		K_3C_{60}	39
		$PbMo_6S_8$	15.2
		NbPS	12
		$LaFeAsO_{1-x}F_x$	26
		$BaFe_2As_2$	23

[2] The Chevrel phases discussed in Section 25.14(d) have some of the highest observed values of H_c.

superconducting below 39 K (see Box 14.5). Then, in 1986, the first **high-temperature superconductor** (HTSC) was discovered. Several materials are now known with T_c well above 77 K, the boiling point of the relatively inexpensive refrigerant liquid nitrogen, and in a few years the maximum T_c in complex copper oxide materials was increased substantially to around 134 K. Hydrogen-rich materials under very high pressures have been shown to become superconducting above 200 K.

Two types of superconductors are known:

- **Type I** superconductors show abrupt loss of superconductivity when an applied magnetic field exceeds a value characteristic of the material.
- **Type II** superconductors, which include high-temperature materials, show a gradual loss of superconductivity above a critical field, denoted H_c.

Figure 30.16 shows that elements that exhibit superconductivity are restricted to certain regions of the periodic table. Note in particular that the ferromagnetic metals Fe, Co, and Ni do not display superconductivity, nor do the majority of the alkali metals or the coinage metals Cu, Ag, and Au.

The first HTSC reported was $La_{1.8}Ba_{0.2}CuO_4$ (T_c = 35 K), which is a member of the solid-solution series $La_{2-x}Ba_xCuO_4$, in which Ba replaces a proportion of the La sites in La_2CuO_4. This material has the K_2NiF_4 structure type (Section 25.11(d), Figure 25.39) with layers of edge-sharing CuO_6 octahedra separated by the La^{3+} and Ba^{2+} cations, although the octahedra are axially elongated by a Jahn–Teller distortion (Section 7.16 and Chapter 20). A similar compound with Sr replacing Ba in this structure type, as in $La_{1.8}Sr_{0.2}CuO_4$ (T_c = 38 K), is also known.

One of the most widely studied HTSC oxide materials, $YBa_2Cu_3O_{7-x}$ (T_c = 93 K; informally this compound is called '123', from the proportions of metal atoms in the compound, or YBCO, pronounced 'ib-co'), has a structure similar to perovskite but with ordered O vacancies. In terms of the structure shown in Fig. 30.17, the stoichiometric $YBa_2Cu_3O_7$ unit cell consists of three simple perovskite cubes stacked vertically with Y and Ba in the A sites of the original perovskite and Cu atoms in the B sites. However, unlike in a true and perfect perovskite structure, the B sites are not surrounded by an octahedron of O atoms: the 123 structure has a large number of vacancies on the O^{2-} sites. As a result, some Cu atoms have five O atom neighbours in a square-pyramidal arrangement, and others have only four, as square-planar CuO_4 units.

Similarly, the Y^{3+} and Ba^{2+} cations on the A-type sites have less than 12-coordination to oxide anions. The compound $YBa_2Cu_3O_7$ readily loses oxygen from some sites within the CuO_4 square planes, forming $YBa_2Cu_3O_{7-x}$ ($0 \leq x \leq 1$), but as x increases above 0.1 the critical temperature drops rapidly from 93 K. A sample of the 123-material made in the laboratory and heated under pure oxygen at 450°C as the final stage of its preparation is typically slightly oxygen-deficient with $x < 0.1$.

If we assign the usual oxidation numbers $N_{ox}(Y) = +3$, $N_{ox}(Ba) = +2$, and $N_{ox}(O) = -2$, then the average oxidation number of copper turns out to be +2.33, so it is inferred that $YBa_2Cu_3O_{7-x}$ is a mixed oxidation state material that contains Cu^{2+} and Cu^{3+}. Note that in $YBa_2Cu_3O_{7-x}$ the material formally contains some Cu^{3+} until x increases above 0.5. An alternative view is that the number of electrons in $YBa_2Cu_3O_{7-x}$ is such that a partially filled band is present: this view is consistent with the high electrical conductivity and metallic behaviour of this oxide at room temperature. If this band can be considered as being constructed from Cu 3d orbitals, then the partial filling is a result of holes in this level (corresponding to Cu^{3+}).

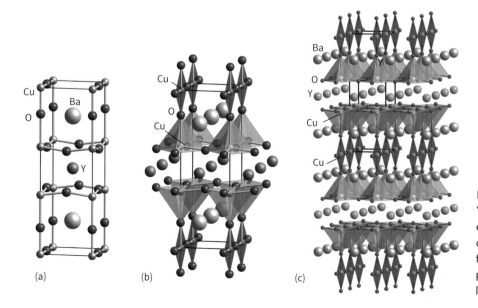

FIGURE 30.17 Structure of the $YBa_2Cu_3O_7$ superconductor: (a) the unit cell; (b) oxygen polyhedra around the copper ions; (c) showing the layers formed from linked CuO_5 square pyramids and chains formed from corner-linked CuO_4 square planes.

The square-planar CuO_4 units in $YBa_2Cu_3O_{7-x}$ are arranged in chains and the CuO_5 units link together to form infinite sheets. The stoichiometry of an infinite sheet of vertex-sharing CuO_4 square planes is CuO_2. The addition of one or two apical O atoms in the cases where the layers are constructed from, respectively, linked square-based pyramids or octahedra, maintains the CuO_2 sheet. This structural feature is also seen in all other oxocuprate HTSCs, and it is a crucial component of the mechanism of superconduction in these materials.

Some HTSC and other cuprate superconducting materials are listed in Table 30.3. All of them may be considered to have at least part of their structure derived from that of perovskite, as a layer of linked CuO_n (n = 4, 5, 6) polyhedra is a section of that structural type. Lying between these cuprate layers (which may include up to six such perovskite-derived CuO_2 sheets) can be a variety of other simple structural units containing s- and p-block metals in combination with oxygen, such as rock-salt and fluorite structures. Thus $Tl_2Ba_2Ca_2Cu_3O_{10}$ can be considered as having three perovskite layers based on Cu, O, and Ca separated by double layers of a rock-salt structure, built from Tl and O; the Ba^{2+} ions lie between the rock-salt and perovskite layers (Fig. 30.18).

The synthesis of high-temperature superconductors has been guided by a variety of qualitative considerations, such as the demonstrated success of the layered structures and of mixed oxidation state Cu in combination with heavy p-block elements. Additional considerations are the radii of ions and their preference for certain coordination environments. Many of these materials are prepared simply by heating an intimate mixture of the metal oxides to 800–900°C in an open alumina crucible. Others, including mercury- and thallium-containing complex copper oxides, require reactions involving the volatile and toxic oxides Tl_2O and HgO; in such cases the reactions are normally carried out in sealed gold or silver tubes (Section 25.9).

There is, as yet, no settled explanation of high-temperature superconductivity in cuprates. It is believed that the movement of pairs of electrons, known as '**Cooper pairs**' and responsible for conventional superconductivity in simple metals and alloys, is also important in the high-temperature superconductor materials. There are currently two theories for superconductivity in cuprates: the **resonating valence bond theory (RVB)**, and **spin fluctuation**. Both theories enable the electron pairing in the CuO_2 sheets that allows electrons to move with zero resistance.

(b) Other superconducting oxides and phases

KEY POINTS Complex metal oxides not containing copper, sulfides, and hydrides can exhibit superconductivity; critical temperatures near and above room temperature have been observed in hydrides under extreme pressures.

The observation of superconductivity in the complex cuprates is unusual in terms of the high critical temperatures reached, but many other oxides and oxide phases demonstrate a transition to zero electrical resistance, albeit normally at considerably lower temperatures. Some compositionally simple examples include phases from the solid solution $Li_{1+x}Ti_{2-x}O_4$ (which adopts the spinel structure and has T_c = 13.7 K for x = 0) and $Na_{0.35}CoO_2 \cdot H_2O$ with $T_c \approx$ <5 K (see also Table 30.3).

The complex bismuth oxides of composition $(K_{0.87}Bi_{0.13})BiO_3$ (for which T_c = 10.2 K) and $(Ba_{0.6}K_{0.4})BiO_3$ (T_c = 30 K), which adopt perovskite structures with Bi as the B-type cation, are among a number of similar bismuthates that exhibit superconductivity. Several complex oxides that adopt the pyrochlore structure (Section 25.11(e), Fig. 25.41) and have composition $M_{2-x}B_2O_{7-x}$, where M is a Group I, Group II, or post-transition-metal cation such as Cs, Ca, or Cd and where B is a heavy d metal, show superconductivity. For example, $Cd_2Re_2O_7$ is superconducting below 1.4 K, and KOs_2O_6 is superconducting below 10 K.

Recently, interest has centred on a different family of superconductors with critical temperatures approaching the best cuprates. These lanthanoid iron arsenic oxides

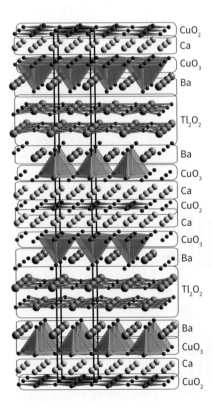

FIGURE 30.18 $Tl_2Ba_3Ca_2Cu_3O_{10}$ structure formed from three oxygen-deficient perovskite layers produced from linked CuO_4 square planes and square-based pyramids separated by Ca on the A-type cation position. Double layers of stoichiometry Tl_2O_2 with rock-salt type arrangements of the Tl and O atoms are interleaved between the multiple perovskite layers.

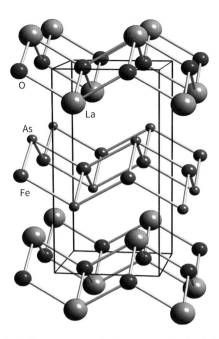

FIGURE 30.19 The structure of the superconductor LaFeAsO, delineating the two types of layer.

(of composition LnFeAs(O,F)$_{1-x}$) were first reported for Ln = La in LaFeAsO$_{1-x}$F$_x$ with T_c = 26 K, and for compositionally similar Pr and Sm compounds, critical temperatures of 52 K and 55 K, respectively, have been achieved. The structures of these compounds are based on alternating layers of composition LnO and FeAs (Fig. 30.19). Related materials that also exhibit superconductivity include LiFeAs (T_c = 18 K) and AFe$_2$As$_2$ (A = Ca, Sr, Ba; T_c up to 38 K).

Elemental hydrogen has been predicted to be a high-temperature superconductor under extremely high pressures, and recent advances in so-called **poly-** or **super-hydride** materials have generated several new high-temperature, superconducting phases. High-temperature superconductivity can be associated with high phonon or vibrational frequencies in a solid, as the resistance to free movement of electrons through a solid is removed when it is coupled with these vibrations.

The presence of light elements, and particularly hydrogen, in compounds leads to high vibrational frequencies, Section 8.5, and thus high superconducting temperatures. However, formation of these superhydrides, which contain species such as H$^-$, H$_3^-$, and H$_2$ molecules, only occurs when simple hydrides, such as H$_2$S and NaH, are subjected to extreme pressures. Thus, when NaH/H$_2$ mixtures are heated to 1850°C under pressures exceeding 30 GPa, NaH$_7$ is formed.

Superconductivity has been found in many superhydride phases that form under these very high pressures, including H$_3$S (T_c = 203 K at 150 GPa), 2H$_2$Se·H$_2$ (T_c = 105 K at 135 GPa), and LaH$_{10}$ (T_c = 250 K at 170 GPa). A report in early 2023 of superconductivity at room temperature (294 K) and a relatively moderate pressure of 1 GPa in LuH$_{3-\delta}$N$_x$ has yet to be fully substantiated.

30.8 Hydrogen storage

H$_2$ may be generated using renewable energy through the electrolysis of water by using solar photovoltaic or wind-turbine power. The storage and distribution of this green hydrogen to end users requires new technologies and materials that overcome difficulties associated with handling large volumes of pure hydrogen gas. For example, distribution of hydrogen gas in pipelines designed for natural gas is problematic due to the high rates of diffusion of the small hydrogen molecule through the pipe walls. Compressed gas cylinders that contain hydrogen at 70 MPa are usually made of steel and are expensive and heavy, though lightweight carbon-fibre tanks are being developed. Solid materials that can store large quantities of hydrogen gas and then release it at near ambient conditions would facilitate a hydrogen energy economy by enabling production, storage, and transportation of green H$_2$ generated when renewable energy is plentiful.

The general principles of hydrogen storage have been introduced in previous chapters (Boxes 11.5 and 13.4). Materials that have been proposed for hydrogen storage include solid metal hydrides and hydride complexes (with reversible uptake of hydrogen), and physisorbed hydrogen in porous solids. The development and application of these material types are discussed further here.

Targets for a hydrogen storage material are whole-system gravimetric capacities of near 6 wt% H$_2$ and volumetric capacities of around 4%. The cost of systems needs to be under $300 per kg of H$_2$, operate at 3–5 kbar pressure, and at temperatures between 40°C and 60°C. A whole system (including the weight of any vessel) gravimetric capacity of 6 wt% will require the enclosed hydrogen storage material to have a gravimetric capacity above 10%.

(a) Hydrides and complex hydrides

KEY POINT Complex metal hydrides such as hydridoaluminates (alanates), amides, and tetrahydridoborates release H$_2$ when heated.

Simple metal hydrides, such as LiH and MgH$_2$, formed by the reaction of the metal with H$_2$(g) could potentially be used as hydrogen storage materials; Box 11.5, Section 11.6, Chapters 12 and 13, and Section 25.15. However, the high temperatures of decomposition of these saline hydrides make them less viable for applications. This section covers compositionally more complex hydrides that often evolve hydrogen at lower temperatures, and it is these materials that are being investigated for hydrogen storage. Complex hydrides include the tetrahydridoaluminates (containing AlH$_4^-$), amides (NH$_2^-$), and boranates (Box 13.1, including tetrahydridoborates containing BH$_4^-$). NaAlH$_4$ and Na$_3$AlH$_6$ have good theoretical hydrogen-storage capacities, but they exhibit poor reversibility in their decomposition reactions which also proceed in stages at different temperatures, which is problematic for applications. The final amounts of hydrogen are evolved only above 400°C (the targets for automotive hydrogen-fuel systems are less than 100°C for release of hydrogen and less than 700 bar for recharging):

$$3\,NaAlH_4(s) \xrightleftharpoons{200°C} Na_3AlH_6(s) + 2\,Al(s) + 3\,H_2(g)$$
yield : 3.6 per cent H_2 by mass

$$2\,Na_3AlH_6(s) \xrightleftharpoons{260°C} 6\,NaH(s) + 2\,Al(s) + 3\,H_2(g)$$
yield : 1.8 per cent H_2 by mass

$$2\,NaH(s) \xrightleftharpoons{425°C} 2\,Na(s) + H_2(g)$$
yield : 2.0 per cent H_2 by mass

Various additives, such as Ti and Zr, have been used to prepare doped materials which, in some cases, have enhanced the rates of hydrogenation and dehydrogenation. Ball-milling to produce smaller particle sizes improves rates of hydrogen evolution at lower temperatures. The lithium hydridoaluminates $LiAlH_4$ and Li_3AlH_6 (Section 14.13) have higher mass percentages of hydrogen than the corresponding sodium compounds and would be attractive for hydrogen storage except for their chemical instability and poor reversibility. $LiAlH_4$ initially decomposes easily, but the resulting products cannot be rehydrogenated back to $LiAlH_4$; furthermore, LiH, one of the initial decomposition products, only loses hydrogen above 680°C.

Lithium tetrahydridoborate, $LiBH_4$ (Section 14.6), has a high theoretical hydrogen storage capacity of 18.5 wt%. It decomposes in a series of stages between 250°C and 500°C to form LiH(s), B(s) and hydrogen, releasing 13.8 wt% as $H_2(g)$; however, the reverse process reforming $LiBH_4$ requires pressures of 150 kbar of H_2. Various systematic studies of the effect of additives and compositional substitutions on the decomposition temperature $LiBH_4$ and reaction reversibility have been undertaken. Addition of many other inorganic materials, including hydrides such as MgH_2, oxides including Fe_2O_3, and metals such as Mg, reduce the initial decomposition temperature to values in the range 70°C to 225°C. $NaBH_4$ (10.4 wt% H) and KBH_4 (7.4 wt% H) have also been studied, but decompose fully only at higher temperatures than $LiBH_4$, and the reverse reaction requires very high $H_2(g)$ pressures. The hydridoberyllate $Li_3Be_2H_7$ has excellent reversibility, but only above 150°C. $Li_4Al_3(BH_4)_{13}$ has a theoretical hydrogen capacity of 17.2 wt% but releases diborane on heating.

Lithium nitride, Li_3N, reacts with hydrogen to form a mixture of $LiNH_2$ and LiH:

$$Li_3N(s) + 2\,H_2(g) \rightarrow LiNH_2(s) + 2\,LiH(s)$$

This mixture evolves hydrogen above 230°C, producing Li_2NH, and has a theoretical hydrogen-storage capacity of 6% H_2 by mass. One problem with nitrides as hydrogen

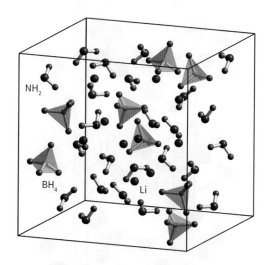

FIGURE 30.20 The structure of $Li_3(NH_2)_2BH_4$.

storage materials is that partial decomposition to produce ammonia can occur; the presence of even low levels of ammonia can interfere with the later uses of the hydrogen released in fuel cells and combustion. Complex lithium amide-boranates, such as $Li_3(NH_2)_2BH_4$ (Fig. 30.20) and $Li_4(NH_2)_2(BH_4)_2$, have been proposed as hydrogen-storage materials and seem to offer reduced evolution of ammonia compared with $LiNH_2$, but with only limited reversibility (Box 11.5).

Related metal-amine complexes—for example $Mg(NH_3)_6Cl_2$—have also been proposed as 'indirect' hydrogen-storage materials (where evolved ammonia might be converted in a secondary reaction to hydrogen), as they have high hydrogen contents. Metal borohydride ammonia complexes of the type $M(BH_4)_n \cdot xNH_3$, M = Li, Mg, Ca, Al, have also been investigated as potential hydrogen storage materials, but they evolve ammonia at low temperatures (30–60°C) in preference to hydrogen.

Mixtures of hydrides and amides show promising properties as hydrogen storage materials. Examples include $LiNH_2 \cdot 2LiH$, $2NaNH_2 \cdot 3MgH_2$, and $3Mg(NH_2)_2 \cdot 8LiH$, which have theoretical H_2 capacities of 12.5, 7.3, and 6.9 wt%, respectively; the onset hydrogen desorption temperatures of these composites are about 170°C, 130°C, and 140°C, respectively. However, the full reversibility of hydrogen uptake and desorption at near ambient conditions required for commercial applications remains poor.

Consideration of the many metal hydrides and complex hydrides proposed for hydrogen storage indicates that finding fully reversible systems that operate efficiently near room temperature over numerous cycles may be problematic. Thermodynamic losses, the structural changes that occur (with associated large changes in material volume), and the potential for irreversible side reactions, which all take place on cycling, are concerns that will have to be addressed to make this type of storage material commercially viable.

(b) Intermetallic materials for hydrogen storage

KEY POINT Intermetallic hydrides release H_2 reversibly near 298 K, but the levels of H_2 incorporated are low.

The hydrides of intermetallic compounds including Laves phases were described in Section 25.15(c). While these materials often show good reversibility for hydrogen uptake and desorption, the wt% of $H_2(g)$ that can be contained in these materials, around 2%, is far lower than the targets needed for applications and commercialization, Table 25.5. Many other intermetallic phases have been investigated, and some of the highest hydrogen wt% values have been found for $TiCr_{1.8}$ (2.4 wt%), $TiMn_{1.4}V_{0.6}$ (3.6 wt%), and $(V_{0.9}Ti_{0.1})_{0.95}Fe_{0.05}$ (3.7 wt%). These materials readily absorb and release hydrogen near room temperature.

(c) Other inorganic hydrogen-storage materials

KEY POINT High-surface-area and porous inorganic compounds can take up high levels of hydrogen gas, but only at very low temperatures.

Physisorption on to the surface of very high-surface-area materials, including any internal pores, offers high hydrogen-storage capacities. However, these high values may be reached only by cooling the system or use of very high pressures, neither of which may be applicable to eventual applications. Some inorganic systems currently being studied include metal alloys that have been templated by using zeolites, inorganic clathrates, or metal organic frameworks (MOFs, Section 29.4), all of which have highly porous structures formed from relatively light elements.

Carbon in its various forms is also being studied as a hydrogen-storage material. Graphite itself in a nanostructure (Section 28.9) adsorbs 7.4 mass percent hydrogen under 1 MPa (10 bar) of hydrogen at 77 K, although more typically activated carbon/graphite materials develop surface monolayers equivalent to 1.5–2.0% hydrogen by mass. Carbon and boron nitride nanotubes (Section 28.8), with their curved surfaces, tend to show enhanced adsorption of hydrogen in comparison with graphite sheets, with reported values of up to 8% by mass at 77 K.

Zeolites (Sections 29.1 and 29.2) have also been proposed as possible hydrogen-storage materials, and many of the different topologies have been studied. The best framework types include faujasites (FAU, zeolites X and Y), zeolite A (LTA), and chabazite, with maximum capacities of around 2.0% hydrogen by mass. The presence of the relatively heavy elements Si and Al in these zeolites limits their weight percent hydrogen uptakes, and other related zeotypes based on the lighter Li, Be, and B are being investigated.

Many MOFs (Sections 25.17(c) and 29.4) can absorb hydrogen at low temperatures (liquid nitrogen temperatures, 77 K) and high pressures, and values nearing 10 wt% have been found in some materials. However, storage capacities near room temperature are typically less than 1 wt% as the strength of the interaction between the MOF and the hydrogen molecules is typically very weak (5–10 kJ mol^{-1}). Methods of increasing the strength of the interaction are being investigated, and include incorporation of metal nanoparticles (Section 28.7(b)) into the MOF pores, or designing the organic linker so that the metal site of the MOF is partially exposed or unsaturated, enabling a direct metal to H_2 coordination (Section 29.4).

30.9 Rechargeable battery materials

The widespread need for mobile power sources in electronic goods and vehicles has resulted in the rapid development of rechargeable Li-ion batteries over the past 40 years. Grid-scale storage of renewable energy using rechargeable batteries is also being studied, where systems based on sodium ions or other mobile cations are being researched. Improving the characteristics of these batteries involves increased specific energy capacity, faster charge and discharge times, and longer lifetimes through numerous charge–discharge cycles. As well as the optimization of the existing inorganic materials used for the battery cathode, anode, and electrolyte, through, for example, nanostructuring (Chapter 28), new materials that offer step changes in battery performance are being sought.

In this section we discuss recent developments in both Li-ion and Na-ion battery materials and with metal–air batteries. The three key elements of an Li-ion rechargeable battery (anode, cathode, and electrolyte)[3] are shown in Figure 25.25: we consider these in turn for Li-ion systems and then for the equivalent rechargeable Na-ion battery.

(a) Lithium ion battery cathodes

KEY POINT The redox chemistry associated with the extraction and insertion of Li$^+$ ions into transition-metal oxide and phosphate structures is exploited in the cathodes of rechargeable batteries.

The existence of complex oxide phases that demonstrate good ionic conductivity associated with the ability to vary the oxidation state of a d-metal ion has led to the development of materials for use as the cathode in rechargeable batteries (Box 12.2 and Section 25.8(d)). Cathode materials that have been commercialized include $LiCoO_2$, with a layer-type structure based on sheets of edge-linked CoO_6 octahedra separated by Li$^+$ ions (Fig. 25.24), and various lithium transition-metal spinels; for example, $LiMn_2O_4$ and $LiNi_{0.5}Co_{0.2}Mn_{0.3}O_2$. In each of these compounds the battery is charged by removing the mobile Li$^+$ ions from the complex metal oxide cathode and discharged through the reverse electrochemical reaction.

---

[3] The term 'cathode' is defined in rechargeable batteries for the direction of charging—where Li$^+$ ions from the electrode are transferred to the electrolyte and electrons move to the external circuit.

TABLE 30.4 Li-ion battery cathode characteristics

Cathode material	Operating Voltage / V	Practical specific capacity / mAhg^{-1}
LiCoO$_2$	3.9–4.5	140
LiMn$_2$O$_4$ (spinel)	3.0	110
LiNi$_{0.5}$Co$_{0.2}$Mn$_{0.3}$O$_2$	3.8	160
LiFePO$_4$	3.5	130
LiFeSO$_4$F	4.3	120
Li$_2$FeSiO$_4$	3.2	190
V$_2$O$_5$	< 3.0	280
Li$_3$V$_2$(PO$_4$)$_3$	3.6–4.2	185

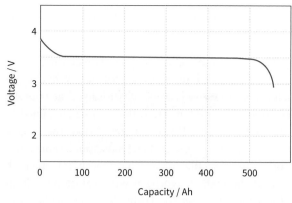
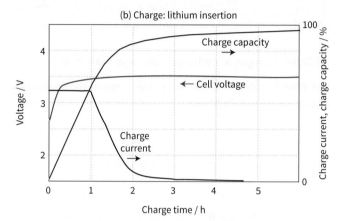

FIGURE 30.21 The (a) discharge and (b) charge characteristics of a typical rechargeable Li-ion battery.

Key parameters in identifying materials for commercial applications are the specific energy capacity of the compound (the stored energy divided by the mass, while sometimes the volumetric capacity is also considered) and the cell voltage. The specific capacity of rechargeable battery cathode materials (mAhg^{-1}) is maximized by using light elements, such as Li and the 3d transition metals (because they are the lowest-density elements with variable oxidation states). While the theoretical specific energy of LiCoO$_2$ is 274 mAhg^{-1}, the usable specific capacity is under 150 mAhg^{-1}, as not all the lithium can be removed from LiCoO$_2$ without degrading the cathode significantly and shortening the battery lifetime. Table 30.4 summarizes the properties of a number of commercial and candidate Li-ion battery cathode materials.

Practical and commercial high cathode capacities are obtained from a large amount of lithium that may be reversibly extracted (for over 1000 discharge/recharge cycles) from the material. Figure 30.21 shows the charge and discharge characteristics of a typical Li-ion rechargeable battery. The cell current needs to be delivered during discharge at a constant and high potential difference (of between 3.3 and 4 V). Charging takes place over a few hours, and in a modern battery around 80% of the battery capacity can be restored at a voltage of 3.3–4.0 V in 1–2 hours.

The cell voltage of a LiCoO$_2$ battery can be related to the high oxidation states of cobalt (+3 and +4) that are involved with lithium extraction from LiCoO$_2$ to give Li$_{1-x}$CoO$_2$. Because cobalt is expensive and fairly toxic, the search continues for oxide cathode materials with improved properties and sustainability over LiCoO$_2$. New materials need to match, or approach closely, the high levels of reversibility found for lithium cobaltate, cell voltages between 3 and 4.5 V, with similar or higher capacities. Partial substitution of cobalt in LiCoO$_2$, producing phases of the composition Li(Co$_{1-x}$M$_x$)O$_2$, M = Ni, Mn, Mg, Fe, and Al, has been found to improve battery cycling and cathode stability. Other layered cathode phases with structures similar to LiCoO$_2$ include LiNiO$_2$ and LiNi$_{0.5}$Co$_{0.2}$Mn$_{0.3}$O$_2$, Fig. 30.22(a).

One important commercialized system is the lithium manganese spinels, including LiMn$_2$O$_4$ and Li(Mn$_{2-x}$M$_x$)O$_4$, M = Al, Ni, x~0.1, Fig. 30.22(b). In this spinel structure, Section 25.11(c), the lithium occupies the tetrahedral sites and can be extracted to form Li$_{1-x}$Mn$_2$O$_4$, 0 < x < 1, on charging the battery. The three-dimensional connectivity of the tetrahedral sites in the spinel structure allows for high lithium mobility. Nanostructured forms of these various complex oxides (Section 28.2(c)), because of their small particle size and short diffusion pathways for Li$^+$ ions, can offer improved reversibility for lithium extraction and insertion.

LiFePO$_4$ shows good characteristics for a cathode material and contains cheap and nontoxic iron. Its structure is that of the natural mineral olivine ((Mg,Fe)$_2$SiO$_4$) and consists of linked FeO$_6$ octahedra and PO$_4$ tetrahedra (Fig. 30.23) surrounding lithium ions that are extracted from the structure during battery charging to produce 'FePO$_4$'. Because of its low electronic conductivity, LiFePO$_4$ is normally used in a small-particle form and coated with conducting graphitic carbon. While LiFePO$_4$ has a lower energy density than LiCoO$_2$, it offers higher peak power ratings, and commercial applications include power tools and electric vehicles,

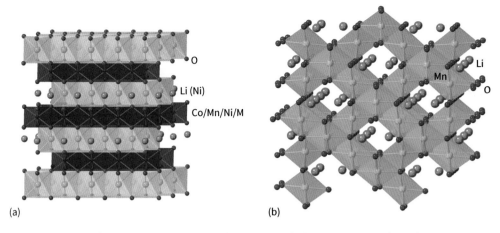

FIGURE 30.22 (a) Layer structure of LiNi$_{0.5}$Co$_{0.2}$Mn$_{0.3}$O$_2$-type cathode materials; (b) the structure of spinel-type cathode materials, LiMn$_2$O$_4$.

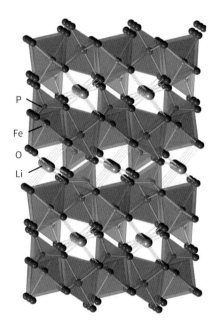

FIGURE 30.23 The structure of LiFePO$_4$ showing the linked FeO$_6$ octahedra and PO$_4$ tetrahedra forming channels containing the Li$^+$ cations.

especially the less expensive vehicles in model ranges. Other similar materials being investigated as rechargeable Li-ion battery cathodes include first-row transition-metal sulfates and fluorides; for example, LiFeSO$_4$F.

(b) Li-ion battery anode materials

KEY POINTS Anodes in rechargeable Li-ion batteries consist of lithium metal, other metals that alloy with lithium, graphite, and related carbon materials, transition-metal oxides, or silicon.

The anode in a rechargeable Li-ion battery can simply be Li metal, which completes the overall cell reaction through the process Li(s) → Li$^+$ + e$^-$ during discharge. Lithium ions then migrate to the cathode through an electrolyte, which in current technologies is typically an anhydrous lithium salt, such as LiPF$_4$ or LiC(SO$_2$CF$_3$)$_3$, dissolved in a polymer; for example, poly(propene carbonate). Solid-state electrolytes are being developed; Section 30.9(c). However, the use of Li metal as the anode has a number of problems associated with its reactivity and volume changes that occur in the cell. Dendrites can grow out from the anode surface through the electrolyte and to the cathode, leading to the battery shorting out, exploding, or setting on fire.

An alternative anode material, now frequently used in rechargeable batteries, is graphitic carbon, which is light and can electrochemically intercalate lithium to form LiC$_6$ (intercalation compounds with graphite were introduced in Section 15.5). As the cell is discharged, Fig. 25.25, Li is transferred from between the carbon layers at the anode and intercalated into the metal oxide at the cathode (and vice versa on charging), with the following overall processes:

$$\text{Li}_y\text{C}_6 + \text{Li}_{1-x}\text{CoO}_2 \underset{\text{charge}}{\overset{\text{discharge}}{\rightleftarrows}} \text{C}_6 + \text{Li}_{1-x+y}\text{CoO}_2$$

However, graphitic carbon has relatively low capacities (under 400 mAhg^{-1}), and some of the safety issues found with Li metal anodes persist, including the growth of lithium metal dendrites.

Other metals or mixtures that form alloys with lithium have been investigated as anode materials, including Al, Sn, Mg, Ag, and Sb. When lithium atoms are formed from Li$^+$ at the metal anode, M, they are incorporated into the metal to form an alloy, MLi$_x$. These metal anode systems produce a higher overall cell voltage than graphite anodes, and have theoretical high capacities (up to 3860 mAhg^{-1}) but are prone to large volume changes during charge and discharge cycles. One of the most promising materials for a Li-ion battery anode is silicon which has a high lithium storage capacity of 3600 mAhg^{-1} (when forming Li$_{15}$Si$_4$ in the completely charged state at 298 K). The volume expansion of a factor of

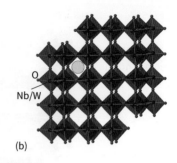

FIGURE 30.24 The structures of (a) TiNb$_2$O$_7$ and (b) Nb$_{16}$W$_5$O$_{55}$ formed from linked MO$_6$ octahedra; sites that incorporate Li$^+$ ions on battery charging are shown with circles.

almost 3 is less than that of many other anode materials, but is still too large for commercial applications. Methods of reducing the volume expansion and improving the electrical conductivity involve the formation of nanoparticulate and nanocomposite forms with carbon.

Other anode systems being researched for lithium batteries include metal oxides; the titanium niobium oxide, TiNb$_2$O$_7$, demonstrates good capacities, similar to Si/C composites, and reasonable voltages (1.6 V vs Li$^+$/Li). The TiNb$_2$O$_7$ structure, Fig. 30.24(a), contains tunnels that can intercalate Li$^+$ ions forming, on complete battery charging, Li$_4$TiNb$_2$O$_7$ (capacity of 385 mAhg^{-1}), and because the lithium ions are incorporated in the tunnels the expansion of the anode is minimal. Nanostructured forms of TiNb$_2$O$_7$ can also be synthesized with improved performance for rapid battery cycling. The niobium tungsten oxide Nb$_{16}$W$_5$O$_{55}$ adopts a Magnéli-type ordered defect ReO$_3$ structure, Section 25.4, Fig. 30.24(b), and shows a low volume change on lithium incorporation and a capacity of around 170 mAhg^{-1}.

(c) Solid-state electrolytes for Li-ion batteries

KEY POINT Room-temperature, fast, lithium-ion conductors are being developed for all-solid-state Li-ion batteries.

The majority of commercial Li-ion batteries use electrolyte pastes, formed of a polymer combined with LiPF$_4$ or LiC(SO$_2$CF$_3$)$_3$ in an organic solvent, to transfer Li$^+$ ions between the anode and cathode. Problems associated with these materials include flammability and reactions with the electrodes leading to a fading of the battery capacity. Solid Li-ion conductor electrolytes are being researched to overcome these problems. The characteristics of solid fast (Li) ion conductors that could be used as an electrolyte have been described in Section 25.7.

Materials being investigated for use as Li-ion battery electrolytes include Li$_{1+x}$Al$_x$Ti$_{2-x}$(PO$_4$)$_3$ (LATP) and Li$_{1+x}$Al$_x$Ge$_{2-x}$(PO$_4$)$_3$ (LAGP) (superionic conductors which adopt the NASICON structure type, Fig. 25.22), Li$_{3x}$La$_{2/3-x}$TiO$_3$ with the perovskite structure, Li$_7$La$_3$Zr$_2$O$_{12}$, and complex metal sulfides such as Li$_{10}$GeP$_2$S$_{12}$. The Li-ion conductivity of these materials at room temperature does not reach the levels required for Li-ion battery applications (>10^{-3} Scm^{-1}), and hybrid composite materials consisting of a fast Li-ion conductor and polymer are the most promising systems for a solid-state electrolyte.

An additional potential application of a fully solid-state battery is that the whole cell could be deposited by using physical and chemical methods (described in Sections 28.3 and 28.4), as extremely thin nanolayers of anode, electrolyte, and cathode. While the stored energy capacity of a very thin, small battery would be limited, direct applications in electronic circuitry as a back-up power supply for sensors have been proposed.

(d) Na$^+$-ion batteries

KEY POINT Cheap rechargeable sodium-ion batteries, for the static storage of grid-scale or household solar energy, would use transition-metal oxides and phosphates as their cathodes.

The advantages of using Li-ion rechargeable batteries for vehicles and portable electronic devices is based on the high specific capacity of light lithium-based materials. For other static energy storage applications, such as the grid-scale storage of energy produced by wind or solar farms, rechargeable batteries based on the much cheaper sodium-ion have been proposed, Box 12.4. Many of the materials being investigated for sodium-ion batteries have parallels to those employed in Li-ion batteries.

Cathode materials being researched and in production in small scale devices include iron phosphates (Na$_2$FePO$_4$F and Na$_3$Fe$_2$(PO$_4$)$_3$), and layered transition-metal oxides—for example, Na$_{0.67}$[Fe$_{0.5}$Mn$_{0.5}$]O$_2$—and the related layered oxides in the general solid solution, Na$_y$[Mn,Ni,Fe,Mg,]O$_2$. Capacities that reach 160 mAhg^{-1} at voltages between 3.2 V and 3.5 V versus Na$^+$/Na(s) have been found but, in general, practical gravimetric energy densities are around half those of Li-ion systems. Sodium ions do not intercalate between the layers of carbon in graphite (as in LiC$_6$) and sodium metal or alternative sodium-rich materials, including the intermetallics Na$_3$Sb, and Na$_{15}$Sn$_4$, are being studied as the anode material. The most commonly used electrolyte is NaPF$_6$ in an organic polymer carbonate solvent.

(e) Metal–air and iron–air batteries

Lithium–air batteries were introduced in Box 12.3 and are an example of a metal–air battery where the anode is a pure metal and the cathode is oxygen in the air. During discharge, the metal is oxidized to metal oxides or hydroxides depending on the nature of the electrolyte. Potential problems with rechargeability in these batteries centre around incomplete discharge at the anode and reactions of the electrolyte with oxidizing species, such as the peroxide and superoxide ions formed at the cathode. Recent advances with lithium–air batteries, including the nanostructuring of a solid electrolyte, show promise in ensuring complete formation of Li_2O at the anode and associated excellent reversibility and attained capacities.

For nonmobile energy-storage applications, alternatives to lithium may be used as the metal anode. Sodium–air batteries offer energy densities of 1600 Wh kg^{-1}, but the stability of the electrolyte over cycling remains a problem. Aluminium–air batteries have high capacities but are not rechargeable.

One of the most promising rechargeable metal–air batteries for low-cost grid-scale energy storage is the iron–air battery, Fig. 30.25. While this battery only develops a potential of 1.28 V and a specific charge capacity of under 300 Ah kg^{-1}, its very low cost, environmentally friendly components, and safe operation make it of interest for large, static energy-storage facilities.

Iron–air batteries can suffer from low charging efficiencies due to hydrogen evolution at the iron electrode, though

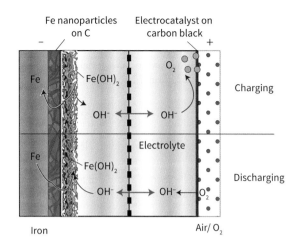

FIGURE 30.25 Schematic of an iron–air battery showing the chemical processes occurring on charge and discharge.

nanostructuring of the iron metal to produce a high surface area alleviates this issue, as does the addition of carbon black. The air cathode must be bifunctional and catalyse both the oxygen reduction reaction (on discharging) and oxygen evolution reaction (on charging). As well as using the traditional electrocatalysts (Section 30.3) for these reactions (Pt, Pd, IrO_2, and RuO_2), the mixed electronic-oxide-ion conductors (Section 25.8(c)) $La_{1-x}Sr_xFeO_{3-\delta}$ (perovskite) and $LaMnO_{3-\delta}$ have been used.

FURTHER READING

D. W. Bruce, D. O'Hare, and R. I. Walton, *Energy Materials (Inorganic Materials Series Book 18)*. John Wiley & Sons (2011).

M. A. Benvenuto, *Chemistry and Energy: From Conventional to Renewable*. De Gruyter (2022).

D. S. Ginley and D. Cahen (eds.), *Fundamentals of Materials for Energy and Environmental Sustainability*. Cambridge University Press (2011).

S. J. Dhoble, N. T. Kalyani, B. Vengadaesvaran, and A. K. Arof (eds.), *Energy Materials: Fundamentals to Applications*. Elsevier (2021).

A. S. Bandarenka, *Energy Materials: A Short Introduction to Functional Materials for Energy Conversion and Storage*. CRC Press (2022).

S. J. Ikhmayies, *Advances in Energy Materials (Advances in Material Research and Technology)*. Springer (2020).

X. Tao, Y. Zhao, S. Wang, C. Li, and R. Li, Recent advances and perspectives for solar-driven water splitting using particulate photocatalysts. *Chem. Soc. Rev.*, 2022, 51, 3561.

Y. Huang, Y. Cheng, and J. Zhang, A review of high density solid hydrogen storage materials by pyrolysis for promising mobile applications. *Industrial & Engineering Chemistry Research*, 2021, 60, 2737–71. DOI: 10.1021/acs.iecr.0c04387.

P. U. Nzereogu, A. D. Omah, F. I. Ezema, E. I. Iwuoha, and A. C. Nwanya, Anode materials for lithium-ion batteries: A review. *Applied Surface Science Advances*, 2022, 9, 100233. DOI: 10.1016/j.apsadv.2022.100233.

EXERCISES

30.1 Explain why ZnO is white, ZnSe is pale yellow, and ZnTe ruby-red.

30.2 Draw a band structure for fluorine-doped tin oxide and explain how this performs as a transparent conducting oxide.

30.3 Explain why the best inorganic pigments involve a charge-transfer band or are mixed valence compounds rather than absorb light based on d–d transitions.

30.4 What is the difference between a direct band gap semiconductor and an indirect one? How does this affect their use in photovoltaic devices?

30.5 Describe the differences in the structures of a-Si and c-Si used as photovoltaics.

30.6 Discuss how the use of Earth-abundant elements influences research on new photovoltaic materials.

30.7 Draw an energy-level diagram to explain how a tandem photocatalyst device can use visible light (wavelength ~500 nm) to split water.

30.8 What are the important thermal and electrical conductivity characteristics of an inorganic material used as a thermoelectric?

30.9 Describe the chemical process happening at the cathode of an SOFC; why is a good electronic and oxide-ion conductor needed for this electrode?

30.10 Describe the differences in the chemical and physical properties of the strong permanent magnets $SmCo_5$ and $Nd_2Fe_{14}B$.

30.11 Why are soft, rather than hard, ferrites used in transformer cores?

30.12 Determine the oxidation state of copper in (a) $La_{1.8}Sr_{0.2}CuO_4$, (b) $YBa_2Cu_3O_{6.5}$, (c) $Hg(II)Ba_2CuO_{4.2}$. Which of these phases might exhibit high-temperature superconductivity?

30.13 Compare alkali and alkaline-earth metal (a) hydridoaluminates and (b) hydridoborates as hydrogen storage materials.

30.14 Calculate the weight percent H_2 stored in $Li_3(NH_2)_2BH_4$ and $Li_4(NH_2)_2(BH_4)_2$ if (a) all the hydrogen is released reversibly and (b) if only the hydridoborate group decomposes reversibly.

30.15 $Li(Ni_{0.8}Co_{0.15}Al_{0.05})O_2$ is being developed as a lithium battery cathode material with a measured energy storage of 190 mAhg^{-1}. What advantages does this material have over $LiCoO_2$?

30.16 Compare the advantages and disadvantages of Li–air batteries versus $LiCoO_2$ cathode-based batteries.

TUTORIAL PROBLEMS

30.1 In his book *Bright Earth: The Invention of Colour* (Vintage 2008), Philip Ball describes how innovation through chemistry has led to the discovery of new pigments and brighter colours than nature provides. Discuss how chemistry has improved (a) white and (b) green pigments since their first use in cave paintings.

30.2 Discuss and compare the chemistries and pigment properties of ultramarine blue, Prussian blue, and copper phthalocyanine blue.

30.3 In 2018, W. Schnick was awarded the Lieberg Medal for his groundbreaking work in the synthesis of inorganic materials for warm-white high-performance LEDs and LED light sources. Describe the compounds synthesized by W. Schnick and his co-workers for this application, and discuss how they have revolutionized LED lighting.

30.4 Describe recent advances towards an LT-SOFC (Low-Temperature Solid Oxide Fuel Cell).

30.5 Discuss the advances in superhydride superconductor chemistry by reference to CaH_6, LuH_3, LaH_{10}, sulfur hydrides, and other related phases.

30.6 $Nd_2Fe_{14}B$ supermagnets are often commercially produced as a two-phase mixture of α-Fe and $Nd_2Fe_{14}B$. Describe the processes that occur when this material undergoes corrosion, and the methods being researched to improve the corrosion resistance.

30.7 Rechargeable batteries based on Mg^{2+} intercalation/deintercalation into metal oxides are being researched. What might be the advantages and disadvantages of a rechargeable battery that uses Mg^{2+} compared to Li^+?

30.8 Calculate the amount of lithium required for all the cars in Europe to be full EVs. Assume a usable average battery capacity of 60 kWh and 250 million vehicles in Europe; use data for a car battery based on $LiCoO_2$ and that in Table 30.4. The amount of lithium in a battery is approximately 0.3 × Ah capacity.

31 Catalysis

Homogeneous catalysis

31.1 **Alkene metathesis**
31.2 **Hydrogenation of alkenes**
31.3 **Hydroformylation**
31.4 **Wacker oxidation of alkenes**
31.5 **Asymmetric oxidations**
31.6 **Palladium-catalysed C–C bond-forming reactions**
31.7 **Methanol carbonylation: ethanoic acid synthesis**

Heterogeneous and nanoparticle catalysis

31.8 **The nature of heterogeneous catalysts**
31.9 **Ammonia synthesis**
31.10 **Sulfur dioxide oxidation**
31.11 **Fischer–Tropsch synthesis**
31.12 **Reactions involving heterogeneous nanoparticle catalysts**
31.13 **Zeolites and microporous structures in heterogeneous catalysis**

Heterogenized homogeneous and hybrid catalysis

31.14 **Alkene oligomerization and polymerization**
31.15 **Tethered catalysts**
31.16 **Biphasic systems**

Further reading

Exercises

Tutorial problems

In this chapter we apply the concepts of organometallic chemistry, coordination chemistry, and materials chemistry, introduced in previous chapters, to discuss the applications of inorganic complexes and solids in catalysis. Catalysts perform a fundamental role across the chemical industry and underpin around 30% of European gross domestic product. Catalysts are involved in the processing of approximately 80% of all manufactured products. There is also the need to optimize current catalyst systems and develop new catalysts that will impact ongoing challenges in clean energy and water, and for sustainable food supplies. Initially we survey a number of commercially important, homogeneously catalysed reactions and discuss proposed mechanisms for the catalysed transformations. A similar theme is then developed in heterogeneous catalysis on nanoparticle surfaces and inside porous materials; we shall see that many parallels exist between homogeneous and heterogeneous catalysis. In both types of catalysis the mechanisms involved are still being investigated, and there is still considerable scope for making new discoveries and finding new applications for catalysts.

A **catalyst** is a substance that increases the rate of a reaction but is not itself consumed. We introduced the general principles of catalysis in Section 2.15, and that section should be revisited before we look at some detailed catalytic processes here. Catalysts are widely used in nature, in industry, and in the laboratory, and it is estimated that they contribute to one-sixth of the value of all manufactured goods in industrialized countries. As shown in Table 31.1, 16 of the top 20 synthetic chemicals in the USA are produced directly or indirectly by catalysis. For example, a key step in the production of a dominant industrial chemical, sulfuric acid, is the catalytic oxidation of SO_2 to SO_3. Ammonia, another chemical essential for industry and agriculture, is produced by the catalytic reduction of N_2 by H_2. Inorganic catalysts are also used for the production of the major organic chemicals and

TABLE 31.1 The top 20 synthetic chemicals in the USA in 2008 (based on mass)

Rank	Chemical	Catalytic process	Rank	Chemical	Catalytic process
1	Sulfuric acid	SO_2 oxidation, heterogeneous	11	Sodium hydroxide	Electrolysis, not catalytic
2	Ethene	Hydrocarbon cracking, heterogeneous	12	Ammonium nitrate	Precursors catalytic
3	Propene	Hydrocarbon cracking, heterogeneous	13	Urea	NH_3 precursor catalytic
3	Polyethene	Polymerization, heterogeneous	14	Ethylbenzene	Alkylation of benzene, homogeneous
5	Chlorine	Electrolysis, not catalytic	15	Styrene	Dehydrogenation of ethylbenzene, heterogeneous
6	Ammonia	$N_2 + H_2$, heterogeneous	16	HCl	Precursors catalytic
7	Phosphoric acid	Not catalytic	17	Cumene	Alkylation of benzene, heterogeneous
8	1,2-Dichloroethane	Ethene + Cl_2, heterogeneous	18	Ethylene oxide	Ethene + O_2, heterogeneous
9	Polypropene	Polymerization, heterogeneous	19	Ammonium sulfate	Precursors catalytic
10	Nitric acid	$NH_3 + O_2$, heterogeneous	20	Sodium carbonate	Not catalytic

Source: Facts & Figures of the Chemical Industry, *Chem. Eng. News*, 2009, **87**, 33.

petroleum products, including fuels, petrochemicals, and polyalkene plastics.

Catalysts play a steadily increasing role in achieving a cleaner environment through, for example, the destruction of pollutants (as with the catalytic converters found on the exhaust systems of non-electric vehicles), the development of better industrial processes that are more efficient with higher product yields and fewer unwanted by-products, and in clean energy generation in fuel cells. Industrially important catalysts are almost invariably inorganic. Enzymes—a class of biochemical catalysts, often with a metal ion at the centre of a complex molecule—are discussed in Chapter 26. Photocatalysts were discussed in detail in Section 30.3.

In addition to the economic importance of catalysts, the chemistry that takes place during catalytic reactions requires investigation in its own right. For example, the subtle interactions between a catalyst and reactants can completely change the outcome of a reaction, and small changes to the composition and structure of the catalyst can markedly affect its effectiveness. The understanding of the mechanisms of catalytic reactions has improved considerably in recent years with the greater availability of isotopically labelled molecules, improved methods for determining reaction rates, improved spectroscopic and diffraction techniques, and much more reliable calculations (at both the microscopic and the macroscopic level).

Homogeneous catalysis

Here we concentrate on some important homogeneous catalytic reactions based on organometallic compounds and coordination complexes. We describe their currently favoured mechanisms, but it should be noted that, as with nearly all mechanistic proposals, catalytic mechanisms are subject to refinement or change as more detailed experimental information becomes available. Unlike simple reactions, a catalytic process frequently contains many steps over which the experimentalist has little control. Moreover, highly reactive intermediates are often present in concentrations too low to be detected spectroscopically. The best attitude to adopt towards these catalytic mechanisms is to learn the pattern of transformations and appreciate their implications but be prepared to accept new mechanisms that might be indicated by future work.

The scope of homogeneous catalysis ranges across hydrogenation, oxidation, and a host of other processes. Often the complexes of all metal atoms in a group will exhibit catalytic activity in a particular reaction, but the 4d-metal complexes are often superior as catalysts to the complexes of their lighter and heavier congeners. In some instances the difference may be associated with the greater substitutional lability of 4d organometallic compounds in comparison with their 3d and 5d analogues. It is often the case that the complexes of costly metals must be used on account of their superior performance compared with the complexes of cheaper metals.

Chapters 23 and 24 described reactions that take place at metal centres: these reactions lie at the heart of the catalytic processes we now consider. In general, we need to invoke a variety of the processes described in those chapters, and it will be useful to review the following sections:

Ligand substitution reactions: Sections 23.1–23.9 and 24.21.

Redox reactions: Sections 23.10–23.12.

Oxidative addition and reductive elimination reactions: Section 24.22.

Migratory insertion reactions: Section 24.24.

1,2-insertions and β-hydride eliminations: Section 24.25.

Together with the direct attack on coordinated ligands, these reaction types (and in some cases their reverse), often in combination, account for the mechanisms of most of the homogeneous catalytic cycles that have been proposed for organic transformations. Other reactions yet to be fully investigated will undoubtedly use other reaction steps.

31.1 Alkene metathesis

KEY POINTS Alkene metathesis reactions are catalysed by homogeneous organometallic complexes that allow considerable control over product distribution; a key step in the reaction mechanism is the dissociation of a ligand from a metal centre to allow an alkene to coordinate.

In an **alkene metathesis** reaction, carbon–carbon double bonds are redistributed, as in the cross-metathesis reaction:

Alkene metathesis was first reported in the 1950s, with poorly defined mixtures of reagents, such as WCl_6/Bu_4Sn and MoO_3/SiO_2, being used to catalyse a number of different reactions (Table 31.2). In recent years a number of newer catalysts have been introduced, and the development of the well-defined ruthenium alkylidene compound (**1**) by Grubbs in 1992 and the molybdenum imido-alkylidene compound by Schrock in 1990 (**2**) were of seminal importance.[1]

[1] The 2005 Nobel Prize was awarded to Robert Grubbs, Yves Chauvin, and Richard Schrock for their work on developing metathesis catalysts.

TABLE 31.2 The scope of the alkene metathesis reaction

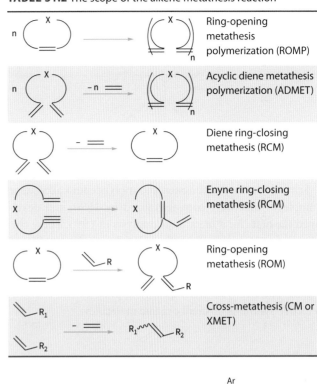

Alkene metathesis reactions proceed through a metallacyclobutane intermediate:

In the case of Grubbs' catalyst it is known that the dissociation of a PCy_3 ligand from the Ru metal centre is crucial in allowing the alkene molecule to coordinate prior to metallacyclobutane formation.

The identification of this mechanism led Grubbs to replace one of the PCy_3 ligands with a bis(mesityl) N-heterocyclic carbene (NHC) ligand, reasoning that the stronger σ-donor and poorer π-acceptor ability of the NHC ligand would both encourage PCy_3 dissociation and stabilize the alkene complex. In a triumph of rational design, the so-called **second-generation Grubbs' catalyst** (**3**) proved to be more active than the original bisphosphine complex. The second-generation Grubbs' catalyst is active in the presence of a large number of different functional groups on substrates, and can be used in many solvent systems. It is commercially available and has been widely used, including in the total synthesis of a number of natural products. In Grubb's third generation catalyst the phosphine ligand is replaced by a heterocycle, such as pyridine, and has a reduced initiation time.

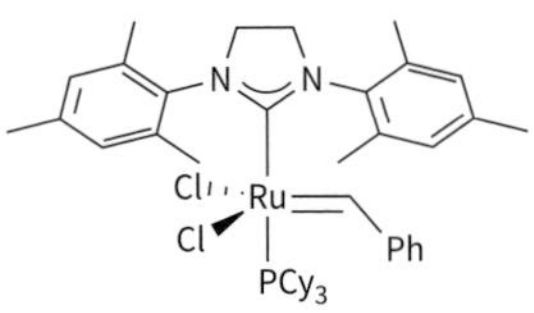

3

Schrock developed tungsten and molybdenum alkylidene complexes that were commercialized by 1990, and developed the first chiral metathesis catalysts in 1993. These chiral catalysts were molybdenum-based complexes (4) and were synthesized to address stereochemical control, known in this context as tacticity (Section 31.14), in ROMP processes. The catalysts were quickly applied to the enantioselective synthesis of small organic molecules.

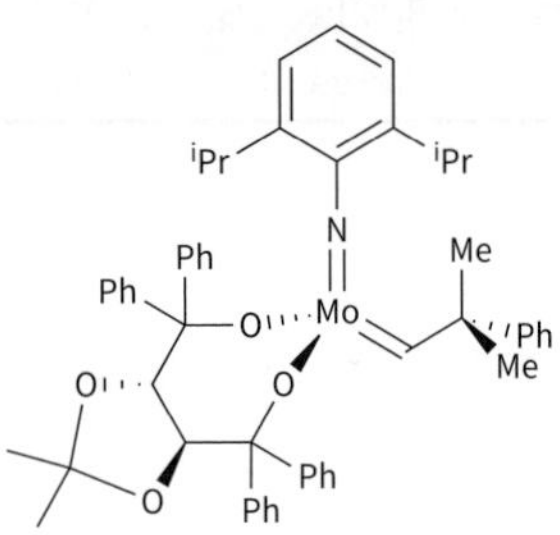

4

The driving force for alkene metathesis reactions varies. For ROM and ROMP (see Table 31.2) it is the release of ring strain from a strained starting material that provides the energy to drive the reaction. For metathesis reactions that result in the generation of ethene (such as RCM or CM) it is the removal of the liberated ethene that can be used to encourage the formation of the desired products. Where there is no clearly identifiable thermodynamically favourable product possible, mixtures of alkenes result, with their relative proportions being determined by the statistical likelihood of their formation.

31.2 Hydrogenation of alkenes

KEY POINTS Wilkinson's catalyst, [RhCl(PPh$_3$)$_3$], and related complexes are used for the hydrogenation of a wide variety of alkenes at pressures of hydrogen close to 1 atm or less; suitable chiral ligands can lead to enantioselective hydrogenations.

The addition of hydrogen to an alkene to form an alkane is favoured thermodynamically ($\Delta_r G^\ominus = -101 \text{ kJ mol}^{-1}$ for the conversion of ethene to ethane). However, the reaction rate is negligible at ordinary conditions in the absence of a catalyst. Efficient homogeneous and heterogeneous catalysts are known for the hydrogenation of alkenes and are used in such diverse areas as the manufacture of nondairy spreads, pharmaceuticals, and petrochemicals.

One of the most studied catalytic systems is the Rh(I) complex [RhCl(PPh$_3$)$_3$], which is often referred to as **Wilkinson's catalyst**. This useful catalyst hydrogenates a wide variety of alkenes and alkynes at pressures of hydrogen close to 1 atm or less at room temperature. The dominant cycle for the hydrogenation of terminal alkenes by Wilkinson's catalyst is shown in Fig. 31.1. It involves the oxidative addition of H$_2$ to the 16-electron complex [RhCl(PPh$_3$)$_3$] (A), to form the 18-electron dihydrido complex (B). The dissociation of a phosphine ligand from (B) results in the formation of the coordinatively unsaturated complex (C), which then forms the alkene complex (D). Hydrogen transfer from the Rh atom in (D) to the coordinated alkene yields a transient 16-electron alkyl complex (E). This complex takes on a phosphine ligand to produce (F), and hydrogen migration to carbon results in the reductive elimination of the alkane and the reformation of (A), which is set to repeat the cycle. A parallel but slower cycle (which is not shown) is known in which the order of H$_2$ and alkene addition is reversed. Another cycle is known, based around the 14-electron intermediate [RhCl(PPh$_3$)$_2$]. Even though there is very little of this species present, it reacts much faster with hydrogen than [RhCl(PPh$_3$)$_3$] and makes a significant contribution to the catalytic cycle. In this cycle, (E) would eliminate alkane directly, regenerating [RhCl(PPh$_3$)$_2$], which rapidly adds H$_2$ to give (C).

Wilkinson's catalyst is highly sensitive to the nature of the phosphine ligand and the alkene substrate. Analogous complexes with alkylphosphine ligands are inactive, presumably because they are more strongly bound to the metal atom and do not readily dissociate. Similarly, the alkene must be

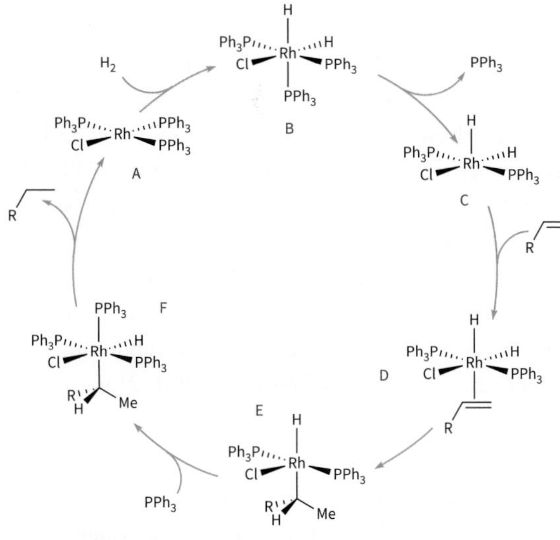

FIGURE 31.1 The catalytic cycle for the hydrogenation of terminal alkenes by Wilkinson's catalyst.

just the right size and shape: highly hindered alkenes or the sterically unencumbered ethene are not hydrogenated by the catalyst, presumably because the sterically crowded alkenes do not coordinate, and ethene forms a stable complex that does not react further. These observations emphasize the point made earlier that a catalytic cycle is usually a delicately poised sequence of reactions, and anything that upsets its flow may block catalysis or alter the mechanism.

Wilkinson's catalyst is used in laboratory-scale organic synthesis and in the production of fine chemicals. Related Rh(I) phosphine catalysts that contain a chiral phosphine ligand have been developed to synthesize optically active products in **enantioselective reactions** (reactions that produce a particular chiral product). The alkene to be hydrogenated must be **prochiral**, which means that it must have a structure that leads to R or S chirality when complexed to the metal. The resulting complex will have two diastereomeric forms, depending on which face of the alkene coordinates to the metal atom. In general, diastereomers have different stabilities and labilities, and in favourable cases one or the other of these effects leads to product enantioselectivity. Enantioselectivities are normally measured in terms of the **enantiomeric excess** (ee), which is defined as the percentage yield of the major enantiomeric product minus the percentage yield of the minor enantiomeric product.

A BRIEF ILLUSTRATION

A reaction that gives 51% of one enantiomer and 49% of another would be described as having an enantiomeric excess of 2%; a reaction that gave 99% of one enantiomer and 1% of another would have ee = 98%.

An enantioselective hydrogenation catalyst containing a chiral phosphine ligand referred to as DiPAMP (**5**) is used to synthesize L-dopa (**6**), a chiral amino acid used to treat Parkinson's disease. An interesting detail of the process is that the minor diastereomer in solution leads to the major product.

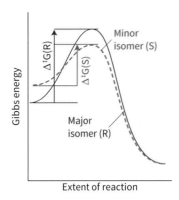

FIGURE 31.2 Kinetically controlled stereoselectivity. Note that $\Delta^\ddagger G_S < \Delta^\ddagger G_{R'}$ so the minor isomer reacts faster than the major isomer.

The explanation of the greater turnover frequency of the minor isomer lies in the difference in activation Gibbs energies (Fig. 31.2). Spurred on by clever ligand design and using a variety of metals, this field has grown rapidly and provides many clinically useful compounds; of particular note are systems derived from ruthenium(II) BINAP (**7**).[2]

7

31.3 Hydroformylation

KEY POINT The mechanism of hydrocarbonylation is thought to involve a pre-equilibrium in which octacarbonyldicobalt combines with hydrogen at high pressure to give a monometallic species that brings about the actual hydrocarbonylation reaction.

In a **hydroformylation reaction**, an alkene, CO, and H_2 react to form an aldehyde containing one more C atom than in the original alkene:

$$RCH=CH_2 + CO + H_2 \rightarrow RCH_2CH_2CHO$$

The term 'hydroformylation' derived from the idea that the product resulted from the addition of methanal (formaldehyde, HCHO) to the alkene, and the name has stuck even though experimental data indicate a different mechanism. A less common but more appropriate name is **hydrocarbonylation**. Both cobalt and rhodium complexes are used

---

[2] Ryoji Noyori and William Knowles were jointly awarded the 2001 Nobel Prize for their work on asymmetric hydrogenation. The prize was shared with Barry Sharpless for his work on asymmetric oxidations (Section 31.5).

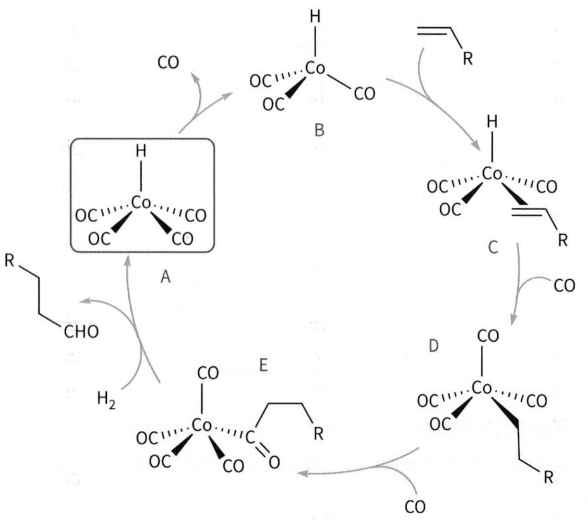

FIGURE 31.3 The catalytic cycle for the hydroformylation of alkenes by a cobalt carbonyl catalyst.

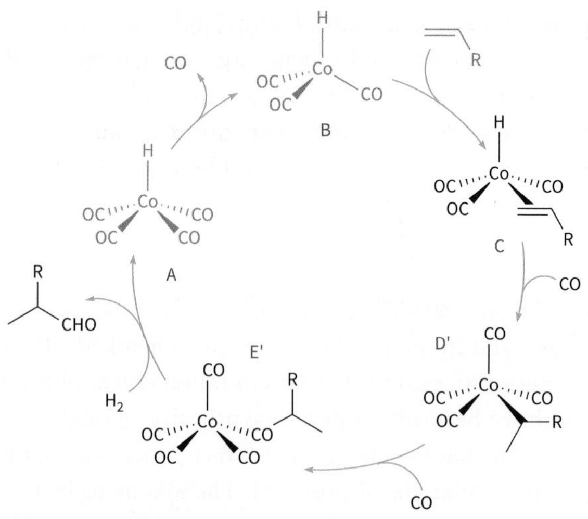

FIGURE 31.4 The formation of branched aldehydes in hydroformylation reactions occurs when the alkyl group is not terminally bound.

as catalysts. Aldehydes produced by hydroformylation are normally reduced to alcohols that are used as solvents and plasticizers, and in the synthesis of detergents. The scale of production is enormous, amounting to millions of tonnes annually.

The general mechanism of cobalt-carbonyl-catalysed hydroformylation was proposed in 1961 by Heck and Breslow, by analogy with reactions familiar from organometallic chemistry (Fig. 31.3). Their general mechanism is still invoked, but has proved difficult to verify in detail. In the proposed mechanism, a pre-equilibrium is established in which octacarbonyldicobalt combines with hydrogen at high pressure to yield the known tetracarbonylhydridocobalt complex (A):

$$[Co_2(CO)_8] + H_2 \rightarrow 2[Co(CO)_4H]$$

This complex, it is proposed, loses CO to produce the coordinatively unsaturated complex $[Co(CO)_3H]$ (B):

$$[Co(CO)_4H] \rightarrow [Co(CO)_3H] + CO$$

It is thought that $[Co(CO)_3H]$ then coordinates an alkene, producing (C) in Fig. 31.4, whereupon the coordinated hydrido ligand migrates onto the alkene, and CO recoordinates. The product at this stage is a normal alkyl complex (D). In the presence of CO at high pressure, (D) undergoes migratory insertion and coordinates another CO, yielding the acyl complex (E), which has been observed by IR spectroscopy under catalytic reaction conditions. The formation of the aldehyde product is thought to occur by attack of either H_2 (as depicted in Fig. 31.3), or the acidic complex $[Co(CO)_4H]$, to yield an aldehyde and generate $[Co(CO)_4H]$, or $[Co_2(CO)_8]$, respectively. Either of these complexes will regenerate the coordinatively unsaturated $[Co(CO)_3H]$.

A significant portion of branched aldehyde is also formed in the cobalt-catalysed hydroformylation. This product may result from a 2-alkylcobalt intermediate formed when reaction of (C) leads to an isomer of (D), with hydrogenation then yielding a branched aldehyde as set out in Fig. 31.4. When the linear aldehyde is required, such as for the synthesis of biodegradable detergents, the isomerization can be suppressed by the addition of an alkylphosphine to the reaction mixture. One plausible explanation is that the replacement of CO by a bulky ligand disfavours the formation of complexes of sterically crowded 2-alkyls:

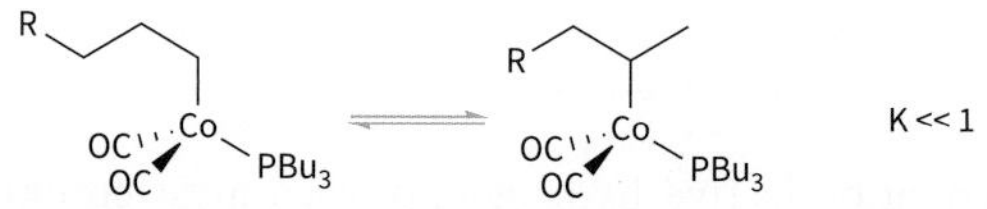

Here again we see an example of the powerful influence of ancillary ligands on catalysis.

EXAMPLE 31.1 Predicting the products from a hydroformylation reaction

Predict the products formed when pent-1-ene reacts with CO and H_2 in the presence of $[Co_2(CO)_8]$. Comment on the effect of adding PMe_3 or PPh_3 to the reaction mixture. How would increasing the CO partial pressure affect the ratio of any linear and branched products?

Answer By analogy with the cycles in Figs. 31.3 and 31.4 we would expect two possible intermediates to be formed following coordination of the alkene and hydrido migration.

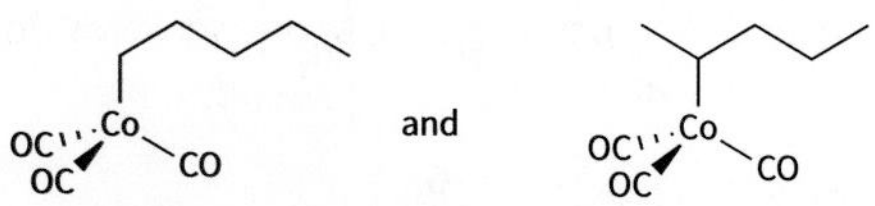

Completion of the catalytic cycles will yield the linear and the branched products, $CH_3CH_2CH_2CH_2CH_2CHO$ and $CH_3CH_2CH_2CH(CHO)CH_3$. The added phosphine would coordinate to the catalysts and the increased steric crowding would inhibit production of the branched product. This effect would be greater for PPh_3 than PMe_3. Increasing the CO pressure would reduce the concentration of the coordinatively unsaturated $[Co(CO)_3H]$ species. This species allows the coordinated alkene to isomerize via β-hydride elimination. Thus, increasing the CO pressure will favour the linear alkene.

Self-test 31.1 Predict the product or products from the hydroformylation of cyclohexene.

Another effective hydroformylation catalyst precursor is $[Rh(CO)H(PPh_3)_3]$ Complex (8), which loses a phosphine ligand to form the coordinatively unsaturated 16-electron complex $[Rh(CO)H(PPh_3)_2]$, which promotes hydroformylation at moderate temperatures and 1 atm. This behaviour contrasts with the cobalt carbonyl catalyst, which typically requires 150°C and 250 atm. The rhodium catalyst is useful in the laboratory as it is effective under convenient conditions. Because it favours linear aldehyde products, it competes with the phosphine-modified cobalt catalyst in industry. The cobalt catalyst is used for synthesis of medium and long-chain aldehydes, and the rhodium catalyst is used for hydroformylation of prop-1-ene.

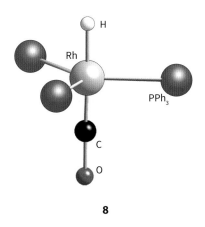

8

EXAMPLE 31.2 Interpreting the influence of chemical variables on a catalytic cycle

An increase in CO partial pressure above a certain threshold decreases the rate of the cobalt-catalysed hydroformylation of 1-pentene. Suggest an interpretation of this observation.

Answer The decrease in rate with increasing partial pressure suggests that CO suppresses the concentration of one of the catalytic species. An increase in CO pressure will lower the concentration of $[Co(CO)_3H]$ in the equilibrium

$$[Co(CO)_4H] \rightleftharpoons [Co(CO)_3H] + CO$$

This type of evidence was used as the basis for postulating the existence of $[Co(CO)_3H]$ as an important intermediate, even though it is not detected spectroscopically in the reaction mixture.

Self-test 31.2 Predict the influence of added triphenylphosphine on the rate of hydroformylation catalysed by $[Rh(CO)H(PPh_3)_3]$.

31.4 Wacker oxidation of alkenes

KEY POINTS The Wacker process is used to produce ethanal from ethene and oxygen; the most successful system uses a palladium catalyst to oxidize the alkene, with the palladium being reoxidized via a secondary copper catalyst.

The **Wacker process** is used primarily to produce ethanal (acetaldehyde) from ethene and oxygen:

$$C_2H_4 + O_2 \rightarrow CH_3CHO \quad \Delta_rG^\ominus = -197\,kJ\,mol^{-1}$$

Its invention at the Wacker Consortium für Elektrochemische Industrie in the late 1950s marked the beginning of an era of production of chemicals from petroleum feedstock. Although the Wacker process is no longer of major industrial concern, it has some interesting mechanistic features that are worth noting.

The actual oxidation of ethene is known to be caused by a palladium(II) salt:

$$C_2H_4 + PdCl_2 + H_2O \rightarrow CH_3CHO + Pd(0) + 2\,HCl$$

The exact nature of the Pd(0) species is unknown, but it is probably present as a mixture of compounds. The slow oxidation of Pd(0) back to Pd(II) by oxygen is catalysed by the addition of Cu(II), which shuttles back and forth to Cu(I):

$$Pd(0) + 2[CuCl_4]^{2-} \rightarrow Pd^{2+} + 2[CuCl_2]^- + 4\,Cl^-$$
$$2[CuCl_2]^- + \tfrac{1}{2}O_2 + 2\,H^+ + 4\,Cl^- \rightarrow 2[CuCl_4]^{2-} + H_2O$$

The overall catalytic cycle is shown in Fig. 31.5. Detailed stereochemical studies on related systems indicate that the

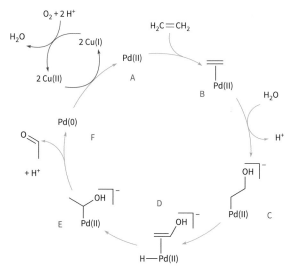

FIGURE 31.5 The catalytic cycle for the palladium-catalysed oxidation of alkenes to aldehydes.

hydration of the alkene–Pd(II) complex (B) occurs by the attack of H₂O from the solution on the coordinated ethene rather than the insertion of coordinated OH. Hydration, to form (C), is followed by two steps that isomerize the coordinated alcohol. First, β-hydrogen elimination occurs with the formation of (D), and then migration of a hydride results in the formation of (E). Elimination of the ethanal and an H⁺ ion then leaves Pd(0), which is converted back to Pd(II) by the auxiliary copper(II)-catalysed air oxidation cycle.

One important observation that the mechanism must account for is that when the reaction is carried out in the presence of D$_2$O, no deuterium is incorporated into the final product. This observation suggests that either intermediate (D) is very short-lived and does not exchange the Pd–H for a Pd–D, or that intermediate (C) rearranges directly to (E).

Alkene ligands coordinated to Pt(II) are also susceptible to nucleophilic attack, but only palladium leads to a successful catalytic system. The principal reason for palladium's unique behaviour appears to be the greater lability of the 4d Pd(II) complexes in comparison with their 5d Pt(II) counterparts. Furthermore, the potential for the oxidation of Pd(0) to Pd(II) is more favourable than for the corresponding Pt couple.

31.5 Asymmetric oxidations

KEY POINT Appropriate chiral ligands can be used in conjunction with d-metal catalysts to induce chirality into oxidation products of organic substrates.

In addition to catalysing reductions, d-metal complexes are also active in oxidations. For example, in the **Sharpless epoxidation**, prop-2-en-1-ol (allyl alcohol) or a derivative is oxidized with *tert*-butylhydroperoxide, in the presence of a Ti catalyst with diethyl tartrate as a chiral ligand, producing an epoxide:

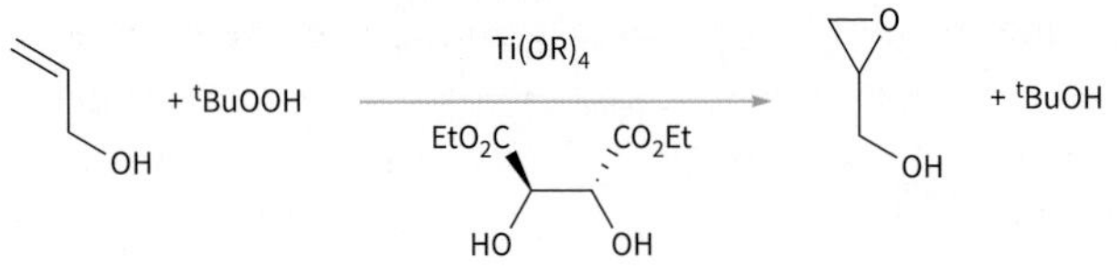

The reaction is thought to go through a transition state in which both the peroxide and the allyl alcohol are coordinated to a Ti atom through their O atoms. Each Ti atom is known to have one diethyl tartrate attached to it, and the chiral environment that the diethyl tartrate produces around the Ti atom is sufficient to differentiate the two prochiral faces of the allyl alcohol. Additional experimental evidence points to a dimeric intermediate (9). Enantiomeric excesses of greater than 98% have been reported for Sharpless epoxidations.

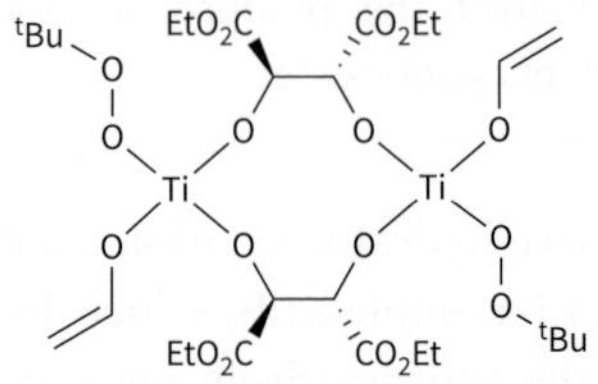

A **Jacobsen oxidation** is a reaction in which the catalyst is an Mn complex of the mixed 2 × N + 2 × O donor ligand known as salen (Fig. 31.6). Hypochlorite ions (ClO⁻) are used to oxidize the Mn(III) complex to an Mn(V) oxide, which is then able to deliver its O atom to an alkene to generate an epoxide (Section 10.8). This oxidation has been used with a wide variety of substrates and routinely delivers enantiomeric excesses greater than 95%. The mechanism of the reaction has not been defined precisely, but proposals include the existence of a dimeric form of the catalyst or a radical oxygen transfer step.

31.6 Palladium-catalysed C–C bond-forming reactions

KEY POINTS A number of palladium-catalysed coupling reactions are known; they all proceed through oxidative addition of reagents at the metal centre, followed by the reductive elimination of the two fragments.

A large number of palladium-catalysed carbon–carbon bond-forming ('coupling') reactions are known. They include

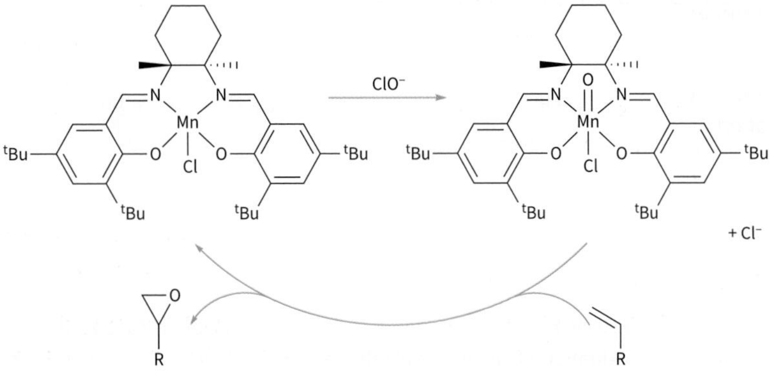

FIGURE 31.6 The Jacobsen oxidation relies on a manganese complex of a salen-based ligand and chlorate(I).

the coupling of a Grignard with an aryl halide and the Heck, Stille, and Suzuki coupling reactions:[3]

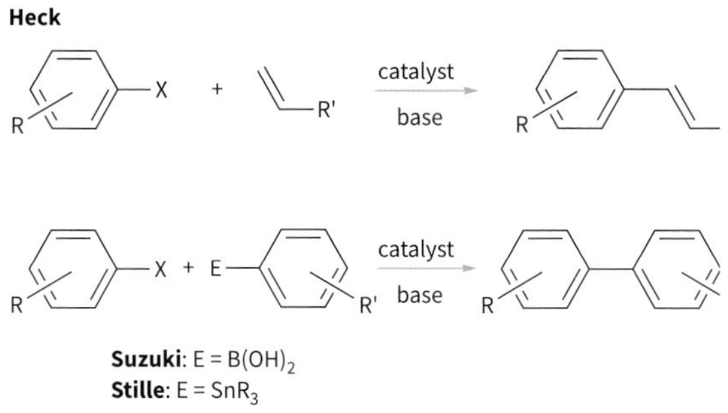

Heck

R—⬡—X + ⫽—R' →(catalyst, base)→ R—⬡—CH=CH—R'

R—⬡—X + E—⬡—R' →(catalyst, base)→ R—⬡—⬡—R'

Suzuki: E = B(OH)$_2$
Stille: E = SnR$_3$

Normally, either a Pd(II) complex, such as [PdCl$_2$(PPh$_3$)$_2$], in the presence of additional phosphine or a Pd(0) compound, such as [Pd(PPh$_3$)$_4$], is used as the catalyst, although many other Pd/ligand combinations are active. The precise reaction pathway is unclear (and probably differs with each Pd/ligand/substrate combination), but it is apparent that all these reactions follow the same general sequence. Figure 31.7 shows an idealized catalytic cycle for the coupling of an ethenyl group with an aryl halide. An initial oxidative addition of an aryl–halogen bond to an unsaturated Pd(0) complex (A) results in a Pd(II) species (B). Coordination of an alkene results in complex (C); 1,2-insertion results in an alkyl complex (D), which can be deprotonated with the loss of the halide to give the organic product attached to the palladium atom (E).

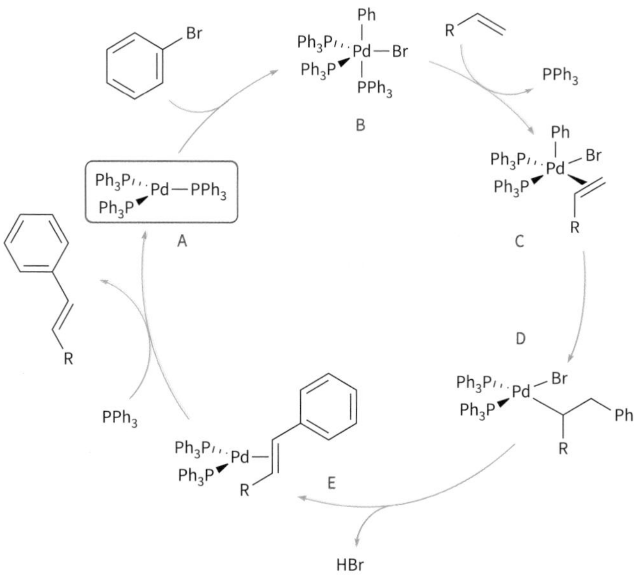

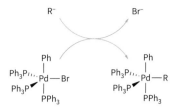

FIGURE 31.8 The exchange of a halide for an organic fragment at a Pd centre can be thought of as nucleophilic displacement.

In other palladium-catalysed coupling reactions, such as that of a Grignard reagent with an aryl halide, initial oxidative addition proceeds as in Fig. 31.7. The second organic group is thought to be introduced with the Grignard reagent behaving as the nucleophilic R$^-$ group displacing the halide at the metal centre in (B), to give two organic fragments attached to the Pd atom, as indicated in Fig. 31.8. These two adjacent fragments can then couple and reductively eliminate to regenerate the starting Pd(0) species (A).

In all palladium-catalysed coupling reactions, it is necessary for the two fragments that are coupling to be *cis* to each other at the metal centre before insertion or reductive elimination can take place; this requirement has led to the use of chelating diphosphines such as dppe (**10**) and the ferrocene derivative (**11**).

Ph$_2$P⌒PPh$_2$ Fc(PPh$_2$)$_2$
10 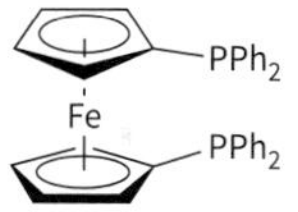   **11**

Palladium-catalysed coupling reactions are tolerant to a wide range of substitution on both fragments and a versatile reaction that takes place at room temperature and in aqueous solution is the Sonogashira coupling:

$$HC\equiv CR + R'X \xrightarrow[base]{Pd(0)/Cu(I)\ catalyst} R'C\equiv CR + HX$$

The two catalysts are a Pd(0) complex, such as [PdCl$_2$(PPh$_3$)$_2$], and a Cu(I) halide. This reaction is used in the synthesis of many pharmaceuticals, including treatments for psoriasis, Parkinson's disease, Tourette's syndrome, and Alzheimer's disease. The Heck coupling is used in the synthesis of steroids, strychnine, and the herbicide Prosulfuron®, which is produced industrially on a large scale. The C–C coupling step in the synthesis is shown below:

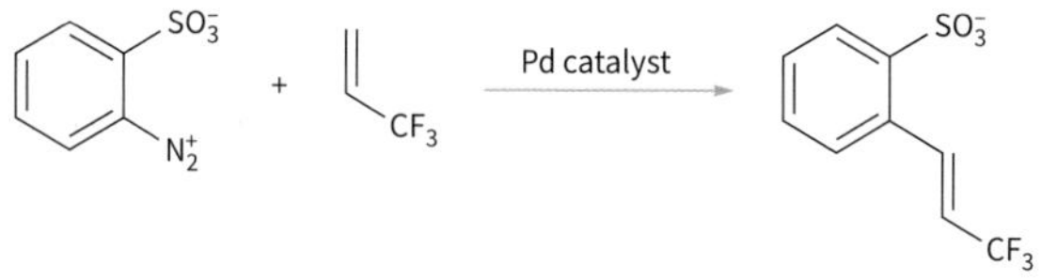

FIGURE 31.7 An idealized catalytic cycle for the coupling of an alk-1-ene to an aryl halide in the Heck reaction.

---
[3] The 2010 Nobel Prize in Chemistry was awarded to Richard Heck, Akira Suzuki and Ei-ichi Negishi for palladium-catalysed coupling reactions.

31.7 Methanol carbonylation: ethanoic acid synthesis

KEY POINT Rhodium and iridium complexes are highly active and selective in the carbonylation of methanol to form acetic acid.

The time-honoured method for synthesizing ethanoic (acetic) acid is by aerobic bacterial action on dilute aqueous ethanol, which produces vinegar. However, this process is uneconomical as a source of concentrated ethanoic acid for industry. A highly successful commercial process is based on the carbonylation of methanol:

$$CH_3OH + CO \rightarrow CH_3COOH$$

The reaction is catalysed by all three members of Group 9 (Co, Rh, and Ir). Originally, a Co complex was used, but then a Rh catalyst developed at Monsanto greatly reduced the cost of the process by allowing lower pressures to be used. As a result, the rhodium-based **Monsanto process** was used throughout the world. Subsequently, British Petroleum developed the **Cativa process**, which uses a promoted Ir catalyst. Both processes are highly selective and generate ethanoic acid of sufficient purity that it can be used in food for humans.

The Monsanto and Cativa processes follow essentially the same reaction sequence, so the rhodium-based cycle described here captures the principal features of the iridium-based process too (Fig. 31.9). Under the conditions used, I⁻ ions that are present react with methanol to set up an appreciable concentration of iodomethane in the first step of the reaction. Starting with the four-coordinate, 16-electron complex $[Rh(CO)_2I_2]^-$ (A), the next step is the oxidative addition of iodomethane to produce the six-coordinate

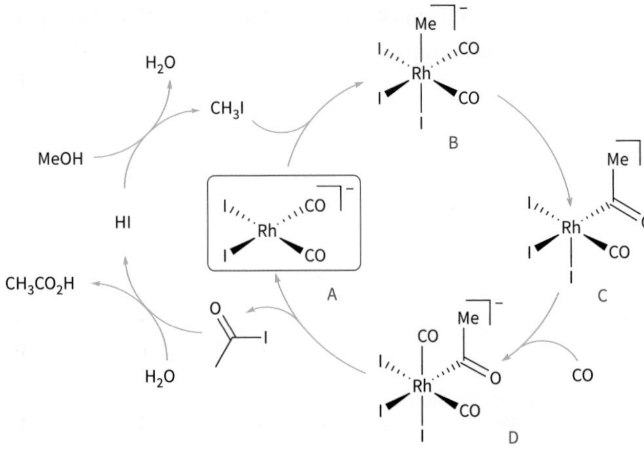

FIGURE 31.9 The catalytic cycle for the formation of ethanoic (acetic) acid with a rhodium-based catalyst. The oxidative addition ste (A ≥ B) is rate-determining.

18-electron complex $[Rh(Me)(CO)_2I_3]^-$ (B). This step is followed by methyl migration, yielding a 16-electron acyl complex (C). Coordination of CO restores an 18-electron complex (D), which is then set to undergo reductive elimination of acetyl iodide with the regeneration of $[Rh(CO)_2I_2]^-$. Water then hydrolyses the acetyl iodide to acetic acid and regenerates HI.

Under normal operating conditions, the rate-determining step for the rhodium-based system is the oxidative addition of iodomethane, whereas for the iridium-based system it is the migration of the methyl group. An important feature is that methyl migration on iridium is favoured by formation of a neutral intermediate, and iodide-accepting promoters help facilitate substitution of an iodide ligand by CO in the Ir analogue of complex (B).

Heterogeneous and nanoparticle catalysis

Numerous important industrial processes are facilitated by heterogeneous catalysis. Practical heterogeneous catalysts are typically high-surface-area solid materials that may contain several different phases and operate at pressures of 1 atm and higher. The high surface areas, either on the external or internal surfaces of a solid, are often generated through the formation of nanoparticles or design of nano- or mesoporous materials. In some cases, the bulk of a high-surface-area single-phase material serves as the catalyst, and is known as a **uniform catalyst**; a simple example is a very finely divided metal, as in 'skeletal nickel'. More often, **multiphasic catalysts** are used, which consist of a high-surface-area material that serves as a support on to which active catalyst nanoparticles are deposited (Fig. 31.10).

In Sections 31.8 and 31.9 we concentrate on the inorganic chemistry involved in reactions on metal nanoparticle

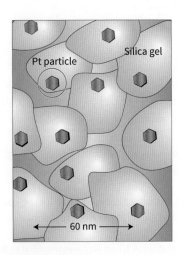

FIGURE 31.10 Schematic diagram of metal nanoparticles supported on a finely divided silica such as silica gel.

surfaces, not the physical chemical aspects of adsorption and reaction. Catalysis on the internal pores and surfaces of microporous zeolites and uniform mesoporous materials is discussed in Section 31.13.

31.8 The nature of heterogeneous catalysts

The basis of catalysis was introduced in Section 2.14, and homogeneous catalysis has been discussed in detail in the earlier sections of this chapter. There are many parallels between the individual reaction steps encountered in heterogeneous and homogeneous catalysis, but we need to consider some additional points.

(a) Surface area and porosity

KEY POINTS Heterogeneous catalysts are typically high-surface-area materials formed from finely divided, sometimes nanoparticulate, substrates which can be coated with nanoparticles of a metal or metal oxide.

An ordinary dense or polycrystalline solid is generally unsuitable as a catalyst because its surface area, where the reactions take place, is quite low. Thus α-alumina, which is a polycrystalline material consisting of particles around 1 μm or greater in dimension, has a low specific surface area (the surface area divided by the mass of the sample), and is used much less as a catalyst support than the nanocrystalline-solid γ-alumina. γ-Alumina can be prepared with small particle size, ~50 nm, and therefore a high specific surface area. The high surface area results from the many small but connected substrate nanoparticles, like those shown in Fig. 31.10 for silica, and a gram or so of a typical catalyst support has a surface area equal to that of a tennis court. Similarly, polycrystalline quartz, particle size ~1 μm, is not used as a catalyst support, but the high-surface-area versions of SiO_2, known as fumed silicas, Section 28.3, are widely used. In a typical heterogeneous catalyst this substrate surface already has a large number of active sites. Nanoparticles of metals (like Pt in Fig. 31.10) or metal oxides can be deposited on the finely divided substrate particles, producing extremely high numbers of catalytically active centres.

Both γ-alumina and high-surface-area silica are metastable nanomaterials, but under ordinary conditions they do not convert to their more stable phases (α-alumina and polycrystalline quartz, respectively). The preparation of γ-alumina with a particle size of around 5 nm involves the dehydration of an aluminium oxide hydroxide at relatively low temperatures:

$$2\,AlO(OH) \xrightarrow{\Delta/450°C} \gamma\text{-}Al_2O_3 + H_2O$$

Similarly, high-surface-area fumed silica is prepared from the acidification of silicates to produce $Si(OH)_4$, which rapidly forms a hydrated silica gel from which much of the adsorbed water can be removed by gentle heating (Section 25.1). When viewed with an electron microscope, the textures of the fumed silica or γ-alumina appear to be that of a rough gravel bed with irregularly shaped voids between the interconnecting nanoparticles (as in Fig. 31.10). Other high-surface-area nanomaterials used as supports in heterogeneous catalysts are TiO_2, Cr_2O_3, ZnO, MgO, and carbon black.

(b) Surface acidic and basic sites

KEY POINT Surface acids and bases are highly active for catalytic reactions such as the dehydration of alcohols and isomerization of alkenes.

When exposed to atmospheric moisture, the surface of γ-alumina is covered with adsorbed water molecules. Dehydration at 100–150°C leads to the desorption of water, but surface OH groups remain and act as weak Brønsted acids:

[diagram: Al–OH, Al–OH, Al–OH → Al–O–Al, Al–OH + H_2O]

At even higher temperatures, adjacent OH groups condense to liberate more H_2O and generate exposed Al^{3+} Lewis acid sites as well as O^{2-} Lewis base sites (**12**). The rigidity of the surface permits the coexistence of these strong Lewis acid and base sites, which would otherwise immediately combine to form Lewis acid–base complexes. Surface acids and bases are highly active for catalytic reactions such as the dehydration of alcohols and isomerization of alkenes; similar Brønsted and Lewis acid sites exist in the interior of many zeolites, Section 25.15 and Chapter 29. Different oxides and their mixtures show variations in surface acidity; thus an SiO_2/TiO_2 mixture is more acidic than SiO_2/Al_2O_3, and promotes different catalytic reactions.

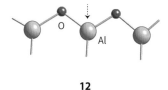

12

(c) Surface metal sites

KEY POINT Metal nanoparticles on ceramic oxide substrates are very active catalysts for a range of reactions.

Metal nanoparticles are often deposited on supports to provide a catalyst. Acidic and basic surfaces, Section 31.8(b), act as useful substrates for depositing other catalytic centres, particularly metal particles. Treatment of γ-alumina with H_2PtCl_6 followed by heating in a reducing environment, produces Pt particles of dimensions 1–50 nm distributed over the alumina surface; see also Section 25.24. A supported metal nanoparticle about 2.5 nm in diameter has about 40% of its atoms on the surface, and the particles are protected from fusing together into bulk metal by their separation. The high proportion of exposed atoms is a great advantage for these small supported particles, particularly for metals such as platinum and the even more expensive rhodium.

Nanoparticles of Pt/Re alloys distributed on the surface of γ-alumina particles are used to interconvert hydrocarbons, and Pt/Rh nanoparticles supported on γ-alumina are used in the catalytic converters of vehicles to promote the combination of O_2 with CO and hydrocarbons to form CO_2 and the reduction of nitrogen oxides to nitrogen (Box 31.1).

The metal atoms on the surface of metal nanoparticles are capable of forming bonds such as M–CO, M–CH_2R, M–H, and M–O (Table 31.3). Often the nature of surface ligands is inferred by comparison of IR spectra with those of organometallic or inorganic complexes. Thus, both terminal and bridging CO groups can be identified on surfaces by IR spectroscopy, and the IR spectra of many hydrocarbon ligands on surfaces are similar to those of discrete organometallic complexes. The case of the N_2 ligand is noteworthy because coordinated N_2 was identified by IR spectroscopy on metal surfaces before dinitrogen complexes (Section 24.17) had been prepared.

(d) Chemisorption, desorption

KEY POINTS Adsorption is essential for heterogeneous catalysis to occur, but must not be so strong that it blocks the catalytic sites and prevents further reaction.

TABLE 31.3 Chemisorbed ligands on surfaces

a Ammonia adsorbed on the Lewis and Al^{3+} sites of γ-alumina.

b,c CO coordinated to platinum metal nanoparticle.

d Hydrogen dissociatively chemisorbed on platinum metal nanoparticle.

e Ethane dissociatively chemisorbed on platinum metal nanoparticle.

f Nitrogen dissociatively chemisorbed on iron metal nanoparticle.

g Hydrogen dissociatively chemisorbed on ZnO.

h Ethene η^2 coordinated to a Pt atom.

i Ethene bonded to two Pt atoms.

j Oxygen bound as a superoxide to a metal nanoparticle surface.

k Oxygen dissociatively chemisorbed on a metal nanoparticle surface.

Adapted from R. L. Burwell, Jr, Heterogeneous catalysis. *Surv. Prog. Chem.*, 1977, **8**, 2.

The adsorption of molecules on the surface of a heterogeneous surfaces often activates molecules just as coordination to a metal centre activates complexes, Sections 24.21–24.26. Likewise, the desorption of product molecules that refreshes the active sites in heterogeneous catalysis is analogous to

---

BOX 31.1 How does a catalytic converter work?

Catalytic convertors are used to reduce toxic emissions from internal combustion engines, which include nitrogen oxides (NO_x), carbon monoxide, and unburnt hydrocarbons (HC). By 2014, when the Euro VI emission standards came into use in Europe, the levels of these compounds in exhaust gases were restricted in passenger vehicle diesel engines to 0.50 (CO), 0.08 (NO_x), and 0.17 (HC+NO_x) g km^{-1}, respectively; these values are similar to those demanded in California. The catalytic converter consists of a honeycomb stainless steel or ceramic structure onto which silica and alumina are deposited, followed by a mixture of platinum, rhodium, and palladium as nanoparticles, with diameters typically between 10 and 50 nm. A three-way converter, used with petrol (gasoline) engines, catalyses the following three reactions:

$$2\,NO_x(g) \rightarrow xO_2(g) + N_2(g)$$
$$2\,CO(g) + O_2(g) \rightarrow 2\,CO_2(g)$$
$$C_xH_{2x+2}(g) + 2xO_2(g) \rightarrow xCO_2(g) + 2xH_2O(g)$$

The first stage of the catalytic converter involves reduction of the NO_x on a reduction catalyst, which consists of a mixture of platinum and rhodium; rhodium is highly reactive towards NO. The second stage involves catalytic oxidation, which removes the unburnt hydrocarbons and carbon monoxide by oxidizing them on the mixed metal platinum/palladium catalyst. A two-way catalytic converter used in conjunction with most diesel engines undertakes only the oxidation reactions (the second and third of those above).

In order to facilitate the near-complete conversion of the gases emerging from the engine, the correct initial air:fuel mixture is used. The ideal stoichiometric air:fuel mixture entering the engine is 14.7:1, so that after combustion the gases entering the catalytic converter contain about 0.5% oxygen. If richer or leaner air:fuel mixtures are used (that is, with lower or higher air contents, respectively), then the level of oxygen entering the exhaust stream may be too high or low for effective operation of the catalytic converter. For these reasons, various metal oxides, particularly Ce_2O_3 and CeO_2, are incorporated into the catalytic coating to store and release oxygen as the oxygen content of the gases emerging from the engine changes.

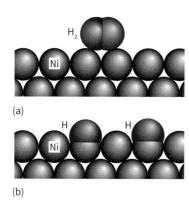

FIGURE 31.11 Schematic representation of (a) physisorption and (b) chemisorption of hydrogen on a nickel metal nanoparticle surface.

the dissociation of a complex in homogeneous catalysis. Before a heterogeneous catalyst is used it is usually 'activated'. Activation is a catch-all term; in some instances it refers to the desorption of adsorbed molecules such as water from the surface, as in the dehydration of γ-alumina. In other cases it refers to the preparation of the active site by a chemical reaction, such as by reduction of metal oxide particles to produce highly active metal nanoparticles.

An activated surface can be characterized by the adsorption of various inert and reactive gases. The adsorption may be either **physisorption**, when no new chemical bond is formed, or **chemisorption**, when surface–adsorbate bonds are formed (Fig. 31.11).

The interaction of small molecules with metal surfaces is similar to their interaction with low-oxidation-state metal complexes. Table 31.4 shows that a wide range of metals chemisorb CO, and that fewer are capable of chemisorbing N_2, just as there is a much wider variety of metals that form carbonyls than form dinitrogen complexes. Furthermore, just as with metal carbonyl complexes, both bridging and terminal CO surface species have been identified by IR spectroscopy. The dissociative chemisorption of H_2 is analogous

TABLE 31.4 The abilities of metals to chemisorb simple gas molecules*

	O_2	C_2H_2	C_2H_4	CO	H_2	CO_2	N_2
Ti, Zr, Hf, V, Ta, Cr, Mo, W, Fe, Ru, Os	+	+	+	+	+	+	+
Ni, Co	+	+	+	+	+	+	+
Rh, Pd, Pt, Ir	+	+	+	+	+	+	–
Mn, Cu	+	+	+	+	±	+	–
Al, Au	+	+	+	+	–	–	–
Na, K	+	+	–	–	–	–	–
Ag, Zn, Cd, In, Si, Ge, Sn, Pb, As, Sb, Bi	+	–	–	–	–	–	–

* Chemisorption is strong (+), weak (±), unobservable (–).
Adapted from G. C. Bond, *Heterogeneous Catalysis*, Oxford University Press (1987).

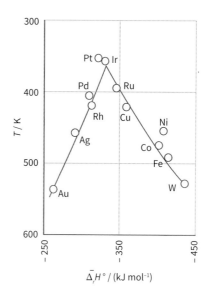

FIGURE 31.12 A 'volcano diagram', in this case the reaction temperature for a set rate of methanoic (formic) acid decomposition, plotted against the stability of the corresponding metal methanoate as judged by its enthalpy of formation. (Based on W. J. M. Rootsaert and W. M. H. Sachtler, *Z. Phyzik. Chem.*, 1960, **26**, 16.)

to the oxidative addition of H_2 to metal complexes (Sections 11.5 and 24.22).

Although adsorption is essential for catalysis to occur, it must not be so strong as to block the active sites and prevent further reaction. This factor is in part responsible for the limited number of metals that are effective catalysts. The catalytic decomposition of methanoic (formic) acid on metal nanoparticle surfaces,

$$HCOOH \xrightarrow{M} CO + H_2O$$

provides a good example of this balance between adsorption and catalytic activity. It is observed that the catalysis is most effective by using metals for which the metal methanoate is of intermediate stability (Fig. 31.12). The plot in Fig. 31.12 is an example of a 'volcano diagram', and is typical of many catalytic reactions. The implication is that the earlier d-block metals form very stable surface compounds (with the reactants or products), whereas the later noble metals such as silver and gold form very weak surface compounds, both of which are detrimental to a catalytic process. Between these extremes, the metals in Groups 8 to 10 have high catalytic activity, especially the platinum metals (Group 10).

(e) Nanoparticle size and shape and catalytic activity

KEY POINT Controlling the morphology of nanoparticles determines the nature of the surface reaction sites and hence catalytic activity.

The active sites of heterogeneous catalysts are not uniform, and many diverse sites are exposed on the surface of a nanocrystalline solid such as γ-alumina or silica gel. Even

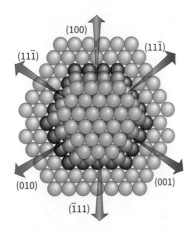

FIGURE 31.13 Some possible metal crystal planes that might be exposed on a metal nanoparticle surface to a reactive gas. The planes labelled [$\bar{1}$ 1 1], [1 $\bar{1}$ 1], etc. are hexagonally close-packed, and the planes represented by [100], [010], etc. have square arrays of atoms.

metal nanoparticles are not usually uniform, though the synthesis methods described in Sections 28.2–28.4 aim to optimize the route to give highly reactive monodispersed (in terms of size and shape) nanoparticles. Without a degree of control in the synthesis process, a nanoparticle crystal typically has more than one type of exposed plane, each with its characteristic pattern of surface atoms (Fig. 31.13). In addition, metal surfaces have irregularities such as steps that expose metal atoms with low coordination numbers (Fig. 31.14). These highly exposed, coordinatively unsaturated sites are often particularly reactive. As a result, the different sites present on the surface may serve different functions and promote a variety of catalytic reaction pathways. The variety of sites thus accounts for the lower selectivity of many heterogeneous catalysts in comparison with their homogeneous analogues.

The control of catalytic nanoparticle size and shape has been accomplished by addition of various reaction pathway-controlling species to the reaction mixture. Platinum nanoparticles, which can form as tetrahedra, cubes, or cuboctahedra, are often obtained by reaction of dihydrogenhexachloroplatinate, H_2PtCl_6, with alcohols. Addition of polymers or branched dendrimers to the reaction mixture prevents nanoparticles aggregating and typically reduces particle sizes from 10 to 1 nm. Addition of Br^- to the reaction mixture directs the formation of nanocubes alone which have enhanced catalytic activity for many reactions, including hydrogenation, Section 31.12(a). It has been proposed that Br^- stabilizes the growth of the (100) planes of the Pt structure, leading to the nanocube form.

(f) Chemisorption and surface migration

KEY POINT Adsorbed atoms and molecules migrate over metal surfaces.

The surface analogue of fluxional mobility in metal clusters is diffusion, and there is abundant evidence for the diffusion of chemisorbed molecules or atoms on metal nanoparticle surfaces. For example, adsorbed H atoms and CO molecules are known to move over the surface of a metal particle. These 2D diffusion pathways generally involve the adsorbed molecules moving through a variety of different coordination sites on the metal surface. So, for example, CO migration can result from a molecule moving between sites interacting with one (terminal CO) and between two and four (bridging CO) metal atoms on the surface. The energy barrier to this process is relatively low (a few tens of kilojoules per mole), and thus migration rates are very high under typical catalytic reaction conditions. This mobility is important in catalytic reactions, as it allows atoms or molecules to find and approach one another rapidly.

31.9 Ammonia synthesis

KEY POINT Catalysts based on iron metal are used for the synthesis of ammonia from nitrogen and hydrogen.

The synthesis of ammonia has already been discussed from several different viewpoints (Sections 16.6 and 27.6). Here we concentrate on details of the catalytic steps. The formation of ammonia is exergonic and exothermic at 25°C, the relevant thermodynamic data being $\Delta_r G^\ominus = -116.5$ kJ mol^{-1}, $\Delta_r H^\ominus = -146.1$ kJ mol^{-1}, and $\Delta_r S^\ominus = -199.4$ J K^{-1} mol^{-1}. The negative entropy of formation reflects the fact that two NH_3 molecules form in place of four reactant molecules.

The great inertness of N_2 (and to a lesser extent H_2) requires that a catalyst be used for the reaction. In the course of developing the original process, Haber, Bosch, and their co-workers investigated the catalytic activity of most of the metals in the periodic table and found that the best are Fe, Ru, and U promoted by small amounts of alumina and potassium salts. Cost and toxicity considerations led to the choice of iron as the basis of the commercial catalyst. The role of the various promoters, particularly K, in the Fe metal

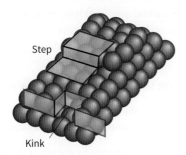

FIGURE 31.14 Schematic representation of surface irregularities, steps, and kinks.

catalyst had been the subject of much scientific research. G. Ertl[4] found that in the presence of potassium, N_2 molecules adsorb more readily on the metal surface and the adsorption enthalpy is made more exothermic by about 12 kJ mol^{-1}, probably as a result of the increased electron-donating abilities of the Fe/K surface. The more strongly adsorbed N_2 molecule is then cleaved more easily in the rate-determining step in the process. The other reactant, H_2, undergoes much more facile dissociation on the metal surface, and a series of insertion reactions between adsorbed species leads to the production of NH_3:

$N_2(g) \longrightarrow N_2(g) \longrightarrow N\ N$

$H_2(g) \longrightarrow H_2(g) \longrightarrow H\ H$

$N + H \longrightarrow NH \longrightarrow NH_2 \longrightarrow NH_3 \longrightarrow NH_3(g)$

Because of the slowness of the N_2 dissociation, even when catalysed, it is necessary to run the industrial process at high temperatures, typically 450°C, in order to achieve a reasonable reaction rate. However, because the reaction is exothermic, the high temperature decreases the equilibrium constant of the reaction. Thus, higher pressures, typically of the order of 100 atm, are used to help favour the formation of product. Fig. 31.15 shows how the theoretical equilibrium yields of ammonia vary with temperature at a number of different pressures. Catalysts operating at room temperature, such as the enzyme nitrogenase (Section 26.14), would give good equilibrium yields of NH_3, even at low pressures, and are much sought after. Catalysts based on Co, Ni, and Ru have all shown promise, but so far none have been commercialized.

The source of hydrogen for most industrial plants synthesising ammonia is methane. When the use of methane is coupled with the high operating temperatures and pressures required, the whole process becomes very energy-intensive, ultimately generating significant CO_2 emissions. Indeed, it has been estimated that 1.2% or all human-derived CO_2 emissions worldwide come from the Haber–Bosch process. The use of renewably generated electricity to produce hydrogen via the electrolysis of water should enable an industrial process to synthesize 'green' ammonia without the carbon footprint associated with the older processes. This route would require significant quantities of pure water, and has yet to be realized on a large scale.

31.10 Sulfur dioxide oxidation

KEY POINT The most widely used catalyst for the oxidation of SO_2 to SO_3 is molten potassium vanadate supported on a high-surface-area silica.

The oxidation of SO_2 to SO_3 is a key step in the production of sulfuric acid (Section 17.11). The reaction of sulfur with oxygen to produce SO_3 gas is exergonic ($\Delta_r G^\ominus = -371$ kJ mol^{-1}) but very slow, and the principal product of combustion is SO_2:

$$S(s) + O_2(g) \rightarrow SO_2(g)$$

The combustion is followed by the catalytic oxidation of SO_2:

$$SO_2(g) + \tfrac{1}{2} O_2(g) \rightarrow SO_3(g)$$

---

[4] Gerhard Ertl was awarded the 2007 Nobel Prize for chemistry for his work on chemical processes at solid surfaces.

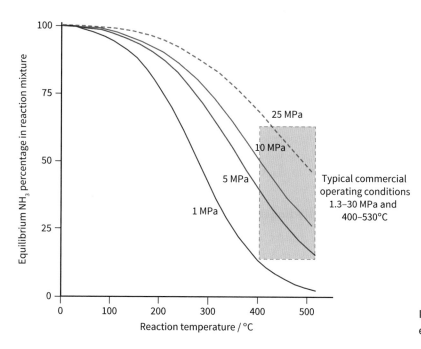

FIGURE 31.15 The relationship between equilibrium ammonia proportion and temperature.

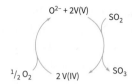

FIGURE 31.16 Cycle showing the key elements involved in the oxidation of SO_2 by V(V) compounds.

This step is also exothermic and thus, as with ammonia synthesis, has a less favourable equilibrium constant at elevated temperatures. The process is therefore generally run in stages. In the first stage, the combustion of sulfur raises the temperature to about 600°C, but by cooling and pressurizing before the catalytic stage, the equilibrium is driven to the right, and high conversion of SO_2 to SO_3 is achieved.

Several quite different catalytic systems have been used to catalyse the combination of SO_2 with O_2. The most widely used catalyst at present is potassium or caesium vanadate molten salt covering a high-surface-area silica. The current view of the mechanism of the reaction, known as the Contact process, is that the rate-determining step is the oxidation of V(IV) to V(V) by O_2 (Fig. 31.16). In the melt, the vanadium and oxide ions are part of a polyoxovanadate complex (Box 20.1), but little is known about the evolution of the oxo species.

31.11 Fischer–Tropsch synthesis

KEY POINT Hydrogen and carbon monoxide can be converted to hydrocarbons and water by reaction over iron or cobalt catalysts.

The conversion of **syngas**, a mixture of H_2 and CO, into hydrocarbons over metal catalysts was first discovered by Franz Fischer and Hans Tropsch at the Kaiser Wilhelm Institute for Coal Research in Müllheim in 1923. In the Fischer–Tropsch reaction, CO reacts with H_2 to produce hydrocarbons, which can be written symbolically as the formation of the chain extender ($-CH_2-$), and water:

$$CO + 2\,H_2 \rightarrow -CH_2- + H_2O$$

The process is exothermic, with $\Delta_r H^\ominus = -165$ kJ mol^{-1}. The product range consists of aliphatic straight-chain hydrocarbons that include methane (CH_4) and ethane, LPG (C_3 to C_4), gasoline (C_5 to C_{12}), diesel (C_{13} to C_{22}), and light and heavy waxes (C_{23} to C_{32} and $> C_{33}$, respectively). Side reactions include the formation of alcohols and other oxygenated products. The distribution of the products depends on the catalyst and the temperature, pressure, and residence time. Typical conditions for the Fischer–Tropsch synthesis are a temperature range of 200–350°C and pressures of 15–40 atm.

There is general agreement that the first stages of the hydrocarbon synthesis involve the adsorption of CO on the metal, followed by its cleavage to give a surface carbide and oxide. Hydrogenation of such species to water surface methyne (CH), methylene (CH_2), and methyl (CH_3) species takes place, but there is still debate on what happens next and how chain growth occurs. One proposal suggests a polymerization of bridging surface $-CH_2-$ groups initiated by a surface $-CH_3$ group. However, the fact that many such species have been isolated and are stable as metal complexes suggests that the mechanism is probably not so simple. Other mechanistic investigations of these processes have suggested an alternative possibility for chain growth in the hydrocarbon synthesis; namely, that it proceeds by combination of surface bridging $-CH_2-$ groups and alkenyl chains (M–CH=CHR), rather than by the combination of alkyl chains (M–CH_2CH_2R) with methylene groups in the surface.

Several catalysts have been used for the Fischer–Tropsch synthesis; the most important are based on Fe and Co. Cobalt catalysts have the advantage of a higher conversion rate and a longer life (of over five years). The cobalt catalysts are in general more reactive for hydrogenation and produce fewer unsaturated hydrocarbons and alcohols than iron catalysts. Iron catalysts have a higher tolerance for sulfur, are cheaper, and produce more alkene products and alcohols. The lifetime of the iron catalysts is short, however, and in commercial installations generally limited to about eight weeks. Additives containing potassium or copper are often added to the catalyst to enhance their activity.

The reaction of CO_2 with renewably produced hydrogen to synthesize hydrocarbons has the potential to provide a route to generating 'green' fuels, though this has yet to be commercially realized.

31.12 Reactions involving heterogeneous nanoparticle catalysts

The high surface area and reactivity of metal nanoparticles distributed on a metal oxide surface or within the pores of a mesoporous solid can be exploited in a wide variety of catalytic reactions, particularly those involving dissociation of gaseous species. The iron catalyst used in the Haber–Bosch process for the reaction of nitrogen with hydrogen (described earlier in Section 31.9) is a highly porous nanoparticulate consisting of a mixture of $Fe_{1-x}O$, Fe_3O_4, and iron metal. Similarly, in the **contact process** (Section 31.10) for oxidizing SO_2 to SO_3, the vanadium oxide catalyst is prepared by deposition on V_2O_5 nanoparticles on an alumina or TiO_2 substrate.

(a) Hydrogenation catalysts

KEY POINTS Alkenes are hydrogenated on supported metal nanoparticles by a process that involves H_2 dissociation and migration of H to an adsorbed ethene molecule. Skeletal or nanoporous nickels can be used to reduce alkanals to alkanols.

A milestone in heterogeneous catalysis was Paul Sabatier's observation in 1890 that nickel catalyses the hydrogenation of alkenes. He was in fact attempting to synthesize [Ni(C$_2$H$_4$)$_4$] in response to Mond, Langer, and Quinke's synthesis of [Ni(CO)$_4$] (Section 24.18). However, when he passed ethene over heated nickel he detected ethane, so he repeated the experiment but included hydrogen with the ethene, whereupon he observed a good yield of ethane.

The hydrogenation of alkenes on supported metal nanoparticles is thought to proceed in a manner very similar to that in metal complexes. As pictured in Fig. 31.17, H$_2$, which is dissociatively chemisorbed on the surface, is thought to migrate to an adsorbed ethene molecule, giving first a surface alkyl and then the saturated hydrocarbon. When ethene is hydrogenated with D$_2$ over platinum, the simple mechanism depicted in Fig. 31.17 indicates that CH$_2$DCH$_2$D should be the product. In fact, a complete range of C$_2$H$_n$D$_{6-n}$ ethane isotopologues is observed. It is for this reason that the central step is written as reversible; the rate of the reverse reaction must be greater than the rate at which the ethane molecule is formed and desorbed in the final step.

One of the most important class of heterogeneous hydrogenation catalysts is the '**skeletal nickels**', sometimes referred to as '**Raney nickel**', which are used for various processes such as conversion of alkanals (aldehydes) to alkanols (alcohols), as in

$$\text{CH}_3\text{CH}_2\text{CH}_2\text{CHO} + \text{H}_2 \rightarrow \text{CH}_3\text{CH}_2\text{CH}_2\text{CH}_2\text{OH}$$

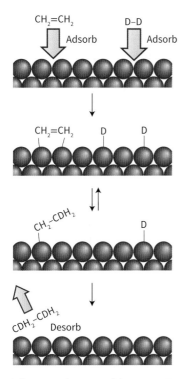

FIGURE 31.17 Schematic diagram of the stages involved in the hydrogenation of ethene by deuterium on a metal nanoparticle surface.

and reduction of alkylchloronitroanilines to the corresponding amines. Skeletal nickels and similar catalytically active metal alloys are produced by preparing a metal alloy such as NiAl at high temperature and then selectively dissolving most of the aluminium by treatment with sodium hydroxide. Other metals, including molybdenum and chromium, can be added to the original alloy and may act as promoters that affect the reactivity and selectivity of a catalyst for certain reactions. The resulting spongy or nanoporous metals are rich in nickel (>90%), and their high surface areas lead to very high catalytic activities. One further application of these catalysts is to the conversion of naturally occurring polyunsaturated fats, which are liquids, to solid polyhydrogenated fats, as in nondairy spreads.

(b) Heterogeneous catalysis using mesoporous-nanoparticle materials

KEY POINT The nanoscale properties of mesoporous materials and metal nanoparticles can be combined for enhanced heterogeneous catalytic activity.

Mesoporous nanomaterials, Chapter 29, also offer routes to functionalize pores to increase catalytic activity and selectivity. They have received much attention as host materials for the inclusion of guests such as organometallic complexes, polymers, d-metal complexes, macromolecules, and optical laser dyes. Mesoporous silicates have large ordered arrays of pores in the range 12–20 nm and generate very high specific surface areas (of over 1000 m^2 g^{-1}). The large pores allow larger molecules to undergo catalytic processes, although their acidity is weak compared with the zeolites. More importantly, other catalytic centres, such as nanoparticles of metals, alloys, and metal oxides including platinum or Pt/Sn, may be deposited within the mesostructured channels, Chapter 29.

As one example, Co nanoparticles deposited on the mesoporous silica support MCM-41 (MCM-41 stands for Mobil Crystalline Material type 41) promote the cycloaddition of alkynes with alkenes and carbon monoxide to produce cyclopentenones. This is an example of the Pauson–Khand reaction in which an alkene, an alkyne, and CO react together to form an unsaturated five-membered cyclic ketone:

31.13 Zeolites and microporous structures in heterogeneous catalysis

KEY POINTS Zeolites act as heterogeneous catalysts and have strongly acidic sites on their internal surfaces that promote reactions such as isomerization via carbonium ions; shape selectivity may arise at various stages of the reaction because of the relative dimensions of the zeolite channels and the reactant, intermediate, and product molecules.

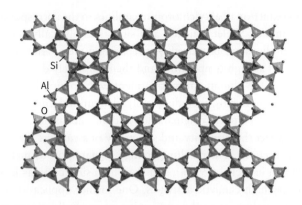

FIGURE 31.18 The zeolite faujasite (also known as zeolite X or Y) framework structure, showing the large pores in which catalytic cracking occurs. Tetrahedra are SiO_4 or AlO_4.

Zeolites (Sections 15.15, 25.16, 29.1, and 29.2) are important heterogeneous catalysts for the interconversion of hydrocarbons and the alkylation of aromatics, as well as in oxidation and reduction. Two commercially important zeolites used for reactions of this type are faujasite (Fig. 31.18), also known as zeolite X or zeolite Y (the X or Y terminology is defined by the Si:Al ratio of the material: X has a higher Al content), and zeolite ZSM-5, an aluminosilicate zeolite with a high Si content. (The catalyst was developed in the research laboratories of Mobil Oil; the initials denote Zeolite Socony-Mobil).

The channels of zeolites X and Y, and ZSM-5 (Fig. 31.19), consist of a three-dimensional maze of intersecting tunnels, and as with other aluminosilicate catalysts, such as clays, the presence of aluminium sites makes the zeolite strongly acidic.

The charge replacement of tetrahedrally coordinated Si(IV) by Al^{3+} requires the presence of an added positive ion in the pores. When this ion is H^+ (Fig. 31.20) the Brønsted acidity of the aluminosilicate can be higher than that of concentrated H_2SO_4 and the material is termed a *superacid* (Section 6.14); the turnover frequency (Section 2.14(c)) for hydrocarbon reactions at these sites can be very high. When the Brønsted acid form of the zeolite is dehydrated, catalytically active Lewis acid sites on aluminium are introduced. Four examples of zeolite-based heterogeneous catalysis are described.

(a) Catalytic cracking by zeolites and other microporous materials

Natural petroleum consists of only about 20% of alkanes suitable for use in petrol (gasoline) and diesel, with chain lengths ranging from C_5H_{12} (pentane) to $C_{12}H_{26}$. Conversion of the higher-molar-mass hydrocarbons to the valuable lighter ones involves not only breaking the C–C bonds, but also structural rearrangement of the hydrocarbons through dehydrogenation, isomerization, and aromatization reactions. All these processes are catalysed by **solid acid zeolite catalysts**. The principal zeolite used for catalytic cracking is zeolite Y, in which the extra-framework H^+ cations have been partially replaced with lanthanoid ions, typically a mixture of La, Ce, and Nd. The mechanism of the catalytic cracking initially involves protonation of the alkane or alkene chain by the Brønsted acid sites in the zeolite pores, followed by cleavage of the C–C bond in the β-position to the C atom carrying the positive charge. For example:

$$RCH_2C^+HCH_2(CH_2)_2R' \rightarrow RCH_2CH=CH_2 + {}^+CH_2CH_2R'$$

(b) Interconversion of aromatics by zeolites

Acidic zeolite catalysts also promote rearrangement reactions via carbonium ions. For example, the isomerization of 1,3-dimethylbenzene to 1,4-dimethylbenzene is believed to occur by the following steps:

Reactions such as dimethylbenzene (xylene) isomerization and methylbenzene (toluene) disproportionation illustrate the selectivity that can be achieved with acidic zeolite catalysis. The shape selectivity of these zeolite catalysts was previously

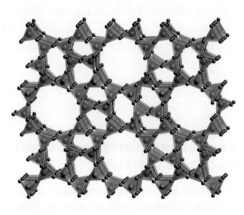

FIGURE 31.19 The zeolite ZSM-5 structure, highlighting the channels along which small molecules may diffuse. Tetrahedra are SiO_4 or AlO_4.

attributed to a variety of processes. In **reactant selectivity** the molecular sieving abilities of zeolites are important, as only molecules of appropriate size and shape can enter the zeolite pores and undergo a reaction. In **product selectivity**, a reaction-product molecule that has dimensions compatible with the channels will diffuse faster, allowing it to escape; molecules that do not fit easily within the channels diffuse slowly and, on account of their long residence in the zeolite, have ample opportunity to be converted to the more mobile isomers that can escape rapidly.

A currently more favoured view of zeolite selectivity is based on **transition-state selectivity**, where the orientation of reactive intermediates within the zeolite channels favours specific products. In the case of dimethylbenzene (xylene) isomerization, the narrower intermediates formed during the generation of 1,4-dialkylbenzene molecules are accommodated better within the pores. Another common reaction in zeolites is the alkylation of aromatics with alkenes.

> **EXAMPLE 31.3** Proposing a mechanism for the alkylation of benzene
>
> In its protonated form, ZSM-5 catalyses the reaction of ethene with benzene to produce ethylbenzene. Write a plausible mechanism for the reaction.
>
> **Answer** We should recall that protonated forms of zeolite catalysts are very strong acids. Therefore, a mechanism involving protonation of the organic species present in the system is the likely pathway. The acidic form of ZSM-5 is strong enough to generate carbocations from aliphatic hydrocarbons, so the initial stage would be
>
> $$CH_2CH_2 + H^+ \rightarrow CH_2CH_3^+$$
>
> The carbocation can attack benzene as a strong electrophile and subsequent deprotonation of the intermediate yields ethylbenzene:
>
> $$CH_2CH_3^+ + C_6H_6 \rightarrow C_6H_5CH_2CH_3 + H^+$$
>
> **Self-test 31.3** A pure silica analogue of ZSM-5 can be prepared. Would you expect this compound to be an active catalyst for benzene alkylation? Explain your reasoning.

The full range of porous inorganic framework materials (Chapter 29) is being investigated as potential heterogeneous catalysts, either in their own right or as hosts for active species. The large pores of pillared clays and MOFs makes it easier for these materials to accommodate additional species and, thereby, sites for heterogeneous catalysis. For example, the d-metal constituents of metal–organic frameworks (MOFs) can take part directly in redox reactions but these materials also have very large pores that can incorporate nanoparticles of metals, such as those of Pt (Section 29.4).

(c) Small-pore zeolite NOx reduction and methanol-to-alkene conversion

Catalysis using small-pore zeolites (with a maximum ring size of 8) have important applications in the reactions of small molecules. Two applications have been commercialized: NOx elimination from vehicle exhaust gases and the conversion of methanol to light alkenes.

The selective catalytic reduction, **SCR** or **deNOx**, process is used to remove NO from lean-burn diesel vehicle exhausts. The catalyst commercialized for this process is Cu-SSZ-13, which is a copper exchanged form of the aluminosilicate zeolite with the chabazite, CHA, structure topology. The catalytic reaction, Fig. 31.21, involves the copper centre reducing NO to N_2 using NH_3 as the reductant and oxygen to reinstate the active catalyst: ammonia is produced in the diesel engine exhaust from urea which is added to the fuel. The overall reaction occurring in the Cu-SSZ-13 catalyst is

$$4\,NH_3 + 4\,NO + O_2 \rightarrow 6\,H_2O + 4\,N_2$$

Methanol can be converted to ethene, propene, and butenes (**MTO**, methanol-to-olefin) using Cu-SAPO-34, a small-pore silico-alumino-phosphate with the chabazite (CHA) zeolite topology. Methanol can be obtained from biomass, shale gas, or coal by using the syngas process (Section 11.4). The catalytic processes in the small-pore zeolite at 350°C involves a series of stages, including the dehydration of two CH_3OH molecules to form dimethyl ether, followed by C–C bond formation.

(d) New directions in heterogeneous catalysis using zeolites and framework materials

The development of solid-phase catalysts is very much a frontier subject for inorganic chemistry with the continuing discovery of compositions for promoting reactions,

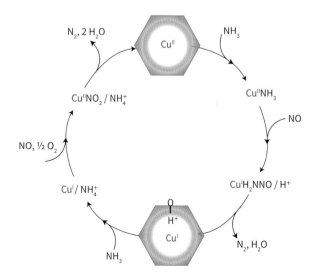

FIGURE 31.21 Catalytic cycle for the selective reduction of NO in diesel exhausts. The 6-ring windows in which the copper ions reside are shown schematically as hexagons.

particularly for petrochemicals. One very active area is the investigation of selective heterogeneous oxidation catalysts, which allow partial oxidation of hydrocarbons to useful intermediates in, for example, the polymer and pharmaceutical industries. Examples of such reactions include alkene epoxidation, aromatic hydroxylation, and ammoxidation (an oxidation in the presence of ammonia that generates nitriles) of alkanes, alkenes, and alkyl aromatics. In all these cases it is desirable to derive the products without complete oxidation of the hydrocarbon to carbon dioxide.

One example where new catalysts are being investigated is in the partial oxidation of benzene to phenol. At present, the three-step **cumene process** produces about 95% of the phenol used in the world and gives propanone (acetone) as a by-product, although the market for acetone is oversupplied from other industrial processes. The cumene process involves three stages: namely, alkylation of benzene with propene to form cumene (a process catalysed by phosphoric acid or aluminium chloride), direct oxidation of cumene to cumene hydroperoxide by using molecular oxygen, and finally, cleavage of cumene hydroperoxide to phenol and acetone, which is catalysed by sulfuric acid. Far better would be a single-stage process that accomplishes the reaction

$$C_6H_6 + \tfrac{1}{2}O_2 \rightarrow C_6H_5OH$$

Examples of catalysts investigated for this and similar processes include iron-containing zeolites with the silicalite framework type (Section 25.16), mixtures of $FeCl_3/SiO_2$, photocatalysts based on $Pt/H_2SO_4/TiO_2$, and various vanadium salts.

Heterogenized homogeneous and hybrid catalysis

Some catalytic systems cannot be defined as specifically homogeneous or heterogeneous. Research has centred on trying to achieve the best of both systems: the high selectivity of homogeneous catalysts coupled with the ease of separation of heterogeneous catalysts from the reaction products. This approach is sometimes referred to as 'heterogenizing' homogeneous catalysts.

31.14 Alkene oligomerization and polymerization

KEY POINTS Ethene can be oligomerized to linear alkenes by homogenous catalysis with a nickel catalyst. Heterogeneous Ziegler–Natta catalysts are used in alkene polymerization; the Cossee–Arlman mechanism describes their function; low molar mass homogeneous catalysts also catalyse the alkene polymerization reaction; considerable control over polymer tacticity is possible with judicious ligand design.

The development of alkene polymerization catalysts in the second half of the twentieth century, producing polymers such as polypropene and polystyrene, ushered in a revolution in construction materials, fabrics, and packaging. Polyalkenes are most often prepared by use of organometallic catalysts. The catalysts used can be homogeneous, heterogenized homogeneous, or heterogeneous; polymerization provides a good example of how homogenous catalysis has influenced the design of industrially important heterogeneous catalysts. All three types of catalyst are discussed in this section.

Ethene is readily available from natural gas and petroleum by steam cracking of heavier hydrocarbons. It can be converted to the much more valuable long-chain alkenes, sometimes still referred to as olefins, by processes such as the Shell Higher Olefin Process (SHOP). SHOP was developed for the conversion of ethene to C_{10}–C_{14} internal alkenes; that is, an alkene in which the double bond is not at the end of the carbon chain. The product alkenes are mainly converted into linear primary alcohols for use in detergents, but they can be modified to obtain linear alkenes of just about any range.

SHOP is a three-step process. The first step is homogeneously catalysed alkene **oligomerization** to form short chains of up to ten monomer units. The catalyst is generated *in situ* from bis(cyclooctadiene)nickel(0) and a bidentate phosphine carboxylate ligand. A nickel hydride complex is generated by displacement of the cycooctadiene by an incoming ethene molecule. An initial hydride shift (1,2-insertion reaction, Section 24.25) is followed by successive alkene migrations to build the oligomer:

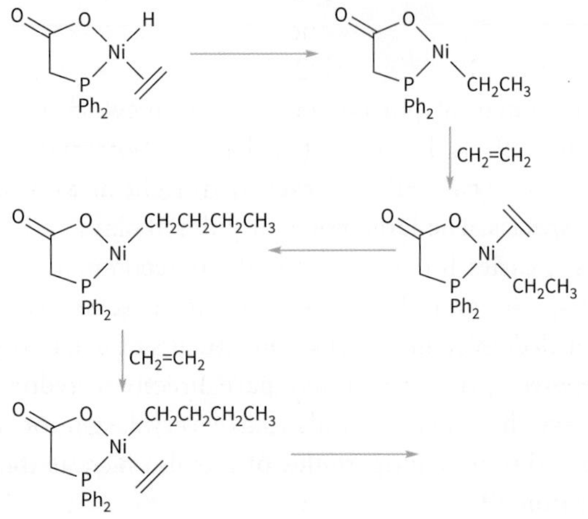

The hydrocarbon chain finally terminates by β-elimination to produce a terminal alkene:

[structure: Ni complex with CH₂CH₂(CH₂CH₂)ₙCH₂CH₃ chain] → [Ni-H complex] + CH₂=CH(CH₂CH₂)nCH₂CH₃

The products are linear α-alkenes (with the double bond between the first and second carbon atoms) with between four and twenty carbon atoms; they are separated by fractional distillation.

Whereas the SHOP system is not very selective (the products have to be separated), some other catalysts are very effective for selective oligomerization. For instance, a homogeneous chromium catalyst generated *in situ* by mixing a Cr(II) or Cr(III) halide with a so-called PNP phosphine (**13**), followed by activation with methylaluminoxane (MAO) under ethene, provides a system that is very active for trimerization of ethene, resulting in 99.9% 1-hexene. This system generates no polymer by-product, which is important in industry, where reactors need to be kept free from solid materials. Changing the PNP ligand can alter the products; $Ph_2PN(iPr)PPh_2$ generates 1-octene and 1-hexene as the major products.

13

In the 1950s, J. P. Hogan and R. L. Banks discovered that chromium oxides supported on silica—a so-called **Philips catalyst**—polymerized alkenes to long-chain polyenes. Also in the 1950s, K. Ziegler, working in Germany, developed a catalyst for ethene polymerization based on a catalyst formed from $TiCl_4$ and $Al(C_2H_5)_3$, and soon thereafter G. Natta in Italy used this type of catalyst for the stereospecific polymerization of propene (see below). Both the **Ziegler–Natta catalysts** and the chromium-based heterogeneous polymerization catalysts are widely used today.

The full details of the mechanism of Ziegler–Natta catalysts are still uncertain, but the **Cossee–Arlman mechanism** is regarded as highly plausible (Fig. 31.22). The catalyst is prepared from $TiCl_4$ and $Al(C_2H_5)_3$, which react to give polymeric $TiCl_3$ mixed with $AlCl_3$ in the form of a fine powder. The alkylaluminium alkylates a Ti atom on the surface of the solid, and an ethene molecule coordinates

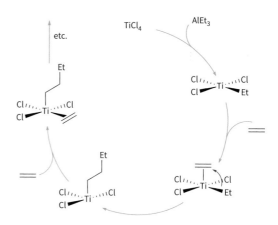

FIGURE 31.22 The Cossee–Arlman mechanism for the catalytic polymerization of ethene. Note that the Ti atoms are not discrete but are part of an extended structure containing bridging chlorides.

to the neighbouring vacant site. In the propagation steps for the polymerization, the coordinated alkene undergoes a migratory insertion reaction. This migration opens up another neighbouring vacancy, and so the reaction can continue and the polymer chain can grow. The release of the polymer from the metal atom occurs by β-hydrogen elimination, and the chain is terminated. Some catalyst remains in the polymer, but the process is so efficient that the amount is negligible.

The proposed mechanism of alkene polymerization on a Philips catalyst involves the initial coordination of one or more alkene molecules to a surface Cr(II) site, followed by rearrangement to metallocycloalkanes on a formally Cr(IV) site. Unlike Ziegler–Natta catalysts, the solid-phase catalyst does not need an alkylating agent to initiate the polymerization reaction; instead, this species is thought to be generated by the metallocycloalkane directly or by formation of an ethenylhydride by cleavage of a C–H bond at the chromium site.

Homogeneous catalysts related to the Philips and Ziegler–Natta catalysts provide additional insight into the course of the reaction and are of considerable industrial significance in their own right, being used commercially for the synthesis of specialized polymers. These **Kaminsky** catalysts utilize metals from Group 4 (Ti, Zr, Hf) and are based on a bis(cyclopentadienyl) metal system: the tilted ring complex $[Zr(\eta^5-Cp)_2(CH_3)L]^+$ (**14**) is a good example. These Group 4 metallocene complexes catalyse alkene polymerization by successive insertion steps that involve prior coordination of the alkene to the electrophilic metal centre. Catalysts of this type are used in the presence of a co-catalyst, the so-called methylaluminoxane (MAO), a poorly defined compound of approximate formula $(MeAlO)_n$, which, among other functions, serves to methylate a starting chloride complex. Kaminsky catalysts can also be supported on silica and are used industrially for the polymerization of α-alkenes and styrene. They are thus examples of **heterogenized homogeneous catalysts**.

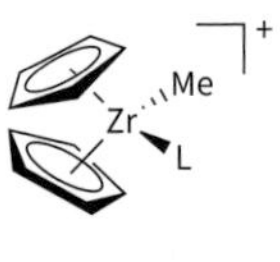

14

Additional complications arise with alkenes other than ethene. We shall discuss only terminal alkenes such as propene and styrene, as these are relatively simple. The first complication to consider arises because the two ends of the alkene molecule are different. In principle, it is possible for the polymer to form with the different ends head-to-head (**15**), head-to-tail (**16**), or randomly.

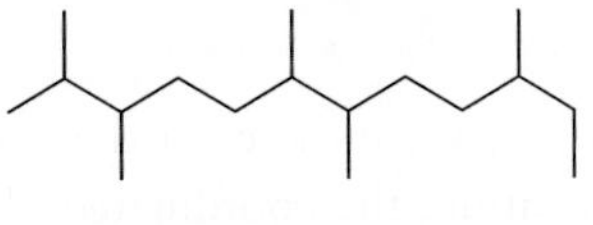

15

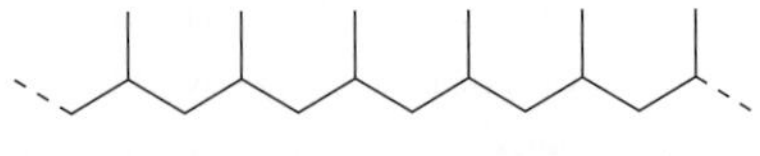

16

Studies on catalysts such as (**14**) show that the growing chain migrates preferentially to the more highly substituted C atom of the alkene, thus giving a polymer chain that contains only head-to-tail orientations:

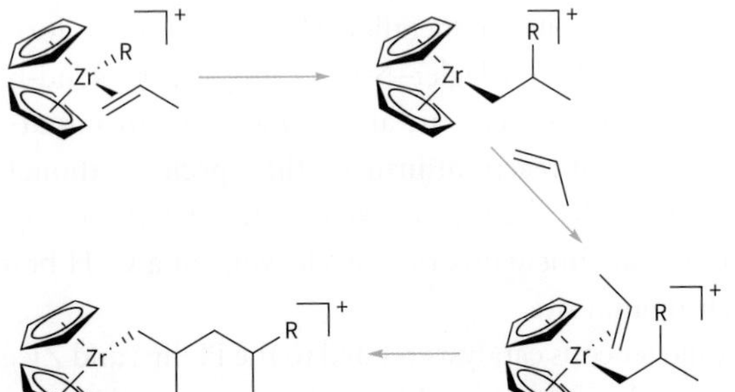

If we consider propene, we can see that the coordinated alkene induces less steric strain if its smaller CH_2 end is pointing into the cleft of the $(Cp)_2Zr$ catalyst (**17**), rather than the larger methyl-substituted end (**18**). The migrating polymer chain is thus adjacent to the methyl-substituted end of the propene molecule, and it is to this methyl-substituted C atom that the chain attaches, giving a head-to-tail sequence to the whole polymer chain.

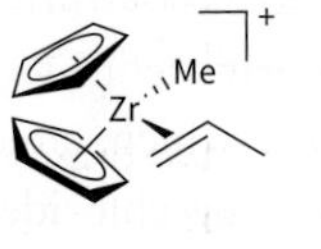

17 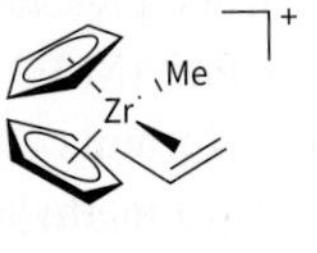   **18**

The second structural modification of polypropene is its **tacticity**, the relative orientations of neighbouring groups in the polymer. In a regular **isotactic** polypropene, all the methyl groups are on the same side of the polymer backbone (**19**). In regular **syndiotactic** polypropene, the orientation of the methyl groups alternates along the polymer chain (**20**). In an **atactic** polypropene, the orientation of neighbouring methyl groups is random (**21**). Control of the tacticity of a polymer is equivalent to controlling the stereospecificity of the reaction steps. The orientation of neighbouring groups is not simply of academic interest because the orientation has a significant effect on the properties of the bulk polymer. For example, the melting points of isotactic, syndiotactic, and atactic polypropene are 165°C, 130°C, and below 120°C, respectively.

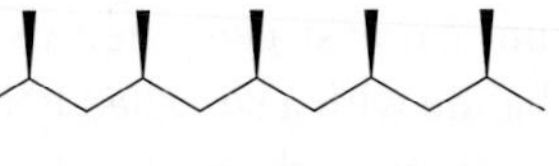

19

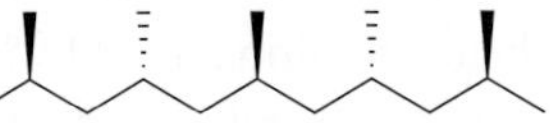

20

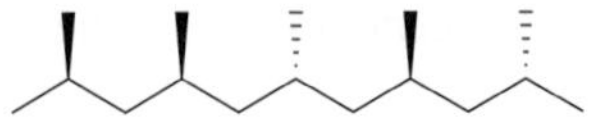

21

It is not possible to control the tacticity of polypropene with a Zr catalyst such as (**17**), and an atactic polymer results. However, with other catalysts it is possible to control the tacticity. The type of catalyst normally used to control the tacticity has a metal atom bonded to two indenyl groups that are linked by a CH_2CH_2 bridge. Reaction of the bis(indenyl) fragment with a metal salt gives rise to three compounds: two enantiomers (**22**) and (**23**), which have C_2 symmetry, and a nonchiral compound (**24**). These compounds are called *ansa*-metallocenes (derived from the Latin for 'handle' and used to indicate a bridge). It is possible to separate the two enantiomers from the nonchiral compounds, and both enantiomers of these *ansa*-metallocenes can catalyse the stereoregular polymerization of propene.

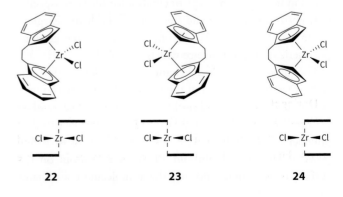

22   **23**   **24**

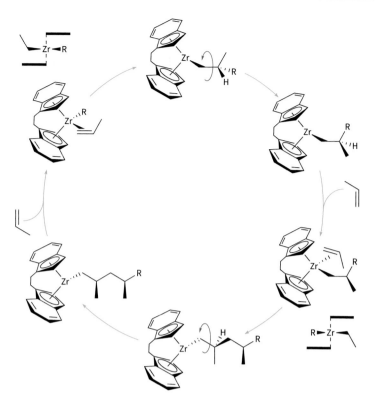

FIGURE 31.23 When propene is polymerized with an indenyl metallocene catalyst, isotactic polypropene results. The zirconium species all have a single positive charge, but here this has been omitted for clarity.

If we now consider the coordination of propene to one of the enantiomeric compounds (**25**) or (**26**) there is a second constraint in addition to the steric factor mentioned above (the CH_2 group pointing towards the cleft). Of the two potential arrangements of the methyl group shown in (**25** and **26**), the latter is disfavoured by a steric interaction with the phenyl ring of the indenyl group.

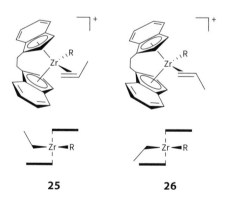

During the polymerization reaction, the R group migrates preferentially to one side of the propene molecule; coordination of another alkene is then followed by migration, and so on. Figure 31.23 shows how an isotactic polypropene then results.

EXAMPLE 31.4 **Controlling the tacticity of polypropene**

Show that polymerization of propene with a catalyst containing CH_2CH_2-linked fluorenyl and cyclopentadienyl groups (**27**) should result in syndiotactic polypropene.

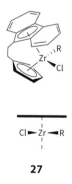

27

Answer We need to consider how the propene reactant will coordinate to the catalyst: in complex (**27**) a coordinated propene will always coordinate with the methyl group pointing away from the fluorenyl and towards the cyclopentadienyl ring. A series of sequential alkene insertions, as outlined in Fig. 31.24, will therefore lead to a product that should be syndiotactic.

Self-test 31.4 Demonstrate that polymerization of propene with a simple $[Zr(Cp)_2Cl_2]$ catalyst would give rise to atactic polypropene.

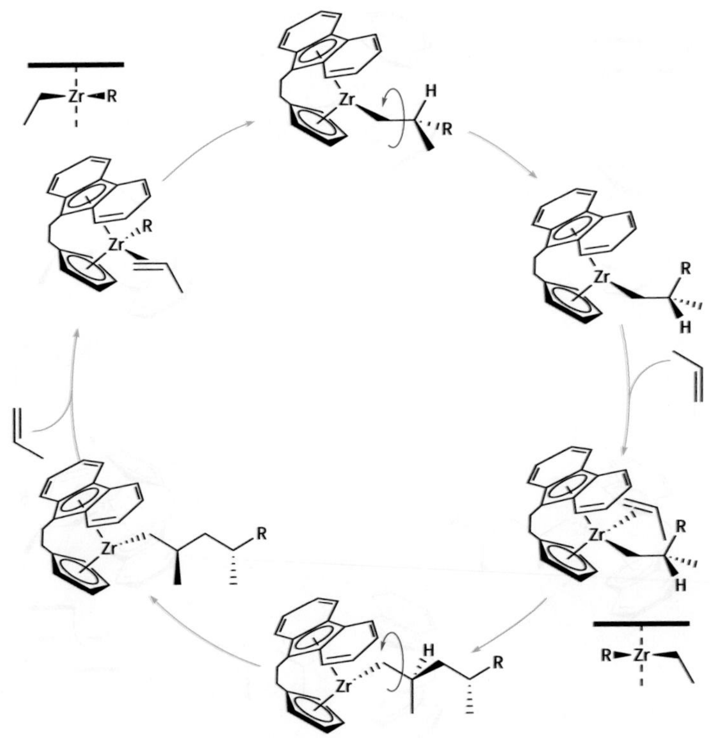

FIGURE 31.24 When a propene is polymerized with a fluorenyl metallocene catalyst, syndiotactic polypropene results. The zirconium species all have a single positive charge, which here is omitted for clarity.

Although the Group 4 metal catalysts are by far the most common for alkene polymerization, active catalysts based on other transition metals and lanthanoids have been reported. Polymerization with Group 5 metal complexes has been limited because of their thermal instability, but careful ligand design has enabled synthesis of catalysts (**28**) that are thermally stable and highly active for ethene polymerization in the presence of organoaluminium chloride co-catalysts. Catalytic activity of the Pd- and Ni-diimine complexes (**29**) is affected by the N-aryl groups of the ligand, and electron-donating groups on the aryl group stabilize the cationic metal centre to give a high-molecular-weight polymer. A variety of neodymium-based catalysts are used in industry for the polymerization of 1,3-butadiene.

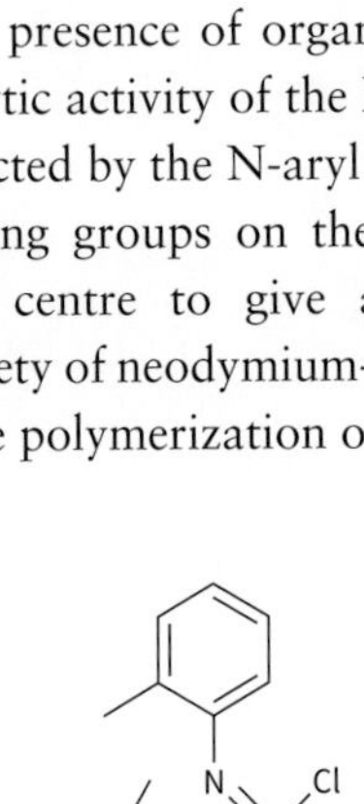

31.15 Tethered catalysts

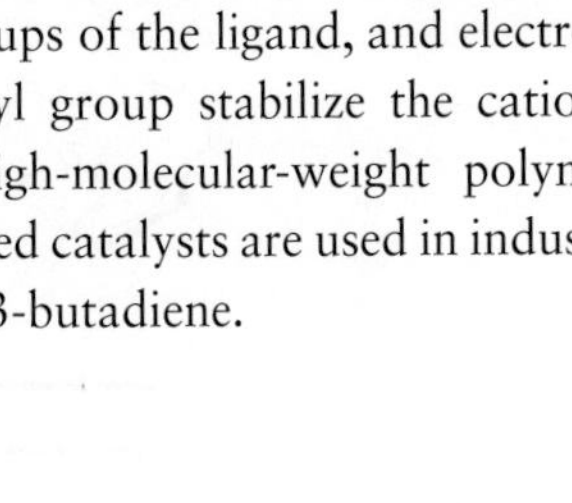

KEY POINT The tethering of a catalyst to a solid support allows easy separation of the catalyst with little loss in catalyst activity.

One popular technique has been the **tethering** of a homogeneous catalyst to a solid support. Thus, a hydrogenation catalyst such as Wilkinson's catalyst can be attached to a silica surface by means of a long hydrocarbon chain:

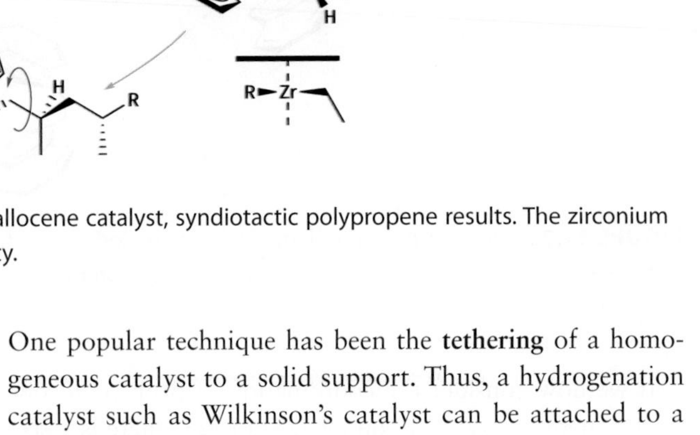

When the silica monolith is immersed in a solvent, the rhodium-based catalytic site behaves as though it is in solution, and reactivity is largely unaffected. Separation of the products from the catalyst simply requires the decanting of the solvent. Commercially available functionalized silica precursors include amino-, acrylate-, allyl-, benzyl-, bromo-, chloro-, cyano-, hydroxy-, iodo-, phenyl-, styryl-, and vinyl-substituted reagents. Relatively simple reactions can result in the synthesis of a whole host of further reagents (such as the phosphine compound in the scheme above). In addition to silica, polystyrene, polyethene, polypropene, and various clays have been used as the solid support and have led to reports of the successful heterogenization of most reactions that rely on soluble metal complexes.

In a recent novel application of tethered catalysis, a dicationic ionic liquid was tethered to superparamagnetic iron oxide nanoparticles as a catalyst for the efficient synthesis of Betti bases (**30**) which are attractive chiral ligands in enantioselective reactions as well as being precursors for the synthesis of molecules being evaluated for their effects on bradycardia and hypotension in humans. Due to the magnetic properties of the catalyst, it can be successfully recovered by a simple external magnet, removing the need for filtration.

30

In some cases the activity of a supported catalyst is greater than that of its unsupported analogue. This improvement normally takes the form of enhanced selectivity brought about by the steric demands of approaching a catalyst constrained to a surface or an increase in catalyst turnover frequencies brought about by protection from the support. Often, however, supported catalysts suffer from catalyst leaching and reduced activity.

31.16 Biphasic systems

KEY POINT Biphasic systems offer another way of combining the selectivity of homogeneous catalysts with the ease of separation of heterogeneous catalysts.

Another popular method of attempting to combine the best of both homogeneous and heterogeneous catalysts has been the use of two liquid phases that are not miscible at room temperature but are miscible at higher temperatures. The existence of immiscible aqueous/organic phases together with a compound that facilitates the transfer of reagents between the two phases will be familiar; two other systems worthy of note are ionic liquids and the 'fluorous biphase' system.

Ionic liquids, introduced in Section 5.15(e), are typically derived from 1,3-dialkylimidazolium cations (**31**) with counter-ions such as PF_6^-, BF_4^-, and $CF_3SO_3^-$. These systems have melting points of less than (and often much less than) 100°C, a very low viscosity, and an effectively zero vapour pressure. If it can be arranged that the catalysts are preferentially soluble in the ionic phase (for instance, by making them ionic), immiscible organic solvents can be used to extract organic products. As an example, consider the hydroformylation of alkenes by an Rh catalyst with phosphine ligands. When the ligand used is triphenylphosphine, the catalyst is extracted from the ionic liquid together with the products. However, when an ionic sulfonated triphenylphosphine is used as the ligand, the catalyst remains in the ionic liquid, and product separation from the catalyst is complete. It should be borne in mind that the ionic liquid phase is not always unreactive, and may induce alternative reactions.

31

The fluorous biphase system, which typically consists of a fluorinated hydrocarbon and a 'normal' organic solvent such as methylbenzene (toluene), offers two principal advantages over aqueous/organic phases: the fluorous phase is unreactive, so sensitive groups (such as hydrolytically unstable groups) are stable in it, and polyfluorinated and hydrocarbon solvents, which are not miscible at room temperatures, become so on heating, giving rise to a genuinely homogeneous system. A catalyst that contains polyfluorinated groups is preferentially retained in the fluorous solvent, with the reactants (and products) preferentially soluble in the hydrocarbon phase. Figure 31.25 indicates the type of sequence that is used.

Separation of the products from the catalyst becomes a trivial matter of decanting one liquid from another. A number of ligand systems that confer fluorous solubility on a catalyst have been developed; typically they are phosphine-based, such as (**32**), (**33**), and (**34**).

32

33

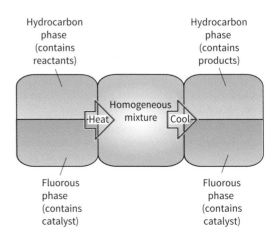

FIGURE 31.25 The sequence of processes that occurs during the use of a fluorous biphase catalyst.

34

Rhodium- and Pd-based catalysts of these ligands have then been prepared. With a fluorous solvent such as perfluoro-1,3-dimethylcyclohexane (**35**), miscibility with an organic phase can be achieved at 70°C, and catalysts have been used in hydrogenation (Section 31.2), hydroformylation (Section 31.3), and hydroboration reactions.

35

More recent developments in this area include the use of gold nanoparticles stabilized by fluorous ligands such as (**36**) and $C_9F_{19}CH_2CH_2$–SH, which were able to catalyse the selective oxidation of alkenes in a biphasic system, with no loss of catalytic activity, even after multiple cycles.

36

FURTHER READING

T. M. Trnka and R. H. Grubbs, The development of L2X2Ru=CHR olefin metathesis catalysts: an organometallic success story. *Acc. Chem. Res.*, 2001, **34**, 18. A review of the development of alkene metathesis catalysts.

P. Espinet and A. M. Echavarren, The mechanisms of the Stille reaction. *Angew. Chem., Int. Ed. Engl.*, 2004, **43**, 4704. A good review of the Stille reaction that touches on the mechanism of all palladium-catalysed coupling reactions.

H. H. Brintzinger, D. Fischer, R. Mülhaupt, B. Rieger, and R. M. Waymouth, Stereospecific olefin polymerization with chiral metallocene catalysts. *Angew. Chem., Int. Ed. Engl.*, 1995, **34**, 1143. A review of the area of control of polymer tacticity.

R. Prins, A. Wang, X. Li, and F. Sapountzi, *Introduction to Heterogeneous Catalysis (Advanced Textbooks in Chemistry)*, 2nd ed. World Scientific Publishing Company (2022).

M. R. Cesario and D. A. de Macedo, *Heterogeneous Catalysis: Materials and Applications*. Elsevier (2022).

J. M. Thomas and W. J. Thomas, *Principles and Practice of Heterogeneous Catalysis*, 2nd ed. Wiley-VCH (2014).

J. R. H. Ross, *Heterogeneous Catalysis: Fundamentals and Applications*. Elsevier (2011).

EXERCISES

31.1 Which of the following constitute genuine examples of catalysis, and which do not? Present your reasoning. (a) The addition of H_2 to C_2H_4 when the mixture is brought into contact with finely divided platinum. (b) The reaction of an H_2/O_2 gas mixture when an electrical arc is struck. (c) The combination of N_2 gas with lithium metal to produce Li_3N, which then reacts with H_2O to produce NH_3 and LiOH.

31.2 Define the terms (a) turnover frequency, (b) selectivity, (c) catalyst, (d) catalytic cycle, (e) catalyst support.

31.3 Classify the following as homogeneous or heterogeneous catalysis, and present your reasoning. (a) The increased rate in the presence of NO(g) of SO_2(g) oxidation by O_2(g) to SO_3(g). (b) The hydrogenation of liquid vegetable oil using a finely divided nickel catalyst. (c) The conversion of an aqueous solution of D-glucose to a D,L mixture catalysed by HCl(aq).

31.4 You are approached by an industrialist with the proposition that you develop catalysts for the following processes at 80°C with no input of electrical energy or electromagnetic radiation:

(a) The splitting of water into H_2 and O_2.
(b) The decomposition of CO_2 into C and O_2.
(c) The combination of N_2 with H_2 to produce NH_3.
(d) The hydrogenation of the double bonds in vegetable oil.

The industrialist's company will build the plant to carry out the process, and the two of you will share equally in the profits. Which of these would be easy to do, which are plausible candidates for investigation, and which are unreasonable? Describe the chemical basis for the decision in each case.

31.5 Addition of PPh_3 to a solution of Wilkinson's catalyst, $[RhCl(PPh_3)_3]$, reduces the turnover frequency for the hydrogenation of propene. Give a plausible mechanistic explanation for this observation.

31.6 The rates of H_2 gas absorption (in dm^3 mol^{-1} s^{-1}) by alkenes catalysed by $[RhCl(PPh_3)_3]$ in benzene at 25°C are: hexene, 2910; *cis*-4-methyl-2-pentene, 990; cyclohexene, 3160; 1-methylcyclo-hexene, 60. Suggest the origin of the trends, and identify the affected reaction step in the proposed mechanism (Fig. 31.1).

31.7 Sketch the catalytic cycle for the production of butanal from prop-1-ene. Identify the step at which selectivity to the *n* or *iso* isomer occurs.

31.8 Draw the catalytic cycle for the Ziegler–Natta polymerization of propene. Explain each of the steps involved, and predict what the physical properties of the polymer produced would be like.

31.9 Infrared spectroscopic investigation of a mixture of CO, H_2, and 1-butene under conditions that bring about hydroformylation indicate the presence of compound (E) in Fig. 31.3 in the reaction mixture. The same reacting mixture in the presence of added tributylphosphine was studied by infrared spectroscopy and neither (E) nor an analogous phosphine-substituted complex was observed. What does the first observation suggest as the rate-limiting reaction in the absence of phosphine? Assuming that the sequence of reactions remains unchanged, what are the possible rate-limiting reactions in the presence of tributylphosphine?

31.10 Show how reaction of MeCOOMe with CO under conditions of the Monsanto ethanoic acid process can lead to ethanoic anhydride.

31.11 Suggest reasons why (a) ring-opening alkene metathesis polymerization and (b) ring-closing metathesis reactions proceed.

31.12 (a) Starting with the alkene complex shown in Fig. 31.5 with *trans*-DHC=CHD in place of C_2H_4, assume that dissolved OH⁻ attacks from the side opposite the metal. Produce a stereochemical drawing of the resulting compound. (b) Assume attack on the coordinated *trans*-DHC=CHD by an OH⁻ ligand coordinated to Pd, and draw the stereochemistry of the resulting compound. (c) Does the stereochemistry differentiate these proposed steps in the Wacker process?

31.13 Why is the platinum/rhodium catalyst in automobile catalytic converters dispersed on the surface of a ceramic rather than used in the form of a thin metal foil?

31.14 Alkanes are observed to exchange hydrogen atoms with deuterium gas over some platinum metal catalysts. When 3,3-dimethylpentane in the presence of D_2 is exposed to a platinum catalyst and the gases are observed before the reaction has proceeded very far, the main product is $CH_3CH_2C(CH_3)_2CD_2CD_3$ plus unreacted 3,3-dimethylpentane. Devise a plausible mechanism to explain this observation.

31.15 The effectiveness of platinum in catalysing the reaction $2\,H^+(aq) + 2\,e^- \rightarrow H_2(g)$ is greatly decreased in the presence of CO. Suggest an explanation.

31.16 Consider the validity of each of the following statements, and provide corrections where required. (a) A catalyst introduces a new reaction pathway with lower enthalpy of activation. (b) As the Gibbs energy is more favourable for a catalytic reaction, yields of the product are increased by catalysis. (c) An example of a homogeneous catalyst is the Ziegler–Natta catalyst made from $TiCl_4(l)$ and $Al(C_2H_5)_3(l)$. (d) Highly favourable Gibbs energies for the attachment of reactants and products to a homogeneous or heterogeneous catalyst are the key to high catalytic activity.

TUTORIAL PROBLEMS

31.1 A catalyst might not just lower the enthalpy of activation, but might make a significant change to the entropy of activation. Discuss this phenomenon. (See A. Haim, *J. Chem., Educ.*, 1989, **66**, 935.)

31.2 When direct evidence for a mechanism is not available, chemists frequently invoke analogies with similar systems. Describe how J. E. Bäckvall, B. Åkermark, and S. O. Ljunggren (*J. Am. Chem. Soc.*, 1979, **101**, 2411) inferred the attack of uncoordinated water on η^2-C_2H_4 in the Wacker process.

31.3 The addition of promoters can further enhance the rate of a catalysed reaction. Describe how the promoters allowed the iridium-based Cativa process to compete with the rhodium-based process in the carbonylation of methanol. (See A. Haynes, P. M. Maitlis, G. E. Morris, G. J. Sunley, H. Adams, and P. W. Badger, *J. Am. Chem. Soc.*, 2004, **126**, 2847.)

31.4 Discuss the advantages and disadvantages of the homogeneous Grubbs-type catalysts, compared with the heterogeneous Schrock-type catalysts, in the alkene metathesis reaction. Consider both small-scale laboratory-based reactions and large-scale industrial processes.

31.5 Whereas many enantioselective catalysts require the precoordination of a substrate, this is not always the case. Use the example of asymmetric epoxidation to demonstrate the validity of this statement, and indicate the advantages that such catalysts might have over catalysts that require precoordination of a substrate. (See M. Palucki, N. S. Finney, P. J. Pospisil, M. L. Güler, T. Ishida, and E. N. Jacobsen, *J. Am. Chem. Soc.*, 1998, **120**, 948.)

31.6 Discuss shape selectivity with respect to catalytic processes involving zeolites.

31.7 Discuss the importance of the Haber–Bosch process in relation to global food supplies. Make reference to estimates of the proportion of worldwide energy consumption used by the process.

31.8 Discuss whether ammonia synthesis can ever become truly sustainable. See S. Ghavam, M. Vahdati, I. A. G. Wilson, and P. Styring, *Front. Energy Res.* 2021, **9**:580808. doi: 10.3389/fenrg.2021.580808.

31.9 Discuss the advantages of a solid support in catalysis by reference to the use of $(Ni(POEt)_3)_4$ in alkene isomerization. (See A. J. Seen, *J. Chem. Educ.*, 2004, **81**, 383, and K. R. Birdwhistell and J. Lanza, *J. Chem. Educ.*, 1997, **74**, 579.)

31.10 J. A. Bota et al. discuss the catalytic conversion of vegetable oils into hydrocarbons suitable for use as biofuels (*Catalysis Today*, 2012, **195**, 1, 59). What are the most important features of catalysts that are used for these reactions? How was the incorporation of transition metals expected to modify catalyst properties? Outline how the modified catalysts were prepared and characterized. What reactions occurred in the reactor in addition to catalytic cracking? Which reactions lead to aromatic products? Which of the modified catalysts produced most coke build-up? Explain why this build-up did not deactivate the catalyst.

31.11 β-Pinene is a major component of natural turpentine, and its polymer is nontoxic and is used in a wide variety of industrial applications such as adhesives, varnishes, food packaging, and chewing gum. Novel catalysts for the polymerization of β-pinene have been produced based on niobium and tantalum pentahalides (M. Hayatifar et al., *Catalysis Today*, 2012, **192**, 1, 177). Which catalyst was the most effective for the polymerization, and why are the reaction conditions so attractive? Describe the nature of the catalysts–β-pinene interaction.

31.12 A. Arbaoui and C. Redshaw (*Polym. Chem.*, 2010, **1**, 801) review catalysts for the synthesis of biodegradable polymers via ring-opening metathesis polymerization. Summarize the need for biodegradable polymers and why new catalysts are required. From the details provided, identify which group of metals produce the most active catalysts with which types of ligands, and illustrate with examples.

Inorganic chemistry in medicine

32

The chemistry of elements in medicine
- 32.1 Inorganic compounds in cancer treatment
- 32.2 Anti-arthritis drugs
- 32.3 Bismuth pharmaceuticals
- 32.4 Lithium in the treatment of bipolar disorders
- 32.5 Involvement of metal complexes in the cause and treatment of malaria
- 32.6 Metal complexes as antiviral agents
- 32.7 Metal drugs that slowly release CO: an agent against post-operative stress
- 32.8 Chelation therapy
- 32.9 Imaging agents
- 32.10 Nanoparticles in directed drug delivery
- 32.11 Outlook

Further reading

Exercises

Tutorial problems

Medicine exploits several elements that are not normally used by biology and commonly considered to be poisons. These elements range across almost the entire periodic table, from lithium to bismuth. Taking a new drug to market is a lengthy and expensive process. Drug discovery is often serendipitous, a classic example being the seemingly unrelated experiments that led to cisplatin (and thereafter) other Pt complexes as cures for many types of cancer. Drugs interfere with biological targets, causing them to be suppressed or destroyed, so a key factor is to ensure that this action is directed selectively to diseased tissue. Metal-containing drugs may undergo numerous chemical changes en route to their molecular targets, making it very difficult to establish how they work. Inorganic complexes may show stereochemical diversity at a single site that is not possible with carbon, leading to important opportunities for drug design. Aside from drugs themselves, inorganic compounds are also important in diagnosis.

The chemistry of elements in medicine

Serendipity has played an important role in drug discovery, with many effective treatments arising from chance discoveries. There appears to be a special role for compounds containing metals that are not otherwise used by biological systems, for example Li, Pt, Au, Ag, Ru, As, and Bi. Many treatments involve killing invader cells be they bacteria, parasites, or cancers, and it should come as no surprise that a compound or a component element that is foreign to a living cell may, if able to penetrate the cell, be very effective in displaying *cytotoxicity*—the property of being lethal to that cell. Penetrating the cell membrane is rarely a specific stage of the attack process, and 'close relatives' (mimics) of an element or molecule often enter the target cell with little difficulty. Then, like the Trojan Horse, the more specific characteristics of an element are brought into play for the 'killing'. Figure 32.1 shows a periodic table that depicts the major 'inorganic' elements responsible for the specific activity of drugs that are either already in clinical practice or under intense investigation.

In addition to interfering chemically with targets in microbes or cancer cells, an increasing range of compounds deliver radioactive nuclei to detect and/or destroy diseased tissue by radiation. Certain elements and their compounds have desirable properties that enable them to replace parts of the body lost to injury or disease. Examples include polyphosphazenes (Box 16.6), which are exploited when biodegradability is important (such as temporary support of tissue repairs), the use of zirconia, ZrO_2, and doped derivatives in dental crowns, and titanium in joint replacements (Box 32.1).

A much-adapted objective in pharmacology is 'to cure the disease without killing the patient'; drugs tend to be

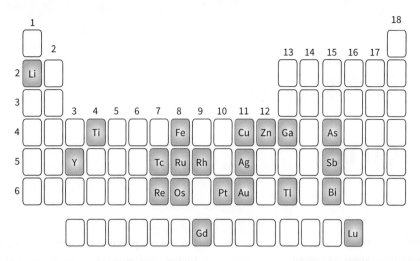

FIGURE 32.1 Periodic table showing the main elements in drugs or diagnostic agents, either in clinical use or attracting strong interest.

BOX 32.1 Using inorganic materials to replace solid tissue. The use of titanium in artificial joints.

Osteoarthritis—a painful condition affecting the joints—is the result of prolonged damage to the protective cartilage lining the ends of bones, arising either from injury or another condition such as rheumatoid arthritis. In most cases, the symptoms can be relieved by changes to lifestyle (exercise, weight loss) or painkillers, but in more serious cases the best remedy is surgery, by which the affected joint is repaired or replaced by an artificial version, known as a prosthesis. Hip and knee replacements are routine hospital procedures.

Titanium is by far the most important material used in artificial implants to replace joint parts or roots of teeth (dental implants). An X-ray photograph of a total knee replacement is shown in Fig. B32.1. The femoral (upper) and tibial (lower) implants are carefully aligned and fixed in place using a cement or (increasingly) a thin layer of porous tantalum.

Titanium has several important properties that enable it to substitute for bone. In the first place it is light, strong, corrosion-resistant, nontoxic, and widely available. More specifically, titanium has excellent biocompatibility (Fig. B32.2). The upper and middle panels show how titanium interacts with bone

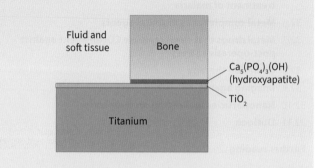

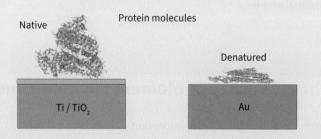

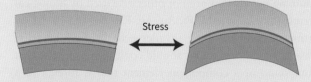

FIGURE B32.2 The biocompatibility of titanium.

tissue, or with soft tissue and body fluids, the crucial factor being that the metal surface comprises an extremely thin layer of TiO_2. In a process known as osseointegration, the oxide coating forms bonds with hydroxyapatite ($Ca_5(PO_4)_3(OH)$), the inorganic component of bone; this in turn ultimately merges

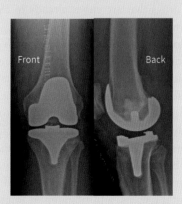

FIGURE B32.1 X-ray photograph of a titanium knee replacement.

with bone itself, which is a composite material. The TiO$_2$ coating is hydrophilic and forms a benign interface with body fluids and soft tissue. Proteins such as serum albumin tend to preserve their native conformation, whereas adsorbed on a more hydrophobic surface, such as gold, they unfold to displace water molecules and present soft donors such as cysteine-S to the metal. The lower panel represents the mechanical compatibility between Ti and bone—the similarity in elasticity (quantified in terms of Young's modulus) ensuring a concordant response when stress is applied.

Similar principles apply wherever inorganic materials are used to replace worn or damaged solid tissue, including teeth, for which ZrO$_2$ has properties making it particularly suitable for dental implants.

'blunt tools', damaging healthy as well as infected tissues. A major challenge for developing 'smart' drugs lies in elucidating the mechanism of action at the molecular level, bearing in mind that the drug that is administered may differ from the molecule that ultimately reacts at the target site, due to it undergoing modification in the body fluids. This gap in knowledge is particularly true for metal complexes, which are usually more susceptible to hydrolysis than organic molecules and may be more easily modified by oxidants or reductants present in the tissue.

The precursor is often referred to as a **prodrug**, and various options for its transformation to the active agent that has reached its final target zone are represented in Fig. 32.2. The coordination shell may remain intact, one or more ligands may be released to expose a binding site, the entire coordination shell may be jettisoned to release an active metal ion, or the metal may be removed to yield an active ligand. Linking one or more biologically active molecules to a scaffold or metal complex is known as **bioconjugation**.

In general, a plausible mechanism of action is proposed based upon an extrapolation of *in vitro* studies, which includes using the analytical strategy of 'metallomics', as explained in Section 26.2. Orally administered drugs are highly desirable because they avoid the trauma and potential hazards of injection; however, they may not pass through the gut wall or survive hydrolytic enzyme action. Inorganic compounds are also used in the diagnosis of disease or damage, a particularly important example being the exploitation of radioactive technetium, an artificial element, to image injured or diseased tissue. Water solubility and O$_2$ stability are important issues, yet organometallic compounds, usually associated with anaerobic organic solvents, are finding increasing applications as drugs. The potency of a particular drug is gauged by its IC$_{50}$ value, which is the molar concentration required to achieve 50% inhibition of a certain biological activity. A drug is often administered as part of a package (a regimen) that includes other agents.

Drug development is a lengthy and costly venture, and a typical timeline is shown in Fig. 32.3. The first stage is discovery, which begins with identifying a *target* compound and then synthesizing and testing this compound along with

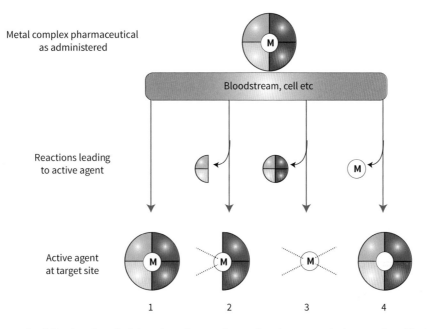

FIGURE 32.2 Possible scenarios following the administration of a metal-complex pharmaceutical to a patient. The final active agent may be (1) the intact complex, (2) a reactive fragment formed upon dissociation of label ligands, (3) the metal ion (ligand 'chaperone' is jettisoned), (4) the ligand (the metal ion 'chaperone' is jettisoned).

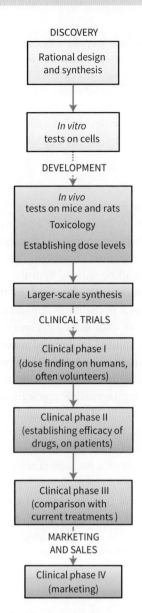

FIGURE 32.3 Timeline for drug development.

closely related analogues. The next stage is development, which investigates whether the compound is safe, and in what form or amount it is likely to be administered to a patient. The third stage is clinical trials, of which there are three scientific phases: in Phase I the new drug is administered to humans, usually healthy volunteers, to determine how it is tolerated and whether it behaves as expected from earlier experiments; in Phase II, the new compound is tested on a small number of patients having the illness, for which the drug is intended to benefit; in Phase III, any encouraging compounds are compared within a much larger patient group, alongside other medicines that are already being used to treat the condition and a dummy substance known as a 'placebo'. If at the end of Phase III a compound has proved successful, a licence application is made and the drug proceeds to Phase IV—the marketing and sales

through which the pharmaceutical company capitalizes on its investment. This chapter describes compounds that are currently approved and marketed in several countries, as well as others that illustrate important strategies despite not being in clinical use.

32.1 Inorganic compounds in cancer treatment

KEY POINTS The great success of cis-$[PtCl_2(NH_3)_2]$ (cisplatin) in treating many cancers is due to its ability to bind to DNA, preventing its replication and halting uncontrolled cell division and proliferation. Other Pt complexes having fewer serious side effects have followed. There is great interest in developing drugs that will destroy only cancerous cells, and some ingenious chemistry is involved.

'Cancer' is a term that covers a large number of different types of the disease, all characterized by the uncontrolled replication of transformed cells that overwhelm the normal operation of the body. Ideally, the principle of the treatment is to apply drugs that destroy malignant cells while leaving healthy cells unharmed. For many years, the main focus has been on anticancer drugs that attack DNA—a rather nonspecific target—with efforts being made to help ensure that cancerous cells rather than healthy cells are preferentially destroyed. Recognition, where successfully achieved, has thus been at the tissue/organ level rather than the molecular level. More recently, research is being directed on specific downstream products, such as RNAs and proteins, that are produced by certain cancer cells and offer more selective targets.

Arsenic-containing compounds (arsenicals) have an important place in medical history. Although toxic and carcinogenic, arsenic was used in ancient Chinese medicine and has long been used to combat cancer and other serious diseases. As described in Chapter 16, arsenic resembles phosphorus, but the main oxidation state is As(III), and being a softer Lewis acid it forms more stable compounds with S-donor ligands. The oxide 'As_2O_3', which consists of As_4O_6 molecules (**1**) is an effective drug against acute promyelocytic leukaemia (APL)—a disease that was once considered to be incurable. The regimen typically consists of As_2O_3 along with all-*trans* retinoic acid, and the five-year survival rate for patients afflicted with APL is 90%. In neutral aqueous solution, As(III) is largely present as $As(OH)_3$. The mechanisms of action of As are complex and multifold, but include (a) inducing cell death (apoptosis) by binding to exposed thiol groups of proteins located in the mitochondrial membranes, (b) contributing to the formation of harmful, reactive oxygen species (commonly abbreviated ROS, see Section 26.10), and (c) interfering with gene expression crucial in cell differentiation.

1 As$_4$O$_6$

The remarkable action of the square-planar d^8 platinum (II) complex cis-[PtCl$_2$(NH$_3$)$_2$] (**2**, known as cisplatin) was discovered in 1964 during examination of the effect of an electric field on the growth of bacteria. The behaviour of a colony of *Escherichia coli* suspended in solution between two platinum electrodes was observed, and it was noted that upon application of an alternating voltage the cells continued to grow in size by forming long filaments, but stopped replicating. The effect was traced to a complex that was formed electrochemically by dissolution of Pt into the electrolyte, which contained NH$_4$Cl. Since then, cisplatin has been highly effective in treating many forms of cancer, particularly testicular cancer for which the success rate approaches 100%. The other geometric isomer, trans-[PtCl$_2$(NH$_3$)$_2$], is relatively ineffective.

2 Cisplatin

The ultimate molecular basis of the chemotherapeutic action of cisplatin and related drugs is the formation of a complex between Pt(II) and DNA. Cisplatin is administered into the bloodstream of the patient, where, because the plasma contains high concentrations of Cl$^-$, it tends to remain as the neutral dichlorido species. The electrical neutrality of the dichlorido complex facilitates its passage through the cell membrane (Fig. 32.4).

Once subjected to the lower Cl$^-$ concentrations inside the cell, the Cl$^-$ ligands are replaced by H$_2$O (or OH$^-$): the cationic species that may result are attracted electrostatically to DNA and ultimately form inner-sphere complexes in which the –Pt(NH$_3$)$_2$ fragment is coordinated by N atoms of the nucleotide bases. Some classic studies have shown that the preferred target is a pair of N atoms on consecutive guanine bases in the same strand. Complexes of the –Pt(NH$_3$)$_2$ fragment with oligonucleotides have been studied by X-ray crystallography and ^{195}Pt-NMR spectrometry (Fig. 32.5). Complexation with Pt causes the helix to kink and partially unwind, rendering the DNA incapable of replication or repair. The distortion also makes the DNA recognizable by 'high mobility group' proteins that bind to bent DNA, thus initiating cell death. Altering the DNA of a cancer cell to render it incapable of replication is now widely accepted as the major mode of action of Pt-based anticancer drugs.

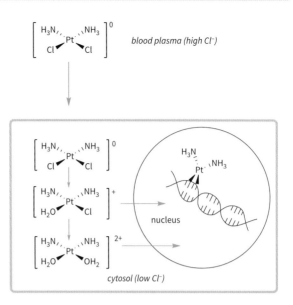

FIGURE 32.4 Mechanism of transport of cisplatin from a patient's bloodstream to the DNA of a cancer cell.

Despite its efficacy, cisplatin has highly undesirable side effects; in particular, it causes serious damage to the kidneys before it is eventually excreted. Great efforts have been made to find Pt complexes that have fewer side-effects. Other examples in clinical use are carboplatin (**3**) and oxaliplatin (**4**), but like cisplatin (and indeed As$_2$O$_3$) these drugs are blunt tools, attacking healthy as well as diseased tissue.

3 Carboplatin   **4** Oxaliplatin

Cancerous tumours are usually hypoxic, meaning that their O$_2$ level is below 400 Pa—much lower than that of normal tissue which lies in the range 3–11 kPa. Hypoxia arises from restricted blood supply as well as the higher metabolic activity of cancer cells. The reducing environment offers the opportunity to extend the prodrug concept to exploit Pt(IV) which, being a d^6 ion, forms six-coordinate octahedral complexes. Having penetrated the cell membrane, the Pt(IV) complex is easily reduced, losing both axial ligands to give the active, square-planar Pt(II) form. The strategy is illustrated by bioconjugate compound **5**, which upon arrival in cancer tissue is reduced to cisplatin and two molecules of (D)-1-methyl tryptophan are released: the latter is an inhibitor of indoleamine-2,3-dioxygenase, an enzyme responsible for enabling cancer cells to evade the immune system, and

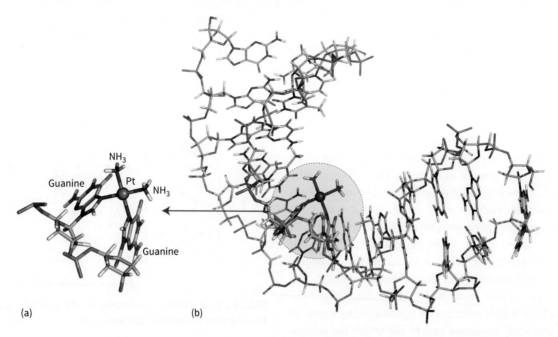

FIGURE 32.5 Structure of a complex formed between the Pt(NH$_3$)$_2$ fragment and two adjacent guanine bases on an oligonucleotide. (a) The square-planar ligand arrangement around the Pt atom. (b) Coordination of Pt causes bending of the DNA helix.

itself an anticancer agent. The Pt(IV) prodrug thus provides two modes of chemotherapeutic action.

5 A bioconjugate Pt(IV) prodrug activated by reduction

The strategy of using Pt(IV) bioconjugates to undertake hypoxia-triggered release of biologically active ligands has wide scope. In another example, the Pt(IV) complex shown in Fig. 32.6 contains two estradiol groups attached to *trans* positions through carboxylate-terminated linker groups. The complex enters breast or ovarian cancer cells and upon reduction forms the active Pt(II) complex and releases two molecules of the estrogen derivative which induces the formation of a protein that inhibits repair of platinated DNA.

Other metals are under intense investigation, expanding the armoury of anticancer drugs by offering alternative modes of cell targeting and different mechanisms of DNA binding or other cytotoxic action. The Ru(III) complex (**6**) known as NAMI-A (new anti-tumour metastasis inhibitor) is effective in killing secondary cancer cells that have spread from the primary tumour—an invasive process known as 'metastasis'. Offering an alternative to Pt-based drugs, NAMI-A has received a great deal of attention. Another Ru(III) complex (**7**) referred to simply as KP1019 is more active against primary tumours. Both of these Ru(III) complexes become active upon reduction after entering the target cell, leading researchers to investigate how Ru(II) complexes may operate.

6 NAMI-A

7 KP1019

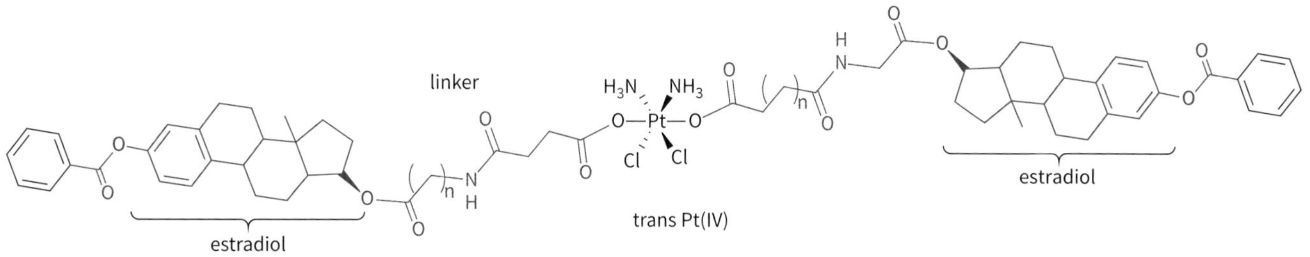

FIGURE 32.6 Example of a Pt(IV) bioconjugate containing two estradiol groups attached to *trans* positions through carboxylate-terminated linker groups.

The half-sandwich Ru(II) arene complex (**8**) forms a complex with guanine-N in a similar way to Pt complexes, but the interaction with DNA is supplemented by intercalation of the biphenyl group within the hydrophobic DNA core as well as hydrogen bonding between guanine and the $-NH_2$ groups of the ethylenediamine ligand. The use of an intercalating polyaromatic ligand to engage in π-stacking interactions with the bases offers improved efficacy.

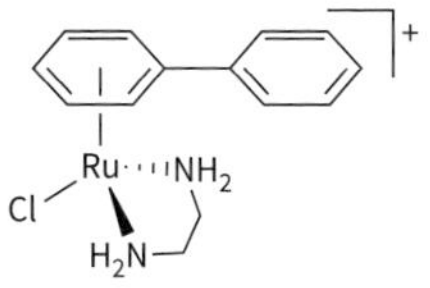

8 Half-sandwich Ru(II) arene complex

The Cu complex (**9**) illustrates a further innovative strategy: one of the ligands is a DNA intercalator, the other is a molecule of suberoylanilide hydroxamic acid (SAHA), a clinically approved cancer drug that inhibits histone deacetylase, a Zn-containing enzyme that modifies the structure of histones. Histones are proteins responsible for 'packaging' DNA in the tidy compact coils found in a chromosome. By inhibiting histone deacetylase, SAHA allows DNA to remain in a partly unwound state that, through increased transcription, corrupts cancer cell metabolism. The unwound state also exposes the DNA to attack by the intercalator.

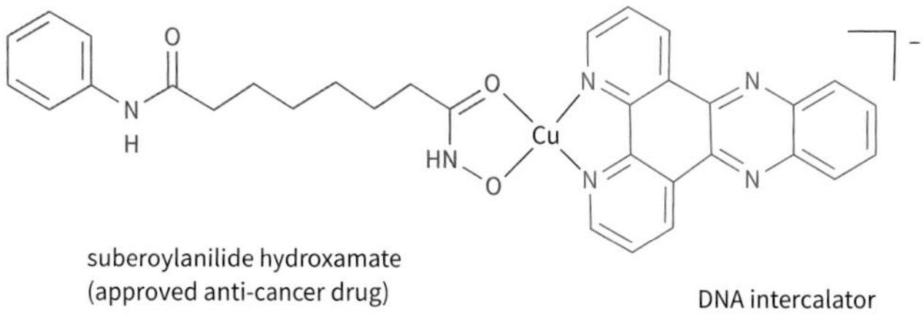

suberoylanilide hydroxamate (approved anti-cancer drug)

DNA intercalator

9 Cu(II) complex combining SAHA and intercalating ligand

Another prodrug strategy is photoactivation, whereby a complex becomes active only upon irradiation. This method of treating cancers is known as **photodynamic therapy**. A Pd porphyrin complex marketed under the name 'Padeliporfin', shown in Fig. 32.7, is used to treat prostate cancer. After injection, the drug binds to the blood plasma carrier protein known as serum albumin. The role of the Pd(II) is to tune the electronic properties of the porphyrin. Excitation by a 735-nm laser directed by optical fibres inserted into the diseased zone (tissue is more transparent to longer wavelength visible light) results in the photocatalytic reduction of O_2 to superoxide and hydroxyl radicals. These reactive species (hydroxyl in particular) cause lethal and highly localized damage to the targeted cancer cells. The photocatalytic system can undergo numerous cycles.

Other examples under investigation include an Rh(III) complex having a light-sensitive diimine ligand (**10**) and the

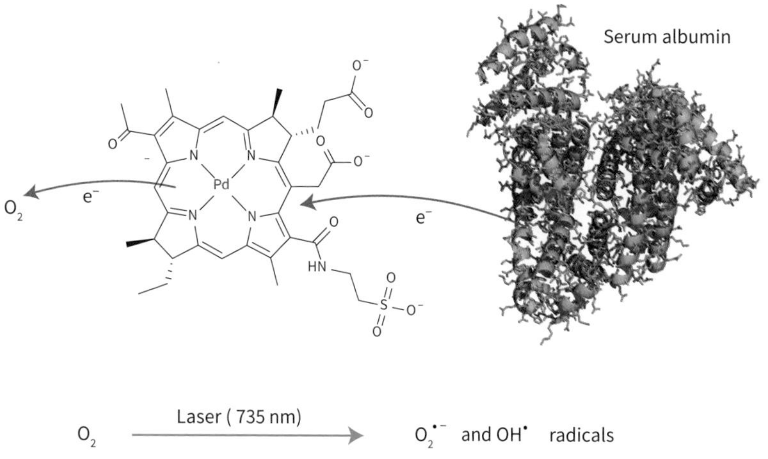

FIGURE 32.7 Photodynamic therapy using a Pd porphyrin complex that is administered to the tumour region and binds to serum albumin. Excitation by a laser produces a powerful oxidant that abstracts electrons from serum albumin. The reduced product, now in its ground state, is a powerful reductant and reacts with O_2 to generate reactive oxygen radicals that attack the local tissue.

10 Photosensitive Rh complex

trans-diazido Pt(IV) complex shown below, which breaks down upon irradiation (intense LMCT; $N_3^- \rightarrow Pt$) to form a Pt(II) complex, releasing highly reactive $N_3 \cdot$ (azidyl) radicals which may attack tissue or rearrange to form N_2 gas.

Surprisingly, analogues in which the two azide ligands are in *cis* positions are much less active: this contrasts with the established *cis* preference for Pt(II) complexes, indicating that other factors, such as photoactivation or the release of $N_3 \cdot$ radicals, may be more important than the nature of the Pt(II) product.

Destroying cancer cells using radiation is known as **radiotherapy**, an important challenge being to ensure that only cancerous tissue is destroyed. Thus, instead of using an external beam of X-rays, the dose is generated internally by **radiopharmaceuticals** introduced to a specific tissue. Radionuclides that decay by beta emission (β^-, electron, or β^+, positron) are most useful, as beta particles travel only a short distance in tissue, thus localizing the lethal dose. In addition, the annihilation of a positron by an electron produces γ photons, the detection of which is exploited to image tissue at high resolution—a technique known as positron emission tomography (PET). Some examples of radionuclides currently in medical use or under investigation are listed in Table 32.1 (and imaging is discussed in Section 32.9). Selection of a particular nuclide depends on the type and energy of the radiation delivered (which affects its penetration), the half-life (which, depending on application, should be in the order of hours or days), the ease of generation from a precursor nuclide, and ultimately its chemical suitability.

> **EXAMPLE 32.1** Explaining why Pt complexes are used as anticancer drugs
>
> What are the chemical properties of Pt complexes that make them suitable in cancer therapy?
>
> **Answer** Both Pt(II) and Pt(IV) form nonlabile complexes that can be crystallized in pure forms. The high affinity for N-donor ligands as provided by DNA bases contrasts with the relatively low affinity for O- and Cl⁻-ligands: the latter are therefore easily displaced during specific stages of the drug's transport to DNA and final complex formation. Under hypoxic conditions that prevail in cancer cells, octahedral Pt(IV) complexes (d^6) are reduced to square-planar Pt(II) (d^8), jettisoning the axial ligands together with their functionalities that may have assisted uptake into the cell, and may themselves be therapeutic agents.
>
> **Self-test 32.1** How can the reduction potentials of Cu–semithiocarbazone complexes be varied and tuned to optimize their activity?

Targeting a particular organ is achieved through bioconjugation, whereby a ligand system affording stable coordination of the radioactive metal is attached to a molecule that is recognized by the desired biological receptor. Conventionally, the bioconjugate is assembled before administering it to the patient. An alternative strategy is to 'synthesize' the drug *in situ*, bringing metal complex radionuclide and bioactive ligand together at the site of action. Here the bioactive ligand may be a monoclonal antibody against the cancer cell, in this case derivatized to form the bioconjugate once the metal or metal complex is present. This treatment is known as **radioimmunotherapy**. These strategies are represented in Fig. 32.8.

An extensive range of bioconjugate complexes (**11**) can be formed that contain semithiocarbazone ligands coordinated

TABLE 32.1 Some metal-atom radionuclides used in radiotherapy and radioimaging

Nuclide	Half-life	Dominant decay mode and energy	Tissue penetration distance (β particle)	Ligand	Application
^{64}Cu	12.7 h	β^+ (0.65 MeV) β^- (0.58 MeV)	1 mm	semithiocarbazone	therapy/imaging
^{67}Cu	61.8 h	β^- (0.56 MeV)	6 mm	semithiocarbazone	therapy
^{68}Ga	67.7 m	β^+ (0.84 MeV)	1 mm	DOTA	imaging/therapy
^{90}Y	64.1 h	β^- (2.3 MeV)	10 mm	DOTA	therapy
^{99m}Tc	6 h	γ (140 keV)		various (Sect. 32.9)	imaging
^{177}Lu	6.7 d	β^- (0.50 MeV) γ (113, 208 keV)	2 mm	DOTA	therapy
^{201}Tl	73 h	γ (80 keV)		aqueous	imaging

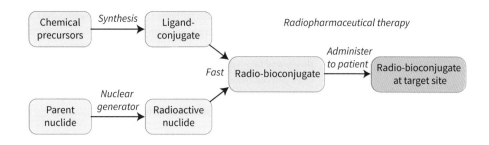

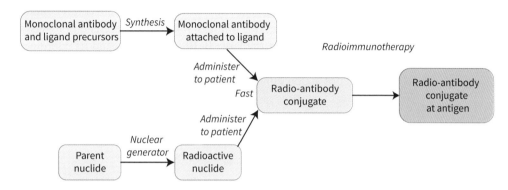

FIGURE 32.8 Principles of radiopharmaceutical therapy and radioimmunotherapy.

to radioactive ^{64}Cu, which decays ($t_{1/2}$ = 12.7 h) by several routes—principally β-decay and positron emission, but also electron capture and gamma radiation. The isotope is usually produced from ^{63}Cu by neutron capture in a reactor, then used immediately. Radiotherapy makes use of the emitted β-particles which have a mean path length of only about 1 mm, ensuring highly specific killing of cells that have taken up ^{64}Cu. Selective delivery is based on the principle that Cu(II) semithiocarbazone complexes are neutral and diffuse into the cancer cell where the hypoxic environment causes reduction to the charged Cu(I) form: the lethal nuclide is thereby trapped. Many variations are possible depending on the nature of the substituents R selected to recognize a particular target.

11 Cu–semithiocarbazone complex

Instead of Cu(II), which forms strong complexes with soft ligands, a hard 3+ metal ion such as a lanthanoid may be used as the radionuclide, exploiting the rapid formation of stable complexes with hard, polydentate ligands. An important class of bioconjugate compounds used to treat cancers of the neuroendrocrine system (the cells of which release many important hormones) comprises the octadentate macrocyclic ligand 2,2′,2″,2‴-(1,4,7,10-tetraazacyclododecane-1,4,7,10-tetrayl)tetraacetic acid (commonly referred to simply as dodecane tetracetic acid and given the acronym 'DOTA') attached to a synthetic analogue of somatostatin (**12**). Somatostatin is a peptide that binds to the somatostatin receptor in neuroendocrine cells and blocks the production of hormones. In a typical procedure, ^{90}Y or ^{177}Lu are generated from suitable precursors (^{90}Sr, ^{176}Yb, or ^{176}Lu, respectively) and the cation reacts with the bioconjugate ligand to form the complex that is injected into the patient.

12 Metal-DOTA bioconjugate with somatostatin analogue

32.2 Anti-arthritis drugs

KEY POINTS Complexes of Au(I) are effective against rheumatoid arthritis. Mechanisms of action probably involve the binding of Au to thiol-containing proteins.

Gold drugs are used in the treatment of rheumatoid arthritis, an autoimmune disease that results when the body's immune system turns on its owner, attacking the tissue around joints.

The inflammation is caused by the action of hydrolytic enzymes in cell compartments known as lysosomes that are associated with the Golgi apparatus (see Fig. 26.1). Commonly administered drugs include sodium aurothiomalate (myochrisin, **13**), sodium aurothioglucose (solganol, **14**; the linkage between units is uncertain), and auranofin (**15**), each of which feature Au(I) with the expected linear coordination (Section 20.11).

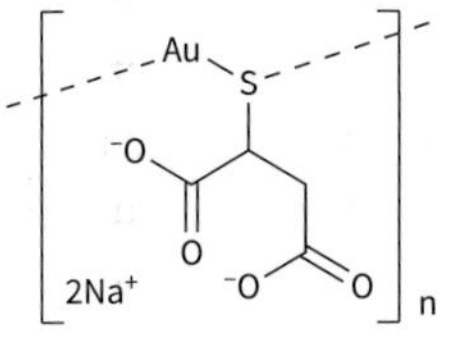

13 Myochrisin

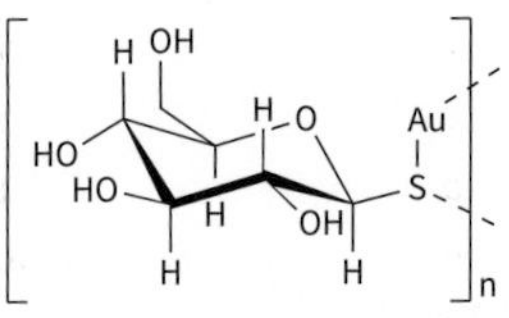

14 Solganol

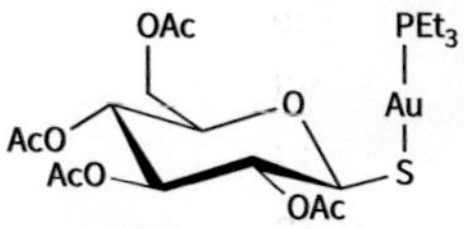

15 Auranofin

Side-effects of Au drugs include skin allergies as well as kidney and gastrointestinal problems. Many Au drugs, including myochrisin and solganol, are water-soluble polymers that are injected into the muscle: they cannot be administered orally because they undergo acid hydrolysis in the stomach. Auranofin is a monomer, due to its single phosphine ligand: it can be given orally, but is reported to be less effective than the injected compounds. Aurofin is also under investigation as an antiviral drug.

The mechanisms by which gold drugs act have long been controversial, and many different protein targets are implicated. As a soft metal ion, Au(I) forms stable complexes with S-donor ligands provided by cysteine or methionine side chains in proteins. The main candidates for therapeutic Au(I) binding and inhibition include several enzymes in which the active site is a cysteine:

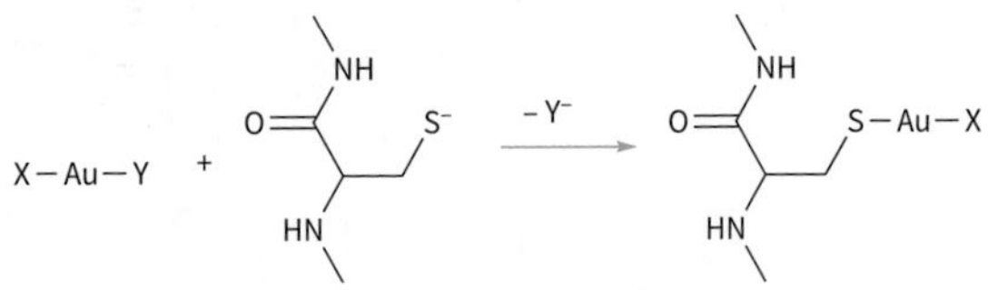

These enzymes include thioredoxin reductase, an enzyme involved in maintaining a constant reducing environment in the cell, and cathepsins, cysteine proteases that are involved in inflammation. Not surprisingly, Au(I) reacts with glutathiol, a tripeptide of glycine, cysteine, and glutamate, that also serves as an important cellular redox buffer.

32.3 Bismuth pharmaceuticals

KEY POINTS Bismuth pharmaceuticals have long been used to combat bacterial infections ranging from common irritations of the eye and conjunctiva to a notorious pathogen, *Helicobacter pylori*, which invades the stomach. When complexed with carboxylate ligands, Bi(III) forms insoluble coatings on the stomach wall from which Bi^{3+} can be released slowly and taken up by bacterial cells. Once inside the cell, Bi(III) interferes with enzyme activities that are essential for the pathogen's survival in the acidic gastric environment. Bismuth is under investigation as a tool for combating microbial resistance.

Compounds of bismuth (Chapter 16), essentially the heaviest element to have a stable nucleus (the half-life of ^{209}Bi is 20×10^{18} y), are familiar antibacterial agents for common health problems. Ointments for eye infections contain bibrocathol (**16**), although this simple representation masks the fact that the large Bi(III) ion will expand its coordination shell, forming insoluble polymers and making its mode of action difficult to establish.

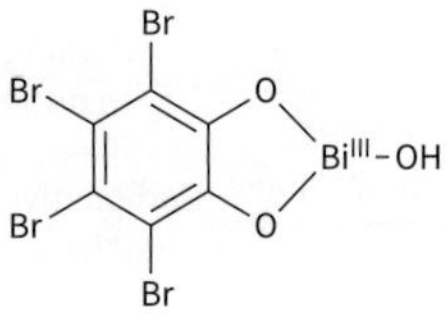

16 Bibrocathol

The notorious bacterium *Helicobacter pylori* is responsible for stomach complaints extending from indigestion to gastric and duodenal ulcers, and it is believed to be a major cause of stomach cancer. Fortunately, Bi(III) pharmaceuticals such as bismuth subsalicylate (BSS), marketed as Pepto-Bismol®, and colloidal bismuth subcitrate (CBS), offer an effective treatment for *H. pylori* infections. The compounds are administered orally, often with a cocktail of antibiotics.

Bismuth(III) is resistant to oxidation or reduction (Bi(V) is powerfully oxidizing), and its aqueous chemistry is dominated by polynuclear insoluble or colloidal Bi–O or Bi–carboxylate species, a property that may be important in retaining Bi(III) in the stomach, where the environment is highly acidic (pH 1.5–3) and rich in organic acids. Species such as binuclear $[Bi_2(cit)_2]^{2-}$ (**17**) and polymeric salicylate sheets based on Bi_2O_2 rings (as present in BSS, **18**) or Bi_6O_7 octahedra (**19**) are taken up in the gastric mucus to form a protective coating on the craters of ulcers, thereby limiting the adherence of *H. pylori*. The structure of BSS (**18**) was finally established in 2022, well over a century after its therapeutic value was noted.

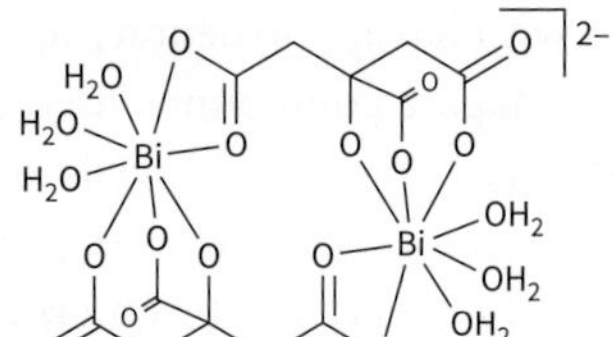

17 $[Bi_2(cit)_2]^{2-}$

18 Bi₂O₂ salicylato complex (BSS)

19 Salicylato complexes based on Bi₆O₄ octahedra

neutralizing stomach acidity by converting urea to ammonia and carbamate. Bismuth inhibits fumarase—an enzyme catalysing the hydration of fumarate to form malate, a step that is essential in energy production.

Bismuth(III) has been shown to inhibit metallo-β-lactamases that contain a pair of Zn^{2+} ions at their active site. Produced by pathogenic bacteria, these enzymes are responsible for hydrolysing the four-membered β-lactam ring in penicillins and cephalosporins—antibiotics that act by inhibiting cell wall synthesis. The growing problem of antibiotic resistance, much of it arising from their overuse in medicine and agriculture, can be traced (in part) to bacteria adapting through mutations to the metallo-β-lactamase gene. Those strains able to produce metallo-β-lactamase variants that are not so effectively inhibited tend to survive and multiply. A crystal structure of β-lactamase shows that a single Bi^{3+} can replace both Zn^{2+} ions (Fig. 32.9). Low levels of Bi(III) appear non-toxic to humans; consequently, the favourable replacement of Zn(II) by Bi(III) in a microbial enzyme, rendering it inactive, offers a potential way to overcome antibiotic resistance that does not require redesigning drugs.

Investigation of the exact speciation in a biological environment is very difficult because Bi(III) lacks convenient spectroscopic characteristics. Further therapeutic action appears to involve bismuth being released slowly from these polymeric storage sites, presumably as soluble species and even including Bi^{3+}(aq) (pK_a approximately 1.5) that may persist under such acidic conditions, and eventually entering the pathogen using the metal-ion uptake system of the bacteria. Once inside the cell, bismuth may target proteins that are essential for the pathogen's survival:

Bi(III) oxido-colloid ⇌ Bi^{3+} → Bi^{3+} - protein
steady release   *uptake by pathogen*

Bismuth(III) forms strong complexes with a wide range of protein ligand donors—particularly O, N, and S—and its potential interactions are investigated by metalloproteomics strategies (Section 26.2). Bismuth interferes with vital metal-ion-binding proteins, including transferrin (see Section 26.6), thus preventing Fe uptake, a histidine-rich protein known as Hpn that transports Ni, the latter being the active metal in hydrogenases (Section 26.13), and urease, an enzyme that makes up about 10% of the total protein in *H. pylori*. Such a high level of urease is essential for

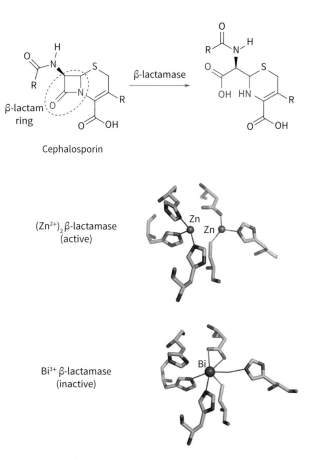

FIGURE 32.9 β-lactamases are produced by bacteria to destroy compounds containing the β-lactam ring, which include penicillins and cephalosporins (shown). The enzymes are inhibited when two Zn^{2+} are replaced by one Bi^{3+}.

32.4 Lithium in the treatment of bipolar disorders

KEY POINTS Lithium has long been used in the treatment of manic depression, yet its mode of action is still uncertain. Several targets are proposed, particularly enzymes that normally use Mg^{2+}, its close diagonal relative in the periodic table.

The lithium aqua ion, administered as its chloride or carbonate salts, is the simplest of drugs and a very effective agent in the treatment of bipolar disorders that are commonly referred to as manic depression and are characterized by severe swings in behaviour. A 'mood stabilizer', Li^+ is known to inhibit or activate specific neurophysiological processes. Its many roles include inhibition of glycogen synthase kinase 3 (GSK-3), an enzyme that is activated during chronic stress; inhibition of inositol monophosphatase (an enzyme needed to produce inositol, a sugar and signalling agent concentrated in the brain); suppressing the activity of dopamine and glutamate receptors that are important in neurotransmission; and activation of receptors for γ-aminobutyric acid, an inhibitory neurotransmitter of the central nervous system.

Despite being in use for more than 50 years, the mechanism of action of Li^+ remains unclear at the molecular level, although a knowledge of inorganic chemistry provides important clues. Lithium is unusual. The Li^+ cation is smaller and more polarizing than its Group 1 counterparts Na^+ (the main extracellular cation) and K^+ (the main intracellular cation), but it shares a close diagonal relationship with Mg^{2+} (Section 10.3), which is closely involved in the metabolism of phosphate and other hard oxygen-donor ligands. Competition with any of these mobile cations is probable.

32.5 Involvement of metal complexes in the cause and treatment of malaria

KEY POINTS Resistance to quinine and related antimalarial drugs is an urgent medical issue worldwide. Malaria is caused by a parasite known as *Plasmodium*: quinine works by interfering with the parasite's ability to deal with haematin, an Fe–porphyrin product of haemoglobin degradation that generates reactive species from O_2. One way of overcoming resistance is to modify the quinine's structure by attaching it to a metal.

Malaria, a major killer affecting half a billion people worldwide and causing over a million deaths per year, is caused by infections of *Plasmodium*, a genus of parasites transmitted to humans by mosquitoes. Once in the bloodstream of its victim, the parasite attacks red blood cells (erythrocytes) to obtain the iron that it needs for survival—a process that requires the degradation of haemoglobin via an Fe(III)–porphyrin intermediate, aquaferriprotoporphyrin IX, commonly known as haematin (**20**).

20 Haematin

It is important for *Plasmodium* to avoid accumulating haematin, which is highly toxic because it catalyses the generation of reactive oxygen species, particularly peroxides, that oxidize lipids, resulting in membrane damage. *Plasmodium* survives by converting haematin into a highly insoluble microcrystalline substance called haemozoin (Fig. 32.10). Antimalarial drugs, the most important of which are based on the quinoline family ('quinine'), interfere with the production and stability of haemozoin, thereby exposing the parasite to self-inflicted Fe overload. Many strains of *Plasmodium* have become resistant to quinolines, and ingenious ways are being devised to find derivatives that can overcome the parasite's immune defences.

An important strategy here is to attach a metal complex to the quinoline group, the bioconjugate having a greater ability to deceive the organism's resistance pathways. Ferroquine (**21**), a bioconjugate of chloroquinine with a ferrocene group, has undergone extensive trials: despite peaks and troughs in interest, it continues to attract attention, first as a component of regimens administered with other antimalarials, and also for its apparent efficacy in the treatment of other diseases, including cancers.

21 Ferroquine

32.6 Metal complexes as antiviral agents

KEY POINTS Some drugs already approved for the treatment of other diseases have antiviral activity, but their modes of action are unclear. The discovery and mechanistic investigation of inorganic complexes that stall the invasion of viruses in highly targeted ways is a great challenge for chemists.

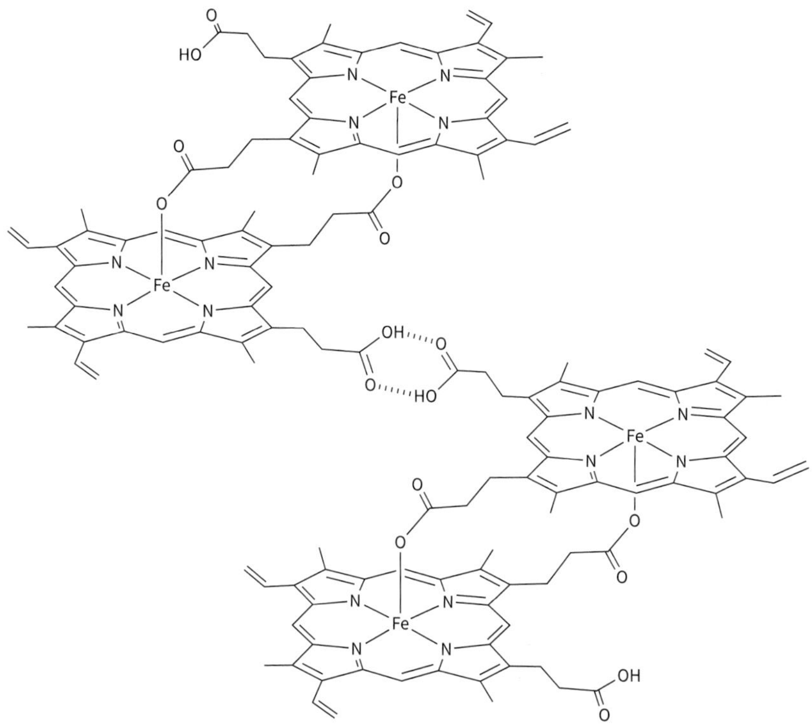

FIGURE 32.10 Haemozoin, an insoluble compound produced by malaria parasites to protect them against the lethal damage caused by haematin.

Viruses are non-living pathogenic particles that invade healthy cells and seize control of their metabolic machinery. The particles consist of nucleic acid (DNA, or RNA in the case of retroviruses) enclosed by a protein in a coating called a capsid. Once inside a cell, the virus imposes its own nucleic acid blueprint and recruits indigenous enzymes to produce what it requires to survive and replicate, quickly overwhelming the host. The ideal target for an antiviral drug is an enzyme that is useful only to the virus.

Metal complexes are being investigated in the search for new antiviral drugs against a host of different diseases, including Human Immunodeficiency Virus (HIV), influenza, and strains of Severe Acute Respiratory Syndrome Covid (SARS-CoV), as well as many others. Auranofin (**15**), normally used to treat rheumatoid arthritis, has been found to show activity against SARS-COV-2, though its mechanism of action remains unclear.

Looking to a more rational approach to discovering new antivirals, two classes of inorganic complexes that have received a good deal of attention are based on Co(III) and Au(III). Although a long way from being used clinically, these examples are interesting because they illustrate how some important principles of inorganic chemistry may be exploited.

A Co(III) complex such as the example shown in Fig. 32.11 may interfere with viral proteins by undergoing redox-assisted ligand exchange to produce a Co(III)-protein complex having impaired function. Once inside a cell, the Co(III) complex can undergo transient reduction to Co(II) which, unlike Co(III), is substitution labile. Upon coordination to a protein functionality, which may be an imidazole-N donor on a histidine sidechain, reoxidation occurs to form the inert product. The challenge for pharmacologists is to target such action specifically towards a viral protein.

The Au(III) complex [Au(bnpy)Cl$_2$] (bnpy = 2-benzylpyridine) has been shown to inactivate transcription factors by unravelling Zn-fingers (see Section 26.15). The mechanism involves reduction to Au(I) and formation of a covalent bond between one of the cysteines and the 2-benzylpyridine ligand that is released (Fig. 32.12). The reaction releases the

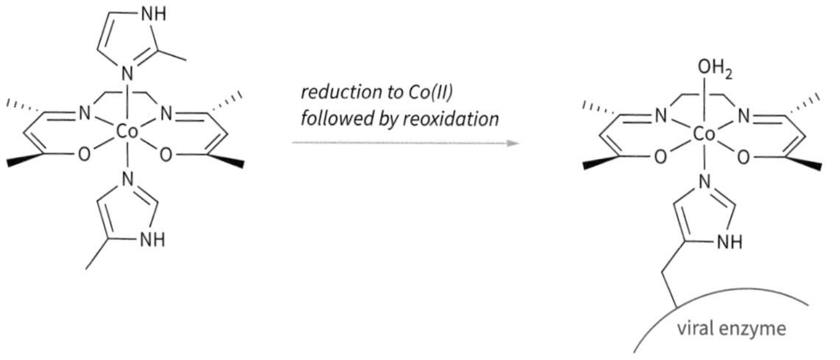

FIGURE 32.11 Redox-assisted formation of a Co(III) complex with a viral protein. An imidazole ligand on the original complex is replaced by a histidine sidechain.

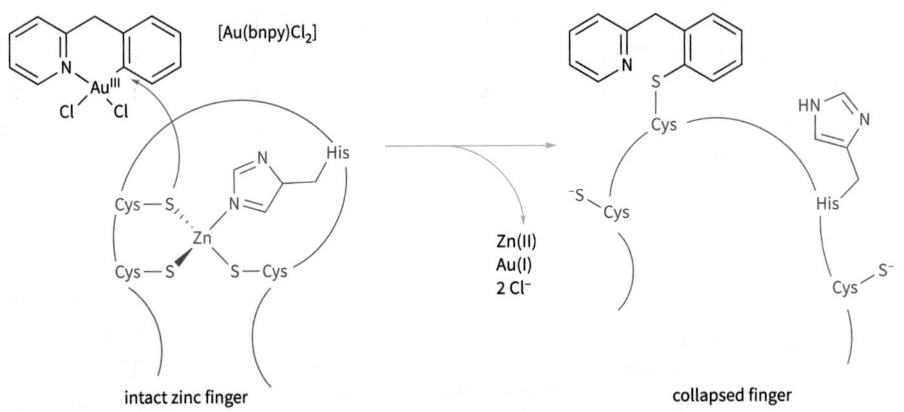

FIGURE 32.12 Mechanism proposed for the unravelling of a Zn finger by an Au(III) complex.

Zn^{2+} and removes the ability of the transcription factor to locate its target on a nucleic acid.

Hepatitis C kills more than 350,000 people every year. One strategy for antiviral therapy is to target specific RNA sequences that are present in the hepatitis C virus but mainly absent from the natural genome of the host that has been infected, thereby supporting the major treatments based on the naturally produced defensive proteins known as interferons. Bioconjugate-based strategies use metal complexes derivatized with peptides that can bind to certain stem-loop motifs of viral RNA.

Derivatives of cyclams—macrocyclic ligands forming strong complexes with d-block metal ions (Section 7.14)—have been under investigation as anti-HIV therapeutics. Entry of the HIV virus into a cell is initiated by an interaction between a glycoprotein on the virus and a receptor protein known as CD4 that is present on the target-cell membrane. The resulting reaction sets off a chain of events that involve other receptor proteins. Cyclams interrupt these events, though it is unclear how they do so. One hypothesis is that cyclams bind free Zn^{2+} and that the resulting macrocyclic complexes form termolecular adducts with cysteine thiolate (RS^-) functionalities present on special protein receptors known as CXCR4. These receptors are required for HIV to enter cells, suggesting that Zn-cyclams interfere with the invasion process. Bicyclams are particularly effective and show promising results when used as bioconjugates with the well-known anti-HIV drug AZT (22).

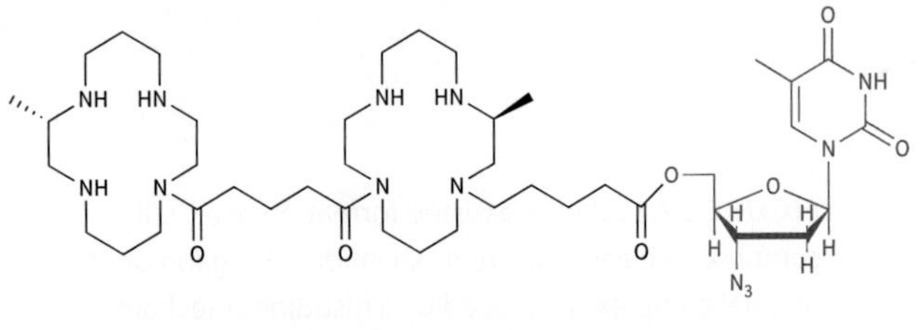

22 Bicyclam–AZT conjugate

32.7 Metal drugs that slowly release CO: an agent against post-operative stress

KEY POINTS Like NO, CO is also a signalling agent, and trace levels are known to be highly beneficial for relieving trauma following an operation. The best way to administer small amounts of CO in a controlled manner is via metal complexes known as CORMs, which release it slowly into the bloodstream. Recent investigations are showing that some CORMs have potent antimicrobial activity.

Carbon monoxide is well known as a very poisonous gas, yet CO has a natural beneficial function as a signalling agent like NO. Trace amounts of CO are continually being released in the body through the degradation of haemoglobin (CO is produced by the action of haem oxygenases on porphyrins). Carbon monoxide is now known to be a vasodilatory and anti-inflammatory agent that can be very useful in combating post-operative trauma. It has been found that rather than administering CO directly to a patient, it can be introduced at a continuous low level through the action of CO-releasing molecules (CORMs). These agents not only serve a cytoprotective role but are also active against pathogenic bacteria such as strains of *E. coli*, *Staphylococcus*, *Pseudomonas*, and *Campylobacter* (a leading cause of gastroenteritis). A much-studied water-soluble CORM is [Ru(CO)$_3$Cl(glycinate)] (23), which releases CO upon reaction with biological ligands such as cysteine.

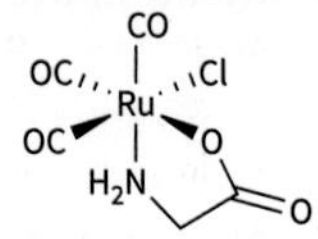

23 [Ru(CO)$_3$Cl(glycinate)]

In order to be most clinically useful it is beneficial to be able to control the rate at which CO is delivered, and attention has turned to encapsulating CORMs in nanostructures known as CO-releasing polymers (CORPs). Passive CO release may be retarded in a water-soluble polymer (24).

Alternatively, CO release can be coupled to an external stimulus, an example being the photoactivatable CORP (**25**) in which dark-stable $Mn(CO)_3$ units are linked by organic ligands to produce an amorphous coordination polymer. Upon UV irradiation, the CORP particles release one CO per Mn unit.

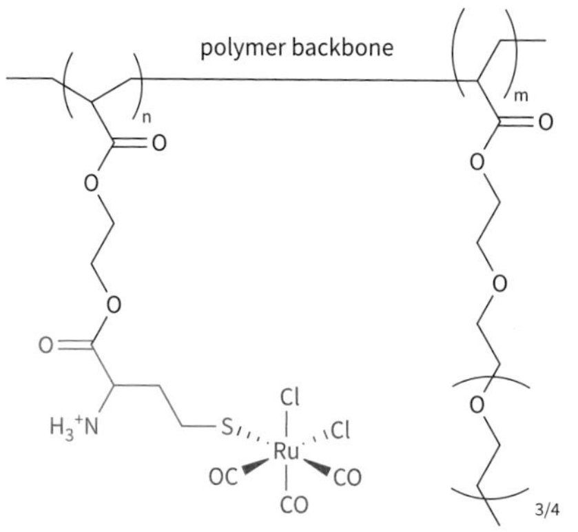

24 A section of a slow-acting water-soluble CO-releasing polymer

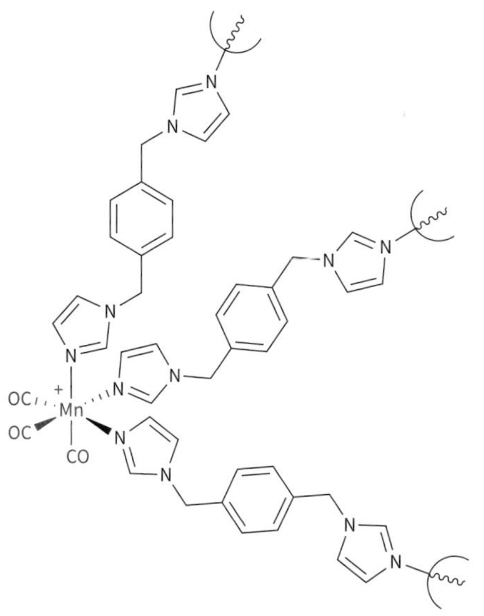

25 A section of a photoactivated CO-releasing polymer

32.8 Chelation therapy

KEY POINTS Despite its unique importance to living organisms, iron is a very toxic element unless complexed strongly by proteins. Uncomplexed Fe species (excess Fe) catalyse reactions such as the Fenton reaction, producing harmful hydroxyl radicals that attack sensitive molecules such as DNA. The treatment of Fe overload involves sequestration of Fe by ligands having structures based upon or inspired by the bacterial ligands known as siderophores.

'Iron overload' is the name given to several serious conditions that affect a large proportion of the world's population. Here we recall that despite its great importance, Fe is potentially a highly toxic element, particularly in its ability to produce harmful radicals by reaction with O_2, and its levels are normally strictly controlled by regulatory systems (Section 26.16). In many groups of people, a genetic disorder results in breakdown of this regulation, the chief culprit being *thalassaemia*, which is endemic in Southeast Asia. One kind of iron overload is caused by an inability of the patient to produce sufficient porphyrin. Other problems are caused by faults in the regulation of Fe levels through ferritin or transferrin production (Section 26.6).

Iron overload is treated by **chelation therapy**, the administration of a ligand to sequester Fe and allow it to be excreted. Desferrioxamine (Desferal, **26**) is a ligand that is similar to the siderophores described in Section 26.6. It is a very successful agent for iron overload, apart from the trauma of its introduction into the body, which involves being connected to an intravenous supply.

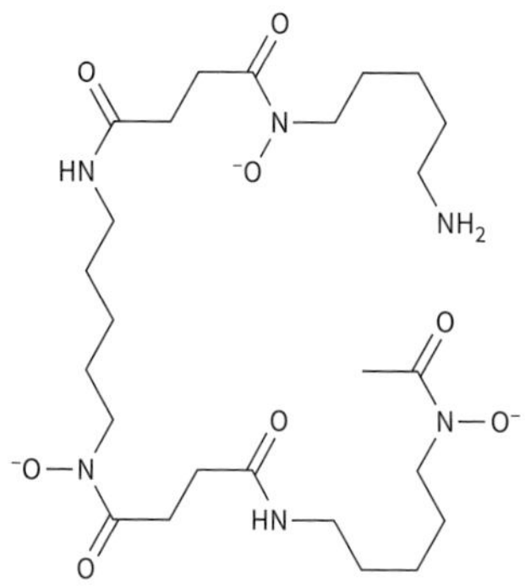

26 Desferal

The latter concern has led to the development of orally administered drugs such as deferasirox (**27**) and deferiprone (**28**). These small lipophilic ligands are able to cross the intestinal wall to enter the bloodstream. A complex of two deferasirox molecules with one Fe(III) is shown (**29**).

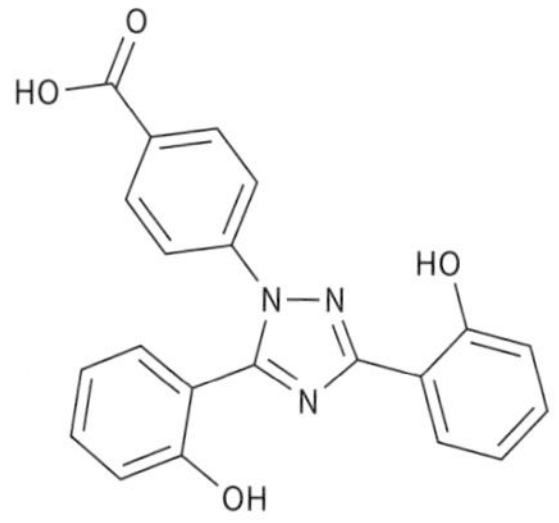

27 Deferasirox

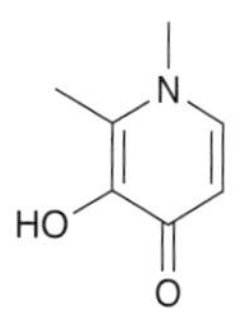

28 Deferiprone

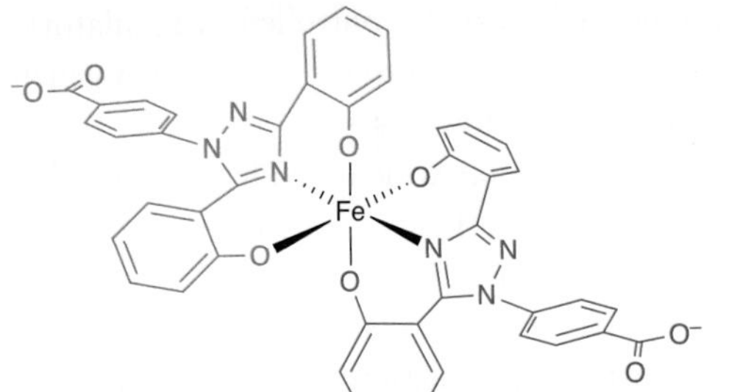

29

A special case of chelation therapy is the treatment of individuals who have been contaminated with plutonium following exposure to nuclear weapons. In its common oxidation states, Pu(IV) and Pu(III) have similar charge densities to Fe(III). Siderophore-like chelating ligands have been developed, such as 3,4,3-LIMACC (**30**), which contains four catechol groups.

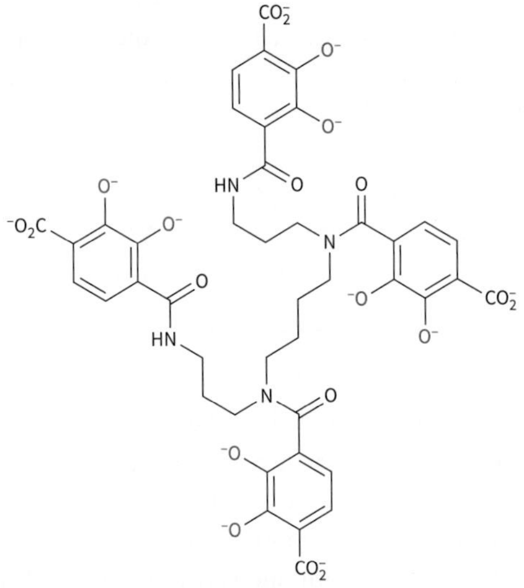

30 3,4,3-LIMACC

32.9 Imaging agents

KEY POINTS Damaged and diseased tissue can be located noninvasively by using compounds that concentrate in that tissue and reveal their location via tomographic scanning, either by interfering with the nuclear relaxation of protons in water or emitting radiation. Particular organs and tissues are targeted according to the ligands that are present.

Complexes of gadolinium(III) (f^7) are used in magnetic resonance imaging (MRI), which has become an important technique in medical diagnosis. Through their effect on the relaxation time of ^{1}H-NMR spectroscopic resonances, Gd(III) complexes are able to enhance the contrasts between different tissues and highlight details such as abnormalities of the blood–brain barrier or cancerous tumours. A number of Gd(III) complexes have been approved for clinical use, each exhibiting different degrees of rejection or retention by certain tissues, as well as stabilities, rates of water exchange, and relaxation parameters.

As expected with all lanthanoids, stable complexes of Gd(III) are based on chelating ligands, particularly those having multiple carboxylate groups (Section 21.7). Of particular importance are complexes formed with the macrocyclic ligand DOTA, mentioned earlier. The MRI agent (**31**) is known clinically as Dotarem. Uncomplexed Gd^{3+} is very toxic, and interest has increasingly focused on extending the range of imaging agents based on DOTA and its derivatives which (through the macrocyclic effect) form more stable complexes than those formed with the linear ligand diethylenetriamine-N,N,N',N''-pentaacetic acid (DTPA).

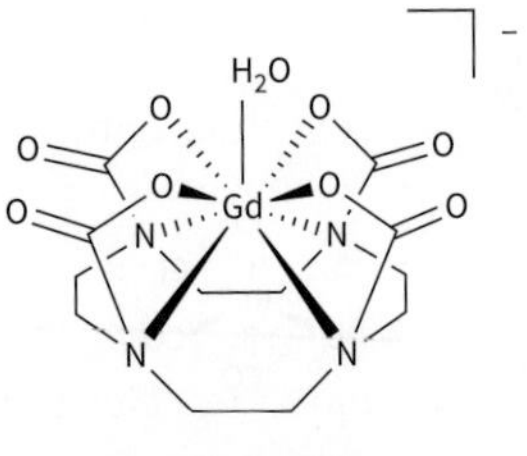

31 Dotarem

Magnevist (**32**)—an MRI agent based on DTPA—has largely been discontinued due to toxicity concerns. Elaborate bioconjugate MRI tracers are being developed that have greater specificity for their targets. An example is the Gd contrast agent EP-210R (Fig. 32.13), which features a Gd^{3+} DOTA complex linked to a peptide that recognizes 'oncoprotein extradomain-B fibronectin' (EDB-FN)—a protein expressed in large amounts by certain cancer cells. Expanded porphyrins ('texaphyrins') in which the macrocycle contains five pyrrole nitrogen donors offer even greater scope for exploiting lanthanoids.

32 Magnevist

Technetium, an artificial element, has found an important use as an imaging agent. The active radionuclide is ^{99m}Tc ('m' denotes metastable), which decays by γ-emission (140 keV) and has a half-life of 6 h. The image is obtained using a scintillation ('gamma') camera. Production of ^{99m}Tc involves bombarding ^{98}Mo with neutrons and separating it as soon as it is formed from the unstable product ^{99}Mo:

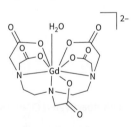

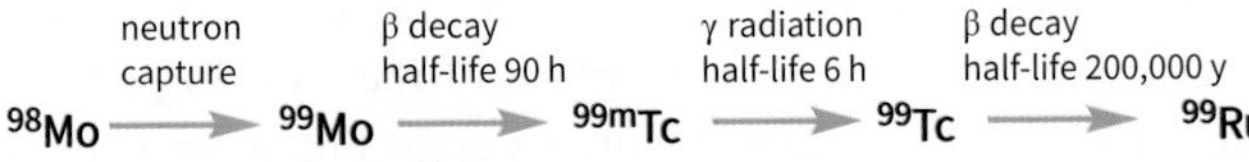

Localized high-energy γ-rays are less harmful to tissue than α- or β-particles. The chemistry of technetium

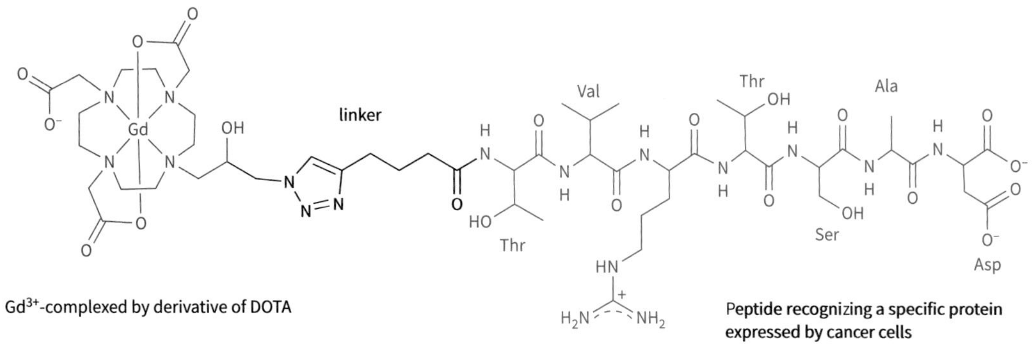

FIGURE 32.13 A bioconjugate Gd contrast agent containing a peptide that recognizes and binds to a protein expressed by certain cancer cells.

resembles manganese except that higher oxidation states are much less oxidizing.

To produce Tc tracers, radioactive $^{99}MoO_4^{2-}$ is passed on to an anion exchange column, where it binds tightly until nuclear decay occurs to give the pertechnate ion $^{99m}TcO_4^-$ and the lower charge causes it to be eluted (Fig. 32.14). The eluate is treated with a reducing agent, usually Sn(II), and the ligands required to convert it into the desired imaging agent. The resulting compound is then administered to the patient at low concentration (~10 nM).

A variety of substitution-inert Tc complexes can be made that, if injected into the patient, rapidly localize in particular tissues and provide information on their status. Complexes have been developed that 'light up' specific organs such as the heart (revealing tissue damage due to a heart attack), kidney (imaging renal function), or bone (revealing cancerous lesions and fracture lines). A good basis for organ targeting appears to be the charge on the complex: cationic complexes target the heart, neutral complexes target the brain, and anionic complexes target the bone and kidney.

Of the different imaging agents, the Tc(I) isonitrile complex $[Tc(CNCH_2C(CH_3)_2OCH_3)_6]^+$ (**33**) is the best established: known as cardiolite, it is widely used as a heart-imaging agent. Cardiolite accumulates in myocardial tissue (heart muscle) but is excreted from the body within two days.

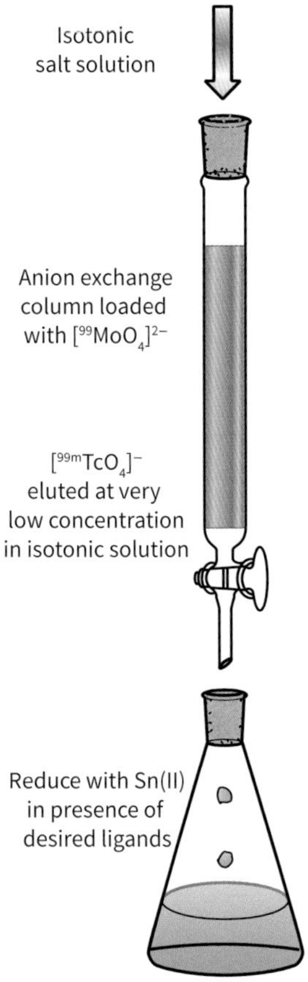

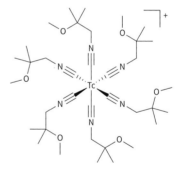

33 Cardiolite

The compound of Tc(V) with mercaptoacetyltriglycine (**34**), known as Tc-MAG3, is used to image kidneys because of its rapid excretion.

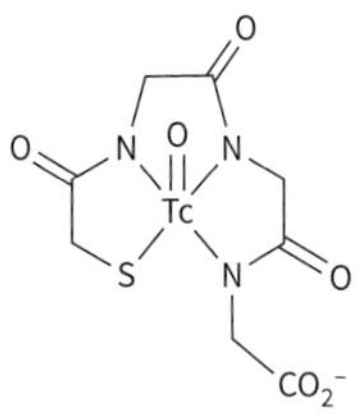

34 Tc-MAG3

Complexes of Tc(VII) with diphosphonate ligands (**35**) are effective in imaging bone: the peripheral hard O-atom

FIGURE 32.14 Principles of preparation of Tc-99m complexes.

donors bind to reactive exposed surface sites, locating stress fractures and other abnormalities.

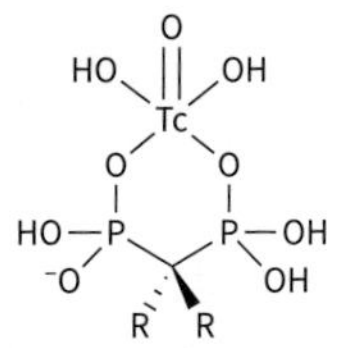

35 Tc(VII) complex with diphosphonate

Brain imaging is carried out with neutral compounds such as Ceretec (**36**).

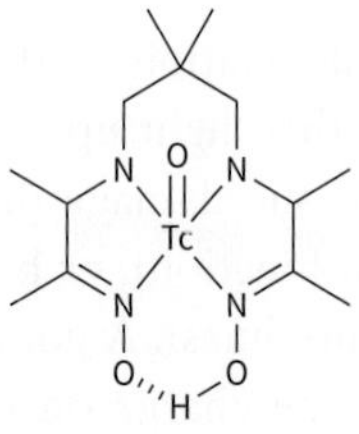

36 Ceretec

Bioconjugates of technetium are being developed to improve specificity further. The compound known as ^{99m}Tc-tilmanocept (Fig. 32.15) is a soluble polysaccharide used to image lymphatic tissue. Each Tc^{3+} cation is coordinated by a DTPA ligand that is linked to the dextrin backbone. A mannose unit, also attached to the backbone (one per Tc(DTPA)), recognizes specific receptors found in lymph nodes.

Gallium-68 is a positron emitter (Table 32.1) and is suitable for imaging by PET. The ^{68}Ga-containing version of (**12**) which contains a somatostatin analogue is used to image neuroendocrine cells.

Thallium-201 (half-life 73 h, Table 32.1) is used as an myocardial imaging agent to observe and diagnose heart malfunction. Injected as a trace amount of ^{201}Tl$^+$(aq) (produced by proton bombardment of ^{203}Tl in a cyclotron), the gamma-emitter is taken up rapidly by muscle cells via transporters that are normally specific for K$^+$, which it closely resembles (Sections 14.1 and 26.4).

32.10 Nanoparticles in directed drug delivery

KEY POINT Drugs can be delivered to specific targets by transport on nanoparticles modified with recognition functionalities.

Nanoparticles offer numerous opportunities as carriers for the selective delivery of a package of drugs or imaging agents into a target zone, usually a solid tumour. Drug carriers usually comprise a protein such as albumin or a specifically tagged liposome (a vesicle) containing crystals of the drug, but inorganic nanomaterials are rapidly emerging as important tools with special advantages, as are smaller cages based on metal–organic frameworks (MOFs) (Section 29.4). The principle of these 'magic warheads' is outlined in Fig. 32.16.

The payload of drugs is released upon reaching the target. The action is triggered by a particular local stimulus, such as a change in pH, exposure to light, or the activity of certain enzymes. Gold nanocages, often coated with a polymer, are promising vehicles, along with mesoporous silica cages and iron-oxide nanoparticles. The light-absorbing properties of Au nanoparticles depend on their size, allowing their application in photothermal therapy.

As well as carrying a lethal payload of drugs, the nanoparticle can also be modified with radioactive nuclei that are able to destroy targeted tissue, an example being ^{177}Lu, a β-particle emitter (Table 32.1) that can be grafted onto Au nanoparticles by modifying them with polymer chains terminating in the Lu analogue of Dotarem (**31**). An important consideration with inorganic nanomaterials is their biocompatibility. Size is an important factor in determining their speed of passage through the kidneys and retention in the bloodstream for a sufficiently long time to locate their target, but biodegradability and long-term toxicity are also relevant.

32.11 Outlook

KEY POINT Future medicine will benefit greatly from the work of scientists with expertise in inorganic chemistry.

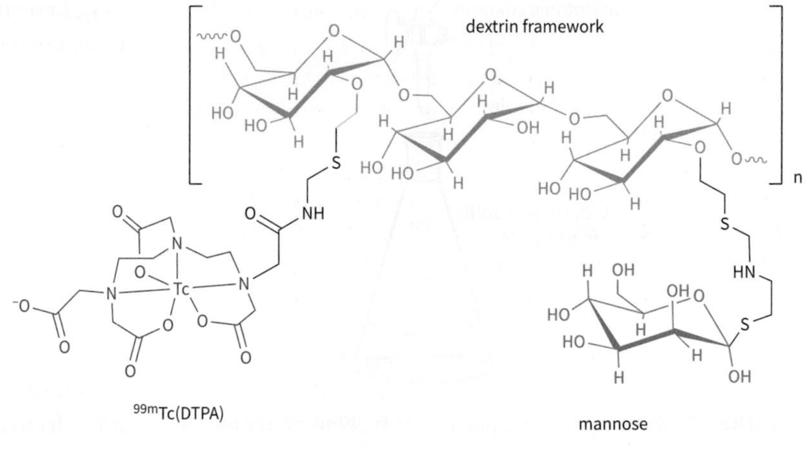

FIGURE 32.15 The structure of 'Tilmanocept', used to image lymphatic tissue. The dextrin backbone is modified by incorporating linkers to Tc(DTPA) complexes and mannose units. The latter recognize specific receptors found at high concentration in lymph nodes.

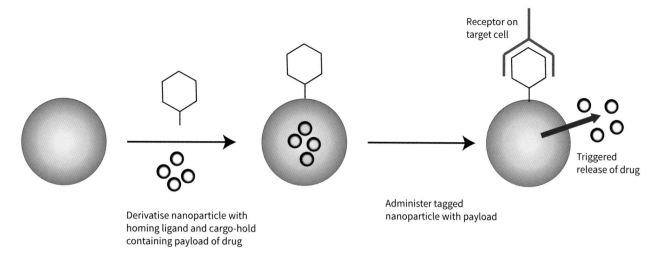

FIGURE 32.16 Targeted delivery of a drug payload by nanoparticles.

The topics outlined in this chapter probably represent only the 'tip of the iceberg' in terms of the opportunities existing for inorganic chemistry to play an important role in medicine. Alongside time-honoured remedies such as Ag compounds used in antibacterial ointments, and necessary 'best option' drugs such as sodium stibogluconate (a complex of antimony(V) used to treat Leishmaniasis, a tropical skin disease), many new medicines are still at early stages of investigation. They range from fluorescent imaging agents for Zn released during a seizure, to metal complexes used in electrochemical sensors.

As stressed in this chapter, a major challenge is in establishing the detailed mechanisms of action of these compounds, the active forms of which may differ greatly from the form in which they are administered.

Advances in inorganic materials and their composites will continue to be an important field as life expectancy increases and natural 'wear and tear' makes greater demands on solid tissue such as bones, joints, and teeth. In addition to technical factors such as biocompatibility, strength, and appearance, the costs of producing the highest-quality materials need to be minimized, ensuring benefits to patients of limited financial means.

FURTHER READING

P. Chellan and P. J. Sadler, The elements of life and medicines. *Phil. Trans. R. Soc. A*, 2015, **373**, 20140182.

N. P. E. Barry and P. J. Sadler, Exploration of the medical periodic table: Towards new targets. *Chem. Commun.*, 2013, **49**, 5106.

E. J. Anthony, E. M. Bolitho, et al., Metallodrugs are unique: Opportunities and challenges of discovery and development. *Chem. Sci.*, 2020, **11**, 12888–917. An excellent and very readable review on progress in different aspects of inorganic therapeutics.

J. Karges, R. W. Stokes, and S. M. Cohen, Metal complexes for therapeutic applications. *Trends in Chemistry*, 2021, **3**, 524–34. A review of current strategies in the search for new drugs featuring metal complexes.

N. P Paul, A. E. Galván, K. Yoshinaga-Sakurai, B. P. Rosen, and M. Yoshinagam, Arsenic in medicine: past, present and future. *Biometals* (2022) https://doi.org/10.1007/s10534-022-00371-y

H. Shi, C. Imberti, and P. J. Sadler, Diazido platinum(IV) complexes for photoactivated anticancer chemotherapy. *Inorg. Chem. Front.*, 2019, **6**, 1623–38.

J. J. Wilson and S. J. Lippard, Synthetic methods for preparation of platinum anticancer complexes. *Chem. Rev.*, 2014, **114**, 4470. A review focusing on strategies for synthesizing Pt anticancer drugs.

Q. Peña, A. Wang, et al., Metallodrugs in cancer nanomedicine. *Chem. Soc. Rev.*, 2022, **51**, 2544–82.

D. M. Griffith, H. Li, M. V. Werrett, P. C. Andrews, and H. Sun, Medicinal chemistry and biomedical applications of bismuth-based compounds and nanoparticles. *Chem. Soc. Rev.*, 2021, **50**, 12037–69.

E. L. M. Ochoa, Lithium as a neuroprotective agent for bipolar disorder: An overview. *Cellular and Molecular Neurobiology*, 2022, **42**, 85–97. https://doi.org/10.1007/s10571-021-01129-9

Angelo Frei, J. Zuegg, et al., Metal complexes as a promising source for new antibiotics. *Chemical Science*, 2020, **11**, 2627–39.

R. E. F. de Paiva, A. Marçal Neto, I. A. Santos, A. C. G. Jardim, P. P. Corbid, and F. R. G. Bergamini, What is holding back the development of antiviral metallodrugs? A literature overview and implications for SARS-CoV-2 therapeutics and future viral outbreaks. *Dalton Trans.*, 2020, **49**, 16004–033.

T. Hanawa, Titanium–tissue interface reaction and its control with surface treatment. *Frontiers in Bioeng. Biotechnol.*, 2019, **7**, 170.

J. Cheng and J. Hu, Recent advances on carbon monoxide releasing molecules for antibacterial applications. *ChemMedChem.*, 2021, **16**, 3628–34. A short review describing the application of CORMs as antimicrobials.

C. Biot, W. Castro, C. Botte, and M. Navarro, The therapeutic potential of metal-based antimalarial agents: implications for the mechanism of action. *Dalton Trans.*, 2012, **41**, 6335.

E. Svensson Grape, V. Booth, M. Nero, T. Willhammar, and A. K. Inge, Structure of the active pharmaceutical ingredient bismuth subsalicylate. *Nature Commun.*, 2022, **13**, 1984.

E. C. Dreaden, A. M. Alkilany, X. Huang, C.J. Murphy, and M. El-Sayed, The golden age: Gold nanoparticles for biomedicine. *Chem. Soc. Rev.*, 2012, **41**, 2740.

D. Papagiannopoulou, Technetium-99m radiochemistry for pharmaceutical applications. *J. Labelled Compounds and Radiopharmaceuticals*, 2017, **60**, 502–520.

Z.-R. Lu, V. Laney, and Y. Li, Targeted contrast agents for magnetic resonance molecular imaging of cancer. *Acc. Chem. Res.*, 2022, **55**, 2833–47.

EXERCISES

32.1 Compounds of Au(III) are under investigation as anticancer drugs. Predict some of the similarities and contrasts with Pt(II) compounds.

32.2 What properties of gadolinium (a) make it suitable for imaging, or (b) may render it so toxic when unbound?

32.3 Although artificial knee and hip joints are made from titanium, the screws and plates that are used to hold bone together after a fracture are usually made from stainless steel. Explain why titanium is not used for this purpose.

32.4 With reference to appropriate sections of this textbook, list some reasons why ZrO_2 is widely used as a material in dental implants (crowns).

32.5 Boranocarbonate ($[H_3BCO_2]^{2-}$) has been considered as a promising CORM candidate. It is stable in alkali but decomposes slowly to release CO once introduced to neutral or mildly acidic solutions such as the bloodstream. Predict the products of decomposition of boranocarbonate and propose a mechanism.

32.6 Identify particular chemical properties of bismuth that suit it for its special role in treating gastric ailments, noting that the stomach environment is highly acidic.

32.7 What properties of technetium make it so useful for imaging tissue?

TUTORIAL PROBLEMS

32.1 Write short essays on selected elements of the periodic table with regard to their medical uses.

32.2 In their article 'Targeting and delivery of platinum-based anticancer drugs' (*Chem. Soc. Rev.* 2012, **42**, 202), X. Wang and Z. Guo review the expanding field of nanoparticle-based drug delivery. Summarize the various ways that metal-containing drugs are attached to nanostructures, and the advantages of modifying a drug in such a way.

32.3 Write an essay on the different ways that metallobioconjugates are being used in medicine.

Resource section 1
Selected ionic radii

Ionic radii are given (in picometres, pm) for the most common oxidation states and coordination geometries. The coordination number is given in parentheses: (4) refers to tetrahedral and (4SP) refers to square-planar. All d-block species are low-spin unless labelled with †, in which case values for high-spin are quoted. Most data are taken from R. D. Shannon, *Acta Crystallogr.*, 1976, **A32**, 751, where values for other coordination geometries can be found. Where Shannon values are not available, Pauling ionic radii are quoted and are indicated by *.

Group 1
- Li+: 59 (4), 76 (6), 92 (8)
- Na+: 99 (4), 102 (6), 118 (8)
- K+: 137 (4), 138 (6), 151 (8)

Group 2
- Be2+: 27 (4), 45 (6)
- Mg2+: 57 (4), 72 (6), 89 (8)
- Ca2+: 100 (6), 112 (8)

Group 3
- Sc3+: 75 (6), 87 (8)

Group 4
- Ti4+: 42 (4), 61 (6), 74 (8)
- Ti3+: 67 (6)
- Ti2+: 86 (6)

Group 5
- V5+: 36 (4), 54 (6)
- V4+: 58 (6), 72 (8)
- V3+: 64 (6)
- V2+: 79 (6)

Group 6
- Cr6+: 26 (4), 44 (6)
- Cr5+: 49 (6)
- Cr4+: 41 (4), 55 (6)
- Cr3+: 62 (6)
- Cr2+: 73 (6)

Group 7
- Mn7+: 25 (4), 46 (6)
- Mn6+: 26 (4)
- Mn5+: 33 (4)
- Mn4+: 37 (4), 53 (6)
- Mn3+: 65 (6)
- Mn2+: 67 (6), 96 (8)

Group 8
- Fe6+: 25 (4)
- Fe4+: 58 (6)
- Fe3+: 49 (4)†, 55 (6), 78 (8)†
- Fe2+: 63 (4)†, 61 (6), 92 (8)†

Group 9
- Co4+: 40 (4), 53 (6)†
- Co3+: 55 (6)
- Co2+: 58 (4)†, 65 (6), 90 (8)

Group 10
- Ni4+: 48 (6)
- Ni3+: 56 (6)
- Ni2+: 55 (4), 69 (6)

Group 11
- Cu3+: 54 (6)
- Cu2+: 57 (4), 73 (6)
- Cu+: 60 (4), 77 (6)

Group 12
- Zn2+: 60 (4), 74 (6), 90 (8)

Group 13
- B3+: 11 (4), 27 (6)
- Al3+: 39 (4), 53 (6)
- Ga3+: 47 (4), 62 (6)

Group 14
- C4+: 15 (4), 16 (6)
- Si4+: 26 (4), 40 (6)
- Ge4+: 39 (4), 53 (6)
- Ge2+: 73 (6)

Group 15
- N3−: 146 (4), 140 (6), 142 (8)
- N3+: 16 (6)
- P5+: 17 (4), 38 (6)
- P3+: 44 (6)
- P3−: 192 (4)
- As5+: 34 (4), 46 (6)
- As3+: 58 (6)

Group 16
- O2−: 138 (4), 140 (6), 142 (8)
- S2−: 184 (6)
- S6+: 12 (4), 29 (6)
- S4+: 37 (6)
- Se2−: 198 (6)
- Se6+: 28 (4), 42 (6)
- Se4+: 50 (6)

Group 17
- F−: 131 (4), 133 (6)
- Cl−: 181 (6)
- Cl7+: 8 (4), 27 (6)
- Br−: 196 (6)
- Br7+: 39 (6)

Group 18
- Ne+: 112*
- Ar+: 154*
- Kr+: 169*

1	2	3	4	5	6	7	8	9	10	11	12	13	14	15	16	17	18
Rb^+	Sr^{2+}	Y^{3+}	Zr^{4+}	Nb^{5+}	Mo^{6+}	Tc^{7+}	Ru^{8+}	Rh^{5+}	Pd^{4+}	Ag^{3+}	Cd^{2+}	In^{3+}	Sn^{4+}	Sb^{5+}	Te^{6+}	I^-	Xe^+
152 (6)	118 (6)	90 (6)	59 (4)	48 (4)	41 (4)	37 (4)	36 (4)	55 (6)	62 (6)	67 (4)	78 (4)	62 (4)	55 (4)	60 (6)	43 (4)	220 (6)	190*
160 (8)	126 (8)	102 (8)	72 (6)	64 (6)	59 (6)	56 (6)				75 (6)	95 (6)	80 (6)	69 (6)		56 (6)		
			84 (8)	74 (8)							110 (8)	92 (8)	81 (8)				
				Nb^{4+}	Mo^{5+}	Tc^{5+}	Ru^{7+}	Rh^{4+}	Pd^{3+}	Ag^{2+}			Sn^{2+}	Sb^{3+}	Te^{4+}	I^{7+}	Xe^{8+}
				68 (6)	46 (4)	60 (6)	38 (4)	60 (6)	76 (6)	79 (4)			102 (6)	76 (6)	66 (4)	42 (4)	40 (4)
				79 (8)	61 (6)					94 (6)					97 (6)	53 (6)	48 (6)
				Nb^{3+}	Mo^{4+}	Tc^{4+}	Ru^{5+}	Rh^{3+}	Pd^{2+}	Ag^+							
				72 (6)	65 (6)	66 (6)	57 (6)	67 (6)	64 (4)	67 (2)							
						95 (8)			86 (6)	100 (4)							
										115 (6)							
					Mo^{3+}		Ru^{4+}		Pd^+								
					69 (6)		62 (6)		59 (2)								
							Ru^{3+}										
							68 (6)										
Cs^+	Ba^{2+}	La^{3+}	Hf^{4+}	Ta^{5+}	W^{6+}	Re^{7+}	Os^{8+}	Ir^{5+}	Pt^{5+}	Au^{5+}	Hg^{2+}	Tl^{3+}	Pb^{4+}	Bi^{5+}	Po^{6+}	At^{7+}	
167 (6)	135 (6)	103 (6)	58 (4)	64 (4)	42 (4)	38 (4)	39 (4)	57 (6)	57 (6)	57 (6)	96 (4)	75 (4)	65 (4)	76 (6)	67 (6)	62 (6)	
174 (8)	142 (8)	116 (8)	71 (6)	74 (8)	60 (6)	53 (6)					102 (6)	89 (6)	78 (6)				
			83 (8)								114 (8)	98 (8)	94 (8)				
				Ta^{4+}	W^{5+}	Re^{6+}	Os^{7+}	Ir^{4+}	Pt^{4+}	Au^{3+}	Hg^+	Tl^+	Pb^{2+}	Bi^{3+}	Po^{4+}		
				68 (6)	62 (6)	55 (6)	53 (6)	63 (6)	63 (6)	68 (4SP)	119 (6)	150 (6)	119 (6)	103 (6)	94 (6)		
										85 (6)		159 (8)	129 (8)	117 (8)	108 (8)		
				Ta^{3+}	W^{4+}	Re^{5+}	Os^{6+}	Ir^{3+}	Pt^{2+}	Au^+							
				72 (6)	66 (6)	58 (6)	55 (6)	68 (6)	60 (4SP)	137 (6)							
									80 (6)								
						Re^{4+}	Os^{5+}										
						63 (6)	58 (6)										
							Os^{4+}										
							63 (6)										
Fr^+	Ra^{2+}	Ac^{3+}															
180 (6)	148 (8)	81 (6)															
	170 (12)																

Lanthanoids

Ce^{4+}	Pr^{4+}	Nd^{3+}	Pm^{3+}	Sm^{3+}	Eu^{3+}	Gd^{3+}	Tb^{4+}	Dy^{3+}	Ho^{3+}	Er^{3+}	Tm^{3+}	Yb^{3+}	Lu^{3+}
87 (6)	85 (6)	98 (6)	97 (6)	96 (6)	95 (6)	94 (6)	76 (6)	91 (6)	90 (6)	89 (6)	88 (6)	87 (6)	86 (6)
97 (8)	96 (8)	111 (8)	109 (8)	108 (8)	107 (8)	105 (8)	88 (8)	103 (8)	102 (8)	100 (8)	99 (8)	99 (8)	98 (8)
Ce^{3+}	Pr^{3+}	Nd^{2+}		Sm^{2+}	Eu^{2+}		Tb^{3+}	Dy^{2+}			Tm^{2+}	Yb^{2+}	
101 (6)	99 (6)	129 (8)		127 (8)	117 (6)		92 (6)	107 (6)			103 (6)	102 (6)	
114 (8)	113 (8)				125 (8)		104 (8)	119 (8)			109 (8)	114 (8)	

Actinoids

Th^{4+}	Pa^{5+}	U^{6+}	Np^{7+}	Pu^{6+}	Am^{4+}	Cm^{4+}	Bk^{4+}	Cf^{4+}	Es	Fm	Md	No^{2+}	Lr
94 (6)	78 (6)	52 (4)	71 (6)	71 (6)	85 (6)	85 (6)	83 (6)	82 (6)				110 (6)	
105 (8)	91 (8)	73 (6)			95 (8)	95 (8)	93 (8)	92 (8)					
		86 (8)											
	Pa^{4+}	U^{5+}	Np^{6+}	Pu^{5+}	Am^{3+}	Cm^{3+}	Bk^{3+}	Cf^{3+}					
	90 (6)	76 (6)	72 (6)	74 (6)	98 (6)	97 (6)	96 (6)	95 (6)					
	101 (8)				109 (8)								

	Pa³⁺	U⁴⁺	Np⁵⁺	Pu⁴⁺	Am²⁺
	104 (6)	89 (6)	75 (6)	86 (6)	126 (8)
		100 (8)	96 (8)		
		U³⁺	Np⁴⁺	Pu³⁺	
		103 (6)	87 (4)	100 (6)	
			98 (6)		
			Np³⁺		
			101 (6)		
			Np²⁺		
			110 (6)		

Resource section 2
Electronic properties of the elements

Ground-state electron configurations of atoms are determined experimentally from spectroscopic and magnetic measurements. The results of these determinations are listed below. They can be rationalized in terms of the building-up principle, in which electrons are added to the available orbitals in a specific order in accord with the Pauli exclusion principle. Some variation in order is encountered in the d and f blocks to accommodate the effects of electron–electron interaction more faithfully. The closed-shell configuration $1s^2$ characteristic of helium is denoted [He], and likewise for the other noble-gas element configurations. The ground-state electron configurations and term symbols listed below have been taken from S. Fraga, J. Karwowski, and K. M. S. Saxena, *Handbook of Atomic Data*. Elsevier (1976).

The first three ionization energies of an element E are the energies required for the following processes:

$$I_1 : E(g) \rightarrow E^+(g) + e^-(g)$$
$$I_2 : E^+(g) \rightarrow E^{2+}(g) + e^-(g)$$
$$I_3 : E^{2+}(g) \rightarrow E^{3+}(g) + e^-(g)$$

The electron affinity E_{ea} is the energy *released* when an electron attaches to a gas-phase atom:

$$E_{ea} : E(g) + e^-(g) \rightarrow E^-(g)$$

The values given here are taken from various sources, particularly C. E. Moore, *Atomic Energy Levels*, NBS Circular 467, Washington (1970), and W. C. Martin, L. Hagan, J. Reader, and J. Sugar, *J. Phys. Chem. Ref. Data*, 1974, **3**, 771. Values for the actinoids are taken from J. J. Katz, G. T. Seaborg, and L. R. Morss (eds.), *The Chemistry of the Actinide Elements*. Chapman & Hall (1986). Electron affinities are from H. Hotop and W.C. Lineberger, *J. Phys. Chem. Ref. Data*, 1985, **14**, 731.

For conversions to kilojoules per mole and reciprocal centimetres, see Resource section 7.

Atom			Ionization energy (eV)			Electron affinity E_{ea} (eV)
			I_1	I_2	I_3	
1	H	$1s^1$	13.60			+0.754
2	He	$1s^2$	24.59	54.51		−0.5
3	Li	[He]$2s^1$	5.320	75.63	122.4	+0.618
4	Be	[He]$2s^2$	9.321	18.21	153.85	≤0
5	B	[He]$2s^2 2p^1$	8.297	25.15	37.93	+0.277
6	C	[He]$2s^2 2p^2$	11.257	24.38	47.88	+1.263
7	N	[He]$2s^2 2p^3$	14.53	29.60	47.44	−0.07
8	O	[He]$2s^2 2p^4$	13.62	35.11	54.93	+1.461
9	F	[He]$2s^2 2p^5$	17.42	34.97	62.70	+3.399
10	Ne	[He]$2s^2 2p^6$	21.56	40.96	63.45	−1.2
11	Na	[Ne]$3s^1$	5.138	47.28	71.63	+0.548
12	Mg	[Ne]$3s^2$	7.642	15.03	80.14	≤0
13	Al	[Ne]$3s^2 3p^1$	5.984	18.83	28.44	+0.441
14	Si	[Ne]$3s^2 3p^2$	8.151	16.34	33.49	+1.385
15	P	[Ne]$3s^2 3p^3$	10.485	19.72	30.18	+0.747
16	S	[Ne]$3s^2 3p^4$	10.360	23.33	34.83	+2.077
17	Cl	[Ne]$3s^2 3p^5$	12.966	23.80	39.65	+3.617
18	Ar	[Ne]$3s^2 3p^6$	15.76	27.62	40.71	−1.0
19	K	[Ar]$4s^1$	4.340	31.62	45.71	+0.502
20	Ca	[Ar]$4s^2$	6.111	11.87	50.89	+0.02
21	Sc	[Ar]$3d^1 4s^2$	6.54	12.80	24.76	
22	Ti	[Ar]$3d^2 4s^2$	6.82	13.58	27.48	
23	V	[Ar]$3d^3 4s^2$	6.74	14.65	29.31	
24	Cr	[Ar]$3d^5 4s^1$	6.764	16.50	30.96	
25	Mn	[Ar]$3d^5 4s^2$	7.435	15.64	33.67	
26	Fe	[Ar]$3d^6 4s^2$	7.869	16.18	30.65	
27	Co	[Ar]$3d^7 4s^2$	7.876	17.06	33.50	
28	Ni	[Ar]$3d^8 4s^2$	7.635	18.17	35.16	
29	Cu	[Ar]$3d^{10} 4s^1$	7.725	20.29	36.84	
30	Zn	[Ar]$3d^{10} 4s^2$	9.393	17.96	39.72	
31	Ga	[Ar]$3d^{10} 4s^2 4p^1$	5.998	20.51	30.71	+0.30
32	Ge	[Ar]$3d^{10} 4s^2 4p^2$	7.898	15.93	34.22	+1.2
33	As	[Ar]$3d^{10} 4s^2 4p^3$	9.814	18.63	28.34	+0.81
34	Se	[Ar]$3d^{10} 4s^2 4p^4$	9.751	21.18	30.82	+2.021
35	Br	[Ar]$3d^{10} 4s^2 4p^5$	11.814	21.80	36.27	+3.365
36	Kr	[Ar]$3d^{10} 4s^2 4p^6$	13.998	24.35	36.95	−1.0

Atom			Ionization energy (eV)			Electron affinity E_{ea} (eV)
			I_1	I_2	I_3	
37	Rb	[Kr]5s^1	4.177	27.28	40.42	+0.486
38	Sr	[Kr]5s^2	5.695	11.03	43.63	+0.05
39	Y	[Kr]4d^{1}5s^2	6.38	12.24	20.52	
40	Zr	[Kr]4d^{2}5s^2	6.84	13.13	22.99	
41	Nb	[Kr]4d^{4}5s^1	6.88	14.32	25.04	
42	Mo	[Kr]4d^{5}5s^1	7.099	16.15	27.16	
43	Tc	[Kr]4d^{5}5s^2	7.28	15.25	29.54	
44	Ru	[Kr]4d^{7}5s^1	7.37	16.76	28.47	
45	Rh	[Kr]4d^{8}5s^1	7.46	18.07	31.06	
46	Pd	[Kr]4d^{10}	8.34	19.43	32.92	
47	Ag	[Kr]4d^{10}5s^1	7.576	21.48	34.83	
48	Cd	[Kr]4d^{10}5s^2	8.992	16.90	37.47	
49	In	[Kr]4d^{10}5s^{2}5p^1	5.786	18.87	28.02	+0.3
50	Sn	[Kr]4d^{10}5s^{2}5p^2	7.344	14.63	30.50	+1.2
51	Sb	[Kr]4d^{10}5s^{2}5p^3	8.640	18.59	25.32	+1.07
52	Te	[Kr]4d^{10}5s^{2}5p^4	9.008	18.60	27.96	+1.971
53	I	[Kr]4d^{10}5s^{2}5p^5	10.45	19.13	33.16	+3.059
54	Xe	[Kr]4d^{10}5s^{2}5p^6	12.130	21.20	32.10	−0.8
55	Cs	[Xe]6s^1	3.894	25.08	35.24	
56	Ba	[Xe]6s^2	5.211	10.00	37.51	
57	La	[Xe]5d^{1}6s^2	5.577	11.06	19.17	
58	Ce	[Xe]4f^{1}5d^{1}6s^2	5.466	10.85	20.20	
59	Pr	[Xe]4f^{3}6s^2	5.421	10.55	21.62	
60	Nd	[Xe]4f^{4}6s^2	5.489	10.73	22.1	
61	Pm	[Xe]4f^{5}6s^2	5.554	10.90	22.28	
62	Sm	[Xe]4f^{6}6s^2	5.631	11.07	23.42	
63	Eu	[Xe]4f^{7}6s^2	5.666	11.24	24.91	
64	Gd	[Xe]4f^{7}5d^{1}6s^2	6.140	12.09	20.62	
65	Tb	[Xe]4f^{9}6s^2	5.851	11.52	21.91	
66	Dy	[Xe]4f^{10}6s^2	5.927	11.67	22.80	
67	Ho	[Xe]4f^{11}6s^2	6.018	11.80	22.84	
68	Er	[Xe]4f^{12}6s^2	6.101	11.93	22.74	
69	Tm	[Xe]4f^{13}6s^2	6.184	12.05	23.68	
70	Yb	[Xe]4f^{14}6s^2	6.254	12.19	25.03	
71	Lu	[Xe]4f^{14}5d^{1}6s^2	5.425	13.89	20.96	
72	Hf	[Xe]4f^{14}5d^{2}6s^2	6.65	14.92	23.32	
73	Ta	[Xe]4f^{14}5d^{3}6s^2	7.89	15.55	21.76	
74	W	[Xe]4f^{14}5d^{4}6s^2	7.89	17.62	23.84	
75	Re	[Xe]4f^{14}5d^{5}6s^2	7.88	13.06	26.01	
76	Os	[Xe]4f^{14}5d^{6}6s^2	8.71	16.58	24.87	
77	Ir	[Xe]4f^{14}5d^{7}6s^2	9.12	17.41	26.95	
78	Pt	[Xe]4f^{14}5d^{9}6s^1	9.02	18.56	29.02	
79	Au	[Xe]4f^{14}5d^{10}6s^1	9.22	20.52	30.05	
80	Hg	[Xe]4f^{14}5d^{10}6s^2	10.44	18.76	34.20	
81	Tl	[Xe]4f^{14}5d^{10}6s^{2}6p^1	6.107	20.43	29.83	
82	Pb	[Xe]4f^{14}5d^{10}6s^{2}6p^2	7.415	15.03	31.94	
83	Bi	[Xe]4f^{14}5d^{10}6s^{2}6p^3	7.289	16.69	25.56	
84	Po	[Xe]4f^{14}5d^{10}6s^{2}6p^4	8.42	18.66	27.98	
85	At	[Xe]4f^{14}5d^{10}6s^{2}6p^5	9.64	16.58	30.06	
86	Rn	[Xe]4f^{14}5d^{10}6s^{2}6p^6	10.75			
87	Fr	[Rn]7s^1	4.15	21.76	32.13	
88	Ra	[Rn]7s^2	5.278	10.15	34.20	
89	Ac	[Rn]6d^{1}7s^2	5.17	11.87	19.69	
90	Th	[Rn]6d^{2}7s^2	6.08	11.89	20.50	
91	Pa	[Rn]5f^{2}6d^{1}7s^2	5.89	11.7	18.8	
92	U	[Rn]5f^{3}6d^{1}7s^2	6.19	10.6	19.1	
93	Np	[Rn]5f^{4}6d^{1}7s^2	6.27	11.7	19.4	
94	Pu	[Rn]5f^{6}7s^2	6.06	11.7	21.8	
95	Am	[Rn]5f^{7}7s^2	5.99	12.0	22.4	
96	Cm	[Rn]5f^{7}6d^{1}7s^2	6.02	12.4	21.2	
97	Bk	[Rn]5f^{9}7s^2	6.23	12.3	22.3	
98	Cf	[Rn]5f^{10}7s^2	6.30	12.5	23.6	
99	Es	[Rn]5f^{11}7s^2	6.42	12.6	24.1	
100	Fm	[Rn]5f^{12}7s^2	6.50	12.7	24.4	
101	Md	[Rn]5f^{13}7s^2	6.58	12.8	25.4	
102	No	[Rn]5f^{14}7s^2	6.65	13.0	27.0	
103	Lr	[Rn]5f^{14}6d^{1}7s^2	4.96	14.8	23.0	
104	Rf	[Rn]5f^{14}6d^{2}7s^2	6.02	14.3	23.8	
105	Db	[Rn]5f^{14}6d^{3}7s^2	6.80	14.0	23.1	
106	Sg	[Rn]5f^{14}6d^{4}7s^2	7.80	17.1	25.8	
107	Bh	[Rn]5f^{14}6d^{5}7s^2	7.70	17.5	26.7	
108	Hs	[Rn]5f^{14}6d^{6}7s^2	7.60	18.2	29.3	

Resource section 3
Standard potentials

The standard potentials quoted here are presented in the form of Latimer diagrams (Section 6.12) and are arranged according to the blocks of the periodic table in the order s, p, d, f. Data and species in parentheses are uncertain. Most of the data, together with occasional corrections, come from A. J. Bard, R. Parsons, and J. Jordan (eds.), *Standard Potentials in Aqueous Solution*. Marcel Dekker (1985). Data for the actinoids are from L. R. Morss, *The Chemistry of the Actinide Elements*, Vol. 2 (ed. J. J. Katz, G. T. Seaborg, and L. R. Morss). Chapman & Hall (1986). The value for $[Ru(bpy)_3]^{3+/2+}$ is from B. Durham, J. L. Walsh, C. L. Carter, and T. J. Meyer, *Inorg. Chem.*, 1980, **19**, 860. Potentials for carbon species and some d-block elements are taken from S. G. Bratsch, *J. Phys. Chem. Ref. Data*, 1989, **18**, 1. For further information on standard potentials of unstable radical species, see D. M. Stanbury, *Adv. Inorg. Chem.*, 1989, **33**, 69. Potentials in the literature are occasionally reported relative to the standard calomel electrode (SCE), and may be converted to the H^+/H_2 scale by adding 0.2412 V. For a detailed discussion of other reference electrodes, see D. J. G. Ives and G. J. Janz, *Reference Electrodes*. Academic Press (1961).

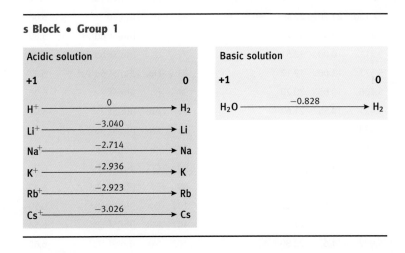

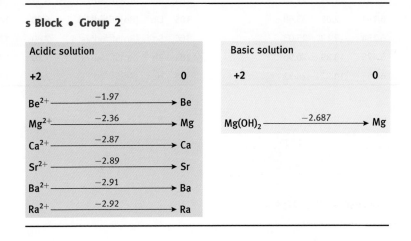

p Block • Group 13

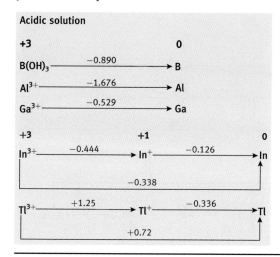

p Block • Group 14

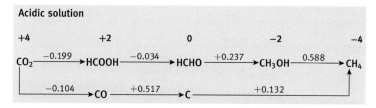

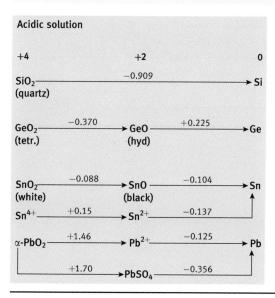
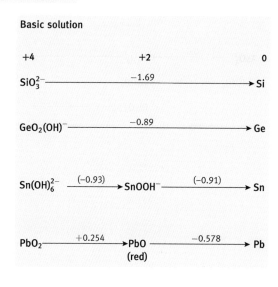

p Block • Group 15

Acidic solution

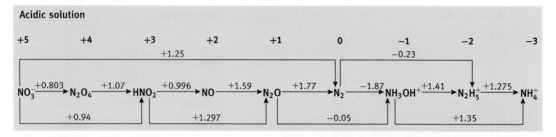

Basic solution

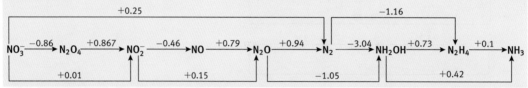

Acidic solution

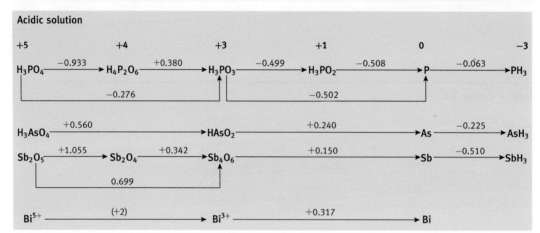

Basic solution

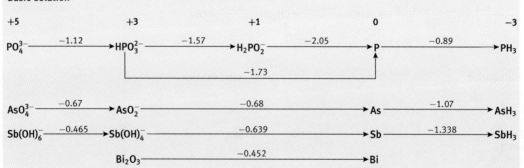

p Block • Group 16

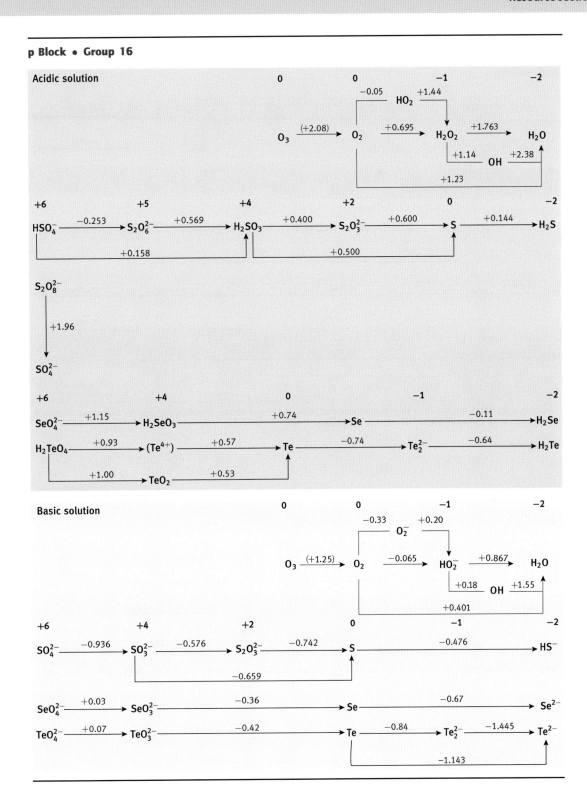

p Block • Group 17

Acidic solution

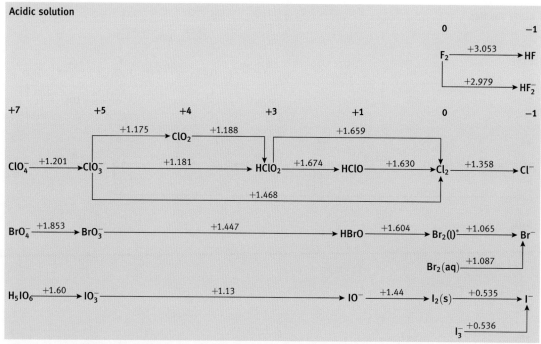

*Bromine is not sufficiently soluble in water at room temperature to achieve unit activity. Therefore, the value for a saturated solution in contact with $Br_2(l)$ should be used in all practical calculations.

Basic solution

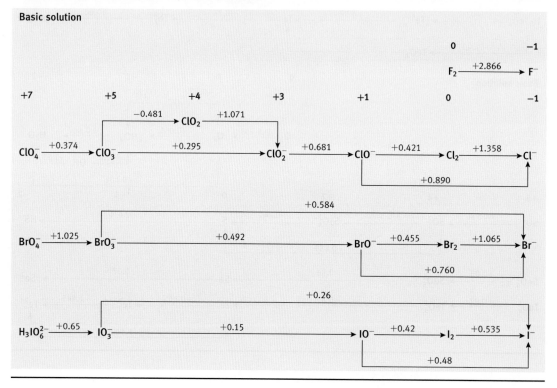

p Block • Group 18

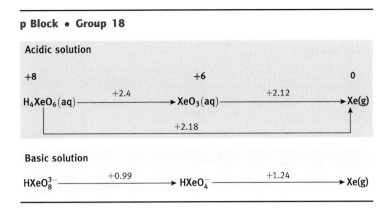

d Block • Group 3

Acidic solution

+3   0

$Sc^{3+} \xrightarrow{-2.03} Sc$

$Y^{3+} \xrightarrow{-2.37} Y$

$La^{3+} \xrightarrow{-2.38} La$

Basic solution

+3   0

$Sc(OH)_3 \xrightarrow{-2.60} Sc$

d Block • Group 4

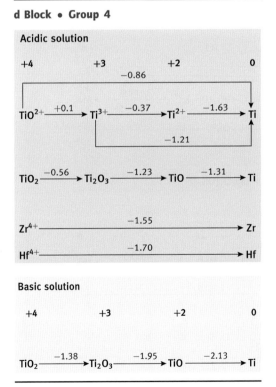

d Block • Group 5

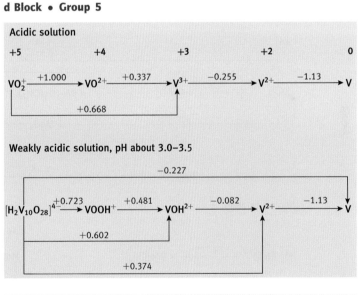

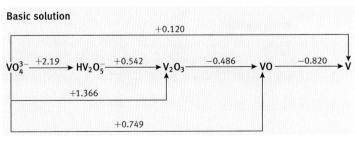

d Block • Group 5 (Continued)

Acidic solution

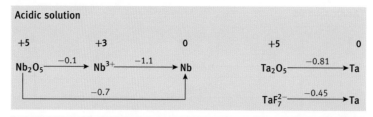

d Block • Group 6

Acidic solution

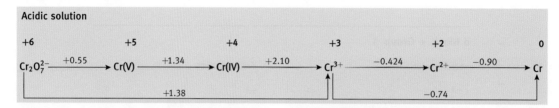

Neutral solution

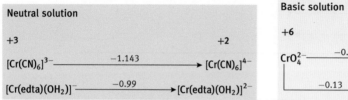

Basic solution

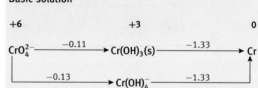

Acidic solution

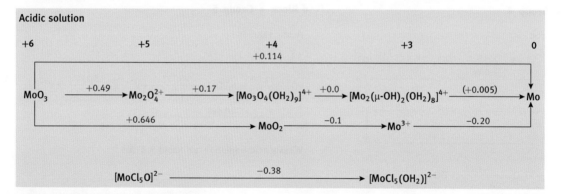

Neutral solution

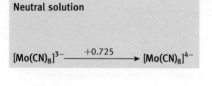

Basic solution

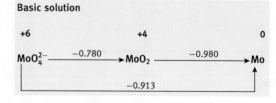

Acidic solution

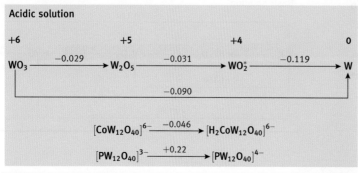

d Block • Group 6 (Continued)

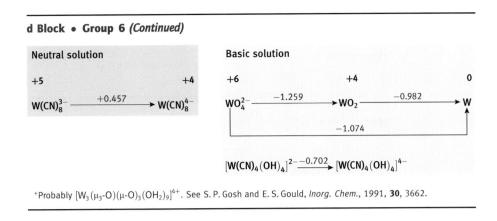

*Probably $[W_3(\mu_3\text{-O})(\mu\text{-O})_3(OH_2)_9]^{4+}$. See S. P. Gosh and E. S. Gould, *Inorg. Chem.*, 1991, **30**, 3662.

d Block • Group 7

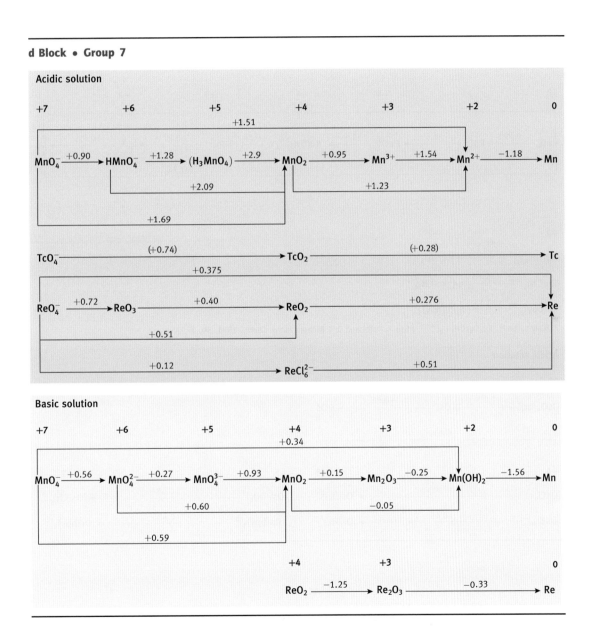

d Block • Group 8

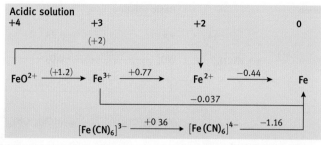

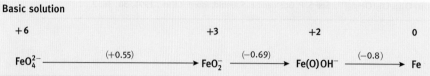

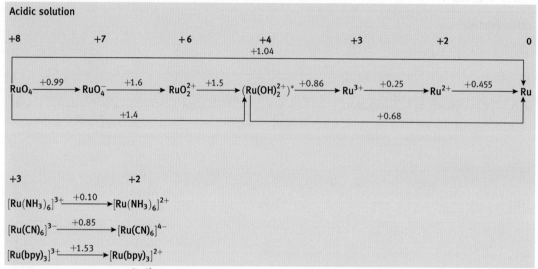

*Likely to be $H_n[Ru_4O_6(OH_2)_{12}]^{(4+n)+}$. See A. Patel and D. T. Richen, *Inorg. Chem.*, 1991, **30**, 3792.

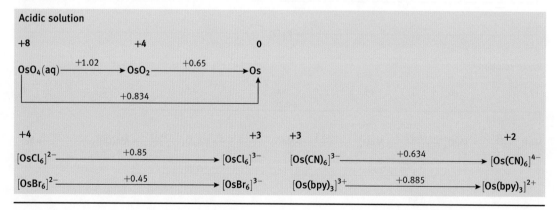

d Block • Group 9

Acidic solution

$$CoO_2 \xrightarrow{+1.4} Co^{3+} \xrightarrow{+1.92} Co^{2+} \xrightarrow{-0.282} Co$$
(+4) (+3) (+2) (0)

Basic solution

$$CoO_2 \xrightarrow{(+0.7)} Co(O)OH \xrightarrow{(-0.22)} Co(OH)_2 \xrightarrow{-0.873} Co$$
(+4) (+3) (+2) (0)

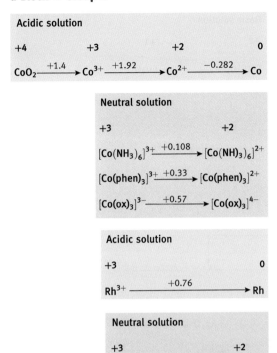

Neutral solution

+3 → +2

$[Co(NH_3)_6]^{3+} \xrightarrow{+0.108} [Co(NH_3)_6]^{2+}$

$[Co(phen)_3]^{3+} \xrightarrow{+0.33} [Co(phen)_3]^{2+}$

$[Co(ox)_3]^{3-} \xrightarrow{+0.57} [Co(ox)_3]^{4-}$

Acidic solution

+3 → 0

$Rh^{3+} \xrightarrow{+0.76} Rh$

Neutral solution

+3 → +2

$[Rh(CN)_6]^{3-} \xrightarrow{+0.9} [Rh(CN)_6]^{4-}$

Acidic solution (Ir)

+4, +3, 0

$IrO_2 \xrightarrow{+0.23} (Ir^{3+}) \xrightarrow{+1.16} Ir$

$IrO_2 \xrightarrow{+0.93} Ir$

$[IrCl_6]^{2-} \xrightarrow{+0.867} [IrCl_6]^{3-} \xrightarrow{+0.86} Ir$

$[IrBr_6]^{2-} \xrightarrow{+0.805} [IrBr_6]^{3-}$

$[IrI_6]^{2-} \xrightarrow{+0.49} [IrI_6]^{3-}$

d Block • Group 10

Acidic solution

+4, +2, 0

$NiO_2 \xrightarrow{+1.59} Ni^{2+} \xrightarrow{-0.257} Ni$

Basic solution

+4, +3, +2, 0

$NiO_2 \xrightarrow{+0.7} NiOOH \xrightarrow{+0.52} Ni(OH)_2 \xrightarrow{-0.72} Ni$

Neutral solution

+2, 0

$[Ni(NH_3)_6]^{2+} \xrightarrow{-0.49} Ni$

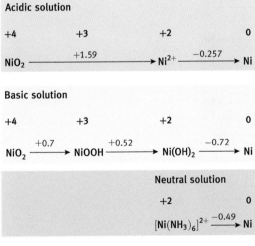

Acidic solution (Pd)

+4, +2, 0

$PdO_2 \xrightarrow{+1.194} Pd^{2+} \xrightarrow{+0.915} Pd$

$[PdCl_6]^{2-} \xrightarrow{+1.47} [PdCl_4]^{2-} \xrightarrow{+0.60} Pd$

$[PdBr_4]^{2-} \xrightarrow{+0.49} Pd$

Basic solution (Pd)

+4, +2, 0

$PdO_2 \xrightarrow{+1.47} PdO \xrightarrow{+0.897} Pd$

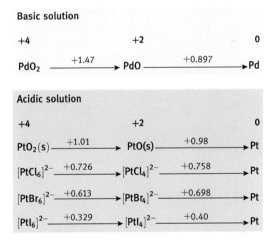

Acidic solution (Pt)

+4, +2, 0

$PtO_2(s) \xrightarrow{+1.01} PtO(s) \xrightarrow{+0.98} Pt$

$[PtCl_6]^{2-} \xrightarrow{+0.726} [PtCl_4]^{2-} \xrightarrow{+0.758} Pt$

$[PtBr_6]^{2-} \xrightarrow{+0.613} [PtBr_4]^{2-} \xrightarrow{+0.698} Pt$

$[PtI_6]^{2-} \xrightarrow{+0.329} [PtI_4]^{2-} \xrightarrow{+0.40} Pt$

d Block • Group 11

Acidic solution

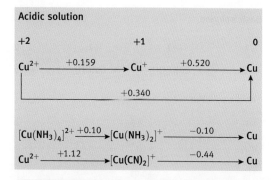

Basic solution

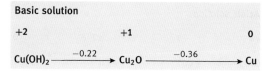

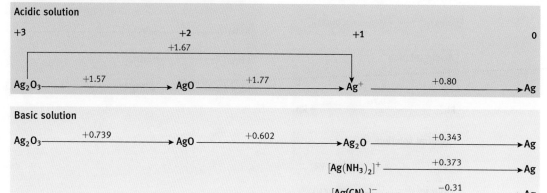

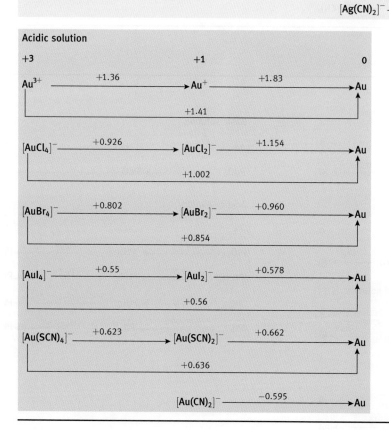

d Block • Group 12

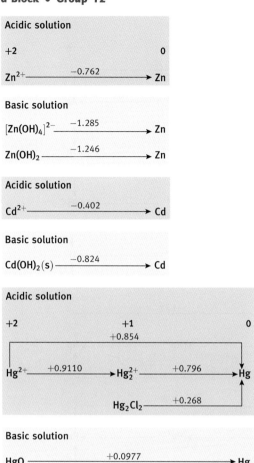

f Block • Lanthanoids

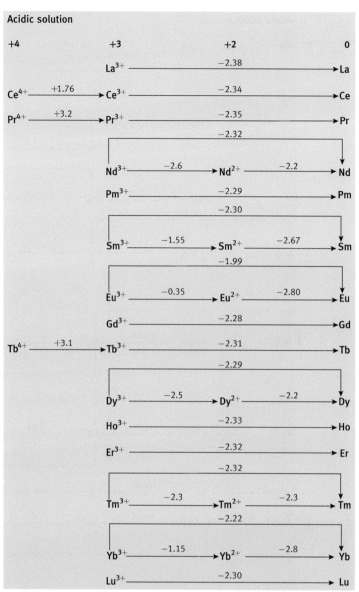

f Block • Actinoids

Acidic solution

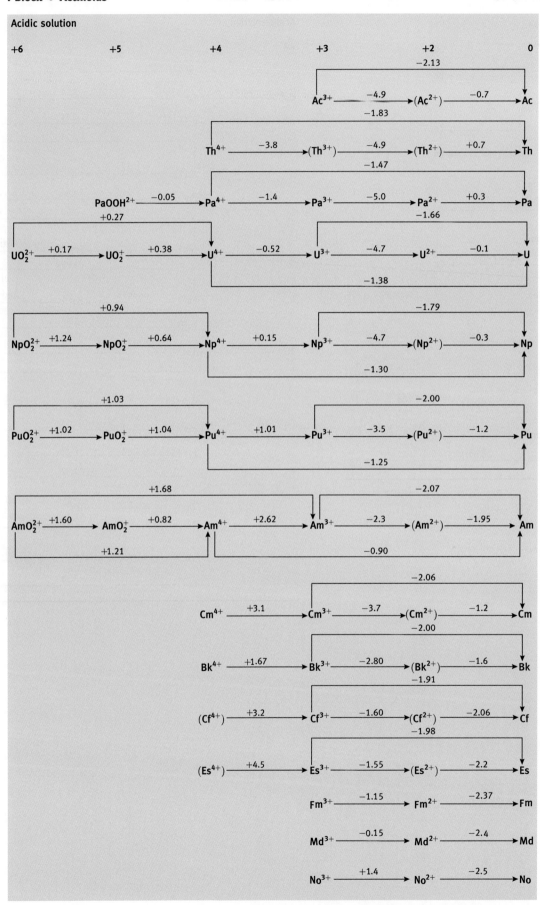

Resource section 4
Character tables

The character tables that follow are for the most common point groups encountered in inorganic chemistry. Each one is labelled with the symbol adopted in the Schoenflies system of nomenclature (such as C_{3v}). Point groups that qualify as crystallographic point groups (because they are also applicable to unit cells) are also labelled with the symbol adopted in the International System (or the Hermann–Mauguin system, such as $2/m$). In the latter system, a number n represents an n-fold axis, and a letter m represents a mirror plane. A diagonal line indicates that a mirror plane lies perpendicular to the symmetry axis, and a bar over the number indicates that the rotation is combined with an inversion.

The symmetry species of the p and d orbitals are shown on the right of the tables. Thus, in C_{2v}, a p_x orbital (which is proportional to x) has B_1 symmetry. The functions x, y, and z also show the transformation properties of translations and of the electric dipole moment. The set of functions that span a degenerate representation (such as x and y, which jointly span E in C_{3v}) are enclosed in parentheses. The transformation properties of rotation are shown by the letter R on the right of the tables. The value of h is the order of the group.

The groups C1, Cs, Ci

C_1 (1)	E			$h=1$
A	1			

$C_s = C_h$ (m)	E	σ_h		$h=2$	
A′	1	1	x, y, R_z	x^2, y^2, z^2, xy	
A″	1	−1	z, R_x, R_y	yz, zx	

$C_i = S_2$ (1)	E	i		$h=2$	
A_g	1	1	R_x, R_y, R_z	$x^2, y^2, z^2, xy, zx, yz$	
A_u	1	−1	x, y, z		

The groups C_n

C_2 (2)	E	C_2		$h=2$	
A	1	1	z, R_z	x^2, y^2, z^2, xy	
B	1	−1	x, y, R_x, R_y	yz, zx	

C_3 (3)	E	C_3	C_3^2	$h=3, \varepsilon = \exp(2\pi i/3)$	
A	1	1	1	z, R_z	$x^2 + y^2, z^2$
E	$\begin{cases}1 \\ 1\end{cases}$	$\begin{matrix}\varepsilon \\ \varepsilon^*\end{matrix}$	$\begin{matrix}\varepsilon^* \\ \varepsilon\end{matrix}$	$(x,y)(R_x, R_y)$	$(x^2-y^2, xy)(yz, zx)$

C_4 (4)	E	C_4	C_2	C_4^3	$h=4$	
A	1	1	1	1	z, R_z	x^2+y^2, z^2
B	1	−1	1	−1		x^2-y^2, xy
E	$\begin{cases}1 \\ 1\end{cases}$	$\begin{matrix}i \\ -i\end{matrix}$	$\begin{matrix}-1 \\ -1\end{matrix}$	$\begin{matrix}-i \\ i\end{matrix}$	$(x,y)(R_x, R_y)$	(yz, zx)

The groups C_{nv}

C_{2v} (2mm)	E	C_2	$\sigma_v(xz)$	$\sigma_v'(yz)$	$h=4$	
A_1	1	1	1	1	z	x^2, y^2, z^2
A_2	1	1	−1	−1	R_z	xy
B_1	1	−1	1	−1	x, R_y	zx
B_2	1	−1	−1	1	y, R_x	yz

C_{3v} (3m)	E	$2C_3$	$3\sigma_v$	$h=6$	
A_1	1	1	1	z	x^2+y^2, z^2
A_2	1	1	−1	R_z	
E	2	−1	0	$(x,y)(R_x, R_y)$	$(x^2-y^2, xy)(zx, yz)$

C_{4v} (4mm)	E	$2C_4$	C_2	$2\sigma_v$	$2\sigma_d$	$h=8$	
A_1	1	1	1	1	1	z	x^2+y^2, z^2
A_2	1	1	1	−1	−1	R_z	
B_1	1	−1	1	1	−1		x^2-y^2
B_2	1	−1	1	−1	1		xy
E	2	0	−2	0	0	$(x,y)(R_x, R_y)$	(zx, yz)

C_{5v}	E	$2C_5$	$2C_5^2$	$5\sigma_v$	$h=10, \alpha=72°$	
A_1	1	1	1	1	z	x^2+y^2, z^2
A_2	1	1	1	−1	R_z	
E_1	2	$2\cos\alpha$	$2\cos 2\alpha$	0	$(x,y)(R_x, R_y)$	(zx, yz)
E_2	2	$2\cos 2\alpha$	$2\cos\alpha$	0		(x^2-y^2, xy)

C_{6v} (6mm)	E	$2C_6$	$2C_3$	C_2	$3\sigma_v$	$3\sigma_d$			h = 12
A_1	1	1	1	1	1	1	z		$x^2 + y^2, z^2$
A_2	1	1	1	1	−1	−1	R_z		
B_1	1	−1	1	−1	1	−1			
B_2	1	−1	1	−1	−1	1			
E_1	2	1	−1	−2	0	0	(x, y)	(R_x, R_y)	(zx, yz)
E_2	2	−1	−1	2	0	0			$(x^2 − y^2, xy)$

$C_{\infty v}$	E	$2C_\phi$	...	$\infty \sigma_v$			h = ∞
$A_1 (\Sigma^+)$	1	1		1	z		$x^2 + y^2, z^2$
$A_2 (\Sigma^-)$	1	1		−1	R_z		
$E_1 (\Pi)$	2	$2\cos\phi$		0	(x,y)	(R_x, R_y)	(zx, yz)
$E_2 (\Delta)$	2	$2\cos 2\phi$		0			$(xy, x^2 − y^2)$

The group C_{2h}

C_{2h}	E	$C_2(z)$	i	σ_h		h = 4
A_g	1	1	1	1	R_z	x^2, y^2, z^2, xy
B_g	1	−1	1	−1	(R_x, R_y)	(zx, yz)
A_u	1	1	−1	−1	z	
B_u	1	−1	−1	1	(x, y)	

The groups D_n

D_2 (222)	E	$C_2(z)$	$C_2(y)$	$C_2(x)$		h = 4
A	1	1	1	1		x^2, y^2, z^2
B_1	1	1	−1	−1	z, R_z	xy
B_2	1	−1	1	−1	y, R_y	zx
B_3	1	−1	−1	1	x, R_x	yz

D_3 (32)	E	$2C_3$	$3C_2$			h = 6
A_1	1	1	1			$x^2 + y^2, z^2$
A_2	1	1	−1	z, R_z		
E	2	−1	0	(x, y) (R_x, R_y)		$(x^2 − y^2, xy)$ (zx, yz)

The groups D_{nh}

D_{2h} (mmm)	E	$C_2(z)$	$C_2(y)$	$C_2(x)$	i	$\sigma(xy)$	$\sigma(xz)$	$\sigma(yz)$		h = 8
A_g	1	1	1	1	1	1	1	1		x^2, y^2, z^2
B_{1g}	1	1	−1	−1	1	1	−1	−1	R_z	xy
B_{2g}	1	−1	1	−1	1	−1	1	−1	R_y	zx
B_{3g}	1	−1	−1	1	1	−1	−1	1	R_x	yz
A_u	1	1	1	1	−1	−1	−1	−1		
B_{1u}	1	1	−1	−1	−1	−1	1	1	z	
B_{2u}	1	−1	1	−1	−1	1	−1	1	y	
B_{3u}	1	−1	−1	1	−1	1	1	−1	x	

D_{3h} ($\bar{6}m2$)	E	$2C_3$	$3C_2$	σ_h	$2S_3$	$3\sigma_v$		h = 12
A'_1	1	1	1	1	1	1		$x^2 + y^2, z^2$
A'_2	1	1	−1	1	1	−1	R_z	
E'	2	−1	0	2	−1	0	(x, y)	$(x^2 − y^2, xy)$
A''_1	1	1	1	−1	−1	−1		
A''_2	1	1	−1	−1	−1	1	z	
E''	2	−1	0	−2	1	0	(R_x, R_y)	(zx, yz)

D_{4h} (4/mmm)	E	$2C_4$	$C_2(=C_4^2)$	$2C_2'$	$2C_2''$	i	$2S_4$	σ_h	$2\sigma v$	$2\sigma_d$		$h=16$
A_{1g}	1	1	1	1	1	1	1	1	1	1		x^2+y^2, z^2
A_{2g}	1	1	1	−1	−1	1	1	1	−1	−1	R_z	
B_{1g}	1	−1	1	1	−1	1	−1	1	1	−1		x^2-y^2
B_{2g}	1	−1	1	−1	1	1	−1	1	−1	1		xy
E_g	2	0	−2	0	0	2	0	−2	0	0	(R_x, R_y)	(zx, yz)
A_{1u}	1	1	1	1	1	−1	−1	−1	−1	−1		
A_{2u}	1	1	1	−1	−1	−1	−1	−1	1	1	z	
B_{1u}	1	−1	1	1	−1	−1	1	−1	−1	1		
B_{2u}	1	−1	1	−1	1	−1	1	−1	1	−1		
E_u	2	0	−2	0	0	−2	0	2	0	0	(x, y)	

D_{5h}	E	$2C_5$	$2C_5^2$	$5C_2$	σ_h	$2S_5$	$2S_5^2$	$5\sigma v$		$h=20, \alpha=72°$
A_1'	1	1	1	1	1	1	1	1		x^2+y^2, z^2
A_2'	1	1	1	−1	1	1	1	−1	R_z	
E_1'	2	$2\cos\alpha$	$2\cos 2\alpha$	0	2	$2\cos\alpha$	$2\cos 2\alpha$	0	(x, y)	
E_2'	2	$2\cos 2\alpha$	$2\cos\alpha$	0	2	$2\cos 2\alpha$	$2\cos\alpha$	0		$(x-y^2, xy)$
A_1''	1	1	1	1	−1	−1	−1	−1		
A_2''	1	1	1	−1	−1	−1	−1	1	z	
E_1''	2	$2\cos\alpha$	$2\cos 2\alpha$	0	−2	$-2\cos\alpha$	$-2\cos 2\alpha$	0	(R_x, R_y)	(zx, yz)
E_2''	2	$2\cos 2\alpha$	$2\cos\alpha$	0	−2	$-2\cos 2\alpha$	$-2\cos\alpha$	0		

D_{6h} (6/mmm)	E	$2C_6$	$2C_3$	C_2	$3C_2'$	$3C_2''$	i	$2S_3$	$2S_6$	σ_h	$3\sigma_d$	$3\sigma v$		$h=24$
A_{1g}	1	1	1	1	1	1	1	1	1	1	1	1		x^2+y^2, z^2
A_{2g}	1	1	1	1	−1	−1	1	1	1	1	−1	−1	R_z	
B_{1g}	1	−1	1	−1	1	−1	1	−1	1	−1	1	−1		
B_{2g}	1	−1	1	−1	−1	1	1	−1	1	−1	−1	1		
E_{1g}	2	1	−1	−2	0	0	2	1	−1	−2	0	0	(R_x, R_y)	(zx, yz)
E_{2g}	2	−1	−1	2	0	0	2	−1	−1	2	0	0		(x^2-y^2, xy)
A_{1u}	1	1	1	1	1	1	−1	−1	−1	−1	−1	−1		
A_{2u}	1	1	1	1	−1	−1	−1	−1	−1	−1	1	1	z	
B_{1u}	1	−1	1	−1	1	−1	−1	1	−1	1	−1	1		
B_{2u}	1	−1	1	−1	−1	1	−1	1	−1	1	1	−1		
E_{1u}	2	1	−1	−2	0	0	−2	−1	1	2	0	0	(x, y)	
E_{2u}	2	−1	−1	2	0	0	−2	1	1	−2	0	0		

$D_{\infty h}$	E	$\infty C_2'$	$2C_\phi$	i	$\infty \sigma_v$	$2S_\phi$	$h = \infty$	
$A_{1g}(\Sigma_g^+)$	1	1	1	1	1	1		x^2+y^2, z^2
$A_{1u}(\Sigma_u^+)$	1	−1	1	−1	1	−1	z	
$A_{2g}(\Sigma_g^-)$	1	−1	1	1	−1	1	R_z	
$A_{2u}(\Sigma_u^-)$	1	1	1	−1	−1	−1		
$E_{1g}(\Pi_g)$	2	0	$2\cos\phi$	2	0	$-2\cos\phi$	(R_x, R_y)	(zx, yz)
$E_{1u}(\Pi_u)$	2	0	$2\cos\phi$	−2	0	$2\cos\phi$	(x, y)	
$E_{2g}(\Delta_g)$	2	0	$2\cos 2\phi$	2	0	$2\cos 2\phi$		(xy, x^2-y^2)
$E_{2u}(\Delta_u)$	2	0	$2\cos 2\phi$	−2	0	$-2\cos 2\phi$		
⋮	⋮	⋮	⋮	⋮	⋮	⋮		

The groups D_{nd}

$D_{2d} = V_d$ (42m)	E	$2S_4$	C_2	$2C_2'$	$2\sigma_d$	$h = 8$	
A_1	1	1	1	1	1		x^2+y^2, z^2
A_2	1	1	1	−1	−1	R_z	
B_1	1	−1	1	1	−1		x^2-y^2
B_2	1	−1	1	−1	1	z	xy
E	2	0	−2	0	0	$(x, y)(R_x, R_y)$	(zx, yz)

D_{3d} (3m)	E	$2C_3$	$3C_2$	i	$2S_6$	$3\sigma_d$	$h = 12$	
A_{1g}	1	1	1	1	1	1		x^2+y^2, z^2
A_{2g}	1	1	−1	1	1	−1	R_z	
E_g	2	−1	0	2	−1	0	(R_x, R_y)	$(x^2-y^2, xy)(zx, yz)$
A_{1u}	1	1	1	−1	−1	−1		
A_{2u}	1	1	−1	−1	−1	1	z	
E_u	2	−1	0	−2	1	0	(x, y)	

D_{4d}	E	$2S_8$	$2C_4$	$2S_8^3$	C_2	$4C_2'$	$4\sigma_d$	$h = 16$	
A_1	1	1	1	1	1	1	1		x^2+y^2, z^2
A_2	1	1	1	1	1	−1	−1	R_z	
B_1	1	−1	1	−1	1	1	−1		
B_2	1	−1	1	−1	1	−1	1	z	
E_1	2	$\sqrt{2}$	0	$-\sqrt{2}$	−2	0	0	(x, y)	
E_2	2	0	−2	0	2	0	0		(x^2-y^2, xy)
E_3	2	$-\sqrt{2}$	0	$\sqrt{2}$	−2	0	0	(R_x, R_y)	(zx, yz)

The cubic groups

T_d (43m)	E	$8C_3$	$3C_2$	$6S_4$	$6\sigma_d$		$h = 24$
A_1	1	1	1	1	1		$x^2 + y^2 + z^2$
A_2	1	1	1	−1	−1		
E	2	−1	2	0	0		$(2z^2 − x^2 − y^2, x^2 − y^2)$
T_1	3	0	−1	1	−1	(R_x, R_y, R_z)	
T_2	3	0	−1	−1	1	(x, y, z)	(xy, yz, zx)

O_h (m3m)	E	$8C_3$	$6C_2$	$6C_4$	$3C_2(=C_4^2)$	i	$6S_4$	$8S_6$	$3\sigma_h$	$6\sigma_v$		$h = 48$
A_{1g}	1	1	1	1	1	1	1	1	1	1		$x^2 + y^2 + z^2$
A_{2g}	1	1	−1	−1	1	1	−1	1	1	−1		
E_g	2	−1	0	0	2	2	0	−1	2	0		$(2z^2 − x^2 − y^2, x^2 − y^2)$
T_{1g}	3	0	−1	1	−1	3	1	0	−1	−1	(R_x, R_y, R_z)	
T_{2g}	3	0	1	−1	−1	3	−1	0	−1	1		(xy, yz, zx)
A_{1u}	1	1	1	1	1	−1	−1	−1	−1	−1		
A_{2u}	1	1	−1	−1	1	−1	1	−1	−1	1		
E_u	2	−1	0	0	2	−2	0	1	−2	0		
T_{1u}	3	0	−1	1	−1	−3	−1	0	1	1	(x, y, z)	
T_{2u}	3	0	1	−1	−1	−3	1	0	1	−1		

The icosahedral group

I	E	$12C_5$	$12C_5^2$	$20C_3$	$15C_2$		$h = 60$
A_1	1	1	1	1	1		$x^2 + y^2 + z^2$
T_1	3	½(1 + √5)	½(1 − √5)	0	−1	$(x, y, z) (R_x, R_y, R_z)$	
T_2	3	½(1 − √5)	½(1 + √5)	0	−1		
G	4	−1	−1	1	0		
H	5	0	0	−1	1		$(2z^2 − x^2 − y^2, x^2 − y^2, xy, yz, zx)$

Resource section 5
Symmetry-adapted orbitals

Table RS5.1 gives the symmetry classes of the s, p, and d orbitals of the central atom of an AB_n molecule of the specified point group. In most cases the z-axis is the principal axis of the molecule; in C_{2v} the x-axis lies perpendicular to the molecular plane.

The orbital diagrams that follow show the linear combinations of atomic orbitals on the peripheral atoms of AB_n molecules of the specified point groups. Where a view from above is shown, the dot representing the central atom is either in the plane of the paper (for the D groups) or above the plane (for the corresponding C groups). Different phases of the atomic orbitals (+1 or −1; amplitudes) are shown by different colours. Where there is a large difference in the magnitudes of the orbital coefficients in a particular combination, the atomic orbitals are drawn large or small to represent their relative contributions to the linear combination. In the case of degenerate linear combinations (those labelled E or T), any linearly independent combination of the degenerate pair is also of suitable symmetry. In practice, these different linear combinations look like the ones shown here, but their nodes are rotated by an arbitrary angle around the z-axis.

Molecular orbitals are formed by combining an orbital of the central atom (as in Table RS5.1) with a linear combination of the same symmetry.

TABLE RS5.1 Symmetry species of orbitals on the central atom

	$D_{\infty h}$	C_{2v}	D_{3h}	C_{3v}	D_{4h}	C_{4v}	D_{5h}	C_{5v}	D_{6h}	C_{6v}	T_d	O_h
s	Σ	A_1	A_1'	A_1	A_{1g}	A_1	A_1'	A_1	A_{1g}	A_1	A_1	A_{1g}
p_x	Π	B_1	E'	E	E_u	E	E_1'	E_1	E_{1u}	E_1	T_2	T_{1u}
p_y	Π	B_2	E'	E	E_u	E	E_1'	E_1	E_{1u}	E_1	T_2	T_{1u}
p_z	Σ	A_1	A_2''	A_1	A_{2u}	A_1	A_2''	A_1	A_{2u}	A_1	T_2	T_{1u}
d_{z^2}	Σ	A_1	A_1'	A_1	A_{1g}	A_1	A_1'	A_1	A_{1g}	A_1	E	E_g
$d_{x^2-y^2}$	Δ	A_1	E'	E	B_{1g}	B_1	E_2'	E_2	E_{2g}	E_2	E	E_g
d_{xy}	Δ	A_2	E'	E	B_{2g}	B_2	E_2'	E_2	E_{2g}	E_2	T_2	T_{2g}
d_{yz}	Π	B_2	E''	E	E_g	E	E_1''	E_1	E_{1g}	E_1	T_2	T_{2g}
d_{zx}	Π	B_1	E''	E	E_g	E	E_1''	E_1	E_{1g}	E_1	T_2	T_{2g}

$D_{\infty h}$	C_{2v}
Σ_g	A_1
Π_g	A_2
Π_u	B_1
Σ_u	B_2
	A_1
	B_2

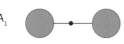
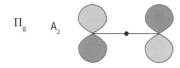
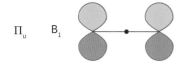

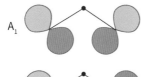
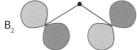

D_{3h}	C_{3v}
A_1'	A_1
A_2'	A_2
E'	E
E'	E
A_2''	A_1
E''	E

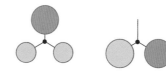
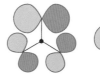

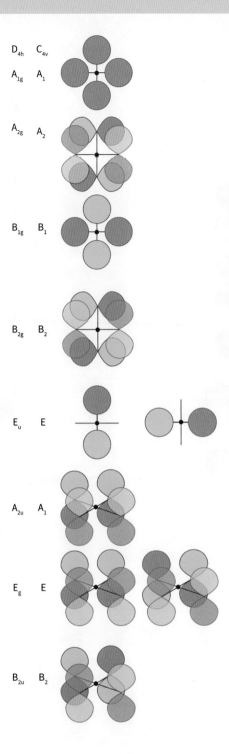

D_{6h}	C_{6v}			
A_{1g}	A_1		E_{2g} $\quad$ E_2	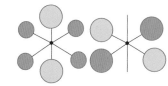
A_{2g}	A_2		A_{2u} $\quad$ A_1	
B_{1u}	B_1		B_{2g} $\quad$ B_1	
B_{2u}	B_2		E_{1g} $\quad$ E_1	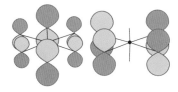
E_{1u}	E_1		E_{2u} $\quad$ E_2	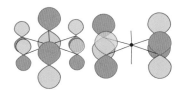

T_d

A_1

T_2

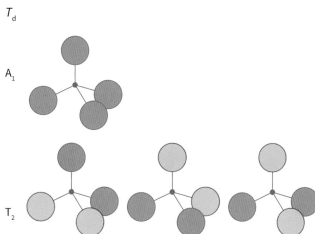

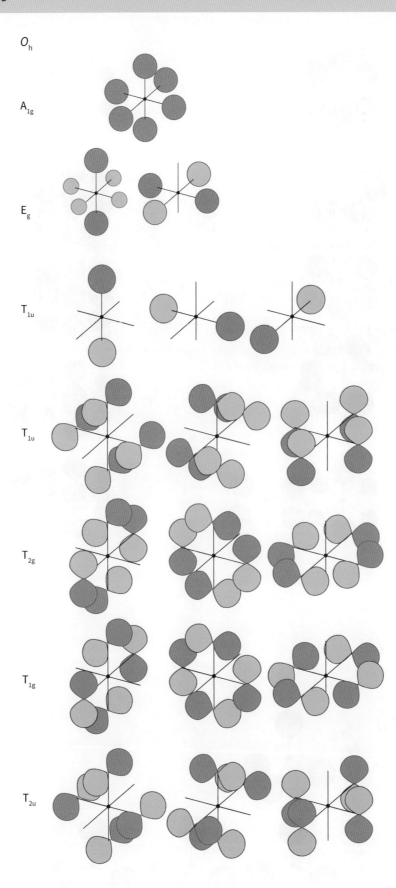

Resource section 6
Tanabe–Sugano diagrams

This section collects together the Tanabe–Sugano diagrams for octahedral complexes with electron configurations d^2 to d^8. The diagrams, which were introduced in Section 22.2, show the dependence of the term energies on ligand-field strength. The term energies E are expressed as the ratio E/B, where B is a Racah parameter, and the ligand-field splitting Δ_O is expressed as Δ_O/B. Terms of different multiplicity are included in the same diagram by making specific, plausible choices about the value of the Racah parameter C, and these choices are given for each diagram. The term energy is always measured from the lowest energy term, and so there are discontinuities of slope where a low-spin term displaces a high-spin term at sufficiently high ligand-field strengths for d^4 to d^8 configurations. Moreover, the noncrossing rule requires terms of the same symmetry to mix rather than to cross, and this mixing accounts for the curved rather than the straight lines in a number of cases. The term labels are those of the point group O_h.

The diagrams were first introduced by Y. Tanabe and S. Sugano, *J. Phys. Soc. Japan*, 1954, **9**, 753. They may be used to find the parameters Δ_O and B by fitting the ratios of the energies of observed transitions to the lines. Alternatively, if the ligand-field parameters are known, then the ligand-field spectra may be predicted.

1. d^2 with $C = 4.428B$

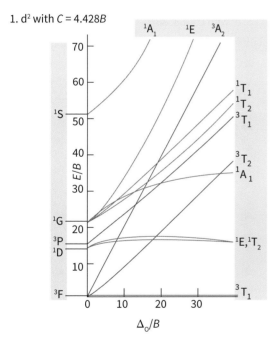

2. d^3 with $C = 4.502B$

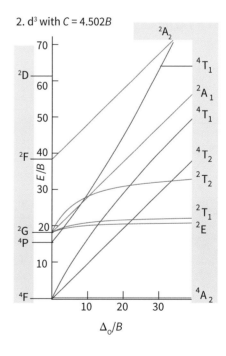

3. d^4 with $C = 4.611B$

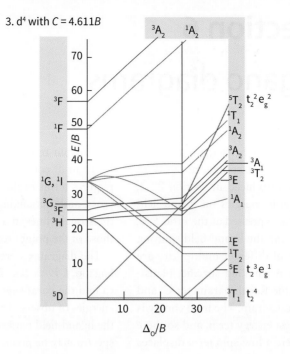

4. d^5 with $C = 4.477B$

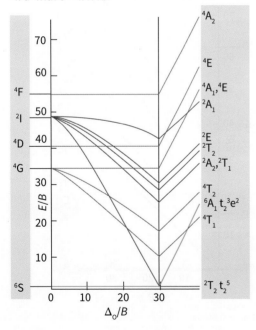

5. d^6 with $C = 4.808B$

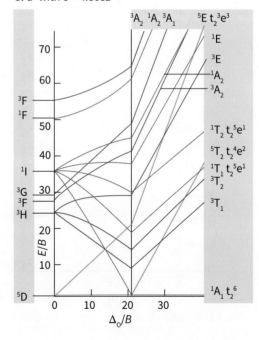

6. d^7 with $C = 4.633B$

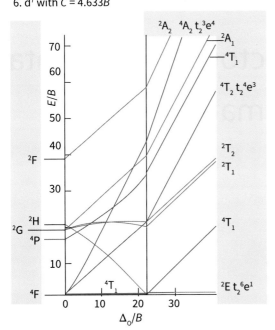

7. d^8 with $C = 4.709B$

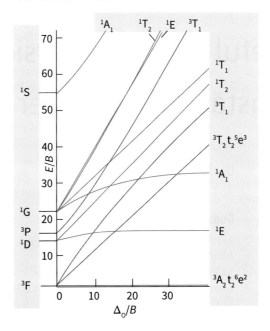

Resource section 7
Useful data: conversion factors, fundamental constants, and other information

Useful relations
At $T = 298.15$ K, $RT = 2.4790$ kJ mol^{-1}, and $RT/F = 25.693$ mV

Conversion factors
1 atm = 101.325 kPa = 760 Torr (exactly)
1 bar = 10^5 Pa
1 eV = $1.602\ 18 \times 10^{-19}$ J = 96.485 kJ mol^{-1} = 8065.5 cm^{-1}
1 cm^{-1} = 1.986×10^{-23} J = 11.96 J mol^{-1} = 0.1240 meV
1 cal = 4.184 J (exactly)
1 D = $3.335\ 64 \times 10^{-30}$ C m
1 T = 10^4 G
1 Å = 100 pm = 10^{-10} m
1 M = 1 mol dm^{-3}

General data and fundamental constants

Quantity	Symbol	Value
Speed of light	c	$2.997\ 925 \times 10^8$ m s^{-1}
Elementary charge	e	$1.602\ 177 \times 10^{-19}$ C
Faraday's constant	$F = N_A e$	$9.648\ 53 \times 10^4$ C mol^{-1}
Boltzmann's constant	k	$1.380\ 65 \times 10^{-23}$ J K^{-1}
		8.6173×10^{-5} eV K^{-1}
Gas constant	$R = N_A k$	$8.314\ 46$ J K^{-1} mol^{-1}
		$8.205\ 74 \times 10^{-2}$ dm^3 atm K^{-1} mol^{-1}
Planck's constant	h	$6.626\ 07 \times 10^{-34}$ J s
	$\hbar = h/2\pi$	$1.054\ 57 \times 10^{-34}$ J s
Avogadro's constant	N_A	$6.022\ 14 \times 10^{23}$ mol^{-1}
Atomic mass constant	m_u	$1.660\ 54 \times 10^{-27}$ kg
Mass of electron	m_e	$9.109\ 38 \times 10^{-31}$ kg
Vacuum permittivity	ε_0	$8.854\ 19 \times 10^{-12}$ J^{-1} C^2 m^{-1}
	$4\pi\varepsilon_0$	$1.112\ 65 \times 10^{-10}$ J^{-1} C^2 m^{-1}
Bohr magneton	$\mu_B = e\hbar/2m_e$	$9.274\ 01 \times 10^{-24}$ J T^{-1}
Nuclear magneton	$\mu_N = e\hbar/2m_p$	$5.050\ 78 \times 10^{-27}$ J T^{-1}
Bohr radius	$a_0 = 4\pi\varepsilon_0 \hbar^2/m_e e^2$	$5.291\ 77 \times 10^{-11}$ m
Rydberg constant	$R_\infty = m_e e^4/8h^3 c \varepsilon_0^2$	$1.097\ 37 \times 10^5$ cm^{-1}

Prefixes

atto	femto	pico	nano	micro	milli	centi	deci	kilo	mega	giga	tera	peta
a	f	p	n	µ	m	c	d	k	M	G	T	P
10^{-18}	10^{-15}	10^{-12}	10^{-9}	10^{-6}	10^{-3}	10^{-2}	10^{-1}	10^{3}	10^{6}	10^{9}	10^{12}	10^{15}

Greek alphabet

A, α	alpha	I, ι	iota	P, ρ	rho
B, β	beta	K, κ	kappa	Σ, σ	sigma
Γ, γ	gamma	Λ, λ	lambda	T, τ	tau
Δ, δ	delta	M, µ	mu	Y, υ	upsilon
E, ε	epsilon	N, ν	nu	Φ, φ	phi
Z, ζ	zeta	Ξ, ξ	xi	X, χ	chi
H, η	eta	O, o	omicron	Ψ, ψ	psi
Θ, θ	theta	Π, π	pi	Ω, ω	omega

Index

absorbance, 271
absorption, 886
   selectivity, 887
   spectroscopy, 269
acceptor band, 735
accident prevention, 856
acid phosphatases, 806
acid rain, 516
acid–base catalysis by enzymes, 801
acid–base discrimination window, 174
acidic oxide, 158
acidic proton, 153
acidification, of the oceans, 146
acidity constant, 147
   Table, 149
aconitase, 807, 834
actinium, 592
actinoid(s) 596
   beyond fermium, 618
   contraction, 615
   bonding in the dioxido unit, 615
   dioxido complexes, 613
   electronic configurations and oxidation states, 613
   electronic spectra, 639
   Frost diagrams, 614
   half-lives, 598
   organometallic chemistry, 616
   sandwich complexes, 617
   synthesis by neutron capture, 597
   synthesis by nuclear fusion, 598
   use of f-orbitals in bonding, 615
activated carbon, 451
activation
   entropy of, 655
   volume of, 655
activator ion, 904
active organic compounds, 896
adenosine diphosphate (ADP), 159
adenosine triphosphate (ATP), 159
aerogels, 744
agostic hydrogen, 695
agrochemically-active compounds, 896
airbag, 481
alkali metal halides
   Born–Haber cycle, 373
   radius ratio, 373

   standard enthalpies of formation, 373
alkali metal(s), 114, 362, 770
alkalide ion, 379
alkaline earth carbonates, 139
alkaline earth metals, 385
alkaline fuel cell (AFC), 190
alkaline phosphatase, 803
alkane, 444
   as a ligand, 695
alkenyl ligand, 686
alkyl ligand, 686
alkyl transfer in biology, 821
alkylammonium ion, 139
alkylidene, 694
alkynyl ligand, 686
allotropes, 314
   group 15 elements, 472
allowed transition, 634
alloy(s), 111, 115, 120, 566, 724, 915, 916
Allred–Rochow electronegativity, 31
allyl ligand, 691
alnico, 730
Altair process, 849
alternation effect, 28, 329
   electronegativity, 329
   p block oxide reactivity, 330
   p block pentahalides, 330
alumina, 435, 937
aluminium
   applications, 416
   extraction, 854
   hydride, 433
   occurrence and recovery, 416
   oxide, 724
   production by electrolysis, 212
   recycling, 417
   trihalide, 434
aluminium-doped zinc oxide (AZO), 901
aluminophosphates (ALPOs), 765, 885
   (Al,P)O_4 frameworks, 882
aluminosilicate(s), 180, 447, 463, 764
Alzheimer's disease, 935
ambidentate, 219
americium, 617
amides, 762, 920
amine oxidases, 811
amine–boranes, 423
amino acid residues, 780
amino acids (modified), 782
amino acids codes (Table), 781

amino acids as ligands, 782
ammonia, 93, 478, 482, 851
   colours, 478
   industrial uses, 478
   liquid, 175
   production by microorganisms, 829
   solvated electrons, 176
   synthesis, 849, 853
ammonium carbonate, 278
ammonium nitrate, 117
ammonolysis of oxides, 757
amorphous, 755
amorphous silicon (a-Si), 906
amphiprotic substance, 146
amphoteric, 158, 318
   beryllium, 402
   oxides, 158, 318
amyloid peptides, 840
Anastas, Paul, 848
Ångstrom (Å), 263
angular nodes, 12, 14
angular wavefunction, 14
anhydrous lithium salts, 923
anhydrous oxides, 158
animal feeds, 886
anion mobility, 740
anion size, 139
anion vacancies, 740
anode, 188
anomalous, first group member, 326
ansa-complexes, 575
antibiotic resistance, 965
antibonding (molecular) orbital, 47, 48
antiferromagnetic, 268, 641
antifluorite, 118, 122
   structure, 760
anti-HIV therapeutics, 968
antimalarial drugs, 966
antimonides, 482
antimony
   applications, 478
   complex for treating tropical skin disease, 973
   pentafluoride, 175, 179
antiperovskites, 752
anti-site defect, 723
anti-static, 900
anti-Stokes lines, 277
anti-Stokes phosphors, 905
antiviral drugs, 967
apatite, 765, 912
apoprotein, 782
apparent quantum efficiency (AQE), 909

applications of IR and Raman spectroscopy, 278
applications of zeolites, 885
aqua acid, 153
aqua acids (strengths of), 154
aqua regia, 489
arachno, 414, 427
area-detector, 266
arene ligand, 690
argentite, 586
argon, 135, 550
   coordination compounds, 559
   uses, 553, 554
Arkel–Ketelaar triangle, 65
Arrhenius, acidity concept, 145
arsabenzene, 497
arsane, 484
arsenic, 471
   applications, 478
   environmental toxicity, 477
   medicine, 958
arsenicals, 478
arsenides, 482
arsole, 497
artificial metalloenzymes, 842
aryl ligand, 686
arylometals, 497
arylxenon(II), 558
asbestos, 458
associative rate determining step, 650
associative activation, 660
associative mechanism, 649
astantine, 525
atactic, 948
atmophiles, 310
atom diffusion, 737
atom economy, 849
atomic absorption spectroscopy (AAS), 294
atomic clock, *see* caesium clock/atomic clock
atomic emission spectroscopy (AES), 294
atomic force microscopy (AFM), 302
atomic layer deposition (ALD), 865, 866
atomic number, 3
atomic orbital, 9
atomic radii, 23, 565
   diagonal relationship, 328
   trends, 310
atom-interchange defect, 723
ATP, *see* adenosine triphosphate (ATP)

attenuated total reflectance (ATR), 277
Aufbau principle, 18
auranofin, 964, 967
austenite, 115, 116
autoionization, 148, 175
autoprotolysis, 148
autoprotolysis constant, 148, 173, 174
auxiliary substances, 851
azide, 480
azimuthal quantum number, 9
azurin, 800

Bailar twist, 663
Balmer series, 7
band, 58
   formation, 58
   gap, 60, 758, 906
   structure, 748, 750
   width, 58
band-gap engineering, 909
banded iron formations, 503
barium, 390
   applications, 391
   extraction, 390
   titanate, 743, 753, 876
Bartlett, Neil, 554
barycentre, 245
base hydrolysis, 661
basic oxide, 158
basicity constant, 148
basis set, 47
basket Fe-porphyrin, 794
bastnaesite ($LnCO_3F$), 597
batteries, lithium, 369
Bayer process, aluminium production, 212
Beer–Lambert law, 272
bent metallocene, 706
benzene ligand, 690
Berry pseudorotation, 232, 663
beryl, 402, 724
beryllates, 402
beryllium, 386
   amphoteric, 402
   anomalous properties, 389
   applications, 390
   extraction, 389
   halides, 393
   occurrence and abundance, 386
   zero valent compounds, 406
beryllocene, 404
bidentate, 219
Big Bang, 3
BIMEVOX, 912
BINAP, 684, 931
binary phases, 118
binding energy, 4
bioconjugation, 957
biodegradable, 855

biohydrogen, 339, 827
bioinformatics, 785
bioinspired catalysis, 840
biological concentrations of elements (Table), 779
biological feedstocks, 854
biological hydrogen cycle, 339
biological standard state, 194
biomass, 854
biomimicry, 853
biomineralization, 837
biominerals, 765
biphasic catalysts, 951
bipolar junction transistor (BJT), 736
BISCO, 587
bismuth, 477
   applications, 478
   cytotoxicity in bacteria, 965
   inhibition of β-lactamases, 965
   pharmaceuticals, 964
   selenide, 911
   small molecule activation, 496
   subcitrate, 964
   subsalicylate (Pepto-Bismol®), 964
   telluride, 911
   vanadate, 909
bismuthates, 918
bismuthides, 482
bismuthole, 497
bite angle, 222
blast furnace, 210
block, 22
blue Cu centre, 799
blue Cu oxidases, 811
blue-emitting LED, 275
body-centred, 103
body-centred cubic, 112
Bohr magneton, 248
Bohr radius, 13
Bohr, Niels, 8
bond dissociation enthalpy, 63
bond enthalpy trends, 319
bond length, 57, 62, 266
bond order, 56
bond strength, 56
bonding, single versus multiple, 327
bonding (molecular) orbital, 47, 48
bones and teeth, 837
boranates, 919
boranes, 417
   higher, 426
borazine, 424
boric acid, 422
borides, 415, 425
Born equation, 153
Born–Haber cycle(s), 131, 134, 137
Born–Landé equation, 141

Born–Mayer equation, 132, 134, 141
borohydride(s), 414, 427
   characteristic reactions, 430
   structural correlations, 429
boron, 410
   allotropes, 411
   anomalous properties, 410
   applications, 416
   clusters, 414, 427
   flame colour, 411
   for cancer treatment, 426
   nitrogen compounds, 423
   occurrence and recovery, 415
   oxide, 412
   oxygen compounds, 422
boron halides, with boron–boron bonds, 421
boron hydride(s), 416
   classification, 427
boron nitride, 412, 423
   applications, 413
boron trihalide(s), 420
   Lewis acidity, 162, 411
   properties, 412
   reactions, 420
borophene, 874
borosilicate (glass), 755, 756
bottom-up approaches, 860
boundary surface, 14, 15
Brackett series, 7
Bragg angle, 264
Bragg's equation, 263
brass, 115, 588, 728, 729
bridging hydrogens, 351
bridging ligand, 681
broad band decoupling, 282
bromine, 525
Brønsted acid, 146, 173
Brønsted base, 146
Brønsted–Lowry concept of acidity, 146, 175
bronze, 115, 444, 566
   molybdenum, tungsten, 575
Bronze Age, 210
Brust–Schiffrin method, 861
builder (in detergents), 888
building-up principle, 18, 52
butadiene ligand, 688

C_{60}, 108
cadmium, 588
   in enzyme, 804
   selenide nanocrystal, 869
   selenide, 902
   sulfide, 862, 902
   telluride, 734, 907
cadmium-chloride structure, 123
cadmium-iodide structure, 123
caesium chloride, 118
caesium chloride structure, 120, 365

caesium clock/atomic clock, 370
calamitic (rod-like) liquid crystal, 773
calciferation, 837
calcite as gravity sensor device (inner ear), 837
calcium, 386
   applications, 391
   cement, 392
   comparison with magnesium in biology, 788
   halides, 395
   hydride, 180
   hydrogenphosphate, 765
   in biology, 838
   in cell signalling, 788
   in muscle contraction (troponin C), 788
   occurrence and abundance, 386
   occurrence and extraction, 390
   phosphate, 837
   pumps, in cell membranes, 788
   signalling proteins, 788
calcium carbonate, 389, 837
   biomineral, 401
   geology, 400
   minerals, 400
   polymorphs, 399
calcium oxide, 396
   in carbon capture, 399
calculation of lattice enthalpies, 132
caliche, 367
californium, 598
   uses, 618
calmodulin, 788
Calvin cycle, 805
cancer treatment, 958
   boron, 426
capped octahedron, 227
capped trigonal prism, 227
carbene
   complex, 700
   Fischer, 694
   ligand, 694
   N-heterocyclic, 694
   Schrock, 694
   singlet, 694
   triplet, 694
carbene transfer catalysis by enzymes, 842
carbides
   interstitial, 462
   metal, 446
   metallic, 460
   metalloid, 460
   saline, 460
carbon, 921
   activated, 451
   allotropes, 442

black, 451, 866, 903, 910
clusters, 449
compounds with oxygen, 455
compounds with sulfur, 455
cycle, fast, 457
fibres, 451
halide reactivity, 454
halides, 453
nanotubes (CNTs), 450, 871, 921
oxides, 445
carbon capture, 164, 456, 475
by calcium oxide, 399
reutilization, 456
storage, 456
carbon dioxide, 455, 850, 854
capture, 164
dehydratase (carbonic anhydrase), 802
fixation by ribulose (mechanism), 806
phase diagram, 177
Lewis acidity, 163
transport in biology, 802, 803
carbon monoxide, 55, 455, 682
dehydrogenases, 827
binding to haemoglobin, 794
detection by bacteria, 835
how organisms deal with it, 827
releasing molecules (CORMs) in medicine, 968
releasing polymers (CORPs), 968
carbonates, 138
carbonic acid, 157
carbonic anhydrase, 802
mechanism, 803
carbonium ion, 180
carbonyl, 276, 682
carbonyl dichloride, 850
carbonyl groups, 279
carbonyl halides, 454
carbonyl stretching frequencies, 683
carbonylation, methanol, 584
carborane(s), 180, 415, 431
carboxylates, 768, 891
carboxypeptidase, 803
Cardiolyte®, 579, 971
cast iron, 116
catalysis, 66, 55, 885, 927
alkene metathesis, 929
ammonia synthesis, 940
asymmetric oxidations, 934
C–C bond-forming reactions, 934
catalytic cracking, 944
ethanoic acid synthesis, 936
Fischer–Tropsch synthesis, 942
heterogeneous, 936
homogeneous, 928

hydroformylation, 931
hydrogenation, 942
hydrogenation of alkenes, 930, 939, 950
oligomerization and polymerization, 946
sulfur dioxide oxidation, 941
Wacker oxidation of alkenes, 933
catalyst, 66, 565, 849, 885, 897, 927
biphasic, 951
multiphasic, 937
support, 937
tethered, 950
catalytic activity, 878
catalytic converters, 938
catalytic cycle, 67
catalytic efficiency, 67
catenand, 243
catenane, 243
cathode, 188
cathodoluminescence, 904
cation size, 139
cationic interhalogens, 538
Cativa process, 583, 936
C-centred, 103
CE process, 300
cell membrane, 777
cell potential, 193
cell voltage, 922
cement, 392, 885
production, 164
centre of inversion, 74
ceramic, 755
Ceretec (brain imaging agent), 971
CFCs, see chlorofluorocarbons (CFCs)
CFSE, see crystal-field stabilization energy (CFSE)
chalcogens, 501
chalcophiles, 310
character table, 79
charge characteristics, 922
charge ordering, 915
charge transfer, 901, 902
charge-transfer complex, 166
charge-transfer reaction, 670
chelate
complex, 221
effect, 221, 241
chelating ligand, 221
chelation therapy, 969
chemical analysis, 294
chemical beam epitaxy (CBE), 865
chemical industry, 847
chemical loop process, 603
chemical looping, 211
chemical properties, relativistic effects, 331

chemical shift, 281
chemical similarities
chlorine and manganese, 336
p block and d block, 336
phosphorus and vanadium, 336
silicon and titanium, 336
sulfur and chromium, 336
chemical vapour deposition (CVD), 863, 865, 871, 873
chemical vapour transport (CVT), 745
chemiosmotic theory, 795
chemisorption, 938
Chevrel phases, 761, 911
Chinese blue, 901
chiral molecule, 81
chiral resolution, 236
chirality, 232, 235
chloralkali, 530
chloralkali industry, 371
chloride process, 848, 853
chlorine, 525
production by electrolysis, 212
chlorofluorocarbons (CFCs), 527, 531, 546, 851
chlorophyll, 403, 783, 819
chloroplasts, 777
CHN analysis, 295
chromium, 573
in biology, 838
terephthalate, MIL-101, 895
chromophore, 870
CIGS (copper indium gallium selenide), 587, 907
cinnabar, 589
CIS (copper indium diselenide), 587
cisplatin, 267, 585, 959
discovery, 959
transport in body, 959
cis–trans isomerism, 230, 233
clathrate(s), 168, 552
group 18559
Claus process, 211, 508
clays, 766, 945
Clebsch–Gordan series, 623
clinoptilolite, 885, 888, 889
clock reaction, 543
close-packed array, 118
close-packed layer, 105
close-packed structure, 104, 105
closo, 414, 427
cloud seeding, 532
cluster, metal, 228, 590, 706
cluster valence electrons (CVE), 708
cobalamin, 583, 783, 821
cobalt, 112, 583
arsenide, skutterudite, 911
blue, 901
enzymes, 821

enzymes, methyl-transfer catalysis, 822
enzymes, radical-based rearrangement reactions, 822
in biology, 839
COD, 688
coenzyme B_{12}, 821
catalysis of radical-based isomerizations, 822
cofactors (of enzymes), 778
coinage metals, 322
colossal magnetoresistance, 580, 914
colour centre, 725
colour wheel, 273
colours of hydrogen, 346
combinations, 277
compartmentalization (within a living cell), 779
complementary colours, 273
complex, 159, 217
16 electron, 678
18 electron, 678
ansa, 575
formation, 161
high-spin, 247
homoleptic, 697
inner-sphere, 218
low-spin, 247
octahedral, 225, 233, 244
outer-sphere, 218
polymetallic, 228
spin crossover, 642
square-planar, 224, 230, 251
square-pyramidal, 232
tetrahedral, 224, 232, 250
trigonal-bipyramidal, 232
trigonal-prismatic, 226
complex hydrides, 762, 919
complex metal sulfides, 924
complex oxides, 749
complexation, influence on reduction potential, 197
comproportionation, 196
computer modelling, 884
conduction band, 61, 900
conductivity, 900
conjugate acid, 147
conjugate base, 147
constructive interference, 9
contact process, 519
Cooper pairs, 918
cooperative magnetism, 913
coordinate bond, 54, 160
coordination compound, 217
group 1 element, 380
coordination isomerism, 229
coordination number, 106, 114, 118, 124, 218
trends, 319
coordination sphere, primary, 218

copernicium, 592
copper, 586
 chromite, 903
 complex containing DNA intercalator (anticancer drug), 961
 electron transfer centres, 799
 enzymes catalysing production of melanine pigments, 818
 in biological electron transfer, 799
 in biology, 839
 indium gallium selenide (CIGS), 587, 907
 regulation in cells, 836
 semithiocarbazone complexes in medicine, 962
 zinc tin sulfide (CZTS), 907
core–sheath nanowires, 872
core–shell composite nanoparticles, 861, 862, 870
correlation energy, 19
corrin, 783
corundum structure, 749
Cossee–Arlman mechanism, 947
COT, 690
Coulomb potential, 141
Coulombic energy, 133
Coulombic repulsion, 64
counting electrons, 679
coupling reactions, 934, 935
covalency, 758
covalent bond, 35
covalent halides, stability, 320
covalent radius, 23, 62
Cp, 692
cross-polarization, 284
crown ethers, 380
cryptand ligands, 380
crystal lattice, 101
crystal structure, 101
crystal systems, 101
crystal-field splitting parameter, 245
crystal-field stabilization energy (CFSE), 246
crystal-field theory, 244
crystalline Si (c-Si), 906
crystallographic shear plane, 732
cube, 227, 228
cubic close-packed, 106
cubic groups, 75, 76
cubic mesoporous (MCM-48), 878
cumene process, 946
cuprate layers, 917
Curie temperature, 640, 753, 913
CVD, see chemical vapour deposition (CVD)

CVE, see cluster valence electrons (CVE)
cyanide ion, 460, 768
cyanobacteria, 791
cyanogen, 460
cyanomethane, 555
cyclam, 968
cyclic voltammetry, 299
cyclobutadiene, 688
cycloheptatriene, 692
cyclopentadiene, 692
cytochrome c, 796
cytochrome c oxidase, 801, 809
 active site of O_2 reduction, 810
 catalytic cycle, 810
cytochrome c peroxidase, 808
cytochrome P450, 812
cytochrome, optimization for electron transfer, 796
cytochromes, 796
cytoplasm, 777
cytotoxicity, 955
Czochralski process, for silicon production, 211

d band, 59
d orbitals, 11
 shape, 15
dative bond, 54, 160
d-block metal, 562
 disulfides, structures, 515
 first row, 327
 oxides, stability, 321
 redox stability, 316
 sulfides, structures, 514
DCD, 687
d–d reaction, 670
d–d transitions, 626, 901
DDT, 851
de Broglie, Louis, 8
de Broglie relation, 268
decomposition temperature, 139
defect pyrochlores, 754
defects, 722, 738, 747
deferasirox (orally administered iron chelator), 969
degenerate orbitals, 10, 80
degradation, 855
delayed reaction, 670
deltahedra, 426
demineralization, 837
density, 112
density of states, 60
deoxymyoglobin, 792
desalination of water, 888
desferrioxamine (Desferal) in chelation therapy, 969
deshielded, 281
designing safer chemicals, 851
desorption, 938

destructive interference, 9
desulfurization of natural gas, 887
deuteration, 279
 effect on physical properties, 345
deuterium, 4, 340
Dewar–Chatt–Duncanson, 687
diagonal relationship, 328
 atomic radii, 328
 beryllium and aluminium, 328
 boron and silicon, 329
 carbon and phosphorus, 329
 lithium and magnesium, 328
 Pauling electronegativity, 328
diamagnetic, 248
diamond, 442, 448
 anvil cells, 745
 synthesis, 449
 thermal conductivity, 448
diastereomer, 236
diborane, 97, 417
 Lewis acidity, 418
dicarbides, 461
dichalcogenide anions, 759
diethanolamine, 851
differential scanning calorimetry, 296
differential thermal analysis, 296
diffraction methods, 263
diffraction pattern(s), 263, 729
diffusion selectivity, 887
dihydrogen, 338
 complexes of d-metals, 357
 ligand, 686
dimethyl carbonate, 850
dimethyl sulfoxide, 173, 175
dinitrogen, 472
 ligand, 696
 oxide, greenhouse gas, 491
 pentoxide, 488
 tetroxide, 177, 488
 trioxide, 488
dioxygen difluoride, 541
dioxygenases, 812
DiPAMP, 931
dipole moment, 887
dip-pen nanolithography (DPN), 866
direct band gap, 906
direct methanol fuel cell (DMFC), 190
directed evolution, 841
discharge (of lithium ion battery), 922
discotic (disc-like) liquid crystal, 773
disilene, 453
dispersion interaction, 134
displacement reaction, 161

displays, 874
disproportionation, 196, 200
dissociation constant, 240
dissociative rate determining step, 651
dissociative mechanism, 649
distortion
 rhombic, 226
 tetragonal, 225, 252
distribution diagram (species as function of pH), 150
disulfurous acid, 519
ditetralenes, 467
ditetralynes, 467
dithionate, 518
dithionic acid, 520
dithionite, 518
dithionous acid, 520
DNA, 167
 information transfer, 833
dodecahedron, 227
dodecane tetracetic acid (DOTA) as ligand, 963
donor–acceptor complex, 166
donor-pair, counting electrons, 679, 680
dopant, 60, 724
dopant band, 735
Dotarem, MRI agent, 970, 972
double bond, 35
double exchange (DE), 915
Down's process, 368
Drago–Wayland equation, 171, 172
Drago–Wayland parameters (Table), 172
drug development, timeline, 958
drug potency (IC_{50} value), 957
drug precursor (prodrug), 957
drug-delivery agents, 896
DSC, see dye solar cells (DSC)
dubnium, 592
dye solar cells (DSC), 671
dynamic Jahn–Teller effect, 253
dysprosium, 914

EAN (effective atomic number) rule, 677
Earth's core, 114
EC process, 300
E-factor, 848
effective mass, 276
effective nuclear charge, 16
efficiency (of a catalyst), 67, 206
effluent, 855
Egyptian blue, 901
Eigen cation, $H_9O_4^+$, 169
Eigen–Wilkins mechanism, 656
einsteinium, characterization of a heavy actinoid complex, 618
electric field gradients, 288

electric vehicles, 914
electride(s), 176, 365
electrocatalysis, 206
  for water oxidation, 506
electrocatalytic cycle, 206
electrochemical extraction of elements, 212
electrochemical sensor, 741
electrochemical series, 192
electrochromism, 757
electrogenic proton pump, 795
electroluminescence, 889
electrolysis, 207
  of water, 195, 348
electrolyte material, 912
electromagnetic, 270
electromagnetic force, 3
electromagnetic spectrum, 7
electromotive force (emf), 189
electron affinity, 29, 551
  trends, 312
electron attachment, 96
electron configuration, 52
electron counting in clusters, 707
electron deficiency, 97, 411
electron domain, 38
electron impact ionization (EI), 292
electron microscopy, 302, 732
electron pair, 54
electron paramagnetic resonance (EPR), 285
electron self-exchange reaction, 666
electron spectroscopy for chemical analysis (ESCA), 289
electron spin, 11
electron spin resonance (ESR), 285
electron transfer, 795
electron transfer in biology, 795
electron tunnelling in biology, 796
electron-deficient molecular hydrides, 351
electron-deficient compounds, 382
electronegativity, 30, 64, 565
  alternation effect, 329
  Allred–Rochow, 31
  Mulliken, 30
  Pauling, 30
  trends, 313
electron-gain enthalpy, 29
electronic conductivity, 59, 739, 913
electronic factor, 667
electronic spectra, actinoid ion, 639
electronic spectroscopy, 270

electron-precise molecular hydrides, 351
electron-rich molecular hydrides, 351
electrons, unpaired, 325
electron-transfer complex, two proteins, 797
electropositive, 728
electrospinning methods, 879
electrospray ionization (ESI), 292
electrostatic force microscopy, 302
electrostatic parameter (ζ), 137, 153
elements, 592
  availability to early life forms, 778
  occurrence, 309
  superheavy (6d), 592
  used in medicine (periodic table), 956
Ellingham diagram, 207
emerald, 724
emission spectroscopy, 269, 274
enantiomer(s), 81, 236
enantiomeric excess, 931
enantiomeric pair, 232
enantioselective reactions, 931
encounter complex, 667
encounter pair, 656
endocytosis, 790
endohedral fullerene–metal complexes, 450
endoplasmic reticulum, 777, 788
energy-dispersive analysis of X-rays (EDAX), 296
energy-dispersive spectroscopy (EDS), 296, 303
energy efficiency, 853, 854
energy flow in metabolism, 159
energy materials, 899
entering group, 647
enterobactin, 216, 789
enthalpy of atomization, 25
enthalpy of formation ionic chlorides, 320
enthalpy of solvation, 152
entropy of activation, 655
epitaxial films, 864
epoxidation, Sharpless, 934
epoxide, 850
equilibrium bond length, 62
Ertl, Gerhard, 941
esterases, 788
eta (η), 681
ethanoic acid synthesis, 936
ethylene oxide, 849
eukaryote, 777
eukaryotic cell, 777
europium, aqueous chemistry, 607
Evans balance method, 298

Evans NMR method, 299
exchange energy, 19, 28, 580, 602
exclusion principle, 48
exclusion rule, 83
exocytosis, 790
extended defects, 722, 731
extended X-ray absorption fine structure (EXAFS), 291
extinction coefficient, 272
extracellular enzymes, 779
extraction, 563
extraction of elements, 207
extrinsic defects, 722, 724
extrinsic semiconducting materials, 734, 901

f block, 595
  first row, 327
  redox stability, 317, 599
f orbitals, shape, 15
fac, 233
face-centred, 103
face-centred cubic (fcc), 106
Fajan's rules, 32
Faraday balance, 298
Faraday effect, 769
fast atom bombardment (FAB), 292
fast carbon cycle, 457
F-centre, 725
Fenton reaction, 510
fermentation, 855, 856
Fermi distribution, 59
Fermi level, 59
ferredoxins, 798
ferrichrome, 789
ferrihydrite, 791
ferrimagnet, 567, 642
ferrite(s), 116, 913
ferritin, 790
ferrocene, 582, 676, 856
ferroelectric, 752
ferromagnet, 640
ferromagnetic, 268, 876
ferromagnetic and antiferromagnetic coupling, 798
ferromagnetic material, 288
ferromagnetic properties, 913
ferroxidase centre (in ferritin), 791
ferryl ($Fe^{IV}=O$), 813
fertilizer, 478, 886
fibres, carbon, 451
figure of merit (zT) (for thermoelectrics), 911
fillers, 879
fingerprint detection, sulfur nitrides, 522
fireworks, 393
first group member, anomalous, 326

first order phase transition, 756
first row
  d block, 327
  f block, 327
first-order spectra, 282
Fischer, Ernst Otto, 676
Fischer carbene, 694, 700
Fischer–Tropsch catalysis by an enzyme, 831
Fischer–Tropsch process, 573
Fischer–Tropsch synthesis, 942
flame atomization, 294
flame emission spectroscopy, 294
flame photometry, 294
flame pyrolysis, 863
flame synthesis, 863
flame tests, alkali metals, 363
flares, 393
FLP, see frustrated Lewis pair (FLP)
fluorescence, 635, 900
fluorescence spectroscopy, 274
fluorescent displays, 869
fluorescent tube lamps, 905
fluoride glasses, 759
fluorides, 758
fluorimetry, 274
fluorine, 139, 525, 551
fluorine-doped tin oxide (FTO), 901
fluorite, 118, 131, 754
fluorite structure, 121, 122, 741
fluorocarbons, 546
fluorophiles, 442
fluorosulfonic acid, 176
fluxional, 270
  metallocenes, 705
fluxionality, 283
fool's gold, 581
forbidden transition, 634
force constant, 276
formal reduction potential, 194
formaldehyde, 851
formamidinium lead iodide, 907
formation constant, 239
  overall, 239
  stepwise, 239
four-circle diffractometer, 266
Fourier transform NMR, 281
Fourier transformation, 277
fractional coordinates, 103
framework electrolytes, 739
framework representation, 464
framework structures, 763, 764
francium, 362
Franck–Condon principle, 630, 634, 666
Frasch process, 508
Frenkel defect, 723
frictional force microscopy, 302

Friedel–Crafts alkylation, 179, 180
frontier orbital, 30, 53
Frost diagram, oxidation state diagram, 200
frustrated Lewis pair (FLP), 171, 419
fuel cell(s), 189, 737, 912
fullerene(s), 108, 449, 559
   endohedral complex, 559
fullerene–metal complexes, 450
fullerides, 769
fumed silica, 867
functionalization of MOFs, 895
Fuoss–Eigen equation, 657

g value, 285
gadolinite, 579
gadolinium, in MRI, 970
gallium, 410
   applications, 416
   arsenide (GaAs), 436, 734, 874, 906
   hydride, 433
   occurrence and recovery, 416
   sulfides, 436
   trihalide, 434
gallium-68 in medical imaging, 972
galvanic cell, 188
gamma-radiation, 287
gamma-rays, 270
garnet structure, 606
gas absorption, 895
gas condensation, 863
gas monitors, 856
gas separation, 895
gas storage, 895
gemstones, 724
general acid catalysis, 178
general base catalysis, 178
geometric isomerism, 229
gerade, 50
germane, 453
germanene, 874
germanium, 441
   halides, 454
   oxides, 459
germylenes, 467
giant magnetoresistance, 914
Gibbs energy, 131
   of solvation, 152
Gillespie–Nyholm rules, 37
glass, 445, 755
glass fibres, 757
GO, *see* graphene oxide (GO)
gold, 112, 586
   complexes as drugs, 963
   extraction using sulfur-cycle bacteria, 212
   nanoparticles, 587, 861, 867
   refinement using cyanide, 212

Goldschmidt classification, 114, 127, 310
Golgi, 777, 964
Gouy balance, 298
gram-negative bacteria, 777
gram-positive bacteria, 777
graphene, 443, 856, 873
graphene oxide (GO), 873
graphite, 442, 448, 921
   intercalation compounds, 461
   pyrolytic, 448
graphitic carbon, 923
Grätzel cell, 671, 863, 902, 908
Great Oxygenation Event, 503, 778
green chemistry, 847
greenhouse effect, 456
greenhouse gas, 452, 849, 851, 854, 855
   dinitrogen oxide, 491
Grignard reagent, 405, 935
Grotthuss mechanism, proton transfer in water, 169
ground state, 15
ground-state configuration, 16, 18
ground term, 624
group 1 element(s), 362
   atomic radius, 364
   binary compounds, 365
   carbides, 377
   carbonates, 376
   compounds of other oxoacids, 377
   coordination compounds, 380
   first ionization energy, 364
   halides, 372
   hydrides, 372
   hydrogencarbonates, 376
   hydroxides, 375
   metal salts, 910
   metallic radius, 364
   nitrides, 377
   organometallic compounds, 381
   oxides, 374
   peroxides, 374
   properties, 363
   solubility and hydration, 378
   sulfides, selenides, and tellurides, 375
   superoxides, 374
group 12/16 semiconductors, 515
group 13 elements, 409
   compounds, 411
   compounds with group 15 elements, 436
   halides, 413
   hydrides, 413
   hydrogen storage, 412
   low oxidation state halides, 434

   metallic character trend, 410
   occurrence and abundance, 410
   occurrence and recovery, 415
   organometallic compounds, 437
   oxides, 413
   oxo compounds, 435
   oxo salts, 413
   selected properties, 410
group 13/15 compounds, 436
group 13/15 semiconductors, 436
group 14 elements, 441
   applications, 443
   bond enthalpies, 444
   compounds with halogens, 453
   compounds with organic substituents, 465
   compounds with nitrogen, 460
   hydrides, 444, 451
   metallic character trend, 441
   occurrence and abundance, 442
   occurrence and recovery, 448
   selected properties, 442
   simple compounds, 444
group 15 elements, 471
   allotropes, 472
   applications, 477
   Frost diagram, 487
   halides, 485
   halogen compounds, 475
   hydrides, 475, 482
   metallic character trend, 472
   occurrence and recovery, 476
   organometallic compounds, 495
   oxides, 475, 492
   oxoanions, 492
   oxohalides, 486
   pentahalides, 485
   selected properties, 472
   simple compounds, 473
   trihalides, 485
group 16 elements, 502
   compounds with hydrogen, 509
   Frost diagram in acidic solution, 510
   halides, 504, 513
   hydrides, 503
   oxides, 515
   oxohalides, 517
   selected properties, 502
   simple compounds, 503
group 17 elements, 525
group 18 elements, 550
   clathrates, 559
   coordination compounds, 559

   hydrides, 557
   occurrence, 551
   reactivity, 551
   uses, 554
group 2 elements, 385
   atomic radius, 387
   biological compounds, 403
   carbides, 398
   carbonates, 398
   decomposition, 399
   catalysis, 405
   compound solubility, 388
   coordination compounds, 403
   flame tests, 387
   halides, 388
   hydrides, 387, 392
   hydrogencarbonates, 398
   hydroxides, 388, 395, 397
   ionization energies, 387
   in biology, 392
   lower oxidation state compounds, 406
   metallic radius, 387
   nitrates, 401
   nitrides, 388, 398
   organometallic compounds, 404
   oxides, 388, 395
   peroxides, 396
   selected properties, 386
   simple compounds, 387
   solubility and hydration, 402
   sulfates, 140, 401
   sulfides, 397
group II/VI semiconductors, 515
group III/V compounds, 436
group oxidation state, 316
group theory, 72
Grubbs' catalyst, 929
guanyl cyclase, 835
gypsum, 401
gyromagnetic ratio, 280

^{1}H NMR chemical shifts, 346
Haber–Bosch process, 478, 482, 581, 853, 940
haematin, toxicity to malaria parasite, 966
haemerythrin, 794
haemocyanin, 793
haemoglobin, 581, 792
   cooperativity, 793
haemozoin, 967
hafnium, 568
half-reaction, 187
halides, 758
   group 1 element, 372
Hall–Héroult process for aluminium production, 212
halogen bonding, 166, 534

halogen oxides, 528, 541
halogenases, 814
halogens, 525
halons, 532
haloperoxidase, 570
Hammett acidity function H_0, 174
hapticity, 681
hard acid, 169
hard ferrites, 913
hard ferromagnet, 641
hard-disk drives (HDDs), 915
hardness
　abrasion, 423
　permanent, 401
　temporary, 388, 401
hardness (chemical consequences), 170
hard/soft classification, 169, 565
　rule, 170
HCFCs, see hydrochlorofluorocarbons (HCFCs)
Heck coupling, 935
hectorite, 766
Heisenberg uncertainty principle, 8
Heisenberg, Werner, 8
Helicobacter pylori, 964
helium, 550
　extraction, 553
　occurrence, 552
　uses, 553
helium-II, 553
hepatitis C, 968
heteroepitaxy, 865
heterogeneous catalysis, 68, 852, 936
heterogeneous nucleation, 864
heterogenized homogeneous catalysts, 946
heteronuclear coupling, 282
heteronuclear diatomic molecule(s), 53, 93
heteropolyoxometallates, 159
hexagonal lamellar mesoporous (MCM-50), 878
hexagonal mesoporous (MCM-41), 878
hexagonally close-packed, 106
HFCs, see hydrofluorocarbons (HFCs)
high anion mobility, 738, 739
high pressure, 745
higher boranes, 426
　characteristic reactions, 430
　synthesis, 429
higher borohydrides, synthesis, 429
higher oxides, 749
highest occupied molecular orbital (HOMO), 53
high-spin complex, 247

high-temperature superconductors, 605, 916
hole, 108
　in close-packed structure, 108
hole size, 128
holoenzyme, 782
homeostasis, 779
homoepitaxy, 865
homogeneous catalysis, 68, 852, 928
homoleptic complex, 697
homolytic dissociation, 350
homonuclear coupling, 282
homonuclear diatomic molecule, 43, 49, 93
Hoppe, Rudolf, 554
host-guest complex, 168
host-guest compound, 888
Hund's rules, 18, 47, 624
hybrid, 891
hybrid nanomaterials, 878
hybrid orbital, 45, 97
hybridization, 45, 97
hybridization schemes, 46
hydrate isomerism, 229
hydration enthalpy, 140
hydrazine, 483
hydride, 340
　acidity, 699
　gap, 355
　ion affinity (HIA), 172
　ligand, 685
　non-classical, 686
hydrides, 919
　boron, 416
　group 1 element, 372
　metallic, 342, 354
　molecular, 342
　saline, 342, 353, 387, 392
hydridicity, 357
hydrido complexes of d-metals, 357
hydridoaluminates (alanates), 762
hydroboration, 418
hydrocarbons, 452
hydrochloric acid, 851
hydrochlorofluorocarbons (HCFCs), 546
hydrofluorocarbons (HFCs), 527, 531, 851
hydroformylation, 852, 931
hydrogen, 4, 114, 919
　absorption, 896
　agostic, 695
　biological cycle, 339
　cation, 340
　colours of, 346
　compounds, 342
　dissociation, 341
　fuel, 341
　intermolecular forces, 340

　isotopes, 340
　molecular cation, 340
　molecule, 42
　nuclear properties, 344
　position in the periodic table, 340
　properties, 340
hydrogen bond, 166, 536
　enthalpies, 353
　energies of, 167
　influence on physical properties, 353
hydrogen carriers, 354
hydrogen compound(s)
　bond energies, 343
　Gibbs energy of formation, 343
　reactivity, 343
　synthesis, 358
hydrogen cyanide, 851
　solid, 167
Hydrogen Evolution Reaction (HER), 348, 349, 350, 908
hydrogen fluoride, 54, 176
　hydrogen bonding in, 167
hydrogen halides, 527
Hydrogen Oxidation Reaction (HOR), 348, 350
hydrogen peroxide, 510
　environmentally friendly bleach, 511
　in biology, 808
hydrogen production, 339, 345
　carbon sources, 347
　laboratory, 347
　using solar energy, 349
　water, 347
hydrogen purifier, 356
hydrogen reactions
　heterolytic dissociation, 350
　homolytic dissociation, 350
　radical chain reactions, 350
hydrogen selenide, 513
hydrogen storage materials, 268, 762, 919
　group 13 element compounds, 412
　magnesium hydride, 393
hydrogen sulfide, 512
hydrogen superoxide, 511
hydrogen telluride, 513
hydrogenases, 827
　with [FeFe] active site, 828
　with [NiFe] active site, 828
hydrogenated amorphous silicon (a-Si:H), 906
hydrogenation catalysts, 942
hydrogenation of alkenes, 930, 939
hydrogenic atoms, 7
hydrometallurgical extraction, 209

hydronium ion, 146
hydrosilylation, 453
hydrothermal reaction, 884
hydrothermal techniques, 746
hydroxoacid, 154
hydroxyl radicals, 910
hydroxylamine, 484
hypercoordinate, 462
hyperfine structure, 286
hypertetrahedral frameworks, 889
hyperthermophile, 778
hypervalence, 45, 95
hypo, 427
hyponitrite, 489
hypophosphite, 492
hypothyroidism, 533
hypoxia, 834

I_2 molecules, 114
ice, structure, 168
icosahedral group, 76
identity operation, 73
ilmenite, 848
image plate, 266
imidazoles, 891
improper rotation, 74
improper-rotation axis, 75
in operando techniques, 262, 265
in situ techniques, 262
incineration, 855
incomplete octet, 411
indirect band gap, 906
indium, 410
　applications, 416
　occurrence and recovery, 416
　sulfides, 436
　trihalide, 434
indium tin oxide (ITO), 435, 900
inert, 647
inert gas rule, 677
inert gases, 550
inert pair effect, 332
　rationalization, 332
infrared (IR) radiation, 270
infrared (IR) spectroscopy, 82, 276, 701
inherent stability (redox), 200
inner-sphere mechanism, 663
inorganic framework chemistry, 767
inorganic liquid crystals, 773
inorganic–organic nanocomposites, 878
insertion, 714
insulator, 61, 733
interband electron transfer, 901
interband transitions, 868
intercalation compound, 186, 448, 760

interchange, 656
interchange mechanism, 649
interference, 45
interference (between wavefunctions), 46
interhalogen, 139, 528, 537, 539, 555
  cationic, 538
intermediate, 649
intermetallic compounds, 728, 729
intermetallic hydrides, 763, 921
International Union of Pure and Applied Chemistry (IUPAC), 21
interstitial carbides, 462
interstitial compounds, 111
interstitial mechanism, 740
interstitial solid solutions, 115, 728
intersystem crossing, 636
intervalence charge transfer, 901
intraband transitions, 870
intrinsic defects, 722
intrinsic semiconductivity, 61, 734
inverse micelle synthesis, 864
inverse spinel structure, 751
inversion (operation), 50, 74
inverted behaviour (electron transfer), 668
iodine, 525, 889
  as a Lewis acid, 166
ion channels, 785
ion cyclotron resonance (ICR) mass spectrometer, 293
ion diffusion, 737, 744
ion exchange, 888, 890
ion-exchange chromatography, 597
  separation of actinoids, 613
  separation of lanthanoids, 606
ion pump (biological membrane), 780
ion transport in biology, 780
ion transporter (biological membrane), 780
ionic bonding, 101, 130
ionic chlorides, enthalpy of formation, 320
ionic conductivity, 737, 739
ionic liquids, 177, 852, 951
ionic model, 117, 127
ionic radii, 23, 127, 128, 129, 758
ionic solids, 117
ionization energy, 26, 96, 551
  trends, 27, 311
ionization isomerism, 229
ionophore, 785
ionothermal reactions, 746

IR active, 277
IR spectroscopy, see infrared (IR) spectroscopy
iridium, 112, 583
iron, 116, 581
  content of rice, 840
  cycle, in biology, 789
  in biology, 839
  metal, 113
  overload, 789
  overload (medical condition), 969
  oxide, 747
  pentacarbonyl, 270, 696
  porphyrin, distal region, 792
  porphyrin, proximal region, 792
  regulatory protein (IRP), 834
  Pourbaix diagram for naturally occurring aqua species, 204
  toxicity, 789
  uptake and storage by organisms, 789
iron(IV) in peroxidases and oxidases, 808
iron(IV) oxo complexes, models for oxygenases, 817
Iron Age, 207, 210, 566
iron–air battery, 925
iron-protein sensors for NO or CO, 835
iron–sulfur centre, 798
  biosynthesis, 798
  in radical SAM enzymes, 823
  in radical-based catalysis, 823
  in the fumarate–nitrate regulator, 834
iron–sulfur clusters, 797
iron–sulfur proteins as sensors, 833
irreducible representation, 79, 85, 91
Irving–Williams series, 253, 813, 832
isocyanate, 851
isolated-island structures, 880
isolobal analogies, 53, 708
isomer shift, 288
isomerism
  coordination, 229
  geometric, 229
  hydrate, 229
  ionization, 229
  linkage, 229
  optical, 232
isoreticular, 894
isosbestic points, 274
isotactic, 948
isotopes, 6, 293
  hydrogen, 340
isotopologue, 345, 943

isotopomer, 276, 345
ITO, 435
IUPAC, 21

Jacobsen oxidation, 934
Jahn–Teller distortion, 226, 252, 575, 580, 588
jj-coupling, 623
Jupiter, 114

Kaminsky catalyst, 947
Kapustinskii equation, 136
Ketelaar triangle, 65, 115, 117, 121, 122, 731
Kevlar®, 167
kinases, 804
kinetic diameter, 887
kinetic isotope effect, 345
Kocheshkov redistribution reaction, 466
Koopmans' theorem, 289
Kroll process, titanium production, 211, 568
krypton, 550
  clathrates, 559
  coordination compounds, 559
  fluorides, 559

L type ligand, 679
labile, 647
lanthanoid-based permanent magnetic materials, 913
lanthanoid binary compounds, metallic vs insulators, 604
lanthanoid contraction, 25, 330, 564, 599
  impact on atomic radii, 330
  impact on densities, 330
  impact on ionic radii, 331
lanthanoid ions as probes for calcium proteins, 788
lanthanoid iron arsenic oxides, 918
lanthanoid nickel borocarbides (superconductors), 605
lanthanoid organometallics as homogeneous catalysts, 611
lanthanoids, 562, 596
  applications, 598
  aqua ions, 606
  as optical or magnetic tracers, 607
  atomic properties (Table), 599
  coordination compounds, 606
  coordination numbers, 601
  crustal abundance, 597
  electronic spectra, 636
  extraction, 597
  f- and d-orbital energies, 603
  halides, 604

  in biology, 840
  in metal–organic frameworks (MOFs), 607
  ligand-field effects, 601
  magnetic properties, 643
  nitrides, 603
  organometallic chemistry, 608
  oxides, 603
  participation of d-orbitals in bonding, 610
  prevalence of 3+ oxidation state, 599
  radii, 599
  redox chemistry, 602
  standard potentials, 602
  standard potentials for the Ln^{3+}/Ln^{2+} couple, 607
  ternary and complex oxides, 605
lanthanum, 566
lanthanum barium copper oxide, 917
Laporte selection rule, 634, 902
Large Hadron Collider, 553
laser pyrolysis, 863
laser-ablation, 863
lasers, 554, 870, 900
lasing effects, 757
Latimer diagram (reduction potential diagram), 199
lattice, 101
  energy, 131, 743
  enthalpy, 131, 132, 134, 138, 140
  parameter, 265, 726
  points, 101, 103, 104, 113, 120
Laves phases, 763
layered double hydroxides, 766
layered metal sulfides, 760
LCA, see life cycle analysis (LCA)
LCAO (linear combination of atomic orbitals), 46
L-dopa, 931
lead, 112
  chromate, 902
  halides, 455
  iodide perovskite, 129, 907
  oxides, 459
  zirconate titanate, 569, 911
leaving group, 647
LEDs, see light-emitting diodes (LEDs)
Lewis acid catalysis, 179
Lewis acidity, 159, 218
  boron trihalide, 411
  compounds of Group 13, 162
  compounds of Group 14, 163
  compounds of Group 15, 165
  compounds of Group 16, 165
  compounds of Group 17, 165

diborane, 418
s-block elements, 162
Lewis acids and bases, classification, 169
Lewis base, 159, 218, 685
Lewis structure, 36
LFAE, see ligand-field activation energy (LFAE)
LFER, see linear free-energy relation (LFER)
LFSE, see ligand-field stabilization energy (LFSE)
Li-ion battery materials, 268, 923
life cycle analysis (LCA), 854
ligand, 217
   ambidentate, 219
   chelating, 221
   chirality, 237
   macrocyclic, 241, 243
   spectator, 658
   strong field, 246
   substitution, 647
   substitution reaction, 710
   tripodal, 225
   weak field, 246
ligand-field activation energy (LFAE), 659
ligand-field splitting parameter, 245
ligand-field stabilization energy (LFSE), 240, 246, 575, 647, 751
ligand-field theory, 244, 254
ligand-field transitions, 626
ligand-to-metal charge-transfer (LMCT) transition, 631, 632
light-emitting diodes (LEDs) 735, 869, 904
lighting 566, 900, 905
light transmission, absorption and emission, 899
Li-ion battery, 737
   anodes, 872
   electrolytes, 924
   materials, 921
limewater, 397
limonite, 585
linear combination of atomic orbitals, 46
linear free-energy relation (LFER), 658
linkage isomerism, 229
linked polyhedral metal centres, 766
lipid bilayer, 777
liquid natural gases, 896
lithium, 362
   alloys, 923
   amide-boranates, 920
   applications, 369

atypical properties, 366
borofluoride, 759
cobalt oxide, 742, 921
diisopropylamide (LDA), 180
iron phosphate, 922
iron sulfate fluoride, 922
manganese spinels, 742, 922
metal, 923
nitride, 180, 378, 757, 920
occurrence and extraction, 367
recycle, 368
tetrahydridoborate, 920
treatment for bipolar disorders, 966
lithium–air batteries, 925
lithium batteries, 369, 741
   primary, 369
   rechargeable, 369
   cathode materials, 752, 921
lithium ion conductors, 924
lithophiles, 310
livermorium, 502
living cell, 777
LMCT, see ligand-to-metal charge-transfer (LMCT)
localization, 96
lodestone, 581
lone pair, 35, 38, 41
Lowenstein's rule, 765
lower oxidation state compounds, group 2 elements, 406
lowest unoccupied molecular orbital (LUMO), 53
low-spin complex, 247
luminescence, 635, 900, 903
   lanthanoid ion, 638
LUMO, see lowest unoccupied molecular orbital (LUMO)
Lyman series, 7
lysosome, 777

macrocycle, 243, 783
macrocyclic effect, 241
Maddrell's salt, 493
Madelung constant, 132, 133, 136
'magic acid', 179
magic-angle spinning (MAS), 284
magnesium, 386
   alloys, 391
   applications, 390
   boride, 425
   carbonate, 138
   chlorophyll, 403
   complex with ATP, 780
   enzyme in CO_2 fixation, 805
   enzymes, 804
   extraction, 389
   halides, 395

hydride, 393, 762
in biology, 838
in endonucleases, 805
occurrence and abundance, 386
magnet, single molecule, 643
magnetic coercivity, 913
magnetic domain, 640
magnetic force microscopy, 302
magnetic properties, 749
magnetic quantum number, 9
magnetic remanence, 913
magnetic resonance, 279
magnetic resonance imaging (MRI), 553, 554
magnetic structure, 268
magnetic susceptibility, 298, 640
magnetite, 837
magnetometry, 248, 298
magnetoresistance, 914
magnetoresistive, 876
magnetotactic bacteria, 837
Magnevist, MRI agent, 970
main group, allotropes, 314
main group metal chalcogenides, 759
malaria, 966
MALDI, see matrix-assisted laser desorption/ionization (MALDI)
manganese, 579
   in biology, 839
   Frost diagram, 203
   Pourbaix diagram, 206
manganites, 914
Marcus cross-relation, 669
Marcus equation, 667
Marcus theory, 796
Marsh test, 478
martensitic stainless steels, 116
mass defect, 292
mass number, 3
mass spectrometry, 292
matrix isolation, 279, 557, 559
matrix-assisted laser desorption/ionization (MALDI), 292
mean bond enthalpy, 63
Meissner effect, 916
melting point, trends, 315
membrane separator, 477
Mendeleev, Dmitri, 20
Menkes disease (Cu deficiency), 836
mer, 233
mercury, 113, 589
meridional, 233
mesogenic liquid crystal, 773
mesoporous materials, 877
   catalysis, 937
mesostructured nanomaterials, 877

metal(s), 111, 916
   amine complexes, 920
   borides, 425
   carbides, 446
   carbonyl basicity, 699
   catalyst, 871
   chalcogenides, 911
   chloride structures, 759
   clusters, 706, 888
   complexes, 894
   covalent oxides, 514
   disulfides, 514
   fluorides, 747
   hydride batteries, 356
   hydrides, 761, 762
   nanoparticles on surfaces, 862
   nitrides, 747
   oxides, 513, 747
   selenides, 515
   sulfide chemistry, 891
   sulfides, 514
   tellurides, 515
metal-complex pharmaceutical (what happens after administration), 957
metal-disulfide structures, 760
metal–metal bond, 575, 590, 672, 706
metal–organic chemical vapour deposition (MOCVD), 865
metal–organic frameworks (MOFs), 746, 767, 768, 891, 921
   applications, 895
metal-oxide fillers, 880
metal-rich chalcogenides, 761
metal-to-ligand charge-transfer transition (MLCT), 631, 633
metallaboranes, 415, 431
metallic bonding, 66, 100
metallic character, trends, 314
   group 15 elements, 472
metallic conductor, 60, 733
metallic elements, 112
metallic hydrides, 342, 354
metallic nanoparticles, 870
metallic radius, 23, 114, 728
metallocene, 702
   bent, 706
   fluxionality, 705
metalloids, 20
metallomics, 778, 957
metalloneurochemistry, 840
metallophosphate systems, 890
metalloproteins, 781
metallothioneins, 833
metastable, 747
metathesis, 161, 929

methane, 452
  clathrates, 168, 452
  monooxygenase, 816
  production by microorganisms, 821
  production in biology (methanogenesis), 824
methanol carbonylation, 936
methides, 461
methylaluminumoxane, MAO, 947
methylammonium lead iodide, 907
methylcobalamin, 821, 822
methylcoenzyme M reductase (Ni), 824
metmyoglobin, 792
microcompartments (of cells), 777
microporous solids, 463, 764, 890
microscopy, 866
microstates, 622
microwaves, 854
midpoint potential, 194
migration, 714
migratory insertion reaction, 714
Minamata, 589
minerals, silicates, 458
minimal basis set, 49
mirror plane, 73
mischmetal, 566, 598
missing xenon, 557
mitochondria, 777, 788
mixed ionic–electronic conductors, 741
mixed oxides, 749
mixed-valence, 798
MLCT, see metal-to-ligand charge-transfer transition (MLCT)
M–M bond, 698
mobile ions, 738
MOF(s), see metal–organic frameworks (MOFs)
molar absorption coefficient, 272
molar absorptivity, 272
molecular absorption, 884
molecular beam epitaxy (MBE), 864
molecular hydrides, 342, 351
  electron-deficient, 351
  electron-precise, 351
  electron-rich, 351
  reactions, 352
molecular inorganic materials, 769
molecular magnets, 772
molecular materials chemistry, 770

molecular orbital (Lewis acid-base), 160
molecular orbital (MO) theory, 46
molecular orbital diagram, 48, 49
molecular orbital energy-level diagram, 48
molecular orbitals, 88
molecular potential energy curve, 43
molecular recognition AFM, 302
molecular shape, 278
molecular sieves, 464, 886, 887
molecular term symbols, 627
molecular vibrations, 82, 276
molybdenum, 573
  in biology, 839
  sulfide, 761
molybdenum and tungsten enzymes, 825
molybdopterin, 783
monazite, 597
Mond, Ludwig, 676
Mond process, 585, 698
monodentate, 219
monooxygenases, 812
monoxides of the 3d metals, 747
Monsanto, 851
Monsanto acetic acid process, 583, 936
montmorillonite, 766, 855
moscovium, 21, 471
Mössbauer spectroscopy, 287
MRI, see magnetic resonance imaging (MRI)
mu (μ), 681
Mulliken electronegativity, 30
Mulliken, Robert, 30
multinuclear NMR, 280
multiple quantum well (MQW), 875
multiple versus single bonds, 327
multiwalled carbon nanotube (MWNT), 450, 871
MXenes, 874
myochrisin, 964
myoglobin, 791

Na ion batteries, 924
  materials, 921
NADH dehydrogenase (Complex I of mitochondria), 798
Nafion®, sodium perfluorosulfonate, 190
nanobuds, 450
nanocarbons, 879
nanomaterial, 860

nanoparticles, 101, 745, 859, 888, 903, 909
  in drug delivery, 972
nanoparticulate TiO_2, 867
nanoscience, 860
nanosized reaction vessels, 864
nanostructures, 866
nanostructuring, 911
nanotechnology, 860
nanotubes, 871, 872
  carbon, 450
  multiwalled, 450
  single-walled, 450
naproxen, 852
NASICON, 738, 739, 924
natural waters, 205
natural zeolites, 885
Nd:YAG, 566
Néel temperature, 641, 890
nematic phase, 773
neodymium iron boride, 598, 913
neon, 550
nephelauxetic parameter, 631
nephelauxetic series, 631
neptunium, 617
  synthesis of complex using a nuclear reaction, 617
Nernst equation, 192, 741
network modifier, 756
neutral-ligand, counting electrons, 679
neutron diffraction, 267
NEXAFS, 291
N-heterocyclic carbene (NHC), 694, 929
  silver complex, 588
nickel, 584
  arsenide, 118
  arsenide structure, 121, 122
  cadmium (NiCad) battery, 189
  in biology, 839
  in urease, 802
nido, 414, 427
nihonium, 21, 409
niobium, 570
  halides, 573
niobium titanium oxide, 924
Nitinol, 730
nitrate, 489
nitrate reductase, 831
nitric acid, 488
nitric oxide, 488
  synthase, 832
  substrate for guanylyl cyclase, 832
nitride(s), 474, 480, 757, 905
  covalent, 480
  interstitial, 480
  nitrido wall, 336
  saline, 480

nitrile dehydratase, 839
nitrite, 489
  role in curing meat, 491
nitrogen, 471
  activation, 479
  applications, 477
  cycle, 474, 828
  dioxide, 488
  distillation, 477
  fixation, 829
  Frost diagram, 201
  halides, 485
  ligand, 479
  monoxide, ligand, 696
  oxides and oxoanions, 476, 487
  trifluoride, 485
nitrogenase, 479, 570, 574, 829, 853
  'P-cluster', 830
  FeMo cofactor (FeMoco), 830
nitrogen-to-ammonia, the Lowe–Thorneley cycle, 831
nitronium, 177, 489
nitrophorin, 832
nitrosium, 489
nitrosyl cation, 489
nitrosyl halides, 486
nitrous acid, 490
nitrous oxide, 488
  reductase, 801, 831
nitryl cation, 489
nitryl halides, 486
NMR spectroscopy, 554, 701
  typical ^{1}H NMR chemical shifts, 346
noble character, 322
noble gas, ligand, 695
noble gas, compounds, 135, 551
nodal planes, 14
nodes, 12
nonaqueous solvents, 173
  Table, 174
non-bonding orbital, 49
non-bonding pair, 35
nonclose-packed, 112
noncrossing rule, 51, 629
nonlabile, 647
nonstoichiometry, 726, 747, 752
normal mode(s), 84, 276
normal spinel structure, 751
NOx removal, 885
n-type semiconductivity, 735, 736, 905, 913
nuclear fission, 5, 612
nuclear frequency factor, 667
nuclear fusion, 5
nuclear magnetic resonance (NMR), 279
nuclear power, 6

nuclear properties, hydrogen, 344
nuclear reactor, 268
nuclear spin characteristics, 280
nuclear waste, 888, 890
nucleon number, 3
nucleophilic discrimination factor, 652
nucleophilic substitution, 648
nucleophilicity, 648
  parameter, 652
nucleosynthesis, 4
nucleus (of a cell), 777

occupation factor of a spinel, 751
occurrence, elements, 309
octahedral complex, 233, 244
octahedral hole, 108, 117, 124
octahedron, 225
octet rule, 35
oganesson, 21, 550, 551
oil refining industry, 848
oleum, 165
oligomerization, 946
one-dimensional control, 870
one-dimensional metals, 770
one-dimensional solid, 59
one-equivalent process, 664
optical activity, 81
optical isomer, 232, 235
orbital angular momentum quantum number, 9
orbital approximation, 15, 46
orbital overlap, 48
order, of group, 79
ores, 563
organelle, 777
organic linker, 894
organoaluminium, 438
organoarsenic compounds, 496
organoboranes, 558
organoboron, 437
organodifluoroboranes, 558
organolithium, 382
organomagnesiation, 405
organometallic compounds
  group 1 element, 381
  group 2 element, 404
  of the d-block, 675
  of the lanthanoids, 608
organophosphines, 485
organophosphorus compounds, 487
organoxenon, 558
Orgel diagram, 628
ORTEP diagrams, 267
orthosilicate, 446
oscillating reaction, 543
osmiridium, 583
osmium, 112, 300, 581
osmotic pressure, 777
Ostwald process, 478, 488

Ostwald ripening, 861
outer-sphere mechanism, 663
overlap, p orbital, 327
overlap of atomic orbitals, 58
overpotential, 190, 206
overtones, 277
oxazaborolidine, 171
oxidant, 186
oxidation, 185
  Jacobsen, 934
oxidation by atmospheric oxygen, 196
oxidation number, 65, 186, 200
oxidation of carbonyls, 699
oxidation state, 139
  stability, 200
  trends, 315
oxidative addition, 686, 712, 936
oxide-fluorides, 759
oxide-ion conductivity, 741
oxide-nitrides, 758, 903, 909
oxides, amphoteric, 318
oxidizing agent, 186
oxido complexes, 335
oxo wall, 334, 565, 813
oxoacid, 154, 155
  structural anomalies, 157
  structures and pK_a values, 155
oxoanions, 158, 764
oxofluorides, 556
oxophiles, 442
oxygen, 290, 505
  abundance, 502
  adaptation in higher animals, 834
  biogenic, 503
  binding curves for myoglobin and haemoglobin, 793
  binding to myoglobin and haemoglobin, 792
  carrier proteins, 791
  difluoride, 528, 541
  evolution during photosynthesis, catalytic cycle, 820
  evolution reaction (OER), 206, 820, 908
  evolving centre (OEC), 819
  Frost diagram, 201
  Pourbaix diagram, 511
  reactions, 505
  rebound mechanism, 813
  reduction reaction (ORR), 206
  sensor, 741
  singlet, 506
  solubility, 505
oxygenases, 812
  dependent on 2-oxoglutarate, 813, 835
  non-haem Fe, 813

  non-haem (table of reactions), 814
oxygen-atom transfer reactions, enthalpies, 827
oxymyoglobin, 792
ozone, 502, 507
  depletion, 851
  hole, 532
  layer, 532
ozonides, 508

^{31}P NMR, 685
p band, 59
p block oxide, reactivity, 330
p orbital, overlap, 327
p orbitals, 11
paint industry, 852
pairing energy, 246
palladium, 584
  complex in photodynamic therapy, 961
paramagnetic, 248
paramagnetism of dioxygen, 52
Parkinson's disease, 935
partially crystalline carbon, 451
Pascal's triangle, 283
Paschen series, 7
passivation, 195
Pauli exclusion principle, 15, 38, 48, 49
Pauli repulsion, 19, 38, 49, 64
Pauling electronegativity, 30, 65
  diagonal relationship, 328
Pauling, Linus, 30, 64
Pauling's rules, 156
Pauson–Khand reaction, 943
pearlescent pigments, 903
Pechini method, 745
Peierls distortion, 771
Peltier effect, 911
penetration, 15
penicillin, biosynthesis of, 815
pentagonal bipyramid, 42, 227
pentaphenylantimony, 497
pentaphenylarsenic, 497
pentlandite, 585
perbromate, 542
perchlorate, 542
period, 22
periodate, 542
periodic table, 20
periplasm, 777
permanent hardness, 401
permanent magnets, 914
permanganate, 579
pernitride anion, 137
perovskite(s), 118, 727, 752, 875, 907, 918
perovskite structure, 124, 911, 912, 924
peroxidase(s), catalytic cycle, 808, 809

peroxide(s), 139, 510
peroxisome, 777
peroxosulfuric acid, 519
perxenate(s), 552, 556
PET, see positron emission tomography (PET)
petroleum, 944
petroleum refining, 887
pH, 147
pharmaceutical industry, 848
pharmaceutically active compounds, 896
phase change materials (PCMs), 910
phase diagrams, 265
phase identification, 265
Philips catalyst, 947
phosgene, 454
phosphate(s), 492, 765
  condensed, 493
  ester hydrolysis by acid phosphatase, 807
  in biology, 780
  in the food industry, 493
phosphazenes, 494
  biomedical applications, 495
phosphides, 481
phosphine, 484
  ligand, 684
phosphite, 492
phosphorescence, 635, 900
phosphoric acid, 159
phosphors, 735, 900, 904
  lanthanoid, 639
phosphorus
  applications, 478
  Latimer diagrams, 492
  oxides, 492
  pentachloride, applications, 486
  plant nutrient, 478
phosphoryl halides, 487
photocatalysts, 758, 849, 897, 908
photochemical reaction, 670
photochromic glasses, 757
photochromism, 757
photodissociation, 555
photodynamic therapy, 961
photoelectron spectroscopy (PES), 49, 93, 289
photoionization, 96
photoluminescence, 904
photo-substitution, 710
photosynthesis, 819
photosynthetic electron-transport chain, 820
photovoltaic (PV) materials, 905
physical properties, effect of deuteration, 345
physical techniques in inorganic chemistry, 262

Index

physical vapour deposition (PVD), 864
physisorption, 921
Pidgeon process, magnesium production, 207
piezoelectric, 569, 753, 911
piezoresistive, 448
pigment(s), 752, 758, 848, 852, 863, 867, 872, 900, 901
pillared clays, 766
plasma synthesis, 863, 864
plaster of Paris, 401
plastocyanin, 800
platinum, 584
   anticancer drugs, 959
   complex with DNA, 960
   metals, 322
platinum(IV) complexes as prodrugs, 960
plumbane, 453
plutonium, 617
   chelation therapy, 970
   environmental hazard, 617
   waste, 198
p-n type junction, 735
pnictogens, 471
point defects, 722
point group, 75
point-group symmetry, 73
polar molecule, 81
polarizability, 32, 135
polarizing ability, 32
pollutants, 908
polonium, 113, 115, 509
polyarsane compounds, 496
polyatomic molecular orbitals, 93
polyatomic molecule, 44, 93
polybromides, 540
polycarbonates, 850
polychlorinated biphenols, 852
polycrystalline, 263
polydentate, 219
polyhalide(s), 166, 539, 554
polyhedral units, 767
polyhydride, 919
polylactic acid, 855
polymer, 850, 852
polymer matrix, 879
polymer nanocomposites (PNCs), 879
polymerase chain reaction, 841
polymorphism, 113, 114, 116
polynitrogen compounds, 480
polyoxoanion, 158
polyoxometallate (POM), 159, 571, 575
polyphosphates, 159
   industrial applications, 494
polyphosphoric acid, 494
polyprotic acid, 150

polytetrafluoroethene (PTFE), 527, 546, 547
polytypes, 105, 112
polytypism, 112
polyurethane, 851
POM, see polyoxometallate (POM)
porosity, 768, 895
porous coordination polymers (PCPs), 891
porphyrin, 783, 850
positive hole, 61
positron emission tomography (PET), 962
post-translational modification, 781
potassium, 362
   channel, 784, 786
   in biology, 780, 838
   nitrate, 137, 910
   tetrafluoridonickelate(II), 753
potential-gated K^+ channel, 785
Pourbaix diagram, 196, 204
powder diffraction, 264
powder diffractometer, 264
powder method, 263
powder X-ray diffraction (PXRD), 265, 755
primitive, 102
primitive cubic, 113
principal axis, 73
principal quantum number, 9, 130
principles of green chemistry, 847
probability density, 9
product selectivity, 945
production, hydrogen, 339
projection operator, 92
prokaryote, 777
prolyl hydroxylases, 834
promotion, 44
prompt reaction, 670
Prosulfuron® (herbicide), 935
protein structure (determination by X-ray diffraction), 784
protein structure representation, 784
proteins, 780
proteome, 781
protiodide, 590
proton, 340
   activity, 147, 151
   affinity (Table), 151
   affinity (effective), 151
   affinity (gas phase), 151
   conductor, 169
   transfer in water, 169
   transfer reaction, 147
proton-coupled electron transfer (PCET), 204

proton-exchange membrane fuel cell (PEMFC), 190
proton-gain enthalpy, 151
proton-gain reaction (cycle), 151
Prussian blue, 768, 902
pseudohalogen, 460, 535
PTFE, see polytetrafluoroethene (PTFE)
PtPOP, 672
p-type semiconductivity, 736, 905
pulsed-laser deposition (PLD), 864, 865
purely siliceous zeolite, 884
purification, 887
pyrite, 515
pyrochlore structure, 754, 918
pyrolusite, 579
pyrolytic graphite, 448
pyrometallurgical process, 207
PZT (lead zirconate titanate), 569

quadruple bond, 577
quadrupolar splitting, 288
quadrupole mass filter, 292
quadrupole moment, 887
quantization, 8
quantum cascade (QC) lasers, 875
quantum confinement, 868
quantum dot(s) (QD), 275, 862, 868
quantum dot (QD) monolayers, 869
quantum dot (QD)-LED, 869
quantum mechanics, 8
quantum numbers, 9
quantum wells, 874
quantum yield, 670
quartz, 445
quasicrystals, 729
quenched, 248
quinol, 559
quintuple bond, 576, 579

Racah parameters, 625
radial distribution function, 13
radial distribution function (Ln vs An), 596
radial node, 12, 597
radical, 902
radical chain reactions hydrogen reactions, 350
radicals, 325
radicals (dimerization of), 53
radio waves, 270
radioimmunotherapy, 962
radionuclides, 888, 890
   in medicine (Table), 962
radiopharmaceuticals, 962

radiotherapy, 962, 963
radium, 385
   applications, 391
   extraction, 390
   radioactivity, 390
radius ratio, 128, 134
   alkali metal halides, 373
radon, 550
   fluorides, 559
   uses, 553
Raman active, 277
Raman spectroscopy, 82, 276
Raney nickel, 943
rare earth elements, 596
Raschig process, 483
rate-determining step (RDS), 67, 649
rate law, 648
rates of diffusion, 738
Ray–Dutt twist, 663
RDS, see rate-determining step (RDS)
reactant selectivity, 945
reaction
   charge transfer, 670
   clock, 543
   coordinate, 66
   d–d, 670
   delayed, 670
   electron self-exchange, 666
   enthalpy, 62
   entropy (from Ellingham diagram), 208
   oscillating, 543
   photochemical, 670
   prompt, 670
reactive oxygen species (ROS), 511, 835, 966
real-time analysis, 855
rechargeable battery, 189, 742
   materials, 921
redox condensation, 709
redox couple, 187
redox reaction, 186, 663
redox reactions, balancing equations, 187
redox reactions, influence of pH, 194
redox stability, 193
   d block, 316
   f block, 317
reducible representation, 85
reducing agent, 186
reductant, 186
reduction, 185, 563
reduction formula, 91
reduction of carbonyls, 699
reduction potentials, 96
reductive carbonylation, 698
reductive elimination, 713
reflection, 263
relative permittivity, 153, 174

relativistic effects, 25, 331, 587, 589, 592
　chemical properties, 331
remediation of nuclear waste, 888
renewable energy, 853
renewable feedstocks, 854
reorganization energy, 667, 796, 798
representation, 91
resonance, 37, 156
resonance hybrid, 37
resonance Raman spectroscopy, 278
resonating valence bond theory (RVB), 918
respiratory chain, 795
resting state (of catalyst), 67
resting state (of enzyme), 785
rhenium, 579
　trioxide structure, 750
rheumatoid arthritis, 963
rhodium, 583
　complex in photodynamic therapy, 962
ribozyme, 780
ribulose 1,5-bisphosphate carboxylase (Rubisco), 805
Rietveld method, 265
ring-opening metathesis polymerization (ROMP), 929
ring-opening metathesis (ROM), 929
ring sizes, 883
ring slip, 691, 711
rocket fuels, 484
rock-salt, 118, 131
　structure, 117, 130, 133, 135, 365
ROM, see ring-opening metathesis (ROM)
ROMP, see ring-opening metathesis polymerization (ROMP)
rotation axis, 73
Roundup® (herbicide), 851
ruby, 724, 750
Ruddlesden–Popper phases, 754
Russell–Saunders coupling, 623
ruthenium, 581
　complexes as anticancer drugs, 960
ruthenocene, 582
rutile, 118, 131, 265
rutile structure, 123, 762
Rydberg constant, 7
Rydberg, Johann, 7

s band, 59
s orbitals, 11
Sabatier principle, 69

S-adenosylmethionine (SAM), 823
safer solvents, 851
safety principle, 856
SALCs, see symmetry-adapted linear combinations (SALCs)
salen, 850
saline hydrides, 342, 353, 387, 392
samarium cobalt, 913
samarocene, 609
sapphire, 724
scanning electron microscopy (SEM), 303, 866
scanning probe microscopy, 302
scanning tunnelling microscopy (STM), 302
$scCO_2$ (supercritical CO_2), 177, 852
scH_2O (supercritical water), 178, 852
scheelite, 574
Schlenk equilibria, 405
Schoenflies symbol, 75
Schottky defect, 722
Schrock carbene, 694
Schrock catalyst, 929
Schrödinger equation, 8
Schrödinger, Erwin, 8
sea urchin, 837
seaborgium, 592
Sea-Nine® (antifouling agent), 855
secondary structure (of a protein), 782
Seebeck effect, 911
selection rules, 633
selective catalyst, 68
selective uptake of metal ions (by organisms), 779
selectivity, 68
　product, 945
　reactant, 945
　transition-state, 945
selenides, 759
selenium, 509
　oxides, 516
　polyanions, 520
　polycations, 521
selenocysteine, 783
self-assembled monolayers (SAMs), 866
self-cleaning windows, 910
SEM, see scanning electron microscopy (SEM)
semiconducting nanoparticles, 868
semiconductivity, 733, 748, 856, 908
semiconductor and oxide nanoparticles, 862

semiconductors, III/V and II/VI types, 906
semimetals, 20, 60
sensitivity of NMR, 279
sensitizer, 904
separation, 886, 887
sequestration of CO_2, 895
Severe Acute Respiratory Syndrome Covid (SARS-CoV), 967
shape-memory alloy (SMA), 730
Sharpless epoxidation, 934
shell, 10
Shell Higher Olefin Process (SHOP), 946
shielded, 281
shielding, 15
shielding constant, 16
Shockley–Queisser (SQ) limit, 906
Si:Al ratios in zeolites, 885
siderophiles, 310
siderophore, 789
Sidgwick, N. V., 677
silane(s), 444, 452
silica, 458, 863, 877, 937
silicate minerals, 159, 458
silicates, 446, 755
siliceous framework, 884
silicides, 463
silicon, 441
　compounds with oxygen, 458
　dioxide, 755
　halides, 454
　halides, Lewis acidity, 164
　in biology, 840
　oxides, 445
　production by Czochralski process, 211
silicon aluminium phosphates (SAPOs), 885
silicone, 466
　polymers, 466
silver, 586
　compounds as antibacterial agents, 973
　halides, 134
　nanoparticles, 867
　tetraiodidomercurate(II), Ag_2HgI_4, 738, 739
silver@zeolite, 889
single bond, 35
single molecule magnets, 643
single-crystal diffraction, 263
single-crystal X-ray diffraction, 266
single-layer nanomaterials, 873
single-molecule magnets, 772
single-walled nanotube (SWNT), 450, 871
singlet oxygen, 506

singlet carbene, 694
singly occupied molecular orbital (SOMO), 53
site-directed mutagenesis, 785
skeletal electrons, 427
Slater's rules, 16
slow-release agents, 897
small-pore zeolites, 885
smart windows, 900
smelting, 207
sodalite, 463
　structure (SOD), 883
sodium, 362
　β-alumina, 738, 739
　azide, 378
　chloride, 725
　hydride, 180
　in biology, 838
　ion batteries, 370
　ion-exchange materials, 371
　iron phosphate, 924
　montmorillonite (Na-MMT), 879
　occurrence and extraction, 367
　nitrate, 910
　pump (Na^+/K^+-ATPase), 786
　solution in liquid ammonia, 379
sodium–sulfur battery, 372
soft acid, 169
soft ferrites, 913
soft ferromagnet, 641
soil conditioners, 886
solar cell, 671, 869, 905
　lead iodide perovskites, 456
　materials, 271
solar thermal collectors, 910
solar-to-hydrogen efficiency (STH), 909
solder, 444
solganol, 964
sol–gel methods, 756
sol–gel processes, 744
solid acids, 180
solid C_{60}, 770
solid electrodes, 738, 739, 741
solid electrolytes, 738, 739, 740
　for Li-ion batteries, 924
solid oxide fuel cell (SOFC), 190, 742, 912
solid solution, 115, 727, 752
solid-state NMR, 284
solid-state superlattices, 875
solubility, 140
solubility and hydration, group 1 element, 378
solubility product, K_{sp}, 198
solution in liquid ammonia, sodium, 379
solvated electrons, 602
solvatochromism, 631

solvent, of crystallization, 223
solvent (as acid and base), 175
solvent levelling, 173
solvents, 851
solvent-system definition, 175
solvothermal reactions and conditions, 746, 893
Sonogashira coupling, 935
spallation, 268
spark-discharge method, 863
specific acid catalysis, 179
specific energy capacity, 922
spectator ligand, 648, 658
spectrochemical series, 245
spectrophotometry, 274
spectroscopic terms, 622, 627
sphalerite (zinc blende, cubic), 118, 120, 588
spin-allowed, 634
spin correlation, 19
spin crossover complex, 249, 642
spin fluctuation, 918
spin-forbidden, 634
spin magnetic quantum number, 11
spin-only magnetism, 248
spin–orbit coupling, 623
spin pairing, 43
spin-restricted, 506
spin selection rules, 634
spin–spin coupling, 282
spin valve, 915
spinel(s), 118, 253, 724, 750, 751, 913
    structure, 125, 901, 918, 922
spintronics, 915
spontaneity (of a chemical reaction), 187
spray pyrolysis, 863
sputtering, 863, 864
square antiprism, 227
square-planar complex, 224, 230, 251, 678
square-pyramidal structure, 42
stability
    covalent halides, 320
    d-block oxides, 321
stabilizers, 861
staging, 448
stainless steel, 115
standard cell potential, 189
standard conditions, 188
standard enthalpy of formation, 132
standard Gibbs energy, 188
standard hydrogen electrode (SHE), 188
standard potential, 186, 188
    factors contributing to, 191
    Table, 192
standard reduction potential, 188

stannane, 453
steel(s), 111, 116, 573, 579, 581, 728
    production using hydrogen, 210
stereochemically-inert lone pair, 42
steric hindrance, 794
steric influence of domains, 38
stibane, 485
stibole, 497
Stille coupling, 935
Stokes lines, 277
Stone Age, 566
stopped-flow technique, 274
stretching wavenumbers, 276
strong base, 149
strong force, 3, 4
strong permanent magnets, 913
strong-field case, 246
strong-field ligand, 246
strontium, 385
    applications, 391
    occurrence and extraction, 390
structural type with radius ratio, 128
structure determination, 266
structure-directing agent (SDA), 746, 882, 884
structure map, 130
subatomic particles, 3
subshell, 10
substitution reaction, 161
substitutional alloy, 115, 728
Sudbury, 585
sulfate, 518
sulfate process, 848, 853
sulfide spinels, 912
sulfides, 759, 903
sulfite, 518
sulfite oxidase, 825
    structure, 826
    active site, 842
sulfite reductase, 842
sulfur, 508
    allotropes, 502, 508
    biology, 512
    dioxide, 165
    dioxide (liquid), 175
    dioxide oxidation, 941
    hexafluoride, 95, 513
    nitrides for fingerprint detection, 522
    occurrence, 502
    oxides, 504, 515
    oxoacids, redox properties, 517
    polyanions, 520
    polycations, 521
    production by Claus process, 211

    properties of allotropes and polymorphs, 509
    trioxide, 165
sulfuric acid, 154, 176, 518
sulfur–nitrogen compounds, 521
sulfurous acid, 157, 516, 519
sunlight, 909
sunscreens, 903
superacid, 165, 179, 944
superbase, 180
superconducting cuprate, 865
superconducting quantum interference device (SQUID), 299
superconduction, 570, 733, 761, 916
superconductor, lanthanoid barium cuprate, 605
supercritical carbon dioxide, 177, 852
supercritical fluid, 177, 852
supercritical water, 178, 852
superexchange, 642, 749
superheavy elements, 592
superhydride, 919
superlattice nitrides, 875
superlattice structures, 875
superoxide, 511
supertetrahedral clusters, 891
surface acids, 180, 937
surface bases, 937
surface enhanced Raman spectroscopy (SERS), 278
surface metals, 937
surface migration, 939
surface plasmons, 870
sustainable development, 847
Suzuki coupling, 935
sweep rate, 300
symmetry, 72
    element, 73
    operation, 73
    species, 79
symmetry-adapted linear combinations (SALCs), 86, 87, 90, 92, 254
synchrotron, 267
    radiation, 267
syndiotactic, 948
synergistic coordination, 790
syngas, 942
'Synroc' ('synthetic rock'), 757
synthesis, 884
    of materials, 743
    of MOFs, 893
    of nanoparticles, 861
synthetic models (of metal cofactors in enzymes), 785
synthetic oxygen carriers, 794

tacticity, 948
Tanabe–Sugano diagram, 629

tandem two-photon device, 909
tantalum, 570
    halides, 573
    nitride, 757
technetium, 6, 579
    in medical imaging, 971
teflate, 517
Teflon®, 546, 547
tellurides, 759
tellurium, 509
    oxides, 516
    polyanions, 520
    polycations, 521
TEM, see transmission electron microscopy (TEM)
template(s), 765, 884
    effect, 242
temporary hardness, 401
tennessine, 21, 525
term symbol, 623, 627
    lanthanoid ion, 637
terms, spectroscopic, 623
ternary phases, 124, 126
tethered catalysts, 950
tetracyanidoplatinate(II) complexes, 771
tetraethyllead, 466
tetrafluoroethane, 527
tetrahalomethane
    properties, 453
    synthesis, 453
tetrahedral complex, 224, 232, 250
tetrahedral hole, 109, 124
tetrahedral oxoanions, 764
tetrahydridoaluminates, 919
tetrahydridoborate(s), 419, 762
tetralenes, 467
tetrathiometallate anions, 515
tetrathionate, 518
thalassemia, 969
thalidomide, 851
thallium, 410
    applications, 416
    in heart imaging, 972
    occurrence and recovery, 416
    substituting for potassium in cells, 787
    sulfides, 436
    trihalide, 434
theoretical design, 884
therapeutic agent, 867
thermal conductivity, 872, 873
    diamond, 448
thermal decomposition, 138
thermal solar energy, 910
thermal stabilities of ionic solids, 138
thermochemical radii, 137
thermoelectric material, 911
thermogravimetric analysis, 296
thermoplastic polymers, 850

thin-film applications, 900
thiocyanate, 171
thiospinels, 760
thiosulfate, 518, 519
thiosulfuric acid, 519
thorium, 554
    compounds, 615
    metal–metal bonds, 617
three-dimensional (3D) supramolecular architectures, 876
thyroid, 533
thyroxine, 533
time-of-flight (TOF) mass spectrometer, 293
timescales, 270
tin, 441
    halides, 455
    halides as Lewis bases, 164
    oxides, 459
tin-doped antimony oxide, 901
titania, 863, 867, 877
titanium, 568
    dioxide, 848, 853, 903, 909
    in artificial joints, 956
    monoxide, 723
    production by Kroll process, 211
    slag, 848
titanosilicate, 890
Tolman cone angle, 659, 685, 710
top-down approaches, 860
touchscreen, 900
    capacitive, 436
Tourette's syndrome, 935
trace elements, 778
trans effect, 653, 711
trans influence, 282, 653
transcription factors, 833
transducers, 911
transferrin receptor, 790
transferrins, 789
transformers, 913
transition
    allowed, 634
    charge transfer (CT), 631
    forbidden, 634
transition glass transition temperature, T_g, 756
transition metal, 562
    chalcogenides, 759
    complexes, 272
transition state, 650
    effect, 653
    selectivity, 945
transition temperature, 114
translation (of the genetic code), 781
transmembrane proton gradient, 779
transmission electron microscopy (TEM), 303

transmittance, 900
transparent materials, 900
transuranium chemistry, 617
tributyltin, 855
tributyltin oxide, 855
tricapped trigonal prism, 228
tridentate, 219
trifluoroselenium cation, 172
trigonal bipyramidal structure, 42
triiodide ion, 166, 539
triiodothyronine, 533
trimesic acid (TMA), 894
triphosphate, 493
triple bond, 35
triplet carbene, 694
tritium, 4, 340
tube furnace, 745
tungsten, 104, 112, 573
    carbide, 728
    in biology, 840
    in formate dehydrogenase, 826
    oxide, 732
turnover frequency (TOF), 67
turnover number (TON), 67
two-dimensional pattern, 101
two-dimensional structure, 873
two-equivalent process, 664
type I and type II superconductors, 917

ultramarine, 272, 902
ultrasound, 854
ultraviolet (UV) radiation, 270
ultraviolet photoelectron spectroscopy, 49, 289
ultraviolet–visible spectroscopy, 270, 275
ungerade, 50
unit cell, 101
    parameters, 102
unpaired electrons, 325
ununoctium, 21
ununpentium, 21
ununseptium, 21
ununtrium, 21
uranium, 554
    compounds, 615
    fission products, 612
    halides, 616
    hexafluoride, 616
    oxides, 616
uranocene, 617
UV photoelectron spectrum, 49, 289
UV photolysis, 557
UV–visible spectroscopy, 270, 275
UV–visible spectrum, 271

vacancy, 722
valence band, 61
valence bond theory, 42
valence shell, 22
valence shell electron pair repulsion (VSEPR), 37, 555
valinomycin, 381, 785
van der Waals interaction, 134
van der Waals radius, 63
vanadium, 570
    as active site of enzyme (bromoperoxidase), 808
    halides, 573
    in nitrogenase, 831
    oxide, 570
    pentoxide, 570
vapour transport crystal growth, 745
Vaska's complex, 584
Vegard's rule, 726
verdigris (basic copper carbonate), 196, 586
vermillion, 589
vibrating sample magnetometer (VSM), 298
vibrational analogy, 90
vibrational displacements, 85
vibrational energy, 276, 911
vibrational modes, 90
vitamin B_{12}, 583
vitreous state, 755, 756
vitrification, 888
volcano plot (diagram), 69, 939
voltammogram, 206
volume of activation, 655
VSB-1 (Versailles–Santa Barbara), 890
VSEPR, *see* valence shell electron pair repulsion (VSEPR)

Wacker process, 933
Wade's rules, 426
    origin, 428
Wade–Mingos–Lauher rules, 708
Wadsley defects, 732
Warner, John, 848
waste prevention, 848
waste products, 848, 855
waste treatment, 852
water, 510
    as a redox agent, 194
    electrolysis, 348
    stability field, 196, 205
water oxidation, electrocatalysts, 506
water splitting, photocatalysis, 908
wavefunction, 8, 42
wavenumber, 276

wave–particle duality, 8
weak base, 149
weak-field case, 246
weak-field ligand, 246
Werner, Alfred, 218, 244
Wilkinson, Geoffrey, 702
Wilkinson's catalyst, 584, 930, 950
Wilson's disease (Cu accumulation), 836
window (in porous structures), 883
wolframite, 574
Wurtz coupling, 405
wurtzite, 588
    hexagonal, 118
    structure, 121

X type, ligand, 679
XANES, 291
X-band spectrometer, 285
xenodeborylation, 558
xenon, 135, 550
    clathrates, 559
    coordination compounds, 558
    fluorides, 551, 554
    hydrides, 552
    insertion compounds, 557
    missing, 557
    oxofluorides, 552, 556
    organo compounds, 558
    oxides, 552, 556
    tetraoxide, 552
    trioxide, 552, 556
    uses, 553, 554
^{129}Xe-NMR spectroscopy, 556
xerogels, 744
X-ray absorption spectroscopy (XAS), 291
X-ray diffraction, 114, 117, 127, 263, 759
X-ray emission lines, 296
X-ray fluorescence (XRF), 296
X-ray free electron lasers (XFEL), 267
X-ray photoelectron spectroscopy (XPS), 289
X-rays, 270

YAG, *see* yttrium aluminium garnet (YAG)
YBCO, *see* yttrium barium copper oxide (YBCO)
YIG, *see* yttrium iron garnet (YIG)
YSZ, *see* yttrium-stabilized zirconia (YSZ)
yttrium, 566
yttrium aluminium garnet (YAG), 566, 606

yttrium barium copper oxide (YBCO), 587, 743, 917
yttrium iron garnet (YIG), 567, 606
yttrium-stabilized zirconia (YSZ), 569, 738, 739, 912

Zachariasen rules, 756
Zeise's salt, 586, 676, 678
zeolites, 180, 447, 463, 746, 764, 882, 921
   applications, 463
   catalysis, 937, 943
   structures, 883
Zeolite Encapsulated (transition) Metal Complex (ZEMC), 889
zeotypes, 882, 889
zero valent compounds
   beryllium, 406
zero-point energy, 277

Ziegler–Natta catalysts, 947
ZIF (zeolitic imidazolate framework), 891
zinc, 588
   enzyme catalysis, 'zinc carbonyl' mechanism, 802
   enzyme catalysis, 'zinc hydroxide' mechanism, 802
   enzymes, 802
   in biology, 839
   in brain tissue, 840
   in protein structure stabilization, 802
   oxide, 736
zinc finger(s), 588, 785, 832, 968
   interacting with a section of DNA, 833
   nucleases, principle, 833
   nucleases, use in gene therapy, 832
zinc-blende structure, 120

zincophosphates, 591
Zintl ion, 437
Zintl phases, 117, 437, 731
   containing alkali metals, 379
zirconia, 741
zirconium, 568
zirconium dioxide in dental implants, 956
Z-scheme, 909
Zundel cation, $H_5O_2^+$, 169

0D nanoparticles, 868
1,1-migratory insertion reaction, 714
1,2-insertion reaction, 715
16 electron complex, 678
18 electron complex, 678
18 electron rule, 677
3c,2e bonding, 382, 426
3d-metal monoxides, 748, 749
4c,2e bonding, 382

α-cage, 464
α-helix, 782
α-hydride elimination, 715
β-sheet, 782
β-hydride elimination, 715
η, 222, 681
η^2-alkene, 687
κ, 221, 222
μ, 222, 681
π acid, 170, 257
π-acceptor ligand, 257
π backbonding, 682
π-base, 257
π bond, 44
π-donor ligand, 257
π orbital, 50
σ bond, 43, 44
σ orbital, 50
σ-bond metathesis, 713